The Photosynthetic Bacteria

The Photosynthetic Bacteria

Edited by
Roderick K. Clayton
Cornell University
Ithaca, New York

and

William R. Sistrom
University of Oregon
Eugene, Oregon

PLENUM PRESS · NEW YORK AND LONDON

Library of Congress Cataloging in Publication Data

Main entry under title:

The Photosynthetic bacteria.

Includes bibliographies and index.
1. Bacteria, Photosynthetic. I. Clayton, Roderick K. II. Sistrom, W. R.
QR88.5.P48   589.9'01'3342   78-2835
ISBN 0-306-31133-X

© 1978 Plenum Press, New York
A Division of Plenum Publishing Corporation
227 West 17th Street, New York, N.Y. 10011

Printed in the United States of America

Contributors

J. Amesz · Department of Biophysics, Huygens Laboratory, University of Leiden, The Netherlands

Assunta Baccarini-Melandri · Institute of Botany, University of Bologna, Bologna, Italy

Margareta Baltscheffsky · Arrhenius Laboratory, Department of Biochemistry, University of Stockholm, Strockholm, Sweden

Robert G. Bartsch · Chemistry Department, University of California, San Diego, California

James R. Bolton · Photochemistry Unit, Department of Chemistry, University of Western Ontario, London, Ontario, Canada

A. Yu. Borisov · Moscow State University, Moscow, USSR

Richard W. Castenholz · Department of Biology, University of Oregon, Eugene, Oregon

B. Chance · Department of Biochemistry and Biophysics, Johnson Research Foundation, University of Pennsylvania, Philadelphia, Pennsylvania

Roderick K. Clayton · Division of Biological Sciences, Cornell University, Ithaca, New York

R. J. Cogdell · Department of Botany, University of Glasgow, Glasgow, Scotland

Prasanta Datta · Department of Biological Chemistry, The University of Michigan, Ann Arbor, Michigan

Gerhart Drews · Institut für Biologie II, Universität Freiburg, Freiburg, West Germany

P. Leslie Dutton · Johnson Research Foundation and Department of Biochemistry and Biophysics, University of Pennsylvania, Philadelphia, Pennsylvania

W. Charles Evans · Department of Biochemistry and Soil Sciences, University College of North Wales, Bangor, Gwynedd, United Kingdom

George Feher · Department of Physics, University of California at San Diego, La Jolla, California

Darrell Fleischman · Charles F. Kettering Research Laboratory, Yellow Springs, Ohio

R. C. Fuller · Department of Biochemistry, University of Massachusetts, Amherst, Massachusetts

Kenneth D. Gibson · Department of Physiological Chemistry, Roche Institute of Molecular Biology, Nutley, New Jersey

Gabriel Gingras · Département de Biochimie, Université de Montréal, Montréal, Québec, Canada

Florian Göbel · Institut für Mikrobiologie der Gesellschaft für Strahlen- und Umweltforschung mbH München in Göttingen, Göttingen, Germany

Ernest D. Gray · Departments of Biochemistry and Pediatrics, University of Minnesota, Minneapolis, Minnesota

O. T. G. Jones · Department of Biochemistry, University of Bristol, Bristol, England

Samuel Kaplan · Department of Microbiology, University of Illinois, Urbana, Illinois

Joseph J. Katz · Chemistry Division, Argonne National Laboratory, Argonne, Illinois

Donald L. Keister · Charles F. Kettering Research Laboratory, Yellow Springs, Ohio

Christine N. Kenyon · Space Sciences Laboratory, University of California, Berkeley, California, and Department of Pharmaceutical Chemistry and Department of Biochemistry and Biophysics, University of California, San Francisco, California

David B. Knaff · Department of Chemistry, Texas Tech University, Lubbock, Texas

J. Lascelles · Department of Bacteriology, University of California, Los Angeles, California

J. S. Leigh, Jr. · Johnson Research Foundation, University of Pennsylvania, Philadelphia, Pennsylvania

Synnøve Liaaen-Jensen · Organic Chemistry Laboratories, Norwegian Institute of Technology, University of Trondheim, Trondheim-NTH, Norway

Barry L. Marrs · Department of Biochemistry, Saint Louis University School of Medicine, Saint Louis, Missouri

D. Mauzerall · The Rockefeller University, New York, New York

Hubert Mayer · Max-Planck-Institut für Immunbiologie, Freiburg, West Germany

Bruno Andrea Melandri · Institute of Botany, University of Bologna, Bologna, Italy

J. M. Merrick · Department of Microbiology, School of Medicine, State University of New York at Buffalo, Buffalo, New York

Robert A. Niederman · Department of Microbiology, Rutgers University, New Brunswick, New Jersey

James R. Norris · Chemistry Division, Argonne National Laboratory, Argonne, Illinois

Melvin Y. Okamura · Department of Physics, University of California at San Diego, La Jolla, California

John M. Olson · Biology Department, Brookhaven National Laboratory, Upton, New York

William W. Parson · Department of Biochemistry, University of Washington, Seattle, Washington

Norbert Pfennig · Institut für Mikrobiologie der Gesellschaft für Strahlen- und Umweltforschung m.b.H., Göttingen, West Germany

Beverly K. Pierson · Department of Biology, University of Puget Sound, Tacoma, Washington

Patricia B. Pinder · Departments of Biochemistry and Microbiology, Dartmouth Medical School, Hanover, New Hampshire

Roger C. Prince · Johnson Research Foundation and Department of Biochemistry and Biophysics, University of Pennsylvania, Philadelphia, Pennsylvania

Charles C. Remsen · Center for Great Lakes Studies, The University of Wisconsin–Milwaukee, Milwaukee, Wisconsin

Karin Schmidt · Institut für Mikrobiologie der Gesellschaft für Strahlen- und Umweltforschung mbH, Göttingen, West Germany

W. R. Sistrom · Department of Biology, University of Oregon, Eugene, Oregon

Lucile Smith · Departments of Biochemistry and Microbiology, Dartmouth Medical School, Hanover, New Hampshire

Gary A. Sojka · Department of Biology, Indiana University, Bloomington, Indiana

Walther Stoeckenius · Cardiovascular Research Institute and Department of Biochemistry and Biophysics, University of California, San Francisco, California; Ames Research Center, NASA, Moffett Field, California

Charles E. Strouse · Department of Chemistry, University of California, Los Angeles, California

J. Philip Thornber · Department of Biology and Molecular Biology Institute, University of California, Los Angeles, California

Terry L. Trosper · Department of Biology and Molecular Biology Institute, University of California, Los Angeles, California

Hans G. Trüper · Institut für Mikrobiologie der Universität Bonn, Bonn, West Germany

R. L. Uffen · Department of Microbiology, Michigan State University, East Lansing, Michigan

Jürgen Weckesser · Institut für Biologie II, Universität Freiburg, Freiburg, West Germany

C. A. Wraight · Department of Physiology and Biophysics and Department of Botany, University of Illinois at Urbana–Champaign, Urbana, Illinois

Duane C. Yoch · Department of Cell Physiology, University of California, Berkeley, California

Kenneth L. Zankel · Martin Marietta Laboratories, Baltimore, Maryland

Preface

Our knowledge of the biology of the photosynthetic bacteria was last summarized in 1963 in the Kettering Symposium entitled *Bacterial Photosynthesis* [H. Gest, A. San Pietro, and L. P. Vernon (eds.), 1963, Antioch Press, Yellow Springs, Ohio]. Subsequently, there has been an enormous increase in our understanding of nearly every facet of the life of these fascinating creatures: development, structure, genetics, regulatory physiology, photochemistry, electron transport, phosphorylation, and many aspects of intermediary metabolism. The new and detailed knowledge of the photochemical mechanism obtained by studying photochemical reaction centers of these bacteria has helped to elucidate the mechanism of green plant photosynthesis as well. In this volume, we attempt to bring these topics together at a level that reflects the forefront of our present knowledge. We believe that the next few years will see an even greater increase in our understanding of the photosynthetic bacteria than the decade and a half since the appearance of *Bacterial Photosynthesis*. Our optimism is founded on two developments. The first is the coming together of the classic approaches of biochemistry and biophysics with those of molecular biology and genetics. The second is the discovery over the past several years of previously unknown kinds of photosynthetic bacteria, which suggests that these organisms are even more diverse than hitherto suspected.

The arrangement of subject matter is shown in the Contents. Inevitably, the treatment of some topics failed to materialize during the editorial process. For example, the protective function of carotenoid pigments is not discussed; for a discussion of this subject, the reader is referred to the review by Krinsky [Krinsky, N. I., 1971, in: *The Carotenoids* (O. Isler, ed.), pp. 669–716, Birkhäuser Verlag, Basel]. Some redundancy can be inferred from the chapter titles, but this redundancy has been allowed to stand so that each chapter can give a coherent presentation from the viewpoint of its author. Thus, for example, Chapters 16–18 are all on the subject of excitation energy transfer, but the different approaches to this topic provide illuminating contrasts.

The Appendix contains a list of the more useful mutant strains of photosynthetic bacteria. We have omitted a list of culture media; the Kettering Symposium contains an excellent list of these. Similarly, there is no compilation of methods for the enrichment and isolation of these bacteria; for such a compilation, the reader is referred to van Niel's review [*Methods in Enzymology* (A. San Pietro, ed.), 1971, Vol. 23A, pp. 3–28, Academic Press, New York].

Concerning our choice of title, many readers will feel that *Photosynthetic Bacteria* is imprecise and that *Phototrophic Bacteria* should have been chosen. Although we appreciate the strength of this argument, we have opted for familiarity rather than precision.

For our understanding of photosynthesis, we owe a great debt to the pioneering work of Prof. C. B. van Niel, whose comparative biochemical studies of photosynthesis generated many valuable insights. Undoubtedly, the most significant of these insights was the appreciation that the primary photochemistry in photosynthesis is a light-driven oxidation–reduction process. We regret that Prof. van Niel was not able to contribute personally to this book; his contributions are implicit in many of the chapters, and we dedicate this volume to him.

R.K.C.
W.R.S.

Contents

CHAPTER 9: **Photosynthetic Apparatus and Cell Membranes of the Green Bacteria**

BEVERLY K. PIERSON AND RICHARD W. CASTENHOLZ

CHAPTER 10: **Metabolism of Reserve Materials**

J. M. MERRICK

PART III: **CHEMISTRY OF CELLULAR COMPONENTS**

CHAPTER 11: **Bacteriochlorophyll and Photosynthetic Evolution**

D. MAUZERALL

CHAPTER 12: **Chemistry of Carotenoid Pigments**

SYNNØVE LIAAEN-JENSEN

PHOTOMETABOLISM: REACTION CENTERS

CHAPTER 23:   **EPR Studies of Primary Events in Bacterial Photosynthesis**

J. S. LEIGH, JR.

PHOTOMETABOLISM: PRIMARY AND SECONDARY ELECTRON
TRANSPORT AND INTERACTIONS WITH PROTONS

CHAPTER 24:   **Protonation and the Reducing Potential of the Primary Electron Acceptor**

ROGER C. PRINCE AND P. LESLIE DUTTON

CHAPTER 25:   **Quinones as Secondary Electron Acceptors**

WILLIAM W. PARSON

CHAPTER 31:   **Coupling Factors**

ASSUNTA BACCARINI-MELANDRI AND BRUNO ANDREA
MELANDRI

CHAPTER 32:   **Reducing Potentials and the Pathway of NAD⁺ Reduction**

DAVID B. KNAFF

CHAPTER 33:   **Oxygen-Linked Electron Transport and Energy Conservation**

LUCILE SMITH AND PATRICIA B. PINDER

PART VI: PERIPHERAL OXIDATIONS AND REDUCTIONS

CHAPTER 34:   **Nitrogen Fixation and Hydrogen Metabolism by
Photosynthetic Bacteria**

DUANE C. YOCH

PART I

Introduction: Physiology, Ecology, and Taxonomy

CHAPTER 1

General Physiology and Ecology of Photosynthetic Bacteria

Norbert Pfennig

1. Common and Differentiating Characteristics of the Groups of Prokaryotic Phototrophs

1.1. Basic Cellular Organization

The phototrophic green and purple bacteria that are the subject of this book share their basic prokaryotic cellular organization with the third group of phototrophic prokaryotes, the cyanobacteria, or blue-green algae (Stanier, 1974a,b). There follows a brief summary of the common features of these groups, which separate them sharply from the eukaryotic phototrophs, after which the typical and differentiating characteristics of each group will be considered: (1) The genophore, consisting of double-stranded DNA fibrils, is embedded directly in the cytoplasm, and lacks a surrounding membrane. (2) The relatively rigid cell walls contain as a structural element the mucopeptide heteropolymer, murein or peptidoglycan, that is characteristic of prokaryotes. (3) Special unit-membrane-bound cytoplasmic organelles are lacking. (4) The ribosomes are of the 70 S type. (5) The cells of several species of all groups are buoyant in water due to unique non-unit-membrane-bound gas vesicles in the cytoplasm (Cohen-Bazire et al., 1969; Walsby, 1972). (6a) Gliding motility is shown by species with filamentous flexible cells. (6b) Characteristic flagella are present in species with

swimming motility (Doetsch, 1971). (7) The ability to use molecular nitrogen as a nitrogen source for growth has been demonstrated in many species of all groups (Stewart, 1973; Keister and Fleischmann, 1973; Stanier, 1974c). (8) The characteristic bacterial viruses, bacteriophages and cyanophages, have been described for both purple bacteria (Bosecker et al., 1972) and cyanobacteria (Padan et al., 1970).

1.2 Major Differences between the Cyanobacteria and the Green and Purple Bacteria

We shall now consider the differentiating physiological and biochemical characteristics between the cyanobacteria, on one hand, and the green and purple bacteria, on the other.

Like the eukaryotic phototrophs, the cyanobacteria carry out oxygenic photosynthesis with two photosystems (Avron, 1967) in which water is the electron donor and oxygen is the ultimate oxidation product. They have chlorophyll a and phycobiliproteins as characteristic pigments, together with β-carotene and zeaxanthin as the most common carotenoids (e.g., Stanier, 1974a). The phototrophic green and purple bacteria, in contrast, carry out an anoxygenic photosynthesis using only one photosystem. They therefore require electron donors of a lower redox potential than water, such as reduced sulfur compounds, molecular hydrogen, or simple organic compounds. Instead of oxygen, the corresponding oxidized products are sul-

Norbert Pfennig · Institut für Mikrobiologie der Gesellschaft für Strahlen- und Umweltforschung m.b.H., Göttingen, West Germany

Table 1. Bacteriochlorophylls of Phototropic Green and Purple Bacteria

Designations[a]	Characteristic absorption maxima of living cells (nm)			
Bacteriochlorophyll a (a_p or a_{Gg})[b]	375,	590,	800–810,	830–890
b	400,	605,	835–850,	1015–1035
c	335,	460,	745–760,	812
d	325,	450,	725–745,	805
e	345,	450–460,	715–725,	805

[a] The designations of Bchl's a–d are given by Jensen *et al.* (1964); that of Bchl e, by Gloe *et al.* (1975).

[b] Bchl a_p contains phytol as the esterifying alcohol; Bchl a_{Gg} contains, instead of phytol, geranylgeraniol (see Katz *et al.*, 1972; Künzler and Pfennig, 1973; Gloe and Pfennig, 1974).

Table 2. Carotenoid Groups of Phototrophic Green and Purple Bacteria[a]

Group	Designation	Major components	Representative species
1	Normal spirilloxanthin series	Lycopene, rhodopin, spirilloxanthin	*Rhodospirillum rubrum, Chromatium vinosum*
		Also: dihydro derivatives of neurosporene and lycopene	*Rhodopseudomonas viridis*
		Also: tetrahydrospirilloxanthin	*Thiocapsa pfennigii*
2	Rhodopinal branch of spirilloxanthin series	Lycopenal, rhodopinal	*Lamprocystis, Thiodictyon, Rhodospirillum tenue*
3	Alternative spirilloxanthin series	Hydroxyneurosporene, spheroidene, spheroidenone	*Rhodopseudomonas sphaeroides, Rp. capsulata, Rp. gelatinosa*
4	Okenone series and R.g.-keto-carotenoids	Okenone	*Chromatium okenii*
		Diketo-tetrahydrospirilloxanthin	*Rhodopseudomonas globiformis*
5	Isorenieratene series	β-Carotene, isorenieratene	*Chlorobium phaeobacteroides*
		γ-Carotene, chlorobactene	*Chlorobium limicola*
		β-Carotene and γ-carotene, γ-carotene-glucoside	*Chloroflexus aurantiacus*

[a] From Schmidt *et al.* (1965), Liaaen Jensen (1967), Pfennig (1967), and Pierson and Castenholz (1974*b*).

fate, protons, and organic compounds and CO_2, respectively. Since the protons formed by the oxidation of molecular hydrogen are necessary for the reduction of CO_2 to cell material, there is no directly observable oxidized product in this case. The photosynthetic pigments are bacteriochlorophylls a, b, c, d, and e (Table 1) and a great variety of carotenoids that have been ordered in five groups (Table 2).

The physiological and ecological consequences of the difference between the oxygenic photosynthesis of the cyanobacteria and the anoxygenic photosynthesis of the green and purple bacteria are profound. However, that an organism possessing two photosystems may, under suitable conditions, grow and carry out photosynthetic carbon dioxide assimilation with one photosystem only was recently demonstrated by Cohen *et al.* (1975). The cyanobacterium *Oscillatoria limnetica*, isolated from the sulfide-rich hypolim-

nion* of a salt lake, appears to be the first phototroph reported to grow, depending on the conditions, with either type of photosynthesis. Growth and CO_2 assimi-

---

* Following are brief definitions of limnological terms used in this chapter: The body of water of almost all lakes is stratified during the summer and consists of three distinct layers, each with different thermal characteristics and specific gravities. Due to wind movements and thermal convection currents, the surface layer, termed the *epilimnion*, is well mixed, aerobic, and of fairly uniform temperature. The water layer below the epilimnion is characterized by a sharp decrease in temperature, and is therefore called the *thermocline, discontinuity layer*, or *metalimnion*. The water mass below the thermocline and extending to the lake bottom is termed the *hypolimnion*. It is characterized by low temperatures (4–8°C), anaerobic conditions, and the presence of sulfide. Lakes with annual complete turnover of the whole water mass are called *holomictic* lakes, whereas those in which only the upper layers turn over are called *meromictic* lakes. The stagnant lower layer of the water mass of meromictic lakes (the hypolimnion) is usually of higher specific gravity (e.g., sea water below fresh water), permanently anaerobic, and rich in sulfides.

lation are shown to proceed at the same rate either without sulfide or, after a short lag period, in the presence of 3–4 mM sulfide. The water-cleaving photosystem II is completely inhibited by sulfide, which at the same time serves as electron donor for photosynthesis with photosystem I alone. This process proved to be unaffected by either DCMU (which is an inhibitor of photosystem II; Padan et al., 1970) or the use of light of the "red drop" region around 700 nm. Sulfide is oxidized to elemental sulfur, which accumulates outside the cells and cannot be further oxidized. Oscillatoria limnetica thus provides an excellent example in one and the same organism for the theory of van Niel (1931), which postulated the equivalence of H_2O and H_2S as photosynthetic electron donors in the CO_2 assimilation of algae and green sulfur bacteria, as well as the equivalence of oxygen and elemental sulfur as the corresponding oxidation products.

Use of universally available water as the electron donor for autotrophic carbon dioxide assimilation became possible with the acquisition of the second photosystem in ancient phototrophic prokaryotes (e.g., cyanobacteria). Photosynthesis thus became sufficiently independent from reduced substances already present in the environment or produced by the metabolic activities of other microorganisms. This independence may well have been a prerequisite for the development of the chloroplast-endosymbiosis of phototrophic prokaryotes in eukaryotic cells, as it has been envisaged at the roots of the evolution of the major algal groups and higher plants (Stanier, 1970, 1974a). Although physiologically plausible, an endosymbiosis of phototrophic green or purple bacteria in a purely fermentative eukaryotic cell would never have gained evolutionary significance after the advent of oxygenic photosynthesis due to the inefficiency of fermentation compared with aerobic respiration. Today, the ecological niches of the anaerobic phototrophs are provided only in aqueous environments, either by sulfur springs or through the activities of other microorganisms the metabolic end products of which became the primary elements for their photometabolism.

1.3. Major Characteristics of Green and Purple Bacteria

Although both green and purple bacteria carry out an anoxygenic photosynthesis and both contain Bchl a_p as one of several pigments, they constitute two remarkably different groups of phototrophic pro-

karyotes on the basis of their major cytological properties. With the exception of the gliding Chloroflexus, all green bacteria are nonmotile. Most purple bacteria, in contrast, are flagellated and exhibit advantageous photo- and chemotactic responses to their environmental conditions. In both groups of the purple bacteria (Chromatiaceae and Rhodospirillaceae), the pigments are located on intracytoplasmic unit membrane systems of various types that are continuous with the cytoplasmic membrane. These membrane systems are structurally comparable to those of the cyanobacteria, the chemoautotrophic nitrifying and methane-oxidizing bacteria. In contrast, the pigments of the green bacteria are located predominantly in special structures, the chlorobium vesicles, which underlie and are firmly attached to the cytoplasmic membrane and presumably form a functional unit with it (Cohen-Bazire et al., 1964; Fowler et al., 1971).

Since comparable structural elements have so far not been found in other prokaryotic or in eukaryotic cells, the chlorobium vesicles—together with the Bchl c, d, or e—readily characterize a particular organism as a member of the green bacteria.

In all groups of the phototrophic bacteria, the pigment content of the cells is regulated as a function of the light intensity. The specific concentrations of the pigments reach maximum values in dim light under anaerobic conditions. Along with Bchl and carotenoids, the amount of pigment-bearing structures in the cells, i.e., the number of chlorobium vesicles per cell in the green bacteria (Holt et al., 1966) and the amount of intracytoplasmic membrane system in the purple bacteria (Cohen-Bazire, 1963; Oelze and Drews, 1972), is regulated. While the extent of this regulation is limited in the strictly anaerobic green and purple sulfur bacteria, an almost complete repression of the intracytoplasmic membrane system could be achieved in several purple nonsulfur bacteria by high light intensities under anaerobic conditions (Cohen-Bazire, 1963). In the latter group of organisms, a corresponding degree of repression is reached by oxygen in strongly aerated aerobic cultures.

Under anaerobic conditions in the dark, all phototrophic bacteria obtain their energy for maintenance by a very slow fermentation of storage polysaccharides (Larsen, 1953; van Gemerden, 1968; Schön, 1969; Gürgün, 1974). A slow fermentation of exogenous substrates, particularly pyruvate, has also been demonstrated in the purple bacteria (Kohlmiller and Gest, 1951; Uffen and Wolfe, 1970; Gürgün, 1974). In the case of Rhodospirillum rubrum, Uffen and Wolfe (1970) showed that increases in both protein and cell number can occur under strictly anaerobic conditions in the dark.

2. Physiological–Ecological Groups of the Green and Purple Bacteria

The green and purple bacteria are considered in four physiological–ecological groups that correspond to the four existing families, if the newly established family Chloroflexaceae (Trüper, 1976) for the gliding filamentous green bacteria is included.

2.1. The Green Bacteria

2.1.1. The Green and Brown Sulfur Bacteria: Chlorobiaceae

(a) General View of the Group. Nine species of this small and surprisingly uniform group of nonmotile obligate anaerobic and phototrophic bacteria have been studied in pure culture and well described. A few more types have been observed in samples from nature and grown in enrichment cultures. On the basis of this knowledge, the following view of the Chlorobiaceae has emerged during the past several years.

Almost all morphological–cytological types that became known as green-colored species have deep-brown-colored counterparts. This correspondence not only covers the gas-vacuole-free and gas-vacuole-containing species, but also includes even the long-known unique ectosymbiontic associations of regularly arranged *Chlorobium* cells around a motile, colorless central bacterium: *Chlorochromatium consortium* is the green form and *Pelochromatium consortium* its brown counterpart (Lauterborn, 1915; Pfennig and Trüper, 1974).

All species of the Chlorobiaceae have in common their general physiology and the fine structure of the photosynthetic apparatus: the cytoplasmic membrane with attached chlorobium vesicles. The green-colored species are regularly characterized by the possession of Bchl c or d (light-harvesting), Bchl a_p (reaction center), and the carotenoids chlorobactene and OH-chlorobactene. In contrast, the brown-colored species regularly contain Bchl e (light-harvesting), Bchl a_p (presumably reaction center) (Gloe *et al.*, 1975), and the carotenoids isorenieratene and β-isorenieratene (Liaaen Jensen, 1965). The latter carotenoids contribute to the brown color of these species and the broader absorption range between 480 and 550 nm. This extended absorption range, in blue-green light is obviously of ecological significance, since the brown species usually occur in greater depth in lakes than the green forms (Trüper and Genovese, 1968).

A requirement for vitamin B_{12} has been established for most species.

(b) Photometabolism of Sulfide and Carbon Sources. All species of the Chlorobiaceae are strictly anaerobic and depend on both hydrogen sulfide and light for their development. Sulfide not only serves as the electron donor for CO_2 assimilation, but also provides the necessary low redox potential for growth, and since an assimilatory sulfate reduction is lacking, it is a source of reduced sulfur for biosynthesis (Lippert, 1967). In the course of CO_2 reduction, sulfide is oxidized to sulfate. Depending on the sulfide concentration and light intensity, elemental sulfur may be formed as an intermediate oxidation product. Globules of elemental sulfur always arise outside the cells, never inside; they can nevertheless be further oxidized to sulfate. Most strains can use molecular hydrogen and some strains thiosulfate as electron donors for CO_2 reduction.

Along with sulfide-dependent CO_2 assimilation, a few simple organic compounds can be photoassimilated by all members of the Chlorobiaceae. Acetate is most effectively used, almost doubling the cell yield; propionate, butyrate, pyruvate, lactate, glutamate, and some amino acids may be used in addition by some strains, although much less effectively (Kelly, 1974). Sadler and Stanier (1960) showed that the amount of acetate assimilated by *Chlorobium* is strictly proportional to either the sulfide or the bicarbonate concentration when either is growth-limiting. The inability to photoassimilate simple organic substrates without sulfide or to use them as electron donors for CO_2 assimilation shows that in difference from the purple bacteria, the Chlorobiaceae have a purely anabolic intermediary carbon metabolism.

Although low activities of the key enzymes of the reductive pentose phosphate cycle were obtained in cell-free extracts of *Chlorobium* strains (Smillie *et al.*, 1962; Tabita *et al.*, 1974; Quandt, 1975) the contribution of this cycle to total CO_2 assimilation appears to be of minor importance in these organisms. This lesser importance is another significant physiological difference between the green and purple bacteria; in the latter, more than 50% of the CO_2 is fixed via this cycle. Ferredoxin-dependent carboxylation reactions catalyzed by pyruvate- and α-keto-glutarate synthase (Buchanan *et al.*, 1967) were shown to occur in green and purple bacteria. These reactions, together with other carboxylic acid cycle enzymes, were demonstrated to be of primary importance for CO_2 fixation in the green bacteria (Buchanan *et al.*, 1972). A reductive carboxylic acid cycle was therefore postulated as the main port of entry of CO_2 in *Chlorobium* (Evans *et al.*, 1966; Sirevag and Ormerod, 1970; Sirevag, 1974). However, no activity of citrate lyase, the key enzyme of this cycle, could be demonstrated in *Chlorobium*. On the contrary, citrate synthase was found to be active (Beuscher and Gottschalk, 1972). The cyclic regeneration of acetyl-CoA in these organisms therefore remains to be elucidated experimentally.

(c) Enrichment Conditions and Habitats. On the basis of their selective advantage in enrichment cultures, two physiological–ecological subgroups have been differentiated among the green and brown sulfur bacteria. The first subgroup includes the green and brown species of the genera *Chlorobium* and *Prosthecochloris*, which are selectively enriched at medium-to-high sulfide concentrations (4–8 mM) and at light intensities around or above light saturation (700–1500 lux). The second subgroup comprises the gas-vacuole-containing species of the genera *Pelodictyon* and *Ancalochloris* and their brown-colored counterparts. These forms outgrow members of the first group at low light intensities (50–100 lux, near the lower limit), low sulfide concentrations (0.4–2 mM), and low incubation temperatures (10–20°C).

Members of the first group inhabit predominantly the upper layers of the sulfide-rich black mud in freshwater and estuarine environments. Natural enrichments of the gas-vacuole-containing species are the green- or reddish-brown-colored blooms in the upper layer of the sulfide-rich hypolimnion of stratified holomictic or meromictic* lakes (Gorlenko and Lebedeva, 1971; Gorlenko, 1972). Apparently, the gas vesicles allow them to adopt this narrowly circumscribed position in the water gradient, where both sulfide from below and light from above are simultaneously available.

In the form of the ectosymbiotic association (consortium) with a motile, colorless central bacterium, the nonmotile green and brown *Chlorobium* cells attain the capacities of the motile purple sulfur bacteria. Both "*Chlorochromatium*" and "*Pelochromatium*" consortia exhibit chemo- and phototactic responses to the environmental conditions when they accumulate in optimum sulfide concentration and light intensity. In enrichment cultures, the consortia have a selective advantage at low sulfide concentration and low light intensity. In nature, they have been found only in freshwater habitats; they may occur both in shallow pools above the mud and in the layers of green or purple sulfur bacteria in stratified lakes.

(d) Position in the Ecosystem and Its Reasons. In stagnant ponds, lakes, or estuarine habitats with vertical gradients of both light (from above) and hydrogen sulfide (from below), the green sulfur bacteria regularly form the lowermost layer of phototrophic organisms in the muddy or sandy sediment or stratified water underneath the layers of purple sulfur bacteria and algae (Baas Becking and Wood, 1955; Fenchel, 1969; Fenchel and Straarup, 1971; Kusnezow and Gorlenko, 1973; Jørgensen and Fenchel, 1974). That they do can be understood when the

* See footnote on p. 4.

following special features of the Chlorobiaceae are considered: (1) The cells are nonmotile. (2) The cells are obligately phototrophic and sulfide-dependent. (3) Photosynthetic electron donor cannot be stored in the form of elemental sulfur inside the cells. (4) Photosynthesis reaches light saturation at lower light intensities (700 lux; Lippert and Pfennig, 1969) than in the purple sulfur bacteria (1000–2000 lux; Trüper and Schlegel, 1964). (5) The spectrum of light absorbed by the Chlorobiaceae (700–760 nm) is not absorbed by algae, cyanobacteria (which absorb below 690 nm), or purple sulfur bacteria (which absorb above 800 nm). Consequently, the uppermost limit of the green bacterial growth layer must coincide with the level of permanent sulfide production in the habitat that is not exhausted even at the time of maximum photosynthetic activity during the daylight period. The lowermost limit of the layer certainly reflects absolute light limitation. In contrast, the motile and sulfur-storing purple sulfur bacteria actively adjust themselves to the diurnal changes in sulfide concentration that they cause by their own photosynthetic activity during daylight: they move downward when sulfide is being consumed during the day, and they rise with the increasing sulfide concentrations in the dark. They are therefore able to grow in the layers of lower and changing sulfide content but higher light intensity above the stable green bacterial zone.

(e) Pronounced Capacity for Syntrophy and Its Functional and Structural Basis. An ecologically significant feature of the Chlorobiaceae is their ability to grow particularly well in syntrophic community with anaerobic organotrophic bacteria. Good examples of this syntrophy are the "*Chlorochromatium*" and "*Pelochromatium*" consortia. The syntrophic capacity, however, is by no means confined to the species that take part in these fairly stable ectosymbiotic associations, but rather is characteristic of the green bacteria in general, including *Chloroflexus aurantiacus*.

It has been shown that *Chlorobium* cultures excrete large amounts of organic substances into the growth medium. Only 75–85% of the radioactivity of $[^{14}C]CO_2$ assimilated was recovered in the cell material (Lippert, 1967), while 15–25% appeared as organic substances in the medium. Under natural conditions in lakes, excreted amounts of more than 30% were found during primary production measurements in *Chlorobium* blooms (Czeczuga and Gradzki, 1973). It is therefore not surprising that sulfate- and sulfur-reducing bacteria grow well in purely mineral media in the presence of Chlorobiaceae; the nutrients required by these organisms are provided by the Chlorobiaceae. While different kinds of syntrophic mixtures grow readily and exhibit an unusual longevity, growth and maintenance of pure cultures of Chlorobiaceae require special care.

Significant for the syntrophic ability is the structural organization of the cell and the particular metabolic functions associated with that organization. The sensitivity of the Chlorobiaceae to all kinds of unfavorable environmental conditions is comparable to that of the the sulfate-reducing and methane-forming bacteria. Important redox enzymes may be poorly protected, as was shown for hydrogenase in *Desulfovibrio gigas* by Bell *et al.* (1974). The photosynthetic apparatus and electron-transport system are confined to the periphery of the cell and do not extend into the central portion, as they do in the purple bacteria. A metabolic counterpart to the interspecies hydrogen transfer exists in the green sulfur bacteria in their interspecies sulfide and sulfur transfer.

A good example of this transfer is provided by the work of Gray *et al.* (1973). The presumed pure culture "*Chloropseudomonas ethylica* 2K" was shown by these authors to be a syntrophic mixture of an unknown ethanol- or acetate-utilizing anaerobic bacterium together with *Chlorobium*. Recently, another culture with the same designation was studied in detail (Pfennig, and Biebl, 1976). The green component proved to be a strain of *Prosthecochloris aestuarii*. The organotrophic companion is a gram-negative, motile, rod-shaped bacterium that oxidizes ethanol and acetate to CO_2, and at the same time reduces elemental sulfur to sulfide.This pink organism, *Desulfuromonas acetoxidans* (Pfennig and Biebl, 1976), shows a pure cytochrome *c* spectrum and is unable to reduce sulfate. It forms robust syntrophic mixtures with any member of the green and brown sulfur bacteria.

That catalytic amounts of sulfide, initially added to the syntrophic mixture together with ethanol or acetate, give rise to extremely dense cultures indicates that in the mixture, the green or brown sulfur bacterium grows exclusively by the oxidation of sulfide to sulfur; the elemental sulfur is redistributed in soluble form in the medium, where it is reduced back to sulfide by the sulfur-reducing companion. No sulfur globules appear and no sulfur is lost from the interspecies cycle by further oxidation to sulfate as long as the *Desulfuromonas* has acetate or ethanol as electron donor. The successful competition of *Desulfuromonas* over the green sulfur bacteria for the excreted elemental sulfur in the medium might be explained by the assumption that this organism has a higher affinity for elemental sulfur than the green sulfur bacteria.

It is not unreasonable to expect the enzymes of the photosynthetic electron-transport system to be located in the cytoplasmic membrane, i.e., to expect oxidation of electron-donor molecules not to take place inside the cell where the chlorobium vesicles are located. The latter appear to serve exclusively as powerful light-harvesting structures with an energy transfer that is directed back to the cytoplasmic membrane. A similar conclusion was reached by Pierson and Castenholz (1974*b*) for the facultatively aerobic green bacterium *Chloroflexus aurantiacus*. In this organism, an intimate contact of respiratory and photosynthetic electron carriers is indicated by the inhibition of oxygen uptake in the light. And since respiratory electron carriers are likely to be located in the cytoplasmic membrane, the authors concluded that "these results support the hypothesis that a portion of the photosystem is located on the cell membrane in *Chloroflexus*."

This view of the functional specialization of cytoplasmic membrane and chlorobium vesicles in photosynthesis is further supported by the analyses of Cruden and Stanier (1970), Fowler *et al.* (1971), Olson *et al.* (1976), and Fuller *et al.* (1976). The results of the latter three groups indicate that the electron-transport system with reaction center Bchl *a*, cytochromes, and carotenoids is located in the metabolically active cytoplasmic membrane, while the chlorobium vesicles, highly enriched with photochemically inactive Bchl *c*, serve a light-harvesting function. This specialization would explain the lack of photophosphorylation activity from isolated chlorobium vesicles (Lippert, 1967). For an evaluation of the special functional contribution of the vesicles, it will be most important to confirm the exact localization of the reaction center Bchl *a*.

2.1.2. The Filamentous Gliding Green Bacteria: Chloroflexaceae (*Chloroflexus aurantiacus*)

Until 1971, it seemed as though the phototrophic green bacteria with their typical chlorobium vesicles would exist only in the form of the nonmotile, physiologically somewhat restricted green and brown sulfur bacteria. The situation changed in 1971 when Pierson and Castenholz first described filamentous phototrophic bacteria of hot springs containing Bchl *c* and *a* in a ratio similar to *Chlorobium*; the γ- and β-carotene that are present are closely related to the carotenoids of *Chlorobium*. Further work on 16 isolates from hot springs in Japan, the United States, Iceland, and New Zealand (Pierson and Castenholz, 1974*a,b*) revealed that these thermophilic bacteria (temperature optimum: 50–60°C) with gliding motility like flexibacteria or Oscillatoriaceae possess typical chlorobium vesicles and carry out, under anaerobic conditions, an anoxygenic photosynthesis. Although *Chloroflexus* may be capable of photoautotrophic growth with sulfide as the electron donor (Madigan and Brock, 1975), it is primarily a photoorganotrophic organism. Yeast extract proved to be the best substrate; it could be replaced in part by glycerol or acetate, and to a lesser extent also by glucose, lactate, pyruvate, or glutamate. Under fully aerobic conditions, Bchl synthesis is repressed; the color of the cultures changes from dull green to orange, and the organism continues to grow chemoorganotrophically by respiration. No growth occurs under anaerobic conditions in the dark.

With this regulatory capacity—to thrive either under anaerobic conditions by anoxygenic photosynthesis or, when air (oxygen) is present, by respiration—*Chloroflexus* represents, among the green bacteria, a counterpart to the purple nonsulfur bacteria.

The natural habitats of *Chloroflexus* are almost all hot-springs areas over the world. When the springs are rich in sulfide, dull green to orange mats of *Chloroflexus* occur upstream and dissociated from cyanobacteria. When the springs have little or no sulfide, *Chloroflexus* forms nearly unispecific orange-colored gel-like mats several millimeters thick. These mats occur either underneath or overlying thinner layers of cyanobacteria, which provide the necessary organic carbon sources for the chemoorganotrophic growth of *Chloroflexus*. Pierson (1973) confirmed this capacity for syntrophic growth with mixed pure cultures. In the presence of *Synechococcus lividus*, *Chloroflexus* grew well on inorganic media, deriving the needed organic nutrients from the cyanobacterium.

2.2. The Purple Bacteria

2.2.1. General View of the Group

From a physiological point of view, the purple bacteria may be considered to represent developments parallel to the green bacteria. But while the latter are structurally conservative, lack flagellar motility, and show an extreme metabolic fastidiousness in both physiological groups, the purple bacteria apparently inherited a cellular organization of greater flexibility. It allowed not only the development of structural varieties and of photo- and chemotactically controlled flagellar motility, but also the acquisition of a greater metabolic versatility, resulting in a broader ecological valence. In addition to Bchl *a* or *b* (three species), a large number of different carotenoid pigments (Table 2, Groups 1–4) give rise to a spectrum of colors ranging from yellow and brown to red and purple-violet. All the different types of photopigment-bearing intracytoplasmic membrane systems originate in and are continuous with the cytoplasmic unit membrane, and may fill almost the entire volume of the cell. As a consequence, the photometabolism is not confined to the cell periphery, but extends into the inner part of the cells; that it does is obvious in the case of the purple sulfur bacteria by the globules of elemental sulfur transiently formed inside the cells. With these sulfur globules, the cells carry with them a large reservoir of photosynthetic electron donor for CO_2 fixation. Flagellar motility together with the capacity for positive and negative photo- and chemotactic responses to environmental conditions allow each individual cell to adjust to optimum growth conditions in a particular habitat (Clayton, 1957; Pfennig, 1967; Sorokin, 1970).

In comparison with the carbon metabolism of the green bacteria, that of the purple bacteria exhibits far greater capacities. In all species studied so far, the reductive pentose phosphate cycle is the major port of entry for CO_2 under photoautotrophic conditions; it also plays an important role under photoorganotrophic conditions (Slater and Morris, 1973), however, and may not be repressed to more than 60% by organic substrates (Hurlbert and Lascelles, 1963). Simple organic compounds may not only be incorporated as in the green sulfur bacteria, but also can be metabolized and serve as sources of reducing power. Substances that are more oxidized than cell material may, in part, become converted to CO_2 and reducing equivalents that are used to assimilate the substrate to cell material. When compounds that are more reduced than cell substance (e.g., higher fatty acids) are photoassimilated, the CO_2 fixation via the reductive pentose phosphate cycle and ferredoxin-dependent carboxylation reactions serves as a sink for the excess hydrogen present in the substrate.

2.2.2. Differentiating Characteristics of Chromatiaceae and Rhodospirillaceae

The differentiation of two physiological groups (the former Athiorhodaceae, now Rhodospirillaceae, and the former Thiorhodaceae, now Chromatiaceae) among the purple bacteria was introduced by Molisch (1907) and is still useful today, particularly from an ecological point of view. Today, the distinction is based on the ability or inability of the cells to use elemental sulfur as an electron donor for phototrophic CO_2 assimilation and to oxidize it to sulfate (Pfennig and Trüper, 1974). The sulfur may be formed either inside or outside the cells as an intermediate oxidation product of sulfide or thiosulfate, or it may be elemental sulfur of any other origin. All members of the purple sulfur bacteria, Chromatiaceae, are able to oxidize elemental sulfur to sulfate. The purple nonsulfur bacteria, Rhodospirillaceae, lack this capacity. When species of this group use sulfide or thiosulfate as an electron donor, they can either oxidize it to sulfate without intermediate formation of elemental sulfur (*Rhodo-*

pseudomonas palustris, Rp. sulfidophila), or oxidize it only to elemental sulfur, which is deposited outside the cells and cannot be oxidized further to sulfate (many species; Hansen and van Gemerden, 1972). There is another difference between the two families in their metabolism of sulfur compounds: the key enzyme of the dissimilatory sulfur metabolism of purple and green sulfur bacteria, adenylylsulfate (APS) reductase, has proved to be absent from all Rhodospirillaceae studied so far (Trüper and Peck, 1970; Hansen and Veldkamp, 1972; Kirchhoff and Trüper, 1974).

It is therefore not surprising that in natural habitats dominated by sulfide from either sulfur springs or the activity of sulfate-reducing bacteria, the green and purple sulfur bacteria have a selective advantage over the predominantly photoorganotrophic Rhodospirillaceae. The same is true, of course, for sulfide-containing enrichment cultures, in which, under specific, experimentally established conditions, certain species can be selectively enriched and isolated (Pfennig, 1967; van Niel, 1971; van Gemerden, 1974). The inability of the Rhodospirillaceae to compete successfully with the sulfur bacteria may also be due to the growth-inhibiting action of sulfide or the different growth factor requirements of both groups. Strains of the sulfur bacteria were found to require only vitamin B_{12} (Pfennig and Lippert, 1966), a growth factor produced in large amounts in their sulfide-rich, highly reduced natural habitats. The Rhodospirillaceae, on the other hand, depend on biotin, *p*-aminobenzoic acid, thiamine, and nicotinic acid; they reach reasonable growth rates only in the presence of simple organic carbon and complex organic nitrogen sources.

In connection with these physiological differences between the two groups of purple bacteria, it is additionally noteworthy that apparent differences also exist in the fine structure. All species of the Chromatiaceae, in which intracellular sulfur globules form, have the same vesicular type of intracytoplasmic membrane system occupying most of the cell volume. Only the Bchl-*b*-containing *Thiocapsa pfennigii* has groups of parallel tubes instead of vesicles. All members of *Ectothiorhodospira*, in which sulfur globules arise outside the cells, form lamellar membrane stacks similar to those present in the brown *Rhodospirilla* of the Rhodospirillaceae. In addition to this type, three more types of membrane systems have been encountered in the Rhodospirillaceae: (1) the vesicular type characteristic of the Chromatiaceae; (2) membranes parallel and underlying the cytoplasmic membrane; and (3) a few tubular intrusions in addition to the cytoplasmic membrane. The Rhodospirillaceae are therefore structurally as well as physiologically the most diverse group of the phototrophic bacteria (Lascelles, 1968; Drews, 1972).

2.2.3. The Purple Sulfur Bacteria: Chromatiaceae

(a) Metabolic Characteristics. The metabolic theme of the green sulfur bacteria—the linkage between the oxidation of sulfide and sulfur to sulfate and the fixation of CO_2 to cell material in an anaerobic, obligately phototrophic metabolism—is repeated in the purple sulfur bacteria. The 26 described species of ten genera differ with respect to their tolerance toward sulfide concentration and light intensity as well as their capacity to grow in the presence or absence of sulfide with simple organic substances. The more fastidious forms are the large *Chromatium* species (*Chr. okenii, Chr. weissei, Chr. warmingii, Chr. buderi*), *Thiospirillum*, and the gas-vacuole-containing genera *Lamprocystis, Thiodictyon,* and *Thiopedia*. These forms are strictly anaerobic and inhibited by oxygen; they depend on sulfide, since they lack an assimilatory sulfate reduction. Acetate and often pyruvate as well are photoassimilated in the presence of sulfide and CO_2; vitamin B_{12} is required as a growth factor. The remaining species of the purple sulfur bacteria (small *Chromatium* species, *Thiocystis, Thiocapsa, Thiosarcina, Amoebobacter,* and *Ectothiorhodospira*) are usually capable of photoassimilating a wider range of organic substrates, including lower fatty acids, dicarboxylic acids, glycerol, and fructose. With the exception of *E. halophila*, which lacks assimilatory sulfate reduction, they can grow in the absence of sulfide if anaerobic reducing conditions are provided (e.g., by addition of ascorbate). Molecular hydrogen can usually serve as a photosynthetic electron donor for CO_2 fixation. The least fastidious species of this group, which is most resistant to organic pollution and oxygen, is *Thiocapsa roseopersicina*; it commonly forms pink to reddish blooms in sewage lagoons. The organism is even able to grow organotrophically under microaerophilic conditions in the dark. Under fully aerobic conditions, growth ceases after a few generations when the photopigments are diluted out, the synthesis of the pigments being repressed by oxygen.

As in the case of the green sulfur bacteria, two main physiological–ecological subgroups have been differentiated among the purple sulfur bacteria on the basis of their selective advantage in enrichment cultures (Pfennig, 1967). The first subgroup includes the motile and nonmotile members of the genera without gas vacuoles, such as *Chromatium, Thiocystis, Thiocapsa, Thiosarcina,* and *Ectothiorhodospira*; these forms have a selective advantage at medium-to-high sulfide concentrations (2–8 mM) and light intensities near light saturation (1000–2000 lux). The second subgroup comprise species that can be selectively enriched only at low sulfide concentrations (0.4–1 mM), nearly growth-limiting light intensities, and temperatures below 20°C. The large *Chromatium* species, *Thiospirillum*, and all gas-vacuole-containing species of the genera *Amoebobacter*, *Lamprocystis*, *Thiodictyon*, and *Thiopedia* belong to this subgroup. Like their green counterparts, these organisms form blooms in the uppermost layer of the sulfide-containing hypolimnion of stratified lakes. They may also be found, however, in the form of purple-violet layers on top of or below decaying leaves in the sulfide-containing clear water of shady forest ponds.

(b) Factors That Affect Cellular Arrangement, Motility, and Diurnal Migrations. A feature of major ecological significance shown by most species of the Chromatiaceae is their capacity to develop either in the form of single cells or in nonmotile cell aggregates of variable size embedded in slime. Adverse environmental conditions to which single cells are susceptible (e.g., air and light) may well be endured by the cells protected within the aggregates. Winogradsky (1888) first described the development of the cell aggregates for many species, using cell material collected in nature and observed in special slide cultures. With pure cultures, the different modes of growth may be obtained experimentally by the variation of sulfide concentration, pH, light regimen, light intensity, and temperature (Pfennig, 1967). At high sulfide concentrations (4–8 mM) and light intensities above saturation, the flagellated forms (*Thiospirillum*, *Chromatium*, *Thiocystis*, *Lamprocystis*) as well as the immotile species grow in the form of nonmotile cell colonies that are surrounded by slime and stick to each other and any other available surface. The cells are tightly packed and appear nearly polygonal to spherical; they are usually filled with large globules of elemental sulfur.

In nature, these growth forms are regularly found in the pink to purple-red blooms of Chromatiaceae in shallow pools and estuarine environments above black mud that are exposed to bright sunlight. Only at low sulfide concentrations (0.4–1 mM) and low light intensities do the cell aggregates start to grow loose, the individual cells of motile species begin to move, and the cell aggregates to disappear. Pure culture studies revealed that the various species differ with respect to the sulfide concentration and light intensity at which the cells are able to grow permanently in the motile or free stage. These studies provide an understanding for the fact that in natural habitats, the same species may be found in different growth stages depending on the local conditions of sulfide concentration and light intensity. In areas with direct sunlight, species of *Chromatium*, *Thiocystis*, or *Thiospirillum* often occur in slimy cell families forming purple-red layers on the bottom of the pond. In duckweed-covered areas of ponds or shady parts of forest ponds, the same organisms are found growing in reddish clouds of swarming individual cells exhibiting characteristic bioconvection patterns (Pfennig, 1962).

When, on a bright day, sulfide in the body of water becomes exhausted by the photosynthetic activity of these organisms, clouds of swarming cells can be seen to move closer and closer to the surface of the mud, and they eventually disappear completely in the upper layers. During the night, the sulfide concentration in the free water increases again, and the swarming purple sulfur bacteria rise. Such diurnal movements have been observed not only in shallow ponds, but also in stratified lakes. According to the measurements of Sorokin (1970), the purple-red layer of motile *Chromatium okenii* cells in the meromictic Lake Belovod showed diurnal vertical displacements of up to 2 m on a bright day. These vertical migrations in correspondence to the local growth conditions clearly show the broader ecological valence of the purple bacteria over the nonmotile Chlorobiaceae, which are rarely observed in mass development above the level of a fairly constant sulfide concentration.

2.2.4. The Purple and Brown Nonsulfur Bacteria: Rhodospirillaceae

(a) Capacity for Chemoorganotrophic Growth in the Dark. The cellular organization of the purple bacteria reaches its greatest structural and metabolic diversity in the physiological group of the purple nonsulfur bacteria. Like *Chloroflexus* among the green bacteria, most species of the Rhodospirillaceae have the capacity to thrive not only as anaerobic phototrophs, but also as facultatively microaerophilic to aerobic chemoorganotrophs. When grown anaerobically in the light, the cells of many species exhibit as much as about half the respiratory activity of fully adapted, aerobically, dark-grown cells. The respiration—not the growth—of such pigmented light-grown cells is partially inhibited by light, indicating that the respiratory and photosynthetic electron-transport systems compete for the electrons made available from the substrate. That many species (e.g., *Rp. capsulata*, *Rp. sphaeroides*, *Rp. gelatinosa*, *Rs. rubrum*) have the capacity for both types of energy generation without much regulatory change shows that these organisms are particularly adapted to thrive in habitats of variable oxygen tensions.

The tolerance for oxygen varies considerably in different species. For prolonged growth in the dark, the brown-colored *Rhodospirillum* species (*Rs. fulvum*, *Rs. molischianum*, and *Rs. photometricum*) depend on extreme microaerophilic conditions (3–6 torr P_{O_2}). Under these conditions, the cells are well pigmented and grow only slowly (38-hr doubling time). At higher oxygen tensions (10 torr P_{O_2}), the formation of the photosynthetic pigments and of electron-transport compounds is repressed. The capacity to form an additional oxidative electron-transport system, characteristic of most other Rhodospirillaceae, is lacking in these species. Instead, the brown-colored Rhodospirillaceae are able to respire only under extreme microaerophilic conditions in the dark, using the existing, but quantitatively slightly modified, electron-transport system (Lehmann, 1975). A similar situation appears to exist in *Rp. viridis*, the green-colored and Bchl-*b*-containing species of the group. Most other species grow well under aerobic conditions in the dark; the cultures are then faintly colored to colorless. Except for strains of *Rp. capsulata*, *Rp. sphaeroides*, and *Rs. rubrum*, the organisms may be injured in strongly aerated media either when small inocula are transferred to media lacking complex organic nutrients or when the carbon and energy source is depleted and growth has ceased. As soon as the partial pressure of oxygen is lowered to a certain level (about 20 torr P_{O_2}), the synthesis of photosynthetic pigments and the intracytoplasmic membrane system resumes, independently of the presence of light (Oelze and Drews, 1972).

(b) Photometabolism of Carbon Compounds and Ecological Niches. The purple nonsulfur bacteria are able to photoassimilate a wide variety of organic compounds; acetate, pyruvate, and the dicarboxylic acids are the best and universally used substrates. Some species also grow well on lower and higher fatty acids, methanol, or ethanol, while others are able to use sugars and sugar alcohols more readily. Complex organic nitrogen sources (e.g., yeast extract) are widely used and give increased growth rates. That most species have a requirement for one or several vitamins is in good agreement with the photoorganotrophic nature of these organisms; such compounds are always present in habitats dominated by the active breakdown of organic matter.

Photoautotrophic growth is shown by many species with molecular hydrogen as the electron donor in the presence of the necessary growth factors (Klemme, 1968). Sulfide and thiosulfate may be used effectively by the two species *Rp. sulfidophila* and *Rp. palustris*.

With the exception of the ability of *Rp. gelatinosa* to liquefy gelatin, the known purple nonsulfur bacteria characteristically lack the capacity to break down organic macromolecules such as starch, cellulose, pectin, chitin, neutral lipids, and proteins. In natural habitats, they therefore depend on the preceding activity of chemoorganotrophic bacteria capable of degrading such macromolecules. This

dependence may be one reason why Rhodospirillaceae are never seen in blooms comparable to those of the green and purple sulfur bacteria (van Niel, 1971); instead, they are regularly found in relatively high proportions together with chemoorganotrophic bacteria (Kaiser, 1966; Biebl, 1973). Particularly when the active breakdown of plant residues results in anaerobic conditions, low-molecular-weight breakdown products are liberated and provide growth conditions for Rhodospirillaceae. Such a situation is likely to occur in the mud of eutrophic ponds and ditches, the littoral zone of lakes, rivers, and the sea, and in all kinds of sewage lagoons. The most commonly encountered species in nature are *Rp. gelatinosa* and *Rp. palustris* (Biebl and Drews, 1969).

3. Position of the Phototrophic Bacteria in the Cycle of Matter

3.1. Ecological Niches in Aquatic Ecosystems and the Anaerobic Cycle of Matter

In the preceding sections, the major metabolic features of all subgroups of the green and purple bacteria were considered in connection with particular habitats that provide the necessary conditions for their development. The ecological niches in which these aquatic organisms compete successfully with other microorganisms are characterized by two features: (1) the presence of light and (2) the microbial breakdown of organic matter under conditions of oxygen limitation or anaerobiosis. In stagnant bodies of water, the decomposition of plant and animal residues proceeds in different stages in vertical zonation accompanied by different redox potentials. The conditions are aerobic at first, and oxygen-consuming mineralization processes are dominant (+450 to +200 mV). As a consequence, in the photosynthetically inactive deeper layers, oxygen gradually becomes exhausted and a fermentative degradation takes place (+200 to +100 mV or less); reduced fermentation products accumulate. At this stage, the facultatively aerobic, photoorganotrophic, purple nonsulfur bacteria find their ecological niche: they photoassimilate the low-molecular-weight breakdown products into cell material. In the absence of light, these products are metabolized by anaerobically respiring bacteria using nitrate and sulfate instead of oxygen as ultimate electron acceptors. With the formation of H_2S by the sulfate-reducing bacteria, the highly reducing, oxygen-free conditions (+100 to −250 mV and lower) under which hydrogen-evolving and methane-forming bacteria develop become established; the latter are able to use CO_2 as an electron acceptor for their metabolism of hydrogen or highly reduced substances (−400 mV). At this end of the chain of anaerobic fermentative and respiratory processes, all further possibilities for microbial energy generation and growth are exhausted. The accumulating products can be metabolized only when either oxygen or light for energy-yielding reactions gains access to the habitat. Anaerobically in the light, phototrophic green and purple sulfur bacteria find their adequate growth conditions without any competitors; they metabolize the available end products, including H_2S, H_2, CO_2, and small amounts of reduced organic substances. The toxic sulfide is already photooxidized to sulfate in the anaerobic part of the water; changes of E_h from −180 to +196 mV were measured (Fenchel, 1969). Carbon dioxide and simple organic substances are photoassimilated into cell material of the phototrophic bacteria.

When the biomass thus formed remains in the anaerobic part of the water, it becomes the substrate of the anaerobic microflora. New cleavage products will be formed that may in part be used by phototrophic bacteria again. These organisms therefore establish a small cycle of matter driven by the energy of light. Under anaerobic conditions, however, various organic compounds cannot be decomposed to products that can be photoassimilated again; these substances accumulate slowly in the mud (Czeczuga and Czerpak, 1968). Due to this incomplete recycling, the anaerobic microflora would gradually die off if its habitat did not regularly receive decomposable organic materials from the aerobic part of the water.

3.2. Contribution to Secondary Production and Sulfide Detoxification

It is important for the productivity of ponds and lakes that the biomass produced by the phototrophic bacteria from nutrients inaccessible under anaerobic conditions contributes significantly to the secondary production of these habitats. Above the layers of phototrophic sulfur bacteria, high population densities of protozoa, rotatoria, cladocera, and copepoda were observed; the intestines of these organisms were filled with the phototrophic bacteria on which they were feeding (Culver and Brunskill, 1969; Sorokin, 1970; Fenchel, 1969; Hayden, 1972). During their diurnal vertical migrations, the zooplankton organisms rise to the fully aerobic parts of the water, where they may become the prey of larger animals and fish. This conversion of substances of the anaerobic sulfide-containing zone into photosynthetically grown cell material that enters the food chain of the aerobic zone is particularly important in permanently stratified meromictic lakes.

The contribution of the primary production by phototrophic sulfur bacteria to the total productivity has been determined in different kinds of lakes. In Table 3, values of milligrams of carbon assimilated under 1 m² of lake surface area are summarized for five lakes of the meromictic type. The contribution of the phototrophic bacteria ranges between 20 and a maximum 85% of the total daily production. These values are interesting not only in view of the food value of the phototrophic bacteria for the secondary production in the lake, but also in consideration of

Table 3. Productivity of Phototrophic Sulfur Bacteria in Meromictic Lakes

Lake	References	Daily productivity rate under 1 m² of lake surface area	
		mg carbon/m² · day	Percentage of total production (phytoplankton and photosynthetic bacteria)
Belovod	Lyalikova (1957)	180	40
	Sorokin (1970)	110	20
Suigetsu	Takahashi and Ichimura (1968)	45	20
Kisaratsu Reservoir	Takahashi and Ichimura (1968)	800	60
Medicine Lake	Hayden (1972)	190	55
Fayetteville Green Lake	Culver and Brunskill (1969)	2470	85

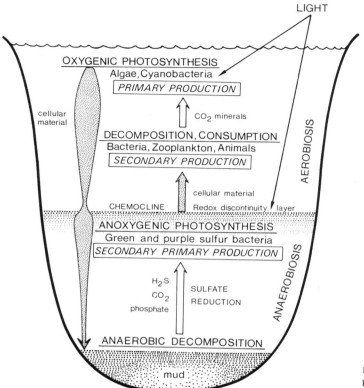

Fig. 1. Production, consumption, and decomposition in aquatic ecosystems with a redox discontinuity layer.

the amount of toxic sulfide that became photooxidized anaerobically in the course of CO_2 assimilation.

Culver and Brunskill (1969) in Green Lake and Hayden (1972) in Medicine Lake obtained carbon assimilation rates of 240 g and 30 g carbon/m² per year, respectively, in the layers of the phototrophic bacteria. When the photosynthetically active areas of these lakes (Green Lake, 0.25 km²; Medicine Lake, 0.125 km²) are taken into account, 60 and 3.75 tons, respectively, of carbon are assimilated in the lake per year. Since the phototrophic sulfur bacteria convert 1.4 g hydrogen sulfide to sulfate for every 1 g carbon assimilated, it can be calculated that in the course of a year, these organisms photooxidize during their growth in the lake 84 tons (Green Lake) and 5.25 tons (Medicine Lake) of toxic sulfide to sulfate.

The different types of production, consumption, and decomposition processes discussed in the preceding sections for aquatic ecosystems with a redox discontinuity layer are schematically represented in Fig. 1. Fenchel (1969) pointed out that such a model may be equally relevant for stagnant lakes, ponds, and marine sediments with important anaerobic decomposition, i.e., on either a large or a very small scale. For the sake of simplicity, the production of

chemoautotrophic sulfur bacteria that was shown to occur at the chemocline (Fenchel, 1969; Sorokin, 1970) was omitted. The phototrophic bacteria compete successfully with them for the available sulfide as long as the light intensity at the chemocline is not the primary growth-limiting factor (Jørgensen and Fenchel, 1974; Pfennig, 1975).

ACKNOWLEDGMENT

I thank Harry A. Douthit for his reading of the manuscript.

4. References

Avron, M., 1967, Mechanism of photoinduced electron transport in isolated chloroplasts, *Curr. Top. Bioenerg.* **2**:1–19.

Baas Becking, L. G. M., and Wood, E. J. F., 1955, Biological processes in the estuarine environment I–II: Ecology of the sulfur cycle, *Proc. K. Ned. Akad. Wet. Ser. B* **58**:160–181.

Bell, R., Le Gall, J., and Peck, H. D., 1974, Evidence for the periplasmic location of hydrogenase in *Desulfovibrio gigas*, *J. Bacteriol.* **120**:994–997.

Beuscher, N., and Gottschalk, G., 1972, Lack of citrate lyase—the key enzyme of the reductive carboxylic acid cycle—in *Chlorobium thiosulfatophilum* and *Rhodospirillum rubrum*, *Z. Naturforsch.* **27b**:967–973.

Biebl, H., 1973, Die Verbreitung der schwefelfreien Purpurbakterien im Plussee und anderen Seen Ostholsteins, Thesis, University of Freiburg.

Biebl, H., and Drews, G., 1969, Das *in vivo* Spektrum als taxonomisches Merkmal bei Untersuchungen zur Verbreitung der Athiorhodaceae, *Zentralbl. Bacteriol. Parasitenkd. Infektionskr. Hyg. Abt. 2* **123**:425–452.

Biebl, H., and Pfennig, N., 1978, Growth yields of green sulfur bacteria in mixed cultures with sulfur and sulfate reducing bacteria, *Arch. Microbiol.* **117**:9–16.

Bosecker, K., Drews, G., and Tauschel, H. D., 1972, Untersuchungen zur Adsorption des Bakteriophagen Rp1 an *Rhodopseudomonas* 1e5, *Arch. Mikrobiol.* **87**:139–148.

Buchanan, B. B., and Sirevag, R., 1976, Ribulose 1,5-diphosphate carboxylase and *Chlorobium thiosulfatophilum*, *Arch. Microbiol.* **109**:15–20.

Buchanan, B. B., Evans, M. C. W., and Arnon, D. I., 1967, Ferredoxin-dependent carbon assimilation in *Rhodospirillum rubrum*, *Arch. Mikrobiol.* **59**:32–40.

Buchanan, B. B., Schurmann, P., and Shaumugam, K. T., 1972, Role of the reductive carboxylic acid cycle in a photosynthetic bacterium lacking ribulose 1,5-diphosphate carboxylase, *Biochim. Biophys. Acta* **283**:136–145.

Clayton, R. K., 1957, Phototaxis of purple bacteria, in: *Encyclopedia of Plant Physiology*, Vol. 17/1 (W. Ruhland, ed.), pp. 371–387, Springer-Verlag, Berlin.

Cohen, Y., Padan, E., and Shilo, M., 1975, Facultative bacteria-like photosynthesis in the blue-green alga *Oscillatoria limnetica*, *J. Bacteriol.* **123**:855–861.

Cohen-Bazire, G., 1963, Some observations on the organization of the photosynthetic apparatus in purple and green bacteria, in: *Bacterial Photosynthesis* (H. Gest, A. San Pietro, and L. P. Vernon, eds.), pp. 89–110, Antioch Press, Yellow Springs, Ohio.

Cohen-Bazire, G., Pfennig, N., and Kunisawa, R., 1964, The fine structure of green bacteria, *J. Cell Biol.* **22**:207–225.

Cohen-Bazire, G., Kunisawa, R., and Pfennig, N., 1969, Comparative study of the structure of gas vacuoles, *J. Bacteriol.* **100**:1049–1061.

Conrad, R., and Schlegel, H. G., 1977, Differential degradation pathways for glucose and fructose in *Rhodopseudomonas capsulata*, *Arch. Microbiol.* **112**:39–48.

Cruden, D., and Stanier, R. Y., 1970, The characterization of chlorobium vesicles and membranes isolated from green bacteria, *Arch. Mikrobiol.* **72**:115–134.

Culver, D. A., and Brunskill, G. J., 1969, Fayetteville Green Lake, New York. V. Studies of primary production and zooplankton in a meromictic lake, *Limnol. Oceanogr.* **14**:862–873.

Czeczuga, B., and Czerpak, R., 1968, Investigations on vegetable pigments in post-glacial bed sediments of lakes, *Hydrologie* **30**:217–231.

Czeczuga, B., and Gradzki, F., 1973, Relationship between extracellular and cellular production in the sulphuric green bacterium *Chlorobium limicola* Nads. as compared to primary production of phytoplankton, *Hydrobiologia* **42**:85–95.

Doetsch, R. N., 1971, Functional aspects of bacterial flagellar motility, *CRC Crit. Rev. Microbiol.*, pp. 73–103.

Drews, G., 1972, Die photosynthetischen Bakterien als Modelle zur Untersuchung von zellulären Differenzierungsprozessen, *Naturwiss. Rundsch.* **25**:213–220.

Dubinina, G. A., and Gorlenko, V. M., 1975, New filamentous photosynthetic green bacteria containing gas vacuoles, *Mikrobiologiya* **44**:511–517.

Evans, M. C. W., Buchanan, B. B., and Arnon, D. I., 1966, A new ferredoxin-dependent carbon reduction cycle in a photosynthetic bacterium, *Proc. Natl. Acad. Sci. U.S.A.* **55**:928–934.

Fenchel, T., 1969, The ecology of marine microbenthos. IV. Structure and function of the benthic ecosystem, *Ophelia* **6**:1–182.

Fenchel, T., and Staarup, B. J., 1971, Vertical distribution of photosynthetic pigments and the penetration of light in marine sediments, *Oikos* **22**:172–182.

Fowler, C. F., Nugent, N. A., and Fuller, R. C., 1971, The isolation and characterization of a photochemically active complex from *Chloropseudomonas ethylica*, *Proc. Natl. Acad. Sci. U.S.A.* **68**:2278–2282.

Fuller, R. C., Boyce, C. D., and Oyewole, S. H., 1976, The location of the photochemical reaction center on the cytoplasmic membrane in the green bacteria, *Abstracts of the International Conference on the Primary Electron Transport and Energy Transduction in Photosynthetic Bacteria*, Brussels.

Gloe, A., and Pfennig, N., 1974, Das Vorkommen von Phytol und Geranylgeraniol in den Bacteriochlorophyllen roter und grüner Schwefelbakterien, *Arch. Mikrobiol.* **96**:93–101.

Gloe, A., Pfennig, N., Brockmann, H., Jr., and Trowitzsch, W., 1975, A new bacteriochlorophyll from brown-colored Chlorobiaceae, *Arch. Microbiol.* **102**:103–109.

Gorlenko, V. M., 1972, A new species of phototrophic brown sulfur bacteria *Pelodictyon phaeum* nov. spec., *Mikrobiologiya* **41**:370–371.

Gorlenko, V. M., 1975, Characteristics of filamentous phototrophic bacteria, *Mikrobiologiya* **44**:756–758.

Gorlenko, V. M., and Lebedeva, E. V., 1971, New green sulfur bacteria with appendages, *Mikrobiologiya* **40**:1035–1039.

Gorlenko, V. M., and Pivovarova, T. A., 1977, On the belonging of blue green alga *Oscillatoria coerulescens* Gicklhorn, 1921, to a new genus of chlorobacteria *Oscillochloris* nov. gen., *Izv. Akad. Nauk SSSR, Ser. Biol., No. 3*, 396–409.

Gorlenko, V. M., Chebotarev, E. N., and Rachalkin, V. I., 1974, Participation of microorganisms in sulfur turnover in Pomyaretskoe Lake, *Mikrobiologiya* **43**:908–913.

Gray, B. H., 1977, Rejection of *Chloropseudomonas ethylica* as a nomina rejicienda. Request for an opinion, *Int. J. Syst. Bacteriol.* **27**:168.

Gray, B. H., Fowler, C. F., Nugent, N. A., Rigopoulos, N., and Fuller, R. C., 1973, Reevaluation of *Chloropseudo-*

monas ethylica strain 2K, *Int. J. Syst. Bacteriol.* **23**:256–264.

Gürgün, V., 1974, Untersuchungen über den anaeroben Dunkelstoffwechsel einiger Arten der phototrophen Purpurbakterien, Thesis, University of Göttingen.

Gürgün, V., Kirchner, G., and Pfennig, N., 1976, Vergärung von Pyruvat durch sieben Arten phototropher Purpurbakterien, *Z. Allg. Mikrobiol.* **16**:573–586.

Hansen, T. A., and Gemerden, H. van, 1972, Sulfide utilization by purple nonsulfur bacteria. *Arch. Mikrobiol.* **86**:49–56.

Hansen, T. A., and Veldkamp, H., 1972, *Rhodopseudomonas sulfidophila, nov. spec.*, a new species of the purple nonsulfur bacteria, *Arch. Mikrobiol.* **92**:45–58.

Hayden, J. F., 1972, A limnological investigation of a meromictic lake (Medicine Lake, South Dakota), Master's thesis, University of South Dakota, Vermillion.

Holt, S. C., Conti, S. F., and Fuller, R. C., 1966, Effect of light intensity on the formation of the photochemical apparatus in the green bacterium *Chloropseudomonas ethylicum, J. Bacteriol.* **91**:349–355.

Hurlbert, R. E., and Lascelles, J., 1963, Ribulose diphosphate carboxylase in Thiorhodaceae, *J. Gen. Microbiol.* **33**:445–458.

Imhoff, J. F., and Trüper, H. G., 1977, *Ectothiorhodospira halochloris* sp. nov., a new extremely halophilic phototrophic bacterium containing bacteriochlorophyll *b*, *Arch. Microbiol.* **114**:115–121.

Jensen, A., Aasmundrud, O., and Eimhjellen, K. E., 1964, Chlorophylls of photosynthetic bacteria, *Biochim. Biophys. Acta* **88**:466–479.

Jørgensen, B. B., and Fenchel, T., 1974, The sulfur cycle of a marine sediment model system, *Mar. Biol.* **24**:189–201.

Kaiser, P., 1966, Ecologie des bacteries photosynthetiques, *Rev. Ecol. Biol. Sol. T* **III**:409–472.

Katz, J. J., Strain, H. H., Harkness, A. L., Studier, M. H., Svec, W. A., Janson, T. R., and Cope, B. T., 1972, Esterifying alcohols in the chlorophylls of purple photosynthetic bacteria. A new chlorophyll, bacteriochlorophyll (gg), all-*trans*-geranyl-geranyl bacteriochlorophyllide *a, J. Am. Chem. Soc.* **94**:7938–7939.

Keister, D. L., and Fleischmann, D. E., 1973, Nitrogen fixation in photosynthetic bacteria, in: *Photophysiology* (A. Giese, ed.), Vol. 8, pp. 157–183, Academic Press, New York.

Kelly, D. P., 1974, Growth and metabolism of the obligate photolithotroph *Chlorobium thiosulfatophilum* in the presence of added organic nutrients, *Arch. Microbiol.* **100**:163–178.

Keppen, O. I., and Gorlenko, V. M., 1975, A new species of purple budding bacteria containing bacteriochlorophyll *b*, *Mikrobiologiya* **44**:258–264.

Kirchoff, J., and Trüper, H. G., 1974, Adenylylsulfate reductase of *Chlorobium limicola, Arch. Microbiol.* **100**:115–120.

Klemme, J., 1968. Untersuchungen zur Photoautotrophie mit molekularem Wasserstoff bei neuisolierten schwefelfreien Purpurbakterien, *Arch. Mikrobiol.* **67**:29–42.

Kohlmiller, E. F., and Gest, H., 1951, A comparative study of the light and dark fermentations of organic acids by *Rhodospirillum rubrum, J. Bacteriol.* **61**:269–282.

Kondratieva, E. N., Zhukov, V. G., Ivanosky, R. N., Petushkova, Yu. P., and Monosov, E. Z., 1976, The capacity of phototrophic sulfur bacterium *Thiocapsa roseopersicina* for chemosynthesis, *Arch. Microbiol.* **108**:287–292.

Künzler, A., and Pfennig, N., 1973, Das Vorkommen von Bacteriochlorophyll a_P and a_{Gg} in Stämmen aller Arten der Rhodospirillaceae, *Arch. Mikrobiol.* **91**:83–86.

Kusnezow, S. I., and Gorlenko, W. M., 1973, Limnologische und mikrobiologische Eigenschaften von Karsteen der A. S. R. Mari, *Arch. Hydrobiol.* **71**:475–486.

Larsen, H., 1953, On the microbiology and biochemistry of the photosynthetic green sulfur bacteria, photosynthetic apparatus, *Adv. Microb. Physiol.* **2**:1.

Larsen, H., 1953, On the microbiology and biochemistry of the photosynthetic green sulfur bacteria, *K. Nor. Vidensk. Selsk. Skr.*, No. 1, 119 pp.

Lascelles, J., 1968, The bacterial photosynthetic apparatus, *Adv. Microb. Physiol.* **2**:1.

Lauterborn, R., 1915, Die sapropelische Lebewelt, *Verh. Dtsch. Naturhist.-Med. Ver. Heidelberg*, N. F. 13, pp. 395–481.

Lehmann, H., 1976, Wachstumsphysiologische Untersuchungen und Charakterisierung des Elektronentransport-Systems fakultativ aerober und microaerober Rhodospirillaceae, Thesis, University of Göttingen.

Liaaen Jensen, S., 1965, Bacterial carotenoids. XVIII. Arylcarotenes from *Phaeobium, Acta Chem. Scand.* **19**:1025–1030.

Liaaen Jensen, S., 1967, Recent advances in the chemistry of natural carotenoids, *Pure Appl. Chem.* **14**:227–244.

Lippert, K. D., 1967, Die Verwertung von molekularem Wasserstoff durch *Chlorobium thiosulfatophilum*, Ph.D. thesis, University of Göttingen.

Lippert, K. D., and Pfennig, N., 1969, Die Verwertung von molekularem Wasserstoff durch *Chlorobium thiosulfatophilum, Arch. Mikrobiol.* **65**:29–47.

Lyalikova, N. N., 1957, A study of the assimilation of free carbon dioxide by purple bacteria in Lake Belovod, *Mikrobiologiya*, **26**:97–103.

Madigan, M. T., and Brock, T. D., 1975, Photosynthetic sulfide oxidation by *Chloroflexus aurantiacus*, a filamentous, photosynthetic gliding bacterium, *J. Bacteriol.* **122**:782–784.

Molisch, H., 1907, *Die Purpurbakterien nach neuen Untersuchungen*, Fischer Verlag, Jena, 95 pp.

Oelze, J., and Drews, G., 1972, Membranes of photosynthetic bacteria, *Biochim. Biophys. Acta* **265**:209–239.

Olson, J. M., Prince, R. C., and Brune, D. C., 1976, The reaction center of green bacteria and its relation to bacteriochlorophyll *a*-proteins and chlorobium vesicles, *Abstracts of the International Conference on the Primary Electron Transport and Energy Transduction in Photosynthetic Bacteria*, Brussels.

Padan, E., Ginzburg, D., and Shilo, M., 1970, The reproductive cycle of cyanophage LPP1-G in *Plectonema boryanum* and its dependence on photosynthesis and respiratory system, *Virology* **40**:514–521.

Pfennig, N., 1962, Beobachtungen über das Schwärmen von *Chromatium okenii, Arch. Mikrobiol.* **42**:90–95.

Pfennig, N., 1967, Photosynthetic bacteria, *Annu. Rev. Microbiol.* **21**:285–324.

Pfennig, N., 1975, The phototrophic bacteria and their role in the sulfur cycle, *Plant Soil* **43**:1–16.

Pfennig, N., 1977, Phototrophic green and purple bacteria: A comparative, systematic survey, *Annu. Rev. Microbiol.* **31**:275–290.

Pfennig, N., 1978, *Rhodocyclus purpureus* gen nov. and sp. nov., a ringshaped, vitamin B_{12}-requiring member of the family Rhodospirillaceae, *Int. J. Syst. Bacteriol.* **28**:283–288.

Pfennig, N., and Biebl, H., 1976, *Desulfuromonas acetoxidans gen. nov.* and *sp. nov.*, a new anaerobic, sulfur-reducing, acetate-oxidizing bacterium, *Arch. Microbiol.* **110**:3–12.

Pfennig, N., and Lippert, K. D., 1966, Über das Vitamin B_{12}-Bedürfnis phototropher Schwefelbakterien, *Arch. Mikrobiol.* **55**:245–256.

Pfennig, N., and Trüper, H. G., 1974, The phototrophic bacteria, in: *Bergey's Manual of Determinative Bacteriology*, 8th Ed. (R. E. Buchanan and N. E. Gibbons, eds.), pp. 24–64, The Williams & Wilkins Co., Baltimore.

Pierson, B. K., 1973, The characterization of gliding filamentous phototrophic bacteria, Ph.D. thesis, University of Oregon, Eugene.

Pierson, B. K., and Castenholz, R. W., 1974a, A phototrophic gliding filamentous bacterium of hot springs, *Chloroflexus aurantiacus, gen.* and *sp. nov.*, *Arch. Microbiol.* **100**:5–24.

Pierson, B. K., and Castenholz, R. W., 1974b, Studies of pigments and growth in *Chloroflexus aurantiacus*, a phototrophic filamentous bacterium, *Arch. Microbiol.* **100**:283–305.

Pivovarova, T. A., and Gorlenko, V. M., 1977, Fine structure of *Chloroflexus aurantiacus var. mesophilus* (nom. prof.) grown in the light under aerobic and anaerobic conditions, *Mikrobiologiya* **46**:329–334.

Puchkova, N. N., and Gorlenko, V. M., 1976, New brown chlorobacterium *Prosthecochloris phaeoasteroidea* nov. sp., *Mikrobiologiya* **45**:655–660.

Puchkova, N. N., Gorlenko, V. M., and Pivovarova, T. A., 1975, A comparative ultrastructural study of vibrioid green sulfur bacteria, *Mikrobiologiya* **44**:108–114.

Quandt, L., 1975, Versuche zum Nachweis von Ribulose-1,5-Diphosphat-Carboxylase in Chlorobium-Arten, Diplomarbeit, University of Göttingen.

Quandt, L., Gottschalk, G., Ziegler, H., and Stichler, W., 1977, Isotope discrimination by photosynthetic bacteria, *FEMS Microbiol. Lett.* **1**:125–128.

Sadler, W. R., and Stanier, R. Y., 1960, The function of acetate in photosynthesis by green bacteria, *Proc. Natl. Acad. Sci. U.S.A.* **46**:1328–1334.

Satoh, O., Hoshino, Y., and Kitamura, H., 1976, *Rhodopseudomonas sphaeroides forma denitrificans*, a denitrifying strain as a subspecies of *Rhodopseudomonas sphaeroides, Arch. Microbiol.* **108**:265–267.

Schmidt, K., Pfennig, N., and Liaaen Jensen, S., 1965, Carotenoids of Thiorhodaceae. IV. The carotenoid composition of 25 pure isolates, *Arch. Mikrobiol.* **52**:132–146.

Schön, G., 1969, Der Einfluss der Reservestoffe auf den ATP-Spiegel in Zellen von *Rhodospirillum rubrum* beim Übergang von aerober zu anaerober Dunkelkultur, *Arch. Mikrobiol.* **68**:40–50.

Sirevag, R., 1974, Further studies on carbon dioxide fixation in *Chlorobium, Arch. Microbiol.* **98**:3–18.

Sirevag, R., and Ormerod, J. G., 1970, Carbon dioxide-fixation in photosynthetic green sulfur bacteria, *Science* **169**:186–188.

Slater, J. H., and Morris, I., 1973, The pathway of carbon dioxide assimilation in *Rhodospirillum rubrum* grown in turbidostat continuous-flow culture, *Arch. Mikrobiol.* **92**:235–244.

Smillie, A. M., Rigopoulos, N., and Kelly, H., 1962, Enzymes of the reductive pentose phosphate cycle in the purple and in the green photosynthetic sulfur bacteria, *Biochim. Biophys. Acta* **56**:612–614.

Sorokin, Yu. I., 1970, Interrelations between sulphur and carbon turnover in meromictic lakes, *Arch. Hydrobiol.* **66**:391–446.

Sorokin, Yu. I., and Donata, N., 1975, On the carbon and sulfur metabolism in the meromictic Lake Faro (Sicily), *Hydrobiologia* **47**:241–252.

Stanier, R. Y., 1970, Some aspects of the biology of cells and their possible evolutionary significance, in: *Twentieth Symposium of the Society of General Microbiology* (H. P. Charles and B. C. J. G. Knight, eds.), pp. 1–38, University Press, Cambridge.

Stanier, R. Y., 1974a, The origins of photosynthesis in eukaryotes, in: *Twenty-fourth Symposium of the Society of General Microbiology* (M. J. Carlile and J. J. Skehel, eds.), pp. 219–240, University Press, Cambridge.

Stanier, R. Y., 1974b, The cyanobacteria, in: *Bergey's Manual of Determinative Bacteriology*, 8th Ed. (R. E. Buchanan and N. E. Gibbons, eds.), p. 22, The Williams & Wilkins Co., Baltimore

Stanier, R. Y., 1974c, The relationship between nitrogen fixation and photosynthesis, *Aust. J. Exp. Biol. Med. Sci.* **52**:3–20.

Stewart, W. D. P., 1973, Nitrogen fixation by photosynthetic microorganisms, *Annu. Rev. Microbiol.* **27**:283–316.

Tabita, F. R., McFadden, B. A., and Pfennig, N., 1974, D-Ribulose-1,5-biphosphate carboxylase in *Chlorobium thiosulfatophilum* Tassajara, *Biochim. Biophys. Acta* **341**:187–194.

Takahashi, M., and Ichimura, S., 1968, Vertical distribution and organic matter production of photosynthetic sulfur bacteria in Japanese lakes, *Limnol. Oceanogr.* **13**:644–655.

Trüper, H. G., 1976, Higher taxa of the phototrophic bacteria: *Chloroflexaceae* fam. nov., a new family for the gliding, filamentous, phototrophic "green" bacteria, *Int. J. Syst. Bacteriol.* **26**:74–75.

Trüper, H. G., and Genovese, S., 1968, Characterization of photosynthetic sulfur bacteria causing red water in Lake Faro (Messina, Sicily), *Limnol. Oceanogr.* **13**:225–232.

Trüper, H. G., and Peck, H. D. Jr., 1970, Formation of adenylylsulfate in phototrophic bacteria, *Arch. Mikrobiol.* **73**:125–142.

Trüper, H. G., and Schlegel, H. G., 1964, Sulfur metabolism in Thiorhodaceae. I. Quantitative measurements on growing cells of *Chromatium okenii, Antonie van Leeuwenhoek; J. Microbiol. Serol.* **30**:225–238.

Uffen, R. L., and Wolfe, R. S., 1970, Anaerobic growth of purple non-sulfur bacteria under dark conditions, *J. Bacteriol.* **104**:462–472.

van Gemerden, H., 1968, On the adenosine triphosphate generation of *Chromatium* in the darkness, *Arch. Mikrobiol.* **64**:118–124.

van Gemerden, H., 1974, Coexistence of organisms competing for the same substrate: An example among the purple sulfur bacteria. *Microb. Ecol.* **1**:104–119.

van Niel, C. B., 1932, On the morphology and physiology of the purple and green sulfur bacteria, *Arch. Mikrobiol.* **3**:1–112.

van Niel, C. B., 1971, Techniques for the enrichment, isolation, and maintenance of the photosynthetic bacteria, in: *Methods in Enzymology*, Vol. 23, Part A (A. San Pietro, ed.), pp. 3–28, Academic Press, New York.

Walsby, A. E., 1972, Structure and function of gas vacuoles, *Bacteriol. Rev.* **36**:1–32.

Winogradsky, S., 1888, *Beiträge zur Morphologie und Physiologie der Bakterien*, Vol. 1, *Zur Morphologie und Physiologie der Schwefelbacterien*, Verlag A. Felix, Leipzig, 120 pp.

Wolfe, R. S., and Pfennig, N., 1977, Reduction of sulfur by spirillum 5175 and syntrophism with *Chlorobium, Appl. Environ. Microbiol.* **33**:427–433.

CHAPTER 2

Taxonomy of the Rhodospirillales

Hans G. Trüper and Norbert Pfennig

1. The Order Rhodospirillales

The order Rhodospirillales is defined as comprising those bacteria that contain bacteriochlorophylls (Bchl's) and carry out an anoxygenic photosynthesis (Pfennig and Trüper, 1971c; 1974; Trüper, 1976). The two suborders of this order are the Rhodospirillineae and the Chlorobiineae.

The suborder Rhodospirillineae comprises those bacteria that contain Bchl a or b; these pigments are always located in intracytoplasmic membrane systems of different types continuous with the cytoplasmic membrane.

This suborder contains the families Rhodospirillaceae and Chromatiaceae.

The suborder Chlorobiineae comprises those phototrophic bacteria that contain Bchl c, d, or e. These pigments are located in non-unit-membrane bound, lens- to cigar-shaped organelles, the chlorobium vesicles, that underlie the cytoplasmic membrane. Besides these pigments, the Chlorobiineae contain Bchl a, usually in small amounts. The photosynthetic reaction centers of these organisms contain only Bchl a.

This suborder contains the families Chlorobiaceae and Chloroflexaceae (Trüper, 1976).

Hans G. Trüper · Institut für Mikrobiologie der Universität Bonn, Bonn, West Germany   Norbert Pfennig · Institut für Mikrobiologie der Gesellschaft für Strahlen- und Umweltforschung m.b.H., Göttingen, West Germany

2. The Suborder Rhodospirillineae

2.1. The Family Rhodospirillaceae

The characteristic property of this family (Pfennig and Trüper, 1971c) is the preference for photoorganoheterotrophic growth. Only two species grow on fairly high concentrations of sulfide in a photolithoautotrophic manner, namely, *Rhodopseudomonas sulfidophila* (Hansen and Veldkamp, 1973) and *Rp. sulfoviridis* (Keppen and Gorlenko, 1975). Some strains of other species are also able to utilize sulfide as an electron donor; however, they do so only at low sulfide concentrations, and preferably in continuous cultures.

None of the Rhodospirillaceae tested so far contains adenylyl sulfate reductase (APS reductase; EC 1.8.99.2). Those growing with sulfide form either elemental sulfur or sulfate or tetrathionate as the end product of sulfide oxidation. All species except *Rhodopseudomonas globiformis* and *Rp. sulfoviridis* can utilize sulfate as sulfur source.

Most species require one or more vitamins.

The abilities to utilize hydrogen gas as an electron donor and to fix molecular nitrogen are rather common in this family.

Five species divide by budding: *Rhodopseudomonas palustris*, *Rp. viridis*, *Rp. acidophila*, *Rp. sulfoviridis*, and *Rhodomicrobium vannielii*; all other species divide by binary fission. In the first three of the species that divide by budding, a daughter cell develops at the end of a short, narrow division tube that arises (at least in the case of *Rp. palustris*) from

Table 1. Properties of the Rhodospirillaceae[a]

Species	Cell shape, width/length (μm)	Intracytoplasmic membrane system	DNA base ratio (mol G + C/dl)	Predominant carotenoids[b]	Color anaerobic culture	Quinones[c]	Aerobic or microaerophilic growth in the dark[d]	Required growth factors[e]
Rhodocyclus purpureus	Half-circle/circle, 0.6–0.7/2.7–5	Tubes	65.3	rl, rh	Purple-violet	Q_{10}	m	Vitamin B_{12} + paba + biotin
Rhodomicrobium vannielii	Ovoid and stalk, 1.0–1.2/2–2.8	Lamellae	61.8–63.8	sp, β-c	Orange-brown	ND	m	None
Rhodopseudomonas								
acidophila	Rod, 1.0–1.3/2–5	Lamellae	62.2–66.8	rh, rg, rlg	Purple-red or orange-brown	ND	m, ae	None
capsulata	Rod/sphere, 0.5–1.2/2–2.5	Vesicles	65.5–66.8	sn, se	Yellow to brown	Q_{10}	ae	Thiamine ± biotin ± niacin
gelatinosa	Rod, 0.4–0.5/1–2	Tubes	70.5–72.4	sn, se	Yellow-brown to pinkish	$Q_8 + MK_8$	ae	Biotin + thiamine
globiformis	Sphere, 1.6–1.8	Vesicles	66.3	kts	Purple-red	ND	m	Biotin + paba
palustris	Rod, 0.6–0.9/1.2–2	Lamellae	64.8–66.3	sp, ly, rh	Red-brown	Q_{10}	ae	Paba ± biotin
sphaeroides	Sphere/ovoid, 0.7/2–2.5	Vesicles	68.4–69.9	sn, se	Green-brown to brown	Q_{10}	ae	Biotin + thiamine + niacin
sulfidophila	Rod/sphere, 0.6–0.9/0.9–2.0	Vesicles	67.0–71.0	sn, se	Yellow-brown to red	ND	ae	Biotin + thiamine + niacin + paba
sulfoviridis	Rod, 0.5–0.9/1.2–2	Lamellae	67.8–68.4	neu, sp	Olive green	ND	m	Biotin + pyridoxine + paba
viridis	Rod, 0.6–0.9/1.2–2	Lamellae	66.3–71.4	2H-neu, 2H-ly	Green	$Q_9 + MK_9$	m	Paba + Biotin
Rhodospirillum								
fulvum	Spiral, 0.5–0.7/3.5	Stacks	64.3–65.3	ly, rh	Brown	$Q_9 + MK_9$	m	Paba
molischianum	Spiral, 0.7–1.0/5–8	Stacks	61.7–64.8	ly, rh	Brown	$Q_9 + MK_9$	m	Amino acids
photometricum	Spiral, 1.2–1.5/7–10	Stacks	65.8	ly, rh	Brown	ND	m	Yeast extract
rubrum	Spiral, 0.8–1.0/7–10	Vesicles	63.8–65.8	sp	Red	$Q_{10} + RK$	ae	Biotin
tenue	Spiral, 0.3–0.5/3–6	Tubes	64.8	ly, rh, rl	Purple-violet or brown-orange	$Q_8 + MK_8$	ae	None

[a] The data were collected from Hansen and Veldkamp (1963), Keppen and Gorlenko (1975), Mandel *et al.* (1971), Maroc *et al.* (1968), Pfennig (1974), Pfennig and Trüper (1971*d*, 1974), and van Niel (1944).

[b] (β-c) β-Carotene (61); (2H-ly) 1,2-dihydrolycopene (69); (2H-neu) 1,2-dihydroneurosporene (71); (kts) diketo-tetrahydrospirilloxanthin (20); (ly) lycopene (67); (neu) neurosporene (70); (rg) rhodopin glucoside (51); (rh) rhodopin (50); (rl) rhodopinal-D glucoside (2); (rlg) rhodopinal-D glucoside (3); (se) spheroidene (28); (sn) spheroidenone (23); (sp) spirilloxanthin (37). The numbers in parentheses are the serial numbers of the carotenoids in Chapter 12, Table 1.

[c] (MK) Menaquinone; (Q) ubiquinone; (RK) rhodoquinone. The subscript numbers indicate isoprenoid chain lengths. (ND) Not determined.

[d] (ae) Aerobic; (m) microaerophilic.

[e] Required by some strains.

the pole opposite that bearing the flagellum. In *Rp. sulfoviridis*, a division tube is not seen; the daughter cell buds directly from the mother cell. In this species, the daughter cell may be either motile or nonmotile and embedded in slime. The division tubes of *Rhodomicrobium vannielii* can arise from both poles of the mother cell; they are considerably longer than in the other species and are occasionally branched. As a

Table 2. Electron Donors and Carbon Sources Used by the Rhodospirillaceae[a]

Donor/source	Rc. purpureus	Rm. vannielii	Rp. acidophila	Rp. capsulata	Rp. gelatinosa	Rp. globiformis	Rp. palustris	Rp. sphaeroides	Rp. sulfidophila	Rp. sulfoviridis	Rp. viridis	Rs. fulvum	Rs. molischianum	Rs. photometricum	Rs. rubrum	Rs. tenue
Acetate	+	+	+	+	+	−	+	+	+	+	+	+	+	+	+	+
Arginine	−	−	−	0	0	−	−	0	−	−	0	−	−	−	+	+
Aspartate	−	−	−	+	0	−	−	0	±	0	0	±	±	±	+	−
Benzoate	+	−	−	−	−	−	+	−	−	−	−	+	−	−	−	−
Butyrate	−	+	+	+	±	−	+	±	+	+	−	+	+	+	+	+
Caproate	+	+	±	+	−	−	+	±	+	0	−	+	+	−	+	+
Caprylate	−	+	−	+	0	−	+	±	+	0	0	+	+	−	0	+
Citrate	−	−	+	−	+	−	−	+	−	−	−	−	−	−	−	−
Ethanol	−	+	+	−	+	+	+	+	±	+	+	+	+	+	+	±
Formate	−	−	±	0	±	−	+	0	+	−	−	0	0	−	−	−
Fructose	−	−	−	+	0	+	−	+	±	+	0	−	−	+	±	−
Fumarate	+	+	+	+	+	+	+	+	0	+	0	+	+	+	+	+
Gluconate	0	0	0	−	0	+	0	+	0	0	0	0	0	0	−	0
Glucose	−	−	±	+	+	+	−	+	+	+	+	±	−	+	−	−
Glutamate	−	−	−	+	0	−	+	0	+	0	+	0	0	−	+	0
Glycerol	−	−	±	−	−	−	+	+	0	+	−	−	−	+	−	−
Glycolate	−	−	±	0	0	−	+	0	0	0	0	0	−	+	0	−
Lactate	−	+	+	+	+	−	+	+	+	+	−	−	±	+	+	+
Malate	+	+	+	+	+	+	+	+	+	+	+	+	+	+	+	+
Malonate	−	+	±	0	0	−	+	0	0	−	0	0	−	0	0	−
Mannitol	−	−	−	−	±	+	−	+	−	−	0	−	+	−	−	−
Methanol	−	±	±	0	±	−	−	0	−	0	0	0	−	−	±	±
Pelargonate	−	−	−	+	0	−	−	±	+	0	0	+	+	±	0	+
Propionate	−	+	+	+	±	−	+	±	+	−	−	+	+	±	+	+
Pyruvate	+	+	+	+	+	±	+	+	+	+	+	+	+	+	+	+
Succinate	−	+	+	+	+	+	+	+	+	+	−	+	+	+	+	+
Tartrate	−	−	±	−	±	+	−	+	−	−	0	0	−	−	−	−
Valerate	−	+	+	+	0	−	+	±	+	−	−	+	+	+	+	+
Yeast extract	−	+	+	+	+	+	+	+	+	0	0	+	+	+	+	+
Casamino acids	−	+	+	+	0	−	+	0	+	+	+	−	−	+	+	+
Hydrogen (H₂)	+	+	+	+	±	0	+	+	+	0	−	+	0	0	+	+
Sulfide	0	(+)	0	(+)	0	0	(+)	0	+	+	0	0	0	0	(+)	0
Thiosulfate	−	−	−	−	−	−	+	−	+	+	−	−	−	−	−	−

[a] The data were collected from Hansen and Veldkamp (1973), Keppen and Gorlenko (1975), Pfennig (1974), and Pfennig and Trüper (1974).

[b] Genera: (Rc.) *Rhodocyclus*; (Rm.) *Rhodomicrobium*; (Rp.) *Rhodopseudomonas*; (Rs.) *Rhodospirillum*. Symbols: + = utilized; ± = utilized by some strains; (+) = utilized at low concentrations, preferably in continuous cultures; − = not utilized; 0 = not tested.

Table 3. Properties of the Chromatiaceae[a]

Species	Cell shape, width/length (μm)[b]	DNA base ratio (mol G + C/dl)[b]	Motility	Slime	Cell aggregates	Gas vacuoles	Predominant carotenoids[c]	Color of cell suspension	Hydrogenase activity[b]
Amoebobacter									
pendens	Sphere, 1.5–2.5	65.3	−	+	−	+	sp	Pink-red	+
roseus	Sphere, 2.0–3.0	64.3	−	+	−	+	sp	Pink-red	+
Chromatium									
buderi	Rod, 3.5–4.5/4.5–9	62.2–62.8	+	−	−	−	rl	Purple-violet	−
gracile	Rod, 1.0–1.3/2–6	68.9–70.4	+	+	−	−	sp, ly, rh	Brown-red	+
minus	Rod, 2.0/2.5–6	52.0–62.2	+	−	−	−	ok	Purple-red	+
minutissimum	Rod, 1.0–1.2/2.0	63.7	+	−	−	−	sp, ly, rh	Brown-red	+
okenii	Rod, 4.5–6.0/8–15	48.0–50.0	+	−	−	−	ok	Purple-red	−
vinosum	Rod, 2.0/2.5–6	61.3–66.3	+	−	−	−	sp, ly, rh	Brown-red	+
violascens	Rod, 2.0/2.5–6	61.8–64.3	+	−	−	−	rl	Brown-red	+
warmingii	Rod, 3.5–4.0/5–11	55.1–60.2	+	−	−	−	rl	Purple-violet	+
weissei	Rod, 3.5–4.0/5–11	48.0–50.0	+	−	−	−	ok	Purple-violet	−
Ectothiorhodospira									
halophila	Spiral, 0.8/5.0	68.4	+	−	−	−	sp	Red	ND
mobilis	Spiral, 0.7–1.0/2–2.6	67.3–69.0	+	−	−	−	sp, rh	Brown-red	+
shaposhnikovii	Spiral, 0.8–0.9/1.5–2.5	62.3	+	−	−	−	sp, rh	Brown-red	+
Lamprocystis									
roseopersicina	Sphere, 3.0–3.5	63.8	+	−	Clumps	+	la, lo	Purple	−

Thiocapsa								
pfennigii	Sphere, 1.2–1.5	69.4–69.9	–	–	–	–	ts	Orange-brown	+
roseopersicina	Sphere, 1.2–3.0	63.3–66.3	–	+	–	–	sp	Pink-red	+
Thiocystis								
gelatinosa	Sphere, 3.0	61.3	+	+	–	–	ok	Purple-red	+
violacea	Sphere, 2.5–3.0	62.8–67.9	+	+	Clumps	–	rl	Purple-violet	+
Thiodictyon								
bacillosum	Rod, 1.5–2.0/3–6	66.3	–	–	Clumps	+	rl, rh	Purple-violet	–
elegans	Rod, 1.5–2.0/3–8	65.3	–	–	Nets	+	rl, rh	Purple-violet	–
Thiopedia								
rosea	Ovoid, 1.0–2.0/1.2/2.5	ND	–	–	Platelets	+	ok	Purple-violet	–
Thiosarcina								
rosea	Sphere, 2.0–3.0	ND	–	+	Packets	–	sp	Pink-red	+
Thiospirillum								
jenense	Spiral, 2.5–4.5/30–40	45.5	+	–	–	–	ly, rh	Orange-brown	–

[a] The data were collected from Mandel *et al.* (1971) and Pfennig and Trüper (1971a,d, 1974).

[b] (ND) Not determined.

[c] (la) Lycopenal (1); (lo) lycopenol (56); (ly) lycopene (67); (ok) okenone (12); (rh) rhodopin (50); (rl) rhodopinal (2); (sp) spirilloxanthin (37); (ts) 3,4,3′,4′-tetrahydro-spirilloxanthin (39). The numbers in parentheses are the serial numbers of the carotenoids in Chapter 12, Table 1.

consequence, extensive networks of cells connected by division tubes can occur.

With the exception of *Rhodomicrobium vannielii*, the motile species have polar or subpolar to lateral flagellation and are either monotrichous or multitrichous. The motile cells of *Rhodomicrobium vannielii* are peritrichously flagellated.

The properties of the 16 species of Rhodospirillaceae are given in Table 1; photosynthetic electron donors and carbon sources are listed in Table 2.

Special properties of individual species are as follows:

1. *Rhodopseudomonas viridis* and *Rp. sulfoviridis* are the only species that contain Bchl *b* instead of Bchl *a*.
2. *Rhodopseudomonas gelatinosa* is the only species able to liquefy gelatin.
3. *Rhodocyclus purpureus* is the only nonmotile species (Pfennig, 1978).
4. *Rhodomicrobium vannielii* is the only species with the ability to produce exosporelike, moderately heat-resistant cysts.
5. The spectrum of whole cells of *Rhodospirillum rubrum* is unique in two respects: (a) a prominent absorption band at 550 nm due to spirilloxanthin and (b) a single, symmetrical long-wavelength absorption band of Bchl *a* at about 885 nm. All other species that contain Bchl *a* show two or more major absorption bands in the region from 800 to 890 nm.

The type genus of the family is *Rhodospirillum*.

2.2. The Family Chromatiaceae

The species of this family (Pfennig and Trüper, 1971*b*) carry out an anoxygenic photosynthesis in which sulfide is oxidized to sulfate via elemental sulfur; the sulfur accumulates in the form of globules inside or outside the cells (the latter mode occurs only in the genus *Ectothiorhodospira*). The ultimate oxidation product of sulfide, sulfur, thiosulfate, or sulfite is always sulfate. APS reductase is present.

All species can photoassimilate a number of simple organic compounds, acetate and pyruvate being the most widely used. When assimilatory sulfate reduction is lacking, organic compounds are used only in the presence of a reduced sulfur compound.

All species multiply by binary fission. Cells of the motile species are polarly flagellated, and are either monotrichous or multitrichous. Several species form characteristic cell aggregates (see Table 3). Due to various cultural or environmental conditions, many species may develop either as single cells or form mostly nonmotile aggregates or clusters of variable size and shape embedded in slime. These unspecific aggregates can confound identification, which accordingly is reliable only by pure culture studies.

The "giants" in this family, *Thiospirillum jenense* and *Chromatium okenii*, *Chr. weissei*, *Chr. warmingii*, and *Chr. buderi*, differ from the other species in size, in their inability to utilize organic compounds other than acetate and pyruvate, in their vitamin B_{12} requirement, and in the lack of assimilatory sulfate reduction. Vitamin B_{12} is also required by *Amoebobacter roseus* and *Ectothiorhodospira mobilis*.

The general properties of the 24 species of Chromatiaceae are given in Table 3.

Special properties of individual genera and species are as follows:

1. *Thiocapsa pfennigii* is the only species that contains Bchl *b*; all other species contain Bchl *a*. In addition, it is the only species in which the intracytoplasmic membrane system has the form of bundles of tubes. All other species except the genus *Ectothiorhodospira* contain membrane vesicles.
2. In the genus *Ectothiorhodospira*, the intracytoplasmic membranes have the form of stacks of membrane disks (thylakoids). Only in this genus are sulfur globules formed outside rather than inside the cells.
3. *Ectothiorhodospira halophila* is extremely halophilic, while moderate salt requirements are typical for *Ectothiorhodospira mobilis*, *Chr. buderi*, and *Chr. gracile*. *Ectothiorhodospira halophila* lacks assimilatory sulfate reduction.

The type genus of the family is *Chromatium*.

3. The Suborder Chlorobiineae

3.1. The Family Chlorobiaceae

The species of this family (Trüper and Pfennig, 1971) carry out an anoxygenic photosynthesis in which reduced sulfur compounds (sulfide, sulfur, thiosulfate) serve as electron donors. All species are strictly anaerobic and obligately phototrophic. Elemental sulfur is formed extracellularly and further oxidized to sulfate. APS reductase is present. Assimilatory sulfate reduction is absent. Some simple organic substrates are assimilated in the presence of sulfide and carbon dioxide. Many strains are also able to utilize molecular hydrogen under these conditions. Many strains require vitamin B_{12}.

Cells possess rigid cell walls and divide by binary fission. In *Pelodictyon clathratiforme*, ternary fission also occurs, leading to the formation of three-dimensional nets.

Table 4. Properties of the Chlorobiaceae[a]

Species	Cell shape, width/length (μm)	DNA base ratio (mol G + C/dl)[b]	Aggregates in pure culture	Gas vacuoles	Color of cell suspension	Predominant Bchl[b]	Predominant carotenoid[c]
Chlorobium							
limicola	Rod, 0.7–1.0/0.9–1.5	51.0–58.1	Chains	—	Green	c or d	chl
vibrioforme	Vibrio, 0.5–0.7/1.0–1.2	52.0–57.1	Spirals	—	Green	d or c	chl
phaeobacteroides	Rod, 0.6–0.8/1.3–2.7	49.0–50.0	Spirals	—	Brown	e	irt
phaeovibrioides	Vibrio, 0.3–0.4/0.7–1.4	52.0–53.0	Spirals	—	Brown	e	irt
Prosthecochloris							
aestuarii	Sphere + prosthecae, 0.5–0.7/1.0–1.2	50.0–56.0	Chains	—	Green	c	chl
Pelodictyon							
luteolum	Ovoid, 0.6–0.9/1.2–2.0	53.5–58.1	Clumps, spheres	+	Green	c or d	chl
clathratiforme	Rod, 0.7–1.2/1.5–2.5	48.5	Nets	+	Green	c or d	chl
phaeum	Vibrio, 0.6–0.9/1.0–2.0	ND	Spirals	+	Brown	e (?)	irt
Clathrochloris							
sulfurica	Sphere, 0.5–1.5	ND	Strings (?)	+	Green	ND	ND
Ancalochloris							
perfilievii	Sphere + prosthecae, 0.5–1.0; prosthecae, up to 2.0	ND	Microcolonies	+	Green	ND	ND

[a] The data were collected from Gloe et al. (1975), Gorlenko (1972), Gorlenko and Lebedeva (1971), Mandel et al. (1971), Pfennig and Trüper (1971d), and Trüper and Pfennig (1971).

[b] (ND) Not determined.

[c] (chl) Chlorobactene (66); (irt) isorenieratene (65). The numbers in parentheses are the serial numbers of the carotenoids in Chapter 12, Table 1.

The main distinguishing properties of the ten species of *Chlorobiaceae* are given in Table 4.

None of the accepted species is motile.

The type genus of the family is *Chlorobium*.

Table 4 does not contain data about the species *Chloropseudomonas ethylica*. According to recent published and unpublished results (Gray *et al.*, 1973; Pfennig and Siefert, unpublished; Pfennig and Biebl, 1976), this organism is apparently a syntrophic mixture of green bacteria and sulfur- and sulfate-reducing bacteria (see Chapter 1). In their report that "*Chloropseudomonas ethylica* strain 2-K" was a mixed culture, Gray *et al.* (1973) identified the green component as *Chlorobium limicola*. This identification was supported by photomicrographs that showed short rods typical of *Chlorobium limicola*. Recently, Pfennig identified the green bacterium in "*Chloropseudomonas ethylica*" strain N2. (M. C. W. Evans, Botany Department, University of London, King's College, London SE24 9JF, England) as *Chlorobium limicola* (Pfennig, unpublished; Pfennig and Biebl, 1976). However, Pfennig identified the green bacterium in "*Chloropseudomonas ethylica*" strain 2-K" (J. M. Olson, Biology Department, Brookhaven National Laboratory, Upton, New York 11973) and in a subculture of this strain (K.-I. Takamiya, Department of Biology, Kyushu University, Fukuoka 812, Japan) as *Prosthecochloris aestuarii*. These findings mean that it is impossible to be certain of the identity of the green bacteria in cultures of "*Chloropseudomonas ethylica*" used in the past.

For the original description of *Chloropseudomonas ethylica*, see Pfennig and Trüper (1974).

Several ectosymbiotic associations (consortia) are known in which a central, chemoorganotrophic, motile bacterium is covered by synchronously dividing Chlorobiaceae: *Chlorochromatium aggregatum*, *Chlorochromatium glebulum*, *Cylindrogloea bacterifera*, and *Pelochromatium roseo-viride* (Gorlenko and Kuznetsov, 1971).

These consortia were originally described as species; however, species names cannot be applied to symbiotic associations (cf. Trüper and Pfennig, 1971; Pfennig and Trüper, 1974).

3.2. The Family Chloroflexaceae

Showing the basic properties of the Chlorobiineae, this family is morphologically similar to the flexibacteria. The filamentous, flexible cells show gliding motility on surfaces.

So far, only one species has been described, *Chloroflexus aurantiacus* (Pierson and Castenholz, 1974). The description is as follows: Flexible filaments 0.6–0.7 μm wide (some strains up to 1.0 μm) and generally 30–100 μm long. Cross walls revealed by electron microscopy occur every 2–6 μm. Shows gliding motility on surfaces. Chlorobium vesicles present. The major pigments are Bchl's *a* and *c*, β-carotene, and γ-carotene. Capable of photolithotrophic growth with CO_2 and sulfide. Better growth occurs photoorganotrophically in the presence of yeast extract and casamino acids, glycerol, or acetate. Under aerobic conditions, can grow chemoorganotrophically. Temperature optimum: 50–60°C, minimum: 30°C, maximum: 70°C; pH optimum, 7.6–8.4; DNA base ratio: 53–55 mol guanine + cytosine/dl. Natural habitat is the outflow of hot sulfur springs.

4. Note Added in Proof

Since the above chapter was composed, the following new (anaerobic) phototrophic bacteria have been described:

Chromatiaceae:

    Ectothiorhodospira halochloris (Imhoff and Trüper, 1977, *Arch. Microbiol.* **114**:115–121)

Chlorobiaceae:

    Chlorobium chlorovibrioides (Gorlenko *et al.*, 1974, *Mikrobiologiya* **43**:908–914)

    Prosthecochloris phaeoasteroides (Puchkova and Gorlenko, 1976, *Mikrobiologiya* **45**:655–659)

Chloroflexaceae:

    Chloroflexus aurantiacus f. mesophilus (Pivovarova and Gorlenko, 1977, *Mikrobiologiya* **46**:329–333)

    Chloronema giganteum (Dubinina and Gorlenko, 1975, *Mikrobiologiya* **44**:511–517)

    Chloronema spiroideum (Dubinina and Gorlenko, 1975, *Mikrobiologiya* **44**:511–517)

    Oscillochloris chrysea (Gorlenko and Pivovarova, 1977, *Izv. Akad. Nauk SSSR, Ser. Biol.* **3**:396–409)

Rhodospirillaceae:

    Rhodocyclus purpureus (Pfennig, 1978, *Int. J. Syst. Bacteriol.* **28**:283–288)

5. References

Gloe, A., Pfennig, N., Brockmann, H., Jr., and Trowitzsch, W., 1975, A new bacteriochlorophyll from brown-colored Chlorobiaceae, *Arch. Microbiol.* **102**:103.

Gorlenko, V. M., 1972, Phototrophic brown sulphur bacteria *Pelodictyon phaeus nov. sp.*, *Mikrobiologiya* **41**:370.

Gorlenko, V. M., and Kuznetsov, S. I., 1971, Vertical distribution of photosynthetic bacteria in Lake Kononer Mariyskoy, ASSR, *Mikrobiologiya* **40**:746.

Gorlenko, V. M., and Lebedeva, E. V., 1971, New green sulphur bacteria with apophyses, *Mikrobiologiya* **40**:1035.

Gray, B. H., Fowler, C. F., Nugent, N. A., Rigopoulos, N., and Fuller, R. C., 1973, Reevaluation of *Chloropseudomonas ethylica* strain 2-K, *Int. J. Syst. Bacteriol.* **23**:256.

Hansen, T. A., and Veldkamp, H., 1973, *Rhodopseudomonas sulfidophila, nov. spec.*, a new species of the purple nonsulfur bacteria, *Arch. Microbiol.* **92**:45.

Keppen, O. I., and Gorlenko, V. M., 1975, A new species of purple budding bacteria containing bacteriochlorophyll *b*, *Mikrobiologiya* **44**:258.

Mandel, M., Leadbetter, E. R., Pfennig, N., and Trüper, H. G., 1971, Deoxyribonucleic acid base compositions of phototrophic bacteria, *Int. J. Syst. Bacteriol.* **21**:222.

Maroc, J., De Klerk, H., and Kamen, M. D., 1968, Quinones of *Athiorhodaceae, Biochim. Biophys. Acta* **162**:621.

Pfennig, N., 1974, *Rhodopseudomonas globiformis, sp. n.*, a new species of the Rhodospirillaceae, *Arch. Microbiol.* **100**:197.

Pfennig, N. and Biebl, H., 1976, *Desulfuromonas acetoxidans gen. nov.* and *sp. nov.*, a new anaerobic, sulfur-reducing, acetate-oxidizing bacterium, *Arch. Microbiol.* **110**:3.

Pfennig, N., and Trüper, H. G., 1971*a*, New nomenclatural combinations in the phototrophic sulfur bacteria, *Int. J. Syst. Bacteriol.* **21**:11.

Pfennig, N., and Trüper, H. G., 1971*b*, Conservation of the family name *Chromatiaceae* Bavendamm 1924 with the type genus *Chromatium* Perty 1852, request for an opinion, *Int. J. Syst. Bacteriol.* **21**:15.

Pfennig, N., and Trüper, H. G., 1971*c*, Higher taxa of the phototrophic bacteria, *Int. J. Syst. Bacteriol.* **21**:17

Pfennig, N., and Trüper, H. G., 1971*d*, Type and neotype strains of the species of phototrophic bacteria maintained in pure culture, *Int. J. Syst. Bacteriol.* **21**:19.

Pfennig, N., and Trüper, H. G., 1974, The phototrophic bacteria, in: *Bergey's Manual of Determinative Bacteriology*, 8th Ed. (R. E. Buchanan and N. E. Gibbons, eds.), pp. 24–64, The Williams & Wilkins Co., Baltimore.

Pierson, B. K., and Castenholz, R. W., 1974, A phototrophic gliding filamentous bacterium of hot springs, *Chloroflexus aurantiacus, gen.* and *sp. nov., Arch. Microbiol.* **100**:5.

Trüper, H. G., 1976, Higher taxa of the phototrophic bacteria: *Chloroflexaceae fam. nov.*, a family for the gliding, filamentous, phototrophic "green" bacteria, *Int. J. Syst. Bacteriol.* **26**:74.

Trüper, H. G., and Pfennig, N., 1971, The family of phototrophic green sulfur bacteria: *Chlorobiaceae* Copeland, the correct family name; rejection of *Chlorobacterium* Lauterborn; and the taxonomic situation of the consortium-forming species. Request for an opinion, *Int. J. Syst. Bacteriol.* **21**:8.

van Niel, C. B., 1944, The culture, general physiology, morphology and classification of the non-sulfur purple and brown bacteria, *Bacteriol. Rev.* **8**:1.

PART II

Structure

CHAPTER 3

Comparative Subcellular Architecture of Photosynthetic Bacteria

Charles C. Remsen

1. Introduction

Attempts to isolate and characterize bacterial photosynthetic membranes began with the work of French (1938), who compared the absorption spectrum of whole cells of *Spirillum rubrum* (*Rhodospirillum rubrum*) with that of methanol extracts of the cells. He noted that the extract did not contain spirilloxanthin, but did contain bacteriochlorophyll (Bchl); however, the peaks had shifted. He concluded from this finding that in whole cells, Bchl is bound to protein, and that the Bchl is extracted by methanol. In 1940, French (1940) further reported that he had obtained the Bchl–protein complex in aqueous extracts of cells, since the absorption spectra of whole cells and of these extracts were the same.

Schachman *et al.* (1952) reported that when cells of *Rs. rubrum* were ground with alumina and ultracentrifuged, a particulate fraction could be isolated that contained all the pigments of the cell. The sedimentation coefficient was 190 S, and the size of the particles, which were vesicular, was estimated to be 110 nm for the flattened disks that were actually observed, or 60 nm for the vesicles. The absorption spectra of the cells and of the particles were identical, and the name *chromatophore* was given to these particles. Shortly after this report, Frenkel (1954) stimulated additional interest in chromatophores by demonstrating that in the presence of ADP, inorganic phosphate, and light, they could carry out photophosphorylation. Thus, particles had been isolated from photosynthetic bacteria that contained all the Bchl and carotenoids of whole cells and that were photochemically active.

Once it became apparent that a photochemical organelle could be isolated from photosynthetic bacteria, numerous studies were begun in an attempt to characterize both the structure and the functions of these particles. It soon became evident that the method of cell disruption and the conditions of isolation of the particles influenced the composition of the "chromatophores." Newton and Newton (1957) used a variety of breakage methods, and after differential centrifugation, succeeded in isolating two fractions, a "chromatophore" fraction (up to 100 nm in diameter) and a "small particle" fraction (around 40 nm in diameter), both of which could carry out photophosphorylation. Other workers used density-gradient centrifugation (Frenkel and Hickman, 1959; Cohen-Bazire and Kunisawa, 1963; Worden and Sistrom, 1964), and were able to isolate two main pigmented bands, which were named the *light* and *heavy* chromatophore fractions. Comparison of the data showed that the "chromatophore" fraction of Newton and Newton (1957) corresponded to the "heavy" chromatophore fraction and the "small particles" to the "light" fraction. Further comparison, together with the finding that the "chromatophore" fraction of Newton and Newton was contaminated with cell-wall material, eventually led investigators to conclude that the chromatophore was the "small particle" of Newton and Newton or the "light" chromatophore of Cohen-Bazire and Kunisawa (1963).

Charles C. Remsen · Center for Great Lakes Studies, The University of Wisconsin–Milwaukee, Milwaukee, Wisconsin 53201

Table 1. Rhodospirillaceae Examined by Thin Section or Freeze–Etching[a]

Species	References
Rhodomicrobium vannielii	Boatman and Douglas (1961), Conti and Hirsh (1965), Trentini and Starr (1967)
Rhodopseudomonas acidophila	Tauschel and Hoeniger (1974)
Rhodopseudomonas capsulata	Drews *et al.* (1969, 1971), Lampe *et al.* (1972)
Rhodopseudomonas gelatinosa	Weckesser *et al.* (1969)
Rhodopseudomonas palustris	Solov'eva and Fedenko (1970), Tauschel and Drews (1967, 1969a–c), Whittenbury and McLee (1967)
Rhodopseudomonas sphaeroides	Peters and Cellarius (1972), Vatter and Wolfe (1958), Giesbrecht and Drews (1966)
Rhodopseudomonas viridis	Whittenbury and McLee (1967)
Rhodospirillum molischianun	Drews (1960), Gibbs *et al.* (1965), Giesbrecht and Drews (1962), Hickman and Frenkel (1965)
Rhodospirillum rubrum	Boatman (1964), Cohen-Bazire and Kunisawa (1963), Drews (1960), Drews and Giesbrecht (1963), Holt and Marr (1965a–c, Ketchum and Holt (1970), Niklowitz and Drews (1955), Oelze *et al.* (1969), Schön and Jank-Ladwig (1972), Schön and Ladwig (1970), Vatter and Wolfe (1958)

[a] As of June 1975.

Table 2. Chromatiaceae Examined by Thin Section or Freeze–Etching[a]

Species	References
Chromatium buderi	Remsen and Trüper (1973)
Chromatium okenii	Kran *et al.* (1963)
Chromatium vinosum	Petrova (1959), Vatter and Wolfe (1958)
Ectothiorhodospira halophila	Raymond and Sistrom (1967)
Ectothiorhodospira mobilis	Remsen *et al.* (1968), Holt *et al.* (1968)
Ectothiorhodospira shaposhnikovii	Cheri *et al.* (1969)
Thiocapsa floridana	Takács and Holt (1971)
Thiocapsa pfennigii	Eimhjellen *et al.* (1967)
Thiocapsa roseopersicina	Bogorov (1974)
Amoebobacter pendens	Cherni *et al.* (1969)

[a] As of June 1975

Table 3. Chlorobiaceae Examined by Thin Section or Freeze–Etching[a]

Species	References
Chlorobium limicola	Cohen-Bazire *et al.* (1964), Vatter and Wolfe (1958)
Chlorobium limicola f. thiosulfatophilum	Cohen-Bazire *et al.* (1964), Solov'eva and Federov (1970), Tomina and Federov (1967)
Chloropseudomonas ethylica	Holt *et al.* (1966)
Pelodictyon clathratiforme	Pfennig and Cohen-Bazire (1967)
Prosthecochloris aestuarii	Gorlenko and Zhilina (1968)

[a] As of June 1975.

Since these findings, it has been shown that chromatophores of all members of the Rhodospirillaceae and Chromatiaceae studied so far are membranous structures. "Chromatophores" of the green sulfur bacteria or Chlorobiaceae, however, were shown to be unique in that they were nonmembranous. To differentiate them from chromatophores, these structures were named *chlorobium vesicles* (Cohen-Bazire, 1963; Cohen-Bazire *et al.*, 1964).

A great many electron micrographs of photosynthetic bacteria are now found in the published literature (Tables 1–3), and from this literature one can get a general idea of the structure of these bacteria.

Most bacterial photosynthetic membranes appear to arise by invagination of the cytoplasmic membrane (see Section 3). In some bacteria, the invaginated membranes pinch off and form detached vesicles; in others, the photosynthetic membranes remain connected to the cytoplasmic membrane. The photosynthetic membranes that remain attached to the cytoplasmic membrane may be tubular invaginations or folded stacks of membranes.

Both vesicles and lamellar stacks of photosynthetic membranes occur in members of the Rhodospirillaceae. Those that have vesicles include *Rs. rubrum* (Vatter and Wolfe, 1958) and *Rhodopseudomonas sphaeroides* (Cohen-Bazire, 1963); most of the other species in this family have lamellar membranes. There are two modes of cell division in this group of bacteria: binary fission and budding. The brown *Rhodospirillum* species *Rs. fulvum*, *Rs. molischianum*, and *Rs. photometricum* (Cohen-Bazire and Sistrom, 1966; Giesbrecht and Drews, 1962) reproduce by binary fission and have cytomembrane stacks randomly arranged near the periphery of the cell. In contrast, the organisms that reproduce by budding, such as *Rhodomicrobium vannielii* (Boatman and Douglas, 1961; Conti and Hirsch, 1965), *Rp. palustris* (Tauschel and Drews, 1967), and *Rp. viridis* (Whittenbury and McLee, 1967), have membranes at only one end of the cell. During budding, the cytomembranes remain with the mother cell, and the daughter cell forms new ones by an invagination of the cytoplasmic membrane.

Thiospirillum (Cohen-Bazire, 1963), *Thiocystis* (Remsen, unpublished), and *Thiocapsa* (Cohen-Bazire, 1963) have vesicular membranes; only a single *Thiocapsa* species (Eimhjellen *et al.*, 1967; Pfennig, 1967) has tubular membranes. Both vesicular and lamellar membranes have been seen in some strains of *Thiocapsa* (Cohen-Bazire, 1963) and *Chromatium* (Cohen-Bazire and Sistrom, 1966; Fuller *et al.*, 1963; Kran *et al.*, 1963). The only Chromatiaceae that have stacks of membranes belong to the genus *Ectothiorhodospira* (Remsen *et al.*, 1968; Raymond and

Sistrom, 1967; Holt *et al.*, 1968). All species in this genus deposit sulfur extracellularly rather than intracellularly, an activity shared by members of the Chlorobiaceae. The bacteria in this latter group, however, lack a true lamellar membrane system, but instead have the unique "chlorobium vesicles" located adjacent and parallel to the plasma membrane.

Since the term *chromatophore* has been defined in different ways by different authors, it is no longer adequate to describe the photosynthetic membranes of the various photosynthetic bacteria. Therefore, in the remainder of this discussion, the term *intracytoplasmic membrane system* will be used for all photosynthetically active membrane structures.

2. Fine Structure of the Photosynthetic Bacteria—Intracytoplasmic Membrane Systems

2.1. The Rhodospirillaceae

The Rhodospirillaceae—or, more familiarly, the purple, nonsulfur bacteria—are for the most part unable to grow well with sulfur compounds as photosynthetic electron donors. Most species are microaerophilic, and it has been demonstrated quite clearly that as the concentration of dissolved oxygen increases, the photopigment content and the intracytoplasmic membrane systems decrease. Cell division generally occurs by binary fission, with the exception of the genus *Rhodomicrobium* and the species *Rp. palustris*, *Rp. viridis*, and *Rp. acidophila*, which divide by budding. Cells of *Rhodomicrobium* are peritrichously flagellated, while the species of *Rhodospirillum* and *Rhodopseudomonas* have polar flagella. Exosporelike heat-resistant bodies have been observed in the genus *Rhodomicrobium*. None of the known species of the Rhodospirillaceae contains gas vacuoles. The types of intracytoplasmic membrane systems that are found in the purple, nonsulfur bacteria range from vesicular and lamellar to tubular systems. Depending on individual interpretations of membrane configurations, the Rhodospirillaceae can be divided into either four (Pfennig, 1967) or five (Oelze and Drews, 1972) groups.

In general, these bacteria can be divided into groups of species by particular types of extensions of the cytoplasmic membrane to form a characteristic form of intracytoplasmic membrane system.

2.1.1. Intracytoplasmic Membrane Systems of Species That Multiply by Binary Cell Division

(a) Vesicular Intracytoplasmic Membrane System. The cytoplasmic membrane differentiates by forming

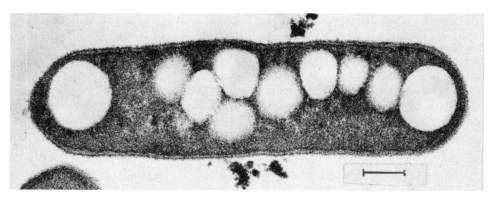

Fig. 1. Section of *Rhodopseudomonas sphaeroides* after semi-aerobic culture in the dark. Cell has a few distinguishable chromatophores, but is packed with numerous poly-β-hydroxybutyrate (PHB) granules. Bar represents 0.2 μm. Reprinted from Peters and Cellarius (1972) by permission.

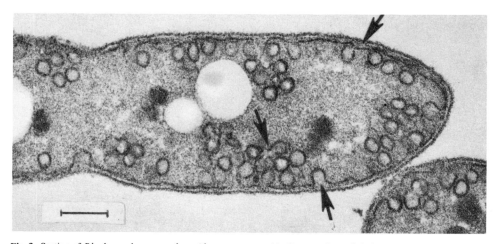

Fig. 2. Section of *Rhodopseudomonas sphaeroides* grown anaerobically at moderate light intensity. Areas of chromatophore continuity with the cytoplasmic membrane and proliferating vesicles are indicated by arrows. Bar represents 0.2 μm. Reprinted from Peters and Cellarius (1972) by permission.

vesicular or tubular invaginations and extends inward in the form of connected vesicles and bulged tubes (Holt and Marr, 1965a–c; Peters and Cellarius, 1972; Drews and Giesbrecht, 1963; Cohen-Bazire and Kunisawa, 1963; Boatman, 1964). Under aerobic conditions, intracytoplasmic membrane development is minimal (Fig. 1), since growth is achieved through oxidative metabolism and the extent of Bchl synthesis is determined by the oxygen partial pressure. When cells are grown anaerobically at moderate light intensities, however, the vesicular intracytoplasmic membrane system becomes well developed (Fig. 2). The Rhodospirillaceae that have this type of intracytoplasmic membrane system include *Rp. capsulata*, *Rp. globiformis*, *Rp. sphaeroides*, and *Rhodospirillum rubrum*.

When these vesicular chromatophores were first observed, it was suggested that they were free cytoplasmic organelles, i.e., that they were discrete and not continuous with the cytoplasmic membrane (Vatter and Wolfe, 1958). However, it was demonstrated quite clearly by subsequent investigators (Cohen-Bazire and Kunisawa, 1963; Drews and Giesbrecht, 1963; Holt and Marr, 1965a–c; Boatman, 1964; Peters and Cellarius, 1972) that this is not the case. The vesicles first appear as simple membrane protrusions, which then invaginate and become constricted, resulting in a stalked, more-or-less spherical vesicle that is open at the cell-wall end (Fig. 3). Finally, these vesicles penetrate more deeply into the cytoplasm of the cell, often assuming a tubular appearance, and proliferate into a branched network (Fig. 4). New protrusions continue to arise at other areas of the cytoplasmic membrane, and the process is repeated.

(b) Single, Small and Irregular Membrane Invaginations. The most weakly developed type of intracytoplasmic membrane system found in the Rhodospirillaceae is that represented by *Rp. gelatinosa*,

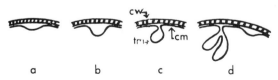

Fig. 3. Schematic illustration of the invagination of cytoplasmic membrane to form thylakoids (chromatophores). (a) Protrusion of cytoplasmic membrane; (b) invagination; (c) constriction at cell-wall end; (d) proliferation and new protrusion. (cm) Cytoplasmic membrane; (cw) cell wall; (tm) thylakoid (chromatophore) membrane. Reprinted from Peters and Cellarius (1972) by permission.

Rhodospirillum tenue, and *Rhodocyclus purpureus*. Few studies have been made on this group of bacteria, a single investigation on the fine structure and taxonomy of *Rp. gelatinosa* (Weckesser *et al.*, 1969) being the major contribution to date. These authors conclude that the weak development of an intracytoplasmic membrane system despite a Bchl content of 10–25 μg Bchl/mg protein suggests that the cytoplasmic membrane may be included in the system of the photosynthetic apparatus.

(c) Short Lamellar Membrane Systems, Usually in the Form of Stacks. Stacks of short lamellae, formed by tubular intrusions of the cytoplasmic membrane, are characteristic of this group of photosynthetic bacteria. These stacks of infolded disks are not parallel to the cytoplasmic membrane, but are at a sharp angle to it. Included in this group are the strictly anaerobic non-sulfur brown *Rhodospirillum* species, *Rs. fulvum*, *Rs. molischianum*, and *Rs. photometricum* (Giesbrecht

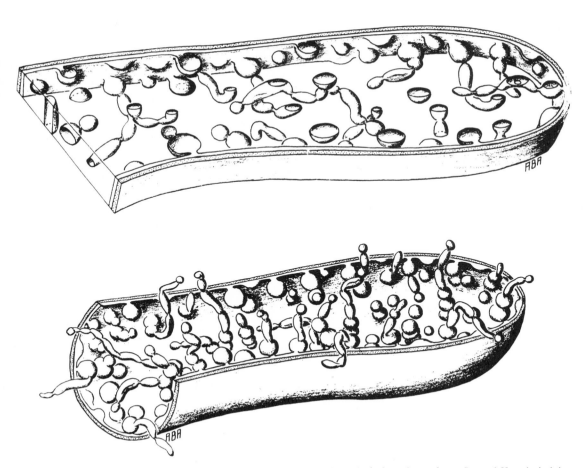

Fig. 4. (top) Composite representation of stereoelectron micrographs of sections of *Rhodospirillum rubrum*. (bottom) Hypothetical three-dimensional representation of the internal membrane system of *Rhodospirillum rubrum*. Reprinted from Holt and Marr (1965a) by permission from the American Society for Microbiology, Bethesda, Maryland.

and Drews, 1962; Gibbs *et al.*, 1965; Hickman and Frenkel, 1965). Hickman and Frenkel (1965) suggest that the photosynthetic lamellae of the bacterium *Rs. molischianum* originate as extensions of the cytoplasmic membrane into the cytoplasm of the cell. The first lamellae extensions are narrow folds and occur independently of one another; they appear to average about 8.0 nm in width, representing one side of the fold in the cytoplasmic membrane, or 16.0 nm when the two sides of the fold are closely appressed. The 16.0-nm lamellae then increase in number and may associate to form larger lamellae. Finally, "... the space within each fold increases, the two appressed regions of the cytoplasmic membrane in each fold separate to form distinct invaginations, and the lamellae observed at this stage are formed by an association of the sides of adjacent invaginations" (Hickman and Frenkel, 1965).

2.1.2. Intracytoplasmic Membrane Systems of Species That Multiply by Budding

This group of bacteria, primarily distinguished by their method of division, can be grouped together (Pfennig, 1967) or separated into three different groups (Oelze and Drews, 1972). The photosynthetic budding bacteria include the genus *Rhodomicrobium* and three species of *Rhodopseudomonas*: *Rp. acidophila*, *Rp. palustris*, and *Rp. viridis*. Evidence to date suggests that the cytoplasmic membrane extends in the form of paired lamellae that are peripherally arranged in many layers parallel to the cytoplasmic membrane. In *Rhodomicrobium vannielii* (Conti and Hirsch, 1965; Boatman and Douglas, 1961), *Rp. palustris* (Cohen-Bazire and Sistrom, 1966), and *Rs. viridis* (Giesbrecht and Drews, 1966), the membrane layers are usually open in one or both ends of the cell. Cross sections of these cells show that the lamellae form more or less open rings, suggesting that as the cells age, the area occupied by the lamellae gradually extends to both sides along the rim, finally closing to form concentric rings.

2.2. The Chromatiaceae

The Chromatiaceae—more familiarly known as the purple sulfur bacteria—can use either reduced carbon or reduced sulfur compounds as electron donors. The usual organic acids function well as reduced carbon, and sulfide or thiosulfate is used as sulfur electron donor. Elemental sulfur accumulates under these conditions as sulfur globules inside the cell; only in one genus, *Ectothiorhodospira*, is sulfur deposited outside the cell. With the exception of *Thiocapsa roseopersicina*, the purple sulfur bacteria are strict anaerobes; *T. roseopersicina* can grow in the dark under microaerophilic conditions. All species multiply by binary fission, and the motile species are polarly flagellated. Depending on the culture conditions or environmental conditions in nature, all species may develop either as single cells or form mostly nonmotile aggregates of variable size and shape embedded in slime (Figs. 5 and 6).

As with the Rhodospirillaceae, the Chromatiaceae can be grouped according to the type of intracytoplasmic membrane system present in the cells.

2.2.1. Vesicular Intracytoplasmic Membrane System

Included in this group are almost all members of the Chromatiaceae. Thus, all species of the genera *Thiospirillum*, *Chromatium*, *Thiocystis*, *Thiosarcina*, *Lamprocystis*, *Thiodictyon*, *Thiopedia*, and *Amoebobacter* have vesicular intracytoplasmic membrane systems. In addition, *Thiocapsa roseopersicina* develops an intracytoplasmic membrane system of the vesicular type.

The cytoplasmic membrane differentiates by forming vesicular or tubular invaginations, and extends inward in the form of connected and bulged tubes. As growth proceeds, the vesicles become tightly packed within the cell. In both sectioned (Fig. 7) and frozen–etched (Fig. 8) preparations, the tight packing of vesicles is easily seen; often it becomes difficult to resolve the individual vesicle membranes. The similarity of this arrangement in various species and genera can be easily demonstrated when one compares the ultrastructure of *Chromatium buderi* (Fig. 9), *Chr. gracile* (Fig. 8), *Chr. violascens* (Fig. 7), *Thiocystis violacea* (Fig. 10), and *Thiopedia rosea* (Fig. 11). When partially lysed cells are fixed, the individual vesicles and their origin from the cytoplasmic membrane can be easily seen (Figs. 12 and 13). The membrane-bound vesicles measure about 30–40 nm in diameter.

In addition to the vesicular intracytoplasmic membranes, extended lamellar membranes are also found dispersed throughout the cell or concentrated around the periphery of the cell (Fig. 10). The presence of these intracytoplasmic membranes is probably related to exposure of the cells to high light intensities or may, in some cases, reflect a general decline in cultural conditions. Fuller *et al.* (1963) found that in *Chr. vinosum* strain D, paired membranes were formed under conditions of high light intensities (7000 ft-c. of incident light), and were observed most frequently in the peripheral areas of the cell. Under "low" light intensities (100 ft-c. of incident light intensities, i.e., about 1000 lux), only the usual vesicular intracytoplasmic membranes were found in *Chr. vinosum*.

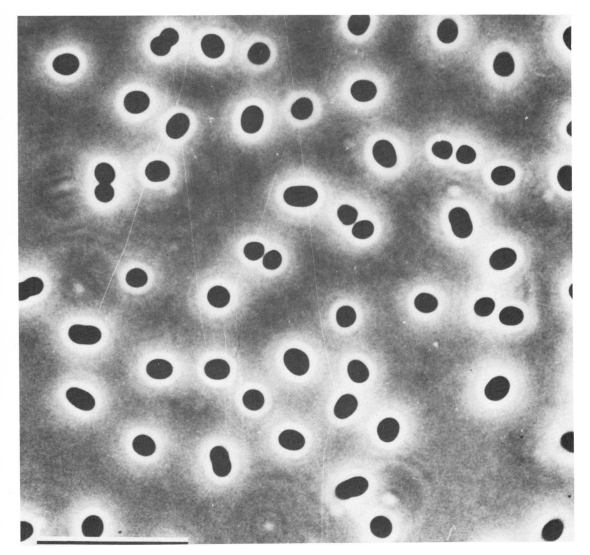

Fig. 5. Phase-contrast micrograph of an India-ink-stained preparation of *Thiocapsa roseopersicina* (*Thiocapsa floridana*). Bar represents 10 μm. Reprinted from Takács and Holt (1971) by permission from Elsevier Publishing Co., Amsterdam, The Netherlands.

Paired lamellar membranes were also found in *Chr. buderi* (Remsen and Trüper, 1973) cells grown at light intensities of 500–1000 lux. Under the conditions of Fuller *et al.* (1963), these might represent low light intensities; however, it was shown by Trüper and Schlegel (1964) that for large-cell *Chromatium* species, e.g., *Chr. buderi* and *Chr. okenii*, 500 to 1000 lux represents optimal to high light intensities.

2.2.2. Tubular Intracytoplasmic Membrane Systems

In 1967, a new species of phototrophic sulfur bacteria was described by Eimhjellen *et al.* (1967).

The organism differed from all other Chromatiaceae so far studied in having a tubular type of intracytoplasmic membrane system. This new organism was given the name *Thiocapsa pfennigii* (*Thiococcus sp.*, Eimhjellen, 1970), and is the single representative of this group. The most conspicuous feature of these cells is the extensive internal membrane system occupying the greater part of the outer circumference of the cell (Fig. 14). Unlike the situation in other photosynthetic bacteria, the sectioned profiles of the membranous structures vary from almost circular to elliptical to very oblong and tubular. The only structure that is compatible with these observations would be a

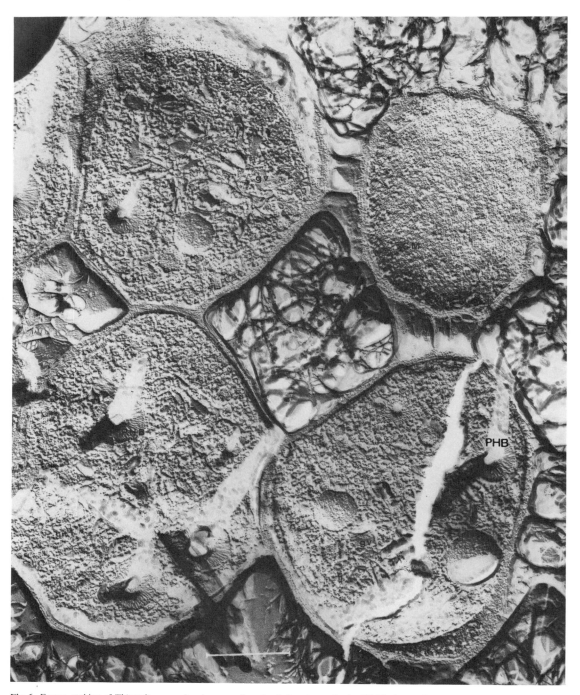

Fig. 6. Freeze–etching of *Thiopedia rosea* showing a number of cells forming a platelet. Within these cells are the vesicular chromatophores. gas vacuoles (gv), and poly-β-hydroxybutyrate granules (PHB). (CM) Cytoplasmic membrane. Bar represents 0.5 μm. From Remsen (unpublished micrograph).

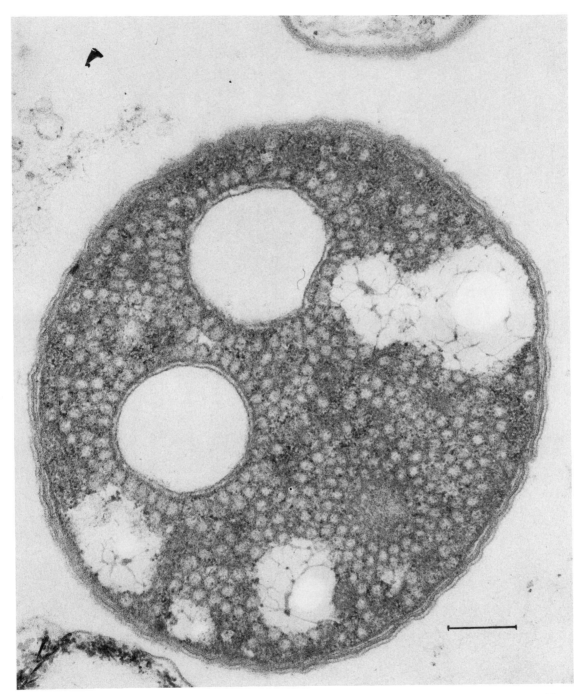

Fig. 7. Section of *Chromatium violascens* showing characteristic packing of chromatophores and sulfur granules. Bar represents 0.2 μm. From Remsen (unpublished micrograph).

Fig. 8. Freeze-etching of *Chromatium gracile* showing cell wall (CW), cytoplasmic membrane (CM), sulfur granules, and densely packed chromatophores. Bar represents 0.5 μm. From Remsen (unpublished micrograph).

membrane system composed of long tubes of even diameter.

On examining these cells after extrusion from a French pressure cell, Eimhjellen *et al.* (1967) found a close similarity in the construction of the tubular inner membranes with the cytoplasmic membrane, and they were able to demonstrate a continuity between the two (Fig. 15). Thus, good evidence was provided indicating that the tubes originated from the ctyoplasmic membrane. Further studies with negatively stained, osmotically shocked cells showed that the tubular membranes were composed of regularly arranged subunits (Fig. 16). The subunits are arranged in a regular hexagonal pattern, the centers of the subunits being approximately 12–13 nm apart. Similar fine structure was observed on the surface of internal membranes in *Rs. rubrum* (Holt and Marr, 1965*b*) and in *Rp. viridis* (Giesbrecht and Drews, 1966).

2.2.3. Short Lamellar Membrane Systems, Usually in the Form of Stacks

Within this group are all species of the genus *Ecto-thiorhodospira*. The number and arrangement of

lamellar stacks appear to differ among species, but the development of this type of intracytoplasmic membrane system appears to be identical (Fig. 17). A significant difference between this group and others in the Chromatiaceae is that all species of *Ectothio-rhodospira* deposit sulfur outside the cell. As pointed out in Section 3, this may be related to the type of membrane system present and whether it always remains connected with the cytoplasmic membrane.

In sections of *E. mobilis* (Remsen *et al.*, 1968; Holt *et al.*, 1968) and *E. halophila* (Raymond and Sistrom, 1967), numerous connections are found between the cytoplasmic membrane and lamellar stacks. In negatively stained preparations of whole cells, the stacks of membranes were always more electron-dense than the surrounding cytoplasm, indicating an opening through the plasma membrane that permitted the entrance of the negative stain.

According to Remsen *et al.* (1968), the cytoplasmic membrane enfolds, forming a tubelike or sac-like primary structure that in turn enfolds and distends, forming disklike structures (Fig. 18). The secondary invaginations must be able to occur at any point on the circumference of the primary structure;

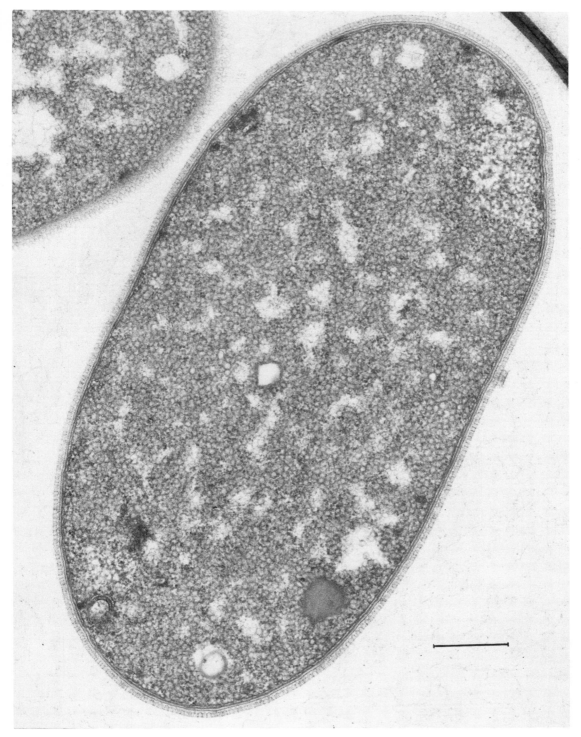

Fig. 9. Section of *Chromatium buderi* showing close packing of chromatophores and dispersed nuclear material. Unique cell wall layer is also shown. Bar represents 0.5 μm. From Remsen (unpublished micrograph).

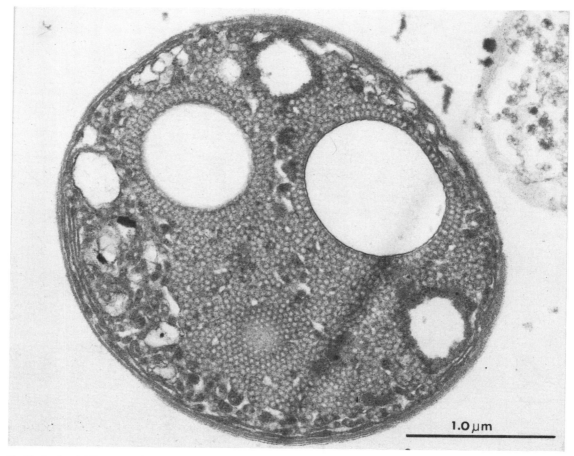

Fig. 10. Section of *Thiocystis violacea* showing packed chromatophores, sulfur granules, and lamellar to tubular membrane characteristic of old cells of this species. Bar represents 1.0 μm. From Remsen (unpublished micrograph).

however, cross sections of the disks illustrate that only a small area of their circumference is attached to the primary structure (e.g., Figs. 7, 13, and 15 in Remsen *et al.*, 1968).

2.3. The Chlorobiaceae

The Chlorobiaceae—or, more familiarly, the green sulfur bacteria—are strictly anaerobic and obligately phototrophic. The family is unique in that they possess organelles, the chlorobium vesicles, that contain the bulk of the photosynthetic pigments. The green sulfur bacteria are nonmotile.

Of all the photosynthetic bacteria, the green sulfur bacteria are the most poorly described and by far the least understood. In terms of their general ultrastructure, very little has been done; perhaps ten studies concerned with the ultrastructure of these extremely interesting bacteria have been published. As a result,

our information on these bacteria is quite limited, and what can be said must be limited in its generality.

In the first survey of the structure of green sulfur bacteria as revealed by electron microscopy of their sections, Vatter and Wolfe (1958) examined a strain of *Chlorobium limicola*. They were unable to detect vesicular elements, which are such conspicuous cytoplasmic constituents of most purple bacteria, but did observe that the cytoplasm contained numerous electron-opaque granules 15–25 nm in diameter, which they interpreted as the "chromatophores" of *Chlorobium limicola*. In retrospect, these granules could be more reasonably interpreted as inclusions of metaphosphate, which are always abundant in the cells of green bacteria grown with an excess of inorganic phosphate in the medium.

Bergeron and Fuller (1961) examined thin sections of a strain of *Chlorobium thiosulfatophilum*—or, more properly, *Chlorobium limicola f. thiosulfato-*

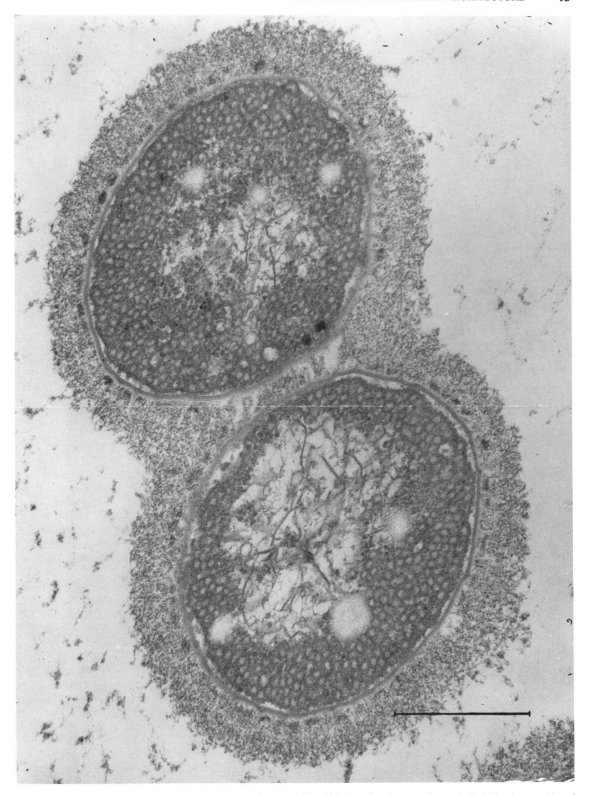

Fig. 11. Section of *Thiopedia rosea* showing the dense capsular material in which the cell packets are often embedded. The dense packing of the chromatophores is also very evident. In the center of the cells are the collapsed gas vacuole membranes. Bar represents 1.0 μm. From Hirsch (unpublished micrograph).

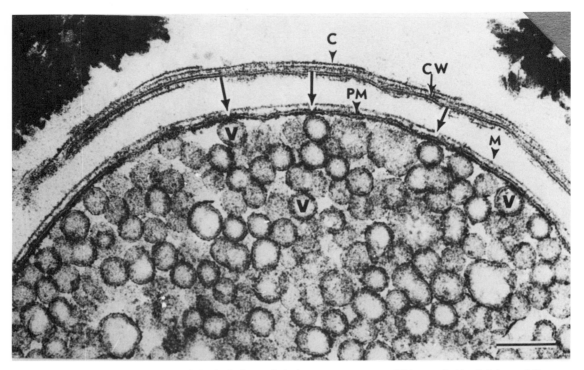

Fig. 12. Section of osmotically shocked and chemically fixed cell of *Thiocapsa roseopersicina* (*Thiocapsa floridana*). Release of ribosomes provides a clearer view not only of the membrane-bound vesicles (V), but also of their relationship to the plasma membrane (arrows). Remnants of the capsule (C), the unit-membrane-like layer of the cell wall (CW), microlayer of the cell wall (M), and the cytoplasmic membrane (PM) are also resolved. Bar represents 100 nm. Reprinted from Takács and Holt (1971) by permission from Elsevier Publishing Co., Amsterdam, The Netherlands.

philum—and were unable to detect either vesicular or lamellar elements in the cytoplasm. Apart from metaphosphate granules, the only cytoplasmic structures they observed were numerous particles 15 nm in diameter.

Even though both these initial studies failed to identify the photosynthetic apparatus of the green sulfur bacteria, they did strongly suggest that green bacteria have a photosynthetic structure different from and considerably simpler in architecture than that of either purple bacteria or blue-green algae.

The first detailed study on the location of the pigment system in cell-free systems was also carried out by Bergeron and Fuller (1961). Using a relatively abrasive method of cell breakage, they found that the bulk of the pigment system in the cell extract was associated with particles about 15 nm in diameter, and difficult to separate from ribosomes. Since the only structural feature observed in thin sections were also 15 nm particles, Bergeron and Fuller made the obvious correlation that the pigments were associated with these observed cellular elements.

Two years later, Cohen-Bazire (1963) reported on some new data that she had obtained in collaboration with Dr. Norbert Pfennig. In contrast to the two previous studies on the fine structure of green sulfur bacteria, they found that the green bacteria they examined, all strains of *Chlorobium*, had a "highly distinctive and complex fine structure, quite unlike that found in purple bacteria—or, indeed, in any other type of bacterium so far studied by modern techniques of electron microscopy."

In addition to a fairly complex cell wall, a rather complex cell membrane system was found, consisting of either one or two unit membranes. Parallel and adjacent to this membrane system was a thinner, electron-dense membrane, approximately 5 nm thick, that appeared to surround a series of "... large, clear, oblong areas which line the cortex of the cytoplasm between the surface membrane system and the ribosomal region." These structures were given the name *chlorobium vesicles* and are relatively large, ranging from 100 to 150 nm long by 30 to 40 nm wide. Further observations on fractions of cell-free extracts indicated that the photosynthetic pigment system is associated primarily, if not entirely, with the chlorobium vesicle fraction (Cohen-Bazire *et al.*, 1964).

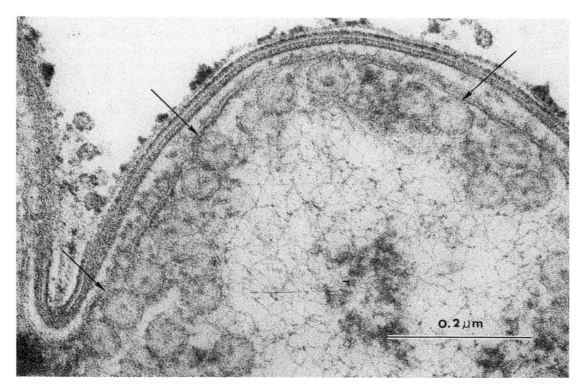

Fig. 13. Section of *Thiocystis violacea* showing association of chromatophore vesicles with the cytoplasmic membrane (arrows). Bar represents 0.2 μm. From Remsen (unpublished micrograph).

Holt *et al.* (1966), examining the ultrastructure of "*Chloropseudomonas ethylica*" (a dubious species),* showed that when cells were broken by ballistic disruption and examined by electron microscopy, vesicles 130–150 nm long and 30–50 nm wide were found to rim the periphery of the cell (Fig. 19). They further found, by studying purified preparations obtained through density gradient centrifugation, that possible interconnections existed between the vesicles and that a macromolecular fine structure might be present on the vesicle "membrane." Of particular interest was that these authors were able to show a separation of the cytoplasmic membrane-bound enzyme succinic dehydrogenase from the chlorophyll-containing vesicle. This seemed to indicate that the photosynthetic vesicles of the green sulfur bacteria may be more highly differentiated from the photosynthetic membrane system of the purple bacteria.

Further studies on the green sulfur bacteria established the presence of the chlorobium vesicles as the photosynthetic apparatus in this group. Tomina and Federov (1967) confirmed their presence in *Chlorobium limicola f. thiosulfatophilum*, and Pfennig

* See Chapter 2 for a discussion of this syntrophic system.

and Cohen-Bazire (1967) showed that they were also present in *Pelodictyon clathratiforme*. Unlike other green bacteria so far studied, *P. clathratiforme* also produces large, electron-transparent, tubular vesicles with conical ends, surrounded by a nonunit membrane 2 nm or less thick. Their average diameter was 75 nm (Pfennig and Cohen-Bazire, 1967). Since they were similar in appearance to the elements observed in sections of filamentous blue-green algae that contain gas vacuoles, they were interpreted as components of a gas vacuole system. In addition to the chlorobium vesicles and gas vacuoles, *P. clathratiforme* also contained large, highly convoluted mesosomes and an intracytoplasmic fibrillar organelle of unknown function. In members of *Chlorobium* and *Pelodictyon*, mesosomal invaginations of the cytoplasmic membrane have been observed that are analogous to structures in numerous nonphotosynthetic bacteria, and the formation of these invaginations appears to be associated with cell division (Cohen-Bazire *et al.*, 1964; Pfennig and Cohen-Bazire, 1967). In general, the cytoplasm and the nuclear region of green bacteria show no structural differences from other prokaryotic microorganisms. *Chlorobium* species differ from *P. clathratiforme* with respect to cell division. *Chloro-*

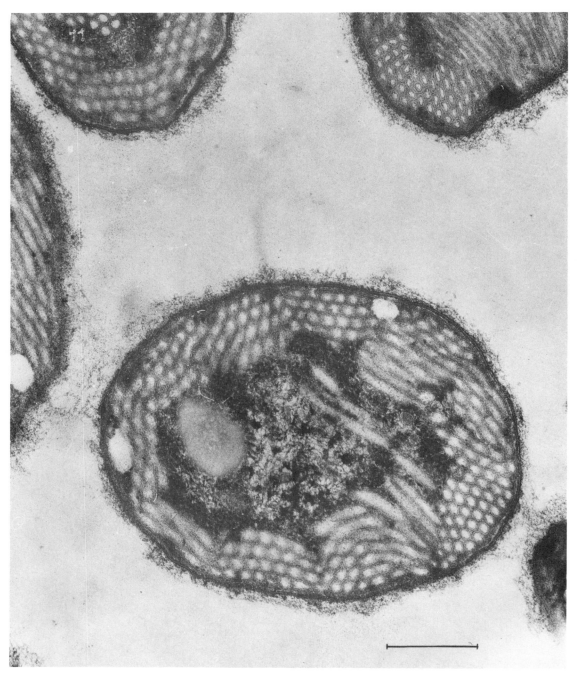

Fig. 14. Section of *Thiocapsa pfennigii* showing the stacking arrangement of tubular membranes around the periphery of the cell. Bar represents 0.5 μm. Reprinted from Eimhjellen *et al.* (1967) by permission from Springer-Verlag, Berlin and New York.

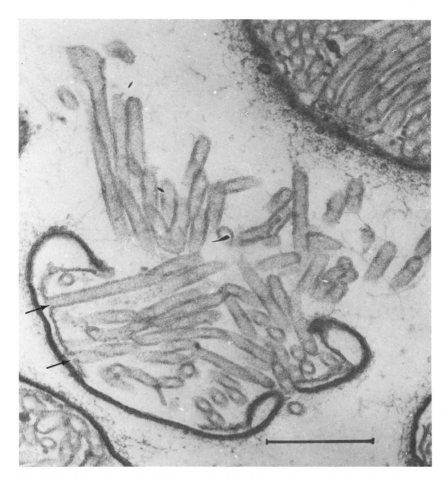

Fig. 15. Detail from thin section of *Thiocapsa pfennigii* cell disrupted by extrusion from French pressure cell. Note association of tubular membrane with the cytoplasmic membrane (arrows). Bar represents 0.5 μm. Reprinted from Eimhjellen *et al.* (1967) by permission from Springer-Verlag, Berlin and New York.

bium species divide by forming a cross wall with subsequent development of two sister cells, whereas the latter organisms are capable of triple division preceded by true ramification of the cell (Pfennig and Cohen-Bazire, 1967).

The ultrastructure of *Prosthecochloris aestuarii* has been studied (Gorlenko and Zhilina, 1968; Remsen and Watson, unpublished), and results of these investigations further confirmed the presence of chlorobium vesicles and established it as the photosynthetic apparatus of green bacteria. *Prosthecochloris aestuarii* is a spherical to ovoid bacterium, forming about 20 prosthecae per cell. These prosthecae range from 0.1 to 0.5 μm long and 0.1 to 0.17 μm wide and are the predominant site of the chlorobium vesicles (Figs. 20 and 21).

Finally, a new genus, *Chloroflexus*, has recently been proposed for the group of gliding, filamentous bacteria that has numerous similarities with the green sulfur bacteria (Pierson and Castenholz, 1974; Madigan and Brock, 1975). The presence of "chlorobium vesicles" and Bchl *c* and the formation of sulfur granules outside the cell support the suggestion that *Chloroflexus* be classified as a member of the Chlorobiaceae, despite its filamentous, gliding nature and its ability to grow aerobically in the dark.

3. Fine Structure of the Photosynthetic Bacteria—Other Structures

3.1. Internal Structures: Storage Granules, Gas Vacuoles

One of the more conspicuous structures found in the photosynthetic sulfur bacteria is the sulfur

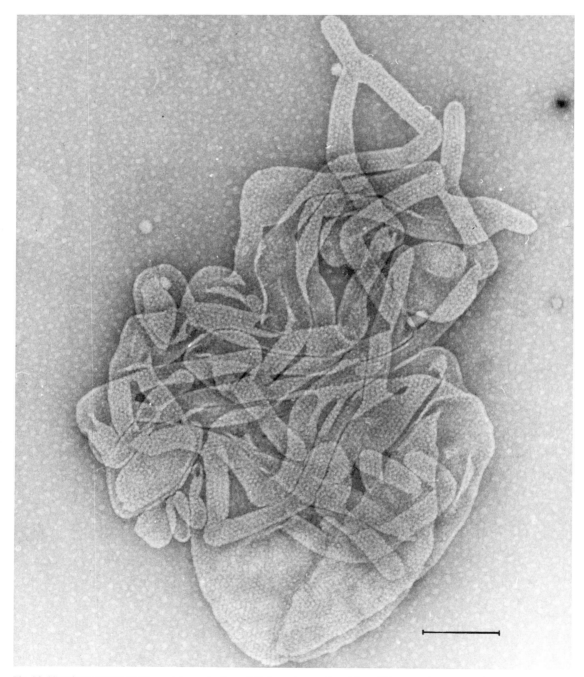

Fig. 16. Negatively stained membrane tubes and a large folded membrane released from *Thiocapsa pfennigii* cells (in stationary phase of growth) by osmotic shock treatment. Bar represents 0.2 μm. Reprinted from Eimhjellen *et al.* (1967) by permission from Springer-Verlag, Berlin and New York.

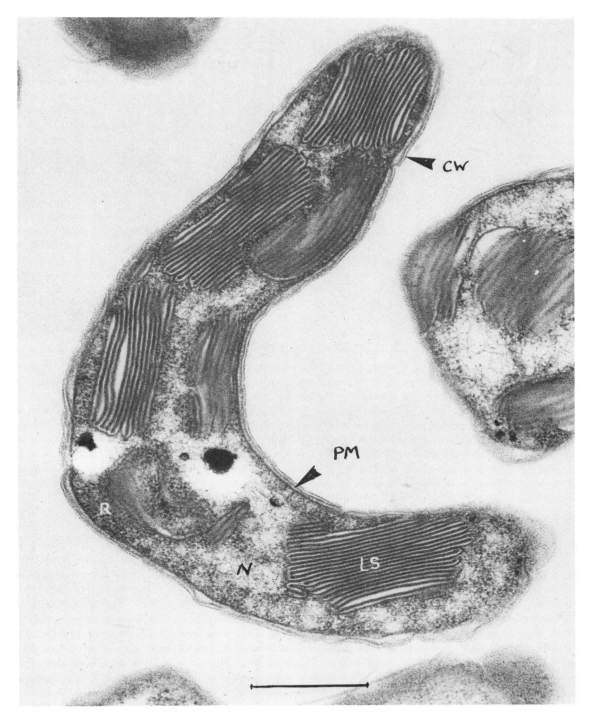

Fig. 17. Section of *Ectothiorhodospira mobilis* showing the general fine structure of the cell. A multilayered cell wall (CW), cytoplasmic membrane (PM), nucleoplasm (N), ribosomes (R), and six to seven lamellar stacks (LS) can be seen. Bar represents 0.5 μm. Reprinted from Remsen *et al.* (1968) by permission from the American Society for Microbiology, Bethesda, Maryland.

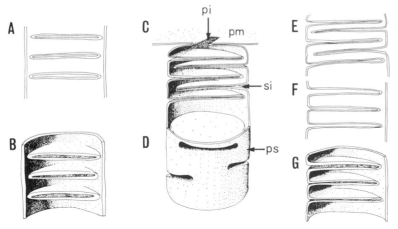

Fig. 18. Reconstruction of various arrangements of the membranes of *Ectothiorhodospira mobilis* showing spatial relationships. Shown are three-dimensional (B–D, G) and two-dimensional (A, E, F) views of this "chloroplastlike" structure. The plasma membrane (pm), primary (pi) and secondary (si) invaginations, and primary structure (ps) are illustrated. Reprinted from Remsen *et al.* (1968) by permission from the American Society for Microbiology, Bethesda, Maryland.

granules. Often, sulfur granules are distributed throughout the cytoplasm and can be up to 1.0 μm in diameter (Fig. 10). In thin section, the sulfur granule is usually seen as an empty vacuole, bounded by a limiting membrane (Figs. 7, 10, 14, and 22). In freeze–etchings, sulfur granules have a somewhat characteristic appearance when the cleavage plane passes through the central portion of the sulfur granule (Figs. 8 and 23). The appearance of sulfur granules in frozen–etched preparations is such that they can be distinguished quite easily from other cellular inclusions such as glycogen, polyphosphate, or PHB granules.

Occasionally, the cleavage plane in freeze–etchings will pass along the outer surfaces of the sulfur granules, giving some insight into the nature of the limiting membrane and the surrounding vesicular intracytoplasmic membrane (Figs. 8 and 23). This relationship of sulfur granules and vesicles is even more clearly demonstrated when sulfur granules are examined by the negative stain technique (Fig. 23) or by examining sections of osmotically lysed cells (Fig. 22). As these micrographs illustrate, there is a very close association between the vesicles and the membrane of the sulfur granule, suggesting that the limiting membrane could be derived from the intracytoplasmic membrane (see also Section 4).

Nicholson and Schmidt (1971), Schmidt *et al.* (1971), and Gonye *et al.* (submitted) described the limiting membrane of sulfur granules as being a unique proteinaceous membrane consisting of 2.5-nm globular components having a molecular weight of 13,500 daltons for *Chr. vinosum* strain D (Schmidt *et al.*, 1971) and 18,500 daltons for *Chr. buderi* (Gonye *et al.*, in prep.), indicating a structure perhaps similar to both the chlorobium vesicle and the gas vacuoles.

According to Pfennig (1967), "The cells of all those strains of purple and green bacteria which can be selectively enriched by their buoyancy, contain highly refractile, irregularly shaped structures which can be identified . . ." as gas vacuoles.

Electron-microscope studies of thin sections of gas-vacuole-containing purple and green sulfur bacteria (Hirsch, unpublished; Pfennig and Cohen-Bazire, 1967) revealed that the fine structure of the gas vacuole is essentially identical in these organisms.

The gas vacuole is resolved by the electron microscope into groups of parallel arrays of cylindrical vesicles with conical ends (Fig. 24). Individual vesicles are about 0.15 μm in diameter and differ markedly in length in different organisms; they are bounded by a membrane that appears as a single, electron-dense layer approximately 2 nm wide. In centrifuged cells in which no gas vacuoles can be detected by the light microscope, the formerly cylindrical vesicles appear to be collapsed (Fig. 11), and a regular array of subunits can be seen.

3.2. Structure and Attachment of Flagella

The fine structure of the polar flagella common to all the photosynthetic bacteria can probably be illustrated by work done with *Ectothiorhodospira mobilis* (Remsen *et al.*, 1968). The flagellum with its basal body is graphically illustrated in Fig. 25, and is generally in agreement with that described for other photosynthetic bacteria (Cohen-Bazire and London, 1967), but is different from that described for petrichously flagellated organisms. The flagella of *E. mobilis* appear to originate from a spherical structure about 50 nm in diameter; this spherical structure seems to be part of a larger structure (Fig. 26) or polar plate. Each flagellum therefore appears to have

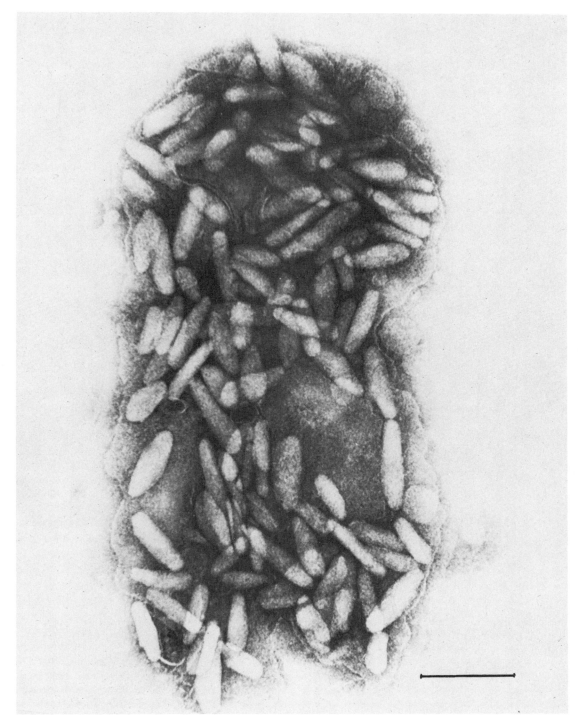

Fig. 19. *"Chloropseudomonas ethylica"* grown at 10 footcandles and ballistically opened. Note the presence of the large 150 × 50 mm vesicles in the opened cell envelope. Negatively stained with 2% phosphotungstic acid. Bar represents 0.2 μm. Reprinted from Holt *et al.* (1966) by permission of the American Society for Microbiology, Bethesda, Maryland.

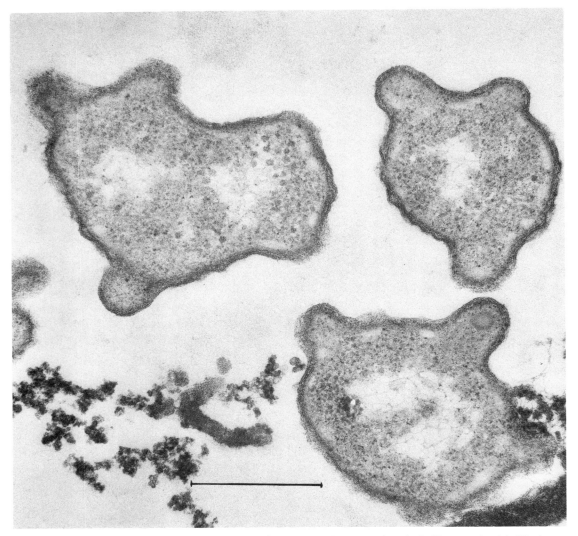

Fig. 20. Section of *Prosthecochloris aestuarii* showing the irregular structure of these green bacteria. In this preparation, it is difficult to see the chlorobium vesicles. Bar represents 0.5 μm. From Remsen and Watson (unpublished micrograph).

originated from a disklike structure or basal body. These basal bodies are kept together by their attachment or association with a larger body or polar plate.

The "ladder-shaped" structure seen in Fig. 8 is similar in shape to the polar membrane seen in sections of *E. mobilis* (Remsen *et al.*, 1968) and *Chr. buderi* (Remsen and Trüper, 1973), and is analogous to the polar membranes described in *Rs. rubrum* (Cohen-Bazire and London, 1967; Hickman and Frenkel, 1965), *Rs. molischianum*, and *Rs. fulvum* (Hickman and Frenkel, 1965) and the polar organelle

described in *Rp. palustris* (Tauschel and Drews, 1969b). The function of the polar membrane has yet to be determined.

Tauschel and Drews (1970) examined the fine structure of the flagellum of *Rp. palustris*. They found that the flagellum consists of two major regions, the flagellar hook (0.24–0.53 μm in length) and the flagellar filament (4.5–5.5 μm in length). Both regions contain a flagellar core (12.5 nm in diameter) that is composed of 2-nm subunits. Comparable data on the flagella of other photosynthetic bacteria are lacking.

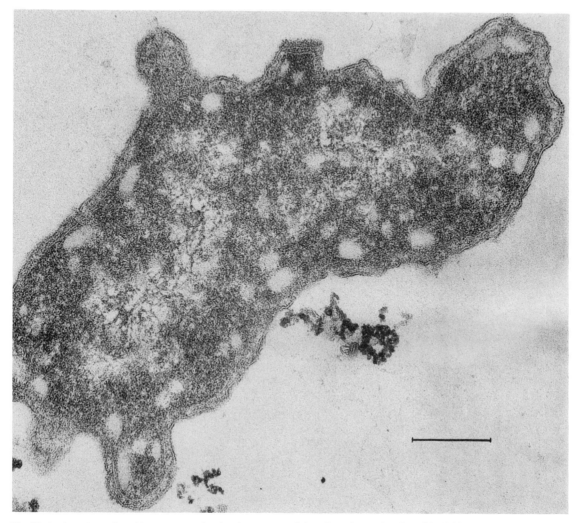

Fig. 21. Section of *Prosthecochloris aestuarii* showing the structure of the cell envelope and a view of the "chlorobium" vesicles arranged along the periphery of the cell. Bar represents 0.2 μm. From Remsen and Watson (unpublished micrograph).

4. Thoughts, Speculations, and Hypotheses

A major point of confusion among morphologists and physiologists working with the photosynthetic bacteria is the question whether all parts of the intracytoplasmic membranes are connected to each other and to the cytoplasmic membrane.

Unfortunately, as bacteriologists, we are biased by the fact that bacteria do not have discrete organelles. Pfennig (1967), in his review of photosynthetic bacteria, accurately reflects this position: "In the eucaryotic phototrophs, the major cellular functions are compartmentalized. The uptake and excretion of substances is regulated by the cytoplasmic membrane;

the respiratory electron transport is located in the mitochondria, and the enzymes and pigments of photosynthetic electron transport are located in the chloroplasts. There is increasing evidence that no such clear differentiation exists in organisms of the procaryotic type. The cytoplasmic membrane of bacteria is functionally more complex than that of eucaryotic organisms, and is potentially able to fulfill all three of the above-mentioned functions."

Stanier and van Niel (1962), in their essay on the concept of a bacterium, state it even more clearly: "Within the enclosing cytoplasmic membrane of the eucaryotic cell, certain smaller structures, which house sub-units of cellular function, are themselves surroun-

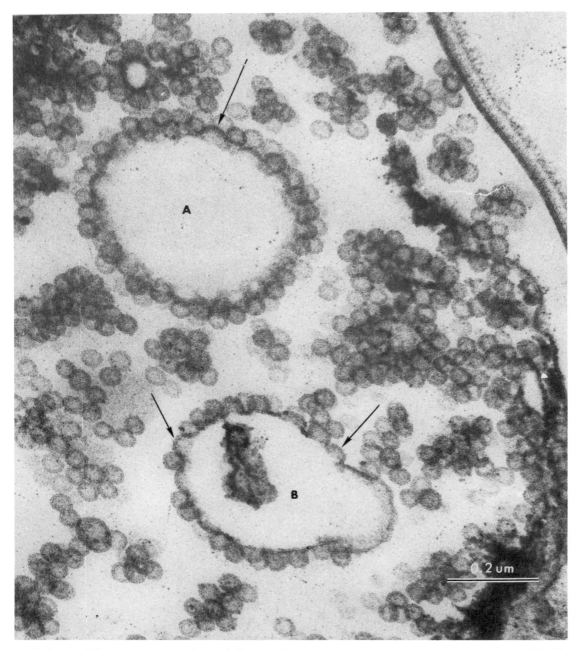

Fig. 22. Section of *Chromatium vinosum* cell, osmotically lysed prior to fixation, showing the association of chromatophores with sulfur granules A and B. Arrows indicate where the chromatophore membrane has coalesced with the sulfur granule membrane. Bar represents 0.2 μm. From Remsen (unpublished data).

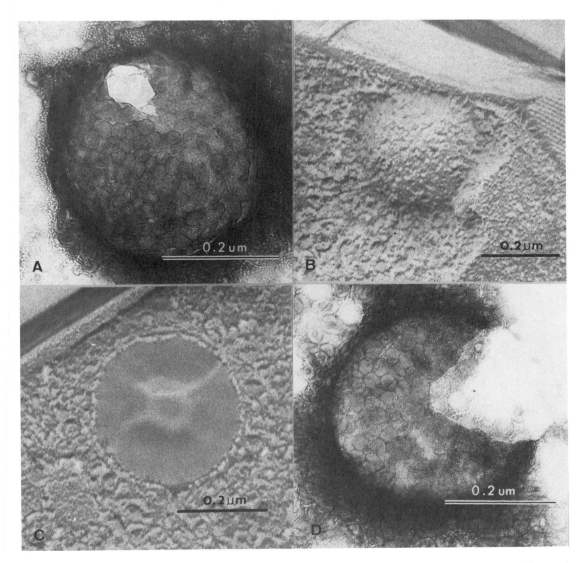

Fig. 23. (A, D) Negatively stained fractions of "purified" sulfur granules showing chromatophores aggregated around the sulfur granule membrane. Bars represent 0.2 μm. From Remsen (unpublished data). (B, C) Freeze–etchings of *Chromatium vinosum* showing (A) the smooth internal structure of a sulfur granule and its coalescing chromatophores, and (B) the outer surface of a sulfur granule with chromatophores apparently associated with the sulfur membrane. Bars represent 0.2 μm. From Remsen (unpublished data).

ded by individual membranes, interposing a barrier between them and other internal regions of the cell. In the procaryotic cell, there is no equivalent structural separation of major sub-units of cellular functions; the cytoplasmic membrane itself is the only major bounding element which can be structurally defined." Even more specifically, they continue "... In the eucaryotic cell, the enzymatic machinery of respiration and of photosynthesis is housed in specific organelles enclosed by membranes, the mito-

chondrion and chloroplast, respectively. Homologous, membrane-bounded organelles responsible for the performance of these two metabolic functions have not been found in the procaryotic cell."

There is a good deal of evidence to support this thesis. Numerous papers and reviews document the fact that the intracytoplasmic membranes of bacteria are continuous with the cytoplasmic membrane. Among photosynthetic bacteria, especially the Rhodospirillaceae, it appears that the intracytoplasmic

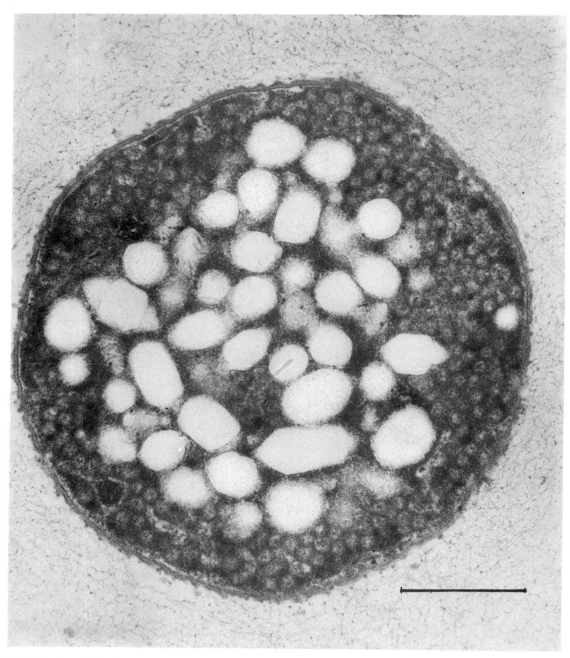

Fig. 24. Section of *Thiopedia rosea* showing arrangement and structure of gas vacuoles usually located in the central portion of the cell. Bar represents 0.5 μm. From Hirsch (unpublished micrograph).

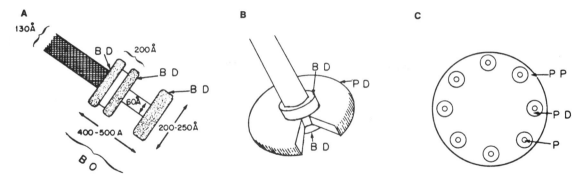

Fig. 25. (A) Diagram illustrating the fine structure of the basal portion of the flagella: basal disks (BD) and basal organelle (BO). (B) Insertion of the flagellum into polar disk (PD); basal disks (BD) appear to "clamp" flagella into polar disk. (C) Relationship of polar disk (PD) to polar plate (PP); flagella enter the polar disk through a central pore (P). Reprinted from Remsen *et al.* (1968) by permission from the American Society for Microbiology, Bethesda, Maryland.

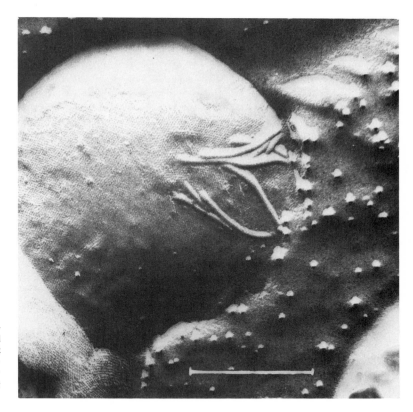

Fig. 26. Freeze–etching of *Ectothiorhodospira mobilis* showing external surface of the cell wall and the polar tuft of flagella. Bar represents 0.5 μm. Reprinted from Remsen *et al.* (1968) by permission from the American Society for Microbiology, Bethesda, Maryland.

photosynthetic membrane system and the cell membrane are a continuum. The situation is less clear in the Chromatiaceae. In at least some of this group, it is possible that vesicles may separate from the cytoplasmic membrane even though, as discussed earlier, they almost certainly arise from it. We should like to speculate that such "free vesicles" can occur and that

they are closely associated with the formation of intracellular sulfur granules. This speculation is based on the following considerations:

All purple sulfur bacteria, with the exception of *Ectothiorhodospira*, accumulate sulfur intracellularly. *Ectothiorhodospira* is also the only genus of the Chromatiaceae in which the intracytoplasmic mem-

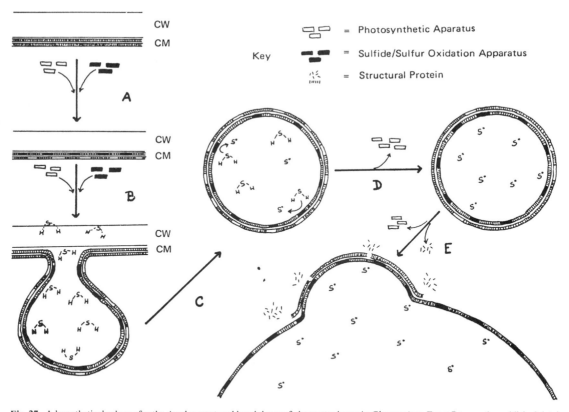

Key

= Photosynthetic Aparatus

= Sulfide/Sulfur Oxidation Apparatus

= Structural Protein

Fig. 27. A hypothetical scheme for the development and breakdown of chromatophores in *Chromatium*. From Remsen (unpublished data).

brane system has been definitely shown to remain attached to the cell membrane. In the other Chromatiaceae, it is not known whether the entire intracytoplasmic membrane system remains continuous with the cell membrane.

The speculation is more directly supported by recent work that we have been doing on sulfur granule genesis in *Chr. buderii* (Gonye *et al.*, submitted). When sulfur granules from this organism are collected and examined, the membrane surrounding the sulfur is found not to have a unit membrane structure and is only 3–4 nm thick. Furthermore, and consistent with previous reports (Schmidt *et al.*, 1971), sodium dodecylsulfate–polyacrylamide gel electrophoresis indicated a single major polypeptide with a molecular weight of approximately 18,500 based on its mobility. By the same procedures, we found that both intracytoplasmic membrane and cell envelope preparations contain large amounts of a protein with a molecular weight of 16,000–18,000. This suggests that the sulfur granule membrane, the intracytoplasmic membrane, and the cell membrane are related in some way. We propose that the sulfur granule membrane may be

derived from the cell membrane via the intracytoplasmic membranes. This idea is illustrated in Fig. 27.

The cytoplasmic membrane differentiates by incorporating the machinery for photosynthesis and sulfide oxidation onto separate halves of the membrane (Step A). As the cytoplasmic membrane continues to differentiate, it begins to invaginate, forming a vesicular structure; when this occurs, sulfide is incorporated into the vesicle (Step B). Once the cytoplasmic membrane has invaginated sufficiently to form a vesicle approximately 50 nm in diameter, the membrane pinches off, forming a discrete vesicle no longer connected to the cytoplasmic membrane and containing sulfide (Step C). As the sulfide is depleted, the vesicle may lose its ability to carry out photosynthesis. This is marked by a loss of protein from the outer half of the intracytoplasmic vesicular membrane (Step D). This is a "small" sulfur granule. There is a tendency for these to coalesce, forming typical large sulfur granules. We assume that during this process the membrane is further modified to the extent that the outer protein portion, and perhaps the phospholipids, break down to low-molecular-weight components, leaving only the inner protein part of the membrane. This is the "membrane" found around isolated sulfur granules (Step E).

ACKNOWLEDGMENTS

This chapter is Contribution No. 170, Center for Great Lakes Studies, The University of Wisconsin— Milwaukee, Milwaukee, Wisconsin 53201.

5. References

Bergeron, J. A., and Fuller. R. C., 1961, The submicroscopic basis of bacterial photosynthesis, in: *Macromolecular Complexes* (M. V. Edds, Jr., ed.), pp. 179–202, Ronald Press, New York.

Boatman, E. S., 1964, Observations on the fine structure of spheroplasts of *Rhodospirillum rubrum*, *J. Cell Biol.* **20**:297.

Boatman, E. S., and Douglas, H. C., 1961, Fine structure of the photosynthetic bacterium *Rhodomicrobium vannieli*, *J. Cell Biol.* **11**:469.

Bogorov, L. V., 1974, Properties of *Thiocapsa roseopersicina*, strain BBS, isolated from a white sea estuary, *Microbiologiya* **43**:275.

Cherni, N. E., Solov'eva, Zh. V., Federov, V. D., and Kondrat'eva, E. N., 1969, Ultrastructure of two species of purple sulfur bacteria, *Mikrobiologiya* **38**:479.

Cohen-Bazire, G., 1963, Some observations on the organization of the photosynthetic apparatus in purple and green bacteria, in: *Bacterial Photosynthesis* (H. Gest, A. San Pietro, and L. P. Vernon, eds.), pp. 89–114, Antioch Press, Yellow Springs, Ohio.

Cohen-Bazire, G., and Kunisawa, R., 1963, The fine structure of *Rhodospirillum rubrum*, *J. Cell Biol.* **16**:401.

Cohen-Bazire, G., and London, J., 1967, Basal organelles of bacterial flagella, *J. Bacteriol.* **94**:458–465.

Cohen-Bazire, G., and Sistrom, W. R., 1966, The procaryotic photosynthetic apparatus, in: *The Chlorophylls* (L. P. Vernon and G. R. Seely, eds.), pp. 313–341, Academic Press, New York.

Cohen-Bazire, G., Pfennig, N., and Kunisawa, R., 1964, The fine structure of green bacteria, *J. Cell Biol.* **22**:207.

Conti, S. F., and Hirsch, P., 1965, Biology of budding bacteria. III. Fine structure of *Rhodomicrobium* and *Hyphomicrobium* spp., *J. Bacteriol.* **89**:503.

Drews, G., 1960, Untersuchungen zur Substruktur der "Chromatophoren" von *Rhodospirillum rubrum* und *Rhodospirillum molischianum*. *Arch. Mikrobiol.* **36**:99.

Drews, G., and Giesbrecht, P., 1963, Zur Morphogenese der Bakterien "Chromatophoren" (-Thylakoide) und zur Synthese des Bakteriochlorophylls bei *Rhodopseudomonas spheroides* und *Rhodospirillum rubrum*, *Zentralbl. Bakteriol. Parasitenkd. Infektionskr. Hyg. Abt. 1: Orig., Reihe A* **190**:508.

Drews, G., and Giesbrecht, P., 1965, Die Thylakoidstrukturen von *Rhodopseudomonas spec.*, *Arch. Mikrobiol.* **52**:242.

Drews, G., Lampe, H.-H., and Ladwig, R., 1969, Die Entwicklung des Photosyntheseapparates in Dunkelkulturen von *Rhodopseudomonas capsulata*, *Arch. Mikrobiol.* **65**:12.

Drews, G., Leutiger, I., and Ladwig, R., 1971, Production of protochlorophyll, protopheophytin and bacteriochlorophyll in the mutant AIa of *Rhodopseudomonas capsulata*, *Arch. Mikrobiol.* **76**:349.

Eimhjellen, K. E., 1970, *Thiocapsa pfennigii* sp. nov., a new species of the phototrophic sulfur bacteria, *Arch. Mikrobiol.* **73**:193–194.

Eimhjellen, K. E., Steensland, H., and Traetteberg, J., 1967, A *Thiococcus* sp. nov. gen., its pigments and internal membrane system, *Arch. Mikrobiol.* **59**:82.

French, C. S., 1938, The chromoproteins of photosynthetic purple bacteria, *Science* **88**:60–62.

French, C. S., 1940, The pigment-protein compound in photosynthetic bacteria. I. The extraction and properties of photosynthesis, *J. Gen. Physiol.* **23**:469–481.

Frenkel, A. W., 1954, Light induced photophosphorylation by cell free preparations of photosynthetic bacteria, *J. Am. Chem. Soc.* **76**:5568–5569.

Frenkel, A. W., and Hickman, D. D., 1959, Structure and photochemical activity of chlorophyll-containing particles from *Rhodospirillum rubrum*, *J. Biophys. Biochem. Cytol.* **6**:285–290.

Fuller, R. C., Conti, S. F., and Mellin, D. B., 1963, The structure of the photosynthetic apparatus in the green and purple sulfur bacteria, in *Bacterial Photosynthesis* (H. Gest, A. San Pietro, and L. P. Vernon, eds.), pp. 71–87, Antioch Press, Yellow Springs, Ohio.

Gibbs, S. P., Sistrom, W. R., and Worden, P. B., 1965, The photosynthetic apparatus of *Rhodospirillum molischianum*, *J. Cell Biol.* **26**:395.

Giesbrecht, P., and Drews, G., 1962, Elektronenmikroskopische Untersuchungen über die Entwicklung der "Chromatophoren" von *Rhodospirillum molischianum* Giesberger, *Arch. Mikrobiol.* **43**:152.

Giesbrecht, P., and Drews, G., 1966, Über die Organisation und die makromolekuläre Architektur der Thylakoide "lebender" Bakterien, *Arch. Mikrobiol.* **54**:297.

Gorlenko, V. M., and Zhilina, T. N., 1968, Study of the ultrastructure of green sulfur bacteria, strain SK-413, *Mikrobiologiya* **37**:1052.

Hickman, D. D., and Frenkel, A. W., 1965, Observations on the structure of *Rhodospirillum molischianum*, *J. Cell Biol.* **25**:261.

Holt, S. C., and Marr, A. G., 1965a, Location of chlorophyll in *Rhodospirillum rubrum*, *J. Bacteriol.* **89**:1402.

Holt, S. C., and Marr, A. G., 1965b, Isolation and purification of the intracytoplasmic membranes of *Rhodospirillum rubrum*, *J. Bacteriol.* **89**:1413.

Holt, S. C., and Marr, A. G., 1965c, Effect of light intensity on the formation of intracytoplasmic membranes in *Rhodospirillum rubrum*, *J. Bacteriol.* **89**:1421.

Holt, S. C., Conti, S. F., and Fuller, R. C., 1966, Effect of light intensity on the formation of the photochemical apparatus in the green bacterium *Chloropseudomonas ethylicum*, *J. Bacteriol.* **91**:349.

Holt, S. C., Trüper, H. C., and Takács, B. J., 1968, Fine structure of *Ectothiorhodospira mobilis* Strain 8113 thylakoids: Chemical fixation and freeze–etching studies, *Arch. Mikrobiol.* **62**:111.

Ketchum, P. A., and Holt, S. C., 1970, Isolation and characterization of the membrane from *Rhodospirillum rubrum*, *Biochim. Biophys. Acta* **196**:125.

Kran, Von G., Schlote, F. W., and Schlegel, H. G., 1963,

Cytologische Untersuchungen an *Chromatium okenii* Perty, *Naturwissenschaften* **50**:128.

Lampe, H. H., Oelze, J., and Drews, G., 1972, Die Fraktionierung des Membransystems von *Rhodopseudomonas capsulata* und seine Morphogenese, *Arch. Mikrobiol.* **83**:78.

Madigan, M. T., and Brock, T. D., 1975, Photosynthetic sulfide oxidation by *Chloroflexus aurantiacus*, a filamentous, photosynthetic, gliding bacterium, *J. Bacteriol.* **122**:782.

Newton, J. W., and Newton, G. A., 1957, Composition of the photoactive subcellular particles from *Chromatium*, *Arch. Biochem. Biophys.* **71**:250–265.

Nicholson, G. I., and Schmidt, G. L., 1971, Structure of the *Chromatium* sulfur particle and its protein membrane, *J. Bacteriol.* **105**:1142.

Niklowitz, W., and Drews, G., 1955, Zur Elektronenmikroskopischen Darstellung der Feinstruktur von *Rhodospirillum rubrum*, *Arch. Mikrobiol.* **23**:123.

Oelze, J., and Drews, G., 1972, Membranes of photosynthetic bacteria, *Biochim. Biophys. Acta* **265**:209.

Oelze, J., Biedermann, M., and Drews, G., 1969, Die Morphogenese des Photosyntheseapparates von *Rhodospirillum rubrum*. I. Die Isolierung und Charakterisierung von zwei Membransystem, *Biochim. Biophys. Acta* **173**:436.

Peters, G. A., and Cellarius, R. A., 1972, Photosynthetic membrane development in *Rhodopseudomonas spheroides*. II. Correlation of pigment incorporation with morphological aspects of thylakoid formation, *Bioenergetics* **3**:345.

Petrova, E. A., 1959, The morphology of purple sulfur bacteria of the genus *Chromatium* in relation to the medium, *Mikrobiologiya* **28**:414.

Pfennig, N., 1967, Photosynthetic bacteria, *Annu. Rev. Microbiol.* **21**:285.

Pfennig, N., and Cohen-Bazire, G., 1967, Some properties of the green bacterium *Pelodictyon clathratiforme*, *Arch. Mikrobiol.* **59**:226.

Pierson, B. K., and Castenholz, R. W., 1974, A phototrophic gliding, filamentous bacterium of hot springs, *Chloroflexus aurantiacus gen.* and *sp. nov.*, *Arch. Mikrobiol.* **100**:5.

Raymond, J. C., and Sistrom, W. R., 1967, The isolation and preliminary characterization of a halophilic photosynthetic bacterium, *Arch. Mikrobiol.* **59**:255.

Remsen, C. C., and Trüper, H. G., 1973, The fine structure of *Chromatium buderi*, *Arch. Mikrobiol.* **90**:269.

Remsen, C. C., Watson, S. W., Waterbury, J. B., and Trüper, H. G., 1968, Fine structure of *Ectothiorhodospira mobilis* Pelsh, *J. Bacteriol.* **95**:2374.

Schachman, H. K., Pardee, A. B., and Stanier, R. Y., 1952, Studies on the molecular organization of microbial cells, *Arch. Biochem.* **38**:213–221.

Schmidt, G. L., Nicholson, G. L., and Kamen, M. D., 1971, Composition of the sulfur particle of *Chromatium vinosum* strain D, *J. Bacteriol.* **105**:1137.

Schön, G., and Jank-Ladwig, R., 1972, Veränderungen des intracytoplasmatischen Membransystem (Thylakoide) von *Rhodospirillum rubrum* unter Stickstoff-limitierung, *Arch. Mikrobiol.* **85**:319.

Schön, G., and Ladwig, R., 1970, Bacteriochlorophyllsynthese und Thylakoidmorphogenese in anaerober Dunkelkultur von *Rhodospirillum rubrum*, *Arch. Mikrobiol.* **74**:356.

Solov'eva, Zh. V., and Fedenko, E. P., 1970, Ultrafine cell structure of the parent strain and the pigmented mutant of *Rhodopseudomonas palustris*, *Microbiologiya* **39**:94.

Solov'eva, Zh. V., and Federov, V. D., 1970, Effect of growth condition on ultrathin structure of *Chlorobium thiosulfatophilum*, *Microbiologiya* **39**:739.

Stanier, R. Y., 1963, The organization of the photosynthetic apparatus in purple bacteria, in: *The General Physiology of Cell Specialization* (D. Mazia and A. Tyler, eds.), pp. 242–252, McGraw-Hill Book Co., New York.

Stanier, R. Y., and van Niel, C. B., 1962, The concept of a bacterium, *Arch. Mikrobiol.* **42**:17

Takács, B. J., and Holt, S. C., 1971, *Thiocapsa floridana*; a cytological, physical and chemical characterization. I. Cytology of whole cells and isolated chromatophore membranes, *Biochim. Biophys. Acta* **233**:258.

Tauschel, H.-D., and Drews, G., 1967, Thylakoidmorphogenese bei *Rhodopseudomonas palustris*, *Arch. Mikrobiol.* **59**:381.

Tauschel, H.-D., and Drews, G., 1969a, Zum Norkommen helixförmig angevordneter Ribosomen bei *Rhodopseudomonas palustris*, *Arch. Mikrobiol.* **64**:377.

Tauschel, H.-D., and Drews, G., 1969b, Der Geisselapparat von *Rhodopseudomonas palustris*. I. Untersuchungen zur Feinstruktur des Polorganells, *Arch. Mikrobiol.* **66**:166.

Tauschel, H.-D., and Drews, G., 1969c, Der Geisselapparat von *Rhodopseudomonas palustris*. II. Entstehung und Feinstruktur der Geissel-Basalkörper, *Arch. Mikrobiol.* **66**:180.

Tauschel, H.-D., and Drews, G., 1970, Der Geisselapparat von *Rhodopseudomonas palustris*. III. Untersuchungen zur Feinstruktur der Geissel, *Cytobiologie Z. Exp. Zellforsch.* **2**:87.

Tauschel, H.-D., and Hoeniger, J. F. M., 1974, The fine structure of *Rhodospirillum acidophila*, *Can. J. Microbiol.* **20**:13.

Tomina, I. V., and Federov, V. D., 1967, Ultra-fine structure of *Chlorobium thiosulfatophilum*, a green sulfur bacterium, *Mikrobiologiya* **36**:663.

Trentini, W. C., and Starr, M. P., 1967, Growth and ultrastructure of *Rhodomicrobium vannielii* as a function of light intensity, *J. Bacteriol.* **93**:1699.

Trüper, H. G., and Schlegel, H. G., 1964, Sulphur metabolism in Thiorhodaceae. I. Quantitative measurements on growing cells of *Chromatium okenii*, *Antonie van Leeuwenhoek*; *J. Microbiol. Serol.* **30**:225–238.

Vatter, A. E., and Wolfe, R. S., 1958, The structure of photosynthetic bacteria, *J. Bacteriol.* **75**:480–488.

Weckesser, J., Drews, G., and Tauschel, H.-D., 1969, Zur Feinstruktur und Taxonomie von *Rhodopseudomonas gelatinosa*, *Arch. Mikrobiol.* **65**:346.

Whittenbury, R., and McLee, A. G., 1967, *Rhodopseudomonas palustris* and *Rh. viridis*—photosynthetic budding bacteria, *Arch. Mikrobiol.* **59**:324.

Worden, P. B., and Sistrom, W. R., 1964, The preparation and properties of bacterial chromatophore fractions, *J. Cell Biol.* **23**:135–150.

CHAPTER 4

Cell Envelopes

Gerhart Drews, Jürgen Weckesser, and Hubert Mayer

1. Functions and Properties

Cells of photosynthetic bacteria are surrounded by
envelopes. The first envelope external to the cytoplasmic membrane is a multilayered structure, the cell
wall. The cell wall is responsible for the shape of the
cell and withstands the osmotic pressure of the
protoplasts. Naked protoplasts are spherical (Boatman, 1964) and burst if not osmotically stabilized.
Photosynthetic bacteria have spherical (*Thiocapsa*),
rod-shaped (*Rhodopseudomonas*), screw-shaped
(*Rhodospirillum*), prosthecate (*Prosthecochloris*), or
stalk-forming, budding cells (*Rhodomicrobium*). Envelopes communicate directly with the surroundings
by means of phage receptors, *O*-antigens, or, generally
speaking, by areas specified by their chemical structure and charge.

External to the typical cell wall, some photosynthetic organisms form regular surface patterns or thick
capsules or thin dissolving slime layers.

2. Fine Structure

2.1. Cell Walls

So far as is known, photosynthetic bacteria are
gram-negative. A typical gram-negative cell wall
consists of a basic electron-dense layer (width 3–8 nm)

Gerhart Drews and Jürgen Weckesser · Institut für Biologie II,
Universität Freiburg **Hubert Mayer** · Max-Planck-Institut für
Immunbiologie, Freiburg, West Germany

and the triple-layered so-called "outer membrane"
(width about 7.5 nm) (Glauert and Thornley, 1969;
Costerton *et al.*, 1974). The electron-dense layer,
which is identical with the murein layer (Fig. 1), can
usually be resolved only with difficulty in thin sections
of photosynthetic bacteria. Either it is closely associated with the outer membrane, so that it is seen as a
thickening of the inner surface of the outer membrane,
or it is seen as an additional layer that follows the
contours of the outer membrane closely.

Apparently all photosynthetic bacteria have an
outer membrane (Fig. 1, OM) in the cell wall that has
the appearance of a unit membrane in thin sections,
with a thickness of 7–8 nm. The outer membrane is
not visible in thin sections of bacteria extracted with
hot phenol/water (Westphal *et al.*, 1952). Partial disappearance of outer cell-wall material has been
observed after extraction with saline (Weckesser *et al.*,
1972*b*). The demonstration of lipopolysaccharides
(LPS) in both kinds of extracts and the reaction of
isolated LPS with antisera prepared in rabbits with
whole bacteria demonstrate that LPS are components
of the outer membrane (Weckesser *et al.*, 1972*a,b*). In
addition to LPS, lipids and proteins were found in the
outer cell-wall layers extractable by saline (Weckesser
et al., 1972*b*).

The surfaces of the photosynthetic bacteria appear
smooth in negative stained and etched preparations
(Fig. 2). The inner surfaces of frozen–etched cell-wall
preparations show a wrinkled fracture face studded
with fine particles, which is apparently a protein layer
within the outer membrane (Fig. 3).

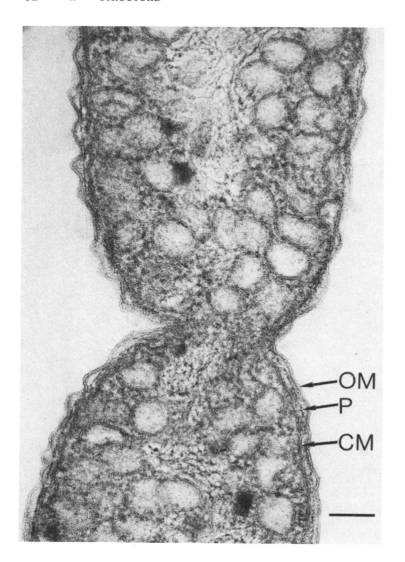

Fig. 1. *Rhodospirillum rubrum*—ultrathin section. (CM) Cytoplasmic membrane; (OM) outer membrane of cell wall; (P) peptidoglycan layer. The bars in this figure and following figures represent 0.1 μm. Micrograph by R. Marx.

2.2. External Envelopes

Structured surface layers, external and additional to the outer membrane, are formed in some gram-negative bacteria (Glauert and Thornley, 1969). Consequently, they cover the O-antigens. In shadowed preparations of replicas or freeze–dried cells of *Rhodospirillum rubrum* and of a *Rhodopseudomonas* strain, regular arrays of particles with center-to-center distances of 16.5 and 13.0 nm, respectively, were described (Salton and Williams, 1954; de Boer and Spit, 1964). *Ectothiorhodospira mobilis* and *Chromatium gracile* have an external layer of spherical particles with a diameter of 5 nm and a center-to-center distance of 8 nm (Remsen *et al.*, 1968, 1970). The surface of *Chr. buderi* appears to consist of tightly packed five-sided, cup-shaped units approximately 3.5 nm in diameter. The wine-glass-like cup was found to be resting on a thin stem, 5 nm wide and 15 nm long. The units are held together at the base of the stem by a thin, dense 7.5-nm layer situated above the outer membrane (Remsen *et al.*, 1970). The cell walls of *Chr. okenii* and *Chr. weissei* (Fig. 4) are covered by a structured layer composed of a hexagonal array of hollow cone-shaped subunits approximately 25 nm long and 13 nm in diameter (Hageage and Gherna, 1971). A surface layer composed of hollow, spherical subunits with a diameter of 10 nm was found in a prosthecate green sulfur bacterium (Fig. 5).

Capsular material was observed on the surfaces of many green sulfur bacteria (Cohen-Bazire *et al.*, 1964;

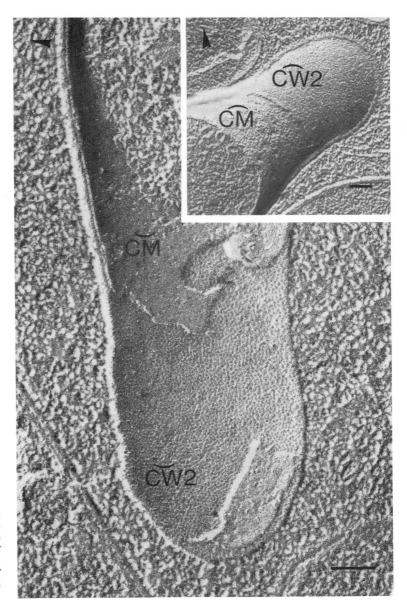

Fig. 2. *Rhodospirillum rubrum*—frozen–etched preparation. Fracture faces exposing cytoplasmic membrane (CM) and cell wall layers. (CW2) Convex and concave fracture faces of a particle layer within the cell wall. The arrow indicates the direction of shadow. Preparation and micrograph by J. Golecki.

Pfennig and Cohen-Bazire, 1967). Some photosynthetic bacteria, such as *Thiocapsa*, produce large amounts of slime in which single cells or cell aggregates are embedded. The slime has a spongy, fine-fibrous structure (Fig. 6).

3. Chemical Composition

Our present knowledge of the chemical composition of cell envelopes of photosynthetic bacteria is restricted to peptidoglycan and LPS. It seems reasonable to assume that outer membrane and external layers have additional macromolecular components like other gram-negative cell walls (Braun and Hantke, 1974; Costerton *et al.*, 1974; Henning *et al.*, 1973; Ames *et al.*, 1974; Randall-Hazelbauer and Schwartz, 1973; Sleytr and Thornley, 1973; Thornley *et al.*, 1974). The data from photosynthetic bacteria are given in Section 3.3.

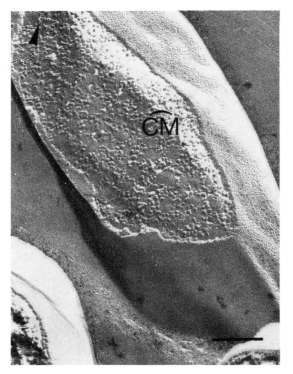

Fig. 3. *Rhodospirillum rubrum*—frozen–etched preparation. The smooth surface of the cell wall is exposed by etching. Preparation and micrograph by J. Golecki.

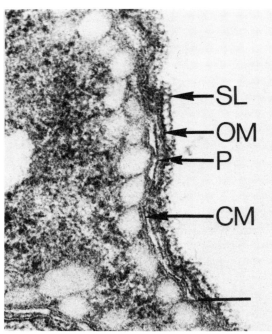

Fig. 5. Cell envelope of a prosthecae-forming green sulfur bacterium—cross-sectioned. (SL) Structured outer layer of cell wall; (OM) outer membrane; (P) peptidoglycan layer; (CM) cytoplasmic membrane. Micrograph by R. Ladwig.

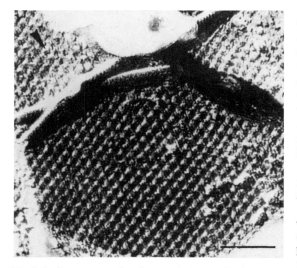

Fig. 4. Surface pattern of a frozen-etched cell of *Chromatium weissii*. Reproduced from Hageage and Gherna (1971) by courtesy of the authors.

3.1. Murein (Peptidoglycan)

In contrast to the variability of peptidoglycan types in gram-positive bacteria, gram-negative cell walls apparently have only one type of peptidoglycan. It contains *N*-acetylmuramic acid, *N*-acetylglucosamine, L-alanine, D-glutamic acid, *m*-diaminopimelic acid, and D-alanine in the molar ratio $1:1:1:1:1:1$ (Schleifer and Kandler, 1972). *N*-Acetyl-muramic acid and *N*-acetylglucosamine are linked together by β 1→ 4-glycosidic linkages. The peptide side chain is bound to the carboxyl group of muramic acid through the amino group of L-alanine (Martin, 1966). Figure 7 illustrates the primary structure and the directly cross-linked variation of the gram-negative peptidoglycan type. The general structure of this peptidoglycan type is discussed by Martin (1966), Ghuysen *et al.* (1968), and Schleifer and Kandler (1972).

The characteristic amino acid pattern of this type was found in the peptidoglycan of *Rs. rubrum* (Newton, 1968). The average extent of peptide cross-linking in bacilliform mutants of this bacterium was 33%, in contrast to 44% in spirilla. A high percentage of cross-linkage was also found in *Spirillum serpens* peptidoglycan by Kolenbrandner and Ensign (1968).

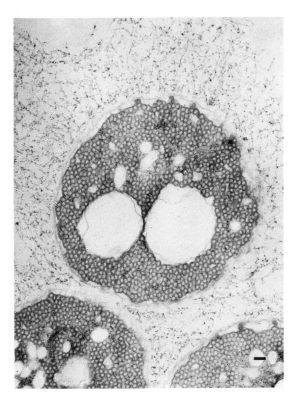

Fig. 6. *Thiocapsa sp.*—cells are embedded in slime. Micrograph by R. Ladwig.

$$-G-M-G-$$
$$\downarrow$$
$$L-Ala$$
$$\downarrow$$
$$D-Glu \overset{\alpha}{\rightarrow} NH_2$$
$$\downarrow \gamma$$
$$m-Dap \overset{\omega}{\longleftarrow} D-Ala$$
$$\downarrow \qquad \uparrow$$
$$D-Ala \qquad m-Dap$$
$$\qquad \uparrow$$
$$NH_2 \overset{\alpha}{\rightarrow} D-Glu$$
$$\uparrow$$
$$L-Ala$$
$$\uparrow$$
$$-G-M-G-$$

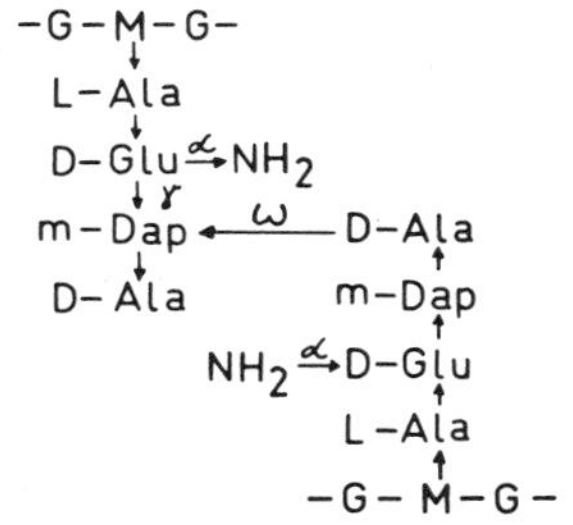

Fig. 7. Peptidoglycan type of gram-negative cell walls.

It was proposed by Newton (1968, 1970, 1971) that interference with the metabolism of D-alanine can alter the extent of cross-linking and in consequence the morphology of *Rs. rubrum*.

3.2. Lipopolysaccharides (LPS)

3.2.1. General Properties

A common feature of gram-negative bacteria is the occurrence of glycoconjugates on the cell envelope (Lüderitz *et al.*, 1968). LPS is the trivial name for these macromolecular substances, which show the same gross chemical composition as the well-investigated enterobacterial LPS. These heteropolymers, also called *O*-antigens, carry the type-specific *O*-antigenicity and form one of the main components of the outer membrane of gram-negative cell walls. LPS of most species exhibit as "endotoxins" more or less pronounced biological activities (lethal toxicity, pyrogenicity, etc.; Lüderitz *et al.*, 1968, 1971).

LPS seem to be linked to the cell wall primarily by noncovalent bonds, i.e., hydrophobic or ionic interac-

tions (Lüderitz *et al.*, 1968; Shands, 1971). It has been suggested that phospholipids and LPS form a lipid bilayer in the outer membrane of the cell wall interspersed with proteins (Forge and Costerton, 1973; Forge *et al.*, 1973; Shands, 1971). Freeze–etching studies have shown that there exists a cleavage plane in the outer membrane (Fig. 2) (Forge *et al.*, 1973; Costerton *et al.*, 1974), indicating a hydrophobic zone within the outer membrane. In consequence, the outer membrane cannot be penetrated by many molecules from the medium inward or from the periplasmatic space outward and contributes to a barrier within the cell envelopes (Costerton *et al.*, 1974). Investigations carried out with S- and R-strains of *Salmonella typhimurium* showed that LPS are important components of this barrier (Schlecht and Westphal, 1970; Costerton *et al.*, 1974). From immunological studies, it is clear that the hydrophilic parts of the LPS, i.e., the *O*-specific chains, reach the surface of the cell.

The chemical investigation of LPS started with enterobacterial strains and was greatly facilitated by previous serological classification of *Salmonella* and other members of the family *Enterobacteriaceae*, as documented in the well-known Kauffmann–White scheme of *Salmonella* (Kauffmann, 1966). It is therefore not surprising that the detailed knowledge accumulated in the past years on the serological reactivity, the chemical composition, and their mutual relationship, as well as on the biosynthesis of LPS and their genetic determination, is generally used as a comparative basis for studies of *O*-antigens of bacteria belonging to other taxonomic groups. The general structure of a *Salmonella* LPS is given in Fig. 8, which shows the three distinct structural subregions that are independently synthesized and that are under separate genetic control (Stocker and Mäkelä, 1971). The lipid moiety, called lipid A, consists of a glucosamine–disaccharide backbone, on which phosphorus and long-chain fatty acids are bound. The fatty acids form amide or ester bonds to the amino sugar. Characteristic constituents of the R-core oligosaccharide (Fig. 8) are 2-keto-3-deoxyoctonate (KDO) and L-glycero-D-mannoheptose. KDO forms the acid-labile binding

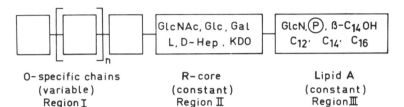

O-specific chains | R-core | Lipid A
(variable) | (constant) | (constant)
Region I | Region II | Region III

Fig. 8. General structure of a *Salmonella* type lipopolysaccharide Lüderitz *et al.* (1971).

between the polysaccharide moiety and lipid A. Lipid A and the R-core show little variation in structure within the Enterobacteriaceae. In contrast, the repeating units of the O-specific chains are distinguished by a high diversity of the strain-specific sugar compositions. S-LPS, formed by smooth wild-type strains, contains all three subregions; R-LPS, formed by rough-mutants, lacks the O-specific chains (Ra) or, in addition, parts of the R-core (Rb–Re). In the following discussion we will use the terminology established for enterobacterial LPS, keeping in mind, however, that at present our knowledge of the structure of LPS of the photosynthetic bacteria is fragmentary.

3.2.2 Extraction Procedures

Of the numerous procedures developed for the extraction of LPS from gram-negative bacteria (Lüderitz *et al.*, 1968), three should be mentioned here, since they have been successfully used for extracting LPS of photosynthetic bacteria.

The most common method is the hot phenol/water procedure (Westphal *et al.*, 1952). LPS are usually extracted into the aqueous phase. With Rhodospirillaceae, however, it was often noticed that LPS were either totally or partly extracted into the phenol phase, e.g., the LPS of *Rp. palustris* strains (Weckesser *et al.*, 1973a). A method described by Galanos *et al.* (1969) is especially suitable for the extraction of lipophilic glycolipids (R-LPS) of enterobacterial R-mutants and involves an extraction of carefully dried bacteria with a mixture of phenol/chloroform/petroleum ether. It was found that this method gives good results with wild-type strains of *Rp. gelatinosa* and *Rs. tenue*; the LPS of these resemble the LPS of enterobacterial R-mutants in their chemical composition.

Treatment of aqueous cell suspensions with chelating agents such as EDTA or with hot saline releases some LPS and proteins from the bacteria (Weckesser *et al.*, 1972b).

Thus, the behavior of LPS in the various extraction procedures provides first information on chemical and physical properties of the extracted macromolecules. A major impediment in the study of LPS is, however,

the marked polydispersity and heterogeneity found in many preparations (Shands, 1971; Nowotny, 1971).

3.2.3. Physical State

In water, LPS form turbid dispersions; the size and shape of the aggregates seem to depend on hydrophobic interactions and on the content of polyvalent cations (putrescine, spermidine, Ca^{2+}, and Mg^{2+}) that are able to cross-link separate chains of polyanionic macromolecules (Galanos and Lüderitz, 1975).

Sections of embedded LPS material show trilaminar structures in the electron microscope (Shands, 1971). It was proposed that LPS in aqueous environment exist as a bilayer of polysaccharide and lipid, with the nonpolar lipids occupying the interior of the bilayers (Burge and Draper, 1967; de Petris, 1967; Shands *et al.*, 1967). A comparison between the morphology of purified LPS preparations and preparations containing also other constituents of the outer membrane suggested, however, that the structure of the outer membrane is determined not only by LPS, but also by the phospholipid–protein constituents.

3.2.4. LPS of Photosynthetic Bacteria

(a) Discovery. The presence of LPS in the cell walls of photosynthetic bacteria was first proposed by Salton (1960) and Newton (1967). They found glucose, fucose, rhamnose, and an unknown sugar in the cell walls or LPS fraction of *Rs. rubrum*. From *Rp. capsulata*, a LPS similar to enterobacterial O-antigens was isolated by Weckesser *et al.* (1972a). This LPS was extractable by the hot phenol/water procedure into the aqueous phase; it was shown to be localized in the outer membrane, and it exhibited O-antigenicity. furthermore, it showed lethal toxicity in mice, although the LD_{50} is considerably higher than that reported for enterobacterial LPS (Weckesser *et al.*, 1972b). It formed aggregates that were sedimentable at 100,000g in the ultracentrifuge and appeared as ribbonlike patterns in the electron microscope. Treatment with deoxycholate resulted in a reversible splitting into smaller units (Weckesser *et al.*, 1972b).

Hydrolysis with 1% acetic acid resulted in the formation of a water-insoluble lipid material (lipid A) and a water-soluble polysaccharide moiety (degraded

Table 1. Sugar Constituents of Lipopolysaccharides of Rhodospirillaceae and Chromatiaceae

Structure	Configuration		Trivial name	Occurrence[a]
Hexoses	Glucose	D		Frequent
	Galactose	D		Frequent
	Galactose, 2-*O*-methyl-	D		*Rp. gelatinosa*
	Galactose, 2,3-di-*O*-methyl-	D		*Rp. gelatinosa*
	Mannose	D		*Rp. viridis, Rp. gelatinosa, Thiocapsa roseopersicina*
	Mannose, 3-*O*-methyl-	D		*Rp. viridis*
6-Deoxyhexoses	Galacto-	D	Fucose	*Rs. tenue*
	Galacto-	L	Fucose	*Rp. gelatinosa*
	Galacto-, 2-*O*-methyl-	L		*Rp. gelatinosa*
	Manno-	L	Rhamnose	Frequent
	Manno-	D	Rhamnose	*Rs. tenue, Rs. rubrum* 2.4.6
	Manno-, 2-*O*-methyl-	D		*Rs. tenue*
	Manno-, 2-*O*-methyl-	L		*Rp. sphaeroides* 28/3A and 28/4, *Thiocapsa roseopersicina*
	Manno-, 3-*O*-methyl-	L	Acofriose	*Rp. capsulata* 37b4, *Rs. rubrum* 2.4.6
	Talo-	D	Talomethylose	*Rp. palustris*
	Talo-	L	Talomethylose	*Rp. sphaeroides* W31
	Talo-, 3-*O*-methyl-	D	Acovenose	*Rp. palustris*
3,6-Dideoxyhexoses	Galacto-	D	Abequose	*Rp. gelatinosa*
	Galacto-	L	Colitose	*Rp. sphaeroides* 413
Pentoses	Arabinose	D		*Rs. tenue*
	Arabinose	L		*Chr. vinosum*
	Xylose	D		*Rp. palustris*
	Xylose, 4-*O*-methyl-	D		*Rp. palustris*
	Xylose, 3-*O*-methyl-	L		*Rp. viridis, Rs. rubrum* W
	Ribose, 3-*O*-methyl-	D		*Chr. vinosum*
Heptoses				
L-glycero	Manno-	D		Frequent
D-glycero	Manno-	D		*Rp. gelatinosa, Rs. tenue*
2-Amino-2-deoxyhexoses	Gluco-	D	Glucosamine	Frequent
	Gluco-, 6-*O*-methyl-	D		*Rp. palustris*
	Galacto-	D	Galactosamine	Frequent
	Manno-	D	Mannosamine	*Chr. vinosum*
2-Amino-2,6-dideoxyhexoses	Gluco-	D	Quinovosamine	*Rp. palustris, Rp. viridis, Rs. tenue, Chr. vinosum*
	Galacto-	?	Fucosamine	*Rs. tenue*
2,3-Diamino-2,3-dideoxyhexoses	Gluco-	D		*Rp. palustris, Rp. viridis*
2-Amino-2-deoxyhexuronic acids	Galacto	?	Galactosamin-uronic acid	*Rp. viridis, Rs. tenue*
2-Keto-3-deoxyoctonic acids	Manno-	D	KDO	Frequent
Sialic acids	Manno-	D	Neuraminic acid	*Rp. capsulata* 37b4

[a] Strain designations are not given for *Rp. gelatinosa*, *Rp. viridis*, *Rp. palustris*, and *Rs. tenue*, since detailed tables of their sugar composition are given in Tables 2, 4, and 6.

Table 2. Sugar Patterns in LPS of Strains of *Rhodopseudomonas palustris* and *Rhodopseudomonas viridis*[a]

Strain	Chemo-type	4-O-Methyl-D-xylose	3-O-Methyl-L-xylose	3-O-Methyl-D-mannose	3-O-Methyl-6-deoxy-D-talose	6-O-Methyl-D-glucosamine	D-Mannose	D-Galactose	D-Glucose	D-Glucosamine	2,3-Diamino-2,3-dideoxy-D-glucose	Quinovosamine	2-Keto-3-deoxyoctonate	L-Glycero-D-mannoheptose	D-Galactosamine	Rhamnose	6-Deoxytalose	Xylose	2-Amino-2-deoxygalacturonic acid	Amphoteric amino sugars I	Amphoteric amino sugars II
Rp. palustris																					
2/2	I	O					●	●	(●)	●	●	●	●	●						O	
K/1	I	O					●	●	(●)	●	●	●	●	●						O	
1e5	II	O			O		●	●	(●)	●	●	●	●	●			O	O			O
8/1	II	O			O		●	●	(●)	●	●	●	●	●			O	O			O
42	III					O	●	●	●	●	●	●	●	●	O	O					
15	III					O	●	●	●	●	●	●	●	●	O	O					
2/2	III					O	●	●	●	●	●	●	●	●	O	O					
1f2	III					O	●	●	●	●	●	●	●	●	O	O					
1a1	III					O	●	●	●	●	●	●	●	●	O	O					
16	III					O	●	●	●	●	●	●	●	●	O	O					
V₂	III					O	●	●	●	●	●	●	●	●	O	O					
E51	III					O	●	●	●	●	●	●	●	●	O	O					
Rp. viridis																					
F			O	O			●	●	●	●		●	●	●					O		
FV 103			O	O			●	●	●	●		●	●	●					O		
FV 104			O	O			●	●	●	●		●	●	●					O		
9350			O	O			●	●	●	●		●	●	●					O		
E1P9.2			O	O			●	●	●	●		●	●	●					O		
GPO1			O	O			●	●	●	●		●	●	●					O		
2450			O	O			●	●	●	●		●	●	●					O		
2750			O	O			●	●	●	●		●	●	●					O		
P2F6			O	O			●		●	●		●	●	●					O		
GP1 P10.4			O	O			●	●	●	●		●	●	●					O		
GRF7			O	O			●	●	●	●		●	●	●					O		

[a] Symbols:
(●) Present in phenol/water extract, but not in phenol/chloroform/petroleum ether extract.
● Sugars present in all strains.
O Strain-specific sugars.

Table 3. Immunological Relationships among Lipid A's from Various Rhodospirillaceae and Enterobacteriaceae[a]

Antiserum against	Indicator antigen (lipid A) *Salmonella typhimurium*	*Rhodopseudomonas gelatinosa*	*Rhodopseudomonas viridis*
Salmonella typhimurium or *minnesota*	+++	+++	—
Rhodopseudomonas gelatinosa	+++	+++	—
Rhodospirillum tenue	+++	+++	—
Rhodopseudomonas viridis	—	—	+++
Rhodopseudomonas palustris	—	—	+++

[a] The reactivity was measured in a passive hemolysis test using alkali-treated lipid A as antigen. Lipid A-antisera were prepared with HOAc-degraded bacteria by intravenous immunization of rabbits (Roppel, 1975; Galanos *et al.*, 1977). Boxed entries indicate homologous titers.

polysaccharide). Chemical analysis of the polysaccharide part revealed the following constituents: D-glucose, D-galactose, L-rhamnose, 3-O-methyl-L-rhamnose (acofriose), 2-keto-3-deoxyoctonate (KDO), and neuraminic acid. A fraction of the degraded polysaccharide including the immunodeterminant group was obtained by gel-chromatography and contained glucose, rhamnose, and acofriose as constituents. The lipid A fraction contained D-glucosamine as sole sugar constituent. It was found to carry ester-linked (lauric and β-hydroxycapric acid) and amide-linked (β-hydroxymyristic acid) fatty acids.

(b) Sugar Constituents. The observation that photosynthetic bacteria, which are quite different in their physiological properties, their natural habitats, and probably their evolution from Enterobacteriaceae, nevertheless contain O-antigenic LPS in their outer cell wall, prompted an extensive investigation of the chemical constituents of LPS from different taxonomic groups of photosynthetic bacteria.

Thus far, studies have been made of the sugar and fatty acid composition of about ten strains of each of the following species: *Rp. palustris*, *Rp. viridis*, *Rp. gelatinosa*, and *Rs. tenue*. In addition, LPS from a limited number of strains of *Rs. rubrum*, *Rs. molischianum*, *Rp. sphaeroides*, *Rp. capsulata*, *Rhodomicrobium vannielii*, and *Chr. vinosum* (Hurlbert *et al.*, 1976; Hurlbert and Hurlbert, 1977) have been isolated and at least partially characterized.

Besides the common constituents of LPS from gram-negative bacteria, a high number of unusual sugars—some of them described for the first time from natural products—were encountered and characterized. Table 1 gives a survey of the constituents hitherto isolated from LPS of photosynthetic bacteria. It is especially noteworthy that O-methyl ethers of

Table 4. Sugar Constituents of the Heteropolysaccharides (Haptens) and Lipopolysaccharides from *Rhodopseudomonas gelatinosa*[a]

Hapten chemotype	Strain	2,3-Di-O-methyl-D-galactose	2-O-Methyl-L-fucose	2-O-Methyl-D-galactose	3,6-Dideoxy-D-galactose (abequose)	Rhamnose	L-Fucose	D-Mannose	D-Galactose	D-Glucose	D-Glucosamine	D-Glycero-D-mannoheptose	2-Keto-3-deoxyoctonate
I	2150, P8P9	O						O	O	O	O		
II	Dr 2		O				O	O	O		O		
III	K 32, P18f3.1			O				O	O	O	O		
IV	39/2				O	O		O					
LPS of all strains											●	●	●

[a] Symbols:
● Sugars present in all strains.
O Strain-specific sugars.

Table 5. Characteristic Constituents of Lipopolysaccharides of Rhodospirillaceae[a]

| Species | P[b] | Main fatty acids | | | | | | | Amino sugar in lipid A | | KDO | Heptoses[d] | |
| | | β-Hydroxy fatty acids | | | Saturated fatty acids | | | | | | | | |
		C$_{10}$—OH	C$_{14}$—OH	C$_{16}$—OH	C$_{12}$	C$_{14}$	C$_{16}$	C$_{18}$	GlcN	DAG[c]		L,D-Hep	D,D-Hep
Rs. molischianum	+		+	+			+		+	−	+	+	−
Rs. rubrum	+		+	+			+		+	−	ND	+	−
Rs. tenue I	++	+				+	+		+	−	+	+	−
Rs. tenue II	++	+				+	+		+	−	+	+	+
Rs. fulvum	+	+	+					+	ND	ND	ND	+	−
Rp. gelatinosa	++	+	+		+	+			+	−	+	−	+
Rp capsulata	++	+	+		+				+	−	+	+	−
Rp. palustris	+	+	+	+			+	+	−	+	+	+	−
Rp. viridis	+	+	+						−	+	+	−	−
Rp. sphaeroides	++	+	+						+	−	ND	ND	ND

[a] (+) Present; (−) tested, but not present; (ND) not determined.
[b] P (phosphorus) values: (++) ⩾ 0.5%; (+) ⩽ 0.5%.
[c] (DAG) 2,3-Diamino-2,3-dideoxy-D-glucose.
[d] (L,D-Hep) L-glycero-D-mannoheptose; (D,D-Hep) D-glycero-D-mannoheptose.

sugars of various sugar classes occur in nearly all species. They must be considered characteristic constituents of photosynthetic prokaryotic LPS (see Tables 1–4), although they do occur in a few LPS of other groups of bacteria (Lüderitz et al., 1971; Mayer et al., 1973a, 1974). Further constituents that seem to be by far more common in these LPS than in entero-bacterial ones are 2-amino-2-deoxyhexuronic acids (aminouronic acids), especially 2-amino-2-deoxy-D-galacturonic acid (Weckesser et al., 1973b, 1977a).

(c) **Fatty Acid Constituents.** The fatty acids of LPS of almost all gram-negative bacteria differ from those of phospholipids of the cell wall or the cytoplasmic membrane in a characteristic way. (Lüderitz et al., 1973; Schröder et al., 1969). Lipid A of Enterobacteriaceae, for example, consists of D-glucosamine (20%), phosphorus (about 2%), and fatty acids (about 60%) such as lauric (C_{12}), myristic (C_{14}), palmitic (C_{16}), D-β-hydroxymyristic acids, and others, which are linked to the OH-groups and the amino group of glucosamine. β-Hydromyristic acid occurs pre-dominantly in an amide-linkage, and is one of the most characteristic markers of enterobacterial lipid A (Rietschel et al., 1972). The fatty acid pattern found in LPS of Rhodospirillaceae (Table 5) is similar to that of Enterobacteriaceae in many respects. There are, however, characteristic dissimilarities such as the replacement of amide-linked β-hydroxymyristic acid by β-hydroxycapric acid in strains of Rp. gelatinosa and Rs. tenue and the lack of significant amounts of ester-linked fatty acids in Rp. viridis. It must be pointed out that the fatty acid spectrum obtained from phenol-soluble LPS may be falsified by fatty acids originating from non-LPS material. β-Hydroxy-myristic acid in Rp. palustris, Rp. viridis, and Rp. capsulata is amide-linked. The optical configuration of β-hydroxymyristic acid from Rp. viridis was shown to be the D-form as in Salmonella LPS (Roppel, 1975).

The lack of ester-linked fatty acids in Rp. viridis is partly explained by the fact that LPS of this species does not contain D-glucosamine in its lipid A moiety, but rather 2,3-diamino-2,3-dideoxy-D-glucose (Rop-pel et al., 1975; Keilich et al., 1976). The additional amino group can also carry an amide-linked fatty acid.

(d) **Species-Specific Patterns.** Rhodopseudomonas palustris is a very common, rod-shaped organo-trophic member of Rhodospirillaceae. The lamellar and branched intracytoplasmic membranes, arranged parallel to the cytoplasmic membrane at one pole of the cell, and the budding type of growth are charac-teristic morphological features (Tauschel and Drews, 1967; Whittenbury and McLee, 1967; Pfennig and Trüper, 1974).

The LPS of 12 strains of Rp. palustris thus far investigated are extracted to the extent of about 90%

into the phenol phase of phenol/water extracts. They are also extractable by phenol/chloroform/petroleum ether, but the LPS is not precipitated by adding water to the phenol layer after removal of chloroform and petroleum ether (Weckesser et al., 1973a). The comparative sugar analysis enabled these 12 strains to be classified into three distinct chemotypes (Weckesser et al., 1973a). This chemical classification was later fully corroborated by serological studies affording three distinct serotypes coinciding with the three chemotypes (Framberg et al., 1974). Table 2 gives the sugar composition and shows those sugars that are common to all strains, and those that are specific for the respective chemo- (sero-) types. Four of the five O-methyl sugars occurring in Rp. palustris and Rp. viridis, and the 2,3-diamino-2,3-dideoxy-D-glucose, were hitherto not reported from bacterial LPS (Mayer et al., 1974; Keilich et al., 1976; Roppel et al., 1975). Amino uronic acids of hitherto unknown configura-tions differing from those having the gluco-, galacto-, manno-, or gulo- configuration were demonstrated in LPS of chemotypes I and II (Table 2).

The data on degraded polysaccharides of Rp. palustris are too preliminary to decide whether a common R-core structure comparable to the entero-bacterial R-cores exists. The extremely complex sugar composition raises the question whether the O-chains of these and other members of Rhodospirillaceae are built by repeating units as is the case in O-chains of Enterobacteriaceae. Since no detailed structural studies have been performed, this important question remains unanswered.

Rhodopseudomonas viridis is morphologically re-lated to Rp. palustris. It shows a similar budding-like mode of cell division (Whittenbury and McLee, 1967), but has Bchl b instead of the Bchl a found in Rp. palustris. It is an obligate photoorganotroph.

The LPS are predominantly extracted into the aqueous phase by hot phenol/water extraction, unlike the phenol-soluble LPS of Rp. palustris. The LPS of all 11 strains studied so far belong to a single chemotype (Type 2), and probably, judged from their pronounced serological cross reactions, also to a single serotype (Weckesser et al., 1974a). Many sugars of the LPS of Rp. viridis are likewise common to all three chemotypes of Rp. palustris (see Table 2). A characteristic difference is, however, the lack of L-glycero-D-mannoheptose (or of any other heptose) in LPS of Rp. viridis. LPS of Rp. viridis have an extremely low phosphorus content (less than 0.2%); negative charges are contributed, however, by 2-keto-3-deoxy-octonic acid (KDO) and by galactosamin-uronic acid, which has been identified in LPS of all strains.

The morphological relationship between *Rp. palustris* and *Rp. viridis* is paralleled by a special type of lipid A hitherto unique to these two species (Mayer *et al.*, 1973*a*). D-Glucosamine, which as a phosphorylated disaccharide forms the backbone of enterobacterial LPS and of the LPS of the other Rhodospirillaceae, is entirely replaced in these two species by 2,3-diamino-2,3-dideoxy-D-glucose, which may formally be considered as a 3-amino derivative of D-glucosamine (Roppel *et al.*, 1975; Keilich *et al.*, 1976). Neither this sugar nor any other member of this sugar class has previously been reported from natural sources.

Aqueous-phase material from *Rp. viridis* was shown to contain only two fatty acids, namely, amide-linked β-hydroxymyristic acid and tetradecenoic acid. The latter can arise artificially from the former by elimination of water under usual hydrolytic conditions (Rietschel *et al.*, 1972; Wilkinson, 1974).

Lipid A antisera obtained by immunizing rabbits with acetic-acid-treated bacteria according to Galanos *et al.* (1971) showed an intense serological cross-reaction between lipid A of *Rp. palustris* and that of *Rp. viridis* (Table 3). According to expectation, no cross-reaction at all was demonstrable between lipid A of *Rp. viridis* and lipid A from other Rhodospirillaceae (Table 3) (Roppel, 1975).

Lipid A is now generally recognized as the active principle of enterobacterial endotoxins (Lüderitz *et al.*, 1971, 1973). It is therefore not unexpected that preliminary results indicate that the lethal toxicity as well as the pyrogenicity of LPS of the two species mentioned above are greatly reduced when compared with enterobacterial LPS (Galanos *et al.*, 1977).

Rhodopseudomonas gelatinosa and *Rhodospirillum tenue* are morphologically related. *Rp. gelatinosa* is a slender, facultative photoorganotrophic member of Rhodospirillaceae. Cells of many strains are long and spirilloid, and similar in appearance to *Rs. tenue*. Both species have a primitive, poorly developed intracytoplasmic membrane system that consists of single and small tubular or lamellar invaginations of the cytoplasmic membrane. *Rp. gelatinosa* tends to form slime and to sediment during the stationary growth phase in voluminous aggregates. It is able to liquefy gelatin and to use citrate or aspartate as sole carbon source.

The LPS of *Rp. gelatinosa* are rather lipophilic and therefore easily extractable by phenol/chloroform/petroleum ether similar to those of enterobacterial R-mutants. LPS of all 12 strains that have been investigated contain only a single neutral sugar constituent identified as D-glycero-D-mannoheptose. This is the 6-epimer of L-glycero-D-mannoheptose,

which is the ubiquitous R-core constituent of enterobacterial LPS (Table 4) (Weckesser *et al.*, 1975*a*). 2-Keto-3-deoxyoctonate forms, as usual, the acid-labile ketosidic linkage between lipid A and the heptose-oligosaccharide. D-Glucosamine represents the only amino sugar, and originates, like its phosphate ester, exclusively from the lipid moiety (Table 4).

The passive hemagglutination test carried out with isolated LPS and antisera against heat-killed bacteria revealed the existence of two distinct serotypes (Weckesser *et al.*, 1975*a*). Serotype I comprises those strains that do not synthesize a haptenlike polysaccharide (see below). "Haptens" were, however, found in strains of serotype II. LPS of serotype I seem to possess a hitherto unidentified acid-labile component, judging from the strong formation of huminlike substances during acid hydrolysis (Weckesser *et al.*, 1975*a*).

Haptens of Rhodopseudomonas gelatinosa. Of the 12 strains of *Rp. gelatinosa* studied so far, 9 were found to produce low-molecular-weight heteropolysaccharides that resemble the *O*-specific haptens of *Salmonella rfa* mutants (Stocker and Mäkelä, 1971). In these mutants, the *rfaL* gene product, a component of the translocase system, is missing. Its function is the effective transfer of the completed *O*-specific chains to their attachment site on the complete R-core. They are accumulated and are found as low-molecular-weight polysaccharides in the aqueous phase of phenol/water extracts of these mutants (Stocker and Mäkelä, 1971). The amounts of strain-specific polysaccharides isolated from *Rp. gelatinosa* varied from 1.2–2.3% in 6 strains to less than 0.2% of the bacterial dry weight in 3 strains. They were extracted from bacteria with phenol/water (following an extraction of the LPS with phenol/chloroform/petroleum ether) and were obtained in the supernatant of the water phase after high-speed centrifugation. Chemical analysis of the polysaccharide after removal of RNA and glucan revealed constituents that are considered as being highly characteristic for *O*-specific chains and not for either *K*- or *M*-antigens (Lüderitz *et al.*, 1968), i.e., *O*-methyl hexoses or 3,6-dideoxyhexoses (abequose) (Weckesser *et al.*, 1975*b*). By analogy, therefore, we designate the polysaccharides of *Rp. gelatinosa* as haptens. Table 4 shows the chemical composition of the four haptens that have been identified.

D-Glycero-D-mannoheptose and KDO, the characteristic components of *Rp. gelatinosa* LPS, as well as fatty acids, were present only in trace amounts or not at all in the polysaccharide fractions. Antisera prepared by intravenous immunization of rabbits with heat-killed cell suspensions react specifically in many cases with the isolated polysaccharides from the respective strains (for a table, see Weckesser et al., 1975b).

It is now well-established that D-glycero-D-manno-heptose is an intermediate in the biosynthesis of L-glycero-D-mannoheptose (Ginsburg et al., 1962; Lehmann et al., 1973), the latter being one of the characteristic common R-core constituents of many gram-negative bacteria. A defined group of Salmonella typhimurium mutants was recently described (Lehmann et al., 1973) that contain both L-glycero-D-manno- and D-glycero-D-mannoheptose in their respective LPS due to defects in the epimerase systems. It was noticed that the few O-chains that could be completed did contain the L-glycero-D-mannoheptose, whereas the D-glycero-D-mannoheptose was highly enriched in R-core stubs. Assuming that the biosynthesis of LPS in Rhodospirillaceae follows a pattern similar to that of Enterobacteriaceae, it is possible that the haptenic heteropolysaccharides of Rp. gelatinosa pile up since they cannot be incorporated due to the "wrong" heptose in the core region. Another explanation, however, is also reasonable, namely, that the so-called O-translocase system,

genetically determined in Salmonella by rfaL and rbfT genes (Stocker and Mäkelä, 1971), has not been developed in these species.

The LPS of Rs. tenue are, like those of Rp. gelatinosa, extractable by phenol/chloroform/petroleum ether due to their rather lipophilic character (Weckesser et al., 1977a). The sugar analyses of the O-antigens from nine strains are summarized in Table 6. Again, a group of sugars is common to all strains, indicating that a common R-core may exist. Other sugars are restricted to only a few strains and occur in small amounts, leading to a number of separate chemotypes.

It is remarkable that four strains also contain, in addition to the ubiquitous L-glycero-D-mannoheptose, D-glycero-D-mannoheptose (Tables 5 and 6), which is the 6-epimer of the former. The mutual occurrence of these two heptoses, which are biogenetically related (Eidels and Osborn, 1974), has been reported from a few enterobacterial S-forms, such as Proteus mirabilis (Bagdian et al., 1966; Kotelko et al., 1965, 1974), and Salmonella R-mutants (Adams et al., 1967; Lehmann et al., 1973). Whether these two heptoses are constit-

Table 6. Sugar Patterns in Lipopolysaccharides of *Rhodospirillum tenue*[a]

Strains	2-O-Methyl-D-rhamnose	D-Rhamnose	D-Fucose	D-Mannose	D-Glycero-D-mannoheptose	L-Glycero-D-mannoheptose	D-Glucose	D-Arabinose	2-Keto-3-deoxyoctonate	D-Glucosamine	D-Galactosamine	Quinovosamine	Fucosamine	2-Amino-2-deoxygalacturonic acid
P16						●	●	●	●	●			O	O
6750, 3760						●	●	●	●	●	O			O
Eu1	O	(O)				●	●	●	●	●				O
GFU			(O)	O		●	●	●	●	●		O		(O)
P₄P₁.₂					O	●	●	●	●	●		(O)		
3661					O	●	●	●	●	●	O			O
2761					O	●	●	●	●	●		O		O
3761		(O)			O	●	●	●	●	●	O			O

[a] Symbols:
(O) Present to less than 1% of LPS dry weight.
● Sugars present in all strains.
O Strain-specific sugars.

uents of two different structural regions (L-glycero-D-mannoheptose in the R-core, the D epimer in the O-chains) or whether they belong to two different LPS species occurring in the same serotype as recently described by Nghiem and Staub (1975) for LPS of *Salmonella zuerich* remains to be elucidated.

D-Arabinose, a pentose so far not encountered in LPS (Lüderitz *et al.*, 1966, 1971), is a characteristic sugar of LPS from all *Rs. tenue* strains. Studies on strains GFU and $P_4P_{1,2}$ showed that arabinose together with glucosamine and phosphorylated glucosamine is part of the lipid A region (Tharanathan *et al.*, 1978a).

Table 3 shows that lipid A antisera prepared with acetic-acid-degraded strains of *Rs. tenue* (strain 2761) and *Rp. gelatinosa* (strain 29/2) cross-react serologically with each other, as well as with similarly prepared antisera against lipid A of *Salmonella minnesota* 1144 and *Salmonella typhimurium* 1135, indicating that structural similarities exist in the lipid A region of these species. No cross-reaction is found with lipid A of *Rp. viridis* or *Rp. palustris*.

(e) **Taxonomic Relevance.** A comparison of the sugar composition of all investigated Rhodospirillaceae strains demonstrates a large diversity of components within a small number of species. Remarkable is the large number of individual O-methyl sugars and 2-amino-hexoses, including 2-amino-2-deoxy-hexuronic acids. The only sugar component found in all LPS samples is 2-keto-3-deoxy-octonate, which is the constituent providing the linkage between the lipid and the polysaccharide moieties in LPS of gram-negative bacteria. Not unexpectedly, in view of the wide variety of components, every species is characterized by a specific pattern of constituents (Tables 2 and 4–6). In addition to the various sugars of the O-chains, the components of the R-core (see Fig. 8) and lipid A, which might be expected to be more conservative, show a remarkable diversity (Table 5). The replacement of glucosamine by 2,3-diamino-2,3-dideoxy-D-glucose in lipid A of *Rp. palustris* and *Rp. viridis* parallels a considerable change in the antigenicity of their respective lipid A (Table 3).

The data presented in this chapter indicate that the basic pattern of LPS structure found in the Enterobacteriaceae seems to occur in the phototrophic bacteria. The same is apparently true of both unicellular and filamentous blue-green algae (Weise *et al.*, 1970; Weckesser *et al.*, 1974b; Katz *et al.*, 1977). In addition, the chemical composition of LPS and their reflection in the serological properties of the respective strains supports the current taxonomy of the Rhodospirillaceae and provides an efficacious tool for further taxonomic work (Weckesser *et al.*, 1977b).

3.3. Other Components

Proteins are regular constituents of gram-negative cell walls (Costerton *et al.*, 1974). Some were shown to be enzymes localized in the periplasmatic space (Machtiger and Fox, 1973; Costerton *et al.*, 1974). Other proteins together with LPS and phospholipids are constituents of the outer membrane of cell walls (Rosenbusch, 1974; Schnaitman, 1974). At present, the structures or functions of very few of these proteins are known (Hantke and Braun, 1973; Braun and Hantke, 1974). Braun and co-workers (Braun, 1973; Braun and Hantke, 1974) described the structure

Table 7. Serological Cross-Reactions between Rabbit Antisera Against Heat-Killed Cells of Rhodospirillaceae and the Lipoprotein of *Escherichia coli*[a] in Passive Hemagglutination[b]

Titer	Rhodospirillaceae						Enterobacteriaceae	
	Rp. gelatinosa	*Rs. rubrum*	*Rs. tenue*	*Rp. capsulata*	*Rp. viridis*	*Rp. palustris*	R^i-mutants	S-forms
5120—								
—	2							
2560—	1						1	
—	2						1	
1280—	6		3				1	
—	4							
640—	2	2	3					
—			1				3	
320—	1		1					
—								
160—			2				2	2
—								
80—								1
—					1			
40—								3
—								
20—					1	1	3	2
—								
10—								2
—						2		
<10—			2		6	12		2

[a] Braun *et al.* (1976).

[b] The figures are numbers of individual antisera. Average titers from parallel passive hemagglutination tests are given. For comparison, R^i-mutants with incomplete R-core and encapsulated or nonencapsulated S-forms of *Esch. coli*, *Salmonella*, and *Shigella* are included (Mayer *et al.*, 1973b).

of a lipoprotein that is covalently bound to the murein-sacculus of *Escherichia coli* B and protrudes into the outer membrane. The same lipoprotein was later also found in a free form in *Esch. coli* (Inouye *et al.*, 1972; Inouye, 1974). At present, it is uncertain whether this lipoprotein occurs in cell walls of other gram-negative bacteria. However, substantial cross-reaction between the isolated lipoprotein of *Esch. coli* B and antisera prepared against heat killed R-mutants of a variety of Enterobacteriaceae (*Salmonella, Shigella, Esch. coli,* and *Citrobacter*) has been reported (Mayer *et al.*, 1973*b*). Similar studies were undertaken with a number of different antisera prepared in the same way against heat-killed Rhodospirillaceae. Table 7 shows that six species of Rhodospirillaceae fall into two groups. The first includes *Rp. gelatinosa, Rs. rubrum,* and *Rs. tenue,* and is characterized by rather high titers against the lipoprotein of *Esch. coli* B (as indicator antigen). The titers are of the same magnitude as those found in antisera against the R-mutants used by Mayer and co-workers (see above). The second group consists of *Rp. capsulata, Rp. viridis,* and *Rp. palustris,* and is characterized by at most very moderate lipoprotein titers. This finding does not necessarily mean that a lipoprotein of the type found in *Esch. coli* B is completely absent in these species; it might also indicate that it is not easily accessible.

Sodium dodecylsulfate–polyacrylamide gel electrophoresis of cell-wall fractions from photosynthetic bacteria showed distinct protein bands. In *Rs. rubrum,* 43,000- and 14,000-dalton proteins dominate (Oelze *et al.*, 1975). In the cell-wall fraction of *Chr. vinosum,* one major protein with a molecular weight of approximately 45,500 daltons was observed (Hurlbert *et al.*, 1974). In contrast, a 60,000-dalton protein dominates in the cell-wall fraction of *Rp. capsulata* (Drews, 1974). Presumably, these proteins are not identical with the lipoprotein described above.

The fatty acid patterns of the cell-wall fraction were shown to be distinctly different from the cytoplasmic-membrane fraction (Drews, 1974; Hurlbert *et al.*, 1974; Schröder *et al.*, 1969). This is due to the enrichment of LPS in the outer membrane, while phospholipids are the major lipid compound in the cytoplasmic membrane.

4. References

Adams, G. A., Quadling, C., and Perry, M. B., 1967, D-Glycero-D-mannoheptose as a component of lipopolysaccharides from gram-negative bacteria, *Can. J. Microbiol.* **13**:1605.

Ames, G. F.-L., Spudich, E. N., and Nikaido, H., 1974, Protein composition of the outer membrane of *Salmonella typhimurium, J. Bacteriol.* **117**:406.

Bagdian, G., Dröge, W., Kotelko, K., Lüderitz, O., and

Westphal, O., 1966, Vorkommen zweier Heptosen in Lipopolysacchariden enterobakterieller Zellwände: L-Glycero- und D-Glycero-D-Mannoheptose, *Biochem. Z.* **344**:197.

Boatman, E. S., 1964, Observations on the fine structure of spheroplasts of *Rhodospirillum rubrum, J. Cell Biol.* **20**:297.

Braun, V., 1973, Molecular organization of the rigid layer and the cell wall of *Escherichia coli. J. Infect. Dis.* **128**:1.

Braun, V., and Hantke, K., 1974, Biochemistry of bacterial cell envelopes, *Annu. Rev. Biochem.* **43**:89.

Braun, V., Bosch, V., Klumpp, E. R., Neff, I., Mayer, H., and Schlecht, S., 1976, Antigenic determinants of murein lipoprotein and its exposure at the surface of Enterobacteriaceae, *Eur. J. Biochem.* **62**:555.

Burge, R. E., and Draper, J. C., 1967, The structure of the cell wall of the Gram-negative bacterium *Proteus vulgaris, J. Mol. Biol.* **28**:205.

Cohen-Bazire, G., Pfennig, N., and Kunisawa, R., 1964, The fine structure of green bacteria, *J. Cell Biol.* **22**:207.

Costerton, J. W., Ingram, J. M., and Cheng, K.-J., 1974, Structure and function of the cell envelope of Gram-negative bacteria, *Bacteriol. Rev.* **38**:87.

de Boer, W. E., and Spit, B. J., 1964, A new type of bacterial cell wall structure revealed by replica technique, *Antonie van Leeuwenhoek; J. Microbiol. Serol.* **30**:239.

de Petris, S., 1967, Ultrastructure of the cell wall of *Escherichia coli* and chemical nature of its constituent layers, *J. Ultrastruct. Res.* **19**:45.

Drews, G., 1974, Composition of a protochlorophyll–protopheophytin complex, excreted by mutant strains of *Rhodopseudomonas capsulata* in comparison with the photosynthetic apparatus, *Arch. Microbiol.* **100**:397.

Eidels, L., and Osborn, M. J., 1974, Phosphoheptose isomerase, first enzyme in the biosynthesis of aldoheptose in *Salmonella typhimurium, J. Biol. Chem.* **249**:5642.

Forge, A., and Costerton, J. W., 1973, Biophysical examination of the cell wall of a gram-negative marine pseudomonad, *Can. J. Microbiol.* **19**:451.

Forge, A., Costerton, J. W., and Kerr, K. A., 1973, Freeze–etching and X-ray diffraction of the isolated double-track layer from the cell wall of a pseudomonad, *J. Bacteriol.* **113**:445.

Framberg, K., Mayer, H., Weckesser, J., and Drews, G., 1974, Serologische Untersuchungen an isolierten Lipopolysacchariden aus *Rhodopseudomonas palustris*-Stämmen, *Arch. Microbiol.* **98**:239.

Galanos, C., and Lüderitz, O., 1975, Electrodialysis of lipopolysaccharides and their conversion to uniform salt forms, *Eur. J. Biochem.* **54**:603.

Galanos, C., Lüderitz, O., and Westphal, O., 1969, A new method for extraction of R-lipopolysaccharides, *Eur. J. Biochem.* **9**:245.

Galanos, C., Lüderitz, O., and Westphal, O., 1971, Preparation and properties of antisera against the lipid-A component of bacterial lipopolysaccharides, *Eur. J. Biochem.* **23**:116.

Galanos, C., Roppel, J., Weckesser, J., Rietschel, E. T., and Mayer, H., 1977, Biological activities of lipopolysaccharides and lipid A from Rhodospirillaceae, *Infect. Immun.* **16**:407.

Ghuysen, J. M., Strominger, J. L., and Tipper, D. J., 1968, Bacterial cell walls, in: *Comprehensive Biochemistry* (M. Florkin and E. H. Stotz, eds.), Vol. 26A, pp. 53–104, Elsevier Publishing Co., Amsterdam and New York.

Ginsburg, V., O'Brien, P. J., and Hall, C. W., 1962, Guanosinediphosphate D-glycero-D-mannoheptose, *J. Biol. Chem.* **237**:497.

Glauert, A. M., and Thornley, M. J., 1969, The topography of the bacterial cell wall, *Annu. Rev. Microbiol.* **23**:159.

Hageage, G. J., and Gherna, R. L., 1971, Surface structure of *Chromatium okenii* and *Chromatium weissii*, *J. Bacteriol.* **106**:687.

Hantke, K., and Braun, V., 1973, Covalent binding of lipid to protein. Diglyceride and amide-linked fatty acid at the *N*-terminal end of the murein-lipoprotein of the *Escherichia coli* outer membrane, *Eur. J. Biochem.* **34**:284.

Henning, U., Höhn, B., and Sonntag, I., 1973, Cell envelope and shape of *Escherichia coli*, *Eur. J. Biochem.* **39**:27.

Hurlbert, R. E., and Hurlbert, I. M., 1977, Biological and physicochemical properties of the lipopolysaccharide of *Chromatium vinosum*, *Infect. Immunity* **16**:983.

Hurlbert, R. E., Golecki, J. R., and Drews, G. G., 1974, Isolation and characterization of *Chromatium vinosum* membranes, *Arch. Microbiol.* **101**:169.

Hurlbert, R. E., Weckesser, J., Mayer, H., and Fromme, I., 1976, Isolation and characterization of the lipopolysaccharide of *Chromatium vinosum*, *Eur. J. Biochem.* **68**:365.

Inouye, M., 1974, A three-dimensional molecular assembly model of a lipoprotein from the *Escherichia coli* outer membrane, *Proc. Natl. Acad. Sci. U.S.A.* **71**:2396.

Inouye, M., Shaw, J., and Shen, C., 1972, The assembly of a structural lipoprotein in the envelope of *Escherichia coli*, *J. Biol. Chem.* **247**:8154.

Katz, A., Weckesser, J., Drews, G., and Mayer, H., 1977, Chemical and biological studies on the lipopolysaccharide (*O*-antigen) of *Anacystis nidulans*, *Arch. Microbiol.* **113**:247.

Kauffmann, F., 1966, *The Bacteriology of Enterobacteriaceae*, Williams & Wilkins Co., Baltimore.

Keilich, G., Roppel, J., and Mayer, H., 1976, Characterization of a diaminohexose (2,3-diamino-2,3-dideoxy-D-glucose) from *Rhodopseudomonas viridis* lipopolysaccharides by circular dichroism, *Carbohydr. Res.* **51**:129.

Kolenbrander, P. E., and Ensign, J. C., 1968, Isolation and chemical structure of the peptidoglycan of *Spirillum serpens* cell walls, *J. Bacteriol.* **95**:201.

Kotelko, K., Lüderitz, O., and Westphal, O., 1965, Vergleichende Untersuchungen von Antigenen von *Proteus mirabilis* und einer stabilen L-Form, *Biochem. Z.* **343**:227.

Kotelko, K., Gromska, W., Papierz, M., Szer, K., Krajewska, D., and Sidorczyk, Z., 1974, The constitution of core in *Proteus* lipopolysaccharides, *J. Hyg. Epidemiol. Microbiol. Immunol.* **18**:405.

Lehmann, V., Hämmerling, G., Nurminen, M., Minner, I., Ruschmann, E., Lüderitz, O., Kuo, T.-T., and Stocker, B. A. D., 1973, A new class of heptose-defective mutants of *Salmonella typhimurium*, *Eur. J. Biochem.* **32**:268.

Lüderitz, O., Staub, A. M., and Westphal, O., 1966, Immunochemistry of *O* and *R* antigens of *Salmonella* and related Enterobacteriaceae, *Bacteriol. Rev.* **30**:192.

Lüderitz, O., Jann, K., and Wheat, R., 1968, Somatic and capsular antigens of Gram-negative bacteria, in: *Comprehensive Biochemistry* (M. Florkin and E. H. Stotz, eds.), Vol. 26A, pp. 105–228, Elsevier Publishing Co., Amsterdam and New York.

Lüderitz, O., Westphal, O., Staub, A. M., and Nikaido, H., 1971, Isolation and chemical and immunochemical characterization of bacterial lipopolysaccharides, in: *Microbial Toxins: A Comprehensive Treatise* (G. Weinbaum, S. Kadis, and S. J. Ajl, eds.), Vol. IV, pp. 369–433, Academic Press, London and New York.

Lüderitz, O., Galanos, C., Lehmann, V., Nurminen, M., Rietschel, E. T., Rosenfelder, G., Simon, M., and Westphal, O., 1973, Lipid A, chemical structure and biological activity, *J. Infect. Dis.* **128**:17.

Machtiger, N. A., and Fox, C. F., 1973, Biochemistry of bacterial membranes, *Annu. Rev. Biochem.* **42**:575.

Martin, H. H., 1966, Biochemistry of bacterial cell walls, *Annu. Rev. Biochem.* **35**:457.

Mayer, H., Weckesser, J., Roppel, J., and Drews, G., 1973*a*, *O*-Antigens of Rhodospirillaceae, in: *Abstracts of the Symposium on Prokaryotic Photosynthetic Organisms*, Freiburg, pp. 190–192.

Mayer, H., Schlecht, S., and Braun, V., 1973*b*, Immunogenicity of lipoprotein (murein-lipoprotein) of Gram-negative bacteria, Joint Meeting of the European Societies for Immunology, Strasbourg.

Mayer, H., Framberg, K., and Weckesser, J., 1974, 6-*O*-Methyl-D-glucosamine in lipopolysaccharides of *Rhodopseudomonas palustris* strains, *Eur. J. Biochem.* **44**:181.

Newton, J. W., 1967, Bacilliform mutants of *Rhodospirillum rubrum*, *Biochim. Biophys. Acta* **114**:633.

Newton, J. W., 1968, Linkages in walls of *Rhodospirillum rubrum* and its bacilliform mutants, *Biochim. Biophys. Acta* **165**:534.

Newton, J. W., 1970, Metabolism of D-alanine in *Rhodospirillum rubrum* and its bacilliform mutants, *Nature (London)* **228**:1100.

Newton, J. W., 1971, *Vibrio* mutants of *Rhodospirillum rubrum*, *Biochim. Biophys. Acta* **244**:478.

Nghiem, H. O., and Staub, J. M., 1975, Molecular immunological heterogeneity of the *Salmonella zuerich* cell-wall polysaccharides, *Carbohydr. Res.* **40**:153.

Nowotny, A., 1971, Chemical and biological heterogeneity of endotoxins, in: *Microbial Toxins: A Comprehensive Treatise* (G. Weinbaum, S. Kadis, and S. J. Ajl, eds.), Vol. IV, pp. 309–329, Academic Press, New York and London.

Oelze, J., Golecki, J. R., Kleinig, H., and Weckesser, J., 1975, Characterization of two cell-envelope fractions from chemotrophically grown *Rhodospirillum rubrum*, *Antonie van Leeuwenhoek; J. Microbiol. Serol.* **41**:273.

Pfennig, N., and Cohen-Bazire, g., 1967, Some properties of the green bacterium *Pelodictyon clathratiforme*, *Arch. Mikrobiol.* **59**:226.

Pfennig, N., and Trüper, H. G., 1974, Phototrophic bacteria, in: *Bergey's Manual of Determinative Bacteriology* (R. E.

Buchanan and M. E. Gibbons, eds.), 8th Ed., pp. 24–75, The Williams & Wilkins Co., Baltimore.

Randall-Hazelbauer, L., and Schwartz, M., 1973, Isolation of the bacteriophage lambda receptor from *E. coli*, *J. Bacteriol.* **116**:1436.

Remsen, C. C., Watson, S. W., Waterbury, J. B., and Trüper, H. G. 1968, Fine structure of *Ectothiorhodospira mobilis*, *J. Bacteriol.* **95**:2374.

Remsen, C. C., Watson, S. W., and Trüper, H. G., 1970, Macromolecular subunits in the walls of marine photosynthetic bacteria, *J. Bacteriol.* **103**:254.

Rietschel, E. T., Gottert, H., Lüderitz, O., and Westphal, O., 1972, Nature and linkages of the fatty acids present in the lipid-A component of Salmonella lipopolysaccharide, *Eur. J. Biochem.* **28**:166.

Roppel, J., 1975, Chemische, immunchemische und biologische Untersuchungen der Lipopolysaccharide von *Rhodopseudomonas viridis* Stämmen, Thesis, University of Freiburg, Germany.

Roppel, J., Mayer, H., and Weckesser, J., 1975, Identification of a 2,3-diamino-2,3-dideoxyhexose in the Lipid A component of lipopolysaccharides of *Rhodopseudomonas viridis* and *Rhodopseudomonas palustris*, *Carbohydr. Res.* **40**:31.

Rosenbusch, J. P., 1974, Characterization of the major envelope protein from *E. coli*, *J. Biol. chem.* **249**:8019.

Salton, M. R. J., 1960, Monosaccharide constituents of the walls of Gram-negative bacteria, *Biochim. biophys. Acta* **45**:364.

Salton, M. R. J., and Williams, R. C., 1954, Electron microscopy of the cell walls of *Bacillus megaterium* and *Rhodospirillum rubrum*, *Biochim. Biophys. Acta* **14**:455.

Schlecht, S., and Westphal, O., 1970, Untersuchungen zur Typisierung von Salmonella-R-Formen 4. Mitteilung: Typisierung von *S. minnesota*-R-Mutanten mittels Antibiotica, *Zentralbl. Bakteriol. Parasitenkd. Infektionskr. Hyg. Abt. 1: Orig.* **213**:356.

Schleifer, K. H., and Kandler, O., 1972, Peptidoglycan types of bacterial cell walls, *Bacteriol. Rev.* **36**:407.

Schnaitman, C. A., 1974, Outer membrane proteins of *Escherichia coli*, *J. Bacteriol.* **118**:442.

Schröder, J., Biedermann, M., and Drews, G., 1969, Die Fettsäuren in ganzen Zellen, Thylakoiden, Lipopolysacchariden von *Rhodospirillum rubrum* und *Rhodopseudomonas capsulata*, *Arch. Mikrobiol.* **66**:273.

Shands, J. W., 1971, The physical structure of bacterial lipopolysaccharides, in: *Microbial Toxins: A Comprehensive Treatise* (G. Weinbaum, S. Kadis, and S. J. Ajl, eds.), Vol. IV, pp. 127–144, Academic Press, New York and London.

Shands, J. W., Graham, J. A., and Nath, K., 1967, The morphologic structure of isolated bacterial lipopolysaccharide, *J. Biol.* **25**:15.

Sleytr, U. B., and Thornley, M. J., 1973, Freeze–etching of the cell-envelope of an Acinetobacter, *J. Bacteriol.* **116**:1383.

Stocker, B. A. D., and Mäkelä, P. H., 1971, Genetic aspects of biosynthesis and structure of Salmonella lipopolysaccharide, in: *Microbial Toxins: A Comprehensive Treatise* (G. Weinbaum, S. Kadis, and S. J. Ajl, eds.), Vol. IV, p. 394, Academic Press, New York and London.

Tauschel, H.-D., and Drews, G., 1967, Thylakoid-morphogenese bei *Rhodopseudomonas palustris*, *Arch. Mikrobiol.* **59**:381.

Tharanathan, R. N., Weckesser, J., and Mayer, H., 1978a, Structural studies on the D-arabinose-containing lipid A from *Rhodospirillum tenue* 2761, *Eur. J. Biochem.* **84**:385.

Tharanathan, R. N., Mayer, H., and Weckesser, J., 1978b, Location of O-methyl sugars in antigenic (lipo-)polysaccharides of photosynthetic bacteria and cyanobacteria. *Biochem. J.* **171**:403.

Thornley, M. J., Thorne, K. J. I., and Glauert, A. M., 1974, Detachment and chemical characterization of the subunits from the surface of an Acinetobacter, *J. Bacteriol.* **118**:654.

Weckesser, J., Drews, G., Fromme, I., 1972a, Chemical analysis of and degradation studies on the cell wall lipopolysaccharide of *Rhodopseudomonas capsulata*, *J. Bacteriol.* **109**:1106.

Weckesser, J., Drews, G., and Ladwig, R., 1972b, Localization and biological and physicochemical properties of the cell wall lipopolysaccharide of *Rhodopseudomonas capsulata*, *J. Bacteriol.* **110**:346.

Weckesser, J., Drews, G., Fromme, I., and Mayer, H., 1973a, Isolation and chemical composition of lipopolysaccharides of *Rhodopseudomonas palustris* strains, *Arch. Mikrobiol.* **92**:123.

Weckesser, J., Mayer, H., and Fromme, I., 1973b, O-Methyl-sugars in lipopolysaccharides of Rhodospirillaceae, *Biochem. J.* **135**:293.

Weckesser, J., Drews, G., Roppel, J., Mayer, H., and Fromme, J., 1974a, The lipopolysaccharides (O-antigens) of *Rhodopseudomonas viridis*, *Arch. Microbiol.* **101**:233.

Weckesser, J., Katz, A., Drews, G., Mayer, H., and Fromme, I., 1974b, Lipopolysaccharide containing L-acofriose in the filamentous blue-green alga *Anabaena variabilis*, *J. Bacteriol.* **120**:672.

Weckesser, J., Mayer, H., Drews, G., and Fromme, I., 1975a, Lipophilic O-antigens containing D-glycero-D-mannoheptose as sole neutral sugar in *Rhodopseudomonas gelatinosa*, *J. Bacteriol.* **123**:449.

Weckesser, J., Mayer, H., Drews, G., and Fromme, I., 1975b, Low molecular weight polysaccharide antigens isolated from *Rhodopseudomonas gelatinosa*, *J. Bacteriol.* **123**:456.

Weckesser, J., Drews, G., Indira, R., and Mayer, H., 1977a, Lipophilic O-antigens in *Rhodospirillum tenue*, *J. Bacteriol.* **130**:629.

Weckesser, J., Drews, G., and Mayer, H., 1977b, Lipopolysaccharide aus Zellwänden photosynthetischer Prokaryonten. Bedeutung für Fragen der Taxonomie und der endotoxischen Aktivität, *Naturw. Rundschau* **30**:360.

Weise, G., Drews, G., Jann, B., and Jann, K., 1970, Identification and analysis of a lipopolysaccharide in cell walls of *Anacystis nidulans*, *Arch. Mikrobiol.* **71**:89.

Westphal, O., Lüderitz, O., and Bister, F., 1952, Über die Extraktion von Bakterien mit Phenol/Wasser, *Z. Naturforsch.* **7b**:148.

Whittenbury, R., and McLee, G. A., 1967, *Rhodopseudomonas palustris* and *R. viridis*—photosynthetic budding bacteria, *Arch. Mikrobiol.* **59**:324.

Wilkinson, S. G., 1974, Artefacts produced by acidic hydrolysis of lipids containing 3-hydroxy alkanoic acids, *J. Lipid Res.* **15**:181.

CHAPTER 5

Isolation and Physicochemical Properties of Membranes from Purple Photosynthetic Bacteria

Robert A. Niederman and Kenneth D. Gibson

1. Historical Background

1.1. Association of the Photosynthetic Pigments with "Soluble" Proteins

Extensive physicochemical studies of the purple photosynthetic bacteria during the past 25 years have firmly established that the photosynthetic pigments of these organisms [both bacteriochlorophyll (Bchl) and carotenoids] are localized in a system of intracytoplasmic membranes. Before this period, however, it was thought that the pigments were cytoplasmically localized in association with soluble components. This assumption was based on shifts in the absorption spectra of the pigments seen when they were extracted into organic solvents. In addition, microscopic examination had suggested that the photopigments were uniformly distributed throughout the cytoplasm. Mechanical disruption of several purple bacteria liberated carotenoid (French, 1940a) and Bchl components (French, 1940b,c; Katz and Wassink, 1939; Wassink *et al.*, 1939) with absorption bands virtually identical to those of the whole cells from which they were derived. Fractionation with ammonium sulfate resulted in the precipitation of the pigmented material (French, 1940b). This led to the view that *in vivo* the

pigments exist as complexes with water-soluble protein components. Although several carotenoid–protein and Bchl–protein complexes have been isolated from these organisms in recent years (see below), it is now clear that these complexes normally exist as integral components of bacterial photosynthetic membranes.

1.2. Discovery of the Particulate Photosynthetic Apparatus

More than a decade after the work of French, Pardee *et al.* (1952) reported that the photosynthetic pigments of *Rhodospirillum rubrum* were actually associated with a discrete particulate fraction. These pigmented particles were observed in extracts of alumina-ground cells, and they sedimented more rapidly in the analytical ultracentrifuge than any other subcellular component. After purification by differential centrifugation, this particulate fraction produced a somewhat diffuse sedimentation boundary with a sedimentation coefficient of approximately 200 S (Schachman *et al.*, 1952). The position of this boundary coincided with that of the pigment. The pigment particles were designated as *chromatophores*. They appeared in electron micrographs as collapsed, disk-shaped structures with a diameter of about 100 nm. On the assumption that the particles were spherical, a minimum value of 30 nm was calculated for their anhydrous diameter, based on their sedimentation characteristics. The absorption bands

Robert A. Niederman · Department of Microbiology, Rutgers University, New Brunswick, New Jersey 08903 Kenneth D. Gibson · Department of Physiological Chemistry, Roche Institute of Molecular Biology, Nutley, New Jersey 07110

observed in the near-IR and visible regions with the isolated chromatophore fraction were identical to those of whole-cell preparations. No chromatophore band was observed in the ultracentrifuge with extracts prepared from cells that were bleached by aerobic growth in the dark.

It was soon demonstrated by Frenkel (1954) that the isolated chromatophore fraction participates directly in photochemical reactions. Such a particulate fraction from phototrophically grown cells catalyzed photophosphorylation of ADP (Frenkel, 1954, 1956) and several photochemical reductions (Kamen and Vernon, 1954; Frenkel, 1958). It thus appeared that isolated chromatophores represented the bacterial photosynthetic apparatus.

The concept that the pigments of purple bacteria were localized in discrete particles received additional support from the electron-microscopic studies of Vatter and Wolfe (1958). In ultrathin sections of *Rs. rubrum, Rhodopseudomonas sphaeroides,* and *Chromatium vinosum* strain D, the cytoplasm appeared to be closely packed with numerous discrete vesicular structures. These structures were surrounded completely by a membrane, and their interiors appeared less dense than the adjacent cytoplasm. No such organized structure was seen in aerobically grown *Rs. rubrum,* and it was concluded that the membrane-bounded particles observed in phototrophically grown cells were identical to the isolated chromatophores of Schachman *et al.* (1952). As will be discussed below, it is still possible that some chromatophores exist as independent structures within the cell, but subsequent ultrastructural and biochemical evidence has suggested that they are derived from and may largely remain attached to the cytoplasmic membrane.

2. Membrane Isolation

2.1. Early Attempts at Chromatophore Isolation

In the studies of Schachman *et al.* (1952), some purification of the chromatophores of *Rs. rubrum* was achieved by differential centrifugation. Newton and Newton (1957) purified chromatophores from *Chr. vinosum* by similar procedures, and demonstrated that aside from the photosynthetic pigments and some carbohydrate, these structures were chiefly composed of protein and phospholipid. Sonic oscillation of the isolated chromatophore fraction produced smaller pigmented particles, which they designated as *chromatophore subunits.* Both types of particles were capable of catalyzing photophosphorylation.

In further studies on the isolation of chromatophores from *Rs. rubrum,* the technique of zone sedimentation in sucrose density gradients was employed by Frenkel and Hickman (1959), Cohen-Bazire and Kunisawa (1960), and Holt and Marr (1965b). Sucrose-gradient ultracentrifugation was also applied to the isolation of chromatophores from *Rp. sphaeroides* (Bull and Lascelles, 1963; Worden and Sistrom, 1964). In general, two bands that contained pigmented material were observed; these bands were designated as *light* and *heavy* chromatophores. On the basis of these results, it was concluded that the photosynthetic apparatus exists in two distinct forms within the cell. The specific Bchl content (μg Bchl/mg protein) was higher in the light chromatophore fraction, and it was thought to represent the purified chromatophore membrane. On the other hand, the heavy fraction was more heterogeneous, and consisted of material that originated from both the chromatophore membrane and the cell envelope.* The existence of a heavy chromatophore fraction containing cell-envelope material seemed to offer direct support for the hypothesis put forward on morphological grounds that chromatophores originate as invaginations of the cytoplasmic membrane (Cohen-Bazire, 1963; Cohen-Bazire and Kunisawa, 1963; Stanier, 1963) and exist *in vivo* as a continuous tubular network attached to the periphery of the cell (Holt and Marr, 1965a). It was thus suggested that the heavy fraction is largely derived form the peripheral membranes and represents a partially differentiated intermediate structure in the formation of the chromatophore membrane (Lascelles, 1968). It is now clear that much of the pigmented material in the heavy fraction resulted from the high ionic strengths of the gradients used, and under carefully controlled conditions, virtually all the Bchl and carotenoids can be isolated in a single zone that bands in the position of light chromatophores (see below).

2.2. A Critical Appraisal of Differential and Zone Sedimentation Techniques

As noted above, initial attempts at chromatophore isolation relied on differential centrifugation to separate this fraction from other subcellular constituents. Indeed, "chromatophores" isolated in this manner still

---

* Because portions of the cytoplasmic membrane in phototrophically grown cells apparently invaginate to form the chromatophore membrane, and, in some cases, additional cytoplasmic membrane is separated by mechanical disruption of the cell, the envelope fraction isolated from phototrophic cells is actually enriched in material of cell-wall origin. In cases in which EDTA–lysozyme is employed, the peptidoglycan is apparently degraded, and the isolated envelope fraction is actually enriched in outer membrane.

constitute the starting material for many biochemical investigations. More recent investigations of such preparations have indicated that they contain quantities of nonchromatophore material far in excess of that present in preparations purified by gel-filtration and density-gradient centrifugation procedures (Fraker and Kaplan, 1971). Since chromatophores sediment more rapidly than ribosomes but more slowly than the cell-envelope fraction, considerable cross-contamination of the chromatophore fraction would be expected. Repeated cycles of differential centrifugation would be necessary to affect further purification.

The rate of sedimentation of a subcellular particle depends on its size, shape, and density, as well as on the density and viscosity of the suspension medium. Particles of different size and density sediment at different rates when a centrifugal field is applied. In an ideal procedure for the fractionation of subcellular particles by differential centrifugation, the initial step is to sediment the heavier particles for a sufficient period to ensure their complete precipitation with minimal contamination by the lighter particles. As pointed out by deDuve and Berthet (1954), unless the sedimentation properties of the unresolved particles are very different, gross contamination of the initial sediment is unavoidable, and purification can be achieved only through careful washing of the individual precipitates. In general, only the fraction that sediments most slowly can be obtained in a satisfactory state of purity. An additional complication is that careful and complete decantation of the supernatant fluid is essential, since many of the lighter particles will be located near the bottom of the tube when complete sedimentation of heavier particles has been achieved. Under these circumstances, the upper portion of the heavy-particle precipitate will be enriched in less dense particles, since density is the sole factor that determines the position of a completely sedimented particle in a precipitate.

The sedimentation rate of a spherical particle in a given medium is proportional to the product $r^2(\rho_p - \rho_m)$, where r is the radius of the particle, ρ_p is the density of the particle, and ρ_m is the density of the medium (deDuve and Berthet, 1954). On this basis, it can be calculated that in an aqueous extract of *Rs. rubrum*, the sedimentation rate of ribosomes is 0.13 times that of the chromatophores when respective diameters of 160 and 800 Å (Holt and Marr, 1965b) and buoyant densities of 1.60 (McCall and Potter, 1973) and 1.19 g/cm³ (Collins and Niederman, 1976b) are employed. Accordingly, at the moment when chromatophores are completely sedimented, they would be contaminated with approximately 20% of the ribosomes initially present in the crude suspension. The cell-envelope fragments from which the chromatophores are separated would be expected to have a

density of about 1.30 g/cm³ (Collins and Niederman, 1976a); in electron micrographs, they appear very heterogeneous in size (Collins and Niederman, 1976b). Thus, substantial contamination of chromatophores by these structures would also be anticipated in a differential centrifugation procedure. These predictions are supported by the studies of Fraker and Kaplan (1971) on the fractionation of subcellular particles in *Rp. sphaeroides*. After three differential centrifugations, the crude chromatophore pellet was shown in electron micrographs to contain considerable quantities of ribosomes and cell-envelope fragments. The result of Fraker and Kaplan (1971), and the simple theoretical considerations advanced above, should be borne in mind when assessing studies of chromatophores that have been purified using differential centrifugation alone; such preparations are likely to be significantly contaminated by ribosomes and fragments of cell envelope.

Many of the problems inherent in the separation of subcellular particles by differential centrifugation are obviated by zone sedimentation in density gradients. In this procedure, the polydisperse suspension is layered on top of a density-gradient-stabilized liquid column, and a high gravitational field is applied. To achieve adequate separation, it is necessary that the density gradient exceed the inverted density gradient produced by the sample (Britten and Roberts, 1960). In addition, the density gradient prevents mixing within the liquid column, and the system is stabilized against convective disturbances by an increase in the density of the gradient in the direction of sedimentation. Separations are thereby achieved on the basis of differences in sedimentation velocity. However, osmotic pressure and viscosity gradients significantly affect separations when the gradients are formed with substances of high osmolarity, such as sucrose. As will be seen below, membrane separations are also markedly affected by the concentration of ionic species in the gradients. The limits of resolution are generally defined by the diffusion coefficient of the individual macrospecies. In dilute solution, the diffusion coefficient of a spherical particle is inversely proportional to the square root of its sedimentation coefficient (deDuve *et al.*, 1959). Accordingly, if one assumes an average $s_{20,w}$ of 250 S for the chromatophores of *Rs. rubrum* (Collins and Niederman, 1976b) and 4 S for a protein, then the diffusion coefficient would be approximately 8 times lower for chromatophores. In practice, rate-zonal sedimentation procedures have been established in which chromatophores form a narrow band that is satisfactorily resolved from ribosomes and a cell-envelope fraction on sucrose density gradients (Niederman, 1974).

Sucrose density gradients have also been employed

for the separation of membranes by isopyknic sedimentation. The relatively small diffusion coefficient for chromatophores permits satisfactory banding in this procedure as well. The sample is also layered onto a preformed gradient, and when equilibrium is achieved, a subcellular particle bands at its buoyant density within the gradient. Under appropriate conditions, several components will band at a position where the density of the gradient equals their equilibrium buoyant density, and provided these differ sufficiently, adequate resolution is attained. As described below, both zone and isopyknic sedimentation have been employed for the complete resolution of membranes of the purple photosynthetic bacteria. Two simple and reproducible procedures developed in this laboratory for the separation of membranes from phototrophically and aerobically grown cells will now be presented. Alternative procedures are discussed and evaluated elsewhere in this chapter.

2.3. Procedure for the Isolation of Chromatophore and Cell-Envelope Fractions

We describe a simple zone sedimentation procedure for the isolation of chromatophore and cell-envelope fractions from phototrophically grown *Rp. sphaeroides* (Niederman and Gibson, 1971; Niederman, 1974) and from concentrated cell suspensions in which chromatophore formation is induced at low aeration (Niederman *et al.*, 1976). All isolation procedures are performed at 0–4°C with 1 mM Tris–HCl buffer (pH 7.5); these ionic conditions minimize nonspecific chromatophore–cell envelope interactions (Niederman, 1974). Harvested cells are washed twice in buffer, and resuspended to a final concentration of 0.2–0.4 g wet wt./ml. After addition of RNase A and DNase I (10 µg of each/ml), the cell suspensions are disrupted by two passages through a French pressure cell operated at 20,000 lb/inch². The resulting extracts are centrifuged at 10,000g for 10 min, and the supernatant fraction is centrifuged at 254,000g for 75 min. The precipitate is resuspended by homogenization in approximately 1 ml buffer for each original gram of wet weight. After a further centrifugation at 27,000g for 10 min to remove aggregated material, the resuspended precipitates (2.5–3.0 ml) are layered on 32-ml linear 5–30% (wt./wt.) sucrose gradients prepared on top of a 4-ml cushion of 60% (wt./wt.) sucrose. All sucrose solutions are prepared in 1 mM Tris–HCl buffer. The gradients are centrifuged at 96,000g for 230 min (11.0 × 10¹⁰ $\omega^2 t$ units) in a swinging-bucket rotor. The gradients are fractionated by upward flow displacement with 65% (wt./wt.) sucrose. The fractions are appropriately diluted in

deionized water, and their absorbance is determined at 260, 280, and 850 nm (the latter represents the near-IR absorption maximum of *Rp. sphaeroides* Bchl *in vivo*). The succinic dehydrogenase (succinate: phenazine methosulfate oxidoreductase, EC 1.3.99.1) and NADH oxidase activities of individual fractions may also be assayed. Unfolded and partially degraded ribosomes band near the top of the gradient, a rather homogeneous chromatophore band appears at a position slightly below the center of the tube, and a photopigment-depleted cell-envelope fraction is found at the interface between the gradient and the cushion. Appropriate fractions can be collected by sedimentation at 254,000g for 75 min after dilution in buffer. They are resuspended in buffer for further analysis. A linear 5–60% (wt./wt.) sucrose gradient has also been used, but the separation between the chromatophore and ribosome fractions is not usually as wide. A reconstruction experiment (Fraker and Kaplan, 1971) suggested that the isolated chromatophores are 95% free from contamination by particulate protein of nonchromatophore origin (Niederman, 1974). Depending on the amount of material needed, smaller swinging-bucket rotors or large-scale zonal rotors may also be used, but when these are centrifuged at higher speeds, the gradient is slightly altered. This may be ascribed to some redistribution of sucrose molecules, since sedimentation of sucrose occurs at gravitational fields in excess of 70,000g (Sykes, 1971). However, satisfactory resolution has been achieved at forces as high as 200,000g.

2.4. Resolution of the Cell Envelope into Cytoplasmic and Outer Membranes

This procedure separates the cytoplasmic and outer membrane from aerobically grown *Rs. rubrum* (Collins and Niederman, 1976a). Cells are washed in 10 mM Tris buffer (pH 8.1), and lysed at room temperature by a modification of the method of Godson and Sinsheimer (1967). The cells are first plasmolyzed in 25% (wt./wt.) sucrose prepared in the same buffer. Lysozyme (EC 3.2.1.17), 0.7 mg/ml; EDTA, 2.0 mg/ml; Brij 58 (polyoxyethylene cetyl ether), 0.65%; and MgCl₂, 11.5 mM, together with a few crystals of DNase I, are sequentially added at 10-min intervals.* All subsequent operations are performed at 0–4°C. Cell debris and remaining whole cells are removed by repeated centrifugation at 750g for 10 min until no further pellet is obtained. Subcellular particles are sedimented at 254,000g for 75 min, and the resulting precipitate is homogenized in

---

* A similar lysis procedure for *Rs. rubrum* was reported by Simon and Siekevitz (1973).

and dialyzed against 10 mM *N*-2-hydroxyethylpiperazine *N'*-2'-ethanesulfonic acid (HEPES) buffer (pH 7.5). The dialyzed samples are layered onto 37-ml 35–55% (wt./wt.) sucrose gradients prepared in the same buffer. Isopyknic sedimentation of membrane fractions is achieved by 13.75-hr centrifugation at 96,000*g*. Gradients are fractionated as described above, and the absorbance of diluted fractions is determined at 235, 260, and 280 nm. The succinic-dehydrogenase-rich cytoplasmic membrane bands near the top of the gradient, while the outer membrane fraction is found near the bottom. Apparently as a result of the viscosity of the gradient, the ribosome fraction cosediments with the cytoplasmic membranes, and it is necessary to separate them largely on the basis of their differences in sedimentation velocity. This is accomplished by zone sedimentation for 2 hr at 200,000*g* on a 13-ml 5–45% (wt./wt.) sucrose gradient prepared in 1 mM Tris (pH 7.5). The membranes sediment more rapidly and band mostly in the bottom half of the tube. The gradients are collected and analyzed as described above, and appropriate fractions are recovered by the procedure described in the previous section.

2.5. Evidence for Separability of Chromatophores from the Peripheral Cytoplasmic Membrane

2.5.1. Association of Bchl with One Type of Structure

As noted above, in early studies on the isolation of membranes from photosynthetic bacteria, the photosynthetic pigments were observed nearly equally distributed between two bands in sucrose density gradients. In contrast, Gibson (1965*a*) reported that 85% or more of the photosynthetic pigments in extracts of *Rp. sphaeroides* were associated with a single, rather homogeneous chromatophore band for which a sedimentation coefficient of 160 S was calculated. Such chromatophore preparations were isolated in high yield by zone sedimentation in cesium chloride gradients. Studies of various types demonstrated that under some conditions, the isolated chromatophores readily formed aggregates (Gibson, 1965*b*). Aggregation was very sensitive to the ionic environment, and was influenced by both ionic strength and the nature of the ion present. Thus, magnesium and other divalent cations caused extensive chromatophore aggregation at ionic strengths in excess of 0.01 μ. On the other hand, with monovalent cations such as cesium, no aggregation was observed. Since high concentrations of ionic species such as magnesium were present in the sucrose gradients used by Cohen-Bazire and Kunisawa (1960), Worden and Sistrom

(1964), and Holt and Marr (1965*b*), it seemed possible that the presence of pigments in two bands largely reflected the ionic environments within the gradients. It was subsequently shown that the separation of chromatophores from the cell envelope in *Rp. sphaeroides* was markedly affected by the presence of ionic species (Niederman and Gibson, 1971). In gradients that were 10 mM with respect to both Tris buffer and magnesium, chromatophores were distributed nearly equally between an upper and a lower pigmented band. About half the chromatophores were released from the lower band when magnesium was excluded from the gradient. In gradients prepared with aqueous sucrose, virtually all the chromatophores were resolved from a lower photosynthetic-pigment-depleted band. Subsequently, the lower band was identified as a Bchl-deficient cell-envelope fraction (Niederman, 1974). Thus, nearly all the photosynthetic pigments could be isolated in a single, discrete chromatophore band; these studies suggested that the photosynthetic apparatus of *Rp. sphaeroides* is localized in membranes largely distinct from the cell wall–cytoplasmic membrane complex.

It thus seemed possible that the heavy chromatophore fraction observed in *Rp. sphaeroides* resulted from magnesium-induced aggregation between light chromatophores and the cell envelope. In further studies (Niederman, 1974), heavy chromatophores were dialyzed against 10 mM Tris buffer and sedimented on sucrose gradients prepared in the absence of divalent cations. An extensive light chromatophore band appeared in the appropriate position, while a photosynthetic-pigment-depleted cell envelope band was observed in the usual position of heavy chromatophores. In addition, isolated light-chromatophore and cell-envelope fractions were combined, dialyzed against Tris buffer containing 10 mM magnesium, and sedimented in a sucrose gradient prepared with the same buffer. The majority of the pigment banded in the usual position of the heavy chromatophore fraction; the reconstituted heavy fraction had a Bchl content characteristic of heavy chromatophores, and after dialysis and zone sedimentation in the absence of magnesium, much of the chromatophore material present could be released from this fraction. Magnesium concentrations of 1.0 mM or higher caused extensive heavy-chromatophore formation. It can be concluded that magnesium causes chromatophores to interact reversibly with the cell envelope. Significant aggregation was also promoted by other divalent cations (Co^{2+} > Mn^{2+} > Ca^{2+} > Mg^{2+}), but aggregation was less extensive with monovalent cations. Thus, the nature of the ions present markedly influences chromatophore–cell envelope interactions. It is possible that in the formation of heavy chromatophores, divalent cations increase the stability of chromatophore–envelope interactions by interchain cross-linking between negative charges on the surface of the two structures.

These resolution and reconstitution studies strongly suggest that the heavy chromatophore fraction of *Rp. sphaeroides* represents an artifact of the ionic conditions employed in previous isolation schemes, and that the chromatophore material present in this fraction is identical to light chromatophores. However, evidence presented by Worden and Sistrom (1964) was consistent with possible intrinsic differences in the chromatophore material present in their light and heavy fractions. The ratio of carotenoids to Bchl and the near-IR absorption spectra of the two fractions from *Rp. sphaeroides* strain Ga differed significantly. However, Niederman and Gibson (1971) reported virtually consistent carotenoid/Bchl ratios for light and heavy chromatophore preparations from wild-type *Rp. sphaeroides*; the ratios were quite similar even when only negligible amounts of chromatophore

material remained bound to the envelope. The *in vivo* spectrum of *Rp. sphaeroides* contains two light-harvesting (LH) Bchl components that together account for more than 95% of the total Bchl; these include B850 with absorption bands at 800 and 850 nm and B875, which absorbs maximally at 875 nm (Sistrom, 1964). Worden and Sistrom (1964) observed that the ratios of absorbance at 800, 850, and 873 nm were consistently different in their heavy and light chromatophore fractions. Specifically, the heavy fraction had a lower relative absorbance, per milligram Bchl, at 850 and 800 nm than the light fraction. These spectral differences could reflect nonrandom binding of a special population of chromatophores to the cell envelope. However, other studies from our laboratory do not support this possibility. A peripheral-membrane-enriched fraction that banded in the usual position of heavy chromatophores exhibited absorption bands at 850 and 800 nm that were diminished by about 20% and 10%, respectively, when compared with those of the usual light chromato-

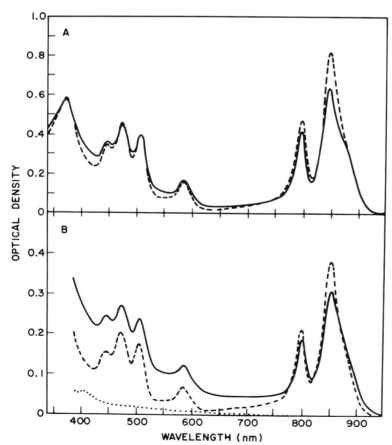

Fig. 1. Comparison of spectra of heavy fractions and chromatophores from *Rhodopseudomonas sphaeroides*. (A) Heavy fraction enriched in cytoplasmic membrane from phototrophically grown cells (Niederman *et al.*, 1972) (——); chromatophores (– – –). The amounts of each fraction were adjusted so that the height of the Soret band at 376 nm was equal in each spectrum. (B) Aggregated heavy fraction prepared by the addition of chromatophores to aerobically grown cells (Niederman *et al.*, 1972) (——); the aerobic cytoplasmic membrane fraction (·····); chromatophores (– – –). Here, the concentrations of the two pigmented fractions were adjusted to give an equal absorbance at 870 nm.

phore fraction (Fig. 1A). Reconstruction experiments indicated that the bulk of the Bchl in this aggregated fraction arose from light chromatophores (Niederman *et al.*, 1972). A near-IR spectrum typical of heavy chromatophores was also observed in a reconstituted heavy fraction (Fig. 1B). This fraction was prepared by the addition of purified light chromatophores to aerobically grown cells, passage of the mixture through the French press, followed by differential and rate-zone sedimentation. The reconstituted fraction had a Bchl content similar to that of the phototrophic peripheral membrane fraction. In addition, the spectrum of chromatophores resolved from a heavy fraction was identical to that of light chromatophores isolated in the usual manner (Niederman *et al.*, 1972). All these results suggest that heavy chromatophores are an artifact that form *in vitro* through random interactions of light chromatophores with the cell envelope. Rather than reflecting an intrinsic difference between two nonoverlapping classes of chromatophores, the spectral differences observed in the heavy fraction result from the effects of extensive aggregation on molecular interactions responsible for the individual absorption bands of the Bchl present in a single type of particle. That the B875 component is particularly sensitive to the state of aggregation of the membrane is suggested by its irreversible conversion to monomeric Bchl when chromatophores are treated with the zwitterionic detergent lauryl dimethyl amine oxide (LDAO) under conditions that do not greatly alter B850 (Clayton and Clayton, 1972). Transformation of the longest-wavelength form of Bchl by detergent results in its inability to accept excitation energy (Bril, 1963), and it has been suggested that the longest-wavelength component represents an aggregated state that is easily disrupted by detergents (Olson and Stanton, 1966).

In further studies of the heavy-chromatophore fraction from *Rs. rubrum*, Ketchum and Holt (1970) isolated a Bchl-depleted heavy band from light-grown cells when magnesium was excluded from their isolation scheme. Additional chromatophores were released when the isolated heavy band was disrupted in a French pressure cell. Oelze *et al.* (1969a) resolved chromatophores from cell-envelope material by zone sedimentation on Ficoll gradients prepared in the absence of magnesium. The results of these investigations suggested that a Bchl-deficient cytoplasmic membrane is conserved during phototrophic growth in *Rs. rubrum*; the content of succinic dehydrogenase and cytochrome in the cell-envelope fraction from *Rp. sphaeroides* grown under similar conditions is also consistent with this possibility (Niederman, 1974). A pigment-depleted heavy band was also observed in magnesium-free sucrose gradients with extracts pre-

pared from *Rs. rubrum* during adaptation to photometabolism (Yamashita and Kamen, 1969). The isolation of membrane fractions was recently investigated in phototrophically grown *Rs. rubrum* after lysis by the EDTA–lysozyme–Brij procedure described in Section 2.4 (Collins and Niederman, 1976b). A quantitative separation of chromatophores from a cell-envelope fraction was achieved by isopyknic sedimentation in density gradients prepared with HEPES-buffered sucrose. The isolated heavy fraction had a low Bchl content, and on this basis was at least 93% free of contamination by chromatophore protein. The Bchl-rich chromatophore fraction was resolved from contaminating ribosomal material by zone sedimentation on sucrose gradients; reconstruction experiments showed that resulting chromatophore fractions are 95% free from particulate protein of nonchromatophore origin. Thus, as observed for *Rp. sphaeroides*, the photosynthetic apparatus of *Rs. rubrum* appears to be housed in a single system of chromatophore membranes separable from the cell-envelope complex.

In summary, earlier reports indicating the existence of two distinct classes of chromatophores in extracts of *Rp. sphaeroides* or *Rs. rubrum* have proved on closer examination to have been mistaken. By ensuring that the ionic strength of the medium used to isolate the different membrane fractions is maintained at a sufficiently low value, it is now possible to isolate from French-pressure-cell extracts of either organism a single major pigmented fraction that contains almost all the photopigments. The pigmented particles correspond to the fraction that was named light chromatophores in the early studies. An additional spin-off, as discussed in the next section, has been the development of methods for partial separation of unpigmented cell envelopes into their cytoplasmic-membrane and cell-wall components.

2.5.2. Separation of Cell-Envelope Membranes

The existence of Bchl-deficient cytoplasmic membrane continuous with intracytoplasmic chromatophore membrane, together with the postulated origin of chromatophores at intrusions of the cytoplasmic membrane, suggests that discrete regions exist within the plasma membrane of the Rhodospirillaceae. Accordingly, peripheral regions of the membrane would be closely apposed to the cell wall and structurally and functionally differentiated from Bchl-containing internal portions (the chromatophore membrane). A critical test of this possibility requires the resolution of the various membranes and the isolation of highly purified preparations of each. To obtain representative preparations of the peripherally localized cytoplasmic membrane, it is first necessary to resolve the cell envelope into its cytoplasmic-membrane and cell-wall (or outer-membrane) components.

(a) Aerobically Grown Cells. Since the purification of cell-envelope membranes is complicated by the presence of chromatophores, envelope resolution is most simply achieved in cells grown at high aeration that are devoid of chromatophore membranes. Through a combination of differential- and zone-sedimentation techniques, Niederman *et al.* (1972) isolated two membrane fractions after French-press disruption of such *Rp. sphaeroides* cells. One fraction was obtained in the form of small membrane fragments that did not sediment during a 90-min centrifugation at 150,000*g*. On the basis of their high succinic dehydrogenase and cytochrome content, chemical composition, polypeptide profile in sodium dodecyl sulfate (SDS)–polyacrylamide gel electrophoresis, and appearance in electron micrographs, they were identified as a fraction enriched in portions of the cytoplasmic membrane. A fraction isolated from the initial 150,000*g* precipitate was identified as cell-wall material depleted in cytoplasmic membrane. More recently, Ding and Kaplan (1976) extensively washed and sonicated a crude envelope fraction obtained from French-press extracts of chemotrophically grown *Rp. sphaeroides*. The envelope was resolved into cytoplasmic-membrane and cell-wall components by isopyknic sedimentation on a discontinuous sucrose gradient. In addition, Guillotin and Reiss-Husson (1975) reported the separation of the cytoplasmic and outer membranes from mechanically disrupted lysozyme–EDTA spheroplasts of aerobic *Rp. sphaeroides*.

A procedure recently developed by Collins and Niederman (1976*a*) for the resolution of surface layers from Brij-58-treated spheroplasts of aerobic *Rs. rubrum* is presented in Section 2.4. The peptidoglycan layer is apparently destroyed by this procedure, and highly purified cytoplasmic- and outer-membrane preparations are obtained. Their appearance in electron micrographs and their polypeptide patterns on SDS–gel electrophoresis suggest that the isolated structures are quite similar to preparations obtained from other gram-negative bacteria. The bulk of the membrane-bound succinic dehydrogenase activity was present in the cytoplasmic-membrane fraction; this fraction was also enriched tenfold in *b*- and *c*-type cytochromes with respect to the outer membrane. A much greater carbohydrate content was observed in the outer membrane fraction, which was also enriched in arachidic acid, a specific component of the lipopolysaccharide of *Rs. rubrum* (Schröder *et al.*, 1969).

Two cell-envelope fractions were also obtained from chemotrophically grown *Rs. rubrum* by Oelze *et al.*

(1975). Cells that were disrupted in a French press were subjected to a series of differential and isopyknic sedimentations. The isolated envelope fractions appeared to be enriched in cytoplasmic membrane and cell wall, respectively. Their appearance in electron micrographs and their fatty acid and overall chemical compositions suggest that, apart from differences in peptidoglycan content, they are quite similar to the preparations obtained in this laboratory. A comparison of their carbohydrate contents and polypeptide profiles, however, suggests that the resolution obtained by Oelze *et al.* (1975) is less complete.

Thus, a number of procedures are now available for the preparation of resolved fractions of the cell-envelope layers of aerobically grown Rhodospirillaceae in varying states of purity. The physicochemical properties of some of these preparations are presented in Section 3.

(b) Phototrophically Grown Cells. The differential centrifugation procedures that resulted in partial resolution of the cytoplasmic membrane and cell wall of aerobic *Rp. sphaeroides* were also employed with light-grown cells (Niederman *et al.*, 1972). Under these circumstances, approximately 80–90% of the Bchl was sedimented. The membrane fragments that did not sediment contained about 50% of the succinic dehydrogenase activity and the remaining Bchl. After sedimentation at 100,000*g* for 18 hr, the aggregated precipitate was placed on sucrose gradients, and small membrane fragments were resolved from contaminating ribosomal material by zone sedimentation. The membranes banded in the position of cell-envelope material, and their specific Bchl content was about 25% that of chromatophores; however, their specific succinic dehydrogenase activity was greater than that of chromatophores. Reconstruction experiments suggested that this fraction consisted of a mixture of non-pigmented small membrane fragments similar to those found in aerobic cells, together with whole or fragmented chromatophores (the near-IR absorption spectrum of this fraction is discussed in Section 2.5.1.). Despite the significant contamination by material derived from chromatophores, this fraction was apparently enriched in Bchl-depleted regions of the cytoplasmic membrane. In pulse-chase studies, it appeared to contain precursors of chromatophore proteins (Gibson *et al.*, 1972).

Further elucidation of the role of the peripheral membrane as a possible precursor of chromatophores necessitates the isolation of pigment-depleted regions in a form free from contamination by whole or broken chromatophores. Since putative pigmented precursors of chromatophores have been identified in *Rp. sphaeroides* during adaptation to phototrophic growth (Shaw and Richards, 1971), additional resolution of

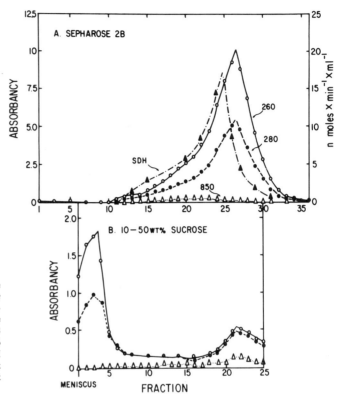

Fig. 2. Isolation of peripheral-cytoplasmic-membrane fraction from phototrophically grown *Rhodopseudomonas sphaeroides*. (A) Chromatography of 150,000g supernatant on Sepharose 2B. (SDH) Succinic dehydrogenase activity. (B) Fractions 19–24 from a Sepharose 2B column were pooled, aggregated, placed on a 10–50% (wt./wt.) sucrose density gradient prepared in 0.01 M HEPES buffer, pH 7.5, and centrifuged for 20 hr at 78,000g. The band near the bottom of the gradient contained highly purified cytoplasmic membrane.

small membrane fragments (Niederman *et al.*, 1972) should permit a critical test of the possibility that such precursors exist under steady-state growth conditions. The further resolution and characterization of small membrane fragments is currently in progress in this laboratory (Parks and Niederman, in prep.). Apparent peripheral-cytoplasmic-membrane preparations nearly free from photosynthetic pigment have been isolated from light-grown cells. These preparations were obtained from French-press extracts by a procedure that involves differential centrifugation, chromatography on Sepharose 2B, ultrafiltration, and isopyknic sedimentation on sucrose gradients. The results of a representative separation are shown in Fig. 2. The 150,000g, 90-min supernatant was placed directly on a Sepharose 2B column and eluted with the phosphate–EDTA buffer of Fraker and Kaplan (1971). The location of membrane fragments is indicated by the distribution of succinic dehydrogenase, which shows maximum activity between fractions 24 and 26 (Fig. 2A). The high ratio of absorbance at 260 nm to that at 280 nm indicates a wide distribution of unfolded ribosomal material in the column. To assure that the membrane fragments thus obtained are relatively free from Bchl, fractions with

the highest succinic dehydrogenase/$A_{850\,nm}$ ratio were pooled, concentrated by ultrafiltration, and sedimented at 100,000g for 18 hr. The pelleted membrane fragments were resolved from contaminating ribosomal material by an isopyknic sedimentation on a HEPES-buffered sucrose gradient (Fig. 2B). The purified membrane fraction forms a band near the bottom of the gradient. The isolated fraction is markedly depleted in photosynthetic pigments; the Bchl contents of several preparations suggest that they are 90–97% free from contamination by chromatophores. Significant amounts of b- and c-type cytochromes and succinic dehydrogenase activity are present in the isolated fraction. In SDS–gel electrophoresis performed by the method of Laemmli (1970), polypeptide profiles are obtained for this material that appear distinct from chromatophore and cell-wall fractions but essentially identical to aerobic cytoplasmic-membrane fractions isolated by this method and that of Ding and Kaplan (1976). The latter technique, however, could not be applied successfully to the isolation of peripheral cytoplasmic membrane free from Bchl from phototrophically grown cells. The membrane fraction isolated by the newly developed procedure is markedly depleted in a major polypeptide band of approximately 40,000 molecular weight. This protein is apparently a component of the outer membrane (Niederman *et al.*, 1972), and may be similar to that observed in *Escherichia coli* (Schnaitman, 1970a) and other gram-negative bacteria (Schnaitman, 1970b). Electron micrographs of thin sections of the membrane fraction reveal typical small vesicular membrane structures identical to cytoplasmic-membrane fractions from aerobically grown cells. These are distinct from isolated-cell-wall preparations, which yield much larger fragments and show typical coiled structures. That the Bchl-depleted cytoplasmic-membrane preparation from phototrophically grown cells originates from peripheral regions of the cell is suggested by the equivalence between the physicochemical properties of this fraction and that derived from chemotrophically grown cells, in which much of the cytoplasmic membrane was observed in close apposition to the cell wall (Brown *et al.*, 1972; Peters and Cellarius, 1972).

2.6. Evaluation of Available Membrane-Isolation Procedures

The choice of membrane-isolation procedures is dictated largely by the kind of end product that is desired. For critical investigations of the structure, localization, and biogenesis of various membrane components, it is necessary to separate the membranes from other cellular constituents as well as from each other. This generally requires a combination of differential and density-gradient centrifugation. Such procedures may result in the loss or inactivation of labile components such as those involved in the photophosphorylation of ADP (Gibson, 1965b,c). Nevertheless, it is now possible to obtain representative membrane preparations for studies on the formation of major chromatophore components and the possible role of other cellular membranes in this process. There follows a critical appraisal of available membrane-isolation techniques, with special emphasis on the extent of resolution and the apparent level of purity of the material that is obtained.

2.6.1. Rhodospirillaceae

As noted already, most of the early studies on the resolution of the membrane components from phototrophically grown Rhodospirillaceae were seriously impeded by the inclusion of high concentrations of magnesium in buffers employed for isolation procedures. Following the studies of Gibson (1965a), in which 85% or more of the Bchl present in magnesium-free extracts of *Rp. sphaeroides* was shown to be localized in a single, discrete chromatophore band that

aggregated readily in the presence of divalent cations, several separations of chromatophores and cell-envelope material from mechanically disrupted extracts were reported. Provided that ionic conditions are carefully controlled, such separations have been obtained by zone sedimentation on gradients of Ficoll or sucrose. Although apparently satisfactory chromatophore preparations have been obtained from *Rs. rubrum* by Ficoll gradient centrifugation (Oelze *et al.*, 1969a), the isolated "cytoplasmic membrane" fraction appeared to be heavily pigmented and contained succinic dehydrogenase activity, a marker that was recently shown to be specific to chromatophore membrane in phototrophically grown *Rs. rubrum* (Collins and Niederman, 1976b). The sucrose-gradient centrifugation procedure of Yamashita and Kamen (1969) was applied to phototrophically adapted *Rs. rubrum*. This resulted in the separation of a heavy band from chromatophores; the latter fraction appeared to be resolved from ribosomes that banded near the top of the gradient. The heavy band was relatively free from 880 nm absorbance, and the majority of the absorbance at this wavelength was associated with the chromatophores. Both the chromatophore and heavy fractions contained NADH oxidase and oxidative phosphorylation activity. Sedimentation on sucrose density gradients in the absence of magnesium was also employed in a scheme for the isolation of chromatophores and a Bchl-depleted heavy fraction from phototrophically grown *Rs. rubrum* by Ketchum and Holt (1970). After the initial gradient centrifugation, however, their chromatophore fraction was contaminated with ribosomes, and the diffused heavy band contained significant quantities of photopigments. The chromatophores were resolved from ribosomes by sucrose-gradient electrophoresis, and the heavy fraction was treated with a French press and subjected to a second isopyknic sedimentation to remove some of the chromatophores. Substantial amounts of Bchl and succinic dehydrogenase activity were retained in the heavy band after treatment by this procedure.

As noted in Section 2.5.1, isopyknic and zone sedimentation of extracts of *Rs. rubrum* lysed by sequential lysozyme, EDTA, and Brij 58 treatment resulted in virtually complete resolution of chromatophores from a diffuse heavy band (Collins and Niederman, 1976b). Although the latter fraction was enriched in outer membrane, its cytochrome and NADH oxidase content suggest that it also contains Bchl-depleted cytoplasmic membrane. The dense fraction obtained from these extracts is consistently more free from contamination with Bchl than that isolated from French-press extracts. In contrast to earlier claims that *Rs. rubrum* spheroplasts do not release chromatophores on osmotic lysis (Tuttle and Gest, 1959), lysis with lysozyme and EDTA alone resulted in the liberation of about 50% of the total Bchl in the form of typical chromatophores (Collins and Niederman, 1976b). Chromatophores from extracts obtained by lysozyme–EDTA lysis in the presence and in the absence of Brij had physicochemical properties similar to those of French-press preparations. Similar near-IR absorption spectra, overall chemical composition, polypeptide patterns in SDS–gel electrophoresis, and equilibrium buoyant densities (see Section 3.4) were observed. The procedure presented for the isolation of chromatophores and cell envelope fractions by a single density-gradient centrifugation (Section 2.3) is also applicable to French-press extracts of *Rs. rubrum*.

With regard to the separation of subcellular particles from *Rp. sphaeroides*, the preparative zone sedimentation techniques described in Section 2.3 result in satisfactory yields of chromatophores and cell-envelope preparations in a high state of purity. Gorchein *et al.* (1968) and Shaw and Richards (1971) also reported sucrose density-gradient centrifugation procedures for extracts of mechanically disrupted *Rp. sphaeroides* prepared in the absence of divalent cations. In the studies of Gorchein *et al.* (1968), a discontinuous sucrose gradient prepared with high concentrations of Tris buffer was employed. Although a light chromatophore band was obtained, a considerable quantity of chromatophore material was lost with the pelleted cell-envelope material. In the procedure of Shaw and Richards (1971), isopyknic centrifugation in a fixed-angle rotor was applied to sonicated extracts prepared from cells undergoing adaptation to phototrophic growth. Although almost no chromatophores were formed during this period, several nonchromatophore membrane fractions were isolated that were designated as precursors of chromatophores on the basis of labeling studies. One of these fractions, however, appeared to be heavily contaminated with ribosomes.

A procedure for the preparation of large quantities of highly purified chromatophores from *Rp. sphaeroides* was developed by Fraker and Kaplan (1971). It was found that chromatophore aggregation could be minimized by performing all purification steps with 0.1 M sodium phosphate buffer (pH 7.6) containing 0.01 M EDTA. French-press extracts were chromatographed on a Sepharose 2B column, and the chromatophore band was further purified by centrifugation on a 20–40% (wt./vol.) linear sucrose gradient. A reconstruction experiment was performed in which [^{14}C]phenylalanine-labeled aerobic cells devoid of chromatophores were mixed with unlabeled phototrophically grown cells. In this manner, it was

shown that the gradient-purified chromatophores contained less than 1% of the nonchromatophore protein present prior to chromatography. The purified chromatophore fraction also appeared homogeneous in the electron microscope. By carefully monitoring the absorbance of individual fractions at 260, 280, and 850 nm, it is possible to isolate chromatophores of satisfactory purity directly from the Sepharose 2B column (Niederman and Segen, unpublished). No increase in the specific Bchl of the isolated chromatophore fraction was observed after sucrose density-gradient centrifugation. However, the cell-envelope fraction that is eluted from the column prior to the chromatophores and purified by centrifugation through sucrose (Huang and Kaplan, 1973a) contained higher levels of photopigment than the envelope fraction isolated directly from sucrose gradients without prior column treatment (Niederman and Segen, unpublished). The Sepharose 2B column chromatographic procedure of Fraker and Kaplan (1971) formed the basis of a procedure for the isolation of small cytoplasmic membrane fragments from *Rp. sphaeroides* (Parks and Niederman, in prep.) described in Section 2.5.2.

In contrast to the variety of procedures developed for the isolation of membranes from *Rs. rubrum* and *Rp. sphaeroides*, only a few reports on the resolution of the membranes of other Rhodospirillaceae have appeared. Lampe *et al.* (1972) recently reported a scheme for the fractionation of the membrane system in *Rp. capsulata* by centrifugation on gradients of Ficoll (density range 1.01–1.07 g/cm³). Three membrane fractions were observed; these fractions were identified on the basis of their relative Bchl content and appearance in electron micrographs. The light fraction was designated as small cytoplasmic membrane vesicles, and was present in extracts of aerobically and phototrophically grown cells. A middle band designated as intracytoplasmic membranes was observed only in aerobically grown cells, while the heavy band contained highly purified chromatophores from phototrophic and semiaerobic cultures. It had been observed previously in thin sections that aerobically grown *Rp. capsulata* contains some tubular intracytoplasmic membrane (Drews *et al.*, 1969b). The results of pulse-chase studies in which phospholipids were labeled with 2-[^{14}C]acetate led to the conclusion that the chromatophores and cytoplasmic membranes form a morphogenetic unit in which the component membranes can be transformed into each other in response to altered oxygen tension or light intensity (cf. Irschik and Oelze, 1973). Definitive identification of the isolated fractions awaits further assessment of their purity and physicochemical characteristics.

The *in vivo* morphology of the internal membranes

of other Rhodospirillaceae from which these structures have been isolated is very different from the afore-named members of this group, in which a vesicular system of chromatophore membranes is observed in thin sections of whole cells (Vatter and Wolf, 1958; Cohen-Bazire and Kunisawa, 1963; Lampe *et al.*, 1972). Gibbs *et al.* (1965) reported a procedure for the isolation of lamellar chromatophore membranes from French-press extracts of *Rs. molischianum*. The chromatophore material was isolated after centri-fugation in a gradient formed with 27% (wt./wt.) rubidium chloride. Although these preparations were apparently free from ribosomes, it is not clear whether the isolated chromatophore membranes had been resolved from cell-envelope material. Hickman *et al.* (1963) reported the isolation of chromatophore material from *Rs. molischianum* in a single zone after sedimentation on discontinuous sucrose density gradients. The branched lamellar chromatophore membrane of *Rp. palustris* (Cohen-Bazire and Sistrom, 1966; Tauschel and Drews, 1967) and the irregular intrusions of the cytoplasmic membrane observed in *Rp. gelatinosa* (Weckesser *et al.*, 1969) were isolated by sucrose-gradient centrifugation of sonic extracts by Hansen and deBoer (1969). With *Rp. gelatinosa*, the latter investigators observed that inclusion of 0.01 M magnesium in the gradients resulted in the appearance of two Bchl-containing bands; only a single zone was seen when magnesium was omitted. The chromatophore material from *Rp. palustris* was located in a heavy band in the presence of magnesium, whereas a single upper band of pigmented material was observed in its absence. These results are in agreement with the magnesium-induced membrane aggregation observed in other species of Rhodospirillaceae (Section 2.5.1). Pigmented mem-branes were also isolated from *Rp. palustris* by flotation on CsCl gradients by Garcia *et al.* (1968). Biedermann and Drews (1968) reported the isolation of the stacked, granalike chromatophore material (Giesbrecht and Drews, 1966) from sonic extracts of *Rp. viridis* by density-gradient centrifugation. More recently, Pucheu *et al.* (1973, 1974) sedimented such extracts on discontinuous sucrose density gradients and isolated a photosynthetic membrane fraction that was sensitive to aggregation by magnesium.

2.6.2. Chromatiaceae

The most extensive investigation of isolation pro-cedures for the membranes of Chromatiaceae was performed in *Thiocapsa floridana* (now *T. roseopersicina*) by Takacs and Holt (1971a). Despite the presence of magnesium, only a single pigmented band was observed in sucrose density gradients in the position of the light chromatophores of Rhodospiril-laceae. Column chromatography of a crude French-press supernatant on Sepharose resulted in the isolation of three membrane fractions. On the basis of electron micrographs of negatively stained pre-parations, these fractions were designated in the order of their elution as fragmented cell wall, purified chromatophores, and small membrane fragments contaminated with ribosomes [cf. Section 2.5.2(b)]. In an alternative preparative procedure, crude pre-parations of subcellular particles were centrifuged on gradients formed with 30% (wt./wt.) rubidium chloride. The bulk of the chromatophore fraction floated to the top of the gradient, whereas the cell-wall fragments and ribosomes formed a pellet. Negatively stained preparations of these chromatophores in-dicated that they were relatively homogeneous in size, with an average diameter of approximately 650 Å; no cell-wall or ribosomal contamination was observed. The purified chromatophore preparation gave a single, symmetrical Schlieren peak in the analytical ultracentrifuge, with a sedimentation coefficient of 150 S when extrapolated to infinite sample dilution (Takacs and Holt, 1971b). The isolated preparations were also capable of catalyzing the photophosphoryla-tion of ADP, and their near-IR and visible absorption spectra were essentially identical to those of whole cells.

In other studies on the resolution of membrane components of Chromatiaceae, two Bchl-containing membrane fractions were isolated from *Chr. vinosum* by Cusanovich and Kamen (1968a). Cells were disrupted at high pressure in a Ribi cell fractionator; a buffer prepared with 0.1 M potassium phosphate was used in the preparative procedures. After removal of ribosomes by centrifugation in 35% (wt./wt.) rubidium chloride, velocity sedimentation in sucrose gradients resolved the membranes into two pigmented bands: a light chromatophore band and a heavy fraction. Sedimentation studies suggested that the heavy par-ticles attach some light chromatophores. Aside from some increases in light scattering and the relative absorbance of the longest-wavelength Bchl band (absorption maximum at 888 nm) observed in the heavy fraction, the near-IR and visible absorption spectra were similar for both fractions. As seen in *Rp. sphaeroides* (Section 2.5.1), the differences in the relative absorbance of bands in the near-IR may reflect the increased state of aggregation of the heavy fraction, rather than an intrinsic diffrence between light and heavy chromatophores of *Chr. vinosum*. In addition to their high Bchl content, the isolated fractions also contained high levels of heme and carotenoids. The ratios of these components to Bchl were quite similar in the light and heavy chromatophores, although they were somewhat variable in the latter fraction, which was thought to arise from the cytoplasmic membrane. A greater relative yield of light chromatophores was obtained from alumina ground cells, even though this was considered to be a gentler means of cell breakage than the Ribi cell fractionator. A sedimentation coefficient of 145 S at infinite dilution was determined for the light particles by analytical ultracentrifugation, and identical results were obtained with both Schlieren and absorption optical systems. Electron micrographs of negatively stained preparations of the light chromatophores indicated that their average diameters were similar to those noted above for *T. floridana* (now *T. roseopersicina*). The light fraction was enriched in lipids, and was obtained reproducibly with a fixed content of photoactive pigments and electron-transfer components.

In a recent report by Hurlbert *et al.* (1974), a total membrane fraction was obtained from *Chr. vinosum* after plasmolysis, treatment with lysozyme and EDTA, and lysis of the resulting spheroplasts by brief sonication or passage through a syringe. The membranes were collected by centrifugation and resolved by isopyknic sedimentation on a discontinuous sucrose gradient into three intracytoplasmic-membrane fractions and a heavy band. The three intracytoplasmic-membrane bands had apparent buoyant densities of 1.11, 1.14, and 1.16 g/cm³, respectively, and contained the major portion of the photosynthetic pigments, with the highest content in the middle band. French-pressure-cell treatment of whole-cell suspensions resulted in a single chromatophore band at the position of the middle band. The heavy fraction from the lysed spheroplasts (ρ = 1.23 g/cm³) was nearly devoid of Bchl, and its polypeptide pattern in SDS–gel electrophoresis and appearance in electron micrographs suggest that it consisted largely of outer membrane. The polypeptide patterns of the pigmented fractions were qualitatively similar, but some differences in the levels of various components were observed. Contamination of the two heavier intracytoplasmic-membrane fractions by the major outer-membrane polypeptide was also observed. These two fractions also contained considerable quantities of diaminopimelic acid, an apparent peptidoglycan component. The lightest intracytoplasmic-membrane fraction and the outer-membrane band were essentially devoid of diaminopimelic acid. Electron micrographs of negatively stained preparations revealed that the pigmented material consisted of large reticular structures with vesicular appendages and flattened tubular extensions. Although the relationship between the isolated intracytoplasmic-membrane fractions and possible pigment-depleted regions of the peripheral cytoplasmic membrane is not clear, this gentle disruptive procedure apparently results in the release of much of the photosynthetic apparatus in a form that closely resembles that in whole cells. A recent electron-microscopic investigation by Oyewole and Holt (1976) indicated that in isolated intracytoplasmic membranes from *Ectothiorhodospira mobilis*, the folded organization of the lamellar membrane stacks had remained intact.

3. Membrane Characterization

3.1. Gross Chemical Composition

The overall chemical composition of representative chromatophore preparations from several of the Rhodospirillaceae is presented in Table 1. The major portion of the dry weight of each of the isolated chromatophore fractions can be accounted for by protein and lipid. Aside from the apparently higher content of carbohydrate in the chromatophores of *Rs. rubrum*, a striking similarity in the levels of the various chemical constituents in the photosynthetic membranes from each organism is observed. With the exception that up to 8% of their dry weight can be accounted for by photosynthetic pigments, the chemical composition of these membrane preparations is typical of that observed for the plasma membranes of a variety of non-photosynthetic bacteria (Salton, 1967; Miura and Mizushima, 1968; Osborn *et al.*, 1972). The variations observed in photopigment contents among several of the chromatophore preparations may have resulted from the effect of different conditions of growth (e.g., light intensity, self-shading, and stage of growth) on the cells from which these preparations were isolated; some of the disparities in Bchl values largely reflect differences in the extinction coefficients employed. The low nucleic acid content of the isolated chromatophores indicates that they are essentially free from contamination by DNA and ribosomes. This is also supported by the fairly close agreement between the levels of total phosphorus and lipid phosphorus in purified chromatophore preparations from *Rp. sphaeroides* (Bull and Lascelles, 1963; Gibson, 1965c; Gorchein *et al.*, 1968). The chemical composition data for chromatophores isolated from Chromatiaceae (Table 2) generally indicate that these preparations differ from those of the Rhodospirillaceae in their high total lipid content. This unusual property is shared with the photosynthetic vesicles isolated from Chlorobiaceae, but much of the total lipid can be accounted for as Bchl *c* in the latter group (Cruden and Stanier, 1970).

The chemical composition of several cell-envelope fractions recently isolated from *Rp. sphaeroides* and

Table 1. Chemical Composition (% Dry Weight) of Chromatophores Isolated from Rhodospirillaceae

Organism	Protein	Total lipid[a]	Phospho-lipid	Total phosphorus	Bchl	Caro-tenoids	Carbo-hydrate	Nucleic acid	References
Rp. sphaeroides	58.4	24.6	—	0.68	2.7[b]	1.0	4.2	0.90	Bull and Lascelles (1963)
	53.0	36.5	—	1.44	7.3	—	3.5	—	Schmitz (1967)
	62.0	26.7	—	1.03	4.0[b]	2.0	4.0	0.10	Gorchein *et al.* (1968)
	64.0	—	25.0	—	4.6	—	0.15	0.38	Fraker and Kaplan (1971)
	63.0	30.0	—	0.93	5.9	1.9	2.3	0.64	Niederman (1974)
Rp. viridis	56.5	30.0	—	0.52	—	—	3.5	—	Drews (1968)
Rs. rubrum	46.5	32.0	—	0.83	4.7	—	14.0	—	Drews (1968); Schröder and Drews (1968)
	55.2	—	15.6	0.62	3.2	—	5.6	0.36	Collins and Niederman (1976b)

[a] Includes photosynthetic pigments.
[b] These values were recalculated with the extinction coefficient of 75 mM^{-1}·cm^{-1} at 770 nm reported by Clayton (1963).

Table 2. Chemical Composition (% Dry Weight) of Chromatophores Isolated from Chromatiaceae

Organism	Protein	Total lipid[a]	Bchl	Carotenoids	Carbohydrate	References
Chromatium vinosum	27.0	67.0	8.3	1.3	0.3	Cusanovich and Kamen (1968a)
Thiocapsa floridana (now *T. roseopersicina*)	49.0	47.0	3.0	—	1.8	Takacs and Holt (1971b)

[a] Includes photosynthetic pigments.

Rs. rubrum is presented in Table 3. The preparations obtained from aerobically grown cells are essentially devoid of Bchl, and those from phototrophic cells are considerably depleted. Aside from the "membrane fragments" of Gorchein *et al.* (1968) that were isolated by differential centrifugation, these preparations are very low in nucleic acid content. The protein/lipid ratio approaching 2:1 observed in the cytoplasmic membrane fractions and the much higher ratio obtained with some aerobic cell-wall and outer-membrane preparations was also observed in the corresponding fractions from other gram-negative bacteria (Miura and Mizushima, 1968; Osborn *et al.*, 1972). With the exception of their low photopigment content, the chemical composition of the various cytoplasmic-membrane fractions is quite similar to that of the chromatophore preparations (see Table 1). The high carbohydrate content of the cell-wall and outer-membrane fractions is consistent with the presence of significant quantities of lipopolysaccharide in these preparations. The fatty acid composition of these fractions from *Rs. rubrum* also supports this possibility (Section 3.3). The diaminopimelate content of several of the cell-envelope fractions suggests that the peptidoglycan has largely fractionated with the outer-membrane-rich preparations, and, accordingly, those isolated from French-press extracts have been designated as cell walls.

It has been suggested that the carbohydrate present in isolated chromatophores arises largely from contamination by cell-wall components (Lascelles, 1968; Oeize and Drews, 1972). An extensive analysis of the carbohydrate present in purified chromatophore preparations from *Rs. rubrum* and *Rp. viridis* was

Table 3. Chemical Composition (% Dry Weight) of Cell-Envelope Fractions

Organism	Protein	Total lipid[a]	Phospho-lipid	Total phosphorus	Bchl	Carbo-hydrate	Nucleic acid	Diamino-pimelate	References
Rp. sphaeroides									
Aerobically grown									
Cytoplasmic-membrane-enriched									
	55.0	23.0	—	1.33	0	16.0	4.0	0.37	Gorchein *et al.* (1968)
	65.0	—	22.0	0.87	0.03	4.2	0.39	0	Parks and Niederman (in prep.)
Cell-wall-enriched									
	67.0	—	26.0	1.05	0	9.1	0.28	8.2	Parks and Niederman (in prep.)
Phototrophically grown									
Cytoplasmic-membrane-enriched									
	65.0	—	25.0	1.01	0.55	4.5	0.38	0	Parks and Niederman (in prep.)
Cell-wall-enriched									
	70.0	—	31.0	1.25	0.55	8.8	0.19	8.8	Parks and Niederman (in prep.)
Rs. rubrum									
Aerobically grown									
Cytoplasmic-membrane-enriched									
	43.0	28.0	21.1	—	—	5.6	—	0.16	Oelze *et al.* (1975)
	60.9	—	33.7	1.35	0.07	4.8	0.17	—	Collins and Niederman (1976*a*)
Cell-wall-enriched									
	49.0	8.8	3.4	—	—	25.0	—	1.6	Oelze *et al.* (1975)
Outer-membrane-enriched									
	70.1	—	9.3	0.37	0.02	27.3	0	—	Collins and Niederman (1976*a*)
Phototrophically grown									
Cell-envelope fraction									
	72.8	—	—	—	0.5	15.1	2.3	—	Collins and Niederman (1975; unpublished)

[a] Includes photosynthetic pigments.

reported by Drews (1968). In the *Rs. rubrum* preparation, 90% of the total sugar content could be accounted for as glucose; the other sugars present were rhamnose and fucose. In the *Rp. viridis* chromatophores, glucose comprised 50% of the sugars present, while the remainder was attributed to nearly equal quantities of galactose, mannose, and rhamnose. After phenol–water extraction of the chromatophore preparations, some of the carbohydrate (expressed as glucose) appeared in the phenol phase with the protein, but the majority was found in the aqueous phase, where lipopolysaccharide normally fractionates. However, several recent observations suggest that both glycolipid and glycoprotein are intrinsic compo-nents of the chromatophore membranes of purple bacteria. Although the chromatophores of this group do not contain the monogalactosyl diglycerides found in the photosynthetic membranes of Chlorobiaceae and all other phototrophs (Cruden and Stanier, 1970), glycolipid was identified as a major component of the purified lipid fraction from the chromatophores of *T. floridana* (now *T. roseopersicina*) (Takacs and Holt, 1971*b*). Both glucosyl and rhamnosyl diglycerides appeared to be present. With regard to the presence of glycoprotein in the chromatophore membrane, a highly purified carotenoid–protein preparation was isolated recently from *Rs. rubrum* (Schwenker and Gingras, 1973); approximately 14% of the dry weight

Table 4. Levels (nmol/mg Protein) of Various Electron-Transfer Components in Isolated-Membrane Preparations

Organism/fraction	Flavin	Heme	Nicotinamide nucleotide	Nonheme iron	Ubiquinone	References
Chr. vinosum						
Chromatophore	0.5	3.0	1.0	26.0	10.0	Newton (1963)
	—	14.9	—	118.0	8.7[a]	Cusanovich and Kamen (1968a)
Rp. sphaeroides						
Chromatophore	—	2.9	—	<5.7	—	Lascelles (1962, 1968)
	—	0.7	—	—	24.0	Guillotin and Reiss-Husson (1975)
Aerobic						
Cytoplasmic-membrane-enriched	—	2.5	—	—	9.6	Guillotin and Reiss-Husson (1975)
Outer-membrane-enriched	—	0	—	—	7.3	Guillotin and Reiss-Husson (1975)
Rs. rubrum						
Aerobic						
Cytoplasmic-membrane-enriched	—	—	—	—	7.2[b]	Oelze *et al.* (1975)
Cell-wall-enriched	—	—	—	—	0.42[b]	Oelze *et al.* (1975)
Cell-envelope	0.6	0.4–0.6	—	0.8	0.30[c]	Taniguchi and Kamen (1965)

[a] Ubiquinone 35 (CoQ_7).   [b] Ubiquinone 10 (CoQ_2).   [c] Ubiquinone 50 (CoQ_{10}).

of the purified material was attributed to carbohydrate (Schwenker *et al.*, 1974). Since the carotenoid–protein component was thought to comprise about 10% of the membrane protein, the carbohydrate present in this glycoprotein could account for a substantial portion of that found in the chromatophores of *Rs. rubrum*. Material that gives a typical periodic acid–Schiff staining reaction also appears to be associated with the most rapidly migrating band after SDS–polyacrylamide gel electrophoresis of *Rp. sphaeroides* and chromatophores (R. L. Hall, personal communication). However, only small quantities of protein-bound hexose were detected in highly purified chromatophores from the latter organism by Fraker and Kaplan (1971).

Other analyses of sugar components were performed on cell–envelope fractions from aerobically grown *Rs. rubrum* (Oelze *et al.*, 1975) and on membrane preparations from *Chr. vinosum* (Hurlbert *et al.*, 1974). Aerobic cell-wall and cytoplasmic-membrane fractions from *Rs. rubrum* contained rhamnose, fucose, and glucose in nearly equal proportions, although the former was greatly enriched in carbohydrate content (Table 3). No ribose was

detected in either of these preparations. Sugars present in the pigmented intracytoplasmic-membrane fractions from *Chr. vinosum* included glucose (2–4% of the dry weight of the membranes), smaller amounts of mannose, and traces of glucosamine. In contrast, carbohydrate represented approximately 20% of the dry weight of the outer membrane; approximately two-thirds of this was accounted for as glucose. In addition, substantial quantities of 3-O-methyl-D-ribose, D-ribose, L-arabinose, D-mannose, and mannosamine, and traces of D-rhamnose, glucosamine, and quinovosamine were also detected in the outer-membrane fraction. An extensive discussion of sugars found in the lipopolysaccharide fractions from purple bacteria is presented in Chapter 4.

The levels of various electron-transfer components have also been estimated in several of the isolated-membrane fractions (Table 4). The content of these components in such preparations does not necessarily reflect that in the intact membranes, since cofactors may be removed during the purification procedures (Oelze and Drews, 1972). In such cases, the purification of the membranes may not result in an increase in the specific activity of the reactions with which these

Table 5. Amino Acid Composition (mol/100 mg) of Isolated Chromatophores

Amino acid	Rhodopseudomonas sphaeroides		Rhodospirillum rubrum (Gorchein et al., 1968)	Thiocapsa floridana (T. roseopersicina) (Takacs and Holt, 1971b)
	(Gorchein et al., 1968)	(Fraker and Kaplan, 1971)		
Ala	18.2	14.7	11.3	14.7
Arg	3.3	3.4	4.4	2.6
Asp	6.7	5.9	8.6	7.3
Cys	—	0.39	—	0.65
Glu	10.8	8.0	10.7	10.2
Gly	9.3	8.1	9.9	7.2
His	2.0	2.1	2.0	2.8
Ile	2.5	4.8	5.4	5.4
Leu	10.0	10.8	9.6	10.6
Lys	4.5	3.9	4.3	4.0
Met	2.7	3.2	2.0	1.7
Phe	3.7	4.8	6.1	6.8
Pro	6.3	5.4	4.0	6.7
Ser	5.5	5.1	6.0	4.0
Thr	6.2	5.9	6.6	5.1
Trp	—	2.2	—	—
Tyr	2.5	3.0	1.8	2.4
Val	5.8	8.5	7.3	8.0
Nonpolar residues[a]	49.2	54.8	45.7	54.6

[a] Calculated according to the classification of Guidotti (1972) in which Ala, Cys, Ile, Leu, Met, Phe, Pro, Trp, and Val are considered as nonpolar residues.

components are associated. Nevertheless, substantial levels of cytochrome-associated heme, nonheme iron, and ubiquinone are found in both chromatophore and cytoplasmic membrane preparations. The presence of significant quantities of such components in isolated chromatophores together with ATPase (Reed and Raveed, 1972) further suggests that these pigmented structures are the site of photophosphorylation and associated electron-transfer reactions.

The results of amino acid analyses of several highly purified chromatophore fractions are presented in Table 5, and in Table 6, the amino acid composition of cell-envelope fractions from Rp. sphaeroides is given. The amino acid composition of the cell-wall fractions apparently reflects not only that of the outer membrane proteins, but also those amino acids present in the peptidoglycan layer. The proportion of nonpolar residues in several of these preparations is unusually high, especially in the chromatophores. By comparison, the erythrocyte membrane contains 45.9 mol nonpolar residues/100 mg (Rosenberg and Guidotti, 1968). The high proportion of apolar aminoacyl residues is consistent with a composition for the major proteins that would permit them to interact within a

lipid bilayer. However, only with a knowledge of the overall structure of these proteins can the nature of their interactions within the membrane be assessed with confidence. Information on their tertiary structure within the membrane will be essential to distinguish whether their apolar residues are localized, as would be expected in specific regions conferring a hydrophobic character to the proteins, or whether they are dispersed throughout the structure, having little influence on the overall properties of the protein. The relative insolubility of the major chromatophore proteins suggests that they possess a largely hydrophobic nature. These considerations are discussed further in the following section.

3.2. Protein and Enzymatic Composition

With the advent of SDS–polyacrylamide gel electrophoretic techniques (Kiehn and Holland, 1968; Weber and Osborn, 1969; Laemmli, 1970), many of the apparent integral membrane proteins of purple bacteria have been resolved and characterized. The most completely characterized include the following chromatophore proteins: the three polypeptide compo-

Table 6. Amino Acid Composition (mol/100 mg) of Cell-Envelope Fractions from *Rhodopseudomonas sphaeroides*

Amino acid	Aerobically grown cells				Phototrophically grown cells		
	Cell envelope	Cytoplasmic-membrane-enriched		Cell-wall-enriched	Cytoplasmic-membrane-enriched	Cell-wall-enriched	
	(Huang and Kaplan, 1973a)	(Gorchein et al., 1968)	(Parks and Niederman, unpublished)	(Parks and Niederman, unpublished)	(Parks and Niederman, unpublished)	(Parks and Niederman, unpublished)	(Gorchein et al., 1968)
Ala	11.2	14.5	18.1	19.7	17.0	19.8	16.0
Arg	4.7	4.8	4.8	2.5	4.5	2.4	3.3
Asp	11.2	10.6	7.5	9.8	8.0	9.0	10.1
Cys	1.7	—	—	—	—	—	—
Glu	8.3	11.6	7.8	9.2	8.1	9.6	11.6
Gly	12.2	9.0	15.5	20.7	14.9	19.1	11.4
His	1.4	1.6	1.3	0.93	1.2	1.0	1.5
Ile	4.0	3.9	3.9	2.2	4.4	2.3	3.6
Leu	6.9	9.9	8.2	4.7	8.1	5.2	6.5
Lys	4.2	4.6	3.7	3.5	3.8	3.6	3.9
Met	3.0	2.3	0.91	0.83	1.4	0.88	2.1
Phe	4.0	3.5	2.8	2.8	2.6	2.8	3.9
Pro	2.7	3.4	5.9	3.1	5.7	3.8	3.4
Ser	5.3	5.3	5.6	5.4	5.7	5.4	5.2
Thr	5.7	5.1	5.3	5.5	5.4	5.5	6.7
Trp	3.2	[a]	[a]	[a]	[a]	[a]	[a]
Tyr	2.9	2.5	1.5	2.3	1.5	2.3	3.1
Val	7.2	7.6	7.4	7.0	7.5	7.2	7.5
Nonpolar residues[b]	43.9	45.1	47.2	40.3	46.7	42.0	43.0

[a] Tryptophan was not determined.

[b] Calculated according to the classification of Guidotti (1972).

nents associated with the photochemical reaction center in *Rp. sphaeroides* (Feher, 1971; Okamura *et al.*, 1974; Clayton and Haselkorn, 1972; Takemoto and Lascelles, 1974); a polypeptide from *Rp. sphaeroides* to which the LH Bchl is attached (Fraker and Kaplan, 1972; Clayton and Clayton, 1972); and a carotenoid–protein complex from *Rs. rubrum* (Schwenker *et al.*, 1974). Table 7 presents apparent-molecular-weight values and the relative proportions of apolar aminoacyl residues for these chromatophore proteins and other membrane proteins isolated from *Rp. sphaeroides* by Huang and Kaplan (1973a,b). For the determination of their molecular weights, the membranes were dissociated in SDS, and the resulting polypeptides were electrophoresed on polyacrylamide gels in the presence of this detergent. As pointed out by Takemoto and Lascelles (1973), the values for the components that migrate most rapidly should be considered as approximations since molecular weights below about 17,000 cannot be determined with accuracy on such gels. Accordingly, these authors assigned a mass of 5000–10,000 daltons to band 15 on the basis of its relative migration in gel electrophoresis performed by the method of Laemmli (1970). Special gel systems such as that of Swank and Munkres (1971) are needed for more precise determinations of molecular weights within this range. When this system was applied to the carotenoid–protein complex of *Rs. rubrum* by Schwenker *et al.* (1974), a value of 11,000 was obtained; this was consistent with a minimum molecular weight of 10,500 that was calculated from amino acid composition data.

Table 7 also indicates that nonpolar aminoacyl residues account for about 47–56% of those found in the major chromatophore proteins. In accordance with

Table 7. Some Membrane Proteins Resolved by Polyacrylamide Gel Electrophoresis

Organism	Designation	Mol. wt.[a]	Nonpolar amino acids (mol/100 mg)[b]	Comments[c]	References
Rp. sphaeroides					
Chromatophore	Band 15	10,000[d]	55.6	LH-Bchl- and phospholipid-associated.	Fraker and Kaplan (1972), Clayton and Clayton (1972)
	RC$_c$	21,000	55.1	RC-Bchl-associated.	Okamura *et al.* (1974), Steiner, L. A., *et al.* (1974)
	RC$_b$	24,000	55.2	RC-Bchl-associated.	
	RC$_a$	28,000[e]	49.3	RC-associated.	
	Band 13	27,000	50.8	May be identical to RC$_a$.	Huang and Kaplan (1973b)
	Band 12	44,000	49.4	Chromatophore-specific.	Huang and Kaplan (1973b)
Aerobic cell envelope	Band 10	12,000	38.3		Huang and Kaplan (1973a)
	Band 9	14,500	37.5		
	Band 8	18,300	38.4		
	Band 5	28,700	37.4		
	Band 3	45,000	45.0		
Rs. rubrum					
Chromatophore	Carotenoprotein	11,000[f]	46.6	Complex also contains spirilloxanthin, phospholipid, and carbohydrate.	Schwenker *et al.* (1974)

[a] Apparent molecular weight as determined by SDS–polyacrylamide gel electrophoresis.
[b] Calculated according to the classification of Guidotti (1972).
[c] (LH) Light-harvesting; (RC) reaction center.
[d] Minimal molecular weight calculated from amino acid composition was 7000; that estimated by sedimentation equilibrium was 15,000 (Fraker and Kaplan, 1972); SDS–polypeptide molecular weight reported by Clayton and Clayton (1972) was 9000.
[e] Apparent SDS–polypeptide molecular weight determined by Clayton and Clayton (1972) for RC$_a$ was 29,000, whereas the weight for RC$_b$ and RC$_c$ were identical to values reported by Okamura *et al.* (1974).
[f] Minimal molecular weight calculated from amino acid composition was 10,500.

the criteria of Guidotti (1972), this would permit them to be classified as "nonpolar membrane proteins." Although the cell envelope of *Rp. sphaeroides* is even more refractory to solubilization than chromatophores (Huang and Kaplan, 1973a), the amino acid composition of the major cell-envelope protein components are unusual for membrane proteins in their relatively low content of nonpolar residues. This may be explained in part by the absence of data on their cysteine and tryptophan contents; these nonpolar amino acids account for about 5% of the aminoacyl residues in the total envelope fraction (see Table 6). In addition, some of these proteins may be amphipathic components of the outer membrane with a portion of their structure exposed at the cell surface.

Estimates of the levels of various protein components in *Rp. sphaeroides* chromatophores indicate that the three reaction center (RC) subunits are present in equimolar quantities (Clayton and Haselkorn, 1972; Okamura *et al.*, 1974) and together account for approximately 25% of the total chromatophore protein; the LH-Bchl-associated protein comprises approximately 40–50% (Fraker and Kaplan, 1972). Immunological evidence (Fraker and Kaplan, 1972; Huang and Kaplan, 1973b) and the absence of these protein components from various photosynthetically incompetent mutants (Takemoto and Lascelles, 1973, 1974) indicate that they are specifically associated with the chromatophore and do not occur in other cellular fractions. Chromatophore-specific protein components have also been distinguished on the basis of solubility in 2-chloroethanol (Fraker and Kaplan, 1971). The RC_b and RC_c subunits of the reaction center have been isolated in a photochemically active form in association with RC Bchl (Feher, 1971; Okamura *et al.*, 1974). The subunit structure of RC preparations isolated form *Rs. rubrum* is very similar to that of *Rp. sphaeroides* (Noël *et al.*, 1972; Oelze and Golecki, 1975); it consists of three polypeptide components with apparent molecular weights of 21,000, 25,500, and 28,000 (van der Rest *et al.*, 1974). Although detergent treatment has yielded highly purified reaction centers only from a carotenoid-less mutant of *Rp. sphaeroides*, such preparations have been obtained from wild-type *Rs. rubrum* (Noël *et al.*, 1972; van der Rest *et al.*, 1974). Photochemically active RC preparations from *Rp. capsulata* also contain three polypeptides; they are present in a molar ratio of 1:1:1, and apparent molecular weights of 20,500, 24,000, and 28,000 were reported for the respective components (Nieth *et al.*, 1975). Even though the amino acid composition of isolated reaction centers from *Rs. rubrum* is quite similar to that of *Rp. sphaeroides*, they do not share antigenic determinants (Steiner, L. A., *et al.*, 1974).

Much of the Bchl and phospholipid originally present in highly purified chromatophores from *Rp. sphaeroides* has been isolated in association with the polypeptide designated as band 15 (Fraker and Kaplan, 1971, 1972). Highly purified preparations contained 59% protein, 6% Bchl, and 35% phospholipid; the NH_2-terminal amino acid of the protein moiety was identified as methionine (Fraker and Kaplan, 1972). On the basis of differences in amino acid composition, antigenic reactivity, and peptide maps, the band 15 protein appeared distinct from two other apparent chromatophore-specific polypeptides designated as bands 12 and 13 (Huang and Kaplan, 1973*b*). When band 15 was cleaved with cyanogen bromide, three oligopeptide components were produced; the sum of their molecular weights determined in SDS–polyacrylamide gel electrophoresis was 8600, whereas a minimum molecular weight of 11,000 was assigned to the protein on the basis of methionine content (Huang and Kaplan, 1973*c*). In the latter study, glycine was unambiguously identified at the *COOH*-terminus of the protein, and part of the primary structure in this region of the molecule was determined.

After exposure of purified *Rp. sphaeroides* chromatophores to 1% LDAO, Clayton and Clayton (1972) isolated a pigment–protein complex with an apparent molecular weight of 9000. The isolated complex had more than 60% of the LH Bchl associated with it, and its specific Bchl content was nearly 1.5 times that of purified chromatophores. Absorption spectra showed the typical B800- and B850-components of LH Bchl plus the carotenoid bands characteristic of spheroidene and spheroidenone. The isolated protein component appears to be identical to that of Fraker and Kaplan (1972), and it is thought to exist in the membrane in an oligomeric form. Other investigators, however, have observed heterogeneity in the low-molecular-weight protein of *Rp. sphaeroides* (Segen and Gison, 1971; Hall *et al.*, 1973; Takemoto and Lascelles, 1974; Niederman *et al.*, 1976), which has been resolved into two components by independent electrophoretic techniques. The LH Bchl of *Rs. rubrum* is apparently associated with a polypeptide of approximately 10,000 molecular weight that is a major component of the pellet that remains after removal of reaction centers from chromatophores with LDAO (van der Rest *et al.*, 1974). Two components with molecular weights of 8500 and 9500 were resolved in such preparations from a carotenoidless mutant of *Rs. rubrum* by Oelze and Golecki (1975).

An oligomeric protein with an apparent subunit molecular weight of 9000 is excreted in association with various Bchl precursors by mutants of *Rp. sphaeroides* with defects in Bchl synthesis (Richards *et al.*, 1975). This apparent LH-Bchl-associated polypeptide was not detected in membranes from several of these mutants by Takemoto and Lascelles (1973). A tetrapyrrole precursor present in the excreted complex from one of the mutants was displaced by free Bchl, which was converted to bacteriopheophytin and bound by the protein; the absorption maxima of the bacteriopheophytin had been shifted largely to the 840 to 865-nm region. It was suggested that the excreted pigment–protein complex contains the protein component associated with LH Bchl in the chromatophore membranes of the wild-type strain, and that this polypeptide normally combines with Bchl as it is incorporated into developing photosynthetic membranes. Shaw (1974) identified an apparent aggregate of this polypeptide component as the major protein of a chromatophore precursor subunit from wild-type *Rp. sphaeroides*. A pigment–protein complex similar to that of the mutants was excreted by wild-type cells under adaptation conditions that resulted in a slowed synthesis of Bchl. As development proceeds, additional chromatophore-protein components are apparently added to the precursor subunit.

As noted above, Schwenker and Gingras (1973) isolated a pigment–protein complex that contains a portion of the spirilloxanthin present in *Rs. rubrum* chromatophores (Table 7). The carotenoprotein was solubilized from the membrane with SDS and purified by ammonium sulfate fractionation and gel filtration.

The isolated complex contained 78% protein, 5.2% spirilloxanthin, 3.4% phospholipid, and 13.5% carbohydrate (Schwenker *et al.*, 1974); the latter accounts for much of the carbohydrate present in the chromatophore (Section 3.1). The purified complex migrated as a single band in SDS–polyacrylamide gel electrophoresis as demonstrated by both Coomassie blue and periodic acid–Schiff stains. Immunodiffusion also indicated that the preparation was highly purified. Spirilloxanthin was noncovalently associated with the polypeptide chain in a 1 : 1 stoichiometry. Lysine was identified at the NH_2-terminus of the protein component. The visible absorption spectrum of the isolated complex showed a single major band at 373 nm that is not observed for spirilloxanthin in chromatophores. This had been reported previously for the carotenoid complexes isolated from *Rs. rubrum* by others (Vernon and Garcia, 1967; Fujimori, 1969; Hall *et al.*, 1973). After extraction of the pigment from the complex into organic solvents, a typical spirilloxanthin spectrum was observed. On the basis of this spectral shift and the visible CD spectrum, Schwenker *et al.* (1974) proposed that in the form in which the complex is isolated, the carotenoid is twisted about its central double bond at such an angle as to break the system of conjugated double bonds, giving rise to two halves with six double bonds each. Since the pigment is associated with a specific chromatophore-protein component after extensive purification, and a complex with the same spectral properties has been isolated by a variety of procedures, the isolated material was thought to represent the carotenoid–protein complex of the intact chromatophore membrane.

As indicated in Table 7, SDS–gel electrophoresis of isolated cell envelopes from *Rp. sphaeroides* suggests that this structure also contains multiple protein components. That this results from a heterogeneity among different protein species was suggested by the amino acid composition and tryptic peptide fingerprints of selected envelope proteins from aerobically grown cells (Huang and Kaplan, 1973a). These authors also observed a similar electrophoretic profile in a cell-envelope fraction from phototrophically grown cells, and a common structure was suggested for the cell envelope. A banding pattern similar to that of the cell envelope was observed for the small fraction of chromatophore protein insoluble in 2-chloroethanol. Although reconstruction experiments suggested that this fraction forms part of the chromatophore structure, from morphological considerations, it would be expected that only the cytoplasmic-membrane moiety of the cell envelope would exist as part of the chromatophore.

Other studies on the protein composition of the cell envelope have indicated that the heterogeneity observed for the major protein species is more pronounced in the cytoplasmic membrane than in the outer envelope membrane. Electrophoresis of envelope fractions isolated from aerobically grown *Rp. sphaeroides* indicated that cytoplasmic-membrane-enriched material contained many bands with nearly equal staining intensities as well as several minor bands (Niederman *et al.*, 1972). An essentially similar profile was observed in a fraction enriched in small fragments of the cytoplasmic membrane from phototrophic cells. In contrast, a cell-wall-enriched fraction from both aerobically and phototrophically grown cells showed one major polypeptide band with an apparent molecular weight of 35,000 that accounted for more than 50% of the stained protein bands. In this regard, the outer membrane of *Rp. sphaeroides* appears to be similar to that of *Esch. coli*, which contains a single major polypeptide of 44,000 molecular weight (Schnaitman, 1970a). Depending on the conditions of solubilization and electrophoresis, the *Esch. coli* band may be resolved into several components (Schnaitman, 1973). A similar overall distribution of polypeptides also prevails for the more highly resolved cell-envelope membranes of aerobically grown *Rs. rubrum* (Collins and Niederman, 1976a). Again, the cytoplasmic membrane contained a variety of bands with similar staining intensities; a major protein band of approximately 40,000 molecular weight was observed with the outer membrane from both aerobically and phototrophically grown cells (Collins and Niederman, 1976a,b). Qualitatively similar results were reported for two cell-envelope fractions from chemotrophically grown *Rs. rubrum* enriched in cytoplasmic membrane and cell wall, respectively (Oelze *et al.*, 1975). In the latter study, gels were also stained with periodic acid–Schiff reagent. No bands were observed with the cytoplasmic-membrane fraction. Although the cell-wall fraction contained a single carbohydrate positive band of 14,000 molecular weight, the major zones of staining did not coincide with any protein band. In recent studies with *Chr. vinosum*, a polypeptide component of approximately 42,000 molecular weight was the major band observed in outer-membrane preparations (Hurlbert *et al.*, 1974).

Although the procedures used for the isolation of membranes may result in the loss of loosely associated cofactors and peripheral protein components, several enzymatic activities have been demonstrated in highly purified chromatophore preparations. These include succinic dehydrogenase, NADH oxidase, and the enzymes that catalyze light-drive phosphorylation of ADP, including the ATPase coupling factor. Both *b*- and *c*-type cytochromes have also been observed. After zone sedimentation of subcellular particles from phototrophically grown *Rp. sphaeroides*, the succinic dehydrogenase activity was distributed in both the chromatophore and cell-envelope fractions; however, the specific activity of the less dense portion of the cell-envelope band was approximately twice that of chromatophores (Niederman, 1974). A fraction enriched in small fragments of the cytoplasmic membrane of phototrophic cells had a specific succinic dehydrogenase activity as much as 5 times that of chromatophores (Niederman *et al.*, 1972). In the latter study, the bulk of the succinic dehydrogenase activity of aerobic cells was localized in a cytoplasmic-membrane-enriched fraction. Thus, in *Rp. sphaeroides*, this enzyme serves as a specific membrane marker and has proved useful for the detection of membrane fractions during isolation procedures.

A very different picture has emerged with regard to the distribution of succinic dehydrogenase in membrane fractions from *Rs. rubrum*. Although virtually all the activity in aerobically grown cells is localized in the cytoplasmic membrane (Oelze *et al.*, 1975; Collins and Niederman, 1976a), in extracts from phototrophically grown cells, the distribution of the enzyme after both isopyknic and rate-zone sedimentation coincided exactly with the 880 nm absorbance. When extracts were prepared by EDTA–lysozyme–Brij lysis, the succinic dehydrogenase activity was found exclusively in chromatophores; no activity was observed in the Bchl-depleted dense fraction in which peripheral cytoplasmic membrane was localized.* It is concluded that in photosynthetically grown *Rs. rubrum*, succinic dehydrogenase is a chromatophore-specific protein and is not associated with the inner cell-envelope membrane as in other gram-negative bacteria (Schnaitman, 1970a). This is supported by studies of Simon and Siekevitz (1973) in which the specific activity of succinic dehydrogenase was much higher in a cell-membrane fraction from aerobically grown *Rs. rubrum* than in a similar fraction from phototrophic cells. The chromatophore fraction isolated by these authors had approximately the same specific activity as the aerobic cell-membrane fraction, but much more than the phototrophic cell-membrane fraction. It was suggested that during adaptation to photometabolism, changes in the composition of the peripheral cytoplasmic membrane occur. On the basis of the essentially constant specific activity of succinic dehydrogenase observed in purified *Rs. rubrum* chromatophores of very different pigment contents (Cohen-Bazire and Kunisawa, 1960), Lascelles (1968)

suggested that this enzyme forms an integral and invariant part of the chromatophore membrane structure. Hatefi *et al.* (1972) solubilized the enzyme from crude chromatophores of *Rs. rubrum* with chaotropic agents. SDS–polyacrylamide gel electrophoresis suggested that purified preparations consist of two protein subunits; the larger subunit had a molecular weight of approximately 60,000 and contained flavin, while the smaller protomer had a molecular weight of approximately 25,000. The purified enzyme was similar to that from beef heart in several of its spectral and enzymatic properties.

Of several enzymes that catalyze electron-transfer reactions, only succinic dehydrogenase activity survived the extensive washing and purification procedures used in the isolation of chromatophore fractions from *Chr. vinosum* (Cusanovich and Kamen, 1968a). Relative to their specific Bchl contents, the heavy band was sixfold enriched in this activity as compared with the light chromatophores, which suggests that the former contained pigment-depleted cytoplasmic membrane. No NADH oxidase activity survived the preparative procedure.

The distribution of NADH oxidase has been investigated extensively by Drews, Oelze, and their collaborators in the membranes of *Rs. rubrum* (Throm *et al.*, 1970; Oelze and Drews, 1970b; Irschik and Oelze, 1973) and *Rp. capsulata* (Lampe and Drews, 1972). On the basis of these studies, NADH oxidation has been employed as a marker for oxidative electron-transfer components, and has provided one of the bases of a model for chromatophore-membrane biogenesis (Oelze and Drews, 1972) in which the activity of such cytoplasmic-membrane components gradually reaches a lowered level in the developing photosynthetic apparatus.

Although it has not been possible to demonstrate significant rates of photophosphorylation in *Rp. sphaeroides* chromatophores (Gibson, 1965b), photochemically active chromatophore preparations can be readily isolated by zone sedimentation of extracts from *Rs. rubrum* (Frenkel and Hickman, 1959; Cohen-Bazire and Kunisawa, 1960). On a Bchl basis, such purified particles were capable of catalyzing a light-dependent phosphorylation of ADP and reduction of NAD in the presence of succinate at rates equivalent to those of crude chromatophore preparations (Frenkel and Hickman, 1959). Similar observations were made with purified chromatophore preparations from *Rs. rubrum* by Geller and Lipmann (1960); in addition, a stimulation of photophosphorylation by phenazine methosulfate and succinate was observed. These findings have established chromatophores (and the subchromatophore particles derived from them) as the site of photochemical reactions in the cell.

---

* In recent experiments in this laboratory (Collins and Niederman, 1976, unpublished), no succinic-dehydrogenase-containing cytoplasmic membrane fragments could be obtained from extracts of phototrophically grown *Rs. rubrum* by chromatography on Sepharose 2B under conditions that would be expected to yield such material from *Rp. sphaeroides* [Section 2.5.2(b)].

Significant rates of photophosphorylation have been demonstrated in highly purified chromatophore preparations from *Rs. molischianum* (Hickman *et al.*, 1963) and *Rp. capsulata* (Lampe and Drews, 1972).

In studies of photophosphorylation by chromatophores from Chromatiaceae, relatively high activities were demonstrated in the extensively purified light-chromatophore fraction of *Chr. vinosum* (Cusanovich and Kamen, 1968b). A cyclic phosphorylation mechanism coupled to a light-driven pathway of electron flow was proposed for the *Chr. vinosum* preparations. Takacs and Holt (1971b) demonstrated photophosphorylation in purified chromatophore preparations form *T. floridana* (now *T. roseopersicina*) and in reassociated chromatophores derived from detergent solubilized preparations.

A factor required for coupling photophosphorylation to light-driven electron flow was removed from a crude *Rp. capsulata* chromatophore preparation by sonic disruption in the presence of EDTA (Baccarini-Melandri *et al.*, 1970). Both photophosphorylation and ATPase activity could be restored by addition of the extracted material to the depleted particles. In a subsequent study, biochemical and immunological evidence suggested that the coupling factor preparations from both aerobically and phototrophically grown cells were fully interchangeable in their stimulation of the respective oxidative and light-driven phosphorylation reactions (Melandri *et al.*, 1971). Particles that contained much of the Mg^{2+}-dependent ATPase activity were removed from purified *Rp. sphaeroides* chromatophores by extraction with Triton X-100 or EDTA (Reed and Raveed, 1972). The isolated particles had an apparent molecular weight of approximately 3.0×10^5, and in electron micrographs of negatively stained preparations, an average diameter of 9 nm was observed. It was suggested that the 9-nm particles are located on the outer surface of the asymmetrical chromatophore membrane. An ATPase coupling factor was also demonstrated in *Rs. rubrum* (Johansson, 1972); it was purified to apparent homogeneity (Berzborn *et al.*, 1975), and the subunit structure of the enzyme was recently described (Johansson and Baltscheffsky, 1975). A more detailed discussion of coupling factors from photosynthetic bacteria and their role in energy conservation is to be found in Chapter 31.

The cytochromes detected in isolated photosynthetic membranes represent a tightly bound (Bartsch, 1971) or entrapped (Prince *et al.*, 1975) chromatophore fraction of the total cytochromes present in the cell. Although only relative concentrations have been reported, b- and c-type cytochromes have been demonstrated in highly purified chromatophore preparations form *Rp. sphaeroides* (Worden

and Sistrom, 1964; Niederman, 1974) and *Rs. rubrum* (Collins and Niederman, 1976b). In the latter case, however, lower levels were observed. In contrast, a cytoplasmic membrane fraction from aerobically grown *Rs. rubrum* contained more than twice the concentration of these components found in the chromatophores from this organism (Collins and Niederman, 1976a). Significant levels of both b- and c-type cytochromes of nonchromatophore origin were present in the cell-envelope fractions from phototrophically grown *Rp. sphaeroides* and *Rs. rubrum*. This is consistent with the conservation of Bchl-deficient plasma membrane under these conditions. As indicated by the heme concentrations of isolated chromatophore preparations (see Table 4), the light-particle fraction from *Chr. vinosum* had a relatively high cytochrome content (Cusanovich and Kamen, 1968a). The observed heme was accounted for by cytochromes c', c-552, and c-555; their respective concentrations were 2.5, 2.9, and 3.9 nmol/mg chromatophore protein. They have been estimated to comprise approximately 9% of the dry weight of the chromatophore. No b-type cytochrome was detected.

Further information on the distribution of cytochromes in photosynthetic bacteria may be found in Chapter 13.

3.3 Lipid Composition

The majority of the lipid of the chromatophore membrane in *Rp. sphaeroides* has been accounted for by phospholipid (Gorchein, 1968c). The major phospholipid components were identified as phosphatidylethanolamine, phosphatidylglycerol, and phosphatidylcholine; they represented 35, 34, and 23%, respectively, of the total lipid phosphorus. The remaining lipid phosphorus was attributed to phosphatidic acid, 3.3%, and a phospholipid tentatively identified as cardiolipin, 3.9%. A similar phospholipid pattern was observed in phototrophically grown cells and a supernatant fraction that apparently contained small fragments of cytoplasmic membrane. An ornithine lipid (Gorchein, 1964) was also detected in chromatophores; an acid hydrolysate of the extracted lipid contained 26 nmol ornithine/mg dry wt. Additional ornithine lipid was thought to be localized in the cytoplasmic membrane (Gorchein, 1968c). In further studies, Gorchein (1968d) identified this component as ornithine esterified with a fatty alcohol derivative and N-acetylated with a C16 or C18 fatty acid. An ornithine lipid was also isolated from *Rs. rubrum* by DePinto (1967).

A cytoplasmic-membrane fraction from aerobically grown *Rp. sphaeroides* contained essentially the same relative proportions of major phospholipids as chromatophores; however, a much lower ornithine lipid concentration was observed (Gorchein, 1968c). The phospholipid composition of cytoplasmic-membrane and cell-wall fractions isolated from aerobically grown *Rs. rubrum* was reported by Oelze *et al.* (1975); each was comprised largely of phosphatidylethanolamine (77 and 65%, respectively). Smaller quantities of phosphatidylglycerol (16 and 22%, respectively), cardiolipin, and lysophosphatidylethanolamine were detected. In the Chromatiaceae, nearly equal quantities of phosphatidylethanolamine and phosphatidylglycerol were observed as the major chromatophore phospholipids of *Chr. vinosum* (Haverkate *et al.*, 1965), while in the chromatophores of *T. floridana* (now *T. roseopersicina*), the major lipid components were lysophosphatidylethanolamine, phosphatidylethanolamine, phosphatidylglycerol (cardiolipin), and glycolipids (Takacs and Holt, 1971b).

In the chromatophores of Rhodospirillaceae, vaccenic acid has been identified as the major fatty acid, while in preparations from *Chr. vinosum*, nearly equal quantities of C18:1, palmitic, and palmitoleic acids were reported (Table 8). Essentially the same fatty acid composition was recently reported for a chromatophore fraction isolated from *Chr. vinosum* after osmotic lysis of EDTA–lysozyme spheroplasts (Hurlbert *et al.*, 1974). Cytoplasmic- and outer-membrane fractions from aerobically grown *Rs. rubrum* differed significantly in their relative contents of several fatty acid components (Table 8). The cytoplasmic membrane contained a greater proportion of unsaturated fatty acids. The outer membrane was enriched fivefold in arachidic acid (20:0), a specific component of *Rs. rubrum* lipopolysaccharide (Schröder *et al.*, 1969). Other differences observed between the fatty acid composition of cytoplasmic and outer membrane apparently result from fatty acids associated with lipopolysaccharide. In this regard, a similar profile of fatty acid moieties was observed for the phospholipids isolated from the cell-wall and cytoplasmic-membrane fractions of *Esch. coli* (White *et al.*, 1972); a major difference was observed only in their contents of palmitic acid.

3.4. Sedimentation Behavior

Most published procedures for isolating chromatophores and other membrane fractions from photosynthetic bacteria rely heavily on the preparative ultracentrifuge. However, with few exceptions, purified membrane fractions are not suitable for study by analytical ultracentrifugation because of their extreme polydispersity and tendency to aggregate. Almost the only information that the ultracentrifuge can supply in such cases is an estimate of density, which can be obtained by isopyknic centrifugation. Estimates of density, usually from isopyknic centrifugation in sucrose gradients, are now becoming customary for all

Table 8. Fatty Acid Composition (% Total Fatty Acids) of Isolated-Membrane Preparations

Fatty acid	Chromatophores				Aerobically grown cells (*Rs. rubrum*)			
	Rs. rubrum[a]	*Rp. capsulata*[a]	*Rp. sphaeroides*[b]	*Chr. vinosum*[c]	Cytoplasmic membrane		Outer membrane	
14:0	2.6	—	—	2.8	2.9[d]	2.5[e]	12.8[d]	12.0[e]
15:0	0.9	—	—	—	[f]	0.2	[f]	0.5
15:1	—	—	—	—	[f]	—	[f]	0.6
16:0	10.4	2.1	4.6	28.5	10.3	13.1	17.4	16.4
16:1	30.8	1.7	2.1	30.3	37.8	32.3	20.4	22.5
17:1	0.2	—	—	—	[f]	0.4	[f]	0.5
18:0	1.4	1.4	9.7	3.1	1.6	1.2	6.9	1.4
18:1	53.3	94.6	76.8	35.3	47.4	50.1	40.5	44.8
19:0	—	—	6.8	—	[f]	—	[f]	—
19:1	—	—	—	—	[f]	—	[f]	0.9
20:0	—	—	—	—	[f]	0.1	[f]	0.5

[a] Schröder *et al.* (1969).
[b] Schmitz (1967).
[c] Haverkate *et al.* (1965); similar values were reported for intracytoplasmic membrane fractions from *Chr. vinosum* by Hurlbert *et al.* (1974).
[d] Oelze *et al.* (1975).
[e] Collins and Niederman (1976a).
[f] Values below 1% were omitted.

subcellular membrane fractions. Densities obtained in this manner for chromatophores from various species of photosynthetic bacteria are typically in the range of 1.14–1.17 g/cm³, i.e., essentially what would be predicted from the chemical composition of these organelles. Fractions that are enriched in cytoplasmic membrane have densities that are almost the same (1.12–1.15 g/cm³); fractions enriched in the outer membrane or cell wall usually show higher values of 1.22–1.25 g/cm³. To a large extent, these differences in density determine the behavior of the fractions in the preparative ultracentrifuge and are responsible for the separations that have been obtained.

When centrifuged in NaBr or CsCl gradients, chromatophores from *Rs. rubrum* band with slightly higher buoyant densities in the range 1.19–1.20 g/cm³ (Ketchum and Holt, 1970; Collins and Niederman, 1976b). In gradients of Ficoll, however, the same chromatophores had a buoyant density of 1.07 g/cm³ (Ketchum and Holt, 1970). the latter authors explained these variations by appealing to a theory concerning the behavior of sealed vesicles in density gradients (Steck *et al.*, 1970); the low density in Ficoll was attributed to impermeability of the chromatophores to this solute, and the increased density in NaBr was taken to indicate that the vesicles were completely permeable to inorganic salts. Solutions of sucrose, which were also considered impermeable to the vesicles, would tend to withdraw water from these structures, thus resulting in smaller volumes and higher densities than in solutions of Ficoll. Equilibrium buoyant densities determined in CsCl gradients for cytoplasmic-membrane preparations from both *Rs. rubrum* (Collins and Niederman, 1976a) and *Rp. sphaeroides* (Ding and Kaplan, 1976) were approximately 1.18 g/cm³, while that for the outer membrane of *Rs. rubrum* was 1.30 g/cm³, and the cell wall of *Rp. sphaeroides* was 1.24 g/cm³. As has been observed for chromatophores, these values are greater than those obtained by equilibrium banding in sucrose gradients; thus, these data suggest that the cell-envelope layers are also isolated largely as closed vesicles that are semipermeable in nature.

Equilibrium banding on CsCl gradients also serves as a means for estimating the homogeneity of isolated chromatophore preparations with respect to their density. This may be assessed by simultaneously labeling cells with an amino acid (e.g., [³H]leucine) and a precursor of Bchl (e.g., [¹⁴C]δ-aminolevulinic acid) and determining the radioactivity of each fraction in the chromatophore band, or as reported recently for *Rs. rubrum*, by determining the $A_{280\,nm}$ and $A_{880\,nm}$ of appropriate fractions (Collins and Niederman, 1976a). Equilibrium centrifugation in CsCl gradients has also provided a method for analyz-

ing the growth of the chromatophore membrane of *Rp. sphaeroides* (Kosakowski and Kaplan, 1974). By such a centrifugation procedure, it was possible to distinguish "old" membrane obtained from cells grown on a medium supplemented with 70% D₂O from "new" membrane obtained from cells shifted to a H₂O-based medium. Under the former conditions, chromatophores were obtained with a density of 1.220–1.230 g/cm³, while those from a H₂O medium had a density of 1.175–1.180 g/cm³. The latter values are in agreement with those of 1.17–1.18 g/cm³ obtained by Gibson (1965a) by plotting viscosity times $s_{20,w}$ against density in a series of moving boundary sedimentation runs.

The one type of membrane subfraction that can be examined profitably in the analytical ultracentrifuge is the vesicular chromatophores characteristic of ·*Rp. sphaeroides*, *Rs. rubrum*, and several Chromatiaceae. There are various reasons for this. In the first place, although chromatophores are by no means monodisperse, they frequently show a range of sizes that approximate a normal distribution with a standard deviation that is small relative to the mean sizes. Thus, a boundary formed by such particles in the analytical ultracentrifuge is intrinsically fairly sharp, and, in addition, is likely to show considerable self-sharpening, especially at concentrations in the range required for observation with Schlieren optics. Finally, because of their size and simple shape, interpretation of sedimentation coefficients of these chromatophores is trivial.

Moving-boundary sedimentation has most frequently been employed as a test of homogeneity of chromatophore preparations. From studies of this type, values of $s_{20,w}$ have been reported for chromatophores of several organisms; these values tend to vary somewhat, depending on the investigator and the optical system employed, and more importantly on whether the value was obtained by extrapolation to infinite dilution. For chromatophores from *Rp. sphaeroides*, reported $s_{20,w}$ values are: 153 S (Worden and Sistrom, 1964) and 160 S (Gibson, 1965a; Niederman and Gibson, 1971); for *Chr. vinosum*, 145 S (Cusanovich and Kamen, 1968a); for *T. floridana* (now *T. roseopersicina*), 150 S (Takacs and Holt, 1971b); and for *Rs. rubrum*, 250 S (Collins and Niederman, 1976b). Conflicting results have been obtained with regard to the dependence of $s_{20,w}$ on concentration. Worden and Sistrom (1964) observed a strong inverse relationship between $s_{20,w}$ and concentration with chromatophores from *Rp. sphaeroides*, whereas both Cusanovich and Kamen (1968a), working with chromatophores from *Chr. vinosum*, and Takacs and Holt (1971b), using chromatophores from *T. floridana* (now *T. roseopersicina*), found that $s_{20,w}$

increased with concentration. Possibly, a rapidly reversible concentration-dependent aggregation might be responsible for the latter behavior.

In a few studies, $s_{20,w}$ values have been combined with other information to yield estimates of the mean chromatophore diameter or additional parameters. By applying the Scheraga–Mandelkern equation to the values at infinite dilution of the sedimentation coefficient $s_{20,w}$, and the intrinsic viscosity, $[\eta]$, as well as the partial specific volume, Cusanovich and Kamen (1968a) calculated a molecular weight of 1.29×10^7 for chromatophores of *Chr. vinosum*; using the same methods, Takacs and Holt (1971b) reported a value of 1.26×10^7 for chromatophores of *T. floridana* (now *T. roseopersicina*). These values refer to unhydrated chromatophores. Diameters of 52.5 and 54.5 nm, respectively, were calculated for the equivalent hydrodynamic spheres. Using a less sophisticated approach, Gibson (1965a) obtained a value of 43.5 nm for the diameter of the equivalent hydrodynamic sphere in the case of chromatophores form *Rp. sphaeroides*. By comparison of this value with the value of 57.0 nm for the diameter of chromatophores measured from electron micrographs of fixed chromatophores, it was deduced that 55% of the weight of the intact chromatophore was due to water.

It should be emphasized that all these calculations may be subject to quite significant errors. Thus, Cusanovich and Kamen (1968a) obtained a value of 0.667 for the partial specific volume of the chromatophores they studied, which, as they point out, is scarcely compatible with the high lipid content of the particles (see Table 2). Takacs and Holt (1971b) did not report the value of $[\eta]$ or of the partial specific volume that they used in their calculations; however, from their values for molecular weight and radius, one can calculate that the partial specific volume of their chromatophores was 0.63. Again, this appears to be a very low value for a structure nearly half the dry weight of which is lipid (see Table 2); it also disagrees with their value of 1.14 g/cm^3 for the density of these chromatophores obtained by isopyknic centrifugation in sucrose gradients. The calculations of both molecular weight and particle diameter depend very strongly on the accuracy of the determination of partial specific volume; hence, the estimates provided by both sets of authors may need revision. Possible errors associated with the methods used by Gibson (1965a) are of a different nature, and involve partly technical limitations and partly the use of salts to vary density. This introduces a third component into the system, and requires a more sophisticated theoretical analysis of the results than was attempted. Here also, a more refined analysis would probably be well worthwhile.

3.5. Electron Microscopy

Electron-microscopic studies of photosynthetic bacteria have been used to advantage to examine the morphology of the intracellular membrane system, to aid in the identification of isolated membrane fragments, and most recently to investigate aspects of the structure of the membranes themselves. Improvements in fixation techniques and resolution, and the introduction of freeze–fracture methods, have led to increased understanding of photosynthetic membranes during the last two decades. Of particular importance have been investigations of whole cells grown under various conditions (see, for example, Cohen-Bazire, 1963; Cohen-Bazire and Kunisawa, 1963; Holt and Marr, 1965c; Peters and Cellarius, 1972); these investigations not only have served to identify the *in vivo* nature of the photosynthetic apparatus, but also remain the most informative technique for studying photosynthetic membranes as they occur in the cell. Since several excellent reviews of these studies have appeared in recent years (Cohen-Bazire and Sistrom, 1966; Lascelles, 1968; Pfennig, 1967; Oelze and Drews, 1972), we will content ourselves in this section with presenting a brief overview of the work and will mention some salient points.

The usefulness of electron-microscopic studies of whole cells was first demonstrated in the classic studies of Cohen-Bazire and her collaborators (Cohen-Bazire, 1963; Cohen-Bazire and Kunisawa, 1963; Cohen-Bazire and Sistrom, 1966). These investigators examined thin sections of *Rs. rubrum* and *Rp. sphaeroides* grown under a variety of conditions, ranging from dark aerobiosis through anaerobiosis under various intensities of illumination. They observed that in cells grown anaerobically at low to moderate light intensities, the cytoplasm contained numerous nearly circular vesicles, bounded with narrow electron-dense membranes surrounding relatively featureless interiors. In cultures grown at different light intensities, there was a good correlation between the average number of vesicles per cell and the concentration of Bchl in harvested cells. In cultures grown aerobically, or anaerobically at high light intensities, very few vesicles were seen, especially in the interior of the cell. These investigators concluded that the membrane-bound vesicles housed the photosynthetic apparatus in the cell.

Subsequent work has completely confirmed these conclusions, and extended them to show that *Rp. capsulata* (Drews *et al.*, 1969b), several species of *Chromatium* (Cohen-Bazire, 1963; Kran *et al.*, 1963; Fuller *et al.*, 1963), *Thiospirillum jenense* (Cohen-Bazire, 1963), and *Thiocapsa floridana* (now *T. roseopersicina*) (Cohen-Bazire, 1963; Takacs and Holt,

1971*a*) contain morphologically similar vesicles, which almost certainly represent the *in vivo* forms of the photosynthetic apparatus of these organisms. The diameters of the vesicles are usually fairly uniform within organisms of a given strain, but vary among species over a range of 30–80 nm. Although most of the vesicles appear as separate entities, more or less frequent examples of partially joined vesicles are seen in most sections. A particularly important feature of all the observations is the occurrence of vesicles, of approximately the same size as the intracytoplasmic vesicles, that appear to be continuous with the cytoplasmic membrane, and resemble invaginations of that membrane. This general pattern of rounded, membrane-bounded vesicles, often joined to each other or to the cytoplasmic membrane, is also consistent with electron micrographs of whole *Rs. rubrum* cells obtained by negative staining (Simon, 1972) or freeze–fracture techniques (Crofts, 1970; Golecki and Oelze, 1975). The freeze–etch studies of Takacs and Holt (1971*a*) also support this possibility in *T. floridana* (now *T. roseopersicina*).

The most generally accepted interpretation of these electron-microscopic data is that *in vivo*, the photosynthetic apparatus of all the species named above is a continuous tubular network of intracytoplasmic membrane enclosing an internal space that is continuous with the periplasmic space, with a very large number of nearly spherical swellings separated from each other by narrow constricted channels (Cohen-Bazire and Kunisawa, 1963; Holt and Marr, 1965*a*). Although several alternative explanations exist (Gibson, 1965*d*; Cusanovich and Kamen, 1968*a*), these are now usually discounted by most investigators. A further conclusion that was drawn by Cohen-Bazire and her collaborators, and that until recently was regarded as proved, was that the photosynthetic apparatus arises by invagination of the cytoplasmic membrane, with the addition of certain components, such as Bchl, that are characteristic of pigmented cells. However, advancing knowledge of the composition of chromatophores and cytoplasmic membranes, as well as direct investigation of chromatophore biogenesis, have now necessitated considerable modification of the original proposal (see Section 4).

The approach taken by Cohen-Bazire and her collaborators has been applied with little modification, other than improvements in technique, to most of the photosynthetic bacteria. The results of these studies have shown that there is remarkable diversity in the morphology of the photosynthetic apparatus in purple bacteria. The nearly spherical vesicular membrane system characteristic of *Rs. rubrum* and *Rp. sphaeroides* is replaced by a system of tubular membranes in *Thiocapsa pfennigii* (Eimhjellen *et al.*,

1967); by sparse irregular membrane invaginations in *Rp. gelatinosa* and *Rs. tenue* (Weckesser *et al.*, 1969; deBoer, 1969); and by various types of stacked, lamellar membrane systems in *Rs. molischianum* and some other species of *Rhodospirillum*, *Rp. palustris*, *Rp. acidophila*, *Rhodomicrobium vannielii*, and *Ectothiorhodospira*. The different forms that can be taken by the photosynthetic apparatus *in vivo* are excellently summarized in Fig. 1 of the review by Oelze and Drews (1972). It is important to note that species that are closely related by most taxonomic criteria can show widely differing morphology in their photosynthetic apparatus; conversely, species that seem not to be very closely related can have photosynthetic apparatuses of remarkably similar form. Presumably the morphology, but not the functional efficiency, of the bacterial photosynthetic apparatus is very sensitive to small variations in some component of these membranes, such as the spectrum of fatty acids in the lipids.

Isolated chromatophores from several purple bacteria have been examined extensively by electron microscopy. To date, almost all the work has been done with organisms the photosynthetic apparatus of which *in vivo* is of the vesicular type, and, in particular, with *Rs. rubrum*, *Rp. sphaeroides*, and various species of *Chromatium*. The earliest and most popular technique for examining chromatophores from these species involved negative staining of isolated membrane fractions. With this technique, the chromatophores appear as flattened disks, the diameters of which are greater and more variable than those of the membrane-bounded vesicles observed in fixed cells. The difference is probably due to the large degree of hydration of isolated chromatophores; on dehydration, the hydrated spheres collapse and spread to a variable degree, leading to the observed profile. Much of the diagnostic power of the electron microscope is lost when chromatophore preparations are examined in this way, since it is difficult to relate the dimensions of the flattened chromatophore disks to those of the membrane-bounded vesicles in whole cells. In an attempt to overcome this difficulty, Gibson (1965*d*) compared fixed preparations of purified chromatophores from *Rp. sphaeroides* with thin sections of whole cells. Measurements of the dimensions of the purified chromatophores and the intracellular vesicles showed that the diameters of both conformed to a normal distribution with a mean of 57 nm and a standard deviation of 3 nm. The study also showed that the purified chromatophore preparation, which accounted for more than 80% of the Bchl in the original culture, was contaminated with other types of membrane to the extent of less than 3%. The methods used by Gibson (1965*d*) for comparing isolated chromatophore preparations with the *in vivo* photosynthetic apparatus are probably too time-consuming for everyday use; however, the study referred to does illustrate the usefulness of electron microscopy in the characterization of such preparations.

Electron micrographs of preparations enriched in cytoplasmic membrane and cell wall or outer membrane, obtained from some of the species mentioned above, reveal structures that are typical for gram-negative bacteria (Oelze *et al.*, 1975; Collins and Niederman, 1976*a*; Ding and Kaplan, 1976). Thus, in thin sections of fixed material, cytoplasmic membrane preparations appear largely as closed vesicles the diameters of which cover a fairly wide range centered around 50 nm in the case of *Rp. sphaeroides* (Parks and Niederman, in prep.) and 40 nm in the case of *Rs. rubrum* (Oelze *et al.*, 1975). On the other hand, fractions enriched in outer membrane or cell walls show characteristic large vesicles and coiled structures in thin sections (Collins and Niederman, 1976*a,b*); when the fractions are negatively stained, occasional disrupted hulls are seen (Ketchum and Holt, 1970; Niederman *et al.*, 1972). Such electron micrographs are essentially identical to similar micrographs of preparations of nonphotosynthetic gram-negative bacteria (Schnaitman, 1970*a*; Osborn *et al.*, 1972). Indeed, it is perhaps worth emphasizing that from a morphological, as well as a chemical, point of view, the outer membranes of the photosynthetic bacteria, whether they are observed *in vivo* or after more or less extensive purification, do not differ significantly from the outer membranes of other gram-negative bacteria, and that this is a further indication of the nonspecialized nature of at least the major part of the envelope membranes.

3.6. Effects of Light Intensity and Stage of Development on Chromatophore Composition

With the availability of membrane-isolation procedures, the effects of light intensity on the composition of the bacterial photosynthetic apparatus became a subject of intensive investigation (for reviews of the earlier studies, see Cohen-Bazire and Sistrom, 1966; Lascelles, 1968). As a result of more recent findings, many of the earlier contradictory observations can now be satisfactorily explained. Ketchum and Holt (1970) observed an essentially constant specific Bchl content in chromatophores purified from steady-state cultures of *Rs. rubrum* that varied in specific growth rate form 0.038 to 0.185 hr^{-1}. This represented a range of light intensities from 100 to 3100 ft-c. However, as much as a twofold increase in specific Bchl content was reported for chromatophores isolated from stationary-phase cells in which significant self-shading had occurred. Thus, the physiological age of such cultures, rather than the intensity of the incident light to which they are exposed, is the critical factor in determining the Bchl content of their chromatophore membranes. These findings suggest that the inverse relationship between light intensity and specific Bchl content observed in steady-state *Rs. rubrum* cells (Holt and Marr, 1965*c*; Ketchum and Holt, 1970) can be accounted for entirely by changes in the overall quantities of chromatophore membrane in which the relative contents of Bchl and protein remain constant. Differences in the amount of chromatophore membrane observed in thin sections of whole cells also support this possibility (Cohen-Bazire and Kunisawa, 1963; Cohen-Bazire, 1963; Holt and Marr, 1965*c*).

In contrast to the observed constancy in specific Bchl content, Ketchum and Holt (1970) reported that the lipid-phosphorus content of purified chromatophores was directly proportional to the specific growth rates of the cultures from which they were isolated. However, data presented by S. Steiner *et al.* (1970) suggest that the Bchl content per unit of lipid phosphorus remains constant in *Rp. capsulata* grown at different light intensities, and in the studies of Lascelles and Szilágyi (1965), essentially parallel increases in both Bchl and lipid phosphorus were observed in low-aeration suspensions of *Rp. sphaeroides* in which chromatophore formation was induced. Similar results were reported for *Rs. rubrum* and *Rp. capsulata* during the induction of chromatophore formation (Schröder and Drews, 1968).

On the basis of their studies on the regulation of Bchl synthesis, Aagaard and Sistrom (1972) provided a possible explanation for the differences in the responses of *Rs. rubrum* and *Rp. sphaeroides* to alterations in light intensity. Although specific growth rates affect the overall Bchl content of the respective organisms similarly (Sistrom, 1962*b*; Holt and Marr, 1965*c*; Cohen-Bazire and Sistrom, 1966), an inverse relationship was observed between the specific Bchl content of *Rp. sphaeroides* chromatophores and the light intensity to which cultures were exposed (Worden and Sistrom, 1964; cf. Gorchein, 1968*a*). It was concluded that in *Rp. sphaeroides*, both the amount of the chromatophore membrane and its specific Bchl content vary as a function of changes in the specific Bchl levels of the cell. Thus, large quantities of chromatophore membrane of high specific Bchl content are found in photosynthetic-pigment-rich cells grown under low-intensity illumination. Aagaard and Sistrom (1972) suggested that the different responses of these two organisms reflect the manner in which the levels of their LH-Bchl components are regulated.

In *Rs. rubrum*, the ratio of the level of the single B880 LH Bchl to that of RC Bchl is essentially invariant and unrelated to the specific Bchl content of the cells. In contrast, levels of the accessory B850 LH-Bchl component in wild-type *Rp. sphaeroides* are independent of RC-Bchl levels and increase as the specific Bchl content of the cells is increased; however, amounts of the B875 LH-Bchl complex are directly related to the levels of RC Bchl in the membrane. If it is assumed that RC Bchl forms an essentially invariant portion of the chromatophore membrane in each of these organisms, then in *Rs. rubrum*, the ratio of total Bchl to chromatophore protein would always remain fixed, while the wide variations observed in the specific Bchl content of *Rp. sphaeroides* chromatophores would be accounted for by differences in the relative amounts of the independently regulated B850 component.* Since *Rs. molischianum* also contains multiple LH-Bchl components (Nelson and Frenkel, 1974), the direct relationship observed between the specific Bchl content of isolated chromatophores and that of whole cells (Gibbs *et al.*, 1965) may be explained in part by regulatory phenomena similar to those of *Rp. sphaeroides*.

Changes in composition have also been observed in chromatophore preparations isolated from facultatively aerobic Rhodospirillaceae in which the formation of the photosynthetic apparatus is induced by decreased oxygen tension or transition from aerobic to phototrophic growth. In photochemically active chromatophores purifed from *Rs. rubrum* that had been transferred from high to low aeration conditions, the specific Bchl content increased in parallel to that of the crude cell extracts (Cohen-Bazire and Kunisawa, 1960). Oelze *et al.* (1969*b*) reported that during adaptation of *Rs. rubrum* to phototrophic growth, three distinct phases of chromatophore membrane formation occurred. In the initial phase (approximately 3–6 hr in duration), no chromatophores could be isolated; this was thought to be a consequence of incorporation of newly synthesized chromatophore-specific material into the preexisting cytoplasmic membrane. In the second stage, chromatophores were isolated in which the specific Bchl content increased in direct proportion to that of the adapting cells. In the final phase, no further increases in the Bchl content of the chromatophores were observed despite a continued accumulation of intracellular pigment.

A correlation between chromatophore morphogenesis and the development of photosynthetic competence suggests that the first of these phases also occurs during repigmentation in *Rp. sphaeroides* at reduced oxygen tensions (Peters and Cellarius, 1972). Although within 60–90 min, photochemical activity as measured by interactions between the photosynthetic pigment and electron-transfer systems approached that observed in fully pigmented cells, no typical chromatophores were observed in thin sections of whole cells until approximately 3 hr after the transition to low aeration. The incorporation of photochemical components into the cytoplasmic membrane during the initial phases of repigmentation is also suggested by recent results of Jones and Plewis (1974). It was possible to reconstitute light-driven electron flow by the addition of purified reaction centers to cytochrome-containing crude membrane preparations from a photosynthesis-deficient mutant of *Rp. sphaeroides* unable to synthesize Bchl.

From estimates of Bchl and protein levels in highly purified chromatophore preparations, it has been suggested that the photosynthetic membranes of *Rp. sphaeroides* have an essentially constant composition after the initial stages of adaptation to phototrophic growth (Huang and Kaplan, 1973*b*). It was concluded that a coordinated synthesis of Bchl and chromatophore-specific proteins occurs, and that these protein components are formed synchronously. The results of an analysis of Bchl fluorescence during induction of chromatophore formation in *Rp. sphaeroides* were interpreted in terms of a synchronous assembly of photosynthetic pigment components (Cellarius and Peters, 1969). In contrast, the recent studies of Niederman *et al.* (1976) with concentrated suspensions of *Rp. sphaeroides* in which photosynthetic-pigment formation was induced at low aeration suggested a noncoordinate assembly of functionally essential and accessory components in the developing photosynthetic apparatus (Fig. 3). Under these conditions, the synthesis of photosynthetic pigments was initiated without a significant lag and an essentially parallel increase in Bchl, and the major carotenoid was observed in the whole cell preparations. However, a computer-assisted analysis of the levels of LH-Bchl complexes in highly purified chromatophore preparations (Section 2.3) indicated that the incorporation of these components follows an asynchronous course (Fig. 3, bottom panel). B875 was preferentially inserted into the chromatophore membrane during the early stages of induction,† and

---

* In confirmation of this hypothesis, Takemoto and Huang Kao (1977) recently reported that in *Rp. sphaeroides* grown at high light intensity, the membrane contains lower amounts of B850-associated polypeptides than in cells grown at low light levels; however, the relative amounts of RC-associated polypeptides were approximately equal regardless of the light intensity at which the cells were grown.

† In a study by Takemoto (1974), a preferential insertion of RC Bchl and the polypeptide components of the photochemical reaction center was demonstrated during the early stages of repigmentation in *Rp. sphaeroides*.

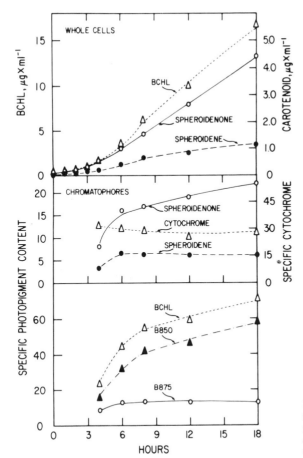

Fig. 3. Levels of various pigmented components in whole-cell and chromatophore preparations during the induction of chromatophore membrane formation in low-aeration suspensions of *Rhodopseudomonas sphaeroides*. Adapted from data of Niederman *et al.* (1976).

thereafter its levels were maintained at a steady state; the three RC-polypeptides (not shown) and *b*- and *c*-type cytochromes were also maintained at essentially constant levels. Several of these components have essential functions in primary photochemical events and in light-driven electron flow. An essential role for the B875 complex is suggested by the fixed stoichiometry of B875 and RC Bchl in cells of varying Bchl content (Aagaard and Sistrom, 1972) and the reduced levels of B875 in mutants of *Rp. sphaeroides* lacking RC Bchl (Sistrom and Clayton, 1964; Sistrom, (1966).* The levels of the accessory LH components spheroidenone and B850, together with the B850-associated polypeptide component, continued to increase in the isolated chromatophores in an essentially coordinated manner. The increases in the specific Bchl content of the chromatophores were accounted

for largely by the elevated concentrations of the B850 component. These findings are consistent with a stepwise assembly mechanism for associated Bchl and protein components, and suggest that separate regulatory mechanisms control the levels of functionally essential and accessory components within the membrane.

A sequential mechanism for the assembly of the chromatophore membrane was also observed recently in low-aeration suspensions of *Rp. capsulata* by Nieth and Drews (1975). Synthesis of RC Bchl preceded that of bulk LH Bchl, and this was correlated with the sequential appearance of their associated polypeptides in the developing membranes. During a 2-hr incubation of low-aeration cell suspensions, the photophosphorylation rate in isolated membranes remained essentially constant on an RC-Bchl basis, whereas on a total-Bchl basis, it decreased markedly. A multistep assembly of the photosynthetic apparatus during the initial phase of repigmentation in *Rs. rubrum* was recently reported by Oelze and Pahlke (1976).

An obligatory coupling between Bchl and protein

---

* Although light-dependent cyclic electron flow was reconstituted by the addition of reaction centers to membrane preparations from a Bchl-less mutant of *Rp. sphaeroides* (Jones and Plewis, 1974), it is not clear whether the efficiency of energy conversion is as great as in LH-Bchl-containing membranes.

synthesis has been demonstrated in the Rhodospirillaceae (Sistrom, 1962a; Bull and Lascelles, 1963; Drews, 1965). It has been proposed that this reflects a specific association between Bchl synthesis and the formation of proteins that accommodate the photopigments within the membrane (Lascelles, 1968). In a later study, Takemoto and Lascelles (1973) demonstrated that in *Rp. sphaeroides*, Bchl formation is tightly coupled to the synthesis of at least four chromatophore-specific proteins; these included LH-Bchl-associated protein and that associated with the photochemical RC (Takemoto and Lascelles, 1974). In the membranes of mutant strains unable to synthesize Bchl, these proteins were either missing or present in significantly reduced levels, which suggests that the completed Bchl molecule is essential for their occurrence in the membrane. The recent density-shift analysis of Kosakowski and Kaplan (1974) is consistent with the possibility that Bchl is interspersed into the membranes of steady-state cells in Bchl–protein complexes.

Several possible mechanisms have been proposed to account for the effects of light intensity and oxygen tension on Bchl (and chromatophore membrane) synthesis (a comprehensive discussion of earlier hypotheses may be found in the review of Oelze and Drews, 1972). More recently, the mechanism by which molecular oxygen reversibly inhibits Bchl synthesis was assessed critically by Marrs and Gest (1973b) in various respiration-deficient mutants of *Rp. capsulata* (Marrs and Gest, 1973a). In a mutant that lacked cytochrome oxidase activity, Bchl formation was repressed by oxygen to levels equivalent to those observed in the parental strain. This does not support the original proposal by Cohen-Bazire *et al.* (1957) in which the oxidation–reduction state of an intermediate electron carrier was thought to regulate pigment synthesis. Bchl synthesis was hypersensitive to oxygen in another mutant that lacked NADH and succinic dehydrogenase activities. It was concluded that molecular oxygen directly inactivates a factor required for Bchl synthesis. It is thought that when the oxygen tension is lowered sufficiently, the factor is reactivated by a flow of electrons diverted from the branched respiratory chain that is derived initially from NADH or succinate. We wish to suggest a modification of the proposed mechanism that may account for the inverse relation between light intensity and the overall Bchl content of whole cells. It is possible that illumination causes a diversion of electrons away from the proposed regulatory factor, since Keister and Yike (1967) showed that light stimulates succinate-linked NAD reduction via a reversal of electron flow. This would result in the observed inhibition of Bchl synthesis as the intensity of the incident illumination is increased. A detailed discussion of the regulation of photosynthetic membrane development is presented in Chapter 43.

4. Speculations on the Nature of *in Vivo* Chromatophore Membrane Structure

In view of the results of numerous morphological and biochemical investigations, the term "chromatophore" as initially applied to the bacterial photosynthetic apparatus by Pardee *et al.* (1952) has undergone considerable modification. Indeed, this designation was abandoned altogether in the review of Oelze and Drews (1972) and replaced by "intracytoplasmic membrane." The latter term is rather vague, however, and has been used to designate numerous different structures in nonphotosynthetic bacteria. Frenkel and Nelson (1971) pointed out that the term chromatophore is also not very specific, and should include a designation that the particles have been derived from photosynthetic bacteria. With these caveats in mind, it is suggested that "chromatophore" be restricted to isolated photosynthetic membranes that have been separated from other cellular material, and that "chromatophore membrane" designate the structure as it exists *in vivo*. As a result of recent studies from this laboratory and others (Section 3.6), it is also possible to provide an operational definition of the chromatophore membrane from a biogenetic standpoint. No chromatophores have been detected by sucrose density-gradient centrifugation of extracts of *Rp. sphaeroides* grown at high aeration (Niederman *et al.*, 1972, 1976), and none has been observed in thin sections of such cells (Peters and Cellarius, 1972; Brown *et al.*, 1972). After transfer to low-aeration conditions, a photosynthetic-pigment-containing band appears in the gradients at the usual position of chromatophores from phototrophically grown cells (Niederman *et al.*, 1976). We believe that the new band from the induced cells represents the developing chromatophore membrane. In the recent report of Huang and Kaplan (1973b), the limits that oxygen tension imposes on chromatophore formation were defined by immunological criteria. That isolated chromatophore preparations are derived from the photosynthetic apparatus is suggested by the results of reconstruction experiments in *Rp. sphaeroides* (Fraker and Kaplan, 1971; Niederman, 1974) and *Rs. rubrum* (Collins and Niederman, 1976b) in which radio-labeled, aerobically grown cells were mixed with phototrophically grown cells, and chromatophores essentially devoid of radioactivity were isolated. In the recent isopyknic banding studies of Kosakowski and Kaplan (1974), cells grown in H_2O- and D_2O-based media were mixed prior to breakage; two discrete bands of chromatophores were observed. These had the same densities as the bands seen in unmixed preparations. Therefore, isolated chromatophore vesicles originate through a local fragmentation and resealing of the chromatophore membrane; they do not represent hybrid structures derived from distant regions of an overall membrane continuum. Taken together, these data suggest that it is possible to draw conclusions about the *in vivo* nature of the bacterial photosynthetic apparatus from the properties of chromatophore preparations that have been isolated in a high state of purity.

The question of the physical continuity of various regions of the chromatophore membrane with the peripheral-cytoplasmic membrane has not been entirely resolved. Since the invaginations of the cytoplasmic membrane observed in thin sections of photosynthetically grown cells (Section 3.5) have suggested that a morphogenetic relationship exists between the cytoplasmic and chromatophore membranes, this question will be approached first from a developmental standpoint. Two possible interpretations for such membrane intrusions were advanced by Drews *et al.* (1969a). In the first model, both chromatophore and cytoplasmic membranes are thought to consist of the same macromolecular components, but their relative proportions differ in the two structures. In the alternative model, the chromatophore membrane, though anchored to the cytoplasmic membrane, is a highly specialized structure and quite different in composition from the peripheral membrane.

The first model was further elaborated by Oelze and Drews (1972). On the basis of their studies in *Rs. rubrum*, they suggested that during the induction of chromatophore formation, the cytoplasmic membrane becomes differentiated through the accretion of specialized components associated with light-dependent reactions. At the point that the rate of synthesis of the differentiating cytoplasmic membrane exceeds that of the cell wall, the cytoplasmic membrane begins to invaginate, which results in the formation of the chromatophore membrane. As development proceeds, differentiation preferentially occurs in the chromatophore membrane, and a gradient of enzyme activity from chromatophore to cytoplasmic membrane becomes established. It was also proposed that the differentiation process is reversible when the photosynthetic apparatus is no longer needed, as, for example, during transition from phototrophic to chemotrophic growth. Under such circumstances, a preferential insertion of respiratory components and specific proteins into the cytoplasmic rather than chromatophore membrane would occur along with a direct shift of chromatophore phospholipid into the peripheral membrane (Oelze and Drews, 1970a).

However, a more recent investigation by Irschik and Oelze (1973) suggests that in photosynthetically grown *Rs. rubrum*, transition from low- to high-intensity illumination results in a structural and functional conservation of the chromatophore membrane. Their results suggested further that no intraconversion of the membranes occurred; the chromatophore and cytoplasmic membrane are formed independently of each other during such shifts in growth conditions. When no longer needed by the cell, the chromatophore membrane becomes diluted out as the cells divide. Since succinic dehydrogenase activity in *Rs. rubrum* is confined to the chromatophore membrane in phototrophically grown cells (Collins and Niederman, 1976b) but is a component of the cytoplasmic membrane during aerobic growth (Collins and Niederman, 1976a), we suggest that a differentiation in both the peripheral and the intracytoplasmic portion of the membrane occurs during the formation of the photosynthetic apparatus.

The alternative model for chromatophore biogenesis presented above is supported by the recent results of studies with *Rp. sphaeroides* by Kaplan and his collaborators (Huang and Kaplan, 1973b; Kosakowski and Kaplan, 1974). In a double-labeling study (Huang and Kaplan, 1973b), it was shown that adaptation to phototrophic growth was accompanied by a preferential synthesis of 2-chloroethanol-soluble proteins specific to the chromatophore [a preferential synthesis of several chromatophore-specific proteins was also demonstrated in low-aeration cell suspensions by Takemoto (1974)]. This fraction accounted for more than 95% of the chromatophore protein during the latter stages of induction. In contrast, 2-chloroethanol-insoluble protein was gradually diluted from the chromatophore membrane as the adaptation process continued; the latter fraction was thought to represent the site of attachment of the chromatophore to the cytoplasmic membrane. It was concluded that although the initial stages of chromatophore formation may occur at the periphery of the cell, the continued growth of the photosynthetic apparatus is independent from the cytoplasmic membrane. In this connection, it was proposed by Segen and Gibson (1971) that the chromatophore membrane consists of several parts, some of which may be synthesized independently of one another.

The density-shift analysis of Kosakowski and Kaplan (1974) in steady-state *Rp. sphaeroides* indicated that newly synthesized phospholipid is interspersed into the growing chromatophore membrane randomly and homogeneously. This is consistent with the fluidity of the phospholipids within the membrane mosaic (Singer and Nicolson, 1972). Surprisingly, the incorporation of protein and Bchl followed a nonhomogeneous course, and it was suggested that pigment and protein are inserted into the growing membrane in the form of a large precursor. It had been suggested previously by labeling experiments that the chromatophores of *Rp. sphaeroides* may be derived from a precursor structure (Shaw and Richards, 1971; Gibson *et al.*, 1972). It is possible that such precursor structures are inserted as conserved units in a relatively fixed position within the membrane. This is supported by the studies of Peters and Cellarius (1972) on repigmentation in *Rp. sphaeroides* under conditions in which the oxygen tension is reduced. Based on the spatial limitations imposed by energy-transfer considerations and because energy transfer and photochemical reaction center activity are observed almost simultaneously with the insertion of photopigments into the membrane (Cellarius and Peters, 1969), it was suggested that pigment molecules are incorporated at discrete sites on the membrane surface. Kosakowski and Kaplan (1974) concluded that the chromatophore membrane is formed by a condensation of newly synthesized material onto the plasma membrane. It was proposed that each cell contains a single condensation site on the peripheral membrane, and that the chromatophore and cytoplasmic membranes are discontinuous with respect to composition. Continued growth was thought to result from homogeneous and non-homogeneous incorporation of new material along replicating and non-replicating faces of the single chromatophore membrane, respectively. However, the results of a recent freeze–etch study in phototrophically grown *Rs. rubrum* by Golecki and Oelze (1975) are not compatible with this model. It was possible to visualize invaginations as protuberances on the concave face of the cytoplasmic membrane and indentations on its convex face. The number of invagination sites was found to increase as the specific Bchl content of the cells was increased. It was concluded that the overall chromatophore-membrane content of the cells is increased through both an elongation of existing invaginations and an increase in the number of such invaginations.

As a result of the electron-microscopic and biochemical studies discussed above, there appears little doubt that at least a portion of the photosynthetic apparatus forms a continuum with the cytoplasmic membrane. However, these investigations have not established firmly that all regions of the chromatophore membrane remain structurally interconnected, and they have left open the possibility that "mature" chromatophore vesicles "bud off" from the membrane continuum. The recent studies of Prince *et al.* (1974, 1975) on the structural and functional asymmetry of the chromatophore membrane were addressed to this crucial question. Monospecific antibodies against the cytochrome c_2 from *Rp. sphaeroides* and *Rp. capsulata* were used to determine the location of this heme protein within the cell. This approach was based on earlier observations by Racker *et al.* (1970) in which the inhibition of respiration by specific antibody was employed in the localization of cytochrome c between the inner and outer membranes of intact mammalian mitochondria. Accordingly, Prince *et al.* (1974, 1975) tested the effect of antibody on the photooxidation of cytochrome c_2 in various cell fractions from these bacteria. Both the photooxidation measurements and reduced vs. oxidized difference spectra were consistent with a similar cytochrome c_2/Bchl content in whole cells and in crude chromatophore preparations isolated from French-press extracts; however, as much as 85% of the cytochrome c_2 was lost during spheroplast formation. In the chromatophore preparations, it was possible to inhibit photooxidation of cytochrome c_2 by antibody only when the membranes were rendered "leaky" with 1% sodium cholate. It was therefore concluded that cytochrome c_2 is specifically localized within the periplasmic space in intact cells, and that in the chromatophores prepared from them, it becomes entrapped inside the resulting vesicles. The loss of cytochrome c_2 on spheroplast formation suggested that the contents of the chromatophore membrane are continuous with the periplasmic fluid, and that chromatophores do not exist as sealed vesicles *in vivo*. Similar conclusions on the location of cytochrome c_2 in *Rp. capsulata* were drawn by Hochman *et al.* (1975). Other antibody experiments by Prince *et al.* (1975) indicated that the coupling factor is localized on the outer surface of isolated chromatophores. This would correspond to the surface facing the cytoplasm in intact cells.

As noted above, approximately 15% of the cytochrome c_2 was not released into solution during the formation of spheroplasts. This residual cytochrome c_2 was photooxidized in a reaction that was not sensitive to antibody, and it was suggested that such cytochrome is protected from release by undigested peptidoglycan (Prince *et al.*, 1974, 1975). An equally plausible explanation is that this fraction of cytochrome c_2 is contained within chromatophores that have formed vesicles *in vivo* and are structurally independent from the cytoplasmic membrane. Some support for this possibility is provided by the release of approximately 50% of the chromatophores after EDTA–lysozyme lysis of *Rs. rubrum* (Collins and Niederman, 1976b). Preliminary studies suggest that they contain slightly elevated levels of cytochromes when compared with chromatophores from French-press extracts. The chromatophores released during osmotic lysis may represent a "free" population within the cell (cf. Gorchein, 1968*b*; Hurlbert *et al.*, 1974). If they had arisen by disruption of the photosynthetic apparatus during the lysis of spheroplasts, it is expected that they would be depleted in cytochrome. It is also possible that such "free" chromatophores exist within each cell in dynamic equilibrium with a continuous chromatophore membrane.

It is clear that despite the rapid advances that have been made in our understanding of the structure and properties of isolated chromatophores, much further work will be necessary to resolve the question of the *in vivo* status of the bacterial photosynthetic apparatus.

5. Note Added in Proof

The LH-Bchl associated protein from the chromatophore membranes of *Rs. rubrum* (Section 3.2) has been characterized further. An organic-solvent-soluble polypeptide was extracted from chromatophores and photoreceptor complexes with chloroform-methanol and purified by gel permeation chromatography (Tonn, S. J., Gogel, G. E., and Loach, P. A., 1977, *Biochemistry* **16**:877). A single polypeptide component with an apparent molecular weight of 12,000 was observed in SDS–polyacrylamide gel electrophoresis of these extracts. A minimal molecular weight of 19,000 was calculated from amino acid analysis on the basis of one histidine residue per polypeptide. The polypeptide contained three methionine residues; cyanogen bromide cleavage yielded four oligopeptides. The *COOH*-terminal amino acid was identified as glycine, but the *NH₂*-terminus appeared to be blocked. The polypeptide accounted for approximately 50% of the chromatophore protein and apparently binds Bchl and carotenoids. In photoreceptor complexes, the extraction procedure resulted in the quantitative removal of protein with an apparent molecular weight of 10,000 in SDS–polyacrylamide gels. In chromatophores, a substantial quantity of protein remained in this portion of the gel. The residual polypeptide band had an amino acid composition different from that of the organic-solvent-soluble protein. An apparent LH complex composed of 66% protein, 5% Bchl, and 29% phospholipid has also been isolated from RC-depleted chromatophores of the *Rs. rubrum* carotenoidless mutant G-9 (Cuendet, P. A., and Zuber, H., 1977, *FEBS Lett.* **79**:96). The protein moiety could be solubilized with chloroform-methanol. When solubilized with LDAO, the near-IR absorption spectrum of the complex was similar to that of RC-depleted chromatophores. After precipitation with ammonium sulfate, a single polypeptide band with an apparent molecular weight of 14,000 was observed in SDS–polyacrylamide gel electrophoresis;

a minimal molecular weight of 14,000 was also calculated from the amino acid composition, while a value of 12,000 was obtained from sedimentation equilibrium analysis.

The isolation of LH-Bchl complexes from *Rp. capsulata* and *Rp. palustris* has also been reported (Drews, G., Feick, R., Schumacher, A., and Firsow, N., 1977, *Proceedings of the Fourth International Congress on Photosynthesis*, p. 83). The isolated B875 complex from each organism contained a single polypeptide component with an apparent molecular weight of 12,000–13,000 in SDS–polyacrylamide gel electrophoresis. A purified B850 complex from *Rp. capsulata* contained three polypeptides with apparent molecular weights of 8000, 10,000, and 14,000. In studies on the synthesis of these complexes in *Rp. palustris*, a marked decrease in the ratio of B850 to RC Bchl was noted after a ten-fold increase in light intensity; however, the relative concentrations of B875 and RC Bchl within the membrane remained constant. These data confirm previous results reported for *Rp. sphaeroides* and *Rp. capsulata* (see Section 3.6). In a further report (Firsow, N. N., and Drews, G., 1977, *Arch. Microbiol.* **115**:299), the synthesis of the Bchl complexes in *Rp. palustris* responded to decreases in oxygen tension and transitions from chemotrophic to phototrophic growth in a manner similar to that observed with the other species. It was also reported that a B850 complex isolated from *Rp. palustris* contained polypeptides with apparent molecular weights of 9000 and 11,000. An isolated B875 complex was obtained in association with RC Bchl and the three characteristic RC polypeptide subunits. The absorption spectra of the isolated LH-Bchl complexes closely resembled those obtained with intact membrane.

Further evidence for the specific localization of succinate dehydrogenase activity in the intracytoplasmic membranes of *Rs. rubrum* (see Section 3.2) has been presented by Oelze and his collaborators. During transitions from low to high intensity illumination in which the content of photosynthetically active chromatophore membranes decreased, the apparent incorporation of succinate dehydrogenase activity and the activity of other succinate-dependent electron transport reactions was confined mainly to the chromatophore membrane (Irschik, H., and Oelze, J., 1976, *Arch. Microbiol.* **109**:307). In contrast, NADH-dependent electron transport activities appeared in both the cytoplasmic and chromatophore membranes. An inhomogeneous distribution of succinate- and NADH-dependent respiratory reactions was also shown to occur within the membranes of *Rs. rubrum* during different stages of aerobic growth (Oelze, J., Golecki, J. R., and Kruczek, J., 1977, *FEMS Microbiol. Lett.* **2**:229). An increased ratio of succinate– to NADH–cytochrome *c* reductase activity was observed in membrane preparations from cells isolated at the end of exponential growth. This was correlated with the appearance of concentric intracytoplasmic membranes in electron micrographs of such cells. In sucrose density gradients, a new band of dense membranes was noted that was enriched in the succinate-dependent activity.

ACKNOWLEDGMENTS

Work in the laboratory of the senior author was supported by Public Health Service grant GM20367 from the National Institute of General Medical Sciences and National Science Foundation grant BMS74-18151. The senior author is the recipient of Public Health Service Research Career Development Award GM00093-03 from the National Institute of General Medical Sciences.

We thank Mary Lynne Perille Collins and Lawrence C. Parks for their critical reading of the manuscript and for their participation in some of the studies reported here. We also thank Ronald D. Poretz for performing amino acid analyses and for useful discussions.

6. References

Aagaard, J., and Sistrom, W. R., 1972, Control of synthesis of reaction center bacteriochlorophyll in photosynthetic bacteria, *Photochem. Photobiol.* **15**:209.

Baccarini-Melandri, A., Gest, H., and San Pietro, A., 1970, A coupling factor in bacterial photophosphorylation, *J. Biol. Chem.* **245**:1224.

Bartsch, R. G., 1971, Cytochromes: bacterial, in: *Methods in Enzymology*, Vol. 23A (A. San Pietro, ed.), pp. 344–363, Academic Press, New York.

Berzborn, R. J., Johansson, B. C., and Baltscheffsky, M., 1975, Immunological and fluorescence studies with the coupling factor ATPase from *Rhodospirillum rubrum*, *Biochim. Biophys. Acta* **396**:360.

Biedermann, M., and Drews, G., 1968, Trennung der Thylakoidbausteine einiger Athiorhodaceae durch Gelelektrophorese, *Arch. Mikrobiol.* **61**:48.

Bril, C., 1963, Studies on bacterial chromatophores. II. Energy transfer and photooxidative bleaching of bacteriochlorophyll in relation to structure in normal and carotenoid-depleted *Chromatium, Biochim. Biophys. Acta* **66**:50.

Britten, R. J., and Roberts, R. B., 1960, High-resolution density gradient sedimentation analysis, *Science* **131**:32.

Brown, A. E., Eiserling, F. A., and Lascelles, J., 1972, Bacteriochlorophyll synthesis and the ultrastructure of wild type and mutant strains of *Rhodopseudomonas sphaeroides, Plant Physiol.* **50**:743.

Bull, M. J., and Lascelles, J., 1963, The association of protein synthesis with the formation of pigments in some photosynthetic bacteria, *Biochem. J.* **87**:15.

Cellarius, R. A., and Peters, G. A., 1969, Photosynthetic membrane development in *Rhodopseudomonas sphaeroides*: Incorporation of bacteriochlorophyll and development of energy transfer and photochemical activity, *Biochim. Biophys. Acta* **189**:234.

Clayton, R. K., 1963, Toward the isolation of a photochemical reaction center in *Rhodopseudomonas sphaeroides, Biochim. Biophys. Acta* **75**:312.

Clayton, R. K., and Clayton, B. J., 1972, Relations between pigments and proteins in the photosynthetic membranes of *Rhodopseudomonas sphaeroides, Biochim. Biophys. Acta* **283**:492.

Clayton, R. K., and Haselkorn, R., 1972, Protein components of bacterial photosynthetic membranes, *J. Mol. Biol.* **68**:97.

Cohen-Bazire, G., 1963, Some observations on the organization of the photosynthetic apparatus in purple and green bacteria, in: *Bacterial Photosynthesis* (H. Gest, A. San Pietro, and L. P. Vernon, eds.), pp. 89–110, Antioch Press, Yellow Springs, Ohio.

Cohen-Bazire, G., and Kunisawa, R., 1960, Some observations on the synthesis and function of the photosynthetic apparatus in *Rhodospirillum rubrum, Proc. Natl. Acad. Sci. U.S.A.* **46**:1543.

Cohen-Bazire, G., and Kunisawa, R., 1963, The fine structure of *Rhodospirillum rubrum, J. Cell Biol.* **16**:401.

Cohen-Bazire, G., and Sistrom, W. R., 1966, The procaryotic photosynthetic apparatus, in: *The Chlorophylls* (G. R. Seeley and L. Vernon, eds.), pp. 313–341, Academic Press, New York.

Cohen-Bazire, G., Sistrom, W. R., and Stanier, R. Y., 1957, Kinetic studies of pigment synthesis by non-sulfur purple bacteria, *J. Cell Comp. Physiol.* **49**:25.

Collins, M. L. P., and Niederman, R. A., 1976a, Membranes of *Rhodospirillum rubrum*: Isolation and physicochemical properties of membranes from aerobically grown cells, *J. Bacteriol.* **126**:1316.

Collins, M. L. P., and Niederman, R. A., 1976b, Membranes of *Rhodospirillum rubrum*: Physicochemical properties of chromatophore fractions isolated from osmotically and mechanically disrupted cells, *J. Bacteriol.* **126**:1326.

Crofts, A. R., 1970, The chromatophore as a functional unit, in: *Electron Transport and Energy Conservation* (J. M. Tager, S. Papa, E. Quagliariello, and E. C. Slater, eds.), pp. 221–228, Adriatica Editrice, Bari.

Cruden, D. L., and Stanier, R. Y., 1970, the characterization of *Chlorobium* vesicles and membranes isolated from green bacteria, *Arch. Mikrobiol.* **72**:115.

Cusanovich, M. A., and Kamen, M. D., 1968a, Light-induced electron transport in *Chromatium* strain D. I. Isolation and characterization of *Chromatium* chromatophores, *Biochim. Biophys. Acta* **153**:376.

Cusanovich, M. A., and Kamen, M. D., 1968b, Light-induced electron transport in *Chromatium* strain D. III. Photophosphorylation by *Chromatium* chromatophores, *Biochim. Biophys. Acta* **153**:418.

deBoer, W. E., 1969, On the ultrastructure of *Rhodopseudomonas gelatinosa* and *Rhodospirillum tenue*, *Antonie van Leeuwenhoek, J. Microbiol. Serol.* **35**:241.

deDuve, C., and Berthet, J., 1954, The use of differential centrifugation in the study of tissue enzymes, *Int. Rev. Cytol.* **3**:225.

deDuve, C., Berthet, J., and Beaufay, H., 1959, Gradient centrifugation of cell particles—theory and applications, *Prog. Biophys.* **9**:325.

DePinto, J. A., 1967, Ornithine-containing lipid in *Rhodospirillum rubrum, Biochim. Biophys. Acta* **144**:113.

Ding, D. H., and Kaplan, S., 1976, Separation of inner and outer membranes of *Rhodopseudomonas sphaeroides, Prep. Biochim.* **6**:61.

Drews, G., 1965, Untersuchungen zur Regulation der Bacteriochlorophyll-Synthese bei *Rhodospirillum rubrum, Arch. Mikrobiol.* **51**:186.

Drews, G., 1968, Nachweis der Zucker in den Thylakoiden von *Rhodospirillum rubrum* und *Rhodopseudomonas viridis, Z. Naturforsch.* **23b**:671.

Drews, G., Biedermann, M., and Oelze, J., 1969*a*, Investigation of the thylakoid morphogenesis in *Rhodospirillum rubrum, Prog. Photosynth. Res.* **1**:204.

Drews, G., Lampe, H.-H., and Ladwig, R., 1969*b*, Die Entwicklung des Photosyntheseapparates in Dunkelkulturen von *Rhodopseudomonas capsulata, Arch. Mikrobiol.* **65**:12.

Eimhjellen, K. E., Steensland, H., and Traetteberg, J., 1967, A *Thiococcus sp. nov. gen.*, its pigments and internal membrane system, *Arch. Mikrobiol.* **59**:82.

Feher, G., 1971, Some chemical and physical properties of a bacterial reaction center particle and its primary photochemical reactants, *Photochem. Photobiol.* **14**:373.

Fraker, P. J., and Kaplan, S., 1971, Isolation and fractionation of the photosynthetic membranous organelles from *Rhodospirillum sphaeroides, J. Bacteriol.* **108**:465.

Fraker, P. J., and Kaplan, S., 1972, Isolation and characterization of a bacteriochlorophyll-containing protein from *Rhodospirillum sphaeroides, J. Biol. Chem.* **247**:2732.

French, C. S., 1940*a*, Absorption spectra of the carotenoids in the red and brown forms of a photosynthetic bacterium, *Bot. Gaz. (Chicago)* **102**:406.

French, C. S., 1940*b*, The pigment–protein compound in photosynthetic bacteria. I. The extraction and properties of photosynthin, *J. Gen. Physiol.* **23**:469.

French, C. S., 1940*c*, The pigment–protein compound in photosynthetic bacteria. II. The absorption curves of photosynthin from several species of bacteria, *J. Gen. Physiol.* **23**:483.

Frenkel, A., 1954, Light induced phosphorylation by cell-free preparations of photosynthetic bacteria, *J. Am. Chem. Soc.* **76**:5568.

Frenkel, A. W., 1956, Phosphorylation of adenine nucleotides by cell-free preparations of purple bacteria, *J. Biol. Chem.* **222**:823.

Frenkel, A. W., 1958, Simultaneous reduction of diphosphopyridine nucleotide and oxidation of reduced flavin mononucleotide by illuminated bacterial chromatophores, *J. Am. Chem. Soc.* **80**:3479.

Frenkel, A. W., and Hickman, D. D., 1959, Structure and photochemical activity of chlorophyll-containing particles from *Rhodospirillum rubrum, J. Biophys. Biochem. Cytol.* **6**:285.

Frenkel, A. W., and Nelson, R. A., 1971, Bacterial chromatophores, in: *Methods in Enzymology*, Vol. 23A (A. San Pietro, ed.), pp. 256–268, Academic Press, New York.

Fujimori, E., 1969, Bacteriochlorophyll pheophytinization in chromatophores and subchromatophores from *Rhodospirillum rubrum, Biochim. Biophys. Acta* **180**:360.

Fuller, R. C., Conti, S. F., and Mellin, D. B., 1963, The structure of the photosynthetic apparatus in the green and purple sulfur bacteria, in: *Bacterial Photosynthesis* (H. Gest, A. San Pietro, and L. P. Vernon, eds.), pp. 71–87, Antioch Press, Yellow Springs, Ohio.

Garcia, A., Vernon, L. P., Ke, B., and Mollenhauer, H., 1968, Some structural and photochemical properties of *Rhodopseudomonas palustris* subchromatophore particles obtained by treatment with Triton X-100, *Biochemistry* **7**:319.

Geller, D. M., and Lipmann, F., 1960, Photophosphorylation in extracts of *Rhodospirillum rubrum, J. Biol. Chem.* **235**:2478.

Gibbs, S. P., Sistrom, W. R., and Worden, P. B., 1965, The photosynthetic apparatus of *Rhodospirillum molischianum, J. Cell Biol.* **26**:395.

Gibson, K. D., 1965*a*, Nature of the insoluble pigmented structures (chromatophores) in extracts and lysates of *Rhodopseudomonas sphaeroides, Biochemistry* **4**:2027.

Gibson, K. D., 1965*b*, Isolation and characterization of chromatophores from *Rhodopseudomonas sphaeroides, Biochemistry* **4**:2042.

Gibson, K. D., 1965*c*, Structure of chromatophores of *Rhodopseudomonas sphaeroides*. Removal of nonpigmented outer layer of lipid, *Biochemistry* **4**:2052.

Gibson, K. D., 1965*d*, Electron microscopy of chromatophores of *Rhodopseudomonas sphaeroides, J. Bacteriol.* **90**:1059.

Gibson, K. D., Segen, B. J., and Niederman, R. A., 1972, Membranes of *Rhodopseudomonas sphaeroides*. II. Precursor–product relations in anaerobically growing cells, *Arch. Biochem. Biophys.* **152**:561.

Giesbrecht, P., and Drews, G., 1966, Über die Organisation und die makromolekulare Architektur der Thylakoide "lebender" Bakterien, *Arch. Mikrobiol.* **54**:297.

Godson, G. N., and Sinsheimer, R. L., 1967, Lysis of *Escherichia coli* with a neutral detergent, *Biochim. Biophys. Acta* **149**:476.

Golecki, J. R., and Oelze, J., 1975, Quantitative determination of cytoplasmic membrane invaginations in phototrophically growing *Rhodospirillum rubrum*. A freeze–etch study, *J. Gen. Microbiol.* **88**:253.

Gorchein, A., 1964, Ornithine in *Rhodopseudomonas sphaeroides, Biochim. Biophys. Acta* **84**:356.

Gorchein, A., 1968*a*, The relation between the pigment content of isolated chromatophores and that of the whole cell in *Rhodopseudomonas sphaeroides, Proc. R. Soc. London Ser. B.* **170**:247.

Gorchein, A., 1968*b*, The nature of the internal fine structure of *Rhodopseudomonas sphaeroides* as determined by the study of cell fragments, *Proc. R. Soc. London Ser. B.* **170**:255.

Gorchein, A., 1968c, The separation and identification of the lipids of *Rhodopseudomonas sphaeroides*, *Proc. R. Soc. London Ser. B.* **170**:279.

Gorchein, A., 1968d, Studies on the structure of an ornithine-containing lipid from non-sulphur purple bacteria, *Biochim. Biophys. Acta* **152**:358.

Gorchein, A., Neuberger, A., and Tait, G. H., 1968, The isolation and characterization of subcellular fractions from pigmented and unpigmented cells of *Rhodopseudomonas sphaeroides*, *Proc. R. Soc. London Ser. B.* **170**:229.

Guidotti, G., 1972, Membrane proteins, *Annu. Rev. Biochem.* **41**:731.

Guillotin, J., and Reiss-Husson, F., 1975, Cytoplasmic and outer membranes separation in *Rhodopseudomonas sphaeroides*, *Arch. Microbiol.* **105**:269.

Hall, R. L., Kung, M. C., Fu, M., Hales, B. J., and Loach, P. A., 1973, Comparison of phototrap complexes from chromatophores of *Rhodospirillum rubrum*, *Rhodopseudomonas sphaeroides*, and the R-26 mutant of *Rhodopseudomonas sphaeroides*, *Photochem. Photobiol.* **18**:505.

Hansen, T. A., and deBoer, W. E., 1969, Sucrose-gradient centrifugation of chromatophorous material from Athiorhodaceae, *Antonie van Leeuwenhoek*; *J. Microbiol. Serol.* **35**:243.

Hatefi, Y., Davis, K. A., Baltscheffsky, H., Baltscheffsky, M., and Johansson, B. C., 1972, Isolation and properties of succinate dehydrogenase from *Rhodospirillum rubrum*, *Arch. Biochem. Biophys.* **152**:613.

Haverkate, F., Teulings, F. A. G., and van Deenen, L. L. M., 1965, Studies on the phospholipids of photosynthetic microorganisms. II, *K. Ned. Akad. Wet. Versl. Gewone Vergad. Afd. Natuurkd.* **68**:154.

Hickman, D. D., Frenkel, A. W., and Cost, K., 1963, Isolation of photochemically active chromatophores from *Rhodospirillum molischianum*, in: *Bacterial Photosynthesis* (H. Gest, A. San Pietro, and L. P. Vernon, eds.), pp. 111–114, Antioch Press, Yellow Springs, Ohio.

Hochman, A., Fridberg, I., and Carmeli, C., 1975, The location and function of cytochrome c_2 in *Rhodopseudomonas capsulata* membranes, *Eur. J. Biochem.* **58**:65.

Holt, S. C., and Marr, A. G., 1965a, Location of chlorophyll in *Rhodospirillum rubrum*, *J. Bacteriol.* **89**:1402.

Holt, S. C., and Marr, A. G., 1965b, Isolation and purification of the intracytoplasmic membranes of *Rhodospirillum rubrum*, *J. Bacteriol.* **89**:1413.

Holt, S. C., and Marr, A. G., 1965c, Effect of light intensity on the formation of intracytoplasmic membranes in *Rhodospirillum rubrum*, *J. Bacteriol.* **89**:1421.

Huang, J. W., and Kaplan, S., 1973a, Membrane proteins of *Rhodopseudomonas sphaeroides*. III. Isolation, purification, and characterization of cell envelope proteins, *Biochim. Biophys. Acta* **307**:301.

Huang, J. W., and Kaplan, S., 1973b, Membrane proteins of *Rhodopseudomonas sphaeroides*. IV. Characterization of chromatophore proteins, *Biochim. Biophys. Acta* **307**:317.

Huang, J. W., and Kaplan, S., 1973c, Membrane proteins of *Rhodopseudomonas sphaeroides*. V. Additional chemical characterization of a pigment-lipid-associated protein

isolated from chromatophores, *Biochim. Biophys. Acta* **307**:332.

Hurlbert, R. E., Golecki, J. R., and Drews, G., 1974, Isolation and characterization of *Chromatium vinosum* membranes, *Arch. Microbiol.* **101**:169.

Irschik, H., and Oelze, J., 1973, Membrane differentiation in phototrophically growing *Rhodospirillum rubrum* during transition from low to high light intensity, *Biochim. Biophys. Acta* **330**:80.

Johansson, B. C., 1972, A. coupling factor from *Rhodospirillum rubrum* chromatophores, *FEBS Lett.* **20**:339.

Johansson, B. C., and Baltscheffsky, M., 1975, On the subunit composition of the coupling factor (ATPase) from *Rhodospirillum rubrum*, *FEBS Lett.* **53**:221.

Jones, O. T. G., and Plewis, K. M., 1974, Reconstitution of light-dependent electron transport in membranes from a bacteriochlorophyll-less mutant of *Rhodopseudomonas sphaeroides*, *Biochim. Biophys. Acta* **357**:204.

Kamen, M. D., and Vernon, L. P., 1954, Enzymatic activities affecting cytochromes in photosynthetic bacteria, *J. Biol. Chem.* **211**:663.

Katz, E., and Wassink, E. C., 1939, Infrared absorption spectra of chlorophyllous pigments in living cells and in extra-cellular states, *Enzymologia (Acta Biocatalytica)* **7**:97.

Keister, D. L., and Yike, N. J., 1967, Energy-linked reactions in photosynthetic bacteria. I. Succinate-linked ATP-driven NAD^+ reduction by *Rhodospirillum rubrum* chromatophores, *Arch. Biochem. Biophys.* **121**:415.

Ketchum, P. A., and Holt, S. C., 1970, Isolation and characterization of the membranes from *Rhodospirillum rubrum*, *Biochim. Biophys. Acta* **196**:141.

Kiehn, E. D., and Holland, J. J., 1968, Multiple protein components of mammalian cell membranes, *Proc. Natl. Acad. Sci. U.S.A.* **61**:1370.

Kosakowski, M. H., and Kaplan, S., 1974, Topology and growth of the intracytoplasmic membrane system of *Rhodopseudomonas sphaeroides*: Protein, chlorophyll, and phospholipid insertion into steady-state anaerobic cells, *J. Bacteriol.* **118**:1144.

Kran, V. G., Schlote, F. W. and Schlegel, H. G., 1963, Cytologische Untersuchungen an *Chromatium okenii* Perty, *Naturwissenschaften* **50**:728.

Laemmli, U. K., 1970, Cleavage of structural proteins during the assembly of the head of bacteriophage T4, *Nature (London)* **227**:680.

Lampe, H.-H., and Drews, G., 1972, Die Differenzierung des Membransystems von *Rhodopseudomonas capsulata* hinsichtlich seiner photosynthetischen und respiratorischen Funktionen, *Arch. Mikrobiol.* **84**:1.

Lampe, H.-H., Oelze, J., and Drews, G., 1972, Die Fraktionierung des Membransystems von *Rhodopseudomonas capsulata* und seine Morphogenese, *Arch. Mikrobiol.* **83**:78.

Lascelles, J., 1962, The chromatophores of photosynthetic bacteria, *J. Gen. Microbiol.* **29**:47.

Lascelles, J., 1968, The bacterial photosynthetic apparatus, in: *Advances in Microbial Physiology*, Vol. 2 (A. H. Rose and J. F. Wilkinson, eds.), pp. 1–42, Academic Press, New York.

Lascelles, J., and Szilágyi, J. F., 1965, Phospholipid synthesis by *Rhodopseudomonas sphaeroides* in relation to the formation of photosynthetic pigments, *J. Gen. Microbiol.* **38**:55.

Marrs, B., and Gest, H., 1973*a*, Genetic mutations affecting the respiratory electron-transport system of the photosynthetic bacterium *Rhodopseudomonas capsulata*, *J. Bacteriol.* **114**:1045.

Marrs, B., and Gest, H., 1973*b*, Regulation of bacteriochlorophyll synthesis by oxygen in respiratory mutants of *Rhodopseudomonas capsulata*, *J. Bacteriol.* **114**:1052.

McCall, J. S., and Potter, B. J., 1973, *Ultracentrifugation*, Williams & Wilkins Co., Baltimore.

Melandri, B. A., Baccarini-Melandri, A., San Pietro, A., and Gest, H., 1971, Interchangeability of phosphorylation coupling factors in photosynthetic and respiratory energy conversion, *Science* **174**:514.

Miura, T., and Mizushima, S., 1968, Separation by density gradient centrifugation of two types of membranes from spheroplast membrane of *Escherichia coli* K12, *Biochim. Biophys. Acta* **150**:159.

Nelson, R. A., and Frenkel, A. W., 1974, Computer-assisted analyses of the near-infrared absorption spectra of chromatophores isolated from *Rhodospirillum molischianum*, *Fed. Proc. Fed. Am. Soc. Exp. Biol.* **33**:1330.

Newton, J. W., 1963, Composition of bacterial chromatophores, in: *Bacterial Photosynthesis* (H. Gest, A. San Pietro, and L. P. Vernon, eds.), pp. 469–474, Antioch Press, Yellow Springs, Ohio.

Newton, J. W., and Newton, G. A., 1957, Composition of the photoactive subcellular particles from *Chromatium*, *Arch. Biochim. Biophys.* **71**:250.

Niederman, R. A., 1974, Membranes of *Rhodopseudomonas sphaeroides*: Interactions of chromatophores with the cell envelope, *J. Bacteriol.* **117**:19.

Niederman, R. A., and Gibson, K. D., 1971, The separation of chromatophores from the cell envelope in *Rhodopseudomonas sphaeroides*, *Prep. Biochem.* **1**:141.

Niederman, R. A., Segen, B. J., and Gibson, K. D., 1972, Membranes of *Rhodopseudomonas sphaeroides*. I. Isolation and characterization of membrane fractions from extracts of aerobically and anaerobically grown cells, *Arch. Biochem. Biophys.* **152**:547.

Niederman, R. A., Mallon, D. E., and Langan, J. J., 1976, Membranes of *Rhodopseudomonas sphaeroides*. IV. Assembly of chromatophores in low-aeration cell suspension, *Biochim. Biophys. Acta* **440**:429.

Nieth, K.-F., and Drews, G., 1975, Formation of reaction centers and light-harvesting bacteriochlorophyll–protein complexes in *Rhodopseudomonas capsulata*, *Arch. Microbiol.* **104**:77.

Nieth, K.-F., Drews, G., and Feick, R., 1975, Photochemical reaction centers from *Rhodopseudomonas capsulata*, *Arch. Microbiol.* **105**:43.

Noël, H., van der Rest, M., and Gingras, G., 1972, Isolation and partial characterization of a P870 reaction center complex from wild type *Rhodospirillum rubrum*, *Biochim. Biophys. Acta* **275**:219.

Oelze, J., and Drews, G., 1970*a*, Der Einfluss der Lichtintensität und der Sauerstoffspannung auf die Differenzierung

der Membranen von *Rhodospirillum rubrum*, *Biochim. Biophys. Acta* **203**:189.

Oelze, J., and Drews, G., 1970*b*, Variations of NADH oxidase activity and bacteriochlorophyll contents during membrane differentiation in *Rhodospirillum rubrum*, *Biochim. Biophys. Acta* **219**:131.

Oelze, J., and Drews, G., 1972, Membranes of photosynthetic bacteria, *Biochim. Biophys. Acta* **265**:209.

Oelze, J., and Golecki, J. R., 1975, Properties of reaction center depleted membranes of *Rhodospirillum rubrum*, *Arch. Microbiol.* **102**:59.

Oelze, J., and Pahlke, W., 1976, The early formation of the photosynthetic apparatus in *Rhodospirillum rubrum*, *Arch. Microbiol.* **108**:281.

Oelze, J., Biedermann, M., and Drews, G. 1969*a*, Die Morphogenese des Photosyntheseapparates von *Rhodospirillum rubrum*. I. Die Isolierung und Charakterisierung von zwei Membransystemen, *Biochim. Biophys. Acta* **173**:436.

Oelze, J., Biedermann, M., Freund-Mölbert, E., and Drews, G., 1969*b*, Bacteriochlorophyllgehalt und Proteinmuster der Thylakoide von *Rhodospirillum rubrum* während der Morphogenese des Photosynthese-Apparates, *Arch. Mikrobiol.* **66**:154.

Oelze, J., Golecki, J. R., Kleinig, H., and Weckesser, J., 1975, Characterization of two cell-envelope fractions from chemotrophically grown *Rhodospirillum rubrum*, *Antonie van Leeuwenhoek*; *J. Microbiol. Serol.* **41**:273.

Okamura, M. Y., Steiner, L. A., and Feher, G., 1974, Characterization of reaction centers from photosynthetic bacteria. I. Subunit structure of the protein mediating the primary photochemistry in *Rhodopseudomonas sphaeroides* R-26, *Biochemistry* **13**:1394.

Olson, J. M., and Stanton, E. K., 1966, Absorption and fluorescence spectra of bacterial chlorophylls *in situ*, in: *The Chlorophylls* (G. R. Seeley and L. Vernon, eds.), pp. 381–398, Academic Press, New York.

Osborn, M. J., Gander, J. E., Parisi, E., and Carson, J., 1972, Mechanism of assembly of the outer membrane of *Salmonella typhimurium*: Isolation and characterization of cytoplasmic and outer membrane, *J. Biol. Chem.* **247**:3962.

Oyewole, S. H., and Holt, S. C., 1976, Structure and composition of intracytoplasmic membranes of *Ectothiorhodospira mobilis*, *Arch. Microbiol.* **107**:167.

Pardee, A. B., Schachman, H. K., and Stanier, R. Y., 1952, Chromatophores of *Rhodospirillum rubrum*, *Nature* (*London*) **169**:282.

Peters, G. A., and Cellarius, R. A., 1972, Photosynthetic membrane development in *Rhodopseudomonas sphaeroides*. II. Correlation of pigment incorporation with morphological aspects of thylakoid formation, *J. Bioenerg.* **3**:345.

Pfennig, N., 1967, Photosynthetic bacteria, *Annu. Rev. Microbiol.* **21**:285.

Prince, R. C., Hauska, G., and Crofts, A. R., 1974, Asymmetry of the bacterial membrane: The localization of cytochrome c_2 in *Rhodopseudomonas sphaeroides*, *Biochem. Soc. Trans.* **2**:534.

Prince, R. C., Baccarini-Melandri, A., Hauska, G. A., Melandri, B. A., and Crofts, A. R., 1975, Asymmetry of an energy transducing membrane: The location of cytochrome c_2 in *Rhodopseudomonas spheroides* and *Rhodopseudomonas capsulata*, *Biochim. Biophys. Acta* **387**:212.

Pucheu, N. L., Kerber, N. L., and Garcia, A. F., 1973, Some environmental factors influencing the state of the membranes isolated by gradient centrifugation from cell-free extracts of *Rhodopseudomonas viridis*, *FEBS Lett.* **33**:119.

Pucheu, N. L., Kerber, N. L., and Garcia, A. F., 1974, Comparative studies on membranes isolated from *Rhodopseudomonas viridis* grown in the presence and in the absence of yeast extract, *Arch. Microbiol.* **101**:259.

Racker, E., Burstein, C., Loyter, A., and Christiansen, R. O., 1970, The sidedness of the inner mitochondrial membrane, in: *Electron Transport and Energy Conservation* (J. M. Tager, S. Papa, E. Quagliariello, and E. C. Slater, eds.), pp. 235–252, Adriatica Editrice, Bari.

Reed, D. W., and Raveed, D., 1972, Some properties of the ATPase from chromatophores of *Rhodopseudomonas sphaeroides* and its structural relationship to the bacteriochlorophyll proteins, *Biochim. Biophys. Acta* **283**:79.

Richards, W. R., Wallace, R. B., Tsao, M. S., and Ho, E. 1975, The nature of a pigment–protein complex excreted from mutants of *Rhodopseudomonas sphaeroides*, *Biochemistry* **14**:5554.

Rosenberg, S. A., and Guidotti, G., 1968, The protein of human erythrocyte membranes. I. Preparation, solubilization, and partial characterization, *J. Biol. Chem.* **243**:1985.

Salton, M. R. J., 1967, Structure and function of bacterial cell membranes, *Annu. Rev. Microbiol.* **21**:417.

Schachman, H. K., Pardee, A. B., and Stanier, R. Y., 1952, Studies on the macromolecular organization of microbial cells, *Arch. Biochem. Biophys.* **38**:245.

Schmitz, R., 1967, Über die Zusammensetzung der pigmenthaltigen strukturen des Pupurbakteriums *Rhodopseudomonas sphaeroides*, *Z. Naturforsch.* **22b**:645.

Schnaitman, C. A., 1970a, Protein composition of the cell wall and cytoplasmic membrane of *Escherichia coli*, *J. Bacteriol.* **104**:890.

Schnaitman, C. A., 1970b, Comparison of the envelope protein compositions of several gram-negative bacteria, *J. Bacteriol.* **104**:1404.

Schnaitman, C. A., 1973, Outer membrane proteins of *Escherichia coli*. I. Effect of preparative conditions on the migration of protein in polyacrylamide gels, *Arch. Biochem. Biophys.* **157**:541.

Schröder, J., and Drews, G., 1968, Quantitative Bestimmung der Fettsäuren von *Rhodospirillum rubrum* und *Rhodopseudomonas capsulata* während der Thylakoidmorphogenese, *Arch. Mikrobiol.* **64**:59.

Schröder, J., Biedermann, M., and Drews, G., 1969, Die Fettsäuren in ganzen Zellen, Thylakoiden und Lipopolysacchariden von *Rhodospirillum rubrum* und *Rhodopseudomonas capsulata*, *Arch. Mikrobiol.* **66**:273.

Schwenker, U., and Gingras, G., 1973, A carotenoprotein from chromatophores of *Rhodospirillum rubrum*, *Biochem. Biophys. Res. Commun.* **51**:94.

Schwenker, U., St.-Onge, M., and Gingras, G., 1974, Chemical and physical properties of a carotenoprotein

from *Rhodospirillum rubrum*, *Biochim. Biophys. Acta* **351**:246.

Segen, B. J., and Gibson, K. D., 1971, Deficiencies of chromatophore proteins in some mutants of *Rhodopseudomonas sphaeroides* with altered carotenoids, *J. Bacteriol.* **105**:701.

Shaw, M. A., 1974, Chromatophore morphogenesis in *Rhodopseudomonas sphaeroides*, Ph.D. thesis, Simon Fraser University, Burnaby, B.C., Canada.

Shaw, M. A., and Richards, W. R., 1971, Evidence for the formation of membranous chromatophore precursor fractions in *Rhodopseudomonas sphaeroides*, *Biochem. Biophys. Res. Commun.* **45**:863.

Simon, S. R., 1972, Preparation and properties of membrane fractions from *Rhodospirillum rubrum*, Ph.D. thesis, Rockefeller University.

Simon, S. R., and Siekevitz, P., 1973, Biochemical properties of purified membrane preparations from *Rhodospirillum rubrum*, in: *Mechanisms in Bioenergetics* (G. F. Azzone, L. Ernster, S. Papa, E. Quagliariello, and N. Siliprandi, eds.), pp. 3–31, Academic Press, New York.

Singer, S. J., and Nicolson, G. L., 1972, The fluid mosaic model of the structure of cell membranes, *Science* **175**:720.

Sistrom, W. R., 1962a, Observations on the relationship between the formation of photopigments and the synthesis of protein in *Rhodopseudomonas sphaeroides*, *J. Gen. Microbiol.* **28**:599.

Sistrom, W. R., 1962b, The kinetics of the synthesis of photopigments in *Rhodopseudomonas sphaeroides*, *J. Gen. Microbiol.* **28**:607.

Sistrom, W. R., 1964, Calculation of the absorption coefficients of the individual components of the spectra of bacteriochlorophyll in chromatophore preparations of *Rhodopseudomonas sphaeroides*, *Biochim. Biophys. Acta* **79**:419.

Sistrom, W. R., 1966, The spectrum of bacteriochlorophyll *in vivo*: Observations on mutants of *Rhodopseudomonas sphaeroides* unable to grow photosynthetically, *Photochem. Photobiol.* **5**:845.

Sistrom, W. R., and Clayton, R. K., 1964, Studies on a mutant of *Rhodopseudomonas sphaeroides* unable to grow photosynthetically, *Biochim. Biophys. Acta* **88**:61.

Stanier, R. Y., 1963, The organization of the photosynthetic apparatus in purple bacteria, in: *General Physiology of Cell Specialization* (D. Mazia and A. Tyler, eds.), pp. 242–252, McGraw-Hill, New York.

Steck, T. L., Straus, J. H., and Wallach, D. F. H., 1970, A model for the behavior of vesicles in density gradients: Implication for fractionation, *Biochim. Biophys. Acta* **203**:385.

Steiner, L. A., Okamura, M. Y., Lopes, A. D., Moskowitz, E., and Feher, G., 1974, Characterization of reaction centers from photosynthetic bacteria. II. Amino acid composition of the reaction center protein and its subunits in *Rhodopseudomonas sphaeroides* R-26, *Biochemistry* **13**:1403.

Steiner, S., Sojka, G. A., Conti, S. F., Gest, H., and Lester, R. L., 1970, Modification of membrane composition in growing photosynthetic bacteria, *Biochim. Biophys. Acta* **203**:571.

Swank, R. T., and Munkres, K. D., 1971, Molecular weight analysis of oligopeptides by electrophoresis in polyacrylamide gel with sodium dodecyl sulfate, *Anal. Biochem.* **39**:462.

Sykes, J., 1971, Centrifugal techniques for the isolation and characterization of sub-cellular components from bacteria, in: *Methods in Microbiology*, Vol. 5B (J. R. Norris and D. W. Ribbons, eds.), pp. 55–207, Academic Press, New York.

Takacs, B. J., and Holt, S. C., 1971*a*, *Thiocapsa floridana*; a cytological, physical and chemical characterization. I. Cytology of whole cells and isolated chromatophore membranes, *Biochim. Biophys. Acta* **233**:258.

Takacs, B. J., and Holt, S. C., 1971*b*, *Thiocapsa floridana*; a cytological, physical and chemical characterization. II. Physical and chemical characteristics of isolated and reconstituted chromatophores, *Biochim. Biophys. Acta* **233**:278.

Takemoto, J., 1974, Kinetics of photosynthetic membrane protein assembly in *Rhodopseudomonas sphaeroides*, *Arch. Biochem. Biophys.* **163**:515.

Takemoto, J., and Huang Kao, M. Y. C., 1977, Effects of incident light levels on photosynthetic membrane polypeptide composition and assembly in *Rhodopseudomonas sphaeroides*, *J. Bacteriol.* **129**:1102.

Takemoto, J., and Lascelles, J., 1973, Coupling between bacteriochlorophyll and membrane protein synthesis in *Rhodopseudomonas sphaeroides*, *Proc. Natl. Acad. Sci. U.S.A.* **70**:799.

Takemoto, J., and Lascelles, J., 1974, Function of membrane proteins coupled to bacteriochlorophyll synthesis. Studies with wild type and mutant strains of *Rhodopseudomonas sphaeroides*, *Arch. Biochem. biophys.* **163**:507.

Taniguchi, S., and Kamen, M. D., 1965, The oxidase system of heterotrophically grown *Rhodospirillum rubrum*, *Biochim. Biophys. Acta* **96**:395.

Tauschel, H.-D., and Drews, G., 1967, Thylakoid Morphogenese bei *Rhodopseudomonas palustris*, *Arch. Mikrobiol.* **59**:381.

Throm, E., Oelze, J., and Drews, G., 1970, The distribution of NADH oxidase in the membrane system of *Rhodospirillum rubrum*, *Arch. Mikrobiol.* **72**:361.

Tuttle, A. L., and Gest, H., 1959, Subcellular particulate systems and the photochemical apparatus of *Rhodospirillum rubrum*, *Proc. Natl. Acad. Sci. U.S.A.* **45**:1261.

van der Rest, M., Noël, H., and Gingras, G., 1974, An immunological and electrophoretic study of *Rhodospirillum rubrum* chromatophore fragments, *Arch. Biochem. Biophys.* **164**:285.

Vatter, A. E., and Wolfe, R. S., 1958, The structure of photosynthetic bacteria, *J. Bacteriol.* **75**:480.

Vernon, L. P., and Garcia, A. F., 1967, Pigment–protein complexes derived from *Rhodospirillum rubrum* chromatophores by enzymatic digestion, *Biochim. Biophys. Acta* **143**:144.

Wassink, E. C., Katz, E., and Dorrestein, R., 1939, Infrared absorption spectra of various strains of purple bacteria, *Enzymologia (Acta Biocatalytica)* **7**:113.

Weber, K., and Osborn, M., 1969, The reliability of molecular weight determinations by dodecyl sulfate–polyacrylamide gel electrophoresis, *J. Biol. Chem.* **244**:4406.

Weckesser, J., Drews, G., and Tauschel, H.-D., 1969, Zur Feinstruktur und Taxonomie von *Rhodopseudomonas gelatinosa*, *Arch. Mikrobiol.* **65**:346.

White, D. A., Lennarz, W. J., and Schnaitman, C. A. 1972, Distribution of lipids in the wall and cytoplasmic membrane subfractions of the cell envelope of *Escherichia coli*, *J. Bacteriol.* **109**:686.

Worden, P. B., and Sistrom, W. R., 1964, The preparation and properties of bacterial chromatophore fractions, *J. Cell Biol.* **23**:135.

Yamashita, J., and Kemen, M. D., 1969, Observations on distribution of NADH oxidase in particles from dark-grown and light-grown *Rhodospirillum rubrum*, *Biochem. Biophys. Res. Commun.* **34**:418.

CHAPTER 6

A Comparative Review of Photochemical Reaction Center Preparations from Photosynthetic Bacteria

Gabriel Gingras

1. Introduction

The isolation of the photochemical reaction center from photosynthetic bacteria had considerable theoretical significance. Its mere possibility imposed new and better-defined boundaries on the structure-and-function models of the photosynthetic unit. It showed that charge separation did not occur at impurity centers in a crystal-like chlorophyll matrix (Calvin, 1956, 1961; Arnold and Sherwood, 1957; Arnold, 1965). It stressed instead that the primary photochemistry occurs within chlorophyll molecules functionally inseparable from a specific protein, which is itself a constituent of a biological membrane, the chromatophore. It thus opened a new era in the study of the first events of photosynthesis, the era of biochemistry and of molecular biology.

This chapter is a survey of various reaction center preparations from photosynthetic bacteria, and of the methods of their isolation. For a detailed account of the chemical and optical properties of reaction centers from *Rhodopseudomonas sphaeroides*, see Chapter 19.

2. Historical Review: Isolation of the Photochemical Reaction Center

Two lines of evidence indicated that the bacterial photochemical reaction center must be of a particulate

nature and could therefore theoretically be isolated. The first line was provided by the observation that most of the light-harvesting or antenna bacteriochlorophyll (Bchl) could be selectively destroyed in several bacterial species while their photochemical reaction centers remained intact. Clayton (1963) achieved that result with a carotenoidless mutant (R-26) of *Rp. sphaeroides* either by pheophytinization *in vivo* or by illuminating isolated chromatophores in the presence of Triton X-100. Independently, Loach *et al.* (1963) and Kuntz *et al.* (1964) obtained a similar effect by treating chromatophores from wild-type strains of *Rhodospirillum rubrum*, *Rp. sphaeroides*, and *Chromatium vinosum* with K_2IrCl_6. Over 95% of the antenna Bchl could be destroyed without impairing the photochemical reaction of P_{870} (Clayton, 1963), alias $P_{0.44}$ (Loach *et al.*, 1963; Kuntz *et al.*, 1964), associated with the reaction center. These observations indicated that the photochemically active Bchl must be in a specialized environment.

The other line of evidence was genetic: a P_{870}-minus mutant of *Rp. sphaeroides* with a normal antenna pigment complement was isolated by Sistrom and Clayton (1964). It was found to be an obligate heterotroph. P_{870} thus appeared to be in close association with a molecular species.

The isolation of particles enriched in photochemical reaction center was a natural outcome of these experiments. This isolation was announced in 1968 by two groups working independently. Reed and Clayton (1968) worked with the carotenoidless mutant R-26 of *Rp. sphaeroides*, and Gingras and Jolchine (1969)

Gabriel Gingras · Département de Biochimie, Université de Montréal, Montréal, Québec, Canada

with the carotenoidless mutant G-9 of *Rs. rubrum*. In both cases, the membranes were solubilized with the detergent Triton X-100 and the particles fractionated by density-gradient centrifugation.

These preparations were of high minimal molecular weights ($\approx 10^6$) and contained cytochromes (Gingras and Jolchine, 1969) and high amounts of lipid, ubiquinone, and other substances (Reed, 1969). They were enriched in photochemical activity with respect to their parent membranes, but obviously could not be said to represent isolated photochemical reaction center. In addition, the methods by which they were obtained were inapplicable even to the wild-type strains of the same species, not to speak of other organisms.

Several techniques were rapidly developed, however, to overcome these drawbacks. These techniques made use of different detergents and of new fractionation methods. The three main detergents that were introduced were the anionic sodium dodecyl sulfate (SDS), the quaternary ammonium salt cetyltrimethylammonium bromide (CTAB), and a nonionic detergent, dodecyldimethylamine *N*-oxide (LDAO).

Using SDS in conjunction with chromatography on hydroxylapatite, Thornber *et al.* (1969) were able to obtain a photochemical reaction center preparation from *Rp. viridis*. A similar technique applied by Thornber (1970) to *Chr. vinosum* led to a P_{870}-enriched particle. Segen and Gibson (1971) used SDS solubilization followed by density-gradient centrifugation to extract a photochemical reaction center fraction from wild-type *Rp. sphaeroides* as well as from some mutants with altered carotenoids. Slooten (1972*a*) solubilized the chromatophores isolated from wild-type *Rp. sphaeroides* with SDS. Fractionation was achieved by differential and by density-gradient centrifugation. After initial lack of success, this technique was also found applicable to *Rs. rubrum* (wild-type) (Slooten, 1972*b*). Smith *et al.* (1972) were also successful in preparing photochemical reaction center from wild-type *Rs. rubrum* using SDS in the presence of a large excess of ascorbate. Purification was achieved by density-gradient centrifugation and by electrophoresis.

The quaternary ammonium salt CTAB was used by Jolchine *et al.* (1969) and by Reiss-Husson and Jolchine (1972) for the extraction and purification of the photochemical reaction center from wild-type (strain Y) *Rp. sphaeroides*. Here again, purification was achieved by density-gradient centrifugation.

The detergent that has proved to have the most general applicability and that has led to the purest and best-characterized preparations is LDAO. Its use was introduced independently by Feher (1971) and by Clayton and Wang (1971) for the extraction of the photochemical reaction center from the R-26 strain (carotenoidless) of *Rp. sphaeroides*. Purification was obtained by ammonium sulfate fractionation and by chromatography on molecular sieves (Clayton and Wang, 1971) or on DEAE–cellulose (Feher, 1971). An analogous method based on the use of the same detergent was developed by Noël *et al.* (1972) for wild-type *Rs. rubrum*. This method, however, was inapplicable to the carotenoidless G-9 mutant of the same organism. Therefore, other procedures were designed (Wang and Clayton, 1973; Okamura *et al.*, 1974; Vadeboncoeur and Gingras, in prep.) that led to purified but unstable preparations except, at least in our hands, when LDAO was replaced with Triton X-100 immediately after the extraction. More recently, Jolchine and Reiss-Husson (1974) succeeded in extracting the photochemical reaction center from wild-type *Rp. sphaeroides* with LDAO. Finally, with this detergent, Lin and Thornber (1975) obtained a greatly purified preparation of photochemical reaction center from *Chr. vinosum*.

A word should be added about some preparations that are, in fact, particles enriched in photochemical reaction center. Among them, one should mention the particles obtained by Ke and Chaney (1971) from carotenoid-deficient *Chr. vinosum* using Triton X-100. Such particles were also obtained from certain Chlorobiaceae by Fowler *et al.* (1971, 1973) and by Olson *et al.* (1973) without detergent solubilization of the membranes (see Chapter 8).

3. Comparison of Photochemical Reaction Center Preparations from Different Photosynthetic Bacteria

A given preparation is defined by three different parameters: the species from which it came, the method of preparation, and the properties of the material thereby obtained. In this section, we shall review and compare various preparations, using these parameters as standards of reference. For the sake of convenience, the subject matter will be grouped according to the bacterial species considered beginning with the wild-type strains. No detailed preparation procedures will be reproduced here, since they can be found in the original literature. Outlines will be given, however, stressing some important points as guidance to the reader. In addition, each preparation will be characterized by a few properties used as criteria of purity. Indeed, purity is the main concern of this section.

This material is summarized in Table 1, to which frequent reference will be made throughout this chapter. Table 1 is not exhaustive; it lists only the properties of the first and of well-purified preparations from each species. Two of these properties pertain to the preparation procedure: the detergent used for solubilizing the membranes and the overall yield of the method. The latter is given as the percentage of the photochemical activity recovered from the starting material (usually chromatophores). The other properties given are properties of the preparations themselves, and are criteria of purity. The first and the most obvious is the absence of antenna pigments. This criterion is generally met, although some preparations may contain residual amounts of antenna Bchl.

The next important property is the centesimal composition for a given dry weight of the preparation. This is determined after removal of the solubilizing detergent. With some detergents such as Triton X-100, this may prove to be a difficult task. Table 1 lists only the centesimal protein content, since protein is by far the most important component (at least 90%) of the well-purified preparations.

Another important criterion is the specific activity, which is usually expressed as the activity per unit protein content. Since the extinction coefficients of the two best-studied photochemical reaction centers are now known (Straley et al., 1973; van der Rest and Gingras, 1974), we will use the parameter "minimal molecular weight" expressed as grams of protein per mole of P_{870}. One mole of P_{870} is defined as one electron-transfer equivalent of material. Each reaction-center particle, containing four molecules of Bchl, effects the transfer of one electron in response to a brief saturating flash of light. All the values reported here for preparations from Rp. sphaeroides and from Rs. rubrum have been normalized to the more recent determinations of $\varepsilon_{870}^{red-ox} = 112$ and $126 \text{ mM}^{-1} \text{ cm}^{-1}$, respectively. Whenever possible, the amount of protein should be evaluated from the amino acid composition. Provided certain precautions are taken (Noël et al., 1972), however, the method of Lowry et al. (1951) is probably accurate within 10% for this protein.

The A_{280}/A_{802} ratio is also proportional to the amount of protein corresponding to a given quantity of photochemical reaction center. The absorption at 280 nm is due to aromatic amino acids and probably also to a nonnegligible contribution from the pigments attached to the protein (Loach et al., 1971). The absorption at 802 nm is proportional to the amount of photochemical reaction center. An equally useful ratio would be A_{280}/A_{870}, provided P_{870} is fully reduced. However, since A_{802} is less sensitive to the redox state of the preparation, it has been more widely used as a reference. Although it is a handy purity check, the A_{280}/A_{802} ratio should not be taken as an absolute characteristic of any given preparation unless the composition of the medium is specified. The latter has

been found to have a considerable effect on the A_{280}/A_{802} ratio in the photochemical reaction center preparation from wild-type Rs. rubrum (Noël and Gingras, in prep.). In 50 mM phosphate buffer (pH 7.0)–0.03% LDAO, $A_{280}/A_{802} = 1.02$. From the same preparation in 10 mM Tris-Cl (pH 8.0)–0.1% LDAO, the ratio is 1.22. If the latter medium also contains 0.1 M NaCl, the ratio becomes about 1.0. These variations can probably be explained by conformational changes in the protein–detergent complex. Another factor is light scattering, which tends to increase this ratio. But in well-purified preparations, this is probably negligible, provided sufficient amounts of detergents are used for solubilization.

Electrophoresis on polyacrylamide gel in the presence of SDS is one of the most sensitive existing methods for determining the purity of a protein. With chromatophore proteins, a few tenths of a microgram of a single component can be detected. Therefore, when a photochemical reaction center preparation shows only the three usual polypeptides with protein loads of 15–50 μg, its purity may be conservatively estimated to be 95%. This is the purity level of preparations described as "Excellent" in the "Purity by electrophoresis" column in Table 1.

3.1. Rhodospirillaceae

3.1.1. *Rhodopseudomonas sphaeroides*

This species has been the source of the best-studied photochemical reaction center. While the purest and best-characterized preparations were obtained from mutant strain R-26, well-purified ones also came from wild-type strains.

(a) Wild-Type. Jolchine et al. (1969) and Reiss-Husson and Jolchine (1972) solubilized various membrane fractions with the detergent CTAB. The purification procedure, which has evolved with time, at first used centrifugation on a layer of sucrose followed by either high-speed centrifugation, ammonium sulfate precipitation, or molecular sieve filtration. The bacteria were grown on either "normal" or "low-iron" medium. Preparations of decreasing minimal molecular weight were obtained from "crude chromatophores," "purified chromatophores," and "light particles," respectively (see Table 2). "Light particles" ("low-iron" bacteria) yielded the purest material, which had an A_{280}/A_{802} ratio of 1.5. These preparations were free from light-harvesting Bchl, but contained cytochromes. One peculiarity of the procedure was that the main purification step with respect to specific activity was achieved before solubilization of the membranes. Starting from "light particles" was a lengthy procedure, however, and the final yield of

Table 1. Purity of Various Photochemical Reaction Center Preparations from Different Bacteria[a]

Organism and references	Protein (% dry wt.)	Minimal mol. wt. × 10⁻³	$\dfrac{A_{280}}{A_{802}}$	Antenna Bchl	Cyto-chromes	Purity by electro-phoresis	Detergent	Yield[b]
Rp. sphaeroides (wild-type)								
Jolchine et al. (1969)	NR	246	NR	None	Present	NR	CTAB	NR
Slooten (1972a)	NR	144[c]	NR	None	Present	NR	Triton	NR
Jolchine and Reiss-Husson (1974, 1975)	80–90	86	1.4	None	Traces	NR	LDAO	~20%[d]
Rp. sphaeroides (R-26) (carot.⁻)								
Reed (1969). Reed et al. (1970)[e]	69	530	NR	None	Present	NR	Triton	50%
Clayton and Wang (1971)	NR	NR	1.3	None	None	Excellent	LDAO	25%
Okamura et al. (1974)	90	90	1.22	None	None	Excellent	LDAO	~30[f]
Rs. rubrum (wild-type)								
Smith et al. (1972)	NR	NR	3.1	Sm. amts.	Present	Contamin.	SDS	NR
Noël et al. (1972)	90	90	1.22	None	None	Excellent	LDAO	35[g], 50%[h]
Rs. rubrum (G-9) (carot.⁻)								
Gingras and Jolchine (1969)	NR	1000[i]	NR	None	Present	NR	Triton	20%
Wang and Clayton (1973)	NR	210	2	None	None	NR	LDAO	NR
Okamura et al. (1974)	NR	NR	1.25	None	None	Excellent	LDAO	NR
Vadeboncoeur and Gingras (unpublished)	NR	85	1.25	None	None	Excellent	LDAO	50%

Rp. capsulata Ala[+] (carot.[−]) Nieth *et al.* (1975)	NR	145	1.25	None	None	Good	LDAO	NR
Rp. gelatinosa (carot.[−]) Clayton and Clayton (1978)	NR	150	1.8	None	None	Good	LDAO	NR
Rp. viridis (wild-type)								
Thornber *et al.* (1969)	NR	NR	3.7[j]	None	Present	"Reason."	SDS	NR
Clayton and Clayton, 1978	NR	190	2.1	None	Present	Good	LDAO	NR
Chr. vinosum (carot. + and −) Lin and Thornber (1975)	NR	NR	1.6	None	Present	Contamin.	LDAO	NR

[a] Abbreviations: (NR) not reported; (sm.amts.) small amounts; (contamin.) contaminated; (reason.) reasonable; (carot.) carotenoid.

[b] Yield expressed as percentage of P_{870} recovered from the starting material.

[c] AUT-RC preparation.

[d] Recalculated from the author's data assuming 30% of the chromatophore proteins are photochemical reaction center protein.

[e] Preparation first described by Reed and Clayton (1968).

[f] Recalculated from the author's data assuming that in chromatophores, 2% of A_{865} is due to P_{870}.

[g] Fractionation with a saturated solution of ammonium sulfate.

[h] Fractionation with solid ammonium sulfate.

[i] Based on total nitrogen content.

[j] A_{280}/A_{830} ratio.

Table 2. Minimal Molecular Weights of Photochemical Reaction Center Preparations from Different Membrane Fractions[a]

Preparation	Normal bacteria	Low-iron bacteria
Crude chromatophores	446,000	926,000
Purified chromatophores	274,000	—
Light particles	219,000	133,000

[a] Recalculated from the data of Reiss-Husson and Jolchine (1972).

photochemical reaction center was poor (Reiss-Husson and Jolchine, 1972).

A more recent version of this method avoids some of these limitations (Jolchine and Reiss-Husson, 1974). After CTAB solubilization of the purified chromatophores and low-speed centrifugation, the crude photochemical reaction center preparation was concentrated by ammonium sulfate precipitation. It was then purified by either of two methods: (1) ammonium sulfate fractionation with decreasing salt concentration according to Feher (1971) or (2) high-speed centrifugation for 24 hr, followed by chromatography on Sepharose 6B.

The minimal molecular weight of the material obtained by method 1 was 102,000; that obtained by method 2 had a minimal molecular weight of 132,000. Method 1 therefore appears to be preferable. In all cases, CTAB must be removed at an early stage of the preparation and replaced by a milder detergent such as Triton X-100 or Brij 35 for maintaining solubility.

The same group (Jolchine and Reiss-Husson, 1974) was also successful with LDAO, which had been shown to be entirely satisfactory for other organisms (Feher, 1971; Clayton and Wang, 1971; Noël et al., 1972), but not for wild-type Rp. sphaeroides. The solubilization was carried out at 26°C in 100 mM phosphate (pH 7.5)–0.25% LDAO. Tris buffer cannot be substituted for phosphate (Reiss-Husson, personal communication). Purification was obtained by methods 1 and 2 described above. Both methods yielded preparations of minimal molecular weights (86,000) comparable to that of other LDAO preparations (see Table 1). The authors state that both preparations are dissociated into three polypeptide bands by SDS–polyacrylamide gel electrophoresis, but they do not use this method as a criterion of purity. Interestingly, both preparations contain high amounts of lipid (50–60% of the dry weight), more than half of which is phospholipid.

The method developed by Slooten (1972a) consists in incubating the chromatophores with 3% SDS in 50 mM Tris (pH 8.0)–10 μM MgCl$_2$ for 2 hr at room temperature. After this time, NaCl was added (0.2 M final concentration), and the suspension was layered on 0.5 M sucrose–0.2 M NaCl and centrifuged for 4 hr at 200,000g. The photochemical reaction center particles in the supernatant were centrifuged again on a linear sucrose density gradient (0.4–2 M). This yielded a so-called "SDS-RC preparation" (minimal molecular weight 244,000), which could be further purified by another sucrose density-gradient centrifugation at alkaline pH, in the presence of 1 M urea and 0.3% Triton X-100. This last step, analogous to a procedure used by Loach et al. (1970) to obtain sub-chromatophore particles, yielded a more purified preparation called "AUT-RC." The latter preparation had a minimal molecular weight of 144,000 and contained 0.25 mol cytochrome, 1.2 mol ubiquinone, and negligible amounts of antenna Bchl. Polyacrylamide gel electrophoresis in the presence of 0.3% SDS and 8 M urea was stated to reveal three polypeptide bands. The presence of possible contaminants was not discussed. The A_{280}/A_{802} could not be measured in the AUT-RC preparation because of the absorption by Triton X-100 in the ultraviolet. For the SDS-RC preparations, this ratio was 2.0 in the best case (recalculated from the author's data). See Table 1 for a summary.

No specific activity was reported for the SDS preparation of Segen and Gibson (1971). Although it was nearly free of antenna Bchl, polyacrylamide gel electrophoresis in the presence of SDS showed it to be considerably contaminated by foreign proteins.

(b) Carotenoid-Deficient Mutant R-26. For historical reasons, this organism has supplied the most extensively studied isolated photochemical reaction center. Two basically similar procedures were developed independently by Clayton and Wang (1971) and by Feher (1971). Clayton and Wang (1971) added 1% LDAO (final concentration) to twice-washed chromatophores ($A_{870} = 50$) resuspended in 10 mM Tris-Cl (pH 7.5). This suspension was centrifuged on a discontinuous sucrose density gradient (0.5 and 1 M) for 90 min at 200,000g.

The uppermost fraction was collected and subjected first to two successive ammonium sulfate precipitations at 50% saturation and then to a fractionation between 30 and 50% saturation. The last floating precipitate was redissolved in 10 mM Tris-Cl (pH 7.5)–0.3% LDAO and purified by molecular-sieve filtration on Sephadex G-200 or on BioGel in the presence of LDAO 0.03%.

The solubilization method employed by Feher (1971) and later by Okamura et al. (1974) was similar to that just described. The chromatophores ($A_{865} = 40$) were solubilized at 4°C in 10 mM Tris-Cl (pH 8.0) containing 1% LDAO. This solution was centrifuged at high speed, and the resulting supernatant was precipitated with ammonium sulfate (22% wt./vol. in H$_2$O) at 20°C. Initially, (Feher, 1971), the floating precipitate was redissolved and purified by successive precipitations with decreasing ammonium sulfate concentrations (18–13% wt./vol.). Alternatively, this could be replaced by adsorption chromatography on DEAE cellulose.

An ingenious modification of this method was introduced by Okamura *et al.* (1974). The crude photochemical reaction center preparation obtained by ammonium sulfate precipitation was resuspended in 10 mM Tris-Cl (pH 8.0) buffer to give $A_{802} = 5$. After clarification, celite and ammonium sulfate (24% wt./vol. final concentration) were added, and the slurry was packed in a column. The protein was eluted with a linear ammonium sulfate gradient of decreasing concentration (24–15% wt./vol.). The eluted protein fraction was dialyzed and chromatographed on DEAE cellulose in Tris-Cl (pH 8.0)–0.1% LDAO buffer.

The preparations described by both groups are free of contaminating antenna pigments. The photochemical reaction center isolated by Clayton and Wang (1971) had a A_{280}/A_{802} ratio of 1.3. Electrophoresis on polyacrylamide gel in the presence of SDS revealed the three polypeptides assigned to the photochemical reaction center without any evidence of noticeable protein contaminants (Clayton and Haselkorn, 1972). The purity of the preparation described by Feher (1971) and by Okamura *et al.* (1974) has been established by a more thorough chemical analysis. Its characteristics are as follows: The protein content accounts for at least 90% of the dry weight. Its A_{280}/A_{802} ratio is 1.22 in 10 mM Tris-Cl (pH 8.0) buffer containing 0.1% LDAO. Polyacrylamide gel electrophoresis in the presence of SDS reveals no components other than the three polypeptides. Its minimal molecular weight is 90,000. This value is based on a protein content obtained from the amino acid composition and on a spectrophotometric assay of P$_{870}$ using the extinction coefficient determined by Straley *et al.* (1973).

3.1.2. *Rhodospirillum rubrum*

(a) Wild-Type. Smith *et al.* (1972) solubilized isolated chromatophores with 1% SDS in the presence of a large excess of ascorbate. Purification was effected by means of sucrose density-gradient centrifugation followed by gel filtration on Sephadex G-100. This preparation contained cytochromes, small amounts of Bchl–protein complex, as well as some colorless proteins, and had an apparent molecular weight of about 100,000. The best A_{280}/A_{802} ratio obtained was 3.1. Electrophoresis on polyacrylamide

gel in the presence of 0.05% SDS revealed several polypeptide bands, one of which (apparent molecular weight 35,000) was photoactive. This band may perhaps be an aggregate of the *b* and *c* polypeptides (the M and L polypeptides in Feher's terminology; Chapter 19) as observed by Okamura *et al.* (1974). This preparation does not appear to meet the criteria for an isolated photochemical reaction center, and is probably more akin to an electron-transport particle (see Table 1).

In the method of Noël *et al.* (1972), twice-washed chromatophores (final $A_{880} = 37.5$) were suspended in 50 mM phosphate buffer (pH 7.0)–0.25% LDAO at 4°C for 1 hr. The LDAO concentration was then brought to 0.1% by dilution and the preparation centrifuged (1.5 hr, 105,000*g*). The suspension was equilibrated at 20°C and subjected to ammonium sulfate fractionation between 35 and 45% saturation. Although the original procedure called for the use of a saturated ammonium sulfate solution, addition of the solid salt (21 and 28%, wt./vol.) may be preferred, since it leads to smaller final volumes and to higher yields. Likewise, in the original procedure, further purification was obtained by a 40-hr dialysis, which resulted in a selective precipitation of the photochemical reaction center through removal of the detergent. Since this step does not produce consistently pure preparations, it has been replaced in our laboratory by DEAE–cellulose chromatography. About 20 mg protein is applied to a 16 × 300 mm column equilibrated with 10 mM Tris-Cl (pH 8.0)–0.1% LDAO. The column is washed with 2 vol. of this buffer and then with at least 6 vol. 25 mM NaCl–10 mM Tris-Cl (pH 8.0)–0.1% LDAO, and the photochemical reaction center is eluted with 125 mM NaCl–10 mM Tris-Cl (pH 8.0)–0.1% LDAO. After elution, the preparation must be rapidly dialyzed or filtered on Sephadex G-25 to replace the last solution with 10 mM Tris-Cl (pH 8.0); otherwise, the combined effects of high concentrations of salt and of detergent cause instability of the preparation. The material may be kept as a precipitate after removal of the detergent by dialysis or in a solution of 10 mM Tris-Cl (pH 8.0)–0.05% LDAO.

After a 190-hr dialysis against H$_2$O to remove most of the detergent, the preparation was found to contain about 90% protein and 7% pigment on a dry weight basis. Using the ε_{868} of 143 mM^{-1}·cm^{-1} determined by van der Rest and Gingras (1974), its minimal molecular weight was estimated to be 90,000. Its A_{280}/A_{802} ratio was 1.22 (Noël and Gingras, in prep.). This preparation is free of antenna Bchl and of cytochromes, and contains only traces of lipid material (Noël *et al.*, 1972). Only the three characteristic polypeptides were revealed by polyacrylamide gel electrophoresis in the presence of SDS in Tris or in phosphate–urea buffer (Noël et al., 1972; van der Rest et al., 1974).

(b) Carotenoidless Strain G-9. Wang and Clayton (1973) solubilized isolated chromatophores of this organism with 0.3% LDAO in 50 mM Tris-Cl (pH 7.4). The suspension was centrifuged on a layer of 0.6 M sucrose, and the resulting supernatant was collected and filtered on Sephadex G-200. The preparation thus obtained was purified severalfold with respect to that of Gingras and Jolchine (1969) (see Table 1). Its A_{280}/A_{802} ratio was about 2. Its minimal molecular weight was about 210,000 as recalculated from the author's data, using the value of 126 mM^{-1}·cm^{-1} for $\varepsilon_{870}^{red-ox}$ as found by van der Rest and Gingras (1974).

Okamura et al. (1974) resuspended isolated chromatophores (final $A_{865} = 40$) in 10 mM phosphate (pH 7.0) containing 0.4% LDAO. The supernatant of a high-speed centrifugation (60 min, 250,000g) was precipitated with 22% (wt./vol.) ammonium sulfate and packed with celite on a column. The column was washed with 1 vol. 23% (wt./vol.) ammonium sulfate in 10 mM phosphate (pH 7.0)–0.04% LDAO and then with 2 vol. 23% (wt./vol.) ammonium sulfate in Tris-Cl (pH 8.0) without LDAO. The photochemical reaction center was eluted from the celite with a decreasing concentration gradient of ammonium sulfate (24–0% wt./vol.) in the last-named buffer.

This material had a A_{280}/A_{802} ratio of 1.25. Polyacrylamide gel electrophoresis in the presence of SDS revealed only the three polypeptides also observed with the wild-type preparation (Noël et al., 1972; van der Rest et al., 1974). The minimum molecular weight was not reported.

Finally, Vadeboncoeur and Gingras (in prep.) have obtained a preparation that appears to be pure by present standards. Moreover, it is stable except under intense light and can be kept for several months in the cold and in the dark. The chromatophore suspension ($A_{870} = 38$) in 25 mM phosphate buffer (pH 7.0) is treated at 4°C with 0.28% (wt./vol.) LDAO. The suspension is then diluted with 2 vol. 60 mM phosphate (pH 7.0)–0.1% (wt./vol.) Triton X-100, and is centrifuged for 90 min at 100,000 g. The supernatant is dialyzed against 10 mM Tris-Cl (pH 7.5) (called "Tris" hereafter), and the protein is adsorbed on DEAE Sephadex A-50. The resin is washed with Tris–100 mM NaCl and eluted with a Tris–1 M NaCl–0.1% (wt./vol.) Triton X-100 solution. After dialysis, the material is adsorbed on a DEAE–cellulose column. The column is washed with a Tris–20 mM NaCl–0.1% Triton solution, and the photochemical reaction center is eluted with Tris–0.1% Triton–125 mM NaCl. The NaCl is dialyzed out, and the

protein is applied on another DEAE–cellulose column. After the column is washed as above, the protein is eluted as a sharp band ($A_{865} = 5$) with Tris–1 M NaCl–0.1% Triton X-100. NaCl is immediately removed by dialysis against Tris. The yield in P_{870} from the chromatophores is typically about 50%.

Polyacrylamide gel electrophoresis in Tris buffer containing 0.1% SDS shows the classic pattern of three polypeptides without evidence of contaminants. The A_{280}/A_{802} ratio is 1.25 (in the absence of Triton X-100). From its $\varepsilon_{865} = 143$ mM^{-1}·cm^{-1} (van der Rest and Gingras, 1974) and from its amino acid composition, this preparation has a minimal molecular weight of 85,000. It contains no cytochromes and only traces of lipids. Replacement of LDAO by Triton X-100 at an early stage of the preparation is important for good stability.

(c) Rhodopseudomonas viridis. This organism differs from other better-known photosynthetic bacteria in that it contains Bchl b and has a near-IR absorption maximum at 1010 nm. A method of preparing its photochemical reaction center was described by Thornber et al. (1969) and by Thornber (1971). In this method, the cells were ruptured by sonic oscillation and treated with SDS. The homogenate was then centrifuged and the supernatant chromatographed on hydroxylapatite columns. The final eluate is the photochemical reaction center preparation. Its characteristic absorption bands at 830 and 960 nm are affected by bright light or oxidizing agents in a manner analogous to P_{800} and P_{870} of the other Rhodospirillaceae.

This preparation apparently contains nonpigmented proteins as well as cytochromes (C-558 and C-553). Its molecular weight is about 110,000. According to Thornber (1971), it should be regarded as an electron-transport particle. It is to be hoped that this particle can be further purified to the stage of an isolated photochemical reaction center.

3.2. Chromatiaceae

3.2.1. Chromatium vinosum

Until recently, it had not proved possible to isolate the photochemical reaction center from this sulfur purple bacterium (Thornber, 1970; Ke and Chaney, 1971). By using the detergent LDAO, however, Lin and Thornber (1975) finally succeeded in obtaining a fraction much purer than had heretofore been available. In their method, the crude chromatophore fraction in 50 mM Tris-Cl (pH 8.0) is treated with 40 g LDAO per mole Bchl for 3 hr at room temperature. The treated material is chromatographed on hydroxylapatite, and the eluate is precipitated with ammonium sulfate. The precipitate is redissolved in 50 mM Tris-Cl (pH 8.0)–0.05% LDAO–0.2 M NaCl and chromatographed on Sephadex 6B. This procedure was found to be successful for both carotenoid-containing and carotenoid-deficient cultures.

The absorption spectrum of this material was very similar to those of analogous preparations from the nonsulfur purple bacteria. For example, the reduced state has the three typical bands at 756, 800, and 870–875 nm. The preparation was found to contain cytochromes C-555 (high potential) and C-553 (low potential). Electrophoresis on polyacrylamide gel in Tris–SDS revealed five polypeptide bands, three of which were prominent. It is a matter of speculation whether the latter are associated with the photochemical reaction center protein (but see Chapter 19, Fig. 3). The specific activity is unknown. Further purification is necessary before quantitative comparison can be made with the photochemical reaction centers from other purple bacteria.

3.3. Chlorobiaceae

The particles obtained by Fowler *et al.* (1971, 1973) and by Olson *et al.* (1973) from *Chloropseudomonas ethylica* and from *Chlorobium limicola* are nearly devoid of chlorobium Bchl *c*, but still contain about 80 Bchl *a* molecules per mole P_{840}. These fractions contain various cytochromes (Gray *et al.*, 1972) and have received considerable spectroscopic attention (Fowler *et al.*, 1973; Olson *et al.*, 1973). They seem to form polymeric associations of similar subunits with a minimum size of 1.5×10^6 (Fowler *et al.*, 1971, 1973). More details on these preparations can be found in Chapter 8.

4. Summary and Conclusions

The discussion that follows summarizes from the author's point of view the advantages and disadvantages of the preparations just reviewed.

Rhodopseudomonas sphaeroides, wild-type. Although the AUT-RC fraction of Slooten (1972*a*) is fairly well purified, only the LDAO preparations obtained by Jolchine and Reiss-Husson (1974) can be claimed to be isolated photochemical reaction center. As judged by its minimal molecular weight, this material is as pure on a protein basis as the best present preparations (see Table 1). It is therefore surprising that its lipid content is as high as 50% of the dry weight. Representative LDAO preparations generally contain only traces of lipid material (Feher, 1971; Noël and Gingras, in prep.; Vadeboncoeur and Gingras, in prep.). The preparations described by

Feher and Okamura (Chapter 19), however, contain firmly bound LDAO up to 33% of the dry weight. The yield of the Jolchine and Reiss-Husson method (20%) is fair, although one disadvantage is that it calls for a purification of the chromatophores before their solubilization.

Rhodopseudomonas sphaeroides, strain R-26. This is the best-studied photochemical reaction center preparation. The methods of Clayton and Wang (1971) and of Feher's laboratory (Feher, 1971; Okamura *et al.*, 1974) seem to produce equally pure material. That obtained by the latter group has been the object of a better-documented chemical analysis; its protein content is in excess of 90%, and its purity is estimated to be over 95%. Both methods have comparable yields of about 30%.

Rhodospirillum rubrum, wild-type. The preparations obtained from this bacterium after solubilization with SDS either have not been characterized (Slooten, 1972*b*) or have not been extensively purified (Smith *et al.*, 1972). Data on the "large fragment" (apparent molecular weight 100,000) obtained by Smith *et al.* (1972) are presented in Table 1. This fragment is obviously contaminated. These authors also obtained a smaller fragment (apparent molecular weight 35,000) after polyacrylamide gel electrophoresis in the presence of 0.05% SDS. This "small fragment" was photoactive and may be compared to a similar aggregate of subunits b and c described by Okamura *et al.* (1974).

The photochemical reaction center preparation obtained by Noël *et al.* (1972) with LDAO contains about 90% protein and 7% pigment (Noël and Gingras, in prep.). Its purity is estimated to be at the 95% level. The yield of the method is high, especially if solid ammonium sulfate is used in the fractionation step (see Table 1).

Rhodospirillum rubrum, strain G9. The preparations obtained by Okamura *et al.* (1974) and by Vadeboncoeur and Gingras appear to have the same level of purity of at least 95%. The latter is probably more stable due to replacement of LDAO by Triton X-100 after the solubilization step. Although its protein/dry weight ratio has not been determined, its phospholipid content is low, i.e., on the order of 1 mol per mol P_{870}. The yield of the preparation procedure is high (50%).

Rhodopseudomonas viridis. Only one preparation from this organism has been described so far (Thornber *et al.*, 1969; Thornber, 1971). It is contaminated by foreign proteins, e.g., cytochromes. Purer and better-characterized preparations should be obtained in the near future.

Chromatium vinosum. The preparation obtained with LDAO is much purer than the membrane fragments derived earlier from this organism. Its A_{280}/A_{802} ratio of 1.6 testifies that it is relatively free of protein contaminants, even if Lin and Thornber (1975) found that it contains cytochromes.

In contrast to the biophysicist, who can single out a specific property (e.g., an optical or electrical signal) of a molecule without physically separating it from its environment, the biochemist must, from the outset of his work, isolate the object of his study. From the biochemist's point of view, no preparation is entirely satisfactory until it can be freed from all foreign elements. This is the attitude that was adopted in this chapter. A perusal of Table 1 will convince the reader that we, as biochemists, have made considerable progress toward this objective since 1968. We have learned to isolate, and, in some cases, have begun to characterize (see Chapter 19), the photochemical reaction center from a few bacterial species. The use of the detergent LDAO has been an important factor in this progress. The next ten years will no doubt witness an extension of these techniques to other organisms. This will open the way to comparative biochemistry and biophysics of the primary reactions of photosynthesis.

5. Note Added in Proof

The following is a brief account of preparations that have been described since this manuscript was completed.

Rhodopseudomonas sphaeroides Y. The LDAO preparation from this strain was delipidated by treatment with sodium deoxycholate followed by chromatography on a Sephadex G-75 column (Jolchine and Reiss-Husson, 1975). The resulting material contained 1.3 μg phospholipid per mg protein. This is equivalent to about 3% by weight, assuming a molecular weight of 100,000 for the photoreaction center.

Rhodopseudomonas capsulata Ala+ (carotenoidless). Prince and Crofts (1973) were the first to describe a preparation of photoreaction center from *Rp. capsulata*. This preparation, obtained by the method of Clayton and Wang (1971), had a spectrum typical of denatured photoreaction center. Although it contained no detectable cytochromes, its purity is difficult to assess from the authors' data. Nieth and Drews (1974) obtained a photoreaction-center-enriched particle by solubilization of the chromatophores with Triton X-100. This preparation was contaminated with foreign proteins and 14–22% of antenna Bchl. Nieth *et al.* (1975) used a modification of the method of Clayton and Wang (1971) in which the membranes solubilized with LDAO were

centrifuged for 11 hr at 200,000g on a 0.2–0.5 M sucrose density gradient. Three bands were thus obtained and the middle one was subjected to ammonium sulfate fractionation and to chromatography on Sephadex G-200. This preparation had an infrared absorption spectrum similar to those from *Rp. sphaeroides* and *Rs. rubrum*. It was free of antenna Bchl and of cytochrome *c*. The three usual polypeptides (20,000, 24,000, and 28,000 apparent molecular weight) were prominent on SDS–polyacrylamide gels after electrophoresis. The ratio A_{280}/A_{802} was routinely 1.3–1.5 and reached 1.25 in the best preparations. Minimal molecular weight varied widely according to how it was calculated. Its most trustworthy value is probably 145,000. This is based on the protein content and an estimated $\varepsilon_{850} = 100,000$ mM^{-1} cm^{-1}.

Rhodopseudomonas gelatinosa, strain EM-1 (carotenoidless). Clayton and Clayton (1978) extracted and purified the photoreaction center from carotenoidless strain EM-1. They first purified the chromatophores by centrifugation on a discontinuous sucrose density gradient before solubilization with 3% LDAO. The extract was purified by two successive chromatographic steps on hydroxylapatite columns. The photoreaction center was free of antenna Bchl and of cytochromes. It had infrared peaks at 799 nm and 850 nm compared to 802 nm and 860–865 nm for photoreaction centers from *Rp. sphaeroides* (R-26) and *Rs. rubrum* (G-9). The ratio of A_{280}/A_{799} was 1.8 in the best preparations. Minimal molecular weight based on the protein content and on the absolute molar extinction coefficient was 150,000. Polyacrylamide gel electrophoresis in the presence of SDS showed the presence of only two polypeptides (apparent molecular weights of 33,000 and 25,000) instead of the three usually found in other Rhodospirillaceae. Attempts to isolate the photoreaction center from wild-type strain TG-9 were not successful.

Rhodospirillum rubrum, strain G-9. Zürrer *et al.* (1977) isolated the photoreaction center from this carotenoidless strain by a modification of the method of Okamura *et al.* (1974). The chromatophores were treated with approx. 0.225% LDAO, then centrifuged for 60 min at 304,000g. The supernatant was concentrated by ultrafiltration on Amicon UM 20E and then purified by chromatography on Sepharose 6B and on DEAE–cellulose. This preparation was nearly free of antenna Bchl and had a ratio of $A_{280}/A_{802} = 1.25$ which is indicative of good purity. Electrophoresis indicated the presence of residual antenna Bchl–protein complex. The cytochrome content and the yield of the procedure were not reported.

Rhodopseudomonas viridis. One of the problems of the original procedure of Thornber *et al.* (Thornber *et al.*, 1969; Thornber, 1971) was that it resulted in preparations which contained high amounts of P_{685} attributed to oxidized antenna Bchl *b*. An improved method was described by Trosper *et al.* (1977) in which sodium dithionite was present from the start of solubilization with SDS to the final purification step. This method resulted in a smaller amount of the P_{685} contaminant. The same result was obtained by Pucheu *et al.* (1976) who introduced LDAO as a solubilizing agent for the membranes of this species. They purified the photoreaction center by one of two procedures. Procedure A was a modification of the method of Okamura *et al.* (1974) with a centrifugation on a linear sucrose density gradient after DEAE–cellulose chromatography. Procedure B involved no ammonium sulfate fractionation but only repeated DEAE–cellulose chromatography followed by centrifugation on a sucrose density gradient. The preparation was free of antenna Bchl *b* and had a ratio of A_{830}/A_{685} of 9 as compared to 1.3 in the original preparation of Thornber *et al.* (1969). The ratio of $A_{830}/A_{965} = 2.3$ (Pucheu *et al.*, 1976) and 2.5 (Trosper *et al.*, 1977) is comparable to the ratio of A_{800}/A_{870} obtained with isolated photoreaction center from other purple bacteria. The preparation of Pucheu *et al.* (1976) contained four polypeptide bands detectable by SDS–polyacrylamide gel electrophoresis. The ratio of A_{280}/A_{830} was 2.5. Neither its minimal molecular weight nor its yield was reported. Although probably purer than the preparation of Thornber *et al.* (1969), it also contained firmly bound cytochromes C-553 and C-558. This also applied to the preparation of Trosper *et al.* (1977). Thornber *et al.* (1977) also used LDAO in a modification of procedure B of Pucheu *et al.* (1976) in which hydroxylapatite replaced DEAE–cellulose for column chromatography. They also made progress in the isolation of the photoreaction center from *Thiocapsa pfennigii*, another Bchl-*b*-containing purple bacterium. Their method made use of LDAO for solubilization followed by chromatography on DEAE–cellulose. The preparation was thermolabile and still contained the 685 nm absorbing pigment.

Clayton and Clayton (1978) described the purest preparation obtained yet from *Rp. viridis*. Purified chromatophores were extracted with 5% LDAO for 5 minutes and the extract was purified by repeated chromatography on hydroxylapatite followed by ammonium sulfate fractionation. This preparation had a A_{280}/A_{830} ratio of 2.1 and a minimal molecular weight of 190,000, based on an extinction coefficient of 123 ± 25 mM^{-1} cm^{-1} at 960 nm, also determined by the authors. This preparation contained about two molecules of cytochrome-558 and three of cytochrome-552 per mole of photoreaction center, in agreement with the values reported by Trosper *et al.* (1977). SDS–polyacrylamide gel electrophoresis showed a pattern of three polypeptides with apparent molecular weights of 41,000, 37,000, and 31,000. It is not known whether any of these bands can be identified with cytochrome-558 or -552.

Chromatiaceae. Ackerson (Feher and Okamura, 1976) modified the method of Lin and Thornber (1975) and obtained an electrophoretically pure preparation of the photoreaction center from *Chromatium vinosum.* Only three polypeptides of apparent molecular weights comprised between 20,000 and 30,000 were apparent on polyacrylamide gels after electrophoresis in the presence of SDS.

Large photosynthetic electron-transport particles obtained from the genus *Chromatium* were studied by several groups. Although they cannot be assimilated to isolated photoreaction center, they are worthy of mention here because of the valuable information they have yielded on the primary process of photosynthesis. Halsey and Byers (1975) studied the polypeptide composition of 700,000-molecular-weight fragments by SDS–polyacrylamide gel electrophoresis. Five polypeptides were found, three of which had apparent molecular weights comparable to those of the triad observed in other photoreaction center preparations. One of these bands ($\approx 45{,}000$) was identified as cytochromes C-553 and C-555. Other particles obtained by methods similar to that of Thornber (1970) were depleted of their contaminating antenna Bchl by extraction with cold acetone (Tiede *et al.*, 1976a,b; van Grondelle *et al.*, 1976; Romijn and Amesz, 1977). Shuvalov and Klimov (1976) obtained a complex ($P_{890}/B_{890} = 1/32$) after solubilization of the chromatophores of *Chromatium minutissimum* and purification by polyacrylamide gel electrophoresis. Antenna Bchl was oxidized selectively by incubation on the complex in ferricyanide.

ACKNOWLEDGMENT

The author is indebted to Dr. Margaret Mamet-Bratley for performing a critical reading of this manuscript.

6. References

Arnold, W., 1965, An electron-hole picture of photosynthesis, *J. Phys. Chem.* **69**:788–791.

Arnold, W., and Sherwood, H. K., 1957, Are chloroplasts semi-conductors?, *Proc. Natl. Acad. Sci. U.S.A.* **43**:105–114.

Calvin, M., 1956, The photosynthetic carbon cycle, in:

Proceedings of the Third International Congress of Biochemistry, Brussels, 1955, pp. 211–335, Academic Press, New York.

Calvin, M., 1961, Quantum conversion in photosynthesis, *J. Theor. Biol.* **2**:258–287.

Clayton, R. K., 1963, Toward the isolation of a photochemical reaction center in *Rhodopseudomonas spheroides*, *Biochim. Biophys. Acta* **75**:312–323.

Clayton, R. K., and Clayton, B. J., 1978*a*, Properties of photochemical reaction centers purified from *Rhodopseudomonas gelatinosa*, *Biochim. Biophys. Acta* **501**:470–477.

Clayton, R. K., and Clayton, B. J., 1978*b*, Molar extinction coefficients and other properties of an improved reaction center preparation from *Rhodopseudomonas viridis*, *Biochim. Biophys. Acta* **501**:478–487.

Clayton, R. K., and Haselkorn, R., 1972, Protein components of bacterial photosynthetic membranes, *J. Mol. Biol.* **68**:97–105.

Clayton, R. K., and Wang, R. T., 1971, Photochemical reaction centers from *Rhodopseudomonas spheroides*, in: *Methods in Enzymology* (S. P. Colowick and N. O. Kaplan, eds.), Vol. 23, pp. 696–704, Academic Press, New York and London.

Feher, G., 1971, Some chemical and physical properties of a bacterial reaction center particle and its primary photochemical reactants, *Photochem. Photobiol.* **14**:373–387.

Feher, G., and Okamura, M. Y., 1976, Reaction centers from *Rhodopseudomonas sphaeroides*, in: *Chlorophyll-Proteins, Reaction Centers and Photosynthetic Membranes, Brookhaven Symp. Biol.* **28**:183–194.

Fowler, C. F., Nugent, N. A., and Fuller, R. C., 1971, The isolation and characterization of a photochemically active complex from *Chloropseudomonas ethylica*, *Proc. Natl. Acad. Sci. U.S.A.* **68**:2278–2282.

Fowler, C. F., Gray, B. H., Nugent, N. A., and Fuller, R. C., 1973, Absorbance and fluorescence properties of the bacteriochlorophyll *a* reaction center complex and bacteriochlorophyll *a* protein in green bacteria, *Biochim. Biophys. Acta* **292**:692–699.

Gingras, G., and Jolchine, G., 1969, Isolation of a P_{870}-enriched particle from *Rhodospirillum rubrum*, in: *Progress in Photosynthesis Research: Proceedings of the International Congress of Photosynthesis Research, Freudenstadt, 1968* (H. Metzner, ed.), Vol. 1, pp. 209–216, H. Laupp, Jr., Tübingen, Germany.

Gray, B. H., Fowler, C. F., Nugent, N. A., and Fuller, R. C., 1972, A reevaluation of the presence of low midpoint potential cytochrome 551.5 in the green photosynthetic bacterium *Chloropseudomonas ethylica*, *Biochem. Biophys. Res. Commun.* **47**:322–327.

Halsey, Y. D., and Byers, B., 1975, A large photoreactive particle from *Chromatium vinosum* chromatophores, *Biochim. Biophys. Acta* **387**:349–367.

Jolchine, G., and Reiss-Husson, F., 1974, Comparative studies on two reaction center preparations from *Rhodopseudomonas spheroides* Y, *FEBS Lett.* **40**:5–8.

Jolchine, G., and Reiss-Husson, F., 1975, Studies on pigments and lipids in *Rhodopseudomonas spheroides* Y reaction centers, *FEBS Lett.* **52**:33–36.

Jolchine, G., Reiss-Husson, F., and Kamen, M. D., 1969, Active center fractionation from *Rhodopseudomonas spheroides* strain Y (De Klerk), *Proc. Natl. Acad. Sci. U.S.A.* **64**:650–654.

Ke, B., and Chaney, T. H., 1971, Spectral and photochemical properties of subchromatophore fractions from carotenoid-deficient *Chromatium* by Triton treatment, *Biochim. Biophys. Acta* **226**:341–353.

Kuntz, I. D., Loach, P. A., and Calvin, M., 1964, Absorption changes in bacterial chromatophores, *Biophys. J.* **4**:227–249.

Lin, L., and Thornber, J. P., 1975, Isolation and partial characterization of the photochemical reaction center of *Chromatium vinosum* (strain D), *Photochem. Photobiol.* **22**:37–40.

Loach, P. A., Androes, G. M., Maksim, A. F., and Calvin, M., 1963, Variation of electron paramagnetic resonance signals of photosynthetic systems with the redox level of their environment, *Photochem. Photobiol.* **2**:443–454.

Loach, P. A., Sekura, D. L., Hadsell, R. M., and Stemer, A., 1970, Quantitative dissolution of the membrane and preparation of photoreceptor subunits from *Rhodopseudomonas spheroides*, *Biochemistry* **9**:724–733.

Loach, P. A., Bambara, R. A., and Ryan, F. J., 1971, Identification of the major ultraviolet absorbance photochanges in photosynthetic systems, *Photochem. Photobiol.* **13**:247–257.

Lowry, O. H., Rosebrough, N. J., Farr, A. L., and Randall, R. J., 1951, Protein measurement with the Folin phenol reagent, *J. Biol. Chem.* **193**:265–275.

Nieth, K. F., and Drews, G., 1974, The protein patterns of intracytoplasmic membranes and reaction center particles isolated from *Rhodopseudomonas capsulata*, *Arch. Microbiol.* **96**:161–174.

Nieth, K. F., Drews, G., and Feick, R., 1975, Photochemical reaction centers from *Rhodopseudomonas capsulata*, *Arch. Microbiol.* **105**:43–45.

Noël, H., van der Rest, M., and Gingras, G., 1972, Isolation and partial characterization of a P_{870} reaction center complex from wild type *Rhodospirillum rubrum*, *Biochim. Biophys. Acta* **275**:219–230.

Okamura, M. Y., Steiner, L. A., and Feher, G., 1974, Characterization of reaction centers from photosynthetic bacteria. I. Subunit structure of the protein mediating the primary photochemistry in *Rhodospseudomonas spheroides* R-26, *Biochemistry* **13**:1394–1402.

Olson, J. M., Philipson, K. D., and Sauer, K., 1973, Circular dichroism and absorption spectra of bacteriochlorophyll–protein and reaction center complexes from *Chlorobium thiosulfatophilum*, *Biochim. Biophys. Acta* **292**:206–217.

Prince, R. C., and Crofts, A. R., 1973, Photochemical reaction centers from *Rhodopseudomonas capsulata* Ala Pho$^+$, *FEBS Lett.* **35**:213–216.

Pucheu, N. L., Kerber, N. L., and Garcia, A. F., 1976, Isolation and purification of reaction center from *Rhodopseudomonas viridis* NHTC 133 by means of LDAO, *Arch. Microbiol.* **109**:301–305.

Reed, D. W., 1969, Isolation and composition of a photosynthetic reaction center complex from *Rhodopseudomonas spheroides*, *J. Biol. Chem.* **244**:4936–4941.

Reed, D. W., and Clayton, R. K., 1968, Isolation of a reaction center fraction from *Rhodopseudomonas spheroides*, *Biochem. Biophys. Res. Commun.* **30**:471–475.

Reed, D. W., Raveed, D., and Israel, H. W., 1970, Functional bacteriochlorophyll–protein complexes from chromatophores of *Rhodopseudomonas spheroides* strain R-26, *Biochim. Biophys. Acta* **223**:281–291.

Reiss-Husson, F., and Jolchine, G., 1972, Purification and properties of a photosynthetic reaction center isolated from various chromatophore fractions of *Rhodopseudomonas spheroides* Y, *Biochim. Biophys. Acta* **256**:440–451.

Romijn, J. C., and Amesz, J., 1977, Purification and photochemical properties of reaction centers of *Chromatium vinosum*. Evidence for the photoreduction of a naphthoquinone, *Biochim. Biophys. Acta* **461**:327–338.

Segen, B. J., and Gibson, K. D., 1971, Deficiencies of chromatophore proteins in some mutants of *Rhodopseudomonas spheroides* with altered carotenoids, *J. Bacteriol.* **105**:701–709.

Shuvalov, V. A., and Klimov, V. V., 1976, The primary photoreactions in the complex cytochrome $P_{890} \cdot P_{760}$ (Bacteriopheophytin$_{760}$) of *Chromatium minutissimum* at low redox potentials, *Biochim. Biophys. Acta* **440**:587–599.

Sistrom, W. R., and Clayton, R. K., 1964, Studies on a mutant of *Rhodopseudomonas spheroides* unable to grow photosynthetically, *Biochim. Biophys. Acta* **88**:61–73.

Slooten, L., 1972a, Reaction center preparations of *Rhodopseudomonas spheroides*, *Biochim. Biophys. Acta* **256**:452–466.

Slooten, L., 1972b, Electron acceptors in reaction center preparations from photosynthetic bacteria, *Biochim. Biophys. Acta* **272**:208–218.

Smith, W. R., Jr., Sybesma, C., and Dus, K., 1972, Isolation and characteristics of small, soluble photoreactive fragments of *Rhodospirillum rubrum*, *Biochim. Biophys. Acta* **267**:609–615.

Straley, S. C., Parson, W. W., Mauzerall, D. C., and Clayton, R. K., 1973, Pigment content and molar extinction coefficient of photochemical reaction centers from *Rhodopseudomonas spheroides*, *Biochim. Biophys. Acta* **305**:597–609.

Thornber, J. P., 1970, Photochemical reactions of purple bacteria as revealed by studies of three spectrally different carotenobacteriochlorophyll–protein complexes from *Chromatium*, strain D, *Biochemistry* **9**:2688–2697.

Thornber, J. P., 1971, The photochemical reaction center of *Rhodopseudomonas viridis*, in: *Methods in Enzymology* (S. P. Colowick and N. O. Kaplan, eds.), Vol. 23, pp. 688–691, Academic Press, New York and London.

Thornber, J. P., Olson, J. M., Williams, D. M., and Clayton, M. L., 1969, Isolation of reaction center of *Rhodopseudomonas viridis*, *Biochim. Biophys. Acta* **172**:351–354.

Thornber, J. P., Dutton, P. L., Fajer, J., Forman, A., Olson, J. M., and Prince, R. C., 1977, Isolated photochemical reaction centers from bacteriochlorophyll *b* containing organisms, *Proceedings of the Fourth International Congress on Photosynthesis*, p. 382, Reading, U.K.

Tiede, D. M., Prince, R. C., and Dutton, P. L., 1976a, EPR and optical spectroscopic properties of the electron carrier intermediate between the reaction center bacteriochlorophylls and the primary acceptor in *Chromatium vinosum*, *Biochim. Biophys. Acta* **449**:447–467.

Tiede, D. M., Prince, R. C., Reed, G. H., and Dutton, P. L., 1976b, EPR properties of the electron carrier intermediate between the reaction center bacteriochlorophylls and the primary acceptor in *Chromatium vinosum*, *FEBS Lett.* **65**:301–304.

Trosper, T. L., Benson, D. L., and Thornber, J. P., 1977, Isolation and spectral characteristics of the photochemical reaction center of *Rhodopseudomonas viridis*, *Biochim. Biophys. Acta* **460**:318–330.

van der Rest, M., and Gingras, G., 1974, The pigment complement of the photosynthetic reaction center from *Rhodospirillum rubrum*, *J. Biol. Chem.* **249**:6446–6453.

van der Rest, M., Noël, H., and Gingras, G., 1974, An immunological and electrophoretic study of *Rhodospirillum rubrum* chromatophore fragments, *Arch. Biochem. Biophys.* **164**:285–292.

van Grondelle, R., Romijn, J. C., and Holmes, N. G., 1976, Photoreaction of the long-wavelength bacteriopheophytin in reaction centers and chromatophores of the photosynthetic bacterium *Chromatium vinosum*, *FEBS Lett.* **72**:187–192.

Wang, R. T., and Clayton, R. K., 1973, Isolation of photochemical reaction centers from a carotenoidless mutant of *Rhodospirillum rubrum*, *Photochem. Photobiol.* **17**:57–61.

Zürrer, H., Snozzi, M., Hanselmann, K., and Bachofen, R., 1977, Localisation of the subunits of the photosynthetic reaction centers in the chromatophore membrane of *Rhodospirillum rubrum*, *Biochim. Biophys. Acta* **460**:273–279.

CHAPTER 7

Bacteriochlorophyll *in Vivo*: Relationship of Spectral Forms to Specific Membrane Components

J. Philip Thornber, Terry L. Trosper, and Charles E. Strouse

1. Introduction

The purple sulfur bacteria (Chromatiaceae) and purple nonsulfur bacteria (Rhodospirillaceae) contain either bacteriochlorophyll (Bchl) *a* or *b* associated with intracytoplasmic membranes. The green sulfur bacteria (Chlorobiaceae) contain one of the classes of pigments now known as Bchl's *c*, *d*, or *e*, as well as 5–10% Bchl *a* (Gloe *et al.*, 1975). A species of phototropic gliding bacteria, *Chloroflexus aurantiacus*, which should probably be assigned to a new family (Pierson and Castenholtz, 1974*a*), also contains Bchl *c* and *a* (Pierson and Castenholtz, 1974*b*). In the green sulfur bacteria and in the gliding bacteria, the photosynthetic pigments, with the possible exception of Bchl *a* (Boyce *et al.*, 1976), are associated with characteristic vesicular structures within the cell (Cohen-Bazire and Sistrom, 1966). Calculations based on the total content of bacterial chlorophylls and the amount of intracellular material with which they are associated have indicated that local concentrations of pigment may exceed 0.05 M (Sauer, 1975). Since the pigment molecules can be extracted into polar organic solvents, they are not strongly bound to the cellular structures with which they are associated. The Bchl molecules do, however, interact with each other or with other

components of their immediate environment *in vivo*, or both, as is shown by differences between the absorption spectra of monomeric solutions and of preparations of cellular material. These differences are most easily observed in the near-IR region (Olson and Stanton, 1966), where pigment monomers have a single large absorption band, and the cellular preparations generally have a more complex spectrum of two or more lower-energy absorption bands.

The precise forms of the IR absorption spectra of the photosynthetic bacteria have been shown to depend on a large variety of parameters, including strain of organism, carbon source, source of reducing power, growth conditions, and age of culture (Biebl and Drews, 1969). This fact was recognized early in the investigation of the state of Bchl *in vivo* (Wassink *et al.*, 1939), but has not always been adequately heeded. Variations in spectra are presumably indicative of alterations in the molecular interactions of Bchl's or in the environments of the chromophores, or in both. Correct specific interpretations will be possible when the molecular organization of the Bchl-containing structures is known. It was originally suggested that each absorption band or spectral form of the pigment corresponded to a particular Bchl–protein complex (Katz, E., and Wassink, 1939; Wassink *et al.*, 1939; French and Young, 1956), the relative amounts of which might vary within and among organisms. Later, after many data on aggregated pigments *in vitro* became available (Krasnovskii *et al.*, 1952; Krasnovskii, 1969; Katz,

J. Philip Thornber and Terry L. Trosper · Department of Biology and Molecular Biology Institute, University of California, Los Angeles, California 90024   Charles E. Strouse · Department of Chemistry, University of California, Los Angeles, California 90024

J. J., and Norris, 1973), the individual absorption bands *in vivo* were attributed to various states of aggregation of the Bchl molecules themselves (Clayton, 1965). In the one case for which molecular structure has been determined, the Bchl *a*–protein of green bacteria (Fenna and Matthews, 1975) [see also Olson (Chapter 8)], both Bchl–Bchl and Bchl–protein interactions are present. It is probable that each will be found to be involved, to varying extents, in the different components of Bchl-containing photosystems.

The Bchl complement of photosynthetic bacteria may be divided functionally into reaction center Bchl's, which are photochemically active, and antenna Bchl's, which absorb light quanta and transfer the absorbed energy to the reaction centers (see Chapters 6 and 15). Many investigators have used several methods to achieve physical separation of these functional classes of pigments, with degree of success dependent on the organism as well as the method used (Section 3) [see also Gingras (Chapter 6)]. However, the fractions obtained usually still display more than one IR band. Interpretations of fractionation studies in terms of the organization of Bchl molecules *in vivo* are discussed in Section 5, following a summary (Sections 2–4) of the spectral data available. Information relevant to explanations of the different spectral forms of Bchl *in vivo* has also been provided by studies of mutant organisms and of cells grown under abnormal conditions.

Recent experiments have demonstrated unequivocally that bacteriophaeophytin *a* is also an integral component of the reaction center of *Rhodopseudomonas sphaeroides* strain R-26 (Reed and Peters, 1972; Straley *et al.*, 1973). Spectral attributes of reaction centers of other purple photosynthetic bacteria have led to the assumption that bacteriophaeophytin *a* or *b* is a normal component of reaction centers.

2. Spectra of Whole Cells and Chromatophores

2.1. Absorption Spectra

Nearly four decades ago, E. Katz and Wassink (1939) and Wassink *et al.* (1939) published spectral studies of some whole purple bacteria. Since then, the near-IR spectra of dozens of species and strains of photosynthetic bacteria have been reported. These data are collected in Table 3, in which the positions of maxima and shoulders are recorded for all species for which spectra appear in the literature. Included are data from earlier reviews of the subject (French and Young, 1956; Clayton, 1963; Olson and Stanton, 1966).

When absorption by whole cells was observed, investigators overcame the problem of light scattered from the turbid suspensions by using opal glass or an integrating sphere, sandwiching a thin layer of suspension between glass plates (Katz, E., and Wassink, 1939), or suspending cells in bovine serum albumin (Schlegel and Pfennig, 1961; Sojka *et al.*, 1970), as was originally suggested by Barer and Joseph (1955) for refractometry. When the pigmented membrane or vesicle fraction of broken cells (chromatophore fraction; Schachman *et al.*, 1952) was shown to have a spectrum indistinguishable from that of whole cells [however, see Goedheer (1972) with respect to green bacteria], most investigators subsequently reported spectra of these fractions, for which scattering effects are usually minimal. Data obtained from both types of samples are included in Table 3.

The IR absorption bands of the purple bacteria are identified by the approximate positions of the band maxima. Up to five bands may be observed, which are conventionally designated B800, B*800, B820, B850, and B870–890 (Duysens, 1952). In this notation, the "B" and "B*" indicate an antenna Bchl form,† and the letter "P" is used to signify spectral types in reaction centers. The organisms usually display one of three typical spectra: one major (B870–890) and one minor band at about 800 nm; three bands (B800, B850, and B870–890), with the lowest energy band frequently a distinct shoulder on the B850 band; or more than three bands (B*800 or B820 or both, in addition to the others) of varying absorbances. In the discussion (Section 5), bacteria exhibiting these spectra will be classified as Group I, II, and III, respectively.

Fewer absorption bands are observed in the spectra of the green sulfur bacteria. In general, a single absorption maximum of the light-harvesting Bchl *c*, *d*, or *e* is observed at wavelengths 70–100 nm greater than that of the monomer absorption, and a much smaller maximum attributed to Bchl *a* is observed at 805–812 nm. The phototrophic gliding bacteria also show a single Bchl *c* band at 740 nm, but in this species, two bands, at 802 and 865 nm, are attributed to Bchl *a* (Pierson and Castenholtz, 1974*a,b*).

More recently, low-temperature spectroscopic techniques have been applied to chromatophore fractions of some species of photosynthetic bacteria (Vredenberg and Amesz, 1966; Litvin and Gulyaev, 1969; Goedheer, 1972; Saunders and Jones, 1974). Results of these investigations are also given in Table 3. Absorption spectra recorded at liquid nitrogen temperature are similar in overall form to room-temperature spectra; however, shoulders may be

---

† B800 and B*800 are used to refer to *two* distinct spectral forms of Bchl with absorption maxima near 800 nm that have been observed in the low-temperature spectra of some bacteria.

sharpened and more clearly resolved at the lower temperature. The lower-energy IR bands of purple bacteria are usually shifted to longer wavelengths by 5–15 nm at liquid nitrogen temperature, whereas the B800 and B820 bands are not so affected (Vredenberg and Amesz, 1967; Goedheer, 1972). Goedheer noticed that band shifts may occur over much smaller temperature ranges and suggested that caution should be exercised in interpreting difference spectra when the sample temperature could change during an experiment. Few low-temperature spectra of green bacteria are available, but Litvin and Gulyaev (1969) reported the resolution of a large number of components in the low-temperature, second-derivative spectrum of *Chloropseudomonas ethylica*[†] (see Table 3).

Several general phenomena become apparent on inspection of the data in Table 3:

1. The absorption maxima *in vivo* are always shifted to longer wavelengths than the position of the near-IR Bchl absorption band in organic solvents.

2. Two or more absorption bands are usually present, although one may be markedly predominant.

3. Different "wild-type" strains of the same species may have somewhat different spectral characteristics.

4. The same strain grown under different conditions may also show spectral variations.

5. Mutants of some species show altered spectra, with a particular band or bands missing or greatly attenuated.

Points (3) and (4) were emphasized by Wassink *et al.* (1939), and were investigated in detail by Biebl and Drews (1969), who suggested that the spectra—although somewhat variable—are generally useful tools for taxonomic investigations. Haskins and Kihara (1967) also used near-IR spectra for identification of common species of purple bacteria. The significance of the number and relative heights of the distinguishable absorption bands for the forms of Bchl *in vivo* will be considered in more detail in Sections 3–5.

2.2. Fluorescence Spectra

Fluorescence spectra of chromatophore fractions from some photosynthetic bacteria were reviewed by Olson and Stanton (1966). In the species investigated, *Rhodospirillum rubrum*, *Chromatium vinosum* strain D, *Rp. viridis*, and *Chlorobium limicola f. thiosulfatophilum*, the fluorescence bands are of lower energy than the longest-wavelength absorption band, and their shape is roughly a mirror image of this band.

Vredenberg and Amesz (1966) reported that absorption in all the IR bands by *Rp. sphaeroides*, *Rp. palustris*, and *Chr. vinosum* strain D is effective in exciting fluorescence of the longest wavelength band. Zankel and Clayton (1969) observed emission from the 800 and 850 nm bands of *Rp. sphaeroides* chromatophores as well as from the 870 nm band. Excitation with wavelengths absorbed by B870 is not noticeably effective in exciting B800 fluorescence, but is efficient in exciting B850 emission. Excitation of B850 does, however, cause approximately 1% of the total emission to come from B800. If B800 and B850 are components of the same pigment–protein complex and B870 resides in a different complex (see Section 5), this fluorescence behavior is not unreasonable.

Low-temperature fluorescence spectra, like those at room temperature, are indicative of emission by the lowest-energy absorption band. They are displaced to longer wavelengths than room-temperature spectra, a phenomenon expected because of shifts of the absorption bands. The extent of the fluorescence band shifts is about equal to that of the absorption bands, indicating that the Stokes shift is not altered at low temperatures (Goedheer, 1972).

Ebrey and Clayton (1969) reported a slight polarization of fluorescence from *Rp. sphaeroides* R-26 chromatophores excited by 760 to 850 nm light. Goedheer (1973) also investigated fluorescence polarization spectra of several photosynthetic bacteria—*Rp. palustris*, *Rs. molischianum*, *Rp. gelatinosa*, *Rp. viridis*, *Rs. rubrum*, *Thiocystis violacea*, and *Chr. vinosum* strain D. In all cases, the most highly polarized emitted light is excited by light of wavelength greater than or equal to 800 nm, regardless of the temperature at which the chromatophores are maintained. The difference in fluorescence polarization for exciting light of 800 and 880 nm is slight, but fluorescence excited with the longer-wavelength light is more polarized in most cases. The maximum polarization observed ($p \approx 0.1$–0.15) is considerably lower than the theoretical maximum possible. Because various lines of evidence indicate that energy transfers do occur and the absorbing molecule is unlikely to be the same as the emitter, the results are interpreted as indicating partial orientation of the neighboring Bchl molecules.

2.3. Circular and Linear Dichroism

Most studies of circular dichroism (CD) of photosynthetic bacteria have been performed on subchromatophore particles (Phillipson and Sauer, 1973a; see Section 5 below). However, Sauer and co-workers reported CD of *Rs. rubrum* and *Rp. sphaeroides* chromatophores, and Steffen and Calvin

---

[†] The validity of this species is in doubt; see Trüper and Pfennig (Chapter 2).

(1970) studied CD of mutant *Rp. sphaeroides* R-26 chromatophores. Double CD components corresponding to the IR absorption bands are detected for all three organisms. The detailed properties are interpreted as indicating interactions among small aggregates of Bchl that are not aligned parallel to each other (Sauer, 1975). Phillipson and Sauer (1973*b*) further showed that the shape of the CD spectrum is not significantly influenced by differential scattering of left and right circularly polarized light by particulate samples. Such scattering effects should be absent from chromatophore preparations of these bacteria, since these organisms do not possess ordered (i.e., lamellar) membrane structures.

Large linear dichroic effects have been observed for magnetically oriented whole cells of *Rp. palustris* and *Rp. viridis* in which the photosynthetic pigments are organized in a lamellar system, whereas cells of *Rp. sphaeroides* in which the pigments are located in vesicles give no linear dichroism signals (Breton, 1974). However, chromatophores of all these bacteria, if physically oriented by deposition on a flat surface, show linear dichroic effects. *Rhodopseudomonas palustris* cells oriented in a flow apparatus also exhibit anisotropic absorption by the Bchl bands (Morita and Miyazaki, 1971). The data obtained have been interpreted as showing that (1) Bchl molecules are partially oriented with respect to the plane of the membranes in which they are bound, and (2) photosynthetic bacteria containing lamellae can be oriented with their lamellar contents in a preferred direction with respect to a magnetic field (Breton, 1974). It should be noted that B800 and B850 are oriented in the same direction and to about the same extent in all cases for which data are given.

3. Spectral Forms of Bacteriochlorophyll in Subchromatophore Fractions

The aims of this section are: (1) to document as completely as possible the spectral forms of Bchl that are present in subchromatophore fractions that have been obtained from any photosynthetic bacterium; (2) to describe, when it is known, the nature of the protein associated with the different spectral forms of antenna and reaction center Bchl; and (3) to indicate that certain spectral forms are always fractionated together, i.e., occur in the same component of the photosynthetic apparatus. Somewhat surprisingly, few types of photosynthetic bacteria have been subjected to fractionation (Table 1) compared with the multitude of organisms the absorption spectra of which have been reported (see Table 3). Generalizations and interpretations that can be made from the assembled

data about the organization of Bchl *in vivo* are discussed in Section 5.

3.1. Purple Sulfur Bacteria

3.1.1. *Chromatium vinosum* Strain D

This bacterium has frequently been selected from the many available to attempt fractionation of the spectral forms of Bchl because it can exhibit more spectral forms (B*800, B800, B820, B850, and B890) than most other bacteria (see Table 3). The possibility that the shorter-wavelength forms could be separated from B890 was indicated by the fluorescence measurements of Bril (1960) on deoxycholate (DOC)-treated chromatophores in which energy transfer from B850 to B890 was affected by the detergent. Clayton (1962) was the first to obtain spectrally different fractions by centrifugation of DOC-treated chromatophores, which gave a pellet that was considerably enriched in 890-nm absorbing material and a supernatant that contained the 800 and 850 nm forms. A much cleaner spectral separation was achieved by Garcia *et al.* (1966*a*), who used the detergent Triton X-100 to solubilize the chromatophore membranes. Sucrose-density centrifugation of the resulting solution yielded two pigmented zones, a heavy (H) and a light (L) band. The L band had absorption maxima of the same height at 802 and 848 nm, whereas the H band, after further treatment with Triton X-100, yielded a fraction containing B890 together with a much smaller absorbance at 800 nm.

Details of the organization of Bchl in cultures of this organism that contained all five spectral forms were obtained by the use of the anionic detergent sodium dodecylsulfate (SDS). Hydroxylapatite chromatography of SDS-dissociated chromatophores enabled isolation of three spectrally different caroteno–bacteriochlorophyll–protein complexes (Thornber, 1970). One of these complexes, Fraction A, was a photochemically active multienzyme component [reaction center components plus light-harvesting B890 (B890/P883 = 40/1) plus cytochromes] with an absorbance maximum at 890 nm (905 nm at 77°K) and a minor absorbance maximum at 800 nm, i.e., a component essentially analogous to the H fraction of Garcia *et al.* (1966*a*). The two other components isolated from the SDS-solubilized chromatophores, Fractions B and C, have the same molecular size, carotenoid composition, and Bchl/carotenoid ratio. They differ in chromatographic behavior, and, more particularly, in their near-IR absorption spectra. Fraction B contains the antenna Bchl forms, B800 and B850, absorbing at 803 and 846 nm (801 and 859 nm at 77°K), whereas Fraction C, containing B*800 and B820, has a maximum at 800 nm and a shoulder at 820 nm (797 and 821 nm peaks at 77°K). Thus, all five spectral forms of antenna Bchl in this organism appear to be accommodated in three spectrally different pigment–protein complexes, the size and protein component(s) of which have been reported (see below).

Lin and Thornber (1975) recently isolated the reaction center of this organism free of all antenna Bchl by column chromatography of lauryl dimethyl amine oxide (LDAO)-solubilized chromatophores. The absorption spectrum of this preparation has the typical triple-peaked spectrum (760, 800, and 875 nm) common to reaction center preparations from the Rhodospirillaceae (Chapter 19). The wavelength of maximum bleaching of the reaction center (875 nm) is shifted slightly from that generally observed in *Chromatium* chromatophores and subchromatophore fractions (Cusanovitch *et al.*, 1968; Thornber, 1970). The contribution of the reaction center components to total cell absorbance in the IR is minor.

Several polypeptides are present in Fraction A (mol. wt. ≈ 500,000) (Thornber, 1970; Erokhin and Moskalenko, 1973; Halsey and Byers, 1975). It appears that one (≈ 12,000 daltons) may be involved in the organization of the B890 pigment molecules; however, this spectral form has not yet been obtained in a complex free of all other Bchl types. The Bchl-containing particle in which B800 and B850 are the only spectral forms (e.g., Fraction B) has a molecular weight of 100,000 (Thornber, 1970); Erokhin and Moskalenko (1973) reported that two polypeptides of 9800 and 7600 daltons occur in this complex in the related species, *Chr. minutissimum*. The polypeptide composition of Fraction C, containing B*800 and B820, appears to be identical to that of Fraction B (L. Lin, personal communication). A triad of polypeptides (cf. Clayton and Haselkorn, 1972) of 20,000–30,000 daltons are most likely associated with the photochemical reaction center in this organism (Halsey and Byers, 1975; Lin and Thornber, 1975) and in *Chr. minutissimum* (Erokhin and Moskalenko, 1973).

Several workers have noted that the B850 spectral form of *Chr. vinosum* can be readily converted *in vitro* to a form absorbing at shorter wavelength (800–830 nm). Thus, it is not surprising to find that B850 and B820 are found in an essentially identical protein complex. Clayton (1962), Bril (1963), and Garcia *et al.* (1966a) found that exposure to light of fractions containing B850 caused this conversion. Later, Suzuki *et al.* (1969) and Erokhin and Sinegub (1970a) found that the addition of several different nonionic and cationic detergents to chromatophores brought about a similar spectral shift that could subsequently be reversed by removal of the detergent. Anionic detergents, however, unless added to give abnormally high final concentrations (25%) (Komen, 1956), had no

effect on the near-IR absorption spectrum. This conversion of B850 to a shorter wavelength can also be brought about in *Chr. minutissimum* chromatophores by treatment with phospholipase A or by heat (Erokhin and Sinegub, 1970b), and by lowering the pH to 2.2–4.5, or by addition of 8–12 M LiCl or 8 M urea (Sinegub and Erokhin, 1971). In the latter three treatments, the spectral shift can be reversed. Most recently, L. Lin (personal communication) has demonstrated reversibility of the conversion of the spectrum of Fraction B into that of Fraction C on pH lowering and subsequent neutralization.

Ke and Chaney (1971) examined *Chr. vinosum* strain D cells grown in the presence of diphenylamine, which greatly reduces the normal carotenoid content of the cells. The H and L bands were still obtained by Triton X-100 treatment, but the L band exhibited an 805 nm peak and an 825 nm shoulder (805 and 837 nm at 77°K). No change in the absorbance of the B890-containing fraction was observed. One-third of the Bchl was obtained as detergent-solubilized pigment with an absorbance maximum at 780 nm; this fraction was not obtained from cells containing a normal complement of carotenoids.

3.1.2. *Thiocapsa roseopersicina* (formerly *Thiocapsa floridana*)

Solubilization of chromatophore membranes by SDS results in the complete loss of the longest-wavelength Bchl (B890) form in this bacterium (Takacs and Holt, 1971). These researchers showed that the B800–B850 material electrophoresed as a single fast-moving band. Removal of the detergent reaggregated this material; however, the 890 nm shoulder did not reappear. The authors suggested that this antenna spectral form represents an aggregated state of the B800–B850 material.

3.2. Purple Nonsulfur Bacteria

3.2.1. *Rhodopseudomonas sphaeroides*

Bril (1958) and Clayton (1962) obtained a partial separation of the antenna Bchl forms by treatment of chromatophores with Triton X-100 and DOC, respectively. Centrifugation of the detergent-treated material yielded a supernatant containing only the B800 and B850 forms, and a pellet in which the B870 form was considerably enriched in comparison to the starting material. From subsequent fractionation studies on the wild-type and mutant organism, cleaner subchromatophore fractions have now been obtained.

The best known of these fractions is the reaction center preparation, the isolation, function, and bio-

Table 1. Summary of Fractionation Studies on Photosynthetic Bacteria

Bchl	Derived fraction			Monomeric size of polypeptide associated with Bchl in fraction
	Name	Size	Carotenoids	
Chromatium vinosum strain D				
B*800, B820 →	Fraction C[a]	100,000[a]	+	Same as Fraction B?
B800, B850 →	Fraction B[a] or L band[b]	100,000[a]	+	7600 and 9800[c]
B890 →	Fraction A[a] or H band[b]	500,000–800,000[a,c,d]	+	≈12,000 with B890[c–e]
Reaction center (RC) Bchl's →	LDAO fraction[e] (760, 800, 875 nm)	—	+	Triad 20,000–30,000[c–e]
Thiocapsa roseopersicina				
B800, B850 →	SDS fraction[f]	—	?	?
B890 →	Altered by SDS[f]			
RC Bchl's				
Rhodopseudomonas sphaeroides, wild-type				
B800, B850 →	LDAO fraction[g]	>100,000[g,h]	+	9000–10,000[g,i]
B870 →	Altered by LDAO[g] or AUT treatment[h]			
RC Bchl's →	RC preparation[j] (757, 803, 866 nm)	120,000[j]	+	22,000, 24,000, and 27,000[k] (strain Y)
Rhodopseudomonas sphaeroides, mutant R26				
B860 →	Triton fraction[l]	?	–	11,000[m]
RC Bchl's →	RC preparation[n] (737, 803, 866 nm)	44,000[o]	–	21,000 and 24,000[o]
Rhodopseudomonas palustris				
B800, B850 →	? L band[p]	?	?	?
B870, RC Bchl's →	H band[p]	?	?	?
Rhodopseudomonas capsulata				
B800, B850, B870 →	??? - - - - - - - - - - - - - -			10,000–11,000[q]
RC Bchl's →	RC preparation[q,r] (750, 798, 850 nm)	?	–	?
Rhodospirillum rubrum				
B880 →	Triton[s] or AUT[h] fractions	100,000[h]	+	19,000[h], ≤10,000[t] with B888
RC Bchl's →	RC preparation[n] (750, 800, 865 nm)	100,000–140,000[u]	+	Triad 20,000–30,000[m,v]

(continued)

Table 1. (continued)

| Bchl | Derived fraction | | | Monomeric size of polypeptide associated with Bchl in fraction |
	Name	Size	Carotenoids	
Rhodopseudomonas viridis				
B1015 ⟶	Altered by SDS[u,w]	25,000[u,w]	?	?
RC Bchl's ⟶	RC preparation[w] (830, 965 nm)	110,000[w]	+	?
Green bacteria				
Bchl *c, d,* or *e* ⟶	?	?	?	?
Bchl *a* (B810) ⇉	Bchl *a*–protein	144,000[x]	–	42,000[x]
RC Bchl's ⟶	Photoreactive particle[y]	1.5 × 10^6 [y]	?	?

[a] Thornber (1970); [b] Garcia *et al.* (1966a); [c] Erokhin and Moskalenko (1973); [d] Halsey and Byers (1975); [e] Lin and Thornber (1975); [f] Takacs and Holt (1971); [g] Clayton and Clayton (1972); [h] Loach *et al.* (1970a); [i] Fraker and Kaplan (1971); [j] Segen and Gibson (1971), Slooten (1972); [k] Jolchine and Reiss-Husson (1974); [l] Reed *et al.* (1970); [m] Clayton and Haselkorn (1972); [n] Numerous papers (see text); [o] Okamura *et al.* (1974); [p] Garcia *et al.* (1968a); [q] Lampe *et al.* (1972), Neith and Drews (1974, 1975); [r] Prince and Crofts (1973); [s] Garcia *et al.* (1966b); [t] van der Rest *et al.* (1974), Oelze and Golecki (1975); [u] Pucheu *et al.* (1974); [v] Smith, W. R. Jr., *et al.* (1972), Noël *et al.* (1972), Wang and Clayton (1973); [w] Thornber *et al.* (1969), Trosper *et al.* (1976); [x] Olson (Chapter 8, this volume); [y] Fowler *et al.* (1971, 1973)

chemical and biophysical characteristics of which are extensively described elsewhere by Gingras (Chapter 6), Feher and Okamura (Chapter 19), and Clayton (Chapter 20). Only the points particularly pertinent to this chapter need be repeated here. The first reaction center preparation from any photosynthetic organism was obtained by treatment of a carotenoidless mutant (R-26) of this bacterium with Triton X-100 (Reed and Clayton, 1968; Reed, 1969); better preparations were obtained subsequently by using LDAO (Feher, 1971; Clayton and Wang, 1971). The reaction center was later isolated from the wild-type organism by use of SDS (Segen and Gibson, 1971), or by SDS plus a modified alkaline urea–Triton (AUT) treatment (Slooten, 1972), and from the carotenoid-containing strain Y by solubilization of the membranes with CTAB (Jolchine *et al.*, 1969). LDAO yields an improved preparation from this strain (Jolchine and Reiss-Husson, 1974). All these preparations have near-IR absorption maxima at 865–867, 802–804 and 756–758 nm that are attributed to Bchl (as P870 and P800) and bacteriopheophytin (756–758 nm), respectively. The contribution of their absorbances to the 700 to 900 spectrum of the whole cells, however, is minor; the major absorbance is derived from antenna Bchl.

A large portion (>60%) of antenna Bchl (B800 and B850) has been isolated free of reaction center pigments but together with unidentified carotenoids in a homogeneous pigment–protein complex (>100,000) daltons) by fractionation of LDAO-treated wild-type chromatophores (Clayton and Clayton, 1972). This method results in a slight shift of the 850 nm peak to 845 nm, which can be reversed by lowering the detergent concentration. When fully denatured, this complex was found to be composed of aggregates of pigments and a 9000-dalton polypeptide. It appears very likely that the B800–B850 component isolated by Clayton and Clayton (1972) is the same entity as the pigment–phospholipid–protein component (Band 15) studied in considerable detail by Kaplan and his colleagues (Fraker and Kaplan, 1971, 1972; Kaplan and Huang, 1973). Unfortunately, the pigment in Band 15 is pheophytinized during the isolation procedure, so the spectral forms of Bchl cannot be identified. SDS treatment of whole chromatophores markedly affects the spectrum of the major (B800+B850) complex (Segen and Gibson, 1971). The light-harvesting component solubilized in this way has very little 800 nm absorbance, a peak at 855 nm, and a shoulder at 870 nm

The other spectral form of antenna Bchl, B870, has not been isolated in a spectrally pure complex. Bril (1958) and Clayton (1962) preserved the 870 nm form in their fractionation studies with DOC and Triton X-100; in neither case, however, was it solubilized by the detergent. Clayton and Clayton (1972) found that this spectral form in wild-type chromatophores was converted by LDAO to one absorbing at 770 nm (monomeric Bchl), although B800 and B850 were not so affected. Loach's AUT treatment causes a similar spectral shift of B870 (Loach *et al.*, 1970a; Hall *et al.*, 1973). On reduction of the concentration of Triton X-100 in an AUT photoreceptor preparation, the 780 nm form changes its absorbance to 870 nm. Loach and co-workers

(Loach *et al.*, 1970a; Hall *et al.*, 1973) suggested that the 870 nm spectral form may be a property of other antenna complexes aggregated in the membranes, and not necessarily a separate complement of Bchl. This suggestion is not inconsistent with the result of Fraker and Kaplan (1971) that all the bacteriopheophytin in their preparation is associated with a single polypeptide zone on polyacrylamide gels. The question whether photosynthetic bacteria contain the longest-wavelength antenna spectral form in a distinct pigment–protein complex will be reconsidered in Section 5.

Reed *et al.* (1970) fractionated the carotenoidless R-26 mutant, which unlike the wild-type organism has only a single absorption maximum in the near IR at 862 nm. Triton solubilization yielded the antenna Bchl free of reaction center pigments in a large Bchl–protein complex, with the wavelength maximum shifted to 850 nm. Clayton and Haselkorn (1972) and Clayton and Clayton (1972) demonstrated that the polypeptide moiety from this complex is of about the same molecular size as that of the (B800+B850)–protein. However, the complex may instead be related to the B870 spectral form of the wild-type organism (see Section 5).

3.2.2. *Rhodopseudomonas palustris* and *Rhodopseudomonas capsulata*

These two bacteria have an IR absorption spectrum almost identical to that of *Rp. sphaeroides*; i.e., they each contain B800, B850, and B870 in very nearly the same relative proportions. The few attempts at fractionation of these two organisms indicate that organization of Bchl in the two bacteria is very similar to that in *Rp. sphaeroides* and *Chr. vinosum* strain D. Thus, Garcia *et al.* (1968a) obtained two bands (H and L) by sucrose density-gradient centrifugation of Triton X-100-solubilized chromatophores of *Rp. palustris*. The L band had equal absorbance at 802 and 857 nm [cf. the (B800+B850)-containing components of *Chr. vinosum* strain D and *Rp. sphaeroides*]. The photochemically active H fraction had a major absorption peak at 873 nm and a much smaller absorbance at 802 nm, which was probably contributed by the P800 of the photochemical reaction center (cf. the published absorption spectra of the H band and Fraction A of *Chr. vinosum* strain D). The photochemical reaction center of a carotenoidless mutant of *Rp. capsulata* has been isolated (Prince and Crofts, 1973; Nieth and Drews, 1974), but the reaction center of *Rp. palustris* has not. The *Rp. capsulata* preparation exhibits three near-IR absorption bands at 855, 798, and 750 nm. The organization of antenna Bchl in *Rp. capsulata* has not been studied directly. It has been deduced that a low-molecular-weight polypeptide of 10,000–11,000 daltons is associated with the light-harvesting pigments in a carotenoidless strain (Lampe *et al.*, 1972; Nieth and Drews, 1974), and that two or more polypeptides of similar molecular weight are associated with the antenna Bchl in the wild-type organism (Nieth and Drews, 1975).

3.2.3. *Rhodospirillum rubrum*

This bacterium contains fewer spectral forms of Bchl than almost all the other photosynthetic bacteria; thus, a less complex organization of Bchl in this organism might be expected. The hypothesis of Sybesma (Sybesma and Fowler, 1968; Sybesma, 1969) that two spectrally different photosystems are present in *Rs. rubrum* would suggest that fractionation of the spectral forms might be possible. Until recently, however, fractionation yielded subchromatophore components with near-IR absorption spectra essentially identical to the spectrum of whole chromatophores (Garcia *et al.*, 1966b; Vernon and Garcia, 1967; Loach *et al.*, 1970b; Hall *et al.*, 1973). Loach and co-workers showed that AUT treatment yields a photoreceptor component with an average particle size of 100,000 daltons that accounts for 100% of the chromatophore pigments. The complex contains about 40 Bchl and 20 carotenoid molecules. The only fraction that has been obtained that differs in absorbance from the spectrum of whole chromatophores is a preparation of the photochemical reaction center that has been isolated free of antenna Bchl from the wild-type organism by use of SDS (Smith, W. R., Jr., *et al.*, 1972) or LDAO (Noël *et al.*, 1972), and from blue-green mutants by Triton X-100 (Gingras and Jolchine, 1969) and by LDAO (Wang and Clayton, 1973; Oelze and Golecki, 1975). All the reaction center preparations exhibit bands at 865, 800, and 750 nm. The 800 and 750 nm absorption of the reaction center probably accounts for the majority of the absorbance at those wavelengths in the intact chromatophores. It thus seems probable that all the antenna Bchl in *Rs. rubrum* is represented by B870; however, no light-harvesting pigment–protein complex has yet been isolated in a spectrally homogeneous state from this bacterium. The work of van der Rest *et al.* (1974) indicates indirectly that a 10,000-dalton polypeptide may be associated with B870 *in vivo*. Oelze and Golecki (1975) present evidence that a 9500- or an 8500-dalton polypeptide or both may perform a similar function in the blue-green mutant strain VI.

The fractionation studies of Garcia *et al.* (1966b) demonstrated the possibility that a bacteriopheophytin *a* component (A_{max} 750 nm) may be a minor component of *Rs. rubrum*; other workers (Gingras and Jolchine, 1969; Wang and Clayton, 1973), however, noted the considerable lability of Bchl in this organism in detergents. Interestingly, the photosynthetically incompetent mutant M46 excretes a bacteriopheophytin–protein–carbohydrate complex into the medium (Schick and Drews, 1969) (see Section 4).

3.2.4. *Rhodopseudomonas viridis*

This bacterium is one of the few isolated thus far that contains Bchl *b*. Its near-IR absorption spectrum is of a simplicity similar to that of *Rs. rubrum*. However, the minor peak at 830 nm and the major antenna Bchl absorbing at 1015 nm have lower energies than the equivalent 800 and 880 nm peaks in *Rs. rubrum*. Olson and Clayton (1966) proposed on the basis of fluorescence and photochemical studies that the 830 nm absorbance was due to a specialized form of Bchl (P830) that was more closely associated with the reaction center, P985 (Holt, A. S., and Clayton, 1965), than the bulk Bchl (B1015). This was confirmed on the isolation (Thornber *et al.*, 1969) of the reaction center (mol. wt. $\approx$ 110,000), which exhibited a photobleachable absorption maximum at 965 nm and a second maximum at 830 nm. The skewed shape of the 830 nm band, which is also observed in spectra of whole cells, may indicate the presence of a bacteriopheophytin in the reaction center.

As in the case of all bacteria discussed above, the very low-energy antenna form, B1015 in *Rp. viridis*, has not been isolated as a pigment–protein complex free of other Bchl types. Addition of SDS (Thornber, 1971; Pucheu *et al.*, 1974), but not of Triton X-100† (Garcia *et al.*, 1968*b*), to broken cells results in an appreciable shift of the wavelength maximum of the antenna Bchl. Intermediate spectral forms appear during this shift; thus, a discrete form at 880 nm is observed prior to conversion to a form absorbing at 810 nm that in turn yields a 685 nm absorbance. The 685 nm absorbing chromophore is probably oxidized Bchl *b*, since Bchl *b* is particularly labile in the presence of oxygen (Baumgarten, 1970). Also, appearance of B685 can be prevented if antioxidants are present in broken-cell suspensions when detergent is added (Trosper *et al.*, 1976). Pucheu *et al.* (1974) isolated the 685 nm spectral form in a pigment–protein complex, whereas Trosper *et al.* (1976) obtained a preparation of the 810 nm absorbing pigment–protein complex. Both preparations have a low molecular weight (<25,000 daltons). No antenna Bchl form other than B1015 has ever been detected in *Rp. viridis*.

† Triton X-100 neither solubilizes nor fractionates the photosynthetic membranes of this bacterium; thus, it is not surprising that the spectral shifts do not occur in the presence of this detergent.

3.3. Green Bacteria

Olson (Chapter 8) describes in considerable detail our total knowledge of the organization of Bchl *a* in this group of photosynthetic bacteria. But for completeness in this chapter, and because more is known about the organization of Bchl in one constituent of the photosynthetic apparatus in these organisms than about any other chlorophyll–protein complex, the salient points bear repetition here.

Green bacteria contain two different types of Bchls: 80–95% of the chlorophyll is represented by Bchl *c* or *d* or *e* (chlorobium chlorophylls), and the remainder by Bchl *a*. Interestingly, the ratio of antenna Bchl to reaction center Bchl in green bacteria is an order of magnitude (1000–1500/l) greater than in the purple bacteria (Fowler *et al.*, 1971). Studies by Sybesma and Olson (1963) showed that light energy absorbed in the antenna by chlorobium chlorophyll is passed to a Bchl *a* form ($A_{max} \sim$ 810 nm), which is contained in the well-characterized Bchl *a*–protein, and from their to the reaction center, P840 (Sybesma and Vredenberg, 1963).

Nothing is known about the organization of the Bchl *c*, *d*, or *e* in the antenna of these organisms except that it can be deduced from the reported Bchl *c*/protein ratio in chlorobium vesicles (Cruden and Stanier, 1970) that very little protein can be associated with these major antenna pigments. This is the only known instance in which protein is not likely to be involved in the organization of photosynthetic pigments. Fractionation studies have been devoted entirely to investigation of the organization of Bchl *a*. It has been known for over a decade that most of the Bchl *a* can be isolated in a water-soluble, crystallizable Bchl *a*–protein complex; no carotenoids are present in the pigment–protein. Recent X-ray diffraction studies (Fenna and Matthews, 1975) on single crystals showed that the complex (mol. wt. 145,000) is composed of three subunits, each of which forms a "string bag" inside which seven Bchl *a* molecules are located. Every Mg atom in the chromophores is coordinated in the fifth but not the sixth position by a water molecule or by an amino acid residue. This Bchl–protein is photochemically inactive.

Active particles (mol. wt. ~ 1.5×10^6) containing the Bchl *a*–protein and the reaction center have been obtained from several green bacteria (Fowler *et al.*, 1971, 1973; Olson *et al.*, 1973). This large complex contains about 80 Bchl *a* molecules per P840. Comparison of its absorption, fluorescence, and CD spectra with those of the Bchl *a*–protein (Fowler *et al.*, 1973; Olson *et al.*, 1973) indicates that the spectra of the Bchl *a*–protein are distinctly different when in the active, larger complex, and that an additional absorption band at 833 nm that is not observed in the purified Bchl–protein is seen in the particle. This absorbance is not due to P840 (Fowler *et al.*, 1973). Recent fractionation studies (Olson *et al.*, 1976) demonstrated that the 833 nm absorbance can be obtained together with the photochemical reaction center in a subfraction of the 1.5×10^6-dalton complex. Thus, the 833 nm absorbing pigment may be a special pigment of the reaction center, and therefore need not result from rearrangement of the Bchl in the Bchl *a*–protein. Nevertheless, there must be some rearrangement of the pigments in the Bchl *a*–protein when it is purified to account for other distinctly different spectral characteristics.

4. Influence of Growth Conditions and Mutations on the Content and Spectral Forms of Bacteriochlorophyll

4.1. Effect of Light Intensity, Oxygen Tension, and Growth Rate on Total Bchl Content of Cells†

Cohen-Bazire and Sistrom (1966) effectively summarized the data showing that the specific Bchl content of a cell is inversely related to the light intensity under anaerobic conditions and to oxygen tension (at lower levels) under aerobic conditions. Thus, cells grown at low light intensities contain more Bchl per cell than those grown at higher intensities (Cohen-Bazire *et al.*, 1957; Fuller *et al.*, 1963; Holt, S. C., *et al.*, 1966). The variation in Bchl content could reflect either (1) a change in the amount per cell of chromatophore material of fixed Bchl content, (2) a change in the Bchl content of a fixed amount of chromatophore material, or (3) a combination of these two cases (Cohen-Bazire and Sistrom, 1966; Worden and Sistrom, 1964). Changes in the Bchl content in *Rp. palustris* (Cohen-Bazire and Sistrom, 1966) and in *Rs. rubrum* during their exponential growth phase (Ketchum and Holt, 1970) are explained by the first possibility. In all other cases studied, including *Rs. rubrum* in the stationary phase of growth (Ketchum and Holt, 1970), changes in Bchl content fit the second possibility.

At any given light intensity, the specific Bchl content of *Rp. sphaeroides* is directly related to growth rate (Cohen-Bazire and Sistrom, 1966). The demonstrated obligatory coupling between Bchl synthesis and protein synthesis [Lascelles, 1959; Sistrom, 1962; Takemoto and Lascelles, 1973, 1974; Takemoto, 1974; Kaplan (Chapter 43); Lascelles (Chapter 42)] almost certainly accounts for this relationship between

† Also see Chapter 44.

growth rate and Bchl content (cf. Cohen-Bazire and Sistrom, 1966). A recent review (Oelze and Drews, 1972) summarizes studies of the relationship of Bchl content to synthesis of intracytoplasmic membranes.

4.2. Effect of Growth Conditions and Mutations on the Spectral Forms of Bchl

4.2.1. *Rhodopseudomonas sphaeroides*

B870 can be enriched with respect to B800 and B850 by growing cells at higher light intensities (see Chapter 44). This has been observed for the wild-type organism (2.4.1) (Aagard and Sistrom, 1972; Cohen-Bazire *et al.*, 1957; Worden and Sistrom, 1964) and for *Rp. sphaeroides* Ga and mutant PM-9 (Crounse *et al.*, 1963). A similar enrichment is observed in cells grown anaerobically in the dark compared with those grown photoheterotrophically (Clayton and Clayton, 1972; Saunders and Jones, 1974). B870 contributes a greater proportion of the near-IR spectrum in *Rp. sphaeroides* strain Y in cells grown in a medium containing a limiting amount of iron (Reiss-Husson *et al.*, 1971), and in wild-type cells in the exponential growth phase compared with those in the stationary phase (Biebl and Drews, 1969; de Klerk *et al.*, 1969).

It appears pertinent that most mutants with altered spectral forms of Bchl are impaired in their ability to synthesize colored carotenoids. Table 2 gives the spectral data and carotenoid content of several *Rp. sphaeroides* mutants. They can be divided into two classes on the basis of the spectral form of Bchl present: (1) those that, like the wild type, contain B800, B850, and B870 [Ga, PM-8, PM-9 (Sistrom and Clayton, 1964; Sistrom, 1966; Crounse *et al.*, 1963), and G (Segen and Gibson, 1971)]; and (2) those with a single major absorption maximum between 860 and 870 nm, which in most cases exhibit a minor absorbance at 800 nm [PM-8:bg-58 (Sistrom and Clayton, 1964), R-22 and R-26 (Crounse *et al.*, 1963), UV-33 (Sistrom *et al.*, 1956), B and B/G (Segen and Gibson, 1971)]. All those classified in the second group have a block in their biosynthetic pathway prior to neurosporene, whereas those in the first group have malfunctions after neurosporene; this observation will be discussed further in Section 5.

4.2.2. *Chromatium vinosum* Strain D

Bergeron and Fuller (1959) observed that cells grown in the presence of diphenylamine, an inhibitor of synthesis of colored carotenoids, had an enhanced content of B800. They concluded initially that carotenoids influenced the near-IR spectrum of this bacterium. Later, Fuller *et al.* (1963) and Garcia *et al.*

Table 2. Spectra and Carotenoid Content of Some *Rhodopseudomonas sphaeroides* Mutants

Strain	Spectrum	Carotenoid content
Wild-type (2.4.1)	B800-B850-B870	Neurosporene, 2-ketospirilloxanthin, spheroidene, and spheroidenone, plus hydroxyl derivatives of last two
GA	Same as wild type[a,b]	Neurosporene, chloroxanthin[a]
PM-8	Same as wild type[a,b]	Neurosporene, chloroxanthin, hydroxylycopene(?)[a]
PM-9	Same as wild type[c]	Phytoene, phytofluene, zeta-carotene, and neurosporene[c]
G	Same as wild type[d]	Two unidentified carotenoids[d]
PM-8:bg 58	862 nm peak; no absorption at 800 nm[a,b]	Lacks colored carotenoids; colorless polyenes may be present[a]
R-22	867 nm peak[e]	Phytoene, phytofluene, and zeta-carotene (trace)[c]
R-26	862 nm peak; minor 800 nm absorption[f]	No polyenes[c]
uv-33	870 nm peak; minor 800 nm absorption[c,g]	Phytoene[c,g]
B/G	870 nm peak; minor 800 nm absorption[d]	Almost no carotenoids present[d]
B (brown)	855 nm peak; minor 800 nm absorption and possibly 870 nm shoulder[d]	Trace of spheroidenonelike carotenoid[d]

[a] Sistrom and Clayton (1964); [b] Sistrom (1966); [c] Crounse *et al.* (1963); [d] Segen and Gibson (1971); [e] Aagard and Sistrom (1972); [f] Reed *et al.* (1970); [g] Sistrom *et al.* (1956).

(1966a) found that the spectrum could be varied considerably by changing the light intensity during growth: at higher intensities, B890 and, to a lesser extent, B850 were enhanced. Fuller *et al.* (1963) decided that their earlier result was better explained by the fact that the experimental cells were grown at lower light intensities than the control cells. This occurred because only the experimental culture flask was wrapped with red cellophane to prevent photo-destruction of diphenylamine. Other workers (Wassink and Kronenberg, 1962; Bril, 1963; Ke and Chaney, 1971) also found that diphenylamine does not affect the near-IR spectrum of *Chr. vinosum* strain D cells. Earlier, Wassink *et al.* (1939) had reported an increased content of B820 in cells grown at lower light intensities. Thus, all the observed changes in the near-IR spectrum of this bacterium can be explained by light-intensity effects.

No pertinent mutants of *Chromatium* have been reported.

4.2.3. *Rhodopseudomonas capsulata*

If wild-type cells of this species are repeatedly subcultured aerobically in the dark, the usual near-IR spectrum is altered. B880 becomes a distinct peak at 875 nm, B850 eventually disappears, and absorption of B800 is markedly decreased (Klemme and Schlegel,

1969; Lien *et al.*, 1973). Drews and co-workers (Drews *et al.*, 1969; Dierstein and Drews, 1974) reported that aerobic light-grown cultures have a spectrum similar to the dark aerobic cultures of Lien *et al.* (1971), with a small band at 800 nm and a major absorbance at 870 nm. Lowering the oxygen tension suddenly to 5 mm Hg causes an increase in the B870 band within 5 min and a shift in its position to 880 nm. The 800 nm band doubles in absorbance within 30 min, but there is no further change in B870; the presence of B850 is not apparent by this time (Drews *et al.*, 1969). Weaver *et al.* (1975) characterized several strains of this species. When strains of the St. Louis type are aerated, they show increased absorption in the carotenoid region and concomitantly develop a distinct shoulder at 870 nm that is not obvious when the cells are grown under strictly anaerobic conditions.

Two types of mutants have been studied. One, Z-1 and strains derived from it, can grow on media containing arsenate. The absorption of B870 is enhanced in these mutants, but only when they are grown at elevated temperatures ($\geqslant 30°C$) and high light intensities (Lien *et al.*, 1973). Mutant Ala⁻ produces neither carotenoids nor Bchl and cannot photosynthesize (Drews *et al.*, 1971). It excretes protochlorophyll– and protopheophytin–protein complexes, the polypeptides of which are not the same as those of the Bchl–proteins in the wild type (Drews, 1974). A photosynthetically competent revertant, Ala[+], produces some Bchl, although it still releases the complexes into the growth medium. The revertant cells have a major absorption band at 872 nm and a minor band at 800 nm, i.e., a spectrum similar to many of the mutants of *Rp. sphaeroides* shown in Table 2. The carotenoid complement of Ala[+] is not known. Mutant strain W4, derived from strain St. Louis of Weaver *et al.* (1975), has a near-IR spectrum like that of Ala[+]. The single large-band maximum is at 875 nm, and a small band is located at 802 nm. The visible spectrum indicates the complete absence of colored carotenoids from this strain.

4.2.4. *Rhodospirillum rubrum*

When *Rs. rubrum* strain S1 is cultured in the dark under rigorously anaerobic conditions, the IR absorption spectrum is indistinguishable from that of photosynthetic cultures (Uffen and Wolfe, 1970). Three-week-old cultures grown in light show increased IR absorbance—the major 880 nm band increases in extinction with respect to the 800 nm band—compared with young cultures (Biebl and Drews, 1969). A concomitant, but smaller, increase occurs in carotenoid absorption.

Many carotenoid mutants of this organism exhibit altered near-IR spectra, as was seen to be the case for *Rp. sphaeroides*. The blue-green mutant G9 was reported to have a spectrum very similar to those of wild-type strains (Wang and Clayton, 1973). Other blue-green mutants (e.g., strain VI), however, have the major band shifted to 875 nm (Oelze and Golecki, 1975). In mutant M_2B, which has no colored carotenoids but which does contain phytoene, the major band is shifted to 870 nm (Kuhn and Holt, 1972). The nonphotosynthetic strain M46 not only has the major band shifted to 872 nm, but also lacks the 800-nm band (Schick and Drews, 1969), phenomena that could be attributed to the absence of reaction center. No IR spectra are reported for mutants FR_1-VI and S1-B4, which also fail to synthesize carotenoids less saturated than phytoene (Maudinas *et al.*, 1974).

The effects of carotenogenesis inhibitors on *Rs. rubrum* have also been investigated. In the presence of 4-phenoxy-2,6-diaminopyridine, cells of strain S1 accumulate phytoene and ζ-carotene. The Bchl spectral forms of these cells are indistinguishable from normally grown cells (Nugent and Fuller, 1967). Diphenylamine causes accumulation of 7,8,11,12-tetrahydrolycopene and 3,4,11′,12′-tetrahydrospheroidene and lesser amounts of spheroidene and 11,12-dihydrospheroidene (Davies *et al.*, 1969), whereas 2-hydroxy-

biphenyl blocks carotenoid synthesis beyond phytoene and phytofluene, the former representing 80% of the carotenoid present (Herber *et al.*, 1972; Maudinas *et al.*, 1973). IR spectra are not reported in these latter cases.

4.2.5. Other Purple Bacteria

In addition to observing pigmentation of *Rp. sphaeroides* and *Rs. rubrum* under strict anaerobic dark conditions, in which growth is slow, Uffen and Wolfe (1970) also cultured *Rp. viridis* and *Rp. palustris* in this manner. The former bacterium has a spectrum identical to that of normal photosynthetic cells, with the exception that a small band at 685 nm is missing from the dark anaerobic culture. Since this band is probably due to oxidized Bchl *b* (Baumgarten, 1970), its absence from cells grown under strict anaerobic conditions would be expected. Anaerobic dark-grown cells of *Rp. palustris* have a spectrum similar to that of photosynthetic cells, but may have a slightly attenuated B880 shoulder.

Biebl and Drews (1969) determined that the relative absorbancies of the IR spectral forms of *Rp. palustris* are particularly sensitive to the nature of the carbon source used in the culture medium. This is not true for any other species of Rhodospirillaceae they investigated. They also reported that in 2-week-old cultures of *Rp. palustris*, the absorption of B800 exceeded that of B850. The relative height of the B870 shoulder also appeared to vary from strain to strain.

Among 33 strains of *Rp. gelatinosa*, Biebl and Drews (1969) distinguished two main types, one of which has a much more distinct B870 shoulder, carotenoid absorption at slightly lower wavelengths, and a much more rapid growth rate than the second type. A wild-type species of *Rp. gelatinosa* grown in the presence of diphenylamine has an unchanged IR absorption spectrum (Crounse *et al.*, 1963).

Rhodopseudomonas viridis can be cultured successfully in the absence of yeast extract if the growth factors *p*-aminobenzoic acid or folic acid are added to the medium (Pucheu *et al.*, 1974). In the absence of these factors, Pucheu *et al.* (1974) also observed cell growth and some formation of photosynthetic pigments. The IR spectra of the two cultures are not different. On a protein or lipid–phosphorus basis, however, the cultures without growth factors contain much less Bchl *b*. Results of gel-electrophoresis experiments on membrane polypeptides from these cultures may be interpreted as indicating a decreased ratio of light-harvesting to reaction center pigment–protein complexes compared with normally grown cells.

Ectothiorhodospira shaposhnikovii will grow aerobically in the dark, the cultures becoming pale rose if they are not agitated (Uspenskaya and Kondrat'eva, 1972). The IR spectrum of these dark grown cells differs from that of the normally cultured photosynthetic cells in that there is relatively less absorption by B850, and the B800 and B890 bands are shifted a few nanometers to the blue in the former cultures.

5. Some Possible Interpretations of Available Data

The results reviewed in Sections 3 and 4, although numerous, appear insufficient to achieve our ultimate goal. At present, it is not possible to describe the IR spectral properties of any one organism quantitatively in terms of specific molecular interactions. When the data are viewed collectively, however, some interpretations and correlations are suggested, which may serve as a broadly applicable basis for the desired explanations. Some points that may be of fundamental importance are discussed in this section. The eventual goal is to obtain an accurate molecular description of the organization of Bchl *in vivo*, which should then lead to a complete understanding of the interrelationship of the spectral and biochemical phenomena observed.

5.1. Molecular Interactions

While the different spectral forms of Bchl have been observed in numerous whole-cell, chromatophore, and Bchl–protein preparations, no completely satisfactory model has been advanced that identifies the interactions responsible for the spectral shifts. It is clear that some component of the immediate environment of the pigment molecule has a profound influence on its absorption spectrum, and that the spectrum thus serves as an indicator of molecular organization in the photosynthetic apparatus. The wide variety of potentially important interactions, however, at present limits the usefulness of this indicator. Once the predominant interactions are identified, if in fact a few interactions are predominant, it is possible that more detailed information about the pigment environment can be derived from a spectrum. Among the interactions that could influence the spectrum are Bchl–Bchl, Bchl–carotenoid, Bchl–lipid, and Bchl–protein interactions. Assignment of the contribution that each makes to the spectral shifts in systems in which several or all these interactions might be found is at present difficult.

5.1.1. Bchl–Bchl Interactions

The large spectral shifts of the B850 and B890 forms of Bchl *a* have often been attributed to Bchl–Bchl interactions, and indeed, 800, 850, and 890 nm absorptions have been observed *in vitro* from hydrated films of Bchl *a* (Rabinowitch, 1956; Reinach *et al.*, 1973). However, the relationship between the interactions in the potentially large aggregates in the films and those in Bchl–protein complexes, which are known to possess a small number of pigment molecules per subunit, is tenuous. Ballschmiter and Katz (1968) reported absorption maxima of Bchl–water micelles at 820, 850, 865, and 895 nm. The presence of water or some other suitable ligand appears to be required for the formation of the long-wavelength spectral forms. The similarity of the red shift for the light-harvesting pigment in green sulfur bacteria to that for the crystalline pigment may be significant. The high specific Bchl content and low protein content of the vesicles that contain the photosynthetic pigments of these bacteria (cf. Cruden and Stanier, 1970) suggest that this pigment may exist in a highly aggregated form, without specific Bchl–protein interactions.

CD spectra give evidence of Bchl–Bchl interactions in several preparations (Sauer, 1975). Reaction center preparations have been thoroughly studied (Phillipson and Sauer, 1973a); in them, strong coupling between the few Bchl molecules present is considered to give rise to exciton states responsible for the spectral features observed (Sauer, 1975). Also, chromatophore preparations show CD features attributable to exciton interactions among antenna Bchl's (Sauer, 1972) (see Section 2.3).

Study of polarization of fluorescence emitted by P870 has led to the suggestions that the P800 transition is not quite parallel to the P870 transition, and that the chromophores responsible are close to each other (Ebrey and Clayton, 1969). In whole chromatophores of several purple bacteria, fluorescence polarization values vary little with exciting wavelengths 800 nm and above (Goedheer, 1973). The polarization value is not large, but is higher in this wavelength range than at shorter wavelengths. The results are indicative of Bchl molecules with some slight relative orientation close enough to interact weakly. The linear dichroism studies (see Section 2) suggest more specifically that B800 and B850 transitions are roughly parallel, at least in bacteria containing lamellae. Whether this is due to two individual chromophores oriented parallel to each other, or to Bchl's interacting to form a new pair of energy levels with parallel transition moments, cannot be decided until the complex containing both the B800 and B850 spectral forms (see Section 3) is more thoroughly analyzed.

5.1.2. Bchl–Protein Interactions

The recent X-ray structural investigation (Fenna and Matthews, 1975) of the green bacterial Bchl *a*–protein complex gives the first detailed picture of the pigment organization and pigment–protein interactions in such complexes. This pigment environment may prove to be characteristic of forms of Bchl that absorb around 800 nm. Care must be exercised, however, in relating this most unusual, easily solubilized protein to the membrane-bound complexes in purple bacteria with 800 nm absorptions. X-ray structural investigations of isolated Bchl–protein complexes that possess lower-energy spectral forms of Bchl could identify structural features responsible for the larger red shifts, but this work is impeded by the difficulty in preparation of crystalline samples of detergent-solubilized proteins.

5.1.3. Bchl–Carotenoid Interactions

Griffiths *et al.* (1955) were the first to gather evidence that Bchl–carotenoid associations might play a significant role in determining the multipeaked near-IR spectra of photosynthetic bacteria. Calvin (1955) specifically suggested that the associations might be via interactions between π-electrons of the two types of pigment molecules. Results of other early work (Clayton and Arnold, 1961) and of some investigations summarized in Section 4 (e.g., Segen and Gibson, 1971) are consistent with this explanation. In these cases, absence of colored carotenoids occurs in conjunction with a reduction in the number of Bchl near-IR spectral forms. The apparent inconsistency of the data on some other "carotenoidless" mutants and on cells grown in the presence of diphenylamine could be attributed to the presence in these organisms of carotenoid precursors that are already highly conjugated polyenes (Davies *et al.*, 1969), if it is assumed that these molecules substitute for the normal carotenoids in the photosynthetic membranes (cf. Fuller and Anderson, 1958). Spectral studies of organisms in which carotenoid synthesis is known to be blocked at more saturated precursors (Maudinas *et al.* 1972, 1974) would aid resolution of this notion.

Carotenoids might also be expected to influence Bchl spectral forms in other equally important, but less direct, ways. Altered carotenoids might cause altered conformations of the polypeptides that form the immediate environments of the Bchl molecules; e.g., the flexibility of the more saturated precursors (Griffiths *et al.*, 1955) might not impose the same structural constraints on associated membrane proteins. Absence of certain carotenoids in mutants may reflect absent or altered enzymatic machinery or regulatory mechanisms, or both, that control the presence and nature of other Bchl neighbor molecules, e.g., polypeptides (see Sistrom, Chapter 44, for further hypotheses concerning this possibility). Adequate experimental data on any one organism, which would be required to verify these proposals, are not yet available.

5.1.4. Bchl–Lipid Interactions

The experimental evidence available on lipids associated with Bchl *in vivo* and with pigment–protein complexes does not facilitate discussion of their role in determining Bchl spectral forms. Phospholipid has been detected in the (B800+B850)–protein complex of *Rp. sphaeroides* (Fraker and Kaplan, 1971). Treatment of *Chr. minutissimum* chromatophores with phospholipase A results in shift of the B850 spectral form to a B830 form (Erokhin and Sinegub, 1970*a*). When *Rp. viridis* is cultured under conditions that reduce the specific Bchl *b* content of the cells, reduced amounts of ornithine lipid, phosphatidyl ethanolamine, and light-harvesting Bchl–protein are present (Pucheu *et al.*, 1974). These data all suggest that lipids present in chromatophore membranes influence the spectral forms of Bchl, but whether this occurs directly via Bchl–lipid interaction, or indirectly, cannot be asserted. Consideration of data obtained with model membranes of Bchl *a* and phospholipids (Hoff, 1974) may not be appropriate, because protein is absent, and the sizes of the Bchl aggregates and the ratio of Bchl to lipid in the models are unlikely to correspond to those in photosynthetic bacteria.

Systematic investigation of alterations in Bchl spectral forms caused by detergent treatments may provide information on the role of lipids in membrane organization. Triton X-100 has been shown to substitute quantitatively for the fatty acid moieties of lamellar lipids in some subchloroplast fractions (Allen, personal communication). The extent to which detergents used to break up chromatophores replace bacterial membrane lipids has not been determined; this might be related to spectral changes observed. The physical and chemical properties of detergents dictate the types of noncovalent bonds attacked and the extent to which they are disrupted. Results that may be amenable to explanations in these terms are the observations that LDAO causes reversible shifts of the B850 maximum 5 nm to the blue in wild-type *Rp. sphaeroides* (Clayton and Clayton, 1972) and of the 862 nm maximum 12 nm to the blue in mutant strain R-26 (Reed *et al.*, 1970); that SDS shifts the antenna spectral form of Bchl *b* in *Rp. viridis* 200 nm to the blue, in a process during which at least two intermediate spectral forms are observed (Trosper *et al.*, 1976); and that SDS causes the (B800+B850) light-harvesting fraction of *Rp. sphaeroides* to change to a B860 form (Segen and Gibson, 1971). The reversibility of these latter two effects has not been investigated. Whether the detergents substitute for lipid in a direct Bchl–lipid interaction, or alter other interactions involving Bchl molecules, needs study.

5.1.5. Influence of Membrane Environment on Molecular Interactions

Some experimental results indicate that insertion of Bchl–protein complexes into a membrane causes slight shifts of the absorption maxima of some of the Bchl forms. For example, when *Rp. sphaeroides* R-26 reaction centers and membrane fragments of non-photosynthetic strain O are reconstituted into photosynthetically competent membranes, the P800 band is shifted from 804 to 806 nm (Jones and Plewis, 1974), and the Bchl a–protein of green bacteria has slightly different absorption maxima in the isolated state from that of aggregates in cell fractions (Olson *et al.*, 1973). The spectral forms of antenna Bchl in the PM-8 *Rp. sphaeroides* mutant are not exactly the same as those in mutants that do contain reaction center (Sistrom, 1966); Segen and Gibson (1971) also suggested that reaction centers affect the antenna spectral forms in the membranes of *Rp. sphaeroides*.

Liquid-nitrogen temperature results in sharpening and better resolution of Bchl spectral forms, as expected, but also effects marked energy changes in the longer-wavelength forms (see Table 3). The extent to which this result is due to the rigidity of the frozen membrane changing molecular interactions has not been considered.

The observations cited above may be due to an influence of the membrane on the conformations of the complexes within it, as well as to weak interactions among chromophores in different complexes. An additional consideration is the location of a particular complex within the membrane. A pigment–protein *in vivo* may be in an environment with a different dielectric constant than when it is purified in aqueous detergent solution. This difference may not only result in energy-level shifts, but might also result in the complex's being inaccessible to certain reagents—as the reaction center appears to be to potassium iridium chloride (Kuntz *et al.*, 1964). Explanations for these phenomena must await descriptions of membrane structure on a molecular level finer than that yet achieved.

5.2. Pigment–Protein Complexes of Photosynthetic Bacteria

Isolation of pigment–protein complexes from bacteria has progressed substantially during the last decade. The aim of such studies has been to obtain for subsequent biochemical and biophysical characterization the greatest structural simplification without significant alteration of the forms in which the pigments function *in vivo*. This goal imposes the requirement of preservation of secondary and tertiary structure of any polypeptides associated with Bchl in these functioning complexes. Surprisingly, detergents, some of which are normally used as strong denaturing agents, have proved satisfactory for the controlled membrane breakdown required. The selection of detergent and the ratio of detergent to membrane are critical factors (cf. Thornber and Olson, 1971). No one detergent is universally useful; e.g., SDS yields several spectrally different fractions representative of the state of Bchl *in vivo* from *Chr. vinosum* strain D (Thornber, 1970), but destroys much of the specific organization of Bchl when applied to *Rp. sphaeroides* membranes (Segen and Gibson, 1971). Again, SDS does not dissociate all the antenna Bchl from the reaction center of *Chromatium*, but LDAO does. LDAO is well suited to the solubilization of the photochemical reaction center from several bacteria, especially mutants lacking colored carotenoids. The reason for its particular suitability and the fact that the so-called "carotenoidless" mutants are more easily fractionated remain to be explained.

5.2.1. Complexes of Purple Bacteria

Some spectral forms of Bchl can be separated from others after dissociation of the photosynthetic apparatus by detergents; two or more spectrally different fractions have been obtained from several purple bacteria (see Table 1). Reed *et al.* (1970) concluded that protein plays a big role in the maintenance of the environment of the Bchl molecules on the basis of their observations that only Bchl and protein are present in the antenna pigment–protein isolated from *Rp. sphaeroides* R-26, and that proteolytic digestion of the complex results in a shift of the spectral form of Bchl present. This conclusion is supported by less direct evidence, e.g., the effect of protein denaturants on the near-IR spectrum of bacterial chromatophores (e.g. Sinegub and Erokhin, 1971), and the demonstration of an obligatory coupling between Bchl synthesis and protein synthesis (see Section 4.1). Therefore, the earlier hypothesis (Katz, E., and Wassink, 1939; Wassink *et al.*, 1939; French and Young, 1956) that each spectral form of Bchl *in vivo* is associated with a specific protein has been shown to be basically correct. Modifications of the original idea imposed by recent data are discussed below.

Similarities in the characteristics of fractions from the different organisms examined make it tempting to hypothesize that analogous pigment–protein complexes occur in all these organisms. Obviously, one of these complexes is the photochemical reaction center, which has an essentially identical spectral and polypeptide composition in all purple bacteria (however, see Clayton and Haselkorn, 1972). We propose that a second complex is a B890–protein.[†] Such a pigment–protein complex is likely to be present in all purple bacteria (see Sections 3 and 4), even though it has never been isolated free of all other Bchl spectral forms. The best preparations of the B890–protein are contaminated mainly by the photochemical reaction center, e.g., Band H (*Chromatium* and *Rp. palustris*), Fraction A (*Chromatium*), and the AUT preparation (*Rs. rubrum*) (Table I). One cannot yet determine whether carotenoids are integral components or whether they indirectly influence the spectral form of Bchl in the complex. Indirect evidence indicates that the size of the polypeptide associated with B890 is small ($\approx$ 12,000 daltons) (see Table 1). Aggregates of this small pigment–protein are postulated to occur in the intact photosynthetic unit as well as in the best preparations of the B890–protein. A third complex, the (B800+B850)–carotenoid–protein, has been isolated from several purple bacteria. Analogous preparations of this pigment–protein from the different strains show two near-IR absorption peaks of roughly equal intensity. The size of the fully denatured protein moiety is similar to, but slightly smaller (8000–12,000 daltons) than, that proposed to be associated with B890. Hence, the homogeneous preparations of B800+B850 complex must be composed of aggregates of the polypeptide and photosynthetic pigments because the size of such fractions is 100,000 daltons or greater (see Table 1). More than one 100,000 unit is probably present in each photosynthetic unit *in vivo* (cf. Thornber, 1970; Clayton and Clayton, 1972). The fourth class of pigment–protein complexes is a (B*800+B820)–carotenoid–protein that occurs in only a few bacteria under certain culture conditions. Apart from its near-IR spectrum, all other characteristics of this complex are identical to those of the (B800+B850)–protein from which it can be formed by a variety of treatments (see Section 3).

It is possible to explain the three spectrally different groups of purple bacteria (see Section 2) by the presence of varying amounts of these four types of pigment–proteins. Group I organisms (e.g., *Rs. rubrum* and *Rp. viridis*) exhibit a single major antenna spectral form and a very much smaller absorbance at 805 nm (*Rs. rubrum*) or 830 nm (*Rp. viridis*), which is probably due to reaction center pigments. These bacteria are assumed to contain only two of the four

pigment–proteins, the reaction center– and B890– proteins. By varying the amounts of (B800+B850)– protein added to these two, the spectra of all group II organisms (e.g., *Rp. sphaeroides*, *Rp. capsulata*, and *Rp. gelatinosa*) can be mimicked. To obtain spectra equivalent to those of group III organisms (e.g., *Chromatium* and *Rp. palustris*), addition of the (B*800+B820)–protein is also required. It is very likely that the precise location of the IR maxima in analogous chlorophyll–proteins will vary slightly and thereby account for the variations in the precise wavelengths of the spectral forms of Bchl in the different purple bacteria. Subchromatophore fractions of group II and III organisms (e.g., Band H and Fraction A of *Chromatium* or *Rp. palustris*) have spectra and bulk Bchl/reaction center ratios very similar to those of whole chromatophores of group I bacteria such as *Rs. rubrum*. Thus, fractionation studies support the hypothesis. Furthermore, such studies support the early suggestion of J. C. Smith and French (1963) of the nonuniform distribution of the spectral forms over the membrane surface. Mutants may totally lack one or more of the four pigment proteins, and therefore belong to a different spectral group than their wild-type organism.

It is of interest that only in organisms of group I is the esterifying alcohol of the porphyrin not solely phytol; the Bchl's in these bacteria are esterified with geranyl–geraniol or phytol (Kuntzler and Pfennig, 1973; Gloe and Pfennig, 1974). The role of the esterifying alcohol in determining the spectral forms has not been elucidated; it has been reported, however, that in *Rs. rubrum* mutant G-9, an organism having 93% geranyl-geraniol, the only chromophore esterified with phytol is the bacteriopheophytin in the reaction center (Walter, 1975).

5.2.2. Factors That Influence the Pigment–Protein Content of Photosynthetic Membranes of Purple Bacteria

Mutant spectra and changes in spectra of wild-type organisms subjected to varied growth conditions can be explained using different combinations of the same four pigment–proteins. Occasionally, a particular pigment–protein of the normal organism is completely missing from the altered organism. Researchers working intensively in this area (Sistrom, 1964, 1966; de Klerk *et al.*, 1969; Aagaard and Sistrom, 1972; Lien *et al.*, 1971, 1973) have come to a similar conclusion [see also Sistrom (Chapter 44)]. Aagaard and Sistrom (1972) found that in *Rp. sphaeroides* cultures of different specific Bchl content, the ratios of B870/reaction center pigments and of B800/B850 are constant. They determined that variations in the ratio of

---

[†] We will use B890 for the longest-wavelength Bchl form, although the precise location of this Bchl type varies between 870 and 890 nm in different species.

(B800+B850)/(B870 + reaction center pigments) are reflected by an altered antenna Bchl/reaction center ratio and account for the variations in spectra observed under different growth conditions (see also Cohen-Bazire and Sistrom, 1966; Sistrom, Chapter 44). Similarly, Lien *et al.* (1971, 1973) propose that two light-harvesting complexes (LH-C) may be present in varying amounts in *Rp. capsulata* under different growth conditions. LH-CI is similar to the B890–protein, and LH-CII is equivalent to the light-harvesting (B800+B850)–protein. Lien *et al.* (1973) proposed that synthesis of the latter is controlled by the amount of energy-rich compounds available. The plasticity of this component is similar to that of the light-harvesting chlorophyll *a/b*–protein of chlorophyll-*b*-containing plants (cf. Thornber, 1975). All purple bacteria studied show similar spectral responses to changes in growth conditions (see Section 4). It is thus proposed that they are all capable of varying the relative amounts of Bchl–protein complexes present. Under optimum growth conditions (actively dividing cells grown in high light or aerobically in the dark), purple bacteria require a small antenna pigment system. The organisms appear to respond to this requirement by reducing the rate of synthesis of their major antenna Bchl–protein; e.g., group I decreases the relative proportion of the B890–protein, group II organisms reduce the rate at which they synthesize the (B800+B850)–protein. Conversely, under stress conditions (limiting light intensities, stationary phase of growth, or dark anaerobiosis), the organism reacts by increasing the relative rate of synthesis of an antenna Bchl–protein (i.e., B890–protein, B800+B850–protein, and B*800+B820–protein in organisms of groups I, II, and III, respectively). Organisms of spectral group I (e.g., *Rs. rubrum*) are restricted in the variety of ways they can respond because they appear to be incapable of synthesizing more than one antenna pigment–protein complex. In our interpretation of observed phenomena, bacteria of the other two spectral groups apparently find it biosynthetically easier and more economical to vary synthesis of antenna Bchl–protein complexes other than the B890–protein.

A point that is pertinent to this discussion and that has been considered by only a few workers in the area is the nature of the antenna Bchl–protein(s) in the spectrally altered carotenoid mutants. Spectral evidence alone cannot unequivocally indicate which antenna complex is present. Since the size of the polypeptide associated with the antenna pigments in the mutant is the same as that of the (B800+B850)–protein (see Section 3.2.1), it might be concluded that a spectrally altered (B800+B850)–protein is present. However, an alternative hypothesis (Aagaard and Sistrom, 1972) that the (B800+B850)–protein is missing from mutants lacking colored carotenoids and that the pigment–protein present is a spectrally altered form of the B890-containing component appears more likely in view of the following considerations: First, it is now believed that the B890–protein complex contains a polypeptide of roughly the same size as that in the (B800+B850)–protein complex (see Section 3). Second, the antenna Bchl/reaction center ratio in wild-type organisms varies with the specific Bchl content, whereas in two carotenoid-deficient mutants (R-22 and R-26 strains of *Rp. sphaeroides*), this ratio [30/1 (Aagaard and Sistrom, 1972) and 35/1 (Reed, 1969), respectively] is constant and equal to the calculated ratio (Aagaard and Sistrom, 1972) for B870/reaction center in the wild-type (Cohen-Bazire and Sistrom, 1966). Finally, B890 in carotenoidless mutants of group I organisms is shifted to shorter wavelengths by the same amount as would be necessary to shift the B890–protein in *Rp. sphaeroides* to give the observed spectrum of mutants lacking colored carotenoids. Identification of the mutant Bchl–protein complex as an altered form of B890– or (B800+B850)–protein— or possibly as a different component not analogous to either of these—cannot be made until all three isolated complexes are subjected to detailed protein chemistry investigations (e.g., fingerprinting, amino acid analysis, terminal amino acid determination).

5.2.3. Use of Bchl–Protein Models for Interpretation of Spectra

The organization of Bchl in the pigment–proteins isolated from the purple bacteria may ultimately be shown to exhibit a few similarities to that in the Bchl *a*–protein of green bacteria (see above and Olson, Chapter 8). The bonds between peptide and Bchl in the purple bacteria must be located internally, where they are protected from disruption by detergents. Each Bchl must be arranged so that it can interact at least weakly with another chromophore, e.g., Bchl or carotenoid (see above), to account for the CD spectral observations, the high efficiency of energy transfer, and the multiplicity of Bchl absorption maxima. Insertion of the Bchl into a "bag" formed by the polypeptide, as occurs in the Bchl *a*–protein of green bacteria, could meet these requirements. Nevertheless, the much larger size of the polypeptide in the latter complex indicates that some basic differences must exist between the supramolecular organization of antenna Bchl *a* in the purple and the green bacteria. Furthermore, we anticipate that hydrophobic amino acid residues must be located on the exterior of the complexes in the purple bacteria to account for their insolubility in water.

Table 3. Near-IR Bchl Spectral Forms in Whole Organisms and Chromatophores

Organism	Spectral forms					Comment	Reference
	B800	B*800	B820	B850	B870–890		
Rhodospirillaceae							
Rhodospirillum rubrum	800m				875–878		Wassink *et al.* (1939)
	795m				875		French and Young (1956)
	≈800m				≈880		Clayton (1963)
	803m				880		Kuntz *et al.* (1964)
	804m				880.4		Olson and Stanton (1966)
	805m				883		Vredenberg and Amesz (1966)
S1	802m				894	Liq. N$_2$	Vredenberg and Amesz (1966)
F	802m				883		Vernon and Garcia (1967)
	807m				885		Schick and Drews (1969)
Six strains	807–808m				881–885		Biebl and Drews (1969)
4	≈805m				880		Goedheer (1972)
4	803m				895	Liq. N$_2$	Goedheer (1972)
BG1	805m				875		Clayton (1966)
M46	—				872		Schick and Drews (1969)
M$_2$B	806m				870		Kuhn and Holt (1972)
G9			—"As wild type"—				Wang and Clayton (1973)
V1	≈805m				≈875		Oelze and Golecki (1975)
Rhodospirillum tenue[a]							
7a/1	802		858	≈890s			Biebl and Drews (1969)
40/7	802		855	873			Biebl and Drews (1969)
11a/6	809		877				Biebl and Drews (1969)
2761	805		852	880s			Pfenning (1969b)
Rhodospirillum fulvum							
1350	807		850	≈880s			Pfennig *et al.* (1965)
Rhodospirillum molischianum	800			890s			Gibbs *et al.* (1965)
Six strains	803–807		853–855	≈880s			Biebl and Drews (1969)
	≈800	824s	848				Goedheer (1972)
	800		860	895		Liq. N$_2$	Goedheer (1972)
Rhodospirillum photometricum	805			881			Trosper (unpublished)

Strain							Reference
Rhodopseudomonas palustris							
EII 5.1.1	802		862.5				Wassink et al. (1939)
EII 5.1.6	802		850s				Wassink et al. (1939)
EII 5.2.1	802		881				Wassink et al. (1939)
2137	804	820s			≈880s		Olson and Stanton (1966)
	804		858m		880		Vredenberg and Amesz (1967)
	804	819	≈850s		895	Liq. N₂	Vredenberg and Amesz (1967)
22 strains	807–810		862	863–878			Biebl and Drews (1969)
	802			860s	≈880		Goedheer (1973)
	800			870s	895	Liq. N₂	Goedheer (1973)
Rhodopseudomonas viridis							
NHTC 133	830				995	Bchl b	A. S. Holt and Clayton (1965)
	835/836				1020	Bchl b	Drews and Giesbrecht (1965)
NHTC 133	≈820				1014	Bchl b	Olson and Stanton (1966)
NHTC 133	830				1015	Bchl b	Garcia et al. (1968b)
Four species	835–837				1017–1020	Bchl b	Biebl and Drews (1969)
	825				1010	Bchl b	Goedheer (1972)
	832				1034	Liq. N₂	Goedheer (1972)
Rhodopseudomonas acidophila							
7050	805			855–860	890s		Pfennig (1969a)
Rhodopseudomonas gelatinosa							
	≈800			≈850	≈890s		Crounse et al. (1963)
	803			859	≈880s		Haskins and Kihara (1967)
Type I (8 strains)	805–810			862–865	≈890s		Biebl and Drews (1969)
Type II (17 strains)	802–805			854–860	880s		Biebl and Drews (1969)
Rhodopseudomonas capsulata							
	806			863	875s	Liq. N₂	Vredenberg and Amesz (1967)
37b4	806			866	882		Vredenberg and Amesz (1967)
	802			857			Haskins and Kihara (1967)
8 strains	805–806			860–866	880s		Biebl and Drews (1969)
23782	800			860	≈880s		Sojka et al. (1970)
Z-1	802			856	880s		Lien et al. (1971)
Ala pho⁻	—No IR bands—						Drews et al. (1971)
Ala pho⁺	798m			872			Nieth and Drews (1974)
St. Louis	802			858	—		Weaver et al. (1975)
	802			858	870s	Aerated	Weaver et al. (1975)
W4	802			858	875		Weaver et al. (1975)

(continued)

Table 3. (continued)

Organism	Spectral forms					Comment	Reference
	B800	B*800	B820	B850	B870–890		
Rhodopseudomonas sphaeroides							
	≈805			855	890		Clayton (1963)
	799			849	≈880s		Kuntz *et al.* (1964)
	≈800			852	≈880		Olson and Stanton (1966)
	801			852	870s		Vredenberg and Amesz (1967)
	801			856	882	Liq. N$_2$	Vredenberg and Amesz (1967)
37/3	810			853	878s		Biebl and Drews (1969)
39/2	808			863s	880		Biebl and Drews (1969)
Five strains	800–803			863–868	880–889		Biebl and Drews (1969)
28/3	803			860	—		Biebl and Drews (1969)
NCIB 8327	800			850	880s		Segen and Gibson (1971)
	800	850		850	880s		Goedheer (1972)
	799			854	886	Liq. N$_2$	Goedheer (1972)
PM-9	800			850	—	Low light	Crounse *et al.* (1963)
PM-9	≈800m			≈850s	870	High light	Crounse *et al.* (1963)
R-26	800m				870		Clayton (1963)
uv-33	≈800m				870		Sistrom and Clayton (1964)
PM-8	≈800			≈850	—		Sistrom and Clayton (1964)
PM-8 : bg 58				≈860	—		Sistrom and Clayton (1964)
B	800m			855	870		Segen and Gibson (1971)
BG	805m	859			880s		Segen and Gibson (1971)
G	800	≈800		850	880s		Segen and Gibson (1971)
Ga	≈800			≈850s	≈870	Low pigment	Aagard and Sistrom (1972)
V-2			—No IR bands—	≈850	≈875s	High pigment	Aagard and Sistrom (1972)
G-VP	801			852	—		Saunders and Jones (1974)
G-VP	801			858	880s	Liq. N$_2$	Saunders and Jones (1974)
G-VP	805			866	—	Aerobic	Saunders and Jones (1974)
G-VP	805			860s	881	Aerobic. liq. N$_2$	Saunders and Jones (1974)
O			—No IR bands—				Jones and Plewis (1974)
Rhodopseudomonas globiformis							
7950	813			862	870s	895s, also	Piennig (1974)
Rhodopseudomonas sulfidophila							
W4	803			855	895s		Hansen and Veldkamp (1973)

							Reference
Rhodomicrobium vannielii	800			≈870	880ms		Clayton (1963)
2/1	807			872			Biebl and Drews (1969)
Chromatiaceae							
Ectothiorhodospira mobilis							
8112	≈805			≈875	900s		Trüper (1968)
Ectothiorhodospira shaposhnikovii	≈795			≈855	≈895s		Uspenskaya and Kondrat'eva (1972)
Ectothiorhodospira halophila	798			≈850s	885		Raymond and Sistrom (1967)
Thiospirillum jenense	≈800		835	850	890s		Schlegel and Pfennig (1961)
Chromatium okenii	800m			850	880s	Enriched	Schlegel and Pfennig (1961)
Chromatium warmingii	≈805			850	880s	Enriched	Schlegel and Pfennig (1961)
Chromatium buderii	≈800s			840	≈890s	Enriched	Trüper and Jannasch (1968)
Chromatium vinosum							
D	803.5			854	895s		Wassink et al. (1939)
D	795			848	885s		French and Young (1956)
D	800			855	890s		Schlegel and Pfennig (1961)
D	805			845s	887		Kuntz et al. (1964)
D	806		820s	850s	889		Olson and Stanton (1966)
D	800s		825s	850s	890		Vredenberg and Amesz (1967)
D	795	810	825	855	905	Liq. N₂	Vredenberg and Amesz (1967)
D	795s	804	825s	850	888		Goedheer (1972)
D	796	805	825s	860	910	Liq. N₂	Goedheer (1972)
D variant	800	808		850	880		Clayton (1962)
Chromatium minutissimum	≈795			≈850	885s		Erokhin and Sinegub (1970b)
	790	805	835s	870	895s	Liq. N₂	Erokhin and Sinegub (1970b)
Chromatium violasens							
3311	810m			850	≈880s		Trüper and Genovese (1968)
Thiocystis violacea	796s	810s		849s	890		Goedheer (1972)
	796	808		866	910	Liq. N₂	Goedheer (1972)
Thiocapsa roseopersicina	797		824s	853	889s		Vredenberg and Amesz (1966)
	797			866	903	Liq. N₂	Vredenberg and Amesz (1966)
9314	≈800			850	890s		Takacs and Holt (1971)
Thiocapsa pfennigii	830		825		≈1020	Bchl *b*	Eimhjellen et al. (1967)
Amoebobacter roseus	802		825	855s	890		Vredenberg and Amesz (1967)
	795		820	855s	905	Liq. N₂	Vredenberg and Amesz (1967)
Amoebobacter sp.	808			≈850	≈895	Enriched	Schlegel and Pfennig (1961)
Rhodothece sp.							
41	790s		830		883m		Verkhoturov et al. (1969)

(continued)

Table 3. (continued)

Organism	Spectral forms					Comment	Reference
	B800	B*800	B820	B850	B870–890		
Chlorobiaceae	Bchl c, d, or e				Bchl a		
Chlorobium limicola f. sp. thiosulfatophilum	740				803		Goedheer (1972)
	744				804, 813s, 824s	Liq. N₂	Goedheer (1972)
L (Lascelles)	725				810	Bchl d, a	Stanier and Smith (1960)
PM (Lascelles)	747				—	Bchl c, a	Stanier and Smith (1960)
6230, Tassajara	756				812	Bchl c, a	Gloe et al. (1975)
Chlorobium vibrioforme							
6030	736				809	Bchl d, a	Gloe et al. (1975)
Chlorobium vibrioforme f. sp. thiosulfatophilum							
1930	727				800	Bchl d, a	Gloe et al. (1975)
Chlorobium phaeobacterioides							
2430	720				808	Bchl e, a	Gloe et al. (1975)
2431	721				808	Bchl e, a	Gloe et al. (1975)
9230	719				805	Bchl e, a	Gloe et al. (1975)
Chlorobium phaeovibrioides							
B1 (Zenitani)	715				806	Bchl e, a	Gloe et al. (1975)
2531	724				805	Bchl e, a	Gloe et al. (1975)
2631	726				807	Bchl e, a	Gloe et al. (1975)
Prosthecochloris aestuarii							
SK413	750				—	Bchl c	Gorlenko (1970)
Chloropseudomonas ethylica[b]	750				—	Bchl c, a	Shaposhnikov et al. (1960)
Pelodictyon luteolum							
2532	725				805	Bchl d, a	Gloe et al. (1975)
Pelodictyon clathratiforme							
1831	750				—	Bchl c, a	Pfennig and Cohen-Bazire (1967)
2730	739				—	Bchl d, a	Pfennig and Cohen-Bazire (1967)
Gliding Bacteria							
Chloroflexus aurantiacus	740				802, 865	Bchl c, a	Pierson and Castenholtz (1974a)

Species are listed in the order in which they appear in *Bergey's Manual of Determinative Bacteriology.* Spectral forms that are present as shoulders are denoted by the suffix *s*; minor bands, by the suffix *m*.

[a] Pfennig (1969b) has identified the *Rhodospirillum* species designated as Group IV by Biebl and Drews (1969) as *Rs. tenue*, and strains in this group are so listed here.

[b] *Chloropseudomonas ethylica* is a mixed culture of a green bacterium and one or more nonphotosynthetic, sulfate-reducing bacteria (Gray et al., 1973). See Chapter 2 for a discussion of the taxonomy of this syntrophic system. Litvin and Gulyaev (1969) reported a complex low-temperature, second-derivative spectrum of this culture with peaks at 688, 705, 715, 741, 762, 782, 790, 801, 811, 823, and 835 nm.

Apart from the Bchl *a*–protein of green bacteria, no well-characterized, naturally occurring complexes are available to serve as models to interpret near-IR spectra in terms of molecular interactions *in vivo*. Thus, the wealth of spectral data available on whole organisms and subchromatophore fractions cannot yet be used directly to describe the state of Bchl *in vivo*. This information will become more useful when further basic analytical data on Bchl–proteins are obtained. Rigorous analyses are required to establish the identity and relative proportions of all the constituents in the available homogeneous pigment–protein complexes. More important will be purification to homogeneity of other pigment–proteins (e.g., B890–protein), and detailed analysis of their protein moieties. Eventually, X-ray analyses must be performed on crystalline complexes, which then might be used as models to describe Bchl environments *in vivo*, if one also considers the effects of supramolecular aggregation on the energy levels of chromophores.

6. Note Added in Proof

Many data pertinent to this review have appeared since the review was completed, for which the reader should consult the publications of serveral recent photosynthesis meetings: *Brookhaven Symposium in Biology*, Vol. 28 (1977); *IV International Congress of Photosynthesis Research*, Reading, England (Sept., 1977); *Ciba Foundation Symposium*, No. 61, London, England (Feb., 1978, to be published by Elsevier). A particularly pertinent article for some of the hypotheses put forward in this article is H. C. Yen and B. Marrs, 1976, *J. Bacteriol.* **126**:619–629.

ACKNOWLEDGMENTS

The preparation of this review was supported by grants from the National Science Foundation (GB 31207 to J. P. T.; BMS 74-12596 to C. E. S.) and from the United States Public Health Service (Molecular Biology Training Grant GM 1531).

7. References

Aagaard, J., and Sistrom, W. R., 1972, Control of synthesis of reaction center bacteriochlorophyll in photosynthetic bacteria, *Photochem. Photobiol.* **15**:209.

Ballschmiter, K., and Katz, J. J., 1968, Long wavelength forms of chlorophyll, *Nature (London)* **220**:1231.

Barer, R., and Joseph, S., 1955, Refractometry of living cells. II. The immersion medium, *Q. Rev. Microscop. Soc.* **96**:1.

Baumgarten, D. L., 1970, M.A. thesis, University of California, Berkeley.

Bergeron, J. A., and Fuller, R. C., 1959, Influence of carotenoids on the infrared spectrum of bacteriochlorophyll in *Chromatium, Nature (London)* **184**:1340.

Biebl, H., and Drews, G., 1969, Das *in vivo*-Spektrum als taxonomisches Merkmal bei Untersuchungen zur Verbreitung von Athiorhodaceae, *Zentralbl. Bakteriol. Parasitenkd. Infektionskr. Hyg. Abt. 2:* **123**:425.

Boyce, C. O. L., Oyewole, S. H., and Fuller, R. C., 1977, Localization of photosynthetic reaction center in *Chlorobium limicola, Brookhaven Symp. Biol.* **28**:365 (abstract).

Breton, J., 1974, The state of chlorophyll and carotenoids *in vivo*. II. A linear dichroism study of pigment orientation in photosynthetic bacteria, *Biochem. Biophys. Res. Commun.* **59**:1011.

Bril, C., 1958, Action of a non-ionic detergent on chromatophores of *Rhodopseudomonas spheroides, Biochim. Biophys. Acta* **29**:458.

Bril, C., 1960, Studies on bacterial chromatophores. I. Reversible disturbance of transfer of electronic energy between bacteriochlorophyll types in *Chromatium, Biochim. Biophys. Acta* **39**:296.

Bril, C., 1963, Studies on bacterial chromatophores. II. Energy transfer and photooxidative bleaching of bacteriochlorophyll in relation to structure in normal and carotenoid-depleted *Chromatium, Biochim. Biophys. Acta* **66**:50.

Calvin, M., 1955; addendum to Griffiths *et al.* (1955): *Nature (London)* **176**:1215.

Clayton, R. K., 1962, Primary reactions in bacterial photosynthesis. I. Nature of light-induced absorption changes in chromatophores; evidence for a special bacteriochlorophyll component, *Photochem. Photobiol.* **1**:201.

Clayton, R. K., 1963, Absorption spectra of photosynthetic bacteria and their chlorophylls, in: *Bacterial Photosynthesis* (H. Gest, A. San Pietro, and L. P. Vernon, eds.), pp. 495–500, Antioch Press, Yellow Springs, Ohio.

Clayton, R. K., 1965, *Molecular Physics in Photosynthesis*, pp. 149–156, Blaisdell, New York.

Clayton, R. K., 1966, Spectroscopic analysis of bacteriochlorophylls *in vitro* and *in vivo, Photochem. Photobiol.* **5**:669.

Clayton, R. K., and Arnold, W., 1961, Absorption spectra of bacterial chromatophores at temperatures from 300°K to 1°K, *Biochim. Biophys. Acta* **48**:319.

Clayton, R. K., and Clayton, B. J., 1972, Relations between pigments and proteins in photosynthetic membranes of *Rhodopseudomonas spheroides, Biochim. Biophys. Acta* **283**:492.

Clayton, R. K., and Haselkorn, R., 1972, Protein components of bacterial photosynthetic membranes, *J. Mol. Biol.* **68**:97.

Clayton, R. K., and Wang, R. T., 1971, Photochemical reaction centers of *Rhodopseudomonas spheroides, Methods Enzymol.* **23**:696.

Cohen-Bazire, G., and Sistrom, W. R., 1966, The prokaryotic photosynthetic apparatus, in: *The Chlorophylls* (L. P. Vernon and G. R. Seely, eds.), pp. 313–341, Academic Press, New York.

Cohen-Bazire, G., Sistrom, W. R., and Stanier, R. Y., 1957, Kinetic studies of pigment synthesis by non-sulfur purple bacteria, *J. Cell. Comp. Physiol.* **49**:25.

Crounse, J., Sistrom, W. R., and Nemser, S., 1963, Carotenoid pigments and the *in vivo* spectrum of bacteriochlorophyll, *Photochem. Photobiol.* **2**:361.

Cruden, D. L., and Stanier, R. Y., 1970, The characterization of *Chlorobium* vesicles and membranes isolated from green bacteria, *Arch. Mikrobiol.* **72**:115.

Cusanovitch, M. A., Bartsch, R. G., and Kamen, M. D., 1968, Light-induced electron transport in *Chromatium* strain D. II. Light-induced absorption changes in *Chromatium* chromatophores, *Biochim. Biophys. Acta* **153**:397.

Davies, B. H., Holmes, E. A., Loeber, D. E., Toube, T. P., and Weedon, B. C. L., 1969, Carotenoids and related compounds: Part XXIII. Occurrence of 7,8,11,12-tetrahydrolycopene, spheroidene, 3,4,11′,12′-tetrahydrospheroidene and 11,12-dihydrospheroidene in *R. rubrum*, *J. Chem. Soc. C:* 1266.

de Klerk, H., Govindjee, Kamen, M. D., and Lavorel, J., 1969, Age and fluorescence characteristics in some species of Athiorhodaceae, *Proc. Natl. Acad. Sci. U.S.A.* **62**:972.

Dierstein, R., and Drews, G., 1974, N-limited continuous culture of *Rps. capsulata* growing photosynthetically or heterotrophically under low O_2 tensions, *Arch. Microbiol.* **99**:117.

Drews, G., 1974, Composition of a protochlorophyll–protophaeophytin complex excreted by mutant strains of *Rps. capsulata*, in comparison with the photosynthetic apparatus, *Arch. Mikrobiol.* **100**:397.

Drews, G., and Giesbrecht, P., 1965, Die Thylakoidstrukturen von *Rhodopseudomonas* spec., *Arch. Mikrobiol.* **52**:242.

Drews, G., Lampe, H.-H., and Ladwig, R., 1969, Die Entwicklung des photosynthetisches Apparatus in Dunkelkulturen von *Rps. capsulata*, *Arch. Mikrobiol.* **65**:12.

Drews, G., Leutiger, I., and Ladwig, R., 1971, Production of protochlorophyll, protophaeophytin and bacteriochlorophyll by the mutant Ala of *Rps. capsulata*, *Arch. Mikrobiol.* **76**:349.

Duysens, L. M. N., 1952, Thesis, University of Utrecht.

Ebrey, T. G., and Clayton, R. K., 1969, Polarization of fluorescence from bacteriochlorophyll in castor oil, in chromatophores and as P870 in photosynthetic reaction centers, *Photochem. Photobiol.* **10**:109.

Eimhjellen, K. E., Steensland, H., and Traetteberg, J., 1967, A *Thiococcus* sp., nov. gen., its pigment and internal membrane system, *Arch. Mikrobiol.* **59**:82.

Erokhin, Yu. E., and Moskalenko, A. A., 1973, Characteristics of proteins, number of chains and molecular weights of the pigment lipo-protein complexes of *Chromatium*, *Dokl. Akad. Nauk SSR Ser. Biol.* (English translation) **212**:429.

Erokhin, Yu. E., and Sinegub, O. A., 1970a, Molecular organization of the pigment system in purple photosynthesizing bacteria, *Mol. Biol. USSR* (English translation) **4**:319.

Erokhin, Yu. E., and Sinegub, O. A., 1970b, Changes in the absorption spectra of *Chromatium* chromatophores by detergents and organic solvents, *Mol. Biol. USSR* (English translation) **4**:437.

Feher, G., 1971, Some chemical and physical properties of a bacterial reaction center particle and its primary photochemical reactants, *Photochem. Photobiol.* **14**:373.

Fenna, R. E., and Matthews, B. W., 1975, Chlorophyll arrangement in a bacteriochlorophyll–protein from *Chlorobium limicola*, *Nature* (*London*) **258**:573.

Fowler, C. F., Nugent, N. A., and Fuller, R. C., 1971, The isolation and characterization of a photochemically active complex from *Chloropseudomonas ethylica*, *Proc. Natl. Acad. Sci. U.S.A.* **68**:2278.

Fowler, C. F., Gray, B. H., Nugent, N. A., and Fuller, R. C., 1973, Absorbance and fluorescence properties of a bacteriochlorophyll *a*–reaction center complex and bacteriochlorophyll *a*–protein in green bacteria, *Biochim. Biophys. Acta* **292**:692.

Fraker, P. J., and Kaplan, S., 1971, Isolation and fractionation of the photosynthetic membranous organelles from *Rhodopseudomonas spheroides*, *J. Bacteriol.* **108**:465.

Fraker, P. J., and Kaplan, S., 1972, Isolation and characterization of a bacteriochlorophyll-containing protein from *Rhodopseudomonas spheroides*, *J. Biol. Chem.* **247**:2732.

French, C. S., and Young, V. M. K., 1956, The absorption, action and fluorescence spectra of photosynthetic pigments in living cells and in solutions, in: *Radiation Biology III* (A. Hollaender, ed.), pp. 343–391, McGraw-Hill, New York.

Fuller, R. C., and Anderson, I. C., 1958, Inhibition of carotenoid synthesis in photosynthetic bacteria—Suppression of carotenoid synthesis and its effect on the activity of photosynthetic bacterial chromatophores, *Nature* (*London*) **181**:252.

Fuller, R. C., Conti, S. F., and Mellin, D. B., 1963, The structure of the photosynthetic apparatus in the green and purple sulfur bacteria, in: *Bacterial Photosynthesis* (H. Gest, A. San Pietro, and L. P. Vernon, eds.), pp. 76–87, Antioch Press, Yellow Springs, Ohio.

Garcia, A., Vernon, L. P., and Mollenhauer, H., 1966a, Properties of *Chromatium* subchromatophore particles obtained by treatment with Triton X-100, *Biochemistry* **5**:2399.

Garcia, A., Vernon, L. P., and Mollenhauer, H., 1966b, Properties of *Rhodospirillum rubrum* subchromatophore particles obtained by treatment with Triton X-100, *Biochemistry* **5**:2408.

Garcia, A., Vernon, L. P., Ke, B., and Mollenhauer, H., 1968a, Some structural and photochemical properties of *Rhodopseudomonas palustris* subchromatophore particles obtained by Triton X-100, *Biochemistry* **7**:319.

Garcia, A., Vernon, L. P., Ke, B., and Mollenhauer, H., 1968b, Some structural and photochemical properties of *Rhodopseudomonas* NHTC 133 subchromatophore particles obtained by treatment with Triton X-100, *Biochemistry* **7**:326.

Gibbs, S. P., Sistrom, W. R., and Worden, P. B., 1965, The photosynthetic apparatus of *Rsp. molischianum*, *J. Cell Biol.* **26**:395.

Gingras, G., and Jolchine, G., 1969, Isolation of a P_{870}-enriched particle from *Rhodospirillum rubrum*, *Prog. Photosyn. Res. Proc. Int. Congr. 1968* **1**:209.

Gloe, A., and Pfennig, N., 1974, Das Vorkommen von Phytol und Geranyl-Geraniol in den Bacteriochlorophyllen roter und grüner Schwefelbacterien, *Arch. Microbiol.* **96**:93.

Gloe, A., Pfennig, N., Brockmann, H., and Trowitzsch, W., 1975, A new bacteriochlorophyll from brown-colored Chlorobiaceae, *Arch. Microbiol.* **102**:103.

Goedheer, J. C., 1972, Temperature dependence of absorption and fluorescence spectra of bacteriochlorophylls *in vivo* and *in vitro*, *Biochim. Biophys. Acta* **275**:169.

Goedheer, J. C., 1973, Fluorescence polarization and pigment orientation in photosynthetic bacteria, *Biochim. Biophys. Acta* **292**:665.

Gorlenko, V. M., 1970, A new photosynthetic green sulfur bacterium *Prothecochloris aestuarii* nov. gen. nov. spec., *Z. Allg. Mikrobiol.* **10**:147.

Gray, B. H., Fowler, C. F., Nugent, N. A., Rigopoulos, N., and Fuller, R. C., 1973, Reevaluation of *Chloropseudomonas ethylica*, *Int. J. Syst. Bacteriol.* **23**:256.

Griffiths, M., Sistrom, W. R., Cohen-Bazire, G., and Stanier, R. Y., 1955, Function of carotenoids in photosynthesis, *Nature (London)* **176**:1211.

Hall, R. C., Kung, M.-C., Fu, M., Hales, B. J., and Loach, P. A., 1973, Comparison of the phototrap complexes from chromatophores of *Rhodospirillum rubrum*, *Rhodopseudomonas spheroides* and the R-26 mutant of *Rhodopseudomonas spheroides*, *Photochem. Photobiol.* **18**:505.

Halsey, Y. D., and Byers, B., 1975, Detergent release and purification from *Chromatium vinosum* chromatophores of a large photoreactive particle retaining secondary electron acceptor, *Biochim. Biophys. Acta* **387**:349.

Hansen, T. A., and Veldkamp, H., 1973, *Rps. sulfidophila*, nov. sp., a new species of the purple nonsulfur bacteria, *Arch. Mikrobiol.* **92**:45.

Haskins, E. F., and Kihara, T., 1967, The use of spectrophotometry in an ecological investigation of the facultatively anaerobic purple photosynthetic bacteria, *Can. J. Microbiol.* **13**:1283.

Herber, R., Maudinas, B., and Villoutreix, J., 1972, Mise en évidence d'isomères trans de phytoène et de phytofluène chez *Rps. sphaeroides*, *Rsp. rubrum* et *Mucor hiamalis*, *C. R. Acad. Sci. Paris Ser. D:* **274**:327.

Hoff, A. J., 1974, The orientation of chlorophyll and bacteriochlorophyll molecules in an oriented lecithin multilayer, *Photochem. Photobiol.* **19**:51.

Holt, A. S., and Clayton, R. K., 1965, Light-induced absorbancy changes in Eimhjellen's *Rhodopseudomonas*, *Photochem. Photobiol.* **4**:829.

Holt, S. C., Conti, S. F., and Fuller, R. C., 1966, Effect of light intensity on the formation of the photochemical apparatus in the green bacterium, *Chloropseudomonas ethylicum*, *J. Bacteriol.* **91**:349.

Jolchine, G., and Reiss-Husson, F., 1974, Comparative studies on two reaction center preparations from *Rhodopseudomonas spheroides* Y, *FEBS Lett.* **40**:5.

Jolchine, G., Reiss-Husson, F., and Kamen, M. D., 1969, Active center fraction from *Rhodopseudomonas spheroides*, strain Y, *Proc. Natl. Acad. Sci. U.S.A.* **64**:650.

Jones, O. T. G., and Plewis, K. M., 1974, Reconstitution of light dependent electron transport in membranes from a bacteriochlorophyll-less mutant of *Rps. sphaeroides*, *Biochim. Biophys. Acta* **357**:204.

Kaplan, S., and Huang, J. W., 1973, Membrane proteins of *Rhodopseudomonas spheroides*. V. Additional characterization of pigment-lipid associated protein from chromatophores, *Biochim. Biophys. Acta* **307**:332.

Katz, E., and Wassink, E. C., 1939, Infrared absorption spectra of chlorophyllous pigments in living cells and in extracellular extracts, *Enzymologia* **7**:97.

Katz, J. J., and Norris, J. R., 1973, Chlorophyll and light energy transduction in photosynthesis, *Curr. Top. Bioenerg.* **5**:41.

Ke, B., and Chaney, T. H., 1971, Spectral and photochemical properties of subchromatophore fractions derived from carotenoid-deficient *Chromatium* by Triton treatment, *Biochim. Biophys. Acta* **226**:341.

Ketchum, P. A., and Holt, S. C., 1970, Isolation and characterization of the membranes of *Rhodospirillum rubrum*, *Biochim. Biophys. Acta* **196**:141.

Klemme, J.-H., and Schelegel, H. G., 1969, Untersuchungen zum Cytochrome-oxydase-System aus Anaerob im Licht und Aerob im Dunkeln gewachsenen Zellen von *Rps. capsulata*, *Arch. Mikrobiol.* **68**:326.

Komen, J. G., 1956, Observations on the infrared absorption spectra of bacteriochlorophyll, *Biochim. Biophys. Acta* **22**:9.

Krasnovskii, A. A., 1969, The principles of light energy conversion in photosynthesis. Photochemistry of chlorophyll and the state of pigments in organisms, *Prog. Photosynthetic Res.* **II**:709.

Krasnovskii, A. A., Voinovskaya, K. K., and Kosobutskaya, L. M., 1952, The nature of the natural state of bacteriochlorophyll in connection with spectral properties of its colloidal solutions and solid films, *Dokl. Akad. Nauk SSSR* **85**:389.

Kuhn, P. J., and Holt, S. C., 1972, Characterization of a blue mutant of *Rsp. rubrum*, *Biochim. Biophys. Acta* **261**:267.

Kuntz, I. D., Loach, P. A., and Calvin, M., 1964, Absorption changes in bacterial chromatophores, *Biophys. J.* **4**:227.

Kuntzler, A., and Pfennig, N., 1973, Das Vorkommen von Bacteriochlorophyll a_p und a_{gg} in Stämmen aller Arten der Rhodospirillaceae, *Arch. Mikrobiol.* **91**:83.

Lampe, H. H., Oelze, J., and Drews, G., 1972, Die Fraktionierung des Membransystems von *Rhodopseudomonas capsulata* und seine Morphogenese, *Arch. Mikrobiol.* **83**:78.

Lascelles, J., 1959, Adaptation to form bacteriochlorophyll in *Rhodopseudomonas spheroides*: Changes in activity of enzymes concerned in pyrrole synthesis, *Biochem. J.* **72**:508.

Lien, S., San Pietro, A., and Gest, H., 1971, Mutational and physiological enhancement of photosynthetic energy conversion in *Rps. capsulata*, *Proc. Natl. Acad. Sci. U.S.A.* **68**:1912.

Lien, S., Gest, H., and San Pietro, A., 1973, Regulation of chlorophyll synthesis in photosynthetic bacteria, *J. Bioenerg.* **4**:423.

Lin, L., and Thornber, J. P., 1975, Isolation and partial characterization of the photochemical reaction center of

Chromatium vinosum (strain D), *Photochem. Photobiol.* **22**:34.

Litvin, F. F. and Gulyaev, B. A., 1969, Resolution of the structure of the absorption spectrum of chlorophyll *a* and its bacterial analogs in the cell by measurement of the second derivative at 20° and −196°, *Dokl. Akad. Nauk. SSSR* **189**:1385.

Loach, P. A., Sekura, D. L., Hadsell, R. M., and Stemer, A., 1970*a*, Quantitative dissolution of the membrane and preparation of photoreceptor subunits from *Rhodopseudomonas spheroides*, *Biochemistry* **9**:724.

Loach, P. A., Hadsell, R. M., Sekura, D. L., and Stemer, A., 1970*b*, Quantitative dissolution of the membrane and preparation of photoreceptor subunits from *Rhodospirillum rubrum*, *Biochemistry* **9**:3127.

Maudinas, B., Herber, R., and Villoutreix, J., 1972, Photoisomerization du Phytoène par la 9-Fluoreuone, *Photochem. Photobiol.* **16**:267.

Maudinas, B., Herber, R., and Villoutreix, J., 1974, Occurrence of transphytoene in microorganisms grown in the absence of carotenogenesis inhibitors, *Biochim. Biophys. Acta* **348**:357.

Morita, S., and Miyazaki, T., 1971, Dichroism of bacteriochlorophyll in cells of the photosynthetic bacterium *Rps. palustris*, *Biochim. Biophys. Acta* **245**:151.

Nieth, K. F., and Drews, G., 1974, The protein patterns of intracytoplasmic membranes and reaction center particles isolated from *Rps. capsulata*, *Arch. Microbiol.* **96**:161.

Nieth, K. F., and Drews, G., 1975, Formation of reaction centers and light harvesting bacteriochlorophyll–protein complexes in *Rps. capsulata*, *Arch. Microbiol.* **104**:77.

Noël, H., Van der Rest, M., and Gingras, G., 1972, Isolation and partial characterization of P870 reaction center complex from wild type *Rhodospirillum rubrum*, *Biochim. Biophys. Acta* **275**:219.

Nugent, N. A., and Fuller, R. C., 1967, Carotenoid biosynthesis in *Rsp. rubrum*: Effect of pteridine inhibitor, *Science* **158**:922.

Oelze, J., and Drews, G., 1972, Membranes of photosynthetic bacteria, *Biochim. Biophys. Acta* **265**:209.

Oelze, J., and Golecki, J. R., 1975, Properties of reaction center depleted membranes of *Rsp. Rubrum*, *Arch. Microbiol.* **102**:59.

Okamura, M. K., Steiner, L. A., and Feher, G., 1974, Characterization of reaction centers from photosynthetic bacteria. I. Subunit structure of the protein mediating the primary photochemistry in *Rhodopseudomonas spheroides*, R26, *Biochemistry* **13**:1394.

Olson, J. M., and Clayton, R. K., 1966, Sensitization of photoreactions in Eimhjellen's *Rhodopseudomonas* by a pigment absorbing at 830 nm, *Photochem. Photobiol.* **5**:655.

Olson, J. M., and Stanton, E. K., 1966, Absorption and fluorescence spectra of bacteriochlorophyll *in situ*, in: *The Chlorophylls* (L. P. Vernon and G. R. Seely, eds.), pp. 381–398, Academic Press, New York.

Olson, J. M., Phillipson, K. D., and Sauer, K., 1973, Circular dichroism and absorption spectra of bacteriochlorophyll–protein and reaction center complexes from *Chlorobium thiosulfatophilum*, *Biochim. Biophys. Acta* **292**:206.

Olson, J. M., Shaw, E. K., and Englberger, F. M., 1976, Comparison of bacteriochlorophyll–proteins from two green bacteria, *Biochem. J.* **159**:769.

Pfennig, N., 1969*a*, *Rps. acidophila*, sp. n., a new species of the budding purple nonsulfur bacteria, *J. Bacteriol.* **99**:597.

Pfennig, N., 1969*b*, *Rsp. tenue*, sp. n., a new species of purple nonsulfur bacteria, *J. Bacteriol.* **99**:619.

Pfennig, N., 1974, *Rps. globiformis*, sp. n., a new species of the Rhodospirillaceae, *Arch. Microbiol.* **100**:197.

Pfennig, N., and Cohen-Bazire, G., 1967, Some properties of the green bacterium *Pelodictyon clathratiforme*, *Arch. Mikrobiol.* **59**:226.

Pfennig, N., Eimhjellen, K. E., and Liaaen-Jensen, S., 1965, A new isolate of the *Rsp. fulvum* group and its photosynthetic pigments, *Arch. Mikrobiol.* **51**:258.

Phillipson, K. D., and Sauer, K., 1973*a*, Comparative study of CD spectra of reaction centers from several photosynthetic bacteria, *Biochemistry* **12**:535.

Phillipson, K. D., and Sauer, K., 1973*b*, Light scattering effects on the CD of chloroplasts, *Biochemistry* **12**:3454.

Pierson, B. K., and Castenholtz, R. W., 1974*a*, A phototropic gliding filamentous bacterium of hot springs, *Chloroflexus aurantiacus*, gen. and sp. nov., *Arch. Microbiol.* **100**:5.

Pierson, B. K., and Castenholtz, R. W., 1974*b*, Studies of pigments and growth in *Chloroflexus aurantiacus*, a phototropic filamentous bacterium, *Arch. Microbiol.* **100**:283.

Prince, R. C., and Crofts, A. R., 1973, Photochemical reaction centers from *Rhodopseudomonas capsulata*, Ala, Pho⁺, *FEBS Lett.* **35**:213.

Pucheu, N., Kerber, N. L., and Garcia, A. F., 1974, Comparative studies on membranes isolated from *Rps. viridis* grown in the presence and absence of yeast extract, *Arch. Microbiol.* **101**:259.

Rabinowitch, E., 1956, *Photosynthesis and Related Processes*, Vol. 2, Part II, Wiley-Interscience, New York.

Raymond, J. C., and Sistrom, W. R., 1967, The isolation and preliminary characterization of a halophilic photosynthetic bacterium, *Arch. Mikrobiol.* **59**:255.

Reed, D. W., 1969, Isolation and composition of a photosynthetic reaction center complex from *Rhodopseudomonas spheroides*, *J. Biol. Chem.* **244**:4936.

Reed, D. W., and Clayton, R. K., 1968, Isolation of a reaction center fraction from *Rhodopseudomonas spheroides*, *Biochem. Biophys. Res. Commun.*, **30**:471.

Reed, D. W., and Peters, G. A., 1972, Characterization of the pigments in the reaction center preparation from *Rhodopseudomonas spheroides*, *J. Biol. Chem.* **247**:7148.

Reed, D. W., Raveed, D., and Israel, H. W., 1970, Functional bacteriochlorophyll–protein complexes from chromatophores of *Rhodopseudomonas spheroides* strain R-26, *Biochim. Biophys. Acta* **223**:281.

Reinach, P., Aubrey, B. B., and Brody, S. S., 1973, Monomolecular films of bacteriochlorophyll and derivatives at an air–water interface. Surface and spectral properties, *Biochim. Biophys. Acta* **314**:360.

Reiss-Husson, F., de Klerk, H., Jolchine, G., Jauneau, E., and Kamen, M. D., 1971, Some effects of iron deficiency on *Rhodopseudomonas spheroides* strain Y, *Biochim. Biophys. Acta* **234**:73.

Sauer, K., 1972, Circular dichroism and optical rotary dispersion of photosynthetic organelles and their component pigments. *Methods Enzymol.* **24**:206.

Sauer, K., 1975, Primary events and the trapping of energy, in: *Bioenergetics of Photosynthesis* (Govindjee, ed.), p. 115, Academic Press, San Francisco.

Saunders, V. A., and Jones, O. T. G., 1974, Adaptations in *Rps. sphaeroides*, *FEBS Lett.* **44**:169.

Schachman, H. K., Pardee, A. B., and Stanier, R. Y., 1952, Studies on the macromolecular organization of microbial cells, *Arch. Biochem. Biophys.* **38**:245.

Schick, J. and Drews, G., 1969, The morphogenesis of bacterial photosynthetic apparatus. III. The features of a phaeophytin–protein–carbohydrate complex excreted by mutant M46 of *Rsp. rubrum*, *Biochim. Biophys. Acta* **183**:215.

Schlegel, H. G., and Pfennig, N., 1961, Die Anreichungskultur einiger Schwefelpurpurbacterien, *Arch. Mikrobiol.* **38**:1.

Segen, B. J., and Gibson, K. D., 1971, Deficiencies of chromatophore proteins in some mutants of *Rps. sphaeroides* with altered carotenoids, *J. Bacteriol.* **105**:701.

Shaposhnikov, V. V., Kondrat'eva, E. N., and Fedorov, V. D., 1960, A new species of green sulfur bacteria, *Nature (London)* **187**:167.

Sinegub, O. A., and Erokhin, Yu. E., 1971, The effect of pH, ionic strength and oxidizing agents on bacteriochlorophyll in *Chromatium minutissimum*, *Mol. Biol. SSSR* (English translation) **5**:380.

Sistrom, W. R., 1962, Observations on the relationship between formation of photopigments and the synthesis of protein in *Rhodopseudomonas spheroides*, *J. Gen. Microbiol.* **28**:599.

Sistrom, W. R., 1964, Calculation of the absorption coefficients of the individual components of the spectra of bacteriochlorophyll in chromatophore preparations of *Rps. sphaeroides*, *Biochim. Biophys. Acta* **79**:419.

Sistrom, W. R., 1966, The spectra of bacteriochlorophyll *in vivo*: Observations on mutants of *Rhodopseudomonas spheroides* unable to grow photosynthetically, *Photochem. Photobiol.* **5**:845.

Sistrom, W. R., and Clayton, R. K., 1964, Studies of a mutant of *Rps. sphaeroides* unable to grow photosynthetically, *Biochim. Biophys. Acta* **88**:61.

Sistrom, W. R., Griffiths, M., and Stanier, R. Y., 1956, The biology of a photosynthetic bacteria which lacks colored carotenoids, *J. Cell. Comp. Physiol.* **48**:473.

Slooten, L., 1972, Reaction center preparations of *Rhodopseudomonas spheroides*: Energy transfer and structure, *Biochim. Biophys. Acta* **256**:452.

Smith, J. H. C., and French, C. S., 1963, The major and accessory pigments in photosynthesis, *Annu. Rev. Plant Physiol.* **14**:181.

Smith, W. R., Jr., Sybesma, C., and Dus, K., 1972, Isolation and characteristics of small, soluble photoreactive fragments of *Rhodospirillum rubrum*, *Biochim. Biophys. Acta* **267**:609.

Sojka, G. A., Freeze, H. H., and Gest, H., 1970, Quantita-

tive estimation of bacteriochlorophyll *in situ*, *Arch. Biochem. Biophys.* **136**:578.

Stanier, R. Y., and Smith, J. H. C., 1960, The chlorophylls of green bacteria, *Biochim. Biophys. Acta* **41**:478.

Steffen, H., and Calvin, M., 1970, The origin of the long wavelength absorption bands in purple bacteria, *Biochem. Biophys. Res. Commun.* **41**:282.

Straley, S. C., Parson, W. W., Mauzerall, D. and Clayton, R. K., 1973, Pigment content and molar extinction coefficients of photochemical reaction centers from *Rps. sphaeroides*, *Biochim. Biophys. Acta* **305**:597.

Suzuki, Y., Morita, S., and Takamiya, A., 1969, Reversible interconversion of B850 and B810 in chromatophores of *Chromatium* D as induced with detergents, *Biochim. Biophys. Acta* **180**:114.

Sybesma, C., 1969, Light-induced reactions of P890 and P800 in the purple photosynthetic bacterium, *Rhodospirillum rubrum*, *Biochim. Biophys. Acta* **172**:177.

Sybesma, C., and Fowler, C. F., 1968, Evidence for two light-driven reactions in the purple photosynthetic bacterium, *Rhodospirillum rubrum*, *Proc. Natl. Acad. Sci. U.S.A.*, **61**:1343.

Sybesma, C., and Olson, J. M., 1963, Transfer of chlorophyll excitation energy in green photosynthetic bacteria, *Proc. Natl. Acad. Sci. U.S.A.* **49**:248.

Sybesma, C., and Vredenberg, W. J., 1963, Evidence for a reaction center P840 in the green photosynthetic bacterium *Chloropseudomonas ethylicum*, *Biochim. Biophys. Acta* **75**:439.

Takacs, B. J., and Holt, S. C., 1971, *Thiocapsa floridana*; a cytological, physical and chemical characterization. I. Cytology of whole cells and isolated chromatophore membranes, *Biochim. Biophys. Acta* **233**:258.

Takemoto, J., 1974, Kinetics of photosynthetic membrane protein assembly in *Rps. sphaeroides*, *Arch. Biochem. Biophys.* **163**:515.

Takemoto, J., and Lascelles, J., 1973, Coupling between bacteriochlorophyll and membrane protein synthesis in *Rps. sphaeroides*, *Proc. Natl. Acad. Sci. U.S.A.* **70**:799.

Takemoto, J., and Lascelles, J., 1974, Function of membrane proteins coupled to bacteriochlorophyll synthesis, *Arch. Biochem. Biophys.* **163**:507.

Thornber, J. P., 1970, Photochemical reactions of purple bacteria as revealed by studies of three spectrally different carotene bacteriochlorophyll–protein complexes isolated from *Chromatium*, strain D, *Biochemistry* **9**:2688.

Thornber, J. P., 1971, The photochemical reaction center of *Rhodopseudomonas viridis*, *Methods Enzymol.* **21**:688.

Thornber, J. P., 1975, Chlorophyll–proteins: Light harvesting and reaction center components in plants, *Annu. Rev. Plant Physiol.* **26**:127.

Thornber, J. P., and Olson, J. M., 1971, Chlorophyll–proteins and reaction center preparations from photosynthetic bacteria, algae and higher plants, *Photochem. Photobiol.* **14**:329.

Thornber, J. P., Olson, J. M., Williams, D. M., and Clayton, M. L., 1969, Isolation of the reaction center of *Rhodopseudomonas viridis*, *Biochim. Biophys. Acta* **172**:351.

Trosper, T. L., Benson, D. L., and Thornber, J. P., 1976, Further studies on the photochemical reaction center of

Rhodopseudomonas viridis, Biochim. Biophys. Acta
460:318.

Trüper, H. G., 1968, *Ectothiorhodospira mobilis* Pelsh, a photosynthetic sulfur bacterium depositing sulfur outside the cells, *J. Bacteriol.* **95**:1910.

Trüper, H. G., and Genovese, S., 1968, Characterization of photosynthetic sulfur bacteria causing red water in Lake Faro (Messina, Sicily), *Limnol. Oceanogr.* **13**:225.

Trüper, H. G., and Jannasch, H. W., 1968, *Chromatium buderi*, nov. sp., eine neue Art der "grossen" *Thiorhodaceae*, **61**:363.

Uffen, R. L., and Wolfe, R. S., 1970, Anaerobic growth of purple nonsulfur bacteria under dark conditions, *J. Bacteriol.* **104**:462.

Uspenskaya, V. E., and Kondrat'eva, E. N., 1972, Growth of the photosynthetic bacteria *Ectothiorhodospira shaposhnikovii* in the dark under aerobic conditions, *Mikrobiologiya* (English translation) **41**:392.

van der Rest, H., Noël, H., and Gingras, G., 1974, An immunological and electrophoretic study of *Rhodospirillum rubrum* chromatophore fragments, *Arch. Biochem. Biophys.* **164**:285.

Verkhoturov, V. N., Kondrat'eva, E. N., Lopanitsina, V. V., and Rubin, A. B., 1969, Spectrophotometric studies of a new strain of purple sulfur bacteria, *Mol. Biol. SSSR* (English translation) **3**:538.

Vernon, L. P., and Garcia, A. F., 1967, Pigment–protein complexes derived from *Rhodospirillum rubrum* chromatophores by enzymic digestion, *Biochim. Biophys. Acta* **143**:144.

Vredenberg, W. J., and Amesz, J., 1966, Absorption bands of bacteriochlorophyll types in purple bacteria and their response to illumination, *Biochim. Biophys. Acta* **126**:244.

Vredenberg, W. J., and Amesz, J., 1967, Absorption characteristics of bacteriochlorophyll types in purple bacteria and efficiency of energy transfer between them, *Brookhaven Symp. Biol.* **19**:49.

Walter, E., 1975, Thesis, Federal Institute of Technology, Zürich, to be published.

Wang, R. T., and Clayton, R. K., 1973, Isolation of photochemical reaction centers from a carotenoidless mutant of *Rhodospirillum rubrum*, *Photochem. Photobiol.* **17**:57.

Wassink, E. C., and Kronenberg, G. H. M., 1962, Strongly carotenoid deficient *Chromatium* strain D cells with "normal" bacteriochlorophyll absorption peaks in the 800–850 nm region, *Nature (London)* **194**:553.

Wassink, E. C., Katz, E., and Dorrestein, R., 1939, Infrared absorption spectra of various strains of purple bacteria, *Enzymologia* **7**:113.

Weaver, P. F., Wall, J. D., and Gest, H., 1975, Characterization of *Rhodopseudomonas capsulata*, *Arch. Microbiol.* **105**:207.

Worden, P. B., and Sistrom, W. R., 1964. The preparation and properties of bacterial chromatophore fractions, *J. Cell Biol.* **23**:135.

Zankel, K. L., and Clayton, R. K., 1969, Uphill energy transfer in a photosynthetic bacterium, *Photochem. Photobiol.* **9**:7.

CHAPTER 8

Bacteriochlorophyll *a*–Proteins from Green Bacteria

John M. Olson

1. Introduction

The existence of bacteriochlorophyll *a* (Bchl *a*) in two strains of green bacteria was discovered accidentally by J. M. Olson and Romano (1962) in following the procedure of Gibson (1961) for extracting cytochromes from green bacteria. This discovery was independently verified by Holt *et al.* (1963) in Ottawa and by Jensen *et al.* (1964) in Trondheim. The existence of a specific Bchl *a*–protein complex in extracts of green bacteria was demonstrated by Sybesma and Olson (1963).

The function of the Bchl *a*–protein *in vivo* is to accept excitation energy from chlorobium chlorophyll (Bchl *c*, *d*, *e*, or *f*), the main light-harvesting pigment of green bacteria (Sybesma and Olson, 1963), and to transfer this excitation energy to the reaction center Bchl, P840 (Sybesma and Vredenberg, 1963, 1964).

Bchl *a*–proteins have been obtained from three strains of green bacteria: *Prosthecochloris aestuarii*, strain 2K; *Chlorobium limicola f. thiosulfatophilum*, strain L; and *Chl. limicola f. thiosulfatophilum*, strain 6230 (Tassajara). Strain 2K grows best in mixed culture with one or more nonphotosynthetic sulfur-reducing bacteria. Such a syntrophic culture has been known as "*Chloropseudomonas ethylica*, strain 2K" (Gray *et al.*, 1973; Olson, J. M., 1973; Shioi *et al.*, 1976; Pfennig and Biebl, 1976). In the extensive literature on the Bchl *a*–protein from "*Chloropseudomonas ethylica*, strain 2K," the actual source of the Bchl *a*–protein was probably *P. aestuarii* 2K (see

Chapter 2). Unless otherwise noted, all studies cited in this chapter used Bchl *a*–protein from *P. aestuarii* 2K grown in mixed culture. The early work on Bchl *a*–protein from *Chl. limicola f. thiosulfatophilum* utilized strain L; the more recent work is based on the Tassajara strain.

2. Physical and Chemical Properties

2.1. Size

2.1.1. Subunit size

The Bchl *a*–protein from strain 2K is a trimer (Fenna *et al.*, 1974). The molecular weight of each subunit should be the sum of the protein and the Bchl only (Thornber and Olson, 1968). When Bchl *a*–protein was denatured by heating at 100°C in 1% sodium dodecyl sulfate and 1% mercaptoethanol, and then electrophoresed on a discontinuous polyacrylamide gel in the presence of marker proteins from phage T7, the weight of the denatured subunit was found to be 40 ± 2 kdalton, as shown in Fig. 1 (Olson, J. M., *et al.*, 1976*b*). (The denatured subunits from strains 2K and Tassajara are the same size.) At the time of the experiment, it was assumed that each denatured subunit had been completely separated from its seven Bchl molecules (mol. wt = 6.4 kdalton). The molecular weight of the complete native subunit was assumed to be approximately 46 kdalton, corresponding to a trimer weight of approximately 140 kdalton. This value is in fair agreement with the determinations based on equilibrium sedimentation and crystal density listed in Table 1.

John M. Olson · Biology Department, Brookhaven National Laboratory, Upton, New York 11973

Table 1. Values for the Molecular Weight of the Bchl *a*–Protein from *Prosthecochloris aestuarii* 2K

Value (kdaltons)	Method	References
137 ± 5	Equilibrium sedimentation, $\bar{V}$ assumed to be 0.74	J. M. Olson *et al.* (1963)
167 ± 17	Equilibrium sedimentation, $\bar{V}$ measured as 0.79	J. M. Olson (1966)
153 ± 23	Crystal density	Fenna *et al.* (1974)
119 ± 3	Chemical composition (corrected for 3 subunits and 21 Bchl molecules)	Thornber and Olson (1968), J. M. Olson (1971)
135	Chemical composition (3 subunits assumed)	J. M. Olson *et al.* (1976*b*)

Table 2. Comparison of Bchl *a*–Proteins: Amino Acid Composition (mol/Σmol)

Amino acid	Strain 2K[a]	Difference (%)	Strain 2K[b]	Difference (%)	Strain Tassajara[b]
Gly	11.26	1	11.43	−8[c]	10.46
Asp	10.91	1	11.01	1	11.14
Val	9.56	0	9.58	12[c]	10.69
Glu	9.23	2	9.39	−10[c]	8.48
Ser	7.53	3	7.77	18[c]	9.13
Ile	6.59	0	6.57	−39[c]	3.98
Ala	6.01	0	6.04	8[c]	6.57
Leu	5.72	−1	5.65	5	5.91
Arg	5.84	−6	5.52	14[c]	6.32
Pro	4.91	4	5.09	−15[c]	5.87
Lys	5.18	−3	5.02	2	5.10
Phe	4.89	−3	4.76	4	4.95
Thr	4.09	−4	3.92	−15[c]	3.34
Tyr	2.66	1	2.70	3	2.78
His	2.26	−7[c]	2.12	0	2.13
Trp	1.74	7[c]	1.88	−30[c]	1.31
Met	0.99	−1	0.98	23[c]	1.21
Cys	0.63	−9[c]	0.58	3	0.60
Total	100.00		100.01		99.97

[a] Data of Thornber and Olson (1968).   [b] Data of J. M. Olson *et al.* (1976*b*).   [c] Significant difference.

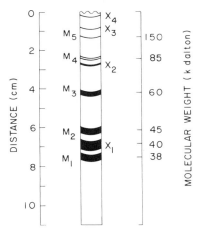

Fig. 1. Electrophoresis of Bchl *a*–protein from strain 2K and marker proteins from bacteriophage T7 (Studier, 1972) on a discontinuous polyacrylamide gel (10%) in 0.1 M phosphate containing 0.1% SDS. Proteins were stained with Coomassie blue. Marker M_1 (38 kdalton) is a major head protein (gene 10); M_2 (45 kdalton) is a minor head protein (gene 8); M_4 (85 kdalton) is a mixture of head protein (gene 15) and tail protein (gene 12); M_5 (150 kdalton) is a head protein (gene 16). Band X_1 corresponds to the Bchl *a*–protein subunit polypeptide. Bands X_2 and X_3 correspond to the dimer and the tetramer, respectively. Band X_4 corresponds to an aggregate of six or more polypeptides.

The molecular weight can be calculated from the amino acid composition (Tables 2 and 3) of the subunit, since Fenna and Matthews (1975, 1977) showed that the subunit contains seven Bchl *a* molecules and at least 354 amino acid residues. The calculated subunit weight for strain 2K is 38.6 kdalton for the protein alone and 45.0 kdalton for both Bchl *a* and protein together. (The corresponding trimer weight of 135 kdalton is listed in Table 1.) The value of 40 kdalton measured by gel electrophoresis is closer to the calculated value without the seven Bchl *a* molecules included. Due to the limits of accuracy and precision in gel electrophoresis, the true molecular weight of the denatured subunit could be anywhere between 38 kdalton and 42 kdalton. it seems reasonable to assume that the denatured subunit has lost its Bchl *a* and therefore has a true molecular weight of about 39 kdalton.

2.1.2. Trimer Weight

For the intact Bchl *a*–protein (trimer) from strain 2K, the effective molecular weight, $M_{eff, 20, w}$, is 36 ± 1 kdalton by equilibrium sedimentation (Olson, J. M., *et*

Table 3. Comparison of Bchl *a*–Proteins: Amino Acid Composition of Subunits (mol/mol) Calculated for 354 Residues/Subunit

| Amino acid | Strain 2K | | Strain Tassajara | | Minimum difference— nearest integer |
	Moles	Nearest integers	Moles	Nearest integers	
Gly[a]	40.5	39–42	37.0	36–38	−1
Asp	39.0	38–40	39.4	38–41	0
Val[a]	33.9	33–35	37.8	37–39	+2
Glu[a]	33.2	32–34	30.0	29–31	−1
Ser 4%	27.5	26–29	32.3	31–34	+2
Ile[a]	23.3	23–24	14.1	14	−9
Ala[a]	21.4	21–22	23.3	23–24	+1
Leu	20.0	19–21	20.9	20–22	0
Arg[a]	19.5	19–20	22.4	22–23	+2
Pro[a] 4%	18.0	17–19	20.8	20–22	+1
Lys	17.8	17–18	18.0	17–18	0
Phe	16.8	16–17	17.5	17–18	0
Thr[a] 4%	13.9	13–14	11.8	11–12	−1
Tyr	9.6	9–10	9.8	10	0
His	7.5	7–8	7.5	7–8	0
Trp[a]	6.6	6–7	4.6	4–5	−1
Met[a]	3.5	3–4	4.3	4	0
Cys	2.0	2	2.1	2	0
Total	354.0		353.6		−13, +8

[a] Significant difference.

al., 1963), and the sedimentation coefficient $s_{20,w}$ is 7.3 S (Olson, J. M., et al., 1969). The value of the actual molecular weight calculated from $M_{eff,20,w}$ and the experimentally determined $\bar{V}$ is 167 ± 17 kdalton (Olson, J. M., 1966), which is in satisfactory agreement with the value of 153 ± 23 kdalton determined by crystal density measurements (Fenna et al., 1974). These values, however, are not in satisfactory agreement with the values of 119 and 135 kdalton based on chemical composition (see Table 1). Some of the obvious discrepancies among these four values could be resolved if the experimental value of $\bar{V}$ used for the 167-kdalton value proved to be too high. If a $\bar{V}$ value (0.74) closer to that expected for a typical protein is used, with the equilibrium sedimentation data, a molecular weight of 137 ± 5 kdalton is obtained. On the other hand, the value of 119 kdalton based on

composition is probably too low, because the calculation was made before the number of residues was known. The best estimate consistent with composition, crystal density, and equilibrium sedimentation is 140 kdalton. The 21 Bchl a molecules account for 19 kdalton, and the protein moiety for 121 kdalton.

2.2. Isoelectric Point

An attempt was made to determine the isoelectric point for each Bchl a–protein by the method of isoelectric focusing (Olson, J. M., et al., 1976b). Since every preparation gave from 5 to 13 bands, it was necessary to graph the distribution pattern for the bands from each protein. (The reason for multiple bands being focused is thought to be heterogeneity in the distribution of amide groups on the surfaces of the

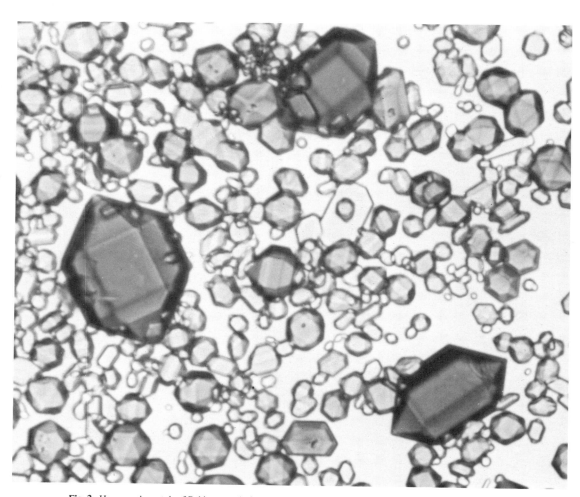

Fig. 2. Hexagonal crystals of Bchl a–protein from strain 2K. The largest crystal is approximately 0.1 mm long.

macromolecules.) For Bchl *a*–protein from strain 2K, there were 17 experiments with 148 bands distributed between pH 5.3 and 7.4. (Material from both pure and mixed cultures was tested.) The distribution was centered at pH 6.0. For Bchl *a*–protein from strain Tassajara, there were 8 experiments with 53 bands distributed between pH 6.1 and 7.8. This distribution was centered at pH 7.0. The clear difference of 1 pH unit between the "average" isoelectric values for the two proteins is consistent with the fact that the Tassajara protein contains at least two more arginine residues and two fewer glutamic acid residues than does the 2K protein.

2.3. Composition

2.3.1. Amino Acid Analyses

(a) Strain 2K. Thornber and Olson (1968) showed that the Bchl *a*–protein from strain 2K consists of Bchl *a* and protein only. The *N*-terminal amino acid is alanine. From the amino acid composition (see Table 2), the molecular weight of the peptide (without amide groups) was calculated to be 33.4 kdalton on the basis of whole numbers of residues for the amino acids in

lowest amount (half cystine, methionine, histidine, and tyrosine). The amino acid composition was determined again with material from both pure and mixed cultures (Olson, J. M., *et al.*, 1976*b*), and earlier results were confirmed except for histidine, tryptophan, and half cystine. (The differences between the two determinations were 7–9% for these residues.) The older value for tryptophan was obtained from a spectrophotometric determination of the molar ratio of Tyr : Trp, whereas the newer value was obtained directly by chromatography. The newer value is more likely to be correct. The reason for the other differences is not known.

(b) Strain Tassajara. The amino acid composition (see Table 2) of the Bchl *a*–protein from strain Tassajara is clearly different from that for the protein from strain 2K. The differences between the two proteins can be calculated on the assumption that each peptide contains at least 354 amino acid residues. A sample calculation for 354 residues (Table 3) indicates that the minimum differences for 354 residues are −1 Gly, +2 Val, −1 Glu, +2 Ser, −9 Ile, +1 Ala, +2 Arg, +1 Pro, −1 Thr, and −1 Trp. Between 8 and 13 residues have to be different in the two proteins.

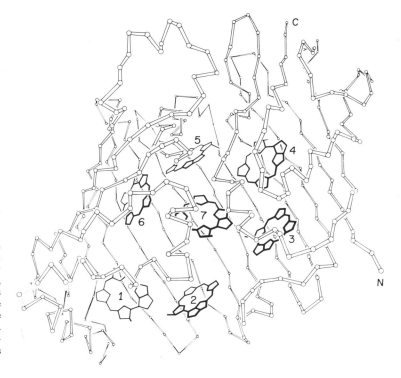

Fig. 3. Subunit of Bchl *a*–protein from strain 2K. This schematic diagram shows the arrangement of the polypeptide backbone and the Bchl core of one subunit. Assumed *α*-carbon positions are indicated by circles. Some connections within the polypeptide chain are uncertain, and particularly ambiguous regions are indicated by broken lines. For clarity, magnesium atoms, Bchl ring substituents, and phytyl chains (except for the first bond) are omitted. The direction of view is from the center of the trimer toward the exterior of the molecule. The threefold axis of the trimer is horizontal. Reproduced from Fenna and Matthews (1977).

2.3.2. Chlorophyll Identification

In strain 2K, the identification of Bchl a as the colored constituent of the Bchl a–protein is certain. When the Bchl is removed from the protein, the absorption spectrum in either is virtually identical to that of Bchl a from *Chromatium vinosum* D (Olson, J. M., *et al.*, 1963). Jensen *et al.* (1964) showed that Bchl a from green bacteria and Bchl a from purple bacteria have the same R_f value in paper chromatography, and Holt *et al.* (1963) established by chemical and physical methods the identity of Bchl a from green bacteria as authentic Bchl a.

2.4. Structure

Fenna and Matthews (1975, 1977) determined the structure of the Bchl a–protein from strain 2K to a nominal resolution of 2.8 Å. The determination was carried out by means of X-ray diffraction of single hexagonal crystals of native Bchl a–protein (see Fig. 2), and of Bchl a–protein to which the following heavy atoms had been bound: mercury, mercury and platinum, platinum, and uranium. The following description of the subunit is adapted from the texts of Fenna and Matthews (1975, 1977).

2.4.1. Subunit

The subunit may be described as a string bag of protein with seven Bchl a molecules inside (see Fig. 3). The polypeptide chain forms a distorted hollow cylinder that contains the Bchl aggregate. One end of the cylinder is closed, while the other end is open, but this opening is covered in the trimer. The wall of the cylinder that faces the outside of the trimer is composed almost entirely of 15 strands of β-sheet, of which 13 are antiparallel. The side of the cylinder in contact with adjacent subunits consists of short lengths of α-helix interspersed with regions of irregular conformation. Tang and Hirs (unpublished) have identified the first 11 amino acid residues from the N-terminus as $NH_2 \cdot Ala \cdot Leu \cdot Phe \cdot Gly \cdot Thr \cdot Lys \cdot Asx \cdot Thr \cdot Thr \cdot Thr \cdot Ala$.

The seven Bchl a molecules occupy the space within an ellipsoid of axial dimensions, $45 \times 35 \times 15$ Å (Fig. 4). The average center-to-center distance between porphine rings is 12 Å for nearest neighbors. Although no porphine ring is exactly parallel to any other, the seven porphine rings do lie close to an "average" plane.* The angles between each individual porphine ring and the "average" plane lie between 10° and 40°. The normal to the "average" plane makes an angle of 62° with the three-fold axis of the trimer, and the center of the Bchl aggregate is about 20 Å from the center of the trimer. In the complete trimer, the Bchl aggregates in each subunit form a triangular funnel around the three-fold axis.

Within each subunit, the phytyl chains lie close together. The phytyl tails of Bchl's 4, 5, and 6 are parallel and form a planar structure between porphine rings 5 and 6 and the β-sheet of the outer wall. The phytyl tails of Bchl's 2, 3, and 7 lie in extended conformation in the inner space between porphine rings, while the tail of Bchl 1 is bent into a U-shaped loop.

There is no evidence for chemical interactions other than hydrophobic associations between neighboring Bchl a molecules. There is, however, solid evidence of interaction between each Bchl molecule and the polypeptide "bag." For each Bchl molecule, a fifth ligand to the Mg atom is indicated by electron density protruding from the center of the porphine ring. In six cases, this density extends as a continuous bridge between the Mg atom and the polypeptide chain. The ligands for Bchl's 3, 4, and 7 are the side chains of three residues (probably histidines) in an α-helix, and the ligand for Bchl 2 appears to be a water molecule. The ligands for Bchl's 1 and 6 are side chains of β-sheet residues that are probably histidines. Bchl 5

---

* The "average" plane is defined as that plane for which the sum of direction cosines with respect to each of the porphine planes is a maximum.

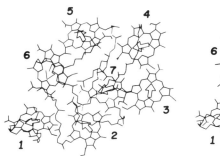 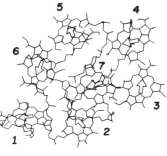

Fig. 4. Stereo drawing of the seven Bchl a molecules inside one subunit. The direction of view is approximately normal to the "average" plane of the Bchl molecules. Each arrow, directed from Ring III to Ring I, is parallel to the Q_y transition moment of each Bchl molecule. Magnesium atoms are omitted. Reproduced from Fenna and Matthews (1975, 1977).

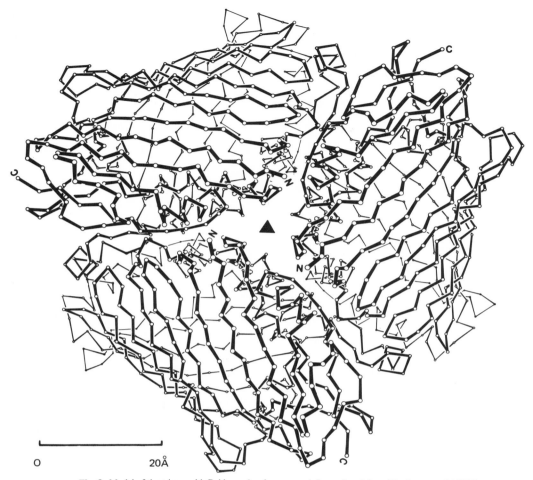

Fig. 5. Model of the trimer with Bchl *a* molecules removed. Reproduced from Matthews *et al.* (1977).

appears to ligand directly to the polypeptide back-bone.

The Bchl *a* ring substituents include a number of potential hydrogen bond acceptors such as oxygen-containing groups. Most of these groups are close to the periphery of the aggregate, and are also close enough to the protein electron density to form hydrogen bonds with suitable donors in the poly-peptide chain. Each Bchl molecule is apparently anchored to the protein through extensive hydrogen bonding and liganding to the Mg atom in addition to hydrophobic interactions through the phytyl tail. The position and orientation of each Bchl molecule are governed mainly by specific interaction with the protein, rather than with other Bchl molecules. Chlorophyll–chlorophyll interactions are limited to hydrophobic interactions between phytyl tails. The Bchl molecules do not occur in regular one- or two-dimensional arrays, as often postulated in models for chlorophyll arrangement *in vivo*.

2.4.2. Trimer

The subunits are tightly packed in the trimer, as shown in Fig. 5. The space occupied by the trimer is roughly that of an oblate ellipsoid of revolution 57 Å along the short axis and 83 Å along the long axis (Fenna *et al.*, 1974). The closest distance between Bchl molecules in adjacent subunits is 24 Å, compared to 12 Å between nearest neighbors within one subunit (Fenna and Matthews, 1975).

2.5. Stability

The Bchl *a*–protein from strain 2K is most stable in solution in the presence of a salt such as NaCl (0.2–1.0 M) and buffered between pH 7 and 8. In this pH range, the protein appears to be unaffected by 5% Triton X-100 (Ghosh and Olson, 1968) or by 8 M urea (Kim and Ke, 1970). Sodium dodecyl sulfate (10–40 mM), however, causes a slow denaturation to a blue intermediate, which eventually becomes pink due to pheophytinization of the Bchl (Ghosh and Olson, 1968; Kim, 1970). The Bchl can be separated quantitatively from the protein by 90% methanol at 40°C, but the complex remains intact in 50% methanol at 4–5°C and in 60% methanol, 30% ethanol, and 10% water at −44°C (Ghosh and Olson, 1968). The complex is stable for 24 hr at pH 7.7 and 40°C; at 72°C, it denatures within an hour. At 20–25°C, the complex is more or less stable between pH 3 and 12, but below pH 3, it is converted to bacteriopheophytin–protein. Below pH 1.5, a blue intermediate form of the Bchl a–protein appears prior to the formation of pheophytin. Similar blue intermediates also form above pH 12.

2.6. Crystal Properties

Bchl a–protein from strain 2K crystallizes in two habits: hexagonal space group $P6_3$ (Olson, J. M., et al., 1969) shown in Fig. 2 or trigonal space group $P3_1$ or $P3_2$ (Fenna et al., 1974). (Bchl a–protein from strain Tassajara forms crystals that appear to be similar to the hexagonal crystals from strain 2K.) Under appropriate conditions, this macromolecule may also form paracrystalline aggregates (Olson, R. A., et al., 1969b; Olson, R. A., 1970).

2.6.1. Hexagonal Crystal

The dimensions of the unit cell in hexagonal crystals are a = b = 112.4 ± 0.4 Å (Fenna et al., 1974) and c = 98.4 ± 0.4 Å (Olson, J. M., et al., 1969). Since each unit cell contains two macromolecules, the symmetry properties of the unit cell require three subunits for each macromolecule. The arrangement of trimers in the hexagonal crystal is shown in Fig. 6. Crystals exhibit a weak birefringence (0.0025 at 550 nm) (Olson, R. A., et al., 1969a). In white light, the polarization color is orange. At room temperature, the absorption spectrum of a crystal is essentially the same as that of a solution. The crystal is weakly dichroic with ratios of 1/(1.21 ± 0.02) at 603 nm and 1/(1.30 ± 0.04) at 809 nm. Selective dispersion of birefringence is negative at 603 nm and positive at 809 nm with respect to the crystal axis.

2.6.2. Trigonal Crystal

Although formation of hexagonal crystals is the usual result of crystallization, trigonal crystals have been obtained on one occasion (Fenna et al., 1974). Cell dimensions are a = b = 83.2 Å and c = 165.8 Å. The molecules are much more densely packed in trigonal crystals than in hexagonal crystals.

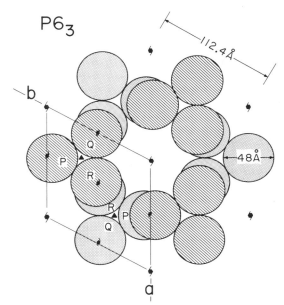

Fig. 6. Arrangement of Bchl a–protein trimers in the hexagonal crystal form seen in projection down the 6_3 axis (Fenna et al., 1974). Each sphere represents one subunit.

2.6.3. Paracrystalline Aggregate

Sometimes Bchl a–protein will come out of solution as a paracrystalline aggregate when additional ammonium sulfate is added to the mother liquor remaining after crystallization (see Section 6) (Olson, R. A., et al., 1969b; Olson, R. A., 1970). The aggregate consists of more or less parallel tubules. Cross sections of tubules in electron micrographs show a hexagonal array of electron dense elements surrounding an electron transparent channel. The proposed packing arrangement of Bchl a protein in the tubules is shown in Fig. 7 compared to the packing in hexagonal crystals. In the tubule model, the channel diameter is 160 Å and the outside diameter (including cusps) is 440 Å. The optical properties of paracrystalline aggregates are similar to those of crystals and solutions. The dichroic ratio at 809 nm is +1.50 with preferential absorption parallel to the filaments in the aggregate. At 603 nm, the dichroic ratio is 1/1.09. Birefringence is of the same order and sign as in hexagonal crystals.

3. Spectral Properties

3.1. Absorption, Circular Dichroism, and Fluorescence at Room Temperature

Absorption spectra for the Bchl a–proteins from both strains in solution at room temperature are

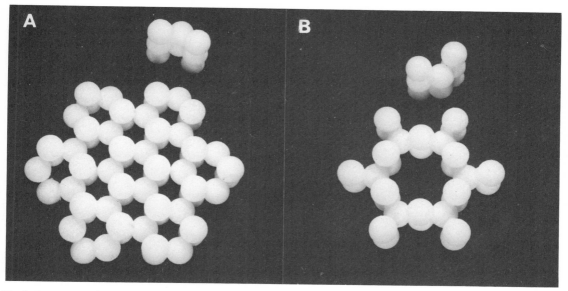

Fig. 7. Arrangement of Bchl *a*–protein macromolecules in the hexagonal crystal lattice (A), and in the proposed configuration for a cross section of a tubule in a paracrystalline aggregate (B) (Olson, R. A., *et al.*, 1969*b*). Each sphere represents one trimer.

shown in Figs. 8 and 9; absorption characteristics (Olson, J. M., 1966; Olson, J. M., *et al.*, 1973) are summarized in Table 4. Circular dichroism (CD) spectra at room temperature are also shown in Fig. 9 (Olson, J. M., *et al.*, 1973). For the Bchl *a*–protein from strain 2K, the value of $\Delta\varepsilon_{370} - \Delta\varepsilon_{400}$ is 50 $\pm$ 5 $M^{-1} \cdot cm^{-1}$ (Olson, J. M., 1971). The CD spectrum in the far red (760–840 nm) can vary depending on pH, temperature, ionic strength, and history of the sample. The same parameters also affect the ab-

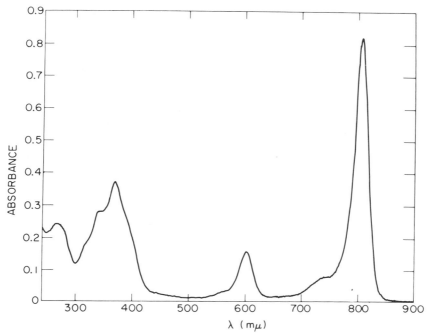

Fig. 8. Absorption spectrum of Bchl *a*-protein (strain 2K) dissolved in 0.25 M NaCl and 20 mM phosphate buffer, pH 7.8. Reproduced from Sybesma and Olson (1963).

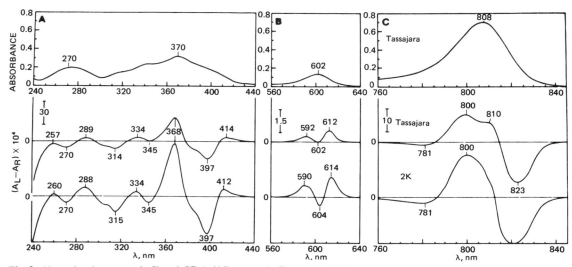

Fig. 9. Absorption (top curve A–C) and CD (middle curve, A–C) spectra of Bchl a–protein from strain Tassajara, and CD spectrum (bottom curve, A–C) of Bchl a–protein from strain 2K. Samples dissolved in 10 mM Tris (pH 8.0), and 0.25 M NaCl; path length, 1.0 cm; temp. 24°C. The figure is composed of spectra from several samples. The absorption spectrum (top curve, A–C) of Bchl a–protein (strain Tassajara) is that of a single solution ($A_{808} = 0.695$). The CD spectrum of Bchl a–protein (strain Tassajara) from 240 to 640 nm (middle curve, A and B) is for a sample with $A_{371} = 0.965$, $A_{602} = 0.4$, and $A_{808} = 2.0$, but for the spectrum from 760 to 840 nm (middle curve, C), another sample ($A_{808} = 1.30$) was used. The CD spectrum of Bchl a–protein (strain 2K) from 240 to 640 nm (bottom curve, A and B) is for a sample with $A_{371} = 1.66$, $A_{603} = 0.72$, and $A_{809} \simeq 3.7$, but for the spectrum from 760 to 840 nm (bottom curve, C), another sample ($A_{809} = 1.51$) was used. Reproduced from J. M. Olson *et al.* (1973).

sorption spectrum in the far red, but to a lesser degree. For quantitative measurement of concentration, the absorbance value at 370.5 nm is recommended. The absorbance ratio $A_{809}/A_{370.5}$ (strain 2K) or $A_{808}/A_{370.5}$ (Tassajara) is a useful criterion of the state of the complex; values of 2.3–2.4 (strain 2K) or 2.2–2.3 (Tassajara) indicate a sample in good condition, whereas values below 2.0 suggest partial denaturation.

The fluorescence emission peak for the Bchl a–protein (strain 2K) at room temperature (20–25°C) is

Table 4. Absorption Characteristics[a] at Room Temperature[b]

	Strain 2K			Strain Tassajara	
λ (nm)	ε(mM^{-1}·cm^{-1})[c]	$\varepsilon_\lambda/\varepsilon_{370.5}$[d]		λ (nm)	$\varepsilon_\lambda/\varepsilon_{370.5}$
267	37	0.55		267	0.57
343	49	0.73		342.5	0.76
370.5	67	1.00		370.5	1.00
603	28.4	0.42		602.5	0.41
745	13.4	0.20		745	0.21
809	154[e]	2.30[e]		808	2.31[e]
		2.1–2.2[f]			2.1–2.2[f]

[a] Samples dissolved in 10 mM Tris (pH 8.0), 0.25 M NaCl.
[b] Data from J. M. Olson (1966) and J. M. Olson and J. A. Cotton (unpublished, 1978).
[c] Values based on Bchl a. Limit of error estimated to be ±4%.
[d] Average values. Standard deviation is about 2%.
[e] Highest values observed.
[f] Typical values depending on pH, temperature, ionic strength, and history of sample.

at 818 nm (Sybesma and Olson, 1963; Olson, J. M., 1966). Optical rotatory dispersion spectra were recorded by Ke (1966).

3.2. Low-Temperature Properties

3.2.1. Absorption and Fluorescence

The differences between the absorption and CD spectra for the two strains of bacteria become more pronounced as the temperature is dropped to approximately 77–100°K. The 809 nm absorption band (strain 2K) is resolved at 77°K into three sharp bands at 825, 813.5, and 805.5 nm, with a broad band at approximately 790 nm, as shown in Fig. 10; the 603 nm band is partially resolved into at least three components, with peaks at 599 and 607 nm and a shoulder at 616 nm. The fluorescence emission peak sharpens and shifts to 831 nm (Olson, J. M., 1971).

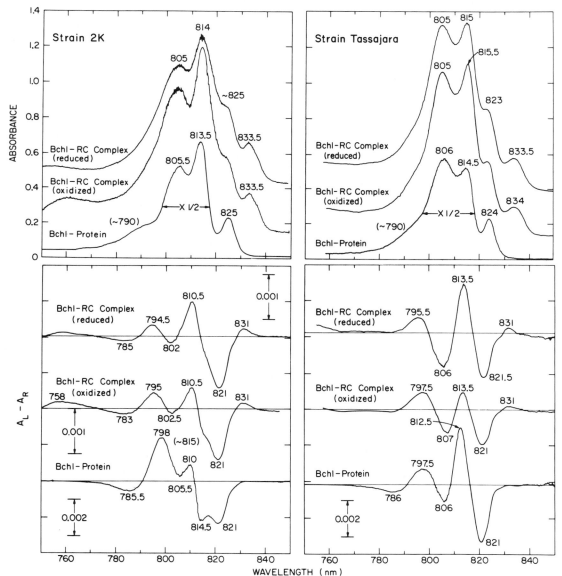

Fig. 10. Absorption and CD spectra of Bchl *a*–protein and Bchl *a*–RC complexes in 50% glycerol at 77°K. The light path is 1.0 mm. Spectra for strain 2K are on the left; spectra for strain Tassajara are on the right. Bchl *a*–proteins were dissolved in 10 mM Tris (pH 8) and approximately 0.1 M NaCl before mixing with 60% glycerol (1:4 by volume). Bchl *a*–RC complexes were dissolved in 10 mM phosphate (pH 7.4) and approximately 35% sucrose. Either sodium ferricyanide (30 mM) or sodium ascorbate (solid) was added to oxidize or reduce the reaction centers before mixing with 60% glycerol (1:4 by volume). Reproduced from J. M. Olson *et al.* (1976a).

The fluorescent yield increases from 0.19 at 293°K to 0.29 at 77°K (Olson, J. M., 1966, 1971).

The 808-nm absorption band (strain Tassajara) is also resolved at 77°K into three sharp bands at 824, 814.5, and 806 nm, with a slight shoulder at approximately 790 nm, as shown in Fig. 10; the relative peak heights are different, however, from those in the spectrum for strain 2K.

3.2.2. Circular Dichroism

The CD spectrum for each strain shows two negative troughs (781 and approximately 823 nm) separated by a positive peak (approximately 800 nm) in the far-red region (see Fig. 9). The spectra look similar but not identical. At 77°K, the similarity between the CD spectra disappears (see Fig. 10). The spectrum for strain 2K shows two peaks and three troughs (Philipson and Sauer, 1972) at 788 (−), 798 (+), 810 (+), 814.5 (−), and 821 (−). Strain Tassajara shows a different pattern: 786 (−), 797.5 (+), 806 (−), 812.5 (+), and 821 (−) (Olson, J. M., *et al.*, 1967*a*).

3.2.3. Exciton Interaction

The multiple components in the far-red absorption and CD spectra for each Bchl *a*–protein are the result of strong exciton interactions between the Bchl molecules in each protein (Philipson and Sauer, 1972). In each subunit of the Bchl *a*–protein from strain 2K, there are seven Bchl molecules spaced 12 Å apart on the average (Fenna and Matthews, 1975, 1977*a*). These seven chromophores are so close together that they act as a single unit for absorbing or emitting photons. In such an array, the Q_y band of Bchl can be split into several components (up to a maximum of seven) in both absorption and rotational strength. The pattern of this "exciton splitting" depends on the positional and directional coordinates of each Bchl molecule in the array. If all coordinates are known, the pattern can be predicted. However, the coordinates cannot be deduced from the pattern alone. From a comparison of the exciton splitting patterns for strains 2K and Tassajara, it is clear that the positional and directional coordinates of the Bchl molecules are not

Table 5. Five Computer-Resolved Components of Absorption and CD Spectra of Bchl *a*–Protein from Strain 2K

λ* and Δλ (nm)			Relative dipole strength	Relative rotational strength	FWHM (nm)[a]	Skew
Abs.	CD	Δ				
Philipson and Sauer (1972)						
792*	787*	+5	13	−8	16.0	0.65
12	13					
804*	800*	+4	37	18	11.9	0.65
9	12					
813*	812*	+1	35	35	7.5	0.75
4	2					
817*	814*	+3	3	−22	8.0	0.65
7	9					
824*	823*	+1	13	−17	9.1	0.82
J. M. Olson et al. (1976a)						
791.1*	791.3*	−0.2	12.0	−5.4	18.7	0.41
14.0	6.8					
805.1*	798.1*	+7.0	40.7	10.6	13.7	0.56
4.9	12.7					
810.0*	810.8*	−0.8	1.8	3.2	5.5	0.52
3.9	2.8					
813.9*	813.6*	+0.3	33.4	−5.3	7.1	0.89
11.2	8.0					
825.1*	821.6*	+3.5	12.0	−6.7	8.4	0.75

[a] Full width at half-maximum for left half of asymmetric Gaussian curve.

exactly the same in the two subunits. Interactions between Bchl molecules in different subunits of the same trimer have been assumed to be negligible, but recent calculations by Fenna and Matthews (1977) suggest that small but significant interactions occur between Bchl's 7 in all three sub-units, and between Bchl 5 of any one subunit and Bchl's 2 and 3 of the appropriate neighboring subunit in the trimer from strain 2K.

3.2.4. Analysis of Spectra

Since there are seven Bchl molecules in each Bchl a–protein subunit from strain 2K (and presumably also from strain Tassajara), it was of interest to find out whether seven components are required to resolve (or deconvolute) the 77°K absorption and CD spectra for the two proteins. From visual inspection of the absorption spectra, four components are obviously required for each protein. From the CD spectra, five components are necessary. If the absorption and CD spectra are analyzed together with the requirement that there be the same number of components in each spectrum and that each pair of components (absorption and CD) share a common bandwidth and skew, a quantitative best fit can be computed for any arbitrary number of components. In this way, Philipson and Sauer (1972) analyzed the low-temperature spectra for the Bchl a–protein from strain 2K in terms of five asymmetric Gaussian components. The results (see Table 5) appeared satisfactory except for the 4- to 5-nm differences between the wavelengths of absorption and CD components in the first two pairs.

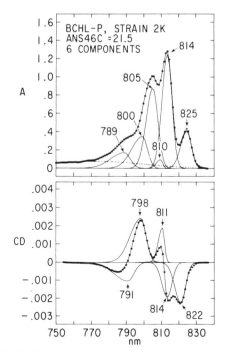

Fig. 11. Six computer-resolved components of absorption and CD spectra of Bchl a–protein from strain 2K. Data points are indicated by squares. The best fit to the data is indicated by the solid curves. Reproduced from J. M. Olson et al. (1976a).

J. M. Olson et al. (1976a) repeated the low-temperature spectroscopy on the protein from strain 2K and analyzed their results with the same computer program GAMET used by Philipson and Sauer

Table 6. Six Computer-Resolved Components of Absorption and CD Spectra of Bchl a–Protein from Strain 2K

λ* and Δλ (nm)			Relative dipole strength	Relative rotational strength	FWHM (nm)[a]	Skew
Abs.	CD	Δ				
789.2*	791.0*	−1.8	9.2	−5.5	18.6	0.43
10.3	7.4					
799.5*	798.4*	+1.1	15.0	10.5	14.3	0.49
5.9	7.0					
805.4*	805.4F*[b]	0.0	28.1	0.1	8.9	0.78
4.7	5.4					
810.1*	810.8*	−0.7	1.6	3.3	5.7	0.48
3.8	2.8					
813.9*	813.6*	+0.3	34.1	−5.3	7.3	0.86
11.2	8.0					
825.1*	821.6*	+3.5	12.0	−6.7	8.4	0.75

[a] Full width at half-maximum for left half of asymmetric Gaussian curve.

[b] The symbol F denotes a fixed parameter.

Table 7. Five Computer-Resolved Components of Absorption and CD Spectra of Bchl *a*–Protein from Strain Tassajara

λ^* and $\Delta\lambda$ (nm)			Relative dipole strength	Relative rotational strength	FWHM (nm)[a]	Skew
Abs.	CD	Δ				
789.4*	789.5*	−0.1	0.5	−4.2	18.2	0.40
11.4	9.7					
800.8*	799.2*	+1.6	26.8	5.9	21.8	0.37
6.2	8.2					
807.0*	807.4*	−0.4	50.1	−6.0	9.5	1.54
8.7	5.1					
815.7*	812.5*	+3.2	13.1	8.9	5.8	0.86
8.3	8.5					
824.0*	821.0*	+3.0	9.4	−7.3	5.5	1.07

[a] Full width at half-maximum for left half of asymmetric Gaussian curve.

(1972). Their analysis in terms of five asymmetric Gaussian components (see Table 5) also contained a major flaw: the 7-nm difference between the wavelengths of the absorption (805.1 nm) and CD (798.1

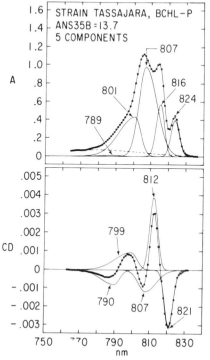

Fig. 12. Five computer-resolved components of absorption and CD spectra of Bchl *a*–protein from strain Tassajara. Data points are indicated by squares. The best fit to the data is indicated by the solid curves. Reproduced from J. M. Olson *et al.* (1976*a*).

nm) components in the second pair. This flaw was rectified by adding a sixth component at 799.5 nm to the absorption spectrum and a sixth component at 805.4 nm to the CD spectrum (see Table 6 and Fig. 11). This addition greatly improved the resolution of the absorption spectrum with negligible effect on the CD spectrum. Six components are required for a satisfactory resolution of the data of J. M. Olson *et al.* (1976*a*), but additional components cannot be justified based on spectral evidence alone.

Low-temperature spectra for the Bchl *a*–protein from strain Tassajara were also analyzed in terms of five asymmetric Gaussian components, as shown in Table 7 and Fig. 12 (Olson, J. M., *et al.*, 1976*a*). This resolution is unsatisfactory because the third component, 807.0 (807.4), has a large skew (1.54) toward the red. The implication is that at least six components might be required to resolve these data in terms of components either symmetrical or skewed toward the blue. Thus, for both Bchl *a*–proteins, the number of real or inferred components in the exciton splitting of the Q_y band is one less than the number of chromophores inside the subunit.

4. Bchl *a*–Reaction Center Complexes

A large complex (mol. wt. > 1.5 million) containing Bchl *a* and photochemical activity was first prepared from strain 2K by Fowler *et al.* (1971). This Bchl *a*–reaction center (RC) complex I contains cytochrome, P840, and carotenoid, but is essentially free of chlorobium chlorophyll (Bchl *c*, *d*, *e*, or *f*). J. M. Olson *et al.* (1973), using strain Tassajara, duplicated the earlier results of Fowler *et al.* (1971). From a comparison of the low-temperature (77–100°K) absorption spectra of the Bchl *a*–protein and the Bchl *a*–RC complex I from either strain of bacteria, it appeared that the Bchl *a*–protein might be a constituent of Bchl *a*–RC complex I (Olson, J. M., *et al.*, 1973; Fowler *et al.*, 1973).

4.1. Low-Temperature Spectra

Comparison of the low-temperature CD spectra for Bchl *a*–protein and Bchl *a*–RC complex I from either strain strengthened the earlier inference based on absorption spectra alone. Complex I from each strain was prepared with the RCs either fully reduced (by ascorbate) or fully oxidized (by ferricyanide). Absorption and CD spectra at 77°K are shown in Fig. 10 (Olson, J. M., *et al.*, 1976a). The absorption spectra for both reduced and oxidized complex I show a pattern of three peaks at 805, 814, and approximately 825 nm (strain 2K) or 805, 815, and approximately 823 nm (Tassajara) quite similar to the pattern for the corresponding Bchl *a*–protein. Likewise, the CD spectra for complex I and the protein from strain 2K show a pattern with peaks and troughs at approximately 784 (−), approximately 796 (+), approximately 804 (−), 810 (+), 814 (−), and 821 nm (−). For strain Tassajara, the CD spectra for complex I and the protein show another pattern, with peaks and troughs at approximately 797 (+), approximately 807 (−), approximately 813 (+), and 821 nm (−). Two new spectral features are unique to complex I: an absorption peak at 834 nm and a positive CD peak at 831 nm. The new features are tentatively ascribed to Bchl *a* (other than P840) in the RC, while the other components are ascribed to Bchl *a*–proteins in complex I.

4.2. Dissociation by Guanidine Hydrochloride

When complex I (Tassajara) is incubated in 2 M guanidine HCl, it dissociates into Bchl *a*–RC complex II (larger component), Bchl *a*–protein (smaller component), and possibly other small proteins (Olson, J. M., *et al.*, 1976c). Complex II, which still contains considerable Bchl *a* as well as carotenoid, cytochromes *b* and *c*, and RC activity, may be completely separated from Bchl *a*–protein by chromatography on Sephadex G-200, Sepharose 4B, or Sepharose CL-4B. Gel electrophoresis of complexes I and II shows that Bchl *a*–protein (39-kdalton polypeptide) is the major component of complex I, but is a lesser component of complex II.

Electron microscopy of positively stained thin sections of complex I and II preparations shows in both cases a population of unit-membrane (thickness = 70–80 Å) vesicles ranging in diameter from 300 to 1000 Å, with most vesicle diameters between 500 and 600 Å (Olson, J. M., and Thornber, 1978). In addition to the vesicles, there also appear some discontinuous sections (400–700 Å) of unit membrane (Olson, J. M., *et al.*, 1977), which are relatively more numerous in preparations of complex I than in complex II. The removal of Bchl *a*–protein by guanidine HCl apparently has no effect on the appearance of the unit membrane.

5. General Remarks

The Bchl *a*–proteins from green bacteria can be models for other chlorophyll–proteins: the chlorophyll *a*–proteins of blue-green algae and chloroplasts and the chlorophyll *a/b*–proteins of chloroplasts (Thornber, 1975). I would predict that most chlorophyll–protein subunits will consist of protein bags with the chlorophyll molecules inside. One possible exception to this generalization may be the chlorophyll–proteins of RCs. In an RC, at least one chlorophyll molecule would need to be exposed to a cytochrome or other electron donor. Likewise, one chlorophyll molecule also needs to be exposed to the primary electron acceptor.

The evolutionary relationships between the various green bacteria can be studied by determining the amino acid sequences of key proteins such as cytochrome and ferredoxin. Cytochrome *c*-555 from *Chl. limicola f. thiosulfatophilum* (strain L or PM) has, for example, only about half its residues in common with cytochrome *c*-555 from *P. aestuarii* 2K (van Beeumen *et al.*, 1976). This supports the morphological evidence that strain 2K belongs to a different species than do strains L, PM, and (by implication) Tassajara. The differences between the Bchl *a*–proteins of strains 2K and Tassajara are consistent with the implications of the cytochrome data, but the amino acid sequences for each Bchl *a*–protein are needed for a really useful contribution to the species question. Bchl *a*–protein is the key protein of the light-harvesting apparatus, and should prove useful in working out family relationships.

ACKNOWLEDGMENTS

This chapter was written under the auspices of the U.S. Energy Research and Development Administration. By acceptance of this chapter, the publisher and/or recipient acknowledges the U.S. Government's right to retain a nonexclusive, royalty-free license in and to any copyright concerning this chapter.

A chapter similar to this one has been written by Fenna and Matthews (1978). I am grateful to Dr. Roger Fenna and Dr. Brian Matthews of the University of Oregon for permission to paraphrase the text of their papers on the structure of the Bchl a–protein from strain 2K (Fenna and Matthews, 1975, 1977). I also thank them for supplying Figs. 3–5 for this chapter.

6. Appendix: Preparation of Bchl a–Proteins*

Production. Prosthecochloris aestuarii 2K and *Chl. limicola f. thiosulfatophilum* 6230 (Tassajara) are grown anaerobically (Bose, 1963; Olson, J. M., *et al.*, 1973) in 20-liter carboys with illumination provided by two 60-W incandescent lamps (18-inch Lumiline type) and culture temperature maintained between 28 and 32°C. The cultures are stirred continually with a magnetic bar. The cells are harvested by precipitation with alum [$KAl(SO_4)_2 \cdot 12H_2O$] added as a saturated solution, about 500 ml per 20 liters of culture. The aggregated cells are collected by low-speed centrifugation. All procedures are carried out at room temperature (20–25°C) except as noted.

Extraction. The packed cells (450 to 500-g wet wt.) or an equivalent amount of frozen cells are suspended in 1 vol. 0.2 M Na_2CO_3. Additional 2 M Na_2CO_3 is added if necessary to give pH 9.5–10, and the mixture is stored at 4–5°C for 18 hr or more. The cell mixture is homogenized for 4 min at low speed in a blender and sonicated for 10 min at 1.2 A in a Raytheon 10-kHz oscillator. The sonicated suspension is centrifuged at 13,700g for 45 min.

Precipitation. Solid ammonium sulfate (30 g/dl) is added to the supernatant solution, which is then stored at 4–5°C for 18 hr or longer. The resulting mixture containing precipitated Bchl a and chlorobium chlorophyll (Bchl c, d, e, or f) is stirred with diatomaceous earth (0.5 kg or more Celite 545), along with sufficient ammonium sulfate solution (35 g/dl, 10 mM Tris, pH 8.0) to make a thick slurry, which should be light gray-green. If too little Celite is used, the green chlorophyll components will be washed off the column. No harm is done if excess Celite is added.

Ammonium Sulfate Chromatography. The slurry is poured into a column (30 × 14 cm) onto a 2.5-cm pad of dry Celite and eluted with a constant gradient of decreasing $(NH_4)_2SO_4$ (35–0 g/dl) in 10 mM Tris, pH 8.0 (approximately 23 liters). Fractions are collected according to the color of the eluate. The first fraction

is colorless; the second is straw-colored (cytochromes; Olson and Shaw, 1969); the third is blue-green (Bchl a–protein) and usually elutes between 12 and 5% $(NH_4)_2SO_4$; the fourth is yellow-green. The blue-green fraction (6–8 liters) is concentrated to about 150 ml by ultrafiltration and then dialyzed at 4–5°C against 10 mM Tris, pH 8.0, to remove $(NH_4)_2SO_4$. (Bchl a–protein from strain Tassajara is dialyzed against 5 mM Tris, pH 8.0.)

DEAE Chromatography. The concentrated salt-free Bchl a–protein is loaded onto a suitably equilibrated DEAE–cellulose (Schleicher & Schuell Type 40) column (60 × 2.5 cm), and is eluted with a constant gradient of increasing NaCl (0–0.25 M) in 10 mM Tris, pH 8.0 (5 mM Tris, pH 8.0, for strain Tassajara). The Bchl a–protein is collected using the spectral criterion $A_{267}/A_{371} < 0.6$. Material having a higher absorbance ratio can be rerun on DEAE–cellulose. By the criterion of polyacrylamide gel electrophoresis, the Bchl a–protein from strain 2K is at least 99% pure, but the Bchl a–protein from strain Tassajara is only 85% pure after DEAE chromatography. Bchl a–protein is stored as a slurry in 30% (wt./vol.) $(NH_4)_2SO_4$ at −10°C.

Crystallization. Bchl a–protein may be crystallized by making the eluted Bchl a–protein solution 1 M in NaCl, concentrating to about 10 mg/ml by ultrafiltration followed by slow dialysis at 4–5°C against 5–10 g $(NH_4)_2SO_4$/dl in 1 M NaCl, 10 mM Tris, pH 8.0. [Strain Tassajara Bchl a–protein is concentrated to about 17 mg/ml and dialyzed against 5–10 g $(NH_4)_2SO_4$/dl in 1 M NaCl, 5 mM Tris, pH 8.0.] Since Bchl a–proteins are slowly oxidized in the presence of light and oxygen, they should be isolated under green light (F40 green fluorescent lamps behind green celluloid or green 2092 Plexiglas) and stored in the dark.

7. Note Added in Proof

7.1. Structure

Five Bchl a molecules (1, 3, 4, 6, and 7) in the 2K subunit appear to have histidine residues as ligands to the Mg atoms. The ligands to Bchl molecules 3 and 7 are adjacent residues occuring in an α-helix which runs along the trimer interface. The ligand to Bchl 4 is a residue about 2–3 residues before the start of this α-helix. The ligands to Bchl's 1 and 6 are about 20 Å apart on two adjacent strands of the β-sheet structure that forms most of the surface exposed to solvent. The ligand to Bchl 5 may be a main chain carbonyl oxygen, and the ligand to Bchl 2 may be a water molecule (unpublished results of B. W. Matthews, R. E. Fenna, M. Bolognesi, and M. Schmid). Atomic coordinates

---

* Written by Elizabeth K. Shaw.

(Ident. Code 1BCL, Protein Data Bank, Brookhaven National Laboratory, Upton, N.Y. 11973) for the seven Bchl *a* molecules inside each subunit have been published (Fenna *et al.*, 1977).

7.2. Paracrystalline Aggregate

Matthews *et al.* (1977) have refined the tubule model (Fig. 7) of R. A. Olson *et al.* (1969*b*) by taking into consideration the trimeric nature of the Bchl *a*–protein and the intermolecular contacts found in both hexagonal and trigonal crystal habits.

7.3. Spectral Properties

In the absorption spectrum of the Tassajara protein (Fig. 9) the fine structure of the 267 nm band is not adequately shown. More recent work (J. M. Olson and J. A. Cotton, unpublished) shows the existence of shoulders at ≈283 and ≈292 nm which are also apparent in the spectrum for the 2K protein (Fig. 8).

When the temperature of the 2K Bchl *a*–protein (dissolved in 4 vol. 75% potassium glycerophosphate, 4 vol. buffer, and one vol. glycerol) is dropped from 77°K to 5°K, the narrow absorption bands at 805.0–805.8, 814.4, and 825.0 nm are further sharpened and slightly shifted to 805.1, 814.9, and 825.5 nm respectively. Shoulders at 793 and 801 nm were located by 4th and 8th derivative spectroscopy, and some evidence was obtained for a small band at 809 nm (Whitten *et al.*, 1978*a*). (To a first approximation these results are consistent with the six components shown in Fig. 11 and Table 6.) The four major bands at 801, 805.1, 814.9, and 825.5 nm are resolved by 8th derivative spectroscopy at all temperatures from 5 to 300°K where they are shifted slightly to 800, 805.0, 814.2, and 823.1 nm. The constancy of the exciton splitting pattern indicates that the conformation of the Bchl *a*–protein trimer remains essentially the same over this temperature range. The most prominent derivative peak (814.2 nm) in the 300°K absorption spectrum has almost the same position (813 nm) and band width (≈12 nm) as the single peak of the linear dichroism spectrum of the protein oriented in an electric field (Whitten *et al.*, 1978*b*). This indicates that the linear dichroism is due to the one or two exciton states that give rise to the 814–815 nm absorption component.

When the Tassajara protein (dissolved in one vol. glycerol and one vol. buffer) is cooled from 77°K to 5°K, the narrow absorption bands (806, 814.5, 824 nm) also are sharpened and slightly shifted to 806, 815.5, and 823 nm respectively. Shoulders at 793 and 802 nm were located by 4th derivative spectroscopy, and some evidence was obtained for a small band at

810 nm (W. B. Whitten, R. M. Pearlstein, and J. M. Olson, unpublished data). [These results indicate one more component (810 nm) than are shown in Fig. 12 and Table 7.] The Q_y bands of *both* Bchl *a*–proteins at 5°K are thus seen to consist of at least six components.

8. References

Bose, S. K., 1963, Media for anaerobic growth of photosynthetic bacteria, in: *Bacterial Photosynthesis* (H. Gest, A. San Pietro, and L. P. Vernon, eds.), pp. 501–510, Antioch Press, Yellow Springs, Ohio.

Fenna, R. E., and Matthews, B. W., 1975, Chlorophyll arrangement in a bacteriochlorophyll protein from *Chlorobium limicola*, *Nature* (*London*) **258**:573–577.

Fenna, R. E., and Matthews, B. W., 1977, Structure of a bacteriochlorophyll *a*–protein from *Prosthecochloris aestuarii*, *Brookhaven Symp. Biol.* **28**:170–182.

Fenna, R. E., and Matthews, B. W., 1978, Bacteriochlorophyll proteins from green photosynthetic bacteria, in: *The Porphyrins* (D. Dolphin, ed.), Academic Press, New York (in press).

Fenna, R. E., Matthews, B. W., Olson, J. M., and Shaw, E. K., 1974, Structure of a bacteriochlorophyll–protein from the green photosynthetic bacterium *Chlorobium limicola*: Crystallographic evidence for a trimer, *J. Mol. Biol.* **84**:231–240.

Fenna, R. E., Ten Eyck, L. F., and Matthews, B. W., 1977, Atomic coordinates for the chlorophyll core of a bacteriochlorophyll *a*–protein from green photosynthetic bacteria, *Biochem. Biophys. Res. Commun.* **75**:751–755.

Fowler, C. F., Nugent, N. A., and Fuller, R. C., 1971, The isolation and characterization of a photochemically active complex from *Chloropseudomonas ethylica*, *Proc. Natl. Acad. Sci. U.S.A.* **68**:2278–2282.

Fowler, C. F., Gray, B. H., Nugent, N. A., and Fuller, R. C., 1973, Absorbance and fluorescence properties of Bchl-*a* RC complex and Bchl *a*–protein in green bacteria, *Biochim. Biophys. Acta* **292**:692–699.

Ghosh, A. K., and Olson, J. M., 1968, Effects of denaturants on the absorption spectrum of the bacteriochlorophyll–protein from the photosynthetic bacterium *Chloropseudomonas ethylicum*, *Biochim. Biophys. Acta* **162**:135–148.

Gibson, J., 1961, Cytochrome pigments from the green photosynthetic bacterium *Chlorobium thiosulphatophilum*, *Biochem. J.* **79**:151–158.

Gray, B. H., Fowler, C. F., Nugent, N. A., Rigopoulos, N., and Fuller, R. C., 1973, A reevaluation of *Chloropseudomonas ethylica* strain 2K, *Int. J. Syst. Bacteriol.* **23**:256–264.

Holt, A. S., Hughes, D. W., Kende, H. J., and Purdie, J. W., 1963, Chlorophylls of green photosynthetic bacteria, *Plant Cell Physiol.* **4**:49–55.

Jensen, A., Aasmundrud, O., and Eimhjellen, K. E., 1964, Chlorophylls of photosynthetic bacteria, *Biochim. Biophys. Acta* **88**:466–479.

Ke, B., 1966, Optical rotatory dispersion of chlorophyll-containing particles from green plants and photosynthetic bacteria, in: *The Chlorophylls* (L. P. Vernon and G. R. Seely, eds.), p. 427, Academic Press, New York.

Kim, Y. D., 1970, The action of detergents on the bacteriochlorophyll–protein complex, *Arch. Biochem. Biophys.* **140**:354–361.

Kim, Y. D., and Ke, B., 1970, Conformational states at acid, neutral, and alkaline pH values, *Arch. Biochem. Biophys.* **140**:341–353.

Matthews, B. W., Fenna, R. E., and Remington, S. J., 1977, An evaluation of electron micrographs of bacteriochlorophyll *a*–protein in terms of the structure determined by x-ray crystallography, *J. Ultrastruct. Res.* **58**:316–330.

Olson, J. M., 1966, Chlorophyll–protein complexes derived from green photosynthetic bacteria, in: *The Chlorophylls* (L. P. Vernon and G. R. Seely, eds.), pp. 413–425, Academic Press, New York.

Olson, J. M., 1971, Bacteriochlorophyll–protein of green photosynthetic bacteria, in: *Methods in Enzymology*, Vol. 23, Part A (A. San Pietro, ed.), pp. 636–639, Academic Press, New York.

Olson, J. M., 1973, Historical note on *Chloropseudomonas ethylica* strain 2K, *Int. J. Syst. Bacteriol.* **23**:265–266.

Olson, J. M., and Romano, C. A., 1962, A new chlorophyll from green bacteria, *Biochim. Biophys. Acta* **59**:726–728.

Olson, J. M., and Shaw, E. K., 1969, Cytochromes from the green photosynthetic bacterium *Chloropseudomonas ethylicum*, *Photosynthetica* **3**:288–290.

Olson, J. M., and Thornber, J. P., 1978, Photosynthetic reaction centers, in: *Membrane Proteins in Energy Transduction* (R. A. Capaldi, ed.), Marcel Dekker, New York.

Olson, J. M., Filmer, D., Radloff, R., Romano, C. A., and Sybesma, C., 1963, The protein–chlorophyll-770 complex from green bacteria, in: *Bacterial Photosynthesis* (H. Gest, A. San Pietro, and L. P. Vernon, eds.), pp. 423–431, Antioch Press, Yellow Springs, Ohio.

Olson, J. M., Koenig, D. F., and Ledbetter, M. C., 1969, A model of the bacteriochlorophyll–protein from green photosynthetic bacteria, *Arch. Biochem. Biophys.* **129**:42–48.

Olson, J. M., Philipson, K. D., and Sauer, K., 1973, Circular dichroism and absorption spectra of bacteriochlorophyll–protein and reaction center complexes from *Chlorobium thiosulfatophilum*, *Biochim. Biophys. Acta* **292**:206–217.

Olson, J. M., Ke, B., and Thompson, K. H. 1976a, Exciton interaction among chlorophyll molecules in bacteriochlorophyll *a*–proteins and bacteriochlorophyll *a*–reaction center complexes from green bacteria, *Biochim. Biophys. Acta* **430**:524–537.

Olson, J. M., Shaw, E. K., and Englberger, F. M., 1976b, Comparison of bacteriochlorophyll *a*–proteins from two green bacteria, *Biochem. J.* **159**:769–774.

Olson, J. M., Giddings, T. H., and Shaw, E. K., 1976c, An enriched reaction center preparation from green photosynthetic bacteria, *Biochim. Biophys. Acta* **449**:197–208.

Olson, J. M., Prince, R. C., and Brune, D. C., 1977, Reaction-center complexes from green bacteria, *Brookhaven Symp. Biol.* **28**:238–245.

Olson, R. A., 1970, Microtubular spherulites: Development and growth in solutions of bacteriochlorophyll protein, *Science* **169**:81–82.

Olson, R. A., Jennings, W. H., and Olson, J. M. 1969a, Chlorophyll orientation in crystals of bacteriochlorophyll–protein from green photosynthetic bacteria, *Arch. Biochem. Biophys.* **129**:30–41.

Olson, R. A., Jennings, W. H., and Hanna, C. H., 1969b, Paracrystalline aggregates of bacteriochlorophyll–protein from green photosynthetic bacteria, *Arch. Biochem. Biophys.* **130**:140–147.

Pfennig, N., and Biebl, H., 1976, *Desulfuromonas acetoxidans* gen. nov. and sp. nov., a new anaerobic, sulfur-reducing, acetate-oxidizing bacterium, *Arch. Microbiol.* **110**:3–12.

Philipson, K. D., and Sauer, K., 1972, Exciton interaction in a bacteriochlorophyll–protein from *Chloropseudomonas ethylica*. Absorption and circular dichroism at 77°K, *Biochemistry* **11**:1880–1885.

Shioi, Y., Takamiya, K., and Nishimura, M., 1976, Isolation and some properties of NAD⁺ reductase of the green photosynthetic bacterium *Prosthecochloris aestuarii*, *J. Biochem. (Tokyo)* **79**:361–371.

Studier, F. W., 1972, Bacteriophage T7, *Science* **176**:367–376.

Sybesma, C., and Olson, J. M., 1963, Transfer of chlorophyll excitation energy in green photosynthetic bacteria, *Proc. Natl. Acad. Sci. U.S.A.* **49**:248–253.

Sybesma, C., and Vredenberg, W. J., 1963, Evidence for a reaction center P840 in the green photosynthetic bacterium *Chloropseudomonas ethylicum*, *Biochim. Biophys. Acta* **75**:439–441.

Sybesma, C., and Vredenberg, W. J., 1964, Kinetics of light-induced cytochrome oxidation and P840 bleaching in green photosynthetic bacteria under various conditions, *Biochim. Biophys. Acta* **88**:205–207.

Thornber, J. P., 1975, Chlorophyll–proteins: Light-harvesting and reaction center components of plants, *Annu. Rev. Plant Physiol.* **26**:127–158.

Thornber, J. P., and Olson, J. M., 1968, The chemical composition of a crystalline bacteriochlorophyll–protein complex isolated from the green bacterium, *Chloropseudomonas ethylicum*, *Biochemistry* **7**:2242–2250.

van Beeumen, J., Ambler, R. P., Meyer, T. E., Kamen, M. D., Olson, J. M., and Shaw, E. K., 1976, Amino acid sequences of cytochromes *c*-555 from two green bacteria of the genus *Chlorobium*, *Biochem. J.* **159**:757–774.

Whitten, W. B., Nairn, J. A., and Pearlstein, R. M., 1978a, Derivative absorption spectroscopy from 5–300K of bacteriochlorophyll–protein from *Prosthecochloris aestuarii*, *Biochim. Biophys. Acta* **503**:251–262.

Whitten, W. B., Pearlstein, R. M., Phares, E. F., and Geacintov, N. E., 1978b, Linear dichroism of electric field oriented bacteriochlorophyll–protein from green photosynthetic bacteria, *Biochim. Biophys. Acta* (in press).

CHAPTER 9

Photosynthetic Apparatus and Cell Membranes of the Green Bacteria

Beverly K. Pierson and Richard W. Castenholz

1. Introduction

During the past two decades, much attention has been given to the photosynthetic apparatus and membranes of the purple bacteria. During the same period, much less attention has been given to the green bacteria. Although some emphasis has been placed on the study of NAD^+ reduction, the pathway of CO_2 reduction, and the nature of the bacteriochlorophyll (Bchl)–protein complex in these bacteria, the fundamental question of the location of their energy-generating system remains unanswered.

Perhaps the reason for the scant attention given to the green bacteria is that they have seemed distantly related to the more familiar photosynthetic organisms (Cruden *et al.*, 1970). Diverse lines of evidence, recently accumulated, suggest that the green bacterial line may not be as remote from other photosynthetic prokaryotes as previously thought (Kenyon and Gray, 1974; Pierson and Castenholz, 1974*a*).

The green bacterial line (Suborder: Chlorobiineae, Family: Chlorobiaceae and Chloroflexaceae) (Trüper, 1976), however, is unique in at least the following respects:

1. Cells lack the intracellular proliferation of unit-membrane structures that bear Bchl in the Chromatiaceae and Rhodospirillaceae, although cells of the green line under proper conditions can have a higher specific content of Bchl.

2. In addition to small amounts of Bchl *a*, they contain larger quantities of chlorobium chlorophylls (Bchl *c*, *d*, or *e*), pigments not found in the purple bacteria.

3. They contain a peripheral layer of small ovoid bodies or sacs termed *chlorobium vesicles*. These structures are surrounded by a 2- to 3-nm-thick envelope that is not a "unit" membrane, and are located adjacent to the cell membrane. Early studies revealed that the majority of the Bchl *c* or *d* appeared to be associated with these vesicles. Thus, the green bacteria appeared to be unique among all phototrophic organisms in possibly having the entire photosynthetic apparatus in something other than a unit-membrane structure. If this were the case, green bacteria would then be unique among all organisms in having an electron-transport-dependent energy-conserving system outside such a membrane.

In this discussion, we shall consider the structure, isolation and composition of the chlorobium vesicles, cell membrane, and mesosomes of the green bacterial group. Also, we shall review what is known of the location and composition of the photosynthetic apparatus.

2. Chlorobium Vesicles

2.1. Structure and Location

Cohen-Bazire (1963) first used the term *chlorobium vesicle* for these structures clearly shown in electron

Beverly K. Pierson · Department of Biology, University of Puget Sound, Tacoma, Washington 98416 **Richard W. Castenholz** · Department of Biology, University of Oregon, Eugene, Oregon 97403

micrographs of three strains of green bacteria. Essentially identical vesicles were subsequently seen in all green bacteria examined (Cohen-Bazire, 1971; Cohen-Bazire *et al.*, 1964; Cohen-Bazire and Sistrom, 1966; Pfennig and Cohen-Bazire, 1967; Holt *et al.*, 1966a; Gorlenko and Zhilina, 1968), and in the recently described *Chloroflexus aurantiacus* (Pierson and Castenholz, 1974a). The vesicles are ovoid or elliptical, and are always very close to or actually appressed to the cell membrane (Fig. 1). Their dimensions are variable and have been reported as 30–40 × 100–150 nm in *Chlorobium limicola* and *Chl. limicola f. thiosulfatophilum* (Cohen-Bazire *et al.*, 1964) and 30 × 100 nm in *Prosthecochloris aestuarii*

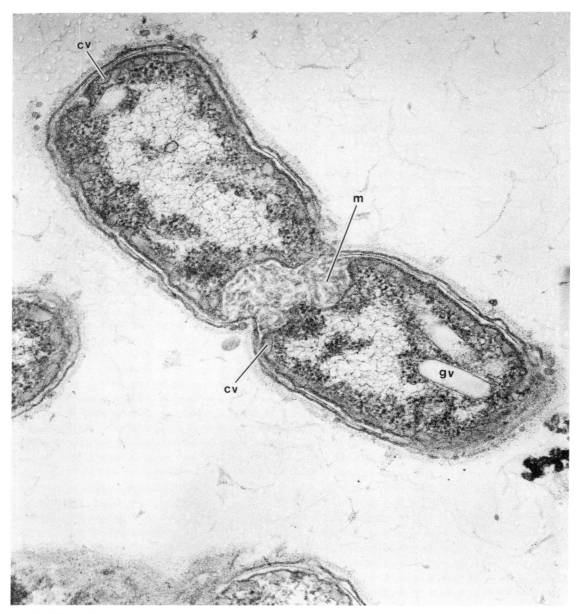

Fig. 1. Dividing cell of *Pelodictyon clathratiforme*, embedded in Maraglass. Chlorobium vesicles (cv) are of medium electron density and peripheral only. A large mesosome (m) occupies the center. Gas vesicles (gv) are also present. Osmium fixation (R.K.) for 2 hr, acetone dehydration. Section was poststained with lead hydroxide. ×80,000, reproduced at 95%. Courtesy of G. Cohen-Bazire, from Fig. 11, Cohen-Bazire (1971), by permission.

(Gorlenko and Zhilina, 1968). In one strain of *Chloroflexus aurantiacus*, the dimensions were 40–90 × 15–20 nm; in two other strains, in which the vesicles are elliptical in cross section, the dimensions were 100 × 50–70 × 20 nm (Pierson and Castenholz, 1974*a*) (Fig. 2). The vesicles were bounded in all cases by a nonunit "membrane" or envelope 2–3 nm thick (Cohen-Bazire and Sistrom, 1966; Pierson and Castenholz, 1974*a*). It appears only as a dark line in electron micrographs (Fig. 2). The appearance of the vesicles changes with different embedding materials (Cohen-Bazire and Sistrom, 1966). Preparations in epon resins appear dark, whereas those embedded in methacrylate are clear and appear empty. In a few cases, a fibrillar array gave some substructure to the vesicles when negatively stained (Holt *et al.*, 1966*b*; Cohen-Bazire and Sistrom, 1966; Cohen-Bazire, 1971). Cohen-Bazire *et al.* (1964) showed that in epon-embedded material, each vesicle was filled with fibrils parallel to the long axis and measuring 1.2–2.0 nm wide. More detail was seen by Cruden and Stanier (1970) in isolated vesicles of *Chl. limicola f. thio-*

sulfatophilum that were of greater than normal width and perhaps broken during isolation. The intravesicular structures appeared as five or six subunits aggregated around a central hole. Each circular aggregate was 9–10 nm in diameter, and vesicles contained 10–25 of them. These substructures remained after the extraction of Bchl's with methanol.

In thin-sectioned cells, the vesicles form a continuous peripheral layer, occupying about 25% of the cell volume in *Chl. limicola f. thiosulfatophilum* (Cohen-Bazire *et al.*, 1974). Cruden and Stanier (1970) suggested 12% as a more accurate estimate in *Chl. limicola f. thiosulfatophilum*. The estimated volume was 9–18% in *Chloroflexus aurantiacus* (Pierson, 1973). In *Prosthecochloris aestuarii*, the elongate prosthecae were also completely lined with vesicles (Fig. 3). Gorlenko and Zhilina (1968) concluded that these arms served to provide extra space for vesicles relative to protoplast volume. However, the area of cell membrane relative to cell volume is also greatly increased by the possession of prosthecae.

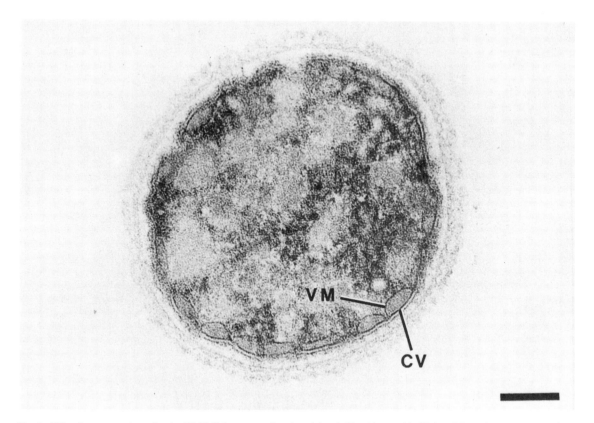

Fig. 2. *Chloroflexus aurantiacus* (strain OK-70-fl) in cross section. A peripheral chlorobium vesicle (CV) and the vesicle "membrane" (VM) are indicated. Glutaraldehyde and osmium fixation. The bar represents 0.1 µm. Reproduced from Fig. 5, Pierson and Castenholz (1974*a*), by permission.

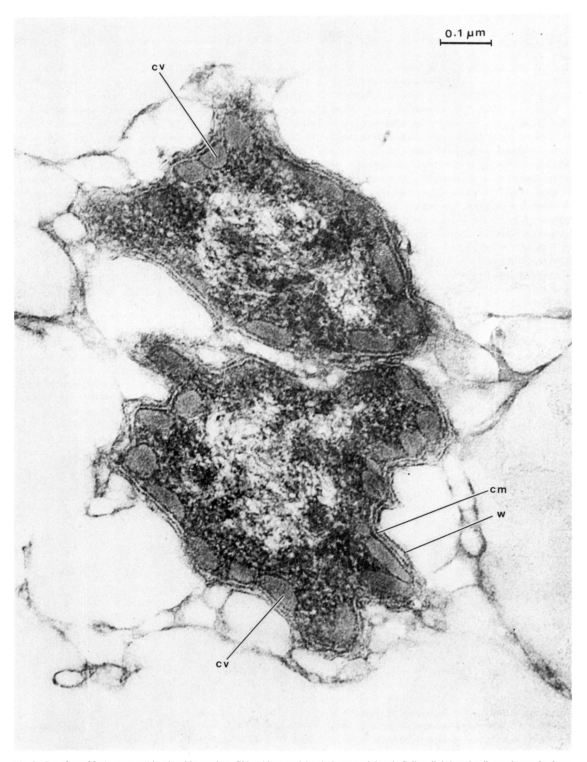

Fig. 3. *Prosthecochloris aestuarii* in ultrathin section. Chlorobium vesicles (cv) are peripheral. Cell wall (w) and cell membrane (cm) are clearly distinguishable. Fixed with osmium, stained with uranyl acetate, and embedded in methacrylate. ×100,000, reproduced at 85%. Courtesy of V. M. Gorlenko, from Fig. 2a, Gorlenko and Zhilina (1968).

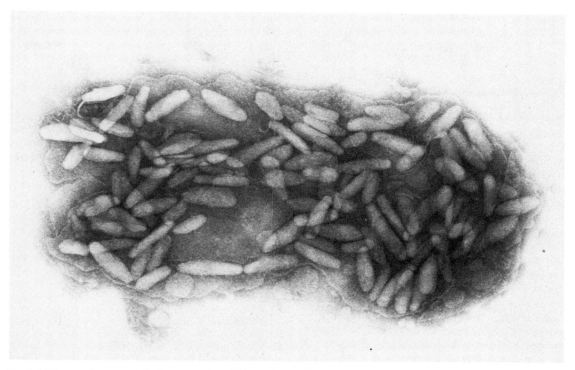

Fig. 4. "*Chloropseudomonas ethylica*" grown at about 100 lux and opened ballistically. Large numbers of chlorobium vesicles are seen in the cell envelope. Negatively stained with 2% phosphotungstic acid. ×90,750. Reproduced courtesy of S. Holt, from Fig. 5, Holt *et al.* (1966*b*), by permission.

Vesicles in the green bacteria may be separated from or appressed to the cell membrane, perhaps even attached. Cruden *et al.* (1970) published micrographs that indicate a definite separation, although small. Others have thought that the vesicles are attached to the cell membrane and even to each other (Gorlenko and Zhilina, 1968; Tomina and Fedorov, 1967; Solov'eva and Fedorov, 1970; Holt *et al.*, 1966*a*; Pierson and Castenholz, 1974*a*). Holt *et al.* (1966*a*) measured the kinetics of vesicle release during cell disruption and concluded that the vesicles were probably bound loosely to the cell membrane or to each other (Fig. 4) (Holt *et al.*, 1966*b*). These results were ambiguous, however, because of possible small breaches in the cell walls that would permit only a slow release of vesicles. Using a similar approach, Sykes *et al.* (1965) compared the kinetics of vesicle release from cells broken by sonication with the release of ribosomes. The two rates were similar, and they concluded that the vesicles were free cytoplasmic elements, not part of some larger structure.

Irregular shrinkages during dehydration and embedding may alter the relationship of vesicles to the membrane, just as distortions between membrane and wall are seen. To our knowledge, no electron-micro-

scopic studies have used freeze–fractured preparations of green bacteria. Even if vesicles are attached, they must have a smaller radius of curvature than the cell. Consequently, only thin sections of a limited area would reveal the attachment. This question may not be easily answered.

In any case, all vesicles are at least closely associated with a piece of cell membrane. No internal chlorobium vesicles have been seen in the green bacterial line.

2.2. Isolation and Purification

Although many components of the green bacterial photosynthetic apparatus have been isolated and characterized, in only a few instances have these components been isolated from a specifically identified subcellular fraction (Cruden and Stanier, 1970; Cruden *et al.*, 1970; Sykes *et al.*, 1965; Fowler *et al.*, 1971; Schmitz, 1967). Isolation procedures for purified vesicle preparations have been described by Cohen-Bazire *et al.* (1964), Cruden and Stanier (1970), Holt *et al.* (1966*a*), and Fowler *et al.* (1971). The procedures used by Cohen-Bazire *et al.* (1964) for cells of *Chl. limicola f. thiosulfatophilum* consisted of differential and density-gradient centrifugation. Cells were broken with a French pressure cell and centrifuged at low speed to remove whole cells. The supernatant was centrifuged for 2 hr at 100,000g. The pellet containing nearly all the Bchl was resuspended in Tris buffer and centrifuged on a linear sucrose gradient (0.5–2.0 M) for 2 hr at 25,000 rpm. A main pigmented fraction was obtained that had twice the specific Bchl content of the whole cells. The criterion for purity of the vesicle fraction was restricted to electron-microscopic verification of the negatively stained preparation. The material was heterogeneous, but did represent some purification of the vesicle fraction. Holt *et al.* (1966*a*) used similar techniques on French-pressure-cell extracts of "*Chloropseudomonas ethylica*" (a dubious species*). Centrifugation on a linear sucrose gradient at 25,000 rpm for 2 hr was followed by density-gradient electrophoresis on sucrose, but the pigmented vesicle fraction was nevertheless contaminated with ribosomes and membranes.

Sykes *et al.* (1965) also attempted to obtain a pure vesicle preparation from *Chl. limicola f. thiosulfatophilum.* They were able to reduce ribosomal contamination, but did not monitor for cell membranes. Schmitz (1967) subjected homogenates of *Chl. limicola f. thiosulfatophilum* prepared with a Hughes press to differential and sucrose density-gradient centrifugation, which resulted in two pigment bands. The lighter fraction had a higher proportion of Bchl to carotenoids than the heavier band. The lighter fraction was considered the vesicle preparation. The purity was assessed under electron microscopy, which disclosed chlorobium vesicles and smaller particles. It was thought to be only slightly contaminated by cell walls, but no direct measure of membrane contamination was reported.

In isolating a photochemically active complex from "*Chloropseudomonas*," Fowler *et al.* (1971) attempted to obtain pure vesicles. They were removed by centrifugation at 40,000g for 60 min and purified by density-gradient centrifugation at 25,000 rpm for 2 hr on a 10–35% sucrose gradient. Fraction 15 of this gradient contained the Bchl c. Bchl a was in an asymmetrical zone associated with it. The authors concluded that this separation was caused by vesicle disruption. On the other hand, the slight separation of pigment bands could have meant different origins. The vesicle fraction was examined by electron microscopy, and some membrane contamination was seen.

A more successful attempt to purify vesicles for

chemical analysis was reported by Cruden and Stanier (1970). Membranes and vesicles were separated from each other, and cross-contamination was monitored for each fraction. Cells were disrupted with a French pressure cell or by sonication and then treated with 5% (vol./vol.) glutaraldehyde to stabilize the vesicles. After removal of glutaraldehyde, an electrophoretic separation of vesicles and membranes from the rest of the particulate fraction was achieved on a linear sucrose gradient. Separation was not complete, but by careful selection of fractions, cross-contamination was minimized. The specific content of Bchl c (vesicles only) and the specific activity of succinic dehydrogenase (membranes only) were used to monitor cross-contamination. Purity was further assessed by electron microscopy, using negative staining. The cleanest vesicle preparation had less than 5% membrane contamination on the basis of succinic dehydrogenase activity. Unfortunately, Bchl a was not assayed.

2.3. Chemical Composition

The first attempt to analyze the contents of isolated chlorobium vesicles was by Cohen-Bazire *et al.* (1964). They found that 87% of the total cell chlorophyll (Bchl c) was in the vesicle fraction in *Chl. limicola f. thiosulfatophilum.* The specific content was 240 μg Bchl c/mg protein, a twofold enrichment over that in the whole cells. Sykes *et al.* (1965) found that the crude vesicle fraction contained 107 μg Bchl c/mg protein in the same organism. They extended their analysis of vesicle composition to include pigment, protein, lipid, RNA, and carbohydrate (Table 1).

Schmitz (1967) reported a very high (2.3 mg Bchl c/mg protein) specific chlorophyll content for his vesicle preparations from *Chl. limicola f. thiosulfatophilum.* He suggested this high value was a result of little protein contamination from sources other than vesicles. Later, Cruden and Stanier (1970) reported 845 μg Bchl c/mg protein in the vesicle fraction from the same species, a ninefold increase in the specific content over that of the crude extract (Table 1). It was concluded that all the Bchl is housed in the vesicles. No attempt was made to measure Bchl a in the vesicles in any of these studies.

The analyses of Cruden and Stanier (1970) also included protein, pigment, lipids, RNA, and carbohydrate content of the vesicles (Table 1). The lipid fraction consisted of both glycolipid and phospholipid with 42 μg carbohydrate/mg lipid and 8.5 μg phosphate/mg lipid, the glycolipids thus being predominant. Thin-layer chromatography revealed that the vesicles contained only one glycolipid, a mono-

---

* See Chapter 2 for a discussion of the taxonomy of this syntrophic system.

Table 1. Chemical Composition of Chlorobium Vesicles[a]

	% Dry weight						
References	Protein	Total lipid	Bchl c	Lipid less Bchl c	RNA	Carbohydrate	Bchl c/protein[b]
Sykes *et al.* (1965)	63.3	25.2	6.8	18.4	4.3	9.0	1/9
Schmitz (1967)	17	48	39	9.4	—	38	2/1
Cruden and Stanier (1970)	32.7	42.7	27.6	15.1	0.4	16.3	1/1.2

[a] The organism was *Chl. limicola f. thiosulfatophilum* in all cases except that of Holt *et al.* (1966a), who used "*Chloropseudomonas*" (see Chapter 2 for a discussion of the taxonomy of this syntrophic system).

[b] Three other studies were concerned with vesicle pigment content relative to protein: Bergeron and Fuller (1961), Bchl c/protein, 1/13; Cohen-Bazire *et al.* (1964), 1/4; Holt *et al.* (1966a), 1/1.2.

galactosyl diglyceride. This type of lipid is not found in purple bacteria, but is found in blue-green bacteria and chloroplasts. The succinic dehydrogenase activity was low (0.010 units/mg protein). It was therefore concluded that contamination of the vesicle fraction by membranes was minimal. The NADH, NADPH-dye reductase activities were higher (0.178 and 0.047 units/mg protein, respectively), but still lower than those in the membrane fraction. All enzyme activities were presumably depressed by the prior glutaraldehyde treatment.

Based on analysis of the photochemically active complex believed to be derived from disrupted vesicles, Fowler *et al.* (1971) concluded that the vesicles contained Bchl c, Bchl a, carotenoid, and cytochrome c-553. Later, Fowler (1974) demonstrated the presence of a b-type cytochrome in the photochemically active complexes from *Chl. limicola* and *Chl. limicola f. thiosulfatophilum*, prepared as described previously. The conclusion as to whether or not any of these components are contained in the chlorobium vesicles depends on the correct assessment of the origin of the photoactive complex.

In summary, the vesicles contain Bchl c, protein, lipid, and carbohydrate. RNA is probably due to ribosomal contamination. If the variation in the Bchl c/protein ratio obtained by different investigators is an indication of differences in purity of the vesicle preparations, then it appears that Schmitz (1967) had the purest preparation. If his estimate of cell-wall contamination ($<3\%$) is correct, the higher carbohydrate content and lower nonchlorophyllous lipid content (Table 1) indicate an even higher proportion of glycolipid to phospholipid in the vesicles than was seen by Cruden and Stanier (1970).

Because of controversy over the origin of the photoactive complex of Fowler *et al.* (1971) and Fowler (1974), we consider the assignment of Bchl a, carotenoid, and b- and c-type cytochromes to the vesicles as uncertain.

2.4. Composition of the Photosynthetic Apparatus

Other investigators have isolated and characterized several components of the photosynthetic apparatus of green bacteria (Knaff and Buchanan, 1975; Shioi *et al.*, 1974). From their data alone, however, it is impossible to determine whether these components are membrane- or vesicle-bound. In such studies, the term *chromatophore* was loosely applied to the photosynthetic structures. Knaff and Buchanan (1975) prepared "chromatophores" by sonication of *Chl. limicola f. thiosulfatophilum* and detected the presence of "membrane-bound" cytochrome b in equimolar concentrations with reaction center Bchl a (P840). Shioi *et al.* (1974), maintaining that their cultures of "*Chloropseudomonas ethylica*" were axenic, proceeded to characterize the electron-transport system. They examined particulate and soluble fractions, again with no attempt to distinguish between vesicles and membranes, and concluded that cytochrome c-555 was weakly bound to the "membranes." Cytochrome c-551, however, was apparently bound to the "membranes" by weaker forces or else was located in the cytosol. They detected no b- or a-type cytochromes or cytochrome c-553. The "photosynthetic apparatus" appeared to be composed of (in number of molecules/pigment P840): Bchl c, 1,530; Bchl a, 89; total cytochrome, 26; total nonheme iron, 410; total acid-extractable flavin, 9. The authors suggested that the most significant difference between the green bacteria and other phototrophs was the much larger "photosynthetic unit" in the green bacteria— approximately 1600 Bchl/P840.

3. Membranes

The cells of green bacteria are clearly bounded by a typical unit membrane about 7–8 nm in thickness that sometimes appears to be continuous with one or more prominent mesosomal-like structures (Cohen-Bazire *et al.*, 1964; Pfennig and Cohen-Bazire, 1967; Gorlenko and Zhilina, 1968; Pierson and Castenholz, 1974*a*). The term *mesosome* has been loosely applied to various intracellular membranous structures in both gram-positive and gram-negative bacteria. It has been suggested by Greenawalt and Whiteside (1975) that "mesosome" should be applied only when specific criteria of origin, location, extrusion, and function are met. In the green bacteria and *Chloroflexus aurantiacus*, the mesosomelike bodies appear to originate from the cell membrane, and often occur at the site of septum formation (see Fig. 1). Since they also look like the mesosomes of gram-positive bacteria, the only criterion remaining unsatisfied is that of extrusion. Mesosomes in *Chloroflexus aurantiacus* frequently appear tubular and vesicular (Pierson and Castenholz, 1974*a*). Mesosomes in green bacteria have not been considered as likely sites for the photosynthetic apparatus (Cohen-Bazire, 1963; Cohen-Bazire *et al.*, 1964).

We are aware of only one study that has sought to purify membranes of green bacteria, that of Cruden and Stanier (1970). The purpose was to characterize both the vesicles and membranes, but not to distinguish cell membrane from the mesosomes. In the purest membrane fraction, the specific content of Bchl *c* was less than 10% of that in the vesicles; this was thought to be due to contamination. The other membrane components were protein, 29%; lipid, 50%; carbohydrate, 9%; and RNA, 0.5%. The membrane fraction was richer in phospholipids and poorer in glycolipids than the vesicles.

The sugar content of membrane lipids was 9.3 and the phosphate content 43.8 µg/mg lipid. The membrane fraction had only one glycolipid. It contained galactose, rhamnose, and a third, unidentified sugar. Enzyme activities were (in units/mg protein): succinic dehydrogenase, 0.220; malic dehydrogenase, 0.015; NADH-dye reductase, 0.225; NADPH-dye reductase, 0.083. All these activities, but particularly those of succinic and malic dehydrogenases, were higher than in the vesicle fraction. It was assumed that succinic dehydrogenase was found in membrane only. If the study had included an assay of Bchl *a*, it would have been interesting to see whether the Bchl *a* distribution paralleled that of Bchl *c* or of succinic dehydrogenase. In gross composition, the membranes appear to be similar to those of purple bacteria, except for the presence of the glycolipid.

4. Physiological Functions Associated with Chlorobium Vesicles and Membranes

The functions probably associated with the photosynthetic apparatus, vesicles, and membranes in green bacteria include the control of vesicle formation and pigment synthesis, respiratory activity, and photosynthetic activity.

4.1. Variations in Vesicle Formation and Pigment Synthesis

The regulation of vesicle formation and pigment synthesis in green bacteria has received little attention. Holt *et al.* (1966*b*) found that the number of vesicles per cell decreased with increasing light intensity. Electron micrographs of thin sections of cells grown at 100 lux had the typical cortical layer of vesicles, each vesicle being approximately 130×50 nm. Micrographs of sectioned cells grown at 10,760 and 107,600 lux showed smaller and fewer vesicles per cell (those present were still peripheral). Similar conclusions were drawn from negatively stained whole cells that had been disrupted ballistically.

As part of the same study, the specific chlorophyll content (Bchl *c* only) was determined for whole cells and for isolated vesicles from cells grown at different light intensities. The data for whole cells from Holt *et al.* (1966*b*) are reproduced in Table 2.

There was about a fivefold decrease in the specific chlorophyll content of whole cells as the light intensity increased from about 100 to 107,000 lux. The authors reported an eightfold decrease in the specific content of the vesicles. However, since the vesicle preparations were so contaminated with cell walls and membrane fragments, the variation in chlorophyll content could be an artifact. It is therefore possible that only the number of vesicles per cell, and not the pigment

Table 2. Specific Bchl *c* Content of "*Chloropseudomonas*" Grown at Various Light Intensities[a]

Approximate light intensity (lux)	Chlorophyll, µg/mg protein (whole cells)	
	Expt. I	Expt. II
100	501	491
1,070	188	162
10,760	117	112
107,600	97	86

[a] Data from Holt *et al.* (1966*b*).

Table 3. Specific Pigment Content of *Chloroflexus aurantiacus* Grown Anaerobically at Various Intensities of Incandescent Light[a]

Light intensity (lux)	Pigment (μg/mg dry wt.)		Ratio Bchl c/a
	Bchl c	Bchl a	
210–270	23.42	2.27	10.3
320–540	16.44	2.03	8.1
1,080	6.67	2.07	3.2
5,400	2.90	1.73	1.7
21,500	1.72	1.09	1.6
54,000	0.63	0.78	0.8

[a] Data from Pierson and Castenholz (1974*b*).

content per vesicle, changed with light intensity. There was no apparent increase in the amount of cell membrane in the more highly pigmented cells.

In *Chloroflexus aurantiacus*, the specific chlorophyll contents of whole cells (μg/mg dry wt.) decreased with increasing light intensity (Pierson and Castenholz, 1974*b*). This was true for both Bchl *c* and *a*, but no electron-microscopic counts of vesicles were made. However, the decrease in specific pigment content of Bchl *c* was greater than for Bchl *a*, resulting in a change in the Bchl *c*/Bchl *a* ratio from 10 at 200–270 lux to 0.8 at 54,000 lux. Table 3 lists pigment contents of *Chloroflexus aurantiacus* at selected intensities. The total Bchl *c* per cell is considerably less than that of the Chlorobiaceae at different light intensities and is more comparable to the specific Bchl *a* content of the Rhodospirillaceae. The growth rate of *Chloroflexus aurantiacus* (doublings/hr) increased with light intensity from 0.03 at less than 1000 lux to 0.3 at 20,000 lux. At higher intensities, the carotenoid-to-chlorophyll ratio increased, and some additional carotenoids were synthesized (Pierson and Castenholz, 1974*b*).

Since the Chlorobiaceae are obligate anaerobes, the specific effect of oxygen on pigment synthesis and vesicle formation has not been analyzed. In contrast, *Chloroflexus aurantiacus* can be grown under fully aerobic conditions. In that case, no vesicles were seen. High oxygen tension completely suppressed synthesis of both Bchl's (Pierson and Castenholz, 1974*b*), but synthesis under microaerobic conditions did occur. The presence of oxygen, however, did not eliminate the synthesis of all types of carotenoids. Two or more were synthesized under fully aerobic conditions (Castenholz, unpublished). Schmidt (1976) also reported the synthesis of carotenoids under aerobic conditions.

Chlorophyll synthesis in *Chloroflexus aurantiacus* appears to be regulated by light and oxygen in a manner similar to that described for the purple nonsulfur bacteria (Cohen-Bazire and Sistrom, 1966). The green bacteria also show a similar response to light intensity. It would be interesting to have a more quantitative analysis of the vesicle number per cell for both types of organisms and to know whether the Bchl *a* content of green bacteria behaves differently from that of Bchl *c* as in *Chloroflexus aurantiacus*.

Although Holt *et al.* (1966*b*) suggested that pigment variations in "*Chloropseudomonas*" could be due to changes in the number of vesicles per cell only and not in the pigment content per vesicle, a more complex situation definitely exists in *Chloroflexus aurantiacus*. Since the two Bchl's do not change at the same rate, a simple explanation involving change in number of vesicles with a constant pigment content cannot apply. The rates of synthesis of the two pigments are not the same after a shift in light intensity (Pierson and Castenholz, 1974*b*). There are two possible explanations: (1) both pigments are housed in vesicles in which relative pigment contents are independently controlled, or (2) one of the pigments is in the vesicle, the other is a part of the cell membrane.

Absolutely nothing is known about the origin, synthesis, growth, or development of the vesicles. Their association with the cell membrane suggests they may be synthesized or assembled by a membrane-associated system.

4.2. Respiratory Activity (Oxygen Consumption)

Although it is known that the purple nonsulfur bacteria are capable of respiratory oxygen consumption, green bacteria (i.e., Chlorobiaceae) are thought to be obligate phototrophs incapable of respiration. Larsen (1953) demonstrated that *Chl. limicola f. thiosulfatophilum* could not use molecular oxygen either for endogenous activity or for the oxidative degradation of external substrate. There was no uptake of oxygen with or without substrate when cultures were exposed to air and 5% CO_2, or to a mixture of nitrogen, 1% O_2, and 1% CO_2. Larsen also concluded that molecular oxygen was toxic.

The relationship of green bacteria to oxygen is not well understood, however, and an examination of the more recent literature on this subject revealed some confusing and intriguing results. Gusev *et al.* (1969) concluded that *Chl. limicola f. thiosulfatophilum* and "*Chloropseudomonas*" were unable to respire after growth ceased in darkness or light in the presence of oxygen. However, cells of *Chl. limicola f. thiosulfatophilum* absorbed oxygen in the presence of acetate.

The nature of this uptake is unknown. In a later publication, Gusev et al. (1970) reported that *Chlorobium* survived 3–4 hr of purging with air in the light.

Shuvalov et al. (1968) reported that light stimulated the absorption of oxygen by *Chl. limicola f. thiosulfatophilum* and "*Chloropseudomonas*," and that this absorption was greater in heat-killed cells. In the light, the rate reached a stationary level, but oxygen absorption declined and disappeared with continued exposure to light. Absorption was intensity-dependent, increasing at least up to 20,000 lux. An action spectrum indicated that absorption was sensitized by short-wavelength (monomeric) forms of Bchl *c* or *d* (Abs 670–680 nm). The authors suggested that this oxygen consumption might be the result of a destructive photodynamic effect or an adaptive removal of oxygen from the environment. There was no indication that the oxygen uptake was associated with a respiratory process.

The question of oxygen metabolism in green bacteria may be even more complex. Gusev and Shenderova (1971) observed light-stimulated uptake of oxygen in "*Chloropseudomonas*" with or without exogenous substrate. They also observed an uptake of oxygen in the dark that was stimulated by malate, succinate, and glucose. The oxygen consumption in the presence of malate was inhibited about 50% by KCN. Using the same culture, Takamiya (1971) demonstrated oxygen uptake and showed that oxygen reversibly inhibited the photooxidation of menaquinone and cytochromes, suggesting the possibility that molecular oxygen may interact with the photosynthetic electron-transport chain in green bacteria. However, since the "*Chloropseudomonas*" used was probably a dual culture of a *Chlorobium* and a sulfate-reducing chemoheterotroph (Gray et al., 1973), results that pertain to this aspect of metabolism are questionable.

Although the Chlorobiaceae cannot grow aerobically, it appears that some kinds of oxygen uptake and perhaps even a type of respiration do exist. If some kind of respiratory metabolism does occur, it is probably associated with membranes, rather than with chlorobium vesicles.

Oxygen is not toxic to *Chloroflexus aurantiacus* (Pierson and Castenholz, 1974b). Vigorous aeration supports growth rates of 0.14–0.22 doubling/hr in chlorophyll-less cells in light or dark using complex organic medium. Oxygen inhibits the synthesis of Bchl's. Cells containing Bchl, however, consume oxygen at high rates in darkness only. Light partially inhibits the uptake of oxygen in pigmented cells, perhaps because of a competitive sharing of some of the same electron carriers by photosynthetic and respiratory processes (Marrs and Gest, 1973; Melandri et al., 1971; Chapter 45). Light-dependent reactions appear to predominate in such competitions. Since all other known electron transport chains are housed in "unit-type" membranes, it is easiest to assume that this is also true in *Chloroflexus aurantiacus*.

4.3. Photosynthetic Activity

The most obvious function of the photosynthetic apparatus of the green bacteria, which may involve both vesicles and cell membrane, is photosynthesis. All the following photosynthetic functions have been found associated with subcellular fractions of green bacteria: photoreduction of ferredoxin, NAD$^+$, and CO$_2$; photooxidation of the reaction center P840 and cytochromes with electron transport; and photophosphorylation. Each of these functions will be briefly reviewed.

4.3.1. Photoreduction of Ferredoxin, NAD$^+$, and CO$_2$

Evans and Buchanan (1965) demonstrated that bacterial particles prepared from *Chl. limicola f. thiosulfatophilum* photoreduced ferredoxin. The particles contained chlorophyll and were prepared from sonified cells by centrifugation and passage through a DEAE–cellulose column. The following were effective as electron donors for photoreduction of ferredoxin: Na$_2$S, 2-mercaptoethanol, cysteine, glutathione, ascorbate–DPIP. The reduced ferredoxin was used directly in CO$_2$ fixation by the ferredoxin-dependent pyruvate synthase reaction. NAD$^+$ was reduced by ferredoxin in *Chl. limicola f. thiosulfatophilum* (Buchanan and Evans, 1969). Similar particles photoreduced ferredoxin in the presence of an electron donor. NAD$^+$ was then reduced when the soluble fraction, which apparently contained the proper enzyme, was added. The rate of photoreduction of NAD$^+$ was dependent on the amount of chlorobium chlorophyll present. The maximum rate observed was 0.6 μmol NAD$^+$ reduced/hr per mg chlorobium chlorophyll. This was close to the observed rate of CO$_2$ assimilation by whole cells (0.7–2.4 μmol CO$_2$/hr per mg chlorobium chlorophyll).

Jones and Whale (1970) were apparently unable to demonstrate a light-induced formation of NAD(P)H by whole cells or chromatophores of *Chl. limicola f. thiosulfatophilum* for technical reasons. They concluded that NADH was probably formed by photoreduction via ferredoxin and a flavoprotein, and that no energy-dependent "reversed electron flow" was involved. They also found a soluble NADH–cytochrome c reductase that resembled the flavoprotein associated with ferredoxin–NAD^+ photoreduction and suggested its participation in a cyclic electron flow.

4.3.2. Electron Transport, Reaction Centers, and the Photosynthetic Unit

Olson and Sybesma (1963) first showed that light-induced oxidations of c-type cytochromes similar to those of purple bacteria occurred in whole cells of green bacteria. Light absorbed by both chlorobium chlorophyll and Bchl a was effective, but only the Bchl a was implicated in actual photochemical reactions, chlorobium chlorophyll behaving as a light-harvesting pigment only. Photooxidations of c-type cytochromes were further characterized by Sybesma (1967) and Meyer et al. (1968), but again with no localization of site.

Takamiya (1971) studied the quinones of whole cells of "*Chloropseudomonas*." Ubiquinone was lacking; menaquinone and chlorobium quinone were present. Light-induced absorbance changes at 270 nm under anaerobic conditions were interpreted to represent the oxidation of 60% of the menaquinone. The molar ratio of menaquinone to chlorobium chlorophyll was $1:50$. "Chromatophores" prepared by sonic treatment did not undergo any photooxidations either of quinones or cytochromes, suggesting that the preparative procedures caused inactivation.

Electron-transport activity in c-type cytochromes has been studied using thiosulfate (Kusai and Yamanaka, 1973a) or sulfide (Kusai and Yamanaka, 1973b) as electron donor. It was suggested that in *Chl. limicola f. thiosulfatophilum* (Kusai and Yamanaka, 1973c), cytochrome c-551 could accept electrons from thiosulfate and transfer them to cytochrome c-555. Alternatively, cytochrome c-553 could accept electrons from sulfide and transfer them to cytochrome c-555, which would then be photooxidized as in the following scheme:

$$S_2O_3^{2-} \rightarrow \text{cyt } c\text{-551} \searrow$$
$$\text{cyt } c\text{-555} \rightarrow \text{RC Bchl} \rightarrow$$
$$HS^- \rightarrow \text{cyt } c\text{-553} \nearrow \qquad \text{Fd} \rightarrow \text{NAD(P)}$$

The earliest photochemically active complex constituting a reaction center was prepared from "*Chloropseudomonas*" by Fowler et al. (1971). They demonstrated photooxidations of reaction center, P840, and cytochrome c-553. The photochemically functional particles were believed to be derived from the chlorobium vesicles and appeared to consist of 4 cytochromes, 80 Bchl a molecules, and 1000–1500 Bchl c molecules per P840.

In a more recent study, Fowler et al. (1973) isolated a reaction center complex from *Chl. limicola* and believed that it was derived from vesicles. In this complex, Bchl c was apparently present in monomeric form and did not transfer excitation energy to Bchl a. The γ-carotene present was also inactive. The Bchl c/Bchl a ratio was $1:5$; the γ-carotene/Bchl a ratio was $1:1$. The identity of an 833 nm component was not determined, but it was apparently different from the reaction center P840 and was not observed in the spectrum of bulk Bchl a–protein. They suggested the existence of an altered Bchl a–protein in the reaction center complex. The purified reaction center complex contained 5–10 cytochromes per P840.

Olson et al. (1973) also isolated from *Chl. limicola f. thiosulfatophilum* a photochemically active reaction center complex that contained Bchl a, P840, cytochromes and carotenoid, but was free of Bchl c or d. The nonphotochemically active Bchl a–protein of green bacteria was shown to be a constituent of this complex and could be completely removed from it, leaving a smaller photochemically active complex of Bchl a, carotenoid, and b- and c-type cytochromes (see Olson, Chapter 8). Preparative procedures did not allow an explicit determination of the origin of the photochemically active complex or of the Bchl–protein in the cells.

Photochemically active particles of uncertain origin from cells of *Chl. limicola* and *Chl. limicola f. thiosulfatophilum* were described by Barsky et al. (1974) and Knaff et al. (1973). The bulk Bchl a/cytochrome c/P840 ratio was $100:1.5:1$ (Barsky et al., 1974). Knaff and Buchanan (1975) reported photochemical electron transport involving a "membrane-bound" cytochrome b in equimolar concentrations with P840 from *Chl. limicola f. thiosulfatophilum*. The photoreduction of cytochrome b was enhanced and the electron transport from Na_2S to NAD was inhibited by antimycin A.

Fowler (1974) found b-type cytochromes in photochemically active complexes of *Chl. limicola* and *Chl. limicola f. thiosulfatophilum*. These were believed to be derived from the vesicles. He reported four of the b- and four of the c-type cytochromes per P840. The b-type cytochrome was bound only to the photoactive particle, which apparently also contained all 80 Bchl a molecules. The b cytochrome underwent reversible photooxidation.

No studies on photosynthetic activity in subcellular fractions of *Chloroflexus aurantiacus* have been reported. A reversible light-induced bleaching at 860 nm was observed in crude cell-free extracts, suggesting the existence of a reaction center Bchl a (Pierson and Castenholz, 1974b).

4.3.3. Photophosphorylation

Since a function of photosynthesis is the making of ATP, one would expect the detection of photophosphorylation in a particular fraction to be a reliable means of identifying the intact photosynthetic apparatus. It is possible to isolate parts of the photosynthetic apparatus capable of electron transport and to lose simultaneously some essential factors for phosphorylation. The detection of photophosphorylation has proved difficult in green bacteria, and the site is unknown.

Williams (1956) observed a light-dependent decrease in phosphate from solution in cell-free sonic preparations of *Chl. limicola*, incubated under N_2. The decrease was comparable to that in similar preparations of *Chr. vinosum*. The uptake of phosphate was inhibited 40–60% under aerobic conditions. In their isolation of photosynthetic particles from *Chl. limicola f. thiosulfatophilum*, Bergeron and Fuller (1961) also observed light-dependent uptake of phosphate in the cell-free extract, but this activity disappeared during the course of fractionation.

Hughes *et al.* (1963) also prepared pigmented particles from *Chl. limicola f. thiosulfatophilum* grown with a decreased polyphosphate content. There was a light-dependent uptake of phosphate from the incubation mixture, presumably because of phosphorylation. The uptake rate was 28 μmol phosphate/mg Bchl *c* per hr. A similar rate of photophosphorylation was affirmed by Fuller *et al.* (1963) for chlorophyll-containing particles from *Chl. limicola f. thiosulfatophilum*. However, Buchanan and Evans (1969) could not demonstrate light-induced ATP synthesis in a particulate fraction that photoreduced NAD^+. Lippert (1967) was unable to detect photophosphorylation in *Chl. limicola f. thiosulfatophilum*. Cells were cultured on a low-phosphate medium and extracts prepared using a French press or sonifier. No phosphorylation was detected in either the crude extract or washed vesicles. Using a similar system, Lippert did measure reasonable rates of photophosphorylation in preparations from *Rhodospirillum rubrum* and *Chromatium*. He also used more delicate means of breaking the cells to try to preserve phosphorylation activity. Cells were treated with lysozyme to make protoplasts, which were subsequently broken by osmotic shock. This treatment did not produce a green chromatophore fraction, as did the previous techniques; instead, it produced a membranous fraction with attached "chromatophores" sedimenting at 16,000*g*. This fraction did not show photophosphorylation either.

Lippert noticed that cell disruption with the French press caused the immediate release of Bchl-*a*-containing particles. The "chromatophores" (presumably the chlorobium vesicles), however, and cell fragments were removed from the crude extract by centrifugation at 45,000*g* for 20 min. This fraction was greatly enriched in Bchl *c*. The yellow supernatant contained particles enriched in Bchl *a* and carotenoids. They were removed by centrifugation at 100,000 *g* for 60 min. Chromatophores prepared by sonic disruption had 40% less Bchl *a* relative to Bchl *c* than whole cells. Further exposure of these chromatophores to ultrasonic frequencies did not release more Bchl *a*. Lippert suggested that the lack of photophosphorylation was due to the loss of a particular Bchl *a* component from the chromatophores. However, the lack of phosphorylation in the lysozyme-treated material was not explained.

The origin of the photophosphorylating particles of Hughes *et al.* (1963) is of considerable interest. There have been isolated from green bacteria subcellular structures that perform the photosynthetic functions of electron transport, but it has not been consistently reported that these fractions function to conserve energy. Coupled with the observations of Lippert on the immediate release of Bchl-*a*-containing particles during cell disruption, these observations indicate that the intact and functional photosynthetic apparatus of green bacteria disintegrates very readily during isolation.

5. An Evaluation of the Organization of the Photosynthetic Apparatus

The location of the complete photosynthetic apparatus in the green bacteria remains uncertain. It seems doubtful that the complete photosynthetic apparatus of green bacteria has even been isolated as an intact structure. That the peripheral oblong bodies called "chlorobium vesicles" contain the accessory pigments is well established. That these bodies also contain the reaction center pigment and associated electron carriers is questionable, primarily because of the difficulty in isolating these bodies free from contaminating membrane fragments. Although it is possible that the pigment-containing bodies of green bacteria do house the entire photosynthetic apparatus, the only solid evidence to support this notion is the presence of the accessory pigments. The problems discussed by Lascelles (1968) for the isolation and accurate analysis of chromatophores have not yet been resolved for green bacteria.

We object to the conclusion that these bodies represent the complete photosynthetic apparatus and the extension of this conclusion to the statement that the green bacteria are therefore unique among all photosynthetic organisms in having a non-membranous photosynthetic apparatus (Oelze and Drews, 1972). Our objections are based on the following observations from the literature cited in this chapter:

1. The vesicle preparations from which reaction centers and electron-transport components have been isolated are known to be contaminated with membrane fragments, thereby introducing the possibility that these components are actually housed in the membrane, not in the vesicles.

2. In the isolation procedures described, it has been repeatedly observed that much of the Bchl a is readily separable from the Bchl c component as soon as the cells are broken, although intact vesicles are still isolated from these preparations.

3. To our knowledge, no one has yet demonstrated photophosphorylation in a preparation of "pure" chlorobium vesicles or even in a preparation of impure vesicles.

We further object to the conclusion that the chlorobium vesicles are the complete photosynthetic apparatus in green bacteria on the basis of the following theoretical argument: No organisms are known in which phosphorylation is coupled to electron transport in anything other than a membranous structure. Evolution is conservative, and all other known electron-transport systems are membrane-bound, vectorial, and closed systems. We believe that it is improbable that the green bacteria, with so many electron carriers and pigments identical to or similar to those of the purple bacteria, would have evolved a unique system for generating ATP that excludes vectorial organization in a unit membrane.

We propose an alternative for the organization of the photosynthetic components in green bacteria. We suggest that the chlorobium vesicles function as bodies of accessory pigment located adjacent to the cell membrane, which is the site of at least the reaction center, electron-transport carriers, and phosphorylation coupling factors. Many observations cited in this chapter agree more with our suggestion. There is a large body of circumstantial evidence that makes the interpretation worthy of consideration. The observations are as follows:

1. Chlorobium vesicles are peripheral in the cell and are always closely associated with, if not directly attached to, the cell membrane. This suggests that being adjacent to or attached to cell membrane is essential to their function. This would be expected if they were accessory pigment bodies funneling excita-tion energy to membrane-bound reaction centers. An alternative reason for this association, however, is the possibility that the vesicles are synthesized by the membrane and remain close to their site of synthesis.

2. The Bchl a is readily dissociated from the accessory Bchl c when cells are broken, suggesting different locations for the pigments. The separation of the two pigments was evident in the very early work of Bergeron and Fuller (1961) before Bchl a was known in green bacteria. In this work, an absorbance maximum at 800 nm was clearly seen in spectra of whole cells of *Chl. limicola f. thiosulfatophilum* (see Fig. 1 of Bergeron and Fuller, 1961). A cleared cell-free extract was prepared, and comparison of the spectrum with that of whole cells showed a change in ratio of the two peaks. More 800 nm pigment (Bchl a) was lost in the pellet than 720 nm pigment (Bchl c or d). Further purification and preparation of a pigmented fraction resulted in two zones on a sucrose gradient: a green zone of "pigmented particles" that was rich in Bchl c and a yellow zone enriched in carotenoids and Bchl a. Although the authors were interested in the separation of the carotenoid from the pigmented particles and suggested that in fact this pigment may not have been part of the particle, they ignored the 800 nm absorbance peak the behavior of which paralleled that of the carotenoid. Such separation has been reported by other authors. Lippert (1967) clearly demonstrated that the breaking of cells released Bchl a, while vesicles appeared to remain intact.

Fowler *et al.* (1971) found the Bchl a in the subcellular fraction of "*Chloropseudomonas ethylica*" associated with but not completely united with the Bchl-c-containing vesicle fraction. They concluded that the Bchl-a-containing particles had partially dissociated from the vesicles, and further suggested that the relative ease with which the Bchl a complex was derived from the intact particle indicated a much different arrangement of the light-harvesting pigment with the reaction center complex than found in other organisms. Their observations could easily be explained if the light-harvesting pigments were in the chlorobium vesicles and the reaction centers in the cell membrane adjacent to the vesicles. One might expect a "loose" attachment of the vesicle to the membrane under these circumstances, which would permit isolation of the two components as a readily dissociable unit.

Olson and others (see Chapter 8) have found an easy separation of the water-soluble Bchl a–protein complex from the Bchl c. An initial separation is achieved simply by centrifuging ruptured cells; the

Bchl *c* is pelleted, and the cytochromes and Bchl *a* remain in the supernatant.

The early work of Bergeron and Fuller (1961) showed electron micrographs of purified pigmented particles (largely lacking Bchl *a*), which were uniformly about 15 nm in diameter and sedimented at 50 S. These particles must have originated by comminution of the larger chlorobium vesicles. With increased purification, the particles aggregated into larger rod-shaped units with dimensions of 100 × 15–40 nm, the dimensions of chlorobium vesicles. Possibly the self-assembly of chlorobium vesicles had occurred. If true, it occurred with most of the Bchl *a* already removed, suggesting that vesicle integrity is not dependent on the presence of Bchl *a*.

3. There exist in green bacteria electron-transport systems that are basically similar to the unit-membrane-bound systems of other photosynthetic organisms. Cytochromes are readily separated together with the Bchl *a* fraction and not the Bchl *c* fraction. The vesicles seem to be poor candidates for the organizational matrix of electron transport, in part because of their relatively low protein and phospholipid content (see Table 1 and Cruden and Stanier, 1970).

4. Photophosphorylation has been difficult to demonstrate in green bacteria, but if this type of energy-conserving system in green bacteria functions at all like that in other organisms (Jagendorf, 1975), then some vectorial organization of electron carriers is required as well as a selectively permeable membrane. The presence of phospholipids in membranes of *Rs. rubrum* was necessary to maintain phosphorylation activity (Razin, 1972). Phospholipids in green bacteria are restricted to the membranes and are absent from the vesicles. We suggest that one reason for the lack of consistent phosphorylation results in green bacteria is the use of chlorobium vesicles to the relative exclusion of the membrane fraction. However, the coupling factor required for phosphorylation in purple bacteria is easily removed from the membranes by sonic oscillation in the presence of EDTA (Melandri *et al.*, 1971). Phosphorylation activity in green bacteria may also be easily lost with a coupling factor during preparative procedures.

5. Observations of the recently described filamentous green bacterium *Chloroflexus aurantiacus* tend to support the suggestion that chlorobium vesicles are sacs of accessory pigments (Pierson and Castenholz, 1974*b*). Both respiratory and photosynthetic activities occur in this organism, and it has been suggested that the electron-transport chains of both systems share some of the same carriers. Respiration can occur in cells lacking or containing chlorobium vesicles. Photo-

synthesis occurs in cells containing vesicles. However, since cells with vesicles respire less in light than darkness, it appears that photosynthesis and respiration may compete for the same electron carriers (Pierson and Castenholz, 1974*b*). Thus, it seems more likely that such carriers are housed in the cell membrane rather than in the vesicles.

It has been suggested by Olson (1970) that the chlorobium chlorophylls evolved more recently than Bchl *a*, and that these pigments were additions to bacteria already containing Bchl *a*. If indeed the archetypal green bacterium had Bchl *a* bound in membranes as in the extant purple bacteria, the simplest evolutionary step would have been the acquisition of an accessory pigment package associated with these membranes, rather than a complete structural reorganization of the whole photosynthetic system. The evolution of this type of system would be similar in principle to the development of biliprotein units (phycobilisomes) in the blue-green bacteria. These are located on the surface of the thylakoid membranes, which contain both light-harvesting and reaction center chlorophyll *a*.

The transfer of excitation energy from light-harvesting Bchl *c*, *d*, or *e* and Bchl *a* to the reaction center poses no problems if the vesicles are indeed attached to the membrane or at least very close to it. In those electron micrographs that indicate a separation of vesicles from membranes, the separation distance does not exceed a few nanometers, which is close enough to permit efficient transfer of excitation energy by resonance between the chlorophyll molecules (Olson and Sybesma, 1963; Borisov and Godik, 1973; Knox, 1975; Fogg *et al.*, 1973).

The organization of the photosynthetic apparatus could be such that the vesicles contain only the accessory chlorobium chlorophylls arranged within some lipid or protein superstructure or in no particular organizational array. Although it is theoretically possible to have a very rapid transfer of light energy throughout such an antenna in the absence of any protein or lipid superstructure (Norris *et al.*, 1975), nothing is known at this time about the molecular interactions within the vesicles.

6. Note Added in Proof

6.1. Ultrastructure of Chlorobium Vesicles and Membranes and Response to Changes in Light Intensity and Oxygen

Several species of bacteria in the Chlorobiineae have been observed recently with electron microscopy and all have chlorobium vesicles similar to those described in this chapter. Three species of vibrioid green bacteria, *Chlorobium vibrioforme*, *Pelodictyon luteolum*, and *Pelodictyon phaeum*, were studied by Puchkova *et al.* (1975). All three species had ovoid cortical vesicles in the range of 70–150 × 30–80 nm and bounded by a non-unit membrane 2–5 nm thick when fixed in glutaraldehyde and osmium or with $KMnO_4$ and embedded in Araldite or Epon. Mesosomes were also observed in these cells as well as invaginations of the plasma membrane associated with electron-transparent cavities. It was suggested that these latter membranous structures represented sites for the excretion of sulfur.

Gorlenko (1975) mentioned the presence of chlorobium vesicles in newly isolated strains of mesophilic *Chloroflexus* species. In a later study on two mesophilic strains of *Chloroflexus aurantiacus*, Pivovarova and Gorlenko (1977) compared the ultrastructure of cells grown in the light (3000 lux) under either aerobic or anaerobic conditions. Cells were fixed in glutaraldehyde and osmium or with 1.5% $KMnO_4$ and embedded in Epon 812. Sections were poststained with uranyl acetate and lead citrate. Strain KN-4 had peripheral chlorobium vesicles bounded by a thin membrane and in contact with the cell membrane. The dimensions of the vesicles were 24–33 × 31–34 nm. Strain BR-1 when grown anaerobically had peripheral vesicles which measured 13–31 × 23–110 nm and appeared as electron-dense bodies when fixed with $KMnO_4$. Vesicles were sometimes observed along the cell septa. Cells of strain BR-1 grown aerobically had very few or no chlorobium vesicles. Such cells did have tubular or vesicular types of mesosomal invaginations of the cell membrane associated with the site of septum formation. Cells of strain BR-1 grown anaerobically had, in addition to the numerous chlorobium vesicles, an extensive vesicular invagination of the cell membrane near the septa. This membrane proliferation was much greater than in aerobically grown cells, occupying up to 23% of the cell volume, and seemed similar to the membranous proliferations in the purple bacteria. No chlorobium vesicles were associated with these invaginated membranes. The authors suggested that these membranes might be associated with the photosynthetic process and might house Bchl *a* as in the purple bacteria. No evidence was presented to support this suggestion.

The chlorobium vesicles in thermophilic strains of *Chloroflexus aurantiacus* have been further characterized by Madigan and Brock (1977). Both fixation (glutaraldehyde and osmium fixation and embedding in Durcupan with poststaining with uranyl acetate and lead citrate) and negative staining (1.5%

phosphotungstic acid) were used. For negative staining the cells were briefly homogenized before being placed on grids to facilitate penetration of the stain. In chemically fixed and sectioned preparations not subjected to homogenization vesicles were located peripherally. In negatively stained homogenized preparations vesicles were observed in the cortical region and also distributed throughout the interior of the cells. In these preparations vesicles appeared as ellipsoidal electron-transparent structures. It is suspected that homogenization is responsible for the distribution of the vesicles in the cell interior. The dimensions of the vesicles were 100–150 × 40–70 nm in strains OK-70-fl and Y-400-fl. The smaller strain had smaller vesicles, 90–95 × 25–35 nm in strain 254-2. The association of smaller vesicles with smaller-diameter filaments was also observed by Pierson and Castenholz (1974*a*). Madigan and Brock (1977) also observed that cells grown in the dark under highly aerated conditions contained no vesicles and that cells of OK-70-fl grown anaerobically either photoheterotrophically or photoautotrophically had identical chlorobium vesicles. It was further observed (in strain OK-70-fl) that when vesicles were released from the cells by homogenization, structural integrity was retained, and the vesicles appeared to remain associated with cell membrane material.

Chlorobium vesicles have also been observed in several recently discovered filamentous bacteria containing Bchl *c* or *d* (Dubinina and Gorlenko, 1975). The filaments were larger than those of *Chloroflexus aurantiacus* and contained gas vesicles. Mesosomes were also observed. The chlorobium vesicles were located peripherally and also in the cytoplasm. The cytoplasmic vesicles, however, were invariably associated with invaginations of the cell membrane. It was suggested by the authors that this internal proliferation of cell membranes with the associated vesicles was a means of increasing the amount of photosynthetic apparatus relative to cell volume in these very large cells (2–2.5 × 3.5–4.5 μm).

Broch-Due *et al.* (1978) studied the effect of light intensity on vesicle formation in *Chlorobium limicola f. thiosulfatophilum*. Cells were grown at 22 lux and 22,000 lux. Vesicles were isolated and compared by sucrose density-gradient centrifugation. It was found that vesicles from cells grown at low light intensity were lower in the gradient than vesicles from cells grown at high light intensity. The conclusion that the vesicles formed at low light intensity were therefore larger was verified with electron microscopy of thin sections. The numbers of vesicles per cell under the two conditions were not determined. It was also observed that the specific chlorophyll content was higher at low light intensities (0.22 mg Bchl d/mg protein at 22 lux and 0.11 mg Bchl d/mg protein at 22,000 lux). The authors claimed there was no difference in the specific content of Bchl a. The *in vivo* absorption maximum for Bchl d was at 742–744 nm at low light intensity and 731–733 nm at high light intensity.

6.2. Molecular Composition of the Photosynthetic Apparatus

Further work on identification of molecular components in the photosynthetic apparatus of green bacteria has revealed the presence of a low-potential component, perhaps the primary electron acceptor in green bacteria. Prince and Olson (1976) found evidence for such an acceptor (with a midpoint potential below −450 mV) in Bchl a reaction-center complexes from *Chlorobium limicola f. thiosulfatophilum*. Such a primary acceptor would be capable of mediating the direct reduction of NAD$^+$ via ferredoxin. Knaff and Malkin (1976) found several membrane-bound iron–sulfur proteins in chromatophores of *Chlorobium limicola f. thiosulfatophilum* which differed from those found in purple bacteria. One of these proteins had a midpoint potential near −550 mV and was thought to be a candidate for the primary electron acceptor although no photoreduction was demonstrated. Jennings and Evans (1977), working with a larger less purified preparation of "photosynthetic membranes" than Prince and Olson's from *Chlorobium limicola f. thiosulfatophilum*, demonstrated the ability of reaction centers to produce a reductant at lower than −500 mV by direct photochemistry.

Olson *et al.* (1976) have reported a more thorough analysis of the composition of Complex I from *Chlorobium limicola f. thiosulfatophilum*. Complex I was found to be a large membranous fragment apparently free of vesicles that could be dissociated into two fractions in the presence of 1.5–2.0 M guanidine·HCl. One of these fractions consisted of some cytochrome c and all of the Bchl a protein. The remaining fraction, named Bchl a reaction center Complex II, was larger than the Bchl a protein and contained Bchl a (P840, B800, B813, B835), cytochrome c, cytochrome b, carotenoid, and bacteriopheophytin c (probably a contaminant). The Bchl a protein was photochemically inactive. Complex II was photochemically active, with photoreduction of cytochrome b and photooxidation of cytochrome c. The ratio of cytochrome c to b in Complex II is 3:1. Two cytochrome c molecules were oxidized for

each P840 oxidation. Not all of the cytochrome c was oxidized, however, and the ratio of total cytochrome c (active and inactive) to P840 was 3–4:1. There were 30–50 chlorophyll molecules (Bchl a) per reaction center.

6.3. Organization of the Photosynthetic Apparatus in Green Bacteria

Some direct evidence has been provided supporting the contention put forth in this chapter that the reaction centers of green bacteria are in the cell membrane and that the chlorobium vesicles house the accessory bacteriochlorophylls. Fuller and Boyce (1976), using sucrose gradients and monitoring for cell wall (muramic acid), Bchl a and reaction centers (absorbance at 810–850 nm), Bchl c (absorbance at 750 nm), and cytoplasmic membrane (tritium-labeled pyridoxal phosphate), found that much of the Bchl a remained associated with the cell envelope fraction rather than with the vesicle fraction.

Although some authors have now indicated that the reaction centers of green bacteria are in the cell membranes (Pfennig, 1977; Fuller and Boyce, 1976; Olson *et al.*, 1976, 1977), and we believe from the review presented here that there is considerable evidence to support this notion, to our knowledge a clearly definitive study that clarifies the total organization of the photosynthetic apparatus in green bacteria has not yet been published. A model for the organization of the photosynthetic apparatus in green bacteria based on the evidence accumulated from several authors has been published, however, by Olson *et al.* (1977). This model shows the accessory chlorobium chlorophylls contained within the chlorobium vesicle bounded by a protein membrane. The vesicles are bound to the cytoplasmic membrane by the Bchl a–protein units. The cytoplasmic membrane contains the reaction centers and associated Bchl a molecules.

7. References

Barsky, E. L., Borisov, A. Yu., Fetisova, Z. G., and Samulov, V. D., 1974, Spectral and energetic characteristics of the photoactive particles obtained from chromatophores of the green bacterium *Chlorobium limicola*, *FEBS Lett.* **42**:275–278.

Bergeron, J. A., and Fuller, R. C., 1961, The photosynthetic macromolecules of *Chlorobium thiosulfatophilum*, in: *Biological Structure and Function*, Vol. II (T. W. Goodwin and O. Lindberg, eds.), pp. 307–324, Academic Press, London and New York.

Borisov, A. Yu., and Godik, V. I., 1973, Excitation energy transfer in photosynthesis, *Biochim. Biophys. Acta* **301**:227–248.

Broch-Due, M., Ormerod, J. G., and Fjerdingen, B. S., 1978, Effect of light intensity on vesicle formation in *Chlorobium, Arch. Microbiol.* **116**:269–274.

Buchanan, B. B., and Evans, M. C. W., 1969, Photoreduction of ferredoxin and its use in $NAD(P)^+$ reduction by a subcellular preparation from the photosynthetic bacterium, *Chlorobium thiosulfatophilum, Biochim. Biophys. Acta* **180**:123–129.

Cohen-Bazire, G., 1963, Some observations on the organization of the photosynthetic apparatus in purple and green bacteria, in: *Bacterial Photosynthesis* (H. Gest, A. San Pietro, and L. P. Vernon, eds.), pp. 89–110, Antioch Press, Yellow Springs, Ohio.

Cohen-Bazire, G., 1971, The photosynthetic apparatus of procaryotic organisms, in: *Biological Ultrastructure: The Origin of Cell Organelles* (P. Harris, ed.), pp. 65–90, Oregon State University Press, Corvallis.

Cohen-Bazire, G., and Sistrom, W. R., 1966, The procaryotic photosynthetic apparatus, in: *The Chlorophylls* (L. P. Vernon and G. R. Seely, eds.), pp. 313–341, Academic Press, London and New York.

Cohen-Bazire, G., Pfennig, N., and Kunisawa, R., 1964, The fine structure of green bacteria, *J. Cell Biol.* **22**:207–225.

Cruden, D. L., and Stanier, R. Y., 1970, The characterization of chlorobium vesicles and membranes isolated from green bacteria, Arch. Mikrobiol. **72**:115–134.

Cruden, D. L., Cohen-Bazire, G., and Stanier, R. Y., 1970, The photosynthetic organelles of green bacteria, *Nature (London)* **228**:1345–1347.

Dubinina, G. A., and Gorlenko, V. M., 1975, New filamentous photosynthetic green bacteria containing gas vacuoles, *Microbiology USSR* (English translation) **44**(3):452–458.

Evans, M. C. W., and Buchanan, B. B., 1965, Photoreduction of ferredoxin and its use in carbon dioxide fixation by a subcellular system from a photosynthetic bacterium, *Proc. Natl. Acad. Sci. U.S.A.* **53**:1420–1425.

Fogg, G. E., Stewart, W. D. P., Fay, P., and Walsby, A. E., 1973, *The Blue-Green Algae,* Academic Press, London and New York.

Fowler, C. F., 1974, Evidence for a cytochrome *b* in green bacteria, *Biochim. Biophys. Acta* **357**:327–331.

Fowler, C. F., Nugent, N. A., and Fuller, R. C., 1971, The isolation and characterization of a photochemically active complex from *Chloropseudomonas ethylica, Proc. Natl. Acad. Sci. U.S.A.* **68**:2278–2282.

Fowler, C. F., Gray, B. H., Nugent, N. A., and Fuller, R. C., 1973, Absorbance and fluorescence properties of the bacteriochlorophyll *a* reaction center complex and bacteriochlorophyll *a* protein in green bacteria, *Biochim. Biophys. Acta* **292**:692–699.

Fuller, R. C., and Boyce, C., 1976, The association of the photochemical reaction center and the cytoplasmic membrane of the green bacteria (abstract), in: *Proceedings of the Second International Symposium on Photosynthetic Procaryotes* (G. A. Codd and W. D. P. Stewart, eds.), Dundee, Scotland.

Fuller, R. C., Conti, S. F., and Mellin, D. B., 1963, The structure of the photosynthetic apparatus in the green and purple sulfur bacteria, in: *Bacterial Photosynthesis* (H. Gest, A. San Pietro, and L. P. Vernon, eds.), pp. 71–87, Antioch Press, Yellow Springs, Ohio.

Gorlenko, V. M., 1975, Characteristics of filamentous phototrophic bacteria from fresh water lakes, *Microbiology USSR* (English translation) **44**(4):682–684.

Gorlenko, V. M., and Zhilina, T. N., 1968, Study of the ultrastructure of green sulfur bacteria, strain SK-413, *Microbiology USSR* (English translation) **27**(6):1052–1056.

Gray, B. H., Fowler, C. F., Nugent, N. A., Rigopoulos, N., and Fuller, R. C., 1973, Reevaluation of *Chloropseudomonas ethylica* strain 2-K, *Int. J. Syst. Bacteriol.* **23**:256–264.

Greenawalt, J. W., and Whiteside, T. L., 1975, Mesosomes: Membranous bacterial organelles, *Bacteriol. Rev.* **39**:405–463.

Gusev, M. V., and Shenderova, L. V., 1971, Effect of light and certain inhibitors on oxygen consumption by photosynthetic bacteria, *Microbiology USSR* (English translation) **40**:638–644.

Gusev, M. V., Shenderova, L. V., and Kondrat'eva, E. N., 1969, Behavior of various species of photosynthesizing bacteria toward molecular oxygen, *Microbiology USSR* (English translation) **38**(5):787–792.

Gusev, M. V., Shenderova, L. V., and Kondrat'eva, E. N., 1970, Influence of the oxygen concentration on the growth and survival of photosynthetic bacteria, *Microbiology USSR* (English translation) **39**(4):562–566.

Holt, S. C., Conti, S. F., and Fuller, R. C., 1966a, Photosynthetic apparatus in the green bacterium *Chloropseudomonas ethylicum, J. Bacteriol.* **91**:311–323.

Holt, S. C., Conti, S. F., and Fuller, R. C., 1966b, Effect of light intensity on the formation of the photochemical apparatus in the green bacterium *Chloropseudomonas ethylicum, J. Bacteriol.* **91**:349–355.

Hughes, D. E., Conti, S. F., and Fuller, R. C., 1963, Inorganic polyphosphate metabolism in *Chlorobium thiosulfatophilum, J. Bacteriol.* **85**:577–584.

Jagendorf, A. T., 1975, Mechanism of photophosphorylation, in: *Bioenergetics of Photosynthesis* (Govindjee, ed.), pp. 413–492, Academic Press, London and New York.

Jennings, J. V., and Evans, M. C. W., 1977, The irreversible photoreduction of a low potential component at low temperatures in a preparation of the green photosynthetic bacterium *Chlorobium thiosulfatophilum, FEBS Lett.* **75**:33–36.

Jones, O. T. G., and Whale, F. R., 1970, The oxidation and reduction of pyridine nucleotides by *Rhodopseudomonas sphaeroides* and *Chlorobium thiosulfatophilum, Arch. Mikrobiol.* **72**:48–59.

Kenyon, C. N., and Gray, A. M., 1974, Preliminary analysis of lipids and fatty acids of green bacteria and *Chloroflexus aurantiacus, J. Bacteriol.* **120**:131–138.

Knaff, D. B., and Buchanan, Bob B., 1975, Cytochrome *b* and photosynthetic sulfur bacteria, *Biochim. Biophys. Acta* **376**:549–560.

Knaff, D. B., and Malkin, R., 1976, Iron–sulfur proteins of the green photosynthetic bacterium *Chlorobium*, *Biochim. Biophys. Acta* **430**:244–252.

Knaff, D. B., Buchanan, Bob B., and Malkin, R., 1973, Effect of oxidation–reduction potential on light-induced cytochrome and bacteriochlorophyll reactions in chromatophores from the photosynthetic green bacterium—*Chlorobium*, *Biochim. Biophys. Acta* **325**:94–101.

Knox, R. S., 1975, Excitation energy transfer and migration: Theoretical considerations, in: *Bioenergetics of Photosynthesis* (Govindjee, ed.), pp. 183–221, Academic Press, London and New York.

Kusai, A., and Yamanaka, T., 1973*a*, A novel function of cytochrome *c* (555, *Chlorobium thiosulfatophilum*) in oxidation of thiosulfate, *Biochem. Biophys. Res. Commun.* **51**:107–112.

Kusai, A., and Yamanaka, T., 1973*b*, Cytochrome *c* (553, *Chlorobium thiosulfatophilum*) is a sulfide–cytochrome *c* reductase, *FEBS Lett.* **34**:325–327.

Kusai, A., and Yamanaka, T., 1973*c*, The oxidation mechanism of thiosulfate and sulfide in *Chlorobium thiosulfatophilum*: Roles of cytochrome *c*-551 and cytochrome *c*-553, *Biochim. Biophys. Acta* **325**:304–314.

Larsen, H., 1953, On the microbiology and biochemistry of the photosynthetic green sulfur bacteria, *K. Nor. Vidensk. Selsk. Skr.* **1**:1–199.

Lascelles, J., 1968, The bacterial photosynthetic apparatus, in: *Advances in Microbial Physiology*, Vol. 2 (A. H. Rose and J. F. Wilkinson, eds.), pp. 1–42, Academic Press, London and New York.

Lippert, K. D., 1967, Die Verwertung von molekularen Wasserstoff durch *Chlorobium thiosulfatophilum*, Ph.D. thesis, University of Göttingen.

Madigan, M. T., and Brock, T. D., 1977, 'Chlorobium-type' vesicles of photosynthetically-grown *Chloroflexus aurantiacus* observed using negative-staining techniques, *J. Gen. Microbiol.* **102**:279–285.

Marrs, B., and Gest, H., 1973, Genetic mutations affecting the respiratory electron-transport system of the photosynthetic bacterium *Rhodopseudomonas capsulata*, *J. Bacteriol.* **114**:1045–1051.

Melandri, B. A., Baccarini-Melandri, A., San Pietro, A., and Gest, H., 1971, Interchangeability of phosphorylation coupling factors in photosynthetic and respiratory energy conversion, *Science* **174**:514–516.

Meyer, T. E., Bartsch, R. G., Cusanovich, M. A., and Mathewson, J. H., 1968, The cytochromes of *Chlorobium thiosulfatophilum*, *Biochim. Biophys. Acta* **153**:854–861.

Norris, J. R., Scheer, H., and Katz, J. J., 1975, Models for antenna and reaction center chlorophylls, *Ann. N. Y. Acad. Sci.* **244**:260–280.

Oelze, J., and Drews, G., 1972, Membranes of photosynthetic bacteria, *Biochim. Biophys. Acta* **265**:209–239.

Olson, J. M., 1970, The evolution of photosynthesis, *Science* **168**:438–446.

Olson, J. M., and Sybesma, C., 1963, Energy transfer and cytochrome oxidation in green bacteria, in: *Bacterial Photosynthesis* (H. Gest, A. San Pietro, and L. P. Vernon, eds.), pp. 413–422, Antioch Press, Yellow Springs, Ohio.

Olson, J. M., Philipson, K. D., and Sauer, K., 1973, Circular dichroism and absorption spectra of bacteriochlorophyll–protein and reaction center complexes from *Chlorobium thiosulfatophilum*, *Biochim. Biophys. Acta* **292**:206–217.

Olson, J. M., Giddings, T. H., and Shaw, E. K., 1976, An enriched reaction center preparation from green photosynthetic bacteria, *Biochim. Biophys. Acta* **449**:197–208.

Olson, J. M., Prince, R. C., and Brune, D. C., 1977, Reaction-center complexes from green bacteria, *Brookhaven Symp. Biol.* **28**:238–246.

Pfennig, N., 1977, Phototrophic green and purple bacteria: a comparative, systematic survey, *Annu. Rev. Microbiol.* **31**:275–290.

Pfennig, N., and Cohen-Bazire, G., 1967, Some properties of the green bacterium *Pelodictyon clathratiforme*, *Arch. Mikrobiol.* **59**:226–236.

Pierson, B. K., 1973, The characterization of gliding filamentous phototrophic bacteria, Ph.D. thesis, University of Oregon, Eugene.

Pierson, B. K., and Castenholz, R. W., 1974*a*, A phototropic gliding filamentous bacterium of hot springs, *Chloroflexus aurantiacus*, gen. and sp. nov., *Arch. Mikrobiol.* **100**:5–24.

Pierson, B. K., and Castenholz, R. W., 1974*b*, Studies of pigments and growth in *Chloroflexus aurantiacus*, a phototrophic filamentous bacterium, *Arch. Mikrobiol.* **100**:283–305.

Pivovarova, T. A., and Gorlenko, V. M., 1977, Fine structure of *Chloroflexus aurantiacus* Var. *mesophilus* (Nom. Prof.): Growth in light under aerobic and anaerobic conditions, *Microbiology USSR* (English translation) **46**(2):276–282.

Prince, R. C., and Olson, J. M., 1976, Some thermodynamic and kinetic properties of the primary photochemical reactants in a complex from a green photosynthetic bacterium, *Biochim. Biophys. Acta* **423**:357–362.

Puchkova, N. N., Gorlenko, V. M., and Pivovarova, T. A., 1975, A comparative ultrastructural study of vibrioid green sulfur bacteria, *Microbiology USSR* (English translation) **44**(1):89–94.

Razin, S., 1972, Reconstitution of biological membranes, *Biochim. Biophys. Acta* **265**:241–296.

Schmidt, K., 1976, Carotenoid glycosides in phototrophic bacteria (abstract), in: *Proceedings of the Second International Symposium on Photosynthetic Procaryotes* (G. A. Codd and W. D. P. Stewart, eds.), pp. 58–60, Dundee, Scotland.

Schmitz, R., 1967, Über die Zusammensetzung der pigmenthaltigen Strukturen aus Prokaryoten. II. Untersuchungen an Chromatophoren von *Chlorobium thiosulfatophilum* Stamm Tassajara, *Arch. Mikrobiol.* **56**:238–247.

Shioi, Y., Takamiya, K., and Nishimura, M., 1974, Studies on electron transfer systems in the green photosynthetic bacterium *Chloropseudomonas ethylica* strain 2-K. II. Composition of pigments and electron transfer systems, *J. Biochem.* **76**:241–250.

Shuvalov, V. A., Kondrat'eva, E. N., and Litvin, F. F., 1968, Photoinduced absorption of oxygen by green photosynthesizing bacteria, *Dokl. Akad. Nauk SSSR* (English translation) **178**(3):711–714.

Solov'eva, Zh. V. and Fedorov, V. D., 1970, Effect of growth conditions on ultrathin structure of *Chlorobium thiosulfatophilum*, *Microbiology USSR* (English translation) **39**(5):739–743.

Sybesma, C., 1967, Light-induced cytochrome reactions in the green photosynthetic bacterium *Chloropseudomonas ethylicum*, *Photochem. Photobiol.* **6**:261–267.

Sykes, J., Gibbon, J. A., and Hoare, D. S., 1965, The macromolecular organization of cell-free extracts of *Chloro-bium thiosulfatophilum* L660, *Biochim. Biophys. Acta* **109**:409–423.

Takamiya, K., 1971, The light-induced oxidation–reduction reactions of menaquinone in intact cells of a green photosynthetic bacterium, *Chloropseudomonas ethylica*, *Biochim. Biophys. Acta* **234**:390–398.

Tomina, Zh. V., and Fedorov, V. D., 1967, Ultrathin structure of the green sulfur bacterium *Chlorobium thiosulfatophilum*, *Microbiology USSR* (English translation) **36**(4):556–559.

Trüper, H., 1976, Higher taxa of the phototrophic bacteria: *Chloroflexaceae* fam. nov., a family for the gliding, filamentous, phototrophic "green" bacteria, *Int. J. Syst. Bacteriol.* **26**:74–75.

Williams, A. M., 1956, Light-induced uptake of inorganic phosphate in cell-free extracts of obligately anaerobic photosynthetic bacteria, *Biochim. Biophys. Acta* **19**:570.

CHAPTER 10

Metabolism of Reserve Materials

J. M. Merrick

1. Introduction

Storage compounds of prokaryotic cells generally include glycogenlike polymers, poly-β-hydroxybutyrate (PHB), and polyphosphates (polyP). These reserve substances are usually accumulated when the supplies of exogenous carbon and energy are abundant and utilized when the supplies are depleted. The synthesis of these substances provides the cell with a mechanism for accumulating and storing carbon and phosphorus in a form that is essentially osmotically inert. Storage of these substances is widespread among the bacteria, and is found in many different morphological and physiological types of gram-negative and gram-positive organisms. There are also examples in which organisms can store more than one type of storage substance, e.g., glycogen and poly-β-hydroxybutyrate by *Rhodospirillum rubrum* (Stanier *et al.*, 1959); glycogen and polyphosphate by *Escherichia coli* (Holme and Palmstierna, 1956; Nesmeyanova *et al.*, 1974). A few bacteria [e.g., *Pseudomonas aeruginosa* (Campbell *et al.*, 1963)] apparently do not synthesize any specific reserve material. Table 1 shows the levels of glycogen, PHB, or polyP during the various stages of growth of several microorganisms. Generally, it may be noted that reserve substances are synthesized during exponential growth, reach maximal levels in the early stationary phase, and decrease during the late stationary phase of growth or during starvation conditions. Several excellent reviews on this topic have appeared (Krebs and Preiss, 1975; Shively, 1974; Dawes and Senior, 1973; Preiss *et al.*, 1973; Preiss, 1969; Doudoroff, 1966; Harold, F. M., 1966; Dawes and Ribbons, 1964).

2. Glycogen

2.1. General Considerations

The accumulation of glycogen by bacterial cells is generally observed when conditions of growth become unbalanced in the presence of an excess of exogenous carbon. This can occur when the supply of a nutrient such as nitrogen (Zevenhuizen, 1966*a*; Holme and Palmstierna, 1956), phosphate or, sulfur (Dicks and Tempest, 1967; Zevenhuizen, 1966*a*) becomes limiting. However, exceptions have been noted. For example, in the case of *Rhodopseudomonas capsulata*, a photosynthetic nonsulfur purple bacterium, glycogen levels are highest during exponential growth (Eidels and Preiss, 1970) (Table 1). This organism synthesizes glycogen in the dark or in the light, although higher levels of the polysaccharide are formed in the light. Environmental factors play a predominant role in determining the type of reserve accumulated by organisms capable of storing alternative compounds. *Rhodospirillum rubrum* cells, for example, can store either PHB or glycogen. Stanier *et al.* (1959) showed that the nature of the carbon source supplied for growth of the organism influences the type of reserve accumulated. Substrates that give rise to pyruvate lead to glycogen formation, while substrates that generate acetyl-CoA without the intermediate formation of pyruvate lead predominantly to PHB.

J. M. Merrick · Department of Microbiology, School of Medicine, State University of New York at Buffalo, Buffalo, New York 14214

Table 1. Levels of Glycogen, PHB, or PolyP During Growth of Various Photosynthetic and Nonphotosynthetic Bacteria

Organism:	Rhodopseudomonas capsulata			Chromatium vinosum			Escherichia coli[a]			Rhodospirillum rubrum		Azotobacter beijerinckii			Chlorobium limicola f. thiosulfatophilum	
Carbon source:	Glucose			CO$_2$			Glucose			Acetate		Glucose			CO$_2$	
Growth phase[b]:	Mid-Exp.	Late Exp.	Stat.	Exp.	Stat.	Starved cells	Exp.	Stat.	Late Stat.	Exp.	Starved cells	Exp.	Early Stat.	Late Stat.	Exp.	Stat.
Glycogen content, mg glucose/g (dry wt.) cells	141	124	97	63	150	c	6	20	11							
PHB content, mg/g (dry wt.) cells										230	30	230	350	160		
PolyP content, μmol/mg N															14	24
Reference	Eidels and Preiss (1970)			Hara et al. (1973)			Preiss (1972)			Stanier et al. (1959)		Senior et al. (1972)			Hughes et al. (1963)	

[a] Glycogen content was reported as mg glucose/g (wet wt.) of cells.
[b] Exp.: exponential; Stat.: stationary.
[c] No analytical data given; in the electron microscope, however, glycogen granules were barely detectable in the cells.

It is generally believed that the ability to accumulate glycogen provides the cell with a mechanism for maintaining cell integrity and viability under conditions of starvation. For example, bacterial glycogen may be used by the cell as a source of maintenance energy (i.e., energy required for such processes as turnover of RNA and protein, motility, osmotic regulation, maintenance of intracellular pH, etc.) during nongrowing conditions. Dawes and Ribbons (1965) showed that degradation of the nitrogenous components of the cell is prevented until the glycogen reserves of the cell are depleted. Other studies have noted the prolonged survival time of glycogen-rich cells over their glycogen-poor counterparts, e.g., *Aerobacter aerogenes*** (Strange *et al.*, 1961), *Esch. coli* (Strange, 1968), *Arthrobacter* species (Zevenhuizen, 1966*a*), and *Streptococcus mitis* (Houte and Jansen, 1970). In the case of *A. aerogenes*, however, the survival time is not affected by the glycogen content in the presence of magnesium ions (Tempest and Strange, 1966). The picture is further complicated by the observation that glycogen-rich *Sarcina lutea* cells die more rapidly under starvation conditions than do cells that are glycogen-deficient (Burleigh and Dawes, 1967).

2.2. Structure

The structures of bacterial glycogens vary somewhat, but are generally similar to those found in eukaryotic cells. Glycogen is made up of D-glucose residues linked by α-1,4-glucosidic bonds. Branching occurs through α-1,6-glucosidic linkages. The average chain length of bacterial glycogens ($\overline{CL}$, the number of glucose residues per mole of nonreducing terminal glucose) falls in the range generally observed for animal and yeast glycogens. This value, which is a measure of the degree of branching, is of the order of 10–14 units (Zevenhuizen, 1966*b*). Some bacterial species [*Arthrobacter* (Zevenhuizen, 1966*b*; Ghosh and Preiss, 1965) and mycobacteria (Antoine and Tepper, 1969)] produce glycogens that are more highly branched with $\overline{CL}$ values between 7 and 9. The purple sulfur bacterium *Chromatium vinosum* strain D synthesizes a glycogen with a $\overline{CL}$ value of 11 (Hara *et al.*, 1973); the glycogen of the blue-green alga *Nostoc muscorum* has a $\overline{CL}$ value of 13 (Chao and Bowen, 1971); the *S. mitis* glycogen has a $\overline{CL}$ of 12 (Builder and Walker, 1970).

Glycogen can be observed to be deposited as nonmembrane-bound granules in a variety of photosyn-

thetic and nonphotosynthetic bacteria, e.g., *Esch. coli* (Shively, 1974), *Chr. vinosum* strain D (Hara *et al.*, 1973), *Rs. rubrum* (Cohen-Bazire and Kunisawa, 1963), and the blue-green algae (Wolk, 1973). The glycogen deposits of *Rs. rubrum* can be seen in Fig. 1 as light areas scattered throughout the cytoplasm. Granules vary in size from about 50 to 150 nm, and are not surrounded by a limiting membrane. In the blue-green algae, the granules are found between the photosynthetic thylakoid membranes and appear as spheres (25–30 nm in diameter), crystals, or rods (Wolk, 1973). Rods found in *Oscillatoria rubescens* are up to 300 nm long and are made up of 70-nm discs with globular subunits and a central pore (Jost, 1965). Isolated granules from *Nostoc muscorum* have average dimensions of 31 nm in width and 65 nm in length and are composed of two equal parts (Chao and Bowen, 1971). Attempts to dissociate the granule into subunits were unsuccessful.

2.3. Biosynthesis

It is now well established that the glycosyl donor for glycogen biosynthesis by prokaryotes is ADP-glucose. Synthesis of ADP-glucose and glycogen proceeds according to the following reaction (Krebs and Preiss, 1975; Preiss, 1969; Greenberg and Preiss, 1965; Shen and Preiss, 1965):

$$\text{ATP} + \text{glucose 1-P} \rightleftharpoons \text{ADP-glucose} + \text{PP}_i\text{*} \tag{1}$$

$$\text{ADP-glucose\dagger} + (1,4\text{-}\alpha\text{-D-glucosyl})_n \rightleftharpoons$$
$$(1,4\text{-}\alpha\text{-D- glucosyl})_{n+1} + \text{ADP} \tag{2}$$

Starch biosynthesis in plants also utilizes ADP-glucose as the glycosyl donor (Recondo and Leloir, 1961). In eukaryotes, however, UDP-glucose is the effective glycosyl donor in glycogen biosynthesis. The synthesis of α-1,6 branch linkages in glycogen by a branching enzyme has been reported in *Arthrobacter globiformis* (Zevenhuizen, 1964) and in *Esch. coli* (Sigal *et al.*, 1965).

It was previously indicated that unbalanced growth conditions such as might exist when there is a limiting supply of nutrient and excess carbon can result in glycogen accumulation. Thus, many bacterial species will accumulate glycogen during the stationary phase of growth. Preiss (1969) demonstrated that both ADP-glucose pyrophosphorylase and glycogen synthase are derepressed at the end of exponential growth. Enzyme levels are six- to tenfold greater than

---

* The correct name for this organism is *Enterobacter aerogenes*. To avoid confusion, however, the name used by the original investigators will be retained.

* ADP-glucose pyrophosphorylase, E.C.2.7.7.27.
† Glycogen 4-α-glucosyl transferase, E.C.2.4.1.21; glycogen synthase.

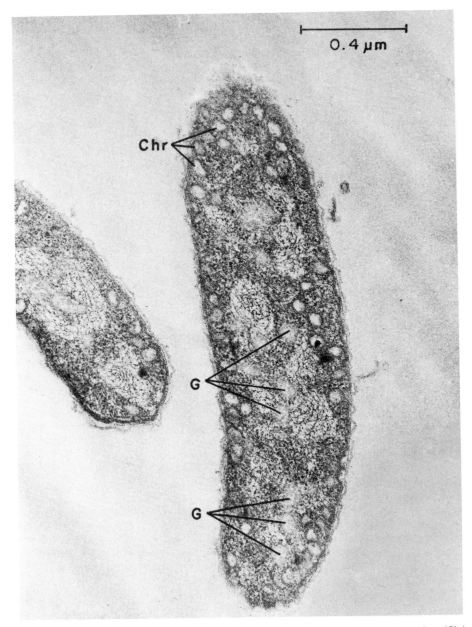

Fig. 1. Section of *Rhodospirillum rubrum* showing glycogen deposits. (G) Glycogen granules; (Chr) chromatophores. Courtesy of G. Stanier, Pasteur Institute, Paris, France.

necessary to account for the observed *in vivo* rate of glycogen accumulation, suggesting that both the activity and the synthesis of the enzymes are controlled. Because the similarity to the derepression of spore formation in the bacilli, Preiss (1969) suggests that both may be survival mechanisms with similar controlling factors. Bacterial extracellular enzyme secretion (Stinson and Merrick, 1974) may also be related to these phenomena, since here, too, exoenzymes are commonly secreted only at the end of exponential growth and presumably provide the cell with survival advantages. The accumulation of glycogen during the late log and stationary phases of growth by *Chr. vinosum* strain D is similar to the pattern observed in other bacteria (Hara *et al.*, 1973). Since the levels of the biosynthetic enzymes were not reported, it is unknown whether they are also derepressed when this organism ceases to grow. *Rhodopseudomonas capsulata* does not, however, follow this general pattern (Eidels and Preiss, 1970). In this case, glycogen accumulation, glycogen synthase, and ADP-glucose pyrophosphorylase activities are at a maximum in either light- or dark-grown cells during the logarithmic phase of growth.

Glycogen synthesis by bacteria is apparently regulated at the level of ADP-glucose synthesis, rather than at the reaction catalyzed by glycogen synthase (Krebs and Preiss, 1975). In eukaryotes, it is the glycogen synthase that is subject to allosteric regulation, and this may reflect the additional role of UDP-glucose in the synthesis of UDP-galactose, UDP-glucuronic acid, and other glycosides. Under these circumstances, the glycogen synthase reaction would be the first specific step in glycogen biosynthesis. ADP-glucose has no apparent function other than its role as precursor in glycogen biosynthesis, and it is the synthesis of this sugar nucleotide that is the first specific step in glycogen biosynthesis by prokaryotes. Allosteric regulation of ADP-glucose pyrophosphorylase would thus act to conserve ATP required for synthesis of ADP-glucose. Generally, ADP-glucose pyrophosphorylase is activated by glycolytic intermediates and inhibited by AMP, ADP, or P_i. Glycogen synthesis can therefore be considered to be under the control of the energy state of the cell whereby a high energy charge (high ATP level relative to total adenine nucleotides) would favor glycogen synthesis. This situation might be expected when growth ceases in the presence of an excess of the carbon and energy source. When the cell is in a low-energy state, e.g., high AMP, glycogen synthesis would be inhibited. In this regard, Shen and Atkinson (1970) showed that ADP-glucose pyrophosphorylase responds sharply to variations in the energy charge.

There appears to be some relationship between the effector that activates the ADP-glucose pyrophosphorylase and the pathway of carbon metabolism by the microorganism (Krebs and Preiss, 1975; Preiss, 1969). The glycolytic pathway of sugar metabolism is utilized by members of the Enterobacteriaceae. Here, it is generally found that ADP-glucose pyrophosphorylase is activated by fructose 1,6-diphosphate (Fru-P_2), NADPH, and pyridoxal phosphate, and inhibited by AMP (Ribereau-Gayon *et al.*, 1971; Preiss *et al.*, 1966). There is evidence suggesting that Fru-P_2 is the effective activator under physiologic conditions (Govons *et al.*, 1973). *Serratia marcescens* appears to be an exception in this group of organisms, since no known metabolites activate this enzyme (Ribereau–Gayon *et al.*, 1971). For the non-

enteric organisms that utilize glycolysis, other variations occur. The enzyme from *Aeromonas formicans* and *Micrococcus lysodeikticus* is activated by either fructose 6-P or Fru-P_2 and inhibited by ADP rather than AMP (Krebs and Preiss, 1975). In the case of *Clostridium pasteurianum*, the enzyme is not activated or inhibited by any of the aforementioned metabolites at physiologic concentrations (Robson *et al.*, 1972). Organisms that utilize the Entner–Doudoroff pathway, e.g., *Agrobacterium tumefaciens*, *Arthrobacter viscosus*, and *Rp. capsulata*, have fructose 6-P and pyruvate as the activators, and Pi, ADP, or AMP as inhibitors (Eidels *et al.*, 1970; Shen and Preiss, 1966). *Rhodospirillum rubrum*, an organism that grows on various organic acids in the light or in the dark, contains an ADP-glucose pyrophosphorylase that is activated only by pyruvate (Furlong and Preiss, 1969), no matter what conditions are used for growth of the organism (Preiss, 1969). No known metabolite inhibits the enzyme. In the case of higher plants and algae, ADP-glucose pyrophosphorylase is activated by 3-phosphoglycerate and inhibited by P_i (Preiss *et al.*, 1973).

Extensive *in vitro* studies of the kinetic properties of ADP-glucose pyrophosphorylase have provided considerable insight into the metabolic regulation of glycogen synthesis (Krebs and Preiss, 1975). In the presence of activators, the apparent affinity of the enzyme for the substrate is increased. Thus, the concentration of the substrates ATP, glucose 1-P, pyrophosphate, and ADP-glucose required to give 50% of the maximal velocity ($s_{0.5}$), is lowered 5- to 15-fold in the presence of the activator. Also, activators may increase the maximal velocity from 2.5- to 60-fold, and, of most importance, they can reverse the effect of the inhibitor. For example, at a Fru-P_2 concentration of 0.06 mM, the K_i for AMP is 3.4 μM. However, at a Fru-P_2 concentration of 1.7 mM, it is 70 μM. When glycogen synthesis is minimal, the relative concentration of activator and inhibitor would maintain ADP-glucose pyrophosphorylase in an inactive state. Either increasing the activator level or decreasing the inhibitor level allows the activity of ADP-glucose pyrophosphorylase to increase, enabling glycogen synthesis to occur. Thus, relative concentrations of activator and inhibitor could modulate glycogen synthesis by affecting the rate of ADP-glucose synthesis. In a culture in which growth is limited because of nutrient depletion, glycogen accumulates if an excess of carbon and energy is available. According to the hypothesis of Preiss and co-workers, glycogen accumulates under these conditions because the levels of ATP and effector molecules such as Fru-P_2 (in the case of *Esch. coli*) rise, increasing the rate of ADP-glucose synthesis and thus the rate of glycogen biosynthesis. Dietzler *et al.* (1973) measured the levels of ATP and Fru-P_2 in *Esch. coli* 4597 (K). When growth ceased because of NH_4^+ depletion, the ATP level increased 50%, but unexpectedly the level of Fru-P_2 decreased 76% from the exponential to the stationary phase of growth. Further studies (Dietzler *et al.*, 1974a) demonstrated that the increase in ATP was at the expense of ADP and AMP, and hence an increase in the adenylate charge was obtained (from 0.74 in the exponential phase to 0.87 in the stationary phase). These workers concluded that the decrease in Fru-P_2 is more than offset by the increase in energy charge, and that Fru-P_2 does contribute to the increased rate of glycogen synthesis observed during nitrogen starvation. This conclusion was supported by studies that demonstrated that the rate of glycogen synthesis was linearly related to the square of the intracellular concentration of Fru-P_2 (Dietzler *et al.*, 1974b).

Studies of *Esch. coli* mutants with altered ADP-glucose pyrophosphorylases have also provided physiological evidence to support the role of activators and inhibitors in the regulation of glycogen biosynthesis. Two mutants, SG5 and CL1136, are able to accumulate considerably more glycogen than their parent because their ADP-glucose pyrophosphorylase has greater affinity for Fru-P_2 as well as a lower affinity for AMP (Preiss *et al.*, 1973). Further, it can be shown that the SG5 enzyme, in the presence of its allosteric activators, is more active at physiologic energy charge (0.8–0.85) than the enzyme from the parent strain (Govons *et al.*, 1973). The studies described above have provided convincing evidence that the allosteric effects observed *in vitro* play a functional role in the regulation of the rate of glycogen synthesis *in vivo*.

Organisms utilizing the Entner–Doudoroff pathway have not been as extensively studied as *Esch. coli*. Here, fructose 6-P is the allosteric activator of ADP-glucose pyrophosphorylase. Preiss (1969) suggested that under limiting growth conditions in the presence of excess carbon, glucose 6-P dehydrogenase would be inhibited by ATP, resulting in an increase in the levels of glucose 6-P and fructose 6-P. ADP-glucose pyrophosphorylase would be activated under these conditions, and therefore, glycogen biosynthesis would take place. The enzyme from *Rp. capsulata* can be activated either by pyruvate or by fructose 6-P, and this may be a reflection of the ability of this photosynthetic organism to grow on either various tricarboxylic acid intermediates or on glucose. *Rhodospirillum rubrum* grows on tricarboxylic acid intermediates in the light or in the dark, but cannot grow

on glucose. ADP-glucose pyrophosphorylase from this organism is activated only by pyruvate. Stanier *et al.* (1959) demonstrated that compounds such as succinate, malate, or pyruvate are assimilated to glycogen by *Rs. rubrum* incubated in the light. From studies with labeled succinate, it was concluded by these workers that hexose units of the polysaccharide were synthesized by the conversion of succinate to pyruvate. Pyruvate was subsequently converted to hexose through a reversal of the glycolytic pathway. Acetate in the presence of CO_2 is also assimilated to glycogen by *Rs. rubrum*. Glycogen synthesis in this case is accounted for by the synthesis of pyruvate from acetyl-CoA and CO_2 according to the following reaction (Buchanan *et al.*, 1967):

$$\text{Acetyl-CoA} + CO_2 + \text{ferredoxin}_{\text{red.}} \rightarrow$$
$$\text{pyruvate} + \text{CoA} + \text{ferredoxin}_{\text{ox.}} \quad (3)$$

Pyruvate thus appears to play a central role in the assimilation of carbon by *Rs. rubrum* (and presumably by *Rp. capsulata*). Since *Rs. rubrum* can convert pyruvate to phosphoenolpyruvate according to the reaction (Buchanan and Evans, 1965)

$$\text{Pyruvate} + \text{ATP} \rightarrow$$
$$\text{phosphoenolypyruvate} + \text{AMP} + P_i \quad (4)$$

it is possible to consider pyruvate as the first glycolytic intermediate in gluconeogenesis by *Rs. rubrum*.

In higher plants and green algae, 3-phosphoglycerate activates ADP-glucose pyrophosphorylase. Here, 3-phosphoglycerate is the first glycolytic intermediate formed as a result of CO_2 fixation, and is converted by the energy and reducing power derived from photosynthesis to hexose-phosphate and starch. Preiss (1969) considers activation of ADP-glucose pyrophosphorylase by 3-phosphoglycerate as "feedforward" activation.

2.4. Breakdown

Glycogen degradation in bacteria has not received as much attention as has its synthesis. Glycogen phosphorylase, which catalyzes the following reaction, has been demonstrated in *Esch. coli* K-12 by Chen and Segel (1968a,b):

$$(\text{Glucosyl})_{n+1} + P_i \rightarrow (\text{glucosyl})_n + \text{glucose 1-P} \quad (5)$$

The synthesis of glycogen phosphorylase is independent of the carbon source used for growth, in contrast to a maltodextrin phosphorylase that is also produced by *Esch. coli* K-12. The latter enzyme is present in glucose-grown cells, but can be induced to higher levels in maltose-grown cells. These phosphorylases differ in substrate specificity. Glycogen phosphorylase prefers glycogen as a substrate, while the maltodextrin phosphorylase is more active with short-chain oligosaccharides. Both enzymes are slightly stimulated by AMP. Glycogen phosphorylase was also competitively inhibited by ADP-glucose, TDP-glucose, and UDP-glucose. Chen and Segel (1968*b*) suggested that intracellular nucleotide glucose compounds regulate glycogen phosphorylase. The levels of these substances would tend to be higher in stationary cells that are accumulating glycogen in the presence of exogenous carbon. Their levels would decrease when the exogenous carbon source is depleted, releasing the inhibition of glycogen phosphorylase and thus permitting glycogen degradation.

A glycogen phosphorylase purified from *Streptococcus salivarius* has also been reported (Khandelwal *et al.*, 1973; Spearman *et al.*, 1973). The enzyme is more active with dextrin, however, than with *Str. salivarius* glycogen. In accord with the work described above, the enzyme is inhibited by ADP-glucose and activated by AMP.

A soluble enzyme (isoamylase) capable of debranching glycogen or amylopectin was reported in *Esch. coli* NCTC 5928 (Palmer *et al.*, 1973) and in other bacterial species [in *Cytophaga* (Gunja-Smith *et al.*, 1970) and in a pseudomonad (Yokobayashi *et al.*, 1970)]. Palmer *et al.* (1973) suggested that this enzyme, the level of which increases more than twofold in the stationary phase of glucose-grown *Esch. coli* NTCC 5928 cells, plays an important role in glycogen breakdown. These workers propose the scheme shown in Fig. 2 for the degradation of bacterial glycogen. The combined actions of isoamylase, glycogen phosphorylase, maltodextrin phosphorylase, and amylomaltase results in the conversion of glycogen to glucose and glucose 1-P, which can then be funneled through glycolysis. Since the reaction catalyzed by isoamylase would be the first specific step

in glycogen degradation, it may be subject to metabolic control. A detailed kinetic analysis of the enzyme, as well as studies with mutants, should provide further insight into the regulation of glycogen breakdown in bacteria.

3. Poly-β-hydroxybutyrate

3.1. General Considerations

Poly-β-hydroxybutyrate (PHB) was initially discovered in *Bacillus megaterium* by Lemoigne (1925). In recent years, it has become apparent that PHB is accumulated by many diverse microorganisms, including both gram-positive and gram-negative genera. It is found in cocci and bacilli, and is not restricted to any particular physiological group. It can be detected in photosynthetic and aerobic groups, and in lithotrophic as well as organotrophic bacteria. A listing of the various microorganisms that can store PHB was published by Dawes and Senior (1973). PHB was also identified in *C. botulinum* type E (Emeruwa and Hawirko, 1973).

The ability of bacterial cells to accumulate massive amounts of PHB—e.g., in *Azotobacter beijerinckii*, the polymer can constitute more than 70% of its total dry weight (Stockdale *et al.*, 1968; Senior *et al.*, 1972)—provides the cells with a reserve of fatty acid carbon in a form that is osmotically inert. Bacteria generally do not store neutral lipids (e.g., triglycerides), and it appears that PHB is the principal lipid reserve of prokaryotic cells. Cultural conditions play a predominant role in PHB accumulation. As in the case of glycogen, PHB is accumulated as a result of nutrient imbalance. Synthesis usually prevails when the medium contains an excess of carbon and energy source but growth is limited due to a deficiency in a required nutrient. For example, nitrogen limitation leads to PHB accumulation in *B. megaterium* (Macrae

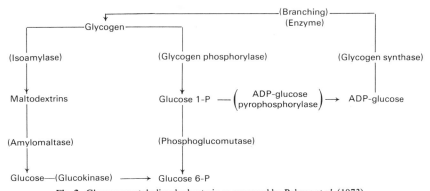

Fig. 2. Glycogen metabolism by bacteria as proposed by Palmer *et al.* (1973).

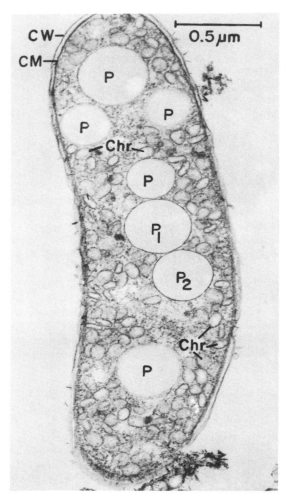

It has been established for a number of organisms that PHB may be used as a substrate for endogenous respiratory activity to provide maintenance energy [*Pseudomonas saccharophilia* (Doudoroff and Stanier, 1959), *B. megaterium* (Macrae and Wilkinson, 1958), *Azotobacter agilis* (Sobek *et al.*, 1966), *H. eutropha* (Schlegel and von Bartha, 1961; Hippe, 1967), *Micrococcus halodenitrificans* (Sierra and Gibbons, 1962), and *Sphaerotilus discophorus* (Stokes and Parson, 1968)]. In some cases, it has been possible to correlate the endogenous utilization of PHB with the survival of the microorganism under starvation conditions. Thus, examples are available in which polymer-rich cells survive for longer periods of times under starvation conditions than do polymer-poor cells [e.g., *M. halodenitrificans* (Sierra and Gibbons, 1962), *H. eutropha* (Hippe, 1967), *S. discophorus* (Stokes and Parson, 1968), and *A. agilis* (Sobek *et al.*, 1966)]. Exceptions

Fig. 3. Section of *Rhodospirillum rubrum* showing PHB granules with a distinct limiting membrane. (P) PHB granules (limiting membrane seen especially clearly in P_1 and P_2); (CW) cell wall; (CM) cell membrane; (Chr) chromatophores. Reproduced from Boatman (1964) courtesy of E. S. Boatman; with permission.

and Wilkinson, 1958), in the chemolithotroph *Hydrogenomonas eutropha* (Schlegel and von Bartha, 1961), and in *Rs. rubrum* (Doudoroff, 1966). Oxygen limitation results in PHB accumulation in the case of *A. beijerinckii* (Senior *et al.*, 1972) and *H. eutropha* (Schuster and Schlegel, 1967). The nature of the exogenous carbon source can also affect PHB production. As discussed earlier (see Section 2.1), *Rs. rubrum* photometabolizes substrates such as lactate, pyruvate, malate, succinate, or CO_2 mainly to glycogen (Stanier *et al.*, 1959). Substrates such as acetate, β-hydroxybutyrate, or butyrate that are metabolized to acetyl-CoA without the intermediate formation of pyruvate lead to PHB accumulation.

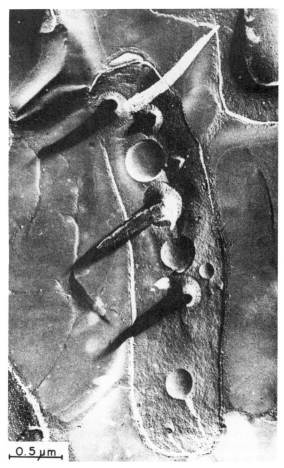

Fig. 4. Intact cell of *Bacillus cereus* showing PHB granules stretched or removed. Reproduced from Dunlop and Robards (1973) courtesy of A. W. Robards; with permission.

have, however, been noted. In the case of *A. agilis*, polymer-rich glucose-grown cells survive longer than polymer-poor cells. Succinate-grown cells, on the other hand, contain less polymer than glucose-grown cells, but degrade their polymer more slowly and are able to survive for longer periods of time. The reasons for this are not apparent, but it has been suggested that the succinate-grown cells metabolize their polymer more efficiently (Sobek *et al.*, 1966). PHB can also be used as an endogenous substrate to support protein synthesis [e.g., *Hydrogenomonas* (Schlegel and von Bartha, 1961) and *Rs. rubrum* (Stanier *et al.*, 1959)]. *Rhodospirillum rubrum* cells, incubated in the light, require CO_2 for the conversion of PHB to carbohydrate and protein. This requirement for CO_2 can now be understood in view of the discovery of the reduced ferredoxin-dependent carboxylation of acetyl-CoA to pyruvate [Reaction (3)].

3.2. Chemical Structure and Composition of Granules

PHB is a linear polyester of D(−)-β-hydroxybutyric acid with the following structure:

$$HO-CH-HCH_2-C-[-O-CH-CH_2-C-]_n-O-CH-CH_2-COOH$$
$$\quad | \qquad\qquad \| \qquad | \qquad\quad \| \qquad\quad |$$
$$\quad CH_3 \qquad\quad O \quad\; CH_3 \qquad O \qquad\; CH_3$$

Lundgren *et al.* (1965) investigated the molecular weights of PHB extracted from a number of different types of bacteria. Polymer molecular weights based on viscometry ranged from 1000 to 250,000. Generally, polymer extracted from bacteria treated with alkaline hypochlorite to destroy other cell constituents gave uniformly lower molecular sizes. Polymer obtained from bacterial species of different genera gave similar X-ray diffractograms and IR spectra, indicating that the same basic molecule is stored in all the bacteria examined. Structural studies with the electron microscope have shown that polymer crystals have a "folded-chain" lamellar morphology (Alper *et al.*, 1963). Based on analysis of an X-ray fiber diagram, Okamura and Marchessault (1967) proposed that the molecule is a compact right-handed helix with a two-fold screw axis along the chain corresponding to two residues per turn. By conformational analysis, a molecular model was suggested by Cornibert and Marchessault (1972).

PHB is stored on the cell as discrete sudanophilic granules that are readily observed in the microscope with phase-contrast or dark-field illumination. The granules are enclosed by a membrane that does not appear to be a typical unit membrane, and depending on the species varies in thickness from 2.0 to 8.0 nm. A limiting membrane surrounding the granules has been observed in thin sections of *Rs. rubrum* (Fig. 3) (Boatman, 1964), in thin sections of frozen and thawed *Bacillus cereus* cells (Pfister and Lundgren, 1964), in the chemolithotrophic bacterium *Ferrobacillus ferrooxidans* (Wang and Lundgren, 1969), in *Caulobacter* (Poindexter, 1964), and in the blue-green alga *Chlorogloea fritschii* (Jensen and Sicko, 1971, 1973).

Using the freeze–etching technique, Dunlop and Robards (1973) examined the appearance of PHB granules from *B. cereus*. After freeze–fracturing, granules in intact cells varied in diameter from 240 to 720 nm and consisted of a central core of diameter between 140 and 370 nm, which occupied less than 50% of the volume of the granule. The central core, surrounded by an outer coat, was stretched on fracturing of the cells. Figure 4 shows PHB granules stretched from the central region, revealing strands in both the stretched part and the fracture surface. In some cases, granules are removed by freeze–fracturing and leave craters behind. Freeze–etching PHB granules isolated by hypochlorite treatment showed a stranded outer surface. In contrast, the outer surfaces of granules isolated from cells subjected to sonic oscillation were covered with particles. Since hypochlorite, but not sonic oscillation, is known to have deleterious effects on polymer granules, these authors suggest that the stranded surface represents an interior layer of the granule and the particulate surface is the outer face of the granule. A schematic representation of an isolated freeze–fractured PHB granule is seen in Fig. 5. The studies of Dunlop and Robards (1973) are of considerable interest, since the division of the granule into an inner core and coat regions would not have been predicted on the basis of chemical analysis. As reported below, the granules from *B. megaterium* consist of 98% PHB. Thus, the inner core and coat regions may represent different physical forms of the polymer.

PHB granules from *B. megaterium* have been extensively purified by differential centrifugation, fractionation in a polymer two-phase system, and density-gradient centrifugation (Griebel *et al.*, 1968). PHB constitutes 97–98% of the dry weight of the granule, the remainder being made up of protein (2%) and lipid (0.5%). These latter substances are presumably components of the membrane of the granule. The lipid fraction contains an unidentified lipid and phosphatidic acid. The procedures used for the purification of granules are relatively mild and result in preparations of "native PHB granules," i.e., granules that are susceptible to enzymatic depolymerization by

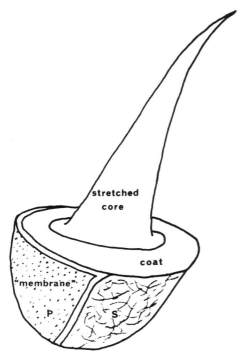

stretched
core

coat

"membrane"

P

S

Fig. 5. Diagram of an isolated freeze–fractured PHB granule. (P) Particulate surface of sonicated specimens; (S) stranded surface of hypochlorite-extracted specimens. Reproduced from Dunlop and Robards (1973) courtesy of A. W. Robards; with permission.

soluble depolymerases and that are associated with the PHB synthetase. These granules are intact, bounded by a membrane, and free from other contaminating cellular components. By application of the carbon-replica technique, Lundgren *et al.* (1964) demonstrated that a discrete membranelike structure encases the granules (Fig. 6). From studies on the enzymatic depolymerization of PHB, Merrick and Doudoroff (1961) noted that a variety of chemical and physical treatments inactivate PHB granules and make them unsuitable as a substrate for the degradative enzymes. Inactivated granules were examined by electron microscopy (Merrick *et al.*, 1965), and in all cases, the distinct morphological appearance of native granules was altered. These changes were characterized mainly by membrane fragmentation, loss of coalescence, and surface alterations.

The size and shape of *B. megaterium* native granules were examined by Ellar *et al.* (1968). Granules are spherical and range in diameter from 200 to 700 nm, with a number-average value of 490 nm. By light-scattering data, the weight-average particle weight was found to be 3.57×10^9 g/mol of particles, indicating that each granule contains at least 10,000

PHB molecules. By use of acetone to disrupt the granule, these workers found that the granule is composed of 10- to 15-nm fibrils involving extended polymeric chains. X-ray diffraction studies indicated that the PHB of the granule was crystalline and had the same unit cell as single crystals.

3.3. Biosynthesis

The synthesis of PHB from D(−)-β-hydroxy-butyryl-CoA was first demonstrated in cell-free extracts of *B. megaterium* KM and *Rs. rubrum* (Merrick and Doudoroff, 1961). In both cases, the PHB synthetase was associated with PHB granules. *Rhodospirillum rubrum* granules also possess an active depolymerase, and were not further studied. A similar enzyme in *A. beijerinckii* (Dawes, 1975) was also reported. Mild alkaline extraction (10m M NaOH) of *B. megaterium* KM PHB granules solubilized approximately 70% of the total protein of the granules (Griebel and Merrick, 1971). Both the solubilized protein fraction and the extracted granule are essentially devoid of PHB synthetase activity unless they are recombined. The solubilized protein fraction was further resolved into two protein components (A-I and A-II) by chromatography on Bio-Gel A-15m. A-I but not A-II could be recombined with the extracted granules to give rise to PHB synthetase activity. Presumably the A-I component functions as the PHB synthetase and the extracted granules as primer in the reconstituted system. Griebel and Merrick (1971) proposed that PHB synthesis proceeds in two partial reactions, with the formation of an acyl-enzyme intermediate in the initial reaction [Reaction (6)]. The acyl-enzyme intermediate is then transferred to a primer acceptor in the second reaction [Reaction (7)]:

D(−)-β-Hydroxybutyryl-CoA + PHB synthetase-SH →

β-hydroxybutyryl-S-synthetase + CoASH   (6)

β-Hydroxybutyryl-S-synthetase + primer →

β-hydroxybutyrate-primer + PHB synthetase-SH   (7)

Ellar *et al.* (1968) proposed that polymer-synthesizing enzymes aggregate into a micellar form. PHB synthesis would then occur within this protein covering, giving rise to PHB fibrils parallel to each other. By extensive coiling and twisting, they would be compacted in the granule. The surface coat in this case is visualized as being primarily composed of polymer-synthesizing molecules.

Two different pathways have been reported for the synthesis of D(−)-β-hydroxybutyryl-CoA. In the case of *Rs. rubrum*, Moskowitz and Merrick (1969)

Fig. 6. Carbon replica of germanium-shadowed native PHB granules from *Bacillus megaterium* showing a defined core and outer membrane layer. Reproduced from Lundgren *et al.* (1964) courtesy of R. Pfister and D. G. Lundgren; with permission.

reported the isolation of two enoyl-CoA hydrases, the combined activity of which resulted in the racemization of L(+)-β-hydroxybutyryl-CoA to D(−)-β-hydroxybutyryl-CoA according to the following series of reactions:

$$\text{L(+)-}\beta\text{-hydroxybutyryl-CoA} \rightleftharpoons \text{crotonyl-CoA} + \underline{\text{H}_2\text{O}}$$
$$(8)$$

$$\text{Crotonyl-CoA} + \text{H}_2\text{O} \rightleftharpoons \text{D(−)-}\beta\text{-hydroxybutyryl-CoA}$$
$$(9)$$

The enoyl-CoA hydrase that reversibly hydrates enoyl-CoA to D(−)-β-hydroxybutyryl-CoA has been extensively purified. It also hydrates crotonyl thioesters of pantetheine and acyl carrier protein, but at approximately 50% of the rate at which it hydrates crotonyl-CoA. It was shown that the product of crotonyl-CoA but not crotonyl-ACP is incorporated into PHB, suggesting that acyl carrier protein thioesters do not play a significant role in PHB synthesis. Ritchie and Dawes (1969) reached a similar conclusion regarding acyl carrier protein derivatives in their studies on PHB biosynthesis in *A. beijerinckii*.

Moskowitz and Merrick (1969) proposed the following pathway for PHB synthesis from acetate by *Rs. rubrum*:

$$\text{Acetate} \rightarrow \text{acetyl-CoA} \rightarrow \text{acetoacetyl-CoA} \rightarrow$$
$$\text{L(+)-}\beta\text{-hydroxybutyryl-CoA} \rightarrow \text{crotonyl-CoA} \rightarrow$$
$$\text{D(−)-}\beta\text{-hydroxybutyryl-CoA} \rightarrow \text{PHB}$$

The enzymes in the early steps of this pathway were reported in *Rs. rubrum* by Eisenberg (1955) (acetyl-CoA kinase) and Stern and Del Campillo (1956) (thiolase, L(+)-β-hydroxybutyryl-CoA dehydrogenase).

The synthesis of D(−)-β-hydroxybutyryl-CoA by *A. beijerinckii* differs from the pathway described for *Rs. rubrum*. This organism possesses an acetoacetyl-CoA reductase that catalyzes the reduction of acetoacetyl-CoA to D(−)-β-hydroxybutyryl-CoA using NADH or NADPH as the electron donor (Ritchie *et al.*, 1971). The reaction rate was fivefold greater with NADPH than with NADH, suggesting that NADPH is the preferred coenzyme. No evidence for the presence of enoyl-CoA hydrases in this organism was found (Dawes and Senior, 1973). Thus, in *A. beijerinckii*, the overall pathway of PHB synthesis appears to be the following reaction sequence:

Acetyl-CoA→acetoacetyl-CoA→

D(−)-β-hydroxybutyryl-CoA→PHB

The β-ketothiolase of *A. beijerinckii* was examined in some detail by Senior and Dawes (1973). The reaction catalyzed by this enzyme is

$$2\ CH_3COSCoA \rightleftharpoons CH_3COCH_2COSCoA + CoA$$

(10)

The condensation reaction is inhibited by CoA, while the reaction in the reverse direction, the cleavage reaction, is substrate-inhibited by acetoacetyl-CoA. A β-ketothiolase with similar properties was also reported for *H. eutropha* by Oeding and Schlegel (1973).

From the kinetic data, both groups of workers proposed that this enzyme, the first key enzyme in PHB biosynthesis, plays an important role in the regulation of PHB metabolism. During exponential growth, acetyl-CoA is mainly funneled through the tricarboxylic acid cycle to generate energy and carbon precursors for biosynthetic reactions. Thus, the intracellular concentration of acetyl-CoA might be expected to be low and that of CoA high as a result of citrate synthase activity. These conditions would favor inhibition of β-ketothiolase and PHB synthesis. When growth slows because of a nutrient deficiency (e.g., a depleted nitrogen source), the acetyl-CoA concentration will increase concomitant with a reduction in free CoA. Under such conditions, β-ketothiolase would be released from inhibition, and PHB would be synthesized. Senior *et al.* (1972) reported that under oxygen-limiting conditions, *A. beijerinckii* accumulates massive amounts of PHB. They suggest that PHB synthesis by oxygen-limited cells provides the cells with a mechanism for reoxidizing excess reducing power, a process that these workers have termed *quasifermentation*. They propose that oxygen-limited cultures accumulate reduced nicotinamide nucleotides that inhibit the activities of citrate synthase and isocitrate dehydrogenase (Senior and Dawes, 1971). Under these conditions, the oxidation of acetyl-CoA through the tricarboxylic acid cycle is prevented, its concentration would thus increase, and it would be channeled into PHB synthesis.

3.4. Degradation

The most detailed study on PHB degradation has been carried out on a cell-free system that utilizes *B. megaterium* KM negative PHB granules as substrate and soluble depolymerizing enzymes from *Rs. rubrum*

(Merrick and Doudoroff, 1964; Merrick and Yu, 1966; Griebel and Merrick, 1971). The requirements for depolymerization include a labile factor associated with the granules and the following soluble components: (1) heat-stable protein factor (activator), (2) a PHB depolymerase, and (3) a hydrolase. The successive action of the activator and the depolymerase on the granules results in the formation of β-hydroxybutyrate as the major product and some dimer (15–20%). The hydrolase hydrolyzes the dimer, as well as the trimer (but not PHB), to monomer. Native PHB granules can be rendered unsuitable as a substrate (denatured) by treatments such as freezing and thawing, extensive proteolytic digestion, repeated centrifugation, and heat, and by treatment with various organic solvents and acids. The denaturation of the granule may be due to damage to the bounding membrane. This conclusion is supported by the observation that denaturing treatments result in morphological alterations that are characterized mainly by membrane fragmentation (Merrick *et al.*, 1965). Griebel and Merrick (1971) proposed that PHB may be hydrolyzable only in a particular conformational state that is maintained by its close association with protein. Alteration of this association may change the physical state of the polymer and result in preparations no longer degradable by the depolymerase.

The action of the heat-stable "activator" precedes that of the depolymerase. The mechanism of activation is poorly understood, but can be simulated by mild tryptic treatment, which might suggest that activator and trypsin were exerting their activity by removal of some protective substance. It has not been possible to demonstrate proteolytic activity with the activator component of *Rs. rubrum*, however, and thus it is possible that activation by activator or by trypsin proceeds by two different mechanisms. Of considerable interest were the experiments of Griebel and Merrick (1971) on the effect of alkaline extraction of *B. megaterium* granules on the depolymerizing system. Extracted granules were much more rapidly hydrolyzed by *Rs. rubrum* extracts than native granules and no longer required pretreatment with activator or trypsin. Readdition of the alkaline extract prevented the direct hydrolysis, and the inhibition was reversed by activator or by trypsin. Griebel and Merrick (1971) concluded that PHB granules are associated with an inhibitor, presumably a protein, that interferes with the hydrolysis of PHB by depolymerase. The inhibitor is destroyed by proteolysis, removed by alkaline extraction, and inactivated by activator in as yet undetermined manner. The relationship of the activator and of the inhibitor of the granule as potential devices by the bacterial cell to regulate PHB degradation cannot be underestimated.

Gavard *et al.* (1966, 1967) isolated a soluble PHB depolymerase system from *B. megaterium*. The enzyme system was partially purified, and two depolymerases were separated. One hydrolyzes native PHB granules or PHB obtained from hypochlorite-digested cells mainly to dimer with some monomer. The other hydrolyzes oligomers to monomer. In *Hydrogenomonas*, Hippe and Schlegel (1967) isolated a depolymerase that degrades native granules to β-hydroxybutyric acid. The properties of the native granules were similar to those previously described for *B. megaterium* KM granules.

As a result of the action of the degradative enzymes, PHB is finally broken down to D(−)-β-hydroxy-butyrate. This substance is further metabolized to acetyl-CoA by the following steps:

1. An NAD-specific D(−)-β-hydroxybutyrate dehydrogenase oxidizes D(−)-β-hydroxybutyrate to acetoacetate. This enzyme has been studied in a number of different organisms [*A. beijerinckii* (Senior and Dawes, 1973), *Azotobacter vinelandii* (Jurtshuk *et al.*, 1968), *Rp. sphaeroides* (Bergmeyer *et al.*, 1967), *Rs. rubrum* (Shuster and Doudoroff, 1962), *Pseudomonas lemoignei* (Delafield *et al.*, 1965), *Hydrogenomonas* (Schindler and Schlegel, 1963)]. Oeding and Schlegel (1973) showed that the *H. eutropha* enzyme is competitively inhibited by NADH, pyruvate, and oxalacetate. In *A. beijerinckii*, NADH, pyruvate, and α-ketoglutarate (Senior and Dawes, 1973) are competitive inhibitors.

2. An acetoacetate:succinyl-CoA transferase carries out the following reaction (Senior and Dawes, 1973):

$$\text{Acetoacetate} + \text{succinyl-CoA} \rightleftarrows \text{succinate} + \text{acetoacetyl-CoA} \quad (11)$$

3. A β-ketothiolase catalyzes the thiolysis of aceto-acetyl-CoA to acetyl-CoA. The latter enzyme oc-cupies a central role in PHB metabolism, since it is the first unique enzyme in PHB synthesis as well as the last enzyme in its degradation.

Dawes and Senior (1973) proposed that polymer degradation might be regulated at the step catalyzed by β-hydroxybutyric dehydrogenase. Here, it is envisaged that the competitive inhibitors of the enzyme, pyruvate and α-ketoglutarate (or oxal-acetate), would be at high concentration during active growth and would inhibit the dehydrogenase, preventing stored PHB from being metabolized. Thus, inhibition of the dehydrogenase would effectively prevent the utilization of PHB under conditions when its need by the cell is minimal.

A scheme for PHB metabolism is illustrated in Fig. 7; it was adapted from Oeding and Schlegel (1973) and Senior and Dawes (1973).

4. Polyphosphate

4.1. General Considerations

The first type of subcellular particle to be recognized in bacteria was the intracellular inclusions that stained metachromatically with basic dyes. These granules have been commonly referred to as *volutin*, *metachromatic*, or *Babes–Ernst* granules, and can be seen by phase-contrast microscopy. It is generally accepted that metachromatic granules are composed of polyphosphate (polyP) (Harold, F. M., 1966). There is at least one report, however, that suggests that the metachromatic granules of *Spirillum volutans* are PHB, not polyP (Martinez, 1963). Many different types of bacteria as well as blue-green algae, yeast, fungi, algae, and protozoa are known to accumulate metachromatic granules. There are also reports of the presence of polyP in higher plants and animals (Harold, F. M., 1966). Compilations of those microorganisms that store polyP have been published by Dawes and Senior (1973) and Kuhl (1960).

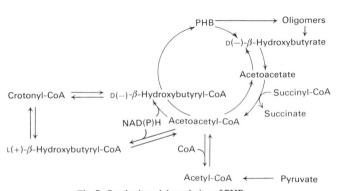

Fig. 7. Synthesis and degradation of PHB.

4.2. Chemical Structure and Composition of Granules

PolyP is a polymer of orthophosphate with phosphoanhydride linkages, and has the following general structure:

$$-O-\overset{\overset{\displaystyle O}{\|}}{\underset{\underset{\displaystyle O^-}{|}}{P}}-O-\overset{\overset{\displaystyle O}{\|}}{\underset{\underset{\displaystyle O^-}{|}}{P}}-\left[O-\overset{\overset{\displaystyle O}{\|}}{\underset{\underset{\displaystyle O^-}{|}}{P}}-\right]_n O-\overset{\overset{\displaystyle O}{\|}}{\underset{\underset{\displaystyle O^-}{|}}{P}}-O^-$$

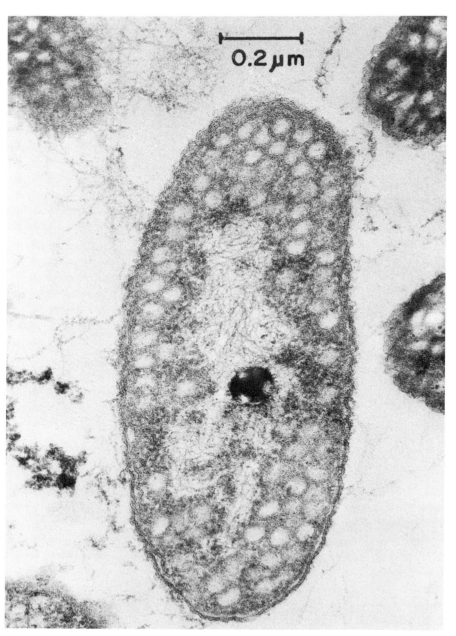

Fig. 8. Polyphosphate granule in *Rhodopseudomonas sphaeroides*. Courtesy of G. Stanier, Pasteur Institute, Paris, France.

These unbranched structures can vary in chain length from 2 (pyrophosphate) to about 10^4. Thermodynamically, these substances can be considered to be high-energy phosphate compounds. The free energy of hydrolysis of the anhydride linkage is about 9 kcal per phosphate bond at pH 5.

PolyP is deposited in the cell as spherical granules that are seen in sections as small electron-opaque bodies that appear to lack a limiting membrane (Fig. 8). F. M. Harold (1966) suggests that volutin granules in the cell could be due to precipitation of polyP as a result of the high ionic strength of the cytoplasm. The structure and composition of native polyP granules are, however, still in doubt. Granules (40–80 nm in diameter) isolated from *M. lysodeikticus* by Friedberg and Avigad (1968) were found to be composed of 24% protein, 30% lipids, 27% polyP, and small amounts of RNA, carbohydrate, and polyvalent cations. It is difficult to assess from their study the degree of contamination of these polyP granules with other cell constituents. Further studies on the structure of polyP granules would be of considerable interest.

4.3. Biosynthesis

There appear to be two pathways for the biosynthesis of polyP: (1) the reaction catalyzed by polyP kinase and (2) the reaction catalyzed by 1,3-diphosphoglycerate-polyP phosphotransferase.

1. PolyP kinase (ATP-polyP phosphotransferase) catalyzes the transfer of the terminal phosphoryl group of ATP to polyP according to the following reaction:

$$\text{ATP} + (P_i)_n \rightleftharpoons \text{ADP} + (P_i)_{n+1} \qquad (12)$$

The enzyme was extensively purified from *Esch. coli* by Kornberg *et al.* (1956). PolyP synthesis is strongly inhibited by ADP due to the reversal of the reaction. The K_m for ADP was 47 μM and for polyP, 26 μM, while that for ATP was 1.4 μM. This observation would suggest that *in vivo*, the reaction would be reversed by low ATP/ADP ratios. PolyP kinase is widely distributed among microorganisms, including the green sulfur bacterium *Chlorobium limicola f. thiosulfatophilum* (Cole and Hughes, 1965). In the case of *A. aerogenes*, mutants that lack polyP kinase are unable to synthesize polyP, indicating that it is the only pathway for biosynthesis of polyP in this organism (Harold, F. M., 1964; Harold, R. L., and Harold, 1963).

2. 1,3-Diphosphoglycerate-polyP phosphotransferase, which was first reported in *Neurospora crassa*

(Kulaev and Bobyk, 1971), catalyzes the following reaction:

$$1,3\text{-Diphosphoglycerate} + (P_i)_n \rightarrow 3\text{-phospho-}$$
$$\text{glycerate} + (P_i)_{n+1} \qquad (13)$$

Escherichia coli, *M. lysodeikticus*, and *Propionibacterium shermanii* possess both the polyP kinase and 1,3-diphosphoglycerate enzymes; however, the former enzyme is usually present at higher activities (three- to tenfold) (Kulaev *et al.*, 1971). The principal enzyme responsible for polyP synthesis in organisms that possess both polyP kinase and 1,3-diphosphoglycerate-polyP phosphotransferase may be resolved by studies with mutants. In some organisms, e.g., *Penicillium chrysogenum* and *N. crassa* (Kulaev *et al.*, 1971; Kulaev and Bobyk, 1971), only the 1,3-diphosphoglycerate enzyme is present, and under these circumstances, this enzyme is presumably the principal pathway for polyP synthesis.

4.4. Degradation

A number of different enzymes that may play a role in polyP degradation have been described in microorganisms:

1. PolyP kinase may play a dual role, functioning in both biosynthesis and degradation of polyP.

2. PolyP-AMP phosphotransferase transfers phosphate from polyP to AMP according to the following reaction:

$$(P_i)_{n+1} + \text{AMP} \rightarrow (P_i)_n + \text{ADP} \qquad (14)$$

Although the enzyme has been reported in *Mycobacterium smegmatis* (Winder and Denneny, 1957) and *Corynebacterium xerosis* (Dirheimer and Ebel, 1965), its physiological role remains in doubt.

3. PolyP glucokinase was discovered in glucose-grown *Mycobacterium phlei*, other mycobacteria, and *Corynebacterium diphtheriae* (Szymona, M., 1962; Szymona, M., *et al.*, 1962). The enzyme has also been reported to be present in a number of other microorganisms (Szymona, O., *et al.*, 1967; Uryson and Kulaev, 1968), but was not detected in *Esch. coli*, *A. aerogenes*, *Aspergillus niger*, or *Rs. rubrum*. The reaction catalyzed is:

$$\text{Glucose} + (P_i)_{n+1} \rightarrow \text{glucose 6-P} + (P_i)_n \qquad (15)$$

The enzyme derived from glucose-grown *M. phlei* phosphorylates glucosamine, but not fructose or mannose. A polyP fructokinase is induced, however, on fructose-grown *M. phlei* cells (Szymona, M., and Szumilo, 1966). The actual function of this enzyme in the degradation of polyP or in the phosphorylation of hexoses is still unclear. The availability of mutants lacking this enzyme would permit a more definite assessment of its physiological role.

4. Polyphosphatases that catalyze the hydrolysis of polyP to inorganic phosphate appear to be widespread in nature. Among the bacteria, the enzymes from *C. xerosis* (Muhammed *et al.*, 1959) and from *A. aerogenes* (Harold, F. M., and Harold, 1965) have been purified. *Aerobacter aerogenes* mutants lacking this enzyme were unable to degrade polyP, suggesting that at least in this organism, it may play a primary role in polyP degradation.

4.5. Accumulation and Utilization

The extrinsic factors that favor polyP accumulation are similar to those observed for other storage substances. Generally, environmental conditions and the physiological state of the cell determine the extent of polyP accumulation and degradation. The early studies of Wilkinson, Duguid, and their associates (Smith *et al.*, 1954; Wilkinson and Duguid, 1960) demonstrated that *A. aerogenes* cells growing exponentially accumulate little polyP. PolyP is accumulated, however, if growth is prevented because of low pH of the culture medium, or a deficiency in a source of nitrogen or sulfur. Magnesium, potassium, phosphate, and a source of energy were also required for polyP deposition. When the cells were transferred to a fresh growth medium, the previously accumulated polyP was degraded. This characteristic behavior is seen with a variety of microorganisms. Some other examples include *C. xerosis* (Hughes and Muhammed, 1962), *Mycobacterium* (Drews, 1960; Mudd *et al.*, 1958), *Hydrogenomonas* (Kaltwasser, 1962), and *Streptococcus* (Tanzer and Krichevsky, 1970). *Micrococcus lysodeikticus* (Friedberg and Avigad, 1968) behaves atypically with regard to polyP accumulation and degradation. This organism synthesizes polyP during the exponential phase and utilizes it during the stationary phase of growth. As indicated earlier, *M. lysodeikticus* deposits polyP granules that may be associated with more complex subcellular structures.

PolyP metabolism has not been extensively studied in the photosynthetic bacteria. A number of scattered reports are available; however, detailed information regarding its overall metabolism, including the molecular machinery involved in its biosynthesis and degradation, is lacking. In a short note, Carr and Sandhu (1966) reported that *Rp. sphaeroides*, grown aerobically or photosynthetically, and incubated under anaerobic conditions in the dark, degrades polyP. On the other hand, photosynthetically grown cells incubated in the light in a basal salts medium without a carbon source accumulated polyP. Thus, a source of

ATP in the absence of carbon source resulted in polyP accumulation, while deprivation of ATP-forming mechanisms resulted in polyP degradation. The obligate phototroph *Chl. limicola f. thiosulfatophilum* (Hughes *et al.*, 1963; Cole and Hughes, 1965) accumulated polyP during both logarithmic and stationary stages of growth. These workers suggest that this may be a reflection of excess ATP production by photophosphorylation under the growth conditions utilized. This organism apparently lacks a polyphosphatase, and polyP degradation appears to be mediated by the successive action of polyP kinase and ATPase. However, further study is required to define this system more clearly. Kulaev *et al.* (1974) observed that *Rs. rubrum* grown in the light accumulated pyrophosphate as well as polyP during various growth stages. PolyP, but not pyrophosphate, was also accumulated by dark-grown cells. The accumulated polyP was successively subfractionated into salt-soluble, alkali-soluble, and hot-perchloric-acid-soluble fractions. These extraction procedures presumably result in polyP preparations of successively increasing molecular weight (Langen *et al.*, 1962; Harold, F. M., 1966). Although all three fractions were observed in the light- or dark-grown cells, the relative concentrations of the salt- and alkali-soluble fractions were different. The salt-soluble polyP was the major polyP in light-grown cells, whereas the alkali-soluble polyP was the predominant polyP in the dark-grown cells. In light-grown cells, the content of the salt-soluble fraction is highest in actively growing cells and decreases during the stationary phase. The content of pyrophosphate in the light-grown cells also followed a similar course. Kulaev *et al.* (1974) suggest that there may be a direct relationship between photophosphorylation and the biosynthesis of salt-soluble polyP, possibly through a pyrophosphate intermediate. In the absence of more definitive data, however, such a conclusion can be considered only tenuous. Further studies in this interesting area are clearly necessary.

In studies with *A. aerogenes*, F. M. Harold (1963, 1964) observed that polyP accumulation due to nutrient imbalance could be divided into two patterns. First, when growth and nucleic acid synthesis are prevented because of the depletion of sulfate or of uracil or amino acids in the case of auxotrophic mutants, polyP is slowly accumulated. If growth is allowed to resume, polyP is degraded rapidly, and phosphate is quantitatively transferred to the nucleic acid fraction. By dissociating cell growth from nucleic acid synthesis with chloramphenicol, F. M. Harold (1965) and F. M. Harold and Harold (1965) observed a dual competitive relationship between nucleic acid and polyP synthesis. Thus, nucleic acid synthesis, possibly by competing for the available ATP, decreased the rate of polyP acumulation and simultaneously stimulated its degradation. The second pattern was observed when phosphate was added to cells that had previously been starved of phosphate. Here, polyP accumulation is rapid and extensive. This phenomenon had been previously observed in yeast, and had been called *Polyphosphate-Überkompensation* by Liss and Langen (1962). F. M. Harold (1966) refers to this as *polyP overplus*.

Some preliminary studies on the regulation of polyP metabolism in *A. aerogenes* were reported by Harold and co-workers (Harold, F. M., 1966). Cells from cultures growing exponentially contain low levels of polyP kinase, polyphosphatase, and alkaline phosphatase. When the cells are subjected to phosphate starvation, all three enzymes are synthesized. The synthesis of these three enzymes appears to be under the control of a common regular gene. Thus, mutants have been isolated in which all three enzymes are either permanently repressed or constitutively derepressed (Harold, F. M., 1964; Harold, F. M., and Harold, 1965). F. M. Harold and Harold (1965) indicated that although these enzymes share a common regulatory gene, the structural genes do not fall within a single operon.

Nesmeyanova *et al.* (1974) recently reported that in *Esch. coli*, phosphate starvation did not induce the synthesis of polyP kinase or 1,3-diphosphoglycerate-polyP phosphotransferase. These results are in contrast to those reported above for *A. aerogenes*. On the other hand, the levels of alkaline phosphatase and polyphosphatase did increase. When phosphate was added to cells previously starved of phosphate, the phosphatases were repressed, but the levels of the polyP-synthesizing enzymes (polyP kinase and 1,3-diphosphoglycerate-polyP phototransferase) increased and polyP was accumulated. The *polyP overplus* phenomenon, however, was not observed.

Although polyP can act as a phosphoryl group donor in the phosphorylation of certain sugars by corynebacteria and mycobacteria, its role as a microbial phosphagen (i.e., to phosphorylate ADP to form ATP) appears to be doubtful. F. M. Harold and Sylvan (1963) demonstrated with *A. aerogenes*, Kaltwasser (1962) with *Hydrogenomonas*, and Liss and Langen (1962) with yeast that polyP accumulation accounts for only a small proportion of the ATP generated by these organisms. These observations argue against the role of polyP as a storage reserve of energy. It appears more likely that its principal physiological role in microorganisms is to function as a phosphorus reserve. Kaltwasser (1962)

proposed that synthesis and degradation of polyP may serve to regulate the inorganic phosphate level of the cell. Unfortunately, however, details on the regulatory aspects of polyP synthesis and degradation are lacking. Further study in this area would provide considerable insight into our understanding of polyP metabolism and its relationship to the overall phosphorus economy of the cell.

ACKNOWLEDGMENTS

I am grateful to the following individuals who generously supplied reprints, illustrations, prepublication material, and micrographs: E. S. Boatman, E. A. Dawes, D. G. Lundgren, R. H. Marchessault, J. Preiss, and G. Stanier. I wish also to thank M. W. Stinson for reading the manuscript and R. Merrick, who provided technical assistance in the preparation of the manuscript.

5. References

Alper, R., Lundgren, D. G., Marchessault, R. H., and Cote, W. A., 1963, Properties of poly-β-hydroxybutyrate. I. General considerations concerning the naturally occurring polymer, *Biopolymers* **1**:545.

Antoine, A. D., and Tepper, B. S., 1969, Characterization of glycogens from mycobacteria, *Arch. Biochem. Biophys.* **134**:207.

Bergmeyer, H. U., Gawehn, K., Klotzsch, H., Krebs, H. A., and Williamson, D. H., 1967, Purification and properties of crystalline 3-hydroxybutyrate dehydrogenase from *Rhodopseudomonas spheroides, Biochem. J.* **102**:423.

Boatman, E. S., 1964, Observations on the fine structure of spheroplasts of *Rhodospirillum rubrum, J. Cell Biol.* **20**:297.

Buchanan, B. B., and Evans, M. C. W., 1965, The synthesis of phosphoenolpyruvate from pyruvate and ATP by extracts of photosynthetic bacteria, *Biochem. Biophys. Res. Commun.* **22**:484.

Buchanan, B. B., Evans, M. C. W., and Arnon, D. I., 1967, Ferredoxin-dependent carbon assimilation in *Rhodospirillum rubrum, Arch. Mikrobiol.* **59**:32.

Builder, J. E., and Walker, G. J., 1970, Metabolism of the reserve polysaccharide of *Streptococcus mitis.* Properties of glycogen synthetase, *Carbohydr. Res.* **14**:35.

Burleigh, I., and Dawes, E. A., 1967, Studies on the endogenous metabolism and senescence of starved *Sarcina lutea, Biochem. J.* **102**:236.

Campbell, J. J. R., Gronlund, A. F., and Duncan, M. G., 1963, Endogenous metabolism of *Pseudomonas, Ann. N. Y. Acad. Sci.* **102**:669.

Carr, N. G., and Sandhu, G. R., 1966, Endogenous metabolism of polyphosphates in two photosynthetic microorganisms, *Biochem. J.* **99**:29P.

Chao, L., and Bowen, C. C., 1971, Purification and

properties of glycogen isolated from a blue-green alga *Nostoc muscorum, J. Bacteriol.* **105**:331.

Chen, G. S., and Segel, I. H., 1968a, *Escherichia coli* polyglucose phosphorylase, *Arch. Biochem. Biophys.* **127**:164.

Chen, G. S., and Segel, I. H., 1968b, Purification and properties of glycogen phosphorylase from *Escherichia coli, Arch. Biochem. Biophys.* **127**:175.

Cohen-Bazire, G., and Kunisawa, R., 1963, The fine structure of *Rhodospirillum rubrum, J. Cell Biol.* **16**:401.

Cole, J. A., and Hughes, D. E., 1965, The metabolism of polyphosphates in *Chlorobium thiosulfatophilum, J. Gen. Microbiol.* **38**:65.

Cornibert, J., and Marchessault, R. H., 1972, Physical properties of poly-β-hydroxybutyrate. IV. Conformational analysis and crystalline structure, *J. Mol. Biol.* **71**:735.

Dawes, E. A., 1975, The role and regulation of poly-β-hydroxybutyrate as a reserve in microorganisms, in: *Proceedings of the International Symposium on Macromolecules*, Rio de Janeiro, July 26–31, 1974 (E. B. Mano, ed.), pp. 433–450, Elsevier Scientific Publishing Co., Amsterdam.

Dawes, E. A., and Ribbons, D., 1964, Some aspects of the endogenous metabolism of bacteria, *Bacteriol. Rev.* **28**:126.

Dawes, E. A., and Ribbons, D., 1965, Studies on the endogenous metabolism of *Escherichia coli, Biochem. J.* **95**:332.

Dawes, E. A., and Senior, P. J., 1973, Energy reserve polymers in micro-organisms, in: *Advances in Microbial Physiology* (A. H. Rose and D. W. Tempest, eds.), Vol. 10, pp. 135–266, Academic Press, New York.

Delafield, F. P., Cooksey, K., and Doudoroff, M., 1965, β-Hydroxybutyrate dehydrogenase and dimer esterase of *Pseudomonas lemoignei, J. Biol. Chem.* **240**:4023.

Dicks, J. W., and Tempest, D. W., 1967, Potassium–ammonium antagonism in polysaccharide synthesis by *Aerobacter aerogenes* NCTC 418, *Biochem. Biophys. Acta* **136**:176.

Dietzler, D. N., Leckie, M. P., and Lais, C. J., 1973, Rates of glycogen synthesis and the cellular levels of ADP and FDP during exponential growth and the nitrogen-limited stationary phase of *Escherichia coli* W4597(K), *Arch. Biochem. Biophys.* **156**:684.

Dietzler, D. N., Lais, C. J., and Leckie, M. P., 1974a, Simultaneous increases of the adenylate energy charge and the rate of glycogen synthesis in nitrogen-starved *Escherichia coli* W597(K), *Arch. Biochem. Biophys.* **160**:14.

Dietzler, D. N., Leckie, M. P., Lais, C. J., and Magnani, J. L., 1974b, Evidence for the allosteric regulation of bacterial glycogen synthesis *in vivo, Arch. Biochem. Biophys.* **162**:602.

Dirheimer, G., and Ebel, J. P., 1965, Characterization d'une polyphosphate-AMP-phosphotransferase dans *Corynebacterium xerosis, C. R. Acad. Sci. Paris* **260**:3787.

Doudoroff, M., 1966, Metabolism of poly-β-hydroxybutyrate in bacteria, in: *Current Aspects of Biochemical Energetics* (N. O. Kaplan and E. P. Kennedy, eds.), pp. 385–400, Academic Press, New York.

Doudoroff, M., and Stanier, R. Y., 1959, Role of poly-β-hydroxybutyric acid in the assimilation of organic carbon by bacteria, *Nature (London)* **183**:1440.

Drews, G., 1960, Untersuchungen zum Polyphosphatstoffwechsel und der Bildung metachromatischer Granula bei *Mycobacterium phlei, Arch. Mikrobiol.* **36**:87.

Dunlop, W. F., and Robards, A. W., 1973, Ultrastructural study of poly-β-hydroxybutyrate granules from *Bacillus cereus, J. Bacteriol.* **114**:1271.

Eidels, L., and Preiss, J., 1970, Carbohydrate metabolism in *Rhodopseudomonas capsulata*: Enzyme titers, glucose metabolism, and polyglucose polymer synthesis, *Arch. Biochem. Biophys.* **140**:75.

Eidels, L., Edelman, P., and Preiss, J., 1970, Biosynthesis of bacterial glycogen. VIII. Activation and inhibition of the adenosine diphosphoglucose pyrophosphorylase of *Rhodopseudomonas capsulata* and of *Agrobacterium tumefaciens, Arch. Biochem. Biophys.* **140**:60.

Eisenberg, M. A., 1955, The acetate-activating enzyme of *Rhodospirillum rubrum, Biochim. Biophys. Acta* **16**:58.

Ellar, D., Lundgren, D. G., Okamura, K., and Marchessault, R. H., 1968, Morphology of poly-β-hydroxybutyrate granules, *J. Mol. Biol.* **35**:489.

Emeruwa, A. C., and Hawirko, R. Z., 1973, Poly-β-hydroxybutyrate metabolism during growth and sporulation of *Clostridium botulinum, J. Bacteriol.* **116**:989.

Friedberg, I., and Avigad, G., 1968, Structures containing polyphosphate in *Micrococcus lysodeikticus, J. Bacteriol.* **96**:544.

Furlong, C. E., and Preiss, J., 1969, Biosynthesis of bacterial glycogen. VII. Purification and properties of the adenosine diphosphoglucose pyrophosphorylase of *Rhodospirillum rubrum, J. Biol. Chem.* **244**:2539.

Gavard, R., Dahinger, A., Hauttecoeur, B., and Raynaud, C., 1966, Degradation du lipide β-hydroxybutyrique par un extrait enzymatique de *Bacillus megaterium*. I. Depolymérase A, *C. R. Acad. Sci. Paris* **263**:1273.

Gavard, R., Raynaud, C., Hauttecoeur, B., and Dahinger, A., 1967, Degradation du lipide β-hydroxybutyrique par un extrait enzymatique de *Bacillus megaterium*. III. Depolymérase B, *C. R. Acad. Sci. Paris* **265**:1557.

Ghosh, H. P., and Preiss, J., 1965, The isolation and characterization of glycogen from *Arthrobacter* sp. NRRLB 1973, *Biochim. Biophys. Acta* **104**:274.

Govons, S., Gentner, N., Greenberg, E., and Preiss, J., 1973, Biosynthesis of bacterial glycogen. XI. Kinetic characterization of an altered adenosine diphosphate glucose synthase from a "glycogen-excess" mutant of *Escherichia coli B, Biol. Chem.* **248**:1731.

Greenberg, E., and Preiss, J., 1965, Biosynthesis of bacterial glycogen. II. Purification and properties of the adenosine diphosphoglucose: glycogen transglucosylase of *Arthrobacter species* NRRL B1973, *J. Biol. Chem.* **240**:2341.

Griebel, R. J., and Merrick, J. M., 1971, Metabolism of poly-β-hydroxybutyrate: Effect of mild alkaline extraction on native poly-β-hydroxybutyrate granules, *J. Bacteriol.* **108**:782.

Griebel, R., Smith, Z., and Merrick, J. M., 1968, Metabolism of poly-β-hydroxybutyrate. I. Purification, composition, and properties of native poly-β-hydroxybutyrate granules from *Bacillus megaterium, Biochemistry* **7**:3676.

Gunja-Smith, Z., Marshal, J. J., Smith, E. E., and Whelan, W. J., 1970, A glycogen-debranching enzyme from *Cytophaga, FEBS Lett.* **12**:96.

Hara, F., Akazawa, T., and Kojima, K., 1973, Glycogen biosynthesis in *Chromatium* strain D. I. Characterization of glycogen, *Plant Cell Physiol.* **14**:737.

Harold, F. M., 1963, Accumulation of inorganic polyphosphate in *Aerobacter aerogenes*. I. Relationship to growth and nucleic acid synthesis, *J. Bacteriol.* **86**:216.

Harold, F. M., 1964, Enzymatic and genetic control of polyphosphate accumulation in *Aerobacter aerogenes, J. Gen. Microbiol.* **35**:81.

Harold, F. M., 1965, Regulatory mechanisms in the metabolism of inorganic polyphosphate in *Aerobacter aerogenes, Colloq. Int. C. N. R. S.* (*Paris*) **124**:307.

Harold, F. M., 1966, Inorganic polyphosphates in biology: Structure, metabolism and function, *Bacteriol. Rev.* **30**:772.

Harold, F. M., and Harold, R. L., 1965, Degradation of inorganic polyphosphate in mutants of *Aerobacter aerogenes, J. Bacteriol.* **89**:1262.

Harold, F. M., and Sylvan, S., 1963, Accumulation of inorganic polyphosphate in *Aerobacter aerogenes*. II. Environmental control and the role of sulfur compounds, *J. Bacteriol.* **86**:222.

Harold, R. L., and Harold, F. M., 1963, Mutants of *Aerobacter aerogenes* blocked in the accumulation of inorganic polyphosphate, *J. Gen. Microbiol.* **31**:241.

Hippe, H., 1967, Abbau und Wiederverwertung von Poly-β-hydroxybuttersäure durch *Hydrogenomonas* H16, *Arch. Mikrobiol.* **56**:248.

Hippe, H., and Schlegel, H. G., 1967, Hydrolyse von PHBS durch intracelluläre Depolymerase von *Hydrogenomonas* H16, *Arch. Mikrobiol.* **56**:278.

Holme, T., and Palmstierna, H., 1956, Changes in glycogen and N-containing compounds in *Escherichia coli* B during growth in deficient media. I. Nitrogen and carbon starvation, *Acta Chem. Scand.* **10**:578.

Houte, J. V., and Jansen, H. M., 1970, Role of glycogen in survival of *Streptococcus mitis, J. Bacteriol.* **101**:1083.

Hughes, D. E., and Muhammed, A., 1962, The metabolism of polyphosphate in bacteria, *Colloq. Int. C. N. R. S.* (*Paris*) **106**:591.

Hughes, D. E., Conti, S. F., and Fuller, R. C., 1963, Inorganic polyphosphate metabolism in *Chlorobium thiosulfatophilum, J. Bacteriol.* **85**:577.

Jensen, T. E., and Sicko, L. M., 1971, Fine structure of poly-β-hydroxybutyric acid granules in a blue-green alga, *Chlorogloea fritschii, J. Bacteriol.* **106**:683.

Jensen, T. E., and Sicko, L. M., 1973, The fine structure of *Chlorogloea fritschii* cultured in sodium acetate enriched medium, *Cytologia* **38**:381.

Jost, M., 1965, Die Ultrastruktur von *Oscillatoria rubescens* D.C., *Arch. Microbiol.* **50**:211.

Jurtshuk, P., Manning, S., and Barrera, C. R., 1968, Isolation and purification of the D(−)-β-hydroxybutyric

dehydrogenase of *Azotobacter vinelandii, Can. J. Microbiol.* **14**:775.

Kaltwasser, H., 1962, Die Rolle der Polyphosphate im Phosphatstoffwechsel eins Knallgasbakteriums (*Hydrogenomonas Stamm* 20), *Arch. Mikrobiol.* **41**:282.

Khandelwal, R. L., Spearman, T. N., and Hamilton, I. R., 1973, Purification and properties of glycogen phosphorylase from *Streptococcus salivarius, Arch. Biochem. Biophys.* **154**:295.

Kornberg, A., Kornberg, S. R., and Simms, E. S., 1956, Metaphosphate synthesis by an enzyme from *Escherichia coli, Biophys. Acta* **26**:215.

Krebs, E. G., and Preiss, J., 1975, Regulatory mechanisms in glycogen metabolism, in: *International Review of Science: Biochemistry* (W. J. Whelan, ed.), Vol. 5, *Biochemistry of Carbohydrates*, Chapt. 7, pp. 337–389, University Park Press, Baltimore.

Kuhl, A., 1960, Die Biologie der kondensierten anorganischen Phosphate, *Ergeb. Biol.* **23**:144.

Kulaev, I. S., and Bobyk, M. A., 1971, Detection of a new enzyme in *Neurospora crassa*, 1,3-diphosphoglycerate: polyphosphate phosphotransferase, *Biochemistry* (*USSR*) (English translation) **36**:356.

Kulaev, I. S., Bobyk, M. A., Nikolaev, N. N., Sergeev, N. S., and Uryson, S. O., 1971, Polyphosphate-synthesizing enzymes in some fungi and bacteria, *Biochemistry* (*USSR*) (English translation) **36**:791.

Kulaev, I. S., Shadi, A., and Mansurova, S. E., 1974, Polyphosphates of phototrophic bacteria *Rhodospirillum rubrum* under different cultivation conditions, *Biochemistry* (*USSR*) (English translation) **39**:656.

Langen, P., Liss, E., and Lohmann, K., 1962, Art, Bildung und Umsatz der Polyphosphat der Hefe, *Colloq. Int. C. N. R. S.* (*Paris*) **106**:603.

Lemoigne, M., 1925, Etudes sur l'autolyse microbienne acidification par formation d'acide β-oxybutyrique, *Ann. Inst. Pasteur Paris* **39**:144.

Liss, E., and Langen, P., 1962, Versuche zur Polyphosphat-Überkompensation in Hefezellen nach Phosphatverarmung, *Arch. Microbiol.* **41**:383.

Lundgren, D. G., Pfister, R. M., and Merrick, J. M., 1964, Structure of poly-β-hydroxybutyric acid (PHB) granules, *J. Gen. Microbiol.* **34**:441.

Lundgren, D. G., Alper, R., Schnaitman, C., and Marchessault, R. H., 1965, Characterization of poly-β-hydroxybutyrate extracted from different bacteria, *J. Bacteriol.* **89**:245.

Macrae, R. M., and Wilkinson, J. F., 1958, The influence of the cultural conditions on poly-β-hydroxybutyrate synthesis in *Bacillus megaterium, Proc. R. Phys. Soc. Edinburgh* **27**:73.

Martinez, R. J., 1963, On the nature of the granules of the genus *Spirillum, Arch. Mikrobiol.* **44**:334.

Merrick, J. M., and Doudoroff, M., 1961, Enzymatic synthesis of poly-β-hydroxybutyric acid in bacteria, *Nature* (*London*) **189**:890.

Merrick, J. M., and Doudoroff, M., 1964, Depolymerization of poly-β-hydroxybutyrate by an intracellular enzyme system, *J. Bacteriol.* **88**:60.

Merrick, J. M., and Yu, C. I., 1966, Purification and

properties of a D(−)-β-hydroxybutyric dimer hydrolase from *Rhodospirillum rubrum, Biochemistry* **5**:3563.

Merrick, J. M., Lundgren, D. G., and Pfister, R. M., 1965, Morphological changes in poly-β-hydroxybutyrate granules associated with decreased susceptibility to enzymatic hydrolysis, *J. Bacteriol.* **89**:234.

Moskowitz, G. J., and Merrick, J. M., 1969, Metabolism of poly-β-hydroxybutyrate. II. Enzymatic synthesis of D(−)-β-hydroxybutyryl coenzyme A by an enoyl hydrase from *Rhodospirillum rubrum, Biochemistry* **8**:2748.

Mudd, S., Yoshida, A., and Koike, M., 1958, Polyphosphate as accumulator of phosphorus and energy, *J. Bacteriol.* **75**:224.

Muhammed, A., Rodgers, A., and Hughes, D. E., 1959, Purification and properties of a polymetaphosphatase from *Corynebacterium xerosis, J. Gen. Microbiol.* **20**:482.

Nesmeyanova, M. A., Dmitriev, A. D., and Kulaev, I. S., 1974, Regulation of the enzymes of phosphorous metabolism and the level of polyphosphate in *Escherichia coli* K-12 by exogenous orthophosphate, *Mikrobiologiya* (English translation) **43**:227.

Oeding, V., and Schlegel, H. G., 1973, β-Ketothiolase from *Hydrogenomonas eutropha* H16 and its significance in the regulation of poly-β-hydroxybutrate metabolism, *Biochem. J.* **134**:239.

Okamura, K., and Marchessault, R. H., 1967, X-ray structure of poly-β-hydroxybutyrate, in: *Conformation of Biopolymers* (B. M. Ramachandran, ed.) Vol. 2, pp. 709–720. Academic Press, New York.

Palmer, T. N., Wöber, G., and Whelan, W. J., 1973, The pathway of exogenous carbohydrate utilization in *Escherichia coli*: A dual function for the enzymes of the maltose operon, *Eur. J. Biochem.* **39**:601.

Pfister, R. M., and Lundgren, D. G., 1964, Electron microscopy of polyribosomes within *Bacillus cereus, J. Bacteriol.* **80**:1119.

Poindexter, J. S., 1964, Biological properties and classification of the *Caulobacter* group, *Bacteriol. Rev.* **28**:231.

Preiss, J., 1969, The regulation of the biosynthesis of α-1,4 glucans in bacteria and plants, in: *Current Topics in Cellular Regulation* (B. L. Horecker and E. R. Stadtman, eds.), Vol. 1, pp. 125–160, Academic Press, New York.

Preiss, J. 1972, Studies on the function of a regulatory site using mutants, *Intra-Sci. Chem. Rep.* **6**:13.

Preiss, J., Shen, L., Greenberg, E., and Gentner, N., 1966, Biosynthesis of bacterial glycogen. IV. Activation and inhibition of the adenosine diphosphate glucose pyrophosphorylase of *Escherichia coli, Biochemistry* **5**:1833.

Preiss, J., Ozbun, J. L., Hawker, J. S., Greenberg, E., and Lammel, C., 1973, ADPG synthetase and ADPG-α-glucan 4-glucosyl transferase: Enzymes involved in bacterial glycogen and plant starch synthesis, *Ann. N. Y. Acad. Sci.* **210**:265.

Recondo, E., and Leloir, L. F., 1961, Adenosine diphosphate glucose and starch biosynthesis, *Biochem. Biophys. Res. Commun.* **6**:85.

Ribereau-Gayon, G., Sabraw, A., Lammel, C., and Preiss, J., 1971, Biosynthesis of bacterial glycogen. IX. Regulatory properties of the adenosine diphosphate glucose pyrophosphorylase of the Enterobacteriaceae, *Arch. Biochem. Biophys.* **142**:675.

Ritchie, G. A. F., and Dawes, E. A., 1969, The non-involvement of acyl-carrier protein in poly-β-hydroxybutyric acid biosynthesis in *Azotobacter beijerinckii, Biochem. J.* **112**:803.

Ritchie, G. A. F., Senior, P. J., and Dawes, E. A., 1971, The purification and characterization of acetoacetyl-coenzyme A reductase from *Azotobacter beijerinckii, Biochem. J.* **121**:309.

Robson, R. L., Robson, R. M., and Morris, J. G., 1972, Regulation of granulose synthesis in *Clostridium pasteurianum, Biochem. J.* **130**:4P.

Schindler, J., and Schlegel, H. G., 1963, D(−)-β-Hydroxybuttersäure Dehydrogenase aus *Hydrogenomonas* H16, *Biochem. Z.* **339**:154.

Schlegel, H. G., and von Bartha, R., 1961, Formation and utilization of poly-β-hydroxybutyric acid by knallgas bacteria (*Hydrogenomonas*), *Nature* (*London*) **191**:463.

Schuster, E., and Schlegel, H. G., 1967, Chemolithotrophes Wachstum von *Hydrogenomonas* H16 in Chemotaten mit elektrolytischer Knallgaserzeugung, *Arch. Microbiol.* **58**:380.

Senior, P. J., and Dawes, E. A., 1971, Poly-β-hydroxybutyrate biosynthesis and the regulation of glucose metabolism in *Azotobacter beijerenckii, Biochem. J.* **125**:55.

Senior, P. J., and Dawes, E. A., 1973, The regulation of poly-β-hydroxybutyrate metabolism in *Azotobacter beijerenckii, Biochem. J.* **134**:225.

Senior, P. J., Beech, G. A. Ritchie, G. A. F., and Dawes, E. A., 1972, The role of oxygen limitation in the formation of poly-β-hydroxybutyrate during batch and continuous culture of *Azotobacter beijerinckii, Biochem. J.* **128**:1193.

Shen, L. C., and Atkinson, D. E., 1970, Regulation of adenosine diphosphate glucose synthase from *Escherichia coli*. Interactions of adenylate energy charge and modifier concentrations, *J. Biol. Chem.* **245**:3996.

Shen, L., and Preiss, J., 1965, Biosynthesis of bacterial glycogen. I. Purification and properties of the adenosine diphosphoglucose pyrophosphorylase of *Arthrobacter* species NRRL B1973, *J. Biol. Chem.* **240**:2334.

Shen, L., and Preiss, J., 1966, Biosynthesis of bacterial glycogen. V. The activation and inhibition of the adenosine diphosphate glucose pyrophosphorylase of *Arthrobacter viscosus* NRRL B1973, *Arch. Biochem. Biophys.* **116**:375.

Shively, J. M., 1974, Inclusion bodies of prokaryotes, *Annu. Rev. Microbiol.* **28**:167.

Shuster, C. W., and Doudoroff, M., 1962, A cold sensitive D(−)-β-hydroxybutyric acid dehydrogenase from *Rhodospirillum rubrum, J. Biol. Chem.* **237**:603.

Sierra, G., and Gibbons, N. E., 1962, Role and oxidation pathway of poly-β-hydroxybutyric acid in *Micrococcus halodenitrificans, Can. J. Microbiol.* **8**:255.

Sigal, N., Cattanéo, J., Chambost, J. P., and Favard, A., 1965, Characterization and partial purification of a branching enzyme from *Escherichia coli, Biochem. Biophys. Res. Commun.* **20**:616.

Smith, I. W., Wilkinson, J. F., and Duguid, J. P., 1954, Volutin production in *Aerobacter aerogenes* due to nutrient imbalance, *J. Bacteriol.* **68**:450.

Sobek, J. M., Charba, J. F., and Forest, W. N., 1966, Endogenous metabolism of *Azotobacter agilis, J. Bacteriol.* **92**:687.

Spearman, R. N., Khandelwal, R. L., and Hamilton, I. R., 1973, Some regulatory properties of glycogen phosphorylase from *Streptococcus salivarius, Arch. Biochem. Biophys.* **154**:306.

Stanier, R. Y., Doudoroff, M., Kunisawa, R., and Contopoulou, R., 1959, The role of organic substrates in bacterial photosynthesis, *Proc. Natl. Acad. Sci. U.S.A.* **45**:1246.

Stern, J. R., and Del Campillo, A., 1956, Enzymes of fatty acid metabolism. II. Properties of crystalline crotonase, *J. Biol. Chem.* **218**:985.

Stinson, M. W., and Merrick, J. M., 1974, Extracellular enzyme secretion by *Pseudomonas lemoignei, J. Bacteriol.* **119**:152.

Stockdale, H., Ribbons, D. W., and Dawes, E. A., 1968, Occurrence of poly-β-hydroxybutyrate in the Azotobacteriaceae, *J. Bacteriol.* **95**:1798.

Stokes, J. L., and Parson, W. L., 1968, Role of poly-β-hydroxybutyrate in survival of *Sphaerotilus discophorous* during starvation, *Can. J. Microbiol.* **14**:785.

Strange, R. E., 1968, Bacterial "glycogen" and survival, *Nature (London)* **220**:606.

Strange, R. E., Dark, F. A., and Ness, A. G., 1961, The survival of stationary phase *Aerobacter aerogenes* stored in aqueous suspension, *J. Gen. Microbiol.* **25**:61.

Szymona, M., 1962, Purification and properties of the new hexokinase utilizing inorganic polyphosphate, *Acta Biochim. Pol.* **9**:165.

Szymona, M., and Szumilo, T., 1966, Adenosine triphosphate and inorganic polyphosphate fructokinases of *Mycobacterium phlei, Acta Biochim. Pol.* **17**:129.

Szymona, M., Szymona, O., and Kulesza, S., 1962, On the occurrence of inorganic polyphosphate hexokinase in some microorganisms, *Acta Microbiol. Pol.* **11**:287.

Szymona, O., Uryson, S. O., and Kulaev, I. S., 1967, Detection of polyphosphate glucokinase in various microorganisms, *Biochemistry (USSR)* (English translation) **32**:408.

Tanzer, J. M., and Krichevsky, M. I., 1970, Polyphosphate formation by caries-conducive *Streptococcus* SL-1, *Biochim. Biophys. Acta* **215**:368.

Tempest, D. W., and Strange, R. E., 1966, Variation in content and distribution of magnesium and its influence on survival in *Aerobacter aerogenes* grown in a chemostat, *J. Gen. Microbiol.* **44**:273.

Uryson, S. O., and Kulaev, I. S., 1968, The presence of polyphosphate glucokinase in bacteria, *Dokl. Biol. Sci.* (English translation) **183**:697.

Wang, W. S., and Lundgren, D. G., 1969, Poly-β-hydroxybutyrate in the chemolithotrophic bacterium *Ferrobacillus ferrooxidans, J. Bacteriol.* **97**:947.

Wilkinson, D. H., and Duguid, J. F., 1960, The influence of cultural conditions on bacterial cytology, *Int. Rev. Cytol.* **9**:1.

Winder, F. G., and Denneny, J. M., 1957, The metabolism of inorganic polyphosphate in mycobacteria, *J. Gen. Microbiol.* **17**:573.

Wolk, C. P., 1973, Physiology and cytological chemistry of blue-green algae, *Bacteriol. Rev.* **37**:32.

Yokobayashi, K., Misaki, A., and Harada, T., 1970, Purification and properties of *Pseudomonas* isoamylase, *Biochim. Biophys. Acta* **212**:458.

Zevenhuizen, L. P. T. M., 1964, Branching enzyme of *Arthrobacter globiformis, Biochim. Biophys. Acta* **81**:608.

Zevenhuizen, L. P. T. M., 1966a, Formation and function of the glycogen-like polysaccharide of *Arthrobacter, Antonie van Leeuwenhoek; J. Microbiol. Serol.* **32**:356.

Zevenhuizen, L. P. T. M., 1966b, Function, structure and metabolism of the intracellular polysaccharide of *Arthrobacter, Meded. Landbouwhogesch. Wageningen,* 66–10.

PART III

Chemistry of Cellular Components

CHAPTER 11

Bacteriochlorophyll and Photosynthetic Evolution

D. Mauzerall

1. Introduction

The porphyrins are clearly the pigments of life. Selected by the fine comb of biological evolution, they have been structurally refined into the awesomely efficient apparatuses of photosynthesis and energy conversion. The understanding of these processes has been one of the major problems of modern science. The fraction of research effort put into these problems has until now been far from proportional to their importance. However, we are beginning not only to know the biochemistry and the physical mechanism of these processes, but also to understand their relationship to the physical and chemical properties of these adaptable molecules. It is the latter aspect of the problem that I will consider in this chapter. I will focus on the particular problem that arises in joining the structure and function of bacteriochlorophyll to its evolutionary position.

2. Properties

Bacteriochlorophyll (Bchl) is a waxy solid that, like chlorophyll (Chl), has not yet been coerced into forming a true three-dimensional crystal. Far less work has been done on Bchl than on the more prevalent Chl's of green plants. Properties of this pigment are to be found in summaries mainly concerned with Chl's: Rabinowitch (1945, 1951, 1956), Smith and Benitez (1955), Vernon and Seely (1966), and Gurinovich et

al. (1968). Like Chl, Bchl is usually isolated with at least 1 mol of a polar molecule, usually water, bound to the magnesium. The blue color of solutions of the pigment is caused by a rather weak band near 590 nm and the tail of the allowed transition at 760 nm. Spectral data for Bchl and bacteriopheophytin (Bph) are collected in Table 1. Analytical methods used in determining the pigment composition of bacterial reaction centers are detailed by Straley et al. (1973). Spectra of aggregated forms of Bchl were determined by Sauer et al. (1966) and Ballschmiter and Katz (1969). Because of relevance to bacterial photosynthesis, the one-electron oxidation–reduction species have been studied in detail. This work is reviewed by Norris and Katz in Chapter 21. The spectrum of the triplet state of Bchl is known (Pekharinen and Linschitz, 1960).

In the early literature, Bchl had the reputation of being extremely unstable. This instability is caused by the ready photo- and autooxidation of the pigment in dilute solution. This behavior is particularly evident in methanol. Concentrated solutions, particularly in clean diethyl ether, are very stable. The impurities and oxidative reactions consume a lesser fraction of the pigment in concentrated solution, and concentration quenching of excited states slows the photochemistry.

Bchl is rather resistant to loss of magnesium to form Bph. The order of increasing stability is porphyrin < chlorin < bacteriochlorin for the magnesium chelates. In ether, about 0.1 M methanesulfonic acid is required to convert Bchl to Bph rapidly and quantitatively.

Bph is much more stable to light than is Bchl, is far less soluble, and is more readily obtained in apparently microcrystalline form. In addition to the Bchl a referred

D. Mauzerall · The Rockefeller University, New York, New York 10021

Table 1. Spectroscopic Data for Bacteriochlorophyll and Bacteriopheophytin

Solvent	λ (nm)	$\varepsilon \times 10^{-4}$ ($M^{-1} \cdot cm^{-1}$)	λ (nm)	$\varepsilon \times 10^{-4}$ ($M^{-1} \cdot cm^{-1}$)	λ (nm)	$\varepsilon \times 10^{-4}$ ($M^{-1} \cdot cm^{-1}$)
			Bacteriochlorophyll			
Ether[a]	770	9.6	573	2.2	357	7.35
Ether[b]	769	9.35	574	2.0	357	7.1
Ether[c]	773	9.1	577	2.1	359	7.35
Acetone[a]	770	6.9	577	1.95	358	6.6
			Bacteriopheophytin			
Ether[c]	749	6.75	523	3.05	358	12.15
Ether[d]	750	6.3	528	3.5	358	14.25
Acetone[e]	745	4.75	522	2.55	356	9.7

[a] Sauer *et al.* (1966); [b] Holt and Jacobs (1954); [c] Smith and Benitez (1955); [d] Weigl (1952); [e] Straley *et al.* (1973). The ε values of references *d* and *e* are of secondary origin.

to above, several other porphyrin-type pigments are found in photosynthetic bacteria. Their spectra and properties were reported by Jensen *et al.* (1964). The structure of Bchl *b* is that of Bchl *a* (Fig. 1), with the ethyl group replaced by an ethylidene group (Scheer *et al.*, 1974). It is of interest because its *in vivo* absorption band near 1 μm is the longest wavelength for any photosynthetic pigment known. The green photosynthetic bacteria have, in addition to Bchl as the photoactive pigment and a small antenna, a large light-gathering antenna of chlorobium Chl's. These

absorb in solution near 650 and 660 nm, but appear to be mixtures of similar compounds (Holt, 1966).

The Bchl pigments are readily isolated by extraction of photosynthetic bacteria and chromatography on sugar. Detailed procedures were given by Strain and Svec (1966).

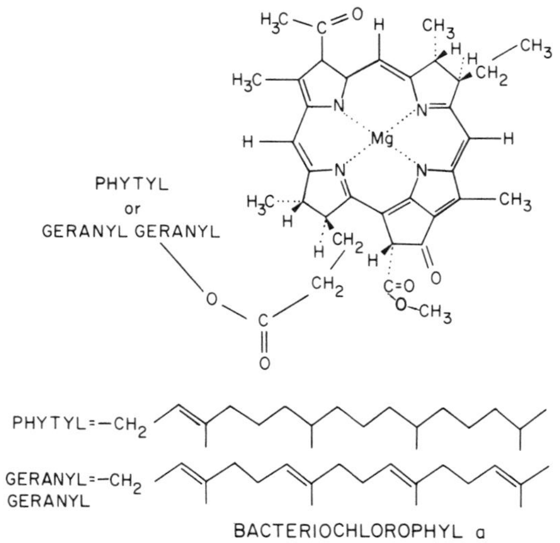

3. Structure and Function

3.1. Conjugated Macrocycle

The structure of Bchl is shown in Fig. 1. It is essentially a conjugated 20-membered macrocycle with four subcycles of five members containing a nitrogen atom. The basic ring plan is that of a porphyrin, which has a four-fold symmetry, rare in organic compounds. The size of the molecule is determined by the requirement that the photosynthetic pigment absorb solar radiation, with a quantum intensity peak near 600 nm. An exercise in quantum mechanics shows the conjugated molecule must be about 15 Å long, or 7 Å in diameter if circular. Since the circular molecules of the right size and shape gain aromatic stability, an effect also explained by quantum mechanics, the basics of the photosynthetic molecule are readily understood. The highly symmetrical porphyrin structure absorbs in the 500 to 600 nm region, but rather weakly. A simple disruption of the four-fold symmetry by reducing one double bond on the outer edge of a pyrrolic ring forms a chlorin, which has a much stronger absorption near 660 nm. One can repeat this trick on an opposite pyrrolic ring and obtain a

Fig. 1. Structure of Bchl *a*.

bacteriochlorin absorbing strongly at 770 nm. These molecules retain the aromatic and the necessary photochemical properties required for the photosynthetic pigment. Reducing an adjacent pyrrolic ring, or interrupting the macrocyclic conjugation by reduction on the meso bridges (phlorins, porphomethenes, and porphyrinogens, Mauzerall and Hong, 1975), leads to unstable molecules or absorption at shorter wavelengths, and thus has not been found useful in evolution.

3.2. Metal Chelate

The metal chelated by the macrocycle, magnesium, is the most electropositive or ionically bound metal that is stable in protonic solvents. These magnesium porphyrin chelates are formed in the laboratory only under anhydrous conditions. The mechanism used by the cell remains a mystery. One easily argues that transition metal chelates, e.g., iron or copper, are useless for photosynthesis, since the key photochemistry is quenched, very likely by internal transitions. Closed-shell, ionically bound metals favor electron donation in the excited state (see Section 4.1). Zinc porphyrins have the same photochemistry as magnesium porphyrins, and the zinc chelates form so easily that zinc (and copper) contaminate samples of free base porphyrins. If the excited states that mediate photochemistry do not involve spin conversion, then the selection of magnesium may have been favored by its effect in disfavoring intersystem crossing. And if the aim is the lowest oxidation potential (which allows the electron to be passed on to the acceptor at the highest reducing potential) combined with stability in water, then magnesium is the minimax solution (Fuhrhop and Mauzerall, 1969) to this problem.

3.3. Side Chains and Biosynthesis

The various side chains of Bchl (Fig. 1) are the residues of the simple biosynthetic path to Bchl (Fig. 2) (also see Chapters 40 and 42). The original ionic acetate and propionate residues are systematically changed to nonpolar residues. The early reactions (Fig. 2) are straightforward decarboxylations, followed by oxidative decarboxylations producing methyl and vinyl groups. One of the latter is reduced to ethyl as in Chl a, but the other (via hydration and oxidation) is converted to an acetyl group. Magnesium is inserted along the way. A complex series of reactions leads to the formation of the "fifth ring" containing the polar carbonyl and carbomethoxy groups and the reactive methylene group. The reduction of the adjacent pyrrolic ring

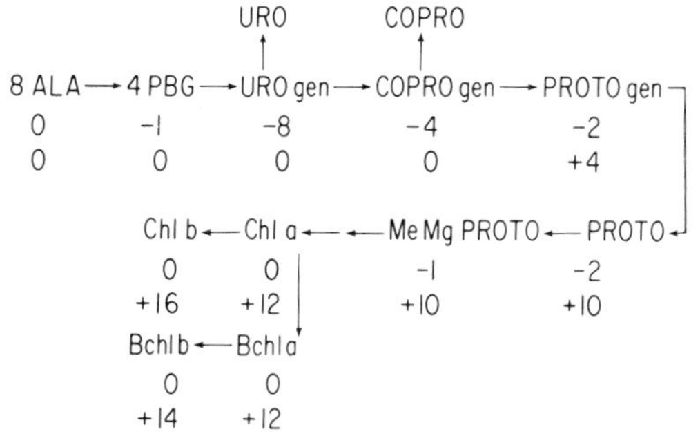

Fig. 2. Outline of the biosynthetic pathway of the Chl's (see Chapter 40 for chemical structures). The first number below each label is the ionic charge per molecule at pH 7; the second number is the equivalent of oxidation relative to the original substrates.

relieves steric crowding, and is possibly the main function of this rather mystifying part of the structure. In the case of Chl a, a variety of suggestions as to the function of the fifth ring have been made (Mauzerall, 1976a). Reduction of the opposite pyrrolic ring leads to the bacteriochlorin structure. Finally, the phytol, or geranylgeranyl (Katz et al., 1972), ester is made, and strongly reinforces the hydrophobic properties of the molecule.

The biosynthesis of Bchl is an excellent example of how an apparently highly complex molecule is assembled from very simple precursor molecules. The symmetric macrocycle is assembled from identical subunits, then the side chains are modified in an increasingly specific manner to lead to the final "complex" product. Our knowledge of the biosynthetic pathway of Bchl owes much to the pioneering work of June Lascelles (1964), and is reviewed in Chapters 40 and 42. The identity of the pathway from δ-aminolevulinic acid (ALA) through protoporphyrin in all cells is a striking example of the unity of nature. At this point, the "iron branch" leads to heme and the "magnesium branch" leads to the Chl's (Granick, 1949). It is believed that the path for Bchl follows that of Chl. This scheme provides an easy explanation of the presence of the "fifth ring", and hence the stepwise reduction of the macrocycle while retaining aromatic and photochemical properties.

4. Photochemistry

4.1. Electron Transfer

Details of the electron-transfer reactions (Mauzerall, 1977a–c) and the photoreactions in membranes (Mauzerall and Hong, 1975) have been reviewed. The Russian work was summarized by Krasnovsky (1958, 1965) and by Chibisov (1969). Our model studies have often used simple porphyrins in place of the Chl or Bchl to achieve quantitative analysis of the reactions. The more complex structures show qualitatively the same reactions. The reactive side chains, e.g., the active methylene in the fifth ring, lead to side reactions that simply complicate the analysis. A function of the protein structure surrounding the Chl or Bchl in reaction centers may well be to prevent these extraneous side reactions. The results of our studies can be briefly summarized: The free base porphyrins or pheophytins are readily reduced both chemically and photochemically (Mauzerall, 1962). In protonic solvents, the reduced free radicals disproportionate to porphyrins and phlorins (Mauzerall and Feher, 1964). The magnesium or zinc chelates are correspondingly easily oxidized (Fuhrhop and Mauzerall, 1969). Although the converse reactions, photooxidation of free base pigments (e.g., Kholmogorov et al., 1969) and photoreduction of the metalloderivatives (Krasnovsky, 1958), can be observed, they usually require chemical oxidants or reductants as partners. A detailed study of the ionic effects on the reaction of triplet zincuroporphyrin with various acceptors in anaerobic, aqueous solution showed several aspects relevant to the hypothesized prebiotic photochemistry (Carapellucci and Mauzerall, 1975). The excited state is a powerful reductant, reducing NAD to the one-electron level ($E_0 \approx -0.7$ V) as rapidly as it reduces ferricyanide ion ($E_0 \approx +0.4$ V). The electron transfer occurs over a long range, through about 10 Å of water separating the donor and acceptor. The reaction can be extremely efficient. For positively charged acceptors, a concentration of 10^{-6} M is sufficient to trap all the negatively charged zincuroporphyrin in the excited state. Extension of these photoreactions to a lipid bilayer system shows the theoretically predicted gains by structural organization of the reaction. Chl, Bchl, or a magnesium or zinc porphyrin is localized in the lipid bilayer, and an acceptor is kept across the interface in the aqueous layer (Hong and Mauzerall, 1974; Mauzerall and Hong, 1975). The vectorial electron transfer produces an electric field across the membrane that is used to monitor the reaction. The electron transfer from the excited state occurs in less than 10^{-7} sec (Hong and Mauzerall, 1976), and thus the reaction occurs in the presence of air. The lifetime of the singlet state is only about 10^{-9} sec, and that of the triplet in the presence of air about 10^{-7} sec. Thus, while the long-lived triplet state ($\approx 10^{-3}$ sec) allows efficient reactions anaerobically, the presence of air requires that the primary photoreactions in photosyn-

thesis be very fast. In fact, they are thought to occur in 10^{-11} sec (Zankel et al., 1968; Rockley et al., 1975; Kaufmann et al., 1975). Thus, these model photoreactions show some essential characteristics of the photosynthetic systems.

4.2. Reaction Centers

Most of our knowledge of the photochemistry of bacterial photosynthesis comes from detailed study of isolated reaction centers. This subject is reviewed in Chapters 19 and 20 and also by Parson and Cogdell (1975). Our analysis of the reaction center from *Rhodopseudomonas sphaeroides* R26 showed that it contains no less than four molecules of Bchl and two molecules of Bph (Straley et al., 1973). The electron-donor unit is a dimer of Bchl (Chapters 19 and 21). The presence of Bph suggests that it is part of the mechanism to efficiently separate charge in the primary act. As I have discussed elsewhere (Mauzerall 1976a, 1977a–c; Mauzerall and Hong, 1975), the mechanism of electron tunneling predicts a sharp maximum in the yield-lifetime product of the charge transfer state vs. distance of separation and orientation of the molecules. Too distant, and the electron does not transfer within the lifetime of the excited state. Too close, and although the electron transfer is extremely rapid, the back-transfer to the ground state becomes faster than the reciprocal lifetime of the excited state. This latter case is called *quenching*, and is the more usual occurrence for reactions in solution. Our reactions with ionically charged molecules of similar sign (Carapellucci and Mauzerall, 1975), or across the lipid water interface (Hong and Mauzerall, 1974), are efficient just because too-close approach is prevented. Evolutionary selection has optimized these distances and orientations in the photosynthetic reaction center. A method to obtain the most distant transfer and thus the longest lifetime of the charge transfer state would be to have the tunneling barrier contain a series of (virtual) states of decreasing energy so as to favor forward transfer. The Bph may fulfill that function, because of the favoring of electron donation by the metallopigment, Bchl, and acceptance by the free base, Bph. It is also possible that the presence of Bph represents a residuum of the postulated primitive photosynthesis wherein the free base pigment oxidized an organic donor, and was itself reduced (see Section 5). It is interesting that in a study of the "greening" of *Rp. sphaeroides*, Cellarius and Peters (1969) observed that a small amount ($\approx 1\%$ of Bchl) of Bph was formed very early in the biosynthesis. This could be the Bph that is built into the reaction center. Some evidence for the involvement of

Bph was observed in picosecond light-pulse experiments on reaction centers (Rockley *et al.*, 1975; Kaufmann *et al.*, 1975). However, the exciting photons are absorbed by a band attributable to Bph, and multiple excitations are highly likely in these ultrafast experiments. My studies on multiple excitations in *Chlorella* showed that these can have large effects on the photosystem (Mauzerall, 1976*b*). Experiments in progress to determine the role of pigment neighbors on these reactions have already shown that electron transfer is highly efficient between excited states of metalloporphyrins (Ballard and Mauzerall, in prep.; Mauzerall, 1977*a*).

5. Biosynthesis and Evolution

5.1. Evolution of Photosynthesis

The origins of photosynthesis have been widely speculated on (Buvet and Ponnamperuma, 1971; Gaffron, 1965; Oparin *et al.*, 1959; Krasnovsky, 1974; Olson, 1970). My approach follows that of Granick (1965), who stressed the prebiotic photochemistry, and the likelihood that photosynthesis was an early rather than a later component of living processes. Krasnovsky (1974) also raised the latter possibility. Granick (1957) even proposed a model using FeS mineral that predated the discovery of bacterial ferredoxin. My stress will be on the photochemical aspects of the evolution of photosynthesis. Since the basic concept has been stated in the context of Chl photosynthesis (Mauzerall, 1973, 1976*a*, 1977*b*), I will present only a brief outline here. Since life originated when the earth's atmosphere was still chemically reducing, the useful photosynthetic reaction would have been the photooxidation of the prevalent reduced organic compounds and the emission of hydrogen. The oxidized organic compounds would be reactive intermediates for biogenesis. Life also originated in the ocean. The first porphyrin on the biosynthetic pathway is uroporphyrin, essentially a colored, carboxylated octopus. In dilute, anaerobic, neutral, aqueous solution, this free base porphyrin readily photooxidizes many organic compounds. Since isomer III of uroporphyrin is the most favored isomer in a random formation, the occurrence of this isomer in biological systems is reasonable (Mauzerall, 1960*a*, *b*). On chelation of a magnesium (or zinc), the favored photoreaction is now the reduction of an acceptor and the oxidation of the pigment. Under reducing conditions, this electron transfer could only be used (as could the previous reaction) to generate ATP in a cyclic reaction. However, once the pigment oxidation

was coupled to the ultimate acceptor, water, oxygen was formed, and the age of modern photosynthesis began. It is the very large gradient of free energy between oxygen and reduced organic compounds (foods or fuels) that allows complex forms of life to exist. The evolutionary advantage of this large gradient highly favored the selection of modern photosynthesis over more passive forms of utilization of light, e.g., formation of ATP by cyclic photoreactions or by the proton pump of bacteriorhodopsin (Chapter 29).

5.2. Bacteriochlorophyll

With this view of photosynthetic evolution in mind, we can pose the problem of Bchl more clearly: did it follow or did it precede Chl in photosynthetic function? Strict adherence to the biosynthetic pathway indicates that it followed Chl, while the usual statements on evolution are that it preceded Chl. Since the latter view is so widely promulgated, I will argue for the more novel former view. Clearly, no strictly logical decision can be made in favor of either view. The novel view, however, correlates chemistry and function, and could possibly lead to an intriguing rearrangement of views on photosynthetic evolution. Let us begin with some *a priori* chemical arguments. The complexity and specificity of the chemical structures of Chl and Bchl strongly argue that they did not arise by prebiotic chemical reactions. Thus, the biosynthetic pathway is very unlikely to have been built backward. In fact, one can pinpoint the position along the biosynthetic chain of these pigments at which simple chemistry stops. It is between copro- and protoporphyrin(ogen). It is at this point that specificity arises: two specific propionic residues are attacked. Moreover, the chemistry is an oxidative decarboxylation to more reactive vinyl groups, quite different from the condensations and simple decarboxylations to more stable species characteristic of all the previous steps in the pathway. The next step, esterification of a specific propionate residue, and the chelation of magnesium are hypothesized (see Section 3.3) to be a turning point in photochemical function, leading finally to the momentous oxidation of water. The formation of the five-membered ring of protochlorophyll is also a strongly oxidizing reaction: six reducing equivalents are lost. It is often stated that Bchl is more reduced that Chl: a tetrahydro- vs. a dihydroporphyrin. In fact, if we consider the total molecules, Chl *a* and Bchl *a* are at exactly the same stoichiometric redox level! The two extra hydrogens on Bchl are balanced by a 2-equivalent oxidation of the (hydrated) vinyl group to an acetyl group in position 2. Thus, the argument that Bchl is more reduced, and therefore more primitive, is on shaky grounds. In fact, the formation of vinyl groups and the fifth ring have cost 10 equivalents of oxidation (Fig. 2), whereas the reduction reactions to Chl *a* (2-ring H and vinyl to ethyl) have returned only 4 reducing equivalents. Thus, the Chl *a* and Bchl *a* structures are 6 equivalents oxidized with respect to URO or COPRO. The oxidation of the methyl group at position 3 to a formyl group in Chl *b* makes it more oxidized by 4 equivalents than Chl *a*. Bchl *b* (Scheer *et al.*, 1974) is 2 equivalents more oxidized than Bchl *a*. Thus, there appears to be an overall trend in the biosynthetic path toward molecules at a higher stoichiometric level of oxidation. I have based the previous discussion on the porphyrins, whereas the biosynthetic intermediates are the porphyrinogens, which are 6 equivalents more reduced than the porphyrins. It is quite possible that the intermediate tetra- (porphomethenes) and dihydro- (phlorin) meso-reduced porphyrins could equilibrate with the ring-reduced species (chlorins and bacteriochlorins). Disproportionation among the meso-reduced species was observed when phlorins were first made photochemically (Mauzerall, 1962). In fact, evidence exists for transformations of simple phlorins to chlorins and vice versa (Whitlock and Oester, 1973) Note that this reinforces the point in Section 5.1 that these early photopigments would be in a highly (stoichiometrically) reduced state compared with modern Chl and Bchl. The recent observation of a bacteriochlorin uroheme as the prosthetic group of a sulfite reductase in a primitive sulfur bacterium (Murphy *et al.*, 1973) gives further support to this view. The structure of the ubiquitous vitamin B_{12}, essentially the cobalt chelate of an alkylated uroporphyrinogen, also suggests the prevalence of these compounds in early biogenesis. Although the actual redox potential for each of the steps discussed above would enter into a strictly quantitative analysis of the problem, all the potentials for the reactions considered above are near one another (slightly reducing in E_0), and thus can be neglected at the present level of discussion.

It is usually assumed that the photosynthetic bacteria preceded the cyanophytes (blue-green algae) in evolution (Brock, 1973; Schopf, 1974). However, reasoning similar to that presented above led Olson (1970) to suggest a common precursor utilizing a pigment similar to Chl *a* for both the photosynthetic bacteria and the cyanophytes. I would go a step further and suggest that the modern photosynthetic bacteria arose from a simplification of a primitive unicellular cyanophyte. Our stress on the very early origins of photosynthetic reactions relieves us from

exclusive reliance on photoheterotrophic reactions at early times. The similarity of photosystem I of cyanophytes to bacterial photosynthesis is as easily explained in this way as in the reverse direction as outlined by Olson (1970). The secondary origin of the Bchl photosynthesis would explain the evolutionary driving force toward an ecological niche (wavelength) not occupied by cyanophytes. As Olson (1970) pointed out, the lowered redox potential of Bchl could prevent the development of an oxygen system. The joining together of a hydrogen photosynthetic system with an oxygen system outlined in Section 5.1 would explain the system I–system II arrangements of the cyanophytes.

The base-pair fractions in the nucleic acids of photosynthetic bacteria overlap with those of unicellular cyanophytes of type 1-A. Three strains of *Chlorobium* have G + C content of 55 ± 3%, while the common photosynthetic bacteria have G + C content of 66 ± 5% (Normore, 1973). The G + C contents of the unicellular cyanophytes of type 1-A have two distribution ranges: 53 ± 5% and 68 ± 3% (Schiff, 1973). These data are consistent with the assumed close relationship of these organisms, and could suggest an evolutionary diversion of the cyanophytes into these bacteria.

The green *Chlorobium* bacteria are an interesting case. In these organisms, a large antenna of chlorobium Chl is connected to a funnel of Bchl protein (Olson, 1971; Olson, Chapter 8), and this in turn to a Bchl reaction center. It is interesting to compare this antenna Chl with the phycobilins, already present in primitive cyanophytes. The latter are formed by oxidative opening of a macrocyclic intermediate (at the protoporphyrin level; see Chapter 40) along the biosynthetic chain to Chl. The *Chlorobium* have adapted an intermediate (Chl) along the path to Bchl for the antenna function. The structural changes in the known chlorobium Chl's (Mathewson *et al.*, 1963; Holt, 1966) can be constructed from Chl *a*: hydration of the vinyl group, decarbomethoxylation of the "fifth ring", and alkylation at the meso bridges. These changes were very likely a response to the Bchl reaction center and funnel, which require a lower-energy antenna than that provided by the phycobilins for efficient energy transfer. As with the phycobilins, a variety of molecular species are found. Whereas the detailed molecular structures may be critical for the photochemical transformation, they are not unique for the simpler function of collecting photons and transmitting excitation.

The amino acid sequences of the ferredoxins and the restricted amino acid composition indicate that they are all elaborations of a common precursor, with those of *Clostridium* being the surviving relic (Hall *et al.*, 1973). The size of the ferredoxins seems to increase with increasing phylogenetic position: *Clostridium* has 55, *Chlorobium* 60, *Rhodospirillum* 66–75, *Chromatium* 81, and spinach 97 residues. It is interesting that the ferredoxin of a *Chlorobium* (Tanaka *et al.*, 1974) is much closer in size and sequence to that of the *Clostridium*. This may indicate the early origin of the green photosynthetic bacteria. It would be most interesting to have data on ferredoxin from a unicellular cyanophyte.

Another possible way to differentiate these various hypotheses is the origin of aminolevulinic acid required for the biosynthesis of all the porphyrin pigments. The photosynthetic bacteria use the glycine–succinate path, as do mammals (Chapter 40). However, the higher plants (Beale and Castelfranco, 1974) and algae (Beale, personal communication) use a different precursor, probably glutamic acid. It will be most interesting to see which path is used by the unicellular cyanophytes. If it is the glycine–succinate path, the intimate relationship of the cyanophytes and the photosynthetic bacteria will be strengthened, while if it is the higher plant path, the argument for a common, but remote, origin will be more likely.

In summary, we can reach no clear decision about the sequence of Chl and Bchl in biosynthesis and photosynthetic evolution. However, the detailed knowledge now becoming available promises some clarification. We can already relate the structure of these pigments to their function in photosynthesis in a reasonably detailed manner. One of the great problems of modern science is in the process of being solved.

6. Note Added in Proof

I wish to dedicate this work to the memory of Dr. S. Granick, who died in April, 1977, and to whom I owe so much.

Since this chapter was written several relevant articles have appeared. The encyclopedic work *The Porphyrins* (D. Dolphin, ed., Academic Press, New York) will be published soon and will contain much relevant work on the properties of Bchl. The chapter by W. A. Svec on preparation and estimation of chlorophylls will contain the newest optical data of the Argonne group on these compounds. Further evidence has accumulated for the intermediate Bph^- in the primary electron-transfer step of reaction centers (J. Fajer *et al.*, 1975, *Proc. Natl. Acad. Sci. USA* **72**:4956–4960; P. L. Dutton *et al.*, 1977, in: *Brookhaven Symposia in Biology*, No. 28, pp. 213–

237; R. C. Prince *et al.*, 1977, *Biochim. Biophys. Acta* **462**:467). Evidence has been found for the presence of cyanophyta fossils in rocks $\approx 3.5 \times 10^9$ years old (A. H. Knoll and E. S. Barghoorn, 1977, *Science* **198**:396). A view of the evolution of Bchl similar in general to that presented in this chapter is to be found in E. Broda (1975, *The Evolution of the Bioenergetic Process*, Pergammon Press, Oxford).

Acknowledgments

The origin of the concepts put forth in this article is to be found in Dr. Granick's writings and in his continually probing questions, from which I have greatly benefited. This manuscript was written at the Marine Biology Laboratory, Woods Hole, while I was an instructor in the Experimental Marine Botany Program. Many of the participants furnished information and criticisms for this paper, and I am indebted to them and to the director, J. Schiff. The financial help of the NSF in the program of Experimental Marine Botany and Grant BMS 74-11747 is also gratefully acknowledged.

7. References

Ballard, G., 1977, Ph.D. thesis, The Rockefeller University, New York.

Ballschmiter, K., and Katz, J. J., 1969, An infrared study of chlorophyll–chlorophyll and chlorophyll–water interactions, *J. Am. Chem. Soc.* **91**:2661.

Beale, S. I., and Castelfranco, P. A., 1974, The biosynthesis of δ-aminolevulinic acid in higher plants, II. Formation of ^{14}C-δ-aminolevulinic acid from labeled precursors in greening plant tissues, *Plant Physiol.* **53**:297.

Brock, T. D., 1973, Evolutionary and ecological aspects of the cyanophytes, in: *The Biology of Blue-Green Algae* (N. G. Carr and B. A. Whitton, eds.), pp. 487–500, University of California Press, Berkeley.

Buvet, R., and Ponnamperuma, C., (eds.), 1971, *Chemical Evolution and the Origin of Life*, Vol. I, *Molecular Evolution*, Elsevier, New York.

Carapellucci, P. A., and Mauzerall, D., 1975, Photosynthesis and porphyrin excited state redox reactions, *Ann. N.Y. Acad. Sci.* **244**:214.

Cellarius, R. A., and Peters, G. A., 1969, Photosynthetic membrane development in *Rhodopseudomonas sphaeroides*: Incorporation of bacteriochlorophyll and development of energy transfer and photochemical activity, *Biochim. Biophys. Acta* **189**:234–249.

Chibisov, A. K., 1969, A flash photolysis study of intermediates in photochemical reactions of chlorophyll, *Photochem. Photobiol.* **10**:331.

Fuhrhop, J. H., and Mauzerall, D., 1969, The one-electron oxidation of metalloporphyrins, *J. Am. Chem. Soc.* **91**:4174.

Gaffron, H., 1965, The role of light in evolution: The transition from a one quantum to a two quanta mechanism, in: *The Origins of Prebiological Systems* (S. W., Fox, ed.), pp. 437–460, Academic Press, New York.

Granick, S., 1949, The structural and functional relationship between heme and chlorophyll, in: *The Harvey Lectures*, pp. 220–245, Charles C. Thomas, Springfield, Illinois.

Granick, S., 1957, Speculations on the origins and evolution of photosynthesis, *Ann. N. Y. Acad. Sci.* **69**:292.

Granick, S., 1965, Evolution of heme and chlorophyll, in: *Evolving Genes and Proteins* (V. Bryson and H. J. Vogel, eds.), pp. 67–88, Academic Press, New York.

Gurinovich, G. P., Sevchenko, A. N., and Solv'ev, K. N., 1968, Spectroscopy of chlorophyll and related compounds, Atomic Energy Commission Translation 7199, Chemistry, Technical Information Division, 4500.

Hall, D.O., Camnack, and Rao, K. K., 1973, The plant ferredoxins and their relationship to the evolution of ferredoxins from primitive life, *Pure Appl. Chem.* **34**:553.

Holt, A. S., 1966, Recently characterized chlorophylls, in: *The Chlorophylls* (L. P. Vernon and G. R. Seely, eds.), pp. 111–118 Academic Press, New York.

Holt, A. S., and Jacobs, E. E., 1954, Spectroscopy of plant pigments. II. Methyl bacteriochlorophillide and bacteriochlorophyll, *Am. J. Bot.* **41**:718.

Hong, F. T., and Mauzerall, D., 1974, Interfacial photoreactions and chemical capacitance in lipid bilayers, *Proc. Natl. Acad. Sci. U.S.A.* **71**:1564.

Hong, F. T., and Mauzerall, D., 1976, Tunable voltage camp method—application to photoelectrical effects in pigmented bilayer lipid membranes, *J. Electrochem. Soc.* **123**:1317.

Jensen, A., Aasmusdrud, O., and Eimhjellen, K. E., 1964, Chlorophylls of photosynthetic bacteria, *Biochim. Biophys. Acta* **88**:466.

Katz, J. J., Strain, H. H., Harkness, A. L. Studier, M. H., Svec, W. A., Janson, T. R., and Cope, B. T., 1972, Esterifying alcohols in the chlorophylls of purple photosynthetic bacteria. A new chlorophyll, bacteriochlorophyll (gg), all-*trans* geranylgeranyl bacteriochlorophyllide a, *J. Am. Chem. Soc.* **94**:7938.

Kaufmann, K. J., Dutton, P. L., Netzel, T. L., Leigh, J. S., and Rentzepis, P. M., 1975, Picosecond kinetics of events leading to reaction center bacteriochlorophyll oxidation, *Science* **188**:1301.

Kholmogorov, V. E., Savel'ev, D. A., and Sidorov, A. N., 1969, Phototransformations of porphyrins in frozen solution, *Biofizika* **14**:414; *Chem. Abstr.* **71**:45995k.

Krasnovsky, A. A., 1958, Reduction photochimique reversible de la chlorophylle et de ses analogues et mécanisme de la photosensibilisation, *J. Chim. Phys.* **55**:968.

Krasnovsky, A. A., 1965, Photochemistry and spectroscopy of chlorophyll, bacteriochlorophyll and bacterioviridin in model systems and photosynthesizing organisms, *Photochem. Photobiol.* **4**:641.

Krasnovsky, A. A., 1974, Chemical evolution of photosynthesis: Models and hypotheses, in: *The Origin of Life and Evolutionary Biochemistry* (K. Dose, S. W. Fox,

G. A. Deborin, and T. E. Povlovskaya, eds.), pp. 233–244, Plenum Press, New York.

Lascelles, J. 1964, *Tetrapyrrole Biosynthesis and Its Regulation*, Benjamin, New York.

Mathewson, J. W., Richards, W. R., and Rapoport, H., 1963, Chlorobium chlorophylls, Nuclear magnetic resonance studies on a chlorobium pheophorbide 660–650, *J. Am. Chem. Soc.* **85**:364.

Mauzerall, D., 1960a, The thermodynamic stability of porphyrinogens, *J. Am. Chem. Soc.* **82**:2601.

Mauzerall, D., 1960b, The condensation of porphobilonogen to uroporphyrinogen, *J. Am. Chem. Soc.* **82**:2605.

Mauzerall, D., 1962, The photoreduction of porphyrins: Structure of the products, *J. Am. Chem. Soc.* **84**:2437.

Mauzerall, D., 1973, Why chlorophyll?, *Ann. N. Y. Acad. Sci.* **206**:483.

Mauzerall, D., 1976a, Chlorophyll and photosynthesis, *Philos. Trans. R. Soc. London Ser. B*: **273**:287.

Mauzerall, D., 1976b, Fluorescence and multiple excitation in photosynthetic systems, *J. Phys. Chem.* **80**:2306.

Mauzerall, D., 1977a, Electron transfer reactions and photoexcited porphyrins, *Brookhaven Symp. Biol.*, No. 28, pp. 64–73.

Mauzerall, D., 1977b, Photoredox reactions of porphyrins and the origins of photosynthesis, in: *Bioorganic Chemistry* (E. Van Tamelen, ed.), Vol. IV, pp. 303–314, Academic Press, New York.

Mauzerall, D., 1977c, Electron transfer photoreactions of porphyrins, in: *The Porphyrins* (D. Dolphin, ed.), Vol. 5, Chapt. 2, pp. 44–80, Academic Press, New York.

Mauzerall, D., and Feher, G., 1964, Optical absorption of the porphyrin free radical formed in a reversible photochemical reaction, *Biochim. Biophys. Acta* **88**:658.

Mauzerall, D., and Hong, F. T., 1975, Photochemistry of porphyrins in membranes and photosynthesis, in: *Porphyrins and Metalloporphyrins* (K. M. Smith, ed.), Chapt. 17, pp. 701–725, Elsevier, Amsterdam.

Murphy, M. J., Siegel, L. M., Kamin, H., and Rosenthal, D., 1973, Reduced nicotinamide adenine dinucleotide phosphate–sulfite reductase of Enterobacteria. II. Identification of a new class of heme prosthetic group: an iron-tetrahydroporphyrin (isobacteriochlorin type) with eight carboxylic acid groups, *J. Biol. Chem.* **248**:2801.

Normore, W. M., 1973, Guanine plus cytosine (GC) composition of the DNA of bacteria, fungi, algae and protozoa, in: *Handbook of Microbiology* (A. I. Laskin and H. W. Lechevalier, eds.), pp. 587–691, CRC Press, Cleveland.

Olson, J. M., 1970, The evolution of photosynthesis, *Science* **168**:438.

Olson, J. M., 1971, Bacteriochlorophyll protein of green photosynthetic bacteria, in: *Methods in Enzymology*, Vol. XXIII, *Photosynthesis, Part A* (A. San Pietro, ed.), pp. 636–644, Academic Press, New York.

Oparin, A. I., Pasynaskii, A. G., Braunstein, A. E., and Pavoloskaya, T. E. (eds.), 1959, *The Origin of Life on the Earth*, Pergamon Press, New York.

Parson, W. W., and Cogdell, R. J., 1975, The primary photochemical reaction of bacterial photosynthesis, *Biochim. Biophys. Acta* **416**:105.

Pekharinen, L., and Linschitz, H. 1960, Studies on metastable states of porphyrins. II. Spectra and decay kinetics of tetraphenylporphine, zinc tetraphenylporphine and bacteriochlorophyll, *J. Am. Chem. Soc.* **82**:2407.

Rabinowitch, E. I., 1945, *Chemistry of Photosynthesis, Chemosynthesis and Related Processes in Vitro and in Vivo*, Vol. I of *Photosynthesis*, Interscience, New York.

Rabinowitch, E. I., 1951, *Spectroscopy and Fluorescence of Photosynthetic Pigments; Kinetics of Photosynthesis*, Vol. II, Part 1, of *Photosynthesis*, Interscience, New York.

Rabinowitch, E. I., 1956, *Kinetics of Photosynthesis* (continued); addenda to Vol. I and Vol. II, part 1, Vol. II, Part 2, of *Photosynthesis*, Interscience, New York.

Rockley, M. G., Windsor, M. W., Cogdell, R. J., and Parson, W. W., 1975, Picosecond detection of an intermediate in the photochemical reaction of bacterial photosynthesis, *Proc. Natl. Acad. Sci. U.S.A.* **72**:2251.

Sauer, K., Smith, J. R. L., and Schutz, A. J., 1966, The dimerization of chlorophyll a, chlorophyll b and bacteriochlorophyll in solution, *J. Am. Chem. Soc.* **88**:2681.

Scheer, H., Svec, W. A., Cope, B. T., Studer, M. H., Scott, R. G., and Katz, J. J., 1974, Structure of bacteriochlorophyll b, *J. Am. Chem. Soc.* **96**:3714.

Schiff, J. A., 1973, The development, inheritance, and origin of the plastid in *Euglena, Adv. Morphorg.* **10**:265.

Schopf, J. W., 1974, Paleobiology of the precambrian: The age of blue-green algae, *Evol. Biol.* **7**:1.

Smith, J. H. C., and Benitez, A., 1955, Chlorophylls: Analysis in plant material, in: *Modern Methods of Plant Analysis* (K. Paech and M. V. Tracey, eds.), Vol. IV, p. 142, Springer-Verlag, Berlin.

Strain, H. H., and Svec, W. A., 1966, Extraction, separation, estimation, and isolation of the chlorophylls, in: *The Chlorophylls* (L. P. Vernon and G. R. Seely, eds.), pp. 21–66, Academic Press, New York.

Straley, S. C., Parson, W. W., Mauzerall, D., and Clayton, R. K., 1973, Pigment content and molar extinction coefficients of photochemical reaction centers from *Rhodopseudomonas sphaeroides, Biochim. Biophys. Acta* **305**:597.

Tanaka, M., Haniu, M., Yasunobu, K. T., Evans, M. C., and Wand Rao, K. K., 1974, Amino acid sequence of ferredoxin from a photosynthetic green bacterium, *Chlorobium limicola, Biochemistry* **13**:2953.

Vernon, L. P., and Seely, G. R. (eds.), 1966, *The Chlorophylls*, Academic Press, New York.

Weigl, J. W., 1952, Concerning the absorption spectrum of bacteriochlorophyll, *J. Am. Chem. Soc.* **74**:999.

Whitlock, H. W., and Oester, M. Y., 1973, Behaviour of di- and tetrahydroporphyrins under alkaline conditions. Direct observation of the chlorin–phlorin equilibrium, *J. Am. Chem. Soc.* **95**:5738.

Zankel, K. L., Reed, D. W., and Clayton, R. K., 1968, Fluorescence and photochemical quenching in photosynthetic reaction centers, *Proc. Natl. Acad. Sci. U.S.A.* **61**:1243.

CHAPTER 12

Chemistry of Carotenoid Pigments

Synnøve Liaaen-Jensen

1. Introduction

The carotenoids of photosynthetic bacteria have been extensively studied since van Niel's and Karrer's pioneering work in the 1930's. Contributions from Goodwin's school in the 1950's and onward stimulated further interest in these pigments. The structures of the carotenoids of photosynthetic bacteria were further studied in Trondheim from the late 1950's. Total synthesis has been performed by Isler's school, Weedon's school, Surmatis and co-workers, and in Trondheim. The spectroscopic properties of these carotenoids have been systematically studied throughout, and this work was extended to ^1H NMR and mass spectroscopy in the 1960's.

The last review restricted to carotenoids of photosynthetic bacteria (Liaaen-Jensen, 1963a) is now out-of-date. Later, more extensive reviews (Liaaen-Jensen, 1965; Liaaen-Jensen and Andrewes, 1972; Goodwin, 1973) have not treated carotenoids of photosynthetic bacteria exhaustively. In the comprehensive monograph by Isler (1971), all chemical evidence available until 1970 is referred to, although not collected under the heading carotenoids of photosynthetic bacteria.

This chapter aims at a compilation and evaluation of the subject (literature survey finished June, 1975), in part to be useful in itself and in part to serve as a guide to the comprehensive collection of data in Isler's

book. By use of cross-references to the valuable compilation by Straub (1971) in Isler's book, references to the original literature prior to 1970 have been greatly reduced.

2. Structures

2.1. Definition and Nomenclature

According to the recently approved rules of carotenoid nomenclature (IUPAC and IUB, 1975), the definition of carotenoids allows more structural variation than previously: Carotenoids are still considered a class of hydrocarbons (carotenes) and their oxygenated derivatives (xanthophylls) consisting of eight isoprenoid units joined in such a manner that the arrangement of isoprenoid units is reversed at the center of the molecule so that the two central methyl groups are in a 1,6-positional relationship and the remaining nonterminal methyl groups in a 1,5-positional relationship. All carotenoids may be formally derived from this basic structure of lycopene (Fig. 1, 67*)—with its central chain of 11 conjugated double bonds—by (1) hydrogenation, (2) dehydrogenation, (3) cyclization, or (4) oxidation, or any combination of these processes. In addition, compounds that arise from certain rearrangements of the skeleton of 67 or by the formal removal of part of this structure are now also classified as carotenoids.

Many carotenoids have trivial names. All

Synnøve Liaaen-Jensen · Organic Chemistry Laboratories, Norwegian Institute of Technology, University of Trondheim, Trondheim-NTH, Norway

*All compounds are numbered in accordance with Table 1.

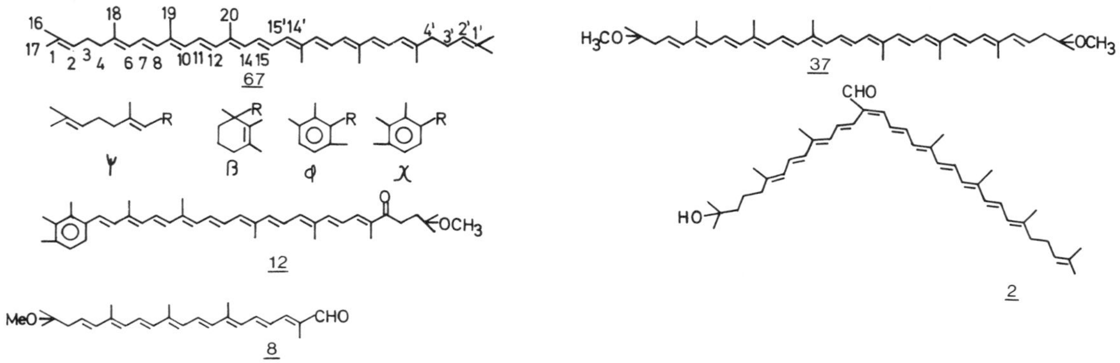

Fig. 1. Illustrations of carotenoid nomenclature.

carotenoids may now be designated by semisystematic names (IUPAC and IUB, 1975) based on the stem name carotene and a double prefix describing the structure of both C_9 end groups by Greek letters. Those of concern for this review are given in Fig. 1. Functional groups present, changes in hydrogenation level, and skeletal modifications are given in a systematic manner, exemplified by spirilloxanthin (*37*; 1,1'-dimethoxy-3,4,3',4'-tetradehydro-1,2,1',2'-tetrahydro-ψ,ψ-carotene), rhodopinal (*2*; 13-*cis*-1-hydroxy-1,2-dihydro-ψ,ψ-caroten-20-al), okenone (*12*; 1'-methoxy-1',2'-dihydro-χ,ψ-catoten-4'-one) and *Thiothece*-425 (*8*; 1-methoxy-1,2-dihydro-12'-apo-ψ,ψ-caroten-12'-al). The use of primed numbers in the "right" half of the structure is illustrated by structure *67* in Fig. 1. The prefix "apo" denotes a formal cleavage of the C_{40}-carbon skeleton (see Fig. 1, structure *8*).

It is recommended that full semisystematic names be given at first mention in a paper, and that introduction of new trivial names be restricted. Simple derivatives of known carotenoids may be named by modification of existing trivial names (e.g., 2,2'-diketosprilloxanthin; *20*). The use of preliminary designations is discussed below.

2.2. Carotenoids of Photosynthetic Bacteria

Of the approximately 350 different carotenoids of natural occurrence for which structures have been suggested (Liaaen-Jensen, 1976), some 78 are synthesized by photosynthetic bacteria. Thus, in the compilation by Straub (1971), 58 such carotenoids were included, plus some that have subsequently been abandoned or revised. Later, 23 additional carotenoids were reported.

These carotenoids are compiled in an abbreviated manner in Table 1 in groups of (1) aldehydes, (2) ketones and esters, (3) methyl ethers, (4) alcohols and glycosides, and (5) hydrocarbons (= carotenes).

Carotenoids containing more than one functional group are listed only in the group of highest priority. Trivial and rational names are included. The full structures may be drawn by means of the abbreviated formulas referring to the end groups and central C_{22}-units depicted in Fig. 2.

The formula numbers in parentheses in Table 1 refer to the numbers used in the compilation by Straub (1971). Through the Straub numbers in the monograph by Isler (1971), rational names and key references to chemical and spectroscopic data for these compounds are readily available. For the more recently reported carotenoids not included in Isler (1971), references to the original literature are included as footnotes *b–e* to Table 1.

2.3. Characteristic Structural Features

Carotenoids of photosynthetic bacteria have rather simple structures. They are generally aliphatic, but some have aromatic or β-end groups. They differ markedly in chromophore length, but chemically have close structural relationships, consistent with biosynthetic pathways involving simple step-reactions (see Chapter 39).

Characteristic functional groups are tertiary hydroxy, glucosyloxy, or methoxy groups (in 1,1'-positions), keto groups (in 2,2',4- or 4'-positions), and cross-conjugated aldehyde or hydroxyl groups in the 20-position, as well as 1,2,3-trimethylphenyl and 1,2,5-trimethylphenyl end groups. Methoxylated carotenoids are not encountered outside photosynthetic bacteria. The same is true for C_{20}-substituted carotenoids.

Structural elements such as allenic or acetylenic bonds, epoxides or furanoxides, nor-carotenoids (carotenoids in which carbon atoms have been eliminated from the skeleton), or C_{45}- or C_{50}-carotenoids are not encountered in photosynthetic bacteria. Only a few

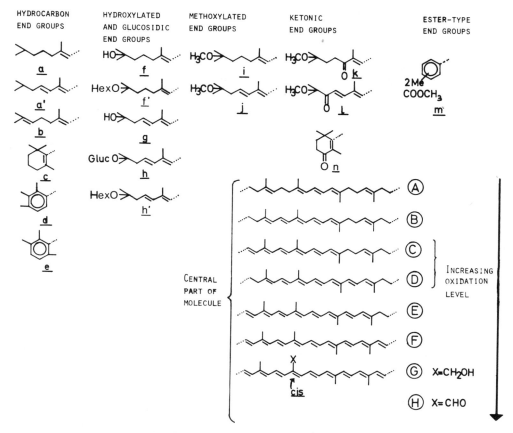

Fig. 2. Central moiety (A–H) and end groups (*a–n*) encountered in carotenoids of photosynthetic bacteria.

Table 1. Carotenoids of Photosynthetic Bacteria[a]

1. Aldehydes

1 Lycopenal; 13-*cis*-ψ,ψ-caroten-20-al; *b*-H-*b*; *143*

2 Rhodopinal; 13-*cis*-1-hydroxy-1,2-dihydro-ψ,ψ-caroten-20-al; *f*-H-*b*; *144*

3 Rhodopinal-D-glucoside; 13-*cis*-1-(β-D-glucopyranosyloxy)-1,2-dihydro-ψ,ψ-caroten-20-al; *h*-H-*b*; *145*

4 Anhydrorhodovibrinal; 12-*cis*-1-methoxy-3,4-didehydro-1,2-dihydro-ψ,ψ-caroten-20-al; *j*-H-*b*; *145a*

5 Methoxylycopenal; 13-*cis*-1-methoxy-1,2-dihydro--ψ,ψ-caroten-20-al; *i*-H-*b*; *146*

6 Tetrahydrospirilloxanthinal; 13-*cis*-1,1dimethoxy-1,2,1′,2′-tetrahydro-ψ,ψ-caroten-20-al; *i*-H-*i*; *147*

7 *Thiothece*-460[b]; 1-methoxy-4-oxo-1,2-dihydro-8′-apo-ψ,ψ-caroten-8-al; *k*-F-CHO

8 *Thiothece*-425[b]; 1-methoxy-1,2-dihydro-12′-apo-ψ,ψ-caroten-12′-al; see Fig. 1

2. Ketones and Esters

9 Echinenone; β,β-caroten-4-one; *n*-F-*c*; *148*

10 4-Keto-γ-carotene; β,ψ-caroten-4-one; *n*-F-b; *151*

11 4-Keto-3′,4′-dehydro-OH-γ-carotene hexoside; 1′-hexosyloxy-3′,4′-didehydro-1′-2′-dihydro-β,ψ-caroten-4-one; *n*-F-*h′*; (cf. *159*)[c]

12 Okenone; 1′-methoxy-1′,2′-dihydro-χ,ψ-caroten-4′-one; *d*-F-*k*; *181*

13 *Thiothece*-474[b]; 1′-methoxy-1′,2′-dihydro-β,ψ-caroten-4′-one; *c*-F-*k*

14 *Thiothece*-484[b]; only partial structure assigned, *m*-F-*k*

(*continued*)

Table 1. (continued)

15 *Thiothece*-OH-484[b]; 1′-hydroxy-1-methoxy-1,2,1′,2′-tetrahydro-ψ,ψ-caroten-4-one; *f*-F-*k*

16 *R.g.* keto-I[d]; 1-methoxy-1,2-dihydro-ψ,ψ-caroten-4-one; *k*-F-*b*

17 *R.g.* keto-II[d]; 1,1′-dimethoxy-1,2,1′,2′-tetrahydro-3′,4′-didehydro-ψ,ψ-caroten-4-one; *k*-F-*j*

18 *R.g.* keto-III[d]; 1,1′-dimethoxy-1,2,1′2′-tetrahydro-ψ,ψ-caroten-4,4′-dione; *k*-F-*k*

19 2-Keto-OH-spirilloxanthin; 1′-hydroxy-1-methoxy-3,4,3′,4-tetrahydro-1,2,1′,2′-tetrahydro-ψ,ψ-caroten-2-one; *l*-F-*g*; *184*

20 2,2′-Diketo-spirilloxanthin; 1,1′-dimethoxy-3,4,3′,4′-tetrahydro-1,2,1′,2′-tetrahydro-ψ,ψ-caroten-2,2′-dione; *l*-F-*l*; *208*

21 *R.g.* keto-VI[d]; -1-methoxy-1,2,7′,8′-tetrahydro-ψ,ψ-caroten-4-one; *k*-D-*b*

22 *R.g.* keto-VII[d]; 1-methoxy-1′-hydroxy-1,2,7′,8′-tetrahydro-ψ,ψ-caroten-4-one; *k*-D-*f*

23 Spheroidenone; 1-methoxy-3,4-didehydro-1,2,7′,8′-tetrahydro-ψ,ψ-caroten-2-one; *l*-E-*b*; *182*

24 OH-Spheroidenone; 1′-hydroxy-1-methoxy-3,4-didehydro-1,2,1′,2′,7′,8′-hexahydro-ψ,ψ-caroten-2-one; *l*-E-*f*; *185*

25 *R.g.* keto-V[d]; 1-methoxy-1,2,7′,8′,11′,12′-hexahydro-ψ,ψ-caroten-4-one; *k*-C-*b*

3. Methyl Ethers (not listed above)

26 Anhydrorhodovibrin; 1-methoxy-3,4-didehydro-1,2-dihydro-ψ,ψ-carotene; *j*-F-*b*; *97*

27 3,4-Dihydroanhydrorhodovibrin; 1-methoxy-1,2-dihydro-ψ,ψ-carotene; *i*-F-*b*; *98*

28 Spheroidene; 1-methoxy-3,4-didehydro-1,2,7′,8′-tetrahydro-ψ,ψ-carotene; *j*-E-*b*; *99*

29 3,4-Dihydrospheroidene; 1-methoxy-1,2,7′,8′-tetrahydro-ψ,ψ-carotene; *i*-E-*b*; *100*

30 11′,12′-Dihydrospheroidene; 1-methoxy-3,4-didehydro-1,2,7′,8′,11′,12′-hexahydro-ψ,ψ-carotene; *j*-C-*b*; *101*

31 3,4,11′,12′-Tetrahydrospheroidene; 1-methoxy-1,2,7′,8′,11′,12′-hexahydro-ψ,ψ-carotene; *i*-C-*b*; *102*

32 Methoxy-phytofluene; 1-methoxy-1,2,7,8,7′,8′,11′,12′-octahydro-ψ,ψ-carotene; *i*-B-*b*; *103*

33 Methoxy-phytoene; 1-methoxy-1,2,7,8,11,12,7′,8′,11′,12′-decahydro-ψ,ψ-carotene; *i*-A-*b*; *104*

34 OH-Spirilloxanthin; 1′-methoxy-3,4,3′,4′-tetradehydro-1,2,1′,2′,-tetrahydro-ψ,ψ-caroten-1-ol; *g*-F-*j*; *105*

35 Rhodovibrin; 1′-methoxy-3′,4′-didehydro-1,2,1′,2′-tetrahydro-ψ,ψ-caroten-1-ol; *f*-F-*j*; *106*

36 OH-Spheroidene; 1′-methoxy-3′,4′-didehydro-1,2,7,8,1′,2′-hexahydro-ψ,ψ-caroten-1-ol; *f*-E-*j*; *107*

37 Spirilloxanthin; 1,1′-dimethoxy-3,4,3′,4′-tetrahydro-1,2,1′,2′-tetrahydro-ψ,ψ-carotene; *j*-F-*j*; *108*

38 3,4-Dihydrospirilloxanthin; 1,1′-dimethoxy-3,4-didehydro-1,2,1′,2′-tetrahydro-ψ,ψ-carotene; *j*-F-*i*; *109*

39 3,4,3′,4′-Tetrahydrospirilloxanthin; 1,1′-dimethoxy-1,2,1′,2′-tetrahydro-ψ,ψ-carotene; *i*-F-*i*; *110*

40 3,4,7,8-Tetrahydrospirilloxanthin; 1,1′-dimethoxy-3,4-didehydro-1,2,1′,2′,7′,8′-hexahydro-ψ,ψ-carotene; *j*-E-*i*; *111*

41 3,4,3′,4′-Hexahydrospirilloxanthin; 1,1′-dimethoxy-1,2,7,8,1′,2′-hexahydro-ψ,ψ-carotene; *i*-F-*i*; *112*

42 Methoxy-OH-phytofluene[e]; 1′-methoxy-1,2,7,8,11,12,1′,2′,7′,8′-decahydro-ψ,ψ-caroten-1-ol; *i*-B-*f*

43 OH-Hexahydrospheroidene[e]; 1′-methoxy-1,2,7,8,11,12,1′,2′-octahydro-ψ,ψ-caroten-1-ol; *i*-C-*f*

44 OH-Tetrahydrospheroidene[e]; 1′-methoxy-1,2,7,8,1′,2′-hexahydro-ψ,ψ-caroten-1-ol; *i*-E-*f*

4. Alcohols and Glycosides (not listed above)

45 OH-γ-Carotene; 1′,2′-dihydro-β,ψ-caroten-1′-ol; *c*-F-*f*; *48*

46 OH-γ-Carotene hexoside; 1′-hexosyloxy-1′,2′-dihydro-β,ψ-carotene; *c*-F-*f′*[c]

47 3′,4′-Dehydro-OH-γ-carotene glucoside; 1′-glucosyloxy-3′,4′-didehydro-1′,2′-dihydro-β,ψ-carotene; *c*-F-*h*[c]

48 OH-Chlorobactene; 1′,2′-dihydro-ϕ,ψ-caroten-1′-ol; *e*-F-*f*; *53*

49 3,4-Dehydrorhodopin; 3,4-didehydro-1,2-dihydro-ψ,ψ-caroten-1-ol; *g*-F-*b*; *55*

50 Rhodopin; 1,2-dihydro-ψ,ψ-caroten-1-ol; *f*-F-*b*; *56*

51 Rhodopin glucoside; 1-(β-D-glucopyranosyloxy)-1,2-dihydro-ψ,ψ-carotene; *h*-F-*b*; *57*

(continued)

Table 1. (continued)

52 Chloroxanthin; 1,2,7′,8′-tetrahydro-ψ,ψ-caroten-1-ol; *f*-E-*b*; *58*
53 OH-Hexahydrolycopene; 1,2,7′,8′,11′,12′-hexahydro-ψ,ψ-caroten-1-ol; *f*-C-*b*; *59*
54 OH-Dihydrophytofluene; 1,2,7,8,11,12,7′,8′-octahydro-ψ,ψ-caroten-1-ol; *b*-B-*f*; *60*
55 OH-Dihydrophytoene; 1,2,7,8,11,12,7′,8′,11′,12′-decahydro-ψ,ψ-caroten-1-ol; *f*-A-*b*; *61*
56 Lycopenol; 13-*cis*-ψ,ψ-caroten-20-ol; *b*-G-*b*; *63*
57 Rhodopinol; 13-*cis*-1,2-dihydro-ψ,ψ-caroten-1,20-diol; *f*-G-*b*; *80*
58 OH-Rhodopin; 1,2,1′,2′-tetrahydro-ψ,ψ-carotene-1,1′-diol; *f*-F-*f*; *81*
59 Di-OH-ζ-carotene; 1,2,7,8,1′,2′,7′,8′-octahydro-ψ,ψ-carotene-1,1′-diol; *f*-D-*f*; *82*
60 OH-ζ-Carotene; 1,2,7′,8′,11′,12′-hexahydro-ψ,ψ-caroten-1-ol; *f*-C-*b*; *59*

5. Hydrocarbons

61 β-Carotene; β,β-carotene; *c*-F-*c*; *3*
62 β-Isorenieratene; β,φ-carotene; *e*-F-*c*; *6*
63 γ-Carotene; β,ψ-carotene; *c*-F-*b*; *8*
64 *retro*-Dehydro-γ-carotene; 4,4′-didehydro-β,ψ-carotene[c,f]
65 Isorenieratene; φ,χ-carotene; *e*-F-*e*; *14*
66 Chlorobactene; φ,ψ-carotene; *e*-F-*b*; *15*
67 Lycopene; ψ,ψ-carotene; *b*-F-*b*; *19*
68 1,2-Dihydro-3,4-dehydrolycopene; 3,4-didehydro-1,2-dihydro-ψ,ψ-carotene; *a′*-F-*b*; *20*
69 1,2-Dihydrolycopene; 1,2-dihydro-ψ,ψ-carotene; *a*-F-*b*; *21*
70 Neurosporene; 7,8-dihydro-ψ,ψ-carotene; *b*-E-*b*; *22*
71 1,2-Dihydroneurosporene; 1,2,7,8-tetrahydroψ,ψ-carotene; *a*-E-*b*; *23*
72 Tetrahydrolycopene; 1,2,1′,2′-tetrahydro-ψ,ψ-carotene; *a*-F-*a*; *24*
73 Asymmetrical ζ-carotene; 7,8,11,12-tetrahydro-ψ,ψ-carotene; *b*-C-*b*; *25*
74 ζ-Carotene; 7,8,7′,8′-tetrahydro-ψ,ψ-carotene; *b*-D-*b*; *26*
75 Hexahydrolycopene; 1,2,7,8,11,12-hexahydro-ψ,ψ-carotene; *b*-C-*a*; *27*
76 Tetrahydroneurosporene; 1,2,7,8,1′,2′-hexahydro-ψ,ψ-carotene; *a*-E-*a*; *28*
77 Hexahydrolycopene; 1,2,7,8,7′,8′-hexahydro-ψ,ψ-carotene; *a*-D-*b*; *29*
78 Phytofluene; 15-*cis*-7,8,11,12,7′,8′-hexahydro-ψ,ψ-carotene; *b*-B-*b*; *30*
79 1,2-Dihydrophytofluene; 1,2,7,8,11,12,7′,8′-octahydro-ψ,ψ-carotene; *a*-B-*b*; *31*
80 Phytoene; 15-*cis*-7,8,11,12,7′,8′,11′,12′-octahydro-ψ,ψ-carotene; *b*-A-*b*; *32*
81 1,2-Dihydrophytoene; 1,2,7,8,11,12,7′,8′,11′,12′-decahydro-ψ,ψ-carotene; *b*-A-*a*; *33*

[a] The entries in this table are: serial number, trivial name, rational name, letter formula (see Fig. 2), and the number assigned to the compound in Isler (1971) (references for compounds not referred to therein are given in footnotes *b–e*).
[b] Andrewes and Liaaen-Jensen (1972); [c] Halfen *et al.* (1972); [d] Schmidt and Liaaen-Jensen (1973); [e] Davies and Than (1974).
[f] The structure of this carotenoid cannot be formed from the elements in Fig. 2; the structure is:

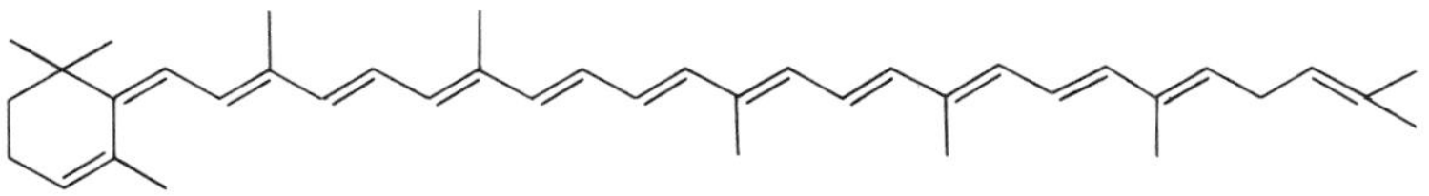

apocarotenoids are reported. Among the aryl carotenoids, no phenolic representatives are known. Esters of alcohols have not been encountered up till now.

Of the approximately 350 naturally occurring carotenoids, about 190 possess 1–6 chiral centers. It is noteworthy that no chiral carotenoids are present in photosynthetic bacteria, except the five glucosidic carotenoids in which the aglycones are achiral.

2.4. Taxonomic and Evolutionary Aspects

The possibility of using carotenoids as taxonomic markers within photosynthetic bacteria was discussed recently by Liaaen-Jensen and Andrewes (1972).

Identification of methoxylated carotenoids, e.g., in lake deposits, readily reveals a prehistory involving photosynthetic bacteria.

From an evolutionary point of view, a few general considerations are evident. The general inability of photosynthetic bacteria to cyclize or to introduce molecular asymmetry (chirality) in their carotenoids places them on a more primitive level than other photosynthetic organisms. Lack of methoxylated carotenoids in other carotenogenic organisms is interesting and raises doubts about any direct evolutionary line from photosynthetic bacteria.

3. Isolation

The carotenoid content of photosynthetic bacteria varies, with 0.1% of the acetone–methanol-extracted dry matter as an average figure. In most cases, the extracts contain rather complex carotenoid mixtures.

Methods for isolation of carotenoids from biological materials were recently reviewed (Liaaen-Jensen, 1971). Detailed, quantitative isolation procedures from

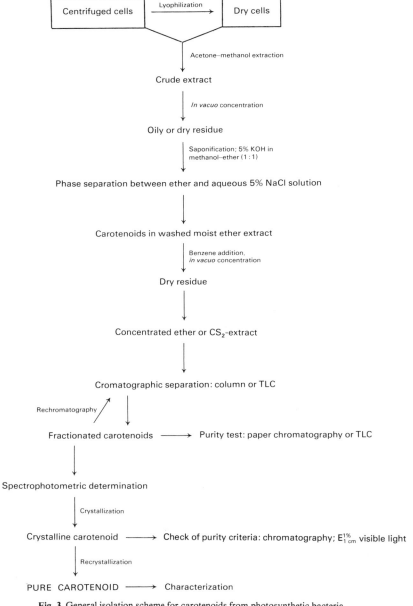

Fig. 3. General isolation scheme for carotenoids from photosynthetic bacteria.

photosynthetic tissues have also been described (Liaaen-Jensen and Jensen, 1971).

Since carotenoids are pigments, visible to the eye, and their concentrations are readily estimated from known, high extinction coefficients in visible light, quantitative aspects of an isolation procedure are easily checked. General precautions, founded on experience, have been detailed (Liaaen-Jensen and Jensen, 1971) and include inert atmosphere (*vacuo*, nitrogen or carbon dioxide), low temperature (−20 to +40°C), darkness (subdued light), and acid- and peroxide-free conditions. These precautions are required due to the polyene nature of carotenoids. A few carotenoids are altered even with weak alkali (esters of carotenols and fucoxanthin-related epoxy ketones). Such carotenoids, however, have not yet been encountered in photosynthetic bacteria.

Solvent extraction of lyophilized or preferably fresh cells is carried out at room temperature. To secure complete extraction, all chlorophylls should also be extracted. Repeated extractions with acetone–methanol mixtures are ideal, acetone extracting the carotenoids and methanol the chlorophylls.

The chlorophylls and colorless lipids are subsequently removed by saponification. Prior to saponification, it is imperative that all acetone be removed, since carotenals, if present, readily form condensation products with acetone in the presence of alkali (Schmidt *et al.*, 1971). Storage of chlorophyll-containing extracts should be avoided, since decomposition of chlorophylls, providing unsaponifiable gray-green products, interferes with the chromatographic separation.

After alkali treatment, the carotenoids are extracted into ether by phase separation. Ether extracts are conveniently dried by azeotropic distillation with benzene (*in vacuo*).

The unsaponifiable matter, containing the carotenoids, is now submitted to chromatographic separation. Usually, other unsaponifiable matter poses no serious contamination problem; photosynthetic bacteria being devoid of steroids. A few exceptions in which colorless impurities have been troublesome are *Rhodopseudomonas gelatinosa* (Liaaen-Jensen, 1963b) and *Rp. palustris* (Liaaen-Jensen, 1962).

Chromatographic separation on columns (Strain, 1958) is still used, but has been largely replaced by thin-layer chromatography (TLC) (Bolliger and König, 1969), mostly on silica gel plates. For separation of complex *cis–trans* isomeric mixtures, kieselguhr-filled circular papers are still superior (Jensen and Liaaen-Jensen, 1959). In the carotenoid field, adsorption chromatography has found much wider application than partition chromatography.

For analytical work, the detailed chromatographic

procedures described elsewhere (Liaaen-Jensen and Jensen, 1971) are applicable to carotenoid mixtures obtained from photosynthetic bacteria, and may readily be adapted to larger-scale preparative work.

Quantitative determination of each component based on electronic spectra in visible light and known extinction coefficients is also described by Liaaen-Jensen and Jensen (1971). Even for preparative work, it is recommended to work quantitatively.

Chromatographically pure (as to carotenoids) fractions containing more than 0.5 mg may provide crystalline products. Fractional crystallization from noncarotenoid contaminants is discussed and reference to original procedures is made in Liaaen-Jensen (1971). The purity of crystalline material must be checked by chromatography (circular paper or TLC) and measurement of extinction coefficients in visible light. A sharp and correct melting point (in an evacuated tube) is a further criterion of purity. Recrystallization from a different solvent pair may be required.

A general isolation scheme for carotenoids from photosynthetic bacteria is given in Fig. 3.

4. Identification and Elucidation of Structure

4.1. General

The first clues to the identification of a carotenoid are the absorption spectrum in visible light, which indicates the nature of the chromophore, and the relative polarity in a known chromotographic system. Measurement of polarities in terms of partition ratios (Petracek and Zechmeister, 1956; Krinsky, 1963) has lost importance.

Mass spectrometry for establishment of molecular weight and characteristic structural features (see below) has become an indispensable tool for identification. There are many examples of misidentifications in the carotenoid field without mass spectrometry. Even unequivocal identification as a carotenoid requires consideration of the fragmentation pattern on electron impact (cf. Andrewes *et al.*, 1973).

Careful cochromatography with authentic material is a necessary requirement for identity, but not a sufficient proof of identity. A collection of reference carotenoids is a necessity and may be obtained by extraction from previously analyzed bacteria.

Even when the precautions mentioned above are taken, *trans–cis* isomerization (see below) inevitably occurs during the isolation procedure. *cis* Isomers usually form discrete zones and have different spectra in visible light. Reversibility to the parent *trans* isomer by iodine catalysis (Zechmeister, 1962) confirms the *cis* nature of such compounds. That a single carotenoid appears as more zones may be misinterpreted as presence of additional carotenoids.

Single carotenoids have already been described too often by many names in the literature, and many compounds believed to represent new structures have subsequently been shown to represent *cis* isomers.

Only after a careful literature survey and thorough characterization including at least iodine-catalyzed stereomutation, mass spectrometry, simple chemical derivatization, and cochromatography with related carotenoids should it be concluded that a carotenoid has not been described previously. Until the structure is elucidated, the unidentified carotenoid is frequently referred to by means of the source and the wavelength of the main absorption maximum in visible light measured in hexane (e.g., *Thiothece*-474; carotenoid with λ_{max} 474 nm obtained from a *Thiothece* species).

On a microscale, preparation and characterization of appropriate chemical derivatives substantiate the identification. Suitable derivatives for carotenoids from photosynthetic bacteria are discussed below.

4.2. Physical Methods

These methods include chromatography, partition behavior, and melting point determinations, already commented on, and recently reviewed elsewhere (Liaaen-Jensen, 1971).

Characterization by spectroscopic methods was recently authoritatively reviewed (Vetter *et al.*, 1971).

4.2.1. Electronic Spectra

The absorption spectrum in visible light gives information about the chromophore present. A large number of chromophores have been described in terms of λ_{max}, spectral fine structure, extinction value, and characteristic solvent shifts (which are especially marked in conjugated ketones). The spectral fine structure has been defined as %III/II, where III represents the optical density of the middle main band and II that of the longest-wavelength main band, with the minimum between these two bands as zero line (Liaaen-Jensen, 1962). The main band (in the visible region) usually comprises three bands. Structural features governing the electronic spectrum have recently been compiled (Vetter *et al.*, 1971). For aliphatic carotenoids from photosynthetic bacteria, the following phenomena are worth remembering: λ_{max} is always bathochromically displaced with increasing conjugated polyene chain. The spectral fine structure is at a maximum for the nonaene chromophore and

drops off with decreasing or increasing polyene chain. Conjugated carbonyl functions decrease the spectral fine structure, particularly in methanolic solution. Carotenoids containing conjugated carbonyl functions further exhibit a bathochromic shift in methanolic relative to petroleum ether solution. The hypsochromic shift caused by complex metal hydride reduction of a conjugated keto group depends on the location of the keto group ($\approx$32 nm in H, $\approx$28 nm in *l*, and $\approx$16 nm in *k*, Fig. 2, in petroleum ether). Cross-conjugated aldehyde functions cause remarkably broad, round-shaped spectra. The bathochromic shift caused by a 1,2,3-trimethylphenyl group corresponds to slightly less than one aliphatic double bond, and that of a 1,2,5-trimethylphenyl group approximately to a *β*-end group.

Extinction coefficients are given in the original literature for the respective crotenoids (cf. Straub, 1971). For analytical work with unknown mixtures, approximate extinction coefficients may be assumed from λ_{max}, which indicates the chromophore. Suitable $E_{1cm}^{1\%}$ values have been recommended (Liaaen-Jensen, 1971).

The effects of introducing *cis* bonds into the polyene chain should be kept in mind: decrease in extinction coefficient, loss in spectral fine structure, hypsochromic shift of λ_{max} by approximately 5 nm for an unhindered *cis* bond, and appearance of a so-called *cis*-peak at a wavelength 142 ± 2 nm less than the longest-wavelength maximum in the *trans* isomer (Zeichmeister, 1962). Entirely aliphatic carotenoids have characteristic, double *cis* peaks. For the reasons stated above, use of approximate extinction coefficients lower than for the all-*trans* isomer is justified.

Because of the high extinction values of carotenoids ($E_{1cm}^{1\%} = 900 - 3600$; $\varepsilon \approx 150,000$), recording of electronic spectra of carotenoids requires only microgram samples. Except for the most saturated members of the series, the UV region gives little structural information.

4.2.2. Infrared Spectra

With micro KBr pellets or microcells, 0.5 mg crystalline carotenoid or less is required. IR spectra generally reveal qualitatively the functional groups present. Characteristic frequencies were recently discussed briefly (Vetter *et al.*, 1971). Conjugated carbonyl functions in carotenoids from photosynthetic bacteria are readily revealed by absorption in the 1660-cm^{-1} region. Tertiary hydroxy groups are indicated by absorption at approximately 3400, 1140, and 905 cm^{-1}; methyl ethers, by absorption at approximately 1080 cm^{-1}; and tetra-substituted phenyl groups with two adjacent free hydrogens, by absorption at 800 cm^{-1}. A rough criterion of purity of the sample is the relative intensity of the absorption at approximately 960 cm^{-1} (strong band for *trans*-disubstituted double bonds).

4.2.3. Proton Magnetic Resonance Spectra

^{13}C NMR spectra for carotenoids characteristic of photosynthetic bacteria are not yet available.

^{1}H NMR spectra of the major compounds are available in the original literature (cf. Straub, 1971).

A general, comprehensive treatment of carotenoid NMR spectra was presented in Vetter *et al.* (1971). For satisfactory documentation of new structures, ^{1}H NMR spectra are required. For the single-scan 60 MHz or 100 MHz spectra hitherto mostly used, 2–8 mg crystalline carotenoid has been required. With modern Fourier-transform instruments, the sample requirements can be greatly reduced. This technique will permit the characterization of minor carotenoids by ^{1}H NMR spectroscopy, provided the pure carotenoid is available.

The methyl region is particularly informative. In carotenoids, the methyl groups generally have no hydrogens at the adjacent carbons, and the methyl groups appear as singlets of different chemical shift revealing different structural surroundings. An exception is the 1,2-dihydro series with end groups *a,a'*, Fig. 2 (Malhotra *et al.*, 1970; Eidem *et al.*, 1976), in which the four *gem.* methyl groups cause a characteristic doublet, caused by spin coupling. Tertiary methoxy groups, of course, show characteristic singlets at lower field, and primary hydroxy functions as in the G-type carotenoids (Fig. 2) show characteristic two-proton singlets, moving to lower fields on acetylation. The

olefinic region can be interpreted in detail only by 220-MHz spectra, or in suitable cases by decoupling techniques.

Interpretation of carotenoid ^{1}H NMR spectra is usually simple and requires only knowledge of simple NMR theory and use of reference data (Vetter *et al.*, 1971; Straub, 1971).

4.2.4. Mass Spectra

For the size of sample required (approximately 20 μg, noncrystalline material), this method gives the most structural information. No identification is considered safe without mass spectrometric examination, and all structures advanced without mass spectrometry can only be considered tentative. A comprehensive review of the application of mass spectrometry to the carotenoid field was given by Vetter *et al.* (1971). For all carotenoids, the molecular ion, even of free glycosides, may be obtained on electron impact. The molecular ion gives the molecular weight of the compound and, by high-precision measurements, the elementary composition as well. By checking characteristic losses from the molecular ion, its identification may be confirmed.

The most characteristic fragmentation of carotenoid pigments is the loss of 92, 106, and 158 mass units from the molecular ion or other fragment ions, caused by the expulsion of toluene, xylene, and dimethylcyclodecapentaene from the polyene chain. By the use of deuterium-labeled carotenoids, the region from which these elimination products come can be deduced exactly (Johansen *et al.*, 1974; Johannes *et al.*, 1974). Substitution of lateral methyl groups as in the G- and H-type carotenoids (Fig. 2) does not prevent this general mode of fragmentation. Increased masses for the expelled fragments reveal the type of substitution

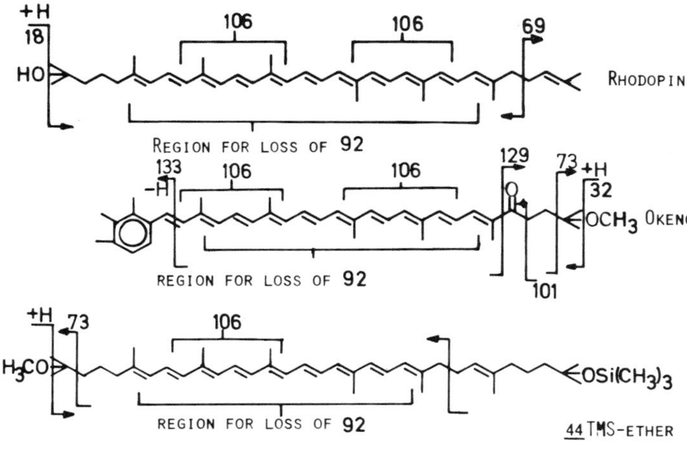

Fig. 4. Examples of characteristic fragmentations of carotenoids on electron impact.

present (Enzell and Liaaen-Jensen, 1971). The intensity ratio for the M-92/M-106 fragment ions may be used to check the chromophore present as a supplement to electronic spectra (Enzell *et al.*, 1968; Francis *et al.*, 1974).

The various end groups encountered in carotenoids of photosynthetic bacteria cause characteristic fragmentations (Enzell *et al.*, 1969). *Bis*-allylic double bonds, as in end group *b*, of A, B, and C (Fig. 2) are readily cleaved; benzylic bonds adjacent to *d* and *e* (Fig. 2) undergo single cleavage with charge retention on the aromatic moiety; and ketones suffer *α*-cleavage (*k* and *l*). Water is eliminated from tertiary alcohols (end groups *f* and *g*) and alcohols from ethers. In-chain cleavages with or without hydrogen transfer are also frequently observed These principles are illustrated for rhodopin (*50*), okenone (*12*), and the trimethylsilyl ether of the monool *44* in Fig. 4. Arrows indicating cleavages point toward the charged fragment. Fragmentation mechanisms are frequently based on paper-chemical rationalizations of observed spectra of known carotenoids. In some cases, however, support from metastable ions and mass spectra of specifically deuterium-labeled compounds is available.

Registration of carotenoid mass spectra requires some training. Conditions should be optimized by exercise with authentic material. Recording conditions are usually stated in the original literature. Carotenoid

glucosides provide diagnostically useful fragmentation patterns after acetylation (Schmidt *et al.*, 1971).

4.3. Chemical Methods

Preparation of chemical derivatives and their subsequent characterization (R_f values, electronic and mass spectra) substantiate any identification or structural conclusions. Such derivatives may readily be prepared on the microscale (20 to 100-*μg* samples). For carotenoids from photosynthetic bacteria, the following reactions, depicted in Fig. 5, are diagnostically useful:

(i) *Silylation of tertiary hydroxy groups* (Fig. 2, end groups *f* and *g*). A general procedure was detailed in Liaaen-Jensen and Jensen (1971). Together with a negative acetylation test under standard conditions (Liaaen-Jensen and Jensen, 1971), this procedure is positive evidence for the presence of a tertiary hydroxy group.

(ii) *Dehydration of tertiary hydroxy groups* with phosphorous oxychloride in pyridine (end group *f*). Procedures are available in Surmatis and Ofner (1963) and Liaaen-Jensen *et al.* (1964).

(iii) *Complex metal hydride reduction of conjugated carbonyl groups* (for end groups) *k* and *l* and transformation of H to G. A general procedure is given in

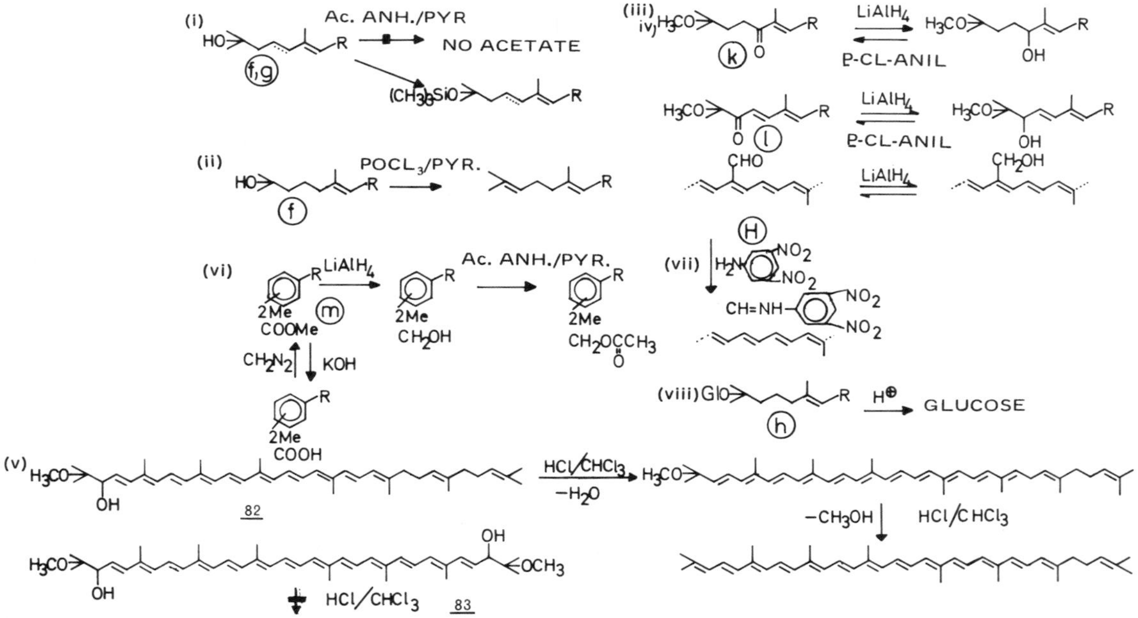

Fig. 5. Useful chemical reactions for the preparation of derivatives of carotenoids from photosynthetic bacteria.

Liaaen-Jensen and Jensen (1971). The reaction causes a hypsochromic shift of the visible spectrum as discussed in Section 4.2. The resulting allylic hydroxy group may be acetylated, oxidized to the original keto or aldehyde group (iv) or submitted to an elimination reaction (v).

(iv) *Oxidation of allylic hydroxy groups* derived from *k* or *l* proceeds smoothly with *p*-chloranil (Liaaen-Jensen, 1965). Oxidation of G to H proceeds with low yield. The oxidation reagents used are O_2 and I_2 (Aasen and Liaaen-Jensen, 1967c) and dicyanodibenzoquinone (Puntervold and Liaaen-Jensen, 1974).

(v) *Elimination of allylic hydroxyl (or methoxyl) groups.* Treatment with 0.03 N HCl in chloroform rapidly eliminates allylic hydroxyl (Entschel and Karrer, 1958), frequently accompanied by allylic rearrangement of the polyene chain in the intermediate carbonium ion, e.g., as for hydride-reduced spheroidenone (*82*) in Fig. 5. The reaction generally results in products of decreased polarity and in extension of the polyene chain. If methoxy groups become allylic during the reaction, elimination of methanol follows. The reaction requires that hydrogen atoms be available for elimination of water or methanol. Thus, 2,2'-dihydroxyspirilloxanthin (*83*) is inert to this reaction despite the presence of allylic hydroxy groups (Liaaen-Jensen, 1963b). G-type carotenoids (Fig. 2) do not provide such elimination

products. A general review of the reaction was presented in Liaaen-Jensen (1971).

(vi) *Carbomethoxy functions*, as in end group *m*, may be saponified to the corresponding carboxylic acid under strongly alkaline conditions (Andrewes and Liaaen-Jensen, 1972). The free carboxylic acid may be reesterified with diazomethane. With $LiAlH_4$, the carbomethoxy group is reduced to a primary alcohol, which may be acetylated or silylated.

(vii) *From H-type carotenoids*, aldehyde derivatives such as oximes and dinitrophenylhydrazones may be formed (Aasen and Liaaen-Jensen, 1967c). Such carotenals also readily undergo the Wittig reaction (Aasen and Liaaen-Jensen, 1967c) and base-catalyzed aldol condensation with acetone (Schmidt *et al.*, 1971).

(viii) *Tertiary glucosides* (end group *h*) may be hydrolyzed under relatively weak acidic conditions and the resulting glucose submitted to paperchromatographic identification (Schmidt *et al.*, 1971). The aglucone is destroyed under these conditions.

Aryl end groups (Fig. 2, *d* and *e*) and methoxy functions in *i*, *j*, and *k* are rather inert to chemical reactions.

Oxidative cleavage of all carotenoids listed in Table 1 is of course possible in principle, but requires larger amounts of sample.

Failure of derivative formation of the types outlined

Table 2. Carotenoids from Photosynthetic Bacteria Prepared by Total Synthesis

1	Lycopenal[a]	*57*	Rhodopinol[a]
2	Rhodopinal[a]	*58*	OH-Rhodopin[h]
12	Okenone[b]	*61*	β-Carotene[k]
20	2,2'-Diketospirilloxanthin[c]	*62*	β-Isorenieratene[l]
23	Spheroidenone[d]	*65*	Isorenieratene[m]
26	Anhydrorhodovibrin[e]	*66*	Chlorobactene[j]
27	3,4-Dihydroanhydrorhodovibrin[f]	*67*	Lycopene[k]
28	Spheroidene[d]	*68*	1,2-Dihydro-3,4-dehydrolycopene[n]
34	OH-Spirilloxanthin[g]	*69*	1,2-Dihydrolycopene[o]
35	Rhodovibrin[e]	*70*	Neurosporene[f]
37	Spirilloxanthin[h]	*71*	1,2-Dihydroneurosporene[n]
39	3,4,3',4'-Tetrahydrospirilloxanthin[i]	*72*	1,2,1',2'-Tetrahydrolycopene[o]
48	OH-Chlorobactene[j]	*73*	Asymmetrical ζ-carotene[p]
49	3,4-Dehydrorhodopin[g]	*74*	ζ-Carotene[p]
50	Rhodopin[j]	*78*	Phytofluene[p]
52	Chloroxanthin[d]	*80*	Phytoene[p]
56	Lycopenol[a]		

[a] Puntervold and Liaaen-Jensen (1974); [b] Aasen and Liaaen-Jensen (1967b); [c] Schwieter *et al.* (1966); [d] Barber *et al.* (1966); [e] Surmatis *et al.* (1961); [f] Kjøsen *et al.* (1971); [g] Schneider and Weedon (1967); [h] Surmatis and Ofner (1963); [i] Aasen and Liaaen-Jensen (1967a); [j] Bonnet *et al.* (1964); [k] several routes—see Mayer and Isler (1971); [l] Cooper *et al.* (1963); [m] Yamaguchi (1959); [n] Eidem *et al.* (1976); [o] Kjøsen and Liaaen-Jensen (1970); [p] Davis *et al.* (1966).

above indicates the absence of particular structural elements required for a reaction to proceed. However, parallel studies with model compounds that undergo the particular reaction are always recommended.

5. Stereochemistry

In the carotenoids of photosynthetic bacteria (Table 1), disregarding the two glucosides *3* and *51*, only *cis–trans* (geometrical isomerism) is encountered, since molecular asymmetry (chiral centers) is absent.

The phenomenon of *cis–trans* isomerism was extensively treated by Zechmeister (1962), and a general review of the stereochemistry of carotenoids was recently given by Weedon (1971).

With the possible exception of phytoene (*80*) and phytofluene (*78*), the carotenoids of photosynthetic bacteria appear to exist in the cells in their all-*trans* form. The H-type (Fig. 2) carotenoids are an exception, which is discussed below. However, the aliphatic carotenoids, frequently with long, conjugated polyene chains, characteristic of photosynthetic bacteria, readily undergo *trans–cis* isomerization during the isolation procedure. Changes of the electronic spectra on introduction of *cis* bonds have already been treated in Section 4.2, as well as iodine-catalyzed reversibility tests used to demonstrate *cis–trans* isomeric relationships.

Generally, the all-*trans* isomer is the thermodynamically most stable one, and the one that is crystallized the most readily and has the highest

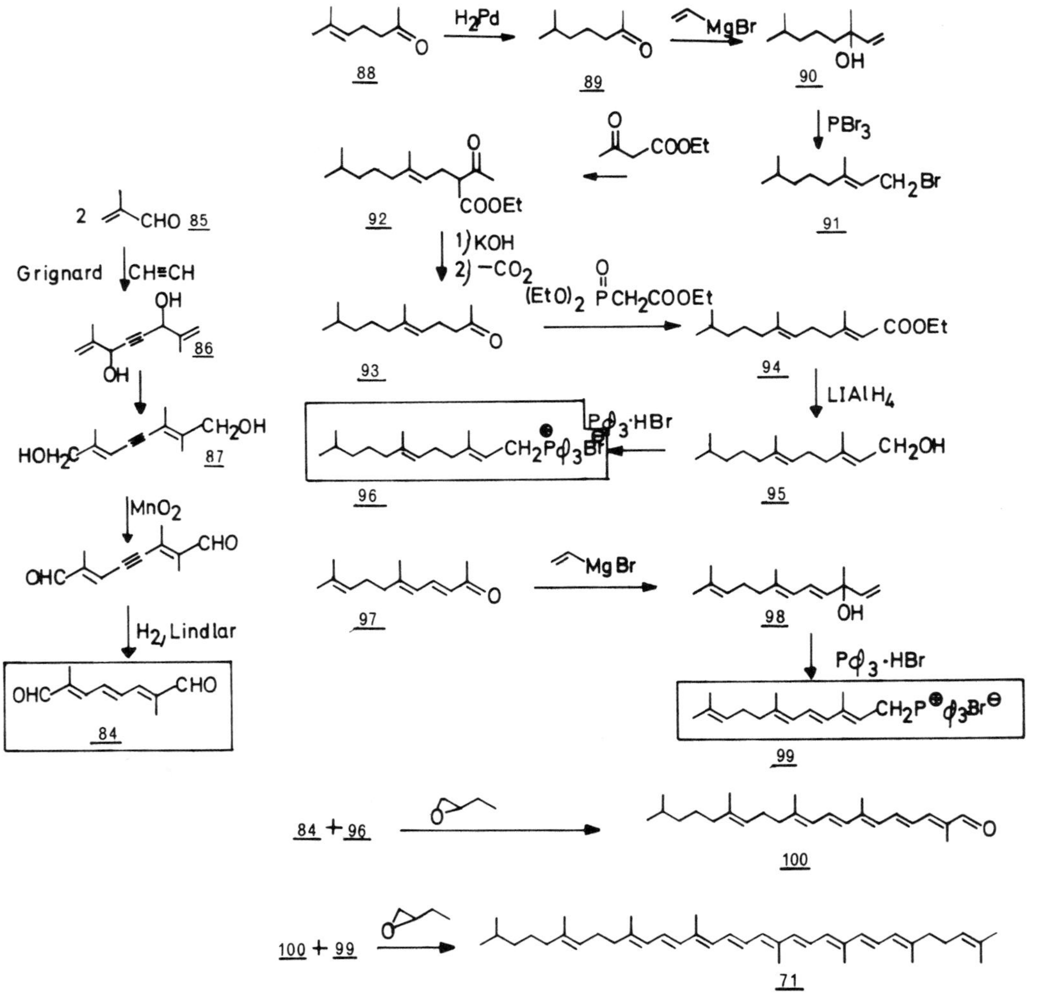

Fig. 6. Example of total synthesis of carotenoids.

melting point. Its absorption spectrum is at the longest wavelength in visible light, and it exhibits the largest spectral fine structure.

The H-type carotenoids (Fig. 2) form an exception with the 13-double bond exclusively in *cis* configuration, an effect caused by the cross-conjugated aldehyde group (Aasen and Liaaen-Jensen, 1967c; Chin and Song, 1974). Also, the G-type carotenoids show a tendency to 13-*cis* configuration, but in these cases, the all-*trans* isomer may also be isolated.

Careful, quantitative study of the composition of the iodine-catalyzed equilibrium mixtures of two carotenoids suspected to be identical may be used as a criterion of identity, although it is not a sufficient proof of identity.

6. Total Synthesis

Total synthesis is, besides X-ray analysis, always the ultimate proof of a structure. In the carotenoids of photosynthetic bacteria, absolute configuration poses no problem for the synthesis design. The major carotenoids of photosynthetic bacteria, representing 33 of 73, have been prepared by total synthesis (Table 2). Frequently, more than one synthesis has been reported; however, only one key reference is included. The serial numbers are those in Table 1.

The general principles used for total synthesis of carotenoids were thoroughly reviewed by Mayer and Isler (1971). A central component of varying size to which the two end groups are attached is frequently used. As an example, Fig. 6 depicts a route for synthesis of 1,2-dihydroneurosporene (*71*; Eidem *et al.*, 1976), involving many of the more common reactions used to build up carotenoids. In this case, the central component was the C_{10}-dialdehyde *84* (Mildner and Weedon, 1953); for its synthesis, methacrolein (*85*) is condensed with acetylene in a Grignard reaction to give the diol *86*, which was converted via the primary dibromide to *87*. Subsequent manganese dioxide oxidation of *87* provided the acetylenic dialdehyde, which was hydrogenated with Lindlar catalyst to the C_{10}-dial *84*; *84* was obtained after *cis*–*trans* isomerization.

Methylheptenone (*88*) was hydrogenated in the presence of palladium to give the saturated ketone *89*. Reaction of *89* with vinyl magnesium bromide in a Grignard reaction gave the tertiary alcohol *90*, which was then converted to the primary allylic bromide *91* using phosphorous tribromide. Condensation of *91* with acetoacetic ester gave the β-keto ester *92*, which after saponification and decarboxylation gave dihydrogeranylacetone (*93*). Horner condensation, suitable for ketones, of *93* with ethyl diethyl-

phosphonoacetate gave the conjugated ester *94*, which was converted to the corresponding allylic alcohol *95* with lithium aluminum hydride and further transformed to the phosphonium salt *96*, providing one desired end group.

Similarly, pseudoionone *97* was converted via the alcohol *98* to the phosphonium salt *99*, representing the other end group.

The dialdehyde *84* was now condensed in a Wittig reaction with the ylid of the phosphonium salt *96*, generated by treatment with butylene oxide, to give the C_{30}-aldehyde *100*. A second Wittig reaction of *100* with the ylid of the phosphonium salt *99* eventually provided the desired carotene (*71*).

7. References

Aasen, A. J., and Liaaen-Jensen, S., 1967a, Bacterial carotenoids XXI. Isolation and synthesis of 3,4,3′,4′-tetrahydrosprilloxanthin, *Acta Chem. Scand.* **21**:371.

Aasen, A. J., and Liaaen-Jensen, S., 1967b, Bacterial carotenoids XXIII. The carotenoids of Thiorhodaceae. 6. Total synthesis of okenone and related compounds, *Acta Chem. Scand.* **21**:970.

Aasen, A. J., and Liaaen-Jensen, S., 1967c, Bacterial carotenoids XXIV. Carotenoids of Thiorhodaceae. 7. Cross-conjugated carotenals, *Acta Chem. Scand.* **21**:2185.

Andrewes, A. G., and Liaaen-Jensen, S., 1972, Bacterial carotenoids XXXVII. Carotenoids of Thiorhodaceae. 10. Structural elucidation of five minor carotenoids from *Thiothece gelatinosa, Acta Chem. Scand.* **26**:2194.

Andrewes, A. G., Hertzberg, S., Liaaen-Jensen, S., and Starr, M. P., 1973, *Xanthomonas* pigments 2. The *Xanthomonas* "carotenoids"—noncarotenoid brominated aryl-polyene esters, *Acta Chem. Scand.* **27**:2383.

Barber, M. S., Jackman, L. M., Manchand, P. S., and Weedon, B. C. L., 1966, Carotenoids and related compounds. Part XVI. Structural and synthetic studies on spirilloxanthin, chloroxanthin, spheroidene and spheroidenone, *J. Chem. Soc. C*, 2166.

Bolliger, H. R., and König, A., 1969, Vitamins including carotenoids, chlorophylls and biologically active quinones, In: *Thin Layer Chromatography, a Laboratory Handbook* (E. Stahl, ed.), pp. 259–311, 2nd Ed., Springer, Heidelberg.

Bonnett, R., Spark, A. A., and Weedon, B. C. L., 1964, Carotenoids and related compounds. Part XII. Syntheses of chlorobactene, "HO-chlorobactene" and rhodopin, *Acta Chem. Scand.* **18**:1739.

Chin, C. A., and Song, P.-S., 1974, Electronic spectra of carotenoids. A theoretical analysis of the electronic spectrum of rhodopinal, *J. Mol. Spectrosc.* **52**:216.

Cooper, R. D. G., Davis, J. B., and Weedon, B. C. L., 1963, Carotenoids and related compounds. Part X. Synthesis of renieratene, isorenieratene, renierapurpurin, and other aryl-polyenes, *J. Chem. Soc.*, 5637.

Davies, B. H., and Than, A., 1974, Monohydroxy caro-tenoids from diphenylamine-inhibited cultures of *Rhodospirillum rubrum*, *Phytochemistry* **13**:209.

Davis, J. B., Jackman, L. M., Siddons, P. T., and Weedon, B. C. L., 1966, Carotenoids and related compounds. Part XV. The structure and synthesis of phytoene, phytofluene, ξ-carotene and neurosporene, *J. Chem. Soc. C*, 2154.

Eidem, A., Buchecker, R., Kjøsen, H., and Liaaen-Jensen, S., 1976, Bacterial carotenoids XLVI. Total syntheses of carotenoids of the 1,2-dihydro series, *Acta Chem. Scand.* **B29**:1015.

Entschel, R., and Karrer, P., 1958, Carotinoidsynthesen XXII. Umsetzungsprodukte des β-Carotins mit Bromsuccinimid (Einführung von Äther- und Hydroxyl-Gruppen in den Kohlenwasserstoff), *Helv. Chim. Acta* **41**:402.

Enzell, C. R., and Liaaen-Jensen, S., 1971, Mass-spectrometric studies of carotenoids. 5. Steric effects in in-chain elimination reactions, *Acta Chem. Scand.* **25**:271.

Enzell, C. R., Francis, G. W., and Liaaen-Jensen, S., 1968, Mass-spectrometric studies of carotenoids. 1. Occurrence and intensity ratios of M-92 and M-106 peaks, *Acta Chem. Scand.* **22**:1054.

Enzell, C. R., Francis, G. W., and Liaaen-Jensen, S., 1969, Mass-spectrometric studies of carotenoids. 2. A survey of fragmentation reactions, *Acta Chem. Scand.* **23**:727.

Francis, G. W., Norgård, S., and Liaaen-Jensen, S., 1974, C_{50}-Carotenoids. 12. Steric effects of the intensity ratios of the M-92/M-106 ions in the mass spectra of carotenoids, *Acta Chem. Scand. B* **28**:244.

Goodwin, T. W., 1973, Microbial carotenoids, In: *Handbook of Microbiology*, Vol. III, *Microbial Products* (A. I. Laskin and H. A. Lechevalier, eds), pp. 75–83, CRC Press, Cleveland.

Halfen, L. N., Pierson, B. K., and Francis, G. W., 1972, Carotenoids of a gliding organism containing bacteriochlorophylls, *Arch. Mikrobiol.* **82**:240.

Isler, O. (ed.), 1971, *Carotenoids*, Birkhäuser, Basel.

IUPAC and IUB, 1975, Nomenclature of carotenoids. Approved 1974, *Biochemistry* **14**:1803.

Jensen, A., and Liaaen-Jensen, S., 1959, Quantitative paper chromatography of carotenoids, *Acta Chem. Scand.* **13**:1863.

Johannes, B., Brzezinka, H., and Budzikiewicz, H., 1974, Massenspektroskopische Fragmentierungsreaktionen. VII. Abbau der Polyenkette bei Carotinoiden, *Org. Mass Spectrom.* **9**:1095.

Johansen, J. E., Eidem, A., and Liaaen-Jensen, S., 1974, Mass spectrometry of carotenoids—In-chain fragmentation of deuterium labelled carotenoids, *Acta Chem. Scand. B* **28**:385.

Kjøsen, H., and Liaaen-Jensen, S., 1970, Bacterial carotenoids XXXV. Total synthesis of 1,2-dihydro and 1,2,1′,2′-tetrahydrolycopene, *Acta Chem. Scand.* **24**:2259.

Kjøsen, H., Liaaen-Jensen, S., and Enzell, C. R., 1971, Mass-spectrometric studies of carotenoids. 4. In-chain elimination reactions, *Acta Chem. Scand.* **25**:85.

Krinsky, N., 1963, A relationship between partition coefficients of carotenoids and their functional groups, *Anal. Biochem.* **6**:293.

Liaaen-Jensen, S., 1962, The constitution of some bacterial carotenoids and their bearing on biosynthetic problems, *K. Nor. Vidensk. Selsk. Skr.*, No. 8 (199 pp.).

Liaaen-Jensen, S., 1963a, Carotenoids of photosynthetic bacteria—distribution, structure and biosynthesis, in: *Bacterial Photosynthesis* (H. Gest, A. San Pietro, and L. P. Vernon, eds.), pp. 19–34, Antioch Press, Yellow Springs, Ohio.

Liaaen-Jensen, S., 1963b, Bacterial carotenoids X. On the constitution of the minor carotenoids of *Rhodopseudomonas*. 1. P518, *Acta Chem. Scand.* **17**:303.

Liaaen-Jensen, S., 1965, Studies on allylic oxidation of carotenoids, *Acta Chem. Scand.* **19**:1166.

Liaaen-Jensen, S., 1971, Isolation, reactions, in: *Carotenoids* (O. Isler, ed.), pp. 61–189, Birkhäuser, Basel.

Liaaen-Jensen, S., 1976, New structures, *J. Pure Appl. Chem.* **47**:129.

Liaaen-Jensen, S., and Andrewes, A. G., 1972, Microbial carotenoids, *Annu. Rev. Microbiol.* **26**:225.

Liaaen-Jensen, S., and Jensen, A., 1971, Quantitative determination of carotenoids in photosynthetic tissues, *Methods Enzymol.* **23**:586.

Liaaen-Jensen, S., Hegge, E., and Jackman, L. M., 1964, Bacterial carotenoids XVIII. The carotenoids of photosynthetic green bacteria, *Acta Chem. Scand.* **18**:1703.

Malhotra, H. C., Britton, G., and Goodwin, T. W., 1970, A novel series of 1,2-dihydro carotenoids, *Int. J. Vit. Res.* **40**:315.

Mayer, H., and Isler, O., 1971, Total synthesis, in: *Carotenoids* (O. Isler, ed.), pp. 325–576, Birkhäuser, Basel.

Mildner, P., and Weedon, B. C. L., 1953, Carotenoids and related compounds. Part II. C_{10} Intermediates for the synthesis of carotenoids, *J. Chem. Soc.*, 3294.

Petracek, F. J., and Zechmeister, L., 1956, Determination of partition coefficients of carotenoids as a tool in pigment analysis, *Anal. Chem.* **28**:1484.

Puntervold, O., and Liaaen-Jensen, S., 1974, Bacterial carotenoids XLV. Synthesis of lycopen-20-al and rhodopin-20(20′)-al, *Acta Chem. Scand. B* **28**:1096.

Schmidt, K., and Liaaen-Jensen, S., 1973, Bacterial carotenoids XLII. New keto-carotenoids from *Rhodopseudomonas globiformis* (Rhodospirillaceae), *Acta Chem. Scand.* **27**:3040.

Schmidt, K., Francis, G. W., and Liaaen-Jensen, S., 1971, Bacterial carotenoids XXXVI. New carotenoid glucosides and remarkable C_{43}-artefacts of cross-conjugated carotenals, *Acta Chem. Scand.* **25**:2476.

Schneider, D. F., and Weedon, B. C. L., 1967, Carotenoids and related compounds. Part XVII. Synthesis of spirilloxanthin, "OH-spirilloxanthin" and 3,4-dehydrorhodopin, *J. Chem. Soc.*, 1686.

Schwieter, U., Rüegg, R., and Isler, O., 1966, Synthesen in der Carotinoid-Reihe. 21. Synthese von 2,2′-Diketospirilloxanthin (P518) und 2,2′-Diketobacterioruberin, *Helv. Chim. Acta* **49**:992.

Strain, H. H., 1958, *Chloroplast Pigments and Chromatographic Analysis*, Pennsylvania State University Press, University Park.

Straub, O., 1971, List of natural carotenoids, in: *Carotenoids* (O. Isler, ed.), pp. 771–850, Birkhäuser, Basel.

Surmatis, J. D., and Ofner, A., 1963, Total synthesis of spirilloxanthin, dehydrolycopene and 1,1'-dihydroxy-1,2,1',2'-tetrahydrolycopene, *J. Org. Chem.* **28**:2735.

Surmatis, J. D., Ofner, A., Gibas, J., and Thommen, R., 1961, Total synthesis of rhodovibrin (P-481) and rhodopin, *J. Org. Chem.* **31**:186.

Vetter, W., Englert, G., Rigassi, N., and Schwieter, U., 1971, Spectroscopic methods, in: *Carotenoids* (O. Isler, ed.), pp. 189–266, Birkhäuser, Basel.

Weedon, B. C. L., 1971, Stereochemistry, in: *Carotenoids* (O. Isler, ed.), pp. 267–324, Birkhäuser, Basel.

Yamaguchi, M., 1959, Total synthesis of isorenieratene, *Bull. Chem. Soc. Jpn.* **32**:1171.

Zechmeister, L., 1962, *Cis–Trans Isomeric Carotenoids, Vitamins A and Arylpolyenes*, Springer-Verlag, Vienna.

CHAPTER 13

Cytochromes

Robert G. Bartsch

1. Introduction

1.1. Background and Objectives

The presence of cytochromes in the photosynthetic bacteria was first recognized by biochemists who proceeded to purify and chemically characterize these heme proteins (Vernon, 1953; Elsden *et al.*, 1953; Kamen and Vernon, 1954, 1955). Subsequently, *in vivo* light-induced redox reactions involving some of these cytochromes were demonstrated (Duysens, 1954; Chance and Smith, 1955) and were seen to be analogous to the similar reactions observed in chloroplasts of green plants (Hill and Scarisbrick, 1951; Davenport and Hill, 1952). The photosynthetic bacteria have since proved to be, on one hand, a prolific source of cytochromes and, on the other, favored objects for study of light-induced electron-transfer reactions.

This discussion will present a description of the variety of cytochromes found in photosynthetic bacteria (see Table 1), with emphasis on the chemical properties of those cytochromes that have been purified from photosynthetically grown cells.

Reviews that are in part concerned with the cytochromes of the photosynthetic bacteria include Bartsch (1968, 1971), Kamen and Horio (1970), Horio and Kamen (1970), Kamen *et al.* (1972), Lemberg and Barrett (1973), Yamanaka and Okunuki (1974), and Kamen (1973).

The photophysiologist's interest in the cytochromes

and related electron-transfer agents of the photosynthetic bacteria has been largely confined to those components that can be implicated in light-driven redox reactions. One theme of this review is to emphasize the occurrence in the bacteria of additional cytochromes of chemical and biological interest. A second theme concerns the recognition of different cytochrome patterns in the various groups of photosynthetic bacteria, patterns that do not strictly correspond to the taxonomic divisions. The third major theme is the comparison of these cytochromes with those of other groups of organisms.

It is accepted that the absorption of a light quantum results in the cyclic passage of an electron from reaction center bacteriochlorophyll (Bchl) to a primary acceptor, quickly followed by transfer of an electron from a ferrocytochrome c to the photooxidized chlorophyll. The cycle is completed by electon transfer from the reduced acceptor via other components, including ubiquinone and cytochrome b, and the reduced state of the cytochrome c is restored. Cytochromes have also been implicated in noncyclic electron transfer from oxidizable substrates.

Based on the types of cytochromes c implicated in the light-driven cyclic electron-transfer process, the photosynthetic bacteria can be divided into several categories. One such category includes those bacteria among the Rhodospirillaceae that contain readily soluble cytochrome c_2. A second category includes several species in which a high-redox-potential cytochrome c-556 and low-redox-potential cytochrome c-552 are tightly bound to the cytoplasmic membrane. Included in this group are *Chromatium vinosum*,

Robert G. Bartsch · Chemistry Department, University of California, San Diego, California 92093

Table 1. Distribution of Cytochromes in Photosynthetic Bacteria^a

| Organism | Soluble cytochromes | | | | | | | | | Membrane-bound cytochromes | | | | |
| | Miscellaneous | | | | | | | | Flavocytochromes | | | | | |
	c_2	c'	c-556	c-554	c_3	"f"	High-spin	Low-spin	c	b-558	c-552, c-556	C (mesoheme)	B (protoheme)	Oxidases
Rs. fulvum														
I	++	++												
II	+													
Rs. molischianum														
I	++	++												
II	+													
Rs. photometricum	++	++										+ (*C*-554)	+	
Rs. rubrum	++	++								+		+	+	+ (o)
Rs. tenue	++	++		+			++ [*C*-552.5(548)]			+		+ (*C*-552)	+	
Rp. acidophila	++	−												
Rp. capsulata	++	++	+	+	+		+ "SHP"					+	+	+ (o?)
Rp. gelatinosa	−	++				+		++ (*c*-551)	++		++			
Rp. palustris														
Str. 37	++	+	+	+	+					+			+	+ (o?)
Str. 6	++	−	+	+	+					+			++	
Str. x	++	++		+						+			+	
Str. 18	++	+	+											
Str. Morita	++	+												

									(C-553)	(B-561)	(o) (a-type)
Rp. sphaeroides	++	++	+	+	+ (SHP)	+				++	+	+ +
Rp. viridis	++	+	−									
Rp. spp., Strains, 2.8.1, 2.10.1	+	+	++								++	
Rm. vannielii	+	−								+ (C-553)	+	+
Chr. vinosum	++	+		+ (b')	+ (c-550)	++			++	++	+	
Thiocapsa pfennigii	+			+ [c-552 (545)]		++			++			
T. roseopersicina				+ (c-550)		+			++	+ (c-552.5)	+	
Chl. limicola f. thiosulfatophilum	++			++ (c-551)		++						
Photosynthetic component of "Chloropseudomonas ethylica"[b]	++											
Chloroflexus aurantiacus										+ (c-554)		

[a] The approximate relative concentrations of those cytochromes that have been isolated or those bound cytochromes that have been detected spectroscopically are indicated. Yields of 1–5 μmol cytochrome heme/kg wet wt. of cells are indicated by (+); greater amounts, by (++). The assured absence of a given component is indicated by (−). No entry indicates that a component was not observed, or in the case of the membrane-bound cytochromes, that no observation has been reported. (?) Indicates uncertain assignment to a given cytochrome class.

[b] See Chapter 2.

Thiocapsa pfennigii, Rhodopseudomonas gelatinosa and *Rp. viridis*. The third category is limited at present to the few species of the Chlorobiaceae that have been examined, as well as *Chloroflexus aurantiacus*, in which membrane-bound cytochromes *c*-553 or *c*-554 are implicated in light-driven redox reactions.

Many of the Rhodospirillaceae are fully capable of aerobic growth in the dark, although some show only low tolerance for oxygen. In these organisms, the same cytochrome c_2 that is implicated in light-driven electron-transfer processes probably also functions in oxidative electron-transfer processes (Taniguchi and Kamen, 1965; Kikuchi *et al.*, 1965). The concentration of some of the cytochromes found in light-grown cells is depressed under aerobic growth conditions (Taniguchi and Kamen, 1965), and in some species, a new aerobic oxidase appears (Kikuchi *et al.*, 1965; Saunders and Jones, 1974) to supplement the cytochrome *o*-type oxidase that seems to be constitutive in many of these bacteria (Klemme and Schlegel, 1969; Connelly *et al.*, 1973) (see also Chapter 33).

Most of the species of Rhodospirillaceae have been surveyed for soluble cytochrome content, as have representatives of several readily grown species of the Chlorobiaceae and Chromatiaceae. The chemical properties of only a few of the membrane-bound cytochromes of any of these bacteria have been determined.

1.2. Nomenclature and Definitions

The general recommendations of the IUB Committee on Cytochrome Nomenclature (Florkin and Stotz, 1965) will be followed, and are summarized here. Heme proteins that can undergo reversible oxidation–reduction reactions involving the ferrous–ferric couple of the prosthetic group are presumed to fulfill an electron-transfer function and are called *cytochromes*. Those cytochromes for which the heme group has been characterized will be designated as cytochrome *a*, *b*, or *c*; cytochromes *a* contain heme *a*, a heme with a formyl side chain; cytochromes *b* contain protoheme; cytochromes *c* contain heme covalently linked to the protein. The hemes of cytochromes *a* and *b* can be readily extracted into acetone or methyl-ethyl ketone acidified to pH 2 with HCl, whereas the heme of the known cytochromes *c* can be split from the protein by disrupting the thioether bonds between cysteine residues of the protein and the substituted vinyl side chains of the heme, which so far has proved to be derived from protoheme. Such splitting can be best accomplished by the action of heavy metal salts at low

pH on the completely denatured cytochrome, e.g., as described by Ambler and Wynn (1973).

The alkaline pyridine ferrohemochrome absorption spectrum provides the simplest means available to distinguish among the types of heme. After the soluble hemes are extracted, the acidified organic solvent is removed by evaporation, the solubilized heme fraction is dissolved in 25% (vol./vol.) pyridine in 0.2 M NaOH, and the absorption spectrum is traced before and after addition of sodium dithionite to reduce the hemes *a* or *b*. The residue from the extraction may be similarly treated to detect the heme *c* content. The *α* peak absorption maximum of the alkaline pyridine ferrohemochrome spectrum of heme *c* is 549–551 nm ($\varepsilon_{mM} \cong 29$); of protoheme (heme *b*), 556–558 nm ($\varepsilon_{mM} \cong 34$)) (see Falk, 1964); and of heme *a*, 580–590 nm ($\varepsilon_{mM} \cong 26$–30) (Lemberg, 1969). Thus, by measurement of the spectrum of the heme derivative, it is feasible to establish the category to which the cytochrome belongs and its concentration. Mixtures of protoheme and *c*-type heme, however, will give intermediate *α* absorption peak maxima, as will the variant *c*-type hemes that are bound by a single thioether bond and thus have a free vinyl group (Pettigrew *et al.*, 1975). This type of variant cytochrome *c* has not been found as yet in the photosynthetic bacteria.

Where appropriate, the wavelength of the reduced *α* peak absorption maximum of a cytochrome will be designated, and the organism in which it occurs will also be specified; thus, cytochrome *c*-554 (*Rp. palustris*). Because of historical precedent, the designations cytochrome c_2 and cytochrome *c'* are retained. The name of the latter cytochrome has undergone numerous changes, but now seems likely to remain as cytochrome *c'*, the designation for predominantly high-spin cytochromes *c*. By analogy, it is appropriate to designate high-spin cytochromes *b* as *b'*.

Cytochromes for which the heme group has not been identified are labeled according to the reduced *α* peak, whether determined from absolute or difference spectra; thus, cytochrome-560 (*Rs. rubrum*).

A much less exacting operational method for designating cytochromes has come into vogue whereby redox substances assumed to be cytochromes of the *b* or *c* class are distinguished by a subscript corresponding to the apparent midpoint redox potential of the cytochrome. The cytochrome has been designated as *c* or *b* depending on whether the maximum absorbance change measured in the redox titration occurs near 550 nm or near 560 nm, respectively; thus, cytochrome b_{-105} of *Rs. rubrum* (Dutton and Jackson, 1972), which might better be designated as cytochrome $B_{-105 \text{ mv}}$, inasmuch as direct chemical identification of the presumed heme has not proved to be feasible, although circumstantial evidence based on the action of inhibitors has been adduced in favor of the identification. That the α absorption bands of c-type and b-type cytochromes overlap in the 550 to 560 nm spectral region further decreases confidence in the accuracy of identification of cytochromes in both difference spectra and redox titration spectra of complex cytochrome mixtures in membrane samples.

2. Methods for Isolation and Characterization

2.1. Purification Methods

The nature, number, and yield of cytochromes recovered from bacteria depend in large degree on the scale of the preparation and on the isolation procedure used. It is therefore appropriate to outline some methods that have been devised.

A general purification procedure that has been successfully applied to many soluble cytochromes of photosynthetic bacteria (Bartsch, 1971) is summarized here:

1. Prepare a cell-free and particle-depleted extract of the desired cells in 0.1 or 0.2 M buffer, pH 7–8.

2. Chromatograph this extract on a DEAE–cellulose column to remove acidic proteins such as bacterial ferredoxin as well as nucleic acid fragments.

3. Desalt the unadsorbed protein by passage through a column of G-25 Sephadex. It may be desirable to buffer the column to limit the extent of nonspecific retardation of cytochromes and to simultaneously adjust the pH of the solution for the next step.

4. Chromatograph the cytochromes on a DEAE–cellulose column to remove and recover acidic proteins. Most unadsorbed cytochromes can be recovered by chromatography on a CM–cellulose column after adjusting the solution to pH 6.

5. Concentrate the recovered cytochrome bands on small DEAE– or CM–cellulose columns after dilution or desalting, and elute with a buffer–salt solution of high ionic strength, such as 0.5 M NaCl.

6. Fractionate with ammonium sulfate to eliminate readily precipitated proteins, which frequently appear to consist of membrane fragments.

7. Chromatograph the redissolved cytochrome fractions on G-75 or G-100 Sephadex columns.

8. The recovered cytochromes can frequently be crystallized from ammonium sulfate solution, or, failing that, repetition of one or more of the chromatography steps can generally produce pure cytochromes.

In several of the operations described above, it is advisable to control the redox state of cytochromes in order to minimize the number of differently charged ionic species that must be separated by ion-exchange chromatography. The high-potential cytochromes can be successfully kept in the reduced state by including a reducing agent such as 1 mM 2-mercaptoethanol or dithiothreitol in the solutions. Alternatively, the cytochromes can be conveniently oxidized by treatment with excess potassium ferricyanide, although reduction by adventitious reductants may occur during subsequent chromatography. The ionic products of the oxidation treatment can be removed from the protein solution by use of a Sephadex G-25 column or more quantitatively by adsorption onto a column of an anion exchanger such as Dowex-1-Cl$^-$, which does not appreciably adsorb large protein molecules. The low-potential cytochromes are best left in their auto-oxidized state.

Among the properties listed in Table 2 are some that are useful for devising a purification strategy for many of the cytochromes listed. These properties include isoelectric point, redox potential, molecular size or weight, and details of actual conditions reported for the chromatography of the cytochromes. Based on experience, it is worth adding the caution that many nuances of technique are not successfully recorded in published procedures, and it is frequently useful to make a few small-scale trial chromatograms to determine proper experimental conditions for use with a given bacterial extract and set of reagents.

An alternate approach useful for recovering large-molecular-weight or polymeric cytochromes that otherwise are lost because they do not chromatograph well on the ion-exchange media is to first concentrate the crude extract either by collecting the fraction that precipitates between 20 and 90% saturated ammonium sulfate or by use of a membrane-type ultrafiltration apparatus. The concentrated material is then chromatographed on a Sephadex G-100 column. The large-molecular-weight zone therefrom is then fractionated by chromatography on hydroxyapatite or by precipitation with ammonium sulfate to obtain proteins such as the high-spin protoheme protein, cytochrome b', from *Chr. vinosum*, a polymeric iron–sulfur protein from *Rp. palustris*, and even cytochrome b-558 from various Rhodospirillaceae (T. E. Meyer, private communication).

2.2. Isolation of Membrane-Bound Cytochromes

Some of the Rhodospirillaceae such as *Rs. rubrum* and *Rp. palustris* appear to contain little cytochrome C intimately bound to the membrane, but all members

Table 2. Properties of the Cytochromes of Photosynthetic Bacteria

Cytochrome	Organism	Spectra[a]		Mol. wt. (kdaltons)[b]	$E_{m,7}$ (mV)	pI	Chromatography conditions[c]	Purity index[d]
		λ	ε_{mM}					
c_2								
	Rs. fulvum[e]							
	I	414.5 549.5 408.5 ox.					35 mM N + 5 mM T, 7.3, ox. (C)	0.11 red.
	II	414.5 549.5 409.5 ox.					24 mM N + above	0.15 red.
	Rs. molischianum[e,f]							
	I	415 549 409 ox.	130 29	10.7_4 (F[g])	381	9.78	27 mM N + above	0.12 red.
	II	415 549 410 ox.	136 29.8	10.5_6 (F[g])	288	9.44	13 mM N + above	0.18 red.
	Rs. photometricum[h]	416 550.5 412 ox.		12 (S)	345		40 mM N + 20 mM T, 7.3, ox. (C)	0.18 red.
	Rs. rubrum[i]	415 550 410 ox.	156 32	12.8_6 (F[j])	307[k] 295[l]	6.2	30 mM T, 8.0, red. (D)	0.18 red.
	Rp. acidophila[e]	416 550.5 410 ox.			370[h]		6 mM N + 5 mM T, 7.3, ox. (C)	0.21 red.
	Rp. capsulata[m]	416 550 410 ox.		12.8 (F[g])	340[n]		40 mM N + 10 mM T, 8.0 (D)	0.17 red.
	Rp. palustris							
	Str. 37[o]	418 551.5 410 ox.	168 27	13.5 (F[g])	330[p] 366[h]	9.7	2 mM P, 6.0, ox. (C)	0.14 red.
	Str. 6[q]	417 551.5 412 ox.		12.8_3 (F[g])	346[h]		15 mM N + 20 mM T, 7.3, ox. (C)	0.17 red.

	Absorption maxima	Mol. wt.		E		Conditions	
Rp. sphaeroides[e]	417, 550, 411 ox.	148, 30.8	14.0_9 (F[g])	340[l], 352[h]	5.5	35 mM N + 20 mM T, 7.3, ox. (D)	0.21 red
Rp. viridis[e]	416, 550.5, 410 ox.		12.3_0 (F[g])	296[h]		5 mM T, 7.3, ox. (C)	0.16 red.
Rm. vannielii[e]	416, 550.5, 409.5 ox.		11.8_0 (F[g])	304[r], 360[h]	7.9	15 mM N + 5 mM T, 7.3, ox. (C)	0.18 red.
Rhodopseudomonas spp., Strains 2.8.1 and 2.10.1[e]	416, 552, 412 ox.					50 mM N + 20 mM T, 7.3, red. (D)	0.18 red.
Rs. fulvum[e]	Not purified					17 mM N + 5 mM T, 7.3, ox. (C)	
Rs. molischianum[e]	426 red., 556, 400 ox., 498, 643		38 (S)		7.2	20 mM T, 7.3 (D)	0.36 ox.
Rs. photometricum[h]	422 red., 549, 391 ox., 495, 635					20 mM T, 7.3 (D)	0.23 ox.
Rs. rubrum[i]	423, 550, 390 ox., 497, 638	98/heme, 25.3, 82, 11, 3	13.5_3 (F[s])	−8	5.6	50 mM T, 8.0 (D)	0.27 ox.
Rs. tenue[e]	427 red., 546, 402 ox., 494, 634					18 mM N + 5 T, 7.3 (C)	0.4 ox.

c'

(continued)

Table 2. (continued)

Cytochrome	Organism	Spectra[a] λ	Spectra[a] ε_{mM}	Mol. wt. (kdaltons)[b]	$E_{m,7}$ (mV)	pI	Chromatography conditions[c]	Purity index[d]
c' (cont.)	Rp. capsulata[m]	425 red., 550, 399 ox., 495, 635		28 (S[q])	0.0	4.7	60 mM N + 10 mM T, 8.0 (D)	0.23 ox.
	Rp. gelatinosa[e]	427 red., 550, 400 ox., 495, 640		14 (A), 27 (H), 35 (S)		9.6	20 mM P, 7.0 (C)	0.24 ox.
	Rp. palustris Strain 37[o]	426, 552, 398 ox., 500, 642	99/heme, 10.3, 85, 11, 3	14 (A[e]), 13.4 (H)	102	9.4	5 mM P, 6.0 (C)	0.21 ox.
	Rp. sphaeroides[e]	425, 546, 400 ox., 496, 633	99, 10.4, 90, 10.5, 3	25 (S)	30[t]	4.9	70 mM N + 20 mM T, 7.3 (D)	0.29 ox.
	Rhodopseudomonas spp., Strain 2.8.1[e]	Not pure		≈15 (S)			50 mM N + 20 mM T, 7.3 (D)	
	Chr. vinosum[u]	426, 547, 399 ox., 495, 634	95/heme, 11, 85, 12, 3	14 (A[e]), 28(H), 37 (S[q])	−5	4.6	100 mM N + 20 mM T, 8.0 (D)	0.33 ox.
	T. pfennigii[v]	426.5 red., 546, 399 ox., 490, 632		14 (A[e]), 11 (P[e])			160 mM N + 20 mM T, 7.3 (D)	0.37 ox.
Miscellaneous high-spin	Rp. sphaeroides "SHP"[e]	429, 548	119	13 (S)	−22[t]		20 mM N + 20 mM T, 7.3 (D)	0.14 ox.

		400 ox. 497 626	171					0.18 red.
c-556	Rp. palustris, Strain 37°	420 556 415 ox.	136 19.9	~12 (S)	230		2 mM P, 6.0, ox. (C)	
	Rhodopseudomonas spp., Strains 2.8.1 and 2.10.1[e]	420 red. 556 411 ox. 506 630					150 mM N + 20 mM T, 7.3 (D)	0.14 red.
c-554	Rp. palustris, Strain 37°	418.5 554 412.5 ox.	163 27	40 (S) 58 (S[c])	−6		50 mM P, 6.0, red. (C)	
	Rp. sphaeroides[e]	419 554 412 ox.	144 25.7	44 (S)	120[w]	4.1	160 mM N + 20 mM T, 7.3, ox. (D)	0.18 red.
c_3	Rs. tenue[e]	417.5 red. 551 408.5 ox.					75 mM N + 5 mM T, 7.3 (C)	0.1 red.
	Rp. palustris, Strain 37°	418.5 551.5 409 ox.		13 (S)	−150	6.1	100 mM N + 20 mM T, 7.3 (D)	0.25 red. (impure)
	Rp. sphaeroides[x]	419 551.5 409 ox.	216/heme 29.5	21 (S[c]) (2 hemes)	−254	4.1	120 mM N + 20 mM T, 7.3 (D)	0.07 red.
"f-like"	Chl. limicola f. thiosulfatophilum[y]	418.5 555 (551) 412.5 ox.	153 21.6[z]	9.4_7 (F[aa])	145[bb]	10.5	20 mM N + 20 mM T, 7.3 (D)	0.16 red.
	Photosynthetic component of "Chloropseudomonas ethylica"[e,cc,ll]	417.5 555 (550) 412 ox.		11.1_5 (F[aa])	103[dd]	4.65	40 mM N + 20 mM T, 7.3 (D)	0.18 red.
	Rp. gelatinosa[e]	417 553 (550) 411 ox.		27 (S)		9.6	10 mM P, 6.0, ox. (C)	

(continued)

Table 2. (continued)

Cytochrome	Organism	Spectra[a] λ	Spectra[a] ε_{mM}	Mol. wt. (kdaltons)[b]	$E_{m,7}$ (mV)	pI	Chromatography conditions[c]	Purity index[d]
"f-like" (cont.)	Chr. vinosum[ee]	417.5 553 (550) 410 ox.	115 16.3	13–50 (H) 19 (S[e])	320	4.4	20 mM T, 7.3 (D)	0.12 red.
	T. pfennigii[v]	417 552 (550) 412 ox.		30 (P) 12/heme (A)			120 mM N + 20 mM T, 7.3 (D)	0.16 red.
Flavocytochrome c	Chl. limicola f. thiosulfatophilum[y]	416.7 553.5 410 ox.	156 30	50 (H) 11/monoheme subunit, 45/flavin subunit (P[ff])	98	6.7	20 mM N + 20 mM T, 7.8 (D)	0.94 ox.
	Chr. vinosum[gg]	416.5 552 410 ox.	310 54	72 (H) 21/diheme subunit, 45/flavin subunit (P[hh])	10	4.5	160 mM N + 20 mM T, 8.0 (D)	0.55 ox.
	T. roseopersicina[m]	416.5 552.5 409 ox.					≈280 mM N + 10 mM T, 7.3 (D)	
Miscellaneous cytochromes c	Rs. tenue[e]	416.5 552.5 (548) 409.5 ox.			405[h]		34 mM N + 5 mM T, 7.3 ox. (C)	0.21 red.
	Rp. gelatinosa[e]	417 551 409 ox.					40 mM N + 10 mM T, 7.3, ox. (C)	0.14 red.
	Chl. limicola f. thiosulfatophilum[v]	416 551 410.5 ox.	114/heme 17.6	45 (H) 60 (S) (≥2 hemes)	135	6.0	40 mM N + 20 mM T, 7.8 (D)	0.39 red. (impure)
	T. pfennigii[v]	415.5 552 (545) 407 ox.		30 (P, S) 12/heme (A)	Low[kk]		220 mM N + 20 mM T, 8.0 (D)	0.2 ox.
b-558	Rs. rubrum[k]	425 557.5 407 ox.		450 (H) 23	−204[kk]	4.6	10 mM P, 7.0 (D)	1.3 ox.

	Absorption maxima (nm)	E_h	Extinction	Preparation	Purity
Rp. palustris, Strain 37[o]	425.5 / 559 / 419 ox.	−200[kk]		40 mM N + 20 mM T, 8.0 (D)	1.3 ox.
Rp. sphaeroides[ii]	426 / 559 / 419 ox.			Ads. CG 50–100 mM acetate, pH 5.0; not ads. pH 5.65	1.2 ox.
b'	438 / 558 / 407 ox.	Low[kk]		40 mM P + 0.2 M N, hydroxyapatite	Not pure
Solubilized bound cytochromes					
Chr. vinosum[q]	≈110 (S)				
c-556, *c*-552 complex, *Chr. vinosum*[jj]	418 Δspec. / 556	$E_h > 100$ mv	325	2% cholate–0.5 M N extr. of acetone powder; 2% cholate +	
	552	$E_h < 0$ mv	8	0.2 M N + 0.1% mercaptoethanol, +50 T, 8.0, hydroxyapatite	
c-553, *Chl. limicola* f. *thiosulfatophilum*[m]	416.5 / 552.5 / 411.5 ox.			2% Triton X-100 extr. of acetone powder, 0.2 N, 20 T, 8.0 (D)	
c-554, *Chloroflexus aurantiacus*[m]	420.5 / 554 / 413.5 ox.			Same as above	

---

[a] The extinction coefficient values are generally determined as absorbance ($A_{\lambda max}$) per millimole heme measured by alkaline pyridine ferrohemochrome analysis. Some spectral data are taken from Bartsch (1963), where representative spectra of several of the cytochromes can be found.

[b] Molecular-weight determination methods: (F) formula weight from amino acid sequence data; (S) molecular sieve size determination; (P) polyacrylamide-gel electrophoresis with sodium dodecyl sulfate; (A) formula weight from amino acid composition.

[c] (T) Tris–HCl buffer and (P) phosphate buffer; (N) NaCl; (D) DEAE–cellulose and (C) CM–cellulose ion exchanger.

[d] The purity index is expressed as the ratio of the uv absorption maximum between 275 and 280 nm to the γ absorption maximum either oxidized (ox.) or reduced (red.).

[e] T. E. Meyer (private communication); [f] Dus *et al.* (1970); [g] R. P. Ambler (private communication); [h] G. W. Pettigrew (private communication); [i] Bartsch *et al.* (1971); [j] Dus *et al.* (1968); [k] Kakuno *et al.* (1973); [l] Dutton and Jackson (1972); [m] R. G. Bartsch (unpublished); [n] Evans and Crofts (1974); [o] R. G. Bartsch and T. Horio (unpublished); [p] Henderson and Nankiville (1966); [q] R. G. Bartsch and T. E. Meyer (unpublished); [r] Morita and Conti (1963); [s] Meyer *et al.* (1975); [t] M. A. Cusanovich (private communication); [u] Bartsch and Kamen (1960); [v] Orlando (1962); [w] Meyer *et al.* (1973); [x] Meyer *et al.* (1971); [y] Meyer *et al.* (1968); [z] van Beeuman and Ambler (1973); [aa] J. Gibson (1961); [bb] Yamanaka and Okunuki (1968); [cc] Olson and Shaw (1969); [dd] Shioi *et al.* (1974); [ee] Cusanovich and Bartsch (1969); [ff] Kusai and Yamanaka (1973); [gg] Bartsch *et al.* (1968); [hh] Kennel and Kamen (1971a, b); [ii] Orlando and Horio (1961); [jj] Kennel (1971); [kk] These low-potential cytochromes reduce very sluggishly with sodium dithionite. [mm] See Chapter 2.

of the group contain appreciable amounts of heme proteins thought to be cytochromes B on the basis of the reduced-minus-oxidized difference spectra. No successful attempt to solubilize these cytochromes B has been reported. Although these cytochromes can be solubilized, as is indicated by experience with reaction center preparations where variable amounts of cytochromes B have been observed, dependent on the type and concentration of detergent used (Reed, 1969), no successful purification of the cytochromes has been reported. My own experience with *Rs. rubrum*, *Rp. capsulata*, *Rp. palustris*, and others indicates that the bound cytochromes B are unstable after solubilization with detergents. The cytochromes disappear throughout purification attempts, and no pure material is ever recovered. A possibly related b-type cytochrome present at high levels in a temperature-sensitive *Rs. rubrum* mutant (Weaver, 1971) is easily extracted, but also is unstable during purification attempts, even in the absence of detergent (Weaver, 1974; R. G. Bartsch, unpublished). The methods such as were applied to the purification of mitochondrial cytochrome c_1 (Yu *et al.*, 1972) and to similar sensitive cytochromes may successfully preserve the bacterial cytochromes B.

Membrane-bound cytochromes C are found in most of the photosynthetic bacteria. Successful isolation of several of the relatively stable cytochromes C has been achieved. The cytochrome c-556–c-552 complex that constitutes about 80% of the total cellular heme of *Chr. vinosum* has been purified (Kennel and Kamen, 1971*a,b*). Subchromatophore Fraction A, which contains the cytochrome c-556–c-552 complex in addition to photosynthetic pigments (Thornber, 1969), proved to be the best starting material. Lipids, which interfere with subsequent chromatographic steps, were removed by extraction with acetone, and the residue was washed with dilute Tris buffer to remove any soluble proteins before the particulate cytochromes were extracted with 2% (wt./vol.) sodium cholate in 0.5 M sodium chloride plus 50 mM Tris buffer, pH 8.0. A high-ionic-strength extracting solution was required for optimum solubilization of the cytochromes. More than one-half of the bound cytochrome was extracted in this manner. Repeated chromatography on hydroxyapatite columns yielded the pure cytochrome c-556–c-552 complex, the properties of which are described in Section 4.2.1.

With the membrane-bound cytochrome c-554 of *Chloroflexus aurantiacus*, it was found that only after the membrane fraction was extracted with cold acetone could any of the cytochrome be solubilized by treatment with 2% (wt./vol.) Triton X-100 in 50 mM Tris buffer, pH 8.0. The resulting extract was chromatographed on DEAE–cellulose to yield nearly pure cytochrome, which appeared to require no supplemental detergent to remain in aqueous solution. A similar though less pure cytochrome was obtained from *Chl. limicola f. thiosulfatophilum* by analogous treatment.

The cytochrome a type oxidase of a *Rs. sphaeroides* strain was solubilized by extracting the membrane fraction from aerobically grown cells with Triton X-100. The cytochrome was partially purified through fractionation by ammonium sulfate precipitation and by chromatography on DEAE–cellulose (Sasaki *et al.*, 1970).

None of the purified cytochromes named above has been successfully reincorporated into depleted membranes to restore their functional activities.

2.3. Measurement of Oxidation–Reduction Potentials

The midpoint oxidation–reduction potential, the voltage at which an electron-transfer agent is half reduced, is indicated by the symbol $E_{m, \text{pH}}$, where the pH of the measurement is indicated in the subscript, and the ambient oxidation–reduction potential of a solution containing redox substances is designated as E_h (Clark, 1960). For meaningful measurement and use of E_m values, the necessity for equilibrium among the components of the reacting system must be assured. As a consequence, the values listed in Table 2 (see Section 4), which were obtained in various ways and with various degrees of experimental care, must be used with caution as to their accuracy. For an example of the problems involved, see the discussion of the E_m value for cytochrome b-558 (*Rs. rubrum*) in Section 4.3.2. The recognized possibility of questionable attainment of equilibria across bacterial membranes was discussed by Dutton and Wilson (1974).

Several procedures for E_m measurements can be listed. These include the method of mixtures (Davenport and Hill, 1952; Hill, 1954; Velick and Strittmatter, 1956); experimental considerations for titration with the ferri–ferrocyanide redox couple, which are also applicable to other redox systems (Hanania *et al.*, 1967); and the electrochemical titration procedures as applied by Loach (1966); Cusanovich *et al.* (1968*a*), and Dutton and Wilson (1974).

2.4. Measurement of Concentrations

It is not easy to determine the concentrations of cytochromes in complex mixtures such as whole cells, membrane fractions, or even cell-free extracts. In Table 1 (see Section 3), this fact is reflected by the very approximate terms in which the cytochrome levels in the various bacteria are reported. The values listed are based largely on yields of cytochromes obtained after initial separation of the proteins.

Several stratagems have been devised to estimate the cytochrome concentrations in natural mixtures. For example, Kakuno et al. (1971) determined the amounts of cytochromes c_2 and c' and of protoheme (cytochrome B) in chromatophores. The cytochrome c_2 was selectively reduced and determined by virtue of its high redox potential and distinctive spectrum. From alkaline pyridine hemochrome measurements, the total heme c, less the cytochrome c_2, gave the cytochrome c' content. The protoheme content was measured separately. Weaver (1974) used a variation of this method in which absorbance measurements of reduced-minus-oxidized cytochrome c' were made at isosbestic points with the otherwise interfering cytochrome c_2 spectrum. This method permitted estimation of both cytochromes in an extract devoid of cytochrome B. These procedures depend on the availability of spectra and extinction coefficients of the purified cytochromes and on the widely separated E_m values of cytochromes c_2 and c', thus allowing cytochrome c_2 to be selectively reduced. Details of the analysis must be worked out for each system studied. Such relatively simple methods are not useful with complex mixtures of cytochromes c with similar spectral and redox properties.

Dutton and Jackson (1972) and Dutton and Wilson (1974) attempted to measure the relative amounts of cytochromes in membrane fractions on the basis of the absorbance changes for each postulated species in spectrophotometric redox titrations. For an accurate measurement from such data, it is necessary to accurately identify the chromophore responsible for an induced absorbance change and to have accurate extinction coefficients for the various cytochrome species assumed to be involved, data that are seldom available. In principle, any method to determine the stoichiometry between the electrons needed for the titration of a redox component and the magnitude of the absorbance changes of that component will yield the needed values directly, but techniques for such measurements with complex mixtures have yet to be developed.

Relatively large variations in reduced-minus-oxidized differential extinction coefficients have been observed. For many cytochromes c, $\Delta\varepsilon_{\mathrm{mM, 550\,nm}} \cong 19\text{--}20$, a value that has frequently been assumed in calculation of cytochrome c concentrations. For cytochrome c-553(550)* (Chr.

vinosum), and for cytochromes with similar spectral properties, $\Delta\varepsilon_{\mathrm{mM, 553\,nm}} \cong 10$ (Cusanovich and Bartsch, 1969). With such differences possible for extinction coefficients, it seems unwarranted to assume values for uncharacterized cytochromes in calculations.

An approach to the measurement of stoichiometries of redox systems is that of Dutton et al. (1975), in which the relative absorbance changes due to photooxidation of excess added cytochromes c_2 and mitochondrial cytochrome c are compared. If equivalent amounts of the cytochromes are oxidized by a reaction center preparation, then the extinction coefficient of the cytochrome c_2 can be calculated relative to the known value of the cytochrome c. Perhaps this principle can be extended to studies of membrane-bound cytochromes.

2.5. Measurement of Miscellaneous Properties

Estimation of molecular weights of the cytochromes can be accomplished by molecular sieve chromatography (Andrews, 1964), by standard ultracentrifuge procedures (Schachman, 1959), or by sodium dodecyl sulfate–polyacrylamide gel electrophoresis of the proteins (Weber and Osborn, 1970). This latter technique is also useful to determine the subunit size of polymeric cytochromes. The ultimate molecular-weight determination is the formula weight calculated from amino acid sequence data. Differences noted in Table 2 among a set of molecular-weight values determined by the different methods indicate nonideal behavior of the protein or the polymeric nature of the soluble cytochrome.

Protein isoelectric pH values can be determined effectively by the electrofocusing technique (Vesterberg and Svensson, 1966), or by polyacrylamide gel electrophoresis using Ampholine (LKB) electrolytes (Righetti and Drysdale, 1971). The electrofocusing technique permits determining if multiple-charged species of a single cytochrome exist in crude extracts as well as in purified preparations (Sletten and Kamen, 1968; Bartsch et al., 1971).

3. Distribution

3.1. Cytochrome Patterns

A summary of the distribution of cytochromes in photosynthetic bacteria is presented in Table 1. The description of individual cytochromes is left for Section 4. From Table 1 it is evident that no fixed pattern of occurrence of heme proteins exists. Variations occur even among strains currently

---

* The number in parentheses, 550 in this case, designates a shoulder on the main α peak.

assigned to the same species. Note, for example, the varieties of cytochromes found among the several *Rp. palustris* strains. The simplest pattern recognized is that of *Rp. acidophila* and *Rhodomicrobium vannielii*, in which the only soluble cytochrome is cytochrome c_2 and the predominant membrane-bound cytochrome appears to be a cytochrome C-554, and that of *Chloroflexus aurantiacus*, in which a tightly bound cytochrome c-554 appears to be the sole cytochrome in the photosynthetically grown organism. With the exceptions noted in Table 1, only cytochromes found in the photosynthetic growth mode are listed.

The yield of soluble cytochromes can vary over a wide range, dependent on the medium and mode of growth, on the culture age, and on the method of extraction and purification, and therefore the relative amounts of cytochromes are presented in approximate terms only. As an example of extreme variation, Hilmer and Gest (private communication) observed that the soluble cytochrome c' content of late log phase *Rp. capsulata* (strain Z-1) varied from 100 μmol/kg dry wt. cells grown in minimal medium at low light intensity and at low temperature (25°C) to 600 μmol/kg dry wt. cells grown at high light intensity and high temperature (39°C) on the same medium supplemented with 1 g/liter glutathione or with glutamate instead of ammonia as nitrogen source. The rationale given was that if the E_h of the medium was maintained at a low level throughout the life of the culture, the cytochrome c' level was enhanced. The cytochrome c_2 content varied inversely with the cytochrome c' level, decreasing by approximately one-half under the second growth conditions. Weaver (1974) observed that the cytochrome c' content of *Rs. rubrum* was decreased at least tenfold in cells grown with succinate rather than with malate as the primary carbon source. Cytochrome c' was found to be almost completely suppressed when *Rs. rubrum* (Taniguchi and Kamen, 1965) or *Rs. sphaeroides* (Kikuchi *et al.*, 1965) was grown aerobically. With limited iron in the medium, the levels of both cytochromes c_2 and c' were severely depressed in *Rp. sphaeroides*, while the membrane-bound cytochromes seemed to be little affected (Agalidis *et al.*, 1974).

3.2. Periplasmic vs. Membrane-Bound Cytochromes

In the Rhodospirillaceae, most of the cytochromes B and variable amounts of cytochrome(s) C are intrinsically bound in the membrane (see Section 4.2.2 for a discussion). The readily extracted cytochromes appear to be located in the periplasmic space of the cell, for at least from *Rp. sphaeroides* and *Rp.*

capsulata, the cytochromes c_2 and c' can be nearly completely extracted by making spheroplasts of the cells (Prince *et al.*, 1975; Hochman and Carmeli, 1974). Nearly quantitative extraction of these cytochromes on a relatively large scale can be made from blue-green mutants of *Rs. rubrum*, *Rp. capsulata*, *Rp. sphaeroides*, and *Rp. palustris* by simply washing the cells with 4 mM EDTA, buffered at pH 7.5, whereas several successive overnight extractions are required with the respective wild-type cells to accomplish incomplete extraction of these cytochromes by this technique (R. G. Bartsch, unpublished). It is not yet known whether the minor soluble *Rhodopseudomonas*-type cytochromes c_3 or c-554 or cytochrome b-558 are extracted under either set of conditions.

It appears that on mechanical disruption of the bacterial cell, as much as one-third of these periplasmic cytochromes becomes encapsulated in the vesicles formed by the resealing of fragments produced by energetic rupture of the membrane. Dutton *et al.* (1975) suggest that chromatophores represent a satisfactory sampling of the native association of the cytochromes together with their natural membrane-bound reaction partners. Most of the entrapped cytochrome c_2 can effectively undergo light-induced electron-transfer reactions, whereas the cytochrome c' remains characteristically inert, or at least optically invisible. However, the stoichiometry between the cytochromes c_2 and c' and the reaction center Bchl and the membrane redox components such as ubiquinone varies considerably, dependent on the organism studied and the extraction process used (e.g., compare Kakuno *et al.*, 1971, and Dutton *et al.*, 1975). In addition, an appreciable portion of the entrapped cytochrome c_2 sometimes fails to undergo rapid light-induced oxidation (Dutton *et al.*, 1975). Prince *et al.* (1975) postulate that the periplasmic cytochromes occupy preferred binding sites relative to the electron-donating components such as NADH-cytochrome c oxidoreductase or the ubiquinone–cytochrome b electron-transferring system and to the electron-accepting components such as reaction center Bchl or the oxidase. There is an obvious analogy between the location of cytochrome c loosely attached to the outer surface of the inner mitochondrial membrane and the location of cytochrome c_2 in the periplasmic space of the bacterial cell.

It is reasonable to postulate the periplasmic location of readily soluble cytochromes of the photosynthetic bacteria, but not all the cytochromes c implicated in rapid light-induced photosynthetic oxidation reactions are so located Several species of photosynthetic bacteria have been recognized in which the cytochrome involved in the light-driven electron-transfer system may be entirely confined within the hydrophobic environment of the membrane. *Chromatium vinosum, Rp. gelatinosa, Thiocapsa pfennigii, Rp. viridis,* and possibly *Chloroflexus aurantiacus* and *Chl. limicola f. thiosulfatophilum* are included in this group. In the first four species, high-potential cytochromes *c*-556 or *c*-557 and low-potential cytochromes *c*-552 are embedded in the bacterial membrane, and, depending on experimental circumstances, one or both of these cytochromes have been shown to take part in light-driven redox reactions (Olson and Chance, 1960a,b; Olson and Nadler, 1965; Cusanovich *et al.*, 1968a,b; Parson and Case, 1970; Case *et al.*, 1970; Dutton *et al.*, 1970). These bound cytochromes *c* have at least in part been solubilized and partially characterized. In several of the Rhodospirillaceae, a relatively abundant bound cytochrome *C*-553 or *C*-554 is seen in reduced-minus-oxidized difference spectra of membrane fractions that have previously been washed with chaotropes or with acetone to remove residual soluble cytochromes. These bound cytochromes have not yet been purified, nor apparently have they been implicated in light-induced redox reactions.

4. Properties

4.1. Soluble Cytochromes *c*

Physical data for the various cytochromes are summarized in Table 2.

4.1.1. Cytochromes c_2

Cytochrome c_2 of *Rs. rubrum* was the first bacterial cytochrome isolated (Vernon, 1953), and was shown to be different in several respects from mitochondrial cytochrome *c* (Elsden *et al.*, 1953; Vernon and Kamen, 1954). Because it is the most completely studied example, it serves as the model for this cytochrome group. Cytochrome c_2 may be defined as that cytochrome found in the Rhodospirillaceae that is structurally similar to, but functionally distinct from, mitochondrial cytochrome *c*. The spectra of these cytochromes have a reduced α peak at 549–552 nm and a γ peak at 416–418 nm. The redox potential ranges from $E_{m,7}$ 290 to 400 mV. With the exception of *Rs. tenue* and *Rp. gelatinosa*, which lack cytochrome c_2, the cytochrome has proved to be the predominant soluble heme protein found in the Rhodospirillaceae. Cytochrome c_2 usually functions as the principal electron donor to reaction center Bchl$^+$ in

photosynthetic cells. *Rhodopseudomonas viridis* may be an exception because the cytochrome *c*-557, *c*-552 pair, but not the cytochrome c_2, has been observed to participate in the light-driven redox reaction (Olson and Nadler, 1965; Case *et al.*, 1970).

As many as seven isoelectric forms of the cytochrome occur in *Rs. rubrum* (Sletten and Kamen, 1968; Bartsch *et al.*, 1971), apparently as artifacts of deamidation (Flatmark, 1966) or of differential ion binding, and are probably not examples of isocytochromes. Similarly, several cytochromes c_2 are also found in *Rp. sphaeroides* (Prince *et al.*, 1974). In *Rs. molischianum*, there occur two cytochromes *c* that have essentially identical absorption spectra but differ in redox potentials, in isoelectric points (Dus *et al.*, 1970; Flatmark *et al.*, 1970), and in amino acid sequences, but contain nearly the same number of residues (R. P. Ambler, private communication). On the basis of closely similar absorption spectra, these two proteins have been designated as isocytochromes c_2 I and II (Flatmark *et al.*, 1970), in analogy to the cytochrome *c* pair found in *Saccharomyces cerevisiae* (Slonimski *et al.*, 1965). It will be interesting to determine whether both bacterial cytochromes are located together within the periplasmic space and whether, despite the approximately 100-mV difference in E_m, both can serve identical functions and therefore be true isocytochromes *c*. A possibly analogous pair of isocytochromes c_2 occurs in *Rs. fulvum* (T. E. Meyer, private communication). To date, only a single species of cytochrome c_2 has been isolated from the other Rhodospirillaceae examined.

From examination of the amino acid sequences and X-ray crystal structures, a strong similarity in structural features is evident among mitochondrial cytochrome *c*, *Paracoccus denitrificans* cytochrome *c*-550, and *Rs. rubrum* cytochrome c_2 (Dickerson and Timkovich, 1975; Timkovich and Dickerson, 1973; Salemme *et al.*, 1973a,b). The characteristic features include a globular molecular shape, a hydrophobic sheath surrounding the heme with the nonpolar amino acid side chains largely within van der Waals contact distance of the heme ring, one edge of the heme (pyrrole ring number 2) near the surface of the molecule, and a group of lysine residues surrounding this surface region. The heme is covalently bound to the peptide chain via two thioether bridges, although a single thioether link proves sufficient in two variant protozoan mitochondrial cytochromes *c* (Pettigrew *et al.*, 1975a). The heme-binding sequence Cys–X–Y–Cys–His is characteristically located within approximately 20 residues of the amino terminal end of the peptide chains. The extraplanar ligands of the heme iron that were predicted to be a histidine nitrogen

(Theorell and Åkeson, 1941) and a methionine sulfur (Harbury *et al.*, 1965) have been established by the various crystallographic structure studies.

From available sequence data, the cytochromes c_2 fall into two structural subgroups (R. P. Ambler, private communication). Cytochromes c_2 from *Rm. vannielii*, *Rp. viridis*, and *Rs. molischianum* are more like mitochondrial cytochrome *c* in chain length and in the number of homologous or invariant residues than are those in the second group, which includes *Rs. rubrum*, *Rp. palustris*, *Rp. sphaeroides*, *Rp. capsulata*, and *Paracoccus denitrificans*. For all the latter cytochromes, several insertions and deletions must be postulated for alignment with the smaller cytochromes. By analogy with the X-ray structures of the *Rs. rubrum* and *Paracoccus* cytochromes, it may be assumed that the extra residues of the larger cytochromes c_2 are all disposed as loops on the surface, relatively removed from the face of the molecules where the heme edge is exposed. In this manner, the basic cytochrome *c* structure is preserved with minimum apparent disturbance by the extra amino acid residues.

An intensive effort is under way to compare the reactivities of the various cytochromes *c* in various test systems in the attempt to deduce the relationship between the more or less subtle structural differences among the cytochromes *c* and the diverse chemical and biochemical reactivities of these cytochromes. From such studies, it is clear that despite the fundamental structural similarity among the cytochromes *c*, large quantitative differences exist in reactivity with various test oxidases and reductases, and therefore enzymatic specificities must be explained on a finer level than is so far evident from the structures available. As an example of such a test system, mitochondrial cytochrome *c* oxidase preparations react about 20 times faster with mitochondrial cytochromes *c* than with cytochromes c_2 (Elsden *et al.*, 1953; Yamanaka, 1972; Davis *et al.*, 1972; B. J. Errede and M. D. Kamen, private communication), but mitochondrial cytochrome *c* reductase preparations (complex II) react only about two times as fast with the mitochondrial cytochromes *c* as with cytochromes c_2 (Davis *et al.*, 1972; B. J. Errede and M. D. Kamen, private communication). Subtle technical problems that need not be discussed here have been found to greatly complicate this type of comparative enzymology (Errede *et al.*, 1976). Intermolecular electron transfer between the cytochrome and reaction partners probably occurs via the heme's pyrrole ring II, which most closely approaches the surface of the cytochrome *c* molecules (Salemme *et al.*, 1973*b*; Dickerson and Timkovich, 1975; Wood

and Cusanovich, 1975). The net positive charge in the vicinity of this access region may be important in the formation of reactive complexes between the cytochrome molecule and oppositely charged active-site areas of the enzymatic reductants or oxidants.

The pH profiles of various reactions involving cytochromes *c* have been studied. One such study involves the effect of pH on the redox potential of the cytochromes. For mitochondrial cytochrome *c*, E_m is independent of pH between pH 2 and pH 8–9, above which the E_m decreases at the rate of 60 mV/pH unit (Rodkey and Ball, 1950; Margalit and Schejter, 1973). It may therefore be assumed that no proton ionization in the physiologic pH range affects the redox potential of that cytochrome. *Rhodopseudomonas capsulata* cytochrome c_2 behaves in a similar manner (G. W. Pettigrew, private communication), but for *Rs. rubrum* c_2, the E_m changes between pH 4 and pH 8 in such a way as to suggest that the ionization of two or more groups affects the heme (Kakuno *et al.*, 1973; Wood and Cusanovich, 1975). For various bacterial cytochromes *c*, the "alkaline pH −60 mV/pH unit slope" begins between pH 6 and 8 (Kakuno *et al.*, 1973; Pettigrew and Schejter, 1974; G. W. Pettigrew, private communication). The suggestion of Salemme *et al.* (1973*b*), that a hydrogen bond network extending from the methionine sulfur that is liganded to the heme to a polar residue at the surface of the cytochrome molecule may provide a mechanism for the redox-dependent ionization, is at best only a partial explanation, even for those cytochromes in which such a network can exist.

From a comparison of the relationship between E_m and pH for a number of cytochromes c_2 (G. W. Pettigrew, private communication), it is apparent that below pH 6, there exists for almost all examples a plateau region over which the E_m value is constant with pH. Consequently, the E_m values of the cytochromes should be compared at a value such as pH 5 within the plateau region, rather than at the more conventionally used pH 7, where there is frequently a relatively steep change in E_m with pH. Perhaps this sensitivity of E_m to pH is somehow exploited in the energy-transduction system of the cell in connection with the light-induced acidification of the cytochrome c_2 side of the chromatophore membrane. It is also observed that in the plateau region, the E_m values of most cytochromes c_2 range between 350 and 400 mV. Only exceptional examples of cytochromes currently classed as cytochromes c_2 have lower E_m values, namely, *Rs. molischianum* iso-2-cytochrome c_2, *Rp. viridis* cytochrome c_2 and *P. denitrificans* cytochrome *c*-550.

The redox properties are also affected by the bound

or free state of the cytochromes. For the *Rp. capsulata* cytochrome c_2 contained in freshly prepared, but not in aged, membrane vesicles, the redox potential has a pH dependence of -60 mV/pH unit between pH 6 and 7, and is independent of pH above pH 7.5, directly opposite to the behavior of the freely soluble cytochrome, perhaps because the equilibrium for redox mediator across the fresh membrane is only slowly established (Evans and Crofts, 1974). The vesicle-trapped cytochrome c_2 of *Rp. sphaeroides* has $E_{m,\,7}$ 295 mV, whereas the free cytochrome has $E_{m,\,7}$ 340 mV (Dutton and Jackson, 1972), but both *Rs. rubrum* (Kakuno *et al.*, 1973) and *Rp. capsulata* (Prince *et al.*, 1975) cytochromes show the same $E_{m,\,7}$ in the bound and free states. Such disparate patterns indicate that several factors must be sought to explain the redox response of cytochromes c_2 to pH.

Prince *et al.* (1974) indicate that in light, the *Rp. sphaeroides* chromatophore interior reaches pH 4.5; i.e., the H^+ concentration on the cytochrome c_2 side of the membrane is about 1000 times greater than in the outside milieu. Therefore, studies on the effect of pH on the redox behavior of the cytochromes have immediate physiological relevance.

Appendix. The pH dependence and possible structural causes of redox potential variations among cytochromes c_2 have been described by Pettigrew *et al.* (1975b, 1978).

4.1.2. Cytochromes c'

Cytochromes c' are predominantly high-spin cytochromes (Ehrenberg and Kamen, 1965; Maltempo *et al.*, 1974) with the heme covalently bound to the peptide chain (Barrett and Kamen, 1961). The spectra as well as the primary architectural features of this group of cytochromes differ from those of the various types of low-spin cytochromes c. The heme binding sequence pattern is like that of low-spin cytochromes c, namely, Cys–X–Y–Cys–His, but is located near the carboxyl end of the protein (Kennel *et al.*, 1972b; Ambler, 1973; Meyer *et al.*, 1975), suggesting that like the low-spin cytochromes c, this His serves as one of the heme extraplanar ligands. As yet, no clue as to the identity of the weak field sixth ligand is available. Whatever this ligand may be, it can be displaced in the reduced cytochrome by carbon monoxide and in either oxidation state by nitric oxide, with characteristic changes in absorption spectra (Taniguchi and Kamen, 1963). Addition of ionic ligands such as cyanide or azide ions causes no changes in absorption spectra, and these agents are assumed to be unreactive with the heme. The cytochromes c' are autooxidizable, with $E_{m,\,7}$ values ranging between -10 and $+100$ mV.

Thus, only uncharged reactants such as O_2, CO, and NO seem to have ready access to the heme of cytochromes c', and the analogy to high-spin heme proteins such as myoglobin (Vernon and Kamen, 1954) is only partial.

The absorption spectra of cytochromes c' undergo reversible changes as a function of pH (Horio and Kamen, 1961; Taniguchi and Kamen, 1963; Cusanovich *et al.*, 1970). Similar spectral changes can be induced by organic solvents that can perturb hydrophobic regions of proteins (Imai *et al.*, 1969). These changes are also reflected in the ESR spectra (Maltempo *et al.*, 1974). Three spectral states are recognized for the oxidized proteins: the high-spin neutral type (I), the possibly mixed-spin intermediate type (II), and the low-spin alkaline type (III) (Imai *et al.*, 1969). At high pH, variously from pH 11 to 14, or at neutral pH in the presence most effectively of *o*-chlorophenol, the spectrum of ferricytochromes c' becomes completely low spin (type III), with visible absorption peaks very similar to those of cytochromes c_2. While no direct physiological significance can yet be attached to studies such as these, they may point to the reason these cytochromes are very difficult to detect *in vivo*.

The absorption spectra of the cytochromes c' have several features that distinguish this cytochrome c class from others. In the oxidized form, there are charge-transfer bands at 490–500 and 630–650 nm and a broad γ absorption maximum centered at 390–400 nm, characteristic of high-spin cytochromes. In the reduced form, there is a poorly defined absorption maximum at 550–565 nm, making for possible confusion with some cytochrome *B* spectra, and the γ peak is typically split, with a maximum near 425 nm and a pronounced shoulder near 430 nm. The spectrum of the carbon monoxide complex has definite low-spin character, with a single γ band at 416 nm and α and β bands that can be confused with those of other heme proteins that can also react with CO, such as cytochrome *o* (Horio and Taylor, 1965). As deduced from kinetic and isotopic analysis of flash-induced dissociation and reassociation of CO with cytochrome c', two CO molecules probably combine with the heme, although only one is liganded to the heme iron with resultant absorption spectrum change (Gibson, Q. H., and Kamen, 1966; Cusanovich and Gibson, 1973). The dissociation constant for CO is considerably higher than for heme proteins such as myoglobin. The site of attachment of the second CO molecule has not been determined.

Most of the cytochromes c' have been isolated in dimeric form, with molecular weights in the range 28–30 kdaltons. These dimers can be dissociated under denaturing conditions (Cusanovich, 1971). From several *Rp. palustris* strains, however, cytochrome *c'* is isolated in monomeric form with molecular weight approximately 13–14 kdaltons (Dus *et al.*, 1967). Amino acid sequence determination of cytochromes *c'* clearly indicate that two identical peptide chains comprise the 28-kdalton forms (Kennel *et al.*, 1972*b*; Ambler, 1973; Meyer *et al.*, 1975). The nature of the binding forces responsible for the dimeric character of the cytochromes *c'* has not been determined, but however these forces act, they have no pronounced effect on the absorption spectra, for the monomeric and dimeric cytochromes *c'* have very similar spectra.

Cytochromes *c'* have been isolated from the Rhodospirillaceae and the Chromatiaceae and from several other types of bacteria, including a denitrifying *Alcaligenes* species (Suzuki and Iwasaki, 1962; Iwasaki and Shidara, 1969; Ambler, 1973), a halotolerant denitrifying *Paracoccus* species (T. E. Meyer, private communication), and *Azotobacter vinelandii* (Yamanaka and Imai, 1972). No common electron-transfer role has yet been discerned for the cytochrome among these various bacteria.

Because of difficulties encountered in observing cytochromes *c'* in intact cells or membrane vesicles, it has been suggested that the cytochrome contained in chromatophores may have spectroscopic properties different from those of the free protein (Kakuno *et al.*, 1971; Weaver, 1974). The ferrocytochrome *c'* retained by *Rs. rubrum* vesicles appears not to react with CO, and the absorption spectra of the reduced and oxidized forms of cytochrome *c'* are effectively masked by more strongly absorbing *b*-type cytochromes (Kakuno *et al.*, 1971). That cytochromes *c'* are retained by various chromatophore preparations has been documented by several types of experiments, and methods have been devised to assess the relative amounts of cytochromes c_2 and *c'* in simple mixtures such as occur in *Rs. rubrum* extracts (Weaver, 1974) or in chromatophores (Kakuno *et al.*, 1971).

With *Rs. rubrum* chromatophores, Kakuno *et al.* (1971) determined that less than one-half of the cytochrome absorbance change at 562 nm was due to reduction of entrapped cytochrome *c'*, and the remainder was due to cytochromes *B* assayed as protoheme. Coincidentally, one-half of the absorbance change, with $E_{m, 7}$ 10–20 mV, was produced by the reaction of succinate with the particles. With NADH, the remaining absorbance change at 562 nm could be produced. On the basis of these observations, the suggestion was made that cytochrome *c'* might be involved in the succinic dehydrogenase electron-transfer system. However, *Rs. rubrum*, when grown with succinate as the sole carbon source, contains very little cytochrome *c'* (Weaver, 1974; R. G. Bartsch, unpublished), and therefore no obligatory connection between the cytochrome and succinic dehydrogenase activity is evident, even though cytochrome *c'* in chromatophores can be readily reduced with added sodium succinate.

The ferricytochrome *c'* ESR spectrum shows high-spin $g \simeq 4.3$–4.9 signals at 100°K (Ehrenberg and Kamen, 1965; Maltempo *et al.*, 1974). Attempts to utilize these signals to detect the cytochrome in chromatophores (Dutton and Leigh, 1973) and in whole cells (Corker and Sharpe, 1975) have been largely frustrated by uncertainties about the differences in signals between the free and the bound cytochrome and by the large temperature sensitivity of the ESR signal, which make it difficult to correlate the *g* values of the free cytochrome with similar ones seen with the particles or cells.

4.1.3. Cytochromes *c*-556

Cytochromes *c*-556 have been isolated from two *Rhodopseudomonas* species. The *Rp. palustris* cytochromes *c*-556 have typical low-spin cytochrome *c* spectra, but with both the *γ* and *α* band maxima shifted to longer wavelengths. The marine *Rhodopseudomonas* species of Elsden (Van Niel strains 2.8.1 and 2.10.1) yield similar cytochromes that have a degree of high-spin character as indicated by the low-intensity absorption band at 630 nm in the oxidized state (T. E. Meyer, private communication). A resemblance to cytochromes *c'* is suggested by the preliminary amino acid sequence data for the *Rp. palustris* cytochrome, which indicate similarity in the heme peptides (R. P. Ambler, private communication). The *Rp. palustris* cytochromes appear to be spectrally unstable, for during storage, the *α* band shifts from 556 nm to 550 nm.

Note Added in Proof. In several marine photosynthetic bacteria, including *Rp. sulfidophila* strains W-4 and W-12, the *Rhodopseudomonas* species mentioned, as well as several incompletely characterized recent isolates (P. Weaver, private communication), the predominant soluble heme protein is cytochrome *c*-556 with properties like those listed in Table 2 for the *Rhodopseudomonas* strains (R. G. Bartsch, unpublished). The possible structural affinity of these cytochromes with cytochromes *c'* is supported by the partial high-spin character of the absorption spectra and by the partial reaction (10–15%) in reduced state with carbon monoxide.

4.1.4. Cytochromes c-554

Cytochromes c-554 have been found in small amounts in several *Rhodopseudomonas* species. These cytochromes c are of relatively large molecular weight and are probably polymeric in nature. The absorption spectra of *Rp. sphaeroides* and *Rp. palustris* cytochromes c-554 resemble *Rp. palustris* cytochrome c-556 in detail, except for the α peak (T. E. Meyer, private communication). Because of the low yields, this cytochrome group has been one of the least studied. It has not been determined whether the soluble cytochrome is related to the membrane-bound cytochrome c-553 or c-554 observed in several of the Rhodospirillaceae.

4.1.5. Miscellaneous High-Spin Cytochromes c

Additional high-spin cytochromes have been isolated in small amounts from various *Rhodopseudomonas* species, but because of inadequate quantities, they have not been characterized. One is a cytochrome designated as "SHP," for *sphaeroides* heme protein (T. E. Meyer, private communication), the reduced γ band of which is of lower intensity than the oxidized band, in contrast to the relative ratio in cytochromes c'. In the reduced state, the cytochrome combines with CO. Another example of this type is a green-colored (in oxidized form) cytochrome from *Rp. capsulata* that behaves somewhat like "SHP" (R. G. Bartsch, unpublished). The possibility exists that these minor components are aberrant denatured forms of another cytochrome; nevertheless, these forms deserve mention and additional study because they are found reproducibly in the bacterial extracts. According to the nomenclature rules, these high-spin cytochromes c should be designated c', but some means must be devised to distinguish them from the quantitatively more significant cytochromes c'.

4.1.6. Cytochromes c₃

Low-potential cytochromes that are found in certain *Rhodospeudomonas* species (Meyer *et al.*, 1971) as well as *Rs. tenue* (T. E. Meyer, private communication) may be related to similar ones found in some blue-green algae and diatoms (Holton and Myers, 1967a,b; Biggins, 1967; Yamanaka *et al.*, 1967). These cytochromes have been labeled cytochromes c_3 on the basis of the redox potentials, $E_{m,\,7} -200 \pm 50$ mV, and spectroscopic characteristics peculiar to cytochromes c_3 isolated from sulfate-reducing bacteria, including the ratio $\gamma_{red}/\gamma_{ox} \simeq 1.5$, rather than $\simeq 1.2$, which obtains for most low-spin cytochromes c. The cytochromes c_3 isolated from photosynthetic

bacteria all have sizes of about 20 kdaltons (molecular sieve chromatography), but the molecular weight per heme ranges from 10 to 14 kdaltons. These probable diheme or dimeric cytochromes c_3 appear to be distinct from the tetraheme and variant triheme cytochromes c_3 of the sulfate reducers (Ambler *et al.*, 1971; Meyer *et al.*, 1971). The nature of at least one of the heme ligands of these cytochromes must differ from those of the high-redox-potential, low-spin cytochromes c and c_2 and from the median-redox-potential, high-spin cytochromes c' to account for the 250- to 600-mV-lower E_m values. A recent suggestion for the three-dimensional structure of sulfate-reducer cytochromes c_3, based on analyses of ESR and NMR spectra, appears to implicate histidine as the extraplanar heme ligands (Dobson *et al.*, 1974).

As yet, no role has been suggested for these very-low-potential cytochromes in the photosynthetic bacteria, and because of the limited amounts purified, detailed structural studies have not yet been conducted.

The cytochrome c_3 isolated from "*Chloropseudomonas ethylica*"* (Olson and Shaw, 1969; Meyer *et al.*, 1971) is derived from the colorless component rather than from the photosynthetic species, and is therefore not related to the cytochromes c_3 of the photosynthetic organisms (Gray *et al.*, 1972). The amino acid sequence of this variant cytochrome is clearly related to that of the *Desulfovibrio* species (Ambler, 1971).

4.1.7. Cytochromes "ƒ"

Cytochromes with spectral similarities to the algal cytochromes ƒ are obtained from several classes of photosynthetic bacteria (Yamanaka and Okunuki, 1968; Cusanovich and Bartsch, 1969; Meyer *et al.*, 1968). There remains to be demonstrated either a close structural or functional similarity among these cytochromes.

The spectra of these cytochromes have low-spin character, but with complex reduced α absorption peaks of low amplitude at 553–555 nm and with shoulders near 550 nm. This low α peak intensity results in the approximately 30% exaggeration of the $\gamma_{red}/\alpha_{red}$ absorption ratio as compared with cytochromes c and c_2.

The amino acid sequences of *Chl. limicola f. thiosulfatophilum* and the green bacterial component of "*Chloropseudomonas ethylica*" cytochromes c-555(550) are closely related (van Beeuman and Ambler, 1973), but show less similarity to the

---

* See Chapter 2 for a discussion of the taxonomy of this syntrophic system.

sequences of algal cytochromes f (Laycock, 1972; Pettigrew, 1974; Ambler and Bartsch, 1975) and still less to *Pseudomonas mendocina* cytochrome c_5 (Ambler and Taylor, 1973). With the exception of the *Chl. limicola* cytochrome, which has 102 residues, the peptide chains of these cytochromes consist of 80–90 residues with the heme-binding sequence Cys–X–Y–Cys–His located very near the amino termini and with an invariant Met–Pro–Ala sequence, which probably provides the Met sixth heme ligand, located about 50 residues farther along the peptide chain.

Two enzymatic methods for reduction of the *Chl. limicola f. thiosulfatophilum* cytochrome c-555(550) are known by which the principal oxidizable growth substrates may be coupled with the electron-transfer system of the cell (Kusai and Yamanaka, 1973). Cytochrome c-551 catalyzes the transfer of electrons from the thiosulfate-cytochrome c-551 oxidoreductase to cytochrome c-555(550), and flavocytochrome c-553 catalyzes the reduction of cytochrome c-555(550) by bisulfide ion.

The light-driven oxidation of cytochromes c-555(550) in the two *Chlorobium* species has not been clearly demonstrated. Unlike the algal and purple photosynthetic bacterial cytochromes f with $E_{m,7} \cong 350$ mV, the *Chlorobium* cytochromes c-555(550) have midpoint redox potentials of about 145 mV (Gibson, J., 1961) and 103 mV (Shioi *et al.*, 1974), only about 100 mV less than the potential of the reaction center Bchl a of the cells. This is perhaps a sufficient difference to permit effective interaction in a light-induced electron-transfer system, as was suggested by Yamanaka and Okunuki (1968). However, with whole cells (Olson and Sybesma, 1963; Sybesma, 1967), and with membrane fragments devoid of the light-harvesting chlorobium chlorophyll, the light-induced difference spectra show an asymmetric α peak at 553 nm and γ peak at 423 nm, with $E_{m,7} \cong 220$ mV. An observation with membrane fragments from *Chl. limicola f. thiosulfatophilum* (Tassajara strain) indicates that the light-reactive cytochrome may have an α maximum at 551 nm (Knaff *et al.*, 1973). Cytochromes c-555(550) might account for some of the absorption spectra observed, but a closer examination is needed. It has been found possible to purify at least a part of the membrane-bound cytochrome c-553. The purified cytochrome does not resemble cytochrome c-555(550), but could conceivably be equivalent to a cytochrome c-553 partly purified by Gibson (1961) (R. G. Bartsch, unpublished).

The *Chl. limicola f. thiosulfatophilum* cytochrome c-555(550) has an anomalous property that distinguishes it dramatically from other bacterial cytochromes c, including the structurally very similar *Chl. limicola* cytochrome c-555(550), for it shows relatively high reactivity with mitochondrial cytochrome c oxidase (Yamanaka and Okunuki, 1968). Both the *Chlorobium* cytochromes are unreactive with the mitochondrial cytochrome c reductase (B. J. Errede and M. D. Kamen, private communication).

Cytochromes c-553(550) that may be analogous to the cytochromes discussed above are isolated in polymeric forms as minor constituents of the cytochrome complements of *Chr. vinosum* (Cusanovich and Bartsch, 1969) and from *Rp. gelatinosa* and *T. pfennigii* (Meyer *et al.*, 1973).

These high-potential cytochromes c have $E_{m,7} \simeq 350$ mV, and the spectra show some similarity to the algal cytochromes f. Acetone extraction promotes the solubilization of the *Chromatium* and the *Rp. gelatinosa* cytochromes, possibly indicating a weakly lipophilic nature. The *Chromatium* cytochrome is isolated as a mixture of polymeric forms and appears to be unstable to prolonged storage at $-20°$C. The degree of aggregation increases during storage and the α absorption maximum changes gradually to 550 nm, in which state it may resemble the minor, uncharacterized cytochrome c-550 found in *Chr. vinosum* extracts.

Any possible involvement of these cytochromes in light-induced oxidation reactions would be completely obscured by the much more abundant membrane-bound cytochrome c-556-7 that participates in the light-induced electron-transfer reactions of these organisms. One can wonder whether there may exist in the purple photosynthetic bacteria cyclic electron-transfer systems that involve these cytochrome-f-like heme proteins.

Note Added in Proof. The so-called soluble cytochromes f from several algal sources have been shown to differ from the characteristically membrane-bound cytochromes f of these organisms (Wood, 1977). Based on this analysis the soluble algal cytochromes are suggested to function instead of plastocyanin as a mobile electron shuttle to membrane-bound reactants. By analogy, the similar cytochromes found in some photosynthetic bacteria may not be immediate participants with reaction center processes, as implied above, and the designation cytochrome f is inappropriate.

From light-induced time-delay difference spectra of *Chr. vinosum* whole cells, Van Grondelle *et al.* (1977) extracted the indication that a cytochrome c-551 ($E_m = 260$ mV) may act as a mobile electron shuttle to reduce photooxidized membrane-bound cytochrome c-555. Perhaps either the cytochrome *c*-553(550) or the cytochrome *c*-550 of *Chr. vinosum* may be this cytochrome.

4.1.8. Cytochrome *c*-551 of Chlorobiaceae

Chlorobium limicola f. thiosulfatophilum also contains cytochrome *c*-551 (Meyer *et al.*, 1968), which has been shown to be the specific electron-transfer intermediary from the thiosulfate–cytochrome *c*-551 oxidoreductase of the cell to cytochrome *c*-555(550) (Kusai and Yamanaka, 1973). This cytochrome *c*-551 has not been observed in "*Chloropseudomonas ethylica.*" Although not completely purified, it has been determined to have a molecular weight of approximately 45 kdaltons and to contain at least two hemes. The cytochrome dissociates into smaller units in sodium dodecyl sulfate–polyacrylamide gel electrophoresis, but the analysis has not been concluded (T. E. Meyer, private communication).

4.1.9. Cytochrome *c*-551 of *Rp. gelatinosa* and Cytochrome *c*-552.5(548) of *Rs. tenue*

Cytochrome *c*-551, the predominant soluble low-spin cytochrome isolated from *Rp. gelatinosa*, and a comparable cytochrome *c*-552.5(548) from *Rs. tenue* (T. E. Meyer, private communication) possess structural affinities with the cytochromes *c*-551 isolated from several denitrifying *Pseudomonas* species and from *Azotobacter vinelandii*, as judged from the amino acid sequences and absorption spectra, but have little resemblance to cytochromes c_2 (R. P. Ambler, private communication). These cytochromes *c*-551 with about 85 residues have the general architectural features of the cytochrome *c* class, including the heme binding sequence Cys–X–Y–Cys–His near the amino terminus, and a Met that serves as the sixth heme ligand, which, however, is located some 20 residues closer to the heme than in the larger cytochromes *c*. Numerous stretches of identical sequence are evident for the several cytochromes. Among the denitrifiers, cytochrome *c*-551 has been implicated as one of the electron-transfer intermediaries between substrate and the cytochrome *cd* nitrite reductase of the *Pseudomonas* species (Yamanaka *et al.*, 1961). No functional studies have been made with these cytochromes.

4.1.10. Cytochrome *c*-552(545) of *T. pfennigii*

Another unique cytochrome among the photosynthetic bacteria is cytochrome *c*-552(545), the most abundant of the soluble cytochromes of *T. pfennigii*

(Meyer *et al.*, 1973). This cytochrome *c* of about 30 kdaltons probably contains two hemes. The cytochrome is only slowly reduced by sodium dithionite, unlike any other cytochrome *c* encountered to date. The spectrum is reminiscent of cytochrome *c* peroxidases isolated from *Pseudomonas aeruginosa* (Soininen and Ellfolk, 1973), as well as from other denitrifying bacteria (Kodama and Mori, 1969; Iwasaki and Matsubara, 1971), but no test for this property has been reported.

Note Added in Proof. A sample of *T. pfennigii* cytochrome *c*-552(545) failed to react as a cytochrome *c* peroxidase when tested vs. *T. pfennigii* ferrocytochrome *c*-552(550) and high-potential iron protein (HiPIP) as well as *Pseudomonas aeruginosa* ferrocytochrome *c*-551. *Ps. aeruginosa* cytochrome *c* peroxidase [cytochrome *c*-557(551)] readily oxidized these ferroproteins. Therefore, despite some similarity in properties, the *T. pfennigii* cytochrome is not a cytochrome *c* peroxidase (R. G. Bartsch, unpublished).

The partial high-spin character of *T. pfennigii* cytochrome *c*-552(545) is indicated by a decrease (about 10%) in intensity of the α, β, and γ bands and a shift of the γ band from 415.5 to 413 nm (R. G. Bartsch, unpublished).

4.1.11. Flavocytochromes *c*

c-Type cytochromes that contain covalently bound flavin have been isolated from *Chr. vinosum* (Bartsch *et al.*, 1968), *Chl. limicola f. sp. thiosulfatophilum* (Meyer *et al.*, 1968), and possibly *Thiocapsa roseopersicina* (Trüper and Rogers, 1971). The first two may act as sulfide dehydrogenases (Kusai and Yamanaka, 1973) because they are reduced by approximately stoichiometric concentrations of bisulfide ion 10–20 times faster than are other cytochromes *c*, and the *Chlorobium* flavocytochrome *c*-553 appears to catalyze the reduction of *Chlorobium* cytochrome *c*-555(550) by bisulfide ion. However, *Chr. vinosum* can grow photoheterotrophically with malate or succinate as the chief oxidizable substrate and with only a small concentration of sulfide ion, or photoautotrophically with hydrogen gas as the oxidizable substrate and no sulfide ion. From all these cultures, the flavocytochrome *c*-552 was obtained in the same yield and with the same properties as that isolated from cells grown photoautotrophically on a sulfide–thiosulfate medium. It would seem that a constitutive cytochrome of this nature might well function as an electron-transfer agent with substrates other than bisulfide ion, but no candidate has yet been identified.

With other properties similar to those of *Chr. vinosum* flavocytochrome *c*-552, it is safe to assume that *T. roseopersicina* flavocytochrome *c*-552.5 is also readily reduced by bisulfide ion. However, the organism contains cytochrome *c*-550, which lacks flavin, but is readily reduced by bisulfide ion with production of elemental sulfur (Trüper *et al.*, 1976). The presence in the same organism of two cytochromes with similar reactivity indicates that flavocytochromes *c* are not uniquely suited as bisulfide ion oxidants.

The photosynthetic component of "*Chloropseudomonas ethylica*" oxidizes sulfide ion and *Rp. palustris* oxidizes thiosulfate ion as oxidizable substrates for growth, and both organisms lack flavocytochrome *c* (T. E. Meyer, private communication). Therefore, photosynthetic bacteria have evolved more than one means to consume these sulfur-containing inorganic substrates.

It was once assumed that the *Chromatium* flavocytochrome *c*-552 corresponded to the cytochrome implicated in noncyclic light-induced electron-transfer reactions in that organism. It is now clear that the cytochrome *c*-552 involved in these reactions is firmly bound in the chromatophore membrane and is not related to the soluble cytochrome, despite spectroscopic and redox potential similarities (see the discussion in Section 4.2.1).

The *Chr. vinosum* flavocytochrome *c*-552 consists of a diheme cytochrome subunit, approximately 21 kdaltons, and a flavoprotein subunit, approximately 45 kdaltons, as determined by sodium dodecyl sulfate–polyacrylamide gel electrophoresis (Kennel, 1971). Similarly, the *Chlorobium* flavocytochrome *c*-553 consists of an 11-kdalton monoheme cytochrome subunit and a 45-kdalton flavoprotein subunit (Yamanaka and Kusai, 1976). The subunits can also be dissociated by several denaturants such as 8 M urea or 4 M guanidinium hydrochloride, by a mercurial such as *p*-mercuribenzene sulfonate, by alkaline pH (pH > 10), and by a freeze–thaw cycle in the presence of 0.4 M sodium perchlorate (Bartsch *et al.*, 1968; R. G. Bartsch and T. E. Meyer, unpublished; Yamanaka and Kusai, 1976). From the dissociated mixtures, the cytochrome *c* subunits can be prepared in relatively pure form. The flavoprotein subunits tend to polymerize and become insoluble and to entrap some of the cytochrome *c*. The polymerization process seems to involve oxidative crosslinking via disulfide bonds, for on reduction with dithiothreitol plus 4 M urea, the insoluble material dissolves. The two *Chromatium* subunits separate

effectively if the dissociated mixture is chromatographed on a G-75 Sephadex column in the presence of 10% (vol./vol.) formic acid (M. A. Cusanovich, private communication). The flavoprotein subunits seem unstable to exposure to air and to light.

The flavin of *Chr. vinosum* flavocytochrome *c*-552 was shown to be FAD (Hendricks and Cronin, 1971). The interpretation that a novel thiohemiacetal linkage exists between the FAD and a cysteine residue of the peptide chain (Walker *et al.*, 1974; Kenney *et al.*, 1974) has been withdrawn (Kenney *et al.*, 1977, see below) in connection with the description of the 8α-S-cysteinyl FAD thioether linkage found in the *Chlorobium* flavocytochrome *c*.

Both the *Chromatium* and the *Chlorobium* flavocytochromes *c* undergo similar reactions with various nucleophilic reagents including CN^-, $S_2O_3^{2-}$, SO_3^{2-}, and several mercaptans (Meyer, 1970; Meyer and Bartsch, 1976; Kusai and Yamanaka, 1973; Yamanaka and Kusai, 1976). These reagents cause the loss of the flavin absorption bands from the oxidized protein with concomitant formation of a long-wavelength charge-transfer absorption band centered near 675 nm. The heme absorption bands are not affected. The original spectrum can be restored by removing the reagents. With several reducing agents, including sodium dithionite, sodium sulfide, and dithiothreitol administered in near-stoichiometric amounts, the transient bleaching of the flavin absorbance and the appearance of the charge-transfer band can be seen during the time required for the heme, and presumably the flavin, to become fully reduced. With all the ligands, full reduction of the flavocytochrome abolishes the long-wavelength absorption band, and indeed, dialysis of the complexed cytochromes under reducing conditions can yield uncomplexed cytochromes.

For the *Chromatium* cytochrome, both the hemes and the flavin appeared to have nearly the same redox potential, $E_{m,7} \simeq 0$ mV, when the redox titration was carried out with sodium dithionite as reductant (W. P. Vorkink and M. A. Cusanovich, private communication). Because sulfite, an oxidation product of dithionite, can react with the flavin, dithionite is not the optimum reductant for redox titrations of these cytochromes. When the titration is conducted using ferrooxalate as reductant, the flavin becomes only partly reduced when complete reduction of the hemes is attained (Bartsch *et al.*, 1968). Similar observations have been made with the *Chlorobium* cytochrome (Meyer *et al.*, 1968), using iron–EDTA mixtures. Because of the overlapping absorbance changes due to the heme and the flavin semiquinone in the 450 to 480 nm region of the spectrum, the redox potential of the flavin has not been determined. It may be guessed that the $E_{m,7}$ value for the semiquinone form is 50–100 mV more negative than that of the heme for both cytochromes.

Unlike the monoheme *Chlorobium* flavocytochrome *c*-553, the diheme *Chr. vinosum* ferroflavocytochrome *c*-552 forms a complex with carbon monoxide, which can be photodissociated by light. The absorption spectrum resembles the CO-complexes of other cytochromes *c* such as cytochrome *c'* (Bartsch, 1963; Bartsch *et al.*, 1968) and denatured mitochondrial cytochrome *c* (Butt and Keilin, 1962). The γ absorbance of these cytochromes is shifted to 414–416 nm and is increased nearly twofold, but the α and β bands are shifted to the red and decreased in intensity relative to the uncomplexed cytochromes. If the average increase in the γ absorbance of the CO complex of ferrocytochromes *c* applies to the flavocytochrome *c*-552, then only one CO may complex per pair of hemes, inasmuch as the γ absorbance is increased only 50%. From the results of a CD study of the cytochrome, it was postulated that one CO may be shared by two hemes located sufficiently close together to permit electronic interaction between them (Bartsch *et al.*, 1968). Based on analysis of optical rotary dispersion spectra in the γ region of the spectrum of the *Chromatium* cytochrome, Yong and King (1970) suggested that the two hemes may occupy different environments, but nevertheless may be sufficiently close together to interact.

The accessibility of CO to the ferroheme implies that unlike cytochromes *c* and c_2, the heme of the native protein is not completely shielded from the solvent.

Like cytochromes c_2 and *c'* in the Rhodospirillaceae, the flavocytochrome *c*-552 that can be trapped in vesicles formed during disruption of the *Chromatium* cells can be released by the action of detergents or chaotropes or after removal of lipids with acetone. Perhaps this cytochrome together with the cytochrome *c'* may be located in the periplasmic space of the cell.

Note Added in Proof. Extracts of *T. roseopersicina* cytochromes contain cytochromes *c'*, *c*-550 (with redox properties somewhat like those of cytochromes c_2), and flavocytochrome *c*-552.5 (R. G. Bartsch, unpublished), as well as the cytochrome *c*-550 of 34-kdalton molecular weight described by Trüper *et al.* (1976).

The flavin peptide of *Chlorobium* flavocytochrome *c* is Val–Thr–Cys(FAD)–Pro–Phe–Ser–Asn, with the FAD linked to the cysteine residue via a thioether bond to the 8α-methylene group of the flavin (Kenney *et al.*, 1977).

4.2. Membrane-Bound Cytochromes *c*

4.2.1. Cytochromes *c*-556, *c*-552 Complex

Two types of membrane-bound cytochromes *c* have been implicated in light-induced electron-transfer reactions in *Chr. vinsoum* (Olson and Chance, 1960*a*,*b*; Parson and Case, 1970; Kennel and Kamen, 1971*a*,*b*; Kennel *et al.*, 1972*a*), in *Rp. gelatinosa* (Dutton, 1971), in *Rp. viridis* (Olson and Nadler, 1965; Case *et al.*, 1970), and in *Thiocapsa pfennigii* (Olson *et al.*, 1969). The light-induced reactions undergone by these high- and low-potential cytochromes have been demonstrated by poising the redox potentials (E_h) of the various membrane preparations between +200 and +400 mV for the cytochromes *c*-556 and −100 and +100 mV for the cytochrome *c*-552. In *Chr. vinosum*, two molecules of cytochrome *c*-556 appear to be associated with each reaction center Bchl *a* (P870) (Parson and Case, 1970). Direct estimates of the cytochrome *c*-552 content seem not to have been made, although from the relative intensities of α band absorbance changes, it has been assumed that the cytochromes occur in the approximate ratio $7:2 = c\text{-}552/c\text{-}556$ (Thornber, 1969). The high-potential cytochrome is considered to be a component of the cyclic photoelectron-transfer system and the low-potential cytochrome to be part of a noncyclic electron-transfer system.

Attempts have been made to characterize these cytochromes, which constitute most of the heme protein content of the bacteria. In *Chr. vinosum* chromatophores, in subchromatophore fraction A (Thornber, 1969), and in the purified cytochrome complex (Kennel and Kamen, 1971*b*), the ratio of α peak absorption changes for cytochromes *c*-556 ($E_{m,7}$ 325 mV) and *c*-552 ($E_{m,7}$ 8 mV) is about 1:3. Because the extinction coefficients have not been deduced for the cytochromes, the relative molar concentrations cannot be specified. The two cytochromes in the purified complex could not be separated, nor was it conclusively shown that the two species had the same or different peptide chains. A comparison by polyacrylamide gel electrophoresis in the presence of sodium dodecyl sulfate showed that the predominant cytochrome band of 45 kdaltons from the purified complex coincided with the sole cytochrome band from the starting subchromatophore fraction A and from chromatophores. The complex also yielded smaller fragments of 23 and 29 kdaltons. From the amino acid and heme composition of the main and subsidiary cytochrome bands, the minimum formula weight appeared to be 11 kdaltons (equivalent to approximately 90 amino acids) per heme, which suggests that the 45-kdalton complex may contain 4 hemes, in the stoichiometry 1 c-556/3 c-552, or, if the extinction of the c-556 α peak is very low, 2 c-556/2 c-552.

The cytochromes c-552 were solubilized from *T. pfennigii* and *Rp. gelatinosa*, but were not appreciably purified (Meyer *et al.*, 1973). The high-potential cytochromes c-556 were not isolated, suggesting that these hemes are more labile than those of the comparable *Chromatium* cytochrome complex, or that they are on a separate peptide chain that was not solubilized. Preliminary SDS–polyacrylamide gel electrophoresis results indicated that the *T. pfennigii* cytochrome c-552 was 70 kdaltons and the *Rp. gelatinosa* cytochrome 10 kdaltons. Other properties were not reported.

There was confusion about the possible identity between the soluble *Chr. vinosum* flavocytochrome c-552 and the low-potential cytochrome c-552 that undergoes very rapid light-induced oxidation in whole cells (Olson and Chance, 1960a,b) even at low temperatues (Vredenberg and Duysens, 1964; Chance and Nishimura, 1960), in chromatophores (Cusanovich *et al.*, 1968b; Parson, 1969; Parson and Case, 1970; Dutton *et al.*, 1970; Takamiya and Nishimura, 1974), and in subchromatophore fractions (Case *et al.*, 1970). From consideration of the amino acid compositions, molecular weights, behavior in SDS–polyacrylamide gel electrophoresis, and hydrophobic vs. water-soluble natures of the cytochromes (Kennel and Kamen, 1971b), and of the very different thermodynamic properties of the two cytochromes (Case and Parson, 1971), it is abundantly clear that these cytochromes are completely different.

4.2.2. Other Membrane-Bound Cytochromes c

The recently devised methods for eliminating the periplasmic cytochromes c by spheroplast preparation or by simple washing with dilute EDTA solution should facilitate the search for additional examples of low concentrations of bound cytochromes c in the photosynthetic bacteria and also facilitate testing the involvement of such bound cytochromes in light-driven electron-transfer reactions by removing background noise caused by the adventitious soluble cytochromes.

Cytochromes C-553-4 that have been observed in reduced-minus-oxidized difference spectra of membrane fractions of various Rhodospirillaceae have not been purified. In one attempt, the abundant cytochrome C-553 of *Rp. sphaeroides* was solubilized from acetone-extracted chromatophores by the action of 2% Triton X-100, but was not successfully purified.

Spectrally similar cytochromes C-553-4 have been solubilized and purified from *Chl. limicola f. thiosulfatophilum* and from *Chloroflexus aurantiacus*, but chemical properties of these cytochromes have not been determined (R. G. Bartsch, unpublished).

4.3. Cytochromes b

4.3.1. Cytochromes b-558

Cytochromes b-558 have been isolated from *Rp. sphaeroides* (Orlando and Horio, 1961), *Rs. rubrum* (Bartsch *et al.*, 1971), *Rp. palustris*, and *Rs. tenue* (R. G. Bartsch, unpublished). For all the examples, the absorption spectra, molecular weights, and redox behavior were very similar to those of the so-called "soluble" cytochrome b_1 obtained from *Escherichia coli* (Deeb and Hager, 1964). The cytochromes were all difficult to reduce with sodium dithionite, a characteristic associated with their polymeric nature (Deeb and Hager, 1964). As much as 60 min was required to effect complete reduction of the cytochromes, although added mediator dyes such as FMN or neutral red increased the reduction rate. Nevertheless, the low redox potentials reported for this group of cytochromes b must be accepted with reservations simply because true thermodynamic equilibria may not have been reached in the reported titrations as a consequence of the remarkably slow reaction rates seen. The analogy between the *Esch. coli* cytochrome b_1 and the Rhodospirillaceae cytochromes b is reinforced by the observation that all the samples exist as polymers approximating the octameric oligomer of the *Esch. coli* cytochrome b (Bartsch *et al.*, 1971).

No functional role can yet be ascribed to the soluble cytochromes b of the photosynthetic bacteria.

4.3.2. Cytochrome b'

A protoheme-containing protein that has catalatic activity, but can also undergo redox reactions like a cytochrome, was purified from extracts of *Chr. vinosum* by R. G. Bartsch and T. E. Meyer (unpublished). The absorption spectra of this cytochrome have high-spin characteristics, namely, charge-transfer bands at 500 and 640 nm in the ferri form and the presence of a broad absorption band at 560 nm in the ferro form. In addition, the γ absorbance of the ferrocytochrome is only about 50% as intense as that of the ferricytochrome. The absorption spectrum is similar to that of horseradish peroxidase. The designation cytochrome b' has been applied to this high-spin protoheme protein. On the basis of spectroscopic differences, cytochrome b' is probably not related to the membrane-bound protoheme-containing cytochrome reported by Knaff and Buchanon (1975) to occur in subchromatophore preparations of *Chr. vinosum*.

4.3.3. Membrane-Bound Cytochromes B

Multiple membrane-bound cytochromes *B* have been identified in all groups of photosynthetic bacteria, Rhodospirillaceae (Baltscheffsky, 1969; Whale and Jones, 1970; Kakuno *et al.*, 1971; Dutton and Jackson, 1972; Connelly *et al.*, 1973), Chromatiaceae (Knaff and Buchanan, 1975) and Chlorobiaceae (Fowler, 1974; Knaff and Buchanan, 1975). In all instances, it has been postulated that one or more of the cytochromes *B* participate in the cyclic light-induced transfer of electrons from a secondary electron acceptor such as ubiquinone to the cytochrome *c* implicated as the immediate electron donor to the reaction center Bchl (Prince and Dutton, 1975; Knaff and Buchanan, 1975). Two or more presumed *B*-type cytochromes have been resolved by the spectrophotometric redox titration method in various Rhodospirillaceae and designated according to the E_m values obtained. Those absorbance changes as a function of imposed E_h that occur in the 560 to 565 nm region of the spectrum are assumed to be related to cytochromes *B*. Further indication of the nature of the cytochrome involved is sometimes given by the effect of electron-transfer inhibitors such as antimycin that block the oxidation of some cytochromes *b* and therefore cause the accumulation of the reduced forms of these possible cytochromes.

No successful purification of any of the cytochromes *B* has been achieved. In the light of this experience, no details of the properties of these important heme proteins could be presented in Table 2.

4.4. Cytochrome Oxidases

Cytochrome *o* is one of the cytochromes *B* (Taniguchi and Kamen, 1965) about which essentially no chemical details are available. The physiological properties of this cytochrome group are discussed by Smith and Pinder in Chapter 33. It is the general practice to label as cytochrome *o* whatever autooxidizable pigment in the cell combines with CO with resultant formation of a spectrum with a γ band near 415 nm and an α band at 560–570 nm, in either absolute or difference spectra. Such a pigment has been detected in all the Rhodospirillaceae examined. The basal level of the cytochrome in photosynthetic

cells is enhanced when the cells are grown aerobically, where the level of cytochromes c', which also react with CO, is repressed (Horio and Taylor, 1965; Sasaki *et al.*, 1970).

Rhodopseudomonas sphaeroides seems to possess an oxidase in addition to cytochrome *o*. Aerobically grown cells of this organism contain a cytochrome a–a_3 oxidase, which, depending on the culture conditions, may be evident only in early log phase cells under conditions of moderate aeration (Kikuchi *et al.*, 1965) or may persist throughout the life of an aerobic culture (Saunders and Jones, 1974). A partially purified cytochrome preparation contains both cytochrome *c*-551 and cytochrome *a* hemes, and the oxidase is about tenfold more reactive with mitochondrial ferrocytochrome *c* than with either *Rs. rubrum* or *Rp. sphaeroides* ferrocytochromes c_2 (Sasaki *et al.*, 1970).

In *Rp. capsulata* (Klemme and Schlegel, 1969; Connelly *et al.*, 1973; Marrs and Gest, 1973) and in *Rp. palustris* (King and Drews, 1975), two terminal oxidative paths seem to occur. Both are inhibited by cyanide ion, but the major one is not inhibited by CO. The latter is evidently a *B*-type cytochrome oxidase that differs from cytochrome *o* (Zannoni *et al.*, 1974) as customarily defined (Castor and Chance, 1959).

5. Concluding Remarks

Sufficient information to correlate the chemical properties of the cytochromes with the physiological roles of these electron-transfer proteins is still largely lacking. Often, it is not possible to correlate substances detected in cells or membranes by redox spectrophotometry with chemically identifiable cytochromes. It is not yet feasible to specify the mechanics of interaction between reduced cytochromes and $BChl^+$ of the reaction center, nor to be certain how many of the cytochromes of a cell participate in the reaction. Indeed, in only a few instances have enzymatic electron-transfer reactions been identified for specific cytochromes.

From results available, it is evident that the cytochromes of photosynthetic bacteria have corollaries in other organisms, and perhaps these cytochromes preserve in some degree the evolutionary history of cytochrome development. The structural relationships between cytochromes c_2 and mitochondrial cytochrome *c*, and between the bacterial *f*-like cytochromes *c* and the algal cytochromes *c*, and the similar properties of the cytochromes *b*-558 and the cytochromes b_1 come to mind.

In view of the relative ease with which many of the

cytochromes of the photosynthetic bacteria can be isolated, these comparative studies can be expected to yield information and concepts applicable to electron-transfer systems of other organisms and modes of life.

ACKNOWLEDGMENTS

I wish to thank colleagues who provided information about their cytochrome studies in advance of publication of their observations, especially Drs. T. E. Meyer and G. W. Pettigrew, who both informed and commented very freely. Many of the observations reported here originated under the aegis of Prof. M. D. Kamen, who began the whole study and still generously supports our efforts. This current effort was supported by grants-in-aid to M. D. Kamen from the National Institutes of Health (GM 18528-11) and from the National Science Foundation (GB36019X).

6. References

Agalidis, I., Jauneau, E., and Reiss-Husson, F., 1974, Influence of iron deficiency on the cytochromes of *Rhodopseudomonas spheroides*, *Eur. J. Biochem.* **47**:573.

Ambler, R. P., 1971, The amino acid sequence of cytochrome c-551.5 (cytochrome c_7) from the green photosynthetic bacterium *Chloropseudomonas ethylica*, *FEBS Lett.* **18**:351.

Ambler, R. P., 1973, The amino acid sequence of cytochrome c' from *Alcaligenes* sp. NC1B11015, *Biochem. J.* **135**:751.

Ambler, R. P., and Bartsch, R. G., 1975, Amino acid sequence similarity between cytochrome f from a blue-green bacterium and algal chloroplasts, *Nature (London)* **253**:285.

Ambler, R. P., and Taylor, E., 1973, Amino acid sequence of cytochrome c_5 from *Pseudomonas mendocina*, *Biochem. J.* **111**:1973.

Ambler, R. P., and Wynn, M., 1973, The amino acid sequences of cytochromes c-551 from three species of *Pseudomonas*, *Biochem. J.* **131**:485.

Ambler, R. P., Bruschi, M., and Le Gall, J., The amino acid sequence of cytochrome c from *Desulfovibrio desulfuricans* (strain El Agheila Z, NCIB 8380), *FEBS Lett.* **18**:347.

Andrews, P., 1964, Estimation of the molecular weight of protein by Sephadex gel-filtration, *Biochem. J.* **91**:222.

Baltscheffsky, M., 1969, Reversed energy conversion reactions of bacterial photophosphorylation, *Arch. Biochem. Biophys.* **133**:46.

Barrett, J., and Kamen, M. D., 1961, On the prosthetic group of an RHP-type haem protein from *Chromatium*, *Biochim. Biophys. Acta* **50**:573.

Bartsch, R. G., 1963, Spectroscopic properties of purified cytochromes of photosynthetic bacteria, in: *Bacterial Photosynthesis* (H. Gest, A. San Pietro, and L. P. Vernon, eds.), pp. 475–494, Antioch Press, Yellow Springs, Ohio.

Bartsch, R. G., 1968, Bacterial cytochromes, *Annu. Rev. Microbiol.* **22**:181.

Bartsch, R. G., 1971, Cytochromes: Bacterial, in: *Methods in Enzymology, Photosynthesis*, Part A (A. San Pietro, ed.), pp. 344–363, Academic Press, New York.

Bartsch, R. G., and Kamen, M. D., 1960, Isolation and properties of two soluble heme proteins in extracts of the photoanaerobe *Chromatium*, *J. Biol. Chem.* **235**:825.

Bartsch, R. G., Meyer, T. E., and Robinson, A. B., 1968, Complex c-type cytochromes with bound flavin, in: *Structure and Function of Cytochromes* (K. Okunuki, M. D. Kamen, and I. Sekuzu, eds.), pp. 443–451, University of Tokyo Press.

Bartsch, R. G., Kakuno, T., Horio, T., and Kamen, M. D., 1971, Preparation and properties of *Rhodospirillum rubrum* cytochromes c_2, cc', and $b_{557.5}$, and flavin mononucleotide protein, *J. Biol. Chem.* **246**:4489.

Biggins, J., 1967, Photosynthetic reactions by lysed protoplasts and particle preparations from the blue-green alga, *Phormidium luridum*, *Plant Physiol.* **42**:1447.

Butt, W. D., and Keilin, D., 1962, Absorption spectra and some other properties of cytochrome c and of its compounds with ligands, *Proc. R. Soc. London* **156b**:429.

Case, G. D., and Parson, W. W., 1971, Thermodynamics of the primary and secondary photochemical reactions in *Chromatium*, *Biochim. Biophys. Acta* **253**:187.

Case, G. D., Parson, W. W., and Thornber, J. P., 1970, Photooxidation of cytochromes in reaction center preparations from *Chromatium* and *Rhodopseudomonas viridis*, *Biochim. Biophys. Acta* **223**:122.

Castor, L. N., and Chance, B., 1959, Photochemical determination of the oxidases of bacteria, *J. Biol. Chem.* **234**:1587.

Chance, B., and Nishimura, M., 1960, On the mechanism of chlorophyll–cytochrome interaction: The temperature insensitivity of light-induced cytochrome oxidation in *Chromatium*, *Proc. Natl. Acad. Sci. U.S.A.* **46**:19.

Chance, B., and Smith, L., 1955, Respiratory pigments of *Rhodospirillum rubrum*, *Nature (London)* **175**:803.

Clark, W. M., 1960, *Oxidation–Reduction Potentials of Organic Systems*, pp. 91–106, The Williams and Wilkins Co., Baltimore.

Connelly, J. L., Jones, O. T. G., Saunders, V. A., and Yates, D. W., 1973, Kinetic and thermodynamic properties of membrane-bound cytochromes of aerobically and photosynthetically grown *Rhodopseudomonas spheroides*, *Biochim. Biophys. Acta* **292**:644.

Corker, G. A., and Sharpe, S. A., 1975, Influence of light on the EPR-detectable electron transport components in whole cell *Rhodospirillum rubrum*, *Photochem. Photobiol.* **21**:49.

Cusanovich, M. A., 1971, Molecular weights of some cytochromes cc', *Biochim. Biophys. Acta* **236**:238.

Cusanovich, M. A., and Bartsch, R. G., 1969, A high potential cytochrome c from *Chromatium* chromatophores, *Biochim. Biophys. Acta* **189**:245.

Cusanovich, M. A., and Gibson, Q. H., 1973, Anomalous ligand binding by a class of high spin c-type cytochromes, *J. Biol. Chem.* **248**:822.

Cusanovich, M. A., Bartsch, R. G., and Kamen, M. D., 1968a, Light-induced absorbance changes in *Chromatium* chromatophores, *Biochim. Biophys. Acta* 153:397.

Cusanovich, M. A., Bartsch, R. G., and Olson, J. M., 1968b, Light-induced reactions in *Chromatium* chromatophores under controlled redox conditions, in: *Comparative Biochemistry and Biophysics of Photosynthesis* (K. Shibata, A. Takamiya, A. T. Jagendorf, and R. C. Fuller, eds.), pp. 186–195, University of Tokyo Press.

Cusanovich, M. A., Tedro, S. M., and Kamen, M. D., 1970, *Pseudomonas denitrificans* cytochrome *cc'*, *Arch. Biochem. Biophys.* 141:557.

Davenport, H. E., and Hill, R., 1952, The preparation and some properties of cytochrome *f*, *Proc. R. Soc. London Ser. B*: 139:327.

Davis, K. A., Hatefi, Y., Salemme, F. R., and Kamen, M. D., 1972, Enzymic redox reactions of cytochromes *c*, *Biochem. Biophys. Res. Commun.* 47:1328.

Deeb, S. S., and Hager, L. P., 1964, Crystalline cytochrome b_1 from *Escherichia coli*, *J. Biol. Chem.* 239:1024.

Dickerson, R. E., and Timkovich, R., 1975, Cytochrome *c*, in: *The Enzymes* (P. D. Boyer, ed.), 3rd Ed., Vol. 11, Chapt. 7, pp. 395–544, Academic Press, New York.

Dobson, C. M., Hoyle, N. J., Geraldes, C. G., Wright, P. E., Williams, R. J. P., Bruschi, M., and LeGall, J., 1974, Outline structure of cytochrome c_3 and consideration of its properties, *Nature (London)* 249:425.

Dus, K., DeKlerk, H., Bartsch, R. G., Horio, T., and Kamen, M. D., 1967, On the monoheme nature of cytochrome *c'* (*Rhodopseudomonas palustris*), *Proc. Natl. Acad. Sci. U.S.A.* 57:367.

Dus, K., Sletten, K., and Kamen, M. D., 1968, Cytochrome c_2 of *Rhodospirillum rubrum*. Complete amino acid sequence and phylogenetic relationships, *J. Biol. Chem.* 243:5507.

Dus, K., Flatmark, T., DeKlerk, H., and Kamen, M. D., 1970, Isolation and chemical properties of two *c*-type cytochromes of *Rhodospirillum molischianum*, *Biochemistry* 9:1984.

Dutton, P. L., 1971, Oxidation–reduction potential dependence of the interaction of cytochromes, bacteriochlorophyll and carotenoids at 77°K in chromatophores of *Chromatium* D and *Rhodopseudomonas gelatinosa*, *Biochim. Biophys. Acta* 226:63.

Dutton, P. L., and Jackson, J. B., 1972, Thermodynamic and kinetic characterization of electron-transfer components *in situ* in *Rhodopseudomonas sphaeroides* and *Rhodospirillum rubrum*, *Eur. J. Biochem.* 30:495.

Dutton, P. L., and Leigh, J. S., 1973, Electron spin resonance characterization of *Chromatium* D hemes, non-heme irons and the components involved in primary photochemistry, *Biochim. Biophys. Acta* 314:178.

Dutton, P. L., and Wilson, D. F., 1974, Redox potentiometry in mitochondrial and photosynthetic bioenergetics, *Biochim. Biophys. Acta* 346:165.

Dutton, P. L., Kihara, T., McCray, J. A., and Thornber, J. P., 1970, Cytochrome C_{553} and bacteriochlorophyll interaction at 77°K in chromatophores and a sub-chromatophore preparation from *Chromatium D*, *Biochim. Biophys. Acta* 205:196.

Dutton, P. L., Petty, K. M., Bonner, H. S., and Morse, S. F., 1975, Cytochrome c_2 and reaction center of *Rhodopseudomonas spheroides* Ga membranes. Extinction coefficients, content, half-reduction potentials, kinetics and electric field alterations, *Biochim. Biophys. Acta* 387:536.

Duysens, L. N. M., 1954, Reversible photo-oxidations of a cytochrome pigment in photosynthesizing *Rhodospirillum rubrum*, *Nature (London)* 173:692.

Ehrenberg, A., and Kamen, M. D., 1965, Magnetic and optical properties of some bacterial haem proteins, *Biochim. Biophys. Acta* 102:333.

Elsden, S., Kamen, M. D., and Vernon, L. P., 1953, A new soluble cytochrome, *J. Am. Chem. Soc.* 75:6347.

Errede, B. J., Haight, G. T., and Kamen, M. D., 1976, On the oxidation of ferrocytochromes by mitochondrial oxidase, *Proc. Natl. Acad. Sci. U.S.A.* 73:113.

Evans, E. H., and Crofts, A. R., 1974, A thermodynamic characterization of the cytochromes of chromatophores from *Rhodopseudomonas capsulata*, *Biochim. Biophys. Acta* 357:78.

Falk, J. E., 1964, *Porphyrins and Metalloporphyrins*, p. 204, Elsevier, Amsterdam.

Flatmark, T., 1966, Some physico-chemical properties of the main subfractions (Cy I–Cy III), *Acta Chem. Scand.* 20:1476.

Flatmark, T., Dus, K., DeKlerk, H., and Kamen, M. D., 1970, Comparative study of physicochemical properties of two *c*-type cytochromes of *Rhodospirillum molischianum*, *Biochemistry* 9:1991.

Florkin, M., and Stotz, E. H., (eds.), 1965, *Comprehensive Biochemistry*, Vol. 13, 2nd Ed., Chapt. 5, Elsevier, Amsterdam.

Fowler, C. G., 1974, Evidence for a cytochrome *b* in green bacteria, *Biochim. Biophys. Acta* 357:327.

Fowler, C. F., Nugent, N. A., and Fuller, R. C., 1974, The isolation and characterization of a photochemically active complex from *Chloropseudomonas ethylica*, *Proc. Natl. Acad. Sci. U.S.A.* 68:2278.

Gibson, J., 1961, Cytochrome pigments from the green photosynthetic bacterium *Chlorobium thiosulfatophilum*, *Biochem. J.* 79:151.

Gibson, Q. H., and Kamen, M. D., 1966, Kinetic analysis of the reactions of cytochrome *cc'* with carbon monoxide, *J. Biol. Chem.* 241:1969.

Gray, B. H., Fowler, C. F., Nugent, N. A., and Fuller, R. L., 1972, A reevaluation of the presence of low midpoint potential cytochrome *c551.5* in the green photosynthetic bacterium *Chloropseudomonas ethylica*, *Biochem. Biophys. Res. Commun.* 47:322.

Hanania, G. H., Irvine, D. H., Eaton, W. A., and George, P., 1967, Thermodynamic aspects of the potassium hexacyanoferrate (III)–(II) system. II. Reduction potential, *J. Phys. Chem.* 71:2022.

Harbury, H. A., Cronin, J. R., Fanger, M. W., Hettinger, T. P., Murphy, A. J., Myer, Y. P., and Vinogradov, S. N., 1965, Complex formation between methionine and a heme peptide from cytochrome *c*, *Proc. Natl. Acad. Sci. U.S.A.* 54:1658.

Henderson, R. W., and Nankiville, D. D., 1966, Electrophoretic and other studies on haem pigments from *Rhodopseudomonas palustris*: Cytochrome 552 and cytochromoid *c*, *Biochem. J.* **98**:587.

Hendriks, R., and Cronin, J. R., 1971, The flavin of *Chromatium* cytochrome *c*-552, *Biochem. Biophys. Res. Commun.* **44**:313.

Hill, R., 1954, The cytochrome *b* component of chloroplasts, *Nature (London)* **174**:501.

Hill, R., and Scarisbrick, R., 1951, The haematin compounds of leaves, *New Phytol.* **50**:98.

Hochman, A., and Carmeli, C., 1974, Photosynthetic electron transport in *Rhodopseudomonas capsulata*, in: *Proceedings of the Third International Congress on Photosynthesis* (M. Avron, ed.), pp. 777–789, Elsevier, Amsterdam.

Holton, R. W., and Myers, J., 1967*a*, Water soluble cytochromes from a blue green alga. I. Extraction, purification and spectral properties of cytochromes *c* (549, 552 and 554, *Anacystis nidulans*), *Biochim. Biophys. Acta* **131**:362.

Holton, R. W., and Myers, J., 1967*b*, Water soluble cytochromes from a blue-green alga. II. Physicochemical properties and quantitative relationships of cytochromes *c* (549, 552 and 554, *Anacystic nidulans*), *Biochim. Biophys. Acta* **131**:375.

Horio, T., and Kamen, M. D., 1961, Preparation and properties of three pure crystalline bacterial haem proteins, *Biochim. Biophys. Acta* **48**:266.

Horio, T., and Kamen, M. D., 1970, Bacterial cytochromes: II. Functional aspects, *Annu. Rev. Microbiol.* **24**:399.

Horio, T., and Taylor, C. P. S., 1965, The photochemical determination of an oxidase of the photoheterotroph *Rhodospirillum rubrum*, and the action spectrum of the inhibition of respiration by light, *J. Biol. Chem.* **240**:1772.

Imai, Y., Imai, K., Sato, R., and Horio, T., 1969, Three spectrally different states of cytochromes *cc'* and *c'* and their interconversion, *J. Biochem.* **65**:225.

Iwasaki, H., and Matsubara, T., 1971, Cytochrome *c*-557(551) and cytochrome *cd* of *Alcaligenes faecalis*, *J. Biochem. (Tokyo)* **69**:847.

Iwasaki, H., and Shidara, S., 1969, Crystallization and some properties of cryptocytochrome *c* from aerobically grown *Pseudomonas denitrificans*, *Plant Cell Physiol.* **10**:291.

Kakuno, T., Bartsch, R. G., Nishikawa, K., and Horio, T., 1971, Redox components associated with chromatophores from *Rhodospirillum rubrum*, *J. Biochem. (Tokyo)* **70**:79.

Kakuno, T., Hosoi, K., Higuti, T., and Horio, T., 1973, Electron and proton transport in *Rhodospirillum rubrum* chromatophores, *J. Biochem. (Tokyo)* **74**:1193.

Kamen, M. D., 1973, Toward a comparative biochemistry of the cytochromes, *Proteins, Nucleic Acids Enzymes (Tokyo)* **18**:753.

Kamen, M. D., and Horio, T., 1970, Bacterial cytochromes: I. Structural aspects, *Annu. Rev. Biochem.* **39**:673.

Kamen, M. D., and Vernon, I. P., 1954, Existence of haem compounds in a photosynthetic obligate anaerobe, *J. Bacteriol.* **67**:617.

Kamen, M. D., and Vernon, L. P., 1955, Comparative studies on bacterial cytochromes, *Biochim. Biophys. Acta* **17**:10.

Kamen, M. D., Dus, K. M., Flatmark, T., and deKlerk, H., 1972, Cytochromes *c*, in: *Electron and Coupled Energy Transfer in Biological Systems*, Vol. 1, Part A (T. E. King and M. Klingenberg, eds.), pp. 243–324, Marcel Dekker, New York.

Kennel, S. J., 1971, Cytochromes and their relation to electron transport in *Chromatium vinosum*, Ph.D. thesis, University of California at San Diego, La Jolla.

Kennel, S. J., and Kamen, M. D., 1971*a*, Iron containing proteins in *Chromatium*. I. Solubilization of membrane-bound cytochrome, *Biochim. Biophys. Acta* **234**:458.

Kennel, S. J., and Kamen, M. D., 1971*b*, Iron containing proteins in *Chromatium*. II. Purification and properties of cholate-solubilized cytochrome complex, *Biochim. Biophys. Acta* **253**:153.

Kennel, S. J., Bartsch, R. G., and Kamen, M. D., 1972*a*, Observations on light-induced oxidation reactions in the electron transport system of *Chromatium*, *Biophys. J.* **12**:882.

Kennel, S. J., Meyer, T. E., Kamen, M. D., and Bartsch, R. G., 1972*b*, On the monoheme character of cytochrome *c'*, *Proc. Natl. Acad. Sci. U.S.A.* **69**:3432.

Kenney, W. C., Edmondson, D. S., and Stinger, T. P., 1974, The covalently bound flavin of *Chromatium* cytochrome c_{552} 2. Sequence of flavin peptides and flavin–tyrosine interaction, *Eur. J. Biochem.* **48**:449.

Kenney, W. C., McIntire, W., and Yamanaka, T., 1977, Structure of the covalently bound flavin of *Chlorobium* cytochrome c_{553}, *Biochim. Biophys. Acta* **483**:467.

Kikuchi, G., Saito, Y., and Motokawa, Y., 1965, On cytochrome oxidase as the terminal oxidase of dark respiration of non-sulfur purple bacteria, *Biochim. Biophys. Acta* **94**:1.

King, M. T., and Drews, G., 1975, The respiratory electron transport system of heterotrophically grown *Rhodopseudomonas palustris*, *Arch. Microbiol.* **102**:219.

Klemme, J. H., and Schlegel, H. G., 1969, Untersuchen zum Cytochrom-oxydase-System aus Anaerob im Licht und Aerob in Dunkeln gewaschenen Zellen von *Rhodopseudomonas capsulata*, *Arch. Mikrobiol.* **68**:326.

Knaff, D. B., and Buchanan, B. B., 1975, Cytochrome *b* and photosynthetic sulfur bacteria, *Biochim. Biophys. Acta* **376**:549.

Knaff, D. B., Buchanan, B. B., and Malkin, R., 1973, Effect of oxidation–reduction potential on light-induced cytochrome and bacteriochlorophyll reactions in chromatophores from the photosynthetic green bacterium *Chlorobium*, *Biochim. Biophys. Acta* **325**:94.

Kodama, T., and Mori, T., 1969, A double peak *c*-type cytochrome, cytochrome *c*-552, 558 of a denitrifying bacterium, *Pseudomonas stutzeri*, *J. Biochem. (Tokyo)* **65**:621.

Kusai, K., and Yamanaka, T., 1973, The oxidation mechanisms of thiosulphate and sulphide in *Chlorobium thiosulphatophilum:* Roles of cytochrome *c*-551 and cytochrome *c*-553, *Biochim. Biophys. Acta* **325**:304.

Laycock, M. V., 1972, The amino acid sequence of cytochrome *c*-553 from the Chrysophycean alga *Monochrysis lutheri, Can. J. Biochem.* **50**:1311.

Lemberg, M. R., 1969, Cytochrome oxidase, *Physiol. Rev.* **49**:63–64.

Lemberg, R., and Barrett, J., 1973, *Cytochromes*, Academic Press, New York.

Loach, P. A., 1966, Primary oxidation–reduction changes during photosynthesis in *Rhodospirillum rubrum, Biochemistry* **5**:592.

Maltempo, M. M., Moss, T. H., and Cusanovich, M. A., 1974, Magnetic studies on the changes in the iron environment in *Chromatium* ferricytochrome *c', Biochim. Biophys. Acta* **342**:290.

Margalit, R., and Schejter, A., 1973, Cytochrome *c*: A thermodynamic study of the relationships among oxidation state, ion-binding and structural parameters 1. The effects of temperature, pH and electrostatic media on the standard redox potential of cytochrome *c, Eur. J. Biochem.* **32**:492.

Marrs, B., and Gest, H., 1973, Genetic mutations affecting the respiratory electron-transport system of the photosynthetic bacterium *Rhodopseudomonas capsulata, J. Bacteriol.* **114**:1045.

Meyer, T. E., 1970, PhD. Thesis, Comparative studies of soluble iron-containing proteins in photosynthetic bacteria and some algae, University of California, La Jolla.

Meyer, T. E., and Bartsch, R. G., 1976, The reactions of flavocytochromes *c* of the phototrophic sulfur bacteria with thiosulfate, sulfite, cyanide and mercaptans, in: *Fifth International Symposium on Flavins and Flavoproteins* (T. P. Singer, ed.), pp. 312–317, Elsevier Scientific Publishing Co., Amsterdam.

Meyer, T. E., Bartsch, R. G., Cusanovich, M. A., and Mathewson, J. H., 1968, The cytochromes of *Chlorobium thiosulfatophilium, Biochim. Biophys. Acta* **153**:854.

Meyer, T. E., Bartsch, R. G., and Kamen, M. D., 1971, A class of electron transfer heme proteins found in both photosynthetic and sulfate-reducing bacteria, *Biochim. Biophys. Acta* **254**:453.

Meyer, T. E., Kennel, S. J., Tedro, S. M., and Kamen, M. D., 1973, Iron protein content of *Thiocapsa pfennigii*, a purple sulfur bacterium of atypical chlorophyll composition, *Biochim. Biophys. Acta* **292**:634.

Meyer, T. E., Ambler, R. P., Bartsch, R. G., and Kamen, M. D., 1975, The amino acid sequence of cytochrome *c'* from the purple photosynthetic bacterium *Rhodospirillum rubrum* S1, *J. Biol. Chem.* **250**:8416.

Morita, S., and Conti, S. F., 1963, Localization and nature of the cytochromes of *Rhodomicrobium vannielii, Arch. Biochem. Biophys.* **100**:302.

Olson, J. M., and Chance, B., 1960a, Oxidation–reduction reactions in the photosynthetic bacterium *Chromatium.* 1. Absorption spectrum changes in whole cells, *Arch. Biochem. Biophys.* **88**:26.

Olson, J. M., and Chance, B., 1960b, Oxidation–reduction reactions in the photosynthetic bacterium *Chromatium.* 2. Dependence of light reactions on intensity of irradiation

and quantum efficiency of cytochrome oxidation, *Arch. Biochem. Biophys.* **88**:40.

Olson, J. M., and Nadler, K. D., 1965, Energy transfer and cytochrome function in a new type of photosynthetic bacterium, *Photochem. Photobiol.* **4**:783.

Olson, J. M., and Shaw, E. K., 1969, Cytochromes from the green photosynthetic bacterium *Chloropseudomonas ethylicum, Photosynthetica* **3**:288.

Olson, J. M., and Sybesma, C., 1963, Energy transfer and cytochrome oxidation in green bacteria, in: *Bacterial Photosynthesis* (H. Gest, A. San Pietro, and L. P. Vernon, eds.), p. 413, Antioch Press, Yellow Springs, Ohio.

Olson, J. M., Carroll, J. W., Clayton, M. L., Gardner, G. M., Linkins, A. E., and Moreth, C. M. C., 1969, Light-induced absorbance changes in a sulfur bacterium containing bacteriochlorophyll *b, Biochim. Biophys. Acta* **172**:338.

Orlando, J. A., 1962, *Rhodopseudomonas spheroides* cytochrome 553, *Biochim. Biophys. Acta* **57**:373.

Orlando, J. A., and Horio, T., 1961, Observations on a *b*-type cytochrome from *Rhodopseudomonas spheroides, Biochim. Biophys. Acta* **50**:367.

Parson, W. W., 1969, Cytochrome photooxidation in *Chromatium* chromatophores. Each P870 oxidizes two cytochrome C_{422} hemes, *Biochim. Biophys. Acta* **189**:397.

Parson, W. W., and Case, G. D., 1970, In *Chromatium*, a single photochemical reaction center oxidizes both cytochrome C_{552} and cytochrome C_{555}, *Biochim. Biophys. Acta* **205**:232.

Pettigrew, G. W., 1974, The purification and amino acid sequence of cytochrome *c*-552 from *Euglena gracilis, Biochem. J.* **139**:449.

Pettigrew, G. W., and Schejter, A., 1974, Conformation changes in cytochrome c_2 from *Rhodospirillum rubrum, FEBS Lett.* **43**:131.

Pettigrew, G. W., Leaver, J., Meyer, T. E., and Pyle, A. P., 1975a, Purification, properties and amino acid sequence of atypical cytochrome *c* from two protozoa, *Euglena gracilis* and *Crithidia oncopelti, Biochem. J.* **147**:291.

Pettigrew, G. W., Meyer, T. E., Bartsch, R. G., and Kamen, M. D., 1975b, pH Dependence of the oxidation-reduction potential of cytochrome c_2, *Biochim. Biophys. Acta* **430**:197.

Pettigrew, G. W., Bartsch, R. G., Meyer, T. E., and Kamen, M. D., 1978, Structural bases for the differences in the redox potentials of the cytochromes c_2, *Biochim. Biophys. Acta* (submitted).

Prince, R. C., and Dutton, P. L., 1975, A kinetic completion of the cyclic photosynthetic electron pathway of *Rhodopseudomonas sphaeroides*: Cytochrome *b*–cytochrome c_2 oxidation–reduction, *Biochim. Biophys. Acta* **387**:609.

Prince, R. C., Cogdell, R. J., and Crofts, A. R., 1974, The photo-oxidation of horse heart cytochrome *c* and native cytochrome c_2 by reaction centres from *Rhodopseudomonas spheroides* R-26, *Biochim. Biophys. Acta* **347**:1.

Prince, R. C., Baccarini-Melandri, A., Hauska, G. A., Melandri, B. A., and Crofts, A. R., 1975, Asymmetry of an energy transducing membrane. The location of cytochrome c_2 in *Rhodopseudomonas spheroides* and *Rhodopseudomonas capsulata*, *Biochim. Biophys. Acta* **387**:212

Reed, D. W., 1969, Isolation and composition of a photosynthetic reaction center complex from *Rhodopseudomonas spheroides*, *J. Biol. Chem.* **244**:4936.

Righetti, P., and Drysdale, J. W., 1971, Isoelectric focussing in polyacrylamide gels, *Biochim. Biophys. Acta* **236**:17.

Rodkey, F. L., and Ball, E. G., 1950, Oxidation–reduction potentials of the cytochrome c system, *J. Biol. Chem.* **182**:17.

Salemme, F. R., Freer, S. T., Xuong, Ng H., Alden, R., and Kraut, J., 1973a, The structure of oxidized cytochrome c_2 of *Rhodospirillum rubrum*, *J. Biol. Chem.* **248**:3910.

Salemme, F. R., Kraut, J., and Kamen, M. D., 1973b, Structural bases for function in cytochromes c, an interpretation of comparative x-ray and biochemical data, *J. Biol. Chem.* **248**:7701.

Sasaki, T., Motokawa, Y., and Kikuchi, G., 1970, Occurrence of both a-type and o-type cytochromes as the functional terminal oxidases in *Rhodopseudomonas spheroides*, *Biochim. Biophys. Acta* **197**:284.

Saunders, V. A., and Jones, O. T. G., 1974, Properties of the cytochrome a-like material developed in the photosynthetic bacterium *Rhodopseudomonas sphaeroides* when grown aerobically, *Biochim. Biophys. Acta* **333**:439.

Schachman, H. J., 1959, *Ultracentrifugation in Biochemistry*, Academic Press, New York.

Shioi, Y., Takamiya, K., and Nishimura, M., 1974, Studies on energy and electron transfer systems in the green photosynthetic bacterium *Chloropseudomonas ethylica* strain 2-K. Composition of pigments and electron transfer systems, *J. Biochem.* (*Tokyo*) **76**:241.

Sletten, K., and Kamen, M. D., 1968, Isoelectric fractionation of soluble hemoprotein from *Rhodospirillum rubrum*, in: *Structure and Function of Cytochromes* (K. Okunuki, M. D. Kamen, and I. Sekuzu, eds.), pp. 422–428, University of Tokyo press.

Slonimski, P. P., Acher, R., Péré, G., Sels, A., and Somlo, M., 1965, in: *Mecanismes de Regulation des Activites Cellulaires chez les Microorganismes*, p. 435, Centre National de la Recherche Scientifique, Paris.

Soininen, R., and Ellfolk, N., 1973, *Pseudomonas* cytochrome c peroxidase V. Absolute spectra of the enzyme and of its compounds with ligands. Inhibition of the enzyme by cyanide and azide, *Acta Chem. Scand.* **24**:35.

Suzuki, H., and Iwasaki, H., 1962, Preparation and properties of crystalline blue proteins and cryptocytochrome c, and a role of copper in denitrifying enzyme from a denitrifying bacterium, *J. Biochem.* (*Tokyo*) **52**:193.

Sybesma, C., 1967, Light-induced cytochrome reactions in the green photosynthetic bacterium *Chloropseudomonas ethylicum*, *Photochem. Photobiol.* **6**:261.

Takamiya, K., and Nishimura, M., 1974, Quantum yield of photooxidation of cytochromes in *Chromatium* chromatophores, *Biochim. Biophys. Acta* **368**:339.

Taniguchi, S., and Kamen, M. D., 1963, On the anomalous

interactions of ligands with *Rhodospirillum* haem protein, *Biochim. Biophys. Acta* **74**:438.

Taniguchi, S., and Kamen, M. D., 1965, The oxidase system of heterotrophically grown *Rhodospirillum rubrum*, *Biochim. Biophys. Acta* **96**:395.

Theorell, H., and Åkeson, Å., 1941, Studies on cytochrome c. III. Titration curves, *J. Am. Chem. Soc.* **63**:1818.

Thornber, J. P., 1969, Fractionation of three spectrally different caroteno–bacteriochlorophyll–protein complexes from *Chromatium* strain D, *Biochemistry* **9**:2688.

Timkovich, R., and Dickerson, R. E., 1973, Recurrence of the cytochrome fold in a nitrate respiring bacterium, *J. Mol. Biol.* **79**:39.

Trüper, H. G., and Rogers, L. A., 1971, Purification and properties of adenyl sulfate reductase from the phototrophic sulfur bacterium, *Thiocapsa roseopersicina*, *J. Bacteriol.* **108**:1112.

Trüper, H. G., Lorenz, C., and Fischer, U., 1976, Metabolism of elemental sulfur by phototrophic sulfur bacteria, in: *Proc. Second International Symp. on Photosynthetic Prokaryotes* (G. A. Codd and W. D. P. Stewart, eds.), pp. 41–42, Dundee, Scotland.

van Beeuman, J., and Ambler, R. F., 1973, Homologies in the sequence of cytochrome c-555 from the green photosynthetic bacteria *Chloropseudomonas ethylica* and *Chlorobium thiosulfatophilum*, *Antonie van Leeuwenhoek; J. Microbiol. Serol.* **39**:355.

Van Grondelle, R., Duysens, L. N. M., Van Der Wel, J. A., and Van Der Wel, H. N., 1977, Function and properties of a soluble c-type cytochrome c-551 in secondary photosynthetic electron transport in whole cells of *Chromatium vinosum* as studied with flash spectroscopy, *Biochim. Biophys. Acta* **461**:188.

Velick, S. F., and Strittmatter, P., 1956, The oxidation–reduction stoichiometry and potential of microsomal cytochrome, *J. Biol. Chem.* **221**:265.

Vernon, L. P., 1953, Cytochrome c content of *Rhodospirillum rubrum*, *Arch. Biochem. Biophys.* **43**:492.

Vernon, L. P., and Kamen, M. D., 1954, Hematin compounds in photosynthetic bacteria, *J. Biol. Chem.* **211**:643.

Vesterberg, O., and Svensson, H., 1966, Isoelectric fractionation, analysis and characterization of ampholytes in natural pH gradients IV. Further studies on the resolving power in connection with separation of myoglobins, *Acta Chem. Scand.* **20**:820.

Vredenberg, W. J., and Duysens, L. M. N., 1964, Light-induced oxidation of cytochromes in photosynthetic bacteria between 20° and −170°, *Biochim. Biophys. Acta* **79**:456.

Walker, W. H., Kenney, W. C., Edmondson, D. G., Singer, T. P., Cronin, J. R., and Hendriks, R., 1974, The covalently bound flavin of *Chromatium* cytochrome c_{552} 1. Evidence for cysteine thiohemiacetal, *Eur. J. Biochem.* **48**:439.

Weaver, P., 1971, Temperature-sensitive mutations of the photosynthetic apparatus of *Rhodospirillum rubrum*, *Proc. Natl. Acad. Sci. U.S.A.* **68**:136.

Weaver, P. F., 1974 Environmental and mutational variations in the photosynthetic apparatus of *Rhodo-*

spirillum rubrum, Ph.D. thesis, University of California, San Diego.

Weber, K., and Osborn, J., 1969, Reliability of molecular weight determinations by dodecylsulfate polyacrylamide gel electrophoresis, *J. biol. Chem.* **244**:4406.

Whale, F. R., and Jones, O. T. G., 1970, The cytochrome system of heterotrophically grown *Rhodopseudomonas spheroides, Biochim. Biophys. Acta* **223**:146.

Wood, F. E., and Cusanovich, M. A., 1975, The reaction of *Rhodospirillum rubrum* cytochrome c_2 with iron hexacyanides, *Bioinorg. Chem.* **4**:337.

Wood, P. M., 1977, The roles of *c*-type cytochromes in algal photosynthesis. Extraction from algae of a cytochrome similar to higher plant cytochrome *f, Eur. J. Biochem.* **72**:605.

Yamanaka, T., 1972, Evolution of cytochrome *c* molecule, *Adv. Biophys.* **3**:227.

Yamanaka, T., and Imai, S., 1972, A cytochrome *c'*-like haemoprotein isolated from *Azotobacter vinelandii, Biochem. Biophys. Res. Commun.* **46**:150.

Yamanaka, T., and Kusai, A., 1976, The function and some molecular features of cytochrome *c*-553 derived from *Chlorobium thiosulfatophilum,* in: *Fifth International Symposium on Flavins and Flavoproteins* (T. P. Singer, ed.), pp. 292–301, Elsevier Scientific Publishing Co., Amsterdam.

Yamanaka, T., and Okunuki, K., 1968, Comparison of *Chlorobium thiosulfatophilum* cytochrome *c*-555 with *c*-type cytochromes derived from algae and nonsulphur purple bacteria, *J. Biochem. (Tokyo)* **63**:341.

Yamanaka, T., and Okunuki, K., 1974, Cytochromes, in: *Microbial Iron Metabolism* (J. Nielands, ed.), pp. 349–400, Academic Press, New York.

Yamanaka, T., Ota, A., and Okunuki, K., 1961, A nitrite reducing system reconstructed with purified cytochrome components of *Pseudomonas aeruginosa, Biochim. Biophys. Acta* **53**:294.

Yamanaka, T., DeKlerk, H., and Kamen, M. D., 1967, Highly purified cytochromes *c* derived from the diatom *Naviculla pelliculosa, Biochim. Biophys. Acta* **143**:416.

Yong, F. C., and King, T. E., 1970, Optical rotatory dispersion of *Chromatium* cytochrome c_{552}, *J. Biol. Chem.* **245**:1331.

Yu, C. A., Yu, L., and King, T. G., 1972, Preparation and properties of cardiac cytochrome c_1, *J. Biol. Chem.* **247**:1012.

Zannoni, D., Baccarini-Melandri, A., Melandri, B. A., Prince, R. C., and Crofts, A. R., 1974, Energy transduction in photosynthetic bacteria. The nature of cytochrome *c* oxidase in the respiratory chain of *Rhodopseudomonas capsulata, FEBS Lett.* **48**:153.

CHAPTER 14

Complex Lipids and Fatty Acids of Photosynthetic Bacteria

Christine N. Kenyon

1. Introduction

The purposes of this chapter are several: (1) to define the complex lipids and fatty acids of particular importance to the photosynthetic bacteria; (2) to describe the methods necessary for adequate identification of the lipids and fatty acids that are present in photosynthetic bacteria; (3) to tabulate the lipid and fatty acid compositions of the photosynthetic bacteria that have been examined, together with strains analyzed and the conditions employed for growth; (4) to compare and evaluate the data in terms of the methods utilized for growth and lipid analyses; (5) to summarize the analytical results and compare the compositions of different bacteria with each other and with other organisms; and (6) to summarize the available data on the mechanisms by which complex lipids and fatty acids are synthesized. It will be made clear that frequently different workers have not utilized comparable methods, and that it is therefore often difficult to compare the results from one strain with those from another. Nevertheless, it is possible to make some significant generalizations concerning the taxonomic and evolutionary implications of the data. Furthermore, the available information on the biosynthesis of lipids and fatty acids in these microorganisms indic-

ates that they will be valuable tools for study of bacterial lipid metabolism and its control in general.

Some of the data on composition and comparative studies have appeared in a number of other reviews, the most recent and relevant being those by Goldfine (1972), O'Leary (1973), Oelze and Drews (1972), Ikawa (1967), and Lennarz (1966). None of the other reviews has been able to cover all the known data on the photosynthetic bacteria. For this reason, this chapter will attempt to be comprehensive in its coverage of the analyses reported in the literature. Neutral lipids will not be discussed in this chapter, with the exception of a few reports on sterols and triterpenes. Poly-β-hydroxybutyrate is covered in Chapter 10.

2. Chemical Nature of Complex Lipids and Fatty Acids

The chemistry of the complex lipids and fatty acids found to occur in photosynthetic bacteria has been thoroughly described elsewhere (Sober, 1970; Goldfine, 1972). The reader is referred to these other sources for details not covered here.

2.1. Complex Lipids

2.1.1. Phospholipids

The basic structure of the majority of the phospholipids found in the photosynthetic bacteria consists of

Christine N. Kenyon · Space Sciences Laboratory, University of California, Berkeley, California 94720, and Department of Parmaceutical Chemistry and Department of Biochemistry and Biophysics, University of California, San Francisco, California 94143

Table 1. Phospholipids of Photosynthetic Bacteria

Abbreviation	Common name	Generic name	−X	Structure of −X
PA	Phosphatidic acid	1,2-Diacyl-sn-glycero-3-phosphoric acid	−H	−H
PG	Phosphatidylglycerol	3-sn-Phosphatidyl-1′-sn-glycerol	L-Glycerol	$-CH_2$ 1′ $HCOH$ 2′ CH_2OH 3′
PE	Phosphatidyl-ethanolamine	3-sn-Phosphatidyl-ethanolamine	Ethanolamine	$-CH_2CH_2NH_2$
PC	Lecithin, phosphatidyl-choline	3-sn-Phosphatidyl-choline	Choline	$-CH_2CH_2\overset{+}{N}(CH_3)_3$

PS	Phosphatidylserine	3-sn-Phosphatidyl-serine	L-Serine	$-CH_2CH-COO^-$
				         $\mid$
				        $\overset{+}{N}H_3$

CL	Cardiolipin, diphos-phatidylglycerol	1′,3′-Di(3-sn-phos-phatidyl)glycerol	PG	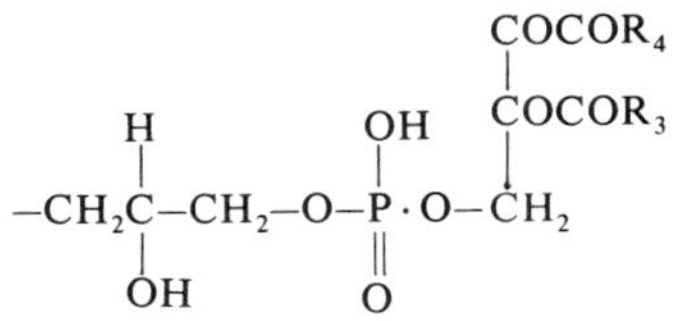 $-CH_2\underset{\underset{OH}{\mid}}{C}-CH_2-O-\underset{\underset{O}{\parallel}}{P}\cdot O-CH_2$ ... COCOR$_4$, COCOR$_3$, OH, H
PI	Phosphatidylinositol	3-sn-Phosphatidyl-inositol	Inositol	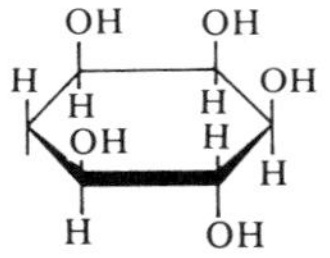 (inositol ring structure with OH groups)
bis-PA	bis-Phosphatidic acid	3-sn-Phosphatidyl-1′-(2′,3′-diacyl-sn-glycerol)	2,3-Diacylglycerol	$-CH_2$ $HCOCOR_3$ H_2COCOR_4

---

an sn-glycerol-3-phosphate backbone to the phosphate of which is attached another group, −X, as summarized in Table 1, and to the other two carbons of which are esterified fatty acid residues (R_1CO_2- and R_2CO_2-):

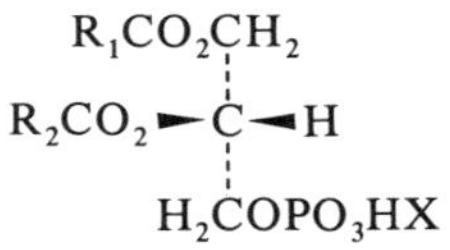

$$R_1CO_2CH_2$$
$$R_2CO_2 \blacktriangleright \overset{\mid}{C} \blacktriangleleft H$$
$$H_2COPO_3HX$$

Phosphatidic acid is defined as 1,2-diacyl-sn-glycerol-3-phosphoric acid.

The plasmalogens are another type of phospholipid that have sometimes been looked for in photosynthetic bacteria. These are derivatives of sn-glycerol-3-phosphate in which the 1-position is substituted with a long-chain alkenyl ether presumably derived from a long-chain aldehyde. The remainder of the molecule is as described for phosphatidic acid and its derivatives (Table 1).

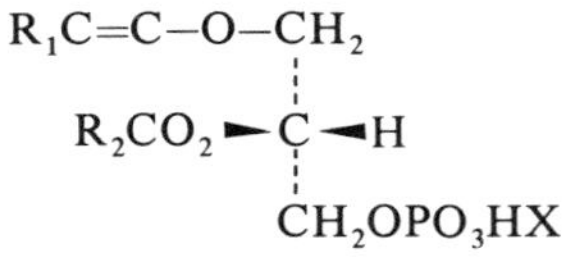

$$R_1C{=}C-O-CH_2$$
$$R_2CO_2 \blacktriangleright \overset{\mid}{C} \blacktriangleleft H$$
$$CH_2OPO_3HX$$

Plasmalogen

2.1.2. Glycolipids

The glycolipids found in the photosynthetic bacteria are glycosyldiglycerides that do not contain phosphorus.

(a) Monogalactolipid (Constantopoulos and Bloch, 1967)

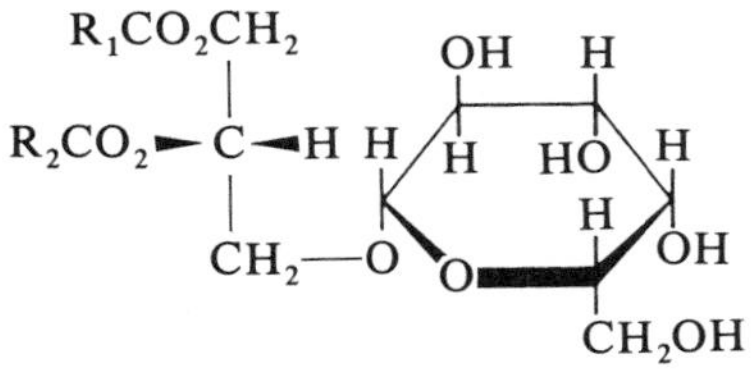

1,2-Diacyl-sn-glycero-3-β-D-galactopyranoside (MGDG, glycolipid I, or monogalactosyldiglyceride)

(b) Glycolipid II (Constantopoulos and Bloch, 1967). This lipid is probably a mixture of diglyceride derivatives containing galactose, rhamnose, and perhaps another relatively nonpolar sugar.

(c) Digalactolipid

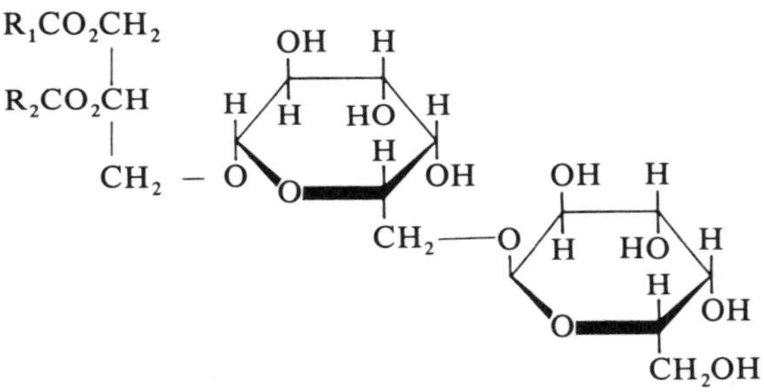

1,2-Diacyl-sn-glycero-3-[β-D-galactopyranosyl-(6→1)-α-D-galactopyranoside]

(DGDG or digalactosyldiglyceride)

This lipid has been found in eukaryotic photosynthetic organisms and cyanobacteria.

(d) Sulfolipid

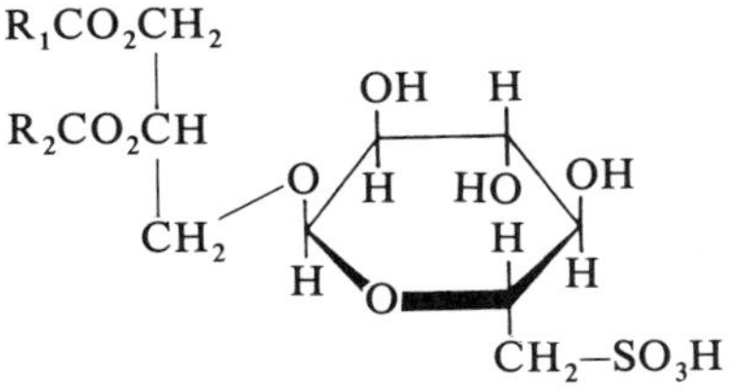

1,2-Diacyl-sn-glycero-3-(6-sulfo-α-D-quinovopyranoside)
(SQDG)

2.1.3. Ornithine Lipids

Three types of ornithine-containing lipids have been reported to occur in species of photosynthetic bacteria.

(a) Ornithine Phosphatidylglycerol (Macfarlane, 1962). This is an ornithine derivative of phosphatidyl-

glycerol in which the ornithine is esterified to one of the free hydroxyl groups of glycerol.

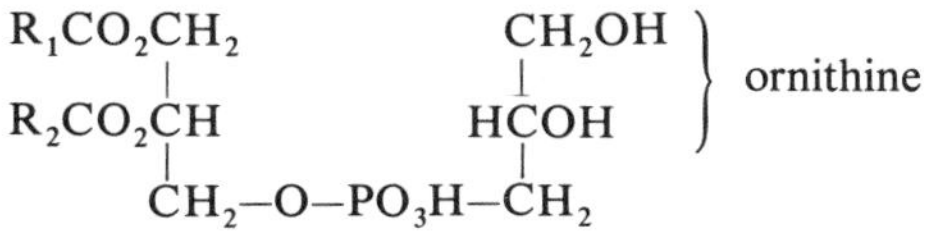

(b) Ornithine Amide I (Gorschein, 1968c). Ornithine amide I is found in *Rhodopseudomonas sphaeroides*. It is an ornithine amide in which a long-chain fatty acid residue forms an amide bond with the α-amino group of ornithine, and an alcohol residue with a cyclopropane group or a methyl branch is esterified to the carboxyl group of ornithine.

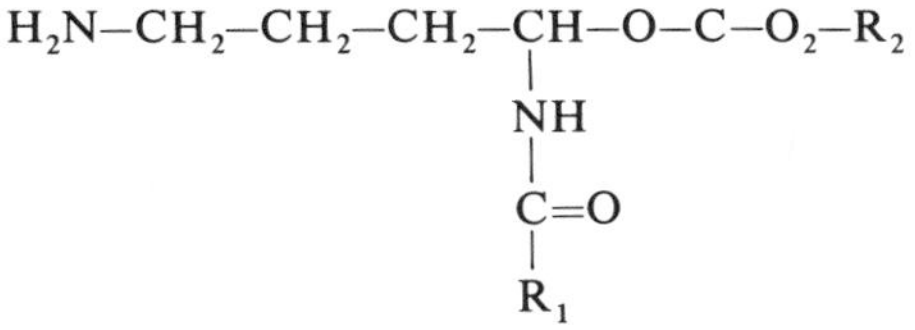

$$H_2N-CH_2-CH_2-CH_2-CH-O-C-O_2-R_2$$

with NH, C=O, R₁ chain shown below.

$-COR_1 = $ a fatty acid residue
$-OR_2 = $ an alcohol residue with a cyclopropane group or a methyl branch

(c) Ornithine Amide II (Brooks and Benson, 1972). The third ornithine lipid, found in *Rhodospirillum rubrum*, is also an ornithine amide in which a long-chain fatty acid esterifies the hydroxyl group of a 3-hydroxy fatty acid, which in turn forms an amide bond with the α-amino group of an ornithine residue.

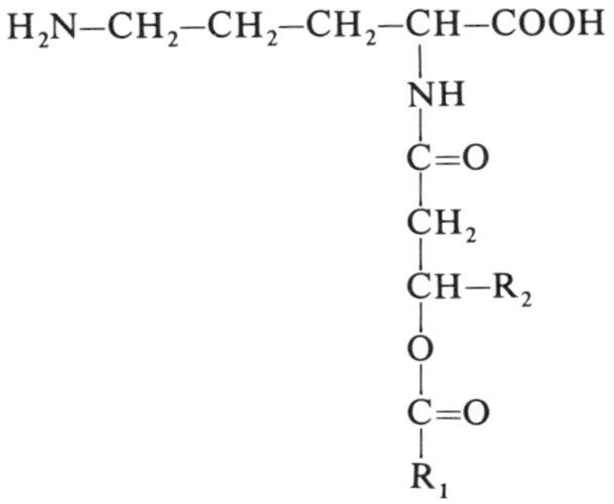

$$H_2N-CH_2-CH_2-CH_2-CH-COOH$$

with NH, C=O, CH₂, CH–R₂, O, C=O, R₁ chain shown below.

$$-O-\overset{O}{\underset{\|}{C}}-R_1 = \text{a long-chain fatty acid residue}$$

$$-\overset{O}{\underset{\|}{C}}-CH_2-CH-R_2 = \text{a 3-hydroxy fatty acid}$$
$$\underset{O-}{|}$$

All these ornithine lipids have free amino groups, and have somewhat similar chromatographic behavior. The first is easily distinguished from the other two by the presence or absence of phosphate.

2.2. Fatty Acids

2.2.1. Straight-Chain Saturated and Unsaturated Fatty Acids

The majority of the fatty acids found in the photosynthetic bacteria are straight-chain saturated and monounsaturated fatty acids, predominantly those 14–18 carbons in length. The major fatty acids are usually those the structures of which are given in Table 2.

2.2.2. Other Fatty Acids

In addition, some photosynthetic bacteria contain hydroxylated fatty acids of 10–14 carbons in chain length, branched-chain fatty acids of 15 or 17 carbons in length, and cyclopropane-group-containing fatty acids with backbones of 16, 18, or 20 carbons. The structures of typical examples of these fatty acid types appear below.

(a) Hydroxylated Fatty Acids

$$CH_3(CH_2)_{10}\overset{\displaystyle H}{\underset{\displaystyle OH}{C}}-CH_2COOH$$

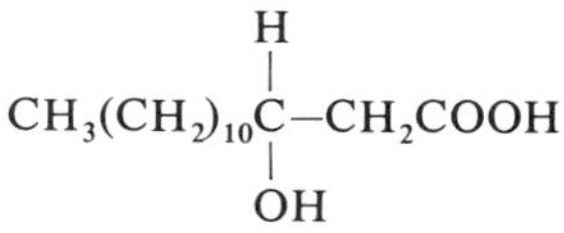

β-Hydroxymyristic acid or 3-hydroxytetradecanoic acid (14:OH)

(b) Branched-Chain Fatty Acids

$$CH_3-\underset{\displaystyle CH_3}{CH}-(CH_2)_{11}COOH$$

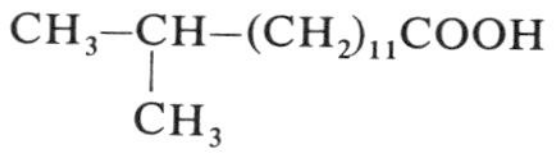

Isopentadecylic acid or 13-methyltetradecanoic acid (15br)

(c) Cyclopropane-Group-Containing Fatty Acids

$$CH_3(CH_2)_7\underset{\displaystyle CH_2}{CH-CH}(CH_2)_7COOH$$

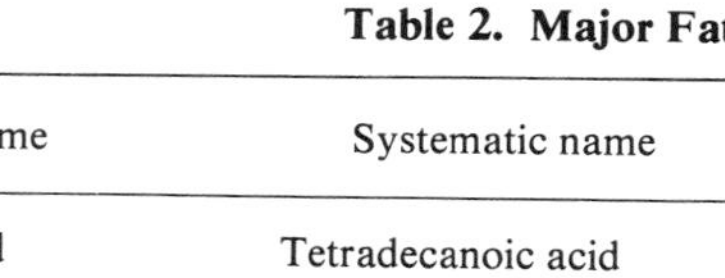

Lactobacillic acid or ω-(2-n-octylcyclopropyl)-octanoic acid (19cy)

3. Distribution of Complex Lipids by Group and Comparison with Other Photosynthetic Groups

3.1. Methods

The lipids of bacteria can be divided into an extractable fraction and a nonextractable fraction. The extractable lipids are those lipids that are extractable from whole cells with lipid solvents such as chloroform–methanol (2 : 1, vol./vol.) without prior chemical treatment of the cells. The extractable lipids of gram-negative bacteria are derived from the cell membrane, intracytoplasmic membranes, if any, and the cell wall. The nonextractable or bound lipids are released from the residue remaining after exhaustive extraction with lipid solvents by treatment with acid or alkali followed by further lipid solvent extraction. If the cells are pretreated with acid or alkali prior to the initial solvent extraction, both types of lipids are extracted.

The only well-characterized bound lipid is the lipopolysaccharide (LPS)-component lipid A of *Escherichia coli* and *Salmonella* spp., which contains phosphate, glucosamine, and characteristic fatty acids including β-hydroxymyristic acid (14 : OH), lauric acid (12 : 0), and myristic acid (14 : 0) (Lüderitz, 1970). The lipid analyses reported below are those of the extractable lipids of the photosynthetic bacteria. In some cases, subcellular fractions have been studied, but in most cases whole cells have been analyzed (see Table 4). The cell envelopes and lipopolysaccharide layers of the photosynthetic bacteria are discussed in Chapter 4.

The methods utilized to study the lipid composition of the photosynthetic bacteria have varied widely from laboratory to laboratory. In reading the following discussion, it would be well to keep in mind a few generalizations. The extraction of lipids by mixtures of chloroform and methanol [chloroform–methanol (2 : 1, vol./vol.) or the Bligh–Dyer method (Bligh and Dyer, 1959)] usually yields the most complete extracts. In the following discussion, the methods for extraction will be described when solvents other than these two systems were used.

Table 2. Major Fatty Acids of Photosynthetic Bacteria

Common name	Systematic name	Formula	Abbreviation
Myristic acid	Tetradecanoic acid	$CH_3(CH_2)_{12}COOH$	14 : 0
Palmitic acid	Hexadecanoic acid	$CH_3(CH_2)_{14}COOH$	16 : 0
Palmitoleic acid	9-Hexadecenoic acid	$CH_3(CH_2)_5CH=CH(CH_2)_7COOH$	16 : 1Δ9
cis-Vaccenic acid	*cis*-11-Octadecenoic acid	$CH_3(CH_2)_5CH=CH(CH_2)_9COOH$	18 : 1Δ11
Oleic acid	*cis*-9-Octadecenoic acid	$CH_3(CH_2)_7CH=CH(CH_2)_7COOH$	18 : 1Δ9

Preliminary examination of the polar lipids by one- or two-dimensional thin-layer chromatography (TLC), or both, before or after column chromatography has almost always been performed. The separated lipids can then be identified by specific staining behavior and comparison of their R_f's with those of known standards, usually run on the same plate. Other methods of identification of polar lipids are necessary before unambiguous identifications of the separated lipids can be made. The products of mild alcoholic base treatment are the deacylated lipid and fatty acids. Separations of the deacylated lipids by several chromatographic systems are usually less ambiguous than similar separations of the total lipids. The deacylated lipid can be hydrolyzed completely, and the relative amounts of its components and of the previously removed fatty acids can be determined chemically by use of specific tests. The total lipid can also be hydrolyzed and analyzed chemically, and various physical methods are available for structure determination. All these methods, including use of several different chromatographic systems at each step, are necessary before conclusive identifications can be made.

The lipids found in each bacterial species will be described in the order in which the species appear in the 8th edition of *Bergey's Manual* (Pfennig and Trüper, 1974).

3.2. Complex Lipid Composition of Whole Cells and Cell Fractions

3.2.1. Rhodospirillaceae

(a) *Rhodospirillum*

Rhodospirillum rubrum: The lipids of *Rs. rubrum* strains have been studied by a large number of groups. The earliest analysis of the lipids of *Rs. rubrum* is that of Benson *et al.* (1959), who isolated the chromatophores of photosynthetically grown cells and extracted the chromatophore lipids. Neither the purity of the chromatophores or lipids nor the completeness of the extraction with chloroform alone was reported. The deacylated lipids were identified by cochromatography on paper with known compounds, by staining behavior, and by neutron activation analysis. PC, PG, PI, and PE were found to be present (Tables 3 and 4, Study 1). It was implied that MGDG, DGDG, and SQDG were also present in chromatophores of *Rs. rubrum*, although the data are not given.

Benson and Strickland (1960) later reported, in addition, the presence of CL in *Rs. rubrum* chromato-

phores, in amounts somewhat less than the amount of PE and PG present (Tables 3 and 4, Study 2). In this study, negligible amounts of PC were found, and the amount of PI found was less than the amounts of PG, PE, and CL. No comment is made on the differences between the findings reported in Benson *et al.* (1959) and in Benson and Strickland (1960). The lipid extraction method was not stated.

Wood *et al.* (1965), Nichols and James (1965), and James and Nichols (1966) reported the first analysis of the whole-cell lipids of *Rs. rubrum* (Tables 3 and 4, Study 4). They employed chloroform–methanol (2 : 1), rather than chloroform alone. The lipids were identified solely by R_f on one- and two-dimensional silicic acid thin-layer chromatograms and by specific staining behavior. No structural or quantitative analyses were performed. The presence of PG, PE, CL, PI, and an ornithine-lipid (suggested to be ornithine-PG) was reported. In apparent contrast to the results of Benson *et al.* (1959), this group did not find SQDG, MGDG, or DGDG in this species. PI was reported in one of the three papers.

Haverkate *et al.* (1965), Haverkate (1965), and van Deenen and Haverkate (1966) analyzed the lipids of whole cells of *Rs. rubrum*, strain 1 (Tables 3 and 4, Study 5). The identity of the lipids was determined both by R_f of ^{32}P-labeled lipids obtained by growth on ^{32}P and by chromatography of deacylated ^{32}P-labeled lipids. The quantities of each lipid found were strikingly different from those found by Benson *et al.* (1959) (Table 4). Moreover, Haverkate *et al.* (1965) found that strain 1 contained less PC and more CL when grown in the light than when grown in the dark. The relative amounts of PG and PE remained the same in the light and in the dark. PI, ornithine-lipid, SQDG, MGDG, and DGDG were not detected in cells grown under either set of conditions.

In 1967, Depinto (1967) reported the presence of an ornithine-containing lipid that did not contain phosphorus in cells of *Rs. rubrum*, S1 (Tables 3 and 4, Study 6) in addition to PG and PE. It is likely that the ornithine-PG reported by Wood *et al.* (1965) is this ornithine amide, since their "ornithine-PG" behaves chromatographically like ornithine amide and its phosphate content was not reported. Authentic ornithine-PG has an R_f on silicic acid TLC similar to that of PC and not similar to that of PE, as does ornithine amide (Houtsmuller and van Deenen, 1963).

The analysis of the lipids of whole cells of *Rs. rubrum* by Hirayama (1968) showed that PG, PE, CL, and ornithine amide were present, while PC, PI, SQDG, MGDG, and DGDG were absent (Tables 3 and 4, Study 7). The lipid characterization by TLC, specific staining behavior, deacylation, and chemical

Table 3. Strains of Bacteria and Methods of Cultivation Used for Analyses Reported in Tables 4 and 6[a]

Genus and species	Study	Strain	Medium Ref. No.[b]	Added carbon compound[c]	Supplement[d]	Gas phase[e]	Light intensity	Temperature (°C)	pH	Growth stage at harvest
Rhodospirillaceae										
Rs. rubrum	1	NR[f]	NR	NR	NR	NR	Light	NR	NR	NR
	2	NR	NR	NR	NR	NR	NR	NR	NR	NR
	3	NR	NR	NR	NR	NR	NR	NR	NR	NR
	4a	NR	1	MG	YE	5% CO$_2$–N$_2$	1 ft. from 100-W lamp	30	6.8	Log phase
	4b	NR	1	MG	YE	Air	Dark	NR	6.8	NR
	5a	1	2,3	P	—	—	25 cm from 40-W lamp	25–30	7.0	3 days
	5b	1	2,3	P	—	Air	Dark	25–30	7.0	3 days
	6a	S1	4	M	CAA	—	Light	NR	6.6–6.8	NR
	6b	S1	4	M	CAA	—	Dark	30	6.6–6.8	NR
	7	NR	4	M	CAA	—	7000–10,000 lux	27	6.6–6.8	Early stationary phase
	8a	F1(FR1)	5	M	YE	—	4000 and 400 lux	NR	6.5–6.8	17 hr
	8b	F1(FR1)	5	M	YE	Air	Dark	30	6.5–6.8	End log
	9	F9	5	M	YE	Semiaerobic	Dark	30	6.5–6.8	48 hr
	10a	S1	1	MG	YE	Air	30 cm from 100-W lamp	24–27	6.8	Late log or early stationary phase
	10b	S1	1	MG	YE	Air	30 cm from 100-W lamp	24–27	6.8	
	10c	S1	1	MG	YE	—	Dark	24–27	6.8	
Rs. molischianum	12	NR	NR	NR	NR	NR	NR	NR	NR	NR
Rs. palustris	13a	2.1.7	1	MG	YE	5% CO$_2$–N$_2$	1 ft. from 100-W lamp	30	6.8	Log
	13b	2.1.7	1	MG	YE	Air	Dark	30	6.8	NR
	14	2.1.7	NR	NR	NR	NR	NR	NR	NR	NR
	14	2.1.23	NR	NR	NR	NR	NR	NR	NR	NR
	15	42	5	M	YE	—	200 ftc.	30	6.5–6.8	End log phase
Rp. viridis	17	F	5	M	YE	—	200 ftc.	30	6.5–6.8	End log phase
	18a	NHTC 133	6	S	YE	—	200 ftc.	30	6.8–7.2	3 days
	18b	NHTC 133	6	S	YE	—	1 ft. from 100-W lamp	30	6.8–7.2	Log phase
Rp. gelatinosa	19a	2.2.13	1	MG	YE	5% CO$_2$–N$_2$	1 ft. from 100-W lamp	30	6.8	Log phase
	19b	2.2.13	1	MG	YE	Air	Dark	30	6.8	3 days
Rp. capsulata	20a	2.3.11	1	SG	YE	5% CO$_2$–N$_2$	1 ft. from 100-W lamp	NR	NR	Log phase
	20b	2.3.11	1	SG	YE	Air	Dark	30	6.8	Log phase
	21	M.P.2	7	Pr	YE	Air	7000–10,000 lux	27	6.8	Early stationary phase
	22a	ATCC 23782	8	M	—	—	>550 ftc.; 40 ftc. or intermittent >550 ftc.	34	6.8	Log phase
	22b	ATCC 23782	8	M	—	Air	Dark	30	6.8	Log phase
	23a	37b4	5	M	YE	Air	4000 and 400 lux	NR	6.8	17 hr
	23b	37b4	5	M	YE	—	Dark	30	6.5–6.8	
	24	37b4	5	M	YE	Air	200 ftc.	30	6.5–6.8	End log phase
	25a	37b4 Ala$^-$ / 37b4 Ala$^+$ r	5	M	±YE	—	>2000 lux	30	6.5–6.8	NR
	25b	37b4 Ala$^-$ / 37b4 Ala$^+$ r	5	M	±YE	Semiaerobic	Dark	30	6.5–6.8	NR

Organism	No.	Strain	Medium ref.[b]	C source[c]	Supplement[d]	Atmosphere[e]	Light	Temp. (°C)	pH	Growth phase
Rp. sphaeroides	26a	NCIB 8253	1	MG	YE	Semiaerobic	200–250 ftc.	32–34	6.8	NR
	b	NCIB 8253	1	MG	YE	Air	Dark	34	6.8	NR
	27a	2.4.1 Ga	9	S	—	N₂	600 ftc.	34	6.8–7.2	24 hr
	b	2.4.1 Ga	9	S	—	Air	Dark	37	6.8	36 hr
	28a	NCIB 8253	1	MG	YE	—	200–250 ftc.	32–34	6.8	End log phase
	b	NCIB 8253	1	MG	YE	Air	Dark	30	6.8	Log phase
	29a	2	1	MG	YE	5% CO₂–N₂	1 ft. from 100-W lamp	30	6.8	NR
	b	2	1	MG	YE	Air	Dark	NR	6.8	3 days
	30a	4	10	M	YE	—	25 cm from 40-W lamp	25–30	6.8	3 days
	b	4	10	M	YE	Air	Dark	25–30	7.4	40–44 hr
	31a	NCIB 8253	11	MG	—	Semiaerobic	60-W lamp	32–34	6.8	24–36 hr
	b	NCIB 8253	11	MG	—	O₂–air	Dark	32	6.8	NR
	32	NR	12	MG	—	CO₂–N₂–H₂	2000 lux	30	6.9	2.5 days
	33	NR	12	MG	—	N₂	3000 lux	30	6.9	NR
	34	NR	NR	NR	NR	—	NR	NR	NR	NR
Rm. vannielii	35	NR	13	L	YE	—	NR	29	7.0	End log phase
Chromatiaceae										
Chr. vinosum	36	D	14	M?	—	—	100-W lamps	25	8.0	6 days
	37	D	NR	NR	NR	NR	NR	NR	NR	NR
	38	D	2, 3	M	—	—	25 cm from 100-W lamp	25–30	7.8	3 days
	39	D	15	—	—	—	100 cm from 120-W lamp	30–32	NR	Early stationary phase
	40	D	16	—	—	—	500 or 100 ftc.	30	NR	Early stationary phase
	41	D	17, 18	M	—	—	25–40-W lamps	25–32	7.8–8.0	Early stationary phase
T. roseopersicina	42	9314	19	—	—	—	300 ftc.	30	6.9–7.1	Log phase
E. halophila	43	SL-1		S	—	—	600–800 ftc.	45	7.6–8.0	Log or late log phase
Chlorobiaceae										
Chl. limicola	44	NR	NR	NR	NR	NR	NR	NR	NR	NR
	45	"*Chloropseudomonas ethylica*"	NR	NR	NR	NR	NR	NR	NR	NR
	46	1230	20 + NaCl	A	—	—	25 ftc.–	25	6.8	Log phase
	47	6330	17 + B₁₂ + Na₂S	A	—	—	700–2000 lux	25–30	6.8	NR
Chl. limicola	48	6230	21	A	—	—	1000 lux	30	6.8	2.5 days
f. thiosulfatophilum	49	6130	20 + Na₂S₂O₃	A	—	—	25 ftc.	25	6.8	Log phase
	50	6230	20 + Na₂S₂O₃	A	—	—	700–2000 lux	25–30	6.8	Log phase
Chl. vibrioforme	51	6030	17 + B₁₂ + Na₂S	A	—	—	700–2000 lux	25–30	6.8	NR
Chl. vibrioforme f. thiosulfatophilum	52	1930	17 + B₁₂ + Na₂S	A	—	—	700–2000 lux	25–30	6.8	NR
Chl. phaeobacterioides	53	2430	17 + B₁₂ + Na₂S	A	—	—	700–2000 lux	25–30	6.8	NR
Chl. phaeovibrioides	54	2631	17 + B₁₂ + Na₂S	A	—	—	700–2000 lux	25–30	6.8	NR
P. luteolum	55	2530	17 + B₁₂ + Na₂S	A	—	—	200–500 lux	16–22	6.8	NR
Chloroflexaceae										
C. aurantiacus	56	OK-70-fl., OH-64-fl., J-10-fl.	22, 23	—	YE	Semiaerobic	170 ftc.	45	8.0 or 7.5–7.6	NR

[a] This table shows details of the experimental materials utilized in each of the studies in Tables 4 and 6. References for the studies will be found in Table 4 (footnote c).

[b] References for media: (1) Lascelles (1959); (2) Bril (1964); (3) Eymers and Wassink (1938); (4) Cohen-Bazire et al. (1957); (5) Drews (1965); (6) Eimhjellen et al. (1963); (7) Kobayashi et al. (1967); (8) Ormerod et al. (1961); (9) Sistrom (1960); (10) Haverkate et al. (1965); (11) Lascelles (1956); (12) Wiessner (1960); (13) Park and Berger (1967a); (14) Newton and Kamen (1956); (15) Fuller (1963); (16) Hurlbert and Lascelles (1963); (17) Pfennig (1965); (18) Takacs and Holt (1971a); (19) Raymond and Sistrom (1967); (20) Pfennig (1961); (21) Pfennig (1962); (22) Van Baalen (1967); (23) Castenholz (1969).

[c] Supplied at substrate level: (A) acetate; (G) glutamate; (L) lactate; (M) malate; (P) peptone; (Pr) propionate; (S) succinate; (—) no organic carbon source.

[d] (YE) Yeast extract; (CAA) casamino acids; (—) no supplement.

[e] (—) No gas phase.

[f] Not reported.

Table 4. Complex Lipid Composition of Photosynthetic Bacteria[a]

Genus and species	Study	Strain	Growth condition (L or D)[b]	Cell fraction	Phospholipids						Ornithine lipids		Glycolipids				Ref. Nos.[c]
					PA	PC	PG	PI	PE	CL	−PG	−amide	MGDG	II	DGDG	SQDG	
Rhodospirillaceae																	
Rs. rubrum	1	NR	L	Chromatophores		30[d]	42	15	12				(+)		(+)	(+)	1
	2	NR	NR	Chromatophores		Tr	+	+	+	+							2
	4	NR	NR	Whole cells			+	+?	+	+	+[e]						3, 4
	5a	1	L	Whole cells		6[d]	29		57	8							5
	b	1	D	Whole cells		12	26		62	+							5
	6a	S1	L	Whole cells					+								6
	b	S1	D	Whole cells					+								6
	7	Wild type	L (N₂)	Whole cells			10[f]		19	5			Tr?				7
	10a	S1	L (air)	Whole cells			+		+	+		17					8
	b	S1	D	Whole cells			+		+	+		+					8
	c	S1		Whole cells			+		+	+		+					8
Rs. molischianum	12	NR	NR	Whole cells		+	+		+	+		+					9, 10
Rp. palustris	13	2.1.7	NR	Whole cells		+	+		+	+	Tr[e]		+				3, 4
	14	2.1.7	NR	Whole cells		+	+		+	+							9, 10
		2.1.23	NR	Whole cells			+			+							9, 10
Rp. viridis	18a	NHTC 133	L + YE	Whole cells		+	+		+	+							11
Rp. gelatinosa	b	NHTC 133	L − YE	Whole cells		+	+		+	−							11
	19	2.2.13	NR	Whole cells		+	+	−	+	−	+[e]						3, 4
Rp. capsulata	20	2.3.11	NR	Whole cells		+	+	−	+	−	+[e]						3, 4
	21	M.P.2	L	Whole cells		11[g]	9[f]		31	1							7
	22a	ATCC 23782	L	Whole cells		23	41		48	8							8
	b	ATCC 23782	D	Whole cells		22	38		38								12
Rp sphaeroides	28a	NCIB 8253	L	TCA precipitate of whole cells	24[g]	22	14		40								13
	b	NCIB 8253	D	TCA precipitate of whole cells	28[g]	20	13		39								13
					24[g]	21	14		41								13
	29	L-70	D	Whole cells		+	+		+		+[e]						13
	30a	2	NR	Whole cells		8[d]	54		36	2							3, 4
	b	4	L	Whole cells		4	54		36	6						+	5
	31a	4	D	Chromatophores		63[h]	80		100							+	14
		NCIB 8253	L	Cell wall	7[g]		−		85[g]			26					14
		NCIB 8253	L	Chromatophores	4[g]	22	36		35	+		15				+	15
		NCIB 8253	L	Cytoplasmic membrane	7[g]	23	34		35	3		+				+	15
		NCIB 8253	L		7[g]	21	34		30	9		+				+	15
	b	NCIB 8253	D	Cytoplasmic membrane and cell wall	4[g]	19	33		45			+				+	15
Rm. vannielii	32	NCIB 8253	D	Whole cells	4[g]	15	39		41			+				+	15
	34	NR	D	Whole cells	9[f]	30			18							3[j]	16
	35	NR	NR	Whole cells	2[d]	27	10		5	7[i]	47	1				Tr	17
		NR	NR	Whole cells													18

No.	Taxon / Fraction	L/D	Strain										Ref.
Chromatiaceae													
	Chr. vinosum												
36	Chromatophores	D							+				19
37	Whole cells	D											20
38	Whole cells	D		53^d	47	—				—			5
39	Whole cells	D		18^g	57		4			—			21
40	Whole cells	L (high)		39^g	50		4	+		3^l	+		22
40	Whole cells	L (low)		36^g	53		5	+		4^l	+		22
42	Whole cells	L	9314	+	+		+	+		+^{i}	+		23
42	Chromatophores	L	9314	+	+		+	+		+^{i}	+		23
43	Crude membranes	L	SL-1	61^g	10			10^m					24
43	Photosynthetic vesicles	L	SL-1	79^g	—			16^m					24
	Thylakoids	L	SL-1	81^g	3			4^m					24
Chlorobiaceae													
44	*Chl. limicola* — Whole cells	NR	NR	+					+^{n}	+^{n}	+^{n}		4
45	Whole cells	NR	"*Chloropseudomonas ethylica*"	+	(+)				+^{n}	+	+	—	9, 10
46	Cell membranes	L	1230							+	+		25
	Photosynthetic vesicles	L	1230								+		25
47	*Chl. limicola* f. *thiosulfatophilum* — Whole cells	L	6330	+			+		+	+	+	+	26
49	Cell membranes	L	6130	+						+	+		25
	Photosynthetic vesicles	L	6130							+			25
	Cell membranes	L	6230	+						+	+		25
	Photosynthetic vesicles	L	6230						+		+		25
50	*Chl. vibrioforme* — Whole cells	L	6230	+	—		+		+	+	+	+	26
51	*Chl. vibrioforme* — Whole cells	L	6030	+	—		+		+	+	+	+	26
52	*Chl. vibrioforme* f. *thiosulfatophilum* — Whole cells	L	1930	+	—		+		+	+	+	+	26
54	*Chl. phaeovibroides* — Whole cells	L	2631	+	—		+			+	+	+	26
55	*P. luteolum* — Whole cells	L	2530	+	—		+			+	+	+	26
Chloroflexaceae													
56	*C. aurantiacus* — Whole cells	L	OK-70-fl.	+	—				+^{o}	+^{o}	+^{o}	+ +	26
	Whole cells	L	OH-64-fl.	+	—				+^{o}	+^{o}	+^{o}	+ +	26
	Whole cells	L	J-10-fl.	+	—				+^{o}	+^{o}	+^{o}	+ +	26

[a] NR: Not reported; +: present; —: absent; Tr: trace.

[b] L: Light; D: dark.

[c] References for studies in Tables 4 and 6: (1) Benson et al. (1959); (2) Benson and Strickland (1960); (3) James and Nichols (1966), Wood et al. (1965); (4) Nichols and James (1965); (5) Haverkate (1965), Haverkate et al. (1965), van Deenen and Haverkate (1966); (6) Depinto (1967); (7) Hirayama (1968); (8) Brooks and Benson (1972); (9) Constantopoulos and Bloch (1967); (10) G. Constantopoulos (personal communication); (11) Pucheu et al. (1974); (12) Steiner et al. (1970b); (13) Lascelles and Szilagyi (1965); (14) Gorschein (1964); (15) Gorschein (1968b); (16) Radunz (1969); (17) D. G. Bishop (personal communication); (18) Park and Berger (1967a); (19) Newton and Newton (1957); (20) Lascelles and Szilagyi, in Erwin and Bloch (1964); (21) Steiner et al. (1969); (22) Steiner et al. (1970a); (23) Takacs and Holt (1971b); (24) J. C. Raymond (personal communication); (25) Cruden and Stanier (1970), Cruden et al. (1970); (26) Kenyon and Gray (1974); (27) Cho and Salton (1964, 1966); (28) Schröder et al. (1969); (29) Oelze and Drews (1970); (30) Weckesser et al. (1973); (31) Weckesser et al. (1974); (32) Drews (1974); (33) Hands and Bartley (1962); (35) Scheuerbrandt and Bloch (1962); (36) Schmitz (1967a); (37) Park and Berger (1967b); (38) Hurlbert et al. (1974); (39) Schmitz (1967b).

[d] Percentage of total phospholipids.

[e] Probably ornithine-amide.

[f] Percentage of total lipids.

[g] Percentage of lipid phosphorus.

[h] Micromoles per milligram dry weight of cell fraction.

[i] bis-PA.

[j] MGDG: monoglucosyldiglyceride; DGDG: (mannosylglucosyl)-diglyceride and (dimannosylglucosyl)-diglyceride.

[k] Percentage of lipid carbon.

[l] Unidentified glycolipid containing glucose and rhamnose.

[m] Phosphate- and ninhydrin-positive, R_f similar to that of PS.

[n] Unidentified galactosyldiglyceride.

[o] Galactose and another sugar present.

analysis should have yielded accurate results. Since a large quantity of cells was extracted, any PC or PI present should have been detected.

Brooks and Benson (1972) further investigated the onithine lipids in *Rs. rubrum* S1 (Tables 3 and 4, Study 10). After rather careful chemical determinations of the components of the major ornithine amide, they proposed an unusual structure in which one residue of fatty acid esterifies the hydroxyl group of a 3-hydroxy fatty acid (mostly 16 : OH), which in turn forms an amide bond with the α-amino group of an ornithine residue [ornithine amide II; see Section 2.1.3(c)]. This lipid is present in approximately equal amounts in anaerobic light-grown cells and in aerobic dark- and aerobic light-grown cells. A second, different ornithine-containing lipid was found in the latter two types of cells only. These authors confirmed the presence of PE, PG, and CL in *Rs. rubrum*, and the absence of PC and aminoacyl PG.

Fiertel and Klein (1959) reported the absence of sterols from the nonsaponifiable lipids of *Rs. rubrum*, and Kamio *et al.* (1969) could find no plasmalogens in this species.

Examination of the reported data on the extractable lipids of whole cells and chromatophores of *Rs. rubrum* reveals that PG and PE were always found to be present when looked for, CL and ornithine amide were always found in the later studies, MGDG is probably not present, and SQDG and DGDG are certainly absent. The reason for the substantial differences with respect to PC and PI content is obscure. The laboratories that investigated the lipids of *Rs. rubrum* utilized five different strains of *Rs. rubrum*, and whether or not any of these is identical is unknown. Furthermore, the growth conditions were often not the same (Table 3). When cells were grown on malate in mineral medium with yeast extract or casamino acids, PC was always absent, while on peptone, PC was present. It is not clear how much difference there would be between different strains grown under identical conditions or in the same strain grown under different conditions. In this regard, it should be mentioned that Haverkate *et al.* (1965) and Brooks and Benson (1972) found no qualitative difference between the lipids of anaerobic, light-grown cells and aerobic dark-grown whole cells (Table 4).

Rhodospirillum molischianum: Constantopoulos and Bloch (1967) and G. Constantopoulos (personal communication) reported the presence of a small amount of a MGDG-like glycolipid in *Rs. molischianum* (Tables 3 and 4, Study 12). This lipid occurs in very much lower concentration than in the higher algae and plants or in green bacteria (see Section 3.3), and has not yet been fully characterized.

(b) *Rhodopseudomonas*

Rhodopseudomonas palustris: Wood *et al.* (1965), Nichols and James (1965), and James and Nichols (1966) showed that PC, PG, PE, CL, and traces of an ornithine lipid were present in *Rp. palustris* 2.1.7 (Tables 3 and 4, Study 13). The methods used were the same as for *Rs. rubrum* [see Section 3.2.1(a)]. No SQDG, MGDG, or DGDG was found. The absence of the glycolipids MGDG and DGDG was confirmed by Constantopoulos and Bloch (1967) (Tables 3 and 4, Study 14).

Aaronson (1964) used two inhibitors, benzmalecene and triparanol, to determine whether sterols are present in *Rp. palustris*. He inferred that since ergosterol, squalene, and oleic acid annul the inhibition of growth of *Rp. palustris* by each of the inhibitors at certain concentrations, sterols probably occur in this microorganism. It seems to this author, however, that the inhibitors could act on the cell in some other way, perhaps at the membrane, that is overcome by oleic acid or by the sterols under the appropriate conditions.

Jensen (1962) isolated paluol, a triterpene, from the nonsaponifiable fraction of both *Rp. palustris* 5 and a strain of *R. palustris* obtained by van Niel from Gaffron. Since this is an isolated report of the occurrence of a triterpene, its significance is unknown.

Rhodopseudomonas viridis: Pucheu *et al.* (1974) found PC, PG, PE, CL, and an undefined ornithine lipid in *Rp. viridis* NHTC 133 grown either with or without yeast extract (Tables 3 and 4, Study 18). It was not clear whether whole cells or membranes were extracted, and lipids were identified on the basis of R_f on TLC and specific staining behavior only. Thus, the full significance of these analyses is not clear.

Rhodopseudomonas gelatinosa: The only analysis of this species is that of Wood *et al.* (1965), Nichols and James (1965), and James and Nichols (1966). They reported the presence of PG, PE, and ornithine lipid in *Rp. gelatinosa* 2.2.13 (Tables 3 and 4, Study 19). This author assumes that the report of PC in this species in one of their publications (James and Nichols, 1966) is an error. No glycolipids, PI, or CL were found. Characterization of the lipids was as described in Section 3.2.1(a) under *Rs. rubrum*.

Rhodopseudomonas capsulata: The earliest analysis of the lipids of this species showed that PC, PG, PE, and an ornithine lipid are present in *Rp. capsulata* 2.3.11, while PI, CL, MGDG, DGDG, and SQDG are absent (Tables 3 and 4, Study 20) (Wood *et al.*, 1965; Nichols and James, 1965; James and Nichols, 1966). Again, this author assumes that the report of SQDG in one of these papers (James and Nichols, 1966) is an error.

Hirayama (1968) found a slightly different composition for strain M.P.2 grown on propionate rather than malate as carbon source (Tables 3 and 4, Study 21). He found PG, PE, and small amounts of CL. SQDG, MGDG, DGDG, PC, PI, and ornithine lipid were absent. The methods used by Hirayama should have detected PC as described in Section 3.2.1(a) under *Rs. rubrum*.

Steiner *et al.* (1970*b*) found PC, PG, and PE in anaerobic light-grown *Rp. capsulata* ATCC 23782 (St. Louis) in better agreement with Wood *et al.* (1965) (Tables 3 and 4, Study 22). This study was one of the few in which growth conditions were carefully controlled. The data are given for lipid-P determined after separation of the deacylated lipids by column chromatography. In contrast to the results cited above for *Rs. rubrum* [Section 3.2.1(a) and Table 4], there was found to be a significant difference in the amount of PE and PG/protein in cells grown anaerobically in the light relative to that in those grown aerobically in the dark. Moreover, the relative proportions of the three phospholipids were found to differ significantly under the two conditions of growth.

Rhodopseudomonas sphaeroides: Lascelles and Szilagyi (1965) studied membrane synthesis in *Rp. sphaeroides* NCIB 8253 and found that the major lipid components of the membranes of cells grown anaerobically in the light were PA, PC, PG, and PE (Tables 3 and 4, Study 28). The lipid composition of cells grown aerobically in the dark was similar, as was that of the nonpigmented mutant of this strain, L-70. The phospholipids were identified by deacylation followed by chromatography.

Wood *et al.* (1965), Nichols and James (1965), and James and Nichols (1966) also reported the presence of PC, PG, and PE, and in addition CL, SQDG, and ornithine-lipid in *Rp. sphaeroides* strain 2 [refer to comments on methods used in Section 3.2.1(a) under *Rs. rubrum*] (Tables 3 and 4, Study 29). They did not mention PA, and MGDG, DGDG, and PI were found to be absent. This was the only strain in which this group found SQDG.

Haverkate *et al.* (1965), Haverkate (1965), and van Deenen and Haverkate (1966) found the same lipids in strain 4 as found by Wood *et al.* (1965), with the exception of ornithine lipid, and in addition reported the relative amounts of each phospholipid (Tables 3 and 4, Study 30). The phospholipid composition of light-, anaerobically grown cells was similar to that of dark-, aerobically grown cells. The amounts of all lipids except PE differ significantly from those of Lascelles and Szilagyi (1965). The methods used are discussed in Section 3.2.1(a) under *Rs. rubrum*.

Gorschein (1964, 1968*a–c*) reported a rather complete analysis of the lipids of whole cells and particulate cell fractions, prepared by differential centrifugation and centrifugation in sucrose gradients, of *Rp. sphaeroides* NCIB 8253 grown both semi-anaerobically in the light and aerobically in the dark (Tables 3 and 4, Study 31). Again PC, PG, and PE were major lipids, PA was found, and several fractions contained CL. Ornithine was a significant component of the lipid fractions. The relative amounts of the different phospholipids were constant from one particulate cell fraction to another with the exception of the cell walls, which did not contain PC or PG. SQDG was detected in whole cells grown either anaerobically or aerobically. The identification of the lipids of *Rp. sphaeroides* by Gorschein was based on R_f, specific staining behavior, and analysis of the products of total and partial hydrolysis of radioactively labeled lipids. Gorschein's determinations suggest strongly that significant amounts of plasmalogens are absent from *Rp. sphaeroides*. The structure and properties of the ornithine amide found were determined by Gorschein (1964, 1968*c*).

The sulfolipid of *Rp. sphaeroides* reported by most of the authors cited above was characterized chemically by Radunz (1969) (Tables 3 and 4, Study 32), who showed that it was the same as the sulfolipid of a number of higher plants, algae, and a cyanobacterium, *Oscillatoria chalybea*. The ether-soluble lipids of *Rp. sphaeroides* were reported by Radunz to consist of 3% sulfolipid.

In summary, the published analyses of *Rp. sphaeroides* suggest that the major lipids of this species are PC, PG, PE, and ornithine amide. Small amounts of SQDG are also present. Reports on the composition of strain NCIB 8253 mention the presence of PA, and CL is also probably present. The ornithine-containing lipid (ornithine amide I) in *Rp. sphaeroides* NCIB 8253 has been well-characterized chemically.

(c) *Rhodomicrobium*

Rhodomicrobium vannielii: A preliminary examination of this species was made in the laboratory of K. Bloch (D. G. Bishop, personal communication). Chloroform–methanol-extracted lipids were separated by column chromatography on silicic acid and identified by staining behavior and R_f on silica gel TLC. Bishop found PA, PE, and PC (Tables 3 and 4, Study 34).

Park and Berger (1967*a*) later analyzed this species more completely (Tables 3 and 4, Study 35). Lyophilized cells were exhaustively extracted with several lipid solvents, including chloroform–methanol. Lipids were fractionated by silicic acid column chromatography and silica gel TLC. Following isolation of the individual lipids, they were characterized chemically by a number of colorimetric and chromatographic methods. The lipids found were PA, PC, PG, PE, bis-PA, ornithine-PG, ornithine amide, and traces of SQDG. No sterols or methyl derivatives of PE were found, and no mention was made of glycosyl diglycerides. Since none of the other Rhodospirillaceae have been shown to contain such large amounts of ornithine-PG and the chromatographic properties of the lipid of Park and Berger (1967a) appear similar to those of the ornithine amide, the component should probably be purified and characterized again. If it is indeed ornithine-PG, it would be most interesting, since *Rm. vannielii* would apparently be unique among the photosynthetic bacteria in containing such large amounts of this lipid.

3.2.2. Chromatiaceae

In contrast to the many analyses of the complex lipids of the Rhodospirillaceae, there have been only a few analyses of the lipids of the Chromatiaceae.

(a) Chromatium vinosum. *Chromatium vinosum* strain D (often referred to merely as *Chromatium strain D*) was first analyzed by Newton and Newton (1957), who separated the chromatophores and extracted their lipids with boiling ethanol and ethanol–ether. They found a molar ratio of glycerol/ethanolamine/P in the lipid fraction of 1 : 1 : 1, and concluded that PE was the only phospholipid present (Tables 3 and 4, Study 36).

In 1964, Erwin and Bloch (1964) reported that Lascelles and Szilagyi found galactolipids (MGDG and DGDG) to be absent from *Chr. vinosum* (Tables 3 and 4, Study 37).

Haverkate *et al.* (1965), Haverkate (1965), and van Deenen and Haverkate (1966) confirmed the presence of PE in and the absence of galactolipids from *Chr. vinosum* (Tables 3 and 4, Study 38). In addition, they showed that PG and CL were present in this species, and that SQDG was absent. The phospholipid composition of chromatophores of *Chr. vinosum* was reported to be similar to that of the whole cells.

Steiner *et al.* (1969) extracted the lipids of *Chr. vinosum* grown anaerobically in the light, and identified them by chromatography and chemical analyses using ^{32}P and ^{14}C as lipid tracers (Tables 3 and 4, Study 39). They found PE and lyso PE, PG, CL, and three unusual glycolipids that contained glucose and mannose: monoglucosyldiglyceride, (mannosylglucosyl)-diglyceride, and (dimannosylglucosyl)-diglyceride. No galactose was found in the hydrolyzed lipid extracts. Glycolipids of this composition have not been reported in photosynthetic bacteria, and indeed are usually characteristic of gram-positive strains

(Shaw and Baddiley, 1968; Shaw, 1970; Goldfine, 1972). When such small amounts of lipids are reported, however, the possibility of artifacts such as contamination of the sample with sugars must be eliminated before drawing too many conclusions. If these glycolipids do exist in *Chr. vinosum*, it will be most interesting to determine their location in the cell. In a later study, Steiner *et al.* (1970a) examined the effect of light intensity on the lipid composition of *Chr. vinosum*. They found that although there was less phospholipid per protein in cells grown at low light intensity relative to those grown at high light intensity, the relative proportions of the lipids remained the same under the two conditions (Tables 3 and 4, Study 40).

(b) Thiocapsa roseopersicina (Thiocapsa floridana). The lipids of whole cells and chromatophores of *T. roseopersicina* strain 9314 were analyzed by Takacs and Holt (1971b) (Tables 3 and 4, Study 42). They found that the total lipid content of chromatophores (47% of the dry weight) was higher than that of whole cells (15% of the dry weight), but that the nature of the lipids found was the same in each cell fraction. PE, PG, CL, and glycolipids containing glucose and rhamnose were found in chloroform–methanol extracts of whole cells and purified chromatophores. The lipids were freed of contaminating material on Sephadex G-25, and were identified after fractionation and TLC using the R_f and colorimetric reactions for identification. These authors found that some glycolipid was lost following dissociation of the chromatophores in SDS and reassociation by dialysis against $MgCl_2$-containing buffer.

(c) Ectothiorhodospira halophila, Strain SL-1. J. C. Raymond (personal communication) analyzed the lipids of crude membranes and of photosynthetic vesicles and thylakoids of this extreme halophile (Tables 3 and 4, Study 43). The lipids appeared to be completely "normal," unlike those of *Halobacterium* as described by Kates *et al.* (1966). In addition to PG, PE, and a ninhydrin-positive phospholipid that has chromatographic and staining behavior similar to ornithine-PG or PS (Table 4), two unknown phospholipids were found. The latter were 17% of the total phospholipids in crude membranes, 19% in photosynthetic vesicles, and 12% in thylakoids, and were not CL, PC, or lysolecithin. The cell fractions appear to be differentiated with respect to lipids, as was found for *Rp. capsulata* (Steiner *et al.*, 1970b) [see Section 3.2.1.(b)]. Raymond reports that the crude membrane fraction had the lowest pigment/protein ratio of all three fractions, and probably contained large amounts of cytoplasmic membranes. The photosynthetic vesicles, five-layered structures made from appressed membranes of the thylakoids, are invaginated photosynthetic membranes. The thylakoids resemble "stacks of coins" and have areas of five-layered structures connected by membranes with a three-layered appearance. The pigment/protein ratio of this fraction is intermediate between that of the vesicles and the crude membrane. The vesicles contained only PG and the amino-lipid. The relative content of PG is higher in the vesicles and thylakoids than in the crude membrane fractions, and the content of the ninhydrin-positive phospholipid is significantly higher in the vesicles than in the thylakoids. The results further suggest that PE is a component of the cytoplasmic membrane, and that the thylakoid fraction probably contains some cytoplasmic membrane, since it contains the cytoplasmic membrane component, PE.

3.2.3. Chlorobiaceae

(a) *Chlorobium limicola*. The Chlorobiaceae are unique with respect to lipid content. The first report of an analysis of the lipids of microorganisms belonging to this family was that of Nichols and James (1965), who reported the presence of galactosyldiglycerides and SQDG in *Chl. limicola* (Tables 3 and 4, Study 44). PG was also implied to be present. Unfortunately, the details of the analysis, the strain employed, and the conditions of growth were not given in this report.

A subsequent analysis of this species was performed by Constantopoulos and Bloch (1967) (Tables 3 and 4, Study 45). Unfortunately, they used "*Chloropseudomonas ethylica*," a species subsequently shown to be a mixed culture of a green bacterium and a sulfate-reducing bacterium.* Despite the dubious origin of the lipids extracted, their careful analysis of the glycolipids was in all likelihood an analysis of the glycolipids of *Chl. limicola*, as proved by later studies. They found that a significant quantity (2.2%) of the total lipids was monogalactolipid. This lipid was shown to be identical with the monogalactolipid found in higher plants and algae. They also found a second, less well characterized glycolipid ("glycolipid II") that contained galactose, rhamnose, and an unidentified sugar that has so far not been found in the glycolipids of other microorganisms. No digalactosyldiglyceride was detected (Table 4). G. Constantopoulos (personal communication) also found PE and PG in "*Chloropseudomonas ethylica*," but because of the rather wide distribution of these lipids among the bacteria, the significance of this finding for *Chl. limicola* was unclear.

Cruden *et al.* (1970) and Cruden and Stanier

* See Chapter 2 for a discussion of the taxonomy of this syntrophic system.

(1970), using pure cultures of *Chl. limicola* strain 1230, showed that the glycolipids of Constantopoulos and Bloch (1967) were indeed characteristic of *Chl. limicola* (Tables 3 and 4, Study 46). Moreover, the MGDG was localized in the photosynthetic vesicles, while glycolipid II was localized in the cell-membrane fraction. The latter was enriched in phospholipid relative to the vesicle fraction.

More recently, we analyzed the composition of the lipids of *Chl. limicola* strain 6330 (Kenyon and Gray, 1974) and found no evidence for PE, which suggests that the PE found by Constantopoulos was indeed derived from the sulfate-reducing bacterium, rather than from *Chl. limicola*. We confirmed the presence of PG, MGDG, and glycolipid II, and found CL and suggestive evidence for SQDG. No PS was detected. There were several unknown lipids in significant amounts. The identifications of phospholipids in our laboratory were based solely on R_f in several solvent systems and staining behavior, since the amount of available material for analysis was limited. Further chemical analyses are clearly indicated. The glycolipids were each isolated, and the component sugars were determined to be the same as those found by Constantopoulos and Bloch (1967) and by Cruden and Stanier (1970).

(b) Other Chlorobiaceae. Analysis of the other type strains of Chlorobiaceae, in parallel with those of *Chl. limicola*, has shown that the other species of this group also contain PG, CL, MGDG, glycolipid II, and SQDG as found in *Chl. limicola* (Cruden *et al.*, 1970; Cruden and Stanier, 1970; Kenyon and Gray, 1974) (Tables 3 and 4, Studies 49–52, 54, and 55). We found that there were some differences in other unknown lipids between the seven type species, but none of the lipids found to be different was available in sufficient quantity to be identified. None of the other species contained PE.

3.2.4. Chloroflexaceae

Finally, we have analyzed the lipids of three strains belonging to the newly described genus *Chloroflexus aurantiacus* (Pierson and Castenholz, 1974; Kenyon and Gray, 1974) (Tables 3 and 4, Study 56). *Chloroflexus aurantiacus*, strains OK-70-fl., OH-64-fl., and J-10-fl., were found to contain PG and PI, but not PE or PS. Glycolipids that behaved chromatographically like those found in higher plants were found, and were shown to contain galactose and possibly a galactose derivative or a degradation product of galactose. Again, a more rigorous chemical analysis of the lipids present is called for.

Table 5. Complex Lipids of Representative Photosynthetic Organisms

Group	Genus and species	PC	PG	PI	PE	CL	Ornithine lipid	MGDG	Glycolipid II	DGDG	SQDG	Ref. Nos.[b]
Higher plants	Spinacea oleracea	+	+	+	+			+	−	+	+	1
Green algae	Chlorella vulgaris	+	+	+	+			+	−	+	+	2
Euglenid	Euglena gracilis	+	+	+	+			+	−	+	+	3
Red algae	Porphyra umbilicalis	+	+	+	+	−		+	−	+	+	4
Yellow algae	Monodus subterraneus	+	+	+	Tr	Tr		+	−	+	+	5
Marine diatoms	Nitzschia thermalis	+	+	+	Tr	Tr		+	−	+	+	6
Freshwater diatoms	Navicula pelliculosa	+	+	+	Tr	Tr		+	−	+	+	6
Cyanobacteria	Synechococcus sp.	−	+	−	−			+	−	+	+	5
PHOTOSYNTHETIC BACTERIA												
Rhodospirillaceae	Rs. rubrum	+?	+	−?	+	+	+	Tr?	−	−	−	c
	Rs. molischianum							+	−	−	−	c
	Rp. palustris	+	+	−	+	+	Tr	−	−	−	−	c
	Rp. viridis	+	+	−	+	+	+		−	−	−	c
	Rp. gelatinosa	−	+	−	+	−	+		−	−	−	c
	Rp. capsulata	+	+	−	+	+	+		−	−	−	c
	Rp. sphaeroides	+	+	−	+	+	+		−	−	−	c
	Rm. vannielii	+	+	−	+	+	−		−	−	−	c
Chromatiaceae	Chr. vinosum		+		+	+		+[d]	−	+[d]	−	c
	T. roseopersicina		+		+	+		+[d]	−	+[d]	−	c
	E. halophila		+		+	+	+?		−	−	−	c
Chlorobiaceae	Chl. limicola		+		−	+		+	+	−	+	c
	Chl. limicola f. thiosulfatophilum		+		−	+		+	+	−	+	c
	Chl. vibrioforme		+		−	+		+	+	−	+	c
	Chl. vibrioforme f. thiosulphatophilum		+		−	+		+	+	−	+	c
	Chl. phaeovibrioides		+		−	+		+	+	−	+	c
	P. luteolum		+		−	+		+	+	−	+	c
Chloroflexaceae	C. aurantiacus		+	+	−			+[e]	−	+[e]	+	c

[a] (+) Present; (−) absent; (blank) not reported; (Tr) trace.
[b] (1) Wintermans (1960), Bishop (1959), Zill and Harmon (1962); (2) James and Nichols (1966), Nichols et al. (1966); (3) Nichols et al. (1966); (4) Davies et al. (1965); (4) Benson and Shibuya (1962); (5) James and Nichols (1962); (6) Kates and Volcani (1966).
[c] See Table 4.
[d] Not galactolipids.
[e] Galactose and another sugar present.

3.3. Comparative Aspects of Complex Lipid Composition

Comparison of the lipid compositions of the various groups of photosynthetic bacteria with each other, with the other major groups of photosynthetic organisms, and with other bacteria reveals that certain lipids are of particular significance to taxonomy or evolution or both (Table 5). Phosphatidylglycerol (PG) has been found in all photosynthetic organisms studied to date. This phospholipid and sometimes its derivative, CL, appear to be the only phospholipids of major quantitative importance in the cyanobacteria (Nichols *et al.*, 1965) and in the Chlorobiaceae. The latter as a group and certain other photosynthetic bacteria (Table 4) are thus an exception to the generalization made by Goldfine (1972) that in gram-negative bacteria, PE and its methylated derivatives are usually quantitatively more important than PG. In all photosynthetic bacteria other than the Chlorobiaceae and *Chloroflexus*; in eukaryotic photosynthetic organisms, except red algae and diatoms; and in nearly all other gram-negative bacteria (Ikawa, 1967; Goldfine, 1972; Oliver and Colwell, 1973), PE is also a major component of the phospholipid fraction. This implies that *Chloroflexus* and the Chlorobiaceae are more closely related to each other and to the cyanobacteria than are the other groups of photosynthetic bacteria.

Generalizations about the distribution of other phospholipids are necessarily less broad. PC is apparently present in most of the Rhodospirillaceae with the possible exception of some strains of *Rs. rubrum* and *Rp. gelatinosa*, but is not reported to be present in the Chromatiaceae, the Chlorobiaceae, or *Chloroflexus*. The rarity of the occurrence of PC in bacteria and the taxonomic significance of its distribution were reviewed by Ikawa (1967) and by Goldfine (1972). This lipid is usually a major component of the membranes of higher organisms, although it is absent from cyanobacteria.

The presence of PI in *Chloroflexus* is of particular interest, because this lipid is usually characteristic only of eukaryotic cells and some gram-positive bacteria (Goldfine, 1972). The only other reports of PI in photosynthetic bacteria are those of Benson *et al.* (1959) and Benson and Strickland (1960) for *Rs. rubrum*. However, until more detailed chemical analyses of the lipids of *Chloroflexus* are performed, and this organism can be grown without yeast extract, the significance of this finding is unclear.

The absence of plasmalogens from all the photosynthetic bacteria in which they have been sought is of particular interest. Goldfine (1972) and Kamio *et al.*

(1969) pointed out that this group of compounds has been found in all strict anaerobes, but not in facultative anaerobes or aerobes. None of the strictly anaerobic photosynthetic bacteria has been shown to contain plasmalogens, but it is still possible that a systematic search will reveal their presence. The effect of such a systematic search on the relationship between plasmalogen occurrence and a phylogenetic tree (Kamio *et al.*, 1969) will be interesting.

The distribution of glycolipids is also significant from a comparative point of view. Only traces of MGDG, if any at all, are found in *Rp. palustris*. Other Rhodospirillaceae contain no glycolipids, with the exception of SQDG, which is found in *Rp. sphaeroides* only. In contrast, *Chr. vinosum* and *Thiocapsa roseopersicina* contain glycolipids, but they are not of the plant or green bacterial (MGDG, DGDG, SQDG, glycolipid II) type. The probable presence of MGDG in the Chlorobiaceae and in *Chloroflexus* suggests that the latter are closely related to each other, as mentioned above, to the cyanobacteria, and to eukaryotic photosynthetic organisms, all of which contain MGDG, DGDG, and SQDG. The glycolipid II of *Chl. limicola* and *Chl. limicola f. thiosulfatophilum* is a cell-membrane component. It seems likely that the non-galactose-containing glycolipids of the other green bacteria and the Chromatiaceae are also likely to be associated with membranes not specifically adapted for photosynthesis. Whether or not the second glycolipid of *Chloroflexus aurantiacus*, which does contain galactose, is DGDG or a similar glycolipid is still not known. If it is DGDG, this would place this organism closer to the cyanobacteria, which also contain DGDG, than any other photosynthetic bacterium.

Extreme halophiles (grown on 25% NaCl) were found by Kates *et al.* (1966) to contain no, or very little, fatty acids in their lipids, and to contain long-chain ethers bound in alkyl ether linkage instead. J. C. Raymond (personal communication) did not find any evidence for alkyl-ether-containing lipids, and found reasonable amounts of esterified fatty acids (see below) in each of the phospholipids of *Ectothiorhodospira halophila* SL-1. The latter strain will not grow at 6% NaCl, the salt concentration used for growth of moderate halophiles shown to contain esterified fatty acids by Kates *et al.* (1966). The lipid analyses of *E. halophila* were performed on cells grown on 20% NaCl, which is approximately the concentration of salt Kates *et al.* (1966) utilized for growth of their extreme halophiles (25% NaCl). *Ectothiorhodospira halophila* grows optimally at 23–24% NaCl. Thus, either the dialkyl phosphatidyl glycerophosphate found in nonphotosynthetic extreme halophiles is not of physiological significance, but only of taxonomic significance, or anaerobic photosynthetic metabolism and its machinery are significantly different from the metabolism of the aerobic halophiles, so that the specialized lipid composition is not necessary. Interestingly, the enzymes of *H. halobium* require high salt for activity, while at least a few of the enzymes of *E. halophila* are inhibited by NaCl (Tabita and McFadden, 1976).

The data in Table 3 point out that there have been relatively few studies that relate the conditions of growth to phospholipid composition. Moreover, different laboratories do not always agree on the effects of different growth conditions. This is possibly true because specific phospholipids are not essential for function of the reaction center (Jolchine and Reiss-Husson, 1975), and species differentiation has involved lipid differentiation. That some phospholipids are essential is emphasized by the finding that phospholipase A causes disruption of photophosphorylation, which is restored by phospholipids (Klemme *et al.*, 1971). Probably, however, there are several different phospholipids that can serve the same structural function. Haverkate *et al.* (1965) reported more PC and less CL in *Rs. rubrum* strain 1 grown aerobically in the dark relative to that in cells grown anaerobically in the light (Table 4, Study 5). Brooks and Benson (1972), however, did not observe this difference for strain S1 (Table 4, Study 10). *Rhodopseudomonas capsulata* ATCC 23782 showed changes similar to *Rs. rubrum* strain 1 (Table 4, Study 22), while a number of strains of *Rp. sphaeroides* did not show such changes, and in fact two strains may have shown opposite changes (Table 4, Studies 28, 30, 31, and 40). It thus appears that if there is differentiation of the membrane on synthesis of bacteriochlorophyll (Bchl) in response to oxygen content of the medium or light, it does not universally involve a differentiation of the membrane with respect to lipids, although in some cases it may (e.g., *Rp. capsulata* ATCC 23782).

The presence or absence of yeast extract from the medium used for growth of *Rp. viridis* NHTC 133 caused no difference in the lipid composition, although the Bchl content of cells grown with yeast extract was considerably higher. This observation, and the lack of qualitative differences reported for dramatic changes in growth conditions [see Section 3.2.1(a)], suggests that the lack of agreement among different groups with respect to the lipid composition is not a reflection of different growth conditions, but is due to use of different strains and methods of analysis.

The picture with respect to lipid composition in subcellular fractions, a different approach to the same question of lipid differentiation, is equally un-

satisfactory. In only a few cases has this question been addressed. Gorschein (1964, 1968b) found all the cell fractions except the cell wall to be similar with respect to lipid composition (Table 4, Study 31). Similarly, *Thiocapsa roseopersicina* 9314 and *Chr. vinosum* chromatophores had lipid compositions similar to those of the whole cells (Table 4, Studies 38 and 42). However, *Ectothiorhodospira halophila* SL-1 showed a lipid distribution among the cell fractions that indicates differentiation (Table 4, Study 43).

4. Distribution of Fatty Acids by Group and Comparison with Other Photosynthetic Groups

4.1. Methods

Fatty acids are usually obtained free after saponification of whole cells, extracted lipids, or other cell fractions and are subsequently methylated. In some cases, they are transmethylated in acidic methanol to yield the methyl esters. The latter procedure is not always quantitative and sometimes leads to modification of existing fatty acids. Following methylation and sometimes purification, the resulting fatty acid methyl esters are eventually separated by gas–liquid chromatography (GLC). For identification of fatty acids, the retention time (RT) of the unknown acid is compared with that of acids of known structure. If only one GLC column and the resulting RTs are used, the apparent identifications can often be erroneous. Other procedures are necessary before accurate identifications can be made. These methods cannot be discussed in detail here, but will be outlined and evaluated briefly. Frequently, the use of both polar and nonpolar column packings will both reveal the presence of fatty acids not visible on either alone and confirm identifications of fatty acids based on behavior on either column. AgNO$_3$-impregnated TLC is useful for the separation of saturated and unsaturated fatty acids, and for separation of different fatty acid isomers. The double bond in unsaturated fatty acids can also be localized by oxidation followed by GLC of the resulting methylated fragments. The chain length of unsaturated fatty acids can be determined by hydrogenation followed by GLC of the resulting saturated fatty acid methyl ester.

In a survey of fatty acid compositions, it is seldom possible to perform all these steps. It should be emphasized that it is not sufficient to identify fatty acids solely on the basis of RT on one GLC column.

The methods used in the studies discussed below varied widely, and will be briefly described where appropriate.

4.2. Fatty Acid Composition of Whole Cells and Chromatophores

4.2.1. Rhodospirillaceae

(a) Rhodospirillum

Rhodospirillum rubrum: Cho and Salton (1964, 1966) compared the fatty acid composition of whole cells and cell envelopes of a strain of *Rs. rubrum* the source of which was not stated (Tables 3 and 6, Study 3). The "suitable" medium used for growth of this strain was also not described. They compared the fatty acids of the lipids of whole cells with those of a cell-envelope fraction, and with those of cell-envelope fractions of other gram-positive and gram-negative bacteria. Unfortunately, the nature and purity of the "envelope" fraction are not stated. The major fatty acids of the whole-cell lipids were found to be 16:0, 16:1, and 18:1 (Table 6). The cell envelopes showed a rather different spectrum and contained more saturated fatty acids (14:0 and 16:0) than did the whole cells, as well as 12% of an unknown (possibly a cyclopropane acid) and 4.3% of 17br, neither of which was detected in significant amounts in whole cells. Since the identifications of the fatty acids were solely by comparison of RTs on polar and nonpolar columns with those of standards, the identity of the latter two acids is speculative.

More conclusive analyses of the cellular fatty acids of *Rs. rubrum* were reported by Wood *et al.* (1965), James and Nichols (1966), Haverkate (1965), Haverkate *et al.* (1965), and van Deenen and Haverkate (1966) (Tables 3 and 6, Studies 4 and 5). Wood *et al.* (1965) and James and Nichols (1966) reported the presence of only saturated and monounsaturated fatty acids in this species (Table 6). The major fatty acids were 16:0, 16:1Δ9, and 18:1Δ11, in confirmation of the analyses of Cho and Salton (1966). More 16:1 and less 16:0 was found in the lipids of dark-grown cells than in those of light-grown cells. No evidence for 16:1Δ3, a major fatty acid of the PG of plants and algae (James and Nichols, 1966), was found, nor were polyunsaturated or cyclopropane fatty acids found. The fatty acids were identified by GLC of the methyl esters of the fatty acids obtained after saponification of the cellular lipids [extracted as described in Section 3.2.1(a)]. The double bond position in the unsaturated fatty acids of *Rs. rubrum* was determined by isolation of individual fatty acids after separation on the GLC column and separation of isomers on AgNO$_3$-impregnated silica gel thin-layer plates. The isomers were further identified by analysis of the products of oxidation of the separated monounsaturated acids.

The results of the fatty acid analysis of a presumably different strain (strain 1) by Haverkate, van Deenen, and co-workers were similar to those of Wood *et al.* (1965) for light-grown cells (Table 6, Study 5) except that more 18:1 was found in the cellular lipids by this group than was found by Wood *et al.* (1965), and dark aerobic cells had a lipid fatty acid composition similar to that of light anaerobic cells (Haverkate, 1965; Haverkate *et al.*, 1965; van Deenen and Haverkate, 1966).

In 1968, Hirayama (1968) reported a more detailed analysis of the lipids of yet another strain of *Rs. rubrum* (wild type) (Tables 3 and 6, Study 7). Again, the major fatty acids of the whole-cell lipids were 16:0, 16:1, and 18:1. He was able to identify by GLC and TLC minor amounts of 15:0, 17:0, 17:1, and 19:1. The individual lipids of this organism were purified [see Section 3.2.1(a)], and their component fatty acids were determined. PG, CL, and PE, which would be expected to be synthesized from the same phosphatidic acid precursor, showed similar fatty acid compositions. Ornithine amide contained more 14:0 and 19:1 and less 16:0 than did the former lipids.

Drews and his group have reported the only analysis of the whole-cell, thylakoid, and lipopolysaccharide (LPS) fatty acid compositions of both anaerobic light-grown and aerobic dark-grown cells of *Rs. rubrum* FR1 (Schröder and Drews, 1968; Schröder *et al.*, 1969; Oelze *et al.*, 1970) (Tables 3 and 6, Study 8). The fatty acid compositions of anaerobic light-grown and aerobic or semiaerobic dark-grown whole cells of the wild type (FR1), of semiaerobic dark-grown cells of several pigment mutants (Oelze *et al.*, 1970), and of thylakoids of the wild type (FR1), were similar to each other and to the fatty acid compositions of the total lipids reported above. In addition, a number of minor fatty acids were detected in these studies. The fatty acids of a Bchl precursor containing complex excreted under semiaerobic dark conditions by a Bchl-less mutant of this strain (F9) were quite different from those of whole cells, although the whole-cell fatty acids of F9, like those of other mutants, were nearly identical to those of FR1 (Oelze and Drews, 1970).

Interestingly, the fatty acids of the LPS of dark- and light-grown cells were similar to each other, but were very different from those of the whole cell (Tables 3 and 6, Study 8) (Schröder *et al.*, 1969). The LPS contained considerably more saturated fatty acids than did the whole cells, the saturated/unsaturated

Table 6. Fatty Acids of Photosynthetic Bacteria

Genus and species	Study	Strain	Growth condition (L or D)	Cell fraction	Lipid fraction	Saturated 14:0	16:0	18:0	14:1	16:1 Δ3 trans	16:1 Δ9	16:1 NR	18:1 Δ9	18:1 Δ11	18:1 NR	PUSFA (18:2)	Present (amount)	Absent	Ref. Nos.
Rhodospirillaceae																			
Rs. rubrum	3	NR	NR	Whole cells	Total lipids	2.0	25.5	1.6				16.4			54.0		Unk	17br	27
	4a	NR	NR	Cell envelope	Total lipids	10.9	48.9	Tr				9.2			14.1		17br(4.3); Unk(12)		27
		NR	L	Whole cells	Total lipids	1.6	16.3	Tr	—		37.6			37.0		—		Cyclopropane acids	3
	b	NR	D	Whole cells	Total lipids	1.6	8.3	Tr	1.4	—	51.0			35.3		—		Cyclopropane acids	3
	5a	1	L	Whole cells	Total lipids	3.0	19.0	+		—		26.1			51.9		15:0(1); 17:0; 17:1(2); 19:1(1)		5
	b	1	D	Whole cells	Total lipids	2.9	17.4	+		—		30.0			49.9		15:0(1); 17:0(1); 17:1(3); 19:1(5)		5
	7	Wild-type	L	Whole cells	Total lipids	1	18	1				37			35		15:0(2); 17:0(2); 17:1(5)		7
		Wild-type	L	Whole cells	Free fatty acids	1	20	1				35			33		15:0(2); 17:0(5); 17:1(2)		7
		Wild-type	L	Whole cells	CL	1	23	2				35			35		15:0(1); 17:1(3)		7
		Wild-type	L	Whole cells	PG	2	34	1				30			30		15:0(1); 17:0(1)		7
		Wild-type	L	Whole cells	PE	1	30	—				31			34		15:0(5); 17:1(2); 19:1(10)		7
		Wild-type	L	Whole cells	Ornithine amide	34	8	1				5			35				7
		Wild-type	L	Whole cells	lysoPE	2	63	2				4			27				7
	8a	F1(FR1)	L	Whole cells	ʃ	3.0	10.5	1.9	Tr			29.9			51.8		10:0; 12:0; 15:0; 19:0; 20:0; 11:1; 17:1(2.0)	14:OH	28
	b	F1(FR1)	D	Whole cells	ʃ	2.9	12.7	1.3	Tr			30.6			52.0		10:0; 12:0; 15:0; 19:0; 20:0; 17:1	11:1; 14:OH	28
	a	F1(FR1)	D	Thylakoids	ʃ	2.6	10.4	1.4	Tr			30.8			53.3		10:0; 12:0; 15:0; 19:0; 20:0; 11:1; 17:1		28
	a	F1(FR1)	L	LPS (water phase)	ʃ	5.3	16.4	10.4	6.2			6.2			19.4		10:0; 12:0(3.5); 15:0(2.1); 19:0(6.2); 20:0(9.6); 11:1(5.0); 17:1(8.7)		28
	b	F1(FR1)	D	LPS (water phase)	ʃ	9.9	22.8	12.3	Tr			6.1			15.4		10:0; 12:0(4.8); 15:0; 19:0(6.5); 20:0(18.8); 11:1; 17:1(3.3)		28
	9	F9	D	Whole cell	ʃ	1.8	17.2	4.0	Tr			27.2			40.3		15:0(1.3); 17:0(1.8); 19:0; 17:1(3.7); 19:1(1.1)	15:1	29
		F9	D	Pigment complex	ʃ	3.4	24.8	18.4	Tr			9.4			23.4		15:0(1.6); 17:0(1.0); 19:0; 17:1(1.5); 19:1(7.1)	15:1	29
Rs. molischianum	10a	S1	NR	Whole cell	Phospholipids	+	+					+			+				8
	a	S1	L	Whole cell	Ornithine amide	+	+					+			+				8
Rp. palustris	12	NR	NR	NR	Total lipids?	Tr	21.9	Tr				29.8			47.1	—	15:OH; 16:OH; 17:OH; Unk		9
	13a	2.1.7	L	Whole cell	Total lipids	12.7	12.7	6.2	—			3.1	72.3			—		14:1	3
	b	2.1.7	D	Whole cell	Total lipids	3.5	3.5	Tr	—			4.8	91.5			—	Unk		3
	14	2.1.7	NR	NR	Total lipids?	Tr	5.12	7.10				1.57h		78.00h		—	Unk(7,91)	12:0	9
		2.1.23	NR	NR	LPS	Tr	6.60	7.80				1.97h		75.00h		—	Unk(8,98)	12:0	9
	15	42	L	LPS ʃ (phenol phase)		Tr	15.9	24.5							8.2		12:0; 14:OH(37.7); 16:OH(7.3); Unk(4.7)		30
Rp. viridis	17	F	L	LPS + phospholipid (phenol phase) ʃ		35.2											14:OH(64.8); 16:OH		31
		F	L	LPS (water phase) ʃ													14:OH(100)		31
	18a	NHTC 133	L + YE	Membranes	Total lipids	3.0	15.0	Tr				18.0			55.0		12:0(1.9); 13:0; 15:0(1.0); 17:0; 19:0; 13:1; 15:1; 17:1; 20:1(1.7)	20:0; 22:0	11
	b	NHTC 133	L — YE	Membranes	Total lipids	1.3	16.0	Tr				8.0			60.0		12:0; 13:0; 15:0; 17:0; 19:0(2.7); 20:0(3.0); 22:0(1.0); 13:1; 15:1; 17:1; 20:1(3.0)		11

Organism	No.	Strain	Config.	Sample	Fraction	Composition values	Fatty acids	Ref.
Rp. gelatinosa	19a	2.2.13	L	Whole cell	Total lipids	2.9; 33.4; Tr; —; 51.0; 0.1; —; 28		3
	b	2.2.13	D	Whole cell	Total lipids	2.2; 13.8; Tr; —; 58.0; Tr; —; 40		3
Rp. capsulata	20a	2.3.11	L	Whole cell	Total lipids	—; 2.3; 4.2; —; 2.4; —; —; 50		3
	b	2.3.11	NR (L?)	Whole cell	PC	Tr; Tr; Tr; 2.ᵃ; 84.0; —; —; 45		3
		2.3.11	NR (L?)	Whole cell	PE	—; 3.0; 4.6; 10.3; 93.6		3
		2.3.11	NR (L?)	Whole cell	PG	—; 3.3; 5.4; 14.5; 82.1		3
	21	M.P.2	L	Whole cell	Total lipids	22.6; 2.4; 6.8; 76.1		3
		M.P.2	L	Whole cell	FFA	6; 10; 2; 66.7	15:0(5); 17:0(14); 17:1(33); 19:1(8) — 19:1	7
		M.P.2	L	Whole cell	PE	—; 25; 3	15:0(5); 17:0(13); 17:1(8) — 19:1	7
		M.P.2	L	Whole cell	PG	2; 6; 4	15:0(4); 17:0(4); 17:1(32) — 19:1	7
	23a	37b4	L	Whole cell	*f*	Tr; 14; 3; 2.6	15:0(3); 17:0(11); 17:1(22) — 19:1	7
	b	37b4	D	Whole cell	*f*	1.9; 1.3; 93.6	10:0; 11:0; 12:0; 17:0; 19:0; 20:0; 12:1; 15:1; 20:1 — 14:OH	28
	b	37b4	D	Whole cell	*f*	2.0; 0.9; 2.4; 94.1	10:0; 11:0; 12:0; 17:0; 19:0; 20:0; 12:1; 15:1; 20:1 — 14:OH	28
	a	37b4	L	Thylakoids	*f*	2.1; 1.4; 1.7; 94.6	10:0; 11:0; 12:0; 17:0; 19:0; 20:0; 12:1; 15:1; 20:1	28
	a	37b4	L	LPS (water phase)	*f*	Tr; 44.2; Tr; 1.1	10:0; 11:0; 12:0; 17:0(0.2); 19:0; 20:0; 12:1(45.2); 15:1(1.3); 20:1(2.5)	28
	b	37b4	D	LPS (water phase)	*f*	Tr; 59.7; Tr; Tr	10:0; 11:0; 12:0; 17:0; 19:0; 20:0; 12:1(33.5); 15:1; 20:1(2.2)	28
	24	37b4	L	LPS	*f*	3; 4	12:0(16); 10:OH(38); 14:OH(33) — 10:1	32
	25	37b4	NR	LPS	*f*	3; 1	12:0(17); 10:OH(41); 14:OH(35) — 10:1; 14:OH	33
		37b4	NR	Pigment complex	*f*	Tr; 24; 50	12:0(3); 10:OH(11); Unk(13) — 10:1; 14:OH	33
		Ala⁺ʳ	NR	LPS	*f*	Tr; 4	12:0(7); 10:OH(50); Unk(40) — 10:1; 14:OH	33
		Ala⁺ʳ	NR	Intracytoplasmic membrane	*f*	16; 52	12:0(5); 10:OH(8); Unk(19) — 10:1; 14:OH	33
		Ala⁻	NR	LPS	*f*	22; 19; 30	10:OH(30) — 12:0; 10:1; 14:OH	33
		Ala⁻	NR	Pigment complex	*f*	17; 8; 47	10:1(7); 10:OH(10); Unk(12) — 12:0; 14:OH	33
Rp. sphaeroides	26a	NCIB 8253	L	Particulate fraction	*f*	6.0; 7.5; 10.0; 2.0; 67; 5	10:0; 12:0(1.5)	34
	b	NCIB 8253	D	Particulate fraction	*f*	8.5; 4.0; 2.0; 2.0; 70; +	10:0(2.0); 12:0(2.5)	34
	27a	2.4.1Ga	D	Particulate fraction	*f*	3; 69	14:1Δ7(6)	35
	b	2.4.1Ga	D	Particulate fraction	*f*	3; 78	14:1Δ7(3)	35
	29a	2	L	Whole cells	Total lipids	1.7; 1.0; 90.8		3
	b	2	D	Whole cells	Total lipids	Tr; Tr; Tr; 99.0		3
	30a	4	L	Whole cells	Total lipids	1.2; 8.9; 11.4; 7.4		5
	b	4	D	Whole cells	Total lipids	0.3; 9.6; 13.0; 3.5		5
		NR	L	Whole cells	Total lipids	Tr; 5.0; 9.7; 2.4; 70.4		36
		NR	L	Thylakoids	Total lipids	Tr; 4.6; 9.7; 2.1; 72.5		36
	33	NR	L	Whole cell	SL	11.0; 23.2; 2.0; 76.6	17:0; Unk(6.4)	16
	32	NR	L	Whole cell	SL	6.25; 3.80; Tr; 76.8	17:0; Unk(6.8)	18, 37
	35	NR	NR	Whole cell	Phospholipids + SL	Tr; —; 62.3; 89.0	17:0(1.5)	18, 37
Rm. vannielii		NR	NR	Whole cell	Neutral lipids	1.75; 5.40; Tr; 1.08; 88.2; 2.70	10:0; 12:0 — 10:0; 12:0; branched and cyclopropane acids	18, 37
		NR	NR	Cell residue after lipid extraction	*f*	+; +; +; +	Branched and cyclopropane acids	18, 37
Chromatiaceae							17:0; 20:0; 22:0; 15:1; 20:1; 22:1; 15br; 17br; 19cy; 21cyʲ	
Chr. vinosum	38	D	L	Whole cell	Total lipids	28.7; +; 32.6; 38.7	12:OH(α); 13:OH(α); 14:OH(α); 14:OH(β); 12:OH(β); 13br:OH(α)	5
		D	L	Whole cell	PG	1.8; 30.6; 2.9; 16.1; 48.6		5
		D	L	Whole cell	PE	1.7; 10.3; 16.1; 15.4; 56.5		5

(continued)

Table 6. (continued)

Saturated[c] group spans 14:0, 16:0, 18:0; Unsaturated[c] group spans 14:1, 16:1 (Δ3 trans, Δ9, NR), 18:1 (Δ9, Δ11, NR), PUSFA (18:2), and Others (Present, Absent).

Genus and species	Study	Strain	Growth condition (L or D)[a,b]	Cell fraction[a]	Lipid fraction	14:0	16:0	18:0	14:1	16:1 Δ3 trans	16:1 Δ9	16:1 NR[a]	18:1 Δ9	18:1 Δ11	18:1 NR[a]	PUSFA (18:2)[i]	Present (amount)[d]	Absent	Ref. Nos.[e]
Chromatiaceae (continued)	41	D	L	Chromatophores envelope	Total lipids	2.8	28.5	3.1		—		35.3			30.3	—	12:0(3.7); 14:OH(11.0)		5
	41	D	L	Total	f	1.0	20.2	1.0	2.8			32.2			28.4				38
		D	L	Intracytoplasmic membrane I	f	1.4	29.8	1.4	—			33.9			33.5			12:0; 14:OH	38
		D	L	Intracytoplasmic membrane II	f	1.0	26.9	1.0	—			36.5			34.5			12:0; 14:OH	38
		D	L	Intracytoplasmic membrane III	f	1.3	26.3	1.3	—			36.9			34.4			12:0; 14:OH	38
T. roseopersicina	42	D	L	Cell wall IV	f	2.0	29.3	—		2.0		23.2			30.3		12:0(4.0); 14:OH(9.1)	18:0	38
		9314	L	Whole cell	f	+	20	+				45			28		10:0; 12:0; 13:0; 13:1		23
		9314	L	Chromatophores	f	+	20	+				45			28		10:1; 12:0; 13:0; 13:1		23
E. halophila	43	SL-1	L	Crude membrane	Amino lipid		13	19				63			1		20:1(2); 19cy(2)		24
		SL-1	L	Crude membrane	Unknown lipid 1		11	19				64			2		20:1(2); 19cy(1)		24
		SL-1	L	Crude membrane	PG		11	18				60			1		20:1(9); 19cy(5)		24
		SL-1	L	Crude membrane	PE		1	16				77			1		20:1(2); 19cy(2)		24
		SL-1	L	Crude membrane	Unknown lipid 2		11	17				64			2		20:1(2); 19cy(1)		24
Chlorobiaceae *Chl. limicol*	47	6330	L	Whole cell	Total lipids	13	17	Tr	2	—		—			57		17cy(3)		26
Chl. limicol f. thiosulfatophilum	48	6230	L	Whole cell	Total lipids	22.4	7.7	Tr	7.1	—		Tr	58				15:0(1.4); 15:1; Unk(2.3)		39
Chl. vibrioforme f. thiosulfatophilum	50	6230	L	Chromatophores	Total lipids	14.6	8.4	1.6	7.1			1.4	60				15:0(2.3); 15:1(1.9); Unk(2.5)		39
	50	6230	L	Whole cell	Total lipids	21	10	Tr	2			1			43		17cy(21)		26
	52	1930	L	Whole cell	Total lipids	12	23	1	Tr			2			52		17cy(3)		26
Chl. phaeobacteroides	53	2430	L	Whole cell	Total lipids	16	15	Tr	Tr			1			64		17cy(1)		26
Chl. phaeovibrioides	54	2631	L	Whole cell	Total lipids	10	29	Tr	Tr			2			51		17cy(2)		26
P. luteolum	55	2530	L	Whole cell	Total lipids	14	21	Tr	1			Tr			47		17cy(11)		26
Chloroflexaceae *Chloroflexus aurantiacus*	56	OH-64-fl	L	Whole cell	f	Tr	12	14	—		4[h]		52[l]			2	15:0; 17:0(3); 19:0(1); 20:0(1); 15:1; 17:1(3); 19:1(3); 20:1(4)		26
		OK-70-fl	L	Whole cell	f	Tr	8	10			3		46			3	15:0(5); 19:0(2); 20:0(1); 15:1; 17:1(7); 19:1(8); 20:1(5)		26
		J-10-fl	L	Whole cell	f	Tr	17	27			2		34			3	15:0; 17:0(3); 19:0(1); 20:0(1); 15:1; 17:1(1); 19:1(2); 20:1(5)		26

[a] (NR) Not reported.
[b] (L) Grown in light; (D) grown in dark; (YE) yeast extract.
[c] Percentage of fatty acids unless otherwise noted; (Tr) trace (<1%); (+) present; (−) absent.
[d] Amount is given in parentheses following acid. (Unk) Unknown.
[e] See Table 4, footnote c.
[f] Whole cells were analyzed without prior lipid extraction.

[g] Growth conditions (whether those of study 10b or 10c in Table 3) of cultures from which phospholipids obtained not specified.
[h] Only one isomer present, position unknown.
[i] Could be 18:2.
[j] All in greater than trace amounts; absolute numbers not reported.
[k] 16:1Δ7 and 16:1Δ9.
[l] Mostly 18:1Δ9; some 18:1Δ7 and 18:1Δ11.

fatty acid ratio (S/US) being 0.18–0.21 for whole cells and thykaloids and 1.18–3.04 for LPS. The major saturated acids of LPS were 14:0, 16:0, 18:0, 19:0, and 20:0, and the major unsaturated acids were 11:1, 16:1, 17:1, and 18:1 in light-grown cells and 16:1, 17:1, and 18:1 in dark-grown cells. These authors were unable to find β-hydroxymyristic acid, which is characteristic of the LPS of the enteric bacteria (see Section 3.1), in whole cells or in isolated LPS of *Rs. rubrum*. Although a rigorous chemical determination of the structures of the fatty acids found was not performed, the paper and gas liquid chromatographic methods employed should have been sufficient for identification of the indicated fatty acids.

Brooks and Benson (1972) confirmed that the major fatty acids of the phospholipids of *Rs. rubrum* S1 were 16:0, 16:1, and 18:1, and that significant amounts of 14:0 were also present. The ornithine amide found in anaerobic grown cells had 14:0, 16:0, 18:1, and an unidentified fatty acid as major fatty acids. This lipid also contained 3-hydroxy-hexadecanoic acid (16:OH) and minor amounts of other hydroxy acids (Tables 3 and 6, Study 10). Again, no β-hydroxymyristic acid was found. The fatty acid methyl esters from the phospholipids were prepared by transesterification in methanol–H_2SO_4.

The fatty acid composition of cellular and lipid fractions of many strains of *Rs. rubrum*, grown under a variety of conditions has been examined. The results of the analyses are remarkably uniform from laboratory to laboratory and strain to strain. The major fatty acids are 16:0, 16:1, and 18:1. When the isomers of the fatty acids were determined, it was found that 16:1 was 16:1Δ9 and 18:1 was 18:1Δ11. The earliest comparison of the cell envelope to the whole cell indicated that there were relatively more saturated fatty acids and less unsaturated fatty acids in the envelope than in the whole cell (Table 6, Study 3). This observation was confirmed by a later study (Table 6, Study 8) of strain FR1, which showed a similar difference between the LPS fatty acids and all other cell fractions. Similarly, two groups have shown that all lipid fractions except the ornithine amide have similar fatty acid compositions in a wild-type strain and in strain S1 (Table 6, Studies 7 and 10). The content of 16:1 in ornithine amide is lower than in other lipids in both strains. Wood *et al.* (1965) and Haverkate *et al.* (1965) do not agree on the changes in fatty acid composition in dark-, aerobic- vs. light-, anaerobic-grown cells (Table 6, Studies 4 and 5). Wood and co-workers found more 16:1 and less 16:0 in dark-grown cells, while Haverkate and co-workers did not. It is not clear whether the changes reported by Wood and co-workers are related to the biochemical

changes or to the change in growth rate, dispersion, temperature, and other factors. Thus, it seems inappropriate to comment further.

16:1Δ3*trans*, cyclopropane acids, β-hydroxymyristic acid, and polyunsaturated fatty acids (PUSFA) are all found to be absent from *Rs. rubrum* when looked for. There may be small amounts of branched-chain acids and other hydroxylated fatty acids, but because they have not always been sought, their occurrence is unknown. Small but significant amounts of odd-chain-length saturated and unsaturated acids are frequently found in cell fractions. In the LPS and ornithine amide, the amounts of odd-chain-length acids are larger.

Rhodospirillum molischianum: Constantopoulos and Bloch (1967) analyzed the fatty acids of *Rs. molischianum* rather carefully. They found no fatty acids with more than one double bond and only one isomer of each unsaturated fatty acid. The position of the double bond was not determined. The major fatty acids were again 16:0, 16:1, and 18:1 (Tables 3 and 6, Study 12). The growth conditions are not stated, and it is only implied that the fatty acids of cellular lipids were studied.

(b) *Rhodopseudomonas*

Rhodopseudomonas palustris: The fatty acids of *Rp. palustris* 2.1.7 and 2.1.23 were analyzed by Wood *et al.* (1965) and by Constantopoulos and Bloch (1967) (Tables 3 and 6, Studies 13 and 14). Both groups found that 18:1 was the major fatty acid in each of what appear to be three different strains. Wood and co-workers found that 18:0 occurred in significant amounts only in anaerobically grown cells of strain 2.1.7. This acid also occurred in significant amounts in both strains analyzed by Constantopoulos and Bloch, who unfortunately did not report growth conditions. The methods of isolation and identification used by these groups were discussed in Section 4.2.1(a).

Weckesser *et al.* (1973) examined the fatty acids of the LPS of 12 strains of *Rp. palustris*. The LPS-containing phenol extract of *Rp. palustris* could not be completely freed of contaminating phospholipid, so the amount of the fatty acids attributable to LPS and to phospholipid is unclear. Although the strains examined fall into several serological groups, their LPS-fatty acid compositions were nearly identical, with the exception of the contents of 18:0 and 18:1, which varied from 0 to 9% of the dry weight of the LPS. The composition of a representative strain is reported in Table 6 (Study 15). As in *Rs. rubrum*, the content of saturated fatty acids is much higher in the LPS than in the extractable lipids. A major portion of the fatty acids of this fraction is 14:OH. The fatty acids were liberated by HCl treatment of the LPS and methylated with HCl–methanol.

Rhodopseudomonas viridis: Weckesser *et al.* (1974) also examined the LPS of a strain closely related to *Rp. palustris, Rp. viridis* F (Tables 3 and 6, Study 17). In this case, the LPS is found mainly in the water phase of phenol/water extracts and is virtually phosphate-free (phospholipid-free). The only fatty acid found in the LPS is 14:OH (Table 6). The phenol phase (phospholipid- and LPS-containing) contained in addition 16:0, and traces of 16:OH (Table 6). In this case also, the LPS fatty acids of a number of strains belonging to one chemotype were analyzed. Since their fatty acid compositions were nearly identical, only one is presented in Table 6. The methods were the same as those used for *Rp. palustris* LPS (see above).

Pucheu *et al.* (1974) analyzed the fatty acids of the extractable lipids of membranes of *Rp. viridis* NHTC 133 obtained by differential centrifugation (Tables 3 and 6, Study 18). The fatty acid methyl esters were prepared in H_2SO_4–methanol and separated by GLC. The means of identification were not stated. They found saturated and monounsaturated fatty acids from 12–22 carbons, with 18:1 being the major fatty acid as in *Rp. palustris*. There was relatively more 16:1 in the membranes of cells grown with yeast extract than in cells grown without yeast extract, in which there was far less Bchl.

Rhodopseudomonas gelatinosa: Wood *et al.* (1965) found that 16:0, 16:1, and 18:1 were the major fatty acids of *Rp. gelatinosa* 2.2.13 (Tables 3 and 6, Study 19). The majority of the 18:1 was *cis*-vaccenic acid (18:1Δ11). There was relatively more 18:1 and relatively less 16:0 in cells grown aerobically in the dark than in cells grown in the light anaerobically. The methods utilized were described in Section 4.2.1(a).

Rhodopseudomonas capsulata: Wood *et al.* (1965) analyzed the fatty acid composition of the lipids of *Rp. capsulata* 2.3.11 grown in the dark and in the light (Tables 3 and 6, Study 20). They determined that the major fatty acid was *cis*-vaccenic acid under both conditions, and a shift from light to dark caused an increase in 16:1 and decreases in 16:0 and 18:0. PC and PE contained almost exclusively 18:1 and 16:1, while PG contained 23% of 16:0 and 67% of 18:1. The significance of the latter distribution is unclear, since the growth conditions (light or dark) for the cells used were not stated. The methods are discussed in Section 4.2.1(a) under *Rs. rubrum*.

Hirayama (1968) also analyzed the fatty acids of the lipid fractions of *Rp. capsulata* M.P.2 (Tables 3 and 6, Study 21). The quantitative and qualitative fatty acid composition he obtained differed significant-

ly from that obtained by Wood *et al.* (1965), although the methods he employed (GLC and TLC) should also have yielded accurate results. He found significant amounts of 15:0, 17:0, 17:1, and 19:1 in the lipids of *R. capsulata* in addition to 16:0 and 18:1. These analyses should clearly be repeated. Despite this confusion, Hirayama and Wood and co-workers agree that PG contains more 16:0 and less 18:1 than does PE. The results contrast strikingly with the similar fatty acid compositions of all phospholipids of *Rs. rubrum* [see Section 4.2.1(a)].

The analysis of *Rp. capsulata* 37b4 grown in the light and the dark by Schröder *et al.* (1969) (Tables 3 and 6, Study 23) agrees well with the analyses of Wood and co-workers except that the small changes with culture conditions were not observed by this group. Schröder and co-workers found that in this species as in *Rs. rubrum*, the fatty acids of the thylakoids were similar to those of whole cells, but fatty acids of the LPS were considerably different. Again, the LPS contains more saturated fatty acids (16:0) and less unsaturated fatty acids (18:1 and 12:1) than do whole cells. In a later paper, Weckesser *et al.* (1972) found a different LPS fatty acid composition for strain 37b4 (Tables 3 and 6, Study 24). There was little 16:0 and a lot of 14:OH and 10:OH reported. They mentioned that this analysis was different from that of Schröder and co-workers, but made no comment on the reason for the difference. Since these analyses came from the same laboratory, the difference is puzzling.

Finally, Drews (1974) again analyzed the fatty acids of the LPS of *Rp. capsulata* strain 37b4. The values obtained were compared with the fatty acids found in an excreted pigment complex of this strain and with fatty acids found in a carotenoid- and Bchl-less mutant of this strain (Ala⁻), as well as with those of a revertant of Ala⁻ that had regained the ability to form Bchl (Ala⁺r) (Tables 3 and 6, Study 25). He found that the LPS fatty acids of the wild type (37b4) were similar to those reported by Weckesser *et al.* (1972). The Bchl–precursor pigment complex of the wild type, in contrast, contained much less hydroxy acids or 12:0 and large amounts of 16:0 and 18:1. β-Hydroxymyristate was also absent from both the membranes and LPS of Ala⁺r and from the pigment complex and LPS of Ala⁻, but large amounts of 10:OH were present in all the fractions. Since the fatty acid compositions of LPS and pigment complex differ, the pigment complex is probably not composed primarily of wall fragments.

Rhodopseudomonas sphaeroides: In 1962, Hands and Bartley (1962) found that a particulate fraction from a semianaerobic light-grown *Rp. sphaeroides*

NCIB 8253 contained 18:1Δ11, 14:0, 16:0, and 18:0 (Tables 3 and 6, Study 26). The fatty acids of aerobic dark cells contained less 18:0. The same year, Scheuerbrandt and Bloch (1962) examined the isomers of the monounsaturated fatty acids of the green mutant 2.4.1 Ga (Tables 3 and 6, Study 27). They found that dark- and light-grown cells had nearly identical unsaturated fatty acid compositions, and that 14:1 was 14:1Δ7, 16:1 was 16:1Δ9, and 18:1 was 18:1Δ11. The analyses of Scheuerbrandt and Bloch (1962) employed rigorous chemical methods for fatty acid identification.

Wood *et al.* (1965) compared the fatty acids of light- and dark-grown cells of *Rp. sphaeroides* 2 and confirmed that the major fatty acid in both cases was *cis*-vaccenic acid, and that there was a little less 18:0 in the dark-grown cells (Tables 3 and 6, Study 29). They did not, however, find significant amounts of any fatty acid other than 18:1 in dark-grown cells, and found only small amounts of 16:0, 18:0, and 16:1 in light-grown cells. As in the other genera they examined, no 16:1Δ3 or polyenoic acids were detected.

Haverkate *et al.* (1965) reported fatty acid analyses of the total extractable lipids of *Rp. sphaeroides* 4 grown in the dark and in the light. The composition of fatty acids was similar under both sets of conditions (Tables 3 and 6, Study 30). Subsequent analyses by Schmitz (1967*a*) of the fatty acids of the lipids of whole cells and thylakoids of a strain of *Rp. sphaeroides* were also similar to the results of the latter workers. 16:0, 18:0, 16:1, and mainly 18:1 were also found in the analysis of the fatty acids of the sulfolipid of *Rp. sphaeroides* by Radunz (1969) (Tables 3 and 6, Study 32). Considerably more 18:0 was present in this fraction than in the others.

The results of analyses of the fatty acids of this species appear to be among the most uniform from group to group and from strain to strain. 18:1 is always the major fatty acid, with smaller amounts of 16:0, 18:0, 16:1. Little change occurs in the fatty acids with change from light to dark. Moreover, the care with which the fatty acids were identified seems to be greater in the case of this species than in the other cases discussed above.

(c) *Rhodomicrobium vannielii.* *Rhodomicrobium vannielii* fatty acids were examined in detail by Park and Berger (1967*a,b*) (Tables 3 and 6, Study 35). They separated the neutral lipids from the phospholipids and sulfolipid before examining the fatty acids of each fraction. They also examined the fatty acids remaining in the cell residue after lipid extraction. Their analysis was careful, and is among the most rigorous chemical analyses of photosynthetic bacterial fatty acids in the literature. Since there are a number

of unusual fatty acids reported, it would be desirable to prove their structure conclusively by methods other than GLC. The findings that the major fatty acid of both lipid fractions is *cis*-vaccenic acid and that branched-chain and cyclopropane fatty acids are absent are similar to findings for the other Rhodospirillaceae. A variety of hitherto undescribed branched-chain and a number of hydroxy fatty acids were found in the cell residue, which suggests that the majority of the reports discussed above are incomplete. It seems inappropriate to comment much further on the distribution of hydroxy and branched-chain acids among photosynthetic bacteria except to mention that hydroxy acids of various structures have been found in the LPS of most of the Rhodospirillaceae. Since there was yeast extract in the medium used for growth, the significance of the presence of 18:2 in *Rm. vannielii* is not clear.

4.2.2. Chromatiaceae

(a) *Chromatium vinosum,* Strain D. There are two reported analyses of the fatty acids of the lipids of *Chr. vinosum* cell fractions (Tables 3 and 6, Studies 38 and 41). Haverkate *et al.* (1965) and Hurlbert *et al.* (1974) agree that 16:0, 16:1, and 18:1 are the most abundant nonpolar fatty acids of this organism. Since Haverkate and co-workers analyzed the extractable lipid-bound fatty acids rather than total envelope fatty acids as reported by Hurlbert and co-workers, they would not be expected to have detected β-hydroxymyristic acid, which was reported by Hurlbert and co-workers to be present in total envelopes and in the cell-wall fraction. The fatty acids of all cell fractions analyzed by either group appear to be otherwise similar. Haverkate and co-workers found that PE and PG differed significantly from the total lipids in content of 16:0, 18:0, 16:1, and 18:1 and from each other in content of 16:0, 18:0, and perhaps 18:1.

(b) *Thiocapsa roseoperisicina.* Takacs and Holt (1971*b*) found that the major fatty acids of both whole cells and chromatophore membranes of *T. roseoperisicina* 9314 were 16:0, 16:1, and 18:1 (Tables 3 and 6, Study 42). Small amounts of 10:0, 12:0, 13:0, 13:1, 14:0, and 18:0 were also present. The authors state that palmitoleate and oleate were the isomers present, but the methods employed would not distinguish 16:1Δ7 and 16:1Δ9 or 18:1Δ9 and 18:1Δ11. The authors were careful to utilize both polar and nonpolar GLC columns, and to hydrogenate the samples in order to prove the chain length of the fatty acids.

(c) *Ectothiorhodospira halophila.* J. C. Raymond (personal communication) analyzed the fatty acids of the lipids of crude membranes of *E. halophila* SL-1 (Tables 3 and 6, Study 43). He found no evidence for ether-linked phytols among the lipids of this species, in contrast to the results of Kates with *Halobacterium* as discussed in Section 3.3. The major fatty acid of all the lipids was 18:1. 16:0 and 18:0 were also present in significant amounts. Moreover, PE had a lower concentration of 16:0 than did the other fractions. The fatty acids were identified by their behavior on GLC by reference to fatty acids thoroughly characterized in our laboratory.

4.2.3. Chlorobiaceae

Schmitz (1967*b*) reported the fatty acid composition of the ether-soluble lipids of whole cells and chlorobium vesicles of *Chl. limicola f. thiosulfatophilum* strain Tassajara (strain 6230; Tables 3 and 6, Study 48). The fatty acids of the whole cell were similar to those of the chlorobium vesicles, except that there was relatively more 14:0 in the whole cells than in the vesicle fraction. 16:1Δ9 was the major fatty acid of both fractions. The position of the double bond in 16:1 was determined by ozonolysis. Although the exact methods used by Schmitz for lipid and fatty acid extraction were not stated, his results agree well with ours for the same strain (Kenyon and Gray, 1974).

Kenyon and Gray (1974) examined the fatty acids of five species of *Chlorobium* and of *Pelodictyon luteolum*. The identification of the fatty acids was based on RTs on GLC as well as on behavior after chemical modification of the fatty acids, and TLC. We found that species belonging to this family uniformly contained large amounts of 14:0, 16:0, 16:1, and, in two cases, 17cy. None of the other species of photosynthetic bacteria contains such a large relative amount of 14:0 and such a low relative amount of 18:1 in the extractable lipids. The isomer of 16:1 is most likely 16:1Δ9, as determined by silver nitrate TLC.

4.2.4. Chloroflexaceae

Whole cells of three strains of *Chloroflexus aurantiacus* were saponified. The major fatty acids of all three strains were 16:0, 18:0, and 18:1. The isomers of each of the unsaturated fatty acids that were present were examined, and multiple isomers of all the unsaturated fatty acids except 17:1 were detected. The finding of 18:2, a fatty acid presumably derived from yeast extract in the growth medium, suggests that some of the minor fatty acids found might also be derived from yeast extract. Until this species can be grown on completely defined medium or appropriate

incorporation studies are performed, this point will be unclear. It should be pointed out that most of the media used for growth of the photosynthetic bacteria contain at least traces of yeast extract, which could contain assimilable fatty acids. When used only in trace amounts, this is most likely not a problem.

4.3. Comparative Aspects of the Fatty Acid Composition

The photosynthetic bacteria as a group are unique among photosynthetic organisms in lacking polyunsaturated 18- or 20-carbon fatty acids and the *trans*-3-hexadecenoic acid (16:1Δ3*trans*) characteristic of the phosphatidylglycerol of eukaryotic photosynthetic organisms. They share the absence of polyunsaturated acids with certain groups of cyanobacteria (mostly unicellulars) and with nearly all other bacteria, and share the lack of *trans*-3-hexadecenoate with all the cyanobacteria examined (Kenyon, 1972; Kenyon *et al.*, 1972; Nichols, 1973). The only cases in which 18:2 has been reported to occur in photosynthetic bacteria are cases in which cells were grown with yeast extract, which contains 18:2, in the medium. The functional significance of this basic division among photosynthetic organisms is unclear, since those cyanobacteria that lack polyunsaturated fatty acids (PUSFA) are capable of the Hill reaction, as are the more highly evolved photosynthetic organisms. The evolutionary significance appears to be that the most primitive organisms, the photosynthetic bacteria and more primitive cyanobacteria, are those that lack PUSFA.

Cyclopropane-group-containing fatty acids are apparently not common among the photosynthetic bacteria, and in fact have been found to be absent from *Rs. rubrum*, *Rm. vannielii*, and *Chloroflexus aurantiacus*. 17cy is a major component of the fatty acids of at least two species of the Chlorobiaceae, and 19cy is present in *Ectothiorhodospira halophila*. Goldfine (1972) pointed out that cyclopropane acids are usually characteristic of gram-negative bacteria belonging to the order Eubacteriales.

The occurrence of odd-chain-length fatty acids in the extractable lipids of photosynthetic bacteria is uncertain. Usually, when reported, they are found in small amounts. Exceptions to this are the analyses of a strain of *Rp. capsulata* and a strain of *Rs. rubrum* by Hirayama (1968) and the reported analysis of *Chloroflexus aurantiacus* (Kenyon and Gray, 1974). Although it is not clear that other authors always looked for these acids, it is likely that if they occurred in large amounts, they would have been reported. Few unknowns appear in published tables in which trace

Table 7. Relative Amounts of the Major Saturated and Unsaturated Fatty Acids Found in Whole Cells of Different Groups of Photosynthetic Bacteria Grown in the Light under Anaerobic Conditions

Organism	Fatty acid[a]				
	14:0	16:0	18:0	16:1	18:1
Rhodospirillaceae					
Rs. rubrum	+	++	+	+++	+++
Rs. molischianum	+	+++		+++	+++
Rp. palustris	+	++	++	+	+++
Rp. gelatinosa	+	+++	+	+++	++
Rp. capsulata	+	+	+	+	+++
Rp. sphaeroides	+	++	++	+	+++
Rm. vannielii	+	++	+	+	+++
Chromatiaceae					
Chr. vinosum	+	+++	+	+++	+++
T. roseopersicina	+	+++	+	+++	+++
E. halophila		++	++	+	+++
Chlorobiaceae					
Chl. limicola	++	++	+	+++	
Chl. limicola f. thiosulfatophilum	+++	++	+	+++	+
Chl. vibrioforme f. thiosulfatophilum	++	+++	+	+++	+
Chl. phaeobacteroides	++	++	+	+++	+
Chl. phaeovibroides	++	+++	+	+++	+
P. luteolum	++	+++	+	+++	+
Chloroflexaceae					
Chloroflexus aurantiacus					
J-10-fl.	+	++	++	+	+++
OK-70-fl.	+	++	++	+	+++
OH-64-fl.	+	++	+++	+	+++

[a] (+) 5% or less; (++) 6–20%; (+++) more than 20%.

amounts of odd-chain-length fatty acids are not reported. It appears that the presence of small but substantial quantities of odd-chain fatty acids is a property that distinguishes *Chloroflexus aurantiacus* and perhaps certain strains of *Rp. capsulata* and *Rs. rubrum*. Drew's group found relatively large amounts of odd-chain-length fatty acids in the LPS of their strain of *Rs. rubrum*, but only trace amounts in the other strains they examined.

The data on the fatty acids of the whole cells or extractable lipids of the strains that have been examined are summarized in Table 7. Some of the choices of category are, of necessity, based on subjective evaluation of the available data. No one fatty acid is a major fatty acid of all groups, although 16:0 occurs in large amounts in most groups. 18:1 is a major acid in all groups with the exception of the Chlorobiaceae. The Chlorobiaceae are also dis-

tinguished from the other groups by the presence of major amounts of 14:0, a minor acid in Rhodospirillaceae, Chromatiaceae, and *Chloroflexus*. This summary table again points out that *Chloroflexus* is more similar in certain propertes (low 14:0 and 16:1, high 18:0 and 18:1) to the Rhodospirillaceae than to the Chlorobiaceae, a point discussed by Pierson and Castenholz (1974), Madigan *et al.* (1974), and Kenyon and Gray (1974).

The occurrence of major amounts of 14:0 is not widespread. Among the eukaryotic photosynthetic organisms, it occurs in fairly large amounts only in certain red and brown algae (Klenk *et al.*, 1963) or in cell fractions of other organisms. More significantly, there are a few typological subgroups of cyanobacteria that are distinguished by large amounts of 14:0, strain cluster 1 of *Synechococcus* (Kenyon, 1972) and a few marine strains (Kenyon and

Waterbury, 1978). The unicellular strains of *Synechococcus*, cluster 1, like the Chlorobiaceae, contain small amounts of PUSFA; the filamentous marine strains contain larger amounts.

My opinion is that the fatty acid composition of organisms is often very useful in determining relatedness of single species or groups of species, and sometimes is even useful in classifying an unknown organism when the fatty acid composition of a certain group is sufficiently distinctive. For example, the marine endospore- and nanocyte-forming cyanobacteria contain nearly exclusively 16-carbon acids and polyunsaturated 16-carbon acids (Kenyon and Waterbury, 1978), the Chlorobiaceae contain a large percentage of 14:0 and a small percentage of 18:1, and *Chloroflexus aurantiacus* contains a high percentage of 18:0. There are very few organisms that contain little 18-carbon or longer fatty acids as do the Chlorobiaceae and certain cyanobacteria.

The significance of the fatty acid composition of different lipids and the occurrence of different fatty acid isomers when it has been examined will be discussed in Sections 5 and 6.

Finally, the functional significance of the fatty acid compositions, if any, is obscure. Oelze and Drews (1972) discuss some aspects of this question. The fatty acid compositions of cells of widely different structures, such as *Rs. rubrum* and *Chr. vinosum*, appear to be grossly similar (Table 7). This again emphasizes the point made earlier that fatty acid compositions are often useful for determining taxonomic relatedness of species, genera, and groups of genera. However, the differing fatty acid compositions of certain organisms with similar metabolism and structure make it rather clear that there are a variety of fatty acids that in combination lead to a physical state of the membrane that facilitates a given reaction sequence.

Certain fatty acids known to be characteristic of higher plants, e.g., α-linolenic acid (18:3α) and *trans*-3-hexadecenoic acid (16:1Δ3*trans*), are known to be localized in the chloroplast fraction of cells. Although few strains have been appropriately fractionated, it appears that the thylakoids, or the intracellular membranes, or the chromatophores contain approximately the same relative amounts of the different fatty acids as do the whole cells (Table 6, Studies 8, 23, 33, 38, 41, 42, and 48). In contrast, the fatty acids of the LPS are usually radically different (Table 6, Studies 3, 8, 15, 17, and 23–25).

As was mentioned for *Rs. rubrum* in Section 4.2.1(a), it was found by Wood *et al.* (1965) that the cellular content of saturated fatty acids usually decreases in dark-, aerobic-grown cells of facultative anaerobes (*Rs. rubrum, Rp. palustris,* and *Rp.*

gelatinosa) when compared with light-, anaerobic-grown cells (Table 6, Studies 4, 13, and 19). Other laboratories, however, have not always confirmed this finding. In *Rs. rubrum* 1, *Rp. capsulata*, and *Rp. sphaeroides*, this change in growth condition causes little or no change in fatty acid composition (Table 6, Studies 5, 20, 23, 26, 29, and 30). When *Rp. viridis* NHTC 133 is grown with yeast extract, a condition in which there is more Bchl than when yeast extract is absent, the membrane lipid content of 16:1 increases (Table 6, Study 18). This change is opposite to the light/dark changes mentioned, and suggests that the fatty acid changes are not controlled by the same mechanism as is Bchl synthesis.

The other sorts of experiments found useful in other bacteria for relating fatty acid composition to function, such as temperature effects (Fulco, 1974), effects of media (Fulco, 1974; Weerkamp and Heinen, 1972) except as noted above, and growth-stage effects (Oliver and Colwell, 1973), have not yet been performed.

5. Biosynthesis of Complex Lipids by Photosynthetic Bacteria

The pathways utilized by most bacteria for the biosynthesis of phospho- and glycolipids are known and have been thoroughly reviewed (Goldfine, 1972; Lennarz, 1966). Very little has been reported, however, concerning the pathways utilized by the photosynthetic bacteria for lipid biosynthesis. A number of groups have studied photosynthetic membrane formation and lipid synthesis *in vivo* in species of Rhodospirillaceae, but there is only one report of *in vitro* studies on the biosynthesis of phospholipids (Lueking and Goldfine, 1975b).

The Rhodospirillaceae, because of their adaptability to growth anaerobically in the light or aerobically in the dark, are excellent subjects for study of photosynthetic membrane formation, and changes in pigment and phospholipid composition during adaptation from high to low aeration or dark to light have frequently been used to trace formation of the pigment-containing structures. From these experiments, some information on the pathways utilized for synthesis of individual phospholipids can also be obtained. Lascelles and Szilagyi (1965) and Gorschein (1968b) used phospholipids and pigments as tracers for chromatophore formation in *Rp. sphaeroides*. Both laboratories found that the relative amounts of phospholipids were constant under high or low aeration (Table 4, Studies 28 and 31), indicating that the chromatophore membranes are not differentiated with respect to the cell membrane. Lascelles and Szilagyi (1965) also observed that PG was labeled with ^{32}P more rapidly than were the other phospholipids of this species, and PC was the least rapidly labeled, with PE and PA labeled at intermediate rates. Since no attempt was made to distinguish *de novo* synthesis from degradation or turnover, these experiments cannot be conclusively interpreted in terms of a pathway. It can be concluded, however, that PG is the most metabolically active phospholipid in *Rp. sphaeroides*, and that PE could be a precursor of PC.

In contrast, Steiner *et al.* (1970*b*) showed that in *Rp. capsulata* ATCC 23782, during adaptation from dark aerobic to a number of anaerobic conditions in the light, the amount of PE and PG/protein increases much more rapidly than that of PC/protein, accounting for the change in phospholipid ratios observed during this adaptation [see Section 3.2.1(b)].

Rhodopseudomonas viridis NHTC 133 also changes its lipid composition in response to a change in growth conditions, in this case the presence or absence of yeast extract in the medium (Pucheu *et al.*, 1974). In the absence of yeast extract (lower Bchl/protein), the ratio of PC/PE increases (PE decreases) and PC/ornithine lipid also increases (ornithine lipid decreases). PC/PG and PC/CL do not change.

The available results can be interpreted in terms of the pathway for biosynthesis of phospholipids expected from studies of other bacteria:

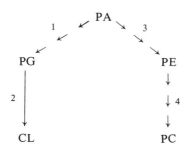

There appear to be four sites of possible control by light intensity, aeration, or medium composition (1–4). The early results of Lascelles and Szilagyi (1965) with *Rp. sphaeroides* separated the pathway to PG from that to PE and PC. Reactions 1 and 3, but not 4, appear to be stimulated during photosynthetic membrane synthesis in *Rp. capsulata* ATCC 23782. PE synthesis (reaction 3) is also stimulated in *Rp. viridis* NHTC 133 when Bchl synthesis is stimulated by the presence of yeast extract. Ornithine lipid synthesis is also stimulated under these conditions. Reactions 1, 2, and 4 proceed at the same relative rates in this species with or without yeast extract. Clearly, further study is necessary, and the availability of *in vitro* systems for

synthesis of phospholipids (see below) will greatly facilitate such studies.

More specific data on the pathways of phospholipid biosynthesis were reported by Gorschein *et al.* (1968*a*), who showed that radioactivity from L-([^{14}C]methyl)methionine was incorporated into the lipid fractions of *Rp. sphaeroides* corresponding to *N*-methylphosphatidylethanolamine, *N,N*-dimethylphosphatidylethanolamine, and PC. They showed that the methyl-PE's were labeled sooner than PC, and that PC eventually had a greater specific activity than the methyl-PE's. These data suggest strongly that the pathway for synthesis of PC is PE → *N*-methyl-PE → *N,N*-dimethyl-PE → PC, as has been found for other bacteria (Goldfine, 1972). *In vitro*, the methylation activity was associated with particulate material from aerobic or anaerobically grown cells, but not specifically with chromatophores. No evidence for incorporation of choline directly could be demonstrated.

Gorschein *et al.* (1968*a*) also showed that [^{14}C]glycine is incorporated *in vivo* into PE, PC, and fatty acids, which suggests that PS may be a precursor of PE, as is found in *Escherichia coli* and other bacteria (Goldfine, 1972). Furthermore, the incorporation of radioactivity *in vivo* from [^{14}C]glycine into PC, but not into PE, was inhibited by L-ethionine, lending further support to the pathway PS → PE → PC.

The only data on synthesis of ornithine amide indicate that [^{14}C]ornithine added to growing cells is incorporated into this fraction (Gorschein, 1968*a*). The phospholipids and ornithine amide of *Rp. sphaeroides* NCIB 8253 are metabolically stable in the dark and in the light (Gorschein *et al.*, 1968*b*; Gorschein, 1968*a*).

In *Rs. rubrum* S1, ornithine amide is also stable, and its synthesis is not related to the Bchl content of the cells (Depinto, 1967). Arginine represses formation of ornithine lipid in this species, and proline and chloramphenicol stimulate. The mechanisms of these effects are not known.

The only *in vitro* study of early steps in biosynthesis of phospholipids in photosynthetic bacteria shows that a particulate fraction of anaerobic light-grown *Rp. sphaeroides* 2.4.1 Ga (green mutant) is capable of transferring fatty acids from acyl-ACP (fatty acyl–*E. coli* acyl carrier protein) to glycerol-3-phosphate to form lysoPA and PA (Lueking and Goldfine, 1975*b*). This is the first specific step in phospholipid synthesis in *E. coli*, and presumably also in *Rp. sphaeroides*. Both palmityl-ACP and oleyl-ACP are good substrates for the initial acylation of glycerol-3-phosphate in *Rp. sphaeroides*. The second acylation is more specific. CoA derivatives of long-chain fatty acids failed to serve as substrates for the sn-glycerol-3-phosphate acyltransferase (EC 2.3.1.15) to any significant extent, although acyl-CoA derivatives are substrates for the comparable reaction in aerobically grown *Esch. coli*.

This report is of particular importance. The only other report of an *in vitro* study of phospholipid synthesis is the synthesis of PC from ([^{14}C]methyl)-methionine by a particulate fraction of *Rp. sphaeroides* mentioned above. More significantly, although ACP has not yet been shown to occur in photosynthetic bacteria, its involvement in lipid biosynthesis is strongly suggested by these results.

6. Biosynthesis of Fatty Acids by Photosynthetic Bacteria

The biosynthesis of fatty acids in bacteria proceeds as in higher organisms by the successive addition of two-carbon units to a growing hydrocarbon chain. The most significant features of the pathway in most bacteria are that it is a soluble, rather than a particulate, system, and utilizes free acyl carrier protein (ACP) as carrier for the growing fatty acid. In *Esch. coli*, the synthesis of unsaturated fatty acids does not involve oxygen, as it does in certain bacteria and eukaryotic cells. It would be expected that the photosynthetic bacteria, all of which are capable of anaerobic growth, would also utilize the anaerobic pathway for synthesis of unsaturated fatty acids. *In vitro* studies confirming this hypothesis have not yet been performed, and in fact the involvement of free ACP in the synthesis of fatty acids by photosynthetic bacteria has not yet been demonstrated.

One method of distinguishing the aerobic and anaerobic pathways for the synthesis of unsaturated fatty acids is by determining the position of the double bond in the end-product fatty acids. Most organisms utilizing the aerobic pathway are found to contain monounsaturated fatty acids with the double bond between carbons 9 and 10. The review by Fulco (1974) summarizes data showing that this is not always the case, but it is true in the vast majority of organisms and in all gram-negative organisms so far examined. On the other hand, the double bond is introduced by β,γ-dehydration at an intermediate stage of elongation of fatty acids in organisms employing the anaerobic pathway. This mechanism leads to unsaturated fatty acids with double bonds at different positions, depending on the length of the fatty acid at the time of β,γ-dehydration, i.e., double bond at $\Delta7$, $\Delta9$, $\Delta11$, and so on. Another way of looking at this is

to note that the position of the double bond is a constant distance from the methyl end of the molecule in fatty acids with different chain lengths, e.g., $14:1\Delta7$, $16:1\Delta9$, $18:1\Delta11$, derived from b,γ-dehydration of β-hydroxydecanoate. Strong suggestive evidence that this was the pathway of fatty acid biosynthesis in *Rp. sphaeroides* 2.4.1 Ga grown in the light anaerobically or in the dark aerobically was presented by Scheuerbrandt and Bloch (1962), who demonstrated that $14:1\Delta7$, $16:1\Delta9$, and $18:1\Delta11$ constituted about 80% of the total long-chain fatty acids in this species. These were the only isomers of these fatty acids found.

Hands and Bartley (1962) reported the presence of both *cis*-vaccenic acid ($18:1\Delta11$) and oleic acid ($18:1\Delta9$) in *Rp. sphaeroides* NCIB 8253. They suggested that a direct desaturation pathway could also exist in these organisms, citing the presence of linoleic acid ($18:2$). It seems likely that the $18:2$ they found was a contaminant of their samples or was incorporated by the bacteria from yeast extract, and that $18:1\Delta9$ is not present, in view of the results of others (Table 6).

Other analytical data supporting the existence of the anaerobic pathway in photosynthetic bacteria are the following. Wood *et al.* (1965) found that $18:1\Delta11$ was the predominant, or only, 18-carbon unsaturated fatty acid in five species of Rhodospirillaceae, the double bond in *Chl. limicola* $16:1$ is between carbons 9 and 10, and multiple isomers of the monoenoic acids of *Chloroflexus aurantiacus* are found (Kenyon and Gray, 1974).

As expected for a pathway similar to that in other organisms in which the fatty acid chain is elongated by successive two-carbon units, 2-[^{14}C]acetate is efficiently utilized for synthesis of both saturated and unsaturated fatty acids by *Rs. rubrum* (Wood *et al.*, 1965; Harris *et al.*, 1965). The incorporation of acetate was greater by cells incubated anaerobically than by cells incubated aerobically, although the relative amounts of saturated and unsaturated fatty acids made were essentially the same in both types of cells. If the aerobic pathway were utilized, under aerobic conditions it would be expected that aeration would affect the relative amounts of saturated and unsaturated fatty acids.

Further evidence for the anaerobic pathway was obtained when Wood *et al.* (1965) examined the utilization of 2-[^{14}C]acetic, 2-[^{14}C]malonic, 1-[^{14}C]octanoic (8:0), 1-[^{14}C]decanoic (10:0), 1-[^{14}C]dodecanoic (12:0), 1-[^{14}C]tetradecanoic (14:0), 1-[^{14}C]hexadecanoic (16:0), and 1-[^{14}C]octadecanoic (18:0) acids for fatty acid biosynthesis by *Rp. capsulata* 2.3.11. When cells were grown anaerobically in the light and incubated aerobically in the light and dark with fatty acids, they took up the fatty acid precursors and converted them to the whole range of bacterial fatty acids. Both 16:0 and 18:0 gave rise to 18:1 (7.3 and 1.2%, respectively, of the total), with little label in shorter acids. The latter result suggests direct desaturation (aerobic pathway). However, *Rp. palustris* also utilized added fatty acids, and the proportions of radioactive 16:0, 16:1, and 18:1 were the same regardless of the fatty acid added. Moreover, the [^{14}C]18:1 made from different labeled fatty acid precursors by either *Rp. capsulata* or *Rp. palustris* showed a random distribution among the first four carbon atoms and a lot of the radioactivity in carbons 5–10, suggesting that there was not direct desaturation of any of the labeled acids.

In summary, the anaerobic pathway is the most likely route for synthesis of unsaturated fatty acids in all the bacteria studied by Wood *et al.* (1965) (*Rp. sphaeroides, Rp. capsulata, Rp. palustris, Rp. gelatinosa*, and *Rs. rubrum*), since 16:1Δ9 and 18:1Δ11 are the major or only monoenes found, aerobic growth of *Rp. capsulata* and *Rp. palustris* does not lead to a different labeling pattern from acetate than does anaerobic growth, and no direct desaturation of long-chain acids is demonstrable in *Rp. capsulata* or *Rp. palustris*.

Both U-[^{14}C]18:1Δ9 and U-[^{14}C]18:2Δ9, 12 are taken up and their label is found in the lipids of *Rp. capsulata* and *Rp. palustris* (Wood *et al.*, 1965). Since the free fatty acids apparently were not removed from the lipid extracts before examination of the labeled fatty acids in these last-mentioned experiments, whether or not added fatty acids are incorporated into lipid is still unknown. However, since the label in isolated fatty acids was the same whether 1-[^{14}C]oleate, U-[^{14}C]oleate, or 2-[^{14}C]acetate was added, it seems likely that all added fatty acids were degraded and resynthesized, and direct incorporation did not occur.

There is little conclusive evidence on the control of the nature and total amount of fatty acids synthesized in the photosynthetic bacteria. It was mentioned above that in some photosynthetic bacteria, the nature of the fatty acids changes in response to change to light, anaerobic conditions from dark, aerobic conditions. The mechanism of this control is unknown. A possible mechanism is suggested by the following arguments: In many bacteria, including *Esch. coli* and *Clostridium* species, the *sn*-glycerol-3-phosphate acyltransferase utilizes acyl-CoA as well as acyl-ACP as substrates for lysoPA and PA synthesis (Goldfine and Ailhaud, 1971; van den Bosch and Vagelos, 1970). They are able to utilize added fatty acids by first activating them

to CoA derivatives (Overath *et al.*, 1969). The product of fatty acid biosynthesis is acyl-ACP. We and others (Kito *et al.*, 1972) have noted that the fatty acid compositions of the phospholipids of *Esch. coli*, which are all derived from PA (PE, PG, CL), are not identical and do not respond identically to change in conditions of growth. This observation suggests that there is either a specificity for fatty acid components during synthesis, or fatty acid exchange in phospholipids, or both. We therefore concluded that there was probably both a *de novo* (ACP-dependent) pathway for lipid biosynthesis through PA and a fatty acid exchange (CoA-dependent) pathway. Lueking and Goldfine (1975*a*) showed that guanosine tetraphosphate has a differential effect on incorporation of –CoA and –ACP derivatives into phosphatidic acid in *Esch. coli*, thus separating the control of the two hypothesized pathways.

Rhodopseudomonas palustris and *Rp. capsulata* are unable to utilize directly added long-chain fatty acids (Wood *et al.*, 1965). Moreover, the particulate acyl transferase of *Rp. sphaeroides* is unable to utilize fatty acyl-CoA derivatives, but utilizes fatty acyl-ACP derivatives efficiently (Lueking and Goldfine, 1975*b*). These observations strongly suggest that *Rp. sphaeroides* has only the *de novo* (ACP-dependent) pathway for PA synthesis found in *Esch. coli*, and is lacking in the CoA-dependent exchange pathway as proposed by Leuking and Goldfine (1975*b*).

Lastly, in *Esch. coli*, adaptation of fatty acid composition can occur both by change in acyltransferase specificity (Sinensky, 1971) and by change in the relative amounts of fatty-acid-synthesizing enzymes in the cell (Cronan, 1974). Control and adaptation of photosynthetic bacterial fatty acid composition could be effected by either or both of these mechanisms.

The control of the extent of fatty acid synthesis in *Rp. capsulata* has been only briefly studied. Mindich (1972, 1973) showed that glycerol auxotrophs of *Rp. capsulata*, like those of *Esch. coli*, do not accumulate free fatty acids when deprived of glycerol. Thus, either fatty acid synthesis is curtailed when lipids are not made, or the fatty acids made are incorporated elsewhere, e.g., into lipid A. Oelze and Drews (1969) suggest that the latter is at least partially true in *Rs. rubrum*, in which there is more cell-wall fatty acid in dark-grown cells than in light-grown cells. However, there is no evidence for incorporation to lipid A in *Esch. coli* during glycerol starvation, and it is known that exogenous unsaturated fatty acids curtail the synthesis of unsaturated fatty acids in *Esch. coli* Silbert *et al.*, 1972). *Rhodopseudomonas capsulata* will be an excellent subject for study of the mechanism of the control of the extent of fatty acid synthesis without the complication of the other control (exchange) mechanisms later in the pathway that occur in *Esch. coli.*

7. Note Added in Proof

Several pertinent articles have appeared since this chapter was completed. Maudinas and Villoutreix (1974) showed evidence for the synthesis of squalene by *Rs. rubrum* S, B₄, and FR VI. The lipids and fatty acids of cell envelope fractions of *Rs. rubrum* were described by Collins and Niederman (1976) and by Oelze *et al.* (1975). The lipids and fatty acids of an additional species of *Ectothiorhodospira* (*E. mobilis*) were reported by Oyewole and Holt (1976). Further chemical characterization of the components of the cell envelope of *Chr. vinosum* was reported by Hurlbert *et al.* (1976), and the distribution of lipids in the intracytoplasmic membrane of *Chr. vinosum* was studied by Shimada and Murata (1976). Rietschel (1976) used the technique of gas–liquid chromatography of diasteriomeric derivatives to determine that the absolute configuration of the 3-hydroxy acids of several photosynthetic bacterial lipopolysaccharides is the D-isomer. Finally, it seems relevant to point out that all fatty acid analyses discussed in this chapter utilized only gas–liquid chromatography, chemical modifications, and thin-layer chromatography. Now that newer techniques of analysis such as coupled gas–liquid chromatography/mass spectrometry (GC/MS) are readily available, it would be appropriate to perform more rigorous total analyses of the fatty acid compositions of the photosynthetic bacteria and other bacteria. Oyewole and Holt (1976) utilized GC/MS to confirm the presence of 19cy in *E. mobilis*. Boon *et al.* (1977) discussed the strength of this technique for analytical studies of a taxonomic nature.

ACKNOWLEDGMENTS

I wish to acknowledge all those from whom I received personal communication of unpublished results and preprints of manuscripts, particularly J. C. Raymond, D. R. Lueking, and H. S. Goldfine. Many helpful suggestions were made by those listed and by G. D. Hegeman and G. L. Kenyon. The author was supported during the majority of the time this chapter was being prepared by U.S. Public Health Service Grant No. AM-13492 from the National Institute of Arthritis, Metabolism and Digestive Diseases, and during a part of the time by U.S. Public Health Service Grant No. HO-7314 from the National Institute of Child Health and Human Development to Dr. George D. Hegeman. I wish to thank Alane M. Gray for exceptionally able technical assistance.

8. References

Aaronson, S., 1964, A role for a sterol and a sterol precursor in the bacterium *Rhodopseudomonas palustris, J. Gen. Microbiol.* **37**:225.

Benson, A. A., and Shibuya, I., 1962, Surfactant lipids, in: *Physiology and Biochemistry of Algae* (R. A. Lewin, ed.), pp. 371–383, Academic Press, New York.

Benson, A. A., and Strickland, E. H., 1960, Plant phospholipids. III. Identification of diphosphatidylglycerol, *Biochim. Biophys. Acta* **41**:328.

Benson, A. A., Wintermans, J. F. G. M., and Wiser, R., 1959, Chloroplast lipids as carbohydrate reservoirs, *Plant Physiol.* **34**:315.

Bishop, N. I., 1959, The reactivity of a naturally occurring quinone (Q-225) in photochemical reactions of isolated chloroplasts, *Proc. Natl. Acad. Sci. U.S.A.* **45**:1696.

Bligh, E. G., and Dyer, W. J., 1959, A rapid method of total lipid extraction and purification, *Can. J. Biochem. Physiol.* **37**:911.

Boon, J. J., de Leeuw, J. W., van den Hoek, G. J., and Vosjan, J. H., 1977, Significance and taxonomic value of iso- and anteisomonoenoic fatty acids and branched β-hydroxy acids in *Desulfovibrio desulfuricans, J. Bacteriol.* **129**:1183.

Bril, C., 1964, Studies on the photosynthetic apparatus of some purple bacteria, Thesis, The State University of Utrecht, Utrecht, The Netherlands.

Brooks, J. L., and Benson, A. A., 1972, Studies on the structure of an ornithine-containing lipid from *Rhodospirillum rubrum, Arch. Biochem. Biophys.* **152**:347.

Castenholz, R. W., 1969, Thermophilic blue-green algae and the thermal environment, *Bacteriol. Rev.* **33**:476.

Cho, K. Y., and Salton, M. R. J., 1964, Fatty acid composition of the lipids of membranes of gram-positive bacteria and "walls" of gram-negative bacteria, *Biochim. Biophys. Acta* **84**:773.

Cho, K. Y., and Salton, M. R. J., 1966, Fatty acid composition of bacterial membrane and wall lipids, *Biochim. Biophys. Acta* **116**:73.

Cohen-Bazire, G., Sistrom, W R., and Stainer, R. Y., 1957, Kinetic studies of pigment synthesis by non-sulfur purple bacteria, *J. Cell. Comp. Physiol.* **49**:25.

Collins, M. L. P., and Niederman, R. A., 1976, Membranes of *Rhodospirillum rubrum*: Isolation and physicochemical properties of membranes from aerobically grown cells, *J. Bacteriol.* **126**:1316.

Constantopoulos, G., and Bloch, K., 1967, Isolation and characterization of glycolipids from some photosynthetic bacteria, *J. Bacteriol.* **93**:1788.

Cronan, J. E., Jr., 1974, Regulation of the fatty acid composition of the membrane phospholipids of *Escherichia coli, Proc. Natl. Acad. Sci. U.S.A.* **71**:3758.

Cruden, D. L., and Stanier, R. Y., 1970, The characterization of chlorobium vesicles and membranes isolated from green bacteria, *Arch. Mirobiol.* **72**:115.

Cruden, D. L., Cohen-Bazire, G., and Stanier, R. Y., 1970, Chlorobium vesicles: The photosynthetic organelles of green bacteria, *Nature (London)* **228**:1345.

Davies, W. H., Mercer, E. I., and Goodwin, T. W., 1965, The occurrence and intracellular distribution of the plant sulfolipid in maize, runner beans, plant tissue cultures, and *Euglena gracilis*, *Phytochemistry* **4**:741.

Depinto, J. A., 1967, Ornithine-containing lipid *Rhodospirillum rubrum*, *Biochim. Biophys. Acta* **144**:113.

Drews, G., 1965, Die Isolierung schwefelfreier Purpurbakterien, *Zentralbl. Bakteriol. Parasitenkd. Infektionskr. Hyg. Abt. 1, Suppl.* **1**:170.

Drews, G., 1974, Composition of a protochlorophyll–protopheophytin complex, excreted by mutant strains of *Rhodopseudomonas capsulata*, in comparison with the photosynthetic apparatus, *Arch. Microbiol.* **100**:397.

Eimjhellen, K. E., Aasmundrud, O., and Jensen, A., 1963, A new bacterial chlorophyll, *Biochem. Biophys. Res. Commun.* **10**:232.

Erwin, J., and Bloch, K., 1964, Biosynthesis of unsaturated fatty acids in microorganisms, *Science* **143**:1006.

Eymers, J. C., and Wassink, E. C., 1938, On the photochemical carbon dioxide assimilation in purple sulphur bacteria, *Enzymologia* **2**:258.

Fiertel, A., and Klein, H. P., 1959, On sterols in bacteria, *J. Bacteriol.* **78**:738.

Fulco, A. J., 1974, Metabolic alterations of fatty acids, *Annu. Rev. Biochem.* **43**:215.

Goldfine, H., 1972, Comparative aspects of bacterial lipids, *Adv. Microb. Physiol.* **8**:1.

Goldfine, H., and Ailhaud, G. P., 1971, Fatty acyl-acyl carrier protein and fatty acyl-CoA as acyl donors in the biosynthesis of phosphatidic acid in *Clostridium butyricum*, *Biochem. Biophys. Res. Commun.* **45**:1127.

Gorschein, A., 1964, Ornithine in *Rhodopseudomonas sphaeroides*, *Biochim. Biophys. Acta* **84**:356.

Gorschein, A., 1968a, Distribution and metabolism of ornithine in *Rhodopseudomonas spheroides*, *Proc. R. Soc. London Ser. B*: **170**:265.

Gorschein, A., 1968b, The separation and identification of the lipids of *Rhodopseudomonas spheroides*, *Proc. R. Soc. London Ser. B*: **170**:279.

Gorschein, A., 1968c, Studies on the structure of an ornithine-containing lipid from nonsulphur purple bacteria, *Biochim. Biophys. Acta* **152**:358.

Gorschein, A., Neuberger, A., and Tait, G. H., 1968a, Incorporation of radioactivity from (Me-¹⁴C)methionine and (2-¹⁴C)glycine into the lipids of *Rhodopseudomonas spheroides*, *Proc. R. Soc. London Ser. B*: **170**:299.

Gorschein, A., Neuberger, A., and Tait, G. H., 1968b, Metabolic turnover of the lipids of *Rhodopseudomonas spheroides*, *Proc. R. Soc. London Ser. B*: **170**:311.

Hands, A. R., and Bartley, W., 1962, The fatty acids of *Rhodopseudomonas* particles, *Biochem. J.* **84**:238.

Harris, R. V., Wood, B. J. B., and James, A. T., 1965, Fatty acid biosynthesis in photosynthetic micro-organisms, *Biochem. J.* **94**:22P.

Haverkate, F., 1965, Phosphatidylglycerol from photosynthetic tissues, Thesis, University of Utrecht.

Haverkate, F., Teulings, F. A. G., and van Deenen, L. L. M., 1965, Studies on the phospholipids of photosynthetic microorganisms II. *K. Ned. Acad. Wet. Proc. Ser. B* **68**:154.

Hirayama, O., 1968, Lipids and lipoprotein complex in photosynthetic tissues. Part IV. Lipids and pigments of photosynthetic bacteria, *Agric. Biol. Chem.* **32**:34.

Houtsmuller, U. M. T., and van Deenen, L. L. M., 1963, Identification of a bacterial phospholipid as an O-ornithine ester of phosphatidyl glycerol, *Biochim. Biophys. Acta* **70**:211.

Hurlbert, R. E., and Lascelles, J., 1963, Ribulose diphosphate carboxylase in thiorhodaceae, *J. Gen. Microbiol.* **33**:445.

Hurlbert, R. E., Golecki, J. R., and Drews, G., 1974, Isolation and characterization of *Chromatium vinosum* membranes, *Arch. Microbiol.* **101**:169.

Hurlbert, R., Weckesser, J., Mayer, H., and Fromme, I., 1976, Isolation and characterization of the lipopolysaccharide of *Chromatium vinosum*, *Eur. J. Biochem.* **68**:365.

Ikawa, M., 1967, Bacterial phosphatides and natural relationships, *Bacteriol. Rev.* **31**:54.

James, A. T., and Nichols, B. W., 1966, Lipids of photosynthetic systems, *Nature (London)* **210**:372.

Jensen, S. L., 1962, The constitution of some bacterial carotenoids and their bearing on biosynthetic problems, *K. Nor. Vidensk. Selsk. Skr.*, No. 8, p. 1.

Jolchine, G., and Reiss-Husson, F., 1975, Studies on pigments and lipids in *Rhodopseudomonas spheroides* Y reaction center, *FEBS Lett.* **52**:33.

Kamio, Y., Kanegasaki, S., and Takahashi, H., 1969, Occurrence of plasmalogens in anaerobic bacteria, *J. Gen. Appl. Microbiol.* **15**:439.

Kates, M., and Volcani, B. E., 1966, Lipid components of diatoms, *Biochim. Biophys. Acta* **116**:264.

Kates, M., Palameta, B., Joo, C. N., Kushner, D. J., and Gibbons, N. E., 1966, Aliphatic diether analogs of glyceride-derived lipids. IV. The occurrence of di-O-dihydrophytylglycerol ether containing lipids in extremely halophilic bacteria, *Biochemistry* **5**:4092.

Kenyon, C. N., 1972, Fatty acid composition of unicellular strains of blue-green algae, *J. Bacteriol.* **109**:827.

Kenyon, C. N., and Gray, A. M., 1974, Preliminary analysis of lipids and fatty acids of green bacteria and *Chloroflexus aurantiacus*, *J. Bacteriol.* **120**:131.

Kenyon, C. N., and Waterbury, J., 1978, Fatty acids belonging to the orders Chamaesiphonales and Pleurocapsales (in prep.).

Kenyon, C. N., Rippka, R., and Stanier, R. Y., 1972, Fatty acid composition and physiological properties of some filamentous blue-green algae, *Arch. Mikrobiol.* **83**:216.

Kito, M., Aibara, S., Kato, M., and Hata, T., 1972, Differences in fatty acid composition among phosphatidylethanolamine, phosphatidylglycerol, and cardiolipin of *Escherichia coli*, *Biochim. Biophys. Acta* **260**:475.

Klemme, B., Klemme, J.-H., and San Pietro, A., 1971,

PPase, ATPase, and photophosphorylation in chromatophores of *Rhodospirillum rubrum*: Inactivation by phospholipase A; reconstitution by phospholipids, *Arch. Biochem. Biophys.* **144**:339.

Klenk, E., Knipprath, W., Eberhagen, D., and Koof, H. P., 1963, Über die ungesättigen Fettsäuren der Fettstoffe von Süsswasser- und Meeresalgen, *Hoppe-Seyler's Z. Physiol. Chem.* **334**:44.

Kobayashi, M., Mochida, K., and Okuda, A., 1967, The amino acid composition of photosynthetic bacterial cells, *Jpn. Soc. Sci. Fish. Bull.* **33**:657.

Lascelles, J., 1956, The synthesis of porphyrins and bacteriochlorophyll by cell suspensions of *Rhodopseudomonas spheroides, Biochem. J.* **62**:78.

Lascelles, J., 1959, Adaptation to form bacteriochlorophyll in *Rhodopseudomonas spheroides*: Changes in activity of enzymes concerned in pyrrole synthesis, *Biochem. J.* **72**:508.

Lascelles, J. and Szilagyi, J. F., 1965, Phospholipid synthesis by *Rhodopseudomonas spheroides* in relation to the formation of photosynthetic pigments, *J. Gen. Microbiol.* **38**:55.

Lennarz, W. J., 1966, Lipid metabolism in the bacteria, *Adv. Lipid Res.* **4**:175.

Lüderitz, O., 1970, Recent results on the biochemistry of the cell wall lipopolysaccharides of *Salmonella* bacteria, *Angew. Chem. Int. Ed. Engl.* **9**:649.

Lueking, D. R., and Goldfine, H., 1975a, The involvement of guanosine 5'-diphosphate-3'-diphosphate in the regulation of phospholipid biosynthesis in *Escherichia coli, J. Biol. Chem.* **250**:4911.

Lueking, D. R., and Goldfine, H., 1975b, sn-Glycerol-3-phosphate acyltransferase activity in particulate preparations from anaerobic, light-grown cells of *Rhodopseudomonas spheroides, J. Biol. Chem.* **250**:853.

Macfarlane, M. G., 1962, Characterization of lipoaminoacids as O-aminoacid esters of phosphatidylglycerol, *Nature (London)* **196**:136.

Madigan, M. T., Petersen, S. R., and Brock, T. D., 1974, Nutritional studies on *Chloroflexus*, a filamentous photosynthetic, gliding bacterium, *Arch. Microbiol.* **100**:97.

Maudinas, B., and Villoutreix, J., 1974, Mise en evidence du squalène chez des bactéries photosynthétiques, *C. R. Acad. Sc. Paris Ser. D* **278**:2995.

Mindich, L., 1972, Control of fatty acid synthesis in bacteria, *J. Bacteriol.* **110**:96.

Mindich, L., 1973, Synthesis and assembly of bacterial membranes, in: *Bacterial Membranes and Walls* (L. Leive, ed.), pp. 1–36, Marcel Dekker, New York.

Newton, J. W., and Kamen, M. D., 1956, *Chromatium* cytochrome, *Biochim. Biophys. Acta* **21**:71.

Newton, J. W., and Newton, G. A., 1957, Composition of the photoactive subcellular particles from *Chromatium, Arch. Biochem. Biophys.* **71**:250.

Nichols, B. W., 1973, Lipid composition and metabolism, in: *The Biology of Blue-Green Algae* (N. G. Carr and B. A. Whitton, eds.), pp. 144–161, University of California Press, Berkeley and Los Angeles.

Nichols, B. W., and James, A. T., 1965, Lipids of photosynthetic tissue, *Biochem. J.* **94**:22P.

Nichols, B. W., Harris, R. V., and James, A. T., 1965, The lipid metabolism of blue-green algae, *Biochem. Biophys. Res. Commun.* **20**:256.

Nichols, B. W., Stubbs, J. M., and James, A. T., 1966, The lipid composition and ultrastructure of normal developing and degenerating chloroplasts, in: *Biochemistry of Chloroplasts*, Vol. II (T. W. Goodwin, ed.), pp. 677–701, Academic Press, London.

Oelze, J., and Drews, G., 1969, Die Morphogenese des Photosyntheseapparates von *Rhodospirillum rubrum*. II. Die Kinetik der Thylakoidsynthese nach Markierung der Membranen mit (2-^{14}C)Azetät, *Biochim. Biophys. Acta* **173**:448.

Oelze, J., and Drews, G., 1970, Die Ausscheidung von partikelgebundenen Bacteriochlorophyllvorstufen durch die Mutante F9 von *Rhodospirillum rubrum, Arch. Mikrobiol.* **73**:19.

Oelze, J., and Drews, G., 1972, Membranes of photosynthetic bacteria, *Biochim. Biophys. Acta* **265**:209.

Oelze, J., Schröder, J., and Drews, G., 1970, Bacteriochlorophyll, fatty acid, and protein synthesis in relation to thylakoid formation in mutant strains of *Rhodospirillum rubrum, J. Bacteriol.* **101**:669.

Oelze, J., Golecki, J. R., Kleinig, H., and Weckesser, J., 1975, Characterization of two cell-envelope fractions from chemotrophically grown *Rhodospirillum rubrum, Ant. van Leewenh.* **41**:273.

O'Leary, W. M., 1973, Lipids, in: *Handbook of Microbiology*, Vol. II, *Microbial Composition* (A. I. Laskin and H. A. Lechevalier, eds.), pp. 243–327, The Chemical Rubber Company, Cleveland, Ohio.

Oliver, J. D., and Colwell, R. R., 1973, Extractable lipids of Gram-negative marine bacteria: Phospholipid composition, *J. Bacteriol.* **114**:897.

Ormerod, J. G., Ormerod, K. S., and Gest, H., 1961, Light-dependent utilization of organic compounds and photoproduction of molecular hydrogen by photosynthetic bacteria; relationships with nitrogen metabolism, *Arch. Biochem. Biophys.* **94**:449.

Overath, P., Pauli, G., and Schairer, H. U., 1969, Fatty acid degradation in *Escherichia coli*. An inducible acyl-CoA synthetase, the mapping of *old*-mutations, and the isolation of regulatory mutants, *Eur. J. Biochem.* **7**:559.

Oyewole, S. H., and Holt, S. C., 1976, Structure and composition of intracytoplasmic membranes of *Ectothiorhodospira mobilis, Arch. Microbiol.* **107**:167.

Park, C.-E., and Berger, L. R., 1967a, Complex lipids of *Rhodomicrobium vannielii, J. Bacteriol.* **93**:221.

Park, C.-E., and Berger, L. R., 1967b, Fatty acids of extractable and bound lipids of *Rhodomicrobium vannielii, J. Bacteriol.* **93**:230.

Pfennig, N., 1961, Eine vollsynthetische Nährlösung zur selektiven Anreicherung einiger Schwefelpurpurbakterien, *Naturwissenschaften* **48**:136.

Pfennig, N., 1962, Beobachtungen über das Schwärmen von *Chromatium okenii, Arch. Mikrobiol.* **42**:90.

Pfennig, N., 1965, Anreichungskultur für rote und grüne Schwefelbakterien, *Zentralbl. Bakteriol. Parasitenkd. Infektionskr. Hyg. Abt. 1, Suppl. 1* **179**:503.

Pfennig, N., and Trüper, H. G., 1974, The photosynthetic

bacteria, in: *Bergey's Manual of Determinative Bacteriology*, 8th Ed. (R. E. Buchanan and N. E. Gibbons, eds.), pp. 24–61, Williams and Wilkins Co., Baltimore.

Pierson, B. K., and Castenholz, R. W., 1974, A phototrophic gliding filamentous bacterium of hot springs, *Chloroflexus aurantiacus*, gen. and sp. nov., *Arch. Microbiol.* **100**:5.

Pucheu, N. L., Kerber, N. L., and Garcia, A. F., 1974, Comparative studies on membranes isolated from *Rhodopseudomonas viridis* grown in the presence and in the absence of yeast extract, *Arch. Microbiol.* **101**:259.

Radunz, A., 1969, Über das Sulfochinovosyl-diacylglycerin aus höheren Pflanzen, Algen, und Purpurbakterien, *Hoppe-Seyler's Z. Physiol. Chem.* **350**:411.

Raymond, J. C., and Sistrom, W. R., 1967, The isolation and preliminary characterization of a halophilic photosynthetic bacterium, *Arch. Mikrobiol.* **59**:255.

Rietschel, E. T., 1976, Absolute configuration of 3-hydroxy fatty acids present in lipopolysaccharides from various bacterial groups, *Eur. J. Biochem.* **64**:423.

Scheuerbrandt, G., and Bloch, K., 1962, Unsaturated fatty acids in microorganisms, *J. Biol. Chem.* **237**:2064.

Schmitz, R., 1967a, Über die Zusammensetzung der pigmenthaltigen Strukturen des Purpurbakteriums *Rhodopseudomonas spheroides*, *Z. Naturforsch.* **22b**:645.

Schmitz, R., 1967b, Über die Zusammensetzung der pigmenthaltigen Strukturen aus Prokaryonten. II. Untersuchungen an Chromatophoren von *Chlorobium thiosulfatophilum* Stamm Tassajara, *Arch. Mikrobiol.* **56**:238.

Schröder, J., and Drews, G., 1968, Quantitative Bestimmung der Fettsäuren von *Rhodospirillum rubrum* und *Rhodopseudomonas capsulata* während der Thylakoidmorphogenese, *Arch. Mikrobiol.* **64**:59.

Schröder, J., Biedermann, M., and Drews, G., 1969, Die Fettsäuren in ganzen Zellen, Thylakoiden un Lipopolysacchariden von *Rhodospirillum rubrum* und *Rhodopseudomonas capsulata*, *Arch. Mikrobiol.* **66**:273.

Shaw, N., 1970, Bacterial glycolipids, *Bacteriol. Rev.* **34**:365.

Shaw, N., and Baddiley, J., 1968, Structure and distribution of glycosyl diglycerides in bacteria, *Nature (London)* **217**:142.

Shimada, K., and Murata, N., 1976, Chemical modification by trinitrobenzenesulfonate of a lipid and proteins of intracytoplasmic membranes isolated from *Chromatium vinosum* and *Azotobacter vinelandii*, *Biochim. Biophys. Acta* **455**:605.

Silbert, D. F., Cohen, M., and Harder, M. E., 1972, The effect of exogenous fatty acids on fatty acid metabolism in *Escherichia coli* K12, *J. Biol. Chem.* **247**:1699.

Sinensky, M., 1971, Temperature control of phospholipid biosynthesis in *Escherichia coli*, *J. Bacteriol.* **106**:449.

Sistrom, W. R., 1960, A requirement for sodium in the growth of *Rhodopseudomonas sphaeroides*, *J. Gen. Microbiol.* **22**:778.

Sober, H. A. (ed.), 1970, *Handbook of Biochemistry: Selected Data for Molecular Biology*, pp. A-15–17, E-1–21, The Chemical Rubber Company, Cleveland, Ohio.

Steiner, S., Conti, S. F., and Lester, R. L., 1969, Separation and identification of the polar lipids of *Chromatium* strain D, *J. Bacteriol.* **98**:10.

Steiner, S., Burnham, J. C., Conti, S. F., and Lester, R. L., 1970a, Polar lipids of *Chromatium* strain D grown at different light intensities, *J. Bacteriol.* **103**:500.

Steiner, S., Sojka, G. A., Conti, S. F., Gest, H., and Lester, R. L., 1970b, Modification of membrane composition in growing photosynthetic bacteria, *Biochim. Biophys. Acta* **203**:571.

Tabita, F. R., and McFadden, B. A., 1976, Molecular and catalytic properties of ribulose 1,5-bisphosphate carboxylase from the photosynthetic extreme halophile *Ectothiorhodospira halophila*, *J. Bacteriol.* **126**:1271.

Takacs, B. J., and Holt, S. C., 1971a, *Thiocapsa floridana*; a cytological, physical, and chemical characterization. I. Cytology of whole cells and isolated chromatophore membranes, *Biochim. Biophys. Acta* **233**:258.

Takacs, B. J., and Holt, S. C., 1971b, *Thiocapsa floridana*; a cytological, physical, and chemical characterization. II. Physical and chemical characteristics of isolated and reconstituted chromatophores, *Biochim. Biophys. Acta* **233**:278.

Van Baalen, C., 1967, Further observations on growth of single cells of coccoid blue-green algae, *J. Phycol.* **3**:154.

van Deenen, L. L. M., and Haverkate, F., 1966, Chemical characterization of phosphatidylglycerol from photosynthetic tissues, in: *Biochemistry of Chloroplasts* (T. W. Goodwin, ed.), pp. 117–131, Academic Press, London and New York.

van den Bosch, H., and Vagelos, P. R., 1970, Fatty acyl-CoA and fatty acyl-acyl carrier protein as acyl donors in the synthesis of lysophosphatidate and phosphatidate in *Escherichia coli*, *Biochim. Biophys. Acta* **218**:233.

Weckesser, J., Drews, G., and Fromme, I., 1972, Chemical analysis of and degradation studies on the cell wall lipopolysaccharide of *Rhodopseudomonas capsulata*, *J. Bacteriol.* **109**:1106.

Weckesser, J., Drews, G., Fromme, I., and Mayer, H., 1973, Isolation and chemical composition of the lipopolysaccharides of *Rhodopseudomonas palustris* strains, *Arch. Mikrobiol.* **92**:123.

Weckesser, J., Drews, G., Roppel, J., Mayer, H., and Fromme, I., 1974, The lipopolysaccharides (*O*-antigens) of *Rhodopseudomonas viridis*, *Arch. Microbiol.* **101**:233.

Weerkamp, A., and Heinen, W., 1972, The effects of nutrients and precursors on the fatty acid composition of two thermophilic bacteria, *Arch. Mikrobiol.* **81**:350.

Wiessner, W., 1960, Wachstum und Stoffwechsel von *Rhodopseudomonas spheroides* in Abhängigkeit von der Versorgung mit Mangan und Eisen, *Flora (Jena)* **149**:1.

Wintermans, J. F. G. M., 1960, Concentrations of phosphatides and glycolipids in leaves and chloroplasts, *Biochim. Biophys. Acta* **44**:49.

Wood, B. J. B., Nichols, B. W., and James, A. T., 1965, The lipids and fatty acid metabolism of photosynthetic bacteria, *Biochim. Biophys. Acta* **106**:261.

Zill, L. P., and Harmon, E. A., 1962, Lipids of photosynthetic tissue. I. Silicic acid chromatography of the lipids from whole leaves and chloroplasts, *Biochim. Biophys. Acta* **57**:573.

PART IV

Photometabolism

CHAPTER 15

Role of the Reaction Center in Photosynthesis

William W. Parson

The concept of a photosynthetic reaction center has its origins in the classic studies by Emerson and Arnold (1932*a,b*) of CO_2 fixation in the green alga *Chlorella pyrenoidosa*. Emerson and Arnold (1932*a*) measured the amount of CO_2 that suspensions of algae could fix when they were exposed to short flashes of light from a neon lamp. To obtain the maximal amount of CO_2 fixation per flash, it was necessary to allow an adequate period of darkness between flashes. Each flash apparently generated products that had to be consumed before the photochemical apparatus could function properly again. Allowing an optimal period between the flashes, Emerson and Arnold (1932*b*) found that the amount of CO_2 that was fixed per flash increased with increasing flash intensity, until it reached a maximum that presumably was set by the concentration of the photosynthetic apparatus itself. Unexpectedly, the maximum amount of CO_2 fixation turned out to be quite small, compared with the chlorophyll content of the algae. One molecule of CO_2 was fixed per 2480 molecules of chlorophyll.

Although the interpretation was not clear at the time, the work of Emerson and Arnold indicated that most of the chlorophyll of the cells does not participate directly in the photochemical reactions of photosynthesis. Instead, the major part of the pigments of photosynthetic cells serves as a light-harvesting antenna, funneling energy from the light that it

absorbs to a special set of reactive molecules. This thesis was developed by Duysens (1951, 1952) in a series of experiments that proved to be germinal for our understanding of bacterial photosynthesis. As Wassink *et al.* (1939) had noted, the bacteriochlorophyll (Bchl) of the photosynthetic bacteria occurs in three or more distinct forms or complexes, with slightly different absorption spectra. Duysens called the three main forms in *Chromatium vinosum* "B800," "B850," and "B890," for their absorption maxima in the near-IR. On examining the fluorescence emission spectrum of *Chr. vinosum* cells, Duysens found that only B890 fluoresced. The fluorescence excitation spectrum, however, indicated that both B800 and B850 were capable of transferring energy efficiently to B890. Even when the cells were excited with blue-green light that was absorbed by carotenoids rather than by any of the Bchl complexes, much of the energy found its way to B890.

To investigate whether the transfer of energy to B890 was an essential step in photosynthesis, Duysens compared the fluorescence excitation spectrum with the action spectrum for photosynthesis. (For convenience, Duysens actually measured phototaxis; earlier workers had shown that the action spectra for phototaxis and photosynthesis were similar.) The two action spectra were essentially indistinguishable. Duysens concluded that the energy of light became available to drive the "chemical" part of photosynthesis only after an array of absorbers had transferred the energy to B890. Similar measurements in

William W. Parson · Department of Biochemistry, University of Washington, Seattle, Washington 98195

algae led to a similar conclusion: light that was absorbed by a variety of pigments contributed to photochemistry only insofar as the energy was transferred to chlorophyll *a*. Duysens suggested that the migration of energy within the light-harvesting array occurred by the phenomenon of inductive resonance that Förster (1951) had recently described.

Searching for clues to the photochemical reactions that must ultimately occur, Duysens built a sensitive split-beam spectrophotometer. The instrument made it possible to measure small differences between the absorption spectra of two turbid suspensions of cells, one of which was illuminated with auxiliary light. Illumination proved to cause two distinctly different sorts of absorbance changes. Judging from its spectrum, one of the two clearly represented the oxidation of a cytochrome. The cytochrome photooxidation could be induced by weak illumination of anaerobic suspensions of *Rhodospirillum rubrum* in the presence of reducing substrates (Duysens, 1954). These observations were extended by Chance and Smith (1955), who found that either illumination or aeration caused the oxidation of several different cytochromes in *Rs. rubrum*. Chance and his colleagues (Chance and Nishimura, 1960; Olson and Chance, 1960*a,b*) continued this work in an extensive study of cytochrome photooxidation in *Chr. vinosum*.

The second type of light-induced absorbance change that Duysens (1952) discovered appeared to involve the Bchl itself. Its main feature was a decrease of about 2% in the absorbance of the cells near 890 nm in *Chr. vinosum*, or 870 nm in *Rs. rubrum*. Accompanying this bleaching, there also occurred a small decrease in the absorbance near 810 nm (Duysens *et al.*, 1956), and small increases near 430, 790 (Duysens, 1953, 1954; Duysens *et al.*, 1956), and 1250 nm (Clayton, 1962*a*). All these absorbance changes reversed rapidly when the exciting light was turned off. Duysens called the component that was responsible for the absorbance changes "P890." Because the spectrum varies somewhat from species to species, other workers have used various other designations, such as P_{870}. For simplicity, the designation P_{870} is useful in a general sense.

Unlike the absorbance changes that indicated cytochrome photooxidation, the absorbance changes due to P_{870} were most pronounced if the cells were illuminated under oxidizing conditions. They were more difficult to detect if the cell suspensions were anaerobic and if they contained reducing substrates (Duysens, 1952, 1953; Duysens *et al.*, 1956). This suggested that the absorbance changes reflected the photooxidation of P_{870} to a form (P_{870}^+) that could be reduced again rapidly in the presence of appropriate

substrates. Goedheer (1958, 1960) and Duysens (1958) strengthened this interpretation with the discovery that they could cause similar absorbance changes by the addition of the relatively mild oxidant potassium ferricyanide, or reverse the absorbance changes with the reductant potassium ferrocyanide. Clayton (1962*a*) and Kuntz *et al.* (1964) confirmed these observations, and extended them to additional species and strains of bacteria. Kuntz *et al.* (1964) induced the absorbance changes by excitation with a short flash, and found that reducing conditions facilitated the reversal of the absorbance changes after the flash. Oxidizing conditions retarded the reversal.

An important stimulus for the development of these ideas came from parallel studies of green plants. Kok (1961) and Witt *et al.* (1961) had detected similar light-induced absorbance changes in the chlorophyll *a* of chloroplasts, and had also interpreted them in terms of a photooxidation.

Early in the studies of both bacteria and chloroplasts, the question arose whether the light-induced absorbance changes reflected a minor alteration of the absorption spectra of many molecules of Bchl or a major alteration of the spectra of a relatively small number of molecules. With the discovery that the absorbance changes could occur in dried films of chromatophores at temperatures as low as 1°K, Arnold and Clayton (1960) at first suggested that light caused the separation of mobile electrons and holes in an extended array of Bchl molecules. The finding of second-order decay kinetics after the illumination appeared to support such an interpretation at the time (Arnold and Clayton, 1960), although later studies indicated that the decay kinetics were actually first-order (Parson, 1967; McElroy *et al.*, 1974). Calvin and Androes (1962) also advanced the idea that light generated mobile electrons and holes, building on the observation that illumination at low temperatures caused the development of ESR signals. But evidence for a different view began to emerge from studies by Clayton (1962*a*, 1963) of a mutant of *Rhodopseudomonas sphaeroides* that was defective in carotenoid biosynthesis, strain R-26.

In strain R-26, as in the wild-type, illumination or gentle chemical oxidation caused an absorbance decrease of about 2% in the main absorption band of Bchl near 870 nm. The Bchl of the carotenoidless strain, however, was exceptionally labile. If an anaerobic culture was allowed to stand in the light for several weeks, a large part of the Bchl was destroyed, or, more accurately, converted to bacteriopheophytin. Suspensions of chromatophores from such aged cells still exhibited absorbance changes when they were illuminated or treated with ferricyanide, but now the bleaching amounted to almost 100% of the absorbance at 870 nm. The spectrum of the absorbance changes caused by light or ferricyanide was essentially the same as it was in fresh cells, but P_{870} had become a major fraction of the Bchl. Kuntz *et al.* (1964) and Clayton (1966*a*) obtained similar results by treating chromatophores with the strong oxidant K_2IrCl_6. The oxidation destroyed most of the Bchl of the chromatophores without damaging their P_{870}. The treated preparations responded to light just as untreated ones did, but the absorbance changes in the IR were now quite dramatic.

Clayton (1962*a*, 1963, 1966*a*) concluded that P_{870} consisted of a small number of molecules of Bchl in a special environment. The results of extraction and analysis of pigments from the aged or oxidized preparations supported the view that the Bchl in P_{870} was no different in structure from the main light-harvesting Bchl (Clayton 1962*a*, 1963, 1966*a*; Clayton and Sistrom, 1966). There must therefore be something about the "special environment" that endowed the Bchl with its photochemical reactivity and its resistance to destruction.

One might raise the question whether the destructive treatments that Clayton and Kuntz and co-workers used could have created new "reaction center" Bchl, rather than simply unmasking a small amount of P_{870}. Clayton and Sistrom (1966) provided evidence against this in a study of a mutant of *Rp. sphaeroides* that was incompetent at photosynthesis, strain PM-8. Chromatophores from strain PM-8 appeared to lack P_{870}, because they did not exhibit any of the absorbance changes that could be induced by light or oxidants in chromatophores from the wild-type. Treatment of the PM-8 chromatophores with K_2IrCl_6 destroyed all their Bchl, without generating anything resembling P_{870}.

Clayton and Sistrom's studies of the nonphotosynthetic *Rp. sphaeroides* strain PM-8 provided a compelling argument that the photooxidation of P_{870} was a key step in bacterial photosynthesis. The mutant was quite capable of synthesizing Bchl and carotenoids, and it grew well aerobically, but photochemically it was inert. Its chromatophores were incapable of conducting any of the light-driven reactions of endogenous or exogenous electron donors and acceptors that chromatophores of the wild-type cells could perform (Sistrom and Clayton, 1964; Clayton *et al.*, 1965). Additional indications of the derangement of photochemical activity in the mutant were apparent from an examination of the fluorescence from the antenna Bchl. In wild-type cells or chromatophores, the yield of fluorescence was quite low, apparently because energy that was captured by the antenna Bchl was directed efficiently into photochemistry at the reaction center (Vredenberg and Duysens, 1963). On continued illumination, the fluorescence yield typically increased, after an induction period that appeared to reflect the saturation of the photochemical apparatus and that paralleled the photooxidation of P_{870}. In strain PM-8, on the other hand, the fluorescence yield was relatively high, right from the start of illumination, and it did not change during continued illumination (Clayton, 1966*b*; Sistrom and Clayton, 1964).

Clearly, the lack of P_{870} in strain PM-8 was accompanied by a catastrophic failure of photochemical activity. Unfortunately, the absence of information on the actual genetic lesion in the mutant prevented the distinction between cause and effect in this relationship from being completely clear. The recent finding (Clayton and Haselkorn, 1972) that PM-8 lacks not just one, but all three, of the polypeptides of the reaction center hints at the possible complexity of the situation.

There were, in fact, reasons to ask whether the photooxidation of P_{870} might be a side reaction that occurred only when another, more fundamental reaction was prevented or saturated. The difficulty was that the photooxidation generally could not be detected in suspensions of intact cells, except after illumination with very intense light. If one illuminated with more moderate intensity, one observed the cytochrome photooxidation instead. Olson (1962) had measured the quantum yield of the cytochrome photooxidation in *Chr. vinosum*, and found it to be very close to 1.0, and there was no convincing evidence that P_{870} played any role in this process.

A kinetic study of Beugeling and Duysens (1966) illustrated the same point. When they illuminated suspensions of *Chr. vinosum* cells in a mixture of glycerol and potassium glycerolphosphate, the cytochrome photooxidation commenced immediately. The oxidation of P_{870}, on the other hand, began only after the cytochrome was largely oxidized. Chromatophores of *Chr. vinosum* also performed cytochrome photooxidation in response to a weak and brief flash, but exhibited P_{870} photooxidation only after prolonged illumination (Clayton, 1962*b*). Similarly, Arnold and Clayton (1960) had found that suspensions of *Rp. sphaeroides* cells in water exhibited only the cytochrome photooxidation. To detect the photooxidation of P_{870}, it was necessary to add an inhibitor that abolished the cytochrome photooxidation, such as hydroxylamine or sodium azide. In *Rs. rubrum*, P_{870} photooxidation could be seen at moderate light intensity if the cells were suspended in NaCl solution, but only at extremely high intensities if the suspension was supplemented with reducing substrates (Duysens, 1952; Duysens et al., 1956). In the presence of reducing substrates, the cytochrome photooxidation required only weak illumination.

Of course, there were two possible interpretations of these observations. First, P_{870} photooxidation could be the "primary" photochemical reaction of bacterial photosynthesis, if one postulated that the cytochromes ordinarily reduced the oxidized P_{870} very rapidly. If this happened, P_{870^+} would not accumulate until the cytochromes had been drained of electrons. Alternatively, the cytochrome photooxidation might be the primary reaction, and the oxidation of P_{870} might be a side reaction of little physiological importance. The demonstration that P_{870} photooxidation could occur at cryogenic temperatures did not settle the question, because Chance and Nishimura (1960) had found that the cytochrome photooxidation also occurred at low temperatures.

To distinguish between the two possibilities, fast kinetic measurements were necessary. If the photooxidation of P_{870} was an intermediate step in the photooxidation of the cytochrome, P_{870^+} should appear very rapidly after a short flash of light, and the reduction of the P_{870^+} should occur simultaneously with the oxidation of the cytochrome.

Studies of the cytochrome oxidation kinetics initially gave misleading results. Using continuous illumination, Chance and Nishimura (1960) observed a half-time of several seconds for the photooxidation of the low-potential cytochrome C552 in Chr. vinosum, both at room temperature and at 77°K. From the observation that the rate did not change with temperature, they concluded that the cytochrome oxidation was a primary photochemical event. If this was correct, the role of Bchl presumably would be that of an electron acceptor, rather than a donor. Vredenberg and Duysens (1964), however, reported that the cytochrome oxidation rate was independent of temperature only if the exciting light was relatively weak. With stronger illumination, the rate decreased with decreasing temperature. The rates that Chance and Nishimura (1960) had measured were evidently limited by the rate of excitation, rather than by the true kinetics of the electron-transfer reaction.

Although Vredenberg and Duysens (1964) concluded correctly that the cytochrome oxidation was not a photochemical event, their measurements also appear to have been limited by something other than the electron-transfer rate. This became clear when Chance and DeVault (1964) used a Q-switched laser flash for excitation, and measured a half-time of 20 μsec—some 10^5 times shorter than the half-times that either Chance and Nishimura (1960) or Vredenberg

and Duysens (1964) had obtained with continuous excitation. In a more complete study two years later, DeVault and Chance (1966) obtained even faster kinetics for cytochrome C552 photooxidation, indicating a half-time of 2 μsec. Later work showed that even this underestimated the rate by a factor of 2 (Parson and Case, 1970; Seibert and DeVault, 1970).

When technical developments allowed fast kinetic measurements of absorbance changes in the near-IR, it became clear that the photooxidation of P_{870} did precede that of the cytochrome (Parson, 1968). In Chr. vinosum chromatphores, P_{870} photooxidation was complete within 0.5 μsec after a flash. The P_{870^+} returned to the reduced state with a half-time of 2 μsec, and the high-potential cytochrome C555 became oxidized simultaneously. Using sufficiently fast and sensitive detection techniques, P_{870} photooxidation could be seen with weak flashes, as well as with strong ones, and the quantum yield appeared to be essentially the same as that of the cytochrome photooxidation (Parson, 1968). Later work (Parson and Case, 1970; Seibert and DeVault, 1970) demonstrated that the kinetics of photooxidation and reduction of P_{870} could be measured spectrophotometrically at wavelengths other than the 882 nm that was used initially. The kinetic correlation between the oxidation of a cytochrome and the reduction of P_{870^+} was also extended to the low-potential cytochrome C552 in Chr. vinosum, both at room temperature (Parson and Case, 1970; Case et al., 1970) and at cryogenic temperatures (Dutton, 1971; Dutton et al., 1971). More recently, Dutton et al. (1975) established a similar kinetic correlation for the oxidation of cytochrome c_2 in Rp. sphaeroides.

Reexamination of the temperature dependence of the cytochrome oxidation kinetics in several different species has shown that the rates do decrease gradually with falling temperature, at least down to a point near 130°K (DeVault and Chance, 1966; DeVault et al., 1967; Kihara and McCray, 1973). Below this point, the rates in some cases become independent of temperature.

Taken together, the studies outlined above leave little doubt that the photooxidation of P_{870} is the primary photochemical reaction of bacterial photosynthesis. The cytochrome oxidation fits into place as a secondary, dark reaction that returns an electron to the reaction center Bchl. These conclusions set the stage for inquiries into the structure of the reaction center complex, the succession of excited states through which P_{870} travels before it expels an electron, and the nature of the acceptor that captures the electron that P_{870} releases. For investigations into these topics, the use of K_2IrCl_6 or photooxidation to free the reaction center from the antenna Bchl has given way to gentler and more discriminating techniques employing detergents, and highly purified preparations have become available from many different species of photosynthetic bacteria.

References

Arnold, W., and Clayton, R. K., 1960, The first step in photosynthesis: Evidence for its electronic nature, *Proc. Natl. Acad. Sci. U.S.A.* **46**:769.

Beugeling, T., and Duysens, L. N. M., 1966, P890 and cytochrome *c* in *Chromatium*, in: *Currents in Photosynthesis* (J. B. Thomas and J. C. Goedheer, eds.), pp. 59–65, Donker, Rotterdam.

Calvin, M., and Androes, G. M., 1962, Primary quantum conversion in photosynthesis, *Science* **138**:867.

Case, G. D., Parson, W. W., and Thornber, J. P., 1970, Photooxidation of cytochromes in reaction center preparations from *Chromatium* and *Rhodopseudomonas viridis*, *Biochim. Biophys. Acta* **223**:122.

Chance, B., and DeVault, D., 1964, On the kinetics and quantum efficiency of the chlorophyll–cytochrome reaction, *Ber. Bunsenges. Phys. Chem.* **68**:722.

Chance, B., and Nishimura, M., 1960, On the mechanism of chlorophyll–cytochrome interaction: The temperature insensitivity of light-induced cytochrome oxidation in *Chromatium*, *Proc. Natl. Acad. Sci. U.S.A.* **46**:19.

Chance, B., and Smith, L., 1955, Respiratory pigments of *Rhodospirillum rubrum*, *Nature (London)* **175**:803.

Clayton, R. K., 1962a, Primary reactions in bacterial photosynthesis. I. The nature of light-induced absorbancy changes in chromatophores; evidence for a special bacteriochlorophyll component, *Photochem. Photobiol.* **1**:201.

Clayton, R. K., 1962b, Primary reactions in bacterial photosynthesis. III. Reactions of carotenoids and cytochromes in illuminated bacterial chromatophores, *Photochem. Photobiol.* **1**:313.

Clayton, R. K., 1963, Toward the isolation of a photochemical reaction center in *Rhodopseudomonas spheroides*, *Biochim. Biophys. Acta* **75**:312.

Clayton, R. K., 1966a, Spectroscopic analysis of bacteriochlorophylls *in vitro* and *in vivo*, *Photochem. Photobiol.* **5**:669.

Clayton, R. K., 1966b, Fluorescence from major and minor bacteriochlorophyll components *in vivo*, *Photochem. Photobiol.* **5**:679.

Clayton, R. K., and Haselkorn, R., 1972, Protein components of bacterial photosynthetic membranes, *J. Mol. Biol.* **68**:97.

Clayton, R. K., and Sistrom, W. R., 1966, An absorption band near 800 mμ associated with P870 in photosynthetic bacteria, *Photochem. Photobiol.* **5**:661.

Clayton, R. K., Sistrom, W. R., and Zaugg, W. S., 1965, The role of a reaction center in photochemical activities

of bacterial chromatophores, *Biochim. Biophys. Acta* **102**:341.

DeVault, D., and Chance, B., 1966, Studies of photosynthesis using a pulsed laser. I. Temperature dependence of cytochrome oxidation rate in *Chromatium*. Evidence for tunneling, *Biophys. J.* **16**:825.

DeVault, D., Parkes, J. H., and Chance, B., 1967, Electron tunnelling in cytochromes, *Nature (London)* **215**:642.

Dutton, P. L., 1971, Oxidation–reduction potential dependence of the interactions of cytochromes, bacteriochlorophyll and carotenoids at 77°K in chromatophores of *Chromatium* D and *Rhodopseudomonas gelatinosa*, *Biochim. Biophys. Acta* **226**:63.

Dutton, P. L., Petty, K. M., Bonner, H. S., and Morse, S. D., 1975, Cytochrome c_2 and reaction center of *Rhodopseudomonas spheroides* Ga membranes: Extinction coefficients, content, half-reduction potentials, kinetics and electric field alterations, *Biochim. Biophys. Acta* **387**:536.

Dutton, P. L., Kihara, T., McCray, J. A., and Thornber, J. P. T., 1971, Cytochrome C_{553} and bacteriochlorophyll interaction at 77°K in chromatophores and a subchromatophore preparation from *Chromatium* D, *Biochim. Biophys. Acta* **226**:81.

Duysens, L. N. M., 1951, Transfer of light energy within the pigment systems present in photosynthesizing cells, *Nature (London)* **168**:548.

Duysens, L. N. M., 1952, Transfer of excitation energy in photosynthesis, Thesis, Utrecht.

Duysens, L. N. M., 1953, Reversible changes in the light absorption of purple bacteria caused by illumination, *Carnegie Inst. Washington Yearb.* **52**:157.

Duysens, L. N. M., 1954, Reversible photo-oxidation of a cytochrome pigment in photosynthesizing *Rhodospirillum rubrum*, *Nature (London)* **173**:692.

Duysens, L. N. M., 1958, The path of light in photosynthesis, in: *The Photochemical Apparatus: Its Structure and Function, Brookhaven Symp. Biol.* **11**:10.

Duysens, L. N. M., Huiskamp, W. J., Vos, J. J., and van der Hart, J. M., 1956, Reversible changes in bacteriochlorophyll in purple bacteria upon illumination, *Biochim. Biophys. Acta* **19**:188.

Emerson, R., and Arnold, W., 1932a, A separation of the reactions in photosynthesis by means of intermittent light, *J. Gen. Physiol.* **15**:391.

Emerson, R., and Arnold, W., 1932b, The photochemical reaction in photosynthesis, *J. Gen. Physiol.* **16**:191.

Förster, T., 1951, *Fluoreszenz organischer Verbindungen*, Vandenhoeck and Ruprecht, Göttingen.

Goedheer, J. C., 1958, Reversible oxidations of pigments in bacterial chromatophores, in: *The Photochemical Apparatus: Its Structure and Function, Brookhaven Symp. Biol.* **11**:325.

Goedheer, J. C., 1960, Spectral and redox properties of bacteriochlorophyll in its natural state, *Biochim. Biophys. Acta* **38**:389.

Kihara, T., and McCray, J. A., 1973, Water and cytochrome oxidation–reduction reactions, *Biochim. Biophys. Acta* **292**:297.

Kok, B., 1961, Partial purification and determination of the oxidation–reduction potential of the photosynthetic chlorophyll complex absorbing at 700 mμ, *Biochim. Biophys. Acta* **48**:527.

Kuntz, I. D., Jr., Loach, P. A., and Calvin, M., 1964, Absorption changes in bacterial chromatophores, *Biophys. J.* **4**:227.

McElroy, J. D., Mauzerall, D. C., and Feher, G., 1974, characterization of primary reactants in bacterial photosynthesis. II. Kinetic studies of the light-induced EPR signal (g = 2.0026) and the optical absorbance changes at cryogenic temperatures, *Biochim. Biophys. Acta* **333**:261.

Olson, J. M., 1962, Quantum efficiency of cytochrome oxidation in a photosynthetic bacterium, *Science* **135**:101.

Olson, J. M., and Chance, B., 1960a, Oxidation–reduction reactions in the photosynthetic bacterium *Chromatium*. I. Absorption changes in whole cells, *Arch. Biochem. Biophys.* **88**:26.

Olson, J. M., and Chance, B., 1960b, Oxidation–reduction reactions in the photosynthetic bacterium *Chromatium*. II. Dependence of light reactions on intensity of irradiation and quantum efficiency of cytochrome oxidation, *Arch. Biochem. Biophys.* **88**:40.

Parson, W. W., 1967, Flash-induced absorbance changes in *Rhodospirillum rubrum* chromatophores, *Biochim. Biophys. Acta* **131**:154.

Parson, W. W., 1968, The role of P$_{870}$ in bacterial photosynthesis, *Biochim. Biophys. Acta* **153**:248.

Parson, W. W., and Case, G. D., 1970, In *Chromatium*, a single photochemical reaction center oxidizes both cytochrome C$_{552}$ and cytochrome C$_{555}$, *Biochim. Biophys. Acta* **205**:232.

Seibert, M., and DeVault, D., 1970, Relations between the laser-induced oxidations of the high and low potential cytochromes of *Chromatium* D, *Biochim. Biophys. Acta* **205**:220.

Sistrom, W. R., and Clayton, R. K., 1964, Studies on a mutant of *Rhodopseudomonas spheroides* unable to grow photosynthetically, *Biochim. Biophys. Acta* **88**:61.

Vredenberg, W. J., and Duysens, L. N. M., 1963, Transfer of energy from bacteriochlorophyll to a reaction centre during bacterial photosynthesis, *Nature* (*London*) **197**:355.

Vredenberg, W. J., and Duysens, L. N. M., 1964, Light-induced oxidation of cytochromes in photosynthetic bacteria between 20 and −170°, *Biochim. Biophys. Acta* **79**:456.

Wassink, E. C., Katz, E., and Dorrestein, R., 1939, Infrared absorption spectra of various strains of purple bacteria, *Enzymologia* **7**:113.

Witt, H. T., Müller, A., and Rumberg, B., 1961, Oxidized cytochrome and chlorophyll in photosynthesis, *Nature* (*London*) **192**:967.

CHAPTER 16

Energy-Migration Mechanisms in Antenna Chlorophylls

A. Yu. Borisov

1. Introduction

The most ancient photochemist on our planet is Nature. Nature started from the very beginning and improved its mastery by the method of trial and error. Photosynthetic organisms are Nature's highest achievements in this field. They can transform solar light into types of energy utilizable in the conditions of our civilization.

What is the efficiency of such processes? In fact, it is rather low—only a small percentage of light energy can eventually be stored in the overall biomass. But the last decade has brought us more encouraging evidence. A great number of studies demonstrated that during primary stages of photosynthesis, electronic excitation induced by light in chlorophyll (Chl) molecules efficiently generates pairs of opposite charges. These charges serve as starting points for a sequence of redox reactions in biomembranes, producing electrochemical gradients and hence an electrochemical type of energy. It is important to note that a considerable part of the light energy absorbed is present in these primary stages. Hence, in our age, the age of energy crisis, it becomes an important task of Science to interrupt the sequence of photosynthesis stages here and to use the preformed redox couple either for the generation of electric current or for the initiation of some energy-consuming chemical reactions providing fuel (e.g., hydrogen from water), useful material, and other needs.

On the other hand, it may appear reasonable to develop artificial light-converting systems using some findings and principles "invented" by Nature in photosynthesis. In fact, photosynthesis has a remarkable advantage—it does not require the highest degree of material purity and crystallinity as various semiconductor photocells do. This significant circumstance may play a decisive role in our future options from a technological point of view. Thus, it is of great importance to elucidate the detailed mechanisms of primary steps in photosynthesis, and to achieve optimum efficiency in solar energy conversion, especially in strong, direct sunlight.

One of the most important steps in the evolution of photosynthesis was the development of molecular arrays in which 100–1000 Chl molecules per electron-transport chain serve as molecular antennae. These antennae increase the solar-energy consumption per chain up to the maximum possible level. And an important question arises here: how is this process realized? How is the energy of electronic excitations induced by light in Chl antennae focused efficiently and collected by the input devices of electron-transport chains, the reaction centers? That is the main problem of the first photophysical stage of photosynthesis, in both plants and bacteria.

So far, much evidence has been accumulated proving that the main principles of these processes are similar for purple bacteria and photosystem I of plants. But purple bacteria (as compared with green bacteria, algae, and higher plants) give many advan-

A. Yu. Borisov · Moscow State University, Moscow, USSR

tages in investigating primary processes. Among these advantages are:

1. Purple bacteria are characterized by the highest content of reaction centers and other components of electron-transport chains with respect to antenna Chl. Hence, the amplitudes and kinetics of their redox and other reactions can be checked with greater precision, and the conclusions obtained are the most reliable.

2. These organisms appear to have only one kind of photochemical system. It is therefore relatively easy to correlate fluorescence data with those obtained for redox transitions of the reaction center to make definite conclusions about the mechanisms of primary processes.

3. Absorption maxima of the main bacteriochlorophyll (Bchl) forms are well resolved. Hence, quantitative measurements of the amounts of respective molecules, and of energy-migration efficiencies between various Bchl forms, as well as some conclusions of the state of Bchl *in vivo*, can be made (see Thornber *et al.*, Chapter 7).

4. The dimensions of the cells are small, and the long-wave absorption bands of their Bchl components are located in the region of 800–900 nm and even up to 1000 nm. That is why of all photosynthetic organisms, they exhibit minimal light scattering. The optical data obtained even with suspensions of intact cells are not very much obscured, and those obtained with their chromatophores and subchromatophore particles are not obscured at all, by scattering.

5. Photosynthesis in purple bacteria and their particles can be saturated by relatively low light intensities. This facilitates their investigation considerably, because saturating light intensities can be obtained easily in any part of the optical region.

Purple bacteria are therefore to be preferred for fundamental research in the field of primary photosynthetic events.

Let us now survey the possible mechanisms of energy migration in antenna Chl, of photosynthetic bacteria.

2. Energy-Migration Mechanisms

It is well established that purple bacteria have on an average 30–200 and green bacteria 1000–1500 light-harvesting Bchl molecules per reaction center. The proportion of light absorbed directly by the reaction center is therefore very small, if not negligible. But the yield of photosynthesis is known to be not small; studies carried out in several laboratories have revealed that the quantum yields of reaction center and cytochrome photooxidation, quinone reduction,

and other processes are within 0.1–1.0 in purple bacteria and their chromatophores. These facts mean that efficient energy migration, delivering light-induced electronic excitations from antenna Bchl to reaction centers, is at work.

The first extensive study of this problem was undertaken by L. N. M. Duysens. In his classic treatise, Duysens (1952) clearly demonstrated that many phenomena, like energy migration from accessory pigments to Chl or between different Chl forms (both in bacteria and plants), can be qualitatively explained in terms of the inductive resonance theory advanced by Förster (1948). It was a very constructive and elucidating approach, the most important in that decade. But these experiments did not prove irrevocably that Förster's mechanism really works in photosynthesis.

Duysens (1952) established for *Chromatium* cells that the yield of fluorescence emitted by antenna Bchl component B890 is the same for exciting light absorbed by Bchl components B800 and B850. This finding was recently confirmed in other works (Goedheer, 1973; Fowler *et al.*, 1973) and in the author's laboratory for *Chromatium minutissimum* and *Ectothiorhodospira shaposhnikovii*. Two important conclusions may be derived from these data:

1. The energy migration from B800 and B850 to B890 proceeds via the first excited singlet level.

2. The quantum yield of energy transfer from B800 and B850 to B890 is nearly equal to unity.

Light saturation of photosynthesis usually brings about two- to threefold increase in the fluorescence of the longest-wavelength Bchl form (Vredenberg and Duysens, 1963; Sybesma and Vredenberg, 1963; Clayton, 1967a,b; Barsky and Borisov, 1971). This means that at least a considerable part of the excitation is delivered to reaction centers via the first singlet-excited level of Bchl. Hence, we shall be concerned in the following discussion only with singlet mechanisms of energy migration. In fact, triplet and semiconductivity energy-transfer mechanisms seem hardly to be realistic in Chl. There is no evidence that the quantum yield of photoionization for complex organic molecules (like Chl's) may be close to unity even in the most condensed state of crystallinity, the more so as most of the Chl or Bchl molecules *in vivo* do not exist as condensed crystals. The semiconductivity mechanism will therefore be ruled out from further considerations. As to exchange-resonance energy transfer, no experimental evidence is as yet available to prove that photoinduced excitations may be delivered from antenna Bchl to reaction centers via the triplet level. On the contrary, the estimated upper limit (Borisov and Godik, 1972) for the quantum yield of triplet formation in some purple bacteria does not exceed 5%.

Now let us specify the essential requirements that excitation-migration mechanisms should meet.

3. Homogeneous Energy Migration

Consider a Chl array containing many traps. On an average, N molecules ($N \gg 1$) contain $N - 1$ light-harvesting or antenna Chl molecules and one trap molecule.

3.1. Slow Inductive Resonance: Förster Type of Energy Migration

This means that the excitation is always localized in a definite molecule, and the jump time ($\Delta\tau_j$), i.e., the mean time interval between two successive jumps of excitation, exceeds 10^{-12} s. Each excitation jump may result either in hitting a trap or in excitation of another antenna Chl molecule, $1/N$ and $1 - 1/N$ being respective probabilities. Evidently, the probability that after m successive jumps electronic excitation will not be trapped amounts to $(1 - 1/N)^m$. Taking into account that

$$(1 - 1/x)^x \rightarrow e^{-1} \qquad \text{if } x \rightarrow \infty$$

the mean number of the jumps, n_m, needed for trapping $1 - e^{-1}$ ($\simeq 63\%$) of the excitation quanta is equal to N. Thus, the following approximation is valid for mean excitation lifetime:

$$\tau_{ex} = \tau_{fl} = n_m \cdot \Delta\tau_j \simeq N \cdot \Delta\tau_j$$

where τ_{fl} is the mean fluorescence lifetime. It is possible that the equation $n_m \simeq 0.72\, N \log N + 0.26\, N$, obtained in works by Pearlstein, Robinson, and Knox (see, e.g., Robinson, 1967) for two-dimensional molecular arrays, is more precise, but the difference is well within the uncertainty in n_m due to real features of Chl complexes *in vivo* (see below).

Equation (1) is valid for arrays in which separate molecules are similarly oriented and located with respect to their neighbors. What troubles may arise with photosynthetic objects?

1. It is hardly possible that Chl *in vivo* forms highly regular arrays. Hence, various Chl molecules may have different numbers of neighbors at different distances. Besides, it is currently accepted that in purple bacteria, the reaction centers contain not one but two photoactive Bchl molecules. The quantity of adjoining molecules may therefore increase, thus reducing the number of excitation jumps needed.

2. There is no evidence that the reaction centers can be regarded as absolute traps. Hence, the trapping efficiency, ϕ_{tr}, may amount to

$$\phi_{tr} = \frac{K_{tr}}{\beta \cdot K_m + K_{tr} + K_\Sigma} \simeq \frac{K_{tr}}{\beta \cdot K_m + K_{tr}} \qquad (2)$$

where K_{tr}, K_m, and K_Σ are the first-order rate constants for, respectively, electronic excitation in the reaction center to be trapped, to migrate back to the antenna Chl, and to be deactivated via trivial molecular mechanisms. The constant β is the mean number of neighboring molecules that could intercept excitation.

Taking into account all the considerations above, the following equations may be advanced:

$$\tau_{fl} \simeq n_m \cdot \Delta\tau_j(\phi_{tr})^{-1} \cdot \gamma \qquad (3a)$$

$$\Delta\tau_j \simeq (\beta \cdot K_m)^{-1} \qquad (3b)$$

The coefficient γ indicates that the precise picture of energy migration may depend on the intermolecular distances and mutual orientations, on the two- (or three-?) dimensional pattern of Chl organization, on accessibility of the reaction centers to excitations induced by light in various Chl domains, as was demonstrated by Goedheer (1957), and on the positions of reaction centers in Chl domains. It appears that the uncertainty may be as much as $2 \geqslant \gamma \geqslant \frac{1}{3}$, but, more likely, $\frac{3}{2} \geqslant \gamma \geqslant \frac{1}{3}$, because (as was suggested by G. Seely) Nature could improve the Chl complex in the course of evolution, thus facilitating energy migration to the reaction center. Facilitation corresponds to $\gamma < 1$.

According to Wraight and Clayton (1972), the quantum yield of reaction center photooxidation amounts to 1.02 ± 0.04, i.e., $\geqslant 0.98$. Consequently

$$\tau_{fl}^{RC} = 1/(K_{tr} + K_\Sigma) \simeq 1/K_{tr} \qquad (3c)$$

where τ_{fl} and τ_{fl}^{RC} are, respectively, the Bchl fluorescence lifetimes for intact cells or chromatophores and reaction center particles.

The experimental approach to the determination of n_m was advanced in the work of Ebrey and Clayton (1969). The fluorescence polarization of Bchl was measured to be $+0.43$ *in vitro* and $+0.08$ in chromatophores of *Rhodopseudomonas sphaeroides* R-26.

If attributed to energy migration in randomly oriented Bchl, this *in vivo* polarization decrease for chromatophores corresponds to 5–8 excitation jumps, which is rather below the figures that may be obtained from equation (3a). It is possible that either an exciton type of energy migration or some degree of regularity in the Bchl complex *in vivo* (or both) is responsible for underestimation of the n_m value derived from fluorescence polarization data. The latter phenomenon was proved to occur in photosystem I of green plants in the work of Junge and Eckhof (1974).

3.2. Exciton Type of Energy Migration

For the exciton type of migration the intermolecular interaction exceeds $kT \simeq 0.026$ eV; excitation is delocalized over a molecular ensemble $\gg N$. For strong intermolecular coupling (W $\gtrsim 0.1$ eV as in ionic crystals and, to some extent, in molecular ones, like benzene), the velocity of exciton propagation increases greatly and may even approach the velocity of light. Hence, in this limit

$$\tau_{ex} = \tau_{fl} \to \tau_{fl}^{RC} \qquad (4)$$

This means that energy trapping is the limiting step.

But it appears doubtful that fairly complicated porphyrin molecules may provide such strong mutual coupling even in the most condensed crystalline state.

3.3. Intermediate Case (Localized Excitons)

In this case, the excitation is spread over (or localized in) a limited group of molecules, several to several hundreds. It exhibits features of both the exciton and the slow inductive resonance types of energy migration.

The energy of molecular interaction is of the order of 10^{-3}–10^{-2} eV (10–100 cm^{-1}), the rate constant of pairwise energy migration, 10^{12}–10^{13} sec^{-1}, being moderately temperature-dependent (Förster, 1948, 1960). This fact is consistent with an early observation (Arnold and Clayton, 1960) that the efficiency of reaction center photooxidation in *Rp. sphaeroides* was not drastically inhibited by lowering the temperature to 1°K. If the exciton extends over several to several dozens of Chl molecules, i.e., $\ll N$, the efficiency of its trapping by the reaction center is still governed by the rate of energy migration, the value of ϕ_{tr}, and the proportion of the reaction centers in an active state, as it is for slow inductive resonance. The random-walk method is then still applicable to the Chl complexes as the mathematical model of energy migration. On the contrary, for excitons covering an ensemble of Chl molecules $\gg N$, the trapping probability depends only on the rate constant of energy trapping in the reaction center and on the proportion of reaction centers in an active state (Paillotin, 1972; Borisov *et al.*, 1974).

4. Heterogeneous Energy Migration

Most purple bacteria have several (usually three) prominent antenna forms of Bchl. This circumstance greatly accelerates energy migration to the longest-wavelength Bchl form that includes the reaction centers. Such heterogeneous energy transfer is easily achieved, as was shown by Heathcote and Clayton (1976). These authors separated antenna particles (containing B800 and B850 Bchl's and no reaction centers) from wild-type *Rp. sphaeroides* and photoactive reaction center particles from their R-26 carotenoidless mutant. After dialysis of their mixture in order to remove detergent, they produced associates with photochemical activity (photooxidation of P870) partially restored. It is doubtful that very specific reconstruction of protein–pigment complexes was achieved. More likely, an unspecific association was sufficient to restore powerful energy migration from antenna pigments to P870. Such unspecificity, provided by using the longest-wave absorbing pigments (Chl's), which are characterized by critical intermolecular distances for energy migration up to 70–80 Å, may be one of the most important achievements of Nature in the course of evolution.

Efficient heterogeneous energy migration can be accounted for in two ways, which are discussed below.

4.1. Decrease in the Number of Energy Jumps

In the work of Borisov and Fetisova (1971), 14 models of two-dimensional pigment complexes were examined by the random-walk method. The probability matrix method was used in computer calculations. Each molecule could deliver its excitation to four neighbors, with probabilities being proportional to the corresponding integrals of the overlapping of the donor fluorescence spectrum with the absorption spectrum of the acceptor. Each pigment domain investigated contained 48 Chl molecules and one trap molecule in the middle. The 48 Chl molecules were subdivided into two, three, or four forms differing in their absorption maxima. For example, 800, 850, and 890 nm maxima were used in three-form models, the corresponding molecules being in ratios 28:12:8, 12:28:8, 16:20:12, 24:16:8, 20:20:8, which are approximately the same as in purple bacteria. The initial spreading of excitation was homogeneous (equal for each Chl molecule). The probability of energy losses in the course of energy migration to an absolute trap was set equal either to 10^{-3} per jump or to zero. The mean numbers of excitation jumps needed for decreases of excitation concentration to levels e^{-1}, e^{-2}, and 0.01 were calculated for each model and compared with those for corresponding homogeneous models. The following conclusions were obtained:

1. The greatest decrease in the number of energy jumps needed for trapping, up to four- to tenfold as

compared with a homogeneous model, was obtained with the models comprising three or four Bchl forms—B800, B850, B890, or B800, B830, B860, B890—forming concentric areas from the border of the domain to the trap inside. The models comprising two forms (B800 + B890) were less effective in focusing excitations to the trap; their decrease of jumps was not more than 2.5 to 4-fold.

2. For the models organized nonconcentrically, the decrease in energy jumps was equal to 2.5- to 4-fold. It thus appears reasonable than $n_m \simeq N/3$ should be used for purple bacteria containing three Bchl forms, such as *Chr. vinosum* strain D, *Chr. minutissimum*, *Rp. sphaeroides*, *Ectothiorhodospira shaposhnikovii*, and some others.

3. The decay of excitation concentration down to the 0.5 level is not exponential, but further decay from e^{-1} to e^{-2} and 0.01 levels is precisely exponential. This means that the rate of quasi-equilibrium state achievement is comparable with the one for energy trapping when $\phi_{tr} \to 1$. On the contrary, a quasi-equilibrium state is established if $\phi_{tr} \ll 1$ (even for $\phi_{tr} \simeq 0.3$–0.2). The experiments conducted by Zankel and Clayton (1969) with *Rp. sphaeroides* proved the ratio of emissions from B800, B850, and B870 to depend on which Bchl form absorbed the light or on the state of the reaction center. This means that the quasi-equilibrium state was not quite achieved, thus favoring the idea that *in vivo*, either $1.0 > \phi_{tr} > 0.5$, or the faster exciton energy migration is at work.

4.2. Paillotin's Model

Paillotin has accommodated the physical theory developed for crystals with a narrow conductance band to the *in vivo* case of relatively strong Chl interaction energy equal to $W \simeq 100$ cm$^{-1} \simeq 0.01$ eV (Paillotin, 1972). This W value was first predicted by Robinson (1967) and then grounded in experiments of Borisov and Godik (1970) for purple bacteria. The "density" of exciton distribution in Chl was introduced by Paillotin in an equilibrium state

$$P(i) = \frac{\exp(-E_i/KT)}{\sum_j^M \exp(-E_j/KT)} \qquad (5)$$

where $P(i)$ is the probability of excitation in the ith molecule, E_i and E_j are the energies of the singlet-excited state for corresponding "i" and "j" molecules, and M is the number of molecules covered by an exciton.

For the case $M > N$, the following formula was advanced:

$$\tau_{fl}^{min} = \tau_0[1 + P(p)(\tau_0/\tau_{fl}^{RC} - 1)]^{-1} \qquad (6)$$

(notation as used in this chapter), where τ_{fl}^{min} is the antenna fluorescence lifetime with the reaction centers in the active state, p denotes a molecule capable of trapping, τ_0 is the intrinsic radiative lifetime of Chl, equal to 20–30 nsec for Bchl a (Zankel *et al.*, 1968; Slooten, 1972) and to 18 nsec for the Bchl c of green bacteria (Fetisova and Borisov, 1973/1974).

The concrete form of equation (5) is very attractive, although it is desirable that the relevance of the model to *in vivo* Chl complexes be proved. Besides, this theory appears to be more appropriate for a nearly homogeneous Chl complex. The energy gap, even between B850 and B870 forms, is equal to 260 cm^{-1}. It means that an exciton cannot be spread over Bchl molecules belonging to both forms because its characteristic coupling energy ($\simeq 100$ cm^{-1}) is smaller. And, indeed, experiments (Zankel and Clayton, 1969) revealed that the ratio of fluorescence emissions coming from B800, B850, and B870 depends on which Bchl form has absorbed the light.

5. Experimental Data

Let us now compare the considerations discussed above with available experimental data, especially those obtained for fast kinetics in nano- and pico-second ranges.

5.1. Rate Constant of Trapping

Formally, the trapping rate is characteristic of the reaction center, because it is doubtful that the value of K_{tr} changes significantly in the course of reaction center separation from the remaining chromatophore material. We must bear in mind that the K_{tr} value *governs* the excitation lifetime (and hence the efficiency of trapping) in all Bchl complexes for excitons, and in the long-wave Bchl for slow inductive resonance as well. Hence, besides the trapping mechanism, this parameter of the reaction center must be considered in close connection with the energy-migration mechanism, as was done in the works of Paillotin (1972) and Borisov *et al.* (1974).

The only experimental data obtained as yet for K_{tr} are indirect estimates of the fluorescence lifetime of P870 Bchl of photoactive reaction center particles. The fluorescence yield (ϕ_{fl}^{RC}) of P870 was measured and the lifetime was obtained, by means of the formula $\tau_{fl}^{RC} = \phi_{fl}^{RC} \cdot \tau_0$, to be equal to 7 psec (Zankel *et al.*,

1968), or 9 psec (Slooten, 1972) for reaction center particles* of *Rp. sphaeroides* R-26.

Several factors could contribute to uncertainty in the value of this most important parameter of phososynthesis:

1. The pigments of the reaction center may form dimer, tetramer, or even hexamer due to specific interactions. This circumstance may substantially change the value of τ_0 as compared with that for a free Bchl molecule.

2. There is no experimental evidence as yet that the well-known equation $\tau = \phi \cdot \tau_0$ is valid in picosecond time intervals after the absorption of light. It may well be that in the course of Franck–Condon reorganization of the structure of excited molecules, the radiative probability varies dramatically. Hence, direct measurements of τ_{fl}^{RC} are urgently needed, especially their correlation with redox states of P870 and the primary electron acceptor.

5.2. Lifetime of Excitation in Chlorophyll and Bacteriochlorophyll Complexes *in Vivo*

Excitation lifetimes have been obtained *via* Chl fluorescence. Beginning from pioneering works (Brody and Rabinowitch, 1957; Dmitrievsky *et al.*, 1957), in more than ten subsequent publications, values of $\tau_{fl} \simeq 0.5$–2.0 nsec were measured for various plant systems with phase fluorometry techniques. Similar data were obtained for purple bacteria ($\tau_{fl} \simeq 1.0$ nsec; Rubin and Osnitskaja, 1963; Govindjee *et al.*, 1972) and green bacteria ($\tau_{fl} \simeq 2.0$ nsec; Müller *et al.*, 1969). The situation seemed to be clear in this field up to 1970, and considerations of the mechanism of *in vivo* energy-transfer naturally led one to the realm of slow inductive resonance. Only Robinson (1967) had predicted an excitonic mechanism of *in vivo* energy migration. His physical intuition prompted him to suggest that the *in vivo* absorption spectra of Chl's manifest relatively strong intermolecular coupling, which corresponds to an interaction energy $\simeq 100$ cm^{-1} and a jump time $\simeq 10^{-13}$ sec. To bring his considerations into agreement with the currently accepted data on Chl fluorescence lifetimes, he had to

conclude that the trapping efficiency of the reaction center is very low, about 10^{-2}.

The situation became ambiguous in 1970, however, when Godik and the author (Borisov and Godik, 1970) advanced new experimental data obtained with the same phase fluorometry techniques. In these measurements, the *mean* lifetime of fluorescence was observed to decrease while the yield increased, as reaction centers were converted from active to inactive states. This contradictory situation was resolved by proposing the existence of two components in the fluorescence. In addition to known fluorescence emission(s) with lifetime(s) of approximately 1 nsec, a new one, a very short-lived "photosynthetic" emission in the picosecond time scale, was thus implicated by measurements with four types of purple bacteria: *Rs. rubrum*, *E. shaposhnikovii*, *Chr. minutissimum*. and *Rp. sphaeroides*. The yield and lifetime of this fast component of fluorescence is governed by the state of the reaction center. The former emission(s) was defined as a background one, because it came from a small portion ($\simeq 5$–10%) of Bchl with approximately constant fluorescence yield ($\phi_{bg} \simeq 5\%$) and lifetime (τ_{bg}). It seems likely that this Bchl represents domains, small groups and even individual molecules, that have lost the ability to deliver their excitations to the reaction center. The τ_{bg} for domains was obtained to be approximately 0.5–1.5 nsec due to the presence of some quenching centers, whereas for small groups and individual Bchl molecules, it may be as high as 2–4 nsec. This "dead" component may be negligible in cells and chromatophores of high photochemical activity.

The most important values of τ_{fl}^{min}, representing the lifetime of excitation in the main part of Bchl with active reaction centers, were deduced for green and purple bacteria to be within 8–50 psec, increasing to not more than 100–150 psec for aerobic conditions with 50–70% of the reaction centers being photooxidized (Borisov and Godik, 1970, 1972).

The method advanced by Borisov and Godik suffers from the following main sources of possible errors in the determination of τ_{fl}^{min}:

1. Possible complex nature of the fluorescence, which may comprise several individual emissions differing in corresponding τ values.

2. Absence of knowledge of the validity of the well-known equation $\tau = \phi \cdot \tau_0$ in the picosecond time scale.

It should be noted that similar values of τ_{fl} were recently obtained for photosystem I of plants: pea chloroplasts, $\tau_{fl} \gtrsim 30$ psec (Borisov and Il'ina, 1973); *Chlorella* cells and spinach chloroplasts, $\tau_{fl} \gtrsim 10$ psec (Seibert and Alfano, 1974). But there is some evidence

---

* New experimental data appeared as this manuscript was being prepared. The trapping time in reaction centers was shown to be <20 psec (Rockley *et al.*, 1975) and $\gtrsim 7$ psec (Kaufmann *et al.*, 1975). Hence, it is either about 7–9 psec, as mentioned in the text, or considerably smaller (down to 10^{-14} sec as in charge-transfer complexes). In the latter case, the reaction center emission is of the delayed type due to reversal of the P^F state. Nevertheless, the trapping constant should be about 10^{-11} sec or somewhat larger for cells and chromatophores with prominent antenna complexes, characterized by moderate excitonic interactions between molecules.

(Campillo *et al.*, 1976) that in fluorescence measurements on the picosecond time scale, using mode-locked lasers, lifetimes may be underestimated as a result of biphotonic interactions that can occur at high intensities of excitation.

6. Discussion

6.1. Energy Migration in Long-Wavelength Bacteriochlorophyll Forms

The foregoing shows a serious dilemma to exist concerning the magnitude of τ_{fl}^{min}. Either it is of the order of 10^{-9} sec, or considerably shorter, 10–50 psec, as was obtained in the works of Borisov and Godik (1970, 1972) and Barsky *et al.* (1974). These contradictory values lead to essentially different conclusions on the mechanism of energy migration. The statement may be illustrated by the following analysis for purple bacteria.

For species with nearly homogeneous Chl, $n_m \simeq N$ should be used, thus providing: *Rp. sphaeroides*, R-26, $n_m \simeq 20$–25; *Rs. rubrum*, $n_m \simeq 200$. For most purple bacteria with three Bchl forms, $n_m = N/3$ was suggested above (Borisov and Fetisova, 1971). This gives $n_m \simeq 50$–80 for *Chr. vinosum* strain D, *Chr. minutissimum*, *Rp. sphaeroides* (wild-type), *E. shaposhnikovii*, and some others. Similar figures may be obtained by using Paillotin's approach. We shall choose "typical" parameters as follows: $n_m \simeq 65$

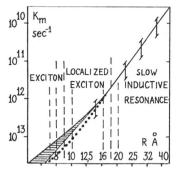

Fig. 1. Rate constant of pairwise energy migration vs. intermolecular distance. The left (above shaded area) and right parts of the curve exhibit $K_m \simeq R^{-3}$ (exciton) and $K_m \simeq R^{-6}$ (slow inductive resonance) dependences, respectively (Förster, 1960). The downward uncertainty in the exciton region (shaded area) is due to possible contributions of high multipole terms to the Chl interaction energy. The lower curve under the shaded area is in close agreement with a recent development of Förster's theory advanced by V. M. Kenkre and R. S. Knox [*Phys. Rev. Lett.* 33:803 (1974)]. According to these authors, the R^{-6} dependence is valid down to jump times $\simeq 10^{-13}$ sec in the exciton region (see the dotted curve).

(mean value for purple bacteria); $\beta = 4$ (assumed); and $\tau_{fl}^{RC} = K_{tr}^{-1} \simeq 10^{11}$ sec (close to the 7–9 psec obtained by Zankel *et al.*, 1968; Slooten, 1972).

By substituting the figures above into equations (3a–c), we obtain

$$\tau_{fl}^{min} = 65 \cdot \frac{1}{4 \cdot K_m} \cdot \frac{10^{11} + 4K_m}{10^{11}} \cdot \gamma \qquad (7)$$

1. $\tau_{fl}^{min} \simeq 10^{-9}$ sec.

For $0.33 \leqslant \gamma \leqslant 1.43$, this gives $7 \cdot 10^9 \gtrsim K_m \gtrsim 5 \cdot 10^{11}$ sec^{-1}; 36 psec $\gtrsim \Delta\tau_j \gtrsim 0.5$ psec; $0.78 \gtrsim \phi_{tr} \gtrsim 0.05$, and, according to T. Förster's equation $K_m = \tau_0^{-1}(R_0/R)^6$ (see Förster, 1948), the mean Bchl intermolecular distances (R) are within $38 \gtrsim R \gtrsim 18$ Å. Everything fits the slow inductive resonance theory.

2. $\tau_{fl}^{min} = 10$–50 psec

Now equation (7) can be solved only for γ values as low as 0.02–0.22, which are well below the region $0.33 \gtrsim \gamma \gtrsim 1.50$ justified in this chapter. This means that the slow inductive resonance is absolutely inconsistent with $\tau_{fl}^{min} \simeq 10$–50 psec, and the only possible conclusion is that some type of exciton energy migration is at work in this case (see Fig. 1 for details).

6.2. Energy Migration in Short-Wavelength Bacteriochlorophyll Forms

The shorter-wavelength positions of B800 and B850 of purple bacteria appear to indicate that the corresponding molecules are less condensed, their pairwise interaction energy being decreased as compared with B870 or B890. The rate of excitation transfer must be, obviously, the lowest in the B800 form. The molecules belonging to B850 and B870 or B890 may be considered as "traps" for excitations originally induced in B800. Taking into account that the amount of B850 + B890 usually exceeds that of B800, it may be concluded that on the average, not more than one or two excitation jumps in the latter form may precede its "trapping."

The experiments conducted by Dr. V. I. Godik with *Chr. minutissimum* and *E. shaposhnikovii* suspensions revealed that the portion of fluorescence emitted by B800 was no more than 0.7% of that from B890 for blue exciting light homogeneously absorbed by all the Bchl molecules under conditions of active photosynthesis. The following equation may reflect these data:

$$7 \times 10^{-3} = \frac{[B800^*]\phi_{800}}{[B890^*]\phi_{890}} \qquad (8)$$

where [B800*] and [B890*] are the quantities of excitation induced in the respective Bchl forms per unit of time, and ϕ_{800} and ϕ_{890} are fluorescence yields of the respective Bchl's.

In addition, the following equations are valid

$$\phi_{800} \simeq \tau_{800} \cdot (\tau_{800}^{i})^{-1} \quad \text{and} \quad \phi_{890} \simeq \tau_{890} \cdot (\tau_{890}^{i})^{-1} \quad (9)$$

where τ_{800} and τ_{890} are the fluorescence lifetimes in corresponding Bchl's, and τ_{800}^{i} and τ_{890}^{i} are the respective intrinsic radiative lifetimes. By combining equations (8) and (9) and taking into account that $\tau_{800}^{i} \simeq \tau_{850}^{i} \simeq \tau_{890}^{i}$ (Borisov and Godik, 1972), we may obtain

$$7 \times 10^{-3} \simeq \frac{[B800^{*}] \cdot \tau_{800}}{[B890^{*}] \cdot \tau_{890}} \simeq \frac{[B800^{*}] \cdot n_{800} \cdot \Delta\tau_{800}}{[B890^{*}] \cdot \tau_{fl}^{min}} \quad (10)$$

where $\Delta\tau_{800}$ and n_{800} correspondingly represent the mean jump time and the number of jumps in B800.

The ratio of B800 to all Bchl's $\simeq 1:3$ for *Chr. minutissimum* and *E. shaposhnikovii*. Consequently, for blue actinic light, $[B800^{*}]:[B890^{*}] \simeq 1:3$ if almost all quanta absorbed by B800 and B850 are delivered to B890. From earlier considerations, $n_m \simeq 1$. All these figures being substituted into equation (10) produce

$$\Delta\tau_{800} \simeq 2 \cdot 10^{-2} \cdot \tau_{fl}^{min}$$

For $\tau_{fl}^{min} \simeq 10^{-9}$ sec, energy migration in all Bchl forms is evidently of the slow inductive resonance type.

For $\tau_{fl}^{min} = 10{-}50$ psec (Borisov and Godik, 1970, 1972), the rate constant of energy migration in B800 (K_{800}^{m}) is equal to

$$K_{800}^{m} = (\Delta\tau_{800})^{-1} \simeq (1{-}5) \cdot 10^{12} \text{ sec}^{-1}$$

Taking account of relative interaction energies, the value of K_{850}^{m} should then be restricted in the following way:

$$K_{800}^{m} \simeq 2 \cdot 10^{12} < K_{850}^{m} < K_{890}^{m} \simeq 10^{13} \text{ sec}^{-1}$$

Similar K_{800}^{m} values may be obtained for *Rp. sphaeroides*, characterized by $F_{800}:F_{870} \simeq 10^{-2}$ under similar conditions (Zankel and Clayton, 1969).

Consequently, the energy migration in B800 and B850 proceeds at rates not much lower than in B870 or B890, in the semiexcitonic fashion.

The situation in green bacteria, in which the Bchl *c*-750/Bchl *a*-800 ratio equals 20–25:1, is quite different. Every 20–25 B750 molecules have on the average one "trapping" molecule of B800, and 40–50 of the latter have, in turn, one P840 trap. Consequently, $n_m \simeq 20{-}25$ and 40–50 should be used for B750 and B800, respectively. These figures provide $K_m \simeq 10^{13}$ sec^{-1} for both Bchl forms if $\tau_{fl}^{min} \simeq 15{-}30$ psec is used as was obtained in the works of Borisov and Godik (1972) and Barsky *et al.* (1974).

This principle of antenna Chl formation seems to be the most advantageous one. If one trap could efficiently serve, for example, 100 molecules and the latter could, in their turn, efficiently gather excitations from approximately 10^{4} shorter-wavelength Chl molecules (etc.?!), the quantum losses would increase proportionally to the net excitation lifetime, which would only double, while the light-harvesting antenna would increase by 100 times. These considerations appear to provide an explanation for the most developed antenna of green bacteria.

6.3. Conclusions

1. The precise mechanism of *in vivo* energy migration was shown to govern strictly the magnitude of the Chl fluorescence lifetime.

2. New experimental data suggest that these lifetimes for various photosynthetic organisms are in a picosecond time scale:

   a. The fluorescence lifetime of Bchl from reaction center particles separated from antenna pigment was first evaluated by Zankel *et al.* (1968) to be about 7×10^{-12} sec, and then by Slooten (1972) to be 9×10^{-12} sec.

   b. The fluorescence lifetimes of antenna Bchl's in cells and chromatophores of several purple bacteria and one green bacterium were first shown by Borisov and Godik (1970, 1972) to be within 10–100 psec in the state of active photosynthesis in contrast to the generally accepted values of about 10^{-9} sec.

3. If these fluorescence lifetime values are accepted, they prove irrevocably that the energy-migration mechanism in antenna Bchl's of photosynthetic bacteria (and photosystem I of plants as well) is of the exciton type at moderate Bchl interaction energies.

On the contrary, the excitation lifetime of reaction centers (7–9 psec) together with antenna fluorescence lifetimes of about 10^{-9} sec irrevocably manifest the slow inductive resonance type of energy migration.

Direct measurements of all the fluorescence lifetimes discussed herein, especially in relation to the proportion of active reaction centers, are therefore urgently needed to solve puzzles offered by the phase fluorometry technique and to arrive at a final understanding of energy-migration mechanisms in photosynthesis.

7. References

Arnold, W. W., and Clayton, R. K., 1960, The first step in photosynthesis: Evidence for its electronic nature, *Proc. Natl. Acad. Sci. U.S.A.* **46**:769–776.

Barsky, E. L., and Borisov, A. Yu., 1971, Determination of the quantum yields of the primary photosynthesis events and the photosynthetic unit types in purple bacteria, *J. Bioenerg.* **2**:275–281.

Barsky, E. L., Borisov, A. Yu., Fetisova, Z. G., and Samuilov, V. D., 1974, Spectral and energetic characteristics of the photoactive particles obtained from chromatophores of the green bacterium *Chlorobium limicola*, *FEBS Lett.* **43**(3):275–278.

Borisov, A. Yu., and Fetisova, Z. G., 1971, A study of resonance energy migration in heterogeneous pigment complexes, *Mol. Biol. (Soviet)* **5**(4):509–517.

Borisov, A., Yu., and Godik, V. I., 1970, Fluorescence lifetime of bacteriochlorophyll and reaction center photooxidation in purple bacterium, *Biochim. Biophys. Acta* **223**:441–443.

Borisov, A., Yu., and Godik, V. I., 1972, Energy transfer in bacterial photosynthesis, *J. Bioenerg.* **3**(3):211–220; **3**(6):515–523.

Borisov, A. Yu., and Il'ina, M. D., 1973, The fluorescence lifetime and energy migration mechanism in photosystem I of plants, *Biochim. Biophys. Acta* **305**:364–373.

Borisov, A., Yu., Godik, V. I., and Fetisova, Z. G., 1974, Determination of the quantum yield of the primary process of photosynthesis. I. Applicability of the technique for various PSU types, *Mol. Biol. (Soviet)* **8**(3):458–466.

Brody, S. S., and Rabinowitch, E., 1957, Excitation lifetime of photosynthetic pigments *in vitro* and *in vivo*, *Science* **125**:555–559.

Campillo, A. J., Shapiro, S. L., Kollman, V. H., Winn, K. R., and Hyer, R. C., 1976, Picosecond exciton annihilation on photosynthetic systems, *Biophys. J.* **16**:93–97.

Clayton, R. K., 1967a, Fluorescence from major and minor bacteriochlorophyll components *in vivo*, *Photochem. Photobiol.* **5**:679–688.

Clayton, R. K., 1967b, Relations between photochemistry and fluorescence in cells and extracts of photosynthetic bacteria, *Photochem. Photobiol.* **5**:807–821.

Dmitrievsky, O. D., Ermolaev, V. L., and Terenin, A. N., 1957, Fluorescence lifetime of chlorophyll *in vivo*, *Dokl. Akad. Nauk SSSR* **114**:75–78.

Duysens, L. N. M., 1952, Transfer of excitation energy in photosynthesis, Thesis, Utrecht.

Ebrey, T. G., and Clayton, R. K., 1969, Polarization of fluorescence from bacteriochlorophyll in castor oil, in chromatophores and as P870 in reaction centers, *Photochem. Photobiol* **10**:109–117.

Fetisova, Z. G., and Borisov, A. Yu., 1973/1974, The intrinsic lifetimes of bacterioviridin-660 and chlorophyll-*a* in different solvents, *J. Photochem.* **2**:151–155.

Förster, T., 1948, Zwischenmolekulare Energiewanderung und Fluoreszenz, *Ann. Phys.* **2**:55–75.

Förster, T., 1960, Excitation transfer, in: *Comparative Effects of Radiation*, pp. 300–341, John Wiley and Sons, New York and London.

Fowler, C. F., Gray, B. H., Nugent, N. A., and Fuller, R. C., 1973, Absorbance and fluorescence properties of the bacteriochlorophyll *a* reaction center complex and bacteriochlorophyll *a*-protein in green bacteria, *Biochim. Biophys. Acta* **292**:692–699.

Goedheer, J. C., 1957, Thesis, Utrecht.

Goedheer, J. C., 1973, Fluorescence polarization and pigment orientation in photosynthetic bacteria, *Biochim Biophys. Acta* **292**:665–676.

Govindjee, Hammond, J. H., and Merkelo, H., 1972, Lifetime of excited states *in vivo*. II. Bacteriochlorophyll in photosynthetic bacteria at room temperature, *Biophys. J.* **12**:809–814.

Heathcote, P., and Clayton, R. K., 1976, Reconstituted energy transfer from antenna pigment–protein to reaction centers isolated from *Rhodopseudomonas sphaeroides*, *Biochim. Biophys. Acta* **459**(3):506–515.

Junge, A., and Eckhof, W., 1974, Photoselection studies on the orientation of chlorophyll a_1 in the functional membrane of photosynthesis, *Biochim. Biophys. Acta* **357**:103–117.

Kaufmann, K. J., Dutton, P. L., Netzel, T. L., Leigh, J. S., and Rentzepis, P. M., 1975, Picosecond kinetics of events leading to reaction center bacteriochlorophyll oxidation, *Science* **188**:1301.

Müller, A., Lumry, R., and Walker, M. S., 1969, Light-intensity dependence of the *in vivo* fluorescence lifetime of chlorophyll, *Photochem. Photobiol.* **9**:113–126.

Paillotin, G., 1972, Transport and capture of electronic excitation energy in the photosynthetic apparatus, *J. Theor. Biol.* **36**:223–235.

Robinson, G. W., 1967, Excitation transfer and trapping in photosynthesis, *Brookhaven Symp. Biol.*, No. 19, pp. 16–48.

Rockley, M. G., Windsor, M. W., Cogdell, R. J., and Parson, W. W., 1975, Picosecond detection of an intermediate in the photochemical reaction of bacterial photosynthesis, *Proc. Natl. Acad. Sci. U.S.A.* **72**:2251.

Rubin, A. B., and Osnitskaja, L. K., 1963, On the interrelation between physiological state of purple bacteria and bacteriochlorophyll fluorescence lifetime, *Mikrobiologiya* **32**:200–203.

Seibert, M., and Alfano, R. R., 1974, Probing photosynthesis on a picosecond time scale, *Biophys. J.* **14**:269–276.

Slooten, L., 1972, Reaction center preparations from *Rp. sphaeroides*. Energy transfer and structure, *Biochim. Biophys. Acta* **256**:452–466.

Sybesma, C., and Vredenberg, W. J., 1963, Evidence for a reaction center P840 in the green photosynthetic bacterium *Ch. ethylicum*, *Biochim. Biophys. Acta* **75**:439–441.

Vredenberg, W. J., and Duysens, L. N. M., 1963, Transfer of energy from bacteriochlorophyll to a reaction center during bacterial photosynthesis, *Nature (London)* **197**:355–357.

Wraight, C. A., and Clayton, R. K., 1972, The absolute quantum efficiency of bacteriochlorophyll photooxidation in reaction centers, *Biochim. Biophys. Acta* **333**:246–260.

Zankel, K. L., and Clayton, R. K., 1969, Uphill energy transfer in a photosynthetic bacterium, *Photochem. Photobiol.* **9**:7–15.

Zankel, K. L., Reed, D. W., and Clayton, R. K., 1968, Fluorescence and photochemical quenching in photosynthetic reaction centers, *Proc. Nat. Acad. Sci. U.S.A.* **61**:1243–1249.

CHAPTER 17

Fluorescence and Energy Transfer

J. Amesz

1. Introduction

It has long been known that the absorption spectra of photosynthetic bacteria are in general more complicated than those of solutions of the combined pigments present in the cell (Wassink *et al.*, 1939; Giesberger, 1947; Clayton, 1964; Olson and Stanton, 1966). This is especially true for the IR spectra of most purple bacteria (Chromatiaceae and Rhodospirillaceae), which usually display several absorption bands in the region 790–890 nm, whereas a solution of bacteriochlorophyll *a* (Bchl *a*) (see Chapter 11) shows only one electronic absorption band in the near-IR region, located near 770 nm. The absorption spectrum of *Chromatium vinosum* strain D shows as many as five bands, situated near 800, 807, 823, 850, and 891 nm; at 100°K, these bands are located at 798, 807, 823, 855, and 905 nm (Vredenberg and Amesz, 1966) (see also Thornber *et al.*, Chapter 7). Other species examined so far show four or less bands. There are several indications that the different bands can be ascribed to the lowest singlet electronic transition of different Bchl *a* molecules. After the approximate location of their IR bands, these Bchl's were designated B800, B850, etc., by Duysens (1952). This designation, with wavelength rounded to the nearest 10 nm, has been widely adopted in the literature for the antenna Bchl's (that are not part of the so-called "reaction center"), and will be used in this chapter.

J. Amesz · Department of Biophysics, Huygens Laboratory University of Leiden, The Netherlands

Like many other porphyrins, Bchl fluoresces on excitation. Therefore, measurements of fluorescence provide an important method to obtain information about the transfer of excitation energy from other pigments to Bchl and between Bchl molecules. Probably the most obvious way to obtain this kind of information is by measurement of excitation spectra of fluorescence, since the shape of the excitation spectrum of a fluorescing pigment provides direct quantitative evidence about transfer to this pigment from other substances that absorb light in the region of measurement. Some 25 years ago, Duysens (1952) applied this method to study energy transfer in various species of purple bacteria. The method has also been used by others (e.g., Goedheer, 1959; 1973; Amesz and Vredenberg, 1966; Clayton and Sistrom, 1966). A second way to obtain information about energy transfer from fluorescence measurements is by determining yields or lifetimes of fluorescence of the energy-transferring pigments. Transfer of excitation energy to another molecule and fluorescence may be viewed as competitive processes leading to de-excitation from the lowest excited (singlet) state, and therefore the fluorescence yield and lifetime will be lowered if transfer of energy to another pigment occurs. Conversely, the fluorescence yield may rise strongly if energy transfer is blocked, e.g., by physical separation of the pigments (Thornber, 1970) or by bleaching of the pigment that acted as energy "sink" (Vredenberg and Duysens, 1963). Energy transfer between identical molecules cannot be observed by this method, but in this case, at least in principle, a third method, viz., measurement of fluorescence polarization, may be used. If the arrangement of energy-transferring molecules is random, as in solution, fluorescence polarization will be lowered by energy transfer. Although a quantitative treatment is complicated (Knox, 1968), the method can then be used to estimate the number of transfers. If this number is large, however, the extent of polarization will be too small to be measured with sufficient precision. For this reason, the method would not appear to be very suitable for photosynthetic bacteria to calculate the number of transfers between pigment molecules (see also Steffen and Calvin, 1971), but it has been used to obtain information about their orientation (Ebrey and Clayton, 1969; Goedheer, 1973; Michel-Villaz, 1976).

This chapter will discuss mainly experimental data obtained with intact cells and preparations of so-called "chromatophores" (vesicles obtained on disrupting the cell). Results obtained with so-called "reaction center preparations" are treated in Chapters 19–23. For a discussion of theoretical aspects of energy transfer in general and in photosynthetic systems, the reader is referred to Förster (1948, 1951), Robinson (1967), Pearlstein (1967), Hoch and Knox (1968), and Knox (1973; 1975). A more detailed discussion of some points to be mentioned here is given by Borisov in Chapter 16 and by Zankel in Chapter 18.

2. Energy Transfer between Antenna Pigments

2.1. Fluorescence Yield and Emission Spectra

For most purple bacteria (a designation that will be reserved here for species containing Bchl a as the only Bchl pigment), the emission spectrum* indicates that nearly all fluorescence excited by visible or near-IR radiation is emitted by the antenna Bchl absorbing at the longest wavelength: B870 or B890, depending on the species (Duysens, 1952; Goedheer, 1972). This point alone gives strong evidence that an efficient transfer of excitation energy occurs from Bchl absorbing at shorter wavelengths (such as B800 or B850) to

these pigments, since the intrinsic fluorescence yield (in the absence of energy transfer) of the other Bchl's a is approximately the same as of B870 or B890. Duysens (1952) observed that a cell-free preparation of *Chr. vinosum* strain D in which B890 had been destroyed by mild heating showed an emission band at 880 nm, apparently due to B850. Bril (1964) obtained comparable results by the action of detergents. Similar observations have been reported for chromatophore preparations from which B890 and the reaction center had been removed by detergent treatment (Thornber, 1970). Untreated chromatophores of *Chr. vinosum* strain D showed mainly emission by B890, with a maximum at 906 nm at 294°K and at 925 nm at 77°K, but a subchromatophore preparation prepared with sodium dodecyl sulfate that did not contain B890 and reaction centers as judged from the absorption and light-induced absorption difference spectrum showed a relatively strong emission band at 859 nm (884 nm at 77°K).

Fluorescence spectra of *Rhodopseudomonas sphaeroides* show, in addition to the band of B870, a fairly strong emission by B850 (Goedheer and van der Tuin, 1967; De Klerk *et al.*, 1969) and even by B800 (Zankel and Clayton, 1969). Emission by B850 was most clearly seen at low intensity of excitation (Zankel and Clayton, 1969), when the yield of B870 fluorescence is relatively low due to trapping of energy by the reaction center (see Section 3). Some emission by B850 could be detected on excitation of B890 with light of 890 or 910 nm, demonstrating so-called "uphill" transfer of energy from B870 to B850, albeit at a lower rate than the "downhill" transfer from B850 to B870. The lower rate of "uphill" transfer is easily explained by the smaller overlap between the emission and absorption bands of the emitter and receptor molecule, respectively, a factor which the rate of energy transfer according to the Förster theory of inductive resonance is proportional (Förster, 1948, 1951). The relative contributions of B850 and B870 to the emission spectrum on excitation of B870 were the same as to that of the so-called "variable fluorescence" (i.e., the increment of fluorescence observed on closing the reaction center traps) on excitation with blue light. This indicates that energy transfer to the traps proceeds via B870. Little if any emission by B850 can be observed in the emission spectra of *Chr. vinosum* strain D and *Rs. molischianum* or in the emission spectrum of *Rp. sphaeroides* at 77°K (Goedheer, 1972). This may be due to a different overlap between the Bchl emission and absorption bands in these species, resulting in lower rates of "uphill" transfer. Emission by B800 in *Rp. sphaeroides* at room temperature was found to be about two orders of

---

* The emission spectrum alone does not allow one to distinguish between emission by B870 or B890 and emission by the reaction center pigments P870 and P890. The fluorescence yield of the latter pigments, however, is very low (Clayton, 1966a; Zankel *et al.*, 1968; Slooten, 1972). The fluorescence yield of B870 or B890 is usually 3–5% in various species of purple bacteria (Wang and Clayton, 1971), depending on wavelength and intensity of excitation; the fluorescence yield of the reaction center pigment is, at least in *Rhodopseudomonas sphaeroides*, 20 times lower under the most favorable conditions, i.e., when the primary electron acceptor is in the reduced state.

magnitude smaller than by B850. No emission was observed on excitation of B890, but transfer from B850 to B800 could be demonstrated (Zankel and Clayton, 1969).

A qualitatively similar reasoning applies to green bacteria, that contain relatively small amounts of Bchl *a*, in addition to much larger amounts of Bchl's *c* or *d* (also called "chlorobium Chl's 660 and 650," respectively; Jensen *et al.*, 1964). The emission spectrum of "*Chloropseudomonas ethylica*" [which later (Gray *et al.*, 1973) turned out to be a mixture of a green photosynthetic bacterium and a colorless motile heterotrophic bacterium*] shows emission bands of Bchl's *a* and *c*, at 814 and 769 nm, respectively, on excitation of the latter pigment (Sybesma and Olson, 1963). A similar spectrum was observed by Krasnovskii *et al.* (1962) for *Chlorobium* sp. From these data, and using certain assumptions, Sybesma and Olson estimated an efficiency of about 33% for energy transfer from Bchl *c* to Bchl *a*. Since energy transfer to

the reaction center probably occurs mainly via Bchl *a*—which is present in much lower concentration than Bchl *c*, and therefore absorbs only a small fraction of the incident light at most wavelengths—this number, if correct, would imply a low efficiency of overall light conversion for this species. As far as we know, direct confirmation of this low yield has not been reported as yet.

The emission spectrum of delayed fluorescence (Fig. 1) showed a relatively lower band of Bchl *c* than that of prompt fluorescence (Clayton, 1965). The same applied to the emission spectrum of the variable fluorescence. It is generally assumed that delayed fluorescence is caused by a reversal of photochemistry (Chapter 27), resulting in a reexcitation of the antenna pigment(s) that normally transfers energy directly to the reaction center, in this case Bchl *a*. The data cited above therefore indicate that energy transfer from Bchl *a* to Bchl *c* is less efficient than vice versa, which can be explained by the locations of the absorption and emission bands of these pigments and their mutual overlap.

* See Chapter 2 for a discussion of the taxonomy of this syntrophic system.

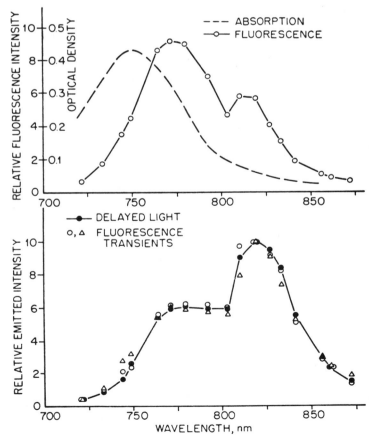

Fig. 1. Emission spectra of prompt and delayed fluorescence of *Chlorobium limicola* ("*Chloropseudomonas ethylica*") The data shown in the lower part of the figure were normalized at 819 nm. Reprinted from Clayton (1965) by permission of the Rockefeller University Press.

2.2. Action Spectra

The action spectra of Bchl fluorescence and of other light-induced processes in purple bacteria are in general agreement with the results discussed above, and show that energy is transferred to B870 or B890. The action spectra of Bchl *a* fluorescence of *Chr. vinosum* strain D, *Rs. rubrum,* and *Rs. molischianum* indicate a rather low efficiency (about 40%) of energy transfer from carotenoids to Bchl (Duysens, 1952), but in *Rp. sphaeroides,* an efficiency of about 90% was observed (Goedheer, 1959). The same can be concluded from action spectra of other photoreactions: phototaxis (Manten, 1948; Duysens, 1952; Clayton, 1953), reduction of intracellular NAD (Amesz, 1963), and inhibition of oxygen uptake (Fork and Goedheer, 1964). For *Rp. viridis,* Olson and Nadler (1965) observed an efficiency of 30% for energy transfer from carotenoids to Bchl *b* (the only Bchl in this species).

Action spectra for B870 or B890 fluorescence indicate that light absorbed by the other antenna Bchl's is transferred to B870 and B890 with high efficiency (Amesz and Vredenberg, 1966; Goedheer, 1973). The spectra suggest that for some pigments (e.g., B850 in *Rp. sphaeroides* and B800 and B810 in *Chr. vinosum* strain D), this transfer may occur with only about 80% efficiency. For *Rp. sphaeroides,* this is in qualitative agreement with the emission spectra measured by Zankel and Clayton (1969) discussed above, but measurements of the fluorescence yield by Wang and Clayton (1971) suggested a higher rather than a lower yield on excitation with light of 850 than with 810 or 870 nm. This applied also to similar measurements with *Chr. vinosum* strain D.

In *Rs. rubrum* and the mutant strain R-26 of *Rp. sphaeroides,* the weak absorption bands at 800 nm are virtually absent in the excitation spectra for fluorescence (Clayton and Sistrom, 1966; Zankel, 1969; Wang and Clayton, 1971; Ebrey, 1971; Goedheer, 1973), although they are present in the action spectra for photochemical reactions (see also Fork and Goedheer, 1964). These bands are, however, at least partly due to reaction center pigments. Apparently, the probability of energy transfer from the reaction center to antenna Bchl is low, even if the reaction center traps are closed (Zankel, 1969; Wang and Clayton, 1971).

3. Energy Transfer between Antenna Bacteriochlorophyll and Reaction Centers

For light to be converted into "useful" energy, the excitation energy must be transferred to the reaction centers. Evidence was presented in the previous

section that in purple bacteria, energy is transferred mainly via the long-wave Bchl, B870 or B890. Although transfer of energy from other Bchl's to the reaction center is favored by the overlap of fluorescence and absorption bands in some cases—e.g., from B800 to the absorption band of the reaction center near 800 nm (due to "P800") and from B850 to the bands near 870 or 890 nm (due to "P870" and "P890", respectively)—substantial transfer of energy does not appear to take place in this way (see also Zankel and Clayton, 1969; Zankel, Chapter 18). The main reason it does not is probably the low concentration of reaction centers compared with that of B870 or B890. In addition, the results of fractionation experiments with *Chr. vinosum* chromatophores suggest that in this species, B890 is more closely associated with the reaction center than the other Bchl's (Thornber, 1970). This does not seem to be the case, however, with all species of purple bacteria (Clayton, 1973). In green bacteria, light-harvesting Bchl *a* seems to be more intimately connected with the reaction center than Bchl *c* (Fowler *et al.,* 1971; Barsky *et al.,* 1974).

For purple bacteria, the evidence discussed in the last paragraph of the previous section indicates that energy transfer from the antenna Bchl to the reaction center is largely unidirectional. This appears not to be true for the Bchl-*b*-containing species *Rp. viridis.* Olson and Clayton (1966) measured action spectra for Bchl fluorescence and for photosynthetic electron transfer in this bacterium. They estimated that transfer of energy from antenna Bchl to the reaction center has an efficiency of only 50–80%, and that the rate of energy transfer from the reaction center to antenna Bchl is considerably higher than in purple bacteria. The reason may be that in this species, energy transfer to the reaction center is an "uphill" process, the antenna pigment absorbing at 1017 and the reaction center pigment at 985 nm.

Vredenberg and Duysens (1963) observed that the fluorescence yield of B890 in intact cells of *Rs. rubrum* is quantitatively related to the oxidation–reduction level of the primary electron donor "P890." Absorption difference spectra brought about by illumination had indicated that the absorption band near 890 nm decreases on oxidation of P890. Assuming that oxidized P890 is completely or nearly completely bleached at this wavelength (experiments with reaction center preparations have later shown that this is indeed the case), so that transfer of energy to oxidized P870 would not occur,* they devised a

---

* Experiments with isolated reaction centers indicate that actually some transfer of energy to a reaction center with oxidized P870 may also occur.

simple model in which the rate of transfer from B890 to the reaction centers was proportional to the concentration of unbleached P870:

$$\frac{d[B^*]}{dt} = k_1 I - (k_2[B^*] + k_3[B^*] + k_4[B^*][P])$$

where I is the light intensity; k_1–k_4 are the "rate constants" for absorption, fluorescence, internal conversion, and energy transfer to the reaction center; and [B*] and [P] are the concentrations of excited B890 and unbleached P890. During illumination, the rate of change of [B*] can be neglected. Therefore, and since the fluorescence yield of B890 is proportional to [B*], the equation can easily be modified into one that predicts a linear relationship between the concentration of P890 (or of oxidized P890) and the reciprocal of the fluorescence yield. A linear relationship was indeed observed with intact cells of *Rs. rubrum*, and subsequently was also found in other species of purple bacteria and in *Rp. viridis* (Clayton, 1966b) and in green bacteria (Sybesma and Vredenberg, 1963). The model of Vredenberg and Duysens implies that the excitation energy of a given antenna Bchl molecule can be transferred to many different reaction centers. [For an analysis of this point, see Clayton (1967).] Such a model was called a "multicentral" or "statistical" model by Borisov and Godik (1973). A linear relationship between the reciprocal of the fluorescence yield and the redox level of P870 or P890 as discussed above has not been observed in all cases. Deviations were observed by Clayton (1966b) in anaerobic cell suspensions of purple bacteria and in chromatophore preparations.

In any simple model of the photosynthetic unit, one should expect the fluorescence lifetime to be proportional to the fluorescence yield. A linear relationship was indeed observed for algae and higher plant chloroplasts (Müller *et al.*, 1969; Tumerman and Sorokin, 1967; Borisov and Godik, 1972a), in which most of the chlorophyll *a* fluorescence is emitted by photosystem II. Rather unexpectedly, however, a different result was obtained for photosynthetic bacteria by Borisov and Godik (1970, 1972a). They observed that the mean lifetime of Bchl fluorescence decreased as the fluorescence yield increased in various species of purple bacteria, including *Rs. rubrum* and *Rp. sphaeroides*, and in "*Chloropseudomonas ethylica*." For a suspension of *Rs. rubrum*, they observed a decrease of the lifetime of fluorescence from 1.6 nsec at low to 1.2 nsec at high intensity of illumination (Fig. 2); for *Rp. sphaeroides*, these lifetimes were 1.2 and 1.0 nsec, respectively. Similar results were obtained with other species and also with a subchromatophore preparation of *Chlorobium*

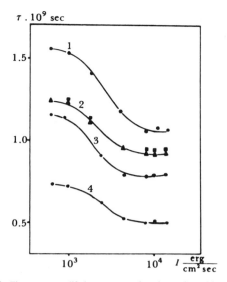

Fig. 2. Fluorescence lifetimes as a function of exciting light intensity for aerobic cell suspensions of photosynthetic bacteria. (1) *Rhodospirillum rubrum*; (2) *Ectothiorhodospira shaposhnikovii* (▲) and *Rhodopseudomonas sphaeroides* (■); (3) "*Chloropseudomonas ethylica*"; (4) *Chlorobium minutissimum*. Reproduced from Borisov and Godik (1972a).

limicola (Barsky *et al.*, 1974). Since lifetimes were measured with a phase fluorometer set at a fixed frequency, they may represent a weighted average for several fluorescing components.

Borisov and co-workers explained their results by assuming that the lifetimes measured were due to two components: a component that shows a very low yield and short lifetime in the presence of open traps, and the yield and lifetime of which increase 10- to 20-fold on closing the traps, and a smaller pool of Bchl with high, invariant fluorescence yield and lifetime. At low intensity of illumination, the lifetimes measured would be due mainly to the last component; at high intensities, the contribution of the other pool would cause a lowering of the "mean" lifetime. For aerobic cells of *Chr. minutissimum* and *E. shaposhnikovii*, a fit with the expected theoretical relationship between fluorescence yield and the proportion of bleached P890 was obtained in this way with a constant "background" of fluorescence that amounted to about 70% at low and 20% of the total fluorescence at high light intensity (Barsky and Borisov, 1971). The fluorescence lifetimes for the "active Bchl," i.e., the pool that is directly involved in energy transfer to the reaction center, were calculated for these bacteria to be about 200 and 500 psec, respectively, at high light intensity and 10 times lower at low light intensities, where the reaction center traps are fully operative. The lifetime of background fluorescence was estimated to be 1.25 and 0.9 nsec, respectively (Borisov and Godik, 1972b).

The short lifetimes of Bchl fluorescence obtained by Borisov and Godik are in agreement with recent reports indicating that in bacteria and bacterial preparations, photochemistry in the reaction centers on excitation with very short flashes occurs within a few picoseconds after the flash (see the review by Parson and Cogdell, 1975). They also explain the discrepancy between the relatively low increase in fluorescence yield and the high efficiency of photochemistry observed for many species of purple bacteria. However, the data of Borisov and co-workers do not agree with results obtained by Govindjee *et al.* (1972) by means of a phase fluorometer operating at higher frequency than that of Borisov. For *Rs. rubrum*, these authors reported lifetimes of 0.4 nsec at low and 1.0 nsec at high light intensity, and similar results were obtained with "*Chloropseudomonas ethylica*" and the Bchl-*b*-containing *Rp. viridis*. It is also not clear why Vredenberg and Duysens (1963) and Clayton (1966b) (see above) obtained a reasonable fit according to the multicentral model between the intensity of fluorescence and the amount of reaction center pigment bleached without applying a correction for background fluorescence. Clearly, further experimentation, preferably with more direct analysis of fluorescence decay curves, will be needed to resolve these contrasting observations.

4. References

References published in 1976 and 1977 were inserted after completion of this paper.

Amesz, J., 1963, Kinetics, quantum requirement and action spectrum of light-induced phosphopyridine nucleotide reduction in *Rhodospirillum rubrum* and *Rhodopseudomonas spheroides, Biochim. Biophys. Acta* **66**:22.

Amesz, J., and Vredenberg, W. J., 1966, Near-infrared action spectra of fluorescence, cytochrome oxidation and shift in carotenoid absorption in purple bacteria, *Biochim. Biophys. Acta* **126**:254.

Barsky, E. L., and Borisov, A. Yu., 1971, Determination of the quantum yields of the primary photosynthesis events and the photosynthetic unit types in purple bacteria, *J. Bioenerg.* **2**:275.

Barsky, E. L., Borisov, A. Yu., Fetisova, Z. G., and Samulov, V. D., 1974, Spectral and energetic characteristics of the photoactive particles from chromatophores of the green bacterium *Chlorobium limicola, FEBS Lett.* **42**:275.

Borisov, A. Yu., and Godik, V. I., 1970, Fluorescence lifetime of bacteriochlorophyll and reaction center photooxidation in a photosynthetic bacterium, *Biochim. Biophys. Acta* **223**:441.

Borisov, A. Yu., and Godik, V. I., 1972a, Energy transfer in bacterial photosynthesis. I. Light intensity dependence of fluorescence lifetimes, *J. Bioenerg.* **3**:211.

Borisov, A. Yu., and Godik, V. I., 1972b, Energy transfer to the reaction centres in bacterial photosynthesis. II. Bacteriochlorophyll fluorescence lifetimes and quantum yields for some purple bacteria, *J. Bioenerg.* **3**:515.

Borisov, A. Yu., and Godik, V. I., 1973, Excitation energy transfer in photosynthesis, *Biochim. Biophys. Acta* **301**:227.

Borisov, A. Yu., Fetisova, Z. G., and Godik, V. I., 1977, Energy transfer in photoactive complexes obtained from green bacterium *Chlorobium limicola, Biochim. Biophys. Acta* **461**:500.

Bril, C., 1964, Studies on the photosynthetic apparatus of some purple bacteria, Thesis, University of Utrecht.

Campillo, A. J., Hyer, R. C., Monger, T. G., Parson, W. W., and Shapiro, S. L., 1977, Light collection and harvesting processes in bacterial photosynthesis investigated on a picosecond time scale, *Proc. Natl. Acad. Sci. U.S.A.* **74**:1997.

Clayton, R. K., 1953, Studies on the phototaxis of *Rhodospirillum rubrum*. I. Action spectrum, growth in green light and WEBER law adherence, *Arch. Mikrobiol.* **19**:107.

Clayton, R. K., 1964, Absorption spectra of photosynthetic bacteria and their chlorophylls, in: *Bacterial Photosynthesis* (H. Gest, A. San Pietro, and L. P. Vernon, eds.), pp. 495–500, Antioch Press, Yellow Springs, Ohio.

Clayton, R. K., 1965, Characteristics of fluorescence and decayed light emission form green photosynthetic bacteria and algae, *J. Gen. Physiol.* **48**:633.

Clayton, R. K., 1966a, Fluorescence from major and minor bacteriochlorophyll components *in vivo, Photochem. Photobiol.* **5**:679.

Clayton, R. K., 1966b, Relations between photochemistry and fluorescence in cells and extracts of photosynthetic bacteria, *Photochem. Photobiol.* **5**:807.

Clayton, R. K., 1967, An analysis of the relations between fluorescence and photochemistry during photosynthesis, *J. Theor. Biol.* **14**:173.

Clayton, R. K., 1973, Primary processes in bacterial photosynthesis, *Annu. Rev. Biophys. Bioeng.* **2**:131.

Clayton, R. K., and Sistrom, W. R., 1966, An absorption band near 800 mμ associated with P870 in photosynthetic bacteria, *Photochem. Photobiol.* **5**:661.

De Klerk, H., Govindjee, Kamen, M. D., and Lavorel, J., 1969, Age and fluorescence characteristics in some species of Athiorhodaceae, *Proc. Natl. Acad. Sci. U.S.A.* **62**:972.

Duysens, L. N. M., 1952, Transfer of excitation energy in photosynthesis, Thesis, University of Utrecht.

Ebrey, T. G., 1971, Anomalous energy transfer behaviour of light absorbed by bacteriochlorophyll in several photosynthetic bacteria, *Biochim. Biophys. Acta* **253**:385.

Ebrey, T. G., and Clayton, R. K., 1969, Polarization of fluorescence from bacteriochlorophyll in castor oil, in chromatophores and as P870 in photosynthetic reaction centers, *Photochem. Photobiol.* **10**:109.

Fork, D. C., and Goedheer, J. C., 1964, Studies on light-induced inhibition of respiration in purple bacteria: Action spectra for *Rhodospirillum rubrum* and *Rhodo-*

pseudomonas spheroides, Biochim. Biophys. Acta **79**:249.

Förster, T., 1948, Zwischenmolekuläre Energiewanderung und Fluoreszenz, *Ann. Phys.* **2**:55.

Förster, T., 1951, *Fluoreszenz organischer Verbindungen*, Vandenhoeck and Ruprecht, Göttingen.

Fowler, I., Nugent, H., and Fuller, G., 1971, The isolation and characterization of a photochemically active complex from *Chloropseudomonas ethylica*, *Proc. Natl. Acad. Sci. U.S.A.* **68**:2278.

Giesberger, G., 1947, Some observations on the culture, physiology and morphology of some brown-red *Rhodospirillum* species, *Antonie van Leeuwenhoek; J. Microbiol. Serol.* **13**:135.

Goedheer, J. C., 1959, Energy transfer between carotenoids and bacteriochlorophyll in chromatophores of purple bacteria, *Biochim. Biophys. Acta* **35**:1.

Goedheer, J. C., 1972, Temperature dependence of absorption and fluorescence spectra of bacteriochlorophylls *in vivo* and *in vitro, Biochim. Biophys. Acta* **275**:169.

Goedheer, J. C., 1973, Fluorescence polarization and pigment orientation in photosynthetic bacteria, *Biochim. Biophys. Acta* **292**:665.

Goedheer, J. C., and van der Tuin, A. K., 1967, Decline of bacteriochlorophyll fluorescence induced by carotenoid absorption, *Biochim. Biophys. Acta* **143**:399.

Govindjee, Hammond, J. H., and Merkelo, H., 1972, Lifetime of the excited state *in vivo*. II. Bacteriochlorophyll in photosynthetic bacteria at room temperature, *Biophys. J.* **12**:809.

Govindjee, Hammond, J. H., Smith, W. R., Govindjee, R., and Merkelo, H., 1975, Lifetime of the excited states *in vivo*. IV. Bacteriochlorophyll and bacteriopheophytin in *Rhodospirillum rubrum, Photosynthetica* **9**:216.

Gray, B. H., Fowler, C. F., Nugent, N. A., Rigopoulos, N., and Fuller, R. C., 1973, Re-evaluation of *Chloropseudomonas ethylica* strain 2-k, *Int. J. Syst. Bacteriol.* **23**:256.

Heathcote, P., and Clayton, R. K., 1977, Reconstituted energy transfer from antenna pigment-protein to reaction centres isolated from *Rhodopseudomonas sphaeroides, Biochim. Biophys. Acta* **459**:506.

Hoch, G., and Knox, R. S., 1968, Primary processes in photosynthesis, in: *Photophysiology* (A. C. Giese, ed.), Vol. III, pp. 225–251, Academic Press, New York and London.

Jensen, A., Aasmundrud, O., and Eimhjellen, K. E., 1964, Chlorophylls of photosynthetic bacteria, *Biochim. Biophys. Acta* **88**:466.

Klimov, V. V., Shuvalov, V. A., Krakhmaleva, I. N., Karapetyan, N. V., and Krasnovsky, A. A., 1976, Bacteriochlorophyll fluorescence changes attributed to the bacteriopheophytin photoreduction in the chromatophores of purple sulfur bacteria (in Russian), *Biokhimiya* **41**:1435.

Knox, R. S., 1968, Theory of concentration quenching by excitation transfer, *Physica* **39**:361.

Knox, R. S., 1973, Transfer of electronic excitation energy in condensed systems, in: *Primary Molecular Events in Photobiology* (A. Checcucci and R. A. Weale, eds.), pp.

45–7.7, Elsevier Scientific Publishing Co., Amsterdam—London—New York.

Knox, R. S. 1975, Excitation energy transfer and migration: theoretical considerations, in *Bioenergetics of Photosynthesis* (Govindjee, ed.), pp. 183–221, Academic Press, New York.

Krasnovskii, A. A., Erokhin, Yu. E., and Yui-Tsun, Kh., 1962, Fluorescence of aggregated forms of bacteriochlorophyll, bacterioviridin and chlorophyll connected with the state of the pigments in photosynthesizing organisms, *Dokl. Akad. Nauk SSSR* **143**:456.

Manten, A., 1948, Phototaxis in the purple bacterium *Rhodospirillum rubrum* and the relation between phototaxis and photosynthesis, *Antonie von Leeuwenhoek; J. Microbiol. Serol.* **14**:65.

Michel-Villaz, M., 1976, Fluorescence polarization: pigment orientation and energy transfer in photosynthetic membranes, *J. Theor. Biol.* **58**:113.

Monger, T. M., and Parson, W. W., 1977, Singlet–triplet fusion in *Rhodopseudomonas sphaeroides* chromatophores. A probe of the organization of the photosynthetic apparatus, *Biochim. Biophys. Acta* **460**:393.

Müller, A., Lumry, R., and Walker, M. S., 1969, Light-intensity dependence of the *in vivo* fluorescence of chlorophyll, *Photochem. Photobiol.* **9**:113.

Olson, J. M., and Clayton, R. K., 1966, Sensitization of photoreactions in Eimhjellen's *Rhodopseudomonas* by a pigment absorbing at 830 mμ, *Photochem. Photobiol.* **5**:655.

Olson, J. M., and Nadler, K. D., 1965, Energy transfer and cytochrome function in a new type of photosynthetic bacterium, *Photochem. Photobiol.* **4**:783.

Olson, J. M., and Stanton, E. K., 1966, Absorption and fluorescence spectra of bacterial chlorophylls *in situ*, in: *The Chlorophylls* (L. P. Vernon and G. Seely, eds.), pp. 381–398, Academic Press, New York and London.

Parson, W. W., and Cogdell, R. J., 1975, The primary photochemical reaction of bacterial photosynthesis, *Biochim. Biophys. Acta* **416**:105.

Paschenko, V. Z., Kononenko, A. A., Protasov, S. P., Rubin, A. B., Rubin, L. B., and Uspenskaya, N. Ya., 1977, Probing the fluorescence emission kinetics of the photosynthetic apparatus of *Rhodopseudomonas sphaeroides* strain 1760–1, on a picosecond pulse fluorometer, *Biochim. Biophys. Acta* **461**:403.

Pearlstein, R. M., 1967, Migration and trapping of excitation quanta, in: *Energy Conversion by the Photosynthetic Apparatus*, pp. 8–15, Biology Department, Brookhaven National Laboratory, Upton, New York.

Robinson, G. W., 1967, Excitation, transfer and trapping in photosynthesis, in: *Energy Conversion by the Photosynthetic Apparatus*, pp. 16–48, Biology Department, Brookhaven National Laboratory, Upton, New York.

Slooten, L., 1972, Reaction center preparations of *Rhodopseudomonas spheroides*: Energy transfer and structure, *Biochim. Biophys. Acta* **256**:452.

Steffen, H., and Calvin, M., 1971, Spectroscopic investigation of the inhibitory effect of fatty acids on photosynthetic systems, *Nature (London) New Biol.* **234**:165.

Sybesma, C., and Olson, J. M., 1963, Transfer of chlorophyll excitation energy in green photosynthetic bacteria, *Proc. Natl. Acad. Sci. U.S.A.* **49**:248.

Sybesma, C., and Vredenberg, W. J., 1963, Evidence for a reaction center P840 in the green photosynthetic bacterium *Chloropseudomonas ethylicum*, *Biochim. Biophys. Acta* **75**:439.

Thornber, J. P., 1970, Photochemical reactions of purple bacteria as revealed by studies of three spectrally different carotenobacteriochlorophyll–protein complexes isolated from *Chromatium*, strain D, *Biochemistry* **9**:2688.

Tumerman, L. A., and Sorokin, E. M., 1967, The photosynthetic unit: A "physical" or a "statistical" concept?, *Mol. Biol. (Moscow)* **1**:628.

Vermeglio, A., and Clayton, R. K. 1976, Orientation of chromophores in reaction centers of *Rhodopseudomonas sphaeroides*. Evidence for two absorption bands of the dimeric primary electron donor, *Biochim. Biophys. Acta* **449**:500.

Vredenberg, W. J., and Amesz, J., 1966, Absorption bands of bacteriochlorophyll types in purple bacteria and their response to illumination, *Biochim. Biophys. Acta* **126**:244.

Vredenberg, W. J., and Duysens, L. N. M., 1963, Transfer of energy from bacteriochlorophyll to a reaction centre during bacterial photosynthesis, *Nature (London)* **193**:355.

Wang, R. T., and Clayton, R. K., 1971, The absolute yield of bacteriochlorophyll fluorescence *in vivo*, *Photochem. Photobiol.* **13**:215.

Wassink, E. C., Katz, E., and Dorrestein, R., 1939, Infrared absorption spectra of various strains of purple bacteria, *Enzymologia* **7**:113.

Zankel, K. L., 1969, Transfer of energy from reaction center to light harvesting chlorophyll in a photosynthetic bacterium, *Photochem. Photobiol.* **10**:259.

Zankel, K. L., and Clayton, R. K., 1969, "Uphill" energy transfer in a photosynthetic bacterium, *Photochem. Photobiol.* **9**:7.

Zankel, K. L., Reed, D. W., and Clayton, R. K., 1968, Fluorescence and photochemical quenching in photosynthetic reaction centers, *Proc. Natl. Acad. Sci. U.S.A.* **61**:1243.

CHAPTER 18

Energy Transfer between Antenna Components and Reaction Centers

Kenneth L. Zankel

1. Introduction

In this chapter, energy flow will be traced through the various antenna bacteriochlorophyll (Bchl) components to the reaction center Bchl. Although laser techniques are beginning to make it possible to trace some of this flow through absorption changes, most of our information, at present, comes from the fluorescence of the various components. The fluorescence yields of these components are used as an indication of how much singlet excitation energy resides in any one pigment component, while the fluorescence lifetimes are used as an indication of the time of residence. This chapter will discuss the relationships between these measurable quantities and possible interpretations will be used to indicate pathways of energy flow and times for such flow to occur.

2. General Background

In the simple case, a component characterized by a molecular absorption band, once excited, will transfer its energy either to fluorescence, to some other component, or to radiationless de-excitation. The sum of the rates for these processes is the rate k of energy removal from the excited component. The time $\tau = 1/k$

Kenneth L. Zankel · Martin Marietta Laboratories, 1450 South Rolling Road, Baltimore, Maryland 21227

is the lifetime of this excitation, or the fluorescence lifetime of the component. The fluorescence yield is determined by the ratio of the rate for emission, k_f, to the total rate k. There is an intrinsic lifetime $\tau_0 = 1/k_f$ that would result if no loss processes were present other than that due to fluorescence. The intrinsic lifetime can be estimated from the absorption spectrum (Strickler and Berg, 1962):

$$\frac{1}{\tau_0} \simeq \frac{2.9 \times 10^{-9}}{\langle \bar{\nu}^{-3} \rangle_{\text{av}}} \, n^2 \int \frac{\varepsilon}{\bar{\nu}} \, d\bar{\nu} \qquad (1)$$

where n is the index of refraction, ε is the molar extinction coefficient (M^{-1} cm^{-1}) at $\bar{\nu}$, and $\bar{\nu}$ is the reciprocal of the wavelength (cm^{-1}). The quantity $\langle \bar{\nu}^{-3} \rangle_{\text{av}}$ is the value of $\bar{\nu}^{-3}$ averaged over the fluorescence band:

$$\langle \bar{\nu}^{-3} \rangle_{\text{av}} = \frac{\int \bar{\nu}^{-3} I_f \, d\bar{\nu}}{\int I_f \bar{\nu}}$$

where I_f is the fluorescence intensity in quanta per unit reciprocal wavelength interval. Equation (1) contains the assumption that the ratios of partition functions for the ground and excited states are unity (for a discussion of this assumption, see, for example, Ross, 1975). For Bchl's in vivo, a value of $\tau_0 \simeq 18$ nsec was obtained by Zankel et al. (1968) based on equation (1) and an assumed index of refraction of 1.33 [a typographical error in that paper led to the omission of the index of refraction in equation (1)]. The measured lifetime is related to the measured fluorescence yield by the expression $\tau = \phi\tau_0$, where ϕ is the fluorescence yield.

When energy flows from an antenna Bchl component to reaction centers in photosynthetically efficient manner, the rate of flow to the reaction centers predominates and $k \simeq k_R$, where k_R is the rate of flow to the reaction centers. Since k_R is usually large compared to k_f, the fluorescence yield is low and the actual lifetime short. As traps become closed (i.e., when reaction center Bchl becomes bleached), the rate of flow from the antenna component to reaction centers decreases and the fluorescence lifetime and yield increase.

As will be seen in examples given later, it is not always useful to consider flow from one pigment system to another without considering the return flow. Singlet excitation energy can flow not only from the more energetic (shorter-wavelength) components to the less energetic (longer-wavelength) ones, but also in the reverse direction. The presence of such "uphill" flow was demonstrated by Zankel and Clayton (1969): light absorbed by an 850 nm Bchl component produced emission from an 800 nm Bchl component in *Rhodopseudomonas sphaeroides*. If the times for flow between components are much shorter than the fluorescence lifetime, equilibrium between the excited states is achieved, and the ratio of the fluorescence yield of a more energetic to that of a less energetic component is proportional to the ratio of the backward to the forward rate of energy flow. Since, as will be discussed below, this ratio is predictable from the absorption spectra, it is possible to determine whether equilibrium is achieved and thus compare transfer rates with fluorescence lifetimes.

When pigment bands are too closely spaced for their fluorescence yields to be separated easily, a method devised by Kennard (1918) can be employed. Kennard showed that for thermal equilibrium in the excited states, the fluorescence spectrum can be predicted from the absorption spectrum by appropriately multiplying the absorption at each wavelength by the Planck radiation equation:

$$I_{\bar{\nu}} \propto \bar{\nu}^3 e^{-h\bar{\nu}c/k'T} \qquad (2)$$

where $I_{\bar{\nu}}$ is the fluorescence intensity per unit wave number interval, c is the velocity of light, h is Planck's constant, k' is Boltzmann's constant, and T is the absolute temperature. The Stokes shift results by virtue of the higher weighting that this radiation formula gives to the longer wavelengths. A simple extension of the derivation of Kennard can be used to show that the procedure is also valid for different absorption bands in equilibrium. When the total absorption of the system is used, the calculations of

predicted emission spectra (being based on thermodynamics) are independent of the mode of energy transfer. Discrepancies between the calculated and measured emission spectra would indicate lack of equilibrium. Such a discrepancy could come about, for example, if there were two components with different absorption spectra not in equilibrium with one another.

For single-antenna Bchl bands, the *in vivo* fluorescence lifetimes of tenths of nanoseconds or more (see, for example, Govindjee *et al.*, 1972), are sufficiently long so that thermal equilibrium within the excited state can occur before emission. The calculations based on such equilibrium are applicable for predicting the emission from each component using its absorption spectrum. This may not hold true for the shorter-lived ($\simeq 10^{-11}$ sec) reaction center Bchl emission (Zankel *et al.*, 1968).

A simple model (Fig. 1) will be used to discuss some of the ramifications of the foregoing discussion. This model consists of a two-component system: pigment band A at higher energy with shorter absorption wavelengths and pigment band B at lower energy with corresponding longer wavelengths. Quanta absorbed by A and B can be passed back and forth between A and B. If the rates k_{AB} and k_{BA} for this transfer are much greater than the rates k_A and k_B for dissipative processes including fluorescence and trapping, equilibrium will be obtained and Equation (2) can be applied to give the total spectrum for emission.[*] Alternatively (for equilibrium), the spectrum is obtained when the ratio k_{AB}/k_{BA} is equated to the ratio of singlet quanta in the two components and the total emission calculated as the sum of the individual emissions. Both procedures must give the same spectrum. The ratio of

---

[*] Note that these are rates and not the usual rate constants used in chemistry. They represent the rate (probability per unit time) of transfer of any given quantum, independent of the number of molecules involved.

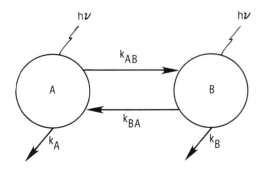

Fig. 1. Schematic for energy transfer between two antenna components.

the rates k_{AB}/k_{BA} depends not only on energy (wavelength) differences between the bands, but also on their relative absorption. If both components are Bchl, the relative absorption is nearly the same as the relative number of molecules, and the emissions are nearly proportional to the number of singlet quanta in each band. Therefore, for a system of two Bchl components, it is not necessary either to know the transfer mechanism or to calculate overlaps of absorption and fluorescence bands to estimate the ratio; the ratio can be estimated from the peaks of the absorption bands and the relative number of molecules in each band:

$$k_{BA}/k_{AB} \simeq N_A\, e^{-hc\bar{v}_A/k'T}/N_B\, e^{-hc\bar{v}_B/k'T} \qquad (3)$$

where N_A is the number of molecules in A, N_B is the number in B, and $\bar{v}_A$ and $\bar{v}_B$ are the reciprocals of their respective peak wavelengths. The same approximate results must be obtained from calculations done this way and calculations based on a specific mechanism, such as that of Förster (1948). An implicit assumption for both equation (3) and equation (2) is that excitation transfer between molecules within each component (A or B) is rapid compared to transfer between components.

3. Transfer between Antenna Bacteriochlorophyll Components

For cases in which the wavelengths of different antenna Bchl bands are far apart, the difference in energy may be sufficiently large so that the "uphill" transfer can be neglected: the fluorescence lifetime of the shorter-wavelength component is then inversely proportional to the sum of the rates for dissipative processes and for transfer to the longer-wavelength components. The fluorescence yield of the shorter-wavelength Bchl is proportional to this lifetime. If absorption by the shorter-wavelength component is very efficient for photosynthesis, the rate of transfer must predominate, and the fluorescence lifetime and yield are measures of the rate of transfer. Since, in this case, insignificant energy flows from the longer- to the shorter-wavelength Bchl, one would not expect the latter's flurescence yield to depend on the states of traps, which communicate only with the longer-wavelength antenna component.

For two closely spaced Bchl bands, the situation is quite different. Energy levels are, in many cases, close enough so that uphill flow is significant. For example, taking into account the energy differences and relative absorptions of the B_{850} and B_{870} antenna Bchl bands of *Rp. sphaeroides*, the rate of downhill flow to the component absorbing at 870 nm is calculated to be

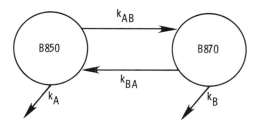

Fig. 2. Schematic for energy transfer between B_{850} and B_{870} in *Rhodopseudomonas sphaeroides*.

only about twice as large as the uphill rate to the component absorbing at 850 nm. Zankel and Clayton (1969) studied chromatophores of this bacterium in detail and found, when exciting at short wavelengths, that the emission spectrum varied with the state of the traps (open vs. closed), and that the fluorescence of B_{870} relative to that of B_{850} was less than calculated on the basis of equilibrium. This discrepancy was even more pronounced when the traps were active (open) (similar unpublished results were obtained by the author for intact bacteria). Let us restrict the discussion for the moment to homogeneous sytems with well-described rates, rather than heterogeneous ones cotaining varying rates or dead* pigments or both. The low fluorescence of B_{870} (relative to B_{850}) then indicates that the transfer rates k_{AB} and k_{BA} (Fig. 2) between B_{850} and B_{870} are not sufficiently fast to maintain B_{850} and B_{870} in equilibrium, i.e., that k_{AB} and k_{BA} are not much faster than the quenching rate k_B from B_{870}. The dependence of this quenching on the state of the traps indicates that most of the energy is transferred from B_{870} to the traps, i.e., that the B_{870} and not the B_{850} molecules are connected to the traps.

When B_{870} was excited directly, the emission spectrum in the steady state did agree with that calculated assuming equilibrium [equation (2)]. This indicated that the rate of dissipation, k_A, was less than the rates of transfer between the two components, k_{AB} and k_{BA}. (It can be shown that in the steady state, when B_{870} is excited, the rate of quenching from B_{870}, k_B, does not enter into the calculations except for establishing the overall level of fluorescence.) As further evidence that energy to the traps flowed primarily through B_{870}, it was shown that the variable part of the fluorescence yield (the change in response to closing the traps) also showed the equilibrium spectrum. This was expected, since it should make no difference whether B_{870} got more energy by direct illumination of B_{870} or by the closing of traps

---

* A dead pigment is a low-absorbing, high-fluorescing one, not connected to the reaction centers. It has been invoked at times to explain the residual fluorescence when traps are open.

connected to B_{870}. Delayed light had the same equilibrium emission spectrum (Zankel, 1969), consistent with the notion that delayed light originates in the traps and is transferred from there to B_{870}.

Since the emission spectrum changes considerably with the state of the traps, the rates of transfer between the antenna components must be the same order of magnitude as the rate to the traps: $k_{AB} \simeq 2k_{BA} \simeq k_B$. The fluorescence yield of *Rp. sphaeroides* is between 2.5 and 5.5%, depending on light intensity (state of traps) and wavelength of excitation (Wang and Clayton, 1971). Assuming an intrinsic fluorescence lifetime of 18 nsec, the lifetimes predicted from these yields would be between 0.45 and 1 nsec (recall that $\tau = \phi\tau_0$). A calculation of the rates of transfer between B_{850} and B_{870} utilizing these results indicates times of transfer in the tenths of nanoseconds.

Rhodopseudomonas sphaeroides does not appear to be unique in its fluorescence and transfer behavior. In the green bacterial consortium "*Chloropseudomonas ethylica*,"* the variable part of the fluorescence also has the same spectral shape as that of delayed light (Clayton, 1965). The spectrum of the variable emission appears to be that expected for equilibrium, considering the differences in energy of the two main Bchl bands and the preponderance of the shorter-wavelength Bchl c over that of the longer-wavelength Bchl a. It appears that, indeed, energy transfer to the traps is from the Bchl a. Since lifetimes of fluorescence for this green bacterium are similar to those for the purple bacterium *Rp. sphaeroides* (Govindjee *et al.*, 1972), the transfer times between the two antenna components must also be similar, in the tenths of nanoseconds.

All the results cited above are applicable either to a separate unit system, in which each trapping center has its own antenna molecules, or one in which antenna molecules are shared by more than one trap. In the case of separate units, the foregoing discussion applies, with Fig. 2 representing each unit and with the total fluorescence yield increasing proportionally to the number of closed traps. The discussion is also applicable for a completely cooperative or "multicentral" system, with Fig. 2 representing the whole system and with a rate constant for the system k_B decreasing as traps become closed. Clayton (1966) showed that in such a multicentral system, the Vredenberg and Duysens (1963) relationship holds; i.e., the change of the inverse of the fluorescence yield is proportional to the change of the number of open traps. Most measurements (comparing absorption to fluorescence changes) indicate that bacteria under

aerobic conditions follow the Vredenberg–Duysens relationship (Vredenberg and Duysens, 1963; Clayton, 1967; Sybesma and Vredenberg, 1963). Contradictory results were obtained, however, by Barsky and Borisov (1971). They observed that the inverse relationship is obtained only after subtracting a considerable portion of dead fluorescence, i.e., fluorescence from pigments that are presumably not associated with photosynthesis. In light of the variability of the absorption spectrum of *Rp. sphaeroides* (Crounse *et al.*, 1963), the emission spectrum of the dead Bchl component would not (*a priori*) be expected to be the same as that of the live (variable) component. When B_{870} is excited, however, the total spectrum is the same as the variable spectrum. Since these common spectra agree with the predicted equilibrium spectrum (Zankel and Clayton, 1969), the simplest interpretation is that the fixed fluorescence contains little "dead" fluorescence. When apparent discrepancies from the Vredenberg–Duysens relationship are observed, they may be due to changes in quenching other than those due to the measured Bchl absorption changes.

4. Transfer from Antenna to Reaction Center Bacteriochlorophyll

The same type of analysis that was presented for transfer between light-harvesting components can be utilized to study the transfer between the light-harvesting Bchl and the reaction center Bchl (Fig. 3). Since reaction center Bchl absorbs at approximately the same wavelengths as the long-wavelength component of the antenna pigments, and since there is usually a preponderance of the antenna pigments, the uphill rate k_{PA} is usually greater than the forward k_{AP}. The reaction center pigment does not trap by virtue of its energy level. On the contrary, energy would tend to reside in the antenna component. Trapping occurs by the quenching of the quanta by the photosynthetic

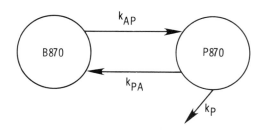

Fig. 3. Schematic for energy transfer between B_{870} and P_{870} in *Rhodopseudomonas sphaeroides*.

process during the fraction of time in which the quanta reside in the reaction center. The observed lifetime is not necessarily the transfer time from antenna to reaction center Bchl. This is true only when the rate for back-transfer k_{PA}, is small compared with the trapping rate k_P in the center ($k_{PA} \ll k_P$) and not much excitation energy returns from the traps. For $k_{PA} \gg k_P$, the energy will pass back and forth between reaction center and antenna Bchl. The observed lifetime will be determined by the fraction of time spent in the reaction center and the time for trapping: $\tau_{obs} = \tau_P k_{PA}/k_{AP}$. The trapping time τ_P is defined here as the time for singlet excitation *in the trap* to be quenched by the photosynthetic process. Since reaction centers are very efficient (Bolton *et al.*, 1969; Wraight and Clayton, 1974), the trapping time approximates the reaction center lifetime.

For chromatophores of a blue-green mutant (R-26) of *Rp. sphaeroides*, there is little antenna pigment absorption at 800 nm, and it is possible to preferentially excite a reaction center Bchl (P_{800}) absorbing at 800 nm. Since reaction centers are relatively nonfluorescent, fluorescence from the chromatophores should serve as a marker for transfer from reaction center to antenna pigments. It was found that such fluorescence was not detectable (Clayton and Sistrom, 1966). This finding suggests that there is little transfer of energy from reaction centers with open traps to antenna Bchl, and that mainly the forward rate k_{AP} determines the lifetime of antenna fluorescence for systems with such open traps. The situation is quite different for *Rp. viridis*, in which, in a similar experiment (Olson and Clayton, 1966), reaction center absorption was as much as two-thirds as effective in exciting fluorescence as antenna absorption. It appears that in this bacterium, the return transfer cannot be neglected. This may account for the 2.5-fold-higher fluorescence lifetimes found (Govindjee *et al.*, 1972) for *Rp. viridis* than for other bacteria.

The fluorescence yield of active and inactive (acceptor-reduced) reaction centers from the blue-green mutant of *Rp. sphaeroides* has been measured and the fluorescence lifetimes estimated. The fluorescence lifetimes so calculated are $(7 \pm 3) \times 10^{-12}$ sec for active traps and 5 times that lifetime for inactive ones (Zankel *et al.*, 1968). Estimates made from fluorescence yields of reaction centers from wild-type *Rp. sphaeroides* give similar results (Slooten, 1973). The calculated rate of transfer from reaction center to antenna pigments for the blue-green mutant of *Rp. sphaeroides* (considering the amounts of pigments in each and the small energy differences between the two) is about 10 times that of the opposing rate. This estimate was made on the basis of the overlap between

antenna and reaction center absorptions and emissions. Possible errors include: (1) differences in the spectra of isolated and *in vivo* reaction centers and (2) differences in the times for reentry to the reaction centers as opposed to that for entering the first time. Although errors due to the latter might alter the numerical values given below, they are not sufficient to alter the general conclusions. The use of the factor of 10 estimated, and an estimate of 0.5 nsec for energy to reach the traps, led to an estimate of 5×10^{-11} sec to return. In that time, most of the energy would have gone to the photosynthetic reactions (assuming a trapping time of 7×10^{-12} sec), and only a small fraction ($<15\%$) would return to the antenna pigments to fluoresce. This small fraction would be difficult to observe in an excitation spectrum, in agreement with the results of Clayton and Sistrom. For inactive chromatophores with unbleached reaction center pigments (accomplished by reducing the photochemical electron acceptor), the rate of reaction center quenching should be comparable to the rate calculated for return to the antenna, and measurable fluorescence is to be expected. The expected experimental result was obtained (Zankel, 1969): reduction of the system allowed the observation of antenna fluorescence induced by P_{800} absorption. In *Rp. viridis*, the reaction center lies at shorter wavelengths than the antenna Bchl (Holt and Clayton, 1965), making the ratio of transfer rate from reaction centers to transfer rate from antenna pigments higher—evidently sufficiently higher that antenna fluorescence arising from energy transferred from the reaction center Bchl is easily detected, even when the traps are open.

For the bacteria studied to date, it appears possible to describe the fluorescence phenomena farily well by considering all photosynthetic units alike, with a single rate constant for each step. Complete homogeneity of the systems may well be an oversimplification. Different units may have different rates of transfer; transfer within absorption components may be such that transfer between two components cannot be described by a simple rate constant. In either case, the constants given above would describe some "average" behavior of the systems.

For a completely homogeneous sytem, the observed lifetime would be expected to increase proportionally to the observed fluorescence yield, while with dead pigment, this is not necessarily so. If there were a considerable dead component, its fluorescence would have its own fixed lifetime. It is possible in such a system (having considerable dead fluorescence) for the average lifetime to decrease as the fluorescence yield increases. This could come about if the fluorescence lifetime of live units with closed traps is less than that of the dead ones. Phase fluorimetric measurements have been made to determine the dependence of the average lifetime on trap state. The results are contradictory: Borisov and Godik (1970, 1972) found that the lifetime in several bacteria, including *Rp. sphaeroides*, decreased as the traps were closed. They interpreted this to signify considerable dead fluorescence. On the other hand, Govindjee *et al.* (1972) consistently found an increase in lifetime as the traps were closed.

The absorption and phase fluorimetric measurements of Barsky and Borisov (1971) and Borisov and Godik (1972) were used to calculate the lifetimes of the dead and live components separately (Borisov and Godik, 1972). The resulting lifetimes for *Rp. sphaeroides* (measured under aerobic conditions) were 0.8 nsec for the dead portion, 7×10^{-12} sec for the active (open trap) live portion, and 7×10^{-11} sec for the inactive (closed trap) live portion. From this, they calculated the fluorescence yields of the individual components. The yields so obtained were about 4 times lower than those obtained in other laboratories (see, for example, Wang and Clayton, 1971).

It can be seen from the preceding discussion that for a photosynthetically efficient system, a measured lifetime of 7×10^{-12} sec implies a much shorter reaction center lifetime, τ_P. Since $k_{PA}/k_{AP} \simeq 10$, the trapping time τ_P implied is less than 7×10^{-13} sec, which is an order of magnitude shorter than the reaction center lifetime of 7×10^{-12} sec estimated from yield measurements of isolated reaction centers. On the other hand (as was shown above), the longer total system lifetimes measured by Govindjee and co-workers are compatible with the estimated reaction center lifetime of 7×10^{-12} sec.

Additional experimental evidence is necessary to definitely resolve the contradiction between the measured lifetime data of Borisov and Godik (1970) and those of Govindjee *et al.* (1972). It appears at the moment that for bacteria under aerobic conditions and for chromatophores, the higher total system lifetimes of Govindjee and co-workers are more readily accommodated to the body of information available from other laboratories.

5. Summary

In concluding this chapter, it must be emphasized that the concepts presented can form, at best, a very incomplete picture of excitation transfer in intact photosynthetic bacteria. Most of the discussion dealt with studies on relatively few species of bacteria and their chromatophores. The complex behavior of bacteria under anaerobic conditions was not discussed. Some aspects of energy transfer, such as transfer between individual molecules within an antenna component and the relationships between the quantum yields for photosynthesis and those for fluorescence, are dealt with in other chapters in this volume.

Most of the available data fit quite well with rather simple interpretations: although a completely homogeneous model is an oversimplification, it does appear to give a reasonable description of most of the experimental evidence, at least for chromatophores. This model indicates that transfer between the long-wavelength antenna Bchl components takes place in tenths of nanoseconds. Transfer from antenna Bchl to reaction center Bchl is channeled through the longest-wavelength antenna Bchl component. Transfer from antenna Bchl to reaction center Bchl also takes place in tenths of nanoseconds. Energy in active reaction centers becomes trapped in about 7 psec. This is sufficiently rapid in *Rp. sphaeroides* so that little ($<15\%$) returns to the antenna Bchl. In *Rp. viridis*, transfer back to the antenna Bchl from reaction center Bchl is rapid enough so that more than half the energy entering the reaction centers is transferred back. The consequences of these results to the possible mechanisms for transfer and energy conversion are discussed elsewhere in this volume.

ACKNOWLEDGMENTS

The author is grateful for the helpful suggestions of Bessel Kok, Richard Radmer, and Fred Fowler. This work was supported in part by a grant from the Energy Research and Development Administration, Contract E(11-1)-3326. This report was also prepared with the support of National Science Foundation Grant No. BMS74-20736. Any opinions, findings, conclusions, or recommendations expressed herein are those of the author and do not necessarily reflect the views of the NSF.

6. References

Barsky, E. L., and Borisov, A. Yu., 1971, Determination of the quantum yields of the primary photosynthesis events and the photosynthetic unit types in purple bacteria, *Bioenergetics* **2**:275.

Bolton, J. R., Clayton, R. K., and Reed, D. W., 1969, An identification of the radical giving rise to the light induced electron spin resonance signal in photosynthetic bacteria, *Photochem. Photobiol.* **9**:204.

Borisov, V. I., and Godik, A. Yu., 1970, Fluorescence lifetime of bacteriochlorophyll and reaction center photooxidation in a photosynthetic bacterium, *Biochim. Biophys. Acta* **223**:441.

Borisov, V. I., and Godik, A. Yu., 1972, Energy transfer to reaction centers in bacterial photosynthesis. II. Bacteriochlorophyll fluorescence lifetimes and quantum yields for some purple bacteria, *Bioenergetics* **3**:313.

Clayton, R. K., 1965, Characteristics of fluorescence and delayed light emission from green photosynthetic bacteria and algae, *J. Gen. Physiol.* **48**:633.

Clayton, R. K., 1966, Relations between photochemistry and fluorescence in cells and extracts of photosynthetic bacteria, *Photochem. Photobiol.* **5**:807.

Clayton, R. K., 1967, An analysis of the relations between fluorescence and photochemistry during photosynthesis, *J. Theor. Biol.* **14**:173.

Clayton, R. K., and Sistrom, W. R., 1966, An absorption band near 800 mμ associated with P870 in photosynthetic bacteria, *Photochem. Photobiol.* **5**:661.

Crounse, J., Sistrom, W. R., and Nemser, S., 1963, Carotenoid pigments and the *in vivo* spectrum of bacteriochlorophyll, *Photochem. Photobiol.* **2**:361.

Förster, T. W., 1948, Zwischenmolekulare Energiewanderung und Fluoreszenz, *Ann. Phys.* **2**:55.

Govindjee, Hammond, J. H., and Merkle, H., 1972, Lifetime of the excited state *in vivo*. II. Bacteriochlorophyll in photosynthetic bacteria at room temperature, *Biophys. J.* **11**:809.

Holt, A. S., and Clayton, R. K., 1965, Light-induced absorbancy changes in Eimhjellen's *Rhodopseudomonas*, *Photochem. Photobiol.* **4**:829.

Kennard, E. H., 1918, On the thermodynamics of fluorescence, *Phys. Rev.* **11**:29.

Olson, J. M., and Clayton, R. K., 1966, Sensitization of photoreactions in Eimhjellen's *Rhodopseudomonas* by a pigment absorbing at 830 mμ, *Photochem. Photobiol.* **5**:655.

Ross, R. T., 1975, Radiative lifetimes and thermodynamic potential of excited states, *Photochem. Photobiol.* **21**:401.

Slooten, L., 1973, Isolation and properties of reaction centers from photosynthetic bacteria, Thesis, University of Leiden, p. 79.

Strickler, S. J., and Berg, R. A., 1962, Relation between absorption intensity and fluorescence lifetime of molecules, *J. Chem. Phys.* **37**:814.

Sybesma, C., and Vredenberg, W. J., 1963, Evidence for a reaction center P840 in the green photosynthetic bacterium *Chloropseudomonas ethylicum, Biochim. Biophys. Acta* **75**:439.

Vredenberg, W. J., and Duysens, L. N. M., 1963, Transfer of energy from bacteriochlorophyll to a reaction center during bacterial photosynthesis, *Nature* (*London*) **197**:355.

Wang, R. T., and Clayton, R. K., 1971, The absolute yield of bacteriochlorophyll fluorescence *in vivo, Photochem. Photobiol.* **13**:215.

Wraight, C. A., and Clayton, R. K., 1974, Absolute quantum efficiency of bacteriochlorophyll photooxidation in reaction centers of *Rhodopseudomonas sphaeroides, Biochim. Biophys. Acta* **333**:246.

Zankel, K. L., 1969, Transfer of energy from reaction center to light harvesting chlorophyll in a photosynthetic bacterium, *Photochem. Photobiol.* **10**:259.

Zankel, K. L., and Clayton, R. K., 1969, "Uphill" energy transfer in a photosynthetic bacterium, *Photochem. Photobiol.* **9**:7.

Zankel, K. L., Reed, D. W., and Clayton, R. K., 1968, Fluorescence and photochemical quenching in photosynthetic reaction centers, *Proc. Natl. Acad. Sci. U.S.A.* **61**:1243.

CHAPTER 19

Chemical Composition and Properties of Reaction Centers

George Feher and Melvin Y. Okamura

1. Introduction

A great deal of progress has been made during the past few years in understanding the primary photochemical events in bacterial photosynthesis. Much of this progress is due to advances made in the isolation, purification, and characterization of the specialized bacteriochlorophyll–protein complex that is the site of the primary photochemistry. This complex is called the *reaction center* (RC),* and will be defined more precisely in Section 2.

Following the initial purification of RC's from *Rhodopseudomonas sphaeroides* R-26 by Reed and Clayton (1968) and from *Rhodospirillum rubrum* G-9 by Gingras and Jolchine (1969), a large number of RC preparations of increasing purity have been obtained from several organisms. A comparative review of these different preparations is given by Gingras in Chapter 6. In this chapter, we shall focus on the detailed chemical composition and properties of the

best characterized of these RC preparations obtained from the carotenoidless mutant, *Rp. sphaeroides* R-26 (Clayton and Wang, 1971; Feher, 1971; Okamura *et al.*, 1974; Steiner *et al.*, 1974*b*). Special emphasis will . be placed on the role of the individual components of the RC in the primary photochemical reaction. More recently, RCs of comparable purity have been obtained from *Rs. rubrum* (Nöel *et al.*, 1972; Okamura *et al.*, 1974) and *Chromatium vinosum* (Lin and Thornber, 1975; Ackerson *et al.*, 1975). Whenever possible, their composition and properties will be compared with those from *Rp. sphaeroides* R-26.

2. Minimum-Size Reaction Center—Criteria for "Native" Primary Photochemical Activity

The term *reaction center* has been used to describe a large variety of particles that contain, in addition to the specialized (Bchl)–protein complex, various amounts of membrane constituents and light-harvesting Bchl's.† Thus, the size and composition of these RC units vary from preparation to preparation. There must be, however, a minimum-size RC unit definable in terms of an operational (functional) criterion, i.e., the smallest entity capable of performing the primary photochemical act. In this review, we shall define primary photochemical activity

---

* Abbreviations used in this chapter: ($A_{800}^{1\,cm}$ V) amount of material in volume V having an absorbance A at 800 nm in a 1-cm path length; (Bchl) bacteriochlorophyll; (Bpheo) bacteriopheophytin; (χ_0) magnetic susceptibility; (CTAB) cetyl trimethylammonium bromide; (ΔE_Q) quadrupole splitting; (EDTA) ethylenediaminetetraacetic acid; (ENDOR) electron nuclear double resonance; (EPR) electron paramagnetic resonance; (I.S.) isomer shift; (LDAO) lauryl dimethylamine oxide; (PMSF) phenyl methyl sulfonyl fluoride; (Q) quinone; (RC) reaction center; (SDS) sodium dodecyl sulfate; (SDS-PAGE) sodium dodecyl sulfate–polyacrylamide gel electrophoresis; (TL buffer) 0.1% LDAO, 0.01 M Tris-HCl, pH 8; (UQ) ubiquinone.

† RC-like particles that contain the light-harvesting Bchl's have been called *photoreceptor units* (Loach, 1970*a*). Particles that contain secondary, tertiary, etc., electron donors and acceptors should more appropriately be called *electron-transport particles*.

George Feher and Melvin Y. Okamura · Department of Physics, University of California at San Diego, La Jolla, California 92093

as the light-induced charge separation between that donor–acceptor pair (referred to as primary) that has been stabilized (with respect to charge recombination) for times of the order of milliseconds. Thus, according to this definition, the transient (intermediate) donor–acceptor states that have recently been observed to occur in the picosecond to nanosecond range (Parson *et al.*, 1975; Rockley *et al.*, 1975; Kaufman *et al.*, 1975; Fajer *et al.*, 1975) are not sufficiently stabilized to be considered as primary photochemical products.

Criteria for "native" primary photochemistry in RCs must be based on measurable quantities that are independent of secondary and tertiary reactants and should be sensitive to the native configuration of the Bchl–protein complex. One such criterion is the kinetics of the low-temperature charge recombination due to the back-reaction between the reduced primary acceptor and the oxidized primary donor observed when the actinic light is turned off. [This reaction is reversible at low temperature and has decay times of about 30 msec in whole bacteria and chromatophores of various photosynthetic bacteria (McElroy *et al.*, 1974).] The use of low (cryogenic) temperatures avoids the possibility of diffusion of exogenous, nonphysiological acceptors or donors during illumination. If, however, these had formed an association with the RCs at room

temperatures (prior to freezing), the low-temperature kinetics would likely be altered. The reason is that the charge-recombination kinetics at low temperatures is believed to be critically dependent on the donor–acceptor distance (McElroy *et al.*, 1974; Hopfield, 1974) and redox properties. Thus, the proper low-temperature kinetics is a necessary condition that the structure of purified RCs is the same as that in whole cells.

There are two accepted assays for testing whether a preparation has photochemical activity. One is the change in the optical spectrum on illumination with actinic light; the most conveniently monitored change is the bleaching of the peak at 865 nm* (see Fig. 1A). The other is the occurrence of the light-induced electron paramagnetic resonance (EPR) signals; the most pronounced and hence the most often monitored is the narrow $g = 2.0026$ signal due to the primary donor (see Fig. 1B) (McElroy, *et al.*, 1974). The low-temperature kinetics of the light-induced optical changes and of the EPR signal have been shown by McElroy *et*

---

* The chemical moiety that gives rise to this peak in *Rp. sphaeroides* is referred to as P_{865}. Since the exact wavelength of the maximum absorbance depends on the bacterial species, some authors have referred to it generically as P_{870}. In *Rp. sphaeroides*, the peak shifts at low temperatures to 890 nm.

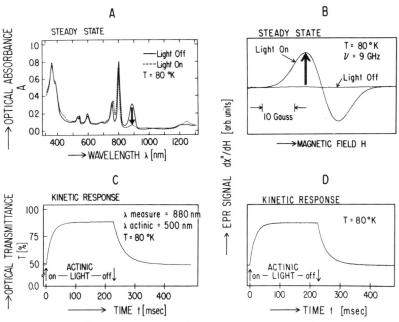

Fig. 1. Assays for primary photochemistry (charge separation) observed at low temperatures in RCs of *Rhodopseudomonas sphaeroides* R-26. (A,B) Steady-state optical absorption and the $g = 2.0026$ EPR signal with and without actinic light; (C,D) kinetic response to turning the actinic light on and off at the position of the arrows. Reproduced (with modifications) from McElroy *et al.* (1974).

al. (1974) to be the same; hence, either can be used to obtain the kinetic parameters. Experimental traces of the low-temperature (80°K) kinetic responses of RCs from *Rp. sphaeroides* R-26 are shown in Figs. 1C,D.

3. Purity of Reaction Centers

An updated procedure for purifying RCs from *Rp. sphaeroides* R-26 is given in Appendix A. It is a modification of the earlier procedures that initiated the use of the detergent lauryl dimethylamine oxide (LDAO) (Feher, 1971; Clayton and Wang, 1971). In preparing the RCs, the ratio of the absorbance at 280 nm to that at 802 nm served as a useful index of relative purity. As contaminating proteins were eliminated, this ratio decreased until a value of 1.22 ± 0.03 was reached. In *Rs. rubrum*, this ratio has been reported to depend on the buffer system and its ionic strength (Gingras, Chapter 6). For instance, when the NaCl concentration in 10 mM Tris-Cl (pH 8.0) buffer was raised to 0.1 M, the ratio decreased from 1.22 to 1.0. No corresponding change with salt concentration was observed in *Rs. sphaeroides* R-26 (Okamura, 1975).

The purity of proteins with respect to other protein contaminants was determined by polyacrylamide gel electrophoresis in the presence of sodium dodecyl sulfate (SDS-PAGE) (Shapiro *et al.*, 1967; Weber and Osborn, 1969). A typical electrophoretogram of RCs is shown in Fig. 2. Three major bands, corresponding

to the subunits of the RC protein (see Section 4), can be seen. Occasionally, minor components that probably represent aggregates of the major components are observed. The results of the electrophoretograms indicate that the pruity of the RCs is at least 95% (Okamura *et al.*, 1974). RCs of similar purity were prepared from wild-type *Rs. rubrum* (Nöel *et al.*, 1972), its carotenoidless mutant, *Rs. rubrum* G-9 (Okamura *et al.*, 1974), and *Chr. vinosum* strain D (Lin and Thornber, 1975; Ackerson *et al.*, 1975).

Immunochemistry offers another sensitive, albeit more qualitative, method for determining the purity of a protein. Antisera were prepared against purified RCs of *Rp. sphaeroides* R-26. When these were reacted with the RCs in double diffusion in agar containing LDAO, a single precipitin band was formed (Okamura *et al.*, 1974; Steiner *et al.*, 1974b). This finding is consistent with the purity of the RC preparation.

4. Structure and Composition of Reaction Center Protein

4.1. Subunit Structure

4.1.1. Dissociation of Reaction Centers by SDS-PAGE

As shown in Fig. 2, RCs of *Rp. sphaeroides* R-26 can be dissociated into three photochemically inactive subunits (Feher *et al.*, 1971; Feher, 1971; Clayton and Haselkorn, 1972). These subunits have been labeled L, M, and H (for light, medium, and heavy) (Okamura *et al.*, 1974). The electrophoretic mobilities of the components in SDS-PAGE (50 mM Tris-Cl buffer, pH 8) correspond to proteins having molecular weights of 21 ± 1, 24 ± 1, and 28 ± 1 kdaltons (see Section 4.5). It should be noted that the electrophoretic pattern depends on the buffer used; e.g., in phosphate buffer (pH 7) in the presence of 0.1% SDS, the three bands were not resolved (Feher, 1971; Reiss-Husson and Jolchine, 1972). The subunits are not derived from each other by enzymatic cleavage, as shown from the amino acid analysis of the subunits (see Section 4.3).

Besides RCs from *Rp. sphaeroides*, those from *Rs. rubrum* (Nöel *et al.*, 1972; Okamura *et al.*, 1974), *Chr. vinosum* (Ackerson *et al.*, 1975), and *Rp. capsulata* (Nieth *et al.*, 1975) all show the presence of only three subunits. The electrophoretic mobilities of the subunits of *Rs. rubrum* and *Chr. vinosum* were found to be a small percentage lower than those of *Rp. sphaeroides* (Fig. 3) (Okamura *et al.*, 1974; Ackerson *et al.*, 1975). Clayton and Haselkorn (1972) prepared

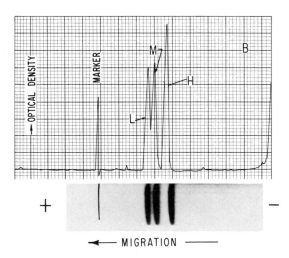

Fig. 2. Subunit structure of RC protein from *Rhodopseudomonas sphaeroides* R-26. Polyacrylamide gel electrophoresis performed in 0.1% SDS, 0.05 M Tris-Hcl (pH 8) shows the presence of three subunits, L, M, and H. Reproduced (with modifications) from Feher (1971) and Feher *et al.* (1971).

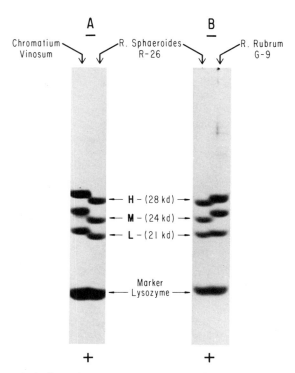

Fig. 3. Comparison of the subunit structure of RCs from *Rhodopseudomonas sphaeroides* R-26 and *Rhodospirillum rubrum* G-9 (Okamura *et al.*, 1974) and *Chromatium vinosum* (Ackerson *et al.*, 1975) by electrophoresis on a split SDS–polyacrylamide gel stained with Coomassie blue. Egg-white lysozyme was used as a marker.

subchromatophore preparations from *Rs. rubrum*, *Rp. capsulata*, *Rp. palustris*, *Rp. gelatinosa*, and *Rp. viridis*. The SDS-PAGE patterns obtained from the preparations of *Rs. rubrum*, *Rp. capsulata*, and *Rp. palustris* suggest the presence of three proteins that are similar in their electrophoretic mobilities to the subunits of RCs from *Rp. sphaeroides* R-26, whereas these protein bands were not evident in the preparations from *Rp. gelatinosa* and *Rp. viridis*. Since the proteins were not purified in this study, a large number of bands was observed. Consequently, the existence of only three subunits was not established with certainty.

4.1.2. Stoichiometry of Subunits

The molar ratios of the subunits (stoichiometry) in RCs have been determined by three independent methods. In the first, the integrated intensity of the protein bands in SDS-PAGE gels stained with Coomassie blue was used to arrive at a provisional stoichiometry of 1 : 1 : 1 (Clayton and Haselkorn, 1972). This method assumes the same binding capacity of the dye as well as equal recovery for all three subunits. For RCs, these assumptions were shown not to be valid (Okamura *et al.*, 1974). The staining efficiency of the H subunit was found to be 1.5 times higher than that of the L subunit, whereas the recovery of the H unit in the gel was only approximately 70% of that of the L subunit. The fortuitous

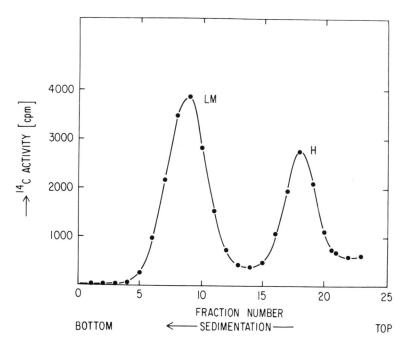

Fig. 4. Sucrose-gradient centrifugation of ^{14}C-labeled RCs in 1% SDS–0.6% LDAO. The faster-sedimenting band contains the two subunits LM, and the slower one contains H. From the ratio of radioactivity (^{14}C) in the two bands, the stoichiometry of the L, M, and H subunits was determined. After Okamura *et al.* (1974).

cancellation of these two effects produced the staining pattern characteristic of an equimolar ratio of subunits.

A second method of determining the stoichiometry utilized radioactively (^{14}C) labeled RCs that were quantitatively dissociated into their LM and H constituents by centrifugation in a sucrose gradient containing SDS and LDAO (see Appendix B, Section B.1). The C^{14} was distributed in two bands, as shown in Fig. 4. The purity of the material in the LM peak and H peak was demonstrated by SDS-PAGE. From the total radioactivity of the LM fractions and H fractions, the stoichiometry of 1:1:1 was unequivocally deduced.

In the third method, a 1:1:1 stoichiometry was obtained from the amino acid composition of the RCs and their three subunits (Steiner et al., 1974b) (see Section 4.3).

4.2. Immunological Analysis

In Section 4.1.1, we discussed the similarities in electrophoretic mobilities between RCs from *Rp. sphaeroides* R-26 and *Rs. rubrum* G-9. A qualitative test of the structural relationship between the RCs of the two bacteria was performed by immunodiffusion (Steiner et al., 1974b). Antisera prepared against each of these RCs precipitated with the homologous antigen, but no cross-reaction could be detected (Fig. 5). Thus, the two proteins are antigenically distinct. Similarly, antisera against R-26 RCs did not react with RCs from *Chr. vinosum* (Ackerson et al., 1975). Clayton and Haselkorn (1972) found that antibodies against RCs from *Rp. sphaeroides* did not react

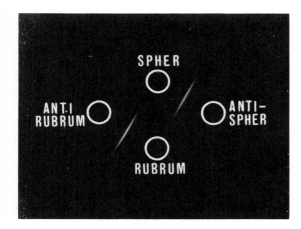

Fig. 5. Immunodiffusion of RCs from *Rhodospirillum rubrum* G-9 and *Rhodopseudomonas sphaeroides* R26 and their antisera. Precipitation occurs only with the homologous antigens. Reproduced from Steiner et al. (1974b).

with chromatophores or LDAO extracts prepared from the following bacteria: *Rs. rubrum*, *Rp. capsulata*, *Rp. palustris*, *Rp. gelatinosa*, and *Rp. viridis*.

Antisera against the H subunit and the LM unit of *Rp. sphaeroides* RCs reacted only with the homologous antigens. Antisera against RCs reacted with both the H subunits and the LM units.

4.3. Amino Acid Composition

A detailed amino acid analysis of RCs of *Rp. sphaeroides* R-26 and their subunits was carried out by Steiner et al. (1974b). As discussed in the last section, RCs can be dissociated into an LM unit and an H subunit by centrifugation in an sucrose gradient, and into three subunits (L, M, and H) by SDS-PAGE. For the determination of the amino acid composition of the subunits, H was prepared by centrifugation and the L and M subunits by gel electrophoresis. The appropriate bands were cut out after the unstained gels were optically scanned, and the protein was eluted electrophoretically from the slices and hydrolyzed. Corrections were made for small amounts of crosscontamination among the subunits as well as traces of some residues found in blank gels. In view of these corrections and the small amount of material eluted, the amino acid composition of the subunits was determined with less precision ($\approx$10%) than that of the intact RC protein (1–2%). The results of the amino acid analysis are shown in Table 1.

The most striking feature of the composition of the RC protein is its high content of nonpolar (hydrophobic) residues. This is consistent with its origin in the bacterial membrane (see Section 4.6.2) and with its solubility properties. Indeed, the protein falls into the category of "integral membrane proteins" as defined by Singer and Nicolson (1972), i.e., proteins that can be dissociated from the membrane only by detergents, or other "vigorous" reagents. In Table 1, the residues have been divided into polar and nonpolar groups. For most water-soluble proteins, the molar percentage of polar residues is in the vicinity of 50%, whereas for integral membrane proteins, it lies between 29 and 50% (Capaldi and Vanderkooi, 1972). Thus, the L subunit belongs to one of the most hydrophobic proteins (29%) reported. The H subunit has the most polar residues of the three subunits. This is consistent with the finding that H is associated with the surface of the membrane (Steiner et al., 1974a; Valkirs et al., 1976) (see also Section 4.6.2).

In view of the possible role of sulfhydryl groups in binding iron (see Section 5.2), special attention was devoted by Steiner et al. (1974b) to the determination of half-cystines. Since the recovery of cysteic acid is not always quantitative, a radiolabeling method util-

Table 1. Amino Acid Composition of Reaction Centers and Their Subunits from *Rhodopseudomonas sphaeroides* R-26 (mol%)[a]

Amino acid	Subunits			Whole RCs (LMH)
	L	M	H	
Polar residues	28.7	29.9	37.9	32.3
Acidic residues	11.1	12.9	16.4	13.61
Asp	6.3	6.4	8.2	6.91
Glu	4.8	6.5	8.2	6.70
Basic residues	8.0	7.0	11.8	8.90
Lys	2.4	0.9	5.1	2.69
His	2.3	2.1	2.3	2.38
Arg	3.3	4.0	4.4	3.83
Hydroxyl residues	9.6	10.0	9.7	9.77
Thr	5.2	4.4	4.7	4.83
Ser	4.4	5.6	5.0	4.94
Nonpolar residues	71.4	70.2	62.1	67.7
Aliphatic	53.9	51.9	54.6	53.57
Gly	12.0	11.7	10.3	10.95
Ala	9.7	9.9	10.5	10.43
Val	6.0	5.5	7.8	6.49
Ile	6.2	5.1	5.4	5.49
Leu	11.2	11.9	9.9	10.84
Pro	5.9	4.7	7.8	6.04
Met	1.6	3.1	2.0	2.62
$\frac{1}{2}$ Cys	1.3	0.0	0.9	0.71
Aromatic	17.5	18.3	7.5	14.13
Tyr	4.3	3.3	2.5	3.11
Phe	7.5	7.7	4.2	6.70
Trp	5.7	7.3	0.8	4.32

[a] Data taken from Steiner *et al.* (1974*b*).

izing RCs containing [^{35}S]cystine was used to determine recovery of cysteic acid after performic acid oxidation. For this purpose, RCs were obtained from bacteria grown in the presence of [^{35}S]cystine. It was found that the M subunit contains no half-cystine and the L and H subunits contain an equal number of half-cystines [(Cys inH)/(Cys inL)] = 0.95 ± 0.05). The absolute number of half-cystines in each subunit is either 2 or 3, the uncertainty being due to the uncertainty in the molecular weight; the value of 3 is considered at present to be the more likely one (see Section 4.5). It has not been determined yet whether the half-cystine residues are present in the reduced (cysteine) or in the oxidized (cystine) form. Preliminary experiments indicate that at least 1 half-cystine of the H subunit binds to *p*-(hydroxymercuri)benzoate in the absence of reducing and denaturing agents (Okamura, 1975).

The presence of half-cystine in the smaller (L) subunit and its absence in the larger (M) subunit prove that the smaller unit cannot be derived from the larger by simple proteolytic cleavage. Similarly, the small amount of tryptophan found in the H subunit eliminates the possibility that the L or M subunits are derived from it by cleavage.

From the amino acid content of the subunits and the intact RC, one can, in principle, determine the stoichiometry of the subunit. For the correct stoichiometry, the average fractional difference between the amount of each residue in the intact RC and the sum of the amounts of that residue in the three subunits should be less than that for other stoichiometries. Using this procedure, Steiner *et al.* (1974*b*) found that the most probable stoichiometry was 1:1:1. This is in accord with the results discussed in Section 4.1.2 although the confidence limit of the assignment, based

Table 2. Comparison of Amino Acid Composition of Reaction Centers from *Rhodospirillum rubrum* G-9 *Chromatium vinosum*, and *Rhodopseudomonas sphaeroides* R-26

Amino acid	*Rs. rubrum* (mol%)[a,c]	*Chr. vinosum* (mol%)[b,c]	*Rp. sphaeroides* (mol%)[a]	*Rs. rubrum* (mol%) / *Rp. sphaeroides* (mol%)	*Chr. vinosum* (mol%) / *Rp. sphaeroides* (mol%)
Lys	3.4	3.0	2.8	1.21	1.07
His	1.9	2.2	2.4	0.79	0.92
Arg	5.3	4.4	4.0	1.33	1.10
Asp	6.9	6.4	7.1	0.97	0.90
Thr	5.8	5.4	4.7	1.23	1.15
Ser	4.9	4.9	4.6	1.06	1.06
Glu	7.2	6.8	6.9	1.04	0.99
Pro	5.5	6.6	6.3	0.87	1.05
Gly	10.0	9.9	11.2	0.89	0.88
Ala	11.2	10.7	10.5	1.07	1.02
Val	6.4	6.5	6.3	1.02	1.03
Met	2.1	2.4	2.6	0.81	0.92
Ile	5.6	6.2	5.2	1.08	1.19
Leu	9.9	9.2	10.9	0.91	0.84
Tyr	3.1	3.7	3.1	1.00	1.19
Phe	5.8	6.7	6.4	0.91	1.05

[a] Data from Steiner *et al.* (1974*b*).

[b] The amino acid analysis was performed by Steiner (1976) on RCs supplied by Ackerson *et al.* (1975).

[c] The values for *Rs. rubrum* are the averages of five, and for *Chr. vinosum* of two, 24-hydrolyses not corrected for partial destruction or incomplete hydrolysis of some residues. Tryptophan and half-cystine were not determined; consequently, the values were normalized to 95 mol% (based on the results of Table 1).

on the amino acid analysis, is considerably lower than that obtained from the radiolabeled subunit structure work (Okamura *et al.*, 1974).

A comparison of the amino acid composition of RCs from *Rp. sphaeroides* R-26, *Rs. rubrum* G-9, and *Chr. vinosum* (D) was made by Steiner and co-workers (Steiner *et al.*, 1974*b*; Steiner, 1976). The results (Table 2) reveal a general resemblance in amino acid composition, although there are significant differences in a number of residues. It would be interesting to determine whether these differences are reflected in all three RC subunits and to explore homologies among species. The latter should aid in understanding the evolution of photosynthetic bacteria.

4.4. Amino Acid Sequence

To sequence proteins, the subunits need to be isolated and purified in milligram quantities. A procedure to do this was developed by Rosen *et al.* (1977*a*,*b*) and is described in Appendix B, Sections B.2 and B.3. Sequence work has been started on these subunits; the preliminary results are as follows (Rosen *et al.*, 1977*a*): The *N*-terminal sequence of the L subunit is: Ala–Leu–Leu–X–Phe–Glu–Arg–Lys–Tyr–Arg–Val–

Pro–Gly–Gly–Thr–Leu–Val–Gly–Gly–Asn–Leu–Phe–Asp–Phe.

In the case of the H and M subunits, only a low yield of *N*-terminal sequence was found. Presumably, their *N*-terminals are partially blocked.

4.5. Molecular Weight

The simplest method for determining the minimum molecular weight of the RC protein is by SDS-PAGE. It should be remembered, however, that the validity of this method has been established only for water-soluble (hydrophilic) proteins (e.g., Weber and Osborn, 1969; Dunker and Rueckert, 1969), and not for highly hydrophobic proteins such as the RC protein. If, for instance, SDS binds preferentially to nonpolar residues, SDS-PAGE would lead to an underestimate of the reported molecular weights. Evidence that this may indeed be the case was provided by Okamura *et al.* (1974), who found a small increase in apparent molecular weight of the subunits with increasing concentration of polyacrylamide gel. Such an increase is consistent with the RC protein binding more SDS than the marker proteins (Banker and Cotman, 1972). The combined molecular weight of the three subunits

Table 3. Molecular-Weight Estimates of Reaction Centers from
Rhodopseudomonas sphaeroides[a]

Method	Molecular weight	References
SDS-PAGE	73,000 ± 2,000[b]	Feher (1971), Okamura *et al.* (1974)
SDS-PAGE	68,000	Clayton and Haselkorn (1972)
From half-cystine	62,000 (4 CyS)	Steiner *et al.* (1974b)
determination	93,000 (6 CyS)	Steiner *et al.* (1974b)
From amino acid determination and extinction coefficient	92,000 ± 5,000	Steiner *et al.* (1974b) Straley *et al.* (1973)
Gel filtration	150,000 ± 15,000[c]	Reiss-Husson and Jolchine (1972)
Sedimentation	153,000	Reiss-Husson and Jolchine (1972)

[a] The last two entries include the weight of six tetrapyrroles ($\approx$6000 mol. wt.); the others represent the molecular weight of the protein only.

[b] The uncertainty represents the estimated statistical error (standard deviation of the mean), not a possible systematic error of the method (see the text).

[c] Subtracting the molecular weight of six tetrapyrroles and the associated detergent (0.5 mg LDAO/mg protein), one obtains a value of 96,000.

determined by SDS-PAGE is 73,000 ± 2000 (Okamura *et al.*, 1974). The quoted uncertainty represents, however, only the statistical error, not a possible systematic error arising from effects discussed above. Different values of molecular weights obtained from several independent experiments are summarized in Table 3.

Another possible way of determining the minimum molecular weights is, in principle, to determine it from the amino acid analysis of the RC protein, i.e., by looking for integral values of residues. Although the composition of the intact RC protein has been obtained with a relatively high degree of precision (Table 1) (Steiner *et al.*, 1974b), with the exception of half-cystines, there are too many residues of each kind to permit estimation of minimum molecular weight. Using half-cystine for the determination of the minimum molecular weight, Steiner *et al.* (1974b) obtained a value of 62,000 assuming 4 half-cystines per RC and 93.000 if the number of half-cystines is assumed to be 6 (note that the number of half-cystines must be even, since two of the subunits contain the same amount of this residue, whereas the third subunit is devoid of cystine).

The subunits are of a more reasonable size for the estimation of the minimum molecular weight from the amino acid composition. Unfortunately, it has been difficult until recently to prepare pure L and M sub-

units in sufficiently large quantities. Consequently, the composition of the subunits has not been determined with sufficient precision to permit a calculation of minimum molecular weight (Steiner *et al.*, 1974b) (see Section 8.1).

The amino acid determination can also be used in conjunction with the molar extinction coefficient to estimate the molecular weight of the RC protein. Steiner *et al.* (1974b) found 0.319 mg amino acid residues in 1 ml of RCs having an optical absorbance $A_{802}^{1\,cm} = 1.00$. Using the molar extinction coefficient $\varepsilon_{802}^{M} = 2.88\ (\pm0.14) \times 10^5$ M^{-1} cm^{-1} (Straley *et al.*, 1973), the calculated molecular weight of the RC protein is: $(mg/A_{802}^{1\,cm}\,V)(\varepsilon_{802}^{M}) = 92,000 \pm 5000$.

Gel filtration and sedimentation experiments performed by Reiss-Husson and Jolchine (1972) on RCs isolated with cetyl trimethylammonium bromide (CTAB) and Triton X-100 from *Rp. sphaeroides* Y indicated a molecular weight of approximately 150,000. Similarly, Noël *et al.* (1972), using gel filtration, determined a molecular weight of 140,000 for RCs isolated with LDAO from *Rs. rubrum*. The discrepancies between these results and those obtained by SDS-PAGE were interpreted as being due to either aggregation (e.g., dimerization) of the RCs or the presence of detergent molecules. We can estimate the effect of detergent on the molecular-weight determination by using the detergent-binding capacity of RCs from *Rp. sphaeroides* R-

26. Ackerson *et al.* (1975) found that approximately 0.5 mg LDAO bound to 1 mg RC protein (see Section 6.6). Subtracting the weight of this amount of detergent and the six tetrapyrrole molecules from 150,000 and 140,000, one arrives at molecular-weight values of 96,000 and 86,000, respectively. This is in fair agreement with the values obtained from the amino acid determinations. It does not seem necessary, therefore, to invoke dimerization of RCs to account for the molecular weight obtained by gel filtration.

In summary, the molecular weight of the RC protein is approximately between 70,000 and 95,000, the higher values being favored at present. A more accurate determination awaits the determination of the primary sequence or a more accurate redetermination of the amino acid composition of the subunits. It would also be desirable to repeat the equilibrium sedimentation experiments with radiolabeled detergents, in particular by using the general method developed by Casassa and Eisenberg (1964) for multicomponent systems.

4.6. Function of the Three Subunits

4.6.1. Are All Three Subunits Required for the Primary Photochemistry?

What role does each of the three RC subunits play in the primary photochemistry? A straightforward way to answer this question would be to determine the effect of removal of each of the subunits on the photochemical activity. This has been partially accomplished with RCs from *Rp. sphaeroides* R-26 (Feher, 1971; Feher *et al.*, 1971; Okamura *et al.*, 1974).

When RCs of *Rp. sphaeroides* R-26 are exposed to relatively mild denaturing conditions (0.1% SDS, 25°C), they split into two components that can be separated by SDS-PAGE (Feher, 1971; Feher *et al.*, 1971). One component contains only the L and M subunits. This component has an optical spectrum similar to that of RCs, and was found to be photochemically active at room temperature (Fig. 6). The other component is the H subunit, which contains no pigment. Okamura *et al.* (1974) subsequently showed that the LM unit could be stabilized by adding small amounts of LDAO to SDS and devised a preparative procedure for isolating LM units involving sucrose-gradient centrifugation. Recently, several different procedures for separating the LM unit have been devised; the most successful of these involves the use of the chaotropic agent LiClO$_4$ (see Appendix B, Sections B.1 and B.2).

The observation of reversible bleaching of P$_{865}$ at room temperature in LM units (see Fig. 6) suggested the possibility that only two subunits are required for

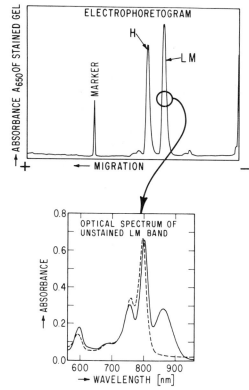

Fig. 6. Top: Electrophoretogram of RCs from *Rhodopseudomonas sphaeroides* R-26 under mild denaturing conditions (0.1% SDS, temp. = 25°C). The RCs split into two major bands, H and LM. Bottom: Optical absorption of the LM band in the gel (——) and demonstration of room temperature photoactivity when illuminated with actinic light (– – –). Reproduced from Feher *et al.* (1971).

photochemical activity. When the more stringent requirement of low-temperature activity (see Section 2) was applied, however, it was found that the results depended critically on the method of preparation of the LM units. When LM units were prepared with SDS and LDAO, only a small fraction of the photochemical activity ($\approx$10%) was retained at cryogenic temperatures. When LiClO$_4$ and LDAO were used instead (see Appendix B, Sections B.1 and B.2), the LM unit exhibited fully reversible photobleaching and EPR changes at low temperatures. However, the decay time, τ_D,* of charge recombination at low temperature was found to be approximately 50 msec, as compared with 30 msec for chromatophores and "native" RCs. This change in kinetics indicates an alteration in the donor–acceptor con-

---

* τ_D is the time required for the EPR signal (or optical transmission) to reach $1/e$ (36.8%) of its maximum value (see Fig. 1). Some authors quote the half-life $\tau_{1/2}$, which is related to τ_D by $\tau_{1/2} = 0.69\tau_D$.

figuration. It should be kept in mind, however, that the tunneling mechanism (McElroy *et al.*, 1974; Hopfield, 1974) is exquisitely sensitive to the donor–acceptor distance (a 5% change in the distance could account for the observation). Thus, the donor–acceptor configuration in the LM unit may be very close to the "native" configuration. Nevertheless, it prevents us from stating categorically at present that LM is the smallest unit with "native" photochemical activity. Whether the H subunit is required or whether a gentler separation technique will suffice to restore the proper kinetics remains to be determined.

4.6.2. Role of the H Subunit—Structural Relationship to the Membrane

As pointed out in the last section, the H subunit contains no pigments or other cofactors (see also Section 5) and plays at most a minor part in the primary photochemistry. We do not know at present the main function of the H subunit. Perhaps it provides binding sites for proteins associated with the electron-transfer chain (e.g., cytochromes) or with the energy transduction (e.g., ATPase). To shed some light on these questions, the spatial arrangement of the subunits was investigated by immunological techniques.

Steiner *et al.* (1974a) prepared antisera against RCs from *Rp. sphaeroides* R-26. This antiserum precipitated with chromatophores from this organism and also from the nonphotosynthesizing mutant, *Rp.*

sphaeroides 8-17, a strain that contains only the H subunit (Takemoto and Lascelles, 1973). When this antiserum was absorbed with membrane fragments of the 8-17 mutant (thereby leaving only antibodies against L and M), it did not precipitate with chromatophores of R-26, although it did still react with RCs. This suggested that particular antigenic sites of the H subunit but not the L and M subunits are exposed on the outer chromatophore membrane.

A more direct method of visualizing RCS by electron microscopy utilizes ferritin-labeled antibodies. This technique was first applied to the RC problem by Reed *et al.* (1975), who prepared a ferritin conjugate of anti-RC. The antibodies did not react with untreated chromatophores of *Rp. sphaeroides* R-26, but did react with chromatophores treated with EDTA, which removes the ATPase from the surface of the chromatophore. This work was extended by Valkirs *et al.* (1976), who prepared antisera against the subunits of H and LM and whole RCs from *Rp. sphaeroides* R-26. When chromatophores were incubated with these antisera in the presence of EDTA and subsequently reacted with ferritin-conjugated goat anti-rabbit IgG, extensive ferritin labeling of the outside of the chromatophores was observed with anti-H and anti-RC (Fig. 7). This provides additional proof that part of the H subunit is exposed on the outside of the chromatophore membrane. In conjunction with the findings of Reed *et al.* (1975), these results suggest that ATPase may be bound to the H subunit.

The experiments discussed above were all performed on chromatophores in which only the outer mem-

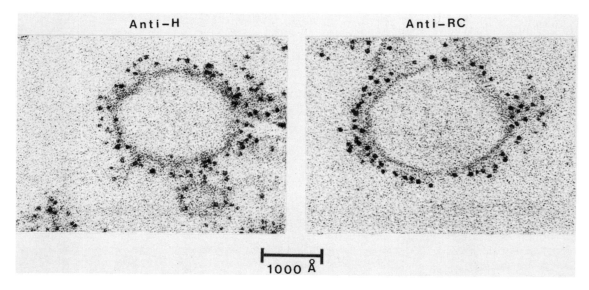

Fig. 7. Electron micrographs of ferritin-labeled *Rhodopseudomonas sphaeroides* chromatophores treated with anti-RC and anti-H sera. These results show that the H subunit is exposed on the outside of the chromatophore membrane. Reproduced from Valkirs *et al.* (1976)

brane was accessible to antibodies. To expose both sides of the membrane, punctured spheroplasts were made and exposed to antibodies (Valkirs *et al.*, 1976; Feher and Okamura, 1977). Preliminary results show that anti-RC and anti-LM label *both* sides of the plasma membrane, whereas H labels only the inside of the plasma membrane, which forms the outside of the chromatophore. Thus, the LM unit seems to span the entire membrane, consistent with the picture that the charge separation occurs across the membrane. The earlier, negative immunological experiments (Steiner *et al.*, 1974*a*) could be explained by postulating that the antisera used were not directed against the particular antigenic sites that are exposed on the surface.

5. Cofactors Important in the Primary Photochemistry

5.1. Tetrapyrrole Pigments: Bacteriochlorophyll and Bacteriopheophytin

5.1.1. Optical Spectroscopy

Purified RCs exhibit a characteristic optical absorption spectrum (see Figs. 1 and 8), indicating the presence of tetrapyrrole pigments [see Norris and Katz,

Chapter 21; Thornber *et al.*, Chapter 7; Mauzerall, Chapter 11; and the reviews by Parson and Cogdell (1975) and Sauer (1974)]. The most striking feature of the spectra are three absorption bands in the near-IR, which, at room temperature, appear at approximately 760, 800, and 865 nm; the absorbance ratios $A_{760}/A_{800}/A_{865}$ are approximately $1:2:1$. In addition, there are absorption peaks of lower amplitude at 590 and 530 nm, as well as a strong Soret peak at 360 nm. If the RC pigments are extracted into an organic solvent, their spectra are markedly altered, particularly in the near-IR (see Fig. 8). This suggests that the optical absorption (and presumably the functional) properties of the RC pigments are due to their special environment and arise from interactions mediated by the RC protein. It has been suggested that the shifts and splitting of the bands in the near-IR arise from interactions between tetrapyrrole molecules. Additional support of the coupling between Bchl molecules is provided by the CD work of Sauer *et al.* (1968) and Reed and Ke (1973). Thus, it may not be meaningful to assign an absorption band to an individual molecule. To a first approximation, however, the bands at 865, 800, and 590 nm have been assigned to Bchl and the bands at 760 and 530 nm to bacteriopheophytin (Bpheo) (see Parson and Cogdell, 1975). At cryogenic temperatures, the bands sharpen and the

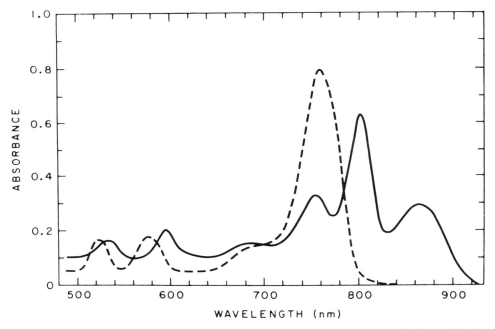

Fig. 8. Absorption spectra of intact RCs from *Rhodopseudomonas sphaeroides* R-26 (———) in 0.01 M Tris-HCl buffer (pH 7.5) containing 0.1% LDAO and a neutral organic extract of the pigments (– – –) at the same concentration in dry petroleum ether. The presence of both Bpheo and Bchl in the preparations is shown by the two absorption maxima between 500 and 600 nm. A blue shift of the absorption maxima in the extract is attributed to a change in the specialized environment of the pigments in the RC. Reproduced from Reed and Peters (1972).

530 nm band is observed to split into two bands (Clayton et al., 1972), and a shoulder is resolved on the 800 nm peak (Feher, 1971). In addition, the long-wavelength absorption band is red-shifted to 890 nm (see Fig. 1).

An indication of the functional role of Bchl in the primary photochemistry is the observation that the long-wavelength absorption band, P_{865}, bleaches on illumination with actinic light (see Fig.1). In addition, the absorption band at 800 nm shifts to shorter wavelengths.* These optical changes were first observed by Duysens (1952) in whole bacteria. Similar optical changes produced by chemical oxidation of chromatophores (Goedheer, 1959) indicate that P_{865} is associated with the primary donor.

Several other observations indicate that P_{865} is the primary donor: (1) The bleaching of P_{865}^+ coincides with the loss of photochemical activity with a midpoint potential of $E_m = +0.44$ V (Loach et al., 1963; Loach, 1966). (2) The quantum yield for P_{865} oxidation is near unity (Loach and Sekura, 1968; Bolton et al., 1969; Wraight and Clayton, 1974). (3) The P_{865} bleaching occurs before cytochrome oxidation (Parson, 1968) and has a rise time of less than 6 psec (Netzel et al., 1973; Rockley et al., 1975). This rapid bleaching of P_{865} is due to photooxidation of the primary donor, as was recently shown by picosecond spectroscopy (Dutton et al., 1975; Moskowitz and Malley, 1978), in which the oxidized donor absorbance at 1250 nm appeared concurrently with the P_{865} bleaching.

Large bathychromic shifts similar to those seen in P_{865} have been observed in $Chl \cdot H_2O$ aggregates, and led Katz and co-workers to postulate a special chlorophyll dimer linked via an H_2O molecule at the photoactive site of electron transfer (Katz and Norris, 1973) (see also Norris and Katz, Chapter 21). A specialized Bchl dimer, having a slightly different geometry, was proposed more recently by Fong (1974).

An early indication that Bpheo may be involved in the photochemistry of RCs came from the observation of spectroscopic changes of the absorption peaks assigned to Bpheo (particularly a red shift of a band at 760 nm). These changes were observed in chromatophores (Loach, 1966) and in RCs (Clayton and Straley, 1972) when the primary acceptor was reduced. Clayton suggested that these changes could be due to an "electrochromic shift" of the Bpheo

absorption due to the presence of a nearby reduced acceptor species (Clayton, 1972).

Recent pulsed laser studies have indicated that Bpheo may play an important role in the primary photochemistry. Parson and collaborators have shown that a transient state, P^F, is involved as an intermediate state in the primary photosynthetic process. P^F was first observed in RCs in which the primary acceptor was reduced; it had a lifetime of approximately 10 nsec (Parson et al., 1975). This state was later observed in RCs in which the primary acceptor was unreduced; in this system, P^F had a rise time of less than 10 psec and a decay time of approximately 200 psec (Rockley et al., 1975; Kauffman et al., 1975). The observation of a rapid ($\approx$3- to 6-psec) appearance of the 1250 nm absorption increase (Dutton et al., 1975; Moskowitz and Malley, 1978), indicative of D^+, showed that P^F must involve an electron transfer from a donor to an acceptor. Furthermore, since P^F was formed in RCs in which the primary acceptor, A, was reduced, a different, transient, intermediate electron acceptor, I, must be involved (i.e., $P^F = D^+I^-$). The subsequent electron transfer from I to A accounts for the rapid (200-psec) decay of P^F in unreduced RCs.

Fajer et al. (1975) postulated that the intermediate acceptor, I, is Bpheo. His conclusion was based on a comparison of the optical spectrum of Bpheo$^-$ with that of P^F and the fact that Bpheo had a redox potential (-0.55 V vs. normal hydrogen electrode) that was more consistent with an intermediate acceptor than did Bchl (-0.85 V).

More recently, several groups of workers have succeeded in trapping the reduced intermediate I$^-$ in chromatophore preparations from Chr. minutissimum (Shuvalov and Klimov, 1976), subchromatophore fragments from Chr. vinosum (Tiede et al., 1976, 1977) and RCs from Rp. sphaeroides R-26 (Okamura et al., 1977a). The trapping was accomplished by illuminating the preparations at low redox potential in the presence of cytochrome. The cytochrome rapidly transfers an electron to the oxidized primary donor, thus converting the short-lived D^+I^- state to a stable DI^- species. The optical spectrum of the trapped I$^-$ shows bleaching of bands of 540 and 760 nm indicative of Bpheo$^-$. In addition, bleaching of bands at 590 and 800 nm was observed, indicating that Bchl is also involved.

5.1.2. EPR Spectroscopy

(a) The Primary Donor. Concomitant with the main optical changes (e.g., bleaching of P_{865}), a narrow, structureless, light-induced EPR signal at $g = 2.0026$ is observed when RCs are illuminated.

---

* Recent studies of linear dichroism in the absorption spectra of RC chromophores suggest that the apparent blue shift of the 800 nm band is in fact due to the bleaching of a band at 815 nm and the appearance of a new band at 790 nm (Vermeglio and Clayton, 1976). The bleachable bands at 865 and 815 nm might reflect the spectrum of a Bchl dimer that functions as photochemical electron donor (see Section 5.1.2).

Such an EPR signal was first observed in green plant preparations by Commoner et al. (1956) and in photosynthetic bacteria by Sogo et al. (1959). This EPR signal is formed with a high quantum yield (Loach and Walsh, 1969; Bolton et al., 1969), and has the same redox behavior (Loach et al., 1963) and rise and decay kinetics as the optical changes at 865 nm (Loach and Sekura, 1967; McElroy et al., 1974). The species giving rise to this EPR signal (D+) has therefore been associated with the oxidized primary donor (P_{865}^+). The g values of this signal and that seen in the monomeric Bchl cation radical are identical (g = 2.0026).

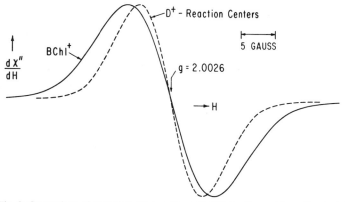

Fig. 9. Comparison of EPR lines from oxidized (Bchl+) *in vitro* and from illuminated RCs of *Rhodopseudomonas sphaeroides* R-26 (D+) (temp. = 77°K, v_e = 9 GHz). The ratio of the line widths is approximately $2^{1/2}$. Reproduced from Feher et al. (1975a).

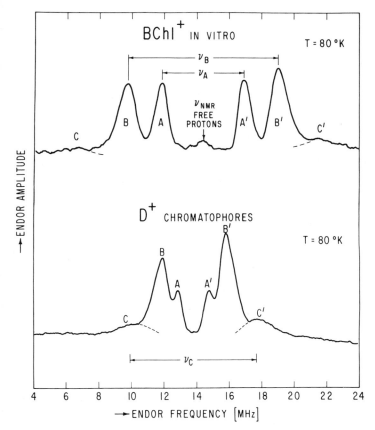

Fig. 10. Comparison of ENDOR spectra from Bchl+ *in vitro* (top) and chromatophores of *Rhodopseudomonas sphaeroides* R-26 (bottom). The frequencies v_A, v_B, and v_C are proportional to the electron spin densities at different protons. Their values in chromatophores are approximately one-half as large as those observed in Bchl+, showing that the unpaired electron is shared between two Bchl molecules. Reproduced from Feher et al. (1975a).

The line width of the D^+-cation radical, however, is narrower than that of $Bchl^+$ by a factor of 1.4 (see Fig. 9) (Norris et al., 1971; McElroy et al., 1972). This difference in line width led Norris and Katz to propose that the primary donor consists of a special pair of strongly interacting Bchl molecules, the odd electron being delocalized over both of them (Norris et al., 1971).

Confirmation of the dimer model came from electron nuclear double resonance (ENDOR) experiments in which the interactions (hyperfine) between the electron spin and the proton nuclei were determined (Feher et al., 1973, 1975a; Norris et al., 1973, 1975). In a dimer, the unpaired electron spends only half its time on each Bchl molecule; consequently, the expected average spin densities and hence hyperfine splittings are one-half as large as those observed in the monomeric Bchl cation radical.* This was indeed found to be the case (Fig. 10).

(b) The Intermediate Acceptor. The EPR signal from the intermediate acceptor, I^-, that was trapped as discussed in the previous section was first observed by Shuvalov and Klimov (1976) and Tiede et al. (1976). To determine its chemical identity, the EPR characteristics of I^- in RCs of Rp. sphaeroides R-26 were

---

* This model implicitly assumes that: (a) the rate of hopping of the electron between the two Bchl's is fast compared to the inverse hyperfine interaction frequency; (b) the two Bchl's are equivalent; and (c) the electronic structure (at least the part responsible for the hyperfine interactions) of the Bchl monomer does not change significantly upon dimerization.

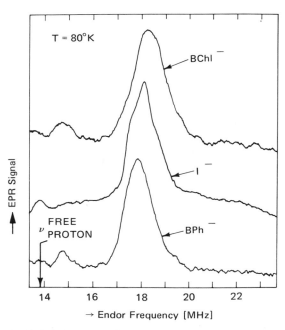

Fig. 12. ENDOR spectra from I^-, $Bchl^-$, and $Bpheo^-$. The preparative procedure is the same as in Fig. 11. Reproduced from Feher et al. (1977).

compared with those obtained from chemically reduced $Bchl^-$ and $Bpheo^-$ (Fajer et al., 1977; Feher et al., 1977). The results (see Fig. 11) show that the g values and line widths of all three species are the same, within experimental accuracy. Similarly, the ENDOR spectra are very similar, as shown in Fig. 12. We conclude, therefore, that I^- is a *monomeric* tetrapyrrole anion, unlike D^+, which is a specialized *Bchl* dimer (compare Figs. 10 and 12). At present, however, it is not possible to differentiate between $Bchl^-$ and $Bpheo^-$.

An interesting difference was observed between the EPR spectra of I^- formed at 200°K in subchromatophore particles of Chr. vinosum (Tiede et al., 1976) and those produced at 300°K† in RCs of Rp. sphaeroides R-26 (Feher et al., 1977). In Rp. sphaeroides, a single resonance was observed (see Figs. 11 and 13b), whereas in Chr. vinosum, two additional broad lines were obtained (Fig. 13a). This doublet structure was postulated to result from the magnetic interaction of I^- with the acceptor, Q^-Fe^{2+} (Tiede et al., 1976) (see Section 5.3.4). The absence of the doublet structure at room temperature was attributed to a thermally activated dark reaction (Okamura et al., 1977a) in which I^- transfers its electron to Q^-, forming the dia-

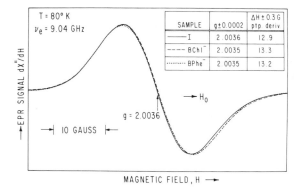

Fig. 11. Comparison of the EPR signal of the intermediate electron acceptor I^- with the signals obtained from $Bchl^-$ and $Bpheo^-$. I^- was produced by illuminating RCs of Rhodopseudomonas sphaeroides R-26 at 25°C in the presence of 0.2 mM cytochrome c, 2 mM $Na_2S_2O_4$, pH 8. $Bchl^-$ and $Bpheo^-$ were photoreduced with Na_2S in pyridine with approximately 2% H_2O. The g values and line widths of all three species are the same within experimental accuracy. Reproduced from Feher et al. (1977).

---

† In Rp. sphaeroides, the cytochrome does not react with D^+ at 200°C; consequently, I^- had to be produced at higher temperatures.

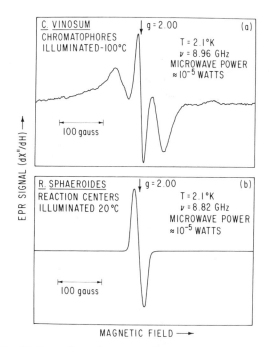

Fig. 13. Comparison of EPR signals from the intermediate acceptor I^- observed in chromatophores of *Chromatium vinosum* (top) and RCs of *Rhodopseudomonas sphaeroides* R-26 (bottom). Chromatophores with an $A_{800}^{1 \text{ cm}}$ of 100 were illuminated in the presence of 2 mM $Na_2S_2O_4$, 50 mM Tris, pH 8. RCs with an $A_{800}^{1 \text{ cm}}$ of 18 were illuminated at 20°C for 5 min in the presence of 0.2 mM cytochrome c and 2 mM $Na_2S_2O_4$, 50 mM Tris, pH 8. Reproduced from Okamura *et al.* (1977).

magnetic species, Q^{2-} (see Section 5.3.1). This process can be represented as follows:

$$DI(Q^-Fe^{2+}) \xrightarrow[\text{cyt}^{2+} \; h\nu \; \text{cyt}^{3+}]{} DI^-(Q^-Fe^{2+}) \longrightarrow$$

$$DI(Q^{2-}Fe^{2+}) \xrightarrow[\text{cyt}^{2+} \; h\nu \; \text{cyt}^{3+}]{} DI^-(Q^{2-}Fe^{2+})$$

The doublet EPR signal results from a magnetic interaction of I^- and (Q^-Fe^{2+}), while the narrow signal is due to $I^-(Q^{2-}Fe^{2+})$.

If the model described above is correct and the kinetic parameters are appropriate, one should observe the doublet signal after a short period of illumination and rapid freezing [i.e., $DI^-(Q^-Fe^{2+})$ can accumulate]. Indeed, the first doublet signals were observed in chromatophores of *Chr. vinosum* under such conditions (Evans *et al.*, 1974). Attempts to observe the doublet in native RCs of *Rp. sphaeroides* R-26 after short illumination times were unsuccessful (Okamura *et al.*, 1977a). When ubiquinone (UQ) was replaced by menaquinone [the primary quinone (Q) of *Chr.*

vinosum], however, a doublet signal was observed after a brief (≈5-sec) illumination. This suggests that the doublet signal is very sensitive to the structure of the primary Q. A systematic analysis of the doublet splittings should give information about the interaction between I^- and the primary acceptor.

5.1.3. Chemical Analysis—Stoichiometry and Extinction Coefficients

Early estimates by Clayton (1966) indicated that the RC was associated with a complex of three Bchl molecules. Subsequently, it was shown from the low-temperature optical absorption spectrum that there are at least 4 pigment molecules associated with the RC (Feher, 1971). Recent experiments in which tetrapyrroles were quantitatively extracted from RC preparations from *Rp. sphaeroides* R-26 (Mauzerall, 1972; Reed and Peters, 1972; Straley *et al.*, 1973), as well as *Rp. sphaeroides* Y (Jolchine and Reiss-Husson, 1975), showed that the RC contains 4 molecules of Bchl and 2 molecules of Bpheo.

The result cited above was obtained from the observation that in RCs the molar ratio Bchl/Bpheo is $2:1$. Thus, the RC must contain $2n$ Bchl molecules and n Bpheo molecules (where n is a small integer). From the measured concentration of tetrapyrrole pigments that were extracted from RCs of a given absorbance, Straley *et al.* (1973) calculated possible molar extinction coefficients assuming different values of n. These were compared with an extinction coefficient obtained by an independent method based on the optical absorbance changes due to cytochrome c oxidation that occurs concomitantly with the optical absorbance changes due to P_{865} reduction. From the known extinction coefficient from cytochrome c oxidation (at $\lambda = 550$ nm) the extinction coefficient of RCs (at $\lambda = 865$ nm) was determined. The best agreement between the two sets of extinction coefficients was obtained for a pigment composition of 4 Bchl's and 2 Bpheo's per RC ($n = 2$). The values of the extinction coefficients were determined to be

$$\varepsilon_{802} = 2.88 \times 10^5 \; (\pm 0.04) \; M^{-1} \cdot cm^{-1}$$

$$\varepsilon_{865} = 1.28 \times 10^5 \; (\pm 0.06) \; M^{-1} \cdot cm^{-1}$$

Reed and Peters (1972) found the same pigment stoichiometry by observing that the 530 nm peak was broadened, indicating that there were 2 Bpheo's and, consequently, 4 Bchl's.

Similar determinations made on *Rs. rubrum* RCs by van der Rest and Gingras (1974) resulted in the same pigment composition, i.e., 4 Bchl's and 2 Bpheo's. Although the pigment composition of *Chr. vinosum* RCs has not been analyzed in detail, the similarity between the optical absorption spectra (Lin and

Thornber, 1975) of *Chr. vinosum* and *Rp. sphaeroides* suggests that their pigment compositions are identical.

The presence of Bpheo in RCs is particularly interesting, since it was not considered to be a normal constituent of photosynthetic bacteria. Its presence was considered as possibly being due to pheophytinization of Bchl during the RC isolation process (Norris *et al.*, 1975). Most workers in the field, however, do not consider this to be the case. The presence of Bpheo in RCs in a precise stoichiometric amount argues that

it is not merely a contaminant. Furthermore, spectroscopic changes have been observed in RCs, indicating a functional role for Bpheo in the primary photochemistry (see Section 5.1.1).

5.1.4. Removal and Replacement of Tetrapyrroles

Exhaustive extraction of RCs with organic solvents removes all the tetrapyrrole pigments [although an as yet unidentified chromophore, which may be a Bchl degradation product, remains (Feher, 1971)]. The

Table 4. Chemical Composition of Reaction Centers from Different Bacterial Species

Cofactor	Structure	Cofactor Composition (mole fraction)		
		Rhodopseudomonas sphaeroides	Rhodospirillum rubrum	Chromatium vinosum
Bacteriochlorophyll	(structure; phytyl or geranylgeranyl)	4[a-d]	4[e]	ND[‡]
Bacteriopheophytin	(structure; phytyl or geranylgeranyl)	2[a-d]	2[e]	ND[‡]
Iron	Fe^{2+} (high spin)	1[†f,g]	1[†h]	1[h]
Ubiquinone	(structure)	1-2[*i-m]	1-2[*h,n]	<0.1[o]
Menaquinone (Vitamin K_2)	(structure)	—	—	1[o]

[a] Mauzerall (1972); [b] Reed and Peters (1972); [c] Straley *et al.* (1973); [d] Jolchine and Reiss-Husson (1975); [e] van der Rest and Gingras (1974); [f] Feher (1971); [g] (Feher *et al.* (1974); [h] Okamura (1975); Feher *et al.* (1972); [j] Slooten (1972a); [k] Jolchine and Reiss-Husson (1974); [l] Prince *et al.* (1974); [m] Okamura *et al.* (1975); [n] Smith *et al.* (1972); [o] Okamura *et al.* (1976).

* There have been reports, however, of less than 1 UQ/RC (Clayton and Yau, 1972; Noël *et al.*, 1972) (see the text).

† Removal of Fe does not abolish primary photochemistry.

‡ ND: Not determined.

resulting extracted protein is inactive. As yet, there has been no success in partially and selectively removing either Bchl or Bpheo.

Loach et al. (1975) have, however, reported experiments that may indicate that chemical exchange between RC-bound Bchl and added Bchl is possible. They observed that the line width of the light-induced EPR signal in chromatophore preparations was decreased in the presence of deuterated Bchl. They proposed a model of the RC in which 4 Bchl's are involved in the primary donor. Although these experiments may be complicated by the oxidation of exogenous Bchl, the possibility of exchanging Bchl molecules offers a promising method for probing the role of Bchl in RCs.

5.1.5. Function of Bacteriochlorophyll and Bacteriopheophytin

The RC contains 4 Bchl and 2 Bpheo molecules (see Table 4). Two of the Bchl molecules are involved in the primary donor D (i.e., P_{865}). The evidence for this assignment is summarized below (for further details, see Norris and Katz, Chapter 21).

(a) Bleaching of the Optical Absorption Band at 865 nm Due to Oxidation of the Primary Donor. This bleaching, as well as the appearance of a band at longer wavelengths (1250 nm), is characteristic of the oxidation of Bchl and has been associated with the primary photoact (Parson, 1968). The difference in the spectrum of P_{865} and Bchl may be due to interactions between Bchl molecules.

(b) Formation of a Narrow EPR Signal Due to Oxidation of the Primary Donor. The g value ($g = 2.0026$) is identical with $Bchl^+$. The line width is indicative of a Bchl dimer.

(c) ENDOR Spectrum of the Primary Donor. The features of the hyperfine splittings of D^+ are similar to $Bchl^+$. The reduction in the magnitude of the hyperfine splitting of D^+ by one-half shows that D^+ is a Bchl dimer.

The functional role of Bpheo is not as well established as that of the Bchl dimer. The presently accepted working hypothesis is that one Bpheo is associated with the intermediate acceptor, I (Fajer et al., 1975). Evidence for this assignment comes from otpical spectroscopy (Rockley et al., 1975; Fajer et al., 1975) and measurements of redox potential (Fajer et al., 1975). The EPR and ENDOR results are consistent with this assignment (Feher et al., 1977), although at present they do not enable one to discriminate between $Bchl^-$ and $Bpheo^-$.

Thus, we can account for 2 of the 4 Bchl molecules and 1 Bpheo molecule. What role, if any, do the other

tetrapyrrole molecules play? We do not know. Perhaps they are involved in some transient electron-transfer role at an earlier stage in the photochemical process, either as part of a larger donor aggregate or as acceptor species.

5.2. Iron

5.2.1. EPR Spectroscopy

The idea that metal ions (besides Mg, which is part of the Bchl molecule) could play a role in the primary photochemistry of RCs was suggested by the observation of a broad, light-induced EPR signal at cryogenic temperatures by McElroy et al. (1970). A typical, up-to-date trace obtained in RCs of Rp. sphaeroides R-26 is shown in Fig. 14. Besides the narrow EPR signal at $g = 2.0026$, a broad signal at $g = 1.8$ is observed. In the original trace, smaller peaks at lower magnetic fields were also seen (Feher, 1971). These have now been attributed to Mn (see Section 5.2.3). The low-temperature (80°K) kinetics of the broad and narrow EPR signals were found to be identical (Feher, 1971;

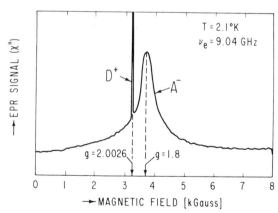

Fig. 14. Light-induced EPR signal from RCs of Rhodopseudomonas sphaeroides R-26. The EPR signal was obtained (2.1°K and 9.04 GHz) with 4 Hz actinic light modulation ($\lambda = 855$ nm, bandwidth = 50 nm, actinic light intensity ≈ 100 mW/cm²), and is proportional to the light-induced absorption (χ''), not to the usual derivative ($d\chi''/dH$) seen with magnetic field modulation (see Fig. 12). The sample was in a flat quartz cell, 50% glycerol (vol. = 0.1 ml, $A_{800}^{1\,cm} = 15$). The narrow signal (D^+) $g = 2.0026$ is due to the oxidized donor. The broad signal (A^-) $g = 1.8$ is due to the reduced acceptor. Similar signals were seen in chromatophores, except that the A^- signal was approximately 20% narrower. The A^- signal in RCs could be narrowed by 10–20% on addition of 10^{-3} M o-phenanthroline. An additional peak has occasionally been observed at $g \approx 2.2$. This peak can be eliminated by using high RC concentrations ($A_{800} > 10$) and 50% glycerol. Other peaks observed earlier at 0.7, 1.6, and 2.8 kG (McElroy et al., 1970; Feher et al., 1971) are due to Mn (see Fig. 16).

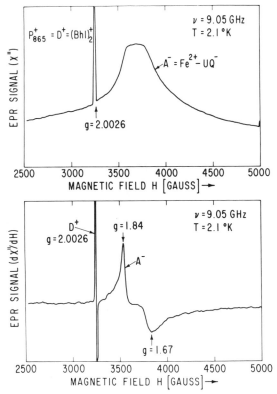

Fig. 15. Narrow $g = 2.0026$ and broad $g = 1.8$ signals observed in RCs of *Rhodopseudomonas sphaeroides* R-26. Conditions the same as in Fig. 14. The top trace shows the EPR absorption obtained with light modulation. The signal is proportional to the susceptibility χ''. The bottom trace was obtained with magnetic field modulation under constant illumination, which gives a signal proportional to the derivative $d\chi''/dH$. Similar signals were observed from RCs of *Rhodospirillum rubrum* and *Chromatium vinosum* with small differences in the shape of the broad signal. Reproduced from Feher *et al.* (1975*b*).

Feher *et al.*, 1974). Both these signals were also observed in chromatophores of *Rp. sphaeroides*, although the line width of the broad signal was consistently found to be approximately 20% less than in RCs (Feher *et al.*, 1975*b*). Addition of 10^{-3} M *o*-phenanthroline to RCs and chromatophores had the effect of narrowing the line widths of the $g = 1.8$ signals by approximately 10–20% (Feher *et al.*, 1975*b*). Dutton and Leigh observed a similar broad signal in chromatophores from *Chr. vinosum*, as well as from RCs of *Rp. sphaeroides* R-26 (Leigh and Dutton, 1972; Dutton and Leigh, 1973; Dutton *et al.*, 1973). The difference in appearance between their spectra and the spectrum seen by McElroy *et al.* (1970) is due to their use of a derivative ($d\chi''/dH$) presentations, as opposed to an absorption presentation (χ'') in the former work (see Fig. 15). Dutton

and Leigh (1973) determined that the redox potential of the broad signal was the same as that expected for the primary acceptor (-50 mV, pH 7.0) in *Rp. sphaeroides* RCs. The broad EPR signal was assigned to the reduced form of the primary acceptor (see Bolton, Chapter 22).

5.2.2. Chemical Analysis

Since a broad EPR signal is often characteristic of transition-metal ions, atomic absorption spectroscopy was used to analyze for various metal ions (Feher, 1971). It was indeed found that purified RCs of *Rp. sphaeroides* R-26 contain approximately 1 Fe/RC. A similar Fe content was found in RCs of *Rs. rubrum* G-9 and *Chr. vinosum* (Okamura, 1975). A puzzling result of the analysis of RCs of *Rp. sphaeroides*, however, was that the Fe content per RC was persistently lower than unity (0.8–0.9). Another indication of the low Fe content was the observation that the Mg/Fe ratio was 5:1 instead of the expected ratio of 4:1 (Feher, 1971).

5.2.3. Replacement of Fe with Mn

The puzzle of the low Fe content was resolved by the observation that Fe could be partially replaced by Mn in RCs of *Rp. sphaeroides* (Feher *et al.*, 1974). This was accomplished by growing the bacteria in a Mn-rich medium. By this expedient, the Mn content could be increased to approximately 0.3 Mn/RC; concomitantly, the Fe content in these RCs dropped to 0.7 Fe/RC. At the other extreme, by using a low-Mn growth medium, it was possible to prepare RCs with

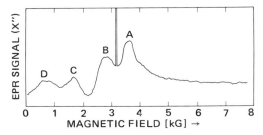

Fig. 16. EPR spectrum from RCs from *Rhodopseudomonas sphaeroides* R-26 containing 0.7 Fe/RC and 0.3 Mn/RC. The bacteria were grown in a medium containing 20 ppm Mn. The spectrum was taken with 4 Hz actinic light modulation (temp. = 2.1°K, $\nu = 8.96$ GHz). The sample absorbance was $A_{800}^{1\,cm} = 5$. Peaks, C, D, and most of B do not appear in samples that contain no Mn. Peak B is also largely due to Mn. In this sample, however, there is a contribution from Fe to peak B that can be eliminated, as was done in the sample of Fig. 14. The decay kinetics of all peaks were found to be identical ($\tau_D = 26 \pm 3$ msec). Reproduced from Feher *et al.* (1974).

0.02 Mn and 0.99 Fe/RC. The complementarity between the Fe and Mn contents indicates that there is a metal-binding site on the RC that can accept either of these transition metals.

Evidence that Mn replaces Fe at a functional site was obtained from EPR and low-temperature kinetics experiments (Feher *et al.*, 1974). In RCs containing Mn, an additional broad light-induced EPR signal was observed at lower magnetic field (see Fig. 16). It is this EPR signal that was superimposed on the originally observed $g = 1.8$ signal, as discussed in Section 5.2.1. The two broad signals due to Mn and Fe are sufficiently well resolved to provide a method of spectroscopically separating RCs containing Mn from those

containing Fe. The low-temperature, light-induced kinetics of the EPR signals were found to be identical for both the Mn and Fe peaks.

5.2.4. Removal of Fe

Although the RCs from *Rp. sphaeroides* R-26 contain approximately 1 Fe/RC, there were several reports of preparations that contained significantly less than stoichiometric amounts of Fe (Reiss-Husson and Jolchine, 1972; Loach and Hall, 1972). The lack of Fe does not prove, however, that it has no functional role in the primary photochemistry, unless evidence is presented showing that the photochemical activity, as

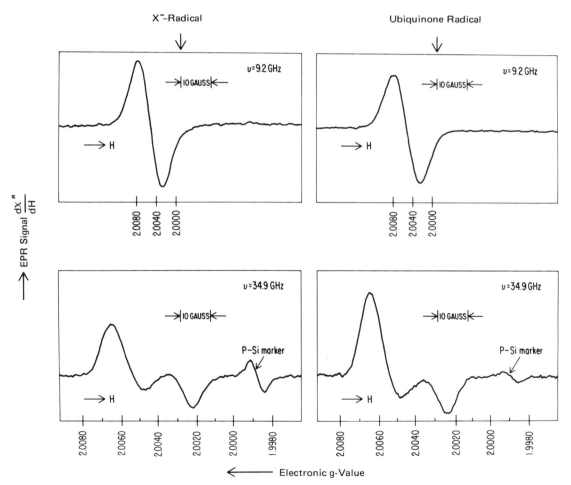

Fig. 17. EPR signal from a reduced acceptor species (X⁻) in Fe-depleted RCs of *Rhodopseudomonas sphaeroides* (left) compared with UQ (Q-50) radical (right) at microwave frequencies of 9.2 and 34.9 GHZ (temp. = 1.3°K). The P(865)⁺ radical has been reduced with cytochrome *c* to reveal the pure spectrum of X⁻. The identical *g*-values and line shapes have been taken as evidence that X⁻ is a UQ radical. Reproduced from Feher *et al.* (1972). The narrow EPR signal is replaced by a broad ($g = 1.8$) signal in Fe-containing RCs (see Figs. 14 and 15).

defined in Section 2, remains unimpaired in Fe-depleted RCs. Such experiments were performed by Loach and Hall (1972) on photoreceptor units from *Rs. rubrum* that contained approximately 0.3 Fe/unit. These preparations were active at cryogenic temperatures and exhibited the same decay kinetics as chromatophores (Loach *et al.*, 1975). In these preparations, a narrow, light-induced EPR signal was observed (at 9 GHz, $\Delta H_{ptp} \equiv 7.0 \pm 0.3$ gauss, $g = 2.0050 \pm 0.0003$). Subsequently, Feher *et al.* (1972) were able to remove Fe from RCs of *Rp. sphaeroides* R-26 by treating them with a mixture of two detergents (0.1% SDS, 0.02% LDAO). (A gentler treatment for removing Fe is given in Appendix C.) These Fe-depleted RCs were photochemically active at cryogenic temperatures and exhibited a narrow EPR signal (at 1.4°K, 9 GHz, $g = 2.0046 \pm 0.0002$, $\Delta H_{ptp} = 8.1 \pm 0.5$ gauss), similar to the one observed by Loach and Hall (1972) in place of the broad ($g = 1.8$) signal observed in Fe-containing preparations. The new signal was identified by EPR spectroscopy at 9 and 35 GHz as being due to a UQ radical (see Fig. 17).

5.2.5. Determination of the Valence of Fe

(a) Mossbauer Spectroscopy. If Fe were the primary acceptor, one would expect its valence to change on reduction, either by actinic light or by a chemical reductant. Mossbauer spectroscopy provides a sensitive technique to determine the valence of Fe, and was applied by Debrunner *et al.* (1975) to RCs prepared from bacteria grown in an ^{57}Fe-enriched medium. Mossbauer spectra were taken before and after reduction with dithionite. The extent of reduction was estimated from the amplitude of the broad $g = 1.8$ signal to be at least 90%. The Mossbauer spectra of nonreduced and reduced RCs show essentially the same isomer shifts (I.S.) and quadrupole splittings (ΔE_Q), indicating that the valence of Fe does not change on reduction of the primary acceptor (see Fig. 18). The values of I.S. and ΔE_Q are characteristic of Fe^{2+} (high-spin) (Brady *et al.*, 1965).

(b) Susceptibility Measurements. The d.c. magnetic susceptibility, χ_0, depends on the spin state of the Fe, which, in turn, depends on its valence. Susceptibility measurements were performed on an RC sample of *Rp. sphaeroides* R-26 in the temperature range of 0.7–200°K (Johnson *et al.*, 1976). From the high-temperature data, a magnetic moment of the Fe of 5.7 ± 0.4 Bohr magnetons was obtained. This is in agreement with observed values of high-spin Fe^{2+} of 5.1–5.7 Bohr magnetons (Figgis and Lewis, 1960). No significant change in χ_0 was observed when the acceptor was chemically reduced. This result corroborates the Mossbauer experiments; i.e., the susceptibility is due predominantly to the high magnetic moment of Fe^{2+}, which does not change during the reduction of the acceptor.

A more detailed analysis of the Mossbauer and susceptibility measurements at different temperatures and magnetic fields should give information concerning the different excited levels of the Fe^{2+} (see Section 8.4).

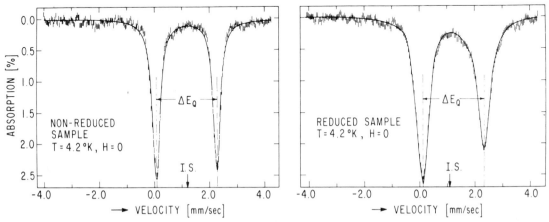

Fig. 18. Mossbauer spectra of native (nonreduced) and reduced RCs from *Rhodopseudomonas sphaeroides* (enriched approximately 80% in Fe^{57}) (temp. = 4.2°K, H = 0, pH 8, 0.1 LDAO). Both spectra were fitted by two Lorentzians (solid line). The separation of quadrupole splittings (ΔE_Q) and isomer shifts (I.S.) is characteristic of Fe^{2+} (high-spin), showing that the Fe does not change valence on reduction of the acceptor. The reduced sample, which was treated with 50 mM NaS_2O_4, exhibited the broad $g = 1.8$ EPR signal and was estimated to be 90% reduced. The broadening and asymmetry of the spectrum of the reduced sample can be accounted for by a magnetic interaction between the Fe and a nearby paramagnetic radical species. Velocity scale is relative to Fe. Reproduced from Debrunner *et al.* (1975).

5.2.6. Evidence Against Fe Alone Being the Primary Acceptor

Although the presence of Fe in RCs (see Table 4), together with the observation of a broad, light-induced EPR signal, suggested that Fe may be the primary acceptor, the evidence presented in the previous sections argues strongly against it. Let us briefly summarize the main arguments.

(a) Retention of Photochemical Activity by Fe-Depleted RCs. Although it may be argued that a nonphysiological acceptor takes over the function of Fe when the latter is removed, the unaltered low-temperature kinetics in Fe-free preparations indicates that the acceptor has not been changed.

(b) Identity of the Low-Temperature Kinetics of Reaction Centers Containing Mn and Fe (see Section 5.2.3). The donor–acceptor charge-recombination time is expected to be strongly dependent on the redox properties of the acceptor. Since Mn and Fe have different redox potentials (Loach, 1970b), one should observe different kinetics in Mn- and Fe-containing RCs if Fe were the primary acceptor. This was not found to be the case.

(c) Unchanged Valence State of Fe on Reduction of the Primary Acceptor [see Section 5.2.5(c)]. This provides the strongest evidence that Fe is not the primary acceptor.

5.2.7. Function of Fe

If Fe is not the primary acceptor, what then is its function? Since most RC preparations contain Fe in stoichiometric quantities, it is reasonable to assume that it does fulfill a specific role. As we shall show in Section 5.3, Fe is in proximity to two UQs that form the primary and secondary electron acceptors. This close association between Fe and UQs leads to the hypothesis that the function of the Fe is to serve as a path for electron transfer between the primary and secondary electron acceptors ("an iron wire"). Such a transfer may be facilitated by a partial delocalization of the wave function of the odd electron of the UQ onto the Fe.

It should be noted that Fe–S proteins (e.g., ferredoxins) are known components of electron-transfer chains. These proteins, however, characteristically contain 2 or more Fe and 2 or more labile sulfides (Palmer, 1973). The chemical composition of the RC (1 Fe, no labile sulfide) shows that it is *not* such an Fe–S protein.

5.3. Quinones

5.3.1. EPR and Optical Spectroscopy

We discussed in Section 5.2.4 that photoreceptor units from *Rs. rubrum* (Loach and Hall, 1972) and

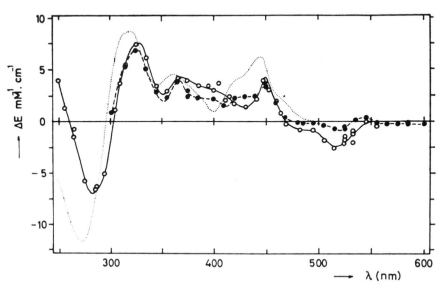

Fig. 19. Optical absorption changes due to reduction of an acceptor in RCs of *Rhodopseudomonas sphaeroides* (O and ●) compared with the optical absorption changes for the formation of ubisemiquinone anion (·······). (O) Obtained from a kinetic analysis of room-temperature light-induced absorbance changes; (●) obtained by illumination in the presence of 10 mM ascorbate to reduce oxidized P_{870}. Taken from Slooten (1972b). Similar changes were seen earlier by Clayton and Straley (1970).

RCs from *Rp. sphaeroides* R-26 (Feher *et al.*, 1972) from which Fe had been removed exhibit an EPR signal that is characteristic of a semiquinone radical (see Fig. 17). The low-temperature kinetics of this light-induced EPR signal made UQ a serious contender for the role of the primary acceptor (see Bolton, Chapter 22).

Independent evidence for the involvement of Qs in RCs of *Rp. sphaeroides* R-26 came from the observation of optical absorbance changes associated with the reduction of an acceptor (Clayton and Straley, 1970, 1972; Slooten, 1972*b*). To separate these optical changes from those due to the oxidation of P_{865}, Clayton and Straley (1970, 1972) added electron donors so that P_{865} remained fully reduced. Slooten (1972*b*) obtained similar optical absorption changes by kinetic analyses of the light-induced optical changes as well as by illumination in the presence of a weak reductant. In Fig. 19, the absorbance changes in RCs of *Rp. sphaeroides* observed by Slooten are compared with the difference spectra for the *in vitro* conversion of Q to semiquinone (Land *et al.*, 1971). The similarity between these optical changes was taken as evidence that the reduced acceptor species is ubisemiquinone (Slooten, 1972*b*).

Evidence that the UQ is the primary acceptor comes from the observation that the optical absorbance changes were unaltered in the presence of *o*-phenanthroline, a known inhibitor of the electron transfer from the primary to the secondary acceptor (Clayton and Straley, 1972).

Additional evidence concerning the role of UQ as a primary *and* secondary electron acceptor has come from recent flash experiments of Vermeglio and Clayton (1977) and Wraight (1977). These authors studied the absorbance changes (ΔA_{450}) due to semiquinones in RCs of *Rp. sphaeroides* after a series of short light flashes. In the absence of exogenous UQ, they found an increase in A_{450}, after the first and third flash, indicating that a ubisemiquinone anion, Q^-, was formed. After the second flash, the absorbance at A_{450} decreased. These experimental results were interpreted by the following reaction scheme:

where D_2 is an exogenous secondary electron donor (e.g., cyt^{2+} or diaminodurene) and Q_1 and Q_2 are the primary and secondary electron acceptors. In the presence of exogenous ubiquinone, the oscillatory behavior of the 450 nm absorbance changes continued for many flashes (Vermeglio, 1977; Wraight, 1977). This oscillatory behavior can be explained if an additional constraint is imposed on the model, namely, that Q_1^- reacts rapidly with the quinone pool but Q_2^- does not. The half-time for electron transfer from Q_1 to Q_2 has been measured to be 200 μsec (Vermeglio and Clayton, 1977).

Flash experiments have been used to measure the EPR spectra of the reduced primary and secondary quinones. Wraight (1977) observed the oscillatory behavior of the $g = 1.8$ signal after odd flashes in the presence of excess quinone, indicating that the secondary quinone was also coupled to the Fe^{2+}. Okamura *et al.* (1978) confirmed this result and showed that the EPR signal of the secondary quinone, while similar to that of the primary quinone (both being centered around $g = 1.8$), had an ≈30% broader line width. The observation of magnetic coupling between both Q_1 and Q_2 and Fe^{2+} is consistent with the iron-wire hypothesis (see Section 5.2.7).

When cytochrome *c* was used as the exogenous electron donor, D, the absorbance changes accompanying the oxidation of cyt^{2+} were used to corroborate this scheme (Okamura *et al.*, 1977*b*). When RCs containing either 2 UQs/RC or 1 UQ/RC were illuminated in the presence of reduced cytochrome *c*, a total of 3.0 or 1.0 cytochrome/RC, respectively, were rapidly oxidized.* The monitoring of the cytochrome oxidation provides a convenient and accurate method of measuring the number of Qs attached to the reaction center (see also Section 5.3.3).

5.3.2. Chemical Analysis

Several workers have determined the UQ content of RCs by chemical methods. The results vary greatly

---

* The cytochrome actually continues to be oxidized at a very slow rate that presumably is determined by the slow loss of electrons from $DQ^-Fe^{2+}Q^{2-}$.

1st flash

$$DQ_1Fe^{2+}Q_2 \xrightarrow{h\nu} D^+Q_1^-Fe^{2+}Q_2 \xrightarrow[\;]{D_2 \quad D_2^+} DQ_1^-Fe^{2+}Q_2 \longrightarrow DQ_1Fe^{2+}Q_2^-$$

2nd flash

$$DQ_1Fe^{2+}Q_2^- \xrightarrow{h\nu} D^+Q_1^-Fe^{2+}Q_2^- \xrightarrow[\;]{D_2 \quad D_2^+} DQ_1^-Fe^{2+}Q_2^- \longrightarrow DQ_1Fe^{2+}Q_2^{2-}$$

3rd flash

$$DQ_1Fe^{2+}Q_2^{2-} \xrightarrow{h\nu} D^+Q_1^-Fe^{2+}Q_2^{2-} \xrightarrow[\;]{D_2 \quad D_2^+} DQ_1Fe^{2+}Q_2^{2-}$$

according to the purity of the preparations and the bacterial species. Relatively impure RCs of *Rp. sphaeroides* were found to contain up to 10 UQs (Reed, 1969; Reiss-Husson and Jolchine, 1972). Most of these probably "carried over" from the chromatophores that are known to contain a large excess of UQ with respect to RCs (Lester and Crane, 1959). Many workers have observed a UQ content of 1–2 UQs/RC in RCs from *Rp. sphaeroides* R-26 (Feher *et al.*, 1972; Prince *et al.*, 1974; Okamura *et al.*, 1975), *Rp. sphaeroides* Y (Jolchine and Reiss-Husson, 1974), *Rp. sphaeroides* wild-type (Slooten, 1972a), *Rs. rubrum* G-9 (Okamura, 1975), and *Rs. rubrum* wild (Smith *et al.*, 1972).

Purified RCs from *Chr. vinosum* have been found to contain no UQ; instead, they contain approximately 1 molecule of menaquinone (vitamin K) (Okamura *et al.*, 1976; Feher and Okamura, 1977; Romijn and Amesz, 1977). This recent finding explains the seemingly puzzling observation that chromatophores that have been extensively extracted to remove all UQ retained the ability to perform the primary photochemistry (Parson and Cogdell, 1975). RCs from *Rp. viridis* have also been reported to contain menaquinone (Pucheu *et al.*, 1976).

On the other hand, several workers have found significantly less than 1 UQ/RC in fully active RC preparations of *Rp. sphaeroides* (Clayton and Yau,

1972) and *Rs. rubrum* (Noël *et al.*, 1972). We do not understand these latter findings, and are at present inclined to ascribe them to an unexplained artifact.

5.3.3. Ubiquinone Removal, Readdition, and Replacement by Other Quinones

When chromatophores and RC preparations were extracted with hydrocarbons such as isooctane, UQ was removed, although the capacity to perform the primary photochemistry was not impaired. Presumably only the UQ associated with secondary electron transfer were involved (Bolton and Cost, 1973; Halsey and Parson, 1974).

Cogdell *et al.* (1974) showed that by extracting RCs from *Rp. sphaeroides* R-26 with isooctane containing a trace amount (0.1%) of methanol, the primary photochemistry was abolished; it could be restored by adding either UQ or naphthaquinone to the depleted RCs.

A more quantitative and thorough investigation of the removal of UQ and reconstitution of RCs with Qs was made by Okamura *et al.* (1975). They developed a method of UQ removal and readdition that avoids the use of organic solvents, thereby reducing the danger of denaturing the reaction centers. They found that UQ could be removed from RCs of *Rp. sphaeroides* R-26 by treatment with varying amounts of

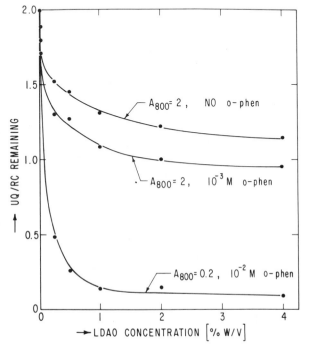

Fig. 20. Removal of UQ from RCs of *Rhodopseudomonas sphaeroides* R-26 with LDAO and *o*-phenanthroline (*o*-phen). One UQ is weakly bound and can be removed by a mild treatment (upper two curves, incubation time 3 hr); the other is more tightly bound (lower curve, incubation time 6 hr). A second incubation in 4% LDAO, 10^{-2} MM *o*-phenanthroline, $A_{800}^{1\ cm} = 0.2$, resulted in less than 0.05 UQ/RC. The binding of UQ to these RCs is shown in Fig. 21. Reproduced from Okamura *et al.* (1975).

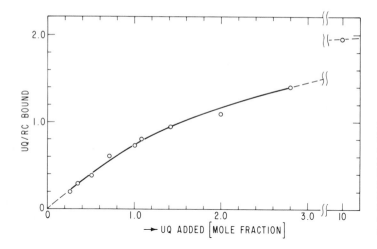

Fig. 21. Binding of UQ (UQ-50) to UQ-depleted RCs of *Rhodopseudomonas sphaeroides* R-26 (see Fig. 20). RCs ($A_{800}^{1\,cm} = 2$) containing less than 0.05 UQ/RC were incubated with ^{14}C-labeled UQ in 1% LDAO for 10 hr at 4°C. Unbound UQ was removed by passage through DEAE cellulose. Up to 2 UQs/RC could be bound per RC when a high concentration of UQ was used. Reproduced from Okamura *et al.* (1975).

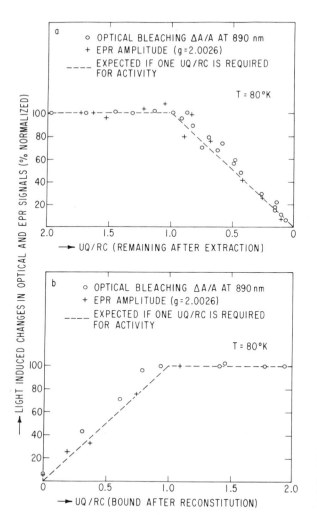

Fig. 22. Relationship between low-temperature photoactivity (normalized to untreated RCs) and the number of UQs per RC. (a) Obtained with RCs exposed to varying concentrations of LDAO and *o*-phenanthroline; (b) obtained with addition of ^{14}C]UQ in varying amounts to RCs that had less than 0.05 UQ/RC. The experimental points fall close to the values (‒‒‒) expected if 1 UQ/RC is required for activity. The Fe content ($\approx$1 Fe/RC) remained unaffected by the extraction procedure. Reproduced from Okamura *et al.* (1975).

LDAO and o-phenanthroline, as shown in Fig. 20. Initially, purified RCs contained close to 2 UQs/RC. One of the UQs could be removed by mild treatment. The ease of removal of this UQ probably accounts for the variable amounts of UQ (between 1 and 2 UQs/RC) observed by various workers. The other UQ was removed under more rigorous conditions. After the treatment, the Fe content of the RCs remained unaltered at 1 Fe/RC. The binding of UQ (UQ-50) to UQ-depleted RCs is shown in Fig. 21. When UQ was added in excess, 2 UQs/RC were bound. The results cited above show the RCs of *Rp. sphaeroides* R-26 contain two UQ-binding sites.

The photochemical activity of untreated, UQ-depleted and reconstituted RCs was measured at cryogenic temperatures, using the criteria discussed in Section 2, i.e., the light-induced change in the optical absorbance of the 890 nm peak (see Fig. 1A) and the amplitudes and low-temperature kinetics of the two light-induced EPR signals at $g = 2.0026$ and $g = 1.8$ (see Figs. 1B and 14). In UQ-depleted RCs, the light-induced optical and EPR signals were virtually eliminated. On addition of UQ, photochemical activity was fully restored. The quantitative relationship between low-temperature photochemical activity and UQ content was determined on RC samples at intermediate stages of UQ removal and reconstitution. The results shown in Fig. 22 show that only one UQ is required for photochemical activity.

Besides UQ, other Qs were used by Okamura *et al.* (1975) to reconstitute UQ-depleted RCs. A list of Qs that were successfully incorporated is shown in Table 5. The two parameters that are believed to reflect the environment of the primary acceptor are the decay time τ_D and the line width of the broad EPR signal. Both of these changed when RCs were reconstituted with Qs other than UQ, thus showing that the Q plays a major role in the primary photochemistry. The results also indicate that although neither the isoprenoid chain nor the methoxy groups of UQ are absolutely required for binding and activity, "native" photochemical activity (according to our definition in Section 2) is obtained only with the physiological acceptor, UQ. It is interesting to note that when RCs from *Rp. sphaeroides* are reconstituted with menaquinone obtained from *Chr. vinosum* RCs, the low-temperature decay kinetics τ_D increases from 29 to 35 msec, whereas in RCs of *Chr. vinosum*, τ_D is 29 msec (McElroy *et al.*, 1974). Evidently, the bacteria evolved in such a manner that their respective RC proteins sterically accommodate a different Q to maintain the same decay time. In addition, since menaquinones (vitamin K) generally have lower redox potentials (by approximately 150 mV) than UQ (Loach, 1970b), the replacement of UQ by menaquinone in *Chr. vinosum* may explain the lower redox potential observed in this species (Parson and Cogdell, 1975).

Table 5. Low-Temperature Decay Kinetics and EPR Line Widths of the Acceptor in Reaction Centers of *Rhodopseudomonas sphaeroides* R-26 Reconstituted with Different Quinones[a]

COMPOUND	STRUCTURE	DECAY TIME τ_D[†] [msec]	LINE WIDTH ΔH [GAUSS][‡]
UBIQUINONE-50	CH₃O CH₃ structure	29±2	600±30
UBIQUINONE-30	CH₃O CH₃ structure	29±2	600±30
PLASTOQUINONE-C	CH₃ structure	32±2	680±40
DUROQUINONE	CH₃ structure	105±10	1000±60
MENADIONE	CH₃ structure	37±3	740±40
PHYLLOQUINONE (VITAMIN K₁)	CH₃ $C_{16}H_{33}$ structure	34±2	760±40
ANTHRAQUINONE	structure	135±10	740±40
UNTREATED REACTION CENTERS		28±2	600±30

[a] UQ-depleted RCs were incubated for 10 hr at 4°C in the presence of different Qs, solubilized in 1% LDAO, pH 8. The EPR samples were prepared in 50% glycerol ($A_{800}^{1 \, cm} = 10$). After Okamura *et al.* (1975) (in the original publication, the protons on the isoprenoid chain were inadvertently omitted).

[†] τ_D is the time required for the narrow EPR signal ($g = 2.0026$) to reach $1/e$ of its amplitude after cessation of illumination (measured at temp. = 80°K).

[‡] ΔH is the full width at half amplitude of the broad $g = 1.8$ signal (see Fig. 13), measured at 2.1°K.

5.3.4. Interaction of Quinone with Fe—The Ferroquinone Complex

It was shown in Section 5.2.4· that RCs devoid of Fe exhibit a narrow EPR signal (see Fig. 17), whereas RCs devoid of UQ give essentially no EPR signal. On the other hand, when both Fe and UQ are present, a broad $g = 1.8$ EPR signal is observed (see Fig. 14). These findings point to the conclusion that both Fe and UQ are responsible for the characteristic broad $g = 1.8$ EPR signal (Okamura *et al.*, 1975). High-spin Fe^{2+}, having an even number of electrons, does not usually give rise to an EPR signal (e.g. Abragam and Bleaney, 1970). It can, however, interact magnetically with the spin of the semiquinone radical and shift and broaden the UQ signal (in this case by two orders of magnitude!). The species resulting from the close association of UQ with Fe was called a *ferroquinone complex* (Okamura *et al.*, 1975). Its existence had been postulated earlier in connection with the identity of the primary acceptor (Feher *et al.*, 1972, 1974; Bolton and Cost, 1973). It should be stressed, however, that the experimental evidence points only to a magnetic complex without implying anything concerning the importance of the complexing with regard to the primary photochemistry. The reason for this caution is that energies involved in chemical reactions are much larger than magnetic energies. Consequently, the finding that the EPR spectrum of UQ is greatly altered by the presence of Fe does not mean that the latter is essential to the primary photochemistry. Indeed, the facts summarized in Section 5.2.6 indicate that Fe plays a relatively minor role in the primary photoact, the electron being located mostly on the UQ.

The foregoing argument also explains the observation that whereas the EPR spectrum due to Q is greatly perturbed by the presence of Fe (compare Figs. 14 and 17), the optical absorbance changes arising from the Q reduction do not seem to be greatly perturbed by the presence of Fe (see Fig. 19). The reason is that the energies required to perturb an optical spectrum are of the order of chemical energies, which are several orders of magnitude larger than magnetic interaction energies. Thus, we conclude that the quinone is weakly coupled to the Fe. The structural relationship between Fe and UQ, as well as the detailed nature of their interaction, is at present not understood. We do know, however, that the primary UQ must be in a unique environment, since available evidence indicates that it acts as a one-electron acceptor (e.g., Parson and Cogdell, 1975), while Q in aqueous solutions act as two-electron acceptors (Morton, 1965). This difference may be due to the absence of protons in the environment of the primary UQ, and is consistent with its being buried in the hydrophobic interior of the protein. Loach *et al.* (1975) also pointed out that such a buried Q may have a very low redox potential. We do now know, however, what effect the Fe or the protein environment has on the redox behavior of the Q.

5.3.5. Function of the Tightly Bound Quinone

Although RCs from several bacterial species have been found to contain 2 Q/RC (see Table 4), several experimental observations presented in the previous sections point to the requirement of only 1 Q as the primary acceptor in RCs from *Rp. sphaeroides*. We summarize here the main evidence for this:

(a) Removal and Readdition of Ubiquinone. The results are presented in Fig. 20; they clearly show that only one UQ, the tightly bound one, is essential for the primary photochemistry.

(b) Replacement of Ubiquinone with Other Quinones. The low-temperature kinetics and line width of the broad, $g = 1.8$, EPR signal are different in RCs that have been reconstituted with other Qs (see Table 5). Such differences are to be expected from a change in the donor–acceptor distance or electronic properties brought about by the differences between the different Qs.

(c) Light-Induced EPR Spectrum of Fe-Depleted Reaction Centers. The spectrum is characteristic of a UQ radical (see Fig. 17); the low-temperature kinetics is the same as observed in chromatophores.

(d) Optical Absorbance Changes. When the acceptor is reduced, optical absorbance changes that are similar to those produced by a one-electron reduction of UQ are observed (see Fig. 19).

5.3.6. Function of the Loosely Bound Quinone

Since only the tightly bound Q is required for the primary photochemistry, it is natural to assume that the loosely bound Q serves as the secondary electron acceptor. Direct evidence that this is indeed the case was given by Halsey and Parson (1974). They showed that the electron-transfer reaction from the primary to the secondary electron acceptor in chromatophores of *Chr. vinosum* was abolished when the loosely bound Q was extracted and restored when Q was added back. More recent evidence that Q serves as the secondary electron acceptor comes from the flash experiments in RCs by Vermeglio (1977) and Vermeglio and Clayton (1977) and Wraight (1977), which were discussed in Section 5.3.1 (for a comprehensive review, see Parson, Chapter 25).

The ease of displacement of the loosely bound UQ by *o*-phenanthroline offers a simple explanation of the inhibitory effect of *o*-phenanthroline on the electron transfer from primary to secondary electron acceptors (Parson and Case, 1970). A possible mechanism of the removal of UQ is the chelation of *o*-phenanthroline to Fe, which may play a role in the attachment of the UQ. This suggests that the secondary (as well as the primary) UQ is in proximity to the Fe. A further interesting feature of this loosely bound UQ is the observation that its removal is facilitated by reducing conditions (Okamura *et al.*, 1975). This is consistent with a mechanism of electron transfer *in vivo* in which the reduced Q is released from the RC to a Q pool.

6. Other Chemical Moieties

6.1. Carotenoids

6.1.1. Chemical Analysis

Several workers have observed that RCs prepared from wild-type photosynthetic bacteria were found to contain carotenoids that varied in different species (Beugeling *et al.*, 1972; Smith *et al.*, 1972; Jolchine and Reiss-Husson, 1974; van der Rest and Gingras, 1974; Lin and Thornber, 1975; Ackerson *et al.*, 1975; Cogdell *et al.*, 1976). In several strains of *Rp. sphaeroides*, the major RC carotenoid was found to be spheroidene (Beugeling *et al.*, 1972; Jolchine and Reiss-Husson, 1974; Cogdell *et al.*, 1976). In *Rp. sphaeroides* Ga, however, it was chloroxanthin (Cogdell *et al.*, 1976), and in *Rs. rubrum* RCs, spirilloxanthin (Smith *et al.*, 1972; van der Rest and Gingras, 1974). Although it is possible that the carotenoid in RC preparations occurs as an impurity, several facts argue against this. First, the amount of carotenoid in highly purified RC preparations was recently determined to be stoichiometric (i.e., carotenoid/RC $\approx$ 1) (van der Rest and Gingras, 1974; Cogdell *et al.*, 1976). Second, the RC carotenoid does not reflect the carotenoid composition of the whole cell. This is particularly true for *Rp. sphaeroides* Ga, in which the RC carotenoid is chloroxanthin, while the major carotenoid in the organism is neurosporene (Cogdell *et al.*, 1976).

6.1.2. Function

One role of the carotenoid is to serve as a light-harvesting pigment. An indication of this role is the observation that light absorbed by the RC carotenoid effectively excites the fluorescence of the RC Bchl (Slooten, 1973; Cogdell *et al.*, 1975).

A second, perhaps more important, function is the protection of RCs from photochemical damage. Cogdell *et al.* (1976) showed that the RC carotenoid effectively quenches the state P^R, presumably a triplet state of the RC Bchl. This triplet-quenching may protect the RC from photooxidation, and may at least partially account for the fact that wild-type photosynthetic bacteria can grow in the presence of light *and* oxygen, while carotenoidless mutants are killed under these conditions (Sistrom *et al.*, 1956).

6.2. Carbohydrates

Since many membrane proteins are also glycoproteins (Steck and Fox, 1972), it was thought possible that RCs also contain carbohydrates. Consequently, RCs were tested for carbohydrate both by the phenol sulfuric acid method and by gas chromatography of the alditol acetate derivates (Okamura *et al.*, 1974). The former revealed less than 0.1% (wt./wt.) neutral sugar. The latter showed less than 0.1% (wt./wt.) sucrose, mannose, galactose, glucosamine, and galactosamine, but showed the presence of 0.2% (wt./wt.) glucose. In view of the results of the phenol sulfuric acid analysis, the presence of the small amount of glucose was probably due to an accidental contamination. It is thus concluded that the RC protein is *not* a glycoprotein.

6.3. Labile Sulfide

The presence of Fe and the observation of a broad $g = 1.8$ EPR signal suggested the possibility of an Fe—S center (Palmer, 1973). Since such centers are characterized by the occurrence of acid-labile sulfide (Malkin, 1973), RCs were treated with acid and analyzed for any released hydrogen sulfide. Olfactory analysis revealed less than 0.01 acid-labile sulfide/RC (Okamura *et al.*, 1974).

6.4. Pteridine

Pteridine has been considered as a possible candidate for the role of the primary acceptor. Reed and Mayne (1971) determined the pteridine content of purified *Rp. sphaeroides* R-26 RCs using a fluorimetric assay, and found it to be less than 0.15 pteridine/RC. This rules pteridine out as the primary acceptor.

6.5. Lipid

Reaction centers are membrane-bound proteins that are soluilized by means of detergent. Thus, they probably exist *in vivo* in close association with phospholipids (Singer and Nicolson, 1972). In the process of detergent solubilization and purification, most, if not all, of the phospholipid is removed from RCs and replaced by detergent. Purified RCs from *Rp. sphaeroides* were reported to contain less than 1 phosphorus/RC by Feher (1971). The transition of the RC from the phospholipid to a detergent environment seems to have no effect on the photochemical activity as indicated by low-temperature kinetic data (McElroy *et al.*, 1974). However, the role of phospholipids in subsequent reactions (e.g., cytochrome oxidation, UQ reduction) has not yet been explored. It is interesting to note that in RCs that still retain some phospholipid (≪1 phospholipid/RC), there is evidence for the occurrence of an unusual ornithine phospholipid (Steiner *et al.*, 1974*b*; Jolchine and Reiss-Husson, 1974).

6.6. Bound Detergent

As mentioned above, RC purification is accomplished by replacing phospholipid molecules associated with RCs *in vivo* with detergent molecules. The binding of detergent to RCs is therefore important not only in interpreting the properties (e.g., molecular weight) of the isolated RC, but also in anticipating further work on the incorporation of RCs into phospholipid vesicles and model membranes. The binding of LDAO to RCs was studied using ^{14}C-labeled LDAO (Ackerson *et al.*, 1975). RCs isolated with ^{14}C-labeled LDAO and dialyzed vs. 0.025% [^{14}C]LDAO have been found to bind 0.15 ± 0.02 mg detergent/A_{800} V. Using the value of 0.319 mg protein/A_{800} V for the RC (Steiner *et al.*, 1974*b*), this corresponds to 0.47 ± 0.06 mg LDAO/mg protein. This bound LDAO may explain the high molecular weight of 150,000 obtained by Reiss-Husson and Jolchine (1972) (see Section 4.3).

The kinetics of removal of LDAO was studied by dialyzing ^{14}C-labeled LDAO–RCs against different buffer solutions at 4°C, pH = 8.0 (Ackerson *et al.*, 1975). When dialyzed against buffer containing no LDAO, free LDAO and some loosely bound LDAO (≈30% of the total bound LDAO) were removed in the first 5–10 hr. After 4–5 days, an additional amount of approximately 20% of bound LDAO was removed, causing the RCs to precipitate. After approximately 30 days, there was still approximately 40% of the initially bound LDAO attached to the RCs. The removal of LDAO was not significantly accelerated by dialyzing against cold LDAO. Thus, a large fraction of the LDAO is very tightly bound to the RCs.

6.7. Cytochromes

Cytochromes are universally found in photosynthetic bacteria (see Bartsch, Chapter 13) and have been shown to be secondary electron donors (Parson, 1968, 1969). RCs can be prepared that contain no cytochrome, showing that cytochrome is not a primary reactant (Feher, 1971) (Gingras, Chapter 6). The interaction of RCs with exogenous cytochrome *c* has been studied by several workers (e.g. Ke *et al.*, 1970; Prince *et al.*, 1974). The electron-donating

capacity of cyt^{2+} has been used in several instances to elucidate the primary photochemistry (see Fig. 17 and Section 5.3.1).

7. Summary and Discussion

In this chapter, we have reviewed our present knowledge of the chemical composition and properties of RCs, placing special emphasis on the role of the individual components in the primary photochemistry. The three bacterial species from which highly purified RCs have been obtained are *Rp. sphaeroides*, *Rs. rubrum*, and *Chr. vinosum*. Of these, RCs from *Rp. sphaeroides* R-26 are the best characterized. We shall briefly summarize in this section the basic facts (referring to *Rp. sphaeroides* R-26 unless specified otherwise), point out unanswered questions, and mention several areas of inquiry that are characterized by almost complete ignorance. The latter includes the three-dimensional structure of the RC unit and the detailed mechanisms responsible for the remarkably high primary quantum conversion efficiency.

RCs prepared by detergent fractionation are composed of a pure* protein to which the primary acceptor and donor are attached as prosthetic groups. Some RCs have carotenoids associated with them; these increase the light-gathering efficiency and protect the reactants from oxidation, but they are not essential for the primary photochemistry. In addition, the RC has detergent molecules attached to it; they are required to keep the RC in solution.

The protein of RCs from *Rp. sphaeroides* R-26, *Rs. rubrum*, and *Chr. vinosum* is composed of three subunits present in equimolar proportions, called L, M, and H. Their combined molecular weight is at present estimated to be between 70 and 95 kdaltons. A more precise determination will require careful sedimentation work and a knowledge of the partial specific volume and the amount of detergent attached. The primary structure (sequence) of the protein has not been determined; the structure, of course, would give the most accurate molecular weight. The amino acid composition of the three subunits has been obtained in RCs from *Rp. sphaeroides*. The most striking feature is the large number of nonpolar residues, the H subunit being the most polar of the three. The amino acid composition of the RCs of the three bacterial species is similar, as is the electrophoretic mobility of their subunits. Antigenically, however, they were found not to be related.

---

* Since the protein has been rather well characterized, the time may be ripe to give it a name. Why not call it *PET* (photoelectron transferase)?!

The question whether all three subunits are essential for "native" primary photochemistry has not been answered unequivocally. Two of the subunits (LM) contain all the pigments and primary reactants and can perform the primary photochemistry at low temperatures. However, the kinetics of the back-reaction at low temperatures (charge recombination) was found to differ slightly from those in chromatophores and RCs containing all three subunits. This may be due either to an inherent conformational change of the LM unit or to a change brought about by the preparative procedure. At any rate, the conformation of the LM unit is unlikely to differ greatly from the *in vivo* configuration.

There are 6 tetrapyrrole pigments (4 Bchl and 2 Bpheo) attached to the RC. Two Bchl's comprise a specialized pair that functions as the primary donor, and one Bpheo has been postulated to function as an intermediate acceptor. The role of the other pigments is not known; perhaps they are also involved in the intermediate steps that intervene between the singlet excited state of Bchl and the appearance of the primary oxidizing and reducing entities. The preferential removal of Bchl and Bpheo and reconstruction offers a promising avenue to explore this question. The spatial arrangement of the pigments has not been determined.

In addition to the tetrapyrrole pigments, there are 2 UQs and 1 Fe in RCs of *Rp. sphaeraides* R-26. The UQ that is more tightly bound to the RC is magnetically coupled to the Fe^{2+} and forms the primary acceptor. The transferred electron in the reduced acceptor is localized mostly on the UQ, which can function as the primary acceptor in the absence of Fe. In *Chr. vinosum*, this "primary" Q is menaquinone; in *Rs. rubrum*, the question of the Q has not yet been definitively resolved. The second, more loosely bound UQ is also magnetically coupled to Fe and is assumed to function as the secondary acceptor. A possible role of the Fe^{2+} is to help shuttle the electron from the primary to the secondary acceptor. Again, little is known about the spatial arrangements of these moieties. By postulating a tunneling mechanism for the charge recombination in the dark, one can make at present only a rough estimate of the donor–acceptor distance from the observed decay kinetics. A more precise knowledge of this distance would be important in understanding the mechanism of charge separation and coupling to the rest of the electron-transfer chain (e.g., how does it occur across the chromatophore membrane?). A promising approach to determine this distance is based on the interaction between the spins located on Fe^{2+} and D^+. By studying the effect of Fe^{2+} on the spin-lattice relaxation time and line width of D^+, this interaction can be determined. Preliminary experiments performed in our laboratory have shown that this is feasible (e.g., the spin-lattice relaxation time of D^+ at cryogenic temperatures is shortened in the presence of Fe^{2+}).

Besides—"because of" might be more appropriate—our ignorance of the structural relationship between the primary reactants and the RC protein, there remain the unanswered questions concerning the detailed mechanisms leading to the charge separation. It is this basic process, which proceeds with such remarkably high quantum efficiency, that distinguishes the operation of the photosynthetic unit from other photochemical reactions in solutions. Each photon that is absorbed produces essentially one charged donor–acceptor pair with a lifetime (milliseconds) that is long in comparison with the coupling time (microseconds) to the electron-transfer chain. This corresponds to a quantum efficiency of unity, whereas most other photochemical reactions proceed with an efficiency one or two orders of magnitude lower. The reason is that in solution chemistry, the excited singlet states decay via wasteful paths and the charged species are not stabilized for long enough times. In RCs, the stabilization seems to be facilitated by a fast removal of the electron from the vicinity of the donor via intermediate states and by interposition of a potential barrier between the spatially separated reactants; the details of this process remain to be elucidated.

Another unresolved problem deals with the thermodynamic efficiency of the energy-transduction process. The photon having an energy of 1.4 eV ($\lambda = 870$ nm) produces an oxidized and reduced species the difference in redox potential of which is commonly assumed to be approximately 0.5 eV. If we assume a maximum efficiency allowable by the laws of thermodynamics of approximately 70% (for a review on this question, see, for instance, Knox, 1969), 0.5 eV of the incident photon is not utilized. Does this mean that nature has picked a process in which half the incident energy is wasted? Or is it, perhaps, the price one pays for the fast charge separation?

In summary, by the use of standard biochemical and physical techniques, the components of RCs have been well characterized. The components, however, define only the boundary conditions within which the primary photochemistry must operate. Their characterization constitutes the first necessary step in understanding the structure–function relationship of the RC. The second, more difficult problem of the detailed organization of the components into a functional unit, as well as the mechanism of action, remain largely unknown, and should provide a fruitful field of endeavor for the future. One should not forget, however, that the

RC represents only a small part of the photosynthetic machinery. Once the RC is understood, the challenge will be to reconstitute larger active units until the whole process of photosynthesis is understood and reproduced in the test tube.

ACKNOWLEDGMENTS

We thank P. G. Debrunner and L. A. Steiner, as well as the members of our group, L. Ackerson, R. Isaacson, D. Rosen, and G. Valkirs, for their permission to quote their results prior to publication. The work cited from our laboratory was supported by grants from the National Science Foundation (BMS-74-21413) and the National Institutes of Health (GM 13191).

Appendix A. Purification of Reaction Centers from *Rhodopseudomonas sphaeroides* R-26

Cells of *Rp. sphaeroides* R-26 are grown anaerobically in a modified Hutner medium that uses succinate as the sole carbon source and is devoid of extraneous sources of Mn ([Mn] <5 parts per billion). Culture bottles (1-liter volume) are inoculated with approximately 50 ml of bacteria that are in the late log phase of growth having an optical absorbance of $A_{865}^{1\,cm} \approx 1$. The sealed bottles are kept in the dark for approximately 10–12 hr and then exposed at 30°C to light (≈ 1 mW/cm^2) for 6 days and harvested in the stationary phase. The yield of cells is about 6 g wet wt. (whole cells)/liter culture. The cells are suspended in 10 mM Tris buffer, pH 8, at a concentration of 100 g wet wt./300 ml and broken by passage through a Sorvall-type RF-1 (Ribi) refrigerated cell fractionator at 20,000 p.s.i. DNAse (1 mg/300 ml) and Ca^{2+} and Mg^{2+} (1 mM) are added, and the mixture is stirred at room temperature for 1 hr, then centrifuged at 250,000g for 1 hr to obtain a pellet of chromatophores that can be stored at −20°C for future use.

A typical preparation starts with 90 g (wet wt.) of chromatophores that are suspended overnight by stirring in 10 mM Tris buffer at 4°C at a concentration of 0.3 g (wet wt.) chromatophores/ml solution. All operations are performed in dim green light. LDAO (30%) is added to a final concentration of 1.2%, and the solution is immediately centrifuged at 250,000g for 1 hr. The supernatant fraction, which contains the RCs, is carefully decanted and brought to room temperature (24 ± 2°C). Concentrated ammonium sulfate is added to the supernatant solution to a final concentration of 25% (wt./vol.). The solution is centrifuged (50,000g, 10 min), and the floating pellet is resuspended in 10 mM Tris, pH 8, to an absorbance of $A_{800}^{1\,cm} = 5$. To the subnatant solution, LDAO (30%) is added to give an additional concentration of 0.6% LDAO. The solution is centrifuged as before (50,000g, 10 min). The floating pellet from the second centrifugation is redissolved in 10 mM Tris ($A_{800}^{1\,cm} = 5$) and combined with the pellet from the first centrifugation. The RC solution is allowed to incubate at room temperature in the dark overnight while stirring. The next day, the solution is clarified by centrifugation at 50,000g for 30 min.

To the supernatant solution, 8 g Celite (Johns Manville) is added (≈ 1 g Celite/100 A_{800} V RC). Concentrated ammonium sulfate containing 0.1% LDAO at pH 8 is added (while the solution is stirred) to a final concentration of 25% (wt./vol.). The resulting slurry is filtered through a funnel supporting a thin bed of 4 g Celite. The material is then washed with 500–1000 ml 25% ammonium sulfate solution in TL buffers until the washings are nearly colorless. The RC–Celite mixture is suspended in 25% ammonium sulfate in TL buffer and packed in a chromatographic column (2.5 cm i.d.) and eluted with ammonium sulfate (25–15%, TL buffer) gradient (300 ml). The RC fraction is dialyzed vs. 0.025% LDAO at 4°C (3 × 6-liter changes). The last dialysis is performed against 0.1% LDAO. In order to prevent possible degradation of the RCs by proteases, all dialyses are performed in the presence of 10^{-4} M PMSF.

A DEAE (Whatman DE 52) column is prewashed with 1 M NaCl, and equilibrated with TL buffer. The RC sample is applied to the column and washed with 0.06 M NaCl in TL buffer, until the RCs begin to elute from the column. The RCs are then removed from the column with 0.12 M NaCl in TL buffer and dialyzed at 4°C vs. 0.025% LDAO 10 mM Tris, pH 8. At this stage, if the RCs are not completely pure, as judged by the ratio $A_{280}/A_{800} > 1.24$, the DEAE step is repeated. The RCs may be concentrated ($A_{800} \approx 30$) by applying them to a small DEAE column and removing them with 1 M NaCl, TL buffer.

A flow chart for this preparative procedure is given in Fig. 23.

Appendix B. Preparative Procedures for the Isolation of Subunits

B.1. Preparation of LM by Sucrose-Gradient Centrifugation

The preparation of the LM unit can be accomplished by sucrose-gradient centrifugation in a solu-

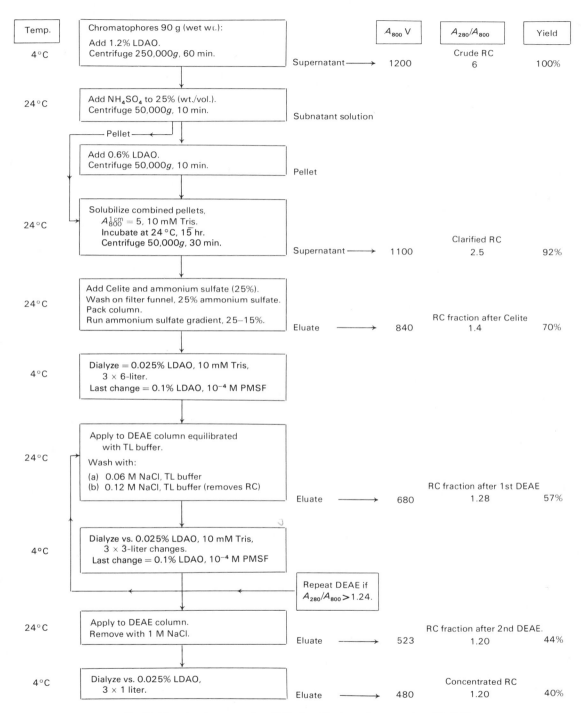

Fig. 23. Flow chart for preparation of RCs from *Rhodopseudomonas sphaeroides* R-26.

tion containing the chaotropic agent LiClO$_4$ (1.0 M), 0.05% LDAO, 50 mM Tris (pH 8.0). In a typical run, 5.3 ml of a sucrose gradient (0–10%) is layered into a centrifuge tube (5.5-ml maximum volume). RCs ($A^{1\,cm}_{800}$ = 10, 0.20 ml), also in 0.05% LDAO, 50 mM Tris (pH 8), are layered onto the top of the gradient. The tube is centrifuged at 65,000 rpm in a Spinco SW65 Ti rotor ($\approx$300,000g) for 20 hr at 20°C.

After centrifugation, the tube is pierced and 0.25-ml fractions are collected. The LM fraction is a blue band that sediments approximately 1.5 cm down the tube. The H fraction remains at the top. The LM prepared by this procedure is relatively pure, with less than 5% H, and approximately 50–60% active.

B.2. Preparation of LM by Precipitation of H

An alternate method of preparing LM that can easily be scaled up to produce large quantities has been developed by Rosen *et al.* (1977).

RCs ($A^{1\,cm}_{800}$ = 10) are heated at 40°C in 0.5 M LiClO$_4$, 10% ethanol, 50 mM CaCl$_2$ (0.025% LDAO, 10 mM Tris, pH 8) for 2 hr. All the H and about 15% of the LM precipitate and are centrifuged out (20 min at 2000g). The supernatant solution is dialyzed at 4°C vs. 0.025% LDAO, 10 mM Tris, pH 8, and contains pure LM (<2% H). The pellet is highly enriched in H,

containing about 85% H and 15% L and M. The purity of the product was assayed by SDS-PAGE. The results, together with a summary of the procedure, are shown in Fig. 24.

B.3. Preparation of L

LM prepared with LiClO$_4$, as discussed above, is concentrated on DEAE-cellulose, eluted with 1 M NaCl (TL buffer) and dialyzed against 0.025% LDAO, 10 mM Tris, pH 8, 1 mM EDTA ($A^{1\,cm}_{800}$ = 40). The sample is purged in a test tube, stoppered with a serum cap, with N$_2$ gas, and SDS is added to 1%. The tube is then evacuated and heated at 65°C for 1 hr. The sample is then passed over a Sepharose 4B column to which *p*-hydroxymercuribenzoate has been bound, using the method of Cuatrecasas and Anfinsen (1971). The unbound fraction contains M as well as a small amount (10%) of L. The material bound on the column is washed with 1% SDS, 100 mM Tris, pH 8 ($\approx$10-column vol.) and eluted with 50 mM cysteine in 1% SDS, 100 mM Tris, pH 8, 1 mM EDTA. The eluate is dialyzed vs. 10 mM Tris, pH 8 (4 hr at 23°C, then overnight at 4°C). The product contains pure L (<2% H,M). The electrophoretograms, together with the procedure, are shown in Fig. 25 (Rosen *et al.*, 1977).

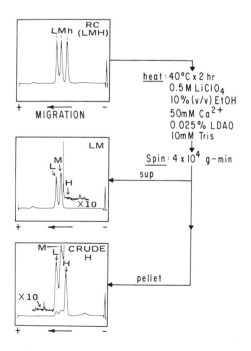

Fig. 24. Flow chart for preparation of LM units, with electrophoretograms of the products.

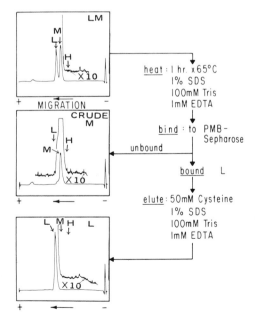

Fig. 25. Flow chart for preparation of L subunits, with electrophoretograms of the products.

Appendix C. Removal of Fe

RCs ($A^{1\,cm}_{800} = 10$) are incubated overnight at 20°C in 1 M LiClO$_4$, ≈0.2 mM o-phenanthroline (saturated solution in LiClO$_4$), 5 mM ascorbate (0.025% LDAO, 50 mM Tris, pH 8). The samples are subsequently centrifuged for 10 min at 2000g to remove precipitated H and dialyzed at 4°C vs. 0.025% LDAO, 10 mM Tris, pH 8. The Fe content was less than 0.05 Fe/RC.

8. Note Added in Proof

8.1. On the Molecular Weight

The amino acid composition of the L subunit was redetermined with sufficiently high accuracy to enable one to use the method of integral residues to determine its minimum molecular weight (see Section 4.5) (D. Rosen, L. A. Steiner, E. Moskowitz, M. Okamura, and G. Feher, to be published). By using a computer program into which the amino acid composition was fed, the molecular weight of the L subunit was determined to be 28,000, as compared with 21,000 obtained from SDS-PAGE. If one assumes the same proportional increase in molecular weights for the M and H subunits, a value of 97,000 for the molecular weight of the whole RC protein results. This probably represents a slight overestimate since the M and H subunits are less hydrophobic and will probably require a smaller correction.

8.2. The Primary Donor, D$^+$, in *Rp. viridis*

Fajer *et al.* (1977*b*) pointed out a seeming inconsistency with the dimer model in this species. They found an EPR line width of 14 Gauss for the monomeric Bchl *b* cation as compared with a width of 12 for D$^+$ in whole cells (McElroy *et al.*, 1972). Thus, the ratio of $\sqrt{2}$ between the two widths that led to the dimer model in the other bacterial species does not apply to *Rp. viridis*. This, however, does not rule out the presence of a dimer species and the anomalous finding may be attributed to the breakdown of the assumptions leading to the $\sqrt{2}$ factor for *Rp. viridis* [see footnote*, p. 362].

8.3. On the Intermediate Acceptor (see Section 5.1.2b)

Several articles have recently appeared on the intermediate acceptor. The consensus at present is that the electron is localized on a pheophytin which interacts with Bchl (for a recent review on the primary photochemical processes, see Holten *et al.*, 1978). EPR

experiments performed on RCs enriched with ^{25}Mg corroborate this view (Norris *et al.*, 1978).

The kinetics of formation of the doublet EPR signal in RCs of *Rp. sphaeroides* that were reconstituted with menaquinone were investigated by measuring the rate constant for the reaction I$^-$Q$^-$ → IQ^{2-}. For menaquinone the rate was found to be 0.2 sec^{-1}, whereas for ubiquinone it was 100 sec^{-1}. This explains the inability to trap the doublet-producing species I$^-$Q$^-$ in native RCs of *Rp. sphaeroides*. The strength of the coupling (as determined from the doublet splitting) and the measured activation energy were used to estimate the electron-transfer rate using the theory for thermally activated tunneling (Hopfield, 1974). Good agreement was obtained between the calculated and experimentally determined rate of 0.2 sec^{-1} (M. Y. Okamura, R. A. Isaacson, and G. Feher, 1978, submitted).

8.4. The Electronic Structure of FE^{2+} and the Fe^{2+}–Quinone Complex

The susceptibility measurements (see Section 5.2.5) have been extended and used to determine the parameters of the five-fold ground state of Fe^{2+}, as well as to explain the g value of the broad g = 1.8 EPR signal (see Fig. 14) (Butler *et al.*, 1978). The low-temperature date of the unreduced RCs were fitted with the spin Hamiltonian.

$$\mathscr{H} = g\beta\bar{H}\cdot\bar{S} + D[S_z^2 - \tfrac{1}{3}S(S+1)] + E(S_x^2 - S_y^2)$$

with D = 6.0 cm^{-1} and E = 2.0 cm^{-1}. Data for reduced RCs were fitted by adding to the above Hamiltonian a Zeeman term for the quinone and an exchange coupling term $-J\bar{S}_1\cdot\bar{S}_2$, where $\bar{S}_1$ and $\bar{S}_2$ are the spins of the quinone and Fe^{2+}, respectively. The best fit was obtained for J ≈ −0.6 cm^{-1}. Calculations using this value of J predict the observed broad g ≈ 1.8 EPR signal. The sign of J shows that in the ground state the spins on the quinone and Fe^{2+} point in opposite directions (antiferromagnetic coupling). The observed increase in the EPR line width of the secondary quinone (see Section 5.3.1) suggests that it has a slightly larger exchange coupling than the primary quinone (Okamura *et al.*, 1978).

8.5. Attachment Site of the Primary Quinone

The technique of photoaffinity labeling (for a review, see Knowles, 1972) was used to determine the subunit that binds the primary quinone (T. Marinetti, M. Okamura, and G. Feher, to be published). RCs depleted of UQ were reconstituted with ^{3}H-labeled 2-azidoanthraquinone and were shown to be photochemically active. The reconstituted RCs were photolyzed with UV light to form a covalent attachment between the quinone and the protein at the quinone-binding site. SDS-PAGE runs of the RCs showed that most of the radioactive quinone was bound to the M subunit, leading to the conclusion that the primary quinone is bound to M.

9. References

Abragam, A., and Bleaney, B., 1970, *Electron Paramagnetic Resonance of Transition Metal Ions*, p. 443, Clarendon Press, Oxford.

Ackerson, L. A., Okamura, M. Y., and Feher, G., 1975 (unpublished).

Banker, G. A., and Cotman, C. W., 1972, Measurement of free electrophoretic mobility and retardation coefficient of protein–sodium dodecyl sulfate complexes by gel electrophoresis, *J. Biol. Chem.* **247**:5856.

Beugeling, T., Slooten, L., and Barelds-van de Beck, P. G. M. M., 1972, Thin-layer chromatography of pigments from reaction center particles of *Rhodopseudomonas spheroides*, *Biochim. Biophys. Acta* **238**:328.

Bolton, J. R., and Cost, K., 1973, Flash photoylsis–electron spin resonance: A kinetic study of endogenous light-induced free radicals in reaction center preparations from *Rhodopseudomonas spheroides*, *Photochem. Photobiol.* **18**:417.

Bolton, J. R., Clayton, R. K., and Reed, D. W., 1969, An identification of the radical giving rise to the light-induced electron spin resonance signal in photosynthetic bacteria, *Photochem. Photobiol.* **9**:209.

Brady, P. R., Duncan, J. E., and Mok, K. F., 1965, Chemical aspects of the Mossbauer effect, *Proc. R. Soc. (London) Ser. A*: **287**:343.

Butler, W. F., Johnston, D. C., Okamura, M. Y., Shore, H. B., and Feher, G., 1978, Magnetic Properties of Reaction Centers from *R. sphaeroides R-26*, *Biophys. J.* **21**, *Abstract* 8a.

Capaldi, R. A., and Vanderkooi, G., 1972, The low polarity of many membrane proteins, *Proc. Natl. Acad. Sci. U.S.A.* **69**:930.

Casassa, E. F., and Eisenberg, H., 1964, Thermal dynamic analysis of multicomponent solutions, *Adv. Protein Chem.* **19**:287.

Clayton, R. K., 1966, Spectroscopic analysis of bacteriochlorophylls *in vitro* and *in vivo*, *Photochem. Photobiol.* **5**:669.

Clayton, R. K., 1972, Physical mechanisms in photosynthesis: Past elucidations and current problems, *Proc. Natl. Acad. Sci. U.S.A.* **69**:44.

Clayton, R. K., and Haselkorn, R., 1972, Protein components of bacterial photosynthetic membranes, *J. Mol. Biol.* **68**:97.

Clayton, R. K., and Straley, S. C., 1970, An optical absorption change that could be due to reduction of the primary photochemical electron acceptor in photosynthetic reaction centers, *Biochem Biophys. Res. Commun.* **39**:1114.

Clayton, R. K., and Straley, S. C., 1972, Photochemical electron transport in photosynthetic reaction centers. IV. Observations related to the reduced photoproducts, *Biophys. J.* **12**:1221.

Clayton, R. K., and Wang, R. T., 1971, Photochemical reaction centers from *Rhodopseudomonas spheroides*, *Methods Enzymol.* **XXIII**(23):696.

Clayton, R. K., and Yau, H. F., 1972, Photochemical electron transport in photosynthetic reaction centers. I. Kinetics of the oxidation and reduction of P_{870} as affected by external factors, *Biophys. J.* **12**:867.

Clayton, R. K., Fleming, H., and Szuts, E. Z., 1972, Photochemical electron transport in photosynthetic reaction centers from *Rhodopseudomonas spheroides*. II. Interaction with external donors and acceptors and a re-evaluation of some spectroscopic data, *Biophys. J.* **12**:46.

Cogdell, R. J., Brune, D. C., and Clayton, R. K., 1974, Effects of extraction and replacement of ubiquinone upon the photochemical activity of reaction centers and chromatophores from *Rhodopseudomonas spheroides*, *FEBS Lett.* **45**:344.

Cogdell, R. J., Monger, T. G., and Parson, W. W., 1975, Carotenoid triplet states in reaction centers from *Rhodopseudomonas sphaeroides* and *Rhodospirillum rubrum*, *Biochim. Biophys. Acta* **408**:189.

Cogdell, R. J., Parson, W. W., and Kerr, M. A., 1976, The type, amount, and energy transfer properties of the carotenoid in reaction centers from *Rhodopseudomonas sphaeroides*, *Biochim. Biophys. Acta* **430**:83.

Commoner, B., Heise, J. J., and Townsend, J., 1956, Light-induced paramagnetism in chloroplasts, *Proc. Natl. Acad. Sci. U.S.A.* **42**:710.

Cuatrecasas, P., and Anfinsen, C. B., 1971, Affinity chromatography, *Methods Enzymol.* **22**:345.

Debrunner, P. G., Schulz, C. E., Feher, G., and Okamura, M. Y., 1975, Mossbauer study of reaction centers from *Rp. spheroides*, *Biophys. Soc. Abstr.* **15**:226.

Dunker, A. K., and Rueckert, R. R., 1969, Observations on molecular weight determinations on polyacrylamide gel, *J. Biol. Chem.* **244**:5074.

Dutton, P. L., and Leigh, J. S., 1973, Direct measurement of the oxidation–reduction midpoint potential of the primary electron acceptor (photoredoxin) in bacterial photosynthesis, *Biophys. Soc. Abstr.* **13**:60.

Dutton, P. L., Leigh, J. S., and Reed, D. W., 1973, Primary events in the photosynthetic reaction centre in *Rhodopseudomonas spheroides* strain R-26: Triplet and oxidized states of bacteriochlorophyll and the identification of the primary electron acceptor, *Biochim. Biophys. Acta* **292**:654.

Dutton, P. L., Kaufmann, K. J., Chance, B., and Rentzepis, P. M., 1975, Picosecond kinetics of the 1250 nm band of the *Rps. sphaeroides* reaction center: The nature of the primary photochemical intermediary state, *FEBS Lett.* **60**:275.

Dutton, P. L., Prince, R. C., Tiede, D. M., Petty, K., Kaufmann, K. J., Netzel, T. L., and Rentzepis, P. M., 1976, Electron transfer in photosynthetic reaction centers, *Brookhaven Symp. Biol.*, No. 28.

Duysens, L. N. M., 1952, Transfer of excitation energy in photosynthesis, Thesis, University of Utrecht.

Evans, M. C. W., Lord, A. V., and Reeves, S. G., 1974, The detection and characterization by EPR spectroscopy of iron–sulphur proteins and other electron-transport components in chromatophores from the purple bacterium *Chromatium*, *Biochem. J.* **138**:177.

Fajer, J., Brune, D. C., Davis, M. S., Forman, A., and Spaulding, L. D., 1975, Primary charge separation in bacterial photosynthesis: Oxidized chlorophylls and reduced pheophytin, *Proc. Natl. Acad. Sci. U.S.A.* **72**:4956.

Fajer, J., Davis, M. S., and Forman, A., 1977a, ENDOR and ESR characteristics of bacteriophytin and bacteriochlorophyll anion radicals, *Biophys. J. Abstr.* **17**:150.

Fajer, J., Davis, M. S., Brune, D. C., Spaulding, L. D., Borg, D. C., and Forman, A., 1977b, *Brookhaven Symp. Biol.*, No. 28, p. 93.

Feher, G., 1971, Some chemical and physical properties of a bacterial reaction center particle and its primary photochemical reactants, *Photochem. Photobiol.* **14**:373.

Feher, G., and Okamura, M. Y., 1977, Reaction centers from *Rhodopseudomonas sphaeroides*, in: *Chlorophyll–Proteins, Reaction Centers, and Photosynthetic Membranes, Brookhaven Symp. Biol.*, No. 28, pp. 183–194.

Feher, G., Okamura, M. Y., Raymond, J. A., and Steiner, L. A., 1971, Subunit structure of reaction centers from *Rhodopseudomonas spheroides*, *Biophys. Soc. Abstr.* **11**:38.

Feher, G., Okamura, M. Y., and McElroy, J. D., 1972, Identification of an electron acceptor in reaction centers of *Rhodopseudomonas spheroides* by EPR spectroscopy, *Biochim. Biophys. Acta* **267**:222.

Feher, G., Hoff, A. J., Isaacson, R. A., and McElroy, J. D., 1973, Investigation of the electronic structure of the primary electron donor in bacterial photosynthesis by the ENDOR technique, *Biophys. J. Abstr.* **13**:61.

Feher, G., Isaacson, R. A., McElroy, J. D., Ackerson, L. C., and Okamura, M. Y., 1974, On the question of the primary acceptor in bacterial photosynthesis: Manganese substituting for iron in reaction centers of *Rhodopseudomonas sphaeroides* R-26, *Biochim. Biophys. Acta* **368**:135.

Feher, G., Hoff, A. J., Isaacson, R. A., and Ackerson, L. C., 1975a, ENDOR experiments on chlorophyll and bacteriochlorophyll *in vitro* and in the photosynthetic unit, *Ann. N. Y. Acad. Sci.* **244**:239.

Feher, G., Isaacson, R. A., and Okamura, M. Y., 1975b (unpublished).

Feher, G., Isaacson, R. A., and Okamura, M. Y., 1977, Comparison of EPR and ENDOR spectra of the transient acceptor in reaction centers of *R. sphaeroides* with those of bacteriochlorophyll and bacteriopheophytin radicals, *Biophys. J. Abstr.* **17**:149.

Figgis, B. N., and Lewis, A. J., 1960, *Modern Coordination Chemistry*, pp. 400–454, John Wiley & Sons, New York.

Fong, F. K., 1974, Molecular basis for the photosynthetic primary process, *Proc. Natl. Acad. Sci. U.S.A.* **71**:3692.

Gingras, G., and Jolchine, G., 1969, Isolation of a P_{870}^--enriched particle from *Rhodospirillum rubrum*, in *Progress in Photosynthesis Research, Proceedings of the International Congress of Photosynthesis Research* (H. Metzner, ed.), Vol. 1, pp. 209–216, Freudenstadt.

Goedheer, J. C., 1959, Reversible oxidations of pigments in bacterial chromatophores, in *The Photochemical Apparatus—Its Structure and Function, Brookhaven Symp. Biol.*, No. 11, pp. 325–331.

Halsey, Y. D., and Parson, W. W., 1974, Identification of ubiquinone as the secondary electron-acceptor in the photosynthetic apparatus of *Chromatium vinosum*, *Biochim. Biophys. Acta* **347**:404.

Holten, D., Windsor, M. W., Parson, W. W., and Thornber, J. P., 1978, *Biochim. Biophys. Acta* **501**:112–126.

Hopfield, J., 1974, Electron transfer between biological molecules by thermally activated tunneling, *Proc. Natl. Acad. Sci. U.S.A.* **71**:3640.

Johnson, D., Okamura, M. Y., and Feher, G., 1976 (unpublished).

Jolchine, G., and Reiss-Husson, F., 1974, Comparative studies on two reaction center preparations from *Rhodopseudomonas spheroides* Y, *FEBS Lett.* **40**:5.

Jolchine, G., and Reiss-Husson, F., 1975, Studies on pigments and lipids in *Rhodopseudomonas spheroides* Y reaction centers, *FEBS Lett.* **52**:33.

Katz, J. J., and Norris, J. R., 1973, Chlorophyll and light energy transduction in photosynthesis, in *Current Topics in Bioenergetics* (D. R. Sanadi and L. Packer, eds.), Vol. 5, pp. 41–75, Academic Press, New York.

Kaufmann, K. J., Dutton, P. L., Netzel, T. L., Leigh, J. S., and Rentzepis, P. M., 1975, Picosecond kinetics of events leading to reaction center bacteriochlorophyll oxidation, *Science* **188**:1301.

Ke, B., Chaney, T. H., and Reed, D. W., 1970, The electrostatic interaction between the reaction center bacteriochlorophyll derived from *Rhodopseudomonas spheroides* and mammalian cytochrome *c* and its effects on light-activated electron transport, *Biochim. Biophys. Acta* **216**:373.

Knowles, J. R., 1972, *Accounts Chem. Res.* **5**:155.

Knox, R. S., 1969, Thermodynamics and the primary processes of photosynthesis, *Biophys. J.* **9**:1351.

Land, E. J., Simic, M., and Swallow, A. J., 1971, Optical absorption spectrum of half-reduced ubiquinone, *Biochim. Biophys. Acta* **226**:239.

Leigh, J. S., and Dutton, P. L., 1972, The primary electron acceptor in photosynthesis, *Biochem. Biophys. Res. Commun.* **46**:414.

Lester, R. L., and Crane, F. L., 1959, The natural occurrence of coenzyme Q and related compounds, *J. Biol. Chem.* **234**:2169.

Lin, L., and Thornber, J. P., 1975, Isolation and partial characterizations of the photochemical reaction center of *Chromatium vinosum* (strain D), *Photochem. Photobiol.* **22**:34.

Loach, P. A., 1966, Primary oxidation–reduction changes during photosynthesis in *Rhodospirillum rubrum*, *Biochemistry* **5**:592.

Loach, P., 1970a, Quantitative dissolution of the membrane and preparation of photoreceptor subunits from *Rhodopseudomonas spheroides*, *Biochemistry* **9**:724.

Loach, P. A., 1970b, Oxidation–reduction potentials, absorbance bands and molar absorbance of compounds used in biochemical studies, in *Handbook of Biochemistry*. 2nd Ed. (H. A. Baker, ed.), pp. J33–J40, Chemical Rubber Co., Cleveland.

Loach, P. A., and Hall, R. L., 1972, The question of the primary electron acceptor in bacterial photosynthesis, *Proc. Natl. Acad. Sci. U.S.A.* **69**:786.

Loach, P. A., and Sekura, D. L., 1967, A comparison of decay kinetics of photoproduced absorbance EPR and luminescence changes in chromatophores of *Rhodospirillum rubrum*, *Photochem. Photobiol.* **6**:381.

Loach, P. A., and Sekura, D., 1968, Primary photochemistry and electron transport in *Rhodospirillum rubrum*, *Biochemistry* **7**:2642.

Loach, P. A., and Walsh, K., 1969, Quantum yield for the photoproduced electron paramagnetic resonance signal in chromatophores from *Rhodospirillum rubrum*, *Biochemistry* **8**:1908.

Loach, P. A., Androes, G. M., Maksim, A. F., and Calvin, M., 1963, Variations of electron paramagnetic resonance signals of photosynthetic systems with the redox level of their environment, *Photochem. Photobiol.* **2**:443.

Loach, P. A., Kung, M., and Hales, B. J., 1975, Characterization of the phototrap in photosynthetic bacteria, *Ann. N. Y. Acad. Sci.* **244**:297.

Malkin, R., 1973, The chemical properties of ferredoxins, in *Iron–Sulfur Proteins* (W. Lovenberg, ed.), Vol. II, Chapt. 1, pp. 1–26, Academic Press, New York.

Mauzerall, D., 1972, Pigment content of purified reaction centers from *Rp. spheroides* mutant R-26 *Fed. Proc. Fed. Am. Soc. Exp. Biol.* **31**:885.

McElroy, J. D., Feher, G., and Mauzerall, D., 1970, Observation of a second light induced EPR signal from reaction centers of photosynthetic bacteria, *Biophys. Soc. Abstr.* **10**:204.

McElroy, J. D., Feher, G., and Mauzerall, D. C., 1972, Characterization of primary reactants in bacterial photosynthesis. I. Comparison of the light-induced EPR signal ($g = 2.0026$) with that of a bacteriochlorophyll radical, *Biochim. Biophys. Acta* **267**:363.

McElroy, J. D., Mauzerall, D. C., and Feher, G., 1974, Characterization of primary reactants in bacterial photosynthesis. II. Kinetic studies of the light-induced signal ($g = 2.0026$) and the optical absorbance changes at cryogenic temperatures, *Biochim. Biophys. Acta* **333**:261.

Morton, R. A., 1965, Introductory account of quinones, in: *Biochemistry of Quinones* (R. A. Morton, ed.), Chapt. 1, pp. 1–21, Academic Press, New York.

Moskowitz, E., and Malley, M. M., 1978, Energy transfer and photo-oxidation kinetics in reaction centers on the picosecond time scale, *Photochem. Photobiol.* **27**:55.

Netzel, T. L., Rentzepis, P. M., and Leigh, J., 1973, Picosecond kinetics of reaction centers containing bacteriochlorophyll, *Science* **182**:238.

Nieth, K. F., Drews, G., and Flick, R., 1975, Photochemical reaction centers from *Rhodopseudomonas capsulata*, *Arch. Microbiol.* **105**:43.

Noël, H., van der Rest, M., and Gingras, G., 1972, Isolation and Partial characterization of a P_{870} reaction center complex from wild type *Rhodospirillum rubrum*, *Biochim. Biophys. Acta* **275**:219.

Norris, J. R., Uphaus, R. A., Crespi, H. L., and Katz, J. J., 1971, Electron spin resonance of chlorophyll and the origin of signal I in photosynthesis, *Proc. Natl. Acad. Sci. U.S.A.* **68**:625.

Norris, J. R., Druyan, M. E., and Katz, J. J., 1973, Electron nuclear double resonance of bacteriochlorophyll free radical *in vitro* and *in vivo*, *J. Am. Chem. Soc.* **95**:1680.

Norris, J. R., Scheer, H., and Katz, J. J., 1975, Models for antenna and reaction center chlorophylls, *Ann. N. Y. Acad. Sci.* **244**:260.

Norris, J., Bowman, M., and Thurnauer, M., 1978, *Biophys. J. Abstr.* **21**:10a.

Okamura, M. Y., 1975 (unpublished).

Okamura, M. Y., Steiner, L. A., and Feher, G., 1974, Characterization of reaction centers from photosynthetic bacteria. I. Subunit structure of the protein mediating the primary photochemistry in *Rhodopseudomonas spheroides* R-26, *Biochemistry* **13**:1394.

Okamura, M. Y., Isaacson, R. A., and Feher, G., 1975, The primary acceptor in bacterial photosynthesis: The obligatory role of ubiquinone in photoactive reaction centers of *Rp. spheroides*, *Proc. Natl. Acad. Sci. U.S.A.* **72**:3491.

Okamura, M. Y., Ackerson, L. C., Isaacson, R. A., Parson, W. W., and Feher, G., 1976, The primary electron acceptor in *Chromatium vinosum* (strain D), *Biophys. Soc. Abstr.* **16**:67.

Okamura, M. Y., Isaacson, R. A., and Feher, G., 1977a, On the trapping of the transient acceptor in reaction centers (RCs) of *R. sphaeroides* R-26, *Biophys. J. Abstr.* **17**:149.

Okamura, M. Y., Isaacson, R. A., and Feher, G., 1977b (unpublished).

Okamura, M. Y., Isaacson, R. A., and Feher, G., 1978, EPR Signals from the primary and secondary quinone in reaction centers of *R. sphaeroides* R-26, *Biophys. J.* **21**, *Abstract* 8a.

Palmer, G., 1973, Current insights into the active center of spinach ferredoxin and other iron–sulfur proteins, in: *Iron–Sulfur Proteins* (W. Lovenberg, ed.), Vol. II, Chapt. 8, pp. 285–325, Academic Press, New York.

Parson, W. W., 1968, The role of P_{870} in bacterial photosynthesis, *Biochim. Biophys. Acta* **153**:248.

Parson, W. W., 1969, The reaction between primary and secondary electron acceptors in bacterial photosynthesis, *Biochim. Biophys. Acta* **189**:384.

Parson, W. W., and Case, G. D., 1970, In *Chromatium*, a single photochemical reaction center oxidizes both cytochrome C_{552} and cytochrome C_{555}, *Biochim. Biophys. Acta* **205**:232.

Parson, W. W., and Cogdell, R., 1975, The primary photochemical reaction of bacterial photosynthesis, *Biochim. Biophys. Acta* **416**:105.

Parson, W. W., Clayton, R. K., and Cogdell, R. J., 1975, Excited states of photosynthetic reaction centers at low redox potentials, *Biochim. Biophys. Acta* **387**:265.

Prince, R. C., Cogdell, R. J., and Crofts, A. R., 1974, The photo-oxidation of horseheart cytochrome *c* and native cytochrome c_2 by reaction centers from *Rhodopseudomonas spheroides* R-26, *Biochim. Biophys. Acta* **347**:1.

Pucheu, N. L., Kerber, N. L., and Garcia, A. F., 1976, Isolation and purification of reaction center from *Rhodopseudomonas viridis* NHTC 133 by means of LDAO, *Arch. Microbiol.* **109**:301.

Reed, D. W., 1969, Isolation and composition of a photosynthetic reaction center complex from *Rhodopseudomonas spheroides*, *J. Biol. Chem.* **244**:4936.

Reed, D. W., and Clayton, R. K., 1968, Isolation of a reaction center fraction from *Rhodopseudomonas sphaeroides*, *Biochem. Biophys. Res. Commun.* **30**:471.

Reed, D. W., and Ke, B., 1973, Spectral properties of reaction center preparations form *Rhodopseudomonas spheroides*, *J. Biol. Chem.* **248**:3041.

Reed, D. W., and Mayne, B. C., 1971, The subcellular localization of the pteridines in strain R-26 of *Rhodopseudomonas spheroides*, *Biochim. Biophys. Acta* **226**:477.

Reed, D. W., and Peters, G. A., 1972, Characterization of the pigments in reaction center preparations from *Rhodopseudomonas spheroides*, *J. Biol. Chem.* **247**:7148.

Reed, D. W., Raveed, D., and Reporter, M., 1975, Localization of photosynthetic reaction centers by antibody binding to chromatophore membranes from *Rhodopseudomonas spheroides* strain R-26, *Biochim. Biophys. Acta* **387**:368.

Reiss-Husson, F., and Jolchine, G., 1972, Purification and properties of a photosynthetic reaction center isolated from various chromatophore fractions of *Rhodopseudomonas spheroides* Y, *Biochim. Biophys. Acta* **256**:440.

Rockley, M. G., Windsor, M. W., Cogdell, R. J., and Parson, W. W., 1975, Picosecond detection of an intermediate in the photochemical reaction of bacterial photosynthesis, *Proc. Natl. Acad. Sci. U.S.A.* **72**:2251.

Romijn, J. C., and Amesz, J., 1977, Purification and photochemical properties of RCs of *Chr. vinosum*. Evidence for the photoreduction of a naphthaquinone, *Biochim. Biophys. Acta* **461**:327.

Rosen, D., Okamura, M. Y., Feher, G., Steiner, L. A., and Walker, J. E., 1977a, Separation and *N*-terminal sequence analysis of the subunits of the reaction center protein from *R. sphaeroides* R-26, *Biophys. J. Abstr.* **17**:67.

Rosen, D., Okamura, M. Y., and Feher, G., 1977b (unpublished).

Sauer, K., 1974, Primary events and the trapping of energy, in: *Bioenergetics of Photosynthesis* (Govindjee, ed.), pp. 115–181, Academic Press, New York.

Sauer, K., Dratz, E. A., and Coyne, L., 1968, Circular dichroism spectra and the molecular arrangement of bacteriochlorophylls in the reaction centers of photosynthetic bacteria, *Proc. Natl. Acad. Sci. U.S.A.* **61**:17.

Shapiro, A. L., Vinuela, E., and Maizel, J. V., 1967, Molecular weight estimation of polypeptized chains by electrophoresis in SDS–polyacrylamide gels, *Biochem. Biophys. Res. Commun.* **28**:815.

Shuvalov, V. A., and Klimov, V. V., 1976, The primary photoreactions in the complex cytochrome P_{890}–P_{760} (bacteriopheophytin$_{760}$) of *Chromatium minutissimum* at low redox potentials, *Biochim. Biophys. Acta* **440**:587.

Singer, S. J., and Nicolson, G. L., 1972, The fluid mosaic model of the structure of cell membranes, *Science* **175**:720.

Sistrom, W. R., Griffiths, M., and Stanier, R. Y., 1956, The biology of a photosynthetic bacterium which lacks colored carotenoids, *J. Cell. Comp. Physiol.* **48**:473.

Slooten, L. 1972a, Reaction center preparations of *Rhodopseudomonas spheroides*: Energy transfer and structure, *Biochim. Biophys. Acta* **256**:452.

Slooten, L., 1972b, Electron acceptors in reaction center preparations from photosynthetic bacteria, *Biochim. Biophys. Acta* **272**:208.

Slooten, L., 1973, Fluorescence excitation spectra and the relative numbers of pigment molecules in reaction centers from *Rhodopseudomonas spheroides*, *Biochim. Biophys. Acta* **314**:15.

Smith, W. R., Jr., Sybesma, C. S., and Dus, K., 1972, Isolation and characteristics of small, soluble photoreactive fragments of *Rhodospirillum rubrum*, *Biochim. Biophys. Acta* **267**:609.

Sogo, P. Jost, M., and Calvin, M., 1959, Evidence for freeradical production in photosynthesizing systems, *Radiat. Res., Suppl. 1*, p. 511.

Steck, T. L., and Fox, C. F., 1972, Membrane proteins, in: *Membrane Molecular Biology* (C. F. Fox and A. D. Keith, eds.), Chapt. 2, pp. 27–75, Sinauer Associates, Stanford, Connecticut.

Steiner, L. A., 1976 (unpublished).

Steiner, L. A., Lopes, A. D., Okamura, M. Y., Ackerson, L. C., and Feher, G., 1974a, On the spatial arrangement of reaction center subunits in the bacterial membrane of *Rhodopseudomonas spheroides*, *Fed. Proc. Fed. Am. Soc. Exp. Biol.* **33**:1461.

Steiner, L. A., Okamura, M. Y., Lopes, A. D., Moskowitz, E., and Feher, G., 1974b, Characterization of reaction centers from photosynthetic bacteria. II. Amino acid composition of the reaction center protein and its subunits in *Rhodopseudomonas spheroides* R-26, *Biochemistry* **13**:1403.

Straley, S. C., Parson, W. W., Mauzerall, D. C., and Clayton, R. K., 1973, Pigment content and molar extinction coefficients of photochemical reaction centers from *Rhodopseudomonas spheroides*, *Biochim. Biophys. Acta* **305**:597.

Takemoto, J., and Lascelles, J., 1973, Coupling between bacteriochlorophyll and membrane protein synthesis in *Rhodopseudomonas spheroides*, *Proc. Natl. Acad. Sci. U.S.A.* **70**:799.

Tiede, D. M., Prince, R. C., Reed, G. H., and Dutton, P. L., 1976, EPR properties of the electron carrier intermediate between the primary reaction center bacteriochlorophylls and the primary acceptor in *Chromatium vinosum*, *FEBS Lett.* **65**:301.

Tiede, D. M., Prince, R. C., and Dutton, P. L., 1977, EPR and optical spectroscopic properties of the electron carrier intermediate between the RC bacteriochlorophylls and the primary acceptor in *C. vinosum*, in: *Chlorophyll–Proteins, Reaction Centers, and Photosynthetic Membranes*, *Brookhaven Symp. Biol.*, No. 28, pp. 213–237.

Valkirs, G., Rosen, D., Tokuyasu, K. T., and Feher, G., 1976, Localization of reaction center protein in chromatophores from *Rhodopseudomonas spheroides* by ferritin labeling, *Biophys. Soc. Abstr.* **16**:223.

van der Rest, M., and Gingras, G., 1974, The pigment complement of the photosynthetic reaction center from *Rhodospirillum rubrum, J. Biol. Chem.* **249**:6446.

Vermeglio, A., 1977, Secondary electron transfer in reaction centers of *Rhodopseudomonas sphaeroides*: Out-of-phase periodicity of two for the formation of ubisemiquinone and fully reduced ubiquinone, *Biochim. Biophys. Acta* **459**:516–524.

Vermeglio, A., and Clayton, R. K., 1976, Orientation of chromophores in reaction centers of *Rhodopseudomonas spheroides*: Evidence for two absorption bands of the dimeric primary electron donor, *Biochim. Biophys. Acta* **449**:500–515.

Vermeglio, A., and Clayton, R. K., 1977, Electron transfer between primary and secondary electron acceptors in reaction centers from *Rhodopseudomonas sphaeroides*, *Biophys. J. Abstr.* **17**:147.

Weber, K., and Osborn, M., 1969, The reliability of molecular weight determinations by dodecyl sulfate–polyacrylamide gel electrophoresis, *J. Biol. Chem.* **244**:4406.

Wraight, C. A. 1977, The primary acceptor of photosynthetic bacterial reaction centers: Direct observation of oscillatory behavior suggesting two closely equivalent species, *Biochim. Biophys. Acta* **459**:525–531.

Wraight, C. A., and Clayton, R. K., 1974, The absolute quantum efficiency of bacteriochlorophyll photooxidation in reaction centers of *Rhodopseudomonas spheroides*, *Biochim. Biophys. Acta* **333**:246.

CHAPTER 20

Physicochemical Mechanisms in Reaction Centers of Photosynthetic Bacteria

Roderick K. Clayton

1. Introduction

This chapter will describe the elucidation of physicochemical mechanisms in the photochemistry of bacterial photosynthesis. We will consider first the earlier studies with cells and chromatophores (membrane fragments) that have implicated certain compounds in the photochemistry, and next the more detailed and informative experiments with isolated photochemical reaction centers. The measurements with reaction centers have again pointed to certain reactants and products, but at a more refined level of understanding. Next, we will review the manifestations and properties of physical states that either mediate the photochemistry or that appear as byproducts: excited singlet, biradical, and triplet states involving bacteriochlorophyll (Bchl) or bacteriopheophytin (Bpheo). This will lead us to an outline of the entire sequence of events that begins with the absorption of light and culminates in the storage of stable chemical energy. The chapter will end with a brief look toward the future.

2. Photochemical Reactants and Products

2.1. Earlier Studies with Intact Cells and Chromatophore Preparations

Beginning with the work of Duysens (1953, 1954) and Chance and Smith (1955), using newly developed

techniques of sensitive differential absorption spectrometry, two types of compounds became implicated in the photochemistry of bacterial photosynthesis: Bchl and cytochromes (for a detailed account of which specific cytochromes are involved, see Chapters 13 and 28). Light-induced absorbance changes in cells and chromatophores, when compared with changes induced through oxidation by ferricyanide, indicated that illumination causes oxidation of both Bchl (Duysens, 1953; Duysens et al., 1956; Goedheer, 1959) and cytochromes (Duysens, 1954; Chance and Smith, 1955; Smith and Ramirez, 1959) in various species of photosynthetic bacteria. The most conspicuous change reflecting Bchl oxidation is a reversible bleaching centered near the long-wave absorption maximum, usually near 870–890 nm. The oxidation of Bchl *in vivo* was seen to follow a one-electron redox titration with a midpoint potential of about +0.5 V (Duysens, 1959; Goedheer, 1959). Of the total Bchl in a cell, only a small fraction, now known to be part of the reaction center (RC), could be oxidized. In chromatophores of photosynthetic bacteria, ubiquinone (UQ) was identified tentatively as a photochemical electron acceptor on the basis of a reversible light-induced bleaching centered near 280 nm (Clayton, 1962a).

The photooxidation of cytochrome in *Chromatium vinosum* was observed to proceed, although irreversibly, at 77°K (Chance and Nishimura, 1960). In chromatophores of *Rhodopseudomonas sphaeroides*, the light-induced absorbance changes suggesting Bchl

Roderick K. Clayton · Division of Biological Sciences, Cornell University, Ithaca, New York 14853

oxidation were seen to occur reversibly at temperatures as low as 1.3°K (Arnold and Clayton, 1960). At room temperature, the oxidation of Bchl was generally not observed unless some cytochrome had previously become oxidized (Duysens et al., 1956; Clayton, 1962b). The kinetics of these reactions in the millisecond time range were compatible with two hypotheses: (1) Oxidized Bchl is a primary photochemical product that in turn oxidizes cytochrome. (2) Oxidized cytochrome is a primary photochemical product; the oxidation of Bchl is a side reaction (perhaps a protective device) that occurs only when reduced cytochrome is not available as a primary reactant. With the advent of measurements using pulsed lasers, with time resolved to less than a microsecond (Chance and DeVault, 1964), the first hypothesis became established (Parson, 1968). In chromatophores of Chr. vinosum, the oxidation of Bchl is complete in much less than 1 μsec, and the time course of the subsequent oxidation of cytochrome mirrors exactly the rereduction of oxidized Bchl, over a period of several microseconds (Parson, 1968). Similar behavior can be seen in other species of bacteria, except that the halftime of the reaction between cytochrome and oxidized Bchl varies from approximately 1 to 30 μsec at room temperature, and from 5 μsec to 2.5 msec or more at 80°K (see Chapter 28). This great range of reaction times might reflect quantum mechanical tunneling of the electron through an energy barrier between the two components (DeVault and Chance, 1966).

The quantum efficiency of this reaction sequence, Bchl oxidation followed by and closely coupled to cytochrome oxidation, could be measured on the basis of the known extinction coefficient of the appropriate cytochrome. Efficiencies were generally estimated to be greater than 90% (Loach and Sekura, 1968; Olson, 1962; Parson, 1968).

One of the earliest applications of electron spin resonance (ESR) spectroscopy in biology was the detection of a light-induced signal in green plant tissues, and then in photosynthetic bacteria. The first signal seen in photosynthetic bacteria (Sogo et al., 1959) had the same kinetics of formation and decay as the oxidation of Bchl observed by optical absorption spectrometry (Bolton et al., 1969; Loach and Sekura, 1967; McElroy et al., 1974). This "$g = 2$" signal is now securely identified with the oxidized Bchl that appears as a primary photochemical product (see Chapter 21).

In summary, the earlier measurements with cells and chromatophores of photosynthetic bacteria gave a picture that has been confirmed more recently through studies of purified RCs: light causes the efficient oxidation of Bchl coupled to reduction of UQ; the oxidized Bchl in turn oxidizes a cytochrome of the c type.

2.2. Photochemical Studies with Reaction Centers

The foregoing reactions, first detected in cells and chromatophores, are exhibited with essentially the same characteristics but with greater clarity by purified photochemical RCs (see also Chapters 19 and 28). In RCs from Rp. sphaeroides, Bchl is photooxidized with quantum efficiency exceeding 98% (Wraight and Clayton, 1974), and the oxidation of added cytochrome can be coupled closely to the reduction of oxidized Bchl (Straley et al., 1973). A distinctive set of optical absorbance changes (Clayton and Straley, 1972; Slooten, 1972b) shows that UQ, a component of the RC, is reduced photochemically to the anionic semiquinone, UQ$\bar{\cdot}$.

The photooxidation of Bchl in the living cell can be ascribed solely to this process in the RCs. A mutant strain of Rp. sphaeroides, strain PM-8, has antenna Bchl and carotenoids, but lacks RCs (Sistrom and Clayton, 1964; Clayton and Haselkorn, 1972). The mutant cannot grow photosynthetically; it shows none of the photochemical reactions associated with photosynthesis (Sistrom and Clayton, 1964; Clayton et al., 1965). We therefore believe that we have identified these RCs correctly as the seat of the photochemical part of bacterial photosynthesis.

RCs isolated from three diverse species of photosynthetic bacteria, Chr. vinosum, Rp. sphaeroides, and Rhodospirillum rubrum, are remarkably similar in their composition and their properties (see Chapters 6 and 19). In each, the fundamental RC particle contains 4 molecules of Bchl a and 2 of Bpheo a bound noncovalently to a protein consisting of three peptides in the range 20–30 kdaltons. Absorption spectra of these RCs are strikingly similar. The RCs of Rp. sphaeroides and Rs. rubrum contain UQ (Feher et al., 1972; Slooten, 1972a) (Chapter 19); that of Chr. vinosum contains menaquinone as an essential component (Okamura et al., 1976). The RC of Rp. viridis (Thornber et al., 1969) is superficially similar, except that the characteristic absorption bands are shifted to greater wavelengths because these RCs contain Bchl b and Bpheo b instead of Bchl a and Bpheo a.

A new light-induced ESR signal could be detected in RCs of Rp. sphaeroides (McElroy et al., 1970) (Chapter 19); subsequently this "$g = 1.82$" signal was found in chromatophores of Chr. vinosum as well as in Rp. sphaeroides (Leigh and Dutton, 1972; Dutton et al., 1973). The kinetic and redox properties of the $g = 1.82$ signal are appropriate for its identification with the photochemical electron acceptor, but the signal does not resemble that of $UQ^{\bar{}}$. Its spectrum suggests nonheme Fe. In RCs or subchromatophore particles depleted of Fe, the $g = 1.82$ signal is replaced by one that does resemble that of $UQ^{\bar{}}$ (Loach and Hall, 1972; Feher *et al.*, 1972). The Fe depletion can be done in such a way that the photochemical electron transfer from Bchl to UQ appears to be unchanged. And if Mn is substituted for Fe, still a third type of ESR signal (with fine structure suggesting Mn) can be seen, while the photochemistry remains unaltered judging from the kinetics of oxidation and rereduction of Bchl (Feher *et al.*, 1974). An economical interpretation is that the $g = 1.82$ signal originates from an unpaired electron in $UQ^{\bar{}}$, but the ESR spectrum of $UQ^{\bar{}}$ is modulated strongly through magnetic interaction with a nearby Fe or Mn atom (Chapters 19 and 22). Any electronic interaction between $UQ^{\bar{}}$ and Fe is not sufficient to affect the optical absorption spectrum of $UQ^{\bar{}}$.

The essentiality of quinone as photochemical electron acceptor is shown most clearly by extraction–reconstitution experiments (Cogdell *et al.*, 1974; Okamura *et al.*, 1975). RCs isolated from *Rp. sphaeroides* contain 2 molecules of UQ/RC particle. One of these is removed easily, either by hydrocarbon solvents or by *o*-phenanthroline in an aqueous detergent environment, and the removal of this one does not affect the photochemical capability of the RC. The second molecule is bound more tightly, but can be removed by hydrocarbon solvents containing methanol (Cogdell *et al.*, 1974) or by higher concentrations of *o*-phenanthroline and detergent (Okamura *et al.*, 1975). Photochemistry stops when the second molecule of UQ is removed, but apparently normal photochemistry is restored if UQ is then added back to the RCs. In the case of *Chr. vinosum*, the "essential" quinone is menaquinone rather than UQ (Okamura *et al.*, 1976), although UQ serves as secondary electron acceptor in *Chr. vinosum* as well as in *Rp. sphaeroides* (Halsey and Parson, 1974). The situation in *Rs. rubrum* seems unsettled; assays have indicated that there is only 1 tightly bound molecule of UQ for every 2 RCs (Morrison *et al.*, 1976).

The second molecule of UQ in RCs of *Rp. sphaeroides* may serve to translate the one-electron reduction of "primary" UQ

$$UQ \xrightarrow{1e^-} UQ^{\bar{}}$$

into a two-electron reduction of a secondary pool of UQ in the chromatophore

$$UQ \xrightarrow{2e^- + 2H^+} UQH_2$$

A sequence of brief flashes of light, applied to RCs with added electron donor and excess UQ, can elicit an oscillating sequence of absorbance changes (Vermeglio, 1977; Wraight, 1977). The pattern indicates that $UQ^{\bar{}}$ appears after flashes 1, 3, 5, ... and disappears after flashes 2, 4, 6, ..., while a new molecule of UQH_2 is formed after every even-numbered flash. A simple view is that after a flash, the first (essential) molecule of UQ transfers its electron quickly to the second. In consecutive flashes, the second molecule of UQ cycles through the states UQ, $UQ^{\bar{}}$, UQ^{2-} or UQ, $UQ^{\bar{}}$, and UQH_2. The UQ^{2-} or UQH_2 discharges its two electrons (and H^+'s) spontaneously into the secondary pool and reverts to UQ.

2.3. Bacteriochlorophyll as Photochemical Electron Donor in Reaction Centers

A single RC particle from *Rp. sphaeroides*, containing 4 molecules of Bchl and 2 of UQ, can transfer just one electron from Bchl to UQ in response to a single saturating laser flash (Straley *et al.*, 1973). The resulting oxidized Bchl must be reduced again before the particle can perform another photochemical electron transfer. Of the 4 Bchl molecules, 2 share the electron donation, becoming a cation radical dimer $(Bchl)_{\frac{1}{2}}^{+}$. This has been shown by analysis of ESR and electron-nuclear double resonance spectra (Feher *et al.*, 1975; Norris *et al.*, 1971, 1973), and is documented in detail in Chapter 21. Studies of Bchl *in vitro* (Katz *et al.*, 1976) suggest that water is an essential component in the structure of this dimer, and it has long been known (Clayton, 1967) that the exhaustive dehydration of chromatophores or RCs causes loss of photochemical activity as well as minor changes in the absorption spectrum of the RC. The changes caused by dehydration are fully reversible by rehydration. No systematic study has been made of this aspect of the functioning of RCs.

Absorption bands attributable to Bchl in RCs have maxima at 803 and 865 nm. The formation of $(Bchl)_2^+$ is attended by loss of the 865 nm absorption band and an apparent shift of the 803 nm band to 798 nm. It appears now that the apparent blue-shift of the 803 nm band is in fact due to the bleaching of a lesser component at 815 nm and the appearance of a new component at 790 nm, while the main band at 803 nm is not changed. Measurements of linear dichroism (polarized absorption) in spatially oriented RCs show that the transition moments of the "bleachable" 815 nm absorption and of the light-induced increase of 790 nm absorption have different orientations (Vermeglio and Clayton, 1976). They would have the same orientation if they were both caused by the blue-shift of a single band. Moreover, the 815 nm component can be seen as a shoulder on the 803 nm band at low temperature. The bleaching of two components at 815 and 865 nm can be due to loss of the absorption bands of the dimer $(Bchl)_2$ on its oxidation, and the new band at 790 nm can be a property of the singly oxidized dimer $(Bchl)_2^+$.

3. Physical States That Mediate the Photochemistry in Reaction Centers

3.1. Fluorescence and Utilization of the Excited Singlet State of Bacteriochlorophyll

The lowest excited singlet state of Bchl in the RC is the immediate precursor of events leading to the photochemistry. This state is formed directly when light is absorbed by the long-wave absorption band of RCs, or indirectly if light is absorbed at shorter wavelengths by Bchl or by accessory pigments. There is no evidence that absorption at shorter wavelengths changes the photochemistry, except for possible energy losses attending conversion to the lowest excited singlet state.

In RCs from *Rp. sphaeroides*, the lowest excited singlet state of Bchl, corresponding to the absorption band at 865 nm, is signaled by fluorescence with a maximum at 900 nm (Zankel *et al.*, 1968). The intensity of this fluorescence measures the excited state population. A quantum of excitation can be dissipated by radiation (giving fluorescence), by radiationless decay to the ground state, or by utilization for photochemistry. If these events are characterized by fixed probabilities per unit time, they can be assigned first-order rate constants k_f, k_d, and k_p, respectively. The quantum yield of fluorescence is then

$$\phi_f = k_f/(k_f + k_d + k_p) \qquad (1)$$

and the photochemical efficiency is

$$\phi_p = k_p/(k_f + k_d + k_p) \qquad (2)$$

It must be understood that this expression for ϕ_p takes into account only the first step in utilizing the lowest excited singlet state; it ignores losses of efficiency at earlier or later stages.

Equations (1) and (2) determine the intensity of fluorescence and the rate of photochemistry in the quasi-steady state achieved during constant illumination. The greater the value of k_p, the less the fluorescence. We can compare the fluorescence of a suspension of RCs in two states: (1) fully active, k_p maximal and ϕ_f minimal; (2) rendered inactive so that $k_p = 0$ and ϕ_f is maximal. If the values of k_f and k_d are not altered when k_p is reduced to zero, the foregoing equations give

$$\phi_p = 1 - (\phi_f^{min})/(\phi_f^{max}) \qquad (3)$$

In fact, we find that the 900 nm fluorescence of RCs from *Rp. sphaeroides* rises three- to fourfold when the

RCs are brought to low redox potential so as to reduce the photochemical electron acceptor (Zankel *et al.*, 1968; Clayton *et al.*, 1972). From equation (3), this would correspond to $\phi_p = 0.7$, but actually $\phi_p = 0.98 \pm 0.04$ (Wraight and Clayton, 1974). This contradiction means either that the inactivation of RCs by exposure to low redox potential does not bring k_p to zero or that the treatment changes the value of k_d or k_f. We have no further information on this point. Since k_p is the rate constant for the *first* step that normally leads to photochemistry, it does not necessarily become zero if a later step (electron transfer to UQ) is blocked. We shall see later that k_p is probably the rate constant for conversion from the excited singlet state to a biradical state denoted P^F. For the present, we can say that the level of fluorescence signals the redox state of the RC, but we have not learned how to fit the fluorescence yield into a simple theory as expressed by equation (3).

Another way to apply knowledge of the fluorescence yield is to predict the lifetime of the excited singlet state. For a given pair of electronic states (e.g., ground and lowest excited singlet) of a molecule, the probability of a radiative downward transition (fluorescence) is proportional to the probability of an upward transition (absorption). The rate constant k_f is therefore proportional to the integrated probability of absorption; i.e., k_f can be computed from the area of the absorption band (see Zankel *et al.*, 1968). The area of the 865 nm absorption band of RCs from *Rp. sphaeroides* predicts that $k_f = 5 \times 10^7$ sec^{-1}. The reciprocal of k_f, denoted τ_0, is the intrinsic lifetime of the excited state; i.e., the lifetime if fluorescence were the only mechanism for leaving the excited state. The actual lifetime τ is less because the excited state can also be quenched by the processes represented by k_d and k_p; we can write

$$\tau = 1/(k_f + k_d + k_p) \qquad (4)$$

and combining this with equation (1),

$$\tau/\tau_0 = k_f/(k_f + k_d + k_p) = \phi_f \qquad (5)$$

In RCs from *Rp. sphaeroides*, ϕ_f has been measured to be 3.5×10^{-4} in the active state, and 10^{-3} when rendered inactive by lowering the redox potential (Zankel *et al.*, 1968). From equation (5), with $\tau_0 = 20$ nsec (the reciprocal of k_f), we compute $\tau = 7$ psec for the excited singlet state in active RCs, and 20 psec at low redox potential. The high photochemical efficiency of active RCs means, by equation 2, that $k_p \simeq k_d$ or k_f. Then, from equation 4, we compute $k_p \gg 1/(7\ \text{psec}) = 1.4 \times 10^{11}$ sec^{-1}. This gives the rate of the first step by which photochemistry is launched from the lowest excited singlet state.

The foregoing argument would have to be modified if a state other than the excited singlet could give appreciable fluorescence by regenerating the excited singlet; we shall return to this point in connection with the state P^F.

The fluorescence of antenna Bchl in chromatophores of photosynthetic bacteria becomes more intense when the RCs become closed (rendered inactive), either by photochemical oxidation of the Bchl dimer electron donor or by reduction of the electron acceptor (Clayton, 1966; Vredenberg and Duysens, 1963). For these situations, we can write rate constants k_f, k_d, and k_p for rates of de-excitation in the antenna, with k_p representing the quenching by the RCs (see Chapters 16–18). It has often been assumed that a closed RC does not quench excitation from the antenna (k_p becoming zero), but this should not be expected. The measurements with purified RCs suggest that an open (active) RC should merely be three times as good a quencher as a closed RC, since τ is computed to be 7 psec in open RCs and 20 psec in closed ones. This prediction has been verified through experiments with *Rp. sphaeroides* in which purified RCs have been recombined with purified antenna Bchl–protein so as to restore energy transfer from the antenna to the RCs (Heathcote and Clayton, 1977). The fluorescence yield in the reconstituted system was about 1.5% with open RCs and 3% with closed RCs; the antenna pigment–protein, treated in the same way except that no RCs were added, showed a fluorescence yield of 9%. Similar values are observed with chromatophores with open and closed RCs, respectively (1 and 3%), and with chromatophores from a mutant that lacks RCs (9%). In the model system with closed RCs, the yield was 3% whether the RCs had been closed through photochemical oxidoreduction or by chemical reduction.

3.2. Triplet State of Bacteriochlorophyll *in Vivo*

The triplet states of chlorophylls have long been regarded as likely intermediates in the photochemistry of photosynthesis because they mediate many photochemical reactions *in vitro*. Until recently, there was only tenuous evidence for the formation of triplet Bchl in illuminated photosynthetic tissues. Convincing evidence for triplet formation in RCs was first obtained by Dutton *et al.* (1972, 1973), who observed a light-induced ESR signal appropriate for triplet Bchl in chromatophores or RCs brought to low redox potential and examined at temperatures below 20°K. The triplet state could not be seen in photochemically active RCs; it was necessary to bring the redox potential below a midpoint of −0.05 V, which is sufficient to reduce the UQ that serves as photochemical electron acceptor. The quantum efficiency of triplet formation in closed RCs was found to be approximately unity, as was the quantum efficiency of

photochemistry in open RCs at the same low temperature (Wraight *et al.*, 1974; Clayton and Yamamoto, 1976).

Optical investigations by Parson *et al.* (1975) then revealed light-induced absorbance changes suggesting the formation of triplet Bchl, at room temperature as well as at low temperature, again in RCs brought to low redox potential but not in photochemically active RCs. This (presumed) triplet state was formed, in RCs from *Rp. sphaeroides*, with quantum efficiency less than 15% at room temperature, becoming approximately 100% at low temperature. The state decayed exponentially in the dark, with a half-time of 6 μsec at room temperature, and 120 μsec at temperatures below 80°K. The low quantum efficiency at room temperature casts some doubt on its role as a photochemical intermediate, even though it could not be observed (conceivably as a result of rapid photochemical utilization) in active RCs.

A peculiarity of this triplet state of Bchl, revealed by the ESR spectrum, is that the unpaired electron spin is strongly polarized relative to magnetic forces in its molecular environment. This suggests that it is formed from a polarized precursor such as a biradical or charge transfer state, not directly from the unpolarized excited singlet state (see Chapter 21). We shall see that this is undoubtedly the case, and that the formation of the triplet state is a side-reaction rather than an essential step in the photochemical transfer of an electron from a Bchl dimer to UQ.

3.3. The Short-Lived Intermediate State P^F

The optical studies of Parson *et al.* (1975), with RCs from *Rp. sphaeroides* at low redox potential, revealed not only the absorbance changes suggesting triplet Bchl, but also a second class of changes indicating the formation of a distinct state of shorter lifetime. The difference spectrum of this second state, denoted P^F (for "fast"), showed a sharp negative band near 540 nm. The P^F state is formed with high quantum efficiency (near unity) at room temperature as well as at temperatures near 20°K. In closed RCs, it decays with half-time about 10 nsec at room temperature, increasing to about 40 nsec at low temperature. It can be seen in RCs that have been treated so as to extract all of the UQ, as well as in RCs at a redox potential low enough to reduce the UQ. Because the P^F state and the triplet state are both formed with quantum efficiency near unity at low temperature, with P^F decaying in 40 nsec and triplet persisting for more than 100 μsec, it could be concluded that the P^F state is a precursor of the triplet state in closed RCs.

The possibility of P^F as a photochemical intermediate was established through experiments in two laboratories (Kaufmann *et al.*, 1975; Rockley *et al.*,

1975) in which P^F was detected and characterized in active RCs. Using absorption spectrometry with mode-locked laser flashes to provide time resolution shorter than 20 psec, these investigators found that P^F is formed within 20 psec after a flash. It decays with half-time about 200 psec, in contrast to the 10-nsec half-time in closed RCs. During the 200-psec decay process, the absorption difference spectrum of P^F gives way to the familiar spectrum of the stable photochemical products, in which the features of $(Bchl)_2^+$ are predominant.*

The two most significant features in the spectrum of P^F are the sharp negative band near 540 nm and a positive band at 1250 nm (Dutton et al., 1975). The former suggests reduction of just one of the two Bpheo molecules in the RC (Fajer et al., 1975; Clayton and Yamamoto, 1976), and the latter is characteristic of $(Bchl)_2^+$. The P^F state is therefore described provisionally as a biradical, a polarized excited state of the type

$$(Bchl)_2^+ \cdots (Bpheo)^-$$

The decay of P^F in active RCs is presumably due to further movement of the electron from $Bpheo^-$ to UQ. This decay is signaled most clearly by disappearance of the negative band at 540 nm. The spectral features of $(Bchl)_2^+$, such as the 1250 nm band, are common to both P^F and the final photochemical product; these features do not change in magnitude during the decay process. This fact helps to establish P^F securely as an intermediate in the photochemistry.

The P^F state has been detected in RCs from *Rs. rubrum* as well as *Rp. sphaeroides* (Cogdell et al., 1975).

3.4. Further Details Concerning the P^F State

No evidence for triplet Bchl has been found in active RCs, even in the picosecond time domain. In closed RCs at low temperature, the P^F state decays to a state resembling triplet Bchl; we can imagine that the electron of $Bpheo^-$ returns to the Bchl dimer and generates a polarized triplet state, a biradical within the Bchl dimer. This manner of genesis of the triplet state explains its high degree of electron spin polarization (see Chapter 21).

If these interpretations are correct, the 1250 nm band of P^F ought to disappear as P^F is converted to triplet at low redox potential, whereas at physiological redox potential, this band persists as a property of the stable $(Bchl)_2^+$. This prediction has not been tested.

RCs prepared from wild-type strains of *Rp.*

sphaeroides or *Rs. rubrum* contain a molecule of a carotenoid pigment. In these RCs at low redox potential, the P^F state is a precursor of triplet carotenoid instead of triplet Bchl (Cogdell et al., 1975).

Having recognized the position of P^F in the photochemical sequence, we can return to the question of fluorescence with a clearer perspective. Does fluorescence come exclusively from the singlet state that is first formed on light absorption, in competition with the conversion of excited singlet to P^F? Or, alternatively, does some of the fluorescence follow the regeneration of excited singlet in a back-reaction from P^F? That the fluorescence rises three- to fourfold on lowering of the redox potential is compatible at least qualitatively with the second alternative, since the lifetime of P^F is increased from 0.2 to 10 nsec on lowering of the redox potential. The first alternative would require the additional assumption that the rate of conversion from singlet to P^F is decreased three- to fourfold when the redox potential is lowered. In that case, the rise time for P^F formation should be about 7 psec in active RCs and about 20 psec in closed RCs of *Rp. sphaeroides*; this point has not been tested critically. The second alternative would predict a component of fluorescence having a lifetime equal to the decay time of P^F; this point also awaits decisive clarification.

If the more intense fluorescence at lower redox potential reflects a greater regeneration of excited singlet from P^F, and if the difference in energy between excited singlet and P^F is appreciable, then the fluorescence at low redox potential should show a temperature dependence that reflects this energy difference. In fact, the intensity of the 900 nm fluorescence from RCs of *Rp. sphaeroides* at low redox potential is independent of temperature, within $\pm 10\%$, between 30 and 180°K (Clayton, 1977). This leaves two possibilities: (1) The fluorescence comes exclusively from the primary excited singlet state, not from P^F by way of a back-reaction to excited singlet, with the further provision that the rate of "singlet → P^F" is independent of temperature. (2) The fluorescence comes from P^F as well as from the primary excited singlet, and the energy of the P^F state is equal to that of the excited singlet state within 10^{-3} eV, or within 0.1% of the excited singlet energy (Clayton, 1977). If we accept the second possibility, we must forego the attractive idea that the conversion to P^F provides stabilization of excitation energy by giving up a fraction of the energy. Instead, we must explain how, without an energy loss, the forward reaction (excited singlet → P^F) is much faster than the reverse ($P^F \to$ excited singlet).

The energy of the excited singlet state in the range

---

* Although UQ· is probably formed at this stage, this has not been established with certainty. The spectrum of UQ^- is masked by the larger features of $(Bchl)_2^+$ except in the UV, where measurements are still lacking in the subnanosecond time domain.

865–900 nm is about 1.4 eV. The redox midpoint potentials of $(Bchl)_2/(Bchl)_2^+$ and $(Bpheo)/(Bpheo)^-$ are about 0.45 and −0.55 V, respectively (Fajer *et al.*, 1975), so the energy of P^F might be estimated at about 0.45 + 0.55, or 1.0 eV, although it could be greater than the sum at the midpoint potentials of the (presumed) separate components.

Recently, it has been found possible to trap a state resembling the reduced side of P^F, with features resembling Bpheo⁻ prominent in its absorption difference spectrum. This state was designated I⁻ by Tiede *et al.* (1976). The trapping of I⁻ has been demonstrated with RCs from photosynthetic bacteria such as *Chr. vinosum* or *Rp. viridis*, in which a cytochrome of the *c* type is strongly coupled as an electron donor to the RC even at low temperature. The technique is to place the material at low redox potential so as to reduce the quinone electron acceptor, and then to expose it to strong illumination at a suitable temperature between 200 and 300°K. The rationale is that illumination causes continuous cycling through the P^F state, and occasionally an electron from cytochrome neutralizes the positive side, $(Bchl)_2^+$, of P^F. This leaves a relatively stable combination of oxidized cytochrome and "trapped" reduced side of P^F. After many cycles, a large proportion of the RCs have been driven into this trapped state. Of course, there is no guarantee that I⁻, the reduced side of P^F accompanied by oxidized cytochrome, is identical to the "normal" reduced side (reduced Bpheo?) accompanied by $(Bchl)_2^+$.

I⁻ has been exhibited at 200°K in RCs from *Chr. vinosum* that retain *c*-type cytochromes (Tiede *et al.*, 1976), and in RCs and chromatophores from *Rp. viridis* at room temperature (Trosper and Benson, 1976). The absorption spectrum of I⁻ in RCs from *Chr. vinosum* has features like those in the P^F state that suggest reduced Bpheo, and other features that are not evident in the P^F state of *Rp. sphaeroides* (P^F has not yet been measured in *Chr. vinosum*). These additional features, a bleaching centered at 595 nm and a combination of bleaching and blue-shift of the 802 nm band, indicate involvement of the 2 Bchl molecules of the RC that are not part of the dimeric electron donor. Perhaps in I⁻ an electron interacts with these 2 Bchl's as well as with a molecule of Bpheo. In P^F, the attraction of the plus charge of $(Bchl)_2^+$ might confine this electron more closely to the Bpheo.

A state with spectral features similar to the I⁻ of *Chr. vinosum* has been trapped in RCs from *Rp. viridis* (which also contain *c*-type cytochromes) by illumination at room temperature in the presence of sodium dithionite as a reducing agent (Trosper and Benson, 1976). Bands at 790 and 540 nm, ascribed to Bpheo *b*, are bleached, and a band at 830 nm, attribut-

able to the Bchl *b* that does not act as photochemical electron donor, shows a combination of bleaching and blue shift.

Assuming that I⁻ represents the reduced part of P^F in *Rp. viridis* as well as in *Chr. vinosum*, and that P^F is a precursor of triplet Bchl *b* in closed (acceptor-reduced) RCs, Prince *et al.* (1976) sought to measure the redox midpoint potential of I/I⁻ by finding potentials below which the triplet cannot be formed. The midpoint was found to be −0.4 V. The absence of light-induced triplet formation at lower potentials was explained by saying that with I (Bpheo *b*?) already in its reduced form, P^F could not be formed. A corresponding estimation of the midpoint potential of I/I⁻ in other species of photosynthetic bacteria has not been made.

4. Summary and a Look to the Future

The foregoing material can be summarized in a scheme that represents a reasonable current view of the photochemistry and subsequent events in bacterial photosynthesis (Fig. 1). The relationship of these events to peripheral electron transfers and to the translocation of protons across the photosynthetic membrane is detailed in Chapters 26 and 28. This outline is consistent with available data on the energetics and kinetics of each step; it accounts for the efficient conversion of light energy to forms that can be used for the formation of ATP and for all the metabolic activities of the living and growing cell.

A major problem for the future is to elucidate in detail the structure of the photosynthetic membrane, including the positions and orientations of all components of the RC, the antenna, peripheral electron and proton carriers, and enzymes of phosphorylation. Paralleling this effort, one might attempt to reassemble the separated components of the photosynthetic membrane so as to restore the significant aspects of its function.

Much of our present knowledge of the mechanism of bacterial photosynthesis has been obtained through experiments with isolated RCs. The insights gained from our understanding of photochemical processes in these RCs might profitably be applied to the study of green plant photosynthesis, in which the details of primary events in the RCs of photosystems I and II remain relatively obscure.

Our picture of how the RC works might also serve as a guide in the construction of a solar cell using chlorophyll as the photosensitizer. One could begin by trying to construct *in vitro* a functioning model of the RC embedded in an artifical membrane.

ABSORPTION AND ENERGY TRANSFER

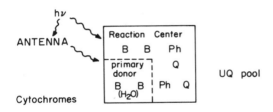

PHOTOCHEMICAL EVENTS

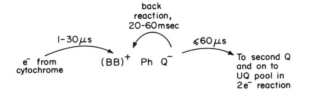

SECONDARY ELECTRON TRANSFERS

Fig. 1. An outline of the early stages in bacterial photosynthesis. (B) Bchl; (Ph) Bpheo; (Q) quinone; (UQ) ubiquinone; (*) the excited singlet state. The characteristic times for secondary electron transfers are taken from the following sources: cytochrome to (BB)$^+$, Chapter 28; Q$^-$ to secondary Qs, Chapter 25; back-reaction, Chapter 19 and Clayton and Yau (1972).

ACKNOWLEDGMENTS

Preparation of this manuscript was supported by Contract No. E(11-1)-3162 with the United States Energy Research and Development Administration and by Grant No. PCM76-10556 from the National Science Foundation.

5. References

Arnold, W., and Clayton, R. K., 1960, The first step in photosynthesis: Evidence for its electronic nature, *Proc. Natl. Acad. Sci. U.S.A.* **46**:769.

Bolton, J. R., Clayton, R. K., and Reed, D. W., 1969, An identification of the radical giving rise to the light-induced electron spin resonance signal in photosynthetic bacteria, *Photochem. Photobiol.* **9**:209.

Chance, B., and DeVault, D., 1964, On the kinetics and quantum efficiency of the chlorophyll–cytochrome reaction, *Ber. Bunsenges. Phys. Chem.* **68**:722.

Chance, B., and Nishimura, M., 1960, On the mechanism of chlorophyl–cytochrome interaction: The temperature insensitivity of light-induced cytochrome oxidation in *Chromatium, Proc. Natl. Acad. Sci. U.S.A.* **46**:19.

Chance, B., and Smith, L., 1955, Respiratory pigments of *Rhodospirillum rubrum, Nature (London)* **175**:803.

Clayton, R. K., 1962a, Evidence for the photochemical reduction of coenzyme Q in chromatophores of photosynthetic bacteria, *Biochem. Biophys. Res. Commun.* **9**:49.

Clayton, R. K., 1962b, Primary reactions in bacterial photosynthesis. III. Reactions of carotenoids and cytochromes in illuminated bacterial chromatophores, *Photochem. Photobiol.* **1**:313.

Clayton, R. K., 1966, Relations between photochemistry and fluorescence in cells and extracts of photosynthetic bacteria, *Photochem. Photobiol.* **5**:807.

Clayton, R. K., 1967, The bacterial photosynthetic reaction center, *Brookhaven Symp. Biol.* **19**:62.

Clayton, R. K., 1977, Fluorescence of bacteriochlorophyll in reaction centers from *Rhodopseudomonas sphaeroides* at low temperatures, *Plant Cell Physiol. Jpn., Special Issue No. 3*:87.

Clayton, R. K., and Haselkorn, R., 1972, Protein components of bacterial photosynthetic membranes, *J. Mol. Biol.* **68**:97.

Clayton, R. K., and Straley, S. C., 1972, Photochemical electron transport in photosynthetic reaction centers. IV. Observations related to the reduced photoproducts, *Biophys. J.* **12**:1221.

Clayton, R. K., and Yamamoto, T., 1976, Photochemical quantum efficiency and absorption spectra of reaction centers from *Rhodopseudomonas sphaeroides* at low temperature, *Photochem. Photobiol.* **24**:67.

Clayton, R. K., and Yau, H. F., 1972, Photochemical electron transport in photosynthetic reaction centers from *Rhodopseudomonas sphaeroides*. I. Kinetics of the oxidation and reduction of P870 as affected by external factors, *Biophys. J.* **12**:867.

Clayton, R. K., Sistrom, W. R., and Zaugg, W., 1965, The role of a reaction center in photochemical activities of bacterial chromatophores, *Biochim. Biophys. Acta* **102**:341.

Clayton, R. K., Fleming, H., and Szuts, E. Z., 1972, Photochemical electron transport in photosynthetic reaction centers from *Rhodopseudomonas sphaeroides*. II. Interaction with external electron donors and acceptors and a reevaluation of some spectroscopic data. *Biophys. J.* **12**:46.

Cogdell, R. J., Brune, C. D., and Clayton, R. K., 1974, Effects of extraction and replacement of ubiquinone upon the photochemical activity of reaction centers and chromatophores from *Rhodopseudomonas sphaeroides*, *FEBS Lett.* **45**:344.

Cogdell, R. J., Monger, T. G., and Parson, W. W., 1975, Carotenoid triplet states in reaction centers from *Rhodopseudomonas sphaeroides* and *Rhodospirillum rubrum*, *Biochim. Biophys. Acta* **408**:189.

DeVault, D., and Chance, B., 1966, Studies of photosynthesis using a pulsed laser. I. Temperature dependence of cytochrome oxidation rate in *Chromatium vinosum*. Evidence for tunneling, *Biophys. J.* **6**:825.

Dutton, P. L., Leigh, J. S., and Seibert, M., 1972, Primary processes in photosynthesis: *In situ* ESR studies on the light induced oxidized and triplet state of reaction center bacteriochlorophyll, *Biochem. Biophys. Res. Commun.* **46**:406.

Dutton, P. L., Leigh, J. S., and Reed, D. W., 1973, Primary events in the photosynthetic reaction centre of *Rhodopseudomonas sphaeroides* strain R-26: Triplet and oxidized states of bacteriochlorophyll and the identification of the primary electron acceptor, *Biochim. Biophys. Acta* **292**:654.

Dutton, P. L., Kaufmann, K. J., Chance, B., and Rentzepis, P. M., 1975, Picosecond kinetics of the 1250 nm band of the *Rhodopseudomonas sphaeroides* reaction center: The nature of the primary photochemical intermediary state, *FEBS Lett.* **60**:275.

Duysens, L. N. M., 1953, Reversible changes in the light absorption of photosynthetic bacteria caused by illumination, *Carnegie Inst. Washington Yearb.* **52**:157.

Duysens, L. N. M., 1954, Reversible photo-oxidation of a cytochrome pigment in photosynthesizing *Rhodospirillum rubrum, Nature* (*London*) **173**:692.

Duysens, L. N. M., 1959, The path of light energy in photosynthesis, *Brookhaven Symp. Biol.* **11**:10.

Duysens, L. N. M., Huiskamp, W. J., Vos, J. J., and van der Hart, J. M., 1956, Reversible changes in bacteriochlorophyll in photosynthetic bacteria upon illumination, *Biochim. Biophys. Acta* **19**:188.

Fajer, J., Brune, D. C., Davis, M. S., Forman, A., and Spaulding, L. D., 1975, Primary charge separation in bacterial photosynthesis: Oxidized chlorophylls and reduced pheophytin, *Proc. Natl. Acad. Sci. U.S.A.* **72**:4956.

Feher, G., Okamura, M. Y., and McElroy, J. D., 1972, Identification of an electron acceptor in reaction centers of *Rhodopseudomonas sphaeroides* by EPR spectroscopy, *Biochim. Biophys. Acta* **267**:222.

Feher, G., Isaacson, R. A., McElroy, J. D., Ackerson, L. C., and Okamura, M. Y., 1974, On the question of the primary electron acceptor in bacterial photosynthesis: Manganese substituting for iron in reaction centers of *Rhodopseudomonas sphaeroides* R-26, *Biochim. Biophys. Acta* **368**:135.

Feher, G., Hoff, A. J., Isaacson, R. A., and Ackerson, L. C., 1975, ENDOR experiments on chlorophyll and bacteriochlorophyll *in vitro* and in the photosynthetic unit, *Ann. N. Y. Acad. Sci.* **244**:239.

Goedheer, J. C., 1959, Reversible oxidation of pigments in bacterial chromatophores, *Brookhaven Symp. Biol.* **11**:325.

Halsey, Y. D., and Parson, W. W., 1974, Identification of ubiquinone as the secondary electron acceptor in the photosynthetic apparatus of *Chromatium vinosum*, *Biochim. Biophys. Acta* **347**:404.

Heathcote, P., and Clayton, R. K., 1977, Reconstituted energy transfer from antenna pigment-protein to reaction centres isolated from *Rhodopseudomonas sphaeroides, Biochim. Biophys. Acta* **459**:506.

Katz, J. J., Oettmeier, W., and Norris, J. R., 1976, Organization of antenna and photo-reaction centre chlorophylls on the molecular level, *Philos. Trans. R. Soc. London Ser. B:* **237**:227.

Kaufmann, K. J., Dutton, P. L., Netzel, T. L., Leigh, J. S., and Rentzepis, P. M., 1975, Picosecond kinetics of events leading to reaction center bacteriochlorophyll oxidation, *Science* **188**:1301.

Leigh, J. S., and Dutton, P. L., 1972, The primary electron acceptor in photosynthesis, *Biochem. Biophys. Res. Commun.* **46**:414.

Loach, P. A., and Hall, R. L., 1972, The question of the primary electron acceptor in bacterial photosynthesis, *Proc. Natl. Acad. Sci. U.S.A.* **69**:786.

Loach, P. A., and Sekura, D. L., 1967, A comparison of decay kinetics of photo-produced absorbance, EPR and luminescence changes in chromatophores of *Rhodospirillum rubrum, Photochem. Photobiol.* **6**:381.

Loach, P. A., and Sekura, D. L., 1968, Primary photochemistry and electron transport in *Rhodospirillum rubrum, Biochemistry* **7**:2642.

McElroy, J. D., Feher, G., and Mauzerall, D. C., 1970, Observation of a second light-induced EPR signal from reaction centers of photosynthetic bacteria, *Biophys. J.* **10**:204a.

McElroy, J. D., Mauzerall, D. C., and Feher, G., 1974, Characterization of primary reactants in bacterial photosynthesis. II. Kinetic studies of the light-induced signal (g = 2.0026) and the optical absorbance changes at cryogenic temperatures, *Biochim. Biophys. Acta* **333**:261.

Morrison, L. E., Runquist, J. A., and Loach, P. A., 1976, Tightly bound ubiquinone and photochemical activity in *Rhodospirillum rubrum, Biophys. J.* **16**:222a.

Norris, J. R., Uphaus, A., Crespi, H. L., and Katz, J. J., 1971, Electron spin resonance of chlorophyll and the origin of signal 1 in photosynthesis, *Proc. Natl. Acad. Sci. U.S.A.* **68**:625.

Norris, J. R., Druyan, M. E., and Katz, J. J., 1973, Electron nuclear double resonance of bacteriochlorophyll free radical *in vitro* and *in vivo, J. Am. Chem. Soc.* **95**:1680.

Okamura, M. Y., Isaacson, R. A., and Feher, G., 1975, The primary acceptor in bacterial photosynthesis: The obligatory role of ubiquinone in photoactive reaction centers of *Rhodopseudomonas sphaeroides, Proc. Natl. Acad. Sci. U.S.A.* **72**:3491.

Okamura, M. Y., Ackerson, L. C., Isaacson, R. A., Parson, W. W., and Feher, G., 1976, The primary electron acceptor in *Chromatium vinosum* (strain D), *Biophys. J.* **16**:223a.

Olson, J. M., 1962, Quantum efficiency of cytochrome oxidation in a photosynthetic bacterium, *Science* **135**:101.

Parson, W. W., 1968, The role of P$_{870}$ in bacterial photosynthesis, *Biochim. Biophys. Acta* **153**:248.

Parson, W. W., Clayton, R. K., and Cogdell, R. J., 1975, Excited states of photosynthetic reaction centers at low redox potentials, *Biochim. Biophys. Acta* **387**:265.

Prince, R. C., Leigh, J. S., and Dutton, P. L., 1976, Thermodynamic properties of the reaction center of *Rhodopseudomonas viridis: In vivo* measurement of the reaction center bacteriochlorophyll—primary acceptor intermediary electron carrier, *Biochim. Biophys. Acta* **440**:662.

Rockley, M. G., Windsor, M. W., Cogdell, R. J., and Parson, W. W., 1975, Picosecond detection of an intermediate in the photochemical reaction of bacterial photosynthesis, *Proc. Natl. Acad. Sci. U.S.A.* **72**:2251.

Sistrom, W. R., and Clayton, R. K., 1964, Studies on a mutant of *Rhodopseudomonas sphaeroides* unable to grow photosynthetically, *Biochim. Biophys. Acta* **88**:61.

Slooten, L., 1972a, Reaction center preparations of *Rhodopseudomonas sphaeroides*: Energy transfer and structure, *Biochim. Biophys. Acta* **256**:452.

Slooten, L., 1972b, Electron acceptors in reaction center preparations from photosynthetic bacteria, *Biochim. Biophys. Acta* **272**:208.

Smith, L., and Ramirez, J., 1959, Reactions of pigments of photosynthetic bacteria following illumination or oxygenation, *Brookhaven Symp. Biol.* **11**:310.

Sogo, P., Jost, M., and Calvin, M., 1959, Evidence for free-radical production in photosynthesizing systems, *Radiat. Res., Suppl. 1,* p. 511.

Straley, S. C., Parson, W. W., Mauzerall, D. C., and Clayton, R. K., 1973, Pigment content and molar extinction coefficients of photochemical reaction centers from *Rhodopseudomonas sphaeroides, Biochim. Biophys. Acta* **305**:597.

Thornber, J. P., Olson, J. M., Williams, D. M., and Clayton, M. L., 1969, Isolation of reaction center of *Rhodopseudomonas viridis, Biochim. Biophys. Acta* **172**:351.

Tiede, D. M., Prince, R. C., and Dutton, P. L., 1976, EPR and optical properties of an electron carrier intermediate in reaction centers of *Chromatium vinosum, Brookhaven Symp. Biol.* **28**:368.

Trosper, T., and Benson, D. L., 1976, Spectral properties of an improved *Rhodopseudomonas viridis* reaction center preparation, *Brookhaven Symp. Biol.* **28**:367.

Vermeglio, A., 1977, Secondary electron transfer in reaction centers of *Rhodopseudomonas sphaeroides*: Out-of-phase periodicity of two for the formation of ubisemiquinone and fully reduced ubiquinone, *Biochim. Biophys. Acta* **459**:516.

Vermeglio, A., and Clayton, R. K., 1976, Orientation of chromophores in reaction centers of *Rhodopseudomonas sphaeroides*: Evidence for two absorption bands of the dimeric primary electron donor, *Biochim. Biophys. Acta* **449**:500.

Vredenberg, W. J., and Duysens, L. N. M., 1963, Transfer of energy from bacteriochlorophyll to a reaction centre during bacterial photosynthesis, *Nature (London)* **197**:355.

Wraight, C. A., 1977, The primary acceptor of photosynthetic bacterial reaction centers: Direct observation of oscillatory behaviour suggesting two closely equivalent species, *Biochim. Biophys. Acta* **459**:525.

Wraight, C. A., and Clayton, R. K., 1974, The absolute quantum efficiency of bacteriochlorophyll photooxidation in reaction centres of *Rhodopseudomonas sphaeroides, Biochim. Biophys. Acta* **333**:246.

Wraight, C. A., Leigh, J. S., Dutton, P. L., and Clayton, R. K., 1974, The triplet state of reaction center bacteriochlorophyll: Determination of a relative quantum yield, *Biochim. Biophys. Acta* **333**:401.

Zankel, K. L., Reed, D. W., and Clayton, R. K., 1968, Fluorescence and photochemical quenching in photosynthetic reaction centers, *Proc. Natl. Acad. Sci. U.S.A.* **61**:1243.

CHAPTER 21

Oxidized Bacteriochlorophyll as Photoproduct

James R. Norris and Joseph J. Katz

1. Introduction

The concept of oxidized chlorophyll (Chl) as the product formed in the primary events of photosynthesis has emerged over the last two decades and provides a key to the understanding of the initial steps of light conversion in photosynthesis. From the first observations by optical (Duysens, 1952; Kok, 1959; Goedheer, 1960) and ESR spectroscopy (Commoner et al., 1956; Sogo et al., 1959), it was suspected that oxidized Chl was produced in the primary step of photosynthesis.

More recently, it has been shown that the primary, light-induced, electron-transfer act of bacterial photosynthesis results in the oxidation of a special pair of bacteriochlorophyll (Bchl) molecules (Katz et al., 1968; Garcia-Morin et al., 1969; Norris et al., 1971, 1972, 1973, 1974, 1975b; McElroy et al., 1972; Feher et al., 1973, 1975; Katz and Norris, 1973). The light energy necessary to drive this initial chemical reaction of photosynthesis is collected by a large group of Chl molecules, which, by virtue of the role they play in harvesting light energy, are often referred to as "antenna" (Emerson and Arnold, 1932a,b; van Niel, 1935) Bchl. Antenna Bchl constitutes the bulk of the Chl present *in vivo*, and is chemically inert, serving only to transmit the energy of the absorbed photon to the photoreactive Chl situated in a reaction center

(RC). This latter species of Chl is therefore generally designated as *reaction* (or *photoreaction*) *center* (Duysens et al., 1956; Clayton, 1962) Bchl. When a photon excites a resting RC, within picoseconds of acquiring the excitation energy, light-energy conversion begins to occur (Parson et al., 1975; Rockley et al., 1975; Kaufmann et al., 1975; Dutton et al., 1975). The excited Chl special pair quickly ejects an electron, leaving the Bchl special pair cation as the primary oxidized photoproduct produced in the course of bacterial photosynthesis. The electron ejected from the special pair is the source of the chemical energy necessary to carry out all the dark chemical reactions taking place in photosynthetic organisms.

The experimental evidence that shows that the primary oxidized photoproduct involves two special Chl molecules acting in concert strongly suggests that these special Chl's constitute the essential part of the primary electron pump of bacterial photosynthesis. Similar observations have been made for green plant photosynthesis (Norris et al., 1971, 1972, 1974, 1975b; Katz and Norris, 1973; Feher et al., 1975). The special pair model for RC Chl thus provides a common basis for the interpretation of the primary events in light conversion in bacterial as well as green plant photosynthesis. It is not surprising, therefore, that considerable experimental work has been carried out to test the validity and scope of the special pair concept. Here we review those studies that involve oxidized Bchl either directly or indirectly, with particular emphasis on the *in vivo* photooxidation.

James R. Norris and Joseph J. Katz · Chemistry Division, Argonne National Laboratory, Argonne, Illinois 60439

2. Bacteriochlorophyll Species

2.1. Antenna Bacteriochlorophyll

Whereas *in vivo* RC Bchl is photoactive, antenna Bchl appears to be photochemically inert, and its properties in our present context can be quickly summarized. We have prepared aggregates of Bchl and water that absorb near 800 and near 850 nm (Fig. 1) and thus mimic the properties of various *in vivo* Bchl antenna species (Katz *et al.*, 1976). A species absorbing at 829 nm can also be prepared. All these aggregates are observed when Bchl is hydrated in aliphatic hydrocarbon solvents under a variety of conditions. Much research has shown that the central Mg of the Chl's with coordination number 4 (Fig. 2) is coordinatively unsaturated and requires an additional, basic (nucleophilic) donor ligand in one or both of the Mg axial positions (Katz, 1968). The requisite axial ligand may be provided either by the oxygen of water or by a donor functional group present in another Bchl (such as the 9-keto C=O or 2-acetyl C=O of Fig. 2b).

The studies that have characterized these interactions, particularly for Chl *a* systems, are very numerous, and the reader is referred to recent reviews (Katz *et al.*, 1976; Katz and Janson, 1973) for a detailed discussion. Of particular interest here is that these *in vitro* antenna preparations exhibit photoreversible ESR signals (Katz *et al.*, 1976). The small ESR signals that can be induced by light in these

Bchl–H_2O adduct species may be attributed either to small amounts of special pair Chl that are present only in very low concentrations or to very small amounts of impurity electron acceptors likely to be present in all such preparations, or to both. Since intentional addition of significant amounts of acceptor molecules does not proportionately increase the ESR photoactivity, it may be suggested that it is the special pair that is essential for photochemistry, and that only small amounts of the special pair are formed in these antenna preparations. The *in vitro* preparations have broad optical spectra compared with the optical spectrum of *in vivo* antenna Bchl, and thus it is not unreasonable that absorption by the P865 special pair is present in the red wing of the *in vitro* 850 nm transition. In any event, these experiments imply that water may play a significant role in the formation of both *in-vivo*-like antenna and photoreactive Bchl. Antenna Chl in green plants appears to consist of Chl–Chl self-aggregates, and thus differs fundamentally from bacterial antenna.

2.2. Reaction Center Bacteriochlorophyll

The most informative and direct means so far employed to establish the nature of oxidized Chl are ESR and electron-nuclear double resonance (ENDOR) spectroscopy. In fact, ESR and ENDOR provide the most direct evidence that oxidation of *in vitro* Chl produces a one-electron oxidation product,

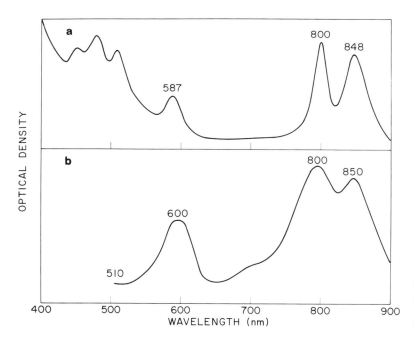

Fig. 1. (a) *Rhodopseudomonas sphaeroides* electronic transition (E.T.) spectrum showing P_{800} and P_{850}; (b) E.T. spectrum of *in vitro* preparation of P_{800} and P_{850}, showing close correlation with the organism *Rp. sphaeroides*.

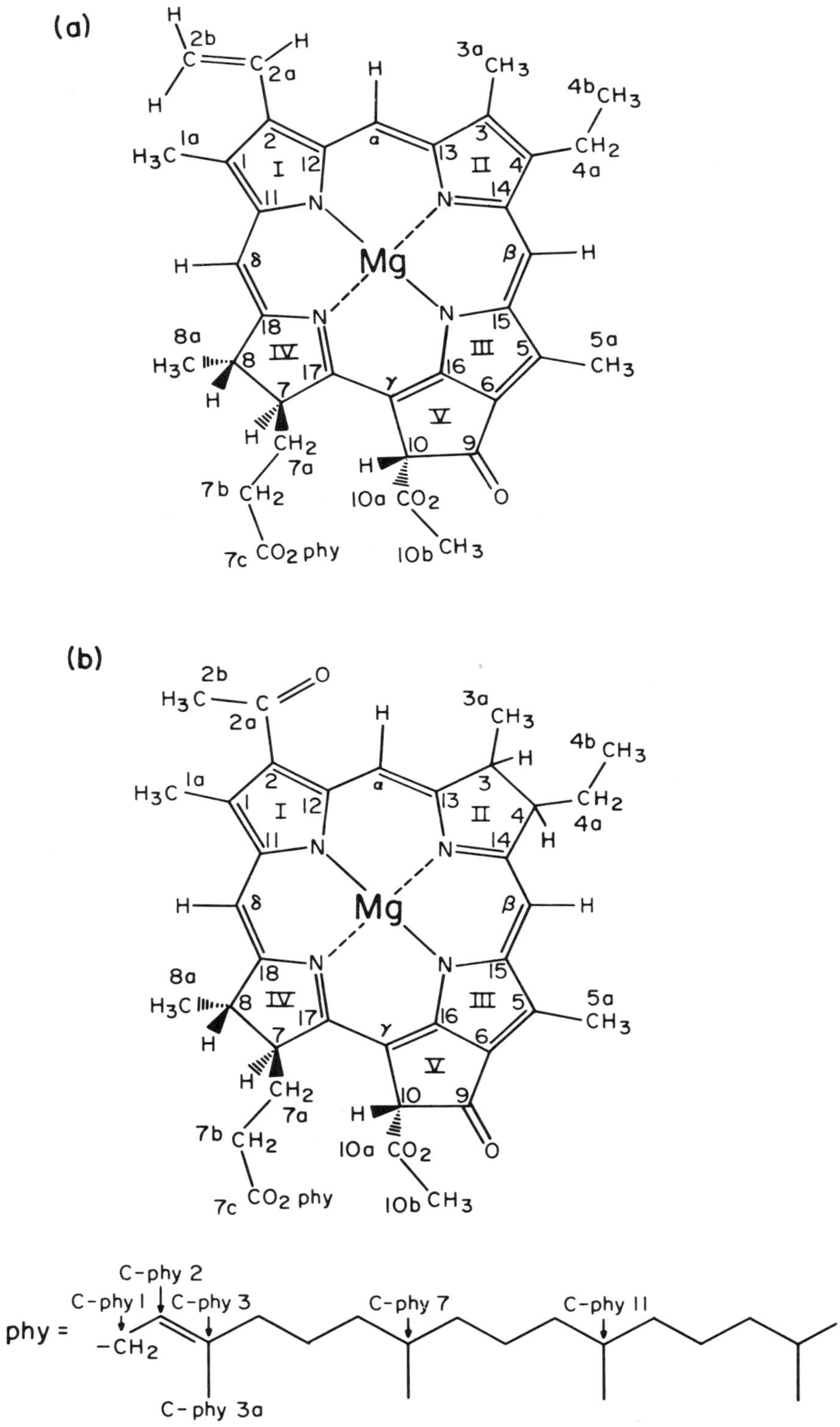

Fig. 2. Structural formulas for Chl *a* (a) and Bchl *a* (b).

and establishes not only the existence but also the nature of oxidized Chl *in vivo*. To obtain a satisfactory description of *in vivo* oxidized Bchl, which is our primary goal, we compare *in vivo* observations by ESR and ENDOR with studies on well-characterized *in vitro* Chl free radical cations. The cations of Chl are most reproducibly generated by "dark" chemical reactions or by electrochemical techniques (Fuhrhop and Mauzerall, 1969; McElroy *et al.*, 1969; Fajer *et al.*, 1970, 1974, Borg *et al.*, 1970), and such systems are easily investigated by ESR, ENDOR and electronic transition (E.T.) spectra, and electrochemical techniques. The cations, prepared in anhydrous solvents and completely free of air, are very stable at low temperature ($<100°$K), and are even sufficiently stable at room temperature to allow detailed characterization by ESR and ENDOR.

The interpretation of magnetic resonance data is facilitated by observations on E.T. spectra; the latter is a convenient way to study photoreactive Bchl, although this spectroscopic technique provides little structural information in comparison with magnetic resonance techniques. Less direct but nevertheless pertinent information is provided by fluorescence, triplet state ESR, and electrochemical measurements of the redox properties of oxidized Chl.

3. Experimental Observation of Photoreactive Bacteriochlorophyll

3.1. Visible Absorption (Electronic Transition) Spectra

One of the first and most important observations to indicate the formation of oxidized Bchl in the primary light conversion steps of photosynthesis is in the optical "bleaching" observed in the E.T. spectra of irradiated photosynthetic bacteria. This type of experiment was first carried out in bacteria by Duysens (1952) and later in green plants by Kok (1959). In bacteria, a complete loss of optical absorption near 870 nm occurs either when the RC Bchl is chemically oxidized in the dark or when the RC is irradiated with actinic light. This change in optical absorption was attributed to the oxidation of an *in vivo* form of Bchl designated as P870, since Bchl extracted from bacteria and dissolved in a polar (basic) solvent absorbs light maximally near 770 nm. It is also true that monomeric Bchl absorbing at 770 nm can be reversibly oxidized in a one-electron process by chemical means, and this form of Bchl also shows essentially complete bleaching of its red band, except that now the bleaching is observed to occur at 770 nm instead of near 870

nm, as is the case for the *in vivo* systems (Goedheer, 1960; Fuhrhop and Mauzerall, 1969; Fajer *et al.*, 1974).

Because none of the pigments that can be extracted from the living organism absorbs near 870 nm, it was assumed, correctly it turns out, that P870 is a Bchl cation, but in a special environment (McElroy *et al.*, 1969; Bolton *et al.*, 1969). We now know that this special environment *in vivo* is produced primarily through various forms of interaction of the Chl's, and that one of these aggregated forms, the special pair, constitutes the primary donor of photosynthesis. The proof for this conclusion is based largely on magnetic resonance spectroscopy.

3.2. ESR Spectroscopy of the Special Pair

3.2.1. Introduction to ESR Spectroscopy

ESR spectroscopy has been an especially powerful tool in the study of photosynthesis because the first relatively stable oxidized product of photosynthesis is a doublet-state free radical. Several very good general treatments of ESR theory and application are now available (Wertz and Bolton, 1972; Swartz *et al.*, 1972).

For Chl cations that contain one unpaired electron per molecule, the major interaction of the unpaired electron (spin angular momentum $\frac{1}{2}$) is with other magnetic nuclei, of which protons (nuclear spin $\frac{1}{2}$) of the Chl macrocycle dominate. Additionally, the magnetic environment of the electron includes ^{14}N (spin 1), ^{13}C (abundance 1.08%; nuclear spin $\frac{1}{2}$), and certain anisotropic effects of the orbital angular momentum of the unpaired electron (i.e., *g*-anisotropy). It should be especially noted that the most important contribution to the environment sensed by the unpaired electron arises from the protons on the Chl macrocycle.

Monomeric Bchl cations (Bchl$^+$)* produced by chemical or electrolytic oxidation give rise to ESR signals typical of aromatic free radicals. From the ESR properties of *in vitro* Bchl$^+$ (Fig. 3) listed in Table 1, it can be seen that Bchl$^+$ has a single, very nearly Gaussian lineshape envelope, with a linewidth, ΔH_{pp} (peak-to-peak first-derivative linewidth), of 13.0 gauss. The spectrum center has essentially the free-electron *g* value of 2.0026 $\pm$ 0.0002 (Norris *et al.*, 1971; McElroy *et al.*, 1969, 1972; Kohl *et al.*, 1965) (the free-electron *g* value is 2.0023). An additional

---

* We write monomer Bchl free radical schematically as Bchl$^+$, but it should be understood that one or two molecules of nucleophilic ligand (solvent) may be present in the Mg axial positions of the Chl.

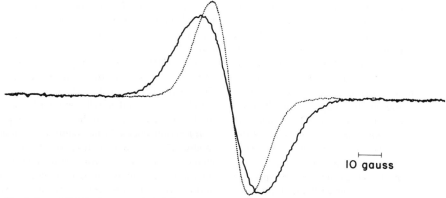

Fig. 3. (———) Oxidized monomeric Bchl‡ monomer *in vitro* generated by oxidation of Bchl in CH$_2$Cl$_2$/MeOH (3:1, vol./vol.) by I$_2$; (......) photoinduced signal generated in *Rhodospirillum rubrum* by red light. Note that both curves have a Gaussian line shape and the same $g = 2.0026$.

observation also noted in Table 1 is that ΔH_{pp} decreases to 5.4 gauss in Bchl deuterated to the extent of 99.5% (Norris *et al.*, 1971; McElroy *et al.*, 1969, 1972; Kohl *et al.*, 1965). Four simple observations on the line shape, line width, line center, and isotope effect on line shape reveal a great deal about Bchl‡.

3.2.2. Basic ESR Properties of Bacteriochlorophyll Cations

(a) Significance of the Free-Electron *g* Value. The symmetrical nature of the Gaussian lineshape allows accurate determination of a *g* value, i.e., the center of

Table 1. ESR Data of *in Vitro* Bacteriochlorophyll Cations

Isotopic composition	Temp. (°K)	Solvent	Oxidant	Line shape[a]	g Value	ΔH_{pp} (gauss)	Ref. No.[b]
^{1}H	77	CH$_3$OH–glycerol (1:1, vol./vol.)	I$_2$	G	2.0025 ± 0.0001	12.8 ± 0.5	1, 2
^{1}H	100	12% CH$_3$OH in CH$_2$Cl$_2$ (vol./vol.)	I$_2$	G	—	13	3
^{1}H	—	CH$_2$Cl$_2$	c	G	2.0025 ± 0.0001	13.2	4
^{1}H	—	CHCl$_3$	c	G	2.0025 ± 0.0001	12.8	4
^{1}H	—	CH$_3$OH	c	G	2.0025 ± 0.0001	12.7	4
^{1}H	2–300	CH$_3$OH–glycerol (1:1, vol./vol.)	I$_2$	G	2.0025 ± 0.0001	13.0 ± 0.2	5
^{1}H	2–300	Same	I$_2$	G	2.0025 ± 0.0001[d]	14.2 ± 0.2[d]	5
^{1}H	300	CH$_3$OH	I$_2$	—	2.0026 ± 0.0002	12.6 ± 5%	6
^{1}H	—	Same	I$_2$	—	2.0025 ± 0.0001	12.8 ± 0.5	7
^{2}H	77	CH$_3$OH–glycerol (1:1, vol./vol.)	I$_2$	G	2.0026 ± 0.0001	5.4 ± 0.2	1, 2
^{2}H	2–300	Same	I$_2$	G	2.0025 ± 0.0001	5.0 ± 0.2	5
^{2}H	2–300	Same	I$_2$	e	2.0026 ± 0.0001[d]	7.1 ± 0.2[d]	5

[a] (G) Gaussian line shape within signal-to-noise.
[b] (1) McElroy *et al.* (1969); (2) Norris *et al.* (1971); (3) Norris *et al.* (1973); (4) Fajer *et al.* (1974); (5) McElroy *et al.* (1972); (6) Fuhrhop and Mauzerall (1969); (7) Loach *et al.* (1971).
[c] Electrochemical, I$_2$, or cation of Zn tetraphenyl porphyrin.
[d] Microwave ESR frequency was near 35 GHz. All other measurements were at 9 GHz. The high frequency increases the ΔH_{pp} because of the small amount of *g*-anisotropy contributing to the line width.
[e] Since this system is fully deuterated and also recorded at 35 GHz, the line width and line shape are more affected by *g*-anisotropy. Thus, there is no longer a purely Gaussian shape due to a 10% asymmetry in peak heights.

the resonance envelope. Since simple aromatic hydrocarbon cation free radicals typically have g values near that of the free-electron g value of 2.0023, we know that Bchl‡ cation free radicals, with $g = 2.0026$, are basically similar to that of an aromatic hydrocarbon. Thus, we know that the six oxygen atoms of Bchl do not interact strongly with the unpaired electron of Bchl‡ (quinone free radicals, in which the unpaired electron is significantly localized near the oxygen atoms, have g values in the vicinity of 2.004). In Bchl‡, therefore, the unpaired electron resides largely on the C–H skeleton of the macrocycle.

(b) **Significance of the Gaussian Line Shape.** The Gaussian line shape is consistent with a high and even delocalization of the unpaired electron over the entire macrocycle. If an unpaired electron with a spin density of approximately 1 interacts with a proton at a single site (i.e., no delocalization) in a π-bond system typical of aromatic free radicals, two distinct peaks separated by approximately 25 gauss would be observed in the ESR spectrum. If the spin density were 0.5 for two equivalent sites, the splitting would be half as large, or 12.5 gauss, and there would now be three peaks in the spectrum. In other words, the strength of the electron–proton interaction measured in gauss is proportional to unpaired-electron spin density. Thus, as more delocalization of the unpaired electron occurs, the weaker the interaction and the more numerous the spectral lines become. Roughly speaking, the sum of the spin densities within a molecule with one unpaired electron must be approximately 1. Hence, for large delocalization, we have small spin densities at many sites and small but quite numerous splittings in the ESR signal. For example, suppose an unpaired electron interacts only with 10 equivalent protons of nuclear spin $\frac{1}{2}$. In this case, the spin density at each proton would be 0.1, and the electron–proton splitting would be 2.5 gauss. To first order, the largest and smallest internal magnetic field that could be experienced by the unpaired electron would be either with all protons pointing with or against the external magnetic field. In contrast, no net internal field is experienced by the electron when exactly half the protons are aligned against the external magnetic field. Since the no-net-field case is 252 times more probable than either extreme internal field case, the center of the resonance pattern is most intense. Likewise, information about the extreme field environment is found in the wings of the ESR signal. Thus, the more uniform and extensive the delocalization of the unpaired spin, the more numerous and smaller become the individual interactions between electron and magnetic nuclei in the environment, and the more closely the envelope of resonances approaches a Gaussian shape. Furthermore, if any one electron–proton interaction becomes significantly greater than the others because of uneven electron-spin delocalization, then the resonance envelope ESR line shape deviates from a Gaussian shape. The Gaussian nature of the line shape can therefore be taken to show that the unpaired electron is interacting essentially simultaneously and equally on the ESR time scale with all the magnetic nuclei in the Chl macrocycle.

(c) **Significance of Line Width and Changes in Line Width.** For the Bchl cation, the Gaussian ESR line shape envelope arises from more than a million combinations of small electron-magnetic nuclei interactions with the unpaired electron. As mentioned in Section 3.2.2(b), the more numerous the interactions, the greater becomes the probability that any given electron will experience a very small net environmental shift. Thus, in systems that contain a delocalized electron, the more intense becomes the center of the Gaussian envelope and the narrower becomes the Gaussian line width.

A rough quantification of the foregoing qualitative argument indicates that approximately 12 protons are interacting with the unpaired electron in monomeric Bchl‡. The square root of the number of nuclei that interact with the unpaired electron is inversely proportional to the overall peak-to-peak line width. If the line width were to decrease by a factor of 2, relative to the monomer free radical, then it would be expected that $(2)^2 \times 12 = 48$ protons, or an aggregate consisting of 4 Bchl molecules, share the unpaired electron. If the line width were narrowed in a particular instance by $\sqrt{2}$, then $(\sqrt{2})^2 \times 12 = 24$ protons would be involved. Thus, for us, the real significance of the ESR line width is in its ability to tell whether a cation free radical *in vivo* arises from Bchl at all, and if it does, how many Bchl molecules are involved in spin-sharing.

(d) **Effect of ^{2}H Substitution for ^{1}H.** We have used protons to discuss the nature of the environmental field because protons account for the largest fraction of the line width in Bchl‡. We know this to be so because there is a decrease in line width by a factor of 2.4 when the ESR signal of ^{2}H-Bchl‡ is observed. A decrease in line width in this case is expected for magnetic interactions of ^{2}H (nuclear spin 1) vs. ^{1}H, because the magnetic moment of ^{1}H is 6.514 times larger than the magnetic moment of ^{2}H. If all the ESR line width resulted from proton–electron interactions, then the Gaussian line width would decrease by the maximum value of 4 (Kohl, 1965). The decrease in line width by a factor of 2.4 tells us that the extensive delocalization of the unpaired spin in the free radical involves hydrogen–electron interactions to the extent that

Table 2. ESR Data of *in Vivo* Photosynthetic Bacterial Systems

Organism	Temp. (°K)	Oxidant	Line shape[a]	g Value	ΔH_{pp} (gauss)	Ref. No.[b]
[1]H-*Rs. rubrum*	288	Actinic light	G	c	9.1 ± 0.5	1, 2
[2]H-*Rs. rubrum*	288	Actinic light	G	c	4.0 ± 0.5	1, 2
[1]H-*Rs. rubrum*	77	Red light	G	2.0026 ± 0.0001	9.5 ± 0.5	2, 3
[2]H-*Rs. rubrum*	77	Red light	G	2.0026 ± 0.0001	4.2 ± 0.5	2, 3
[1]H-*Rs. rubrum*	300	Red light	G	c	9.2 ± 0.6	2, 4
[1]H-*Rp. sphaeroides*[d]	c	Red light	G	2.0025 ± 0.0002	9.6 ± 0.2	2, 5
[1]H-*Rs. rubrum*	2–300	Red light	G	2.0026	9.4 ± 0.2	6
[1]H-*Rs. rubrum*	2–300	Red light	G	2.0026	10.3 ± 0.2^e	6
[2]H-*Rs. rubrum*	2–300	Red light	f	2.0027	5.9 ± 0.3^e	6
[1]H-*Rp. sphaeroides* R-26	2–300	Red light	G	c	9.3 ± 0.2	6
[1]H-*Rp. sphaeroides* R-26	2–300	Red light	G	c	10.2 ± 0.2^e	6
[1]H-*Rp. sphaeroides* R-26[g]	2–300	Red light	G	c	9.8 ± 0.2	6
[1]H-*Rp. sphaeroides* R-26[g]	2–300	Red light	G	c	10.7 ± 0.2^e	6
[1]H-*Rp. sphaeroides* R-26[g]	2–300	$K_3Fe(CN)_6$	G	c	9.8 ± 0.2	6
[1]H-*Rp. sphaeroides* R-26[g]	2–300	$K_3Fe(CN)_6$	G	c	10.8 ± 0.2^e	6
[1]H-*Chromatium* strain D[h]	2–300	Red light	G	c	10.1 ± 0.3	6
[1]H-*Chromatium* strain D[h]	2–300	Red light	G	c	10.7 ± 0.3^e	6
[1]H-*Rp. viridis*[i]	2–300	Red light	G	c	12.2 ± 0.5	6
[1]H-*Rp. viridis*[i]	2–300	Red light	G	c	12 ± 1^e	6
[1]H-*Rp. viridis*[i]	2–300	$K_3Fe(CN)_6$	G	c	12.0 ± 0.2	6
[1]H-*Rp. viridis*[i]	2–300	$K_3Fe(CN)_6$	G	c	12.6 ± 0.2^e	6

[a] G indicates Gaussian line shape.

[b] (1) Kohl *et al.* (1965); (2) Norris *et al.* (1971); (3) McElroy *et al.* (1969); (4) P. A. Loach (private communication); (5) Bolton *et al.* (1969); (6) McElroy *et al.* (1972).

[c] Not determined or not available.

[d] RC preparation, whole cells, and chromatophores.

[e] 35 GHz microwave ESR frequency. All others at approximately 9 GHz.

[f] In the case of ^{2}H samples at 35 GHz, the line shape is significantly affected by *g*-anisotropy such that a 10% asymmetry in peak heights occurs. The line shape was fitted with two Gaussians.

[g] Reaction centers.

[h] Chromatophores.

[i] This system contains Bchl *b*. That the line width of this organism is significantly greater than that of others probably indicates that Bchl *b* is involved.

approximately 85% of the ESR line width is determined by proton–electron interactions.

3.2.3. *In Vivo* ESR Spectroscopy of the Special Pair

(a) Range of Application of ESR Spectroscopy to *in Vivo* Systems. We have gone into some detail in Section 3.2.2. to give a physical interpretation of four ESR signal properties primarily because (1) ESR is the most revealing probe of the nature of the oxidized species formed in the primary act of photosynthesis—it can be applied to organisms from ambient temperature to liquid helium temperatures and can be applied equally well to both intact, living organisms and RC preparations; and (2) with the exception of ENDOR, ESR yields the most detailed information available on cations of Bchl both *in vivo* and *in vitro*.

It turns out experimentally that all four of the ESR parameters discussed in Section 3.2.2 remain essentially unchanged from room temperature to 1.5°K (except for very slight changes probably due to slowing down the rotation of methyl groups at very low temperatures), and thus it is important to understand explicitly what the ESR of the *in vivo* doublet state reveals. The invariance of the ESR properties with temperature, organism, and sample preparation gives confidence that ESR and ENDOR studies at low temperature with RC preparations provide data relevant to the natural process of *in vivo* photosynthesis in intact organisms.

(b) ESR Spectroscopy *in Vivo*: The Special Pair. In Table 2 are listed four pertinent ESR properties for several *in vivo* systems. As is true for *in vitro* monomeric data, the ESR signals of *in vivo* systems exhibit Gaussian line shape envelopes, and a *g* value of

2.0026, very near that of the free electron. In fully deuterated photosynthetic bacteria, the ESR line width is decreased by a factor of approximately 2.4 on deuteration. Particularly striking and important is a reduction in ESR line width of no less than 40% for the *in vivo* signal as compared with the *in vitro* monomeric Bchl$^+$ signal. Three of the four ESR properties for both the *in vivo* and the *in vitro* systems are identical, and only the line widths are different. From the qualitative discussion given previously, it is therefore expected that the unpaired electron in the *in vivo* system is more delocalized than in *in vitro* Bchl$^+$.

We have previously developed a general quantitative treatment of the effects of spin delocalization on Gaussian line shapes. When extensive delocalization of an unpaired electron that interacts with many nuclei occurs, line-narrowing occurs to the extent

$$\Delta H_N = (1/\sqrt{N})\Delta H_M \qquad (1)$$

where ΔH_N is the peak-to-peak line width of an unpaired spin developed over an aggregate of size N (an N-mer), in which the electron is shared equally; and ΔH_M is the line width of the monomer free radical.

It has long been recognized that the *in vivo* line width of the presumed Bchl$^+$ cation is distinctly narrower than that of monomer Bchl$^+$ prepared in the laboratory (McElroy *et al.*, 1969). Although the *in vivo* ESR line is approximately 40% narrower than monomer Bchl$^+$, the *in vivo* signal was attributed to monomer Bchl$^+$, and an otherwise unspecified "biological environment" was invoked to account for the line width difference. We can now ask whether a process of spin delocalization might also account for the *in vivo* line shape. In Table 3, ESR data for a variety of photosynthetic organisms with various isotopic compositions are compared, and we can ask

Table 3. Comparison of *in Vivo* and *in Vitro* ESR Line Widths

Organism[a]	Isotopic composition	ΔH Observed	ΔH Predicted[b]	Ratio[c]	Aggregation No.[d]
Rs. rubrum	^{1}H	9.3	9.1	1.02	1.9
Rs. rubrum	^{2}H	4.0	3.7	1.08	1.7
Rp. sphaeroides	^{1}H	9.6	9.10	1.05	1.8
Rp. sphaeroides R-26	^{1}H	9.6	9.10	1.05	1.8
Chromatium strain D	^{1}H	10.1	9.10	1.11	1.6
Rp. viridis	^{1}H	12.1	e	e	e

[a] Average values for all 9 GHz systems of Table 2.
[b] Calculated from $\Delta H_1/\sqrt{N}$, where $N = 2$. $\Delta H_1 = 12.86$ for ^{1}H and $\Delta H_1 = 5.2$ for ^{2}H. These ΔH values are average values obtained from Table 1.
[c] Ratio of ΔH(observed)/ΔH(predicted).
[d] Aggregate number determined by $(\Delta H_1/\Delta H_{obs})^2$.
[e] The ΔH_{pp} of *in vitro* Bchl b^+ is not yet well determined.

whether there is a value for N in equation 1 that would account for the narrowing of the *in vivo* signal. From the two right-hand columns in Table 3 it can be seen that a value $N = 2$ accounts with remarkable precision for the *in vivo* line width.

The reduction in line width observed *in vivo* can be rationalized on the basis of the *in vitro* ESR studies of monomeric Bchl‡ if it is assumed that a special pair of Bchl molecules share the unpaired electron in the cation free radical that is formed on light irradiation, and that the *in vivo* signal should be assigned to a cation containing two Bchl molecules. Since it is the 865 nm band of Bchl *in vivo* that bleaches on oxidation, the simplest correlation of optical and ESR data equates P_{865} with the Bchl special pair. Since the ESR data hold from *in vivo* ambient temperatures to 1.5°K, we can reasonably assume that interpretation of the primary electron donor in bacterial photosynthesis in terms of a special pair of Bchl molecules is valid over the entire range of temperatures, organisms, and active center preparations.

We designate the two Bchl molecules constituting P_{865} as a *special pair* rather than a dimer. Chl dimers, trimers, and oligomers are formed by *direct* donor–acceptor interactions of Chl molecules involving the central Mg atom of one Chl molecule and an oxygen function of another (Katz *et al.*, 1976; Katz, 1968; Katz and Janson, 1973; Ballschmiter and Katz, 1969) (the Ring V keto C=O in Chl a, the Ring V keto C=O and the Ring I acetyl C=O functions in Bchl a). By analogy with photoactive Chl a– and Bchl-a–water adducts that can be prepared in the laboratory, it is reasonable to suppose that in the special pair, the Bchl molecules are linked by water: Bchl–H_2O–Bchl or Chl–H_2O–Chl. For laboratory Chl a–water adducts, the analytical evidence (Ballschmiter and Katz, 1969) indicates that 1 water molecule is sufficient to cross-link 2 Chl a molecules. For Bchl–water adducts, no analytical data are as yet available. For the *in vivo* special pair, the simplest hypothesis is that the special pair is linked by at least 1 water molecule, and that the function of the water, in addition to imposing a geometry conducive to charge separation, is to balance charge separation in the special pair. A role

for water in the Bchl$_{sp}$ is supported by the observation that desiccation of several photosynthetic bacteria results in loss of photochemical activity (at least as seen on a time scale of more than 1 msec) (Clayton, 1966). Desiccation also decreases the optical density at 865 nm, increases absorption at 790 nm, and causes a blue shift of the 597 nm band (R. K. Clayton, private communication). All this is compatible with the presence of water in the RC. Since dehydration would also be expected to affect the conformation of protein to which the Chl may be attached, additional evidence would be desirable. Although direct experimental evidence that firmly establishes the presence of water in the special pair *in vivo* is still lacking, the evidence that two Chl molecules are involved is compelling.

3.3. ENDOR Spectroscopy of Oxidized Bacteriochlorophyll

3.3.1. Introduction to ENDOR Spectroscopy

ENDOR spectroscopy, a technique introduced by Feher (1956), provides by far the most detailed information so far available on the nature of both the *in vivo* and *in vitro* cations of Bchl. ENDOR is much more revealing than ordinary ESR, and it is limited only in that it has as yet been applied to photosynthetic organisms only at low temperatures.

We begin by describing some of the limitations of conventional ESR that can be mitigated by ENDOR. In an ESR experiment, we observe the interaction of an unpaired electron with many nuclei. We illustrate the interaction between an unpaired electron and 10 protons on the Chl macrocycle in Fig. 4. Note that there are 10 interactions felt by the unpaired electron, but only 1 interaction experienced by an individual proton. As discussed in Section 3.2.2(b), the 10 interactions of the nuclei generate many ESR transitions (ranging from a minimum of 11 separate resonances to a maximum of 1024). Since the difference in the many environments sampled by the electron is smaller than the transition line widths, only a single envelope that includes a large number of interactions is observable. A typical ENDOR spectrum is obtained by

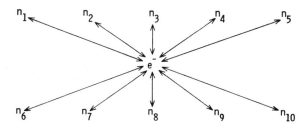

Fig. 4. Interaction of 10 nuclei $n_{i=1,10}$ with one electron. The double-headed arrows represent the hyperfine interactions designated by A's in the text. Note that no nuclear–nuclear interactions are included, and that one nucleus interacts only with one electron. The electron interacts, however, with all 10 nuclei simultaneously.

monitoring an ESR resonance (≈9 GHz) as a function of radiofrequency (1–30 MHz for the Chl's) at a constant resonant magnetic field. Such a spectrum can be described as a nuclear magnetic resonance spectrum of a free radical, using an ESR transition as a detection system. In an ESR experiment, resonant energy is supplied to flip unpaired electrons, which have a large intrinsic magnetic spin moment, and thus a large amount of energy is required to effect the transition. In ENDOR, a much smaller resonant energy is sufficient to flip the heavy nuclei. Furthermore, each nucleus, by virtue of its own "localization," interacts only with the electron, and only two nuclear resonance transitions occur. These two transitions correspond to flipping the nuclear moment in the external magnetic field perturbed by the two different energy states of a single electron (an electron with its magnetic moment parallel or opposed to the external magnetic field). The magnitude of the electron–proton interaction is the same as in an ESR experiment, but the number of interactions is greatly reduced. For the

case illustrated in Fig. 4, the minimum number of lines in an ENDOR spectrum is 2 (in the case in which all nuclei experience an identical interaction), and the maximum number of lines is 20 (all protons nonequivalent). Compare the 2–20 lines of the ENDOR spectrum with the 11–1024 lines of ESR, and the power of ENDOR is readily appreciated. The resolution in an ENDOR experiment is significantly higher than in ESR, because the number of lines is greatly reduced while the size and energy widths of the electron proton interactions remain approximately unchanged. Because the Chl's are so unsymmetrical in structure, a large number of distinct interactions is expected in ESR, but only a small number of interactions in the ENDOR. ENDOR thus has the ability to map the location of the unpaired electron over the cation Chl macrocycle. We indicated above that the strength of the electron–nuclear interaction is dependent on the spin density (i.e., concentration or effective lifetime at a particular site) of the electron at a nucleus of interest. Most important for our present

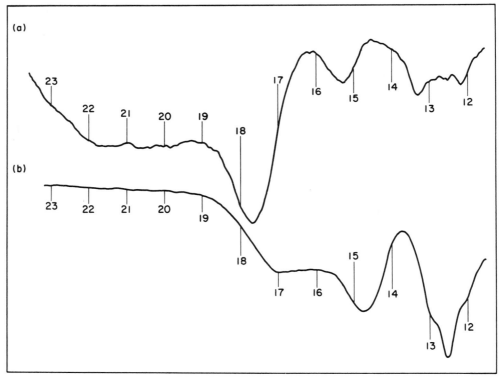

Fig. 5. (a) Bchl a (extracted from *Rhodospirillum rubrum*) oxidized *in vitro* in CD_2Cl_2/CD_3OD (3:1, vol./vol.) by I_2, temp. ≈ 15°K. Following nomenclature developed for Chl a ENDOR spectra (see Fig. 6), the Bchl a ENDOR peak near 17.7 MHz is associated with A_3 and the peak near 15.3 MHz with A_1. (b) *Rhodospirillum rubrum* oxidized by $K_3Fe(CN)_6$, temp. ≈ 15°K. In this case, A_3 is near 14.8 MHz and A_1 is observable only by deconvolution. In both (a) and (b) only the high-frequency half of the ENDOR spectrum is shown.

purposes is that ENDOR can provide a stringent test for the special pair model, by making it possible to compare in detail the spin distribution of the cation free radicals *in vitro* and *in vivo*. First, the electron–proton interactions are measured *in vitro*. For an unpaired spin delocalized over 2 Bchl molecules, such as we have postulated for the *in vivo* special pair, the interactions should be one-half the *in vitro* interaction, because the concentration (spin density) of electron at any given site in the molecule has been reduced by a factor of 2 by delocalization over the special pair. These interactions are known as electron–nuclear hyperfine constants, and their strength is measured in units of magnetic field (Gauss) or frequency (MHz) for the characteristic resonances of each interaction. Each proton–electron interaction results in two resonances (or peaks) in the ENDOR spectrum. The splitting (difference in energy measured in gauss or MHz) between these two peaks is the proton–electron hyperfine coupling constant A_i (see Fig. 5). This coupling constant is proportional to unpaired electron spin density at or near the interacting proton. This relationship, in its simplest form, is

$$A_i = Q\rho_i$$

where A_i is the ith hyperfine coupling constant, ρ_i is the corresponding unpaired electron spin density, and Q is a constant that is approximately 25 Gauss or 70 MHz (McConnell, 1956; Bersohn, 1956; Weissman, 1956; McConnell and Chesnut, 1958; Wertz and Bolton, 1972).

To bring the ENDOR technique to bear on the problem of the *in vivo* special pair, we must first show that the *in vivo* free radical has hyperfine coupling constants one-half those in the *in vitro* Bchl[‡]. Second, we must show that a decrease by one-half is not merely a coincidence, and we can do this by showing that we are actually comparing the coupling constants of the same protons in both the *in vitro* and *in vivo* cases. That is to say, the coupling constants must be assigned. To determine whether the spectra show a change in the *in vitro* vs. *in vivo* coupling constants merely requires a comparison between the two sets of ENDOR spectra. To make specific assignments for the hyperfine interactions, however, we must make a reasonable assignment of a particular resonance in an ENDOR spectrum to a particular proton or group of protons, and this we can do by a process of selective deuteration. Deuterium, as noted, has a much smaller magnetic moment than does ^{1}H. In an ENDOR spectrum, the effect of deuterium substitution for ^{1}H results in the disappearance of a resonance peak, for deuterium is essentially invisible to an ENDOR spectrometer set up to detect protons. Thus, by

Table 4. Comparison of *in Vitro* and *in Vivo* ENDOR Data

Protons[a]	Ref.	Hyperfine coupling constants (MHz)		Aggregation number[b]
		Bchl a[‡]	*In vivo*	
(α, β, δ, 10)	c	c	c	c
	d	1.4	0.8	1.7
1a	e	5.0 ± 0.1	2.0 ± 0.1	2.5
	d	5.32	2.2	2.4
5a	e	9.2 ± 0.2	4.2 ± 0.2	2.2
	d	9.8	4.7	2.1
(7, 8, 3, 4)	e	16	8	2.0
	d	14	7	2.0
Average				2.1 ± 0.3

[a] Proton numbering from Fig. 2b. Parentheses indicate that the coupling constant arises from some or all of the indicated groups.
[b] The aggregation number is the ratio of coupling constants *in vitro* to coupling constants *in vivo*.
[c] Not available.
[d] From Norris *et al.* (1974). *In vitro* Bchl a[‡] was produced by I_2 in 12% CH_3OH/CH_2Cl_2 (vol./vol.). *In vivo* signal was produced by $K_3Fe(CN)_6$ in whole cells of *Rs. rubrum*. Temperature, 100°K.
[e] From Feher *et al.* (1975). Bchl a[‡] was produced by I_2 in CH_3OH–glycerol (1 : 1, vol./vol.). *In vivo* signal was produced by light and freezing in *Rp. sphaeroides* R-26 chromatophores. Temperature, 80°K.

inserting deuterium into specific sites of the Bchl *in vitro* and *in vivo*, we can determine the origin of peaks in the ENDOR spectrum in both cases.

In practice, ENDOR spectroscopy is nonlinear, and this means that relative peak intensities do not accurately represent the relative populations of nuclei giving rise to the peaks. Intensities depend on temperature, solvent, instrument settings, and instrument design, making intensity comparisons among spectra recorded in different laboratories difficult. These are additional reasons to use deuterium replacement to ascertain the origin of ENDOR spectral peaks.

3.3.2. ENDOR Data Relevant to the Special Pair

The ENDOR spectra of *in vivo* and *in vitro* systems are shown in Fig. 5, from which proton–electron coupling constants can be extracted and aggregation numbers calculated (Table 4). We define an aggregation number N_j as the ratio of the hyperfine coupling constant of a particular proton in a Bchl[‡] monomer relative to its value in a system in which the unpaired spin is delocalized

$$N_j = A_{1j}/A_{Nj}$$

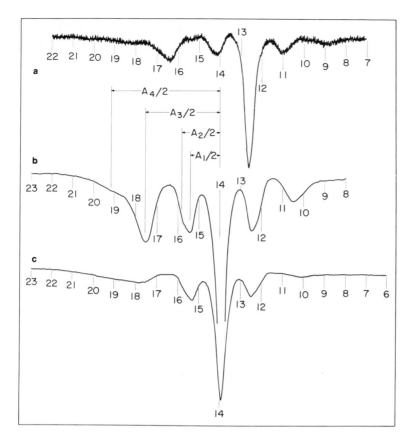

Fig. 6. ENDOR spectra. All species in CH_2Cl_2/CH_3OH (3:1, vol./vol.) oxidized by I_2 at approximately 100°K. (a) Chl a; (b) pyromethylchlorophyllide a; (c) 5-C[^{2}H]$_3$-pyromethylchlorophyllide a.

where A_{1j} is the hfi for a specific proton in a monomer and A_{Nj} is the hfi for the same nucleus when the spin is delocalized. N_j then gives an experimental value for the number of Bchl molecules over which the spin is delocalized. The special pair concept requires all aggregation numbers to be near the value of 2. It can be seen from Table 4 that all the *in vivo* aggregation numbers determined by ENDOR are indeed remarkably close to the average value of 2.

As indicated previously, the comparison between *in vitro* and *in vivo* hyperfine coupling constants would be much more satisfying if the *in vivo* and *in vitro* hyperfine splittings could be assigned independently. We have carried out many selective deuteration experiments, primarily on Chl a derivatives, to make the *in vitro* coupling assignments (Scheer *et al.*, 1977). For example, we have employed pyromethyl chlorophyllide a as a model compound for Chl a. The ENDOR spectra of Chl a, methylchlorophyllide a, and 5-C[^{2}H]$_3$-methylchlorophyllide are shown in Fig. 6. Comparison of Figs. 6a and b shows that pyrochlorophyllide a is a good model compound for Chl a, since the ENDOR spectra of the two are very nearly

identical. In Fig. 6c, the ENDOR resonance near 17.5 MHz (labeled A_3) and 10.5 MHz diminishes tremendously when the 5-methyl is deuterated. Thus, in Chl a, we know the 5-methyl hyperfine coupling constant. We can also demonstrate that the origin of the A_1 and A_3 of Bchl a systems is due to the 1a and 5a methyl groups (Norris *et al.*, 1973, 1974, 1975b; Feher *et al.*, 1975). To accomplish this, we have also selectively introduced deuterium into Bchl *in vivo* by growing *Rhodospirillum rubrum* in ^2H_2O plus succinic acid-$^1H_4$ as a carbon source. This nutrient medium is known to produce Bchl containing ^2H at all the methine positions and at positions 3, 4, 7, and 8. In contrast, *Rs. rubrum* grown in ^1H_2O on succinic acid-$^2H_4$ contains Bchl with ^1H at positions 3, 4, 7, and 8 and the methine protons (Dougherty *et al.*, 1966; Katz *et al.*, 1966). In both cases, the 1a- and 5a-CH_3 groups contain both ^1H and ^2H. With the orgnisms grown in ^2H_2O, the 1a- and 5a-methyl groups dominate the ENDOR spectrum, and thus the assignments are easily made. In Fig. 7, *in vitro* Bchl biosynthetically deuterated except in the 1a- and 5a-methyl groups still exhibits a peak near 19 MHz and 16.7 MHz, and thus

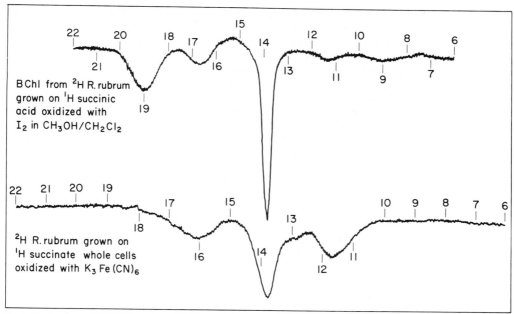

Fig. 7. Comparison of ENDOR spectra of Bchl containing ENDOR-active protons only at the 1-CH$_3$ and 5-CH$_3$ for both *in vivo* and *in vitro* systems at 100°K. *In vitro*: A_3 near 19 MHz and 9 MHz; A_1 near 16.7 MHz and 11.3 MHz. *In vivo*: A_3 near 16.0 and 11.8 MHz; A_1 near 14.7 and 12.9 MHz.

these peaks must be due to the 5a- and 1a-methyl groups.* Likewise, the *in vivo* ENDOR spectrum of Fig. 7 still has a resonance at 16 MHz and a shoulder near 13 MHz that again must arise from the 1a- and 5a-methyl groups (Norris *et al.*, 1973, 1974, 1975*b*; Feher *et al.*, 1975). Feher *et al.* (1975) also carried out such experiments and came to similar conclusions. Because it is known that in monomeric Chl$^+$ the larger coupling constant A_3 is associated with the 5a-methyl group, we assume that A_1 originates from the 1a-methyl group and that the larger coupling constant A_3 belongs to the 5a-methyl groups for Bchl$^+$ both *in vivo* and *in vitro*. Agreement between the experimental findings (Norris *et al.*, 1974, 1975*b*; Feher *et al.*, 1975) and theoretical calculations is excellent, and all experiments with isotopically substituted Bchl are highly compatible with the *in vitro* Chl assignments.

Additional support for the assignment of the A_1 and A_3 peaks to methyl groups has been provided by experiments at cryogenic temperatures, at which the rotation of the methyl group is frozen out. As long as the methyl groups are free to rotate, only a single ENDOR peak is observable. At temperatures near

---

* In this system grown in ^2H_2O with succinic acid-$^1H_4$ as a carbon substrate, ^1H exists in sites other than the 1a- and 5a-CH$_3$ groups; however, these additional sites cannot be expected to give rise to significant ENDOR coupling constants, and thus for ENDOR purposes, ^1H exists only in the 1a- and 5a-CH$_3$ positions.

that of liquid helium, three peaks emerge, one for each proton of the methyl group (Feher *et al.*, 1975).

The rather surprising capability to selectively deuterate *in vivo* Bchl greatly reduces the uncertainty in the comparisons of *in vitro* and *in vivo* data, and provides a firm experimental base for the special pair as the primary electron donor in bacterial photosynthesis.

3.4. Triplet State of Bacteriochlorophyll by ESR Spectroscopy

3.4.1. Primary Events

Direct observation of the doublet state of oxidized Bchl free radical in the RC is by no means the only source of information about the primary donor. Triplet states can also be observed, and as in the case of the doublet state, comparison between *in vivo* and *in vitro* triplets yields valuable insights into the nature of the primary donor.

In Section 1, it was pointed out that the photoreaction center can function only if the special pair is in a neutral state, and that the normal course of events following excitation results in a stable cation of the Bchl special pair. Work at low temperatures has shown that this cation pair is stable for at least tens of milliseconds (McElroy *et al.*, 1969, 1974; Loach and

Sekura, 1967), and thus the free radical is easily observed by standard ESR techniques. When the RC of an organism is fully reduced, however, as, for example, by addition of sodium dithionite, the cation (doublet) signal is no longer observed by ESR. Instead, an ESR triplet signal with highly unusual characteristics is observed (Dutton *et al.*, 1972; Wraight *et al.*, 1974; Leigh and Dutton, 1974; Uphaus *et al.*, 1974; Norris *et al.*, 1975a; Thurnauer *et al.*, 1975). Within a few picoseconds after light irradiation (Parson *et al.*, 1975; Rockley *et al.*, 1975; Kaufmann *et al.*, 1975; Netzel *et al.*, 1973; Leigh *et al.*, 1974; Parson and Cogdell, 1975; Kaufmann and Rentzepis, 1975), P_{865} is bleached, and changes in the optical spectrum are consistent with the formation of Bchl$^+$ (Rockley *et al.*, 1975; Dutton *et al.*, 1975; Fajer *et al.*, 1975). At first this may appear strange, since the ESR no longer observes a doublet ESR signal that is attributable to *in vivo* oxidized Bchl free radical. However, the optical detection of Bchl cation at times less than 10 nsec can be shown to be entirely consistent with, if not demanded by, the observed triplet-state ESR spectrum.

Picosecond optical spectroscopy has revealed the presence of at least two states in the course of the primary events of photosynthesis. A very short-lived intermediate state exists under all conditions of electron-transfer activity. If the acceptor (presumably a quinone–Fe complex) is not bleached by chemical reduction or removal and thus is ready to receive an electron, this state disappears with a lifetime of a few hundred picoseconds. On the other hand, if the usual acceptor molecule is already fully reduced, the intermediate state lasts for less than 10 nsec, and appears to give rise to a new, much longer lived state. This longer-lived state has optical properties (including a several-hundred-microsecond lifetime) characteristic of a triplet state. The interpretation of these valuable optical experiments follows from an analysis of the triplet ESR data.

3.4.2. Introduction to Triplet ESR Spectroscopy

The triplet state of Bchl is most directly observable by ESR techniques, for the spectra confirm the formation of a triplet state. Maximum resolution is obtainable only in very rigid media, in which molecular motions are prevented, as in Chl's in a frozen medium, preferably at temperatures as low as possible. The *in vivo* signals are likewise best observed at temperatures near liquid helium, in the range 1.8–15°K. Triplet ESR spectra contain important new information not available from doublet-state ESR, ENDOR, or E.T. spectra. The ESR and ENDOR

spectra of the doublet state revealed the existence of the special pair *in vivo*, but the data gave no information on the geometry of the special pair. This is because the doublet-state magnetic resonance spectra are dominated by electron–nuclear magnetic interactions that have a very short range. In principle, an ENDOR study could determine the geometry of a bond in or near the macrocycle, e.g., the average position of the protons of the 5-CH$_3$ group with respect to the macrocycle. However, the nature of the extremely short-range interactions in the doublet state severely limits inferences that can be made on the geometry of the special pair.

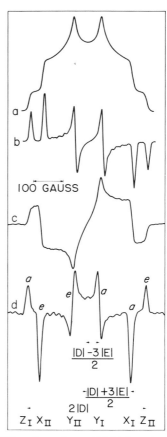

Fig. 8. (a) ESR spectrum of normal triplet shown in straight absorption mode. (b) ESR spectrum of normal triplet shown in first derivative absorption mode. (c) ESR spectrum of highly unusual electron spin polarized triplet spectra obtained for "blocked" photosynthetic bacteria in straight "absorption" mode. Compare with Fig. 8a. Note that downward peaks are actually from emission of microwaves. (d) ESR spectrum of blocked photosynthetic bacteria in first derivative mode. The peaks are labeled *a* for absorption and *e* for emission. Compare with Fig. 8b. Also, the zero field splittings D and E are labeled, as well as the canonical transitions X_i, Y_i, and Z_i where i = I or II (see Tables 5 and 6).

The triplet state contributes two new pieces of information about the RC Chl. First, it can tell the size and shape of the "box" that contains the triplet, especially if the box contains more than one Chl molecule. Second, the triplet-state spectrum obtained at low temperature retains a "memory" of the state it existed in at birth, and thus the triplet contains information about the earliest stages of the primary events associated with the light conversion event. To understand how this new information can be extracted, let us briefly explore the origin of triplet ESR spectra.

A molecule in the triplet state contains two unpaired electrons that split into three energy levels in a high magnetic field in an ESR experiment (the external ESR field). The typical ESR triplet spectrum of a solid solution is shown in Fig. 8. Just as the Bchl doublet spectrum is an envelope of resonances, so is the triplet spectrum. In the ESR of triplets, however, the line-broadening is primarily produced by magnetic dipole interactions between the two unpaired electrons in the triplet state. Each electron of the triplet experiences a magnetic field from its triplet partner. Since the electron has a magnetic moment more than three orders of magnitude larger than a proton, the triplet spectrum is orders of magnitude broader than a typical, organic doublet ESR signal. Thus, the interaction is much longer-ranged than nuclear–electron interactions.

In the first derivative presentation, the Chl triplet spectrum exhibits six peaks that allow determination of the zero field splitting (ZFS) parameters D and E (Fig. 9). Figure 9 labels these transitions according to a rectangular coordinate system within the Chl macrocycle. The z-axis *in vitro* is perpendicular to the plane of the Chl macrocycle.

Of relevance to this discussion is the qualitative meaning of the ZFS. The D parameter is related to the size of the triplet "box"; a large D implies a small box, or conversely, a small D, a large box. The E parameter is related to the shape of the box. If E is zero, the box has axial or higher symmetry. If E is large, the box deviates significantly from axial symmetry. (D is also related to the shape of the box, since D vanishes if the box has cubic or higher symmetry.)

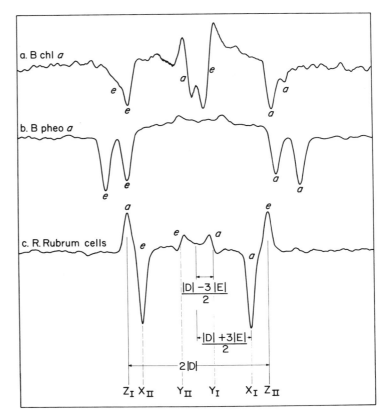

Fig. 9. Comparison of triplets of *in vitro* Bchl *a* (a), Bpheo *a* (b), and *Rhodospirillum rubrum* (c). The peaks are labeled *e* for emission and *a* for absorption. The ZFS parameters D and E are also labeled, along with the X, Y, and Z peaks for both transitions I and II (see Tables 5 and 6).

3.4.3. Interpretation of Triplet ZFS Data

Triplet ZFS data for Bchl systems are shown in Table 5. Attention is directed to the reduction in the ZFS parameters of triplets situated in bacterial photoreaction centers relative to monomeric Bchl, which strongly implies participation by more than one Bchl molecule in the genesis of the triplet signal. The optical data, which cover a time domain from a few picoseconds to milliseconds, are best interpreted with the special pair as the origin of the triplet. Reductions in ZFS parameters have been observed in other systems, and are caused by a sharing of the triplet excitation over two electron states (DeGroot *et al.*, 1969; Schwoerer and Wolf, 1967), or in donor–acceptor systems that form triplet charge-transfer complexes (Hayashi *et al.*, 1969; Hayashi and Nagakura, 1970; Möhwald and Sackmann, 1974). In the latter case, D and E are reduced by the same fraction of charge-transfer character in the complex, and thus a very nearly constant D/E ratio is maintained.

A description that combines the rapid transfer (on the ESR time scale) of triplet excitation between two suitably oriented Bchl molecules with a small amount of charge-transfer character can explain the ZFS results in Table 5. If the reduction in the *in vivo* ZFS is interpreted on this basis, the Chl macrocycles in the special pair are maintained parallel, as suggested in various models (Katz and Norris, 1973; Ballschmiter and Katz, 1969; Fong, 1974a,b, 1975). A rotation of approximately 40° about the z-axis (the axis perpendicular to the plane of macrocycle) yields the average D/E ratio observed in Table 5. Incorporating the appropriate charge-transfer percentage into the rotated structure then gives all ZFS values reported in Table 5. Note that oxidized Bchl is formally a part of such a charge-transfer description.

Table 5. ZFS and Electron Spin Polarization for *in Vitro* and *in Vivo* Bacteriochlorophyll Triplets

Species	ZFS[a]		Polarization[b]					
	$\|D\|$(cm^{-1})	$\|E\|$(cm^{-1})	Z_I	X_{II}	Y_{II}	Y_I	X_I	Z_{II}
In vitro[c]								
Bchl a[d]	0.0224	0.0055	*e*	*e*	*e*	*a*	*a*	*a*
Bpheo a[d]	0.0256	0.0045	*e*	*e*	—	—	*a*	*a*
Bchl b[e]	0.0212	0.0055	*e*	*e*	*e*	*a*	*a*	*a*
Bpheo b[c]	0.0249	0.0050	*e*	*e*	*e*	*a*	*a*	*a*
In vivo								
[1]H-*Rs. rubrum*[d]	0.0185	0.0033	*a*	*e*	*e*	*a*	*a*	*e*
[2]H-*Rs. rubrum*[d]	0.0185	0.0034	*a*	*e*	*e*	*a*	*a*	*e*
[1]H-*Rp. sphaeroides*[d]	0.0182	0.0035	*a*	*e*	*e*	*a*	*a*	*e*
[2]H-*Rp. sphaeroides*[d]	0.0183	0.0032	*a*	*e*	*e*	*a*	*a*	*e*
[1]H-*Rp. palustris*[d]	0.0182	0.0035	*a*	*e*	*e*	*a*	*a*	*e*
[2]H-*Rp. palustris*[d]	0.0184	0.0031	*a*	*e*	*e*	*a*	*a*	*e*
[1]H-*Rp. gelatinosa*[d]	0.0184	0.0028	*a*	*e*	*e*	*a*	*a*	*e*
[1]H-*Rp. sphaeroides* R-26[f,g]	0.0183	0.0031	*a*	*e*	*e*	*a*	*a*	*e*
[1]H-*Rp. sphaeroides* Ga[f,h]	0.0185	0.0031	*a*	*e*	*e*	*a*	*a*	*e*
[1]H *Chromatilum* D[f,i]	0.0177	0.0034	*a*	*e*	*e*	*a*	*a*	*e*
[1]H-*Rs. rubrum* G9[f,i]	0.0185	0.0031	*a*	*e*	*e*	*a*	*a*	*e*
[1]H-*Chromatium* D[f,h]	0.0178	0.0033	*a*	*e*	*e*	*a*	*a*	*e*
[1]H-*Rp. sphaeroides* Ga[f,h]	0.0185	0.0031	*a*	*e*	*e*	*a*	*a*	*e*
[1]H-*Rp. gelatinosa*[f,i]	0.0186	0.0027	*a*	*e*	*e*	*a*	*a*	*e*
[1]H-*Rp. sphaeroides* R-26[f,i]	0.0186	0.0031	*a*	*e*	*e*	*a*	*a*	*e*

[a] Obtained near 5°K.
[b] See Fig. 8 for transition definitions. Note that in these columns, *a* and *e* denote *absorption* and *emission*, as in Fig. 8.
[c] Solvent of [²H]toluene with 10% [²H]pyridine; all ¹H species.
[d] Data from Thurnauer *et al.* (1975); *in vivo* systems are whole cells.
[e] Data from Thurnauer and Norris (1977).
[f] Data from Prince *et al.* (1976).
[g] RC preparations.
[h] Chromatophores.
[i] Whole cells.

3.4.4. Triplet Polarization Effects and Their Interpretation

The memory aspect of ESR triplet spectroscopy is directly related to the oxidation of Bchl special pairs in the primary event. At low temperatures, most photo-excited triplet states exhibit emission or enhanced absorption. The physics of the birth and decay of triplets optically generated is sufficiently well understood to predict the consequences for all combinations of non-Boltzmann populations of randomly oriented triplet molecules produced by normal intersystem crossing (ISC). A few of these combinations are listed in Table 6 along with the expected memory patterns. Such memory patterns can explain all known *in vitro* Chl spectra as well as any other triplet spectrum except that obtained from "blocked" photosynthetic bacteria RCs.

The *in vivo* triplet spectrum is unusual because only the center T_0 level of the triplet is born with population no matter which molecular axis is along the external magnetic field. This is most unusual because in normal triplets, ISC always occurs to the upper and lower levels for at least some orientations. The *in vivo* triplet data can be easily explained if one assumes chemistry has occurred. By chemistry, we mean that the electron has been transferred, producing some

Table 6. Comparison of Polarization of Bacteria with Normal Polarizations

Relative populations at zero field[a]			Polarization at canonical orientation					
P_x	P_y	P_z	Z_I	X_{II}	Y_{II}	Y_I	X_I	Z_{II}
1	0	0	*e*	*e*	*a*	*e*	*a*	*a*
0	1	0	*e*	*a*	*e*	*a*	*e*	*a*
0	0	1	*a*	*a*	*a*	*e*	*e*	*e*
1	1	0	*e*	*e*	*e*	*a*	*a*	*a*
1	0	1	*a*	*e*	*a*	*e*	*a*	*e*
0	1	1	*a*	*a*	*e*	*a*	*e*	*e*
2	1	0	*e*	*e*	—	—	*a*	*a*
2	0	1	—	*e*	*a*	*e*	*a*	—
1	2	0	*e*	—	*e*	*a*	—	*a*
0	2	1	—	*a*	*e*	*a*	*e*	—
1	0	2	*a*	—	*a*	*e*	—	*e*
0	1	2	*a*	*a*	—	—	*e*	*e*
Bacteria			*a*	*e*	*e*	*a*	*a*	*e*

[a] $P_i \propto p_i / k_i$ where p_i represents a populating rate constant and k_i represents the depopulating rate constant of level i = x, y, z.

Bchl[+]. The "ejected" electron is on a molecule near Bchl[+], such as Bchl or bacteriopheophytin (Bpheo). The latter explanation is favored by some (Dutton *et al.*, 1975; Fajer *et al.*, 1975); i.e., $Bchl_{sp}^{+}$ Bpheo[−], where the subscript sp indicates that the special pair is an intermediate state in the photosynthetic pathway. Two unpaired electrons relatively well separated are produced in such a complex. When electrons are sufficiently far apart and then recombine to form an excited triplet state, the mechanism predicts that essentially only the center triplet level T_0 will be populated (Dutton *et al.*, 1973; Uphaus *et al.*, 1974; Norris *et al.*, 1975a; Thurnauer *et al.*, 1975; Abragam and Bleaney, 1970; Itoh *et al.*, 1969). Thus, such a recombination is a plausible mechanism to explain the triplet polarization pattern. Thus, if the acceptor complex, probably a quinone–iron (Q–Fe) complex, is blocked, chemistry still takes place, and the Bchl special pair cation is still formed, but the chemical state lasts only for about 10 nsec, then decaying to a triplet state. This 10 nsec is too short a time to observe $Bchl_{sp}^{+}$ using ordinary ESR techniques. If the Q–Fe acceptor complex is not blocked by reduction, again $Bchl_{sp}^{+}$ is formed, and the ejected electron gets even farther away from the special pair. This state can last for milliseconds without back-reacting. If a back-reaction occurs, the system now either has insufficient energy to produce a triplet state or produces triplets so slowly that the triplet is unobservable. It is noteworthy that all explanations of the primary events postulate the formation of oxidized Chl within the special pair on a very short time scale. We discuss some possible mechanisms of the primary photochemistry in the special pair and the significance of the triplet state data in Section 4.

3.5. Redox Properties of Bacteriochlorophyll

Knowledge of the redox properties of Bchl and Bpheo is essential when considering the chain of electron-transfer reactions in photosynthesis following conversion of red light quanta into reducing power. As expected, the Chl's are relatively easily oxidized or reduced. Table 7 lists the midpoint redox potential, E_m, for several *in vitro* Chl species, as well as for the *in vivo* photooxidation of Bchl. *In vitro* cyclic voltammetry, controlled potential electrolysis, and coulometry indicate that one-electron reversible oxidation (or reduction) occurs with the Chl's (Fuhrhop and Mauzerall, 1969; McElroy *et al.*, 1969; Fajer *et al.*, 1970, 1974; Borg *et al.*, 1970; Kuntz *et al.*, 1964). The oxidation (or reduction) products are indicated to be cations (or anions) by their reversible cyclic voltammetry patterns. Furthermore, it has been shown that the one-electron oxidation product of Bchl

Table 7. Bacteriochlorophyll *a* and Bacteriopheophytin *a* Redox Potentials (mV)[a]

	Bpheo (CH$_2$Cl$_2$)	Bchl (CH$_2$Cl$_2$)	(Bchl)$_2$ P$_{870}$
For cation formation	+960[b] (+720)	+640[c] (+400)	+450[d] (+210)
For anion formation	−580[c] (−820)	−860[c] (−1100)	Not determined

[a] Values in parentheses are referenced to an aqueous saturated calomel electrode (sce). Other values are referenced to the normal hydrogen electrode (NHE) using the equation E_m (sce) = E_m (NHE) − 240 mV.

[b] From Fajer *et al.* (1974).

[c] From Fajer *et al.* (1975).

[d] From Kuntz *et al.* (1964) and Loach *et al.* (1963).

migrates to the cathode, which confirms the assignment of oxidized Bchl to be a cation (Fajer, private communication).

In general, the Chl's are easier to oxidize than the corresponding pheophytins. Likewise, the pheophytins are easier to reduce than the corresponding Chl's. The redox potentials of the Chl's depend to a small extent on solvent, but again probably not enough to account for the lower redox E_m and Bchl$_{P_{870}}$ vs. monomeric Bchl. We believe that the diminished redox properties of P$_{870}$ are most likely a consequence of special pair formation *in vivo*.

4. Summary and Conclusions

In all photosynthetic organisms so far studied, the evidence is good that a special pair of Chl molecules constitutes the primary electron donor in photosynthesis. This generalization appears equally valid both for green, oxygen-producing plants and for purple photosynthetic bacteria. We can now briefly consider how the special pair may function and its relationship to the electron acceptors and donors that constitute the photoreaction center. The photoreaction scheme we will discuss is shown in Fig. 10, which also provides a key to the nomenclature employed below.

The main pathway of photosynthesis in Fig. 10 is indicated by double-line arrows (⇒), whereas a side path of photosynthesis is indicated by single-line arrows (→). The components we consider are the special pair, written schematically as [Bchl Bchl], and two electron acceptors designated P and X. Since it is customary to refer to the primary electron acceptor as X, we will consider P to be a preprimary or transient electron acceptor. We further assume that P is Bpheo, and that X is a [Q–Fe] complex. The light-induced reaction occurs from the normal resting condition of photosynthesis, indicated as state 1 in Fig. 10. Within picoseconds after light absorption has occurred, oxidized Bchl$^+$ is observed, in agreement with the reaction course 1 ⇒ 2 ⇒ 3. The reaction 3 ⇒ 4 occurs in the normal course of events of photosynthesis, and results in a further spatial separation between the oxidized Chl special pair and subsequent reduced acceptors in the electron-transport chain. It is the electron on X$^-$ that ultimately provides reducing capacity for chemical reactions. Insertion of an electron from a donor (a cytochrome, not shown) replenishes the electron deficiency in the cation radical 4, and the reaction 4 ⇒ 1 restores the electron pump to its initial state. State 4 is the source of the photoinduced special pair cation doublet signal observed by ESR.

In some instances, the reaction 3 ⇒ 4 may be prevented either by prior reduction of X to X$^-$ or because

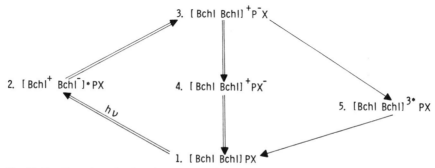

Fig. 10. Reaction scheme for the primary events of photosynthesis. (⇒) Main pathway; (→) side pathway that leads to triplet formation (state 5). States 2 and 3 are also called P^F. State 5 is also referred to as P^R. State 4 gives rise to an ESR doublet signal characteristic of the special pair cation. State 5 gives rise to a triplet signal characteristic of the special pair and must be formed from a radical pair state.

X is missing. Thus, when $3 \Rightarrow 4$ is blocked, a new reaction $3 \rightarrow 5$ occurs. In state 3, also called P^F (Parson *et al.*, 1975), an electron-transfer reaction has occurred. Since the normal electron transfer to X is blocked, the electron transfer reverses as described in reaction $3 \rightarrow 5$, forming the triplet state 5. State 5, also called P^R (Parson *et al.*, 1975), is the origin of the triplet signal that can be observed by ESR. Transition $5 \rightarrow 1$ is reported to occur in approximately 500 μsec at $2°K$ [magnetic resonance measurements in *Rs. rubrum* (Clarke *et al.*, 1975)] and 120 μsec at approximately $15°K$ [optical measurements in *Rp. sphaeroides* (Parson and Cogdell, 1975)]. Because of the radical pair nature of state 2 or state 3, which precedes the formation of state 5, the triplet state acquires the unusual electron spin polarization discussed in Section 3.4.4. (Thurnauer *et al.*, 1975).

Some time ago, Katz (1971) proposed that the initial electron-transfer reaction occurred in the Chl special pair as indicated by state 2 of Fig. 10. Such an electron transfer may be energetically feasible because of the structure of the special pair. Since we believe that the special pair is in part held together by keto

$$
\begin{array}{c}
\text{H} \\
| \\
\text{C} = \text{O} \cdots \text{H} \cdots \text{O} \cdots \text{Mg}
\end{array}
$$

[Ring V keto C=O of one Bchl and water coordinated to the Mg of the other (see Fig. 2b)], it is possible that the energy required to produce the initial electron transfer is significantly reduced by a rapid internal proton transfer that helps neutralize the electrostatic charge separation. Optical evidence supporting the presence of such species has recently been obtained for RC preparations (Rockley *et al.*, 1975; Kaufmann *et al.*, 1975; Dutton *et al.*, 1975). Optical evidence on this point in intact organisms or in chromatophores is not yet available. The optical transients observed by picosecond laser spectroscopy of RC preparations have been interpreted in terms of a cation of Bchl and a Bchl anion or a Bpheo anion. For illustrative purposes in Fig. 10, we have included Bpheo (designated P) in the electron-transport scheme (Thurnauer *et al.*, 1975; Dutton *et al.*, 1975; Fajer *et al.*, 1975). Even if P is an essential part of bacterial photosynthesis, the initial electron transfer may still occur within the special pair as shown by $1 \Rightarrow 2$. State 2 will likely have an effective lifetime of only a few picoseconds, and thus is difficult to detect.

As we have already mentioned, the manner in which the *in vivo* special pair is constructed may well be critical to its ability to function. A variation of the special pair structure has been suggested that has C_2 symmetry and in which the two Chl molecules are held together by two water molecules hydrogen-bonded to the two Ring V (see Fig. 2b) carbomethoxy groups (Fong, 1974a,b, 1975). Although water is one of the most reasonable means of joining two Chl's to form a special pair, no *in vivo* evidence bears directly on this question. In the absence of *in vivo* structural data, an equally valid argument can be made for a special pair that has "approximate symmetry" or no symmetry at all. Even if C_2 symmetry is present in the special pair, it most probably does not occur by interactions involving the Ring V carbomethoxy as proposed by Fong (1974a,b, 1975), for the keto C=O function of Ring V is from all evidence a far better donor group than are the ester C=O groups. The question of the symmetry of the special pair cannot yet be answered, since the ESR, ENDOR, and optical data can be interpreted quite independently of *exact* symmetry arguments.

The requirement for water in the special pair is by no means unequivocal. Although Clayton (1966) showed that certain bacteria lose photochemical activity upon desiccation, conformational changes in protein affecting the relative positions of the two Bchl molecules in the special pair, or the distances between $Bchl_{sp}$ and the components of the electron acceptor chain, could also be affected by desiccation. Moreover, water is not the only ligand that can bond two Chl molecules together to form aggregated species the optical properties of which mimic those of *in vivo* P_{870}. Alcohols (ROH) and amines (RNH_2) can also coordinate to Mg and form hydrogen bonds, and in the laboratory can be used to produce highly red-shifted Chl species (Shipman *et al.*, 1976). This suggests that protein side-chain functions containing $-OH$ or $-NH_2$ groups may serve to decrease the energy requirement for electron transfer, by neutralizing electrostatic charge separation as mentioned above. For example, $[(Bchl^+OR^-)(Bchl^-H^+)]$ is lower in energy than $[Bchl^+Bchl^-]$, and thus arguments based on *in vitro* redox potentials for the formation of $Bchl^+$ and $Bchl^-$ are not necessarily valid for determining whether a single 870 nm photon can produce such a state. The use of serine or lysine or other donor functions in protein side-chains has the additional advantage of providing a role for protein in the genesis of the special pair and in the primary act of photosynthesis.

Thus, the exact structure of the special pair *in vivo* and its mode of operation are not completely established. On the other hand, oxidized Bchl in the form of the special pair as the primary donor unit appears well established, and thus provides a focal point for future investigations of the primary events of bacterial photosynthesis.

ACKNOWLEDGMENT

Work performed under the auspices of the U.S. Energy Research and Development Administration.

5. References

Abragram, A., and Bleaney, B., 1970, *Electron Paramagnetic Resonance of Transition Ions*, pp. 502–514, Clarendon Press, Oxford.

Ballschmiter, K., and Katz, J. J., 1969, An infrared study of chlorophyll–chlorophyll and chlorophyll–water interactions, *J. Am. Chem. Soc.* **91**:2661.

Bersohn, R., 1956, Proton hyperfine interactions in semiquinone ions, *J. Chem. Phys.* **24**:1066.

Bolton, J. R., Clayton, R. K., and Reed, D. W., 1969, An identification of the radical giving rise to the light-induced electron spin resonance signal in photosynthetic bacteria, *Photochem. Photobiol.* **9**:209.

Borg, D. C., Fajer, J., Felton, R. H., and Dolphin, D., 1970 The π-cation radical of chlorophyll *a*, *Proc. Natl. Acad. Sci. U.S.A.* **67**:813.

Clarke, R. H., Connors, R. E., Norris, J. R., and Thurnauer, M. C., 1975, Optically detected zero-field magnetic resonance studies of the photoexcited triplet state of the photosynthetic bacterium *Rhodospirillum rubrum*, *J. Am. Chem. Soc.* **97**:7178.

Clayton, R. K., 1962, Primary reactions in bacterial photosynthesis. I. The nature of light-induced absorbancy changes in chromatophores; evidence for a special bacteriochlorophyll component, *Photochem. Photobiol.* **1**:201.

Clayton, R. K., 1966, Fluorescence from major and minor bacteriochlorophyll components *in vivo*, *Photochem. Photobiol.* **5**:679.

Commoner, B., Heise, J. J., and Townsend, J., 1956, Light-induced paramagnetism in chloroplasts, *Proc. Natl. Acad. Sci. U.S.A.* **42**:710.

DeGroot, M. S., Hesselmann, I. A. M., and Van der Waals, J. H., 1969, Paramagnetic resonance in phosphorescent aromatic hydrocarbons. VI. Mesitylene in β-trimethyborazole, *Mol. Phys.* **16**:61.

Dougherty, R. C., Crespi, H. L., Strain, H. H., and Katz, J. J., 1966, Nuclear magnetic resonance studies of plant biosynthesis: Bacteriochlorophyll, *J. Am. Chem. Soc.* **88**:2854.

Dutton, P. L., Leigh, J. S., and Siebert, M., 1972, Primary processes in photosynthesis: *In situ* ESR studies on the light induced oxidized and triplet state of reaction center bacteriochlorophyll, *Biochem. Biophys. Res. Commun.* **46**:406

Dutton, P. L., Leigh, J. S., and Reed, D. W., 1973, Primary events in the photosynthetic reaction center from *Rhodopseudomonas spheroides* strain R26: Triplet and oxidized states of bacteriochlorophyll and the identification of the primary electron acceptor, *Biochim. Biophys. Acta* **292**:654.

Dutton, P. L., Kaufmann, K. J., Chance, B., and Rentzepis, P. M., 1975, Picosecond kinetics of the 1250 nm band of the *Rp. sphaeroides* reaction center: The nature of the primary photochemical intermediary state, *FEBS Lett.* **60**:275.

Duysens, L. N. M., 1952, Transfer of excitation energy in photosynthesis, Thesis, State University of Utrecht, The Netherlands.

Duysens, L. N. M., Huiskamp, W. J., Vas, J. J., and Van der Hart, J. M., 1956, Reversible changes in bacteriochlorophyll in purple bacteria on illumination, *Biochim. Biophys. Acta* **19**:188.

Emerson, R., and Arnold, W., 1932a, The separation of the reactions in photosynthesis by means of intermittent light, *J. Gen. Physiol.* **15**:391.

Emerson, R., and Arnold, W., 1932b, The photochemical reaction in photosynthesis, *J. Gen. Physiol.* **16**:191.

Fajer, J., Borg, D. C., Forman, A., Dolphin, D., and Felton, R. H., 1970, π-Cation radicals and dications of metalloporphyrins, *J. Am. Chem. Soc.* **92**:3451.

Fajer, J., Borg, D. C., Forman, A., Felton, R. H., Dolphin, D., and Vegh, L., 1974, The cation radicals of free base and zinc bacteriochlorin, bacteriochlorophyll and bacteriopheophytin, *Proc. Natl. Acad. Sci. U.S.A.* **71**:994.

Fajer, J., Brune, D. C., Davis, M. S., Forman, A., and Spaulding, L. D., 1975, Primary charge separation in bacterial photosynthesis: Oxidized chlorophylls and reduced pheophytin, *Proc. Natl. Acad. Sci. U.S.A.* **72**:4956.

Feher, G., 1956, Observation of nuclear magnetic resonances via the electron spin resonance line, *Phys. Rev.* **103**:834.

Feher, G., Hoff, A. J., Isaacson, R. A., and McElroy, J. D., 1973, Investigation of the electronic structure of the primary electron donor in bacterial photosynthesis by the ENDOR technique, *Biophys. Soc. Abstr.* **13**:61a.

Feher, G., Hoff, A. J., Isaacson, R. A., and Ackerson, L. C., 1975, ENDOR experiments on chlorophyll and bacteriochlorophyll *in vitro* and in the photosynthetic unit, *Ann. N. Y. Acad. Sci.* **244**:239.

Fong, F. K., 1974a, Energy upconversion theory of the primary photochemical reaction in plant photosynthesis, *J. Theor. Biol.* **46**:407.

Fong, F. K., 1974b, Molecular basis for the photosynthesis primary process, *Proc. Natl. Acad. Sci. U.S.A.* **71**:3692.

Fong, F. K., 1975, Molecular symmetry and exciton interaction in photosynthetic primary events, *Appl. Phys.* **6**:151.

Fuhrhop, J.-H., and Mauzerall, D., 1969, The one-electron oxidation of metalloporphyrins, *J. Am. Chem. Soc.* **91**:4174.

Garcia-Morin, M., Uphaus, R. A., Norris, J. R., and Katz, J. J., 1969, Interpretation of chlorophyll electron spin resonance spectra, *J. Phys. Chem.* **73**:1066.

Goedheer, J. C., 1960, Spectral and redox properties of bacteriochlorophyll in its natural state, *Biochim. Biophys. Acta* **38**:389.

Hayashi, H., and Nagakura, S., 1970, The ESR and phosphorescence spectra of some dicyanobenzene complexes with methyl-substituted benzenes, *Mol. Phys.* **19**:45.

Hayashi, H., Iwata, S., and Nagakura, S., 1969, ESR spectra of the charge-transfer triplet states of some molecular complexes, *J. Chem. Phys.* **50**:993.

Itoh, K., Hayashi, H., and Nagakura, S., 1969, Determination of the singlet–triplet separation of a weakly interacting radical pair from the ESR spectrum, *Mol. Phys.* **17**:561.

Katz, J. J., 1968, Coordination properties of magnesium in chlorophyll from IR and NMR spectra, *Dev. Appl. Spectrosc.* **6**:201.

Katz, J. J., 1971, Conference on Primary Photochemistry of Photosynthesis, Argonne National Laboratory, Argonne, Illinois.

Katz, J. J., and Janson, T. R., 1973, Chlorophyll–chlorophyll interaction from ^{1}H and ^{13}C nuclear magnetic resonance spectroscopy, *Ann. N. Y. Acad. Sci.* **206**:579.

Katz, J. J., and Norris, J. R., 1973, Chlorophyll and light energy transduction in photosynthesis, *Curr. Top. Bioenerg.* **5**:41.

Katz, J. J., Dougherty, R. C., Crespi, H. L., and Strain, H. H., 1966, Nuclear magnetic resonance studies of plant biosynthesis: A bacteriochlorophyll isotope mirror experiment, *J. Am. Chem. Soc.* **88**:2856.

Katz, J. J., Ballschmiter, K., Garcia-Morin, M., Strain, H. H., and Uphaus, R. A., 1968, Electron paramagnetic resonances of chlorophyll–water aggregates, *Proc. Natl. Acad. Sci. U.S.A.* **60**:100.

Katz, J. J., Oettmeier, W., and Norris, J. R., 1976, Organization of antenna and photo-reaction centre chlorophylls on the molecular level, *Philos. Trans. R. Soc. London Ser B* **237**:227.

Kaufmann, K. J., and Rentzepis, P. M., 1975, Picosecond spectroscopy in chemistry and biology, *Acc. Chem. Res.* **8**:407.

Kaufmann, K. J., Dutton, P. L., Netzel, T. L., Leigh, J. S., and Rentzepis, P. M., 1975, Picosecond kinetics of events leading to reaction center bacteriochlorophyll oxidation, *Science* **188**:1301.

Kohl, D. H., 1965, Studies of photosynthesis in intact cells by electron spin resonance, Thesis, Washington University, St. Louis, Missouri.

Kohl, D. H., Townsend, J., Commoner, B., Crespi, H. L., Dougherty, R. C., and Katz, J. J., 1965, Effects of isotopic substitution on electron spin resonance signals in photosynthetic organisms, *Nature (London)* **206**:1105.

Kok, B., 1959, Light-induced absorption changes in photosynthetic organisms. II. A split-beam difference spectrometer, *Plant Physiol.* **34**:184.

Kuntz, I.D., Loach, P. A., and Calvin, M., 1964, Absorption changes in bacterial chromatophores, *Biophys. J.* **4**:227.

Leigh, J. S., and Dutton, P. L., 1974, Reaction center bacteriochlorophyll triplet states: Redox potential dependence and kinetics, *Biochim. Biophys. Acta* **357**:67.

Leigh, J. S., Netzel, T. L., Dutton, P. L., and Rentzepis, P. M., 1974, Primary events in photosynthesis: Picosecond kinetics of carotenoid bandshifts in *Rhodopseudomonas spheroides* chromatophores, *FEBS Lett.* **48**:136.

Loach, P. A., and Sekura, D. L., 1967, Comparison of decay kinetics of photo-produced absorbance, EPR, and luminescence changes in chromatophores of *Rhodospirillum rubrum, Photochem. Photobiol.* **6**:381.

Loach, P. A., Androes, G. M., Maksim, A. F., and Calvin, M., 1963, Variation in electron paramagnetic resonance signals of photosynthetic systems with the redox level of their environment, *Photochem. Photobiol.* **2**:443.

Loach, P. A., Bambara, R. A., and Ryan, F. J., 1971, Identification of the major ultraviolet absorbance photochanges in photosynthetic systems, *Photochem. Photobiol.* **13**:247.

McConnell, H. M., 1956, Indirect hyperfine interaction in the paramagnetic resonance spectra of aromatic free radicals, *J. Chem. Phys.* **24**:764.

McConnell, H. M., and Chesnut, D. B., 1958, Theory of isotopic hyperfine interactions in π-electron radicals, *J. Chem. Phys.* **28**:107.

McElroy, J. D., Feher, G., and Mauzerall, D. C., 1969, On the nature of the free radical formed during the primary process of bacterial photosynthesis, *Biochim. Biophys. Acta* **172**:180.

McElroy, J. D., Feher, G., and Mauzerall, D. C., 1972, Characterization of primary reactants in bacterial photosynthesis. I. Comparison of the light-induced EPR signal ($g = 2.0026$) with that of a bacteriochlorophyll radical, *Biochim. Biophys. Acta* **267**:363.

McElroy, J. D., Mauzerall, D. C., and Feher, G., 1974, Characterization of primary reactants in bacterial photosynthesis. II. Kineic studies of the light-induced EPR signal ($g = 2.0026$) and the optical absorbance changes at cryogenic temperatures, *Biochim. Biophys. Acta* **333**:261.

Möhwald, H., and Sackmann, E., 1974, Mobile and trapped triplet states in single crystals of charge transfer complexes, *Z. Naturforsch.* **299**:1216.

Netzell, T. L., Leigh, J.S., and Rentzepis, P. M., 1973, Picosecond kinetics of reaction centers containing bacteriochlorophyll, *Science* **182**:238.

Norris, J. R., Uphaus, R. A., Crespi, H. L., and Katz, J. J., 1971, Electron spin resonance of chlorophyll and the origin of signal I in photosynthesis, *Proc. Natl. Acad. Sci. U.S.A.* **68**:625.

Norris, J. R., Uphaus, R. A., and Katz, J. J., 1972, Electron spin resonance in ^{13}C-labeled chlorophyll and ^{13}C-labeled algae, *Biochim. Biophys. Acta* **275**:161.

Norris, J. R., Druyan, M. E., and Katz, J. J., 1973, Electron nuclear double resonance of bacteriochlorophyll free radical *in vitro* and *in vivo, J. Am. Chem. Soc.* **95**:1680.

Norris, J. R., Scheer, H., Druyan, M. E., and Katz, J. J., 1974, An electron-nuclear double resonance (ENDOR) study of the special pair model for photoreactive chlorophyll in photosynthesis, *Proc. Natl. Acad. Sci. U.S.A.* **71**:4897.

Norris, J. R., Uphaus, R. A., and Katz, J. J., 1975a, ESR of triplet states of chlorophylls a, b, c_1, c_2 and bacteriochlorophyll a: Applications of ZFS and electron spin polarizations to photosynthesis, *Chem. Phys. Lett.* **31**:157.

Norris, J. R., Scheer, H., and Katz, J. J., 1975b, Models for antenna and reaction center chlorophylls, *Ann. N. Y. Acad. Sci.* **244**:261.

Parson, W. W., and Cogdell, R. J., 1975, The primary photochemical reaction of bacterial photosynthesis, *Biochim. Biophys. Acta* **416**:105.

Parson, W. W., Clayton, R. K., and Cogdell, R. J., 1975, Excited states of photosynthetic reaction centers at low redox potentials, *Biochim. Biophys. Acta* **387**:268.

Prince, R. C., Leigh, J. S., Jr., and Dutton, P. L., 1976, Thermodynamic properties of the reaction center of *Rhodopseudomonas viridis: In vivo* measurement of the reaction center bacteriochlorophyll–primary acceptor intermediary electron carrier, *Biochim. Biophys. Acta* **440**:622.

Rockley, M. G., Windsor, M. W., Cogdell, R. J., and Parson, W. W., 1975, Picosecond detection of an intermediate in the photochemical reaction of bacterial photosynthesis, *Proc. Natl. Acad. Sci. U.S.A.* **72**:2251.

Scheer, H., Katz, J. J., and Norris, J. R., 1977, Proton–electron hyperfine coupling constants of the chlorophyll *a* cation radical by ENDOR spectroscopy, *J. Am. Chem. Soc.* **99**:1372.

Schwoerer, M., and Wolf, H. C., 1967, ESR study of naphthalene-d_8-naphthalene-h_8 mixed crystals in the triplet state, *Mol. Crystallogr.* **3**:177.

Shipman, L. L., Cotton, T. M., Norris, J. R., and Katz, J. J., 1976, New proposal for structure of special pair chlorophyll, *Proc. Natl. Acad. Sci. U.S.A.* **73**:1791.

Sogo, P., Jost, M., and Calvin, M., 1959, Free radical production in photosynthesizing systems, *Radiat. Res. Suppl.* **1**:511.

Swartz, H. M., Bolton, J. R., and Borg, D. C., 1972, *Biological Applications of Electron Spin Resonance*, Wiley-Interscience, New York.

Thurnauer, M. C., and Norris, J. R., 1977, The ordering of the zero field triplet spin sublevels in the chlorophylls: A magnetophotoselection study, *Chem. Phys. Lett.* **47**:100.

Thurnauer, M. C., Katz, J. J., and Norris, J. R., 1975, The triplet state in bacterial photosynthesis: Possible mechanisms of the primary photo-act, *Proc. Natl. Acad. Sci. U.S.A.* **72**:3270.

Uphaus, R. A., Norris, J. R., and Katz, J. J., 1974, Triplet states in photosynthesis, *Biochem. Biophys. Res. Commun.* **61**:1057.

van Niel, C. B., 1935, Photosynthesis of bacteria, *Quant. Biol.* **3**:138.

Weissman, S. I., 1956, Isotropic hyperfine interactions in aromatic free radicals, *J. Chem. Phys.* **25**:890.

Wertz, J. E., and Bolton, J. R., 1972, *Electron Spin Resonance Elementary Theory and Practical Applications*, McGraw-Hill, New York.

Wraight, C. A., Leigh, J. S., Dutton, P. L., and Clayton, R. K., 1974, The triplet state of reaction center bacteriochlorophyll: Determination of a relative quantum yield, *Biochim. Biophys. Acta* **333**:401.

CHAPTER 22

Primary Electron Acceptors

James R. Bolton

1. Introduction

It is now well established that the primary photochemical event in bacterial photosynthesis involves the transfer of an electron from a special bacteriochlorophyll (Bchl) species (usually called P870*) to an electron acceptor (see Chapters 15 and 19–21 for a full discussion of the evidence behind this statement). This primary electron transfer is thought to occur within a membrane-bound reaction center (RC) that may be a single protein or a complex of protein subunits (see Chapters 6 and 19).

The purpose of this chapter is to discuss what is known or can be inferred concerning the nature, function, and identity of the primary electron acceptor. I shall discuss the electron transfer process itself, the known physical characteristics of the acceptor, the evidence concerning its identity (particularly the possibilities of iron or ubiquinone or both), electron transfer out of the primary acceptor, and finally some speculation on the direction in which future research is likely to go. Several recent reviews have also treated this subject (Parson, 1974a,b; Parson and Cogdell, 1975; Sauer, 1974).

---

* P870 will be used throughout this chapter to designate the Bchl component that becomes oxidized in the primary electron transfer. The number 870 refers to the λ_{max} for this spectral bleaching, but it varies somewhat from species to species.

James R. Bolton · Photochemistry Unit, Department of Chemistry, University of Western Ontario, London, Ontario, Canada N6A 5B7. This chapter is Publication No. 152 from the Photochemistry Unit.

There is some question as to the exact definition of the "primary" acceptor. I have arbitrarily chosen the following definition: "The primary acceptor is that species that has accepted the electron ejected from the primary donor after a time of the order of 100 nsec has passed." This time limitation precludes possible transient intermediate states such as a charge-transfer triplet state of the Bchl dimer (Dutton et al., 1972; Wraight et al., 1974) or the P^F state detected by picosecond spectroscopy (Rockley et al., 1975; Parson et al., 1975; Kaufmann et al., 1975). These are very interesting observations and will help a great deal in the understanding of the photochemical electron-transfer process, but in this review I will concentrate on the longer-lived species that has traditionally been called the "primary" acceptor.

2. Electron-Transfer Processes

2.1. The Primary Photochemical Electron Transfer

Clayton (1962a) was the first to propose that the primary electron transfer occurs from a "special" form of Bchl (which is now designated as P870). This electron transfer occurs with virtually zero activation energy, since it is still observed at 1°K (Arnold and Clayton, 1960) and is completed in a time less than 0.2 μsec (Parson, 1968). It also occurs with a very high efficiency. Many quantum yield determinations have been made,* most giving values between 0.5 and 1.0 electrons per photon, but the most recent determination is probably the most reliable, that of Wraight and Clayton (1974), who obtained 1.02 ± 0.04 electrons per photon in a RC preparation from the R-26 mutant of *Rhodopseudomonas sphaeroides*. Barskii *et al.* (1974), using an entirely different method, reported a quantum yield of 0.97 ± 0.03 electron per photon in chromatophores of *Rhodospirillum rubrum*. Thus, it appears that the primary electron-transfer process occurs at an efficiency of about 96–100%. We will represent this process as follows:

$$P870-X \xrightarrow{h\nu} P870^+-X^-$$

where X is the primary electron acceptor. Once this electron-transfer process has occurred, there are three reactions that can occur:

1. A secondary donor molecule can reduce $P870^+$ to P870:

$$D-P870^+-X^- \rightarrow D^+-P870-X^-$$

2. The electron on X can be transferred to secondary acceptor(s):

$$P870^+-X^--Y \rightarrow P870^+-X-Y^-$$

3. A back-reaction can occur, returning the electron to $P870^+$:

$$P870^+-X^- \rightarrow P870-X$$

If (1) or (2), or both, do not occur fairly rapidly, then (3) always occurs.

2.2. Back-Electron Transfer

The back-electron transfer (process 3) is a remarkable chemical reaction. It occurs at temperatures as low as 1°K, as evidenced by the reversibility of P870 photooxidation (Arnold and Clayton, 1960) and the reversibility of the light-induced ESR signal ascribed to $P870^+$ (McElroy *et al.*, 1969). The decay follows first-order kinetics, and the rate constant (≈ 30 sec^{-1}) is virtually temperature-independent from 1 to 100°K (McElroy *et al.*, 1969, 1974). Above approximately 150°K, the rate constant *decreases* with increasing temperature (Parson, 1967; Hsi and Bolton, 1974; Loach *et al.*, 1975). Clayton and Yau (1972) found that the decay rate constant slows sharply between 190 and 230°K, but the 85% glycerol solution they used may have affected their results (Hales, 1976). This behavior has been interpreted in terms of a quantum mechanical electron tunneling

model (McElroy *et al.*, 1969) in which the barrier thickness is estimated to be 30–40 Å (McElroy *et al.*, 1974; Hsi and Bolton, 1974; Hales, 1976). The inverse Arrhenius temperature dependence has been interpreted in terms of an expansion of the matrix causing the distance between $P870^+$ and X^- to increase slightly, thus increasing the tunneling time. A sophisticated calculation by Hopfield (1974) indicates, however, that the tunneling model may require a barrier distance of less than 10 Å. Such a distance would not be compatible with the fact that there is no observed influence of the paramagnetism of X^- on the ESR signal of $P870^+$. A dipole–dipole interaction calculation indicates that such an effect would be observable if the distance between X^- and $P870^+$ were 15 Å or less. Clearly, more theoretical work is required, especially since Hopfield's calculation does not account for the observed temperature dependence. Pending further theoretical work, this reviewer will assume that the distance between X and P870 is approximately 30–40 Å, since this distance is a reasonable one in view of the known dimensions of photosynthetic membranes and the characteristics of the RC protein.

2.3. Electron Transfer to Secondary Acceptors

Under physiological conditions, electron transfer from primary to secondary acceptors (process 2) is usually quite fast. Using a double laser flash technique, Parson (1969a) found that the half-time in *Chromatium vinosum* strain D is approximately 60 μsec at pH 7 with an activation energy of approximately 35 kJ mol^{-1}. The rate varies with pH. The reaction is blocked almost completely by certain inhibitors such as *o*-phenanthroline (Parson and Case, 1970; Clayton *et al.*, 1972b; Hsi and Bolton, 1974; Halsey and Parson, 1974). In *Rhodopseudomonas viridis*, the secondary electron-transfer rate appears to be faster, since the half-time is approximately 6 μsec at pH 7 (Carithers and Parson, 1975). Further information concerning the secondary acceptor can be found in Chapter 25.

3. Physical Characteristics of the Primary Electron Acceptor

3.1. Electron Capacity

There is now good evidence that the primary acceptor can accept only one electron. For example, under *low* redox potentials ($E_m^\circ < -100$ mV) where the primary electron acceptor has undergone a one-electron chemical reduction, the primary electron-

---

* Consult Wraight and Clayton (1974) for references to earlier quantum yield determinations. The first determination was by Loach and Sekura (1968); they obtained 0.95 ± 0.05.

transfer process does not occur (Dutton *et al.*, 1973*a,b*). Also, when two actinic laser flashes are administered with a delay time less than approximately 30 μsec (so that secondary electron transfer does not occur), then the second flash does not cause a significant amount of P870 photooxidation (Parson, 1969*b*). Quantitative measurement of cytochrome *c* photooxidation also indicates a capacity of one electron (Clayton *et al.*, 1972*a*).

3.2. Redox Potential

Most attempts to measure the redox potential of the primary acceptor are based on a technique in which some property of the primary electron-transfer components (e.g., optical bleaching, ESR signal, fluorescence) is monitored as the equilibrium redox potential of the medium is varied by using chemical redox buffers. Reported values (collected in Table 1) are fairly

Table 1. Oxidation–Reduction Midpoint Potentials Associated with the Primary Electron Acceptor in Photosynthetic Bacteria

Species	Preparation[a]	E_m° (mV)	pH	pH Dependence[a]	n	Method[b]	References
Rs. rubrum	Chrm	−44	7.4	ND	1	A	Kuntz *et al.* (1964)
	Chrm	−22	7.6	ND	2	A	Loach (1966)
	Cells	−18	7.3	ND	4	G	Loach (1966)
	Cells	−145	8.0	ND	1	D	Cramer (1969)
	Chrm	−140	7.4	ND	1	C	Dutton and Baltscheffsky (1972)
	Chrm	−145	8.0	ND	1	C	Dutton and Jackson (1972)
	Chrm	−120	7.2	ND	1	B, C	Leigh and Dutton (1974)
Rp. sphaeroides	Chrm	−18	7.6	ND	2	A	Loach (1966)
	Cells	−65, −90 (hysteresis)	8.0	ND	1	D	Cramer (1969)
	RC	−30	7.5	ND	2	H	Nicolson and Clayton (1969)
	RC	−50	7.5	Ind.	1	D	Reed *et al.* (1969)
	Chrm	−25	7.0	ND	1	A	Dutton and Jackson (1972)
	Chrm	−80	7.5	ND	1	C	Dutton and Jackson (1972)
	RC	−45	7.2	Ind.	1	C, B, F	Dutton *et al.* (1973*b*)
	Chrm	−50	7.4	−60 mV/pH	1	C, B	Dutton *et al.* (1973*b*)
	Chrm	−20	7.0	−60 mV/pH	1	A, E	Jackson *et al.* (1973)
	Chrm	−50	7.2	ND	1	B, C	Leigh and Dutton (1974)
Chr. vinosum strain D	Chrm	−135	7.5	ND	2	A	Cusanovich *et al.* (1968)
	Cells	−160	8.0	ND	1	D	Cramer (1969)
	Chrm	−130[c]	7.7	−40 mV/pH	1	E	Case and Parson (1971)
	Chrm	−135	7.4	ND	1	A	Dutton (1971)
	Chrm and SCP	−135	7.0	ND	1	E	Dutton *et al.* (1971)
	Chrm	−134	7.4	ND	1	E	Seibert and De Vault (1971)
	Chrm	−127	7.4	ND	1	A	Seibert *et al.* (1971)
	Chrm	−133	7.4	ND	1	E	Seibert *et al.* (1971)
	SCP	−120	7.4	ND	1	F	Dutton and Leigh (1973)
	Chrm	−110	7.0	−60 mV/pH	1	A, E	Jackson *et al.* (1973)
	Chrm	−135	8.0	ND	1	F	Evans *et al.* (1974)
	Chrm	−120	7.2	ND	1	B, C	Leigh and Dutton (1974)
Rp. gelatinosa	Chrm	−140	7.4	ND	1	A	Dutton (1971)
Chl. limicola f. thio-sulfatophilum	Chrm	−130	8.5	Ind	1	A, E	Knaff *et al.* (1973)
Rp. viridis	Chrm	−110	7.5	−30 mV/pH	1	E	Cogdell and Crofts (1972)

[a] (Chrm) Chromatophores; (RC) reaction center protein preparation; (SCP) subchromatophore particles; (ND) not determined.

[b] Methods: (A) light-induced absorbance changes in the IR; (B) ESR signal of Bchl triplet state; (C) light-induced ESR signal due to P870[+]; (D) fluorescence intensity; (E) cytochrome photooxidation following a flash at low temperature; (F) light-induced ESR signal at $g = 1.82$; (G) carotenoid band shifts; (H) kinetics of P870[+] reduction in the presence of redox dyes.

[c] Measured E_m as a function of temperature.

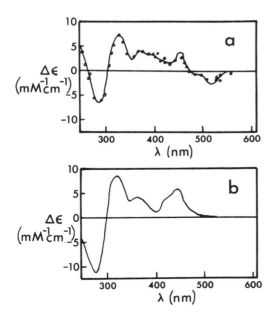

Fig. 1. (a) Optical difference spectrum of $(X^- - X)$ in RC particles from *Rhodopseudomonas sphaeroides*. Data from Slooten (1972*b*). (b) Optical difference spectrum of $Q^- - Q$ where Q is UQ-10 as determined by pulse radiolysis in methanol. Data from Bensasson and Land (1973).

consistent among a variety of techniques, but they do depend on the assumption that the primary acceptor is in thermodynamic equilibrium with the medium. That similar results are obtained for forward and backward titrations in most reported measurements is a reasonable indication that equilibrium has been achieved. Loach *et al.* (1975) reported that the redox potential of the primary acceptor may be as low as -320 mV in *Rs. rubrum*, but Parson and Cogdell (1975) and Prince and Dutton (1976) criticized these measurements on the basis that hysteresis in forward and backward titrations suggests that thermodynamic equilibrium had not been achieved. Much of the species variation in E_m° values, at least in chromatophores, may be due to the pK of the primary acceptor. Prince and Dutton (1976) showed that the E_m° is constant for pH values above the pK. These values are -160 mV in *Chr. vinosum* strain D, -200 mV in *Rs. rubrum*, and -180 mV in *Rp. sphaeroides* chromatophores. They argue that these are the operational potentials, since the rate of electron transfer to secondary acceptors is much greater than the rate of protonation of the primary acceptor. The problem of why the E_m°'s for RC preparations are independent of pH and yet have more positive values is still unresolved. It may be that the disruption necessary to solubilize these membrane proteins alters the environment sufficiently to shift the

redox potential of the primary acceptor. Electron transfer to secondary acceptors appears to involve a simultaneous proton uptake (Cogdell *et al.*, 1973*a,b*).

3.3. Optical Absorption Changes

It is quite difficult to separate the optical difference spectrum of $(X^- - X)$ from that of $(P870^+ - P870)$ or that due to reduction of secondary acceptors. By using chemical reductants to keep P870 in a reduced state and by subtracting effects due to reduction of secondary acceptors, Slooten (1972*b*) obtained the difference spectrum shown in Fig. 1a. These results are in essential agreement with the earlier work of Clayton and Straley (1970, 1972).

3.4. ESR Spectrum

Feher (1971) was the first to report an ESR spectrum due to the X^-. His spectrum was obtained using a light-modulation technique that results in an absorption-mode ESR signal. Leigh and Dutton (1972) subsequently obtained an ESR signal in the first derivative mode. Major g components are at 1.82 and 1.68 and a broad component around 2.0. A first-derivative light-minus-dark ESR spectrum of X^- is shown in Fig. 2. Feher *et al.* (1974) showed that their

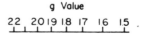

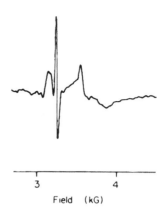

Fig. 2. ESR light-minus-dark difference spectrum of reaction centers from *Rhodopseudomonas sphaeroides* R-26 mutant. Spectrum taken from Dutton *et al.* (1973*a*). The spectrum was run at 10°K and 9.5 GHz microwave frequency. The spike at $g = 2.00$ is due to P870$^+$, and is strongly saturated by the microwave power used (50 mW).

absorption-mode ESR signal is compatible with that shown in Fig. 2. The ESR spectrum of X^- is very temperature-dependent, and disappears above approximately 20°K due to a very short spin-lattice relaxation time at higher temperatures that makes the spectrum too broad to detect. Under conditions where iron has been extracted, the spectrum shown in Fig. 2 is replaced by a much narrower spectrum with a single line at $g = 2.0050$ (Loach and Hall, 1972; Feher *et al.*, 1972). In contrast to the spectrum of X^-, which can be detected only at temperatures of less than 15°K, this new spectrum can be seen at room temperature (Loach and Hall, 1972; Bolton and Cost, 1973).

The ESR spectrum shown in Fig. 2 can be observed either by chemical reduction with $Na_2S_2O_4$ or by light. The main evidence that indicates that this ESR spectrum is due to X^-, however, is that the kinetics of formation and decay are identical with those of P870, at both $g = 1.68$ and $g = 1.82$ (Dutton *et al.*, 1973a). In addition, in experiments with chromatophores from *Chr. vinosum* strain D, Dutton and Leigh (Leigh and Dutton, 1972; Dutton and Leigh, 1973) showed that if cytochrome C-552 was reduced when the chromatophores were frozen, the broad ESR signal shown in Fig. 2 was produced permanently on illumination and did not decay in the dark. If it is accepted that only one electron per RC is transferred by light at low temperatures, then this evidence provides a very strong base for ascribing the ESR spectrum in Fig. 2 to that of X^-, the reduced primary acceptor.

4. Identity of the Primary Electron Acceptor

Now that RC protein has been isolated and chemically characterized, the number of possibilities for the identity of X is severely limited. Only two appear at least partially compatible with the characteristics listed in Section 3. These are iron and ubiquinone, which are each present at a level of approximately 1 molecule/RC protein (Feher, 1971; Beugeling *et al.*, 1972; Feher *et al.*, 1974; Prince and Crofts, 1973; Reiss-Husson and Jolchine, 1972; Slooten, 1972a).

4.1. Evidence For and Against the Premise That Iron Is the Primary Acceptor

If one accepts that the ESR spectrum shown in Fig. 2 is due to that of X^-, then one must conclude that the molecular entity giving rise to this ESR spectrum contains a paramagnetic transition metal ion. This conclusion arises from the fact that the g factors deviate markedly from the free-electron value of approximately 2.00. Chemical analysis shows that Fe

is the only transition metal present in significant amounts (Feher, 1971; Feher *et al.*, 1972; Loach and Hall, 1972).

Hsi and Bolton (1974) observed that *o*-phenanthroline forms a specific complex with RC particles and causes an inhibition of secondary electron transfer. Since *o*-phenanthroline is known to chelate with Fe, this is indirect evidence that Fe is involved. Also, *o*-phenanthroline shifts the redox potential of the primary acceptor in *Rp. viridis* (Cogdell and Crofts, 1972), *Rp. sphaeroides*, and *Chr. vinosum* strain D (Jackson *et al.*, 1973; Dutton *et al.*, 1973b; Evans *et al.*, 1974).

Nevertheless, there exists strong evidence against Fe being the primary acceptor (at least by itself). Loach and Hall (1972) were able to extract over 90% of the Fe from a subchromatophore preparation from *Rs. rubrum* without greatly affecting the level of P870 photooxidation. Feher *et al.* (1972) obtained similar results from RC protein preparations from *Rp. sphaeroides*. Okamura *et al.* (1974) found that the RC protein from *Rp. sphaeroides* R-26 mutant can be resolved into three subunits, L, M, and H. Two of these subunits (LM) form a complex that is still photoactive but appears to contain no Fe. Also, Feher *et al.* (1974) were able to substitute Mn for Fe in their RC protein preparations without significantly changing the kinetics or level of P870 photooxidation. The ESR spectrum of X^- did change, however, indicating an involvement of the transition metal ion in the structure of the primary acceptor. In an addendum to their paper, Feher *et al.* (1974) report that P. Debrunner has measured the Mossbauer spectrum of ^{57}Fe-enriched reaction centers from *Rp. sphaeroides*. The results are consistent with a model in which Fe is in a paramagnetic 2+ state and in which *the valence of the iron does not change* on chemical reduction with $Na_2S_2O_4$. These experiments can be criticized on the basis that (1) reduction of the RC was chemical rather than photochemical, and (2) the "reduced" RCs were mailed to another laboratory for measurements, and there was no evidence that the state of reduction was checked after transit. These experiments are important and need to be repeated. If the results can be confirmed, then Fe must be ruled out as a direct candidate for the primary electron acceptor.

4.2. Evidence For and Against the Premise That Ubiquinone Is the Primary Electron Acceptor

Lester and Crane (1959) were the first to observe that photosynthetic bacteria contain a high level of ubiquinone (UQ). Clayton (1962b) observed a light-induced absorbance change at approximately 270 nm, and proposed that it was due to the photoreduction of

UQ. He was the first to propose that the primary electron acceptor is UQ. Since then, much attention has been paid to the light-induced UV absorption changes (Beugeling, 1968; Ke, 1969; Ke *et al.*, 1968, 1973; Reed *et al.*, 1969). Perhaps the best optical evidence implicating UQ is the work by Clayton and Straley (1972) and by Slooten (1972*b*) on the optical difference spectrum of the primary acceptor and the work by Bensasson and Land (1973) on the UQ–ubisemiquinone optical difference spectrum. The optical difference spectrum of $(X^- - X)$ is shown in Fig. 1a and that of $(UQ^- - UQ)$ in Fig. 1b. The similarity, not only qualitatively but quantitatively, is striking.

Further evidence for UQ is provided by ESR studies. As previously mentioned, Feher *et al.* (1972) and Loach and Hall (1972) observed that extraction of Fe did not affect the photooxidation of P870. When Fe was extracted, however, a new ESR signal appeared with $g = 2.0050$ and a line width of approximately 7 G at 9 GHz. By comparing the *in vivo* spectra with model spectra of the anion radical of UQ at 9 and 35 GHz, Feher *et al.* (1972) provided strong evidence that the new ESR signal was that of UQ^-. Bolton and Cost (1973) observed the same ESR spectrum, but at room temperature, using a flash photolysis technique. On extraction with isooctane (which removes most of the UQ), the new spectrum disappeared, but returned again when UQ was added back. This signal, however, represented only about 15% of that due to $P870^+$.

Finally, if UQ is the primary electron acceptor, then complete removal of UQ should inhibit the photooxidation of P870. This has indeed been observed by Cogdell *et al.* (1974) using isooctane with a trace of methanol to extract the UQ from RCs of *Rp. sphaeroides* R-26 mutant. Probably the best evidence to date for UQ being the primary acceptor is some recent work by Okamura *et al.* (1975). They found that 2 UQ molecules are normally bound to each RC protein of *Rp. sphaeroides*; one is easily removed, one is more tightly bound. They found that the photochemical activity of the preparation was removed quantitatively in direct correlation with the removal of the tightly bound UQ. Furthermore, a full restoral of activity was achieved with readdition of UQ.

Despite the strong evidence for UQ cited above, there is some evidence against. Ke *et al.* (1973) observed that in subchromatophore particles from *Chr. vinosum* strain D, extensive extraction of UQ with isooctane did not change the optical difference spectrum between 240 and 950 nm. Takamiya and Takamiya (1970) observed that even after over 99% of the quinones had been extracted, the optical difference spectrum was little changed and nearly 50%

of the level of optical absorption change remained. Takamiya and Nishimura (1975) recently reported that extensive extraction of UQ does inhibit the photooxidation of cytochrome C-552, but not of cytochrome C-555. They propose that there are two types of RCs in *Chr. vinosum* strain D, one with UQ as the primary acceptor, one with some other substance as the primary acceptor. This proposal is not in agreement, however, with the laser flash studies of Parson and Case (1970).

Nöel *et al.* (1972) isolated from wild-type *Rs. rubrum* a RC particle that reportedly contains less than 0.3 mol UQ/P870, yet is still fully photochemically active.

Halsey and Parson (1974) observed that extraction of UQ from chromatophores of *Chr. vinosum* strain D destroys their ability to perform photochemistry on the second of two closely spaced laser flashes without affecting photochemistry on the first flash. They propose that UQ functions solely as a secondary electron acceptor. Further to this, Okamura *et al.* (1976) showed that all the UQ in *Chr. vinosum* strain D can be removed without affecting the photochemical activity; what remains as the essential component is menaquinone. When menaquinone is removed, photochemistry ceases; when it is added back, photochemistry returns.

Finally, Loach *et al.* (1971) claim that most of the optical density changes in the UV can be accounted for quantitatively by the changes involved in the oxidation of P870.

The results of Cogdell *et al.* (1974) and of Okamura *et al.* (1975) may be reconciled with those of Ke *et al.* (1973), Takamiya and Takamiya (1970), Takamiya and Nishimura (1975), and Halsey and Parson (1974) if menaquinone acts as the primary acceptor in *Chr. vinosum* strain D as Okamura *et al.* (1976) proposed.

4.3. Perhaps the Primary Acceptor Is a Complex of Iron and Ubiquinone?

So far, the evidence for and against both Fe and UQ seems to be evenly balanced, so one way out of the paradox is to assume that both are involved in the primary electron acceptor. Bolton *et al.* (1969) were the first to allude to this possibility when they commented on the fact that (at that time) no ESR signal had been observed for the primary acceptor. They pointed out that if the UQ anion radical were formed, then an ESR signal should be detectable. Bolton and Cost (1973) extended the hypothesis by suggesting that UQ could be the primary acceptor if a paramagnetic ion such as Fe^{2+} or Fe^{3+} were nearby to broaden out the ESR spectrum. This would explain why extraction of Fe has little effect on the photooxidation of P870 but reveals the ESR spectrum of UQ^- when no Fe is present to broaden the spectrum. Feher *et al.* (1972) also proposed a similar Fe–UQ complex, but one in which the electron is predominantly on Fe. They recently revised this hypothesis (Feher *et al.*, 1974) in light of the Mossbauer experiments cited above.

Finally, it is interesting to note that the ESR spectrum recently reported for the primary acceptor in photosystem I of green plants (McIntosh *et al.*, 1975; Evans *et al.*, 1975) is very similar to that of the primary acceptor of bacterial photosynthesis, which may indicate some similarity of structure. It is difficult, however, to explain the large difference in redox potential between the two systems ($E_m^\circ \approx -600$ mV in photosystem I vs. ≈ -200 mV for bacteria).

5. Conclusions and a Look to the Future

In this review, I have attempted to survey the results relating to the primary electron acceptor of bacterial photosynthesis. The evidence for and against both Fe and UQ as candidates appears to be contradictory; hence, the most plausible hypothesis is that the primary acceptor is a *complex* of UQ and Fe in which UQ accommodates most of the added electron. A paramagnetic Fe entity must be closely associated with the UQ to explain the very broad and anisotropic ESR spectrum. I must emphasize, however, that at this point such a model has not been proved, and should be considered only as a working hypothesis.

Further research in this area should concentrate on studies in which a number of physical parameters (e.g., optical absorption changes, redox level, ESR spectra) can be monitored as UQ or Fe or both are progressively extracted. This work should be done for a number of different species to reveal possible species variations, as between *Rp. sphaeroides* and *Chr. vinosum*. Good assay techniques will be needed to obtain reliable measurements of the important ratio of UQ to P870.

6. References

References dated 1976 and 1977 have been added since this chapter was written and hence are not referred to in the text.

Arnold, W., and Clayton, R. K., 1960, The first step in photosynthesis: Evidence for its electronic nature, *Proc. Natl. Acad. Sci. U.S.A.* **46**:769.

Barskii, E. L., Bosisov, A. Yu., Godik, V. I., and Il'ina, M. I., 1974, Determination of the quantum yield of primary energy transformation during photosynthesis. II. Two modifications of the method, *Mol. Biol. (Moscow)* **8**:927.

Bensasson, R., and Land, E. J., 1973, Optical and kinetic properties of semireduced plastoquinone and ubiquinone: Electron acceptors in photosynthesis, *Biochim. Biophys. Acta* **325**:175.

Beugeling, T., 1968, Photochemical activities of $K_3Fe(CN)_6$ treated chromatophores from *Rhodospirillum rubrum*, *Biochim. Biophys. Acta* **153**:143.

Beugeling, T., Slooten, L., and Barelds-van-de-Beak, P. G. M. M., 1972, Thin layer chromatography of pigments from reaction center particles of *Rhodopseudomonas sphaeroides*, *Biochim. Biophys. Acta* **283**:328.

Bolton, J. R., and Cost, K., 1973, Flash photolysis–electron spin resonance: A kinetic study of endogenous light-induced free radicals in reaction center preparations from *Rhodopseudomonas sphaeroides*, *Photochem. Photobiol.* **18**:417.

Bolton, J. R., Clayton, R. K., and Reed, D. W., 1969, An identification of the radical giving rise to the light-induced ESR signal in photosynthetic bacteria, *Photochem. Photobiol.* **9**:209.

Carithers, R. P., and Parson, W. W., 1975, Delayed fluorescence from *Rhodopseudomonas viridis* following single flashes, *Biochim. Biophys. Acta* **387**:194.

Case, G. D., and Parson, W. W., 1971, Thermodynamics of the primary and secondary photochemical reactions in *Chromatium, Biochim. Biophys. Acta* **253**:187.

Chamorovsky, S. K., Remennikov, S. M., Kononenko, A. A., Venediktov, P. S., and Rubin, A. B., 1976, New experimental approach to the estimation of rate of electron transfer from the primary to secondary acceptors in the photosynthetic electron transport chain of purple bacteria, *Biochim. Biophys. Acta* **430**:62.

Clayton, R. K., 1962a, Primary reactions in bacterial photosynthesis. I. The nature of light-induced absorbancy changes in chromatophores; evidence for a special bacteriochlorophyll component, *Photochem. Photobiol.* **1**:201.

Clayton, R. K., 1962b, Evidence for the photochemical reduction of coenzyme Q in chromatophores of photosynthetic bacteria, *Biochem. Biophys. Res. Commun.* **9**:49.

Clayton, R. K., and Straley, S. C., 1970, An optical absorption change that could be due to reduction of the primary photochemical electron acceptor in photosynthetic reaction centers, *Biochem. Biophys. Res. Commun.* **39**:1114.

Clayton, R. K., and Straley, S. C., 1972, Photochemical electron transport in photosynthetic reaction centers. IV. observations related to the reduced photoproducts, *Biophys. J.* **12**:1221.

Clayton, R. K., and Yau, H. F., 1972, Photochemical electron transport in photosynthetic reaction centers from *Rhodopseudomonas sphaeroides*. I. Kinetics of the oxidation and reduction of P870 as affected by external factors, *Biophys. J.* **12**:867.

Clayton, R. K., Fleming, H., and Szuts, E. Z., 1972a, Photochemical electron transport in photosynthetic reaction centers from *Rhodopseudomonas sphaeroides*. II. Interaction with external electron donors and acceptors and a re-evaluation of some spectroscopic data, *Biophys. J.* **12**:46.

Clayton, R. K., Szuts, E. Z., and Fleming, H., 1972*b*, Photosynthetic electron transport in photosynthetic reaction centers from *Rhodopseudomonas sphaeroides*. III. Effects of *o*-phenanthroline and other chemicals, *Biophys. J.* **12**:64.

Cogdell, R. J., and Crofts, A. R., 1972, Some observations on the primary acceptor of *Rhodopseudomonas viridis*, *FEBS Lett.* **27**:176.

Cogdell, R. J., Jackson, J. B., and Crofts, A. R., 1973*a*, The effect of redox potential on the coupling between rapid hydrogen-ion binding and electron transport in chromatophores from *Rhodopseudomonas sphaeroides*, *J. Bioenerg.* **4**:211.

Cogdell, R. J., Prince, R. C., and Crofts, A. R., 1973*b*, Light-induced proton uptake catalyzed by photochemical reaction centers from *Rhodopseudomonas sphaeroides*, strain R26, *FEBS Lett.* **35**:204.

Cogdell, R. J., Brune, D. C., and Clayton, R. K., 1974, Effects of extraction and replacement of ubiquinone upon the photochemical activity of reaction centres and chromatophores from *Rhodopseudomonas sphaeroides*, *FEBS Lett.* **45**:344.

Corker, G. A., 1976, A survey of EPR investigations of bacterial photosynthesis, *Photochem. Photobiol.* **24**:617.

Cramer, W. A., 1969, Low potential titration of the fluorescence yield changes in photosynthetic bacteria, *Biochim. Biophys. Acta* **189**:54.

Cusanovich, M. A., Bartsch, R. G., and Kamen, M. D., 1968, Light-induced electron transport in *Chromatium vinosum* strain D. II. Light-induced absorbance changes in *Chromatium* chromatophores, *Biochim. Biophys. Acta* **153**:397.

Dutton, P. L., 1971, Oxidation–reduction potential dependence of the interaction of cytochromes. Bacteriochlorophyll and carotenoids at $77^\circ K$ in chromatophores of *Chromatium vinosum* strain D and *Rhodopseudomonas gelatinosa*, *Biochim. Biophys. Acta* **226**:63.

Dutton, P. L., and Baltscheffsky, M., 1972, Oxidation–reduction potential dependence of pyrophosphate-induced cytochrome and bacteriochlorophyll reactions in *Rhodospirillum rubrum*, *Biochim. Biophys. Acta* **267**:172.

Dutton, P. L., and Jackson, J. B., 1972, Thermodynamic and kinetic characterization of electron-transfer components *in situ* in *Rhodopseudomonas sphaeroides* and *Rhodospirillum rubrum.*, *Eur. J. Biochem.* **30**:495.

Dutton, P. L., and Leigh, J. S., 1973, Electron spin resonance characterization of *Chromatium vinosum*, strain D hemes, non-heme irons and the components involved in primary photochemistry, *Biochim. Biophys. Acta* **314**:178.

Dutton, P. L., Kihara, T., McCray, J. A., and Thornber, J. P., 1971, Cytochrome C553 and bacteriochlorophyll interaction at $77^\circ K$ in chromatophores and a subchromatophore preparation from *Chromatium vinosum* strain D, *Biochim. Biophys. Acta* **226**:81.

Dutton, P. L., Leigh, J. S., and Seibert, M., 1972, Primary

processes in photosynthesis: *In situ* electron spin resonance studies on the light-induced oxidized and triplet state of reaction center bacteriochlorophyll, *Biochem. Biophys. Res.Commun.* **46**:406.

Dutton, P. L., Leigh, J. S., and Reed, D. W., 1973*a*, Primary events in the photosynthetic reaction center from *Rhodopseudomonas sphaeroides*, strain R26. Triplet and oxidized states of bacteriochlorophyll and the identification of the primary electron acceptor, *Biochim. Biophys. Acta* **293**:654.

Dutton, P. L., Leigh, J. S., and Wraight, C. A., 1973*b*, Direct measurement of the midpoint potential of the primary electron acceptor in *Rhodopseudomonas sphaeroides in situ* and in the isolated state: Some relationships with pH and *o*-phenanthroline, *FEBS Lett.* **36**:169.

Dutton, P. L., Kaufmann, K. J., Chance, B., and Rentzepis, P. M., 1975, Picosecond kinetics of the 1250 nm band of the *Rhodopseudomonas sphaeroides* reaction center: the nature of the primary photochemical intermediary state, *FEBS Lett.* **60**:275.

Evans, M. C. W., Lord, A. V., and Reeves, S. G., 1974, The detection and characterization by electron paramagnetic resonance spectroscopy of iron–sulphur proteins and other electron-transport components in chromatophores from the purple bacterium *Chromatium vinosum*, strain D, *Biochem. J.* **138**:177.

Evans, M. C. W., Sihra, C. K., Bolton, J. R., and Cammack, R., 1975, The primary electron acceptor complex of photosystem I in spinach chloroplasts: Identification of a component undergoing reversible photoreduction at cryogenic temperatures, *Nature (London)* **256**:668.

Fajer, J., Brune, D. C., Davis, M. S., Forman, A., and Spaulding, L. D., 1975, Primary charge separation in bacterial photosynthesis: oxidized chlorophylls and reduced pheophytin, *Proc. Natl. Acad. Sci. U.S.A.* **72**:4956.

Feher, G., 1971, Some chemical and physical properties of a bacterial reaction center particle and its primary photochemical reactants, *Photochem. Photobiol.* **14**:373.

Feher, G., Okamura, M. Y., and McElroy, J. D., 1972, Identification of an electron acceptor in reaction centers of *Rhodopseudomonas sphaeroides* by electron paramagnetic resonance, *Biochim. Biophys. Acta* **267**:222.

Feher, G., Isaacson, R. A., McElroy, J. D., Ackerson, L. C., and Okamura, M. Y., 1974, On the question of the primary acceptor in bacterial photosynthesis: Manganese substituting for iron in reaction centres of *Rhodopseudomonas sphaeroides*, R-26 strain, *Biochim. Biophys. Acta* **368**:135.

Hales, B. J., 1976, Temperature dependency of the rate of electron transport as a monitor of protein motion, *Biophys. J.* **16**:471.

Halsey, Y. D., and Parson, W. W., 1974, Identification of ubiquinone as the secondary electron acceptor in the photosynthetic apparatus of *Chromatium vinosum*, strain D., *Biochim. Biophys. Acta* **347**:404.

Hoff, A. J., Gast, P., and Romijn, J. C., 1977, Time-resolved ESR and chemically induced dynamic electron polarization of the primary reaction in a reaction center particle of *Rhodopseudomonas sphaeroides* wild type at low temperatures, *FEBS Lett.* **73**:185.

Hopfield, J. J., 1974, Electron transfer between biological molecules by thermally activated tunneling, *Proc. Natl. Acad. Sci. U.S.A.* **71**:3640.

Hsi, E. S. P., and Bolton, J. R., 1974, Flash photolysis–electron spin resonance study of the effect of *o*-phenanthroline and temperature on the decay time of the electron spin resonance signal B1 in reaction center preparations and chromatophores of mutant and wild strains of *Rhodospirillum rubrum* and *Rhodopseudomonas sphaeroides*, *Biochim. Biophys. Acta* **347**:126.

Jackson, J. B., Cogdell, R. J., and Crofts, A. R., 1973, Some effects of *o*-phenanthroline on electron transport in chromatophores from photosynthetic bacteria, *Biochim. Biophys. Acta* **292**:218.

Jennings, J. V., and Evans, M. C. W., 1977, The irreversible photoreduction of a low potential component at low temperatures in a preparation of the green photosynthetic bacterium *Chlorobium thiosulphatophilum*, *FEBS Lett.* **75**:33.

Kaufmann, K. J., Dutton, P. L., Netzel, J. L., Leigh, J. S., and Rentzepis, P. M., 1975, Picosecond kinetics of events leading to reaction center bacteriochlorophyll oxidation, *Science* **188**:1301.

Kaufmann, K. J., Petty, K. M., Dutton, P. L., and Rentzepis, P. M., 1976, Picosecond kinetics in reaction centers of *Rhodopseudomonas sphaeroides* and the effects of ubiquinone extraction and reconstitution, *Biochem. Biophys. Research Commun.* **70**:839.

Ke, B., 1969, Nature of the primary electron acceptor in bacterial photosynthesis, *Biochim. Biophys. Acta* **172**:583.

Ke, B., Vernon, L. P., Garcia, A., and Ngo, E., 1968, Coupled photooxidation of bacteriochlorophyll P890 and photoreduction of ubiquinone in a photochemically active subchromatophore particle derived from *Chromatium vinosum*, strain D., *Biochemistry* **7**:311.

Ke, B., Garcia, A. F., and Vernon, L. P., 1973, Light-induced absorption changes in *Chromatium* subchromatophore particles exhaustively extracted with nonpolar solvents, *Biochim. Biophys. Acta* **292**:226.

Klimov, V. V., Shuvalov, V. A., Krakhmaleva, I. N., Karapetyan, N. V., and Krasnovsky, A. A., 1976, Bacteriochlorophyll fluorescence changes related to the bacteriopheophytin photoreduction in the chromatophores of purple sulfur bacteria, *Biokhimiya* **41**:1435.

Knaff, D. B., Buchanan, B., and Malkin, R., 1973, Effect of redox potential on light-induced cytochrome and bacteriochlorophyll reactions in chromatophores from the photosynthetic green bacterium *Chlorobium*, *Biochim. Biophys. Acta* **325**:94.

Kuntz, I. D., Loach, P. A., and Calvin, M., 1964, Absorption changes in bacterial chromatophores, *Biophys. J.* **4**:227.

Leigh, J. S., and Dutton, P. L., 1972, The primary electron acceptor in photosynthesis, *Biochem. Biophys. Res. Commun.* **46**:414.

Leigh, J. S., and Dutton, P. L., 1974, Reaction center bacteriochlorophyll triplet states: Redox potential dependence and kinetics, *Biochim. Biophys. Acta* **357**:67.

Lester, R. L., and Crane, F. L., 1959, The natural occurrence of coenzyme Q and related compounds, *J. Biol. Chem.* **234**:2169.

Loach, P. A., 1966, Primary oxidation–reduction changes during photosynthesis in *Rhodospirillum rubrum*, *Biochemistry* **5**:592.

Loach, P. A., 1977, Yearly review: Primary photochemistry in photosynthesis, *Photochem. Photobiol.* **26**:87.

Loach, P. A., and Hall, R. L., 1972, The question of the primary electron acceptor in bacterial photosynthesis, *Proc. Natl. Acad. Sci. U.S.A.* **69**:786.

Loach, P. A., and Sekura, D. L., 1968, Primary photochemistry and electron transport in *Rhodospirillum rubrum*, *Biochemistry* **7**:2642.

Loach, P. A., Bambara, R. A., and Ryan, F. J., 1971, Identification of the major ultraviolet absorbance photochanges in photosynthetic systems, *Photochem. Photobiol.* **13**:247.

Loach, P. A., Kung, M., and Hales, B. J., 1975, Characterization of the phototrap in photosynthetic bacteria, *Ann. N. Y. Acad. Sci.* **244**:297.

McElroy, J. D., Feher, G., and Mauzerall, D. C., 1969, On the nature of the free radical formed during the primary process of bacterial photosynthesis, *Biochim. Biophys. Acta* **172**:180.

McElroy, J. D., Mauzerall, D. C., and Feher, G., 1974, Characterization of primary reactants in bacterial photosynthesis. II. Kinetic studies of the light-induced electron paramagnetic resonance signal ($g = 2.0026$) and the optical absorbance changes at cryogenic temperatures, *Biochim. Biophys. Acta* **333**:261.

McIntosh, A. R., Chu, M., and Bolton, J. R., 1975, Flash photolysis–electron spin resonance studies of the electron acceptor species at low temperatures in photosystem I of spinach subchloroplast particles, *Biochim. Biophys. Acta* **376**:308.

Netzel, T. L., Rentzepis, P. M., Tiede, D. M., Prince, R. C., and Dutton, P. L., 1977, Effect of reduction of the reaction center intermediate upon the picosecond oxidation reaction of the bacteriochlorophyll dimer in *Chromatium vinosum* and *Rhodopseudomonas viridis*, *Biochim. Biophys. Acta* **460**:467.

Nicolson, G. L., and Clayton, R. K., 1969, The reducing potential of the bacterial photosynthetic reaction center, *Photochem. Photobiol.* **9**:395.

Nöel, H., Von der Rest, M., and Gingras, G., 1972, Isolation and partial characterization of a P870 reaction center complex from wild type *Rhodospirillum rubrum*, *Biochim. Biophys. Acta* **275**:219.

Okamura, M. Y., Steiner, L. A., and Feher, G., 1974, Characterization of reaction centers from photosynthetic bacteria, I. Subunit structures of the protein mediating the primary photochemistry in *Rhodopseudomonas sphaeroides*, R-26 strain, *Biochemistry* **13**:1394.

Okamura, M. Y., Isaacson, R. A., and Feher, G., 1975, Primary acceptor in bacterial photosynthesis: Obligatory role of ubiquinone in photoactive reaction centres of

Rhodopseudomonas sphaeroides, Proc. Natl. Acad. Sci. U.S.A. **72**:3491.

Okamura, M. Y., Ackerson, L. C., Isaacson, R. A., Parson, W. W., and Feher, G., 1976, The primary electron acceptor in *Chromatium vinosum* (strain D), *Biophys. J.* **16**:223a.

Parson, W. W., 1967, Flash-induced absorbance changes in *Rhodospirillum rubrum* chromatophores, *Biochim. Biophys. Acta* **131**:154.

Parson, W. W., 1968, The role of P870 in bacterial photosynthesis, *Biochim. Biophys. Acta* **153**:248.

Parson, W. W., 1969a, The reaction between primary and secondary electron acceptors in bacterial photosynthesis, *Biochim. Biophys. Acta* **189**: 384.

Parson, W. W., 1969b, Cytochrome photooxidations in *Chromatium* chromatophores: Each P870 oxidizes two cytochrome C$_{422}$ hemes, *Biochim. Biophys. Acta* **189**:397.

Parson, W. W., 1974a, Bacterial photosynthesis, *Annu. Rev. Microbiol.* **28**:41.

Parson, W. W., 1974b, Rapid reactions in photobiology, in: *Chemical and Biological Applications of Lasers* (C. B. Moore, ed.), Academic Press, New York.

Parson, W. W., and Case, G. D., 1970, In *Chromatium* a single photochemical reaction center oxidizes both cytochrome C$_{552}$ and cytochrome C$_{555}$, *Biochim. Biophys. Acta* **205**:232.

Parson, W. W., and Cogdell, R. J., 1975, The primary photochemical reaction of bacterial photosynthesis, *Biochim. Biophys. Acta* **416**:105.

Parson, W. W., Clayton, R. K., and Cogdell, R. J., 1975, Excited states of photosynthetic reaction centers at low redox potentials, *Biochim. Biophys. Acta* **387**:265.

Petty, K. M., and Dutton, P. L., 1976, Properties of the flash-induced proton binding encountered in membranes of *Rhodopseudomonas sphaeroides:* A functional pK on the ubisemiquinone?, *Arch. Biochem. Biophys.* **172**: 335.

Prince, R. C., and Crofts, A. R., 1973, Photochemical reaction centers from *Rhodopseudomonas capsulata* ALA PHO$^+$, *FEBS. Lett.* **35**:213.

Prince, R. C., and Dutton, P. L., 1976, The primary acceptor of bacterial photosynthesis: Its operating midpoint potential?, *Arch. Biochem. Biophys.* **172**:329.

Prince, R. C., and Thornber, J. P., 1977, A novel electron paramagnetic resonance signal associated with the "primary" electron acceptor in isolated photochemical reaction centers of *Rhodospirillum rubrum*, *FEBS Lett.* **81**:233.

Prince, R. C., Leigh, J. S., and Dutton, P. L., 1976, Thermodynamic properties of the reaction center of *Rhodopseudomonas viridis. In vivo* measurement of the reaction center bacteriochlorophyll–primary acceptor intermediary electron carrier, *Biochim. Biophys. Acta* **440**:622.

Reed, D. W., Zankel, K. L., and Clayton, R. K., 1969, The effect of redox potential on P870 fluorescence in reaction centers from *Rhodopseudomonas sphaeroides, Proc. Natl. Acad. Sci. U.S.A.* **63**:42.

Reiss-Husson, F., and Jolchine, G., 1972, Purification and

properties of a photosynthetic reaction center isolated from various chromatophore fractions of *Rhodopseudomonas sphaeroides, Biochim. Biophys. Acta* **256**:440.

Rockley, M. G., Windsor, M. W., Cogdell, R. J., and Parson, W. W., 1975, Picosecond detection of an intermediate in the photochemical reaction of bacterial photosynthesis, *Proc. Natl. Acad. Sci. U.S.A.* **72**:2251.

Romijn, J. C., and Amesz, J., 1976, Photochemical activities of reaction centers from *Rhodopseudomonas sphaeroides* at low temperature and in the presence of chaotropic agents, *Biochim. Biophys. Acta* **423**:164.

Romijn, J. C., and Amesz, J., 1977, Purification and photochemical properties of reaction centers of *Chromatium vinosum:* Evidence for the photoreduction of a naphthoquinone, *Biochim. Biophys. Acta* **461**:327.

Sauer, K., 1974, Primary events and the trapping of energy, in: *Bioenergetics of Photosynthesis* (Govindjee, ed.), pp. 115–181, Academic Press, New York.

Seibert, M., and De Vault, D., 1971, Photosynthetic reaction center transients P$_{435}$ and P$_{424}$ in *Chromatium vinosum*, strain D, *Biochim. Biophys. Acta* **253**:396.

Seibert, M., Dutton, P. L., and De Vault, D., 1971, A low-potential photosystem in *Chromatium vinosum*, strain D, *Biochim. Biophys. Acta* **226**:189.

Shuvalov, V. A., Khakhmaleva, I. N., and Klimov, V. V., 1976, Photooxidation of P-960 and photoreduction of P-800 (bacteriopheophytin b-800) in reaction centers from *Rhodopseudomonas viridis, Biochim. Biophys. Acta* **449**:597.

Slooten, L., 1972a, Reaction center preparations of *Rhodopseudomonas sphaeroides*: Energy transfer and structure, *Biochim. Biophys. Acta* **256**:452.

Slooten, L., 1972b, Electron acceptors in reaction center preparations from photosynthetic bacteria, *Biochim. Biophys. Acta* **275**:208.

Takamiya, K.-I., and Nishimura, M., 1975a, Nature of photochemical reactions in chromatophores of *Chromatium vinosum*, strain D. III. Heterogeneity of the photosynthetic units, *Biochim. Biophys. Acta* **396**:93.

Takamiya, K.-I., and Nishimura, M., 1975b, Dual roles of ubiquinone as primary and secondary electron acceptors in light-induced electron transfer in chromatophores of *Chromatium vinosum*, strain D, *Plant and Cell Physiol.* **16**:1061.

Takamiya, K.-I., and Takamiya, A., 1970, Nature of photochemical reactions in chromatophores of *Chromatium vinosum*, strain D. I. Effects of P890 and ubiquinone in chromatophores of *Chromatium* D, *Biochim. Biophys. Acta* **205**:72.

Tiede, D. M., Prince, R. C., and Dutton, P. L., 1976a, EPR and optical spectroscopic properties of the electron carrier intermediate between the reaction center bacteriochlorophylls and the primary acceptor in *Chromatium vinosum, Biochim. Biophys. Acta* **449**:447.

Tiede, D. M., Prince, R. C., Reed, G. H., and Dutton, P. L., 1976b, EPR properties of the electron carrier intermediate between the reaction center bacteriochlorophylls and the primary acceptor in *Chromatium vinosum, FEBS Lett.* **65**:301.

van Grondelle, R., Romijn, J. C., and Holmes, N. G.,

1976, Photoreduction of the long wavelength bacterio-pheophytin in reaction centers and chromatophores of the photosynthetic bacterium *Chromatium vinosum* strain D, *FEBS Lett.* **72**:187.

Vermeglio, A., 1977, Secondary electron transfer in reaction centers of *Rhodopseudomonas sphaeroides*. Out-of-phase periodicity of two for the formation of ubisemiquinone and fully reduced ubiquinone, *Biochim. Biophys. Acta* **459**:516.

Vermeglio, A., and Clayton, R. K., 1977, Kinetics of electron transfer between the primary and secondary electron acceptor in reaction centers from *Rhodopseudomonas sphaeroides, Biochim. Biophys. Acta* **461**:159.

Wraight, C. A., 1977, Electron acceptors of photosynthetic bacterial reaction centers. Direct observation of oscillatory behavior suggesting two closely equivalent ubiquinones, *Biochim. Biophys. Acta* **459**:525.

Wraight, C. A., and Clayton, R. K., 1973, The absolute quantum efficiency of bacteriochlorophyll photo-oxidation in reaction centers of *Rhodopseudomonas sphaeroides, Biochim. Biophys. Acta* **333**:246.

Wraight, C. A., Leigh, J. S., Dutton, P. L., and Clayton, R. K., 1974, The triplet state of reaction centre bacterio-chlorophyll: Determination of a relative quantum yield, *Biochim. Biophys. Acta* **333**:401.

Wraight, C. A., Cogdell, R. J., and Clayton, R. K., 1975, Some experiments on the primary electron acceptor in reaction centers from *Rhodopseudomonas sphaeroides, Biochim. Biophys. Acta* **396**:242.

CHAPTER 23

EPR Studies of Primary Events in Bacterial Photosynthesis

J. S. Leigh, Jr.

1. Introduction

Elucidation of many electron-transfer events in photosynthesis has been made possible by the use of electron paramagnetic resonance (EPR) spectroscopy.* Since early electron-transfer steps involve almost exclusively single-electron events, the resulting changes are readily studied by EPR. This chapter is an attempt to survey those EPR studies that seem most relevant to the description of early events in bacterial photosynthesis. Emphasis has been placed on current views and problems, and no attempt has been made to provide a comprehensive review of the field.

The first application of EPR spectroscopy to photosynthetic studies came in the middle 1950s (Commoner *et al.*, 1956, 1957; Sogo *et al.*, 1959) when a photoinduced "free radical" was observed at $g = 2.0026$. This signal is now identified with oxidized reaction center bacteriochlorophyll (B^+B).

Loach *et al.* (1963) introduced redox potentiometry to photosynthetic EPR studies and found that the $g = 2$ EPR signal had the same oxidation–reduction mid-

point potential as P870 (Duysens *et al.*, 1956) and suggested their identity.

During the 1960s, the identification of the photo-induced $g = 2$ signal with the oxidation of reaction center bacteriochlorophyll (RC Bchl) was further strengthened by many groups (Kuntz *et al.*, 1964; Beinert and Kok, 1964; Ruby *et al.*, 1964; McElroy *et al.*, 1969; Bolton *et al.*, 1969). Little was known, however, about the detailed structure of the RC Bchl except that it was almost certainly an "aggregated" state of Bchl (cf. Smith and French, 1963; Clayton, 1966; Dratz *et al.*, 1966). Many workers favored a model that was akin to semiconductor physics with "conduction bands," "traps," and so on. A minor revolution took place with the striking demonstration by Reed and Clayton (1968) that RCs were actually physical realities—specific proteins containing Bchl as well as Fe, ubiquinone (UQ), and bacteriopheophytin (Bpheo).

2. "Primary" Components

The generally accepted complement of RC components includes: four Bchl's and two Bpheo's (Reed and Peters, 1972; Straley *et al.*, 1973), two UQs† (Okamura *et al.*, 1975), and a single Fe atom (Feher,

---

* Readers unfamiliar with EPR technique will find numerous excellent texts available, some of which are listed in the References. For clear and easy reading, the small monograph by Feher (1970) is highly recommended. Comparable clarity and a bit wider scope are available with Carrington and McLachlan (1967)

† *Chromatium vinosum* (Okamura *et al.*, 1976) and probably *Rp. viridis* (Garcia *et al.*, 1968) have been shown to contain one tightly bound menaquinone molecule replacing one of the two UQs.

J. S. Leigh, Jr. · Johnson Research Foundation, University of Pennsylvania, Philadelphia, Pennsylvania 19174

1971). Since some photosynthetic bacteria also have two molecules of cytochrome c that can donate electrons to the RC Bchl at cryogenic temperatures, it is necessary to consider them as well when describing early steps of RC electron transfer.

3. Reaction Center Bacteriochlorophyll

Using purified RCs from *Rhodopseudomonas sphaeroides* (R-26 mutant), Feher and co-workers (McElroy *et al.*, 1969, 1972; Feher, 1971) carefully compared the EPR properties of light-induced $g = 2$ signals with that of isolated oxidized Bchl a. By the use of isotopically substituted samples and measurement at two microwave frequencies, they obtained values for the various components that make up the linewidth. With *Rhodospirillum rubrum*, values of 8.8, 3.0, and 1.2 gauss for the contributions of proton hyperfine, nitrogen hyperfine, and g-value anisotropy, respectively, were obtained. The squares of the various contributions add to comprise the square of the composite Gaussian line (9.4 gauss). These values may be compared with 12.4, 3.7, and 1.4 gauss, respectively, for the contributions to the linewidth in isolated Bchl. The ratio of proton hyperfine splittings is 1.4. Norris *et al.* (1971) pointed out that if the unpaired electron in the oxidized RC was shared *equally** between two Bchl molecules, the contribution to the Gaussian envelope of unresolved hyperfine splittings would show a decrease by a factor of $\sqrt{2}$. This hypothesis was strongly supported by subsequent ENDOR measurements (Feher *et al.*, 1973, 1975; Norris *et al.*, 1973). It thus seems reasonable to designate the oxidized RC Bchl as $B^{+}B$. However, *Rp. viridis*, which contains Bchl b, may not exactly agree with this picture (Prince *et al.*, 1976). Oxidized extracted Bchl^{+} b shows a linewidth of 14 gauss (Fajer *et al.*, 1975), while the photoinduced $g = 2$ signal in *Rp. viridis* is 12 gauss wide (McElroy *et al.*, 1972; Prince *et al.*, 1976). This is perhaps indicative of unequal sharing between a "special pair." Additional ENDOR studies should be illuminating.

RCs may be incorporated into oriented phospholipid multilayers. A preliminary study of the orientation dependence of the photoinduced $B^{+}B$ signal shows approximately 4 gauss of variation with angle, the maximum g-value being observed when the plane of the "membranes" is oriented perpendicular to the magnetic field (J. S. Leigh, J. Pachence, and J. K. Blasie, unpublished). In addition, the linewidth and lineshape are orientation-dependent, with the largest linewidth observed with the plane perpendicular to the magnetic field. These results are consistent with the Bchl planes being oriented perpendicular to the membrane plane. Optical dichroism studies by Vermiglio and Clayton (1976) are in agreement with this conclusion. Detailed structural interpretation of this kind of result must await more information on the orientation of the g-tensor within the oxidized "special pair."

Oxidized RC Bchl in *Rp. sphaeroides* has a spin-lattice relaxation time (T_1) of approximately 100 μsec at 77°K, which shows an approximate $1/T^2$ temperature dependence (McElroy *et al.*, 1974). This perhaps indicates that spin-lattice relaxation takes place via a Raman mechanism (cf. Abragam and Bleaney, 1970). No obvious effects of spin–spin interaction with the reduced acceptor or with RC cytochromes have yet been detected. This would seem to indicate >20 Å separation between centers. A hint of possible weak spin–spin interaction between the reduced acceptor, X^-, and photoinduced B^+B may have been observed by McElroy *et al.* (1974), who noted difficulty in exact fitting of kinetic EPR recovery rates at temperatures below 4°C. It is to be hoped that further studies will quantitate this observation. With the addition of paramagnetic probes (e.g., Gd^{3+}, Gd EDTA$^-$, Ni^{2+}), it is sometimes possible to determine the intramembrane location of EPR-visible species. Dipolar interactions cause both lineshape alterations (e.g., broadening) and relaxation time effects (Case and Leigh, 1976). Lack of any observable effects of added Gd^{3+} on the B^+B signal in *Chromatium vinosum* chromatophores has been interpreted as indicating that B^+B is within the membrane.

In addition to the much-studied RC Bchl radical, RC UQ radicals have also been observed. In modified (Fe-free) RCs, Loach and Hall (1972) and Feher *et al.* (1972) observed photoinduced quinone radicals. Bolton and Cost (1973) also reported transient quinone radical signals from chromatophores that were exposed to short light flashes; these may represent the signal from secondary electron acceptors. Flavin radicals have not yet been unequivocally identified in photosynthetic bacteria.

4. The "Acceptor"

Illumination of *Rp. sphaeroides* RCs in the proper redox state (400 mV $> {\sim}E_h > {\sim}0$ mV) at cryogenic temperatures results in the formation of a broad signal

---

* It is not inconsistent with this hypothesis that the unpaired electron may be shared between more than two molecules of chlorophyll in an *unequal* manner and result in a 1.4-fold linewidth reduction. For example, if the unpaired spin were shared among three chlorophylls in a $1:4:1$ ratio, the narrowing would be $\sqrt{2}$, or if shared over four chlorophylls in, say, $1:1:1:6.5$ or $1:1:2:7.5$ ratios.

centered at $g = 1.8$ as well as the more easily observed $g = 2.00$ signal (Feher, 1970, 1971). The signal may also be formed by chemical reduction (Leigh and Dutton, 1972) and displays a midpoint oxidation–reduction potential appropriate for identification as the traditional "primary electron acceptor," X (Dutton and Leigh, 1973; Dutton *et al.*, 1973*b*). Similar signals have been observed in a variety of photosynthetic bacteria, including *Chr. vinosum* (Leigh and Dutton, 1972; Evans *et al.*, 1974), *Rp. viridis* (Prince *et al.*, 1976), and *Rp. capsulata* (Prince *et al.*, 1974). The extremely low g value of the signal and the absence of a $g = 1.8$ signal in RCs from which the Fe has been removed (Loach and Hall, 1972; Feher *et al.*, 1972, 1973) implicate Fe as the (partial) source of this signal. Replacement of RC Fe with Mn results in a very different EPR signal for X^- (Feher *et al.*, 1974). Removal of the RC quinones (Qs) has also been shown to prevent formation of a *photoinduced* $g = 1.8$ signal (Okamura *et al.*, 1975). Removal and replacement of the RC Qs seems to alter somewhat the lineshape of the X^- signal. Reconstitution with various kinds of Qs alters both the linewidth and dark decay kinetics (Okamura *et al.*, 1975).

From the pH dependence of the midpoint potential of this signal in chromatophores, it may be inferred that when chemically reduced (at neutral pH), X^- binds a proton to become XH (Dutton *et al.*, 1973*a*). However, no significant difference in lineshape has been noted between EPR signals from chemically reduced XH and photoinduced X^-. Addition of *o*-phenanthroline causes a shift in midpoint potential without much change in lineshape. This shift in apparent midpoint is actually a shift of the pK of the reduced species X^- (Prince and Dutton, 1976; Prince *et al.*, 1976). It is thus unlikely that *o*-phenanthroline interacts directly with X, but rather likely that it displaces the loosely bound Q from its normal position (cf. Okamura *et al.*, 1975). With both Fe and Q appearing to play a significant role in the integrity of the $g = 1.8$ species, the existence of an Fe–Q complex has been proposed (Feher *et al.*, 1972; Bolton and Cost, 1973). This has led to interest in the "question" whether in the reduced state (X^-), the electron is "primarily" on Q or on Fe.

A central g value of 1.8 is difficult to understand if the model for reduced "photoredoxin" is that of a semiquinone radical exchange-coupled to a high-spin ferrous ion. For example, in the case of "strong" antiferromagnetic coupling (e.g., Gibson *et al.*, 1966; Johnson *et al.*, 1971) leading to spin states of $S = \frac{3}{2}$ and $S = \frac{5}{2}$ with $S = \frac{3}{2}$ lying lowest in energy, the g values are expected to be

$$g = \tfrac{6}{5} g^{Fe} - \tfrac{1}{5} g^{Q\cdot}$$

That is, the g value will be less than 2.0 only if the g value of ferrous iron is less than 2 (since the g value of $Q\cdot$ cannot differ appreciably from 2.0). High-spin Fe(II) normally has g values greater than 2.0 (Abragam and Bleaney, 1970), but this may be somewhat misleading, since the RC Fe is in an unusual context. Possibility a mixture of zero-field splitting and exchange will provide the answer.

It has been suggested (W. Blumberg, personal communication; Dutton *et al.*, 1973*a*) that a high-spin ($S = \frac{3}{2}$) Fe(I) with four strong (S^-?) equatorial ligands and two weaker (O^-?) axial ligands could give rise to the observed spectrum. Resonance would be observed from the lowest Kramers doublet. The suggestion was based on a computer analysis designed to calculate interactions among the t_{2g} states of Fe(III), and thus is applicable only qualitatively to Fe(I), since some interaction with the e_g states should also be considered.

The zero-field splitting [or possibly the exchange coupling in the case of the exchange-coupled Q^-Fe(II) model] may be estimated by assuming that the major relaxation process at temperatures greater than approximately $4°K$ is via a two-phonon Orbach process (Orbach, 1961). A logarithmic plot of the linewidth of the $g = 1.8$ signal vs. $1/T$ shows a slope appropriate to a splitting of approximately 20 cm^{-1}. This suggestion is further strengthened by consideration of the integrated intensity of the $g = 1.8$ signal as a function of temperature. Below about $8°K$, the signal intensity approximates Curie law ($1/T$) behavior; at higher temperatures, however, the integrated intensity decreases faster than $1/T$ as the thermal population of the excited state becomes appreciable. The rate of decrease is again consistent with a zero-field splitting of approximately 20 cm^{-1} (± 5 cm^{-1}) (J. Salerno, D. Tiede, and J. S. Leigh, unpublished). No resonance has yet been observed from the excited state(s).

Mossbauer absorption spectra from ^{57}Fe-enriched RCs have been observed (Debrunner *et al.*, 1975; A. J. Bearden, P. L. Dutton and J. S. Leigh, unpublished). With RCs in the oxidized state, a single spectrum is seen with an isomer shift of 1.1 mm/sec (relative to Fe metal) and a quadrupole splitting of 2.1 mm/sec. Reduction of the RC causes a change in isomer shift of less than 0.2 mm/sec, and a small decrease in quadrupole splitting is observed. These results are strong evidence that the oxidized state is essentially high-spin ferrous [Fe(II), $S = 2$]. The small changes observed on reduction are unfortunately somewhat ambiguous. Reduction of the iron to high-spin Fe(I) might be expected to result in a small increase in isomer shift of approximately 0.2 mm/sec

(Walker *et al.*, 1964), but an increase of $4s$ electron contribution of 10–20% (covalency) could completely nullify this expected shift [as occurs in $Fe^{III}(CN)_6$ reduction]. Thus, the Mossbauer observations do not seem inconsistent with either the Q-radical–ferrous-iron hypothesis or the supposition that the reduced state of the acceptor is describable essentially as an Fe(I) species. Assuming that only one Q acts as an Fe ligand, a structure composed of a mixture of "resonance states" is, of course, the most general proposal:

$$Fe^I-Q \leftrightharpoons Fe^{II}-Q^{\underline{.}} \leftrightharpoons Fe^{III}-Q^=$$

Consideration of the EPR spectral properties would seem to give emphasis to the first such structure, with the Mossbauer results favoring the second as do the effects of Q and Fe extraction and replacement. More detailed interpretation must await further spectroscopic and structural studies. The temperature dependence of light-induced changes in magnetic susceptibility, for example, would be very helpful. Chemically Fe seems to have two major functions, to prevent more than one electron being accepted, and to facilitate transfer to secondary Q.

Paramagnetic probe studies have shown that the X^- species is located within 20 Å of the outside surface of the chromatophore membrane (in *Chr. vinosum*) (Case and Leigh, 1974). This result agrees with immunological studies on the arrangement of RC subunits in the chromatophore membrane (in *Rp. sphaeroides*) (Steiner *et al.*, 1974).

5. The Intermediate Acceptor

Picosecond spectroscopy (Netzel *et al.*, 1973; Kaufmann *et al.*, 1975; Rockley *et al.*, 1975; Dutton *et al.*, 1975) on *Rp. sphaeroides* RCs has shown that the oxidation of RC Bchl is accomplished in less than 10 psec. However, reduction on the "primary" acceptor, X, is not detected until approximately 150 psec later. Thus, the traditional primary acceptor is not primary! During the intervening time period, the electron must reside on an intermediate electron acceptor, I (probably Bpheo). One form of the reduced intermediate, I^-, shows weak exchange-coupling to the reduced electron acceptor, X^- ($J \sim 6 \times 10^{-3}$ cm^{-1}), in good agreement with the 150-psec electron-transfer time (Tiede *et al.*, 1976). The lineshape of the exchange-coupled pair is well fit by computer simulation (J. Salerno, personal communication). This magnitude of exchange coupling is consistent with a distance of approximately 8 Å between I and X^- (Hopfield, 1974). At higher temperatures (temp. >

$10°K$), where occupation of the excited state of X^- is appreciable, an unsplit singlet appears in the center of the split doublet.

6. Reaction Center Triplets

A RC Bchl triplet signal has been observed in photosynthetic bacteria, in preparations ranging from whole cells to RCs (Dutton *et al.*, 1972; Leigh and Dutton, 1974; Thurnauer *et al.*, 1975; Prince *et al.*, 1976). EPR spectra of organic molecules in triplet states ($S = 1$) have been studied for many years (cf. Hutchison and Mangum, 1958). Six lines due to anisotropic spin–spin interaction are usually observable, symmetrically displaced around $g = 2.00$ in a characteristic pattern (Kottis and Lefebvre, 1964; Wasserman *et al.*, 1964). The line positions are described by the zero-field splitting parameters usually given as D and E. At very low temperatures where electron-spin relaxation becomes quite slow, differential entry and exit rates into/from the three magnetic levels often leads to non-Boltzmann distribution of spins within the triplet state. This phenomenon is described as (photoinduced) spin polarization, and leads to enhanced EPR absorption and even EPR emission signals (Schwoerer and Sixl, 1968). Spin-polarized RC triplet EPR signals are seen under conditions where electron transfer to X^- is blocked by prior chemical reduction (Dutton *et al.*, 1972; Leigh and Dutton, 1974; Thurnauer *et al.*, 1975). Table I gives a compilation of zero field splitting parameters from a representative list of bacteria. It may be noted that the D and E values are highly specific for a given kind of bacterium, and might even be of some taxonomic value. There appears to be a rough correlation between the zero field splitting parameters and the long-wavelength absorption band of the RC "special pair"; i.e., smaller D values are observed in bacteria with longer-wave optical absorbance. The magnitude of the zero field splitting parameters is proportional to the average reciprocal cube of the distance between the two spins of the triplet state. The values of D shown in Table I are appropriate for a root-mean-reciprocal-cube separation of approximately 4 Å (Leigh and Dutton, 1974). This distance value is appreciably greater than that for monomeric Bchl, and thus provides support for the "special pair" (dimer) model (Norris *et al.*, 1971).

From rapid kinetic optical studies, it appears likely that the RC triplet is formed as a result of recombination of the electron from the $B^+B^-I^-$ "intermediate" species, leading to a B^+B^-I state (Kaufmann *et al.*, 1975; Prince *et al.*, 1976. Return to a "singlet"

Table 1. Zero-Field Splitting Parameters

Species	Preparation	D (cm^{-1}) ($\times 10^4$)	E (cm^{-1}) ($\times 10^4$)
Rp. viridis			
Strain HNTC 133	Cells	157	37
	Chromatophores	158	38
Chr. vinosum			
Strain D	Cells	177	34
	Chromatophores	178	33
	Fraction A	177	33
Rp. sphaeroides			
Strain 2.4.1	Cells	185	31
Strain Ga	Cells	186	31
	Chromatophores	185	31
Strain R-26	Cells	186	31
	Chromatophores	186	31
	RC	183	31
Rp. capsulata			
Strain St. Louis	Cells	183	31
Strain SB 25	Cells	184	31
Strain BY 761	Cells	183	31
Rp. gelatinosa			
Strain I	Cells	186	27
Rp. palustris			
Strain 2.1.6	Cells	184	34
Rs. rubrum			
Strain S 1	Cells	186	34
Strain G 9	Cells	185	33

B$^+$B$^-$ would lead to rapid reaction fluorescence. Return into the spin-polarized triplet state leads to a longer-lived optical state that Parson *et al.* (1975) termed P^R ($t \approx 120$ μsec at $<150°$K). The EPR signal from the spin-polarized triplet has an apparent lifetime of approximately 6 μsec, with highly damped oscillatory kinetics. This complex decay most likely represents spin-lattice relaxation among the triplet levels, since strong temperature dependence of this apparent lifetime has been observed (Leigh and Dutton, 1974). Prince *et al.* (1976) showed that chemical reduction of I prior to illumination prevents the formation of the spin-polarized triplet state. It thus appears that the triplet is not formed on the "forward" pathway of light-induced electron transfer, but rather as a result of an abortive return pathway that occurs on photoactivation when the acceptor X$^-$ is already reduced. Orientation studies with the triplet should help to determine the relative orientation of the RC Bchl's with respect to the "plane" of the chromatophore membranes.

7. Cytochromes

Very few cytochromes in photosynthetic bacteria have yet been studied by EPR techniques. At low temperatures, EPR signals from *Chr. vinosum* in the $g = 3$ region, which correspond to cytochromes c_{553} and c_{555}, have been observed (Leigh and Dutton, 1972; Dutton and Leigh, 1973).

Cytochrome c_{553} is active as an irreversible electron donor to the oxidized RC Bchl at liquid helium temperatures in a time of approximately 2 msec. The midpoint potential measured for c_{553} by low-temperature EPR ($+10$ mV) is in excellent agreement with that measured spectrophotometrically at room temperature. Preliminary studies with oriented *Chr. vinosum* chromatophore films indicate that the heme planes of c_{553} and c_{555} are approximately parallel and perpendicular to the membrane plane, respectively (J. S. Leigh and D. Tiede, unpublished).

The absence of observable cytochrome signals in other photosynthetic bacteria is somewhat puzzling. In *Rp. sphaeroides* and *Rs. rubrum*, for instance, cytochrome c_2 has not yet been detected in low-temperature EPR studies of either whole cells or chromatophores, even though the isolated cytochromes display perfectly "normal" low-spin heme resonances with g values of approximately 3.0, 2.2, and 1.5 (Leigh and Dutton, unpublished).

Cytochrome c' has also been detected in *Chr. vinosum* and *Rp. capsulata* (Dutton and Leigh, 1974; Prince *et al.*, 1974). Cytochrome c' was shown by Maltempo *et al.* (1974) to be describable (at least partially) as an intermediate spin ($S = \frac{3}{2}$) heme protein in some forms. In this respect, it bears a striking resemblance to horseradish peroxidase (Leigh *et al.*, 1975). In fact, it has been possible to show that EPR spectral changes may be readily observed in cytochrome c' (in *Chr. vinosum* chromatophores) on addition of typical peroxidase "substrates" such as resorcinol and aminotriazole (Leigh, unpublished). Further work will be needed to determine whether cytochrome c' might possibly function as a peroxidase in photosynthetic bacteria.

8. Oriented Multilayers

EPR spectra from oriented multilayer samples have not yet been used extensively to obtain structural information from photosynthetic systems. It seems likely that this approach will yield some interesting information, so a few short notes here about the technique may be useful.

Membranous samples, say chromatophores or RCs

in phospholipids, may be oriented by controlled drying on planar substrates (cf. Coleman *et al.*, 1969; Dupont *et al.*, 1973). If placed in the EPR cavity in such a way that the oriented plane is normal to the direction of the magnetic field, then all disorder (in the plane of the oriented multilayer) is about an axis parallel to the magnetic field direction. In this case, what is essentially a single-crystal EPR spectrum is obtained. That is, since EPR spectral parameters are invariant to rotation about an axis parallel to the magnetic field direction, there is no effect from disorder in the plane of the multilayer. Thus, one observes a single EPR transition at a *g* value determined by the orientation of the molecular axes with respect to the membrane normal; call it g_n. Then,

$$g_n^2 = g_x^2 \sin^2 \theta \cos^2\phi + g_y^2 \sin^2 \theta \sin^2 \phi + g_z^2 \cos^2 \theta$$

where g_x, g_y, and g_z are the usual *g*-value principal values ($g_x < g_y < g_z$) and θ and ϕ are the azimuthal and polar angles of the membrane normal with respect to the *g* value (molecule fixed) coordinate system. If the *g* tensor has axial symmetry ($g_x = g_y$), then the value of θ may be immediately calculated. In the more general case of a "rhombic" *g* tensor ($g_x \neq g_y \neq g_z$), it is necessary to also measure the EPR spectrum with the membrane plane parallel to the magnetic field direction in order to calculate the two angles θ and ϕ. With the membrane plane parallel to the magnetic field, disorder in the multilayer will lead to a range of *g* values being observed (a partial "powder pattern"). The EPR spectrum will stretch between two extreme *g* values, say, g_{max} and g_{min}. With the value of g_n measured with the multilayer normal to the magnetic field direction, one may then calculate the angles θ and ϕ from the following equations:

$$\sin^2\theta = \frac{g_z^2(g_z^2 - g_n^2) + g_x^2 g_y^2 - \frac{1}{4}[(g_x^2 + g_y^2 + g_z^2 - g_n^2)^2 - (g_{max}^2 - g_{min}^2)^2]}{(g_z^2 - g_x^2)(g_z^2 - g_y^2)}$$

$$\sin^2\phi = \frac{(g_z^2 - g_x^2)\sin^2\theta - (g_z^2 - g_n^2)}{(g_y^2 - g_x^2)\sin^2\theta}$$

In the event that out of plane disorder is large enough to make application of the above equations invalid (not an unusual occurrence) the simple strategy of rotating the multilayer and taking EPR spectra at, say, every 10° between 0 and $\pi/2$ will lead to clear maxima in the amplitude of an observed signal at the principal *g* values when the multilayer is tilted by an angle θ_i. Since "θ_i" can then be measured for each of the principal *g* value directions, the orientation of the EPR center is thus determined. Computer simulation of spectra is easily accomplished to confirm the results in detail.

Potentially of even more importance is the measurement of directionality of spin dipolar interactions in oriented systems. Then not only may orientations of the components with respect to the membrane be observed, but their distance apart and the direction angles from one another can also be determined.

9. References

Abragam, A., and Bleaney, B., 1970, *Electron Paramagnetic Resonance of Transition Ions*, Clarendon Press, Oxford.

Beinert, H., and Kok, B., 1964, An attempt at quantitation of the sharp light-induced electron paramagnetic resonance signal in photosynthetic materials, *Biochim. Biophys. Acta* **88**:278–288.

Bolton, J. R., and Cost, K., 1973, Flash photolysis–electron spin resonance: A kinetic study of endogenous light-induced free radicals in reaction center preparations from *Rhodopseudomonas sphaeroides, Photochem. Photobiol.* **18**:417–421.

Bolton, J. R., Clayton, R. K., and Reed, D. W., 1969, An identification of the radical giving rise to the light-induced ESR signal in photosynthetic bacteria, *Photochem. Photobiol.* **9**:209–218.

Carrington, A., and McLachlan, A. D., 1967, *Introduction to Magnetic Resonance with Applications to Chemistry and Chemical Physics*, Harper and Row, New York.

Case, G. D., and Leigh, J. S., 1974, Intramembrane location of photosynthetic electron carriers in *Chromatium* revealed by EPR interactions with gadolinium (III), 11th Rare Earth Research Conference, Traverse City, Michigan.

Case, G. D., and Leigh, J. S., 1976, Intramitochondrial position of cytochrome haem groups determined by dipolar interactions with paramagnetic cations, *Biochem. J.* **160**:769–783.

Clayton, R. K., 1966, Physical processes involving chlorophyll *in vivo*, in: *The Chlorophylls* (L. P. Vernon and G. R. Seely, eds.), pp. 609–641, Academic Press, New York.

Coleman, R., Finean, J. B., and Thompson, J. E., 1969, Structural and functional modifications induced in muscle microsomes by trypsin, *Biochim. Biophys. Acta* **173**:51–61.

Commoner, B., Heise, J. J., and Townsend, J., 1956, Light induced paramagnetism in chloroplasts, *Proc. Natl. Acad. Sci. U.S.A.* **42**:710–718.

Commoner, B., Heise, J., Lippincott, B., Norberg, R., Passonneau, J., and Townsend, J., 1957, Biological activity of free radicals, *Science* **126**:57–63.

Debrunner, P. G., Schultz, C. E., Feher, G., and Okamura, M. Y., 1975, Mossbauer study of reaction centers from *Rp. sphaeroides, Biophys. J.* **15**:226a.

Dratz, E. A., Schultz, A. J., and Sauer, K., 1966, Chlorophyll–chlorophyll interactions, *Brookhaven Symp. Biol.* **19**:303–318.

Dupont, Y., Harrison, S. C., and Hasselbach, W., 1973, Molecular organization in the sarcoplasmic reticulum membrane studied by X-ray diffraction, *Nature (London)* **244**:555–558.

Dutton, P. L., and Leigh, J. S., 1973, Electron spin resonance characterization of *chromatium* D hemes, nonheme irons and the components involved in primary photochemistry, *Biochim. Biophys. Acta* **314**:178–190.

Dutton, P. L., Leigh, J. S., and Seibert, M., 1972, Primary processes in photosynthesis: In situ ESR studies on the light-induced oxidized and triplet state of reaction center bacteriochlorophyll, *Biochem. Biophys. Res. Commun.* **46**:406–413.

Dutton, P. L., Leigh, J. S., and Reed, E. W., 1973*a*, Primary events in the photosynthetic reaction center from *Rhodopseudomonas spheroides* strain R26: Triplet and oxidized states of bacteriochlorophyll and the identification of the primary electron acceptor, *Biochim. Biophys. Acta* **292**:654–664.

Dutton, P. L., Leigh, J. S., and Wraight, C. A., 1973*b*, Direct measurement of the midpoint potential of the primary electron acceptor in *Rhodopseudomonas sphaeroides in situ* and in the isolated state: Some relationships with pH and *o*-phenanthroline, *FEBS Lett.* **36**:169–173.

Dutton, P. L., Kaufmann, K. J., Chance, B., and Rentzepis, P. M., 1975, Picosecond kinetics of the 1250 nm band of the *Rp. sphaeroides* reaction center: The nature of the primary photochemical intermediary state, *FEBS Lett.* **60**:275–280.

Duysens, L. N. M., Husikamp, W. J., Vos, J. J., and van der Hart, J. M., 1956, Reversible changes in bacteriochlorophyll in purple bacteria upon illumination, *Biochim. Biophys. Acta* **19**:188–190.

Evans, M. C. W., Lord, A. V., and Reeves, S. G., 1974, The detection and characterization by electron-paramagnetic-resonance spectroscopy of iron–sulphur proteins and the electron transport components in chromatophores from the purple bacterium *Chromatium*, *Biochem. J.* **138**:177–183.

Fajer, J., Brune, D. C., Davis, M. S., Forman, A., and Spaulding, L. D., 1975, Primary charge separation in bacterial photosynthesis: Oxidized chlorophylls and reduced pheophytin, *Proc. Natl. Acad. Sci. U.S.A.* **72**:4956–4960.

Feher, G., 1970, *EPR with Application to Selected Problems in Biology*, Gordon and Breach, New York.

Feher, G., 1971, Some chemical and physical properties of a bacterial reaction center particle and its primary photochemical reactants, *Photochem. Photobiol.* **14**:373–387.

Feher, G., Okamura, M. Y., and McElroy, J. D., 1972, Identification of an electron acceptor in reaction centers of *Rhodopseudomonas sphaeroides* by EPR spectroscopy, *Biochim. Biophys. Acta* **267**:222–226.

Feher, G., Hoff, A. J., Isaacson, R. A., and McElroy, J. D., 1973, Investigation of the electronic structure of the primary electron donor in bacterial photosynthesis by the ENDOR technique, *Biophys. J.* **13**:61a.

Feher, G., Isaacson, R. A., McElroy, J. D., Ackerson, L. C., and Okamura, M. Y., 1974, On the question of the primary acceptor in bacterial photosynthesis: Manganese substituting for iron in reaction centers of *Rhodo-*

pseudomonas sphaeroides R-26, *Biochem. Biophys. Acta* **368**:135–139.

Feher, G., Hoff, A. J., Isaacson, R. A., and Ackerson, L. C., 1975, ENDOR experiments on chlorophyll and bacteriochlorophyll *in vitro* and in the photosynthetic unit, *Ann. N. Y. Acad. Sci.* **244**:239–259.

Garcia, A., Vernon, L. P., Ke, B., and Mollenhauer, H., 1968, Some structural and photochemical properties of *Rhodopseudomonas* species NHTC 133 subchromatophore particles obtained by treatment with Triton X-100, *Biochemistry* **7**:326–332.

Gibson, J. F., Hall, D. O., Thornley, J. H. M., and Whatley, F. R., 1966, The iron complex in spinach ferredoxin, *Proc. Natl. Acad. Sci. U.S.A.* **56**:987–990.

Hopfield, J. J., 1974, Electron transfer between biological molecules by thermally activated tunneling, *Proc. Natl. Acad. Sci. U.S.A.* **71**:3640–3644.

Hutchison, C. A., and Mangum, B. W., 1958, Paramagnetic resonance absorption of naphthalene in its phosphorescent state, *J. Chem. Phys.* **29**:952–953.

Johnson, C. E., Cammack, R., Rao, K. K., and Hall, D. O., 1971, The interpretation of the EPR and Mossbauer spectra of two-iron, one-electron, iron–sulphur proteins, *Biochem. Biophys. Res. Commun.* **43**:564–571.

Kaufmann, K. J., Dutton, P. L., Netzel, T. L., Leigh, J. S., and Rentzepis, P. M., 1975, Picosecond kinetics of events leading to reaction center bacteriochlorophyll oxidation, *Science* **188**:1301–1304.

Kottis, P., and Lefebvre, R., 1964, Calculation of the electron spin resonance line shape of randomly oriented molecules in a triplet state. II. Correlation of the spectrum with the zero-field splittings. Introduction of an orientation-dependent linewidth, *J. Chem. Phys.* **41**:379–393.

Kuntz, I. D., Loach, P. A., and Calvin, M., 1964, Absorption changes in bacterial chromatophores, *Biophys. J.* **4**:227–249.

Leigh, J. S., 1970, ESR rigid-lattice line shape in a system of two interacting spins, *J. Chem. Phys.* **52**:2608–2612.

Leigh, J. S., and Dutton, P. L., 1972, The primary electron acceptor in photosynthesis, *Biochem. Biophys. Res. Commun.* **46**:414–421.

Leigh, J. S., and Dutton, P. L., 1974, Reaction center bacteriochlorophyll triplet states: Redox potential dependence and kinetics, *Biochim. Biophys. Acta* **357**:67–77.

Leigh, J. S., Maltempo, M. M., Ohlsson, P. I., and Paul, K. G., 1975, Optical, NMR and EPR properties of horseradish peroxidase and its donor complexes, *FEBS Lett.* **51**:304–308.

Loach, P. A., and Hall, R. L., 1972, The question of the primary electron acceptor in bacterial photosynthesis, *Proc. Natl. Acad. Sci. U.S.A.* **69**:786–790.

Loach, P. A., Androes, G. M., Maksim, A. F., and Calvin, M., 1963, Variation in electron paramagnetic resonance signals of photosynthetic systems with the redox level of their environment, *Photochem. Photobiol.* **2**:443–454.

Maltempo, M. M., Moss, T. H., and Cusanovich, M. A., 1974, Magnetic studies on the changes in the iron environment in *Chromatium* ferricytochrome *c'*, *Biochim. Biophys. Acta* **342**:290–305.

McElroy, J. D., Feher, G., and Mauzerall, D. C., 1969, On the nature of the free radical formed during the primary process of bacterial photosynthesis, *Biochim. Biophys. Acta* **172**:180–183.

McElroy, J. D., Feher, G., and Mauzerall, D. C., 1972, Characterization of primary reactants in bacterial photosynthesis. I. Comparison of the light-induced EPR signal (g = 2.0026) with that of a bacteriochlorophyll radical, *Biochim. Biophys. Acta* **267**:363–374.

McElroy, J. D., Mauzerall, D. C., and Feher, G., 1974, Characterization of primary reductants in bacterial photosynthesis II. Kinetic studies of the light-induced EPR signal (g = 2.0026) and the optical absorbance changes at cryogenic temperatures, *Biochim. Biophys. Acta* **333**:261–277.

Netzel, T. L., Rentzepis, P. M., and Leigh, J. S., 1973, Picosecond kinetics of reaction center bacteriochlorophyll excitation, *Science* **182**:238–241.

Norris, J. R., Uphaus, R. A., Crespi, H. L., and Katz, J. J., 1971, Electron spin resonance of chlorophyll and the origin of signal I in photosynthesis, *Proc. Natl. Acad. Sci. U.S.A.* **68**:625–628.

Norris, J. R., Druyan, M. E., and Katz, J. J., 1973, Electron nuclear double resonance of bacteriochlorophyll free radical *in vitro* and *in vivo*, *J. Am. Chem. Soc.* **95**:1680–1682.

Okamura, M. Y., Isaacson, R. A., and Feher, G., 1975, Primary acceptor in bacterial photosynthesis: Obligatory role of ubiquinone in photoactive reaction centers of *Rhodopseudomonas sphaeroides*, *Proc. Natl. Acad. Sci. U.S.A.* **72**:3491–3495.

Okamura, M. Y., Ackerson, L. C., Isaacson, R. A., Parson, W. W., and Feher, G., 1976, The primary electron acceptor in *Chromatium vinosum* (strain D), *Biophys. J.* **16**:223.

Orbach, R., 1961, Spin-lattice relaxation in rare-earth salts, *Proc. R. Soc. London Ser. A*: **264**:458–484.

Parson, W. W., and Monger, T. G., 1976, Interrelationships among excited states in bacterial reaction centers, *Brookhaven Symp. Biol.* **28**:195–212.

Parson, W. W., Clayton, R. K., and Cogdell, R. C., 1975, Excited states of photosynthetic reaction centers at low redox potentials, *Biochim. Biophys. Acta* **387**:265–278.

Prince, R. C., and Dutton, P. L., 1976, The primary acceptor of bacterial photosynthesis: Its operational midpoint potential, *Arch. Biochem. Biophys.* **172**:329–334.

Prince, R. C., Leigh, J. S., and Dutton, P. L., 1974, An electron-spin-resonance characterization of *Rhodopseudomonas capsulata*, *Biochem. Soc. Trans* **2**:950–953.

Prince, R. C., Leigh, J. S., and Dutton, P. L., 1976, Thermodynamic properties of the reaction center of *Rhodopseudomonas viridis*: *In vivo* measurement of the reaction center bacteriochlorophyll-primary acceptor intermediary electron carrier, *Biochim. Biophys. Acta* **440**:622–636.

Reed, D. W., and Clayton, R. K., 1968, Isolation of a reaction center fraction from *Rhodopseudomonas sphaeroides*, *Biochem. Biophys. Res. Commun.* **30**:471–475.

Reed, D. W., and Peters, G. A., 1972, Characterization of the pigments in reaction center preparations from *Rhodopseudomonas sphaeroides*, *J. Biol. Chem.* **247**:7148–7152.

Rockley, M. G., Windsor, M. W., Cogdell, R. J., and Parson, W. W., 1975, Picosecond detection of an intermediate in the photochemical reaction of bacterial photosynthesis, *Proc. Natl. Acad. Sci. U.S.A.* **72**:2251–2255.

Schepler, K. L., Dunham, W. R., Sands, R. H., Fee, J. A., and Abeles, R. H., 1975, A physical explanation of the EPR spectrum observed during catalysis by enzymes utilizing coenzyme B_{12}, *Biochim. Biophys. Acta* **397**:510–518.

Schwoerer, M., and Sixl, H., 1968, Optical spin polarization in the triplet state of naphthalene, *Chem. Phys. Lett.* **2**:14–19.

Smith, J. H. C., and French, C. S., 1963, The major and accessory pigments in photosynthesis, *Annu. Rev. Plant Physiol.* **14**:181–224.

Sogo, P., Jost, M., and Calvin, M., 1959, Evidence for free radical production in photosynthesizing systems, *Radiat. Res. Suppl.* 1, pp. 551–518.

Steiner, L. A., Lopes, A. D., Okamura, M. Y., Ackerson, L. C., and Feher, G., 1974, On the spatial arrangement of reaction center subunits in the bacterial membrane of *Rhodopseudomonas sphaeroides*, *Fed. Proc. Fed. Am. Soc. Exp. Biol.* **33**:1461.

Straley, S. C., Parson, W. W., Mauzerall, D. C., and Clayton, R. K., 1973, Pigment content and molar extinction coefficients of photochemical reaction centers from *Rhodopseudomonas sphaeroides*, *Biochim. Biophys. Acta* **305**:597–609.

Thurnauer, M. C., Katz, J. J., and Norris, J. R., 1975, The triplet state in bacterial photosynthesis: Possible mechanisms of the primary photo-act, *Proc. Natl. Acad. Sci. U.S.A.* **72**:3270–3274.

Tiede, D. M., Prince, R. C., Reed, G. H., and Dutton, P. L., 1976, EPR properties of the electron carrier intermediate between the reaction center bacteriochlorophylls and the primary acceptor in *Chromatium vinosum*, *FEBS Lett.* **65**:301–304.

Vermiglio, A., and Clayton, R. K., 1976, Orientation of chromophores in reaction centers of *Rhodopseudomonas sphaeroides*, evidence for two absorption bands of the dimeric primary electron donor, *Biochim. Biophys. Acta* **449**:500–515.

Walker, L. R., Wertheim, G. K., and Jaccarino, V., 1961, Interpretation of the ^{57}Fe isomer shift, *Phys. Rev. Lett.* **6**:101.

Walker, L. R., Wertheim, G. K., and Jaccarino, V., 1964, in: *Mossbauer Effect: Principles and Applications* (G. K. Wertheim, ed.), p. 54, Academic Press, New York.

Wasserman, E., Snyder, L. C., and Yager, W. A., 1964, ESR of the triplet states of randomly oriented molecules, *J. Chem. Phys.* **41**:1763–1772.

CHAPTER 24

Protonation and the Reducing Potential of the Primary Electron Acceptor

Roger C. Prince and P. Leslie Dutton

1. Introduction

The primary electron-transfer reactions of bacterial photosynthesis occur within the reaction center (RC) bacteriochlorophyll (Bchl) protein. Following the absorption of light energy, the RC primary donor (a Bchl dimer, termed P) becomes oxidized, while the primary acceptor becomes reduced (the primary acceptor is probably an iron–quinone compound, but because a complete characterization is not available, it is usually designated X). Further details of the possible chemical natures of these primary reactants are discussed more fully elsewhere in this book; this chapter will examine their electrochemical properties.

It is now realized that at least two "primary" reactions occur in the RC Bchl/bacteriopheophytin (Bpheo) complement before the reduction of X. Recent evidence (Dutton et al., 1975, 1976) indicates that the electron is ejected from the Bchl dimer of the RC within 10 psec of a flash of light, and yet does not arrive at the primary acceptor until some 100–200 psec later (Kaufmann et al., 1975; Rockley et al., 1975). During this time, it is assumed to reside on what must be an electronegative intermediary electron carrier (I), which Fajer et al. (1975) suggested may be Bpheo. If the forward progress of the electron from I to the primary acceptor (X) is blocked, however, the lifetime of the reduced intermediate (I^-) is only approximately 10

nsec (Parson et al., 1975) and the electron returns to P^+. The system then returns to the ground state, probably by way of a triplet state of the Bchl.

The primary acceptor (X) is thus the agent that confers stability on the early events of bacterial photosynthesis; once the electron arrives on X, the resulting P^+IX^- state has an intrinsic lifetime of some tens of milliseconds before the electron on X^- returns to P^+. This lifetime is suitably in excess of the times for the delivery of an electron to P^+ by cytochrome c oxidation (10^{-7}–10^{-4} sec half-time, depending on the species) or from X^- to secondary acceptors (10^{-5}–10^{-4} sec half-time), resulting in the almost unit quantum yield for the light reaction as determined for RC or cytochrome c oxidation (Wraight and Clayton, 1973; Loach and Sekura, 1968; Parson, 1968; Seibert and DeVault, 1970; Parson and Case, 1970).

This chapter will be concerned chiefly with the electrochemical properties of the primary acceptor that allow it to fulfill this role of ensuring not only the essentially irreversible light reaction, but also the efficient reduction of the cyclic electron-transfer system. It will compare the redox properties of the reactants that are functional in the time scale of light-driven electron flow with those obtained by equilibrium techniques. It will be concerned mainly with the properties of the primary acceptors of the purple bacteria (Rhodospirillaceae and Chromatiaceae), which seem to be very similar, but will also discuss recent work on the primary acceptors of the green bacteria (Chlorobiaceae), which seem to be very different, perhaps

Roger C. Prince and P. Leslie Dutton · Johnson Research Foundation and Department of Biochemistry and Biophysics, University of Pennsylvania, Philadelphia, Pennsylvania 19104

more akin to the acceptor of green plant photosystem I. The redox properties of the accompanying primary donors (P) are also briefly reviewed.

2. Oxidation–Reduction Midpoint Potentials of the Primary Reactants as Measured at Equilibrium

As we have already mentioned, the primary light reaction of bacterial photosynthesis involves the oxidation of the primary donor (P) and, via the intermediate (I), the reduction of the primary acceptor (X). Clearly, this cannot occur either if the primary acceptor is reduced or if the primary donor is oxidized, prior to light activation. Thus, photochemistry leading to the stable formation of P^+ can occur only within a range of environmental redox potentials defined by the midpoint potentials (E_m) of the primary donor and the primary acceptor. Many workers have exploited this behavior, usually using redox potentiometry combined with simultaneous optical or ESR spectroscopy, to measure the E_m values of P and X in a variety of organisms. This technique was recently reviewed by Dutton and Wilson (1974), and is also dealt with in Chapters 28 and 32. One point that should be emphasized is that the successful use of the technique can measure only *equilibrium* values of redox couples, which thereby include all the thermodynamically favorable physical events coupled to the oxidation or reduction reaction; these may include conformational changes, charge or ionic effects, and protonation reactions. Such coupled reactions may occur at rates that are many orders of magnitude slower than the intrinsic rates of electron transfer between the redox component and its electron donors and acceptors. In a kinetic situation with an appropriate redox potential driving force, a component may be reduced or oxidized, ignoring the coupled events that would occur at equilibrium. If such considerations applied to a component under study, the half-reduction midpoint potential obtained at equilibrium would have little relevance to the value that is functional on the time scale of physiological electron flow. This consideration is especially relevant with respect to the light reaction, which is very rapid and has sufficient energy to drive the initial reactions to near completion; as we shall see, it has been experimentally proved to be of importance in the case of the primary acceptor.

Table 1 collects the equilibrium midpoint potentials of the RC Bchl, P, of those bacteria for which they have been determined. The E_m value is very similar in a wide range of bacteria, although the value for *Chlorobium limicola f. thiosulfatophilum* is con-

siderably lower than values for the Rhodospirillaceae and Chromatiaceae. In *Rhodospirillum rubrum* (Loach *et al.*, 1963), *Rhodopseudomonas sphaeroides* (Jackson *et al.*, 1973), *Chromatium vinosum* (Case and Parson, 1971), and *Rp. viridis* (Prince *et al.*, 1976), the measured midpoint potential of the couples are essentially *independent* of pH, indicating that the oxidation of P involves only an electron at physiological values of pH:

$$P^+ + e^- \rightleftharpoons P$$

The midpoint potential of the primary acceptor has proved more controversial, and Table 2 presents values reported in the literature. The values reported for different species vary over a wide range, and in those cases in which it has been investigated, the midpoint potential varies with pH. We have recently found that chromatophores are stable over a much wider range of pH than is usually considered; indeed, they are stable from pH 4.7 to 11.0 as judged by a variety of criteria including cytochrome c photooxidation and reduction, cytochrome b reactions, and the carotenoid bandshift. Figure 1 shows a typical titration of the primary acceptor of *Rp. sphaeroides* at pH 11.0; the titration was completely reversible, independent of mediator concentration, and gave similar results when either cytochrome c_2 photooxidation, RC photooxidation, or the carotenoid bandshift was used as a monitor of photochemistry. This titration, like others at lower pH values, displays an $n = 1$ value, indicating a one-electron oxidation–reduction reaction. The E_m values obtained from a number of similar titrations over a wide range of pH are shown in Fig. 2A, which also includes the earlier data of Jackson *et al.* (1973) and Dutton *et al.* (1973). At physiological values of pH, the midpoint potential varies with pH by −60 mV/pH unit, implying that in the physiological pH range, the equilibrium reduction of the primary acceptor not only involves the receipt of an electron, but also requires the accompanying coupled binding of a proton, as follows:

$$X + e^- + H^+ \rightleftharpoons XH$$

At higher values of pH, however, the midpoint potential becomes essentially independent of pH, indicating that the equilibrium reduction of the primary acceptor now involves only an electron:

$$X + e^- \rightleftharpoons X^-$$

The intersection of the E_m/pH line of −60 mV/pH unit with that of 0 mV/pH unit identifies a pK of the reduced form of the primary acceptor, in this case at approximately pH 9.8.

$$XH \rightleftharpoons X^- + H^+$$

Table 1. Equilibrium Oxidation–Reduction Midpoint Potentials of the Reaction Center Bacteriochlorophyll

Organism	Midpoint potential (mV)	pH	Method[a]	pH dependency	References
Rp. sphaeroides	+450	7	RC changes, chromatophores	ND	Dutton and Jackson (1972)
	+440	7	RC changes, chromatophores	0 mV/pH (pH 6–8)	Jackson et al. (1973)
Rp. capsulata	+440	7	RC changes, chromatophores	ND	E. H. Evans and Crofts (1974)
Rp. gelatinosa	+440	7.4	RC changes, chromatophores, 77°K	ND	Dutton (1971)
Rp. viridis	+390 +450	7	RC changes, subchromatophore particles	0 mV/pH	Thornber and Olson (1971)
	+480	7.8	Delayed fluorescence, chromatophores	ND	Carithers and Parson (1975)
	+500	8	RC changes, chromatophores	0 mV/pH (pH 5.3–10.3)	Prince et al. (1976)
Rs. rubrum	+440	7	RC changes (EPR), chromatophores	0 mV/pH (pH 5–10)	Loach et al. (1963)
	+439	7.4	RC changes, chromatophores	ND	Kuntz et al. (1964)
	+440	7.4	RC changes, chromatophores	ND	Remennikov et al. (1975)
Chr. vinosum	+490	7.5	RC changes, chromatophores	ND	Cusanovich et al. (1968)
	+475	7.4	RC changes, chromatophores, 77°K	ND	Dutton (1971)
	+486	7.4	RC changes, chromatophores, 77°K	ND	Seibert and DeVault (1971)
	+490	7.7	RC changes, chromatophores	−23 mV/pH (pH 6–9)	Case and Parson (1971)
E. shaposhnikovii	+390	7.4	RC changes, chromatophores	ND	Remennikov et al. (1975)
"Chloropseudomonas ethylica 2K"[c]	+240	7.4	RC changes, Bchl–RC complex[d]	ND	Fowler et al. (1971)
Chl. limicola f. thiosulfatophilum	+330	6.5	RC changes, "chromatophores"[e]	ND	Knaff et al. (1973)
	+250	6.8	RC changes, Bchl–RC complex[d]	ND	Prince and Olson (1976)

[a] All redox potentiometry was carried out at room temperature, and in most cases photochemistry was assayed optically at this temperature. Exceptions are noted.

[b] ND: Not determined.

[c] "Chloropseudomonas ethylica 2K" is now recognized as a symbiotic association of a colorless organism with a green sulfur bacterium that has been variously classified as *Chlorobium limicola* (Gray et al., 1972, 1973) or *Prosthecochloris aestuarii* (Shioi et al., 1976). See Chapter 2.

[d] The Bchl–RC complexes of the Chlorobiaceae are more akin to chromatophores of the Rhodospirillaceae than to RCs from this group. The Bchl–RC complexes are large, and contain many components in addition to the RC itself [Fowler et al. (1971), Prince and Olson (1976), Olson et al. (1976)].

[e] "Chromatophores" from the Chlorobiaceae are considerably larger than chromatophores from the Rhodospirillaceae, containing in addition the chlorobium chlorophyll complex peculiar to this group [see Fowler et al. (1971)].

Table 2. Equilibrium Oxidation–Reduction Midpoint Potentials of the Primary Acceptor

Organism	Midpoint potential (mV)	pH	Method[a]	pH dependency	References
Rp. sphaeroides	−25	7.3	Cyt c_2 photooxidation, whole cells	ND	Loach (1966)
	−65 or −90 (hysteresis)	8.0	Fluorescence yield, whole cells[c]	ND	Cramer (1969)
	−20	7.0	Cyt c_2 photooxidation and carotenoid bandshift, chromatophores	ND	Dutton and Jackson (1972)
	−20	7.0	Cyt c_2 photo-oxidation, chromatophores	−60 mV/pH (pH 6–8)	Jackson *et al.* (1973)
	−20	7.0	RC photooxidation (EPR), 10°K, chromatophores	−60 mV/pH (pH 6–8)	Dutton *et al.* (1973)
Rp. capsulata	−25	7.0	Cyt c_2 photooxidation, chromatophores	−60 mV/pH (pH 6.5–8)	E. H. Evans and Crofts (1974)
	−30	7.0	RC photooxidation (EPR), 10°K, chromatophores	ND	Prince *et al.* (1974)
Rp. gelatinosa	−140	7.4	RC photooxidation (77°K), chromatophores	ND	Dutton (1971)
Rp. viridis	−95	7.0	Cyt c_2 photooxidation, chromatophores	−30 mV/pH (pH 6–9)	Cogdell and Crofts (1972)
Rs. rubrum	−44	7.4	RC changes, chromatophores	ND	Kuntz *et al.* (1964)
	−22	7.62	RC changes, chromatophores	ND	Loach (1966)
	−145	8.0	Fluorescence yield, whole cells[c]	ND	Cramer (1969)
	−140	7.4	RC changes (EPR), 10°K, chromatophores	ND	Dutton and Baltscheffsky (1972)
Chr. vinosum	−130	7.5	Cyt c photooxidation, chromatophores	ND	Cusanovich *et al.* (1968)
	−160	8.0	Fluorescence yield, whole cells[c]	ND	Cramer (1969)
	−134	7.4	Cyt c and RC photooxidation, 77°K, chromatophores	ND	Dutton (1971)
	−129	7.4	Cyt c and RC photooxidation, 77°K, chromatophores	ND	Seibert *et al.* (1971)
	−130	7.7	Cyt c photooxidation, chromatophores	−40 mV/pH (pH 6–9)	Case and Parson (1971)

(continued)

Table 2. (continued)

Organism	Midpoint potential (mV)	pH	Method[a]	pH dependency	References
	−134	7.4	RC changes, 77°K, chromatophores	ND	Seibert and DeVault (1971)
	−100	7.0	RC changes, chromatophores	−60 mV/pH (pH 6–8)	Jackson et al. (1973)
	−120	7.4	g = 1.82 signal (EPR), 8°K, chromatophores	ND	Dutton and Leigh (1973)
	−135	8.0	g = 1.82 signal (EPR), 8°K, chromatophores	−98 mV/pH[d] (pH 6.5–8)	M. C. W. Evans et al. (1974)
Chl. limicola f. thiosulfatophilum	−130	6.5	Cyt c photooxidation, "chromatophores;"[e]	0 mV/pH (pH 6.5–8.5)	Knaff et al. (1973)
	~−540	10.3	Cyt c photooxidation, Bchl–RC complex[f]	Apparently independent	Olson et al. (1976)

[a] All redox potentiometry was carried out at room temperature, and in most cases, photochemistry was assayed using optical spectroscopy at this temperature. Exceptions are noted. (Cyt) Cytochrome.

[b] ND: Not determined.

[c] The work of Cramer (1969) used cells harvested in the stationary growth phase. All other work appears to have used cells or chromatophores obtained during the exponential growth phase.

[d] The pH dependency reported by M. C. W. Evans et al. (1974) is appreciably greater than the −60 mV/pH unit expected. However, their experiments were performed only at pH 6.5 and 8.

[e] "Chromatophores" from the Chlorobiaceae are considerably larger than those from the Rhodospirillaceae, containing in addition the chlorobium chlorophyll–protein complex peculiar to this group [see Fowler et al. (1971)]. The measurements of Knaff et al. (1973) used continuous illumination, which, as they pointed out, could lead to erroneous results. In the light of more recent data (Prince and Olson. 1976; Olson et al., 1976), this was probably the case.

[f] The Bchl–RC complex of the Chlorobiaceae is more akin to a chromatophore from the Rhodospirillaceae than a RC from this group [see Fowler et al. (1971)].

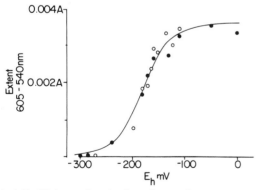

Fig. 1 Equilibrium redox titration of the primary acceptor of *Rhodopseudomonas sphaeroides* Ga at pH 11. Chromatophores (20.4 M Bchl) were suspended in 20 mM glycine, 100 mM KCl, pH 11.0, with 7 μM of each of the following redox mediators; 2,3,5,6-tetramethyl phenylenediamine, phenazine methosulfate, phenazine ethosulfate, pyocyanine, and 2-OH-1,4-naphthaquinone. The addition of 100 μM each of methyl and benzyl viologens had no effect on the titrations. The titrations were performed as described by Dutton and Wilson (1974), the equilibrium midpoint potential of the primary acceptor being measured indirectly from the amplitude of photochemistry following a "single turnover" flash of light as a function of the redox potential measured before the flash. (●) Reductive titration; (○) oxidative titration.

Very similar results have been obtained with chromatophores of *Rs. rubrum* (Fig. 2B), *Chr. vinosum* (Fig. 2C), and *Rp. viridis* (Fig. 2D). These figures also include the data of other workers, and the agreement of data from the several sources and techniques emphasizes the repeatability of this method. Interestingly, although the pK of the reduced form of the primary acceptor is at approximately pH 7.8 in *Rp. viridis*, pH 8.0 in *Chr. vinosum*, pH 8.8 in *Rs. rubrum*, and pH 9.8 in *Rp. sphaeroides*, the equilibrium midpoint potentials above the pKs are very similar, being −150 mV in *Rp. viridis*, −160 mV in *Chr. vinosum*, −180 mV in *Rp. sphaeroides*, and −200 mV in *Rs. rubrum*. In addition, the primary acceptors of *Rp. sphaeroides*, *Chr. vinosum*, and *Rp. viridis* (Dutton et al., 1973; Dutton and Leigh, 1973; Prince et al., 1976) have similar EPR spectra, although we have not yet detected this signal in *Rs. rubrum*. With the midpoint potential of the unprotonated couple, this may represent a universal property of the primary acceptor in the Rhodospirillaceae and Chromatiaceae. It is noteworthy that Figs. 2C and D resolve the anomalous E_m/pH relationships of −40 mV/pH unit and −30 mV/pH unit found by

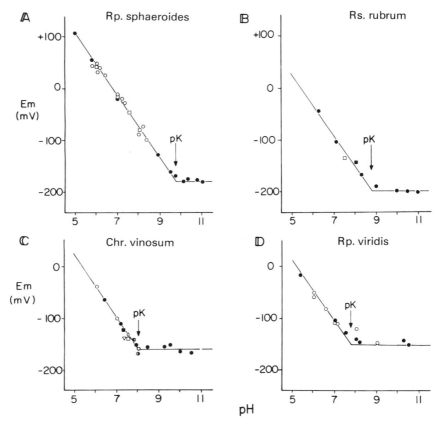

Fig. 2. Equilibrium midpoint potentials of the primary acceptors of *Rhodopseudomonas sphaeroides*, *Rhodospirillum rubrum*, *Chromatium vinosum*, and *Rp. viridis* as functions of pH. The midpoints were determined in experiments similar to those in Fig. 1. The chromatophores were allowed to equilibrate with the pH buffer for 20–30 min prior to each experiment, which each took about an hour. The pH of the mixture was measured before and after each experiment; the value never varied by more than ± 0.1 pH unit. (A) *Rp. sphaeroides*—the points are from three sources: (●) Prince and Dutton (1976); (○) Dutton *et al.* (1973); (□) Jackson *et al.* (1973). (B) *Rs. rubrum*—the points are from three sources: (●) Prince and Dutton (1976); (□) Dutton and Baltscheffsky (1972); (■) Cramer (1969). (C) *Chr. vinosum*—the points are from several sources: (●) Prince and Dutton (1976); (○) Jackson *et al.* (1973); (◇) Cusanovich *et al.* (1968); (□) Dutton (1971); (▽) Seibert *et al.* (1971); (◑) Cramer (1969); (◐) Case and Parson (1971); (■) Dutton and Leigh (1973). (D) *Rp. viridis*—the points are from two sources; (○) Cogdell and Crofts (1972); (●) Prince *et al.* (1976).

Case and Parson (1971) and Cogdell and Crofts (1972) in *Chr. vinosum* and *Rp. viridis*, because these workers averaged their data between pH 6 and 9, and hence over the pKs revealed in Fig. 2.

3. The Nature of the Primary Acceptor Redox Couple during Light-Induced Electron Flow

Although the equilibrium reduction of the primary acceptor at physiological pH involves a proton, there is no indication that this proton is involved in the time scale of light-induced electron flow. The light-induced reduction of the primary acceptor is a very rapid reaction ($t_{1/2}$ 100–200 psec, Kaufmann *et al.*, 1975; Rockley *et al.*, 1975), and its subsequent reoxidation and the reduction of ubiquinone (UQ) occur in the 50- to 400-μsec range at physiological pH (Halsey and Parson, 1974; Petty and Dutton, 1976a). This latter reaction is sufficiently rapid that the protonation of the reduced primary acceptor does not happen if secondary electron flow to UQ occurs. Under these conditions, a proton is taken by the chromatophore from its external aqueous environment (Cogdell *et al.*, 1972; Halsey and Parson, 1974; Petty and Dutton, 1976a), but this is associated with the reduction of the secondary UQ, not the primary acceptor. This means that the kinetically important midpoint couple of the primary acceptor is X/X$^-$, not the X/XH couple measured at equilibrium. As such, the *operational* midpoint potential will be measured at equilibrium beyond the pK of the reduced form (i.e., -175 ± 25 mV, depending on species*). This is substantially more negative than the values measured at equilibrium at physiological pH (see Table 2).

One might also ask the nature of the operating primary acceptor redox couple. It is generally accepted that the native primary acceptor is a quinone (Q)– iron complex (Bolton and Cost, 1973): UQ–Fe in *Rp. sphaeroides* (Okamura *et al.*, 1975) and menaquinone–Fe in *Chr. vinosum* (Okamura *et al.*, 1976). There is evidence that when the primary acceptor is reduced, the electron resides mainly on the Q, since the redox state of the Fe does not appear to change during the reduction of X (Debrunner *et al.*, 1975), and the Fe can even be replaced by Mn (Feher *et al.*, 1974). Since redox titrations of the primary acceptor, such as that in Fig. 1, routinely indicate that only one electron is involved in the reduction of the primary acceptor, regardless of whether there is an accompanying proton, the evidence suggests that the redox couple of the primary acceptor involves a stable semiquinone, and that the role of Fe may be to stabilize this species. On the time scale of electron flow, the semiquinone radical would be the anionic form (Q$^{\bar{\cdot}}$). The reason this chemical species should have a different pK in the different bacterial species is not clear. It mighty reflect fundamentally different properties of the Qs of various species, or perhaps different local environments of the primary acceptors in the various organisms. Fromhertz and Masters (1974) demonstrated that the pK of an indicator dye may be shifted as much as plus or minus 4 pH units by the presence of nearby charged groups, and such considerations might well apply to the primary acceptor.

4. Functional Importance of the Unprotonated Reduced Primary Acceptor

The role of the primary acceptor (X) in the primary events of the light reaction is to effectively stabilize the energy of the absorbed photon as redox potential free energy and energy stored in the physical

separation of charge. As such, its reduction by the primary donor must be an essentially irreversible reaction under physiological conditions. The identification of the unprotonated primary acceptor redox couple as the functional couple in these events allows a thermodynamic understanding of how this is achieved. Figure 3 shows the midpoint potentials of some of the known components of the light reaction and associated cyclic electron-flow system of *Rp. sphaeroides*. An incident photon of 870-nm wavelength represents an energy of 1.42 eV, and if this photon is absorbed by the RC Bchl dimer (P), it excites it to the lowest excited singlet state. This then reduces the intermediate (I), identified by picosecond spectroscopy, within 10 psec. Fajer *et al.* (1975). suggested that I may be a single molecule of Bpheo, which in a variety of organic solvents has an E_m of -550 mV in the monomeric form. The difference in midpoint potential between this and the P$^+$/P couple (see Table 1) is approximately 1 V, which, as Fajer *et al.* (1975) pointed out, would represent about 70% of the incident light energy captured as redox potential chemical energy. The reduced intermediate in turn reduces the primary acceptor, again releasing a substantial amount of chemical free energy (if E_m of I/I$^-$ = -550 mV, the ΔE_m between I/I$^-$ and X/X$^-$ would be 370 mV), which would certainly ensure that the reaction is driven essentially to completion. With such a large free-energy drop from the excited singlet state of the Bchl dimer, through I to X, the reaction is virtually irreversible (half-time of forward reaction $\approx$ 200 psec; of reverse reaction, tens of milliseconds), and the reduced primary acceptor is thus a relatively stable species. *In vivo*, the reverse reaction is rendered even less probable by the much more rapid re-reduction of P$^+$ by c-type cytochromes (half-time varying between 100 nsec and 300 μsec in different species).

The unprotonated primary acceptor thus has no choice but to reduce the secondary acceptor UQ. There is some evidence (Petty and Dutton, 1967a) (see also Chapter 28) that this involves the reduction of UQ to the protonated ubisemiquinone (i.e., to UQ·H), at least at pH 7. The midpoint potential of the UQ/protonated ubisemiquinone couple is unknown, but an estimate on thermodynamic grounds would be -25 to -125 mV. The reduced secondary acceptor then reduces a cytochrome b. The midpoint potential of this cytochrome, like the primary acceptor, varies by -60 mV/pH unit in the physiological pH range (Petty and Dutton, 1976b), but in contrast to the primary acceptor, the reduction of the cytochrome b does appear to involve a proton on the time scale of electron flow (Petty and Dutton, 1976b) (see also Chapter 28). The functional b-cytochrome couple is thus the same as that measured at equilibrium (i.e., cyt

---

* It is important to note that this value of -175 ± 25 mV is effectively the upper limit for the operating midpoint potential. It has accounted for the lack of protonation in the physiological time range, but if other events, such as ionic changes or conformational changes that do not occur on the functional time scale, are still occurring at equilibrium beyond the pK, the operating midpoint potential may be still lower.

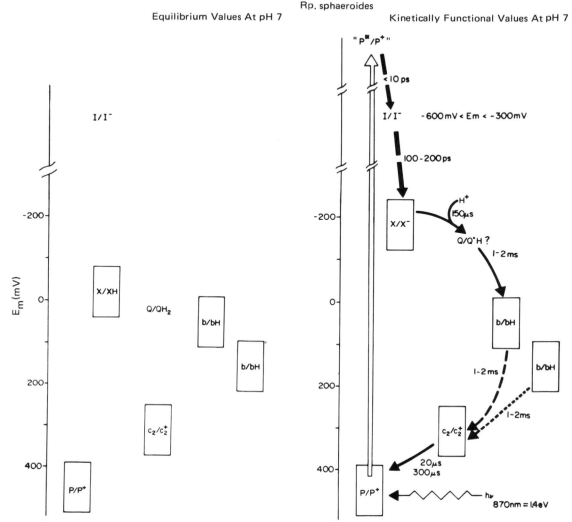

Fig. 3. Equilibrium and kinetically functional midpoint potentials of components involved in the cyclic electron-flow system of *Rhodopseudomonas sphaeroides*. The early events in the cycle are discussed in the text; the cytochrome reactions are discussed more fully in Chapter 28.

b/cyt b H) at pH 7, where the midpoint potential is +50 mV. The ΔE_m between it and the primary acceptor is thus 230 mV, which is enough to again drive the reduction effectively to completion.

The midpoint potentials of the electron-transfer components between P and cytochrome b are thus appropriate to ensure than each photon absorbed by the RC is effectively utilized and its energy converted to usable chemical–potential free energy. In addition to this redox free energy of 630 mV [the difference in midpoint potential (ΔE_m) between the P$^+$/P and X/X$^-$ couples], there has also been a charge separation, since an electron has moved from P to X.

It seems probable that the free energy available in the I-to-X redox span drives this physical separation of the positive and negative charges involved in the final P$^+$IX$^-$ state. At the chromatophore membrane level, it appears that the RCs of many species are oriented such that X is close enough to the outer aqueous interface to donate an electron to the secondary UQ, which then takes up a proton from the outer aqueous phase. In *Rp. sphaeroides*, this has been measured to be 1.0 H$^+$/RC per turnover (Petty and Dutton, 1976a) indicating that all electrons arrive at the outer surface. On the other hand, cytochrome c_2, which serves to re-reduce P$^+$, resides at the inner membrane–aqueous

Table 3. Redox Spans of the Light Reaction in *Rhodopseudomonas sphaeroides*, *Rhodospirillum rubrum*, *Chromatium vinosum*, and *Rhodopseudomonas viridis*

Organism	E_m (mV)		
	P/P^{+a}	X$^-$/X^a	ΔE_m (mV)
Rp. sphaeroides	+450	−180	630
Rs. rubrum	+450	−200	650
Chr. vinosum	+490	−160	650
Rp. viridis	+500	−150	650

[a] P/P$^+$ represents the RC Bchl couple; X$^-$/X represents the unprotonated primary acceptor couple as measured at equilibrium above the pK of the reduced form. The values are taken from Table 1 and Fig. 2.

interface (Prince *et al.*, 1975), and carotenoid band-shift studies (Jackson and Dutton, 1973) suggest that P occupies a place within the membrane dielectric. Thus, at least in *Rp. sphaeroides*, the RC, and its immediate electron donors and acceptors, spans the chromatophore membrane, and is thus electrogenic. During steady-state illumination, the membrane potential across the chromatophore membrane has been reported to be some 200 mV [inside *minus* outside (Jackson and Crofts, 1969; Casadio *et al.*, 1974)]. The large ΔE_m between that predicted for the I/I$^-$ and X/X$^-$ couples would appear to be sufficient to sustain normal forward operation of the light reaction against such a gradient without loss of quantum efficiency, especially if the primary light reaction operates across only a part of the membrane dielectric, and hence against only a part of the membrane potential. Following single turnover activation, when there is little or no preexisting membrane potential, the light reaction would generate the membrane potential, and it is conceivable that this would give the bacterium a substantial extra amount of biochemically usable energy per photon, in addition to the 630 mV available in the P-to-X redox span.

It is noteworthy that the redox span between the reaction center P$^+$/P couple and the primary acceptor X/X$^-$ couple is essentially identical in all the four species shown in Fig. 2 (see Table 3), even though *Rp. viridis* contains Bchl *b* (Eimhjellen *et al.*, 1963), while the others contain Bchl *a*. While the long-wavelength maximum of the RC Bchl *a* is around 870 nm, the maximum of Bchl *b* in the *Rp. viridis* RC is at approximately 985 nm (Holt and Clayton, 1965), indicating that the energy of the absorbed photon is only 1.29 eV in *Rp. viridis*, while it is 1.42 eV in those species containing Bchl *a*. If the quantum efficiency in *Rp. viridis* is the same as in *Rp. sphaeroides*

(1.02 ± 0.04, Wraight and Clayton, 1973), this might imply that the E_m of the I/I$^-$ couple in *Rp. viridis* would be appreciably more positive than it is in the Bchl-*a*-containing species. Recent work on *Rp. viridis* (Prince *et al.*, 1976) measured the E_m of the I/I$^-$ couple as −400 mV, which is in line with such predictions. Interestingly, the ΔE_m between the P$^+$/P and I/I$^-$ redox couples is 900 mV, which is approximately 70% of the energy available in the incident photon; three somewhat different approaches to calculating the maximal efficiency of the conversion of radiant energy to chemical free energy predict an approximate 70% conversion efficiency (Duysens, 1958; Ross and Calvin, 1967; Knox, 1969; Crofts *et al.*, 1971).

To return to the operating potential of the primary acceptor, alternative suggestions have been made to assign a lower midpoint potential to the primary acceptor than that measured at equilibrium at pH 7. A recent report by Loach *et al.* (1975) suggested that the equilibrium-measured midpoint potential of the "true" primary acceptor of *Rs. rubrum* and *Rp. sphaeroides* is as low as −370 mV at pH 7, but that this can be measured only under totally dark conditions. Even then, only one turnover of photochemistry can occur at such low potentials, due to a presumed conformational change preventing reoxidation of the primary acceptor. Because, in the view of these authors, photochemistry is irreversible if the secondary acceptor is reduced, titrations that employ light in the measuring beam of the spectrophotometer measure only the secondary acceptor, which has the E_m that previous workers (see Table 2) have attributed to the primary acceptor.

Their suggestion may be contrasted with ours in that they envisage that the primary acceptor has a low midpoint potential that functions in a simple equilibrium manner, while we suggest that the kinetically active low midpoint potential primary acceptor couple is significantly different from the protonated couple measured at equilibrium at physiological values of pH. There is, however, some uncertainty regarding the redox dyes used by Loach *et al.* (1975); in view of the high midpoint potentials of their mediators [apart from the oxidation products of dithionite itself, none has an E_m of lower than −46 mV at pH 7.0 (see Clark, 1960)] and their poor mediation characteristics (see also Clark, 1960), it is possible that their data may be the result of poor mediation between the primary acceptor and the added redox indicators (see Chapter 32; Parson and Cogdell, 1975), rather than a result of the suggested conformational change. Indeed, when high concentrations of flavin mononucleotide, or low concentrations of viologen dyes, were used as mediators (Loach, 1976), the low potential activity rapidly disappeared (Loach, 1976), suggesting that perhaps mediation did occur with these dyes. Furthermore, the problem of measuring-beam illumination or even background room light as actinic sources does not seem to be important under typical experimental conditions. In *Chr. vinosum*, the light-induced oxidation of cytochrome c_{553} is irreversible at 77°K, so if this cytochrome is reduced prior to activation at low temperatures, only one turnover of the RC can occur before it assumes an incompetent cytochrome c_{553}^+–P–X$^-$ state (see Chapter 28). Nevertheless, titrations measuring reactions at 77°K give similar results to experiments at room temperature, at which light-induced cytochrome c_{553} oxidation is reversible (see Table 2).

5. Inhibition by *o*-Phenanthroline

The identification of the unprotonated primary acceptor as the active species in electron flow also suggests how some inhibitors, such as 1,10-diazaphenanthrene (*o*-phenanthroline), may act. *o*-Phenanthroline is a well-known inhibitor between the primary and secondary acceptors (Parson and Case, 1970; Jackson *et al.*, 1973), although its mode of action is obscure. Jackson *et al.* (1973) demonstrated that it caused an apparent positive shift in the equilibrium midpoint potential of the primary acceptor in both *Rp. sphaeroides* and *Chr. vinosum*, and a similar result was obtained with *Rp. viridis* (Cogdell and Crofts, 1972). Figure 4 shows that this effect originates in a shift of the pK of the reduced form to a more alkaline value. A simple explanation for this would be that *o*-phenanthroline assists the protonation of the primary acceptor at alkaline pH, and it is tempting to speculate (Prince and Dutton, 1976) that the mode of operation lies in the proximity of protonated *o*-

phenanthroline (pK$_{aq}$ ≈ 4.9, Oettmeier and Grewe, 1974), to the primary acceptor, which might be able to interact with X$^-$ much more rapidly than a free proton could. If this were faster than the rate of reduction of UQ, the functional primary acceptor couple would be X/X$^-$-phenanthroline-H$^+$, which might have an appreciably higher midpoint potential than the X/X$^-$ couple. The extent of secondary UQ reduction would then be drastically reduced. If the protonated form of *o*-phenanthroline is the active species, it would explain the high concentrations of the inhibitor required for inhibition. If the pK in the membrane is approximately the same as that in aqueous solution, 2 mM *o*-phenanthroline at pH 7 would be only 20 μM protonated *o*-phenanthroline.

6. Photoreduction of Viologen Dyes and the Possibility of Direct Photoreduction of NAD$^+$

The much lower operational midpoint potential of the primary acceptor (particularly for *Rp. sphaeroides* and *Rs. rubrum*), compared with the equilibrium value usually considered, would also increase the likelihood of the primary acceptor being able to reduce low-potential (<−300 mV) redox dyes such as the viologens (Govindjee *et al.*, 1974; Silberstein and Gromet-Elhanan, 1974) and NAD$^+$ (Govindjee and Sybesma, 1972) under rather specialized nonphysiological conditions. The reduction appears to happen only when the cyclic electron-transport chain is almost completely filled by electrons from a reducing agent, such as ascorbate plus dichlorophenol indophenol (E_m + 217 mV; pH 7.0), in the light. Under these conditions, the reoxidation of the photoreduced primary acceptor is prevented by the prior reduction of the secondary acceptor (Bales and Vernon, 1963). We may consider two alternatives for the fate of the light-produced X$^-$ with respect to viologens or NAD$^+$ and protons. The first entails reaction with the viologens or NAD$^+$ to approach a redox equilibrium between X/X$^-$ and the viologens. The second involves the protonation of X$^-$, which under these conditions can be considered as irreversible in the absence of an oxidant for the XH. If the first situation predominates completely, the final amount of viologen or NAD$^+$ reduced will be governed by its E_m and its experimental concentration, assuming excess ascorbate. If the rate of protonation is significant with respect to the rate of viologen or NAD$^+$ reduction, the termination of the latter reaction will be a measure of the competing protonation rate of X$^-$. Although a rapid reaction between the primary acceptor of green

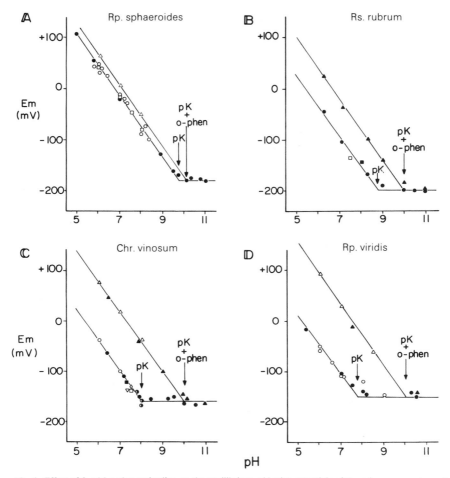

Fig. 4. Effect of 2 mM o-phenanthroline on the equilibrium midpoint potentials of the primary acceptors of *Rhodopseudomonas sphaeroides*, *Rhodospirillum rubrum*, *Chromatium vinosum*, and *Rp. viridis*. The figure includes the data of Fig. 2, with the following additional data: in A: (△) Jackson *et al.* (1973; in B: (▲) Prince and Dutton (1976); in C: Jackson *et al.* (1973), (▲) Prince and Dutton (1976); in D: (△) Cogdell and Crofts (1972), (▲) Prince *et al.* (1976). All these additional points were obtained in the presence of 2 mM o-phenanthroline. We have extrapolated the data in Fig. 4A to show a pK in the presence of o-phenanthroline by analogy with the three other figures.

plant photosystem I and added viologens has been reported (Ke, 1973), a similar rapid interaction in photosynthetic bacteria remains to be established. The data of Govindjee *et al.* (1974) and Silberstein and Gromet-Elhanan (1974) suggest that the interaction is rather slow. Similarly, although Govindjee and Sybesma (1972) reported the reduction of NAD^+ under these conditions, the reaction was not very rapid, and it is generally recognized that the physiological mechanism for the reduction of pyridine nucleotides in the Rhodospirillaceae and the Chromatiaceae is energy-dependent reversed electron flow (Keister and Yike, 1967a,b; Jackson and Crofts, 1968; Jones, C. W., and Vernon, 1969; Jones,

O. T. G., and Whale, 1970; Jones, O. T. G., and Saunders, 1972) in a manner analogous to the reaction that can occur in mitochondria (Chance and Hollunger, 1961). There is little evidence that the direct photoreduction of NAD^+ (in a manner analogous to the photoreduction of $NADP^+$ by green plant photosystem I—see Vishniac and Ochoa, 1951; Hill and Bendall, 1960), ever occurs under physiological conditions in the purple bacteria.

The situation in the green sulfur bacteria may, however, be very different. In this group, there is a considerable body of evidence that the reduction of NAD^+ can occur directly in the light (Evans, M. C. W., and Buchanan, 1965; Buchanan and Evans,

1969; Jones, O. T. G., and Whale, 1970), albeit at a rather low rate. This reaction is independent of inhibitors of reverse electron flow (Evans, M. C. W., 1969) and requires ferredoxin (Buchanan and Evans, 1969). *Chlorobium limicola f. thiosulfatophilum* ferredoxin is unstable (Evans, M. C.W., and Buchanan, 1965). and its redox potential is unknown, but the reduction of NAD$^+$ occurs in the presence of *Chr. vinosum* ferredoxin (Evans, M. C. W., and Buchanan, 1965), and this has a midpoint potential of -490 mV (Bachofen and Arnon, 1966). Recent data show that dithionite cannot reduce the primary acceptor of the Bchl–RC complex from *Chl. limicola f. thiosulfatophilum* at pH 6.8, indicating that the primary acceptor has a midpoint potential substantially below -450 mV (Prince and Olson, 1976). Titrations at higher pH indicate that it lies close to -540 mV (Olson *et al.*, 1976). This value would be sufficiently negative to be thermodynamically capable of effective reduction of *Chr. vinosum* ferredoxin, and is also in the same region as values reported for the primary acceptor of green plant photosystem I (Ke *et al.*, 1973; Lozier and Butler, 1973). It is substantially lower than the value reported by Knaff *et al.* (1973) for *Chl. limicola f. thiosulfatophilum* vesicles, but these workers used continuous actinic illumination rather than single turnover activation, and did point out that their value might have been erroneously high. More recently, Knaff and Malkin (1976) identified an Fe–S protein in their preparation with a ferredoxin-like signal ($g = 1.94$) and an apparent equilibrium midpoint potential at pH 10 of -550 mV. Although they were unable to demonstrate the reduction of this component at cryogenic temperatures, it is possible that it may be the primary acceptor in this organism, and this is discussed more fully in Chapter 32.

The possibility of more than one functional type of RC in photosynthetic bacteria has been suggested several times, and was recently reviewed by Parson (1974). There is little evidence for such suggestions; in the few cases in which optical changes have been shown to occur at potentials where the primary acceptor was chemically reduced (e.g., Seibert *et al.*, 1971), the changes may be ascribed to light-induced triplet states of Bchl or carotenoids (for a review, see Parson and Cogdell, 1975). The very low potential photochemistry monitored in *Rp. gelatinosa* (Dutton, 1971) at 77°K may have been due to this, or perhaps to electron flow from X$^-$ to low-potential redox dyes.

7. Conclusion

In summary, we have discussed the functional oxidation–reduction midpoint potential of the primary

acceptor in several members of the Rhodospirillaceae and Chromatiaceae. In each case, the properties of the primary acceptor are very similar, and it seems probable that they are general features in these photosynthetic bacteria. The operational redox couple is substantially more negative than that measured at equilibrium, and provides a larger electrochemical potential driving force for electrons entering the cyclic electron-transfer cycle than is usually considered; this redox potential free energy is enough to effectively completely reduce the secondary acceptor, but at the same time, the operational couple is also sufficiently electropositive that its reduction by the photoexcited Bchl (via Bpheo) is effectively irreversible.

However, although the identification of the unprotonated primary acceptor redox couple as the operational couple has resolved several former anomalies, such as species differences and the reduction of low-potential dyes, it has not explained the behavior of the primary acceptor of isolated RCs. In *Rp. sphaeroides*, the E_m of this component has been shown to be apparently independent of pH, with a value of approximately -50 mV (Reed *et al.*, 1969; Dutton *et al.*, 1973). Further investigations on the isolated system are clearly needed to understand this difference.

The primary acceptor of the Chlorobiaceae appears to be quite distinct from that in the purple bacteria, and may perhaps be similar to the primary acceptor of green plant photosystem I.

8. References

Bachofen, R., and Arnon, D. I., 1966, Crystalline ferredoxin from the photosynthetic bacterium *Chr. vinosum*, *Biochim. Biophys. Acta* **120**:259–265.

Bales, H., and Vernon, L. P., 1963, Effect of reduced 2,6-dichlorophenolindophenol upon the light induced absorbancy changes in *Rs. rubrum* chromatophores; a coupled reduction of ubiquinone in the photosynthetic bacteria (H. Gest, A. San Pietro, and L. P. Vernon, eds.), pp. 269–274, Antioch Press, Yellow Springs, Ohio.

Bolton, J. R., and Cost, K., 1973, Flash photolysis EPR; a kinetic study of endogenous light-induced free radicals in reaction center preparations from *Rp. sphaeroides*, *Photochem. Photobiol.* **18**:417–421.

Buchanan, B. B., and Evans, M. C. W., 1969, Photoreduction of ferredoxin and its use in NAD (P)$^+$ reduction by a subcellular preparation from the photosynthetic bacterium *Chl. limicola f. thiosulfatophilum*, *Biochim. Biophys. Acta* **180**:123–129.

Carithers, R. P., and Parson, W. W., 1975, Delayed fluorescence from *Rp. viridis* following single flashes, *Biochim. Biophys. Acta* **387**:194–211.

Casadio, R., Baccarini-Melandri, A., and Melandri, B. A., 1974, On the determination of the transmembrane pH difference in bacterial chromatophores using 9-amino acridine, *Eur. J. Biochem.* **47**:121–128.

Case, G. D., and Parson, W. W. 1971, Thermodynamics of the primary and secondary photochemical reactions in *Chr. vinosum, Biochim. Biophys. Acta* **253**:187–202.

Chance, B., and Hollunger, G., 1961, The interaction of energy and electron transfer reactions in mitochondria. I. General properties and nature of the products of succinate-linked reduction of pyridine nucleotides, *J. Biol. Chem.* **236**:1534–1543.

Clark, W. M., 1960, *Oxidation–Reduction Potentials of Organic Systems*, Williams and Wilkins, Balitmore.

Cogdell, R. J., and Crofts, A. R., 1972, Some observations on the primary acceptor of *Rp. viridis, FEBS Lett.* **27**:176–178.

Cogdell, R. J., Jackson, J. B., and Crofts, A. R., 1972, The effect of redox potential on the coupling between rapid hydrogen-ion binding and electron transport in chromatophores of *Rp. sphaeroides, Bioenergetics* **4**:211–227.

Cramer, W. A., 1969, Low potential titration of the fluorescence yield changes in photosynthetic bacteria, *Biochim. Biophys. Acta* **189**:54–59.

Crofts, A. R., Wraight, C. A., and Fleischmann, D. E., 1971, Energy conservation in the photochemical reactions of photosynthesis and its relationship to delayed fluorescence, *FEBS Lett.* **15**:89–100.

Cusanovich, M. A., Bartsch, R. G., and Kamen, M. D., 1968, Light-induced electron transport in *Chr. vinosum.* II. Light-induced absorbance changes in *Chr. vinosum* chromatophores, *Biochim. Biophys. Acta* **153**:397–417.

Debrunner, P. G., Schulz, C. E., Feher, G., and Okamura, M. Y., 1975, Mossbauer study of reaction centers from *Rp. sphaeroides, Biophys. J.* **15**:226a.

Dutton, P. L., 1971, Oxidation–reduction potential dependence of the interaction of cytochromes, bacteriochlorophyll and carotenoids at 77°K in chromatophores of *Chromatium* D and *Rp. gelatinosa, Biochim. Biophys. Acta* **226**:63–80.

Dutton, P. L., and Baltscheffsky, M., 1972, Oxidation–reduction potential dependence of pyrophosphate-induced cytochrome and bacteriochlorophyll reactions in *Rs. rubrum, Biochim. Biophys. Acta* **267**:172–178.

Dutton, P. L., and Jackson, J. B., 1972, Thermodynamic and kinetic characterization of electron transport components *in situ* in *Rp. sphaeroides* and *Rs. rubrum, Eur. J. Biochem.* **30**:495–510.

Dutton, P. L., and Leigh, J. S., 1973, Electron spin resonance characterization of *Chr. vinosum* hemes, non-heme irons and the components involved in primary photochemistry, *Biochim. Biophys. Acta* **314**:178–190.

Dutton, P. L., and Wilson, D. F., 1974, Redox potentiometry in mitochondrial and photosynthetic bioenergetics, *Biochim. Biophys. Acta* **346**:165–212.

Dutton, P. L., Leigh, J. S., and Wraight, C. A., 1973, Direct measurement of the midpoint potential of the primary acceptor in *Rp. sphaeroides in situ* and in the isolated state: Some relationships with pH and *o*-phenanthroline, *FEBS Lett.* **36**:169–173.

Dutton, P. L., Kauffmann, K. J., Chance, B., and Rentzepis, P. M., 1975, Picosecond kinetics of the 1250 nm band *Rp.*

sphaeroides reaction centers; the nature of the primary photochemical intermediate, *FEBS Lett.* **60**:275–280.

Dutton P. L., Prince, R. C., Tiede, D. M., Petty, K. M., Kaufmann, K. J., Netzel, T. L., and Rentzepis, P. M., 1976, Electron transfer in the photosynthetic reaction center, *Brookhaven Symp. Biol.* **28**:213–237.

Duysens, L. N. M., 1958, The path of light energy in photosynthesis, *Brookhaven Symp. Biol.* **11**:10–23.

Eimhjellen, K. E., Aasmundrud, O., and Jensen, A., 1963, A new bacterial chlorophyll, *Biochem. Biophys. Res. Commun.* **10**:232–236.

Evans, E. H., and Crofts, A. R., 1974, *In situ* Characterization of photosynthetic electron transport in *Rp. capsulata, Biochim. Biophys. Acta* **357**:89–102.

Evans M. C. W., 1969, Ferredoxin NAD⁺ reductase and the photoreduction of NAD⁺ by *Chl. limicola f. thiosulfatophilum*, in: *Progress in Photosynthesis Research* (H. Metzner, ed.), Vol. III, pp. 1474–1475, International Union of Biological Sciences, Tübingen, West Germany.

Evans, M. C. W., and Buchanan, B. B., 1965, Photoreduction of ferredoxin and its use in CO₂ fixation by a subcellular system from a photosynthetic bacterium, *Proc. Natl. Acad. Sci. U.S.A.* **53**:1420–1425.

Evans, M. C. W., Lord, A. V., and Reeves, S. G., 1974, The detection and characterization by electron paramagnetic resonance spectroscopy of iron–sulfur proteins and other electron-transport components in chromatophores from the purple bacterium *Chr. vinosum, Biochem. J.* **138**:177–183.

Fajer, J., Brune, D. C., Davis, M. S., Forman, A., and Spaulding, L. D., 1975, Primary charge separation in bacterial photosynthesis: Oxidized chlorophylls and reduced pheophytin, *Proc. Natl. Acad. Sci. U.S.A.* **72**:4956–4960.

Feher, G., Isaacson, R. A., McElroy, J. D., Ackerson, L. C., and Okamura, M. Y., 1974, On the question of the primary acceptor in bacterial photosynthesis: Manganese substituting for iron in reaction centers of *Rp. sphaeroides, Biochim. Biophys. Acta* **368**:135–139.

Fowler, C. F., Nugent, N. A., and Fuller, R. C., 1971, The isolation and characterization of a photochemically active complex from *Cps. ethylica, Proc. Natl Acad. Sci. U.S.A.* **68**:2278–2282.

Fromhertz, P., and Masters, P., 1974, Interfacial pH at electrically charged lipid monolayers investigated by the lipoid pH-indicator method, *Biochim. Biophys. Acta* **356**:270–275.

Govindjee, R., and Sybesma, C., 1972, The photoreduction of NAD⁺ by chromatophore fractions from *Rs. rubrum, Biophys. J.* **12**:897–908.

Govindjee, R. Smith, W. R., and Govindjee, 1974, Interaction of viologen dyes with chromatophores and reaction center preparations from *Rs. rubrum, Biophys. J.* **20**:191–199.

Gray, B. H., Fowler, C. F., Nugent, N. A., and Fuller, R. C., 1972, A re-evaluation of the presence of low midpoint potential cytochrome 551.5 in the green photosynthetic bacterium *Cps. ethylica, Biochem. Biophys. Res. Commun.* **47**:322–327.

Gray, B. H., Fowler, C. F., Nugent, N. A., Rigopoulos, N.,

and Fuller, R. C., 1973, Re-evaluation of *Cps. ethylica* strain 2-K, *Int. J. Syst. Bacteriol.* **23**:256–264.

Halsey, Y. D., and Parson, W. W., 1974, Identification of ubiquinone as the secondary electron acceptor in the photosynthetic apparatus of *Chr. vinosum*, *Biochim. Biophys. Acta* **347**:401–416.

Hill, R., and Bendall, F., 1960, Function of two cytochrome components in chloroplasts; a working hypothesis, *Nature (London)* **186**:136–137.

Holt, A. S., and Clayton, R. K., 1965, Light induced absorbancy changes in Eimhjellens' *Rhodopseudomonas*, *Photochem. Photobiol.* **4**:829–831.

Jackson, J. B., and Crofts, A. R., 1968, Energy linked reduction of NAD in cells of *Rs. rubrum*, *Biochem. Biophys. Res. Commun.* **32**:908–915.

Jackson, J. B., and Crofts, A. R., 1969, The high energy state in chromatophores from *Rp. sphaeroides*, *FEBS Lett.* **4**:185–189.

Jackson, J. B., and Dutton, P. L., 1973, The kinetic and redox potentiometric resolution of the carotenoid band-shift in *Rp. sphaeroides* chromatophores: Their relationship to electric field alterations in electron transport and energy coupling, *Biochim. Biophys. Acta* **325**:102–113.

Jackson, J. B., Cogdell, R. J., and Crofts, A. R., 1973, Some effects of o-phenanthroline on electron transport in chromatophores from photosynthetic bacteria, *Biochim. Biophys. Acta* **290**:218–225.

Jones, C. W., and Vernon, L. P., 1969, *Biochim. Biophys. Acta* **180**:149–164.

Jones, O. T. G., and Saunders, V. A., 1972, Energy linked electron transfer reactions in *Rp. viridis*, *Biochim. Biophys. Acta* **275**:427–436.

Jones, O. T. G., and Whale, F. R., 1970, The oxidation and reduction of pyridine nucleotides by *Rp. sphaeroides* and *Chl. limicola f. thiosulfatophilum*, *Arch. Mikrobiol.* **72**:48–59.

Kaufmann, K., Dutton, P. L., Netzel, T. L., Leigh, J. S., and Rentzepis, P. M., 1975, Picosecond kinetics of events leading to reaction center bacteriochlorophyll oxidation, *Science* **188**:1301–1304.

Ke, B., 1973, The primary electron acceptor of photosystem I, *Biochim. Biophys. Acta* **301**:1–33.

Ke, B. Hansen, R. E., and Beinert, H., 1973, Oxidation–reduction potentials of bound iron–sulfur proteins of photosystem I, *Proc. Natl. Acad. Sci. U.S.A.* **70**:2941–2945.

Keister, D. L., and Yike, N. J., 1967a, Energy linked reactions in photosynthetic bacteria. I. Succinate linked ATP driven NAD$^+$ reduction by *Rs. rubrum* chromatophores, *Arch. Biochem. Biophys.* **121**:415–422.

Keister, D. L., and Yike, N. J., 1967b, Energy linked reactions in photosynthetic bacteria. II. The energy dependent reduction of NADP$^+$ by NADH in chromatophores of *Rs. rubrum*, *Biochemistry* **6**:3847–3857.

Knaff, D. B., and Malkin, R., 1976, Iron–sulfur proteins of the green photosynthetic bacterium *Chl. limicola f. thiosulfatophilum*, *Biochim. Biophys. Acta* **430**:244–252.

Knaff, D. B., Buchanan, B. B., and Malkin, R., 1973, Effect of oxidation–reduction potential on light-induced cytochrome and bacteriochlorophyll reactions in chromato-

phores from the photosynthetic green bacterium *Chl. limicola f. thiosulfatophilum*, *Biochim. Biophys. Acta* **325**:94–101.

Knox, R. T., 1969, Thermodynamics and the primary processes of photosynthesis, *Biophys. J.* **9**:1351–1362.

Kuntz, I. D., Loach, P. A., and Calvin, M., 1964, Absorption changes in bacterial chromatophores, *Biophys. J.* **4**:228–249.

Loach, P. A., 1966, Primary oxidation–reduction changes during photosynthesis in *Rs. rubrum*, *Biochemistry* **5**:592–600.

Loach, P. A., 1976, Chemical properties of the phototrap in photosynthetic bacteria, *Progress in Bio-organic Chemistry* (T. Kaiser and F. Kezdy, eds.), Vol. 4, pp. 89–92, John Wiley & Sons, New York.

Loach, P. A., and Sekura, D. L., 1968, Primary photochemistry and electron transport in *Rs. rubrum*, *Biochemistry* **7**:2642–2649.

Loach, P. A., Androes, G. M., Maksim, A. F., and Calvin, M., 1963, Variation in EPR signals of photosynthetic systems with the redox level of their environment, *Photochem. Photobiol.* **2**:443–454.

Loach, P. A., Kung, M. C., and Hales, B. J., 1975, Characterization of the phototrap in photosynthetic bacteria, *Ann. N. Y. Acad. Sci.* **244**:297–319.

Lozier, R. H., and Butler, W. L., 1973, Redox titration of the primary electron acceptor of photosystem I in spinach chloroplasts, *Biochim. Biophys. Acta* **333**:460–464.

Oettmeier, W., and Grewe, R., 1974, Inhibition of photosynthetic electron transport by azaphenanthrenes, *Z. Naturforsch.* **29c**:545–551.

Okamura, M. Y., Isaacson, R. A., and Feher, G., 1975, Primary acceptor in bacterial photosynthesis: Obligatory role of ubiquinone in photoactive reaction centers of *Rp. sphaeroides*, *Proc. Natl. Acad. Sci. U.S.A.* **72**:3491–3495.

Okamura, M. Y., Ackerson, L. C., Isaacson, R. A., Parson, W. W., and Feher, G., 1976, The primary acceptor in *Chr. vinosum* strain D, *Biophys. J.* **16**:233a.

Olson, J. M., Prince, R. C., and Brune, D. C., 1976, Reaction-center complexes from green bacteria, *Brookhaven Symp. Biol.* **28**:238–246.

Parson, W. W., 1968, The role of P870 in bacterial photosynthesis, *Biochim. Biophys. Acta* **153**:248–259.

Parson, W. W., 1974, Bacterial photosynthesis, *Annu. Rev. Microbiol.* **28**:41–59.

Parson, W. W., and Case, G. D., 1970, In *Chr. vinosum*, a single photochemical reaction center oxidizes both cytochrome c_{552} and cytochrome c_{555}, *Biochim. Biophys. Acta* **205**:232–245.

Parson, W. W., and Cogdell, R. J., 1975, The primary photochemical reactions of bacterial photosynthesis, *Biochim. Biophys. Acta* **416**:105–149.

Parson, W. W., Clayton, R. K., and Cogdell, R. J., 1975, Excited states of photosynthetic reaction centers at low redox potentials, *Biochim. Biophys. Acta* **387**:265–278.

Petty, K. M., and Dutton, P. L., 1976a, Properties of the flash-induced proton binding reaction of *Rp. sphaeroides* membranes: A functional pK on the ubisemiquinone, *Arch Biochem. Biophys.* **172**:335–345.

Petty, K. M., and Dutton, P. L., 1976*b*, Ubiquinone–cytochrome *b* electron and proton transfer: A functional pK on cytochrome b_{50} in *Rp. sphaeroides* membranes, *Arch. Biochem. Biophys.* **172**:346–353.

Prince, R. C., and Dutton, P. L., 1976, The primary acceptor of bacterial photosynthesis; its operating midpoint potential?, *Arch. Biochem. Biophys.* **172**:329–334.

Prince, R. C., and Olson, J. M., 1976, Some thermodynamic and kinetic properties of the primary photochemical reactants in a complex from a green photosynthetic bacterium, *Biochim. Biophys. Acta* **423**:357–362.

Prince, R. C., Leigh, J. S., and Dutton, P. L., 1974, An electron spin resonance characterization of *Rp. capsulata*, *Biochem. Soc. Trans.* **2**:950–953.

Prince, R. C., Baccarini-Melandri, A., Hauska, G. A., Melandri, B. A., and Crofts, A. R., 1975, Asymmetry of an energy transducing membrane; the location of cytochrome c_2 in *Rp. sphaeroides* and *Rp. capsulata*, *Biochim. Biophys. Acta* **387**:212–227.

Prince, R. C., Leigh, J. S., and Dutton, P. L., 1976, Thermodynamic properties of the reaction center of *Rp. viridis*: *In vivo* measurement of the reaction center bacteriochlorophyll–primary acceptor intermediary electron carrier, *Biochim. Biophys. Acta* **440**:622–636.

Reed, D. W., Zankel, K. L., and Clayton, R. K., 1969, The effect of redox potential on P870 fluorescence in reaction centers from *Rp. sphaeroides*, *Proc. Natl. Acad. Sci. U.S.A.* **63**:42–46.

Remennikov, S. M., Chamorovsky, S. K., Kononenko, A. A., Venediktov, P. S., and Rubin, A. B., 1975, High potential oxidation–reduction titration of absorbance changes induced by pulsed laser and continuous light in chromatophores of photosynthesizing bacteria *Rs. rubrum* and *E. shaposhnikovii*, *Stud. Biophys.* **51**:1–13.

Rockley, M. G., Windsor, M. W., Cogdell, R. J., and Parson, W. W., 1975, Picosecond detection of an intermediate in the photochemical reaction of bacterial photosynthesis, *Proc. Natl. Acad. Sci. U.S.A.* **72**:2251–2255.

Ross, R. T., and Calvin, M., 1967, Thermodynamics of light emission and free energy storage in photosynthesis, *Biophys. J.* **7**:595–614.

Seibert, M., and Devault, D., 1970, Relations between the laser-induced oxidations of the high and low potential cytochromes of *Chr. vinosum*, *Biochim. Biophys. Acta* **205**:220–231.

Seibert, M., and DeVault, D., 1971, Photosynthetic reaction center transients, P435 and P424, in *Chr. vinosum*, *Biochim. Biophys. Acta* **253**:396–411.

Seibert, M., Dutton, P. L., and DeVault, D., 1971, A low potential photosystem in *Chr. vinosum*, *Biochim. Biophys. Acta* **226**:189–192.

Shioi, Y., Takamiya, K., and Nishimura, M., 1976, Isolation and some properties of NAD$^+$ reductase of the green photosynthetic bacterium *Prosthecochloris aestuarii*, *J. Biochem.* **79**:361–371.

Silberstein, B. R., and Gromet-Elhanan, Z., 1974, P430, a possible primary electron acceptor in *Rs. rubrum*, *FEBS Lett.* **42**:141–144.

Thornber, J. P., and Olson, J. M., 1971, Chlorophyll proteins and reaction center preparations from photosynthetic bacteria, algae and higher plants, *Photochem. Photobiol.* **14**:329–341.

Vishniac, W., and Ochoa, S., 1951, Photochemical reduction of pyridine nucleotides by spinach grana and coupled carbon dioxide fixation, *Nature (London)* **167**:768–769.

Wraight, C. A., and Clayton, R. K., 1973, The absolute quantum efficiency of bacteriochlorophyll photooxidation in reaction centers of *Rp. sphaeroides*, *Biochim. Biophys. Acta* **333**:246–260.

CHAPTER 25

Quinones as Secondary Electron Acceptors

William W. Parson

1. Distribution and Amounts of Quinones in Photosynthetic Bacteria

The photosynthetic bacteria are richly endowed with quinones (Qs). Lester and Crane (1959) were probably the first to appreciate this, when they included *Rhodospirillum rubrum* and *Chromatium vinosum* in a survey on the natural occurrence of ubiquinone (UQ). Indications of the special importance of UQ in photosynthetic metabolism emerged shortly thereafter, with the finding of Geller (1962) and Sugimura and Rudney (1962) that the UQ content of *Rs. rubrum* was considerably higher in cells growing photosynthetically than it was in cells growing aerobically. This must have been a striking finding at the time, because the evidence was already persuasive that UQ played an essential role in respiratory electron transport and oxidative phosphorylation. Carr and Excell (1965) extended Geller's and Sigimura and Rudney's observations to other species of Rhodospirillaceae, and discovered that the UQ content also depended on the nature of the organic acids in the culture medium.

Table 1 collects information on the amounts of UQ and related Qs that have been found in photosynthetically grown cells of various species. The ratio of the amount of UQ to that of bacteriochlorophyll (Bchl) is typically on the order of 1:4, although in *Rs. rubrum*, it can exceed 1:2. Assuming an average photosynthetic unit size of about 100 Bchl molecules per reaction center (RC), this means that the photosynthetic bacteria typically contain 25 or more molecules of UQ per RC.

The length of the multiprenyl (isoprenoid) side chain of the Q varies from species to species (Table 1). Maroc *et al.* (1968) pointed out that those species that contain Q-7, Q-8, or Q-9 also contain menaquinone (vitamin K_2) (MK) with a side chain of the same length (MK-7, MK-8, or MK-9). The amount of MK present is typically about 30% that of UQ. Species that contain Q-10, on the other hand, appear to lack MK. The only other Q that has been found in species of this group is rhodoquinone, an aminoquinone (Moore and Folkers, 1965) that forms metabolically from UQ in *Rs. rubrum* (Parson and Rudney, 1965). Rhodoquinone has not been found in any species other than *Rs. rubrum*, but in that species, it is present in an amount that is about 20% that of UQ (Okayama *et al.*, 1968). The Chlorobiaceae lack benzoquinones altogether, but they do contain MKs, including chlorobiumquinone. This is a MK with an additional keto group in the isoprenoid side chain (Frydman and Rapaport, 1963; Powls *et al.*, 1968).

One can find information on the methodology of extraction, identification, and analysis of various Qs in the articles by Barr and Crane (1971), Crane and Barr (1971), Dunphy and Brodie (1971), and Mayer and Isler (1971). For recommendations on nomenclature, see IUPAC-IUB (1975). Amesz (1973) provided a useful review of the role of plastoquinones in chloroplast photosynthesis.

William W. Parson · Department of Biochemistry, University of Washington, Seattle, Washington 98195

Table 1. Quinones in Photosynthetic Bacteria

Organism	Prenylogue	UQ/Bchl	MK/Bchl	References
Chr. vinosum	7	0.15	0.056	Takamiya *et al.*(1967)
		0.21	0.075	Fuller *et al.* (1961)
		0.16	Not detected[a]	Halsey and Parson (1974)
		0.026	Not reported	Cusanovich and Kamen (1968)
Chl. limicola				
f. *thiosulfatophilum*	7	Lacking	0.070	Powls and Redfearn (1969)
"*Chloropseudomonas ethylica*"[b]	7	Lacking	0.050	Takamiya *et al.* (1967)
		—	0.070	Powls and Redfearn (1969)
Rp. gelatinosa	8	0.29	0.086	Maroc *et al.* (1968)
		0.10	Not reported	Carr and Exell (1965)
Rs. molischianum	9	0.11	0.028	Maroc *et al.* (1968)
Rs. fulvum	9	0.13	0.031	Maroc *et al.* (1968)
Rs. rubrum	10	0.49	Lacking	Maroc *et al.* (1968)
		0.35	Lacking	Carr and Exell (1965)
		0.33	Lacking	Takamiya *et al.* (1967)
		0.30	Lacking	Noël *et al.* (1972)
Rs. rubrum				
Strain G-9 (carotenoidless)	10	0.56	Lacking	Okayama *et al.* (1968)
Rp. sphaeroides	10	0.23	Lacking	Maroc *et al.* (1968)
		0.14	Lacking	Carr and Exell (1965)
		0.48	Lacking	Takamiya *et al.* (1967)
Rp. capsulata	10	0.22	Lacking	Maroc *et al.* (1968)
		0.26	Lacking	Carr and Exell (1965)
Rp. palustris	10	0.22	Lacking	Maroc *et al.* (1968)
		0.013	Lacking	Carr and Exell (1965)
		0.23	Lacking	Takamiya *et al.* (1967)
Rm. vannielii	10	0.28	Lacking	Carr and Exell (1965)

[a] MK was subsequently found in chromatophores prepared as described in this work and extracted so as to remove UQ (Okamura *et al.*, 1976). It is not clear why Halsey and Parson did not detect it in the intact chromatophores.

[b] "*Chloropseudomonas ethylica*" is a symbiotic association of a green photosynthetic bacterium and a nonphotosynthetic organism (see Chapter 2). Whether the latter makes a significant contribution to the Q content is not clear.

2. Identification of Ubiquinone as a Secondary Electron Acceptor

2.1. Double-Flash Experiments

When suspensions of cells or chromatophores are excited with a single, short flash of light, P_{870} releases an electron, reducing the primary acceptor (Chapters 15 and 19–22). In species that contain membrane-bound *c*-type cytochromes, the P_{870} radical cation (P_{870}^+) rapidly extracts an electron from one of the cytochromes, so that P_{870} returns to the reduced state within a few microseconds (Chapters 15 and 28). Although P_{870} should be able to undergo photooxidation again at this point, the photochemical apparatus as a whole is not prepared to function. A second flash arriving shortly after the first does not cause the photooxidation of P_{870}, or of additional cytochromes (Parson, 1969). Instead, the energy of the second flash is diverted into fluorescence, and presumably into heat. The yield of fluorescence on the second flash is about 4 times as great as it is on the first, as high as it is if photochemistry is blocked by chemical reduction of the primary electron acceptor.

These observations most likely mean that photochemistry on the second flash depends primarily on the redox state of the primary acceptor (Parson, 1969). Evidently, the acceptor is capable of handling only one electron at a time, and the passage of the electron to a secondary acceptor is a comparatively slow process. If this is correct, a study of the kinetics with which the system regains photochemical activity provides a measure of the rate of the secondary electron-transfer reaction. In *Chr. vinosum* chromatophores, at pH 7 and at room temperature, the recovery is complete by 1 msec, and halfway complete in about 60 μsec (Parson, 1969). The recovery does not follow first-order kinetics, but rather appears to be second-order in the amount of the primary acceptor that remains reduced. The speed of the recovery increases with decreasing pH, though not in strict proportion to the proton concentration. It increases with temperature, having an activation energy of approximately 8 kcal mol^{-1} (Parson, 1969). The recovery kinetics are slightly slower in whole cells of *Chr. vinosum* (Parson and Case, 1970). In *Rp. viridis* chromatophores, the recovery kinetics are also approximately second-order, and also depend on pH, but the reaction is about 10 times faster than it is in *Chr. vinosum* (Carithers and Parson, 1975). In chromatophores of *Ectothiorhodospira shaposhnikovii*, the kinetics are first-order, with a half-time of 100 μsec at pH 7 (Chamorovsky et al., 1976).

If the two excitation flashes are spaced between 1 and 10 msec apart, the amount of photochemistry that occurs on the second flash can provide a measure of the integrity or capacity of the secondary electron acceptor. Using this assay, Case et al. (1970) found that the secondary electron-transfer reaction was missing or severely inhibited in a subchromatophore preparation from *Chr. vinosum* ("Fraction A"). However, another preparation (Fraction A′), which was obtained with a different combination of detergents, retained photochemical activity on the second flash (Case, 1972; Halsey and Byers, 1975). The amounts of UQ in the two preparations differed, being about 5 times greater in Fraction A′ than in Fraction A.

2.2. Extraction and Reconstitution

When *Chr. vinosum* chromatophores were lyophilized and extracted with light petroleum so as to deplete them of UQ, they lost almost entirely the ability to perform photochemistry on the second flash (Halsey and Parson, 1974). The extraction did not affect photochemical activity on the first flash. [Okamura et al. (1976) subsequently found that the extracted chromatophores contained approximately 1 equivalent of a MK. As discussed in Chapter 19, the MK would appear to account for the retention of the primary photochemical reaction.] Reconstitution of the extracted chromatophores with the lipid extract or with pure UQ or phylloquinone (vitamin K$_1$) restored the photochemical activity on the second flash completely. To identify the active principle in the extract, the extract was fractionated by thin-layer chromatography. A single fraction accounted for the activity,

and it contained essentially pure UQ (Halsey and Parson, 1974). The extraction of UQ from *Rp. viridis* chromatophores causes a similar inhibition of photochemical activity on a second flash (Carithers and Parson, 1975), but reconstitution has not been studied in this case.

Several unnatural Qs such as menadione, and redox dyes such as *N*-methylphenazonium methosulfate (PMS) or methylene blue, are able to replace UQ in restoring secondary acceptor capacity to extracted chromatophores, Fraction A, or other subchromatophore preparations (Nicholson and Clayton, 1969; Case et al., 1970; Halsey and Parson, 1974). With these artificial acceptors, however, the kinetics of removal of electrons from the primary acceptor are generally much slower than they are with UQ.

The extraction of UQ also destroys the ability of chromatophores to perform photophosphorylation (Horio et al., 1968; Okayama et al., 1968; Yamamoto et al., 1970) and to catalyze the reduction of exogenous cytochrome c$_2$ by succinate (Higuti et al., 1975). Photophosphorylation can be restored along with electron transfer, by reconstituting the extracted chromatophores with UQ. Photophosphorylation in the reconstituted chromatophores retains its sensitivity to inhibition by antimycin. Rhodoquinone is much less effective than UQ in restoring photophosphorylation, but it does afford a partial restoration, if the photophosphylation is measured with PMS as a cofactor (Okayama et al., 1968).

The extraction experiments show that UQ or a related Q is essential for the completion of the secondary electron-transfer reaction. Although this could mean that UQ is required for the activity of some other electron carrier, the simplest interpretation is that UQ itself is the secondary acceptor. This conclusion would agree with the evidence that is discussed below that the illumination of cells or chromatophores can cause the reduction of the endogenous UQ (see Section 3.1).

The loss of the secondary electron-transfer reaction actually does not proceed in parallel with the fraction of the UQ pool that is removed by the extraction. One can remove at least 75% of the UQ without impairing the ability of chromatophores to perform photochemistry on the second flash, and 90% of the UQ must be extracted to reduce photochemical activity by 50% (Halsey and Parson, 1974). Takamiya and Takamiya (1970) reached a similar conclusion in a study of the effects of UQ extraction on P$_{870}$ photooxidation, as measured during continuous illumination of *Chr. vinosum* chromatophores. These results are consistent with the relatively large amount of UQ that occurs in chromatophores (Section 1). A small part of the UQ pool might be sufficient to allow rapid removal of electrons from the primary acceptor. It is possible that the UQ pool is heterogeneous, and that the fraction that is extracted most easily does not participate directly in the secondary electron-transfer reaction.

King and Drews (1973) suggested that the binding or interaction sites for UQ in *Rp. palustris* are heterogeneous. They found that the extraction of UQ from the membranes of aerobically grown cells inactivated the oxidation of both NADH and succinate. Reconstitution with Q-4, Q-6, Q-8, or Q-10 restored succinate oxidase activity. To restore NADH oxidation, however, it was necessary to use Q-10; the shorter-chain UQ prenylogues were comparatively ineffective. This suggests that the site at which UQ interacts with NADH dehydrogeanse is structurally different from the site at which it interacts with succinic dehdyrogenase. Electron transfer between the two sites must occur despite the differences, because either NADH or succinate can reduce at least 80% of the UQ in the membranes (King and Drews, 1973). Studies by Norling *et al.* (1974) gave a similar picture of the UQ pool in mitochondria.

Sections 3.1 and 4 discuss additional indications that the UQ that participates in the secondary reaction is distinct from the UQ of the bulk pool.

2.3. Inhibition of the Secondary Reaction

In addition to the extraction of UQ, there are other ways to block the secondary electron-transfer reaction. One is simply to lower the temperature. If *Chr. vinosum* chromatophores are excited with a single flash at 77°K, the P_{870} that is photooxidized can remove an electron from the low-potential cytochrome C552 and return to the reduced state with a half-time of 2 msec. At this temperature, a second flash does not cause reoxidation of the P_{870}, even if one allows many minutes to elapse between the flashes (Dutton *et al.*, 1971). The complete failure of the secondary reaction at very low temperatures is consistent with the temperature dependence that has been measured at higher temperatures (Parson, 1969; Chamorovsky *et al.*, 1976) (see also Section 2.1).

The secondary electron-transfer reaction is also sensitive to several different inhibitors. The first of these to be recognized was *o*-phenanthroline (1,10-diazaphenanthrene). At a concentration of 1 mM, *o*-phenanthroline has no effect on the ability of chromatophores to perform the photooxidation of P_{870} or cytochrome on a single flash, but it greatly decreases the photochemical activity on a second flash 1 msec after the first (Parson and Case, 1970). Other agents that act similarly include piericidin A, and

certain quinoline-quinones with bulky alkyl side chains (Halsey and Parson, 1974; Carithers and Parson, 1975). 2,5-Dibromo-3-methyl-6-isopropyl-*p*-benzoquinone also inhibits (K. Petty and P. L. Dutton, personal communication). The most potent agent known, 5-*n*-pentadecyl-6-hydroxy-4,7-dioxobenzothiazole, gives half-maximal inhibition at concentrations between 5 and 50 μM in *Chr. vinosum* or *Rp. viridis* chromatophores; it is substantially less effective in *Rp. sphaeroides* (Carithers and Parson, 1976).

o-Phenanthroline and the other inhibitors act on chromatophores that have been extracted and reconstituted with UQ just as they do on native chromatophores (Halsey and Parson, 1974). They generally do not inhibit electron transfer to artificial acceptors such as PMS or methylene blue (Case *et al.*, 1970).

The sensitivity of the secondary electron-transfer reaction to *o*-phenanthroline, piericidin A, and the substituted quinolinequinones points out a similarity of the secondary reaction to the reactions of NADH dehydrogenase and succinic dehydrogenase in mitochondria and aerobic microorganisms. The dehydrogenases, which pass electrons to UQ by way of nonheme Fe, are sensitive to many of the same inhibitors (Hatefi *et al.*, 1969; Jeng *et al.*, 1968; Gutman *et al.*, 1970; Porter *et al.*, 1971, 1973; Loschen and Azzi, 1974). The presence of nonheme Fe in the photosynthetic RC (Chapter 19) underscores the similarity between the systems. However, not all agents that block NADH dehydrogenase are effective against the photosynthetic reaction; amytal is not inhibitory (Halsey and Parson, 1974).

The mechanism by which any of the inhibitors blocks the secondary electron-transfer reaction is not clear. Piericidin A and the inhibitory Qs bear some structural resemblance to UQ, but there is no evidence that the resemblance forms the basis of their inhibitory activity. Certain substituted benzoquinones that resemble UQ even more closely do not inhibit the secondary reaction *Chr. vinosum* (Halsey and Parson, 1974), although they do inhibit electron transport in chloroplasts and mitochondria. [Some of these probably act between cytochromes *b* and *c* in mitochondria, rather than between nonheme Fe and Q (Phelps and Crane, 1975)].

It is tempting to assume that the inhibitory effect of *o*-phenanthroline results from chelation of the nonheme Fe atom of the RC. In agreement with this, the treatment of isolated RCs with *o*-phenanthroline in the presence of chaotropic agents like $LiClO_4$ causes the extraction of the Fe (M. Y. Okamura, personal communication). In the presence of relatively high concentrations of detergents (but the absence of chaotropes), *o*-phenanthroline causes the extraction of UQ from the RCs (Okamura *et al.*, 1975). These facts suggest that *o*-phenanthroline disrupts a complex of Fe and UQ that exists in the RC. Sections 2.5 and 3.1 describe additional evidence that the Fe plays a critical role in the secondary electron-transfer reaction. It is perhaps surprising, therefore, that many other Fe chelators are not inhibitory. These include 8-hydroxyquinoline sulfonate, diethyldithiocarbamate, α,α'-(ethylenedinitrilo)-di-*o*-cresol, 2,2'-biquinoline, oxalate, salicylaldoxime, 2,9-dimethyl-4,7-diphenyl-1,10-diazaphenanthrene sulfonate, pyrocatechol-3,5-disulfonate, 4,7-diphenyl-1,10-diazaphenanthrene sulfonate, ethylenediamine-*N,N,N',N'*-tetraacetate, ethylene glycol-bis-(β-aminoethylether)-*N,N,N',N'*-tetraacetate, and α,α'-bipyridyl (Clayton *et al.*, 1972*b*; Parson, unpublished). Bathophenanthroline (4,7-diphenyl-1,10-diazaphenanthrene) does inhibit, but not as effectively as *o*-phenanthroline (Clayton *et al.*, 1972*b*). R. K. Clayton (personal communication) has pointed out that an inhibitor might require a certain level of solubility in both water and lipid solvents, in order to find its way to the region of the primary electron acceptor. Improper solubility properties could account for the inactivity of some of the agents listed above. In chloroplasts, may different substituted azaphenanthrenes and diazaphenanthrenes inhibit secondary electron transfer at System II, but there is no obvious correlation between the effectiveness of different derivatives as inhibitors and their effectiveness as Fe^{II} chelators (Satoh, 1974; Oettmeier and Grewe, 1974). 1-Azaphenanthrene, for example, is a relatively good inhibitor in chloroplasts, although it is not a chelator at all.

The addition of *o*-phenanthroline causes a positive shift in the apparent $E_{m,7}$ of the primary electron acceptor, both in photosynthetic bacteria and in chloroplasts (Cogdell and Crofts, 1972; Jackson *et al.*, 1973; Evans *et al.*, 1974; Knaff, 1975), and it is possible that this effect underlies the inhibition of the secondary reaction. Prince and Dutton (1976) (see also Prince and Dutton, Chapter 24) showed that the change in the E_m results from an increase in the pK_a of the primary acceptor. They suggest that *o*-phenanthroline acts by transferring a proton to the acceptor, following the primary electron-transfer reaction. The resulting rise in the E_m could render the primary acceptor thermodynamically unable to reduce the secondary acceptor. This hypothesis leaves unexplained why *o*-phenanthroline inhibits the secondary reaction in purified RCs, in which the E_m of the primary acceptor is independent of pH, and in which *o*-phenanthroline has little or no effect on the E_m (Dutton *et al.*, 1973).

2.4. Competition between the Secondary Reaction and Reversal of the Primary Reaction

Blocking the transfer of an electron from the primary acceptor to the secondary acceptor increases the possibility that the electron will return to P_{870} in a reversal of the primary reaction. There are at least two mechanisms by which such a back-reaction can occur. One mechanism, a true reversal of the photochemical reaction, returns P_{870} to an excited singlet state, and results in delayed fluorescence (Fleischman, 1974; Fleischman, Chapter 27; Carithers and Parson, 1975, 1976). The other mechanism bypasses the excited singlet state, and appears to proceed by electron tunneling (Parson, 1967; McElroy *et al.*, 1974; Clayton and Yau, 1972; Carithers and Parson, 1975; Hsi and Bolton, 1974). The tunneling back-reaction, which is quantitatively the more important process, gives a first-order decay of P_{870}^+ with a half-time of 1–60 msec, depending on the species. The kinetics are almost independent of temperature, the reaction being even somewhat faster at cryogenic temperatures than it is at room temperature.

Both types of back-reaction are so slow as to be insignificant compared with the rate at which cytochromes can reduce P_{870}^+ and the rate of transfer of electrons to the secondary acceptor. They become important, however, under conditions that prevent the normal forward reactions. If one selects conditions that prevent the cytochromes from reducing P_{870}^+, the probability of both types of back-reaction increases enormously on lowering of the temperature, or on the addition of *o*-phenanthroline or one of the other inhibitors of the secondary reaction (Fleischman and Clayton, 1968; Clayton *et al.*, 1972*b*; Fleischman, 1974; Halsey and Parson, 1974; Hsi and Bolton, 1974; Carithers and Parson, 1975; Chamorovsky *et al.*, 1976). The extraction of UQ from chromatophores or subchromatophore preparations has just the same effect (Ke *et al.*, 1973; Clayton and Yau, 1972; Halsey and Parson, 1974; Carithers and Parson, 1975; Cogdell *et al.*, 1974). Again, reconstitution of the extracted material with UQ reverses the effect of the extraction, preventing the back-reactions from occurring. Following the reconstitution with UQ, the back-reaction can be induced once again by adding an inhibitor of the secondary reaction.

A puzzling observation that remains to be explained is that *o*-phenanthroline is much more effective than piericidin A in promoting delayed fluorescence in *Rp. sphaeroides* chromatophores, but only slightly more effective in promoting the tunneling back-reaction (Carithers and Parson, 1976). This discrepancy is not observed in *Rp. viridis* (Carithers and Parson, 1975).

The double-flash technique described in Section 2.1 requires that the reduction of P_{870}^+ occur within a few microseconds after the first flash. In *Rs. rubrum* and *Rp. sphaeroides*, the cytochrome *c* that reacts with P_{870}^+ is not tightly bound to the chromatophore membrane, and the cytochrome is extensively solubilized during the preparation of chromatophores. The reduction of P_{870}^+ in these chromatophores is too slow for the double-flash method to be applicable. An alternative technique that is useful in such cases is to measure the effectiveness with which the secondary electron-transfer reaction competes with the back-reaction. If one knows the rate constant for the back-reaction, measurements of the fraction of the P_{870}^+ that decays by that path allow one to calculate the rate constant for the secondary reaction. Chamorovsky *et al.* (1976) used this technique to determine the kinetics of the secondary reaction in *E. shaposhnikovii* and *Rs. rubrum*, at temperatures between 150 and 270°K. The apparent half-time of the secondary reaction is on the order of 300 msec at the lower end of this temperature range and 2 msec at the upper end. The activation energies are 12.4 kcal mol^{-1} in *E. shaposhnikovii* and 9.9 kcal mol^{-1} in *Rs. rubrum*. Extrapolating to 300°K gives a half-time of about 150 μsec in the former species and 270 μsec in the latter. The value for *E. shaposhnikovii* agrees well with the 100-μsec half-time that the same authors measured using the double-flash technique.

2.5. Ubiquinone as a Secondary Acceptor in Purified Reaction Centers—Role of Fe in the Secondary Reaction

RCs that are purified from *Rp. sphaeroides* with the use of lauryldimethylamine oxide contain only approximately 1–2 mol UQ/mol P_{870} (Chapter 19), but the pool of UQ can be enlarged by the addition of exogenous UQ. If one illuminates such preparations with continuous light in the presence of excess ferrous cytochrome *c*, the amount of the cytochrome that undergoes photooxidation varies in proportion to the size of the UQ pool (Clayton *et al.*, 1972a,b). The stoichiometry of the cytochrome photooxidation, relative to the UQ content, indicates that the UQ must be fully reduced to the dihydroquinone, rather than stopping at the semiquinone. Direct measurements of the UV absorbance changes that result from the reduction of the UQ show that the expected amount of UQ is, in fact, reduced (Clayton *et al.*, 1972a).

RC preparations made from the same species but with Triton X-100 as the detergent contain a larger amount of endogenous UQ, generally 5–10 mol/mol P_{870}. Again, the amount of cytochrome *c* that such preparations are capable of photooxidizing varies with the size of the UQ pool, and the stoichiometry implies that the UQ is reduced to the level of the dihydroquinone (Clayton *et al.*, 1972a).

The large-scale reduction of UQ that occurs in either of these preparations of RCs appears to involve the natural secondary electron-transfer reaction, in addition to the primary photochemical reaction, because it is sensitive to inhibition by *o*-phenanthroline. In the presence of 1 mM *o*-phenanthroline, the RCs can oxidize rapidly only approximately 1 mol cytochrome *c*/mol P_{870} (Clayton *et al.*, 1972b).

Clayton *et al.* (1972a,b) used the fluorescence from P_{870} in another approach to studying the transfer of electrons from the primary acceptor to UQ. The yield of fluorescence from P_{870} is very low, as long as P_{870} is able to enter into the photochemical electron-transfer reaction. If th photooxidation of P_{870} is blocked by the prior reduction of the primary electron acceptor, the fluorescence yield increases by a factor of about 3. The fluorescence yield can thus serve as a measure of the ability of secondary electron acceptors to keep the primary acceptor in the oxidized state. If isolated RCs from *Rp. sphaeroides* are illuminated in the presence of *o*-phenanthroline and excess reduced cytochrome, the fluorescence yield rises abruptly after the absorption of approximately 1 quantum of light per RC. In the absence of *o*-phenanthroline, the increase in fluorescence occurs more slowly, after a lag period that varies with the UQ content of the RCs. A quantitative analysis of the lag again supports the conclusion that the UQ must be reduced to the dihydroquinone (Clayton *et al.*, 1972a,b).

Studies of purified RCs have shown that the Fe of the reaction center plays an important role in the secondary electron-transfer reaction. Loach and Hall (1972) and Feher *et al.* (1972, 1974) found that the Fe can be removed or replaced by Mn without interfering with the primary electron transfer reaction. Okamura *et al.* (1975) suggested, therefore, that the role of the Fe might be to facilitate the transfer of electrons from the primary acceptor to a secondary acceptor (see Chapter 19). In agreement with this suggestion, RCs that have been depleted of Fe are unable to carry out the secondary reaction (R. E. Blankenship and W. W. Parson, unpublished). In the presence of reductants that reduce P_{870}^+ rapidly after flash excitation, the P_{870} of Fe-depleted RCs is photochemically inactive on a second flash, even in the presence of excess UQ. In the absence of reductants, the P_{870}^+ in the depleted RCs decays with the kinetics that are characteristic of the tunneling back-reaction, again even in the presence of excess UQ. In both of these regards, the Fe-depleted reaction centers behave like RCs that contain only a single Q, or that have been treated with o-phenanthroline.

Exactly how the Fe participates in the electron-transfer reaction is not yet clear. The Fe is ferrous in dark-adapted RCs, and its redox state appears not to change when the primary acceptor is reduced with dithionite (Debrunner et al., 1975).

3. Absorbance Changes Accompanying Light-Induced Oxidation and Reduction of Quinones

3.1. Photoreduction: Formation and Dismutation of Semiquinones

In solution, UQ has an absorption band at 275 nm. This band bleaches and is replaced by a weaker band near 290 nm when the Q is reduced to either the semiquinone or the dihydroquinone. The semiquinone also has absorption bands near 320 and 450 nm that vary in their relative intensities, depending on whether or not the semiquinone is protonated (Bensasson and Land, 1973). Illumination of cell suspensions, chromatophores, or RC preparations from photosynthetic bacteria causes absorbance changes in the regions of all these bands. Some of these absorbance changes are associated with the reduction of a Q in the primary photochemical reaction (see Chapters 19 and 22). The subsequent transfer of an electron from the primary acceptor to the secondary one appears to cause only very small absorbance changes, in agreement with the view that the secondary acceptor is also a quinone.

Ke (1969) investigated the speed with which an absorbance decrease occurred at 280 nm, following excitation of *Chr. vinosum* chromatophores with a short flash. He found that the absorbance change was essentially complete within the 0.5-μsec limit of resolution of his detection equipment. This is considerably less than the half-time of 60 μsec that appears to be required for the movement of an electron from the primary electron acceptor to a secondary acceptor (see Section 2.1). Ke (1969) was not able to detect the occurrence of any further absorbance changes at 280 nm over the next 100 μsec following the initial absorbance decrease. It seems likely that the fast absorbance decrease that he measured was due largely to the primary photochemical reaction, and that any subsequent absorbance changes that accom-

panied the movement of an electron to the secondary acceptor were too small for him to detect.

The reduction of the primary electron acceptor in isolated RCs also causes absorbance increases near 320 and 450 nm that are similar to those that accompany the reduction of UQ to the anionic semiquinone (Clayton and Straley, 1970, 1972; Slooten, 1972) (see also Chapter 19). These absorbance changes can be seen most clearly after illumination in the presence of reductants that keep P_{870} fully reduced. Like the absorbance decrease at 275 nm, the absorbance increase at 450 nm does not relax when an electron acceptor moves from the primary acceptor to a secondary one. Following the excitation of reaction centers with a single flash, the 450 nm absorption band can persist for several minutes, even if the RCs contain excess UQ (Wraight et al, 1975; Wraight, 1977b; Vermeglio, 1977). But if the RCs are excited by a second flash, the 450 nm absorbance band that was generated by the first flash disappears abruptly (Wraight, 1977b; Vermeglio, 1977; Vermeglio, and Clayton, 1977). The band reappears if one gives a third flash, and continues to oscillate with a periodicity of two on subsequent flashes. Similar oscillations occur in an ESR signal that appears to reflect a ubi-semiquinone radical interacting with the Fe of the RC (Wraight, 1977b). The oscillations do not occur if the secondary reaction is blocked with o-phenanthroline, and they do not occur if the UQ content of the RCs is reduced to 1 mol/mol P_{870}.

These observations suggest that the UQ that is reduced in the secondary reaction following the first flash remains in the form of an anionic semiquinone. Its absorption spectrum would thus be similar to that of the reduced primary acceptor. When the primary reaction is repeated on the second flash, generating another semiquinone, the two semiquinones dismutate rapidly, giving the fully reduced and fully oxidized quinones (Vermeglio, 1977; Vermeglio and Clayton, 1977; Wraight, 1977b). In agreement with this interpretation, Vermeglio (1977) found that the absorbance decrease at 275 nm does not disappear on the even-numbered flashes, but instead becomes progressively larger in amplitude on each flash. One expects the 275-nm absorption band of UQ to bleach when either the semiquinone or the fully reduced quinone is formed, whereas the 450 nm band occurs only in the semiquinone.

The reduction of the primary or secondary quinones also causes small shifts in the 540 and 760 nm absorption bands of the bacteriopheophytin (Bpheo) in the RC (Clayton and Straley, 1970, 1972; Vermeglio and Clayton, 1977). These appear to reflect electrostatic interactions of the Bpheo with the semiquinones.

Vermeglio and Clayton (1977) found that the magnitudes and shapes of the spectral changes are somewhat different, depending on whether it is the primary or secondary Q that is reduced. As indicated above, the spectrum of the absorbance changes accompanying reduction of the primary Q can be obtained by excitation with single flashes in the presence of *o*-phenanthroline, and that associated with the secondary Q from similar measurements in the absence of *o*-phenanthroline. Vermeglio and Clayton (1977) used the change in the spectrum in the Bpheo bands to measure the rate of the secondary electron-transfer reaction. They found that the spectrum changed from one form to the other with a half-time of 200 μsec, following excitation of *Rp. sphaeroides* RCs with a single flash in the absence of *o*-phenanthroline. The rate appeared to be essentially the same after the second flash as it was after the first. [Triple-flash experiments in *Chr. vinosum* chromatophores are in agreement with this conclusion (Case and Parson, 1971).]

The measurements by Vermeglio and Clayton (1977) of the Bpheo absorption band shifts suggest that the semiquinone that is generated in the secondary electron-transfer reaction remains closely associated with the RC, rather than migrating into the bulk pool of Q. The semiquinone would have an electrostatic effect on the Bpheo only if the two molecules were reasonably close together. Measurements of the ESR spectrum of the semiquinone support this view: the secondary semiquinone appears to interact magnetically with the Fe of the RC in much the same way that the primary semiquinone does (Wraight, 1977b). In accord with this conclusion, Okamura *et al.* (1975) found that isolated RCs from *Rp. sphaeroides* bind 2 mol UQ per mol P_{870}, one of which is relatively easily removed by treatment with *o*-phenanthroline and detergents.

Vermeglio and Clayton (1977) found that the dismutation of the two semiquinones, as reflected by the disappearance of the 450 nm absorption band on the second flash, occurs at the same rate as the secondary electron-transfer reaction, as reflected by the change in the electrostatic effect on the Bpheo. The transfer of electrons to the larger pool of UQ would appear to follow the dismutation, or to occur simultaneously with it. If isolated RCs are supplemented with a large amount of UQ, oscillations in the semiquinone concentration can continue for as many as 20 flashes, while the pool of UQ becomes increasingly reduced (Vermeglio, 1977).

Continuous illumination of cells or chromatophores can cause a similar photoreduction of the bulk pool of endogenous UQ (Vernon, 1968). This process requires repeated cycling of the phosotynthetic apparatus and the presence of an appropriate source of electrons, and it is seen most clearly if one adjusts conditions so that the Q pool is largely in the oxidized state in the dark. One can study the change in the redox state of the Q pool either by direct absorbance measurements *in situ* (Bales and Vernon, 1963; Vernon, 1963; Takamiya and Takamiya, 1967, 1969a), or by rapid extraction of the quinone into an organic solvent and analysis of the extract (Takamiya and Takamiya, 1969b). Among the conditions that are capable of eliciting the photoreduction are the illumination of anaerobic chromatophores of *Rs. rubrum* in the presence of reduced 2,6-dichlorophenolindophenol (Bales and Vernon, 1963) and the illumination of aerobic cells of *Chr. vinosum* in the presence of $Na_2S_2O_3$ (Takamiya and Takamiya, 1967, 1969b).

It is tempting to assume that the Q pool in chromatophores provides a channel for the transfer of electrons from the region of one RC to that of another. If this is the case, the migration of reducing equivalents through the pool might allow semiquinones that are generated at separate RCs to dismutate rapidly even after a single flash. Oscillations like those that occur in isolated RCs would then not be seen. There is as yet no firm evidence on this point, but we shall return to it in Section 5. (See also Section 6.)

3.2. Photooxidation

If anaerobic suspensions of intact cells contain adequate levels of a reducing substrate such as succinate or malate, the endogenous Qs are largely reduced (Sugimura and Okabe, 1962; Redfearn, 1967; Takamiya and Takamiya, 1969b). Under these conditions, illumination causes the Q pool to become substantially more oxidized. Again, the shift in redox state can be detected either by rapid extraction techniques (Redfearn, 1967; Takamiya and Takamiya, 1969b) or by direct measurement of absorbance changes in the cell suspension (Parson, 1967; Takamiya and Takamiya, 1967, 1969b). Using rapid extraction, Takamiya and Takamiya (1969b) found that the endogenous UQ of an anaerobic *Chr. vinosum* cell suspension changed from 24% oxidized to 78% oxidized on illumination. In aerobic suspensions supplied with malate, 82% of the UQ was in the oxidized state in the dark, and light caused only a small additional oxidation. Using direct absorbance measurements, Takamiya (1971) obtained similar results in "*Chloropseudomonas ethylica*."* Under anaerobic conditions, light caused increases in optical absorbance between

---

* See Chapter 2 for a discussion of the taxonomy of this syntrophic system.

260 and 290 nm that Takamiya (1971) interpreted as reflecting the oxidation of MK-7, the predominant Q in this mixture of symbionts. Takamiya calculated that illumination caused the oxidation of about 60% of the total MK-7 pool. Under aerobic conditions, the absorbance changes were in the same direction, but were much smaller.

Like the photoreduction of the Q pool, the photo-oxidation of endogenous Qs is a relatively slow process. During continuous illumination, the absorbance changes that reflect the oxidation generally require about 1 min to reach completion (Parson, 1967; Takamiya and Takamiya, 1967; Takamiya, 1971). A lag period of 10–20 sec sometimes precedes the onset of the oxidation.

It seems likely that the photooxidation of endogenous Qs results from an energy-linked reverse electron flow to pyridine nucleotides. NADH can feed electrons into the Q pool through one or more dehydrogenases (King and Drews, 1973; Nisimoto et al., 1973), and the flow of electrons in the opposite direction could, in principle, be driven by ATP or a chemiosmotic potential that is generated in the light. Because the $E_{m,7}$ of the endogenous UQ is near 0 mV (see Section 4), the reduction of NADH ($E_{m,7} = -0.32$ V) would require the expenditure of approximately 14 kcal/mol. Chapter 32 discusses the photoreduction of NADH in greater detail.

4. Redox Titrations of the Secondary Acceptor

By titrating with mixtures of succinate and fumarate, and determining the redox state of the Q spectrophotometrically, Kakuno et al. (1973) measured an $E_{m,7}$ of +2 mV for the endogenous Q-10 of Rs. rubrum. The reduction was a two-electron process, coupled to the uptake of one proton per electron. Using a polarographic technique, Erabi et al. (1975) measured an $E_{m,7}$ of +50 mV; they identified the polarographic wave due to UQ by the fact that it disappeared if the chromatophores were extracted with i-octane and reappeared when they were reconstituted with Q-10. Urban and Klingenberg (1969) reported an $E_{m,7}$ of +50 mV for the Q-10 of beef heart mitochondria.

The polarographic half-wave potential of UQ in aqueous ethanol was reported to be +126 mV by Moret et al. (1961) and +43 mV by Erabi et al. (1975). [Other workers have pointed out complications in polarographic measurements of this type (O'Brien and Olver, 1969).] Rhodoquinone and the menaquinones have $E_{m,7}$ values that are lower than

that of UQ. Wagner et al. (1974) gave values near −80 mV for various MKs in solution; Erabi et al. (1975) gave −63 mV for rhodoquinone in aqueous ethanol and −30 mV for rhodoquinone bound to chromatophores.

In nonpolar, aprotic solvents, the one-electron reduction of UQ to the anionic semiquinone occurs reversibly at a potential of approximately −0.6 V with respect to the standard hydrogen electrode (Marcus and Hawley, 1971). This is considerably more negative than the apparent E_m of the primary electron acceptor (see Chapters 22 and 24). Environmental effects must influence the E_m significantly in vivo. [The two-electron reduction potentials mentioned in the preceding paragraphs are the means of the E_m's for the two one-electron steps; the difference between the E_m's of the two steps is related to the dismutation equilibrium constant. Mitchell (1976) provides further discussion of these relationships.]

There have been several attempts to distinguish absorbance changes due to the secondary electron-transfer reaction from those due to the primary reaction by examining differences in the way they depend on the ambient redox potential. If the E_m of the secondary acceptor is more positive than that of the primary acceptor, lowering the redox potential might block the secondary electron-transfer reaction before it blocks the primary reaction. Because redox titrations of this type require the establishment of redox equilibrium prior to the illumination of the sample, they probe only equilibrium states of the system. If the secondary acceptor is UQ, the redox titration could reflect the overall two-electron reduction to the dihydroquinone, rather than the one-electron reduction of the anionic semiquinone, which is probably the initial product in the light-driven reactions.

In his study of Chr. vinosum chromatophores (Section 3.1), Ke (1969) reported that the relatively mild reductant ascorbate prevented the occurrence of the flash-induced absorbance changes in the UV, without preventing the photooxidation of P_{870}, as measured from IR absorbance changes. Later, however, Ke et al. (1973) attributed these results to the use of different conditions for the two types of measurements. The UV absorbance measurements had involved repeated excitation of the chromatophores, with what could have been insufficient time for recovery between excitations. Allowing a longer recovery period, Ke et al. (1973) found that the UV absorbance changes did in fact occur in the presence of ascorbate. This is what one would expect of absorbance changes that are due mainly to the primary photochemical reaction (see Section 3.1).

In another study, Reed et al. (1969) found that

lowering the ambient redox potential through the region from +50 to −50 mV prevented the UV absorbance changes that resulted from continuous illumination of *Rp. sphaeroides* RCs. The $E_{m, 7.5}$ of the titration was approximately 0 mV. Fluorescence measurements showed that the photooxidation of P_{870}, in contrast, still occurred at lower redox potentials, until it finally declined with an E_m of approximately −50 mV. Whether the two types of measurements were strictly comparable in these experiments is not entirely clear; the experiments deserve repeating.

Case and Parson (1971) attempted to titrate the secondary acceptor in *Chr. vinosum* chromatophores by measuring the effect of redox potential in double-flash experiments. Under certain conditions, the amount of photochemistry that occurs on the second of two flashes declines with an apparent $E_{m, 7.7}$ of approximately −100 mV. This value is about 30 mV more positive than the E_m with which photochemical activity on the first flash declines. The two redox titrations exhibit similar, though not quite identical, dependences on temperature, pH, and ionic strength (Case and Parson, 1971, 1973).

The apparent $E_{m, 7.7}$ of −100 mV was obtained by titrating chromatophores in the presence of a collection of redox buffers that did not include PMS. If 100 μM PMS was included in the redox buffers, photochemical activity on the second flash titrated with a much more positive apparent $E_{m, 7}$ of +80 mV (Case and Parson, 1971). It is not clear whether this means that PMS perturbs the titration, or that the secondary acceptor is titrated properly only in the presence of PMS. PMS does not affect redox titrations of the other known electron carriers in *Chr. vinosum* chromatophores.

To calculate a true E_m from an apparent one, it is necessary to have information on the size of the secondary electron acceptor pool. If the pool is relatively large, the reduction of a major part of the pool might have little effect on the capacity of the pool to oxidize the primary acceptor, just as does the extraction of a large part of the UQ (Section 2.2). The apparent E_m of the secondary acceptor could therefore be considerably more negative than the true E_m. A possible source of information on the pool size is the amount of photochemistry that occurs on the second flash, when the flashes are spaced relatively far apart (10 msec), and when the redox potential is sufficiently positive so that the secondary acceptor pool is fully oxidized before the flashes. Under these conditions, the amount of cytochrome photooxidation that occurs on the second flash is about 75% of that which occurs on the first flash. Case and Parson (1971) interpreted this to mean that the primary acceptor does not become

completely reoxidized in the period between the flashes. Assuming that the primary and secondary acceptors attain electrochemical equilibrium in this period, the extent of reoxidation of the primary acceptor will depend on the pool sizes, as well as on the equilibrium constant for the reaction between the two acceptors (i.e., the difference between the true E_m values). Using a simple model, Case and Parson (1971) derived theoretical curves for the amounts of photochemistry that would occur on two flashes, both at high redox potentials and through the redox titrations of the two acceptors, depending on the equilibrium constant and the pool sizes (see also Case, 1972). The redox titrations in the absence of PMS were fit best on the assumptions that the secondary acceptor was a one-electron carrier with an E_m about 33 mV more positive than that of the primary acceptor, and that the pool size of the secondary acceptor was twice that of the primary acceptor. This would suggest that the UQ that participates in the secondary reaction is quite distinct from the main UQ pool of the chromatophores.

The assumptions that underlie this treatment were called into question by later studies of chromatophores that had been extracted and reconstituted with various Qs (Halsey and Parson, 1974). At high redox potentials, the amount of photochemistry that occurred on the second a pair of widely separated flashes proved to be much the same, whether UQ or phylloquinone was used for the reconstitution. Because the $E_{m, 7}$'s of the two Qs differ by approximately 100 mV, the E_m of the secondary acceptor evidently does not play a decisive role in determining the amount of photochemistry that occurs on the second flash. The relationship between the apparent redox titrations of the secondary acceptor and the true E_m of the acceptor thus remains unclear.

For a final puzzle concerning the redox properties of the secondary electron acceptor, we might return to Table 1 and ask why some species of photosynthetic bacteria enjoy the presence of both UQ and MK, while others contain only one type of Q. J. M. Olson (personal communication) suggested that MK is required for growth under strongly reducing conditions. Because its E_m is substantially lower than that of Q, MK might remain functional at redox potentials that are low enough so that Q is almost completely reduced. As yet, however, there is no evidence that MK acts as a secondary electron acceptor at low potentials, or for that matter, at any potential. In *Chr. vinosum* cell suspensions, the kinetics of the secondary reaction are slower at low redox potentials than they are at higher potentials (Parson and Case, 1970), but it is not clear whether this is due to the operation of different acceptors, or simply to a decrease in the availability of oxidized UQ at low potentials. The role of MK needs further exploration.

5. Proton Uptake Linked to the Secondary Reaction

In both the presence and the absence of PMS, the apparent E_m of the secondary acceptor in *Chr. vinosum* chromatophores depends on the pH. The E_m decreases by about 30 mV/pH unit in the absence of PMS, and by about 60 mV in its presnece (Case and Parson, 1971). The latter value is the expected one for the uptake of one proton per electron on reduction, and would be consistent with the reduction of UQ to the dihydroquinone. The binding of a proton on the reduction of the secondary acceptor would fit well with the observation that the rate of the secondary electron-transfer reaction increases with decreasing pH (see Section 2.1).

As Chapters 24 and 26 discuss, proton uptake can be measured directly, after flash excitation of chromatophores. The proton-binding is largely (though not completely) inhibited by *o*-phenanthroline or by the extraction of UQ, suggesting that it is linked to the secondary electron-transfer reaction (Cogdell *et al.*, 1972; Halsey and Parson, 1974; Carithers and Parson, 1975). Its dependence on temperature and pH are similar to those of the secondary reaction (Petty and Dutton, 1976). In *Rp. sphaeroides* chromatophores, the proton uptake involves a component with an apparent $E_{m,7}$ of +5 mV (Cogdell *et al.*, 1972); the uptake decreases as one lowers the ambient potential through this region. This is close to the $E_{m,7}$ values that Kakuno *et al.* (1973) and Erabi *et al.* (1975) assigned to UQ pool in *Rs. rubrum* chromatophores. (One should bear in mind the distinction between real and apparent E_m values; if the proton-binding component has a large pool size, proton uptake measured with single flashes might not titrate in parallel with the titration of the pool. In addition, components that are required for the proton uptake to occur are not necessarily the species that actually bind the protons.) The apparent pK of the proton-binding component is 8.5 when the neighboring P_{870} is in the reduced state, but it drops to 7.5 if the P_{870} is oxidized (Petty and Dutton, 1976). Petty and Dutton (1976) point out that the pK values are reasonably close to the value of 6.5 that Land and Swallow (1970) gives for ubisemiquinone in aqueous ethanol. They interpret the influence of the redox state of P_{870} as a coulombic effect of the P_{870}^+ cation.

The kinetics of the proton uptake following flash excitation differ significantly from the kinetics that have been measured for the secondary electron-transfer reaction (see Section 2.1). The proton uptake is kinetically first-order. At pH 7, its half-time is 140 μsec in *Chr. vinosum* (Halsey and Parson, 1974), 150 μsec in *Rp. sphaeroides* (Petty and Dutton, 1976), and 510 μsec in *Rp. viridis* (Carithers and Parson, 1975). Half-times do not serve to characterize the secondary electron-transfer reaction adequately, because the process is not cleanly first-order, but the electron transfer appears to be decidedly faster than the proton uptake in *Chr. vinosum* and *Rp. viridis*.

To explain the kinetic discrepancies, Petty and Dutton (1976) suggested that the UQ molecule that first removes electrons from the primary acceptor is buried within the RC complex, so that is is not accessible for protonation. The proton uptake would follow the movement of an electron from this UQ into a pool of UQ that has access to the solution on the outside of the chromatophore membrane.

Additional insight into the kinetics of the proton uptake is provided by work on isolated RCs. As Section 3.1 discusses, the UQ that is reduced in the secondary reaction following the excitation of RCs with a single flash appears to remain as an unprotonated semiquinone. In agreement with this, proton uptake from the solution does not occur after a single flash (Wraight *et al.*, 1975; Wraight, 1977a). Proton uptake does occur on a second flash, however, and on subsequent flashes the proton uptake continues to oscillate with a periodicity of two (Wraight, 1977a). Note that the oscillations in proton uptake are out of phase with the oscillations in the 450 nm absorption band or the ESR signal reflecting the presence of the anionic semiquinone. Proton uptake is apparently linked, not to the secondary electron-transfer reaction itself, but to the dismutation that occurs on the even-numbered flashes. However, the transfer of reducing equivalents from the Qs of the RC to the UQ pool also appears to accompany the dismutation (see Section 3.1). It is therefore not clear whether the proton uptake reflects the dismutation reaction *per se*, or the reduction of UQ in the bulk pool.

Unlike isolated RCs, chromatophores do take up protons after excitation with a single flash. One could explain this on the assumption that in chromatophores, semiquinones generated in separate RCs can interact so that the dismutation can occur after one flash. The additional steps that are necessary for this interaction could explain the slow kinetics of the proton uptake, although one might expect that the proton uptake would then be kinetically second order.

The reduction of the secondary acceptor evidently does not result in the binding or release of any ions other than protons, because the reduction does not cause a change in the net electric charge on the chromatophore membrane. This conclusion follows from the observation that the apparent E_m of the acceptor (as measured in the absence of PMS) is independent of ionic strength (Case and Parson, 1973).

6. Note Added in Proof

Two recent reports (Y. Barouch and R. K. Clayton, 1977, *Biochim. Biophys. Acta* **462**:785, and B. G. de Grooth, J. C. Romijn, and M. P. J. Pulles, 1977, *Proc. 4th Int. Congr. Photosynth.*, Reading, England) describe oscillations in quinone reduction and proton uptake in intact chromatophores.

ACKNOWLEDGMENTS

I am indebted to R. E. Blankenship, R. K. Clayton, R. J. Cogdell, P. L. Dutton, and C. A. Wraight for helpful discussion, and to the National Science Foundation for financial support.

7. References

Amesz, J., 1973, The function of plastoquinone in photosynthetic electron transport, *Biochim. Biophys. Acta* **301**:35.

Bales, H., and Vernon, L. P., 1963, Effect of reduced 2,6-dichlorophenolindophenol upon the light-induced absorbancy changes in *Rhodospirillum rubrum* chromatophores: A coupled reduction of ubiquinone, in: *Bacterial Photosynthesis* (H. Gest, A. San Pietro, and L. P. Vernon, eds.). 269–274, Antioch Press, Yellow Springs, Ohio.

Barr, R., and Crane, F. L., 1972, Quinones in algae and higher plants, in: *Methods in Enzymology*, Vol. XXIIIa (A. San Pietro, ed.), pp. 372–408, Academic Press, New York.

Bensasson, R., and Land, E. J., 1973, Optical and kinetic properties of semireduced plastoquinone and ubiquinone: Electron acceptors in photosynthesis, *Biochim. Biophys. Acta* **325**:175.

Carithers, R. P., and Parson, W. W., 1975, Delayed fluorescence from *Rhodopseudomonas viridis* following single flashes, *Biochim. Biophys. Acta* **387**:194.

Carithers, R. P., and Parson, W. W., 1976, Delayed fluorescence from *Rhodopseudomonas sphaeroides* following single flashes, *Biochim. Biophys. Acta* **440**:215.

Carr, N. G., and Exell, G., 1965, Ubiquinone concentrations in Athiorhodaceae grown under various environmental conditions, *Biochem. J.* **96**:688.

Case, G. D., 1972, Free energy capture and photosynthetic electron transport in *Chromatium*, Ph.D. Thesis, University of Washington, p. 49.

Case, G. D., and Parson, W. W., 1971, Thermodynamics of the primary and secondary photochemical reactions in *Chromatium, Biochim. Biophys. Acta* **253**:187.

Case, G. D., and Parson, W. W., 1973, Redistribution of electric charge accompanying photosynthetic electron transport in *Chromatium, Biochim. Biophys. Acta* **292**:677.

Case, G. D., Parson, W. W., and Thornber, J. P., 1970, Photooxidation of cytochromes in reaction center preparations from *Chromatium* and *Rhodopseudomonas viridis, Biochim. Biophys. Acta* **223**:122.

Chamorovsky, S. K., Remennikov, S. M., Kononenko, A. A., Venediktov, P. S., and Rubin, A. B., 1976, New experimental approach to the estimation of rate of electron transfer from the primary to secondary acceptors in the photosynthtic bacteria, *Biochim. Biophys. Acta* **430**:62.

Clayton, R. K., and Straley, S. C., 1970, An optical absorption change that could be due to reduction of the primary photochemical electron acceptor in photosynthetic reaction centers, *Biochem. Biophys. Res. Commun.* **39**:1114.

Clayton, R. K., and Straley, S. C., 1972, Photochemical electron transport in photosynthetic reaction centers. IV. Observations related to the reduced photoproducts, *Biophys. J.* **12**:1221.

Clayton, R. K., and Yau, H. F., 1972, Photochemical electron transport in photosynthetic reaction centers from *Rhodopseudomonas spheroides*. I. Kinetics of the oxidation and reduction of P-870 as affected by external factors, *Biophys. J.* **12**:867.

Clayton, R. K., Fleming, H., and Szuts, E. Z., 1972a, Photochemical electron transport in photosynthetic reaction centers from *Rhodopseudomonas spheroides*. II. Interaction with external electron donors and acceptors and a re-evaluation of some spectroscopic data, *Biophys. J.* **12**:46.

Clayton, R. K., Szuts, E. Z., and Fleming, H., 1972b, Photochemical electron transport in photosynthetic reaction centers from *Rhodopseudomonas spheroides*. III. Effects of orthophenanthroline and other chemicals, *Biophys. J.* **12**:64.

Cogdell, R. J., and Crofts, A. R., 1972, Some observations on the primary acceptor of *Rhodopseudomonas viridis, FEBS Lett.* **27**:176.

Cogdell, R. J., Jackson, J. B., and Crofts, A. R., 1972, The effect of redox potential on the coupling between rapid hydrogen-ion binding and electron transport in chromatophores from *Rhodopseudomonas spheroides, Bioenergetics* **4**:413.

Cogdell, R. J. Brune, D. C., and Clayton, R. K., 1974, Effects of extraction and replacement of ubiquinone upon the photochemical activity of reaction centers and chromatophores from *Rhodopseudomonas spheroides, FEBS Lett.* **45**:344.

Crane, F. L., and Barr, R., 1971, Determination of ubiquinones, in: *Methods in Enzymology*, Vol. XVIIIc (D. B. McCormick and L. D. Wright, eds.), pp. 137–165, Academic Press, New York.

Cusanovich, M. A., and Kamen, M. D., 1968, Light-induced electron transport in *Chromatium* chromatophores, *Biochim. Biophys. Acta* **153**:376.

Debrunner, P. G., Schultz, C. E., Feher, G., and Okamura, M. Y., 1975, Mossbauer study of reaction centers from *R. spheroides, Biophys. J.* **15**:226a (abstract).

Dunphy, P. J., and Brodie, A. F., 1971, The structure and function of quinones in respiratory metabolism, in: *Methods in Enzymology*, Vol. XVIIIc (D. B. McCormick and L. D. Wright, eds.), pp. 407–461, Academic Press, New York.

Dutton, P. L., Kihara, T., McCray, J. A., and Thornber, J. P., 1971, Cytochrome C_{553} and bacteriochlorophyll interaction at 77°K in chromatophores and a subchromatophore preparation from *Chromatium D, Biochim. Biophys. Acta* **226**:81.

Dutton, P. L., Leigh, J. S., and Wraight, C. A., 1973, Direct measurement of the midpoint potential of the primary electron acceptor in *Rps. spheroides in situ* and in the isolated state: Some relationships with pH and *o*-phenanthroline, *FEBS Lett.* **36**:169.

Erabi, T., Higuti, T., Kanuko, T., Yamashita, J., Tanaka, M., and Horio, T., 1975, Polarographic studies on ubiquinone-10 and rhodoquinone bound with chromatophores from *Rhodospirillum rubrum, J. Biochem. (Tokyo)* **78**:795.

Evans, M. C. W., Lord, A. V., and Reeves, S. G., 1974, The detection and characterization by electron-paramagnetic-resonance spectroscopy of iron–sulphur proteins and other electron-transport compounds in chromatophores from the purple bacterium *Chromatium, Biochem. J.* **138**:177.

Feher, G., Isaacson, R. A., McElroy, J. D. Ackerson, L. C., and Okamura, M. Y., 1974, On the question of the primary acceptor in bacterial photosynthesis: Manganese substituting for iron in reaction centers of *Rhodopseudomonas spheroides* R-26, *Biochim. Biophys. Acta* **368**:135.

Feher, G., Okamura, M. Y., and McElroy, J. D., 1972, Identification of an electron acceptor in reaction centers of *Rhodopseudomonas sphaeroides* by EPR spectroscopy, *Biochim. Biophys. Acta* **267**:222.

Fleischman, D. E., 1974, Delayed fluorescence and the reversal of primary photochemistry in *Rhodopseudomonas viridis, Photochem. Photobiol.* **19**:59.

Fleischman, D. E., and Clayton, R. K., 1968, The effect of phosphorylation uncouplers and electron transport inhibitors upon spectral shifts and delayed light emission of photosynthetic bacteria. *Photochem. Photobiol.* **8**:287.

Frydman, B., and Rapoport, H., 1963, Non-chlorophyllous pigments of *Chlorobium thiosulfatophilum*, chlorobiumquinone, *J. Am. Chem. Soc.* **85**:823.

Fuller, R. C., Smillie, R. M., Rigopoulos, N., and Yount, V., 1961, Comparative studies of some quinones in photosynthetic systems, *Arch. Biochem. Biophys.* **95**:197.

Geller, D. M., 1962, Oxidative phosphorylation in extracts of *Rhodospirillum rubrum, J. Biol. Chem.* **237**:2947.

Gutman, M., Singer, T. P., Beinert, H., and Casida, J. E., 1970, Reaction sites of rotenone, piericidin A, and amytal in relation to the nonheme iron components of NADH dehydrogenase, *Proc. Natl. Acad. Sci. U.S.A.* **65**:763.

Halsey, Y. D., and Byers, B., 1975, A large photoactive particle from *Chromatium vinosum* chromatophores, *Biochim. Biophys. Acta* **387**:349.

Halsey, Y. D., and Parson, W. W., 1974, Identification of ubiquinone as the secondary electron acceptor in the photosynthetic apparatus of *Chromatium vinosum, Biochim. Biophys. Acta* **347**:404.

Hatefi, Y., Stempel, K. E., and Hanstein, W. G., 1969, Inhibitors and activators of the mitochondrial reduced diphosphopyridine nucleotide dehydrogenase, *J. Biol. Chem.* **244**:2358.

Higuti, T., Erabi, T., Kakuno, T., and Horio, T., 1975, Role of ubiquinone-10 in electron transport system of chromatophores from *Rhodospirillum rubrum, J. Biochem. (Tokyo)* **78**:51.

Horio, T., Nishikawa, K., Okayama, S., Horiuti, Y., Yamamoto, N., and Kakutani, Y., 1968, The requirement of ubiquinone-10 for an ATP-forming system and an ATPase system of chromatophores from *Rhodospirillum rubrum, Biochim. Biophys. Acta* **153**:913.

Hsi, E. S. P., and Bolton, J. R., 1974, Flash photolysis–electron spin resonance study of the effect of *o*-phenanthroline and temperature on the decay time of the ESR signal B1 in reaction-center preparations and chromatophores of mutant and wild strains of *Rhodopseudomonas sphaeroides* and *Rhodospirillum rubrum, Biochim. Biophys. Acta* **347**:126.

IUPAC–IUB, 1975, Nomenclature on quinones with isoprenoid side chains, *Biochim. Biophys. Acta* **387**:397.

Jackson, J. B., Cogdell, R. J., and Crofts, A. R., 1973, Some effects of *o*-phenanthroline on electron transport in chromatophores from photosynthetic bacteria, *Biochim. Biophys. Acta* **292**:218.

Jeng, M., Hall, C., Crane, F. L., Takahashi, N., Tamura, S., and Folkers, K., 1968, Inhibition of mitochondrial electron transport by piericidin A and related compounds, *Biochemistry* **7**:1311.

Kakuno, T., Hosoi, K., Higuti, T., and Horio, T., 1973, Electron and proton transport in *Rhodospirillum rubrum* chromatophores, *J. Biochem.* **74**:1193.

Ke, B., 1969, Nature of the primary electron acceptor in bacterial photosynthesis, *Biochim. Biophys. Acta* **172**:583.

Ke, B., Garcia, A. F., and Vernon, L. P., 1973, Light-induced absorption changes in *Chromatium* subchromatophore particles exhaustively extracted with nonpolar solvents, *Biochim. Biophys. Acta* **292**:226.

King, M. T., and Drews, G., 1973, The function and localization of ubiquinone in the NADH and succinate oxidase systems of *Rhodopseudomonas palustris, Biochim. Biophys. Acta* **305**:230.

Knaff, D. B., 1975, The effect of *o*-phenanthroline on the midpoint potential of the primary electron acceptor of photosystem II, *Biochim. Biophys. Acta* **376**:583.

Land, E. J., and Swallow, A. J., 1970, One-electron reactions in biochemical systems as studied by pulse radiolysis. III. Ubiquinone, *J. Biol. Chem.* **245**:1890.

Lester, R. L., and Crane, F. L., 1959, The natural occurrence of coenzyme Q and related compounds, *J. Biol. Chem.* **234**:2169.

Loach, P. A., and Hall, R. L., 1972, The question of the primary electron acceptor in bacterial photosynthesis, *Proc. Natl. Acad. Sci. U.S.A.* **69**:786.

Loschen, G., and Azzi, A., 1974, Dibromothymoquinone: A new inhibitor of mitochondrial electron transport at the level of ubiquinone, *FEBS Lett.* **41**:115.

Marcus, M. F., and Hawley, M. D., 1971, Electrochemical studies of the redox behavior of ubiquinone-1, *Biochim. Biophys. Acta* **226**:234.

Maroc, J., de Klerk, H., and Kamen, M. D., 1968, Quinones of Athiorhodaceae, *Biochim. Biophys. Acta* **162**:621.

Mayer, H., and Isler, O., 1971, Isolation of vitamins K, in: *Methods in Enzymology*, Vol. XVIIIc (D. B. McCormick and L. D. Wright, eds.), pp. 469–491, Academic Press, New York.

McElroy, J. D., Mauzerall, D. C., and Feher, G., 1974, Characterization of primary reactants in bacterial photosynthesis. II. Kinetic studies of the light-induced EPR signal ($g = 2.0026$) and the optical absorbance changes at cryogenic temperatures, *Biochim. Biophys. Acta* **333**:261.

Mitchell, P., 1976, Possible molecular mechanisms of the protonmotive function of cytochrome systems, *J. Theor. Biol.* **62**:327.

Moore, H. W., and Folkers, K., 1965, Coenzyme Q. LXII. Structure and synthesis of rhodoquinone, a natural aminoquinone of the coenzyme Q group, *J. Am. Chem. Soc.* **87**:1409.

Moret, V., Pinamonti, S., and Fornasari, E., 1961, Polarographic study on the redox potential of ubiquinones, *Biochim. Biophys. Acta* **54**:381.

Nicholson, G., and Clayton, R. K., 1969, The reducing potential of the bacterial photosynthetic reaction center, *Photochem. Photobiol.* **9**:395.

Nisimoto, Y., Kakuno, T., Yamashita, J., and Horio, T., 1973, Two different NADH dehydrogenases in respiration of *Rhodospirillum rubrum* chromatophores, *J. Biochem.* **74**:1205.

Noël, H., van der Rest, M., and Gingras, G., 1972, Isolation and partial characterization of a P_{870} reaction center complex from wild type *Rhodospirillum rubrum*, *Biochim. Biophys. Acta* **275**:219.

Norling, B., Glazek, E., Nelson, B. D., and Ernster, L., 1974, Studies with ubiquinone-depleted submitochondrial particles. Quantitative incorporation of small amounts of ubiquinone and its effects on the NADH and succinate oxidase activities, *Eur. J. Biochem.* **47**:475.

O'Brien, F. L., and Olver, J. W., 1969, Electrochemical study of ubiquinone-6 in aqueous methanol, *Anal. Chem.* **41**:1810.

Oettmeier, W., and Grewe, R., 1974, Inhibition of photosynthetic electron transport by azaphenanthrenes, *Z. Naturforsch.* **29c**:545.

Okamura, M. Y., Isaacson, R. A., and Feher, G., 1975, Primary acceptor in bacterial photosynthesis: Obligatory role of ubiquinone in photoactive reaction centers of *Rhodopseudomonas spheroides*, *Proc. Natl. Acad. Sci. U.S.A.* **72**:3491.

Okamura, M. Y., Ackerson, L. C., Isaacson, R. A., Parson, W. W., and Feher, G., 1976, The primary electron acceptor in *Chromatium vinosum* (strain D), *Biophys. J.* **16**:223A (abstract).

Okayama, S., Yamamoto, N., Nishikawa, K., and Horio, T., 1968, Roles of ubiquinone-10 and rhodoquinone in photosynthetic formation of adenosine triphosphate by chromatophores from *Rhodospirillum rubrum*, *J. Biol. Chem.* **243**:2995.

Parson, W. W., 1967, Observations on the changes in ultraviolet absorbance caused by succinate and light in *Rhodospirillum rubrum*, *Biochim. Biophys. Acta* **143**:263.

Parson, W. W., 1969, The reaction between primary and secondary electron acceptors in bacterial photosynthesis, *Biochim. Biophys. Acta* **189**:384.

Parson, W. W., and Case, G. D., 1970, In *Chromatium*, a single photochemical reaction center oxidizes both cytochrome C_{552} and cytochrome C_{555}, *Biochim. Biophys. Acta* **205**:232.

Parson, W. W., and Rudney, H., 1965, The biosynthesis of ubiquinone and rhodoquinone from *p*-hydroxybenzoate and *p*-hydroxybenzaldehyde in *Rhodospirillum rubrum*, *J. Biol. Chem.* **240**:1855.

Petty, K. M., and Dutton, P. L., 1976, Properties of the flash-induced proton binding encountered in membranes of *Rhodopseudomonas sphaeroides*: A functional pK on the ubisemiquinone? *Arch. Biochem. Biophys.* **172**:335.

Phelps, D. C., and Crane, F. L., 1975, Inhibition of mitochondrial electron transport by hydroxysubstituted 1,4-quinones, *Biochemistry* **14**:116.

Porter, T. H., Skelton, F. S., and Folkers, K., 1971, Synthesis of new 5,8-quinolinequinones as inhibitors of coenzyme Q and as antimalarials, *J. Med. Chem.* **14**:1029.

Porter, T. H., Bowman, C. M., and Folkers, K., 1973, Antimetabolites of coenzyme Q. 16. New antimalarial curative activity, *J. Med. Chem.* **16**:1314.

Powls, R., and Redfearn, E. R., 1969, Quinones of the Chlorobacteriaceae. Properties and possible function, *Biochim. Biophys. Acta* **172**:429.

Powls, R., Redfearn, E. R., and Trippett, S., 1968, The structure of chlorobiumquinone, *Biochim. Biophys. Acta* **33**:408.

Prince, R., and Dutton, P. L., 1976, The primary acceptor of bacterial photosynthesis: Its operating midpoint potential?, *Arch. Biochem. Biophys.* **172**:329.

Redfearn, E. R., 1967, Redox reactions of ubiquinone in *Rhodospirillum rubrum*, *Biochim. Biophys. Acta* **131**:218.

Reed, D. W., Zankel, K. L., and Clayton, R. K., 1969, The effect of redox potential on P_{870} fluorescence in reaction centers from *Rhodopseudomonas spheroides*, *Proc. Natl. Acad. Sci. U.S.A.* **63**:42.

Satoh, K., 1974, Action of some derivatives of 1,10-phenanthroline on electron transport in chloroplasts, *Biochim. Biophys. Acta* **333**:127.

Slooten, L., 1972, Electron acceptors in reaction center preparations from photosynthetic bacteria, *Biochim. Biophys. Acta* **275**:208.

Sugimura, T., and Okabe, K., 1962, The reduction of

ubiquinone (coenzyme Q) in chromatophores of *Rhodospirillum rubrum* by succinate, *J. Biochem.* **52**:235.

Sugimura, T., and Rudney, H., 1962, The effect of aerobiosis and diphenylamine on the content of ubiquinone in *Rhodospirillum rubrum, Biochim. Biophys. Acta* **62**:167.

Takamiya, K., 1971, The light-induced oxidation–reduction reactions of menaquinone in intact cells of a green photosynthetic bacterium, *Chloropseudomonas ethylica, Biochim. Biophys. Acta* **234**:390.

Takamiya, K., and Takamiya, A., 1967, Light-induced reactions of ubiquinone in photosynthetic bacterium, *Chromatium* D, *Plant Cell Physiol.* **8**:719.

Takamiya, K., and Takamiya, A., 1969a, Light-induced reactions of ubiquinone in photosynthetic bacterium, *Chromatium* D. II. Effects of inhibitors and other experimental conditions, *Plant Cell Physiol.* **10**:113.

Takamiya, K., and Takamiya, A., 1969b, Light-induced reactions of ubiquinone in photosynthetic bacterium, *Chromatium* D. III. Oxidation–reduction state of ubiquinone in intact cells of *Chromatium* D, *Plant Cell Physiol.* **10**:363.

Takamiya, K., and Takamiya, A., 1970, Nature of photochemical reactions in chromatophores of *Chromatium* D. I. Effects of isooctane extraction on the photochemical reactions of P_{890} and ubiquinone in chromatophores of *Chromatium* D, *Biochim. Biophys. Acta* **205**:72.

Takamiya, K., Nishimura, M., and Takamiya, A., 1967, Distribution of quinones in some photosynthetic bacteria and algae, *Plant Cell Physiol.* **8**:79.

Urban, P. F., and Klingenberg, M., 1969, On the redox potentials of ubiquinone and cytochrome *b* in the respiratory chain, *Eur. J. Biochem.* **9**:519.

Vermeglio, A., 1977, Secondary electron transfer in reaction centers of *Rhodopseudomonas sphaeroides*: Out-of-phase periodicity of two for the formation of ubisemiquinone and fully reduced ubiquinone, *Biochim. Biophys. Acta* **459**:516.

Vermeglio, A., and Clayton, R. K., 1977, Kinetics of electron transfer between the primary and the secondary electron acceptor in reaction centers from *Rhodopseudomonas sphaeroides Biochim. Biophys. Acta* **461**:159.

Vernon, L. P., 1963, Photooxidation and photoreduction reactions catalyzed by chromatophores of purple photosynthetic bacteria, in: *Bacterial Photosynthesis* (H. Gest, A. San Pietro, and L. P. Vernon, eds.), pp. 235–274, Antioch Press, Yellow Springs, Ohio.

Vernon, L. P., 1968, Photochemical and electron transport reactions of bacterial photosynthesis, *Bacteriol. Rev.* **32**:243.

Wagner, G. C., Kassner, R. J., and Kamen, M. D., 1974, Redox potentials of certain vitamins K: Implications for a role in sulfite reduction by obligately anaerobic bacteria, *Proc. Natl. Acad. Sci. U.S.A.* **71**:253.

Wraight, C. A., 1977a, A two-electron gate between the primary and secondary electron transfer processes of photosynthesis, *Bull. Am. Phys. Soc.* **22**:JI-10 (abstract).

Wraight, C. A., 1977b, Electron acceptors of photosynthetic bacterial reaction centers: Direct observation of oscillatory behavior suggesting two closely equivalent ubiquinones, *Biochim. Biophys. Acta* **459**:525.

Wraight, C. A., Cogdell, R. J., and Clayton, R. K., 1975, Some experiments on the primary electron acceptor in reaction centers from *Rhodopseudomonas sphaeroides, Biochim. Biophys. Acta* **396**:242.

Yamamoto, N., Hatekeyama, H., Nishikawa, K., and Horio, T., 1970, The function of ubiquinone-10 both in the electron transport system and in the energy conservation system of chromatophores from *Rhodospirillum rubrum, J. Biochem.* **67**:587.

CHAPTER 26

Ion Transport and Electrochemical Gradients in Photosynthetic Bacteria

C. A. Wraight, R. J. Cogdell, and B. Chance

1. Introduction

The primary source of energy in biological processes is that of oxidoreductive (redox) reactions in which electrons or hydrogen atoms are passed from one molecule to another with an overall drop in free energy. Mechanisms have evolved whereby the reaction pathway of a potentially free-energy-yielding reaction is constrained so as to be obligately couple to another reaction that, when considered alone, is endergonic. The free energy of one reaction is converted into that of another, and the overall (and, in general, the standard) free-energy change in the complete "coupled" process is thus small; the free energy of the exergonic, driving reaction is said to be conserved.

Free energy is available from the environment mainly through redox reactions, but metabolic free energy has many chemical and physical forms. Energy conservation must therefore allow energy transductions in which different free-energy forms are mechanistically linked. Although the essential coupling mechanisms of some comparatively minor transduction processes are known—e.g., in glycolysis—the major energy-conserving and -transducing mechanisms of the electron-transport pathways of respiration and photosynthesis have remained obscure, and attempts to apply the principles observed in the simpler soluble systems have not enjoyed much success. The general principles of electron-transport pathways were put forward many years ago by Keilin and have been impressively developed and extended by many otners since then. Much information has been gained on the nature and general classes of the electron carriers—cytochromes, flavins, iron–sulfur compounds, and quinones—and the positions of many components have been worked out by an enormous variety of techniques. However, progress on the nature of the coupling reactions and the components involved has been slow and painstaking. No functional chemical intermediate has been discovered and, instead, a growing body of evidence has emerged that localized and delocalized potentials and conformational interactions play a major role in the energy-coupling process.

The failure of more conventional approaches has led to an increased attention to the energetics of coupling. Work on photosynthetic organisms has contributed very significantly to a more widespread appreciation of the energetic requirements of the coupling process and, in particular, to the study of electrical and chemical gradients and their energetic contributions. This chapter reviews our current understanding of these topics in photosynthetic bacteria.

C. A. Wraight · Department of Physiology and Biophysics and Department of Botany, University of Illinois at Urbana–Champaign, Urbana, Illinois 61801 **R. J. Cogdell** · Department of Botany, University of Glasgow, Glasgow G12 8QQ, Scotland **B. Chance** · Department of Biochemistry and Biophysics, Johnson Research Foundation, University of Pennsylvania, Philadelphia, Pennsylvania 19174

2. An Outline of the Chemical and Chemiosmotic Coupling Hypotheses

2.1. Apology

Three main classes of coupling mechanisms can be somewhat arbitrarily defined, as described by the mode of energy transferral: chemical, conformational, and active-proton. Chemical coupling, which is represented by a large number of more or less detailed refinements on the classic hypothesis due to Slater (1953, 1971), is outlined in Section 2.2.

Conformational coupling has had a checkered history, with many transient proponents. It has suffered from a lack of experimentally testable tenets and from a high-energy intermediate that is as elusive as the chemical one. It is by no means discounted,

however, and serious formulations continue to be developed (Boyer, 1965, 1967; Boyer et al., 1973; Slater, 1972; Chance et al., 1970c).

Active-proton coupling is represented by two distinct approaches: the intramembrane active-proton hypothesis (Williams, 1961, 1962, 1969) and the transmembrane active-proton or chemiosmotic hypothesis (Mitchell, 1961, 1966, 1968).

The chemical and chemiosmotic hypotheses have been conceptually developed to a much greater extent that any of the other serious contenders, and the major points of contention for all coupling hypotheses are suitably highlighted by comparison of these two. A brief description of these theories is therefore presented here. In particular, it may be noted that the essential features of the chemical and chemiosmotic mechanisms seem to be a matter of emphasis—the

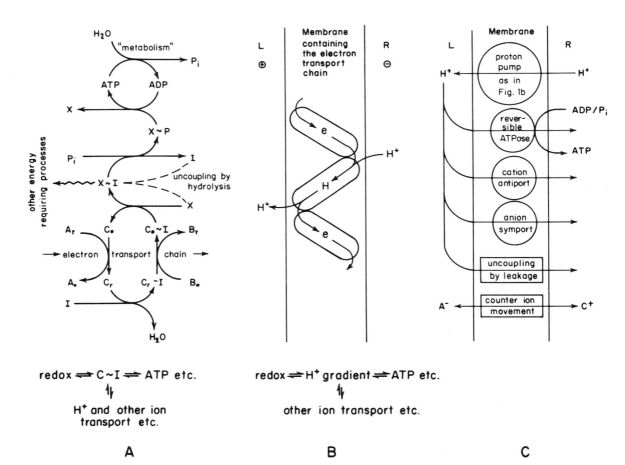

Fig. 1. Schematic representations of chemical (A) and chemiosmotic (B, C) coupling hypothesis. (A) A chemical pathway coupling the electron-transport chain to ATP synthesis and other energy-conserving and -dissipating processes; (B) a chain of alternating hydrogen and electron carriers arranged as a proton pump; (C) the proton pump coupled to proton-translocating utilization and dissipative pathways.

chemical mechanisms propose the interaction of chemical species with the redox couples, while the chemiosmotic mechanism rests on the interaction of charges with the redox couples and an essential directionality of these interactions.

It is worth pointing out, perhaps, that when reduced to bare formalism, all coupling hypotheses are equivalent and are described by thermodynamic necessities. This is clearly demonstrated by the work of DeVault (1971) (see also Dutton and Wilson, 1974), who has developed a general thermodynamic description of energy-conserving redox cycles that is clearly applicable to any of the categories of coupling mechanisms.

2.2. Hypotheses

2.2.1. Chemical

A coupling mechanism must serve two purposes: (1) it must store free energy at least to the same extent as the overall process, and (2) it must offer some entity that can interact with both the initial and the final form of the energy. By analogy with known mechanisms of "substrate-level phosphorylation" (e.g., in glycolysis and in the Krebs cycle), a chemical-coupling hypothesis for oxidative and photosynthetic phosphorylation proposed a chemical intermediate linking the oxidoreduction and hydrodehydration free-energy forms (Slater, 1953, 1971; Chance, 1972). This intermediate, C, is essentially a redox component of the respiratory chain (cytochrome, flavin, or whatever) that can, with another entity, I, undergo a condensation reaction in both its oxidized and reduced states. I is *not* an active redox component. The free-energy change of the condensation reaction is greater in one state than in the other (C is, after all, a different chemical species in the two states), and thus, by alternate oxidation and reduction, a "high-energy" ($\sim$) condensation bond can be formed and then transferred eventually to ADP plus phosphate to make ATP—as shown diagrammatically in Fig. 1A. Completely equivalent schemes can be devised for either C_0–I or C_r–I as the high-energy form. The extra free energy of the high-energy form relative to the other is directly related to and provided by the free energy of the oxidation–reduction span $A \rightarrow C \rightarrow B$.

The high-energy coupling intermediate is $C \sim I$. The simplest scheme would have phosphate as I undergoing direct condensation with C, but experiments reveal the need for a "nonphosphorylated intermediate." $X \sim I$, which is not *a priori* necessary, is some mechanistic entity of the enzymic ATP-synthetic apparatus needed in any coupling mechanism to

account for certain experimental observations on exchange reactions catalyzed by *in vivo* systems. Other energy-linked phenomena, such as active transport, are also driven by enzymic machinery utilizing the energy of the high-energy intermediate—just as is ATP synthesis. The essential feature of the chemical coupling theory is that a chemical intermediate, $C \sim I$, involving a redox component of the electron-transport chain condensed to a component of the transduction system, is the common precursor of all other energy-requiring processes (see Fig. 1A).

2.2.2. Chemiosmotic

Having its origins in earlier ideas of ion transport (Lundegårdh, 1945; Robertson, 1951, 1960; Davies, 1951), the chemiosmotic hypothesis supposes that the high-energy intermediate is an electrochemical activity gradient of protons generated and maintained across the closed membrane of the electron-transport system (Mitchell, 1961, 1966, 1968). The unique feature of Mitchell's hypothesis is the manner in which this intermediate is produced by a chain designed to result in an anisotropic transfer of H^+ from one side of the membrane to the other. This is shown in Fig. 1B. It involves the alternation of hydrogen (specifically protonated on reduction) and electron (not protonated) redox carriers back and forth across the membrane to form a redox "loop". The protons are driven across by the redox reactions, and thus the H^+ gradient contains the free energy of these reactions. For the H^+ gradient to serve as a transduction mechanism, the membrane must be intrinsically impermeable to protons except via subsequent energy-utilizing pathways, the enzymically controlled mechanisms of which must be specifically and obligately coupled to the vectorial dissipation of the H^+ gradient. Thus, the ATP-synthesizing apparatus (or reversible ATPase, as it is called) must be proton-translocating, as must various ion-transport pathways (Fig. 1C). That this is so for the ATPase of chloroplasts, mitochondria, and chromatophores is strongly supported by a number of reports of ATP synthesis driven by artificially induced pH gradients (Jagendorf and Uribe, 1966; Reid *et al.*, 1966; Leiser and Gromet-Elhanan, 1974) and by the occurrence of H^+ translocation during ATP hydrolysis (Mitchell and Moyle, 1965; Crofts, 1968; Moyle and Mitchell, 1973). A large number of proton-coupled and membrane-potential-driven ion-transport systems have been demonstrated in mitochondria and in bacteria (Mitchell, 1963, 1967; Mitchell and Moyle, 1965; Chappell and Crofts, 1966; Chappell and Haarhoff, 1967; Lehninger *et al.*, West, 1970; Kaback, 1974; Slayman and Slayman, 1974; Klingenberg, 1975). In addition, it has been shown that isolated mitochondrial ATPase (coupling factor, F_1, plus a "hydrophobic protein" component) incorporated into the membranes of artificial phospholipid vesicles can synthesize ATP in response to a proton gradient (Racker and Stoeckenius, 1974) (see also Chapter 29).

The active transport of protons across the membrane is envisaged as electrogenic, since protons are charged, and the resulting gradient is thus fully electrochemical, having both electrical and chemical contributions. The difference in electrochemical activity of protons across the membrane (the electrochemical potential gradient, $\Delta\bar{\mu}_{H^+}$, or "protonmotive force," Δp) can be written in terms of the separate electrical and chemical components

$$\Delta\bar{\mu}_{H^+} = \Delta p = \Delta\psi - Z \cdot \Delta pH$$

where $\Delta\psi$ and ΔpH are the electrical potential and pH gradients across the membrane and $Z = 2.3\ RT/F$ is approximately 60 mV at room temperature.

The relative contributions of the two components to the total potential depends on secondary processes of co- and counterion movement, which will tend to neutralize the electrical contribution without initially affecting the chemical (ΔpH) component, and exchange diffusion of protons for other equivalently charged ions, which will affect the chemical component without altering the electrical one. Mechanisms that facilitate the net movement of protons themselves across the membrane will dissipate the whole complement of stored energy. Such pathways, of course, are normally the energy-utilizing processes such as ATP synthesis, but they may also be artificially induced by the addition of compounds known as *uncoupling agents* or by making the membrane generally leaky to ions by osmotic rupture or treatment with detergents.

As originally proposed, the essential feature of the chemiosmotic hypothesis is that the high-energy coupling intermediate is an electrochemical gradient of protons maintained across the closed membrane of the vesicular electron-transport system and generated in the manner diagrammed in Fig. 1B. One might broaden this description by allowing formation of the gradient by any vectorial activity of the electron-transport chain (e.g., Skulachev, 1971, 1972, 1974) provided that coupling to driven reactions such as ATP synthesis and ion transport is achieved via a proton gradient between the bulk aqueous phases on opposite sides of the membrane and that no *direct* mechanical or chemical interactions are necessary in the transfer of free energy. This would still distinguish it from the hypothesis of Williams (1961, 1962, 1969), which invokes a localized, intramembrane proton

potential. However, since this is likely to be in equilibrium with a bulk-phase, transmembrane proton gradient to account simply for many ion-transport phenomena, the distinction is largely lost at the macroscopic level.

2.3. Nature of Ionophorous Antibiotics

Artificial induction of selective permeability and ion exchange, achieved by means of the so-called "ionophorous antibiotics" (Pressman et al., 1967; Lardy et al., 1967), has been of enormous use in probing energy-transduction processes. One class of antibiotics of significance here is exemplified by nigericin, which will perform an electroneutral cation exchange (e.g., K^+/H^+) according to the combined concentration gradients, but is not responsive to or affected by an electrical potential gradient.

A second important class is represented by valinomycin, which can mediate net movement of certain cations (e.g., K^+, Rb^+) according to both electrical and concentration components of an electrochemical gradient. In response to an electrogenic proton pump, valinomycin will allow the electrophoretic transport of potassium, for example, down the electrical gradient, dissipating the transmembrane potential but not directly affecting the H^+-concentration gradient (ΔpH).

Some note should be made of the polarities of these cation movements relative to the polarity of an electrogenic proton pump as primary motivator. If the pump is operating from right to left, as in Fig. 1B, then valinomycin will induce electrophoretic movement of K^+ down the electrical potential gradient, from left to right. Collapsing the membrane potential will partially decrease the magnitude of the proton-motive force, relieving some of the back pressure on the pumping mechanism and allowing a stimulated activity of the pump until, ideally, the loss of $\Delta\psi$ is compensated for by an increased ΔpH. Nigericin, on the other hand, will cause the movement of K^+, in electroneutral exchange for H^+, driven by the H^+-concentration gradient. Thus, K^+ will move from right to left as H^+ ions come back from left to right. Net proton transport will be inhibited, and ΔpH will be decreased, but a roughly equivalent increase in $\Delta\psi$ should take its place through a temporary increase in the turnover of the pump.

More detailed descriptions of the behavior of these and many other related antibiotics may be found in the literature (Pressman et al., 1967; Lardy et al., 1967; Mueller and Rudin, 1969; Crofts and Jackson, 1970; Henderson et al., 1969; Haydon and Hladky, 1972; McLaughlin and Eisenberg, 1975).

3. Osmotic and Electrical Gradients in Photosynthetic Bacteria

3.1. Ion Movements

3.1.1. pH Measurements

Light-induced pH changes in bacterial suspensions were first observed by Baltscheffsky and von Stedingk (1966), using a chromatophore preparation of *Rhodospirillum rubrum*. Using a pH-sensitive glass electrode, alkalinization of the suspension occurred on illumination, indicating the consumption of H^+ ions. Bacterial chromatophores are known to be roughly spherical vesicles, approximately 600 Å in diameter (Crofts, 1970; Saphon *et al.*, 1975a), and, as had previously been observed for chloroplast suspensions (Jagendorf and Hind, 1963; Neumann and Jagendorf, 1964), the extent of the H^+ disappearance in the steady state was rather large (0.5–1.0 H^+/Bchl compared to 3–6 H^+/Chl in chloroplasts) and invited interpretation in terms of net proton uptake into the interior of the chromatophore vesicles. This is strongly supported by the fact that whole cells pump protons *into* the external medium (Scholes *et al.*, 1969), implying that chromatophores are "inside out" with respect to whole cells, much as submitochondrial particles are inside out with respect to whole mitochondria (Mitchell and Moyle, 1965). Thus, the protons are envisioned as moving from the aqueous phase on one side of the membrane to that on the other.

The notion of proton transport across the membrane was also supported by the effects of some ionophorous antibiotics on the pH change. Valinomycin in the presence of potassium salts stimulated both the extent and the initial rate of the pH increase, while gramicidin inhibited them (Baltscheffsky and von Stedingk, 1966). The effect of the gramicidin, which is in marked contrast to that previously observed in mitochondria (Chappell and Crofts, 1965), was viewed as anomalous, but is almost certainly due to limitations on the reservoir of K^+ imposed by the opposite polarity of the two systems—chromatophores take up protons and mitochondria extrude them—and, indeed, preincubation of the chromatophores in high-K^+ medium does allow gramicidin to stimulate H^+ uptake relative to that in low-K^+ (see Jackson *et al.*, 1968, Fig. 6).

3.1.2. H^+ and K^+ Ion Movements and Estimates of ΔpH and $\Delta\psi$

Measurement of external ion-concentration changes in chromatophore suspensions was extended to both H^+ and K^+ movement (Jackson *et al.*, 1968). The effects of valinomycin and nigericin on proton uptake by *Rs. rubrum* chromatophores could be readily interpreted in terms of an inwardly directed, light-induced electrogenic proton pump on the basis of known properties of the antibiotics in inducing passive ion movements in artificial and biological membrane systems.

Briefly, valinomycin stimulated H^+ uptake and K^+ extrusion, indicating an electrophoretic mobility of the K^+ in response to, and thus collapsing, a membrane potential generated by the proton pump. Nigericin, on the other hand, inhibited the pH change and stimulated a K^+ *uptake*, presumably by mediating an exchange of H^+ for K^+. These authors supported the view that the light-induced proton gradient was an energetic precursor of phosphorylation as described by the chemiosmotic hypothesis. That the electrical and osmotic components of the proton gradient could be separated in this way explained the observed insensitivity of phosphorylation to either one of these antibiotics alone, since the loss of one component is compensated for by a roughly equivalent increase in the other. In addition, the polarity of the proton pump, as in the case of gramicidin described above, limits the efficacy of valinomycin via the internal K^+ reservoir. Uncoupling, which requires the complete dissipation of the proton-motive force, occurred in the presence of both valinomycin and nigericin, permitting a rapid cycling of K^+ against H^+ (Jackson *et al.*, 1968; Thore *et al.*, 1968).

Nishimura and Pressman (1969) arrived independently at similar conclusions regarding the electrogenicity of the proton pump, but, finding no synergistic inhibition of the initial rate of proton uptake by valinomycin and nigericin, they concluded that the proton gradient was not a precursor of phosphorylation which was affected by a synergistic action. At the onset of illumination, however, there would initially be no back-pressure of a proton gradient on the pump, and it is only because the electrical capacity of the system is so small that an effect of valinomycin on the initial rate is seen at all; i.e., with respect to the response time of the electrode measurements, the membrane-potential component builds up very rapidly. This effect with valinomycin having been eliminated, no further effect of nigericin is to be expected on the rate.

By varying the external K^+ concentration and observing the H^+ and K^+ movements on addition of nigericin to chromatophores in the dark and in the light, Jackson *et al.* (1968) could deduce a probable value for the contribution of the H^+-concentration gradient (Δp) to the total protonmotive force (Δp). The value obtained was approximately 1 pH unit. The effect of valinomycin on H^+ movements at varying external K^+ concentrations could be used to give a value for $\Delta\psi$, but these workers found that even 100 mM external KCl did not prevent additional H^+ uptake on addition of valinomycin in the light. Using their value of 4 mM for the internal K^+ concentration, one could only say that $\Delta\psi$ is greater than about 90 mV. In subsequent studies on *Rhodopseudomonas sphaeroides* using a range of external KCl concentrations, the concentration required to just abolish the extra H^+ uptake was obtained by extrapolation (Crofts and Jackson, 1970). The light-induced membrane potential in the steady state was estimated to be 200–210 mV at pH 7–8.

3.1.3. General Ion-Permeability Properties of Chromatophore Membranes

As expected of a biological membrane involved in energy coupling, the chromatophore membrane is impermeable to most charged species, including protons, alkali and alkaline earth metal cations, amino acids, chloride, phosphate, and many other commonly occurring cations and anions. Sulfate, however, appears to be permeable enough to allow good uncoupling by ammonium sulfate or nigericin plus potassium sulfate (Montal *et al.*, 1970). Permeation of iodide was indicated by a marked stimulation of both the extent and the initial rate of H^+-ion uptake in continuous light by chromatophores of *Rp. sphaeroides* (Crofts and Jackson, 1970). Simultaneously, the steady-state extent of the carotenoid bandshift was diminished, although the initial extent and kinetics were unaffected. Photophosphorylation was inhibited at somewhat higher concentrations of iodide. Comparable effects could be seen with KCl in the presence of low concentrations of the detergent Triton X-114. A similar effect of nitrate on the carotenoid shift was observed by Jackson and Crofts (1971), but others found that nitrate and iodide, as well as chloride, bicarbonate, and acetate, are not permeable enough to elicit significant uncoupling as the ammonium salts (Montal *et al.*, 1970; Briller and Gromet-Elhanan, 1970).

A number of physiologically less likely ionic species have proved useful in studies of oxidative and photosynthetic energy coupling, precisely because of their high permeabilities. An important class of these is the uncoupling agents. These compounds are generally lipid-soluble weak acids, membrane-permeable in both the protonated and unprotonated forms, e.g., dinitrophenol (DNP), pentachlorophenol (PCP), 2,4,6-trichlorocarbonylcyanide phenylhydrazone (CCCP).

3.1.4. Ionophores and Photophosphorylation

Photophosphorylation by bacterial chromatophores is only partially sensitive to valinomycin (Baltscheffsky and Arwidsson, 1962; Jackson *et al.*, 1968; Gromet-Elhanan, 1970). In *Rs. rubrum* chromatophores, which are inhibited by antimycin A or 8-hydroxy-quinoline-N-oxide (HOQNO), electron transport, phosphorylation (Baltscheffsky and Arwidsson, 1962), and H^+ uptake (von Stedingk, 1967), are restored by N-methylphenazonium methosulfate (PMS). The restored phosphorylation is completely insensitive to valinomycin (Baltscheffsky and Arwidsson, 1962; Baltscheffsky and von Stedingk, 1966). Such observations and others (Gromet-Elhanan, 1970; Shavit *et al.*, 1968) have led to suggestions of two fundamentally different sites of energy conversion. However, since Nishimura (1962) showed that PMS bypasses a rate-limiting step in the electron-transport scheme of *Rs. rubrum*, it seems probable that the anomalies of valinomycin sensitivity can be explained by the expected replacement of $\Delta\psi$ by a larger ΔpH. Rate limitation of electron transport by the low internal pH would prevent complete compensation of $\Delta\psi$ by ΔpH in the normal system, but not in the PMS-restored system, where the sensitive step is bypassed. Control of electron transport by the internal pH has been established in chloroplasts (Bamberger *et al.*, 1973; Wraight *et al.*, 1972), and a similar control may be postulated for chromatophores of photosynthetic bacteria.

3.1.5. Initial Rate Measurements

A strictly coupled and stoichiometric relation between H^+ movements and electron transport is mandatory for the chemiosmotic hypothesis in the limit of perfect coupling, but the rather slow response time of glass electrodes did not permit comparison of simultaneous measurements. Improvement in the time resolution of pH measurements was achieved by the use of pH-indicator dyes to follow the events spectrophotometrically (Chance and Mela, 1966; Chance *et al.*, 1966). The pH-indicator dye bromthymol blue (BTB) was found to give responses opposite to those exhibited by a glass electrode, and was taken to be responding to the internal vesicular pH. Such data seemed to confirm that H^+ ions were indeed translocated by the vesicles across the limiting membrane. In both mitochondria (Chance and Mela, 1966) and chromatophores of *Rs. rubrum* (Chance *et al.*, 1966), however, the apparent rates of internal pH change were far slower than the observed rates of electron transport, and were nonstoichiometric with respect to the cytochrome turnover (less than 1 H^+ per 4 cytochrome *c*).

Subsequent studies on BTB showed that it was actually responding to some energized state of the vesicle—perhaps a conformational change in the membrane (Cost and Frenkel, 1967) or the development of a membrane potential (Mitchell *et al.*, 1968; Jackson and Crofts, 1969*a*)—and was probably not inside the vesicle, but dissolved in the membrane itself. Furthermore, the slowness of the effect was due to a rate limitation in the response of the indicator to this state (Jackson and Crofts, 1969*a*). The use of pH indicators has been indispensable, however, in kinetic and quantitative studies of H^+ uptake and binding reactions. Bromcresol purple, which is bound to chromatophores to a much lesser extent than BTB, gave a reliable measure of the *external* pH change, and rapid H^+ uptake by *Rs. rubrum* chromatophores could be seen at the onset of continuous illumination (Jackson and Crofts, 1969*a*). After only 100 msec or so, this was abruptly curtailed to a lower rate, which continued for some seconds. The initial rate was at least as fast as the initial rate of electron transport, and in the presence of valinomycin, which stimulated the H^+ uptake and prevented the subsequent sharp falloff, was 2–3 times faster.

3.2. Electric Fields and Pigmented Bandshifts

3.2.1. General

Reversible red shifts in the absorption spectra of carotenoids, induced either by light or oxygenation, were first observed in *Rs. rubrum* (Chance and Smith, 1955) and in *Rp. sphaeroides* (Chance, 1958). Since then, they have been observed in a number of species of photosynthetic bacteria; a typical light-induced difference spectrum of the carotenoid bandshift in chromatophores from *Rp. sphaeroides* is shown in Fig. 2. Each of the major carotenoid absorption peaks is shifted to the red. Early studies seem to have focused on an absorption increase seen at 435 nm, which was of unknown and complex origin (Duysens *et al.*, 1956; Smith and Ramirez, 1959; Smith *et al.*, 1960; Geller, 1967). This absorption change was observed to be somewhat diminished in the presence of ADP and phosphate, and a similar effect on the carotenoid absorption changes was reported. The distinction in the early literature between the 435 nm absorption increase and the other spectral alterations due to carotenoids is not always clear, and the effects of phosphorylating conditions were associated clearly only with the 435 nm change. As originally suggested (Duysens *et al.*, 1956), this is largely due to the appearance of oxidized reaction center (RC) bacterio-

chlorophyll (Bchl), as is clear from the studies of Geller (1967), although there is some contribution from cytochrome *b* and carotenoids.

Although it was clear that carotenoids were not essential to the photosynthetic process, since mutants totally devoid of carotenoids grew satisfactorily provided they were not exposed to air and light simultaneously (Griffiths *et al.*, 1955; Sistrom *et al.*, 1956), carotenoids drew considerable attention for two reasons. First, Arnold and Clayton (1960) found that air-dried chromatophores of *Rp. sphaeroides* still exhibited the light-induced carotenoid spectrum at 1°K, although Okade *et al.* (1970), using an aqueous chromatophore suspension, could observe the photo-oxidation of Bchl only at liquid nitrogen temperature. A quite complicated temperature dependence of the carotenoid bandshifts was found in *Rp. gelatinosa*, with the effect disappearing at about −30°C and reappearing at lower temperatures (Dutton, 1971), and reversible light-induced carotenoid shifts at 77°K are seen in *Rp. gelatinosa* but not in *Rp. palustris* or *Chromatium vinosum* (Kihara and Dutton, 1970). The degree of reversibility in the dark was dependent on the extent of aeration, i.e., redox potential.

Second—the onset of the bandshifts was extremely rapid—faster than any observed electron-transport event including the fastest cytochrome oxidations (Smith and Ramirez, 1960). This surprising result was partly resolved when it was estimated that the carotenoid bandshifts were produced with a quantum yield of at least 3 (Amesz and Vredenberg, 1966). Largely because of significant displacement of the null points of the difference spectrum relative to the peaks of the absolute spectrum, the light-induced spectrum was considered to be the result of a red shift of a few nanometers by a small percentage of the total carotenoid complement, rather than a much smaller shift by all the carotenoids. The usual complexity of the carotenoid complement weakens the authority of such an estimate, but the same conclusion was recently drawn from studies on a mutant of *Rp. sphaeroides* (G1C) that contains only one significant (≈97%) carotenoid component, neurosporene (Crofts *et al.*, 1974). By curve-fitting, 7–11% of the carotenoids were found to undergo a red shift of 7 nm during continuous illumination, in good agreement with the earlier estimates. The dark decay of the carotenoid bandshift is accelerated by the uncoupling agent, FCCP; the initial rate of onset in the light is not affected. The implied energy dependence and the high quantum yield led to the suggestion that the carotenoids were responding to some physical stimulus such as a conformational change (Vredenberg *et al.*, 1965; Amesz and Vredenberg, 1966).

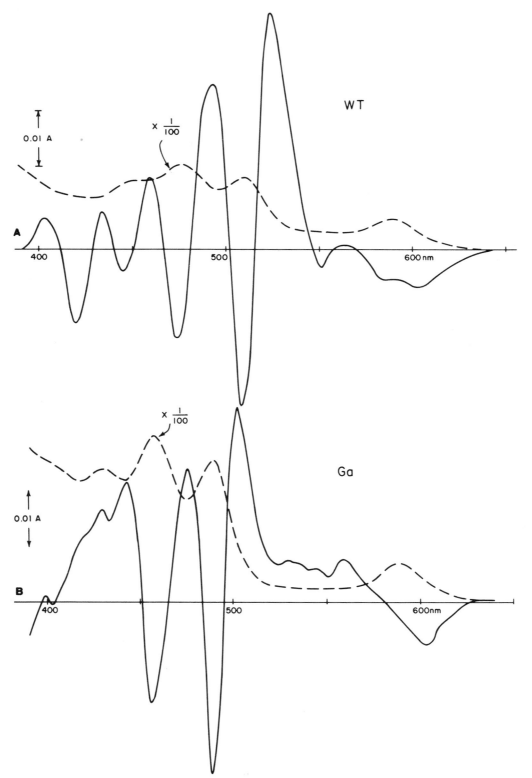

Fig. 2. Light-induced difference spectra in chromatophores of *Rhodopseudomonas sphaeroides* wild-type (A) and green mutant Ga (B). The absorption spectra are also shown, reduced 100-fold.

Studies on the effects of various inhibitors and uncouplers of electron transport and phosphorylation on the carotenoid bandshifts and on delayed light emission in *Rp. sphaeroides* indicated that both phenomena were intimately related to the high-energy state of phosphorylation (Fleischman and Clayton, 1968). Various possibilities of direct interaction with components of the electron-transport chain and indirect physical effects such as membrane potentials and pH gradients were considered, but the great speed of the carotenoid shift was seemingly not compatible with the rather slow kinetics of proton uptake.

Carotenoid bandshifts can be induced in *Rs. rubrum* chromatophores by addition of ATP or pyrophosphate (Baltscheffsky, 1967), and a similar effect of pyrophosphate was seen in *Rp. sphaeroides* (Sherman and Clayton, 1972). In *Rs. rubrum*, pyrophosphate was in fact more efficient than ATP, in keeping with the relative abilities in energy-linked functions such as reversed electron transport. It was suggested that the carotenoids responded to some energy-linked conformational change of the membrane (Baltscheffsky, 1969), and, indeed, early studies had already indicated a dependence on the integrity of the membrane, since the carotenoid absorbance changes were not seen in detergent-treated bacteria (Clayton, 1962).

3.2.2. The Carotenoid Bandshift as Electric-Field Indicator

(a) Bandshifts in Plants. A light-induced absorbance change at about 520 nm was first described by Duysens (1954) in algae, and has since been widely investigated. The absence of this signal in a photosynthetically competent, pale-green mutant of *Chlamydomonas* indicated the nonessential nature of the response (Chance and Sager, 1975) and lent some support to the supposition that carotenoids were involved in the generation of the absorbance change (Chance and Strehler, 1957). In extensive studies by Witt *et al.* (1965), it was attributed to an uncertain activity of Chl *b*, and there has in fact been some controversy over the identity of the species responsible for the absorption spectrum, which has contributions in the blue and the red in addition to the most prominent peak at about 520 nm (Witt, 1967; Fork and Amesz, 1967). It now seems clear that although Chl *b* is involved, as indeed are almost all the pigments in the membrane (Amesz and Visser, 1971), the major contributors to the 520 nm peak are indeed the carotenoids.

Junge and Witt (1968) found that the 520 nm absorbance change, which was rapidly developed after a laser flash, decayed in the dark at a rate that was highly sensitive to various treatments of the chloroplasts that might be expected to damage the thylakoid membrane. In addition, the decay was greatly accelerated by the ionophorous antibiotic gramicidin at concentrations of the order of 1 molecule gramicidin/10^5 Chl, which is about the size of a thylakoid. These observations led them to suggest that the 520 nm absorbancy change was a response to an electric field or membrane potential generated in the light and dissipated by electrophoretic ion movements across the thylakoid membrane (Junge and Witt, 1968; Witt, 1971). However, the kinetic responses to continuous illumination and the light-intensity dependence of the 520 nm absorption change are complex. At least part of the total response is sensitive to oxygen and exhibits a high-intensity light dependence (Chance and Strehler, 1957). This behavior is more indicative of carotenoid triplet formation, and this has been demonstrated in flash-excitation studies (Zieger *et al.*, 1961; Hildreth *et al.*, 1966; Mathis, 1966; Mathis and Galmiche, 1967; Wolff and Witt, 1969). At intensities that are not supersaturating and at times longer than a few microseconds, however, there is little likelihood of confusion between the carotenoid triplet and the bandshifts.

(b) Bandshifts in Bacteria. Following the inference of an electrogenic proton pump from the H$^+$ and K$^+$ movements in *Rs. rubrum* (Jackson *et al.*, 1968), an explanation of the energy-linked bacterial carotenoid bandshifts was sought in terms of an electric-field indicator as proposed for the 515 nm absorbance change in chloroplasts. The carotenoid bandshifts in *Rp. sphaeroides* chromatophores were found to be inducible in the dark by diffusion potentials generated by the addition of KCl in the presence of valinomycin or of HCl in the presence of an uncoupler (FCCP) (Jackson and Crofts, 1969b). Such diffusion potentials would be expected to be positive inside. Alternatively, addition of valinomycin to chromatophores (which contain some internal K$^+$) suspended in a medium lacking KCl, or the addition of KOH in the presence of uncoupler, induced carotenoid shifts in the opposite direction; potentials generated by these methods would be expected to be negative inside. The full difference spectra generated by the first method (positive inside) were identical to that of the light-induced response, while those of the second method were "mirror images." The size of the carotenoid absorbance change at a fixed wavelength was proportional to the logarithm of the external KCl concentration, as expected if the absorbance change were a linear function of the membrane potential. A log-linear voltage–concentration relationship is in agreement with theoretical expectation if the valinomycin-induced permeability to K$^+$ is much greater than any intrinsic permeability to a possible co-ion (e.g., Cl⁻), which in turn implies that the induced diffusion potential is most likely equivalent to 60 mV per decade of KCl concentration. This allows a tentative calibration of the carotenoid response and an estimation of the light-induced change. The initial peak of the carotenoid absorbance response to continuous illumination after a long dark period corresponded to a potential of over 400 mV; this subsequently decayed in the light to a steady value of about 200 mV, in good agreement with the earlier estimate of the steady-state potential obtained from the effect of valinomycin and K⁺ on light-induced H⁺ movements (Crofts and Jackson, 1970).

The involvement of an uncoupler (FCCP) in mediating similar acid-induced carotenoid shifts, an "energy-linked" response, is of particular note since it is quite incompatible with classic notions of uncoupling activity. KCl-induced shifts in the presence of valinomycin were inhibited by FCCP.

3.2.3. Decay Kinetics of the Bandshifts

The use of flash illumination greatly facilitated the study of the decay kinetics of the carotenoid shift. After a single turnover flash, the carotenoid absorbance change decays slowly in a polyphasic manner. An approximate half-time of 1 or 2 sec should be obtained in well-coupled preparations, but the exact time is highly dependent on preparation conditions, age of culture, and any treatments that affect the integrity of biological membranes.

As originally observed by Amesz and Vredenberg (1966) and Fleischman and Clayton (1968), the decay is very sensitive to uncoupling agents. Uncoupling concentrations of FCCP greatly accelerate the decay of the laser-flash-induced absorbance change in chromatophores of *Rs. rubrum* (Baltscheffsky, 1969), *Rp. sphaeroides* (Crofts and Jackson, 1970) and *Chr. vinosum* (Case and Parson, 1973), but do not affect the extent. The decay rate is also markedly stimulated by very low concentrations of valinomycin in the presence of potassium (Nishimura, 1970). The degree of acceleration is a linear function of the FCCP or valinomycin concentration over at least three orders of magnitude (Jackson and Crofts, 1971; Cogdell et al., 1972; Case and Parson, 1973). In the presence of accelerating agents, the decay becomes monophasic (Nishimura, 1970; Jackson and Crofts, 1971). It was recently shown that the minimum effective concentration of valinomycin is equivalent to about one molecule per chromatophore (Saphon et al., 1975a).

In the absence of exogenous agents, the slow decay of the carotenoid bandshift in *Rp. sphaeroides* exactly parallels the decay of a flash-induced H⁺-uptake measured with a pH-indicating dye (Cogdell et al., 1972; Saphon et al., 1975b). Addition of FCCP causes an equivalent acceleration of the decay of the pH change and the carotenoid bandshift (Cogdell et al., 1972). Electrophoretic movement of protons is thus the major intrinsic cause of collapse of the membrane potential in coupled chromatophores, reflecting a high degree of impermeability to a wide range of anionic and cationic species used as osmotic support, including, for example, Cl⁻, K⁺, Na⁺, and choline. The electrophoretic nature of the proton movement is indicated by the effect of valinomycin plus K⁺, which stimulates the dark decay of the carotenoid bandshift but retards the decay of the pH shift (Saphon et al., 1975b). A similar effect would be expected with permeating anions such as NO₃⁻ and I⁻. Conversely, nigericin greatly stimulates the decay of the flash-induced H⁺-binding, but does not affect the carotenoid bandshift (Saphon et al., 1975b).

3.2.4. The Electrogenic Photoact

(a) Rise Time of the Pigment Bandshifts. The rapidity of the carotenoid response on illumination (Smith and Ramirez, 1960) could not be meaningfully explored until the advent of flash spectrophotometric techniques. The rise time of the response in *Rs. rubrum* following a saturating, single-turnover flash is faster than 2 μsec (Nishimura, 1970), and in *Rp. sphaeroides* is complete in less than 100 nsec (Jackson and Crofts, 1971). Similarly, in chloroplasts, the rise time of the 515—nm shift has been found to be less than 20 nsec under conditions designed to avoid interference from carotenoid triplet formation (Wolff et al., 1969). It seems, therefore, that in all cases, the rise time of the light-induced pigment bandshifts is faster than can be resolved by conventional techniques. Recent attempts to resolve the rise time in the subnanosecond time domain (Leigh et al., 1974) have run afoul of interference from the triplet state. This problem is most apparent with high-intensity illumination and may well preclude any true resolution with intense flash activation (Wolff and Witt, 1969; Cogdell et al., 1975; Bensasson et al., 1976).

The great rapidity of appearance of bandshifts has been taken as indicating that an electric field is generated by the photoact itself in a process of charge separation, arranged vectorially across the membrane of the chromatophore or thylakoid, which leaves the primary donor oxidized (e.g., P⁺) on one side and the primary acceptor reduced (e.g., X⁻) on the other (see Fig. 7) (Wolff et al., 1969).

Although the precise rise time of the bandshifts may be somewhat indeterminate due to interference from the carotenoid triplet state, determination of the amplitude of the bandshift absorption changes is unlikely to be affected, since the triplet state undoubtedly decays in no more than a few microseconds (Zieger *et al.*, 1961; Hildreth *et al.*, 1966; Mathis, 1966; Mathis and Galmiche, 1967), whereas one of the characteristic features of the bandshifts is their relative stability on a time scale of at least several hundred milliseconds.

(b) Magnitude of the Primary Electrical Event. In general, added uncouplers, antibiotics, or inhibitors such as antimycin A have little effect on the extent of the rapid, laser-induced (single-turnover) bandshift. The maximum extent is about 25–35% that of the light-on spike of the continuous light response (Jackson and Crofts, 1971). From the earlier voltage calibration (Jackson and Crofts, 1969b), this implies 100–140 mV per turnover. Assuming an average chromatophore diameter of 60 nm (Crofts, 1970; Saphon *et al.*, 1975a) and a plausible range of membrane capacitance (e.g., from Mueller and Rudin, 1969), 20–100 electronic charges would have to be transported across the membrane per saturating flash per chromatophore (Jackson and Crofts, 1971). A recent estimate of the number of Bchl's per vesicle of about 4800 (Saphon *et al.*, 1975a), together with a value of the bulk Bchl/RC ratio of about 100–200 gleaned from later work of Crofts and co-workers, yields 25–50 as the number of RCs per chromatophore. This gives gratifying order-of-magnitude agreement with the calculated number of charges transported and to the notion of an electrogenic photoprocess.*

3.2.5. Bacteriochlorophyll Bandshifts

(a) The Membrane Potential. The spectral response of carotenoids to salt-induced diffusion potentials is clearly a bulk pigment phenomenon, and in fact is not limited to these pigments. In plants, many "electrochromic" absorbance changes have been observed. Apart from the 515 nm component, the most notable are those of various forms of Chl's *a* and *b* (Fork and Amesz, 1967; Emrich *et al.*, 1969; Amesz and Visser, 1971).

In bacteria, Bchl absorption spectrum shifts have been noted in the literature almost as long as those of the carotenoids. A red shift of an absorption band at 860 nm was first observed in *Rp. sphaeroides*, and it was speculated that a photochemical oxidation–reduction reaction was responsible (Clayton, 1963). The notion of an alternative photoreaction to the established one of P870 has also been suggested to account

---

* In some species of photosynthetic bacteria, the unit size can vary widely under different growth conditions. For this reason, it would be of great help in comparing data if RC/bulk Bchl ratios were quoted in research communications as a general practice.

---

for similar absorption changes in *Chr. vinosum* (Cusanovitch *et al.*, 1968; Morita, 1968; Schmidt, G. L., and Kamen, 1971) and in *Rs. rubrum* (Sybesma and Fowler, 1968; Sybesma, 1969). Vredenberg and co-workers, however, observed similar bandshifts in a variety of species, and noted that the derived absolute spectra were those of the light-harvesting pigments (Vredenberg *et al.*, 1965; Vredenberg and Amesz, 1967). They calculated that a large fraction of the total pigment participated in the shift, and that as many as 10–15 molecules responded to the absorption of a single quantum. They therefore concluded that it was indeed light-harvesting, bulk Bchl that was responsible for the absorption changes. The IR absorption changes are very similar to the carotenoid bandshifts with respect to both kinetics and sensitivity to uncoupling agents and inhibitors (Amesz and Vredenberg, 1966; Fleischman and Clayton, 1968). Bchl bandshifts in *Rp. sphaeroides* can be induced by potassium pulses in the presence of valinomycin in just the same way as the carotenoid bandshifts (Crofts *et al.*, 1971a), and the same is true for *Chr. vinosum* (Case and Parson, 1973), seemingly dispelling any possibility that the Bchl response in this species is a manifestation of a second photochemical reaction.

(b) The pH Gradient. Although the Bchl bandshifts seem to be of the same general origin as those of the carotenoids, certain differences are apparent. In *Chr. vinosum*, a calibration of Bchl shifts gives a different value for the light-induced potential than does the carotenoid calibration (G. D. Case, unpublished). Also, Bchl bandshifts can be induced by K^+ pulses in the presence of nigericin or by Ca^{2+} in the presence of the antibiotic A23187 (G. D. Case, unpublished). Both these ionophores are known to mediate a neutral cation–proton exchange. Thus, the bandshifts in this case are in response to a pH gradient rather than a membrane potential; the carotenoids do not respond under these conditions. Sherman and Cohen (1972) demonstrated a quenching by the pH gradient of Bchl fluorescence in *Rp. sphaeroides* chromatophores, and a similar effect is known in chloroplasts (Wraight and Crofts, 1970). A general sensitivity of Chl optical parameters to ΔpH may therefore be indicated. The ΔpH bandshifts display a different, roughly mirror-image spectrum to that of the $\Delta\psi$ (K^+/valinomycin) bandshifts. Taking the ΔpH response into account, the discrepancy between the light-induced Bchl and carotenoid bandshifts can be largely explained (G. D. Case, personal communication).

3.2.6. Nature of the Electric-Field Response

We will consider, at this point, what physically realistic mechanisms could account for the observed relationship between the pigment bandshifts and a transmembrane potential. Jackson and Crofts (1969b) measured absorption changes induced by diffusion potentials of up to 200 mV and estimated that light-induced potentials were as high as 400 mV. These figures correspond to electric-field strengths of $5 \cdot 10^5$ $V \cdot cm^{-1}$ across a membrane of, say, 5-nm thickness. Although this value is large, dielectric breakdown of plant cell membranes does not occur until potentials of 600–1000 mV are applied, the critical value decreasing with increasing temperature (Zimmerman et al., 1974).

There are a number of ways in which an electric field may alter the absorption spectrum of a chromophore. These may be roughly divided into two categories:

1. Indirect effects mediated via the environment, changes in rotation of the chromophore and restrictions on chromophore mobility, conformational or isomer interconversion by conformational changes in the suspending matrix, alterations in the dielectric constant of the medium, electroconstriction, and others.

2. Direct interaction between the electric field and the dipolar character (permanent or induced) of the ground and/or excited states of the absorber.

The great rapidity of onset of the in vivo spectral alterations and occurrence at liquid helium temperatures may rule out some of the first type, but certainly not all. It is also tempting to eliminate many of the second type—such as changes in chemical or physicochemical equilibria between different absorbing forms and effects dependent on molecular reorientation in the field. All these possibilities may contribute to the longer-term and steady-state changes, but even contributions to the rapid and low-temperature effects cannot be rigorously excluded, especially since it is well established that certain types of isomerization can occur at low temperature, e.g., in the visual process (Hubbard and Kropf, 1958; Yoshizawa and Wald, 1963; Rosenfeld et al., 1977).

Although the rapidity and temperature independence of the response cannot really exclude any of the possible mechanisms, intrinsic electrochromism has received by far the most attention. Initially, this may have been merely because it is amenable to theoretical description, but recent evidence suggests that it may even give a fairly adequate account of the phenomenon!

Electrochromism is related to the Stark effect, the splitting of spectral lines of atoms and small molecules in the gas phase by an electric field, but is a somewhat broader term describing electrostatic influences on the less resolved absorption and emission spectra of large molecules in condensed phases

(Platt, 1961). It has been widely adopted as a basis for interpretation of the in vivo bandshifts (Schmidt, S., et al., 1971a,b). We can define two main types:

1. A direct effect on the energy states of a molecule due to differential dipolar character of the initial and final (Franck–Condon) states of an electronic transition, giving rise to a bandshift in absorption and emission spectra.

2. A direct effect on the transition moment of a molecule, resulting in changes in absorption intensity for a particular transition.

A fairly clear description of these effects is given by Liptay (1969) and by Reich and Schmidt (1972) and S. Schmidt and Reich (1972a,b), and a more obtuse treatment by Labhardt (1967). A simpler, hand-waving view is given in the Appendix.

3.2.7. Degree of the Electric-Field Dependence

(a) Expectations. A simple model for electrochromism, as outlined in the Appendix, which ignores electric-field dependence of the *intensity* of absorption, predicts a progressive wavelength shift with increasing electric field—a continuum of spectra. A light-minus-dark difference spectrum should therefore show no true isosbestic point, the null point as well as the extent of the absorbance change varying with the light-induced electric field. The degree of the electric-field dependence depends on the nature of the interaction with the electric field, whether it is through a difference in *permanent* dipoles in the ground and excited states (linear dependence) or *induced* dipoles (quadratic dependence) (see the Appendix).

A two-state system, on the other hand, such as a voltage-dependent equilibrium between two isomers or a conformational change in the membrane, could exhibit a true isosbestic point, and only the extent of the absorbance changes would vary with electric field. Again, the degree of the dependence is a function of the nature of the interaction with the field.

(b) In Vivo and in Vitro Measurements on the Extent of the Absorption Change. Carotenoids are usually symmetrical or nearly so, having a center of symmetry in the conjugated double-bond system. They thus lack a significant permanent dipole in the ground state, and, barring gross conformational changes on excitation, the same is probably true of the excited states. Electrochromism is only expected, therefore, via a change in polarizability. This expectation is borne out by studies on electrochromism of lutein in cadmium arachidate multilayers (Schmidt, S., et al., 1971a,b; Schmidt, S., and Reich, 1972b). The observed absorbance changes were distinctly quadratic in field strength and closely resembled the first

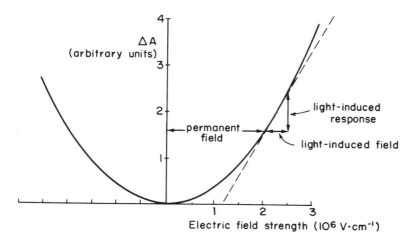

Fig. 3. Schematic representation of the quadratic field dependence of electrochromic absorption changes and the linearizing effect of a large permanent field.

derivative of the absorption spectrum. Both linear and quadratic effects were observed in monolayers of Chl *a* and Chl *b* (Kleuser and Bücher, 1969).

In vivo measurements, both in plants (Strichartz and Chance, 1972; Reinwald *et al.*, 1968; Schliephake *et al.*, 1968; Witt and Zickler, 1974) and in bacteria (Jackson and Crofts, 1969*b*; Evans and Crofts, 1974), strongly indicate a linear dependence of the absorption change on field strength (e.g., diffusion potential or estimated light-induced potential). The apparent linearity of the *in vivo* absorbance changes has been interpreted as indicating that there is a large permanent field in the photosynthetic membrane due to fixed charges (Schmidt, S., *et al.*, 1971*a,b*; Schmidt, S., and Reich, 1972*b*). A permanent field was also suggested to account for the "mirror-image" difference spectra seen with diffusion potentials of opposite polarity (Jackson and Crofts, 1969*b*) [see Section 3.2.2(b)]. The effect of a large permanent field is to shift the field dependence of the bandshifts into a quasilinear behavior, since the applied field (light- or salt-induced) is about an order of magnitude smaller than the underlying permanent one (Fig. 3).

Observations on salt-induced bandshifts in deionized (zero-net-charge) chromatophores indicate that there is no great change in fixed-charge asymmetry across the membrane at the isoelectric point (G. D. Case, unpublished). The behavior of spin labels at different depths in phospholipid bilayers has been taken to indicate that there are large (10^6–10^8 V·cm^{-1}) *short-range* (i.e., dipole) fields due to fixed charges on the membrane surface (Seelig *et al.*, 1972). This conclusion has been challenged, however, and the effect on spin labels interpreted as due to partial penetration by water of the membrane lipid structure. Calculations on a model basis showed that although large field gradients of the size envisaged by Seelig and

co-workers may exist, they would not account for the observed ESR phenomena (Griffith *et al.*, 1974)

(c) *In Vivo* **Measurements of the Wavelength Shift.** Although it might be argued that the linear dependence of the absorption change on field strength is not sufficiently accurate to warrant a firm conclusion, or that the range of membrane potentials covered was not great enough to reveal a departure from linearity, the *magnitude* of the wavelength shift is also not consistent with a simple electrochromic effect. Even for the large changes in polarizability observed for long, conjugated molecules like carotenoids, the expected bandshift in a field of 10^5 V·cm^{-1} or so is less than 0.05 nm, two orders of magnitude less than those reported (see the Appendix).

In view of these rather large *in vivo* effects, Crofts *et al.* (1974) attempted to measure the field dependence of the actual wavelength shift by examining the isosbestic point of the difference spectrum. Using a mutant of *Rp. sphaeroides*, G1C, containing only a single carotenoid component, the supposed field was varied roughly two- to threefold by a series of flashes. After the first flash, the isosbestic point was about 2.5 nm removed from the absorbance peak in the dark, corresponding to an apparent bandshift of 5 nm. In subsequent flashes, however, although the field should have been increasing, the isosbestic point moved only a further 0.4 nm. This behavior is not consistent with either a linear or a quadratic effect on the bulk pigment. More recently, even this small progressive shift was attributed to light-induced light-scattering changes (Crofts *et al.*, 1975*b*).

3.2.8. Orientation of the Carotenoids *in Vivo*

The conclusions of Crofts *et al.* (1974, 1975*b*) that a small fraction of the carotenoids undergoes a large immediate shift and subsequently displays little or no field dependence are based on the fact that curve analysis of the absolute absorption spectrum did not reveal significant spectral heterogeneity, and thus the initial bandshift was measured from the position of the main absorbance peak. It seems rather plausible, however, that there is heterogeneity in the single carotenoid population with respect to environment and orientation, and that this could well be missed in an analysis of Gaussian components. Breton and co-workers (Breton *et al.*, 1973; Breton and Mathis, 1974) observed strong linear dichroism of the 515 nm component in chloroplasts, indicating an average orientation of the carotenoids largely in the plane of the thylakoid membrane. In chromatophores, the carotenoids have an average orientation at least 45° to the plane of the membrane, which may account for the larger effects seen in bacteria (Breton, 1974). Since the change in polarizability is expected to be collinear with the long axis of a carotenoid, those molecules oriented in the plane of the membrane are not expected to respond significantly to a transverse field. The fraction of molecules that do respond (Amesz and Vredenberg, 1966; Crofts *et al.*, 1974) may correspond to those that are favorably oriented out of the plane. These molecules would also feel any permanent dipole field arising from fixed charges on the surface of the membrane most strongly, which would necessarily result in a permanent bandshift giving rise to a distribution of spectral forms. Indeed, a permanent field somewhat greater than 10^6 V·cm^{-1} would produce a bandshift of a few nanometers for the $\Delta\alpha$ values observed in carotenoids. Such heterogeneity would be sufficient to account for the anomalously large initial displacement of the isosbestic point in the *in vivo* carotenoid difference spectrum, which would thus be apparent rather than real.

S. Schmidt *et al.* (1971a,b) constructed thin films of natural pigments (Chl and carotenoid mixtures) and simulated (apart from the quadratic rather than linear field dependence) the *in vivo* difference spectrum with applied fields of about 10^5 V·cm^{-1}. Best fits, however, were obtained with pigment constitutions rather different from the *in vivo* ratios. This may reflect different degrees of orientation of the various pigments.

3.2.9. Magnitude of the Wavelength Shift

As noted above, the magnitude of the *in vivo* wavelength shift is not consistent with a simple induced dipole electrochromism in a field of about 10^5 V·cm^{-1}. The notion of a permanent field, however, in addition to linearizing the response, also helps rationalize the magnitude. The induced bandshift is then given by:

$$\Delta\lambda = \frac{\lambda^2}{hc} \cdot \Delta\tilde{\alpha}(\mathbf{E_0} + \mathbf{E})\mathbf{E}$$

where $\mathbf{E_0}$ and $\mathbf{E}$ are the permanent and induced fields, respectively. for $\Delta\tilde{\alpha} = 1000 \times 10^{-24}$ cm^3, $E_0 = 2 \cdot 10^6$ V·cm^{-1}, and $E = 2 \cdot 10^5$ V·cm^{-1}, the expected shift is about 0.5 nm, which may be compared with the small (0.4-nm) progressive shift in isosbestic point seen by Crofts *et al.* (1974).

3.2.10. Effect of Hypochromism

Although the explanations of the bandshift anomalies discussed above are physically quite plausible, they are speculative, and it must be remembered that up to now we have neglected the probable influence of the field on the absorption intensity. As can be seen in Fig. 4, the simultaneous occurrence of hypochromism and a small red shift is a further possible cause for a significant displacement of the isosbestic point to the red. It also produces a displacement of the spectrum below the zero line, in qualitative agreement with the *in vivo* spectra, which are not exactly symmetrical. It can be readily seen from a rough sketch that a hypochromic absorption change of about the same magnitude as the bandshift-related absorption changes will produce a red shift in isosbestic point of 10–15% of the half-width of the absorption band.* Crofts *et al.* (1974) observed an initial isosbestic displacement of about 2.5 nm for an absorption band of 20 nm half-width (by curve analysis).

* Full width at half-maximum.

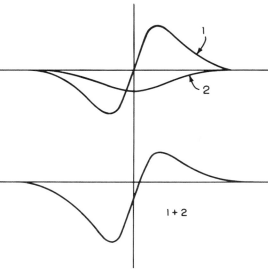

Fig. 4. Effect of hypochromism on a bandshift difference spectrum.

As a significant refinement of the electrochromic model, this point was recently considered in theoretical detail (Conjeaud and Michel-Villaz, 1976), and it was concluded that hypochromism, together with the notion of a permanent field electrochromism, adequately accounts for the characteristics of the *in vivo* bandshifts. Interestingly, the combined effects of hypochromism and bandshift were shown to result in an isosbestic point that does *not* shift progressively with increasing field. Thus, the small or even negligible shift observed by Crofts *et al.* (1974, 1975*b*) may not be at variance with the electrochromic model, albeit one that is more complex. For this reason, we have not developed any of the possible two state models the major attraction of which is that they account for a true isosbestic point. They are, in any case, conceptually quite simple.

3.2.11. Kinetics of Onset of the Bandshifts

(a) Preliminaries. The carotenoid bandshift in *Rp. sphaeroides* exhibits several phases of appearance on illumination by single-turnover, multiple-turnover, and continuous illumination. Apart from the rapid phase seen with the single-turnover flash, two other rise phases are seen with a multiple-turnover xenon flash (Jackson and Crofts, 1971). One of these phases is sensitive to antimycin A and was interpreted as indicating a second electrogenic site in the dark electron-transfer chain in the region of a cytochrome *b*. Conclusive analysis is complicated, however, by the multiple-turnover nature of the flash.

With continuous illumination after a long dark period, the bandshift rises rapidly to a maximum and then decays in the light to a lower, steady-state level. This response was calibrated in terms of millivolts of

membrane potential by Jackson and Crofts (1969*b*). The declining phase is kinetically very similar to the light-induced proton uptake and suggests that during this time, part of the initial membrane potential is converted into a more extensive pH gradient by electrophoretic movement of ions (Crofts and Jackson, 1970). The electrophoretic nature of this phase is strongly supported by the effects of permanent anions, which accelerate and stimulate the extent of the decline, lowering the steady-state membrane potential and increasing the pH-gradient (Crofts and Jackson, 1970; Jackson and Crofts, 1971).

(b) Multiple Phases following a Single Turnover. Clarification of the flash-induced kinetics of the carotenoid absorbance change was obtained using controlled redox potentials to define the conditions more precisely (Jackson and Dutton, 1973; Dutton *et al.*, 1975). In response to a single turnover flash, a very rapid phase of development (phase I) is found to correlate precisely with the functional redox potential range of the primary photoact. It thus disappears either as the primary electron acceptor is chemically reduced at low potentials ($E_{m, 7.0} \simeq -15$ mV) or as the primary donor is oxidized at high potentials ($E_{m, 7.0} = +450$ mV) (see Fig. 5).

In a titration from the high-potential end of photochemical activity, a second phase (phase II) becomes apparent as the redox potential is lowered, corresponding to the chemical reduction of cytochrome c_2 ($E_{m, 7.0} = +295$ mV). The correspondence is kinetic as well as potentiometric, both the light-induced oxidation of cytochrome c and phase II of the carotenoid shift occurring with a half-time of about 150 μsec. Recently, the oxidation of cytochrome c in *Rp. sphaeroides* was found to be biphasic, with a rapid component, $t_{1/2} \leqslant 30$ μsec, and a variable slower

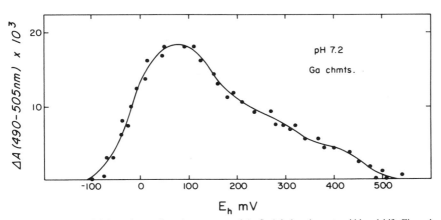

Fig. 5. Redox-potential dependence of maximum extent of the flash-induced carotenoid bandshift. Figure by courtesy of P. L. Dutton.

component, $t_{1/2} \simeq 0.3–8$ msec (Dutton *et al.*, 1975). Similarly, phase II of the carotenoid absorbance change can also be resolved into two components (IIa and IIb) corresponding precisely to the two phases of cytochrome oxidation (Prince and Dutton, unpublished) (see also Dutton and Prince, Chapter 28). There is thus little doubt that oxidation of cytochrome c_2 by the RC is causally related to a portion of the carotenoid bandshift.

A third, much slower phase (phase III) becomes apparent over the same potential range as phase II, with a half-time of a few milliseconds. It is further enhanced and also accelerated at a potential of $+150$ mV, apparently corresponding to the titration of a cytochrome b ($E_{m,7.0} = +155$ mV). Unlike phases I and II, phase III is inhibited by antimycin A (Fig. 6).

All three phases exhibit the same spectrum and the same sensitivity to uncouplers and antibiotics; the relative contributions of the three phases to the maximal change are $0.5 : 0.5 : 1.0$.

(c) Bandshifts and Cytochrome Oxidation. Phase II of the carotenoid bandshift is not readily incorporated into any notions of a separate electrogenic site in view of its precise correspondence with cytochrome c_2 oxidation by P$^+$. The initial charge separation between P and X had previously been envisioned as extending

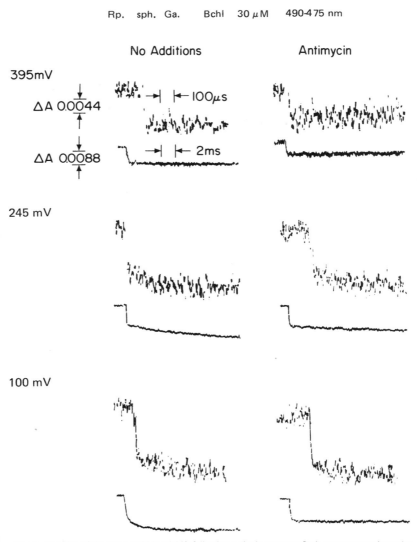

Fig. 6. Kinetics of the carotenoid bandshift following a single-turnover flash at representative redox potentials, and the effect of antimycin A. Figure by courtesy of P. L. Dutton.

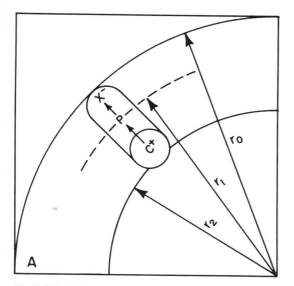 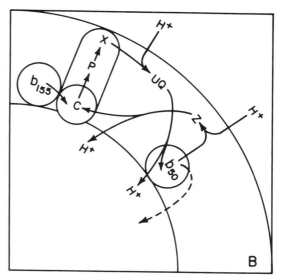

Fig. 7. (A) A RC–cytochrome c_2 arrangement across the coupling membrane permitting progressive charge separation following the electrogenic photoact. (B). An electron-transport chain incorporating the hypothetical component, Z, into a second proton-translocating loop.

right across the membrane, which was equated with a capacitor. Jackson and Dutton (1973) suggested placing P in the dielectric matrix of the membrane, implying that the full span of the membrane is involved only after subsequent oxidation of cytochrome c_2, which is known to lie at the aqueous interface of the inner membrane surface (Prince *et al.*, 1975). (Fig. 7B). This is equivalent to increasing the dielectric spacing of the capacitor, which requires energy (the redox span between P and cytochrome c_2) and causes an increase in the voltage drop across the dielectric. In a parallel-plate capacitor model, the electric field ($E = V/x$) remains constant, but would now encompass a larger volume of the membrane and thus could affect more chromophores. As Jackson and Dutton pointed out, however, the geometry of chromatophores is suited more to a spherical capacitor model with internal and external radii of about 250 and 300 Å, respectively. In this case, the electric field at any point in the dielectric is given by $Q/4\pi\varepsilon_0 r^2$, and charge transfer to cytochrome c_2 would cause an increase in the average electric field as well as affect a larger volume. The increased field strength is due to the greater charge density on the inner capacitor plate as the internal radius decreases (Fig. 7B).

Qualitatively similar results were reported for the carotenoid and Bchl bandshifts in *Chr. vinosum* (Case and Parson, 1973). In this species, however, the major contribution to the bandshifts is potentiometrically related to the oxidation of cytochrome c_{555} ($E_{m,7.0} \simeq$

+340 mV). At potentials below about +50 mV, cytochrome c_{552} ($E_{m,7.0} \simeq$ +10 mV) is titrated in and kinetically largely supersedes cytochrome c_{555} as a secondary donor, resulting in a loss of most of the bandshift amplitude. What little remains can be ascribed to the small success cytochrome c_{555} has as a donor in competition with cytochrome c_{552}. At high potentials, cytochrome c_{555} becomes titrated out due to prior chemical oxidation and the bandshifts also diminish, but in this case, the Bchl bandshifts disappear completely, even though P870 is still photochemically active, whereas the carotenoids are decreased only to a third of their maximum size.

In terms of the spherical capacitor model, these results would indicate that the primary charge separation is barely vectorial, both P and X lying close to the plane of the membrane and to one side, with cytochrome c_{555} on the opposite side. Most of the vectorial charge displacement would therefore be associated with the oxidation of cytochrome c_{555}. A corollary is that cytochrome c_{552} should most probably be on the opposite side of the membrane to cytochrome c_{555} and on the same side as P and X. How this might be arranged as a topological reality is left to the imagination at the present time!

(d) A Second Electrogenic Process. Phase III of Jackson and Dutton (1973) can be more or less equated with the antimycin-A-sensitive, slow phase of Jackson and Crofts (1971). An explanation for this phase could lie in associating it with the oxidation of a cytochrome *b* situated still further to the inside of the vesicle than cytochrome c_2 (cf. phase II above). However, this would have to be reconciled with the known interaction between cytochrome c_2 and the aqueous phase (Prince *et al.*, 1975; Dutton *et al.*, 1975) (see also Dutton and Prince, Chapter 28). An alternative explanation, first offered by Jackson and Crofts (1971), is that of a second electrogenic coupling site in the overall electron-transport scheme, the first being the light reaction itself. This site was suggested to be of the general chemiosmotic type described in Section 2.2.2 and shown in Fig. 1B. The sensitivity to antimycin A implicates a *b*-type cytochrome in the process, and the enhancement of the bandshift in the region of +150 mV might suggest cytochrome b_{155} as the active species. The cytochrome *b* involved in normal cyclic electron flow seems to be cytochrome b_{50}, however, whereas cytochrome b_{155} is too slowly reduced (after rapid oxidation) to participate in the cycle (Dutton and Jackson, 1972). It might be noted, in addition, that the midpoint of the carotenoid enhancement may not be exactly coincident with that of cytochrome b_{155}, but the precision of current measurements does not permit a firm conclusion.

A complete, chemiosmotic coupling site would require both an electron carrier and a hydrogen carrier (Mitchell, 1966). The electron carrier could be identified with a cytochrome *b*, but there is no suitable candidate for a hydrogen carrier known in this region of the electron-transport chain. An unknown hydrogen carrier, Z, has therefore been postulated to act between cytochrome b_{50} and cytochrome c_2 (Jackson and Crofts, 1971). Z, with $E_{m, 7.0} = +150$ mV, is deemed responsible for the enhancement of the millisecond phase of the carotenoid bandshift (phase III) in the potential region of +150 mV. It is not at all obvious, however, how reduction of the high-potential, hydrogen-carrying component of a redox loop can stimulate activity in the low-potential, electron-carrying arm actually responsible for the electrogenicity. One would, instead, expect the carotenoid enhancement to titrate in with the electron carrier, cytochrome b_{50} (Fig. 7B). Nevertheless, titrations of the carotenoid bandshift at different pH's show that the enhancement region does exhibit a pH-dependent midpoint potential, compatible with its identity as a hydrogen carrier (—60 mV/pH unit) (see Dutton and Prince, Chapter 28, Fig. 3). Actually, this does not distinguish it from cytochrome b_{155}, which, in equilibrium titrations at least, exhibits the same pH dependence (Dutton *et al.*, 1974; Petty and Dutton, 1976*b*). Neither does it address the problem of *how* a hydrogen carrier stimulates the electrogenic site. The nature of this second electrogenic site will be returned to later.

3.2.12. Delocalization of the Field

Placing the initial charge separation within the dielectric matrix of the membrane, rather than right across it, raises the question when and if delocalization of the electric field occurs. A delocalized field, or membrane potential, is necessary if subsequent coupling to ATP synthesis and ion fluxes is to occur in a chemiosmotic manner, but is not a requirement of some other, more conformationally oriented, coupling models. The existence of salt-induced bandshifts certainly seems to show that the pigments can respond to delocalized fields, and the close equivalence between these and the light-induced spectra suggests that the field is delocalized in the steady state (Jackson and Crofts, 1969*b*). Furthermore, the sensitivity of the entire absorption change to ionophorous antibiotics at concentrations as low as one molecule per vesicle (Saphon *et al.*, 1975*a*) is also strong support for delocalization in the millisecond time range.

On the other hand, the light-induced field is clearly localized initially and must become delocalized subsequently. In the steady state, this is associated with protolytic reactions binding protons at the outer surface and releasing them at the inner surface (see Section 3.3.2), so that the net charges of the electrogenic step are transferred to the bulk aqueous phases. At shorter times, however, and in particular when the secondary donor (cytochrome *c*) is inoperative due to prior oxidation, the charges are still associated with molecular sites in the membrane, and delocalization must occur through induced charge redistribution in the aqueous phases. In chloroplasts, where the net proton movements to the bulk phases are slow (Ausländer and Junge, 1974), such charge redistribution was measured with electrodes following a nonsaturating flash (Fowler and Kok, 1974; Witt and Zickler, 1973) and was found to occur over at least the area of a thykaloid and probably an entire grana stack in less than 10 μsec. Measurement of the time required in bacterial chromatophores was not possible, probably because it was too fast (C. F. Fowler, personal communication). In a saturating flash, redistribution would need to occur only between adjacent RCs, a much smaller area, and the necessary time would be much less. Thus, one can conclude that the delocalization should occur in considerably less than 1 μsec.

As yet, no kinetic characteristics of the bandshift have been observed to correspond with the delocalization process. However, the existence of the bandshifts at low temperatures (Arnold and Clayton, 1960; Kihara and Dutton, 1970) is suggestive of stable, localized field effects, although delocalization by capacitive coupling (dipole rearrangement) or proton conduction in ice cannot be ruled out.

Fast enthalpy and volume changes following flash-excitation of *Chr. vinosum* chromatophores show a kinetic correspondence with the carotenoid bandshift (Callis *et al.*, 1972). Interpretation of the volume changes in terms of electroconstriction of the membrane revealed significant discrepancies with the predictions of a simple membrane capacitor model involving a delocalized field. The large volume decrease could be explained if the charge separation generated a localized electric field that was not compensated by ion exchange or redistribution for at least several milliseconds. This would imply considerable electrostatic shielding of the charges from the aqueous phase. However, a variety of conformational changes and bond rearrangements could also account for the volume decrease.

The complete loss of Bchl bandshifts at high redox potentials in *Chr. vinosum*, compared with the partial loss of the carotenoid shifts (Case and Parson, 1973), is also difficult to reconcile with the notion of a fully delocalized electric field.

3.2.13. Other Probes of the Membrane Potential

The carotenoid bandshift provides an extremely rapid and sensitive readout of energization of photosynthetic coupling membranes. The nature of this energization can be closely correlated with the onset of an electrical field gradient, as we have shown in the preceding sections. Ultimately, however, this correlation should rest on experiment rather than agreement with a semiquantitative theoretical description. The electrostatic measurements described above provide a strong experimental basis for rapid charge redistribution.

Other less direct techniques have also contributed experimental correlations. Foremost among these is the use of extrinsic optical probes such as anilino naphthalene sulfonate (ANS) (Azzi *et al.*, 1969, 1971; Gromet-Elhanan, 1972; Chance, 1974; Azzi, 1975), auromine-*O* (Kobayashi and Nishimura, 1972), and cyanine and oxanol dyes (Chance and Baltscheffsky, 1975; Chance *et al.*, 1974). In general, these probes have provided confirmatory data, but these data have frequently been more indirect than originally supposed. ANS undergoes very marked fluorescence yield and spectral alterations in response to the polarity of its environment. The current view of *in vivo* responses is that ANS responds to and interacts with the surface charge of a membrane. Similarly, merocyanines respond to polarity changes by a variety of mechanisms, such as dimerization and solvatochromism,

many of which involve translational mobility of the probe. Consequently, response times of these probes are generally slower than that of the carotenoid bandshift, and this has hindered any firm inferences as to the localized or delocalized nature of the membrane potential (Chance *et al.*, 1974).

More than 300 merocyanines and related dyes have been tested in the squid giant axon for suitable correlation with electrode determinations of the electrical properties (Cohen *et al.*, 1974), and many have been found to be accurate indicators of steady-state potentials; however, they are universally slower than the electrodes in response to fast transients such as action potentials. More recently, a new class of dye, the rhodanine-merocyanines, has shown some promise of rapid response in nerve preparations. Successful determinations of steady-state cellular and mitochondrial membrane potentials have been reported using cyanine dyes (Hoffman and Laris, 1974; Laris *et al.*, 1975).

Another method of determining the membrane potential is through the distribution of permeant ions. Distribution of potassium in the presence of valinomycin has been used in mitochondria (Mitchell and Moyle, 1969), and related studies on chromatophores are described in Section 3.1.2. Permeant anions have also been used, and an elegant albeit rather slowly responding technique was developed by Skulachev and co-workers (Isaev *et al.*, 1970). This employs phenyldicarbaundecaborane (PCB⁻) as a permeant anion, and redistribution is determined by measuring the external concentration with a PCB⁻–phospholipid electrode. This technique is therefore indirect, but has been used to show that chromatophores of *Rs. rubrum* generate a membrane potential in the light and in the dark on addition of ATP or reversal of the energy-linked transhydrogenase (Isaev *et al.*, 1970; Ostroumov *et al.*, 1973; Barsky *et al.*, 1975).

3.2.14. Anomalous Effects on the Bandshifts

The induction of Bchl bandshifts by salt-induced ΔpH generation (K⁺/nigericin or Ca²⁺/A23187) as described in Section 3.2.5(b) cannot be accounted for by the simple electrochromic model. Other anomalous origins of bandshifts are known—e.g., through the addition of certain organic solvents (W. K. Cheng and G. D. Case, unpublished). Both red and blue shifts could be generated with different solvent additions. Light-induced bandshifts were also markedly affected. The effects are multifarious and diverse, and have no obvious unifying explanation other than unspecified solvatochromic interactions and influences on the permanent field. Indeed, the phenomenology of bandshifts is so diverse that it has been suggested that even sneezing can induce them.

The acceleration of the decay of the 515 nm bandshift in chloroplasts by very low concentrations of gramicidin was probably the seminal observation leading to the electric-field hypothesis of these absorbance changes (Junge and Witt, 1968). In bacterial preparations, however, titration with gramicidin at equally low concentrations causes a progressive attenuation of the light-induced carotenoid shift (W. W. Parson, unpublished). This appears not to be due to instrumental loss of a very rapidly decaying portion of the absorbance change, but to be a genuine diminution of the initial extent. A similar effect was recently observed for valinomycin (J. B. Jackson and P. L. Dutton, personal communication) in addition to the well-known acceleration of the decay rate (Nishimura, 1970; Jackson and Crofts, 1971). These effects on the amplitude of the signal bring to mind the large-scale effects that these and other agents have on the fluidity and order characteristics of lipid bilayers revealed by NMR (Hsu and Chan, 1973) and ESR (M. Adamich and S. P. Van, unpublished). Valinomycin, for example, was found to affect the packing of the polar head groups at concentrations less than 1 valinomycin/1000 phospholipids and to inhibit the mobility of the hydrocarbon chains at somewhat higher levels (Hsu and Chan, 1973).

Also unaccounted for by a simple model is the lack of salt-jump or pyrophosphate-induced carotenoid bandshifts in a mutant of *Rp. sphaeroides* lacking RCs but with an apparently normal complement of bulk pigments (Sherman and Clayton, 1972). A necessary interaction between the carotenoids and the RC complex was inferred.

Although it may be possible to account for all these effects on an electrochromic basis by such devices as changes in dielectric constant of the membrane or freezing of chromophores in unfavorable orientations, they also provide some impetus to explore possibilities for conformational origins of the bandshifts. The close relationship between electro- and solvatochromism (Liptay, 1969) implies that a conformational involvement need not be severely at odds with the generally satisfactory electrochromic model.

3.3. Proton Uptake and Electron Transport

3.3.1. Flash-Induced H⁺ Uptake

(a) Initial Observations of Single-Turnover Kinetics. The measurement of external H^+-ion changes under continuous illumination at rates comparable with electron transport (Jackson and Crofts, 1969a)

was promptly followed by flash kinetic studies. Flash-induced H^+-ion changes were first measured in chromatophores of *Chr. vinosum*, at pH 6.3, using the pH indicator bromcresol purple (BCP) (Chance et al., 1970a,b). A short flash elicited a rapid disappearance of protons with a half-time of about 400 μsec and a stoichiometry of 1 H^+/100 Bchl, comparable to the amounts of individual electron-transport components. The measured rate of H^+ disappearance could not be correlated with any known electron-transfer step in this species. Oxidation of the primary acceptor under equivalent conditions takes 30 μsec and may exhibit different pH and temperature dependences (Parson, 1969), while the re-reduction of cytochrome c_{555}, the secondary donor, takes several tens of milliseconds. Later measurements showed the half-time of the H^+-binding reaction to be 130–140 μsec at room temperature (Callis et al., 1972; Halsey and Parson, 1974), but this is still too slow to equate with the X → Y reaction.

Chance et al. (1970a) offered two explanations of the rapid pH increase. The first proposed an unidentified redox component that, on reduction via the secondary electron acceptor in a time of about 400 μsec, would bind a proton. The second interpretation suggested that an electron-transfer event, such as the primary-to-secondary-acceptor step, induced a pK change in the membrane in a time of about 400 μsec resulting in H^+-binding. In this case, the protonatable group would not be associated with an active redox component, and the model was described as a "membrane-Bohr" effect by analogy with the Bohr effect in hemoglobin.

A similar H^+-binding reaction has been observed in *Rp. sphaeroides*, *Rp. capsulata*, and *Rs. rubrum*. Early studies on *Rp. sphaeroides* demonstrated a half-time for proton-binding on laser activation of about 300 μsec and a stoichiometry of 1 H^+/170 Bchl at pH 6.3 (Crofts et al., 1971b; Cogdell et al., 1972). Addition of valinomycin caused a 60% increase in the extent as well as a reported 70% stimulation in the initial rate. The kinetics became distinctly biphasic with a previously undetected slow phase ($t_{1/2}$ of a few milliseconds). An effect of valinomycin on the initial rate for a single turnover, is anomalous, however, and is very likely due to the increased *extent* coupled with a response limitation in the apparatus, since the reported time constant was 100 μsec. Recent data indicate that the actual half-time of the rapid H^+ uptake at pH 6.3 is about 120 μsec (Petty and Dutton, 1967a) which is suggestively similar to the lower values reported for *Chr. vinosum*.

(b) Multiple-Turnover Kinetics. The use of multiple-turnover flashes has tended to add only confusion to the complex H^+-uptake phenomenology. Chance *et al.* (1970*a*) found, however, that a multiple-turnover flash caused a 2.5- to 3-fold greater proton uptake in *Chr. vinosum* than did a single-turnover flash. This is consistent with the known rapid turnover of the primary electron acceptor and donor. The greater extent of binding was independent of the actual halfwidth of the broad flash from 0.2 to 2 msec, suggesting that a secondary electron pool of about 2 equivalents was involved that was isolated from the rest of the electron-transport chain by a rate-limiting step of a few milliseconds. This is in agreement with a later estimate of a secondary-electron-acceptor pool (Y) of 2 equivalents (Case and Parson, 1971); the validity of the assumptions necessary for this estimate, however, has been questioned (Halsey and Parson, 1974).

(c) Decay Kinetics of the H^+ Uptake. The rate of onset of the rapid H^+-binding is unaffected by various additions of uncouplers and antibiotics, but the slow-rise kinetics and subsequent decay rate are altered. In the absence of any such reagents, the flash-induced pH change decays slowly with an approximate half-time of several seconds; addition of valinomycin slows down the decay, while uncouplers accelerate it (Chance *et al.*, 1970*a*). The rate of decay of the rapid H^+-binding is a roughly linear function of FCCP concentration, and the rate is essentially identical to that for the decay of the carotenoid bandshift over a wide range of FCCP concentrations (Cogdell *et al.*, 1972). Accepting the carotenoid bandshift as an indicator of a membrane potential and the mode of action of valinomycin as a K^+ uncoupler, these observations are good evidence for the electrogenicity of the proton-uptake mechanism and the electrophoretic nature of subsequent ion movements. Valinomycin has also been found to slow the rate of H^+ decay induced by FCCP (Petty and Dutton, 1976*a*), showing that the action of FCCP is as a true uncoupler acting on both the osmotic and the electrical components of the proton pump.

3.3.2. Relationship between Flash-Induced H^+-Binding and Steady-State H^+ Uptake

In addition to the effects of uncouplers and antibiotics on the rapid proton binding, Cogdell and Crofts (1974) made a good case for the elemental relationship between the flash-induced H^+-binding and steady-state H^+ uptake. Using repetitive flashes at intervals of a few seconds, they showed that a flash-induced steady-state proton uptake occurs in which each flash turnover is identical, kinetically and stoichiometrically, to a single rapid binding reaction. Thus, the observed binding reaction is but the first step

of a transmembrane proton pump. More recently, release of protons inside the chromatophore was observed directly by means of a pH indicator trapped within the vesicles (Petty and Dutton, 1976*b*).

3.3.3. Redox-Potential Characteristics of the Flash-Induced H^+ Uptake

(a) Redox-Potential Dependence. Initial studies on the redox-potential dependence of the H^+-binding reaction showed it to appear roughly in coincidence with photochemistry at high potentials and to attenuate at the low-potential end with $E_{m,7.5}$ about +5 mV (Cogdell *et al.*, 1972). A small enhancement was seen over the region of titration of cytochrome c_2 ($E_{m,7.2} = +293$ mV). In the presence of valinomycin, a general enhancement was seen, largely attributable to the appearance of a slow phase of H^+-binding.

More recent studies have shown that even in the absence of valinomycin, the picture is complicated by more than one component, but that only one component is seen in the presence of antimycin A (Petty *et al.*, 1977). Under these conditions, the low-potential attenuation occurs with $E_{m,7.0}$ approximately +85 mV, considerably higher than that reported earlier.

Since the midpoint potential of the primary electron acceptor in *Rp. sphaeroides* is about -15 mV at pH 7.0 (Reed *et al.*, 1969; Dutton *et al.*, 1973*b*; Prince and Dutton, 1976), the low-potential attenuation of the antimycin-insensitive H^+-binding can be attributed to the titration of a secondary acceptor system. This is consistent with the considerable, although not always complete, inhibition of the H^+-binding by *o*-phenanthroline and insensitivity to antimycin A, i.e., after the primary acceptor and up to and including cytochrome *b*.

The extent of H^+-binding during this phase is very close to 1 H^+/e (Petty and Dutton, 1976*a*). In the absence of antimycin, the extent increases to approximately 1.6 H^+/e, indicating the involvement of a second component that, however, is not readily distinguishable kinetically from the first, and is probably complete within a few hundred microseconds. In the absence of antimycin, H^+ uptake exhibits an apparent $E_{m,7.5}$ of +5 mV (Cogdell *et al.*, 1972) due to a contribution from this second phase of binding, which seems to occur throughout the potential range of photochemical activity, attenuating only as the primary acceptor is reduced or the primary donor oxidized (Petty *et al.*, 1977).

(b) pH Dependence of the Low-Potential Attenuation of H^+-Binding. If the component responsible for the low-potential attenuation of H^+-binding were itself the binding site, then the midpoint potential for attenuation should be pH-dependent. A dependence of −60 mV/pH unit over a narrow pH range was reported for *Rp. sphaeroides* for the complex situation in the absence of antimycin A (Cogdell *et al.*, 1972). The single, antimycin-insensitive component also has a −60 mV/pH unit dependence at low pH values, indicating 1 H^+ bound per electron, but it becomes pH-insensitive above about pH 7.5; i.e., it exhibits a pK on the reduced form (Petty *et al.*, 1977). It corresponds to a one-electron process throughout the whole pH range. When H^+-binding is measured during a series of flashes separated by a few milliseconds, proton uptake on the first flash exhibits a pK, but that on the second does not (Petty *et al.*, 1977).

A preliminary determination of the midpoint potential for H^+-binding in *Chr. vinosum* indicated an $E_{m, 6.9}$ of −70 mV (Cogdell *et al.*, 1972). This was considered compatible with the value for the secondary electron acceptor ($E_{m, 7.7} = −90$ mV) reported to show a pH dependence of −38 mV/pH unit (Case and Parson, 1971). The physical significance of such a dependence, indicating about 0.6 H^+ bound per electron, is obscure, and it seems possible that the reported value is due to inadvertent averaging over a pH range including a pK of the reduced form. This was recently shown to be the case (Prince and Dutton, 1976) for the equally anomalous −36 mV/pH unit dependence reported for the primary acceptor in *Chr. vinosum* (Case and Parson, 1971). The primary acceptor is now known to have a pK at about pH 8, and a similar value may be inferred for the secondary acceptor. This would be in rather good agreement with that directly observed in *Rp. sphaeroides* for rapid H^+-binding.

3.3.4. Multiplicity of H^+-Binding Sites

Following a multiple-turnover flash, redox-potential titrations of H^+-binding revealed an enhancement with apparent $E_{m, 7.5}$ of +130 mV (Cogdell *et al.*, 1972), which was further stimulated by valinomycin. Comparison with the enhancement of the carotenoid bandshift over the same redox-potential region ($E_{m, 7.5} = +150$ mV, Jackson and Dutton, 1973) led to the suggestion that both phenomena were expressions of a second electrogenic, proton-translocating site or loop situated between cytochrome *b* ($E_{m, 7.2} = +50$ mV) and cytochrome c_2 ($E_{m, 7.0} = +295$ mV) and capitalizing on the energy available from this redox span. An energy-conservation site in this region is also indicated by the acceleration of electron flow between these two components by valinomycin and by uncouplers (Prince and Dutton, 1975; Crofts *et al.*, 1974). How-

ever, although enhancement of both the H^+ uptake and the carotenoid bandshift in this potential region is due to slow phases, only the carotenoid response is seen after a single-turnover flash in the absence of valinomycin.

In the presence of valinomycin, proton-binding following a single-turnover flash is stimulated largely due to the appearance of a new, millisecond binding phase (Chance *et al.*, 1970a; Crofts *et al.*, 1971b; Cogdell *et al.*, 1972). The fast phase, however, is also enhanced—presumably an effect on the antimycin-sensitive fast component. Since the new slow phase is equivalent to about 1 H^+/e, it appears that as many as three protons may be taken up following a single-turnover excitation, but the precise stoichiometry is still uncertain, and the "second" proton may in fact be taken up in a biphasic manner. In view of the recent finding of H^+/e ratios of 4 in mitochondria (Brand *et al.*, 1976a,b), however, the possibility of "superstoichiometries," once thought laid to rest, must again be considered, and the bacterial system may prove a very useful testing ground for this question.[*]

3.3.5. Dependence of the Rate of H^+-Binding on Physical Parameters

(a) Temperature. A preliminary report on the temperature dependence of H^+-binding in *Chr. vinosum*, following a multiple-turnover flash, indicated an apparent activation energy of 2.0 kcal · mol⁻¹ (Chance *et al.*, 1970a), significantly different from that reported for the X → Y reaction (8.3 kcal · mol⁻¹) (Parson, 1969), which, however, was measured with single-turnover excitation. In *Rp. sphaeroides*, the apparent activation energy for the antimycin-insensitive, rapid H^+-binding is 10.5 kcal · mol⁻¹ at both pH 6.0 and 8.0 (Petty and Dutton, 1976a).

(b) Viscosity. Glycerol (50%) has no effect on either the rate of rapid H^+-binding in *Rp. sphaeroides* or its activation energy (Petty and Dutton, 1976a).

(c) pH. The kinetics of H^+-binding in *Chr. vinosum* was found to be almost independent of pH over a limited range studied (pH 5.1–7.1) (Chance *et al.*, 1970a). A similar conclusion was reached for *Rp. sphaeroides* (Cogdell *et al.*, 1972), and even over a much wider range of pH only a small dependence was observed (Petty and Dutton, 1976a).

One may conclude from the lack of effect of viscosity and pH on the *rate* of proton uptake that the

---

[*] Recent work has shown the maximum stoichiometry in *Rp. sphaeroides* to be 2 H^+ per electron, and "superstoichiometries" do not appear to arise in this system (Dutton *et al.*, 1977; Petty *et al.*, 1977).

reaction is not limited by the availability of protons in the aqueous phase, but, presumably, by that of the electron [see Section 3.3.8(b)].

3.3.6. Functional pH Range of H⁺ Uptake

Although the rate of H⁺-binding is rather independent of pH, the *extent* of H⁺-binding in *Rp. sphaeroides* shows a marked dependence (Cogdell and Crofts, 1971; Cogdell *et al.*, 1972). Although probably related, this pK determination is not necessarily identical to that observed in the pH dependence of the midpoint-potential described above [see Section 3.3.3(b)]. The value of the pK for the extent of H⁺-binding is dependent on the oxidation state of the primary electron donor (P/P⁺). At high redox potentials, when cytochrome c_2 is oxidized before excitation, the primary donor remains in the oxidized state following activation, and the proton-binding reaction shows a pK of 7.5. At lower potentials P⁺ is partially rereduced by cytochrome c_2 in a time (first half-time $\simeq$ 30 μsec) somewhat faster than the H⁺-binding process ($t_{1/2} \simeq$ 150 μsec), and the pK is shifted up to 8.5. Thus, possibly the proximity of the positive charge on the oxidized donor decreases the H⁺ affinity of an acid–base group associated with the reduced form of a secondary acceptor. The rereduction of the primary donor, however, is not complete in this time range, but so far no reflection of the biphasic cytochrome c_2 oxidation kinetics has been detected in the proton-binding (Petty and Dutton, 1976a).

The redox-potential dependence of the pK for H⁺-binding can account for the small stimulation of uptake observed at pH 7.5 in titrations through the midpoint-potential region of cytochrome c_2 (Cogdell *et al.*, 1972).

3.3.7. Ubiquinone and the Rapid H⁺ Uptake

(a) The Primary and Secondary Acceptors. Quinones (Qs) have long been known in biological oxidation–reduction processes (Crane *et al.*, 1960; Redfearn and Pumphrey, 1960). In bacterial photosynthesis, ubiquinone-10 (UQ-10), which is present in large amounts [10–20 mol/mol RC (Carr and Exell, 1965; Peters and Cellarius, 1972)], has been implicated in both primary and secondary acceptor roles (Cogdell *et al.*, 1974) (see Chapter 25 for a survey of Qs in photosynthetic bacteria). The primary electron acceptor in RCs from *Rp. sphaeroides* was recently identified as UQ-10 (Okamura *et al.*, 1975), while that in *Chr. vinosum* appears to be menaquinone (Okamura *et al.*, 1976). In *Rp. sphaeroides*,

the primary acceptor Q is much more tightly bound than that of the secondary acceptor pool and, judging by its involvement in the primary events and its magnetic coupling to an iron atom, is to some extent in a specialized environment (Okamura *et al.*, 1975; Feher *et al.*, 1974) (see also Chapter 19). Recent results suggest that two Qs are normally associated with isolated RCs (Okamura *et al.*, 1975), and that they are almost equivalent with respect to their interaction with the iron atom (Wraight, 1977a).

The general occurence of excess UQ in isolated RCs and its activity as a secondary acceptor in such preparations suggests that it is, in fact, also the secondary acceptor component. This is supported by extraction studies on chromatophores (Halsey and Parson, 1974; Cogdell *et al.*, 1974). Removal of UQ from *Chr. vinosum* also inhibited proton uptake to about the same extent as did *o*-phenanthroline (Halsey and Parson, 1974). Apart from this, evidence linking UQ directly with H⁺-binding is circumstantial. Secondary acceptors are discussed in more detail by Parson in Chapter 25.

(b) Reaction Center Studies. Light-induced H⁺ uptake by RCs isolated from a blue-green mutant of *Rp. sphaeroides* can be observed only when supplemented with an electron donor, such as cytochrome *c*, and in the presence of a hydrogen-carrying acceptor, such as Q (Cogdell *et al.*, 1973). Proton uptake is stoichiometric with the quantity of secondary acceptor reduced according to the overall equation

$$2 \text{ cyt } c^{II} + Q + 2 H^+ \rightarrow 2 \text{ cyt } c^{III} + QH_2$$

With isolated RCs, this process is not vectorial in nature, and so far, successful reconstitution of H⁺-pumping activities using phospholipid vesicles has not been reported. The fully reduced Q is assumed to form by disproportionation. However, no H⁺ uptake is observed after the first flash of a series, but only after subsequent ones (Wraight *et al.*, 1975). This is consistent with the identification, by optical and ESR methods, of the reduced primary electron acceptor as an anionic ubisemiquinone in RCs from both blue-green mutant and wild-type *Rp. sphaeroides* (Slooten, 1972; Clayton and Straley, 1970, 1972; Feher *et al.*, 1972; Okamura *et al.*, 1975). A similar component was demonstrated optically in RCs from *Rs. rubrum* (Slooten, 1972). The similarity of the low-temperature ESR spectra of the primary acceptors of these and other species of photosynthetic bacteria suggests similar chemical identities throughout (Leigh and Dutton, 1972; Dutton *et al.*, 1973a). Disproportionation to give the fully reduced Q must therefore occur only at the secondary acceptor level. The anionic semiquinone form of the primary acceptor in isolated RCs is extremely stable, lasting for several minutes after flash-induced formation even in the presence of excess Q as secondary acceptor (Wraight et al., 1975). Provided such secondary acceptors are present, however, the presence of the semiquinone does not preclude further photochemical turnover; in other words, the RCs are not "closed" by a one-electron reduction of the supposed primary acceptor (Wraight et al., 1975). More recent results now show that the semiquinone formed by a single flash is destroyed by a subsequent flash presumably forming the fully reduced Q, and that the semiquinone is formed and destroyed on alternate odd and even flashes in a series showing a periodicity of two. This behavior is exhibited by both the optical and ESR ($g = 1.82$) manifestations of the semiquinone, and it is suggested that two closely equivalent Qs are normally active in the primary acceptor region of RCs (Wraight, 1977a). Proton uptake by RCs also oscillates with a period of two, as would be expected if the fully reduced, protonated Q were being formed only on even-numbered flashes (Wraight, 1977b, and unpublished). Generation of the fully reduced Q on even-numbered flashes of a series is also indicated by absorbance changes at 270 nm (Vermeglio, 1977).

(c) Midpoint Potentials in Vitro and in Vivo. The standard redox potential for UQ in aqueous ethanol has been reported as +0.542 V (Morton et al., 1958) and the polarographic half-wave potential at pH 7.4 as +98 mV (Moret et al., 1961), but others have found it to be indeterminate (O'Brien and Olver, 1969).

The measured midpoint potential of the fast H^+-binding component in Rp. sphaeroides ($E_{m,7.0} \simeq +85$ mV, Petty et al., 1977) would appear to be in reasonably good agreement with that reported for the mitochondrial Q pool ($E_{m,7.0} = +66$ mV, Urban and Klingenberg, 1969). Any significant pool size would lead to underestimation of a kinetically determined value such as the photosynthetic one.

3.3.8. Protonation State and Operating Midpoint Potentials

(a) The Primary Acceptor. The anionic nature of the RC semiquinone and the lack of proton-binding in the absence of secondary acceptors is consistent with the pH independence of the primary acceptor in RC preparations (Reed et al., 1969; Dutton et al., 1973b). This is in contrast, however, to the situation in chromatophores, in which the primary acceptor exhibits a −60 mV/pH unit dependence in many species (Jackson et al., 1973; Dutton et al., 1973b; Prince and Dutton, 1976). Since the rapid, antimycin-insensitive H^+-binding in chromatophores, as in RCs,

is associated with the reduction of a secondary acceptor, the pH dependence of the primary acceptor would appear to represent a comparatively slow H^+-binding, evident in equilibrium measurements but not kinetically functional on the time scale of electron-transfer events (Dutton et al., 1973b). Recent work on the primary acceptor in situ has shown that the pH dependence of the midpoint potential reveals a pK above which the potential is pH-independent. Interestingly, this limiting potential is close to −180 mV for all species studied so far despite considerable differences in the value of the pK (Prince and Dutton, 1976) (see also Prince and Dutton, Chapter 24). The kinetic nonfunctionality of the pH dependence of the primary acceptor in situ means that this component is operating with a midpoint potential governed by the pK of the reduced form, i.e., −180 mV.

The lack of pH dependence in isolated RCs even in equilibrium titrations raises the possibility that the in situ dependence is due to a membrane-Bohr effect of the type originally suggested by Chance et al. (1970a). That the E_m in RCs is −50 mV rather than the expected −180 mV is an anomaly that is at present unresolved.

(b) Kinetic Discrepancy between H^+-binding and the Secondary Acceptor Reaction. The supposition that the rate of H^+ uptake is limited by the appearance of the electron in some state of "availability" is in agreement with the conclusion of Chance et al. (1970a) that the H^+-binding in Chr. vinosum (in 150–400 μsec) does not correlate kinetically with the reduction of the secondary acceptor (Y), which occurs in about 30 μsec (Parson, 1969). Studies on the kinetics of this secondary electron-transfer reaction in Chr. vinosum have been possible because the rereduction of P^+ by cytochrome c-555 is extremely fast and thus the turnover time of the photoreaction at times longer than a few microseconds is governed by reoxidation of the primary acceptor (X). In Rp. sphaeroides, re-reduction of P^+ is slower and biphasic, even the fast phase taking some 30 μsec (Dutton et al., 1975). This has generally inhibited attempts to measure the reoxidation rate of X^-. However, the effect of double flashes on the carotenoid bandshift was reported by Crofts et al. (1971b). A portion of the photochemical activity recovered in less than 35 μsec while the rest was very much slower. This is strikingly reminiscent of the cytochrome c_2 oxidation kinetics, and suggests that the rate limitation observed was due to the re-reduction of P^+. If this were the case, then the reoxidation rate of X^- could be at least as fast as 30 μsec and the H^+-binding rate in Rp. sphaeroides would also show a discrepancy with the rate of the $X \rightarrow Y$ reaction.

(c) The Secondary Acceptor. A kinetic discrepancy between the X → Y reaction and the rapid H⁺-binding suggests that the reduced secondary acceptor is also *not immediately* amenable to protonation and initially is also operating with a midpoint potential governed by its pK. In *Rp. sphaeroides*, the sudden plethora of pK's associated with H⁺ uptake [see Section 3.3.3(b) and 3.3.6] make it difficult to decide, in the absence of a definite model, which one to choose. The most appropriate one seems to be that actually determined from the midpoint-potential titrations, which give a limiting midpoint potential of +45 mV above pH 7.5 (Petty *et al.*, 1977). As noted above, however, this pK is associated with H⁺-binding only on the first flash. It should be remembered that in the case of component pools of significant size, flash-activated determinations may underestimate the true midpoint potential of the chemical species. We might therefore expect that the secondary acceptor would display an equilibrium midpoint potential somewhat more positive than the kinetic determinations reveal.

3.3.9. Kinetic Barrier to Protonation and Quinone Mobility

A plausible candidate for the secondary acceptor is UQ. As we shall discuss below, however, the chemical nature of the redox couple involved is unknown and is probably critical. For the time being, therefore, we shall refer to the species only as "oxidized" and "reduced" Q, with additional usage of the term "anionic" to indicate that the reduced form is unprotonated.

It is known that protonation reactions in solution are diffusion-limited (Ilgenfritz, 1966), and one might expect protonation of the reduced Q to follow the arrival of the electron rather closely. In chloroplasts, proton uptake occurs in about 60 msec and is rate-limited by a diffusion barrier (Ausländer and Junge, 1974). Addition of lipid-soluble proton carriers (i.e., uncoupling agents) or apparent removal of the diffusion barrier by sand grinding accelerates the uptake considerably. In bacteria, in which the normal rate of H⁺-binding is considerably faster, no evidence for a partial diffusion barrier is forthcoming. No significant acceleration of binding is observed with increased H⁺ concentration, increased buffering power, presence of uncoupling agents, or after sand grinding (Petty and Dutton, 1976*a*; P. L. Dutton, personal communication). It thus appears that the secondary anionic Q is completely inaccessible at first, and that no protonation occurs until the electron becomes available to the aqueous phase.

Since Qs are widely thought of as lipid-soluble, small molecules, the 150- to 400-μsec lag between reduction of the secondary Q and protonation may seem rather large. Measurements of fluorescence quenching of probes by endogenous Qs in natural and artificial membranes have yielded quenching constants close to the diffusion-limited value (Chance *et al.*, 1975). This is, however, more a measure of the mobility of the probe than of the Q. In fact, naturally occurring Qs are not small; the isoprenoid side chain of UQ-10 is about 45 Å long and, even allowing for extensive folding, cannot be considered small relative to the membrane dimensions. It is very likely that the mobility of the individual Q molecules is rather limited, as was recently suggested for UQ in *Esch. coli* (Stroobant and Kaback, 1975). The kinetic barrier to protonation may therefore correspond to the time necessary for the electron to escape from the "woodwork" of the RC complex by diffusion of the initial anionic Q, or, perhaps, by the operation of a "bucket brigade" of UQ molecules. The usual association of 2 molecules of UQ with isolated RCs, one of which is of the loosely bound, secondary variety (Clayton *et al.*, 1972; Okamura *et al.*, 1975; Cogdell *et al.*, 1974), may provide some support for the notion of a transiently trapped secondary acceptor, inaccessible to protonation by association with the RC complex.

3.3.10. Quinone Pools

In *Rp. sphaeroides*, there is no evidence for a small, potentiometrically distinct pool of secondary acceptor; thus, the mobility of the UQs must be sufficient to render them all equivalent on the time scale of redox potentiometric titrations (minutes to hours). In *Chr. vinosum*, however, there are indications that the secondary acceptor is a small pool of approximately 2 equivalents, with an apparent $E_{m,7.7}$ of −90 mV (Case and Parson, 1971). Conceivably, this may correspond to one or more UQs permanently trapped in the more extensive "woodwork" of the RC complex of this species (Thornber, 1970). Extraction of UQ did not affect the X → Y reaction until the last 2 molecules were removed (Halsey and Parson, 1974). Further work is required on this point and on the pH dependence of this component and of the H⁺-binding in this species.

3.3.11. Completion of the Proton Translocation

It is well established that the rapid H⁺-binding is but the first step is net H⁺ uptake (Cogdell and Crofts, 1974) and that proton release on the inside of the chromatophore vesicle does occur (Petty and Dutton, 1976*b*). Thus, the reduced secondary acceptor (UQ?)

must somehow release its associated proton on the other side of the membrane. It is an almost inescapable conclusion that a redox-driven proton-translocating process is involved, which can be completed by a subsequent electron transfer step. The kinetics of oxidation of the reduced UQ have received little attention so far, but it is possible that the oxidant is cytochrome b_{50} (Petty and Dutton, 1976b).

Traditionally, cytochromes have been viewed as electron carriers, in which case oxidation of a reduced, protonated Q (e.g., QH_2) by cytochrome b_{50} would allow the release of protons on the inside of the vesicle. However, many cytochromes, particularly those of the b-variety, have revealed pH-dependent midpoint potentials in equilibrium redox titrations. This raises the distinct possibility that a cytochrome b, for example, could act as a hydrogen carrier in a redox loop. The significance of the proton-binding capacity of cytochromes on the time scale of electron-transport events is largely unknown, but the pH dependence of the midpoint potential of cytochrome b_{50} was recently shown to be kinetically functional in cyclic electron flow in Rp. sphaeroides (Petty and Dutton, 1976b). This is revealed by the effect of pH on proton release, which does not necessarily occur on oxidation of the reduced Q by cytochrome b_{50}. Internal H^+ release occurs at this step only above pH 7.4, while at lower pH's H^+ release occurs only on oxidation of the cytochrome b by the next component. It is inferred that cytochrome b_{50} is a hydrogen carrier at low pH's, but not at higher values, exhibiting a pK on the reduced form at pH 7.4 in agreement with equilibrium titrations of this species. This does not preclude cytochrome b_{50} from performing an electrogenic electron transfer, for example, to Z, as depicted in Fig. 7B, since the proton could be released inwardly while the electron is transferred out across the membrane, but it does render the whole situation more equivocal.

3.3.12. Models (Wampeters, Foma and Granfalloons)*

An electrogenic site in the cyclic electron-transport chain is evidenced by the antimycin-A-sensitive slow phase of the carotenoid bandshift and the enhancement of this phase in the potential region of +150 mV. Such a site (see Fig. 7B) could also be responsible for the antimycin-A-sensitive uptake of a second proton following a flash. There are, however, many obser-

* "A wampeter is an object around which the lives of many otherwise unrelated people may revolve ... Foma are harmless untruths, intended to comfort simple souls ... A granfalloon is a proud and meaningless association of human beings." From *Wampeters, Foma and Granfalloons* by Kurt Vonnegut, Jr. (Delta Book published by Dell Publishing Co., Inc., New York).

vations that are difficult to reconcile with common organizational concepts of the electron-transport chain. Further difficulties are encountered in the kinetics of electron transport among the components themselves (see Dutton and Prince, Chapter 28). Outstanding difficulties include the "activation" of the carotenoid bandshift by a component (Z) that behaves like a hydrogen carrier, the existence of approximately two *rapid* (<msec) H^+-binding reactions (one of which is antimycin-A-sensitive), and a third, slow H^+-binding that perhaps complements the slow carotenoid bandshift but is seen *only* in the presence of valinomycin.

To date, no scheme adequately accounts for the phenomenology. The traditional models such as that shown in Fig. 7B can be readily criticized and dissected. Rather than elaborate on such schemes, we would like to introduce, briefly a novel concept of component interactions recently suggested for mitochondrial electron transport (Mitchell, 1975a,b) in which Q assumes the role of both Q and Z in Fig. 7B.

Two distinct formulations are possible, and these are incorporated into a feasible photosynthetic electron-transport chain in Fig. 8. One aspect is immediately clear—as pointed out by Mitchell (1975b), the system is "marvelously compact." This, however, is its only certain asset! Suitable identities of the cytochromes b_1 and b_2 are discussed in some depth in Chapter 28 by Dutton and Prince. There is a tendency to rule out cytochrome b_{155} because its oxidation following a single flash is not sensitive to uncoupling agents, and its re-reduction rate appears to be too slow to permit a function in the cyclic electron chain. Oxidation of cytochrome b_{50}, on the other hand, is coupled, and its turnover kinetics are compatible with a role in the cyclic system. For the sake of argument, Dutton and Prince (Chapter 28) equate b_1 with another b cytochrome, b_{-90}, that has not been widely considered in the context of photosynthetic electron transport for lack of data, and b_2 with cytochrome b_{50}. This choice achieves a redox potential span consistent with an energy-conservation site, and, as described in their chapter, avoids certain complications in the kinetics of cytochrome b_{50}. It is also consistent with the observation in Rs. rubrum that reversed electron transport driven by hydrolysis of pyrophosphate causes reduction of a low-potential b-type cytochrome ($E_{m,7} \simeq -105$ mV) (Dutton and Baltscheffsky, 1972). The choice of cytochrome b_{50} at the high-potential end of this redox span places it, however, in the outer part of the membrane, which is not readily reconciled with the valinomycin stimulation of electron transport between cytochromes b_{50} and c_2 or with the elegant interpretation of

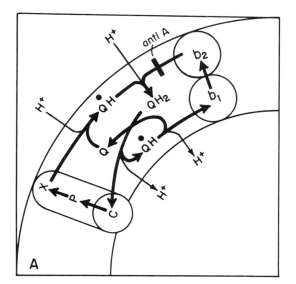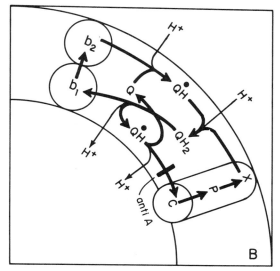

Fig. 8. Two possible arrangements of the proton-motive Q-cycle incorporating a thermodynamically unstable semiquinone (A) and a stable semiquinone (B). The identities of cytochromes b_1 and b_2 are discussed in the text. Possible sites of action of antimycin A are marked.

the effect of the pK of cytochrome b_{50} on internal H^+ release (see Section 3.3.11).

In the schemes of Fig. 8, the redox reactions of the Q are separated into two one-electron processes. In Fig 8A, the first reduction step (Q/Q̇H) has a lower midpoint potential (more negative) than the second step (Q̇H/QH$_2$). The semiquinone is thus inherently unstable, and an equilibrium titration of a Q pool would be expected to yield a single two-electron reduction curve. Such a situation is compatible with the absence, so far, of an observable free-radical ESR signal from a Q pool, but is in apparent conflict with the strongly $n = 1$ titrations for the proton-binding component, attributed to ubisemiquinone (Petty *et al.*, 1977).

In Fig. 8B, the alternate order of reduction potentials is utilized. Here, the Q/Q̇H step is less strongly reducing than the second, Q̇H/QH$_2$, and the system is characterized by two distinct one-electron processes with a stable semiquinone. This, of course, is not in agreement with the lack of observable semiquinone in photosynthetic and oxidative electron-transport systems, but does provide a component for the $n = 1$ titration of rapid H^+-binding.

In both cases, the semiquinone is restricted in its transmembrane interaction, so that mobility across the membrane involves only Q and QH$_2$. The overall operation is that of a hydrogen cycle within an electron cycle with two protons translocated per complete revolution. The possible occurrence of three H^+-binding reactions is unaccounted for by any current model that is not unreasonably elaborate. It is possible, of course, that one of these H^+-uptake

processes is not involved in the repetitive cycles of the chain but represents a binding site of slow turnover.

The action of antimycin A is also not easily incorporated into these schemes. Although steady-state inhibition could be readily achieved by blocks at the positions indicated in Fig. 8, the very tight control exhibited by antimycin A (inhibition of both cyto-chrome c_2 reduction and b oxidation on the first turn-over) is hard to account for without a second site of action, especially at extremes of the redox-potential scale. This is a matter more readily accounted for by traditional schemes in which cytochrome b is oxidized directly by cytochrome c_2. Alternatively, involvement of another carrier between these components would require action of antimycin A directly on this un-known intermediate, preventing both its oxidation and its reduction. Difficulties such as these, and related problems of the rates of oxidation and reduction of the b-cytochromes, can no doubt be overcome by alter-native schemes. Some variations have already ap-peared (Crofts *et al.*, 1975a; Kröger, 1976), and it seems probable that the renewed emphasis on the multiplicity of Q oxidation and protonation states will provide fertile ground for reexamination and ex-planation of previously puzzling phenomena.

Finally, we would like to draw attention to a particularly perplexing aspect of all current schemes in the chemiosmotic mode. Proton uptake, both in the presence and in the absence of antimycin A, exhibits a pK between pH 7.5 and 8.5 (Petty and Dutton, 1976a). At sufficiently high pH's no proton uptake is observed, at least on the first flash. If the photoelectron were to traverse the electron-transport chain unprotonated, eventually to return to cytochrome c_2, the light-induced membrane potential should be short-circuited and the carotenoid bandshift should decay in the turnover time of cyclic electron flow, which appears to be not slower than a few hundredths of a second. Under such conditions, however, the carotenoid bandshift is as stable as at lower pH's, where H^+ uptake is maximal (see Dutton and Prince, Chapter 28, Fig. 28). This is a conundrum indeed, if, in fact, the electron has made a complete circuit of the electron-transport chain. Although kinetic matching of cytochrome b_{50} oxidation and c_2 reduction was demonstrated in the uncoupled state (Prince and Dutton, 1975), discrepancies exist in the coupled state, and, in any case, identification of the electron arriving at cytochrome b_{50} with that entering the H^+-binding pool is only speculative.

3.3.13. Energetics of Coupling

The free energy stored in transmembrane electrical and osmotic gradients is crucial to the chemiosmotic coupling hypothesis, and much effort has been put into estimates and measurements of this parameter. Necessarily, the free energy of the electrochemical proton gradient must be sufficient to account for observed phosphorylation activities. Chemiosmotic coupling assumes a reversible interaction between the redox-driven and ATPase-driven proton pumps, and at equilibrium, the two driving forces will be equal and opposite. If leak pathways are minimal, this will be approximately true in the steady state also. Thus, the proton-motive force or electrochemical proton gradient must be at least as great as the energy stored in the prevailing phosphate potential modified by a stoichiometric factor. If the ATPase mechanism involves n protons per ATP, the free energies are related as

$$(1/n)\Delta G_{ATP} = \Delta \mu_{H^+} = \Delta p$$

Stoichiometries of 1–4 H^+/ATP have been suggested for various ATPases (Mitchell, 1966; Schröder et al., 1971; Rumberg and Schröder, 1972). Direct measurements have provided a value of 2 H^+/ATP for the mitochondrial coupling factor (Moyle and Mitchell, 1973), and measurements on chromatophores of Rp. sphaeroides have yielded a similar value (Jackson et al., 1975).

Measurements of photophosphorylation by chromatophores of Rp. sphaeroides show that a steady-state phosphate potential of some 15 kcal·mol⁻¹ can be achieved (Crofts and Jackson, 1970). This is equivalent to 0.65 eV·mol⁻¹, and with a stoichiometric factor of 2 H^+/ATP, indicates that a

proton-motive force of more than 300 mV is required. As we have seen in Sections 3.1.2 and 3.2.2(b), the carotenoid bandshift and ion movements suggest an electrical potential of up to 430 mV at the start of continuous illumination. This declines to a steady value of 200 mV at the same time as a pH gradient of about 1 pH unit (60 mV) is developing (Jackson and Crofts, 1971). Thus, the total proton-motive force attains values of 260–430 mV. Simultaneous measurements of phosphate potential, ΔpH, and $\Delta \psi$ in chromatophores of Rp. capsulata tended to confirm the energetic sufficiency of the proton-motive force (Casadio et al., 1974). The carotenoid bandshift was used to estimate $\Delta \psi$, while ΔpH was measured from the light-induced quenching of 9-amino acridine (Deamer et al., 1972). Under all conditions tested, the value of the electrochemical proton gradient was equal to or greater than the phosphate potential, assuming 2 H^+/ATP. Values of Δp greater than 400 mV were recorded. Measurements of Δp were also reported for Rs. rubrum using the distribution of [¹⁴C]CNS⁻ (thiocyanate, a penetrating anion) and [¹⁴C]methyl-amine as indicators of $\Delta \psi$ and ΔpH, respectively (Schuldiner et al., 1974). Values for Δp were only about 200 mV, but the phosphate potential was not recorded, and the overall impression was similar to that seen in Rp. capsulata.

Clearly, there is a large amount of energy available in the H^+ gradient, and it is very probably sufficient to account for ATP synthesis. Such a correlation, however, does not distinguish the chemiosmotic hypothesis, since most coupling hypotheses do assume equilibration among the various energy-conserving pathways. Of some critical interest in this respect is a study on flash-yields of ATP synthesis in Rp. sphaeroides (Saphon et al., 1975a,b; Jackson et al., 1975).

The decay of the flash-induced carotenoid bandshift was shown to be sensitive to the presence of phosphorylation conditions, in confirmation of much earlier reports (Smith and Baltscheffsky, 1959; Smith and Ramirez, 1960). However, only some 10–20% of the amplitude of the bandshift was responsive. No evidence could be found for a critical electrical potential, as was suggested for chloroplasts (Junge, 1970). This leaves two major possibilities; (1) either the chromatophore population is heterogeneous and only about 10% is capable of phosphorylation, or (2) there is some unidentified control on ATP synthesis other than via the energy supplied by the proton-motive force.

The first has the attraction of being simple, but it is not really consistent with electron-microscopic evidence for a considerable abundance of coupling factors (Reed and Raveed, 1972), and inhibitor studies indicate an average of at least 15 active ATPase per chromatophore (Jackson *et al.*, 1975). Furthermore, whole cells show a distinction between fast and slow phases of decay of the carotenoid bandshift, although unequivocal correlation with phosphorylation has not been demonstrated (Jackson *et al.*, 1975). The second possibility implies a regulatory interaction between the ATPase and the electron-transport chain, with the energy supplied by the electrochemical proton gradient. There are no indications of what this control might be, but it is of interest to note that Reed *et al.* (1975) recently showed that RCs *in situ* in chromatophores are masked from specific antibodies unless the coupling factor is removed. This may be no more than a physical obstruction, but it could also be an indication of a mechanistically significant association between the two.

3.4. Conclusions

Recent work has provided strong evidence for the electrochromic nature of the carotenoid and chlorophyll bandshifts seen in photosynthetic vesicles, and there seems, at present, no real need to invoke other models to account for the data. There are anomalies still unaccounted for, however, and it would be rash to conclude that electrochromism is proved as the sole mechanism of the electric-field-indicating absorption changes. Similarly, the energetics of conservation are adequately accounted for by the H^+ electrochemical gradient in the few cases in which sufficient data are available to attempt a judgment.

The least satisfactory realm is, perhaps not surprisingly, at the mechanistic level. Our improved understanding of the energetics of coupling processes has led fairly recently to a renewed interest in and invigorated approach to questions of mechanism, and new results are appearing almost too fast to assimilate. Strong evidence now exists for a proton-pumping activity of the electron-transport chain, and some localization and identity of the components involved has been achieved. However, current knowledge of the protonation behavior of the secondary acceptor pool (UQ?) and of cytochrome b_{50} has raised a number of intriguing questions concerning proton translocation and, in particular, the nature of the second coupling site. The current pace of events and the state of expectant turmoil lead us to the optimistic suggestion that a significant understanding may be soon within our reach. Such an event can be expected to have far-reaching and exciting influences on many related areas of membrane-component research.

4. Appendix

4.1. Electrochromic Responses

(a) Bandshifts. The total dipole moment of a molecule in an electric field, E, is given by

$$\mu^E = \mu^0 + a \cdot E$$

where μ^E and μ^0 are the dipole moments in the presence and in the absence of the field and a is the polarizability tensor. The energy of interaction with the field is

$$U = -\int \mu^E \cdot dE$$

The energy of a dipolar molecular system, i, in an electric field is thus

$$U_i^E = U_i^0 - (\mu_i^0 + \tfrac{1}{2} a_i E) E$$

where U_i^E and U_i^0 are the energies in the presence and in the absence of the field. Consequently, if the ground and excited states of the molecule differ in dipolar character, the energy gap between them will change in an electric field. This change in energy is seen as a change in the frequency of absorption such that

$$\Delta v = -\frac{1}{h} [(\mu_e^0 - \mu_g^0) + \tfrac{1}{2}(a_e - a_g) E] E \qquad (3)$$

where Δv and h are the frequency shift and Planck's constant, respectively, and subscripts e and g refer to the excited and ground states. For an absorption band with absorption maximum at λ, the wavelength shift is given by

$$\Delta\lambda = \frac{\lambda^2}{hc} (\Delta\mu^0 + \tfrac{1}{2}\Delta a \cdot E)E \qquad (4)$$

The direction and magnitude of the shift depend on the field strength, the *change* in permanent dipole moment ($\Delta\mu^0$) and polarizability (Δa), and the orientation of these changes in molecular dipole moment with respect to the electric field vector. The dependence on orientation can be more explicitly stated as follows:

$$\Delta\lambda = \frac{\lambda^2}{hc} (\Delta\mu \cdot E \cdot \cos\theta + \tfrac{1}{2}\Delta a \cdot E^2 \cdot \cos^2\phi) \qquad (5)$$

where θ is the angle between the electric field and the change in permanent dipole, ϕ is the angle between the electric field and the change in polarizability, and c is the speed of light.

Bandshifts of a molecular species in homogeneous and isotropic dispersion may result from two types of interaction with an electric field. Initially on application of a field, the difference in permanent dipole moment between ground and excited states will cause equivalent red and blue shifts of the absorption band; averaging over $\cos\theta$, the first term in equation

(5) goes to zero. Although there is no shift in the wavelength maximum, there is a symmetrical broadening of the absorption band. In free solution, any dipolar character of the ground state will cause subsequent orientation of the absorbers in the field; this will result in some asymmetry of the broadening, and consequently an observable shift in the position of the absorption peak (this effect is not included in the equations above). If, however, the absorbing molecules have some fixed orientation prior to the application of the field, as may be the case *in vivo*, then a shift in peak position will be immediately apparent.

Interaction with the field through a change in polarizability on excitation will also induce an immediate bandshift, the *direction* of which depends only on the sign of the polarizability change and not on the orientation of the molecule. The magnitude of the shift is proportional to the square of the field strength resolved in the direction of the appropriate polarizability component.

(b) Magnitude of Electrochromic Bandshifts. A rough guide to the magnitudes of these effects may be useful. Electrochromism can be observed quite easily for changes in dipole moment down to a few tenths of a debye ($D = 10^{-18}$ esu·cm); the shifts are, however, small. For a field strength of 2.5×10^5 V·cm^{-1} and $\Delta\mu^0 = 10$ D, an appreciable size, the expected maximum wavelength shift for an absorption band at 500 nm is about 1 nm. Changes in dipole moment of 1–10 D are common, although much larger values are known.

The order of magnitude for molecular polarizabilities is conveniently taken from the volume of the molecule. Carotenoids might thus be expected to exhibit values of a few hundred cubic angstroms. The linear geometry and conjugated bond system suggest that the actual values might be somewhat higher. This is borne out by experiment. Measurements of the *change* in polarizability on excitation of crocetin (9 conjugated bonds) and bixin (11 conjugated bonds) dimethyl esters have yielded values of 460–780 and 930–1560 $\times 10^{-24}$ cm^3, respectively (Labhardt, 1967). S. Schmidt and Reich (1972*b*) determined $\Delta\alpha_{11}$ (the component parallel to the long axis of the molecule) for lutein (10 conjugated bonds) to be 910×10^{-24} cm^3. Even such large values, however, under the same conditions as used above for the permanent dipole interaction, yield bandshifts of less than 0.05 nm.

(c) Hypo- and Hyperchromism. Analogously to the dipole moments of the ground and excited states in an electric field, the transition moment may be written as

$$\mu_{ge}^E = \mu_{ge}^0 + \mathbf{a}_{ge} \cdot \mathbf{E}$$

where μ_{ge}^E and μ_{ge}^0 are the transition moments in the presence and in the absence of an electric field, $\mathbf{E}$, and $\mathbf{a}_{ge}$ is the transition polarizability tensor. The transition moment can therefore be influenced by an electric field, with the result that the intensity of absorption is also field-dependent. This will be significant whenever $\mathbf{a}_{ge}$ is significant relative to μ_{ge}^0, and can occasionally produce absorption changes of the same magnitude as those due to bandshifts (Liptay, 1969).

Although it is more than likely that this effect contributes to the *in vivo* absorbance changes, it has generally been neglected in discussions of their origin. This is probably because the general shape of the absorbance changes clearly indicates a bandshift. As discussed in the main text, however, inclusion of changes in absorption intensity may account for some

Table 1. Spectral Alterations Due to an Electric Field[a]

Effect	Relevant parameters
Shift in chemical equilibrium	Dipole moments and polarizabilities of ground states of isomers, etc.
Field-induced orientation (dichroism)	Dipole moment and anisotropy of polarizability of ground state: $\mu^E = \mu^0 + \mathbf{a} \cdot \mathbf{E}$
Bandshifts and broadening	Change in dipole moment and polarizability between ground and excited states: $\Delta\mu^E = \mu_e^E - \mu_g^E = (\mu_e^0 - \mu_g^0) + (\mathbf{a}_e - \mathbf{a}_g)\mathbf{E}$
Changes in absorption intensity	Transition moment dipole: $\mu_{ge}^E = \mu_{ge}^0 + \mathbf{a}_{ge} \cdot \mathbf{E}$

[a]All these effects can occur simultaneously.

of the anomalies encountered in a purely bandshift description.

In Table 1, we have summarized briefly the different effects of an electrical field together with the relevant parameters.

4.2. Measurement of Electrochromic Shifts

Electrochromic shifts in peak position are expected, in general, to be small and difficult to observe directly. However, the small displacement relative to the reference position can give rise to significant changes in extinction at the flanks of the absorption band, where the spectrum is changing rapidly with wavelength. For small shifts, the difference spectrum closely approximates the first derivative of the absorption spectrum (Fig. 9). This was first shown by Scheraga (1961), using a Taylor series expansion to give the shifted spectrum:

$$\Delta\varepsilon(\lambda) = \varepsilon(\lambda - \Delta\lambda) - \varepsilon(\lambda) = \sum_{n=1}^{\infty} \frac{(-\Delta\lambda)^n}{n!} \cdot \frac{\partial^n \varepsilon}{\partial\lambda^n} \quad (6)$$

Hence, neglecting $n > 1$

$$\Delta A(\lambda) = C \cdot \Delta\varepsilon(\lambda) = -C\Delta\lambda \cdot \frac{\partial\varepsilon}{\partial\lambda} \quad (7)$$

where $\Delta A(\lambda)$ is the observed absorption change and $\Delta\varepsilon(\lambda)$ the molar extinction coefficient at wavelength λ, and C is the molar concentration of chromophores actually shifted.

For small shifts, therefore, the difference spectrum will have positive and negative wings with an isosbestic point close to the wavelength maximum of the absorption spectrum (see Fig. 9). From this, it is possible to determine only the product $C\Delta\lambda$. For larger shifts, it becomes apparent that the isosbestic wavelength is actually at $\lambda = \lambda_{max} + \frac{1}{2}\Delta\lambda$, allowing independent determination of C and $\Delta\lambda$. The use of the first derivative to accurately describe the shape of the spectrum, however, is limited to shifts of less than about 5 nm (Fisher *et al.*, 1969).

4.3. Degree of the Electric-Field Dependence

Equation 5 above indicates that the extent of the wavelength shift can depend on the magnitude of the electric field in both a linear and a quadratic fashion, corresponding to interaction through the permanent and induced dipole moments, respectively. Since the shifts are usually studied indirectly via the magnitude of the absorbance changes, rather than direct observation of either the isosbestic point or the position of the absorbance maximum, it is instructive to see in what manner the absorbance changes reflect the bandshift.

The field dependence of the difference spectrum can be readily obtained by substituting equation (5) for the wavelength shift into equation (6) to give

$$\Delta\varepsilon = \frac{\lambda^2}{hc} \Delta\mu E \cdot \cos\theta \cdot \frac{\partial\varepsilon}{\partial\lambda} + \frac{1}{2}\frac{\lambda^2}{hc} \cdot \Delta\alpha E^2 \cdot \cos^2\phi \frac{\partial\varepsilon}{\partial\lambda}$$

$$+ \frac{1}{2}\frac{\lambda^4}{h^2c^2} \cdot \Delta\mu^2 \cdot E^2 \cdot \cos^2\theta \frac{\partial^2\varepsilon}{\partial\lambda^2} + \cdots$$

The first and third terms on the right are both due to the change in permanent dipole moment. The first term, linear with respect to the field strength and proportional to the first derivative of the absorption spectrum, arises from unidirectional shifts in oriented molecules. In isotropic solution, it averages to zero, leaving the third term, which corresponds to the band-broadening described above, which is quadratic in field strength and proportional to the second derivative of the absorption band. The second term is the first-order contribution from the induced dipole moment; it is quadratic in field strength and proportional to the first derivative of the absorbance spectrum, and does not average out in isotropic suspensions.

Further aspects of electrochromism are described and discussed in the main text.

5. Note Added in Proof

Since the completion of the literature survey for this chapter a number of significant publications have appeared which bear note in this addendum. In brief, recent work on electrochromism has strongly and gratifyingly confirmed the conclusions and suppositions of Section 3.2. Recent developments in H^+-binding have also been forthcoming and serve to clarify some of the confusing issues described in Section 3.3; however, in this respect we are still far from a mechanistic understanding of the H^+-pumps and, in particular, the nature of the second electrogenic site.

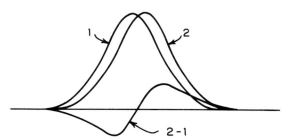

Fig. 9. Generation of "first-derivative-type" bandshift difference spectrum.

5.1. The Electrochromic Nature of the Light-Induced Bulk Pigment Responses

The studies started by Crofts and co-workers (Holmes and Crofts, 1977a,b) were aimed at correlating the amplitudes of carotenoid absorbance changes with the bandshift proper, determined by the position of the isosbestic points. The anomalous behavior could be explained, as described in Sections 3.2.7–3.2.9, by the supposition that the carotenoids were heterogeneous and that the electrochromic response was elicited in a pool with absorbance maxima already red-shifted by, for example, a permanent field. Recent studies by two groups have strongly confirmed this view for the bulk pigment responses in *Rp. sphaeroides* (de Grooth and Amesz, 1977a,b; Symons et al., 1977) and in higher plants (Amesz and de Grooth, 1976).

An elegant study by Takamiya and Dutton (1977), using the carotenoid bandshift to determine the membrane potential, has shown the electrogenic nature of the cyt c_2 oxidation step and has semiquantitatively confirmed the position of P within the membrane [Section 3.2.11c].

5.2. Local Field Effects

Small carotenoid absorbance changes remain in the presence of gramicidin and appear to be due to a red shift of a minor carotenoid pool (de Grooth and Amesz, 1977a). Similar absorbance changes can be seen in isolated reaction centers (Heathcote et al., 1977; Cogdell et al., 1977). These bandshifts appear to occur in response to local electric fields associated with both the oxidized primary donor and the reduced primary acceptor.

5.3. The Origin of the Induced Dipole Moment

Formally the induced dipole moment in the polarizable carotenoids has been ascribed to a large permanent field of the order of $2 \cdot 10^6$ V·m^{-1} (see Sections 3.2.7–3.2.9). Reich and co-workers have attempted to give a more molecular description of this field and have suggested that specific, asymmetric associations between certain carotenoids and Bchl provide a strong permanent polarization due to the electron affinity of the Mg-atom, equivalent to a local electric field. This effect is clearly seen *in vitro* (Reich and Sewe, 1977; Sewe and Reich, 1977a) and serves to rationalize some interesting *in vivo* data of Okada and Takamiya (1970) as well as the more recent problem of the permanent field effect (Sewe and Reich, 1977b). The agreement between the *in vivo* electrochromic absorbance changes in chloroplasts and in *in vitro* pigment mixtures has been much improved by the use of oriented layers and can now be considered excellent (Reich et al., 1976).

The function of the carotenoid:Bchl association is probably to provide an efficient route for deexcitation of chlorophyll-triplet excited states via torsional modes of the carotenoids or through excited-state-fusion processes. These have recently been studied both in reaction centers and bulk Bchl by Parson and co-workers (Parson and Monger, 1976; Monger and Parson, 1977).

5.4. H$^+$-Binding Studies

Recent publications have somewhat clarified the confusion surrounding the uptake of H$^+$-ions (Petty et al., 1977) and the nature of the component, Z (Prince and Dutton, 1977; Crofts et al., 1977; see also Chapter 28, Section 7). New results have also been forthcoming on the oscillatory phenomena of the quinone aceptor complex demonstrating the role of the Fe-atom and of H$^+$-binding in the primary-to-secondary-acceptor electron transfer (Wraight, 1977c). Oscillations of the semiquinone and of H$^+$-binding have also been observed in chromatophores with some, as yet uncertain, redox potential dependence (Barouch and Clayton, 1977; de Grooth et al., 1977).

ACKNOWLEDGMENTS

Many people have contributed helpful comments during the preparation of this paper, including, in particular, Drs. P. L. Dutton and W. K. Cheng. We would also like to thank Drs. P. L. Dutton, G. D. Case, A. R. Crofts, and M. Michel-Villaz for access to unpublished material and preprints of manuscripts in preparation or in press. Preparation of this chapter was funded by NSF grant BMS 75-03127 to C. A. Wraight.

6. References

Amesz, J., and de Grooth, B. G., 1976, Photosynthetic electron transport and electrochromic effects at sub-zero temperatures, *Biochim. Biophys. Acta* **440**:301–313.

Amesz, J., and Visser, J. W. M., 1971, Light-induced shifts in pigmented absorption in green, red and blue-green algae, *Biochim. Biophys. Acta* **234**:62–69.

Amesz, J., and Vredenberg, W. J., 1966, Absorbancy changes of photosynthetic pigments in various purple bacteria, in: *Current in Photosynthesis* (J. B. Thomas and J. C. Goedheer, eds.), pp. 75–83, Ad. Donker, Rotterdam.

Arnold, W. and Clayton, R. K., 1960, The first step in photosynthesis: Evidence for its electronic nature, *Proc. Natl. Acad. Sci. U.S.A.* **46**:769–776.

Ausländer, W., and Junge, W., 1974, The electric generator in the photosynthesis of green plants. II. Kinetic correlation between protolytic reactions and redox reactions, *Biochim. Biophys. Acta* **357**:285–298.

Azzi, A., 1975, The application of fluorescent probes in membrane studies, *Q. Rev. Biophys.* **8**:237–316.

Azzi, A., Chance, B., Radda, G. K., and Lee, C. P., 1969, A fluorescent probe for energy dependent structural changes in fragmented membranes, *Proc. Natl. Acad. Sci. U.S.A.* **62**:612–619.

Azzi, A., Baltscheffsky, M., Baltscheffsky, H., and Vanio, H., 1971, Energy-linked changes of the membrane of *Rhodospirillum rubrum* chromatophores detected by the fluorescent probe 8-anilinonaphthalene-1-sulfonic acid, *FEBS Lett.* **17**:49–52.

Baltscheffsky, M., 1967, Inorganic pyrophosphate as an energy donor in photosynthetic and respiratory electron transport phosphorylation systems, *Biochem. Biophys. Res. Commun.* **28**:270–276.

Baltscheffsky, M., 1969, Energy conversion-linked changes of carotenoid absorbance in *Rhodospirillum rubrum* chromatophores, *Arch. Biochem. Biophys.* **130**:646–652.

Baltscheffsky, H., and Arwidsson, B., 1962, Evidence for two phosphorylation sites in bacterial cyclic photophosphorylation, *Biochim. Biophys. Acta* **65**:425–428.

Baltscheffsky, H., and von Stedingk, L.-V., 1966, Energy transfer from two coupling sites in bacterial photophosphorylation, in: *Currents in Photosynthesis* (J. B. Thomas and J. C. Goedheer, eds,), pp. 253–261, Ad. Donker, Rotterdam.

Bamberger, E. S., Rottenberg, H., and Avron, M., 1973, Internal pH, ΔpH and the kinetics of electron transport in chloroplasts, *Eur. J. Biochem.* **34**:557–563.

Barsky, E. L., Bonch-Osmolovskaya, E. A., Ostroumov, S. A., Sumuilov, V. D., and Skulachev, V. P., 1975, A study on the membrane potential and pH gradient in chromatophores and intact cells of photosynthetic bacteria, *Biochim. Biophys. Acta* **387**:388–395.

Barouch, Y., and Clayton, R. K., 1977, Ubiquinone reduction and proton uptake by chromatophores of *Rhodopseudomonas sphaeroides* R-26: Periodicity of two in consecutive flashes, *Biochim. Biophys. Acta* **462**.

Bensasson, R., Land, E. J., and Maudinas, B., 1976, Triplet states of carotenoids from photosynthetic bacteria studied by nanosecond ultraviolet and electron pulse spectroscopy, *Photochem. Photobiol.* **23**:189–193.

Boyer, P. D., 1965, Carboxyl activation as a possible common reaction in substrate-level and oxidative phosphorylation and in muscle contraction, in: *Oxidases and Related Redox Systems* (T. E. King, H. S. Mason and M.

Morrison, eds.), Vol. 2, pp. 994–1008. John Wiley & Sons, New York.

Boyer, P. D., 1967, ¹⁸O and related exchanges in enzymic formation and utilization of nucleotide triphosphates, *Curr. Top. Bioenerg.* **2**:99–149.

Boyer, P. D., Cross, R. L., and Momsen, W., 1973, A new concept for energy coupling in oxidative phosphorylation based on a molecular explanation of the oxygen exchange reactions, *Proc. Natl. Acad. Sci. U.S.A.* **70**:2837–2839.

Brand, M. D., Chen, C.-H., and Lehninger, A. L., 1976a, Stoichiometry of H⁺-ejection during respiration dependent accumulation of Ca^{2+} by rat liver mitochondria, *J. Biol. Chem.* **251**:968–974.

Brand, M. D., Reynafarje, B., and Lehninger, A. L., 1976b, Stoichiometric relationship between energy-dependent proton ejection and electron transport in mitochondria, *Proc. Natl. Acad. Sci. U.S.A.* **73**:437–441.

Breton, J., 1974, The state of chlorophyll and carotenoid *in vivo*. II. A linear dichroism study of pigment orientation in photosynthetic bacteria, *Biochem. Biophys. Res. Commun.* **59**:1011–1017.

Breton, J., and Mathis, P., 1974, Polarization of the 515 nm effect in chloroplasts oriented by a magnetic field, *Biochem. Biophys. Res. Commun.* **58**:1071–1078.

Breton, J., Michel-Villaz, M., and Paillotin, G., 1973, Orientation of pigments and structural proteins in the photosynthetic membrane of spinach chloroplasts: A linear dichroism study, *Biochim. Biophys. Acta* **314**:42–56.

Briller, S., and Gromet-Elhanan, Z., 1970, Effect of ammonium salts, amines and antibiotics on proton uptake and photophosphorylation in *Rhodospirillum rubrum* chromatophores, *Biochim. Biophys. Acta* **205**:263–272.

Callis, J. B., Parson, W. W., and Gouterman, M., 1972, Fast changes in enthalpy and volume on flash excitation of *Chromatium* chromatophores, *Biochim. Biophys. Acta* **267**:348–362.

Carr, N. G., and Exell, G., 1965, Ubiquinone concentrations in Athiorhodaceae grown under various environmental conditions, *Biochem. J.* **96**:688–692.

Casadio, R., Baccarini-Melandri, A., Zannoni, D., and Melandri, B. A., 1974, Electrochemical proton gradient and phosphate potential in bacterial chromatophores, *FEBS Lett.* **49**:203–207.

Case, G. D., and Parson, W. W., 1971, Thermodynamics of the primary and secondary photochemical reactions in *Chromatium, Biochim. Biophys. Acta* **253**:187–202.

Case, G. D., and Parson, W. W., 1973, Shifts of bacteriochlorophyll and carotenoid absorption bands linked to cytochrome c_{555} photooxidation in *Chromatium, Biochim. Biophys. Acta* **325**:441–453.

Chance, B., 1958, Oxygen-linked absorbancy changes in photosynthetic cells, *Brookhaven Symp. Biol.* **11**:74–86.

Chance, B., 1965, Steady state and kinetic responses of ubiquinone in phosphorylating mitochondria, in: *Biochemistry of the Quinones* (R. A. Morton, ed.), pp. 459–501, Academic Press, New York.

Chance, B., 1972, The nature of electron transfer and energy coupling reactions, *FEBS Lett.* **23**:3–20.

Chance, B., 1974, Deep and shallow probes of natural and artificial membranes, in: *Proceedings of the Fourth International Biophysics Congress—Moscow* L. Kayushin, ed.), pp. 911–923, USSR Academy of Sciences, Publishing House, Moscow.

Chance, B., and Baltscheffsky, M., 1975, Carotenoid and merocyanine probes in chromatophore membranes, in *Biomembranes* (H. Eisenberg, E. Katchalski-Katzir, and L. A. Manson, eds.), Vol. 7, pp. 33–55, Plenum, New York.

Chance, B., and Mela, L., 1966, Proton movements in mitochondrial membranes, *Nature (London)* **212:**372–376.

Chance, B., and Sager, R., 1957, Oxygen and light-induced oxidations of cytochrome, flavoprotein and pyridine nucleotide in a *Chlamydomonas* mutant, *Plant Physiol.* **32:**548–561.

Chance, B., and Smith, L., 1955, Respiratory pigments of *Rs. rubrum, Nature (London)* **175:**803–806.

Chance, B., and Strehler, B., 1957, Effects of oxygen and red light upon the absorption of visible light in green plants, Plant Physiol. **32:**536–538.

Chance, B., Nishimura, M., Avron, M., and Baltscheffsky, M., 1966, Light-induced intravesicular pH-changes in *Rhodospirillum rubrum* chromatophores, *Arch. Biochem. Biophys.* **117:**158–166.

Chance, B., Crofts, A. R., Nishimura, M., and Price, B., 1970*a*, Fast membrane H$^+$-binding in the light-activated state of *Chromatium* chromatophores, *Eur. J. Biochem.* **13:**364–374.

Chance, B., McCray, J. A., and Bunkenburg, J., 1970*b*, Fast spectrophotometric measurement H$^+$-changes in *Chromatium* chromatophores activated by a liquid dye laser, *Nature (London)* **225:**705–708.

Chance, B., Lee, C. P., Lee, I. Y., Ohnishi, T., and Higgins, J., 1970*c*, Electron transport and energy coupling, in *Electron Transport and Energy Conservation* (J. M. Tager, S. Papa, E. Quagliariello, and E. C. Slater, eds.), pp. 29–59, Adriatica Editrice, Bari, Italy.

Chance, B., Baltscheffsky, M., Vanderkooi, J., and Cheng, W., 1974, Localized and delocalized potentials in biological membranes, in: *Perspectives in Membrane Biology* (S. Estrado-O and C. Gitler, eds.), pp. 329–369, Academic Press, New York.

Chance, B., Erecinska, M., and Radda, G. K., 1975, 12-(9-anthroyl) stearic acid, a fluorescent probe for the ubiquinone region of the mitochondrial membrane, *Eur. J. Biochem.* **54:**521–529.

Chappell, J. B., and Crofts, A. R., 1965, Gramicidin and ion transport in isolated liver mitochondria, *Biochem. J.* **95:**393–402.

Chappell, J. B., and Crofts, A. R., 1965, Gramicidin and ion reversible volume changes in isolated mitochondria, in: *Regulation of Metabolic Processes in Mitochondria* (J. M. Tager, S. Papa, E. Quagliariello, and E. C. Slater, eds.), BBA Library, Vol. 7, pp. 293–314, Elsevier, Amsterdam.

Chappell, J. B., and Haarhoff, K. N., 1967, The penetration of the mitochondrial membrane by anions and cations, in: *Biochemistry of Mitochondria* (E. C. Slater, Z. Kaniuga, and L. Wojtczak, eds.), pp. 75–91, Academic Press, London.

Clayton, R. K., 1962, Primary reactions in bacterial photosynthesis. III. Reactions of carotenoids and cytochromes in illuminated bacterial chromatophores, *Photochem. Photobiol.* **1:**313–323.

Clayton, R. K., 1963, Two light reactions of bacteriochlorophyll *in vivo, Proc. Natl. Acad. Sci. U.S.A.* **50:**583–587.

Clayton, R. K., and Straley, S. C., 1970, An optical absorption change that could be due to reduction of the primary photochemical electron acceptor in photosynthetic reaction centers, *Biochem. Biophys. Res. Commun.* **39:**1114–1119.

Clayton, R. K., and Straley, S. C., 1972, Photochemical electron transport in photosynthetic reaction centers. IV. Observations related to the reduced photoproducts, *Biophys. J.* **12:**1221–1234.

Clayton, R. K., Fleming, H., and Szuts, E. Z., 1972, Photochemical electron transport in photosynthetic reaction centers from *Rp. sphaeroides*. II. Interaction with external electron donors and acceptors and a re-evaluation of some spectroscopic data, *Biophys. J.* **12:**46–63.

Cogdell, R. J., and Crofts, A. R., 1971, The effect of antimycin A and 1:10-phenanthroline on rapid H$^+$-uptake by chromatophores from *Rhodopseudomonas sphaeroides*, in: *Proceedings of the Second International Congress on Photosynthesis* (G. Forti, M. Avron, and B. A. Melandri, eds.), Vol. 2, pp. 977–983, Dr. W. Junk, The Hague.

Cogdell, R. J., and Crofts, A. R., 1974, H$^+$-uptake by chromatophores from *Rp. sphaeroides*. The relation between rapid H$^+$-uptake and the H$^+$-pump, *Biochim. Biophys. Acta* **347:**264–272.

Cogdell, R. J., Jackson, J. B., and Crofts, A. R., 1972, The effect of redox potential on the coupling between rapid hydrogen-ion binding and electron transport in chromatophores from *Rhodopseudomonas sphaeroides*, *Bioenergetics* **4:**413–429.

Cogdell, R. J., Prince, R. C., and Crofts, A. R., 1973, Light-induced H$^+$-uptake catalyzed by photochemical reaction centers from *Rp. sphaeroides* R26, *FEBS Lett.* **35:**204–208.

Cogdell, R. J., Brune, D. C., and Clayton, R. K., 1974, Effects of extraction and replacement of ubiquinone upon photochemical activity of reaction centers and chromatophores from *Rp. sphaeroides*, *FEBS Lett.* **45:**344–347.

Cogdell, R. J., Monger, T. G., and Parson, W. W., 1975, Carotenoid triplet states in reaction centers from *Rhodopseudomonas sphaeroides* and *Rhodospirillum rubrum*, *Biochim. Biophys. Acta* **408:**189–199.

Cogdell, R. J., Celis, S., Celis, H., and Crofts, A. R., 1977, Reaction center carotenoid band shifts, *FEBS Lett.* **80:**190–194.

Cohen, L. B., Salzberg, B. M., Davila, H. V., Ross, W. N., Landowne, D., Waggoner, A. S., and Wang, C. H., 1974, Changes in axon fluorescence during activity: Molecular probes of membrane potential, *J. Membrane Biol.* **19:**1–36.

Conjeaud, H., and Michel-Villaz, M., 1976, Photoinduced electric field effect on the optical properties of photosynthetic membranes, *J. Theor. Biol.* **62:**1–16.

Cost, K., and Frenkel, A. W., 1967, Light-induced interactions of *Rhodospirillum rubrum* chromatophores with bromthymol blue, *Biochemistry* 6:663–667.

Crane, F. L., Ehrlich, B., and Kegel, L. P., 1960, Plastoquinone reduction in illuminated chloroplasts, *Biochim. Biophys. Res. Commun.* 3:37–40.

Crofts, A. R., 1968, Ammonium ion uptake by chloroplasts and the high energy state, in: *Regulatory Functions of Biological Membranes* (J. Järnefelt, ed.), pp. 247–263, Elsevier, Amsterdam.

Crofts, A. R., 1970, The chromatophore as a functional unit, in: *Electron Transport and Energy Conservation* (J. M. Tager, S. Papa, E. Quagliariello, and E. C. Slater, eds.), pp. 221–228, Adriatica Editrice, Bari, Italy.

Crofts, A. R., 1974, The electron transport system as a H^+-pump in photosynthetic bacteria, in: *Perspectives in Membrane Biology* (S. Estrada-O and C. Gitler, eds.), pp. 373–412, Academic Press, New York.

Crofts, A. R., and Jackson, J. B., 1970, The high-energy state in chromatophores from photosynthetic bacteria, in: *Electron Transport and Energy Conservation* (J. M. Tager, S. Papa, E. Quagliariello, and E. C. Slater, eds.), pp. 383–408, Adriatica Editrice, Bari, Italy.

Crofts, A. R., Jackson, J. B., Evans, E. H., and Cogdell, R. J., 1971a, The high-energy state in chloroplasts and chromatophores, in: *Proceedings of the Second International Congress on Photosynthesis* (G. Forti, M. Avron, and B. A. Melandri, eds.), Vol. 2, pp. 873–902, Dr. W. Junk, The Hague.

Crofts, A. R., Cogdell, R. J., and Jackson, J. B., 1971b, The mechanism of H^+-uptake in *Rhodopseudomonas sphaeroides*, in: *Energy Transduction in Respiration and Photosynthesis* (E. Quagliariello, S. Papa, and C. S. Rossi, eds.), 883–901, Adriatica Editrice, Bari, Italy.

Crofts, A. R., Prince, R. C., Holmes, N. G., and Crowther, D., 1974, Electrogenic electron transport and the carotenoid change in photosynthetic bacteria, in: *Proceedings of the Third International Congress on Photosynthesis* (M. Avron, ed.), pp. 1131–1146, Elsevier Scientific, Amsterdam.

Crofts, A. R., Crowther, D., and Tierney, G. V., 1975a, Electrogenic electron transport in photosynthetic bacteria, in: *Electron Transport and Oxidative Phosphorylation* (E. Quagliariello, S. Papa, F. Palmieri, E. C. Slater, and N. Siliprandi, eds.), pp. 233–241, North-Holland/American Elsevier, Amsterdam and New York.

Crofts, A. R., Holmes, N. G., and Crowther, D., 1975b, Chemiosmotic coupling in photosynthetic bacteria, in: *Proceedings of the Tenth FEBS Meeting*, Vol. 40, *Electron Transport Systems* (P. Desnuelle and A. M. Michelson, eds.), pp. 287–304, North-Holland, Amsterdam.

Crofts, A. R., Crowther, D., Boyer, J., and Tierney, G. V., 1977, Electron transport through the antimycin sensitive site in *Rhodopseudomonas capsulata*, in: *Structure and Function of Energy-Transducing Membranes* (K. van Dam and B. F. van Gelder, eds.), Elsevier-North Holland Biomedical Press, Amsterdam.

Cusanovitch, M. A., Bartsch, R. G., and Kamen, M. D., 1968, Light-induced electron transport in *Chromatium* strain D. II. Light-induced absorbance changes in

Chromatium chromatophores, *Biochim. Biophys. Acta* 153:397–417.

Davies, R. E., 1951, The mechanism of hydrochloric acid production by the stomach, *Biol. Rev.* 26:87–120.

Deamer, D. W., Prince, R. C., and Crofts, A. R., 1972, The response of fluorescent amines to pH gradients across liposome membranes, *Biochim. Biophys. Acta* 274:323–335.

de Groth, B. G., and Amesz, J., 1977a, Electrochromic absorbance changes of photosynthetic pigments in *Rhodopseudomonas sphaeroides*. I. Stimulation by secondary electron transport at low temperature, *Biochim. Biophys. Acta* 462:237–246.

de Groth, B. G., and Amesz, J., 1977b, Electrochromic absorbance changes of photosynthetic pigments in *Rhodopseudomonas sphaeroides*. II. Analysis of the band shifts of carotenoid and bacteriochlorophyll, *Biochim. Biophys. Acta* 462:247–258.

de Groth, B. G., Romijn, J. C., and Pulles, M. P. J., 1977, Oscillating absorbance changes in purple bacteria, *Abstracts of the Fourth International Congress on Photosynthesis*, Reading, England.

DeVault, D., 1971, Energy transduction in electron transport, *Biochim. Biophys. Acta* 225:193–199.

Dutton, P. L., 1971, Oxidation–reduction potential dependence of the interaction of cytochromes, bacteriochlorophyll and carotenoids at 77°K in chromatophores of *Chromatium* D and *Rp. gelatinosa*, *Biochim. Biophys. Acta* 226:63–80.

Dutton, P. L., and Baltscheffsky, M., 1972, Oxidation–reduction potential dependence of pyrophosphate-induced cytochrome and bacteriochlorophyll reactions in *Rs. rubrum*, *Biochim. Biophys. Acta* 267:172–178.

Dutton, P. L., and Wilson, D. F., 1974, Redox potentiometry in mitochondrial and photosynthetic bioenergetics, *Biochim. Biophys. Acta* 346:165–212.

Dutton, P. L., Leigh, J. S., and Reed, D. W., 1973a, Primary events in the photosynthetic reaction center from *Rp. sphaeroides* strain R26: Triplet and oxidized states of bacteriochlorophyll and the identification of the primary electron acceptor, *Biochim. Biophys. Acta* 292:654–666.

Dutton, P. L., Leigh, J. S., and Wraight, C. A., 1973b, Direct measurements of the midpoint potential of the primary electron acceptor *in situ* and in the isolated state: Some relationships with pH and o-phenanthroline, *FEBS Lett.* 36:169–173.

Dutton, P. L., Morse, S. D., and Wong, A. M., 1974, The thermodynamic and kinetic relationship of the electron and proton in redox transitions, in: *Dynamics of Energy-Transducing Membranes* (L. Ernster, R. W. Estabrook, and E. C. Slater, eds.), pp. 233–241, Elsevier Scientific, Amsterdam.

Dutton, P. L., Petty, K. M., Bonner, H. S., and Morse, S. D., 1975, Cytochrome c_2 and reaction center of *Rhodopseudomonas sphaeroides* Ga membranes. Extinction coefficients, content, half-reduction potentials, kinetics and electric field alterations, *Biochim. Biophys. Acta* 387:536–556.

Dutton, P. L., Prince, R. C., Petty, K. M., and van den Berg, W. H., 1977, Single turnover electron and proton stoichiometry, kinetics and thermodynamics in ubi-quinone—cytochromes b–c_2 oxido-reductase, 61st Annual Meeting, Chicago, *Fed. Proc. Fed. Am. Soc. Exp. Biol.* **36** (abstract 2390).

Duysens, L. N. M., 1954, Reversible changes in the absorption spectrum of *Chlorella* upon irradiation, *Science* **120**:353–354.

Duysens, L. N. M., Huiskamp, W. J., Vos, J. J., and van der Hart, J. M., 1956, Reversible changes in bacteriochlorophyll in purple bacteria upon illumination, *Biochim. Biophys. Acta* **19**:188–190.

Emrich, H. M., Junge, W., and Witt, H. T., 1969, Further evidence for an optical response of chloroplast bulk pigments to a light induced electrical field in photosynthesis, *Z. Naturforsch.* **24b**:1144–1146.

Evans, E. H., and Crofts, A. R., 1974, The relationship between delayed fluorescence and the carotenoid shift in chromatophores from *Rhodopseudomonas capsulata*, *Biochim. Biophys. Acta* **333**:44–51.

Feher, G., Okamura, M. Y., and McElroy, J. D., 1972, Identification of an electron acceptor in reaction centers of *Rp. sphaeroides* by EPR spectroscopy, *Biochim. Biophys. Acta* **267**:222–226.

Feher, G., Isaacson, R. A., McElroy, J. D., Ackerson, L. C., and Okamura, M. Y., 1974, On the question of the primary acceptor in bacterial photosynthesis: Manganese substituting for iron in reaction centers of *Rp. sphaeroides* R26, *Biochim. Biophys. Acta* **368**:135–139.

Fisher, H. F., Adija, D. L., and Cross, D. G., 1969, Dehydrogenase-reduced coenzyme difference spectra, their resolution and relationship to the stereospecificity of hydrogen transfer, *Biochemistry* **11**:4424–4430.

Fleischman, D. E., and Clayton, R. K., 1968, The effect of phosphorylation uncouplers and electron transport inhibitors upon spectral shifts and delayed light emission of photosynthetic bacteria, *Photochem. Photobiol.* **8**:287–298.

Fork, D. C., and Amesz, J., 1967, Light-induced shifts in the absorption spectrum of carotenoids in red, brown and yellow-green algae and in a barley mutant, *Carnegie Inst. Washington Yearb.* **66**:160–165.

Fowler, C. F., and Kok, B., 1974, Direct observation of a light-induced electric field in chloroplasts, *Biochim. Biophys. Acta* **357**:308–318.

Geller, D. M., 1967, Correlation of light-induced absorbance changes with photophosphorylation in *Rhodospirillum rubrum* extracts, *J. Biol. Chem.* **242**:40–46.

Griffith, O. H., Dehlinger, P. J., and Van, S. P., 1974, Shape of the hydrophobic barrier of phospholipid bilayers (evidence for water penetration in biological membranes), *J. Membrane Biol.* **15**:159–192.

Griffiths, M., Sistrom, W. R., Cohen-Bazire, G., and Stanier, R. Y., 1955, Function of carotenoids in photosynthesis, *Nature (London)* **176**:1211–1214.

Gromet-Elhanan, Z., 1970, Differences in sensitivity to valinomycin and nonactin of various photophosphorylating and photoreducing systems of *Rhodospirillum rubrum* chromatophores, *Biochim. Biophys. Acta* **223**:174–182.

Gromet-Elhanan, Z., 1972, Changes in the fluorescence of atebrin and of analinonaphthalene sulfonate reflecting two different light-induced processes in *Rhodospirillum rubrum* chromatophores, *Eur. J. Biochem.* **25**:84–88.

Halsey, Y. D., and Parson, W. W., 1974, Identification of ubiquinone in the secondary electron acceptor in the photosynthetic apparatus of *Chromatium vinosum*, *Biochim. Biophys. Acta* **347**:404–416.

Haydon, D. A., and Hladky, S. B., 1972, Ion transport across thin lipid membranes: A critical discussion of mechanisms in selected systems, *Q. Rev. Biophys.* **5**:187–282.

Heathcote, P., Vermeglio, A., and Clayton, R. K., 1977, The carotenoid band shift in reaction centers from *Rhodopseudomonas sphaeroides, Biochim. Biophys. Acta* **461**:358–364.

Henderson, P. J. F., McGivan, J. D., and Chappell, J. B., 1969, The action of certain antibiotics on mitochondrial, erythrocyte and artificial phospholipid membranes. The role of induced proton permeability, *Biochem. J.* **111**:521–535.

Hildreth, W. W., Avron, M., and Chance, B., 1966, Laser activation of rapid absorption changes in spinach chloroplasts and *Chlorella, Plant Physiol.* **41**:983–991.

Hoffman, J., and Laris, P. C., 1974, Determination of membrane potentials in human and *Amphiuma* red blood cells by means of a fluorescent probe, *J.. Physiol.* **239**:519–552.

Holmes, N. G., and Crofts, A. R., 1977a, The carotenoid shift in *Rhodopseudomonas sphaeroides*: The flash induced change, *Biochim. Biophys. Acta* **459**:492–505.

Holmes, N. G., and Crofts, A. R., 1977b, The carotenoid shift in *Rhodopseudomonas sphaeroides*: Change induced under continuous illumination, *Biochim. Biophys. Acta* **461**:141–150.

Hsu, M., and Chan, S. I., 1973, Nuclear magnetic resonance studies of the interaction of valinomycin with unsonicated lecithin bilayers, *Biochemistry* **12**:3872–3876.

Hubbard, R., and Kropf, A., 1958, The action of light on rhodopsin, *Proc. Natl. Acad. Sci. U.S.A.* **44**:130–139.

Ilgenfritz, G., 1966, Chemical relaxation in strong electric fields, Dissertation, University of Göttingen.

Isaev, P. I., Liberman, E. A., Samuilov, V. D., Skulachev, V. P., and Tsofina, L. M., 1970, Conversion of biomembrane-produced energy into electric form. III Chromatophores of *Rhodospirillum rubrum, Biochim. Biophys. Acta* **216**:22–29.

Jackson, J. B., and Crofts, A. R., 1969a, Bromothymol blue and bromocresol purple as indicators of pH-changes in chromatophores of *Rhodospirillum rubrum*, *Eur. J. Biochem.* **10**:226–237.

Jackson, J. B., and Crofts, A. R., 1969b, The high energy state in chromatophores from *Rhodospeudomonas sphaeroides, FEBS Lett.* **4**:185–189.

Jackson, J. B., and Crofts, A. R., 1971, The kinetics of light-induced carotenoid changes in *Rhodospeudomonas sphaeroides* and their relation to electrical field generation across the chromatophore membrane, *Eur. J. Biochem.* **18**:120–130.

Jackson, J. B., and Dutton, P. L., 1973, The kinetic and redox potentiometric resolution of the carotenoid shifts in

Rhodopseudomonas sphaeroides chromatophores: Their relationship to electric field alterations in electron transport and energy coupling, *Biochim. Biophys. Acta* **325**:102–113.

Jackson, J. B., Crofts, A. R., and von Stedingk, L.-V., 1968, Ion-transport induced by light and antibiotics in chromatophores from *Rhodospirillum rubrum, Eur. J. Biochem.* **6**:41–54.

Jackson, J. B., Cogdell, R. J., and Crofts, A. R., 1973, Some effects of *o*-phenanthroline on electron transport in chromatophores from photosynthetic bacteria, *Biochim Biophys. Acta* **292**:218–225.

Jackson, J. B., Saphon, S., and Witt, H. T., 1975, The extent of the stimulated electrical potential decay under phosphorylating conditions and the H^+/ATP ratio in *Rhodopseudomonas sphaeroides* chromatophores following short flash excitation, *Biochim. Biophys. Acta* **408**:83–92.

Jagendorf, A. T., and Hind, G., 1963, Studies on the mechanism of photophosphorylation, in: *Photosynthetic Mechanisms of Green Plants*, National Academy of Sciences–National Research Council Publication No. 1145, pp. 599–610.

Jagendorf, A. T., and Uribe, E., 1966, ATP formation caused by an acid–base transition of spinach chloroplasts, *Proc. Natl. Acad. Sci. U.S.A.* **55**:170–177.

Junge, W., 1970, The critical electric potential difference for photophosphorylation, *Eur. J. Biochem.* **14**:582–592.

Junge, W., and Witt, H. T., 1968, On the ion transport system of photosynthesis. Investigations on a molecular level, *Z. Naturforsch.* **23b**:244–254.

Kaback, H. R., 19,4, Transport studies in bacterial membrane vesicles, *Science* **186**:882–892.

Kihara, T., and Dutton, P. L., 1970, Light-induced reactions in photosynthetic bacteria. I. Reactions in whole cells and in cell-free extracts at liquid nitrogen temperatures, *Biochim. Biophys. Acta* **205**:196–204.

Kleuser, D., and Bücher, H., 1969, Electrochromic effects of a chlorophyll *a* and a chlorophyll *b* monolayer, *Z. Naturforsch.* **24b**:1371–1374.

Klingenberg, M., 1975, Energetic aspects of transport of ADP and ATP through the mitochondrial membrane, in: *Energy Transformation in Biological Systems*, Ciba Found. Symp. (New Ser.) **31**:105–121.

Kobayashi, Y., and Nishimura, M., 1972, Fluorescence change of auromine-*O* bound to chromatophores of *Rhodospirillum rubrum*—Analysis in connection to ionic environment and ion transport, *J. Biochem.* **71**:275–284.

Kröger, A., 1976, The interaction of the radicals of ubiquinone in mitochondrial electron transport, *FEBS Lett.* **65**:278–280.

Labhart, J., 1967, Electrochromism, in: *Advances in Chemical Physics* (I. Prigogine, ed.), Vol. XIII, pp. 179–204, Interscience, New York.

Lardy, H. A., Graven, S. N., and Estrado-O, S., 1967, Specific induction and inhibition of cation and anion transport in mitochondria, *Fed. Proc. Fed. Am. Soc. Exp. Biol.* **26**:1355–1359.

Laris, P. C., Bahr, D. P., and Chaffee, R. R. J., 1975,

Membrane potentials in mitochondrial preparations as measured by means of a cyanine dye, *Biochim. Biophys. Acta* **376**:415–425.

Lehninger, A. L., Carafoli, E., and Rossi, C. S., 1967, Energy-linked ion movements in mitochondrial systems, *Adv. Enzymol.* **29**:259–320.

Leigh, J. S., and Dutton, P. L., 1972, The primary electron acceptor in photosynthesis, *Biochim. Biophys. Res. Commun.* **46**:414–421.

Leigh, J. S., Netzel, T. L., Dutton, P. L., and Rentzepis, P. M., 1974, Primary events in photosynthesis: Picosecond kinetics of carotenoid bandshifts in *Rhodopseudomonas sphaeroides* chromatophores, *FEBS Lett.* **48**:136–140.

Leiser, M., and Gromet-Elhanan, Z., 1974, Demonstration of acid–base phosphorylation in chromatophores in the presence of a K^+-diffusion potential, *FEBS Lett.* **43**:267–270.

Liptay, W., 1969, Electrochromism and solvatochromism, *Angew. Chem. Int. Ed. Engl.* **8**:177–188.

Lundegårdh,H., 1945, Absorption, transport and exudation of inorganic ions by the roots, *Ark. Bot.* **32A**:1–139.

Mathis, P., 1966, Variation d'absorption de courte durée, induite dans une suspension de chloroplastes par une éclair laser, *C. R. Acad. Sci. Paris* **263**:1770–1772.

Mathis, P., and Galmiche, J. M., 1967, Action des gaz paramagnétiques sur un état transitoire induit par un éclair laser dans une suspension de chloroplastes, *C. R. Acad. Sci. Paris* **264**:1903–1906.

McLaughlin, S., and Eisenberg, M., 1975, Antibiotics and membrane biology, *Annu. Rev. Biophys. Bioeng.* **5**:335–366.

Mitchell, P., 1961, Coupling of phosphorylation to electron and hydrogen transfer by a chemiosmotic type of mechanism, *Nature (London)* **191**:144–148.

Mitchell, P., 1963, The chemical asymmetry of membrane transport processes, in: *Cell Interface Reactions*, pp. 33–50, Scholar's Library, New York.

Mitchell, P., 1966, Chemiosmotic coupling in oxidative and photosynthetic phosphorylation, *Biol. Rev.* **41**:445–502.

Mitchell, P., 1967, Proton-translocation phosphorylation in mitochondria, chloroplasts and bacteria: Natural fuel cells and solar cells, *Fed. Proc. Fed. Am. Soc. Exp. Biol.* **26**:1370–1379.

Mitchell, P., 1968, *Chemiosmotic Coupling and Energy Transduction*, Glynn Research, Bodmin, England.

Mitchell, P., 1975a, Proton-motive redox mechanism of the cytochrome b–c_1 complex in the respiratory chain: Proton-motive ubiquinone cycle, *FEBS Lett.* **56**:1–6.

Mitchell, P., 1975b, The proton-motive Q-cycle: A general formulation *FEBS Lett.* **59**:137–139.

Mitchell, P., and Moyle, J., 1965, Stoichiometry of proton translocation through the respiratory chain and adenosine triphosphatase system of rat-liver mitochondria, *Nature (London)* **208**:147–151.

Mitchell, P., and Moyle, J., 1969, Estimation of membrane potential and pH difference across the cristae membrane of rat liver mitochondria, *Eur. J. Biochem.* **7**:471–484.

Mitchell, P., Moyle, J., and Smith, L., 1968, Bromthymol blue as a pH-indicator in mitochondrial suspensions, *Eur. J. Biochem.* **4**:9–19.

Monger, T. G., and Parson, W. W., 1977, Singlet–triplet fusion in *Rhodopseudomonas sphaeroides* chromatophores: A probe of the organisation of the photosynthetic apparatus, *Biochim. Biophys. Acta* **460**:393–407.

Montal, M., Nishimura, M., and Chance, B., 1970, Uncoupling and charge transfer in bacterial chromatophores, *Biochim. Biophys. Acta* **223**:183–188.

Moret, V., Pinamonti, S., and Fornasari, E., 1961, Polarographic study on the redox potential of ubiquinones, *Biochim. Biophys. Acta* **54**:381–383.

Morita, S., 1968, Evidence for three photochemical systems in *Chromatium* D, *Biochim. Biophys. Acta* **153**:241–247.

Morton, R. A., Gloor, V., Schindler, O., Wilson, G. M., Chopard-dit-Jean, L. H., Hemming, F. W., Isler, O., Leat, W. M. F., Pennock, J. F., Rüegg, R., Schwieter, V., and Wiss, O., 1958, Die Struktur des Ubichinons aus Schweinherzen, *Helv. Chim. Acta* **4**:2343–2357.

Moyle, J., and Mitchell, P., 1973, Proton translocation quotient for the adenosine triphosphatase of rat liver mitochondria, *FEBS Lett.* **30**:317–320.

Mueller, P., and Rudin, D. O., 1969, Translocators in bimolecular lipid membranes: Their role in dissipative and conservative bioenergy transductions, *Curr. Top. Bioenerg.* **3**:157–249.

Neumann, J. S., and Jagendorf, A. T., 1964, Light-induced pH-changes related to phosphorylation by chloroplasts, *Arch. Biochem. Biophys.* **107**:109–119.

Nishimura, M., 1962, Studies on bacterial photophosphorylation. II. Effects of reagents and temperature on light-induced and dark phases of photophosphorylation in *Rhodospirillum rubrum* chromatophores, *Biochim. Biophys. Acta* **57**:96–103.

Nishimura, M., 1970, The sizes of the photosynthetic energy-transducing units in purple bacteria determined by single flash yield, titration by antibiotics and carotenoid absorption bandshift, *Biochim. Biophys. Acta* **197**:69–77.

Nishimura, M., and Pressman, B., 1969, Effects of ionophorous antibiotics on the light-induced internal and external H⁺-ion changes and phosphorylation in bacterial chromatophores, *Biochemistry* **8**:1360–1370.

O'Brien, F. L., and Olver, J. W., 1969, Electrochemical study of ubiquinone-6 in aqueous methanol, *Anal. Chem.* **41**:1810–1813.

Okada, M., and Takamiya, A., 1970, Ferricyanide-induced absorbance changes of carotenoid in chromatophores of *Rhodopseudomonas sphaeroides* correlated to the oxidation of bacteriochlorophyll 885, *Plant Cell Physiol.* **11**:713–721.

Okada, M., Murata, N., and Takamiya, A., 1970, Light-induced absorbance change of carotenoid in chromatophores of photosynthetic bacterium, *Rhodopseudomonas sphaeroides*, *Plant Cell Physiol.* **11**:519–530.

Okamura, M. Y., Isaacson, R. A., and Feher, G., 1975, Primary acceptor in bacterial photosynthesis: Obligatory role of ubiquinone in photoactive reaction centers of *Rp. sphaeroides*, *Proc. Natl. Acad. Sci. U.S.A.* **72**:3491–3495.

Okamura, M. Y., Ackerson, L. C., Isaacson, R. A., Parson, W. W., and Feher, G., 1976, The primary electron acceptor in *Chromatium vinosum* (strain D), *Biophys. J.* **16**:223a.

Ostroumov, S. A., Samuilov, V. D., and Skulachev, V. P., 1973, Transhydrogenase-induced responses of carotenoids, bacteriochlorophyll and penetrating anions in *Rhodospirillum rubrum* chromatophores, *FEBS Lett.* **31**:27–30.

Parson, W. W., 1969, The reaction between primary and secondary electron acceptors in bacterial photosynthesis, *Biochim. Biophys. Acta* **189**:384–396.

Parson, W. W., and Monger, T. G., 1976, Interrelationships among excited states in bacterial reaction centers, *Brookhaven Symp. Biol.* **28**:195–212.

Peters, G. A., and Cellarius, R. A., 1972, The ubiquinone homologue of the green mutant of *Rp. sphaeroides*, *Biochim. Biophys. Acta* **256**:544–547.

Petty, K. M., and Dutton, P. L., 1976a, Properties of the flash-induced proton binding encountered in membranes of *Rhodopseudomonas sphaeroides*: A functional pK on the ubisemiquinone? *Arch. Biochem. Biophys.* **172**:335–345.

Petty, K. M., and Dutton, P. L., 1976b, Ubiquinone–cytochrome b electron and proton transfer: A functional pK on cytochrome b_{50} in *Rp. sphaeroides* membranes, *Arch. Biochem. Biophys.* **172**:346–353.

Petty, K. M., Jackson, J. B., and Dutton, P. L., 1977, Kinetics and stoichiometry of proton binding in *Rhodopseudomonas sphaeroides* chromatophores, *FEBS Lett.* **84**:299–303.

Platt, J. R., 1961, Electrochromism, a possible change in colour producible in dyes by an electric field, *J. Chem. Phys.* **34**:862–863.

Pressman, B. C., Harris, E. J., Jagger, W. S., and Johnson, J. H., 1967, Antibiotic mediated transport of alkali ions across lipid membranes, *Proc. Natl. Acad. Sci. U.S.A.* **58**:1949–1955.

Prince, R. C., and Dutton, P. L., 1975, A kinetic completion of the cyclic photosynthetic electron pathway of *Rhodopseudomonas sphaeroides*: Cytochrome b–cytochrome c_2 oxidation–reduction, *Biochim. Biophys. Acta* **387**:609–613.

Prince, R. C., and Dutton, P. L., 1976, The primary acceptor of bacterial photosynthesis: Its operating midpoint potential?, *Arch. Biochem. Biophys.* **172**:329–334.

Prince, R. C., and Dutton, P. L., 1977, Single and multiple turnover reactions in the ubiquinone-cytochrome b-c_2 oxidoreductase of *Rhodopseudomonas sphaeroides*: The physical chemistry of the major electron donor to cytochrome c_2 and its coupled reactions, *Biochim. Biophys. Acta* **462**:731–747.

Prince, R. C., Baccarini-Melandri, A., Hauska, G. A., Melandri, B. A., and Crofts, A. R., 1975, Asymmetry of an energy transducing membrane: The localisation of cytochrome c_2 in *Rhodopseudomonas sphaeroides* and *Rhodopseudomonas capsulata*, *Biochim. Biophys. Acta* **387**:212–227.

Racker, E., and Stoeckenius, W., 1974, Reconstitution of purple membrane vesicles catalyzing light-driven proton-uptake and adenosine triphosphate formation, *J. Biol. Chem.* **249**:662–663.

Redfearn, E. R., and Pumphrey, A. M., 1960, Oxidation-reduction levels of ubiquinone (Coenzyme Q) in different metabolic states of rat liver mitochondria, *Biochem. Biophys. Res. Commun.* **3**:650–653.

Reed, D. W., and Raveed, D., 1972, Some properties of the ATPase from chromatophores of *Rp. sphaeroides* and its structural relationship to the bacteriochlorophyll proteins, *Biochim. Biophys. Acta* **283**:79–91.

Reed, D. W., Zankel, K. L., and Clayton, R. K., 1969, The effect of redox potential of P_{870} fluorescence in reaction centers from *Rp. sphaeroides*, *Proc. Natl. Acad. Sci. U.S.A.* **63**:42–46.

Reed, D. W., Raveed, D., and Reporter, M., 1975, Localisation of photosynthetic reaction centers by antibody binding to chromatophore membranes, *Biochim. Biophys. Acta* **387**:368–378.

Reich, R., and Schmidt, S., 1972, Über den Einfluss elektrischer Felder auf das Absorptionsspektrum von Farbstoffmolekülen in Lipidschichten. I. Theorie, *Ber. Bunsenges. Phys. Chem.* **76**:589–598.

Reich, R., and Sewe, K.-U., 1977, The effects of molecular polarisation on the electrochromism of carotenoids. I. The influence of a carboxylic group, *Photochem. Photobiol.* **26**:11–17

Reich, R., Scheerer, R., Sewe, K.-U., and Witt, H. T., 1976, The effect of electric fields on the absorption spectrum of dye molecules in lipid layers. V. Refined analysis of the field-indicating absorption changes in photosynthetic membranes by comparison with electrochromic measurements *in vitro*, *Biochim. Biophys. Acta* **449**:295–294.

Reid, R. A., Moyle, J., and Mitchell, P., 1966, Synthesis of adenosine triphosphate by a proton motive force in rat liver mitochondria, *Nature (London)* **212**:257–258.

Reinwald, E., Stiehl, H. H., and Rumberg, B., 1968, Correlation between plastoquinone reduction, field formation and proton translocation in photosynthesis, *Z. Naturforsch.* **23b**:1616–1617.

Robertson, R. N., 1951, Mechanism of absorption and transport of inorganic nutrients in plants, *Annu. Rev. Plant Physiol.* **2**:1–24.

Robertson, R. N., 1960, Ion transport and respiration, *Biol. Rev.* **35**:231–264.

Rosenfeld, T., Honig, B., Ottolenghi, M. Hurley, J., and Ebrey, T. G., 1977, *Cis–trans* isomerization in the photochemistry of vision, *Pure Appl. Chem.* **49**:341–351.

Rumberg, B., and Schröder, H., 1972, Ion transfer and phosphorylation, in: *Proceedings of the Sixth International Congress on Photobiology—Bochum* (G. O. Schenck, ed.), Abstract 036, Deutsche Gesellschaft für Lichtforschung, Frankfurt.

Saphon, S., Jackson, J. B., Lerbs, V., and Witt, H. T., 1975a, The functional unit of electrical events and phosphoryl-ation in chromatophores from *Rhodopseudomonas sphaeroides*, *Biochim. Biophys. Acta* **408**:58–66.

Saphon, S., Jackson, J. B., and Witt, H. T., 1975b, Electrical potential changes, H^+-translocation and phosphorylation induced by short flash excitation in *Rhodopseudomonas sphaeroides* chromatophores, *Biochim. Biophys. Acta* **408**:67–82.

Scheraga, H. A., 1961, *Protein Structure*, Academic Press, New York.

Schliephake, W., Junge, W., and Witt. H. T., 1968, Correlation between field formation, proton translocation and the light reactions in photosynthesis, *Z. Naturforsch.* **23b**:1571–1578.

Schmidt, G. L., and Kamen, M. D., 1971, Redox properties of the "P836" pigment complex of *Chromatium*, *Biochim. Biophys. Acta* **234**:70–72.

Schmidt, S., and Reich, R., 1972a, Über den Einfluss elektrischer Felder auf das Absorptionsspektrum von Farbstoffmolekülen in Lipidschichten. II. Messungen and Rhodamin B, *Ber. Bunsenges. Phys. Chem.* **76**:599–602.

Schmidt, S., and Reich, R., 1972b, Über den Einfluss elektrischer Felder auf das Absorptionsspektrum von Farbstoffmolekülen in Lipidschichten. III. Elektrochromie eines Carotinoids (Lutein), *Ber. Bunsenges. Phys. Chem.* **76**:1202–1208.

Schmidt, S., Reich, R., and Witt, H. T., 1971a, Electrochromism of chlorophylls and carotenoids in multilayers and in chloroplasts, *Naturwissenschaften* **8**:414–429.

Schmidt, S., Reich, R., and Witt, H. T., 1971b, Electrochromic measurements *in vitro* as a test for the interpretation of field indicating absorption changes in photosynthesis, in: *Proceedings of the Second International Congress on Photosynthesis* (G. Forti, M. Avron, and B. A. Melandri, eds.), Vol. 2, pp. 1087–1095, Dr. W. Junk, The Hague.

Scholes, P., Mitchell, P., and Moyle, J., 1969, The polarity of proton translocation in some photosynthetic microorganisms, *Eur. J. Biochem.* **8**:450–454.

Schröder, H., Mühle, H., and Rumberg, B., 1971, Relationship between ion transport phenomena and phosphorylation in chloroplasts, in *Proceedings of the Second International Congress on Photosynthesis* (G. Forti, M. Avron, and A. Melandri, eds.), Vol. 2, pp. 919–930, Dr. W. Junk, The Hague.

Schuldiner, S., Padan, E., Rottenberg, H., Gromet-Elhanan, Z., and Avron, M., 1974, ΔpH and membrane potential in bacterial chromatophores, *FEBS Lett.* **49**:174–177.

Seelig, J., Limacher, L., and Bader, P., 1972, Molecular architecture of liquid crystalline bilayers, *J. Am. Chem. Soc.* **94**:6364–6371.

Sewe, K.-U., and Reich, R., 1977a, The effect of molecular polarisation on the electrochromism of carotenoids. II. Lutein–chlorophyll complexes: The origin of the field-indicating absorption change at 520 nm in the membranes of photosynthesis, *Z. Naturforsch.* **32c**:161–171.

Sewe, K.-U., and Reich, R., 1977b, Influence of the chlorophylls on the electrochromism of carotenoids in the membranes of photosynthesis, *FEBS Lett.* **80**:30–34.

Shavit, N., Thore, A., Keister, D. L., and San Pietro, A., 1968, Inhibition by nigericin of the light-induced pH changes in *Rhodospirillum rubrum* chromatophores, *Proc. Natl. Acad. Sci. U.S.A.* **59**:917–922.

Sherman, L. A., and Clayton, R. K., 1972, The lack of carotenoid bandshifts in a non-photosynthetic, reaction centerless mutant of *Rhodopseudomonas sphaeroides*, *FEBS Lett.* **22**:127–132.

Sherman, L. A., and Cohen, W. S., 1972, Proton uptake and quenching of bacteriochlorophyll fluorescence in *Rhodopseudomonas sphaeroides*, *Biochim. Biophys. Acta* **283**:54–66.

Sistrom, W. R., Griffiths, M., and Stanier, R. Y., 1956, The biology of a photosynthetic bacterium which lacks coloured carotenoids, *J. Cell. Comp. Physiol.* **48**:473–515.

Skulachev, V. P., 1971, Energy transformation in the respiratory chain, *Curr. Top. Bioenerg.* **4**:127–190.

Skulachev, V. P., 1972, Solution of the problem of energy coupling in terms of chemiosmotic theory, *Bioenergetics* **3**:25–38.

Skulachev, V. P., 1974, Enzymic generators of membrane potential in mitochondria, *Ann. N. Y., Acad. Sci.* **227**:188–202.

Slater, E. C., 1953, Mechanism of phosphorylation in the respiratory chain, *Nature (London)* **172**:975–982.

Slater, E. C., 1971, The coupling between energy-yielding and energy-utilizing reactions in mitochondria, *Q. Rev. Biophys.* **4**:35–71.

Slater, E. C., 1972, Mechanism of energy conservation in *Mitochondria/Biomembranes* (S. G. van den Bergh, P. Borst, L. L. N. van Deenen, J. C. Riemersma, E. C. Slater, and J. M. Tager, eds.), *Proceedings of the Eighth FEBS Meeting—Amsterdam*, Vol. 28, pp. 133–146, North-Holland/American Elsevier, Amsterdam.

Slayman, C. L., and Slayman, C. W., 1974, Depolarization of the plasma membrane of *Neurospora* during active transport of glucose: Evidence for a proton-dependent cotransport system, *Proc. Natl. Acad. Sci. U.S.A.* **71**:1935–1939.

Slooten, L., 1972, Electron acceptors in reaction center preparations from photosynthetic bacteria, *Biochim. Biophys. Acta* **275**:208–218.

Smith, L., and Baltscheffsky, M., 1959, Respiration and light-induced phosphorylation in extracts of *Rhodospirillum rubrum*, *J. Biol. Chem.* **234**:1575–1579.

Smith, L., and Ramirez, J., 1959, Absorption spectrum changes in photosynthetic bacteria following illumination or oxygenation, *Arch. Biochem. Biophys.* **79**:233–244.

Smith, L., and Ramirez, J., 1960, Reactions of carotenoid pigments in photosynthetic bacteria, *J. Biol. Chem.* **235**:219–225.

Smith, L., Baltscheffsky, M., and Olson, J. M., 1960, Absorption spectrum changes observed on illumination of aerobic suspensions of photosynthetic bacteria, *J. Biol. Chem.* **235**:213–218.

Strichartz, G. R., and Chance, B., 1972, Absorbance changes at 520 nm caused by salt addition to chloroplast suspensions in the dark, *Biochim. Biophys. Acta* **256**:71–84.

Stroobant, P., and Kaback, H. R., 1975, Ubiquinone-mediated coupling of NADH dehydrogenase to active transport in membrane vesicles from *E. coli*, *Proc. Natl. Acad. Sci. U.S.A.* **72**:3970–3974.

Sybesma, C., 1969, Light-induced reactions of P890 and P800 in the purple photosynthetic bacterium, *Rs. rubrum*, *Biochim. Biophys. Acta* **172**:177–179.

Sybesma, C., and Fowler, C. F., 1968, Evidence for two light-driven reactions in the purple photosynthetic bacterium, *Rs. rubrum*, *Proc. Natl. Acad. Sci. U.S.A.* **61**:1343–1348.

Symons, M., Swysen, C., and Sybesma, C., 1977, The light-induced carotenoid absorbance changes in *Rhodopseudomonas sphaeroides*: An analysis and interpretation of the band shifts, *Biochim. Biophys. Acta* **462**:706–717.

Takamiya, K., and Dutton, P. L., 1977, The influence of transmembrane potentials on the redox equilibrium between cytochrome c_2 and the reaction center in *Rhodopseudomonas sphaeroides* chromatophores, *Photochem. Photobiol.* **80**:279–284.

Thore, A., Keister, D. L., Shavit, N., and San Pietro, A., 1968, Effects of antibiotics on ion transport and photophosphorylation in *Rhodospirillum rubrum* chromatophores, *Biochemistry* **7**:3499–3506.

Thornber, J. P., 1970, Photochemical reactions of purple bacteria as revealed by studies of three spectrally different carotenobacteriochlorophyll–protein complexes isolated from *Chromatium*, strain D, *Biochemistry* **9**:2688–2698.

Urban, P. F., and Klingenberg, M., 1969, On the redox potentials of ubiquinone and cytochrome *b* in the respiratory chain, *Eur. J. Biochem.* **9**:519–525.

Vermeglio, A., 1977, Secondary electron transfer in reaction centers of *Rhodopseudomonas sphaeroides*: Out-of-phase periodicity of two for the formation of ubisemiquinone and fully reduced quinone, *Biochem. Biophys. Acta* **459**:516–524.

von Stedingk, L.-V., 1967, Light-induced reversible pH-changes in chromatophores from *Rhodospirillum rubrum*, *Arch. Biochem. Biophys.* **120**:537–541.

Vredenberg, W. J., and Amesz, J., 1967, Absorption bands of bacteriochlorophyll types in purple bacteria and their response to illumination, *Biochim. Biophys. Acta* **126**:244–253.

Vredenberg, W. J., Amesz, J., and Duysens, L. N. M., 1965, Light-induced spectral shifts in bacteriochlorophyll and carotenoid absorption in purple bacteria, *Biochem. Biophys. Res. Commun.* **18**:435–439.

Walker, D. A., and Crofts, A. R., 1970, Photosynthesis, *Annu. Rev. Biochem.* **39**:389–428.

West, I. C., 1970, Lactose transport coupled to proton movements in *E. coli*, *Biochem. Biophys. Res. Commun.* **41**:655–661.

Williams, R. J. P., 1961, Possible functions of chains of catalysts, *J. Theor. Biol.* **1**:1–17.

Williams, R. J. P., 1962, Possible functions of chains of catalysts II, *J. Theor. Biol.* **3**:209–229.

Williams, R. J. P., 1969, Electron transfer and energy conservation, *Curr. Top. Bioenerg.* **3**:79–156.

Witt, H. T., 1967, On the analysis of photosynthesis by the pulse techniques in the 10^{-1} to 10^{-8} second range, in: *Fast Reactions and Primary Processes in Chemical Kinetics*

(S. Claesson, ed.), Nobel Symposium V, pp. 261–310, Almquist and Wiksell/Interscience, Stockholm, and London.

Witt, H. T., 1971, Coupling of quanta, electrons, fields, ions and phosphorylation in the functional membrane of photosynthesis, *Q. Rev. Biophys.* **4**:365–477.

Witt, H. T. and Zickler, A., 1973, Electrical evidence for the field indicating absorption change in bioenergetic membranes, *FEBS Lett.* **37**:307–310.

Witt, H. T., and Zickler, A., 1974, Vectorial electron flow across the thylakoid membrane. Further evidence by kinetic measurements with an electrochromic and electrical method, *FEBS Lett.* **39**:205–208.

Witt, H. T. Rumberg, B., Schmidt-Mende, P., Siggel, V., Skerra, B., Vater, J., and Weikard, J., 1965, On the analysis of photosynthesis by flashlight techniques, *Angew. Chem. Int. Ed. Engl.* **4**:799–812.

Wolff, C., and Witt, H. T., 1969, On metastable states of carotenoids in primary events of photosynthesis, *Z. Naturforsch.* **24b**:1031–1037.

Wolff, C., Buchwald, H. E., Rüppel, H., Witt, K., and Witt, H. T., 1969, Rise time of the light induced electrical field across the function membrane of photosynthesis, *Z. Naturforsch.* **24b**:1038–1041.

Wraight, C. A., 1977a, Electron acceptors of photosynthetic bacterial reaction centers: Direct observation of oscillatory behavior suggesting two closely equivalent ubiquinones, *Biochim. Biophys. Acta* **459**:525–531.

Wraight, C. A., 1977b, A two-electron gate between the primary and secondary electron transport processes of photosynthesis, *Bull. Am. Phys. Soc.* **22**(1):JI-10 (abstract).

Wraight, C. A., 1977c, The role of iron in the electron acceptor region of reaction centres of photosynthetic bacteria, *Abstracts of the Fourth International Congress on Photosynthesis*, Reading, England (*FEBS Lett.* in press, 1978).

Wraight, C. A., and Crofts, A. R., 1970, Energy-dependent quenching of chlorophyll *a* fluorescence in isolated chloroplasts, *Eur. J. Biochem.* **17**:319–327.

Wraight, C. A., Kraan, G. P. B., and Gerrits, N. M., 1972, The pH dependence of delayed and prompt fluorescence in uncoupled chloroplasts, *Biochim. Biophys. Acta* **283**:259–267.

Wraight, C. A., Cogdell, R. J., and Clayton, R. K., 1975, Some experiments on the primary electron acceptor in reaction centres from *Rp. sphaeroides*, *Biochim. Biophys. Acta* **296**:242–249.

Yoshizawa, T., and Wald, G., 1963, Pre-lumirhodopsin and the bleaching of visual pigments, *Nature (London)* **197**:1279–1286

Zieger, G., Müller, A., and Witt, H. T., 1961, Über eine photochemische Reaktion bei der Photosynthese: Tieftemperaturmessungen and Löschung durch paramagnetische Gase, *Z. Phys. Chem. (N.F.)* **29**:13–24.

Zimmerman, U., Pilwat, G., and Riemann, F., 1974, Dielectric breakdown of cell membranes, *Biophys. J.* **14**:881–900.

CHAPTER 27

Delayed Fluorescence and Chemiluminescence

Darrell Fleischman

1. Introduction

All photosynthetic organisms that have been examined re-emit light for many minutes after they have been illuminated. Such delayed light emission was first observed by Strehler and Arnold (1951) in experiments with spinach chloroplasts. The luminescence of photosynthetic organisms, in particular of higher plants, is the subject of an excellent recent review by Lavorel (1975).

Arnold and Thompson (1955) found that the purple photosynthetic bacteria *Rhodospirillum rubrum*, *Rhodopseudomonas gelatinosa*, and *Rp. palustris* emit delayed light that is substantially weaker than that of higher plants and algae. Preliminary action spectra and emission spectra suggested that the effective exciting light was absorbed by bacteriochlorophyll (Bchl), and that the light was emitted from Bchl excited singlet states that had been generated after the illumination had ended. These observations were confirmed and extended by Goedheer (1962, 1963).

Strong evidence that bacterial delayed fluorescence requires functional photosynthetic reaction centers (RCs) and not just the presence of antenna Bchl was furnished by Clayton and Bertsch (1965). They found that a mutant of *Rp. sphaeroides* (PM-8) that lacks functional RCs emits no measurable delayed fluorescence. The mutant nevertheless contains a normal complement of Bchl, and in fact displays a higher prompt fluorescence yield than its parent, wild-type, strain Ga.

The first truly detailed examination of the delayed fluorescence of a photosynthetic bacterium was performed by Clayton (1965) with the green photosynthetic bacterium "*Chloropseudomonas ethylica*."[*] Clayton found that *Chlp. ethylica* cells emit delayed fluorescence only under anaerobic conditions.

Chloropseudomonas ethylica contains Bchl c (λ_{max} = 750 nm *in vivo*) and Bchl a (λ_{max} = 810 nm *in vivo*) in 70 : 1 ratio. A minor Bchl component absorbing at 840 nm (P840) can be reversibly photobleached and is believed to function as the primary electron donor, as does the P870 complex in the purple bacteria. Light energy absorbed by Bchl c is funneled by resonance transfer to the Bchl a and finally to P840. Clayton found that light emission from Bchl c predominates in the spectrum of the prompt fluorescence emitted at the start of illumination of anaerobic intact cells. But most of the delayed fluorescence, measured at 3 or 11 msec after excitation, is emitted by the Bchl a (Fig. 1). This behavior would be expected if the excitations responsible for delayed fluorescence originate at the photochemical RC, P840, and migrate into the light-harvesting Bchl aggregate before being emitted as fluorescence photons.

Darrell Fleischman · Charles F. Kettering Research Laboratory, Yellow Springs, Ohio 45387

[*] "*Chloropseudomonas ethylica*" refers to a dubious species (see Chapter 2).

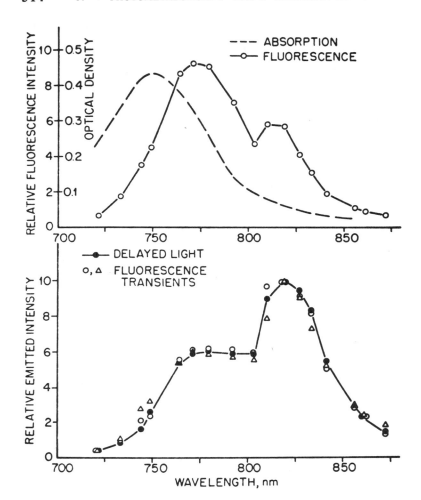

Fig. 1. Spectra of absorption, prompt fluorescence, and delayed fluorescence for suspensions of *Chloropseudomonas ethylica* cells. The upper curves show optical density and the initial value of the prompt fluorescence. The lower curve (●) shows the spectrum of delayed fluorescence and the varying (transient) part of the prompt fluorescence (O, △). The data for delayed and variable prompt fluorescence were normalized so as to have the same value at 819 nm. Reproduced from Clayton (1965).

2. Luminescence and the Reversal of the Primary Light Reaction

2.1. Variable Prompt Fluorescence

The prompt fluorescence of photosynthetic systems generally includes a constant component ("dead" fluorescence), presumably from chlorophylls (Chl's) that do not transfer energy directly or indirectly to RCs, and a component that varies during the course of illumination, presumably in response to the changes in the redox states of the primary electron donor and acceptor. Clayton (1965) noticed a number of similarities between the delayed fluorescence and the variable part of the prompt fluorescence of anaerobic *Chlp. ethylica* cells. Of particular interest, the emission spectrum of variable prompt fluorescence is identical to that of delayed fluorescence, and differs from that of dead prompt fluorescence (Fig. 1). Among other

possibilities, Clayton suggested that variable prompt fluorescence might in fact be a fast component of delayed fluorescence. Until quite recently, however, evidence has seemed to favor an interpretation of fluorescence yield variations in terms of competition between photochemistry and the emission of fluorescence photons (Clayton, 1967). More recent measurements of the quantum yields of photochemistry and of singlet and triplet formation in photosynthetic RCs cannot easily be interpreted in this way (Wraight *et al.*, 1974).

There is increasing evidence that the first photochemical step in bacterial photosynthesis involves a separation of charges within a strongly interacting complex consisting of 4 Bchl and 2 bacteriopheophytin (Bpheo) molecules (Straley *et al.*, 1973) (see also Chapter 19). An electron is believed to be transferred from this complex to the primary electron acceptor within about 150 psec in *Rp. sphaeroides* RCs (Kaufmann *et al.*, 1975). If the primary acceptor is already reduced, the charges appear to recombine within the donor complex in less than 6 μsec (Parson et al., 1975). (Current ideas about the earliest photochemical events are discussed in detail in Chapters 19–21.)

Parson et al. (1975) suggested that such charge recombination within the primary donor complex might sometimes form singlet excited Bchl. The variable fluorescence of RCs could then in reality be delayed fluorescence generated by this mechanism. If this is indeed true, a knowledge of the properties of this luminescence (e.g., quantum yield, temperature dependence, response to electrical fields) could be of vital importance in efforts to understand slower luminescence phenomena. It is quite possible that a final step—perhaps singlet generation accompanying charge recombination within the P870 complex—is common to all components of delayed fluorescence and to all the known forms of chemiluminescence.

2.2. Chlorophyll Chemiluminescence *in Vitro*

Excited states of Chl can indeed be formed in electron-transfer reactions. Goedheer and Vegt (1962) reported that Chl *a* and Bchl in methanolic solutions can be oxidized with ferric ion and reduced to their original spectroscopic states with ferrous ion. Light emission follows the addition of the reductant, and the spectrum of the emitted light is similar to that of Chl fluorescence.

Recent work by Norris and Katz and their associates (see Chapter 21) suggests that Chl–H_2O aggregates in hydrocarbon solvents form a more realistic model of the RC. Reversible light-induced spin production has been observed in such solutions. It would be of interest to know whether light emission accompanies electron transfers or charge recombinations in such systems.

2.3. 0.5-msec Delayed Fluorescence

Recent experiments by Carithers and Parson (1975) have provided the most convincing support thus far for the idea that delayed fluorescence may result from a reversal of the primary charge separation. Their strategy was to prevent secondary electron transfers by chemical means. The properties of the delayed fluorescence that is emitted as electrons return from the primary acceptor to the primary donor following flash excitation could then be observed. Using chromatophores from the photosynthetic bacterium *Rp. viridis*, they blocked electron transfer from the primary acceptor to secondary acceptors with *o*-phenanthroline. The magnitudes of the flash-induced bleaching of cytochrome C558 and P985 (the primary donor complex of *Rp. viridis*) and the intensity of delayed fluorescence were then measured as the redox potential was varied between 200 and 500 mV.

2.3.1. Correlation of Delayed Fluorescence with the Number of P985⁺X⁻ Pairs

A delayed fluorescence component that decays in about 0.5 msec appeared when the redox potential was such that P985 was reduced and cytochrome C558 was oxidized at the time of the exciting flash (Fig. 2). Under these conditions, P985 donates an electron to the primary acceptor (X) during the flash, and cannot in turn immediately accept an electron from the cytochrome. After the flash, the electron returns from the reduced primary acceptor to the oxidized P985. Carithers and Parson (1975) found that the decay of delayed fluorescence precisely parallels the reappearance of P985. Similar results were obtained when electron transfer to secondary acceptors was prevented by extracting lyophilized chromatophores with petroleum ether. This treatment is believed to

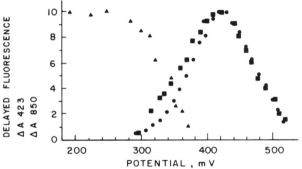

Fig. 2. Potentiometric titrations of cytochrome C-558, P985, and 0.5-msec delayed fluorescence of *Rhodopseudomonas viridis*. Cytochrome oxidation (▲) was measured at 432 nm following an Xe flash in the near-IR (Corning 2600 filter). P985 photooxidation (■) was measured at 850 nm following a blue-green Xe flash (Dchott BG-18 filter). Delayed fluorescence (●) was measured at 1 msec following a Xe flash in the near-IR. Titrations were performed at 20°C using small additions of potassium ferricyanide (6 mg/ml), and a saturated calomel–Pt electrode pair continuously monitored the solution potential. The full scale (10) on the graph represents $2.1 \cdot 10^{-3}$ ΔA (■); $5.3 \cdot 10^{-3}$ ΔA (▲) and 4.1 V (●). pH 7.8. *o*-Phenanthroline, 0.4 mM (▲) or 1 mM (■, ●). Reproduced from Carithers and Parson (1975).

remove the secondary acceptor, presumably a quinone.

When electron transfer to secondary acceptors was not prevented, the delayed fluorescence decayed much more rapidly. The decay half-time of about 24 μsec corresponds closely to the half-time of electron transfer from the primary acceptor to the secondary acceptor.

Thus, under these conditions in *Rp. viridis*, there is now direct evidence that the delayed fluorescence decay kinetics matches the kinetics of the disappearance of P985+X− pairs, whether the pairs disappear by charge recombination or as a result of secondary electron transfers.

2.3.2. Influence of the Prompt Fluorescence Yield

The experiments cited above indicate that for the 0.5-msec component of delayed fluorescence in *Rp. viridis*, the intensity L, in photons emitted per second, obeys the relationship

$$L = C_1[P985^+X^-] \qquad (1)$$

where C_1 is a constant the nature of which can now be considered.

The delayed fluorescence emission spectrum coincides with the prompt fluorescence emission spectrum, confirming that in *Rp. viridis*, as in *Chlp. ethylica*, the light emission accompanies the lowest excited singlet-to-ground transition in the antenna Bchl. One would therefore expect factors that influence the yield of prompt fluorescence to affect the delayed fluorescence yield in a similar way. But experimentally, L depends on the first power of [P985+X−]; i.e., within experimental error, C_1 is not a function of [P985+X−]. The delayed fluorescence yield does not depend on the part of the prompt fluorescence yield that varies with the state of the RCs. This is perhaps not surprising if delayed fluorescence excitations originate by charge recombination in the RCs. Each photosynthetic unit in which a delayed fluorescence excitation has just arisen will of necessity have its RC in the P985X ("open") state. Presumably, then, C_1 includes only the prompt fluorescence yield, ϕ_0, of photosynthetic units having open RCs, so that

$$L = C_2\phi_0[P985^+X^-] \qquad (2)$$

where C_2 is a constant.

The influence of prompt fluorescence yield is discussed in detail in the review by Lavorel (1975), in which the effect of excitation transfer between photosynthetic units is considered.

2.3.3. Rate and Yield of Excited-Singlet Formation

The rate at which the delayed fluorescence mechanism generates singlet excitons in the antenna Chl can be estimated by comparing the intensity of delayed fluorescence with the intensity of the prompt fluorescence that results when the Chl absorbs photons directly at a known rate. Using this method, Carithers and Parson (1975) found the apparent first-order rate constant for the production of singlet excitons in the antenna Chl from P985+X− to be 3.7 sec−1. Since this is more than two orders of magnitude smaller than the rate constant for the net back-reaction, it is clear that the back-reaction P985+X− → P985X usually occurs by a pathway that does not involve the generation of excited singlets.

The 3.7 sec−1 rate constant is that for the generation of excited singlets in the antenna Bchl, not that for the generation of excited P985 (P985*). An excitation originating in P985 is more likely to be photochemically trapped than is one originating in the antenna Chl. The rate constant for the reaction P985+X− → P985*X must therefore be somewhat greater. Since excitation migration is likely to occur more rapidly than electron transfer, one might guess that the rate constant for P985* generation may be no more than an order of magnitude greater than 4 sec−1.

2.3.4. Temperature Dependence of Delayed Fluorescence

Carithers and Parson (1975) found that the temperature dependence of the initial intensity of delayed fluorescence conforms to a linear Arrhenius plot, yielding an activation energy of 12.5 kcal/mole at pH 8. In contrast, the rate of the net back-reaction is almost temperature-independent. In our own laboratory, we have observed the 0.5-msec delayed fluorescence component in *Rp. viridis* at temperatures ranging from +40°C to −130°C, and find the decay kinetics to vary only slightly (Fleischman, unpublished).

In summary, the RC relaxation appears to proceed by two routes, both kinetically first-order in [P985+X−]:

(a) $P985^+X^- \xrightarrow{k_n} P985X$

$k_n = 10^3 \text{ sec}^{-1}$

and

(b) $P985^+X^- + Bchl \xrightarrow{k_r} P985X + Bchl^*$

$k_r = 6 \times 10^9 \exp(-12.5 \text{ kcal}/RT)$,

where Bchl* represents an excited singlet in the antenna Bchl.

Combining equation 2 with the expression for k_r, we may now write for the delayed fluorescence intensity

$$L = 6 \times 10^9 \phi_0 [\text{P985}^+\text{X}^-] \exp(-12.5 \text{ kcal}/RT) \quad (3)$$

2.3.5. Thermodynamics of the Primary Light Reaction

Since primary photochemistry can proceed at liquid helium temperature, there is unlikely to be an activation barrier for the reaction P985$^+$X$^-$ → P985*X. Therefore, unless charge recombination within the primary donor complex is rate-limiting, the activation energy for delayed fluorescence should represent the energy difference between the states P985$^+$X$^-$ and P985*X.

From the absorption spectrum of P985, Carithers and Parson (1975) found that the energy difference between P985X and P985*X is about 28.6 kcal/mol. Then, if the energy difference between P985*X and P985$^+$X$^-$ is 12.5 kcal/mol (the activation energy of delayed fluorescence), the energy stored in the primary reaction P985X → P985$^+$X$^-$ is about 16.6 kcal/mol. The free-energy difference between P985X and P985$^+$X$^-$, estimated from potentiometric titration of P985 and X, is approximately 14.0 kcal/mol. Using the relationship $\Delta G^\circ = \Delta H^\circ - T\Delta S^\circ$, it may be calculated that the primary light reaction in *Rp. viridis* is accompanied by an entropy increase of about 8.8 cal/mol degree.

This result is in sharp contrast to the results of calorimetric and potentiometric experiments performed with *Chromatium vinosum* (Callis *et al.*, 1972). These experiments demonstrated that in *Chr. vinosum* chromatophores, most of the free energy in the primary reaction is stored in the form of an entropy decrease. The reason for the apparent discrepancy is not yet clear. Carithers and Parson (1975) suggest that the reaction P985$^+$X$^-$ → P985*X may occur in two steps, the rate-limiting step having the smaller activation energy. The observation that the midpoint potential of P985 in *Rp. viridis* is much less temperature-dependent than that of P870 in *Chr. vinosum* (Carithers and Parson, 1975) is nevertheless consistent with the interpretation that the primary light reaction in *Rp. viridis* may indeed store free energy primarily in the form of enthalpy.

2.4. 7-msec Delayed Fluorescence

Experiments quite similar to those described in the preceding section were performed in our laboratory,

also with *Rp. viridis* chromatophores. Our strategy, however, was to slow secondary electron transfers by the use of low temperatures. In the ensuing discussion, we consider secondary, tertiary, and quaternary electron acceptors Y, Z, and B, in addition to the primary acceptor X.

P985 can be photooxidized reversibly at temperatures as low as 78°K (Fleischman and Cooke, 1971). After a flash of light sufficient for more than three turnovers of the RC, starting with cytochrome 558 in its reduced state, the photooxidized P985 recovers in less than 0.1 sec. We guessed that the P985 recovery was due to the return of an electron to P985$^+$ from the primary electron acceptor, and predicted that the electron transfer would be accompanied by delayed fluorescence. The delayed fluorescence was found in subsequent experiments (Fleischman, 1974).

This delayed fluorescence has the following properties:

1. The delayed fluorescence decay parallels P985 recovery, but with a half-time of about 7 msec (Fig. 3). The evidence that the 0.5-msec P985 recovery cor-

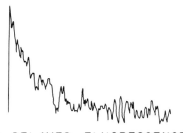

a. DELAYED FLUORESCENCE

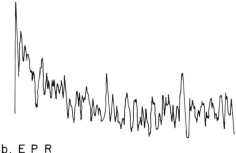

b. EPR

├────┤
10 msec

Fig. 3. Comparison of dark decay kinetics of delayed fluorescence and EPR absorption of *Rhodopseudomonas viridis* chromatophores at −40°C. (a) Delayed fluorescence. Approximately 500 traces from each of two samples were summed with a TMC Computer of Average Transients. (b) EPR. Ten traces with each of 12 samples were summed. Reproduced from Fleischman (1974).

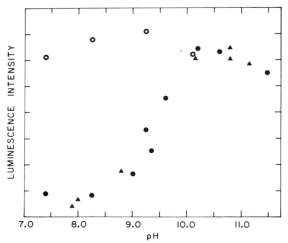

Fig. 4. Dependence on pH of the initial intensity of 0.5-msec delayed fluorescence of *Rhodopseudomonas viridis* chromatophores, and dependence of the initial intensity of acid–base luminescence on the final pH. (○) Delayed fluorescence in the presence of 1 mM *o*-phenanthroline; (●) delayed fluorescence in the absence of *o*-phenanthroline; (▲) acid–base luminescence. Delayed fluorescence and acid–base luminescence intensities were arbitrarily normalized.

responds to the back-reaction between the primary photoproducts is convincing; perhaps the 7-msec process involves electron transfer from a more peripheral electron acceptor to P985+.

2. The 7-msec delayed fluorescence and P985 recovery disappear as electrons move on from the putative secondary (or tertiary, etc.) acceptor along the electron-transport chain. The disappearance of 7-msec delayed fluorescence parallels the disappearance of P985+ · · · Y− (or P985+ · · · Z−, etc.) pairs by this process over a wide temperature range.

3. The net rate of the P985+ · · · Y− → P985 · · · Y reaction varies only slightly with temperature. Apparently, electrons can move to P985+ from either X− or Y− by quantum mechanical tunneling. The

Arrhenius activation energy of 7-msec delayed fluorescence is about 4 kcal/mol.

4. The first-order rate constant for the formation of excited singlets in the antenna Bchl is 1×10^{-3} sec^{-1} at $-23°C$. Again, the temperature-independent, dark recombination route predominates. Only about 0.03% of the Y− → P985+ electron transfers are accompanied by excited singlet formation.

Most of the differences between the 0.5-msec and 7-msec delayed fluorescence components are consistent with the conjecture that the former involves primary acceptor and the latter involves secondary acceptor. The low activation energy of the 7-msec component is surprising, however. Some possible explanations will be considered in the following discussion of the effects of membrane potential and pH.

2.5. Effect of pH

Delayed fluorescence decaying in 0.5 msec can be observed with *Rp. viridis* chromatophores even in the absence of *o*-phenanthroline, if the redox potential is such that cytochrome C558 is initially reduced. Its intensity is strongly pH-dependent (Fig. 4). A simple interpretation would be that the quaternary electron acceptor cannot accept an electron if an acidic group with a pK of 9.5 is ionized. Then, during the course of illumination, the third electron to be transferred would be unable to proceed beyond X−.

If chromatophores are illuminated at neutral pH and the pH is then increased, light is emitted (Fleischman, 1967, 1969). After illumination, the capacity for acid–base luminescence decays in parallel with the recovery of P985. The dependence of the intensity of acid–base luminescence on the final pH is also indicated in Fig. 4. This behavior would be consistent with the following scheme, in which BH and B− are protonated and nonprotonated forms of the quaternary acceptor:

$$C558^{+2}C558^{+2}P985XYZ(BH) \xrightarrow{3h\nu} C558^{+3}C558^{+3}P895^+XY^-Z^-(BH)^-$$

$$\downarrow OH^- \qquad\qquad\qquad OH^-\downarrow$$

$$C558^{+2}C558^{+2}P985XYZ(B^-) + H_2O \xrightarrow{3h\nu} C558^{+3}C558^{+3}P985^+X^-Y^-Z^-(B^-) + H_2O$$

$$\downarrow$$

$$C558^{+3}C558^{+3}P985^*XY^-Z^-(B^-)$$

The decay rate of 0.5-msec delayed fluorescence at high pH is somewhat faster and much less temperature-dependent than that seen at low pH in the presence of o-phenanthroline. One notable difference between the two experimental conditions is that the secondary acceptor is presumably reduced in the former instance, but not in the latter. Van Gorkom and Donze (1973) made the interesting suggestion that repulsion between multiple charges accumulated in the donor or acceptor pools could encourage the back-reaction and stimulate delayed fluorescence. Such repulsion between reduced acceptors may well be responsible for the low activation energy of 7-msec delayed fluorescence too, since three electrons were transferred in these experiments.

Discussion of the extensive studies of chemically and physically induced luminescence in higher plants, in which many of the phenomena were discovered, is beyond the scope of this chapter. For a recent review of this field, see Fleischman and Mayne (1973).

2.6. Glow Curves

During prolonged illumination of *Rp. viridis* chromatophores at low temperatures, the 7-msec component of delayed fluorescence fades as electrons move forward from Y^- along the electron-transport chain. After such an illumination is ended, P985 no longer recovers in 7 msec. Instead, the recovery occurs in two first-order phases of equal amplitude. The Arrhenius activation energy for the faster phase is about 7 kcal/mol (Fleischman and Cooke, 1971). If, after illumination, the chromatophores are warmed in the dark at a constant rate, light is emitted in a single glow peak (Fleischman, 1971a; see also Arnold and Sherwood, 1959). If partial recovery of P985 is allowed in the interval between the end of illumination and the start of heating, the intensity of the glow peak is directly proportional to the amount of P985 that has yet to recover in the faster phase at the time light emission begins. As before, the intensity of light emission seems to be directly proportional to the number of $P870^+$-reduced acceptor pairs. For the glow curve, the acceptor involved is presumably subsequent to X and Y in the electron-transport chain.

In our original report (Fleischman, 1971a), and in several discussions of glow curves obtained with higher plants, attempts were made to calculate the trap depth of the electrons responsible for the glow peaks. This was approached by employing the rate of heating and the temperature at which the glow intensity is maximal. However, this approach involves the assumption that the rate of light emission is pro-

portional to the rate of electron-untrapping over a broad temperature range. This assumption is clearly not valid for *Rp. viridis*, in which light emission and net untrapping may have quite different temperature dependences.

Anaerobic intact *Rp. viridis* cells exhibit a glow peak that is similar to that obtained with chromatophores. Aerobic cells display a second broad glow peak. Its source is not yet known.

We were unable to observe glow curves from *Rs. rubrum* cells or chromatophores.

2.7. Luminescence Induced by Ferricyanide, Hydrosulfite, and Oxygen

Chromatophores of *Rp. viridis* that have recently been illuminated will re-emit light if a solution of ferricyanide is injected into the suspension (Fleischman, 1967, 1969). Ferrocyanide is ineffective. The illumination must be sufficient to bring about the photooxidation of cytochrome C558, but not to bleach P985. The capacity for ferricyanide-induced luminescence decays after the illumination, in parallel with the recovery of the photooxidized cytochrome C558. A plausible interpretation is that the ferricyanide oxidizes the P985. The electron that had been transferred from cytochrome C558 to an acceptor during the illumination can then return to $P985^+$, and in so doing generate singlet excitons. The kinetics of the light emission is similar to the kinetics of the faster phase of P985 recovery, which we believe to be responsible for the glow curve, and the same electron acceptor may be involved.

If hydrosulfite is added to an *Rp. viridis* chromatophore suspension, and an excess of ferricyanide is then injected, light is emitted even if the chromatophores have not been exposed to light. The intensity and kinetics of the light emission are similar to those of the ferricyanide-induced luminescence obtained with pre-illuminated chromatophores.

When a suspension of *Rp. viridis* chromatophores is treated with hydrosulfite, illuminated, and ferricyanide is injected, extraordinarily bright light emission occurs. The decay of the luminescence is complete within a fraction of a second.

On addition of a hydrosulfite solution to a pre-illuminated suspension of *Rp. viridis* chromatophores, light is emitted in two waves. The kinetics of the light emission varies with the time between the illumination and the hydrosulfite injection.

Light is emitted when oxygen is injected into a pre-illuminated suspension of intact *Rp. viridis* cells. The intensity of the emitted light is enhanced in the presence of aliphatic alcohols, and is attenuated by phosphorylation uncouplers.

Ferricyanide-induced luminescence can be obtained with chromatophores from other bacteria such as *Chr. vinosum*, but it is very weak, and the capacity for luminescence decays within a few seconds after the preillumination ends.

As far as I am aware, chemiluminescence experiments with photosynthetic bacteria have been performed only in our own laboratory. Far more questions have been raised than have yet been answered.

For example, the intense, fast luminescence that accompanies the injection of ferricyanide into chromatophore suspensions that have been illuminated in the presence of hydrosulfite may result from the transfer of electrons from X^- to ferricyanide-oxidized P985. If so, the illumination requirement may imply that X is physically inaccessible to hydrosulfite and can only be photoreduced. Preliminary measurements of the effect of hydrosulfite and light on the prompt fluorescence yield in very dim light support this idea.

Ferricyanide-induced luminescence is inhibited by *o*-phenanthroline, as would be expected. But it is also inhibited by antimycin A or 2-*n*-heptyl-4-hydroxy-quinoline-*N*-oxide. P985 is likely to be near the inner surface of the chromatophore membrane, and therefore inaccessible to ferricyanide. This raises the question whether electrons flow from P985 through the antimycin-A-sensitive site to accessible electron carriers on the exterior surface of the chromatophore.

3. Influence of Membrane Potential

Experiments of Mayne (1967) implicating the "high-energy phosphorylation intermediate" in the delayed fluorescence mechanism of chloroplasts led us to look for similar effects in bacteria. We found that phosphorylation uncouplers do indeed eliminate the 4-msec delayed fluorescence of *Rp. sphaeroides* chromatophores (Fleischman and Clayton, 1968). In addition, the delayed fluorescence of *Rs. rubrum* chromatophores is attenuated while ADP is being photophosphorylated, but can be restored if ATP formation is blocked near the terminal step with an energy-transfer inhibitor such as oligomycin. Similar observations were reported by Kononenko *et al.* (1974).

We also examined the response of the light-induced red shifts of the carotenoid and Bchl absorption bands to electron-transport inhibitors and phosphorylation uncouplers. Their response was very similar to that of delayed fluorescence. Subsequent studies strongly support the concept that these absorption changes are a response to electrical fields within or across the

chromatophore membranes, generated by electron transfer between asymmetrically arranged carriers. Such concepts are discussed more fully by Wraight and co-workers in Chapter 26. Noticing the similar behavior of delayed fluorescence and the band shifts, Crofts suggested that membrane potential may influence delayed fluorescence. Drawing on ideas that had been advanced by Mitchell (1966), Witt *et al.* (1969), and others, he presented the following argument (Crofts *et al.*, 1971).

Comparison of the direction of the light-induced carotenoid bandshift with the direction in which the bands shift in response to an artificially created diffusion potential indicates that the light-generated membrane potential in chromatophores is inside-positive. In some bacteria, a red shift of the carotenoid absorption bands also accompanies the primary electron transfer (P870*X → P870$^+$X$^-$), measured with fast instrumentation or at very low temperatures. Crofts *et al.* (1971) suggested that P870 lies near the inner surface of the chromatophore membrane and X lies near the outer surface. The primary electron transfer would then create dipoles with components normal to the plane of the membrane. The electrical field of the dipoles would be responsible for the associated red shift of the carotenoid absorption bands. An inside-positive membrane potential would then enhance delayed fluorescence by pulling electrons back across the membrane from X^- to P870$^+$, generating singlet excitons during the charge recombination. I suggested that an electrical field would have the effect of lowering the activation energy required for the charge recombination (Fleischman, 1971*b*). Equation 2 could then be expanded thus:

$$L = C_3\phi_0[P^+X^-] \exp\left[-(E^{\ddagger} - \psi)/kT\right] \quad (4)$$

where C_3 is a constant that includes the entropy of activation for the light emission, $E^{\ddagger}$ is the energy of activation, and ψ is the membrane potential.

Much earlier, we had found that preilluminated chromatophore suspensions would emit light if they were quickly mixed with concentrated salt solutions (Fleischman, 1967, 1969). Work with chloroplasts has since shown that the salt enhances delayed fluorescence by creating a cation diffusion potential the magnitude of which is given by the Goldman equation

$$\psi = \frac{RT}{F} \ln \frac{\sum_j PC_j[C_j]_0 + \sum_j PA_j[A_j]_i}{\sum_j PC_j[C_j]_i + \sum_j PA_j[A_j]_0} \quad (5)$$

where F is the Faraday constant, PC_j and PA_j are the permeability coefficients of the *j*th cation and anion, respectively, and $[C_j]_0$, $[A_j]_0$, $[C_j]_i$ and $[A_j]$ are the concentrations of the *j*th cation and anion in the outer phase and the inner phase, respectively (Miles and Jagendorf, 1969; Barber and Varley, 1971). The sum is taken over all ions present. A discussion of diffusion potentials is included in the review by Lavorel (1975).

Kraan et al. (1970) reported that the intensity of the light emitted when KCl is injected into a suspension of preilluminated chloroplasts that have been treated with valinomycin is directly proportional to the amount of KCl injected. Valinomycin increases the K^+ permeability coefficient. This has the result that the concentrations of other ions can be ignored in equation 5, so that ψ is proportional to $\ln[K^+]_0$, and the luminescence intensity is indeed proportional to $[K^+]_0$. We repeated the experiment with chromatophores of Chr. vinosum, and again observed a direct proportionality between the concentration of injected K^+ and the luminescence intensity (Fig. 5).

Sherman (1972) furnished additional support for an association between delayed fluorescence intensity and membrane potential. He found that when cyclic electron transport is stimulated with diaminodurene, delayed fluorescence and the magnitude of the carotenoid band shift in Rp. sphaeroides are dramatically enhanced. A similar association between delayed fluorescence intensity and the carotenoid shift was reported by Crofts et al. (1972), who found that the rise kinetics of both phenomena respond in a similar way to electron-transport inhibitors and phosphorylation uncouplers.

Exactly how membrane potential influences delayed fluorescence remains to be determined. Our original suggestion that the electrical field lowers the activation energy for the $P870^+X^- \rightarrow P870X$ reaction was clearly wrong, since that process requires no activation energy. Perhaps the membrane potential does furnish activation energy for the competing $P870^+X^- \rightarrow P870^*X$ pathway, and so increases the probability of exciton-generating charge recombinations. Recent developments suggest, however, that this analysis may be too simple.

Delayed fluorescence of Rp. viridis is almost unaffected by phosphorylation uncouplers (Carithers and Parson, 1975), so membrane potential may not be an absolute requirement for bacterial delayed fluorescence.

4. Note Added in Proof

It is now generally accepted that an intermediate electron carrier (I), probably a pheophytin, mediates electron transfer from the excited primary-donor complex (P) to the "primary" acceptor (X). Luminescence thought to result from charge recombination between P^+ and I^-, decaying in less than 10 nsec, has been reported (Shuvalov and Klimov, 1975; Klimov et al., 1977; Godik and Borisov, 1977). The latter authors furnish new data about the relation between delayed fluorescence and variable prompt fluorescence.

Delayed fluorescence and thermoluminescence believed to be emitted from a protoporphyrin rather than from chlorophyll have been observed (Carithers and Parson, 1976; Govindjee et al., 1977).

Delayed fluorescence decay after single flashes has been studied with Rp. sphaeroides (Carithers and Parson, 1976) and with Chr. vinosum (Arata et al., 1977).

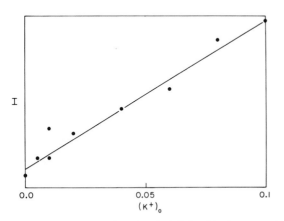

Fig. 5. Dependence of the intensity of salt-induced luminescence of *Chromatium* chromatophores on the concentration of K^+ injected. Chromatophores were suspended in a solution containing 0.025 M Tricine, pH 8.0, 1 µg valinomycin/ml and 0.5 M choline chloride. Then, 2-ml portions of the suspension were illuminated for 0.5 sec with light from a tungsten halogen lamp (0.20 µW/cm²) filtered through 2 cm of water and a red (Corning 2-64) filter. At 1 sec later, 1 ml of a solution containing KCl and choline chloride (0.5 M total salt concentration) was injected with a hypodermic syringe. Reproduced from Fleischman (1971b).

5. References

Arata, H., Takamiya, K.-I., and Nishimura, M., 1977, Delayed fluorescence from bacteriochlorophyll in *Chromatium vinosum* chromatophores, *Biochim. Biophys. Acta* **459**:36.

Arnold, W., and Sherwood, H., 1959, Energy storage in chloroplasts, *J. Phys. Chem.* **63**:2.

Arnold, W., and Thompson, J., 1955, Delayed light production by blue-green algae, red algae and purple bacteria, *J. Gen. Physiol.* **39**:311.

Barber, J., and Varley, W. J., 1971, Stimulation of delayed light emission by salt gradients and estimation of the relative ionic permeabilities of the thylakoid membranes, *J. Exp. Bot.* **23**:216.

Callis, J. B., Parson, W. W., and Gouterman, M., 1972, Fast changes of enthalpy and volume on excitation of *Chromatium* chromatophores, *Biochim. Biophys. Acta* **267**:348.

Carithers, R. P., and Parson, W. W., 1975, Delayed fluorescence from *Rhodopseudomonas viridis* following single flashes, *Biochim. Biophys. Acta* **387**:194.

Carithers, R. P., and Parson, W. W., 1976, Delayed fluorescence from *Rhodopseudomonas sphaeroides* following single flashes, *Biochim. Biophys. Acta* **440**:215.

Clayton, R. K., 1965, Characteristics of fluorescence and delayed light emission from green photosynthetic bacteria and algae, *J. Gen. Physiol.* **48**:633.

Clayton, R. K., 1967, The bacterial photosynthetic reaction center, *Brookhaven Symp. Biol.* **19**:460.

Clayton, R. K., and Bertsch, W. F., 1965, Absence of delayed light emission in the millisecond time range from a mutant of *Rhodopseudomonas spheroides* which lacks functioning photosynthetic reaction centers, *Biochem. Biophys. Res. Commun.* **18**:415.

Crofts, A. R., Wraight, C. A., and Fleischman, D. E., 1971, Energy conservation in the photochemical reactions of photosynthesis and its relation to delayed fluorescence, *FEBS Lett.* **15**:89.

Crofts, A. R., Jackson, J. B., Evans, E. W., and Cogdell, R. J., 1972, The high energy state in chloroplasts and chromatophores, in: *Proceedings of the IInd International Congress on Photosynthesis Research* (G. Forti, M. Avron, and A. Melandri, eds.), pp. 873–902, W. Junk, The Hague.

Fleischman, D. E., 1967, Chemiluminescence of bacteria chromatophores, *Abstr. Biophys. Soc., 11th Annu. Meet.*:WE7.

Fleischman, D. E., 1969, Chemiluminescence in photosynthetic bacteria, in: *Progress in Photosynthesis Research* (H. Metzner, ed.), pp. 952–955, H. Laupp, Tübingen, West Germany.

Fleischman, D. E., 1971a, Glow curves from photosynthetic bacteria, *Photochem. Photobiol.* **14**:65.

Fleischman, D. E., 1971b, Luminescence in photosynthetic bacteria, *Photochem. Photobiol.* **14**:277.

Fleischman, D. E., 1974, Delayed fluorescence and the reversal of primary photochemistry in *Rhodopseudomonas viridis*, *Photochem. Photobiol.* **19**:59.

Fleischman, D. E., and Clayton, R. K., 1968, The effect of phosphorylation uncouplers and electron transport inhibitors upon spectral shifts and delayed light emission of photosynthetic bacteria, *Photochem. Photobiol.* **8**:287.

Fleischman, D. E., and Cooke, J. A., 1971, Electron transport in *Rhodopseudomonas viridis* at low temperatures, *Photochem. Photobiol.* **14**:71.

Fleischman, D. E., and Mayne, B. C., 1973, Chemically and physically induced luminescence as a probe of photosynthetic mechanisms, *Curr. Top. Bioenerg.* **5**:77.

Godik, V. I., and Borisov, A. Yu., 1977, Excitation trapping by different states of photosynthetic reaction centres, *FEBS Lett.* **15**:89.

Goedheer, J. C., 1962, Afterglow of chlorophyll *in vivo* and photosynthesis, *Biochim. Biophys. Acta* **64**:294.

Goedheer, J. C., 1963, A cooperation of two pigment systems and respiration in photosynthetic luminescence, *Biochim. Biophys. Acta* **66**:61.

Goedheer, J. C., and Vegt, G. R., 1962, Chemiluminescence of chlorophyll *a* and bacteriochlorophyll in relation to redox reactions, *Nature (London)* **193**:875.

Govindjee, Desai, T. S., Tatake, V. G., and Sane, P. V., 1977, A new glow peak in *Rhodopseudomonas sphaeroides*, *Photochem. Photobiol.* **25**:119.

Kaufmann, K. J., Dutton, P. L., Netzel, T. L., Leigh, J. S., and Rentzepis, P. M., 1975, Picosecond kinetics of events leading to reaction center bacteriochlorophyll oxidation, *Science* **188**:1301.

Klimov, V. V., Shuvalov, V. A., Krakhmaleva, I. N., Klevanik, A. V., and Krasnovskii, A. S., 1977, Photoreduction of bacteriopheophytin *b* in the primary light reaction of the chromatophores of *Rhodopseudomonas viridis*, *Biokhimiya* **42**:398.

Kononenko, A. A., Venediktov, P. S., Chemeris, Y. K., Adamova, N. P., and Rubin, A. B., 1974, Relation between electron transport-linked processes and delayed luminescence in photosynthesizing purple bacteria, *Photosynthetica* **8**:176.

Kraan, G. P. B., Amesz, J., Velthuys, B. R., and Steemers, R. G., 1970, Studies on the mechanism of delayed and stimulated delayed fluorescence of chloroplasts, *Biochim. Biophys. Acta* **223**:129.

Lavorel, J., 1975, Luminescence, in: *Bioenergetics of Photosynthesis* (Govindjee, ed.), pp. 233–317, Academic Press, New York.

Mayne, B. C., 1967, The effect of inhibitors on the delayed light emission of chloroplasts, *Photochem. Photobiol.* **6**:189.

Miles, C. D., and Jagendorf, A. T., 1969, Ionic and pH transitions triggering chloroplast post-illumination luminescence, *Arch. Biochem. Biophys.* **129**:711.

Mitchell, P., 1966, Chemiosmotic coupling in oxidative and photosynthetic phosphorylation, *Biol. Rev. Cambridge Philos. Soc.* **41**:445.

Parson, W. W., Clayton, R. K., and Cogdell, R. J., 1975, Excited states of photosynthetic reaction centers at low redox potentials, *Biochim. Biophys. Acta* **387**:265.

Sherman, L. A., 1972, The effect of diaminodurene on the delayed light and the carotenoid band shift in *Rhodopseudomonas spheroides*, *Biochim. Biophys. Acta* **283**:67.

Shuvalov, V. A., and Klimov, V. V., 1976, The primary photoreactions in the complex cytochrome-P890·P760 (bacteriopheophytin$_{760}$) of *Chromatium minutissimum* at low redox potentials, *Biochim. Biophys. Acta* **440**:587.

Straley, S. C., Parson, W. W., Mauzerall, D. C., and Clayton, R. K., 1973, Pigment content and molar extinction coefficients of photochemical reaction centers from *Rhodopseudomonas spheroides*, *Biochim. Biophys. Acta* **305**:597.

Strehler, B. L., and Arnold, W., 1951, Light production by green plants, *J. Gen. Physiol.* **34**:809.

Van Gorkom, H. J., and Donze, M., 1973, Charge accumulation in the reaction center of photosystem 2, *Photochem. Photobiol.* **17**:333.

Witt, H. T., Rumberg, B., Junge, W., Döring, G., Stiehl, J., Weikard, J., and Wolff, C., 1969, Evidence for coupling of electron transfer, field changes, proton translocation and phosphorylation in photosynthesis, in: *Progress in Photosynthesis Research* (H. Metzner, ed.), pp. 1361–1373, H. Laupp, Tübingen, West Germany.

Wraight, C. A., Leigh, J. S., Dutton, P. L., and Clayton, R. K., The triplet state of reaction center bacteriophyll: Determination of a relative quantum yield, *Biochim. Biophys. Acta* **333**:401.

CHAPTER 28

Reaction-Center-Driven Cytochrome Interactions in Electron and Proton Translocation and Energy Coupling

P. Leslie Dutton and Roger C. Prince

1. Introduction

This chapter deals with functional aspects of cytochromes handling electrons in light-driven cyclic and substrate-linked electron flow. Our attention will be chiefly confined to those cytochromes that can be recognized as principal components of reaction center (RC)-associated electron-transfer systems. With this restriction, the considerable number of different cytochromes associated with photosynthetic bacteria (see Chapter 13) is greatly diminished. Furthermore, several common features of physical chemistry and function emerge from the diversity of photosynthetic bacteria that have been studied.

Section 2 presents a current view of cytochromes that have been resolved *in vivo* and that are likely to be present in numbers equal to or greater than the RC itself. This will be done on a thermodynamic basis, relating them to other redox components, with a discussion of various pertinent physicochemical properties determined at equilibrium. It will also examine their location with respect to the cytoplasmic membrane. In Sections 3–6, these data and the kinetics of cytochrome reactions are combined to describe current attitudes concerning how cytochromes operate

to transfer electrons, and their role in energy-conversion processes.

2. Cytochromes of Various Photosynthetic Bacteria

2.1. Half-Reduction Potentials and Relationships with Other Electron-Transfer Components

2.1.1. Rhodospirillaceae

Figure 1 shows the redox components that have been resolved in the functional cytoplasmic membranes (chromatophores) of *Rhodopseudomonas sphaeroides*. The components are presented vertically on a redox potential scale related to the standard hydrogen electrode (E_h) at pH 7.0; the boxes span the redox potential ranges required at equilibrium to take the components from 91% oxidized to 91% reduced according to the Nernst equation. The lateral positioning of the components does not necessarily imply any particular kinetic relationship; where this is known, it will be dealt with later. The enclosed area in this and succeeding figures represents, however, those components associated with the functionally active RC complexes that can be isolated following well-defined detergent fractionation of chromatophores. In Fig. 1,

P. Leslie Dutton and Roger C. Prince · Johnson Research Foundation and Department of Biochemistry and Biophysics, University of Pennsylvania, Philadelphia, Pennsylvania 19104

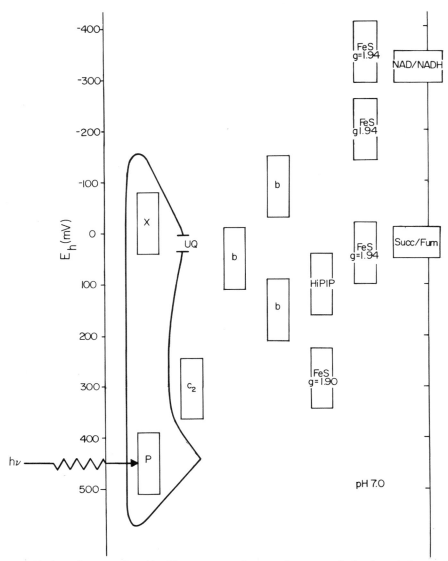

Fig. 1. *Rhodopseudomonas sphaeroides.* The components shown are those present in the chromatophore, and the midpoint potentials of which are known. Each box spans the potential range over which the component goes from 91% reduced to 91% oxidized according to the Peters–Nernst equation $E_h = E_m + RT/n_i^- \ln [\text{Ox}]/[\text{Red}]$ where E_h is the ambient redox potential; E_m is the midpoint potential at the pH under consideration; R, T, and F are the gas constant, absolute temperature, and the Faraday constant, respectively; n is the number of electrons transferred during oxidation or reduction; and [Ox] and [Red] denote the concentrations of the oxidized and reduced form of the couple, respectively. The values for P, X, and the cytochromes are from Dutton and Jackson (1972); for the iron–sulfur (Fe–S) centers, from Prince *et al.* (1975*b*); and for High Potential Iron Protein (HiPIP), from Ingledew and Prince (1977). We have included ubiquinone (UQ) because there is strong circumstantial evidence that it is involved in electron flow (Cogdell *et al.*, 1972, 1973; Petty and Dutton, 1976*a*), but its *in vivo* midpoint potential and *n*-value are not yet clear (but see Takamiya *et al.*, 1978). The UQ analogue present is UQ-10 (Peters and Cellarius, 1972). We have omitted those components that have been identified only in a soluble form, including soluble cytochrome c_2 ($E_{m,7} = +340$ mV, Dutton and Jackson, 1972; Prince *et al.*, 1975*a*; Dutton *et al.*, 1975), cytochrome c' ($E_{m,7} = +5$ mV, Dutton, *et al.*, 1974), cytochrome c_{554} ($E_{m,7} = +120$ mV, Orlando, 1962), cytochrome c_3 ($E_{m,7} = +254$ mV, Meyer *et al.*, 1971), and a soluble cytochrome b (Orlando and Horio, 1961). Under high-oxygen-tension conditions, the synthesis of the RC (enclosed area) is repressed, and it is replaced by a classic cytochrome oxidase containing cytochromes a and a_3 ($E_{m,7} = +200$ and $+375$ mV, Saunders and Jones, 1974). Two extra b-type cytochromes are also produced under such conditions ($E_{m,7} = +390$ and $+255$ mV; Saunders and Jones, 1975) in addition to the three other b-type cytochromes and cytochrome c_2 shown above (Connelly *et al.*, 1973); an additional cytochrome c ($E_{m,7} = +120$ mV; Saunders and Jones, 1974) is also found at very low concentrations ($\approx 10\%$ of the level of cytochrome c_2), and a similar cytochrome at similar low levels has also been detected in anaerobically grown cells (Dutton, P. L., unpublished).

the enclosed area represents RCs prepared using lauryl dimethylamine-N-oxide (LDAO) (Clayton and Wang, 1971; Feher, 1971). It includes the photo-oxidizable RC bacteriochlorophyll (Bchl), designated "P" (see Chapters 19–21), and the primary electron acceptor, most often anonymously called "X," although it is currently considered to be a quinone–iron (Q–Fe) complex (see Chapters 19 and 22). The

ubiquinone (UQ) present is UQ-10 (Carr and Exell, 1965; Peters and Cellarius, 1972), and is found in a 10- to 20-fold excess over the RC itself. It is not unusual (Cogdell *et al.*, 1973) for isolated RCs to contain one or more UQ molecules as functional secondary acceptors, and for this reason, we place the UQ in contact with the RC protein in Fig. 1.

Three electrochemically distinct cytochromes b are

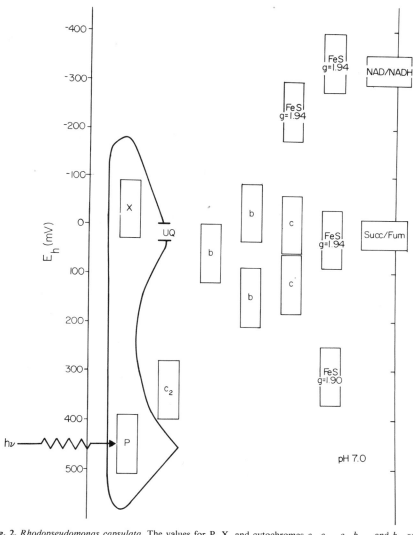

Fig. 2. *Rhodopseudomonas capsulata.* The values for P, X, and cytochromes c_2, c_{120}, c_0, b_{-25}, and b_{50} are from E. H. Evans and Crofts (1974a,b); for cytochrome b_{155}, from P. L. Dutton and R. C. Prince (unpublished); and for the Fe–S centers, from Prince *et al.* (1974c). The UQ is shown by analogy with *Rp. sphaeroides*; the form present is UQ-10 (Carr and Exell, 1965). We have omitted those components known only in a soluble form, including cytochrome c_2 ($E_{m.7} = +340$ mV, Evans, E. H., and Crofts, 1974b) and cytochromes c', c_{554}, and c_3 (Prince, R. C., and Hauska, G. A., unpublished). Under high-oxygen-tension conditions, the synthesis of the RC (enclosed area) is repressed, and it is replaced by an oxidase that apparently contains two cytochromes b (Zannoni *et al.*, 1974), and under these conditions, cytochrome c_2 and b_{50} are also present (Zannoni *et al.*, 1974).

shown. Their identification comes from the approximate location of their reduced-minus-oxidized difference spectra in the 560 nm region. In the absence of completely reliable, and suitably different, α-band absorptions, they have been operationally designated b_{-90}, b_{50}, and b_{155} (Dutton and Jackson, 1972), the subscripts representing their half-reduction potentials at neutral pH. Their respective α-bands (λ_{max}) are at 564–565, 560, and approximately 558–559 nm, and they are typically present at pH 7.0 in a ratio of roughly 1:3:2. Single-turnover flash experiments indicate that cytochromes b_{50} and b_{155} are involved in electron transfer at the major level, becoming reduced or oxidized, respectively, in the millisecond time scale

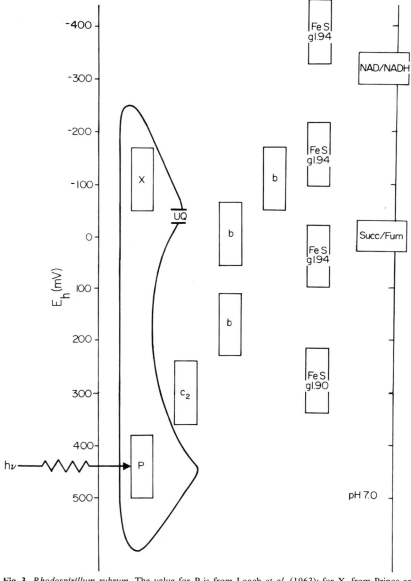

Fig. 3. *Rhodospirillum rubrum.* The value for P is from Loach *et al.* (1963); for X, from Prince and Dutton (1976a); and for the cytochromes, from Dutton and Jackson (1972). The cytochrome c ($E_{m,7}$ = +295 mV) has been designated c_2 by analogy with *Rp. sphaeroides* (Prince *et al.*, 1975a). The UQ present is UQ-10 (Carr and Exell, 1965). The values for the Fe–S centers are from Yoch and Knaff (personal communication), and the soluble cytochromes c_2 and c' ($E_{m,7}$ = +320 mV and −8 mV, Bartsch, 1971) have been omitted.

following a flash (Dutton and Jackson, 1972; Prince and Dutton, 1975; Petty and Dutton, 1976b). Cytochrome b_{-90}, which appears to have an α-band λ_{max} at significantly longer wavelength (Dutton and Jackson, 1972; Jones, O. T. G., 1969), has not been recognized on a single-turnover basis, although it has been shown to respond promptly to steady-state illumination under certain conditions (Jones, O. T. G., 1969), and can be observed on a seconds-or-less time scale.

The other major cytochrome shown has a reduced-minus-oxidized absorbance maximum at 551 nm and is cytochrome c_2. This cytochrome was previously (Dutton and Jackson, 1972) designated cytochrome c_{295} (again from its E_m at pH 7.0), but recent antibody studies (Prince et al., 1975a) provide good evidence that it is the well-known cytochrome c_2 in the membrane-associated state (see Section 3). The half-reduction potential measured in vivo is some 50 mV

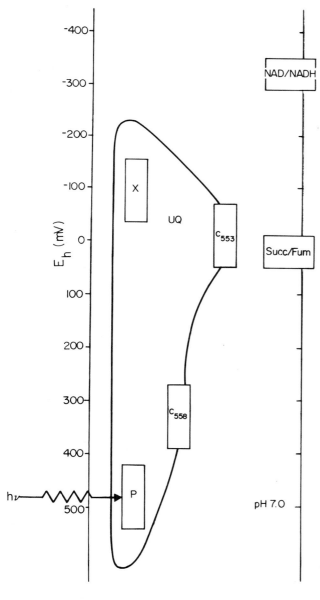

Fig. 4. *Rhodopseudomonas viridis.* The value for P is from Carithers and Parson (1975); for X, from Cogdell and Crofts (1972); and for the cytochromes, from Thornber and Olson (1971). The presence of UQ as the secondary acceptor is indicated by the data of Carithers and Parson (1975), but the analogue present is unknown. We have omitted the soluble ascorbate-reducible cytochrome $c_{551.5}$ (c_2) (Olson and Nadler, 1965). The enclosed area contains the fraction isolated using detergents (Thornber and Olson, 1971).

more negative than cytochrome c_2 in the isolated aqueous-solvated or membrane dissociated state, which has an E_m value of 345 mV (pH 7.0) (Dutton *et al.*, 1975; Prince *et al.*, 1975*a*) in suspending media of similar ionic strength. We shall come back to this in Section 3.

Five iron–sulfur (Fe–S) proteins are listed that are readily resolvable *in vivo* (Prince *et al.*, 1975*b*). The most electropositive has an E_m and EPR lineshape ($g_y = 1.90$) in the reduced form similar to that discovered by Rieske *et al.* (1964) in the mitochondrial b–c_1 complex of the respiratory chain. There is also a high-potential iron protein (HiPIP) that has properties very similar to those of the one found in beef heart mitochondria (Ingledew and Ohnishi, 1975). The other

three are more typical $g_y = 1.94$ components; we do not rule out the possibility of there being more (see Ohnishi, 1973). So far, no single-turnover flash work has been done on the Fe–S proteins, although the $g = 1.90$ "Rieske" Fe–S protein was demonstrated to undergo oxidation in chromatophores after 10 sec of illumination in the presence of redox mediators (Prince *et al.*, 1975*b*). Much more work is needed in this area.

Figure 1 is restricted to components under consideration as important or major parts of the photosynthetic bacterial electron-transfer system, as are Figs. 2 and 7. Omitted are any components that from current evidence are very unlikely candidates for any major role in photosynthetic electron-transfer processes. Also omitted are redox species that in our

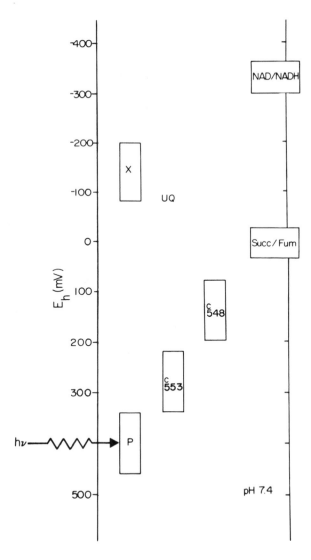

Fig. 5. *Rhodopseudomonas gelatinosa.* The values are from Dutton (1971). UQ is present (Carr and Exell, 1965), and the analogue present is UQ-8. We have omitted the variety of soluble c-type cytochromes noted by Bartsch (1971) and the soluble high-potential iron protein ($E_{m,7} = +330$ mV, DeKlerk and Kamen, 1966). It should be noted that the a-band λ_{max} of the c-type cytochromes was measured at 77°K.

opinion have not been convincingly detected in the functional chromatophore or measured therein, or components that, relative to the RC, are present at levels too low to be considered at this time for a major functional role. With these restrictions, we hope to usefully minimize the impression of awe that is often conveyed to the reader by authors of general reviews of cytochromes when the point has come to discuss function! The omitted components are, however, listed in the figure captions.

Figures 2 and 3 show similar arrangements for two other members of the Rhodospirillaceae, *Rp. cap-*

sulata (Fig. 2) and *Rhodospirillum rubrum* (Fig. 3). The overall similarities of Figs. 1–3 can be readily appreciated. Although studied in less depth on a general basis, a departure from the component patterns of the three species named above is recognizable in other members of the Rhodospirillaceae. Figures 4 and 5 show what is known from *in vivo* studies on *Rp. viridis* (Fig. 4) and *Rp. gelatinosa* (Fig. 5). In these latter species, in addition to the high-potential cytochrome *c*, analogous to cytochrome c_2 above, there is a second, lower-potential *c*-type cytochrome in chromatophores, and both are capable of

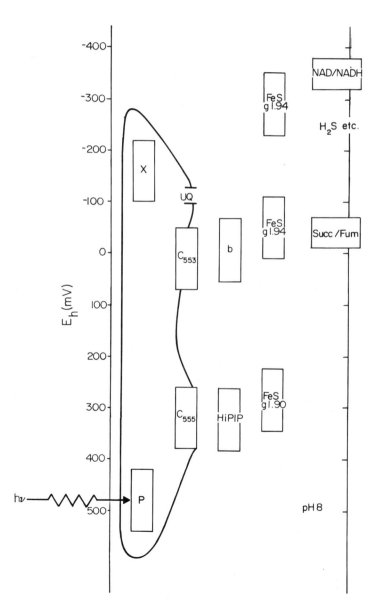

Fig. 6. *Chromatium vinosum.* The value for P is from Cusanovich *et al.* (1968); for X, from Prince and Dutton (1976*a*); and for the two *c*-type cytochromes, from Case and Parson (1971). The "double-flash" data of Case and Parson (1971) suggest that the midpoint of the secondary acceptor at pH 8 is −120 mV, and that there is a pK on the reduced form of the secondary acceptor at pH 8 (cf. the primary acceptor, see Chapter 24; Prince and Dutton, 1976*a*). Halsey and Parson (1974) identified this secondary acceptor as UQ. The values for the Fe–S centers and HiPIP are from Dutton and Leigh (1973) and M. C. W. Evans *et al.* (1974), and for cytochrome *b*, from Knaff and Buchanan (1975). The figure does not include components known only in a soluble form, including cytochromes c_{553} and *c'* ($E_{m,7} = +10$ mV and −5 mV, Bartsch and Kamen, 1960), cytochrome c_{556} ($E_{m,7} = +325$ mV, Kennel and Kamen, 1971) and soluble HiPIP ($E_{m,7} = +350$ mV, Bartsch, 1963). The enclosed area contains the components of "Fraction A" (Thornber, 1970).

very rapid direct electron transfer to the RC P$^+$ (Case *et al.*, 1970; Cogdell and Crofts, 1972; Dutton, 1971). In *Rp. viridis* (Fig. 4), the two cytochromes are integral parts of the detergent-isolated RC-enriched subchromatophore complex (Thornber *et al.*, 1969; Case *et al.*, 1970). Unfortunately, redox potentiometry has not been investigated in the other members of the Rhodospirillaceae, but as we shall discuss later, the identification of both high- and low-potential *c*-type cytochromes donating to the same RC may be a common feature in this family, with the exception of *Rp. sphaeroides*, *Rp. capsulata*, and *Rs. rubrum*. Also in contrast to *Rp. sphaeroides*, *Rp. capsulata*, and *Rs. rubrum*, the existence of *b*-type cytochromes in the other purple nonsulfur bacteria on a general basis remains to be firmly established. In *Rp. viridis*, alkaline pyridine hemochromogen analysis has not thus far revealed protoheme IX, indicative of cytochrome *b* (Jones, O. T. G., and Saunders, 1972), although Olson and Nadler (1965) provided spectroscopic evidence for a cytochrome with an α-band λ_{max} at 560 nm, and O. T. G. Jones and Saunders (1972) demonstrated that antimycin inhibited cyclic electron flow in this species.

2.1.2. Chromatiaceae

Figure 6 displays the components of *Chromatium vinosum*. Two different cytochromes of the *c*-type, low-potential cytochrome c_{553} and high-potential cytochrome c_{555}, can serve to supply electrons rapidly to the RC P$^+$ following flash activation, and as with *Rp. viridis*, both are integral parts of the chromatophore membrane and isolated subchromatophore preparations. Sodium dodecyl sulfate (SDS) (Thornber, 1970) and Triton X-100 (Garcia *et al.*, 1966) solubilize cytochrome *c*–RC subchromatophore preparations in which the components of the light reaction and both cytochromes c_{553} and c_{555} retain their *in vivo* E_m values and RC-induced kinetics. The cytochromes can be "separated" from the antenna and RC Bchl complements by acetone powder–cholate treatment of the SDS particles (Kennel and Kamen, 1971); they remain associated in a single complex, and both retain their *in vivo* E_m values.

Much less is known about the *c*-type cytochromes of other members of the Chromatiaceae. *Ectothiorhodospira shaposhnikovii*, previously known as *Rhodopseudomonas* sp. (Rubin *et al.*, 1969) but now assigned to the genus *Ectothiorhodospira* (Trüper, 1968; Rubin *et al.*, 1970), was reported to have both high- and low-potential cytochromes *c* (Rubin *et al.*, 1969, 1970), and the data of Kihara and Chance (1969) suggest that the same may also be true for *Chr.*

violascens and *Thiocapsa roseopersicina* (see Section 3.2). Recent work on *T. pfennigii* [originally known as a *Thiococcus* sp. (Eimhjellen *et al.*, 1967)] has shown that this organism also possesses both high- and low-potential *c*-type cytochromes (Prince, 1978).

After years of negative indications of a cytochrome *b* at functional levels, recent investigations by Knaff and Buchanan (1975) demonstrated the functional presence of a cytochrome *b* in *Chr. vinosum* membranes (α-band λ_{max} 560 nm), and supported it by alkaline pyridine hemochromogen analysis of the protoheme IX. *In vivo* electron paramagnetic resonance (EPR) analysis reveals several chromatophore-associated Fe–S proteins including the Rieske *g* = 1.90 species and the much-studied HiPIP protein in the electropositive redox regions, and more typical *g* = 1.94 proteins at lower potentials (Dutton and Leigh, 1973; Evans, M. C. W., *et al.*, 1974). Steady-state illumination has been reported to induce the photooxidation of both HiPIP (on a seconds time scale, redox dyes present; Dutton and Leigh, 1973) and the *g* = 1.90 protein (3 min illumination, redox dyes absent; Evans, M. C. W., *et al.*, 1974), but more exact pulsed work on these components is needed to establish whether their rates of oxidation and reduction are commensurate with a role in cyclic or noncyclic electron flow, or in both. None of these Fe–S centers has been detected (Dutton and Leigh, 1973) in the SDS subchromatophore particle of Thornber (1970).

2.1.3. Chlorobiaceae

The functional cytochromes of the green sulfur bacteria have not yet been extensively investigated, and the available data are confused because there is some controversy regarding the taxonomy of this group. This is particularly relevant in the case of "*Chloropseudomonas ethylica* 2K," which is now recognized as a symbiotic association of a colorless bacterium with a green photosynthetic organism (see Olson, 1973) that has been variously identified as *Chlorobium limicola* (Gray *et al.*, 1972, 1973) and *Prosthecochloris aestuarii* (Shioi *et al.*, 1976).* The only other green bacterium that has been extensively studied is *Chlorobium limicola f. thiosulfatophilum*.

The pigment composition of these organisms is markedly different from that of the purple bacteria, for, in addition to Bchl *a*, the green bacteria also possess large amounts of chlorobium Chl (Bchl's *c* and *d*), the photosynthetic unit containing approximately 80 Bchl *a* molecules and 1000–1500 chlorobium Chl's/RC (Fowler *et al.*, 1971). This has

---

* See Chapter 2 for a discussion of the taxonomy of this syntrophic system.

made it difficult to observe small light-induced optical changes because of the large background absorbance of the vesicles. It is possible, however, to remove the chlorobium Chl by mild procedures without detergents to yield what have become known as Bchl–RC

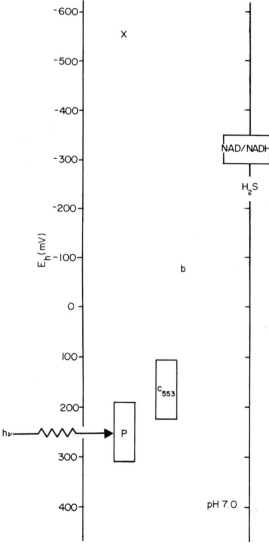

Fig. 7. *Chlorobium limicola f. thiosulfatophilum.* The figure represents the data of Prince and Olson (1976), Fowler *et al.* (1971), Fowler (1974), and Olson *et al.* (1976) obtained with the Bchl–RC complex prepared essentially by the method of Fowler *et al.* (1971). The $E_{m,7}$ of cytochrome b is between 0 and −150 mV (Fowler, 1974). We have not included the soluble c-type cytochromes (c_{551}, $E_{m,7} = +135$ mV; c_{553}, $E_{m,7} = +98$ mV; c_{555}, $E_{m,7} = +145$ mV) purified by Meyer *et al.* (1968) or the data of Knaff *et al.* (1973) and Knaff and Buchanan (1975) obtained with a preparation still containing Bchl's c and d (*Chlorobium* Chl's).

complexes, which are similar in pigment composition to chromatophores from purple bacteria (Fowler *et al.*, 1971).

Figure 7 shows the components known to be present in the "Bchl–RC-complex" of *Chl. limicola f. thiosulfatophilum* (Prince and Olson, 1976). Very similar values for the RC Bchl and cytochrome c_{553} of "*Chlp. ethylica*" were obtained by Fowler *et al.* (1971). A cytochrome b (α-band $\lambda_{max} = 564$ nm) is also present (Fowler, 1974), with a midpoint potential between 0 and −150 mV at pH 7.4. Knaff *et al.* (1973) examined vesicles from *Chl. limicola f. thiosulfatophilum* that still contained the *Chlorobium* Chl, and reported somewhat higher midpoint potential values for both the RC P ($E_{m,7} = +330$ mV) and cytochrome c_{553} ($E_{m,7} = +220$ mV); the reasons for these differences remain unclear. They also reported a much higher value for the primary acceptor ($E_{m,7} = −130$ mV), but this was probably due to the use of continuous actinic illumination instead of single-turnover activation (see Knaff *et al.*, 1973; Prince and Olson, 1976; Olson *et al.*, 1976). Knaff and Buchanan (1975) also detected in their preparation a functional cytochrome b ($E_{m,7.2} = −90$ mV) that could be rapidly reduced in the light.

2.2. A Simple View of Major Photosynthetic Electron-Flow Patterns

This section serves to summarize the various redox components discussed above, and to draw attention to those carriers that have been established as major parts of RC-activated electron flow, either on a single-turnover basis or by light-induced work at low temperatures.

Three kinetic patterns of electron flow are emerging from the photosynthetic bacteria, typified by *Rp. sphaeroides* (Fig. 8A), *Chr. vinosum* (Fig. 8B), and *Chl. limicola f. thiosulfatophilum* (Fig. 8C). The dotted lines represent possible links to substrate (e.g., sulfide, thiosulfate, and succinate) electrons to replenish the cyclic system and for pyridine nucleotide reduction. In the purple bacteria, the reduction of pyridine nucleotides is by energy-linked reversed electron flow (Keister and Yike, 1967a,b; Jackson and Crofts, 1968; Jones, C. W., and Vernon, 1969; Jones, O. T. G., and Whale, 1970; Jones, O. T. G., and Saunders, 1972) (see also Chapter 32), which may be accomplished by the net reverse operation of a coupling mechanism similar to the mitochondrial "Site I," consisting chiefly of Fe–S centers (Ohnishi, 1973). In the green bacteria, there is evidence that the midpoint potential of the primary acceptor is sufficiently low that the reduction of NAD$^+$ can occur directly via the

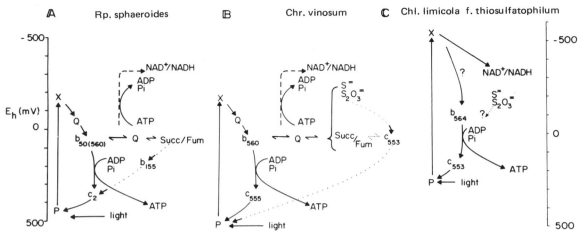

Fig. 8. Photosynthetic electron-flow pathways in photosynthetic bacteria. (——) Cyclic electron flow, indicating the net direction of flow (→); (. . . .) noncyclic, presumably substrate-linked electron flow; (————) energy-linked pathways that operated predominantly in a "reverse" manner. The multiple presentation of quinones (Q) represents different forms (redox states or electrochemical species) of the carrier; this subject is discussed later.

light-induced reduction of ferredoxin, as shown in Fig. 8C (Evans, M. C. W., and Buchanan, 1965; Buchanan and Evans, 1969; Jones, O. T. G., and Whale, 1970; Prince and Olson, 1976) (see also Chapter 32). It is perhaps noteworthy that the light reactions of the photosynthetic bacteria bear striking resemblances to those in green plants. For example, the photosynthetic electron pathway of *Chl. limicola f. thiosulfatophilum* appears to be very similar to photosystem I, while in the purple bacteria, the primary donors resemble the primary donor of photosystem I, and the primary acceptors (see Chapter 22) resemble the acceptor of photosystem II (Knaff, 1975). The possible evolutionary implications of these similarities seem worthy of further consideration.

Little is known of how substrate electrons feed into the photosynthetic system of the green bacteria, but in the purple bacteria one can discern two pathways (Figs. 8A and 8B) that differ with respect to the point at which what might be regarded as "noncyclic electron flow" converges with the cyclic system. In *Chr. vinosum*, studies (Parson and Case, 1970; Seibert and DeVault, 1970; Case *et al.*, 1970) with short single-turnover flashes of light have provided convincing evidence that both the high- and low-potential cytochromes *c* operate to share the same RC P. Cytochrome c_{555}, a very probable part of the energy-coupled (ATP-producing) cyclic system, reduces P$^+$ in approximately 2 μsec half-time, while cytochrome c_{553}, considered to be part of the substrate-linked noncyclic system, reduces P$^+$ in approximately 1 μsec half-time. If both are reduced prior to the flash, the latter reaction is dominant in reducing P$^+$. Work with *Rp.*

viridis (Case *et al.*, 1970), *Ectothiorhodospira shaposhnikovii* (Rubin *et al.*, 1969, 1970), and *T. pfennigii* (Olson *et al.*, 1969; Meyer *et al.*, 1973; Prince, 1978), and low-temperature studies on *Rp. gelatinosa* (Dutton, 1971) indicate a similar situation in these species, and in other bacterial species from the work of Kihara and Chance (1969), who examined the low-temperature (80°K) oxidation of cytochromes in 11 different species of Chromatiaceae and Rhodospirillaceae. In all cases, the rates of oxidation of the lower-potential cytochromes appear to be faster than those of the higher-potential cytochromes. This subject is discussed further in Section 3.

In contrast to the situation discussed above in which noncyclic and cyclic electron-delivery systems converge at the RC, in *Rp. sphaeroides* (Fig. 8A) the convergence seems to be one step back, with two *b*-type cytochromes interacting with the same cytochrome c_2 complement (Dutton and Jackson, 1972; Prince and Dutton, 1975). Cytochrome b_{50} (cyclic, energy-linked) is capable of donating an electron to flash-oxidized cytochrome c_2 in energetically uncoupled situations with a half-time of 1–2 msec, although under more energetically demanding conditions it is considerably slower (see later). Cytochrome b_{155}, which may be part of the noncyclic substrate-linked system, appears to donate to flash-oxidized cytochrome c_2 in about 2 msec, a rate that is apparently independent of the energy status of the chromatophore. We anticipate that *Rp. capsulata* and *Rs. rubrum* may fall into the same category as *Rp. sphaeroides*; they both have multiple cytochromes *b* and the soluble cytochrome c_2 as sole apparent physiological electron donor to P+, and, unlike the other photosynthetic bacterial species, do not seem capable of supporting cytochrome c oxidation at 80°K. To date, this latter property seems to be general with respect to the low-potential noncyclic cytochromes c of the *Chr. vinosum* category (see Section 3).

These differences in the electron-transfer patterns of the purple bacteria may, however, be functionally of little consequence. All the bacteria have an energy-coupled cycle involving, at the least, UQ, a cytochrome b, and a high-potential cytochrome c, and all appear to have a converging non-energy-coupled electron-supply system to a point electropositive of the coupling mechanism. However, while it is clear that in all cases the cyclic system is geared to the ultimate production of ATP, and is probably in contact with "noncyclic" electrons at a point negative of the coupling mechanism for cyclic replenishment and reversed electron flow to pyridine nucleotides, we do not yet fully understand the role of the noncyclic electron-supply system in the economy of the cell, or why this electron flow is not coupled to ATP synthesis. For example, in *Chr. vinosum*, the noncyclic ferrocytochrome c_{553} delivers an electron to P+ over an apparently uninterrupted redox potential drop of approximately 500 mV. However, what might at first glance seem to be a totally counterproductive competition with the cycle is probably rendered insignificant because the rate of re-reduction of the cytochrome by substrates is in the hundreds-of-milliseconds range (Seibert and DeVault, 1970), compared with the 1- to 10-msec turnover time of the cycle under low-energy conditions. The functional significance may lie in the fact that under high-energy conditions (e.g., high ATP/ADP · P$_i$ ratios, large ion gradients), the electropositive end of the cycle will tend to become oxidized (reverse electron flow; slow apparent cycle turnover time), so that the non-cyclic system may operate to provide a steady supply of substrate electrons delivered to oxidized P (P+) independent of the energy status of the chromatophore. Are we forced to consider that the cooperative effects of too many P+ species in the intracytoplasmic membrane are not good for the bacterium?

2.3. Oxidation–Reduction of Cytochromes and the Thermodynamic Requirement for the Proton

So far, we have dealt with cytochromes as electron carriers. It is now a well-established fact that protons are rapidly bound by chromatophores following a single-turnover flash (Chance *et al.*, 1970) (see also Chapter 26). As will be discussed later, they are most probably incorporated into the cyclic electron-transfer system by UQ (Cogdell *et al.*, 1972, 1973) in a time and with a stoichiometry commensurate with electron-transfer rates (Cogdell and Crofts, 1974; Petty and Dutton, 1976a). The fate of these protons, and of those initially contained in substrates, may rest on the proton-carrying properties of the cytochromes. The most common experimental criterion for deciding that the oxidation–reduction of a redox couple does or does not involve the release–binding of a proton comes from the equilibrium measurement of half-reduction potentials of the couple at various pH values over a suitable pH range. The absence of a pH dependency indicates that the net exchange of protons does not occur in the reaction; a pH dependency of −59 mV/pH unit indicates that one proton per electron is involved, as follows:

$$\text{Oxidized} + e^- + H^+ \rightleftharpoons \text{reduced H}$$

Most of the systematic work done on the proton requirement of cytochromes and their companion redox couples has been with *Rp. sphaeroides* and *Chr. vinosum*, and Fig. 9A shows the pH dependency of the half-reduction potentials of the cytochromes of *Rp. sphaeroides* chromatophores. In the left panel, the E_m values (at the various pH values) were measured by "dark" equilibrium techniques by simply relating the measured redox potential to simultaneously assayed absorbance changes at 560–540 nm for the cytochromes b (Petty and Dutton, 1976b) and at 551–540 nm for cytochrome c_2 (Dutton *et al.*, 1975). In the right panel are E_m/pH relationships of cytochrome b_{50} and cytochrome c_2 determined in a different way: the individual E_m values were obtained by assaying a kinetic response as a function of the E_h measured at equilibrium just before activation. In these cases, the extent of flash-induced rapid reduction of cytochrome b_{50} (Petty and Dutton, 1976b) or oxidation of cytochrome c_2 was assayed in chromatophores in the presence of antimycin A (Prince and Dutton, 1977a). These latter measurements are not merely a confirmation of the "dark" titrations; they are important because they uniquely identify the components by their functional involvement in rapid electron transfer, whereas the "dark" titrations make no distinction between the functional and nonfunctional cytochromes. Flash-activated measurements on cytochrome b_{-90} have not been done because its reactions after a single turnover have not yet been reported; there is a problem in that its E_m is more negative than the apparent equilibrium E_m of the primary acceptor itself (see Chapter 24). Cytochrome b_{155} has not been measured directly, although examination of the re-reduction kinetics of cytochrome c_2 (no antimycin) following a single flash indicates a −59 mV/pH unit

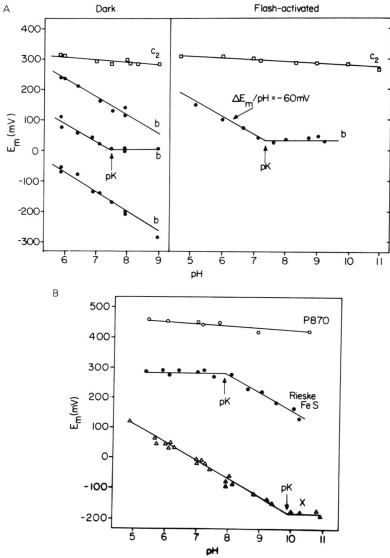

Fig. 9. Midpoint potentials of electron carriers of *Rhodopseudomonas sphaeroides* as functions of pH. This figure combines the data of Dutton *et al.* (1974), Petty and Dutton (1976a), Prince *et al.* (1975b), and Prince and Dutton (1976a, 1977a). Full descriptions of the interpretation of such data as indicating pK's on the oxidized or reduced forms of the couples involved can be found in Dutton and Wilson (1974) and Clark (1960).

relationship similar to that encountered by standard equilibrium techniques. Figure 9A shows that the midpoint potential of cytochrome c_2 *in vivo* has little dependency on pH, indicating that protons are not involved in its oxidation–reduction reaction. On the other hand, the data for cytochromes b_{155} and b_{-90} indicate that their oxidation–reduction reactions involve the coupled release and uptake of a proton (see Dutton and Wilson, 1974), at least at equilibrium.

Cytochrome b_{50} has a pK on its reduced form at pH 7.4, such that at pH values below this, the redox reaction is

$$\text{cyt. } b_{\text{oxidized}} + e^- + H^+ \rightleftharpoons \text{cyt. } b H_{\text{reduced}}$$

while at pH values above pH 7.4, the redox reaction is

$$\text{cyt. } b_{\text{oxidized}} + e^- \rightleftharpoons \text{cyt. } b^{(-)}_{\text{reduced}}$$

This pK is seen in both equilibrium and flash-induced titrations (Fig. 9A) (Petty and Dutton, 1976b), and there is reason to believe that the pK is truly functional on the time scale of electron flow (see Section 4 and Petty and Dutton, 1976b).

Figure 9B shows the E_m/pH relationships of some other electron carriers found in chromatophores of *Rp. sphaeroides*. The RC P does not change its midpoint potential with changing pH, while the Rieske $g = 1.90$ Fe–S center and the primary acceptor both do, at least over some range of pH. The operating midpoint potential of the primary acceptor is dealt with in Chapter 24; here, it will suffice to say that there is evidence that it operates above its reduced pK, because no proton is taken up on the time scale of light-induced electron flow. The operational E_m is thus -180 mV.

In contrast to the data in Figs. 9A and B, which suggest pKs on the reduced forms of cytochrome b_{50} and the primary acceptor, data on the midpoint potential of the Rieske $g = 1.90$ Fe–S center indicate the existence of a pK on the oxidized form of this electron carrier at pH 8, so that at pH values below 8, the redox reaction is

$$[\text{FeS}_{ox}]\text{H}^+ + e^- \rightleftharpoons [\text{FeS}_{red}]\text{H}$$

while at pH values above pH 8, the reaction is

$$[\text{FeS}_{ox}] + e^- + \text{H}^+ \rightleftharpoons [\text{FeS}_{red}]\text{H}$$

As yet, however, there is no firm evidence that the Rieske $g = 1.90$ Fe–S center is involved in cyclic electron flow, and its role is unclear.

The midpoint potential of UQ is a complicated topic that has not yet been fully resolved; it could well have multiple midpoint potentials, both at equilibrium and on a kinetic time scale. A major problem is that it can be measured only indirectly (e.g., by flash-induced proton uptake) (Cogdell *et al.*, 1972; Petty and Dutton, 1976a), which involves an assumption that what is being measured is indeed a property of UQ. The general consensus is currently that UQ is at least involved in the primary and secondary acceptors in the cycle; in the latter role, it may operate as the UQ/protonated semiquinone couple (Petty and Dutton, 1976a). Further considerations on the possible multiple roles of UQ in the cycle are considered in Section 5.

The extensive thermodynamic studies of Case and Parson (1971) on *Chr. vinosum* revealed that the membrane-associated cytochromes c_{553} ($E_m = +10$ mV, pH 7.0; noncyclic system) and c_{555} ($E_m + 350$ mV, pH 7.0; cyclic system) have E_m/pH unit relationships of less than 5 mV/pH unit from pH 5 to 9, and hence their oxidation–reduction does not involve a proton. The same non-proton-involved behavior at

near-neutral pH would appear to hold for the Rieske Fe–S center if we take the similar E_m values obtained at pH 7.2 ($E_m = +280$ mV, Dutton and Leigh, 1973) and pH 8.0 ($E_m = +285$ mV, Evans, M. C. W., *et al.*, 1974), although preliminary titrations (Prince, R. C., and Knaff, D. B., unpublished) suggest a pK on the oxidized form at approximately pH 8 in this organism as well. The redox reactions of cytochrome b_{560} ($E_m = -5$ mV at pH 8.0, -70 mV at pH 9.0; Knaff and Buchanan, 1975) appear likely to involve the proton in the physiological pH range, although it will be important to examine this cytochrome over a wider range of pH.

Some useful similarities thus exist between *Chr. vinosum* and *Rp. sphaeroides*: P and the cyclic cytochromes c (c_{555} and c_2) do not appear to be capable of acting as electron-transfer-coupled proton carriers, while the respective cyclic-system cytochromes b (b_{560} of *Chr. vinosum* and b_{50} of *Rp. sphaeroides*, also α-band λ_{max} 560 nm) are, with possible qualifications due to pKs, potential proton carriers in the neutral and acid pH range. In addition, with the appropriate arguments (see Chapter 24), there is reason to believe that the primary acceptors of *Chr. vinosum* and *Rp. sphaeroides* may, for kinetic reasons, ignore any thermodynamic proton-carrying indications and not bind a proton following reduction during light-induced electron flow. If this is so, the operative couple is X/X$^-$, the E_m in both cases being approximately -180 mV throughout the physiological pH range.

We do recognize further differences, however, in what we identify as the noncyclic systems of these bacteria. Not only do they employ different cytochrome types with different E_m values and different *in vivo* oxidants, but also one is clearly not a proton carrier (c_{553}), while the oxidation–reduction of the other (b_{155}) appears to involve coupled proton release–binding.

2.4. Current Information Concerning Membrane Location

The functional position of components with respect to the membrane, and the propensity to carry or not carry protons coupled to electron flow, are factors of prime importance in considerations of the translocation of electrons and protons within and across the chromatophore membrane. In·contrast to the steadily mounting thermodynamic data on the electron carriers present in chromatophores, information concerning their location with respect to the membrane is fragmentary and in many cases very tentative. To our knowledge, the only study on the membrane location of a cytochrome in photosynthetic bacteria—beyond an identification as "cytoplasmic" or "membrane-bound"—that rests on direct evidence is the immunological localization of cytochrome c_2 in *Rp. sphaeroides* and *Rp. capsulata* (Prince *et al.*, 1975a). In these species, the buffer-soluble cytochrome c_2 is merely a soluble form of the chromatophore-bound cytochrome, and *in vivo* it is localized in the intact cell between the cell wall and the cell membrane in the periplasmic space, as shown diagrammatically in Fig. 10. It had been suggested that a membrane-bound cytochrome c was the electron donor to P, with cytochrome c_2 acting merely as a "pool." Immunological evidence shows, however, that the immediate donor to P is immunologically indistinguishable from soluble cytochrome c_2, and removal of the cell wall to form spheroplasts with exposure to moderate ionic strength permits the release of cytochrome c_2, leaving P without an electron donor (Prince *et al.*, 1975a). In addition, rapid kinetic analysis (Dutton *et al.*, 1975) demonstrated a strict two-cytochromes-c_2-to-one-P stoichiometry, with no evidence of a "pool." This subject will be discussed more fully in the next section.

Immunological evidence has also been presented that suggests that part of the RC protein (specifically the H subunit) is exposed to the outer aqueous–membrane interface of chromatophores (Steiner *et al.*,

1974; Reed *et al.*, 1975; Chapter 19), while the functional oxidizing part of this protein is near the inner aqueous–membrane interface (Prince *et al.*, 1974b, 1975a). Note that the inside of a chromatophore corresponds topologically to the periplasmic space of an intact cell (Fig. 10).

Further evidence that the RC spans the chromatophore membrane comes from the fact that rapid proton uptake associated with the reduction of the secondary acceptor, which is probably UQ (Cogdell *et al.*, 1972, 1973; Halsey and Parson, 1974), occurs with a stoichiometry of 1.0 ± 0.1 H$^+$ for *every* electron delivered from P, via X, to UQ (Petty and Dutton, 1976a).

Interpretations based on the assumption that the carotenoid bandshift (see Chapter 26) responds solely to electric-field alterations across the chromatophore membrane as a result of flash-induced electron transfer also yield results in keeping with the evidence cited above. In *Rp. sphaeroides* chromatophores, such assumed electric fields can be associated both kinetically and thermodynamically with electron transfer from P to X (forming P$^+$X$^-$) and the subsequent re-reduction of P$^+$ by cytochrome c_2 (forming c_2^+PX$^-$) (Jackson and Dutton, 1973; Dutton *et al.*, 1975). The carotenoid bandshift enhancement (approximately twofold) associated with the latter reaction would

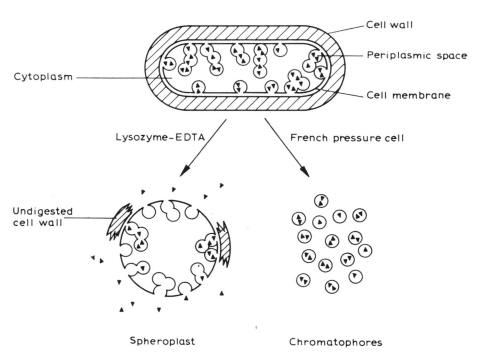

Fig. 10. Location of cytochrome c_2 (▲) in *Rhodopseudomonas sphaeroides* membranes.

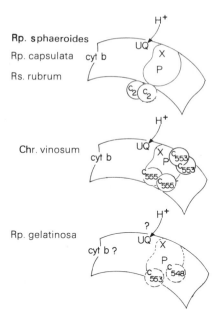

Fig. 11. Possible arrangement of the RC and associated cytochromes *c* in the chromatophore membrane.

suggest that P is in fact somewhere near the middle of the membrane dielectric, so that the reduction of P^+ by cytochrome c_2 in effect pulls the positive charge farther toward the inner side of the membrane, thereby increasing the membrane potential. The lack of an enhancement as X^- reduces UQ (the ensuing protonation effectively delocalizing the negative charge into the outer aqueous phase) could be similarly interpreted as indicating that X^- is quite close to the low dielectric outer surface, so that no further electric-field-generating charge separation could occur during the protonation of UQ.

The information from these varied pieces of evidence is combined in Fig. 11A; the location of cytochrome *b* will be discussed in Section 4. What evidence there is available for *Rp. capsulata* and *Rs. rubrum* suggests that these two bacteria may share the same basic pattern.

The intramembrane location of electron-transport components seems to be somewhat different in other photosynthetic bacteria. In *Chr. vinosum* and *Rp. viridis*, the cytochrome *c* complement is tightly integrated with the RC protein (Thornber, 1970; Case *et al.*, 1970) and is separated only with difficulty (Kennel and Kamen, 1971). In these cases, simple techniques of washing are inapplicable, and the production of antibodies for immunological approaches is fraught with problems of isolation and purity. In all species studied, however, flash activation elicits the rapid

binding of protons at the outer chromatophore membrane surface, consistent with UQ being on the outer side in all cases.

A novel, and perhaps more general, approach to positional determination was recently introduced by Case and Leigh (1974). This approach exploits the paramagnetic resonance properties of redox components, and has been used to study *Chr. vinosum*, in which a large number of EPR-detectable compounds have been resolved (Dutton and Leigh, 1973; Evans, M. C. W., *et al.*, 1974). Case and Leigh's approach utilizes the effects of paramagnetic dipole interactions of the trivalent gadolinium cation (Gd^{3+}) with the membrane-associated paramagnetic redox components. Magnetic effects are apparent only when the interacting species are closer than 10–15 Å, since dipolar splitting is proportional to $r^{-1/3}$ or $r^{-1/6}$, depending on the relative spin-lattice relaxation times of the redox component and the added Gd^{3+} (Slichter, 1963; Leigh, 1970). In chromatophores, Gd^{3+} affected cytochrome c_{553} and the primary acceptor (as seen by its EPR spectrum in the reduced form, $g_y = 1.82$), so these were assigned locations near the outer side of the chromatophore, within 10 Å of the aqueous interface. In contrast, there was no effect on P or cytochrome c_{555}, so these were assigned a position nearer the inner surface of the membrane. Similarly, the $g = 1.94$ Fe–S center [E_m (pH 8.0) $= -50$ mV] was placed close to the outer side, while the $g = 1.90$ "Rieske center" [E_m (pH 8) $= +280$ mV] was placed on the inner side. This promising technique is still in its infancy, and the reader is referred to the original paper (Case and Leigh, 1974) for the uncertainties and assumptions involved. Nevertheless, the results are in remarkable agreement with information for *Chr. vinosum* gleaned by inference from its flash-induced carotenoid bandshift (Case and Parson, 1973). In contrast to the situation in *Rp. sphaeroides*, there is only a small carotenoid bandshift associated with the primary charge separation. However, there is a four-fold increase in the extent of the shift when the P^+ is re-reduced by c_{555} (electrons from the cyclic pathway), but none at all if P^+ is re-reduced by cytochrome c_{553} (electrons from substrates). With the same assumptions about the carotenoid bandshift as were used in the *Rp. sphaeroides* case, the results would imply that X was toward the outer side of the chromatophore membrane, with P a small distance toward the center of the dielectric; cytochrome c_{555} would be toward the inner aqueous interface, and cytochrome c_{553} on the outer side. The combined information is illustrated diagrammatically in Fig. 11B.

Similar interpretations of the carotenoid bandshift can also be made for *Rp. gelatinosa* (Dutton, 1971).

In this case, the carotenoid bandshift enhancement, over that elicited by P^+X^- alone, was observed with either cytochrome c_{548} or c_{553} acting as electron donors to P^+. In this case, the experiments were performed at 77°K, which explains the short wavelengths of the α-band maxima. Interestingly, both cytochromes can be photooxidized at 77°K. The enhancement was 1.5-fold for cytochrome c_{548} and twofold for cytochrome c_{553}. (This enhancement is a low value, since only 50% of the cytochrome c_{553} population is capable of reducing P^+ at 80°K; see Section 3.) With the assumption that X is on the outside, P would be toward the middle, cytochrome c_{548} next, and cytochrome c_{553} nearer still to the inner side, as is shown diagrammatically in Fig. 11C.

Whether the many assumptions made in the various approaches for localizing membrane components in the membrane are valid, only future experiments will reveal. At present, the importance of the interpretations lies in the fact that the unambiguous location of a component represents a formidable task, and as many different experimental approaches should be used as possible. Individually, the results should perhaps be taken with a grain of salt, but together they do provide a corroborative picture that may serve as a useful working guide in this complicated and important area of research.

3. Oxidation and Reduction of c-Type Cytochromes

3.1. Some Historical and Introductory Notes

The first *in vivo* spectroscopic observations of a cytochrome c oxidation elicited by light in photosynthetic bacteria were made by Vernon and Kamen (1953) with chromatophores of *Rs. rubrum*, and by Duysens (1954) and Chance and Smith (1955) with *Rs. rubrum* cells. Since that time, the reaction has been demonstrated in every known photosynthetic bacterium and has occupied the time of researchers in many laboratories. It was accepted at an early point, although not proved, that a close linkage was likely to exist between cytochromes of the c type and the light-excited Bchl. Support for the notion was provided in *Chr. vinosum* when the reaction was demonstrated to occur rapidly and efficiently at liquid nitrogen temperatures (Chance and Nishimura, 1960). Further support came from quantum yield values of near unity for light-induced oxidation of cytochrome c_2 in *Rs. rubrum* (Loach and Sekura, 1968) and cytochromes c_{553} and c_{555} in *Chr. vinosum* (Parson, 1968; Parson and Case, 1970; Seibert and DeVault, 1970). The

most compelling evidence for the direct kinetic coupling of cytochrome c and RC P came, however, from work done with *Chr. vinosum* by Parson (1969). This elegant spectrophotometric study produced the exact matching of the time course of cytochrome c_{555} oxidation with the re-reduction of laser-generated P^+. Further use of the technique led to the demonstration that both cytochromes c_{555} and c_{553} shared the same RC (Parson and Case, 1970; Seibert and DeVault, 1970) and resolved the controversy of multiple light reactions in this organism (cf. Morita, 1968; Cusanovich *et al.*, 1968). It was subsequently shown that there were in fact two equivalent cytochrome c_{555} hemes (Parson, 1969; Case and Parson, 1971), and two similarly equivalent cytochrome c_{553} hemes (Dutton, 1971; Tiede *et al.*, 1976), per RC.

The kinetics of light-induced cytochrome c_2 oxidation in *Rp. sphaeroides*, *Rs. rubrum*, and *Rp. capsulata* have proved more difficult to study (Parson, 1968; Loach and Sekura, 1968; Dutton and Jackson, 1972; Jackson and Dutton, 1973; Evans, E. H., and Crofts, 1974*b*), and it is only recently that experimentally observed multiple phases of cytochrome c_2 oxidation and RC P^+ re-reduction have been satisfactorily matched and explained (Dutton *et al.*, 1975). As was found with the analogous high-potential cytochrome c_{555} of *Chr. vinosum*, two kinetically and thermodynamically equivalent cytochrome c_2 molecules are bound to a single RC (Jackson and Dutton, 1973; Dutton *et al.*, 1975). However, the damage-prone properties of the electrostatic cytochrome c_2/RC binding of *Rp. sphaeroides*, *Rs. rubrum*, and *Rp. capsulata*, while contributing to the early problems, have aided the easy isolation of cytochrome-c_2-free RCs (Reed and Clayton, 1968; Clayton and Wang, 1971; Feher, 1971; Wang and Clayton, 1973; Prince and Crofts, 1973), and the isolated RC from *Rp. sphaeroides* has provided a valuable vehicle for gaining an insight into the nature of the interaction of cytochromes c and c_2 with the RC in model systems (Ke *et al.*, 1970; Prince *et al.*, 1974*a*) (see also Section 3.3).

In this section, we shall discuss the thermodynamic and kinetic aspects of electron transfer from cytochromes c to the RC; we shall also discuss what can be deduced regarding the structural factors that may govern this reaction.

3.2. Low-Temperature Cytochrome c Oxidation by the Reaction Center

It has long been known that the primary photochemical electron transfer in the RC proceeds rapidly (<1 nsec), unimpaired, and reversibly at liquid nitrogen temperatures and below. High activation energies prevent the electron going beyond X into the cycle, and X^- re-reduces P^+ in the dark in a temperature-insensitive reaction ($t_{1/2}$ 20 msec, although this time apparently varies ± 10 msec in different species):

$$P^+ \xrightarrow[< 1\,\text{nsec}]{hv} X^- \qquad (1)$$

$$P \xleftarrow[\sim 20\,\text{msec}]{\text{dark}} X \qquad (2)$$

In many cases, ferrocytochrome c is able to donate an electron to the RC P^+, and if this reaction is irreversible, an electron is stranded on X^-:

$$\text{Ferrocyt. } c \qquad P^+ \xrightarrow[< 1\,\text{nsec}]{hv} X^- \qquad (3)$$

$$\text{Ferricyt. } c \xrightarrow[\mu-\text{msec}]{} P \qquad X^- \qquad (4)$$

The initial discovery of Chance and Nishimura (1960) showed that *Chr. vinosum* was capable of supporting the oxidation of a low-potential cytochrome c (c_{553}) at 80°K. At the time, however, the effect did not seem to be general, since the same was not true for high-potential cytochrome c_{555} in *Chr. vinosum* or for cytochrome c_2 in *Rp. sphaeroides* or *Rs. rubrum* (Vredenberg and Duysens, 1964). The later and monumental efforts of Kihara and Chance (1969) confirmed that *Rp. sphaeroides* (two strains) and *Rs. rubrum* (17 strains!), as well as *Rp. capsulata*, were incapable of supporting c-type cytochrome photo-oxidation at liquid nitrogen temperatures. These three species, however, were more the exception than the rule; every other species they examined showed efficient and rapid low-temperature capabilities at cryogenic temperatures. These species were *Chr. vinosum* (two different strains), *Chr. violascens*, and *Thiocapsa roseopersicina* (then known as *T. floridana*) of the Chromatiaceae family, and *Rp. gelatinosa*, *Rp. viridis*, *Rhodomicrobium vannielii*, *Rm.* sp. CK, and *Rp.* sp. NW of the Rhodospirillaceae family. Thus, the ability to support low-temperature oxidation had correlation neither with taxonomic type nor with the bacterium's obligatory or facultative anaerobic character. If, however, we examine the environmental conditions required to see low-temperature oxidation in the aforenamed species (Kihara *et al.*, 1969; Kihara and Dutton, 1970; Dutton *et al.*, 1970, 1971; Dutton, 1971), a usefully simplifying pattern is revealed. It seems apparent (with the exceptions of *Rp. sphaeroides*, *Rp. capsulata*, and *Rs. rubrum*, and also *Chl. limicola f. thiosulfato-philum*; see Sybesma and Vredenberg, 1964) that the establishment of anaerobic conditions (ambient reduc-

ing conditions) before freezing the sample of cells or chromatophores invariably allows an *irreversible* light-induced cytochrome c-oxidation. Mild aeration of the sample sufficient to preoxidize (probably indirectly) the low-potential cytochrome c before freezing abolishes the irreversible low-temperature reaction, and although in several species cytochrome c oxidation is still evident, it is now reversible. The findings cited above are consistent with the following tentative conclusions: (1) The low-potential cytochrome c is generally capable of rapid *irreversible* electron transfer to the light-generated P^+ at low temperatures. In this regard, the absence of any low-temperature cytochrome oxidation in *Rp. sphaeroides*, *Rp. capsulata*, *Rs. rubrum*, and *Chl. limicola f. thiosulfatophilum* is in line with these species not possessing the low-potential cytochrome. (2) In some cases, the high-potential cytochrome c appears able to perform 80°K electron transfer to P^+, and where evident, the reaction is partly reversible. Further details supporting and amplifying these identifications are presented below.

3.2.1. Low-Potential Cytochrome c Oxidation

Quantitative redox potential studies using controlled redox potential poising at room temperature, with examination of light-induced reactions at 80°K, have been done on *Chr. vinosum* and *Rp. gelatinosa* (Dutton *et al.*, 1971; Dutton, 1971). The results confirm that the low-potential cytochromes undergo rapid irreversible oxidation at 80°K, providing the expected E_m (pH 7.4) value of $+10$ mV for cytochrome c_{553} of *Chr. vinosum* and $+128$ mV for cytochrome c_{548} of *Rp. gelatinosa* (the low-temperature reduced-minus-oxidized α-band was at 548 nm; the γ-band was at 410 nm). In both cases, the reaction was measured to be rapid ($t_{1/2}$ of 2.5 msec and 10 μsec, respectively, although a more recent averaged number for what is likely to be the latter reaction is 4.6 μsec; Kihara and McCray, 1973) and complete, all the P^+ generated in the light being reduced by the associated cytochrome. It would appear that these low-potential, presumed substrate-linked, cytochromes are so arranged that they are capable of electron transfer down to very low temperatures without any impairment or dissociation of the reactants as the temperature is lowered. In fact, in the case of cytochrome c_{553} of *Chr. vinosum*, there is reason to believe that either of two cytochrome c_{553} hemes is (at least down to 77°K) equally capable of delivering an electron to the single oxidizing equivalent of the RC P^+ (Dutton, 1971).

The activation energy for the oxidation of cytochrome c_{553} of *Chr. vinosum* is 3.5 kcal/mol from 300°K ($t_{1/2}$ of 1 μsec) to 120°K ($t_{1/2}$ of 2.5 msec), but below 120°K, the reaction becomes essentially independent of temperature (DeVault and Chance, 1966). This is true of whole cells, chromatophores (DeVault and Chance, 1966), and detergent-treated subchromatophore particles (Dutton *et al.*, 1971), and has been the basis for electron tunneling considerations between the cytochrome and the RC P$^+$ (DeVault and Chance, 1966; Hopfield, 1974). Similar results were presented for cells of *Rp.* sp. NW and *Rp. gelatinosa* (Kihara and McCray, 1973). Activation energies of 2.1 and 2.3 kcal/mol, respectively, were measured from 300 to 150°K, and approximately 0 kcal/mol below 150°K. The reaction half-times at 80°K were some thousandfold faster than those of *Chr. vinosum*: approximately 0.7 μsec at 300°K and 4.6 μsec at 80°K for *Rp. gelatinosa*, and 0.55 μsec at 300°K and 4.2 μsec at 80°K for *Rp.* sp. NW. Kihara and Chance (1969) discussed how such large differences in rate may be caused by quite small variations in distance or tunneling barrier height between cytochrome *c* and RC P in different bacterial species. It is worth mentioning here that Kihara and McCray (1973) suggested that their measurements on cytochrome photooxidations in *Rp. gelatinosa* and *Rp.* sp. NW involved the high-potential cytochrome *c*, since they used aerobic cells. There are, however, two reasons to believe that in fact they were observing the oxidation of the *low-potential* cytochrome. The first is that Kihara and Chance (1969) reported that aeration of *Rp.* sp. NW abolished low-temperature cytochrome oxidation; i.e., the high-potential cytochrome cannot donate to P$^+$ at cryogenic temperatures. The second is that the half-times of photooxidation reported by Kihara and McCray (1973) are similar to those obtained for the respective low-potential cytochrome oxidations (Kihara and Chance, 1969; Kihara and Dutton, 1970; Dutton, 1971).

It is beyond the scope of this chapter to go into the interpretations of the biphasic Arrhenius plot of the temperature dependency of low-potential cytochrome *c* photooxidation (DeVault and Chance, 1966; Hopfield, 1974), but it is of interest to note that the recent work of Parson *et al.* (1975) found similar biphasic Arrhenius plots for the relaxation of flash-induced intermediary states within the RC Bchl and bacteriopheophytin components.

Kihara and McCray (1973) thoroughly studied the effects of water in the cytochrome *c* photooxidation reaction, showing that lyophilization abolished cytochrome oxidation without affecting the light reaction itself. They also substituted D$_2$O for H$_2$O and showed a $\sqrt{2}$ isotope effect on the rate of cytochrome oxidation in both *Rp. gelatinosa* and *Rp.* sp. NW. This

effect was seen down to 80°K, with no effect on the activation energies. The same factor of approximate $\sqrt{2}$ slowing was observed at room temperature in the re-reduction of the cytochrome, and they also demonstrated a similar isotope effect in mitochondrial *c* and *b* redox reactions. They concluded from electron tunneling considerations in the temperature-insensitive range of cytochrome *c* oxidation that the mass of the hydrogen atom was directly involved in the rate-limiting step, and suggested that water played a direct mechanistic role in electron-transport reactions.

The irreversibility observed in most cases for the cytochrome oxidation reaction is worth considering on a simple chemical equilibrium basis. In *Chr. vinosum*, for example, the midpoint potential difference between cytochrome c_{553} and P is large (i.e., ~480 mV; see Fig. 6), and the reaction from P to c_{553} would be highly unfavored. If we assume that following illumination [equations (3) and (4)], the *overall* back reaction

$$c_{553}^+ \leftarrow P \leftarrow X^- \qquad (5)$$

is made strongly favorable by the larger potential difference between X and P (~650 mV) in the *potentially* rapid X → P electron-transfer reaction ($t_{1/2}$ 20 msec), then the cytochrome c_{553} re-reduction is feasible in the dark. However, inspection of the partial reaction

$$c_{553} + P^+ \xrightarrow[k_{-1}]{k_1} c_{553}^+ + P \qquad (6)$$

using the expression*

$$E_{m(P)} - E_{m(c)} = (RT/nF) \ln (k_1/k_{-1}) \qquad (7)$$

reveals at room temperature, for a k_1 of 0.7 × 10^6 sec^{-1} ($t_{1/2}$ ~1 μsec), a k_{-1} rate of 0.7 × 10^{-2} sec^{-1} ($t_{1/2}$ 100 sec); this, of course, will not be evident physiologically, since X is rapidly oxidized by secondary acceptors and cytochrome c_{553} is re-reduced (10 sec^{-1}) by substrate; these components would have to be isolated in a detergent preparation to test this back-reaction sequence. However, at 80°K, using a forward-rate constant k_1 of 3 × 10^2 sec^{-1} (DeVault and Chance, 1966) ($t_{1/2}$ 2.5 msec), the k_{-1} back rate is 3 × 10^{-28} sec^{-1} ($t_{1/2}$ 2 × 10^{27} sec; cf. the age of the Universe, of approximately 1.4 × 10^{17} sec), and clearly the back-reaction will not be observed! Similar considerations with *Rp. gelatinosa*, in which the E_m difference between P and cytochrome c_{548} is less (270 mV; see Fig. 5) and the forward rate k_1 ($t_{1/2}$ at 80°K ~5 μsec) is faster, still put the back-reaction out of experimental

---

* See the Fig. 1 caption for a definition of the term (RT/nF).

reach at this temperature. These considerations do not include the possibility that the room-temperature ΔE_m values may not be applicable to prevailing electron affinities at low temperatures. Such alterations could convert a reaction that is thermodynamically favorable at low temperatures, and vice versa. For a full under-temperatures, and vice versa. For a full understanding, thermodynamic alterations of the components following light activation also need to be considered. In regard to the former point, the most comprehensive studies on the temperature dependence of E_m values have been on cytochrome c_{555}, P, and X of *Chr. vinosum* (Case and Parson, 1971); from 40 to 0°C, the values changed linearly by approximately −1.6, −1.2, and −2.6 mV/°C, respectively. Unfortunately, the X couple measured was the X/XH couple, not the functional X/X⁻ couple (Prince and Dutton, 1976a; Chapter 24), and cytochrome c_{553} was not measured. For illustrative purposes, however, if a similar differential exists between cytochrome c_{553} and X as for cytochrome c_{555} and X, if the temperature profile of X/X⁻ is similar to X/XH, and if the linearity is maintained to lower temperatures, then X would be significantly more positive than cytochrome c_{553} at 80°K. This being so, an electron on X, once driven there by light, would not even try to go back to an oxidized cytochrome c_{553} by any means.

Of special interest to these considerations is the partial reversibility (reduction) of high-potential cytochrome c oxidation in some photosynthetic bacteria at low temperatures. In *Rp. gelatinosa* and *Rp. palustris* at 80°K, the re-reduction half-time is 2 or 3 sec (Kihara and Chance, 1969; Dutton *et al.*, 1970; Dutton, 1971). Simple equilibrium considerations as discussed above would predict that the 80°K back-reaction rate k_{-1} driven by the electron on X⁻ (the X-to-P electron transfer being intrinsically 12-msec $t_{1/2}$) from P to the high-potential ferricytochrome c_{553} of *Rp. gelatinosa* ($k_1 \approx 10^4$ sec⁻¹; $t_{1/2} \approx 60$ μsec; midpoint potential difference at room temperature = 120 mV) would be over an hour, and would be quite temperature-sensitive. In fact, the activation energy between 295 and 80°K was reported to be 0.1–0.5 kcal/mol. Does this indicate a direct electron transfer from X⁻ to the ferricytochrome c_{553} ("tunneling"), or have the relative electron affinities of cytochrome c_{553} and P (i.e., the midpoint potential difference) come closer together, making the overall back-reaction feasible on a seconds time range? Perhaps the partial reversibility is an expression of how close they are? Further effects of temperature on the forward reaction of the high-potential cytochrome c oxidation in photosynthetic bacteria are dealt with in the next section.

3.2.2. High-Potential Cytochrome c Oxidation

In those species that do exhibit a significant high-potential cytochrome c oxidation at 80°K [notably *Rp. palustris* (Kihara and Chance, 1969) and *Rp. gelatinosa* (Dutton *et al.*, 1970; Dutton, 1971)], the reaction is rapid; e.g., in *Rp. gelatinosa*, cytochrome c_{553} oxidation has a half-time of approximately 60 μsec at 80°K. In the quantitative measurements (Dutton, 1971) done on the high-potential cytochrome c_{553} ($E_m = +280$ mV; pH 7.4), however, it is clear that a large portion (about half) of the cytochrome c–RC units have become incapable of detectably performing the reaction at 80°K, even with prolonged steady-state illumination. At higher temperatures, it is apparent that more *Rp. gelatinosa* cytochrome c_{553} is oxidized on illumination (approximately 1.5 times as much at 200 as at 80°K; Dutton, 1971).

A more numerical indication of the apparent loss of competence as a function of temperature is exemplified by flash-activated cytochrome c_{555} oxidation in *Chr. vinosum*. This cytochrome oxidation has an activation energy of approximately 2.5 kcal/mol at temperatures above 270°K, and the half-time for the oxidation is approximately 2 μsec at 300°K. If the activation energy were the only temperature-dependent effect, the cytochrome c_{555} oxidation would be expected to be readily visible at 80°K in the millisecond time range. As the temperature is lowered, however, apart from the expected minor rate effect, the extent of the reaction decreases dramatically (Seibert, 1971) until it has apparently disappeared between 210 and 170°K [by a linear extrapolation of Seibert's single-turnover data; in fact, some 5% of the initial amount is still observable with steady-state illumination at 80°K (Dutton, 1971)]. In this case, and in that of cytochrome c_2 in *Rp. sphaeroides*, *Rp. capsulata*, and *Rs. rubrum* and cytochrome c_{553} in *Chl. limicola f. thiosulfatophilum*, a total oxidative incompetence has been reported at 240°K (Vredenberg and Duysens, 1964; Sybesma and Vredenberg, 1964). There are several possibilities that could explain this marked effect of temperature. In the electrostatically linked cases, the trivial explanation that low temperature leads to the observed dissociation as a result of the build-up of local high salt concentrations might seem applicable. However, such a problem ought not to affect *Chr. vinosum* and the many other species in which the high-potential cytochromes are apparently hydrophobically bonded to the RC. An alternative explanation might be that lowering the temperature changes the thermodynamic properties of the reactants so that the E_m of the cytochrome becomes more positive than that of the RC P, but extrapolation of the data of Case and Parson (1971) suggests that this will not happen unless a radical change in the temperature differential occurs at lower temperatures, at least in *Chr. vinosum*.

A third proposal, which is more attractive experimentally for reasons that will become apparent in the next section, is that the physical relationship of the RC to the high-potential cytochromes *c* is significantly different, for functional reasons, from its relationship to the low-potential cytochromes. In contrast to the latter case, in which the cytochromes always seem to be in close physical association with the RC, it is possible that the high-potential cytochromes exist in more than one state with respect to the RC; one might be considered to be a functionally "close" configuration, while another would be a "distant" location, with respect to the RC. We imagine that the proposed "close–distant" reaction represents a two-state equilibrium in which (1) only the "close" configuration is competent to transfer an electron to the RC P$^+$ and (2) with decreasing temperature, the incompetent "distant" configuration becomes increasingly favored. The tentative recognition of this two-state equilibrium in the high-potential cytochromes *c* may have relevance to two aspects of their function. The first relates to their position in the cyclic electron-transfer system involved in the rapid transfer of electrons from the membrane-bound "reductase" to the RC. If electron entry and exit into the cytochrome *c* are strongly favored at one part or in one orientation of the molecule (for discussions, see Chance and Williams, 1956; Margoliash *et al.*, 1970; Davis *et al.*, 1972; Smith *et al.*, 1973; Nicholls, 1974), e.g., through the heme cleft, then the cytochrome may be forced to undergo conformational alterations or physical movement with respect to its reductase and RC, and produce a measurable "close–distant" relationship with each. The second relates to the low-potential-cytochrome-*c*-containing species, in which the oxidation of the low-potential cytochrome dominates over the oxidation of the high-potential cytochrome should they both be reduced at the time of activation. It appears clear in the case of *Chr. vinosum* that the dominance goes further than a probability based on the faster rate (×2) of oxidation of the low-potential cytochrome (Case *et al.*, 1970; Seibert and DeVault, 1970), and it is not unreasonable to consider that the reduced state of the low-potential cytochrome may induce the high-potential cytochrome into its RC "distant" configuration. CD studies on the isolated cytochromes c_{555} and c_{553} indicate some interaction between the cytochromes and their states of reduction (Kennel and Kamen, 1971). Further evidence for the

two-state reductase–oxidase dynamic equilibrium of the high-potential cyclic cytochromes *c* is presented in the next section.

3.3. Molecular Aspects of Cytochrome *c* Oxidation and Reduction at Physiological Temperatures

In *Rp. sphaeroides*, the kinetics of cytochrome c_2 oxidation following a single-turnover flash is biphasic (Dutton and Jackson, 1972; Dutton *et al.*, 1975), and the same kinetics are seen for each of the two cytochrome hemes associated with the RC (Dutton *et al.*, (1975). Figure 12 shows the oxidation kinetics of the individually measured cytochromes (Dutton *et al.*, 1975): there is a fast phase ($t_{1/2} \approx 30$ μsec), and a second variable slow phase can be controlled at the preparative stage of the chromatophores; it depends on the ionic strength of the medium in which the cells are broken (low ionic strength media yield a slower second phase), but is not alterable experimentally by ionic strength or pH after this step. Breakage in buffered 100 mM KCl usually provides a second slow phase in the 300-μsec half-time range. The biphasic kinetics and the two cytochromes c_2 per RC protein are also apparent in the whole cell, in which the slow-phase half-time is about 1–2 msec, indicating that the biphasic kinetics measured in the chromatophores are a physiological situation. This information has been interpreted (Dutton, 1974; Dutton *et al.*, 1975) as indicating that each cytochrome c_2 operates in a dynamic two-stage equilibrium with a binding site on or near the RC protein. Thus, at the instant of the single-turnover generation of P$^+$, those cytochromes of the population in the "close" position undergo the 30-μsec half-time oxidation; those found in the "distant" configuration are not oxidized until the dynamics of equilibrium brings them to the "close" position, where

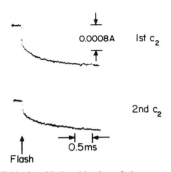

Fig. 12. Individual oxidation kinetics of the two cytochrome c_2 molecules per RC in *Rhodopseudomonas sphaeroides*. The experimental procedures used to obtain these data are described in Dutton *et al.* (1975).

they are oxidized, reducing the waiting P^+. The low-ionic-strength-induced slowing of the slow phase may then come from hindrance by other proteins retained inside the chromatophore during breakage, which slow the "distant"–"close" transitions.

A hint that the proposed "close"–"distant" transitions are not an "on"–"completely off" relationship comes from dark redox titrations with *Rp. sphaeroides* chromatophores (Dutton *et al.*, 1975). As mentioned in Section 3.1, the binding of cytochrome c_2 to the RC in the chromatophore is very damage-prone; in the membrane-associated state, the E_m of cytochrome c_2 is 295 mV, and in the isolated "free" state, it is 345 mV. Mechanical breakage of the cells normally produces chromatophores with 20–30% of the RCs devoid of cytochrome c_2, although, depending on the method of breakage, this value can be increased to over 50%; the remaining RCs, however, have the full functional complement of two cytochrome c_2 molecules. Dark redox titrations done simply by measuring the absorbance change at 550–540 nm as a function of E_h reveal two electrochemical components of cytochrome c_2, one with a midpoint of +340 mV similar to that for the free cytochrome, in the proportions of the functional to damaged cytochrome c_2–RC units (Dutton *et al.*, 1975). Such a resolved redox titration implies that the free cytochrome is not in rapid equilibrium with the membrane-bound form. This in turn indicates that the two kinetic phases of oxidation of the functional c_2 (E_m + 295 mV) do not describe a "close"–"completely off" exchange situation, but represent perhaps two configurations of a bound state.

Further support for the existence of a clearly defined two-state equilibrium comes from studies with model systems of isolated RCs incorporated into phospholipid vesicles with externally added cytochrome c_2 (Dutton *et al.*, 1976). One such experiment is shown in Fig. 13; the experimental details are in the caption. This experiment, although preliminary, clearly reveals a two-state reaction of the cytochrome c_2 with the RC. Again, the fast phase of the oxidation of cytochrome c_2 can be interpreted as representing cytochrome c_2 "close" or bound to the RC protein at the time of activation; the half-time of this phase (≈ 25 μsec) is apparently independent of cytochrome concentration, which is appropriate for the oxidation occurring rapidly in a complex. The half-time of the fast oxidation of cytochrome c_2 is similar to that obtained by Ke *et al.* (1970) for the oxidation of mammalian cytochrome c by RCs. Scatchard-type plots, taking the fast phase of oxidation as being a measure of how many cytochromes are bound in the close position per added cytochrome c_2, give an estimated apparent dissociation constant (K_D) of about 4×10^{-6} M. Further detailed work (not shown) on this system studying the dependence of the rate of the slow phase on cytochrome c_2 and RC concentrations and ratios has revealed the following information. With cytochrome c_2/RC ratios of less than unity, the rate of cytochrome c_2 oxidation (millisecond) also seems independent of reactant concentration. Scatchard-type plots, taking the amount of cytochrome c_2 that undergoes oxidation with this millisecond half-time as indicating binding in a distant configuration, have provided us with the apparent K_D of the "distant" cytochrome c_2–RC interaction; this is much tighter than the "close" configuration, and falls in the range 6–60×10^{-9} M. At cytochrome c_2/RC ratios greater than unity, this slow phase of oxidation sharply becomes dependent on the cytochrome c_2 concentration, as can be seen in Fig. 13, indicative perhaps of multiple "distant" cytochrome c_2–RC configurations. It is at ratios of 1 and above that the fast, "close" phase becomes clearly apparent, consistent with the increased probability that at least one of the c_2 complements of each RC is "close" at the time of flash activation. In summary, it appears that there is a fairly tight binding of cytochrome c_2 to the RC, but in a configuration ("distant") that does not permit microsecond electron transfer. The numbers described above are sensitive to, among other things, the presence of residual detergents, and there is much more work to be done.

The demonstration of the model interaction is of interest, however, apart from roughly mimicking the *in vivo* oxidation kinetics; the net charge at pH 8.0 of the RC (Ke *et al.*, 1970; Prince *et al.*, 1974a) and cytochrome c_2 (Bartsch, 1971) is negative, so general nonspecific electrostatic interactions would not be expected. Thus, it is possible that a specific positive region or regions (e.g., lysine residues at the cytochrome c_2 heme edge, Salemme *et al.*, 1973) on either or both of the molecules provide the electrostatic attraction for the binding.

The transition between the "close" and "distant" cytochrome–RC states could be a physical change of binding sites or a conformational change at one binding site. If the model has a general applicability, then those hydrophobically linked high-potential cytochromes that are not readily dissociable from the RC probably fall into the latter category, as was visualized for cytochrome c_{555} of *Chr. vinosum* by Chance and Williams (1956) and Chance *et al.* (1967) several years ago.

As yet, we cannot be certain of the physiological significance of the RC "distant" binding; there are several possibilities. As indicated above, however, it does not seem to be physically off the RC altogether and free in solution, either *in vivo* or in the model system. An attractive possibility is that *in vivo* there may be a RC binding site that holds the cytochrome in a position functionally close to the reductase. The slow phase of cytochrome c_2 oxidation, which we have already discussed as a transition from functionally "distant" to functionally "close" to the RC P, would thus also represent a transition from functionally "close" to functionally "distant" from the reductase (R), as shown below. The reader is also referred to Section 7 for further developments of the model.

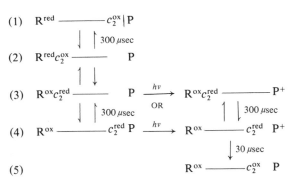

(1) $R^{red} \underline{\hspace{3cm}} c_2^{ox} | P$

(2) $R^{red} c_2^{ox} \underline{\hspace{3cm}} P$ $\Big\uparrow\Big\downarrow\ 300\,\mu sec$

(3) $R^{ox} c_2^{red} \underline{\hspace{2cm}} P \xrightarrow{h\nu} R^{ox} c_2^{red} \underline{\hspace{2cm}} P^+$

OR

(4) $R^{ox} \underline{\hspace{3cm}} c_2^{red}\,P \xrightarrow{h\nu} R^{ox} \underline{\hspace{2cm}} c_2^{red}\ P^+$

(5) $R^{ox} \underline{\hspace{3cm}} c_2^{ox}\ \ P$

Biphasic kinetics are also apparent in the oxidation of the cytochrome c_2 complements of *Rp. capsulata* (Evans, E. H., and Crofts, 1974*b*) and *Rs. rubrum* (Prince, R. C., Baltscheffsky, M., and Dutton, P. L., unpublished) and of the cytochrome c of *Chl. limicola f. thiosulfatophilum* (Prince and Olson, 1976). All have a fast phase, representing about 50% of the total change, and a readily discernible slow phase; the rates in *Rp. capsulata* and *Rs. rubrum* are very similar to those in *Rp. sphaeroides*, while in *Chl. limicola f. thiosulfatophilum* the two phases are significantly faster, with half-times of less than 5 μsec and approximately 50 μsec. All three species also have two molecules of cytochrome (which appear to be kinetically and thermodynamically equivalent) associated with each RC (Prince and Dutton, 1977*a*; Prince and Olson, 1976), so the *in vivo* equilibrium constant for the reductase "close" cytochrome to the RC "close" form would have to be about 0.25 to account for the approximately equal contributions of the fast and slow phases of cytochrome oxidation (Dutton *et al.*, 1975; Prince and Olson, 1976) observed when both cytochromes were reduced before the flash.

The cytochromes c_2 of *Rs. rubrum* and *Rp. sphaeroides* have *in vivo* midpoint potentials some 50 mV more negative than the free *in vivo* forms (Dutton and Jackson, 1972; Prince *et al.*, 1975*a*). One explanation of these shifts would be that they reflect an

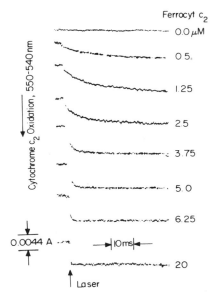

Ferrocyt c_2

0.0 μM
0.5
1.25
2.5
3.75
5.0
6.25
20

Cytochrome c_2 Oxidation, 550–540 nm

0.0044 A |10ms|

↑ Laser

Fig. 13. *In vitro* reconstitution of the biphasic cytochrome c_2 oxidation kinetics observed *in vivo*. RCs, prepared by the method of Clayton and Wang (1971), were incorporated into phospholipid vesicles by a modification of the cholate dialysis method of Kagawa and Racker (1971). The RCs are treated with dithionite to reduce the LDAO detergent in the presence of cholate. The reduction product is not a strong detergent, and the RCs are left suspended in the cholate solution. The phospholipid vesicles are then added (molar excess of lecithin/RCs of 100:1), and the suspension is dialyzed overnight against 10 mM Tris-Cl (pH 8.0). UQ-10 was present in the vesicles (lecithin/UQ = 10:1). Cytochrome c_2 from *Rp. sphaeroides* was added in the amounts shown. The RC concentration shown was 0.7 μM, but some 40% of these did not react at all with external cytochrome, and presumably oriented toward the inside of the vesicles. Activation was provided by a 20-nsec ruby laser pulse.

eightfold preferential binding of the ferricytochrome c_2 over the reduced form to the membranes, but since the cytochromes c_2 of *Rp. capsulata* have kinetics of oxidation similar to those of *Rp. sphaeroides* and *Rs. rubrum*, but do not show the midpoint potential shift (Evans, E. H., and Crofts, 1974*a,b*), the shift in midpoint potential is not a necessary prerequisite for the distant binding site. It should be emphasized, however, that the absence of a shift implies only that the oxidized and reduced forms are equally bound to the membrane, and does not preclude such a binding.

4. Cytochrome *b* Electron- and Proton-Transfer Reactions

Rhodopseudomonas sphaeroides, and to a lesser extent *Rp. capsulata* and *Rs. rubrum*, are the only bacteria in which extensive thermodynamic and flash-induced kinetic work has been done on the cytochromes *b*. The thermodynamic requirement for coupled protonation–deprotonation in the reduction and oxidation of the cytochromes b was described in Section 2.4, and simplistic considerations of their role in electron transport were presented in Section 2.2. In comparison with the cytochrome c_2–RC interaction, however, little is known regarding the electron donors and acceptors that interact with the cytochromes b. Complications have arisen at the experimental level because of their multiplicity, and because they are several steps away from the RC.

There is a history, particularly in mitochondrial and plant photosynthesis studies, of assigning to at least one of the cytochromes b a direct chemical role in energy coupling, in which electron transfer driven through the cytochrome b by the approximately 300-mV redox potential drop to cytochrome c or f generates a "high-energy" form of the cytochrome b that can directly drive the phosphorylation of ADP (for reviews, see Chance, 1972; Dutton and Wilson, 1974). In contrast, the "chemiosmotic" views of Mitchell (1961, 1972, 1975a,b) led to an emphasis being placed on the physical location of the cytochromes b in the membrane, and their involvement with transmembrane electron and proton translocation; in this capacity, they would be sensitive to the energetic status of events across the membrane as a whole. This section describes current developments in studies on the problems of cytochromes b in electron and proton transfer and energy coupling, principally considering the electron-transport system of *Rp. sphaeroides*.

4.1. Ubiquinone–Cytochrome b_{50} Reactions

4.1.1. Proposed Location in Cyclic Electron Transport

Figure 14 shows a simplified step-by-step scheme of the events following a single-turnover flash that lead to the reduction of cytochrome b_{50} (Figs. 14A,B,D,E) and those that lead to the oxidation of one of the cytochrome c_2 hemes (Figs. 14A–C). Experiments representative of Fig. 14 are shown in Fig. 15; all were carried out in the presence of antimycin to prevent electrons from rapidly leaving cytochrome b. Figure 15A shows the binding of 1.0 H^+/e^- ($t_{1/2}$ 150 μsec) to the outside of the chromatophore, a binding that has been suggested to be coupled to the reduction of UQ to the protonated semiquinone (Petty and Dutton. 1976a):

$$UQ + e^- + H^+ \rightleftharpoons UQ\cdot H$$

Figure 15B shows the subsequent reduction of cytochrome b_{50} in approximately 1.5 msec. and Fig.

15C shows that the reduction of the cytochrome b is essentially complete after the first turnover, although two turnovers are required to oxidize the cytochrome c_2 fully (Fig. 15D). Even though the cytochrome b_{50} is essentially fully reduced by the first flash, however, subsequent flashes continue to cause proton uptake stoichiometric with the number of electrons delivered through the primary acceptor to the UQ complement (Petty and Dutton, 1976a), indicating that although there may be only one cytochrome b_{50} heme per cycle, there is multiple electron and proton capacity in the UQ complement before the cytochrome b.

4.1.2. Cytochrome b_{50} and the Fate of the Proton Bound by Ubiquinone

The thermodynamic evidence presented in Fig. 9 suggests that ferrocytochrome b_{50} has a pK at pH 7.4 (see Section 2.3), implying that the reduction of the cytochrome at pH much less than 7.4 will involve a proton, while reduction at pH much greater than 7.4 will not. It has also been established that the agent responsible for the rapid proton binding (ubisemiquinone?) discussed in Section 4.1.1 has an apparent pK at pH 8.5 (Petty and Dutton, 1976a). Thus, on a kinetic timescale following a single-turnover flash, we might expect the following ubisemiquinone–cytochrome b_{50} electron and proton transfers:

(1) At high pH values (pH 9.0):

$$UQ^{\bar{\cdot}} + \text{ferri } b_{50} \rightleftharpoons UQ + \text{ferro } b_{50}^{(-)}$$

(2) Between pH 8.0 and 8.5:

$$UQ\cdot H + \text{ferri } b_{50} \rightleftharpoons UQ + \text{ferro } b_{50}^{(-)} + H^+$$

(3) At pH 6.5 and below:

$$UQ\cdot H + \text{ferri } b_{50} \rightleftharpoons UQ + \text{ferro } b_{50}H$$

Under the pH conditions of case (2), the proton is released from the redox system before, or simultaneously with, ferricytochrome b_{50} reduction, while in case (3), the proton would not be released from the redox system until some point after the reoxidation of the ferrocytochrome b_{50}. In coupled chromatophores, under either condition, the proton does not reappear in the external aqueous phase for hundreds of milliseconds, but uncoupling the membrane (e.g., with >1 μM FCCP) stimulates this proton efflux into the 1- or 2-msec time range. Since antimycin dramatically affects the reoxidation of cytochrome b_{50}, the release of the proton from the redox system would be expected to be antimycin-*sensitive* under the condition of case (3), but antimycin-*insensitive* in case (2). Petty and Dutton (1976b) tested this by monitoring the rates of external efflux of the proton in the presence of anti-

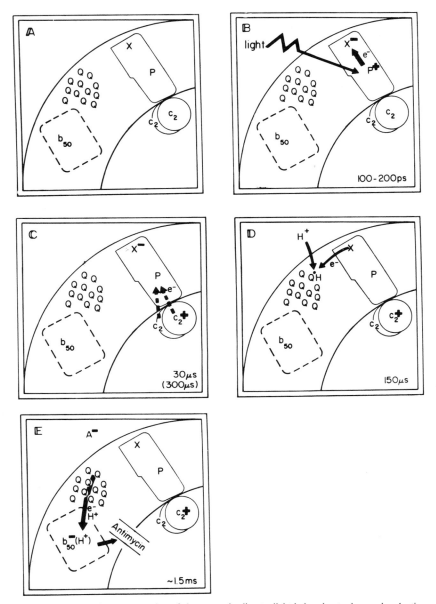

Fig. 14. Diagrammatic representation of the events leading to light-induced cytochrome b reduction, cytochrome c_2 oxidation, and proton uptake in *Rhodopseudomonas sphaeroides*. The times given are for the reactions at pH 7.0 and an E_h of $+200$ mV, so that both cytochrome c_2 molecules are reduced before activation. For the E_h and pH dependencies of the reactions, see Section 4.1.3.

mycin and uncoupler as a function of pH (Fig. 16). Uncoupler would not be expected to stimulate the rate of proton efflux from within the chromatophore membrane if the proton remains bound to the ferrocytochrome b_{50}, until the cytochrome's slow oxidation around the antimycin block as in case (3). In contrast, in case (2), even with antimycin present, uncoupler *would* be expected to stimulate the external efflux of any protons liberated within the chromatophore from the redox system *prior to* the reduction of ferricytochrome b_{50}. Thus, performing this over a range of pH values, the percentage of the uncoupler-stimulatable rapid proton efflux would follow a Henderson–Hasselbalch curve for the pK of cyto-

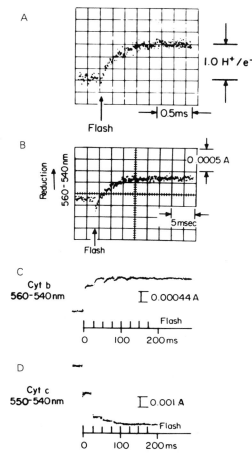

Fig. 15. Flash-induced proton uptake, cytochrome *b* reduction, and cytochrome c_2 oxidation in *Rhodopseudomonas sphaeroides*. In each experiment, the chromatophores were suspended (Bchl about 20 μM for A, C and D; 13 μM for B) in 100 mM KCl alone for A, and with 20 mM morpholinopropane sulfonate for the others, under anaerobic conditions; E_h approximately 220 mV and pH 7.0. In all cases, antimycin (2 μM) was present. In A, 50 μM phenol red was present as the pH indicator, and measurements were taken at 585 nm with a single-wavelength spectrophotometer. In B–D, measurements were made with a dual-wavelength spectrophotometer. Flash activation was with a 90% saturating Xe flash or train of such flashes spaced 24 msec apart as indicated. For further details, see Petty and Dutton (1976a,b) and Dutton *et al.* (1975).

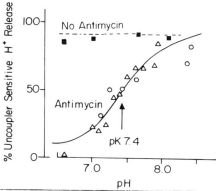

Fig. 16. Functionality of the pK on the reduced form of cytochrome b_{50} in *Rhodopseudomonas sphaeroides*. The experimental details for this experiment can be found in Petty and Dutton (1976b). The rate of proton release was monitored by the decay of the proton uptake as monitored in Fig. 15, using either phenol red ($\triangle$) or cresol red ($\circ$) as indicator dyes. Where indicated, antimycin was added to 2 μM, and 1 μM FCCP (carbonyl cyanide *p*-trifluoromethoxyphenylhydrazone) was also present.

These experiments not only demonstrate that the pK of the ferrocytochrome b_{50} is functional on a time scale of electron flow, but also reveal that the release point of the proton is sufficiently deep within the chromatophore membrane that the diffusional processes for proton efflux can be stimulated by three orders of magnitude by the addition of lipophilic proton carriers (uncouplers).

4.1.3. Brief Notes on Some Mainly Unexplained Influences on Flash-Induced Cytochrome b_{50} Reduction

In this section, we shall briefly mention some factors that influence the rate or extent of cytochrome b_{50} reduction after a single-turnover flash. In most cases, these effects are only poorly understood, but we include them in order to present as complete a picture as possible, for any final model of electron transport must presumably account for them all.

(a) Effects of the State of Reduction of the Cycle and of Ambient pH. At an ambient potential of +200 mV at pH 7 (i.e., with cytochrome c_2 reduced before the flash), the reduction of cytochrome b_{50} in *Rp. sphaeroides* in the presence of antimycin has a half-time of 1–2 msec (see Fig. 14). However, at the same pH, but at +380 mV, where cytochrome c_2 is oxidized before the flash and so is unable to reduce the flash-induced P+, the half-time is about 25 msec (Petty and Dutton, 1976b). Although other factors may be involved, it appears that this effect (see Fig. 17A) is dependent on the state of reduction of cytochrome c_2 before the flash (and hence of the reaction center P+ after the flash).

chrome b_{50}. Figure 16 shows that this is indeed the case. In the absence of antimycin, the efflux of the proton is stimulated by uncouplers at all values of pH, consistent with the normal release of the proton occurring relatively rapidly whether before or after cytochrome b_{50}. The point of proton release when the proton has gone onto ferrocytochrome b_{50} is at present unknown; the latest point in the cycle at which it could occur would be at cytochrome c_2, since this does not require a proton for its reduction (see Section 2.3).

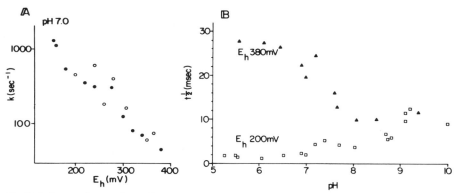

Fig. 17. Influence of externally applied conditions on cytochrome b_{50} reduction in the presence of antimycin in *Rhodopseudomonas sphaeroides*. (A) Rate constant for cytochrome b_{50} reduction, assuming this to be a pseudo-first-order reaction, plotted against ambient redox potential at pH 7. The two symbols represent two different preparations of chromatophores. (B) Plot of the half-time of cytochrome b_{50} reduction at two potentials as a function of ambient pH: (□) $E_h = +200$ mV; (▲) $E_h = +380$ mV. Antimycin, 2 μM, was present in all experiments. More experimental details are described by Petty and Dutton (1976b).

Figure 17B shows that as the pH of the chromatophores is raised through the pK of cytochrome b_{50} at an ambient potential of +200 mV, there is some slowing of cytochrome b_{50} reduction. In contrast, at a potential of +380 mV, the rate becomes faster until at pH values between pH 8 and 9, there is no difference in the rate of reduction at the two potentials (Fig. 17B). At these pH values, neither the UQ nor the cytochrome b_{50} requires a proton for its equilibrium reduction, but whether this has any relation to the rate of cytochrome b_{50} reduction is unclear.

(b) Antimycin. Antimycin is a well-known inhibitor of electron flow through cytochromes *b* (for a review, see Slater, 1973). In *Rp. sphaeroides*, it appears to inhibit the oxidation of both cytochromes b_{50} and b_{155} (Dutton and Jackson, 1972), and the same seems to be true for *Rp. capsulata* (Evans, E. H., and Crofts, 1974b; Prince, R. C., and Dutton, P. L., unpublished) and *Rs. rubrum* (Nishimura, 1963). Antimycin also prevents the oxidation of the cytochromes *b* of *Chr. vinosum* and *Chl. limicola f. thiosulfatophilum* (Knaff and Buchanan, 1975). In addition to the effect of antimycin on the oxidation of cytochromes *b*, however, it may also modify their reduction. A computer analysis of Fig. 18 (Prince and Dutton, 1975) suggests that the addition of 2 μM antimycin slows the rate of reduction of cytochrome b_{50} after a single flash by a factor of approximately 2 (i.e., from ≈1 to ≈2 msec; Pring, M., Prince, R. C., and Dutton, P. L., unpublished), and a similar effect can perhaps be seen in Fig. 19. Antimycin apparently has no effect on the rate of the preceding proton uptake (Petty and Dutton, 1976a).

(c) Other Inhibitors. Several other inhibitors act at or near cytochrome b_{50}. Dibromothymoquinone (DBMIB) was introduced by Trebst *et al.* (1970) as a plastoquinone antagonist in chloroplasts, and it also has inhibitory effects on chromatophores (Baltscheffsky, 1974; Petty, K. M., and Dutton, P. L., unpublished). Concentrations of about 75 μM completely inhibit cytochrome *b* reduction in *Rp. sphaeroides*, but inhibit only 50% of the proton uptake. The rate of cytochrome *b* reduction is unaffected. Similar effects are seen with 7-*N*-heptyl 6-hydroxy-5,8-quinoline quinone; at 30 μM concentrations, the reduction of cytochrome *b* was completely inhibited, while proton uptake was only 50% inhibited. Again, submaximal inhibitory concentrations did not affect the rate of cytochrome *b* reduction. On the other hand, 2 mM *o*-phenanthroline (1,10-diazaphenanthrene) inhibits both the rate and the extent of proton uptake. These inhibitors deserve further studies.

(d) Temperature. The activation energy of flash-activated cytochrome b_{50} reduction in the presence of antimycin at pH 7.0 is 8–10 kcal/mol between 4 and 50°C (Petty, K. M., and Dutton, P. L., unpublished).

4.2. Cytochrome b_{50} to Cytochrome c_2 Electron Transfer under Uncoupled Conditions

Electron flow from cytochrome b_{50} to cytochrome c_2 is perhaps the least-understood portion of the cyclic electron-transport system. As we shall discuss in the next section, electron flow through this region is very sensitive to the energy status of the chromatophore. In this section, we shall discuss only the situation in the uncoupled chromatophore (i.e., in the presence of 10

μM FCCP). This allows repetitive flash activation and signal-averaging techniques without electron flow being limited by energy feedback from the coupled energy-conserving system.

4.2.1. Kinetic Matching of Cytochrome b_{50} and Cytochrome c_2 Reactions

The rationale of the experiments described in this section is as follows: *Rp. sphaeroides* seemingly possesses two *b*-type cytochromes that are capable of reducing ferricytochrome c_2: cytochromes b_{50} (cyclic) and b_{155} ("substrate-linked"). However, although cytochrome b_{155} is oxidized following a single-turnover flash ($t_{1/2} \approx 2$ msec), its re-reduction takes several seconds (Dutton and Jackson, 1972) (see also Fig. 26). Thus, as we discussed earlier, it is not generally considered as part of the cyclic system, and is not significantly involved in electron flow induced by repetitive flashes of light within the seconds time scale.

It seems clear, therefore, that under repetitively pulsed conditions after the first flash, cytochrome b_{50}

is the only *b*-type cytochrome *demonstrably* involved in the cyclic system. Looking at the system from the standpoint of cytochrome c_2 following the first flash, oxidized cytochrome c_2 may receive an electron from cytochrome b_{155}, but after subsequent flashes, it receives electrons only via the cyclic chain through UQ and cytochrome b_{50}. Figure 18A shows the kinetic responses of cytochromes b_{50} and c_2 when subjected to a train of pulses of light, each long enough to cause a single turnover of the RC. The flashes are separated by a dark period of 40 msec, which is long enough to permit the pulsed cytochrome and other chromatophore reactions such as proton uptake-efflux and carotenoid bandshifts to relax completely. Figure 18 shows that cytochrome b_{50} undergoes a transient reduction after each flash, followed by a reoxidation that is complete by 12 msec. Under the same conditions, cytochrome c_2 is rapidly oxidized after each flash, and is subsequently re-reduced. The addition of antimycin has a marked effect on these kinetics, an inhibition that is seen at the first flash after its addition. Antimycin greatly inhibits the rate of oxidation of cytochrome *b* (Fig. 18B), and in its presence, the

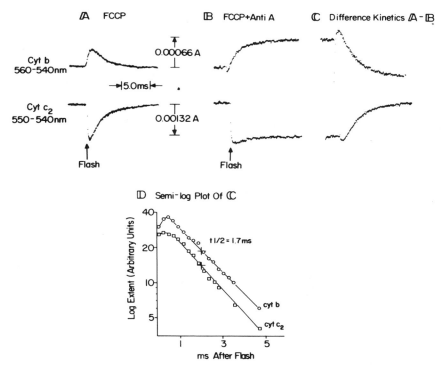

Fig. 18. Pulsed kinetics of cytochromes *b* and c_2 in *Rhodopseudomonas sphaeroides*. The figure shows the kinetics of cytochromes *b* and c_2 in the uncoupled state (+10 μM FCCP). (A, B) Traces averaged from those for 128 Xe flashes for cytochrome *b* and 64 Xe flashes for cytochrome c_2, the flashes being separated by 40 msec (A) and 2 min (B) to allow the reactions to relax between the flashes. (C) Difference kinetics of the traces in (A) and (B). (D) A semilog plot of (C). The E_h was +95 mV. Further experimental details are described by Prince and Dutton (1975).

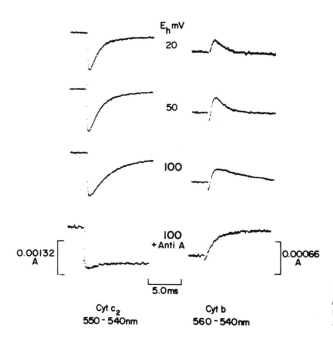

Fig. 19. Pulsed kinetics of cytochromes b and c_2 in *Rhodopseudomonas sphaeroides* as a function of ambient redox potential. These traces are equivalent to those in Figs. 18A and B, except that the ambient potential was as indicated.

extent of cytochrome b reduction is greatly enhanced because normally the reoxidation is intrinsically competitive with the reduction. A similar effect is not seen in the case of cytochrome c_2 (Fig. 18B), since its oxidation rate is sufficiently faster than its re-reduction rate that the two can be resolved in the uninhibited state. Figure 18C shows the difference kinetics recorded in the presence and in the absence of antimycin, with a semilog plot of the difference kinetics shown in Fig. 18D. The resolved kinetics of the two traces are almost identical; i.e., the kinetics of the re-oxidation of cytochrome b are the same as those of the re-reduction of cytochrome c_2.

Figure 19 shows that although the kinetic rates can be altered by varying the ambient potential over the range 20–160 mV, the kinetic matching is maintained; possible reasons for this variance will be discussed later. A similar kinetic matching of cytochromes b_{50} and c_2 is also seen in *Rp. capsulata* (Fig. 20). It is important to add that the rapid kinetic matching of cytochromes b and c_2 can be observed at all values of pH between 5 and 11, but only at potentials at which cytochrome b_{155} is reduced prior to activation. This too will be discussed more fully later.

Thus, in the uncoupled state, electrons are seen to leave cytochrome b as they are seen to arrive at cytochrome c_2. Although this is support for a direct coupling of the two cytochromes, it does not necessarily rule out the possibility of other components that are kinetically active in the cycle between the two cytochromes. However, we can define at least two conditions that such intermediates must fulfill: (1) they must have intrinsically very rapid rates of reduction and oxidation, at least under uncoupled conditions, so that these are not rate-limiting reactions in electron flow between the cytochromes; and (2) antimycin must act *on* the intermediary electron carriers in such a way that both the oxidation of cytochrome b and the reduction of cytochrome c are inhibited at once. The possible existence of intermediary components will be discussed further in the next section.

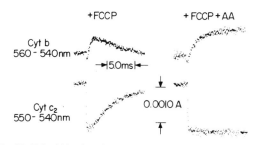

Fig. 20. Pulsed kinetics of cytochromes b and c_2 in *Rhodopseudomonas capsulata*. This experiment was similar to those in Figs. 18 and 19, except that chromatophores from the blue-green mutant of *Rp. capsulata* Ala pho + were used (Bchl = 22.9 μM). The E_h was +100 mV.

4.2.2. A Further Examination of Electron Donors to Cytochrome c_2

Thus far, evidence suggests that two different b-type cytochromes (b_{50} and b_{155}) could be direct electron donors to cytochrome c_2 in a simple convergence of cyclic and substrate-linked pathways. In this section, we shall discuss experiments that exploit the different E_m-pH dependencies of these two cytochromes b (see Fig. 8 and Section 2.3). The pK at pH 7.4 on cytochrome b_{50} means that the midpoint potential of this cytochrome is never lower than approximately +20 mV, while the absence of a pK on ferrocytochrome b_{155} means that at equilibrium, this cytochrome can become more negative than cytochrome b_{50} at high pH. This in turn allows us to observe the flash-activated behavior of the cycle when cytochrome b_{50} is reduced while cytochrome b_{155} is oxidized before activation. If cytochrome b_{50} can donate electrons directly to ferricytochrome c_2, then at high pH this should be observable without contributions from the oxidation of cytochrome b_{155}. To clarify the experiments, the procedure is first described at pH 7.

Figure 21 shows the re-reduction of cytochrome c_2 after a single flash at various ambient potentials. As the potential is lowered, the rate of re-reduction speeds up, until at 20 mV, it has a half-time of about 1 msec. The re-reduction always appears complex, however, not monophasic, so to present the data on a redox potential scale, we have plotted the amount of cytochrome c_2 remaining oxidized 5 msec after the flash. This is shown in Fig. 22, in which the heavy line represents the total extent of the flash-induced cytochrome c_2 photooxidation (i.e., the amount that would be seen in the presence of antimycin), and the symbols represent the amount still oxidized after 5 msec. Thus, at an E_h of 200 mV, little of the cytochrome has been re-reduced by 5 msec, but as the potential is lowered, more and more is rapidly reduced, so that at an E_h of 75 mV, effectively all the cytochrome is re-reduced after 5 msec.

The simplest explanation of these results would be that at high potentials (>200 mV), there are no reduced components able to donate electrons promptly to cytochrome c_2, so the rate of re-reduction reflects the rate of electron flow around a mainly oxidized cycle. At a potential of 100 mV, cytochrome b_{155} would be reduced prior to the flash and so be immediately available for the re-reduction of cytochrome c_2 following a flash, and at 20 mV, cytochrome b_{50} would also be reduced, and would compete with cytochrome b_{155} to reduce ferricytochrome c_2,

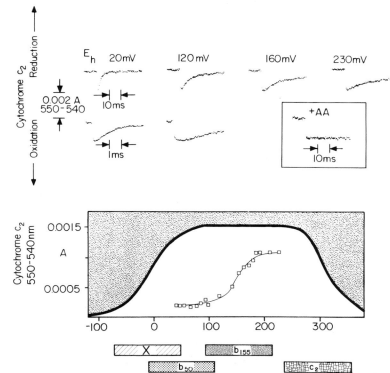

Fig. 21. Re-reduction kinetics of cytochrome c_2 in uncoupled *Rhodopseudomonas sphaeroides* chromatophores. Chromatophores from *Rp. sphaeroides* Ga (Bchl = 21 μM) were suspended at pH 7 in the presence of 10 μM uncoupler, and the re-reduction of cytochrome c_2 was monitored at a variety of ambient redox potentials. The inset shows the kinetics at any potential in the presence of 2 μM antimycin. The traces are averages of 32 Xe flashes separated by 2 min.

Fig. 22. Re-reduction kinetics of cytochrome c_2 in uncoupled *Rhodopseudomonas sphaeroides* chromatophores. This figure shows the results of an experiment similar to that in Fig. 21 performed at pH 7.1. The heavy line represents the total amount of cytochrome c_2 oxidized after a single flash, while the symbols represent the amount of cytochrome remaining oxidized 5 msec ($\square$) after the flash. The shaded rectangles under the curves in this figure and in Figs. 23 and 28 denote the E_h ranges in which the components of the cyclic system go from 91% oxidized to 91% reduced before the flash.

which would thus be reduced even more rapidly. Although for simplicity this is not shown in Fig. 22, it does indeed happen; cytochrome c_2 reduction speeds up by a factor of 1.5 to 2 if cytochrome b_{50} is reduced before activation in addition to cytochrome b_{155}. Such a model would not require any intermediary carriers in the cycle between cytochrome b and c_2. However, the situation is not as simple as this.

Figure 23 shows a series of redox titrations of the re-reduction of cytochrome c_2 at a variety of pH values between pH 6 and 11, and under each titration are shown the equilibrium midpoint potentials of components of the cycle. At pH values below 9, at which point the two cytochromes b have equal midpoint potentials, results qualitatively similar to those at pH 7 are obtained.

At pH values above 10, however, where cytochrome b_{50} can be chemically reduced before the flash with cytochrome b_{155} remaining oxidized, the pattern changes. At such high pH values, the presence of reduced cytochrome b_{50} alone is not enough for a rapid reduction of ferricytochrome c_2; it is apparent that *both* cytochromes must be reduced for this to occur. This result is strong evidence against our facile model of two cytochromes b simply sharing cytochrome c_2 as oxidant. It can be seen from the high pH titrations that the slow reaction at intermediate potentials is not a result of incompetence or damage induced by the high pH conditions, since when both cytochromes are reduced before activation, a rapid reduction can occur. Furthermore, a normal chromatophore integrity at the high pH values is

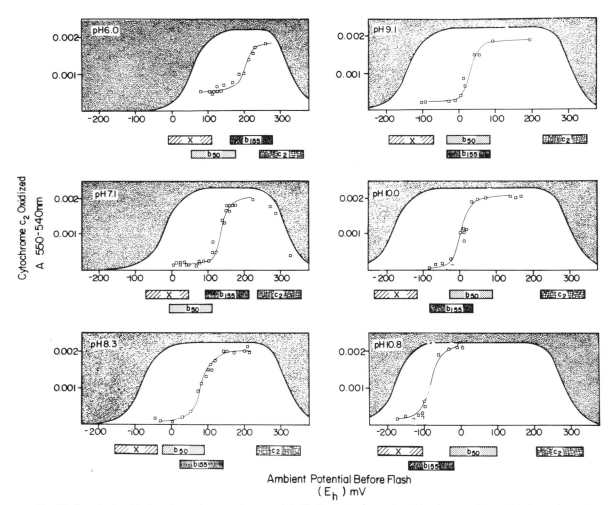

Fig. 23. Re-reduction kinetics of cytochrome c_2 in uncoupled *Rhodopseudomonas sphaeroides* chromatophores. This figure shows the results of several experiments performed as in Figs. 21 and 22 at a variety of pH values.

indicated by the flash-induced carotenoid bandshifts in the absence of uncouplers, which show the usual slow decay characteristics of a coupled membrane.

A major conclusion from these experiments is that more than one component must be reduced before activation if rapid multiturnover electron flow is to occur. One may be identified as cytochrome b_{50}; the other may be simply cytochrome b_{155} that has to donate an electron into the cycle before maximum rates of electron flow can be attained. Alternatively, although it complicates matters, it is also useful at this stage to consider that a third component might exist with electrochemical properties essentially indistinguishable from those of cytochrome b_{155}. As such, it would be a proton carrier from pH 6 to 11, and have an E_m at pH 7 of approximately +150 mV. Its postulated role would be as an intermediate in the cycle, operating between cytochromes b_{50} and c_2. Further considerations on the existence of this hypothetical intermediate will emerge in later sections.

4.3. Ubiquinone–Cytochrome b–Cytochrome c_2 Reactions in the Coupled State

The last section described the kinetics of electron transfers from cytochromes b_{50} to c_2 in situations in which the energy feedback of the coupled "high-energy state" was largely removed by added uncoupler. This section discusses the reaction under energy-coupled conditions.

4.3.1. Pulse-Activated Reactions of Cytochrome c_2 in the Coupled State

Figure 24 contrasts the behavior of the cytochromes c_2 of *Rp. sphaeroides* under coupled and uncoupled conditions at various values of ambient potential. The following points seem clear:

1. At potentials above +200 mV, where no immediate donors to cytochrome c_2 are reduced prior to the flash, the two cytochromes c_2 are readily "flash-oxidized" with little re-reduction in the 25 msec between flashes. Once the cytochromes are fully oxidized, the dark re-reduction rate is in the seconds time range. Uncoupler has the relatively minor effect of stimulating the rate of re-reduction less than tenfold.

2. At potentials that prereduce cytochrome b_{155} (or the hypothetical cytochrome b_{50}–c_2 intermediate or both) before activation, most of the cytochrome c_2 is rapidly re-reduced after the first flash, and is partially rapidly re-reduced after the second flash. The extent of rapid re-reduction continues to decrease after the third and fourth flashes, however, and from there on (pulsed steady-state) the reactions are the same after every flash, with little tendency for re-reduction between the flashes. The time course of cytochrome c_2 re-reduction, after this pulsed steady-state is achieved, is very dependent on the degree of coupling of the chromatophores, but is usually in the seconds time range. An acute sensitivity to uncouplers is encountered in this potential range, as discussed in Section

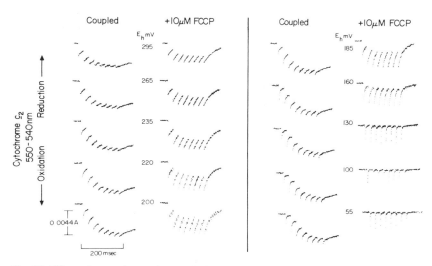

Fig. 24. Effects of uncoupler on the re-reduction kinetics of cytochrome c_2 in *Rhodopseudomonas sphaeroides*. This experiment was similar to that in Fig. 21, except that the responses of cytochrome c_2 to a train of 8 pulses of light (Xe flashes, 25 msec apart) were monitored in the presence or absence of uncoupler. The traces are an average of 16 separate flash trains, each separated by 2 min. A control experiment in the absence of redox mediators gave qualitatively similar results, although in this case the ambient potential could not be quantified. At all potentials and in the presence or absence of uncoupler, the addition of 2 μM antimycin converted the kinetics to those shown in Fig. 15D.

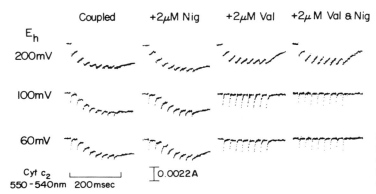

Fig. 25. Effects of nigericin and valinomycin on the re-reduction kinetics of cytochrome c_2. This experiment was identical to that in Fig. 24, except that nigericin or valinomycin or both were added instead of FCCP.

4.2.2. Uncouplers stimulate the re-reduction rate of cytochrome c_2 by a thousandfold; preparative improvements have increased this electron-transport coupling ratio much beyond that first reported by Jackson and Dutton (1973). Figure 24 clearly shows that this dramatic sensititivity to uncoupling occurs over the range of ambient potentials in which cytochrome b_{155} or the postulated cytochrome $b-c_2$ intermediate is reduced before activation.

3. In the coupled state, both cytochrome c_2 hemes become effectively completely oxidized after four flashes or less, irrespective of the prior state of reduction of the cycle and the prior reduction of cytochromes b_{155} or b_{50}.

4. The effects of uncoupler shown in Fig. 24 can be mimicked by valinomycin (Jackson and Dutton, 1973), but not by nigericin (Fig. 25). This is in keeping with the coupling effect on cytochrome c_2 re-reduction being an electrical constraint, since both uncoupler and valinomycin would dissipate a membrane potential, while nigericin would not (Jackson *et*

al., 1968). Crofts *et al.* (1974) reported that, in the presence of valinomycin, a few seconds of illumination to build up a maximum ΔpH (instead of mainly $\Delta\psi$ as in the case after the first few flashes, presumably because of buffering inside the chromatophore) allowed cytochrome c_2 to achieve its coupled, essentially fully oxidized state. Thus, the energy feedback on electron transfer can be interpreted on a chemiosmotic basis as being derived from transmembrane $\Delta\psi$ or ΔpH expressions of the proton-motive force.

5. The re-reduction of ferricytochrome c_2 after the first flash is not markedly stimulated by uncoupler. This is consistent with the re-reduction coming from cytochrome b_{155}, which is apparently not involved in the coupling mechanism. Even if cytochrome b_{155} did not exist, however, the absence of uncoupler stimulation after the first flash could also be interpreted as reflecting the absence of a significant energy feedback after only one turnover.

6. The overall behavior of cytochrome c_2 can be interpreted in the classic view of Chance and Williams

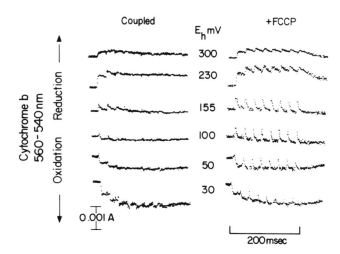

Fig. 26. Effects of uncoupler on the kinetics of cytochrome b reactions in *Rhodopseudomonas sphaeroides*. This experiment was similar to that in Fig. 24, except that cytochrome b reactions were monitored instead of cytochrome c_2.

(1956) as indicating that the cytochrome is at the electropositive end of a coupling mechanism. Appropriately, the RC P behaves similarly (Petty and Dutton, 1976a). In *Rs. rubrum*, the effects of pyrophosphate or ATP have shown that ferrocytochrome c_2 undergoes an energy-linked oxidation consistent with such a proposed position (Baltscheffsky, 1967).

4.3.2. Pulse-Activated Redox States of Cytochromes *b*

The kinetics of the cytochromes *b* in the coupled state are not as straightforward as those of cytochrome c_2. Spectrophotometric analysis in *Rp. sphaeroides* is often complicated by the opposing absorbance changes of what are apparently cytochrome b_{50} reduction and cytochrome b_{155} oxidation (Dutton and Jackson, 1972). Figure 26 is analogous to Fig. 24, and the reactions can be interpreted as follows:

1. With chromatophores poised at +200 mV and above, so that both cytochromes b_{50} and b_{155} are oxidized prior to activation, the clear flash-induced reduction of cytochrome b_{50} ($t_{1/2}$ 1–2 msec) can be observed. As was encountered with cytochrome c_2 re-reduction, the addition of uncoupler stimulates cytochrome b_{50} reoxidation by less than tenfold.

2. At +100 mV, at which cytochrome b_{155} is reduced prior to activation, very little is seen to happen after a flash of light. Dutton and Jackson (1972) interpreted this as the canceling out of absorbance changes due to the reduction of cytochrome b_{50} and the oxidation of cytochrome b_{155}. At potentials low enough to reduce both cytochromes *b* before activation (traces at +50 and +30 mV), the net absorbance decrease after the first flash was interpreted by Dutton and Jackson (1972) as cytochrome b_{155} oxidation, although it would now appear that some contribution from cytochrome b_{50} is also likely, since there is some influence of reduced cytochrome b_{50} on the rate of cytochrome c_2 re-reduction (see Section 4.2.1). The further net oxidation of *b*-type cytochrome after subsequent flashes may prove to be predominantly cytochrome b_{50} becoming oxidized.

3. The reactions of cytochrome b_{50}, like those of cytochrome c_2, become increasingly sensitive to uncoupler as the redox potential before activation is lowered from +200 mV. Figure 26 suggests that both the oxidation and the reduction of cytochrome b_{50} are stimulated by uncoupler. The evidence for the stimulated reduction is the absorbance increase "spike" after the first flash, seen in the +155- and +100-mV traces, indicating that the reduction of cytochromes b_{50} is faster than that recorded at higher potentials and now faster than the opposing approximate 2-msec oxidation half-time of cytochrome b_{155}.

The stimulation of the rate of re-oxidation can be seen in all the traces, but particularly that at +50 mV. Clearly, much more work is needed to resolve the various reactions of the cytochromes *b*.

4. The traces recorded at an ambient potential of +30 mV indicate that cytochrome b_{50} tends to become oxidized under *both* coupled and uncoupled conditions, apparently even if it was reduced prior to activation. Thus, in the repetitively pulsed activated state, cytochrome b_{50} apparently always goes transiently reduced (see Section 4.2.1). This is not easily explicable in terms of a simple "crossover" notion (Chance and Williams, 1956). In simplistic terms, it suggests that the *reduction* of cytochrome b_{50} is not a favored state, even under coupled conditions. This is clearly another problem that needs much greater study.

4.3.3. Carotenoid Bandshift Coupled to Events in the Ubiquinone–Cytochromes *b*–c_2 Region

There are three distinct kinetic phases of the single-turnover-flash-induced carotenoid bandshift (see Section 2.4 and Chapter 26) in *Rp. sphaeroides* (Jackson and Dutton, 1973) and *Rp. capsulata* (Evans, E. H., 1973). The first two (I and II) are in the nanosecond-to-microsecond time range, and appear to be in response to electron transfer from P to X in the RC and from cytochrome c_2 to P+. The third phase is equal to the sum of the first two, and is in the millisecond time range. Its formation requires that a redox component ($E_m = +150$ mV, pH 7.0) be reduced before activation, and is completely inhibited by antimycin.

Two main possibilities for the origin of phase III of the carotenoid bandshift were considered by Jackson and Dutton (1973), and these are shown in Figs. 27A and C. One possibility (Fig. 27A) was that it arose as cytochrome b_{155} reduced cytochrome c_2. The kinetics, E_h dependency, and antimycin sensitivity were consistent with this, but, in retrospect, the position of cytochrome c_2 *within* the membrane so that there was dielectric between cytochromes b_{155} and c_2 seems unlikely (see Section 2.4) unless, as shown in Fig. 27B, cytochrome c_2 occupies an aqueous exposed pocket in the membrane that permits cytochrome b_{155} to lie at the membrane–aqueous interface proper, nearer than cytochrome c_2 to the center of the chromatophore. The second possibility considered by Jackson and Dutton (1973) was based on the early chemiosmotic principles of Mitchell (1961, 1966), and was also favored by Cogdell *et al.* (1972). Again, however, an anomalous position was given to cytochrome b_{155} in that it was placed within the cycle, in contrast to the

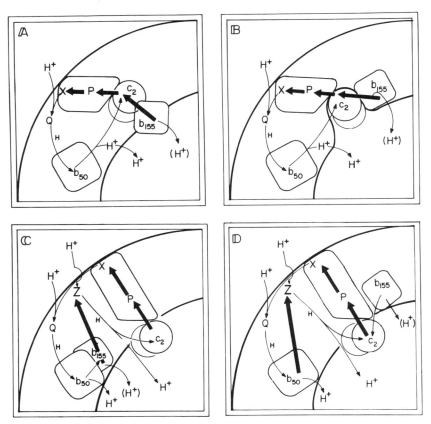

Fig. 27. Models for electron flow in *Rhodopseudomonas sphaeroides* to explain the carotenoid bandshift data of Jackson and Dutton (1973). The heavy arrows represent electron translocations that are considered to cause electric-field alterations across the membrane, and result in a carotenoid bandshift. In addition to A and C, the schemes presented by Dutton and Jackson (1973), B is an alternative to A that attempts to accommodate the finding of Prince *et al.* (1975a) that cytochrome c_2 resides on the inner membrane–aqueous interface. In models of the A or B kind, the proton released on cytochrome b_{50} reduction or oxidation could be involved in intramembrane or transmembrane reactions. In C, cytochrome b_{155} was envisaged as being in the cycle and involved in the generation of the carotenoid bandshift. D is an alternative to C suggesting that cytochrome b_{155} has nothing to do with the generation of the bandshift, which instead originates from a transmembrane reduction of Z by cytochrome b_{50}. An additional possibility based on C is that cytochrome b_{155} does react with Z, but in a noncyclic manner. In such a case, cytochrome b_{50} might react directly with Z on the second turnover, or directly with cytochrome c_2. The protons enclosed in parentheses (H⁺) indicate proton releases that would occur in the models only on the first turnover. The small "hydrogens" (H) placed alongside some arrows indicate that both an electron and a proton are transferred in these reactions between the components at the extremities of the arrows.

previous interpretation of Dutton and Jackson (1972) and considerations throughout this chapter. In this model (Fig. 27C), cytochrome b_{155} is reduced by cytochrome b_{50} and in turn reduces the hypothetical electron/proton intermediary carrier (introduced in the previous section), designated Z, located on the other side of the membrane. In this model, it is this latter reduction that generates the third phase of the carotenoid bandshift. An alternative (Fig. 27D) would have cytochrome b_{50} reducing Z directly, with cyto-

chrome b_{155} involved neither in the cycle nor in the carotenoid bandshift.

All the models shown in Fig. 27 have assumed that all three phases of the carotenoid bandshift reflect partial or full transmembrane charge-separation events. Although there are reasons to consider that the shifts do originate in electric-field alterations, it is not possible with our current knowledge to rule out other physical sources that may account wholly or partially for the shifts (see Chapter 26). For the considerations

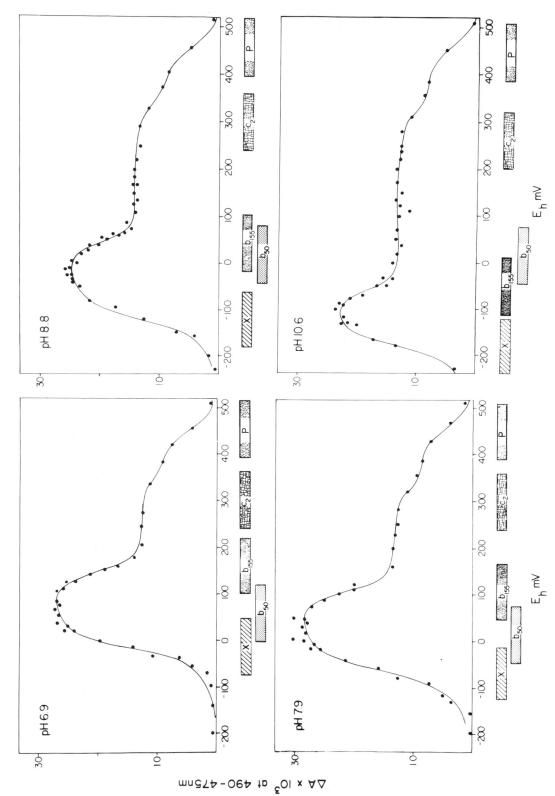

Fig. 28. Extent of the carotenoid bandshift in *Rhodopseudomonas sphaeroides* generated after a single flash, as a function of ambient potential and pH. The extent of the carotenoid shift was monitored 5 msec after a single-turnover flash of light.

in this chapter, however, it is germane to know what kind of reactions they are associated with. The first two phases seem clearly assignable (for whatever reasons) to the P → X and cytochrome c_2 → P^+ electron transfers. The third phase is more difficult.

We have recently done more work on the origins of the third phase of the carotenoid bandshift, again exploiting the different E_h/pH relationships of the cytochromes b.

Figure 28 shows the extent of the carotenoid bandshift at various values of E_h and pH. At each pH, the three phases of the carotenoid bandshift can be clearly identified: phase I, associated with P → X, around +450 mV; phase II, associated with cytochrome c_2→ P^+, below +300 mV; phase III, at lower values of ambient potential. The appearance of the flash-induced third phase clearly has no apparent relationship with the prior state of reduction of cytochrome b_{50}, but always approximates the expected Nernst curve for cytochrome b_{155}. Additional support for the idea that phase III originates in cytochrome b_{155} oxidation comes from the similar time course of the two reactions (although more exact measurements are now needed), and from the observation that both occur predominantly only after the first flash (Dutton and Jackson, 1972). With the introduction of the hypothetical intermediate Z, however, alternative origins, emanating from events in the cyclic system, deserve further consideration when the workings of the system are viewed as a whole. In weighing the evidence for the shift coming from the involvement of Z, it is apparent from Fig. 28 that the published schemes (Figs. 27A and C) are inadequate to explain the data fully without further elaborations on the properties of Z. The inadequacies and elaborations are dealt with in the next section.

5. The Wider Picture: Some Alternatives and Their Problems

The possibilities represented in Fig. 27, used as models to discuss electron-transport-related carotenoid bandshifts, are also required to explain the energy-coupled workings of cyclic electron transfer in *Rp. sphaeroides* on a more comprehensive level. All incorporate the findings on cytochrome c_2 → P → X → UQ → cytochrome b_{50} as covered in Sections 3 and 4.2. The model of Fig. 27B (rejecting Fig. 27A on the grounds that cytochrome c_2 does not occupy an intramembrane position) presents cytochrome b_{50} interacting directly with cytochrome c_2. In addition to the proton-translocating part of the cycle, this model leaves open the possibility that close-order (chemical)

transformation may occur associated with the cytochrome b and possibly its proton, including intramembrane proton release (see Williams, 1969). Figures 27C and D are simple chemiosmotic schemes, portraying two electrogenic and two proton-translocating steps in the whole cycle. An important experimental issue regarding these schemes is how they might serve to explain the rather dramatic differences between coupled and uncoupled electron flow experienced by cytochromes b_{50} and c_2, and why the component with E_m +150 mV at pH 7 (cytochrome b_{155} or Z?) exerts such a profound influence on the reactions. It is worth remembering that, in contrast to the reactions of cytochromes b_{50} and c_2, the effect of uncouplers on stimulating the decay of the carotenoid bandshift, or stimulating the reappearance of the protons taken up on the flash-induced reduction of the secondary acceptor, is independent of the state of reduction of the cycle before activation. Thus, while at low potentials ($E_h \approx$ +100 mV, pH 7) in the presence of uncoupler, the relaxation of cytochromes b_{50} and c_2 reactions, the carotenoid bandshift, and proton uptake take place within milliseconds, at higher potentials ($E_h \approx$ +200 mV, pH 7), the carotenoid bandshift and proton uptake relax in milliseconds while the cytochrome b_{50} remains oxidized for hundreds of milliseconds after the flash. Under these high-potential conditions, at least, the uncoupler-stimulated relaxation of the pulsed redox changes of the cytochromes cannot be coupled to the relaxation of proton uptake or the carotenoid shift; this is similar to the situation (except for proton release under certain understood conditions; see Section 4.1.2) at any poised redox potential in the presence of antimycin.

Two considerations regarding the stimulation of the cycle by the prior reduction of cytochrome b_{155} or the intermediary carrier Z could be as follows: Based on Fig. 27D, and shown with necessary modifications in Fig. 29A, is a model developing the nature of Z. This requires that Z exist in more than one reduction state, such that cytochrome c_2 can react only with ZH_2 specifically, while cytochrome b_{50} will react only with $Z^·H$ ($Z/Z^·H$ couple $E_m \approx$ +150 mV, pH 7.0). In such a case, the prior reduction of Z to $Z^·H$ would be mandatory for effective and rapid electron flow from cytochrome b_{50} to cytochrome c_2, and only under such conditions would uncoupler stimulate electron transport. If Z were in the fully oxidized form, the rate-limiting step in electron flow would be its chemically unfavored reduction by cytochrome b_{50}, and not the energy "back-pressure" of the coupled state. In this model, the third phase of the carotenoid bandshift would be generated as the *flash*-reduced cytochrome b_{50} ($t_{1/2}$ 1–2 msec) *promptly reduced* $Z^·H$ to ZH_2. One

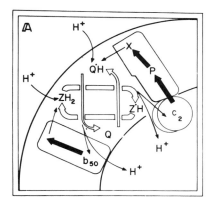 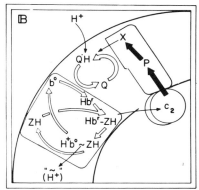

Fig. 29. Developments of the models in Fig. 27. In this figure, the redox states of the proposed hydrogen carrier "Z" are described in redox cycles or chemical sequences. Model A assumes that the component Z can exist in three separate forms, Z, $Z^{\cdot}H$, and ZH_2, with efficient cyclic electron flow occurring only via the $Z^{\cdot}H/ZH_2$ couple. This model is "chemiosmotic" in character, and the three electron-translocating steps are shown as heavy arrows. Its main feature is the reaction of cytochrome b_{50} with $Z^{\cdot}H$ to form, after protonation, ZH_2. Model B assigns the role of ligand to ZH, but this model could also be extended to allow Z to function as a redox-driven hydrogen carrier as well as a ligand. The nature of the "energy currency" in this "chemical" model could be the ferricytochrome $b_{50} \sim ZH$ bond energy, or the associated proton on b_{50} and/or Z if Z is also involved in redox-driven proton exchange.

requirement for this model, assuming that every step is thermodynamically favorable, would be that the E_m of the $Z^{\cdot}H/ZH_2$ couple be appropriately more positive than cytochrome b_{50}, and hence not very much more negative than that of $Z/Z^{\cdot}H$ ($E_m = +150$ mV at pH 7). As such, even though the equilibrium E_m values of the component create what might be regarded as a thermodynamic "tight squeeze" [especially if the reduction of $Z^{\cdot}H$ by cytochrome b_{50} is trans-membrane (electrogenic) and has to work against the energetically opposing transmembrane potential of about +100 mV (Jackson and Crofts, 1969) created by the first electrogenic $c_2^+ \rightarrow X^-$ step], this model might explain the experimental data; it could also work if cytochrome b_{155} re-reduced cytochrome c_2 on the first turnover in a side reaction. However, testing this model further, at high pH, the "tight squeeze" can be made more extreme, and here this model becomes strained. At pH 10.8, the presumed $Z/Z^{\cdot}H$ couple has an E_m of approximately −60 mV, and that of $Z^{\cdot}H/ZH_2$ will be even more negative, while the E_m of cytochrome b_{50} remains at +20 mV. Under these conditions, the cytochrome $b_{50} \rightarrow Z^{\cdot}H$ reaction would be unfavorable, not only with respect to an existing +100 mV $\Delta\psi$, but also on a redox basis by larger than $\Delta E_m = -75$ mV. To make the overall cytochrome $b_{50} \rightarrow$ cytochrome c_2 reaction work, the large free energy available from the ZH_2/cytochrome c_2 reaction would have to be used very efficiently and rapidly. If cytochrome b_{155} reduced cytochrome c_2 after the first flash,

it would be difficult to ascribe the carotenoid band-shift to the reduction of $Z^{\cdot}H$ by cytochrome b_{50}, since this would be unfavorable without the redox free-energy "pull" from ZH_2 to cytochrome c_2.

Figure 29B is an alternative suggestion that formally ascribes the role of "ligand" to ZH, showing that the same processes described above can also be viewed as a specific chemical interaction of ferrocytochrome b_{50} with a reduced form of Z (e.g., $Z^{\cdot}H$). This model also explains the requirement for a prior reduction of Z before activation, if ferricytochrome c_2 can react only with the ferrocytochrome $b_{50}-Z^{\cdot}H$ complex, and $Q^{\cdot}H$ can react only with ferricyto-chrome b_{50}. In connection with the apparent direct effects of prereduced cytochrome b_{50} on ferricyto-chrome c_2 re-reduction, the preformed ferrocyto-chrome $b_{50}-Z^{\cdot}H$ would provide an immediate reduc-tant instead of the 1- to 2-msec delay required to reduce cytochrome b_{50} if it were oxidized prior to the flash. It could be argued that the same attendant limitations that apply to Fig. 29A also apply to 29B, if the third phase of the carotenoid shift arose from conformational or local charge alterations con-comitant with the formation of ferricytochrome $b_{50} \sim Z^{\cdot}H$ from ferrocytochrome $b_{50}-Z^{\cdot}H$, although unlike the chemiosmotic model (Fig. 29A), there is no obligatory requirement for a second electrogenic step, and the third phase could indeed be simply due to the oxidation of cytochrome b_{155}. The formal "chemical" explanation for uncoupling in this model would be directed at the relaxation of the ferricytochrome b_{50} ~Z·H state, in competition with the transfer of its energy to the ATPase, although there is no reason to dismiss the possibility that the mechanism is coupled to the ATPase via a nonredox carrier or "free" proton. The scheme could easily be written to accommodate two H⁺'s bound from the outside per cyclic turnover.

Other more elaborate schemes can also be drawn, but in general they go too far beyond the data to be useful. There are two that deserve comment, however, because they have many of the characteristics of our separate considerations of UQ and Z as hydrogen carriers, and because they provide us with a reference point to schemes that are mechanistically largely hypothetical which were derived mainly from steady-state experiments with the mitochondrial respiratory chain. Mitchell (1975a,b) suggested that the multiple reduction states of UQ allow its operation in a four-step "motion" involving reaction both before and after cytochrome b_{50}, so that it accounts not only for the functions thus far ascribed to ubiquinone, but also for Z.

Figure 30 shows two extremes of the schemes formally adapted for the light-driven system of *Rp. sphaeroides*. Two major differences from our earlier schemes are apparent: (1) cytochrome b_{50}, analogous to mitochondrial cytochrome b_{560} ($E_m = +40$ mV, pH 7), is placed nearer the outer side of the chromatophore; and (2) cytochrome b_{-90}, analogous to mitochondrial cytochrome b_{565} ($E_m = -30$ mV, pH 7), is introduced into the cycle and placed on the inner side of the chromatophore. Figure 30A shows a model in which the E_m of Q/Q·H is more positive than the E_m of Q·H/QH₂, so that the semiquinone is a favored thermodynamic species. In Fig. 30B, the E_m of Q/Q·H is more negative than that of Q·H/QH₂, so a stable semiquinone is not favored. This being the case, the detection at equilibrium of significant semiquinone free radicals would not be expected. Our considerations so far (Petty and Dutton, 1976a,b; Prince and Dutton,

1976a, and in this chapter) are based on this premise. Although the systems are still cyclic in nature, the reactions between the cytochromes b, UQ, and cytochrome c_2 are an interlocking network of events in which altering the characteristics of one part of the network will affect the rest.

In both cases, the second electrogenic step (third phase of the carotenoid bandshift) is presented as occurring between cytochromes b_{-90} and b_{50}, and ignores the possibility that it is associated with the oxidation of cytochrome b_{155}. This being so, it is entirely reliant on a prior reduced state of UQ, and on the flash-generation of a Q species of negative enough potential to reduce cytochrome b_{-90}. This could be either QH₂ generated from Q·H reduction by the primary acceptor (Fig. 30A) or Q·H generated by the oxidation of QH₂ by ferricytochrome c_2 (Fig. 30B). Having the second electrogenic step before cytochrome b_{50} readily accommodates the high pH data, because the ΔE_m between the two b-type cytochromes becomes larger, making the reaction more favorable.

Evidence against the stable semiquinone model (Fig. 30A) comes from the failure to detect any EPR signals ascribable to a stable semiquinone. We have examined concentrated suspensions of *Rp. sphaeroides* chromatophores at ambient and liquid helium temperatures under conditions of E_h and pH that might be expected to maximize ubisemiquinone or Z·H concentrations (Petty, K. M., Prince, R. C., and Dutton, P. L., unpublished). Compared with the light-induced free radical of P870⁺, signals ascribable to a normal ubisemiquinone free radical are negligible (similar efforts by Dr. J. R. Bolton have similarly failed to reveal a semiquinone free-radical signal; personal communication). To accommodate this deficiency, arguments would have to be made involving spin orbit coupling between adjacent closely aligned ubisemiquinones, leading to modified and perhaps broadened magnetic resonances that would be difficult to detect.

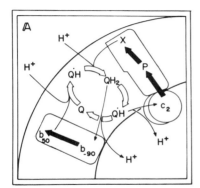 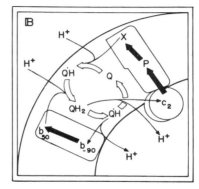

Fig. 30. Proton-motive Q-cycle models. Model A assumes that the semiquinone of UQ is a stable entity; Model B does not. Further details are discussed in the text.

In both models, we must assume that cytochrome b_{-90} oxidation is as fast as its reduction, since the latter has not been detected following flash activation. Such a light-induced reduction of cytochrome b_{-90} might be expected only in the presence of excess reductant when cytochrome b_{50} and ubiquinone were both fully reduced. This has indeed been seen by O. T. G. Jones (1969) using continuous illumination in the presence of succinate. If cytochrome b_{-90} oxidation is as fast as its reduction, and if the third phase of the carotenoid bandshift is generated during the reduction of cytochrome b_{50}, then the rate-limiting step for the bandshift ($t_{1/2}$ 1.2 msec at pH 7) is the reduction of cytochrome b_{-90}. In the model of Fig. 30A, this would be effected by the electronegative QH_2 produced by X^- reduction of $Q \cdot H$. If we take the proton-binding reaction ($t_{1/2}$ 150 μsec at pH 7.0) as representing the reduction of $Q \cdot H$ to QH_2, there is ample time for QH_2 to reduce cytochrome b_{-90}. In the model of Fig. 30B, the reductant would be the electronegative $Q \cdot H$ produced by the ferricytochrome c_2 oxidation of QH_2. Given this, we would have to predict that the ferricytochrome c_2 reduction would take no longer than a 1- to 2-msec half-time, and that all other reactions, as far as cytochrome b_{50} reduction, are not rate-limiting. In contrast to the model of Fig. 30A, the model of Fig. 30B cannot tolerate cytochrome b_{155} as the sole reductant for cytochrome c_2 after the first turnover; for this model, the whole involvement of cytochrome b_{155} would have to be reappraised.

A further distinction between the two models is the predicted n value for the agent responsible for ferricytochrome c_2 reduction and for the cause of the carotenoid bandshift. In Fig. 30A, it would be unity (as it would also be for the models in Figs. 27 and 29), while Fig. 30B predicts a value of 2, in the absence of cytochrome b_{155}. Experiments to date have produced complex re-reduction kinetic patterns for cytochrome c_2 (Figs. 22 and 23) and n values tend to be nearer 2 than 1 (Figs. 22 and 23). Since the writing of this chapter further work has established the n value to be very close to 2.0 (see Section 7). This number is consistent with the above-discussed *lack* of stable semiquinone-type EPR signals and is strong evidence against the schemes of Figs. 27 and 30A, in which cytochrome b_{155}, Q, and Z are presented as $n = 1$ redox agents. Figure 29B is, like many "chemical" schemes, nonspecific and could be written with the cytochrome c_2 donor being $n = 2$. Figure 29A can accommodate Z being an $n = 2$ agent and, as pointed out by Dr. A. R. Crofts (personal discussions), thereby escapes the problems of the "thermodynamic squeeze" associated with the scheme with Z having a stable $Z \cdot H$.

It is also pertinent to ask how the two models in Figs. 30B and 29A (with Z as an $n = 2$ agent) explain the results of Section 4.1, which dealt with $UQ \cdot H$–cytochrome b_{50} reactions in the presence of antimycin. Antimycin prevents the re-reduction of ferricytochrome c_2; in Fig. 30B, this would automatically prevent the reduction of cytochrome b_{50} via cytochrome b_{-90}. The $UQ \cdot H$ produced via X^- after the light reaction would then perhaps reduce cytochrome b_{50} directly, without the generation of the third phase of the carotenoid bandshift. A similar electron-flow situation might also exist at high potentials in the absence of antimycin, when there is no UQH_2 available prior to activation to rapidly reduce ferricytochrome c_2 and cytochrome b_{-90}. $UQ \cdot H$ might then reduce cytochrome b_{50} directly, again without the generation of a carotenoid shift. In such a model, the pathway for the reduction of cytochrome b_{50} at high potentials and in the antimycin-inhibited state would be essentially different from that in the uninhibited low-potential cycle, and the rate of the reaction might be expected to be significantly altered (see Section 4.1.3). In Fig. 29A the prevention by antimycin of cytochrome c_2 reduction by ZH_2 would also prevent the reoxidation of cytochrome b_{50} and eliminate carotenoid bandshift phase III. With and without antimycin at high or low potentials, $UQ \cdot H$ (generated by X^-) would be the agent responsible for the reduction of cytochrome b_{50}.

With the exception of the models in Figs. 27B and 29B, all the models have a second proton-binding reaction that would be expected to be antimycin-sensitive. Such a binding was reported in the millisecond time range (Cogdell *et al.*, 1972), but only in the presence of valinomycin; Crofts *et al.* (1975) offered some perhaps *ad hoc* explanation for this based on models similar to that in Fig. 29A. In contrast, by counting H^+'s bound per single-turnover flash, we (Petty, *et al.*, 1977) demonstrated that there is approximately 0.6 H^+/e^- bound in addition to that (1.0 H^+/e^-) associated with the X-induced reduction of UQ. This is *without* the aid of valinomycin, and its half-time is on the order of hundreds of microseconds. However, we can also confirm that valinomycin promotes H^+ bound per single turnover. The midpoints of the couples binding these antimycin-sensitive protons are not yet known, but more intensive studies should add a further important experimental vehicle for the elucidation of events in the cycle occurring between cytochromes b and c_2.

A key feature on the UQ → cytochromes b → Z → cytochrome c_2 sequence is the dramatic and multiple effects of antimycin. Just where this inhibitor acts to eliminate cytochrome c_2 reduction, cytochrome b_{50}

oxidation, the third phase of the carotenoid bandshift, and an H^+-binding reaction is an important question. The reaction it inhibits (from our current knowledge it could be one of several) is clearly energetically important to the function of a large part of the system.

6. Concluding Remarks

In the development of the simple and familiar "electron-transfer pathway" models of Fig. 8 to the more elaborate models of Figs. 27, 29, and 30, some doubt has been cast on some of the "certainties" presented in the original simplistic models. For some steps, such as the cytochrome c_2 to RC reaction, our increased knowledge has taken us from the first tentative arrow drawn between the components into the realms of mechanistic understanding; other steps, however, have become less clear. The later schemes force us to review some earlier conclusions in a new light. For example, the existence of a cytochrome b_{155} → cytochrome c_2 reaction is a vital issue in some of the schemes. Has cytochrome b_{155} been satisfactorily proved as a kinetically active part of the cycle? The evidence for its involvement rests on the apparently long re-reduction time of what is identified as its flash-induced oxidation (Dutton and Jackson, 1972). If it is involved, is cytochrome c_2 its oxidant, or are there other possibilities, such as a direct reduction of P^+ in RCs that have lost their cytochromes c_2 (see Dutton and Jackson, 1972; Evans, E. H., and Crofts, 1974b)? Or are the b-type cytochromes more intimately associated with UQ than is usually considered? Or again, is the stoichiometry of cytochromes $b_{155}:c_2$ much less than unity, so that although this can act as a pathway for substrate electrons into the cycle, it never dangerously competes with the cycle itself? Much now hangs on quantitation of the reactions of the cytochromes b, and a further characterization of electron donors to ferricytochrome c_2, particularly with regard to their n values. The binding site (or sites) of antimycin with respect to inhibition also needs careful attention, as do some more quantitative correlations of electrons moving around the cycle with proton translocation and ATP synthesis.

Nevertheless, the pathways and mechanisms of the cycle are gradually unfolding, and we feel optimistic that this progress will continue at a rapid rate!

7. Addendum

Since the preparation of this chapter, several developments have taken place, some of them as a direct consequence of questions raised during the writing of the chapter. We shall very briefly summarize the more straightforward ones:

1. Using chromatophores with a partially depleted cytochrome c_2 complement and variable flash intensities for activation, we have found that the extent of cytochrome oxidation is simply proportional to the extent of P^+ formation; those RC that have cytochrome c_2 have two (Prince et al., 1978). This evidence, in addition to the statistical nature of cytochrome c_2 oxidation (Dutton et al., 1975; Prince and Dutton, 1977a), and the resolution of two cytochrome c_2 populations within the chromatophore by slow dark redox titrations (Dutton et al., 1975), strongly suggests that the cytochrome c_2 is enslaved to the RC on a timescale much beyond that relevant to function. In spite of the fact that some can come loose "irreversibly" during chromatophore preparation, there is good reason to believe that the natural association is essentially permanent, as is more readily appreciated for the hydrophobically linked analogues in many other species. Even in the model system (Fig. 13), binding can be inferred by analyzing the kinetics as two first-order processes (see Section 3.3). Direct measurement of the binding equilibrium by the centrifugation of mammalian cytochrome c with dispersions of RC incorporated in egg lecithin have yielded a dissociation constant near 10^{-5} M and suggested binding sites for four cytochromes c (J. M. Pachence, unpublished observation). The oxidation kinetics in the dispersion are roughly similar to those of Fig. 13, obtained with cytochrome c_2. We have also analyzed the oxidation kinetics of cytochrome c in the *centrifuged pellet* (J. M. Pachence, unpublished observation). Although the effective concentrations of the reactants are more than a thousandfold greater than *in vivo*, the kinetics are not proportionally different, providing further evidence that the cytochrome is bound, most probably on the RC protein itself. Remarkably, at a cytochrome c:RC ratio of 2, the oxidation kinetics in the pellet are similar to the *in vivo* "biphasic" kinetics of cytochrome c_2 (Fig. 12).

Further detailed analysis of experiments similar to that of Fig. 13 has shown that the time-course of cytochrome c_2 oxidation can be accounted for not only on a first-order basis, but also on a second-order basis. [It was noted (Dutton et al., 1976) that the half-time of the "slow phase" became slightly slower as the cytochrome c_2:RC ratio was increased to unity, and then became faster as the ratio was increased further. We have since realized that this is a characteristic of a second-order process.] If the reaction is second-order, the apparent second-order rate constant of 10^9 M^{-1} sec^{-1} is high, which is appropriate for a reaction occurring from a "bound" state.

Our view of "close" and "distant" configurations, both *in vivo* and *in vitro*, has not changed, although we now extend the view to the notion that there might not be distinct binding sites, but rather an interaction with an area(s) on the RC on which an electron can be passed rapidly from ferrocytochrome to P^+. In such a case, the cytochrome might be mobile on an area on the RC (perhaps large enough to accommodate four cytochromes, although only inhabited by two).

2. The redox agent that reduces flash-generated ferricytochrome c_2 ($E_m = +150$ mV, pH 7.0) appears to have an n value of 2 (Prince and Dutton, 1976*b*, 1977*b*).

3. The redox agent ($E_m = +150$, pH 7.0) that when reduced before activation leads to the formation of the carotenoid bandshift phase III also has an n value of 2.

4. The "second" proton bound to the membrane following a flash, which was first recognized by Cogdell *et al.* (1972), has been found not to require valinomycin to promote the binding, as was originally thought (Petty *et al.*, 1977). The half-time of this "second" antimycin-sensitive binding can be the same as the "first" antimycin-insensitive proton binding. We call the protons H_{II}^+ and H_I^+, respectively.

5. Both H_I^+ and H_{II}^+ (i.e., 2 H^+/e^-) are bound following every single turnover flash for at least eight turnovers and probably many more. For a full 2.0 stoichiometry beyond three or four turnovers, valinomycin is required to collapse opposing electrical constraints. As stated in the main text, antimycin permits only 1.0 H^+ (H_I^+) bound following each turnover. The agent that binds H_I^+ is considered (see the main text) to be UQ on the outside of the membrane, which is reduced to Q^- by the RC UQ complement, and is subsequently protonated to form $Q^.H$. The agent that binds H_{II}^+ is unknown (unless Q^- is doubly protonated to form QH_2^-!). Since H_{II}^+ can be bound faster than cytochrome b_{50} reduction, this is not thought to be directly responsible, although an indirect interaction is possible (H_{II}^+ is antimycin-sensitive). There is evidence that there is also interaction (source unknown) between cytochrome b_{50} and H_I^+; it seems that the prior state of reduction of cytochrome b_{50} established at equilibrium can prevent H_I^+ binding. From these data, we conclude that the redox dependence of flash-induced H^+ binding does not have any apparent bearing on the state of reduction of the component that actually binds the H^+.

6. Antimycin stops cytochrome c_2 re-reduction on the inner side of the chromatophore, the carotenoid bandshift phase III (the electron transfer from cytochrome b_{-90} to b_{50} across the membrane?) (see the main text), and H_{II}^+ binding on the outside of the membrane at the first turnover of the cycle. Antimycin titrations eliminate all three events concomitantly at a titer of one molecule of antimycin per RC (i.e., per electron-transport system).

7. Taking the foregoing information into account, we currently consider the outer Q complement to have an E_m for the $Q/Q^.H$ couple in the -100- to -180-mV range (-180 mV is the operating E_m of the primary acceptor QFe designated X in the text; see Chapter 24). At equilibrium, however, 17–20 of this Q complement have an n value of 2 and an $E_{m,7}$ of $+90$ mV. Thus if the $Q/Q^.H$ couple is, say, -150 mV then the $Q^.H/QH_2$ couple is $+330$ mV! (Takamiya *et al.*, 1978). Much can be made of this new information.

8. The $+150$ mV ($n = 2$) ZH_2/Z component is thought to be a Q. There is no direct experimental evidence for this, although Q extraction eliminates cytochrome c_2 reduction (Baccarini-Melandri and Melandri, 1977). Our evidence (Prince *et al.*, 1978) suggests that if it is a Q, then it is a single Q in a protein matrix that controls the Q reaction behavior and specificity. Details of the partial reactions are still lacking.

9. It has been established that there is one Z per system (actually 0.8 Z/RC). The reaction of ZH_2 with ferricytochrome c_2 is second-order. It seems clear that an oxidized cytochrome c_2, firmly associated with a reaction center protein, is capable of being reduced by any of at least 4 reductase-bound ZH_2 (Prince *et al.*, 1978). At this point the reader is referred to Petty *et al.* (1977) and Prince and Dutton (1977*b*) for further discussions.

8. References

Baccarini-Melandri, A., and Melandri, A. B., 1977, A role for ubiquinone-10 in the b-c_2 segment of the photosynthetic bacterial electron transport chain?, *FEBS Lett.* **80**:459–464.

Baltscheffsky, M., 1967, Inorganic pyrophosphate as an energy donor in photosynthetic and respiratory electron transport phosphorylation systems, *Biochem. Biophys. Res. Commun.* **28**:270–276.

Baltscheffsky, M., 1974, The effect of dibromothymoquinone on light induced reactions in chromatophores from *R. rubrum*, in: *Dynamics of Energy Transducing Membranes* (L. Ernster, R. W. Estabrook, and E. C. Slater, eds.), pp. 365–376, Elsevier, Amsterdam.

Bartsch, R. G., 1963, Non-heme iron proteins and *C. vinosum* iron protein, in: *Bacterial Photosynthesis* (H.

Gest, A. San Pietro, and L. P. Vernon, eds.), pp. 315–326, Antioch Press, Yellow Springs, Ohio.

Bartsch, R. G., 1971, Cytochromes: Bacterial, in: *Methods in Enzymology* (A. San Pietro, ed.), pp. 344–363, Academic Press, New York.

Bartsch, R. G., and Kamen, M. D., 1960, Isolation and properties of two soluble heme proteins in extracts from the photo-anaerobe *C. vinosum, J. Biol. Chem.* **235**:825–831.

Buchanan, B. B., and Evans, M. C. W., 1969, Photo reduction of· ferredoxin and its use in NAD(P)⁺ reduction by a subcellular preparation from the photosynthetic bacterium *Chl. limicola f. thiosulfatophilum, Biochim. Biophys. Acta* **180**:123–129.

Carithers, R. P., and Parson, W. W., 1975, Delayed fluorescence from *Rps viridis* using single flashes, *Biochim. Biophys. Acta* **387**:194–211.

Carr, N. G., and Exell, G., 1965, Ubiquinone concentrations in Athiorhodaceae grown under various environmental conditions, *Biochem. J.* **96**:688–692.

Case, G. D., and Leigh, J. S., 1974, Intramembrane location of photosynthetic electron carriers in *C. vinosum* revealed by EPR interactions with gadolinium, in: *Proceedings of the 11th Rare Earth Research Conference*, Traverse City, Michigan, October 7–10.

Case, G. D., and Parson, W. W., 1971, Thermodynamics of the primary and secondary photochemical reactions in *C. vinosum, Biochim. Biophys. Acta* **253**:187–202.

Case, G. D., and Parson, W. W., 1973, Shifts of bacteriochlorophyll and carotenoid absorption bands linked to cytochrome c_{555} photo-oxidation in *Chr. vinosum, Biochim. Biophys. Acta* **325**:441–453.

Case, G. D., Parson, W. W., and Thornber, J. P., 1970, Photo-oxidation of cytochromes in reaction center preparations from *C. vinosum* and *Rps. viridis, Biochim. Biophys. Acta* **223**:122–128.

Chance, B., 1972, The nature of electron transfer and energy coupling reactions, *FEBS Lett.* **23**:3–20.

Chance, B., and Nishimura, M., 1960, On the mechanism of chlorophyll–cytochrome interaction: The temperature insensitivity of light induced cytochrome oxidation in *C. vinosum, Proc. Natl. Acad. Sci. U.S.A.* **46**:19–24.

Chance, B., and Smith, L., 1955, Respiratory pigments of *R. rubrum, Nature (London)* **175**:803–806.

Chance, B., and Williams, G. R., 1956, The respiratory chain and oxidative phosphorylation, in: *Advances in Enzymology and Related Subjects* (F. F. Nord, ed.), Vol. 17, pp. 65–134, Interscience Publishers, New York.

Chance, B., DeVault, D., Legallais, V., Mela, L., and Yonetani, T., 1967, Kinetics of electron transfer reactions in biological systems, in: *Nobel Symposium 5: Fast Processes in Chemical Kinetics* (S. Claesson, ed.), pp. 437–468, Interscience, Publishers, New York.

Chance, B., Crofts, A. R., Nishimura, M., and Price, B., 1970, Fast membrane H⁺ binding in the light activated state of *C. vinosum* chromatophores, *Eur. J. Biochem.* **13**:364–374.

Clark, W. M., 1960, *Oxidation–Reduction Potentials of Organic Systems*, Williams and Wilkins, Baltimore.

Clayton, R. K., and Wang, R. T., 1971, Photochemical

reaction centers from *Rps. sphaeroides*, in: *Methods in Enzymology* (A. San Pietro, ed.), Vol. 23A, pp. 696–704, Academic Press, New York.

Cogdell, R. J., and Crofts, A. R., 1972, Some observations on the primary acceptor of *Rp. viridis, FEBS Lett.* **27**:176–178.

Cogdell, R. J., and Crofts, A. R., 1974, H⁺ uptake by chromatophores from *Rps. sphaeroides*. The relation between rapid H⁺ uptake and the H⁺ pump, *Biochim. Biophys. Acta* **347**:264–272.

Cogdell, R. J., Jackson, J. B., and Crofts, A. R., 1972, The effect of redox potential on the coupling between rapid hydrogen ion binding and electron transport in chromatophores from *Rps. sphaeroides, Bioenergetics* **4**:211–227.

Cogdell, R. J., Prince, R. C., and Crofts, A. R., 1973, Light induced H⁺ uptake by photochemical reaction centers from *Rps. sphaeroides* R-26, *FEBS Lett.* **35**:204–208.

Connelly, J. L., Jones, O. T. G., Saunders, V. A., and Yates, D. W., 1973, Kinetic and thermodynamic properties of membrane bound cytochromes of aerobically and photosynthetically grown *Rps. sphaeroides, Biochim. Biophys. Acta* **292**:644–653.

Crofts, A. R., Prince, R. C., Holmes, N. G., and Crowther, D., 1974, Electrogenic electron transport and the carotenoid change in photosynthetic bacteria, in: *Proceedings of the Third International Congress on Photosynthesis* (M. Avron, ed), pp. 1131–1146, Elsevier, Amsterdam.

Crofts, A. R., Crowther, D., and Tierney, G. V., 1975, Electrogenic electron transport in photosynthetic bacteria, in: *Electron Transport and Oxidative Phosphorylation* (E. Quagliariello, S. Papa, F. Palmieri, E. C. Slater, and N. Silliprandi, eds.), pp. 233–241, North Holland/American Elsevier, Amsterdam/New York.

Cusanovich, M. A., Bartsch, R. G., and Kamen, M. D., 1968, Light induced electron transport in *C. vinosum.* II. Light induced absorbance changes in *Chr. vinosum* chromatophores, *Biochim. Biophys. Acta* **153**:397–417.

Davis, K. A., Hatefi, Y., Salemme, F. R., and Kamen, M. D., 1972, Enzymic redox reactions of cytochromes *c, Biochem. Biophys. Res. Commun.* **49**:1329–1335.

DeKlerk, H., and Kamen, M. D., 1966, A high potential non-heme iron protein from the facultative photoheterotrophe *Rps. gelatinosa, Biochim. Biophys. Acta* **112**:175–178.

DeVault, D., and Chance, B., 1966, Studies of photosynthesis using a pulsed laser. I. Temperature dependence of cytochrome oxidation rate in *Chr. vinosum*. Evidence for tunnelling, *Biophys. J.* **6**:825–847.

Dutton, P. L., 1971, Oxidation–reduction potential dependence of the interactions of cytochromes, bacteriochlorophyll and carotenoids at 77°K in chromatophores of *C. vinosum* and *Rps. gelatinosa, Biochim. Biophys. Acta* **226**:63–80.

Dutton, P. L., 1974, Cytochrome c_2–P870 membrane interactions in *Rps. sphaeroides, Fed. Proc. Fed. Am. Soc. Exp. Biol.* **33** (abstract 1346).

Dutton, P. L., and Baltscheffsky, M., 1972, Oxidation–reduction potential dependence of pyrophosphate induced cytochrome and bacteriochlorophyll reactions in *R. rubrum, Biochim. Biophys. Acta* **267**:172–178.

Dutton, P. L., and Jackson, J. B., 1972, Thermodynamic and kinetic characterization of electron transport components *in situ* in *Rps. sphaeroides* and *R. rubrum*, *Eur. J. Biochem.* **30**:495–510.

Dutton, P. L., and Leigh, J. S., 1973, ESR characterization of *C. vinosum* hemes, non-heme irons and the components involved in the primary photochemistry, *Biochim. Biophys. Acta* **314**:178–190.

Dutton, P. L., and Wilson, D. F., 1974, Redox potentiometry in mitochondrial and photosynthetic bioenergetics, *Biochim. Biophys. Acta* **346**:165–212.

Dutton, P. L., Kihara, T., and Chance, B., 1970, Early reactions in photosynthetic energy conservation. The photo-oxidation at liquid nitrogen temperatures of two cytochromes in chromatophores of *Rps. gelatinosa*, *Arch. Biochem. Biophys.* **139**:236–240.

Dutton, P. L., Kihara, T., McCray, J. A., and Thornber, J. P., 1971, Cytochrome c_{553} and bacteriochlorophyll interaction at 77°K in chromatophores and a sub-chromatophore preparation in *C. vinosum*, *Biochim. Biophys. Acta* **226**:81–87.

Dutton, P. L., Morse, S. D., and Wong, A. M., 1974, The thermodynamic and kinetic relationship of the electron and proton in redox titrations, in: *Dynamics of Energy Transducing Systems* (L. Ernster, R. W. Estabrook, and E. C. Slater, eds.), pp. 233–241, Elsevier, Amsterdam.

Dutton, P. L., Petty, K. M., Bonner, H. S., and Morse, S. D., 1975, Cytochrome c_2 and reaction center of *Rps. sphaeroides* Ga membranes. Extinction coefficients, content, half-reduction potentials, kinetics and electric field alterations, *Biochim. Biophys. Acta* **387**:536–556.

Dutton, P. L., Petty, K. M., and Prince, R. C., 1976, *Rps. sphaeroides* cytochrome c_2-reaction center interaction: *In vitro* reconstitution of *in vivo* kinetics, *Fed. Proc. Fed. Am. Soc. Exp. Biol.* **35**:1597.

Duysens, L. N. M., 1954, Reversible photo-oxidation of a cytochrome pigment in photosynthesizing *R. rubrum*, *Nature (London)* **173**:692–693.

Eimhjellen, K. E., Steensland, H., and Traetteberg, J., 1967, A *Thiococcus* sp. nov. gen., its pigments and internal membrane system, *Arch. Mikrobiol.* **59**:82–92.

Evans, E. H., 1973, A thermodynamic characterization of *Rps. capsulata*, Ph.D. thesis, University of Bristol.

Evans, E. H., and Crofts, A. R., 1974a, A thermodynamic characterization of the cytochromes of chromatophores of *Rps. capsulata*, *Biochim. Biophys. Acta* **357**:78–88.

Evans, E. H., and Crofts, A. R., 1974b, In situ characterization of photosynthetic electron flow in *Rps. capsulata*, *Biochim. Biophys. Acta* **357**:89–102.

Evans, M. C. W., and Buchanan, B. B., 1965, Photoreduction of ferredoxin and its use in CO_2 fixation by a subcellular system from a photosynthetic bacterium, *Proc. Natl. Acad. Sci. U.S.A.* **53**:1420–1425.

Evans, M. C. W., Lord, A. V., and Reeves, S. G., 1974, The detection and characterization by ESR spectroscopy of iron–sulfur proteins and other electron transport components in chromatophores from the purple bacterium *C. vinosum*, *Biochim. J.* **138**:177–183.

Feher, G., 1971, Some chemical and physical properties of a bacterial reaction center particle and its primary photochemical reactants, *Photochem. Photobiol.* **14**:373–387.

Fowler, C. F., 1974, Evidence for a cytochrome *b* in green bacteria, *Biochim. Biophys. Acta* **357**:327–331.

Fowler, C. F., Nugent, N. A., and Fuller, R. C., 1971, The isolation and characterization of a photochemically active complex from *Cps. ethylica*, *Proc. Natl. Acad. Sci. U.S.A.* **68**:2278–2282.

Garcia, A., Vernon, L. P., and Mollenhauer, H., 1966, Properties of *C. vinosum* sub-chromatophore particles obtained by treatment with Triton X-100, *Biochemistry* **5**:2399–2407.

Gray, B. H., Fowler, C. F., Nugent, N. A., and Fuller, R. C., 1972, A reevaluation of the presence of low midpoint potential cytochrome 551.5 in the green photosynthetic bacterium, *Cps. ethylica*, *Biochem. Biophys. Res. Commun.* **47**:322–327.

Gray, B. H., Fowler, C. F., Nugent, N. A., Rigopoulos, N., and Fuller, R. C., 1973, Reevaluation of *Cps. ethylica* strain 2-K, *Int. J. Syst. Bacteriol.* **23**:256–264.

Halsey, Y. D., and Parson, W. W., 1974, Identification of ubiquinone on the secondary electron acceptor in the photosynthetic apparatus of *C. vinosum*, *Biochim. Biophys. Acta* **347**:404–416.

Hopfield, J. J., 1974, Electron transfer between biological molecules by thermally activated tunnelling, *Proc. Natl. Acad. Sci. U.S.A.* **71**:3640–3644.

Ingledew, W. J., and Ohnishi, T., 1975, Properties of the S-3 iron–sulfur center of succinate dehydrogenase in the intact respiratory chain of beef heart mitochondria, *FEBS Lett.* **54**:167–171.

Ingledew, W. J., and Prince, R. C., 1977, Thermodynamic resolution of the iron–sulfur centers of the succinic dehydrogenase of *Rps. sphaeroides*, *Arch. Biochem. Biophys.* **178**:303–307.

Jackson, J. B., and Crofts, A. R., 1968, Energy linked reduction of NAD^+ in cells of *Rs. rubrum*, *Biochem. Biophys. Res. Commun.* **32**:908–915.

Jackson, J. B., and Crofts, A. R., 1969, The high energy state in chromatophores from *Rps. sphaeroides*, *FEBS Lett.* **4**:185–189.

Jackson, J. B., and Dutton, P. L., 1973, The kinetic and redox potentiometric resolution of the carotenoid shifts in *Rps. sphaeroides* chromatophores: Their relationship to electric field alterations in electron transport and energy coupling, *Biochim. Biophys. Acta* **325**:102–113.

Jackson, J. B., Crofts, A. R., and von Stedingk, L. V., 1968, Ion transport induced by light and antibiotics in chromatophores from *R. rubrum*, *Eur. J. Biochem.* **6**:41–54.

Jones, C. W., and Vernon, L. P., 1969, NAD^+ photoreduction in *R. rubrum* chromatophores, *Biochim. Biophys. Acta* **180**:149–164.

Jones, O. T. G., 1969, Multiple light induced reactions of cytochromes *b* and *c* in *Rps. sphaeroides*, *Biochem. J.* **114**:793–799.

Jones, O. T. G., and Saunders, V. A., 1972, Energy linked electron transfer in *Rps. viridis*, *Biochim. Biophys. Acta* **275**:427–436.

Jones, O. T. G., and Whale, F. R., 1970, The oxidation and reduction of pyridine nucleotides in *Rps. sphaeroides* and *Chl. limicola f. thiosulfatophilum*, *Arch. Mikrobiol.* **72**:48–59.

Kagawa, Y., and Racker, E., 1971, Partial resolution of the enzymes catalysing oxidative phosphorylation. XXV. Reconstitution of vesicles catalysing $^{32}P_1$–ATP exchange, *J. Biol. Chem.* **246**:5477–5487.

Ke, B., Chaney, T. H., and Reed, D. W., 1970, The electrostatic interaction between the reaction center bacteriochlorophyll derived from *Rps. sphaeroides* and mammalian cytochrome *c* and its effects on light activated electron transport, *Biochim. Biophys. Acta* **216**:373–383.

Keister, D. L., and Yike, N. J., 1967a, Energy linked reactions in photosynthetic bacteria. I. Succinate linked ATP driven NAD$^+$ reduction by *R. rubrum* chromatophores, *Arch. Biochem. Biophys.* **121**:415–422.

Keister, D. L., and Yike, N. J., 1967b, Energy linked reactions in photosynthetic bacteria. II. The energy dependent reduction of NADP$^+$ by NADH in chromatophores of *R. rubrum*, *Biochemistry* **6**:3847–3857.

Kennel, S. J., and Kamen, M. D., 1971, Iron-containing proteins in *C. vinosum*. II. Purification and properties of cholate-solubilized cytochrome complex, *Biochim. Biophys. Acta* **253**:153–156.

Kihara, T., and Chance, B., 1969, Cytochrome photooxidation at liquid nitrogen temperatures in photosynthetic bacteria, *Biochim. Biophys. Acta* **189**:116–124.

Kihara, T., and Dutton, P. L., 1970, Light induced reactions of photosynthetic bacteria. I. Reactions in whole cells and in cell free extracts at liquid nitrogen temperatures, *Biochim. Biophys. Acta* **205**:196–204.

Kihara, T., and McCray, J. A., 1973, Water and cytochrome oxidation–reduction reactions, *Biochim. Biophys. Acta* **292**:297–309.

Kihara, T., Dutton, P. L., and Chance, B., 1969, Low temperature light induced cytochrome oxidation in photosynthetic bacteria, Third International Congress on Biophysics, Cambridge, Massachusetts, abstract IJ7.

Knaff, D. B., 1975, The effect of *o*-phenanthroline on the midpoint potential of the primary electron acceptor of photosystem II, *Biochim. Biophys. Acta* **376**:583–587.

Knaff, D. B., and Buchanan, B. B., 1975, Cytochromes *b* and photosynthetic sulfur bacteria, *Biochim. Biophys. Acta* **376**:549–560.

Knaff, D. B., Buchanan, B. B., and Malkin, R., 1973, Effect of oxidation–reduction potential on light-induced cytochrome and bacteriochlorophyll reactions in chromatophores from the photosynthetic green bacterium *Chl. limicola f. thiosulfatophilum*, *Biochim. Biophys. Acta* **325**:94–101.

Leigh, J. S., 1970, ESR rigid-lattice line shape in a system of two interacting spins, *J. Chem. Phys.* **52**:2608–2612.

Loach, P. A., and Sekura, D. L., 1968, Primary photochemistry and electron transport in *R. rubrum*, *Biochemistry* **7**:2642–2649.

Loach, P. A., Androes, G. M., Maksim, A. F., and Calvin, M., 1963, Variation in EPR signals of photosynthetic

systems with the redox level of their environment, *Photochem. Photobiol.* **2**:443–454.

Margoliash, E., Ferguson-Miller, S., Tulloss, J., Kang, C. H., Feinberg, B. A., Brautigan, D. L., and Morrison, M., 1970, Separate intermolecular pathways for reduction and oxidation of cytochrome *c* in electron transport reactions, *Proc. Natl. Acad. Sci. U.S.A.* **70**:3245–3249.

Meyer, T. E., Bartsch, R. G., Cusanovich, M. A., and Mathewson, J. H., 1968, The cytochromes of *Chl. limicola f. thiosulfatophilum*, *Biochim. Biophys. Acta* **153**:854–861.

Meyer, T. E., Bartsch, R. G., and Kamen, M. D., 1971, Cytochrome c_3; a class of electron transfer heme proteins found in both photosynthetic and sulfate-reducing bacteria, *Biochim. Biophys. Acta* **245**:453–464.

Meyer, T. E., Kennel, S. J., Tedro, S. M., and Kamen, M. D., 1973, Iron protein content of *T. pfennigii*, a purple sulfur bacterium of atypical chlorophyll composition, *Biochim. Biophys. Acta* **292**:634–643.

Mitchell, P., 1961, Coupling of phosphorylation to electron and hydrogen transfer by a chemiosmotic type of mechanism, *Nature (London)* **191**:144–148.

Mitchell, P., 1966, *Chemiosmotic Coupling in Oxidative and Photosynthetic Phosphorylation*, Glynn Research, Bodmin, England.

Mitchell, P., 1972, Structural and functional organization of energy-transducing membranes and their ion-conducting properties, in: *Mitochondria/Biomembranes* (S. G. Van der Bergh, P. Borst, L. L. M. Van Deenen, J. C. Riemersma, E. C. Slater, and J. M. Tager, eds.), Vol. 28, pp. 353–372, North Holland/American Elsevier, Amsterdam/New York.

Mitchell, P., 1975a, Proton motive redox mechanism of the cytochrome *b*–*c₁* complex in the respiratory chain: Proton motive ubiquinone cycle, *FEBS Lett.* **56**: 1–6.

Mitchell, P., 1975b, The proton motive Q cycle: A general formulation, *FEBS Lett.* **59**:137–139.

Morita, S., 1968, Evidence for three photochemical systems in *C. vinosum*, *Biochim. Biophys. Acta* **153**:241–247.

Nicholls, P., 1974, Cytochrome *c* binding to enzymes and membranes, *Biochim. Biophys. Acta* **346**:261–310.

Nishimura, M., 1963, Studies on electron transport systems in photosynthetic bacteria. II. The effect of HOQNO and antimycin on the photosynthetic and respiratory electron transport system, *Biochim. Biophys. Acta* **66**:17–21.

Ohnishi, T., 1973, Mechanism of electron transport and energy conservation in the Site I region of the respiratory chain, *Biochim. Biophys. Acta* **301**:105–128.

Olson, J. M., 1973, Historical note on *Cps. ethylica* strain 2-K, *Int. J. Syst. Bacteriol.* **23**:265–266.

Olson, J. M., and Nadler, K. D., 1965, Energy transfer and cytochrome function in a new type of photosynthetic bacterium, *Photochem. Photobiol.* **4**:783–791.

Olson, J. M., Carroll, J. W., Clayton, M. L., Gardner, G. M., Linkins, A. E., and Moreth, C. M. C., 1969, Light induced absorbance changes of cytochromes and carotenoids in a sulfur bacterium containing bacteriochlorophyll *b*, *Biochim. Biophys. Acta* **172**:338–339.

Olson, J. M., Prince, R. C., and Brune, D. C., 1976, Reaction center complexes from green bacteria, *Brookhaven Symp. Biol.* **28**:238–246.

Orlando, J. A., 1962, *Rps. sphaeroides* cytochrome c_{553}, *Biochim. Biophys. Acta* **57**:373–375.

Orlando, J. A., and Horio, T., 1961, Observations on a b-type cytochrome from *Rps. sphaeroides, Biochim. Biophys. Acta* **50**:367–369.

Parson, W. W., 1968, The role of P870 in bacterial photosynthesis, *Biochim. Biophys. Acta* **153**:248–259.

Parson, W. W., 1969, Cytochrome photo-oxidation in *C. vinosum* chromatophores. Each P870 oxidizes two cytochrome c_{422} hemes, *Biochim. Biophys. Acta* **189**:397–403.

Parson, W. W., and Case, G. D., 1970, In *C. vinosum*, a single photochemical reaction center oxidizes both cytochrome c_{552} and cytochrome c_{555}, *Biochim. Biophys. Acta* **205**:232–245.

Parson, W. W., Clayton, R. K., and Cogdell, R. J., 1975, Excited states of photosynthetic reaction centers at low redox potentials, *Biochim. Biophys. Acta* **387**:265–278.

Peters, G. A., and Cellarius, R. A. 1972, The ubiquinone homologue of the green mutant of *Rps. sphaeroides, Biochim. Biophys. Acta* **256**:544–547.

Petty, K. M., and Dutton, P. L., 1976a, Properties of the flash-induced proton binding reaction of *Rps. sphaeroides* membranes; a functional pK on the ubisemiquinone?, *Arch. Biochem. Biophys.* **172**:335–345.

Petty, K. M., and Dutton, P. L., 1976b, Ubiquinone–cytochrome b electron and proton transfer: A functional pK on cytochrome b_{50} in *Rps. sphaeroides* membranes, *Arch. Biochem. biophys.* **172**:346–353.

Petty, K. M., Jackson, J. B., and Dutton, P. L., 1977, Kinetics and stoichiometry of proton binding in *Rhodopseudomonas sphaeroides* chromatophores, *FEBS Lett.* **84**:299–303.

Prince, R. C., 1978, The reaction center and associated cytochromes of *T. pfennigii*: Their thermodynamic and spectroscopic properties, and their possible location within the photosynthetic membrane, *Biochim. Biophys. Acta* **501**:195–207.

Prince, R. C., and Crofts, A. R., 1973, Photochemical reaction centers from *Rps. capsulata* Ala pho +, *FEBS Lett.* **35**:213–216.

Prince, R. C., and Dutton, P. L., 1975, A kinetic completion of the cyclic photosynthetic electron pathway of *Rps. sphaeroids*: Cytochrome b–cytochrome c_2 oxidation–reduction, *Biochim. Biophys. Acta* **287**:609–613.

Prince, R. C., and Dutton, P. L., 1976a, The primary acceptor of bacterial photosynthesis: Its operating midpoint potential?, *Arch. Biochem. Biophys.* **172**:329–334.

Prince, R. C., and Dutton, P. L., 1976b, The cyclic electron transport system of *Rps. sphaeroides*, in: *Abstracts of the International Conference on the Primary Electron Transport and Energy Transduction in Photosynthetic Bacteria*, Vrije Universiteit Brussel, TB4.

Prince, R. C., and Dutton, P. L., 1977a, The pH dependence of the oxidation–reduction midpoint potential of cytochromes c_2 *in vivo*, *Biochim. Biophys. Acta* **459**:573–577.

Prince, R. C., and Dutton, P. L., 1977b, Single and multiple turnover reactions in the ubiquinone–cytochrome b–c_2

oxidoreductase of *Rhodopseudomonas sphaeroides*: The physical chemistry of the major donor to cytochrome c_2 and its coupled reactions, *Biochim. Biophys. Acta* **462**:731–747.

Prince, R. C., and Olson, J. M., 1976, Some thermodynamic and kinetic properties of the primary photochemical reactants in a complex from a green photosynthetic bacterium, *Biochim. Biophys. Acta* **432**:357–362.

Prince, R. C., Cogdell, R. J., and Crofts, A. R., 1974a, The photo-oxidation of horse heart cytochrome c and native cytochrome c_2 by reaction centers from *Rps. sphaeroides* R-26, *Biochim. Biophys. Acta* **347**:1–13.

Prince, R. C., Hauska, G., and Crofts, A. R., 1974b, Asymmetry of the bacterial membrane: The localization of cytochrome c_2 in *Rps. sphaeroides, Biochem. Soc. Trans.* **2**:534–537.

Prince, R. C., Leigh, J. S., and Dutton, P. L., 1974c, An ESR characterization of *Rps. capsulata, Biochem. Soc. Trans.* **2**:950–953.

Prince, R. C., Baccarini-Melandri, A., Hauska, G. A., Melandri, B. A., and Crofts, A. R., 1975a, Asymmetry of an energy transducing membrane. The location of cytochrome c_2 in *Rps. sphaeroides* and *Rps. capsulata, Biochim. Biophys. Acta* **387**:212–227.

Prince, R. C., Lindsay, J. G., and Dutton, P. L., 1975b, The Rieske iron–sulfur center in mitochondrial and photosynthetic systems: E_m/pH relationships, *FEBS Lett.* **51**:108–111.

Prince, R. C., Bashford, C. L., and Dutton, P. L., 1978, Interactions of cytochrome c_2 with its oxidase and reductase *in vivo*; organization dynamics, *Fed. Proc.* **37**:1519.

Reed, D. W., and Clayton, R. K., 1968, Isolation of a reaction center fraction from *Rps. sphaeroides, Biochem. Biophys. Res. Commun.* **30**:471–475.

Reed, D. W., Raveed, D., and Reporter, M., 1975, Localization of photosynthetic reaction centers by antibody binding to chromatophore membranes of *Rps. sphaeroides* R-26, *Biochim. Biophys. Acta* **387**:368–378.

Rieske, J. S., Hansen, R. E., and Zaugg, W. S., 1964, Studies on the electron transfer system. LVIII. Properties of a new oxidation–reduction component of the respiratory chain as studied by EPR, *J. Biol. Chem.* **239**:3017–3022.

Rubin, L. B., Rubin, A. B., Dubrovin, V. N., and Shvinka, Y. E., 1969, Two systems of primary reactions in the photosynthesis of the purple bacterium *Rhodopseudomonas* sp., *Mol. Biol. (USSR)* **3**:552–558.

Rubin, L. B., Dubrovin, V. N., Adanova, R. P., and Shvinka, Y. E., 1970, The kinetics of photo-induced transformation of cytochromes of *E. shaposhnikovii, Microbiology (USSR)* **39**:229–232.

Salemme, F. R., Freer, S. T., Xuong, Ng. H., Alden, R. A., and Kraut, J., 1973, The structure of oxidized cytochrome c_2 from *R. rubrum, J. Biol. Chem.* **248**:3910–3921.

Saunders, V. A., and Jones, O. T. G., 1974, Properties of the cytochrome a-like material developed in the photosynthetic bacterium *Rps. sphaeroides* when grown aerobically, *Biochim. Biophys. Acta* **333**:439–445.

Saunders, V. A., and Jones, O. T. G., 1975, Detection of two further b-type cytochromes in *Rps. sphaeroides, Biochim. Biophys. Acta* **396**:220–228.

Seibert, M., 1971, Spectral, kinetic and potentiometric studies of the laser induced primary photochemical reactions in the photosynthetic bacterium *C. vinosum,* Ph.D. thesis, University of Pennsylvania, Philadelphia.

Seibert, M., and DeVault, D., 1970, Relationship between the laser-induced oxidation of the high and low potential cytochromes of *C. vinosum, Biochim. Biophys. Acta* **205**:222–231.

Shioi, Y., Takamiya, K. I., and Nishimura, M., 1976, Isolation and some properties of NAD$^+$ reductase of the green photosynthetic bacterium *Prosthecochloris aestuarii, J. Biochem.* **79**:361–371.

Slater, E. C., 1973, The mechanism of action of the respiratory inhibitor antimycin, *Biochim. Biophys. Acta* **301**:129–154.

Slichter, C. P., 1963, *Principles of Magnetic Resonance,* Harper and Row, New York.

Smith, L., Davies, H. C., Reichlin, M., and Margoliash, E., 1973, Separate oxidase and reductase reaction sites on cytochrome *c* demonstrated with purified site-specific antibodies, *J. Biol. Chem.* **248**:237–243.

Steiner, L. A., Okamura, M. Y., Lopes, A. D., Moskowitz, E., and Feher, G., 1974, Characterization of reaction centers from photosynthetic bacteria. II. Amino acid composition of the reaction center protein and its subunits in *Rps. sphaeroides* R-26, *Biochemistry* **13**:1403–1410.

Sybesma, C., and Vredenberg, W. J., 1964, Kinetics of light-induced cytochrome oxidation and P840 bleaching in green photosynthetic bacteria under various conditions, *Biochim. Biophys. Acta* **88**:205–207.

Takamiya, K., Nishimura, M., and Dutton, P. L., 1978, Ubiquinone in *Rhodopseudomonas sphaeroides, Fed. Proc.* **37**:1519.

Thornber, J. P., 1970, Photochemical reactions of purple bacteria as revealed by three spectrally different

caroteno–bacteriochlorophyll–protein complexes isolated from *C. vinosum, Biochemistry* **9**:2688–2698.

Thornber, J. P., and Olson, J. M., 1971, Chlorophyll proteins and reaction center preparations from photosynthetic bacteria, algae and higher plants, *Photochem. Photobiol.* **14**:329–341.

Thornber, J. P., Olson, J. M., Williams, D. M., and Clayton, M. L., 1969, Isolation of the reaction center from *Rps. viridis, Biochim. Biophys. Acta* **172**:351–354.

Tiede, D. M., Prince, R. C., and Dutton, P. L., 1976, EPR and optical spectroscopic properties of the electron carrier intermediate between the reaction center bacteriochlorophylls and the primary acceptor in *Chr. vinosum, Biochim. Biophys. Acta* **449**:447–467.

Trebst, A., Harth, E., and Draber, W., 1970, On a new inhibitor of photosynthetic electron transport in isolated chloroplasts, *Z. Naturforsch.* **25b**:1157–1159.

Trüper, H. G., 1968, *E. mobilis* Pelsh, a photosynthetic sulfur bacterium depositing sulfur outside the cells, *J. Bacteriol.* **95**:1910–1920.

Vernon, L. P., and Kamen, M. D., 1953, Studies on the metabolism of photosynthetic bacteria. XV. Photo autoxidation of cytochrome *c* in extracts of *R. rubrum, Arch. Biochem. Biophys.* **44**:298–311.

Vredenberg, W. J., and Duysens, L. N. M., 1964, Light induced oxidation of cytochromes in photosynthetic bacteria between 20° and −170°, *Biochim. Biophys. Acta* **79**:456–463.

Wang, R. T., and Clayton, R. K., 1973, Isolation of photochemical reaction centers from a carotenoidless mutant of *R. rubrum, Photochem. Photobiol.* **17**:57–61.

Williams, R. J. P., 1969, Electron transfer and energy conservation, *Curr. Top. Bioenerg.* **3**:79–156.

Zannoni, D., Baccarini-Melandri, A., Melandri, B. A., Evans, E. H., Prince, R. C., and Crofts, A. R., 1974, Energy transduction in photosynthetic bacteria. The nature of the cytochrome *c* oxidase in the respiratory chain of *Rps. capsulata, FEBS Lett.* **48**:152–155.

CHAPTER 29

Bacteriorhodopsin

Walther Stoeckenius

1. General Physiology and Biochemistry of Halobacteria

The purple membrane (p.m.)* with its pigment bacteriorhodopsin (bR) is the light-energy-transducing organelle recently discovered in *Halobacterium halobium* (Stoeckenius and Kunau, 1968; Oesterhelt and Stoeckenius, 1973). Halobacteria require high concentrations of NaCl for growth and maintenance of structure; the optimal concentration for growth is approximately 25% and the minimal concentration approximately 15% NaCl. The most intensively studied species of the genus *Halobacterium* are *H. halobium, H. cutirubrum,* and *H. salinarium,* but even these are poorly characterized, and *H. cutirubrum* and *H. salinarium* are considered to be identical by some. Several investigators have used strains isolated from natural sources that have not been classified further. Halobacteria occur in natural salt lakes and salterns,

* Abbreviations in this chapter: (DDA⁺) dibenzyldimethyl ammonium; (CCCP) ketomalononitrile 3-chlorophenylhydrazone (carbonyl-cyanide 3-chlorophenylhydrazone); (DCCD) *N,N'*-dicyclohexylcarbodiimide; (FCCP) ketomalononitrile 4-trifluoromethoxyphenylhydrazone (carbonyl-cyanide 4-trifluoromethoxyphenylhydrazone); (DNP) 2,6-diiodo-4-nitrophenol; (bR) bacteriorhodopsin; (p.m.) purple membrane; (KCN) potassium cyanide; (PMA) phenyl mercuric acetate; (SDS–PAGE) sodium dodecyl sulfate–polyacrylamide gel electrophoresis.

Walther Stoeckenius · Cardiovascular Research Institute and Department of Biochemistry and Biophysics, University of California, San Francisco, California 94143; Ames Research Center, NASA, Moffett Field, California 94035

and first attracted attention because they spoil salted fish and hides prepared with solar salt. [For general reviews on the biochemistry and physiology of halobacteria, see Larsen (1963, 1967), Brown (1964), and Gibbons (1974).]

Typically, the cells of *H. halobium* and *H. salinarium* appear as motile rods 0.6 μm thick and 4–6 μm long with polar flagella; however, shape and size vary considerably with growth conditions. They contain high concentrations of carotenoids, mainly bacterioruberin (Kelly and Liaaen-Jensen, 1967; Kushwaha *et al.,* 1974), which are responsible for the reddish color of cultures and also of their natural habitats when they are present in large numbers. *Halobacterium halobium* strains may also form intracellular gas vacuoles, which allow the cells to float to the surface in liquid media and may afford the cells an opportunity to select to some extent their environmental conditions with respect to oxygen tension and solar-radiation density. Experimental support for such a function has been obtained for the very similar gas vacuoles of some blue-green algae (Walsby, 1972) (see also Chapter 1). Most of the work reported here has been carried out with the spontaneous mutant R_1 of *H. halobium* NRC 34020, which has lost the ability to form gas vacuoles. We prefer this mutant because the presence of gas vacuoles interferes with the isolation of the p.m. and with spectroscopic measurement on whole cells. Even better suited for absorption spectroscopy of the p.m. *in vivo* are derivatives of strain R_1, such as R_1L_3 (our unpublished results) and R_1M_1 (Oesterhelt *et al.,* 1973), which contain much-reduced amounts of bacterioruberin.

While the morphology of halobacteria is similar to that of nonhalophilic bacteria, their chemistry shows some very peculiar features. It is not known, however, how these features are related to the extreme environment of the cells. The cell wall does not contain diaminopimelic or muramic acid. It consists mainly of protein with a rather small carbohydrate content (Mescher and Strominger, 1976). The lipids of the cell membrane are all isoprene derivatives. The phospholipids, which comprise 70% of the total lipid, and the glycolipids all contain dihydrophytanol-glycerol diether as the hydrophobic moiety. Most of the phospholipid is the diether analogue of phosphatidyl glycerol phosphate, and part of the glycolipids carry a sulfate ester on the terminal sugar (Kates, 1972). In addition to bacterioruberin, squalenes and vitamin MK8 have been found as the main neutral lipids (Kushwaha et al., 1974). The sulfoglycolipids appear to occur preferentially or even exclusively in the p.m. (Kushwaha et al., 1975). The cell membrane also carries a respiratory chain, which has been investigated by Lanyi (1968, 1972) and Cheah (1969, 1970). No intracellular membranes have been found except for the gas vacuole membranes, which do not contain any lipid (Stoeckenius and Kunau, 1968).

The proteins of the cell, with the exception of the p.m. and the gas vacuole proteins, show an unusually high concentration of dicarboxylic amino acids. This feature may be responsible at least in part for the disintegration of the cells in low salt concentrations (Soo-Hoo and Brown, 1967; Stoeckenius and Kunau, 1968; Kushner et al., 1964; Onishi and Kushner, 1966). This view, however, does not account for the specific sodium requirement of the cells, and as Lanyi (1974) points out, conformational changes in the proteins take place even at salt dilutions at which the electrostatic charges of the carboxyl groups should still be fully shielded by counterions. Most of the enzymes isolated from halobacteria require high salt concentrations for maximum activity and are irreversibly denatured at salt concentrations below 1.0 M (Lanyi, 1974).

The base composition of halobacteria DNA is not unusual; it has G + C content of 66–68 mol% on density gradients, however, a minor component with 57–60% G + C separates from the main band (Joshi et al., 1963; Moore and McCarthy, 1969). The function of this satellite band is unknown.

Halobacteria have been classified as chemoorganotrophs, relying on respiration as their main energy source; no fermentative pathway has been found in *H. halobium*, *H. cutirubrum*, or *H. salinarium*.* The main substrates for respiration appear to be amino acids; acetate and glycerol are also used (Katznelson and Robinson, 1956). *Halo-*

bacterium halobium also reduces nitrate, nitrite does not accumulate, and the role of this pathway in the energy metabolism of *H. halobium* is not known (A. Danon and W. Stoeckenius, unpublished). Halobacteria do not contain chlorophyll, but a halophilic chlorophyll-containing prokaryote, *Ectothiorhodospira halophila*, has been isolated from the same environment in which halobacteria occur (Raymond and Sistrom, 1967).

Few studies on the permeability of halobacteria have been published. It is generally agreed that the internal salt concentration is as high as or higher than that of the growth medium and that the main internal cation is K^+ (Ginzburg et al., 1970; Christian and Waltho, 1962). It has been reported that practically all the internal K^+ exists in a bound form, and that the cell membrane is permeable to solutes up to a molecular weight of 40,000 (Ginzburg, 1969). This claim is incompatible with most observations by others (Lanyi and Silverman, 1972; Bakker et al., 1976), including our own studies to be discussed below. It is difficult to account for the results of the Ginzburgs. They were obtained mostly with a *Halobacterium* species isolated by the authors and not used by others, but this is not likely to be the reason. It appears possible that their determination of the interior volume of the cells was in error, because the marker for the exterior volume was bound to the cells or because of osmotic effects on the cells.

2. Growth Conditions and Isolation Techniques for Purple Membrane

Laboratory growth conditions and isolation procedures for cell-envelope components including the p.m. were described by Stoeckenius and Kunau (1968) and Oesterhelt and Stoeckenius (1974). Use is made of the cell lysis at low salt concentrations and the difference in the susceptibility to disintegration at low Mg^{2+} concentrations of the cell wall, the p.m., and the rest of the surface membrane. The fractions are separated by differential and density-gradient centrifugation. Their different pigment contents serve as convenient visual markers. The p.m. stands out because of its deep purple color. The high concentrations of bacterioruberin and related pigments give the remainder of the surface membrane an

---

* It was recently pointed out, however, that isolation procedures used for these and most other strains may have selected against carbohydrate-fermenting strains, and a number of carbohydrate-utilizing strains have been isolated recently (Tomlinson and Hochstein, 1972). We have not been able to isolate p.m. from these strains (W. Stoeckenius, unpublished).

orange-red color, and it is usually referred to as the *red membrane*. It is not a homogeneous preparation, however, and may be further subfractionated into membrane fragments of different composition.

It should be pointed out that the use of very low ionic strength media in this technique inactivates enzymes, and the report of no enzyme activity in a fraction may be misleading. The cells are, however, very susceptible to a number of lytic agents such as cholate or deoxycholate at concentrations that leave the p.m. and many enzyme activities essentially intact. Any search for enzyme activity should be carried out on fractions obtained by such techniques, but no detailed protocol has so far been published.

The p.m. was first discovered in *H. halobium* (Stoeckenius and Rowen, 1967; Stoeckenius and Kunau, 1968), and most investigations have been carried out with this species and its mutants. Purple membrane has also been found in many other halobacteria isolates (Kushwaha *et al.*, 1974; Stoeckenius, unpublished), but a careful analysis of such p.m. preparations and an attempt to compare them with *H. halobium* p.m. has apparently been carried out only

with *H. cutirubrum* (Kushwaha and Kates, 1973; Kushwaha *et al.*, 1975; Kushwaha *et al.*, 1976). Work in Russia initiated by Skulachev has mainly used p.m. preparations from a locally isolated *Halobacterium* species (Kayushin *et al.*, 1974). No apparent differences between the *H. halobium* preparations and those of other halobacteria species have been reported so far.

Halobacterium halobium cells synthesize p.m. in substantial amounts only when grown under suboptimal concentrations of oxygen and irradiated with visible light (Oesterhelt and Stoeckenius, 1973). When growing cultures are transferred from aerobic to anaerobic conditions, growth ceases, and only the p.m. concentration of the cells increases sharply. In some instances, increases in cell numbers from four to six times have been observed under anaerobic conditions (Oesterhelt and Krippahl, 1973; Danon and Stoeckenius, unpublished). The extent to which growth under anaerobic conditions is possible is probably determined by the composition of the growth medium. No detailed studies on the requirements for maximal p.m. synthesis have been reported.

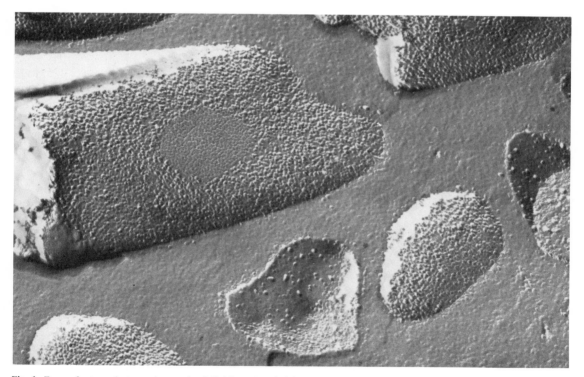

Fig. 1. Freeze–fracture electron micrograph of *Halobacterium halobium* cells containing only a small amount of p.m. The cytoplasmic fracture face of the membrane (A-face) is seen in the upper left quadrant. It contains a small patch of p.m., which can be distinguished from the surrounding red membrane by its smoother surface and a more regular particle pattern. The concave outer fracture faces (B-face) of the red membrane from two cells are seen in the lower right quadrant near the middle and the right edge of the picture. Magnification 84,000×.

3. Morphology

The p.m. is part of the cell membrane of *H. halobium*. In freeze–fracture electron micrographs, it appears as round or oval patches on the membrane fracture face and shows a regular pattern clearly different from the irregular particle distribution seen on the remainder of the membrane (Fig. 1). The p.m. patches are coplanar with the rest of the surface membrane. Light-diffraction analysis of the electron micrographs reveals the presence of a hexagonal crystal lattice in the patches. The centers of the hexagons are

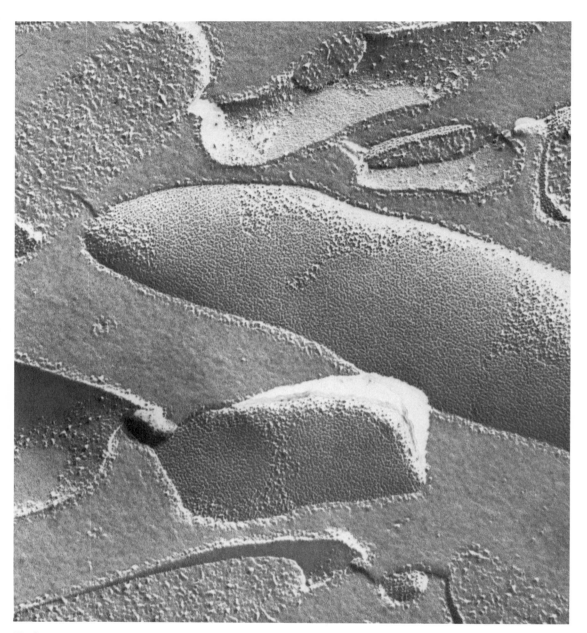

Fig. 2. Freeze–fracture electron micrograph of *Halobacterium halobium* cells containing high concentrations of p.m. The largest fracture face seen (A-face) is mainly occupied by p.m.; so is the smaller A-face directly below it. Above, below, and to the left are B-faces of other cells that show the outer fracture faces of p.m. patches distinguished from the surrounding red membrane by their complete lack of particles. Magnification 82,000×.

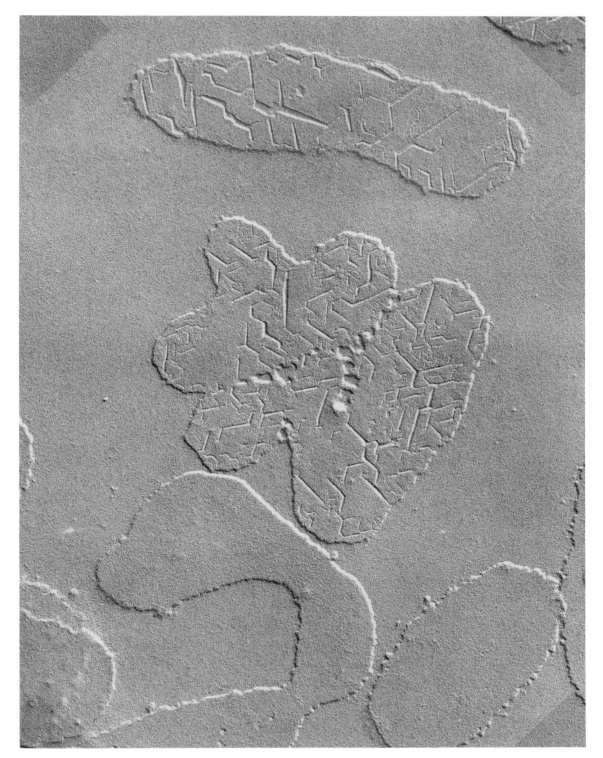

Fig. 3. Electron micrograph of isolated p.m. prepared by drying from a suspension and shadowing with platinum–carbon. The difference between the two surfaces is clearly visible. The exposed outer surface cracks along lattice planes on drying; the inner surface appears smooth. Magnification 69,120×.

63 Å apart (Blaurock and Stoeckenius, 1971). The patches increase in size and number with increasing p.m. concentration. They are roughly circular when they are small, and do not exceed a diameter of approximately 0.6 μm. Larger patches are elongated in the direction of the long axis of the rod-shaped cells. In cells containing high concentrations of p.m., the patches may occupy more than 50% of the total surface (Fig. 2). Usually, freeze–fracturing cleaves cells along an interior plane (Branton, 1969). The two

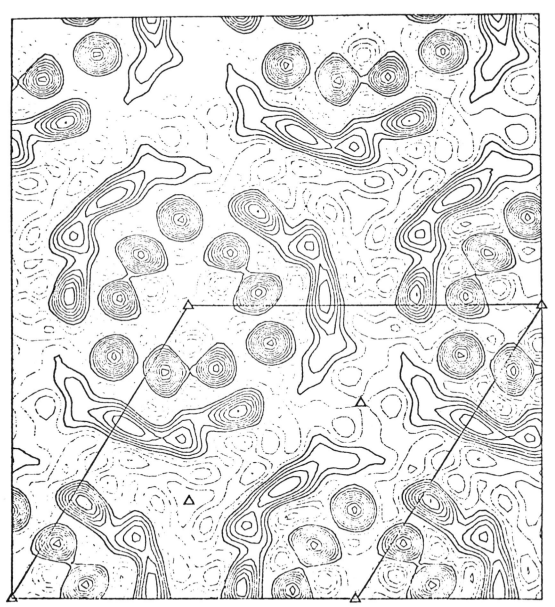

Fig. 4. Structure of the p.m. at 7 Å resolution projected onto the plane of the membrane. The figure is a reconstruction of the repeating unit from an electron micrograph of a single crystal patch taken at minimal beam exposure. The unit cell containing three molecules of bR is outlined in the bottom half of the figure. The triangles indicate three-fold axes of rotation. One molecule of bR presumably comprises two adjacent and approximately parallel rows of α helices; three of these double rows are grouped around the three-fold axes at the corners of the unit cell. The inner row of each pair contains three α-helices, which are seen essentially end-on; the outer row contains four, which are inclined at increasing angles to the plane of the membrane from one end of the row to the other. The spaces around the bR molecules are presumably filled by lipid molecules in a bilayer arrangement. Reproduced from Unwin and Henderson (1975).

complementary fracture faces differ in appearance, reflecting the asymmetry of the membrane. Protruding particles on one face should correspond to depressions on the other. Because of possible distortions during freeze–fracture and the limited resolution of the replication technique, the correspondence is usually not perfect, depressions generally being less clearly seen than protruding particles. In the p.m. patches, the cytoplasmic fracture face (A-face) appears rougher than the very smooth outer fracture face (B-face). This affords the opportunity to recognize the orientation of the p.m. in cell-envelope preparations and reconstituted model systems (Stoeckenius, 1976; Stoeckenius et al., 1976).

The surface of the p.m. is observed in isolated p.m. fractions by freeze–etching and by negative staining or shadowing of p.m. dried from suspension. In these preparations, the patches are preserved essentially intact and appear as round or oval sheets of the same dimensions seen in intact cells. The hexagonal lattice is much more difficult to demonstrate in these preparations because both surfaces show little profile. The external surface usually develops cracks along lattice planes on drying, however, whereas the inner surface appears perfectly smooth (Fig. 3).

The same hexagonal lattice seen in the electron micrographs is also revealed in X-ray diffraction patterns (Blaurock and Stoeckenius, 1971; Blaurock, 1975; Henderson, 1975). Patterns from intact cells and envelope preparations are weak because of diffuse scattering contributions from cell wall and other parts of the surface membrane and because of the high salt content of the suspending medium. Isolated p.m. fractions, however, give very high resolution patterns with sharp reflections out to a spacing of approximately 3.5 Å. in addition to the in-plane hexagonal lattice reflections, membrane profile diffraction of lower resolution is also recognizable. The reconstructed membrane profile shows a membrane thickness of 48 Å with two asymmetric electron-density peaks near the membrane surfaces. In contrast to profiles of other membranes, including the rest of the surface membrane of *H. halobium*, the central trough in the p.m. profile has a relatively high electron density, well above the density of the surrounding water. [The rest of the surface membrane shows an unusual feature: a continuous layer of protein approximately 40 Å wide is present on its outer surface, but is absent on the p.m. (Blaurock *et al.*, 1976).]

The hexagonal lattice of the p.m. has plane group P_3, containing three bR molecules per unit cell (see below). With a newly developed electron-microscopic technique, Unwin and Henderson (1975) recently revealed details of the protein structure (Fig. 4). They

first obtained a two-dimensional Fourier synthesis of its structure at 7 Å resolution and recently completed a three-dimensional map (Henderson and Unwin, 1975). It clearly shows three rows of three α-helices, each oriented at right angles to the plane of the membrane and grouped around a three-fold axis. A second outer row of four α-helices runs parallel to each of the inner rows, but the four helices in this row are inclined at increasing angles toward the plane of the membrane. The α-helices apparently span the central hydrophobic part of the membrane. These results are in good agreement with our CD spectra, which indicate an α-helix content in the protein of 70% or more (Yuen-Wen Tseng and W. Stoeckenius, unpublished).

From these investigations, a picture of the p.m. structure emerges. The bR molecules span the membrane and are all oriented in one direction. They form a very regular hexagonal lattice one molecule thick. One patch thus is a single two-dimensional crystal. The lipid fills the spaces around the bR molecules with its hydrophilic head groups oriented outward. When the membrane is cleaved during freeze–fracturing, the outer half of this discontinuous lipid layer is peeled off, while the bR molecules remain with the inner membrane lamella. As will be seen below, this arrangement fulfills the general structural requirements for the function of the p.m.

4. Chemistry

The p.m. contains approximately 75% protein and approximately 25% lipid (Stoeckenius and Kunau, 1968; Oesterhelt and Stoeckenius, 1971; Kushwaha and Kates, 1973). The main lipid is the diether analogue of phosphatidyl glycerol phosphate. Its hydrophobic moiety consists of two dihydrophytanol residues; i.e., it is the same as the main lipid of the rest of the surface membrane. Of the other cell-membrane lipids, only the sulfolipids appear to be specific for the p.m., and bacterioruberin is absent from it (Kushwaha et al., 1975). The small and somewhat variable quantities found probably have to be explained by the presence of residual red membrane in the p.m. fraction. In a typical preparation, this contamination amounts to less than 0.5% of residual surface membrane. Electron micrographs indicate that the residual red membrane is attached to the periphery of the p.m. patches.

In sodium dodecyl sulfate–polyacrylamide gel electrophoresis (SDS–PAGE) of p.m. dissolved in SDS or guanidine hydrochloride, only one protein

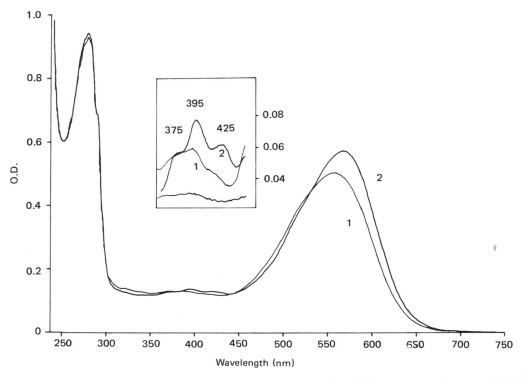

Fig. 5. Absorption spectra of the isolated p.m. suspended in water. Curve 1 is the dark-adapted preparation, curve 2 the light-adapted. The inset shows the region around 400 nm at higher resolution. The numbers above the traces indicate the positions of the absorbance peaks in nanometers for curve 2; the lowest trace is the base line.

band has been found. Amino acid analysis shows a high concentration of hydrophobic residues and one mole of histidine in 26,000 g protein; no cysteine was found (Oesterhelt and Stoeckenius, 1971).

The p.m. owes its color to a broad absorption band around 570 nm, which in a clean fraction has an absorption more than half as high as the maximum of the protein absorption band around 280 nm; minor bands are present at 425, 390, and 375 nm (Fig. 5). The major bands can be used as a rough indicator for the purity of the fraction; the 280:570 nm O.D. ratio should be 1.7 or less in a good preparation.* The absorption bands are due to the protein. They show only minor changes when virtually all the lipid is replaced by detergent and the extracted lipid is colorless. A molar absorption coefficient ε of 54,000 M cm^{-1} has been determined based on a conversion of retinal to the oxime and extraction from the denatured protein (Oesterhelt and Stoeckenius, 1971).

The visible absorption band is due to retinal bound as a Schiff base to an ε-amino group of a lysine residue in the protein (Oesterhelt and Stoeckenius, 1971; Bridgen and Walker, 1976; Schreckenback and Oesterhelt, 1976) (Fig. 6). The 380 nm absorption band of the retinal is red-shifted nearly 200 nm in the retinal protein complex; this is due partly to protonation of the Schiff base and partly to interaction with hydrophobic amino acid residues of the protein (Lewis et al., 1974). The protonated Schiff base is quite stable in the native protein and does not react with small water-soluble molecules such as hydroxylamine and NaBH$_4$ unless the preparation is

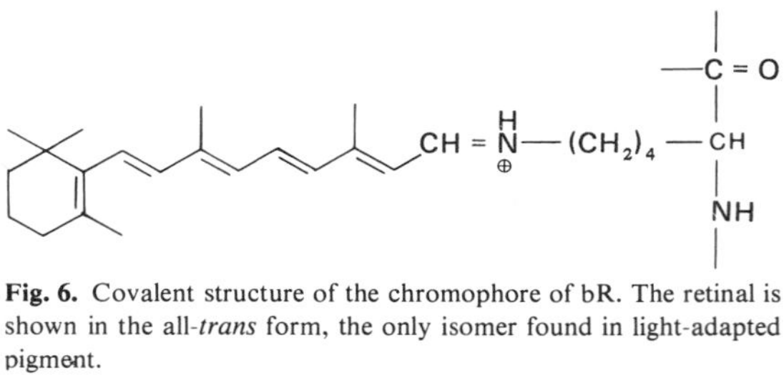

Fig. 6. Covalent structure of the chromophore of bR. The retinal is shown in the all-*trans* form, the only isomer found in light-adapted pigment.

* Note that because of light scattering in p.m. suspensions, this value also depends on the optics of the spectrophotometer used.

exposed to light. When the protein is denatured, the absorption maximum shifts to 370 nm, typical for a retinylidene protein, and the Schiff base hydrolyzes slowly at neutral or slightly alkaline pH and rapidly at acid pH (Oesterhelt and Stoeckenius, 1971). In strong light, the chromoprotein undergoes a cyclic photo-reaction (to be discussed below), during which the Schiff base becomes more accessible. This reaction has been used to remove the retinal without a large disturbance of the protein structure (Oesterhelt et al., 1974). The resulting apoprotein can be reconstituted with either 13-cis or all-trans retinal (Oesterhelt and Schuhmann, 1974). The protein thus very strongly resembles the visual pigments of animals, except for the difference in the retinal isomers active in reconstitution and the cyclic character of its photoreaction.

The native membrane exists in two forms with slightly differing absorption spectra (see Fig. 5). When the membrane is exposed to daylight or the usual artificial light of the laboratory, its absorption maximum is found at 568–570 nm. After prolonged incubation in the dark, it shifts to 560 nm with a slight loss of absorbance; illumination with light absorbed by the purple band reverses the shift, and absorbance decreases. We call these the *light-adapted* and *dark-adapted* forms of bR, bR_{570}^{LA} and bR_{560}^{DA}. Only all-*trans* retinal has been extracted from bR_{570}^{LA}, whereas mixtures of 13-*cis* and all-*trans* retinal have been obtained from bR_{560}^{DA} (Oesterhelt et al., 1973; Pettei et al., 1977), but one report claims that the chromophore of bR^{DA} contains 13-*cis* retinal only (Jan, 1975).

Some of the data given here must be considered tentative. A reevaluation of the molar absorption coefficient has yielded ε of 63,000 M cm^{-1} and is now generally used (Oesterhelt and Hess, 1973; our unpublished results). It is based on retinal extraction and determination according to Futterman and Saslaw (1961). Recent molecular-weight determinations of the protein by SDS–PAGE yielded values near 20,000 (Kushwaha et al., 1975). The best-supported value at present appears to be 25,500 $\pm$ 1000 from the work of Bridgen and Walker (1976). A substantial fraction of the asparagine and glutamine residues (Oesterhelt and Stoeckenius, 1971) may actually be aspartate and glutamate, and the histidyl content was recently found to be consistently slightly below rather than above 0.5 mol% in one laboratory and above 1.0 mol% in another. None of these differences should critically affect the other observations and conclusions reported here. It seems best at present to await the completion of the amino acid sequence analysis, which is presently being undertaken in several laboratories (Keefer and Bradshaw, 1976; Abdulaev et al., 1976), before a reevaluation and recalculation of the data is attempted.

In the meanwhile, the morphological and spectroscopic studies offer a sufficiently firm basis for further exploration of p.m. function.

5. Photophysiology

The first photoreactions of intact cells, which we observed, were phototactic responses (Stoeckenius, unpublished). *Halobacterium halobium* cells typically swim in a straight line. When the intensity of illumination is suddenly decreased in the red part of the spectrum or increased in the blue, the cells stop and then swim off in the opposite direction. The long-wavelength response is present only in cells that contain p.m., whereas the blue response is always observed. This investigation has since been considerably extended, but only short reports have appeared (Forster and Berg, personal communication; Dencher, 1974). The action spectrum for the long-wavelength response closely fits the absorption spectrum of isolated p.m., whereas the blue response has an action spectrum that could be attributed to a carotenoid or flavoprotein (Hildebrand and Dencher, 1975).

These observations clearly established a function for the p.m. in intact cells. In view of the large amounts of p.m. present, however, it seemed unlikely that this would be the only function; unicellular organisms typically contain only very low concentrations of sensory photoreceptors. The high concentration of bR suggested to us, rather, an energy-transducing function, which should manifest itself in influences of light on the ATP concentration in cells. These were indeed observed (Danon and Stoeckenius, 1974). When the cells are suspended in a salt solution without nutrients, they continue to respire on endogenous substrate for hours, and ATP levels remain high. When such cell suspensions are made anaerobic in the dark, the intracellular ATP level drops within a few minutes to values near 30% of the initial and then continues to decrease much more slowly. The cells apparently conserve a low ATP level by reducing energy consumption. In p.m.-containing cells, irradiation with visible light restores the ATP level within 1–2 min, typically to a value slightly higher than the initial aerobic dark level. This level is maintained as long as the illumination continues. Turning the light off causes an equally rapid return to the 30% level. Aeration of the suspension in the dark has essentially the same effect as light, but the steady-state ATP level typically is slightly lower (Fig. 7). A rough action spectrum of the light response corresponds closely to the absorption spectrum of the

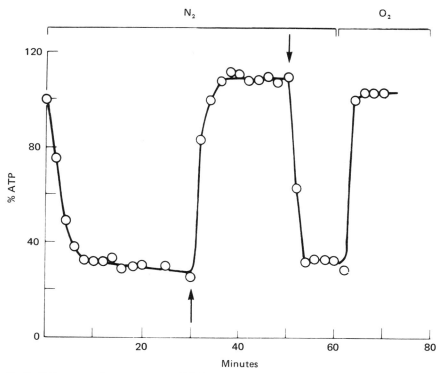

Fig. 7. ATP concentration as percentage of initial concentration in a suspension of *Halobacterium halobium* cells in basal salt solution. At zero time, the suspension was made anaerobic by bubbling with nitrogen (N_2). The upward-pointing arrow indicates the beginning, the downward pointing arrow the end, of illumination with white light. The O_2 at the top of the diagram indicates the time when nitrogen was replaced by air to again render the suspension aerobic.

p.m., and cells that do not contain p.m. show no effect of light on ATP content. The ATP increase caused by light or by oxygen is inhibited by uncouplers of oxidative phosphorylation and photophosphorylation such as DNP, CCCP, FCCP, and 1719, but only the respiratory effect is sensitive to the electron-transport inhibitors KCN, PMA, azide, and others. Both effects are abolished by inhibitors of the membrane-bound ATPases DCCD and Dio-9 (Danon and Stoeckenius, 1974).

These results strongly suggest that both light absorbed by the p.m. and electron transport through the respiratory chain can generate high energy intermediate(s) or membrane state(s) that drive ATP synthesis in the cell, and that ATP synthesis is mediated by membrane-bound ATPase(s). A membrane-bound ATPase in *H. halobium* was first described by De *et al.* (1966). Their results have recently been confirmed and considerably extended by others (D. Oesterhelt, personal communication; H. Baltscheffsky, personal communication; B. May, personal communication) (see also Chapter 30,

Section 3.3). Provided the high-energy intermediate or state is the same for both the light and the O_2 effect, light would be expected to inhibit respiration. This inhibition has been demonstrated by Oesterhelt and Stoeckenius (1973), Oesterhelt and Krippahl (1973), and Bogomolni *et al.* (1976). Quantitatively, 36 quanta absorbed by the p.m. are equivalent to the reduction of one molecule of oxygen (Oesterhelt and Krippahl, 1973). With the high concentrations of bR present in *H. halobium*, this value indicates that photophosphorylation can play a significant role in the cells' energy metabolism, which is also obvious from the fact that strong light generates ATP levels that are as high as or higher than those generated by respiration.*

---

* So far, the possibility that light reduces ATP consumption rather than drives its synthesis has not been rigorously excluded. A reduction of ATP consumption in the light is unlikely, however, because light can raise ATP levels even after prolonged starvation, when ATP synthesis driven by other energy sources should be minimal, and ATP levels appear to remain high in anaerobic illuminated cells for very long times. The model experiments discussed below also argue strongly against such an explanation.

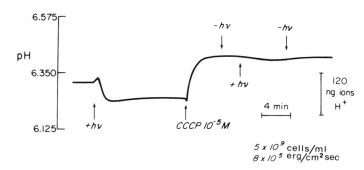

Fig. 8. Light response of a cell suspension in basal salt under anaerobic conditions. When the light is switched on, a transient inflow of protons occurs, recorded as an alkalinization of the medium. When it subsides, the continuing proton ejection by the p.m. results in a sustained acidification of the medium. The proton gradient across the membrane is collapsed completely by the addition of the uncoupler CCCP. Subsequent changes in illumination have no effect on the pH of the medium. Note that the pH after CCCP addition is higher than the initial pH in the dark, indicating that a proton gradient was already present at the beginning of the experiment. For further explanation, see the text.

The coupling between respiratory and light-energy transduction presumably occurs via the high-energy intermediate or state, because it is sensitive to uncouplers.

This raises the question of the identity of the common high-energy intermediate or state. Mitchell's chemiosmotic theory of energy transduction identifies the high-energy state as an electrochemical gradient, generated by the ejection of protons (Mitchell, 1972). Accordingly, light and respiration should both cause an ejection of protons from the cells. This can be checked easily by measuring pH changes in cell suspensions. We have shown that illumination causes ejection of protons from p.m.-containing cells against a potential or concentration gradient or both (Oesterhelt and Stoeckenius, 1973); so does respiration (Belyakova et al., 1975; Bogomolni et al., 1976). Uncouplers, which, according to Mitchell, function as proton carriers through the membrane and therefore collapse the gradient, prevent or reverse the light-induced pH change and relieve the light-induced inhibition of respiration in *H. halobium* cell suspensions (Belyakova et al., 1975; Bogomolni et al., 1976). These observations indicate that light absorbed by the p.m. may cause an ejection of protons from the cell and generate an electrochemical gradient across the cell membrane that in turn can drive ATP synthesis through a chemiosmotic mechanism. In other words, they suggest that the p.m. functions as a light-driven proton pump. Further and better evidence for this assumption will be presented below when the isolated membrane and reconstituted model systems are discussed.

Proton ejection from the cells is usually measured with a pH electrode as a sustained decrease of the pH of the suspending medium. In most instances, the response of the cells to illumination is complex; there is an initial transient increase in the pH of the medium before the lower pH is reached (Fig. 8). Cessation of illumination typically causes a small further acidification before the pH returns to the original dark level (Oesterhelt and Stoeckenius, 1973; Bogomolni et al., 1976). The degree of the transient alkalinization at the beginning and of the transient acidification at the end of the illumination depend on light intensity, p.m. content of the cells, their metabolic state, and the initial pH of the cell suspension. The transient responses appear to dominate at high pH values, low light intensity, and low p.m. content. They practically disappear below pH 5.5.

We view these transient changes as expressions of control functions of the cell, which regulate energy utilization (Bogomolni et al., 1976). The initial alkalinization indicates the light-induced relaxation of a preexisting electrochemical gradient with the same sign as the gradient generated by respiration or light. This interpretation is based on the following observations: (1) Low concentrations of uncouplers that increase the permeability of the membrane for protons will, when added in the dark, cause alkalinization of the medium, presumably because they disperse the preexisting H^+ gradient. A following illumination causes no further alkalinization but an immediate acidification of the medium. (2) Nigericin has essentially the same effect, only the initial alkalinization is more pronounced. This is explained by the action of nigericin as an H^+-K^+ exchange carrier and the high intracellular K^+ concentration. (3) Illumination at low light intensity under proper conditions can cause a net alkalinization only. A subsequent increase in light intensity results in an immediate acidification. (4) Prolonged starvation of aerobic cells in the dark abolishes the initial alkalinization, presumably because in starved cells the preexisting gradient has been dissipated. Adding substrate in the presence of O_2 reestablishes the initial inflow, because it allows the cell to again acquire a gradient. Alternatively, light can be used to first form the gradient, which is then transiently relaxed by a second illumination.

The initial phase of the pH response can be decomposed into a transient inflow of protons and a constant outflow that is linearly dependent on light intensity; the transient inflow may be measurably delayed at low light intensity or acid pH values. The initial inflow apparently occurs through the ATPase. It is abolished by DCCD, a specific inhibitor of the ATPase. When ATP synthesis is followed at high time resolution and correlated with the pH changes in the medium, it is seen that the increase in intracellular ATP concentration parallels the initial inflow of protons. When the maximal intracellular level is obtained, ATP concentration remains high, while the inflow of protons is reduced to a minimal value. The ATP level is presumably limited by the intracellular availability of ADP, while the inflow through the ATPase is reduced to the value determined by intracellular ATP hydrolysis. The continued proton ejection by the p.m. now results in an acidification of the medium, and the new steady-state value of the pH is determined by this ejection of protons, the residual flow through the ATPase, and other gradient-dispersing processes including a passive leak (Bogomolni et al., 1976). It is obvious from the extent of the pH change in the medium that other ions must also be moving across the membrane in response to the potential generated by the p.m. These ions have not been identified so far.

This explanation of the complex pH response of cells assumes that the ATPase allows the inflow of protons only above a limiting value of the proton-motive force. The electrochemical gradient in the energy-deprived cell is apparently maintained just below this critical value. The small increase necessary to activate the ATPase is often not detectable as a pH change in the outside medium, and at high pH, the inflow of protons after activation may be further enhanced by the consumption of protons in the synthesis of ATP. We cannot say at present whether the total proton-motive force or one component of the electrochemical gradient determines the activation of the ATPase. The absence of the initial alkalinization at low pH values argues that either a high intracellular pH or a pH gradient across the membrane activates the ATPase. However, a definitive answer will be possible only when intracellular pH values and the membrane potential are also measured.*

This explains the pH response of cells except for the transient acidification that occurs when the illumination is suddenly terminated. This acidification is

apparently also caused by the ATPase, because it is even more sensitive to the action of DCCD than the initial inflow. This additional ejection of protons also correlates with a short period of accelerated ATP hydrolysis at the end of illumination. It may reflect the time necessary for inactivation of ATPase, which had been activated by the light-induced gradient or pH changes.

Observations that confirm the basic conclusions drawn from observations of intact cells have been obtained with envelope preparations and reconstituted model systems. Envelope preparations that do not phosphorylate and do not maintain a gradient across the membrane in the dark respond to light with an acidification of the medium. They do not show the transient alkalinization at the beginning and acidification at the end of illumination. Essentially the same is true for the still-simpler model systems consisting of isolated p.m. vesicles. The isolated p.m. typically consists of membrane sheets. These can be converted into vesicles maintaining a preferential orientation of the bR by addition of either synthetic or natural phospholipids (Racker and Stoeckenius, 1974; Racker and Hinkle, 1974). Lipids without net charge are preferred, and a substantial amount of the native membrane lipids is exchanged for the added lipids. The most active preparations have been obtained with phosphatidyl choline, and the resulting vesicles contain approximately 10% more lipid than the native membrane. Electron microscopy shows that the preferential orientation of bR in these vesicle preparations is the opposite of that in intact cells (Fig. 9) (Stoeckenius, 1976; S.-B. Hwang and W. Stoeckenius, 1977; Stoeckenius et al., 1976). They would thus be expected to transport protons into the interior volume when illuminated, and this is indeed observed. The pH response is uncoupler-sensitive and the translocation electrogenic (Racker and Hinkle, 1974). Similar results were obtained by a group in Russia (Kayushin and Skulachev, 1974).

The vesicle system has been further developed by the additional incorporation of beef heart mitochondrial fractions or bacterial ATPase. This model system shows light-driven ATP synthesis, and there can be little doubt that the light-energy transduction is coupled to ATP formation through the proton gradient (Racker and Stoeckenius, 1974; Yoshida et al., 1975).

These experiments with reconstituted vesicles eliminate most of the complexity observed in intact cells and unequivocally demonstrate that light absorbed by bR drives the translocation of protons across the p.m., and that the resulting electrochemical gradient is determined by the orientation of the bR

---

* A different explanation for the pH response was offered by Oesterhelt (1975). He interprets the initial inflow of protons solely on the basis of proton consumption during ATP synthesis. In his view, the synthesis is driven by the membrane potential generated by the electrogenic p.m. action. An additional inflow must then be assumed to occur via an electroneutral symport and antiport of protons with other ions.

across the membrane. However, size variations of the lipid vesicles, their small interior volume, and their rather high permeability to anions render this model system unsuitable for many quantitative investigations. Planar lipid films that separate either two aqueous phases or an aqueous and a hydrocarbon phase are being developed (Drachev *et al.*, 1974*a*,*b*; Yaguzhinsky *et al.*, 1976; Hwang *et al.*, 1976). Unfortunately, the concentration and degree of orientation of bR in these planar model systems have not yet been controlled, and it can be concluded only that bR can translocate protons against substantial con-

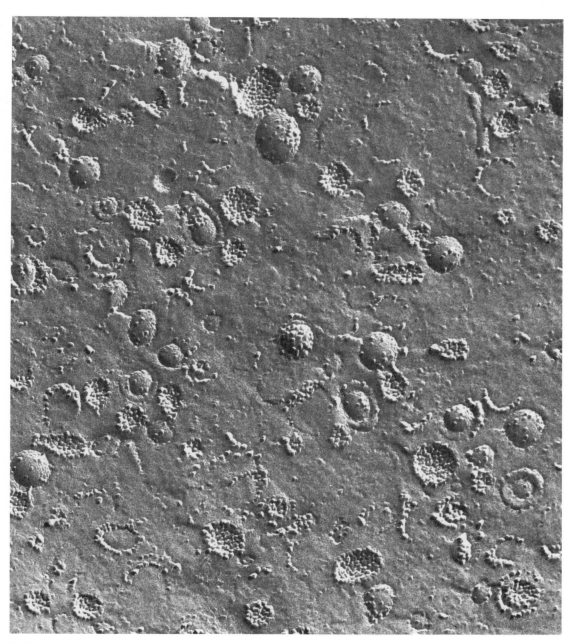

Fig. 9. Freeze–fracture electron micrograph of a vesicle preparation prepared from isolated membrane by adding phosphatidylcholine. Most of the particles are seen on the concave fracture faces (B-faces), indicating that the orientation of bR in the vesicle walls is inside-out as compared with its orientation in intact cells. Magnification 133,300×.

centration gradients or potentials of at least 150 mV or both (Drachev *et al.*, 1974a,b; Yaguzhinsky *et al.*, 1976). The development of almost perfectly oriented planar films with known amounts of p.m. per unit area may overcome these limitations (Hwang *et al.*, 1976; Stoeckenius *et al.*, 1976).

Cell-envelope preparations are intermediate in complexity between intact cells and the p.m. vesicles just described. They may be prepared by osmotic shock, sonication, freezing and thawing, or other means of mechanically breaking cells, with subsequent differential centrifugation to remove the cell contents (Stoeckenius and Rowen, 1967; MacDonald and Lanyi, 1975; Oesterhelt and Stoeckenius, 1974; Kanner and Racker, 1975). The cell wall is lost to different degrees in these preparations, and a small amount of cytoplasmic constituents may be retained. We term these preparations *envelopes* to distinguish them from the lipid vesicles. Envelopes are usually prepared so that they maintain the same orientation of the membrane as intact cells. A convenient indicator of leaky and/or inverted envelopes is menadione reductase activity (MacDonald and Lanyi, 1975).

In the light, proton gradients of the same sign as observed in whole cells are established in these envelope preparations. No transient initial inflows or additional outflows at the end of illumination are observed unless differences in the composition of the internal and external medium are preestablished by loading the envelopes with Na^+ (Lanyi and MacDonald, 1976). This confirms the conclusions from our observations on intact cells that the p.m. acts as a light-driven proton pump, translocating positive charge from the cytoplasmic space into the medium. In a series of publications, Lanyi and his collaborators have recently shown that in these envelope preparations, DDA^+ uptake occurs in the light, and that Na^+ is ejected from the envelopes by a Na^+/H^+ exchange. The uptake of amino acids in turn is driven by the resulting Na^+ gradient (MacDonald and Lanyi, 1975, 1976; Renthal and Lanyi, 1976; Lanyi *et al.*, 1976a, b; Lanyi and MacDonald, 1976).

6. Photochemistry

If the p.m. functions as a light-driven proton pump, the pigment must undergo a cyclic photoreaction, and

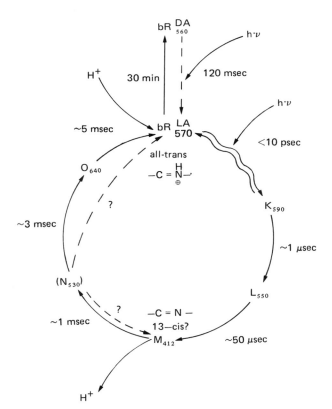

Fig. 10. Tentative scheme for the photoreaction cycle of light-adapted bR. The wavy lines indicate photoreactions; the dashed lines, alternative pathways. For further explanation, see the text.

one would expect to see light-induced changes in the absorption spectrum indicating the presence of intermediates in the cycle. The reaction $bR_{560}^{DA} \rightleftharpoons bR_{570}^{LA}$ (see Section 4) is too slow to account for the efficient proton translocation observed. Under natural conditions and in the laboratory, when p.m.-mediated proton translocation occurs, the pigment exists predominantly as the bR_{570}^{LA} complex, and all the following observations refer to this form unless otherwise stated. For the sake of simplicity, the notation LA is usually omitted, unless it is necessary to emphasize its difference from the dark-adapted complex. Intermediates in the cycle are denoted by letters in alphabetical order beginning with bK_{590}, in which the subscript indicates the maximum of the calculated absorption spectrum for the intermediate complex.†

The reaction cycle of bR_{570} is fast and the photosteady-state concentration of intermediates usually limited by the available light intensity. To observe the absorbance changes that occur on light exposure, it is therefore necessary to use flash spectroscopy with high time resolution or to slow down the reaction by reducing the temperature or by other means. Low-temperature spectroscopy also allows high steady-state concentration of intermediates to be obtained because their rise and decay times are affected differently; in most cases, the decay is more temperature-sensitive than the rise. This is not necessarily true, however, for all intermediates. A third possibility is the use of ether or detergents, which have similar effects on the kinetics of the cycle and greatly delay late reactions (Oesterhelt and Hess, 1973).

Several papers on the photoreaction cycle have been published (Lozier *et al.*, 1975; Stoeckenius and Lozier, 1974; Chance *et al.*, 1975; Dencher and Wilms, 1975; Kaufmann *et al.*,1976; Kung *et al.*, 1975; Lozier *et al.*, 1976; Sherman and Caplan, 1975; Sherman *et al.*, 1976; Slifkin and Caplan, 1975). The basic spectroscopic observations are in reasonable agreement. All interpretations of the data involve some

† This notation arose because the first intermediate observed apparently corresponded to metarhodopsin, and it was termed bM_{412}. As other intermediates were found, they were given letters according to their occurrence before or after bM_{412} in the cycle. Naturally, different investigators use different designations for the intermediate complexes. In some instances, the wavelength at which the individual complexes are observed, i.e., a wavelength close to the maximum of the *difference* spectrum between bR_{570} and the respective complex, has been used. This practice is confusing and should be discontinued. A designation by λ_{max} of the complexes alone is also not satisfactory, because the calculated absorption spectra are not accurate enough and their maxima also show shifts with different conditions, e.g., temperature. We therefore use the letters to unambiguously designate the intermediate in question.

assumptions that have not been verified, however, and these assumptions cause considerable discrepancies in the descriptions of the cycle by the different groups. We initially assumed a linear pathway and first-order reaction kinetics (Lozier *et al.*, 1975). Recently, we have obtained results that seem to require either a branched pathway or a backreaction in a linear pathway. From the number of first-order components observed in the absorbance changes at different times and wavelengths, we conclude that under physiological conditions, at least five intermediates can be observed in the cycle (Fig. 10) (Lozier and Niederberger, 1976).

The first intermediate in the cycle must be the excited state bR_{570}^*. It has not been observed unambiguously. The first observed intermediate is bK_{590} (Lozier *et al.*, 1975). It arises in 15 psec or less (Kaufmann *et al.*, 1976), is formed even at 77° K, and is stable at this temperature. The $bR_{570} \rightleftharpoons bK_{590}$ reaction is photoreversible. Different photosteady-state mixtures of bR_{570} and bK_{590} can be obtained by illumination with different wavelengths at this temperature, but the concentration of bK_{590} never exceeds approximately 50% because the absorption bands overlap. bR_{570} can be restored by illumination of bK_{590} with light of wavelength 635 nm or longer. The formation of bK_{590} or presumably bR_{570}^* is the only required photoreaction in the cycle.

At room temperature and in the dark, bK_{590} decays with a half-time of approximately 1.2 μsec to form bL_{550}, which in turn decays within approximately 30 μsec to bM_{412}. bL_{550} is relatively stable at temperatures below $-140°C$ and bM_{412} below $-60°C$. The rise and decay times of these three intermediates are sufficiently different so that they reach high concentration during the reaction cycle. The absorption spectrum of bM_{412} also shows apparently no overlap with the 570 nm band of the bR_{570} spectrum, and the decrease of 570 nm absorbance when the 412 nm absorbance is at its maximum has been used to calculate the absorption spectra of other intermediates from the recorded difference spectra. This introduces some error because at no time in the reaction cycle is all the cycling pigment actually present as bM_{412}, and the intermediates on both sides of bM_{412} apparently do absorb significantly at 570 nm. When the spectra of the intermediates are calculated from the low-temperature data, the bM_{412} spectrum shows a strong secondary absorption maximum at approximately 550 nm.

The later intermediates have considerably longer and very similar rise and decay times. They also show a stronger temperature dependence of their kinetics than the early intermediates. They are therefore more difficult to identify, and their sequence in the cycle is less certain. A suspected intermediate bN_{530} appears as a second peak in the calculated spectrum of bM_{412} (Lozier *et al.*, 1975). Its existence and position in the cycle, however, have not been unequivocally established. The last intermediate complex in the cycle appears to be bO_{640}; it is clearly observed only at higher temperatures, and its absorption spectrum has therefore been calculated from data obtained at 40°C. All these late intermediates appear and decay with time constants in the range of a few milliseconds. Oesterhelt *et al.* (1975) recently reported that the re-formation of bR_{570} from bM_{412} is accelerated by light absorption in bM_{412}. This is of interest for an elucidation of the mechanism of the photoreaction cycle, but probably of little significance physiolocically because the frequency of cycling is usually limited by light intensity available for absorption by the 570 nm band, and the intensity of blue light is very low in the natural environment.

In the p.m., a photoreaction cycle is therefore completed in a few milliseconds at room temperature and considerably faster at higher temperature. At physiological temperatures, the frequency of cycling can reach 200–500 sec^{-1}. The data given are determined from suspensions of the isolated p.m. in H_2O (Lozier *et al.*, 1975). Qualitative results indicate that the same reaction sequence also occurs in intact cells and envelope and vesicle preparations. In the isolated membrane, the cycle kinetics show little dependence on salt concentrations or pH values between 4 and 7. At lower pH, the pigment is reversibly converted to an inactive form bR_{605}^{acid} with a pK of 3.2 (Oesterhelt and Stoeckenius, 1971). In water, bleaching occurs at pH higher than 12.5 in the dark and is at least partially reversible. This is not true at high salt concentrations, and recent experiments on isolated p.m. in "basal salt" (growth medium without nutrients) and on whole cells indicate a considerably longer cycling time at pH 7.5 or higher (Bogomolni *et al.*, 1976). It is likely that in intact cells, the electrochemical gradient across the membrane also influences the cycle kinetics.

The spectral changes indicate changes in the interaction of retinal with the protein during the cycle and at extreme pH values. No definitive evidence for isomerization of retinal around double bonds during the photoreaction cycle has been obtained so far. Only all-*trans* retinal has been extracted from bR_{570}^{LA} (Oesterhelt *et al.*, 1973; Jan, 1975); recent results from the extraction of intermediates indicate, however, that a transient formation of 13-*cis* retinal may occur during the cycle (Pettei *et al.*, 1977). A deprotonation of the Schiff base in bM_{412} was detected with resonance Raman spectroscopy (Lewis *et al.*, 1974), and it could account at least partially for the large blue shift occurring in the bL → bM reaction. The proton of the

Schiff base is exchanged for a deuteron when the pigment cycles in D_2O (Lewis *et al.*, 1974). The exchange also occurs in the dark, but is detectably slower. These observations are of significance for an interpretation of the proton translocation across the membrane occurring during the cycle and will be discussed below.

As one would expect from the pumping function, changes in the protonation of the membrane during the cycle are also observed with electrodes or indicator dyes in suspensions of the isolated p.m. These changes were first detected in ether-modified preparations (Oesterhelt and Hess, 1973), in which the late reactions are approximately 1000 times slower. bM_{412} therefore accumulates in continuous light, and an acidification of the medium is observed. The pH of the medium returns to its original value when the light is turned off, and the pigment returns to its bR_{570} state. The more relevant kinetics of the unmodified p.m. can be followed by flash spectroscopy when suitable pH indicators are added to the membrane suspensions. Difference between kinetic traces from identical suspensions with and without added indicator yield the time-resolved absorbance changes of the indicator. Calibration of the indicator absorbance with acid pulses yields the amount of protons released (Lozier *et al.*, 1976). It is found that the protons appear in the medium with a time constant slightly longer than the rise of bM_{412}, while they are taken up synchronously with the re-formation of bR_{570}. One proton is exchanged per bR molecule per cycle. These observations are in apparent contradiction to the results obtained by Oesterhelt and Hess (1973) with the ether-modified membrane and by Chance *et al.* (1975) at low temperature in the presence of organic solvents, in which the appearance of the proton was found to correspond to the rise time of bM_{412}. We shall see, however, that the spectral and pH changes are not necessarily tightly coupled, and that the different results need not be contradictory (see Section 7).

If these pH changes in the medium reflect the proton-pumping of the p.m., the protons must be released on the exterior side of the membrane and taken up on the cytoplasmic side, and a mechanism must exist for the translocation of protons through the membrane. Because a proton is released before a proton is taken up by the membrane and the two events can be time-resolved, this postulate can be tested. Single-turnover flashes in the presence of suitable pH indicators show that in envelope preparations, in which the p.m. is oriented in the same direction as in intact cells, i.e., in which protons are translocated from the interior to the exterior, the appearance of protons on the outside is only slightly slower than the rise of bM_{412} and is considerably faster than the decay of

bO_{640}. In lipid vesicle preparations, in which the pumping direction and the orientation of the pigment are the opposite, the alkalinization of the medium occurs with the longer time constant corresponding to the reformation of bR_{570}. In both cases, the decay of the pH signal is many times longer than the cycle time, because it requires back-diffusion of the protons across the vesicle wall. The results therefore show that protons are released on the outer surface of the p.m. and taken up on the cytoplasmic surface as expected (Lozier et al., 1976).

The quantum efficiency of the cycle was first determined in the ether-modified system and found to be 0.79 (Oesterhelt and Hess, 1973). Because one proton is translocated per cycle, this should also be the quantum efficiency for pumping. We have determined it in whole cells and obtained values of 0.7–0.4, depending on the conditions of the measurement. The most reliable value appears to be 0.63 ± 0.06 (Bogomolni et al., in prep.). The determinations are complicated by the transient inflows and the lack of data on pH gradients and transmembrane potentials. Also, the initial rates of proton ejection can be measured only with relatively low accuracy. In view of these limitations, the good agreement of the values is gratifying. However, preliminary measurements of the quantum efficiency for cycling in our hands have consistently yielded values below 0.4 (Lozier, unpublished). The discrepancy is so far unresolved.

All data presented so far refer to the photoreaction cycle of bR_{570}^{LA}; bR_{560}^{DA} also undergoes a photoreaction cycle. Its exploration is complicated by the possibility that bR_{560}^{DA} may constitute a mixture of two different pigments, one containing all-trans retinal, the other 13-cis retinal. The all-trans pigment would be identical to bR_{570}^{LA}, whereas the 13-cis pigment would undergo a different photoreaction cycle with partial conversion to the all-trans pigment (Dencher et al., 1976; Tokunaga et al., personal communication). Earlier observation indicating a photocycle for bR_{560}^{DA} similar to that of bR_{570}^{LA} (Lozier et al., 1975) may be attributable to the fraction of all-trans pigment contained in the dark-adapted p.m.

7. General Discussion

Obviously, halobacteria constitute a new class of photosynthetic organisms, if one is willing to accept the use of light energy to synthesize ATP as a sufficient definition for photosynthesis. In contrast to all other known photosynthetic organisms, they convert light energy to chemical energy with a rhodopsinlike pigment rather than chlorophyll. The efficiency of energy conversion is probably lower than in chlorophyll-based photosynthesis. Preliminary estimates indicate that three protons flowing back through the membrane are necessary to synthesize one molecule of ATP (Bogomolni et al., 1976). With a quantum yield of 0.63 for proton ejection, this amounts to more than 3 quanta/ATP, as compared with 1 or 0.5 for other photosynthetic bacteria. Halobacteria also lack the efficient light-capturing apparatus of the antenna chlorophyll, and the carotenoid pigment in H. halobium cannot transfer light energy to bR (Danon and Stoeckenius, 1974). In addition, halobacteria synthesize their light-capturing membrane at the expense of other surface membrane, and have not evolved a mechanism to increase membrane area by infoldings of the surface membrane. We have been unable to obtain continuous growth of H. halobium under strictly anaerobic conditions in either complex or synthetic media. Purple-membrane-mediated photosynthesis thus appears to be an auxiliary mechanism to satisfy the energy requirements of the cells when O_2 concentrations drop below a critical level. Such conditions can easily arise in their natural environments, in which the salt concentration is near or at saturation and temperatures in strong sunlight may rise well above 30°C. Brine pools show little convection, and the cell density may be high. We found approximately 10^8 cells/ml in the salterns in San Francisco Bay.* It may also be significant that halobacteria are found to grow either on the bodies or in the eyes of salted fish (pink eye), where they have access to light.

The obvious evolutionary advantage of p.m.-mediated light-energy transduction is that synthesis of only one protein and its chromophore enables the cells to use an additional energy source. Also, ATP synthesis is probably not the only process driven by the light-induced electrochemical gradient. We have made observations that suggest that in H. halobium, as in other prokaryotes, locomotion may be driven directly by the gradient rather than through ATP hydrolysis and, at least in envelope preparations, amino acid and potassium uptake and Na^+ extrusion are directly driven by the gradient (MacDonald and Lanyi, 1975; Kanner and Racker, 1975). The relatively high K^+ permeability of H. halobium cells observed earlier (H. Passow and W. Stoeckenius, unpublished) and its relatively low sensitivity to valinomycin (Danon and Stoeckenius, 1974) are arguments that the same K^+-uptake mechanism may be operating in intact cells as well. The p.m. would thus directly energize several vital processes: it would allow the

---

* It should be pointed out, however, that halophilic algae may also grow in the same environment and may produce enough O_2 to keep the O_2 concentration near air saturation during the day. This is the case in the saltern that we investigated.

cells to maintain their high internal K^+ concentration, take up nutrients, and through the phototactic response allow them to search for a favorable environment.

The p.m. is of special interest in photosynthesis and in biological energy transduction; it offers a system to test different theories of energy transduction and to study the structure and function of an ion pump at much higher resolution than is possible in any other system. Because electrogenic proton translocation is a direct result of light absorption in bR, we are led to the conclusion that a chemiosmotic mechanism is involved. There can be little doubt that the coupling to ATP synthesis occurs via the electrochemical gradient. Most of the bR, because it forms large two-dimensional crystals in the surface membrane, cannot be in contact with the membrane-bound ATPase, which is apparently located in the rest of the surface membrane. There is so far no evidence for an efficient energy transfer within the crystal of bR, even though exciton interaction between adjacent bR molecules has been observed (Bauer *et al.*, 1976; Heyn *et al.*, 1975; Becher and Ebrey, 1976). An efficient direct interaction between bR and the ATPase therefore appears to be excluded. The model experiments with lipid vesicles containing p.m. and beef heart "coupling factors" further confirm this conclusion (Racker and Stoeckenius, 1974). However, the p.m. also offers clear evidence that an energized state of the membrane is obligatorily involved in energy transduction and precedes the generation of the electrochemical gradient. Because only the first step in the photoreaction cycle requires light, the first intermediate must contain all the free energy required for completion of the cycle. We are therefore clearly dealing with a high-energy intermediate. The first intermediate (after the excited state) occurs long before the proton is released, and its rise time of less than 15 psec precludes any large conformational changes. An electron redistribution and small conformational change are probably the only events possible. It should be pointed out, however, that the lifetime of the early intermediates K and L is very short compared with the following reactions in the cycle and other reactions in the cell, and that they rapidly generate the electrochemical gradient, which has a lifetime approximately 10^6 times as long as that of K. What is more important is that the use of the energy in the gradient can apparently be controlled by the cell to some extent, which does not appear to be true for the early intermediates. The exact nature of the high-energy intermediate(s) is an exciting question that may be solved with existing techniques.

The structural studies on the p.m. appear to preclude the existence of large aqueous channels through the membrane, which anyway would not be a suitable pathway for proton translocation because the energy of a proton diffusing through an aqueous medium over relatively long distances would be dispersed in thermal collisions. The rigid lattice structure of the p.m. also makes large movements of individual protonated groups on bR highly unlikely. The only alternative models for the proton transport appear to be hydrogen carriers or a chain of proton-exchanging groups spanning the membrane. There is no evidence for redox reactions in the p.m., but there exist in the protein (and also in the lipid) many groups that can act as proton acceptors or donors. A proton translocation based on a chain of such groups therefore appears by far the most likely mechanism for the function of the p.m. In principle, a light-induced reversible change in the pK of one group in the chain could drive the translocation of protons against an electrochemical gradient across the membrane, provided back-reactions are prevented and the proton exchange is thus forced to proceed in one direction along the chain. In such a mechanism, each group should undergo a transient protonation change during the cycle.

One group that fits this description is the Schiff base of the retinal; it is protonated in the bR_{570} complex and deprotonated in the bM_{412} intermediate. That it exchanges its proton for a deuteron faster in the light than in the dark also agrees with a proton-translocating role in the cycle. Because the Schiff base is part of the chromophore and because the energy input occurs at the chromophore, it may also undergo a change in pK. Large pK changes are known to occur on excitation of molecules (Weller, 1961); so are conformational changes in excited molecules. Present evidence, however, favors a pK change leading to deprotonation of the Schiff base, which occurs relatively late in the photoreaction cycle and may be coupled to the conformational change. If the conformational change also varies the distance between the Schiff base and proton acceptor and donor groups, it would also impart the required vectorial character of the proton exchange.

Conformational changes in the chromophore may, but need not, involve isomerization around C=C double bonds or single bond rotations, or concerted actions of two or more single or double bonds could effect rapid conformational changes (Warshel, 1976). Recent results (Pettei *et al.*, 1977) contradict earlier observations (Jan, 1975) and indicate that an all-*trans* to 13-*cis* isomerization may occur during the cycle. Oesterhelt (1974) pointed out the possible occurrence of concerted bond rotations in the lysine residue and the 13-*cis* double bond to explain the change from all-*trans* to 13-*cis* retinal in the $bR_{570}^{LA} \rightleftharpoons bR_{560}^{DA}$ reaction that apparently occurs without large conformational change of the protein.

Such a model can also explain why the appearance and disappearance of protons in the medium need not be tightly coupled to the spectral changes seen in the intermediates, which at least in part reflect the protonation changes in the chromophore. If other proton-transferring groups are inserted between the Schiff base and the surface of the membrane, their kinetics may determine the rate of translocation.

The model for the proton translocation is probably too simple, and may be entirely wrong. However, it is plausible, can be tested, and is at least consistent with the existing data.

Note Added in Proof. This chapter was written in the summer of 1975 and partially updated in March, 1977. Significant new results have been published since that time, which could not be incorporated. A more complete and detailed review has been submitted for publication in *Biochim. Biophys. Acta* early in 1978.

ACKNOWLEDGMENT

This work was supported by NASA Grant NSG-7151 and NHLI Program Project Grant HL-06285. I thank Roberto Bogomolni and Richard Lozier for critical discussion and reading of the manuscript.

8. References

Abdulaev, N. G., Lobanov, N. A., Kiselev, A. V., and Ovchinnikov, Yu. A., 1976, The structural investigation of bacteriorhodopsin, *Abstracts of USSR–USA Symposium on Chemistry and Physics of Proteins*, Riga, August 4–8, 1976, p. 4.

Bakker, E. P., Rottenberg, H., and Caplan, S. R., 1976, An estimation of the light-induced electrochemical potential difference of protons across the membrane of *Halobacterium halobium*, *Biochim. Biophys. Acta* **440**:557.

Bauer, P.-J., Dencher, N. A., and Heyn, M. P., 1976, Evidence for chromophore–chromophore interactions in the purple membrane from reconstitution experiments of the chromophore-free membrane, *Biophys. Struct. Mechanism* **2**: 79.

Becher, B., and Ebrey, T. G., 1976, Evidence for chromophore–chromophore (exciton) interaction in the purple membrane of *Halobacterium halobium*, *Biophys. J.* **16**:100a (abstract TH-AM-G7).

Belyakova, T. N., Kadzyauskas, Yu. P., Skulachev, V. I., Smirnova, I. A., Chekulayeva, L. N., and Yasaytis, A. A., 1975, Generation of electrochemical potential of H⁺ ions and photophosphorylation in *Halobacterium halobium* cells, *Dokl. Akad. Nauk SSSR* **223**: 483.

Blaurock, A. E., 1975, Bacteriorhodopsin: A trans-membrane pump containing α-helix, *J. Mol. Biol.* **93**: 139.

Blaurock, A. E., and Stoeckenius, W., 1971, Structure of the purple membrane, *Nature (London) New Biol.* **233**: 152.

Blaurock, A. E., Stoeckenius, W., Oesterhelt, D., and Scherphof, G. L., 1976, Structure of the cell envelope of *Halobacterium halobium*, *J. Cell Biol.* **71**: 1.

Bogomolni, R. A., Baker, R. A., Lozier, R. H., and Stoeckenius, W., 1976, Light-driven proton translocations in *Halobacterium halobium*, *Biochim. Biophys. Acta* **440**: 68.

Branton, D., 1969, Membrane structure, *Annu. Rev. Plant Physiol.* **20**: 209.

Bridgen, J., and Walker, I. D., 1976, Photoreceptor protein from the purple membrane of *Halobacterium halobium*: Molecular weight and retinal binding site, *Biochemistry* **15**: 792.

Brown, A. D., 1964, Aspects of bacterial response to the ionic environment, *Bacteriol. Rev.* **28**: 296.

Chance, B., Porte, M., Hess, B., and Oesterhelt, D., 1975, Low temperature kinetics of H⁺ changes of bacteriorhodopsin, *Biophys. J.* **15**: 913.

Cheah, K. S., 1969, Properties of electron transport particles from *Halobacterium cutirubrum*. The respiratory chain system, *Biochim. Biophys. Acta* **180**: 320.

Cheah, K. S., 1970, Properties of the membrane-bound respiratory chain system of *Halobacterium salinarium*, *Biochim. Biophys. Acta* **216**: 43.

Christian, J. H. B., and Waltho, J. A., 1962, Solute concentrations within cells of halophilic and non-halophilic bacteria, *Biochim. Biophys. Acta* **65**: 506.

Danon, A., and Stoeckenius, W., 1974, Photophosphorylation in *Halobacterium halobium*, *Proc. Natl. Acad. Sci. U.S.A.* **71**: 1234.

De, K., Passow, H., Stoeckenius, W., and White, M., 1966, Some properties of a potassium-sodium-magnesium activated membrane trinucleotidase of *Halobacterium halobium*, *Pfluegers Arch. Gesamte Physiol. Menschen Tiere* **289**: S.R. 15 (abstract).

Dencher, N., 1974, Functions of bacteriorhodopsin, in: *Biochemistry of Sensory Functions* (L. Jaenicke, ed.), pp. 161–163, Springer-Verlag, Berlin.

Dencher, N., and Wilms, M., 1975, Flash photometric experiments on the photochemical cycle of bacteriorhodopsin, *Biophys. Struct. Mechanism* **1**: 259.

Dencher, N. A., Rafferty, C. N., and Sperling, W., 1976, Photochemistry and dark equilibrium of 13-*cis* and *trans*-bacteriorhodopsin, EMBO Workshop on Transduction Mechanism of Photoreceptors, October 4–8, 1976, Jülich, Abstracts, p. 54.

Drachev, L. A., Jasaitis, A. A., Kaulen, A. D., Kondrashin, A. A., Liberman, E. A., Nemecek, I. B., Ostroumov, S. A., Semenov, A. Yu., and Skulachev, V. P., 1974a, Direct measurement of electric current generation by cytochrome oxidase, H⁺-ATPase and bacteriorhodopsin, *Nature (London)* **249**: 321.

Drachev, L. A., Kaulen, A. D., Ostroumov, S. A., and Skulachev, V. P., 1974b, Electrogenesis by bacteriorhodopsin incorporated in a planar phospholipid membrane, *FEBS Lett.* **39**: 43.

Futterman, S., and Saslaw, L. D., 1961, The estimation of

vitamin A aldehyde with thiobarbituric acid, *J. Biol. Chem.* **236**:1652.

Gibbons, N. E., 1974, Halobacteriaceae, in: *Bergey's Manual of Determinative Bacteriology*, 8th Ed. (R. E. Buchanan and N. E. Gibbons, eds.), pp. 269–272, Williams & Wilkins, Baltimore.

Ginzburg, M., 1969, The unusual membrane permeability of two halophilic unicellular organisms, *Biochim. Biophys. Acta* **173**:370.

Ginzburg, M., Sachs, L., and Ginzburg, B. Z., 1970, Ion metabolism in a *Halobacterium*. I. Influence of age of culture on intracellular concentrations, *J. Gen. Physiol.* **55**:187.

Henderson, R., 1975, The structure of the purple membrane from *Halobacterium halobium*: Analysis of the X-ray diffraction pattern, *J. Mol. Biol.* **93**:123.

Henderson, R., and Unwin, P. N. T., 1975, Three-dimensional model of purple membrane obtained by electron microscopy, *Nature (London)* **257**:28.

Heyn, M. P., Bauer, P.-J., and Dencher, N. A., 1975, A natural CD label to probe the structure of the purple membrane from *Halobacterium halobium* by means of exciton coupling effects, *Biochem. Biophys. Res. Commun.* **67**:897.

Hildebrand, E., and Dencher, N., 1975, Two photosystems controlling behavioral responses of *Halobacterium halobium*, *Nature (London)* **257**:46.

Hwang, S.-B., and Stoeckenius, W., 1977, Purple membrane vesicles. Morphology and proton translocation, *J. Membrane Biol.* **33**:325.

Hwang, S.-B., Korenbrot, J. I., and Stoeckenius, W., 1976, Light-dependent proton transport by bacteriorhodopsin incorporated in an interface film, *Prog. Clin. Biol. Res.* **15**:81.

Jan, L.-Y., 1975, The isomeric configuration of the bacteriorhodopsin chromophore, *Vision Res.* **15**:1081.

Joshi, J. G., Guild, W. R., and Handler, P., 1963, The presence of two species of DNA in halobacteria, *J. Mol. Biol.* **6**:34.

Kanner, B. I., and Racker, E., 1975, Light-dependent proton and rubidium translocation in membrane vesicles from *Halobacterium halobium*, *Biochem. Biophys. Res. Commun.* **64**:1054.

Kates, M., 1972, Ether-linked lipids in extremely halophilic bacteria, in: *Ether Lipids: Chemistry and Biology* (F. L. Snyder, ed.), pp. 351–398, Academic Press, New York and London.

Katznelson, H., and Robinson, J., 1956, Observations on the respiratory activity of certain obligately halophilic bacteria with high salt requirements, *J. Bacteriol.* **71**:244.

Kaufmann, K. J., Rentzepis, P. M., Stoeckenius, W., and Lewis, A., 1976, Primary photochemical processes in bacteriorhodopsin, *Biochem. Biophys. Res. Commun.* **68**:1109.

Kayushin, L. P., and Skulachev, V. P., 1974, Bacteriorhodopsin as an electrogenic proton pump: Reconstitution of bacteriorhodopsin proteoliposomes generating $\Delta\psi$ and ΔpH, *FEBS Lett.* **39**:39.

Kayushin, L. P., Sibeldina, L. A., Lasareva, A. B.,

Vsevolodov, N. N., Kostikov, A. C., Richireva, G. T., and Chekulaeva, L. N., 1974, Membrane protein bacteriorhodopsin from halophilic bacteria, *Stud. Biophys. (Berlin)* **42**:71.

Keefer, L. M., and Bradshaw, R. A., 1977, Structural studies on *Halobacterium halobium* bacteriorhodopsin, *Fed. Proc. Fed. Am. Soc. Exp. Biol.* **36**:799.

Kelly, M., and Liaaen-Jensen, S., 1967, Bacterial carotenoids, XXVI. C_{50}-carotenoids. 2. Bacterioruberin, *Acta Chem. Scand.* **21**:2578.

Kung, M. C., DeVault, D., Hess, B., and Oesterhelt, D., 1975, Photolysis of bacteriorhodopsin, *Biophys. J.* **15**:907.

Kushner, D. J., Bayley, S. T., Boring, J., Kates, M., and Gibbons, N. E., 1964, Morphological and chemical properties of cell-envelopes of the extreme halophile, *Halobacterium cutirubrum*, *Can. J. Microbiol.* **10**:483.

Kushwaha, S. C., and Kates, M., 1973, Isolation and identification of "bacteriorhodopsin" and minor C_{40}-carotenoids in *Halobacterium cutirubrum*, *Biochim. Biophys. Acta* **316**:235.

Kushwaha, S. C., Gochnauer, M. B., Kushner, D. J., and Kates, M., 1974, Pigments and isoprenoid compounds in extremely and moderately halophilic bacteria, *Can. J. Microbiol.* **20**:241.

Kushwaha, S. C., Kates, M., and Martin, W. G., 1975, Characterization and composition of the purple and red membrane from *Halobacterium cutirubrum*, *Can. J. Biochem.* **53**:284.

Kushwaha, S. C., Kates, M., and Stoeckenius, W., 1976, Comparison of purple membrane from *Halobacterium cutirubrum* and *Halobacterium halobium*, *Biochim. Biophys. Acta* **426**:703.

Lanyi, J. K., 1968, Studies of the electron transport chain of extremely halophilic bacteria. I. Spectrophotometric identification of the cytochromes of *Halobacterium cutirubrum*, *Arch. Biochem. Biophys.* **128**:716.

Lanyi, J. K., 1972, Studies of the electron transport chain of extremely halophilic bacteria. VIII. Respiration-dependent detergent dissolution of cell envelopes, *Biochim. Biophys. Acta* **282**:439.

Lanyi, J. K., 1974, Salt-dependent properties of proteins from extremely halophilic bacteria, *Bacteriol. Rev.* **38**:272.

Lanyi, J. K., and MacDonald, R. E., 1976, Existence of electrogenic hydrogen ion/sodium ion antiport in *Halobacterium halobium* cell envelope vesicles, *Biochemistry* **15**:4608–4614.

Lanyi, J. K., and Silverman, M. P., 1972, The state of binding of intracellular K^+ in *Halobacterium cutirubrum*, *Can. J. Microbiol.* **18**:993.

Lanyi, J. K., Yearwood-Drayton, V., and MacDonald, R. E., 1976a, Light-induced glutamate transport in *Halobacterium halobium* envelope vesicles. I. Kinetics of the light-dependent and the sodium-gradient dependent uptake, *Biochemistry* **15**:1595.

Lanyi, J. K., Renthal, R., and MacDonald, R. E., 1976b, Light-induced glutamate transport in *Halobacterium halobium* envelope vesicles. II. Evidence that the driving force is a light-dependent sodium gradient, *Biochemistry* **15**:1603.

Larsen, H., 1963, Halophilism, in: *The Bacteria* (I. C. Gunsalus and R. Y. Stanier, eds.), Vol. IV, pp. 297–342, Academic Press, New York and London.

Larsen, H., 1967, Biochemical aspects of extreme halophilism, *Adv. Microb. Physiol.* 1:97.

Lewis, A., Spoonhower, J., Bogomolni, R. A., Lozier, R. H., and Stoeckenius, W., 1974, Tunable laser resonance Raman spectroscopy of bacteriorhodopsin, *Proc. Natl. Acad. Sci. U.S.A.* **71**:4462.

Lozier, R. H., and Niederberger, W., 1976, The photochemical cycle of bacteriorhodopsin, *Fed. Proc. Fed. Am. Soc. Exp. Biol.* **36**:1805.

Lozier, R. H., Bogomolni, R. A., and Stoeckenius, W., 1975, Bacteriorhodopsin: A light-driven proton pump in *Halobacterium halobium*, *Biophys. J.* **15**:955.

Lozier, R. H., Niederberger, W., Bogomolni, R. A. Hwang, S.-B., and Stoeckenius, W., 1976, Kinetics and stoichiometry of light-induced proton release and uptake from purple membrane fragments, *Halobacterium halobium* cell envelopes, and phospholipid vesicles containing oriented purple membrane, *Biochim. Biophys. Acta* **440**:575.

MacDonald, R. E., and Lanyi, J. K., 1975, Light-induced leucine transport in *Halobacterium halobium* envelope vesicles: A chemiosmotic system, *Biochemistry* **14**:2882.

MacDonald, R. E., and Lanyi, J. K., 1976, Light-activated amino acid transport in *Halobacterium halobium* envelope vesicles, *Fed. Proc. Fed. Am. Soc. Exp. Biol.* **36**:1828.

Mescher, M. F., and Strominger, J. L., 1976, Purification and characterization of a prokaryotic glycoprotein from the cell envelope of *Halobacterium salinarium*, *J. Biol. Chem.* **251**:2005.

Mitchell, P., 1972, Chemiosmotic coupling in energy transduction: A logical development of biochemical knowledge, *J. Bioenerg.* **3**:5.

Moore, R. L., and McCarthy, B. J., 1969, Characterization of the deoxyribonucleic acid of various strains of halophilic bacteria, *J. Bacteriol.* **99**:248.

Oesterhelt, D., 1974, Vergleichende Aspekte der Photorezeption von Retinal-proteinkomplexen, in: *Biochemistry of Sensory Functions* (L. Jaenicke, ed.), pp. 55–77, Springer-Verlag, Berlin and Heidelberg.

Oesterhelt, D., 1975, The purple membrane of *Halobacterium halobium*: A new system for light energy conversion, in: *Energy Transformation in Biological Systems, Ciba Found. Symp. (New Ser.)* **31**:147–167.

Oesterhelt, D., and Hess, B., 1973, Reversible photolysis of the purple complex in the purple membrane of *Halobacterium halobium*, *Eur. J. Biochem.* **37**:316.

Oesterhelt, D., and Krippahl, G., 1973, Light inhibition of respiration in *Halobacterium halobium*, *FEBS Lett.* **36**:72.

Oesterhelt, D., and Schuhmann, L., 1974, Reconstitution of bacteriorhodopsin, *FEBS Lett.* **44**:262.

Oesterhelt, D., and Stoeckenius, W., 1971, Rhodopsin-like protein from the purple membrane of *Halobacterium halobium*, *Nature (London) New Biol.* **233**:149.

Oesterhelt, D., and Stoeckenius, W., 1973, Functions of a new photoreceptor membrane, *Proc. Natl. Acad. Sci. U.S.A.* **70**:2853.

Oesterhelt, D., and Stoeckenius, W., 1974, Isolation of the cell membrane of *Halobacterium halobium* and its fractionation into red and purple membrane, in: *Methods in Enzymology* (S. Fleischer and L. Packer, eds.), Vol. XXXI, *Biomembranes*, Part A, pp. 667–678, Academic Press, New York.

Oesterhelt, D., Meentzen, M., and Schuhmann, L., 1973, Reversible dissociation of the purple complex in bacteriorhodopsin and identification of 13-*cis* and all-*trans*-retinal as its chromophores, *Eur. J. Biochem.* **40**:453.

Oesterhelt, D., Schuhmann, L., and Gruber, H., 1974, Light-dependent reaction of bacteriorhodopsin with hydroxylamine in cell suspensions of *Halobacterium halobium*: Demonstration of an apo-membrane, *FEBS Lett.* **44**:257.

Oesterhelt, D., Hartmann, R., Fischer, U., Michel, H., and Schreckenbach, Th., 1975, Biochemistry of a light-driven proton pump: Bacteriorhodopsin, in: *Proceedings of the Tenth FEBS Meeting*, pp. 239–251, *Enzymes/Electron Transport Systems*, Vol. 40, North-Holland/American Elsevier, Amsterdam/New York.

Onishi, H., and Kushner, D. J., 1966, Mechanism of dissolution of envelopes of the extreme halophile *Halobacterium cutirubrum*, *J. Bacteriol.* **91**:646.

Pettei, M. J., Yudd, A. P., Nakanishi, V., Henselman, R., and Stoeckenius, W., 1977, Identification of retinal isomers isolated from bacteriorhodopsin, *Biochemistry* **16**:1955–1959.

Racker, E., and Hinkle, P. C., 1974, Effect of temperature on the function of a proton pump, *J. Membrane Biol.* **17**:181.

Racker, E., and Stoeckenius, W., 1974, Reconstitution of purple membrane vesicles catalyzing light-driven proton uptake and adenosine triphosphate formation, *J. Biol. Chem.* **249**:662.

Raymond, J. C., and Sistrom, W. R., 1967, The isolation and preliminary characterization of a halophilic photosynthetic bacterium, *Arch. Mikrobiol.* **59**:255.

Renthal, R., and Lanyi, J. K., 1976, Light-induced membrane potential and pH gradient in *Halobacterium halobium* envelope vesicles, *Biochemistry* **15**:2136.

Schreckenbach, T., and Oesterhelt, D., 1976, Photochemical and chemical studies on the chromophore of bacteriorhodopsin, *Fed. Proc. Fed. Am. Soc. Exp. Biol.* **36**:1810.

Sherman, W. V., and Caplan, S. R., 1975, Arrhenius parameters of phototransients in *Halobacterium halobium* in physiological conditions, *Nature (London)* **258**:766.

Sherman, W. V., Slifkin, M. A., and Caplan, S. R., 1976, Kinetic studies of phototransients in bacteriorhodopsin, *Biochim. Biophys. Acta* **423**:238.

Slifkin, M. A., and Caplan, S. R., 1975, Modulation excitation spectrophotometry of purple membrane of *Halobacterium halobium*, *Nature (London)* **253**:56.

Soo-Hoo, T. S., and Brown, A. D., 1967, A basis of the specific sodium requirement for morphological integrity of *Halobacterium halobium*, *Biochim. Biophys. Acta* **135**:164.

Stoeckenius, W., 1976, Structure of biological membranes. Bacteriorhodopsin and the purple membrane, in: *Proceedings of a Conference on Biological and Artificial Membranes and Desalination of Water*, Pontifical Academy of Sciences, Rome, April 14–19, 1975; reprinted from *Ex Aedibvs Academicis in Civitate Vaticana*, 1976 (R. Passino, ed.), pp. 65–77.

Stoeckenius, W., and Kunau, W. H., 1968, Further characterization of particulate fractions from lysed cell envelopes of *Halobacterium halobium* and isolation of gas vacuole membranes, *J. Cell Biol.* **38**:337.

Stoeckenius, S., and Lozier, R. H., 1974, Light energy conversion in *Halobacterium halobium*, *J. Supramol. Struct.* **2**:769.

Stoeckenius, W., and Rowen, R., 1967, A morphological study of *Halobacterium halobium* and its lysis in media of low salt concentration, *J. Cell Biol.* **34**:365.

Stoeckenius, W., Hwang, S.-B., and Korenbrot, J., 1976, Proton translocation by bacteriorhodopsin in model systems, *Nobel Symp. 34*: *The Structure of Biological Membranes*, June 7–11, 1976; reprinted from *Structure of Biological Membranes*, (S. Abrahamsson and I. Pascher, eds.), Plenum Press, New York.

Tomlinson, G. A., and Hochstein, L. I., 1972, Isolation of carbohydrate-metabolizing, extremely halophilic bacteria, *Can. J. Microbiol.* **18**:698.

Unwin, P. N. T., and Henderson, R., 1975, Molecular structure determination by electron microscopy of unstained crystalline specimens, *J. Mol. Biol.* **94**:425.

Walsby, A. E., 1972, Structure and function of gas vacuoles, *Bacteriol. Rev.* **36**:1.

Warshel, A., 1976, Bicycle-pedal model for the first step in the vision process, *Nature* (*London*) **260**:679.

Weller, A., 1961, Fast reactions of excited molecules, *Prog. React. Kinet.* **1**:187.

Yaguzhinsky, L. S., Boguslavsky, L. I. Volkov, A. G., and Rakhmaninova, A. B., 1976, Synthesis of ATP coupled with action of membrane protonic pumps at the octane-water interface, *Nature* (*London*) **259**:494.

Yoshida, M., Sone, N., Hirata, H., Kagawa, Y., Takeuchi, Y., and Ohno, K., 1975, ATP synthesis catalyzed by purified DCCD-sensitive ATPase incorporated into reconstituted purple membrane vesicles, *Biochem. Biophys. Res. Commun.* **67**:1295.

PART V

Phosphorylation

CHAPTER 30

Photosynthetic Phosphorylation

Margareta Baltscheffsky

1. Historical Introduction

About 15 years after the discovery of electron-transport-coupled phosphorylation in animal tissue by Belitzer and Tsibakova (1939), the corresponding reactions in photosynthetic systems were reported to occur. In 1954, Arnon *et al.* (1954) found that chloroplasts from spinach could catalyze the formation of ATP from ADP and P_i with light as the only source of energy. Shortly thereafter, Frenkel (1954) reported that cell-free extracts from *Rhodospirillum rubrum* under illumination, but not in the dark, had the ability to form significant amounts of ATP. In contrast to the chloroplast system, no exogenous electron carrier was necessary. All that was needed in the bacterial system was Mg^{2+} ions, ADP, and P_i. This short communication from Frenkel was a source of inspiration for a great number of scientists to probe into the nature of the photosynthetic phosphorylation reactions.

In looking back over the early literature on the subject, there are some historical circumstances that we may consider. For example, it was not known at first whether the photophosphorylation was electron-transport-linked or of the substrate-level type. So the initial work following Frenkel's first report concentrated on some very basic questions regarding the overall nature of the phosphorylation system. Today, we know that the photophosphorylation in bacteria is indeed an electron-transport-linked reaction. As is obvious from this chapter and other chapters in this book, we now know a great deal about the energy-

transducing system that converts the solar energy into chemical-energy-rich bonds or an energy-rich state, the utilization of which will provide the necessary energy input for biosynthetic and other energy-requiring reactions. The general picture that has evolved is that the mechanism for photophosphorylation is in principle very similar, if not identical, to its counterpart, oxidative phosphorylation in aerobic organisms. Progress today proceeds more or less in parallel in the closely related areas of photophosphorylation and oxidative phosphorylation.

2. Hypotheses on Mechanisms for Photophosphorylation of ADP to ATP

The three well-known current hypotheses for electron-transport-linked photophosphorylation are (1) the classic chemical intermediate; (2) the chemiosmotic; and (3) the conformational change hypothesis. Currently, much active interest is centered on the two latter hypotheses. Since the chemiosmotic theory is well described elsewhere in this book (see Chapter 26), we will concentrate briefly here on the conformational change theory as it has been developed by Boyer and co-workers (Boyer, 1965, 1974; Boyer *et al.*, 1973; Cross and Boyer, 1975). Originally, Boyer (1965) suggested that energy transformation in coupling processes may essentially involve protein conformational changes. Recently, the theory has been extended to include a radically new concept for energy coupling: "In oxidative phosphorylation, ATP synthesized from ADP and P_i at the catalytic coupling site by reversal of hydrolysis is released to the medium

Margareta Baltscheffsky · Arrhenius Laboratory, Department of Biochemistry, University of Stockholm, Fack, S-10691 Stockholm, Sweden

by an energy-driven conformational change" (Boyer, 1974). In other words, the energy-requiring step is not the formation of tightly bound ATP, at a hydrophobic catalytic site, but rather the release of the newly formed ATP from the protein through a change in the protein conformation (Fig. 1).

There are several attractive features of this concept. One is that there is no need of a "classic" covalent energy-rich phosphorylated intermediate, long sought for but never found. Another is that a hydrophobic ATP-binding site would provide a simple explanation of how the hydrolytic reaction (ATP + H_2O → ADP + P_i) may be functioning in the direction of synthesis rather than hydrolysis.

Lutz *et al.* (1974) recently proposed a mechanism for bacterial photophoshorylation that is similar to the

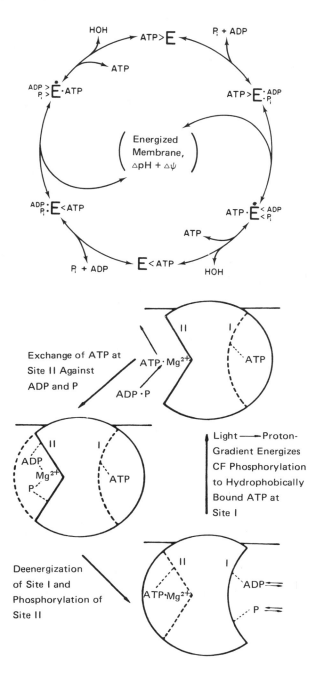

Light ⟶ Proton-Gradient Energizes CF Phosphorylation to Hydrophobically Bound ATP at Site I

Fig. 1. Two proposed mechanisms for ATP synthesis in electron-transport-linked phosphorylation. *Top:* A dual-site mechanism for ATP synthesis as proposed by Boyer (1977). Two catalytic sites are in conformational interaction such that the presence of ADP and P_i at one site allows an energy-requiring conformational change that promotes the binding and activation of P_i while ATP-binding at the second site is decreased, so that ATP bound at that site will be released to the medium. Both sites are equally competent to form ATP from added ADP and P_i. *Bottom:* A two-site mechanism for ATP synthesis in bacterial chromatophores as proposed by Bachofen *et al.* (1974). At site I, in a hydrophobic environment, energization causes adsorbed ADP and P_i to form bound ATP. Through a conformational change, site II becomes accessible to water, and acid-soluble ATP formed in a previous cycle will exchange with ADP and P_i. The ATP is released as soluble ATP, and ADP, P_i and Mg^{2+} can be adsorbed at site II. New acid-soluble ATP is formed at site II concomitant with the hydrolysis of bound ATP at site I. Energization at site I has been transferred to site II, and site I is now accessible to H_2O, ADP, and P_i. Re-energization of the coupling factor excludes water from site I and favors again the formation of firmly bound ATP at this site.

Boyer hypothesis in that it also has firmly bound ATP as the primary product. The following step, however, involves hydrolysis rather than release, of the firmly bound ATP. The energy made available, by hydrolysis of ATP, is then utilized in the formation of soluble ATP. This suggested mechanism involves two sites on the coupling enzyme, one of which alternates, through a conformational change, between a hydrophobic and a hydrophilic conformation. The experimental evidence supporting these mechanisms will be further discussed in the next section.

Since hypotheses on conformational change mechanisms are centered on the final reaction steps in ATP synthesis, we are still left with the unanswered question how the energy generated in electron transport will be transmitted into a protein conformational change. Recent experiments by Wikström (1976) may point toward a solution to this problem. In experiments with mitochondria, he found that induction of a diffusion potential will cause a conformational change in the cytochrome oxidase complex, as well as a probable conformational change in the ATPase moiety, envisaged as a change in aurovertin fluorescence. Thus, it is possible that a combination of the Mitchell and Boyer hypotheses may turn out to take us one step further toward a satisfactory explanation for the mechanism of electron-transport-coupled phosphorylation.

3. Photophosphorylation

3.1. Recent Work on the Mechanism of Light-Induced ATP Synthesis

3.1.1. Membrane-Bound Nucleotides

The role of membrane-bound adenine nucleotides, ADP and ATP, has recently come into focus in connection with the Boyer hypothesis of conformational coupling. An early observation of Eisenhardt and Rosenthal (1968) was that a small portion of $AT^{32}P$ was formed even in the presence of 2,4-dinitrophenol when mitochondria were incubated with $^{32}P_i$, ADP, substrate, and O_2. Boyer et al. (1973) confirmed this and considered the labeling to be due to $P_i \rightleftharpoons ATP$ exchange with membrane-bound ATP. They extended the studies to the exchange reactions $P_i \rightleftharpoons H_2O$, $P_i \rightleftharpoons ATP$, and $ATP \rightleftharpoons H_2O$, and found that only the $P_i \rightleftharpoons H_2O$ exchange is relatively insensitive to uncouplers. Furthermore, the $P_i \rightleftharpoons H_2O$ exchange is inhibited by oligomycin, as is the $P_i \rightleftharpoons ATP$ exchange, indicating the close link between this reaction and the phosphorylation system.

In chromatophores from *Rs. rubrum*, Yamamoto et al. (1972) found firmly bound ADP that is rapidly phosphorylated to ATP. This phosphorylation is oligomycin-insensitive, but uncoupler- and antimycin-sensitive. The binding of this type of ADP is prevented by oligomycin, as is the phosphorylation of added ADP. The phosphorylation of the firmly bound ADP was found to be very rapid; it is complete in 0.2 sec even at 0°C. It is also interesting to note that the detergent Triton X-100 inhibits the phosphorylation of added ADP completely, but the rapid phosphorylation of bound ADP not at all, indicating that the integrity of a membranous structure is more important for what would be the secondary reaction. The amount of firmly bound ADP was determined to be about 0.02 ADP per bacteriochlorophyll (Bchl). This would correspond to less than one ADP per reaction center (RC) complex, assuming a proportion of antenna chlorophyll (Chl) to RC Chl of 30 : 1.

Lutz et al. (1974), Bachofen et al. (1974, 1975), and Beyeler et al. (1975) thoroughly investigated the fate of firmly bound ATP in chromatophores. The relationship between the firmly and loosely bound ATP is not simply that of a direct γ-P transfer, since the loosely bound ATP becomes equally labeled in β- and γ-phosphate groups. They also confirmed, by comparing normal chromatophores with coupling-factor-depleted and with reconstituted chromatophores, that it is likely that the coupling factor ATPase is responsible for the binding of ADP and ATP (Lutz et al., 1974).

Thus, the firmly bound ATP serves as an energy donor rather than as a direct phosphate donor for soluble and loosely associated ADP. This mechanism is compatible with the demonstration by Keister and Minton (1971a) and by M. Baltscheffsky and Baltscheffsky (1972) that PP_i can serve as an energy source but not a phosphate source for ATP formation in the dark from ADP + P_i. The Bachofen group has postulated a mechanism for the last steps of ATP synthesis that is similar, but not identical to that of Boyer. It involves two sites for ATP-binding on the coupling factor ATPase with an energy-dependent first step and is depicted in Fig. 1 (Lutz et al., 1974).

In agreement with Boyer, they assume a conformational change in the transition between firmly bound ATP and soluble ATP. One main difference in the hypotheses is that Boyer considers the release of ATP as the only energy-requiring step, whereas the Bachofen group assumes that energization via a proton gradient is prior to phosphorylation of firmly bound ATP.

No data on the effect of uncouplers or oligomycin appear to have been published yet in connection with the studies on chromatophores by the Bachofen group. Yamamoto *et al.* (1972) did find, however, that the rapid phosphorylation of the firmly bound ADP was as sensitive to uncouplers as the phosphorylation of soluble ADP. On the other hand, Bachofen *et al.* (1974) found that in chloroplasts, also containing a fraction of firmly bound ATP, which could be extracted only with the aid of detergents, the labeling of this bound ATP with $^{32}P_i$ was much less sensitive to desaspidin than the $^{32}P_i$-labeling of soluble ATP.

Very recent data by D. J. Smith *et al.* (1976) show that in spinach chloroplasts, the appearance of γ-labeled $^{32}P_i$-ATP from added ADP and $^{32}P_i$ after an acid–base transition occurs after less than 10 msec. Furthermore, the rate of subsequent formation of γ-$^{32}P_i$-ATP is as fast as the rate that is observed under steady-state photophosphorylation. Their results also indicate that $^{32}P_i$-labeling of bound ADP and ATP occurs at lower rates than the phosphorylation of added ADP. One may hope that similar kinetic studies will soon be performed with bacterial chromatophores, to evaluate the role of bound nucleotides in this system.

3.1.2. Photophosphorylation Following Single Flashes

Of major interest for the mechanism of photophosphorylation is the formation of ATP resulting from single flashes, causing only one turnover of the cyclic electron-transport system. The importance of this approach is obvious, and it constitutes an advantage of photosynthetic systems over many other multienzyme reactions in which a single turnover with saturating conditions may be impossible to obtain. On the other hand, a single light flash may be regarded as a perturbation of the dark steady-state conditions, and perhaps tells us what occurs in the transition from dark to light steady state, rather than what occurs in the conditions under a physiological light steady state. Nevertheless, the information obtained by studying the effects of single flashes is extremely valuable, and sometimes seems to be the only way to obtain a detailed picture of the sequence of events.

By applying this technique to the phosphorylating system in bacteria, Nishimura (1962a) was able to show that photophosphorylation in chromatophores from *Rs. rubrum* could be separated in light and dark phases. The yield of ATP after a single flash required a dark period after the flash to become maximal. This dark period was directly proportional to the intensity of the flashing light. At saturating intensity, the half-life of the dark phase was 19.5 msec (Nishimura, 1962c).

The amount of ATP thus formed in the dark period following a saturating single flash was determined to be 0.047 ATP/Bchl per flash (0.4 msec) (Nishimura, 1962b). This value was not corrected for adenylate kinase activity, which usually is present in *Rs. rubrum* chromatophores even when they are extensively washed.

Very recently, the phosphorylation following single flashes was reinvestigated with chromatophores of *Rp. sphaeroides* in Witt's laboratory (Saphon *et al.*, 1975a,b; Jackson *et al.*, 1975), with special emphasis on the relative contribution of H^+ gradient and membrane potential to ATP formation. The short duration of the flash in their system (20 μsec) gives strictly only one turnover, in contrast to the earlier situation in which the longer duration of the flash made it possible that more than one turnover could have taken place.

In their system, the ionophore valinomycin inhibits totally, and this inhibition is parallel with the stimulation of the decay of the membrane potential, measured as the decay of the carotenoid absorbance change. Equally interesting is their finding that a concentration of antimycin that completely inhibits phosphorylation under continuous illumination lowers the flash yield of ATP by only about 20%. These results are interpreted as strong evidence for direct participation of the electic potential in the formation of ATP after a single flash. That the system is different from that under continuous illumination is obvious, since the effect of the applied inhibitors is so different. The flash yield of ATP with these short flashes is considerably lower than with the longer flashes (1/60) used by Nishimura (1962a), also considering the relative time proportion of respective flashes (1/25), but the phosphorylating capacity of the chromatophore preparations under continuous illumination is also lower (1/16 to 1/4) than that of the chromatophores used by Nishimura. This low yield would indicate, however, that only a very limited number of ATPase molecules actually are activated by these short flashes.

Removal of the coupling factor ATPase abolishes the stimulation of membrane potential decay by ADP + P, relating this stimulated decay intimately to the ATP synthesis (Jackson *et al.*, 1975). In this connection, it should be mentioned that in *Rs. rubrum* chromatophores, removal of the coupling factor specifically removes one kinetic phase of the light-induced carotenoid absorbance change—the slower phase (Baltscheffsky, M., 1974, 1976). Whether this portion of the change also reflects the membrane potential is a question that is under debate (Jackson *et al.*, 1975; Baltscheffsky, M., 1971, 1974, 1976).

The comparative lack of inhibition by antimycin of flash-induced phosphorylation (Saphon *et al.*, 1975*a*) indicates that the energy generated by electron transport between cytochromes *b* and c_2 contributes very little to this ATP synthesis, and that the energy is derived mainly from the primary steps: cytochrome c_2 → P870 → X. It is not clear what the effect of uncouplers on the ATP synthesis in this system is. This would be important to know, since it is well known that uncouplers cause a rapid decay of the carotenoid absorbance change, which is presumed to indicate the transmembrane potential. The experiments by the Witt group undoubtedly provide strong evidence for the participation of a localized or membrane potential in generation of the high-energy state. In recent work in this laboratory (Lundin *et al.*, 1977) on flash-induced phosphorylation in *Rs. rubrum* chromatophores, it has been possible to measure the phosphorylation after each flash (1-msec duration) separately. The uncoupler carbonylcyanide-*p*-trifluoromethoxy-phenylhydrazone (FCCP) causes a total inhibition in the same concentration range as that which inhibits continuous photophosphorylation. The total inhibition by valinomycin found by Saphon *et al.* (1975*a*) has been confirmed, provided the flashes are spaced at least 15 sec apart. More frequent flashes will successively restore part of the phosphorylation, thus converting the system to conditions prevailing under continuous illumination, in which the inhibition by valinomycin is low. It is evident that the potential-driven flash-induced ATP synthesis is different from the steady-state ATP synthesis under continuous illumination. There, the contribution of the membrane potential is small (Melandri *et al.*, 1974), but may be compensated for by an increased H⁺ gradient. The main question—whether the potential in conjunction with the H⁺ gradient is the direct driving force for ATP synthesis or whether the high-energy state can be generated through an equilibrium with a side reaction—is still unresolved.

3.1.3. $\Delta\psi$ and ΔpH and Their Relationship to ATP Synthesis

In contrast to the case in chloroplasts, in chromatophores, a pH gradient alone will not drive any appreciable ATP formation in the dark. Leiser and Gromet-Elhanan (1974) showed that even a ΔpH of 3.5 in an acid–base jump yields very little ATP unless the system is supplied with a valinomycin K⁺-induced diffusion potential in the base stage. On the other hand, the diffusion potential combined with a ΔpH of only 0.5 unit also yields very little ATP, suggesting that both the $\Delta\psi$ and the ΔpH are necessary for energization, leading to ATP synthesis.

The relative contribution to ATP synthesis under continuous illumination by membrane potential and H⁺ gradient in chromatophores from *Rp. capsulata* was recently studied by Melandri *et al.* (1974). It was found that the relationship between the H⁺ gradient and the potential is inverse, so that ionophores that decrease one parameter increase the other, keeping the total proton-motive force constant (Jackson *et al.*, 1968; Melandri *et al.*, 1974). When the total proton-motive force is decreased, phosphorylation is inhibited. This is found both with the uncoupler FCCP and with the combination of valinomycin and nigericin (Thore *et al.*, 1968).

The results are not quite consistent with the Mitchell hypothesis, however, as Melandri *et al.* (1974) also point out. With FCCP, there is a parallel decrease of $\Delta\psi$, ΔpH, and phosphorylation with increasing concentrations of the uncoupler. With the ionophores, where increasing concentrations of nigericin are added in the presence of a constant amount of valinomycin, there is a marked stimulation of $\Delta\psi$ at low nigericin concentration that is greater than the decrease in ΔpH and thus causes an increase in the total proton-motive force. Phosphorylation in this case is already 50% inhibited, whereas the proton-motive force is slightly stimulated and ΔpH is only 25% inhibited. With FCCP, the phosphorylation is 50% inhibited when the ΔpH is about 10% inhibited. The authors suggest that this inconsistency with the chemiosmotic model may be due to a rate-limiting reaction in the ATP-synthetase complex. It seems unlikely, however, that this would be a rate-limiting step in photophosphorylation, since PMS, through accelerating electron transport or increasing the ΔpH (Trebst, 1976), usually greatly increases the rate of photophosphorylation in chromatophores. Instead, these results may be a reflection of an unwarranted use of the total light-induced carotenoid absorbance change as a measure of $\Delta\psi$. If only part of the change, e.g., the fast kinetic phase, reflects the light-induced $\Delta\psi$, a closer investigation of the relative proportions of the different kinetic phases, when the carotenoid change is enhanced by the presence of nigericin, may shed some light on this question. Also, Gromet-Elhanan (1971) presented strong evidence that the ΔpH under continuous illumination is not decreased by phosphorylating conditions. Neither do conditions that reduce ΔpH appreciably decrease the rate of photophosphorylation. It appears from results published by the Witt group that no increase of the carotenoid change in the presence of nigericin is seen after single short flashes (Saphon *et al.*, 1975*b*), which would indicate that the increase observed by others under continuous illumination (Nishimura, 1970; Melandri *et al.*, 1974) belongs mainly to the slower kinetic phase.

3.1.4. The Coupling-Site Problem

The original proposal of two coupling sites was based on inhibition data with valinomycin (Baltscheffsky, H., 1961; Baltscheffsky, H., and Arwidsson, 1962), which inhibited phosphorylation in *Rs. rubrum* chromatophores about 50%. This inhibition was not evident when phenazine methosulfate (PMS) was used as an artificial electron donor. Subsequent work (Jackson *et al.*, 1968; Gromet-Elhanan, 1969) showed that the inhibition with valinomycin under different conditions is variable. It is consistent, though, that the PMS system is not affected even at high concentrations of the inhibitor.

This lack of inhibition, or bypass around an inhibition site, by PMS is also evident in the case of electron-transport inhibitors such as antimycin, 2-heptyl-4-hydroxyquinoline-*N*-oxide (HOQNO), or dibromothymoquinone (DBMIB). The basic rate of photophosphorylation in *Rs. rubrum* chromatophores is usually accelerated in the presence of PMS, and this accelerated rate is not inhibited by the electron-transport inhibitors. This has been interpreted as a bypass being formed by PMS around one coupling site (Baltscheffsky, H., and Baltscheffsky, 1958; Geller and Lipman, 1960, Baltscheffsky, M., 1974). It was recently found by Knaff and Buchanan (1974) that *Chromatium* chromatophores also follow this pattern.

By studying the steady-state redox changes of cytochromes b and c_2 when phosphorylating conditions are induced, M. Baltscheffsky (1967b) identified a crossover point between these cytochromes. In analogy with the situation in animal mitochondria (Chance *et al.*, 1955), this region has been proposed as a site of coupling for ATP formation. This should then be the site that is bypassed in the PMS system. No further sites of crossover in the cyclic electron transport have yet been identified. With the chemiosmotic reasoning, a second site would not be necessary to account for the PMS bypass, since the increased extent of the H^+ gradient formed in the light in the presence of this compound could account for the increased rate of ATP formation. Trebst (1974) postulated that in chloroplasts, the quinone pool physiologically acts as the translocating proton shuttle, and Hauska (1972) suggested that PMS in chloroplasts constitutes an artificial proton translocating loop. These postulates, although proposed for chloroplasts, can be of equal significance in the case of chromatophores. Consistent with this view is that whereas valinomycin alone has no effect on photophosphorylation mediated by PMS, the combination of valinomycin and nigericin causes a total uncoupling of the PMS system (Jackson *et al.*,

1968). On the other hand, some support for the concept of two sites in photophosphorylation, at least one of which is electrogenic and more sensitive to valinomycin than the other, can be found in the recent papers by Jackson *et al.* (1975) and Saphon *et al.* (1975a,b). As was mentioned above, these authors find that phosphorylation after single short flashes causing only one turnover is totally inhibited by valinomycin and only marginally (20%) inhibited by antimycin. The site operative in these experiments apparently depends very little on electron flow in the cytochrome region, and can be visualized as being situated outside this region.

3.2. Photophosphorylation to PP_i

In 1966, H. Baltscheffsky *et al.* (1966) discovered that chromatophores from *Rs. rubrum* could produce inorganic pyrophosphate (PP_i) as an alternative end product to ATP in photophosphorylation. They illuminated chromatophores in the presence of P_i but without the phosphate acceptor, ADP, and found a significant esterification of P_i to PP_i. This reaction is electron-transport-dependent, as was shown by the inhibition with antimycin, and it is abolished by uncouplers, but unlike the phosphorylation to ATP, it is not inhibited by the energy-transfer inhibitor oligomycin (Baltscheffsky, H., and von Stedingk, 1966). On the contrary, PP_i synthesis is markedly stimulated by oligomycin, indicating that this inhibitor prevents an "energy leak" via the ATP-synthesis pathway (Baltscheffsky, H., and von Stedingk, 1966). The PP_i synthesis apparently utilizes the same coupling site on the electron-transport chain as the ATP synthesis, since reversed flow may be induced in the dark by either compound, resulting in the reduction of cytochrome b and the oxidation of cytochrome c_2 (Baltscheffsky, M., 1968). Of the other energy-transfer inhibitors, dicyclohexylcarbodiimide (DCCD) has little effect on PP_i synthesis, whereas Dio-9 does inhibit PP_i formation (Guillory and Fisher, 1972).

It was suggested by M. Baltscheffsky (1967a) that the PP_i-synthetase is a different enzyme from the ATP-synthetase. This suggestion was based on the additivity found in the reversed-flow experiments; the subsequent addition of the other energizing compound always resulted in a further energy-linked cytochrome redox change. Fisher and Guillory (1969) came to the same conclusion using LiCl-treated chromatophores—which showed no ATP-linked reactions but retained the PP_i-linked reactions—and comparing them with butanol-treated chromatophores, in which mainly the PP_i pathway was impaired. The demonstration that PP_i could be used as an energy source for

ATP formation (Keister and Minton, 1971a,b; Baltscheffsky, M., and Baltscheffsky, 1972) also strongly indicated the interaction of two separate enzymes. Johansson (1975) further substantiated the concept of two different terminal enzymes by showing that an antibody against the purified ATPase that completely inhibited ATP synthesis as well as other energy-linked reactions dependent on ATP did not affect the PP_i-linked reactions to any significant degree. Thus, the collected evidence pointing to a separate PP_i-ase is strong. This enzyme, however, is very firmly membrane-bound, and has so far resisted considerable efforts to achieve its isolation.

The photophosphorylation to PP_i shows a different light-saturation pattern than that to ATP. Guillory and Fisher (1972) showed that PP_i formation saturates at a light intensity that is less than one-tenth of that required for saturation of the ATP synthesis. Interestingly enough, at the low saturation intensity for PP_i formation, the ratio of the rates of PP_i to ATP synthesis was as high as 1, whereas at higher intensities, it dropped to 0.25. As Guillory and Fisher point out, it is at low light intensity that *Rs. rubrum* usually grows in nature. This may indicate that the alternative pathway to PP_i is of considerable physiological significance.

It seems that a systematic comparative investigation of photophosphorylation to PP_i in a variety of species of photosynthetic bacteria may well show that the PP_i-linked energy metabolism may be fairly common. So far, there are indications of this pathway in the following species: *Rs. rubrum* S-1 and G-9 (Baltscheffsky, H. *et al.*, 1966); *Rp. palustris* (Knobloch, 1975); *Rp. sphaeroides* (Sherman and Clayton, 1972); and *Rp. viridis* (Jones, O. T. G., and Saunders, 1972).

3.3. Photophosphorylation in *Halobacterium halobium*

One of the most interesting new developments in the field of photophosphorylation is the work by Stoeckenius, Oesterhelt, and their co-workers on photophosphorylation in the halophilic bacterium *Halobacterium halobium* (Oesterhelt and Stoeckenius, 1973; Lozier *et al.*, 1975; Chapter 29). The energy-conversion system in light in this bacterium resides in the purple membrane patches, which contain a rhodopsin-protein. Through a conformational change, this structure has the ability to translocate protons and thus generate a proton gradient over the cell membrane. The proton gradient formed in the light is then proposed to drive ATP synthesis in the same membrane, which is also responsible for the oxidative phosphorylation under dark aerobic growth.

That the light-induced proton gradient generated in the purple patches is able to drive ATP synthesis was elegantly shown by Racker and Stoeckenius (1974). They incorporated isolated purple membranes from *Halobacterium* into phospholipid vesicles, together with the ATPase coupling factor and the "hydrophobic protein" fraction from mitochondria, and were able to show a net synthesis of ATP in this reconstituted system. Very recently, Yoshida *et al.* (1975) further elaborated this work and showed that the hydrophobic protein fraction from mitochondria can be substituted for by the DCCD-sensitive coupling factor from the thermophilic bacterium PS3, thus eliminating the possibility that constituents of the mitochondrial hydrophobic protein other than the coupling factor were a prerequisite for ATP formation. The rate of ATP produced in this system was estimated to be 5.3 molecules of ATP per molecule of coupling factor per minute. The requirement for ATP formation was an electrochemical potential of protons above 200 mV. In whole cells of *Halobacterium halobium*, Danon and Stoeckenius (1974) demonstrated a net increase of ATP when anaerobic cells were illuminated. This increase was inhibited by Dio-9, but not by antimycin A. Neither was it stimulated by PMS. The cells cannot grow anaerobically in the light, however, and thus the physiological role of the "photophosphorylation" may be limited to an accessory function.

4. Partial Reactions of Photophosphorylation

Several dark reactions in all likelihood reflect the final steps of ATP and PP_i synthesis in chromatophores and the reversibility of these reactions. These are the ATPase and PP_i-ase, the ATP $\rightleftharpoons$ $^{32}P_i$ exchange and ADP $\leftrightarrow$ ATP exchange, the PP_i $\rightleftharpoons$ $^{32}P_i$ exchange, and the PP_i-driven ATP synthesis. The last two reactions have so far been found only in chromatophores from *Rs. rubrum*.

4.1. ATPase

This reaction is due to the reversal of the last step in ATP synthesis. ATPase in chromatophores was first reported by Frenkel (1956), and has subsequently been rather extensively investigated by several investigators (Bose and Gest, 1965; Baltscheffsky, M., *et al.*, 1966; Horio *et al.*, 1968). The identity of the activity in the chromatophore membrane with a "coupling-factor" enzyme was conclusively established when Baccarini-Melandri *et al.* (1970) were able to resolve, in chromatophores from *Rp. capsulata*, the coupling factor from the membrane and, with the factor, the ATPase activity. Also in *Rs. rubrum* (Johansson, 1972), *Rp. sphaeroides* (Reed and Raveed, 1972), and *Chromatium vinosum* (Gepstein and Carmeli, 1974), the ATPase activity has been shown to be associated with the "coupling-factor" enzyme, and it is fairly safe to assume that this is the case in all photosynthetic bacteria. Since the coupling factors are treated in Chapter 31, only some significant properties of the enzyme when membrane-bound will be mentioned here.

Uncouplers generally stimulate the activity of the ATPase greatly, up to about 100%, with maximum stimulation in the same concentration range of uncoupler that causes maximum inhibition of photophosphorylation. One exception is the ionophore gramicidin, which, although it is not a true uncoupler in the classic sense, has an uncoupling effect on electron transport and phosphorylation, but inhibits the ATPase reaction. (Baltscheffsky, M., *et al.*, 1966). Other inhibitors of ATPase are oligomycin, DCCD, BAL (British Anti-Lewisite), and mercaptoethanol, the two latter suggesting a role for SH groups in the active site of the enzyme (Bose and Gest, 1965). Horiuti *et al.* (1968) showed a strong dependence of the ATPase reaction on the redox conditions. Many of these results are compatible with current postulates for the ATPase in mitochondria functioning as a proton conductor, the activity of which is controlled by the proton gradient across the energy-transducing membrane.

4.2. PP$_i$-ase

In *Rs. rubrum* (Baltscheffsky, M., *et al.*, 1966; Baltscheffsky, M., 1968) has also been found to be an active, membrane-bound PP$_i$-ase that in many respects resembles the ATPase reaction; it requires Mg^{2+} ions, and it is stimulated by uncouplers, although never to the same extent as the ATPase, and also by gramicidin. In contrast to the ATPase, it is not strongly inhibited by oligomycin (Baltscheffsky, M., *et al.*, 1966). That the PP$_i$-ase is a separate enzyme from the ATPase is evident, from, among other results described in this review, our results (Baltscheffsky, M., 1976), which show that the ATPase can be almost completely removed from chromatophores without inhibiting their PP$_i$-ase activity.

Since the PP$_i$-ase activity, in analogy with the ATPase activity, is considered as being the reversal of the PP$_i$-synthesizing reaction, it is interesting to note that the activity of this enzyme in *Rs. rubrum* usually is higher than that of the ATPase. Only when both activities are maximally stimulated by uncouplers are the rates about equal. This higher rate of hydrolysis

by the PP$_i$-ase is reflected in the faster energization with PP$_i$ than with ATP (Baltscheffsky, M., 1969b), but here the discrepancy is about tenfold. This will be further discussed in Section 7.4. Coupled to the hydrolysis of both ATP and PP$_i$ is the inward translocation of protons (Moyle *et al.*, 1972), rendering the interior of the chromatophore positively charged as compared with the outside.

4.3. ATP ↔ ^{32}P$_i$ Exchange

This reaction, which consists of an exchange of the γ-phosphate of ATP with free ^{32}P$_i$ in the reaction medium, was first demonstrated in chromatophores by Horio *et al.* (1965). Like its counterpart in animal mitochondria, it is inhibited by uncouplers but not by electron-transport inhibitors. Dio-9 partly inhibits this reaction in chromatophores (Fisher and Guillory, 1967) from both *Rs. rubrum* and *Chr. vinosum* (Hochman and Carmeli, 1973). Johansson (1975) showed that it is very likely that the coupling-factor protein catalyzes this exchange reaction since the reaction is inhibited by the antiserum against the coupling factor.

4.4. PP$_i$ ↔ ^{32}P$_i$ Exchange

Keister and Minton (1971b) found a PP$_i$ ↔ ^{32}P$_i$ exchange, corresponding to the ATP ↔ ^{32}P$_i$ exchange, to take place in *Rs. rubrum* chromatophores. As could be expected, this reaction is inhibited, not by oligomycin, but instead by fluoride. A relationship between this reaction and the energy-dependent formation of PP$_i$ is shown by the inhibition with the uncouplers *m*-chlorocarbonylcyanidephenylhydrazone (CCCP) and S-13. The rate of PP$_i$ ↔ ^{32}P$_i$ exchange is considerably lower than that of the ATP ↔ ^{32}P$_i$ exchange activity, being 5 μmol/mg Bchl per hr, as compared with 25 μmol/mg Bchl per hr for the ATP ↔ ^{32}P$_i$ exchange.

4.5. ADP ↔ ATP Exchange

A dark ADP ↔ ATP exchange in chromatophores from *Rs. rubrum* was described by Horio *et al.* (1964, 1965). The reaction does not appear to be closely related to the terminal reactions of ATP synthesis, since uncouplers do not affect it at all. In fact, the only compound having an inhibitory effect is PP$_i$, but since the PP$_i$ concentration in the reported experiments is equal to that of Mg^{2+}, which is necessary for the exchange reaction, this inhibition is probably due to the binding of all Mg^{2+} to PP$_i$.

4.6. PP$_i$-Driven ATP Synthesis

The property of some photosynthetic bacteria of forming PP$_i$ coupled to light-induced electron transport has already been described. The presence of two independent terminal enzymes in energy conservation coupled to the same electron-transport system offers a unique opportunity to study energy transfer on a level close to the terminal reactions. It was shown by Keister and Minton (1971*a,b*) and by M. Baltscheffsky and Baltscheffsky (1972) that PP$_i$ could serve as an energy donor, but not a phosphate donor, in a dark synthesis of ATP. Keister and Minton examined the reaction biochemically, and showed conclusively that ATP was formed from ADP and P$_i$ in this reaction. The demonstration by M. Baltscheffsky and Baltscheffsky consisted of spectrophotometric evidence that the same type of control as is exerted by addition of ADP and P$_i$ on the high-energy state when induced by light is found in the dark when addition of PP$_i$ has generated the high-energy state. The investigation by Keister and Minton also showed that the reaction is inhibited by uncouplers and by oligomycin, but not by the electron-transport inhibitors antimycin A and HOQNO. The pH optimum for the reaction is around 9 (Keister and Minton, 1971*b*); the ADP concentration for half-maximum rate is 5 μM, which is slightly lower than that required for photophosphorylation. The minimum P$_i$ concentration required for this reaction, when the P$_i$ was derived from hydrolyzed PP$_i$, was 0.25 mM, which is considerably lower than that required for maximum photophosphorylation (Horio *et al.*, 1966). This may indicate that the P$_i$ emanating from hydrolyzed PP$_i$ for some time is more readily available as bound to the enzyme or membrane. The stoichiometry is 9–12 PP$_i$ hydrolyzed per ATP formed. Interestingly enough, this stoichiometry is not greatly altered by the simultaneous occurrence of another energy-requiring reaction, e.g., the energy-linked transhydrogenase or succinate-linked pyridine reduction (Keister and Minton, 1971*b*). In this connection, it is interesting to note that the stoichiometry for H$^+$ translocation driven by PP$_i$-ase is 0.5 H$^+$/PP$_i$ (Moyle *et al.*, 1972). Thus, if the ATP synthesis is driven by electrogenic H$^+$ translocation, one would not expect a better stoichiometry than 4 PP$_i$ per ATP synthesized. That the PP$_i$ requirement is even higher, and that the simultaneous proceeding of other energy-requiring reactions does not alter this requirement, may indicate that not all the energy liberated by the hydrolysis of PP$_i$ can be directly utilized for ATP synthesis.

5. Effects of Inhibitors

Since there does not seem to exist any collected information on the action of various inhibitors on bacterial photophosphorylation, this subject will be treated in some detail, in an attempt to provide such information. It is hoped that this treatment will be helpful, especially for those who are newcomers to the field of photophosphorylation.

5.1. Electron-Transport Inhibitors

Many of the inhibitors known to inhibit mitochondrial electron transport and phosphorylation are also effective in bacterial photophosphorylation. The variety is limited, however, to those known to inhibit the mitochondrial system in the region of coupling site II. Inhibitors of cytochrome oxidase, such as KCN or NaN$_3$, have no effect on the bacterial system. Neither do inhibitors of the mitochondrial electron transport around site I such as rotenone have any appreciable effect on the cyclic electron transport in photosynthetic bacteria, even if they do inhibit other electron-transport reactions in these organisms (Thore *et al.*, 1968; Keister and Minton, 1969).

The first reports on inhibition of photophosphorylation appeared in 1956. Geller and Lipmann (1960) reported the inhibition by antimycin A; L. Smith and Baltscheffsky (1956, 1959) found the same effect by HOQNO. The latter authors were also able to correlate the inhibited state with the spectrophotometric observation of an increased light-induced absorption band around 430 nm.

This increased extent of light-induced reduction of cytochrome *b* is also seen with inhibition by antimycin A. Furthermore, Petty and Dutton (1976) recently showed in elegant experiments that the inhibition by antimycin A results in the retention of a membrane-bound proton, which in the uninhibited state is released inside the chromatophore vesicle. Antimycin A inhibition of photophosphorylation has been reported in chromatophores from several species of Rhodospirillaceae, and recently (Knaff and Buchanan, 1974) also in *Chr. vinosum*. The concentration of antimycin A necessary for maximal inhibition has been estimated to be close to five molecules of inhibitor for each chromatophore (Yamamoto *et al.*, 1972). This value can be compared with the concentration of cytochrome *b*, which has also been estimated to be five molecules (Kakuno *et al.*, 1971) per chromatophore. Both estimates were carried out with *Rs. rubrum* chromatophores.

These data correspond closely to the situation in submitochondrial particles, in which Brandon *et al.*

(1972) reported that (1) the stoichiometry of antimycin A to cytochrome b is approximately equal to 1, and (2) antimycin A is ten times more efficient than HOQNO.

In animal mitochondria, a protein of low molecular weight has been identified as the antimycin-binding protein (Das Gupta and Rieske, 1973).

2-Hydroxy-31-ω-cyclohexyloctyl-1,4-naphthoquinone (HHNQ) was reported by Thore *et al.* (1968) to inhibit electron transport and photophosphorylation in *Rs. rubrum* chromatophores at a site before antimycin A.

Another inhibitor of bacterial photophosphorylation on the level of electron transport is the quinone antagonist dibromothymoquinone (DBMIB). M. Baltscheffsky (1974) showed that this inhibitor, known to inhibit quinone oxidation in chloroplasts (Böhme *et al.*, 1971), is also effective in *Rs. rubrum* chromatophores, and that its site of inhibition appears to be parallel to that of antimycin A, since there is a strong synergistic effect on the light-induced oxidation and dark re-reduction of cytochrome c_2 when both inhibitors are present simultaneously. The inhibitory concentration of DBMIB is nearly stoichiometric with the endogenous concentration of ubiquinone in the chromatophore suspension, indicating that the site of action in chromatophores also is at the ubiquinone level.

The inhibition of phosphorylation by antimycin, HOQNO, and DBMIB can be bypassed in *Rs. rubrum* chromatophores by PMS (Geller and Lipmann, 1960; Baltscheffsky, H., and Baltscheffsky, 1958; Baltscheffsky, M., 1974). This ability of PMS to create a bypass around the inhibition site may be related to lipophilic properties of PMS, which, coupled to the electron transport, can transport protons across a lipid barrier in a membrane (Trebst, 1974). The sulfonated analogue of PMS, being hydrophilic, does not function as a mediator in photophosphorylation (Baltscheffsky, M., unpublished). It has also been found that the PMS bypass, when studied at the level of electron transport, apparently does not function in chromatophores from *Rp. sphaeroides* (Prince and Dutton, personal communication).

5.2. Uncouplers

It can generally be said that agents that have an uncoupling effect in other energy-transducing membranes have the same effect in bacterial chromatophores. One notable exception to this is 2,4-dinitrophenol (DNP), which has very poor effect on chromatophores from *Rs. rubrum* except at very high concentration (1 mM causes less than 50% inhibition).

The DNP analogue octyl-DNP, being more strongly lipophilic, does uncouple at normal concentrations. Most uncouplers are weak acids with lipophilic properties and render both natural and artificial membranes permeable to protons. This property is considered, by chemiosmotic reasoning, to be the basis for the inhibitory action, since it would prevent the formation of a pH gradient across the membrane. Consequently, uncouplers inhibit the light-induced H^+ uptake in chromatophores. Table 1 presents a variety of uncouplers that have been found effective in bacterial chromatophores. The interesting work by Hanstein and Hatefi (1974) pointing to the existence of a specific uncoupler-binding protein in the mitochondrial membrane may in the near future lead to a more detailed knowledge about the uncoupling mechanism.

5.3. Ionophores

The groups of antibiotics that have the property of increasing the permeability of biological and artificial membranes to certain cations have been termed *ionophores*. The effects of many of these ionophores have also been extensively studied in bacterial chromatophores. Here, we will dwell only on their inhibitory effects on photophosphorylation, since their other properties as well as specificities are well described elsewhere in this book.

The ionophores have been classified into two groups: the valinomycin type, which forms charged complexes with alkali ions, and the nigericin type, which forms charge-compensated complexes. Both types of complexes are lipophilic (Pressman *et al.*, 1967). Valinomycin, one of the most-used ionophores, creates a specific K^+ permeability in membranes. Valinomycin in the presence of K^+ has some inhibitory effect on photophosphorylation measured under continuous illumination. When it was first found (Baltscheffsky, H., *et al.*, 1960, 1961) to be an inhibitor in chromatophores, the maximum inhibition was consistently found to be around 50%, and furthermore, the PMS-stimulated phosphorylation was not sensitive to valinocycin. Other authors found the maximal inhibition to be less than 50% (Jackson *et al.*, 1968; Thore *et al.*, 1968). Interestingly enough, in their recent study on photophosphorylation after single flashes, Saphon *et al.* (1975a) reported a complete inhibition of ATP synthesis by valinomycin.

Gramicidin, an ionophore of the valinomycin type that transports protons as well as K^+ and Na^+, is a potent inhibitor of photophosphorylation and is similar in its action to an uncoupler.

Nigericin, an ionophore of the type that forms an

Table 1. Some Inhibitors of Photophosphorylation[a]

Inhibitor[b]	Type of inhibition	Concentration (μM)	
		At 50% inhibition	At maximum inhibition
Antimycin A	Electron transport	2×10^{-8}	4×10^{-7} (in *Chromatium*, 2×10^{-5})
DBMIB	Electron transport	5×10^{-6}	1.5×10^{-5}
HHNQ	Electron transport		3.3×10^{-7}
HOQNO	Electron transport	6×10^{-8}	10^{-6}
SN 5949	Electron transport	$\approx 10^{-7}$	10^{-6}
PMA	Electron transport	10^{-4} (45%)	
CCCP	Uncoupler	2×10^{-6}	2×10^{-5}
FCCP	Uncoupler	6×10^{-7}	3×10^{-6}
DNP	Uncoupler	2×10^{-3}	10^{-2}
Desaspidin	Uncoupler	$\approx 10^{-7}$	10^{-6}
Dicoumarol	Uncoupler	5×10^{-5}	2×10^{-4}
PCP	Uncoupler		10^{-5}
S XIII	Uncoupler	7.5×10^{-8}	1.6×10^{-7}
Aurovertin	Energy transfer		
DCCD	Energy transfer	3×10^{-6}	
Dio-9	Energy transfer	8 μg/ml	32 μg/ml
Oligomycin	Energy transfer	10^{-6}	8×10^{-6}
Phlorizin	Energy transfer	1.7×10^{-3}	7.5×10^{-3}
Gramicidin	Ionophore	7×10^{-8}	5×10^{-7}
Valinomycin + KCl	Ionophore	3.3×10^{-8}	
Valinomycin + nigericin (1.5×10^{-7} M)	Ionophore	10^{-7}	10^{-6}

[a] Most of the data in this table were obtained with *Rs. rubrum* chromatophores.
[b] Abbreviations not defined in the text: (PMA) phenylmercuric acetate; (PCP) pentachlorophenol; (S XIII) 5-chloro-3,4-butyl,2'-chloro,4'-NO$_2$-salicylanilide.

uncharged complex with K^+ and transports this ion in exchange for H^+, has in itself no inhibitory effect on photophosphorylation (Thore *et al.*, 1968). The combination of nigericin and valinomycin, however, inhibits completely (Jackson *et al.*, 1968). As a rule, the combination of an ionophore from the valinomycin group with one from the nigericin group will abolish photophosphorylation even if neither of them has any effect by itself. This was shown (Thore *et al.*, 1968) to be true for the combination of valinomycin or nonactin with dianemycin and monensin A, as well as with nigericin. It was suggested that the combined action of both types of ionophores induce an energy-dissipating cyclic transport of K^+ or H^+ across the chromatophore membrane.

Light-induced PP$_i$ formation shows an inhibition pattern with ionophores similar to that of ATP formation: valinomycin alone inhibits up to 50%, and valinomycin plus nigericin give a total inhibition (99%) (Guillory and Fisher, 1972).

5.4. Energy-Transfer Inhibitors

Four inhibitors of photophosphorylation have been found to inhibit close to or at the final reaction in ATP synthesis. These inhibitors are the antibiotics oligomycin, aurovertin, Dio-9, and DCCD. With the exception of aurovertin, they all are inhibitors of the ATPase reaction as well as of photophosphorylation.

Oligomycin (Lardy *et al.*, 1958). aurovertin, and DCCD (Beechey *et al.*, 1967) were first found to inhibit oxidative phosphorylation in animal mitochondria. Aurovertin has been shown to bind stoichiometrically to F$_1$ of submitochondrial particles (Lardy and Lin, 1969). It inhibits ATP synthesis, but not the hydrolysis (Lardy, 1961; Lardy *et al.*, 1964), nor the ATP-induced manifestations of reversed phosphorylation (Lee and Ernster, 1968).

In chromatophores, the effects of aurovertin appear to be similar: photophosphorylation but not ATPase is inhibited; ATP-induced reduction in the dark of cytochrome *b* is unaffected by the inhibitor (Baltscheffsky, M., and Baltscheffsky, H., unpublished).

Oligomycin was found to be a strong inhibitor of oxidative phosphorylation in mitochondria (Lardy *et al.*, 1958) and submitochondrial particles. The antibiotic also inhibits the ATPase reaction as well as ATP-induced reversed phosphorylation reactions (Lee and Ernster, 1968). The rate of electron transport is considerably slowed down by oligomycin, and "respiratory control" can be induced by the antibiotic. This effect is reversed by uncouplers or valinomycin (Lardy, 1961).

Two proteins isolated from mitochondrial membranes have been assigned roles in connection with oligomycin inhibition. One is the oligomycin-sensitivity-conferring protein (OSCP), which has been suggested to constitute the "stalk" that connects the coupling factor ATPase (F_1) complex with the membrane (MacLennan and Tzagoloff, 1968). The other is the "DCCD-binding protein" (see below). This protein has been suggested to bind OSCP and also to be the binding site of oligomycin (Ernster *et al.*, 1974).

Oligomycin is an inhibitor of photophosphorylation to ATP (Baltscheffsky, H., and Baltscheffsky, 1960) and of the membrane-bound ATPase in chromatophores (Baltscheffsky, M., *et al.*, 1966). It does not, however, inhibit photosynthetic PP_i formation, but rather stimulates this reaction (Baltscheffsky, H., and von Stedingk, 1966), nor does oligomycin inhibit membrane-bound PP_i-ase (Baltscheffsky, M., *et al.*, 1966). No studies on the exact mode of action of oligomycin have yet been reported to have been performed with chromatophores.

DCCD was found to be a potent inhibitor of both mitochondrial oxidative phosphorylation and ATPase (Beechey *et al.*, 1967). The pattern of inhibition is very similar to that of oligomycin. DCCD appears to enter in a covalent binding with a protein that has been isolated from mitochondria (Cattell *et al.*, 1971). It is a highly hydrophobic protein with a molecular weight of about 10,000.

DCCD inhibits photophosphorylation to ATP in chromatophores, but, like oligomycin, it does not significantly inhibit photophosphorylation to PP_i, according to Guillory and Fisher (1972). In contrast to the case with oligomycin, however, these authors did not find any stimulation of PP_i synthesis with

DCCD. In fact, they obtained a slight (10–15%) inhibition with this compound.

As mentioned above, Dio-9 was first found to inhibit oxidative phosphorylation and ATPase in rat liver mitochondria, by Guillory (1964). Dio-9 releases respiration in mitochondria, but only in the absence of phosphate; in the presence of phosphate (or arsenate), the effect of Dio-9 is to inhibit respiration.

In chromatophores, Fisher and Guillory (1967) showed that photophosphorylation to ATP, ATPase, and the energy-driven transhydrogenase, driven by ATP, PP_i, or light, are strongly inhibited by Dio-9. Guillory and Fisher (1972) also showed that photosynthetic PP_i synthesis and pyrophosphatase activity are similarly inhibited by the antibiotic. The demonstration that both ATP- and PP_i-linked reactions are inhibited clearly points to a different mode of action than that of oligomycin.

6. Electron-Transport Control

Since electron transport in photosynthetic bacteria is a cyclic process, direct measurements of the rate of electron transport under an illuminated steady state are impossible to perform. Changes in electron-transport rates can, however, be seen as redox changes of components in the electron-transport chain, especially of components located close to a rate-limiting step in the sequence. The phenomenon of electron-transport control is a fundamental characteristic in oxidative phosphorylation in mitochondria from higher organisms. Chance and co-workers (Chance and Williams, 1955; Chance *et al.*, 1955) investigated the relationship between ADP-induced stimulation of respiration and the steady-state oxidation–reduction changes that concomitantly occur at the level of individual electron carriers. They also formulated the crossover theorem for the identification of coupling sites based on the reversible state 4 to state 3 transition. In photosynthetic systems, West and Wiskich (1968) first demonstrated electron-transport control in the noncyclic electron transport of spinach chloroplasts; the rate reversibly increased on addition of low concentrations of ADP. In chromatophores, we (Baltscheffsky, M., and Baltscheffsky, 1972) showed that ADP added to illuminated *Rs. rubrum* chromatophores induced a specific reversible oxidation–reduction cycle of cytochrome *b* and a corresponding reduction–oxidation cycle of cytochrome c_2. Furthermore, addition of uncoupler caused an irreversible oxidation of photoreduced cytochrome *b*. Applying the crossover theorem, these changes were identified as manifestations of electron-transport control, thus extending this regulatory property to the level of bacteria. Additional support for this conclusion was found in the work by Thore *et al.* (1969), who found a stimulation of the rate of NADH oxidation in light in the presence of uncoupler that could best be explained by assuming electron-transport control at a site in the cyclic photoinduced electron transport.

The mechanism of electron-transport control is not yet fully understood in any system.

In the electron-transport chains of mitochondria and of chloroplasts, the influence of the proton electrochemical gradient $\Delta\tilde{\mu}_H$ and its components, ΔpH and $\Delta\psi$, as controlling factors has been investigated. Padan and Rottenberg (1973) found that in mitochondria, the proton electrochemical gradient $\tilde{\mu}_H$ controls the rate of uncoupled electron transport, but that under coupled conditions, the phosphorylation reaction directly controls the rate of respiration. Bamberger *et al.* (1973) concluded that the ΔpH as well as the pH of the inner thylakoid space and the pH of the medium are the factors that control the space and the pH of the medium are the factors that control the rate of electron transport in chloroplasts.

In bacterial chromatophores, although the relative proportions of ΔpH and $\Delta\psi$ in the high-energy state are known (Casadio *et al.*, 1974; Schuldiner *et al.*, 1974), no investigation has yet directly related these parameters to electron-transport control. As mentioned above, it was shown by M. Baltscheffsky and Baltscheffsky (1972) that addition of FCCP under illumination causes an oxidation of photoreduced cytochrome *b*. Also, Prince and Dutton (1975) showed that under uncoupled conditions (in the presence of FCCP), there is no restriction on the electron flow between cytochromes *b* and c_2 in chromatophores from *Rp. capsulata* and *Rp. sphaeroides*. These data can be related to the decrease in $\Delta\tilde{\mu}_H$ under the same conditions reported by Schuldiner *et al.* (1974). There also appears to be a possibility of direct control by the phosphorylating system, since the removal of the coupling factor protein causes a reversible loss of control (Johansson *et al.*, 1972).

7. Reversed Reactions of Photophosphorylation

7.1. ATP- and PP$_i$-Induced Changes in the Redox State of Cytochromes

It was first demonstrated in rat liver mitochondria that ATP, through the reversibility of electron-transport-linked phosphorylation, could affect the redox state of cytochromes in the electron-transport chain (Klingenberg and Schollmeyer, 1961).

In bacterial chromatophores, this manifestation of reversal of photophosphorylation was demonstrated in chromatophores from *Rs. rubrum*, in which it was first found that ATP or PP$_i$, when added in the dark, caused the reduction of an endogenous cytochrome (Baltscheffsky, M., *et al.*, 1966), later identified as cytochrome *b* (Baltscheffsky, M., 1968; Dutton and Baltscheffsky, 1972). The PP$_i$-induced reaction was readily reversible, and was shown to follow the hydrolysis of PP$_i$ (Baltscheffsky, M., 1969a). When induced by ATP, the reversibility due to hydrolysis was very slow, although addition of uncouplers caused immediate reversal of the changes induced by either ATP or PP$_i$. Furthermore, only the ATP-induced change was inhibited by oligomycin, whereas both the ATP- and the PP$_i$-induced reductions were inhibited by antimycin and uncouplers (Baltscheffsky, M., 1967a, 1969b).

Concomitant with the reduction of cytochrome *b*, an oxidation of cytochrome c_2 occurred with the same characteristics as the reduction of cytochrome *b*. Through this reversed electron transport, a coupling site could be identified between these two cytochromes (Baltscheffsky, M., 1967b). The same coupling site was later confirmed in an investigation of the forward reaction (Baltscheffsky, M., and Baltscheffsky, 1972).

In chromatophores from *Rhodopseudomonas viridis*, addition of ATP in the dark causes the oxidation of two *c*-type cytochromes, c_{553} and c_{558}, as was shown by O. T. G. Jones and Saunders (1972). It is not clear what the electron acceptor is in this reaction, since these chromatophores are devoid of *b*-type cytochromes. The oxidation of the two *c*-type cytochromes indicates that both are situated on the oxidizing side of a coupling site. Generally, it appears that this method could be a useful tool for locating coupling sites in cyclic photophosphorylation in well-coupled chromatophores.

7.2. Energy-Linked Transhydrogenase

The presence of an energy-linked transhydrogenase was simultaneously demonstrated by Keister and Yike in *Rs. rubrum* chromatophores (Keister and Yike, 1966, 1967b) and by Orlando in chromatophores from *Rp. sphaeroides* (Orlando *et al.*, 1966; Orlando, 1968). The reaction, which is most active when driven by light, can also be driven in the dark by added ATP or PP$_i$. When driven by ATP, the reaction is sensitive to oligomycin, and with either ATP or PP_i as energy donor, uncouplers abolish the transhydrogenase activity. Electron-transport inhibitors do not inhibit the reaction when driven by high-energy phosphates. The stoichiometry of NADPH formed to ATP utilized in *Rs. rubrum* approaches 1:1. The mechanism of the energy-linked transhydrogenase has been extensively investigated in submitochondrial particles. The participation of –SH groups in the bacterial system was suggested by Orlando (1970). The *Rs. rubrum* transhydrogenase, when driven by light, shows a light saturation lower than that of photophosphorylation (Keister and Yike, 1967b).

An energy-linked transhydrogenase driven by light or ATP has also been reported to occur in chromatophores from *Rp. viridis* (Jones, O. T. G., and Saunders, 1972). Here, too, the reaction was, as could be expected, inhibited by uncouplers, and when driven by ATP, was also inhibited by oligomycin. No mention was made as to whether PP_i could also serve in these chromatophores as an energy source for the transhydrogenase, but since NAD⁺ reduction in the dark with succinate as electron donor could utilize PP_i (cf. below) as energy donor, one may assume that this is the case.

7.3. Energy-Linked NAD⁺ Reduction

In their early abstract, Löw and Alm (1964) reported on the occurrence of an ATP-driven, oligomycin-sensitive reduction of NAD⁺ coupled to the oxidation of succinate in *Rs. rubrum* chromatophores. No further experiments were reported from these authors, but their work was confirmed and extended by Keister and Yike (1967a) and Keister and Minton (1969) in a thorough investigation of the reaction. They investigated the effect of a wide variety of uncouplers and other inhibitors on both the light-driven and the ATP- and PP_i-driven reaction. Frenkel (1958) had demonstrated the inhibition of light-supported NAD⁺ reduction by phosphorylating conditions. This inhibition was shown to be completely reversed by oligomycin (Keister and Yike, 1967a). Keister and Minton (1969) reached the conclusion that all of the reaction, whatever the energy source, is driven by reversed electron flow in *Rs. rubrum* chromatophores. That this could be the case was first suggested by Bose and Gest (1962), and it was further substantiated by C. W. Jones and Vernon (1969). They showed that not only with succinate but also with ascorbate-TMPD as electron donors, the light-driven reaction is inhibited by active photophos-

phorylation. The ATPase activity of the chromatophores was significantly stimulated by the NAD-reducing system. This provided a possibility of measuring the stoichiometry for the ATP-driven reaction. ATP consumed per NAD reduced was determined to be 1.8 with succinate as electron donor and 5.2 with ascorbate-TMPD as a donor system.

O. T. G. Jones and Saunders (1972) determined the energy requirement for the light-induced NAD⁺ reduction in *Rp. viridis* chromatophores to form NADH from the succinate/fumarate couple to be 9.4 kcal. If one extrapolates between the *Rs. rubrum* and the *Rp. viridis* systems, one may calculate that the ATP-driven reaction proceeds with about 65% efficiency (assuming the $\Delta G°$ of ATP hydrolysis to be around 8 kcal).

The finding by Keister and Minton (1969) that rotenone inhibits the light-induced reduction also strongly supports the concept that reversed electron flow is the mechanism for photosynthetic NAD⁺ reduction, since rotenone only marginally inhibits photophosphorylation, but is a potent inhibitor of the NADH oxidase in *Rs. rubrum* chromatophores. One possible exception to this is *Chlorobium limicola f. thiosulfatophilum*, in which Prince and Olson (1976) recently demonstrated that the midpoint potential of the primary acceptor is at least as low as −450 mV, which would make a direct photosynthetic reduction of NAD feasible.

Comparing the light-saturation curves of several energy-linked reactions in *Rs. rubrum* chromatophores, one may put these reactions into two groups, one of which shows saturation or near-saturation at 1–2 × 10⁴, the other at 5–10 × 10⁴, ergs/cm² pe sec. In the former group we find, in addition to the energy-linked transhydrogenase, photophosphorylation to PP_i (Guillory and Fisher, 1972), the fast (but not the slower) kinetic phase of the light-induced carotenoid absorbance change, and the light-induced bandshift of the potential indicating probe merocyanin V (Baltscheffsky, M., 1976). In the group with the higher requirement are photophosphorylation to ATP, light-driven reduction of NAD⁺ with succinate as electron donor (Keister and Minton, 1969), the rate of light-induced H⁺ uptake, and the total (fast and slow) carotenoid bandshift (Baltscheffsky, M., 1976). All these measurements were made with continuous illumination. These collected data may point to a closer dependence of the "low-light" reactions on a localized or membrane potential, whereas the "high-light" reactions require a higher rate of electron transport in which the rate of H⁺ uptake and formation of pH gradient may be the limiting factor for maximum energization.

7.4. ATP- and PP$_i$-Induced Carotenoid Absorbance Change

The energy-linked nature of the carotenoid absorbance change was first indicated in an early observation that the extent of the light-induced carotenoid change decreased under phosphorylating conditions (Smith, L., et al., 1960). This first indication of a link between the energy-conversion system and the carotenoid absorbance changes in bacterial chromatophores was later extensively elaborated by many workers (Baltscheffsky, M., 1967a, 1969a; Fleischman and Clayton, 1968; Jackson and Crofts, 1969, 1971; Nishimura, 1970; Saphon et al., 1975a,b; Jackson et al., 1975). Here the absorbance changes in the dark, induced by energy-rich phosphates, will be emphasized.

The final products of photophosphorylation, ATP and PP$_i$, will, when added to Rs. rubrum chromatophores in the dark, induce changes in carotenoid absorbance that are qualitatively identical to the light-induced absorbance changes (Baltscheffsky, M., 1967a, 1969a). PP$_i$ has also been shown to induce a similar carotenoid change in chromatophores from Rp. sphaeroides Ga (Sherman and Clayton, 1972).

When induced by PP$_i$ or ATP, the carotenoid change is abolished by uncouplers, but not inhibited by electron-transport inhibitors such as antimycin or HOQNO (Baltscheffsky, M., 1969a). Oligomycin prevents the ATP-induced change, but has no, or a slightly stimulatory, effect on the PP$_i$-induced change, in accordance with the differential effect of this antibiotic on the ATP-synthetase and PP$_i$-synthetase reactions. Dio-9, on the other hand, inhibits both the ATP- and the PP$_i$-induced carotenoid change. Kinetically, there is about a tenfold difference in the rate of rise of the change; when induced by PP$_i$, the half-rise time is 300 msec; when induced by ATP, the half-time is about 3 sec (Baltscheffsky, M., 1969a). Also, the extent of the change is about twice as large when PP$_i$ is energizing. Both the faster kinetics and the larger extent are probably related to the fact that the rate of the PP$_i$-ase reaction usually is twice that of the ATPase in Rs. rubrum chromatophores. There is a slight additive effect in the extent when both PP$_i$ and ATP are added (Baltscheffsky, M., 1967a). Jackson and Crofts (1969, 1971) demonstrated that a diffusion potential caused by K$^+$ pulses in the presence of valinomycin induced a carotenoid absorbance change in the dark in chromatophores from Rp. sphaeroides. The same result was obtained with H$^+$ pulses in the presence of uncouplers. This indicates that carotenoid absorbance changes in the dark may occur in response to a membrane potential. This concept was recently strengthened by the report by Barsky et al. (1975) that the PP$_i$-induced change in Rs. rubrum chromatophores is inhibited by the permeant anion SCN$^-$, which is known to inhibit the light-induced membrane potential in these chromatophores (Schuldiner et al., 1974). The work by Sherman and Clayton (1972) strongly indicates that the carotenoids responding in the dark are those closely associated with the RC complex.

8. Epilogue

The fascinating and challenging field of bacterial photophosphorylation today is understood as well, or as poorly, as its counterparts, oxidative phosphorylation in mitochondria and photophosphorylation in plant material. Together with the latter it constitutes the only known instance in which light energy is directly converted into chemical-energy-rich compounds. The continued study of these reactions and the structural architecture of the membranes that support them may yield a more complete understanding not only of the reactions per se, but also of electron-transport-linked phosphorylation reactions in general. A detailed knowledge about photophosphorylation may also serve as a source of information and inspiration for the practical use of solar energy, a problem that in the last few years has become increasingly urgent.

9. References

Arnon, D. I., Allen, M. B., and Whatley, F. R., 1954, Photosynthesis by isolated chloroplasts, *Nature* (*London*) **174**:394.

Baccarini–Melandri, A., Gest, H., and San Pietro, A., 1970, A coupling factor in bacterial photophosphorylation, *J. Biol. Chem.* **245**:1224.

Bachofen, R., Beyeler, W., Dahl, J. S., Lutz, H. U., and Pflugshaupt, C., 1974, Synthesis of free ATP from firmly bound ATP in photosynthetic membranes, in: *Membrane Proteins in Transport and Phosphorylation* (G. F. Azzone, M. E. Klingenberg, E. Quagliariello, and N. Siliprandi, eds.), pp. 61–72, North–Holland, Amsterdam.

Bachofen, R. Beyeler, W., and Pflugshaupt, C., 1975, Firmly bound nucleotides in photosynthetic energy transduction, in: *Electron Transfer Chains and Oxidative Phosphorylation* (E. Quagliariello, S. Papa, F. Palmieri, E. C. Slater, and N. Siliprandi, eds), pp. 167–172, North-Holland, Amsterdam.

Baltscheffsky, H., 1961, Electron transport and phosphorylation in light-induced phosphorylation, *Biol. Struct. Function* **11**:431.

Baltscheffsky, H., and Arwidsson, B., 1962, Evidence for two phosphorylation sites in bacterial cyclic photophosphorylation, *Biochim. Biophys. Acta* **65**:425.

Baltscheffsky, H., and Baltscheffsky, M., 1958, On light-induced phosphorylation in *Rhodospirillum rubrum*, *Acta Chem. Scand.* **12**:1333.

Baltscheffsky, H., and Baltscheffsky, M., 1960, Inhibitor studies on light-induced phosphorylation in extracts of *Rhodospirillum rubrum*, *Acta Chem. Scand.* **14**:257.

Baltscheffsky, H., and von Stedingk, L.-V.,1966, Bacterial photophosphorylation in the absence of added nucleotide: A second intermediate stage of energy transfer in light-induced formation of ATP, *Biochem. Biophys. Res. Commun.* **22**:722.

Baltscheffsky, H., Baltscheffsky, M., and Arwidsson, B., 1960, On electron transport and phosphorylation in plant and bacterial light-induced phosphorylation, *Acta Chem. Scand.* **14**:1844.

Baltscheffsky, H., von Stedingk, L.-V., Heldt, M.-W., and Klingenberg, M., 1966, Inorganic pyrophosphate: Formation in bacterial photophosphorylation, *Science* **153**:1120.

Baltscheffsky, M., 1967*a*, Inorganic pyrophosphate and ATP as energy donors in chromatophores from *Rhodospirillum rubrum*, *Nature* (*London*) **216**:241.

Baltscheffsky, M., 1967*b*, Inorganic pyrophosphate as an energy donor in photosynthetic and respiratory electron transport phosphorylation systems, *Biochem. Biophys. Res. Commun.* **28**:270.

Baltscheffsky, M., 1968, Inorganic pyrophosphate as energy donor in photosynthetic and respiratory structures, in: *Regulatory Functions of Biological Membranes* (J. Järnefelt, ed.), *B.B.A. Libr.* **11**:277.

Baltscheffsky, M., 1969*a*, Energy conversion-linked changes of carotenoid absorbance in *Rhodospirillum rubrum* chromatophores, *Arch. Biochem. Biophys.* **130**:646.

Baltscheffsky, M., 1969*b*, Reversed energy conversion reaction of bacterial photophosphorylation, *Arch. Biochem. Biophys.* **133**:46.

Baltscheffsky, M., 1971, Carotenoids as endogenous indicators of the energized state in chromatophores, in: *Energy Transduction in Respiration and Photosynthesis* (E. Quagliariello, S. Papa, and C. S. Rossi, eds.), pp. 639–648, Adriatica Editrice, Bari, Italy.

Baltscheffsky, M., 1974, The effect of dibromothymoquinone on light induced reactions in chromatophores from *Rhodospirillum rubrum*, in: *Proceedings of the Third International Congress on Photosynthesis* (M. Avron, ed.), pp. 799–806, Elsevier, Amsterdam.

Baltscheffsky, M., 1976, Energy transduction in the chromatophore membrane, in: *Structure of Biological Membranes* (S. Abrahamson and I. Pascher, eds.), pp. 41–62, Plenum, London.

Baltscheffsky, M., and Baltscheffsky, H., 1972, Coupling and control at the cytochrome level of bacterial photosynthetic electron transport, in: *Oxidation Reduction Enzymes* (Å. Åkeson and A. Ehrenberg, eds.), pp. 257–262, Pergamon Press, Oxford and New York.

Baltscheffsky, M., Baltscheffsky, H., and von Stedingk, L.-V., 1966, Light-induced energy conversion and the

inorganic pyrophosphatase reaction in chromatophores from *Rhodospirillum rubrum*, *Brookhaven Symp. Biol.* **19**:246.

Bamberger, E. S., Rottenberg, H., and Avron, M., 1973, Internal pH, ΔpH, and the kinetics of electron transport in chloroplasts, *Eur. J. Biochem.* **34**:557.

Barsky, E. L., Bonch-Osmolovskaya, E. A., Ostroumova, S. A., Samuilov, V. D., and Skulachev, V. P., 1975, A study on the membrane potential and pH gradient in chromatophores and intact cells of photosynthetic bacteria, *Biochim. Biophys. Acta* **387**:388.

Beechey, R. B., Roberton, A. M., Holloway, C. T., and Knight, I. G., 1967, The properties of dicyclohexylcarbodiimide as an inhibitor of oxidative phosphorylation, *Biochemistry* **6**:3867.

Belitzer, V. A., and Tsibakova, E. T., 1939, *Biokhimiya* **4**:516.

Beyeler, W., Lutz, H. O., and Bachofen, R., 1975, Membrane-bound phosphate as driving force for ATP synthesis in chromatophores of *Rhodospirillum rubrum*, *Bioenergetics* **6**:233.

Böhme, H., Reimer, S., and Trebst, A., 1971, The effect of dibromothymoquinone, an antagonist of plastoquinone, on non cyclic and cyclic electron flow systems in isolated chloroplasts, *Z. Naturforsch.* **26b**:341.

Bose, S. K., and Gest, H., 1962, Hydrogenase and light-stimulated electron transfer reactions in photosynthetic bacteria, *Nature* (*London*) **195**:1168.

Bose, S. K., and Gest, H., 1965, Properties of adenosine triphosphatase in a photosynthetic bacterium. *Biochim. Biophys. Acta* **96**:159.

Boyer, P. D., 1965, Carboxyl activation as a possible common reaction in substrate-level and oxidative phosphorylation and in muscle contraction, in: *Oxidases and Related Redox Systems*, Vol. 2 (T. E. King, H. S. Mason, and M. Morrison, eds.), pp. 994–1008, John Wiley and Sons, New York.

Boyer, P. D., 1974, Conformational coupling in biological energy transductions, in: *Dynamics of Energy-Transducing Membranes* (L. Ernster, R. W. Estabrook, and E. C. Slater, eds.), pp. 289–301, Elsevier, Amsterdam.

Boyer, P. D., 1977, Conformational coupling in oxidative phosphorylation and photophosphorylation, *Trends in Biochem. Sci.* **2**:38.

Boyer, P. D., Cross, R. L., and Momsen, W., 1973, A new concept for energy coupling in oxidative phosphorylation based on a molecular explanation of the oxygen exchange reactions, *Proc. Natl. Acad. Sci. U.S.A.* **70**:2837.

Brandon, Y. R., Brocklehurst, J. R., and Lee, C. P., 1972, Effect of antimycin A and 2-heptyl-4-hydroxyquinoline-N-oxide on the respiratory chain of submitochondrial particles of beef heart, *Biochemistry* **11**:1150.

Casadio, R., Baccarini-Melandri, A., Zannoni, D., and Melandri, B. A., 1974, Electrochemical proton gradient and phosphate potential in bacterial chromatophores, *FEBS Lett.* **49**:203.

Cattell, K. J., Lindop, C. R., Knight, I. G., and Beechey, R. B., 1971, The identification of the site of action of

N,N'-dicyclohexylcarbodiimide as a proteolipid in mitochondrial membranes, *Biochem. J.* **125**:169.

Chance, B., and Williams, G. R., 1955, Respiratory enzymes in oxidative phosphorylation. III. The steady state, *J. Biol. Chem.* **217**:409.

Chance, B., Williams, G. R., Holmes, W. F., and Higgins, J., 1955, Respiratory enzymes in oxidative phosphorylation. V. A mechanism for oxidative phosphorylation, *J. Biol. Chem.* **217**:439.

Cross, R. L., and Boyer, P. D., 1975, The rapid labeling of adenosine triphosphate by ^{32}P-labeled inorganic phosphate and the exchange of phosphate oxygens as related to conformational coupling in oxidative phosphorylation, *Biochemistry* **14**:392.

Danon, A., and Stoeckenius, W., 1974, Photophosphorylation in *Halobacterium halobium*, *Proc. Natl. Acad. Sci. U.S.A.* **71**:1234.

Das Gupta, V., and Rieske, J. S., 1973, Identification of a protein component of the antimycin-binding site of the respiratory chain by photoaffinity labeling, *Biochem. Biophys. Res. Commun.* **54**:1247.

Dutton, P. L., and Baltscheffsky, M., 1972, Oxidation–reduction potential dependence of pyrophosphate-induced cytochrome and bacteriochlorophyll reactions in *Rhodospirillum rubrum*, *Biochim. Biophys. Acta* **267**:172.

Eisenhardt, R. H., and Rosenthal, O., 1968, Studies on energy transfer in mitochondrial oxidative phosphorylation. III. On the interaction of adenosine diphosphate with high-energy intermediates, *Biochemistry* **7**:1327.

Ernster, L., Nordenbrand, K., Chude, O., and Juntti, K., 1974, Relationship of components of the ATPase system to the oligomycin-induced respiratory control of submitochondrial particles, in: *Membrane Proteins in Transport and Phosphorylation* (G. F. Azzone, M. E. Klingenberg, E. Quagliariello, and N. Siliprandi, eds.), pp. 29–41, North-Holland, Amsterdam.

Fisher, R. R., and Guillory, R. J., 1967, Inhibition of the energy conservation reactions of *Rhodospirillum rubrum* by Dio-9, *Biochim. Biophys. Acta* **143**:654.

Fisher, R. R., and Guillory, R. J., 1969, Partial resolution of energy-linked reactions in *Rhodospirillum rubrum* chromatophores, *FEBS Lett.* **3**:27.

Fleischman, D. E., and Clayton, R. K., 1968, The effect of phosphorylation uncouplers and electron transport inhibitors upon spectral shifts and delayed light emission of photosynthetic bacteria, *Photochem. Photobiol.* **8**:287.

Frenkel, A. W., 1954, Light induced phosphorylation by cell-free preparations of photosynthetic bacteria, *J. Am. Chem. Soc.* **76**:55.

Frenkel, A. W., 1956, Photophosphorylation of adenine nucleotides by cell-free preparations of purple bacteria, *J. Biol. Chem.* **222**:823.

Frenkel, A. W., 1958, Light-induced reactions of chromatophores of *Rhodospirillum rubrum*, *Brookhaven Symp. Biol.* **11**:276.

Geller, D. M., and Lipmann, F., 1960, Photophosphorylation in extract of *Rhodospirillum rubrum*, *J. Biol. Chem.* **235**:2478.

Gepstein, A., and Carmeli, C., 1974, Properties of adenosine triphosphatase in chromatophores and in coupling factor

from the photosynthetic bacteria *Chromatium* strain D, *Eur. J. Biochem.* **44**:593.

Gromet-Elhanan, Z., 1969, Inhibitors of photophosphorylation and photoreduction by chromatophores from *Rhodospirillum rubrum*, *Arch, Biochem. Biophys.* **131**:299.

Gromet-Elhanan, Z., 1971, Relationship between light-induced quenching of atebrin fluorescence and ATP formation in *Rhodospirillum rubrum* chromatophores, *FEBS Lett.* **13**:124.

Guillory, R. J., 1964, The action of Dio-9: An inhibitor and an uncoupler of oxidative phosphorylation, *Biochim. Biophys. Acta* **89**:197.

Guillory, R. J., and Fisher, R. R., 1972, Studies on the light-dependent synthesis of inorganic pyrophosphate by *Rhodospirillum rubrum* chromatophores, *Biochem. J.* **129**:471.

Hanstein, W. G., and Hatefi, Y., 1974, Characterization and localization of mitochondrial uncoupler binding sites with an uncoupler capable of photoaffinity labeling, *J. Biol. Chem.* **249**:1356.

Hauska, G., 1972, Lipophilicity and catalysis of photophosphorylation. I. Sulfonated phenazonium compounds are ineffective in mediating cyclic photophosphorylation in photosystem-I-subchloroplast vesicles, *FEBS Lett.* **28**:217–220.

Hochman, A., and Carmeli, C., 1973, ATPase and ATP–Pi exchange activities in *Chromatium* strain D chromatophores, *Photosynthetica* **7**:238.

Horio, T., Nishikawa, K., and Yamashita, J., 1964, Adenosine diphosphate–adenosine triphosphate exchange reaction with chromatophores from *Rhodospirillum rubrum*, *J. Biochem.* **55**:327.

Horio, R., Nishikawa, K., Katsumata, M., and Yamashita, J., 1965, Possible partial reactions of the photophosphorylation process in chromatophores from *Rhodospirillum rubrum*, *Biochim. Biophys. Acta* **94**:371.

Horio, T., von Stedingk, L.-V., and Baltscheffsky, H., 1966, Photophosphorylation in presence and absence of added adenosine diphosphate in chromatophores from *Rhodospirillum rubrum*, *Acta Chem. Scand.* **20**:1.

Horio, T., Nishikawa, K., Okayama, S., Horiuti, Y., Yamamoto, N., and Kakutani, Y., 1968, The requirement of ubiquinone-10 for an ATP-forming system and an ATPase system of chromatophores from *Rhodospirillum rubrum*, *Biochim. Biophys. Acta* **153**:913.

Horiuti, Y., Nishikawa, K., and Horio, T., 1968, Oxidation–reduction potential-dependent adenosine triphosphatase activity of chromatophores from *Rhodospirillum rubrum*, *J. Biochem.* **64**:577.

Jackson, J. B., and Crofts, A. R., 1969, The high energy state in chromatophores from *Rhodopseudomonas spheroides*, *FEBS Lett.* **4**:185.

Jackson, J. B., and Crofts, A. R., 1971, The kinetics of light induced carotenoid changes in *Rhodopseudomonas spheroides* and their relation to electriocal field generation across the chromatophore membrane, *Eur. J. Biochem.* **18**:120.

Jackson, J. B., Crofts, A. R., and von Stedingk, L.-V., 1968, Ion transport induced by light and antibiotics in chromatophores from *Rhodospirillum rubrum, Eur. J. Biochem.* **6**:41.

Jackson, J. B., Saphon, S., and Witt, H. T., 1975, The extent of the stimulated electric potential decay under phosphorylating conditions and the H⁺/ATP ratio in *Rhodopseudomonas spheroides* chromatophores following short flash excitation, *Biochim. Biophys. Acta* **408**:83.

Johansson, B. C., 1972, A coupling factor from *Rhodospirillum rubrum* chromatophores, *FEBS Lett.* **20**:339.

Johansson, B. C., 1975, Partial resolution of the energy transfer system in chromatophores from *Rhodospirillum rubrum*. Purification and characterization of the "coupling factor" ATPase, Doctoral thesis, University of Stockholm, Sweden.

Johansson, B. C., Baltscheffsky, M., and Baltscheffsky, H., 1972, Coupling factor capabilities with chromatophore fragments from *Rhodospirillum rubrum*, in: *Proceedings of the Second International Congress on Photosynthesis Research* (G. Forti, M. Avron, and A. Melandri, eds.), pp. 1203–1209, Junk, The Hague.

Jones, C. W., and Vernon, L. P., 1969, Nicotine–adenine dinucleotide photoreduction in *Rhodospirillum rubrum* chromatophores, *Biochim. Biophys. Acta* **180**:149.

Jones, O. T. G., and Saunders, V. A., 1972, Energy-linked electron-transfer reactions in *Rhodopseudomonas viridis*, *Biochim. Biophys. Acta* **275**:427.

Kakuno, T., Bartsch, R. G., Nishikawa, K., and Horio, T., 1971, Redox components associated withy chromatophores from *Rhodospirillum rubrum*, *J. Bioehem.* **70**:79.

Keister, D. L., and Minton, N. J., 1969, Energy-linked reactions in photosynthetic bacteria. III. Further studies on energy-linked nicotinamide–adenine dinucleotide reduction by *Rhodospirillum rubrum* chromatophores, *Biochemistry* **8**:167.

Keister, D. L., and Minton, N. J., 1971a, ATP synthesis driven by inorganic pyrophosphate in *Rhodospirillum rubrum* chromatophores, *Biochem. Biophys. Res. Commun.* **42**:932.

Keister, D. L., and Minton, N. J., 1971b, Energy-linked reactions in photosynthetic bacteria. VI. Inorganic pyrophosphate-driven ATP synthesis in *Rhodospirillum rubrum*, *Arch. Biochem. Biophys.* **147**:330.

Keister, D. L., and Yike, N. J., 1966, Studies on an energy-linked pyridine nucleotide transhydrogenase in photosynthetic bacteria. I. Demonstration of the reaction in *Rhodospirillum rubrum*, *Biochem. Biophys. Res. Commun.* **24**:519.

Keister, D. L., and Yike, N. J., 1967a, Energy-linked reactions in photosynthetic bacteria. I. Succinate-linked ATP-driven NAD⁺ reduction by *Rhodospirillum rubrum* chromatophores, *Arch. Biochem. Biophys.* **121**:415.

Keister, D. L., and Yike, N. J., 1967b, Energy-linked reactions in photosynthetic bacteria. II. The energy-dependent reduction of oxidized nicotinamide–adenine dinucleotide phosphate by chromatophores of *Rhodospirillum rubrum*, *Biochemistry* **6**:3847.

Klingenberg, M., and Schollmayer, D., 1961, Zur Reversibilität der oxydativen Phosphorylierung: Der Einfluss von Adenosintriphosphat auf die Atmungskette in atmenden Mitochondrien, *Biochem. Z.* **335**:231.

Knaff, D. B., and Buchanan, B. B., 1975, Cytochromes *b* and photosynthetic sulfur bacteria, *Biochim. Biophys. Acta* **376**:549.

Knobloch, K., 1975, Energy-linked pyridine nucleotide transhydrogenase activity in photosynthetically grown *Rhodopseudomonas palustris, Z. Naturforsch.* **30c**:771.

Lardy, H., 1961, Reactions involved in oxidative phosphorylation as disclosed by studies with antibiotics, in: *Biological Structure and Function* (T. W. Goodwin and O. Lindberg, eds.), pp. 265–267, Academic Press, London.

Lardy, H. A., and Lin, C. H. C., 1969, in: Inhibition of mitochondrial oxidative phosphorylation by aurovertin, *Inhibitors: Tools for Cell Research* (T. Bücher and H. Sies, eds.), pp. 279–281, Springer-Verlag, New York.

Lardy, H. A., Johnson, D., and McMurray, W. C., 1958, Antibiotics as tools for metabolic studies. I. A survey of toxic antibiotics in respiratory, phosphorylative and glycolytic systems, *Arch. Biochem. Biophys.* **78**:587.

Lardy, H. A., Connelly, J. L., and Johnson, D., 1964, Antibiotics as tools for metabolic studies. II. Inhibition of phosphoryl transfer in mitochondria by oligomycin and aurovertin, *Biochemistry* **3**:1961.

Lee, C. P., and Ernster, L., 1968, Studies of the energy-transfer system of submitochondrial particles. 2. Effects of oligomycin and aurovertin, *Eur. J. Biochem.* **3**:391.

Leiser, M., and Gromet-Elhanan, Z., 1974, Demonstration of acid–base phosphorylation in chromatophores in the presence of a K⁺ diffusion potential, *FEBS Lett.* **43**:267.

Löw, H., and Alm, B., 1964, Reversed electron transport in photophosphorylative particles from *Rhodospirillum rubrum* in the dark, *Abstracts of the First FEBS Meeting*, p. 68.

Lozier, R. H., Bogomolni, R. A., and Stoeckenius, W., 1975, Bacteriorhodopsin: A light-driven proton pump in *Halobacterium halobium, Biophys. J.* **15**:955.

Lundin, A., Thore, A., and Baltscheffsky, M., 1977, Sensitive measurement of flash induced photophosphorylation in bacterial chromatophores by firefly luciferase, *FEBS Lett.* **79**:73.

Lutz, H. U., Dahl, J. S., and Bachofen, R., 1974, Synthesis of free ATP from membrane-bound ATP in chromatophores of *Rhodospirillum rubrum, Biochim. Biophys. Acta* **347**:359.

MacLennan, D. H., and Tzagoloff, A., 1968, Studies on the mitochondrial adenosine triphosphatase system. IV. Purification and characterization of the oligomycin sensitivity conferring protein, *Biochemistry* **7**:1603.

Melandri, B. A., Zannoni, D., Casadio, R., and Baccarini-Melandri, A., 1974, Energy conservation and transduction in photosynthesis and respiration of facultative photosynthetic bacteria, in: *Proceedings of the Third International Congress on Photosynthesis* (M. Avron, ed.), pp. 1147–1162, Elsevier, Amsterdam.

Moyle, J., Mitchell, R., and Mitchell, P., 1972, Proton translocating pyrophosphatase of *Rhodospirillum rubrum*, *FEBS Lett.* **23**:233.

Nishimura, M., 1962a, Studies on bacterial photophosphorylation. I. Kinetics of photophosphorylation in

Rhodospirillum rubrum chromatophores by flashing light, *Biochim. Biophys. Acta* **57**:88.

Nishimura, M., 1962*b*, Studies on bacterial photophosphorylation. II. Effects of reagents and temperature on light-induced and dark-phases of photophosphorylation in *Rhodospirillum rubrum* chromatophores, *Biochim. Biophys. Acta* **57**:96.

Nishimura, M., 1962*c*, Studies on bacterial photophosphorylation. IV. On the maximum amount of delayed photophosphorylation induced by a single flash, *Biochim. Biophys. Acta* **59**:183.

Nishimura, M., 1970, The sizes of the photosynthetic energy-transducing units in purple bacteria determined by single flash yield, titration by antiobiotics and carotenoid absorption and band shift, *Biochim. Biophys. Acta* **197**:69.

Oesterhelt, D., and Stoeckenius, W., 1973, Functions of a new photoreceptor membrane, *Proc. Natl. Acad. Sci. U.S.A.* **70**:2853.

Orlando, J. A., 1968, On the light-dependent reduction of nicotinamide adenine nucleotide phosphate by chromatophores of *Rhodopseudomonas spheroides*, *Arch. Biochem. Biophys.* **124**:413.

Orlando, J. A., 1970, Involvement of sulfhydryl groups in light-dependent transhydrogenase of *Rhodopseudomonas spheroides*, *Arch. Biochem. Biophys.* **141**:111.

Orlando, J. A., Sabo, D., and Curnyn, C., 1966, Photoreduction of pyridine nucleotide by subcellular preparations from *Rhodopseudomonas sphaeroides*, *Plant Physiol.* **41**:937.

Padan, E., and Rottenberg, H., 1973, Respiratory control and the proton electrochemical gradient, *Eur. J. Biochem.* **40**:431.

Petty, K. M., and Dutton, P. L., 1976, Ubiquinone–cytochrome *b* electron and proton transfer: A functional pK on cytochrome b_{50} in *Rhodopseudomonas sphaeroides* membranes, *Arch. Biochem. Biophys.* **172**:346.

Pressman, B., Harris, E. J., Jagger, W. S., and Johnson, J. H., 1967, Antibiotic-mediated transport of alkali ions across lipid barriers, *Proc. Natl. Acad. Sci. U.S.A.* **58**:1949.

Prince, R. C., and Dutton, P. L., 1975, A kinetic completion of the cyclic photosynthetic electron pathway of *Rhodopseudomonas sphaeroides*: Cytochrome *b*–cytochrome c_2 oxidation–reduction, *Biochim. Biophys. Acta* **387**:609.

Prince, R. C., and Olson, J. M., 1976, Some thermodynamic and kinetic properties of the primary photochemical reactants in a complex from a green photosynthetic bacterium, *Biochim. Biophys. Acta* **423**:357.

Racker, E., and Stoeckenius, W., 1974, Reconstitution of purple membrane vesicles catalyzing light-driven proton uptake and adenosine triphosphate formation, *J. Biol. Chem.* **249**:662.

Reed, D. W., and Raveed, D., 1972, Some properties of the ATPase from chromatophores of *Rhodopseudomonas sphaeroides* and its structural relationship to the bacteriochlorophyll proteins, *Biochim. Biophys. Acta* **283**:79.

Saphon, S., Jackson, J. B., Lerbs, V., and Witt, H. T., 1975*a*, The functional unit of electrical events and phosphoryla-

tion in chromatophores from *Rhodopseudomonas sphaeroides*, *Biochim. Biophys. Acta* **408**:58.

Saphon, S., Jackson, J. B., and Witt, H. T., 1975*b*, Electrical potential changes, H^+ translocation and phosphorylation induced by short flash excitation in *Rhodopseudomonas sphaeroides* chromatophores, *Biochim. Biophys. Acta* **408**:67.

Scholes, P., Mitchell, P., and Moyle, J., 1969, The polarity of proton translocation in some photosynthetic microorganisms, *Eur. J. Biochem.* **8**:450.

Schuldiner, S., Padan, E., Rottenberg, H., Gromet-Elhanan, Z., and Avron, M., 1974, ΔpH and membrane potential in bacterial chromatophores, *FEBS Lett.* **49**:174.

Sherman, L. A., and Clayton, R. K., 1972, The lack of carotenoid band shifts in a non-photosynthetic, reaction centerless mutant of *Rhodopseudomonas sphaeroides*, *FEBS Lett.* **22**:127.

Smith, D. J., Stokes, B. O., and Boyer, P. D., 1976, Probes of initial phosphorylation events in ATP synthesis by chloroplasts, *J. Biol. Chem.* **251**:4165.

Smith, L., and Baltscheffsky, M., 1956, Respiration and phosphorylation in extracts of *Rhodospirillum rubrum*, *Fed. Proc. Fed. Am. Soc. Exp. Biol.* **15**:357.

Smith, L., and Baltscheffsky, M., 1959, Respiration and light-induced phosphorylation in extracts of *Rhodospirillum rubrum*, *Biochim. J.* **234**:1575.

Smith, L., Baltscheffsky, M., and Olson, J. M., 1960, Absorption spectrum changes observed on illumination of aerobic suspensions of photosynthetic bacteria, *J. Biol. Chem.* **235**:213.

Thore, A., Keister, D. L. Shavit, N., and San Pietro, A., 1968, Effects of antibiotics on ion transport and photophosphorylation in *Rhodospirillum rubrum* chromatophores, *Biochemistry* **7**:3499.

Thore, A., Keister, D. L., and San Pietro, A., 1969, Studies on the respiratory system of aerobically (dark) and anaerobically (light) grown *Rhodospirillum rubrum*, *Arch. Mikrobiol.* **67**:378.

Trebst, A., 1974, Energy conservation in photosynthetic electron transport of chloroplasts, *Annu. Rev. Plant Physiol.* **25**:423.

Trebst, A., 1976, Artificial energy conservation in bacterial photosynthetic electron transport, *Z. Naturforsch.* **31c**:152.

West, K. R., and Wiskich, J. T., 1968, Photosynthetic control by isolated pea chloroplasts, *Biochem. J.* **109**:527.

Wikström, M., and Saari, T., 1976, Conformational changes in cytochrome aa_3 and ATP synthetase of the mitochondrial membrane and their role in mitochondrial energy transduction, *Mol. Cell. Biochem.* **11**:17.

Yamamoto, N., Yoshimura, S., Higuti, T., Nishikawa, K., and Horio, T., 1972, Role of bound ADP in photosynthetic ATP formation by chromatophores from *Rhodospirillum rubrum*, *J. Biochem.* **72**:1397.

Yoshida, M., Sone, N. Hirata, H., Kagawa, Y., Takeuchi, Y., and Ohno, K., 1975, ATP synthesis catalyzed by purified DCCD-sensitive ATPase incorporated into reconstituted purple membrane vesicles, *Biochem. Biophys. Res. Commun.* **67**:1295.

CHAPTER 31

Coupling Factors

Assunta Baccarini-Melandri and Bruno Andrea Melandri

1. Introduction

Coupling factors were originally defined as factors, generally proteins, that were involved in the synthesis of ATP coupled to electron transfer but did not catalyze oxidation–reduction reactions (cf. Racker, 1970). Until recently, the attention of investigators has been focused mainly on coupling factors from mitochondria or from chloroplasts of higher plants; a large body of research in this field has demonstrated that coupling factors are probably parts or subunits of a transmembrane coupling enzyme, endowed with ATPase activity, and possibly capable of coupling ATP hydrolysis to translocation of protons across the membranes (for reviews, see Racker, 1970; Senior, 1973; Penefsky, 1974; Pedersen, 1975; Jagendorf, 1975). Growing evidence in this direction is now accumulating also for prokaryotic organisms (for reviews, see Abrams and Smith, 1974; Baltscheffsky, H., and Baltscheffsky, 1974).

Although the analogies in the mechanism of ATP synthesis in photophosphorylation by photosynthetic organelles of higher plants or by chromatophores from photosynthetic bacteria are quite evident, studies aimed at resolution and reconstitution of bacterial photophosphorylation were not undertaken until 1970, when the first successful studies in this area were reported for *Rhodopseudomonas capsulata* (Baccarini-Melandri *et al.*, 1970). In this study and

in subsequent studies, the identity of the coupling factor with an enzyme endowed with ATPase activity and possessing many characteristics previously known in coupling factors from eukaryotic cells was firmly established. Since then, several other kinds of photosynthetic bacterial systems have been investigated.

Facultative photosynthetic bacteria (members of the Rhodospirillaceae family) are unique organisms, since, in addition to being able to synthesize ATP by light energy, they have also developed the ability to produce ATP by substrate oxidation (Geller, 1962) and in certain cases by substrate fermentation (Uffen and Wolfe, 1970; Schön and Biedermann, 1972).

Although extensive morphological modifications of the unique membrane system of Rhodospirillaceae accompany the shift from aerobic dark conditions to anaerobic light (for a review, see Oelze and Drews, 1972), the process of photosynthetic differentiation can be envisaged as consisting essentially in the insertion of an active photochemical system in a pre-existing oxidoreduction chain (Jones and Plewis, 1974; Garcia *et al.*, 1974, 1975). On this basis, the coupling enzyme in oxidative and photosynthetic phosphorylation would be expected to be identical in these organisms: the successful resolution of the phosphorylating membrane of some Rhodospirillaceae has indeed confirmed this to be the case (Melandri *et al.*, 1971a; Baccarini-Melandri and Melandri, 1972a; Lien and Gest, 1973; Johansson, 1975), and has provided further support for the view of the existence of a substantial unity between photosynthetic and oxidative ATP production.

Assunta Baccarini-Melandri and Bruno Andrea Melandri
Institute of Botany, University of Bologna, Bologna, Italy

2. Structure and Localization of Coupling Factors

As early as 1964, Löw and Afzelius (1964) reported that the chromatophores from *Rhodospirillum rubrum* were lined with knoblike particles of 12-nm size, very similar to those shown previously by Fernandez Moran (1962) in submitochondrial particles.

At that time, these structures were interpreted as being part of the respiratory electron-transport chain, but Löw and Afzelius suggested that "these stalked knobs represent an important biological unit common to many membranous systems . . . ." Subsequently, it was clearly demonstrated by Kagawa and Racker (1966) that these knobs, observed in submitochondrial particles, were the morphological representation of the coupling ATPase.

Electron micrographs of negatively stained chromatophores from *Rp. sphaeroides* show the presence of these structures on the outer surface of the membrane (Reed and Raveed, 1972). Washings with EDTA or Triton X-100 solutions remove the ATPase activity from the membranes with concomitant release of the knobs, suggesting that the role of these structures in chromatophores is the same as in submitochondrial particles. More recently, a specific antibody prepared against coupling factor ATPase from *Rp. capsulata* was utilized for studies on the localization of the coupling enzyme (Prince *et al.*, 1975). This antibody agglutinates chromatophores but not spheroplasts prepared from EDTA–lysozyme treated cells. In intact cells or spheroplasts, therefore, this enzyme faces the cytoplasm, while in chromatophores, it faces the outside. This reversal is to be expected if isolated chromatophores are formed by a simple pinching-off of invaginations of the cell membrane. Similar results were described by Johansson (1975) for *Rs. rubrum*. The change in orientation of the membrane surfaces is confirmed by the observation that the direction of light-induced proton fluxes in whole cells is opposite to that in chromatophore preparations (von Stedingk and Baltscheffsky, 1966; Scholes *et al.*, 1969).

3. Solubilization, Purification, and Molecular Properties of Coupling Factors

The reversible resolution of the phosphorylating system from photosynthetic bacteria has been achieved so far in three members of the Rhodospirillaceae family: *Rp. capsulata* (Baccarini-Melandri *et al.*, 1970), *Rs. rubrum* (Johansson, 1972), and *Rp. sphaeroides* (Saphon *et al.*, 1975), and one from Chromatiaceae, *Chromatium vinosum* strain D (Hochman and Carmeli, 1971).

The methods employed for the solubilization of coupling factor of photosynthetic bacteria are analogous to those used for mitochondria and chloroplasts; though basically similar for all the genera, they differ in the harshness necessary for obtaining a sufficient resolution of coupling factor from the membrane, a requirement that is related to the stability of the membrane–protein complex. For example, incubation with a low-ionic-strength buffer is sufficient to remove about 90% of the phosphorylation activity from *Chromatium* membranes (Hochman and Carmeli, 1971), but sonication in the presence of 1 mM EDTA is necessary to resolve *Rp. capsulata* or *Rs. rubrum* chromatophores (Baccarini-Melandri *et al.*, 1970; Johansson, 1972); the latter require, however, a longer and more intense sonication. An alternative method that has been successfully used with membranes from Rhodospirillaceae is the removal of coupling factor by washing with high-ionic-strength solutions containing 2 M LiCl or NaBr (Fisher and Guillory, 1969; Gromet-Elhanan, 1974; Melandri *et al.*, 1974); by this technique, a high degree of decoupling (more than 95%) can generally be obtained.

In all these instances, suitable control experiments should be performed to assure that no extensive damage of the electron-transport chain has been caused by the decoupling procedure. These controls are not easily performed in a cyclic electron-transport system, like that operating in bacterial photosynthesis; some indication can be obtained, however, by examining light-induced energy-requiring processes, other than photophosphorylation, which are usually affected only marginally by the detachment of the coupling factor (see below).

Procedures for extensive purification of coupling proteins solubilized from *Rp. capsulata* and *Rs. rubrum* membranes have been described (Baccarini-Melandri and Melandri, 1971; Johansson *et al.*, 1973); purification of coupling factor from *Chromatium* is at present under active investigation (C. Carmeli, personal communication), but no details on this work have yet appeared in the literature. The isolation of coupling factors from Rhodospirillaceae presents experimental difficulties due to the great lability of the enzyme in its soluble form. ATP, at millimolar concentrations, has been shown to be absolutely necessary throughout all stages of purification to preserve the recoupling activity. Despite this precaution, the yields thus far obtained have always been rather low. Aqueous extracts of acetone powders of coupled chromatophores have been used as starting material for purification. In principle, however, a better source for purifying these proteins should be the supernatant obtained after decoupling

Table 1. Fluorescence Ratio and Degree of Purification of Coupling Factor from *Rhodospirillum rubrum*[a]

Step	Total protein (mg)	Specific ATPase activity (μmol hr^{-1} mg^{-1})	Total ATPase activity (μmol hr^{-1})	Fluorescence ratio (nm) 300/350	310/350
1	960	0.5	480	0.2	0.5
2	320	5.2	1.670	0.3	0.6
3	16	300	4.800	0.6	0.9
4	3	640	1.920	2.6	3.5
5	0.9	900	810	5.3	7.8

[a] Slightly modified from Berzborn *et al.* (1975). Purification steps: (1) crude extract; (2) 30–55% ammonium sulfate fraction; (3) eluate from Sepharose 6B; (4) eluate from DEAE Sephadex A50; (5) fraction from sucrose density gradient.

phosphorylating membranes by washing or sonicating in the presence of EDTA; by these methods, in fact, the coupling factor should be solubilized under the mildest possible conditions. Recently, coupling factor from *Rs. rubrum* has also been partially purified from the supernatant obtained after washing chromatophores in the presence of 2 M LiCl and 4 mM ATP (Binder and Gromet Elhanan, 1975), by far the most effective method for decoupling.

The techniques used in the purification include salt fractionation, filtration through agarose gels, membrane ultrafiltration, gradient centrifugation, and ion-exchange chromatography. The preparations obtained are extensively, but not always completely, purified. The most extensive purification (Table 1) has been obtained so far for *Rs. rubrum* coupling factor, which has been prepared as an apparently homogeneous protein (Berzborn *et al.*, 1975); in this instance, the identification of the fractions containing coupling factors has been greatly facilitated by the observation that for chloroplast CF$_1$ (Lien and Racker, 1971), the most active fractions show the highest ratio of fluorescence emission at 300 nm to that at 350 nm due to a lack or a very low content of tryptophan. From this preparation, a monospecific antibody, yielding a single precipitin band arc on immunoelectrophoresis, has been obtained (Berzborn *et al.*, 1975; Johansson, 1975). However, the low yield and the lability of this preparation make its utilization in further studies extremely difficult.

The proteins isolated from Rhodospirillaceae are multicomponent complexes with molecular weights of about 280,000–350,000 daltons (Melandri *et al.*, 1971*b*; Johansson *et al.*, 1973); they appear to be composed of several subunits that can be resolved on

sodium dodecyl sulfate (SDS)–polyacrylamide gel electrophoresis (Johansson and Baltscheffsky, 1975).

The apparent molecular weights of the five subunits, as judged from the mobility in SDS electrophoresis are: 54,000 (α), 50,000 (β), 32,000 (γ), 13,000 (δ), and 7500 (ε). The pattern obtained in the case of *Rs. rubrum* coupling factor is therefore very similar to the patterns present in other eukaryotic or prokaryotic ATPases (see, for example, the review by Pedersen, 1975). This identity in structure is one example of the substantial unity in structure and function of phosphorylation coupling factors existing in all living organisms. Other examples of such a coincidence of properties will be described in the following paragraphs.

It has been suggested that in chloroplast CF$_1$, the lowest-molecular-weight subunit (ε) is identical to an endogenous peptide inhibitor of ATPase activity (Nelson *et al.*, 1972*b*), analogous to that demonstrated in mitochondria (Pullman and Monroy, 1963; see, however, Senior, 1973). Direct evidence of such an inhibitor in *Rs. rubrum* coupling factor is not yet available, but the large increase in total units of ATPase during the purification of this protein (see Table 1) strongly suggests the presence of an inhibitor in the crude acetone powder extract. This should therefore not be coincident with the ε subunit, still present in a homogeneous enzyme preparation endowed with a very high ATPase activity.

4. ATPase Activity of the Membrane-Bound and Soluble Coupling Factors

All coupling factors so far studied in photosynthetic bacteria are endowed with an ATPase activity both when bound to the membrane and as solubilized preparations.

Membrane preparations from photosynthetic bacteria exhibit a Mg^{2+}-dependent ATPase, which is sensitive to energy-transfer inhibitors (Bose and Gest, 1965; Horio et al., 1971; Melandri and Baccarini-Melandri, 1972); oligomycin and phlorizine are ineffective, however, in Chromatium (Gepshtein and Carmeli, 1974). The degree of oligomycin sensitivity may be only partial: in Rp. capsulata, for instance, oligomycin inhibits ATPase activity, when measured in the dark, by only about 50–60% (Melandri and Baccarini-Melandri, 1972). The oligomycin-insensitive activity also appears to be associated with the coupling factor, since studies on the properties of the ATPase, when titrated with a specific antibody or in partially decoupled membranes, demonstrated that the degree of inhibition by oligomycin remains constant, whatever the amount of active enzyme attached on the membrane (Melandri and Baccarini-Melandri, 1972).

Membrane-bound ATPase is less active in the presence of divalent cations other than Mg^{2+}. The cations are effective in the following order: Mn^{2+}, Ca^{2+}, Co^{2+}, Ni^{2+}, Fe^{2+}. It is likely that all these metals act merely as substitutes for Mg^{2+} on the ATPase active site as metal–ATP complexes; the same order of effectiveness is generally observed for other eukaryotic coupling factors. A notable exception should be made, however, for Ca^{2+}, which often appears to be associated with a hydrolytic activity characteristic of modified or partially solubilized enzymes (Johansson et al., 1973).

In several eukaryotic or prokaryotic systems, membrane-bound ATPase activity can be considerably increased by proteolytic digestion of the membrane, usually by trypsin (e.g., Vambutas and Racker, 1965; Eilerman et al., 1971); in chloroplasts, this treatment seems to produce the detachment or the destruction of a peptide ATPase inhibitor (Nelson et al., 1972b). In photosynthetic bacteria, a similar treatment was found to be successful in Chromatium, in which both Ca^{2+} and Mg^{2+}-dependent ATPase could be enhanced five- to eight-fold by tryptic digestion (Gepshtein and Carmeli, 1974); in Rp. capsulata, incubation with trypsin leads only to a slight increase of the activity and subsequently to inactivation of the enzyme (A. Baccarini-Melandri, unpublished); it should be noted, however, that in native membranes from Rp. capsulata, ATPase is already considerably more active than in Chromatium.

If the characteristics of membrane-bound ATPase are rather uniform in different bacterial strains, a much more varied spectrum of properties has been demonstrated as far as the soluble enzymes are concerned. This variability of characteristics is related to the type of decoupling procedure, to the buffers used, to the presence of thiol reagents, and to other factors, and therefore testifies to the rather artifactual nature of the ATPase activity of the soluble enzyme. The most evident differences are observed in the cation requirement for the activity. Preparations of soluble enzyme from Rp. capsulata show a very weak (maximum 10 μmol/h per mg protein) ATPase activity that is dependent on Mg^{2+}, but completely inactive in the presence of Ca^{2+} (Melandri and Baccarini-Melandri, 1972). In contrast, coupling factor from Rs. rubrum prepared from an acetone powder extract exhibits a very active Ca^+-dependent activity (Johansson et al., 1973); Mg^{2+} cannot act as cofactor for this activity, but rather Mg–ATP inhibits the activity as a competitor of Ca–ATP, which is the only metal–ATP complex active as substrate. A Mg^{2+}-dependent soluble ATPase has been reported to be present, however, in the supernatant obtained by sonication of Rs. rubrum chromatophores in the presence of EDTA and dithiothreitol (Konings and Guillory, 1973). In Chromatium, both Ca^{2+}- and Mg^{2+}-dependent activities are present in the crude solubilized enzyme. Both activities are enhanced by tryptic digestion, but the Ca^{2+}-dependent ATPase is increased much more than the Mg^{2+}-dependent one (Gepshtein and Carmeli, 1974).

In Rp. capsulata and Rs. rubrum, the sensitivity of ATPase activity toward oligomycin, which is completely lost on solubilization and purification of the enzyme, is fully restored following reconstitution of the membrane-bound activity (Melandri and Baccarini-Melandri, 1972; Johansson et al., 1973). Sensitivity to oligomycin therefore does not appear to be associated with the EDTA-extractable protein, but requires additional components, probably still present on the decoupled membrane, that must interact with the enzyme in a supermolecular arrangement. Again, this situation reflects the analogy in the organization of bacterial ATP synthetase with that from eukaryotic systems; in mitochondrial or chloroplast enzymes, the inhibition of ATPase by energy-transfer inhibitors (except for aurovertin and azide) requires the interaction of F_1 or CF_1 with specific sites (or components) on the membrane (Pedersen, 1975; Jagendorf, 1975). The presence on decoupled membranes from Rp. capsulata of sites binding oligomycin is suggested by several observations: (1) Oligomycin can stimulate with the same effectiveness light-induced proton uptake in coupled and decoupled chromatophores (Melandri et al., 1970) [unlike chloroplasts (McCarty and Racker, 1967), H^+ uptake is not impaired by solubilization of the coupling factor]. (2) Membranes unable to bind the coupling factor and to confer to it the sensitivity to oligomycin can be prepared by washing LiCl-extracted particles with NH_4OH at pH 9.2. (3) A crude factor, partially purified by salt fractionation from the ammonia supernatant, can partially restore oligomycin sensitivity in ammonia-extracted membranes (Melandri *et al.*, 1974). In *Rs. rubrum*, the properties of oligomycin-sensitive ATPase are very similar; this aspect has not been studied in detail, however, and therefore a complete comparison between the two systems is at present unfeasible.

In addition to the sensitivity toward oligomycin, other properties of *Rs. rubrum* ATPase are modulated by the binding of the protein to the membrane; Ca^{2+}-dependent ATPase of the soluble enzyme becomes partially masked, and its disappearance is paralleled by the appearance of Mg^{2+}-dependent activity (Johansson *et al.*, 1973). Moreover, this Mg^{2+}–ATPase activity is sensitive to digestion with phospholipase A (Klemme, B., *et al.*, 1971), in contrast to the complete insensitivity of soluble Ca^{2+}–ATPase. With some analogy, in *Chr. vinosum* strain D, solubilized crude coupling factor and membrane-bound ATPase differ distinctly in their requirements for optimal concentrations of Ca^{2+} or Mg^{2+} ions and ATP (Gepshtein and Carmeli, 1974; Gepshtein *et al.*, 1975).

These phenomena, observed in different systems, are all examples of so-called "allotopic" properties of the energy-transducing ATPases (Racker, 1970). Rather than an "allosteric" effect of the membrane on the enzyme, these changes in properties probably indicate that the mechanism of ATP hydrolysis, coupled to energization of the membrane, involves components of the ATPase that are intrinsic and extrinsic proteins of the membrane; the mechanism may be quite distinct from that catalyzed by the solubilizable ATPase, which is clearly an extrinsic membrane enzyme.

5. Membrane-Bound Pyrophosphatase from *Rhodospirillum rubrum*

Light-induced synthesis of pyrophosphate (PP_i) in chromatophores from *Rs. rubrum* was first discovered by H. Baltscheffsky and von Stedingk (1966): this activity, at first considered as a side reaction of membrane-bound ATPase, was subsequently recognized to be due to a quite distinct, oligomycin-insensitive, enzyme. Several experiments favor this view: Fisher and Guillory (1969) succeeded in selectively removing ATPase or pyrophosphatase (PPase) with 2 M LiCl or 3.1% butanol, respectively,

leaving the alternate activity unaffected; monospecific antisera against ATPase could inhibit ATP-linked but not PP_i-linked energization of the membrane (Johansson, 1975).

Moreover, ATP synthesis driven by PP_i hydrolysis was demonstrated by Keister and Minton (1971); since no incorporation into ATP of labeled phosphate from $^{32}PP_i$ could be demonstrated, it was concluded that the two reactions were catalyzed by two different enzymatic systems coupled only through the "high-energy state" of the membrane. At variance with these results, PP_i-induced reduction of cytochrome *b* was restored in EDTA-sonicated chromatophores by the crude sonic fluid (Johansson *et al.*, 1972); this discrepancy could be due to excessive damage of the membrane during decoupling and to the use of a crude coupling factor preparation for reconstitution.

Membrane-bound PPase is dependent on the phospholipids of the membrane, as is membrane-bound ATPase (Klemme, B., *et al.*, 1971). No reversible resolution of this enzyme, which is clearly quite distinct from the cytoplasmic PPase, present in the same organism (Klemme, J.-H., and Gest, 1971), has yet been achieved.

Several energy-linked reactions dependent on PP_i hydrolysis have been demonstrated: energy-linked transhydrogenase (Keister and Yike, 1967*a*), cytochrome reduction (Baltscheffsky, M., 1967), succinate-linked NAD^+ reduction (Keister and Yike, 1967*b*), proton uptake (Moyle *et al.*, 1972), and energy-linked carotenoid bandshift (Baltscheffsky, M., 1969). All these data clearly indicate that PPase can mediate energy transduction between PP_i synthesis (or hydrolysis) and other energy-generating (or energy-requiring) reactions of the membrane. The catalytic and possibly physiological function of membrane-bound PPase therefore appears to be equivalent to that of a conventional coupling factor; the reversible resolution of this peculiar enzyme is therefore of great interest, and should be the object of careful investigation.

6. Reconstitution of Phosphorylation; Energy-Conserving Properties of the Membrane

Coupling factors, by definition, are detected by their ability to restore ATP synthesis coupled to electron transport in depleted membranes. Highly purified coupling factors from photosynthetic bacteria can reconstitute photophosphorylation with considerable effectiveness. Typically, 100–200 μg of purified protein from *Rp. capsulata* are saturating in assays containing about 50 μg bacteriochlorophyll (Bchl) (Baccarini-Melandri and Melandri, 1971); this quantity would correspond to about 1 mol of coupling factor per mole of reaction center, if a size of the photosynthetic unit on the order of 100 molecules of Bchl is assumed. In parallel with photophosphorylation, other partial activities such as ATP–^{32}P$_i$ exchange and oligomycin-sensitive and -insensitive ATPase are reconstituted (Baccarini-Melandri and Melandri, 1971) (Fig. 1).

In *Rp. capsulata*, oxidative phosphorylation can be restored by addition of the photosynthetic coupling factor to decoupled respiratory membranes prepared from aerobically grown cells (Melandri *et al.*, 1971a); conversely, a coupling factor able to restore photophosphorylation and oxidative phosphorylation can be purified from aerobic cells (Baccarini-Melandri and Melandri, 1972a,b) (Fig. 2). Since the chemical, enzymatic, and immunological properties of the two preparations are indistinguishable (Melandri *et al.*, 1971a; Baccarini-Melandri and Melandri, 1972a,b;

Lien and Gest, 1973; Johansson, 1975), it is nearly certain that ATP synthesis in photosynthesis and respiration is catalyzed by the same protein; this conclusion is in agreement with the dual function of the membrane of facultative photosynthetic bacteria in respiration and photosynthesis, and with the lack of any compartmentalization in prokaryotic cells.

Considerable information has been accumulated on the role of bacterial coupling factors in energy conservation and transduction; these studies have been facilitated by the peculiar characteristics of these bacterial systems, in which the nearly complete detachment of the coupling factor from the membrane affects only marginally the energy-conserving properties of the membrane itself. Several light-induced energy-dependent processes are retained in decoupled bacterial membranes of Rhodospirillaceae; these include proton uptake (Melandri *et al.*, 1970), quenching of fluorescence of atebrine and 9-amino acridine (Melandri *et al.*, 1972b; Gromet-Elhanan, 1974), enhancement of fluorescence of 8-anilino naphthalene 1-

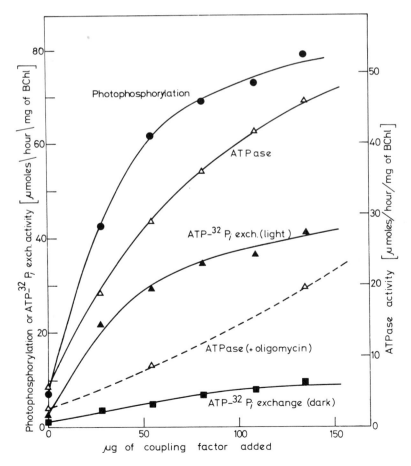

Fig. 1. Restoration of photophosphorylation, ATPase, and ATP–^{32}P exchange in uncoupled membranes (42 μg Bchl) by purified coupling factor. Reproduced from Baccarini-Melandri and Melandri (1971) with kind permission from Academic Press, Inc.

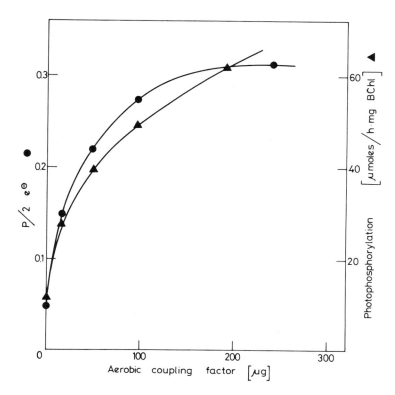

Fig. 2. Effect of purified coupling factor on photophosphorylation and NADH-dependent oxidative phosphorylation. Reproduced from Baccarini-Melandri and Melandri (1972a) with kind permision from Elsevier–North-Holland, Amsterdam.

sulfonic acid (this last phenomenon was, however, decreased by about 50% in *Rs. rubrum*) (Johansson *et al.*, 1972; Gromet-Elhanan, 1974), and carotenoid bandshift (Saphon *et al.*, 1975). The presence of these activities in membranes depleted of coupling factor has been taken as a proof that the decoupling procedure does not impair the cyclic electron-transport system, and that only a marginal structural role of the coupling factor for the integrity of the membrane has to be postulated in bacterial photosynthetic systems.

Conversely, ATP-linked energy-requiring reactions, such as ATP-induced quenching of atebrine fluorescence (Melandri *et al.*, 1972b), ATP-induced reduction of cytochrome *b*, and carotenoid bandshift (Johansson *et al.*, 1972), are strictly dependent on the presence of coupling factor, and are inhibited specifically by antibodies prepared against purified ATPase (Johansson, 1975). In addition, the ADP-induced transient oxidation of cytochrome *b* under steady-state illumination was demonstrated to be dependent on the presence of ATPase (Johansson *et al.*, 1972; Johansson, 1975). The data shown in Fig. 3 are an example of this situation. In these experiments, the quenching of atebrine fluorescence has been taken as an indicator of membrane energization (possibly as a probe of transmembrane pH gradients). It is evident

that the presence of coupling factor on the membranes, whether native or reconstituted (Figs. 3A and C), is essential for ATP-induced energization in the dark, while the light-induced quenching is present in membranes deprived of ATPase (Fig. 3B).

Although light-induced energization is present in membranes deprived of coupling factor, several effects of the presence of ATPase on parameters related to the energized state have been described. Thus, stimulation of light-induced proton uptake by addition of ADP and arsenate was shown to be strictly related to the presence of the membrane-bound ATPase in *Rp. capsulata*; this stimulation could be simulated by oligomycin in both coupled and decoupled membranes (Melandri *et al.*, 1970). The fast component of the biphasic decay of carotenoid shift induced by a single-turnover flash is stimulated by ADP and P_i; also, this effect is dependent on the presence of coupling factor and is inhibited by energy-transfer inhibitors such as oligomycin, venturicidin, and aurovertin (Saphon *et al.*, 1975). Most of these phenomena have been interpreted as due to the influence of coupling factor on the proton conductivity of the membrane and specifically to proton fluxes through the ATPase itself (or through its hydrophobic components still present in the decoupled membranes),

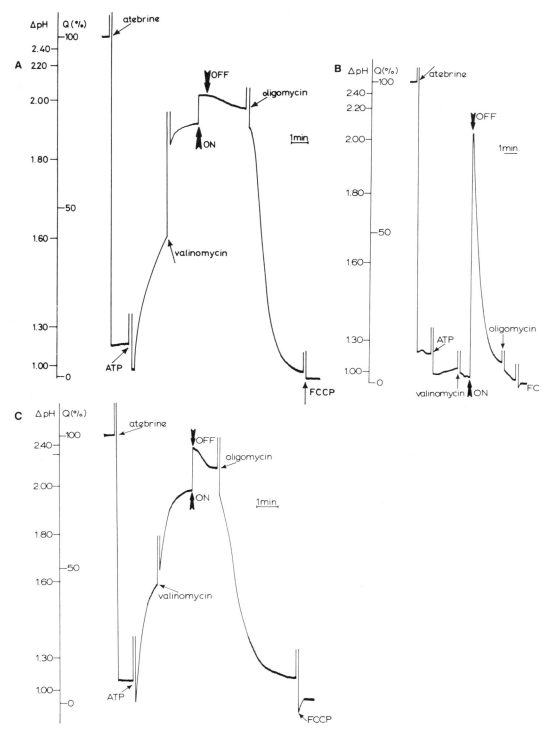

Fig. 3. Quenching of atebrine fluorescence by ATP and light in photosynthetic membranes from *Rhodopseudomonas capsulata*. (A) Untreated particles (25 μg Bchl/assay); (B) EDTA-sonicated particles (30 μg Bchl/assay); (C) EDTA-sonicated particles (30 μg Bchl/assay) reconstituted with 22 μg coupling factor protein. The quenching [Q(%)] is expressed as the percentage decrease of fluorescence on energization in comparison with the fluorescence observed in the uncoupled condition (plus FCCP). The pH scale was calculated assuming that atebrine distributed as an ideal diamine with pKs of 7.9 and 10.1 with an assumed internal volume of 100 μl/mg Bchl. Reproduced from Melandri *et al.* (1972*b*) with the kind permission of Elsevier–North-Holland, Amsterdam.

visualized as a proton translocator, as predicted by the chemiosmotic coupling theory (Mitchell, 1968). In line with this hypothesis, the independence of the electron-transport-linked and ATP-linked energization of the membrane can be quite clearly demonstrated in Rhodospirillaceae, since in this system a clear-cut physical separation of these two processes can be achieved.

7. Modulation of Membrane-Bound ATPase

The activity of coupling factor ATPase is modified by interactions with substrates and with the electron-transport chain. In *Rp. capsulata*, the organism most extensively studied so far in this respect, these properties resemble in many instances analogous phenomena observed either in mitochondrial F_1 or in chloroplast CF_1; again, these observations speak in favor of a substantial constancy, during evolution, of the properties of coupling factor proteins, and are in agreement with the dual role of the coupling factor in this bacterium in both aerobic and photosynthetic metabolism.

The first phenomenon of this type discovered in chromatophores from *Rp. capsulata* was a considerable increase in the ATPase activity caused by illumination of chromatophores (Melandri *et al.*, 1972a). This activation of the enzyme (about two to three times the dark activity) is readily reversible and appears to be related to some interaction of the enzyme with cyclic electron flow, since it is completely inhibited by antimycin A. The relevance in energy transduction of the activity elicited by illumination is attested to by its complete sensitivity to the energy-transfer inhibitors oligomycin and aurovertin.

All uncouplers, when added at concentrations that completely inhibit phosphorylation, also inhibit the activity of ATPase that is sensitive to oligomycin (Baccarini-Melandri *et al.*, 1975a). This effect, which can be observed both in the light and in the dark suggests that the active state of the enzyme is some-what dependent on the energized state of the membrane; in turn, the energized state can be induced by photosynthetic electron flow (activation by light) or in the dark by hydrolysis of ATP. The lower level of activity observed in the dark is compatible with the lower level of energization (a lower proton-motive force) produced by ATP hydrolysis. ATP hydrolysis can be expected to give a lower proton-motive force because the activity of the ATPase is less than the rate of electron transport. Uncouplers, at concentrations not completely inhibitory for photophosphorylation, stimulate ATPase of photosynthetic bacteria as in

other energy-transducing systems, but at higher concentrations, the inhibitory effect, due to the inactivation of the energy-transducing ATPase, becomes predominant. Activation of ATPase by light can also be observed in *Rs. rubrum* chromatophores in the presence of concentrations of uncouplers insufficient to inhibit photophosphorylation completely; also in this system, an excess of uncouplers results in the inhibition of the activity (B. C. Jackson, personal communication). These properties of the coupling ATPase in Rhodospirillaceae are comparable to those of the analogous enzyme from higher plant chloroplasts (Petrack and Lipmann, 1961; Hoch and Martin, 1963); in this system, in fact, Mg^{2+}–ATPase is active only after illumination in the presence of thiol reagents (light-triggered ATPase), and the active state is destabilized by uncouplers (Carmeli, 1969). It is therefore likely that these phenomena reflect a regulatory mechanism specifically present in many, if not all, photosynthetic membranes.

Additional changes in activity, induced by the substrates of photophosphorylation, are superimposed on the activation by light in *Rp. capsulata* (Melandri *et al.*, 1972a), and also in *Rs. rubrum* (B. C. Jackson, personal communication). ADP, at very low (micromolar) concentrations, inhibits the activation by light. On the other hand, 1 mM phosphate, and also arsenate, increases the ATPase activity in the dark by about two- to threefold (Baccarini-Melandri *et al.*, 1975b); again, the activity elicited by phosphate is relevant for energy transduction, since a parallel increase of energy-requiring processes is observed in the presence of this anion, and the induced activity is sensitive to oligomycin and aurovertin. The "allosteric" interaction of ATPase with ADP and phosphate (or arsenate) is not unique to bacterial systems; in higher-plant chloroplasts, ADP destabilizes and phosphate stabilizes the light-triggered state of the enzyme (Carmeli and Lifshitz, 1972). In liver mitochondria, phosphate and arsenate stimulate the ATPase, increasing the V_{max} of the enzyme, and temporarily stabilize a state of the ATPase that is rather insensitive to competitive inhibition by ADP (Mitchell and Moyle, 1970); moreover, a low ADP/ATP ratio, or alternatively, energization due to substrate oxidation, was recently shown to promote the release of the peptide ATPase inhibitor from beef heart submitochondrial particles (van de Stadt *et al.*, 1973).

Is there any conformational change in the structure of the enzyme related to the modulation of the activity? In *Rp. capsulata*, as in spinach chloroplast (McCarty *et al.*, 1972), such a possibility is suggested by the observation that a partial inhibition of ATPase and photophosphorylation by the thiol reagent *N*-

ethylmaleimide (NEM) is observed only if the membranes are exposed to the inhibitor in the light (Baccarini-Melandri et al., 1975a). ADP, phosphate, and uncouplers protect the enzyme against photoinactivation by NEM, suggesting some correlation between the attack by the inhibitor and the modulation of the enzymatic activity by light. The simplest interpretation of these phenomena assumes that on illumination in the absence of phosphorylation substrates, SH groups, important for the catalytic activity, become exposed and susceptible to the attack by NEM. Experiments by McCarty and Fagan (1973) in chloroplasts demonstrated that the inhibition corresponds to a specific labeling by [^{14}C]NEM of the γ subunit of CF_1. More direct evidence of a conformational change of chloroplast coupling factor has been obtained especially in Jagendorf's laboratory (Ryrie and Jagendorf, 1971, 1972), with the demonstration that nonexchangeable tritium, derived from ^3H_2O, is trapped into CF_1 on membrane energization.

There is evidence that coupling factors from photosynthetic bacteria have multiple binding sites for adenine nucleotides, as do those from other organisms (Roy and Moudrianakis, 1971a,b; Nelson et al., 1972a; Catterall and Pedersen, 1972; Harris et al., 1973). In addition to the obvious interaction of solubilized coupling factor with ATP, evident from the protective effect exerted by this nucleotide in all systems tested so far, the presence of firmly bound ATP and ADP in chromatophores from Rs. rubrum was clearly established by Lutz et al. (1974) (see also Bachofen et al., 1974). Of these nucleotides firmly bound to the membrane, some are released by perchloric acid extraction, and the rest (mainly ATP) are completely extracted by SDS; only the SDS-soluble ATP appears to be related to the presence of coupling factor on the membrane and to illumination. Evidence that the acid-soluble ATP could arise from soluble ADP and phosphate and that this synthesis could be concomitant with the hydrolysis of the SDS-extractable ATP has been also presented. Lutz et al. (1974) suggested, therefore, that the hydrolysis of bound ATP could be the energy source for the synthesis of free ATP (acid-soluble), and that this "mechanism is compatible with the concept of conformational energy conservation" (Boyer et al., 1973), possibly through conformational changes of the coupling factor.

Indirect evidence of multiple interactions of ADP with the ATPase of Rp. capsulata is also available; ADP inhibits light activation of ATPase with an apparent affinity constant comparable to the K_m of photophosphorylation (10 μM), and it is, in these actions, much more effective than in the competitive inhibition of ATPase, for which the apparent K_i is

0.26 mM (Melandri et al., 1972a). These data are very reminiscent of similar results obtained in higher plant chloroplasts (Carmeli and Lifshitz, 1972).

Recent hypotheses (Boyer et al., 1973; Slater et al., 1974; Bachofen et al., 1974) propose that all these phenomena are related to the mechanism of ATP synthesis, which is visualized as associated with cyclic changes in the conformation of the ATPase, induced by electron flow and by adenine nucleotides, which cause specific changes in the affinity of the enzyme for ADP and ATP. The data so far accumulated on the interaction of adenine nucleotides with the ATPase in different systems are generally consistent with such a mechanism, although not sufficient to prove it. The importance of such studies for the elucidation of the mechanism of energy transduction mediated by the coupling factor is obvious. It should be noted, however, that such studies may not be conclusive for the elucidation of the mechanism of coupling between electron transport, proton fluxes, and phosphorylation, which still remains a major challenge for future research.

On the basis of the chemiosmotic hypothesis (Mitchell, 1968), decisive evidence on this point could be obtained with the demonstration that transmembrane proton translocation is an intrinsic property of the ATPase complex. Rather conclusive evidence for this has been obtained for mitochondria (Kagawa and Racker, 1971; Racker and Kandrach, 1971; Racker and Stoeckenius, 1974), although some of the membrane proteins used were still not purified. Analogous experiments in bacterial photophosphorylation are not feasible at present, since no data on the hydrophobic components of the ATPase complex are yet available. Studies along this line therefore represent the next decisive step for the comprehension of the function and structure of energy-transducing ATPase in bacterial photosynthesis.

8. Note Added in Proof

Since the manuscript of this review was completed, substantial advances have been made in the studies of the coupling factor from photosynthetic bacteria, especially from Rs. rubrum.

New methods for the purification of Rs. rubrum coupling factor have been described: the modifications proposed include innovative procedures for the dislocation of the enzyme from the membrane, such as extraction with chloroform (Webster et al., 1977) or passage of membranes dialyzed against low-ionic-strength buffer through French press (Lucke and Klemme, 1976), and new approaches for the purification of the extracted ATPase. Among these the most promising appears to be affinity chromatography utilizing adenine nucleotide analogues attached to Sepharose gels (Webster *et al.*, 1977).

These new purification procedures seem to have produced a substantial improvement in the stability and the yield of the enzyme, although with chloroform extraction an incomplete F_1 preparation, deprived of two subunits and unable to restore photophosphorylation, has been obtained (Webster *et al.*, 1977).

Some efforts have also been devoted to the isolation, by detergent extraction, of an oligomycin-sensitive ATPase complex (OS-ATPase) from photosynthetic bacteria, i.e., a preparation including the intrinsic part of the enzyme, which confers sensitivity to oligomycin. Two laboratories have successfully isolated this enzyme from *Rs. rubrum* (Oren and Gromet-Elhanan, 1977; Müller and Baltscheffsky, 1978); the OS-ATPase complex appears to be composed of 9 to 10 subunits, in strict analogy with similar preparations from respiratory systems.

Considerable progress has been achieved in our understanding of the regulatory properties of membrane-bound ATPase of *Rs. rubrum* (Webster *et al.*, 1977). Evidence has been presented which suggests a regulatory interaction of divalent cations, free Mg^{2+} being a very effective inhibitor of ATPase in the presence of high concentrations of uncouplers. The characteristic behavior of solubilized ATPase, which is very active with Ca^{2+} and is inhibited competitively by Mg^{2+}, can also be induced in membrane-bound ATPase by extensive uncoupling, indicating that the change in cation specificity of the solubilized enzyme upon dislocation from the membrane is not due to allotopy, but rather is an intrinsic property of the enzyme and is related to the degree of coupling. A series of anions, of which the most effective is sulfite, but which include also phosphate, reverts the enzyme in extensively uncoupled membranes to the form normally found in coupled chromatophores.

These studies confirm the extremely complex behavior of the coupling ATPase of photosynthetic bacteria, which interacts and is modulated by substrates, free cations, and anions, and whose properties are dictated by the level of the membrane high-energy state.

ACKNOWLEDGMENTS

The authors would like to thank Drs. R. Berzborn, A. Binder, C. Carmeli, B. Jackson, and B. C. Johansson for making available to us preprints of their manuscripts.

Thanks are also due to Consiglio Nazionale delle Ricerche (Italy) for general support.

9. References

Abrams, A., and Smith, J. B., 1974, Bacterial membrane ATPase, in: *The Enzymes* (P. D. Boyer, ed.), Vol. X, pp. 395–429, Academic Press, New York.

Baccarini-Melandri, A., and Melandri, B. A., 1971, Partial resolution of the photophosphorylating system of *Rhodopseudomonas capsulata*, in: *Methods in Enzymology* (A. San Pietro, ed.), Vol. XXIII, pp. 556–561, Academic Press, New York and London.

Baccarini-Melandri, A., and Melandri, B. A., 1972a, Energy transduction in photosynthetic bacteria. III. Coincidence of coupling factor of photosynthesis and respiration in *Rhodopseudomonas capsulata*, *FEBS Lett.* **21**:131.

Baccarini-Melandri, A., and Melandri, B. A., 1972b, Energy transduction in photosynthetic bacteria. II. Coupling factors from aerobically and photosynthetically grown cells of *Rhodopseudomonas capsulata*, in: *Proceedings of the Second International Congress on Photosynthesis* (G. Forti, M. Avron, and B. A. Melandri, eds.), pp. 1185–1193, W. Junk, The Hague.

Baccarini-Melandri, A., Gest, H., and San Pietro, A., 1970, A coupling factor in bacterial photophosphorylation, *J. Biol. Chem.* **245**:1224.

Baccarini-Melandri, A., Fabbri, E., Firstater, E., and Melandri, B. A., 1975a, Energy transduction in photosynthetic bacteria. VII. Inhibition of the coupling ATPase by *N*-ethylmaleimide related to the energized state of the membrane, *Biochim. Biophys. Acta* **376**:72.

Baccarini-Melandri, A., Fabbri, E., and Melandri, B. A., 1975b, Energy transduction in photosynthetic bacteria. VIII. Activation of the energy transducing ATPase by inorganic phosphate, *Biochim. Biophys. Acta* **376**:82.

Bachofen, R., Beyeler, W., Dahl, J. S., Lutz, H. U., and Pflugshaupt, C., 1974, Synthesis of free ATP from firmly bound ATP in photosynthetic membranes, in: *Membrane Proteins in Transport and Phosphorylation* (G. F. Azzone, M. E. Klingenberg, E. Quagliariello, and N. Siliprandi, eds.), pp. 61–72, North-Holland, Amsterdam.

Baltscheffsky, H., and Baltscheffsky, M., 1974, Electron transport phosphorylation, in: *Annual Review of Biochemistry* (E. E. Snell, ed.), pp. 871–897, Annual Reviews, Palo Alto, California.

Baltscheffsky, H., and von Stedingk, L. V., 1966, Bacterial photophosphorylation in the absence of added nucleotide. A second intermediate stage of energy transfer in light induced formation of ATP, *Biochem. Biophys. Res. Commun.* **22**:722.

Baltscheffsky, M., 1967, Inorganic pyrophosphate as an energy donor in photosynthetic and respiratory electron transport phosphorylation systems, *Biochem. Biophys. Res. Commun.* **28**:270.

Baltscheffsky, M., 1969, Energy conversion-linked changes of carotenoid absorbance in: *Rhodospirillum rubrum* chromatophores, *Arch. Biochem. Biophys.* **130**:646.

Berzborn, R. J., Johansson, B. C., and Baltscheffsky, M., 1975, Immunological and fluorescence studies with the coupling factor ATPase from *Rhodospirillum rubrum*, *Biochim. Biophys. Acta* **396**:360.

Binder, A., and Gromet-Elhanan, Z., 1975, Depletion and reconstitution of photophosphorylation in chromatophore membranes of *Rhodospirillum rubrum*, in: *Proceedings of the Third International Congress on Photosynthesis* (M. Avron, ed.), pp. 1163–1170, Elsevier, Amsterdam.

Bose, S. K., and Gest, H., 1965, Properties of adenosine triphosphatase in a photosynthetic bacterium, *Biochim. Biophys. Acta* **96**:159.

Boyer, P. D., Cross, R. L., and Momsen, W., 1973, A new concept for energy coupling in oxidative phosphorylation based on a molecular explanation of the oxygen exchange reactions, *Proc. Natl. Acad. Sci. U.S.A.* **70**:2837.

Carmeli, C., 1969, Properties of ATPase in chloroplasts, *Biochim. Biophys. Acta* **189**:256.

Carmeli, C., and Lifshitz, Y., 1972, Effects of P_i and ADP on ATPase activity in chloroplasts, *Biochim. Biophys. Acta* **267**:86.

Catterall, W. A., and Pedersen, P. L., 1972, Adenosine triphosphatase from rat liver mitochondria. II. Interaction with adenosine diphosphate, *J. Biol. Chem.* **247**:7969.

Edwards, P. A., and Jackson, J. B., 1976, The control of adenosine triphosphatase of *Rhodospirillum rubrum* chromatophores by divalent cations and the membrane high-energy state, *Eur. J. Biochem.* **62**:7.

Eilermann, L. J. M., Pandit-Hovenkamp, H. G., Van der Meer-van Buren, M., Kolk, A. H. J., and Veeenstra, M., 1971, Oxidative phosphorylation in *Azotobacter vinelandii*. Effect of inhibitors and uncouplers on P/o ratio, trypsin-induced ATPase and ADP-stimulated respiration, *Biochim. Biophys. Acta* **245**:305.

Fernandez Moran, H., 1962, Cell membrane ultrastructure: Low-temperature electron microscopy and X-ray diffraction studies of lipoprotein components in lamellar systems, *Circulation* **26**:1039.

Fisher, R. R., and Guillory, R. J., 1969, Partial resolution of energy-linked reactions in *Rhodospirillum rubrum* chromatophores, *FEBS Lett.* **3**:27.

Garcia, A. F., Drews, G., and Kamen, M. D., 1974, On reconstitution of bacterial photophosphorylation *in vitro*, *Proc. Natl. Acad. Sci. U.S.A.* **71**:4213.

Garcia, A. F., Drews, G., and Kamen, M. D., 1975, Electron transport in an *in vitro*-reconstituted bacterial photophosphorylating system, *Biochim. Biophys. Acta* **387**:129.

Geller, D. M., 1962, Oxidative phosphorylation in extracts of *Rhodospirillum rubrum*, *J. Biol. Chem.* **237**:2947.

Gepshtein, A., and Carmeli, C., 1974, Properties of adenosinetriphosphatase in chromatophores and in coupling factor from the photosynthetic bacterium, *Chromatium*, strain D, *Eur. J. Biochem.* **44**:593.

Gepshtein, A., Hochman, Y., and Carmeli, C., 1975, Effect of the interaction between cation–ATP complexes and free cations on ATPase complexes and free cations on ATPase activity in *Chromatium*, strain D chromatophores, in: *Proceedings of the Third International Con-*

gress on Photosynthesis (M. Avron, ed.), pp. 1189–1197, Elsevier, Amsterdam.

Gromet-Elhanan, Z., 1974, Role of photophosphorylation coupling factor in energy conversion by depleted chromatophores of *Rhodospirillum rubrum*, *J. Biol. Chem.* **249**:2522.

Harris, D. A., Rosing, J., van de Stadt, R. J., and Slater, E. C., 1973, Tight binding of adenine nucleotides to beef-heart mitochondrial ATPase, *Biochim. Biophys. Acta* **314**:149.

Hoch, H., and Martin, I., 1963, Photo-potentiation of adenosine triphosphate hydrolysis, *Biochem. Biophys. Res. Commun.* **12**:223.

Hochman, A., and Carmeli, C., 1971, A coupling factor from *Chromatium* strain D chromatophores, *FEBS Lett.* **13**:36.

Horio, T., Nishikawa, K., and Horiuti, Y., 1971, Adenosine triphosphatase: Bacterial, in: *Methods in Enzymology* (A. San Pietro, ed.), Vol. XXIII, pp. 650–654, Academic Press, New York and London.

Jagendorf, A. T., 1975, Mechanism of photophosphorylation. in: *Bioenergetics of Photosynthesis* (Govindjee, ed.), pp. 413–492, Academic Press, New York—San Francisco—London.

Johansson, B. C., 1972, A coupling factor from *Rhodospirillum rubrum* chromatophores, *FEBS Lett.* **20**:339.

Johansson, B. C., 1975, Partial resolution of the energy transfer system in chromatophores from *Rhodospirillum rubrum*. Purification and characterization of the "coupling factor" ATPase, Ph.D. thesis, University of Stockholm.

Johansson, B. C., and Baltscheffsky, M., 1975, On the subunit composition of the coupling factor (ATPase) from *Rhodospirillum rubrum*, *FEBS Lett.* **53**:221.

Johansson, B. C., Baltscheffsky, M., and Baltscheffsky, H., 1972, Coupling factor capabilities with chromatophore fragments from *Rhodospirillum rubrum*, in: *Proceedings of the Second International Congress on Photosynthesis* (G. Forti, M. Avron, and B. A. Melandri, eds.), pp. 1203–1209, W. Junk, The Hague.

Johansson, B. C., Baltscheffsky, M., Baltscheffsky, H., Baccarini-Melandri, A., and Melandri, B. A., 1973, Purification and properties of a coupling factor (Ca^{2+}-dependent adenosine triphosphatase) from *Rhodospirillum rubrum*, *Eur. J. Biochem.* **40**:109.

Jones, O. T. G., and Plewis, K. M., 1974, Reconstitution of light dependent electron transport in membranes from a bacteriochlorophyll-less mutant of *Rhodopseudomonas sphaeroides*, *Biochim. Biophys. Acta* **357**:204.

Kagawa, Y., and Racker, E., 1966, Partial resolution of the enzymes catalyzing oxidative phosphorylation. X. Correlation of morphology and function in submitochondrial particles, *J. Biol. Chem.* **241**:2475.

Kagawa, Y., and Racker, E., 1971, Partial resolution of the enzymes catalyzing oxidative phosphorylation. XXV. Reconstitution of vesicles catalyzing $^{32}P_i$–adenosine triphosphate exchange, *J. Biol. Chem.* **246**:5477.

Keister, D. L., and Minton, N. J., 1971, Energy linked reactions in photosynthetic bacteria. VI. Inorganic pyrophosphate-driven ATP synthesis in *Rhodospirillum rubrum*, *Arch. Biochem. Biophys.* **147**:330.

Keister, D. L., and Yike, N. J., 1967*a*, Energy linked reactions in photosynthetic bacteria. II. The energy dependent reduction of oxidized nicotinamide–adenine dinucleotide phosphate by reduced nicotinamide–adenine dinucleotide in chromatophores of *Rhodospirillum rubrum*, *Biochemistry* **6**:3847.

Keister, D. L., and Yike, N. J., 1967*b*, Energy linked reactions in photosynthetic bacteria. I. Succinate-linked ATP-driven NAD^+ reduction by *Rhodospirillum rubrum* chromatophores, *Arch. Biochem. Biophys.* **121**:415.

Klemme, J.-H., and Gest, H., 1971, Regulation of the cytoplasmic inorganic pyrophosphatase of *Rhodospirillum rubrum*, *Eur. J. Biochem.* **22**:529.

Klemme, B., Klemme, J.-H., and San Pietro, A., 1971, PPase, ATPase and photophosphorylation in chromatophores of *Rhodospirillum rubrum*: Inactivation by phospholipase A; reconstitution by phospholipids, *Arch. Biochim. Biophys.* **144**:339.

Konings, A. N. T., and Guillory, R. J., 1973, Resolution of enzymes catalyzing transhydrogenase. IV. Reconstitution of adenosine triphosphate driven transhydrogenation in depleted chromatophores of *Rhodospirillum rubrum* by the transhydrogenase factor and a soluble oligomycin-insensitive Mg^{++}-adenosine triphosphatase, *J. Biol. Chem.* **248**:1045.

Lien, S., and Gest, H., 1973, On the identity of photo- and oxidative phosphorylation coupling factors in *Rhodopseudomonas capsulata*, *Arch. Biochem. Biophys.* **159**:730.

Lien, S., and Racker, E., 1971, Preparation and assay of chloroplast coupling factor CF_1, in: *Methods in Enzymology* (A. San Pietro, ed.), Vol. XXIII, pp. 547–555, Academic Press, New York and London.

Löw, H., and Afzelius, 1964, Subunits of the chromatophore membranes in *Rhodospirillum rubrum*, *Exp. Cell Res.* **35**:431.

Lucke, F.-K., and Klemme, J.-H., 1976, Coupling factor adenosine-5′-triphosphate from *Rhodospirillum rubrum*: a simple and rapid procedure for its purification, *Z. Naturforsch.* **31c**:272.

Lutz, H. U., Dahl, J. S., and Bachofen, R., 1974, Synthesis of free ATP from membrane bound ATP in chromatophores of *Rhodospirillum rubrum*, *Biochim. Biophys. Acta* **347**:359.

McCarty, R. E., and Fagan, J., 1973, Light stimulated incorporation of *N*-ethylmaleimide into coupling factor 1 in spinach chloroplasts, *Biochemistry* **12**:1503.

McCarty, R. E., and Racker, E., 1967, Effect of a coupling factor and its antiserum on photophosphorylation and hydrogen ion transport, in: *Energy Conservation by the Photosynthetic Apparatus*, pp. 202–212, Brookhaven National Laboratories, Upton, New York.

McCarty, R. E., Pittman, P. R., and Tsuchiya, Y., 1972, Light-dependent inhibition of photophosphorylation by *N*-ethylmaleimide, *J. Biol. Chem.* **247**:3048.

Melandri, B. A., and Baccarini-Melandri, A., 1972, Energy transduction in photosynthetioc bacteria. I. Properties of solubilized and reconstituted ATPase in *Rhodo-*

pseudomonas capsulata photosynthetic membranes, in: *Proceedings of the Second International Congress on Photosynthesis* (G. Forti, M. Avron, and B. A. Melandri, eds.), pp. 1169–1183. W. Junk, The Hague

Melandri, B. A., Baccarini-Melandri, A., San Pietro, A., and Gest, H., 1970, Role of phosphorylation coupling factor in light-dependent proton translocation by *Rhodopseudomonas capsulata* membrane preparations, *Proc. Natl. Acad. Sci. U.S.A.* **67**:477.

Melandri, B. A., Baccarini-Melandri, A., San Pietro, A., and Gest, H., 1971*a*, Interchangeability of phosphorylation coupling factors in photosynthetic and respiratory energy conversion, *Science* **174**:514.

Melandri, B. A., Baccarini-Melandri, A., Gest, H., and San Pietro, A., 1971*b*, Studies on resolution of the photophosphorylating system of the photosynthetic bacterium *Rhodopseudomonas capsulata*, in: *Energy Transduction in Respiration and Photosynthesis* (E. Quagliariello, S. Papa, and C. S. Rossi, eds.), pp. 593–608, Adriatica Editrice, Bari, Italy.

Melandri, B. A., Baccarini-Melandri, A., and Fabbri, E., 1972*a*, Energy transduction in photosynthetic bacteria. IV. Light-dependent ATPase in photosynthetic membranes from *Rhodopseudomonas capsulata*, *Biochem. Biophys. Acta* **275**:383.

Melandri, B. A., Baccarini-Melandri, A., Crofts, A. R., and Cogdell, R. J., 1972*b*, Energy transduction in photosynthetic bacteria, V. Role of coupling factor ATPase in energy conversion as revealed by light or ATP-induced quenching of atebrine fluorescence, *FEBS Lett.* **24**:141.

Melandri, B. A., Fabbri, E., Firstater, E., and Baccarini-Melandri, A., 1974, Allotopic properties and energy dependent conformational changes of bacterial ATPase, in: *Membrane Proteins in Transport and Phosphorylation* (G. F. Azzone, M. E. Klingenberg, E. Quagliariello, and N. Siliprandi, eds.), pp. 55–60, North-Holland, Amsterdam.

Mitchell, P., 1968, *Chemiosmotic Coupling and Energy Transduction*, Glynn Research, Ltd., Bodmin, England.

Mitchell, P., and Moyle, J., 1970, Aurovertin sensitive phosphate activation of mitochondrial adenosine triphosphatase, *FEBS Lett.* **9**:305.

Moyle, J., Mitchell, R., and Mitchell, P., 1972, Proton translocating pyrophosphatase of *Rhodospirillum rubrum*, *FEBS Lett.* **23**:233.

Müller, H. W., and Baltscheffsky, M., 1978, On the oligomycin sensitivity and subunit composition of the ATPase complex from *Rhodospirillum rubrum*, *FEBS Lett.* (in press).

Nelson, N., Nelson, H., and Racker, E., 1972*a*, Partial resolution of the enzymes catalyzing photophosphorylation. XI. Magnesium adenosine triphosphatase properties of heat-activated coupling factor 1 from chloroplasts, *J. Biol. Chem.* **247**:6506.

Nelson, N., Nelson, H., and Racker, E., 1972*b*, Partial resolution of the enzymes catalyzing photophosphorylation. XII. Purification and properties of an inhibitor isolated from chloroplast coupling factor 1, *J. Biol. Chem.* **247**:7657.

Oelze, J., and Drews, G., 1972, Membranes of photosynthetic bacteria, *Biochim. Biophys. Acta* **265**:209.

Oren, R., and Gromet-Elhanan, Z., 1977, Coupling factor adenosine triphosphatase complex from *Rhodospirillum rubrum*: Isolation of an oligomycin-sensitive Ca^{++}, Mg^{++}-ATPase, *FEBS Lett.* **79**:147.

Pedersen, P. L., 1975, Mitochondrial adenosine triphosphatase, *Bioenergetics* **6**:243.

Penefsky, H. S., 1974, Mitochondrial and chloroplast ATPases, in: *The Enzymes* (P. D. Boyer, ed.), Vol. X, pp. 375–394, Academic Press, New York and London.

Petrack, B., and Lipmann, F., 1961, Photophosphorylation and photohydrolysis in cell-free preparations of blue-green algae, in: *Light and Life* (W. D. MacElroy and H. B. Glass, eds.), pp. 621–630, The Johns Hopkins University Press, Baltimore.

Prince, R. C., Baccarini-Melandri, A., Hauska, G. A., Melandri, B. A., and Crofts, A. R., 1975, Asymmetry of an energy transducing membrane: The location of cytochrome c_2 in *Rhodopseudomonas spheroides* and *Rhodopseudomonas capsulata*, *Biochim. Biophys. Acta* **387**:212.

Pullman, H. E., and Monroy, G. C., 1963, A naturally occurring inhibitor of mitochondrial adenosine triphosphatase, *J. Biol. Chem.* **238**:3762.

Racker, E., 1970, Function and structure of the inner membrane of mitochondria and Chloroplasts, in: *Membranes of Mitochondria and Chloroplasts* (E . Racker, ed.), pp. 127–171, Van Nostrand Reinhold, New York.

Racker, E., and Kandrack, A., 1971, Reconstitution of the third site of oxidative phosphorylation, *J. Biol. Chem.* **246**:7069.

Racker, E., and Stoeckenius, W., 1974, Reconstitution of purple membrane vesicles catalyzing light driven proton uptake and adenosine triphosphate formation, *J. Biol. Chem.* **249**:662.

Reed, D. W., and Raveed, D., 1972, Some properties of the ATPase from chromatophores of *Rhodopseudomonas spheroides* and its structural relationship to the bacteriochlorophyll proteins, *Biochim. Biophys. Acta* **282**:79.

Roy, H., and Moudrianakis, E. N., 1971*a*, Interactions between ADP and the coupling factor of photophosphorylation, *Proc. Natl. Acad. Sci. U.S.A.* **68**:464.

Roy, H., and Moudrianakis, E. N., 1971*b*, Synthesis and discharge of the coupling factor adenosine diphosphate complex in spinach chloroplast lamellae, *Proc. Natl. Acad. Sci. U.S.A.* **68**:2720.

Ryrie, I. J., and Jagendorf, A. T., 1971, An energy linked conformational change in the coupling factor protein in chloroplasts. Studies with hydrogen exchange, *J. Biol. Chem.* **246**:3771.

Ryrie, I. J., and Jagendorf, A. T., 1972, Correlation between a conformational change in the coupling factor protein and the high energy state of chloroplasts, *J. Biol. Chem.* **247**:4453.

Saphon, S., Jackson, J. B., and Witt, H. T., 1975, Electrical potential changes, H^+ translocation and phosphorylation induced by short flash excitation in *Rhodopseudomonas sphaeroides* chromatophores, *Biochim. Biophys. Acta* **408**:67.

Scholes, P., Mitchell, P., and Moyle, J., 1969, The polarity of proton translocation in some photosynthetic microorganisms, *Eur. J. Biochem.* **8**:450.

Schön, G., and Biedermann, M., 1972, Bildung flüchtiger Säuren bei der Vergärung von Pyruvat und Fructose in anaerober Dunkelkultur von *Rhodospirillum rubrum*, *Arch. Mikrobiol.* **85**:77.

Senior, A. E., 1973, The structure of mitochondrial ATPase, *Biochim. Biophys. Acta* **301**:249.

Slater, E. C., Rosing, J., Harris, D. A., van de Stadt, R. J., and Kemp, A., 1974, The identification of functional ATPase in energy-transducing membranes, in: *Membrane Proteins in Transport and Phosphorylation* (G. F. Azzone, M. E. Klingenberg, E. Quagliariello, and N. Siliprandi, eds.), pp. 137–147, North-Holland, Amsterdam.

Uffen, R. L., and Wolfe, R. S., 1970, Anaerobic growth of purple non-sulfur bacteria under dark conditions, *J. Bacteriol.* **104**:462.

Vambutas, V. K., and Racker, E., 1965, Partial resolution of the enzymes catalyzing photophosphorylation. I. Stimulation of photophosphorylation by a preparation of a latent, Ca^{++} dependent adenosine triphosphatase from chloroplasts, *J. Biol. Chem.* **240**:2660.

van de Stadt, R. J., De Boer, B. L., and van Dam, K., 1973, The interaction between the mitochondrial ATPase (F_1) and the ATPase inhibitor, *Biochim. Biophys. Acta* **292**:338.

von Stedingk, L.-V., and Baltscheffsky, H., 1966, The light-induced reversible pH change in chromatophores from *Rhodospirillum rubrum*, *Arch. Biochem. Biophys.* **117**:400.

Webster, G. D., Edwards, P. A., and Jackson, J. B., 1977, Interconversion of two kinetically distinct states of the membrane bound and solubilized H^+-translocating ATPase from *Rhodospirillum rubrum*, *FEBS Lett.* **76**:29.

CHAPTER 32

Reducing Potentials and the Pathway of NAD$^+$ Reduction

David B. Knaff

1. The Problem of NAD$^+$ Reduction

The mechanism by which photosynthetic bacteria reduce pyridine nucleotide has been one of the central problems in research on these organisms. Early models for the mechanism of NAD$^+$ reduction by photosynthetic bacteria naturally centered on the well-understood pathway by which pyridine nucleotide is photoreduced in algae and higher plants (Arnon, 1967). The primary light reaction of plant photosystem I produces directly in the light a stable reductant with a midpoint oxidation–reduction potential (E_m) of —530 mV (Lozier and Butler, 1974; Ke, 1974). This membrane-bound reductant, which appears to be an iron–sulfur protein (Bearden and Malkin, 1972a,b; Ke et al., 1973; Evans, M. C. W., et al., 1974a) reduces pyridine nucleotide in a subsequent series of dark reactions that involve intermediate electron transfer to ferredoxin (a soluble iron–sulfur protein) and to a flavoprotein.

One indication that pyridine nucleotide photoreduction by photosynthetic bacteria might proceed by a mechanism different than the one described above for plants came from investigations of the effect of oxidation–reduction potential on photosynthetic electron transport in purple bacteria from the families Rhodospirillaceae and Chromatiaceae. Experiments with several purple photosynthetic bacteria led to the conclusion that the primary electron acceptors had midpoint potentials between 0 and —100 mV at pH 7

David B. Knaff · Department of Chemistry, Texas Tech University, Lubbock, Texas 79409

(see Parson and Cogdell, 1975) (see also Chapter 22). Thus, the reductant generated in the light in these bacteria, unlike in plant photosystem I, is a weaker reductant than is pyridine nucleotide, and the reduction of NAD$^+$ by the photoreduced acceptor would appear to require some additional input of energy. Bose and Gest (1962) had suggested that ATP (or some high-energy precursor) generated by light-driven cyclic electron flow could provide this additional energy, and Keister and Yike (1967) obtained convincing evidence for a pathway of energy-dependent reverse electron transport from succinate ($E'_m = +30$ mV) to NAD.

In contrast to the situation described above for purple bacteria, *Chlorobium limicola f. thiosulfatophilum* (a bacterium from the third family of photosynthetic bacteria, the Chlorobiaceae, or green sulfur bacteria) appears to utilize light to generate directly a reductant sufficiently electronegative to permit the subsequent thermodynamically favorable reduction of NAD$^+$ without any additional input of energy from ATP or its precursors (Evans, M. C. W., 1969; Prince and Olson, 1976; Knaff and Malkin, 1976).

The two different mechanisms for NAD$^+$ photoreduction by photosynthetic bacteria reflect a difference in midpoint potential of primary electron acceptors in the purple and the green bacteria. Evaluation of the feasibility of these mechanisms is not possible without reliable values for the midpoint potential of the primary electron acceptor, the strongest stable reductant generated in the light. This chapter will therefore begin with a brief discussion of the techniques used in the determination of the midpoint potentials of the primary acceptors in photosynthetic bacteria.

2. Experimental Measurement of Oxidation–Reduction Potentials

2.1. General Considerations

A detailed theoretical and practical guide to the determination of oxidation–reduction midpoint potentials, which are a measure of the affinity of an electron carrier for electrons, will be found in the monograph by Clark (1960). A recent review by Dutton and Wilson (1974) and Chapter 28 in this volume, by Dutton and Prince, describe the applications of potentiometry to the study of membrane-bound biological electron-transport systems. The treatment presented below will therefore be relatively simple; readers desiring more detailed discussions are directed to these other sources.

A determination of the oxidation–reduction midpoint potential (E_m) requires the ability to monitor the oxidation state of the carrier as a function of the ambient oxidation potential (E_h). The potential of a sample (in aqueous medium) containing the carriers under study is measured with an inert metal electrode (usually Pt). Because the electron carriers are usually membrane-bound and are often inaccessible to direct interaction with the metal electrode, oxidation–reduction mediators must be added to ensure efficient equilibration between the metal electrode and the electron carrier under investigation. Added mediators must not only equilibrate rapidly with both the electrode and the carrier, but must also meet several other criteria: (1) they cannot induce any modifications in the oxidation–reduction properties of the carrier under study; (2) they must adequately span the oxidation–reduction range covered in the experiment (a reasonable guideline is that one-electron-carrying mediators are useful within a region of ± 60 mV of their midpoint potentials and two-electron-carrying mediators are useful within ± 30 mV of their midpoints); and (3) they cannot obscure the parameters being used to monitor the oxidation state of the electron carrier under investigation.

Oxidation–reduction titrations must always be scrutinized to ascertain whether they have been performed under true equilibrium conditions. The usual minimum criteria are: (1) the measured potential and the oxidation state of the carrier must respond and equilibrate rapidly when either oxidant or reductant is added; (2) the midpoint potential determined should be independent of the concentration of the mediator; and (3) most important, the titration must be reversible, i.e., it should exhibit identical characteristics in both the oxidative and the reductive directions.

2.2. Determination of Midpoint Potentials for Primary Electron Acceptors

The general discussion of midpoint potential determination in the preceding section contains an implication that parameters are available for monitoring the oxidation state of the carrier under investigation. For the primary electron acceptor of photosynthetic bacteria, no such parameter was available for a long time because the identity of the primary electron acceptor was unknown. Recently, an electron paramagnetic resonance (EPR) signal with g values of 1.82 and 1.62 (at liquid-helium temperature) was associated with the reduced form of the acceptor in the Rhodospirillaceae and the Chromatiaceae. This signal led to the direct titration of the primary acceptor (Dutton et al., 1973; Evans, M. C. W., et al., 1974b).

Most determinations of the midpoint potential of the acceptor have involved indirect techniques. One such indirect approach utilizes the effect of oxidation–reduction potential established during a dark incubation period on the subsequent light-induced electron transfer. The rationale for this approach is that chemical reduction of the primary acceptor prior to illumination will block subsequent photochemistry. With no acceptor site available, the electron cannot leave the reaction center bacteriochlorophyll on excitation. Photooxidation of the reaction center bacteriochlorophyll itself and oxidation by the oxidized bacteriochlorophyll of c-type cytochromes have been widely used to monitor photochemical activity. In early investigations (Kuntz et al., 1964; Loach, 1966; Cusanovich et al., 1968), steady-state illumination at room temperature and equipment with relatively slow time-response were used. Such measurements, under conditions that permit multiple turnovers of the electron-transfer chain, do not conclusively establish that the component that stops photochemistry on reduction is the primary electron acceptor. More recent experiments, similar in concept to those described above but using short, "single-turnover" flashes and rapidly responding detectors (Case and Parson, 1971; Jackson et al., 1973), make it possible to conclude with more certainty that the loss of photochemical activity on reduction results from the reduction of the primary acceptor. Similarly, samples poised at defined oxidation–reduction potentials at room temperature can be transferred in the dark to low-temperature environments and examined for photochemical activity at temperatures between 1 and 80°K. At these temperatures, electron transport appears to be confined to the primary light reaction and to the oxidation of secondary electron donors, and there is less chance that reduction of secondary electron acceptors can result in the loss of photochemical activity (Dutton, 1971; Dutton *et al.*, 1972, 1973; Dutton and Leigh, 1971).

Another approach to the determination of the midpoint potential of photosynthetic bacteria involves the fluorescence technique. The rationale behind this method is that the blocking of photochemistry by reduction of the primary electron acceptor eliminates a pathway that competes with fluorescence for excited singlet energy and results in higher fluorescence yield (Clayton, 1972). This technique can be applied to the Rhodospirillaceae and Chromatiaceae (Reed *et al.*, 1969; Cramer, 1969). A conceptually similar technique involves monitoring the low-temperature EPR signal of the bacteriochlorophyll triplet state that is formed when photochemistry is blocked at negative oxidation–reduction potentials (Dutton *et al.*, 1972, 1973).

3. Midpoint Potential of Primary Electron Acceptors

3.1. Purple Bacteria (Chromatiaceae and Rhodospirillaceae)

Table 1 gives the midpoint potentials (subject to the qualifications discussed in the preceding section) of the primary electron acceptors in chromatophores of one species of Chromatiaceae (*Chromatium vinosum*) and five species of Rhodospirillaceae. For some of the organisms studied, several methods have been used, and there has been gratifying agreement in the results obtained by the different techniques. It is apparent from Table 1 that all the purple photosynthetic bacteria that have been studied contain primary electron acceptors that are considerably more electropositive than is the NAD⁺/NADH couple ($E_m = -320$ mV).

It should be noted that most of the bacteria exhibited pH-dependent primary-acceptor midpoint potentials, with the E_m becoming 60 mV more negative

Table 1. Midpoint Oxidation–Reduction Potentials of the Primary Electron Acceptors in Purple Photosynthetic Bacteria

Species	E_m (mV)	pH of determination	pH dependence	References
Rs. rubrum	−145	8.0	—	Cramer (1969)
	−22	7.6	—	Loach (1966)
	−75	7.0	−59 mV/pH	Prince and Dutton (1976)
Rp. sphaeroides	−65 to −95	8.0	—	Cramer (1969)
	−60	7.65	—	Loach (1966)
	−20	7.0	−59 mV/pH	Jackson *et al.* (1973)
	−20	7.0	−59 mV/pH	Dutton *et al.* (1973)
	−15	7.0	−59 mV/pH	Prince and Dutton (1976)
Rp. gelatinosa	−140	7.4	—	Dutton (1971)
Rp. capsulata	−25	7.0	−59 mV/pH	E. H. Evans and Crofts (1974)
Rp. viridis	−95	7.0	−30 mV/pH	Cogdell and Crofts (1972)
Chr. vinosum	−160	8.0	—	Cramer (1969)
	−130	7.5	—	Cusanovich *et al.* (1968)
	−135	7.7	—	Dutton (1971)
	−130	7.7	−40 mV/pH	Case and Parson (1971)
	−100	7.0	−59 mV/pH	Jackson *et al.* (1973)
	−135	8.0	−98 mV/pH	M. C. W. Evans *et al.* (1974*a*)
	−100	7.0	−59 mV/pH	Prince and Dutton (1976)

per each increase of pH unit. The results of kinetic measurements have suggested that the photoreduced primary electron acceptor is oxidized more rapidly than it can be protonated. Thus, the primary acceptor couple functions effectively at a midpoint potential equivalent to that above the pK of the reduced acceptor. The significance of this phenomenon is explored in depth in Chapter 24, by Prince and Dutton, in which the conclusion is drawn that the functional midpoint potentials for the primary acceptors in the Chromatiaceae and the Rhodospirillaceae lie between -160 mV and -200 mV—values that are still 120 mV more electropositive than the value for the $NAD^+/NADH$ couple. These findings make it unlikely that NAD^+ is reduced directly in these organisms.

There have been reports (Silberstein and Gromet-Elhanan, 1974; Govindjee et al., 1974; Loach et al., 1975) that *Rhodospirillum rubrum* may have a primary electron with a midpoint potential considerably more electronegative than the values given in Table 1. These reports are discussed in Chapter 24 and in a recent review by Parson and Cogdell (1975), but do not appear to provide conclusive evidence for the more negative midpoint potentials.

The data in Table 1 are based on experiments with time-resolutions of 50 nsec or slower. Recently, techniques have become available for studying electron transfer in reaction center preparations of *Rhodopseudomonas sphaeroides* in the picosecond time domain (Kaufmann et al., 1975; Rockley et al., 1975; Dutton et al., 1975). These data suggest that an intermediate state that contains the oxidized reaction center bacteriochlorophyll and a reduced species (probably bacteriopheophytin) is formed in less than 10 psec. The acceptor species listed in Table 1 appear to be reduced in a secondary step in 150–250 psec (Kaufmann et al., 1975; Rockley et al., 1975). The species initially photoreduced in 10 psec appears to have a midpoint potential significantly more negative than -420 mV in *Rp. sphaeroides* (Dutton et al., 1975) and equal to -400 mV (at pH 10.8) in *Rp. viridis* (Prince et al., 1976). However, the short lifetime [250 psec when electron transfer is possible and 10 nsec when photochemistry is blocked (Parson et al., 1975)] of this intermediate probably precludes any direct participation in NAD^+ photoreduction. All photochemistry appears to be rapidly channeled from this intermediate into the nonchlorophyll acceptors, and the most reducing species formed with a lifetime long enough to be involved in the "chemistry" of NAD^+ reduction still appears to be the species, the oxidation–reduction properties of which are described in Table 1.

3.2. Green Sulfur Bacteria (Chlorobiaceae)

In the first investigation of the effect of oxidation–reduction potential on the photochemical activity of membrane fragments from a green bacterium (*Chlorobium limicola f. thiosulfatophilum*), it was found that both cytochrome c photooxidation and reaction center bacteriochlorophyll (P840) photooxidation ceased when a component with an E_m of -130 mV became reduced (Knaff et al., 1973). These measurements were made with a slow-response instrument under steady-state illumination. Such measurements are susceptible to errors because the photooxidized membrane-bound electron carriers can be rapidly reduced by the added oxidation–reduction mediators. In addition, in steady-state measurements, the primary electron acceptor cannot be distinguished conclusively from a secondary acceptor. Thus, these measurements can indicate only the electropositive limit for the E_m of the primary acceptor (see Section 2.2), and the midpoint potential of this bacterium could be more electronegative than -130 mV. Evidence for a more electronegative primary acceptor in *Chl. limicola f. thiosulfatophilum* was recently obtained by Prince and Olson (1976), who showed that the magnitude of the rapid cytochrome c photooxidation in reaction center complexes (driven by a single-turnover flash) did not diminish as the potential of the medium was lowered from $+100$ mV to -450 mV (cf. Chapter 24). More extensive titrations (Olson et al., 1976) showed the primary acceptor to have a midpoint potential of -540 mV.

Other evidence is consistent with the presence of a low-potential primary electron acceptor in the Chlorobiaceae. A recent investigation of the iron–sulfur protein content of chromatophores from *Chl. limicola f. thiosulfatophilum* (Knaff and Malkin, 1976) revealed the presence of a membrane-bound iron–sulfur protein (characterized by an EPR signal in the reduced state at $g = 1.94$) with a midpoint potential near -550 mV, identical to E_m for the primary acceptor. The EPR spectrum of this membrane-bound iron–sulfur protein is shown in Fig. 1. Observations on the potential dependence of the photoreduction of a component with an EPR signal at $g = 1.90$ in the reduced state are also consistent with $E_m = -550$ mV for the primary acceptor in green bacteria (Jennings and Evans, 1977). Further supportive evidence for a different mechanism of NAD^+ reduction in the green bacteria, as compared with the purple bacteria, comes from the observation that representative species of both the Chromatiaceae (*Chr. vinosum*) and the Rhodospirillaceae (*Rs. rubrum*) that have primary acceptors that function

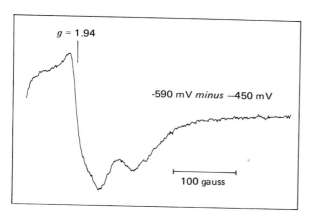

Fig. 1. EPR spectrum of a low-potential, membrane-bound iron–sulfur protein in chromatophores from *Chlorobium limicola f. thiosulfatophilum*. Reproduced from Knaff and Malkin (1976).

near −180 mV contain no low-potential ($E_m <$ −390 mV) iron–sulfur proteins (Knaff and Malkin, 1976).

4. Energy-Linked vs. Direct NAD⁺ Photoreduction

4.1. Purple Bacteria

4.1.1. Occurrence and Enzyme Requirements

NAD⁺ photoreduction has been observed with whole cells of bacteria from both the Chromatiaceae (Olson *et al.*, 1959; Olson and Amesz, 1960) and the Rhodospirillaceae (Olson and Amesz, 1960; Amesz, 1963). Although there are reports of NAD⁺ photoreduction with succinate as the electron donor in chromatophores from *Chr. vinosum* (Hood, 1964; Hinkson, 1965), most studies of NAD⁺ photoreduction in cell-free preparations of purple bacteria have been conducted with Rhodospirillaceae species because of the higher rates obtained. It should be mentioned that in addition to the NAD⁺ photoreduction observed with *Chr. vinosum* chromatophores, it has been shown that the soluble fraction of *Chr. vinosum* catalyzes NAD⁺ reduction in the dark with H₂ as the electron donor. This H₂-linked reduction of NAD⁺ involves the reduction of ferredoxin by H₂ (via hydrogenase) and subsequent reduction of an NAD⁺-specific reductase by the reduced ferredoxin (Buchanan *et al.*, 1964; Weaver *et al.*, 1965; Buchanan and Bachofen, 1968).

NAD⁺ photoreduction by chromatophores from *Rs. rubrum* has been known for some time (Frenkel, 1958; Vernon and Ash, 1958; Nozaki *et al.*, 1961) and utilizes a wide variety of electron donors, including succinate. Similar NAD⁺ photoreduction has been

observed with other Rhodospirillaceae species: *Rp. sphaeroides* (Orlando *et al.*, 1966); *Rp. capsulata* (Klemme and Schlegel, 1967), and *Rp. viridis* (Jones, O. T. G., and Saunders, 1972). The photoreduction occurs at higher rates when NAD⁺ rather than NADP⁺ is the electron acceptor. All the electron carriers necessary for NAD⁺ photoreduction appear to be membrane-bound, since NAD⁺ photoreduction occurs at good rates with washed chromatophores and is not stimulated by the addition of soluble proteins, such as ferredoxin (Nozaki *et al.*, 1961; Trebst *et al.*, 1967).

4.1.2. Energy-Linked NAD⁺ Reduction

As mentioned in Sections 1 and 3.1, the purple photosynthetic bacteria do not appear to be able to generate directly a more electronegative reductant than NADH in the light. It can thus be predicted that some additional input of energy will be required for efficient NAD⁺ reduction. In 1962, Bose and Gest (1962) suggested that NAD⁺ photoreduction in the purple bacteria proceeded by ATP-driven reverse-electron transport from succinate. According to this proposal, the role of light would be confined to the production of ATP via coupled cyclic electron flow. The ATP would then be used to supply the energy for the thermodynamically unfavorable reduction of NAD⁺ by succinate in a manner similar to the ATP-dependent reduction of NAD⁺ by succinate known to occur in mitochondria (Chance and Hollunger, 1960). The first concrete evidence for reverse-electron transport as the mechanism for NAD⁺ photoreduction came from experiments with *Rs. rubrum* by Keister and Yike (1967), who found that NAD⁺ reduction by succinate could be catalyzed in the dark by chromatophores of *Rs. rubrum* if ATP or pyrophosphate were added as a source of energy. ATP-dependent succinate-linked NAD⁺ reduction was also observed in *Rp. viridis* (Jones, O. T. G., and Saunders, 1972) and in *Rp. capsulata* (Klemme, 1969). The rate of ATP-dependent succinate-linked NAD$^+$ reduction was comparable to the rate of NAD$^+$ photoreduction with succinate as the donor (Jones O. T. G., and Saunders, 1972). NAD$^+$ reduction in the dark with ATP as the energy source was inhibited by uncoupler, by rotenone, and by oligomycin. The mode of action of uncouplers and of rotenone will be discussed below. Oligomycin inhibits the hydrolysis of ATP that is coupled to the generation of the high-energy intermediate utilized for reverse electron transport. Inhibition by oligomycin is specific for ATP as the energy source, since no oligomycin inhibition is observed when pyrophosphate is the energy source (Keister and Yike, 1967).

Evidence for the role of energy-dependent reverse-electron transfer in the photoreduction of NAD$^+$ by Rhodospirillaceae came from the observation (Keister and Yike, 1967; Jones, C. W., and Vernon, 1969; Klemme, 1969; Jones, O. T. G., and Saunders, 1972) that agents that inhibited the ability of the chromatophores to form ATP in the light (photophosphorylation) also inhibited succinate-linked NAD$^+$ photoreduction (see Table 2). These agents fall into two categories: (1) uncouplers and (2) inhibitors of cyclic electron flow. Inhibition of cyclic electron transport by inhibitors such as antimycin A and 2-heptyl-4-hydroxyquinoline-*N*-oxide (HOQNO)* (Nozaki *et al.*, 1961; Klemme and Schlegel, 1967; Keister and Yike, 1967; Jones, C. W., and Vernon, 1969; Jones, O. T. G., and Saunders, 1972) blocks the energy-releasing electron-transfer steps that allow the formation of the high-energy precursors of ATP needed to supply the energy for reverse-electron transport to NAD$^+$. [Antimycin A and HOQNO inhibit cyclic electron transport by stopping electron flow from cytochrome *b* to cytochrome *c* (Nishimura, 1963).] These agents do not affect ATP-driven NAD$^+$ reduction in the dark, however, because the high-energy intermediate arises from ATP hydrolysis, and cyclic electron flow is not required.

Uncouplers (agents that block energy transduction but not cyclic electron transport) such as carbonylcyanide *p*-trifluoromethoxyphenylhydrazone (FCCP) dissipate the high-energy intermediate directly and thereby inhibit succinate-linked NAD$^+$ reduction, whether driven by ATP in the dark or by an ATP precursor generated by cyclic electron flow in the light.

The observation that oligomycin, which inhibits ATP synthesis but not the formation of high-energy precursors, does not inhibit succinate-linked NAD$^+$ photoreduction (Keister and Yike, 1967; Jones, O. T. G., and Saunders, 1972) implies that it is the high-energy precursor, rather than ATP itself, that is utilized for NAD$^+$ reduction. The identity of this high-energy ATP precursor is not known, but it may involve an ion gradient, as indicated by the complete inhibition of succinate-linked NAD$^+$ photoreduction on the addition of the ionophore antibiotics valinomycin plus nigericin to chromatophores (Jones, O. T. G., and Saunders, 1972; Gromet-Elhanan, 1970).

Further support for the idea that succinate-linked NAD$^+$ photoreduction in these bacteria requires a high-energy precursor of ATP came from the observation (Keister and Yike, 1967; Jones, C. W., and Vernon, 1969; Jones, O. T. G., and Saunders, 1972) that the addition of ADP plus phosphate, which consumes the high-energy intermediate by allowing ATP synthesis, inhibits NAD$^+$ reduction. This inhibition of NAD$^+$ reduction by ADP plus phosphate was blocked, as expected, when ATP synthesis was inhibited by oligomycin. Inhibition of NAD$^+$ reduction by uncouplers or by ADP plus phosphate would not be expected if NAD$^+$ reduction occurred by a mechanism that involved a photoreduced primary acceptor more electronegative than the NAD/NADH couple. In fact, such inhibition is not observed in pyridine nucleotide photoreduction by plants (Krogmann *et al.*, 1959), where the primary acceptor has a midpoint potential < -530 mV.

An interesting observation (see Table 2) is that the energy-dependent, succinate-linked NAD$^+$ reduction

Table 2. Inhibition Studies on Succinate-Linked NAD$^+$ Reduction by the Particulate Fraction of *Rhodopseudomonas viridis*[a]

Addition	Rate of NAD$^+$ reduction (% of control)	
	Light	ATP
None	100	100
FCCP (1.6 μM)	5	5
Antimycin A (1.6 μM)	0	90
HOQNO (33 μM)	40	90
Rotenone (10 μM)	20	30
Oligomycin (4 μg/ml)	112	5
ADP (0.3 mM) + P$_i$ (5 mM)	50	—

[a] From O. T. G. Jones and Saunders (1972).

---

* Abbreviations in this chapter: (FCCP) carbonylcyanide *p*-trifluoromethoxyphenylhydrazone; (HOQNO) 2-heptyl-4-hydroxyquinoline-*N*-oxide; (CCCP) carbonylcyanide *m*-chlorophenylhydrazone; (DPIP) 2,6-dichlorophenolindophenol; (TMPD) *N*-tetramethyl-*p*-phenylenediamine.

in chromatophores from the Rhodospirillaceae species driven either by light or by ATP in the dark is extremely sensitive to rotenone (Keister and Yike, 1967; Klemme, 1969; Gromet-Elhanan, 1969; Jones, C. W., and Saunders, 1972). NAD^+ reduction is also inhibited by amytal (Keister and Yike, 1967; Klemme and Schlegel, 1967) and by piericidin A (Klemme, 1969). These compounds are powerful inhibitors of mitochondrial NADH dehydrogenase, and all appear to act at the same site in the mitochondrial system (Horgan *et al.*, 1968). Rotenone and related compounds cannot be inhibiting NAD^+ photoreduction in the Rhodospirillaceae by acting as uncouplers or as inhibitors of cyclic electron flow because these compounds do not inhibit cyclic photophosphorylation (Keister and Yike, 1967; Gromet-Elhanan, 1969; Klemme, 1969; Jones, C. W., and Saunders, 1972). Presumably, rotenone and related compounds inhibit NAD^+ photoreduction directly at the level of the NAD^+-reducing enzyme. This enzyme appears to be related to the NADH dehydrogenase of mitochondria on the basis of the similar patterns of inhibitor sensitivities. This similarity is of particular interest because the mitochondrial NADH dehydrogenase is known to be capable of catalyzing energy-dependent, reverse electron transport from succinate to NAD (Chance and Hollunger, 1960). Further evidence in support of possible similarities between the NAD^+-reducing enzymes in the Rhodospirillaceae and mitochondrial NADH dehydrogenase comes from the detection (Prince *et al.*, 1974, 1975) in chromatophores isolated from several Rhodospirillaceae species of iron–sulfur proteins with oxidation–reduction midpoint potentials similar to those of iron–sulfur proteins associated with the mitochondrial NADH and succinate dehydrogenase complexes (Ohnishi, 1974).

Succinate can be replaced as the electron donor for NAD^+ photoreduction in chromatophores from Rhodospirillaceae species by a variety of nonphysiological substances, such as 2,6-dichlorophenolindophenol (DPIP) and N-tetramethyl-p-phenylenediamine (TMPD). NAD^+ photoreduction with DPIP or TMPD as the electron donor differs from the succinate-linked NAD^+ photoreduction in that it is insensitive to antimycin A and to HOQNO (Nozaki *et al.*, 1961; Hinkson, 1965; Trebst *et al.*, 1967; Jones, C. W., and Vernon, 1969; Gromet-Elhanan, 1969; Klemme, 1969). In *Rs. rubrum*, however, DPIP and TMPD also catalyze antimycin-A- and HOQNO-insensitive cyclic photophosphorylation (Nozaki *et al.*, 1961; Jones, C. W., and Vernon, 1969; Gromet-Elhanan, 1969; Trebst *et al.*, 1967; Klemme, 1969). Thus, in the presence of these nonphysiological cofactors, ATP (or a high-energy precursor) generated by an artificial antimycin-A- and HOQNO-insensitive cyclic pathway would still be available to drive energy-dependent reverse-electron transfer. [A possible exception could be *Rp. capsulata* (Klemme, 1969), in which no antimycin-A-insensitive phosphorylation could be demonstrated in the presence of DPIP.] The rates of this DPIP-mediated cyclic phosphorylation could be rapid enough to permit the simultaneous production of ATP (from ADP plus phosphate) and reduction of NAD^+, as observed by Nozaki *et al.* (1961).

Convincing evidence does exist for the conclusion that NAD^+ photoreduction with reduced DPIP or TMPD as the electron donor proceeds through the same energy-dependent pathway as does succinate-linked NAD^+ reduction. Regardless of which electron donor is used, NAD^+ photoreduction is rotenone-sensitive (Keister and Minton, 1969; Klemme, 1969), and is inhibited by the addition of ADP plus phosphate (Jones, C. W., and Vernon, 1969; Gromet-Elhanan, 1969). Furthermore, NAD^+ photoreduction is inhibited by the addition of uncouplers, regardless of the electron donor (Hinkson, 1965; Keister and Minton, 1969; Jones, C. W., and Vernon, 1969; Klemme, 1969; Gromet-Elhanan, 1970). This observation seems more important than some of the differences observed in the uncoupler concentrations required for inhibition when different donors were used (Hinkson, 1965; Klemme, 1969). Because uncouplers, at low concentrations, act only to dissipate high-energy precursors of ATP and have no direct effect on electron transfer, it seems reasonable to conclude that NAD^+ photoreduction in the Rhodospirillaceae does not proceed by the direct generation in the light of a reductant more electronegative than NAD^+, but rather proceeds via energy-dependent reverse-electron transfer.

4.2. Green Sulfur Bacteria

4.2.1. Occurrence and Enzyme Requirements

Cell-free preparations from the Chlorobiaceae have been known for some time to be capable of the photoreduction of low-potential acceptors. The first demonstration of this capability was the photoreduction of ferredoxin (coupled to the reductive carboxylations of acetyl CoA and succinyl CoA) by particles from *Chl. limicola f. thiosulfatophilum* (Buchanan and Evans, 1965; Evans, M. C. W., and Buchanan, 1965). Both the native ferredoxin and the ferredoxin from *Chr. vinosum* were photoreduced. Because the ferredoxin from *Chr. vinosum* is known to have a midpoint oxidation–reduction potential of −490 mV (Ke *et al.*,

1974), its photoreduction was consistent with the idea that this green bacterium was capable of generating an extremely electronegative reductant in the light. On the basis of the evidence available at the time, however, a reverse electron-transport mechanism could not be ruled out (but see below).

It was subsequently demonstrated that the photoreduction of ferredoxin by the particles from *Chl. limicola f. thiosulfatophilum* could be coupled to NAD^+ reduction in the presence of a soluble protein that was replaceable by the flavoprotein, ferredoxin:pyridine nucleotide reductase from spinach (Buchanan and Evans, 1969). [This NAD^+ photoreduction was absolutely dependent on the presence of ferredoxin and the reductase (Buchanan and Evans, 1969; Evans, M. C. W., 1969; Knaff and Buchanan, 1975).] A similar requirement for soluble ferredoxin and an additional soluble enzyme had previously been reported for the dark reduction of NAD^+ by H_2 with extracts from the same bacterium (Weaver *et al.*, 1965). Photoreduction of pyridine nucleotide and pyridine nucleotide reduction by H_2 in the dark were approximately twice as fast with NAD^+ as the acceptor as with $NADP^+$ as the acceptor. Isolation and purification of the ferredoxin:pyridine nucleotide reductase (Kusai and Yamanaka, 1973a) from *Chl. limicola f. thiosulfatophilum* showed that the enzyme was a flavoprotein with FAD as the prosthetic group. The reductase enzyme appears to be able to reduce both pyridine nucleotides, with the relative rates of reduction depending on the nucleotide concentration (Kusai and Yamanaka, 1973a).

Several compounds could serve as electron donors for NAD^+ photoreduction by particles from *Chl. limicola f. thiosulfatophilum*, including 2-mercaptoethanol, sodium sulfide, succinate, and reduced DPIP (Buchanan and Evans, 1969; Evans, M. C. W., 1969; Knaff and Buchanan, 1975). Sulfide and mercaptoethanol were two to four times as effective as the other donors (Evans, M. C. W., 1969). The rates of NAD^+ photoreduction with the most effective donors are low in comparison with those obtained with chromatophores from Rhodospirillaceae species, but are comparable to the rates of CO_2 fixation by whole cells of *Chl. limicola f. thiosulfatophilum* (Buchanan and Evans, 1969). Cells from this bacterium are known to use sulfide as an electron donor for the generation of reducing power during growth, and so the NAD^+ photoreduction with sulfide as the electron donor presumably reflects a physiological process. The mechanism of sulfide oxidation was investigated by Kusai and Yamanaka (1973b), who found that a protein containing both flavin and heme *c* functioned as a sulfide:cytochrome *c* reductase. Thiosulfate, the

other natural electron donor utilized by the cells, does not function in the cell-free system. A possible reason for this may be the solubility of the thiosulfate:cytochrome *c* reductase (Kusai and Yamanaka, 1973b), the enzyme that transfers electrons from thiosulfate to the membrane-bound electron carriers.

As may be seen from the data presented above, pyridine nucleotide photoreduction by particles isolated from a Chlorobiaceae species differs from that observed with chromatophores isolated from the Rhodospirillaceae species and resembles plant photosystem I in the requirement of the addition of a low-potential ferredoxin and a flavoprotein ferredoxin–pyridine nucleotide reductase. NAD^+ photoreduction in *Chl. limicola f. thiosulfatophilum*, unlike the reaction in the Rhodospirillaceae, does not appear to involve an enzyme of the mitochondrial $NADH^+$ dehydrogenase type, as indicated by the observation (see Table 3) that NAD^+ photoreduction is not inhibited by either rotenone or amytal (Evans, M. C. W., 1969). NAD^+ photoreduction with the nonphysiological electron donor 2-mercaptoethanol is not inhibited by antimycin A or HOQNO (Evans, M. C. W., 1969; Knaff and Buchanan, 1975), but NAD^+ photoreduction with sulfide as the donor is inhibited (Knaff and Buchanan, 1975). This pattern of inhibition has been interpreted in terms of an electron-transfer step from a *b*-type cytochrome to a *c*-type cytochrome in the sulfide to NAD^+ pathway, but not in the 2-mercaptoethanol to NAD^+ pathway (Knaff and Buchanan, 1975).

4.2.2. Evidence for Direct Photoreduction

NAD^+ photoreduction by particles from *Chl. limicola f. thiosulfatophilum* differs from that catalyzed by chromatophores from the purple bacteria in that it is unaffected by high concentrations of

Table 3. Effect of Inhibitors on NAD^+ Photoreduction by Particles from *Chlorobium limicola*[a]

Inhibitor	NAD^+ reduced (μmol)
None	1.64
Antimycin A (40 μg/ml)	1.62
Dicoumarol (5×10^{-4} M)	1.22
Rotenone (10^{-5} M)	1.60
Amytal (10^{-3} M)	1.64

[a] From M. C. W. Evans (1969). The electron donor was 2-mercaptoethanol.

dicoumarol (see Table 3) (Evans, M. C. W., 1969), gramicin D, desaspidin, or carbonylcyanide *m*-chlorophenylhydrazone (CCCP) (Knaff and Buchanan, 1975), agents that act as uncouplers in other species. It has not been possible to test the efficacy of these agents as uncouplers directly because particles isolated from *Chl. limicola f. thiosulfatophilum* do not catalyze photophosphorylation *in vitro* (Buchanan and Evans, 1969). The uncoupler-insensitivity of NAD⁺ photoreduction in the Chlorobiaceae suggests that neither ATP nor a high-energy ATP precursor is required for NAD⁺ reduction, and that NAD⁺ photoreduction does not proceed via energy-dependent reverse-electron transport. The uncoupler data, taken in conjunction with the evidence for a primary electron acceptor in *Chl. limicola f. thiosulfatophilum* that has a midpoint potential equal to −540 mV (see Section 3.2), are consistent with a pathway of NAD⁺ reduction in the Chlorobiaceae that produces a strong reductant in the light followed by energetically favorable reduction of a soluble ferredoxin and subsequent reduction of NAD⁺ in a reaction catalyzed by a flavoprotein ferredoxin : NAD⁺ reductase.

5. Conclusion

Pyridine nucleotide photoreduction by photosynthetic bacteria proceeds by two fundamentally diffe-

rent mechanisms (see Fig. 2). In the purple bacteria (Chromatiaceae and Rhodospirillaceae), the first reduced species formed in the light that is stable for longer than 10 nsec is considerably more electropositive than is the NAD⁺/NADH couple. The role of light in NAD photoreduction by these organisms appears to be confined to producing a high-energy precursor of ATP through coupling to light-driven cyclic electron flow. This high-energy precursor is used to power thermodynamically unfavorable electron flow from a donor such as succinate to NAD⁺. The evidence for such an energy-dependent, reverse-electron-flow mechanism in the purple bacteria was obtained largely in studies with cell-free systems. The observation that uncouplers and inhibitors of cyclic electron flow completely inhibit NAD⁺ photoreduction in intact cells (Jackson and Crofts, 1968) suggests, however, that NAD⁺ photoreduction *in vivo* proceeds via an energy-dependent pathway. The molecular mechanism for the reverse-electron transfer is not known.

In the Chlorobiaceae, light produces directly a strong reductant (possibly a membrane-bound iron–sulfur protein with E_m −540 mV). NAD⁺ reduction (mediated by soluble ferredoxin and flavoprotein intermediates) is thermodynamically favorable and requires no additional energy input from ATP. The many resemblances between the mechanisms of pyridine

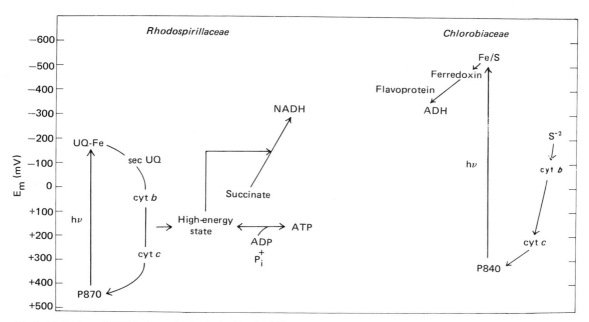

Fig. 2. Mechanism of NAD⁺ photoreduction in the Rhodospirillaceae and the Chlorobiaceae. P870 and P840 are the reaction center bacteriochlorophylls. UQ · Fe is the ubiquinone–iron complex thought to function as the primary electron acceptor in purple bacteria; Fe/S is the iron–sulfur protein at the primary electron acceptor site of green bacteria.

nucleotide reduction in the Chlorobiaceae and in plant photosystem I suggest the possibility that the primary acceptor site of plant photosystem I may have evolved from that in the Chlorobiaceae.

6. References

Amesz, J., 1963, Kinetics, quantum requirement and action spectrum of light-induced pyridine nucleotide reduction in *R. rubrum* and *Rps. sphaeroides*, *Biochim. Biophys. Acta* **66**:22–36.

Arnon, D. I., 1967, Photosynthetic activity of isolated chloroplasts, *Physiol. Rev.* **47**:317–358.

Bearden, A. J., and Malkin, R., 1972a, The bound ferredoxin of chloroplasts: A role as the primary electron acceptor of Photosystem I, *Biochem. Biophys. Res. Commun.* **46**:1299–1305.

Bearden, A. J., and Malkin, R., 1972b, Quantitative EPR studies of the primary reactions of Photosystem I in chloroplasts, *Biochim. Biophys. Acta* **283**:456–468.

Bose, S. K., and Gest, H., 1962, Electron transport systems in purple bacteria, *Nature (London)* **195**:1168–1171.

Buchanan, B. B., and Bachofen, R., 1968, Ferredoxin-dependent reduction of NAD(P) with hydrogen gas by subcellular preparations from the photosynthetic bacterium, *Chromatium*, *Biochim. Biophys. Acta* **162**:607–610.

Buchanan, B. B., and Evans, M. C. W., 1965, The synthesis of α-ketoglutarate from succinate and CO_2 by a subcellular preparation of a photosynthetic bacterium, *Proc. Natl. Acad. Sci. U.S.A.* **54**:1212–1218.

Buchanan, B. B., and Evans, M. C. W., 1969, Photoreduction of ferredoxin and its use in NAD(P)$^+$ reduction by a subcellular preparation from the photosynthetic bacterium *Chlorobium thiosulfatophilum*, *Biochim. Biophys. Acta* **180**:123–129.

Buchanan, B. B., Bachofen, R., and Arnon, D. I., 1964, Role of ferredoxin in the reductive assimilation of CO_2 and acetate by extracts of the photosynthetic bacterium, *Chromatium*, *Proc. Natl. Acad. Sci. U.S.A.* **52**:839–847.

Case, G. D., and Parson, W. W., 1971, Thermodynamics of the primary and secondary photochemical reactions in *Chromatium*, *Biochim. Biophys. Acta* **253**:187–202.

Chance, B. C., and Hollunger, G., 1960, Energy-linked reduction of mitochondrial pyridine nucleotides, *Nature (London)* **185**:666–672.

Clark, W. M., 1960, *Oxidation–Reduction Potentials of Organic Systems*, Williams and Wilkins, Baltimore.

Clayton, R. K., 1972, Physical mechanisms in photosynthesis: Past elucidations and current problems, *Proc. Natl. Acad. Sci. U.S.A.* **69**:44–49.

Cogdell, R. J., and Crofts, A. R., 1972, Some observations on the primary acceptor of *Rps. viridis*, *FEBS Lett.* **27**:176–178.

Cramer, W. A., 1969, Low potential titration of the fluorescence yield changes in photosynthetic bacteria, *Biochim. Biophys. Acta* **189**:54–59.

Cusanovich, M. A., Bartsch, R. G., and Kamen, M. D., 1968, Light-induced electron transport in *Chromatium* Strain D. II. Light-induced absorbance changes in *Chromatium* chromatophores, *Biochim. Biophys. Acta* **153**:397–417.

Dutton, P. L., 1971, Oxidation–reduction potential dependence of the interaction of cytochromes, bacteriochlorophyll and carotenoids at 77°K in chromatophores of *Chromatium* D and *Rhodopseudomonas gelatinosa*, *Biochim. Biophys. Acta* **226**:63–80.

Dutton, P. L., and Leigh, J. S., 1971, ESR characterization of *Chromatium* D hemes, nonheme irons, and the components involved in primary photochemistry, *Biochim. Biophys. Acta* **314**:178–190.

Dutton, P. L., and Wilson, D. F., 1974, Redox potentiometry in mitochondrial and photosynthetic bioenergetics, *Biochim. Biophys. Acta* **346**:165–212.

Dutton, P. L., Leigh, J. S., and Seibert, M., 1972, Primary processes in photosynthesis: *In situ* ESR studies of light induced oxidized and triplet state of reaction center bacteriochlorophyll, *Biochim. Biophys. Res. Commun.* **46**:406–413.

Dutton, P. L., Leigh, J. S., and Wraight, C. A., 1973, Direct measurement of the midpoint potential of the primary electron acceptor in *Rhodopseudomonas sphaeroides in situ* and in the isolated state: Some relationships with pH and *o*-phenanthroline, *FEBS Lett.* **36**:169–173.

Dutton, P. L., Kaufmann, K. J., Chance, B., and Rentzepis, P. M., 1975, Picosecond kinetics of the 1250 nm band of the *Rhodopseudomonas sphaeroides* reaction center: The nature of the primary photochemical intermediary state, *FEBS Lett.* **60**:275–280.

Evans, E. H., and Crofts, A. R., 1974, *In situ* characterisation of photosynthetic electron transport in *Rps. capsulata*, *Biochim. Biophys. Acta* **357**:189–202.

Evans, M. C. W., 1969, Ferredoxin:NAD reductase and the photoreduction of NAD by *Chlorobium thiosulfatophilum*, in: *Progress in Photosynthesis Research* (H. Metzner, ed.), Vol. 3, pp. 1474–1475, Laupp, Tübingen.

Evans, M. C. W., and Buchanan, B. B., 1965, Photoreduction of ferredoxin and its use in CO_2 fixation by a subcellular system from a photosynthetic bacterium, *Proc. Natl. Acad. Sci. U.S.A.* **53**:1420–1425.

Evans, M. C. W., Lord, A. V., and Reeves, S. G., 1974a, The detection and characterization by EPR spectroscopy of iron–sulfur proteins and other electron-transport components in chromatophores from the purple bacterium *Chromatium*, *Biochem. J.* **138**:177–183.

Evans, M. C. W., Reeves, S. G., and Cammack, R., 1974b, Determination of the oxidation–reduction potential of the bound iron–sulfur proteins of the primary electron acceptor complex of Photosystem I in spinach chloroplasts, *FEBS Lett.* **49**:111–114.

Frenkel, A. W., 1958, Simultaneous reduction of DPN and oxidation of reduced FMN by illuminated bacterial chromatophores, *J. Am. Chem. Soc.* **80**:3479–3480.

Govindjee, R., Smith, W. R., and Govindjee, 1974, Interaction of viologen dyes with chromatophores and reaction center preparations from *Rhodospirillum rubrum*, *Photochem. Photobiol.* **20**:191–199.

Gromet-Elhanan, Z., 1969, Inhibitors of photophosphorylation and photoreduction by chromatophores from *Rhodospirillum rubrum*, *Arch. Biochem. Biophys.* **131**:299–305.

Gromet-Elhanan, Z., 1970, Differences in sensitivity to valinomycin and nonaction of various photophosphorylating and photoreducing systems of *Rhodospirillum rubrum* chromatophores, *Biochim. Biophys. Acta* **223**:174–182.

Hinkson, J. W., 1965, NAD photoreduction with *Chromatium* and *Rhodospirillum rubrum*, *Arch. Biochem. Biophys.* **112**:478–487.

Hood, S. L., 1964, Photoreduction of NAD by a cell-free system from *Chromatium*, *Biochim. Biophys. Acta* **88**:461–465.

Horgan, D. J., Singer, T. P., and Casida, J. E., 1968, Studies on the respiratory chain-linked NADH dehydrogenase. XIII. Binding sites of rotenone, piericidin A and amytal in the respiratory chain, *J. Biol. Chem.* **243**:834–843.

Jackson, J. B., and Crofts, A. R., 1968, Energy-linked reduction of nicotinamide adenine dinucleotides in cells of *Rhodospirillum rubrum*, *Biochem. Biophys. Res. Commun.* **32**:908–915.

Jackson, J. B., Cogdell, R. J., and Crofts, A. R., 1973, Some effects of *o*-phenanthroline on electron transport in chromatophores from photosynthetic bacteria, *Biochem. Biophys. Acta* **292**:218–275.

Jennings, V. V., and Evans, M. C. W., 1977, The irreversible photoreduction of a low potential component at low temperatures in a preparation of the green photosynthetic bacterium *Chlorobium thiosulphatophilum*, *FEBS Lett.* **75**:33–36.

Jones, C. W., and Vernon, L. P., 1969, NAD photoreduction in *Rhodospirillum rubrum* chromatophores, *Biochim. Biophys. Acta* **180**:149–164.

Jones, O. T. G., and Saunders, V. A., 1972, Energy-linked electron transfer reactions in *Rps. viridis*, *Biochim. Biophys. Acta* **275**:427–436.

Kaufmann, K. J., Dutton, P. L., Netzel, T. L., Leigh, J. S., and Rentzepis, P. M., 1975, Picosecond kinetics of events leading to reaction center bacteriochlorophyll oxidation, *Science* **188**:1301–1304.

Ke, B., 1974, Some comments on the present status of the primary electron acceptor of Photosystem I, in: *Proceedings of the Third International Congress on Photosynthesis* (M. Avron, ed.), pp. 373–382, Elsevier, Amsterdam.

Ke, B., Hansen, R. E., and Beinert, H., 1973, Oxidation–reduction potentials of bound iron–sulfur proteins of Photosystem I, *Proc. Natl. Acad. Sci. U.S.A.* **70**:2941–2945.

Ke, B., Bulen, W. A. Shaw, E. R., and Breeze, R. H., 1974, Determination of oxidation–reduction potentials by spectropolarimetric titration: Application to several iron–sulfur proteins, *Arch. Biochem. Biophys.* **162**:304–309.

Keister, D. L., and Minton, N. J., 1969, Energy-linked reactions in photosynthetic bacteria. III. Further studies on energy-linked NAD reduction by *Rhodospirillum rubrum* chromatophores, *Biochemistry* **8**:167–173.

Keister, D. L., and Yike, N. J., 1967, Energy-linked reactions in photosynthetic bacteria. I. Succinate-linked ATP-driven NAD⁺ reduction by *Rhodospirillum rubrum* chromatophores, *Arch. Biochem. Biophys.* **121**:415–422.

Klemme, J. H., 1969, Studies on the mechanism of NAD-photoreduction by chromatophores of the facultative phototroph, *Rhodopseudomonas capsulata*, *Z. Naturforsch.* **24b**:67–76.

Klemme, J. H., and Schlegel, H. G., 1967, Photoreduktion von Pyridinnucleotid durch Chromatophoren aus *Rhodopseudomonas capsulata* mit molekularen Wasserstoff, *Arch. Mikrobiol.* **59**:185–196.

Knaff, D. B., and Buchanan, B. B., 1975, Cytochrome *b* and photosynthetic sulfur bacteria, *Biochim. Biophys. Acta* **376**:549–560.

Knaff, D. B., and Malkin, R., 1976, Iron–sulfur proteins of the green photosynthetic bacterium *Chlorobium*, *Biochim. Biophys. Acta* **430**:244–252.

Knaff, D. B., Buchanan, B. B., and Malkin, R., 1973, Effect of oxidation–reduction potential on light-induced cytochrome and bacteriochlorophyll reactions in chromatophores from the photosynthetic green bacterium *Chlorobium*, *Biochim. Biophys. Acta* **325**:94–101.

Krogmann, D. W., Jagendorf, A. T., and Avron, M., 1959, Uncouplers of spinach chloroplast photosynthetic phosphorylation, *Plant Physiol.* **34**:272–277.

Kuntz, I. D., Loach, P. A., and Calvin, M., 1964, Absorbance changes in bacterial chromatophores, *Biophys. J.* **4**:227–244.

Kusai, A., and Yamanaka, T., 1973a, An NAD(P) reductase derived from *Chlorobium thiosulfatophilum*: Purification and some properties, *Biochim. Biophys. Acta* **292**:621–633.

Kusai, A., and Yamanaka, T., 1973b, The oxidation mechanisms of thiosulfate and sulfide in *Chlorobium thiosulfatophilum*. Roles of cytochrome *c*-551 and cytochrome *c*-553, *Biochim. Biophys. Acta* **325**:304–314.

Loach, P. A., 1966, Primary oxidation–reduction changes during photosynthesis in *Rhodospirillum rubrum*, *Biochemistry* **5**:592–600.

Loach, P. A., Kung, M., and Hales, B. J., 1975, Characterization of the phototrap in photosynthetic bacteria, *Ann. N. Y. Acad. Sci.* **244**:297–319.

Lozier, R. H., and Butler, W. L., 1974, Redox titration of the primary electron acceptor of Photosystem I in spinach chloroplasts, *Biochim. Biophys. Acta* **333**:460–464.

Nishimura, M., 1963, Studies on the electron-transfer systems in photosynthetic bacteria. II. The effect of HOQNO and antimycin A on the photosynthetic and respiratory electron-transfer systems, *Biochim. Biophys. Acta* **66**:17–21.

Nozaki, M., Tagawa, K., and Arnon, D. I., 1961, Noncyclic photophosphorylation in photosynthetic bacteria, *Proc. Natl. Acad. Sci. U.S.A.* **47**:1334–1340.

Ohnishi, T., 1974, Mechanism of electron transport and energy conservation in the Site I region of the respiratory chain, *Biochim. Biophys. Acta* **301**:105–128.

Olson, J. M., and Amesz, J., 1960, Action spectrum for fluorescence excitation of pyridine nucleotide in photosynthetic bacteria and algae, *Biochim. Biophys. Acta* **37**:14–24.

Olson, J. M., Prince, R. C., and Brune, D. C., 1976, Reaction center complexes from green bacteria, *Brookhaven Symp. Biol.* **28**:238–246.

Olson, J. M., Duysens, L. N. M., and Kronenberg, O. H., 1959, Spectrofluorometry of pyridine nucleotide reactions in *Chromatium*, *Biochim. Biophys. Acta* **36**:125–131.

Orlando, J. A., Sabo, D., and Curnyn, C., 1966, Photoreduction of pyridine nucleotides by subcellular preparations from *Rhodopseudomonas sphaeroides*, *Plant Physiol.* **41**:937–948.

Parson, W. W., and Cogdell, R. J., 1975, The primary photochemical reaction of bacterial photosynthesis, *Biochim. Biophys. Acta* **416**:105–149.

Parson, W. W., Clayton, R. K., and Cogdell, R. J., 1975, Excited states of photosynthetic reaction centers at low redox potentials, *Biochim. Biophys. Acta* **387**:265–278.

Prince, R. C., and Dutton, P. L., 1976, The primary acceptor of bacterial photosynthesis: Its operating midpoint potential, *Arch. Biochem. Biophys.* **172**:329–334.

Prince, R. C., and Olson, J. M., 1976, Some thermodynamic and kinetic properties of the primary photochemical reactants in a complex from a green photosynthetic bacterium, *Biochim. Biophys. Acta* **423**:357–362.

Prince, R. C., Leigh, J. S., and Dutton, P. L., 1974, An ESR characterization of *Rhodopseudomonas capsulata*, *Biochem. Soc. Trans.* **2**:950–953.

Prince, R. C., Lindsay, J. G., and Dutton, P. L., 1975, The Rieske iron–sulfur center in mitochondrial and photosynthetic systems: E_m/pH relationships, *FEBS Lett.* **5**:108–111.

Prince, R. C., Leigh, J. S., and Dutton, P. L., 1976, Thermodynamic properties of the reaction center of *Rhodopseudomonas viridis*, *Biochim. Biophys. Acta* **440**:662–636.

Reed, D. W., Zankel, K. L., and Clayton, R. K., 1969, The effect of redox potential on P870 fluorescence in reaction centers from *Rhodopseudomonas sphaeroides*, *Proc. Natl. Acad. Sci. U.S.A.* **63**:42–46.

Rockley, M. G., Windsor, M. W., Cogdell, R. J., and Parson, W. W., 1975, Picosecond detection of an intermediate in the photochemical reaction of bacterial photosynthesis, *Proc. Natl. Acad. Sci. U.S.A.* **72**:2251–2255.

Silberstein, B. R., and Gromet-Elhanan, Z., 1974, P430, a possible primary electron acceptor in *Rhodospirillum rubrum*, *FEBS Lett.* **42**:141–144.

Trebst, A., Pistorius, E., and Baltscheffsky, H., 1967, *p*-Phenylenediamines as electron donors for photosynthetic pyridine nucleotide reduction in chromatophores from *Rhodospirillum rubrum*, *Biochim. Biophys. Acta* **143**:257–260.

Vernon, L. P., and Ash, O. K., 1959, The photoreduction of pyridine nucleotides by illuminated chromatophores of *Rhodospirillum rubrum* in the presence of succinate, *J. Biol. Chem.* **234**:1878–1882.

Weaver, P., Tinker, K., and Valentine, R. C., 1965, Ferredoxin-linked DPN reduction, by the photosynthetic bacteria *Chromatium* and *Chlorobium*, *Biochem. Biophys. Res. Commun.* **21**:195–201.

CHAPTER 33

Oxygen-Linked Electron Transport and Energy Conservation

Lucile Smith and Patricia B. Pinder

1. Dark Aerobic Electron-Transport System

It has been known for many years that some species of photosynthetic bacteria are versatile in that they can obtain energy from oxidation–reduction reactions that are either linked to the oxidation of substrates by O_2 or initiated on illumination. Some of these facultative bacteria are capable of high rates of aerobic dark growth, and in the presence of sufficient O_2, the photosynthetic pigments, bacteriochlorophyll (Bchl) and carotenoids, are not synthesized. Bacteria grown aerobically in the dark and those grown anaerobically in the light are capable of respiration in the dark with a number of substrates. For example, the respiration of dark-grown *Rhodospirillum rubrum* is stimulated on addition of succinate, fumarate, lactate, citrate, or glucose (Taniguchi and Kamen, 1965).

Measurements of absorption spectrum changes showed that cytochromes were rapidly oxidized on addition of O_2 to an anaerobic suspension of intact cells of light-grown *Rs. rubrum*; then rapid reduction followed exhaustion of O_2 by cellular respiration (Chance and Smith, 1955; Smith and Ramirez, 1959). Cytochromes were also oxidized on illumination, but they differed qualitatively and quantitatively from those involved in reactions with O_2. Spectrophotometric measurements with intact dark-grown cells of

Rs. rubrum also showed evidence of oxidation and reduction of cytochromes (Taniguchi and Kamen, 1965).

Most of the investigations of the respiratory-chain system of Rhodospirillaceae have been made with broken-cell suspensions or with insoluble membranes isolated from these extracts. The following sections will describe these studies. Unfortunately for this chapter, the dark respiratory-chain–phosphorylating systems have received relatively sparse attention in comparison with the large amount of research into the photosynthetic apparatus.

As a basis for discussion, the following brief scheme is the one usually cited as typical of the respiratory-chain–phosphorylating system of both eukaryotic cells and many species of bacteria:

$$\text{Substrates} \rightarrow \begin{array}{c} \text{Individual} \\ \text{flavoprotein} \\ \text{dehydrogenases} \\ \text{(Fe–S)} \end{array} \rightarrow \text{Quinones} \rightarrow \begin{array}{c} \text{Cytochrome} \\ \text{chain} \\ \text{including} \\ \text{oxidase(s)} \\ \text{(Fe–S)} \end{array} \rightarrow O_2$$

$$\text{2–3 ATP}$$

1.1 Cellular Localization of the Electron-Transport Chains

Respiratory-chain activity is associated with the mixture of membrane fragments obtained on centrifugation of broken-cell extracts of the photosynthetic

Lucile Smith and Patricia B. Pinder · Departments of Biochemistry and Microbiology, Dartmouth Medical School, Hanover, New Hampshire 03755

bacteria. Sometimes a problem in assessing studies on structure and function is a lack of knowledge of the origin of the membrane fractions isolated.

The only membrane evident in electron micrographs of Rhodospirillaceae grown in plentiful O_2 is the cytoplasmic membrane underlying the cell wall; this is a source of fragments obtained from ruptured cells. Bacteria grown anaerobically or in low O_2 concentration synthesize the cytoplasmic membrane plus internal pigmented membranes, which may originate by extension and modification of the cytoplasmic membrane (Oelze and Drews, 1972; Huang and Kaplan, 1973a); broken-cell extracts of these bacteria contain membrane fragments derived from both. An important question here is: do functionally different membranes develop? Some success has been reported in the separation of two kinds of membranes from anaerobically light-grown cells, which appear to correspond to the cytoplasmic membrane (with and without wall attached) and to the intracytoplasmic membranes (e.g., Oelze and Drews, 1972; Huang and Kaplan, 1973a; Simon and Siekevitz, 1973), but these have not been well characterized in all reports. The two kinds of membranes are reported to have different lipid and protein subunit content, as well as different enzymatic activities (Oelze and Drews, 1972; Huang and Kaplan, 1973a,b), although they may share some similar protein subunits (Oelze and Drews, 1969, 1970). Gel electrophoresis and immunological studies showed the envelope* membranes from cells grown anaerobically in the light to be similar to those from aerobically dark-grown cells (Huang and Kaplan, 1973a,b).

The pigmented membrane fraction easily isolated from anaerobically grown bacteria is referred to as *chromatophores*. All agree that it contains fragments of intracytoplasmic membranes, which bear the systems for light-induced electron transport and ATP synthesis. The fragments derived primarily from the cytoplasmic membrane contain cytochromes and may be enriched in dark respiratory-chain activity (Throm *et al.*, 1970; Niederman *et al.*, 1972; Niederman, 1974), although it is puzzling that the succinate dehydrogenase seems to be concentrated in the chromatophore fraction of *Rs. rubrum* (Simon and Siekevitz, 1973), but in the cytoplasmic membrane fraction of *Rhodopseudomonas sphaeroides* (Niederman, 1974).

---

* Huang and Kaplan define "cell envelope" as cell wall and cell membrane. Fragments of the external membrane surrounding the cell wall may well be present in "envelope" preparations, but in most contexts this is unimportant, because the external membrane does not contain enzymes or coenzymes. Such contamination could be significant, however, when specific proteins are being characterized, as in immunological experiments.

A problem that arises in experiments with membrane fractions from anaerobically light-grown bacteria is that under some conditions (1 mM or higher Mg^{2+} ions), the chromatophores can reversibly aggregate with the cytoplasmic membrane fragments; thus, the fractions obtained on centrifugation would have the properties of both membranes (Niederman, 1974). The fractionation of membrane fragments from broken-cell extracts has been found to have a critical dependence on the experimental conditions, particularly the ionic content of the suspending medium (Oelze *et al.*, 1969; Ketchum and Holt, 1970). Also, the resealed vesicles of cytoplasmic membrane may entrap some chromatophores (Simon and Siekevitz, 1973). All membrane fragments of sufficient size form vesicles, some inside-out and some right-side out in relation to the orientation in the intact cell.

It will be apparent throughout this chapter that a better understanding of the localization of all electron-transport–phosphorylating systems would help in yielding answers to such questions as: (1) To what extent, if any, do the O_2-linked and the light-induced systems overlap or cross-react? (2) Does the photosynthetic system catalyze only cyclic photophosphorylation, or is a noncyclic system also involved? (3) In relation to this question, do the membrane-bound NADH and succinate dehydrogenases interact with the light-induced system?

1.2. Activities Observed in Cell-Free Preparations

1.2.1. Respiratory Activities

Membrane fragments isolated from dark-grown photosynthetic bacteria have been shown to catalyze the oxidation of NADH and succinate with molecular oxygen, and partial reactions of these oxidation sequences. Table 1 summarizes these enzymatic activities that have been detected in membranes of *Rp. capsulata* (Baccarini-Melandri *et al.*, 1973; Klemme and Schlegel, 1969), *Rs. rubrum* (Thore *et al.*, 1969; Taniguchi and Kamen, 1965; Geller, 1963), *Rp. sphaeroides* (Whale and Jones, 1970; Connelly *et al.*, 1973; Motokawa and Kikuchi, 1966), and *Rp. palustris* (King and Drews, 1973). In addition, membrane preparations have been shown to catalyze transhydrogenase (Geller, 1962), lactate dehydrogenase (Klemme and Schlegel, 1969), and aldose dehydrogenase (Niederpruem and Doudoroff, 1965).

Similar studies have been made using particles prepared from light-grown *Rp. capsulata* (Klemme and Schlegel, 1969; Marrs *et al.*, 1972), *Rs. rubrum* (Jones, C. W., and Vernon, 1969; Throm *et al.*, 1970),

Table 1. Respiratory Activities Observed in Membrane Particles of Dark-Grown Photosynthetic Bacteria

	Species			
Activity[a]	*Rp. capsulata*	*Rs. rubrum*	*Rp. sphaeroides*	*Rp. palustris*
NADH oxidase	Yes	Yes	Yes	Yes
NADH-cytochrome *c* reductase		Yes	Yes	Yes
NADH-fumarate reductase	Yes	Yes		
NADH-dehydrogenase	Yes	Yes		Yes
NADH-UQ reductase	Yes			
Succinic oxidase	Yes	Yes	Yes	Yes
Succinic-cytochrome *c* reductase	Yes	Yes	Yes	Yes
Succinic dehydrogenase	Yes	Yes		Yes
Cytochrome *c* oxidase	Yes	Yes	Yes	Yes

[a] *Oxidase* refers to the complete enzyme system coupled to the reduction of oxygen; *dehydrogenase* specifies the single enzyme that removes electrons from the designated substrate.

and *Rp. sphaeroides* (Connelly *et al.*, 1973). Although these preparations include cellular components in addition to those found in the cell membranes of dark-grown cells, the respiratory activities are, in general, the same ones present in dark-grown cells. When quantitative comparisons have been made, only NADH oxidase activity was found to be significantly higher on a protein basis in particles from the dark aerobically grown cells (Thore *et al.*, 1969; Yamashita and Kamen, 1969; Oelze and Drews, 1970).

1.2.2. Reversed Electron Transport

It is clear from the array of activities that has been shown to be catalyzed by membranes isolated from Rhodospirillaceae that the respiratory system is similar to that of mitochondria. The bacterial system also resembles mitochondria in that certain oxidation–reduction spans within the complex can be reversed if energy is supplied. For example, the reduction of NADH by succinate is catalyzed by "chromatophore" preparations of *Rs. rubrum* if light or ATP is supplied. The properties of this reaction, including the sensitivity to electron-transport inhibitors and uncoupling agents (Jones, C. W., and Vernon, 1969; Keister and Yike, 1967a; Keister and Minton, 1969a), are consistent with the assumption that in the dark, ATP (or an alternate high-energy substance) gives rise to a high-energy intermediate or state that provides the energy necessary for the reduction of NAD. The stoichiometry obtained experimentally was 1.8 ATP utilized per NAD reduced (Jones, C. W., and Vernon, 1969). Compounds known to inhibit NADH dehydro-

genase, such as rotenone or amytal (Keister and Minton, 1969a; Keister and Yike, 1967a), and uncouplers of phosphorylation, especially carbonyl cyanide *m*-chlorophenylhydrazone (CCCP) (Jones, C. W., and Vernon, 1969), desaspidin, gramicidin, and dicoumarol (Keister and Yike, 1967a), and the energy-transport inhibitor oligomycin (Keister and Yike, 1967a) all inhibit the ATP-requiring NADH formation.

A similar dark ATP-driven reduction of NAD at the expense of succinate was demonstrated in *Rp. capsulata* by Klemme (1969b). This system, too, is inhibited by rotenone, CCCP, 2,4-dinitrophenol (DNP), and oligomycin (Klemme, 1969a,b).

Presumably, when light is used as the energy source for the reaction, the same high-energy intermediate or state is formed by light-induced cyclic electron transport as is formed from ATP in the dark. The rate of photoreduction of NAD by succinate was found to be faster than the rate of the ATP-driven reaction in the same preparation (Jones, C. W., and Vernon, 1969; Klemme, 1969b; Keister and Yike, 1967a; Keister and Minton, 1969a,b).

ATP-driven NAD reduction was also reported in particles from *Rs. rubrum* grown aerobically in the dark (Keister and Minton, 1969b), giving evidence that reversed electron transport can be catalyzed by the dark respiratory-chain system.

1.2.3. Transhydrogenase

Although "chromatophore" preparations from light-grown *Rs. rubrum* catalyze a slow reduction of

NADP by NADH in the absence of an exogenous energy source (Nozaki *et al.*, 1963), this reaction was shown to be stimulated severalfold when energy is supplied as ATP, PP$_i$, or light (Keister and Yike, 1966, 1967*b*; Keister and Minton, 1969*a*). The effect of several uncouplers and inhibitors led to the conclusion that energy for the transhydrogenase reaction could be provided by a high-energy intermediate or state derived from ATP with a stoichiometry of approximately one ATP utilized per NADPH formed. The chromatophore system was resolved into a soluble factor and a membrane fraction, both required for maximal transhydrogenase activity (Fisher and Guillory, 1971*a*). The reaction was found to be a direct H transfer with the same stereospecificity as that of the mitochondrial reaction (Fisher and Guillory, 1971*b*). Transhydrogenase activity has also been demonstrated in "chromatophore" preparations of *Rp. sphaeroides* with properties similar to the activity in *Rs. rubrum*, except that the soluble factor apparently does not participate in the ATP-driven reaction (Orlando, 1970). The only indication we could find that transhydrogenase is a component of the dark electron-transport chain of Rhodospirillaceae was a statement by Keister and Yike (1967*b*) that the rate of the ATP-driven transhydrogenase reaction in *Rs. rubrum* was the same on a protein basis for both the dark- and the light-grown cells.

1.2.4. Inhibitors

Many of the substances known to be inhibitory to electron transport in mitochondria have also been tested and found to be inhibitory to respiratory activities catalyzed by membrane preparations of dark-grown photosynthetic bacteria. Inhibition by rotenone and by cyanide has been demonstrated in *Rp. capsulata* (Baccarini-Melandri *et al.*, 1973; Klemme and Schlegel, 1969), in *Rp. sphaeroides* (Whale and Jones, 1970; Motokawa and Kikuchi, 1966), and in *Rp. palustris* (King and Drews, 1973), and inhibition by these substances and by amytal has been observed in *Rs. rubrum* (Thore *et al.*, 1969; Boll, 1969; Taniguchi and Kamen, 1965). *Rhodospirillum rubrum* and *Rp. sphaeroides* also show inhibition by CO (Taniguchi and Kamen, 1965; Motokawa and Kikuchi, 1966).

However, 2*n*-heptyl-4-hydroxyquinoline-*N*-oxide (HOQNO) and antimycin A, which inhibit the mitochondrial respiratory chain between cytochromes *b* and *c*, have been shown to be less effective inhibitors in these bacterial systems. Results have been variable, and even when high inhibitor concentrations have been used, the inhibition was incomplete (Taniguchi and Kamen, 1965; Whale and Jones, 1970; Simon and Siekevitz, 1973).

The electron-transport systems of most nonphotosynthetic bacteria are insensitive to inhibition by antimycin A, but HOQNO is usually an effective inhibitor. Interestingly, concentrations of HOQNO and antimycin A that do not affect the dark respiratory-chain system of *Rs. rubrum* and *Rp. sphaeroides* will completely inhibit the light-induced ATP formation in these bacteria (Smith and Baltscheffsky, 1959; Ramirez and Smith, 1968; Nishimura, 1963).

1.3. Components of the Respiratory-Chain System

The chemical nature of the dehydrogenases, quinones, cytochromes, and iron–sulfur proteins is described in Chapter 13 and elsewhere. Here, we shall discuss only their function in electron transport. Table 2, the data of Taniguchi and Kamen (1965), gives an illustration of the distribution of redox pigments in the membranes of dark-grown *Rs. rubrum*.

1.3.1. Dehydrogenases

Hatefi *et al.* (1972) isolated and purified the succinate dehydrogenase from chromatophores of light-grown *Rs. rubrum* and found it to resemble the mammalian dehydrogenase in all properties tested. Most remarkably, the dehydrogenase from *Rs. rubrum*

Table 2. Composition of Membrane Fragments of Dark Aerobically Grown *Rhodospirillum rubrum*[a]

Components	Amount (mμmol/mg protein)
Cytochromes	
a-type	Undetectable
b-type (protoheme)	0.30–0.46
c$_2$	0.10–0.15
RHP (cyt *c'*)	Undetectable (≪0.005)
CO-binding pigment	0.20–0.28
Flavin	
Acid-extractable	0.37
Trypsin-and-acid-extractable	0.60
Ubiquinone-10	0.30
Nonheme iron	0.80
Bacteriochlorophyll	0.10–0.50
Lipid	28% (% dry wt.)

[a] Data from Taniguchi and Kamen (1965).

was able to cross-react with the bovine mitochondrial system in reconstitution experiments. The dehydrogenase has not been purified from dark-grown cells.

NADH dehydrogenases have been isolated from membranes of light-grown *Rs. rubrum* (Boll, 1969; Horio *et al.*, 1969), but they may have been altered on removal from the membrane. The activities observed could not account for the NADH oxidase activity of the extract from which they were isolated. Two NADH dehydrogenases that reacted with different electron acceptors were obtained. Antibodies to each of the two dehydrogenases could inhibit the NADH oxidase about 50%, and the inhibitory effects were partially additive (Nisimoto *et al.*, 1973), suggesting two pathways involving NADH in light-grown cells. Similar experiments were not reported with dark-grown *Rs. rubrum*.

1.3.2. Quinones

Ubiquinone-10 (UQ) is found in membranes of both dark- and light-grown *Rs. rubrum*, but the concentration is higher in the latter (Lester and Crane, 1959). Redox changes of UQ in cells or chromatophores of *Rs. rubrum* at rates comparable to the respiration rate (Redfearn, 1967; Sugimura and Okabe, 1962), and decrease in the NADH oxidase activitiy on destruction of the quinone by UV irradiation (Taniguchi and Kamen, 1965; Yamamoto *et al.*, 1970), are compatible with the involvement of UQ in the dark respiratory chain. Similarly, depletion of UQ by pentane extraction or UV irradiation of dark-grown *Rp. palustris* markedly decreased the NADH and succinate oxidase activities of the membranes (King, and Drews, 1973). Differing requirements for the restoration of NADH or succinate oxidase activity to the depleted membranes on adding UQ suggest separate pools of the quinone with different sites of localization in the two systems. The membrane-bound UQ was reduced to the same extent with either substrate, however, indicating that the two pools can interact and exchange electrons. Similar suggestions have been made concerning the NADH and succinate oxidase systems of mitochondria (Lenaz *et al.*, 1971). The decrease in the activity of succinate dehydrogenase on depletion of membrane UQ is also observed with the mitochondrial dehydrogenase (Rossi *et al.*, 1970).

1.3.3. Cytochromes

Table 3 lists accumulated data on the cytochrome content of three species of Rhodospirillaceae that have been grown aerobically in the dark or anaerobically under illumination (for a description of these cytochromes, see Chapter 13). Both *b*- and *c*-type cytochromes are synthesized under both growth conditions. One problem in relating their presence to function, however, is that synthesis of some cytochromes that are part of the photosynthetic electron-

Table 3. Cytochromes Present in Dark- and Light-Grown Rhodospirillaceae[a]

Species	Grown in dark with O_2	Grown anaerobically in light
Rp. sphaeroides	2 or 3 *b*-type 3 *c*-type, one soluble *o* *aa₃* *c'*, soluble	2 or 3 *b*-type 2 *c*-type, one soluble *o*? *c'*, soluble
Rs. rubrum	2 *b*, one is *o*-type *c*? *c₂*, not oxidized enzymatically	3 *b*, one probably *o*-type *c*? *c₂*, oxidized by light *c'*, soluble *c₃*
Rp. capsulata	2 or 3 *b* *c*	2 *b* 3 *c*-type, two soluble *c'*, soluble

[a] See the text for references to these observations.

transport chain continues during dark aerobic growth. For example, both soluble and membrane-associated cytochrome c_2 is found in dark aerobically grown Rs. rubrum, but this cytochrome remains largely reduced in the presence of O_2; thus, although present, it does not participate in the dark respiratory chain (Taniguchi and Kamen, 1965). Similarly, dark aerobically grown Rp. sphaeroides synthesizes a b-type cytochrome that is mainly necessary in photosynthetic electron transport (Connelly et al., 1973).

Like other cells, the Rhodospirillaceae synthesize multiple b-type cytochromes, the relative quantities varying with different growth conditions. The individual b cytochromes are revealed by potentiometric measurements (Connelly et al., 1973; Dutton and Jackson, 1972; Chapter 28), by kinetic studies (Connelly et al., 1973), and by low-temperature absorption spectra (Whale and Jones, 1970). At least two b cytochromes of both dark- and light-grown Rs. rubrum and Rp. sphaeroides are rapidly oxidized and reduced on addition or exhaustion of O_2, indicating participation in the dark respiratory-chain system (Taniguchi and Kamen, 1965; Connelly et al., 1973). One of the b cytochromes may be the cytochrome oxidase known as cytochrome o (see below).

The functioning of c-type cytochromes in the dark respiratory chain of these bacteria is more difficult to assess. Evidence for a c-type cytochrome reducible by substrate and rapidly oxidized by O_2 was seen in dark-grown Rs. rubrum, Rp. sphaeroides, or Rp. capsulata in some experiments (Smith and Ramirez, 1959; Jones, O. T. G., and Plewis, 1974), but not in others (Taniguchi and Kamen, 1965). Also, a relatively high proportion of the cytochrome c of photosynthetically grown Rs. rubrum remains reduced in the presence of O_2 (Chance and Smith, 1955). Measurements of redox potential and of kinetics of oxidation by O_2 with Rp. sphaeroides grown in light or dark gave evidence of only one c-type cytochrome, which had a midpoint potential ($+295$ mV) lower than that of purified cytochrome c_2, but higher than that of typical respiratory-chain-linked cytochromes c (Connelly et al., 1973; Dutton and Jackson, 1972). Other workers have seen evidence for smaller amounts of additional c-type cytochromes (Saunders and Jones, 1974). The function of the soluble c-type cytochromes found in broken-cell extracts of both dark- and light-grown Rp. sphaeroides (Sasaki et al., 1970; Whale and Jones, 1970; Orlando, 1962) has not been demonstrated. Recent studies with antibodies to cytochrome c_2 of Rp. sphaeroides suggest that this cytochrome exists in soluble form in light-grown cells within the periplasmic space, which extends into the intracytoplasmic membranes (Prince et al., 1975). As these

membranes are ruptured, some cytochrome c would then be enclosed within the membrane vesicles formed. Localization within dark-grown cells was not investigated.

Rhodopseudomonas sphaeroides is different from Rs. rubrum and Rp. capsulata in that it can also synthesize a-type cytochrome during dark aerobic growth (Connelly et al., 1973; Saunders and Jones, 1974). Spectrophotometric observations of the reaction with CO (Whale and Jones, 1970) and potentiometric titration (Saunders and Jones, 1974) indicate that this is the terminal oxidase of these cells. The two components seen with midpoint potentials of $+375$ and $+200$ mV correspond well to the cytochromes $a + a_3$ of the mammalian cytochrome oxidase complex measured under the same conditions (Dutton et al., 1970). Growth of the bacteria in a medium with no added copper led to a decreased content of a-type cytochromes, pointing to the participation of copper in the oxidase of dark-grown Rp. sphaeroides, as in the yeast and mammalian oxidases (Saunders and Jones, 1974). Reaction of dark-grown Rp. sphaeroides with CO also gives evidence of cytochrome o (Whale and Jones, 1970), which is a b-type cytochrome that functions as an oxidase in numerous bacterial species. Carbon monoxide action spectra* of the oxidases have not been measured; these spectra would be the best evidence for the identity of an oxidase.

In contrast, Rp. sphaeroides does not synthesize an a-type cytochrome when grown anaerobically in the light. The terminal oxidase of these cells may be cytochrome o; the partially purified oxidase has properties similar to that of Rs. rubrum.

The terminal oxidase of dark aerobically grown Rs. rubrum is cytochrome o, as demonstrated by its CO action spectrum (Taniguchi and Kamen, 1965; Horio and Taylor, 1965). But the presence of cytochrome o in light-grown cells has not been ascertained because the light-reactive photosynthetic pigments present make it difficult to obtain action spectra and because the cytochrome c' also present exhibits a CO absorption spectrum similar to that of cytochrome o. The midpoint potential of cytochrome c' is not appropriate for an oxidase (Taniguchi and Kamen, 1965); its function in photosynthetic bacteria remains unknown.

Rhodopseudomonas capsulata does not synthesize a-type cytochromes in either dark or light (Zannoni et al., 1974; Klemme and Schlegel, 1969), and the nature of the oxidase is not clear. A b-type cytochrome with a high positive midpoint potential ($+410$ mV) (Zannoni

---

* Action spectra for the light-driven release of bound CO, measured by the relief of inhibition.

et al., 1974) was found to be lacking in a mutant deficient in cytochrome oxidase activity. The cytochrome does not combine with CO, however, and the respiration is relatively CO-insensitive; thus, it is not a typical *o*-type cytochrome.

The oxidases of *Rp. sphaeroides, Rs. rubrum*, and *Rp. capsulata* extracts will oxidize soluble mammalian *c* cytochrome at relatively low rates (Sasaki *et al.*, 1970; Klemme and Schlegel, 1969); the oxidation of bacterial cytochrome c_2 by *Rs. rubrum* extracts is even slower (Smith, 1959). Since none of these *c* cytochromes is the natural substrate, this kind of assay is really not a fair measure of the oxidase activity.

Differences in terminal oxidase do not explain the higher NADH oxidase activity in dark-grown Rhodospirillaceae as compared with light-grown cells. The *b*- and *c*-type cytochromes of the membranes of light-grown *Rp. sphaeroides* were oxidized more rapidly by O_2 than were those of dark-grown cells, even though only the latter synthesize the $a+a_3$-type oxidase (Connelly *et al.*, 1973).

1.3.4. Iron–Sulfur Proteins

Although membranes of dark aerobically grown *Rs. rubrum* contain iron–sulfur proteins (0.8 nmol/mg protein) (Taniguchi and Kamen, 1965), we found no investigations of their participation in the electron-transport chain. "Chromatophores" from light-grown *Rs. rubrum* have some iron–sulfur proteins similar to those of the mitochondrial respiratory chain (Dutton and Wilson, 1974).

1.4. Evidence for Multiple or Branched-Chain Pathways

As pictured in Section 1, the respiratory-chain system is usually represented as hydrogen and electron transport from the membrane-bound dehydrogenases via quinone into a common cytochrome chain to O_2. There are, however, a few data that do not fit this simple picture (Norling *et al.*, 1972; Smith *et al.*, 1974). Among the photosynthetic bacteria, there is considerable evidence for multiple or branched pathways.

Evidence of separate pathways from succinate and from NADH in dark-grown Rhodospirillaceae and also for more than one pathway from NADH (Baccarini-Melandri *et al.*, 1973) is provided by differential sensitivities to typical inhibitors of the cytochrome chain (Baccarini-Melandri *et al.*, 1973; Oelze and Kamen, 1975; Thore *et al.*, 1969; Whale and Jones 1970), as well as by different extents of reduction of cytochromes with the two substrates (Klemme and Schlegel, 1969). Different pools of UQ have been suggested for the NADH and succinate oxidase systems (see Section 1.3.2). In addition, the data imply some interaction between the pathways by means of a common component, but this interaction may be slow. The fraction of the NADH oxidase activity of *Rp. capsulata* that is sensitive to antimycin A and to low concentrations of KCN appeared to be coupled to phosphorylation of ADP, while the insensitive part was not, as tested by the quenching of atebrine fluorescence (Baccarini-Melandri *et al.*, 1973), a method sometimes used to demonstrate "energization" in membrane systems (Azzi *et al.*, 1971).

Similar experimentation pointed to differences in the NADH and succinate oxidase systems of light-grown bacteria (Marrs and Gest, 1973; Evans and Crofts, 1974; Horio *et al.*, 1969). These data are even more difficult to interpret, since it is not known whether the dehydrogenases react with the photosynthetic electron-transport chain. Both wild-type and respiratory mutants of *Rp. capsulata* deficient in cytochrome *c* oxidase activity could oxidize NADH and succinate with oxygen (Marrs and Gest, 1973), leading to the postulate of two pathways to a common oxidase, only one of which contains cytochrome *c*. The two paths were thought to "cross-link" at the level of UQ-cytochrome *b*, one pathway being "preferred" by NADH, the other by succinate.

The reduction of different amounts of cytochromes with NADH and succinate appears to rule out pathways including an identical cytochrome chain, unless some of the membranes lost dehydrogenases during preparation. The suggested "cross-linking" between two pathways and the "preferential channeling" (Marrs and Gest, 1973) were not described in detail, and are difficult to envisage in reference to the structure of the membrane-bound systems. Even if the implied cross-reaction via some membrane component is slow compared with electron transport in the main pathways, the cytochromes should all eventually be reduced with excess of the substrate used. The existence of completely separate non-interacting pathways is difficult to rule out. In the suspensions of membrane fragments, soluble cytochromes are present that may bind to the membranes (Smith and Minnaert, 1965). Also, it is conceivable that some of the particle-bound redox pigments could interact to some extent, depending on their location on the surface of the membrane vesicles.

1.5. Summary of Respiratory Chain: Comparison with Nonphotosynthetic Bacteria

The data described above reveal a strong resemblance between the respiratory-chain systems of the Rhodospirillaceae and those of nonphotosynthetic bacteria. Similar components comprise them all, and the variability seen with different growth conditions is also evident in many bacterial species (Smith, 1968; Gel'man et al., 1967). A well-studied example of this variability is the change in content of five separate dehydrogenases and six cytochromes when *Hemophilus parainfluenzae* are subjected to differing growth conditions (Smith, 1968). As with the photosynthetic bacteria (Oelze and Drews, 1970; Sasaki et al., 1970), the concentration of O_2 during growth appears to be of prime importance. An interesting analogy to the effect of O_2 concentration on the synthesis of *a*-type cytochromes by *Rp. sphaeroides* is seen in *Micrococcus denitrificans*, which can obtain energy for growth from electron-transport reactions leading to the reduction of O_2 or nitrate. The synthesis of *a*-type cytochromes is strongly suppressed by growth in the absence of O_2 and derepressed in the presence of O_2; synthesis of *c*- and *b*-type cytochromes is also affected (Scholes and Smith, 1968; Sapshead and Wimpenny, 1972).

Another similarity between photosynthetic and nonphotosynthetic bacteria is the presence of multiple oxidases (Smith, 1968; Gel'man et al., 1967). The combination of *o*- and *a*-type is not unusual, nor is the variation of the relative levels with changes in growth conditions (Broberg and Smith, 1967; Smith, 1968).

If the dark respiratory chain of *Rs. rubrum* actually does contain only *b*- and *o*-type cytochromes, it is unique; most have both *b*- and *c*-type, and many also synthesize *a*-type (Smith, 1968; Gel'man et al., 1967).

2. Energy Conservation Associated with Dark Respiration

Many properties of the dark oxidative phosphorylation system of Rhodospirillaceae resemble those of the mammalian system and of nonphotosynthetic bacteria. Intact photosynthetically grown *Rs. rubrum* can synthesize ATP aerobically in the dark with a rate and a P:O ratio (ratio of molecules of ATP synthesized to atoms of oxygen reduced) comparable to those seen in intact nonphotosynthetic bacteria (Ramirez and Smith, 1968; Welsch and Smith, 1969). The P:O ratio can be as high as 2.8 with β-hydroxybutyrate as substrate (Ramirez and Smith, 1968). Respiratory-chain-linked phos-

phorylation of added ADP can also be demonstrated in cell-free extracts of either dark- or light-grown Rhodospirillaceae, but the P:O ratios are much lower: 0.13–0.64 with broken-cell extracts oxidizing different substrates (Geller, 1962) and 0.17–0.5 for membranes oxidizing NADH or succinate (Taniguchi and Kamen, 1965; Klemme and Schlegel, 1969; Baccarini-Melandri et al., 1973; Yamashita et al., 1967). That the membrane-bound systems do not function as well after rupture of the cells is entirely analogous to the situation seen with broken-cell suspensions of numerous nonphotosynthetic bacteria (Gel'man et al., 1967). One possibility is that ADP and phosphate are not accessible to the proper reaction sites when membrane vesicles are formed; the bacterial membrane does not have an ADP–ATP transporter.

There is disagreement on whether oxidative phosphorylation in extracts of photosynthetic bacteria is tightly obligately coupled, so that the respiratory rate would be controlled by simultaneous phosphorylation. In some experiments, no change in respiration rate was observed on addition of ADP plus phosphate or uncoupler (Melandri et al., 1971; Geller, 1962); in others, a marked stimulation was seen (Yamashita et al., 1967).

Oxidative phosphorylation by membranes of both dark- and light-grown *Rs. rubrum* and *Rp. capsulata* is inhibited by typical "uncouplers" such as DNP or CCCP or by energy-transfer inhibitors such as oligomycin (Geller, 1962; Baccarini-Melandri et al., 1973; Ramirez and Smith, 1968; Klemme and Schlegel, 1969), but photophosphorylation is less sensitive to DNP (Geller, 1962). When different segments of the electron-transport chain were tested by use of inhibitors and added electron donors or acceptors, there was evidence for at least two sites for ATP synthesis coupled to the respiratory chain of membranes of dark-grown *Rp. capsulata* (Baccarini-Melandri et al., 1973).

Protein complexes with ATPase activity, so-called "coupling factors," have been isolated from membranes of both dark- and light-grown *Rp. capsulata* and from "chromatophores" of *Rs. rubrum* (Baccarini-Melandri and Melandri, 1972; Melandri et al., 1971; Johansson and Baltscheffsky, 1975), which could restore phosphorylating activity to membranes depleted of the factor by sonication in the presence of EDTA. Most interesting, the purified coupling factors from dark- or light-grown bacteria were indistinguishable in electrophoretic pattern and in restoring oxidative or photosynthetic phosphorylation to depleted membranes of dark- or light-grown cells. The conclusion that the same coupling factor catalyzes both dark oxidative and photosynthetic phosphorylation is strengthened by the observation that antibodies elicited to one factor will inhibit both reactions (Melandri *et al.*, 1971). The purified coupling factor from *Rs. rubrum* chromatophores resembles that of mammalian mitochondria in molecular weight and in number of peptide subunits (Johansson and Baltscheffsky, 1975).

The change of fluorescence of bound dyes such as 8-anilino-naphthalene-1-sulfonic acid on "energization" of membranes (initiation of energy-producing reactions) is suggested to result from conformational transitions or polarization changes (Chance and Mukai, 1971). Such changes are seen on addition of ADP plus phosphate or of succinate to *Rs. rubrum* chromatophores (Vainio *et al.*, 1972); the change is rather slow, however, with a half-time around 30 sec. The change resulting from succinate addition was inhibited by uncouplers. The significance of the "energization" resulting from succinate addition to "chromatophores," if they catalyze only cyclic photophosphorylation, was not discussed, nor was evidence for the presence or absence of the dark respiratory chain cited.

The mechanism of phosphorylation coupled to the energy conserved in O_2-linked or light-induced oxidation–reduction reactions is discussed in Chapters 26 and 30. Whatever the mechanism, the accompanying shifts in protons and other ions seem to be reflected in changes in the absorption spectrum of the membrane-bound carotenoids. These may result from changes of electric-field interactions (Jackson and Crofts, 1969; Yamashita *et al.*, 1967), and can also be induced when a membrane potential is generated in other ways, such as on addition of ATP (Baltscheffsky, 1969). In this respect, it is interesting that similar carotenoid absorption spectrum changes were observed when an anaerobic suspension of intact light-grown *Rs. rubrum* was either oxygenated or illuminated (Smith and Ramirez, 1959), but no further change resulted from illumination of aerobic cells (Smith *et al.*, 1960). The carotenoid bandshift was also seen on illumination, but not on oxygenation, of a broken-cell extract of *Rs. rubrum* that showed photosynthetic, but not dark oxidative, phosphorylation (Smith and Baltscheffsky, 1959).

Proton efflux into the medium results on either illumination of light-grown *Rs. rubrum* and *Rp. sphaeroides* or oxygenation of anaerobic suspensions of dark-grown cells (Scholes *et al.*, 1969) (see Chapters 26 and 30). In contrast, illumination of "chromatophores" stimulated inward translocation of protons, indicating that the chromatophore vesicles are inside-out with respect to the membrane of intact cells.

Section 1.4 described the postulation of "branched" dark electron-transport chains, one arm of which is not coupled to a phosphorylating system. In fact, this was offered as an explanation for the relatively low P:O ratios of broken-cell extracts. Actually, the P:O ratios are no lower than those of broken-cell extracts of nonphotosynthetic bacteria, and the P:O ratio found with intact cells is high (Ramirez and Smith, 1968). The suggestion of an alternate nonphosphorylating pathway is reminiscent of *Azotobacter*, in which a nonphosphorylating chain seems to serve the function of reducing the O_2 concentration in the area of the O_2-sensitive nitrogenase (Dalton and Postgate, 1969). *Rhodospirillum rubrum* can also synthesize a nitrogen-fixing system, and thus might be comparable to *Azotobacter* in this respect.

3. Relationship between Light-Induced Electron Transport and the O_2-Linked Respiratory Chain

The interesting and still puzzling question is the relationship between the O_2-linked respiratory chain and that catalyzing light-induced electron transport in photosynthetically grown bacteria.

It has long been known (van Niel, 1941) that illumination can reversibly inhibit the respiration of light-grown Rhodospirillaceae to variable extents, depending on growth conditions, time of growth, or "aging" of the cells (Horio and Kamen, 1962). Data with broken-cell extracts are discordant, partial inhibition being seen in some experiments (Katoh, 1961; Thore *et al.*, 1969; Jones, C. W., and Vernon, 1969), and no or slight stimulation in others (Geller, 1962; Smith, 1959). The action spectrum for the inhibitory effect with intact *Rs. rubrum* or *Rp. sphaeroides* corresponds to the absorption spectrum of the photosynthetic apparatus (Horio and Taylor, 1965; Fork and Goedheer, 1964), but light saturation occurs at a lower intensity than that required to saturate photosynthesis (Fork and Goedheer, 1964). In addition, after illumination, respiration may be stimulated for a period of minutes (Ramirez and Smith, 1968).

Two kinds of proposals have been offered to explain the light-induced inhibition of respiration:

First, one complex electron-transport system catalyzes both O_2-linked and light-induced electron transport, and this includes a span common to both pathways (Nishimura and Chance, 1963; Dutton and Wilson, 1974; Horio and Kamen, 1962). Rapid oxidation of these components on illumination would prevent their reaction with O_2. In support of this proposal is the evidence that the two systems contain similar *b*- and *c*-type cytochromes and coupling factors (see Sections 1.3.3 and 2) and the observation that the same shift in carotenoid absorption spectrum is seen on oxygenation as well as on illumination of an anaerobic suspension of *Rs. rubrum*. The ATP produced on either oxygenation or illumination could, however, give rise to the carotenoid shift (see Section 2). The cytochromes $a+a_3$ synthesized in the dark and the Bchl synthesized in the light by *Rp. sphaeroides* could be considered alternate electron acceptors (Dutton and Wilson, 1974), but light-grown *Rp. sphaeroides* can respire in the absence of cytochromes $a+a_3$ (see Section 1.3.3). The proposal of one complex system requires that all components be localized on the same membrane.

Second, there are two separate electron-transport–phosphorylating systems that can interact either via some common membrane-bound component or with some soluble intermediate of the two systems (e.g., Vernon, 1968; Keister and Yike, 1967*a*; Thore *et al.*, 1969; Ramirez and Smith, 1968). Good evidence for this proposal is the marked difference in sensitivitiy of the dark O_2-linked and the light-induced systems to a number of typical inhibitors of electron transport or phosphorylation (see Sections 1.2 and 2). Also, membranes from dark- and light-grown *Rs. rubrum* differ in composition (Geller, 1962) (see also Table 3), and there has been success in the separation of two kinds of membranes from light-grown Rhodospirillaceae with some different activities (see Section 1.1). Evidence has been cited for two different pools of UQ in the respiratory chain and in the photosynthetic electron-transport chain of *Rs. rubrum* showing different absorption spectra (Parson, 1967).

Some postulates of two separate electron-transport–phosphorylating systems suggest that they interact at the level of UQ-cytochrome *b*. However, the two UQ pools postulated in *Rs. rubrum* did not seem to communicate rapidly (Parson, 1967). This proposal would require close association of the two systems on the same membrane, since UQ is insoluble in aqueous media and remains firmly membrane-bound. A remaining possibility for interaction of the two systems is by means of a common pool of soluble metabolites, such as ADP plus phosphate, for which the two systems could compete. If the respiratory chain is obligately coupled to phosphorylation, a lack of phosphate acceptor (utilized on illumination) would be inhibitory. Evidence in favor of this explanation is the prevention of the light inhibition of respiration of intact *Rs. rubrum* (Ramirez and Smith, 1968) or of chromatophores (Thore *et al.*, 1969; Fork and Goedheer, 1964) by inhibitors or uncouplers of light-induced ATP synthesis. There are, however, conflicting data (Katoh, 1961).

The available data cannot definitely distinguish between these two currently advanced hypotheses. In addition, there are other data that are difficult to explain by either. For example, the oxidation of additional cytochrome *c* on illumination after oxygenation of *Rs. rubrum* cells or membranes (Chance and Smith, 1955; Smith and Ramirez, 1959; Smith and Baltscheffsky, 1959) is evidence for an electron-transport chain oxidized on illumination in addition to one oxidized either on illumination or on addition of O_2. In accord with this is the observation that higher levels of ATP are attained in intact *Rs. rubrum* on illumination than on oxygenation in the dark, under conditions in which the dark anaerobic utilization of ATP is very low (Ramirez and Smith, 1968; Welsch and Smith, 1969). Also, methylphenazonium methosulfate inhibits the dark oxidative phosphorylation and part of the light-induced ATP synthesis (Ramirez and Smith, 1968). The rate of light-induced synthesis of ATP was found to be independent of temperature between 4° and 35°C, while the rate of O_2-linked phosphorylation decreased with lowering of temperature (Welsch and Smith, 1969). In fact, at 4°C, biphasic kinetics of ATP synthesis were observed. The simplest explanation for these observations seems to be the existence of different electron-transport–phosphorylation systems localized separately within the cells. One system can respond only to illumination; another responds either to oxygenation or to illumination. On oxygenation, presumably only the fraction of cellular ADP accessible to the latter system will be phosphorylated.

It is conceivable that there is a morphological basis for the functionally different membrane-bound systems observed. For example, as internal membranes are synthesized by the introduction of new components into the existing cytoplasmic membranes (Kosakowski and Kaplan, 1974) (see Section 1.1), it is possible that new components might exist in close association with variable amounts of the oxygen-activated system, permitting interaction between them. As development of the internal membrane proceeds by extension into the cytoplasm, however, it contains the photosynthetic electron-transport components that are at some distance from the original cytoplasmic membrane and are reactive only to light. Not only the original cytoplasmic membrane but also the more extensive internal membrane would possess coupling sites for the available ADP.

This postulate would explain a number of things: (1) the variability of the extent of inhibition of respiration by light with growth conditions of the bacteria; (2) the different extents of reaction of c-type cytochrome with O_2 observed in "dark-grown" *Rs. rubrum* (see Section 1.3.3); (3) apparent multiple pathways with different inhibitor sensitivities; (4) the difficulty experienced in clear-cut separation of different purified membrane fragments; (5) the suggestion of two types of reaction center in *Rs. rubrum* (Frenkel, 1970). Also in agreement with this proposal is the reconstitution of light-induced electron transport in cytoplasmic membranes of Bchl-less mutants of *Rp. sphaeroides* and *Rp. capsulata* by the addition of purified reaction centers (Jones, O. T. G., and Plewis, 1974; Garcia *et al.*, 1974).

The increase in respiration rate that can follow after a period of illumination of intact *Rs. rubrum* does not seem to be related to the above-discussed inhibition of respiration on illumination. Since it can be of several minutes' duration, it most likely results from the accumulation of an oxidizable metabolite, e.g., NADH, produced on illumination.

We should like to end this chapter by emphasizing again the importance of establishing the localization of the different activities and components of the electron-transport chains within the cells. If our postulate is valid, doing so may be experimentally even more difficult than it now appears.

4. References

Azzi, A., Fabbro, A., Santato, M., and Gherardini, P. L., 1971, Energy transduction in mitochondrial fragments: Interaction of the membrane with acridine dyes, *Eur. J. Biochem.* **21**:404.

Baccarini-Melandri, A., and Melandri, B. A., 1972, Energy transduction in photosynthetic bacteria. III. Coincidence of coupling factor of photosynthesis and respiration in *R. capsulata*, *FEBS Lett.* **21**:131.

Baccarini-Melandri, A., Zannoni, D., and Melandri, B. A., 1973, Energy transduction in photosynthetic bacteria. VI. Respiratory sites of energy conservation in membranes from dark-grown cells of *R. capsulata*, *Biochim. Biophys. Acta* **314**:298.

Baltscheffsky, M., 1969, Energy conversion-linked changes of carotenoid absorbance in *Rs. rubrum* chromatophores, *Arch. Biochem. Biophys.* **130**:646.

Boll, M., 1969, Oxidation of reduced NAD in *R. rubrum*. III. Properties of a NADH dehydrogenase solubilized from electron transport particles, *Arch. Mikrobiol.* **69**:301.

Broberg, P. L., and Smith, L., 1967, The cytochrome system of *Bacillus megaterium* KM: The presence and some properties of two CO-binding cytochromes, *Biochim. Biophys. Acta* **131**:479.

Chance, B., and Mukai, Y., 1971, Anilino-1,8-naphthalene sulfonate and bromthymol blue responses to membrane energization, in: *Probes of Structure and Function of Macromolecules and Membranes* (B. Chance, C.-P. Lee, and J. K. Blasie, eds.), Vol. I, *Probes and Membrane Function*, pp. 239–244, Academic Press, New York.

Chance, B., and Smith, L., 1955, Respiratory pigments of *R. rubrum*, *Nature (London)* **175**:803.

Connelly, J. L., Jones, O. T. G., Saunders, V. A., and Yates, D. W., 1973, Kinetic and thermodynamic properties of membrane-bound cytochromes of aerobically and photosynthetically grown *R. spheroides*, *Biochim. Biophys. Acta* **292**:644.

Dalton, H., and Postgate, J. R., 1969, Effect of oxygen on growth of *Azotobacter chroococcum* in batch and continuous cultures, *J. Gen. Microbiol.* **54**:463.

Dutton, P. L., and Jackson, J. B., 1972, Thermodynamic and kinetic characterization of electron-transfer components *in situ* in *R. spheroides* and *Rs. rubrum*, *Eur. J. Biochem.* **30**:495.

Dutton, P. L., and Wilson, D. F., 1974, Redox potentiometry in mitochondrial and photosynthetic bioenergetics, *Biochim. Biophys. Acta* **346**:165.

Dutton, P. L., Wilson, D. F., and Lee, C.-P., 1970, Oxidation–reduction potentials of cytochromes in mitochondria, *Biochemistry* **9**:5077.

Evans, E. H., and Crofts, A. R., 1974, A thermodynamic characterisation of the cytochromes of chromatophores from *R. capsulata*, *Biochim. Biophys. Acta* **357**:78.

Fisher, R. R., and Guillory, R. J., 1971a, Resolution of enzymes catalyzing energy-linked transhydrogenation. II. Interaction of transhydrogenase factor with the *R. rubrum* chromatophore membrane, *J. Biol. Chem.* **246**:4679.

Fisher, R. R., and Guillory, R. J., 1971b, Resolution of enzymes catalyzing energy-linked transhydrogenation. III. Preparation and properties of *R. rubrum* transhydrogenase factor, *J. Biol. Chem.* **246**:4687.

Fork, D. C., and Goedheer, J. C., 1964, Studies on light-induced inhibition of respiration in purple bacteria: Action spectra for *R. rubrum* and *R. spheroides*, *Biochim. Biophys. Acta* **79**:249.

Frenkel, A. W., 1970, Multiplicity of electron transport reactions in bacterial photosynthesis, *Biol. Rev. Cambridge Philos. Soc.* **45**:569.

Garcia, A. F., Drews, G., and Kamen, M. D., 1974, On reconstitution of bacterial photosynthetic phosphorylation *in vitro*, *Proc. Natl. Acad. Sci. U.S.A.* **71**:4213.

Geller, D. M., 1962, Oxidative phosphorylation in extracts of *R. rubrum*, *J. Biol. Chem.* **237**:2947.

Geller, D. M., 1963, Some observations concerning the purification and properties of the aerobic phosphorylation system of *R. rubrum* extracts, in: *Bacterial Photosynthesis* (H. Gest, A. San Pietro, and L. P. Vernon, eds.), pp. 161–174, Antioch Press, Yellow Springs, Ohio.

Gel'man, N. S., Lukoyanova, M. A., and Ostrovskii, D. N., 1967, *Respiration and Phosphorylation of Bacteria*, p. 171, Plenum Press, New York.

Hatefi, Y., Davis, K. A., Baltscheffsky, H., Baltscheffsky, M., and Johansson, B. C., 1972, Isolation and properties of succinate dehydrogenase from *R. rubrum*, *Arch. Biochem. Biophys.* **152**:613.

Horio, T., and Kamen, M. D., 1962, Observations on the respiratory system of *Rs. rubrum, Biochemistry* **1**:1141.

Horio, T., and Taylor, C. P. S., 1965, The photochemical determination of an oxidase of the photoheterotroph, *R. rubrum*, and the action spectrum of the inhibition of respiration by light, *J. Biol. Chem.* **240**:1772.

Horio, T., Bartsch, R. G., Kakuno, T., and Kamen, M. D., 1969, Two reduced NAD dehydrogenases from the photosynthetic bacterium *R. rubrum*, *J. Biol. Chem.* **244**:5899.

Huang, J. W., and Kaplan, S., 1973a, Membrane proteins of *R. spheroides*. III. Isolation, purification and characterization of cell envelope proteins, *Biochim. Biophys. Acta* **307**:301.

Huang, J. W., and Kaplan, S., 1973b, Membrane proteins of *R. spheroides*. IV. Characterization of chromatophore proteins, *Biochim. Biophys. Acta* **307**:317.

Jackson, J. B., and Crofts, A. R., 1969, The high energy state in chromatophores from *R. spheroides, FEBS Lett.* **4**:185.

Johansson, B. C., and Baltscheffsky, M., 1975, On the subunit composition of the coupling factor (ATPase) from *R. rubrum, FEBS Lett.* **53**:221.

Jones, C. W., and Vernon, L. P., 1969, NAD photoreduction in *R. rubrum* chromatophores, *Biochim. Biophys. Acta* **180**:149.

Jones, O. T. G., and Plewis, K. M., 1974, Reconstitution of light-dependent electron transport in membranes from a bacteriochlorophyll-less mutant of *R. spheroides, Biochim. Biophys. Acta* **357**:204.

Katoh, S., 1961, Inhibitory effect of light on oxygen-uptake by cell-free extracts and particulate fractions of *R. palustris, J. Biochem.* (*Tokyo*) **49**:126.

Keister, D. L., and Minton, N. J., 1969a, Energy-linked reactions in photosynthetic bacteria. III. Further studies on energy-linked NAD reduction by *R. rubrum* chromatophores, *Biochemistry* **8**:167.

Keister, D. L., and Minton, N. J., 1969b, Energy-linked reactions in photosynthetic bacteria. IV. Interaction of the photochemical and respiratory systems of *R. rubrum*, in: *Progress in Photosynthesis Research* (H. Metzner, ed.), Vol. III, pp. 1299–1305, H. Metzner, Tübingen.

Keister, D. L., and Yike, N. J., 1966, Studies on an energy-linked pyridine nucleotide transhydrogenase in photosynthetic bacteria. I. Demonstration of the reaction in *R. rubrum, Biochem. Biophys. Res. Commun.* **24**:519.

Keister, D. L., and Yike, N. J., 1967a, Energy-linked reactions in photosynthetic bacteria. I. Succinate-linked ATP-driven NAD$^+$ reduction by *R. rubrum* chromatophores, *Arch. Biochem. Biophys.* **121**:415.

Keister, D. L., and Yike, N. J., 1967b, Energy-linked reactions in photosynthetic bacteria. II. The energy-dependent reduction of oxidized NADP by reduced NAD in chromatophores of *Rs. rubrum, Biochemistry* **6**:3847.

Ketchum, P. A., and Holt, S. C., 1970, Isolation and characterization of the membranes from *R. rubrum, Biochim. Biophys. Acta* **196**:141.

King, M. T., and Drews, G., 1973, The function and localization of ubiquinone in the NADH and succinate oxidase systems of *R. palustris, Biochim. Biophys. Acta* **305**:230.

Klemme, J.-H., 1969a, Hydrogenase and photosynthetic electron transport in chromatophores from the facultative phototroph, *R. capsulata*, in: *Progress in Photosynthesis Research* (H. Metzner, ed.), Vol. III, pp. 1492–1503, H. Metzner, Tübingen.

Klemme, J.-H., 1969b, Studies on the mechanism of NAD-photoreduction by chromatophores of the facultative phototroph, *R. capsulata, Z. Naturforsch. Teil B* **24**:67.

Klemme, J.-H., and Schlegel, H. G., 1969, Untersuchungen zum Cytochrom-Oxydase-System aus anaerob im Licht und aerob im Dunkeln gewachsenen Zellen von *R. capsulata, Arch. Mikrobiol.* **68**:326.

Kosakowski, M. H., and Kaplan, S., 1974, Topology and growth of the intracytoplasmic membrane system of *R. spheroides*: Protein, chlorophyll, and phospholipid insertion into steady-state anaerobic cells, *J. Bacteriol.* **118**:1144.

Lenaz, G., Castelli, A., Littarru, G. P., Bertoli, E., and Folkers, K., 1971, Specificity of lipids and coenzyme Q in mitochondrial NADH and succin-oxidase of beef heart and *S. cerevisiae, Arch. Biochem. Biophys.* **142**:407.

Lester, R. L., and Crane, F. L., 1959, The natural occurrence of coenzyme Q and related compounds, *J. Biol. Chem.* **234**:2169.

Marrs, B., and Gest, H., 1973, Genetic mutations affecting the respiratory electron-transport system of the photosynthetic bacterium, *R. capsulata, J. Bacteriol.* **114**:1045.

Marrs, B., Stahl, C. L., Lien, S., and Gest, H., 1972, Biochemical physiology of a respiration-deficient mutant of the photosynthetic bacterium, *R. capsulata, Proc. Natl. Acad. Sci. U.S.A.* **69**:916.

Melandri, B. A., Baccarini-Melandri, A., San Pietro, A., and Gest, H., 1971, Interchangeability of phosphorylation coupling factors in photosynthetic and respiratory energy conversion, *Science* **174**:514.

Motokawa, Y., and Kikuchi, G., 1966, Cytochrome systems in dark-aerobically grown *R. spheroides, Biochim. Biophys. Acta* **120**:274.

Niederman, R. A., 1974, Membranes of *R. spheroides*: Interactions of chromatophores with the cell envelope, *J. Bacteriol.* **117**:19.

Niederman, R. A., Segen, B. J., and Gibson, K. D., 1972, Membranes of *R. spheroides*. I. Isolation and characterization of membrane fractions from extracts of aerobically and anaerobically grown cells, *Arch. Biochem. Biophys.* **152**:547.

Niederpruem, D. J., and Doudoroff, M., 1965, Cofactor-dependent aldose dehydrogenase of *R. spheroides, J. Bacteriol.* **89**:697.

Nishimura, M., 1963, Studies on the electron-transfer systems in photosynthetic bacteria. II. The effect of heptylhydroxyquinoline-*N*-oxide and antimycin A on the photosynthetic and respiratory electron-transfer systems, *Biochim. Biophys. Acta* **66**:17.

Nishimura, M., and Chance, B., 1963, Studies on the electron-transfer systems in photosynthetic bacteria. I. The light-induced absorption-spectrum changes and the effect of phenylmercuric acetate, *Biochim. Biophys. Acta* **66**:1.

Nisimoto, Y., Kakuno, T., Yamashita, J., and Horio, T., 1973, Two different NADH dehydrogenases in respiration of *R. rubrum* chromatophores, *J. Biochem. (Tokyo)* **74**:1205.

Norling, B., Nelson, B. D., Nordenbrand, K., and Ernster, L., 1972, Evidence for the occurrence in submitochondrial particles of a dual respiratory chain containing different forms of cytochrome *b*, *Biochim. Biophys. Acta* **275**:18.

Nozaki, M., Tagawa, K., and Arnon, D. I., 1963, Metabolism of photosynthetic bacteria. II. Certain aspects of cyclic and noncyclic photophosphorylation in *R. rubrum*, in: *Bacterial Photosynthesis* (H. Gest, A. San Pietro, and L. P. Vernon, eds.), pp. 175–194, Antioch Press, Yellow Springs, Ohio.

Oelze, J., and Drews, G., 1969, Untersuchungen über die Membran-Proteine bei *R. rubrum*, *Arch. Mikrobiol.* **69**:12.

Oelze, J., and Drews, G., 1970, Variations of NADH oxidase activity and bacteriochlorophyll content during membrane differentiation in *R. rubrum*, *Biochim. Biophys. Acta* **219**:131.

Oelze, J., and Drews, G., 1972, Membranes of photosynthetic bacteria, *Biochim. Biophys. Acta* **265**:209.

Oelze, J., and Kamen, M. D., 1975, Separation of respiratory reactions in *R. rubrum*: Inhibition studies with 2-hydroxydiphenyl, *Biochim. Biophys. Acta* **387**:1.

Oelze, J., Biedermann, M., and Drews, G., 1969, Die Morphogenese des Photosynthesapparates von *R. rubrum* I. Die Isolierung und Charakterisierung von zwei Membransystemen, *Biochim. Biophys. Acta* **173**:436.

Orlando, J. A., 1962, *R. spheroides*-cytochrome-553, *Biochim. Biophys. Acta* **57**:373.

Orlando, J. A., 1970, Involvement of sulfhydryl groups in light-dependent transhydrogenase of *R. spheroides*, *Arch. Biochem. Biophys.* **141**:111.

Parson, W. W., 1967, Observations on the changes in ultraviolet absorbance caused by succinate and light in *R. rubrum*, *Biochim. Biophys. Acta* **143**:263.

Prince, R. C., Baccarini-Melandri, A., Hauska, G. A., Melandri, B. A., and Crofts, A. R., 1975, Asymmetry of an energy transducing membrane: The location of cytochrome *c₂* in *R. spheroides* and *R. capsulata*, *Biochim. Biophys. Acta* **387**:212.

Ramirez, J., and Smith, L., 1968, Synthesis of ATP in intact cells of *R. rubrum* and *R. spheroides* on oxygenation or illumination, *Biochim. Biophys. Acta* **153**:466.

Redfearn, E. R., 1967, Redox reactions of ubiquinone in *R. rubrum*, *Biochim. Biophys. Acta* **131**:218.

Rossi, E., Norling, B., Persson, B., and Ernster, L., 1970, Studies with ubiquinone-depleted submitochondrial particles: Effects of extraction and reincorporation of ubiquinone on the kinetics of succinate dehydrogenase, *Eur. J. Biochem.* **16**:508.

Sapshead, L. M., and Wimpenny, J. W. T., 1972, The influence of oxygen and nitrate on the formation of the cytochrome pigments of the aerobic and anaerobic respiratory chain of *Micrococcus denitrificans*, *Biochim. Biophys. Acta* **267**:388.

Sasaki, T., Motokawa, Y., and Kikuchi, G., 1970, Occurrence of both *a*-type and *o*-type cytochromes as the functional terminal oxidases in *R. spheroides*, *Biochim. Biophys. Acta* **197**:284.

Saunders, V. A., and Jones, O. T. G., 1974, Properties of the cytochrome *a*-like material developed in the photosynthetic bacterium *R. spheroides* when grown aerobically, *Biochim. Biophys. Acta* **333**:439.

Scholes, P. B., and Smith, L., 1968, Composition and properties of the membrane-bound respiratory chain system of *Micrococcus denitrificans*, *Biochim. Biophys. Acta* **153**:363.

Scholes, P., Mitchell, P., and Moyle, J., 1969, The polarity of proton translocation in some photosynthetic microorganisms, *Eur. J. Biochem.* **8**:450.

Simon, S. R., and Siekevitz, P., 1973, Biochemical properties of purified membrane preparations from *R. rubrum*, in: *Mechanisms in Bioenergetics* (G. F. Azzone, L. Ernster, S. Papa, E. Quagliariello, and N. Siliprandi, eds.), pp. 3–31, Academic Press, New York.

Smith, L., 1959, Reactions of *R. rubrum* extract with cytochrome *c* and cytochrome *c₂*, *J. Biol. Chem.* **234**:1571.

Smith, L., 1968, The respiratory chain system of bacteria, in: *Biological Oxidations* (T. P. Singer, ed.), pp. 55–122, Interscience, New York.

Smith, L., and Baltscheffsky, M., 1959, Respiration and light-induced phosphorylation in extracts of *R. rubrum*, *J. Biol. Chem.* **234**:1575.

Smith, L., and Minnaert, K., 1965, Interaction of macroions with the respiratory chain system of mitochondria and heart-muscle particles, *Biochim. Biophys. Acta* **105**:1.

Smith, L., and Ramirez, J., 1959, Absorption spectrum changes in photosynthetic bacteria following illumination or oxygenation, *Arch. Biochem. Biophys.* **79**:233.

Smith, L., Baltscheffsky, M., and Olson, J. M., 1960, Absorption spectrum changes observed on illumination of aerobic suspensions of photosynthetic bacteria, *J. Biol. Chem.* **235**:213.

Smith, L., Davies, H. C., and Nava, M., 1974, Reactions of cytochrome *c* in the respiratory chain system, in: *Dynamics of Energy-Transducing Membranes* (L. Ernster, R. W. Estabrook, and E. C. Slater, eds.), pp. 51–59, Elsevier, Amsterdam.

Sugimura, T., and Okabe, K., 1962, The reduction of ubiquinone (coenzyme Q) in chromatophores of *R. rubrum* by succinate, *J. Biochem. (Tokyo)* **52**:235.

Taniguchi, S., and Kamen, M. D., 1965, The oxidase system of heterotrophically grown *R. rubrum*, *Biochim. Biophys. Acta* **96**:395.

Thore, A., Keister, D. L., and San Pietro, A., 1969, Studies on the respiratory system of aerobically (dark) and anaerobically (light) grown *R. rubrum*, *Arch. Mikrobiol.* **67**:378.

Throm, E., Oelze, J., and Drews, G., 1970, The distribution of NADH oxidase in the membrane system of *R. rubrum*, *Arch. Mikrobiol.* **72**:361.

Vainio, H., Baltscheffsky, M., Baltscheffsky, H., and Azzi, A., 1972, Energy-dependent changes in membranes of *R. rubrum* chromatophores as measured by 8-anilino-naphthalene-1-sulfonic acid, *Eur. J. Biochem.* **30**:301.

van Niel, C. B., 1941, The bacterial photosyntheses and their importance for the general problem of photosynthesis, *Adv. Enzymol.* **1**:263.

Vernon, L. P., 1968, Photochemical and electron transport reactions of bacterial photosynthesis, *Bacteriol. Rev.* **32**:243.

Welsch, F., and Smith, L., 1969, Kinetics of synthesis and utilization of ATP by intact cells of *R. rubrum, Biochemistry* **8**:3403.

Whale, F. R., and Jones, O. T. G., 1970, The cytochrome system of heterotrophically grown *R. spheroides, Biochim. Biophys. Acta* **223**:146.

Yamamoto, N., Hatakeyama, H., Nishikawa, K., and Horio, T., 1970, The function of ubiquinone-10 both in the electron transport system and in the energy conservation system of chromatophores from *R. rubrum, J. Biochem. (Tokyo)* **67**:587.

Yamashita, J., and Kamen, M. D., 1969, Observations on distribution of NADH oxidase in particles from dark-grown and light-grown *R. rubrum, Biochem. Biophys. Res. Commun.* **34**:418.

Yamashita, J., Yoshimura, S., Matuo, Y., and Horio, T., 1967, Relation between photosynthetic and oxidative phosphorylation in chromatophores from light-grown cells of *R. rubrum, Biochim. Biophys. Acta* **143**:154.

Zannoni, D., Baccarini-Melandri, A., Melandri, B. A., Evans, E. H., Prince, R. C., and Crofts, A. R., 1974, Energy transduction in photosynthetic bacteria: The nature of cytochrome *c* oxidase in the respiratory chain of *R. capsulata, FEBS Lett.* **48**:152.

PART VI

Peripheral Oxidations and Reductions

CHAPTER 34

Nitrogen Fixation and Hydrogen Metabolism by Photosynthetic Bacteria

Duane C. Yoch

1. Introduction

Photosynthetic bacteria are one of several major groups of free-living bacteria that are capable of enzymatically reducing atmospheric nitrogen to ammonia. Other major groups of bacteria that can use N_2 gas as their only source of nitrogen for biosynthetic processes are the fermentative bacteria, such as *Clostridium pasteurianum*; obligate aerobes, such as *Azotobacter vinelandii*; and coliforms, such as *Klebsiella pneumoniae*. Species of the genus *Rhizobium* infect the roots of leguminous plants, where they fix nitrogen in a symbiotic relationship with the plant. Several other bacterial species fix nitrogen in a symbiotic relationship with nonleguminous plants. The blue-green algae are the only other group of organisms that have been shown to fix nitrogen, and the process therefore appears to be limited to prokaryotic microorganisms.

The extent to which biological nitrogen fixation provides "fixed" metabolizable nitrogen to the biosphere is difficult to assess, but it has been estimated that 63–174 million metric tons of nitrogen are fixed per year by the biological process (Delwiche, 1970; Hardy and Holsten, 1972). Industrial nitrogen fixation is believed by these authors to account for an additional 30 million metric tons of fixed nitrogen being added to the earth each year. A nitrogen balance is maintained because approximately 80 million metric

tons of nitrogen are lost to the atmosphere each year through the process of denitrification.

Because research on nitrogen fixation and hydrogen metabolism by photosynthetic bacteria has been relatively limited, most of the older literature on these topics will be covered in this review. In addition, an attempt will be made to relate the large body of information on the photoevolution of hydrogen to the known properties of the hydrogenase and nitrogenase enzymes. Earlier reviews by Gest (1972), Benemann and Valentine (1972), Stewart (1973), Dalton (1974), and Yoch and Arnon (1974) have touched on various aspects of this problem.

2. Nitrogen Fixation by Photosynthetic Bacteria

2.1. Discovery

The observation by Gest and Kamen (1949a) that both N_2 and NH_4^+ inhibited the photoevolution of H_2 by the photosynthetic bacterium *Rhodospirillum rubrum* suggested similarities to nitrogen-fixing bacteria, in which a relationship between hydrogen metabolism and nitrogen fixation had been established by P. W. Wilson and his associates (Wilson, 1940; Lee and Wilson, 1943). The uptake of $^{15}N_2$ in the light by *Rs. rubrum* demonstrated unequivocally that it was also capable of nitrogen fixation (Kamen and Gest, 1949). The observation that *Rs. rubrum* fixed nitrogen was confirmed at Wisconsin by Lindstrom *et al.*

Duane C. Yoch · Department of Cell Physiology, University of California, Berkeley, California 94720

(1949). Thus, *Rs. rubrum* became the first bacterium to be firmly established as a nitrogen-fixer since *Azotobacter* about 50 years earlier.

After these initial observations in *Rs. rubrum*, nitrogen fixation was also found to occur in a number of other purple nonsulfur bacteria, including *Rhodopseudomonas sphaeroides*, *Rp. capsulata*, *Rp. gelatinosa*, and *Rp. palustris* (Lindstrom *et al.*, 1951; Okuda *et al.*, 1960); in a species of the purple sulfur bacterium *Chromatium*; and in the green sulfur bacterium *Chlorobium limicola f. thiosulfatophilum* (Lindstrom *et al.*, 1949, 1950). In addition, a strain isolated in Russia, *Ectothiorhodospira shaposhnikovii* (Zakhvataeva *et al.*, 1970), and a purple-nonsulfur-like bacterium, *Rhodomicrobium vannielii* (Lindstrom *et al.*, 1951), were also shown to fix nitrogen.

As pointed out by Stewart (1973), it has not been established that all photosynthetic bacteria fix nitrogen; it does appear, however, that all the strains of Rhodospirillaceae that were properly tested were capable of this process. Swoager and Lindstrom (1971) showed that of a large number of isolates of this purple nonsulfur group taken from two central Pennsylvania lakes, all were capable of N_2 fixation. Preliminary evidence indicates that certain members of the "giant" Chromatiaceae (e.g., *Chr. okenii*, *Chr. buberi*, and *Chr. warmingii*) do not fix N_2, in contrast to several species (*Chr. vinosum* and *Chr. violascens*) that do (Trüper and Jannasch, 1968).

2.2. Physiology of Nitrogen Fixation

2.2.1. Nitrogen Fixation in the Light

Photosynthetic bacteria fix N_2 at optimal rates only under anaerobic conditions in the light (Kamen and Gest, 1949; Lindstrom *et al.*, 1949, 1951). The close association between N_2 fixation and photosynthesis was observed more precisely by Pratt and Frenkel (1959) and by Paschinger (1974) during continuous monitoring of N_2 fixation with a recording mass spectrometer that showed that $^{15}N_2$ uptake by *Rs. rubrum* stopped immediately and completely when the light was turned off. Using manometric techniques, Arnon *et al.* (1961) obtained similar results with a *Chromatium* species.

2.2.2. Nitrogen Fixation in the Dark

Although N_2 fixation proceeds optimally in the light, low but significant levels of N_2 appear to be fixed under dark, anaerobic conditions by *Rs. rubrum* (Lindstrom *et al.*, 1951) and a *Chromatium* species (Newton and Wilson, 1953). In an experiment with an alternating light–dark–light pattern, Bennett *et al.* (1964) also found that N_2 uptake by *Chr. vinosum* Strain D did not stop immediately when the light was turned off; N_2 continued to be taken up at a decreasing rate for about 30 min. In an almost identical experiment, Schick (1971a) reported the same results with *Rs. rubrum*. Furthermore, Benemann (personal communication) observed that in the dark, *Rs. rubrum* cells could sustain acetylene reduction (a measure of nitrogenase activity) for periods of up to 24 hr at rates 5% of those observed in the light. The small but significant fixation of N_2 in the dark was ascribed by Schick (1971a) to an accumulation of photosynthetic intermediates in the light. Apparently, during longer periods of N_2 fixation in the dark, these organisms are capable of drawing on their carbohydrate reserves through fermentative processes to supply the ATP and reductant required for nitrogenase activity (see Chapter 46 for details of the fermentative metabolism of photosynthetic bacteria).

Nitrogen fixation by photosynthetic bacteria is completely inhibited by O_2. This is perhaps not unexpected because (1) the nitrogenase enzyme is irreversibly inhibited by O_2 and (2) the strong reductant (reduced ferredoxin) required for nitrogenase activity is easily oxidized by O_2.

2.2.3. Ammonia as a Product and an Inhibitor

Exposure of *Rs. rubrum*, *Chromatium* species, and *Chl. limicola f. thiosulfatophilum* to $^{15}N_2$ showed that the products with the greatest ^{15}N enrichment were NH_4^+ and glutamic acid (Wall *et al.*, 1952). When these cells were exposed to $^{15}NH_4^+$, they also showed high levels of ^{15}N-labeled glutamic acid. It was concluded by these workers that like the nonphotosynthetic nitrogen fixers, these organisms reduced N_2 to NH_4^+, and that this was followed by reductive amination of α-ketoglutaric acid to glutamic acid.

The addition of NH_4Cl to N_2-fixing cultures of *Rs. rubrum* inhibited nitrogen fixation (Kamen and Gest, 1949; Gest and Kamen, 1949b; Lindstrom *et al.*, 1949), as it does in nonphotosynthetic bacteria (Wilson *et al.*, 1943). An analysis of this inhibitory effect has been interpreted to mean that nitrogenase is an inducible enzyme the synthesis of which is repressed by NH_4^+ (Strandberg and Wilson, 1968). Continuous cultures of *Rs. rubrum* ATCC 11170 supplied with limiting concentrations of NH_4^+ gave cell extracts that were more active in N_2 fixation than were extracts from cells cultured on N_2, indicating that the N_2-fixation process produced enough NH_4^+ to partially repress the synthesis of nitrogenase (Munson and Burris, 1969).

Because the inhibition of N_2 fixation by NH_4^+ in *Rs. rubrum* was very rapid (Schick, 1971a), Stewart (1973) suggested that NH_4^+ not only represses nitrogenase synthesis, but might also function as a cell toxin or an inhibitor of cyclic photophosphorylation. This hypothesis seems unlikely, because *Rs. rubrum* grows quite well in the presence of high concentrations of NH_4^+.

Nagatani *et al.* (1971) proposed that under N_2-fixing conditions, bacteria (including photosynthetic bacteria) incorporate NH_4^+ into glutamate via a new pathway recently discovered by Tempest *et al.* (1970). Ammonia is believed to be incorporated by a pathway that involves glutamine synthetase (GS) in combination with glutamate synthase (GOGAT) and α-ketoglutarate (α-KG),

$$\text{Ammonia} + \text{glutamate} \xrightarrow{\text{GS}}$$
$$\text{glutamine} \xrightarrow[\alpha\text{-KG}]{\text{GOGAT}} 2 \text{ glutamate} \quad (1)$$

rather than by the conventional pathway involving glutamate dehydrogenase (GDH):

$$\text{Ammonia} + \alpha\text{-ketoglutarate} \xrightarrow{\text{GDH}}$$
$$\text{glutamate} \xrightarrow[NH_4^+]{\text{GS}} \text{glutamine} \quad (2)$$

The primary evidence for this proposal was that extracts from a N_2-fixing culture of *Klebsiella pneumoniae* had very low levels of glutamate dehydrogenase, whereas GOGAT, the key enzyme of the new pathway, was found to be at high levels.

Although both *Chr. vinosum* Strain D and *Rs. rubrum* had high levels of GS activity, the level of this enzyme in *Chr. vinosum* appeared not to be affected by the nitrogen substrate (N_2 or NH_4^+) (Nagatani *et al.*, 1971). In addition to GOGAT activity, these workers found GS activity in *Chr. vinosum* extracts. Glutamine synthetase has a high affinity (low K_m) for NH_4^+ that is believed to provide non-N_2-fixing cells with a means of scavenging NH_4^+ from a nitrogen-deficient environment or to allow N_2-fixing cells to keep the internal level of the newly synthesized NH_4^+ low, thereby preventing any repression of the nitrogenase enzyme; GS is also thought to have a key regulatory function in N_2 metabolism (Shanmugam and Valentine, 1975).

2.2.4. Accessory Electron Donors

Reducing power and ATP are both required for biosynthetic processes, such as CO_2 reduction and N_2 fixation. To generate the low-potential reductant

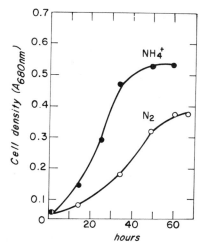

Fig. 1. Effect of nitrogen metabolism on the growth yield of *Rhodopseudomonas palustris*. Reproduced from Yoch *et al.* (1968).

(pyridine nucleotides and ferredoxin) for these processes, photosynthetic bacteria require an oxidizable substrate as a source of electrons; a net input of electrons is not required for ATP synthesis (cyclic photophosphorylation). Depending on the species of photosynthetic bacteria, either organic acids, such as acetate or malate, or reduced sulfur compounds, such as thiosulfate or sulfide, can be oxidized as a source of electrons for these processes. Arnon *et al.* (1961) showed that N_2 fixation by *Chr. vinosum* was strictly dependent not only on light but also on thiosulfate as an accessory electron donor. Thiosulfate can also serve as an electron donor for N_2 fixation by the purple nonsulfur bacterium *Rp. palustris* (Yoch *et al.*, 1968). A differential requirement for reducing power that depended on the type of nitrogen metabolism (N_2 fixation or NH_4^+ assimilation) could be demonstrated with this organism (Fig. 1). With growth-limiting concentrations of pyruvate (5 mM) as the accessory electron donor, *Rp. palustris* achieved a 30% higher cell density when supplied with NH_4^+ rather than N_2 gas as a source of nitrogen, thus demonstrating an increased requirement for reductant during N_2 fixation.

Organic acids not only serve as a source of electrons for the reduction of N_2 to NH_4^+ by photosynthetic bacteria, but also serve as NH_4^+ acceptors. Without suitable carbon skeletons to accept NH_4^+, N_2 fixation in *Chr. vinosum* (Arnon *et al.*, 1961) and *Rs. rubrum* (Schick, 1971c) presumably was inhibited by the accumulated excess NH_4^+.

3. Photoevolution of Hydrogen

3.1. Conditions That Affect the Photoevolution of Hydrogen

Gest and Kamen (1949a,b) were the first investigators to observe the photoevolution of H_2 in photosynthetic bacteria. (The production of H_2 in the dark by photosynthetic bacteria by "conventional" mechanisms is discussed in Section 5.2.) Light-dependent H_2 evolution had previously been observed in the green alga *Scenedesmus* by Gaffron and Rubin (1942), and was thought to be the result of the photodecomposition of water. Hydrogen evolution by algae, however, occurs under very different conditions and in quantities two orders of magnitude lower than that observed in the photosynthetic bacteria.

A number of factors, including cultural conditions, have been shown to affect the photoevolution of H_2 by photosynthetic bacteria. When *Rs. rubrum* was grown photosynthetically on synthetic medium containing as a carbon source either malate, fumarate, or pyruvate (citric acid cycle intermediates) and either N_2, glutamate, or aspartate as a nitrogen source, the culture not only evolved CO_2, but also exhibited a vigorous evolution of H_2 gas (Gest and Kamen, 1949a,b; Gest *et al.*, 1950; Ormerod *et al.*, 1961). The photoevolution of H_2 was completely inhibited by NH_4^+, high concentrations of yeast extract or peptone (Gest and Kamen, 1949b), or amino acids such as alanine and serine that are readily decomposed to release NH_4^+ (Bregoff and Kamen, 1952). If, however, NH_4^+ was maintained at a very low but constant level in the culture (growth in a chemostat), the photoevolution of H_2 was readily observed (Ormerod *et al.*, 1961). The gas phase of the cell suspension also had a significant effect on the photoproduction of H_2. When the gas phase was helium, hydrogen, or argon, H_2 was readily evolved in the light; however, under an atmosphere of N_2, H_2 evolution was halted abruptly (Gest and Kamen, 1949a). Finally, there were indications that the enzyme system responsible for the photoevolution of H_2 was sensitive to oxygen, because aeration of the culture caused a pronounced lag in the photoevolution of H_2 (Ormerod *et al.*, 1961).

The inhibitory effect of NH_4^+ on the photoevolution of H_2 (Gest *et al.*, 1950) suggested that it repressed the synthesis of the H_2-evolving system. Ormerod *et al.* (1961) tested this hypothesis by transferring cultures of *Rs. rubrum* from NH_4^+-containing media, in which there was no H_2 evolution, to nitrogen-free media (in which H_2 evolution was normally observed), and showed that (1) the addition of chloramphenicol (an inhibitor of protein synthesis) to the medium

permanently inhibited photoevolution of H_2; (2) the addition of glutamate caused a considerable acceleration in the rate of H_2 evolution, presumably by providing amino acids for synthesis of the enzyme; and (3) the addition of amino acid analogues to the nitrogen-free medium increased the lag period before the onset of H_2 evolution. These observations were consistent with the idea expressed by Ormerod *et al.* (1961) that NH_4^+ repressed the synthesis of one or more of the enzymes specifically required for light-dependent H_2 production.

During the course of these early studies on the photoevolution of H_2, it seemed that CO_2, which evolved simultaneously with H_2, played an integral part in this phenomenon (Gest and Kamen, 1949a,b; Gest *et al.*, 1950). Furthermore, because of an apparent requirement for CO_2 for the photoevolution of H_2 from organic acid electron donors by *Rp. gelatinosa* (Siegel and Kamen, 1951) and *Chr. vinosum* Strain D (Newton and Wilson, 1953), these investigators concluded that CO_2 was of critical importance for the photoevolution of H_2. Newton and Wilson (1953) proposed that CO_2 condenses with a C_4 compound to form α-ketoglutarate, which could serve as a source of H_2. Although CO_2 seemed to be required for the photoevolution of H_2, Siegel and Kamen (1951) recognized that H_2 evolution should be possible from all metabolizable substrates.

3.2. Early Hypotheses on the Relationship between Hydrogen Evolution and Nitrogen Fixation

In 1950, Gest *et al.* (1950) proposed a generalized scheme to explain the relationship between N_2 fixation and the photoevolution of H_2 by photosynthetic bacteria. This scheme envisioned that these two processes were linked to both the photolysis of water and the oxidation of organic acids. The photolysis of water was thought to yield oxidizing equivalents (YOH) and reducing equivalent (XH). Organic acids (e.g., malate, fumarate) would be used by the cell to reduce the oxidizing equivalents, and the reducing equivalents would be either liberated as H_2 gas or used in the reduction of N_2 or NH_4^+. The inhibition by N_2 or NH_4^+ of the photoevolution of H_2 was believed to be the result of the reducing equivalent being used in the reduction of N_2 and the subsequent reductive ammination and thereby becoming less available for evolution as molecular H_2. Evidence that seemed to support this hypothesis was seen in the vigorous photoevolution of H_2 by cultures of photosynthetic bacteria grown on amino acids, such as glutamate or aspartate, as the nitrogen source. Hydrogen evolution was believed to be possible under these conditions, because protein synthesis took place without the incorporation of N_2 or NH_4 and a surplus of reducing equivalents was generated and subsequently released as H_2 gas.

In a later refinement of the water-splitting hypothesis, Gest et al. (1956) proposed that electrons of low oxidation–reduction (redox) potential or H_2 resulting from the photolysis of water was transferred into a "pool" where the reducing power could eventually be used for N_2 fixation, pyridine nucleotide reduction, or H_2 evolution via hydrogenase. The influence of NH_4^+ and N_2 on the photoevolution of H_2 was still believed to be a competition between the H_2-activation process and the conversion of N_2 to amino acid nitrogen.

On the basis of their determinations of the stoichiometries, Bregoff and Kamen (1952) proposed that the photoevolution of H_2 by photosynthetic bacteria could be accounted for by the dissimilation of organic acids and could be represented by the following equation:

$$C_4H_6O_5 + H_2O \rightarrow C_2H_4O_2 + 2\,CO_2 + 2\,H_2 \quad (3)$$

The basis of the inhibitory effect of N_2 and NH_4^+ on the photoevolution of H_2 was believed to be similar to that proposed previously by Gest et al. (1950).

Another view of the role of light in N_2 fixation and H_2 evolution was proposed by Losada et al. (1961) and Arnon et al. (1961), who demonstrated for the first time that an inorganic electron donor, thiosulfate, could provide the reducing equivalents necessary for the photoevolution of H_2 (and light-dependent N_2 fixation). There was no CO_2 requirement for H_2 evolution from thiosulfate, and the "CO_2 effect" previously observed (Siegel and Kamen, 1951; Newton and Wilson, 1953) was relegated to a secondary role. The organism used in these studies by Losada et al. (1961) was a Chromatium species.

The photoevolution of H_2 from thiosulfate by Chr. vinosum was shown to respond to the conditions of light or dark, the gas phase, and the N_2 source in the same way as did H_2 evolution from Rs. rubrum with organic H_2 donors. These observations suggested that regardless of the electron donor, the enzyme system responsible for the photoevolution of H_2 was similar in all photosynthetic bacteria (Losada et al., 1961). Arnon et al. (1961) concluded from these studies that "Chromatium uses light energy to raise electrons, supplied by thiosulfate to chlorophyll, via cytochromes, to a reducing potential at least equal to that of molecular hydrogen." This view was consistent with the electron-flow theory this group had proposed earlier for plant photosynthesis (Arnon, 1961).

Although a close relationship between light-dependent H_2 evolution and N_2 fixation was recognized, all the workers in this field prior to 1962 believed that the photoevolution of H_2 was caused by the activity of hydrogenase (Gest et al., 1950; Gest and Kamen, 1960; Arnon et al., 1961; Ormerod et al., 1961; Losada et al., 1961). In 1962, the true relationship betwen N_2 fixation and the photoevolution of H_2 by photosynthetic bacteria was surmised by Ormerod and Gest (1962), who reasoned that "the repression of synthesis of the hydrogen-evolving system in Rs. rubrum by ammonia is possibly another example of 'feedback repression.' Since ammonia can be assumed to be the end product of N_2 fixation, this interpretation would suggest that both H_2 production and N_2 activation in Rs. rubrum are catalyzed by the same enzyme or enzyme complex" [emphasis added]. This proposal was made just as Carnahan, Mortenson, and co-workers at Du Pont were achieving for the first time an active and reproducible nitrogenase activity in cell-free preparations of Clostridium pasteurianum (Carnahan et al., 1960a,b).

3.3. ATP-Dependent Hydrogen Evolution from Nitrogenase in Relation to the Photoevolution of Hydrogen

The demonstration of N_2 fixation in cell-free extracts and its requirements for reductant and ATP (Mortenson, 1964; Hardy and D'Eustachio, 1964) was soon extended to the photosynthetic bacterium Rs. rubrum (Schneider et al., 1960; Bulen et al., 1965a). In Rs. rubrum, Bulen et al. (1965a) further showed that the cofactors required for N_2 fixation (a low-potential reductant and ATP) were also required for a new type of hydrogenase activity found in extracts of N_2-fixing organisms. This new type of "ATP-dependent hydrogenase" was shown to be distinct from the "conventional" hydrogenases of heterotrophic organisms that evolve H_2 in the dark in that (1) it was not influenced by the partial pressure of H_2; (2) it was not reversible; (3) it was insensitive to carbon monoxide (CO), a common inhibitor of hydrogenase; and (4) it was found only in extracts of cells grown on N_2 (Bulen et al., 1965a). Furthermore, it was shown that on purification, the nitrogenase activity and ATP-dependent hydrogenase activity closely paralleled one another (see the review by Hardy and Knight, 1967). The conclusion Bulen et al. (1965b) drew from experiments on ATP-dependent H_2 evolution by extracts of Rs. rubrum, Azotobacter vinelandii, and Cl. pasteurianum was that "both N_2 reduction and ATP-dependent H_2 evolution reactions are catalyzed by the same enzyme, namely, nitrogenase."

Table 1. Conditions That Affect the Photoevolution of Hydrogen by Photosynthetic Bacteria Now Known To Have Comparable Effects on Nitrogenase Activity (ATP-Dependent Hydrogen Evolution)

Condition[a]	Photoevolution of hydrogen[a]	Nitrogenase activity (ATP-dependent hydrogen evolution)[a]
Light and an accessory electron donor	Both required[1]	Light and electrons are equivalent to ATP and low-potential reductant (reduced ferredoxin); both required for nitrogenase activity[2]
Excess NH_4^+ in growth medium	Inhibition[3]	Inhibition (synthesis of nitrogenase repressed by NH_4^+)[4]
Low steady-state level of NH_4^+	No effect[5]	No effect (nitrogenase not repressed by NH_4^+ below a concentration of 10^{-6} M)[6]
Presence of oxygen	Inhibition[5]	Inhibition (nitrogenase extremely sensitive to O_2)[7]
Nitrogen starvation or cells grown on N_2 or glutamate	High activity[8]	High activity (nitrogenase derepressed under these conditions)[9]
Cells transferred from NH_4^+ to N_2- or N-free medium:		
a. Effect of adding chloramphenicol	a. Inhibition[5]	a. Inhibition of nitrogenase synthesis[10]
b. Effect of adding glutamate	b. Stimulation[5]	b. Stimulation of nitrogenase synthesis[11]
Presence of carbon monoxide[12]	No effect[12]	No effect[13]
Molybdenum in growth medium	Essential[14]	Essential (nitrogenase is a Mo-containing protein)[15]
Addition of phosphorylation inhibitors	Inhibition[16]	Inhibition (nitrogenase activity requires ATP)[17]

[a] References: (1) Kamen and Gest (1949), Arnon *et al.* (1961), Schick (1971*a*); (2) Mortenson *et al.* (1963), D'Eustachio and Hardy (1964), Mortenson (1964); (3) Gest and Kamen (1949*a*); (4) Wilson *et al.* (1943), Kamen and Gest (1949), Strandberg and Wilson (1968); (5) Ormerod *et al.* (1961); (6) Munson and Burris (1969); (7) Carnahan *et al.* (1960*a,b*), Burns and Bulen (1966); (8) Gest and Kamen (1949*a,b*), Gest *et al.* (1950); (9) Strandberg and Wilson (1968), Burns and Bulen (1966), Yoch and Arnon (1975); (10) Lindsay (1963); (11) Yoch and Pengra (1966); (12) Winter and Arnon (1970); (13) see Wilson (1940) and Bulen *et al.* (1965*b*); (14) Paschinger (1974); (15) Bortels (1930), Esposito and Wilson (1956), Bulen and LeComte (1966); (16) see Gest (1972); (17) Schick (1971*c*).

That this ATP-dependent hydrogenase was found only in extracts of N_2-fixing cells was in agreement with earlier observations that *Rs. rubrum* photo-evolved H_2 only when grown on N_2 or on limiting concentrations of NH_4^+. Furthermore, the light requirement for H_2 evolution by photosynthetic bacteria could be explained by the need for ATP and reducing power. Conditions previously observed to affect both the activity and the development of the system that photoevolves H_2 in photosynthetic bacteria are now seen to be consistent with those conditions required for the activity and development of ATP-dependent H_2 evolution by nitrogenase (Table 1). Thus, it is apparent that the photoevolution of H_2 by photosynthetic bacteria is in fact a manifestation of the ATP-dependent, H_2-evolving activity of the nitrogenase enzyme. Gest (1972) proposed that the photoevolution of H_2 is a means by which photosynthetic bacteria regulate their intracellular levels of ATP and reducing power during periods when synthetic processes are not occurring. It remains to be determined, however, whether nitrogenase would serve such a regulatory

function in nature because the high concentration of N_2 in the biosphere would probably inhibit H_2 evolution and, furthermore, the NH_4^+ that is sometimes present in the environment of the cell would repress the nitrogenase and make this regulatory device unreliable.

4. Biochemistry of Nitrogen Fixation

4.1. Prerequisites for Nitrogen Fixation

4.1.1. Preparation of Active Extracts

The report by Carnahan *et al.* (1960*b*) that extracts of *Cl. pasteurianum* supported high (and reproducible) rates of $^{15}N_2$ incorporation represented a major breakthrough, because this accomplishment had eluded workers in the field for 20 years (see the review by Burris, 1974). Their success was due to (1) anaerobic conditions in all phases of the operation; (2)

keeping cells and cell extracts at room temperature—one of the components is cold-labile; and (3) supplying the extracts with pyruvate at high concentrations (100–150 mM) to serve as a source of ATP and reductant for nitrogenase (see Section 4.2.1).

In experiments using similar techniques of breaking the cells and handling the extracts, Schneider *et al.* (1960) confirmed the observations of Carnahan *et al.* (1960*b*) and extended them to include *Rs. rubrum.* These extracts were reported to be equally active with and without pyruvate as a substrate, and light had no effect. Subsequent attempts to reproduce these results with *Rs. rubrum* have been unsuccessful; however, when the extracts were supplied with the known cofactors for N_2 fixation (ATP and reductant), good activity was observed (Bulen *et al.*, 1965*a*). Success with *Rs. rubrum* extracts seems to be dependent on controlling the pH of the extract (Burns and Bulen, 1966; Yoch and Arnon, 1975). When *Rs. rubrum* cells are broken, the pH drops rather quickly, to about 5.0. This is probably because storage products, such as poly-β-hydroxybutyrate, are being degraded to the level of organic acids. Munson and Burris (1969) obtained active extracts by growing *Rs. rubrum* on low concentrations of NH_4^+ rather than on N_2 (nitrogenase is derepressed when NH_4^+ is kept at a low level). Control of pH was not a problem in extracts of *Chr. vinosum* Strain D, but it was necessary to wash the exogenous substrate from the cells before disruption to prevent the extracts incorporating the newly synthesized NH_4^+ into amino acids, thus resulting in low or no apparent nitrogenase activity (Winter and Arnon, 1970) when measured by the synthesis of NH_4^+.

4.1.2. Reductant and ATP

In N_2-fixing extracts of *Cl. pasteurianum*, pyruvate was shown to have a dual function in that it provided nitrogenase with both reducing power (electrons) and ATP via a highly active phosphoroclastic system (Mortenson *et al.*, 1963; D'Eustachio and Hardy, 1964; Mortenson, 1964). The reducing power of pyruvate was not coupled to nitrogenase directly, but only through an electron carrier such as ferredoxin (Mortenson *et al.*, 1963) or, in iron-starved cells, through flavodoxin (Knight *et al.*, 1966). All extracts tested, including those of photosynthetic bacteria, require a strong reductant and ATP for nitrogenase activity. Although a clostridial-type phosphoroclastic cleavage of pyruvate has been reported for *Chr. vinosum* Strain D (Bennett *et al.*, 1964), pyruvate supported nitrogenase activity in extracts of this organism has yet to be demonstrated. Whereas the

products of pyruvate oxidation appeared to be those of a phosphoroclastic-type reaction, it must be pointed out that no evidence was presented to link the phosphoroclastic reaction of *Chromatium*, like its counterpart in clostridia, to the reduction of ferredoxin.

Pyruvate-supported nitrogenase activity was reported in extracts of "*Chloropseudomonas ethylica*" Strain N2 (Evans and Smith, 1971); this organism, however, was found to be a mixed culture of a green photosynthetic bacterium and an unidentified nonphotosynthetic species (Gray *et al.*, 1972, 1973).* Thus, the possibility of a pyruvate-linked reduction of ferredoxin by the nonphotosynthetic contaminant cannot be discounted.

Nitrogen fixation in cell extracts does not depend on pyruvate—any enzyme system capable of reducing ferredoxin (or flavodoxin) will serve as a source of reducing power. In these photosynthetic microbes, ATP would almost certainly be provided by cyclic photophosphorylation. In this regard, chromatophores have been shown to provide ATP for nitrogenase activity by extracts of *Chr. vinosum* Strain D (Yoch and Arnon, 1970). (For *in vitro* experiments, ATP is best supplied by an ATP-regenerating system, such as creatine phosphate, creatine phosphokinase, and ATP; such a system keeps the level of ADP, a potent inhibitor of nitrogenase, from rising.)

Reducing power for nitrogenase activity in extracts of *Chr. vinosum* Strain D was generated from H_2 coupled to either the native hydrogenase or clostridial hydrogenase (Winter and Arnon, 1970), presumably through ferredoxin, as ferredoxin-linked hydrogenase activity was demonstrated in extracts of *Chr. vinosum* Strain D by Weaver *et al.* (1965). Nitrogenase activity in "*Chloropseudomonas ethylica*" was coupled to ferredoxin photochemically reduced by a particulate fraction of another green sulfur bacterium, *Chl. limicola f. thiosulfatophilum* (Evans and Smith, 1971). Since cell-free extracts of *Chl. limicola f. thiosulfatophilum* (strain Tassajara) photoreduce ferredoxin (Evans and Buchanan, 1965), this is probably the source of reducing power for the nitrogenase of this organism. Finally, ferredoxins from *Chr. vinosum* Strain D (Yoch and Arnon, 1970) and *Rs. rubrum* (Yoch and Arnon, 1975) were reduced artificially by spinach chloroplasts† and coupled to the nitrogenases of these purple photosynthetic bacteria.

---

* See Chapter 2 for a discussion of the taxonomy of this syntrophic system.

† The "chloroplast–nitrogenase" system [reduced dye → chloroplasts → ferredoxin → nitrogenase] has been useful in identifying the electron carriers of a number of N_2-fixing organisms (Yoch, 1977). Unlike dithionite, chloroplasts couple reducing power to nitrogenase only through intermediate electron carriers.

Using sodium dithionite as a reductant for *Azotobacter vinelandii* and *Rs. rubrum* nitrogenases, Bulen *et al.* (1965a) showed that electron carriers were not necessary to couple reducing power to nitrogenase, and that the enzyme could be reduced directly by strong chemical reductants.*

4.1.3. Electron Carriers

Little is known about how photosynthetic bacteria generate reducing power of low redox potential (approximately −400 mV) for nitrogenase reduction (see Section 4.3). It is well known, however, that ferredoxin (or flavodoxin) is the electron carrier that couples this low-potential reducing power to the nitrogenase enzyme, and that all photosynthetic bacteria contain ferredoxin (see the reviews by Buchanan and Arnon, 1970; Yoch and Valentine, 1972).

Ferredoxin was first reported in *Rs. rubrum* by Tagawa and Arnon (1962). In a later study, Shanmugam *et al.* (1972) found that *Rs. rubrum* had two types of ferredoxin (called FdI and FdII), and that both types were present in cells grown phototrophically, whereas only one type (FdII) was found in cells grown chemotrophically in the dark. This discovery was confirmed and extended by Yoch *et al.* (1975), who found that FdI was an 8 Fe–8 S ferredoxin ($M_r = 9000$) and that FdII was a 4 Fe–4 S ferredoxin ($M_r = 15,000$). Both FdI and FdII coupled reducing power to the nitrogenase of *Rs. rubrum* when artificially reduced by illuminated chloroplasts (Yoch and Arnon, 1975). On a molar basis, FdI was found to be three to four times more effective as a reductant than FdII. It may be that FdI is involved in the nitrogenase reaction *in vivo* and FdII serves other ferredoxin-linked enzymes in the cell. Ferredoxins have also been isolated and characterized from *Chr. vinosum* Strain D (Tagawa and Arnon, 1962), *Chl. limicola f. thiosulfatophilum* (Rao *et al.*, 1969), *Rp. sphaeroides* (Yoch *et al.*, 1975), and "*Chloropseudomonas ethylica*" (Rao *et al.*, 1969).

When grown photosynthetically on an iron-deficient medium, *Rs. rubrum*, like many nonphotosynthetic bacteria, produces a flavodoxin (Cusanovich and Edmondson, 1971) that presumably functions in the cell in place of the ferredoxin(s) under conditions of iron starvation. In its general features, this FMN-containing electron transport protein ($M_r = 23,000$) resembles other flavodoxins. Flavodoxins have not yet been reported from members of the families Chromatiaceae and Chlorobiaceae.

In conclusion, the prerequisite for nitrogenase activity in extracts of photosynthetic bacteria (as in extracts of all nitrogen-fixing bacteria) are: (1) anaerobic conditions in all phases of the operation; (2) a source of ATP, either generated endogenously or supplied as an ATP-regenerating system; and (3) a low-potential reductant, either reduced ferredoxin or flavodoxin, or a nonphysiological chemical reductant, such as sodium dithionite, which reduces the nitrogenase directly.

4.2. The Enzyme

Nitrogenase is the name historically given to an enzyme system that reduces N_2 to NH_4^+. The metal-containing nature of this enzyme system was suspected for many years (Wilson and Burris, 1947) prior to the demonstration by Bulen and LeComte (1966) that *Azotobacter vinelandii* nitrogenase consists of two metal-containing components, one a Mo–Fe protein and the other an Fe protein (neither of which has enzymatic activity by itself). Acid-labile sulfur was later found to be associated with both protein fractions (Mortenson *et al.*, 1967).

Since 1966, nitrogenases from such diverse groups as obligate aerobes, fermentative bacteria, legume root nodule bacteroids, and photosynthetic bacteria have been purified to varying degrees, and all proved to be remarkably similar in composition, size, and substrate–inhibitor specificity. Among the photosynthetic bacteria, two nitrogenase protein components have been separated and partially purified from "*Chloropseudomonas ethylica*" (Evans, *et al.*, 1971; Smith, R. V., *et al.*, 1971), *Rs. rubrum* (Biggins *et al.*, 1971; Nordlund and Baltscheffsky, 1973), and *Chr. vinosum* (Winter and Ober, 1973; Evans, *et al.*, 1973). Evans *et al.* (1973) purified to homogeneity the Mo–Fe protein of *Chr. vinosum* strain D.

Nitrogenase is a multielectron-transfer enzyme (6 e^- per N_2 reduced), and the Fe and Mo are believed to play an essential role in the electron-transfer process (Brintzinger, 1966; Hardy *et al.*, 1971). Because these transition metals are paramagnetic, electron paramagnetic resonance (EPR) spectroscopy is an ideal tool for studying the mechanism of electron transfer in this complicated enzyme. The appearance in the Fe protein of an EPR signal at $g = 1.94$ on reduction and its subsequent disappearance on reoxidation by substrate provides direct evidence that Fe is involved in the reduction of N_2 to NH_4^+. [The EPR signal at $g =$

---

*The discovery that an artificial chemical reductant such as dithionite could be used to reduce nitrogenase directly was one of the most significant methodological advances in N_2-fixation research. With dithionite as a reductant, it became possible first to purify and then to characterize nitrogenase without the interference of an enzymatic electron-generating system. Furthermore, the use of dithionite permitted study of nitrogenase in bacteria, in which the natural reducing system had not yet been identified.

1.94 has been well characterized as originating from a nonheme type of Fe in the reduced state (see the review by Beinert, 1965).] The Fe protein from a photosynthetic organism has not been purified, but those from other N_2-fixing bacteria produce EPR signals at $g = 2.05$, 1.94, and 1.86.

The purified Mo–Fe protein of *Chr. vinosum*, when reduced with dithionite, showed low-temperature EPR signals at $g = 4.3$, 3.68, and 2.01 that are similar to those reported earlier for *Azotobacter vinelandii* (Hardy *et al.*, 1971; Orme-Johnson *et al.*, 1972), *Cl. pasteurianum* (Palmer *et al.*, 1972; Orme-Johnson *et al.*, 1972), and *Klebsiella pneumoniae* (Eady *et al.*, 1972). Although the Mo–Fe protein undergoes changes in its EPR spectrum on reduction and re-oxidation, Evans *et al.* (1973) were unable to observe any changes in the EPR spectrum of the *Chr. vinosum* Mo–Fe protein after substitution of ^{95}Mo for the naturally occurring isotope. Significant changes would be expected in the EPR spectrum of any Mo-containing paramagnetic center after such a substitution. Although Mo has been shown to be essential for the growth of microorganisms on N_2 (Bortels, 1930; Esposito and Wilson, 1956; Pengra and Wilson, 1959) and for the synthesis of the Mo–Fe protein component of nitrogenase (Nagatani and Brill, 1974), and appears to be involved in the photoevolution of H_2 by *Rs. rubrum* (Paschinger, 1974), there is as yet no evidence that Mo is directly involved in the electron-transfer reactions of nitrogenase.

From EPR analyses (Orme-Johnson *et al.*, 1972; Zumft *et al.*, 1973; Mortenson *et al.*, 1973) and the substrate and Mg · ATP-binding studies of Bui and Mortenson (1968), the following chain of events is believed to occur in the nitrogenase enzyme during N_2 fixation: (1) the reduced Fe protein binds Mg · ATP, which has the effect of lowering the oxidation–reduction potential of the Fe protein; (2) the reduced Fe protein–Mg · ATP complex reduces the Mo–Fe protein; and (3) the reduced Mo–Fe protein is re-oxidized by the substrate (the reduced Fe protein–Mg · ATP complex is also reoxidized, but presumably through the Mo–Fe protein). The electron-transport scheme has been represented as follows:

$$\text{Reductant} \xrightarrow{e^-} \text{Fe protein} \xrightarrow[\substack{\text{ATP} \quad \text{ADP} + P_i \\ \rightarrow}]{e^-}$$

$$\text{Mo–Fe protein} \xrightarrow{e^-} N_2 \text{ (or } C_2H_2) \quad (4)$$

The observations by EPR spectroscopy of B. E. Smith *et al.* (1973) with *Kl. pneumoniae* nitrogenase differ somewhat from those presented here, but the electron-transport scheme is believed to be the same as that shown above.

A redox titration of the Mo–Fe protein from *Chr. vinosum* Strain D showed two centers with different midpoint potentials, one with $E_m = -60$ mV and the other with $E_m = -260$ mV (Albrecht and Evans, 1973). The relationship of these two paramagnetic centers to enzyme activity is not clear at this time; Zumft *et al.* (1974) suggested from work with *Cl. pasteurianum* Mo–Fe protein that the center with the more positive potential represents the active form of the protein and that the second center, of more negative potential, is from an inactive form of the protein. The reduction of substrate by *Chr. vinosum* nitro-

Table 2. Molecular Weight and Composition of the Two Protein Components of Nitrogenase

| Composition | Clostridium pasteurianum[a] and Klebsiella pneumoniae[b] | | Chromatium vinosum[c] |
	Fe protein	Mo–Fe protein	Mo–Fe protein
Molecular weight	47,000–55,000	212,000–220,000	≈220,000
Subunits	2	4	d
Metal content (g-atom/mol)			
Mo	None	2	2
Fe	3–4	18–24	19
Acid-labile sulfur	4	24	15

[a] Data from Zumft and Mortenson (1975). [b] Data from Eady *et al.* (1972). [c] Data from Evans *et al.* (1973).
[d] Not determined.

genase was shown to require an ambient potential below −450 mV; presumably the Fe protein will not function as an electron carrier above this potential (Evans and Albrecht, 1974).

The similarity of the nitrogenase components from *Cl. pasteurianum* (Zumft and Mortenson, 1975), *Kl. pneumoniae* (Eady *et al.*, 1972), and *Chr. vinosum* (Evans *et al.*, 1973) is shown in Table 2. The Fe protein has a molecular weight of about 55,000, with two subunits of about equal size and 4 Fe (nonheme iron) and 4 S^{2-} (acid-labile sulfur) groups per molecule. The Mo–Fe protein has a molecular weight of about 220,000, four subunits, and about 20 Fe and S^{2-} groups and 1 or 2 atoms of Mo per molecule. The similarity in composition of the Mo–Fe protein from *Chr. vinosum* (Evans *et al.*, 1973) to the proteins of *Kl. pneumoniae* and *Cl. pasteurianum* is also seen in Table 2.

4.2.1. Substrates and Inhibitors

Nitrogenase is highly versatile in its substrate-reducing capacity (see Hardy and Burns, 1968) in that it reduces a number of triple-bonded molecules:

$$N_2 \xrightarrow{6\,e^-} 2\,NH_4^+$$

$$N_3^- \xrightarrow{2\,e^-} NH_4^+ + N_2$$

$$N_2O \xrightarrow{2\,e^-} H_2O + N_2 \qquad (5)$$

$$HCN \xrightarrow[\text{or } 4\,e^-]{6\,e^-} CH_4 + NH_4^+ \text{ or } CH_3NH_2$$

$$C_2H_2 \xrightarrow{2\,e^-} C_2H_4$$

In addition, protons (H^+) are reduced to H_2 (ATP-dependent H_2 evolution). Extracts of *Rs. rubrum* were shown to reduce cyanide, azide, and isocyanide (Munson and Burris, 1969), as well as N_2, acetylene, and protons. Many of the molecules now known to be reduced by nitrogenase had for years been known as inhibitors of N_2 fixation. The discovery by Schöllhorn and Burris (1966) and Dilworth (1966) that nitrogenase reduced acetylene (C_2H_2) to ethylene (C_2H_4) with the same requirements for reductant and ATP as for N_2 reduction led to the development of a highly sensitive, rapid, and inexpensive assay for nitrogenase activity, and represented a significant advance in N_2-fixation methodology* (Hardy *et al.*, 1968).

---

* Prior to the early 1940's, N_2 fixation was assayed by the Kjeldahl method, which permitted detection of about 10^{-6} M nitrogen. When the stable isotope of nitrogen, ^{15}N, became available, Burris *et al.* (1943) developed procedures that permitted detection of increases of 10^{-9} M nitrogen, but the assay was expensive in terms of equipment and time. With the acetylene–ethylene assay, ethylene concentrations as low as 10^{-12} M can be detected quickly and inexpensively by flame-ionization gas chromatography.

4.3. Relationship between Photosynthesis and Nitrogen Fixation

The basic relationship between N_2 fixation and photosynthesis (CO_2 fixation) is that both require energy (ATP) and a low-potential reductant. Therefore, in photosynthetic bacteria one might expect both processes to use the "primary" products (ATP and reductant) of the light reaction, since both CO_2 and N_2 fixation respond directly to light. Although this assumption seems simple enough, it has yet to be demonstrated convincingly that light energy can be used by chromatophores of photosynthetic bacteria to generate both reductant and ATP for nitrogenase activity.

Evidence that the ATP required for N_2 fixation by photosynthetic bacteria can be photochemically generated was provided by the experiments of Yoch and Arnon (1970) on N_2 fixation by extracts of *Chr. vinosum* Strain D in which ATP was supplied by cyclic photophosphorylation in the chromatophores. In these experiments, however, it was necessary to supply an exogenous reductant. Photophosphorylation has not yet been demonstrated in extracts of green sulfur bacteria, but there is no reason to suspect that this process does not provide the ATP required for N_2 fixation and the other biosynthetic processes of these bacteria.

Although there seems to be little doubt that light energy is used directly to supply the nitrogenase with ATP (via cyclic photophosphorylation), such a direct involvement of light in providing a reductant for N_2 fixation has not been demonstrated and is still a matter of speculation. Regardless how the reductant is generated for N_2 fixation, there seems to be little doubt that ferredoxin (or flavodoxin) is the actual reducing agent, as it is for nitrogenases from all other sources (see the review by Yoch, 1978). In photosynthetic bacteria, ferredoxin could be reduced in one of several ways: (1) it could be photochemically reduced as the primary (or secondary) electron acceptor for the bacterial light reaction; (2) it might be reduced by NADH (generated photochemically by ATP-driven reverse electron flow from carbon substrates such as succinate and coupled to ferredoxin by NADH-ferredoxin oxidoreductase; or (3) it could be coupled to the dark oxidation of a substrate such as H_2 or pyruvate by reactions similar to those of the clostridia.

To consider the last possibility first, Bennett *et al.* (1964) proposed that *Chromatium* generates reduced ferredoxin for N_2 fixation by a clostridial-type phosphoroclastic reaction and uses light energy only for ATP synthesis. It might be argued, however, that an organism that uses reduced ferredoxin to synthesize pyruvate from acetyl-CoA and CO_2 (Bachofen et al., 1964) would not use pyruvate (in a back-reaction) to reduce ferredoxin for N_2 fixation.

In considering a light-generated reductant, a noncyclic flow of electrons to the primary acceptor, analogous to noncyclic electron flow in chloroplasts of higher plants, was proposed by Arnon et al. (1961) as a mechanism for providing a reductant for reactions such as CO_2 fixation, pyridine nucleotide reduction, and N_2 fixation in photosynthetic bacteria. When ferredoxin was discovered, it was proposed that this Fe–S protein was the primary electron acceptor in the bacterial noncyclic electron pathway (Tagawa and Arnon, 1962). In the green sulfur consortium "*Chloropseudomonas ethylica*," Evans and Smith (1971) were able to show that extracts could generate reducing power for nitrogenase in a ferredoxin-dependent reaction; however, the photoreduction of ferredoxin in extracts of purple bacteria has yet to be demonstrated.

Although there is no evidence for the direct reduction of ferredoxin by chromatophores of purple bacteria, there has been progress in the past several years in determining the electron-transport constituents of the chromatophores, the potential of the primary reaction, and the nature of the primary electron acceptor. In addition to *b*- and *c*-type cytochromes, four Fe–S centers have been detected by EPR spectroscopy in chromatophores of *Chr. vinosum* (Dutton and Leigh, 1973; Evans et al., 1974), *Rp. sphaeroides*, and *Rp. capsulata* (Prince et al., 1974, 1975). Because none of these Fe–S proteins has a redox potential more negative than −350 mV, none is believed to be a ferredoxin. This view is supported by the observation of Yoch et al. (1975) that the ferredoxins of *Rs. rubrum* are soluble and not membrane-bound. Titrations of the primary photochemical reaction in bacteria indicate a potential of −130 mV, implying that this is the potential of the primary electron acceptor in bacterial photosynthesis (see the review by Dutton and Wilson, 1974, and Chapter 21). A component with a *g*-value of 1.82 (in the reduced form) found in both *Chr. vinosum* (Leigh and Dutton, 1972) and *Rp. sphaeroides* strain R-26 (Dutton et al., 1973) was shown to have a potential of −130 mV, and is now believed to be the primary electron acceptor of the bacterial light reaction. This component was named *photoredoxin* by Dutton and Leigh (1973). Although the potential of photoredoxin at pH 7.0 is −130 mV, the potential is pH-dependent [for each increase of 1 pH unit, the potential of the component becomes 60 mV more negative (Dutton et al., 1973; Jackson et al., 1973; Evans et al., 1974)]. Since the photoreduction of the acceptor *in vivo* appears not to be accompanied by a proton uptake

(Cogdell et al., 1973), it may in fact operate at a potential much lower than −130 mV, perhaps at potentials low enough to reduce ferredoxin. At present, the primary electron acceptor in photosynthesis of purple bacteria is thought to be some type of iron–quinone complex (see Chapters 21 and 22 for detailed discussions of this topic).

It has been proposed (see the review by Gest, 1972) that ATP generated by cyclic photophosphorylation drives the photochemical reduction of NAD by succinate. The first step of this reaction sequence might be represented by (see Chapter 32):

$$\text{Succinate} \xrightarrow[e^-]{\text{ATP}} \text{NADH} \qquad (6)$$

For the reduction of ferredoxin, the second step would also be thermodynamically unfavorable,

$$\text{NADH} \xrightarrow[\text{reductase}]{e^-} \text{ferredoxin}_{\text{red}} \qquad (7)$$

but might occur if the NADH/NAD ratio were high.

Chlorobium vesicles from *Chl. limicola* (unlike chromatophores of purple bacteria) can photoreduce ferredoxin (Evans and Buchanan, 1965; Evans and Smith, 1971). The reaction center bacteriochlorophyll of this organism also has a midpoint potential of −130 mV (Knaff et al., 1973); it has not been determined whether this potential is pH-dependent, as it is in the purple bacteria. Because chlorobium vesicles of green bacteria can reduce ferredoxin *in vitro*, it is obvious that the ability to couple reducing power between the vesicles and the soluble ferredoxin is not completely lost when the cells are broken; in cell extracts of green bacteria, therefore, light-generated reducing power can be coupled to nitrogenase.

5. Hydrogen Metabolism

Until it was realized several years ago that the photoevolution of H_2 by photosynthetic bacteria was a function of the nitrogenase enzyme (see Section 3), H_2 metabolism by this group of microorganisms appeared to be a very complicated phenomenon. Two separate types of H_2 metabolism are now distinguished: (1) photoevolution of H_2 from the nitrogenase enzyme, which is not sensitive to CO (this type of "hydrogenase" activity is found only in nitrogen-starved cells or in those cells cultured on amino acids such as glutamate or aspartate as a nitrogen source; in both situations, nitrogenase is derepressed and synthesized by the cell); and (2) H_2 uptake and evolution by hydrogenase, which is strongly inhibited by CO. Photosynthetic bacteria containing nitrogenase may evolve H_2 from both this enzyme (ATP-dependent H_2 evolution) and hydrogenase, but cells cultured on NH_4^+ will only evolve H_2 by the "conventional" hydrogenase. In evaluating the earlier data in the literature on H_2 metabolism, it is therefore most useful to first determine the nitrogen substrate on which the cells were cultured.

5.1. Growth on Hydrogen

Hydrogen metabolism has been recognized as an important function of the photosynthetic bacteria for more than 40 years. Roelofsen (1935) was the first to observe that H_2 can serve as a source of reducing power for the photoreduction of CO_2 by purple sulfur bacteria, some of which (such as *Chromatium*) could even grow on a mixture of H_2 and CO_2 (Roelofsen, 1935; Gaffron, 1935). Similarly, Larsen (1952) showed that *Chl. limicola f. thiosulfatophilum* could grow on H_2 and CO_2. Thiele (1966) reported that those purple sulfur bacteria (see Chapter 2, Table 3) and green sulfur bacteria that can grow photoautotrophically with H_2 as electron donor possess an active hydrogenase. Because *Chlorobium* species do not reduce sulfate, a small amount of sulfide was required in the medium for amino acid biosynthesis.

The photoautotrophic growth of *Rs. rubrum* strain S1 on a strictly synthetic medium with H_2 as the accessory electron donor was demonstrated by Ormerod and Gest (1962) and confirmed by Anderson and Fuller (1967) and Buchanan *et al.* (1967). Other purple nonsulfur bacteria that grow photoautotrophically with H_2 as an electron donor are *Rp. palustris* (Qadri and Hoare, 1968), *Rp. capsulata* (Klemme and Schlegel, 1967), and *Rp. gelatinosa* (Wertlieb and Vishniac, 1967). The reduction of CO_2 by H_2 seems to be controlled by the availability of organic substrates, since H_2 and CO_2 uptake does not occur in *Rs. rubrum* cultures until the organic substrate is depleted (Schick, 1971*b*).

Preliminary evidence suggests that H_2 will also function as the reductant for N_2 fixation in the dark by extracts of *Chr. vinosum* (Arnon *et al.*, 1961). The use of H_2 as an electron donor for either CO_2 reduction or N_2 fixation suggests a functional hydrogenase and a suitable electron acceptor that will couple to the enzyme in question. The hydrogenases and their associated electron carriers are discussed below.

5.2. Dark Fermentative Hydrogen Evolution

Hydrogen evolution in the dark has been reported in both sulfur and nonsulfur purple bacteria metabolizing organic compounds, such as formate, glucose,

glycerol, glycerophosphate, and pyruvate, and is most probably associated with the breakdown of formic acid (see the review by Gest and Kamen, 1960). Hydrogen evolution from these substrates is not inhibited by N_2 or NH_4^+, and is therefore distinct from light-dependent H_2 evolution by these microorganisms (see Section 3). Cell suspensions of *Rs. rubrum* were shown to decompose formate to CO_2 and H_2 (Kohlmiller and Gest, 1951; Bennett *et al.*, 1964), which is indicative of formate hydrogenlyase activity. Formate hydrogenlyase (formate dehydrogenase plus hydrogenase) activity was demonstrated in extracts of *Rp. palustris*, an organism that also evolves H_2 and CO_2 from formate (Qadri and Hoare, 1968). Although Nakamura (1941) reported the dark anaerobic fermentation of formate to CO_2 and H_2 by *Chr. minutissimum*, neither cell suspensions nor extracts of *Chr. vinosum* Strain D were able to degrade formate in this manner (Bennett *et al.*, 1964). Whereas *Chr. vinosum* does not degrade formate, its breakdown of pyruvate in the dark to equimolar amounts of H_2, CO_2, and acetyl-CoA suggests a phosphoroclastic system (pyruvate dehydrogenase plus hydrogenase) (Bennett *et al.*, 1964).

The dark, anaerobic growth on pyruvate of selected strains of *Rs. rubrum* with a subsequent evolution of H_2 and CO_2 was reported by Uffen (1973), Schön and Biederman (1973), and Schön and Voelskow (1976). Fermentation balances suggested that pyruvate was degraded to acetate and formate by pyruvate-formate lyase (Schön and Biederman, 1973). After pyruvate had been degraded to acetate and formate, the cultures [or cell-free extracts (Jungerman and Schön, 1974)] further degraded the formate to CO_2 and H_2 by the action of the enzyme formate hydrogenlyase. Because this gas evolution followed by several hours the appearance of formate, it suggested that the enzymes responsible were induced by formate (Schön and Voelskow, 1976).

5.3. Hydrogenase

5.3.1. Purple Nonsulfur Bacteria

Cell-free hydrogenase activity in photosynthetic bacteria was first demonstrated by Gest (1952), who measured H_2 uptake in the dark by crude extracts of *Rs. rubrum* with methylene blue as the electron acceptor. This type of hydrogenase activity, unlike the photoevolution of H_2, was present in cells cultured on either NH_4^+, glutamate, or N_2 (Stiffler and Gest, 1954; Gest *et al.*, 1956), suggesting no relationship to N metabolism. This activity was associated with the chromatophores of *Rs. rubrum* and appeared to be un-affected by metal chelating agents (Bose and Gest, 1962). Only after treatment of the chromatophores with deoxycholate could H$_2$-hydrogenase activity be coupled to NAD reduction, and then only in the presence of FMN and an NADH-trapping system (Bose and Gest, 1962).

Untreated chromatophores of *Rp. capsulata* were shown by Klemme and Schlegel (1967) to be capable of photoreducing NAD with H$_2$ as the electron donor. This activity was not enhanced by the addition of ferredoxin. Unlike the H$_2$ uptake coupled to dye reduction by *Rs. rubrum* extracts (Bose and Gest, 1962), the photoreduction of NAD by *Rp. capsulata* was strongly inhibited by the metal chelator *o*-phenanthroline. Although the redox potential of the H$_2$/H$^+$ couple ($E'_0 = -420$ mV) is low enough to reduce NAD ($E'_0 = -320$ mV), the reaction is nevertheless light-dependent, suggesting that the natural electron acceptor in the chromatophore has a more positive potential than that of NAD. This appears to be verified by the fact that ATP is required to couple the reaction in the dark (Klemme, 1969). The effect of uncouplers and oligomycin on the reduction of NAD by H$_2$ suggests an energy-linked reverse electron flow from H$_2$ to NAD.

The hydrogenase of *Rp. palustris* (H$_2$ uptake coupled to the reduction of methylene blue or *p*-benzoquinone) is also localized in the chromatophores (Izawa, 1962). This enzyme, which is strongly inhibited by CO (Izawa, 1962), is believed to be involved in formate metabolism by *Rp. palustris* as part of the formate hydrogenlyase system (Qadri and Hoare, 1968). Although the hydrogenase from *Rp. palustris* coupled reducing power to artificial acceptors, pyridine nucleotide reduction coupled to hydrogenase has yet to be demonstrated in this organism, either in the light or in the dark.

Hydrogenase in the Rhodospirillaceae appears to be a chromatophore-bound enzyme that, because of its sensitivity to inhibitors such as *o*-phenanthroline and *p*-chloromercuribenzoate, may be an Fe–S protein similar to that of the clostridia. Although H$_2$ serves as an accessory electron donor in this group of organisms, there still is no knowledge of the electron acceptor in the chromatophores that is first reduced by the hydrogenase.

5.3.2. Purple Sulfur Bacteria

Crude extracts of *Chr. vinosum* Strain D were shown to reduce both NAD and NADP with H$_2$ in the presence of benzyl viologen (Arnon, 1961). Although both pyridine nucleotides were reduced, activity was highest with NAD; light had no effect on the rate of this reduction.

When supplied with dithionite-reduced methyl viologen as a source of reducing power, *Chr. vinosum* Strain D extracts were shown to evolve H$_2$ in a CO-sensitive reaction (Winter and Arnon, 1970). The photoevolution of H$_2$ by *Chr. vinosum* (now believed to be a function of the nitrogenase enzyme) was also said to be sensitive to CO (Arnon, 1961); Winter and Arnon (1970) later showed, however, that evolution of H$_2$ from nitrogenase is not inhibited by CO.

After the relationship among hydrogenase, ferredoxin, and pyridine nucleotide reduction was shown in *Clostridium* (Valentine *et al.*, 1962), it was demonstrated that the viologen dyes originally required for NAD reduction by H$_2$ in *Chr. vinosum* extracts (Arnon, 1961) could be replaced by ferredoxin (Buchanan *et al.*, 1964; Weaver *et al.*, 1965). Weaver *et al.* (1965) showed that the hydrogenase of *Chr. vinosum* was primarily a soluble protein, whereas Buchanan and Bachofen (1968) found hydrogenase activity not only in the soluble fraction, but also in a particulate (chromatophore) fraction. Reconstitution of the "chromatophore" and soluble fractions was reported to have a stimulatory (greater than additive) effect on ferredoxin-dependent NAD reduction by H$_2$ (Buchanan and Bachofen, 1968).

Feigenblum and Krasna (1970) also separated *Chr. vinosum* Strain D extracts into soluble and particulate (chromatophore) fractions, and they found that ferredoxin-dependent NAD reduction by H$_2$ was associated with the particulate fraction. The soluble fraction, while showing no ferredoxin-dependent NAD reduction, did catalyze a benzyl-viologen-dependent reduction of NAD with H$_2$ as the electron donor. The "particulate" hydrogenase of *Chr. vinosum* Strain D was solubilized either by treating chromatophores with Triton X-100 or deoxycholate or by sonicating whole cells in the presence of one of these detergents (Feigenblum and Krasna, 1970). During the purification, the solubilized hydrogenase not only lost its ability to reduce NAD in the ferredoxin-dependent reaction (which might be explained by the loss of the ferredoxin-NADH oxidoreductase during purification of the hydrogenase), but also was reported to have lost its ferredoxin-reducing activity (Feigenblum and Krasna, 1970; Gitlitz and Krasna, 1975); only the tritium-exchange reaction increased in specific activity during purification. This characteristic (the loss of ferredoxin-reducing capability), if true, makes *Chromatium* hydrogenase unlike that of *Cl. pasteurianum* Strain W5 (Nakos and Mortenson, 1971; Chen and Mortenson, 1974). The loss of ferredoxin-reducing capacity and the diminished benzyl-viologen-reducing capacity of the purified hydrogenase relative to the tritium exchange reaction were ascribed to the loss of "unknown factors present in the crude extract which facilitate this reaction" (Gitlitz and Krasna, 1975).

Purified to homogeneity, the *Chromatium* hydrogenase was shown to be an Fe–S protein with four moles each of nonheme iron and acid-labile sulfur per mole of protein (Gitlitz and Krasna, 1975). The molecular weight of the hydrogenase is about 100,000; the protein consists of two subunits of about 50,000 daltons each. The optical and EPR spectra are consistent with the hydrogenase being an Fe–S protein.

A most astonishing feature of the *Chromatium* hydrogenase is its apparent lack of oxygen lability. The enzyme was purified without the necessity of any anaerobic precautions with a yield (based on the tritium exchange reaction) of 23%. If the *Chromatium* hydrogenase is still in its "natural" state after purification by this procedure, it is again very different from the oxygen-labile clostridial hydrogenase.

5.3.3. Green Sulfur Bacteria

Although various species of *Chlorobium* (and *Pelodictyon*) were reported to photoevolve H_2 in cultures devoid of NH_4^+ (unpublished data of N. Pfennig, cited by Pfennig, 1967), extracts of *Chl. limicola f. thiosulfatophilum* did not take up H_2 to reduce NAD in the dark (Weaver *et al.*, 1965). Although this failure might suggest the absence of hydrogenase in this organism, it must be remembered that *Chlorobium* species grow on H_2 and CO_2, indicating that they must have a hydrogenase to extract electrons from H_2.

There seems to be a significant difference between the hydrogenase of purple nonsulfur and that of purple sulfur bacteria. Whereas the hydrogenases of *Rs. rubrum* and *Rp. capsulata* reduce NAD photochemically (at least in the latter species) in ferredoxin-independent reactions, the hydrogenase of *Chromatium*, like that of the anaerobic fermentative bacteria, is a ferredoxin-linked enzyme and is not light-activated. The significance of these apparent differences is not understood. Whether the hydrogenase of the green sulfur bacteria resembles that of *Chromatium*, which is ferredoxin-linked, or those of *Rs. rubrum* and *Rp. capsulata*, which are linked to unknown electron carriers in the chromatophore, remains to be determined.

6. Inorganic Nitrogen Metabolism (Other Than N_2 and NH_4^+)

All photosynthetic bacteria grow on media containing NH_4^+ (van Niel, 1931, 1944), and most that have been tested appear to fix N_2, but little is known about nitrate metabolism by this group of microorganisms. Whereas there have been occasional reports that purple bacteria will grow on nitrate and can reduce both NO_3^- and NO_2^-, *Chromatium* and various *Rhodopseudomonas* species are reported not to utilize these nitrogenous compounds (see the review by Kondrat'eva, 1963). There are no reports of green sulfur bacteria utilizing either nitrate or nitrite.

Taniguchi and Kamen (1963) reported, however, that *Rs. rubrum* (S1) could be adapted to photosynthetic growth by assimilatory reduction of NO_3^- if the cultures were previously maintained through several transfers in media containing NO_3^- as the only source of nitrogen. Nitrate-supported growth was reported to be slow compared with growth on NH_4^+; however, a considerable amount of nitrate reductase could be measured in cell-free extracts from these NO_3^--grown cells. Reduced pyridine nucleotide appeared to be the physiological electron donor for this enzyme. Attempts by Taniguchi and Kamen (1963) to replace oxygen with nitrate (as a measure of dissimilatory nitrate reductase activity) either in the light or in the dark were unsuccessful with both *Rs. rubrum* and *Rp. sphaeroides*. Although nitrate reductase in *Rs. rubrum* was adaptive, *Rp. sphaeroides* appeared to contain a constitutive nitrate reductase of low activity. The addition of NO_3^- to the growth medium, however, did not enhance this activity, and it was concluded that this weak residual constitutive activity in *Rp. sphaeroides* did not represent a true nitrate reductase, but was an artifact of no physiological significance.

Rhodospirillum rubrum strain S1 was reported by Katoh (1963) to grow either aerobically in the dark or photosynthetically with nitrate as the sole source of nitrogen, but only after the culture had been "trained" to utilize this nitrogen compound. Furthermore, these cultures grew under dark–anaerobic conditions with nitrate as a terminal electron acceptor in an energy-yielding, respiratory-type process.

Nitrate reductase activity was also shown by Katoh (1963) to be associated with the chromatophores. The enzyme utilized NADH or reduced methyl viologen as an electron donor, and is believed to contain a *b*-type cytochrome, flavin, and nonheme iron. The same enzyme is thought to be responsible for both the dissimilatory (nitrate respiration) and assimilatory (growth on nitrate) reduction of nitrate (Katoh, 1963).

Assimilatory nitrate and nitrite reduction was recently observed with a strain of *Rp. sphaeroides* (W. R. Sistrom, personal communication).

Nitrite reductase activity was catalyzed by unwashed chromatophores of *Rs. rubrum* prepared only from cells adapted to photoassimilation of nitrate (Taniguchi and Kamen, 1963). This reductase activity was doubled when the reaction mixture was illuminated, suggesting that the reductant was a product of photosynthesis, perhaps NADH.

7. Concluding Remarks

The processes discussed in this review—N_2 fixation, the photoevolution of H_2, and hydrogenase activity by photosynthetic bacteria—are all intimately associated with bacterial photosynthesis because of their need for light-generated, low-potential reducing power, i.e., reduced ferredoxin. The question of primary interest (which must be addressed in future research) therefore seems to be: how is ferredoxin, which is not a chromatophore component and which has a potential of about −400 mV, reduced by light energy? There seem to be three possibilities: First, it has been proposed that ATP-driven reverse electron flow generates a stable reductant (NADH), and it is possible that a high NADH/NAD ratio might couple electron flow to ferredoxin via a NADH-ferredoxin oxidoreductase. Secondly, that the potential of the primary acceptor in the chromatophore is pH-dependent suggests that it may operate at a potential low enough to reduce ferredoxin. Finally, it cannot as yet be excluded that the primary acceptor (with a low redox potential), capable of reducing ferredoxin, has not yet been identified.

That there are no reports of light-driven, ferredoxin-dependent reactions in purple photosynthetic bacteria lends credence to arguments that ferredoxin reduction in photosynthetic bacteria takes place by a fermentative type of mechanism. Other than the H_2-hydrogenase reduction of ferredoxin in *Chromatium* extracts, however, there have been no reports of substrate-coupled ferredoxin reduction in photosynthetic bacteria. In the opinion of this reviewer, the mechanism of ferredoxin reduction by photosynthetic bacteria is one of the fundamental questions that must be answered to fully understand the nitrogen and hydrogen metabolism of these organisms.

8. Note Added in Proof

Recent findings by Ludden and Burris (1976) and Nordlund *et al.* (1977) have demonstrated that nitrogenase from *Rs. rubrum*, unlike other nitrogenases, requires a chromatophore-bound protein activating factor (AF) and Mn^{2+} for initiation of enzymatic activity. AF was shown to operate only on the Fe protein and, once activated, only the Mo–Fe protein, ATP, and reductant were required for enzymatic activity. Yoch *et al.* (1978) observed that these requirements for activation of *Rs. rubrum* nitrogenase were not general but were linked to the nitrogen nutrition of the cells from which the enzyme was isolated. Activity of nitrogenase isolated from cells grown with glutamate or N_2 was greatly stimulated by AF and Mn^{2+} but nitrogenase from nitrogen-starved cells was active without these cofactors. These observations, along with molecular weight differences in the nitrogenase complex, which depended on the nitrogen substrate of the cell, led Yoch *et al.* (1978) to suggest the existence of two forms of nitrogenase: the always active form (N_2ase A) from nitrogen-starved cells and the regulated form (N_2ase R) from glutamate- or N_2-grown cells. It appears that, when grown on nitrogen substrates which allow derepression of nitrogenase, *Rs. rubrum* can regulate the turnover of this enzyme (N_2ase R) when it finds itself with sufficient nitrogen for growth without having to resort to repression of the system (also see Neilson and Nordlund, 1975). On the other hand, nitrogen-starved cells produce an enzyme (N_2ase A) which is always active and appears not to be subject to regulation. The detailed mechanism of *Rs. rubrum* nitrogenase regulation at enzyme level is under investigation in several laboratories.

A recent development in study of hydrogenases has been the solubilization (by detergent) and isolation of a membrane-bound iron–sulfur-containing hydrogenase from *Rs. rubrum* (Adams and Hall, 1977). Although the endogenous electron carrier for this enzyme was not reported, two candidates exist in ferredoxins III and IV, which were also solubilized from *Rs. rubrum* chromatophores by detergent treatment (Yoch *et al.*, 1977).

9. References

Adams, M. W. W., and Hall, D. O., 1977, Isolation of the membrane-bound hydrogenase from *Rhodospirillum rubrum*, *Biochem. Biophys. Res. Commun.* **77**:730–737.

Albrecht, S. L., and Evans, M. C. W., 1973, Measurement of the oxidation reduction potential of the EPR detectable active centre of the molybdenum iron protein of *Chromatium* nitrogenase, *Biochem. Biophys. Res. Commun.* **55**:1009–1014.

Anderson, L., and Fuller, R. C., 1967, Photosynthesis in

Rhodospirillum rubrum. I. Autotrophic carbon dioxide fixation, *Plant Physiol.* **42**:487–490.

Arnon, D. I., 1961, Cell-free photosynthesis and the energy conversion process, in: *Light and Life* (W. D. McElroy and B. Glass, eds.), pp. 489–566, The Johns Hopkins University Press, Baltimore, Maryland.

Arnon, D. I., Losada, M., Nozaki, M., and Tagawa, K., 1961, Photoproduction of hydrogen, photofixation of nitrogen and a unified concept of photosynthesis, *Nature (London)* **190**:601–610.

Bachofen, R., Buchanan, B. B., and Arnon, D. I., 1964, Ferredoxin as a reductant in pyruvate synthesis by a bacterial extract, *Proc. Natl. Acad. Sci. U.S.A.* **51**:690–694.

Beinert, H., 1965, EPR spectroscopy in the detection, study, and identification of protein-bound non-heme iron, in: *Non-heme Iron Proteins: Role in Energy Conversion* (A. San Pietro, ed.), pp. 23–42, The Antioch Press, Yellow Springs, Ohio.

Benemann, J. R., and Valentine, R. C., 1972, The pathways of nitrogen fixation, *Adv. Microb. Physiol.* **8**:59–104.

Bennett, R., Rigopoulos, N., and Fuller, R. C., 1964, The pyruvate phosphoroclastic reaction and light-dependent nitrogen fixation in bacterial photosynthesis, *Proc. Natl. Acad. Sci. U.S.A.* **52**:762–768.

Biggins, D. R., Kelley, M., and Postgate, J. R., 1971, Resolution of nitrogenase of *Mycobacterium flavum* 301 into two components and cross reaction with nitrogenase components from other bacteria, *Eur. J. Biochem.* **20**:140–143.

Bortels, H., 1930, Molybdän als Katalysator bei der biologischen Stickstoffbindung, *Arch. Mikrobiol.* **1**:333–342.

Bose, S. K., and Gest, H., 1962, Electron transport systems in purple bacteria. Hydrogenase and light-stimulated electron transfer reactions in photosynthetic bacteria, *Nature (London)* **195**:1168–1172.

Bregoff, H. M., and Kamen, M. D., 1952, Studies on the metabolism of photosynthetic bacteria. XIV. Quantitative relations between malate dissimilation, photoproduction of hydrogen, and nitrogen metabolism in *Rhodospirillum rubrum*, *Arch. Biochem. Biophys.* **36**:202–220.

Brintzinger, H., 1966, Formation of ammonia by insertion of molecular nitrogen into metal-hydride bonds. III. Considerations on the properties of enzymatic nitrogen-fixing systems and proposal of a general mechanism, *Biochemistry* **5**:3947–3950.

Buchanan, B. B., and Arnon, D. I., 1970, Ferredoxins: Chemistry and function in photosynthesis, nitrogen fixation, and fermentative metabolism, *Adv. Enzymol.* **33**:119–176.

Buchanan, B. B., and Bachofen, R., 1968, Ferredoxin-dependent reduction of nicotinamide-adenine dinucleotides with hydrogen gas by subcellular preparations from the photosynthetic bacterium, *Chromatium*, *Biochim. Biophys. Acta* **162**:607–610.

Buchanan, B. B., Bachofen, R., and Arnon, D. I., 1964, Role of ferredoxin in the reductive assimilation of CO_2 and acetate by extracts of the photosynthetic bacterium, *Chromatium*, *Proc. Natl. Acad. Sci. U.S.A.* **52**:839–847.

Buchanan, B. B., Evans, M. C. W., and Arnon, D. I., 1967, Ferredoxin-dependent carbon assimilation in *Rhodospirillum rubrum*, *Arch. Mikrobiol.* **59**:32–40.

Bui, P. T., and Mortenson, L. E., 1968, Mechanism of the enzymatic reduction of N_2: The binding of adenosine 5'-triphosphate and cyanide to the N_2-reducing system, *Proc. Natl. Acad. Sci. U.S.A.* **61**:1021–1027.

Bulen, W. A., and LeComte, J. R., 1966, The nitrogenase system from *Azotobacter*: Two enzyme requirement for N_2 reduction, ATP-dependent H_2 evolution, and ATP hydrolysis, *Proc. Natl. Acad. Sci. U.S.A.* **56**:979–986.

Bulen, W. A., Burns, R. C., and LeComte, J. R., 1965a, Nitrogen fixation: Hydrosulfite as electron donor with cell-free preparations of *Azotobacter vinelandii* and *Rhodospirillum rubrum*, *Proc. Natl. Acad. Sci. U.S.A.* **53**:532–539.

Bulen, W. A., LeComte, J. R., Burns, R. C., and Hinkson, J., 1965b, Nitrogen fixation studies with aerobic and photosynthetic bacteria, in: *Non-heme Iron Proteins: Role in Energy Conversion* (A. San Pietro, ed.), pp. 261–287, Antioch Press, Yellow Springs, Ohio.

Burns, R. C., and Bulen, W. A., 1966, A procedure for the preparation of extracts from *Rhodospirillum rubrum* catalyzing N_2 reduction and ATP-dependent H_2 evolution, *Arch. Biochem. Biophys.* **113**:461–463.

Burris, R. H., 1974, Biological nitrogen fixation, 1924–1974, *Plant Physiol.* **54**:443–449.

Burris, R. H., Eppling, F. J., Wahlin, H. B., and Wilson, P. W., 1943, Detection of nitrogen fixation with isotopic nitrogen, *J. Biol. Chem.* **148**:349–357.

Carnahan, J. E., Mortenson, L. E., Mower, H. F., and Castle, J. E., 1960a, Nitrogen fixation in cell-free extracts of *Clostridium pasteurianum*, *Biochim. Biophys. Acta* **38**:188–189.

Carnahan, J. E., Mortenson, L. E., Mower, H. F., and Castle, J. E., 1960b, Nitrogen fixation in cell-free extracts of *Clostridium pasteurianum*, *Biochim. Biophys. Acta* **44**:520–535.

Chen, J.-S., and Mortenson, L. E., 1974, Purification and properties of hydrogenase from *Clostridium pasteurianum*, *Biochim. Biophys. Acta* **371**:283–298.

Cogdell, R. J., Jackson, J. B., and Crofts, A. R., 1973, The effect of redox potential on the coupling between rapid hydrogen-ion binding and electron transport in chromatophores from *Rhodopseudomonas sphaeroides*, *J. Bioenerg.* **4**:211–227.

Cusanovich, M. A., and Edmondson, D. E., 1971, The isolation and characterization of *Rhodospirillum rubrum* flavodoxin, *Biochem. Biophys. Res. Commun.* **45**:327–336.

Dalton, H., 1974, Fixation of dinitrogen by free-living microorganisms, *Crit. Rev. Microbiol.* **3**:183–220.

Delwiche, C. C., 1970, The nitrogen cycle, *Sci. Am.* **223**:136–147.

D'Eustachio, A. J., and Hardy, R. W. F., 1964, Reductants and electron transport in nitrogen fixation, *Biochem. Biophys. Res. Commun.* **15**:319–323.

Dilworth, M., 1966, Acetylene reduction by nitrogen-fixing preparations from *Clostridium pasteurianum*, *Biochim. Biophys. Acta* **127**:285–294.

Dutton, P. L., and Leigh, J. S., 1973, Electron spin resonance characterization of *Chromatium* D hemes, non-heme irons and the components involved in primary photochemistry, *Biochim. Biophys. Acta* **314**:178–190.

Dutton, P. L., and Wilson, D. F., 1974, Redox potentiometry in mitochondrial and photosynthetic bioenergetics, *Biochim. Biophys. Acta* **346**:165–212.

Dutton, P. L., Leigh, J. S., Jr., and Reed, D. W., 1973, Primary events in the photosynthetic reaction centre from *Rhodopseudomonas sphaeroides* R26: Triplet and oxidized states of the bacteriochlorophyll and the identification of the primary electron acceptor, *Biochim. Biophys. Acta* **292**:654–664.

Eady, R. R., Smith, B. E., Cook, K. A., and Postgate, J. R., 1972, Nitrogenase of *Klebsiella pneumoniae*, purification and properties of the component proteins, *Biochem. J.* **128**:655–675.

Esposito, R. G., and Wilson, P. W., 1956, Trace metal requirements of *Azotobacter*, *Proc. Soc. Exp. Biol. Med.* **93**:564–568.

Evans, M. C. W., and Albrecht, S. L., 1974, Determination of the applied oxidation–reduction potential required for substrate reduction by *Chromatium* nitrogenase, *Biochem. Biophys. Res. Commun.* **61**:1187–1192.

Evans, M. C. W., and Buchanan, B. B., 1965, Photoreduction of ferredoxin and its use in carbon dioxide fixation by a subcellular system from a photosynthetic bacterium, *Proc. Natl. Acad. Sci. U.S.A.* **53**:1420–1425.

Evans, M. C. W., and Smith, R. V., 1971, Nitrogen fixation by the green photosynthetic bacterium *Chloropseudomonas ethylicum*, *J. Gen. Microbiol.* **65**:95–98.

Evans, M. C. W., Telfer, A., Cammack, R., and Smith, R. V., 1971, EPR studies of nitrogenase: ATP dependent oxidation of fraction 1 protein by cyanide, *FEBS Lett.* **15**:317–319.

Evans, M. C. W., Telfer, A., and Smith, R. V., 1973, The purification and some properties of the molybdenum–iron protein of *Chromatium* nitrogenase, *Biochim. Biophys. Acta* **310**:344–352.

Evans, M. C. W., Lord, A. V., and Reeves, S. G., 1974, The detection and characterization of electron-paramagnetic-resonance spectroscopy of iron–sulfur proteins and other electron-transport components in chromatophores of the purple bacterium *Chromatium*, *Biochem. J.* **138**:177–183.

Feigenblum, E., and Krasna, A. I., 1970, Solubilization and properties of the hydrogenase of *Chromatium*, *Biochim. Biophys. Acta* **198**:157–164.

Gaffron, H., 1935, Über den Stoffwechsel der Purpurbakterien, *Biochem. Z.* **275**:301–319.

Gaffron, H., and Rubin, J., 1942, Fermentative and photochemical production of hydrogen in algae, *J. Gen. Physiol.* **26**:219–240.

Gest, H., 1952, Properties of cell-free hydrogenases of *Escherichia coli* and *Rhodospirillum rubrum*, *J. Bacteriol.* **63**:111–121.

Gest, H., 1972, Energy conversion and generation of reducing power in bacterial photosynthesis, *Adv. Microb. Physiol.* **7**:243–278.

Gest, H., and Kamen, M. D., 1949*a*, Photoproduction of molecular hydrogen by *Rhodospirillum rubrum*, *Science* **109**:558–559.

Gest, H., and Kamen, M. D., 1949*b*, Studies on the metabolism of photosynthetic bacteria. IV. Photochemical production of molecular hydrogen by growing cultures of photosynthetic bacteria, *J. Bacteriol.* **58**:239–244.

Gest, H., and Kamen, M. D., 1960, The photosynthetic bacteria, *Encycl. Plant Physiol.* **2**:568–612.

Gest, H., Kamen, M. D., and Bregoff, H. M., 1950, Studies on the metabolism of photosynthetic bacteria. V. Photoproduction of hydrogen and nitrogen fixation by *Rhodospirillum rubrum*, *J. Biol. Chem.* **182**:153–170.

Gest, H., Judis, J., and Peck, H. E., Jr., 1956, Reduction of molecular nitrogen and relationships with photosynthesis and hydrogen metabolism, in: *Inorganic Nitrogen Metabolism* (W. D. McElroy and B. Glass, eds.), pp. 298–315, The Johns Hopkins University Press, Baltimore, Maryland.

Gitlitz, P. H., and Krasna, A. I., 1975, Structural and catalytic properties of hydrogenase from *Chromatium*, *Biochemistry* **14**:2561–2568.

Gray, B. H., Fowler, C. F., Nugent, N. A., Rigopoulos, N., and Fuller, R. C., 1972, *Chloropseudomonas ethylica* strain 2-K reevaluated, *Abstr. Annu. Meet. Am. Soc. Microbiol.*, p. 156.

Gray, B. H., Fowler, C. F., Nugent, N. A., Rigopoulos, N., and Fuller, R. C., 1973, Reevaluation of *Chloropseudomonas ethylica* strain 2-K, *Int. J. Syst. Bacteriol.* **23**:256–264.

Hardy, R. W. F., and Burns, R. C., 1968, Biological nitrogen fixation, *Annu. Rev. Biochem.* **37**:331–338.

Hardy, R. W. F., and D'Eustachio, A. J., 1964, The dual role of pyruvate and the energy requirement in nitrogen fixation, *Biochem. Biophys. Res. Commun.* **15**:314–318.

Hardy, R. W. F., and Holsten, R. D., 1972, Global nitrogen cycling: Pools, evolution, transformations, transfers, quantitation and research needs, in: *Symposium Proceedings Environmental Protection Agency, The Aquatic Environment: Microbial Transformations and Water Quality Management Implications* (R. K. Ballantine and L.J. Guarraia, eds.), pp. 87–132, Washington, D.C.

Hardy, R. W. F., and Knight, E., Jr., 1967, The biochemistry and postulated mechanisms of nitrogen fixation, in: *Progress in Phytochemistry* (L. Reinhold, ed.), pp. 387–469, Wiley, London.

Hardy, R. W. F., Holsten, R. D., Jackson, E. K., and Burns, R. C., 1968, The acetylene–ethylene assay for N_2 fixation: Laboratory and field evaluation, *Plant Physiol.* **43**:1185–1207.

Hardy, R. W. F., Burns, R. C., and Parshall, G. W., 1971, The biochemistry of N_2 fixation, *Adv. Chem. Ser., No. 100, Bio-organic Chemistry*, pp. 219–247.

Izawa, S., 1962, Hydrogenase reactions in *Rhodopseudomonas palustris*, *Plant Cell Physiol.* **3**:23–42.

Jackson, J. B., Cogdell, R. J., and Crofts, A. R., 1973, Some effects of *o*-phenanthroline on electron transport in chromatophores from photosynthetic bacteria, *Biochim. Biophys. Acta* **292**:218–225.

Jungermann, K., and Schön, G., 1974, Pyruvate formate

lyase in *Rhodospirillum rubrum* Ha adapted to anaerobic dark conditions, *Arch. Microbiol.* **99**:109–116.

Kamen, M. D., and Gest, H., 1949, Evidence for a nitrogenase system in the photosynthetic bacterium *Rhodospirillum rubrum*, *Science* **109**:560.

Katoh, T., 1963, Nitrate reductase in photosynthetic bacterium, *Rhodospirillum rubrum*. Purification and properties of nitrate reductase in nitrate-adapted cells, *Plant Cell Physiol.* **4**:13–28.

Klemme, J.-H., 1969, Studies on the mechanism of NAD-photoreduction by chromatophores of the facultative phototroph, *Rhodopseudomonas capsulata*, *Z. Naturforsch.* **24B**:67–76.

Klemme, J.-H., and Schlegel, H. G., 1967, Photoreduktion von Pyridinnucleotide durch Chromatophoren aus *Rhodopseudomonas capsulata* mit molekularen Wasserstoff, *Arch. Mikrobiol.* **59**:185–196.

Knaff, D. B., Buchanan, B. B., and Malkin, R., 1973, Effect of oxidation–reduction potential on light-induced cytochrome and bacteriochlorophyll reactions in chromatophores from the photosynthetic green bacterium *Chlorobium*, *Biochim. Biophys. Acta* **325**:94–101.

Knight, E., Jr., D'Eustachio, A. J., and Hardy, R. W. F., 1966, Flavodoxin: A flavoprotein with ferredoxin activity from *Clostridium pasteurianum*, *Biochim. Biophys. Acta* **113**:626–628.

Kohlmiller, E. F., Jr., and Gest, H., 1951, Comparative study of the light and dark fermentations of organic acids by *Rhodospirillum rubrum*, *J. Bacteriol.* **61**:269–282.

Kondrat'eva, E. N., 1963, *Photosynthetic Bacteria* (translated from the Russian), Israel Program for Scientific Translation, Jerusalem, 243 pp.

Larsen, H., 1952, On the culture and general physiology of the green sulfur bacteria, *J. Bacteriol.* **64**:187–196.

Lee, S. B., and Wilson, P. W., 1943, Hydrogenase and nitrogen fixation by *Azotobacter*, *J. Biol. Chem.* **151**:377–385.

Leigh, J. S., and Dutton, P. L., 1972, The primary electron acceptor in photosynthesis, *Biochem. Biophys. Res. Commun.* **46**:414–421.

Lindsay, H. L., 1963, Physiological studies of nitrogen fixation by cells and cell-free extracts of *Aerobacter aerogenes*, Ph.D. dissertation, University of Wisconsin, Madison, Wisconsin (*Diss. Abstr.* **24**:2222).

Lindstrom, E. S., Burris, R. H., and Wilson, P. W., 1949, Nitrogen fixation by photosynthetic bacteria, *J. Bacteriol.* **58**:313–316.

Lindstrom, E. S., Trove, S. R., and Wilson, P. W., 1950, Nitrogen fixation by the green and purple sulfur bacteria, *Science* **112**:197–198.

Lindstrom, E. S., Lewis, S. M., and Pinsky, M. J., 1951, Nitrogen fixation and hydrogenase in various bacterial species, *J. Bacteriol.* **61**:481–487.

Losada, M., Nozaki, M., and Arnon, D. I., 1961, Photoproduction of molecular hydrogen from thiosulfate by *Chromatium* cells, in: *Light and Life* (W. D. McElroy and B. Glass, eds.), pp. 570–575, The Johns Hopkins University Press, Baltimore, Maryland.

Ludden, P. W., and Burris, R. H., 1976, Activating factor

for the iron-protein of nitrogenase from *Rhodospirillum rubrum*, *Science* **194**:424–427.

Mortenson, L. E., 1964, Ferredoxin and ATP, requirements for nitrogen fixation in cell-free extracts of *Clostridium pasteurianum*, *Proc. Natl. Acad. Sci. U.S.A.* **52**:272–279.

Mortenson, L. E., Valentine, R. C., and Carnahan, J. E., 1963, Ferredoxin in the phosphoroclastic reaction of pyruvic acid and its relation to nitrogen fixation in *Clostridium pasteurianum*, *J. Biol. Chem.* **238**:794–800.

Mortenson, L. E., Morris, J. A., and Jeng, D. Y., 1967, Purification, metal composition and properties of molybdoferredoxin and azoferredoxin, two of the components of the nitrogen-fixing system of *Clostridium pasteurianum*, *Biochim. Biophys. Acta* **141**:516–522.

Mortenson, L. E., Zumft, W. G., and Palmer, G., 1973, Electron paramagnetic resonance studies on nitrogenase. III. Function of magnesium adenosine 5'-triphosphate and adenosine 5'-diphosphate in catalysis by nitrogenase, *Biochim. Biophys. Acta* **292**:422–435.

Munson, T. P., and Burris, R. H., 1969, Nitrogen fixation by *Rhodospirillum rubrum* grown in nitrogen-limited continuous culture, *J. Bacteriol.* **97**:1093–1098.

Nagatani, H. H., and Brill, W. J., 1974, Nitrogenase V. The effects of Mo, W and V on the synthesis of nitrogenase components in *Azotobacter vinelandii*, *Biochim. Biophys. Acta* **362**:160–166.

Nagatani, H., Shimizu, M., and Valentine, R. C., 1971, The mechanism of ammonia assimilation in nitrogen fixing bacteria, *Arch Mikrobiol.* **79**:164–175.

Nakamura, H., 1941, Weitere Untersuchungen über die bakterielle Photosynthese. Beiträge zur Stoffwechselphysiologie der Purpurbakterien. VI. *Acta Phytochim.* (*Tokyo*) **12**:43–64; cited by Gest and Kamen (1960).

Nakos, G., and Mortenson, L. E., 1971, Purification and properties of hydrogenase, an iron–sulfur protein, from *Clostridium pasteurianum* W$_5$, *Biochim. Biophys. Acta* **227**:576–583.

Neilson, A. H., and Nordlund, S., 1975, Regulation of nitrogenase synthesis in intact cells of *Rhodospirillum rubrum*: Inactivation of nitrogen fixation by ammonia, L-glutamine and L-asparagine, *J. Gen. Microbiol.* **91**:53–62.

Newton, J. W., and Wilson, P. W., 1953, Nitrogen fixation and photoproduction of molecular hydrogen by Thiorhodaceae, *Antonie van Leeuwenhoek*; *J. Microbiol. Serol.* **19**:71–77.

Nordlund, S., and Baltscheffsky, H., 1973, Two nitrogenase components from *Rhodospirillum rubrum*, *Abstracts of the Ninth International Congress on Biochemistry*, Stockholm, (edited by the International Union of Biochemistry), p. 240, Aktiebolaget Egnellska Boktryckeriet, Stockholm.

Nordlund, S., Eriksson, U., and Baltscheffsky, H., 1977, Necessity of a membrane component for nitrogenase activity in *Rhodospirillum rubrum*, *Biochim. Biophys. Acta* **462**:187–195.

Okuda, A., Yamaguchi, M., and Kobayashi, M., 1960, Nitrogen-fixing microorganisms in paddy soils. IV. Nitrogen fixation in mixed culture of photosynthetic bacteria (*Rhodopseudomonas capsulatus*) under various conditions, *Soil Plant Food* (*Tokyo*) **5**:73–77.

Orme-Johnson, W. H., Hamilton, W. D., Ljones, T., Tso, M.-Y. W., Burris, R. H., Shah, V. K., and Brill, W. J., 1972, Electron paramagnetic resonance of nitrogenase and nitrogenase components from *Clostridium pasteurianum* W5 and *Azotobacter vinelandii* OP, *Proc. Natl. Acad. Sci. U.S.A.* **69**:3142–3145.

Ormerod, J. G., and Gest, H., 1962, Symposium on metabolism of inorganic compounds. IV. Hydrogen photosynthesis and alternative metabolic pathways in photosynthetic bacteria. *Bacteriol. Rev.* **26**:51–66.

Ormerod, J. G., Ormerod, K. S., and Gest, H., 1961, Light-dependent utilization of organic compounds and photoproduction of molecular hydrogen by photosynthetic bacteria; relationships with nitrogen metabolism, *Arch. Biochem. Biophys.* **94**:449–463.

Palmer, G., Multani, J. S., Cretney, W. C., Zumft, W. G., and Mortenson, L. E., 1972, Electron paramagnetic resonance studies on nitrogenase. I. The properties of molybdoferredoxin and azoferredoxin, *Arch. Biochem. Biophys.* **153**:325–332.

Paschinger, H., 1974, A changed nitrogenase activity in *Rhodospirillum rubrum* after substitution of tungsten for molybdenum, *Arch Mikrobiol.* **101**:379–389.

Pengra, R. M., and Wilson, P. W., 1959, Trace metal requirements of *Aerobacter aerogenes* for assimilation of molecular nitrogen, *Proc. Soc. Exp. Biol. Med.* **100**:436–439.

Pfennig, N., 1967, Photosynthetic bacteria, *Annu. Rev. Microbiol.* **21**:285–324.

Pratt, D. C., and Frenkel, A. W., 1959, Studies on nitrogen fixation and photosynthesis of *Rhodospirillum rubrum*, *J. Plant Physiol.* **34**:333–337.

Prince, R. C., Leigh, J. S., Jr., and Dutton, P. L., 1974, An electron-spin-resonance characterization of *Rhodopseudomonas capsulata*, *Biochem. Soc. Trans.* **2**:950–953.

Prince, R. C., Lindsay, J. G., and Dutton, P. L., 1975, The Rieske iron–sulfur center in mitochondrial and photosynthetic systems: E_m/pH relationships, *FEBS Lett.* **51**:108–111.

Qadri, S. M. H., and Hoare, D. S., 1968, Formic hydrogenlyase and the photoassimilation of formate by a strain of *Rhodopseudomonas palustris*, *J. Bacteriol.* **95**:2344–2357.

Rao, K. K., Matsubara, H., Buchanan, B. B., and Evans, M. C. W., 1969, Amino acid composition and terminal sequences of ferredoxins from two photosynthetic bacteria, *J. Bacteriol.* **100**:1411–1412.

Roelofsen, P. A., 1935, On photosynthesis of the Thiorhodaceae, Doctoral thesis, University of Utrecht, Utrecht, The Netherlands.

Schick, H.-J., 1971a, Substrate and light dependent fixation of molecular nitrogen in *Rhodospirillum rubrum*, *Arch. Mikrobiol.* **75**:89–101.

Schick, H.-J., 1971b, Interrelationship of nitrogen fixation, hydrogen evolution and photoreduction in *Rhodospirillum rubrum*, *Arch. Mikrobiol.* **75**:102–109.

Schick, H.-J., 1971c, Regulation of photoreduction in *Rhodospirillum rubrum* by ammonia, *Arch. Mikrobiol.* **75**:110–120.

Schneider, K. C., Bradbeer, C., Singh, R. N., Wang, L. C., Wilson, P. W., and Burris, R. H., 1960, Nitrogen fixation by cell-free preparations from microorganisms, *Proc. Natl. Acad. Sci. U.S.A.* **46**:726–733.

Schöllhorn, R., and Burris, R. H., 1966, Study of intermediates in nitrogen fixation, *Fed. Proc. Fed. Am. Soc. Exp. Biol.* **24**:710.

Schön, G., and Biedermann, M., 1973, Growth and adaptive hydrogen production of *Rhodospirillum rubrum* (F_1) in anaerobic dark cultures, *Biochim. Biophys. Acta* **304**:65–75.

Schön, G., and Voelskow, H., 1976, Pyruvate fermentation in *Rhodospirillum rubrum* after transfer from aerobic to anaerobic conditions in the dark, *Arch. Microbiol.* **107**:87–92.

Shanmugam, K. T., and Valentine, R. C., 1975, Microbial production of ammonium ion from nitrogen, *Proc. Natl. Acad. Sci. U.S.A.* **72**:136–139.

Shanmugam, K. T., Buchanan, B. B., and Arnon, D. I., 1972, Ferredoxin in light- and dark-grown photosynthetic cells with special reference to *Rhodospirillum rubrum*, *Biochim. Biophys. Acta* **256**:477–486.

Siegel, J. M., and Kamen, M. D., 1951, Studies on the metabolism of photosynthetic bacteria. VII. Comparative studies on the photoproduction of H_2 by *Rhodopseudomonas gelatinosa* and *Rhodospirillum rubrum*, *J. Bacteriol.* **61**:215–228.

Smith, B. E., Lowe, D. J., and Bray, R. C., 1973, Studies by electron paramagnetic resonance on the catalytic mechanism of nitrogenase, *Biochem. J.* **135**:331–341.

Smith, R. V., Telfer, A., and Evans, M. C. W., 1971, Complementary functioning of nitrogenase components from a blue-green alga and a photosynthetic bacterium, *J. Bacteriol.* **107**:574–575.

Stewart, W. D. P., 1973, Nitrogen fixation by photosynthetic microorganisms, *Annu. Rev. Microbiol.* **27**:283–316.

Stiffler, H. J., and Gest, H., 1954, Effects of light intensity and nitrogen growth source on hydrogen metabolism in *Rhodospirillum rubrum*, *Science* **120**:1024–1026.

Strandberg, G. W., and Wilson, P. W., 1968, Formation of the nitrogen-fixing enzyme system in *Azotobacter vinelandii*, *Can. J. Microbiol.* **14**:25–31.

Swoager, W. C., and Lindstrom, E. S., 1971, Isolation and counting of Athiorhodaceae with membrane filters, *Appl. Microbiol.* **22**:683–687.

Tagawa, K., and Arnon, D. I., 1962, Ferredoxins as electron carriers in photosynthesis and in the biological production and consumption of hydrogen gas, *Nature* (*London*) **195**:537–543.

Taniguchi, S., and Kamen, M. D., 1963, On the nitrate metabolism of facultative photoheterotrophs, in: *Studies on Microalgae and Photosynthetic Bacteria* (edited by the Japanese Society of Plant Physiologists), pp. 465–484, The University of Tokyo Press, Tokyo.

Tempest, D. W., Meers, J. L., and Brown, C. M., 1970, Synthesis of glutamate in *Aerobacter aerogenes* by a hitherto unknown route, *Biochem. J.* **117**:405–407.

Thiele, H. H., 1966, Wachstumsphysiologische Untersuchungen an Thiorhodaceae; Wasserstoff-Donatoren und Sulfatreduktion, Doctoral thesis, University of Göttingen; cited by Pfennig (1967).

Trüper, H. G., and Jannasch, H. W., 1968, *Chromatium buberi* nov. spec., eine neue Art der "grossen" Thiorhodaceae, *Arch. Mikrobiol.* **61**:363–372.

Uffen, R. L., 1973, Growth properties of *Rhodospirillum rubrum* mutants and fermentation of pyruvate in anaerobic, dark conditions, *J. Bacteriol.* **116**:874–884.

Valentine, R. C., Brill, W. J., and Wolfe, R. S., 1962, Role of ferredoxin in pyridine nucleotide reduction, *Proc. Natl. Acad. Sci. U.S.A.* **48**:1856–1860.

van Niel, C. B., 1931, On the morphology and physiology of the purple and green sulfur bacteria, *Arch. Mikrobiol.* **3**:1–112.

van Niel, C. B., 1944, The culture, general physiology, morphology and classification of the non-sulfur purple and brown bacteria, *Bacteriol. Rev.* **8**:1–118.

Wall, J. S. Wagenknecht, A. C., Newton, J. W., and Burris, R. H., 1952, Comparison of the metabolism of ammonia and molecular nitrogen in photosynthesizing bacteria, *J. Bacteriol.* **63**:563–573.

Weaver, P., Tinker, K., and Valentine, R. C., 1965, Ferredoxin linked DPN reduction by the photosynthetic bacteria *Chromatium* and *Chlorobium*, *Biochem. Biophys. Res. Commun.* **21**:195–201.

Wertlieb, D., and Vishniac, W., 1967, Methane utilization by a strain of *Rhodopseudomonas gelatinosa*, *J. Bacteriol.* **93**:1722–1724.

Wilson, P. W., 1940, *The Biochemistry of Nitrogen Fixation*, University of Wisconsin Press, Madison.

Wilson, P. W., 1958, Asymbiotic nitrogen fixation, *Encycl. Plant Physiol.* **8**:9–47.

Wilson, P. W., and Burris, R. H., 1947, The mechanism of biological nitrogen fixation, *Bacteriol. Rev.* **11**:41–73.

Wilson, P. W., Hull, J. F., and Burris, R. H., 1943, Competition between free and combined nitrogen in the nutrition of *Azotobacter*, *Proc. Natl. Acad. Sci. U.S.A.* **29**:289–294.

Winter, H. C., and Arnon, D. I., 1970, The nitrogen fixation system of photosynthetic bacteria. I. Preparation and properties of a cell-free extract from *Chromatium*, *Biochim. Biophys. Acta* **197**:170–179.

Winter, H. C., and Ober, J. A., 1973, Isolation of particulate nitrogenase from *Chromatium* Strain D, *Plant Cell Physiol.* **14**:769–773.

Yoch, D. C., 1978, Electron transport systems coupled to nitrogenase, in: *A Treatise on Dinitrogen* (N$_2$) *Fixation*

(R. C. Burns and R. W. F. Hardy, eds.), John Wiley & Sons, New York (in press).

Yoch, D. C., and Arnon, D. I., 1970, The nitrogen fixation system of photosynthetic bacteria. II. *Chromatium* nitrogenase activity linked to photochemically generated assimilatory power, *Biochim. Biophys. Acta* **197**:180–184.

Yoch, D. C., and Arnon, D. I., 1974, Biological nitrogen fixation by photosynthetic bacteria, in: *The Biology of Nitrogen Fixation* (A. Quispel, ed.), pp. 687–695, North-Holland, Amsterdam.

Yoch, D. C., and Arnon, D. I., 1975, Comparison of two ferredoxins from *Rhodospirillum rubrum* as electron carriers for the native nitrogenase, *J. Bacteriol.* **121**:743–745.

Yoch, D. C., and Pengra, R. M., 1966, Effect of amino acids on the nitrogenase system of *Klebsiella pneumoniae*, *J. Bacteriol.* **92**:618–622.

Yoch, D. C., and Valentine, R. C., 1972, Ferredoxins and flavodoxins of bacteria, *Annu. Rev. Microbiol.* **26**:139–162.

Yoch, D. C., Mortensen, R. F., and Lindstrom, E. S., 1968, Thiosulfate photooxidation and nitrogen fixation in *Rhodopseudomonas palustris*, *Bacteriol. Proc.*, p. 133.

Yoch, D. C., Sweeney, W. V., and Arnon, D. I., 1975, Characterization of two soluble feredoxins as distinct from bound iron–sulfur proteins in the photosynthetic bacterium *Rhodospirillum rubrum*, *J. Biol. Chem.* **250**:8330–8336.

Yoch, D. C., Carithers, R. P., and Arnon, D. I., 1977, Isolation and characterization of bound iron–sulfur proteins from bacterial photosynthetic membranes. I. Ferredoxins III and IV from *Rhodospirillum rubrum* chromatophores, *J. Biol. Chem.* **252**:7453–7460.

Yoch, D. C., Carithers, R. P., and Arnon, D. I., 1978, Different forms of nitrogenase from *Rhosospirillum rubrum*, *Bacteriol. Proc.* (in press).

Zakhvataeva, N. V., Malofeeva, I. V., and Kondrat'eva, E. N., 1970, Nitrogen fixation capacity of photosynthesizing bacteria, *Mikrobiologiya* **39**:661–666.

Zumft, W. G., and Mortensen, L. E., 1975, The nitrogen-fixing complex of bacteria, *Biochim. Biophys. Acta* **416**:1–50.

Zumft, W. G., Palmer, G., and Mortensen, L. E., 1973, Electron paramagnetic resonance studies on nitrogenase. II. Interaction of adenosine 5'-triphosphate with azotoferredoxin, *Biochim. Biophys. Acta* **292**:413–421.

Zumft, W. G., Mortensen, L. E., and Palmer, G., 1974, Electron-paramagnetic-resonance studies on nitrogenase, *Eur. J. Biochem.* **46**:525–535.

CHAPTER 35

Sulfur Metabolism

Hans G. Trüper

1. Introduction

Anoxygenic photosynthesis requires electron donors other than water. Sulfur compounds at oxidation levels below that of sulfate are commonly used by phototrophic bacteria as electron donors. On the other hand, phototrophic bacteria, like nonphototrophic bacteria, fungi, and plants, require sulfur for the synthesis of sulfur-containing cell constituents. Although the pathways serving these two purposes may be partly identical, it is useful to differentiate between *dissimilatory* sulfur metabolism, i.e., *electron*-supplying reactions with high turnover rates, and *assimilatory* sulfur metabolism, i.e., *sulfur*-supplying reactions with low turnover rates.

2. Dissimilatory Sulfur Metabolism

2.1. Oxidation of Sulfide, Elemental Sulfur, and Sulfite

The oxidation of sulfide by Chromatiaceae and Chlorobiaceae proceeds via the intermediary stage of elemental sulfur and is in strict stoichiometrical relationship with the fixation of carbon dioxide (van Niel, 1931; Trüper, 1964a):

$$2\,H_2S + CO_2 \xrightarrow{\text{light}} 2\,S + H_2O + \langle CH_2O \rangle \qquad (1)$$

Hans G. Trüper · Institut für Mikrobiologie der Universität Bonn, Meckenheimer Allee 168, D-53 Bonn, West Germany

The coupling of sulfur metabolism and carbon assimilation shows that the former is regulated by the demand for electrons by the latter.

The elemental sulfur appears in the form of globules inside or outside the bacterial cells. In the light microscope, the globules are readily identified by their typical birefringence, which is due to their globular shape, not to crystalline structure (Trüper and Hathaway, 1967). While the Chlorobiaceae and the species of the genus *Ectothiorhodospira* deposit elemental sulfur outside the cells, the Chromatiaceae (except *Ectothiorhodospira*) form intracellular globules. In thin sections viewed under the electron microscope, the sulfur globules (or droplets; never granules!) are surrounded by a single nonunit membrane (Kran *et al.*, 1963; Schmidt, G. L., and Kamen, 1970). Analysis of sulfur globules from *Chromatium vinosum* showed that the surrounding membrane (Schmidt, G. L., *et al.*, 1971; Nicholson and Schmidt, 1971) consists entirely of a protein with a molecular weight of approximately 13,500 daltons. The membrane appears to be built up from globular subunits that have a diameter of 2.5 nm. A similar type of membrane has been found to surround gas vacuoles in cyanobacteria and halobacteria. It could be assumed that the membrane surrounding the elemental sulfur globule is itself the location of the enzymes involved in the formation and the metabolism of the sulfur. But on the basis of the results obtained by Nicolson and Schmidt (1971), namely, that the membrane is a unimolecular protein layer, G. L. Schmidt *et al.* (1971) rather assume that the protein acts primarily as a barrier to separate the sulfur from the interior of the cell, and perhaps to provide binding sites for the enzymes of sulfur metabolism.

From electron-microscopic studies, Puchkova *et al.* (1974) concluded that the elemental sulfur deposited outside the cells of Chlorobiaceae (they investigated *Chlorobium vibrioforme*, *Pelodictyon luteolum*, and *Pelodictyon phaeum*) is "liberated from the cells by means of sacs, or invaginations formed by the cytoplasmic membrane." More detailed studies should decide whether the "sacs" consist of a unimolecular protein layer or a unit membrane derived from the cytoplasmic membrane. Still, some doubt will remain because the sulfur globules seen microscopically in cultures of Chlorobiaceae usually have diameters of the same magnitude as the cells themselves. One should be able to watch such a spectacularly active excretion of sulfur globules under the light microscope. So far, this activity has not been reported. On the other hand, if sulfur globules of a rather small size, surrounded by sacs, are excreted, they would have to grow to the usually observed size outside the cell; i.e., if they were membrane-bound (unit or nonunit), this membrane would have to be ruptured. In any event, detailed study of the cell walls of Chlorobiaceae appears to be necessary.

The intracellular elemental sulfur is, in its "liquid" modification, also called "plastic sulfur" (Hageage *et al.*, 1970), and stays in this form as long as it is not subjected to dryness. On being dried, however, the S_8 rings of the sulfur, which are radially arranged, change their angular positions into the corresponding interplanar *d*-distance, thus forming a previously unknown allotrope of sulfur (Hageage *et al.*, 1970). This unstable form finally transforms into the orthorhombic crystalline structure that is the most stable modification of sulfur at room temperature and atmospheric pressure (Trüper and Hathaway, 1967). This form was described earlier for the colorless sulfur bacterium *Thiovulum majus* (La Rivière, 1963).

Further oxidation of elemental sulfur by whole cells leads to the formation of sulfate:

$$2\,S + 3\,CO_2 + 5\,H_2O \xrightarrow{\text{light}} 2\,H_2SO_4 + 3\,\langle CH_2O \rangle \quad (2)$$

The carbon dioxide fixation rates with sulfide and with sulfur are identical (Larsen, 1953; Trüper, 1964*b*). All species of the Chromatiaceae and the Chlorobiaceae are able to oxidize elemental sulfur added to the medium as powdered "flower of sulfur."

The summarizing stoichiometric equation of sulfide oxidation to sulfate is

$$2\,H_2S + 4\,CO_2 + 4\,H_2O \xrightarrow{\text{light}} 2\,H_2SO_4 + 4\,\langle CH_2O \rangle \quad (3)$$

Thus, complete oxidation of sulfide to sulfate provides eight electrons for photosynthesis per molecule of sulfate formed. It has to be emphasized here that equations 1 and 2 do not follow each other in strict mechanisms, but simply depict the stoichiometric relationship in a photolithoautotrophically growing culture between sulfur turnover and carbon dioxide fixation. The kinetics of these transformations in *Chr. okenii* was studied in short-term experiments using a specially constructed reaction vessel (Trüper, 1964*a,b*), as well as in growing cells in batch culture (Trüper and Schlegel, 1964). It was found that the rate of carbon dioxide fixation limits the turnover of sulfide and elemental sulfur to sulfate. The steps shown by equations 1 and 2 do not follow each other in strict succession; sulfate formation begins while sulfide is still being oxidized. This overlapping was also observed with *Thiocapsa roseopersicina* (Trüper, unpublished). On the other hand, in *Ectothiorhodospira mobilis*, sulfide oxidation and extracellular sulfur oxidation occur in strict succession (Trüper and Bürfent, unpublished). The results obtained with *Chr. okenii* were confirmed in studies with different strains of *Chr. vinosum* (van Gemerden, 1968*a,b*; Schmidt, G. L., and Kamen, 1970).

Van Gemerden and Jannasch (1971) applied continuous-culture techniques in their study of sulfur turnover in *Chr. vinosum*. They found a value of 7×10^{-6} M for the saturation constant, K_s, for the transformation of sulfide to sulfur. The further transformation of sulfur to sulfate is limited by the surface area of the intracellular sulfur globules, rather than by the concentration of sulfur. Since the solubility of sulfur in water is extremely low, van Gemerden and Jannasch (1971) had to express K_s for sulfur oxidation in terms of surface area of globules per unit volume. They found a value of 550 cm²/liter.

The sulfide-oxidation rate of a given species is of great importance for the ecological performance of phototrophic sulfur bacteria (van Gemerden, 1974). van Gemerden grew the large *Chr. weissei* together with the small *Chr. vinosum* in a chemostat, after he had determined the specific sulfide oxidation rates as 0.0415 and 0.0185 mM sulfide/mg cell nitrogen per hour, respectively, and the maximum specific growth rates as 0.040 and 0.117/hr, respectively. These data suggested that *Chr. weissei* is able to maintain itself in mixed culture by its higher sulfide oxidation rate, while *Chr. vinosum* does so by its higher specific growth rate. Cultivated together with *Chr. vinosum* in mixed culture in the chemostat under a light regimen of 4 hr light and 8 hr darkness, *Chr. weissei* formed a constant 70% of the total biomass. Even under a regimen of equal light and dark periods, it formed 37% of the total biomass. Van Gemerden concluded that "a high sulfide oxidation rate is a powerful weapon for those organisms unable to compete successfully by means of high specific growth rates."

The enzymology of sulfide and elemental sulfur transformations has been only partly elucidated. Theoretically, there are two possible roles for elemental sulfur: (1) either elemental sulfur is a true intermediate, in which case there should be an enzyme that oxidizes elemental sulfur to sulfite; or (2) it lies on a side path, in which case it would have to be reduced back to sulfide, which would then be oxidized to sulfite by a reversed sulfite reductase. The net yield of electrons for photosynthesis is identical in either case. The enzymes responsible for the reactions leading from sulfide to sulfite are not known.

During dark–anaerobic metabolism, cells of the Chromatiaceae excrete sulfide, which originates from the intracellular sulfur globules. This was found for *Chr. vinosum* (Hendley, 1955) and *Chr. okenii* (Trüper, 1964*a*). Van Gemerden (1967, 1968*b*) demonstrated that in *Chr. vinosum*, there exists a strict stoichiometric coupling between sulfide formation from sulfur and transformation of polyglucose into poly-β-hydroxybutyrate under dark–anaerobic conditions. This means that elemental storage sulfur under these circumstances serves as a sink for electrons, the sulfide is excreted, and ATP is formed via glycolysis. Thus, an electron flow from NAD(P)H toward elemental sulfur must be possible. An interesting finding was reported by Paschinger *et al.* (1974), who incubated *Chl. limicola f. thiosulfatophilum* (strain Tassajara) under CO_2-free hydrogen (or argon) with elemental sulfur in the light. They observed a simultaneous formation of sulfide and sulfate, and considered this a disproportionation of sulfur. Their finding might also show, however, that the oxidation of elemental sulfur to sulfate, at least in the strain or species studied, could possibly proceed via an initial reduction to sulfide.

Possibly the sulfide-oxidizing enzyme is particle-bound, since Evans and Buchanan (1965) found that a particulate fraction of *Chl. limicola f. thiosulfatophilum* utilized sulfide as the electron donor for the light-dependent reductive carboxylation of acetyl-CoA to pyruvate. Kusai and Yamanaka (1973*a,c*) reported that cytochrome *c*-553 of *Chl. limicola f. thiosulfatophilum* has the properties of a sulfide cytochrome *c* reductase. This cytochrome is a *c*-type cytochrome with covalently bound flavin (Bartsch *et al.*, 1968). It was found to catalyze the reduction of *c*-type cytochromes by sulfide. The reaction is strongly inhibited by cyanide, which probably combines with the flavin moiety of the cytochrome. Of the cytochromes of *Chl. limicola* thus far known, cytochrome *c*-555 is ob-

viously the genuine electron acceptor, thus constituting an electron-transport chain from sulfide via cytochrome *c*-553 to cytochrome *c*-555 and, *in vivo*, to the reaction center bacteriochlorophyll. Kusai and Yamanaka (1973*a,c*) did not explain, however, whether sulfide is oxidized during this reaction to elemental sulfur, sulfite, thiosulfate, or still another intermediate. Further, the significance of the catalytic capacity of *Chlorobium* cytochrome *c*-553 has to be proved, since it is well known that *c*-type cytochromes are reducible by hydrogen sulfide. Cytochromes *c* of several phototrophic bacteria including Rhodospirillaceae are reducible by sulfide (Fischer and Trüper, unpublished).

It was shown by Thiele (1966) that not only the Chlorobiaceae but also all strains of Chromatiaceae tested are able to utilize elemental sulfur added to the medium. This fact leads to the assumption that species depositing sulfur outside the cells during sulfide oxidation and species storing sulfur intracellularly possess similar enzyme systems of sulfide oxidation and sulfur utilization.

In this connection, it might be interesting to mention that the members of the genus *Ectothiorhodospira*, which prefer more alkaline pH values than the other Chromatiaceae, form intermediate polysulfides when they are grown in sulfide-containing media. At first, the cultures become yellow but stay translucent; then they become whitish-opaque and sulfur droplets appear outside the cells, while the clear yellow color of the medium disappears. This phenomenon does not appear in cultures of Chlorobiaceae when they are grown near their pH optimum of 6.5–6.8. Polysulfides (sulfanes) are not stable at acid pH values. Those Chromatiaceae that form intracellular sulfur globules do not utilize extracellularly offered polysulfides.

Sulfite appears to be a true intermediate of sulfur oxidation in phototrophic sulfur bacteria, since it is utilized as the photosynthetic electron donor by many strains. Thiele (1966, 1968*b*) found that phototrophic bacteria do not contain a sulfite oxidase (sulfite: oxygen oxidoreductase, EC 1.8.3.1.)

The sulfite-oxidizing enzyme was expected to be adenylyl sulfate (APS) reductase (AMP, sulfite: ferricyanide oxidoreductase, EC 1.8.99.2), since Peck (1966) found APS formation by chromatophores of *Chr. vinosum*, and Thiele (1966) independently found reduction of $Fe(CN)_6^{3-}$ in cell-free extracts in the presence of AMP plus sulfite. The formation of APS was verified in extracts of all strains of Chromatiaceae and Chlorobiaceae tested so far (Trüper and Peck, 1970):

$$AMP + SO_3^{2-} \rightarrow APS + 2\,e^- \qquad (4)$$

The enzyme activity is localized in a particulate fraction, the chromatophores. Depending on species and strain, it is more or less easily leached into the soluble protein fraction. APS reductase was purified from *Thiocapsa roseoperisicina* and its properties characterized by Trüper and Rogers (1971). In addition to the flavin found in the APS reductases of sulfate-reducing bacteria (Peck *et al.*, 1965) and thiobacilli (Bowen *et al.*, 1966; Lyric and Suzuki, 1970), the enzyme from *T. roseopersicina* contains heme groups. The molecular weight of the enzyme is 180,000 daltons. Per molecule, it contains one flavin and two heme groups; further, nonheme iron and labile sulfide were found. Since the enzyme reacts with cytochrome *c* (of *Candida krusei*) as electron acceptor, it is assumed that the electron flow within the APS reductase molecule proceeds from sulfite via enzyme-bound FAD to enzyme-bound hemes and then to the extramolecular electron acceptor, cytochrome *c*. The enzyme also reacts with ferricyanide, which probably receives the electrons from the FAD directly. The purified APS reductase of *Chl. limicola* (Kirchhoff and Trüper, 1974) differs from that of *T. roseopersicina* with respect to molecular composition as well as to reactivity with cytochrome *c*: this enzyme is free of heme groups and does not react with cytochrome *c* (from either horse heart or *C. krusei*). Other properties such as molecular weight (210,000 daltons), flavin content, K_m values, and pH optima do not differ significantly. Thus, the APS reductase of *Chl. limicola* is more similar to the APS reductases of sulfate-reducing bacteria and thiobacilli.

The transformation of APS to sulfate is catalyzed in an energy-conserving step by the enzyme ADP sulfurylase (ADP:sulfate adenylyl transferase, EC 2.7.7.5), as indicated by the work of Thiele (1966, 1968b):

$$APS + P_i \rightarrow ADP + SO_4^{2-} \qquad (5)$$

In combination with adenylate kinase (ATP:AMP phosphotransferase, EC 2.7.4.3), phototrophic sulfur bacteria have in this reaction a possibility of gaining one mole of ATP per two moles of sulfate formed. Whether this ATP, produced by substrate phosphorylation, is of quantitative importance as compared with the amounts provided by photophosphorylation remains to be studied.

A differentiation between Chromatiaceae and Rhodospirillaceae on the basis of sulfide utilization alone is no longer useful, since Hansen and van Gemerden (1972) reported sulfide utilization by several strains of Rhodospirillaceae. They showed that *Rp. capsulata* can tolerate sulfide concentrations to the same extent as observed in *Chr. vinosum*, while

Rp. sphaeroides, *Rp. palustris* and *Rhodospirillum rubrum* strains tolerate about one-fifth the concentration tolerated by *Chr. vinosum*. Addition of small amounts of yeast extract generally increased sulfide tolerance. Hansen (1974) was able to grow these Rhodospirillaceae species at low sulfide concentrations in batch cultures as well as in chemostats ("sulfidostats"), and found that *Rs. rubrum* (strain SMG 107), *Rp. capsulata* (strain SMG 155), and *Rp. sphaeroides* (strain SMG 158) oxidized sulfide no further than to the level of extracellular elemental sulfur, while *Rp. palustris* converted sulfide into sulfate without intermediate accumulation of elemental sulfur or other reduced sulfur compounds. In 1973, Hansen and Veldkamp (1973) described the new species *Rp. sulfidophila*, which can tolerate sulfide to the same extent as ordinary Chromatiaceae, but does not form elemental sulfur or other reduced sulfur compounds while oxidizing sulfide to sulfate. This species grows well in batch cultures. Like the other Rhodospirillaceae (Trüper and Peck, 1970), *Rp. sulfidophila* does not contain adenylylsulfate reductase (Hansen and Veldkamp, 1973; Hansen, 1974). The existence of species lacking APS reductase but able to oxidize sulfide to sulfate without intermediary sulfur formation proves that different pathways of dissimilatory sulfide oxidation exist within the phototrophic bacteria.

Hansen (1974) further studied *Rhodomicrobium vannielii* (strain ATCC 17100) with respect to sulfide utilization. The ability of this species to utilize sulfide as a photosynthetic electron donor had already been mentioned by van Niel (1963). Growing with sulfide in batch culture, this strain transformed 65–80% of the sulfide into thiosulfate, further small amounts of polythionates were found, and some sulfur globules were microscopically visible in the medium. Sulfate was not detected (Hansen 1974). When the same strain was grown in a chemostat, Hansen (1974) found that 86% of the sulfide was transformed into tetrathionate; in addition, small amounts of thiosulfate were found, while sulfate, again, was not detected.

Thus, in each case, sulfide can be oxidized to tetrathionate. From inorganic chemistry, it is known that tetrathionate reacts relatively with sulfide, leading to a formation of thiosulfate and elemental sulfur:

$$4\,S^{2-} + 6\,H_2O \rightarrow S_4O^{2-} + 12\,H^+ + 18\,e^- \qquad (6)$$

$$S_4O_6^{2-} + S^{2-} \rightarrow 2\,S_2O_3^{2-} + S \qquad (7)$$

Because the sulfide concentration at the steady state within the chemostat cultures is extremely low, Reaction 7 cannot proceed in the way it does in batch cultures, and the major sulfide oxidation product becomes tetrathionate. Nakamura (1937) had already observed sulfide and thiosulfate utilization by "*Rhodospirillum giganteum*" (cells 9–10 μm long, isolated from a warm sulfur spring). The sulfide was oxidized to elemental sulfur globules that "covered the cells." It is not possible these 40 years later, however, to determine whether Nakamura was dealing with an *Ectothiorhodospira* or a *Rhodospirillum* species and whether his cultures were completely free of oxygen.

2.2. Oxidation of Thiosulfate and Tetrathionate

The majority of the species of the Chromatiaceae and two subspecies (*formae*) of the genus *Chlorobium* are able to utilize thiosulfate as the photosynthetic electron donor. The large-cell species *Chr. okenii*, *Chr. weissei*, *Chr. warmingii*, *Chr. buderi*, and *Thiospirillum jenense* are not capable of thiosulfate utilization (Trüper and Schlegel, 1964; Thiele, 1966; Trüper and Jannasch, 1968). Of the Rhodospirillaceae, *Rp. palustris* (van Niel, 1944; Rolls and Lindstrom, 1967b) and *Rp. sulfidophila* (Hansen and Veldkamp, 1973) can grow photolithoautotrophically with thiosulfate as the sole electron donor.

In analogy to sulfide oxidation, van Niel (1931, 1936) established the overall stoichiometry of thiosulfate oxidation by *Chr. vinosum* as follows:

$$2\,Na_2S_2O_3 + 4\,CO_2 + 6\,H_2O \xrightarrow{\text{light}}$$
$$2\,H_2SO_4 + 2\,Na_2SO_4 + 4\,\langle CH_2O \rangle \quad (8)$$

The same overall results were obtained by Larsen (1952, 1953) with *Chl. limicola f. thiosulfatophilum*. Several investigators, working with *Chr. vinosum* or similar species (van Niel, 1936; Eymers and Wassink, 1938; Bregoff and Kamen, 1952; Petrova, 1959; Losada *et al.*, 1961; Smith, 1965; Trüper and Pfennig, 1966), observed a transient intracellular storage of elemental sulfur during thiosulfate oxidation, which may be expressed as follows:

$$2\,Na_2S_2O_3 + CO_2 + H_2O \xrightarrow{\text{light}}$$
$$2\,S + 2\,Na_2SO_4 + \langle CH_2O \rangle \quad (9)$$

$$2\,S + 3\,CO_2 + 5\,H_2O \xrightarrow{\text{light}}$$
$$2\,H_2SO_4 + 3\,\langle CH_2O \rangle \quad (10)$$

Comparison of these equations with those of sulfide oxidation shows that for the fixation of a certain amount of CO_2 the amounts of sulfide or thiosulfate are equimolar. The CO_2-fixation rate of *T. roseopersicina* is identical whether the cells are growing with sulfide or thiosulfate (Trüper and Pfennig, 1966).

As equation 8 shows, the oxidation of thiosulfate to sulfate provides eight electrons per molecule for photosynthesis. The two sulfur atoms in the thiosulfate molecule represent a HS– (sulfane) group and a –SO_3H (sulfone) group, and therefore they have to be dealt with by the enzymatic apparatus in different ways. The intermediate storage of elemental sulfur by Chromatiaceae cells growing at the expense of thiosulfate suggested a splitting of the thiosulfate molecule between the sulfane and sulfone groups. Only the sulfane group is rapidly transformed into elemental storage sulfur, whereas the sulfone group is transformed into sulfate and immediately excreted (Smith, 1965; Trüper and Pfennig, 1966). This process could be expressed as follows:

$$HS–S^*O_3H \longrightarrow HS^- \quad + \quad HS^*O_3^-$$
$$\downarrow 2e^- \qquad \downarrow 2e^- \qquad \downarrow 2e^-$$
$$H^+ + S^0 \qquad H^+ + S^*O_4^{2-} \quad (11)$$

Since the ability to oxidize sulfide and sulfite is a general feature in this group of bacteria, the ability to utilize thiosulfate is expressed by the possession of one or two enzymes, namely, the thiosulfate-splitting enzyme and a thiosulfate permease.

The nature of the thiosulfate-splitting enzyme was only recently resolved, due to the long-standing lack of adequate testing methods. The detection of the enzyme rhodanese (thiosulfate: cyanide sulfur-transferase, EC 2.8.1.1) in *Chr. vinosum* by Smith and Lascelles (1966) first led to the conclusion that this enzyme is responsible for thiosulfate cleavage. Rhodanese was also found in *Rp. sphaeroides*, *Rs. rubrum* (Smith and Lascelles, 1966), *Rp. capsulata*, *Rp. palustris*, *Ectothiorhodospira mobilis*, and *Chl. limicola f. thiosulfatophilum* (Yoch and Lindstrom, 1971). The toxicity of cyanide renders it unlikely, however, that the splitting enzyme is rhodanese. The development of a useful method to study thiosulfate cleavage in phototrophic bacteria by Hashwa and Pfennig (1972) allowed detailed enzymological studies. Hashwa (1972), working with *Chr. minus*, *Chr. vinosum*, and *Rp. palustris*, found that in these species, thiosulfate utilization is constitutive and not inducible or dependent on preculture conditions. Extracts of all three species contain rhodanese and thiosulfate reductase activities. These activities are constitutive; they are not affected by growth on thiosulfate. They reside in the soluble fractions. After 90-fold purification, rhodanese showed a molecular weight of 45,000 daltons. The enzyme is not affected by oxygen, but is heat-labile. A 200-fold purification of thiosulfate reductase showed a molecular weight of 98,000 daltons. The enzyme is negatively affected by oxygen, but is heat-stable. As electron donors, reduced benzyl viologen, cysteine, mercaptoethanol, reduced glutathione ($K_m = 1.6 \cdot 10^{-2}$ M), and dihydrolipoate ($K_m = 1.25 \cdot 10^{-3}$ M) were active. Possibly the latter compound or reduced lipoamide is the *in vivo* electron donor. Combined addition of electron donors results in a lower reaction rate.

Much and Hurlbert (1971), working with *Chr. vinosum*, reported partly similar results in a preliminary communication; however, they apparently considered GSH as the natural electron donor of thiosulfate reductase. Glutathione reductases [NAD(P)H: oxidized-glutathione oxidoreductase, EC 1.6.4.2] of phototrophic bacteria that could serve as the producer of GSH by the reaction

$$\text{GS-SG} + \text{NAD(P)H}_2 \rightarrow 2\,\text{GSH} + \text{NAD(P)} \quad (12)$$

have so far been purified only from *Chr. vinosum* (Chung and Hurlbert, 1971) and *Rs. rubrum* (Boll, 1969).

For the green sulfur bacteria, little is known about the enzymatic fate of thiosulfate. To our knowledge, there are no reports in the literature that prove the occurrence of elemental sulfur during thiosulfate oxidation by *Chl. limicola f. thiosulfatophilum* or *Chl. vibrioforme f. thiosulfatophilum*; careful study of the media used—when authors reported formation of elemental sulfur—always revealed that besides thiosulfate, sulfide was also included (e.g., cf. Kelly, 1974).

Larsen (1952) found complete oxidation to sulfate in whole cells, and by manometric experiments could not detect an intermediate accumulation of polythionates in the medium. He pointed out, however, that this finding "does not exclude the possibility that tetrathionate is indeed an intermediate product in this oxidation." Larsen (1952) also found that *Chl. limicola f. thiosulfatophilum* readily utilizes tetrathionate, but not trithionate, sulfite, or dithionate. Thus, Larsen excluded the latter three compounds as possible intermediates of sulfate from tetrathionate.

Mathewson *et al.* (1968) found that crude fractions of *Chl. limicola f. thiosulfatophilum* catalyzed the reduction of cytochrome *c*-551. A protein fraction obtained by ammonium precipitation catalyzed ferricyanide oxidation of thiosulfate. These findings led the authors to the conclusion that the organism studied contained a cytochrome *c*-551:thiosulfate oxidoreductase. Tetrathionate formation, however, was not experimentally proved. Kusai and Yamanaka (1973*b,c*) purified a thiosulfate cytochrome *c* reductase from *Chl. limicola f. thiosulfatophilum*. This enzyme catalyses the reduction of *Chlorobium* cytochrome *c*-551 by thiosulfate. Reduced cytochrome *c*-

551 then reduces cytochrome *c*-555. In addition, cytochrome *c*-555 was found to stimulate the reduction of cytochrome *c*-551 by thiosulfate (Kusai and Yamanaka, 1973*c*).

The formation of tetrathionate from thiosulfate was found by Smith (1966) working with *Chr. vinosum*. Smith found that sulfate was progressively replaced by tetrathionate as the end product of thiosulfate oxidation by whole cells when the pH was decreased from 7.3 to 6.25. Within this range, tetrathionate itself was not further oxidized. Smith purified the enzyme that catalyzed thiosulfate oxidation to tetrathionate, using ferricyanide as the electron acceptor. The pH optimum of the enzyme is 5.0. With whole cells, tetrathionate inhibited growth and the complete oxidation of thiosulfate. At pH 6.75, at which sulfur accumulated during thiosulfate utilization by the splitting enzyme and sulfate and tetrathionate were formed, added tetrathionate inhibited the accumulations of elemental sulfur, but not the production of tetrathionate from thiosulfate. Also, the oxidation of elemental sulfur to sulfate was not inhibited by tetrathionate. Thus, depending on the environmental conditions, thiosulfate may be metabolized by *Chr. vinosum* via either of two pathways:

$$
\begin{array}{c}
\text{S}_2\text{O}_3^{2-}
\begin{array}{l}
\overset{\text{pH 7.3}}{\nearrow} \ \text{S}^0 + \text{SO}_3^{2-} \rightarrow \text{SO}_4^{2-} \\[2ex]
\underset{\text{pH 6.25}}{\searrow} \ \text{S}_4\text{O}_6^{2-}
\end{array}
\end{array}
\quad (13)
$$

Rolls and Lindstrom (1967*a*) found that thiosulfate utilization in *Rp. palustris* is inducible by thiosulfate. As an oxidation product, they detected tetrathionate. Subsequently, Knobloch *et al.* (1971) found a thiosulfate:cytochrome *c* oxidoreductase in a cytochrome-containing fraction isolated from thiosulfate-growing *Rp. palustris*. The pH optimum of this enzymatic activity was at 8.0.

The influence of atmospheric oxygen on sulfur metabolism of the Chromatiaceae, which certainly is of ecological importance, has so far been studied by three investigators. Breuker (1964) found that when cells of *Chr. vinosum* rich in storage sulfur were aerated in the light or dark, they oxidized the elemental sulfur to sulfate. Addition of carbon dioxide to the air even increased the rate of sulfate formation. In contrast, Hurlbert (1967) found that in *Chr. vinosum*, sulfate formation from intracellular storage sulfur was inhibited by oxygen, while thiosulfate disappearance and elemental sulfur formation from the sulfane group of thiosulfate were not.

Gorlenko (1974), using *Amoebobacter roseus*, studied thiosulfate turnover in the dark under micro-aerophilic conditions and observed that the organism was able to grow slowly under these circumstances. The thiosulfate was oxidized to sulfate with an intermediary appearance of intracellular elemental sulfur. Similar results were reported for sulfide in the dark (Gorlenko, 1974). The relevance of these reactions under the influence of oxygen remains to be elucidated.

2.3. Fractionation of Sulfur Isotopes

Like the sulfate-reducing bacteria (*Desulfovibrio* and *Desulfotomaculum* species) and the chemolithotrophic sulfur-oxidizing bacteria (*Thiobacillus, Beggiatoa*, and similar forms), the phototrophic sulfur bacteria participate in the natural mass transformations of sulfur compounds. Therefore, there should be measurable fractionations of stable sulfur isotopes (^{32}S and ^{34}S) during these transformations, which are essentially performed by enzymes. While there are many data available for fractionation rates by sulfate-reducing bacteria, only a few measurements have been done with the sulfide-oxidizing phototrophic bacteria. Fractionation is expressed by the $\delta^{34}S$ value given in parts per thousand:

$$\delta^{34}S(\%_0) = \frac{\left(\dfrac{^{34}S}{^{32}S}\right)_{sample} - \left(\dfrac{^{34}S}{^{32}S}\right)_{standard}}{\left(\dfrac{^{34}S}{^{32}S}\right)_{standard}} \cdot 10^3 \quad (14)$$

As a standard, the sulfur in the mineral troilit of Cañon Diablo, U.S.A., having a ($^{34}S/^{32}S$) of 22.22, is used, $\delta^{34}S = 0$. Negative δ-values then indicate an increase in ^{32}S during a certain process, while positive δ-values demonstrate an increase in ^{34}S.

Generally, in enzymatic reactions, the lighter isotope ^{32}S is preferred; i.e., the reaction product is enriched in ^{32}S and has a negative δ-value.

For *Chr. vinosum*, Kaplan and Rittenberg (1964) found that the intracellular storage sulfur showed a δ of -10.0 parts per thousand against the sulfide it originated from; i.e., ^{32}S is preferred. The sulfate formed, however, showed a much smaller degree of fractionation, namely, $+0.5$ to -2.9 parts per thousand against the sulfide. This finding led Kaplan and Rittenberg to the assumption that the pathway of sulfide oxidation to sulfate does not proceed via elemental sulfur. On the other hand, measurements of Mekht'eva and Kondrat'eva (1966) with *Ectothiorhodospira shaposhnikovii* (at that time still named *Rhodopseudomonas* sp.) resulted in δ-values of -1.0 parts per thousand for elemental sulfur formed from

sulfide and $+2.05$ parts per thousand for sulfate formed from sulfide. The remaining sulfide had a δ-value of -2.65 parts per thousand. A study of green sulfur bacteria (Kondrat'eva *et al.*, 1966) showed that the sulfide residue had δ-values down to -2.35 parts per thousand. The significance of these data, which are in contrast to those obtained for *Chr. vinosum* by Kaplan and Rittenberg (1964), remains to be elucidated.

It may be mentioned here, for comparison, that nonenzymatic inorganic oxidation of H_2S to sulfate results in $\delta = +68$ parts per thousand (against H_2S) at $25°C$, and in $\delta = +48$ parts per thousand (against H_2S) at $100°C$.

3. Assimilatory Sulfur Metabolism

The photoorganotrophic Rhodospirillaceae depend on assimilatory sulfate reduction for the synthesis of sulfur-containing cellular components such as the amino acids cysteine, cystine, and methionine as well as such sulfur compounds as lipoate, biotin, thiamin, and CoA. One species, *Rp. globiformis*, was shown to require thiosulfate for this purpose during photoorganotrophic growth (Pfennig, 1974).

The Chromatiaceae and Chlorobiaceae, however, which usually live in sulfide-containing habitats, would waste energy if they utilized sulfate for amino acid synthesis instead of relying on the readily available sulfide.

The utilization of reduced sulfur, sulfide, and thiosulfate as a source of cell sulfur by members of the Chromatiaceae was first suggested by Hurlbert and Lascelles (1963). When *Chr. okenii* was incubated with radioactively labeled sulfate, the cells did not incorporate any radioactivity (Trüper and Schlegel, 1964). Thiele (1966) studied a considerable number of strains and found that the species *Chr. okenii, Chr. weissei, Chr. warmingii, Chr. minus, Amoebobacter pendens* (syn. *Rhodothece pendens*), and *Amoebobacter roseus* (syn. *Rhodothece conspicua*) are incapable of assimilatory sulfate reduction. Also, *Chr. buderi* (Trüper and Jannasch, 1968) and some strains of the small-celled *Chromatium* and *Thiocapsa* species were not able to utilize sulfate as a source of cell sulfur (Thiele, 1966).

A study by Lippert (1967) of the utilization of molecular hydrogen by a number of *Chlorobium* strains revealed that none of them was able to reduce sulfate to the level of amino acid sulfur, i.e., of sulfhydryl groups.

The *Chlorobium* strains did, however, assimilate cysteine, but not methionine, cystine, cysteic acid, thioglycolate, thioacetamide, or sulfite (Lippert, 1967;

Lippert and Pfennig, 1969), when they were incubated with hydrogen as the photosynthetic electron donor. Cysteine sulfur was found to be transferred into methionine sulfur. Also, *Chr. vinosum* and *T. roseopersicina* strains (Thiele, 1966, 1968a) are able to grow with cysteine, though not with methionine, as the source of sulfur, as long as they are supplied with molecular hydrogen, acetate, or fructose as the photosynthetic electron donor. In these Chromatiaceae, too, the transfer of cysteine sulfur into methionine sulfur was shown.

Although cysteine is utilized as a source of cell sulfur by Chromatiaceae (Thiele, 1966) and Chlorobiaceae, in the latter family, it does not indefinitely sustain growth. Lippert (1967) found that growth ceased after several generations. Therefore, cysteine cannot be considered as a sufficient sulfur source for *Chlorobium*. On the other hand, since *Chlorobium* is unable to assimilate sulfate, Lippert (1967) concluded that some unknown essential sulfur-containing cellular component at the oxidation level of sulfate cannot be synthesized under hydrogen plus cysteine. Such compounds could be sulfate-esters or sulfonates. The most probable compound of this kind is a sulfolipid, first isolated by Benson et al. (1959) from green algae, *Rs. rubrum*, and higher plants, which might be 1-*O*-(β-6'-deoxy-aldohexopyranosyl-6'-sulfonic acid)-3-*O*-oleoyl-glycerol:

Such a sulfolipid was also found by Wood et al. (1965) in *Rp. sphaeroides* (see Chapter 14). It is possible that sulfolipids are of importance in the molecular fine structure of the photosynthetic apparatus.

Little is known about the pathway of sulfate reduction in the Rhodospirillaceae and members of the Chromatiaceae. Ibanez and Lindstrom (1959, 1962) studied the reduction of sulfate by whole cells and chromatophores of *Rs. rubrum*. The chromatophores were able to synthesize 3'-phospho-adenylyl sulfate (PAPS) either in the light or on the addition of ATP in the dark. Light was required for the reduction of sulfate to a volatile form, probably sulfite, and it enhanced the incorporation of radioactive sulfate sulfur into cysteine and cysteic acid of the chromatophores. The authors doubt, however, that this incor-

poration was due to net synthesis, and assume that it arose from exchange reactions with reduced sulfur under the conditions employed. The ratio of activation to reduction was about 100:1, which, together with the fact that exogenous ATP inhibited reduction to volatile sulfur in the light, led to the conclusion that PAPS might not be the *in vivo* substrate of sulfate reduction by chromatophores.

Thiele (1966, 1968a) studied the distribution of radioactive label in hydrolyzed *T. roseopersicina* and *Chr. vinosum* cells after growth with ^{35}S sulfate and found that cystine and methionine (and methionine sulfoxide) were labeled. This observation may be considered as evidence for the formation of the sulfur-containing amino acids from sulfate by these organisms. Cells grown photoheterotrophically with sulfate as sulfur source possess considerable APS reductase activity (up to half the activity of sulfide-grown cells). That they do suggests that APS reductase, not the reduction of PAPS, may be responsible for assimilatory sulfate reduction in Chromatiaceae (Thiele, 1968b; Trüper and Peck, 1970).

Only recently, indications of the presence of a sulfite reductase (hydrogen sulfide:acceptor oxidoreductase, EC 1.8.99.1) in phototrophic bacteria were found in screening experiments by Peck et al. (1974). A methyl-viologen-dependent reduction of sulfite to sulfide was shown to occur in cell-free extracts of *Rp. viridis*, *Rp. gelatinosa*, *Rs. rubrum*, *Rm. vannielii*, *Chr. vinosum*, and *Chl. limicola f. thiosulfatophilum*. In *Chr. vinosum*, the enzyme was present in autotrophically grown cells, as well as in those grown heterotrophically. The occurrence of sulfite reductase in the strictly photolithoautotroph *Ch. limicola*, which is not capable of assimilatory sulfate reduction (Lippert, 1967), as well as in photolithoautotrophically grown *Chr. vinosum*, must be taken as an indication of a possible function of sulfite reductase in dissimilatory sulfur metabolism, i.e., in the oxidation of sulfide (or sulfur) to sulfite. Peck et al. (1974), however, explain the presence of sulfite reductase in photolithoautotrophically grown *Chr. vinosum* by assuming that the enzyme is not subject to feedback control. They further state that the activities they found were too low to allow a mass transformation in a dissimilatory pathway. More work is necessary to resolve this question, and also to show whether the sulfite reductases of phototrophic bacteria contain the new siroheme of Murphy and Siegel (1973), which so far has been found in all purified sulfite reductases.

The pathway of cysteine synthesis from sulfide was studied in some phototrophic bacteria by Chambers and Trudinger (1971). In *Chl. limicola f. thiosulfatophilum* and in *Rp. sphaeroides*, cysteine synthesis from sulfide occurs via serine acetyltransferase (acetyl-CoA:L-serine *O*-acetyltransferase, EC 2.3.1.30) and *O*-acetylserine sulfhydrylase [*O*-acetyl-L-serine acetate lyase (adding hydrogen sulfide), EC 4.2.99.8]:

$$\text{L-Serine} + \text{acetyl} - \text{CoA} \rightarrow$$
$$O\text{-acetylserine} + \text{CoA} \quad (15)$$

$$O\text{-Acetylserine} + H_2S \rightarrow \text{cysteine} + \text{acetate} \quad (16)$$

This pathway was also found in *Rs. rubrum, Rs. fulvum, Rp. palustris, Rp. gelatinosa, Rp. globiformis, Chr. vinosum, T. roseopersicina, Ectothiorhodospira mobilis,* and *Chl. vibrioforme f. thiosulfatophilum* (Hensel and Trüper, 1976). The enzyme *S*-sulfocysteine synthase, which catalyzes a condensation of serine and thiosulfate, was not found in these bacteria (Chambers and Trudinger, 1971).

With respect to methionine biosynthesis in phototrophic bacteria, relatively little is yet known. Cauthen *et al.* (1967) found a vitamin-B_{12}-dependent methionine synthase (N^5-methyltetrahydrofolate: homocysteine B_{12}-methyltransferase, EC 2.1.1.13) in *Rp. sphaeroides.* This enzyme was also found in and partially purified from *Chr. vinosum* (strain D) and *Rs. rubrum* by Ohmori and Fukui (1974). Thus, as was to be expected, the sulfur moiety of the molecule is carried from homocysteine to methionine.

4. Conclusion

Seen as a whole, sulfur metabolism in the phototrophic bacteria is far from being completely understood. With the exception of such basic morphological phenomena as elemental sulfur formation, sulfide disappearance, and sulfate formation, all our present knowledge in this field has been obtained in the past 15 years. The rapid development in biochemical techniques will certainly allow considerable progress in the near future.

5. Note Added in Proof

Since the completion of this chapter the last sentence of its Conclusion has come true. Our knowledge about sulfur metabolism of the phototrophic bacteria has increased considerably since 1975 (Trüper, 1975).

5.1. Dissimilatory Sulfur Metabolism

Experiments with whole cells showed that *Chl. limicola f. thiosulfatophilum* oxidizes sulfide to sulfur and sulfur to sulfate in two strictly separate phases (Lorenz and Trüper, unpublished). Besides elemental sulfur, Schedel (1977) noticed a massive appearance of thiosulfate as an intermediate during this process; only after the complete oxidation of the sulfide were thiosulfate and, at a slower rate, elemental sulfur oxidized to sulfate. The oxidation of elemental sulfur to sulfate occurred without formation of free intermediates such as tetrathionate or other polythionates. While *Chl. limicola f. thiosulfatophilum* does not form elemental sulfur during thiosulfate oxidation (Trüper *et al.,* 1976), *Chl. vibrioforme f. thiosulfatophilum* does (Trüper, unpublished). This explains the contradictory results reported in the literature about growth on thiosulfate by "*Chl. thiosulfatophilum.*"

Photooxidation of sulfide by the newly described thermophilic gliding filamentous bacterium *Chloroflexus aurantiacus* was reported by Castenholz (1973) as well as by Madigan and Brock (1975). The organism oxidizes sulfide anaerobically under photoautotrophic or photoheterotrophic growth conditions and deposits elemental sulfur outside the cells.

As shown by Cohen *et al.* (1975a,b) the cyanobacterium *Oscillatoria limnetica* is capable of anoxygenic photosynthesis under anaerobic conditions using sulfide as the photosynthetic electron donor. The stoichiometric ratio between Na_2S consumed and CO_2 fixed of 2:1 indicates that equation 1 (of this chapter) is fulfilled, i.e., that elemental sulfur is the product of this reaction.

The disproportionation of elemental sulfur by whole cells of Chlorobiaceae as reported by Paschinger *et al.* (1974) was further investigated (Lorenz and Trüper, unpublished). It became clear that this capability is restricted to the thiosulfate-utilizing *formae* of *Chl. limicola* and *Chl. vibrioforme,* while other Chlorobiaceae as well as *Chr. vinosum* and *Ectothiorhodospira mobilis* are incapable of sulfur disproportionation. The reaction(s) occur(s) only anaerobically in the light in the absence of CO_2 and only as long as H_2S is purged from the suspension by argon or nitrogen. The stoichiometry was determined as

$$4\,S + 3H_2O \rightarrow 2H_2S + H_2S_2O_3$$

It was shown by proper analysis that sulfate is not formed during this process. Very small but measurable amounts of sulfite indicated that the thiosulfate may be formed by nonenzymatic condensation of sulfite with sulfur (Trüper *et al.,* 1976).

The oxidation of sulfide to elemental sulfur in *T. roseopersicina* occurs by reduction of cytochrome *c*-550, so far the only cytochrome in the soluble fraction of this organism (Fischer and Trüper, 1977). The reaction is apparently nonenzymatic because the purified cytochrome functions even after 5 min heating at 100°C. Petushkova and Ivanovskii (1976*a*) found that whole cells of *T. roseopersicina* were able to oxidize sulfide after similar heating periods and postulated that the sulfide-oxidizing system is located in the soluble fraction and is thermostabile. Obviously, cytochrome *c*-550 is this "system." While this cytochrome forms elemental sulfur from sulfide the purified flavocytochromes *c*-552 of *Chr. vinosum* and *c*-553 of *Chl. limicola* do not form sulfur but thiosulfate (Fischer, 1977). Knaff and Buchanan (1975) studied photoreduction of cytochrome *b* in phototrophic sulfur bacteria and suggested a function of *b*-type cytochromes in noncyclic electron flow from Na$_2$S to NADP in *Chl. limicola f. thiosulfatophilum*.

A siroheme-containing sulfite reductase was proven to occur in *Chr. vinosum* (Schedel, 1977) and subsequently purified to homogeneity from sulfide/CO$_2$-grown cells. The molecular weight is 280,000 daltons. The complex enzyme consists of eight subunits (four of 37,000; four of 42,000 daltons). It contains 4 siroheme groups, about 48 iron/sulfur groups, no flavin, and reacts with viologen electron donors, but not with NAD(P)H. It is suggested that this enzyme, which resembles that of *Thiobacillus denitrificans* in several properties, is responsible for the oxidation of elemental sulfur to sulfite in purple sulfur bacteria (Schedel and Trüper, unpublished).

Kondrat'eva *et al.* (1975) found that *T. roseopersicina* is able to grow chemolithoautotrophically under aerobic conditions in the dark. This surprising finding points toward a closer relationship between phototrophic sulfur metabolism and that of the thiobacilli than expected earlier. Under the above conditions *T. roseopersicina* oxidizes sulfide to sulfate via elemental sulfur. Further confirming studies by Petushkova and Ivanovskii (1976*a,b*) and Ivanovskii and Petushkova (1976) revealed that during chemolithotrophic growth a sulfite:cytochrome-*c* oxidoreductase participated in sulfite oxidation, while during photolithotrophic growth APS reductase is the more important enzyme. An AMP-independent sulfite-oxidase functioning with ferricyanide and producing sulfate from sulfite was purified from *Chr. vinosum* (Brückenhaus and Trüper, unpublished).

Hansen (1976) reported aerobic oxidation by *Rp. sulfidophila* of thiosulfate, but chemolithotrophy has not been demonstrated so far.

The enzymology of thiosulfate-splitting was further studied by Hashwa (1975) in *Chr. vinosum*, *Chr. minus*, and *Rp. palustris* and by Petushkova and Ivanovskii (1976*c*) in *T. roseopersicina*: all organisms contain thiosulfate reductase as well as rhodanese. Hashwa partially purified these enzymes. Petushkova and Ivanovskii showed that thiosulfate-splitting enzyme activities increased under chemolithotrophic conditions.

The sulfur isotope fractionation studies of Mekht'eva and Kondrat'eva (1966) with *Ectothiorhodospira shaposhnikovii* were resumed and principally confirmed by Ivanov *et al.* (1976).

5.2. Assimilatory Sulfur Metabolism

The enzymes ATP sulfurylase and APS kinase are present in *Rp. capsulata*, *Rp. sulfidophila* and *Rp. viridis* (Cooper and Trüper, unpublished). A. Schmidt and Trüper (1977) tested extracts of 14 species of phototrophic bacteria for their ability to form volatile sulfur compounds from APS and PAPS. The *Rhodospirillum* species tested showed marked activities with both APS and PAPS while the *Rhodopseudomonas* species apparently preferred PAPS. *T. roseopersicina* and *Chr. vinosum* exhibited strongest activities with APS, whereas *Chl. limicola* had equally high activity with PAPS. Subsequently A. Schmidt (1977*a*) purified and characterized the responsible enzyme from *Rs. rubrum*: an APS sulfotransferase with a requirement for thiols. This enzyme resembles the respective plant enzymes. APS sulfotransferase was also found in *Rs. tenue* (Imhoff and Trüper, unpublished). On the other hand, *Rp. sulfidophila* and *Rp. viridis* reduce PAPS in a reaction requiring a heat-resistant protein of low molecular weight, whereas they do not reduce APS. These activities are all located in the soluble fraction of the cells (Imhoff and Trüper, unpublished). These results indicate a dichotomy in the pathway of assimilatory sulfate reduction occurring in the family of the Rhodospirillaceae. A similar dichotomy exists in the cyanobacteria. In a first comparative study of assimilatory sulfate reduction in these bacteria, A. Schmidt (1977*b*) showed that *Plectonema* extracts reduced APS, while those of *Spirulina*, *Synechococcus*, and *Synechocystis* reduced PAPS. The synthesis of PAPS in *Anabaena* has been demonstrated by Sawhney and Nicholas (1976).

In addition to the species mentioned in the chapter above, *O*-acetylserine sulfhydrylase was also found in *Rs. tenue*, *Rs. photometricum*, *Rp. capsulata*, *Rp. sulfidophila*, and *Thiocapsa pfennigii*. In extracts of

Rs. tenue, Rs. gelatinosa, and the Chromatiaceae studied, *O*-acetylserine also reacts with thiosulfate instead of hydrogen sulfide, thus indicating the presence of *S*-sulfocysteine synthase activity (Hensel and Trüper, 1976). Meanwhile it was possible to purify and separate the two activities from *Rs. tenue* (Hensel and Trüper, unpublished).

An interesting new aspect of sulfur metabolism has been introduced by Sandy *et al.* (1975), who showed that 5′-aminolevulinate synthetase, the enzyme initiating tetrapyrrole synthesis in *Rp. sphaeroides*, is activated by trisulfides of the type GSSSG and CySSSCy. Oxygenation of anaerobic cultures effected a decrease in cellular trisulfides and a subsequent inactivation of the enzyme.

6. References

Bartsch, R. G., Meyer, T. E., and Robinson, A. B., 1968, Complex *c*-type cytochromes with bound flavin, in: *Structure and Function of Cytochromes* (K. Okunuki, M. D. Kamen, and I. Sekuzu, eds.), pp. 443–451, University of Tokyo Press and University of Park Press, Tokyo.

Benson, A. A., Daniel, H., and Wiser, R., 1959, A sulfolipid in plants, *Proc. Natl. Acad. Sci. U.S.A.* **45**:1582.

Boll, M., 1969, Glutathione reductase from *Rhodospirillum rubrum*, *Arch. Mikrobiol.* **66**:374.

Bowen, T. J., Happold, F. C., and Taylor, B. F., 1966, Studies on adenosine-5′-phosphosulphate reductase from *Thiobacillus denitrificans*, *Biochim. Biophys. Acta* **118**:566.

Bregoff, H. M., and Kamen, M. D., 1952, Photohydrogen production in *Chromatium, J. Bacteriol.* **63**:147.

Breuker, E., 1964, Die Verwertung von intrazellulärem Schwefel durch *Chromatium vinosum* im aeroben und anaeroben Licht- und Dunkelstoffwechsel, *Zentralbl. Bakteriol. Parasitenkd. Infektionskr. Hyg. Abt.* 2 **118**:561.

Castenholz, R. W., 1973, The possible photosynthetic use of sulfide by the filamentous phototrophic bacteria of hot springs, *Limnol. Oceanogr.* **18**:863.

Cauthen, S. E., Pattison, J. R., and Lascelles, J., 1967, Vitamin B_{12} in photosynthetic bacteria and methionine synthesis by *Rhodopseudomonas spheroides, Biochem. J.* **102**:774.

Chambers, L. A., and Trudinger, P. A., 1971, Cysteine and *S*-sulphocysteine biosynthesis in bacteria, *Arch. Mikrobiol.* **77**:165.

Chung, Y. C., and Hurlbert, R. E., 1971, Characterization of glutathione reductase of *Chromatium* strain D, *Bacteriol. Proc.* **1971**:144.

Cohen, Y., Jørgensen, B. B., Padan, E., and Shilo, M., 1975a, Sulphide-dependent anoxygenic photosynthesis in the cyanobacterium *Oscillatoria limnetica*, *Nature* **257**:489.

Cohen, Y., Padan, E., and Shilo, M., 1975b, Facultative anoxygenic photosynthesis in the cyanobacterium *Oscillatoria limnetica*, *J. Bacteriol.* **123**:855.

Evans, M. C. W., and Buchanan, B. B., 1965, Photoreduction of ferredoxin and its use in carbon dioxide fixation by a subcellular system from a photosynthetic bacterium, *Proc. Natl. Acad. Sci. U.S.A.* **53**:1420.

Eymers, J. G., and Wassink, E. C., 1938, On the photochemical carbon dioxide assimilation by purple sulphur bacteria, *Enzymologia* **2**:258.

Fischer, U., 1977, Die Rolle von Cytochromen im Schwefelstoffwechsel phototropher Schwefelbakterien, Doctoral thesis, University of Bonn, Germany.

Fischer, U., and Trüper, H. G., 1977, Cytochrome *c*-550 of *Thiocapsa roseopersicina*: properties and reduction by sulfide, *FEMS Lett.* **1**:87.

Gorlenko, V. M., 1974, Oxidation of thiosulphate by *Amoebobacter roseus* in the darkness under microaerophilic conditions, *Mikrobiologiya* **43**:729.

Hageage, G. J., Jr., Eanes, E. D., and Gherna, R. L., 1970, X-ray diffraction studies of the sulfur globules accumulated by *Chromatium* species, *J. Bacteriol.* **101**:464.

Hansen, T. A., 1974, Sulfide als electronendonor voor Rhodospirillaceae, Doctoral thesis, University of Groningen, The Netherlands.

Hansen, T. A., 1976, Some aspects of the oxidation of reduced sulfur compounds by *Rhodopseudomonas sulfidophila, Proc. 2nd Internat. Symp. Photosynthetic Prokaryotes* (G. A. Codd and W. D. P. Stewart, eds.), pp. 41–42, Dundee.

Hansen, T. A., and van Gemerden, H., 1972, Sulfide utilization by purple nonsulfur bacteria, *Arch. Mikrobiol.* **86**:49.

Hansen, T. A., and Veldkamp, H., 1973, *Rhodopseudomonas sulfidophila, nov. spec.*, a new species of the purple nonsulfur bacteria, *Arch. Mikrobiol.* **92**:45.

Hashwa, F., 1972, Die enzymatische Thiosulfatspaltung bei phototrophen Bakterien, Doctoral thesis, University of Göttingen, Germany.

Hashwa, F., 1975, Thiosulfate metabolism in some red phototrophic bacteria, *Plant Soil* **43**:41.

Hashwa, F., and Pfennig, N., 1972, The reductive enzymatic cleavage of thiosulfate: Methods and application, *Arch. Mikrobiol.* **81**:36.

Hendley, D. D., 1955, Endogenous fermentation in Thiorhodaceae, *J. Bacteriol.* **70**:625.

Hensel, G., and Trüper, H., 1976, Cysteine and *S*-sulfocysteine biosynthesis in phototrophic bacteria, *Arch. Microbiol.* **109**:101.

Hurlbert, R. E., 1967, Effect of oxygen on viability and substrate utilization in *Chromatium, J. Bacteriol.* **93**:1346.

Hurlbert, R. E., and Lascelles, J., 1963, Ribulose diphosphate carboxylase in Thiorhodaceae, *J. Gen. Microbiol.* **33**:445.

Ibanez, M. L., and Lindstrom, E. S., 1959, Photochemical sulfate reduction by *Rhodospirillum rubrum, Biochem. Biophys. Res. Commun.* **1**:224.

Ibanez, M. L., and Lindstrom, E. S., 1962, Metabolism of sulfate by the chromatophore of *Rhodospirillum, J. Bacteriol.* **84**:451.

Ivanov, M. V., Gogotova, G. I., Matrosov, A. G., and Zyakun, A. M., 1976, Fractionation of sulfur isotopes by photosynthetic sulfur bacteria *Ectothiorhodospira shaposhnikovii*, *Mikrobiologiya* 45:757.

Ivanovskii, R. N., and Petushkova, Y. P., 1976, Effect of growth conditions on substrate phosphorylation during sulfite oxidation in *Thiocapsa roseopersicina*, *Mikrobiologiya* 45:1102.

Kaplan, I. R., and Rittenberg, S. C., 1964, Microbiological fractionation of sulphur isotopes, *J. Gen. Microbiol.* 34:195.

Kelly, D. P., 1974, Growth and metabolism of the obligate photolithotroph *Chlorobium thiosulfatophilum* in the presence of added organic nutrients, *Arch. Microbiol.* 100:163.

Kirchhoff, J., and Trüper, H. G., 1974, Adenylylsulfate reductase of *Chlorobium limicola*, *Arch. Microbiol.* 100:115.

Knaff, D. B., and Buchanan, B. B., 1975, Cytochrome *b* and photosynthetic sulfur bacteria, *Biochim. Biophys. Acta* 376:549.

Knobloch, K., Eley, J., and Aleem, M. I. H., 1971, Thiosulfate-linked ATP-dependent NAD⁺ reduction in *Rhodopseudomonas palustris*, *Arch. Mikrobiol.* 80:97.

Kondrat'eva, E. N., Mekh'eva, V. L., and Sumarokova, R. S., 1966, On the direction of the isotope effect in the first steps of sulfide oxidation by purple bacteria, *Vestn. Mosk. Univ. Ser. VI (Biol. Soil Sci.)* 5:45.

Kondrat'eva, E. N., Petushkova, Y. P., and Zhukov, V. G., 1975, Growth and oxidation of sulfur compounds by *Thiocapsa roseopersicina* in the darkness, *Mikrobiologiya* 44:389.

Kran, G., Schlote, F. W., and Schlegel, H. G., 1963, Cytologische Untersuchungen an *Chromatium okenii* Perty, *Naturwissenschaften* 50:728.

Kusai, A., and Yamanaka, T., 1973a, Cytochrome *c* (553, *Chlorobium thiosulfatophilum*) is a sulphide-cytochrome *c* reductase, *FEBS Lett.* 34:235.

Kusai, A., and Yamanaka, T., 1973b, A novel function of cytochrome *c* (555, *Chlorobium thiosulfatophilum*) in oxidation of thiosulphate, *Biochem. Biophys. Res. Comm.* 51:107.

Kusai, A., and Yamanaka, T., 1973c, The oxidation mechanisms of thiosulphate and sulphide in *Chlorobium thiosulfatophilum*: Roles of cytochrome *c*-551 and cytochrome *c*-553, *Biochim. Biophys. Acta* 325:304.

La Rivière, J. W. M., 1963, Cultivation and properties of *Thiovulum majus* Hinze, in: *Symposium on Marine Microbiology* (C. H. Oppenheimer, ed.), pp. 61–67, Charles C. Thomas, Springfield, Illinois.

Larsen, H., 1952, On the culture and general physiology of the green sulfur bacteria, *J. Bacteriol.* 64:187.

Larsen, H., 1953, On the microbiology and biochemistry of the photosynthetic green sulfur bacteria, *K. Nor. Vidensk. Selsk. Skr.* 1:1.

Lippert, K. D., 1967, Die Verwertung von molekulärem Wasserstoff durch *Chlorobium thiosulfatophilum*, Doctoral thesis, University of Göttingen, Germany.

Lippert, K. D., and Pfennig, N. 1969, Die Verwertung von molekulärem Wasserstoff durch *Chlorobium thiosulfatophilum*: Wachstum und CO₂-Fixierung, *Arch. Mikrobiol.* 65:29.

Losada, M., Nozaki, M., and Arnon, D. I., 1961, Photoproduction of molecular hydrogen from thiosulfate by *Chromatium* cells, in: *A Symposium on Light and Life* (W. D. McElroy and B. Glass, eds.), pp. 570–575, The Johns Hopkins University Press, Baltimore.

Lyric, R. M., and Suzuki, I., 1970, Enzymes involved in the metabolism of thiosulfate by *Thiobacillus thioparus*. II. Properties of adenosine-5'-phosphosulfate reductase, *Can. J. Biochem.* 48:344.

Madigan, M. T., and Brock, T. D., 1975, Photosynthetic sulfide oxidation by *Chloroflexus aurantiacus*, a filamentous photosynthetic gliding bacterium, *J. Bacteriol.* 122:782.

Mathewson, J. H., Burger, L. J., and Millstone, H. G., 1968, Cytochrome *c*-551:thiosulfate oxidoreductase from *Chlorobium thiosulfatophilum*, *Fed. Proc. Fed. Am. Soc. Exp. Biol.* 27:774.

Mekh'eva, V. L., and Kondrat'eva, E. N., 1966, Fractionation of stable sulfur isotopes by the purple photosynthetic sulfur bacteria *Rhodopseudomonas* sp., *Dokl. Akad. Nauk SSSR* 166:465.

Much, A. M., and Hurlbert, R. E., 1971, Characterization of thiosulfate reductase of *Chromatium* strain D, *Bacteriol. Proc.* 1971:149.

Murphy, M. J., and Siegel, L. M., 1973, Siroheme and sirohydrochlorin. The basis for a new type of porphyrinrelated prosthetic group common to both assimilatory and dissimilatory sulfite reductases, *J. Biol. Chem.* 248:6911.

Nakamura, H., 1937, Über die Kohlensäureassimilation von *Rhodospirillum giganteum*, *Acta Phytochim.* 9:231.

Nicolson, G. L., and Schmidt, G. L., 1971, Structure of the *Chromatium* sulfur particle and its protein membrane, *J. Bacteriol.* 105:1142.

Ohmori, H., and Fukui, S., 1974, Vitamin B₁₂-dependent methionine synthetase in photosynthetic bacteria: Partial purification and properties, *Agric. Biol. Chem. (Tokyo)* 38:1317.

Paschinger, H., Paschinger, J., and Gaffron, H., 1974, Photochemical disproportionation of sulfur into sulfide and sulfate by *Chlorobium limicola thiosulfatophilum*, *Arch. Microbiol.* 96:341.

Peck, H. D., Jr., 1966, Some evolutionary aspects of inorganic sulfur metabolism, in: *Lecture Series on Theoretical and Applied Aspects of Modern Microbiology*, pp. 1–22, University of Maryland, College Park.

Peck, H. D., Jr., Deacon, T. E., and Davidson, J. T., 1965, Studies on adenosine 5'-phosphosulfate reductase from *Desulfovibrio desulfuricans* and *Thiobacillus thioparus*, *Biochim. Biophys. Acta* 96:429.

Peck, H. D., Jr., Tedro, S., and Kamen, M. D., 1974, Sulfite reductase activity in extracts of various photosynthetic bacteria, *Proc. Natl. Acad. Sci. U.S.A.* 71:2404.

Petrova, E. A., 1959, The morphology of sulfur purple bacteria of the genus *Chromatium* as a function of the composition of the medium, *Mikrobiologiya* 28:814.

Petushkova, Y. P., and Ivanovskii, R. N., 1976a, Respiration of *Thiocapsa roseopersicina*, *Mikrobiologiya* 45:9.

Petushkova, Y. P., and Ivanovskii, R. N., 1976*b*, Oxidation of sulfite by *Thiocapsa roseopersicina*, *Mikrobiologiya* **45**:592.

Petushkova, Y. P., and Ivanovskii, R. N., 1976*c*, Enzymes participating in thiosulfate metabolism in *Thiocapsa roseopersicina* during its growth under various conditions, *Mikrobiologiya* **45**:960.

Pfennig, N., 1974, *Rhodopseudomonas globiformis*, sp.n., a new species of the *Rhodospirillaceae*, *Arch. Microbiol.* **100**:197.

Puchkova, N. N., Gorlenko, V. M., and Pivovarova, T. A., 1974, Comparative study of fine structure of vibrioid green sulphur bacteria, *Mikrobiologiya* **44**:108.

Rolls, J. P., and Lindstrom, E. S., 1967*a*, Induction of a thiosulfate-oxidizing enzyme in *Rhodopseudomonas palustris*, *J. Bacteriol.* **94**:784.

Rolls, J. P., and Lindstrom, E. S., 1967*b*, Effect of thiosulfate on the photosynthetic growth of *Rhodopseudomonas palustris*, *J. Bacteriol.* **94**:860.

Sandy, J. D., Davies, R. C., and Neuberger, A., 1975, Control of 5'-aminolevulinate synthetase activity in *Rhodopseudomonas spheroides*. A role for trisulphides, *Biochem. J.* **150**:245.

Sawhney, S. K., and Nicholas, D. J. D., 1976, *Planta* **132**:189.

Schedel, M., 1977, Untersuchungen zur anaeroben Oxidation reduzierter Schwefelverbindungen durch *Thiobacillus denitrificans*, *Chromatium vinosum*, und *Chlorobium limicola*, Doctoral thesis, University of Bonn, Germany.

Schmidt, A., 1977*a*, Adenosine-5'-phosphosulfate (APS) as sulfate donor for assimilatory sulfate reduction in *Rhodospirillum rubrum*, *Arch. Microbiol.* **112**:263.

Schmidt, A., 1977*b*, Assimilatory sulfate reduction via 3'-phosphoadenosine-5'-phosphosulfate (PAPS) and adenosine-5'-phosphosulfate in blue-green algae, *FEMS Lett.* **1**:137.

Schmidt, A., and Trüper, H. G., 1977, Reduction of adenylsulfate and 3'-phosphoadenylsulfate in phototrophic bacteria, *Experientia* **33**:1008.

Schmidt, G. L., and Kamen, M. D., 1970, Variable cellular composition of *Chromatium* in growing cultures, *Arch. Mikrobiol.* **73**:1.

Schmidt, G. L., Nicolson, G. L., and Kamen, M. D., 1971, Composition of the sulfur particle of *Chromatium vinosum* strain D, *J. Bacteriol.* **105**:1137.

Smith, A. J., 1965, The discriminative oxidation of the sulphur atoms of thiosulphate by a photosynthetic sulphur bacterium—*Chromatium* strain D, *Biochem. J.* **94**:27P.

Smith, A. J., 1966, The role of tetrathionate in the oxidation of thiosulphate by *Chromatium* sp. strain D, *J. Gen. Microbiol.* **42**:371.

Smith, A. J., and Lascelles, J., 1966, Thiosulphate metabolism and rhodanese in *Chromatium* sp. strain D, *J. Gen. Microbiol.* **42**:357.

Thiele, H. H., 1966, Wachstumsphysiologische Untersuchungen an *Thiorhodaceae*: Wasserstoff-Donatoren und Sulfatreduktion, Doctoral thesis, University of Göttingen, Germany.

Thiele, H. H., 1968*a*, Sulfur metabolism in Thiorhodaceae.
IV. Assimilatory reduction of sulfate in *Thiocapsa floridana* and *Chromatium* species, *Antonie van Leeuwenhoek; J. Microbiol. Serol.* **34**:341.

Thiele, H. H., 1968*b*, Sulfur metabolism in Thiorhodaceae. V. Enzymes of sulfur metabolism in *Thiocapsa floridana* and *Chromatium* species, *Antonie van Leeuwenhoek; J. Microbiol. Serol* **34**:350.

Trüper, H. G., 1964*a*, Sulphur metabolism in Thiorhodaceae. II. Stoichiometric relationship of CO_2 fixation to oxidation of hydrogen sulphide and intracellular sulphur in *Chromatium okenii*, *Antonie van Leeuwenhoek; J. Microbiol. Serol.* **30**:385.

Trüper, H. G., 1964*b*, CO_2-Fixierung und Intermediärstoffwechsel bei *Chromatium okenii* Perty, *Arch. Mikrobiol.* **49**:23.

Trüper, H. G., 1975, The enzymology of sulfur metabolism in phototrophic bacteria—A review, *Plant Soil* **43**:29.

Trüper, H. G., and Hathaway, J. C., 1967, Orthorhombic sulfur formed by photosynthetic sulfur bacteria, *Nature (London)* **215**:435.

Trüper, H. G., and Jannasch, H. W., 1968, *Chromatium buderi* nov. spec., eine neue Art der "grossen" Thiorhodaceae, *Arch. Mikrobiol.* **61**:363.

Trüper, H. G., and Peck, H. D., Jr., 1970, Formation of adenylyl sulfate in phototrophic bacteria, *Arch. Mikrobiol.* **73**:125.

Trüper, H. G., and Pfennig, N., 1966, Sulphur metabolism in Thiorhodaceae. III. Storage and turnover of thiosulphate sulfur in *Thiocapsa floridana* and *Chromatium* species, *Antonie van Leeuwenhoek; J. Microbiol. Serol.* **32**:261.

Trüper, H. G., and Rogers, L. A., 1971, Purification and properties of adenylyl sulfate reductase from the phototrophic sulfur bacterium, *Thiocapsa roseopersicina*, *J. Bacteriol.* **108**:1112.

Trüper, H. G., and Schlegel, H. G., 1964, Sulphur metabolism in Thiorhodaceae. I. Quantitative measurements on growing cells of *Chromatium okenii*, *Antonie van Leeuwenhoek; J. Microbiol. Serol.* **30**:225.

Trüper, H. G., Lorenz, C., and Fischer, U., 1976, Metabolism of elemental sulfur by phototrophic sulfur bacteria, *Proc. 2nd Internat. Symp. Photosynthetic Prokaryotes* (G. A. Codd and W. D. P. Stewart, eds.), pp. 41–42, Dundee.

van Gemerden, H., 1967, On the bacterial sulfur cycle of inland waters, Doctoral thesis, University of Leiden, The Netherlands.

van Gemerden, H., 1968*a*, Utilization of reducing power in growing cultures of *Chromatium*, *Arch. Mikrobiol.* **64**:111.

van Gemerden, H., 1968*b*, On the ATP generation by *Chromatium* in darkness, *Arch. Mikrobiol.* **64**:118.

van Gemerden, H., 1974, Coexistence of organisms competing for the same substrate: An example among the purple sulfur bacteria, *Microb. Ecol.* **1**:19.

van Gemerden, H., and Jannasch, H. W., 1971, Continuous culture of Thiorhodaceae: Sulfide and sulfur limited growth of *Chromatium vinosum*, *Arch. Mikrobiol.* **79**:345.

van Niel, C. B., 1931, On the morphology and physiology of the purple and green sulphur bacteria, *Arch. Mikrobiol.* **3**:1.

van Niel, C. B., 1936, On the metabolism of the Thiorhodaceae, *Arch. Mikrobiol.* **7**:323.

van Niel, C. B., 1944, The culture, general physiology, morphology and classification of the non-sulfur purple and brown bacteria, *Bacteriol. Rev.* **8**:1.

van Niel, C. B., 1963, A brief survey of the photosynthetic bacteria, in: *Bacterial Photosynthesis* (H. Gest, A. San Pietro, and L. P. Vernon, eds.), pp. 459–467, The Antioch Press, Yellow Springs, Ohio.

Wood, B. J. B., Nichols, B. W., and James, A. T., 1965, The lipids and fatty acid metabolism of photosynthetic bacteria, *Biochim. Biophys. Acta* **106**:261.

Yoch, D. C., and Lindstrom, E. S., 1971, Survey of the photosynthetic bacteria for rhodanese (thiosulfate:cyanide sulfur transferase) activity, *J. Bacteriol.* **106**:700.

CHAPTER 36

Photosynthetic Carbon Metabolism in the Green and Purple Bacteria

R. C. Fuller

1. Introduction

Although carbon metabolism in the photosynthetic bacteria has been of interest to many investigators since the turn of the century, the turning point that led to our current understanding of the biochemistry of CO_2 fixation and reduction came in the 1930's with the publication of the classic papers by van Niel (1930, 1931, 1935). These papers, of course, not only demonstrated the biochemistry of the photosynthetic bacteria themselves, but also, by comparing bacterial photosynthesis with higher plant and algal photosynthesis, led to our current level of understanding of the photosynthetic mechanism as a whole. van Niel expressed the analogy with two simple equations.

For higher plants:

$$CO_2 + H_2O \xrightarrow[\text{Chl}]{\text{light}} (CH_2O) + H_2O + O_2$$

For bacteria:

$$CO_2 + H_2S \xrightarrow[\text{Bchl}]{\text{light}} (CH_2O) + H_2O + 2\,S$$

He then put forward the unitary concept for all photosynthesis as

$$CO_2 + 2\,H_2A \xrightarrow{\text{light}} CH_2O + H_2O + 2\,A$$

This comparative biochemical reasoning, of course, suggested not only that the O_2 evolved by plants came from water, not CO_2, but also that the synthesis of cell components came from a reduction of the CO_2 molecule by hydrogen having as its source water in plants and reduced inorganic or organic compounds in bacteria.

It can therefore be stated that real research on the biochemistry of carbon metabolism had its start at this time. The discoveries and controversies were active and intensive over the next decades with the advent of the use of radioisotopes, the techniques of enzymology, and modern molecular biology. It is not the purpose of this chapter to review this fascinating and often bitterly argumentative history of that biochemistry. However, the pioneers in the study of the path of carbon in photosynthesis—Calvin, Benson, Gaffron, Warburg, Arnon, Gibbs, and many others—kept the field a lively and exciting one, and the controversies themselves have led today to *almost* total agreement on the complete mechanism.

The study of the path of carbon in bacterial photosynthesis, although more limited in scope and volume, was vital to our understanding of photosynthesis in general not only because it had its origin in van Niel's hypothesis, but also because of the nature of the organisms themselves. They not only fix and reduce CO_2, but will photoassimilate organic compounds, anaerobically, and some will grow aerobically in the

R. C. Fuller · Department of Biochemistry, University of Massachusetts, Amherst, Massachusetts 01003 dark, and so on; in other words, they show tight metabolic control and consequently enzymatic regulation. Some will grow anaerobically in the dark (see Chapter 46). The ability of the investigator to utilize these "switch" mechanisms in studying particularly the enzymology of carbon metabolism greatly reinforced and often proved or disproved conclusions of modes of metabolic pathways derived from isotope studies.

The main purpose of this chapter will be to summarize the many excellent reviews on the subject and concentrate on the enzymology, isotope pathways, and control-mechanism data that are available on the carbon assimilatory pathways in photosynthetic bacteria. These pathways offer opportunities for future biochemical studies that may indeed be unique, and understanding these pathways may be of great importance for the potential use of solar energy.

Although, because of the nature and scope of this

book, carbon metabolism in higher plants, algae, and blue-green bacteria will not be discussed, many of the conclusions, observations, analogies, and suppositions are dependent on the comparative and evolutionary information available in the published work on these other organisms. The reader is therefore referred to the many excellent reviews on these organisms, especially the works of Walker and Crofts (1970), Hatch and Slack (1970), Smith et al. (1967), Black (1973), and Zellich (1975).

The main achievements in the study of carbon photoassimilation in bacteria at various stages of our understanding of the process are covered in a series of reviews over a period of recent years. Any serious student of the subject of bacterial photosynthesis in general and carbon metabolism in photosynthetic bacteria in particular should refer to the comprehensive reviews by van Niel (1941, 1944, 1954, 1957), Gest

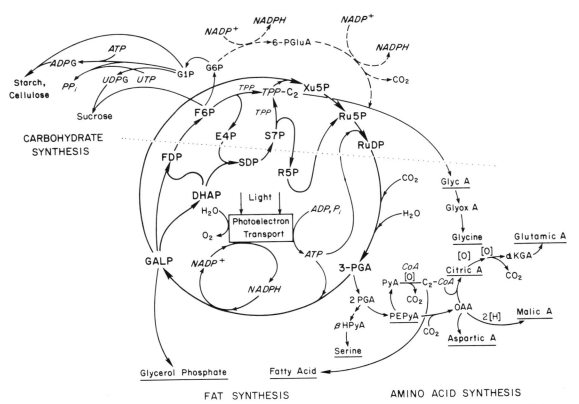

Fig. 1. Pentose reductive or Calvin cycle. This presentation shows the regenerative cycle and its associations with various biosynthetic pathways. Reproduced from Bassham and Kirk (1969). Nonstandard abbreviations: (G1P, G6P) glucose 1- or 6-phosphate; (6-PGluA) 6-phosphogluconic acid; (FDP) fructose diphosphate; (F6P) fructose 6-phosphate; (TPP) thiamine pyrophosphate; (Xu5P) xylose 5-phosphate; (SDP) sedoheptulose diphosphate; (S7P) sedoheptulose 7-phosphate; (R5P) ribose 5-phosphate; (DHAP) dihydroxyacetone phosphate; (GALP) glyceraldehyde phosphate; (PyA) pyruvic acid; (PEPyA) phosphoenolpyruvic acid; (αKGA) α-ketoglutaric acid; (OAA) oxaloacetic acid; (βHPyA) β-hydroxypyruvic acid; (Ru5P) ribulose 5-phosphate; (RuDP) ribulose diphosphate.

(1951), Gest and Kamen (1960), Stanier (1961), Ormerod and Gest (1962), Elsden (1962), Kondrat'eva (1965, 1974), Pfennig (1967), Fuller (1969, 1971), Pfennig and Trüper (1974), and Wiessner (1970). This author will lean heavily on the accumulated material available in these works and often refer to these more comprehensive discussions to emphasize the most recent events in the field.

Although there are more than 50 species of photosynthetic bacteria known and described, detailed studies on the photoassimilation of carbon have been carried on in only a few. Because of the diverse nature of these bacteria, however, it is assumed that the presence and functioning of the pentose reductive (Calvin) cycle is a universal and ancient metabolic cycle in these organisms. Figure 1 depicts the Calvin cycle as it operates in green plants, showing particularly the relationship of the cycle to various biosynthetic mechanisms. The diversity of variations ancillary to the cycle is as great as the diversity of the organisms themselves, as described in several recent reviews (Elsden, 1962; Pfennig, 1967; Fuller, 1969; Tabita *et al.*, 1974; McFadden and Tabita, 1974; Kondrat'eva, 1974).

2. Pentose Reductive (Calvin) Cycle in Phototrophic Bacteria

2.1. Metabolic and Kinetic Evidence in the Obligate and Facultative Purple and Green Bacteria

The following discussion on the path of carbon in photosynthesis will emphasize two major mechanisms of CO_2 fixation that relate to photosynthesis: (1) the regenerative pentose reductive or Calvin cycle, which not only fixes and reduces CO_2, but also, by a cyclic mechanism, regenerates the carbon substrate for the carboxylation of ribulose diphosphate (RuDP); and (2) additional CO_2 fixation reactions that add to cellular carbon, but use as substrates already reduced compounds that are direct products of the Calvin cycle. It is increasingly apparent that individual carboxylations and even additional cycles may function as additive metabolic sequences for cellular synthesis, but are, in turn, dependent on the Calvin cycle intermediates as substrates.

A similar situation was reported by Kornberg in heterotrophic microorganisms. He coined the word *anaplerotic* to describe sequences that essentially replenish lost carbon in such metabolic cycles as the Krebs cycle. When organisms growing on three- and

two-carbon compounds lose an equivalent number of carbon atoms in oxidative metabolism, anaplerotic sequences replenish the cycle and allow cellular synthesis to continue.

Although reactions associated with the Calvin cycle are not truly anaplerotic, in that the Calvin cycle is a regenerative additive cycle in itself, they allow the cell to go about its own particular metabolic business in a much more efficient and directly productive manner. We shall refer to these reactions as *ancillary* carboxylations to differentiate between them and the regenerative reactions of the Calvin cycle.

With some question still open in the green bacteria, the major evidence from many different photosynthetic bacteria as studied in many laboratories indicates that in the photosynthetic bacteria, the regenerative Calvin cycle is the route of CO_2 fixation and reduction.

The establishment of any metabolic pathway or cycle that functions *in vivo* depends on evidence or observations obtained from several diverse experimental approaches, among which are: metabolic need for the pathway; presence of reasonable amounts of the required enzymes; kinetic analysis of the series of reactions, usually using isotopes; and, probably most important, the regulation of the enzymes in question. Any single one of these lines of evidence is not sufficient proof for the function of a metabolic pathway in any system. As mentioned earlier, the photosynthetic bacteria offer excellent experimental material for these approaches, especially the use of $^{14}CO_2$ and the rapid "switch" mechanisms using light and growth conditions as regulators. The one experimental approach in which the evidence for the mechanism of the path of carbon is weaker in bacteria than in plants is the careful chemical degradation of the intermediate labeled compounds that was done in higher plants by Bassham and Calvin (1957) and Gibbs and Kandler (1957). As will be shown in the following pages, however, sufficient evidence has been accumulated to show that the Calvin cycle is the main regenerative pathway of carbon metabolism in at least the modern counterparts of these most primitive photosynthetic cells (Tabita and McFadden, 1974a, b).

2.2. Evidence for the Pentose Reductive Cycle in Chromatiaceae

Several genera of the Chromatiaceae have been investigated relative to the path of CO_2 fixation and reduction in photosynthesis. All show strong evidence for the *in vivo* functioning of the Calvin cycle. Several of the studies meet all the criteria stated in Section 2.1 for the demonstration of the function of the cycle in photosynthesis. The early work on *Chromatium vinosum* strain D gave the first evidence in bacterial systems, using $^{14}CO_2$ fixation kinetics and enzyme control experiments, that the main pathway of CO_2 fixation is the Calvin cycle. This work also showed that ancillary carboxylations play a major role in the total carbon assimilation under photoautotrophic conditions. After all, the major role of either CO_2 fixation or the photoassimilation of organic compounds in bacteria is very different from that of higher plants or algae. Although photosynthetic bacteria do store polysaccharide products such as glycogen (Hara *et al.*, 1973) or poly-β-hydroxybutyrate, the main metabolic role and need for photosynthesis is *not* the synthesis of energy storage compounds, but rather the rapid production of metabolites needed to synthesize new proteins, lipids, and other substances and get on with the important business of cell development and reproduction. The schemes and evidence presented below, therefore, demonstrate the highly selected and biochemically adapted mechanisms that have evolved and are utilized by these organisms for the reproduction of their cellular components in their own particular environments.

2.2.1. *Chromatium vinosum* Strain D

Short-term $^{14}CO_2$-fixation experiments using autotrophically grown *Chr. vinosum* strain D show a decreasing percentage of the total ^{14}C incorporation into phosphoglyceric acid (PGA), indicating the primary carboxylation of RuDP, and the usual increasing percentage of ^{14}C accumulation in secondary products of CO_2 fixation (Fuller *et al.*, 1961). In these experiments, such a "negative slope" of incorporation was also noted for aspartic acid, and this negative slope appeared to be the result of an active ancillary carboxylation of phosphoenolpyruvate to form oxaloacetate and its subsequent transamination to aspartate. This is typical of the specialized ancillary carboxylations that allow this organism to rapidly incorporate substantial amounts of carbon into amino acids and fatty acids for the synthesis of products used primarily for rapid growth and metabolism. *Chromatium vinosum* strain D has a large pool of aspartate as a basic "carbon source" for such reactions. Other organisms use different carboxylation reactions (Buchanan *et al.*, 1972) for the same purpose.

In addition, all the key enzymes of the Calvin cycle have been demonstrated in *Chr. vinosum* strain D (Fuller *et al.*, 1961; Smillie *et al.*, 1962). RuDP carboxylase is also under metabolic control, being repressed 50–90% when the cells are grown under

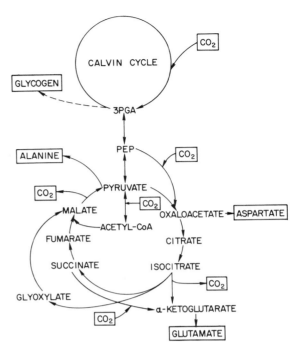

Fig. 2. Path of carbon in *Chromatium vinosum* strain D. Adapted from Fuller *et al.* (1961).

photoheterotrophic conditions. In this species, ATP-producing steps involving two enzymes of the citric acid cycle, malic dehydrogenase and α-ketoglutaric oxidase, are absent. In addition, when grown on acetate, the cells possess isocitric lyase and a "modified" glyoxalate cycle. Losada *et al.* (1960) and Arnon *et al.* (1961) also demonstrated the importance of further carboxylations in this organism.

The data cited above suggested the scheme shown in Fig. 2 for the path of carbon in photosynthesis in *Chr. vinosum*, dependent on the Calvin cycle for the regeneration of the CO_2 acceptor, and *additional* carboxylation reactions for the specialized metabolic needs of this organism.

It is important to note that in this and other metabolic pathways in the photosynthetic bacteria, all these potentially competitive carboxylations and synthetic steps cannot, and do not, operate simultaneously. For instance, as indicated in Fig. 2, the glyxoylate cycle occurs only when cells are grown on acetate as sole source of carbon when the Calvin cycle enzymes are repressed and not needed for cell synthesis. The control mechanisms for these various reactions are virtually unknown, and offer a fertile and fruitful field for research.

2.2.2. Further Variation in Carbon Assimilation in Chromatiaceae

Two other organisms in this family have been thoroughly investigated for their modes of carbon assimilation, namely, *Chr. okenii* (Pfennig and Trüper, 1974) and *Ectothiorhodospira shaposhnikovii* (Firsov *et al.*, 1974). In these organisms, as in *Chr. vinosum*, the enzymes of the Calvin cycle are under metabolic control. Unlike *Chr. vinosum*, however, they have a normal glyoxylate cycle—including malic dehydrogenase—that is derepressed when the cells are grown on acetate. They also do not have α-ketoglutarate-dehydrogenase—and thus lack a complete tricarboxylic acid cycle.

2.3. Evidence for the Pentose Reductive Cycle in Rhodospirillaceae

There has been a much larger volume of work on the carbon metabolism of this family than in all other photosynthetic bacteria. It will therefore be impossible to cover all the experimental work carried out by workers in the field who have contributed so much to the total information. The author does not apologize for omissions, but is overwhelmed by the volume of data. The major emphasis will be placed on the work with *Rhodospirillum rubrum* and *Rhodopseudomonas palustris*, in which the experimental evidence most completely fulfills the criteria for the presence or absence of specific metabolic pathways, e.g., kinetic studies, the presence of enzymes, and regulation of the

mode of the path of carbon in bacterial photosynthesis. The following reviews and articles are responsible for and additive to the data reported here: Glover *et al.* (1952), Hoare (1963), Anderson and Fuller (1967a–d), Stoppani *et al.* (1955), Lascelles (1960), Stokes and Hoare (1969).

2.3.1. Evidence for the Path of Carbon in *Rhodospirillum rubrum*

Evidence using ^{14}C-incorporation studies indicates that the primary product of autotrophic CO_2 fixation in *Rs. rubrum* is 3-PGA. Glutamic acid is also a major "early" labeled compound. The early labeling of glutamic acid is due either to the α-ketoglutarate synthetase described by Buchanan *et al.* (1972) or to an exchange reaction with the six-carbon carboxyl group of glutamic acid.

Both the presence of enzymes of the Calvin cycle and their regulation indicate that the path of carbon in *Rs. rubrum* during autotrophic metabolism involves CO_2 fixation via the Calvin cycle and again numerous ancillary carboxylations.

Table 1 shows the level of various enzymes related to the Calvin cycle when cells are grown under both photoautotrophic conditions (in the light with CO_2 as the sole source of carbon) and during the photoassimilation of acetate and malate, as well as in aerobic dark grown cells.

These enzyme levels cause very different $^{14}CO_2$-incorporation pathways in the cell (Anderson and Fuller, 1967a–c). Thus, it has been demonstrated that

Table 1. Levels of Enzymes of the Reductive Pentose Phosphate Cycle and of the Embden–Meyerhof–Parnas Pathway in *Rhodospirillum rubrum*[a]

Enzymes	Autotrophic, H$_2$	Phototrophic, acetate	Phototrophic, L-malate	Dark chemotrophic, L-malate
	Conditions of growth and hydrogen donors			
Ribulose 1,5-diP carboxylase[b]	240	60	7	4
Ribulose 5-P kinase	34	33	12[c]	0[c]
Ribose 5-P isomerase	40	77	51	66
Glyceraldehyde 3-P dehydrogenase	49	36	11	12
Aldolase	9.4	7.6	9.2	11
Fructose 1,6-diPase (neutral)	0	18	2	2
Fructose 1,6-diPase (alkaline)	27	18	1	1
Enolase	80	110	96	140

[a] Data from Anderson and Fuller (1967c). Specific activities (μmol substrate consumed/min per g protein) are based on protein in the crude extract.
[b] Phosphoglycerate formed (μmol).
[c] ATPase interferes.

greatly different kinetic data related to CO_2-fixation patterns are obtained when cells are grown under different environmental conditions.

Rhodospirillum rubrum can grow aerobically in the dark; this implies that the tricarboxylic acid cycle is functioning in these conditions. It is interesting that the cycle can also function for biosynthetic purposes under anaerobic conditions. Thus, the situation here is very different than in *Chromatium*, in which key enzymes of the citric acid cycle are missing and a modified glyoxylate cycle is induced when cells are grown anaerobically in the light on acetate.

It thus seems clear that in *Rs. rubrum*, the Calvin cycle functions as the primary regenerative route of CO_2 incorporation and reduction. Subsidiary carbon metabolism, including a complete tricarboxylic acid cycle, fits the metabolic needs of this facultative photosynthetic bacterium.

2.3.2. Path of Carbon in *Rhodopseudomonas palustris*

Carbon metabolism in *Rp. palustris* was thoroughly studied in a series of papers by Doman, Tchernadev, and Krasilinikova, as reported by Kondrat'eva (1974). The assimilation of CO_2 proceeds through the Calvin cycle and is supplemented by carboxylation of phosphoenolpyruvate. The experimental results were based both on the use of $^{14}CO_2$ and on enzyme studies. Like *Rs. rubrum*, this organism is a facultative aerobe that will grow on acetate in the light, and under aerobic

conditions in the dark. In both organisms, tricarboxylic cycle intermediates are labeled from ^{14}C acetate in short times. In *Rp. palustris*, $^{14}CO_2$ fixation is stimulated by acetate—but only in the light. We can conclude from these data that both *Rs. rubrum* and *Rp. palustris* utilize acetate, as a carbon source as well as reducing power. Acetate assimilation is stimulated by carboxylations of a reductive nature similar to those in green bacteria, which will be discussed later in this chapter. It has been shown, however, that all the enzymes of the glyoxylate cycle are present in *Rp. palustris*, as well as a complete tricarboxylic cycle. In addition, there is an active ancillary carboxylation of phosphoenolpyruvate. Neither the PEP carboxylase nor the glyoxylate cycle plays any such role in *Rs. rubrum* (Anderson and Fuller, 1967c). If *Rs. rubrum* grows on acetate as sole carbon source, it is converted to poly-β-hydroxybutyrate and stored. In the presence of CO_2, acetate is incorporated into tricarboxylic acid intermediates. The path of both CO_2 fixations and acetate photoassimilation look very different in the two closely related organisms.

The path of carbon in *Rp. palustris* as proposed by Kondrat'eva (1974) is shown in Fig. 3.

The comparison of the carbon metabolism of these two Rhodospirillaceae points out emphatically not only the difference in the biochemical pathways and enzymatic machinery that run them, but also the important role in the growth conditions and the variety of experimentation that must be carried out to arrive at a definitive biochemical answer to the path of carbon assimilation in these closely related cells.

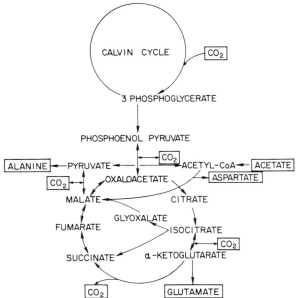

Fig. 3. Path of carbon in *Rhodopseudomonas palustris*. Adapted from Kondrat'eva (1974).

2.4. Path of Carbon and Carbon Metabolism in the Chlorobiaceae

Knowledge of the path of CO_2 fixation, reduction, and utilization in this family of photosynthetic bacteria is in a much less definitive state than in the other two families of photosynthetic bacteria. This is due to a number of factors. On the whole, these organisms are both very different from their purple cousins. Their major (95%) light-harvesting pigment is bacterio-chlorophyll (Bchl) *c*, which is contained in non-unit-membrane chlorobium vesicles. The reaction center contains Bchl *a*. There is no differentiation of the cyto-plasmic membrane to form intracytoplasmic mem-brane or chromatophores. Photochemistry occurs at a much lower redox potential than in the purple bac-teria (Fowler *et al.*, 1971). With the exception of one newly discovered genus, *Chloroflexis*, all are obligately anaerobic photoautotrophs. The genus of gliding green photosynthetic bacteria *Chloroflexis* is the only facul-tative representative in the Chlorobiaceae (Pierson and Castenholtz, 1974) yet described, and although its nutrition is well worked out, little has been done on the study of its CO_2 fixation. Further study with this facultative form should help clarify the paths of carbon in the green bacteria.

The first work on carbon assimilation in the green bacteria was reported by Larsen (1951, 1952). On the basis of $^{14}CO_2$-fixation experiments, he proposed a carboxylation of propionate to succinate. This, of course, was prior to the "establishment" of the Calvin cycle and rather prophetic of the controversies to come in the study of the path of carbon in bacterial photosynthesis, particularly in the green bacteria, in which alternative pathways of CO_2 fixation have been proposed (Buchanan *et al.*, 1972).

The first evidence for the presence of the enzymes of the Calvin cycle was described by Smillie and Fuller (1960) and Smillie *et al.* (1962). At that time, it was noted that the specific activity of the RuDP carboxyl-ase was low compared with that of other photo-synthetic organisms, but that all the necessary enzymes were there. It should also be noted that the experiments were done using *Chlorobium limicola f. thiosulfatophilum*, which was perhaps not identical with strains used later. Sadler and Stanier (1960) and Callely *et al.* (1962) studied the assimilation of acetate by the green bacteria. Sadler and Stanier showed clearly that acetate could be incorporated into cellular components in the light in the green bacteria; unlike the situation in the purple bacteria, however, this in-corporation was dependent on the presence of CO_2. Callely and co-workers showed the presence of the key enzymes of the citric acid cycle, and suggested a mechanism for CO_2 assimilation that was not cyclic but dependent on acetate plus CO_2. Evans *et al.* (1966) proposed, on the basis of enzymatic data, that *Chl. limicola f. thiosulfatophilum* was present in a new reduction cycle, dependent on reduced ferredoxin. They named this the *reductive carboxylic acid cycle* (Fig. 4).

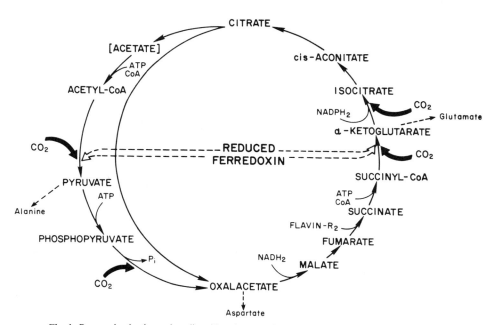

Fig. 4. Proposed reductive carboxylic acid or Arnon cycle. Reproduced from Evans *et al.* (1966).

One of the principal bases for concluding that this cycle was the major pathway for the incorporation of CO_2 was that in *Chl. limicola f. thiosulfatophilum* strain Tassajara, there was no detectable RuDP carboxylase. Subsequently, Gray *et al.* (1972) were unable to detect RuDP carboxylase in *Chloropseudomonas ethylica*—a dubious species (Gray *et al.*, 1971) consisting of a green photosynthetic bacterium and a sulfate-reducing organism living syntro-phically.* RuDP carboxylase had been described in another strain of *Chl. limicola f. thiosulfatophilum* (formerly strain L).

The $^{14}CO_2$-fixation experiments performed by Buchanan *et al.* (1972) and Sirevåg and Ormerod (1970*a,b*) indicated that $^{14}CO_2$ is incorporated

* See Chapter 2 for a discussion of the taxonomy of this syntrophic system.

Table 2. Carboxylation Reaction

Enzyme		Reaction
Systematic name and number (code)	Trivial name	
3-Phospho-D-glycerate-carboxylase (dimerizing); EC 4.1.1.39	Ribulose-diphosphate carboxylase (carboxydismutase)	D-Ribulose-1,5-diphosphate + CO_2 → 2(3-phosphoglycerate)
Orthophosphate : oxaloacetate carboxylase (phosphorylating); EC 4.1.1.31	Phosphopyruvate carboxylase	Phosphoenolpyruvate + H_2O + CO_2 → oxyalacetate + orthophosphate
GTP : oxaloacetate carboxylase (transphosphorylating); EC 4.1.1.32	Phosphopyruvate carboxylase	Phosphoenolpyruvate + CO_2 + GDP (or ADP, IDP) → oxaloacetate + GTP (or ATP, ITP)
Pyrophosphate : oxaloacetate carboxylase (phosphorylating); EC 4.1.1.38	Phosphopyruvate carboxylase	Phosphoenolpyruvate + CO_2 + orthophosphate → oxaloacetate + pyrophosphate
Pyruvate : carbon dioxide lyase (ADP); EC 6.4.1.1	Pyruvate carboxylase	Pyruvate + CO_2 + ATP → oxalo-acetate + ADP + orthophosphate
Propionyl-CoA : carbon dioxide-lyase (ADP); EC 6.4.1.3	Propionyl-CoA carboxylase	Propionyl-CoA + ATP + H_2O → methylmalonyl-CoA + ADP + orthophosphate
L-Malate : NADP-oxidoreductase (decarboxylating); EC 1.1.1.40	Malate dehydrogenase (decarboxyl-ating), "malic enzyme"	Pyruvate + CO_2 + NADPH → malate + $NADP^+$
threo-D$_s$-Isocitrate : NADP-oxido-reductase (decarboxylating); EC 1.1.1.42	Isocitrate dehydrogenase (NADP)	α-Ketoglutarate + CO_2 + NADPH → isocitrate + $NADP^+$
None[c]	Pyruvate synthase	Acetyl-CoA + CO_2 + ferredoxin (reduced) → pyruvate + CoA + ferredoxin
None[c]	α-Ketobutyrate synthase	Propionyl-CoA + CO_2 + ferredoxin (reduced) → α-ketobutyrate + CoA + ferredoxin
None[c]	α-Ketoglutarate synthase	Succinyl-CoA + CO_2 + ferredoxin (reduced) → α-ketoglutarate + CoA + ferredoxin
None[c]	Phenylpyruvate synthase	Phenylacetyl-CoA + CO_2 + ferredoxin (reduced) → phenylpyruvate + CoA + ferredoxin

[a] Translated and adapted from Kondrat'eva (1974).   [b] (*Ect.*) *Ectothiorhodospira*.   [c] No enzyme reaction established.

primarily into amino acids, particularly glutamic acid in *Chlorobium*. Sirevåg (1974), using carefully controlled concentrations of CO_2 during both growth and $^{14}CO_2$-fixation experiments, also concluded that the reductive carboxylic acid cycle is functional in this organism. These authors were unable to detect ribose-5-PO_4 kinase, and were able to detect only trace amounts of RuDP carboxylase in *Chl limicola f. thiosulfatophilum* strain NCIB 8346.

Subsequently, evidence was published showing the presence of several key enzymes of the reductive carboxylic acid cycle in both *Chr. vinosum* and *Rs. rubrum* (Buchanan *et al.*, 1967). It is difficult to understand the metabolic need for the reductive carboxylic acid cycle in the purple bacteria when the Calvin cycle is under tight metabolic control and the key enzymes of that cycle are derepressed under conditions of photoautotrophic growth.

in Photosynthetic Bacteria[a]

Bacteria		
Rhodospirillaceae	Chromatiaceae	Chlorobiaceae
Rs. rubrum, Rp. palustris, Rp. sphaeroides, Rm. vanielii	*Thiocapsa roseopersicina, Chr. vinosum, Chr. okenii, Thiopedia sp., Ect. shaposhnikovii,[b] Thiospirillum sp.*	*Chl. limicola f. thiosulfatophilum*
Rs. rubrum, Rp. palustris	*Chr. vinosum, Chr. okenii, Ect. shaposhnikovii[b]*	*Chl. limicola f. thiosulfatophilum*
Rs. rubrum, Rp. palustris	*Ect. shaposhnikovii[b]*	
Rs. rubrum	*Chr. vinosum, Chr. okenii*	
Rp. sphaeroides, Rs. rubrum		
Rs. rubrum	*Ect. shaposhnikovii[b]*	
	Chr. vinosum	
Rs. rubrum, Rp. palustris	*Chr. vinosum, Chr. okenii, Ect. shaposhnikovii,[b] Thiocapsa roseopersicina*	*Chl. limicola f. thiosulfatophilum*
Rs. rubrum	*Chr. vinosum, Ect. shaposhnikovii,[b] Thiocapsa roseopersicina*	*Chl. limicola f. thiosulfatophilum*
	Chr. vinosum	
Rs. rubrum	*Ect. shaposhnikovii[b]*	*Chl. limicola f. thiosulfatophilum*
	Chr. vinosum	*Chl. limicola f. thiosulfatophilum*

One of the major arguments against the existence of the reductive carboxylic acid cycle in the green bacteria was based on the results of the carefully done experiments of Hoare and Gibson (1964), which showed that the terminal carboxyl carbon of glutamate derived from the carboxyl group of added ^{14}C-labeled acetate during photosynthesis. This finding is incompatible with the operation of the cycle as *either* an oxidative *or* a reductive cycle. In addition, the levels of activity of the key ferredoxin-dependent carboxylases were extremely low compared with those of other enzymes of the cycle.

Besides the evidence against the reductive carboxylic acid cycle cited above, two recent discoveries have cast grave doubts on the necessity for and the functioning of this cycle as the regenerative CO_2-fixing mechanism in any photosynthetic bacterium. Tabita *et al.* (1974) succeeded in isolating and characterizing RuDP carboxylase from *Chl. limicola f. thiosulfatophilum* strain Tassajara. It is relatively unstable—which could account for the low activity, or lack of activity, in the hands of other workers. Thus, the key enzymes of the Calvin cycle are present in at least several species of the green bacteria.

In the functioning of the tricarboxylic acid cycle, citrate synthesis (via citrate synthase) and citrate splitting (via citrate lyase) are irreversible reactions. Thus, the presence or absence of either enzyme will determine the direction (oxidative or reductive) of the cycle, assuming that all other enzymes are present. Citrate synthetase is present in purple and green bacteria. Citrate lyase has been found only in *Rp. gelatinosa*, one of the few photosynthetic bacteria that can utilize citric acid as a carbon source (Schaab *et al.*, 1972). Beuscher and Gottschalk (1972) clearly demonstrated that whereas citrate synthetase is present in *Rs. rubrum* and *Chl. limicola f. thiosulfatophilum*, citrate lyase *is not detectable in either organism*. A reductive carboxylic cycle as described for these two organisms therefore *cannot* function as suggested.

We thus are left with two conflicting views of CO_2 assimilation in photosynthetic bacteria: (1) that autotrophic photoassimilation of CO_2 occurs via the Calvin cycle, and that the enzymes of the tricarboxylic acid cycle and various ferredoxin-dependent carboxylations are ancillary and anaplerotic (Fuller, 1971); or (2) that two carbon cycles coexist in the purple photosynthetic bacteria: the Calvin cycle and the reductive carboxylic acid cycle. In the green bacteria, some strains may have the key enzymes of the Calvin cycle and others may not. If it is finally shown that the reported presence of RuDP carboxylase is not universal, some new mechanism of CO_2 fixation must be proposed.

In summary, these observations (presence of RuDP carboxylase and absence of citrate lyase in the green bacteria) along with the other negative evidence make it extremely unlikely that the reductive carboxylic acid cycle is the regenerative photosynthetic carbon cycle of these or other photosynthetic bacteria described in this chapter.

However, the early evidence, in both the purple sulfur and nonsulfur bacteria and in the green bacteria, of the rapid labeling of amino acids, particularly the recent reports by Buchanan (1973) and Sirevåg (1974), points out the importance to the overall carbon economy of the cell of the many carboxylations associated with the tricarboxylic-acid-cycle enzymes, and of the ferredoxin-dependent reductive carboxylations in photosynthetic cells. It is well documented that many photosynthetic bacteria store polysaccharides. When growing in the light, however, they must rapidly synthesize proteins and lipids for growth, and unlike photosynthetic eukaryotes, cannot be dependent on stored products for growth in the dark. It is not surprising that these light-dependent ancillary carboxylations or even possibly additional cycles should play such a dominant role in bacterial photosynthesis. They may be ancillary to and dependent on the Calvin cycle, but they are vital to the successful photoautotrophic growth of these cells. Perhaps in primitive forms that no longer exist or have not yet been identified, what are now ancillary carboxylations were once parts of a regenerative cycle. The preponderance of evidence in the photosynthetic bacteria strongly suggests that the pentose reductive or Calvin cycle is the major regenerative metabolic route of CO_2 reduction in the light.

Again, the wide diversity and the importance of ancillary carboxylations are shown as summarized by Kondrat'eva (1974) in Table 2.

3. Ribulose Diphosphatase Carboxylase

RuDP carboxylase obviously plays a key role in bacterial as well as in plant photosynthesis. It is an enzyme that is under metabolic control and plays a vital role in the control of the functioning of the Calvin cycle in bacterial photosynthesis. The presence or absence of its activity has been used to argue the mechanism of CO_2 fixation and reduction. The enzyme has apparently functioned since very ancient times, and one of the most interesting aspects of it as a protein is the wide diversity of forms in which it appears, in both photosynthetic and chemoautotrophic bacteria. Clearly, this molecular diversity in photosynthetic bacteria is related to the antiquity of bacterial photosynthesis. Anderson *et al.* (1968) and

Anderson and Fuller (1969) first described the division of RuDP carboxylase into three molecular-size categories in photosynthetic bacteria: (1) large (mol. wt. $\approx$ 500,000); (2) intermediate (mol. wt. $\approx$ 360,000); (3) small (mol. wt. $\approx$ 120,000).

Since that report, papers from the laboratories of McFadden (McFadden, 1973, 1974; Tabita and McFadden, 1974a,b; McFadden and Tabita, 1974) and of Akazawa (Akazawa et al., 1970, 1972; Takabe and Akazawa, 1975a,b) have confirmed this division and given more information on this unique protein, which acts in photosynthetic bacteria, in algae, and in plants as both a carboxylase and an oxygenase. Andrews et al. (1973) showed that the 2,3-enediol of RuDP can be cleaved by oxygen addition as well as by CO_2 addition. The two proposed reactions are shown in Fig. 5.

The carboxylase-oxygenase of the purple sulfur bacterium Chr. vinosum is made up of eight small subunits with molecular weights of approximately 15,000 and eight large subunits with molecular weight of approximately 55,000. In contrast, the enzymes from the green bacteria and most of the facultative purple nonsulfur bacteria have molecular weights of approximately 350,000 and are made up of six large subunits.

Only Rs. rubrum contains a carboxylase with a molecular weight of approximately 115,000 and has the "small" enzyme made up of two large subunits. The evolution of this diversity in unclear—but in any event, the enzyme is ancient enough to have accumulated considerable molecular diversity without affecting its catalytic activity.

The Calvin cycle is also the primary route of CO_2 fixation in the chemolithotrophic bacteria. RuDP carboxylase also occurs in at least two forms in these organisms. Data showing some of the diverse molecular properties of the carboxylase in the photosynthetic bacteria, the cyanobacteria, and the chemolithotrophic bacteria from several laboratories are summarized in Table 3.

Besides the molecular diversity of the carboxylase enzyme in prokaryotes, it may also be associated with intracellular structure. Recently, the work of Shively et al. (1970, 1973a,b) and Purohit et al. (1976) has demonstrated the presence of polyhedral inclusion bodies in several thiobacilli. These investigators have isolated the polyhydral bodies and shown them to be rich in RuDP carboxylase. Also, when cells are grown under heterotrophic conditions, they do not produce carboxylase, and polyhydral bodies are absent. These

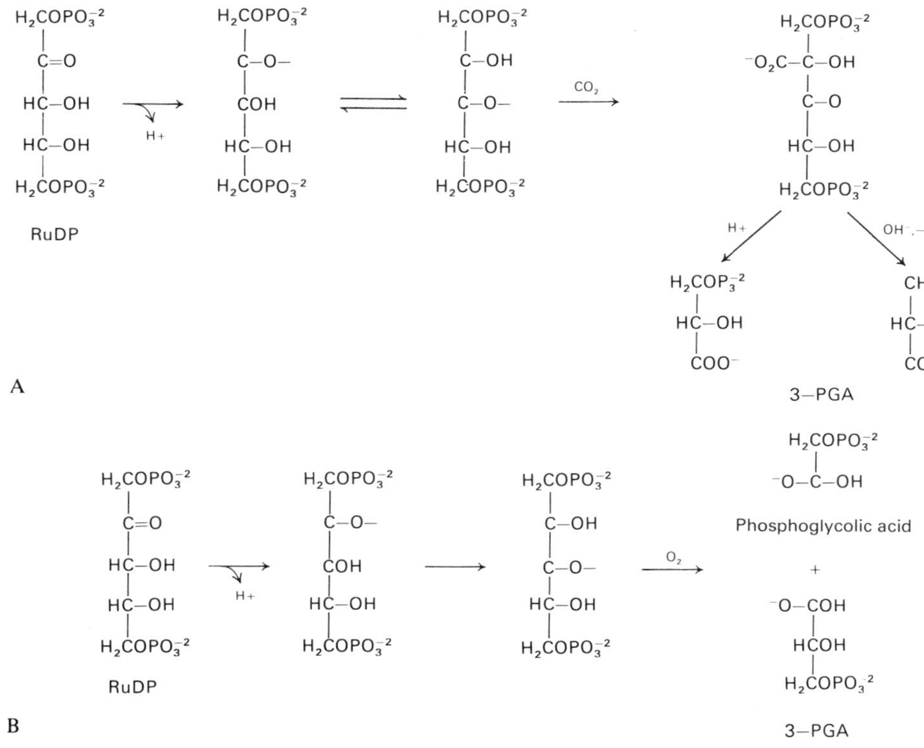

Fig. 5. Carboxylation (A) and oxygenation (B) of ribulose diphosphate.

data certainly indicate that RuDP carboxylase may be organized as a subcellular inclusion or part of an organized structure in the chemolithotrophic bacteria. Whether such structures exist in the photosynthetic microorganisms is yet to be determined.

In an interesting series of experiments, Takabe and Akazawa (1975a,b) showed that SH groups are present at the active site of both the carboxylase and oxygenase activities, and that evidence indicates that both reactions occur at the same site on the protein. This is certainly one of the few enzymes, if not the only enzyme, showing oxygenase activity that does not involve the direct participation of a metal ion, again showing the unique properties of this enzyme in nature. The function of RuDP carboxylase as an oxygenase, especially in the obligate anaerobe *Chromatium*, is not understood. In plants, the phosphoglycolate formed by the oxygenase reaction is the source of glycolic acid, which is the substrate for photorespiration. Glycolate formation and excretion was demonstrated in *Rs. rubrum* (Anderson and Fuller, 1969; Codd and Smith, 1974) and in *Chromatium* (Asami and Akazawa, 1974), and in the latter organism, it occurs only when the cells are placed in an aerobic

Table 3. Molecular Weights and Subunits of Various Ribulose Diphosphate Carboxylases in Photosynthetic and Chemolithotrophic Bacteria[a]

Approximate molecular weight of native enzyme	Organism	Mol. wt. of subunits	Subunits/ molecule	References
Small				
120,000	*Rs. rubrum*	56,000	2 (large)	Tabita and McFadden (1974a)
120,000	*Rs. rubrum*	—	—	Anderson and Fuller (1967a)
86,000	*Rs. rubrum*	—	—	Akazawa et al. (1970)
Intermediate				
360,000	*Rp. palustris*	—	—	Anderson and Fuller (1967a)
360,000	*Rp. sphaeroides*	—	—	Anderson and Fuller (1967a)
360,000	*Rp. sphaeroides*	—	—	Akazawa et al. (1970)
360,000	*Chl. limicola f. thiosulfatophilum*	54,000	6 (large)	Tabita et al. (1974)
360,000	*Thiobacillus denitrificans*	—	—	Tabita et al. (1974)
Large				
550,000	*Chr. vinosum*	57,000 12,000	8 (large) 8 (small)	Akazawa et al. (1972)
660,000	*Chr. vinosum*	—	—	Anderson and Fuller (1967a)
660,000	*Anacystis nidulans*	—	—	Anderson and Fuller (1967a)
18 S rel.	*Plectonema boryanum*	—	—	Kieras and Hazelkorn (1968)
550,000	*Hydrogenomonas facilis*	—	—	Kuehn and McFadden (1969)
520,000	*Hydrogenomonas eutropha*	—	8 (large) 8 (small)	Kuehn and McFadden (1969)
	Thiobacillus novellus		8 (large) 8 (small)	McFadden and Tabita (1974)

[a] Adapted and summarized from McFadden (1974).

environment, but glycolate appears to be formed by a transketolase-type reaction from pentose and hexose (Asimi and Akazawa, 1975), not from the oxygenation of RuDP as suggested in higher plants.

ACKNOWLEDGMENT

The author wishes to thank Professor E. N. Kondrat'eva of Moscow State University for generously allowing the use of several of her excellent summary tables and figures, which until now have been available only in Russian.

4. References

Akazawa, T., Sygiuama, T., and Kotaoka, H., 1970, Further studies on ribulose-1,5-diphosphate carboxylase from *Rhodopseudomonas sphaeroides* and *Rhodospirillum rubrum*, *Plant Cell Physiol.* **11**:541.

Akazawa, T., Kondo, H., Shimazue, T., Nishimura, M., and Sygiuama, T., 1972, Further studies on ribulose 1,5-diphosphate carboxylase from *Chromatium* Strain D, *Biochemistry* **11**:1298.

Anderson, L., and Fuller, R. C., 1967a, The rapid appearance of glycolate during photosynthesis in *Rhodospirillum rubrum*, *Biochim. Biophys. Acta* **131**:198.

Anderson, L., and Fuller, R. C., 1967b, Photosynthesis in *Rhodospirillum rubrum*. I. Autotrophic carbon dioxide fixation, *Plant Physiol.* **42**:487.

Anderson, L., and Fuller, R. C., 1967c, Photosynthesis in *Rhodospirillum rubrum*. II. Photoheterotrophic carbon dioxide fixation, *Plant Physiol.* **42**:491.

Anderson, L., and Fuller, R. C., 1967d, Photosynthesis in *Rhodospirillum rubrum*. III. Metabolic control of reductive pentose phosphate and tricarboxylic acid cycle enzymes, *Plant Physiol.* **42**:497.

Anderson, L., and Fuller, R. C., 1969, Photosynthesis in *Rhodospirillum rubrum*. IV. Isolation and characterization of ribulose-1,5-diphosphate carboxylase, *J. Biol. Chem.* **244**:3105.

Anderson, L., Price, G. B., and Fuller, R. C., 1968, Molecular diversity of the ribulose 1,5-diP carboxylase from photosynthetic microorganisms, *Science* **616**:482.

Andrews, T. J., Lorimer, G. H., and Tolbert, N. E., 1973, Ribulose diphosphate oxygenase I. Synthesis of phosphoglycolate by fraction-1 protein, *Biochemistry* **12**:11.

Arnon, D. I., Das, U. S. R., and Anderson, J., 1961, Photo-assimilation of CO_2 and acetate into protein and other insoluble cell constituents, *Fed. Proc. Fed. Am. Soc. Exp. Biol.* **20**:83.

Asami, S., and Akazawa, T., 1974, Oxidative formation of glycolic acid in photosynthesizing cells of *Chromatium*, *Plant Cell Physiol.* **15**:571.

Asami, S., and Akazawa, T., 1975, Biosynthetic mechanism of glycolate in *Chromatium*. I. Glycolate pathway, *Plant Cell Physiol.* **16**:631.

Bassham, A., and Calvin, M., 1957, *The Path of Carbon in Photosynthesis*, Prentice–Hall, Englewood Cliffs, New Jersey.

Bassham, J. A., and Kirk, M., 1969. Dynamic metabolic regulation of the photosynthetic carbon reduction cycle, in: *Comparative Biochemistry of Photosynthesis* (K. Shibata, A. Takamiya, A. T. Jagendorf, and R. C. Fuller, eds.), p. 365, University of Tokyo Press.

Beuscher, N., and Gottschalk, C., 1972, Lack of citrate ligase—the key enzyme of the reductive carboxylic acid cycle—in *Chlorobium thiosulfatophilum* and *Rhodospirillum rubrum*, *Z. Naturforsch.* **27b**:965.

Black, C. C., 1973, Photosynthetic carbon fixation in relation to net CO_2 uptake, *Annu. Rev. Plant Physiol.* **24**:253.

Buchanan, B. B., Evans, M. C. W., and Arnon, D. I., 1967, The reductive carboxylic acid cycle in green bacteria, *Arch. Mikrobiol.* **59**:32.

Buchanan, B. B., Shürmann, P., and Shanmugan, R. T., 1972, Role of the reductive carboxylic acid cycle in a photosynthetic bacterium lacking ribulose-1,5-diphosphate carboxylase, *Biochim. Biophys. Acta* **283**:136.

Callely, A. G., Rigopoulos, N., and Fuller, R. C., 1968, The assimilation of carbon by *Chloropseudomonas ethylicum*, *Biochem. J.* **106**:615.

Codd, E. A., and Smith, B. M., 1974, Glycolate formation and excretion by the purple photosynthetic bacterium *Rhodospirillum rubrum*, *FEBS Lett.* **48**:105.

Elsden, S. R., 1962, Photosynthesis and lithotrophic carbon dioxide fixation, in: *The Bacteria*, Vol. III (I. C. Gunsalus and R. Y. Stanier, eds.), p. 1, Academic Press, New York.

Evans, M. C. W., Buchanan, B. B., and Arnon, D. I., 1966, A new ferredoxin-dependent carbon reduction cycle in a photosynthetic bacterium, *Proc. Natl. Acad. Sci. U.S.A.* **55**:928.

Firsov, N. N., Chernjadiev, I., Ivanovsky, R. N., Kondrat'eva, E. N., Vdovina, N. V., and Doman, N. G., 1974, Pathways of assimilation of carbon dioxide by *Ectothiorhodospira shaposhnikovii*, *Mikrobiologiya* **43**:214.

Fowler, C. F., Nugent, N. A., and Fuller, R. C., 1971, The isolation and characterization of a photochemically active complex from *Chloropseudomonas ethylica*, *Proc. Natl. Acad. Sci. U.S.A.* **68**:2278–2282.

Fuller, R. C., 1971, The evolution of photosynthetic carbon metabolism: The role of sequences ancillary to the Calvin cycle, in: *Biochemical Evolution and the Origin of Life*, p. 259, North-Holland, Amsterdam.

Fuller, R. C., Smillie, R. M., Sisler, F. C., and Kornberg, H. L., 1961, Carbon Metabolism in *Chromatium*, *J. Biol. Chem.* **236**:2140.

Gest, H., 1951, Metabolic patterns in photosynthetic bacteria, *Bacteriol. Rev.* **15**:183.

Gest, H., and Kamen, M. D., 1960, The photosynthetic bacteria, in: *Bacterial Photosynthesis*, *Handbuch der Pflanzenphysiologie*, Vol. 5 (W. Ruhland, ed.), p. 568, Springer-Verlag, Berlin.

Gibbs, M., and Kandler, O., 1957, Asymmetric distribution of C^{14} in sugars formed in photosynthesis, *Proc. Natl. Acad. Sci. U.S.A.* **43**:446.

Glover, J., Kamen, M. D., and van Genderenen, H., 1952, Studies on the metabolism of photosynthetic bacteria. XII. Comparative light and dark metabolism of acetate and carbonate by *Rhodospirillum rubrum*, *Arch. Biochem. Biophys.* **35**:384.

Gray, B. H., Fowler, C. W., Nugent, N. A., Rigopoulos, N., and Fuller, R. C., 1972, *Chloropseudomonas ethylica* strain 2-K reevaluated, *Abstr. Annu. Meet. Am. Soc. Microbiol*, p. 156.

Hara, F., Akazawa, T., and Kojima, K., 1973, Glycogen biosynthesis in *Chromatium* strain D, *Plant Cell Physiol.* **14**:737.

Hatch, M. D., and Slack, C. R., 1970, Photosynthetic CO_2-fixation pathways, *Annu. Rev. Plant Physiol.* **15**:141.

Hoare, D. S., 1963, The photoassimilation of acetate by *Rhodospirillum rubrum*, *Biochem. J.,* **87**:284.

Hoare, D. S., and Gibson, J., 1964, Photoassimilation of acetate and the biosynthesis of amino acids by *Chlorobium thiosulfatophilum*, *Biochem. J.* **91**:546.

Kieras, F. J., and Hazelkorn, R., 1968, Properties of ribulose-1,5-diphosphate carboxylase from Chinese cabbage and photosynthetic microorganisms, *Plant Physiol.* **43**:1264.

Kondrat'eva, E. N. (1963), translated 1965, *Photosynthetic Bacteria*, Academy of Sciences of the USSR, Moscow, 1963, and Israel Program for Scientific Translations, IPST CAT 1214.

Kondrat'eva, E. N., 1974, Carbon metabolism in phototrophic bacteria, *Usp. Mikrobiol.* **9**:44.

Kuehn, G. D., and McFadden, B. A., 1969, Properties of ribulose diphosphate carboxylase from *Hydrogenomonas eutropha* and *Hydrogenomonas facilis*, *Biochemistry* **8**:2403.

Larsen, H., 1951, Photosynthesis of succinic acid by *Chlorobium thiosulfatophilum*, *J. Biol. Chem.* **193**:167.

Larsen, H., 1952, On the culture and general physiology of the green sulfur bacteria, *J. Bacteriol.* **64**:187.

Lascelles, J., 1960, The formation of ribulose-1,5-diphosphate carboxylase in growing cultures of Athiorhodaceae, *J. Gen. Microbiol.* **23**:496.

Losada, M., Trebst, A. V., Ogata, S., and Arnon, D. I., 1960, Equivalence of light and adenosine triphosphate in bacterial photosynthesis, *Nature (London)* **186**:753.

McFadden, B. A., 1973, Autotrophic CO_2 assimilation and the evolution of ribulose diphosphate carboxylase, *Bacteriol. Rev.* **37**:289.

McFadden, B. A., 1974, The oxygenase activity of D-ribulose-1,5-diphosphate carboxylase from *Rhodospirillum rubrum*, *Biochem. Biophys. Res. Commun.* **60**:312.

McFadden, B. A., and Tabita, F. R., 1974, D-ribulose-1,5-diphosphate carboxylase and the evolution of autotrophy, *Biosystems* **6**:193.

Ormerod, J. G., and Gest, H., 1962, Hydrogen photosynthesis and alternative metabolic pathways in photosynthetic bacteria, *Bacteriol. Rev.* **26**:51.

Pfennig, N., 1967, Photosynthetic bacteria, *Annu. Rev. Microbiol.* **21**:285.

Pfennig, N., and Trüper, H. G., 1974, The phototrophic bacteria, in: *Bergey's Manual of Determinative Bacteriology*, 8th Ed. (R. E. Buchanan and N. E. Gibbons, eds.), Williams and Wilkins, Baltimore.

Pierson, B. K., and Castenholz, R. W., 1974, A photosynthetic gliding filamentous bacterium of hot springs, *Chloroflexus aurantecaecus* gen. and sp. nov., *Arch. Mikrobiol.* **100**:5.

Purohit, K., McFadden, B. A., and Shaykh, M. M., 1976, D-Ribulose-1,5-disphosphate carboxylase and polyhedral inclusion bodies in *Thiobacillus intermedius*, *J. Bacteriol.* **127**:516.

Sadler, W. R., and Stanier, R. Y., 1960, The function of acetate in photosynthesis by green bacteria, *Proc. Natl. Acad. Sci. U.S.A.* **146**:1328.

Schaab, C., Giffhorn, F., Schobert, S., Pfennig, N., and Gottschalk, G., 1972, Phototrophic growth of *Rhodopseudomonas gelatinosa* on citrate; accumulation and subsequent utilization of cleavage products, *Z. Naturforsch.* **27**:962.

Shively, J. M., Decker, G. L., and Greenawalt, J. W., 1970, Comparative ultrastructure of *Thiobacilli*, *J. Bacteriol.* **101**:618.

Shively, J. M., Ball, F. C., Brown, D H., and Sanders, R. E., 1973a, Functional organelles in procaryotes: Polyhedral inclusions (carboxysomes) in *Thiobacillus neapolitanus*, *Science* **182**:584.

Shively, J. M., Ball, F. C., and Kline, B. W., 1973b, Electron microscopy of the carboxysomes (polyhedral bodies) of *Thiobacillus neapolitanus*, *J. Bacteriol.* **116**:1405.

Sirevåg, R., 1974, Further studies on carbon dioxide fixation in *Chlorobium*, *Arch. Microbiol.* **98**:3.

Sirevåg, R., and Ormerod, J. G., 1970a, Carbon dioxide fixation in green sulfur bacteria, *Biochem. J.* **120**:399.

Sirevåg, R., and Ormerod, J. G., 1970b, Carbon dioxide fixation in green photosynthetic bacteria, *Science* **169**:186.

Smillie, R M., and Fuller, R. C., 1960, Further observations on glyceraldehyde 3-phosphate dehydrogenase in plants and photosynthetic bacteria, *Biochem. Biophys. Res. Commun.* **3**:368.

Smillie, R. M., Rigopoulos, N., and Kelly, H., 1962, Enzymes of the reductive pentose phosphate cycle in the purple and in the green photosynthetic sulfur bacteria, *Biochim. Biophys. Acta* **56**:612.

Smith, A. J., London, J., and Stanier, R. Y., 1967, Biochemical basis of obligate autotrophy in blue-green algae and thiobacilli, *J. Bacteriol.* **94**:772.

Stanier, R. Y., 1961, Photosynthetic mechanisms in bacteria and plants; development of a unitary concept, *Bacteriol. Rev.* **25**:1.

Stokes, J. F., and Hoare, D. S., 1969, Reductive pentose cycle and formate assimilation in *Rhodopseudomonas palustris*, *J. Bacteriol.* **100**:890.

Stoppani, A. O. M., Fuller, R. C., and Calvin, M., 1955, Carbon dioxide fixation by *Rhodopseudomonas capsulatus*, *J. Bacteriol.* **69**:491.

Tabita, F. R., and McFadden, B. A., 1974a, D-Ribulose-1,5-diphosphate carboxylase from *Rhodospirillum rubrum*. I. Levels, purification and effect of metallic ions. II. Quaternary structure, composition and catalytic and immunological properties, *J. Biol. Chem.* **249**:3453.

Tabita, F. R., and McFadden, B. A., 1974b, One-step isolation of microbial ribulose-1,5-diphosphate carboxylase, *Arch. Microbiol.* **99**:231.

Tabita, F. R., McFadden, B. A., and Pfennig, N., 1974, D-Ribulose-1,5-diphosphate carboxylase in *Chlorobium thiosulfatophilum* Tassajara, *Biochim. Biophys. Acta* **341**:187.

Takabe, T., and Akazawa, T., 1975a, Further studies on the subunit structure of *Chromatium* ribulose-1,5-diphosphate carboxylase, *Biochemistry* **14**:46.

Takabe, T., and Akazawa, T., 1975b, The role of sulfhydryl groups in the ribulose-1,5-diphosphate carboxylase and oxygenase reactions, *Arch. Biochem. Biophys.* **169**:686.

van Niel, C. B., 1930, Photosynthesis in bacteria, in: *Contribution to Marine Biology*, p. 161, Stanford University Press, Stanford, California.

van Niel, C. B., 1931, On the morphology and physiology of the purple and green sulfur bacteria, *Arch. Mikrobiol.* **3**:138.

van Niel, C. B., 1935, Photosynthesis of bacteria, *Cold Spring Harbor Symp. Quant. Biol.* **3**:138.

van Niel, C. B., 1941, Bacterial photosyntheses and their importance for the general problems of photosynthesis, in: *Advances in Enzymology*, Vol. VI (F. F. Nord, ed.), p. 263, John Wiley & Sons, New York—London—Sydney.

van Niel, C. B., 1944, The culture, general physiology, morphology and classification of the non-sulfur purple and brown bacteria, *Bacteriol. Rev.* **8**:1.

van Niel, C. B., 1954, The chromatographic and photosynthetic bacteria, *Annu. Rev. Microbiol.* **8**:105.

van Niel, C. B., 1957, Order I. *Pseudomonadales orla—*Jensen, 1921, in: *Bergey's Manual of Determinative Bacteriology*, 7th ed., p. 35, Williams and Wilkins, Baltimore.

Walker, D. A., and Crofts, A. R., 1970, Photosynthesis, *Annu. Rev. Biochem.* **39**:389.

Wiessner, W., 1970, Bacterial photosynthesis, in: *Photobiology of Microorganisms* (P. Halldal, ed.), pp. 95–135, Wiley-Interscience, London and New York.

Zellich, I., 1975, Pathways of carbon fixation in green plants, *Annu. Rev. Biochem.* **44**:123.

CHAPTER 37

Metabolism of Nonaromatic Organic Compounds

Gary A. Sojka

1. Introduction

1.1. Previous Reviews and Scope of Material to Be Covered

This chapter deals with the utilization of nonaromatic carbon compounds by photosynthetic bacteria under anaerobic–light and aerobic–dark conditions, an area that has been under investigation for the past 45 years. The last extensive review of this subject (Gest and Kamen, 1960) provides a thorough coverage of the field up to the time of its publication. The subject was again briefly treated by Pfennig (1967) in his general review on the photosynthetic bacteria.

Until recent years, the primary purposes for conducting studies on utilization of carbon sources by the photosynthetic bacteria centered on three primary problems. The utilization of organic carbon for cellular synthesis had taxonomic significance. In addition, it was essential to ascertain which, if any, of the photosynthetic bacteria were obligate autotrophs. Finally, there was a great deal of interest concerning the role played by organic compounds in the mechanisms of bacterial photosynthesis.

1.2. Role of Organic Compounds in Bacterial Photosynthesis

From his classic studies on the Chlorobiaceae and Chromatiaceae, van Niel (1941) developed a unified theory suggesting that the cleavage of water to an oxidized and a reduced moiety was the central photochemical event in all photosyntheses, whether bacterial or green plant. In this scheme, the reduced moiety reduces CO_2 to the level of cell material, while the oxidized moiety either gives rise to molecular oxygen, as in green plants, or oxidizes the reduced compound necessarily present in the culture medium of the photosynthetic bacteria.

The work of Muller (1933) demonstrated that organic compounds could often be substituted for the usual inorganic, reduced-sulfur compounds in anaerobic, photosynthetic cultures of Chromatiaceae. Carbon dioxide was either evolved or consumed, depending on the oxidation level of the organic substrate. Though he conceded the possibility of direct assimilation of the organic substrate under these conditions, van Niel (1941) interpreted these findings in light of his unifying hypothesis. He argued that the photochemically produced oxidant oxidizes the organic substrate to CO_2, which is then reduced to cell material with the photochemically generated reductant.

At nearly the same time that Muller was studying the Chromatiaceae, Gaffron (1933, 1935) initiated his

Gary A. Sojka · Department of Biology, Indiana University, Bloomington, Indiana 47401

investigation of the role of organic substrates in the growth of the Rhodospirillaceae. These organisms, unlike the Chlorobiaceae and Chromatiaceae, ordinarily require organic substrates for anaerobic photosynthetic growth. Gaffron found, as had Muller, that when the organic substrate was more oxidized than the resulting cell material, CO_2 was evolved, whereas growth on organic compounds more reduced than cell material resulted in a net consumption of CO_2. In these studies, Gaffron discovered what appeared to be the reductive photometabolism of acetate with hydrogen in the absence of CO_2. His interpretation of this finding was that the carbon of the acetate was being directly assimilated into cell material without first being converted to CO_2. Van Niel (1936), however, chose to interpret these results in another way, feeling that the organic substrate functioned primarily as a photoreductant for CO_2. He reasoned that CO_2 could be derived from the organic substrate and subsequently incorporated into cell material by the photosynthetic CO_2-fixation mechanism. This view was seemingly supported by the description by Foster (1940) of the incomplete light-dependent oxidation of isopropanol accompanied by reduction of CO_2 to cell material. In this situation, isopropanol was quantitatively converted to acetone with simultaneous CO_2 reduction according to the following equation:

$$2\ CH_3CHOHCH_3 + CO_2 \xrightarrow{\text{light}}$$

$$\underset{\substack{\text{cell} \\ \text{material}}}{(CH_2O)} + 2\ CH_3COCH_3 + H_2O$$

In this instance, it appeared clear that the organic compound (isopropanol) was acting exclusively as a hydrogen donor for the photochemical reduction of CO_2. The finding by Gaffron (1935) that members of the Rhodospirillaceae can grow autotrophically on CO_2 with hydrogen gas as a reductant also seemed to lend support to the van Niel hypothesis that the role of organic substrates in bacterial photosynthesis was primarily one of hydrogen donor for the carbon fixation reactions that were presumed to be essential to the mechanism of photosynthesis in both bacteria and plants. Van Niel's view of the role of carbon sources as reductants in bacterial photosynthesis prevailed until the advent of experiments with ^{14}C-labeled substrates and $^{14}CO_2$.

When *Rhodospirillum rubrum* was grown anaerobically in the light, in the presence either of $^{14}CO_2$ and unlabeled acetate or of ^{14}C-labeled acetate and unlabeled CO_2, it was found that ^{14}C was incorporated into cellular proteins (Cutinelli *et al.*, 1950, 1951). These observations were difficult to reconcile with van

Niel's hypothesis in light of the observations by Glover and Kamen (1951) and by Ormerod (1956) that CO_2 fixation in *Rs. rubrum* was suppressed in the presence of acetate. In the same paper, Ormerod showed that resting cells of *Rs. rubrum* photometabolizing butyrate in the presence of $^{14}CO_2$ completely consumed the butyrate while incorporating $^{14}CO_2$ into cells. Since the amount of $^{14}CO_2$ incorporated coincided exactly with the manometrically determined CO_2 uptake, these results suggested that butyrate carbon was being incorporated into cells without first being converted to CO_2. Studies by Elsden and Ormerod (1956) with the metabolic inhibitor fluoroacetate strongly suggested that although the tricarboxylic acid cycle functioned in the dark–aerobic metabolism of a variety of organic compounds in *Rs. rubrum*, several of these substrates (propionate, succinate, fumarate, malate) were not oxidized to CO_2 by this pathway under anaerobic conditions in the light. At least in the cases of propionate, succinate, fumarate, and malate, therefore, it did not appear that van Niel's theory of complete oxidation of the organic substrate for generation of reducing power was correct. As a result of these experiments, the role played by organic substrates in bacterial photosynthesis remained in doubt.

Conclusive evidence for the direct photoassimilation of organic carbon sources was finally provided by Stanier *et al.* (1959) with the demonstration of incorporation of exogenously supplied carbon into one of two primary reserve materials when the rate of uptake of carbon source exceeded the rate of cellular synthesis. Organic substrates metabolized directly to acetyl subunits (i.e., acetate and butyrate) yielded mostly poly-β-hydroxybutyric acid, while substrates metabolized to pyruvate (i.e., succinate, malate, propionate) contributed primarily to the formation of polysaccharide reserve material.

It is now generally conceded that the primary function of organic substrates in the photosynthesis performed by species of the Chromatiaceae and Rhodospirillaceae is to serve as readily assimilable carbon source as suggested by Gaffron, and not exclusively as a source of reducing power for CO_2 fixation as postulated by van Niel. Nonetheless, it must be pointed out that organic substrates do provide reducing power for photosynthetic CO_2 fixation that may be of qualitative, if not quantitative, significance for photosynthetic growth of these organisms. Also, special cases were reported in which organic substrates, such as certain alcohols, served exclusively as photoreductants for CO_2 fixation (Foster, 1940, 1944). Siegel and Kamen (1950), however, were unable to successfully repeat Foster's experiments because the strains previously used had lost the ability to metabolize isopropanol. A delightful personal account of the so-called "van Niel–Gaffron controversy" is provided in Gaffron's opening address to the 1963 Kettering Symposium on Bacterial Photosynthesis (Gaffron, 1963).

2. Utilization of Organic Compounds by the Green Sulfur Bacteria

2.1. Physiological Basis of Obligate Autotrophy

Species of the family Chlorobiaceae are described as obligately anaerobic and absolutely dependent for growth on light, CO_2, and either an oxidizable inorganic sulfur compound or hydrogen. In other words, these organisms can be considered obligate photoautotrophs in the sense that CO_2 fixation is required for their growth; yet there is considerable evidence for utilization of simple organic compounds by these organisms. In no case, however, can organic compounds relieve the requirement for CO_2 and an inorganic reductant.

The metabolic basis for the condition of obligate autotrophy (in both phototrophs and chemotrophs) has been under investigation for many years. The chemolithotrophic bacterium *Thiobacillus thiooxidans* was shown to be unable to grow on glucose due to the accumulation of toxic products in the medium when it was supplied with such an organic substrate (Borichewski and Umbreit, 1966). The biochemical basis of obligate autotrophy in several blue-green algae and chemolithotrophic bacteria was traced to multiple enzymatic deficiencies (Smith *et al.*, 1967). The obligate autotrophs in this study were shown to have an incomplete tricarboxylic acid cycle. α-Ketoglutarate dehydrogenase was absent, and the organisms exhibited only low levels of malic and succinic dehydrogenase. Such enzymatic deficiencies would preclude the cyclic operation of the tricarboxylic cycle, but would still permit the function of related anabolic pathways. There are examples of obligately photolithotrophic blue-green algae that do not appear to have the metabolic deficiencies just described. Carr (1973) recently suggested that these organisms may be obligate phototrophs simply because they lack the metabolic flexibility to respond to exogenous organic substrates. These organisms therefore remain committed to CO_2 assimilation even in the presence of a preformed source of reduced carbon molecules. Their metabolic pathways of general metabolism and biosynthesis lack mechanisms for control of enzyme synthesis; the same complement of enzymes appear to be formed regardless of whether organic substrates are present or absent.

2.2. Incorporation of Organic Compounds by Members of the Family Chlorobiaceae

It has been repeatedly shown that members of the genus *Chlorobium* are unable to grow at the expense of organic compounds supplied in the culture medium (van Niel, 1931; 1953; Sadler and Stanier, 1960). Sadler and Stanier (1960), however, showed that small but significant increases in growth yield were obtained when a low concentration of either propionate, pyruvate, lactate, glucose, glutamate, or, especially, acetate was added to the culture medium of *Chlorobium limicola*. In the case of acetate, it could be demonstrated that the carbon was incorporated into cell material, with approximately 93% incorporated after 6 days of growth. The amount of acetate carbon assimilated by these cells was strictly dependent, however, on the amount of sulfide and CO_2 furnished in the medium. This finding clearly shows that this organism, even when it obtains carbon from acetate, is still an obligate photoautotroph in the sense that light-driven CO_2 fixation was an absolute requirement for its growth. Sadler and Stanier concluded that *Chl. limicola* lacks the enzymatic capability for acetate oxidation, thus precluding the use of acetate for the generation of either CO_2 or reducing power. Use of acetate is therefore dependent on provision of reducing power and CO_2 from other sources. Pfennig (1967) fully confirmed the observations of Sadler and Stanier with 15 different strains of the genus *Chlorobium*.

Work on acetate photoassimilation was extended to *Chl. limicola f. thiosulfatophilum* by Hoare and Gibson (1964). They also found that acetate incorporation was dependent on bicarbonate and an external source of reducing power. Carbon from ^{14}C-labeled acetate was found in all major fractions of the cells. No poly-β-hydroxybutyrate could be detected in *Chl. limicola f. thiosulfatophilum* even under growth conditions in the presence of acetate that were sufficient to promote synthesis of this storage polymer in *Rs. rubrum*. Both [1-14C]acetate and [2-14C]acetate were incorporated into the amino acids of these cells.

Regulatory aspects of the intermediary metabolism of *Chl. limicola f. thiosulfatophilum* were recently investigated by Kelly (1974), who showed photoassimilation of a variety of organic compounds, none of which was capable of supporting growth by itself. At moderate concentrations (e.g., 1 mM), several of these organic compounds proved to be inhibitory to growth. Among the organic compounds causing inhibition were L-methionine, L-tyrosine, L-tryptophan, L-threonine, L-serine, glycine, and formate. In the case of several of the inhibitory amino acids, the inhibition was overcome by the addition of other amino acids. If the inhibition was due to "autostarvation" (caused by feedback inhibition of an early primary enzyme in a branched biosynthetic pathway), it would be reasonable to assume that the addition of the amino acid the synthesis of which was being curtailed would relieve the growth inhibition. Though the results are not conclusive, Kelly's experiments suggest that at least in several cases, the inhibition is due to such an "autostarvation." For example, threonine, one of the amino acids that cause growth inhibition, is an effective inhibitor of aspartokinase both by itself and in concert with lysine. Addition of isoleucine to the medium completely alleviated growth inhibition caused by threonine. These results suggest that threonine may have been inhibiting growth by causing an isoleucine starvation as the result of inhibition of aspartokinase. It is interesting to note that analogous results were obtained with the nonsulfur purple bacterium *Rhodopseudomonas palustris* (Yen and Gest, 1974). Kelly concludes from his studies that *Chl. limicola f. thiosulfatophilum* is probably similar to most other microorganisms in its general intermediary metabolism and associated control mechanisms. It appears, therefore, that species of the genus *Chlorobium* are unusual only in that they possess what appears to be a purely anabolic intermediary carbon metabolism that results in their characteristic anaerobic photolithotrophy.

2.3. Carbon Metabolism of Members of the Genus *Chloroflexus*

Members of the recently described family Chloroflexaceae closely resemble the Chlorobiaceae in DNA base ratios, in the presence of "chlorobium vesicles" rich in bacteriochlorophyll (Bchl) *c* (Pierson and Castenholz, 1974), and in the synthesis of characteristic glycolipids (Kenyon and Gray, 1974) (see Chapters 2, 9, and 14). Despite these similarities, *Chloroflexus aurantiacus* differs from species of the family Chlorobiaceae in nutrition and physiology. Unlike the Chlorobiaceae, this organism can grow at the expense of a wide variety of organic compounds in the light (Madigan *et al.*, 1974). Several tricarboxylic acid cycle intermediates, short-chain alcohols, amino acids, hexoses, and pentoses support growth of *Chloroflexus* in the light, and many of the same compounds can be utilized by these organisms under aerobic–dark conditions. It is interesting to speculate on the evolutionary significance of these bacteria, which closely resemble the Chlorobiaceae in so many ways, yet exhibit an extreme nutritional diversity. Could such diversity of metabolism and nutrition have resulted from a comparatively small number of enzymatic changes? The answer to this question obviously awaits further investigation of the intermediary metabolism of this interesting new group of organisms.

3. Carbon Metabolism of Members of the Family Chromatiaceae

3.1. Range of Organic Carbon Sources Utilized

The purple sulfur bacteria are able to grow anaerobically in the light either at the expense of organic compounds or by autotrophically fixing CO_2, employing reduced sulfur compounds or H_2 as electron donors (van Niel, 1936, 1944; Muller, 1933). Most of these organisms are unable to grow as chemotrophs on organic compounds in the dark, and are therefore considered as obligate phototrophs. *Thiocapsa roseopersicina* can, however, grow microaerophilically in the dark.

A number of carbon sources, including alcohols, fatty acids, amino acids, and carbohydrates, are utilized by species of *Chromatium* in the presence of CO_2 (Muller, 1933; Gaffron, 1934). It is probable, however, that CO_2 is not required for utilization of many of these carbon sources. As originally pointed out by Muller (1933), utilization of substrates more oxidized than cell material results in a net CO_2 evolution, while growth on substrates more reduced than cell material requires CO_2 fixation coupled to substrate oxidation. [A good approximation of cell material in these organisms is $C_5H_8O_2N$ (van Gemerden, 1968b)]. It seems, therefore, that the role of CO_2 in cultures of purple sulfur bacteria growing on utilizable organic carbon sources is one of an "electron sink" providing an outlet for removal of excess reducing power (Gaffron, 1935; Stanier, 1961).

3.2. Metabolism of Acetate by Purple Sulfur Bacteria

Losada *et al.* (1960) reported growth of an unspecified strain of *Chr. vinosum* on acetate in the absence of added CO_2. Using ^{14}C-labeled acetate, they showed that no significant incorporation of acetate carbon occurred in the dark, while in the light, acetate was directly photoassimilated into cell material. A significant portion of the assimilated acetate carbon was detected in amino acids, with glutamic acid being the most extensively labeled. These authors presented evidence that the primary reaction in acetate incorporation is the ATP-dependent formation of acetyl-CoA, which once formed could condense with a variety of other molecules leading to synthesis of cell material. The condensation of acetyl-CoA and pyruvate led to the formation of citramalate, which this organism converted to glutamic acid by a reversal of the glutamate degradative pathway found in *Clostridium tetanomorphum* (Barker, 1961). Losada *et al.* (1960) also presented evidence for the condensation of acetyl-CoA and oxalacetate to form citrate in the citrate synthesis reaction of the tricarboxylic acid cycle.

When grown on acetate as a carbon source, *Chromatium vinosum* strain D possessed both isocitrate lyase and malate synthase, two key enzymes of the glyoxylate cycle (Fuller *et al.*, 1961). Isocitrate lyase appeared to be inducible, exhibiting high specific activity in cells grown on acetate but not in cells grown on malate or under autotrophic conditions, whereas malate synthetase appeared to be constitutive. The glyoxylate cycle in this organism was shown to be unusual due to the lack of malate dehydrogenase; this apparent block in the cycle was bypassed, however, by two additional reactions. Cultures of *Chr. vinosum* strain D were shown to catalyze an NADP-dependent oxidative decarboxylation of malate to pyruvate. This "malic enzyme" reaction, when combined with the ATP-dependent pyruvate carboxylation reaction demonstrable in extracts of these cells, resulted in the formation of oxalacetate from malate:

$$\text{L-Malate} + \text{NADP}^+ \longrightarrow$$
$$\text{pyruvate} + CO_2 + \text{NADPH} + \text{H}^+$$

$$\text{Pyruvate} + CO_2 \xrightarrow{\text{ATP}} \text{oxalacetate}$$

These reactions, along with the citrate synthase reaction demonstrated by Losada *et al.* (1960), make possible the operation of a complete glyoxylate cycle in cells growing on acetate.

Levels of ribulose-diphosphate carboxylase were examined in *Chr. vinosum* strain D and *Thiopedia* strain PM (which was later shown to be a strain of *Thiocapsa roseopersicina*) grown both autotrophically and in the presence of organic substrates (Hurlbert and Lascelles, 1963). As might be expected, autotrophically growing cells had the highest levels of the carboxylase. Though growth on organic substrates lowered the level of the enzyme present in the cells tested, ribulose-diphosphate carboxylase activity was always demonstrable at significant levels in extracts from cells grown with sources of reduced carbon. The authors interpret these findings in light of the postulated dual role for the carboxylase and the associated enzymes of the reductive pentose cycle. They point out not only that the reductive pentose cycle has to provide the mechanism of CO_2 assimilation under autotrophic conditions, but also that this CO_2 fixation can furnish an outlet for excess reducing power formed during substrate oxidation.

3.3. Mobilization of Carbon Storage Materials in the Dark

Species of Chromatiaceae face an interesting ecological problem due to their obligately phototrophic nature. It appears that these organisms are capable of forming significant amounts of ATP only while they are illuminated. How, then, do these organisms satisfy their energy demands for mobility, cell maintenance, and other processes at night? An interesting series of papers by van Gemerden attempts to answer this and other questions (van Gemerden, 1968a–c). *Chromatium vinosum* strain 6412 produced glucose-containing carbohydrate storage products in the light when provided with CO_2 and H_2S. In experiments in which the cells were exposed to alternating light and dark periods of many hours' duration, it could be shown that this carbohydrate storage product accumulated in the light and disappeared in the dark. In similar experiments, it was discovered that disappearance of the carbohydrate storage product coincided with synthesis of poly-β-hydroxybutyrate. An analogous relationship between sulfur and hydrogen sulfide was also observed in these experiments; in the dark, hydrogen sulfide was produced apparently at the expense of sulfur.

Van Gemerden's interpretation of these results is that in darkness, the carbohydrate storage product is broken down via the Embden–Mayerhof pathway of glucose catabolism to pyruvate, resulting in a gain of ATP through substrate-level phosphorylation. Pyruvate is then converted to poly-β-hydroxybutyrate via acetyl-CoA and acetoacetyl-CoA with the concomitant loss of CO_2. An argument is presented that formation of poly-β-hydroxybutyrate in this fashion does not consume ATP. The electrons released after degradation of the carbohydrate storage product and synthesis of poly-β-hydroxybutyrate are thought to be passed to sulfur through an NAD–NADH-mediated reaction to form hydrogen sulfide.

This scheme provides an explanation for the appearance of CO_2 and sulfide in cultures of *Chromatium* sp. incubated in the dark (Trüper and Schlegel, 1964). It also supplies a metabolic rationale for the disappearance of the carbohydrate storage product and the concomitant synthesis of poly-β-hydroxybutyrate in darkness. Finally, it provides a mechanism for energy release from storage products formed in the light that does not require respiratory metabolism in the dark.

4. Utilization of Organic Compounds by the Nonsulfur Purple Bacteria

4.1. Range of Organic Carbon Sources Utilized

Species of the family Rhodospirillaceae are characterized by the ability to grow on a wide array of different organic carbon sources. Fatty acids, amino acids, alcohols, tricarboxylic acid cycle intermediates, sugars, one-carbon compounds, acetate, and pyruvate have been reported to support the growth of various species in this group. A number of the species (most prominently members of the genus *Rhodopseudomonas*, but also some species of *Rhodospirillum*) are capable of carrying out nonphotosynthetic respiratory metabolism under aerobic–dark conditions. The tricarboxylic acid cycle is thought to be the pathway of terminal substrate oxidation in the dark (Elsden, 1962). This cycle probably also operates anaerobically in the light in most of these organisms (Ormerod and Gest, 1962). It is also possible to grow many members of this group autotrophically using a suitable inorganic reductant such as H_2 (Klemme, 1968).

Organisms with the capacity to metabolize and assimilate a variety of organic substrates are generally equipped with a set of inducible catabolic pathways. This seems to be the case with the nonsulfur purple bacteria as well, with the caution that the word *catabolism* has a somewhat restricted meaning when applied to the anaerobic, light-dependent assimilation of a given carbon source. Under such conditions, no energy is derived from the "catabolism" of an organic substrate. The same pathways of carbon-source breakdown found under photosynthetic conditions, however, are often found to operate in these organisms grown heterotrophically in the dark, where it must be agreed that catabolism, in the full sense of the word, is taking place.

4.2. Utilization of Glucose and Fructose

Rhodopseudomonas sphaeroides was reported to grow on glucose as a carbon source (van Niel, 1944). It was later shown that this organism grew only poorly on glucose, fructose, or mannose (Szymona and Doudoroff, 1960), growth on these substrates resulting

in the accumulation of acidic products including 2-keto-3-deoxy gluconic acid. Though the Embden–Meyerhof pathway was found to be constitutive in this organism, it seemed to be of little consequence for catabolism of glucose, since the key enzyme, aldolase, possessed very low activity. Mutants of *Rp. sphaeroides* were derived that grew much better on glucose than the wild type. These mutants all acquired the enzyme phosphogluconic acid dehydrase. Such mutants metabolized glucose primarily via the pathway that has now become known as the Entner–Doudoroff pathway, though they retained limited ability to dissimilate glucose via the Embden–Meyerhof route. The closely related organism *Rp. capsulata* grows well on glucose. The primary route of glucose dissimilation in *Rp. capsulata* occurs by an inducible Entner–Doudoroff pathway (Eidels, 1969). Growth on glucose as the sole carbon source increased glucose-6-P dehydrogenase 5-fold and 6-phosphogluconate dehydrase 15-fold, and increased the level of the constitutive 2-keto-3-deoxy-6-phosphogluconic acid aldolase 2-fold. It thus appears that glucose induces the synthesis of the Entner–Doudoroff pathway. Cell extracts of *Rp. capsulata* possessed only very low levels of phosphofructokinase and 6-phosphogluconate dehydrogenase, suggesting that the Embden–Meyerhof and hexosemonophosphate shunt pathways of glucose breakdown can be of little quantitative significance in cells growing on glucose (Eidels and Preiss, 1970).

Rhodopseudomonas capsulata is unusual in that it metabolizes glucose via the inducible Entner–Doudoroff pathway, but it metabolizes fructose via the Embden–Meyerhof pathway, which is induced by this sugar (Conrad and Schlegel, 1974). Most bacteria attack these two substrates with the same pathway, be it Embden–Meyerhof or Entner–Doudoroff. The initial step in the dissimilation of fructose in this organism is catalyzed by 1-phosphofructokinase. The involvement of this enzyme in fructose metabolism seems reasonable, since it is known that several photosynthetic bacteria transport fructose into the cell with a phosphotransferase-like system that forms fructose-1-P (Saier *et al.*, 1971). This transport system is composed of two membrane-associated proteins that catalyze the transfer of phosphate from phosphoenolpyruvate to the 1-hydroxyl of fructose during the process of translocation of fructose into the cell. It differs from the phosphotransferase system of other bacteria in that both proteins are membrane-associated under physiological conditions, and there is no evidence for a low-molecular-weight phosphate carrier protein analogous to the HPr of other systems (Roseman, 1969).

4.3. Metabolism of Glutamate and Citrate

Rhodospirillum rubrum can grow both anaerobically in the light and aerobically in the dark on either L- or D-glutamate. Studies employing ^{14}C-labeled glutamate and metabolic inhibitors suggested that *Rs. rubrum* catabolizes glutamate primarily through the tricarboxylic acid cycle (Gibson and Wang, 1968). There is enzymological evidence, however, for the existence of a coenzyme-B$_{12}$-dependent glutamate mutase in extracts of *Rs. rubrum* and *Rp. sphaeroides* (Ohmori *et al.*, 1974), and evidence was obtained that at least some of the glutamate supplied as carbon source is metabolized via the pathway

Glutamate → β-methylaspartate → mesaconate → → citramalate → acetyl-CoA + pyruvate

which is known to occur in other bacteria (Barker, 1961).

Most photosynthetic bacteria are unable to utilize citrate as a carbon source; *Rp. gelatinosa*, however, was shown to grow anaerobically in the light with citrate as sole carbon source (Weckesser *et al.*, 1969). Schaab *et al.* (1972) were able to demonstrate limited growth of *Rp. gelatinosa* on citrate under anaerobic–dark conditions as well. Citrate lyase, the enzyme responsible for citrate degradation, is apparently inactive under aerobic conditions, resulting in the inability of this organism to utilize citrate aerobically. The inducible citrate lyase degrades citrate in a rather uncontrolled manner, resulting in the accumulation of degradation products, acetate and malate, in the culture medium (Schaab *et al.*, 1972). A short lag period follows complete consumption of citrate, after which growth resumes at the expense of the malate and acetate that accumulated during citrate degradation. This lag period, though not always observed, is apparently due to the time required for synthesis of enzymes involved in a utilization of the degradation products. Malate was preferentially consumed; no significant amount of acetate was taken up until after the complete removal of malte from the medium.

The citrate lyase of *Rp. gelatinosa* is rapidly inactivated after exhaustion of citrate from the medium (Giffhorn *et al.*, 1972). This inactivation is essential for efficient utilization of the degradation products, since citrate must be formed by citrate synthase during growth on malate and acetate. Simultaneous activity of citrate synthase and citrate lyase would result in a futile cycling of two-carbon and four-carbon molecules. Citrate lyase from *Rp. gelatinosa* was purified to homogeneity and its physical properties and subunit composition studied by Beuscher *et al.* (1974). The analogous enzyme from *Klebsiella aerogenes* loses activity when it is deacetylated (Buckel *et al.*, 1971),

suggesting a similar mechanism of inactivation of the enzyme from *Rp. gelatinosa*. There are, however, significant differences in the structures of the citrate lyases of these organisms (Beuscher *et al.*, 1974). The mechanism of citrate lyase inactivation following exhaustion of citrate from the culture medium is being actively investigated by G. Gottschalk and his colleagues.

4.4. Glycerol Utilization by Nonsulfur Purple Bacteria

Glycerol serves as a carbon source for *Rp. palustris* and *Rp. sphaeroides*, but does not support aerobic or photosynthetic growth of either *Rp. capsulata* or *Rp. gelatinosa* (van Niel, 1944). *Rhodopseudomonas sphaeroides* degrades glycerol by first forming glycerol-3-phosphate, which is subsequently converted to dihydroxyacetone phosphate (Pike and Sojka, 1975). These reactions are catalyzed by a soluble glycerokinase and a particulate glycerophosphate dehydrogenase in an inducible system similar to the glycerol dissimilation pathway of other bacteria (Koch *et al.*, 1964).

When this organism is grown on a medium containing both glycerol and malate, *Rp. sphaeroides* first consumes malate, during which time the enzymes of glycerol metabolism are repressed. Following malate exhaustion, there is a growth lag during which the glycerol-metabolizing enzymes are simultaneously derepressed. After several hours, growth resumes on glycerol. The diauxic growth curve accompanied by enzyme repression and derepression suggests that the enzymes of glycerol metabolism in *Rp. sphaeroides* wild-type are subject to catabolite repression by malate (Pike and Sojka, 1975).

A spontaneous mutant of *Rp. capsulata* was isolated that has gained the ability to use glycerol as a carbon source in the presence of CO_2, either aerobically in the dark or anaerobically in the light (Lueking *et al.*, 1973). The mutant possesses the same enzymes of glycerol catabolism found in *Rp. sphaeroides*, i.e., soluble glycerokinase and particulate glycerophosphate dehydrogenase. Neither of these enzymes is present in wild-type *Rp. capsulata*. Unlike *Rp. sphaeroides*, the glycerol-utilizing mutant of *Rp. capsulata* synthesizes the glycerol-metabolizing enzymes constitutively. When the mutant is grown in the presence of glycerol and malate, the enzymes are present at a significant level throughout the growth curve, and no diauxic lag is observed. Malate is consumed first, and on its exhaustion, growth immediately continues at the expense of glycerol.

4.5. Growth on Acetate

It has been known for many years that *Rs. rubrum* can incorporate carbon from acetate into cell material (Cutinelli *et al.*, 1951; Glover *et al.*, 1952). The work of Stanier *et al.* (1959) clearly showed that acetate carbon was directly assimilated into the storage polymer, poly-β-hydroxybutyrate. In the presence of added CO_2, acetate carbon is found in polysaccharide, whereas without added CO_2, the acetate carbon is predominantly converted to poly-β-hydroxybutyrate. Unlike many other organisms that utilize acetate as a sole carbon source, *Rs. rubrum* does not possess an active glyoxylate cycle. The enzyme isocitrate lyase, which is essential for the operation of the glyoxylate cycle, is not present in *Rs. rubrum* (Kornberg and Lascelles, 1960; Albers and Gottschalk, 1976). The mechanism by which *Rs. rubrum* is able to grow photosynthetically on acetate as a sole source of carbon in the absence of a functional glyoxylate cycle remains unsolved despite intensive investigation (Olsen, 1967).

Kornberg and Lascelles (1960) examined various species of the family Rhodospirillaceae with regard to the presence or absence of an active glyoxylate cycle. When grown on acetate or butyrate, *Rp. palustris* and *Rp. capsulata* formed significant levels of isocitrate lyase, in contrast to *Rs. rubrum* and *Rp. sphaeroides*, which did not form this enzyme under any of the conditions tested. Kornberg and Lascelles found that malate synthetase was constitutive in all four of the organisms tested.

It was recently reported (Albers and Gottschalk, 1976) that *Rhodomicrobium vannielii* and *Rp. palustris* are constitutive for both isocitrate lyase and malate synthetase, whereas *Rp. sphaeroides* and *Rp. capsulata* possess malate synthetase, but not isocitrate lyase. The reasons for these apparent discrepancies concerning *Rp. palustris* and *Rp. capsulata* are not immediately apparent. Certainly minor strain differences could account for the variation in regulation reported for *Rp. palustris*; the complete lack of isocitrate lyase reported for *Rp. capsulata*, however, is harder to understand. Nielsen and Sojka (1975) recently showed that although wild type *Rp. capsulata* does not produce isocitrate lyase, a spontaneously arising, acetate-utilizing mutant does. On depletion of acetate from the medium, this enzyme undergoes a rapid, energy-dependent inactivation. Readdition of acetate to the culture medium causes a rapid recovery of enzyme activity that is energy-dependent and sensitive to protein synthesis inhibition.

4.6. Utilization of Alcohols, D- and L-Malate, and One-Carbon Compounds

The nonsulfur purple bacteria often inhabit anaerobic niches where fermentation products of other organisms may accumulate. *Rhodospirillum rubrum* was shown to possess an inducible alcohol dehydrogenase (Chaudhary, 1970) that could facilitate the use of some of these fermentation by-products. Growth of *Rs. rubrum* in an ethanol medium increased the activity of the alcohol dehydrogenase 20-fold over the level found in malate-grown cells. The *Rs. rubrum* enzyme is NAD-dependent; yet NADP will substitute for the preferred pyridine nucleotide. *Rhodomicrobium vannielii*, on the other hand, has a strictly NADP-dependent alcohol dehydrogenase (Sandhu and Carr, 1970).

Malic acid has been a very popular carbon source for culture of nonsulfur purple bacteria in the laboratory. A number of commonly used media for these organisms employ malate as the sole carbon source (Bose, 1963). L-Malate is probably metabolized via the tricarboxylic acid cycle through the action of NAD-linked malic dehydrogenase, which is quite active in extracts of these cells. This enzyme is specific for the L-isomer of malate. Many culture media contain D,L-malate, however. Growth–yield data presented by Cohen-Bazire *et al.* (1957) showed that *Rp. sphaeroides* can utilize both isomers of malate. Stahl and Sojka (1973) observed that *Rp. capsulata* grew on D-malate with a growth rate and cell yield nearly equal to that on L-malate under photosynthetic and heterotrophic conditions. No malate racemase activity could be detected in extracts of these cells. D-Malic enzymes that decarboxylate D-malate to pyruvate have been reported for several bacteria (Stern and Hegre, 1966; Stern and O'Brien, 1969). Enzyme systems capable of cleaving either L- or D-malate to acetyl-CoA and glyoxylate have also been described (Stern, 1963; Tuboi and Kikuchi, 1965). As yet, the mechanism of D-malate utilization in the photosynthetic bacteria is not known with certainty.

Though many members of the Rhodospirillaceae grow well on CO_2 when a suitable reductant is present, there have not been many reports of growth of these organisms on other one-carbon compounds. One reason for this apparent lack of ability to utilize such compounds was recently pointed out by Quayle and Pfennig (1975), who were able to obtain growth of a number of strains of *Rp. acidophila*, *Rm. vannielii*, *Rp. gelatinosa*, *Rs. rubrum*, and *Rs. tenue* on methanol. The pH optimum for growth on methanol differed significantly from the optimum for other carbon sources. It was also necessary to add bicarbonate to these cultures to obtain good growth, and yeast extract and casamino acids (in low concentration) had marked stimulatory effects on some of the cultures. Such observations serve to illustrate that many compounds that fail to support growth under one set of conditions might prove to be excellent substrates under different conditions. In these experiments, no methane utilization was detected, nor could any of the cultures grow on methylamine. *Rhodopseudomonas acidophila* was able to grow on formate with added bicarbonate. *Rhodopseudomonas palustris*, which also grows on formate, was shown to have an inducible formic dehydrogenase and a particulate hydrogenase (Qadri and Hoare, 1968). These authors suggest that growth on formate is essentially autotrophic, since the hydrogen lyase system cleaves formate to CO_2 and H_2, the substrates required for autotrophic fixation. Yoch and Lindstrom (1967) also found that *Rp. palustris* could cleave formate to CO_2, and on the basis of their labeling data, they concluded that the CO_2 thus formed is fixed via the reductive carboxylic acid cycle (Evans *et al.*, 1966).

Wertlieb and Vishniac (1967) attempted to grow *Rp. gelatinosa* on methane. It could be concluded that some methane carbon was incorporated into cellular material, and that methane was being oxidized to CO_2; it was not certain, however, that methane was functioning as the sole electron donor in this system. The growth yield of *Rp. galatinosa* on methane was low. It does appear that some species of *Rhodopseudomonas* are capable of growing on carbon monoxide as a sole carbon source anaerobically in the light. Hirsch (1968) isolated, on CO as a carbon source, a strain of nonsulfur purple bacteria that he classified in the genus *Rhodopseudomonas*, and Uffen (1975, personal communication) also reported the isolation of a similar organism from anaerobic mud.

4.7. Regulation of the Intermediary Metabolism of the Nonsulfur Purple Bacteria

Of all the photosynthetic bacteria, the nonsulfur purple bacteria have probably been the most extensively investigated in regard to the control of their intermediary metabolism. For example, we can look at one organism, *Rp. capsulata*, and find that studies have been carried out on regulation of several key enzymes involved in determining the metabolic flow of carbon compounds. Citrate synthase can be considered the first enzyme of tricarboxylic acid cycle,

since it catalyzes the irreversible condensation of acetyl-CoA with oxaloacetate to form citrate. Eidels and Preiss (1970) showed that the amount of this enzyme in *Rp. capsulata* is dependent on the growth conditions. The level of the enzyme is affected by carbon source and also by the mode of growth, whether aerobic–dark or anaerobic–photosynthetic. For any given carbon source, the specific activity of the enzyme was always at least two-fold greater under aerobic conditions, a finding consistent with the concept that the tricarboxylic acid cycle is more active under respiratory conditions than under anaerobic conditions. Though the level of this enzyme can be altered, the specific activity is always quite high, even under the least favorable conditions. This suggests the necessity for additional control mechanisms at the level of enzyme action. Such control exists in the form of inhibition caused by NADH. This inhibition is never complete, and can be entirely reversed by AMP. This type of regulatory system makes metabolic sense for cultures growing in a dark, respiratory mode in which the tricarboxylic acid cycle is the primary route of oxidation of carbon sources. NADH formed in the cycle can be reoxidized by the respiratory electron-transport system, resulting in the generation of ATP. NADH would therefore represent "potential chemical energy," while AMP would signal a depleted cellular energy state. A cell with low levels of NADH would have its tricarboxylic acid cycle stimulated, resulting in increased generation of reducing equivalents and potential chemical energy. Should AMP levels increase, due to either increased biosynthetic demands or an inadequate rate of ATP formation, NADH inhibition would be alleviated, thus stimulating the flow of carbon through the cycle.

Pyruvate kinase, which catalyzes the transfer of phosphate from phosphoenolpyruvate to ADP, forming ATP and pyruvate, was also shown to be a regulated enzyme in *Rp. capsulata* (Klemme, 1974). It therefore represents a critical point of interaction between the energy metabolism and the carbon metabolism of these cells. The kinase is formed constitutively, but its activity is regulated by several effectors; ATP inhibits and AMP activates. One of the tricarboxylic acid cycle intermediates, fumarate, is also a potent inhibitor of the enzyme; this inhibition can also be alleviated by AMP. It appears that in *Rp. capsulata*, pyruvate kinase is functioning as an amphibolic enzyme poised to favor anabolic functions; i.e., fumarate can be considered an end product of a pathway beginning with conversion of PEP to pyruvate. Increased levels of this "end product" inhibit the enzyme; yet should energy become limiting (i.e., when the level of AMP increases), the inhibition is relieved and more ATP is generated.

5. Summary Comments

The metabolism of nonaromatic organic compounds by photosynthetic bacteria has been under investigation for more than 40 years, with the emphasis shifting from time to time. The early experiments on carbon assimilation were designed to probe the mysteries of the fundamental photosynthetic process. The outcome of these investigations, of course, was a profound advance in the understanding of the role played by CO_2 in photosynthesis.

Another area of major concern has been the question of obligate autotrophy and obligate phototrophy. Though it appears that there may be no single metabolic property that can account for this nutritional pattern in all obligate autotrophs, the reasons for this behavior in several of the autotrophs have now been well explained in metabolic terms. An additional consequence of these studies has been an increased understanding of the intermediary metabolism of a number of the photosynthetic bacteria.

It is now generally conceded that the mechanisms of incorporation of organic carbon sources, by photosynthetic bacteria capable of utilizing such compounds, are similar to those found in typical, nonphotosynthetic heterotrophs. There are numerous reports of these bacteria attacking carbon sources by the same metabolic routes found in most other organisms. Work in this area will likely continue because of interest in the relationship of the physiology of these organisms to their ecology and evolution.

The photosynthetic bacteria (particularly those capable of both phototrophic and respiratory growth) offer unique systems for study of certain aspects of bioenergetics and cellular regulations. A basic understanding of their carbon metabolism is, of course, a minimum requirement for a full exploitation of these organisms in such research areas. The emergence of manageable genetic systems in the nonsulfur purple bacteria should greatly facilitate relevant investigations. It seems apparent that continued work on carbon-source utilization and metabolism should effectively complement ongoing research into such aspects of the biology of photosynthetic bacteria as membrane genesis and organization and regulation of the formation of the photosynthetic apparatus.

6. References

Albers, H., and Gottschalk, G., 1976, Acetate metabolism in *Rhodopseudomonas gelatinosa* and several other Rhodospirillaceae, *Arch. Microbiol.* **111**:45.

Barker, H. A., 1961, Fermentations of nitrogenous organic compounds, in: *The Bacteria*, Vol. II (I. C. Gunsalus and R. Y. Stanier, eds.), pp. 151–207, Academic Press, New York.

Beuscher, N., Mayer, F., and Gottschalk, G., 1974, Citrate lyase from *Rhodopseudomonas gelatinosa*: Purification, electron microscopy and subunit structure, *Arch. Microbiol.* **100**:307.

Borichewski, R. M., and Umbreit, W. W., 1966, Growth of *Thiobacillus thiooxidans* on glucose, *Arch. Biochem. Biophys.* **116**:97.

Bose, S. K., 1963, Media for anaerobic growth of photosynthetic bacteria, in: *Bacterial Photosynthesis* (H. Gest, A. San Pietro, and L. P. Vernon, eds.), pp. 501–510, Antioch Press, Yellow Springs, Ohio.

Buckel, W., Buschmeier, V., and Eggerer, H., 1971, Der Wirkungsmechanismus der Citrat-Lyase aus *Klebsiella aerogenes*, *Hoppe-Seyler's Z. Physiol. Chem.* **352**:1195.

Carr, N. G., 1973, Metabolic control and autotrophic physiology, in: *The Biology of Blue-Green Algae* (N. G. Carr and B. A. Whitton, eds.), pp. 39–65, Blackwell, Oxford.

Chaudhary, A., 1970, Studies on the regulation of enzyme activities in *Rhodospirillum rubrum*, Ph.D. thesis, University of Minnesota, Minneapolis.

Cohen-Bazire, G., Sistrom, W. R., and Stanier, R. Y., 1957, Kinetic studies of pigment synthesis by non-sulfur purple bacteria, *J. Cell. Comp. Physiol.* **49**:25.

Conrad, R., and Schlegel, H. G., 1974, Different pathways for fructose and glucose utilization in *Rhodopseudomonas capsulata* and demonstration of 1-phosphofructokinase in photosynthetic bacteria, *Biochim. Biophys. Acta* **358**:221.

Cutinelli, C., Ehrensvärd, G., and Reio, L., 1950, Acetic acid metabolism in *Rhodospirillum rubrum* under anaerobic conditions, *Ark. Kemi* **2**:357.

Cutinelli, C., Ehrensvärd, G., Reio, L., Saluste, E., and Stjernholm, R., 1951, Acetic acid metabolism in *Rhodospirillum rubrum* under anaerobic conditions II, *Ark. Kemi* **3**:315.

Eidels, L., 1969, Regulation of carbohydrate metabolism in *Rhodopseudomonas capsulata*, Ph.D. thesis, University of California, Davis.

Eidels, L., and Preiss, J., 1970, Citrate synthase: A regulatory enzyme from *Rhodopseudomonas capsulata*, *J. Biol. Chem.* **243**:2937.

Elsden, S. R., 1962, Photosynthesis and lithotrophic carbon dioxide fixation, in: *The Bacteria*, Vol. III (I. C. Gunsalus and R. Y. Stanier, eds.), pp. 1–40, Academic Press, New York.

Elsden, S. R., and Ormerod, J. G., 1956, The effect of monofluoroacetate on the metabolism of *Rhodospirillum rubrum*, *Biochem. J.* **63**:691.

Evans, M. C. W., Buchanan, B. B., and Arnon, D. I., 1966, A new ferredoxin-dependent carbon reduction cycle in a photosynthetic bacterium, *Proc. Natl. Acad. Sci. U.S.A.* **55**:928.

Foster, J. W., 1940, The role of organic substrates in photosynthesis of purple bacteria, *J. Gen. Physiol.* **24**:123.

Foster, J. W., 1944, Oxidation of alcohols by non-sulfur photosynthetic bacteria, *J. Bacteriol.* **47**:355.

Fuller, R. C., Smillie, R. M., Sisler, E. C., and Kornberg, H. L., 1961, Carbon metabolism in *Chromatium, J. Biol. Chem.* **236**:2140.

Gaffron, H., 1933, Über den Stoffwechsel der Schwefelfreien Purpurbakterien, *Biochem. Z.* **260**:1.

Gaffron, H., 1934, Über die Lohlensaure-Assimilation der roten Schwefelbakterien. I, *Biochem. Z.* **269**:447.

Gaffron, H., 1935, Über den Stoffwechsel der Purpurbakterien. II, *Biochem. Z.* **275**:301.

Gaffron, H., 1963, van Niel's theory: thirty years after, in: *Bacterial Photosynthesis* (H. Gest, A. San Pietro, and L. P. Vernon, eds.), pp. 3–14, Antioch Press, Yellow Springs, Ohio.

Gest, H., and Kamen, M. D., 1960, The photosynthetic bacteria, in: *Encyclopedia of Plant Physiology* (W. Ruhland, ed.), pp. 568–612, Springer-Verlag, Berlin—Gottingen—Heidelberg.

Gibson, M. S., and Wang, C. H., 1968, Utilization of fructose and glutamate by *Rhodospirillum rubrum, Can. J. Microbiol.* **14**:493.

Giffhorn, F., Beuscher, N., and Gottschalk, G., 1972, Regulation of citrate lyase activity in *Rhodopseudomonas gelatinosa, Biochem. Biophys. Res. Commun.* **49**:467.

Glover, J., and Kamen, M. D., 1951, Observations on the simultaneous metabolism of acetate and carbon dioxide by resting cell suspensions of *Rhodospirillum rubrum, Fed. Proc. Fed. Am. Soc. Exp. Biol.* **10**:190.

Glover, J. Kamen, M. D., and van Gemerden, H., 1952, Studies on the metabolism of photosynthetic bacteria. XII. Comparative light and dark metabolism of acetate and carbonate by *Rhodospirillum rubrum, Arch. Biochem. Biophys.* **35**:384.

Hirsch, P., 1968, Photosynthetic bacterium growing under carbon monoxide, *Nature (London)* **217**:555.

Hoare, D. S., and Gibson, J., 1964, Photoassimilation of acetate and the biosynthesis of amino acids by *Chlorobium thiosulphatophilum, Biochem. J.* **91**:546.

Hurlbert, R. E., and Lascelles, J., 1963, Ribulose diphosphate carboxylase in Thiorhodaceae, *J. Gen. Microbiol.* **33**:445.

Kelly, D. P., 1974, Growth and metabolism of the obligate photolithotroph *Chlorobium thiosulfatophilum* in the presence of added organic nutrients, *Arch. Microbiol.* **100**:163.

Kenyon, C. N., and Gray, A. M., 1974, Preliminary analysis of lipids and fatty acids of green bacteria and *Chloroflexus aurantiacus, J. Bacteriol.* **120**:131.

Klemme, J.-H., 1968, Untersuchungen zur Photoautotrophie mit molekularem Wasserstoff bei neuisolierten schwefelfreien Purpurbakterien, *Arch. Mikrobiol.* **64**:29.

Klemme, J.-H., 1974, Modulation by fumarate of a P_i-insensitive pyruvate kinase from *Rhodopseudomonas capsulata, Arch. Microbiol.* **100**:57.

Koch. J. P., Hayashı, S., and Lin, E. C. C., 1964, The control of dissimilation of glycerol and L-α-glycerophosphate in *Escherichia coli, J. Biol. Chem.* **239**:3106.

Kornberg, H. L., and Lascelles, J., 1960, The formation of isocitratase by the Athiorhodaceae, *J. Gen. Microbiol.* **23**:511.

Larsen, H., 1953, On the microbiology and biochemistry of the photosynthetic green sulfur bacteria, *Kl. Nor. Vidensk. Selsk. Skr.*, p. 1.

Losada, M., Trebst, A. V., Ogata, S., and Arnon, D. I., 1960, Equivalence of light and adenosine triphosphate in bacterial photosynthesis, *Nature (London)* **186**:753.

Leuking, D., Tokuhisa, D., and Sojka, G., 1973, Glycerol assimilation by a mutant of *Rhodopseudomonas capsulata, J. Bacteriol.* **115**:897.

Madigan, M. T., Petersen, S. R., and Brock, T. D., 1974, Nutritional studies on *Chloroflexus*, a filamentous photosynthetic, gliding bacterium, *Arch. Microbiol.* **100**:97.

Muller, F. M., 1933, On the metabolism of the purple sulfur bacteria in organic media, *Arch. Mikrobiol.* **4**:131.

Nielsen, A. M., and Sojka, G. A., 1975, Regulatory aspects of catabolism in *Rhodopseudomonas capsulata, Abstr. Annu. Meet. Am. Soc. Microbiol.* **1975**:176.

Ohmori, H., Ishitani, H., Sato, K., Shimizu, S., and Fukui, S., 1974, Metabolism of glutamate in purple nonsulfur bacteria: Participation of vitamin B_{12}, *Agric. Biol. Chem.* **38**:359.

Olsen, I., 1967, Studies on acetate and propionate metabolism in *Rhodospirillum rubrum*: Isolation of propionyl coenzyme A carboxylase, Ph.D. thesis, State University of New York at Buffalo.

Ormerod, J. G., 1956, The use of radioactive carbon dioxide in the measurement of carbon dioxide fixation in *Rhodospirillum rubrum, Biochem. J.* **64**:373.

Ormerod, J. G., and Gest, H., 1962, Symposium on metabolism of inorganic compounds. IV. Hydrogen photosynthesis and alternative metabolic pathways in photosynthetic bacteria, *Bacteriol. Rev.* **26**:51.

Pfennig, N., 1967, Photosynthetic bacteria, *Annu. Rev. Microbiol.* **27**:285.

Pierson, B. K., and Castenholz, R. W., 1974, A phototrophic gliding filamentous bacterium of hot springs, *Chloroflexus aurantiacus* gen. and sp. nov., *Arch. Microbiol.* **100**:5.

Pike, L., and Sojka, G. A., 1975, Glycerol dissimilation in *Rhodopseudomonas sphaeroides, J. Bacteriol.* **124**:1101.

Qadri, S. M. H., and Hoare, D. S., 1968, Formic hydrogenlyase and the photoassimilation of formate by a strain of *Rhodopseudomonas palustris, J. Bacteriol.* **95**:2344.

Quayle, J. R., and Pfennig, N., 1975, Utilization of methanol by Rhodospirillaceae, *Arch. Microbiol.* **102**:193.

Roseman, S. J., 1969, The transport of carbohydrates by a bacterial phosphotransferase system, *J. Gen. Physiol.* **54**:138.

Sadler, W. R., and Stanier, R. Y., 1960, The function of acetate in photosynthesis by green bacteria, *Proc. Natl. Acad. Sci. U.S.A.* **46**:1328.

Saier, M. H., Jr., Feucht, B. U., and Roseman, S., 1971, Phosphoenolpyruvate-dependent fructose phosphorylation in photosynthetic bacteria, *J. Biol. Chem.* **246**:7819.

Sandhu, G. R., and Carr, N. G., 1970, A novel alcohol dehydrogenase present in *Rhodomicrobium vannielii, Arch. Mikrobiol.* **70**:340

Schaab, C., Giffhorn, F., Schoberth, S., Pfennig, N., and Gottschalk, G., 1972, Phototrophic growth of *Rhodopseudomonas gelatinosa* on citrate: Accumulation and subsequent utilization of cleavage products, *Z. Naturforsch.* **27b**:962.

Siegel, J. M., and Kamen, M. D., 1950, Studies on the metabolism of photosynthetic bacteria. VI. Metabolism of isopropanol by a new strain of *Rhodopseudomonas gelatinosa, J. Bacteriol.* **59**:693.

Smith, A. J., London, J., and Stanier, R. Y., 1967, Biochemical basis of obligate autotrophy in blue-green algae and thiobacilli, *J. Bacteriol.* **94**:972.

Stahl, C. L., and Sojka, G. A., 1973, Growth of *Rhodopseudomonas capsulata* on L- and D-malic acid, *Biochim. Biophys. Acta* **297**:241.

Stanier, R. Y., 1961, Photosynthetic mechanisms in bacteria and plants: Development of a unitary concept, *Bacteriol. Rev.* **25**:1.

Stanier, R. Y., Doudoroff, M., Kunisawa, R., and Contopoulou, R., 1959, The role of organic substrates in bacterial photosynthesis, *Proc. Natl. Acad. Sci. U.S.A.* **45**:1246.

Stern, J. R., 1963, Enzymic activation and cleavage of D- and L-malate, *Biochim. Biophys. Acta* **69**:435.

Stern, J. R., and Hegre, C. S., 1966, Inducible D-malic enzyme in *Escherichia coli, Nature* (London) **212**:1611.

Stern, J. R., and O'Brien, R. W., 1969, Oxidation of D-malic and β-alkylmalic acids by wild type and mutant strains of *Salmonella typhimurium* and by *Aerobacter aerogenes, J. Bacteriol.* **98**:147.

Szymona, M., and Doudoroff, M., 1960, Carbohydrate metabolism in *Rhodopseudomonas sphaeroides, J. Gen. Microbiol.* **22**:167.

Trüper, H. G., and Schlegel, H. G., 1964, Sulphur metabolism in Thiorhodaceae. I. Quantitative measurements on growing cells of *Chromatium okenii, Antonie von Leeuwenhoek; J. Microbiol. Serol.* **30**:225.

Tuboi, S., and Kikuchi, G., 1965, Enzymic cleavage of malyl-coenzyme A into acetyl-coenzyme A and glyoxylic acid, *Biochim. Biophys. Acta* **96**:148.

van Gemerden, H., 1968a, Growth measurements of *Chromatium* cultures, *Arch. Mikrobiol.* **64**:103.

van Gemerden, H., 1968b, Utilization of reducing power in growing cultures of *Chromatium, Arch. Mikrobiol.* **64**:111.

van Gemerden, H., 1968c, On the ATP generation by *Chromatium* in darkness, *Arch. Mikrobiol.* **64**:118.

van Niel, C. B., 1931, On the morphology and physiology of the purple and green sulfur bacteria, *Arch. Mikrobiol.* **3**:1.

van Niel, C. B., 1936, On the metabolism of the Thiorhodaceae, *Arch. Mikrobiol.* **7**:323.

van Niel, C. B., 1941, The bacterial photosyntheses and their importance for the general problems of photosynthesis, *Adv. Enzymol.* **1**:263.

van Niel, C. B., 1944, The culture, general physiology, morphology, and classification of the non-sulfur purple and brown bacteria, *Bacteriol. Rev.* **8**:1.

Weckesser, J., Drews, G., and Tauschel, H.-D., 1969, Zur Feinstruktur und Taxonomie von *Rhodopseudomonas gelatinosa, Arch. Mikrobiol.* **65**:346.

Wertlieb, D., and Vishniac, W., 1967, Methane utilization by a strain of *Rhodopseudomonas gelatinosa, J. Bacteriol.* **93**:1722.

Yen, H., and Gest, H., 1974, Regulation of biosynthesis of aspartate family amino acids in the photosynthetic bacterium *Rhodopseudomonas palustris, Arch. Microbiol.* **101**:187.

Yoch, D. C., and Lindstrom, E. S., 1967, Photosynthetic conversion of formate and CO_2 to glutamate by *Rhodopseudomonas palustris, Biochem. Biophys. Res. Commun.* **28**:65.

CHAPTER 38

Metabolism of Aromatic Compounds by Rhodospirillaceae

P. Leslie Dutton and W. Charles Evans

1. Introduction

Several species of the purple nonsulfur Rhodospirillaceae family are able to grow at the expense of simple aromatic compounds as sole carbon source. As is the case with aliphatic organic acids as growth substrates, the bacteria of this family are able to grow on aromatic compounds both anaerobically in the light by photosynthetic means and aerobically in the dark by respiration.

The photosynthetic and respiratory ways of growth present the bacteria with two starkly different conditions for the tough task of dissimilating the stable aromatic ring structure. This chapter describes the equally different methods utilized for aromatic breakdown by Rhodospirillaceae growing under photosynthetic and respiratory conditions. We also make brief comparisons of the metabolic pathways employed by Rhodospirillaceae with those used by other microorganisms.

2. General Observations of Growth on Aromatic Substrates

Gaffron (1941) showed the quantitative conversion of phenylpropionate to benzoate by *Rhodovibrio*.

Scher (1961) and Scher and Proctor (1960) isolated several strains of photosynthetic bacteria, namely *Rhodopseudomonas palustris*, *Rp. capsulata*, *Rp. sphaeroides*, *Rp. gelatinosa*, and *Rhodospirillum rubrum*, that were shown to grow well anaerobically in the light with benzoate as carbon source. In 1965, *Rs. fulvum*, an obligate phototroph, was isolated from a benzoate enrichment culture (Pfennig *et al.*, 1965). In the same year, Leadbetter and Hawk (1965) found that although *Rp. palustris* could grow anaerobically in the light on benzoate, it was unable to do so aerobically in the dark. 4-Hydroxybenzoate, however, supported growth under both conditions. Table 1 summarizes what is known to date for *Rp. palustris*, and includes some other aromatic substrates tested by Dutton and Evans (1969). Also included are some very recent data of Whittle *et al.* (1976) and Evans (1977) for *Rp. gelatinosa* and *Rp. capsulata* that expand the list to include 1,3,5-trihydroxybenzene (phloroglucinol) and 2,4,6-trihydroxybenzoate.

3. Aerobic Metabolism of Aromatic Compounds

There is, of course, a large literature on oxygen-linked reactions that lead to aromatic ring hydroxylation as a preparation for ring fission. The mono- and dioxygenases are textbook commodities. In the former case, one of the oxygen's two atoms appears incorporated into the aromatic ring as a hydroxyl; in the latter, both atoms of the oxygen molecule are incor-

P. Leslie Dutton · Johnson Research Foundation and Department of Biochemistry and Biophysics, University of Pennsylvania School of Medicine, Philadelphia, Pennsylvania 19104
W. Charles Evans · Department of Biochemistry and Soil Sciences, Memorial Building, University College of North Wales, Bangor, Gwynedd LL57 2UW, United Kingdom

Table 1. Growth Behavior of *Rhodopseudomonas palustris*, *Rhodopseudomonas gelatinosa*, and *Rhodopseudomonas capsulata* with Some Aromatic Substrates[a]

| Substrate | Rp. palustris | | Rp. gelatinosa and Rp. capsulata |
	Photosynthetic growth	Aerobic growth	Photosynthetic growth
Benzoate	+	−	+
3-Hydroxybenzoate	+	−	+
2-Hydroxybenzoate	NT	−	NT
4-Hydroxybenzoate	+	+	+
3,4-Dihydroxybenzoate	−	+	−
Other dihydroxybenzoates	NT	NT	−
2,4,6-Trihydroxybenzoate	NT	NT	+
1,2-Dihydroxybenzoate	−	−	NT
1,3,5-Trihydroxybenzene	+	NT	+

[a] Symbols: (+) growth; (−) no growth; (NT) not tested.

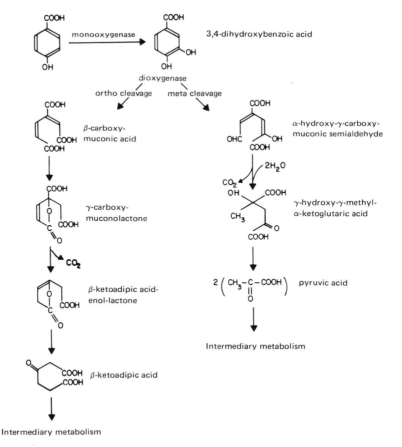

Fig. 1. An example of aromatic metabolism in bacteria under aerobic conditions.

porated into the ring, forming an *ortho-* dihydroxylated compound. Once formed, the dihydroxylated compound is further oxygenated via a dioxygenase, which leads to the opening of the aromatic ring. An example starting with 4-hydroxybenzoate is shown in Fig. 1. This compound is hydroxylated by mono-oxygenase to form 3,4-dihydroxybenzoate (protocatechuate). At this point, there are two possibilities for the ring-cleavage reaction. One uses the protocatechuate-3,4-dioxygenase leading to ring cleavage *between* the two hydroxyl groups, and is commonly called an *ortho-* cleavage mechanism. The other employs protocatechuate-4,5-dioxygenase to break the ring in the adjacent 4,5-position, and is termed a *meta*-cleavage mechanism. The former yields the tricarboxylic β-carboxymuconate, which is further broken down to β-ketoadipate. The latter mechanism, which is less common, forms a muconic semialdehyde that is further broken down to pyruvate. Beyond this point, the breakdown products become part of intermediary metabolism. For the reader interested in bacterial aromatic metabolism in general, four recent articles can serve as source references (Evans, 1976; Dagley, 1976; Trudgill, 1976; Evans, 1977).

Rhodopseudomonas palustris grown aerobically on 4-hydroxybenzoate exclusively utilizes the *meta*-cleavage mechanism, ultimately yielding pyruvate (Hegeman, 1967; Dutton and Evans, 1967, 1969). The pathway was determined by using standard manometric techniques, measuring oxygen consumption by whole cells and cell-free extracts. It is possible to apply this indirect method because the enzymes of the pathway are not constitutive; they are sequentially induced by the growth substrate and metabolites, and display high specificity for the inducing metabolic intermediates. Thus, prompt oxygen uptake is encountered only in the presence of growth substrate and the chemical intermediates of the breakdown sequence. The pathway for 4-hydroxybenzoate breakdown was confirmed by isolation of the intermediates in crude enzyme preparations acting on protocatechuate (Dutton and Evans, 1967, 1969) or γ-carboxy-α-hydroxymuconic semialdehyde (Hegeman, 1967).

4. Anaerobic Photosynthetic Metabolism of Aromatic Compounds

4.1. Early Work

Manometric techniques are not applicable for the study of the breakdown of aromatic substrates by cells grown anaerobically in the light, and from what we

know now about the nature of the photosynthetic bacterial light reaction, this is not surprising. Earlier attempts to elucidate the anaerobic breakdown sequence of aromatic compounds, however, were based on a literal interpretation of van Niel's general schemes (van Niel, 1941) for photosynthetic bacteria and plants. In these schemes, one of the products of the light reaction was a strong oxidant ["OH"], which in plants was converted into molecular oxygen; in bacteria, this proposed light-induced "bound oxygen" was used to oxidize substrates. Thus, in this early work, it was considered that the light-induced oxidant and molecular oxygen were "equivalent," and that the anaerobic photosynthetic breakdown pathways might be the same as those familiar in aerobic breakdown mechanisms. This idea initially received some claims of support from manometric experiments (Scher and Proctor, 1960; Proctor and Scher, 1960). These experiments, however, could not be repeated. Subsequent work by Leadbetter and Hawk (1965) using similar manometric techniques showed that cells of *Rp. palustris* grown anaerobically in the light on 4-hydroxybenzoate failed to respire significantly beyond the endogenous rate following addition of 4-hydroxybenzoate or 3,4-dihydroxybenzoate, the first intermediate of the aerobic pathway. Dutton and Evans (1967) agreed with Leadbetter and Hawk, and extended the observations to *Rp. palustris* grown photosynthetically on benzoate or 3-hydroxybenzoate; neither of these growth substrates was able to promote any respiratory activity in the cells under

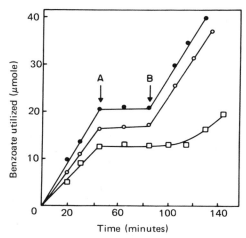

Fig. 2. Effects of air, light, and darkness on benzoate photometabolism in *Rhodopseudomonas palustris* cells. The initial conditions were anaerobic with illumination. In the period between the arrows (A—B), the conditions were air–light (○), air–dark (●), and anaerobic–dark (□).

aerobic conditions. The conclusion from these experiments was that oxygen and any light-generated oxidant were not equivalent.

In fact, Fig. 2 shows directly the dramatic and deleterious effect that oxygen has on the photometabolism of benzoate by photosynthetically grown cells: oxygen totally inhibits benzoate utilization in the light or dark (Dutton and Evans, 1969). Return to anaerobic conditions in the light permits benzoate utilization to proceed again promptly and unimpaired. The obligatory light requirement by the anaerobic breakdown system is also shown in Fig. 2. It is not understood, however, why there is a lag period after return to the illuminated state; there is no lag period if cells are incubated in the dark for the same period without benzoate.

4.2. Pathway of Anaerobic Benzoate Photometabolism

The oxygen inhibition and the obligatory dependence on light were heralds to the finding that photosynthetic aromatic breakdown sequence was a reductive process. Guyer and Hegeman (1969) concluded this from mutant studies with *Rp. palustris*, and suggested that the initial reactions of light-driven benzoate breakdown were reductive. They further suggested that cyclohexanecarboxylate and cyclohex-1-enecarboxylate might be involved. At the same time,

Dutton and Evans (1968, 1969), using ^{14}C-labeled benzoate, determined the key intermediates of the anaerobic photosynthetic metabolism of benzoate in *Rp. palustris*. This was done by incubating unlabeled test intermediates with whole-cell suspensions actively photometabolizing ^{14}C-labeled benzoate. After an appropriate time of incubation, the test compound was extracted and examined for any isotope exchange. The results from this work (Dutton and Evans, 1968, 1969) revealed the pathway for anaerobic benzoate photometabolism in *Rp. palustris* shown in Fig. 3. All the intermediates presented in Fig. 3 became labeled during the incubation treatment. Thus, all seven carbons of benzoate remain together as far as pimelate.

The pathway can be considered in two parts. The first part is the reduction of the aromatic ring; the second part is a sequence of reactions similar to those well known for the β-oxidation of fatty acids.

4.2.1. Reductive Part

We cannot be certain of the precise sequence of aromatic ring reduction from the isotope-exchange experiments. Both cyclohexanecarboxylate and cyclohex-1-enecarboxylate become labeled. Interpreting these results at the simplest level, we might consider that the aromatic ring may become *fully* reduced with the incorporation of six hydrogen

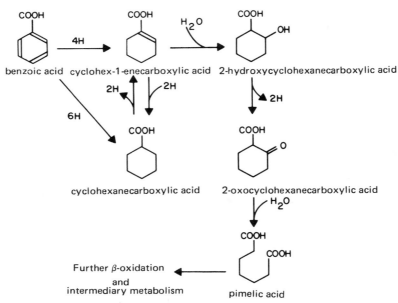

Fig. 3. Anaerobic photometabolism of benzoate by *Rhodopseudomonas palustris*.

equivalents to form cyclohexanecarboxylate. In this case, the subsequent reactions; starting from a fully saturated alicyclic acid, would be classically β-oxidative in nature, with the first reaction being the specific dehydrogenation of cyclohexanecarboxylate in the 1,2-position to produce cyclohex-1-enecarboxylate. Alternatively, however, it is possible that four hydrogen equivalents are incorporated into the aromatic ring to form cyclohex-1-enecarboxylate more directly. In this case, it would have to be proposed that the labeling in the cyclohexanecarboxylate is the result of some nonspecificity in the process.

The only other reduced benzoate compounds to be tested by Dutton and Evans (1969) were cyclohex-2-enecarboxylate and cyclohex-3-enecarboxylate. The former showed some labeling after incubation, while the latter did not. The failure to label the cyclohex-3-enecarboxylate may indicate that the 3,4-positions are obligatorily reduced first, irrespective of what happens next. If this is true, the first intermediate of the breakdown pathway would be cyclohex-1,5-dienecarboxylate. In either of the two possibilities discussed in the previous paragraph, the cyclohex-2-enecarboxylate has to be considered to be a side product.

4.2.2. The β-Oxidative Part and Ring Cleavage

The β-oxidative sequence involves alternate dehydrogenation and hydration. However, because in this case the substrate is a ring compound, instead of the usual release of a two-carbon fragment (acetyl-CoA) at the end of the β-oxidative sequence, β-oxidation leads to breakage of the ring in the 1,2-position.

4.2.3. Cell-Free Preparations and the Role of Coenzyme A and Benzoyl-Phosphate

The early attempts to prepare a functional cell-free extract of cells grown photosynthetically on benzoate failed. However, a useful contributing reason for the failure was revealed by Dutton and Evans (1970). Cell-free extracts, which included the chromatophore particles, were shown to be capable of inhibiting benzoate photometabolism even when added to whole-cell suspensions. The inhibiting material was found to be associated with the particulate chromatophore fraction, and was determined to be composed of long-chain fatty acids. The principal constituent of the fatty acid complement ($\approx75\%$) was determined to be octadeca-11-enoic acid (vaccenic acid). It was concluded that these long-chain fatty acids must be released during and after the breakage of the cells, since almost no free fatty acids were evident in the whole cell.

It is likely that the fatty acids are released by the action of lipases on membrane phospholipids and remain associated with the chromatophore membrane. On addition of chromatophores to whole-cell suspensions, however, it is clear that the fatty acids can be absorbed by, and incorporated into, whole cells. It was subsequently found (Dutton and Evans, 1970) that any of the monocarboxylic acid homologues (but not their methyl esters), even down to propionate, are, in the 1- to 10-mM range, capable of more than 80% inhibition of the rate of benzoate utilization in whole cells. It was also reported that with limiting amounts of fatty acids added or if experimental analysis was pursued long enough, benzoate photometabolism would be seen to resume. The conclusion from these experiments was that the fatty acids can successfully compete as substrates for enzymes or cofactors involved in the benzoate photometabolism. It was suggested (Dutton and Evans, 1970) that the intermediates of the benzoate photometabolism may, like those of the β-oxidation of fatty acids, exist as the CoA esters. Further circumstantial evidence that this could be the case came from the report of Babior and Bloch (1966), who showed that the intermediates of the liver-extract-catalyzed aromatization of cyclohexanecarboxylate (via cyclohex-1-enecarboxylate) operated as CoA esters; significantly, they found that this process was sensitive to octanoate.

Recent work by Whittle et al. (1976) showed that extracts from Rp. palustris cells grown photosynthetically on benzoate are capable of catalyzing the thioesterification of benzoate to form benzoyl-CoA:

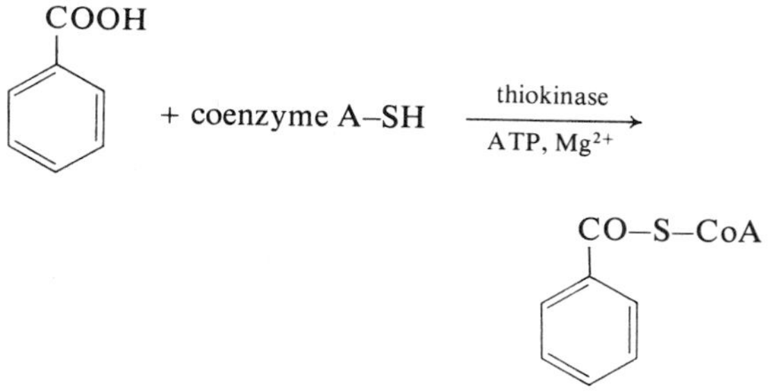

This was done in the dark in a cell-free extract that had been treated with bovine serum albumin to remove the competing fatty acids. The cofactors were ATP and Mg^{2+}, the familiar requirements of the fatty acid thiokinase enzymes.

Further to this, Whittle et al. (1976) successfully obtained a cell-free system capable of converting cyclohex-1-enecarboxylate to pimelate. Again, there is the requirement for CoA, ATP, and Mg^{2+}, but in addition, NAD^+ is an obligatory cofactor. The reaction proceeds in the dark under anaerobic or aerobic conditions.

The basic pathway of Fig. 3 was written more completely by Whittle *et al.* (1976) as follows:

$$\text{Benzoate} \xrightarrow[\text{(1)}]{\text{CoA, ATP}} \text{benzoyl-CoA} \xrightarrow[\text{(2)}]{\text{6H}}$$

$$\text{cyclohexanoyl-CoA} \xrightarrow[\text{(3)}]{\text{FAD}^+}$$

$$\text{cyclohex-1-enoyl-CoA} \xrightarrow[\text{(4)}]{\text{H}_2\text{O}}$$

$$\text{2-hydroxycyclohexanoyl-CoA} \xrightarrow[\text{(5)}]{\text{NAD}^+}$$

$$\text{2-oxocyclohexanoyl-CoA} \xrightarrow[\text{(6)}]{\text{CoA, ATP}} \text{pimelyl-di-CoA}$$

Although steps 1, 4, 5, and 6 have been established in cell-free systems, and as a whole show that the scheme proposed by Dutton and Evans (1968, 1969, 1970) and Guyer and Hegeman (1969) is essentially correct, the following points need to be considered:

1. The demonstration of the cell-free-extract-catalyzed CoA esterification of benzoate is not to be taken as *proof* that this is the first reaction of the pathway, since there is room for doubt about the specificity of some of the enzymes. It has still to be established that the reduction of benzoyl-CoA is possible in cell-free extracts, and that the esterification is obligatory for the reduction (see below). This caution is based on the observations (Dutton and Evans, 1967, 1969) that there is a lack of specificity of *Rp. palustris* cells grown photosynthetically on various benzoic acids. Cells grown on benzoate, for example, could utilize 2-hydroxy-, 3-hydroxy-, and 4-hydroxybenzoates without a detectable lag period; similarly, cells grown on 3-hydroxybenzoate could metabolize benzoate just as well as the growth substrate. Thus, it is possible that the thiokinase that was demonstrated to esterify benzoate can esterify a range of simple benzoic, cyclohexene, and cyclohexane acid derivatives.

2. The proposed FAD-linked step 3 has not yet been demonstrated in cell-free extracts, and the assignment is an assumption based on the known flavoprotein character of the α,β-acyldehydrogenases of fatty acid β-oxidation. It is worth repeating, however, that if the benzoate reduction mechanism could go *directly* to the cyclohex-1-enecarboxylate and not via

cyclohexanecarboxylate, this proposed dehydrogenation step would be redundant as a major part of the pathway.

3. Evidence for the further esterification in step 6 to produce the di-CoA ester of pimelate has not been proved. The reaction is in keeping, however, with expectations of a typical fatty acid β-oxidation mechanism. In this case, because the β-oxidation leads to ring cleavage, both CoA groups are on the same dicarboxylic acid. If the reaction in step 6 is valid, the hydrolytic cleavage proposed in the original pathway (Fig. 3) will become one that involves a thiolytic cleavage.

In contrast to the β-oxidation part of the pathway, it is in the reductive part of the anaerobic benzoate metabolism that much has still to be learned. A very recent experiment of Lunt and Evans (unpublished) gave results that are very significant in this connection. Their experiment is described below.

Cell-free extract (sonicated, 15,000g supernatant, i.e., containing chromatophores) of *Rp. palustris* grown photosynthetically on benzoate, when incubated anaerobically in the light with benzoyl-phosphate caused the disappearance of the aromatic absorption; neither benzoate nor benzoyl-CoA was metabolized in this system. To establish which component of the cell-free extract was responsible, it was separated into the highly pigmented chromatophore fraction (centrifugation at 100,000g for 60 min) and the clear supernatant. Incubation mixtures contained this supernatant *or* washed chromatophores (5 ml), CoA (1 μM), benzoyl-phosphate (50 μM), benzoate (50 μM), and benzoyl-CoA (30 μM), respectively, in separate reaction flasks, and tris-HCl buffer (100 μM, pH 8) in a total volume of 100 ml. A stream of nitrogen (oxygen-free) was passed through the flasks, which were then incubated at 30°C in an illuminated Warburg bath; a duplicate set of chromatophore-containing incubation mixtures was shielded from light in the same bath. Samples were taken periodically and substrate concentrations followed by measuring their characteristic aromatic ultraviolet absorption maxima. The results are shown in Table 2. Over a period of 150 min, the concentration of benzoyl-phosphate had decreased by about 70% during incubation with chromatophores anaerobically in the light; hardly any change occurred in the dark or with the supernatant in the light. No alterations in the concentration of benzoate or benzoyl-CoA occurred in any of the incubation mixtures.

This experiment provides strong evidence for the presence of light-dependent, membrane-bound proton-translocating redox system in the chromatophores

Table 2. Anaerobic Photoreduction of Benzoyl-Phosphate by *Rhodopseudomonas palustris* Chromatophores in the Presence of Coenzyme A

| Time (min) | Benzoyl-phosphate (μM)[a] | | | Benzoate (μM)[a] | | | Benzoyl-CoA (μM)[a] | | |
| | Supernatant | | Chromatophores | Supernatant | Chromatophores | | Supernatant | Chromatophores | |
	L	D	L	L	D	L	L	D	L
0	50	47	50	50	50	50	30	30	30
45	48	45	36	49	49	49	28	27	29
90	48	44	25	49	48	49	27	29	29
115	47	44	20	49	48	49	27	29	29
150	45	43	16	49	49	48	27	26	27

[a] Substrate concentrations (μM) were calculated from λ_{max} at 274 nm (benzoyl-phosphate), 275 nm (benzoate), and 262 nm (benzoyl-CoA). (L) Light; (D) dark.

capable of reducing benzoyl-phosphate; it also affords a subcellular system for further studies.

4.3. Photometabolism of Phloroglucinol by *Rhodopseudomonas gelatinosa*

Whittle *et al.* (1976) reported that *Rp. gelatinosa* is able to grow anaerobically in the light on phloroglucinol (1,3,5-trihydroxybenzene). In contrast to the difficulties encountered in preparing *Rp. palustris* cell-free extracts capable of benzoate reduction, they successfully prepared cell-free extracts from *Rp. gelatinosa* capable of the dark reduction of phloroglucinol to dihydrophloroglucinol:

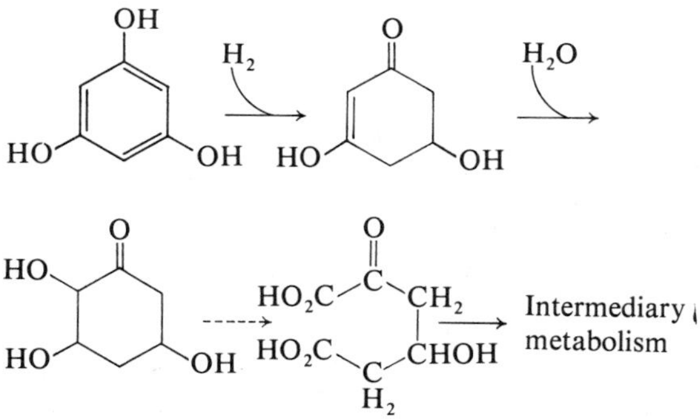

The cell-free system contained EDTA together with cysteine and $NADPH(H^+)$ as reductants. The reactions beyond this point, including the ring-cleavage mechanism, remain to be revealed, although there is evidence for the production of aliphatic acids, one of them probably 2-oxo-4-hydroxyadipate (from mass spectrometry).

5. Some Comparisons

In the early studies, when the anaerobic breakdown pathway was thought to be oxidative, importance was attached to the fact that the bacteria in question were photosynthetic and the possibility that some product derived from the light reaction might be involved. Even after the pathway had been determined to be reductive, there was still a tendency to overvalue the possibility that the reductive forces for benzoate reduction might be special to photosynthetic bacteria. Recent studies have shown that this is not the case, because growth with aromatic compounds as sole carbon source has been reported in denitrifying bacteria. Like *Rp. palustris*, a *Pseudomonas* sp. (Taylor *et al.*, 1970) and a *Moraxella* sp. (Williams and Evans, 1975) are capable of growth at the expense of aromatic compounds both aerobically and anaerobically. Under aerobic conditions, the terminal respiratory oxidant is oxygen, while under anaerobic conditions, the terminal oxidant is nitrate.

Under aerobic conditions, these bacteria utilize oxygen and the metabolic pathways shown in Fig. 1; the *Pseudomonas* sp. has been shown to metabolize benzoate and 4-hydroxybenzoate via the *meta*-cleavage route, while the *Moraxella* sp. was reported to decarboxylate and hydroxylate the benzoate to form catechol (1,2-dihydroxybenzene), with subsequent breakage of the aromatic ring by the *ortho*-cleavage mechanism.

As was found for *Rp. palustris*, *Moraxella* sp. in the absence of oxygen and the presence of nitrate suppresses the enzymes of the aerobic aromatic breakdown pathway and assumes a reductive mechanism. The pathway of anaerobic benzoate metabolism in

Moraxella (Williams and Evans, 1975) driven by nitrate respiration is remarkably similar to that employed by *Rp. palustris*. There is a divergence, however, at 2-oxocyclohexanecarboxylate. Instead of proceeding with the cleavage of 2-oxocyclohexane-carboxylate to form pimelate, decarboxylation occurs, and the ultimate product of the cleavage is adipate.

Another instance of the operation of the reductive pathway is the methanogenic fermentation of aromatic compounds by a "consortium" of bacteria occurring in anaerobic environments—e.g., the rumen and intestinal contents of animals, stagnant waters, and sewage sludge—from anaerobic digesters that, in the absence of O_2, light, NO_3^- or SO_4^{2-}, produce methane and CO_2 from benzoate and many other aromatic substrates (see Evans, 1976, 1977).

In general, it seems likely that the initial outlay of energy for the early endothermic reactions of aromatic breakdown in photosynthetic and denitrifying bacteria and for their methanogenic fermentation comes from ATP generated in the normal events of electron transfer and metabolism. The provision of the necessary reduced cofactors probably originates in ATP-mediated reactions such as reversed electron flow to the nicotinamide adenine nucleotides. In addition, ATP is evidently used directly in esterification reactions, such as those involving CoA, geared to "activate" the aromatic acid, possibly before reduction and more certainly before ring cleavage.

6. References

Babior, B. M., and Bloch, K., 1966, Aromatization of cyclohexanecarboxylic acid, *J. Biol. Chem.* **241**:3643–3651.

Dagley, S., 1976, Microbial catabolism and the carbon cycle, *Biochem. Soc. Trans.* **4**:455–458.

Dutton, P. L., and Evans, W. C., 1967, Dissimilation of aromatic substrates by *Rhodopseudomonas palustris*, *Biochem. J.* **104**:30P.

Dutton, P. L., and Evans, W. C., 1968, The photometabolism of benzoic Acid by *Rhodopseudomonas palustris*, *Biochem. J.* **107**:28P.

Dutton, P. L., and Evans, W. C., 1969, The metabolism of aromatic compounds by *Rhodopseudomonas palustris*: A new reductive method of aromatic ring metabolism, *Biochem. J.* **113**:525–536.

Dutton, P. L., and Evans, W. C., 1970, Inhibition of aromatic photometabolism in *Rhodopseudomonas*

palustris by fatty acids, *Arch. Biochem. Biophys.* **136**:228–232.

Evans, W. C., 1976, Anaerobic and aerobic environments of microbiological catabolism, *Biochem. Soc. Trans.* **4**:452–455.

Evans, W. C., 1977, Biochemistry of the bacterial catabolism of aromatic compounds in aerobic environments, *Nature* **270**:17–22.

Frank, J., and Gaffron, H., 1941, Photosynthesis: facts and interpretations, *Adv. Enzymol.* **1**:200–262.

Gaffron, H., 1941, Unpublished result; cited in Frank and Gaffron (1941).

Guyer, M., and Hegeman, G. D., 1969, Evidence for a reductive pathway for the anaerobic metabolism of benzoate, *J. Bacteriol.* **99**:906–907.

Hegeman, G. D., 1967, The metabolism of *p*-hydroxybenzoate by *Rhodopseudomonas palustris* and its regulation, *Arch. Mikrobiol.* **59**:143–148.

Leadbetter, E. R., and Hawk, A., 1965, Aromatic acid utilization by Athiorhodaceae, *J. Appl. Bacteriol.* **27**:448.

Pfennig, N., Eimhjellen, K. E., and Jensen, S. L., 1965, A new isolate of the *Rhodospirillum fulvum* group and its photosynthetic pigments, *Arch. Mikrobiol.* **51**:258–266.

Proctor, M. H., and Scher, S., 1960, Decomposition of benzoate by a photosynthetic bacterium, *Biochem. J.* **76**:338.

Scher, S., 1961, Photometabolism of benzoic acid by Athiorhodaceae, in: *Proceedings of the Third International Congress of Photobiology*, Copenhagen (B. C. Christiansen and B. Buchmann, eds.), pp. 583–585, Elsevier, Amsterdam.

Scher, S., and Proctor, M. H., 1960, Studies with photosynthetic bacteria: Anaerobic oxidation of aromatic compounds, in: *Comparative Biochemistry of Photoreactive Pigments*, Vol. 25 (M. B. Allen, ed.), pp. 387–393, Academic Press, New York.

Taylor, B. F., Campbell, W. L., and Chinoy, I., 1970, Anaerobic degradation of the benzene nucleus by a facultative anaerobic microorganism, *J. Bacteriol.* **102**:430–436.

Trudgill, P. W., 1976, Microbial degradation of alicyclic and heterocyclic compounds, *Biochem. Soc. Trans.* **4**:458–463.

van Niel, C. B., 1941, The bacterial photosyntheses, *Adv. Enzymol.* **1**:263–328.

Whittle, P. J., Lunt, D. O., and Evans, W. C., 1976, Anaerobic photometabolism of aromatic compounds by *Rhodopseudomonas* sp., *Biochem. Soc. Trans.* **4**:490–491.

Williams, R. J., and Evans, W. C., 1975, The metabolism of benzoate by *Moraxella* species through anaerobic nitrate respiration: Evidence for a reductive pathway, *Biochem. J.* **148**:1–10.

PART VII

Biosynthesis

CHAPTER 39

Biosynthesis of Carotenoids

Karin Schmidt

1. Introduction

Photosynthetic bacteria synthesize many different types of carotenoids, the chemical structures of which differ from those found in algae, fungi, higher plants, and nonphotosynthetic bacteria. These carotenoids are characterized by the following properties (see also Chapter 12):

1. They most often occur as aliphatic compounds.
2. They often contain tertiary hydroxyl and methoxyl groups.
3. There are frequently double bonds in the C-3,4 position.
4. Oxo-groups in conjugation with the polyene chain are found attached to C-2 (aerobic conditions) and to C-4 (anaerobic conditions).
5. Aldehyde groups are sometimes present at the C-20 position.
6. Cyclic carotenoids commonly have aromatic rings, e.g., either 1,2,5-trimethylphenyl or 1,2,3-trimethylphenyl end groups.

During the past 20 years, the carotenoid composition of many strains and species of photosynthetic bacteria has been analyzed. The results of

these studies suggested four main pathways of carotenoid biosynthesis:

1a. Spirilloxanthin (via rhodopin)
  b. Rhodopinal (plus spirilloxanthin via rhodopin)
2. Spheroidene (plus spirilloxanthin via chloroxanthin)
3a. Okenone
  b. Oxo-carotenoids of *Rhodopseudomonas globiformis* (*R.g.*-keto carotenoids)
4a. Isorenieratene
  b. Chlorobactene
  c. γ- and β-Carotene

Regarding the distribution of these carotenoids within the taxonomical groups of photosynthetic bacteria (Table 1), several generalizations can be made. Carotenoids of the spirilloxanthin series (including lycopene and rhodopin) occur in most of the investigated species. Spirilloxanthin, rhodopin, and their intermediates are found in a large number of species of both Rhodospirillaceae and Chromatiaceae (*Rhodospirillum, Rhodopseudomonas, Rhodomicrobium, Chromatium, Thiocapsa, Ectothiorhodospira, Amoebobacter*). There are a few types of carotenoids, however, that seem to be limited to special taxonomic or physiological groups of photosynthetic bacteria. First, spheroidene and its specific biosynthetic pathway have been detected in only three species of *Rhodopseudomonas*: *Rp. sphaeroides, Rp. gelatinosa,* and *Rp. capsulata.* Second, aromatic carotenoids are synthesized only by strictly anaerobic species, which

Karin Schmidt · Institut für Mikrobiologie der Gesellschaft für Strahlen- und Umweltforschung mbH, Göttingen, West Germany

Table 1. Distribution of Carotenoids within Species of Photosynthetic Bacteria

Name of pathway	Number of species			
	Rhodospirillaceae	Chromatiaceae	Chlorobiaceae	Chloroflexaceae
1. Spirilloxanthin (via rhodopin)				
a. Normal spirilloxanthin series	8	10		
b. Rhodopinal series	4	7		
2. Spheroidene	3			
3. Okenone				
a. Okenone		5		
b. *R.g.*-keto carotenoids	1			
4. Isorenieratene				
a. Isorenieratene			2	
b. Chlorobactene			6	
c. *γ*- and *β*-Carotene				1

need reduced sulfur compounds for their growth (okenone in a number of Chromatiaceae, isorenieratene and chlorobactene in *Chlorobium* species). Third, rhodopinal and related cross-conjugated carotenals are formed by anaerobic strains of photosynthetic bacteria only (species of Chromatiaceae, some strains of *Rhodospirillum tenue, Rhodocyclus purpureus*, and anaerobic strains of *Rp. acidophila*).

2. General Pathways

2.1. Formation of the First C_{40}-Hydrocarbon

Carotenoids are usually tetraterpenoids consisting of eight isoprene units. Isoprene is synthesized from acetyl-CoA via β-hydroxy-β-methyl-glutaryl-CoA and mevalonic acid (C_6), as was established in the late 1950's. Decarboxylation of mevalonic acid pyrophosphate yields a C_5-compound, isopentenyl pyrophosphate, four molecules of which combine to yield a C_{20}-intermediate, geranyl-geranyl pyrophosphate. In a tail-to-tail condensation of two geranyl-geranyl pyrophosphate molecules, the basic C_{40}-skeleton is formed. It is now believed that the first C_{40}-intermediate is phytoene. Much of the information that has been obtained about the sequence of reactions (e.g., from labeling and stereochemical experiments and enzyme and cofactor requirement studies) is discussed extensively in several review articles (Goodwin, 1971; Britton, 1971; Davies, 1975). There is still some discussion about the configuration of phytoene, i.e.,

whether it is synthesized in the 1 S-*cis* or all-*trans* form, but there is no doubt that all the more unsaturated carotenoids normally occur as the all-*trans* isomers.

2.2. Desaturation of Phytoene

Although it is not clear yet whether phytoene itself is involved in the biosynthesis of more unsaturated carotenoids, phytoene has been described as a precursor of carotenoids in photosynthetic bacteria by several authors (Liaaen-Jensen *et al.*, 1958, 1961; Liaaen-Jensen, 1963a; Davies, 1970a,b; Malhotra *et al.*, 1970a,b).

The mode of desaturation was first suggested by Porter and Lincoln (1950) from results with mutant strains of tomatoes. They concluded that phytoene is desaturated via phytofluene, ζ-carotene (a symmetrical heptaene, 7,8,7',8'-tetrahydrolycopene), and neurosporene to give lycopene (Fig. 1). In photosynthetic bacteria, however, the sequence of desaturation differs at the heptaene level. Studies with extracts of *Rs. rubrum, Rp. sphaeroides*, and *Rp. globiformis* (Davies, 1970a,b; Schmidt and Liaaen-Jensen, 1973) showed that in purple bacteria, the unsymmetrical isomer of ζ-carotene, 7,8,11,12-tetrahydrolycopene, is the heptaene intermediate. It is very likely that all carotenoids characteristic of photosynthetic bacteria can be regarded as derivatives of either neurosporene or lycopene. There is little experimental evidence for suggesting biosynthetic pathways, but there have been a few kinetic studies of changes in carotenoid com-

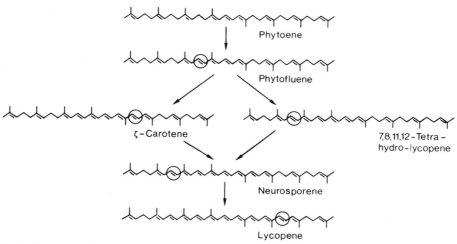

Fig. 1. Desaturation of phytoene to lycopene in higher plants and photosynthetic bacteria. The double bond that appears after each step is circled.

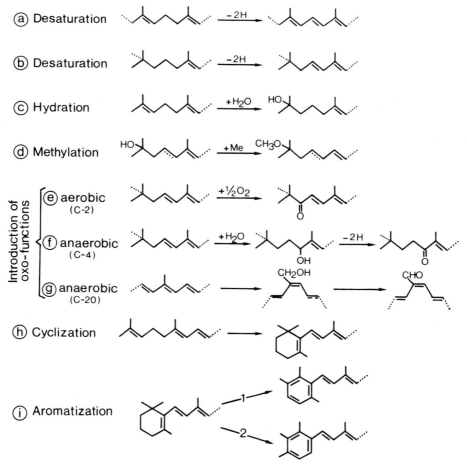

Fig. 2. Types of reactions in the biosynthesis of characteristic carotenoids of photosynthetic bacteria. The letters in Figs. 3–8 and 11 identify these reactions.

position during the synthesis of photopigments in *Rs. rubrum*, *Rp. gelatinosa*, and *Rp. sphaeroides* (Liaaen-Jensen, *et al.*, 1958; Eimhjellen and Liaaen-Jensen, 1964; Cohen-Bazire *et al.*, 1957; Singh *et al.*, 1973a,b). From these studies and from a consideration of the chemical structures of the carotenoids found, a regular sequence of reactions (see Fig. 2) leading from lycopene to spirilloxanthin (see Fig. 3) or from neurosporene to spheroidene (see Fig. 5) was proposed. This sequence includes the successive reaction of: (1) hydration of the terminal isopropylidene group (Fig. 2, reaction type *c*); (2) desaturation at C-3,4 (reaction type *b*); and (3) methylation of the tertiary hydroxyl group (reaction type *d*). It is believed that as a general rule, the uninhibited biosynthetic pathways of most of the carotenoids of photosynthetic bacteria include this sequence of reactions. This hypothesis, together with quantitative comparison of the carotenoid contents of many species and strains, made it possible to suggest specific pathways for the formation of the individual pigments.

All proposed reaction types involved in the synthesis of carotenoids in photosynthetic bacteria are summarized in Fig. 2. Six of them (*a–f*) were pub-lished by Liaaen-Jensen *et al.* (1961) and Liaaen-Jensen (1963a). Following the discovery of more new chemical structures, this scheme was enlarged (Liaaen-Jensen, 1966; Britton, 1971; Schmidt and Liaaen-Jensen, 1973). It was demonstrated by Singh *et al.* (1973a) that the methyl residue in the methoxyl groups arises from *S*-adenosylmethionine. There is nothing known about the mechanisms of the introduction of oxo-groups under anaerobic conditions at the C-4 and C-20 positions. Also, the details of the aromatization step are not yet clear, although a mechanism for this reaction was considered by Liaaen-Jensen (1966). Moshier and Chapman (1973) demonstrated that the aromatic ring in chlorobactene is isoprenoid in origin and that its formation involves the migration of one of the methyl groups. This has not been proved for the synthesis of okenone so far.

3. Special Pathways

3.1. Biosynthesis of Spirilloxanthin via Rhodopin

The so-called "normal spirilloxanthin series" has been the best-known pathway for many years. It

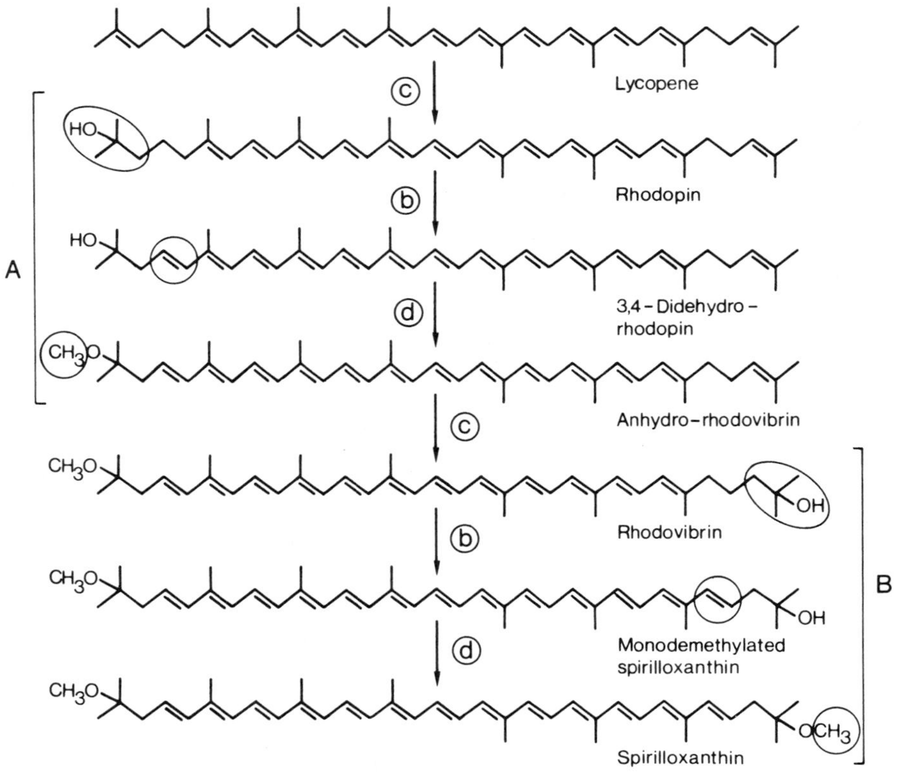

Fig. 3. Postulated scheme of spirilloxanthin biosynthesis (via rhodopin). The new function introduced at each step is circled.

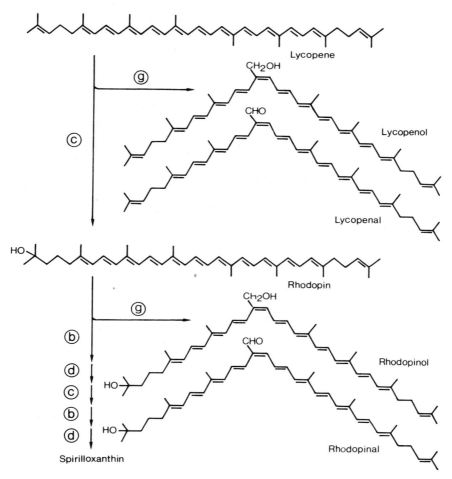

Fig. 4. Postulated scheme of biosynthesis of cross-conjugated carotenals.

functions in a large number of species of purple bacteria (see Table 1). Spirilloxanthin is a symmetrical compound containing methoxyl groups at C-1 and C-1' and additional double bonds in the C-3,4 and C-3',4' positions (Fig. 3).

The synthesis includes several of the reactions described above, which occur sequentially, first on the one half of the molecule (Fig. 3, A) and then on the second half (Fig. 3, B). The intermediates thus are rhodopin, 3,4-dehydrorhodopin, anhydrorhodovibrin, rhodovibrin, and monodemethylated spirilloxanthin. This sequence of reactions and intermediates was first established by Liaaen-Jensen *et al.* (1958, 1961) for *Rs. rubrum*. From analytical data on carotenoid composition in purple bacteria, it is clear that the same pathway occurs in many species. Some of these organisms are unable to perform all reactions on both halves of the molecule, however, and consequently

accumulate intermediates rather than the end product spirilloxanthin (Volk and Pennington, 1950; Goodwin, 1956; Goodwin and Land 1956; Liaaen-Jensen *et al.*, 1961; Conti and Benedict, 1962; Schmidt, 1963; Ryvarden and Liaaen-Jensen, 1964; Pfennig *et al.*, 1965; Schmidt *et al.*, 1965; Schmidt, 1971; Schmidt and Trüper, 1971). Rhodopin is the main carotenoid in many species (*Chromatium vinosum, Chr. violascens,* some strains of *Rp. palustris, Rp. acidophila, Rhodomicrobium vannielii*). Rhodovibrin is the main carotenoid in others (some strains of *Rp. palustris* and *Rs. photometricum*). There are several organisms in which synthesis stops at the level of rhodopin and that therefore accumulate mainly rhodopin together with lycopene in varying proportions (*Thiospirillum jenense, Chr. vinosum* strain 6411, *Rs. molischianum, Rs. fulvum*). In these organisms, dihydroxylycopene is frequently present in considerable amounts. In those strains in which not rhodopin, but a component in the spirilloxanthin or rhodopinal pathway, is the end product, dihydroxylycopene can almost always be detected. This fact may indicate that the enzymes for hydration reactions (reaction type *c*) are much more active than those that catalyze the methylation steps (reaction types *b* and *d*).

3.2. Biosynthesis of Cross-Conjugated Carotenals

The synthesis of rhodopinal ("warmingone series": Schmidt *et al.*, 1965; Pfennig *et al.*, 1968) can be considered to be a branch of the normal spirilloxanthin biosynthetic pathway. Rhodopinal probably derives from rhodopin, which is converted into its oxo-form by replacing the C-20 methyl group by an aldehyde group. The exact mechanism of this oxidation is not yet understood, but it seems very likely that rhodopinol, which is found in most of the pigment extracts, is an intermediate (Fig. 4). Rhodopinal is synthesized only under strictly anaerobic growth conditions. Often, it is found together with spirilloxanthin and its precursors (*Thiocystis violacea*, *Chr. violascens* strain 6111, and some *Rp. acidophila* strains). Mostly, however, spirilloxanthin intermediates occurring later than rhodopin are not found (*Chr. warmingii, Chr. violascens* strain 1313, *Thiodictyon, Rs. tenue, Rhodocyclus purpureus*, one strain of *Rp. palustris*). Especially if some spirilloxanthin is produced, the level of rhodopin in rhodopinal-forming bacteria is remarkably high compared with the amounts of other components of the biosynthetic pathway. A small amount of lycopenal and sometimes lycopenol can be detected in cells that synthesize rhodopinal. Only one strain of *Lamprocystis* has lycopenal as a main carotenoid. Whether lycopenal is a precursor of rhodopinal has not been confirmed. Francis and Liaaen-Jensen (1970) postulated a branched path for each of the derivatives of cross-conjugated carotenals:

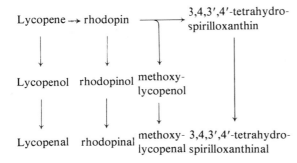

Methoxylycopenal was found in some strains of *Rp. acidophila* (Schmidt, 1971; Schmidt *et al.*, 1971), and

the occurrence of 3,4,3',4'-tetrahydrospirilloxanthinal is so far limited to *Thiocapsa pfennigii* (formerly *Thiococcus* sp. strain *RG3*: Eimhjellen *et al.*, 1967; Aasen and Liaaen-Jensen, 1967; Pfennig *et al.*, 1968).

3.3. Biosynthesis of Spheroidene

In spheroidene-forming organisms, rhodopin is not found, but spirilloxanthin may occur in small quantities. These findings suggest the possibility of an alternative pathway of spirilloxanthin synthesis (Liaaen-Jensen *et al.*, 1961; Liaaen-Jensen, 1963a,b; Eimhjellen and Liaaen-Jensen, 1964). In this case, the reactions for methoxylation of carbon-1 (Fig. 2, reaction types *c*, *b*, and *d*) take place at the level of neurosporene and result in the formation of spheroidene (Fig. 5). In strains of *Rp. capsulata*, spheroidene is the main carotenoid. All other components of the pathway are present only in much smaller amounts, and spirilloxanthin is entirely missing. This indicates that the desaturation step in the 7,8-position is blocked, and hydroxyspheroidene is the end product of synthesis. In cells of *Rp. sphaeroides* and *Rp. gelatinosa*, hydroxyspheroidene can be dehydrogenated in C-7,8, and thus is converted to rhodovibrin, monodemethylated spirilloxanthin, and finally to spirilloxanthin. None of the latter three compounds reaches a high level, but hydroxyspheroidene rather than spheroidene is the main component in many of the investigated strains.

From inhibition experiments with diphenylamine (DPA), which is known to inhibit desaturation in carotenoids, it appears that cells of *Rs. rubrum* are also capable of performing the sequence of methoxylation reactions on the more saturated precursors of lycopene, e.g., phytoene, phytofluene, 7,8,11,12-tetrahydrolycopene, and neurosporene (Davies, 1970a,b; Malhotra *et al.*, 1969, 1970a,b; Davies and Aung Than, 1974). But the methoxylated and hydroxylated compounds are different from those occurring in the normal spirilloxanthin pathway, and very likely their synthesis is caused by the inhibitor. Conclusions from DPA-inhibition experiments should be made with great caution.

If oxygen is introduced into a suspension of *Rp. sphaeroides*, *Rp. gelatinosa*, or *Rp. capsulata*, the methoxylated carotenoids with double bonds at C-3,4 are readily converted into their 2-oxo-derivatives (Goodwin, 1956; Cohen-Bazire *et al.*, 1957; Schneour, 1962a; Eimhjellen and Liaaen-Jensen, 1964). The oxygen of the oxo-group derives directly from the atmosphere, as was shown in experiments with $^{18}O_2$ by Schneour (1962b). The reaction is carried out by a highly active enzyme system (Schneour,

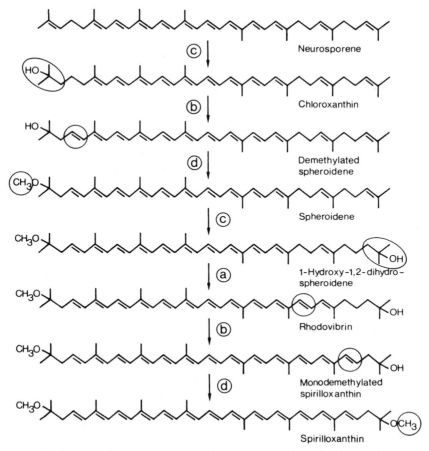

Fig. 5. Postulated alternative scheme for spirilloxanthin biosynthesis (via spheroidene).

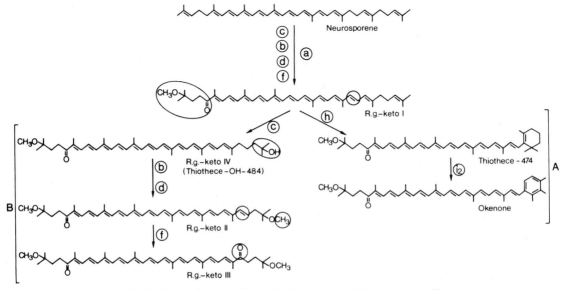

Fig. 6. Postulated scheme of biosynthesis of okenone and *R.g.*-keto carotenoids.

1962a). Spheroidene, hydroxyspheroidene, and rhodovibrin form mono-oxo-derivatives. Spirilloxanthin occurs in its 2,2'-di-oxo-form.

No strain of the species that synthesize the carotenoids of the normal spirilloxanthin series is known to be capable of oxidizing the methoxy-3,4-dehydro members of that pathway. The only organisms provided with this special oxygenating enzyme system so far described are *Rp. sphaeroides*, *Rp. gelatinosa*, and *Rp. capsulata*, indicating that its occurrence is probably linked to the presence of the spheroidene pathway.

3.4. Biosynthesis of Okenone and Oxo-Carotenoids of *Rhodopseudomonas globiformis*

Okenone is a carotenoid with one aromatic end group and with one aliphatic end that is substituted with a tertiary methoxyl group at C-1 and an oxo-function in the C-4 position. It is synthesized only by a number of species of Chromatiaceae. Probable precursors and intermediates in its biosynthetic pathway are found only as minor carotenoids, if at all. So far, very little work has been done on the identification of probable precursors (Schmidt *et al.*, 1963). DPA-inhibited suspensions of *Chr. okenii* contained the normal desaturation series of carotenoids from phytoene to neurosporene. The presence of a very small amount of lycopene is suspected, but has not been rigorously proved. Analyses of the minor carotenoids in extracts of *Thiocystis gelatinosa* (formerly *Thiothece*: Pfennig *et al.*, 1968; Andrewes and Liaaen-Jensen, 1972) revealed some new compounds, which suggest that the aromatic ring is formed via the β-ionone ring precursor, *Thiothece*-474 (Fig. 6, A), and that there probably exists a close relationship to the synthesis of the oxo-carotenoids found in *Rp. globiformis* (Schmidt and Liaaen-Jensen, 1973) (Fig. 6, B). A common compound is the pigment *Thiothece*-OH-484 (or *R.g.*-keto IV), present as about 5% of the total carotenoid content in *Thiocystis* cells. This compound is a side product rather than an intermediate in okenone synthesis.

In *Rp. globiformis* extracts, a series of new aliphatic methoxylated oxo-carotenoids were detected that have not been found in any other organisms before (Schmidt and Liaaen-Jensen, 1973). 7,8,11,12-Tetrahydrolycopene and neurosporene were always present, but lycopene was not detected. Instead, the methoxylated 4-oxo-compound, *R.g.*-keto I, could be isolated from the mixture in small amounts. The sequence of the reactions characteristic of the normal spirilloxanthin series occurs to the second half of the *R.g.*-keto I molecule to yield *R.g.*-keto II (Fig. 6, B). Finally, an oxo-group is introduced in the C-4 position, which results in the symmetrical end product *R.g.*-keto III.

The occurrence of *Thiothece*-OH-484 in both *Thiocystis gelatinosa* and *Rp. globiformis*, as well as the identity of the structure of the aliphatic end groups of okenone and *R.g.*-keto carotenoids, suggest that synthesis of both types of carotenoids utilizes the same precursors (Fig. 6). Partial inhibition of ring formation in cells of the okenone-synthesizing species *Chr. minus* by adding nicotine to the medium resulted in the accumulation of *R.g.*-ketos I–IV in addition to okenone (Schmidt, unpublished). Contrary to the oxidation of spheroidene (Fig. 2, reaction type *e*), the introduction of oxo-functions at C-4 (Figs. 2 and 6, reaction type *f*) in these carotenoids is strictly anaerobic. Nothing is known about the mechanism. It is likely to involve successive hydration and dehydrogenation reactions (Liaaen-Jensen, 1963a) (Fig. 2), but the corresponding carotenols have never been isolated. In contrast, the intermediates lycopenol and rhodopinol have been isolated in rhodopinal-synthesizing organisms.

3.5. Biosynthesis of Isorenieratene and Chlorobactene

Aromatic carotenoids with 1,2,5-trimethylphenyl end groups are found within the photosynthetic bacteria only in the Chlorobiaceae. With respect to their carotenoid biosynthesis, the Chlorobiaceae can be divided into two main groups: the brown species, which are able to cyclize both ends of the carotenoid molecule (isorenieratene synthesis, Fig. 7, A) and the green sulfur bacteria, which cyclize one end only (chlorobactene synthesis, Fig. 7, B).

In green *Chlorobium* species, all compounds of the Porter–Lincoln series (see Fig. 1) and their hydroxy-derivatives were detected, including rhodopin (no exact analysis of the structure of the heptaene intermediate has yet been made). There is always a considerable quantity of γ-carotene present, showing that chlorobactene is formed via this carotenoid. In general, however, chlorobactene is the predominant product (70–90%: Liaaen-Jensen *et al.*, 1964; Schmidt and Schiburr, 1970).

Brown Chlorobiaceae (formerly *Phaeobium*) produce isorenieratene as the main carotenoid pigment, but almost the same amount of β-isorenieratene is accumulated. There is also some β-carotene and chlorobactene present (Liaaen-Jensen, 1965). These data allow the conclusion that both pathways are very closely related, isorenieratene being probably the final product of the pathway.

Moshier and Chapman (1973) demonstrated that the aromatic ring in chlorobactene originates from an isoprenoid precursor, and that a migration of one

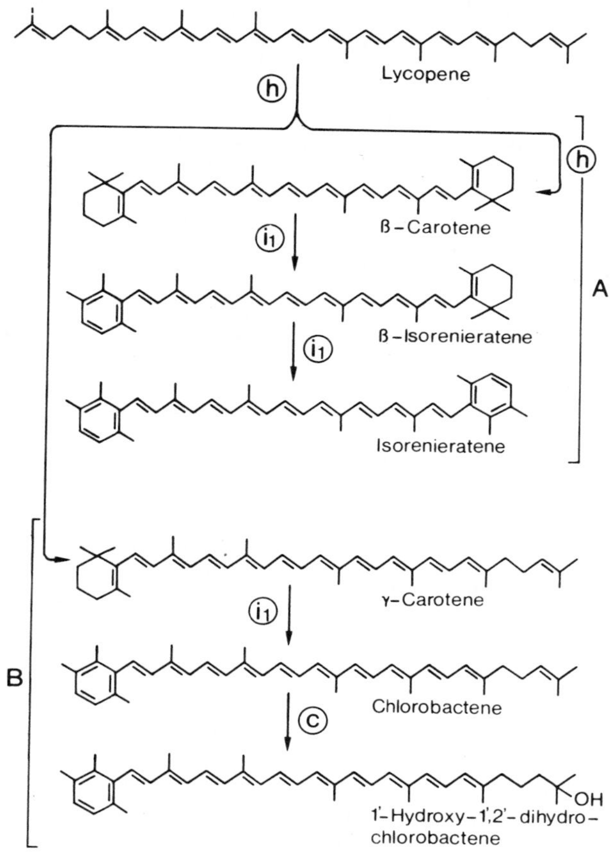

Fig. 7. Postulated scheme of biosynthesis of isorenieratene and chlorobactene.

methyl group from the C-1 to the C-2 position takes place during aromatization (Fig. 2, reaction types h and i_1). There is no evidence as to whether lycopene or neurosporene is the direct precursor of the cyclic carotenoids.

3.6. Biosynthesis of γ-Carotene and β-Carotene and Their Oxo-Derivatives

γ-Carotene and β-carotene are not typical end products of carotenoid biosynthesis in the photosynthetic bacteria. These pigments are characteristic of cyanobacteria, algae, fungi, higher plants, and nonphotosynthetic bacteria. There are, however, a few species of photosynthetic bacteria that produce a relatively small amount of β-carotene (*Rhodomicrobium vannielii*: Volk and Pennington, 1950; Conti and Benedict, 1962; Ryvarden and Liaaen-Jensen, 1964; and some strains of *Rp. acidophila*:

Schmidt, 1971; Schmidt *et al.*, 1971). In the gliding filamentous bacterium *Chloroflexus aurantiacus*, on the other hand, γ- and β-carotene are the main carotenoid pigments (Fig. 8). Oxo-groups are introduced into the C-4 position of the β-ionone ring under aerobic growth conditions, and glucosides of γ-carotene derivatives are also present (Halfen *et al.*, 1972). The 4-oxo-β-rings are typical of the pigments of cyanobacteria, whereas the tertiary hydroxyl group of 1'-hydroxy-1',2'-dihydro-γ-carotene is characteristic of the photosynthetic bacteria. *Chloroflexus aurantiacus* contains bacteriochlorophylls *a* and *c*, and physiological relationships to the Chlorobiaceae are also indicated. Pierson and Castenholz (1974) therefore suggested that this new species is related to the Chlorobiaceae. It may be possible to regard the biosynthesis of γ- and β-carotene in this species as an incomplete isorenieratene pathway. This possibility is discussed more fully in Section 4.1.

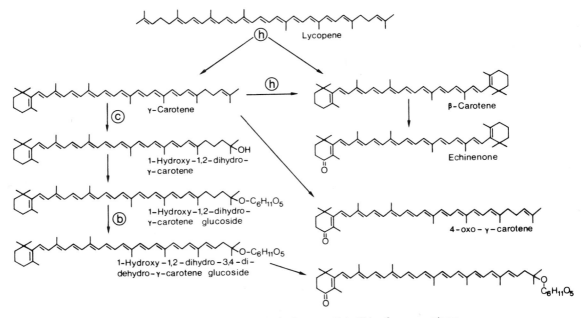

Fig. 8. Postulated scheme of biosynthesis of carotenoids in *Chloroflexus aurantiacus*.

3.7. Biosynthesis of Carotenoid Glycosides

The occurrence of carotenoid glycosides in photosynthetic bacteria is very unusual. They are known to be present in many pigmented nonphotosynthetic bacteria and in cyanobacteria. In all known cases, the hexosyl portion forms a glycosidic linkage with the hydroxyl group of the carotenoid. The known glucosides of rhodopin, rhodopinal, and γ-carotene derivatives are all of the β-type. The first photosynthetic bacterium in which rhodopinal glucoside or rhodopin glucoside (Fig. 9) or both were found in large quantities (30–60% of the total carotenoid content) was *Rp. acidophila* (Schmidt, 1971; Schmidt *et al.*, 1971). Halfen *et al.* (1972) reported that glucosides of γ-carotene and its derivatives (see Fig. 8) constitute up to 34% of the total carotenoids in *Chloroflexus aurantiacus*. Probably more species of photosynthetic bacteria synthesize carotenoid glycosides, although not in such large proportions. In extracts of *Rs. fulvum*, a rhodopin glucoside was found in amounts representing 6–10% of the total carotenoids (Schmidt, unpublished).

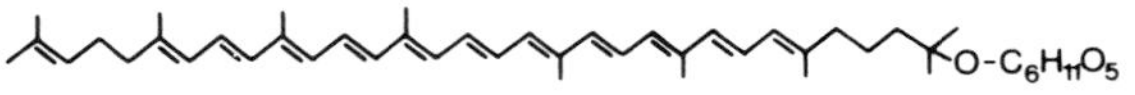

Fig. 9. Structure of rhodopin glucoside.

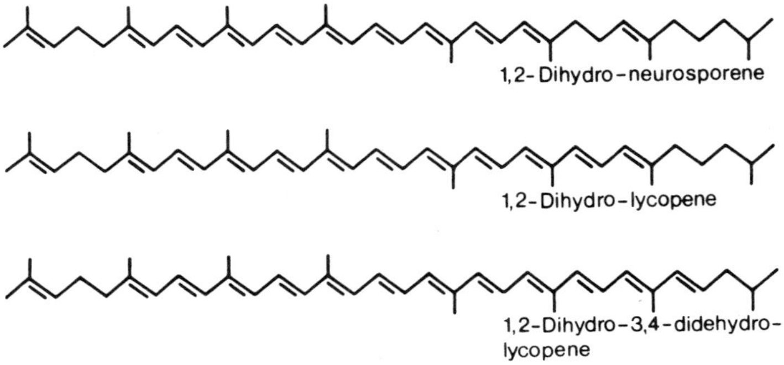

Fig. 10. Structures of the main carotenoids of *Rhodopseudomonas viridis*.

3.8. Biosynthesis of 1,2-Dihydrocarotenoids

Detailed studies on the carotenoid pigments of *Rp. viridis* revealed a series of new compounds that have never been found in other organisms (Malhotra *et al.*, 1970*c,d*; Britton, 1971). Characteristic of these special carotenoids is the saturated terminal isoprenoid unit either on one or on both halves of the molecule. It seems that the cells are able to perform this saturation reaction on the carotenoids at all levels of synthesis, beginning with phytoene and including lycopene. 1,2-Dihydroneurosporene, 1,2-dihydrolycopene, and 1,2-dihydro-3,4-didehydrolycopene are the main carotenoids (Fig. 10). It was also found that the heptaene of the desaturation sequence is a mixture of the symmetrical compound ζ-carotene and 7,8,11,12-tetrahydrolycopene, its unsymmetrical isomer.

4. Regulation

4.1. Regulation of the Different Pathways

So far, no work has been done at the enzymic level concerning the synthesis of carotenoids in photosynthetic bacteria. All suggestions about the sequence of specific reactions are based on chemical structures and properties of the pigments, and on some studies on endogenous carotenoid synthesis. For this purpose, in most cases, strains of Rhodospirillaceae were grown in the presence of an inhibitor of normal carotenoid synthesis such as DPA or nicotine (Goodwin and

Osman, 1954; Goodwin, 1956; Liaaen-Jensen *et al.*, 1958, 1961; Davies, 1970*a*; McDermott *et al.*, 1973). When the inhibitor is removed and the cells are resuspended in buffer and brought back to anaerobic conditions in the light, the sequence of components of the pathway can be postulated by following the successive appearance and disappearance of the individual carotenoids. Suggestions concerning carotenoid biosynthesis can also be based on a comparative quantitative analysis of the carotenoids present in all available strains of one species. Also, incorporation experiments would yield clear evidence, but hardly any work has been done using isotopes to determine kinetic parameters.

To understand the biosynthesis of individual carotenoids from a common C_{40}-precursor, it is necessary to consider the carotenoid compound as a molecule consisting of two halves that originates from the tail-to-tail condensation of two identical C_{20} units (see Section 2.1). It is possible that the enzymes responsible for the substitution and cyclization reactions distinguish between the halves. Liaaen-Jensen *et al.* (1961) were the first to explain the accumulation of either rhodopin, anhydrorhodovibrin, rhodovibrin, or spirilloxanthin in various strains of Rhodospirillaceae by the probable existence of two different enzymes, catalyzing the dehydration reactions (type *b*) specifically on either half of the molecule. A "half-molecule model" was developed for zeaxanthin biosynthesis of *Flavobacterium* by McDermott *et al.* (1974). These authors postulate an enzyme complex containing two reaction centers, each of which is

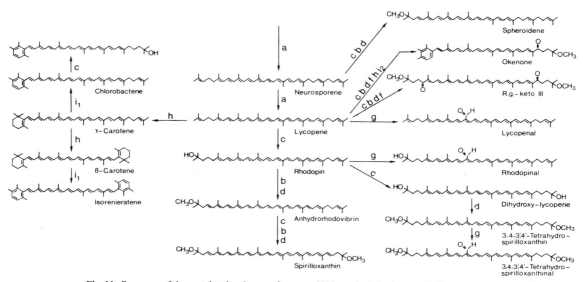

Fig. 11. Summary of the postulated pathways of carotenoid biosynthesis in photosynthetic bacteria.

Tables 2. Rhodospirillaceae: Spirilloxanthin Series*

Organism	Strain	Neurosporene (70)	Lycopene (67)	Rhodopin (50)	Anhydrorhodovibrin (26)	Rhodovibrin (35)	OH-Spirilloxanthin (34)	Spirilloxanthin (37)	OH-Rhodopin (58)	Unidentified polar compounds	β-Carotene (61)	Ref. No.[a]
Rhodospirillum rubrum	NR[b]				≈5 (spanning)			95				1
	1.1.1				≈5 (spanning)			95				2
	S 1			1	2	6		91				3
	9960						11	89				3
	3860					2	39	57				3
	Ha			12	5	11		69				3
	2461				13	8	2	77				3
	6461		1	3	30	29	1	45				3
	1785-1			5	7	29	7	51				3
Rhodospirillum molischianum	Delft		25	75								4
	3660		20	60						11		3
	N131	1	18	77						3		3
Rhodospirillum photometricum	Delft		7	28	20	45						4
Rhodospirillum fulvum	KK		51	48						1		5
	2860		19	73						2	6	3
	1360		15	77						2	6	3
	B 147	1	17	78						1	3	3
	10060	1	16	75						3	5	3
Rhodopseudomonas palustris	8252		14		16	23		59				4
	8289		15	19	16	14	16	21				4
	ATH 2.1.6		2			3	73	15	7			3
	1850		4	40	10		12	34				3
Rhodomicrobium vannielii	Douglas		25	65				8		1	2	6
	Douglas		7	81				12			1	6
	Douglas		9	75	1	2		6		6		6

Carotenoids (% of total)

* Tables 2–10 comprise a presentation of quantitative data on the carotenoid composition of the phototrophic bacteria; see Section 5, p. 747.

[a] References: (1) Goodwin and Sissins (1955); (2) Polgar *et al.* (1944); (3) K. Schmidt (unpublished); (4) Goodwin (1956); (5) Pfennig *et al.* (1965); (6) Ryvarden and Liaaen-Jensen (1964).

[b] (NR) Not recorded.

Table 3. Rhodospirillaceae: Rhodopinal Series

Organism	Strain	Lycopene (67)	Rhodopin (50)	Anhydrorhodovibrin (26)	Rhodovibrin (35)	OH-Spirilloxanthin (34)	Spirilloxanthin (37)	OH-Rhodopin (58)	Lycopenal (1)	Rhodopinol (57)	Rhodopinal (2)	Rhodopin glucoside (51)	Rhodopinal glucoside (3)	β-Carotene (61)	Ref. No.[a]
Rhodospirillum tenue	Eu 1[b]	41	17	20	6	2	14								1
	2761[b]	37	16	32	7		8								1
	3661	12	23							10	5	50			2
	3760	9	28							11	8	44			2
	3761	76	5							1	2	16			2
	P 16	5	18							10	8	59			2
Rhodocyclus purpureus	6770	2	22							7	6	63			2
Rhodopseudomonas acidophila	7050	5	21				5				2	25	40	0.5	2
	3251	13	47			1	1	1	1	1	5	10	20		2
	3750	4	16	1	4		4			1	3	28	38	0.4	2
	7150	4	29				6	1				59			2
	7250	26	37	9	4	4	15	1							2
	7550	8	42				8					40			2
	7750	8	46				15	1				30			2
	2751	8	42			1	3	5				40			2

[a] References: (1) K. Schmidt (unpublished); (2) Schmidt (1971).
[b] This strain is listed in this table even though it accumulates pigments of the spirilloxanthin series, rather than of the rhodopinal series. This exception is explained in Section 4.1.

responsible for the substitutions on one half of lycopene.

These concepts can form the basis for an explanation of the presence, in photosynthetic bacteria, of such a great variety of carotenoid compositions and individual pathways, based mainly on a limited number of reaction types (see Fig. 2). They can also account for a probable relationship among the different pathways (Fig. 11). The spirilloxanthin series is regarded as the central pathway, and its sequence of reaction steps should be considered as the basic sequence characteristic of phototrophic bacteria: (1) hydration (reaction type *c*); (2) desaturation at C-3,4 (reaction type *b*); (3) methylation of the hydroxyl group (reaction type *d*) (Figs. 2 and 3). Enzyme systems performing different reactions may compete with the normal spirilloxanthin synthesis, however, thus leading to deviation of the pathways and the production of the different individual carotenoids.

The only species known to possess an enzyme system that catalyzes the methoxylation of both halves of the molecule very quickly is *Rs. rubrum*. This leads to the formation of spirilloxanthin as about 90–100% of the total carotenoid in the cells. All other species that synthesize the carotenoids of this series contain much less of the end product spirilloxanthin. Many of the purple bacteria have rhodopin, anhydrorhodovibrin, or rhodovibrin as a predominant carotenoid. A few species (*Thiospirillum jenense, Chr. vinosum* strain 6411) do not synthesize any of the compounds following rhodopin, and some (*Rs. fulvum, Rs. molischianum*, Pfennig *et al.*, 1965), to a limited extent, introduce a second hydroxyl group into rhodopin to give the symmetrical compound 1,1′-dihydroxy-1,2,1′,2′-dihydrolycopene.

Table 4. Rhodospirillaceae: Spheroidene Series

Organism	Strain	Neurosporene (70)	Chloroxanthin (52)	Spheroidene (28)	OH-Spheroidene (36)	Spheroidenone (23)	OH-Spheroidenone (24)	Anhydrorhodovibrin (26)	Rhodovibrin (35)	OH-Spirilloxanthin (34)	Spirilloxanthin (37)	2,2′-Diketospirilloxanthin (20)	Ref. No.[a]
Rhodopseudomonas	NR[b]			65		35							1
sphaeroides	NR[b]			70		30							2
	ATH 2.4.1			90		10							3
	1760-1	2		90	4						4		4
Rhodopseudomonas	NR[b]	1	3	27	42	2	5	4	1	1	14		5
gelatinosa	ATH 2.2.1			18	69						13		6
	8290			15	68						18		6
	8290			10	64	8					16	1	7
Rhodopseudomonas	8254			83	4	11	2						7
capsulata	Kb 1	1		87	5	4	2				1		4

[a] References: (1) van Niel (1947); (2) Goodwin *et al.* (1955); (3) Schneour (1962a); (4) K. Schmidt (unpublished); (5) Liaaen-Jensen (1963a); (6) Eimhjellen and Liaaen-Jensen (1964); (7) Goodwin (1956).

[b] (NR) Not recorded.

These results indicate that reaction types *b–d* (Figs. 2 and 3) function well mostly only on one half of the C_{40}-skeleton, completing the methoxylation only at C-1. This results in the accumulation of anhydrorhodovibrin. When the C-1′ position is hydroxylated readily, but further enzyme activities are low, mainly rhodovibrin is formed. In cells that accumulate rhodopin (or rhodovibrin), the desaturation step (reaction type *b*) may be blocked (or limiting), which perhaps prevents the methylation of the hydroxyl

Table 5. Rhodospirillaceae: *R.g.*-Keto Carotenoids[a]

Organism	Strain	Asymmetrical ζ-carotene (73)	Neurosporene (70)	*Thiothece* OH-484 (15)	*R.g.*-keto II (17)	*R.g.*-keto III (18)
Rhodopseudomonas globiformis	7950	3		5	80	12

[a] Data from Schmidt and Liaaen-Jensen (1973).

Table 6. Chromatiaceae: Spirilloxanthin Series

Organism	Strain	Carotenoids (% of total)								Ref. No.[a]
		Neurosporene (70)	Lycopene (67)	Rhodopin (50)	3,4-Dehydrorodopin (49)	Anhydrorhodovibrin (26)	Rhodovibrin (35)	OH-Spirilloxanthin (34)	Spirilloxanthin (37)	
Chromatium vinosum	D		7	48	1	13	4	1	27	1
	1312		13	4		27			55	1
	1611		6	76		2			15	1
	6411	12	66	23						1
	6412		15	52		11			16	1
	2811		6	62	10	5		2	14	1
Thiocapsa roseopersicina	Lascelles		1	3		1	1	1	93	1
	1711								100	1
	1813			4		1	3	2	90	1
	6311		1	4			4	2	89	1
	6612		1	3		3	2	1	91	1
Amoebacter roseus	6611					5			95	1
Amoebobacter pendens	1314			4		11		3	82	1
Ectothiorhodospira mobilis	8112		6	19			9	5	61	2
	8113		4	13			6	4	73	2
	8115		5	19			2		72	2
	8815		5	10			2	6	76	2
Ectothiorhodospira halophila	SL 1						1	4	88	2
Ectothiorhodospira shaposhnikovii	N1		7	18			3	2	70	2
Thiospirillum jenense	1112		12	88						1

[a] References: (1) Schmidt *et al.* (1965); (2) Schmidt and Trüper (1971).

groups. Although it is known from DPA-inhibition experiments that *Rs. rubrum* and *Rp. sphaeroides* can methoxylate all the more saturated precursors of lycopene without preceding desaturation in the C-3,4 position, there is only one species known that can synthesize 3,4,3′,4′-tetrahydrospirilloxanthin under normal growth conditions (*Thiocapsa pfennigii*: Aasen and Liaaen-Jensen, 1967). This bacterium has probably lost completely the ability to perform reaction type *b*.

During rhodopinal synthesis (see Figs. 4 and 11), the introduction of the oxo-function at C-20 is due to a very active enzyme system that prevents or reduces the methoxylation reactions that give rise to spirilloxanthin. Some strains, which normally belong to the group of rhodopinal-synthesizing bacteria, obviously have lost the enzyme system that catalyzes reaction type *g*. From these cells, only members of the normal spirilloxanthin series can be isolated, but they accumulate rhodopin and some 1,1′-dihydroxy-1,2,1′,2′-

Table 7. Chromatiaceae: Rhodopinal Series

Organism	Strain	Carotenoids (% of total)												Ref. No.[a]
		Chloroxanthin (52)	Lycopene (67)	Rhodopin (50)	Anhydrorhodovibrin (26)	Rhodovibrin (35)	Spirilloxanthin (37)	Lycopenol (56)	Lycopenal (1)	Rhodopinol (57)	Rhodopinal (2)	3,4,3',4'-Tetrahydrospirilloxanthin (39)	Tetrahydrospirilloxanthinal (6)	
Chromatium warmingii	1113		3	13					1	10	73			1
	6512		13	17					1	21	48			1
	1311	9	7	15					5	16	48			1
Cromatium violascens	Lascelles[b]		8	77		2	13							1
	6111		8	48	5				1	18	20			1
	6111		5	37			11		1	17	29			1
	1313		3	10					5	16	48			1
Thiocystis violacea	2311		2	12					5	18	63			1
	2311		4	20			1		4	36	36			1
Thiodictyon bacillosum	1814		4	8					3	4	81			1
Thiodictyon elegans	1511		12	12						14	46	16		1
	3011		4	23					2	15	54			2
Lamprocystis roseopersicina	3012		4					27	66	3				2
Thiocapsa pfennigii	Nidelven											95	5	3

[a] References: (1) Schmidt *et al.* (1965); (2) Pfennig *et al.* (1968); (3) Eimhjellen *et al.* (1967).

[b] This strain is listed in this table even though it accumulates pigments of the spirilloxanthin series, rather than of the rhodopinal series. This exception is explained in Section 4.1.

tetrahydrolycopene, rather than spirilloxanthin. This indicates that there is not only a competition of reaction types *g* with *b* and *d*, but also different activities of the enzyme systems performing the individual reactions.

In the biosynthesis of spheroidene (see Figs. 5 and 11), the reaction sequence of methoxylation on one half of the neurosporene molecule precedes the desaturation at C-7,8. In most cases, the spheroidene formed is then first hydroxylated before the desaturation reactions (types *a* and *b*) can take place to give rhodovibrin, monodemethylated spirilloxanthin, and finally spirilloxanthin. These compounds are present only as minor carotenoids, and hydroxyspheroidene is the main pigment. This may indicate

that the enzyme activities for reactions on the second half of the molecule, including the desaturation at C-7',8', are greatly reduced. Thus, no lycopene and rhodopin can be formed because all the neurosporene is readily converted to spheroidene.

The formation of the typical okenone arrangement (see Figs. 6 and 11) during the synthesis of okenone and *R.g.*-keto carotenoids (1-methoxy-1,2-didehydro-4-keto-; reaction types *c*, *b*, *d*, and *f*) happens so quickly to one half of the molecule that no early intermediates can be found. For the formation of *R.g.*-keto carotenoids, the same sequence of reaction types must occur to the other half of the molecule as well to give the symmetrical end product (see Fig. 6, B). Enzyme activities here are much lower. Consequently, some of

Table 8. Chromatiaceae: Okenone Series

Organism	Strain	Thiothece-474 (13)	Thiothece-484 (14)	Thiothece-OH-484 (15)	Okenone (12)	Ref. No.[a]
Chromatium okenii	1111				100	1
	6511				100	1
	1811				100	1
Chromatium weissei	1812				100	1
	6211				100	1
	2111				100	1
Chromatium minus	1211				100	1
Thiocystis gelatinosa	2611	3	5	10	82	2
	2611	1	1	4	94	3

[a] References: (1) Schmidt *et al.* (1965); (2) Pfennig *et al.* (1968); (3) Andrewes and Liaaen-Jensen (1972).

the intermediates are accumulated. The hydroxy compound *R.g.*-keto IV is rapidly converted to *R.g.*-keto II, which is the main product of this type of biosynthesis. The introduction of a second oxo-group in the C-4′ position is obviously a slow reaction, perhaps indicating that the affinity of the second half of the carotenoid molecule for the enzyme complex is not very strong. Unlike okenone-synthesizing bacteria, *Rp. globiformis* has not developed the capability of cyclizing the carotenoid precursors (reaction type *h*).

Ring formation and the subsequent aromatization (reaction types i_1 and i_2) seem to be limited to species of Chlorobiaceae and Chromatiaceae the growth of which is dependent on the presence of reduced sulfur compounds. The enzyme system responsible for the formation of the 1,2,3-trimethylphenyl end group in okenone is of a very high activity. Under normal growth conditions, the ring formation precedes the methoxylating reaction steps that could occur to the second half of the molecule as in *Rp. globiformis*. In most of the okenone-synthesizing cells, hardly any compound other than okenone is found. The occurrence of Thiothece-OH-484 in *Thiocystis gelatinosa* extracts shows, however, that there still exists an enzyme system that promotes the formation of carotenoids of the *R.g.* keto-series in okenone-containing strains as well. Partial inhibition of the ring-formation steps by nicotine supports this hypothesis: reaction types *c, b, d,* and *f* are then carried out on the second half of the *R.g.*-keto I molecule, instead of reaction types *h* and i_2 (see Figs. 6 and 11), giving rise to the formation of *R.g.*-keto carotenoids.

The relationship between the carotenoids in brown and green species of Chlorobiaceae is obvious. The deviation of the individual pathways can be related to differences in enzyme affinity for the carotenoid precursors as well. It seems that green species have lost the ability to cyclize the aliphatic end of γ-carotene (see Figs. 7, B, and 11, reaction types *h* and i_1). Instead, these organisms have developed a high activity of the hydration step (reaction type *c*). Thus, hydroxy-compounds of all members of the biosynthetic path are found in pigment extracts of green *Chlorobium* strains. Brown species, on the contrary, possess a very active cyclizing system (reaction type *h*), but aromatization of the second half of β-carotene (reaction type i_1) is somewhat slower, resulting in the accumulation of β-isorenieratene (Figs. 7, A, and 11).

Table 9. Chlorobiaceae: Chlorobactene and Isorenieratene Series

Organism	Strain	Carotenoids (% of total)														Ref. No.[b]
		Phytoene (80)	Phytofluene (78)	ζ-Carotene (74)	Neurosporene (70)	Lycopene (67)	Rhodopin (50)	Unidentified compound[a]	Pro-γ-carotene-like compound	γ-Carotene (63)	Chlorobactene (66)	OH-Chlorobactene (48)	β-Carotene (61)	β-Isorenieratene (62)	Isorenieratene (65)	
Chlorobium limicola	L1							1			97	2				1
f. thiosulfatophilum	6430			3	3						93	1				2
	6130	33							2	10	55					2
	1630									4	94	2				2
	6230	1	1					1	2	18	74					2
	6131	2	1			2	2			4	86	1				2
	2230	4	2		2	3			1		81	5				2
	2030	1	1		11				1	6	82	1				2
	1430	8	1		1					15	72	1				2
	3331				1	9				2	87	1				2
Chlorobium	8346		15	2	5	1	2	2		2	66	7				2
vibrioforme	8327		1	3	3						90					2
f. thiosulfatophilum	1930	2	1		4	7					77	12				2
Chlorobium	DSIR							1			96	3				1
vibrioforme	6030					3	1	1			90	5				2
	2630					1		1			92	4				2
	6132	0.5	0.5			17	1	4	5	5	62	3				2
	9930					1	1	1			93	2				2
Pelodictyon	2530	1						1	9	11	72	1				2
luteolum	2532					1	1				92	2				2
Chlorobium	2531												1	33	66	3
phaeovibrioides	2631										6		7	38	49	3
	2632										6		4	47	43	3
Chlorobium phaeobacterioides	2430										5		1	14	80	3

[a] Absorption maxima: 415 nm, 432 nm, and 455 nm.
[b] References: (1) Liaaen-Jensen *et al.* (1964); (2) Schmidt and Schiburr (1970); (3) Liaaen-Jensen (1965).

Cells of the new species *Chloroflexus aurantiacus* are not able to perform aromatization steps typical of Chlorobiaceae. That they are not is not surprising, because the growth of *Chloroflexus aurantiacus* is not dependent on reduced sulfur compounds. Carotenoid synthesis by this species therefore results in the accumulation of γ- and β-carotene and their derivatives. Like green *Chlorobium* strains, *Chloroflexus aurantiacus* also tends to hydroxylate carbon-1 of the aliphatic ends.

Table 10. Chloroflexaceae: γ- and β-Carotene

Strain	Carotenoids (% of total)									Ref. No.[a]
	γ-Carotene (63)	retro-Dehydro-γ-carotene (64)	OH-γ-Carotene (45)	OH-γ-Carotene glucoside (46)	3',4'-Didehydro-OH-γ-carotene-glucoside (47)	4-Oxo-3',4'-didehydro-OH-γ-carotene glucoside (11)	4-Oxo-γ-carotene (10)	β-Carotene (61)	Echinenone (9)	
Y-400	58	1	5	11			2	23		1
244-3	57	1	3	12			0.5	23		1
OK-70-fl	50	1	4	17				29		1
J-10-fl[b]	23	2	4	6	23	4	5	28	4	2

[a] References: (1) K. Schmidt (unpublished); (2) Halfen *et al.* (1972).

[b] The higher percentage of oxygenated carotenoids and the increase in the amount of β-carotene compared with γ-carotene in strain J-10-fl may be due to different culture conditions.

It is remarkable that none of the Chlorobiaceae has developed the capability of methylating the tertiary hydroxyl group (including reaction types *b* and *c*), which is the most characteristic reaction type of purple bacteria.

4.2. Regulation by Environmental Factors

Carotenoid synthesis is greatly influenced by light and oxygen. Also, different kinds of nutrients may change the composition of carotenoids in the cells of some photosynthetic bacteria.

From experiments with facultative anaerobic Rhodospirillaceae, it is clear that light intensities and oxygen concentrations regulate mainly the synthesis of photosynthetic membranes. Because these membranes are the site of location of all photopigments, including carotenoids, the control of membrane synthesis causes the regulation of pigment synthesis (Cohen-Bazire and Kunisawa, 1960; Cohen-Bazire, 1963; Lascelles, 1968). Facultatively aerobic strains of *Rp. acidophila* and *Chloroflexus aurantiacus*, however, synthesize carotenoids on a large scale under aerobic growth conditions in the dark, as well as in the light. In the case of *Rp. acidophila*, the carotenoid formed is rhodopin glucoside, which under these circumstances may be localized in the cytoplasmic membrane and act as a protector against photooxidation (Schmidt, unpublished). Cells that are brought back to conditions of photosynthesis start synthesizing carotenoids of the normal spirilloxanthin series at a rate that parallels bacteriochlorophyll and thylakoid membrane synthesis.

Carotenoid composition can also be changed as a result of nutrient supply. That it can be was shown for the biosynthesis of rhodopinal in *Rhodopseudomonas acidophila* strain 7050 (Heinemeyer and Schmidt, unpublished). With fumarate and malate as a carbon source, rhodopin and its glucoside are the predominant carotenoids, whereas pyruvate and succinate promote rhodopinal synthesis at the expense of rhodopin content. Also, the source of nitrogen and the light intensity influence the rhodopin/rhodopinal ratio. These phenomena also await experimental analysis on the other special pathways of carotenoid biosynthesis.

5. Quantitative Data on the Carotenoids of Phototrophic Bacteria

Most of the available quantitative data on the carotenoid composition of the phototrophic bacteria are presented in Tables 2–10. The tables are arranged as follows:

Table 2. Rhodospirillaceae: Spirilloxanthin Series
Table 3. Rhodospirillaceae: Rhodopinal Series
Table 4. Rhodospirillaceae: Spheroidene Series
Table 5. Rhodospirillaceae: *R.g.*-Keto Carotenoids
Table 6. Chromatiaceae: Spirilloxanthin Series

Table 7. Chromatiaceae: Rhodopinal Series

Table 8. Chromatiaceae: Okenone Series

Table 9. Chlorobiaceae: Chlorobactene and Isorenieratene Series

Table 10. Chloroflexaceae: γ- and β-Carotene

Notes:

1. Because of lack of data, the following organisms are not included: *Rp. sulfidophila, Rp. sulfoviridis, Rp. viridis, Chr. buderi, Chr. gracile, Chr. minutissimum, Thiopedia rosea, Thiosarcina rosea, Ancalochloris perfilievii, Clathrochloris sulfurica, Pelodictyon clathratiforme, P. phaeum,* and *Prosthecochloris aestuarii.*

2. Organisms are named in accordance with the taxonomy and nomenclature outlined in Chapter 2. Occasionally, the names are not those used by the original investigators; in such cases, the strain designations provide adequate identification.

3. Carotenoids present in trace amounts (less than 0.4% of total carotenoids) are not included.

4. The numbers in parentheses following the names of the carotenoids are the serial numbers of the compounds in Chapter 12, Table 1. This table gives, for each pigment, its rational name and an abbreviated structural formula.

5. Attention is drawn to the variations in amounts of pigments among different strains of the same species and also among different batches of the same strain (probably because of variations in culture conditions).

6. Data from cultures older than 6 days are not included because of the possibility of artifacts in such moribund material.

6. Conclusion

The great variety of chemical structures of carotenoids that result from the many different biosynthetic reaction types shows that photosynthetic bacteria developed highly specific enzyme systems during evolution. Enzymic and genetic experiments are required for an adequate understanding of the physiological background of such elaborate syntheses.

ACKNOWLEDGMENTS

I wish to thank Dr. G. Britton, Dr. B. H. Davies, Professor H. Douthit, and Professor S. Liaaen-Jensen for reading the manuscript.

7. References

Aasen, A. J., and Liaaen-Jensen, S., 1967, Bacterial carotenoids. XXI. Synthesis of 3,4,3',4'-tetrahydro spirilloxanthin, *Acta Chem. Scand.* **21**:371.

Andrewes, A. G., and Liaaen-Jensen, S., 1972, Bacterial carotenoids. XXXVII. Carotenoids of Thiorhodaceae. 9. Structural elucidation of five minor carotenoids from *Thiothece gelatinosa, Acta Chem. Scand.* **26**:2194.

Britton, G., 1971, General aspects of carotenoid biosynthesis, in: *Aspects of Terpenoid Chemistry and Biochemistry* (T. W. Goodwin, ed.), pp. 255–289, Academic Press, London and New York.

Cohen-Bazire, G., 1963, Some observations on the organization of the photosynthetic apparatus in purple and green bacteria, in: *Bacterial Photosynthesis* (H. Gest, A. San Pietro, and L. P. Vernon, eds.), pp. 89–110, Antioch Press, Yellow Springs, Ohio.

Cohen-Bazire, G., and Kunisawa, R., 1960, Some observations on the synthesis and function of the photosynthetic apparatus in *Rhodospirillum rubrum, Proc. Natl. Acad. Sci. U.S.A.* **46**:1543.

Cohen-Bazire, G., Sistrom, W. R., and Stanier, R. Y., 1957, Kinetic studies of pigment synthesis by non-sulfur purple bacteria, *J. Cell. Comp. Physiol.* **49**:25.

Conti, S. F., and Benedict, C. R., 1962, Carotenoids of *Rhodomicrobium vannielii, J. Bacteriol.* **83**:929.

Davies, B. H., 1970a, A novel sequence for phytoene dehydrogenation in *Rhodospirillum rubrum, Biochem. J.* **116**:93.

Davies, B. H., 1970b, Alternative pathways of spirilloxanthin biosynthesis in *Rhodospirillum rubrum, Biochem. J.* **116**:101

Davies, B. H., 1975, Carotenoids—Aspects of biosynthesis and enzymology, *Ber. Dtsch. Bot. Ges.* **88**:7.

Davies, B. H., and Aung Than, 1974, Monohydroxycarotenoids from diphenylamine-inhibited cultures of *Rhodospirillum rubrum, Phytochemistry* **13**:209.

Eimhjellen, K. E., and Liaaen-Jensen, S., 1964, The biosynthesis of carotenoids in *Rhodopseudomonas gelatinosa, Biochim. Biophys. Acta* **82**:21.

Eimhjellen, K. E., Steensland, H., and Traetteberg, J., 1967, A *Thiococcus sp. nov. gen.,* its pigments and internal membrane system, *Arch. Mikrobiol.* **59**:82.

Francis, G. W., and Liaaen-Jensen, S., 1970, Bacterial carotenoids. XXXIII. Carotenoids of Thiorhodaceae. 9. The structures of the carotenoids of the rhodopinal series, *Acta Chem. Scand.* **24**:2705.

Goodwin, T. W., 1956, The carotenoids of photosynthetic bacteria. II. The carotenoids of a number of non-sulfur purple bacteria (Athiorhodaceae), *Arch. Mikrobiol.* **24**:313.

Goodwin, T. W., 1971, Biosynthesis, in: *Carotenoids* (O. Isler, ed.), pp. 577–636, Birkhäuser Verlag, Basel.

Goodwin, T. W., and Land, D. G., 1956, The carotenoids of photosynthetic bacteria. I. The nature of the carotenoid pigments in a halophilic photosynthetic sulfur bacterium (*Chromatium spec.), Arch. Mikrobiol.* **24**:305.

Goodwin, T. W., and Osman, H. G., 1954, Studies on carotenogenesis: Spirilloxanthin synthesis by washed cells of *Rhodospirillum rubrum*, *Biochem. J.* **56**:222.

Goodwin, T. W., and Sissins, M. E., 1955, Changes in carotenoid synthesis in *Rhodospirillum rubrum* during growth, *Biochem. J.* **61**:xiii.

Goodwin, T. W., Land, D. G., and Osman, H. G., 1955, Studies on carotenogenesis. 14. Carotenoid synthesis in the photosynthetic bacterium *Rhodopseudomonas spheroides*, *Biochem. J.* **59**:491.

Halfen, L. N., Pierson, B. K., and Francis, G. W., 1972, Carotenoids of a gliding organism containing bacteriochlorophylls, *Arch. Mikrobiol.* **82**:240.

Lascelles, J., 1968, The bacterial photosynthetic apparatus, *Adv. Microb. Physiol.* **2**:1.

Liaaen-Jensen, S., 1963a, Carotenoids of photosynthetic bacteria—distribution, structure, and biosynthesis, in: *Bacterial Photosynthesis* (H. Gest, A. San Pietro, and L. P. Vernon, eds.), pp. 19–34, Antioch Press, Yellow Springs, Ohio.

Liaaen-Jensen, S., 1963b, Bacterial carotenoids: The constitution of the minor carotenoids of *Rhodopseudomonas*. 3. OH-Y, *Acta Chem. Scand.* **17**:500.

Liaaen-Jensen, S., 1965, Bacterial carotenoids. XVIII. Arylcarotenes from *Phaeobium*, *Acta Chem. Scand.* **19**:1025.

Liaaen-Jensen, S., 1966, Recent studies on the structure and distribution of carotenoids in photosynthetic bacteria, in: *Biochemistry of Chloroplasts* (T. W. Goodwin, ed.), pp. 437–441, Academic Press, London and New York.

Liaaen-Jensen, S., Cohen-Bazire, G., Nakayama, T. O. M., and Stanier, R. Y., 1958, The path of carotenoid synthesis in a photosynthetic bacterium, *Biochim. Biophys. Acta* **29**:477.

Liaaen-Jensen, S., Cohen-Bazire, G., and Stanier, R. Y., 1961, Biosynthesis of carotenoids in purple bacteria: A reevaluation based on considerations of chemical structure, *Nature (London)* **192**:1168.

Liaaen-Jensen, S., Hegge, E., and Jackman, L. M., 1964, Bacterial carotenoids. XVII. The carotenoids of photosynthetic green bacteria, *Acta Chem. Scand.* **18**:1703.

Malhotra, H. G., Britton, G., and Goodwin, T. W., 1969, The identification of spheroidene and hydroxy spheroidene in diphenylamine-inhibited cultures of *Rhodospirillum rubrum*, *Phytochemistry* **8**:1047.

Malhotra, H. G., Britton, G., and Goodwin, T. W., 1970a, The mono- and dimethoxy-carotenoids of diphenylamine-inhibited cultures of *Rhodospirillum rubrum*, *Phytochemistry* **9**:2369.

Malhotra, H. G., Britton, G., and Goodwin, T. W., 1970b, The occurrence of hydroxy-derivatives of phytoene and phytofluene in diphenylamine-inhibited cultures of *Rhodospirillum rubrum*, *FEBS Lett.* **6**:334.

Malhotra, H. G., Britton, G., and Goodwin, T. W., 1970c, Occurrence of 1,2-dihydro carotenoids in *Rhodopseudomonas viridis*, *J. Chem. Soc. D* **1970**:127.

Malhotra, H. G., Britton, G., and Goodwin, T. W., 1970d, A novel series of 1,2-dihydro carotenoids, *Int. J. Vit. Res.* **40**:315.

McDermott, J. C. B., Ben-Aziz, A., Singh, R. K., Britton, G., and Goodwin, T. W., 1973, Recent studies of carotenoid biosynthesis in bacteria, *Pure Appl. Chem.* **35**:29.

McDermott, J. C. B., Brown, D. J., Britton, G., and Goodwin, T. W., 1974, Alternative pathways of zeaxanthin biosynthesis in a *Flavobacterium* sp.: Experiments with nicotine as inhibitor, *Biochem. J.* **144**:231.

Moshier, S. E., and Chapman, D. J., 1973, Biosynthetic studies on aromatic carotenoids. Biosynthesis of chlorobactene, *Biochem. J.* **136**:395.

Pfennig, N., Eimhjellen, K. E., and Liaaen-Jensen, S., 1965, A new isolate of *Rhodospirillum fulvum* group and its photosynthetic pigments, *Arch. Mikrobiol.* **51**:258.

Pfennig, N., Markham, M. C., and Liaaen-Jensen, S., 1968, Carotenoids of Thiorhodaceae. 8. Isolation and characterization of *Thiothece, Lamprocystis,* and *Thiodictyon* strains and their carotenoid pigments, *Arch. Mikrobiol.* **62**:178.

Pierson, B. K., and Castenholz, R. W., 1974, A phototrophic gliding filamentous bacterium of hot springs, *Chloroflexus aurantiacus*, gen. and sp. nov., *Arch. Mikrobiol.* **100**:5.

Polgár, A., van Niel, C. B., and Zeichmeister, L., 1944, Studies on the pigments of the purple bacteria. II. A spectroscopic and stereochemical investigation of spirilloxanthin, *Arch. Biochem.* **5**:243.

Porter, J. W., and Lincoln, R. E., 1950, Lycopersicon selections containing a high content of carotenes and colourless polyenes. II. The mechanism of carotene biosynthesis, *Arch. Biochem.* **27**:390.

Ryvarden, L., and Liaaen-Jensen, S., 1964, Bacterial carotenoids. XIV. The carotenoids of *Rhodomicrobium vannielii*, *Acta Chem. Scand.* **18**:643.

Schmidt, K., 1963, Die Carotinoide der Thiorhodaceae. II. Carotinoidzusammensetzung von *Thiospirillum jenense* Winogradsky und *Chromatium vinosum* Winogradsky, *Arch. Mikrobiol.* **46**:127.

Schmidt, K., 1971, Carotenoids of purple nonsulfur bacteria. Composition and biosynthesis of the carotenoids of some strains of *Rhodopseudomonas acidophila, Rhodospirillum tenue,* and *Rhodocyclus purpureus, Arch. Mikrobiol.* **77**:231.

Schmidt, K., and Liaaen-Jensen, S., 1973, Bacterial carotenoids. XLII. New keto-carotenoids from *Rhodopseudomonas globiformis* (Rhodospirillaceae), *Acta Chem. Scand.* **27**:3040.

Schmidt, K., and Schiburr, R., 1970, Die Carotinoide der grünen Schwefelbakterien: Carotinoidzusammensetzung in 18 Stämmen, *Arch. Mikrobiol.* **74**:350.

Schmidt, K., and Trüper, H. G., 1971, Carotenoid composition in the genus *Ectothiorhodospira* Pelsch, *Arch. Mikrobiol.* **80**:38.

Schmidt, K., Liaaen-Jensen, S., and Schlegel, H. G., 1963, Die Carotinoide der Thiorhodaceae. I. Okenone als Hauptcarotinoid von *Chromatium okenii* Perty, *Arch. Mikrobiol.* **46**:117.

Schmidt, K., Pfennig, N., and Liaaen-Jensen, S., 1965, Carotenoids of Thiorhodaceae. IV. The carotenoid composition of 25 pure isolates, *Arch. Mikrobiol.* **52**:132.

Schmidt, K., Francis, G. W., and Liaaen-Jensen, S., 1971, Bacterial carotenoids. XXXVI. Remarkable C_{43}-carotenoid artefacts of cross-conjugated carotenals and new carotenoid glucosides from Athiorhodaceae spp., *Acta Chem. Scand.* **25**:2476.

Schneour, E. A., 1962a, Carotenoid pigment conversion in *Rhodopseudomonas spheroides, Biochim. Biophys. Acta* **62**:534.

Schneour, E. A., 1962b, The source of oxygen in *Rhodopseudomonas spheroides* carotenoid pigment conversion, *Biochim. Biophys. Acta* **65**:510.

Singh, R. K., Britton, G., and Goodwin, T. W., 1973a, Carotenoid biosynthesis in *Rhodopseudomonas spheroides*. S-Adenosylmethionine as the methylating agent in the bio-synthesis of spheroidene and spheroidenone, *Biochem. J.* **136**:413.

Singh, R. K., Ben-Aziz, A., Britton, G., and Goodwin, T. W., 1973b, Biosynthesis of spheroidene and hydroxy-spheroidene in *Rhodopseudomonas* species: Experiments with nicotine as inhibitor, *Biochem. J.* **132**:649.

van Niel, C. B., 1947, Studies on the pigments of the purple bacteria. III. The yellow and red pigments of *Rhodopseudomonas spheroides, Antonie van Leeuwenhoek; J. Serol. Microbiol.* **12**:156.

Volk, W. A., and Pennington, D., 1950, The pigments of the photosynthetic bacterium *Rhodomicrobium vannielii, J. Bacteriol.* **59**:169.

CHAPTER 40

Biosynthesis of Porphyrins, Hemes, and Chlorophylls

O. T. G. Jones

1. Introduction

1.1. Nature of the Tetrapyrrole Pigments Found in the Photosynthetic Bacteria

Although many photosynthetic bacteria, particularly members of the Rhodospirillaceae, excrete metal-free porphyrins into the medium during growth, it is unlikely that these porphyrins serve a biological function in the organism. They arise as intermediates, or as oxidation products of intermediates, in the biosynthesis of the functional metal complexes of tetrapyrrole pigments, the chlorophylls (magnesium complexes), the heme pigments (iron complexes), and vitamin B_{12} (a cobalt complex). Bacteriopheophytin (bacteriochlorophyll a lacking magnesium), a component of bacterial reaction centers, may be the only metal-free tetrapyrrole to have a definite, if undefined, biological function.

Several different chlorophyll (Chl) structures are found in photosynthetic bacteria, and only in the case of bacteriochlorophyll (Bchl) a (Fig. 1) is there substantial evidence for the biosynthetic route. As described elsewhere in this volume Bchl a is present in the Chromatiaceae and Rhodospirillaceae and is also found as a relatively minor, but important, component of the photosynthetic apparatus of the Chlorobiaceae, in which the main light-harvesting Chl's are the

chlorobium Chl's (see Chapter 9). Variants of Bchl a were recently found in *Rhodospirillum rubrum*, in which the esterifying alcohol is not phytol but *trans-trans*-geranylgeraniol (Katz *et al.*, 1972; Brockman *et al.*, 1973), and Kunzler and Pfennig (1973) showed that in *Rs. photometricum*, a mixture of bacterio-chlorophyllide a esters of *trans-trans*-geranylgeraniol and of phytol is present. Bchl b, found in *Rhodo-pseudomonas viridis*, is similar in structure to Bchl a (Fig. 1) (Scheer *et al.*, 1974), but the unsaturated substituent at position 4 causes a pronounced shift of the intense absorption band in the far-red region to longer wavelengths. A summary of the distribution of Chl's within the photosynthetic bacteria is given in Table 1.

Hemes are present in photosynthetic bacteria as the prosthetic groups of cytochromes, catalase, and peroxidase. Three hemes are commonly found (Fig. 2): protoheme, which is the prosthetic group of b-type cytochromes, of catalase, and of some peroxidases; heme c, which is the prosthetic group of cytochromes of the c-type, including cytochrome c, is covalently linked to protein; and heme a, which is believed to be the prosthetic group of the cytochrome oxidase developed by *Rp. sphaeroides* when grown with vigorous aeration (Saunders and Jones, 1974). The carbon-monoxide-binding pigment cytochrome o, which acts as an oxidase in some of the Rhodo-spirillaceae, is likely to have a protoheme prosthetic group (see Lemberg and Barrett, 1973). Little is known of the biosynthesis of hemes other than proto-heme by photosynthetic bacteria, although evidence from other organisms suggests that it is likely that

O. T. G. Jones · Department of Biochemistry, University of Bristol, Bristol, England BS8 1TD

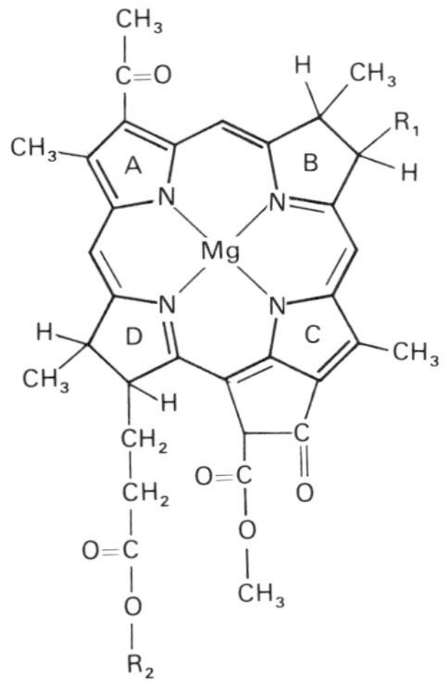

Fig. 1. Bacteriochlorophyll structure. The structures of bacteriochlorophylls a and b are as follows:
Bacteriochlorophyll a: $R_1 = -H$; $R_2 =$ phytyl:

$$\left(-O-CH_2-CH=C-(CH_2)_3-CH-(CH_2)_3-CH-(CH_2)_3-CH \right)$$

Bacteriochlorophyll a_{gg}: $R_1 = -H$; $R_2 =$ geranylgeranyl:

$$\left(-O-CH_2-CH=C-(CH_2)_2-CH=C-(CH_2)_2CH=C-(CH_2)_2-CH=C \right)$$

Bacteriochlorophyll b: $R_1 = =CH-CH_3$; $R_2 =$ phytyl

protoheme is a common intermediate in the biosynthesis of all hemes. Organisms that require protoheme for growth appear to incorporate that heme into cytochrome c or heme a.

1.2. Proposed Biosynthetic Interrelationships of Hemes and Chlorophylls in the Photosynthetic Bacteria

In *Rp. sphaeroides*, the organism in which Bchl a biosynthesis has been most extensively studied, it is likely that Bchl a and protoheme are synthesized by pathways that have the same intermediates as far as protoporphyrin (for a review of this early work, see Lascelles, 1964). Vitamin B_{12} branches from this biosynthetic path at a point before the formation of protoporphyrin. An outline of this biosynthetic scheme is given in Fig. 3, where it can be seen that when magnesium is inserted into the protoporphyrin ring, the magnesium protoporphyrin that is formed is converted into Bchl a in a further series of reactions. If iron is inserted into protoporphyrin, the product, protoheme, is further modified and coordinated to proteins leading to the formation of different hemeproteins. Protoporphyrin is at the branch point in heme and Bchl synthesis. In the following section, the properties of the individual enzymes that are involved

Table 1. Distribution of Bacteriochlorophylls of Photosynthetic Bacteria

Bchl	Group or species	References
Bchl a	Chromatiaceae Rhodospirillaceae Chlorobiaceae[1]	Fischer and Stern (1940)
Bchl a_{gg}	*Rs. rubrum* strains, *Rs. photometricum*	Katz *et al.* (1972) Kunzler and Pfennig (1973)
Bchl b	*Rs. viridis*	Jensen *et al.* (1964), Scheer *et al.* (974)
Bchl c[2]	*Chl. limicola f. thiosulfatophilum* strain 6230 and other *Chlorobium* spp.	Holt *et al.* (1966), Stanier and Smith (1960)
Bchl d[2]	*Chl. vibrioforme f. thiosulfatophilum* and other *Chlorobium* spp.	Purdie and Holt (1965), Stanier and Smith (1960)
Bchl e[2]	*Chl. phaeobacteroides, Chl. phaeovibriodes*	Gloe *et al.* (1975)

[1] Bchl a is a minor component in most species; the amount varies in different species, but may approach 5% of the total Chl. In *Chloroflexus aurantiacus*, however, the amounts of Bchl a and c may approach equality under certain conditions.

[2] Bchl c (sometimes called chlorobium Chl 660), Bchl d (sometimes called chlorobium Chl 650), and Bchl e are mixtures of homologues.

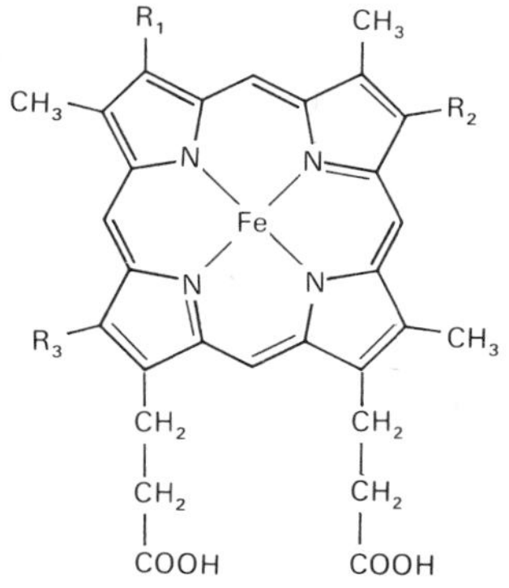

Fig. 2. Heme structure. Structures of hemes found in photosynthetic bacteria are as follows:

Protoheme—the prosthetic group of cytochromes b, cytochromes o, catalase, and peroxidase:

$$R_1 = -CH=CH_2; R_2 = -CH=CH_2; R_3 = =CH_3$$

Heme c—the prosthetic group of cytochrome c; the heme is covalently linked to the apoprotein through thioether bonds:

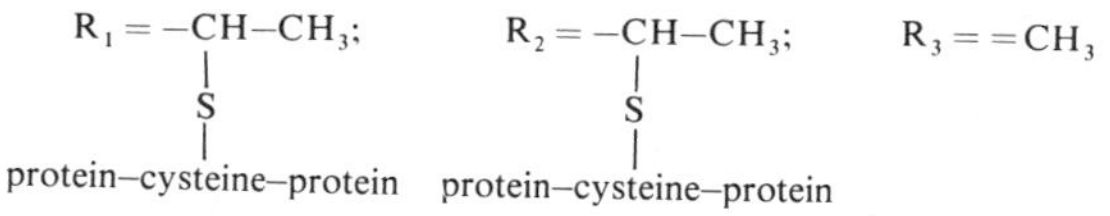

$$R_1 = -CH-CH_3; \qquad R_2 = -CH-CH_3; \qquad R_3 = =CH_3$$

protein–cysteine–protein   protein–cysteine–protein

Heme a—the prosthetic group of cytochromes a and a_3:

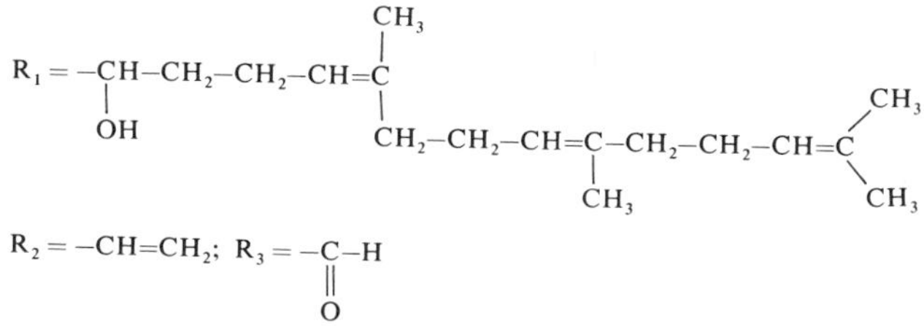

$$R_1 = -CH-CH_2-CH_2-CH=C$$

with OH and CH_3 substituents, and side chain $CH_2-CH_2-CH=C-CH_2-CH_2-CH=C$ bearing CH_3 groups

$$R_2 = -CH=CH_2; R_3 = -C-H$$

with $=O$

in the biosynthesis of protoporphyrin will be discussed in some detail. It must be pointed out that most of the available information about these reactions has been obtained from studies using the Rhodospirillaceae; the pathways in other bacteria have not been fully investigated and may differ.

2. Route to Protoporphyrin

2.1. Formation of 5-Aminolevulinate

It was shown by the outstanding work of Shemin and his collaborators (for an early review, see Shemin, 1956) that in mammalian and avian systems, the carbon atoms of glycine and acetate were incorporated into specific positions in the ring of protoheme. The labeling pattern was best explained by assuming that acetate had first to enter the tricarboxylate cycle to form an asymmetric four-carbon molecule, which then reacted with glycine to give the porphyrin precursor molecule. The asymmetrical four-carbon molecule was found to be succinyl-CoA (Kikuchi et al., 1958; Gibson et al., 1958), and the porphyrin precursor molecule was found to be 5-aminolevulinate (ALA) (Shemin and Russell, 1953; Neuberger and Scott, 1953). The reaction was catalyzed by the enzyme 5-aminolevulinate synthetase (succinyl-CoA-glycine succinyl transferase, E.C. 2.3.1.37). At the time the enzyme from mammalian and avian tissue, concerned in the formation of protoheme, was being characterized, similar activity was found in the photosynthetic bacterium Rp. sphaeroides (Kikuchi et al., 1958; Gibson, 1958), in which it

appeared to be involved in the formation of both heme and Bchl a. In both photosynthetic bacteria and animal tissues, ALA-synthetase is a regulatory enzyme that increases in cellular concentration under conditions of adaptation to increased production of tetrapyrrole pigments. Granick and Sassa (1971) reviewed the central role of ALA-synthetase in heme and Chl synthesis, and Lascelles (Chapter 42) considers more specifically its role in the regulation of Bchl synthesis. In summary, when Rp. sphaeroides is transferred from aerobic conditons, in which the cellular concentration of Bchl a is very low, to anaerobic conditions, in which the specific Bchl content starts to rise (Cohen-Bazire et al., 1957), then this rise is preceded by an increase in the specific activity of ALA-synthetase. When the cells are once more aerated, then the activity of ALA-synthetase diminishes. Not only is the rate of synthesis of the enzyme affected by changes in growth condition, but also the activity of the enzyme is moderated by feedback inhibition from protoheme (Burnham and Lascelles, 1963), by the cellular concentration of polysulfides (Wider de Xifra et al., 1976), and by the cellular concentration of ATP (Fanica-Gaignier and Clement-Metral, 1971).

The reaction catalyzed by ALA-synthetase is shown in Fig. 4. The enzyme is soluble and requires pyridoxal phosphate as a cofactor. Shemin (1957) and Neuberger (1961) proposed that the first step in the reaction sequence is the condensation of glycine with pyridoxal phosphate to form a Schiff base, forming a stable carbanion, with loss of a proton by the fixed glycine. The carbanion condenses with the electrophilic carbonyl carbon atom of succinyl-CoA, yielding

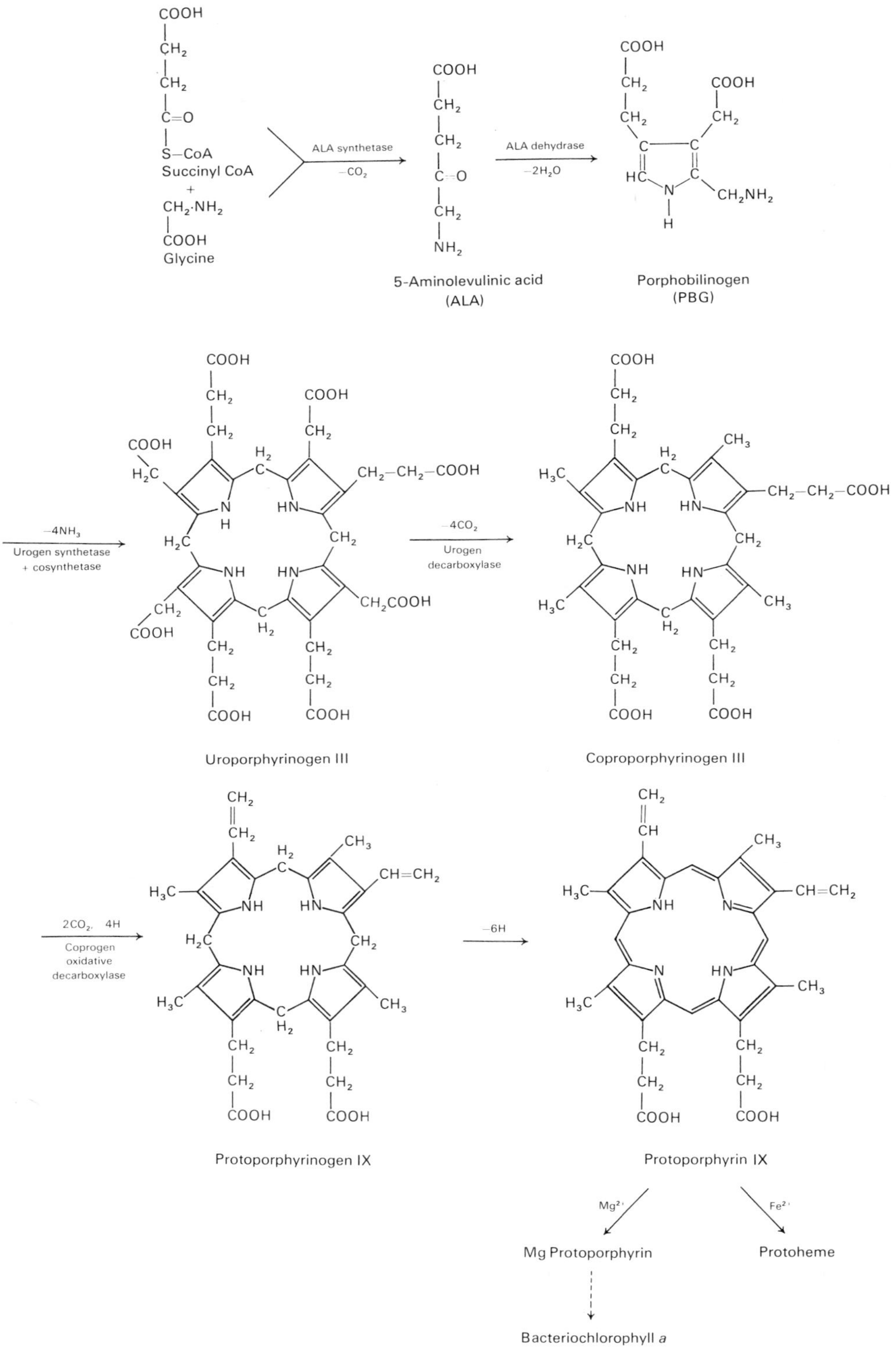

Fig. 3. Biosynthetic sequence leading to the formation of protoporphyrin and showing the branch points at which the routes to bacteriochlorophyll *a* and protoheme diverge.

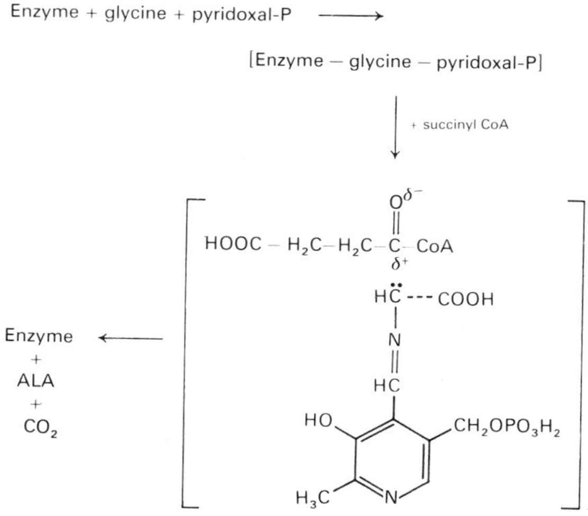

Enzyme + glycine + pyridoxal-P ⟶

[Enzyme — glycine — pyridoxal-P]

+ succinyl CoA

Enzyme
+
ALA ⟵
+
CO_2

Fig. 4. A proposed mechanism for the formation of δ-aminolevulinate by the enzyme ALA-synthetase (Shemin, 1957).

α-amino-β-oxoadipic acid, which decarboxylates spontaneously. The detailed stereochemistry of these reactions was recently characterized by Akhtar et al. (1976), who showed that the *pro* R hydrogen atom of glycine is lost from C-2 of glycine before the reaction with succinyl-CoA.

The succinyl-CoA required for ALA-synthetase in *Rp. sphaeroides* may arise from the tricarboxylate cycle, which is present even when the organism is grown under anaerobic conditions, or from the activity of the enzyme succinyl thiokinase, which generates succinyl-CoA from succinate, CoA, and ATP. Since this latter enzyme is very active in extracts from *Rp. sphaeroides*, there is no requirement for the addition of preformed succinyl-CoA in the assay of ALA-synthetase in crude extracts; sufficient is generated from added succinate, CoA, and ATP.

The purification of ALA-synthetase from *Rp. sphaeroides* has been carried out by several groups of workers, but the properties of the purified enzyme vary in complexity. Warnick and Burnham (1971) appeared to find one major form of the enzyme present after 1300-fold purification. There was one major band on disk electrophoresis, with two minor bands suggested to be caused by aggregation, since they were not found in the presence of sodium dodecyl sulfate (SDS). The molecular weight of the enzyme was 57,000±5000 in the presence or absence of SDS, and it did not give cooperative kinetics when the concentrations of glycine ($K_m = 10$ mM) and succinyl-CoA ($K_m = 25$ μM) were varied, so that its behavior differs from that of many regulatory enzymes. The

purified enzyme was inhibited by protoheme (50% inhibition at around 4 μM). Very similar properties of purified ALA-synthetase of *Rp. sphaeroides* were reported by Yubisui and Yoneyama (1972), who found that the enzyme was more sensitive to inhibition by protoheme (50% inhibition at 0.4 μM), and that it was inhibited by magnesium protoporphyrin (50% inhibition at 2 μM). Both groups reported beneficial effects of thiol reagents on the activity and stability of the enzyme.

Tuboi et al. (1970) presented a somewhat different picture of the ALA-synthetase of *Rp. sphaeroides*. They separated two fractions, I and II, by chromatography on DEAE–cellulose. Since these two forms of the enzyme were not separated by chromatography on Sephadex G-200, they suggest that they are of very similar molecular weights (70,000–80,000). Fraction I was found to exist in two different forms, active and inactive; the inactive form could be made active by overnight dialysis against phosphate buffer containing 0.01 M mercaptoethanol. The pH optimum of fraction I was 7.4, and of fraction II, 7.8, but the K_m values for glycine (5 mM) and succinyl-CoA (5 μM) were identical in the two forms. In a later paper (Tuboi et al., 1970b), it was shown that the responses of fraction I and II enzymes to changes in growth conditions were different. Both were repressed in high oxygen; fraction I enzyme was induced by a reduction of oxygen tension alone, while fraction II was induced by a reduction in oxygen tension only when this was accompanied by illumination of the culture. The inactive form of Fraction I enzyme had an apparent molecular weight of 100,000 (determined by gel filtration), and the molecular weight of the active form was 80,000 (Tuboi and Hayasaka, 1972a,b), although these authors believe that the apparent change in molecular weight may not be due to the loss of a peptide component, but to a conformational change following activation causing a different behavior on molecular sieving. In their long and detailed investigation of the natural activator of ALA-synthetase of *Rp. sphaeroides*, Neuberger and his associates have found that this molecule may be a low-molecular-weight trisulfide. Cysteine trisulfide caused the conversion of the inactive form of fraction I enzyme to its active form (for a review of this work, see Wider de Xifra et al., 1976).

In another report of the purification of ALA-synthetase of *Rp. sphaeroides* (Fanica-Gaignier and Clement-Metral, 1971, 1973a), two fractions were obtained on DEAE–Sephadex columns. Fraction I was later purified 4200-fold and fraction II, 2000-fold. Each appeared homogeneous in disk electrophoresis and had the same molecular weight (about 100,000). Each contained one molar equivalent of pyridoxal phosphate linked at neutral pH as a substituted aldamine derivative with a free amino group of the protein. The two enzymes differed in their isoelectric point (pI was 5.1 for fraction I and 6.0 for fraction II) and in their content of —SH groups (1 per molar equivalent of fraction I, 7 per molar equivalent of fraction II). The activity of both fractions was 75% inhibited by 1 mM ATP, but ADP and AMP were almost without effect on either enzyme. The kinetic properties of fraction I ALA-synthetase were examined in some detail by Fanica-Gaigner and Clement-Metral (1973b). They suggested a mechanism whereby glycine binds to the enzyme first, and 5-amino-levulinate dissociates last. ATP, GTP, and pyro-phosphate inhibited the enzyme, competitively with glycine and noncompetitively with succinyl-CoA. With the use of ^{32}P-labeled ATP, it was found that one molecule of nucleotide was bound per molecule of enzyme and that after binding, one —SH group was masked. Pretreatment of the enzyme with a thiol reagent (p-hydroxymercuribenzoate) prevented bind-ing of ATP as well as inhibited the enzyme, and the authors suggest that ATP inhibits ALA-synthetase by acting on a —SH group of the active center, near or at the glycine-binding site.

2.1.1. Alternative Routes to 5-Aminolevulinate

It is possible for ALA to be formed by homogenates of *Rp. sphaeroides* by a transaminase reaction, involving L-alanine and γ,δ-dioxovalerate (Neuberger and Turner, 1963), a reaction with an equilibrium favoring the formation of ALA. Like ALA-synthetase, it requires free thiol groups and pyridoxal phosphate. Such a route involving transaminase may be important in some groups of photosynthetic bacteria, since it is important to realize that ALA-synthetase has so far been assayed in relatively few photosynthetic bacteria. There is increasing evidence that in green plants, ALA is not formed from glycine plus succinyl-CoA by the "usual" ALA-synthetase, but is formed via a five-carbon intermediate (Beale and Castelfranco, 1974a,b; Beale et al., 1975). When [1-^{14}C]glutamate is supplied as substrate, label is efficiently incorporated into C-5 of ALA. This would not happen if the glutamate were first converted, via the tricarboxylate cycle, to the activated four-carbon intermediate, succinyl-CoA. Schemes for the biosynthesis of the postulated intermediate γ,δ-dioxovalerate are shown in Fig. 5, taken from the article by Tait (1968). It is possible that when the mechanisms of ALA synthesis have been determined in a wider variety of organisms, new ideas about the evolutionary relationships of different groups of organisms will emerge.

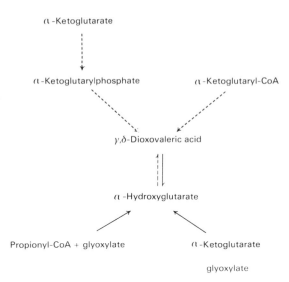

Fig. 5. Possible pathways for the biosynthesis of γ,δ-dioxovaleric acid. See Tait (1968) for a discussion of these routes, which are based on known interconversions of analogous molecules.

2.2. 5-Aminolevulinate Dehydratase

Porphobilinogen (PBG), the monopyrrole precursor of all tetrapyrroles, is formed by the condensation of two molecules of ALA with the loss of two molecules of water (Fig. 6). This reaction is catalyzed by the enzyme ALA-dehydratase (E.C. 4.2.1.24.), a soluble enzyme that has been detected in plants, animals, and bacteria, including the photosynthetic bacteria (Neuberger et al., 1956), and it has been extensively purified from a variety of organisms (Shemin, 1972). Although all preparations of the enzyme so far examined in detail have very similar molecular weights, they appear to belong to different groups, one of which has a requirement for activation by metallic

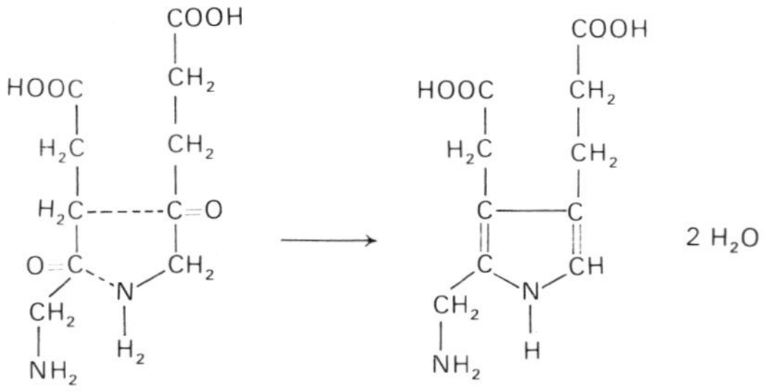

Fig. 6. Scheme for the synthesis of porphobilinogen from two molecules of δ-aminolevulinate.

cations. The ALA-dehydratase of eukaryotic cells does not readily demonstrate a metal activation, but is inhibited by EDTA. Perhaps the most striking effects of metal activation were noted by Shemin and his colleagues (Nandi *et al.*, 1968; Nandi and Shemin, 1968a,b) in their studies of the purification and properties of ALA-dehydratase from *Rp. sphaeroides*. A stimulatory effect of K⁺ on ALA-dehydratase of *Rp. sphaeroides* had been briefly noted by Burnham and Lascelles (1963), but Shemin's group greatly extended this work and showed that in the absence of potassium ions, a plot of the initial reaction rate against ALA concentration yielded a sigmoid curve; in the presence of 0.05 M K⁺, a hyperbolic saturation curve was found. At low substrate concentration (about 4 mM), little activity was found in the absence of K⁺. The effect was not specific to K⁺; Li⁺, NH₄⁺, Rb⁺, and Mg²⁺ were also effective. The addition of Na⁺ also activates the enzyme, but the kinetics are different and the final maximum activity obtained was about only one-half that found with other alkali metals. Centrifugation of the enzyme in a sucrose gradient showed that in the absence of K⁺, it sedimented in a single band the molecular weight of which was calculated to be 250,000. The addition of K⁺ to the sucrose gradient resulted in the formation of an equilibrium mixture of the 250,000-molecular-weight species with other forms of higher molecular weight, calculated at about 500,000 and 750,000. In the presence of 1 M urea, or following dilution of the enzyme solution, the 250,000-molecular-weight species dissociated into subunits of molecular weight about 100,000–130,000. In a subsequent paper, van Heyningen and Shemin (1971) described the electrophoresis of the purified enzyme in the presence of SDS, and found essentially only one species with a molecular weight of 40,000±10%. This low-molecular-weight species bound ¹⁴C-labeled ALA in a specific reaction. Gel filtration chromatography of the purified enzyme showed one species of molecular weight about 240,000 (oligomer) and some material of molecular weight several million. In the presence of K⁺, dioligomers and possibly trioligomers were found, as well as very high-molecular-weight material. They concluded that the ALA-dehydratase of *Rp. sphaeroides* exists as a hexamer of molecular weight about 240,000 daltons, which can dissociate into monomeric units and can aggregate to material having higher molecular weight. It is interesting to compare these results with those obtained for purified ALA-dehydratase of beef liver (Cheh and Neilands, 1973), in which five or six atoms of zinc were required to be present per 280,000-molecular-weight enzyme. These authors suggest that the enzyme contains one Zn²⁺ per

two subunits of molecular weight 36,000, and eight subunits per molecule. Shemin (1975, 1976) found that the crystalline beef liver ALA-dehydratase has a molecular weight of 285,000 and eight subunits of molecular weight 35,000 and 44 Å diameter. There are four to six atoms of zinc per mole of enzyme. These crystals exhibit dihedral symmetry. Binding studies with labeled substrate suggest that there are four active sites per molecule; i.e., it exhibits half-site reactivity.

The ALA-dehydratase of *Rp. capsulata* was purified 400-fold and compared with the enzyme from *Rp. sphaeroides* by Nandi and Shemin (1973). Like the *Rp. sphaeroides* enzyme, its molecular weight was about 260,000 when determined by sedimentation in a sucrose gradient. Unlike the *Rp. sphaeroides* enzyme, a pure dialyzed preparation is active without the addition of K⁺, Na⁺, Mg²⁺ ions, and has normal Michaelis–Menten kinetics in their absence. The enzyme from both these photosynthetic bacteria is insensitive to the chelating agent EDTA over the range 3.3–10 mM (unlike the enzyme from mammalian sources). The purified enzyme from *Rp. sphaeroides* requires thiol compounds for activity, whereas the *Rp. capsulata* is less responsive to thiols. Both enzymes are sensitive to —SH inhibitors, but the *Rp. capsulata* enzyme is relatively insensitive to inhibition by the possible end-product inhibitor protoheme (9% inhibition at 50 μM), whereas the *Rp. sphaeroides* enzyme is 96% inhibited at 16 μM protoheme. These differences in sensitivity to protoheme inhibition and response to metal ions may reflect completely different methods of regulating tetrapyrrole synthesis in these two closely related photosynthetic bacteria. Both enzymes are present within the cells at lower levels when Bchl synthesis rates are low, such as under conditions of intense illumination or aeration.

The mechanism of the ALA-dehydratase reaction was examined by Shemin and his collaborators, using the purified enzyme from *Rp. sphaeroides*. Preliminary results with the enzyme from other sources suggests that at least the early steps in the reaction are the same. The first step is the formation of a Schiff base between the substrate and an ε-NH₂ of a lysine residue on the enzyme substrate. If this enzyme Schiff base is reduced by the addition of sodium borohydride, it forms a stable secondary amine derivative that inactivates the enzyme. This reaction has been used to fix the number of active sites per mole of enzyme. When levulinic acid was added as a substrate, instead of ALA, it was unable to form a pyrrole, but was able to form a Schiff base with the enzyme. Indeed, levulinic acid is a good inhibitor of ALA-dehydratase. When mixtures of ALA and levulinic acid were

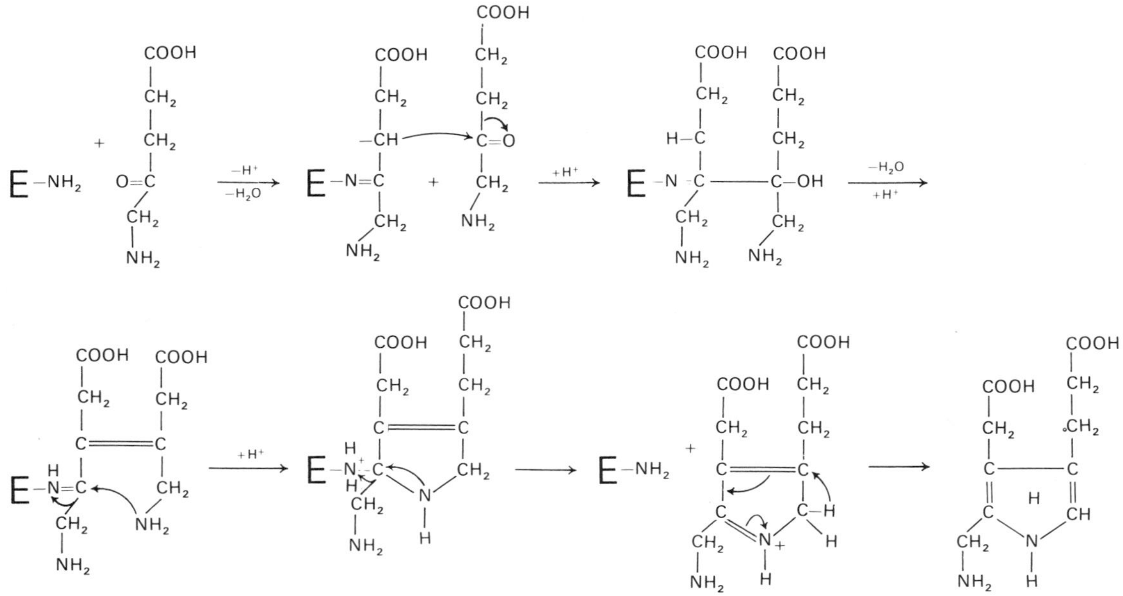

Fig. 7. Outline of the mechanism of porphobilinogen synthesis from two molecules of δ-aminolevulinate, catalyzed by the enzyme (E) ALA-synthetase of *Rhodopseudomonas sphaeroides*. After Nandi and Shemin (1968b).

incubated with ALA-dehydratase, heterologous pyrroles were formed. These observations have been used by Shemin to suggest a mechanism for the ALA-dehydratase reaction. First, a Schiff base is formed with one molecule of substrate (ALA I), resulting in the formation of a stabilized carbanion that can make a nucleophilic attack on the carbonyl carbon atom of a second molecule of ALA (ALA II). Following an aldol condensation, the enzyme catalyzes the elimination of water between the carbon bearing the newly formed hydroxyl group and the adjacent carbon atom on ALA I. A nucleophilic attack of the amino group of ALA molecule II on the highly reactive carbon atom of molecule I in Schiff base linkage with the enzyme causes a transamination and formation of a pyrrole (Fig. 7).

ALA-dehydratase from *Rp. sphaeroides* was immobilized on columns of Sepharose by Gurne and Shemin (1973) and shown to retain activity for up to a month. This will not only make possible the large-scale enzymic production of PBG, avoiding the problem of product inhibition (the reaction halts when PBG concentration reaches 2 mM), but also facilitate further work on the mechanism of the enzyme.

2.3. Formation of Uroporphyrinogen III

The first tetrapyrrole intermediate in the biosynthesis of hemes and Chl's, uroporphyrinogen III, is formed from PBG. The Roman numeral III is added as a suffix to indicate the specific isomer, of the four possible isomers that could arise biologically by the reaction of four molecules of PBG. (See Fig. 8 and the discussion of porphyrin isomerism by Aronoff, 1975.) The symmetrical tetrapyrrole uroporphyrinogen I and its derived porphyrin, uroporphyrin I, are sometimes found in mammals suffering from the disease congenital porphyria. Isomer I is formed by a spontaneous, nonenzymic, head-to-tail reaction of PBG molecules. Uroporphyrinogen I is sometimes formed when tissue or microbial extracts are incubated with PBG. The enzymic production of isomer I is not a reaction that normally proceeds to a significant extent within intact cells or tissues, since this isomer is not believed to be an intermediate in the production of uroporphyrinogen III, the natural isomer. The usual synthetic sequence from PBG to uroporphyrinogen III calls for the simultaneous action of the two enzymes PBG deaminase and uroporphyrinogen III cosynthase. In the presence of PBG deaminase alone, PBG is converted to uroporphyrinogen I, the symmetrical isomer formed by head-to-tail condensation of four molecules of PBG with the elimination of a molecule of ammonia at each condensation, followed by cyclization of the open-chain tetrapyrrole. The enzymic formation of isomer I without isomer III was clearly shown to take place when broken-cell preparations of *Chlorella* were heat-treated at 55°C for

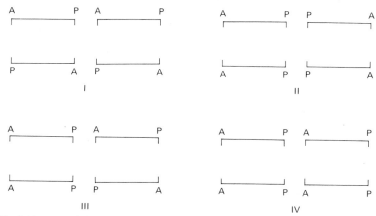

Fig. 8. Representation of the four possible isomers of uroporphyrinogen. (A) —CH$_2$—COOH; (P) —CH$_2$—CH$_2$—COOH. Only the outer edges of the tetrapyrrole rings are shown.

30 min before incubation with PBG; this treatment appeared to inactivate the second enzyme, uroporphyrinogen III cosynthase (Bogorad and Granick, 1953). Later, Bogorad (1958a,b) was able to characterize some of the properties of the two enzymes concerned. PBG deaminase was partly purified from extracts of spinach leaf acetone powder, and uroporphyrinogen III cosynthase was partly purified from extracts of wheat germ. Soon afterwards, two enzymes with the properties of deaminase and cosynthase were partly purified from extracts of thawed and frozen *Rp. sphaeroides* (Heath and Hoare, 1959; Hoare and Heath, 1959). Indeed, the list of mammalian, plant, and microbial tissues from which these enzymes have been extracted is now lengthy, and in every case, heme-, or Chl-synthesizing systems appear to require the mediation of two enzymes in the conversion of PBG to uroporphyrinogen III.

The uroporphyrinogen I synthase purified from spinach leaf has a molecular weight of 40,000 daltons whem measured by analytical centrifugation, gel filtration, or by SDS–polyacrylamide gel electrophoresis. The isoelectric point of this protein was 4.2–4.5. Purified uroporphyrinogen III cosynthase from wheat germ has a molecular weight of 62,000 daltons, determined by gel filtration, and confirmed by analytical centrifugation. The isoelectric point of this protein was 6.1–6.7 (Higuchi and Bogorad, 1975).

The isolated cosynthetase will not catalyze the synthesis of porphyrins from PBG, so it acts possibly by using as a substrate a product of PBG deaminase or, alternatively, by modifying the conformation of the synthase so that its product is a different isomer. Since the cosynthase will not convert preformed uroporphyrinogen I to uroporphyrinogen III, it is possible that some earlier product of PBG deaminase is utilized, and the identity of this product has been actively sought for a number of years. The cosynthase does not cause the isomerization of PBG to form a product that can later be used by PBG deaminase (Bogorad, 1958b).

The mechanism of this unusual reaction, or series of reactions, leading to uroporphyrinogen III formation has been noteworthy for a great deal of speculation (e.g., Margoliash, 1961; Mathewson and Corwin, 1961; Cornford, 1964; Dalton and Dougherty, 1969; Llambias and Battle, 1971; Frydman, R. B., et al., 1971; Davies and Neuberger, 1973; Frydman, B., et al., 1976), and is still unresolved. A number of experimental difficulties combine to make progress slow. A particular problem is the reactivity of PBG and of dipyrrylmethanes; even in the absence of enzymes, these tend to form tetrapyrroles at measurable rates in aqueous solution, and control experiments must be performed with great care.

Bogorad (1963) showed that the formation of porphyrinogens, but not the rate of consumption of PBG, by spinach leaf enzymes was inhibited by the addition of ammonia or hydroxylamine at concentrations near 0.2 M. It was found (Pluscec and Bogorad, 1970) that a number of compounds accumulated under these conditions, including the dipyrryl methane shown in Fig. 9. This compound, when purified, could not be converted to porphyrinogens by spinach enzymes, but when PBG was added to the incubation mixture, the dipyrrylmethane was incorporated into porphyrinogens. This finding suggests that tetrapyrroles may be formed by the sequential addition of PBG molecules to a growing chain of polypyrroles, and not by the reaction of two dipyrrole intermediates.

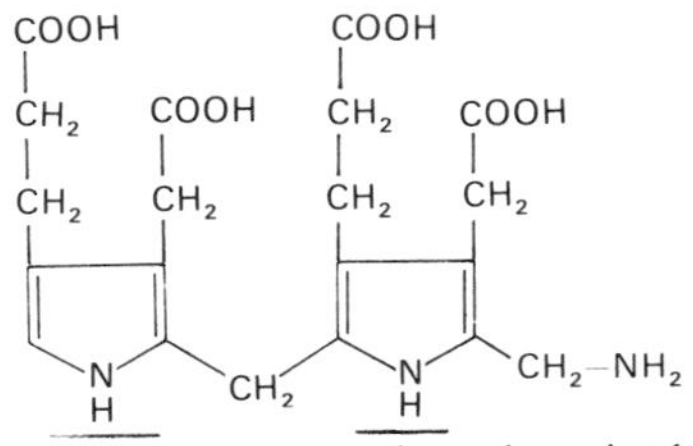

Fig. 9. A dipyrrylmethane that accumulates when spinach leaf PBG deaminase is treated with NH_3 or NH_2OH, at about 0.2 M, to inhibit uroporphyrinogen synthesis. From Pluscec and Bogorad (1970).

Ammonia or hydroxylamine may act by competing for a binding site on the deaminase carrying the growing chain. Among the accumulated intermediates in ammonia- or hydroxylamine-inhibited PBG deaminase of spinach, an uncyclized tetrapyrrole-methane was found (Radmer and Bogorad, 1972). As would be expected, this compound gives a high rate of conversion to uroporphyrinogen I by spontaneous ring closure. It was not converted to uroporphyrinogen III by the cosynthetase, nor was the rate of cyclization increased by PBG deaminase. It is therefore unlikely to be an intermediate in the normal, uninhibited reaction. This highlights another problem: the "intermediates" that accumulate in the presence of inhibitors may never exist under normal conditions. They could arise from an unnatural reaction between a true intermediate that has been displaced from the enzyme and PBG. Free intermediates are not normally liberated into the medium, and the linking of the four PBG units may take place at the active site of the enzyme complex without the release of an intermediate.

A number of potential intermediates in the biosynthesis of uroporphyrinogen III were synthesized and tested by Frydman and his collaborators (see Frydman, B., *et al.*, (1975). They reported that one dipyrrylmethane, 2-aminomethyl-3,4'-(β-carboxyethyl)-4,3'-carboxymethyldipyrrylmethane (see Fig. 10), was incorporated into uroporphyrinogen III.* To accomplish this reaction, it was necessary to have present a mixture of the dipyrrylmethane, PBG, and the two enzymes PBG deaminase and uroporphyrinogen III cosynthase (Frydman, R. B., *et al.*, 1972). More recently, the synthesis of uroporphyrinogen I was shown to proceed using the "symmetrical" tripyrrane and PBG as substrates (Frydman, B., *et al.*, 1975). The relatively high rate of non-enzymic synthesis of uroporphyrinogen III (70–75% of the catalyzed rates) from the synthetic "asymmetrical" polymers of PBG ensures that their role as

intermediates in the normal pathway of uroporphyrinogen III biosynthesis will not be considered as finally established.

Frydman considers that the cosynthase enters into association with the deaminase and acts as a "specifier protein" of the latter, changing the mode of PBG condensation on the enzyme surface. He has found that the deaminase could be attached to an inert Sepharose support, and that by incubating the immobilized enzyme with cosynthase and PBG, the synthesis of uroporphyrinogen III was achieved. When the deaminase–Sepharose complex was packed in a chromatographic column and a cosynthase preparation was filtered through it, most of the cosynthase activity was retained on the column. If the cosynthase was filtered through a control Sepharose column, none of the activity was retained. The column with the two immobilized enzymes catalyzed production of uroporphyrinogen III from added PBG (Frydman, R. B., and Feinstein, 1974). These experiments demonstrated that an association between the two enzymes existed even in the absence of PBG. The formation of an association between purified wheat germ synthase and purified wheat germ cosynthase was investigated by Higuchi and Bogorad (1975). The sedimentation of uroporphyrinogen I synthase or the cosynthase in sucrose gradients was unaffected by the presence of PBG. Yet the presence of PBG in the sucrose gradient affected the sedimentation of a mixture of uroporphyrinogen I synthase and cosynthase, and apparently caused some of the synthase to move farther down the gradient than when PBG is absent. This suggests that at least in the presence of PBG, a complex is formed between the two enzymes. This complex did not appear to be present in gradients run in the absence of PBG.

Carbon-13 NMR spectroscopy has been used as a tool for examining biosynthetic reactions and has been applied to the problem of the synthesis of uroporphyrinogen III. The specifically labeled [2,11-

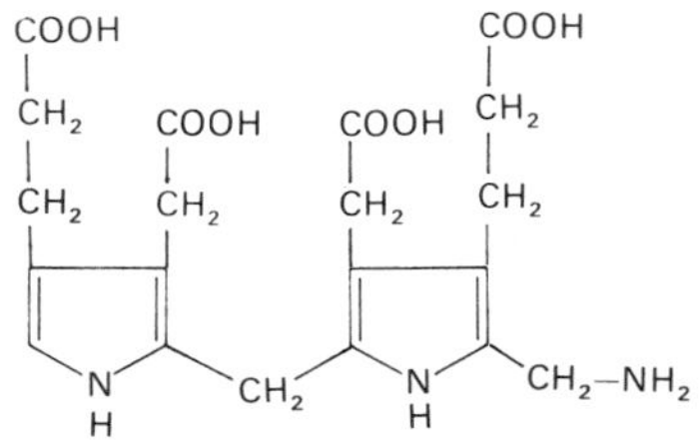

Fig. 10. An "unsymmetrical" dipyrrylmethane, 2-aminomethyl-3,4'-(β-carboxyethyl)-4,3'-carboxymethyldipyrrylmethane, synthesized by R. B. Frydman *et al.* (1972) that may act as a precursor of uroporphyrinogen isomer III.

---

* This is a dipyrrylmethane in which one PBG may be considered to be reversed, as in Ring D (see Fig. 8) of uroporphyrinogen III.

[13C]PBG (Fig. 11) was synthesized enzymatically from 5-amino [5-13C]levulinic acid. A purified uroporphyrinogen I synthetase from spinach leaf converted this labeled molecule to uroporphyrinogen I with the expected symmetrical distribution of label (Fig. 11). When an unfractionated enzyme system from avian erythrocytes was supplied with the labeled PBG, uroporphyrinogen III was formed, and the pattern of labeling in this product was as shown in Fig. 11. A closely similar pattern of labeling was obtained using an enzyme system from *Euglena gracilis* (Battersby *et al.*, 1973). These results suggest that the PBG unit for ring D undergoes intramolecular rearrangement during the biosynthesis of the type III macrocycle, and that PBG units are incorporated intact into rings A, B, and C (see also Battersby and McDonald, 1976).

Uroporphyrinogen III synthesis by the photosynthetic bacterium *Rp. sphaeroides* also required the action of two enzymes, PBG deaminase and uroporphyrinogen III cosynthetase, and these enzymes were partly purified (Heath and Hoare, 1959; Hoare and Heath, 1959). The properties of a 700-fold-purified

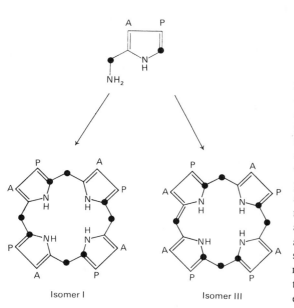

Fig. 11. Biosynthesis of isomer I and III uroporphyrinogens from PBG labeled with 13C at carbons 2 and 11. (A) —CH2—COOH; (P) —CH2—CH2—COOH. The PBG was incubated with purified uroporphyrinogen I synthetase to yield isomer I, or with a preparation from avian erythrocytes to yield isomer III. The label was located by 13C NMR spectroscopy (Battersby *et al.*, 1973). The experiment showed that during the biosynthesis of type III isomers, the PBG unit forming ring D of the tetrapyrrole undergoes *intramolecular* rearrangement.

PBG deaminase of *Rp. sphaeroides* were later described by Jordan and Shemin (1973). This purified enzyme appeared as a single band on analytical polyacrylamide gel electrophoresis. Electrophoresis in the presence of SDS also revealed a single sharp band, corresponding to a molecular weight of about 36,000. Molecular-weight determination by means of gel filtration or sucrose density-gradient centrifugation also gave values close to 36,000. There was no evidence for any species of a higher molecular weight, and it appears that this enzyme exists as a single polypeptide chain, molecular weight 36,000, that catalyzes the polymerization of PBG to yield uroporphyrinogen I with a molecular weight of about 840. Jordan and Shemin speculated that the enzyme contains a single catalytic site that is involved four times during the synthesis of one molecule of the tetrapyrrole. The purified enzyme was rapidly inactivated below pH 6.0 and at low ionic strength. It was sensitive to inhibition by sulfhydryl reagents. The K_m for PBG was 13 μM, which means that under physiological conditions, the concentration of 5-aminolevulinic acid and PBG would be very low.

A purified preparation of PBG deaminase from *Rp. sphaeroides* was also described by Davies and Neuberger (1973). It had a molecular weight of 36,000 daltons in the presence or absence of SDS, but the K_m for PBG (40 μM) was a little higher than reported by Jordan and Shemin (1973). Davies and Neuberger used their purified enzyme in studies on the enzymatic mechanism of uroporphyrinogen I synthesis by *Rp. sphaeroides*. At pH 8.2, the pH normally used for assay, the enzyme consumed PBG and formed uroporphyrinogen I without the accumulation of intermediates. In the presence of hydroxylamine, ammonia, or methoxyamine, the production of porphyrinogen was inhibited, and the enzyme formed liner polypyrroles instead; the rate of consumption of PBG was almost unaltered. The polypyrroles formed porphyrins nonenzymically in a cyclization reaction that was accompanied by the release of the inhibitory amine, and their behavior was consistent with their possession of a tetrapyrrolic structure, probably a pyrrylmethane, that was not cyclized. Once released from the enzyme, the tetrapyrrole was not a substrate for either of the uroporphyrin-synthesizing enzymes. It is suggested that the enzymic cyclization step is more susceptible to interference by the bases than are the earlier condensation steps, and that cyclization involves electrophilic attack by a —CH$_2^+$ group of the pyrrole unit at one end of the open-chain tetrapyrrole on the unsubstituted α-position of the pyrrole unit at the other end. The bases may act by competing for the —CH$_2^+$ group.

2.4. Formation of Coproporphyrinogen III

The conversion of uroporphyrinogen III to coproporphyrinogen III involves the loss of four molecules of carbon dioxide from the carboxyl groups of the acetic acid substituents of the uroporphyrinogen, catalyzed by the enzyme uroporphyrinogen decarboxylase (E.C. 4.1.1.37). Preparations from rabbit reticulocytes, partly purified by zone electrophoresis, carried out this reaction, and it was found that apparent intermediates accumulated in the incubation mixture. These were identified as porphyrinogens with seven, six, or five carboxylic acid groups (Mauzerall and Granick, 1958), and suggested that the enzyme acts by removing carbon dioxide in a series of sequential reactions, possibly at the same active site. A partly purified uroporphyrinogen decarboxylase from *Rp. sphaeroides* had very similar properties (Hoare and Heath, 1959), with chromatographic evidence for the formation of hepta-, hexa-, and pentacarboxylic porphyrinogens as intermediates. The formation of these porphyrinogens as intermediates in all heme-synthesizing systems is suggested by the common occurrence of their derived porphyrins in blood, urine, or feces of mammals suffering diseases of porphyrin metabolism and as excretory products in the supernatant medium of microorganisms (e.g., Eales *et al.*, 1975; Brouillet *et al.*, 1975). Garcia *et al.* (1973) obtained evidence of production of intermediates by the uroporphyrinogen decarboxylase of avian erythrocytes that suggests that the removal of the first carboxyl group proceeds much

more readily than the elimination of the next three. NMR spectroscopy of the intermediate porphyriogens excreted in porphyria has enabled the carboxyl-group substituents to be localized, and indicates that the order of removal of the carboxyl groups in mammalian systems is probably first ring D, then ring A, then ring B, then ring C (Jackson *et al.*, 1975; Stoll *et al.*, 1973).

The partly purified decarboxylase from either rabbit reticulocytes or *Rp. sphaeroides* appeared to have essential sulfhydryl groups and was not isomer-specific. Coproporphyrinogen was formed from the four isomers, uroporphyrinogen I, II, III, and IV, with relative rates in the order III > IV > II > I. This lack of specificity explains the excretion of coproporphyrin I by mammals suffering from certain types of porphyria, where there has been a production of uroporphyrinogen I. Uroporphyrins themselves are not substrates for uroporphyrinogen decarboxylase.

2.5. Formation of Protoporphyrinogen IX

The conversion of coproporphyrinogen III to protoporphyrinogen IX is an oxidative decarboxylation leading to the formation of two vinyl groups from the propionic acid substituents at positions 2 and 4 of the tetrapyrrole ring. The enzyme that catalyzes this reaction is coproporphyrinogen oxidative decarboxylase (E.C. 1.3.3.3). In mammals, it is apparently localized in the mitochondria (Sano and Granick, 1961), and in all aerobic organisms, it has an absolute requirement for molecular oxygen. No other electron

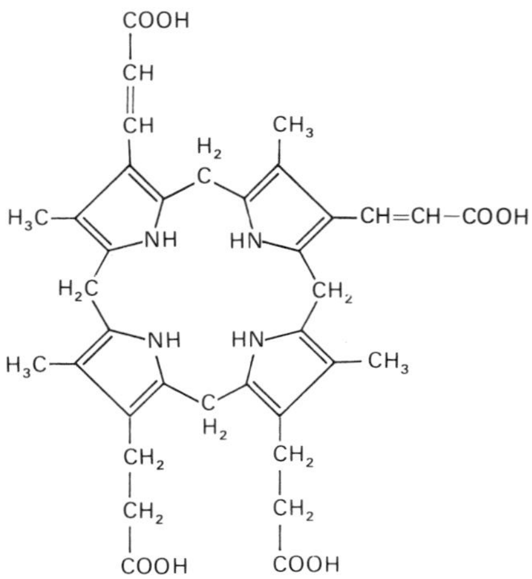

2,4-Diacrylic
Deuteroporphyrinogen IX

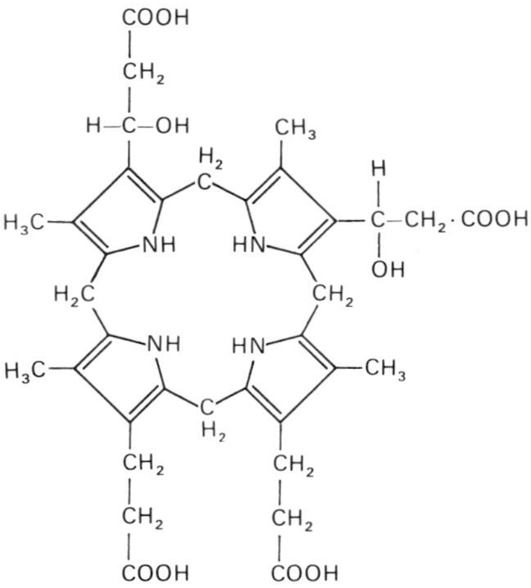

2,4 bis(β-Hydroxypropionic acid)
Deuteroporphyrinogen IX

Fig. 12. Structures of two porphyrinogens synthesized by Sano (1966) and tested as intermediates in the coproporphyrinogen oxidative decarboxylase reaction of bovine liver mitochondria.

acceptor could substitute for oxygen. Using a similar mitochondrial preparation, Porra and Falk (1964) showed that the product of the reaction was protoporphyrinogen and not protoporphyrin, and that although the best substrate was isomer III of coproporphyrinogen, low rates of conversion of coproporphyrinogen IV to a protoporphyrinogen isomer (not IX) could be detected. Coproporphyrinogen isomers I and II were not utilized by this enzyme. Sano and Granick (1961) detected the formation of an intermediate tetrapyrrole in the reaction, with one vinyl group and three carboxylic acid groups, and suggested a sequential mechanism. This type of reaction mechanism is supported by the observations of Porra and Falk (1964) that a synthetic porphyrinogen with three propionic acid substituents, deuteroporphyrinogen IX, 4-propionic acid, acts as a substrate, and of Kennedy et al. (1970) that a 2-vinyl, 4-propionic acid tricarboxylic acid porphyrin (harderoporphyrin) accumulates in considerable quantities in the harderian gland of the rat. Further insight into the mechanism of coproporphyrinogen oxidative decarboxylase of liver mitochondria was obtained by Sano (1966). He synthesized a number of potential intermediates in the reaction (see Fig. 12) and tested them as substrates for the mitochondrial enzyme. Protoporphyrin(ogen) IX was formed from β-hydroxypropionic acid deuteroporphyrinogen IX even in the absence of oxygen from the incubation. It had been shown previously that another possible intermediate, trans-2,4-diacrylicdeuteroporphyrinogen was not utilized (Sano and Granick, 1961). Sano proposed a mechanism for the side-chain alteration whereby oxygen causes a β-oxidation of the propionic acid side chain, followed by dehydration and decarboxylation, yielding a vinyl group, thus:

$$R \cdot CH_2 \cdot CH_2 \cdot COOH \xrightarrow{O} R \cdot \underset{\underset{OH}{|}}{\overset{\overset{H}{|}}{C}} - CH_2 \cdot COOH \rightarrow$$

$$R \cdot CH = CH_2 + H_2O + CO_2$$

No acrylic acid intermediates were thought to be formed.

The reaction was also studied by isotope labeling experiments (Battersby et al., 1972a,b). Two samples of PBG were prepared. One (1) was labeled with deuterium in the methylene group of the proprionic acid side chain next to the proprionate carboxyl, and the other (2) was labeled with deuterium in the methylene group next to the pyrrole nucleus, thus:

$$(1)\ R - CH_2 \cdot CD_2 \cdot COOH$$

or

$$(2)\ R - CD_2 \cdot CH_2 \cdot COOH$$

When (1) was incubated with a crude extract from Euglena gracilis, protoporphyrin IX containing eight deuterium atoms was recovered. When (2) was incubated, the protoporphyrin IX that was recovered contained six deuterium atoms. These results showed that acrylic acids are unlikely to be involved in the conversion of coproporphyrinogen III to protoporphyrin IX, and are in agreement with the mechanism proposed by Sano.

By treating crude extracts of Euglena gracilis with some possible intermediates between coproporphyrinogen III and protoporphyrinogen IX labeled with tritium, Cavaleiro et al. (1974) were able to show that 2-vinyl, 4-propionyldeuteroporphyrinogen IX was a much more effective precursor of porphyrinogen than 4-vinyl, 2-propionyldeuteroporphyrinogen IX. They suggest that the results are best explained if, in the biosynthesis of porphyrinogen from uroporphyrinogen, the 2-propionic acid is converted into vinyl before the 4-substituent is modified. Essentially the same results were briefly reported by Jackson and Games (1975) using hemolysates of avian erythrocytes in place of Euglena gracilis extracts.

Many microorganisms, including the photosynthetic bacteria, form tetrapyrrole pigments when they are grown in conditions in which oxygen is completely absent. The mechanism of the coproporphyrinogen oxidative decarboxylase must be different in these organisms. Mori and Sano (1968, 1972) found, however, that extracts of anaerobically grown Chromatium vinosum did not catalyze the conversion of coproporphyrinogen to protoporphyrinogen when the incubations were performed anaerobically. Under aerobic conditions, the reaction proceeded smoothly. A wide range of electron acceptors and activators were added in unsuccessful attempts to catalyze the anaerobic reaction. The enzyme was purified 23-fold and found to have a K_m for coproporphyrinogen of about 35 μM and to be unaffected by added Fe^{2+} or by iron chelators, unlike the beef liver enzyme. The pH optimum of 6.4 was more acidic than reported for the mammalian liver enzyme (Sano and Granick, 1961).

In contrast with these results are the experiments of Tait (1969, 1972) with extracts from anaerobically grown Rp. sphaeroides. A crude extract catalyzed the conversion of coproporphyrinogen into protoporphyrinogen under either aerobic or anaerobic conditons. The anaerobic activity was observed only when Mg^{2+}, ATP, and L-methionine or S-adenosylmethionine were included in the incubation mixture. In air, the reaction required the addition of cyanide, possibly to prevent the preferential use of oxygen by the respiratory chain of the particles. The pH optima of the aerobic and the anaerobic reactions differed, and the anaerobic reaction, but not the aerobic reaction, was inhibited by low concentrations of 1,10-phenanthroline, α,α'-bipyridyl, riboflavin, FMN, and FAD. These inhibitory effects of iron-chelating agents are consistent with the observations of Lascelles (1956) that in iron deficiency, the cells of *Rp. sphaeroides* excrete coproporphyrin under anaerobic growth conditions, but not under aerobic conditions. All the aerobic activity of disrupted cells was found in the soluble fraction of the crude extract; activity in anaerobic incubations was detected only when the soluble and the particulate fractions were combined. Since the anaerobic reaction is not catalyzed by extracts from *Rp. sphaeroides* that have been grown aerobically, it is clear that this organism has two different enzymes concerned in protoporphyrin formation from coproporphyrinogen, the anaerobic enzyme being formed only when the organism is grown under anaerobic conditions.

If low-molecular-weight materials are removed from crude extracts of anaerobically grown *Rp. sphaeroides*, the enzyme that catalyzes the anaerobic reaction becomes unstable and also needs the addition of nicotinamide nucleotides, in addition to Mg^{2+}, ATP, and methionine, for protoporphyrin to be formed in an anaerobic incubation. NAD^+, $NADP^+$, NADH, or NADPH is an effective supplement, but the greatest stimulation of activity is given by mixtures of NADPH and NAD^+ or NADH and $NADP^+$. The effectiveness of the reduced pyridine nucleotides is surprising if it is assumed that the pyridine nucleotides are required as electron acceptors to permit the reaction to proceed in the absence of oxygen. Tait suggests that *S*-adenosylmethionine acts as a cofactor in this reaction, and that methionine is active only after being converted to *S*-adenosylmethionine. Possibly the *S*-adenosylmethionine is an allosteric effector, and Tait points out that this compound occupies an important position in Bchl synthesis in *Rp. sphaeroides*. In addition to being involved in the methylation of magnesium protoporphyrin (see Section 4.2), it plays some part in controlling the insertion of magnesium into protoporphyrin (see Section 4.2).

Tait (1972) also showed that crude extracts of *Chr. vinosum* exhibited coproporphyrinogenase activity under anaerobic conditions, in the presence of *S*-adenosylmethionine or ATP plus methionine. The addition of NADH plus $NADP^+$ stimulated the reaction, which appears closely similar to that catalyzed by extracts of anaerobically grown *Rp. sphaeroides*. This result is in contrast with the report of Mori and Sano (1972), who added the ATP, Mg^{2+}, and *S*-adenosylmethionine to their incubations of *Chr. vinosum* extracts, but apparently not in combination with pyridine nucleotides.

It is of some interest to compare these results with those of Poulson and Polglase (1974), who described the properties of a 150-fold-purified coproporphyrinogen oxidative decarboxylase obtained from yeast mitochondria. This enzyme worked under aerobic or anaerobic conditions. Aerobically, it needed no cofactors; under anaerobic conditions, there was an absolute requirement for a divalent metal ion, such as Fe^{2+} or Mn^{2+}, together with ATP and L-methionine and NAD^+ or $NADP^+$ as electron acceptor. Anaerobic activity was inhibited by 5 mM GSH or L-ethionine, which did not affect aerobic activity. In this case, however, the same enzyme appears to catalyze activity in either aerobic or anaerobic incubations, and the specific activities were roughly equal when the enzyme was assayed aerobically or anaerobically.

2.6. Conversion of Protoporphyrinogen IX to Protoporphyrin IX

The enzyme protoporphyrinogen oxidase is assumed to be necessary to catalyze the final step in protoporphyrin biosynthesis, although under aerobic conditions, this reaction will proceed spontaneously. Sano and Granick (1961) and Porra and Falk (1964) suggested that an enzyme participated in this reaction, and recently, Poulson and Polglase (1975) described the solubilization and partial purification of protoporphyrinogen oxidase from the mitochondrial membrane fraction of *Saccharomyces cerevisiae*. The enzyme did not catalyze the oxidation of coproporphyrinogen III or uroporphyrinogen III. The molecular weight of the enzyme was $180,000 \pm 18,000$, and the K_m for protoporphyrinogen was approximately 4.8 μM. It appeared sensitive to inhibition by heme or hemin (50% inhibition at about 20 μM).

There is no report of a similar enzyme in the photosynthetic bacteria, but evidence using doubly labeled porphyrinogens suggests that the conversion of protoporphyrinogen to protoporphyrin is catalyzed by hemolysates of chicken erythrocytes, and is a stereospecific process (Jackson *et al.*, 1973).

3. Formation of Hemes. Ferrochelatase: Enzymic Insertion of Fe^{2+} into Porphyrins

The name *ferrochelatase* (protoheme ferro-lyase, E.C. 4.99.1.1) was given to the enzyme, first detected in erythropoietic tissue (e.g., Krueger *et al.*, 1956), that

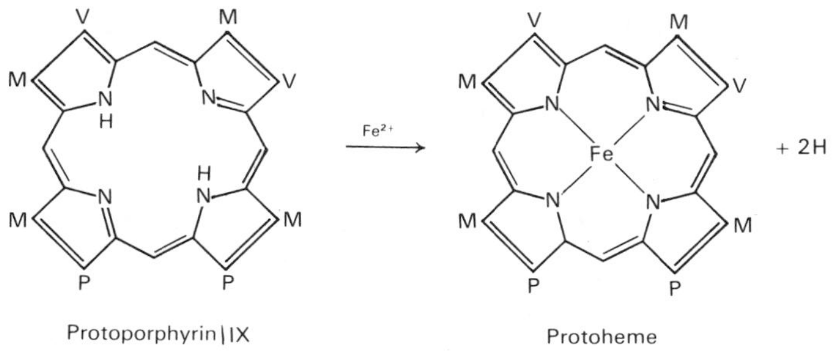

Fig. 13. Representation of the ferrochelatase reaction. (M) $-CH_3$; (V) $-CH=CH_2$; (P) $-CH_2-CH_2-COOH$.

catalyzes the insertion of ferrous ions into proto-porphyrin to form protoheme (Fig. 13). The enzyme is widely distributed, being found not only in such tissues as bone marrow and reticulocytes, but also in organisms and tissues that do not necessarily form hemoglobin. In most eukaryotic cells, ferrochelatase is an insoluble enzyme associated with the mito-chondrial membrane, but in higher plants, ferro-chelatase is also present in both mitochondria and chloroplasts (Porra and Lascelles, 1968; Little and Jones, 1976). In prokaryotic cells, the enzyme is usually attached to the cytoplasmic membrane. The protoheme that is formed may serve directly as the prosthetic group of the b-type cytochromes or, after modification and combination with the appropriate protein, act as the prosthetic group of c- and a-type cytochromes.

Brief preliminary work showed that the fer-rochelatase of *Chr. vinosum* strain D was similar in general properties to the ferrochelatase of nonphoto-synthetic bacteria and of mitochondria (Porra and Jones, 1963), but more detailed analysis of a ferro-chelatase from a photosynthetic bacterium has been carried out using extracts of *Rp. sphaeroides*. The metal specificity of *Rp. sphaeroides* ferrochelatase is not restricted to Fe^{2+}; the membrane fraction from photosynthetically grown *Rp. sphaeroides* catalyzes the insertion of Co^{2+} and Zn^{2+} into protoporphyrin at rates equal to, or greater than, the rate of Fe^{2+} insertion (Johnson and Jones, 1964; Neuberger and Tait, 1964). This broad metal specificity could possibly be due to the presence of more than one metal-chelating enzyme, reflecting the requirement for the synthesis of three different groups of metallo-porphyrins—vitamin B_{12}, Chl's, and hemes—but that the metal specificity of rat liver mitochondrial ferro-chelatase (Jones, M. S., and Jones, 1969) is closely similar to that of *Rp. sphaeroides* makes such an explanation unlikely. Since the ions Co^{2+}, Fe^{2+}, Zn^{2+},

Ni^{2+}, Cu^{2+}, and Mn^{2+} were mutually competitive as metal substrates for the membrane-bound ferro-chelatase of *Rp. sphaeroides* (Jones, M. S., and Jones, 1970), it is probable that only one metal chelatase is being assayed under the conditions used in these experiments, and that this chelatase is not specific for Fe^{2+}, but can utilize other divalent transition metals.

Although much of the ferrochelatase activity of sonicated or French-press-disrupted cells of *Rp. sphaeroides* is found in the particulate fraction, some activity is recovered in the supernatant after centrifugation at $100,000g$ for 2 h. This soluble ferro-chelatase differs in its specificity for porphyrins, but not for metals, from the particulate enzyme. The soluble enzyme utilizes only the "nonbiological" porphyrins, deuteroporphyrin and hematoporphyrin; the particulate enzyme utilizes deutero-, hemato-, meso-, and protoporphyrin (Jones, M. S., and Jones, 1970). The amount of soluble ferrochelatase in *Rp. sphaeroides* extracts is variable, and its relationship to other known ferrochelatases is uncertain; it is possibly a denatured form of the membrane-bound enzyme. The K_m for deuteroporphyrin for the particulate enzyme has been accurately determined as 21.3 μM; the K_m for protoporphyrin is a little lower.

Studies of the kinetic behavior of the ferrochelatase of *Rp. sphaeroides* (Jones, M. S., and Jones, 1970) showed that binding one substrate (e.g., metal ion) did not affect the binding of the second substrate (e.g., porphyrin), and suggested that the reaction has a sequential mechanism. This means that in enzymic metalloporphyrin formation, proton dissociation from the central nitrogen atoms of the porphyrin does not precede reaction with the metal cation. The chelatase reaction of *Rp. sphaeroides* was inhibited by free protoheme (50% inhibition around 8 μM for normal assay conditions) or by magnesium protoporphyrin (50% inhibition at around 5 μM), and these effects may be important in ensuring the balanced production of hemes and Chl's in an organism in which the ratio of these two components varies hugely in response to changes in growth conditions (see Chapter 42 for a discussion of this problem).

The iron substrate for ferrochelatase is the rather unstable Fe^{2+} ion, which is readily autoxidized; Fe^{3+} is not incorporated directly (Porra and Jones, 1963). There is no evidence that the photosynthetic bacteria secrete any iron-scavenging chelators for extracting iron from the medium, but membrane preparations of *Rp. sphaeroides* have the capacity to reduce Fe^{3+} to Fe^{2+} when an oxidizable substrate such as NADH or succinate is supplied. It is apparent that some reduced component of the membrane-bound electron-transport system is able to use Fe^{3+} as an electron acceptor. So, heme synthesis can take place, even under aerobic conditions in which Fe^{2+} is rapidly autoxidized to Fe^{3+}, as long as the oxidizable substrates are provided to regenerate Fe^{2+}. Because of the instability of Fe^{2+}, the K_m for this substrate is not easily measured. The K_m for Co^{2+} was 6.13 μM.

The rate of enzymic formation of metallo-porphyrins by particulate preparations from *Rp. sphaeroides* is markedly reduced following removal of lipid from the membranes by extraction with organic solvents or detergents. The addition of crude lipid fractions partly restored activity. Phosphatidic acid stimulated the particulate ferrochelatase, and a number of organic solvents, e.g., a saturated solution of diethyl ether, also stimulated this activity severalfold. (Mazanowska *et al.*, 1966). It seems possible that the enzyme is a lipoprotein in which the substrate-binding is facilitated by the addition of organic solvents.

The synthetic route to the heme of cytochrome *c* and the heme *a* of the cytochrome oxidase formed by aerated cultures of *Rp. sphaeroides* has not been established. There is evidence from experiments in which labeled protoheme was supplied to heme-requiring organisms that heme *a* is formed by direct modification of precursor protoheme (Sinclair *et al.*, 1967), and that cytochrome *c* is also formed from precursor protoheme in *Spirillum itersonii* and *Physarum polycephalum* (Lascelles *et al.*, 1969; Colleran and Jones, 1973). A likely sequence of intermediates is shown in Fig. 14.

4. Biosynthetic Route from Protoporphyrin to Bacteriochlorophyll *a*

4.1. Introduction

An important difference between Chl synthesis in photosynthetic bacteria and that in higher plants is that in bacterial Chl biosynthesis, there is no evidence for an obligatory light-dependent reaction in the sequence. All the facultative photoheterotrophic photosynthetic bacteria are capable of Chl synthesis in the dark, although this property has not been clearly established for obligate photoheterotrophs or photoautotrophs. Much of the available information about Bchl biosynthesis has been obtained by using members of the family Rhodospirillaceae. Many of these organisms have a vigorous aerobic metabolism and can be grown either photosynthetically, under anaerobic conditions, or aerobically, in the dark if necessary. In the case of *Rp. sphaeroides*, the rate of Chl synthesis is regulated by the oxygen concentration of the medium (Cohen-Bazire *et al.*, 1957) (see also Lascelles, Chapter 42, Section 4), as well as by other factors, such as light intensity. Cultures of *Rp. sphaeroides* are commonly used in the study of Bchl formation because of the ease with which Bchl synthesis can be induced or suppressed in this organism and because mutants blocked in Bchl synthesis can be readily obtained and maintained in dark aerobic culture.

Many of the intermediates in the biosynthesis of Bchl *a* have not been identified unambiguously, and so a number of gaps exist in present representations of the synthetic pathway. Several difficulties have contributed to this uncertainty. One has been the difficulty of synthesis and instability of the many possible intermediates, although recent advances in synthetic techniques have partly solved this problem (e.g., Kenner *et al.*, 1974). Another difficulty has been the repeated failure to obtain measurable, consistent rates of utilization of probable intermediates by disrupted-cell preparations of photosynthetic bacteria. This has led to a common belief that free magnesium tetrapyrroles are not intermediates, but are active only after

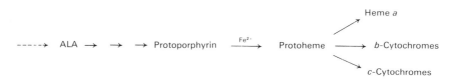

Fig. 14. The likely route to the heme prosthetic groups of cytochromes of photosynthetic bacteria.

complexing with some specific protein, by analogy with the protochlorophyllide holochrome involved in Chl synthesis in higher plants (cf. Smith, J. H. C., 1960). Much of the available information about Bchl *a* biosynthesis has been obtained from the identification of the pigments that accumulate in mutant bacteria that are incapable of Bchl synthesis or in bacteria in which Bchl synthesis has been partly blocked by the addition of inhibitors. A scheme arranging such pigments in a likely biosynthetic sequence is given in Fig. 15, but requires supporting evidence from other types of experimentation before it can be given unreserved acceptance. The available evidence is reviewed in more detail below.

4.2. Protoporphyrin to Magnesium Protoporphyrin Monomethyl Ester

Magnesium protoporphyrin was identified as an intermediate in the biosynthesis of Chl *a* in *Chlorella*, since it accumulated in Chl-less mutants (Granick, 1948). Subsequently, Granick (1961) identified magnesium protoporphyrin monomethyl ester as a product that accumulated in other Chl-less mutants of this organism. These mutant products are most easily explained if the early steps in the Chl pathway in higher plants are:

Protoporphyrin → magnesium protoporphyrin → magnesium protoporphyrin monomethyl ester

Subsequent evidence on the nature of the methylating enzyme in plants has supported this scheme (Radmer and Bogorad, 1967). There is considerable evidence that this pathway is the same in *Rp. sphaeroides* as in *Chlorella*.

Wild-type cells of *Rp. sphaeroides* excreted magnesium protoporphyrin monomethyl ester into their growth medium (Jones, O. T. G., 1963*a*), even under conditions of normal photosynthetic growth. The production of this pigment was increased under conditions in which the iron concentration of the medium was restricted. Suspensions of *Rp. capsulata* also excreted magnesium protoporphyrin monomethyl ester into the medium in an iron-dependent fashion, with an optimum iron requirement around 0.02 mM (Cooper, 1963). Higher iron concentrations suppressed production of the pigment, although more Bchl was then produced, suggesting that iron is required in the later stages of Bchl production. The synthesis of magnesium protoporphyrin monomethyl ester could be increased 10- or 20-fold by the addition of 0.2% of the detergent Tween 80 to the medium. The addition of methionine to the suspending medium also increased the extent of magnesium protoporphyrin

monomethyl ester production. This last effect is readily explained from what is known of the properties of the enzyme from *Rp. sphaeroides* that methylates magnesium protoporphyrin to form the monomethyl ester. This enzyme, described by Gibson *et al.* (1963), was given the name *S*-adenosylmethionine-magnesium protoporphyrin methyl transferase (E.C.2.1.1.11). It catalyzes the transfer of the methyl group from *S*-adenosylmethionine to magnesium protoporphyrin. The activity is found only in the membrane fraction prepared from photosynthetically grown cells. It is present at a low specific activity in cells of *Rp. sphaeroides* grown in the dark with high aeration, but the enzyme activity increases with the onset of anaerobic conditions and Bchl *a* synthesis (Gorchein *et al.*, 1968). Zinc protoporphyrin, magnesium mesoporphyrin, and magnesium deuteroporphyrin were substrates for the enzyme, as well as magnesium porphyrin, but protoporphyrin itself was a poor substrate. This specificity supports the view that magnesium incorporation precedes the methylation step. The enzyme was inhibited by *S*-adenosyl ethionine, which helps to explain the effect of ethionine inhibition of Bchl biosynthesis (Gibson *et al.*, 1962).

Although magnesium protoporphyrin is almost certainly an intermediate in Bchl synthesis, its formation by extracts of photosynthetic bacteria has yet to be demonstrated. The enzymic incorporation of magnesium into protoporphyrin *in vitro* has been unsuccessfully attempted by many groups of research workers, using material from higher plants as well as from bacteria. The ease with which contaminant Zn^{2+} ions are incorporated (see Section 3.1) into protoporphyrin by the enzyme ferrochelatase during attempts at Mg^{2+} incorporation is probably responsible for the few reported measurements of magnesium chelatase activity, since the spectroscopic properties of zinc and magnesium porphyrins are similar. Another approach was successfully used by Gorchein (1972), who attempted indirectly to measure magnesium chelatase activity in whole cells of *Rp. sphaeroides*. The cells were grown photosynthetically in semianaerobic conditions, harvested by centrifugation, and suspended in distilled water in the light under anaerobic conditions. They were supplied with a solution of protoporphyrin dissolved in a crude extract of lipids prepared from photosynthetically grown *Rp. sphaeroides*. After a 90-min incubation, the cells were extracted with ammoniacal acetone, and any accumulated porphyrins were separated and determined. It was found that considerable quantities of magnesium protoporphyrin monomethyl ester were formed by the treated cells. The reaction did not require added Mg^{2+}. Indeed, high concentrations of added Mg^{2+} inhibited the reaction. Presumably, the whole cells contain sufficient Mg^{2+} for the synthesis of this metalloporphyrin. No unesterified magnesium protoporphyrin was found, and the reaction was inhibited by ethionine, an inhibitor of the methylation reaction. This caused Gorchein to suggest that insertion of magnesium into protoporphyrin is obligatorily coupled to the methylation reaction, possibly by a multienzyme complex.

The chelatase reaction of the whole cells was inhibited by the presence of partial pressures of oxygen greater than 15%, and cells grown under pure oxygen lacked the enzyme. This oxygen sensitivity of the chelatase may be important in regulating Bchl synthesis, since Bchl biosynthesis in *Rp. sphaeroides* is suppressed by oxygenation. Although spheroplasts prepared by treating *Rp. sphaeroides* cells with lysozyme retained activity, the disruption of the photosynthetic cells, even by the gentlest techniques, caused a loss of chelatase activity (Gorchein, 1973), and this has prevented detailed studies of the reaction mechanism. An interesting observation made by Gorchein in the course of these experiments is that if isotopically labeled protoporphyrin and magnesium protoporphyrin monomethyl ester were added to whole-cell suspensions of *Rp. sphaeroides*, they were incorporated into Bchl; this suggests that other postulated intermediates in Bchl synthesis could be tested in this relatively simple system.

4.3. Intermediates between Magnesium Protoporphyrin Monomethyl Ester and Bacteriochlorophyll *a*

A number of pigments from inhibited cultures or mutants of *Rp. sphaeroides* have been isolated and characterized and would fit logically on a route between magnesium protoporphyrin monomethyl ester and Bchl (see Fig. 15). The sequence in which the pigments are placed is based on the assumption that chlorophyllide *a*, an intermediate in higher-plant Chl *a* synthesis, is also an intermediate in Bchl synthesis, and that it is formed in the photosynthetic bacteria by a pathway with the same intermediates as in higher plants. The two immediate precursors of chlorophyllide *a* are then likely to be protochlorophyllide (magnesium 2-vinyl, 4-ethyl pheoporphyrin a_5 monomethyl ester) and magnesium 2,4-divinylpheoporphyrin a_5 monomethyl ester (see Fig. 15). Intermediates in the formation of the isocyclic ring have not been found in the photosynthetic bacteria.

Magnesium 2,4-divinylpheoporphyrin a_5 monomethyl ester was first identified as a pigment that

accumulated in cultures of *Rp. sphaeroides* in which Bchl synthesis was partly inhibited by treatment with 8-hydroxyquinoline (Jones, O. T. G., 1963*b,c*). What is apparently the same pigment, sometimes called bacterial protochlorophyll, is also produced by various mutants of *Rp. sphaeroides* (e.g., Griffiths, M., 1962; Lascelles, 1966*a*; Saunders, 1973), of *Rs. rubrum* (Oelze *et al.*, 1970), and of *Rp. palustris* (Krasnovskii *et al.*, 1970). In the last case, the authors state that this magnesium pheoporphyrin is present both in the phytylated and the unphytylated form. The significance of such a finding is not easily assessed, since evidence from other mutants indicates that in *Rp. sphaeroides*, at least, phytylation is the last step in Bchl synthesis (e.g., Richards and Lascelles, 1969; Brown and Lascelles, 1972), as well as in Chl *a* synthesis in higher plants (e.g., Griffiths, W. T., 1974*b*, 1975). The phytylated forms may not be true intermediates.

Excretion of pigments from Bchl-less mutants was enhanced when the cells were grown in conditions of low aeration and also by the addition of 0.2% Tween 80 to the medium. The magnesium divinylpheoporphyrin a_5 was excreted into the medium in association with protein, and within the mutant cells of *Rp. sphaeroides* is attached to the cytoplasmic membrane (Jones, unpublished). Pradel and Clement-Metral (1975) found that a mutant strain of *Rp. sphaeroides* excreted magnesium 2,4-divinylpheoporphyrin a_5 monomethyl ester bound to a macromolecular complex composed of 49% protein, 44% lipids, and 7% pigment, with traces of sugar. The complex exhibited different spectral maxima at different ionic strengths and pH conditions. The complex absorbed at 636 nm, had a molecular weight of 130,000 daltons, and contained 16 pigment molecules. In many of its properties, it resembled the protochlorophyllide holochrome of higher plants.

Drews *et al.* (1971) described a mutant of *Rp. capsulata* that was unable to synthesize Bchl, but that produced protochlorophyll, i.e., a phytylated pigment, together with its magnesium-free derivative, protopheophytin; both pigments were excreted into the medium. The rate of pigment formation by the mutant was dependent on the oxygen tension: in strictly aerobic cultures, no pigment was detectable, whereas under semianaerobic conditions, pigment was formed, and some of the pigment was found in the intracellular membrane of the mutant. The protopheophytin and protochlorophyll found in the medium appear to be excreted by *Rp. capsulata* in a complex consisting of 38% protein, 20% pigment, 13% fatty acids, and 15.5% sugars. The molecular weights of proteins from the pigment complex, determined by

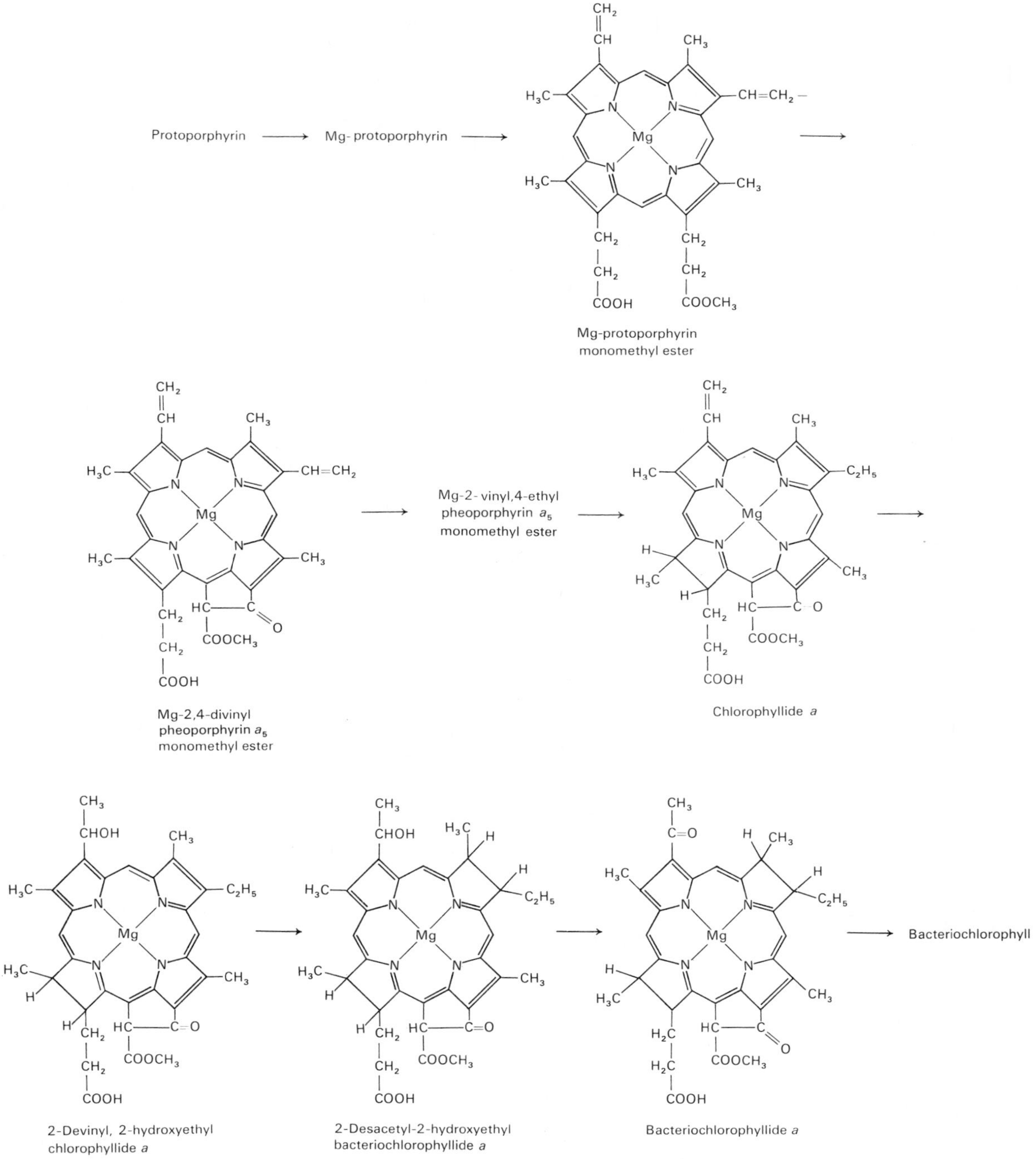

Fig. 15. A proposed biosynthetic route from protoporphyrin to bacteriochlorophyll a.

polyacrylamide gel electrophoresis, did not correspond with those of membrane proteins.

As discussed above, complexes of magnesium porphyrins with protein have been found in a number of mutant photosynthetic bacteria, and these observations may indicate that the later stages of Bchl synthesis involved bound and not free pigments. It may be pointed out, however, that magnesium 2,4-divinyl-pheoporphyrin a_5 monomethyl ester produced by a mutant of *Rp. sphaeroides*, extracted free from any protein and purified by repeated thin-layer chromatography, could be used as an intermediate in Chl synthesis. This pure pigment, when dissolved in dilute cholate, was transformed with high efficiency into chlorophyllide a by flash-illuminated etioplast membranes of barley when supplied with NADPH as reductant (Griffiths, W. T., and Jones, 1975). Protochlorophyllide was an intermediate in the reaction, and pure protochlorophyllide could also be added to the plastid membranes and subsequently converted to chlorophyllide. These experiments have shown that protein-free pigments can be used as intermediates in Chl a synthesis. In higher plants, the specific holochrome protein that catalyzes the light-dependent reduction of protochlorophyllide has the capacity to recycle following a light flash, and an analogous situation may obtain in the photosynthetic bacteria even if the reductive step is not light-dependent.

The conversion of magnesium 2,4-divinylpheoporphyrin a_5 monomethyl ester to protochlorophyllide is a reductive step (calling for the reduction of a vinyl group), and so is the conversion of protochlorophyllide to chlorophyllide a, a reaction that results in the desaturation of a double bond in ring D. In higher plants, both these reactions are driven by NADPH (Griffiths, W. T., and Jones, 1975), and W. T. Griffiths (1974a) has evidence that the availability of NADPH may limit the rate of Chl synthesis in higher plants. Even in gymnosperms, in which Chl a synthesis takes place in seedlings, in the dark, NADPH is utilized for reduction of magnesium 2,4-divinylpheoporphyrin a_5 monomethyl ester and of protochlorophyllide by the plastids from dark-grown seedlings (Griffiths and Mapleston, unpublished). It is tempting to speculate that similar reductive reactions are involved in Bchl synthesis, and that under aerobic conditions, the intracellular concentration of reduced pyridine nucleotides may fall to a level inadequate to maintain Bchl production.

The pigments accumulated by Bchl-less mutants of photosynthetic bacteria sometimes lack magnesium; they are pheoporphyrins or pheophorbides. Since this loss of magnesium can occur readily under mildly acidic conditions, it is assumed that the "natural"

intermediate is the magnesium complex, and that magnesium-free pigments are degradation products. Thus, chlorophyllide a is included in the biosynthetic sequence (see Fig. 15), although pheophorbide a is the compound that has been isolated from the medium of *Rp. sphaeroides* (Jones, O. T. G., 1964; Richards and Lascelles, 1969). Krasnovskii *et al.* (1970) found that a pigment with the spectroscopic and chromatographic properties of Chl a can be detected in mutants of *Rp. capsulata*; i.e., there is some evidence that phytylation can take place at or before this intermediate stage in *Rp. capsulata*.

Magnesium 2-devinyl, 2-hydroxyethyl chlorophyllide a and its magnesium-free derivative have been detected in mutants of *Rp. sphaeroides* (Jones, O. T. G., 1964; Lascelles, 1966b; Richards and Lascelles, 1969), and possibly in *Rp. palustris* (Krasnovskii *et al.*, 1970). Since the excretion of pigment into the medium was increased in the presence of Tween 80, Lascelles (1966a) suggested that during growth, it was normally bound to lipid or protein with the cells. She found that over 80% of the pigment was retained within the bacteria, in the absence of Tween 80, and that it was largely bound to particulate material. After disruption of the cells, treatment of the membranes with Tween 80 caused the liberation of some of the pigment.

Two modifications of the tetrapyrrole structure are necessary for the conversion of magnesium 2-devinyl, 2-hydroxyethyl chlorophyllide a to bacteriochlorophyllide a: the hydroxyethyl substituent at position 2 must be dehydrogenated to form the 2-acetyl group, and a double bond in ring B of the tetrapyrrole must be reduced to give the bacteriochlorin ring structure. Since Richards and Lascelles (1969) identified 2-desacetyl, 2-α-hydroxyethylbacteriochlorophyllide a as an excreted product in a Bchl-less mutant of *Rp. sphaeroides*, it must be assumed that reduction of ring B precedes formation of the acetyl group (see Fig. 15). In the same communication, Richards and Lascelles (1969) described the identification of bacteriochlorophyllide a produced by a mutant of *Rp. sphaeroides* that is unable to synthesize Bchl a, and suggested that the addition of phytol is the final step in the biosynthetic process.

The identification of postulated early intermediates in the phytylated form conflicts with a simple unbranched path such as shown in Fig. 15, and the reported isolation from a mutant of *Rp. sphaeroides* of a pigment identified as 2-desacetyl-2-vinylbacteriopheophorbide a (Pudek and Richards, 1975) cannot be reconciled with this scheme. This pigment was also excreted as a pigment–protein complex. The magnesium derivative of the pheophorbide (Fig. 16) cannot be placed on the biosynthetic pathway il-

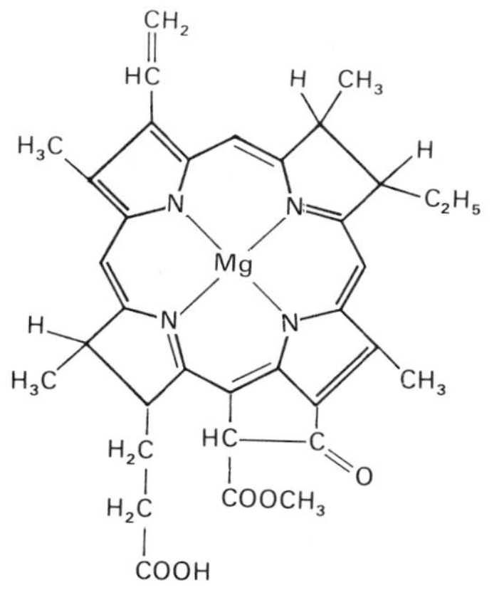

Fig. 16. 2-Desacetyl-2-vinylbacteriochlorophyllide *a*.

lustrated in Fig. 15, and indeed Pudek and Richards drew up a theoretical "metabolic grid" in which the necessary side-chain modifications are postulated to occur while the tetrapyrrole nucleus is at the porphyrin, chlorin, or tetrahydroporphyrin level, more or less at random, giving perhaps three different routes from magnesium protoporphyrin to Bchl. It is apparent that these problems will be resolved only when convincing evidence is available that any postulated intermediate can be incorporated into Bchl or can undergo appropriate transformation reactions *in vitro*.

The discovery of bacteriochlorophyllide *a* accumulations in Bchl-less mutants of *Rp. sphaeroides* (Richards and Lascelles, 1969) is in accord with the sequence of events that is believed to occur in Chl *a* synthesis in higher plants, in which phytylation is the ultimate event in the synthetic sequence. Indeed, Brown and Lascelles (1972) showed that phytol was not synthesized by a variety of mutant strains of *Rp. sphaeroides* that were blocked at various stages of Bchl synthesis, and they suggested that phytol synthesis was tightly coupled to the synthesis of the pyrrole component of Bchl.

5. Biosynthesis of Chlorobium Chlorophylls

The chlorobium chlorophylls are the characteristic pigments of the Chlorobiaceae. Each species contains a mixture of homologues that is called Bchl *c* (or chlorobium Chl 660), a mixture called Bchl *d* (or chlorobium Chl 650), or a mixture called Bchl *e*. Analysis of pheophorbides obtained from chlorobium Chl's by demetallation and hydrolysis revealed the presence of a number of homologues of 2-devinyl-2-(1-hydroxyethyl) pyropheophorbide *a*, as indicated in

Fig. 17; in some cases, the identity of the homologues has not been established beyond doubt. All the Chl's are esters of all-*trans* farnesol. Bchl's *c* and *e* have a *meso* alkyl substituent, and Bchl *e* has a formyl substituent in place of a methyl at position 3.

The determination of the biosynthetic route to the homologous mixture of Chl's found in Chlorobiaceae (see Fig. 17 and Table 1) is made even more difficult by the consistent presence in these organisms of small amounts of Bchl *a* as a component of the photosynthetic reaction center (Jensen *et al.*, 1964). All members of the Chlorobiaceae must possess enzymes necessary for Bchl synthesis, and some of these enzymes may catalyze reactions in common with those necessary for the biosynthesis of the appropriate chlorobium Chl. An example will serve to illustrate this point: crude extracts of *Chlorobium limicola* have been shown to contain the enzyme S-adenosylmethionine magnesium protoporphyrin methyltransferase, with a specific activity approaching that found in crude extracts of *Rp. sphaeroides* or *Rs. rubrum* (Jones, O. T. G., 1968), and it is not unreasonable to suggest that magnesium protoporphyrin monomethyl ester is an intermediate in the synthesis of chlorobium Chl. It is equally possible, however, that the methyl transferase is present only to synthesize the Bchl *a*, present as about 5% of the total Chl. This comment must be borne in mind in considering the early evidence about the biosynthesis of chlorobium Chl's. Thus, Uspenskaya and Kondrat'eva (1964) and Richards and Rapoport (1966) showed that *Chl. limicola f. thiosulpatophilum* excretes uroporphyrins I and III and coproporphyrin I and III, together with porphyrins with two, three, five, six, and seven carboxyl groups and a number of tetracarboxylic porphyrins. No porphyrins with *meso* alkylation were excreted. Godnev *et al.* (1966) showed that the same organisms excreted magnesium protoporphyrin, and Richards and Rapoport (1967) described conditions, in media supplemented with organic substrates plus ethionine, in which these bacteria excreted high concentrations of protoporphyrin monomethyl ester and magnesium protoporphyrin monomethyl ester. On the basis of these observations, Richards and Rapoport suggested that the chlorobium Chl's −650 and −660 were formed with magnesium protoporphyrin monomethyl ester as an intermediate, with the addition of "extra" carbon atom side chains taking place at a later stage, perhaps even by modification of Bchl *a* itself. The loss of the 10-carbomethoxy group was proposed to be a late stage in the process. In agreement with this, Godnev *et al.* (1966) found that the addition of protoporphyrin IX stimulated Chl synthesis by Chlorobiaceae.

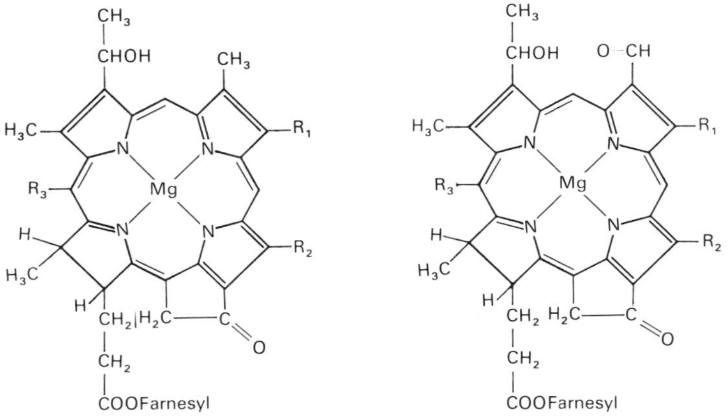

Bacteriochlorophylls *c* and *d*
(known also as chlorobium
chlorophylls 660 and 650)

Bacteriochlorophyll *e*

Fig. 17. Structures of bacteriochlorophylls *c*, *d*, and *e*. These are also known as the chlorobium chlorophylls. Bacteriochlorophyll *c* is a mixture of at least six homologues: R_1 = isobutyl, *n*-propyl, or ethyl; R_2 = ethyl or methyl; R_3 = methyl (and possibly ethyl; see Kenner *et al.*, 1976). Bacteriochlorophyll *d* is a mixture of six homologues: R_1 = isobutyl, *n*-propyl, or ethyl; R_2 = ethyl or methyl; R_3 = H. Bacteriochlorophyll *e* is a mixture of homologues: R_1 = ethyl, *n*-propyl, or isobutyl; R_2 = ethyl; R_3 = methyl.

An alternative approach to the study of chlorobium Chl biosynthesis has been initiated at Liverpool University by Kenner and his collaborators, who have made extensive studies on the NMR spectroscopy of the chlorobium Chl's 660 (Bchl *c*). In particular, they have been able to identify the resonances from the α, β, γ, and δ carbons, confirming that the δ-carbon bears the methyl group (Smith, K. M., and Unsworth, 1975). Bchl *c* differs from Bchl *a* in possessing a δ-methyl substituent, extra methyl units attached to the 4- and 5-side chains, and the hydrated vinyl group at carbon 2, as well as in lacking the methoxy carbonyl at C-10. On the basis of the ease of comparable reactions in model systems, Kenner *et al.* (1976) pointed out that the methylation of the 5-methyl group in protoporphyrin IX, to give a 5-ethyl group, is an unlikely reaction, and that the methylation of side chains would be easier if they were activated toward electrophilic attack. Such activation is available in the side chains of probable biosynthetic precursors, such as uroporphyrinogen III.

After addition of [^{13}C]L-methionine to cultures of *Chloropseudomonas ethylica* (this is a mixed species the classification of which has been somewhat uncertain; see Chapter 2), Kenner *et al.* (1976) extracted and purified the pheophorbides (660) and recorded the NMR spectra. They found considerable enhancement of the resonances attributed to the δ-methyl group and of the methyl group of the 5-ethyl function. Further examination of the ^{13}C-enriched and natural abundance spectra showed that two other signals were significantly enhanced. These correspond with resonances assigned to the terminal methyl in the *n*-propyl group and to the methyl in the isobutyl group. Thus, L-methionine is identified as the biosynthetic origin of the terminal methyl groups in the 4-*n*-propyl, 4-isobutyl, and 5-ethyl functions and the δ-*meso*-methyl for the (660) Chl's from *Chloropseudomonas ethylica*.

In preliminary experiments, Kenner *et al.* (1976) found that sonicated cells of *Chloropseudomonas ethylica* incorporated counts from isotopically labeled porphyrinogens into the chlorobium Chl's. ^{14}C-Labeled uroporphyrinogen III gave incorporations ranging from 2 to 5.4%, whereas tritium-labeled coproporphyrinogen III and protoporphyrinogen IX gave incorporations of 0.4 and 0.3%, respectively. These results are in accord with a biosynthetic scheme whereby the biosynthesis of chlorobium Chl's branches from "normal" Chl synthesis somewhere between uroporphyrinogen III and coproporphyrinogen III. Since the loss of four molecules of carbon dioxide from uroporphyrinogen III in the formation of coproporphyrin III is known to occur in a stepwise manner (see Section 2.4), this branching need not occur at uroporphyrinogen III, and indeed Kenner and co-workers favor a pentacarboxylic porphyrinogen as the key intermediate. The use of ^{13}C NMR spectroscopy should enable such a scheme to be rigorously tested when synthesis of the chlorobium Chl's can be achieved consistently *in vitro*.

6. Formation of the Esters of the Bacterial Chlorophylls

Many higher plants have been found to contain the enzyme chlorophyllase, which catalyzes the hydrolysis of Chl *a* or *b* with the release of phytol, and it has frequently been suggested that this enzyme is involved in the final stage of Chl synthesis, the formation of phytol esters (for a review of some of this literature, see Jones, O. T. G., 1976). The evidence is not completely convincing, since because of the insolubility of the reactants in aqueous buffers, the reaction is measured in very unphysiological media, containing high concentrations of acetone or detergent, and it is frequently found that the rate of the reaction in the forward direction (toward Chl synthesis) is negligible. Similar problems have been found with the chlorophyllase of photosynthetic bacteria, and Uspenskaya (1972) pointed out that in *Rp. palsustris*, there is no correlation between chlorophyllase and Bchl content. The enzyme was present in dark-grown cells and in Bchl-less mutants. Even more disturbing was the observation that chlorophyllase activity (with Bchl *a* as substrate) was present in acetone powders of two nonphotosynthetic bacteria. It is a possibility, then, that chlorophyllase is not involved in biosynthetic reactions, and that Chl or Bchl formation involves some other enzyme or process. It has been suggested that phytol must be activated before esterification will proceed smoothly, and Watts and Kekwick (1974) have evidence that ^{14}C-labeled phytol pyrophosphate is incorporated into Chl by homogenates of leaves of the bean *Phaseolus vulgaris*. As these authors point, out, however, it is possible that the phytol pyrophosphate undergoes hydrolysis by a phosphatase in the crude tissue extract before incorporation of phytol into Chl. Alternatively, it is possible that the chlorophyllide (or bacteriochlorophyllide) is esterified with a precursor of phytol, and that the later stages of biosynthesis do not involve free intermediates. It is believed that in higher plants, phytol is synthesized in the following sequence of reactions (after Goodwin, 1967):

Acetate → mevalonate → isopentenylpyrophosphate → farnesylpyrophosphate → geranylgeranylpyrophosphate → phytol

A protochlorophyllide ester isolated from dark-grown barley leaves contains geranylgeraniol as esterifying alcohol (Liljenberg, 1974).* This finding accords with a postulate that the pyrrole component of

---

* Rüdiger *et al.* (1977) have demonstrated the esterification of chlorophyllide by geranylpyrophosphate catalyzed by extracts from maize shoots.

Chl's may be esterified with a phytol precursor. In the photosynthetic bacteria, the esterifying alcohol of Bchl *a* may be geranylgeraniol in *Rs. rubrum* (Katz *et al.*, 1972; Bockman *et al.*, 1973) and in *Rp. photometricum* (Kunzler and Pfennig, 1973), or farnesol in the chlorobium Chl's of Chlorobiaceae (see Holt *et al.*, 1966). This supports the view that the biosynthetic sequence to phytol in the bacteria is the same as that given above, and also shows that phytol precursors can be esterified with the pyrrole component of bacterial Chl's. Brown and Lascelles (1972) found that phytol synthesis by *Rp. sphaeroides* does not take place in the absence of the synthesis of the pyrrole component of Bchl *a*, which indicates that the two biosynthetic pathways are intimately associated.

7. References

Akhtar, M., Abboud, M. M., Barnard, G., Jordan, P., and Zaman, Z., 1976, Mechanism and stereochemistry of enzymic reactions involved in porphyrin biosynthesis, *Philos. Trans. R. Soc. London Ser. B* **273**:117.

Aronoff, S., 1975, The number of biologically possible porphyrin isomers, *Ann. N. Y. Acad. Sci.* **244**:327.

Battersby, A. R., and McDonald, E., 1976, Biosynthesis of porphyrins and corrins, *Philos. Trans. R. Soc. London Ser. B* **273**:161.

Battersby, A. R., Staunton, J., and Wightman, R. H., 1972a, Biosynthesis of protoporphyrin IX from coproporphyrinogen III, *J. Chem. Soc. Chem. Commun.* **1972**:1118.

Battersby, A. R., Baldas, J., Collins, J., Grayson, D. H., James, K. J., and McDonald, E., 1972b, Mechanism of biosynthesis of the vinyl groups of Protoporphyrin IX, *J. Chem. Soc. Chem. Commun.* **1972**:1265.

Battersby, A. R., Hunt, E., and McDonald, E., 1973, Biosynthesis of type III porphyrins: Nature of the rearrangement process, *J. Chem. Soc. Chem. Commun.* **1973**:442.

Beale, S. I., and Castelfranco, P. A., 1974a, Accumulation of δ-aminolevulinic acid in greening plant tissues, *Plant Physiol.* **53**:291.

Beale, S. I., and Castelfranco, P. A., 1974b, Formation of ^{14}C-δ-aminolevulinic acid from labelled precursors in greening plant tissues, *Plant Physiol.* **53**:297.

Beale, S. I., Gough, S. P., and Granick, S., 1975, Biosynthesis of δ-aminolevulinic acid from the intact carbon skeleton of glutamic acid in greening barley, *Proc. Natl. Acad. Sci. U.S.A.* **72**:2719.

Bogorad, L., 1958a, The enzymatic synthesis of porphyrins from porphobilinogen. I. Uroporphyrin I, *J. Biol. Chem.* **233**:501.

Bogorad, L., 1958b, The enzymatic synthesis of porphyrins from porphobilinogen. II. Uroporphyrin III, *J. Biol. Chem.* **233**:501.

Bogorad, L., 1963, Enzymatic mechanisms in porphyrin synthesis: Possible enzymatic blocks in porphyrins, *Ann. N. Y. Acad. Sci.* **104**:676.

Bogorad, L., and Granick, S., 1953, The enzymatic synthesis of porphyrins from porphobilinogen, *Proc. Natl. Acad. Sci. U.S.A.* **39**:1176.

Brockmann, H., Knobloch, G., Schweer, I., and Trowitzsch, W. 1973, Die Alkoholkomponente des Bacteriochlorophyll *a* aus *Rhodospirillum rubrum*, *Arch. Mikrobiol.* **90**:161.

Brouillet, N., Arselin-De Chateaubodeau, G., and Volland, C. 1975, Studies on protoporphyrin biosynthetic pathway in *Saccharomyces cerevisiae*: Characterisation of the tetrapyrrole intermediates, *Biochemie* **57**:647.

Brown, A. E., and Lascelles, J., 1972, Phytol and bacteriochlorophyll synthesis in *Rhodopseudomonas spheroides*, *Plant Physiol.* **50**:747.

Burnham, B. F., and Lascelles, J., 1963, Control of porphyrin biosynthesis through a negative feedback mechanism, *Biochem. J.* **87**:462.

Cavaleiro, J. A. S., Kenner, G. W., and Smith, K. M., 1974, Pyrroles and related compounds. XXXII. Biosynthesis of protoporphyrin IX from coproporphyrinogen III, *J. Chem. Soc. Perkin. Trans. 1* **1974**:1188.

Cheh, A., and Neilands, J. B., 1973, Zinc, an essential metal ion for beef-liver δ-aminolevulinate dehydratase, *Biochem. Biophys. Res. Commun.* **55**:1060.

Cohen-Bazire, G., Sistrom, W. R., and Stanier, R. Y., 1957, Kinetic studies of pigment synthesis by non-sulfur purple bacteria, *J. Cell. Comp. Physiol.* **49**:25.

Colleran, E. M., and Jones, O. T. G., 1973, Studies on the biosynthesis of cytochrome *c*, *Biochem. J.* **134**:89.

Cooper, R., 1963, Biosynthesis of coproporphyrinogen, magnesium protoporphyrin monomethyl ester and bacteriochlorophyll by *Rhodopseudomonas capsulata*, *Biochem. J.* **89**:100.

Cornford, P., 1964, Transformation of porphobilinogen into porphyrins by preparations from human erythrocytes, *Biochem. J.* **91**:64.

Dalton, J., and Dougherty, R. C., 1969, Formation of the macrocyclic ring in tetrapyrrole biosynthesis, *Nature (London)* **223**:1151.

Davies, R. C., and Neuberger, A., 1973, Polypyrroles formed from porphobilinogen and amines by uroporphyrinogen synthetase of *Rhodopseudomonas sphaeroides*, *Biochem. J.* **133**:471.

Drews, G., Leutiger, I., and Ladwig, R., 1971, Production of protochlorophyll, protophaeophytin and bacteriochlorophyll by the mutant A1 *a* of *Rhodopseudomonas capsulata*, *Arch. Mikrobiol.* **76**:349.

Eales, L., Grosser, Y., and Sears, W. G., 1975, The clinical biochemistry of the human hepatocutaneous porphyrias in the light of recent studies of newly identified intermediates and porphyrin derivatives, *Ann. N. Y. Acad. Sci.* **244**:441.

Fanica-Gaignier, M., and Clement-Metral, J. D., 1971, ATP inhibition of aminolevulinate (ALA) synthetase activity in *Rhodopseudomonas spheroides*, *Biochem. Biophys. Res. Commun.* **44**:192.

Fanica-Gaignier, M., and Clement-Metral, J. D., 1973a, 5-Aminolevulinic acid synthetase of *Rhodopseudomonas spheroides*, *Eur. J. Biochem.* **40**:13.

Fanica-Gaignier, M., and Clement-Metral, J. D., 1973b, Cellular compartmentation of two species of ALA synthetase, *Biochem. Biophys. Res. Commun.* **55**:610.

Fischer, H., and Stern, A., 1940, Bacteriochlorophyll und seine Derivate, in: *Die Chemie des Pyrrols*, Vol. II, Part 2 (H. Fischer and H. Orth, eds.), pp. 305–315, Akademische Verlagsgesellschaft mbH, Leipzig.

Frydman, B., Frydman, R. B., Valasinas, A., Levy, S., and Feinstein, G., 1975, The mechanism of uroporphyrinogen biosynthesis, *Ann. N. Y. Acad. Sci.* **244**:371.

Frydman, B., Frydman, R. B., Valasinas, A., Levy, E. S., and Feinstein, G., 1976, Biosynthesis of uroporphyrinogens from porphobilinogen: Mechanism and the nature of the process, *Phil. Trans. R. Soc. London Ser. B* **273**:137.

Frydman, R. B., and Feinstein, G., 1974. Studies on porphobilinogen deaminase and uroporphyrinogen III cosynthase from human erythrocytes, *Biochem. Biophys. Acta* **350**:358.

Frydman, R. B., Reil, S., and Frydman, B., 1971, Relation between structure and reactivity in porphobilinogen and related pyrroles, *Biochemistry* **10**:1154.

Frydman, R. B., Valasinas, A., Rapoport, H., and Frydman, B., 1972, The enzymatic incorporation of a dipyrrylmethane into uroporphyrinogen III, *FEBS Lett.* **25**:309.

Garcia, R. C., San Martin De Viale, L. C., Tomio, J. M., and Grinstein, M., 1973, Porphyrin biosynthesis: X-porphyrinogen carboxy-lyase from avian erythrocytes—Further properties, *Biochem. Biophys. Acta* **309**:203.

Gibson, K. D., 1958, Biosynthesis of δ-aminolevulinic acid by extracts of *Rp. sphaeroides*, *Biochim. Biophys. Acta* **28**:451.

Gibson, K. D., Laver, W. G., and Neuberger, A., 1958, The formation of δ-aminolevulinic acid from glycıne and succinyl Coenzyme A by particles from chicken erythrocytes, *Biochem. J.* **70**:71.

Gibson, K. D., Neuberger, A., and Tait, G. H., 1962, Studies on the biosynthesis of porphyrin and bacteriochlorophyll by *Rhodopseudomonas sphaeroides*. The effects of ethionine and threonine, *Biochem. J.* **83**:550.

Gibson, K. D., Neuberger, A., and Tait, G. H., 1963, S-adenosyl methionine-magnesium protoporphyrin methyl transferase, *Biochem. J.* **88**:325.

Gloe, A., Pfennig, N., Brockmann, H., and Trowitzsch, W., 1975, A new bacteriochlorophyll from brown-colored Chlorobiaceae, *Arch. Microbiol.* **102**:103.

Godnev, T. N., Kondrat'eva, E. N., and Uspenskaya, V. E., 1966, Possible pathways of biosynthesis of bacterioviridine (Chlorobium chlorophyll), *Izv. Akad. Nauk SSSR Ser. Biol.* **31**:525.

Goodwin, T. W., 1967, Terpenoids and chloroplast development, in: *Biochemistry of Chloroplasts*, Vol. II (T. W. Goodwin, ed.), pp. 721–733, Academic Press, New York.

Gorchein, A., 1972, Magnesium protoporphyrin chelatase activity in *Rhodopseudomonas sphaeroides*. Studies with whole cells, *Biochem. J.* **127**:97.

Gorchein, A., 1973, Control of magnesium-protoporphyrin chelatase activity in *Rhodopseudomonas sphaeroides*. Role of light, oxygen and electron and energy transfer, *Biochem. J.* **134**:833.

Gorchein, A., Neuberger, A., and Tait, G., 1968, Adaptation of *Rhodopseudomonas sphaeroides*, *Proc. R. Soc. London Ser. B* **171**:111.

Granick, S., 1948, Magnesium protoporphyrin as a precursor of chlorophyll in *Chlorella*, *J. Biol. Chem.* **175**:333.

Granick, S., 1961, Magnesium protoporphyrin monoester and protoporphyrin monoethyl ester in chlorophyll biosynthesis, *J. Biol. Chem.* **236**:1168.

Granick, S., and Gassman, M., 1970, Rapid regeneration of protochlorophyllide, *Plant Physiol.* **45**:201.

Granick, S., and Sassa, S., 1971, δ-Aminolevulinic acid synthetase and the control of heme and chlorophyll synthesis, in: *Metabolic Regulation*, Vol. 5 (H. J. Vogel, ed.), pp. 77–141, Academic Press, New York and London.

Griffiths, M., 1962, Further mutational changes in the photosynthetic pigment system of *Rhodopseudomonas sphaeroides*, *J. Gen. Microbiol.* **27**:427.

Griffiths, W. T., 1974a, Source of reducing equivalents for the *in vitro* synthesis of chlorophyll from protochlorophyll, *FEBS Lett.* **46**:301.

Griffiths, W. T., 1974b, Protochlorophyll and protochlorophyllide as precursors for chlorophyll synthesis *in vitro*, *Febs Lett.* **49**:196.

Griffiths, W. T., 1975, Some observations of chlorophyll(ide) synthesis by isolated etioplasts, *Biochem. J.* **146**:17.

Griffiths, W. T., and Jones, O. T. G., 1975, Magnesium 2,4-divinylphaeoporphyrin a_5 as a substrate for chlorophyll biosynthesis *in vitro*, *FEBS Lett.* **50**:355.

Gurne, D., and Shemin, D., 1973, Synthesis of the pyrrole porphobilinogen by Sepharose-linked δ-aminolevulinic acid dehydratase, *Science* **180**:1188.

Heath, H., and Hoare, D. S., 1959, The biosynthesis of porphyrins from porphobilinogen by *Rhodopseudomonas sphaeroides*, *Biochem. J.* **72**:14.

Higuchi, M., and Bogorad, L., 1975, The purification and properties of uroporphyrinogen I synthases and uroporphyrinogen III cosynthase. Interactions between the enzymes, *Ann. N. Y. Acad. Sci.* **244**:401.

Hoare, D. S., and Heath, H., 1959, The biosynthesis of porphyrins from porphobilinogen by *Rhodopseudomonas sphaeroides*. 2. The partial purification and some properties of porphobilinogen deaminase and uroporphyrinogen decarboxylase, *Biochem. J.* **73**:679.

Holt, A. S., Purdie, J. W., and Wasley, J. W. F., 1966, Structures of chlorobium chlorophylls (660), *Can. J. Chem.* **44**:88.

Jackson, A. H., and Games, D. E., 1975, The later stages of porphyrin biosynthesis, *Ann. N. Y. Acad. Sci.* **244**:591.

Jackson, A. H., Games, D. E., Couch, P., Jackson, J. R., Belcher, R. V., and Smith, S. G., 1973, Conversion of coproporphyrinogen III to protoporphyrin IX, *Z. Physiol. Chem.* **354**:865.

Jackson, A. H., Sancovich, H. A., Ferramola, A. M., Evans, N., Games, D. E., Matlin, S. A., Elder, G. H., and Smith, S. G., 1975, Macrocyclic intermediates in the biosynthesis of porphyrins, *Philos. Trans. R. Soc. London Ser. B* **273**:119.

Jensen, A., Aasmundrud, O., and Eimhjellen, K. E., 1964, Chlorophylls of photosynthetic bacteria, *Biochim. Biophys. Acta* **88**:466.

Johnson, A., and Jones, O. T. G., 1964, Enzymic formation of hemes and other metalloporphyrins, *Biochim. Biophys. Acta* **93**:171.

Jones, M. S., and Jones, O. T. G., 1969, The structural organization of heme synthesis in rat liver mitochondria, *Biochem. J.* **113**:507.

Jones, M. S., and Jones, O. T. G., 1970, Ferrochelatase of *Rhodopseudomonas sphaeroides*, *Biochem. J.* **119**:453.

Jones, O. T. G., 1963a, The production of magnesium protoporphyrin monomethyl ester by *Rhodopseudomonas sphaeroides*, *Biochem. J.* **86**:429.

Jones, O. T. G., 1963b, The inhibition of bacteriochlorophyll synthesis in *Rhodopseudomonas sphaeroides* by 8-hydroxyquinoline, *Biochem. J.* **88**:335.

Jones, O. T. G., 1963c, Magnesium 2,4-divinyl phaeoporphyrin a_5 monomethyl ester, a protochlorophyll-like pigment produced by *Rhodopseudomonas sphaeroides*, *Biochem. J.* **89**:182.

Jones, O. T. G., 1964, Studies on the structure of a pigment related to chlorophyll *a* produced by *Rhodopseudomonas sphaeroides*, *Biochem. J.* **91**:572.

Jones, O. T. G., 1968, Chlorophyll biosynthesis, in: *Porphyrins and Related Compounds*, *Biochem. Soc. Symp.*, No. 28 (T. W. Goodwin, ed.), pp. 131–145, Academic Press, London and New York.

Jones, O. T. G., 1976, Chlorophyll *a* biosynthesis, *Philos. Trans. R. Soc. London Ser. B* **273**:207.

Jordan, P. M., and Shemin, D., 1973, Purification and properties of uroporphyrinogen I synthetase from *Rhodopseudomonas sphaeroides*, *J. Biol. Chem.* **248**:1019.

Katz, J. J., Strain, H. H., Harkness, A. L., Studier, M. H., Svec, W. A., Janson, T. R., and Cope, B. T., 1972, Esterifying alcohols in the chlorophylls of purple photosynthetic bacteria. A new chlorophyll bacteriochlorophyll (*gg*), all-*trans* geranylgeranyl bacteriochlorophyllide *a*, *J. Am. Chem. Soc.* **94**:7938.

Kennedy, G. Y., Jackson, A. H., Kenner, G. W., and Suckling, C. J., 1970, Isolation, structure and synthesis of a tricarboxylic porphyrin from the harderian gland of the rat, *FEBS Lett.* **6**:9.

Kenner, G. W., McCombie, S. W., and Smith, K. M., 1974, Pyrroles and related compounds. Part XXX. Cyclisation of porphyrin β-keto-esters to phaeoporphyrins, *J. Chem. Soc. Perkin Trans. 1* **1974**:527.

Kenner, G. W., Rimmer, J., Smith, K. M., and Unsworth, J. F., 1976, Studies on the biosynthesis of the Chlorobium chlorophylls, *Philos. Trans. R. Soc. London Ser. B* **273**:255.

Kikuchi, G., Kumar, A., Talmage, P., and Shemin, D., 1958, The enzymatic synthesis of δ-aminolevulinic acid, *J. Biol. Chem.* **233**:1214.

Krasnovskii, A. A., Fedenko, E. P., Lang, F., and Kondrat'eva, E. N., 1970, Spectrofluorimetry of pigments of the original strain of *Rhodopseudomonas palustris* and of its protochlorophyll mutants, *Dokl. Akad. Nauk SSSR* **190**:218.

Krueger, R. C., Melnick, I., and Klein, J. R., 1956, Formation of heme by broken-cell preparations of duck erythrocytes, *Arch. Biochem. Biophys.* **64**:302.

Kunzler, A., and Pfennig, N., 1973, Das Vorkommen von Bacteriochlorophyll a_p und a_{Gg} in Stammen aller Arten der Rhodospirillaceae, *Arch. Mikrobiol.* **91**:83.

Lascelles, J., 1956, The synthesis of porphyrins and bacteriochlorophyll by cell suspensions of *Rhodopseudomonas sphaeroides*, *Biochem. J.* **62**:78.

Lascelles, J., 1964, *Tetrapyrrole Biosynthesis and Its Regulation*, W. A. Benjamin, New York and Amsterdam.

Lascelles, J., 1966a, The accumulation of bacteriochlorophyll precursors by mutant and wild-type strains of *Rhodopseudomonas sphaeroides*, *Biochem. J.* **100**:175.

Lascelles, J., 1966b, The regulation and synthesis of iron and magnesium tetrapyrroles: Observations with mutant strains of *Rhodopseudomonas sphaeroides*, *Biochem. J.* **100**:184.

Lascelles, J., Rittenberg, B., and Clark-Walker, G. D., 1969, Growth and cytochrome synthesis in a hemin-requiring mutant of *Spirillum itersonii*, *J. Bacteriol.* **97**:455.

Lemberg, R., and Barrett, J., 1973, *Cytochromes*, pp. 226–233, Academic Press, London and New York.

Liljenberg, C., 1974, Characterisation and properties of a protochlorophyllide ester in leaves of dark grown barley with geranyl-geraniol as esterifying alcohol, *Physiol. Plant.* **32**:208.

Little, H. N., and Jones, O. T. G., 1976, The subcellular localization and properties of the ferrochelatase of etiolated barley, *Biochem. J.* **156**:309.

Llambias, E. B. C., and Battle, A. M. C., 1971, Studies on the porphobilinogen deaminase-uroporphyrinogen cosynthetase system of cultured soya-bean cells, *Biochem. J.* **121**:327.

Margoliash, E., 1961, Porphyrins and hemoproteins, *Annu. Rev. Biochem.* **30**:549.

Mathewson, J. H., and Corwin, A. H., 1961, Biosynthesis of pyrrole pigments: A mechanism for porphobilinogen polymerization, *J. Am. Chem. Soc.* **83**:135.

Mauzerall, D., and Granick, S., 1958, Porphyrin biosynthesis in erythrocytes. III. Uroporphyrinogen and its decarboxylase, *J. Biol. Chem.* **232**:1141.

Mazanowska, A. M., Neuberger, A., and Tait, G. H., 1966, Effect of lipids and organic solvents on the enzymic formation of zinc protoporphyrin and heme, *Biochem. J.* **98**:117.

Mori, M., and Sano, S., 1968, Protoporphyrin formation from coproporphyrinogen III by Chromatium cell extracts, *Biochem. Biophys. Res. Commun.* **32**:610.

Mori, M., and Sano, S., 1972, Studies on the formation of protoporphyrin IX by anaerobic bacteria, *Biochim. Biophys. Acta* **264**:252.

Nandi, D. L., and Shemin, D., 1968a, δ-Aminolevulinic acid dehydratase of *Rhodopseudomonas sphaeroides*. Association to polymers and dissociation to subunits, *J. Biol. Chem.* **243**:1231

Nandi, D. L., and Shemin, D., 1968b, δ-Aminolevulinic acid dehydratase of *Rhodopseudomonas sphaeroides*. III. Mechanism of porphobilinogen synthesis, *J. Biol. Chem.* **243**:1236.

Nandi, D. L., and Shemin, D., 1973, δ-Aminolevulinic acid dehydratase of *Rhodopseudomonas capsulata*, *Arch. Biochem. Biophys.* **158**:305.

Nandi, D. L., Baker-Cohen, K. F., and Shemin, D., 1968, δ-Aminolevulinic acid dehydratase of *Rhodopseudomonas sphaeroides*. I. Isolation and properties, *J. Biol. Chem.* **243**:1224.

Neuberger, A., 1961, Aspects of metabolism of glycine and of porphyrins, *Biochem. J.* **78**:1.

Neuberger, A., and Scott, J. J., 1953, Aminolevulinic acid and porphyrin biosynthesis, *Nature (London)* **172**:1093.

Neuberger, A., and Tait, G. H., 1964, Studies on the biosynthesis of porphyrin and bacteriochlorophyll by *Rhodopseudomonas sphaeroides*. 5. Zinc-protoporphyrin chelatase, *Biochem. J.* **90**:607.

Neuberger, A., and Turner, J. M., 1963, γ, δ-Dioxovalerate aminotransferase activity in *Rhodopseudomonas sphaeroides*, *Biochim. Biophys. Acta* **67**:342.

Neuberger, A., Scott, J. J., and Shuster, L., 1956, Synthesis and metabolism of some substances related to δ-aminolevulinic acid, *Biochem. J.* **64**:137.

Oelze, J., Schroeder, J., and Drews, G., 1970, Bacteriochlorophyll, fatty acid and protein synthesis in relation to thylakoid formation in mutant strains of *Rhodospirillum rubrum*, *J. Bacteriol.* **101**:669.

Pluscec, J., and Bogorad, L., 1970, A dipyrrylmethane intermediate in the enzymatic synthesis of uroporphyrinogen, *Biochemistry* **9**:4736.

Porra, R. J., and Falk, J. E., 1964, The enzymic conversion of coproporphyrinogen III into Protoporphyrin IX, *Biochem. J.* **90**:69.

Porra, R. J., and Jones, O. T. G., 1963, Studies on ferrochelatase. 2. An investigation into the role of ferrochelatase in the biosynthesis of various heme prosthetic groups, *Biochem. J.* **87**:186.

Porra, R. J., and Lascelles, J., 1968, Studies on ferrochelatase: The enzymic formation of heme in proplastids, chloroplasts and plant mitochondria, *Biochem. J.* **108**:343.

Poulson, R., and Polglase, W. J., 1974, Aerobic and anaerobic coproporphyrinogenase activities in extracts from *Saccharomyces cerevisiae*: Purification and characterisation, *J. Biol. Chem.* **249**:6367.

Poulson, R., and Polglase, W. J., 1975. The enzymic conversion of protoporphyrinogen IX to protoporphyrin IX, *J. Biol. Chem.* **250**:1269.

Pradel, J., and Clement-Metral, J. D., 1975. A 4-vinylprotochlorophyllide complex as a model of the spectral forms of protochlorophyllide *in vivo*, *Abstr. 10th FEBS Meeting, Paris*, p. 1210.

Pudek, M. R., and Richards, W. R., 1975, A possible alternate pathway of bacteriochlorophyll biosynthesis in a mutant of *Rhodopseudomonas sphaeroides*, *Biochemistry* **14**:3132.

Purdie, J. W., and Holt, A. S., 1965. Structures of Chlorobium chlorophylls (650), *Can. J. Chem.* **43**:3347.

Radmer, R. J., and Bogorad, L., 1967, S-Adenosyl-L-methionine-magnesium protoporphyrin methyltransferase, an enzyme in the biosynthetic pathway of chlorophyll in *Zea Mays*, *Plant Physiol.* **42**:463.

Radmer, R. J., and Bogorad, L., 1972, A tetrapyrryl-methane intermediate in the enzymatic synthesis of uroporphyrinogen, *Biochemistry* **11**:904.

Richards, W. R., and Lascelles, J., 1969, The biosynthesis of bacteriochlorophyll. The characterisation of the latter stage intermediates from mutants of *Rhodopseudomonas sphaeroides*, *Biochemistry* **8**:3473.

Richards, W. R., and Rapoport, H., 1966, The biosynthesis of Chlorobium chlorophylls 660: The isolation and purification of porphyrins from *Chlorobium thiosulfatophilum* 660, *Biochemistry* **5**:1079.

Richards, W. R., and Rapoport, H., 1967, The production of magnesium protoporphyrin monomethyl ester, bacteriochlorophyll and chlorobium phaeoporphyrins by *Chlorobium thiosulfatophilum-660*, *Biochemistry* **6**:3830.

Rüdiger, W., Hedden, P., Köst, H.-P., and Chapman, D. J., 1977, Esterification of chlorophyllide by geranylpyrophosphate in a cell-free system from maize shoots, *Biochem. Biophys. Res. Commun.* **74**:1268.

Sano, S., 1966, 2,4-bis-(β-Hydroxpropionic acid) deuteroporphyrinogen IX, a possible intermediate between coproporphyrinogen III and protoporphyrin IX, *J. Biol. Chem.* **241**:5276.

Sano, S., and Granick, S., 1961, Mitochondrial coproporphyrinogen oxidase and protoporphyrin formation, *J. Biol. Chem.* **236**:1173.

Saunders, V. A., 1973, Electron transfer reactions in photosynthetic bacteria, Ph.D. dissertation, University of Bristol.

Saunders, V. A., and Jones, O. T. G., 1974, Properties of the cytochrome *a*-like material developed in the photosynthetic bacterium *Rhodopseudomonas sphaeroides* when grown aerobically, *Biochim. Biophys. Acta* **333**:439.

Scheer, H., Svec, W. A., Cope, B. T., Studier, M. H., Scott, R. G., and Katz, J. J., 1974, Structure of bacteriochlorophyll *b*, *J. Am. Chem. Soc.* **96**:3714.

Shemin, D., 1957, The biosynthesis of porphyrins, *Ergeb. Physiol.* **49**:299.

Shemin, D., 1972, δ-Aminolevulinic acid dehydratase, in: *The Enzymes*, 3rd Ed., Vol. 7 (P. D. Boyer, ed.), pp. 323–337, Academic Press, New York and London.

Shemin, D., 1975, Porphyrin biosynthesis: Some particular approaches, *Ann. N. Y. Acad. Sci.* **244**:348.

Shemin, D., 1976, 5-Aminolaevulinic acid dehydratase: Structure, function, and mechanism, *Philos, Trans. R. Soc. London Ser. B* **273**:109.

Shemin, D., and Russell, C. S., 1953, δ-Aminolevulinic acid, its role in the biosynthesis of porphyrins and purines, *J. Am. Chem. Soc.* **75**:4873.

Sinclair, P., White, D. C., and Barrett, J., 1967, The conversion of protoheme to heme *a* in *Staphylococcus*, *Biochim. Biophys. Acta* **143**:427.

Smith, J. H. C., 1960, Protochlorophyll transformations, in: *Comparative Biochemistry of Photoreactive Systems* (M. B. Allen, ed.), pp. 257–277, Academic Press, New York.

Smith, K. M., and Unsworth, J. F., 1975, The nuclear magnetic resonance spectra of porphyrins. IX. Carbon-13 nuclear magnetic resonance spectra of some chlorins and other chlorophyll degradation products, *Tetrahedron* **31**:367.

Stanier, R. Y., and Smith, J. H. C., 1960, The chlorophylls of green bacteria, *Biochim. Biophys. Acta* **41**:478.

Stoll, M. S, Elder, G. H., Games, D. E., O'Hanlon, P., Millington, D. S., and Jackson, A. H., 1973, Isocoproporphyrin: Nuclear magnetic resonance and mass-spectral methods for the determination of porphyrin structure, *Biochem. J.* **131**:429.

Tait, G. H., 1968, General aspects of heme synthesis, in: *Porphyrins and Related Compounds* (T. W. Goodwin, ed.), pp. 19–34. Academic Press, New York and London.

Tait, G. H., 1969, Coproporphyrinogenase activity in extracts from *Rhodopseudomonas sphaeroides*, *Biochem. Biophys. Res. Commun.* **37**:116.

Tait, G. H., 1972, Coproporphyrinogenase activities in extracts of *Rhodopseudomonas sphaeroides* and *Chromatium* Strain D, *Biochem. J.* **128**:1159.

Tuboi, S., and Hayasaka, S. 1972*a*, Control of δ-aminolevulinate synthetase activity in *Rhodopseudomonas sphaeroides*, *Arch. Biochem. Biophys.* **150**:690.

Tuboi, S., and Hayasaka, S., 1972*b*, Partial purification of an enzyme which is required for the conversion of the inactive form to the active form of δ-aminolevulinate synthetase from *Rhodopseudomonas sphaeroides*, *J. Biochem. (Tokyo)* **72**:219.

Tuboi, S., Kim, H. J., and Kikuchi, G., 1970*a*, Occurrence and properties of two types of δ-aminolevulinic acid synthetase in *Rhodopseudomonas sphaeroides*, *Arch. Biochem. Biophys.* **138**:147.

Tuboi, S., Kim, H. J., and Kikuchi, G., 1970*b*, Differential induction of fraction I and fraction II of δ-aminolevulinic acid synthetase in *Rhodopseudomonas sphaeroides*, *Arch. Biochem. Biophys.* **138**:155.

Uspenskaya, V. E., 1972, Bacteriochlorophyll synthesis and chlorophyllase activity of a non-sulfur purple bacteria, *Izv. Akad. Nauk SSSR Ser. Biol.* **6**:882.

Uspenskaya, V. E., and Kondrat'eva, E. N., 1964, Formation of free porphyrins by green photosynthetic bacteria, *Doklad. Akad. Nauk SSSR* **157**:678.

van Heyningen, S., and Shemin, D., 1971, Quaternary structure of δ-aminolevulinate dehydratase from *Rhodopseudomonas sphaeroides*, *Biochemistry* **10**:4676.

Warnick, G. R., and Burnham, B. F., 1971, Regulation of porphyrin synthesis. Purification and characterisation of δ-aminolevulinic acid synthetase, *J. Biol. Chem.* **246**:6880.

Watts, R. B., and Kekwick, R. G. O., 1974, Factors affecting the formation of phytol and its incorporation into chlorophyll by homogenates of the leaves of the French bean *Phaseolus vulgaris*, *Arch. Biochem. Biophys.* **160**:469.

Wider de Xifra, E. A., Sandy, J. D., Davies, R. C., and Neuberger, A., 1976, Control of 5-aminolaevulinate synthetase activity in *Rhodopseudomonas sphaeroides*, *Philos. Trans. R. Soc. London Ser. B* **273**:79.

Yubisui, T., and Yoneyama, Y., 1972, δ-Aminolevulinic acid synthetase of *Rhodopseudomonas sphaeroides*: Purification and properties of the enzyme, *Arch. Biochem. Biophys.* **160**:77.

CHAPTER 41

Biosynthesis of Amino Acids

Prasanta Datta

1. Introduction

Photosynthetic bacteria derive their energy from light quanta and convert it into the energy-rich compound ATP for driving metabolic reactions. Most of these organisms can grow photoautotrophically using CO_2 as the sole source of carbon; many can also utilize such simple carbon compounds as malate, glycerol, or acetate heterotrophically. Although some strains require a number of vitamins for growth, these bacteria can synthesize all other cellular components *de novo*, including the common amino acids. What follows is a brief review of the biosynthesis of various amino acids in photosynthetic bacteria.

According to their metabolic interrelationships, all 20 amino acids can be grouped into six different classes (see Table 1). Genetic and biochemical studies using a variety of microorganisms, especially the coliaerogenes group of bacteria, have elucidated the biosynthetic pathways of these amino acids. Although photosynthetic bacteria have not been widely used for such studies, the limited data available indicate that without exception, the same sequences of reactions are found in these organisms for synthesis of these amino acids. Significant differences exist between these two classes of bacteria, however, in the mechanisms of end-product control of the individual biosynthetic enzymes, as well as in the overall regulatory patterns. Regulation of synthesis of the aspartate family of amino acids (see Section 5) is a particularly suitable example to illustrate this point.

Prasanta Datta · Department of Biological Chemistry, The University of Michigan, Ann Arbor, Michigan 48109

2. Biosynthesis of the Glutamate Family of Amino Acids

2.1. Glutamate and Glutamine

Three separate reactions by which synthesis of glutamate can occur are known (Fig. 1). One involves transamination between α-ketoglutarate and an amino acid such as aspartate, yielding glutamate and oxaloacetate (reaction 1). Synthesis of glutamate can also proceed by direct amination of α-ketoglutarate by NH_3 catalyzed by the pyridine-nucleotide-dependent enzyme glutamate dehydrogenase (reaction 2). A newly discovered enzyme, glutamate synthase (reaction 3), can form two moles of glutamate from one mole each of α-ketoglutarate and glutamine. All these schemes require the participation of α-ketoglutarate, an intermediate of the citric acid cycle. In photosynthetic bacteria, all these reactions appear to function for glutamate biosynthesis.

Using cell-free extracts of *Rhodospirillum rubrum* grown photosynthetically in malate–glutamate medium, Hug and Werkman (1957) detected the formation of glutamate via the transamination reaction between α-ketoglutarate and ten amino acids including aspartate, alanine, histidine, leucine, isoleucine, valine, and the aromatic amino acids. Proline and hydroxyproline were not active as amino donors, and methionine and lysine were weakly active. Dark-grown cells also exhibited transaminase activity. Three of these glutamate-forming reactions, including the aspartate–α-ketoglutarate system, were freely reversible. In *Chromatium vinosum* strain D, formation of

Table 1. Classification of Amino Acids According to Their Metabolic Interrelationships

Group	Family	Amino acids
I	Glutamate	Glutamate, glutamine, proline, arginine
II	Serine	Serine, glycine, cysteine
III	Aromatic amino acids	Phenylalanine, tyrosine, tryptophan
IV	Histidine	
V	Aspartate	Aspartate, asparagine, threonine, lysine, methionine
VI	Pyruvate	Alanine, isoleucine, valine, leucine

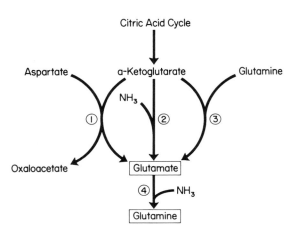

Fig. 1. Synthesis of glutamate and glutamine. Enzymes: (1) aspartate-α-ketoglutarate aminotransferase; (2) glutamate dehydrogenase; (3) glutamate synthase; (4) glutamine synthetase.

During longer exposures, lasting from 14 to 120 sec, glutamate and the citric acid cycle intermediates were labeled; the amount of radioactivity incorporated into these intermediates was always lower, however, than that found in glutamate. Hoare (1963) concluded that glutamate was not formed from citrate via the citric acid cycle. Shigesada *et al.* (1966) and Slater and Morris (1973) confirmed the results obtained by Hoare (1963); additionally, they showed that during dark aerobic metabolism of acetate in *Rs. rubrum*, the labeling pattern of glutamate was consistent with the mechanism involving α-ketoglutarate coming from the citric acid cycle. It is interesting to note that Bachofen and Neeracher (1968) purified an NAD-specific glutamate dehydrogenase from *Rs. rubrum* grown photoheterotrophically.

In *Rhodopseudomonas sphaeroides*, *Rp. capsulata*, and *Chr. vinosum* strain D, the major product of photoassimilation of acetate was also glutamate (Hoare, 1963; Losada *et al.*, 1960). Incubation of washed-cell suspensions of *Chlorobium limicola f. thiosulfatophilum* with [1-^{14}C]- and [2-^{14}C]acetate resulted in labeling of all amino acids (Hoare and Gibson, 1964). No clear indication of metabolic interrelationships among the different amino acids could be deduced from these results.

The formation of glutamate via the glutamate synthase reaction (Fig. 1, reaction 3) was recently uncovered in *Rs. rubrum*, *Chl. limicola f. thiosulfatophilum*, and *Chr. vinosum* (Nagatani *et al.*, 1971). When Slater and Morris considered the labeling patterns of glutamate in *Rs. rubrum* during photoassimilation of acetate as well as during dark aerobic growth, and the discovery of glutamate synthase in cells grown photosynthetically, they speculated that in *Rs. rubrum*, glutamate is synthesized via glutamate dehydrogenase in the dark-grown cells, whereas cells grown photosynthetically synthesize glutamate by the glutamate synthase reaction (Slater and Morris, 1974).

The synthesis of glutamine is catalyzed by glutamine synthetase according to the following reaction (see Fig. 1):

$$\text{Glutamate} + NH_3 + ATP \xrightarrow{\text{Mg}^{2+}}$$

$$\text{Glutamine} + ADP + P_i$$

Extracts of *Rs. rubrum* grown photosynthetically in malate–glutamate medium showed glutamine synthetase activity (Hubbard and Stadtman, 1967; Nagatani *et al.*, 1971), and partially purified enzyme obtained by pH precipitation and acetone fractionation exhibited cumulative feedback inhibition by various end products (Hubbard and Stadtman, 1967).

glutamate by a transamination reaction involving aspartate and α-ketoglutarate was also observed (Wagner *et al.*, 1973).

Although the transaminase reactions detected in cell-free extracts of *Rs. rubrum* appear to suggest the existence of reaction 1 (Fig. 1), formation of glutamate from α-ketoglutarate by either reaction 1 or reaction 2 by cells growing phototrophically could not be firmly established. Based on the time course of photoassimilation of [^{14}C]acetate by *Rs. rubrum*, Hoare (1963) observed that during a 3-sec pulse the major radioactive compound was glutamate, whereas citrate, succinate, fumarate, and malate remained unlabeled.

2.2. Proline

There is a close metabolic relationship among glutamate, proline, and arginine; the latter two amino acids are synthesized from glutamate. A variety of experimental data obtained with different microorganisms supports the following sequence of reactions for proline biosynthesis:

L-Glutamate → L-Glutamate-γ-semialdehyde →

L-Δ^1Pyrroline-5-carboxylate → L-Proline

Coleman (1958) reported that washed suspensions of *Rs. rubrum* cells in phosphate buffer can metabolize [^{14}C]glutamate; some of the radioactivity appeared in proline. These data are not very convincing, however, since glutamate used in these experiments was uniformly labeled, and evolution of $^{14}CO_2$ from glutamate and reassimilation of $^{14}CO_2$ could not be completely eliminated (see, however, Section 6).

2.3. Arginine

Synthesis of arginine from glutamate occurs in several steps (Fig. 2). Although most of the reactions involved in arginine biosynthesis are catalyzed by the same enzymes in all bacteria, in some members of the Enterobacteriaceae (including *Escherichia coli, Aerobacter aerogenes, Salmonella typhimurium,* and *Serratia marcescens*), ornithine is formed by simple deacetylation of N-acetylornithine (Fig. 2, enzyme 5B) whereas in several species of *Pseudomonas, Micrococcus,* and *Streptomyces*, ornithine is formed by the transacetylase reaction (Fig. 2, enzyme 5A), in which the N-acetyl group is transferred to glutamate to form N-acetylglutamate. A survey of several photosynthetic bacteria and blue-green algae (Hoare and Hoare, 1966) demonstrated the existence of two enzymes of the arginine pathway, N-acetylglutamate phosphokinase (enzyme 2) and ornithine acetyltransferase (enzyme 5A). The presence of ornithine transcarbamylase (enzyme 6), which catalyzes the formation of citrulline from ornithine, was also detected in extracts of *Rp. sphaeroides* (Ferretti and Gray, 1968). These data support the concept of arginine biosynthesis from glutamate via the transacetylase reaction with regeneration of N-acetylglutamate.

N-Acetylglutamate phosphokinase activity isolated from a variety of photosynthetic bacteria is subject to feedback inhibition by arginine (Hoare and Hoare, 1966); this finding also supports the transacetylase route of ornithine synthesis. The rationale could be presented as follows: For control of arginine biosynthesis, inhibition by arginine of the activity of the first enzyme, N-acetylglutamate synthetase, which catalyzes the formation of N-acetylglutamate, would not be very effective, since the N-acetylglutamate produced in the transacetylase step could be used for arginine synthesis. Accordingly, in those organisms in which transacetylase is the preferred enzyme for deacetylation of N-acetylornithine, the most effective control scheme for arginine biosynthesis would be feedback regulation of the second enzyme, N-acetylglutamate phosphokinase. This notion is consistent with the finding that the coliaerogenes group of bacteria convert N-acetylornithine to ornithine by the enzyme N-acetylornithinase (enzyme 5B), and the enzyme subject to feedback inhibition by arginine is, as expected, the first enzyme, N-acetylglutamate synthetase (Udaka, 1966).

3. Biosynthesis of Serine, Glycine, and Cysteine

3.1. Serine

Two distinct pathways are known for the biosynthesis of serine from D-glycerate, one involving phosphorylated, the other nonphosphorylated, intermediates (Fig. 3). The first enzyme specific for serine biosynthesis via the phosphorylated pathway is 3-phosphoglycerate dehydrogenase, which converts 3-phosphoglycerate to 3-phosphohydroxypyruvate, the precursor of phosphoserine. A specific phosphoserine phosphatase has been demonstrated for the dephosphorylation reaction to yield serine. The nonphosphorylated pathway involves the formation of serine via 3-hydroxypyruvate.

In mammalian cells, the presence of enzymes 4 and 5 as well as A and B (see Fig. 3) indicates that both pathways coexist (Walsh and Sallach, 1966). In plants and microorganisms, the phosphorylated pathway occurs predominantly, although the enzymes of the nonphosphorylated sequence are known to exist in *Esch. coli* (Blatt *et al.*, 1966) and in *Pseudomonas* sp. AM1 (Large and Quayle, 1963). Extracts of *Rp. capsulata* (Schmidt and Sojka, 1973a) catalyze all the reactions of both pathways. In cells grown photosynthetically with malate as the carbon source, the activities of the phosphorylated pathway enzymes, especially 3-phosphoglycerate dehydrogenase (enzyme 3), were substantially higher than the analogous enzymes of the nonphosphorylated sequence. The specific activity of 3-phosphoglycerate dehydrogenase declined markedly in cells from the late exponential phase; cells grown in low light intensity or on serine as the sole carbon source had low specific activities of the enzymes of the phosphorylated pathway (Schmidt and Sojka, 1973b).

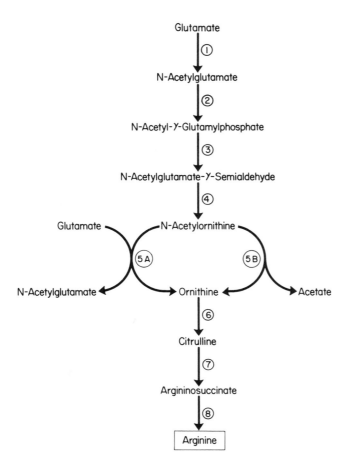

Glutamate

①

N-Acetylglutamate

②

N-Acetyl-γ-Glutamylphosphate

③

N-Acetylglutamate-γ-Semialdehyde

④

Glutamate → N-Acetylornithine

(5A)   (5B)

N-Acetylglutamate ←   → Ornithine ←   → Acetate

⑥

Citrulline

⑦

Argininosuccinate

⑧

Arginine

Fig. 2. Synthesis of arginine from glutamate. Enzymes: (1) *N*-acetylglutamate synthetase; (2) *N*-acetylglutamate phosphokinase; (3) *N*-acetylglutamate semialdehyde dehydrogenase; (4) *N*-acetylornithine Δ-transaminase; (5A) ornithine acetyltransferase; (5B) *N*-acetylornithinase; (6) ornithine transcarbamylase; (7) argininosuccinate synthetase; (8) argininosuccinase.

Beremand and Sojka (1977) obtained serine–glycine auxotrophs of *Rp. capsulata* that were able to grow on serine or glycine and had lesions either in enzyme 3 or in enzyme 5 (see Fig. 3), but not in the enzymes of the nonphosphorylated sequence. When revertants of these mutants were isolated, the reversion to prototrophy was always accompanied by restoration of the defective enzymes. Investigation of the regulation of serine biosynthesis shows that the activities of enzyme 3 and enzyme 5 were inhibited by the end product, L-serine, although the inhibitory effect of L-serine was much more pronounced on 3-phosphoglycerate dehydrogenase (Schmidt and Sojka, 1973a). Thus, the normal patterns of repression of enzyme synthesis and feedback inhibition of the first specific enzyme of the phosphorylated pathway strengthen the conclusion that in *Rp. capsulata*, the sequence of reactions for serine biosynthesis involves phosphorylated intermediates.

3.2. Glycine

The major pathway of glycine synthesis in many organisms involves aldol cleavage of serine, as catalyzed by serine hydroxymethyltransferase:

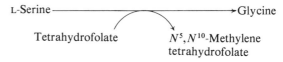

L-Serine ⟶ Glycine

Tetrahydrofolate N^5,N^{10}-Methylene tetrahydrofolate

In extracts of *Rp. capsulata*, the specific activity of this enzyme was high and catalyzed the tetrahydrofolate-dependent formation of glycine from serine. Synthesis of serine from glycine plus formaldehyde was also observed (Schmidt and Sojka, 1973a).

That enzymatic lesions in the phosphorylated pathway for serine biosynthesis resulted in serine–glycine auxotrophy (see Beremand and Sojka, 1977) leads to the conclusion that in *Rp. capsulata*, serine is the pre-

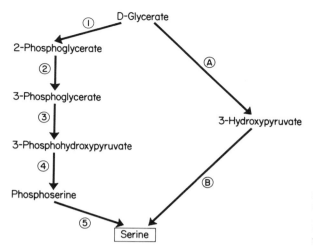

Fig. 3. Pathways of serine biosynthesis. Enzymes: (1) D-glycerate kinase; (2) glycerate phosphomutase; (3) 3-phosphoglycerate dehydrogenase; (4) phosphoserine-(α-ketoglutarate)aminotransferase; (5) phosphoserine phosphatase; (A) D-glycerate dehydrogenase; (B) serine-(pyruvate) aminotransferase.

cursor of glycine. These authors reported that addition of either glycolate or glyoxylate relieved the requirement for serine in serine–glycine auxotrophs, suggesting a separate route for serine formation. It is possible that in this organism, glycine may be formed by a transamination reaction between glutamate and glyoxylate (Campbell, 1956; Nakada, 1964; Cossin and Sinha, 1965); subsequent conversion of glycine to serine would satisfy serine requirement. Tsuiki and Kikuchi (1962) examined the photometabolism of glycine by *Rp. sphaeroides* and concluded that the major route of glycine metabolism proceeds through glyoxylate and malate.

3.3. Cysteine

Genetic and enzymatic studies of cysteine auxotrophs of *Sal. typhimurium* (Dreyfuss and Monty, 1963) and of *Esch. coli* (Jones-Mortimer, 1968) have elucidated the pathway for reduction of sulfate to sulfide. Further studies revealed that the precursor of the carbon skeleton of cysteine is serine, and that the formation of cysteine is accomplished in two steps (Kredich and Tomkins, 1966):

$$\text{L-Serine} + \text{acetyl-CoA} \rightarrow O\text{-acetyl serine} + \text{CoA} \quad (1)$$

$$O\text{-Acetyl serine} + \text{H}_2\text{S} \rightarrow \text{L-cysteine} + \text{H}_2\text{O} + \text{acetate} \quad (2)$$

There is very little experimental evidence that indicates that biosynthesis of cysteine in photosynthetic bacteria follows the scheme for the coliaerogenes group of bacteria, or, for that matter, any other route.

4. Biosynthesis of Aromatic Amino Acids and Histidine

4.1. Phenylalanine, Tyrosine, and Tryptophan

The enzymatic steps and the intermediates involved in the biosynthesis of phenylalanine, tyrosine, and tryptophan in a variety of bacteria have been well characterized (Gibson, F., and Pittard, 1968). The incorporation of radioactivity from [^{14}C]acetate or [^{14}C]bicarbonate into phenylalanine by *Chl. limicola f. thiosulfatophilum* (Hoare and Gibson, 1964), and into tyrosine by *Rs. rubrum* (Ehrensvard and Reio, 1953), is consistent with the biosynthetic scheme (Fig. 4) in which the side chains of these amino acids are derived intact from phosphoenolpyruvate. It was reported by Allison and Robinson (1967), however, that cultures of *Chr. vinosum* strain D and *Rs. rubrum* incorporate radioactivity from [^{14}C]phenylacetate into phenylalanine during photosynthetic growth, but not during aerobic growth in the dark. Furthermore, incorporation of radioactivity from phenylacetate into phenylalanine by *Chr. vinosum* strain D was decreased in the presence either of added phenylalanine or of unlabeled phenylacetate; *de novo* synthesis of phenylalanine from radioactive bicarbonate was also reduced in the presence of phenylacetate. Phenylpyruvate did not affect biosynthesis of phenylalanine either from phenylacetate or from bicarbonate (Allison and Robinson, 1967). The authors proposed that synthesis of phenylalanine from phenylacetate in photosynthetic bacteria may require carboxylation of an intermediate followed by a transamination reaction similar

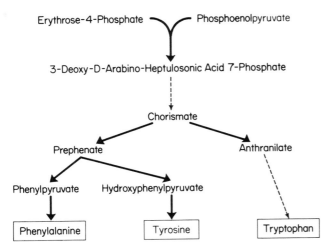

Fig. 4. Synthesis of phenylalanine, tyrosine, and tryptophan.

to that proposed for certain anaerobic bacteria from rumen (Allison, 1965).

The mechanisms of biosynthesis of tyrosine and tryptophan in photosynthetic bacteria remain unexplored.

4.2. Histidine

Although a wealth of information is available on the biosynthesis of histidine as well as on the regulation of synthesis and the activity of various enzymes, especially in *Esch. coli* and *Sal. typhimurium* (for a review, see Truffa-Bachi and Cohen, 1973), practically nothing is known for the photosynthetic bacteria in this regard.

5. Biosynthesis of the Aspartate Family of Amino Acids

5.1. Homoserine and Threonine

In bacteria, aspartic acid provides the carbon skeleton for the synthesis of lysine, methionine, threonine, and isoleucine (Fig. 5). In the initial step, aspartic acid is phosphorylated to aspartyl β-phosphate by the enzyme aspartokinase, which requires ATP and a divalent cation. The phosphorylated product is subsequently dephosphorylated and reduced to yield aspartate β-semialdehyde; both these reactions are presumably catalyzed by the enzyme aspartate β-semialdehyde dehydrogenase in the presence of a reduced pyridine nucleotide. The third enzyme in the sequence, homoserine dehydrogenase, which is also a pyridine-nucleotide-linked enzyme, converts aspartate

β-semialdehyde to homoserine. Two enzymes, homoserine kinase and threonine synthetase, are required for the conversion of homoserine to threonine. A list of these enzymes is given in Table 2.

The existence of reactions 1–4 (Fig. 5 and Table 2) has been extensively documented for a variety of photosynthetic bacteria by identification and characterization of the specific enzymes and from studies of feedback regulation both *in vitro* and *in vivo*. For example, extracts of *Rs. rubrum* (Sturani *et al.*, 1963; Datta and Gest, 1964a, 1965; Datta, 1969), of *Rs. tenue* (Robert-Gero *et al.*, 1972), and of three species of *Rhodopseudomonas*, *Rp. sphaeroides* (Gibson, K. D., *et al.*, 1962; Datta and Prakash, 1966; Datta, 1969; Datta *et al.*, 1973), *Rp. capsulata* (Datta and Gest, 1964a,b; Datta *et al.*, 1973), and *Rp. palustris* (Yen and Gest, 1974), show aspartokinase and homoserine dehyderogenase activities. In addition, aspartokinase activity has been detected in *Rs. molischianum*, *Rs. fulvum*, and *Rp. gelatinosa* (Robert-Gero *et al.*, 1972); extracts of *Rs. tenue* (Robert-Gero *et al.*, 1972) and *Rp. sphaeroides* (Gibson, K. D., *et al.*, 1962) also have the enzyme aspartate β-semialdehyde dehydrogenase. Homoserine kinase activity has been observed in *Rp. palustris*; the enzyme activity is sensitive to threonine and isoleucine inhibition (Yen and Gest, 1974).

With the exception of the aspartokinase from *Rp. sphaeroides* and the homoserine dehydrogenases of *Rs. rubrum* and *Rp. capsulata*, none of the enzymes has been purified to a significant extent. Using a 240-fold-purified preparation of aspartokinase from *Rp. sphaeroides*, Datta and Prakash (1966) established the absolute requirement for ATP and a divalent cation, and showed stimulation of enzyme activity by K⁺,

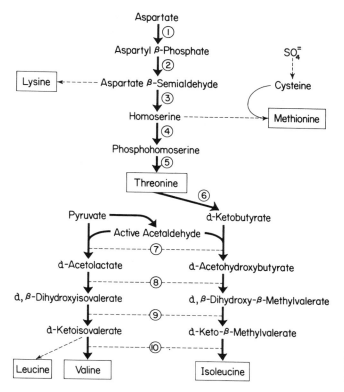

Fig. 5. Synthesis of the aspartate family and the branched-chain amino acids. See Table 2 for the names of the enzymes.

Table 2. Enzymes of the Aspartate Family and Branched-Chain Amino Acids

Step[a]	Enzyme	Substrate
1	Aspartokinase	L-Aspartate
2	Aspartate β-semialdehyde dehydrogenase	Aspartyl β-phosphate
3	Homoserine dehydrogenase	Aspartate β-semialdehyde
4	Homoserine kinase	Homoserine
5	Threonine synthetase	Phosphohomoserine
6	Threonine deaminase	Threonine
7	α-Acetohydroxyacid synthetase	Pyruvate, α-ketobutyrate
8	Reductoisomerase	α-Acetolactate, α-acetohydroxybutyrate
9	Dihydroxyacid dehydratase	α,β-Dihydroxyisovalerate, α,β-dihydroxy-β-methylvalerate
10	Transaminase	α-Ketoisovalerate, α-keto β-methylvalerate

[a] The numbers are the same as those in Figs. 5 and 6.

Cs⁺, or Rb⁺. The K⁺ increased the maximum velocity and facilitated binding of more than one molecule of ATP. The enzyme activity was strongly inhibited by aspartate β-semialdehyde, but not by any amino acid end product (Datta and Prakash, 1966).

Homoserine dehydrogenase was obtained in homogeneous form from *Rs. rubrum* by Datta (1970). The protein has a molecular weight of 110,000 and is made up of two subunits of 55,000 (Epstein, 1976). L-Threonine is a strong feedback inhibitor and induces aggregation of the protein (Datta *et al.*, 1964; Datta and Gest, 1965; Datta, 1970). A detailed study of the effects of threonine, KCl, and pH on the enzyme activity and the state of aggregation of the protein led to the conclusion (Datta and Epstein, 1973) that the primary event in the feedback inhibition of enzyme activity is an intramolecular conformational change in the protein, and that intermolecular interactions leading to enzyme aggregation follow this conformational alteration in the enzyme molecule.

In contrast to the *Rs. rubrum* enzyme, experiments with a 90% pure preparation of homoserine dehydrogenase from *Rp. capsulata* (Datta, unpublished) revealed that this enzyme was inhibited by L-threonine, but did not show a threonine-dependent association–dissociation reaction (Datta *et al.*, 1973).

The multiple patterns of feedback regulation of these reactions are discussed in Section 7.

5.2. Lysine

The synthesis of lysine in bacteria begins with the condensation of aspartate β-semialdehyde with pyruvate to form 2,3-dihydrodipicolinate (Yugari and Gilvarg, 1965), which in turn is converted to lysine via a series of reactions involving diaminopimelate as a metabolic intermediate. This sequence of reactions is referred to as the diaminopimelate pathway. Certain forms of lower and higher fungi, including yeast, utilize a separate pathway for lysine biosynthesis designated as the α-aminoadipate pathway (Rodwell, 1969); the initial reaction involves the condensation of α-ketoglutarate with acetyl-CoA to form homocitrate, and one of the intermediates of the pathway is α-aminoadipate.

Two types of experimental data clearly show that in photosynthetic bacteria, lysine biosynthesis proceeds through the diaminopimelate pathway. Vogel (1965), using radiolabeled aspartate, showed that in *Rp. sphaeroides*, the distribution of radioactivity followed the pattern predicted by this sequence of reactions. The patterns of feedback control by lysine of the early enzymes of the aspartate pathway (see Section 7) are also compatible with the conclusion that aspartate is the precursor of lysine.

5.3. Methionine

As shown in Fig. 5, the amino acid homoserine is a metabolic precursor of methionine in many organisms (Greenberg, 1969). K. D. Gibson *et al.* (1962) showed that photosynthetically grown *Rp. sphaeroides* incorporates radioactivity from [^{14}C]aspartate into methionine; addition of threonine to the medium containing [^{14}C]aspartate decreased the amount of radioactivity in methionine, presumably due to inhibition of homoserine dehydrogenase activity (Fig. 5, enzyme 3) by threonine. Extracts of *Rp. sphaeroides* also catalyze a cobalamin-dependent methylation of homocysteine (Cauthen *et al.*, 1967) by a mechanism

similar to that described for *Esch. coli*. Although there is a lack of detailed enzymatic and genetic studies of the individual steps involved in methionine biosynthesis in photosynthetic bacteria, these results and the regulatory interactions of this amino acid with the enzymes of the aspartate pathway observed thus far clearly imply that the pathway for the synthesis of methionine is similar, if not identical, to that proposed for other bacterial species.

6. Biosynthesis of Isoleucine, Valine, and Leucine

6.1. Isoleucine and Valine

The synthesis of the two branched-chain amino acids isoleucine and valine involves several homologous intermediates; some enzymes also have dual substrate specificities in that they catalyze the conversion of these homologous substrates (Umbarger, 1969).

Formation of isoleucine from threonine proceeds in five enzymatic steps (see Fig. 5 and Table 2), the first reaction unique to this series being the deamination of threonine to α-ketobutyrate by the enzyme threonine deaminase (enzyme 6). The keto acid formed in this reaction is ultimately converted to isoleucine. A part of the side chain of isoleucine is provided by α-hydroxyethyl-2-thiamine pyrophosphate (also known as "active acetaldehyde") generated from the reaction of pyruvate with thiamine pyrophosphate followed by decarboxylation of pyruvate; in *Esch. coli*, this condensation reaction is catalyzed by a pyruvate oxidase. Condensation of "active acetaldehyde" with α-ketobutyrate to form α-acetohydroxybutyrate is mediated by α-acetohydroxyacid synthetase (enzyme 7). The synthesis of isoleucine proceeds through reductive isomerization of α-acetohydroxybutyrate, dehydration of the dihydroxy acid to an α-keto acid, and finally transamination of the α-keto acid to form isoleucine.

In a sequence of reactions paralleling the synthesis of isoleucine from α-ketobutyrate, pyruvate is converted to valine (see Fig. 5). Although the intermediates are chemically different (bearing either a methyl or an ethyl side chain), the same four enzymes are shared for the conversion of the α-keto acids (pyruvate and α-ketobutyrate) to their corresponding amino acids. In *Esch. coli*, there appears to be a second transaminase specific only for valine biosynthesis (Umbarger, 1969).

In photosynthetic bacteria, the existence of threonine deaminase, the first enzyme specific for isoleucine synthesis, has been well established. The enzyme was purified to homogeneity from *Rs. rubrum* by Feldberg and Datta (1971). It has a tetrameric structure of 180,000 daltons with subunits of 46,000. Four moles of pyridoxal 5′-phosphate are tightly bound per mole of enzyme. The enzyme displays a normal Michaelis–Menten substrate saturation curve. High concentrations of isoleucine inhibited enzyme activity somewhat, but only at low levels of substrate; the inhibition was weak (Hughes *et al.*, 1964; Ning and Gest, 1966; Datta *et al.*, 1973) as compared with all other biosynthetic deaminases with the single exception of that obtained from *Rp. palustris* (Yen and Gest, 1974). The 1400-fold-purified threonine deaminase from *Rp. sphaeroides* (Datta, 1966) shows homotropic interactions with threonine and binds valine at two sites. The enzyme is strongly inhibited by isoleucine. An apparent Michaelis constant of 1.6 μM for pyridoxal phosphate was calculated. The threonine deaminase isolated from *Rp. capsulata* (Hughes *et al.*, 1964; Datta, unpublished) has properties similar to the enzyme obtained from *Rp. sphaeroides*. A comparison of the specific activity of the enzyme in dialyzed cell-free extracts from three nonsulfur purple photosynthetic bacteria (Yen and Gest, 1974) showed that *Rp. capsulata* had the highest enzyme level; the deaminases from *Rp. palustris* and *Rs. rubrum* were only 10% as active as that of *Rp. capsulata*. As mentioned above, theonine deaminase of *Rp. palustris* was not inhibited by isoleucine even at a concentration of 10 mM under a variety of assay conditions.

The four enzymes common to the isoleucine–valine biosynthetic pathways (Fig. 5, enzymes 7–10) were detected in extracts of *Rp. sphaeroides*. Barritt (1971) isolated a number of isoleucine–valine auxotrophs and observed that these mutants had lesions either in enzyme 7 or in enzyme 9. In this organism, no significant repression or derepression of threonine deaminase was observed. Valine, as expected, inhibited the activity of α-acetohydroxyacid synthetase. In a somewhat detailed study, Wixom *et al.* (1971) observed the activities of three enzymes, reductoisomerase, dihydroxyacid dehydratase, and transaminase (see Table 2), in photosynthetically grown *Rp. sphaeroides* and *Rs. rubrum*. Several of these enzymatic activities were also demonstrated in extracts from aerobic, dark-grown cells. Extracts of *Chr. vinosum* strain D also catalyzed the dehydratase reaction. The ratio of the rate of dehydratase activity with dihydroxyisovalerate as the substrate over that with dihydroxy-β-methylvalerate was higher in *Rp. sphaeroides* and *Rs. rubrum*, and lower in *Chr. vinosum*. Valine was a potent feedback inhibitor of enzyme 7 in *Chr. vinosum* (Wagner *et al.*, 1973). The transaminase activity from *Rp. sphaeroides* required pyridoxal phosphate and utilized several amino donors; L-glutamate was most effective for the amination of α-ketoisovalerate. These results strongly suggest that the biosynthetic pathways for the synthesis of amino acids of the aspartate family and for isoleucine and valine in photosynthetic bacteria are identical to that proposed for the coliaerogenes group of bacteria.

Some results from the author's laboratory (Dungan and Datta, unpublished) obtained with *Rp. sphaeroides* suggest that an alternate pathway of isoleucine biosynthesis from glutamate may be operative in this organism. *Rhodopseudomonas sphaeroides* strain Ga is defective in carotene biosynthesis (Griffiths and Stanier, 1956). By direct enzyme assay under a variety of experimental conditions, as well as

Table 3. Labeling of L-Isoleucine from [^{14}C]Glutamate in *Rhodopseudomonas sphaeroides*[a]

Strain	Addition	[^{14}C]Isoleucine/ [^{14}C]proline
Rp. sphaeroides		
Wild-type	[^{14}C]Glutamate	0.41
Strain Ga	[^{14}C]Glutamate	0.70
	[^{14}C]Glutamate + 1 mM β-methylaspartate	0.50
	[^{14}C]Glutamate + 0.2 mM α-ketobutyrate	0.37

[a] Washed cells from exponentially growing cultures were incubated with [^{14}C]glutamate, and the radioactive proteins were isolated and hydrolyzed in 6 N HCl. The amino acids were separated by two-dimensional paper chromatography, and spots corresponding to proline and isoleucine were cut out and counted for radioactivity. All results are expressed as the ratio of [^{14}C]isoleucine over [^{14}C]proline. Since proline is also synthesized from glutamate, the data presented this way eliminate unavoidable variables from one experiment to another in terms of labeling efficiency.

by examination of the incorporation of radioactivity into isoleucine from [^{14}C]threonine using exponentially growing cultures, it was shown that strain Ga has less than 0.1% of threonine deaminase activity as compared with the wild-type *Rp. sphaeroides*. Nevertheless, strain Ga does not require isoleucine for growth. Barker *et al.* (1964) reported that in *Clostridium tetanomorphum*, the enzyme glutamate mutase catalyzes the transformation of glutamate to β-methyl-aspartate. Abramsky and Shemin (1965) found that proteins isolated from *Esch. coli* grown on [^{14}C]methyl-labeled β-methylaspartate contained isoleucine labeled in the methyl group; in addition, these authors showed that the requirement of isoleucine in an isoleucine-requiring mutant of *Esch. coli* could be satisfied by β-methylaspartate or by α-ketobutyrate. Table 3 shows that with *Rp. sphaeroides* strain Ga, a significant amount of radioactivity appeared in isoleucine when grown in the presence of [^{14}C]glutamate; unlabeled β-methylaspartate and α-ketobutyrate reduced the amount of radioactivity incorporated into isoleucine. Furthermore, growth experiments demonstrated that strain Ga grew slowly on ammonium sulfate as compared with glutamate as the nitrogen source; significant increases in the growth rates were observed when media containing ammonium sulfate were supplemented with β-methylaspartate or α-keto-butyrate (Dungan and Datta, unpublished). These data suggest that glutamate may serve as a precursor of isoleucine, presumably via a series of reactions as shown below (cf. Abramsky and Shemin, 1965):

Glutamate → β-methylaspartate → methyloxaloacetate

Threonine → α-ketobutyrate ----→ |isoleucine

It was shown by Ohmori *et al.* (1971) that both *Rp. sphaeroides* and *Rs. rubrum* have vitamin-B$_{12}$-dependent glutamate mutase activity.

6.2. Leucine

The third branched-chain amino acid, leucine, is synthesized from α-ketoisovalerate (see Fig. 5). The initial step is the condensation of the α-ketoacid with acetyl-CoA to form α-isopropylmalate; the enzyme isopropylmalate synthetase is the first unique enzyme of the leucine biosynthetic branch and is subject to feedback inhibition by leucine (Kohlhaw *et al.*, 1969). Rearrangement of α-isopropylmalate to its β-isomer followed by decarboxylation yields α-ketoisocaproate, which is converted to leucine by a leucine-specific transaminase (Umbarger, 1969).

The pathway for the biosynthesis of leucine in

photosynthetic bacteria remains to be established. In *Chr. vinosum* strain D, leucine inhibits the activity of α-isopropylmalate synthetase (Wagner *et al.*, 1973).

7. Regulation of the Aspartate Family and the Branched-Chain Amino Acids

7.1. Control of the Early Enzymes

From the biosynthetic scheme shown in Fig. 5, it is apparent that the first two enzymes, aspartokinase and aspartate β-semialdehyde dehydrogenase, are common for the synthesis of lysine, methionine, threonine, and isoleucine. The third enzyme, homoserine dehydrogenase, is shared for the synthesis of three amino acid end products. Enzymes 4–6 are required for the synthesis of threonine and isoleucine, whereas enzyme 6 is unique for isoleucine biosynthesis. Considering these metabolic interrelationships, it is not surprising that a high degree of complexity exists for the control of the early enzymes of the pathway.

In *Esch. coli* (Cohen, 1969) and *Sal. typhimurium*

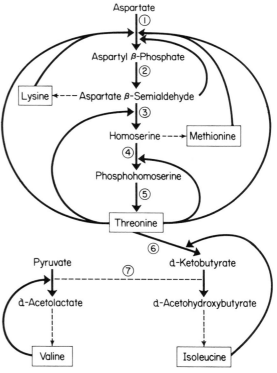

Fig. 6. Multiple regulatory patterns of the aspartate family and the branched-chain amino acids. See Table 2 for the names of the enzymes.

Table 4. Multiple Regulatory Patterns of the Enzymes of the Aspartate Pathway in Several Photosynthetic Bacteria

Organism	Aspartokinase		Homoserine dehydrogenase		Threonine deaminase	Regulatory pattern
	Inhibition by[a]	Reversal by[a]	Inhibition by[a]	Reversal by[a]	Inhibition by[a]	
Rs. rubrum	Thr	Ile, Met	Thr	Ile, Met	None	Compensatory feedback
Rs. tenue	Thr + Lys	Ile, Met				
	Thr	None				
	Lys	Ile, Met	NA	NA	NA	Compensatory feedback?
Rp. palustris	Thr + Lys	Ile				
	Thr + Met	None				
	Thr + Lys	Ile, Met	Thr	Ile, Met	None	Compensatory feedback
	Thr + Met	None				
Rp. capsulata	Thr + Lys	None	Thr	None	Ile	Concerted feedback
Rp. sphaeroides	ASA	None	Thr	None	Ile	Sequential feedback

[a] The usual abbreviations for the amino acids are used. (ASA) Aspartate β-semialdehyde; (NA) data not available.

(Cafferata and Freundlich, 1969), multiple forms of aspartokinase and homoserine dehydrogenase, each regulated by individual end products, have been found. This regulatory mechanism, known as the *isoenzymic pattern*, appears to be the principal mode of control in many coliform bacteria (Cohen, 1969; Datta *et al.*, 1973). In photosynthetic organisms, none of these enzymes is found in more than one form, and, depending on the source of the enzyme, the regulatory interactions between end-product metabolites and the early enzymes are quite different and sometimes quite complex.

In its simplest case, the aspartokinase activity of *Rp. capsulata* is subject to *concerted feedback* inhibition by threonine plus lysine (see Fig. 6); individually, threonine and lysine have no significant effect (Datta and Gest, 1964*a,b*). No other combination of amino acids inhibits enzyme activity. Several variations of this basic scheme of concerted feedback have been observed in three other photosynthetic bacteria (see Table 4). In addition to the concerted action of threonine plus lysine, as seen in *Rp. capsulata* as well as in *Rs. rubrum* (see Datta *et al.*, 1973), a combination of threonine and methionine also inhibits aspartokinase activities from *Rs. tenue* (Robert-Gero *et al.*, 1972) and from *Rp. palustris* (Yen and Gest, 1974). Threo-

nine alone also inhibits enzyme activities from *Rs. rubrum* (Datta and Gest, 1964*a*) and *Rs. tenue* (Robert-Gero *et al.*, 1972), whereas lysine by itself inhibits the activity of only the *Rs. tenue* enzyme. An interesting feature of the aspartokinases isolated from these organisms is that with certain exceptions, inhibition caused by the individual amino acids or by the threonine-plus-lysine combination can be reversed by either isoleucine or methionine or both; in no case, however, was the inhibitory effect of threonine plus methionine compensated (Table 4). Additionally, inhibition of homoserine dehydrogenase of *Rs. rubrum* and *Rp. palustris* by threonine is reversed by isoleucine and methionine (Datta, 1969; Yen and Gest, 1974). In contrast to all other bacterial threonine deaminases examined thus far, the enzymes from *Rs. rubrum* and *Rp. palustris* are not sensitive to isoleucine inhibition. Thus, the overall regulatory pattern in these organisms involves feedback inhibition by one end product or by a combination of end products that can be compensated by a different end product of the branched biosynthetic pathway. This pattern is designated as *compensatory feedback* control (see Datta, 1969; Datta *et al.*, 1973).

A completely distinct control mechanism in which a branch-point intermediate, and not biosynthetic end products, controls the activity of aspartokinase was reported in *Rp. sphaeroides* (Datta and Prakash, 1966). In this organism, the activities of threonine deaminase (Fig. 6, enzyme 6), homoserine dehydrogenase (enzyme 3), and aspartokinase (enzyme 1) are inhibited by isoleucine, threonine, and aspartate β-semialdehyde, respectively. The overall control pattern is a sequential event of progressively lowering the concentration of the earlier metabolites by the negative cascading effect (see Fig. 6). This pattern of *sequential feedback* has not been seen with any other photosynthetic bacteria.

As mentioned earlier, valine inhibits the activity of α-acetohydroxyacid synthetase of *Chr. vinosum* strain D (Wagner *et al.*, 1973) and of *Rp. sphaeroides* (Wixom *et al.*, 1971), thereby controlling its own synthesis from pyruvate.

In contrast to the coliform bacteria, repression of enzyme synthesis, in general, does not appear to play a major role in the regulation of amino acid biosynthetic enzymes in photosynthetic bacteria, although methionine is known to repress aspartokinase and homoserine dehydrogenases in *Rp. sphaeroides* (Datta, 1969), and threonine or a combination of threonine and lysine reduced the specific activity of *Rs. tenue* aspartokinase by about 50% (Robert-Gero *et al.*, 1972).

7.2. Physiological Implications and Evolutionary Significance

Research carried out thus far indicates that multiple overall control patterns exist for the regulation of the early enzymes of the aspartate family of amino acids. In the enteric group of bacteria, isoenzymic control seems to be the dominant pattern, whereas in photosynthetic organisms, even within a single genus, distinct regulatory schemes including concerted feedback, compensatory feedback, and sequential feedback controls are operative (Datta, 1969; Datta *et al.*, 1973). Many of the *in vitro* observations are compatible with the results obtained *in vivo* by cultivating strains under simulated conditions of excess end products (Burlant *et al.*, 1965; Datta, 1969). It is noteworthy that those organisms that utilize compensatory feedback control for the regulation of the early enzymes also have feedback-insensitive threonine deaminase. It is reasonable to assume that each pattern is adequate for a normal flow of metabolites from aspartate to various amino acid end products, and that each regulatory mechanism is peculiarly suited to the physiology and metabolism of the organism.

From the evolutionary viewpoint, it is interesting that several photosynthetic bacteria that are morphologically and physiologically closely related exhibit such divergent control patterns, whereas common control mechanisms, namely, concerted feedback and sequential feedback, are shared by distinct groups of organisms such as *Bacillus* species, *Pseudomonads*, and *Rp. capsulata* in independent biosynthetic pathways. In view of these findings, it is highly likely that if a given control pattern is indeed a highly efficient one, it may be preserved during evolution and would be found in a variety of biosynthetic pathways. Thus, phylogenetic relationships may not indicate similarity in control patterns, and diversity in control mechanisms may not suggest independent evolutionary origin (Datta, 1969; Datta *et al.*, 1973).

ACKNOWLEDGMENTS

The work performed in the author's laboratory was supported in part by grants from the National Science Foundation and the National Institutes of Health, and by the Rackham School of Graduate Studies of the University of Michigan.

8. References

Abramsky, T., and Shemin, D., 1965, The formation of isoleucine from β-methylaspartic acid in *Escherichia coli* W, *J. Biol. Chem.* **240**:2971.

Allison, M. J., 1965, Phenylalanine biosynthesis from phenylacetic acid by anaerobic bacteria from the rumen, *Biochem. Biophys. Res. Commun.* **18**:30.

Allison, M. J., and Robinson, I. M., 1967, Biosynthesis of phenylalanine from phenylacetate by *Chromatium* and *Rhodospirillum rubrum*, *J. Bacteriol.* **93**:1269.

Bachofen, R., and Neeracher, H., 1968, Glutamatdehydrogenase im photosynthetischen Bakterium *Rhodospirillum rubrum*, *Arch. Microbiol.* **60**:235.

Barker, H. A., Rooze, V., Suzuki, F., and Iodice, A. A., 1964, The glutamate mutase system: Assays and properties, *J. Biol. Chem.* **239**:3260.

Barritt, G. J., 1971, Biosynthesis of isoleucine and valine in *Rhodopseudomonas sphaeroides*: Regulation of threonine deaminase activity, *J. Bacteriol.* **105**:718.

Beremand, P., and Sojka, G. A., 1977, Mutational analysis of serine–glycine biosynthesis in *Rhodopseudomonas capsulata*, *J. Bacteriol.* **130**:532.

Blatt, L., Dorer, F. E., and Sallach, H. J., 1966, Occurrence of hydroxypyruvate-L-glutamate transaminase in *Escherichia coli* and its separation from hydroxypyruvatephosphate-L-glutamate transaminase, *J. Bacteriol.* **92**:668.

Burlant, L., Datta, P., and Gest, H., 1965, Control of enzyme activity in growing bacterial cells by concerted feedback inhibition, *Science* **148**:1351.

Cafferata, R. L., and Freundlich, M., 1969, Evidence for a methionine-controlled homoserine dehydrogenase in *Salmonella typhimurium*, *J. Bacteriol.* **97**:193.

Campbell, L. L., 1956, Transamination of amino acids with glyoxylic acid in bacterial extracts, *J. Bacteriol.* **71**:81.

Cauthen, S. E., Pattison, J. R., and Lascelles, J., 1967, Vitamin B$_{12}$ in photosynthetic bacteria and methionine synthesis by *Rhosopseudomonas sphaeroides, Biochem. J.* **102**:774.

Cohen, G. N., 1969, The aspartokinases and homoserine dehydrogenases of *Escherichia coli, Curr. Top. Cell. Regul.* **1**:183.

Coleman, G. S., 1958, The incorporation of amino acid carbon by *Rhodospirillum rubrum, Biochim. Biophys. Acta* **30**:549.

Cossin, E. A., and Sinha, S. K., 1965, Occurrence and properties of L-amino acid: 2-glyoxylate aminotransferase in plants, *Can. J. Biochem.* **43**:495.

Datta, P., 1966, Purification and feedback control of threonine deaminase activity of *Rhodopseudomonas sphaeroides, J. Biol. Chem.* **241**:5836.

Datta, P., 1969, Regulation of branched biosynthetic pathways in bacteria, *Science* **165**:556.

Datta, P., 1970, Homoserine dehydrogenase of *Rhodospirillum rubrum*: Purification and characterization of the enzyme, *J. Biol. Chem.* **245**:5779.

Datta, P., and Epstein, C. C., 1973, Homoserine dehydrogenase of *Rhodospirillum rubrum*: Enzyme polymerization in the presence and absence of threonine, *Biochemistry* **12**:3888.

Datta, P., and Gest, H., 1964a, Alternative patterns of end product control in biosynthesis of amino acids of the aspartic family, *Nature (London)* **203**:1259.

Datta, P., and Gest, H., 1964b, Control of enzyme activity by concerted feedback inhibition, *Proc. Natl. Acad. Sci. U.S.A.* **52**:1004.

Datta, P., and Gest, H., 1965, Homoserine dehydrogenase of *Rhodospirillum rubrum*: Purification, properties, and feedback control of activity, *J. Biol. Chem.* **240**:3023.

Datta, P., and Prakash, L., 1966, Aspartokinase of *Rhodopseudomonas sphaeroides*: Regulation of enzyme activity by aspartate β-semialdehyde, *J. Biol. Chem.* **241**:5827.

Datta, P., Gest, H., and Segal, H. L., 1964, Effects of feedback modifiers on the state of aggregation of homoserine dehydrogenase of *Rhodospirillum rubrum, Proc. Natl. Acad. Sci. U.S.A.* **51**:125.

Datta, P., Dungan, S. M., and Feldberg, R. S., 1973, Regulation of amino acid biosynthesis of the aspartate pathway in different microorganisms, in: *Genetics of Industrial Microorganisms* (Z. Vanek, Z. Hostalek, and J. Cudlin, eds.), pp. 177–193, Academia, Prague.

Dreyfuss, J., and Monty, K. J., 1963, The biochemical characterization of cysteine-requiring mutants of *Salmonella typhimurium, J. Biol. Chem.* **238**:1019.

Ehrensvard, G., and Reio, L., 1953, The formation of tyrosine in *Rhodospirillum rubrum* grown on acetate and carbon dioxide, isotope labelled, under anaerobic conditions, *Ark. Kemi* **5**:327.

Epstein, C. C., 1976, Subunit structure and conformational changes of homoserine dehydrogenase of *Rhodospirillum rubrum*, Ph.D. dissertation, The University of Michigan, Ann Arbor.

Feldberg, R. S., and Datta, P., 1971, L-Threonine deaminase of *Rhodospirillum rubrum*: Purification and characterization, *Eur. J. Biochem.* **21**:438.

Ferretti, J. J., and Gray, E. D., 1968, Enzyme and nucleic acid formation during synchronous growth of *Rhodopseudomonas sphaeroides, J. Bacteriol.* **95**:1400.

Gibson, F., and Pittard, J., 1968, Pathways of biosynthesis of aromatic amino acids and vitamins and their control in microorganisms, *Bacteriol. Rev.* **32**:465.

Gibson, K. D., Neuberger, A., and Tait, G. H., 1962, Studies on the biosynthesis of porphyrin and bacteriochlorophyll by *Rhodopseudomonas sphaeroides, Biochem. J.,* **84**:483.

Greenberg, D. M., 1969, Biosynthesis of amino acids and related compounds, in: *Metabolic Pathways*, Vol. 3 (D. M. Greenberg, ed.), pp. 259–267, Academic Press, New York.

Griffiths, M., and Stanier, R. Y., 1956, Some mutational changes in the photosynthetic pigment system of *Rhodopseudomonas sphaeroides, J. Gen. Microbiol.* **14**:698.

Hoare, D. S., 1963, The photoassimilation of acetate by *Rhodospirillum rubrum, Biochem. J.* **87**:284.

Hoare, D. S., and Gibson, J., 1964, Photoassimilation of acetate and the biosynthesis of amino acids by *Chlorobium thiosulfatophilum, Biochem. J.,* **91**:546.

Hoare, D. S., and Hoare, S. L., 1966, Feedback regulation of arginine biosynthesis in blue-green algae and photosynthetic bacteria, *J. Bacteriol.* **92**:375.

Hubbard, J. S., and Stadtman, E. R., 1967, Regulation of glutamine synthetase. II. Patterns of feedback inhibition in microorganisms, *J. Bacteriol.* **93**:1045.

Hug, D. H., and Werkman, C. H., 1957, Transamination in *Rhodospirillum rubrum, Arch. Biochem. Biophys.* **72**:369.

Hughes, M., Brenneman, C., and Gest, H., 1964, Feedback sensitivity of threonine deaminases in two species of photosynthetic bacteria, *J. Bacteriol.* **88**:1201.

Jones-Mortimer, M. C., 1968, Positive control of sulphate reduction in *Escherichia coli*: Isolation, characterization and mapping of cysteineless mutants of *E. coli* K12, *Biochem. J.,* **110**:589.

Kohlhaw, G., Leary, T. R., and Umbarger, H. E., 1969, α-Isopropylmalate synthase from *Salmonella typhimurium*: Purification and properties, *J. Biol. Chem.* **244**:2218.

Kredich, N. M., and Tomkins, G. M., 1966, The enzymic synthesis of L-cysteine in *Escherichia coli* and *Salmonella typhimurium, J. Biol. Chem.* **241**:4955.

Large, P. J., and Quayle, J. R., 1963, Microbial growth on C$_1$ compounds 5: Enzyme activities in extracts of *Pseudomonas* AM1, *Biochem. J.* **87**:386.

Losada, M., Trebst, A. V., Ogata, S., and Arnon, D. I., 1960, Equivalence of light and adenosine triphosphate in bacterial photosynthesis, *Nature (London)* **186**:753.

Nagatani, H., Schimizu, M., and Valentine, R. C., 1971, The mechanism of ammonia assimilation in nitrogen fixing bacteria, *Arch. Microbiol.* **79**:164.

Nakada, H. I., 1964, Glutamic-glycine transaminase from rat liver, *J. Biol. Chem.* **239**:468.

Ning, C., and Gest, H., 1966, Regulation of isoleucine biosynthesis in the photosynthetic bacterium *Rhodospirillum rubrum, Proc. Natl. Acad. Sci. U.S.A.* **56**:1823.

Ohmori, H., Ishitani, H., Sato, K., Shimizu, S., and Fukui, S., 1971, Vitamin B_{12} dependent glutamate mutase activity in photosynthetic bacteria, *Biochem. Biophys. Res. Commun.* **43**:156.

Robert-Gero, M., LeBorgne, L., and Cohen, G. N., 1972, Concerted feedback inhibition of the aspartokinase of *Rhodospirillum tenue* by threonine and methionine: A novel pattern, *J. Bacteriol.* **112**:251.

Rodwell, V. W., 1969, Biosynthesis of amino acids and related compounds, in: *Metabolic Pathways*, Vol. 3 (D. M. Greenberg, ed.), pp. 334–349, Academic Press, New York.

Schmidt, L. S., and Sojka, G. A., 1973a, Enzymes of serine biosynthesis in *Rhodopseudomonas capsulata*, *Arch. Biochem. Biophys.* **159**:475.

Schmidt, L. S., and Sojka, G. A., 1973b, Regulation of serine biosynthesis in *Rhodopseudomonas capsulata*, *Abstr. Annu. Meet. Am. Soc. Microbiol.*, New York City, p. 142.

Shigesada, K., Hidaka, K., Katsuki, H., and Tanaka, S., 1966, Biosynthesis of glutamate in photosynthetic bacteria, *Biochim. Biophys. Acta* **112**:182.

Slater, J. H., and Morris, I., 1973, The pathway of carbon dioxide assimilation in *Rhodospirillum rubrum* grown in turbidostat continuous flow culture, *Arch. Microbiol.* **92**:235.

Slater, J. H., and Morris, I., 1974, Light-dependent synthesis of glutamate in *Rhodospirillum rubrum*.—Physiological evidence for ammonia assimilation via the glutamine synthetase and glutamine: 2-Oxoglutarate aminotransferase system, *Arch. Microbiol.* **95**:337.

Sturani, E., Datta, P., Hughes, M., and Gest, H., 1963, Regulation of enzyme activity by specific reversal of feedback inhibition, *Science* **141**:1053.

Truffa-Bachi, P., and Cohen, G. N., 1973, Some aspects of amino acid biosynthesis in microorganisms, *Annu. Rev. Biochem.* **42**:113.

Tsuiki, S., and Kikuchi, G., 1962, Catabolism of glycine in *Rhodopseudomonas sphaeroides*, *Biochim. Biophys. Acta* **64**:514.

Udaka, S., 1966, Pathway-specific pattern of control of arginine biosynthesis in bacteria, *J. Bacteriol.* **91**:617.

Umbarger, H. E., 1969, Regulation of amino acid metabolism, *Annu. Rev. Biochem.* **38**:323.

Vogel, H. J., 1965, Lysine biosynthesis and evolution, in: *Evolving Genes and Proteins* (V. Bryson and H. J. Vogel, eds.), pp. 25–40, Academic Press, New York.

Wagner, B. J., Miovic, M. L., and Gibson, J., 1973, Utilization of amino acids by *Chromatium* sp. strain D, *Arch. Microbiol.* **91**:255.

Walsh, D. A., and Sallach, H. J., 1966, Comparative studies on the pathways for serine biosynthesis in animal tissues, *J. Biol. Chem.* **241**:4068.

Wixom, R. L., Heinemann, M. A., Semeraro, R. J., and Joseph, A. A., 1971, Studies in valine biosynthesis. IX. The enzymes in photosynthetic and autotrophic bacteria, *Biochim. Biophys. Acta* **244**:532.

Yen, H.-C., and Gest, H., 1974, Regulation of biosynthesis of aspartate family amino acids in the photosynthetic bacterium *Rhodopseudomonas palustris*, *Arch. Microbiol.* **101**:187.

Yugari, Y., and Gilvarg, C., 1965, The condensation step in diaminopimelate synthesis, *J. Biol. Chem.* **240**:4710.

PART VIII

Physiology

CHAPTER 42

Regulation of Pyrrole Synthesis

J. Lascelles

1. Introduction

The photosynthetic bacteria may be unique in their ability to form the three functional types of metal pyrroles found in nature. Not only do they contain the iron and magnesium derivatives, but also many are rich in corrinoids. The latter substances have not yet been found in other forms of photosynthetic cells, including the blue-green bacteria.

This versatility in biosynthetic capability is combined with subtle regulatory systems that ensure the marshaling of common intermediates into the appropriate channels to maintain vastly different levels of end product. For example, photosynthetically grown cells of *Rhodopseudomonas sphaeroides* contain bacteriochlorophyll (Bchl), hemes, and corrinoids in approximate concentrations of 25, 0.3, and 0.07 nmol/mg dry wt.

The efficiency of control is even more apparent in the ability of the bacteria to adjust the concentration of Bchl without interfering with the quantitatively minor branches and without wastage of common intermediates. This aspect of regulation in seen in the response of organisms such as *Rp. sphaeroides* to changes in light intensity or to aeration (Cohen-Bazire *et al.*, 1957). Bchl synthesis can be halted completely and abruptly, but synthesis of hemes and corrinoids continues without interruption, and without accumulation of intermediates. Failure in the normal mechanisms of "fine" control is manifested by porphyrins in the cultures. This phenomenon was originally observed by van Niel (1944); his description of distinguishing characteristics for Athiorhodaceae included an extracellular pigment with absorption maxima corresponding to coproporphyrin.

A thorough understanding of the regulatory mechanisms requires knowledge of the enzymic steps backed by genetic analysis. As described in other chapters, knowledge of the enzymes is fragmentary, and a genetic system has only recently been discovered. Consequently, much of this review will consider regulatory mechanisms based on circumstantial or indirect evidence obtained mostly with *Rp. sphaeroides*. The first assumption is that the path of pyrrole synthesis, with its Co, Fe, and Mg branches, is as outlined in Fig. 1 (see Chapter 41 for further details). The steps of the Mg path in particular should be treated skeptically; they are based almost entirely on the pattern of Mg pyrroles that accumulate in cultures of mutants blocked in Bchl synthesis.

One striking feature of the overall path is the number of steps involving oxidation or reduction. A particular lack at the enzymic level is knowledge of the physiological H-donors or acceptors in these steps.

Having made apologies, the reviewer will now attempt to fit together the few facts, hoping that the empty acres will encourage future experimenters. It may be assumed that most observations apply to *Rp. sphaeroides* unless stated otherwise.

2. Control of Enzyme Activity

Synthesis of Bchl can be reduced or completely abolished by raising the light intensity or by aeration.

J. Lascelles · Department of Bacteriology, University of California, Los Angeles, California 90024

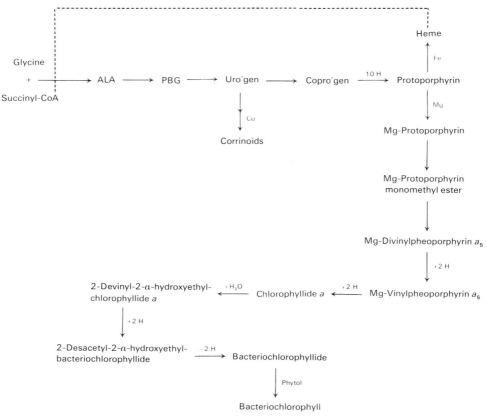

Fig. 1. Outline of pyrrole pathway in *Rhodopseudomonas sphaeroides*. (Uro'gen) Uroporphyrinogen III; (Copro'gen) coproporphyrinogen III.

The response is immediate, yet intermediates do not accumulate even transiently (Cohen-Bazire *et al.*, 1957). This suggests efficient mechanisms for "fine control" of enzyme activity with the focal point at the first step. δ-Aminolevulinate (ALA) synthase is the rate-limiting enzyme for the entire path, at least in *Rp. sphaeroides*. That it is becomes evident when the synthase is bypassed by addition of ALA to cells; rapid accumulation of porphobilinogen (PBG) and a variety of pyrroles occurs (Lascelles, 1956, 1966a,b; Richards and Rapoport, 1966).

ALA synthase (E.C. 2.3.1.37) has now been demonstrated in many photosynthetic bacteria, despite previous problems in detection (Burnham, 1970). There is no firm reason to suspect alternative routes for synthesis of ALA in the bacteria. This contrasts with eukaryotic photosynthetic organisms, in which classic ALA synthase has yet to be shown. instead, there is good evidence that higher plants form ALA by a novel route, deriving from a C5-dicarboxylic acid (Beale and Castelfranco, 1974a,b).

2.1. Regulation of δ-Aminolevulinate Synthase Activity

ALA synthase has been purified from *Rp. sphaeroides* in several laboratories (Lascelles, 1975). The procedure of Warnick and Burnham (1971) provided the most active preparations; the enzyme has an estimated molecular weight of about 60,000, with no evidence of subunit structure. All preparations of the synthase have an exceptionally high K_m for glycine (see Table 1); this observation should be remembered when considering *in vivo* regulation of pyrrole synthesis.

Inhibition of the enzyme by heme is now well documented, but there are also reports of inhibition by Mg-protoporphyrin and protoporphyrin (Yubisui and Yoneyama, 1972) and by ATP (Fanica-Gaignier and Clement-Metral, 1973a,b). Also, activation of the enzyme by trisulfides was recently discovered (Neuberger *et al.*, 1973a,b). All these observations may be significant to the problem of *in vivo* control of

Table 1. K_m Values for Enzymes of Pyrrole Synthesis in *Rhodopseudomonas sphaeroides*

Enzyme	Substrate	K_m (mM)	References
ALA synthase	Glycine + succinyl-CoA	10	Warnick and Burnham (1971)
		0.025	
	Glycine + succinyl-CoA	44	Fanica-Gaignier and Clement-
		0.068	Metral (1973a)
ALA dehydrase	ALA	0.5	Nandi et al. (1968)
Uroporphyrinogen I synthase	PBG	0.01	Jordan and Shemin (1973)
Coproporphyrinogen oxidase	Coproporphyrinogen III	0.03–0.05	Tait (1972)
Mg-Protoporphyrin	Mg-protoporphyrin +	0.04	Gibson et al. (1963)
methyltransferase	S-adenosylmethionine	0.055	

the pyrrole path, and they will now be considered in turn.

2.1.1. Inhibition by Heme

Observations of inhibition of ALA synthase by heme were made originally with crude extracts and partially purified preparations from *Rp. sphaeroides* (Burnham and Lascelles, 1963). Since then, the enzyme has been purified from several sources, including liver, and found to be sensitive to heme (Burnham, 1970; Granick and Sassa, 1971; Whiting and Elliott, 1972).

The enzyme purified from *Rp. sphaeroides* by Warnick and Burnham (1971) is inhibited by low concentrations of heme (13% by 0.1 μM and 57% by 5 μM). There is no evidence of an allosteric type of response. The inhibition is readily reversed by dilution, however, indicating that enzyme-bound heme is in equilibrium with free heme.

It is not difficult to rationalize the *in vitro* effects of heme in physiological terms for organisms forming

iron-porphyrins as major pyrroles. Rather more faith is required for belief in the relevance of heme inhibition in organisms with a predominance of Mg pyrroles.

The model outlined in Fig. 2 provides a working hypothesis, for which there is some support. Competition between the Fe and Mg branches for protoporphyrin is central to the model. Inhibition of the Mg branch would favor diversion to heme; the intracellular concentration would consequently rise to the critical level needed to inhibit ALA synthase. The evidence to support this proposal is based largely on observations with wild-type and mutant strains of *Rp. sphaeroides*, and concerns overproduction of intermediates in relation to heme synthesis.

In the wild-type, iron deficiency reduces synthesis of Bchl and causes the accumulation of porphyrins in the medium (Lascelles, 1956). The quantity of porphyrin formed under these conditions exceeds by at least tenfold the amount of Bchl formed by normal cells, indicating failure of a feedback type of control.

The fundamental cause of overproduction may now be attributed to ALA synthase acting without the

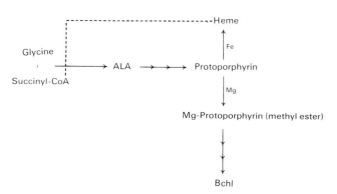

Fig. 2. Model for regulation of pyrrole formation by heme.

Table 2. Pyrrole and Heme Formation by Wild-Type and Mutant *Rhodopseudomonas sphaeroides* Strains[a]

		Pyrrole formed in 4 hr as ALA equivalents (nmol/ml)[c]				
Strain	Enzymic block(s)[b]	Heme	PBG	Coproporphyrin	Mg-protoporphyrin (ester)	Bchl
Wild-type	—	6.7	ND	ND	ND	96
2-33	Coproporphyrinogen to protoporphyrin Mg path	0.7	124	124	ND	ND
2-33R	Mg path	7.4	ND	4	80	6
6-6	Uroporphyrinogen synthase; Mg path	12.8	62	ND	ND	ND
6-6R	Uroporphyrinogen synthase	2.6	200	ND	ND	48

[a] Data from Hatch and Lascelles (1972). Concentrated cell suspensions (2 mg dry wt./ml) were incubated with low aeration in a complete medium supplemented with glycine and succinate.

[b] The enzymic lesions have not been established by direct assay for uroporphyrinogen synthase.

[c] ND: Not detectable.

restraint imposed by heme. Key evidence in favor of this is:

1. The synthase is rate-limiting for heme synthesis. Addition of ALA to cells raises heme synthesis by tenfold or more, and also causes porphyrins and other intermediates to accumulate (Lascelles, 1966a,b; Lascelles and Hatch, 1969).

2. Addition of o-phenanthroline to cells incubated with glycine and succinate blocks heme synthesis, and accumulation of coproporphyrin occurs (Lascelles and Hatch, 1969).

3. A mutant, strain 6-6, with a partial block in uroporphyrinogen I synthase accumulates PBG on inhibition of heme synthesis by o-phenanthroline (Hatch and Lascelles, 1972).

4. Termination of Bchl synthesis by aeration or by addition of inhibitors of protein synthesis is accompanied by increased heme synthesis. Coproporphyrin accumulates when such synthesis is prevented by o-phenanthroline (Lascelles and Hatch, 1969; Hatch and Lascelles, 1972).

5. Accumulation of intermediates by certain mutants is correlated inversely with their ability to form heme (Table 2).

The evidence, though indirect, is consistent with a critical role for heme. More compelling evidence for the validity of the model requires data about the pool concentration of protoporphyrin and of heme in response to various perturbations. Most important, information is needed about the respective Fe and Mg chelating enzymes, such as turnover number and K_m for protoporphyrin.

2.1.2. Protoporphyrin and Mg-Protoporphyrin as Inhibitors

Purified ALA synthase from photosynthetically grown *Rp. sphaeroides* is inhibited by protoporphyrin and Mg-protoporphyrin (Yubisui and Yoneyama, 1972). The Mg derivative at 12 μM concentration gave 83% inhibition. Such strong inhibition invites speculation about regulation. Mg-Protoporphyrin is more easily rationalized than is heme as a feedback regulator of ALA synthase. Also, formation of Mg-protoporphyrin is rate-limited by ALA synthase, since the Mg-pyrrole or its methyl ester accumulates in cells incubated with ALA (Lascelles, 1966a,b; Lascelles and Altshuler, 1969).

Clearly, information is needed about the pool levels of the various putative inhibitors under various conditions of repression and derepression of Bchl synthesis. Such information should be possible to acquire with techniques for extraction of the pyrroles and sensitive methods for measurement by fluorescence spectrophotometry. Such methods are now being developed for analysis of intermediates in chlorophyll synthesis by higher plants (Rebeiz et al., 1975).

2.1.3. Adenine Nucleotides as Effectors

Purified preparations of ALA synthase from *Rp. sphaeroides* strain Y are inhibited strongly by ATP; 70–90% inhibition was observed with concentrations of 1 mM (Fanica/Gaignier and Clement-Metral, 1973b). This observation supports the view that ATP has a role in controlling Bchl synthesis in growing cells

(Sojka and Gest, 1968). Alternatively, the ratio of adenine nucleotides ("energy charge") might also be applied (Zilinsky *et al.*, 1971).

Efforts to correlate adenine nucleotide levels with Bchl synthesis in growing cells have not provided convincing evidence either for or against the notion.

Pool levels of the adenine nucleotides in relation to Bchl synthesis have been measured in growing cultures of *Rp. sphaeroides*, *Rhodospirillum rubrum*, and *Chromatium vinosum* strain D (Fanica-Gaignier *et al.*, 1971; Oelze and Kamen, 1971; Schmidt and Kamen, 1971. In all cases, complex fluctuations were observed, but ATP rather than the other nucleotides was concluded to be important in regulating pigment synthesis. In *Chr. vinosum*, for example, the period of most active Bchl synthesis corresponded to the lowest concentration of ATP in the cell (Schmidt and Kamen, 1971). The rate of pigment synthesis decreased when the cellular ATP reached its maximum level. These experiments were with cultures growing autotrophically with sulfide. In contrast, the observations of Miović and Gibson (1973) provide no support for ATP in a direct regulatory role. Their measurements were in cultures growing photoheterotrophically under varying light intensities. The concentration of ATP was higher in cells growing in dim light than in bright light (0.57 mM and 0.12 mM, respectively). Thus, in these experiments, the higher levels of ATP were found in cells forming Bchl at the maximum rate. Also, changes in pigment synthesis in response to shifts to low or high light occurred without disturbance of adenine nucleotide levels.

2.1.4. Activation by Trisulfides

Assay of ALA synthase activity in crude extracts of bacteria can be a frustrating experience (Burnham, 1970). The trend of the observations has pointed to redox-sensitive modifiers of enzyme activity, which are influenced by the growth conditions. For instance, ALA synthase activity in cultures of *Rp. sphaeroides*

grown photosynthetically declines on transfer to aerobic conditions or to high light intensities (Higuchi *et al.*, 1968; Marriott *et al.*, 1969). Spontaneous activation can occur on storage of extracts (Marriott *et al.*, 1969). Such activation is increased by aeration or by treatment with ferricyanide or hydrogen peroxide (Warnick and Burnham, 1971).

Purification of the enzyme from photosynthetically grown *Rp. sphaeroides* has provided evidence of high- and low-activity forms, designated as *a* and *b*, respectively. Hayasaka and Tuboi (1974) showed interconversion of the two forms by treatment with cystine (*b* to *a*) or thiols (*a* to *b*), and these interconversions were enzyme-dependent.

Recent work from Neuberger's group may have clarified some of these mysteries. Endogenous activators of the synthase were isolated from *Rp. sphaeroides* and identified as cystine trisulfide and the mixed trisulfides of cystine and glutathione (Davies *et al.*, 1973; Sandy *et al.*, 1975). The trisulfides act at micromolar concentrations, and are thought to act as regulators of ALA synthase *in vivo* (Fig. 3). The concentration of trisulfide may be mediated by thiol–disulfide couples, in turn linked to the respiratory or photosynthetic electron-transport chain.

Critical evidence in favor of the scheme includes:

1. Oxygenation of photosynthetically grown cells results in rapid loss of ALA synthase activity, and is accompanied by a decline in trisulfide (from 20–30 to less than 0.5 nmol/g dry wt.).

2. Forms *a* and *b* of ALA synthase can be separated by chromatography of extracts on DEAE-Sephadex. Treatment with cystine trisulfide converts *b* to *a*, but does not affect the activity of the *a* form.

3. The two forms are found in extracts from photosynthetically grown cells. Oxygenation of such cells results in disappearance of *a* and accumulation of the *b* form.

This work is clearly of great importance in considerations of the overall regulation of pyrrole

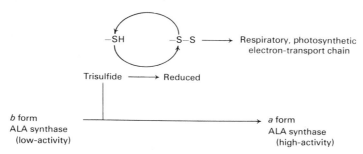

Fig. 3. Model illustrating regulation of ALA synthase activity by trisulfide activation.

path. The trisulfides could act rapidly and in response to environmental changes by modifying the activity of the synthase. Much remains to be established. For instance, the mechanism of activation of the synthase is unknown; is it reversible? How are the trisulfides formed, and how are they rendered inoperative by treatment with thiols? How are they removed by aeration? Cystathionine may be a precursor, since preparations of cystathionase from *Rp. sphaeroides* have been found to convert the disulfide to cystine trisulfide (A. Neuberger, personal communication).

2.1.5. Enzyme Multiplicity

Distinct forms of ALA synthase under independent control would be a satisfactory solution to the problem of controlling the branched pathway.

Two fractions with enzyme activity were reported during purification of ALA synthase from *Rp. sphaeroides* (Tuboi and Hayasaka, 1972; Fanica-Gaignier and Clement-Metral, 1973a). The enzymes purified by the French workers are similar in molecular weight, but differ in isoelectric point and in other properties. At present, it seems that these observations may be attributable to the inter-convertible forms *a* and *b* of the enzyme, discussed in Section 2.1.4.

The idea of enzyme multiplicity would be more attractive if physiological correlates could be established. Different degrees of repression and derepression in response to oxygen pressure might be expected. Differing sensitivity to inhibition by heme and Mg-pyrroles is another property to be examined.

The most convincing proof of the physiological significance of isoenzymes is to isolate mutants lacking one or another form. A mutant strain of *Rp. sphaeroides* (strain H5) lacking ALA synthase activity was described by Lascelles and Altshuler (1969). It requires ALA for heme and for Bchl synthesis, and provides no support for the possibility of isoenzymes. In the absence of a genetic system to determine whether a single mutational event is responsible for lack of enzyme activity, however, the question is still wide open. Also, mutant H5 has several other curious characteristics, indicating multiple genetic lesions. For instance, unlike the parent strain, it does not accumulate massive amounts of porphyrins from ALA, probably due to a defect in transport of ALA.

2.2. Other Enzymes of Biosynthetic Pathways

The characteristics of other enzymes of pyrrole biosynthesis should be considered in relation to their possible significance in control. Some of these enzymes have been extensively purified, but there is lamentable ignorance of many key steps.

2.2.1. δ-Aminolevulinate Dehydratase

ALA dehydratase from *Rp. sphaeroides* was studied in exquisite detail by Shemin and his group, and some of its characteristics could support a regulatory role. First, it is a complex molecule of molecular weight 240,000, and is probably composed of six identical subunits (van Heyningen and Shemin, 1971). Second, it exhibits allosteric responses to K^+ and other monovalent cationic activators (Nandi *et al.*, 1968). Third, it is inhibited strongly by hemin and protoporphyrin (Nandi *et al.*, 1968). These properties may be red herrings; there is no physiological evidence nor is there an obvious teleological reason for control by the dehydratase. The purified enzyme from *Rp. capsulata* does not have the suggestive properties of that from *Rp. sphaeroides*: it is not inhibited by hemin or protoporphyrin, it exhibits simple Michaelis–Menten kinetics, and it does not require activation by metal cations (Nandi and Shemin, 1973).

2.2.2. Uroporphyrinogen I Synthase

Uroporphyrinogen I synthase was purified from *Rp. sphaeroides* to virtual homogeneity by Jordan and Shemin (1973). Its properties do not invite speculations on a role in regulation. An interesting characteristic, however, is the low K_m for PBG relative to the K_m of ALA dehydratase (see Table 1).

2.2.3. Coproporphyrinogen III to Protoporphyrin

Coproporphyrin III is the predominant porphyrin accumulated in most of the bacterial porphyrias (Lascelles, 1964). In *Rp. sphaeroides*, for example, this porphyrin accounts for about 95% of the total formed by iron-deficient cells (Lascelles, 1956). A similar pattern of porphyrin accumulation occurs in response to methionine deficiency (Gibson *et al.*, 1962; Lessie and Sistrom, 1964; Lascelles, 1966a).

Apparently, the enzyme system responsible for converting coproporphyrinogen III to protoporphyrin IX is a critically weak component in the pathway, and it is pertinent to consider what is known in general about this step (Fig. 4).

The overall reaction involves (1) oxidative decarboxylation of the propionic acid side chains at positions 2 and 4 of the tetrapyrrole nucleus to give protoporphyrinogen IX and (2) oxidation of the porphyrinogen to porphyrin. In all, ten hydrogen atoms must be eliminated. The mechanism of decarboxylation is still speculative (Mori and Sano, 1972; Tait, 1972). Conversion of protoporphyrinogen IX to protoporphyrin may proceed via tetrahydroprotoporphyrin IX (P-503); this compound was identified in

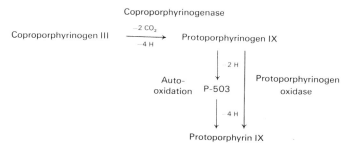

Fig. 4. Reactions catalyzed by coproporphyrinogenase and protoporphyrinogen oxidase.

some bacterial cultures and in anaerobic yeast (Poulson and Polgase, 1973).

The overall reaction in yeast involves two enzymes, which were recently purified from mitochondrial preparations. Coproporphyrinogenase catalyzes the oxidative decarboxylation, and is active anaerobically in the presence of appropriate cofactors (Poulson and Polgase, 1974a). Protoporphyrinogen oxidase is located in the membrane fraction, and is apparently linked obligatorily to oxygen (Poulson and Polgase, 1975; Poulson, 1976).

Oxidation of coproporphyrinogen III to protoporphyrin was demonstrated in extracts of many bacteria in the presence of oxygen (Jacobs, 1974). The anaerobic conversion of protoporphyrinogen to protoporphyrin coupled to the fumarate or nitrate reductase systems has been recently demonstrated in extracts of *Escherichia coli* (Jacobs and Jacobs, 1976). The oxygen-dependent activity was studied in partially purified preparations from *Chr. vinosum* and *Rp. sphaeroides* (Mori and Sano, 1972; Tait, 1972). It is located in the cytoplasmic fraction, and has no demonstrable requirement for cofactors, nor is it inhibited by o-phenanthroline and other metal chelators.

Oxygen is obviously not the physiological oxidant for the enzyme in photosynthetic bacteria growing anaerobically. The demonstration of anaerobic activity *in vitro*, however, was achieved only recently (Tait, 1969, 1972). Failures by such experts as Mori and Sano (1972) may be attributable to the complexity and lability of the anaerobic system. Preparations from *Rp. sphaeroides* and *Chr. vinosum* require Mg^{2+}, S-adenosylmethionine, and reductant provided by NADH and NADP or by succinate. Both the membrane and cytoplasmic fractions of the cell are needed. Iron proteins may also participate, since the anaerobic system is inhibited by o-phenanthroline and other chelators of Fe^{2+}.

The participation of iron (proteins) and of S-adenosylmethionine may account for the accumula-tion of coproporphyrin under conditions of iron or methionine deficiency.

The coproporphyrinogen oxidation system from yeast has many similarities to that from the photosynthetic bacteria (Poulson and Polgase, 1974a, 1975). The anaerobic conversion of coproporphyrinogen III to protoporphyrinogen is activated by Fe^{2+} and requires S-adenosylmethionine and NADP. The overall conversion of coproporphyrinogen to protoporphyrin seems also to be a critical control point in yeast, since repression of heme synthesis by glucose results in the accumulation of P-503 (Poulson and Polgase, 1974b).

In the photosynthetic bacteria, knowledge of the H-acceptors and their recycling is clearly critical to understanding the system. Reverse electron transfer, linked to the energy-linked transhydrogenase, is one possibility. The driving force could be energized intermediates derived from the photosynthetic electron-transport system. If it is, the activity of the system *in vivo* might be expected to respond to light- or oxygen-induced changes in pool levels of ATP and reduced/oxidized ratios of nicotinamide coenzymes.

2.2.4. Incorporation of Iron and Magnesium

The enzymes concerned with incorporating Fe and Mg into protoporphyrin are likely to be the most critical in controlling the flow of the common intermediate into the two branches. Apparently, both systems are sensitive to oxygen, but incorporation of Mg is far more susceptible than that of Fe. That it is has been shown with wild-type and mutant strains of *Rp. sphaeroides* forming heme and Mg-protoporphyrin monomethyl ester with ALA as precursor, under semianaerobic conditions in the light. On aeration, formation of Mg-pyrroles ceases abruptly, whereas heme synthesis continues, though at a reduced rate (Lascelles and Altshuler, 1969; Lascelles and Wertlieb, 1971).

The little that is known of the *in vitro* properties of the enzyme systems suggests a critical role for oxygen in controlling the branch-point enzymes; the effect could be direct or indirect via appropriate redox systems.

(a) Ferrochelatase (Protoheme Ferrolyase, E.C. 4.99.1.1). Ferrochelatase activity is readily demonstrable in the membrane fraction of *Rp. sphaeroides*. Incorporation of ferrous iron into protoporphyrin occurs, but incorporation of Mg has not been observed. Ferrochelatase does not seem to be highly specific for Fe or protoporphyrin, however, and most of the kinetic data concern incorporation of cobalt into deuteroporphyrin (Jones and Jones, 1970). The "model" system is inhibited noncompetitively by heme and by Mg-protoporphyrin.

Ferrochelatase in the photosynthetic bacteria requires further study under conditions more related to the *in vivo* situation. It is usually assayed under strictly anaerobic conditions in the presence of thiols to maintain iron in the ferrous state, but activity with NADH or succinate has been observed (Jones and Jones, 1970). Provision of iron in the ferrous state via the respiratory–photosynthetic electron-transport system could be important in regulating heme and hemoprotein synthesis, and this aspect of heme formation in photosynthetic bacteria deserves more attention.

(b) Incorporation of Magnesium into Protoporphyrin. Incorporation of Mg was observed in whole cells of *Rp. sphaeroides* and *Rs. rubrum*, but this ability is lost on cell disruption (Gorchein, 1972, 1973). Photosynthetically grown *Rp. sphaeroides* incorporated Mg into exogenous protoporphyrin, but only in the presence of EDTA or EGTA. Presumably, the chelators weaken the outer membrane by complexing Mg and Ca and permit passage of protoporphyrin. The product of the reaction is the monomethyl ester of Mg-protoporphyrin, and the incorporation seems to be obligatorily coupled to the methylation step. Cells deprived of methionine have only low activity; also, *S*-adenosylmethionine restores activity to cells treated with pyrophosphate, an inhibitor of *S*-adenosylmethionine synthetase.

The sensitivity to oxygen of the Mg-incorporation system is most pertinent to the problem of regulation. The reaction with *Rp. sphaeroides* is inhibited by pressures of oxygen above 15% (114 mm Hg). Yet, it is dependent on energy-coupling, which can be provided by the respiratory or the photosynthetic electron-transport system. This is evident from the effects of specific inhibitors such as antimycin and oligomycin. It is not possible to interpret the significance of the energy-coupling in the whole-cell system. The rate-limiting step appears to be in the formation of Mg-protoporphyrin, since exogenous Mg-protoporphyrin is rapidly converted to the methyl ester (Gorchein, 1973). One possibility is that the methylation step is required to "pull" the initial incorporation reaction, and ATP may be required to provide an adequate supply of *S*-adenosylmethionine.

3. Kinetic Properties of Enzymes in Relation to Regulation

The kinetic properties of key enzymes in biosynthetic pathways may provide effective means of regulation, without any necessity of invoking the sophistications of negative and positive modifiers of enzyme activity. The relative affinities of enzymes for their substrates could be critical, particularly if these are central metabolites, such as reduced (or oxidized) cofactors or Krebs' cycle intermediates. The K_m's and turnover numbers of enzymes competing for a common substrate could also be crucial in channeling it toward the appropriate end product.

The apparent K_m values for enzymes purified from *Rp. sphaeroides* are shown in Table 1. The high K_m of ALA synthase for glycine raises the question whether the *in vivo* concentration of glycine becomes a factor in limiting its activity. This does not normally seem to be so, however, since the output of "ALA units" can be increased severalfold by interference with heme synthesis (see Table 2). The K_m of ALA dehydratase is high in relation to the values for subsequent enzymes of the path. This disparity may ensure rapid transition of PBG to functional pyrroles, without accumulation of potentially toxic porphyrinogens and porphyrins.

Once again, we must deplore the lack of knowledge of the kinetic properties of the enzymes concerned with incorporation of Fe and Mg.

4. Inhibition of Bacteriochlorophyll Synthesis by Light and Oxygen

The photosynthetic bacteria regulate Bchl synthesis to meet the requirements imposed by the environment with amazing efficiency. Under anaerobic conditions, the light intensity determines the quantity of pigment formed in the cell, and even the proportion of reaction-center to light-harvesting forms of Bchl (Aagaard and Sistrom, 1972). Good housekeeping is particularly apparent in the facultatively aerobic species, in which Bchl synthesis is suppressed by oxygen. This inhibition may create problems for those bacteria that are solely dependent on their photosynthetic apparatus for energy. Growth of these bacteria is inhibited by oxygen, but this inhibition may not necessarily be due to oxygen toxicity *per se*. In *Rs. molischianum*, at least, the primary effect of oxygen is to prevent Bchl synthesis, and hence the cell is deprived of its sole source of metabolic energy (Sistrom, 1965).

The immediate inhibition that occurs on increasing the oxygen pressure or the light intensity points directly to inhibition of enzyme(s) activity. This raises the question of the nature of the inhibitor substance and its locus of action. The original proposal of Cohen-Bazire *et al.* (1957) is still the framework for assembly of scattered facts. They proposed a regulator substance the concentration of which was determined by the state of oxidation of the electron-transport chain. This proposal was slightly modified by Marrs and Gest (1973) on the basis of observations with respiratory mutants of *Rp. capsulata*. They envisage positive control of Bchl synthesis by a factor that is inactivated directly by oxygen. This notion is supported by the observations of oxygen inhibition with mutants lacking in several components of the respiratory chain.

Both proposals envisage an unspecified effector (negative or positive) acting directly on an unspecified area of Bchl synthesis. This reviewer is unable to fill in the specifications. Possibly incorporation of Mg into protoporphyrin is the prime locus. Whole-cell work indicates that this reaction is extremely sensitive to oxygen. It should not be forgotten, however, that the formation of protoporphyrin with its complexities is a critical area of control.

The nature of the effector is completely nebulous. The recent work on the trisulfide activator of ALA synthase encourages notions of redox-sensitive sulfur compounds acting at other loci in the biosynthetic pathway.

5. Repression of Enzyme Formation

The lack of knowledge of regulation at the level of enzyme formation is not surprising in view of the lack of satisfactory assays for so many key steps. This ignorance is compounded by the absence of systems for genetic analysis.

The only evidence for regulation by repression of enzyme synthesis is from *Rp. sphaeroides* (Lascelles, 1975). Strong aeration represses ALA synthase, Mg-protoporphyrin methyl transferase, and the system for incorporation of Mg into protoporphyrin. In each case, the effect of oxygen appears to be due primarily to the prevention of enzyme synthesis, rather than to the presence of inhibitors. Thus, development of these

enzyme activities on transfer of cells to conditions of low aeration is dependent on *de novo* protein synthesis (Lascelles, 1959; Marriott *et al.*, 1969; Davies *et al.*, 1973; Gorchein, 1973). Whether or not these enzymes are also repressed by high light intensity is unclear.

In such well-regulated creatures as *Rp. sphaeroides*, it is to be expected that repression of all the enzymes of the Mg pathway occurs under high aeration, but this can be only an assumption until these enzymes can be assayed. Mutant strains of *Rp. sphaeroides* have been isolated that give support to some type of coordinated regulation (Lascelles and Wertlieb, 1971). These mutants form Bchl under high aeration, and in each case, the methyl transferase is not repressed by oxygen. The mechanism of repression is completely obscure. Questions include: Are regulatory genes involved? Are low-molecular-weight effectors involved in the activation or inactivation of genes or gene products? Are these effectors similar to the substances proposed in the fine control of the biosynthetic enzymes? Progress in this problem is dependent on the development of systems for genetic analysis of mutants, and *Rp. capsulata* is now providing hope.

6. The Cobalt Branch

Vitamin B_{12} derivatives (corrinoids) have been found in sulfur and nonsulfur photosynthetic bacteria, in quantities that vary from organism to organism (Toohey, 1966, 1971). For example, light-grown *Rs. rubrum* contains less than one-tenth the concentration found in *Rp. sphaeroides* (Table 3). Such determinations have been made by assay with vitamin-B_{12}-auxotrophs of *Escherichia coli*, and do not always reflect the actual capacity of the organism to form the corrinoid ring. Many photosynthetic bacteria also form descobalto corrinoids, possibly because of a deficiency in "cobalt chelatase." Such derivatives were discovered by Toohey (1965) in *Chr. vinosum*. The Co-free compounds comprise at least 99% of the total corrinoid formed by this organism, whether or not Co is added to the medium. Both descobalto and normal corrinoids are found in *Rs. rubrum*, but the former comprises more than 90% of the total (Toohey, 1966, 1971).

The vitamin-B_{12}-dependent pathway for methionine biosynthesis has been shown in *Rp. sphaeroides* and *Rs. rubrum* (Cauthen *et al.*, 1967; Ohmori *et al.*, 1974), indicating at least one function for the corrinoids.

Knowledge of the pathway of vitamin B_{12} synthesis

Table 3. Concentration of Corrinoids in Photosynthetic Bacteria

Type of assay	Organism	Corrinoid (nmol/g dry wt.)	Ref. No.[a]
Esch. coli mutant	*Rp. sphaeroides*	68	1
		86	2
	Rs. rubrum	1	1
		5	3
	Chl. limicola *f. thiosulfatophilum*	25	1
Total, including descobalto corrinoid	*Rp. palustris*[b]	10	4
	Rs. rubrum[b]	75	4
	Chr. vinosum strain D[b]	50	4

[a] References: (1) Cauthen *et al.* (1967); (2) Hayashi and Kamikubo (1970); (3) Ohmori *et al.* (1974); (4) Toohey (1966).

[b] Descobalto forms estimated to be about 50% of total in *Rp. palustris*, more than 90% in *Rs. rubrum*, and at least 99% in *Chr. vinosum*.

is fragmentary, with particularly large gaps in the early stages (Scott, 1974; Friedman, 1975; Scott *et al.*, 1974). Insertion of Co probably occurs as the last step, after completion of the corrinoid ring with its amino-isopropanolphosphoribose benzimidazole side chain (Scott *et al.*, 1974). The accumulation of Co-containing precursor corrinoids in propionic acid bacteria and certain *Clostridia* suggests, however, that "cobalt-chelatase" may lack specificity (Friedman, 1975).

Regulation of vitamin B_{12} synthesis in photosynthetic and other bacteria is completely mysterious. The descobalto forms found in some photosynthetic bacteria appear to arise from lack of chelatase, rather than from failure of control mechanisms. These compounds do not accumulate in the abnormally high amounts to be expected of a classic case of failure of feedback control (Table 3).

On quantitative grounds, the drain of the Co branch on the common part of the pyrrole pathway is negligible. The Co branch presumably faces strong competition from the Fe and Mg branches for the common intermediate, presumed to be uroporphyrinogen III. The pool of this intermediate may always be maintained at a level in excess of that needed for maximum activity of the branch-point enzyme of the Co path. There is no information on pool levels of intermediates in photosynthetic bacteria growing normally. Uroporphyrin III is a common component of the mixture of porphyrins formed under abnormal conditions by photosynthetic bacteria, including *Rhodopseudomonas* and *Chlorobium* species (Lascelles, 1964; Richards and Rapoport, 1966).

7. Regulation of Chlorophyll Synthesis in Plants

Formation of ALA is the critical step in the regulation of chlorophyll (Chl) synthesis in eukaryotic plants, but the responsible enzyme has not yet been established (Beale and Castelfranco, 1974a,b). Addition of ALA to developing cotyledons and to algae causes protochlorophyllide to accumulate, and this is largely transformed to chlorophyllide *a* on appropriate illumination (Granick and Sassa, 1971; Rebeiz and Castelfranco, 1973). These observations indicate that the enzymes beyond ALA are not rate-limiting and normally operate below their maximum capacity.

Control of ALA formation apparently involves Fe, possibly in the form of heme. Duggan and Gassman (1974) found etiolated bean leaves to accumulate Mg-porphyrins on treatment with *o*-phenanthroline and other chelators of Fe. Simultaneous administration of levulinic acid (to inhibit the ALA dehydrase) resulted in the accumulation of ALA. These observations support the idea of feedback control by heme of the enzyme forming ALA in a similar fashion to that proposed for bacteria. Castelfranco and Jones (1975) propose diversion of protoporphyrin to heme when the light-dependent reduction of protochlorophyllide is prevented; heme regulates production of ALA by feedback inhibition. The enzyme responsible for ALA formation is still elusive. Isotopic evidence supports glutamate or α-ketoglutarate as a precursor (Beale *et al.*, 1975; Weinstein and Castelfranco, 1977). Clearly, knowledge of this enzyme is vital.

The   conversion   of   protochlorophyllide   to

chlorophyllide is another critical control point in Chl synthesis in higher plants. Many algae, like photosynthetic bacteria, can perform this step in the dark, implying that the required reductant is produced by dark metabolism. Such algae also have the light-dependent system, since mutants have been isolated that form Chl only in the light. This reductive step could be critically responsive to environmental changes such as light intensity, and may be central to regulating the entire path, as proposed by Castelfranco and Jones (1975). Recently, Wang and co-workers (1977) have suggested that protochlorophyllide regulates chlorophyll synthesis by feedback inhibition at the stage of ALA formation and also at the stage of insertion of Mg into protoporphyrin. The evidence is based upon patterns of porphyrins formed by Mendelian mutants of *Chlamydomonas*.

There is apparently a complex system of controls in the region of Mg-protoporphyrin synthesis (Gough, 1972; Von Wettstein *et al.*, 1974). Nuclear gene mutants of barley have been analyzed for their ability to form various porphyrins and Mg-porphyrins in response to ALA. Several structural genes may be involved in determining the conversion of protoporphyrin to Mg-protoporphyrin. Regulatory genes may also participate, as suggested by recent work with *Chlamydomonas* mutants (Wang *et al.*, 1975). Many notions have been put forward, including: (1) channeling of enzymes from ALA to protoporphyrin and from protoporphyrin to protochlorophyllide; (2) processing of Mg for incorporation into protoporphyrin; and (3) carrier (binding) proteins for protoporphyrin and other intermediates.

Support or demolition of these ideas will depend on *in vitro* studies of the enzymic steps. *In vitro* systems from isolated chloroplasts and etioplasts that incorporate radioactive ALA into protochlorophyllide and Chl have been described (Rebeiz and Castelfranco, 1973; Griffiths, 1975a,b; Rebeiz *et al.*, 1975). The yields are small in comparison with the quantities of porphyrins and Mg porphyrins formed in these systems, but they offer hope for the future. Magnesium protoporphyrin chelatase activity has been demonstrated in plastid preparations from etiolated and greening cotyledons (Smith and Rebeiz, 1977). Net formation of protochlorophyllide from magnesium protoporphyrin monoester and from protoporphyrin has also been achieved (Mattheis and Rebeiz, 1977a,b).

8. Concluding Remarks

The versatility of photosynthetic bacteria in forming Fe-, Co-, and Mg-pyrroles is matched by their efficiency in regulation. The production of ALA is rate-limiting for the overall pathway; this seems to be general in all cells and tissues capable of forming pyrroles. Control at this stage has the obvious advantage of preventing wastage of carbon and nitrogen precursors. Possibly of more importance to photosynthetic organisms, the curb on ALA production prevents accumulation of porphyrins with their potential to damage cells through photodynamic action.

It seems that photosynthetic bacteria form ALA from glycine and succinyl-CoA catalyzed by ALA synthase. *In vitro* evidence, supported by physiological data, indicates that the enzyme is regulated by heme through feedback inhibition, and also by intracellular trisulfide activators. The intracellular level of heme may be determined by competition between the Fe and Mg branches for the common intermediate, protoporphyrin. The positive response to trisulfides could be critical in tuning the enzyme to redox changes in the environment.

The ability of photosynthetic bacteria to regulate pyrrole synthesis in response to oxygen pressure or light intensity is spectacular, but the mechanisms are obscure. The enzymes concerned in incorporation of Fe and Mg into protoporphyrin are apparently critical, but knowledge about them is lacking. Many of the steps in the overall pathway involve oxidation or reduction, but the nature of the oxidants or reductants is unknown. Some of these steps could be critical in determining responses to the environment. The conversion of coproporphyrinogen III to protoporphyrin, requiring overall removal of ten hydrogen atoms, is a likely focal point. Also, it is prone to failure, as shown by the predominance of coproporphyrin in bacterial porphyrias.

Progress depends on advances at the enzymic and genetic levels. Systems from expanding leaf tissue seem to be hopeful in enzymic studies, and "gene transfer factor" from *Rp. capsulata* may be a tool for understanding regulation at the genetic level.

9. References

Aagaard, J., and Sistrom, W. R., 1972, Control of reaction center bacteriochlorophyll in photosynthetic bacteria, *Photochem. Photobiol.* **15**:209.

Beale, S. I., and Castelfranco, P. A., 1974a, Biosynthesis of δ-aminolevulinic acid in higher plants. I. Accumulation of δ-aminolevulinic acid in greening plant tissues, *Plant Physiol.* **53**:291.

Beale, S. I., and Castelfranco, P. A., 1974b, Biosynthesis of δ-aminolevulinic acid in higher plants. II. Formation of ^{14}C-δ-aminolevulinic acid from labelled precursors in greening plant tissues, *Plant Physiol.* **53**:297.

Beale, S. I., Gough, S. P., and Granick, S., 1975, Biosynthesis of δ-aminolevulinic acid from the intact carbon skeleton and glutamic acid in greening barley, *Proc. Natl. Acad. Sci. U.S.A.* **72**:2719.

Burnham, B. F., 1970, δ-Aminolevulinic acid synthase, in: *Methods in Enzymology*, Vol. XVIIA (H. Tabor and C. W. Tabor, eds.), pp. 195–200, Academiic Press, New York.

Burnham, B. F., and Lascelles, J., 1963, Control of porphyrin biosynthesis through a negative feedback mechanism, *Biochem. J.* **87**:462.

Castelfranco, P. A., and Jones, O. T. G., 1975, Protoheme turnover and chlorophyll synthesis in greening barley tissue, *Plant Physiol.* **55**:485.

Cauthen, S. E., Pattison, J. R., and Lascelles, J., 1967, Vitamin B_{12} in photosynthetic bacteria and methionine synthesis in *Rhodopseudomonas spheroides, Biochem. J.* **102**:774.

Cohen-Bazire, G., Sistrom, W. R., and Stanier, R. Y., 1957, Kinetic studies of pigment synthesis by non-sulfur purple bacteria, *J. Cell. Comp. Physiol.* **49**:25.

Davies, R. C., Gorchein, A., Neuberger, A., Sandy, J. D., and Tait, G. H., 1973, Biosynthesis of bacteriochlorophyll, *Nature (London)* **245**:15.

Duggan, J., and Gassman, M., 1974, Induction of porphyrin synthesis in etiolated bean leaves by chelators of iron, *Plant Physiol.* **53**:206.

Fanica-Gaignier, M., and Clement-Metral, J. D., 1973a, δ-Aminolevulinate synthetase from *R. spheroides* Y: Purification and some properties, *Eur. J. Biochem.* **40**:13.

Fanica-Gaignier, M., and Clement-Metral, J. D., 1973b, δ-Aminolevulinate synthetase from *R. spheroides* Y. Mechanism and inhibition by ATP, *Eur. J. Biochem.* **40**:19.

Fanica-GIgnier, M., Clement-Metral, J., and Kamen, M. D., 1971, Adenine nucleotide levels and photopigment synthesis in a growing photosynthetic bacterium, *Biochim. Biophys. Acta* **226**:135.

Friedman, H., 1975, Biosynthesis of corrinoids, in: *Cobalamin, Biochemistry and Pathophysiology* (B. M. Babior, ed.), pp. 75–109, Wiley-Interscience, New York.

Gibson, K. D., Neuberger, A., and Tait, G. H., 1962, Studies on the biosynthesis of porphyrin and bacteriochlorophyll by *Rhodopseudomonas spheroides*. 2. The effects of ethionine and threonine, *Biochem. J.* **83**:550.

Gibson, K. D., Neuberger, A., and Tait, G. H., 1963, Studies on the biosynthesis of porphyrin and bacteriochlorophyll by *Rhodopseudomonas spheroides*. 4. S-Adenosylmethionine-magnesium protoporphyrin methyl transferase, *Biochem. J.* **88**:325.

Gorchein, A., 1972, Magnesium protoporphyrin chelatase activity in *Rhodopseudomonas spheroides*: Studies with whole cells, *Biochem. J.* **127**:97.

Gorchein, A., 1973, Control of magnesium protoporphyrin chelatase activity in *Rhodopseudomonas spheroides*: Role of light, oxygen, electron and energy transfer, *Biochem. J.* **134**:833.

Gough, S., 1972, Defective synthesis of porphyrins in barley plastids caused by mutations in plastid genes, *Biochim. Biophys. Acta* **286**:36.

Granick, S., and Sassa, S., 1971, δ-Aminolevulinic acid synthetase and the control of heme and chlorophyll synthesis, in: *Metabolic Regulation* (H. J. Vogel, ed.), pp. 74–141, Academic Press, New York.

Griffiths, W. T., 1975a, Some observations on chlorophyll(ide) synthesis by isolated etioplasts, *Biochem. J.* **146**:17.

Griffiths, W. T., 1975b, Characterization of the terminal stages of chlorophyll(ide) synthesis in etioplast membrane preparations, *Biochem. J.* **152**:623.

Hatch, T. P., and Lascelles, J., 1972, Accumulation of porphobilinogen and other pyrroles by mutant and wild type *Rhodopseudomonas spheroides, Arch. Biochem. Biophys.* **150**:147.

Hayasaka, S., and Tuboi, S., 1974, Control of δ-aminolevulinate synthetase activity in *Rhodopseudomonas spheroides*. III. Partial purification of the Fraction I activating enzyme and the occurrence of two forms of Fraction II, *J. Biochem. (Tokyo)* **76**:157.

Hayashi, M., and Kamikubo, T., 1970, Isolation of 5,6-dimethylbenzimidazolyl cobamide coenzyme from *Rhodopseudomonas spheroides, FEBS Lett.* **10**:249.

Higuchi, M., Ohba, T., Sakai, H., Kurashima, Y., and Kikuchi, G., 1968, Change of metabolic stability of ALA synthetase in *R. spheroides* cells probably related to changes of intracellular oxidation reduction state, *J. Biochem. (Tokyo)* **64**:795.

Jacobs, N. J., 1974, Biosynthesis of heme, in: *Microbial Iron Metabolism* (J. B. Neilands, ed.), pp. 125–148, Academic Press, New York.

Jacobs, N. J., and Jacobs, J. M., 1976, Nitrate, fumarate, and oxygen as electron acceptors for a late step in microbial heme synthesis, *Biochim. Biophys. Acta* **449**:1.

Jones, M. S., and Jones, O. T. G., 1970, Ferrochelatase of *Rhodopseudomonas spheroides, Biochem. J.* **119**:453.

Jordan, P. M., and Shemin, D., 1973, Purification and properties of uroporphyrinogen I synthetase from *Rhodopseudomonas spheroides, J. Biol. Chem.* **248**:1019.

Lascelles, J., 1956, The synthesis of porphyrins and bacteriochlorophyll by cell suspensions of *Rhodopseudomonas spheroides, Biochem. J.* **62**:78.

Lascelles, J., 1959, Adaptation to form bacteriochlorophyll in *Rhodopseudomonas spheroides*: Changes in activity of enzymes concerned in pyrrole synthesis, *Biochem. J.* **72**:508.

Lascelles, J., 1964, *Tetrapyrrole Synthesis and Its Regulation*, W. A. Benjamin, New York.

Lascelles, J., 1966a, The accumulation of bacteriochlorophyll precursors by mutant and wild type strains of *Rhodopseudomonas spheroides, Biochem. J.* **100**:175.

Lascelles, J., 1966b, The regulation of synthesis of iron and magnesium tetrapyrroles: Observations with mutant strains of *Rhodopseudomonas spheroides, Biochem. J.* **100**:184.

Lascelles, J., 1975, The regulation of heme and chlorophyll synthesis in bacteria, *Ann. N. Y. Acad. Sci.* **244**:334

Lascelles, J., and Altshuler, T., 1969, Mutant strains of *Rhodopseudomonas spheroides* lacking δ-aminolevulinate synthase: Growth, heme and bacteriochlorophyll synthesis, *J. Bacteriol.* **98**:721.

Lascelles, J., and Hatch, T. P., 1969, Bacteriochlorophyll and heme synthesis in *Rhodopseudomonas spheroides*: Possible role of heme in regulation of the branched biosynthetic pathway, *J. Bacteriol.* **98**:712.

Lascelles, J., and Wertlieb, D., 1971, Mutant strains of *Rhodopseudomonas spheroides* which form bacteriochlorophyll in the dark: Growth characteristics and enzymic activities, *Biochim. Biophys. Acta* **226**:328.

Lessie, T. G., and Sistrom, W. R., 1964, Control of porphyrin synthesis in *Rhodopseudomonas spheroides*, *Biochim. Biophys. Acta* **86**:250.

Marriott, J., Neuberger, A., and Tait, G. H., 1969, Control of δ-aminolaevulate synthetase activity in *Rhodopseudomonas spheroides*, *Biochem. J.* **111**:385.

Marrs, B., and Gest, H., 1973, Regulation of bacteriochlorophyll synthesis by oxygen in respiratory mutants of *Rhodopseudomonas capsulata*, *J. Bacteriol.* **114**:1052.

Mattheis, J. R., and Rebeiz, C. A., 1977a, Chloroplast biogenesis: Net synthesis of protochlorophyllide from magnesium–protoporphyrin monoester by developing chloroplasts, *J. Biol. Chem.* **252**:4022.

Mattheis, J. R., and Rebeiz, C. A., 1977b, Chloroplast biogenesis: Net synthesis of protochlorophyllide from protoporphyrin IX by developing chloroplasts, *J. Biol. Chem.* **252**:8347.

Miović, M. L., and Gibson, J., 1973, Nucleotide pools and adenylate energy charge in balanced and unbalanced growth of *Chromatium*, *J. Bacteriol.* **114**:86.

Mori, M., and Sano, S., 1972, Studies on the formation of protoporphyrin IX by anaerobic bacteria, *Biochim. Biophys. Acta* **264**:252.

Nandi, D. L., and Shemin, D., 1973, δ-Aminolevulinic acid dehydratase of *Rhodopseudomonas capsulata*, *Arch. Biochem. Biophys.* **158**:305.

Nandi, D. L., Baker-Cohen, K. F., and Shemin, D., 1968, δ-Aminolevulinic acid dehydratase of *Rhodopseudomonas spheroides*. I. Isolation and properties, *J. Biol. Chem.* **243**:1224.

Neuberger, A., Sandy, J. D., and Tait, G. H., 1973a, Control of 5-aminolaevulinate synthetase activity in *Rhodopseudomonas spheroides*: The involvement of sulphur metabolism, *Biochem. J.* **136**:477.

Neuberger, A., Sandy, J. D., and Tait, G. H., 1973b, Control of 5-aminolaevulinate synthetase activity in *Rhodopseudomonas spheroides*: The purification and properties of an endogenous activator of the enzyme, *Biochem. J.* **136**:491.

Oelze, J., and Kamen, M. D., 1971, Adenosine triphosphate cellular levels in *Rhodospirillum rubrum* during transition from aerobic to anaerobic metabolism, *Biochim. Biophys. Acta* **234**:137.

Ohmori, H., Nakatani, K., Shimizu, S., and Fukui, S., 1974, Correlation between the level of vitamin B_{12}-dependent methionine synthetase and intracellular concentration of vitamin B_{12} in some bacteria, *Eur. J. Biochem.* **47**:207.

Poulson, R., 1976, The enzymic conversion of protoporphyrinogen IX to protoporphyrin IX in mammalian mitochondria, *J. Biol. Chem.* **251**:3730.

Poulson, R., and Polgase, W. J., 1973, Evidence for the identification of P-503 with prototetrahydroporphyrin IX, *Biochim. Biophys. Acta* **329**:256.

Poulson, R., and Polgase, W. J., 1974a, Aerobic and anaerobic coproporphyrinogenase activities in extracts from *Saccharomyces cerevisiae*: Purification and characterization, *J. Biol. Chem.* **249**:6367.

Poulson, R., and Polgase, W. J., 1974b, Site of glucose repression of heme biosynthesis, *FEBS Lett.* **40**:258.

Poulson, R., and Polgase, W. J., 1975, The enzymic conversion of protoporphyrinogen IX to protoporphyrin. IX. Protoporphyrinogen oxidase activity in mitochondrial extracts of *Saccharomyces cerevisiae*, *J. Biol. Chem.* **250**:1269.

Rebeiz, C. A., and Castelfranco, P. A., 1973, Protochlorophyll and chlorophyll biosynthesis in cell-free systems from higher plants, *Annu. Rev. Plant Physiol.* **24**:129.

Rebeiz, C. A., Smith, B. B., Mattheis, J. R., Rebeiz, C. C., and Dayton, D. F., 1975, Chloroplast biogenesis: Biosynthesis and accumulation of Mg-protoporphyrin IX monoester and other metalloporphyrins by isolated etioplasts and developing chloroplasts, *Arch. Biochem. Biophys.* **167**:351.

Richards, W. R., and Rapoport, H., 1966, The biosynthesis of chlorobium chlorophylls-660: The isolation and purification of porphyrins from *Chlorobium thiosulfatophilum*-660, *Biochemistry* **5**:1079.

Sandy, J. D., Davies, R. C., and Neuberger, A., 1975, Control of 5-aminolaevulinate synthetase activity in *Rhodopseudomonas spheroides*: A role for trisulphides, *Biochem. J.* **150**:245.

Schmidt, G. L., and Kamen, M. D., 1971, Control of chlorophyll synthesis in *Chromatium vinosum*, *Arch. Mikrobiol.* **76**:51.

Scott, A. I., 1974, Biosynthesis of natural products, *Science* **184**:760.

Scott, A. I., Townsend, C. A., Okada, K., and Kajiwara, M., 1974, Biosynthesis of corrins. I. Experiments with [^{14}C]porphobilinogen and [^{14}C]uroporphyrinogens, *J. Am. Chem. Soc.* **96**:8054.

Sistrom, W. R., 1965, Effect of oxygen on growth and synthesis of bacteriochlorophyll in *Rhodospirillum molischianum*, *J. Bacteriol.* **89**:403.

Smith, B. B., and Rebeiz, C. A., 1977, Chloroplast biogenesis: Detection of Mg-protoporphyrin chelatase *in vitro*, *Arch. Biochem. Biophys.* **180**:178.

Sojka, G. A., and Gest, H., 1968, Integration of energy conversion and biosynthesis in the photosynthetic bacterium *Rhodopseudomonas capsulata*, *Proc. Natl. Acad. Sci. U.S.A.* **61**:1486.

Tait, G. H., 1969, Coproporphyrinogenase activity in extracts from *Rhodopseudomonas spheroides*, *Biochem. Biophys. Res. Commun.* **37**:116.

Tait, G. H., 1972, Coproporphyrinogenase activities in extracts of *Rhodopseudomonas spheroides* and *Chromatium* Strain D, *Biochem. J.* **128**:1159.

Toohey, J. I., 1965, A vitamin B_{12} compound containing no cobalt, *Proc. Natl. Acad. Sci. U.S.A.* **54**:934.

Toohey, J. I., 1966, Cobalt-free corrinoid compounds from photosynthetic bacteria, *Fed. Proc. Fed. Am. Soc. Exp. Biol.* **25**:1628.

Toohey, J. I., 1971, Purification of descobaltcorrins from photosynthetic bacteria, in: *Methods in Enzymology*, Vol. XVIIIC (D. B. McCormick and L. D. Wright, eds.), pp. 71–75, Academic Press, New York.

Tuboi, S., and Hayasaka, S., 1972, Control of δ-aminolevulinate synthetase activity in *Rhodopseudomonas spheroides*. II. Requirement for a disulfide compound for the conversion of the inactive form of Fraction 1 to the active form, *Arch. Biochem. Biophys.* **150**:690.

van Heyningen, S., and Shemin, D., 1971, Quaternary structure of δ-aminolevulinate dehydratase from *Rhodopseudomonas spheroides*, *Biochemistry* **10**:4676.

van Niel, C. B., 1944, The culture, general physiology, morphology and classification of the non-sulfur purple and brown bacteria, *Bacteriol. Rev.* **8**:1.

Von Wettstein, D., Kahn, A., Nielsen, O. F., and Gough, S., 1974, Genetic regulation of chlorophyll synthesis analyzed with mutants in barley, *Science* **184**:800.

Wang, W.-Y., Wang, W. L., Boynton, J. E., and Gillham, N. W., 1975, Genetic control of chlorophyll biosynthesis in *Chlamydomonas*: Analyses of mutants at two loci mediating the conversion of protoporphyrin-IX to Mg-protoporphyrin, *J. Cell. Biol.* **63**:806.

Wang, W.-Y., Boynton, J. E., and Gillham, N. W., 1977, Genetic control of chlorophyll biosynthesis: Effect of increased δ-aminolevulinic acid synthesis on the phenotype of the y-1 mutant of *Chlamydomonas*, *Mol. Gen. Genet.* **152**:7.

Warnick, G. R., and Burnham, B. F., 1971, Regulation of porphyrin biosynthesis: Purification and characterization of δ-aminolevulinic acid synthase, *J. Biol. Chem.* **246**:6880.

Weinstein, J. D., and Castelfranco, P. A., 1977, Protoporphyrin IX biosynthesis from glutamate in isolated greening chloroplasts, *Arch. Biochem. Biophys.* **178**:671.

Whiting, M. J., and Elliott, W. H., 1972, Purification and properties of solubilized mitochondrial δ-aminolevulinic acid synthetase and comparison with the cytosol enzyme, *J. Biol. Chem.* **247**:6818.

Yubisui, T., and Yoneyama, Y., 1972, δ-Aminolevulinic acid synthetase of *Rhodopseudomonas spheroides*: Purification and properties of the enzyme, *Arch. Biochem. Biophys.* **150**:77.

Zilinsky, J. W., Sojka, G. A., and Gest, H., 1971, Energy charge regulation in photosynthetic bacteria, *Biochem. Biophys. Res. Commun.* **42**:955.

CHAPTER 43

Control and Kinetics of Photosynthetic Membrane Development

Samuel Kaplan

1. Introduction

1.1. Intracytoplasmic Membrane Development

Specific aspects of cellular metabolism that must be coordinated and regulated to produce the orderly formation of intracytoplasmic membrane (ICM) are: major precursor pathways for the formation of bacteriochlorophyll (Bchl) (Oelze and Drews, 1972; Lascelles, 1963) (see Chapter 40), carotenoids (Oelze and Drews, 1972; Jensen, 1963) (see Chapter 39), and fatty acids (see Table III of Oelze and Drews, 1972; Schröder *et al.*, 1969) (see Chapter 14). Macromolecular components consisting of six to eight major proteins (Fraker and Kaplan, 1971; Biederman, 1971; Oelze *et al.*, 1970; Biederman and Drews, 1968) the proportions of which may vary under physiological conditions, several different phospholipids (see Table II of Oelze and Drews, 1972; Gorchein, 1968*c*) (see Chapter 14), as well as lesser amounts of several proteins (Nieth and Drews, 1974; Takemoto and Lascelles, 1973; Segen and Gibson, 1971; Fraker and Kaplan, 1971) (see Chapters 5, 6, and 13), and small molecules, such as ubiquinone (Oelze and Drews, 1972) (see Chapter 13), must be formed under specific environmental conditions.

Isolation of the ICM (the term *chromatophores* refers to the comminuted ICM isolated following cell

disruption) from several species of photosynthetic bacteria reveals two major groups of Bchl–protein complexes comprising more than 80% of the total ICM protein. There is a reaction center (RC) protein (see Chapters 6 and 19) containing RC Bchl and consisting of three separable polypeptides (Oelze and Golecki, 1975; Nieth and Drews, 1974; Okamura *et al.*, 1974; Hall *et al.*, 1973; Jolchine and Reiss-Husson, 1972; Clayton and Haselkorn, 1972; Clayton and Clayton, 1972; Feher, 1971; Fraker and Kaplan, 1971) and one or more light-harvesting (LH) complexes, LH Bchl polypeptides (Oelze and Golecki, 1975; Nieth and Drews, 1974; Hall *et al.*, 1973; Clayton and Haselkorn, 1972; Clayton and Clayton, 1972; Fraker and Kaplan, 1971, 1972) (see Chapters 5 and 7). Olson *et al.* (1963) (see Chapter 8) purified a Bchl *a*–polypeptide complex from *Chlorobium limicola*. In *Rhodospirillum rubrum*, there is a specific carotenoid–polypeptide complex (Schwenker and Gingrass, 1973).

In addition to the structural complexity outlined above, the functional relationships among the Bchl–polypeptide complexes have been extensively documented (Hall *et al.*, 1973; Table V of Oelze and Drews, 1972; Aagaard and Sistrom, 1972; Reed and Peters, 1972; Lien *et al.*, 1971; Cellarius and Peters, 1969; Sistrom, 1966; Clayton *et al.*, 1965; Sistrom and Clayton, 1964; Clayton, 1963). The physiological importance of both the RC Bchl proteins and the LH Bchl proteins has been demonstrated by

Samuel Kaplan · Department of Microbiology, University of Illinois, Urbana, Illinois 61801

Clayton *et al.* (1965), and the ratio of the LH Bchl to the RC complex defines the size of the photosynthetic unit (Aargaard and Sistrom, 1972).

Presumably, these molecular species, once formed, require no further information, other than preexisting membrane, for the formation of functional ICM. We should not, however, dismiss the possibility that "processing activities" may supply information for and be involved in the proper assembly of the ICM.

1.2. Control

For the facultatively photosynthetic bacteria, control describes the ability of these cells to either increase, decrease, or maintain a constant differential rate of ICM formation as the result of changes in growth conditions (Oelze and Drews, 1972; Cohen-Bazire *et al.*, 1957). When the bacteria are growing aerobically in either light or dark, precursor pathways and the biosynthesis of macromolecules destined for ICM development are inhibited. Either the complete removal of oxygen or lowering of the pO$_2$ below threshold levels that are species-specific induces the active cellular biosynthesis of the ICM; this process as well as the regulatory events involved are referred to as *induction*. Conversely, the reintroduction of oxygen above threshold levels results in the cessation of synthesis of the ICM for as long as the pO$_2$ levels remain above threshold values; this process and attendant regulatory phenomena are referred to as *inhibition*.

Even when growing phototrophically, in the absence of oxygen, these organisms as well as the obligate phototrophs are able to adjust the level of ICM according to the light intensity and substrate conditions of the medium. When cells growing in steady state are subjected to increases in the available light intensity, the differential rate of ICM biosynthesis is reduced and may become zero. The cellular concentration of ICM accordingly decreases; eventually, it approaches the value characteristic of the new light intensity; the differential rate then increases until both it and the ICM concentration have achieved new steady-state levels. We shall refer to this response as *repression*. Conversely, a decrease in available light intensity results in a rapid increase in the differential rate of ICM synthesis, which gradually returns to a value commensurate with the new steady-state cellular concentration of ICM. This form of regulation will be referred to as *derepression*. Clearly, although there are strong similarities between inhibition and repression on the one hand and induction and derepression on the other, there are also critical differences; because of this as well as for increased clarity, a distinction in the nomenclature is advisable. Finally, the maintenance of

any of an infinite number of possible steady-state ICM concentrations involves the coordinate supply of precursors for ICM formation together with the myriad of other requirements for cell growth. It is suggested that a critical balance between these "competing" cellular requirements will determine the ultimate growth rate of the culture.

In many photosynthetic bacteria, the composition of the ICM can be altered both qualitatively and quantitatively, depending on the conditions of growth; in view of the substructure of the ICM discussed above, the complexity of ICM development and regulation is readily apparent.

2. Early Observations

2.1. Bacteriochlorophyll Response to Stimuli

To define the basic regulatory parameters that govern the ability of photosynthetic bacteria to adapt to phototrophic growth and to adjust to changes in light intensity, it is necessary to adopt some measure of these adjustments. The most obvious measure is the Bchl content. A great deal of work has been done on the effect of various changes in the environment on the Bchl content, and the reader is referred to the following general references for detailed literature citations: Lascelles (1968, 1973), Oelze and Drews (1972), and Cohen-Bazire and Sistrom (1966).

One of the earliest and most informative studies (Cohen-Bazire *et al.*, 1957), employing *Rhodopseudomonas sphaeroides*, demonstrated that cells previously adapted to anaerobic dim light continued to grow without lag and at rates proportional to light intensity, when shifted to higher light intensities. Cells not adapted to photosynthetic growth, when placed under such conditions, displayed a lag before resumption of growth. The length of the lag was generally inversely related to the light intensity, and the subsequent growth rate was directly related to light intensity.

The differential rates of pigment synthesis were found to vary when cells were shifted between different light intensities. For example, in cells undergoing a "shift-up" in light intensity, the growth rate following the shift increased immediately, and pigment synthesis ceased until the specific pigment content for the new light intensity was reached. On the other hand, cells undergoing a "shift-down" to the same intermediate light intensity demonstrated a transient growth lag (suggesting that light-gathering ability was causing the growth limitation), eventually resuming growth at a lower rate. Pigment synthesis increased until the new steady-state growth rate and specific pigment content were reached. These were identical to those attained by the culture that underwent a "shift-up" in light intensity.

When air was introduced into an initially anaerobic culture (the light intensity remaining constant), Bchl synthesis ceased, and after approximately one doubling, the specific pigment content had fallen by approximately 50%. During subsequent growth and in less than one culture doubling following the restoration of anaerobic conditions, the pigment content per cell was restored to nearly normal levels. It was subsequently demonstrated that Bchl synthesis could take place in the dark under low oxygen (Lascelles, 1960), or even under anaerobic, dark conditions (Uffen and Wolfe, 1970; Schön and Ladwig, 1970). That there is an apparent oxygen threshold in regulating pigment production was suggested by Cohen-Bazire and Sistrom (1966), who showed that the rate of pigment synthesis increases as the oxygen partial pressure decreases. For purposes of this discussion, aerobic growth implies oxygen tensions at which ICM development is inhibited and semiaerobic refers to oxygen tensions below which ICM synthesis can take place. If we make the assumption that "dark" conditions are potentially capable of maximum induction of ICM formation, then an observation made by Cohen-Bazire and Kunisawa (1960) can be useful for demonstrating the reciprocity in the interaction of light and oxygen in regulating ICM synthesis. They observed that in a culture grown under semiaerobic, dark conditions, the differential rate of Bchl formation was the same as that observed in a culture growing anaerobically under 2000 ft-c. illumination. In general, for any light intensity, the maximum level of the specific Bchl content for that light intensity can be expressed only if the culture is anaerobic; intermediate levels of expression occur between threshold oxygen levels and anaerobiosis. Conversely, for any oxygen tension below threshold values, light intensity will modulate the specific Bchl content achieved. [See Section 2.2.2(b) for the proposed interaction of light and oxygen on ICM biosynthesis.]

Cohen-Bazire et al. (1957) suggested that "receptors" for both light and oxygen, although different, might be expected to "transmit" their effects through a common cellular system, namely, the electron-transport chain (ETC). Whereas the highly reduced state of some hypothetical carrier of the ETC favored pigment synthesis, inhibition of pigment synthesis was favored by the oxidized state of the carrier. Irrespective of the presence or absence of light, oxygen would block pigment synthesis. In the absence of oxygen, the relative oxidation level of the "carrier" would be related in some manner to both the light intensity and the specific Bchl content. According to the original proposal of Cohen-Bazire et al. (1957), as the specific pigment content of the cell increased, the differential rate of pigment synthesis would be expected to decrease. In contrast, it was found (Sistrom, 1962b) that despite a threefold increase in the specific pigment content of an experimental culture, the differential rate of pigment synthesis remained constant as measured by radioactive phenylalanine incorporation. It was further demonstrated that when a series of cultures of differing initial specific pigment content were "down-shifted" in light intensity, the subsequent differential rates of pigment synthesis were greater the lower the initial specific pigment content. Such a result would be predicted from the hypothesis of Cohen-Bazire et al. (1957).

At the time, Sistrom (1962b) concluded that the earlier hypthesis (Cohen-Bazire et al., 1957) should be modified; Sistrom suggested that the concentration of either the reduced or the oxidized form of the "regulator" and not the ratio of the two forms changed with light intensity, and that the initial rate of pigment production (following a shift-down in light intensity) would therefore be dependent on the previous growth rate of the culture.

By regulating the growth rate of Rp. sphaeroides in chemostat cultures under constant light intensity (Cohen-Bazire and Sistrom, 1966), Sistrom was able to regulate the specific pigment content of the cells as a function of growth rate, and therefore concluded that an obligatory coupling between the Bchl content and growth rate existed. It should be pointed out that succinate availability was used to limit the rate of growth of the chemostat cultures, and complications could have arisen because of the role of succinate both as a precursor for Bchl formation and its more general role as an exogenous electron donor. One parameter that will vary directly as the growth rate changes and that in turn would be expected to affect the subsequent rate of pigment synthesis would be the ribosome content of the cells.

Because a "shift-down" in light intensity led to an increase in the differential rate of pigment synthesis, the question was raised whether there is an increase in the concentration of pigment in the ICM or whether an increase in the amount of ICM accompanies an increase in the specific cellular pigment content. We are now aware that both these changes occur.

2.2. Pigment and Macromolecule Biosynthesis

Because the early investigations of French (1938) and Schachman et al. (1952) demonstrated the particulate nature of Bchl, it was consistent that the observations relating to Bchl synthesis be extended to the macromolecular components of the membrane system in which the Bchl was housed.

2.2.1. Lipid Synthesis

It was shown (Lascelles and Szilagyi, 1965) that in *Rp. sphaeroides* following a shift from aerobic to semi-aerobic growth, an increase in the differential rate of Bchl synthesis was accompanied by a twofold increase in the differential rate of phospholipid synthesis. Steady-state phototrophic cells had a 50–70% increase in phospholipid content relative to chemotrophically grown cells. Gorchein (1968b) observed that the content of lipid ornithine in chromatophores was fourfold higher than that observed for cell membrane (CM). Since there was also an increase in total membrane protein in phototrophic cells, Gorchein concluded that increased membrane synthesis accompanied Bchl synthesis. These conclusions were supported by other investigations (Ketchum and Holt, 1968; Steiner *et al.*, 1970).

Schröder and Drews (1968) extended these studies with *Rs. rubrum* and *Rp. capsulata* when they demonstrated that during induction of ICM in dark grown cells, a 30% increase in the fatty acid content was accompanied by both a 40% increase in the phospholipid content of the cells and an increase in the surface area of the ICM. In cells of *Rs. rubrum* undergoing either induction to high light anaerobic conditions or a "down-shift" in light intensity (Schröder and Drews, 1969), no increased differential rate of fatty acid or phospholipid biosynthesis was observed. It was suggested that in the latter instance, lipid constituents of the ICM were derived from the outer membrane portion of the cell, and therefore no net synthesis occurred.

2.2.2. Protein Synthesis

(a) Qualitative Studies. Several independent lines of evidence have made it clear that the regulatory mechanisms required to describe the adjustments in cellular pigment content involved more than just the regulation of the biosynthetic pathway for Bchl synthesis. Lascelles (1959) observed that the addition to photosynthetic cultures of either chloramphenicol (CAP), *p*-fluorophenylalanine (PFA), or 8-azaguanine (8-AG) caused the abrupt cessation of Bchl synthesis. A similar coupling was also reported by others (Yamashita and Kamen, 1969; Gibson *et al.*, 1962a,b; Sistrom, 1962a).

A more detailed description of the dependence of Bchl synthesis on continued protein synthesis was provided by Bull and Lascelles (1963) and Higuchi *et al.* (1965), who suggested that pigment synthesis might be obligatorily linked to the formation of part of or all the ICM system. By delaying the addition of inhibitors until after derepression of the enzymes for Bchl biosynthesis had taken place, Bull and Lascelles (1963) demonstrated the complete inhibition of Bchl biosynthesis despite only a 25% decrease in the early enzymes of the pathway for Bchl formation. Higuchi *et al.* (1965) also concluded that Bchl synthesis was dependent on concomitant protein synthesis, and that the effect of inhibitors on specific enzymes in the Bchl biosynthetic pathway was not sufficient to account for the cessation of Bchl formation.

(b) Growth Rate and Pigment Content. Although the evidence has indicated an inverse relationship between the pigment content and the growth rate, decreasing the growth rate either by addition of inhibitors or by alterations in the composition of the medium results in lower rather than higher specific pigment contents (Bull and Lascelles, 1963; Gibson *et al.*, 1962a,b). Schön and Ladwig (1970) observed that when cells of *Rs. rubrum* were placed under dark, anaerobic conditions, substrates normally providing the highest growth rates under conditions of growth permitted relatively low specific pigment contents. On the other hand, pyruvate and fructose, normally only weakly stimulatory with respect to growth, promoted the highest specific pigment content. Although there is a direct relationship between growth rate and light intensity, Sistrom (Cohen-Bazire and Sistrom, 1966) showed that this relationship can be uncoupled.

Very recently, Dierstein and Drews (1974) and Drews *et al.* (1974), employing chemostat cultures of *Rp. capsulata*, were able to demonstrate an increased specific Bchl content in both cells and isolated membranes as the growth rate of nitrogen-limited cultures increased (see Table 1). In cells grown semi-anaerobically in the light, the specific Bchl contents of cells did not increase in the same proportion to changes in the growth rate as did the specific Bchl content of the ICM. On the other hand, in cells grown anaerobically in the light, the specific Bchl contents of both the cells and the ICM were constant regardless of growth rate.

In an effort to relate these observations, the model presented in Fig. 16 has been formulated. This model is used to explain the interactions between carbon source, light intensity, oxygen tension, and growth rate in the regulation of ICM biosynthesis.

Let us assume that a corepressor of ICM biosynthesis can be generated via a pathway involving carbon source, light energy, and oxygen tension. For

Table 1. Steady-State Concentrations of Bacteriochlorophyll and Proteins of the Photosynthetic Apparatus in Continuous Cultures of *Rhodopseudomonas capsulata* Strain 37b4[a]

Culture conditions	Growth rate (μm/hr)	Bacteriochlorophyll content			RC Bchl		LH protein/ RC protein (relative units)
		μg/mg cell protein	μg/mg membrane protein	mol total Bchl/ mol RC Bchl	μg/mg membrane protein	ng/mg cell protein	
Dark, aerobically $pO_2 = 5$ mm Hg							
Nitrogen-limited	0.042	4.8	19.6	113	0.17	42	0.26
Nitrogen-limited	0.083	14.5	41.5	278	0.15	52	0.59
Nitrogen-limited	0.125	18.7	56.0	380	0.15	49	0.95
Dark, aerobically $pO_2 = 5$ mm Hg							
Turbidostat	90% μ_{max}	24.2	79.5	443	0.18	54	1.02
Light, anaerobically Nitrogen-limited	0.062	8.2	35.0	165	0.21	50	0.42
	0.125	9.1	39.5	167	0.23	58	0.44

[a] From Dierstein and Drews (unpublished). Generously provided by G. Drews.

the repression system to be active, corepressor in combination with a protein portion, aporepressor, gives rise to the active repressor, i.e., holorepressor. First, the direct interaction of RC Bchl with light (as opposed to the indirect interaction of RC Bchl with light via LH Bchl) is effective in producing corepressor. Conversely, light energy reaching RC Bchl by way of LH Bchl is not only ineffective in generating corepressor, but also is actually counterproductive for corepressor formation. Second, the interaction of oxygen with an "effector molecule" (Marrs and Gest, 1973*b*) is active in corepressor synthesis, and the site of oxygen involvement is metabolically closer to the formation of corepressor than is the interaction of light with RC Bchl. Finally, the carbon source is capable of influencing the level of corepressor, but its involvement in corepressor formation is metabolically more distant than either light or oxygen, and normally, either light or oxygen is present together with carbon source. The immediate metabolic pathway leading to the synthesis of corepressor need not involve either the ETC or the concentration of adenylate nucleotides.

In steady-state cells growing anaerobically in high light, corepressor concentrations will be high, and the cells will be restricted to the synthesis of low levels of ICM, which will be enriched in RC Bchl relative to LH Bchl (B850) due to the greater affinity of the repressor for those regulatory sites involved in LH Bchl synthesis.

Under these conditions, the high steady-state level of corepressor derives from the assumption that was made above, namely, that only the direct interaction of light with RC Bchl gives rise to corepressor. The experiments of Lien *et al.* (1971) employing mutant strain Z1 of *Rp. capsulata*, which is resistant to arsenate, support this interpretation. This organism has increased levels of RC Bchl relative to LH Bchl compared with the parental strain grown under similar cultural conditions. Additionally, strain Z1 has 25% less ICM than the parental strain. This is what might be expected if the interaction of RC Bchl with light resulted in the formation of corepressor. It should be noted that the growth rates of the mutant and parent were similar. Conversely, steady-state cells growing in dim light have lower corepressor levels due to the greatly decreased interaction of light with RC Bchl as a result of the greatly increased interaction of the limiting light with LH Bchl.

Under conditions of changing light intensity (non-steady-state) and employing the model presented in Fig. 16, the results discussed below would be suggested. It must be kept in mind, however, that the regulatory system is best investigated under conditions in which the level of corepressor is responsive to changes in light intensity (see the discussion below). The lower the initial Bchl content of the cells (high light, and a low LH Bchl/RC Bchl ratio), the greater will be the initial differential rate of ICM formation, when such cells are shifted to lower light intensities. Immediately following a shift from high to low light, the absolute level of the interaction of RC Bchl with light will decrease, resulting in a decreased formation of corepressor. The magnitude of these changes in co-repressor levels will be inversely related not to the absolute change in the light intensity, but instead to the actual initial and final light intensities when these are either nonsaturating or not limiting, respectively, with respect to corepressor formation. This prediction derives from the model presented in Fig. 16, since it takes into account both the changing LH Bchl/RC Bchl ratio and the change in the absolute level of RC Bchl as the light intensity used for cell growth changes. This is important, because in dim light the cells may contain a higher concentration of RC Bchl than when grown in high light, but the relative absence of LH Bchl in high light cells increases the effectiveness of RC Bchl in generating corepressor.

Again, it is important to restate the assumption that only the interaction of light with RC Bchl results in co-repressor formation and interaction of light with LH Bchl represents a relative interference with co-repressor formation. In effect, this model, and the example just illustrated, resolve the paradox reported by Sistrom (1962b) (see Section 2.1). We conclude that the differential rate of pigment synthesis for any culture, providing we are in a range of light intensities in which changes are reflected in changes in co-repressor concentration, is dependent both on its initial specific pigment content (which determines its ability to form corepressor) and on the light intensity to which it is subsequently exposed (which determines the initial relative change in its ability to form co-repressor).

Similarly, a "shift-up" in light intensity would result in an immediate absolute increase in the interaction of RC Bchl with light, although the proportion of light interacting with LH Bchl and RC Bchl would not immediately change. The net effect of an increase in the interaction of RC Bchl with light would be increased corepressor synthesis, resulting in a rapid shut-down in LH Bchl synthesis followed by a decline or shut-down in the synthesis of RC Bchl. The precise fate of RC Bchl synthesis would depend on the initial and final light intensities within the limits already described. Roughly, however, the model says that the degree of repression will be greater the higher the initial specific Bchl content of the cells, and will also be greater the higher the new light intensity.

Further, a plot of the LH Bchl/RC Bchl ratio in steady-state cells as a function of light intensity will be sigmoidal. At high light intensities, corepressor concentrations will be saturating, and slight changes in

light intensity will not be reflected in significant changes in corepressor levels. Slight changes in co-repressor levels at high light intensities are more likely to effect RC Bchl levels because of the higher affinity of repressor for the LH Bchl pathway. As the light intensity employed for growth decreases below saturating levels (as regards corepressor synthesis), the LH Bchl/RC Bchl ratio will increase, not only because of the inability of the repressor to saturate the control elements involved in LH Bchl synthesis, but also because LH Bchl acts as a positive regulator of its own synthesis, by interfering indirectly with the formation of corepressor (see the discussion of autogenous gene control in Goldberger, 1974).

At the very high LH Bchl/RC Bchl ratios found in dim-light-grown cells, further decreases in the light intensity will not be reflected in significant changes in the LH Bchl/RC Bchl ratio. The reason is that maximum formation of both components is taking place due to minimal corepressor levels and limitations in other cellular activities; e.g., RNA polymerase, enzyme turnover values, protein synthesis, and other activities become predominant.

Repressor interaction with RC Bchl and LH Bchl synthesis is used in its broadest sense, and is intended to cover the spectrum of regulatory events that are known to take place in other biological systems. Such interactions must include transcriptional control involving RC and LH mRNA synthesis. Because of the rapidity of the responses that are observed, translational control of RC and LH proteins seems inescapable, as well as enzyme regulation in the case of Bchl and phospholipid synthesis. In these latter cases, co-repressor alone may be effective, with holorepressor active only at the transcriptional level.

Further, the model shows that depending on such cultural conditions as carbon source, light intensity, oxygen levels below threshold, and growth rate, it is possible to manipulate the ICM levels in almost any manner. For example, at oxygen levels below threshold, the pO_2 can be used to determine the specific cellular content of the ICM such that the cellular response normally associated with a particular light intensity under anaerobiosis can be overridden by the presence of low levels of oxygen.

Likewise, by making the concentration of apore-pressor directly proportional to the growth rate of the culture, and the corepressor concentration inversely proportional to growth rate, due to its descruction by turnover or further metabolism, it is possible to manipulate a culture experimentally so that ICM formation may be expressed as a direct function of growth rate when the inductive capacity of the culture for ICM formation is directly coupled to growth, e.g., when ICM production is independent of light intensity, as in the experiments of Dierstein and Drews (1974) [see Section 3.4.2(b)].

On the other hand, when light intensity is the agent that controls ICM formation, then the typical inverse relationship between growth rate and ICM is observed.

3. Intracytoplasmic Membrane Development

3.1. Induction of Intracytoplasmic Membrane Development

3.1.1. Intracytoplasmic Membrane and Cell Membrane Continuity

The physical continuity between the ICM and CM has been demonstrated electron-microscopically (Golecki and Oelze, 1975; Cohen-Bazire and Kunisawa, 1963; Cohen-Bazire, 1963; Fuller *et al.*, 1963; Giesbrecht and Drews, 1962; Drews, 1960) and biochemically (Gorchein *et al.*, 1968a; Holt and Marr, 1965a; Boatman, 1964; Tuttle and Gest, 1959), and Oelze and Drews (1972) provided a model for ICM formation from the CM. It is provocative, however, to consider the alternative that the continuity observed between the two membrane systems arises from fusion of the two systems (Niederman, personal communication, 1975; Niederman and Gibson, 1971; Gibson, 1965a,b). The observation that numerous invaginations of the CM are continuous with the ICM may imply that these invaginations result from the fusion of otherwise CM-free ICM with the CM when the two membrane systems are in proximity. If this is true, then the ICM may in fact arise from the CM at only a very few sites. In the extreme, the ICM might be formed totally free of the CM, with subsequent fusion taking place. In the case of the green bacteria, the intracytoplasmic vesicles, which are probably not analogous to the ICM, are free of the CM (Cruden and Stanier, 1970; Holt *et al.*, 1966) (see Chapter 9).

Because of the proximity of the two membrane systems and their tendency to aggregate, a variety of methods have been developed to purify isolated ICM (chromatophores) (Niederman and Gibson, 1971; Fraker and Kaplan, 1971; Oelze *et al.*, 1969b; Gorchein *et al.*, 1968a; Holt and Marr, 1965b; Gibson, 1965a; Worden and Sistrom, 1964) and to separate chromatophores from CM (Niederman, 1974; Huang and Kaplan, 1973a; Niederman *et al.*, 1972; Oelze *et al.*, 1969b; Oelze and Drews, 1969a). The difficulties encountered in obtaining pure preparations have been amply documented (Niederman,

1974; Niederman and Gibson, 1971; Fraker and Kaplan, 1971), and two crucial separations are difficult to make, namely chromatophores from CM derived from cells grown in high light intensities and CM from chromatophores derived from cells grown in low light intensities.

3.1.2. Kinetics of Intracytoplasmic Membrane Induction

(a) Tracer Studies. Early studies (Oelze and Drews, 1969b; Gorchein *et al.*, 1968b) led to the conclusion that the ICM was not formed *de novo*, but remained a part of the CM both physically and chemically. By prelabeling the lipids of aerobically grown *Rp. sphaeroides* or *Rs. rubrum* and then placing the cells under semiaerobic conditions in the light, it was shown that lipids previously formed were metabolically stable and the specific activity of both ICM and CM membrane lipids were identical. It was also concluded that the rate of increase of each membrane system was the same.

Contradictory evidence, although supporting physical continuity between the two membrane systems, indicated chemical discontinuity. Phospholipids, although at present thought to be freely diffusible in membranes, were found to be distinct qualitatively or quantitatively or both in the two kinds of membrane. Steiner *et al.* (1970) found that the lipid composition of the ICM differed significantly from that of cells lacking ICM, and therefore suggested that the two membrane systems, although continuous, may be chemically distinct. Gorchein (1968b) concluded that the ornithine lipid in *Rp. sphaeroides* could undergo transfer from the CM to the ICM during induction and development; however, ornithine lipid appeared to be in greater concentration in the ICM than in the CM. This finding would also support the conclusion of Steiner *et al.* (1970).

Although these investigations were conducted with different species of photosynthetic bacteria, there is clearly a discrepancy. On the one hand, Gorchein *et al.* (1968b) and Oelze and Drews (1969b) observed the movement of lipid between the two membrane systems, and on the other, Steiner *et al.* (1970) and Gorchein (1968b) reported a "chemical" separation of the two systems. One would expect that if the ICM and CM were continuous membrane systems, the lipids, no matter which system they might be incorporated into, should rapidly equilibrate (Singer and Nicolson, 1972; Kornberg and McConnell, 1971). On the other hand, proteins may or may not be expected to equilibrate between the two membrane systems. Yamashita and Kamen (1969) observed that during induction, new protein went exclusively into ICM.

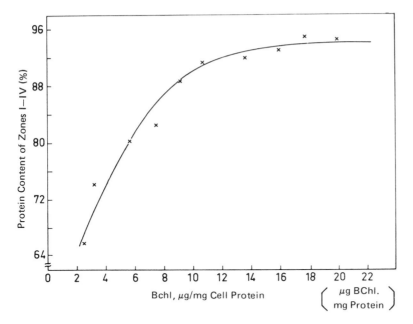

Fig. 1. Relationship between the specific Bchl content of cells grown anaerobically in the light and the level of specific ICM polypeptides associated with photosynthetic growth and identified by PAGE of crude chromatophore preparations. Zones I–IV refer to those regions of the polyacrylamide gel profile found only in photosynthetically growing cells and associated with the ICM. The amount of material in each zone was determined by densitometer tracings of the amino-schwarz stained gels and is expressed as a percentage of the total ICM protein. Reprinted with the kind permission of Springer-Verlag Publishers and Dr. J. Oelze from Oelze *et al.* (1969a).

(b) Protein Studies. Oelze *et al.* (1969a) subdivided the induction process in *Rs. rubrum* into three phases by determining the specific Bchl content of both cells and isolated chromatophores, by observation of polyacrylamide gel electrophoresis (PAGE) patterns (Biederman and Drews, 1968), and by electron microscopy. In the first phase, lasting 2 hr, Bchl synthesis was observed without finding evidence, either chemical or visual, for the formation of ICM. Between 2 and 6 hr postinduction, they observed formation of ICM and the presence of proteins characteristic of the ICM. During this phase, the specific Bchl content of both chromatophores and cells increased. A plot (Fig. 1) of the amount of protein characteristic of the chromatophores (Biederman and Drews, 1968) vs. the specific Bchl content revealed an initial change in the composition of the ICM, followed by an increase in the amount of ICM per cell. Electron micrographs revealed that during the second phase of induction, the diameter of the isolated chromatophores was smaller than that observed for chromatophores derived from steady-state cells (see also Drews and Lampe, 1971).

An elegant investigation of the kinetics of ICM-specific protein formation during induction in *Rp. sphaeroides* was recently conducted by Takemoto (1974). Figure 2 shows that following the incubation of aerobically grown cells under semiaerobic conditions in the dark, there is a coordinate synthesis of the three RC polypeptides, bands, 9, 10, and 11 (28,000 daltons, 23,000 daltons, and 21,000 daltons, respectively), as measured by a pulse of radioactivity added at the beginning of semiaerobic incubation. During the

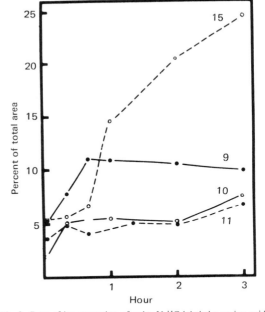

Fig. 2. Rate of incorporation of pulse U-^{14}C-labeled L-amino acids into ICM membrane proteins during incubation of cell suspensions under conditions that induce the formation of ICM. Each point was determined from the areas under the peaks of polypeptide profiles and presented as the percentage of the total area of the gel. The area under each peak was determined by densitometer tracings of SDS-PAGE autoradiograms of the purified membranes. Approximately 100,000 cpm were applied to each gel. The points were plotted at times corresponding to the beginning of their respective labeling periods. (●———●) Band 9; (○———○) band 10; (●--●) band 11; (○--○) band 15. Reprinted with the kind permission of Academic Press, Inc., and Dr. J. Takemoto from Takemoto (1974).

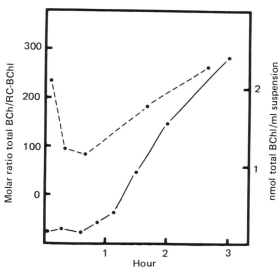

Fig. 3. Changes in the size of the photosynthetic unit during incubation of the cell suspensions. (●———●) Size of the photosynthetic unit calculated from the data presented in Fig. 2; (●———●) Bchl content of whole cells in suspension. The photosynthetic unit is defined as the molar ratio of total Bchl to RC Bchl. The amount of RC Bchl was determined from light-induced absorbance changes at 872 nm and calculated assuming a millimolar extinction value of 100 at 872 nm. Total Bchl was determined in acetone–methanol (7:2, vol/vol) extracts of either whole cells or crude extracts employing a millimolar extinction of 76 at 770 nm. Reprinted with the kind permission of Academic Press, Inc., and Dr. J. Takemoto from Takemoto (1974).

protein synthesis precedes LH protein formation (Niederman *et al.*, 1976), the kinetics of formation of which is sigmoidal (see also Fig. 2), suggesting that this species may regulate its own synthesis (Goldberger, 1974) [see also Section 2.2.2(b)].

Our laboratory has employed a two-dimensional PAGE analysis coupled with specific immune precipitation in the second dimension in order to follow chromatophore-specific protein synthesis after induction. The earliest polypeptide they were able to detect had a molecular weight of 41,000–44,000; this was followed by the appearance of RC polypeptides and finally LH polypeptides. A polypeptide of this size (41,000–44,000) has been observed in many of the ICM preparations reported (Oelze and Golecki, 1975; Nieth and Drews, 1974; Takemoto and Lascelles, 1973; Segen and Gibson, 1971; Fraker and Kaplan, 1971). Its function is unknown, but it is interesting that a mutant of *Rp. capsulata* that is unable to grow anaerobically and is designated Ala⁻ (Drews, 1974; Drews *et al.*, 1971; Schick and Drews, 1969) was found to secrete a protein–pigment complex, and the molecular weight of the protein was estimated to be 45,000.

(c) Precursor Studies. Shaw and Richards (1971*a,b*)

first 40 min of induction, the rate of synthesis of band 9 is higher than that of band 15, with bands 10 and 11 demonstrating this relationship for only 20 min. Thereafter, the rate of synthesis of the LH polypeptides remain relatively constant, although bands 10 and 11 display a slight increase between 2 and 3 hr. These changes are accompanied by a change in the size of the photosynthetic unit (Lien *et al.*, 1973; Aagaard and Sistrom, 1972), as shown in Fig. 3. Antisera (Huang and Kaplan, 1973*b*) prepared against either RC proteins or LH proteins qualitatively identified their presence within several hours following induction in cells incubated under anaerobic conditions. Similar observations were made by Oelze *et al.* (1969*b*) and Oelze and Drews (1970*a*), employing *Rs. rubrum*, and by Cellarius and Peters (1969), employing *Rp. sphaeroides*.

Niederman *et al.* (1976) extended the results of Cellarius and Peters (1969) by following the appearance of different spectral species characteristic of the LH Bchl protein complexes in *Rp. sphaeroides* undergoing induction. Figure 4 illustrates both the kinetics of induction of the LH components and the changing composition of the photosynthetic membrane. RC

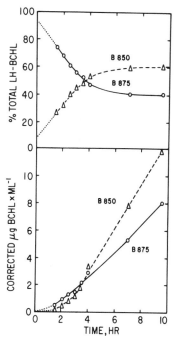

Fig. 4. Relative LH Bchl concentrations and kinetics of B850 and B875 syntheses in whole-cell suspensions. Absorbance at 850 and 875 nm corrected by the method of Crounse *et al.* (1963). Reprinted with the kind permission of Dr. R. A. Niederman.

Table 2. Density of Membrane Fractions from Chemotrophic and Phototrophic *Rhodopseudomonas sphaeroides*

Fraction	Growth medium[a]	Density
Envelope (chemotrophic/phototrophic)	D_2O	1.260
Envelope (chemotrophic/phototrophic)	H_2O	1.205
Chromatophores	D_2O	1.225
Chromatophores	H_2O	1.175

[a] See Kosakowski and Kaplan (1974) for experimental details.

followed the movement of labeled proline through various membrane fractions when the label was added to cells of *Rp. sphaeroides* at the onset of ICM induction. They interpreted their results as demonstrating that proline flowed from a "prephore" fraction (with vesicles similar in size to the chromatophores) to the "prochromatophore" and finally to the chromatophore fraction. This interpretation was based largely on the amount of radioactivity present in various fractions the resolution of which on the gradients was incomplete. Also, an examination of the specific radioactivity of the various fractions makes their conclusions less clear. In particular, the prephore fraction is heavily contaminated with ribosomelike particles, and since two-thirds of the label taken up by the cells (within 5 min) behaves as a pulse of radioactivity, it might be expected to flow through the ribosomes to cell protein.

In our laboratory, we have attempted to determine the relationship during induction between the CM and ICM by following the fate of [³H]leucine, which had been added to *Rp. sphaeroides* growing in H_2O-based medium under anaerobic light conditions, in cells previously grown aerobically in 70% D_2O-containing medium supplemented with [¹⁴C]leucine. They were able to separate (Kosakowski and Kaplan, 1974) "old" aerobic membrane–envelope complex from "new" ICM membrane by centrifugation of isolated membranes on CsCl density gradients (see Table 2 for the densities of the various membrane fractions). The earliest chromatophores, as determined spectrally, appeared after 4 hr; they were at the position of full "light" density, 1.175 g/cc, and although there was a small amount of "old" protein in this region, the ratio of the specific activity of "new" to that of "old" protein was over 300-fold greater than the corresponding ratio for the region of the gradient occupied by "old"

membrane–envelope complex, 1.260 g/cc. Similar results were obtained employing [³H]δ-aminolevulinic acid (δ-ALA) labeling for "new" Bchl with [¹⁴C]leucine labeling of "old" protein. These results imply that although the two membrane systems are physically continuous, they are chemically distinct, a conclusion similar to that made by Steiner *et al.* (1970). If the phospholipids of the two membrane systems were free to undergo rapid lateral movement (Singer and Nicolson, 1972; Kornberg and McConnell, 1971), then we would have expected the "new" chromatophores, based on the composition and calculated contribution of the CM to band at a more dense position, approximately 1.210 g/cc.

Golecki and Oelze (1975) quantified the extent of invagination of the developing ICM during induction in *Rs. rubrum*. Table 3 reveals that no invaginations were observed until 6 hr following induction, and while the specific Bchl content of the cells increased over 20-fold over the next 54 hr, the number of invaginations increased only 4-fold in the same time. Even during the period between 6 and 12 hr, the increase in specific Bchl content was 50% greater than the increase in the number of invaginations. We also know (Oelze *et al.*, 1969a) that between 12 and 24 hr, additional increases in the specific Bchl content of the cells must be due to increases in the development of the ICM. This is particularly evident between 23 and 60 hr, when the

Table 3. Bacteriochlorophyll Contents and Number of Cytoplasmic Membrane Invaginations in *Rhodospirillum rubrum*[a]

Culture conditions[b]	Bchl (μm/mg protein)	Invaginations/μm^2 membrane surface[c]
Aerobic, dark	ND[d]	0
Anaerobic, light (3 hr)	0.6	0
Anaerobic, light (6 hr)	1.7	16.7 ± 8.2
Anaerobic, light (12 hr)	5.7	37.3 ± 11.0
Anaerobic, light (23 hr)	20.6	52.4 ± 9.2
Anaerobic, light (60 hr)	45.4	63.1 ± 6.3

[a] Reprinted with the kind permission of Cambridge University Press and Drs. Golecki and J. Oelze from Golecki and Oelze (1975).
[b] The times after transfer from chemotrophic to phototrophic conditions are given in parentheses.
[c] Significance of the values is based on a 99% confidence level.
[d] ND: Not detectable.

number of invaginations remains roughly constant, but the Bchl content of the cells doubles. During at least the first 3 hr, Golecki and Oelze (1975) concluded, Bchl is incorporated into the CM, since no invaginations are observed. If this incorporation is site-specific, then this region would represent a potential invagination that is not observed due to its small size. Obviously, until more is known, we have a semantic problem when attempting to define CM and ICM at early stages of induction.

3.1.3. Control of Intracytoplasmic Membrane Induction

As previously noted (Cohen-Bazire *et al.*, 1957), the development of the ICM can proceed under semi-aerobic or anaerobic conditions. For *Rp. sphaeroides*, it was determined that at oxygen tensions below 5% oxygen at one atmosphere (38 mm Hg), ICM-specific proteins could be observed, and that the lower the oxygen partial pressure, the greater the level of ICM-specific polypeptides (Huang and Kaplan, 1973*b*). In *Rp. capsulata*, although ICM development responds quantitatively as in *Rp. sphaeroides* to decrease in oxygen partial pressure, the regulatory mechanism is considerably less sensitive to the presence of oxygen. Even at high oxygen partial pressures (60–70 mm Hg), significant levels of ICM are formed (Drews *et al.*, 1969).

That a unique regulator substance or "governor" exists for the induction of ICM development is best illustrated by the isolation of mutants of *Rp.*

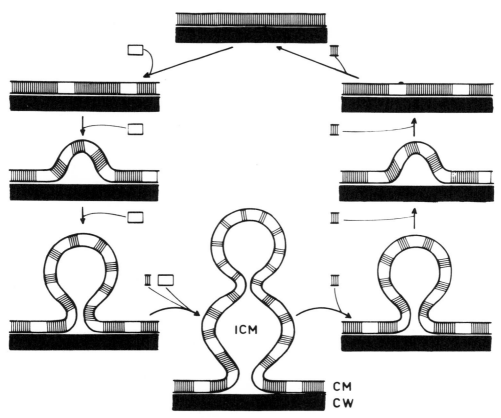

Fig. 5. Hypothetical scheme for the development and dedifferentiation of the photosynthetically active intracytoplasmic membrane system in *Rhodospirillum rubrum*. (☐) All components essential for the formation of the photosynthetic apparatus; (▥) components of the cytoplasmic membrane of aerobically dark-grown cells lacking in any photosynthetic activity. During phototrophic growth, the rate of synthesis of CM constituents is lower than the rate of synthesis of ICM components, and the result is an invagination of the CM, forming the ICM. During adaptation to aerobic growth, the reverse process occurs, resulting in the gradual disappearance of the invagination. (CW) Cell wall; (CM) cytoplasmic membrane; (ICM) intracytoplasmic membrane. For a detailed description, see Oelze and Drews (1972). Reprinted with the kind permission of the ASP Biological and Medical Press and Drs. J. Oelze and G. Drews from Oelze and Drews (1972).

sphaeroides (Lascelles and Wertlieb, 1971; Kaplan, unpublished). Mutant TA-R, isolated by Lascelles and Wertlieb, forms a photochemically competent ICM in oxygen tensions sufficient to repress ICM formation in the wild type; in the absence of oxygen, the mutant responds normally to changes in light intensity.

According to the model of Oelze and Drews (1972) for ICM development (Fig. 5), anaerobiosis induces the formation of those components of the ICM that mediate the light-dependent reactions of the ICM. These light-specific components are not found in the CM of aerobically growing cells (Takemoto and Lascelles, 1973; Huang and Kaplan, 1973a,b; Oelze and Drews, 1970a; Oelze et al., 1970). These protein components, together with phospholipids and pigments, are added to the CM, and it is at these sites on the CM that an increase in the differential rate of ICM synthesis takes place. Whether or not the initial sites of insertion of ICM-specific components into the CM are unique is unknown, but data do exist (Golecki and Oelze, 1975) (see also Section 3.4.1) that suggest that such sites are restricted in number. It would be interesting to determine the relationship between the number of steady-state invaginations and light intensity.

Oelze and Drews (1970b) further suggested that as the invagination of the ICM forms, components of the aerobic respiratory chain normally found in the CM continue to be inserted into the ICM, and that only the increased differential rate of synthesis of ICM-specific proteins causes respiratory chain activity in the ICM to remain low. That aerobic CM polypeptides are present in highly purified chromatophores to approximately 3–5% has been confirmed (Huang and Kaplan, 1973a,b; Fraker and Kaplan, 1971; Oelze et al., 1970; Throm et al., 1970). Conversely, Huang and Kaplan (1973a,b) suggested, based on their inability to demonstrate the appearance of new CM polypeptides in the chromatophore fraction following ICM induction (see Table 3 of Huang and Kaplan, 1973b), that CM polypeptides normally observed are derived from chromatophore vesicles formed at the juncture of the ICM and the CM. This would imply heterogeneity within the chromatophore vesicles with respect to CM-

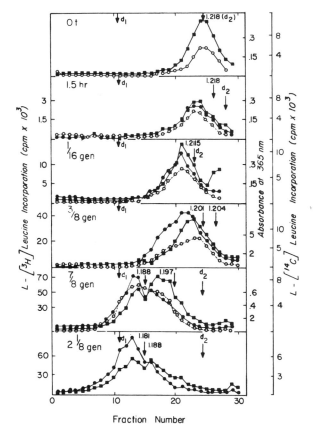

Fig. 6. (O—O—O) A_{365}; (■—■—■) [^{14}C]leucine; (●—●—●) [^{3}H]leucine; (d_1) position of full light chromatophores, 1.175 g/cm^3; (d_2) position of full heavy chromatophores, 1.218 g/cm^3. Cells were grown anaerobically in the light (2000 ft-c.) in medium A of Sistrom containing 70% D_2O and supplemented with [^{14}C]leucine. At a cell density of 2– 3×10^8/ml, the medium was changed to include 100% H_2O and [^{3}H]leucine. The light intensity was reduced to 60 ft-c. and a sample, designated 0_t was removed. Subsequent samples were removed at 1.5 hr following the shift (no increase in cell growth as measured turbidimetrically), and at $\frac{1}{16}$, $\frac{3}{8}$, $\frac{7}{8}$, and $2\frac{1}{8}$ cell doublings as measured turbidimetrically. The samples were harvested and disrupted in the French pressure cell, and chromatophores were isolated by differential centrifugation. Samples were loaded onto 6-ml preformed CsCl density gradients and centrifuged to equilibrium in a Spinco 40 rotor at 32,500 rpm for 24 hr at 20°C. Following centrifugation, each sample was analyzed for OD$_{365}$ nm, [^{3}H], [^{14}C], and refractive index. See Kosakowski and Kaplan (1974) for experimental details.

like activities. The model of Oelze and Drews (1972) would predict a more homogeneous distribution of CM-like proteins within the chromatophore population. Very recent experiments by Jones and Pleivis (1974) with *Rp. sphaeroides* and by Garcia *et al.* (1974) with *Rs. rubrum* clearly suggest a very close functional relationship between the photochemical apparatus on the one hand and the respiratory ETC on the other.

3.2. Derepression of Intracytoplasmic Membrane Development

During a "shift-down" in light intensity of photosynthetic cultures, the requirement for additional photosynthetic units could be accommodated in several ways: (1) new invaginations could arise from the CM; (2) the composition of preexisting ICM could be altered, increasing the amount of ICM-specific proteins per unit membrane; (3) an extension of the preexisting ICM could take place; or (4) a combination of any of these could occur. All these phenomena are actually observed during induction of ICM development (Golecki and Oelze, 1975). It is only during induction, however, that no preexisting ICM need be present, and therefore induction should be treated as distinct from derepression until more is known about both processes at the molecular level. Oelze and Drews (1969a, 1970a) showed that in *Rs. rubrum*, following a shift from 4000 lux to 400 lux, there is a transfer of lipid from the CM to the ICM. We know that during induction, there is also a change in the specific Bchl content of the ICM (Oelze *et al.*, 1969a) that presumably is accompanied by a change in protein composition as well as by an increase in the size of the photosynthetic unit as the light intensity decreases (Aagaard and Sistrom, 1972). The third way, an extension of previously formed invagination into the interior of the cell, may play a role during induction (Golecki and Oelze, 1975) and also during derepression of ICM development.

In our laboratory, we grew cells of *Rp. sphaeroides* in D_2O-containing medium supplemented with a [^{14}C]amino acid in high light and shifted them to low light, at the same time removing the D_2O and ^{14}C label and replacing them with H_2O and the [^3H]amino acid in order to follow the distribution of "old" and "new" ICM (for experimental details, see Kosakowski and Kaplan, 1974). Immediately following the "downshift," growth ceased for approximately 2 hr and protein synthesis was only 15% of its preshift rate, with membrane protein made almost exlusively (Kosakowski and Kaplan, unpublished; Eccleston and Gray, 1973). From Fig. 6, we observe that two populations of chromatophores exist following the

"shift-down." This is first clearly evident at one-sixteenth culture doubling, when the specific activity of the ^3H label exceeds that of the ^{14}C at a density less than 1.2115. The "heavy" population is predominantly "old" protein, and it proceeds to get "lighter" (incorporate new material) precisely as a function of cell growth. The second population of chromatophores is "lighter" and contains a higher proportion of "new" protein, and although it proceeds to get "lighter" with time following the resumption of growth, it gets lighter at a decreasing rate, until the two chromatophore populations are one. These determinations were made by resolving the curves for [^3H], [^{14}C], and OD_{365} using a curve analyzer and assuming a normal distribution for each component.

According to our present experimental evidence, these observations are compatible with model 3 (extension of the preexisting ICM). We would, however, expect the new, "lighter" population of chromatophores to move to a lower density as a function of the growth rate of the culture (see Section 3.4.3). That they incorporate new material at an ever-decreasing rate when compared with growth of the culture is explicable if we assume that the "new" membrane is initially enriched in lipid due to an initial severe limitation in protein synthesis following the downshift due to a lower ATP flux. As the synthetic ability of the cells is restored, this lipid-enriched ICM gradually regains its steady-state composition, and therefore its density continues to decrease, but at a decreasing rate. The "old" membrane replicates as described for cells in the steady state (see Section 3.4.3), i.e., as predicted from the growth rate of the culture. Clearly, the main difference between these two chromatophore populations is that one is derived from "steady-state" cells and the other from cells undergoing a change in cultural conditions.

We have observed that a "shift-down" in light intensity also results in a cessation of ribosome biosynthesis (Eccleston and Gray, 1973), derepressed formation of LH Bchl protein synthesis, and continued synthesis of RC units. Taking these results together, we would expect to see an increase in the size of the photosynthetic unit (Aagaard and Sistrom, 1972), elongation of the ICM into the interior of the cell, an increase in the ratio of RC units to respiratory chain activity, and increased template for increased ICM formation.

Depending on the initial and final steady states to be achieved, factors such as the rate of precursor formation, protein synthetic ability, amount and composition of preexisting ICM, and availability of ATP will all interact to determine the kinetics of ICM synthesis (see Sistrom, 1962b).

3.3. Repression of Intracytoplasmic Membrane Development

3.3.1. Inhibition

When anaerobic cells of *Rs. rubrum* are shifted to aerobic conditions, both Bchl and ICM-specific protein synthesis cease, and the NADH oxidase activity of the ICM increases together with the synthesis of CM-like proteins (Oelze and Drews, 1970*b*). There is also a transfer of phospholipid from the ICM to the CM (Oelze and Drews, 1970*a*). Since the level of CM-markers in the purified chromatophores increased, Oelze and Drews (1972) suggested a reversible modification of the CM and ICM to explain the processes that take place following induction or inhibition of ICM development (see Fig. 5). In fact, however, the two events differ significantly. During induction, there is an increase in the differential rate of ICM-specific proteins incorporated at "unique" regions on the CM, but the specific synthetic rate of CM protein is not affected. During the shift back to aerobic growth, there is no evidence to suggest an increase in the differential rate of insertion of CM proteins into the ICM, but instead ICM-specific protein synthesis ceases and CM protein synthesis merely continues (Oelze and Drews, 1970*b*). Although the net effect is a reversible process, the forward and reverse mechanisms appear to be quite different.

3.3.2. Repression

The results reported by Irschik and Oelze (1973) are important because they differentiate clearly between the cessation of Bchl synthesis promoted by oxygen (inhibition) and that promoted under "shift-up" conditions (repression).

After a "shift-up" in light intensity, Bchl synthesis ceases until the new steady-state level of Bchl is attained, but proteins characteristic of the RC (Oelze and Golecki, 1975) of the ICM continue to be formed at what may be a reduced rate, although the major protein of the ICM associated with Bchl synthesis is not made. On the resumption of Bchl synthesis, therefore, the ICM will have a lower Bchl/RC Bchl ratio, and as a result, the size of the photosynthetic unit would be expected to fall. Further, as the cells approach steady-state photosynthetic growth, we would expect the size of the photosynthetic unit to increase once again to its preshift condition. Nor is there a transfer of phospholipid from the ICM to the CM; instead, both membrane fractions incorporate precursors to the same extent, but independently.

It was also observed that the level of NADH oxidase activity of the ICM remained relatively constant, further confirming the previous observation that the level of CM proteins in the ICM is constant regardless of the light intensity at which the cells were grown.

Finally, we have observed that ribosomal RNA synthesis is derepressed, resulting in an increase in the ribosome content (protein synthetic capacity) of the cells, and therefore increased growth rate.

3.4. Steady-State Synthesis of Intracytoplasmic Membrane

During steady-state growth of photosynthetic cultures, the level of ICM is maintained as long as the environmental and physiological parameters remain constant.

This observation raised a number of important questions: (1) Does the composition of the ICM vary as a function of light intensity? (2) What determines the steady-state concentration of the ICM? (3) How do light intensity and growth rate interact to regulate the final steady-state concentration of the ICM? (4) How is the ICM replicated? (5) How is the ICM partitioned to daughter cells?

Clearly, there are no completely satisfactory answers to any of these questions; in fact, our previous discussions indicated that many of the observations were contradictory. Recent results, however, have shed some light on these questions, and have more sharply defined the controversies that do exist.

3.4.1. Intracytoplasmic Membrane Content

Cohen-Bazire and Kunisawa (1960) concluded that in *Rs. rubrum*, the changes observed in the specific pigment content of cells growing phototrophically could be accounted for by changes in the specific pigment content of the chromatophores. Worden and Sistrom (1964) suggested that both the specific Bchl content of the ICM and the amount of ICM could account for changes in the specific Bchl content of *Rp. sphaeroides*.

Other investigations yielded different results. These investigations (Gorchein, 1968*a*; Holt and Marr, 1965*b,c*) demonstrated that the specific Bchl content of isolated chromatophores derived from steady-state grown cells was invariant, and as the specific Bchl content of the cell increased or decreased, it was only the amount of chromatophore material that changed.

Electron-microscopic studies also revealed a relationship between pigment content and ICM (Gibbs *et al.*, 1965; Cohen-Bazire and Kunisawa, 1963; Cohen-Bazire, 1963). Only Gibbs *et al.* (1965), with *Rs.*

molischianum, attempted to provide a rigorous quantitative estimate of the crucial relationship between the extent of ICM development and specific cellular Bchl content. They observed that the ratio of total ICM to Bchl decreased as the light intensity during growth was increased. On the other hand, the ratio of total membrane (ICM plus CM) to Bchl remained constant over a fivefold range in Bchl content. Overall, the volume of the ICM increased by a factor of 15. There is valid evidence demonstrating that in anaerobic, light-grown cultures, the CM is depleted in Bchl (Huang and Kaplan, 1973a; Niederman et al., 1972; Niederman and Gibson, 1971; Oelze and Drews, 1970b; Throm et al., 1970), but whether this means CM-resident Bchl or chromatophore-contaminated CM is not clear. When Gibbs and co-workers isolated chromatophores, however, they found that the specific Bchl content of the membrane increased as the specific Bchl content of the cells increased.

The apparent paradox revealed by Gibbs et al. (1965) was explicable if it was assumed that the ICM in high-light-grown cells contained significantly more protein per volume membrane than in ICM derived from low-light-grown cells. With the techniques recently developed, it is possible to determine both the qualitative and the quantitative composition of the chromatophores derived from cells grown at different light intensities. Finally, although Gibbs and co-workers rejected the possibility of contamination of the chromatophore preparations with other cellular components, recent evidence (Niederman, 1974; Niederman and Gibson, 1971; Fraker and Kaplan, 1971) makes such a conclusion less likely.

From these conflicting results, we conclude that both a change in the specific Bchl content of the ICM and a change in the amount of ICM accompany changes in the specific Bchl content of *Rp. sphaeroides* growing in steady states under different environmental conditions. On the other hand, *Rs. rubrum*, under similar experimental conditions, shows only a change in the amount of ICM.

3.4.2. Intracytoplasmic Membrane Composition

(a) Light and Oxygen. Aagaard and Sistrom (1972), employing extracts derived from steady-state cultures of *Rp. sphaeroides* and *Rs. rubrum*, examined the relationship between the total Bchl content of either extracts or isolated membranes and the amount of RC Bchl. They showed that as the specific Bchl content of the cells increased (lower growth rate), the ratio of Bchl to RC Bchl increased (this ratio refers to the size of the photosynthetic unit). Dierstein and Drews (1974) and Drews et al. (1974) achieved identical

results in chemostat cultures of *Rp. capsulata* (see Table 1), in which photosynthetic membrane development was gratuitous for cell growth (dark, pO_2 = 5 mm Hg) and the growth rate was limited by the availability of nitrogen. In both laboratories, it was also determined that as the total Bchl content of the cultures increased, the ratio of RC Bchl to cell protein remained relatively constant, although Aagaard and Sistrom (1972) showed that at very low specific Bchl contents, the ratio of RC Bchl to cell protein increases. Dierstein and Drews made the further important observation that the specific RC Bchl content per unit ICM protein remained constant, while the specific Bchl content per unit membrane protein increased at a decreasing rate to some limiting value. In *Rs. rubrum* (Golecki and Oelze, 1975), although no data are presented, we might suggest that as the specific Bchl content of the cells increased, both the specific Bchl content of the membranes increased to some limiting value, and at the same time the amount of ICM also increased.

From these studies and those of Lien et al. (1971, 1973), we can conclude that (1) the density of RC units per unit ICM protein is constant irrespective of the light intensity or growth rate, and is regulated inversely with changes in the oxygen tension, at least during induction; and (2) the density of the bulk LH Bchl protein can vary over wide limits, and its formation is particularly sensitive to oxygen and light intensity.

(b) Growth and Composition. The conditions employed by Dierstein and Drews (semiaerobic, dark) were gratuitous for ICM development and therefore independent of the energy requirements of the cell. The only apparent limitation on ICM synthesis was the ability of the cell to supply precursors for Bchl and ICM formation. If we assume that "dark" microaerophilic conditions maximally potentiate cells for bulk ICM synthesis due to the absence of the interaction of Bchl and light [see Section 2.2.2(b)], then as the growth rate increased due to an increased availability of nitrogen, two events would be taking place simultaneously. These events would lead to the relationship observed by Dierstein and Drews between growth rate and ICM levels. First, increased availability of nitrogen would lead to an increased availability of precursors for ICM synthesis as well as cell growth. Second, increased cell growth would lead to an increased metabolic consumption of corepressor [see Section 2.2.2(b)], which is already at very low levels, and would thereby permit the maximum potential for ICM development (dark) to be expressed to an ever-increasing extent as the metabolism of corepressor increased (with increasing growth rate).

When the cells were grown anaerobically in the light under nitrogen limitation, although increased growth rate provided the biosynthetic capacity for ICM synthesis, it was the light intensity that governed the degree of expression of ICM development, and as observed, the level of ICM was independent of the growth rate. Therefore, it appears that ICM levels can be related to growth rate only under specific physiological conditions in which either the inductive capacity or biosynthetic capacity for ICM formation becomes directly linked to growth. However, the general observation that ICM levels are inversely related to growth rate usually prevails. [See Section 2.2.2(b) for a model for the regulation of ICM formation.]

Since in the steady state "new" ICM biosynthesis takes place over the whole of the "old" ICM (see below), regulation probably exists at the level of precursor supply rather than template availability, because if this were not so, then the amount of template would be expected to increase exponentially during a cell cycle. The final cellular concentration of ICM will therefore be a balance between precursor formation and cell division.

An interesting interpretation that can be made from the observations of Aagaard and Sistrom (1972) regarding "heavy" and "light" chromatophores (Worden and Sistrom, 1964) is that there appears to be a gradient of changing composition over the length of the ICM, with membrane more interior (with reference to the CM) having photosynthetic units of increased size relative to the photosynthetic units of ICM more peripherally located (near the CM) (see Table 4 of Aagaard and Sistrom, 1972). Similarly, it might be expected that a gradient of opposite polarity could exist (and was suggested to exist by Huang and Kaplan, 1973a,b) with respect to respiratory chain activity associated with the ICM (Niederman *et al.*, 1972; Niederman and Gibson, 1971; Oelze and Drews, 1970b; Throm *et al.*, 1970). The density of respiratory chain activity over the ICM may determine the photochemical efficiency of the ICM *in situ*, with peripheral ICM (high light) having a higher density of ETC and intracellular ICM (low light) possessing a lower density of ETC.

3.4.3. Replication and Partitioning of the Intracytoplasmic Membrane

Kosakowski and Kaplan (1974) showed that the precursors destined for "new" ICM synthesis were inserted into preexisting ICM in a random, homogeneous fashion (see Fig. 7). They were able to distinguish between "old" and "new" ICM by growing cells in 70% heavy water for many generations ("old"), followed by a shift to ordinary water ("new"). Analyses of the ICM were performed following cell breakage, on CsCl density gradients employing fixed-angle rotors. Specific macromolecules could also be followed by labeling "D_2O-grown" cultures with a particular ^{14}C compound and "H_2O-grown" cultures with the same compound labeled with ^3H. When the distribution of either "old" or "new" protein or Bchl was followed in the growing ICM, it was observed that "old" material was always slightly displaced to the heavy side, and "new" material was always slightly displaced to the light side, of the bulk ICM. This displacement was never great, and it was maintained

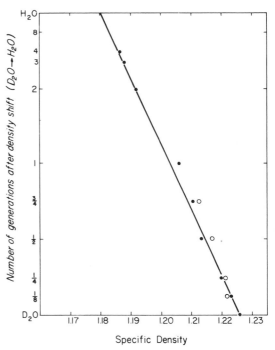

Fig. 7. Specific density of chromatophore material as a function of cell growth after the shift from 70% D_2O medium to H_2O medium. The line represents a theoretical distribution expected for a random distribution of incorporated precursors. (○) Values obtained from the experiment represented in Figs. 3 and 4 of Kosakowski and Kaplan (1974); (●) values from a different experiment. Density values are taken from the OD_{365} profile. Specifically, cells were grown for more than 10 generations in 70% D_2O medium, always maintaining a cell density between 1.5 and 5×10^8 cells/ml. Samples were removed at the appropriate time. Cells were harvested and disrupted, and chromatophores were isolated by differential centrifugation. Samples were loaded onto preformed, linear CsCl gradients and centrifuged to equilibrium in a Spinco 40 rotor for 24 hr at 20°C. Following centrifugation, samples were analyzed for OD_{365} nm, [3H], [^{14}C], and refractive index. Reprinted with the kind permission of the American Society for Microbiology from Kosakowski and Kaplan (1974).

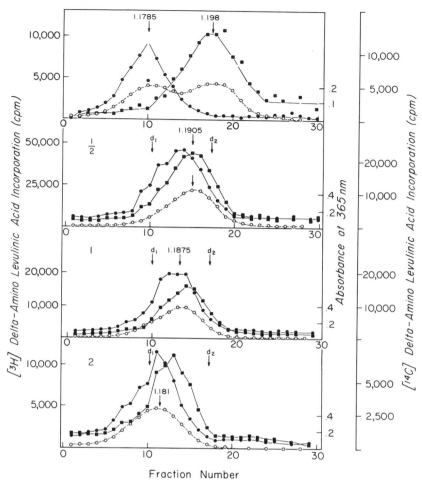

Fig. 8. Equilibrium density centrifugation of chromatophores from a density-shift experiment employing labeled δ-ALA as a marker for chlorophyll. (d_1) Position of light chromatophore material, 1.1785; (d_2) position of heavy chromatophore material, 1.198. (O) Optical density; (●) [³H]; (■) [¹⁴C]δ-ALA, respectively. The number of cell divisions following the shift to H₂O medium is shown in the upper left hand corner of each panel. The top panel is the control resulting from a mixture of "heavy" and "light" cells prior to breakage. See Figs. 6 and 7 and the text for experimental details. Reprinted with the kind permission of the American Society for Microbiology from Kosakowski and Kaplan (1974).

through several cell divisions (see Fig. 8). "Old" and "new" lipid shows no such displacement either from one another or from the bulk ICM (Fig. 9). That Bchl shows the same distribution in the ICM as protein, and not lipid, strongly indicates that Bchl exists as a Bchl–protein complex *in vivo*, and that "new" Bchl goes with "new" protein and "old" Bchl goes with "old" protein (Aagaard and Sistrom, 1972) and no interchange between Bchl–protein complexes occurs. The mere fact that such a displacement is observed and is maintained can be taken as proof against any rapid, lateral movement of Bchl–protein complexes in the membrane, and therefore these complexes are prob-

ably not inserted into the membrane in a zone of synthesis.

Several alternative explanations were offered by Kosakowski and Kaplan that might account for the displacement of "old" and "new" Bchl–protein within the membrane. New material may be inserted into old membrane in a random, nonhomogeneous fashion, or the material may be inserted as relatively large, preformed aggregates. A third possibility extended by Kosakowski and Kaplan is depicted in Fig. 10; this model proposes the replication of preexisting invaginations. The model was developed to account for the results that they observed, in particular the parti-

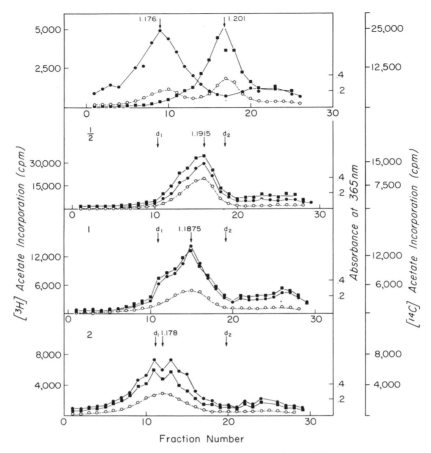

Fig. 9. Equilibrium density centrifugation of chromatophores from a density-shift experiment employing labeled acetate as a marker for lipid synthesis. (◯) Optical density; (●) [³H]acetate; (■) [¹⁴C]acetate. [d_1 (1.176)] Position of light chromatophores; [d_2 (1.201)] position of heavy chromatophores. The designation in the upper left-hand corner indicates the number of generations incubated in H_2O medium. The top panel is a control representing a mixture of D_2O- and H_2O-grown cells. See Figs. 6 and 7 and the text for experimental details. Reprinted with the kind permission of the American Society for Microbiology from Kosakowski and Kaplan (1974).

tioning of ICM into daughter cells; it is not intended as a morphological description of existing structures. Basically, there are two ways of conserving the number of invaginations per cell (Golecki and Oelze, 1975)—by the formation of totally new invaginations or by the replication of preexisting ones. If "new" invaginations were to arise, they should have been observed in the gradients as a second population of chromatophores containing a disproportionate amount of new material and therefore less dense than the old ICM. Likewise, preexisting ICM would be expected to remain disproportionately "old." The slight displacement observed is not sufficient to provide for totally new invaginations. The model of Kosakowski and Kaplan could be modified, however,

to include the existence of numerous replicating invaginations, not just a few, since the results of Golecki and Oelze (1975) demonstrate the apparent random distribution of invaginations over the surface of the CM. Yet if the number of invaginations per daughter cell is conserved (Golecki and Oelze, 1975), how can this be rationalized in the light of what is actually observed?

3.4.4. Precursor–Product Analyses of Intracytoplasmic Membrane Proteins

If "new" proteins are randomly intercalated into "old" ICM during steady-state formation of the ICM,

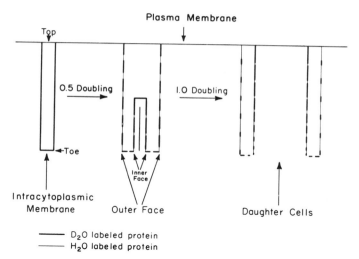

Fig. 10. Model of chromatophore growth. Starting with full, "heavy" intracytoplasmic membrane on the left, new material is incorporated into both outer faces of the intracytoplasmic membrane. As new material is incorporated, however, old material is displaced upward, back toward the plasma membrane. At 0.5 doubling following (center) the shift from D_2O- to H_2O-based media, the outer replicating faces of the intracytoplasmic membrane display a random homogeneous incorporation of "old" and "new" material, resulting in a nonrandom fixation of "old" and "new" protein into the inner nonreplicating faces, which display a gradient of distribution of "old" and "new" material. At the extreme right are depicted the results after 1.0 doubling after the shift in density. The outer faces of the intracytoplasmic membrane are located at the extreme left- and right-hand sides of the daughter membrane, and the "old" and "new" materials show a random, homogeneous distribution. The two new faces (innermost) of the daughter membranes show a gradient of "old" and "new" material, with the most conserved material being closer to the plasma membrane. The heavy and light lines are used to indicate the distribution of density labels, not to suggest absolute amounts of any component. No precise morphological structures are implied in this model. Reprinted with the kind permission of the American Society for Microbiology from Kosakowski and Kaplan (1974).

then in what form and how are these proteins incorporated?

Gibson et al. (1972) followed the fate of a pulse of radioactive amino acids added to steady-state cells of Rp. sphaeroides growing under semi-aerobic conditions in the light. By isolating chromatophores, chromatophore–envelope complex, envelope, and enriched CM (small membrane fraction; Niederman et al., 1972), they were in a position to follow the precursor–product relationships among these fractions. The envelope fraction is the native complex consisting of CM, murein layer, and outer membrane. Enriched CM refers to any fraction in which the CM represents a greater proportion of the total of those three components than is found in the native complex.

A kinetic analysis of the change in specific radioactivity present in the various fractions indicated that label left the small-membrane fraction and entered the chromatophore fraction as a function of time (see Fig. 11). The steady-state specific radioactivity of each fraction was nearly identical. To determine whether the small-membrane fraction served as a precursor for the ICM, the kinetics of labeling of specific chromatophore proteins was followed, and some of these results are presented in Figs. 12 and 13.

Band M (Niederman et al., 1972) is found in small membranes isolated from aerobically as well as from anaerobically grown cells, and is also present in chromatophores. Band L is not found in small membranes derived from aerobically grown cells. A comparison of the data reveals that M is formed approximately three times more rapidly than L and

five times more rapidly than E, C, I, or K. On the other hand, M leaves the small membrane fraction about one-half to one-third as fast as L. If the loss in activity from band M were corrected for the 30% contamination of small membrane with chromatophore material, however, radioactivity would be lost from M some two- to threefold faster. These results would be more compatible with the gain in specific activities observed. It should be pointed out, however, that the loss of activity in one fraction and the gain in another do not prove that in fact a relationship between the two fractions exists.

It is difficult to reconcile these observations with those of Kosakowski and Kaplan (1974). If material destined to reside in the ICM is first incorporated into the CM, this would imply growth of the ICM within a zone adjacent to the CM. This interpretation is not compatible with the density distribution observed for "old" and "new" ICM and for the displacement of "old" and "new" protein from one another. Another possibility is that the small-membrane fragment from anaerobic cells does not have the same origin as does small-membrane fragment from aerobic cells, but instead, these are particulate aggregates of ICM-precursor material formed by the condensation of nascent proteins (Kosakowski and Kaplan, 1974).

Although we have been considering the questions proposed in the beginning of this section as though a unitary regulatory system existed for all the photosynthetic bacteria, this is not likely to be true. We might expect to find that the regulatory mechanisms, although generally applicable, are modified according to the unique requirements of a particular organism. For example, in *Rs. rubrum*, there appears to be no change in the size of the photosynthetic unit (Aagaard and Sistrom, 1972) like that demonstrated for *Rp. sphaeroides*. In *Rp. capsulata* (Drews *et al.*, 1969), it appears that although the synthesis of LH Bchl–protein is sensitive to oxygen as it is in *Rp. sphaeroides* (Lien *et al.*, 1973), the response of RC Bchl–protein

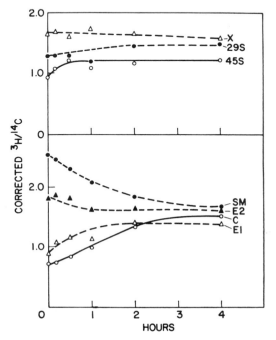

Fig. 11. Corrected specific radioactivity of various cell fractions. Two parallel cultures were grown under semianaerobic conditions in the light. One culture was supplemented with a [^{14}C]amino acid. After 15 hr, during which the dry weight of both cultures increased from 0.015 mg/ml to 0.2 mg/ml, the culture containing ^{14}C was cooled on ice. At this time, a pulse of [^{3}H]amino acid was added to the other culture. After 3 min, an aliquot was removed and poured onto crushed ice. A chase of the appropriate amino acid was added to the remainder of this culture. Further aliquots were removed as a function of time and mixed with an aliquot of the ^{14}C culture. The cells were harvested and disrupted in the French pressure cell. All particulate fractions were isolated by differential and sucrose gradient centrifugation. Specific radioactivities of the various fractions were calculated from the ratio ^{3}H cpm/^{14}C cpm. The numbers were normalized employing the ratio of the dry weights of the ^{14}C culture and the ^{3}H culture. It was also necessary to correct for the dilution of the radioactivity of each fraction caused by growth of the cultures. The resulting number is the corrected specific activity. The pulse and chanse were with phenylalanine. Top: Specific activities of 29 S ribosomes (●——●), 45 S ribosomes (○——○), and fraction X (unfolded ribosomes) (△---△). Bottom: Specific activities of small membranes (●——●), chromatophores (○——○), fraction E1 (bound chromatophores) (△---△), and fraction E2 (cell envelopes) (▲---▲). Reprinted with the kind permission of Academic Press, Inc., and Drs. K. Gibson and R. N. Niederman from Gibson *et al.* (1972).

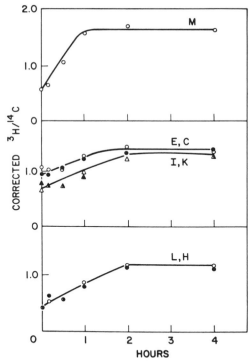

Fig. 12. Corrected specific radioactivity of chromatophore proteins. Methods are the same as for those reported for Fig. 11 except that the appropriate membrane fractions were dissolved in SDS and subjected to PAGE. The gels were fractionated and the radioactivity in each fraction determined. Specific activities were calculated from peaks, separated by SDS gel electrophoresis, and expressed in arbitrary units. Letters refer to the bands identified by Segen and Gibson (1971). Also see the text for details. Middle panel: (●) band C; (○) band E; (▲) band I; (△) band K. Bottom panel: (●) band H; (○) band L. Reprinted with the kind permission of Academic Press, Inc., and Drs. K. Gibson and R. A Niederman from Gibson *et al.* (1972).

synthesis to oxygen tension is sluggish and therefore unlike that observed for *Rp. sphaeroides* (Huang and Kaplan, 1973*b*).

3.5. Mutants and Intracytoplasmic-Membrane-Specific Protein Synthesis

As in other developmental systems, mutant organisms have proved invaluable in the study of ICM development. The isolation by Sistrom and Clayton (1964) of strain PM-8 of *Rp. sphaeroides*, defective in RC Bchl synthesis (Clayton *et al.*, 1965), led Aagaard and Sistrom (1972) to infer at least two separate control units for ICM development (Lien *et al.*, 1973). Takemoto (1974) and Langan and Niederman (personal communication) quantitatively demonstrated the existence of these two control units.

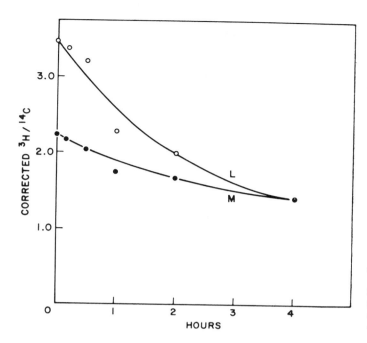

Fig. 13. Corrected specific radioactivity of small membrane proteins. For details, see the Figs. 11 and 12 captions. (L, M) Proteins that migrate like bands L and M of chromatophores. Also see the text for details. Reprinted with the kind permission of Academic Press, Inc., and Drs. K. Gibson and R. A. Niederman from Gibson et al. (1972).

Oelze et al. (1970) divided their mutants of Rs. rubrum into two groups based on coloration, just as did Segen and Gibson (1971) working with Rp. sphaeroides. In both investigations, efforts were made to correlate the presence of specific protein bands (by PAGE) with a specific phenotype. Takemoto and Lascelles, (1973, 1974) extended these and other observations (Hatch and Lascelles, 1972; Brown et al., 1972; Lascelles and Altshuler, 1967). Mutant strains blocked beyond δ-ALA synthesis had neither RC nor LH polypeptides. Mutants lacking RC proteins but possessing LH protein displayed no photophosphorylating activity, and their respiratory activity was not inhibited by illumination (Oelze and Weaver, 1971). All mutants produced normal or near-normal levels of ribulose diphosphate carboxylase activity, and when LH protein was made, near-normal levels of all pigments were produced. Mutants unable to produce δ-ALA, when grown semiaerobically in the absence of δ-ALA, produced small amounts of RC polypeptide but no LH polypeptide(s). When δ-ALA was added, all ICM polypeptides were made. Mutant TA-R, producing pigment in the presence of oxygen, produces a normal pattern of ICM polypeptides under aerobic growth.

Brown and Lascelles (1972) revealed that strains blocked in the phytolation step of Bchl biosynthesis did not make ICM, and mutants blocked in Bchl synthesis prior to the phytolation step did not make the phytol side chain. Similarly, if protein synthesis is blocked, then both branches (pyrrole and hydro-

carbon) of the pathway for Bchl synthesis cease to function. It must be noted, however, that in nearly all cases, the possibility of extensive genetic damage cannot be ruled out due to the absence of a genetic system.

4. Effect of Light and Oxygen

4.1. General Considerations

It is essential to understand how light or oxygen or both serve to regulate the development of the ICM as described. With what systems (receptors) of the cell do these agents interact, what biochemical stimuli (transmitters) are generated, and what molecular interactions (effectors) are responsible for the regulation of ICM development? The first of these considerations has been detailed elsewhere (Gest, 1972; Cohen-Bazire et al., 1957).

4.2. ATP and the Regulation of Intracytoplasmic Membrane Development

Because the intracellular concentration of adenylate nucleotides is not very high, the utilization of ATP must be intimately linked with its synthesis, and therefore, because the levels of adenylate nucleotides (see Gest, 1972) are capable of undergoing continuous change, adenylate nucleotides become prime candidates for effecting regulation of ICM synthesis.

We may hypothesize that high light intensity, by producing relatively high ATP levels, causes a decrease in the availability of precursors to the ICM, resulting in low ICM levels, and conversely for low light intensity. The addition of low levels of ATP to cultures of *Rp. capsulata* growing anaerobically under low light intensity resulted in a better than 50% reduction in the specific Bchl content of the cells (Gest, 1972; Gest *et al.*, 1971; Zilinsky *et al.*, 1971). These experiments would have been more interesting if gratuitous conditions had been used. Additional data employing an arsenate-resistant mutant Z-1 of *Rp. capsulata* (Lien *et al.*, 1971, 1973; Gest *et al.*, 1971) were obtained, and it was concluded that the synthesis of LH Bchl–protein was more susceptible to the "energy state" of the cells, whether this "state" be regulated by oxygen or light. Similarly, Oelze and Kamen (1971) and Schmidt and Kamen (1971) observed a correlation between ATP levels and Bchl synthesis. Only ATP, however, not total adenylate nucleotides, was considered to be involved.

The sensitivity of the LH Bchl–protein complex formation to ATP control could be inferred from the 50–70% inhibition of ALA-synthase activity using either 300-fold purified ALA-synthase or whole cells (Fanica-Gaignier, 1971; Fanica-Gaignier and Clement-Metral, 1971). During repression or de-repression of Bchl synthesis, or both, the changes in ATP levels were paralleled by changes in enzyme activity, with ATP levels responding more rapidly. The addition of δ-ALA prevented changes in Bchl synthesis from taking place. Detailed enzymological studies on ALA-synthase have revealed both active and inactive forms of the enzyme (Tuboi and Hayasaka, 1971; Tuboi *et al.*, 1969, 1970), and the existence of a small molecule regulator has also been suggested (Davies *et al.*, 1973; Tuboi and Hayasaka, 1972).

Miović and Gibson (1973), on the other hand, convincingly demonstrated that neither ATP levels nor energy charge was involved in the regulation of the synthesis of the ICM. Employing *Chromatium vinosum* strain D, they showed that changes in the rates of synthesis of ICM due to changes in light intensity, exposure to intermittent light, or the presence of arsenate in the growth medium could not be correlated with either the ATP levels or the concentration of adenylate nucleotides.

4.3. Electron-Transport Chain and the Regulation of Intracytoplasmic Membrane Development

The original hypothesis proposed by Cohen-Bazire *et al.* (1957) implicated the oxidation–reduction state of some carrier in the ETC as mediating the control of Bchl synthesis by either light or oxygen or both. Similarly, as the intracellular level of ATP is regulated by oxygen, so is the $NADP^+/NADPH$ ratio (Schön, 1971).

Sistrom (1963), employing inhibitors of the ETC, concluded that regulation of Bchl synthesis was mediated at the level of reduced pyridine nucleotides. Schön and Drews (1968) demonstrated that neither the redox state of cytochrome *b* nor that of cytochrome c_2 could be correlated with the regulation of Bchl synthesis.

On the basis of results with mutant strains of *Rp. capsulata* unable to grow aerobically, Marrs and Gest (1973a,b) suggested that the original model of Cohen-Bazire *et al.* (1957) should be modified. Their model is presented in Fig. 14. Mutant M5 demonstrates an almost total inactivation of the terminal respiratory oxidase activity, whereas mutant M2 is almost completely lacking in both NADH dehydrogenase and succinate dehydrogenase activities of the ETC. When

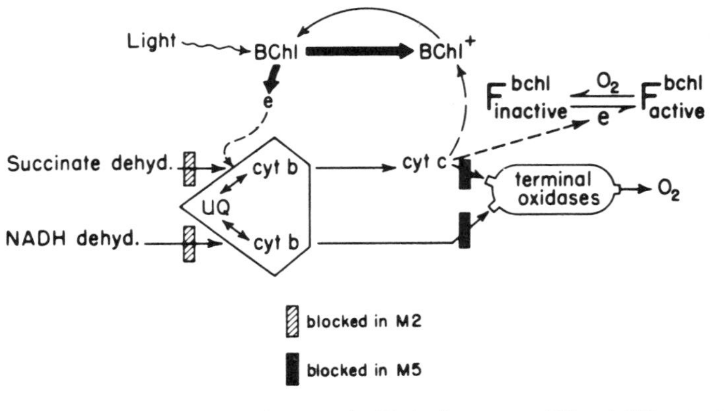

Fig. 14. Scheme for regulation of Bchl synthesis by molecular oxygen in *Rhodopseudomonas capsulata*. Both the respiratory and the light-driven (cyclic) electron-flow systems are indicated (for a discussion of details, see Marrs and Gest, 1973a). (Bchl) Bacteriochlorophyll; (UQ) ubiquinone; (cyt) cytochrome; (e) electrons. Although electrons from NADH and succinate are preferentially channeled to some extent, cytochrome *c* is reducible by both substrates (Segen and Gibson, 1971). Evidence for locations of the mutational blocks in strains M2 and M5 is given in Segen and Gibson (1971). F^{bchl} represents a postulated O_2-sensitive factor required for Bchl synthesis. The equilibrium between active and inactive forms of F^{bchl} is thought to be shifted toward F^{bchl}_{active} by flow of electrons from a region of the electron-transport system between the blocks in mutants M2 and M5; cytochrome *c* is provisionally indicated as the mediator. Reprinted with the kind permission of the American Society for Microbiology and Dr. B. Marrs from Marrs and Gest (1973b).

cultures of the mutant strains and the strain from which they were derived were grown anaerobically in the light on malate-supplemented medium and 2% oxygen was introduced, the following results were observed: The parental strain, after an inhibition of 1–2 hr, resumed Bchl synthesis. Mutant M2 showed an almost complete inhibition of Bchl synthesis for more than 8 hr, whereas M5 showed a 2- to 4-hr inhibition. followed by resumption of Bchl synthesis. Although the model of Cohen-Bazire *et al.* (1957) accommodates the complete inhibition of Bchl synthesis observed with M2, the transient inhibition of Bchl synthesis observed with M5 strongly suggests that oxygen interacts directly with some compound to regulate Bchl synthesis. Because of the gradual release of the inhibition of Bchl synthesis observed with M5, the model further predicts that electron flow through the ETC can be siphoned off at the level of cytochrome *c* to reactivate the oxygen-inactivated effector of Bchl synthesis. As Marrs and Gest (1973*b*) point out, it is the rather normal behavior of mutant M5, despite the presence of the block in the terminal oxidase, that led them to postulate the direct interaction of oxygen with an effector molecule. Because these experiments were performed in the light using photosynthetically grown cells, however, light-induced inhibition of respiration (Oelze and Weaver, 1971) may make the parent strain physiologically equivalent to M5, at least with respect to the control of Bchl synthesis. It would be extremely interesting to perform these experiments in the dark after the introduction of oxygen and to follow the initial rates of Bchl synthesis, if possible.

Experiments reported by Wittenberg and Sistrom (1971) also implicate the "state" of the ETC in the regulation of ICM development. Following anaerobic growth, mutant S37 undergoes approximately three cell doublings before growth and oxygen uptake cease and Bchl synthesis recommences. If these cells are placed under low oxygen so that Bchl synthesis is not inhibited, they retain the ability to grow aerobically. It was concluded that a functional ETC could be formed only when there was concurrent synthesis of Bchl. Since it was observed that heme synthesis drops dramatically on transfer of these cells to aerobic conditions, a reasonable explanation would be to assume that when there is active Bchl synthesis, the derepressed synthesis of δ-ALA overcomes an incomplete block in heme formation due to a defect in the pathway for heme biosynthesis beyond the formation of δ-ALA. We can conclude that as the ratio of reduced to oxidized pyridine nucleotides increases under aerobic conditions, Bchl synthesis resumes. This would fit the proposal made by Sistrom (1963), but is not necessarily compatible with the model of Marrs

and Gest (1973*b*). The data of Oelze and Weaver (1971), however, would be compatible with such a model.

Since mutant TA-R (Lascelles and Wertlieb, 1971) responds normally to changes in light intensity, but is insensitive to regulation by oxygen, these two stimuli must act at different sites or at different levels within the same site.

Finally, Eccleston and Gray (1973) observed a correlation between the intracellular concentration of guanosine tetraphosphate (ppGpp) and the transition from one light intensity to another. As in most of these studies, however, cause and effect are indistinguishable.

5. Conclusions on the Regulation of Intracytoplasmic Membrane Development

Figure 15 is a schematic representation of the events that seem to take place during the regulation of ICM development based on the results discussed above as well as a certain liberal interpretation of these data. The major features of this representation are as follows:

1. All components are subject to oxygen control, since the introduction of oxygen above threshold levels results in the complete inhibition of ICM synthesis. Oxygen control is referred to as *governor control* with respect to ICM development, because it is the only condition that can totally repress ICM synthesis.

2. With the exception of the RC units, all other

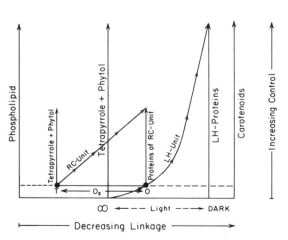

Fig. 15. Schematic representation of major control points affecting the synthesis of the ICM. See the discussion in Section 5 for a full description of the scheme.

components are subject to wide variation in their levels as a function of control by light intensity. This is certainly true for LH complex and carotenoids, and seems to be a reasonable assumption when considering phospholipid synthesis, although the evidence is sparse. This is referred to as *modulator control*.

3. Below a particular threshold for oxygen (which is species-dependent), there is an inverse relationship between the pO_2 and the specific cellular level of RC units. This level is not affected by light intensity. At less then threshold levels of oxygen, the formation of LH complex may be more sensitive to pO_2 than is the synthesis of RC complex. Once induced, under anaerobic conditions, the rate of appearance of LH complex seems to be related to light intensity in a sigmoidal fashion. At very high light intensities, an increase in the proportion of LH complex occurs as the result of relatively small decreases in light intensity. Eventually, a light intensity is reached at which the relationship appears roughly linear, and finally, at very low light intensities, little additional response is noted in terms of additional LH-complex synthesis.

4. There exists a hierarchy of controls, with "governor control" being dominant to "modulator control," so that at light intensities normally leading to high ICM concentrations, oxygen levels exist that will restrict ICM development either partially or entirely.

5. Diagonals represent major precursor pools the syntheses of which are obligately coupled. The presence of two semiindependent vertical lines determining pyrrole and phytol synthesis implies a regulatory division because the Bchl requirements for LH–protein on the one hand and RC–protein on the other appear to be independently regulated. It is also assumed that either the end product of each pathway is a powerful feedback inhibitor of its own synthesis or each is a powerful activator for the synthesis of the other. These assumptions are implied from the data of Brown and Lascelles (1972). The absence of one or both of the major protein components of the ICM therefore results in the immediate cessation of both tetrapyrrole and phytol synthesis. Additionally, a reactant or product of either tetrapyrrole or phytol biosynthesis may serve as a corepressor for transcriptional or translational regulation of ICM protein synthesis. Superimposed on these two levels of control, additional inhibitors or activators of enzyme activities may operate.

6. The scheme also reveals a decreasing order of linkage among activities associated with ICM development. A single mutational event may or may not result

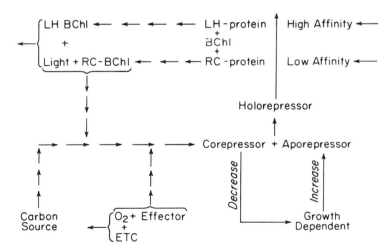

Fig. 16. Proposed scheme for the regulation of ICM development by carbon source, light intensity, oxygen tension, and growth rate. The pathways for the synthesis of LH Bchl–protein and RC Bchl–protein imply all those enzymatic and molecular interactions necessary to form Bchl and proteins. Phospholipid synthesis, carotenoid synthesis, and small-molecule synthesis are implied as being subject to either primary regulation by the repressor system or secondary regulation resulting from the effects of the repressor system at another level. Low affinity is distinguished from high affinity under conditions in which a given repressor concentration is just sufficient to "shutdown" one pathway, while leaving the other operative. The corepressor is imagined to be a small-molecule component of the repressor system that is necessary for repressor activity. The level of repressor will depend primarily on the level of corepressor, since it is the level of corepressor that is regulated by regulating the pathway for corepressor production. The carbon source, through a series of intermediate reactions, can influence the level of corepressor. Similarly, the interaction of RC Bchl–protein with light directly affects the level of corepressor through a series of intermediate steps. Finally, at oxygen levels above threshold values, the interaction of oxygen with an "effector" is transmitted to the pathway for corepressor synthesis. In the case of regulation by light, LH Bchl–protein, by competing with RC Bchl–protein, is acting to inhibit the formation of corepressor. Similarly, at oxygen levels below threshold, the interaction of oxygen with the ETC prevents the interaction of oxygen with effector, thereby causing a decrease in corepressor levels. Since corepressor is a "small molecule," it is suggested that this molecule is subject to degradation or turnover as a function of growth rate. Likewise, the formation of aporepressor increases as the growth rate increases. The balance between corepressor turnover and synthesis and aporepressor synthesis is clearly interactive and dependent on all the parameters that affect growth rate. Finally, regulation through active repressor (or corepressor acting alone, under certain conditions) may be effective at the level(s) of transcription, posttranscription, translation, and the regulation of enzyme activity. See Section 2.2.2(b) of the text for a discussion.

in an observed pleiotropic effect, depending on its position along the horizontal axis.

7. Although little is known about phospholipid synthesis and its relationship to the other components of the ICM, it seems reasonable that any alteration in lipid synthesis would affect total ICM production. The observations of Lien *et al.* (1973) that ICM synthesis in *Rp. capsulata* is sensitive to temperature may reflect a temperature-induced change in the physical state of the ICM lipids.

8. By adjusting the coordinates with respect to some relative concentration of oxygen or light intensity, it is possible to either change or leave unchanged the threshold effect of oxygen or the size of the photosynthetic unit.

In turn, the scheme presented in Fig. 16 and discussed in detail in Section 2.2.2(b) is a plausible model, based on the information available, to explain how light, oxygen, carbon source, and growth rate may interact to give rise to the observations represented in Fig. 15.

From these discussions, it is clear that our image of ICM morphogenesis and control has now emerged to the level of sophistication at which detailed molecular considerations of these processes may now be examined, even though there are considerable areas of uncertainty as well as others in which we suffer from a paucity of experimental evidence. Nevertheless, the last few years have witnessed significant progress in developing the approaches by which we may fill in the details of this fascinating system. It is hoped that this discussion has highlighted what we do not know as well as our achievements, as well as made clear that a system of nomenclature has been organized so that we may all communicate more effectively.

Addendum. Following the completion of this review, experiments were performed in our laboratory by Drs. Lueking and Fraley and by Dr. Wraight of the Department of Botany at the University of Illinois that suggest a very complex control process for the synthesis and assembly of the ICM. Figure 17 is a composite representing many individual experiments. Time and space considerations prevent us from including all these data. Therefore, the abscissas and ordinates for the individual panels do not necessarily bear a precise relationship to one another, since they have been taken from different experiments.

Phototrophic cultures grown under saturating light intensities were synchronized by employing the conditions of Cutler and Evans (1966) for *Escherichia coli* and earlier used by Ferretti and Gray (1967) with *Rp. sphaeroides.* Panels a–c demonstrate a continuous increase in turbidity, and a discontinuous increase in

cell number and DNA, respectively. In panel d, the results of a density-transfer experiment (discussed earlier in Section 3.4) employing synchronous cell cultures are graphed. Note that the log of the density has been plotted vs. time. Since we are performing a density-transfer experiment, we anticipate that the ICM will become less dense as some function of cell growth, and since growth is exponential, the density of the ICM must be followed as an exponential function. The results clearly demonstrate that the ICM undergoes a discontinuous change in density relative to the position of the culture in its stage of synchrony. As previously discussed (see Section 3.4), when employing asynchronous cultures of *Rp. sphaeroides,* Kosakowski and Kaplan (1974) observed that the ICM became polydisperse following density-transfer experiment. Clearly, this polydispersity could have resulted from cell-cycle-dependent heterogeneity associated with the replicative status of each cell, and therefore the point of replication of the ICM, within the cell cycle.

In panel e, we observe that the net amount of lipid phosphorus also accumulates discontinuously in synchronized cell populations. Finally, in panel f, the rate of accumulation of protein into the ICM as measured by the incorporation of radioactive amino acid into the particulate protein per milliliter of culture per 15 min is also observed to increase discontinuously in synchronous cell cultures. We believe the stepwise increase in the rate of protein incorporation in the particulate fraction is due to a gene dosage effect.

The results shown in panels e and f and those of numerous other experiments lead us to suggest that the ratio of protein to lipid in the ICM is not constant during the cell cycle. Both total protein and individual ICM-specific polypeptides are incorporated into the ICM throughout the division cycle (results for individual ICM polypeptides not presented here). The size of the photosynthetic unit as well as the ratio of RC units to cytochrome *c* remain constant (Wraight *et al.,* 1978; Lueking *et al.,* 1978; Fraley *et al.,* 1978). In contrast, there is little or no net incorporation of lipid P during part of the division cycle; during this period, the ratio of protein to lipid P in the ICM must increase. At some point, lipid P begins to be synthesized, so that the original protein/lipid ratio is restored and the total ICM surface is doubled. Finally (Wraight *et al.,* 1978), we have also observed that the activity of succinic dehydrogenase and NADH oxidase in the total particulate fraction also increases discontinuously. At this time, we cannot say whether this increase reflects an increase in protein or just enzyme activity. It is interesting to speculate that perhaps integral proteins are continuously incorporated into the

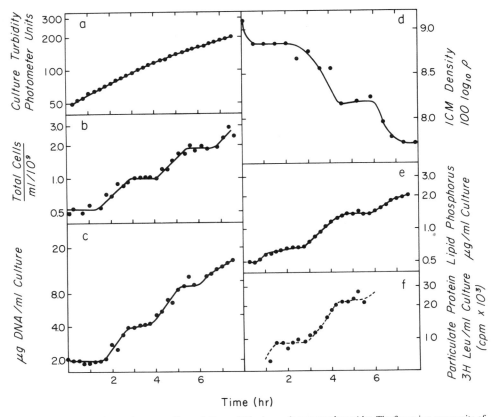

Fig. 17. ICM synthesis in synchronous cell populations of *Rhodopseudomonas sphaeroides*. The figure is a composite of several representative experiments with synchronized cell populations. *Rhodopseudomonas sphaeroides* strain 2.4.1 or a leucine, methionine auxotroph was grown phototrophically and synchronized by the method of Cutler and Evans (1966). (a) Culture turbidity determined with a Klett–Summerson colorimeter with a No. 66 filter. (b) Total number of cells per milliliter determined with a Petroff–Hausser counting chamber. (c) Amount of DNA per milliliter of culture measured by the method of Burton (1956). (d) Results of a density-transfer experiment with cells that had been synchronized in D_2O-based medium. ICM densities were determined as described by Kosakowski and Kaplan (1974). (e) Amount of lipid phosphorus per milliliter of culture. Lipids were extracted by the method of Bligh and Dyer (1959), and total phosphorus in the lipid extract was measured as described by Bartlett (1959). (f) Level of incorporation of [³H]leucine into particulate protein per milliliter of culture per 15 min.

membrane, whereas peripheral proteins are continuously incorporated in relation to the surface area of the membrane and are therefore linked to phospholipid accumulation.

Clearly, investigations pertaining to the synthesis and regulation of the ICM are moving rapidly. The next few years should witness remarkable progress in this exceedingly interesting area.

6. Note Added in Proof

Recent experiments in our laboratory employing the experimental approach described under Fig. 17 have revealed the following sequence of events in the formation of the ICM in synchronous cell cultures of *Rp. sphaeroides*. During the cell cycle the bulk of

the individual ICM polypeptides are synthesized and inserted into the ICM at constant rate and in constant proportions. Bulk Bchl and carotenoids as well as RC Bchl and cytochromes *b* and *c* are similarly continuously formed and inserted into the ICM. During this portion of the cell cycle no net phospholipid accumulation occurs into the ICM. As a result of these molecular events the composition of ICM goes from 65% protein/35% phospholipid to 85% protein/15% phospholipid. These changes have been measured directly and are further substantiated by the increase in ICM density as a function of the stage in the division cycle from which the samples are removed. Just prior to cell division there is a rapid and large increase in the rate of phospholipid synthesis which results in the net accumulation of phospholipid into the ICM restoring the membrane to its original protein/phospholipid composition and which is accompanied by a decrease in the density of the ICM vesicles. The data reveal that phospholipid synthesis and not turnover is being regulated and is responsible for the absence of phospholipid insertion into the ICM during the first two-thirds of the growth cycle and its ultimate accumulation over the last one-third of the growth cycle. Therefore, in any single cell in an exponential culture the amount of ICM protein and chlorophyll is increasing, while the amount of actual ICM (surface area, yield of vesicles, etc.) is remaining constant, until just prior to cell division. Why and how the cell imposes a cell-cycle-specific control of phospholipid biosynthesis during ICM biogenesis remains to be determined. However, these events may be significant in partitioning the ICM to daughter cells following division.

7. References

Aagaard, J., and Sistrom, W. R., 1972, Control of synthesis of reaction center bacteriochlorophyll in photosynthetic bacteria, *Photochem. Photobiol.* **15**:209.

Bartlett, G. R., 1959, Phosphorus assay in column chromatography, *J. Biol. Chem.* **234**:466.

Biederman, M., 1971, Einwirkung von Detergenzien auf die Thylakoide von *Rhodospirillum rubrum*, *Arch. Microbiol.* **75**:171.

Biederman, M., and Drews, G., 1968, Trennung der Thylakoidbausteine einiger Athiorhodaceae durch Gelelektrophorese, *Arch. Mikrobiol.* **61**:48.

Bligh, E. G., and Dyer, W. J., 1959, A rapid method of total lipid extraction and purification, *Can. J. Biochem. Physiol.* **37**:911.

Boatman, E. S., 1964, Observations on the fine structure of spheroplasts of *Rhodospirillum rubrum*, *J. Cell. Biol.* **20**:297.

Brown, A. E., and Lascelles, J., 1972, Phytol and bacteriochlorophyll synthesis in *Rhodopseudomonas sphaeroides*, *Plant Physiol.* **50**:747.

Brown, A. E., Eiserling, F. A., and Lascelles, J., 1972, Bacteriochlorophyll synthesis and the ultrastructure of wild type and mutant strains of *Rhodopseudomonas spheroides*, *Plant Physiol.* **50**:743.

Bull, M. J., and Lascelles, J., 1963, The association of protein synthesis with the formation of pigments in some photosynthetic bacteria, *Biochem. J.* **87**:15.

Burton, K., 1956, A study of the conditions and mechanism of the diphenylamine reaction for the colorimetric estimation of deoxyribonucleic acid, *Biochem. J.* **62**:315.

Cellarius, R. A., and Peters, G. A., 1969, Photosynthetic membrane development in *Rhodopseudomonas spheroides*: Incorporation of bacteriochlorophyll and development of energy transfer and photochemical activity, *Biochim. Biophys. Acta* **189**:234.

Clayton, R., 1963, Photochemical reaction centers in photosynthetic tissue, in: *Bacterial Photosynthesis* (H. Gest, A. San Pietro, and L. P. Vernon, eds.), pp. 377–395, Antioch Press, Yellow Springs, Ohio.

Clayton, R. K., and Clayton, B. J., 1972, Relations between pigments and proteins in the photosynthetic membranes of *Rhodopseudomonas spheroides*, *Biochim. Biophys. Acta* **283**:492.

Clayton, R. K., and Haselkorn, R., 1972, Protein components of bacterial photosynthetic membranes, *J. Mol. Biol.* **68**:97.

Clayton, R. K., Sistrom, W. R., and Zaugg, W. S., 1965, The role of reaction center in photochemical activities of bacterial chromatophores, *Biochim. Biophys. Acta* **102**:341.

Cohen-Bazire, G., 1963, Some observations on the organization of the photosynthetic apparatus in purple and green bacteria, in: *Bacterial Photosynthesis* (H. Gest, A. San Pietro, and L. P. Vernon, eds.), pp. 89–110, Antioch Press, Yellow Springs, Ohio.

Cohen-Bazire, G., and Kunisawa, R., 1960, Some observations on the synthesis and function of the photosynthetic apparatus in *Rhosospirillum rubrum*, *Proc. Natl. Acad. Sci. U.S.A.* **46**:1543.

Cohen-Bazire, G., and Kunisawa, R., 1963, The fine structure of *Rhodospirillum rubrum*, *J. Cell Biol.* **16**:401.

Cohen-Bazire, G., and Sistrom, W. R., 1966, The procaryotic photosynthetic apparatus, in: *The Chlorophylls* (L. P. Vernon and G. R. Seely, eds.), pp. 313–341, Academic Press, New York.

Cohen-Bazire, G., Sistrom, W. R., and Stanier, R. Y., 1957, Kinetic studies of pigment synthesis by non-sulfur purple bacteria, *J. Cell. Comp. Physiol.* **49**:25.

Crounse, J., Sistrom, W. R., and Nemser, G., 1963, Carotenoid pigments and the *in vivo* spectrum of bacteriochlorophyll, *Photochem. Photobiol.* **2**:361.

Cruden, D. L., and Stanier, R. Y., 1970, The characterization of chlorobium vesicles and membranes isolated from green bacteria, *Arch. Mikrobiol.* **72**:115.

Cutler, G. R., and Evans, J. E., 1966, Synchronization of bacteria by a stationary-phase method, *J. Bacteriol.* **91**:469.

Davies, R. C., Gorchein, A., Neuberger, A., Sandy, J. D., and Tait, G. H., 1973, Biosynthesis of bacteriochlorophyll, *Nature (London)* **245**:15.

Dierstein, R., and Drews, G., 1974, Nitrogen-limited continuous culture of *Rhodopseudomonas capsulata* growing photosynthetically or heterotrophically under low oxygen tensions, *Arch. Microbiol.* **99**:117.

Drews, G., 1960, Untersuchungen zur Substruktur der "Chromataphoren" von *Rhodospirillum rubrum* und *Rhodospirillum molischianum*, *Arch. Mikrobiol.* **36**:99.

Drews, G., 1974, Composition of a protochlorophyll–protopheophytin complex, excreted by mutant strains of *Rhodopseudomonas capsulata*, in comparison with the photosynthetic apparatus, *Arch. Microbiol.* **100**:397.

Drews, G., and Lampe, H. H., 1971, The differentiation of the intracytoplasmic membrane system of *Rhodopseudomonas capsulata*, in: *IInd International Congress on Photosynthesis, Stresa* (G. Forti, M. Avron, and A.

Melandri, eds.), pp. 2715–2719, Dr. W. Junks, The Hague.

Drews, G., Lampe, H. H., and Ladwig, R., 1969, Die Entwicklung des Photosyntheseapparates in Dunkelkulturen von *Rhodopseudomonas capsulata*, *Arch. Mikrobiol.* **65**:12.

Drews, G., Leutiger, I., and Ladwig, R., 1971, Production of protochlorophyll, protopheophytin, and bacteriochlorophyll by the mutant Ala of *Rhodopseudomonas capsulata*, *Arch. Mikrobiol.* **76**:349.

Drews, G., Dierstein, R., and Nieth, K. F., 1974, The formation of the photosynthetic apparatus in cells of *Rhodopseudomonas capsulata*, in: *Third International Congress of Photosynthesis*, Rehovot (M. Avron, ed.), pp. 2–6, Elsevier, Amsterdam.

Eccleston, E. D., and Gray, E. D., 1973, Variations in ppGpp levels in *Rhodopseudomonas spheroides* during adaptation to decreased light intensity, *Biochem. Biophys. Res. Commun.* **54**:1370.

Fanica-Gaignier, M., 1971, Control of bacteriochlorophyll synthesis by adenine nucleotides in *Rhodopseudomonas spheroides* Y, in: *IInd International Congress on Photosynthesis, Stresa* (G. Forti, M. Avron, and A. Melandri, eds.), pp. 2721–2726, Dr. W. Junks, The Hague.

Fanica-Gaignier, M., and Clement-Metral, J. D., 1971, ATP inhibition of aminolevulinate (ALA) synthetase activity in *Rhodopseudomonas spheroides* Y, *Biochem. Biophys. Res. Commun.* **44**:192.

Feher, G., 1971, Some chemical and physical properties of a bacterial reaction center particle and its primary photochemical reactants, *Photochem. Photobiol.* **14**:373.

Ferretti, J. J., and Gray, E. D., 1967, Control of enzyme synthesis during adaptation in synchronously dividing populations of *Rhodopseudomonas spheroides*, *Biochem. Biophys. Res. Commun.* **29**:501.

Fraker, P., and Kaplan, S., 1971, The isolation and fractionation of the photosynthetic membranous organelles from *Rhodopseudomonas spheroides*, *J. Bacteriol.* **108**:465.

Fraker, P., and Kaplan, S., 1972, Isolation and characterization of a bacteriochlorophyll-containing protein from *Rhodopseudomonas spheroides*, *J. Biol. Chem.* **247**:2732.

Fraley, R. T., Lueking, D. R., and Kaplan, S., 1978, Intracytoplasmic membrane synthesis in synchronous cell populations of *Rhodopseudomonas sphaeroides*: Polypeptide insertion into growing membrane, *J. Biol. Chem.* **253**:458.

French, C. S., 1938, The chromoproteins of photosynthetic purple bacteria, *Science* **88**:60.

Fuller, R. C., Conti., S. F., and Mellin, D. B., 1963, The structure of the photosynthetic apparatus in the green and purple sulfur bacteria, in: *Bacterial Photosynthesis* (H. Gest, A. San Pietro, and L. P. Vernon, eds.), pp. 71–87, Antioch Press, Yellow Springs, Ohio.

Garcia, A. F., Drews, G., and Kamen, M. D., 1974, On reconstitution of bacterial photophosphorylation *in vitro*, *Proc. Natl. Acad. Sci. U.S.A.* **71**:4213.

Gest, H., 1972, Energy conversion and generation of reducing power in bacterial photosynthesis, in: *Advances in Microbial Physiology* (A. H. Rose and D. W. Tempest, eds.), pp. 243–282, Academic Press, New York.

Gest, H., Lien, S., Zilensky, J. W., and San Pietro, A., 1971, Regulation of metabolic processes in photosynthetic bacteria; the adenylate energy charge control model, in: *IInd International Congress on Photosynthesis, Stresa* (G. Forti, M. Avron, and A. Melandri, eds.), pp. 2701–2713, Dr. W. Junks, The Hague.

Gibbs, S. P., Sistrom, W. R., and Worden, P. B., 1965, The photosynthetic apparatus of *Rhodospirillum rubrum*, *J. Cell Biol.* **26**:395.

Gibson, K. D., 1965a, Nature of the insoluble pigmented structures (chromatophores) in extracts and lysates of *Rhodopseudomonas spheroides*, *Biochemistry* **4**:2027.

Gibson, K. D., 1965b, Isolation and characterization of chromatophores from *Rhodopseudomonas spheroides*, *Biochemistry* **4**:2042.

Gibson, K. D., Newberger, A., and Tait, G. H., 1962a, Studies on the biosynthesis of porphyrin and bacteriochlorophyll by *Rhodopseudomonas spheroides*. 1. The effect of growth conditions, *Biochem. J.* **83**:539.

Gibson, K. D., Neuberger, A., and Tait, G. H., 1962b, Studies on the biosynthesis of porphyrin and bacteriochlorophyll by *Rhodopseudomonas spheroides*. 2. The effects of ethionine and threonine, *Biochem. J.* **83**:550.

Gibson, K. D., Segen, B. J., and Niederman, R. A., 1972, Membranes of *Rhodopseudomonas spheroides*. II. Precursor–product relations in anaerobically growing cell, *Arch. Biochem. Biophys.* **152**:561–568.

Giesbrecht, P., and Drews, G., 1962, Elektronemikroskopische Untersuchungen über die Entwicklung der "Chromatophoren" von *Rhodospirillum molischianum* Giesberger, *Arch. Mikrobiol.* **43**:152.

Goldberger, R. F., 1974, Autogenous regulation of gene expression, *Science* **183**:810.

Golecki, J. R., and Oelze, J., 1975, Quantitative determination of cytoplasmic membrane invaginations in phototrophically growing *Rhodospirillum rubrum*: A freeze–etch study, *J. Gen. Microbiol.* **88**:253.

Gorchein, A., 1968a, The relation between the pigment content of isolated chromatophores and that of the whole cell in *Rhodopseudomonas spheroides*, *Proc. R. Soc. London Ser. B.* **170**:247.

Gorchein, A., 1968b, Distribution and metabolism of ornithine in *Rhodopseudomonas spheroides*, *Proc. R. Soc. London Ser. B.* **170**:265.

Gorchein, A., 1968c, The separation and identification of the lipids of *Rhodopseudomonas spheroides*, *Proc. R. Soc. London Ser. B.* **170**:279.

Gorchein, A., Neuberger, A., and Tait, G. H., 1968a, The isolation and characterization of subcellular fractions from pigmented and unpigmented cells of *Rhodopseudomonas spheroides*, *Proc. R. Soc. London Ser. B.* **170**:229.

Gorchein, A., Neuberger, A., and Tait, G. H., 1968b, Metabolic turnover of lipids of *Rhodopseudomonas spheroides*, *Proc. R. Soc. London Ser. B.* **170**:311.

Hall, R. L., Kung, M. C., Fu, M., Hales, B. J., and Loach, P. A., 1973, Comparison of phototrap complexes from chromatophores of *Rhodospirillum rubrum*, *Rhodopseudomonas spheroides*, and the R-26 mutant of *Rhodopseudomonas spheroides*, *Photochem. Photobiol.* **18**:505.

Hatch, T., and Lascelles, J., 1972, Accumulation of porphobilinogen and other pyrroles by mutant and wild type *Rhodopseudomonas spheroides*: Regulation by heme, *Arch. Biochem. Biophys.* **150**:147.

Higuchi, M., Gato, K., Fujimoto, M., Namiki, O., and Kikuchi, G., 1965, Effect of inhibitors of nucleic acid and protein synthesis on the induced synthesis of bacteriochlorophyll and δ-aminolevulinic acid synthetase by *Rhodopseudomonas spheroides*, *Biochim. Biophys. Acta* **95**:94.

Holt, S. C., and Marr, A. G., 1965*a*, Localization of chlorophyll in *Rhodospirillum rubrum*, *J. Bacteriol.* **89**:1402.

Holt, S. C., and Marr, A. G., 1965*b*, Isolation and purification of the intracytoplasmic membrane of *Rhodospirillum rubrum*, *J. Bacteriol.* **89**:1413.

Holt, S. C., and Marr, A. G., 1965*c*, Effect of light intensity on the formation of intracytoplasmic membrane in *Rhodospirillum rubrum*, *J. Bacteriol.* **89**:1421.

Holt, S. C., Conti, S. F., and Fuller, R. C., 1966, Photosynthetic apparatus in the green bacterium *Chloropseudomonas ethylicum*, *J. Bacteriol.* **91**:311.

Huang, J., and Kaplan, S., 1973*a*, Membrane proteins of *Rhodopseudomonas spheroides*. III. Isolation, purification and characterization of cytoplasmic membrane proteins, *Biochim. Biophys. Acta* **307**:301.

Huang, J., and Kaplan, S., 1973*b*, Membrane proteins of *Rhodopseudomonas spheroides*. IV. Characterization of chromatophore proteins, *Biochim. Biophys. Acta* **307**:317.

Irschik, H., and Oelze, J., 1973, Membrane differentiation in phototrophically growing *Rhodospirillum rubrum* during transition from low to high light intensity, *Biochim. Biophys. Acta* **330**:80.

Jensen, S. L., 1963, Carotenoids of photosynthetic bacteria–distribution, structure and biosynthesis, in: *Bacterial Photosynthesis* (H. Gest, A. San Pietro, and L. P. Vernon, eds.), pp. 19–34, Antioch Press, Yellow Springs, Ohio.

Jolchine, G., and Reiss-Husson, F., 1972, Proteins of *Rp. spheroides* Y reaction centers: Gel electrophoresis and electrofocusing studies, *Biochem. Biophys. Res. Commun.* **48**:333.

Jones, O. T. G., and Pleivis, K. M., 1974, Reconstitution of light-dependent electron transport in membranes from a bacteriochlorophyll-less mutant of *Rhodopseudomonas sphaeroides*, *Biochim. Biophys. Acta* **357**:204.

Ketchum, P. A., and Holt, S. C., 1968, Isolation and characterization of the membranes from *Rhodospirillum rubrum*, *Biochim. Biophys. Acta* **196**:141.

Kornberg, R. D., and McConnell, H. M., 1971, Lateral diffusion of phospholipids in a vesicle membrane, *Proc. Natl. Acad. Sci. U.S.A.* **68**:2564.

Kosakowski, M. H., and Kaplan, S., 1974, Topology and growth of the intracytoplasmic membrane system of *Rhodopseudomonas spheroides*: Protein, chlorophyll and phospholipid insertion into steady state anaerobic cells, *J. Bacteriol.* **118**:1144–1157.

Lascelles, J., 1959, Adaptation to form bacteriochlorophyll in *Rhodopseudomonas spheroides*: Changes in activity of enzymes concerned in pyrrole synthesis, *Biochem. J.* **72**:508.

Lascelles, J., 1960, The synthesis of enzymes concerned in bacteriochlorophyll formation in growing cultures of *Rhodopseudomonas spheroides*, *J. Gen. Microbiol.* **23**:487.

Lascelles, J., 1963, Tetrapyrroles in photosynthetic bacteria, in: *Bacterial Photosynthesis* (H. Gest, A. San Pietro, and L. P. Vernon, eds.), pp. 35–52, Antioch Press, Yellow Springs, Ohio.

Lascelles, J., 1968, The bacterial photosynthetic apparatus, *Adv. Microb. Physiol.* **2**:1.

Lascelles, J., 1973, Microbial photosynthesis, in: *Benchmark Papers in Microbiology* (W. W. Umbreit, ed.), pp. 1–401, Dowden, Hutchison and Ross, Stroudsburg, Pennsylvania.

Lascelles, J., and Altshuler, A., 1967, Some properties of mutant strains of *Rhodopseudomonas spheroides* which do not form bacteriochlorophyll, *Arch. Mikrobiol.* **59**:204.

Lascelles, J., and Szilágyi, J. F., 1965, Phospholipid synthesis by *Rhodopseudomonas spheroides* in relation to the formation of photosynthetic pigments, *J. Gen. Microbiol.* **38**:55.

Lascelles, J., and Wertlieb, D., 1971, Mutant strains of *Rhodopseudomonas spheroides* which form photosynthetic pigments aerobically in the dark, *Biochim. Biophys. Acta* **226**:328.

Lien, S., San Pietro, A., and Gest, H., 1971, Mutational and physiological enhancement of photosynthetic energy conversion in *Rhodopseudomonas capsulata*, *Proc. Natl. Acad. Sci. U.S.A.* **68**:1912.

Lien, S., Gest, H., and San Pietro, A., 1973, Regulation of chlorophyll synthesis in photosynthetic bacteria, *Bioenergetics* **4**:423.

Lueking, D. R., Fraley, R. T., and Kaplan, S., 1978, Intracytoplasmic membrane synthesis in synchronous cell populations of *Rhodopseudomonas sphaeroides*: Fate of "old" and "new" membrane, *J. Biol. Chem.* **253**:451.

Marrs, B., and Gest, H., 1973*a*, Genetic mutations affecting the respiratory electron—transport system of the photosynthetic bacterium *Rhodopseudomonas capsulata*, *J. Bacteriol.* **114**:1045.

Marrs, B., and Gest, H., 1973*b*, Regulation of bacteriochlorophyll synthesis by oxygen in respiratory mutants of *Rhodopseudomonas capsulata*, *J. Bacteriol.* **114**:1052–1057.

Miović, M. L., and Gibson, J., 1973, Nucleotide pools and adenylate energy charge in balanced and unbalanced growth of Chromatium, *J. Bacteriol.* **114**:86.

Niederman, R. A., 1974, Membranes of *Rhodopseudomonas spheroides*: Interactions of chromatophores with the cell envelope, *J. Bacteriol.* **117**:19.

Niederman, R. A., and Gibson, K. D., 1971, The separation of chromatophores from the cell envelope in *Rhodopseudomonas spheroides*, *Prep. Biochem.* **1**:141.

Niederman, R. A., Segen, B. J., and Gibson, K. D., 1972, Membranes of *Rhodopseudomonas spheroides*. 1. Isolation and characterization of membrane fractions from extracts of aerobically and anaerobically grown cells, *Arch. Biochem. Biophys.* **152**:547.

Niederman, J. A., Mallon, D. E., and Langan, J. E., 1976, Membranes of *Rhodopseudomonas sphaeroides.* IV. Assembly of chromatophores in low aeration cell suspensions, *Biochim. Biophys. Acta* **440**:429.

Nieth, K. F., and Drews, G., 1974, The protein patterns of intracytoplasmic membranes and reaction center particles isolated from *Rhodopseudomonas capsulata, Arch. Mikrobiol.* **96**:161.

Oelze, J., and Drews, G., 1969a, Untersuchungen über die Membran-Proteine bei *Rhodospirillum rubrum, Arch. Mikrobiol.* **69**:12.

Oelze, J., and Drews, G., 1969b, Die Morphogenese des Photosynthese-Apparates von *Rhodospirillum rubrum.* II. Die Kinetik der Thylakoidsynthese nach Markierung der Membranen mit [2-^{14}C]Azetat, *Biochim. Biophys. Acta* **173**:448.

Oelze, J., and Drews, G., 1970a, Der Einfluss der Lichtintensität und der Sauerstoffspannung auf die Differenzierung der Membranen von *Rhodospirillum rubrum, Biochem. Biophys. Acta* **203**:189.

Oelze, J. V., and Drews, G., 1970b, Variations of NADH oxidase activity and bacteriochlorophyll contents during membrane differentiation in *Rhodospirillum rubrum, Biochim. Biophys. Acta* **219**:131.

Oelze, J., and Drews, G., 1972, Membranes of photosynthetic bacteria, *Biochim. Biophys. Acta* **265**:209.

Oelze, J., and Golecki, J. R., 1975, Properties of reaction center depleted membranes of *Rhodospirillum rubrum, Arch. Microbiol.* **102**:59.

Oelze, J., and Kamen, M. D., 1971, Adenosine triphosphate cellular levels in *Rhodospirillum rubrum* during transition from aerobic to anaerobic metabolism, *Biochim. Biophys. Acta* **234**:137.

Oelze, J., and Weaver, P., 1971, The adjustment of photosynthetically grown cells of *Rhodospirillum rubrum* to aerobic light conditions, *Arch. Mikrobiol.* **79**:108.

Oelze, J., Biedermann, M., Freund-Mölbert, E., and Drews, G., 1969a, Bacteriochlorophyllgehalt und Proteinmuster der Thylakoids von *Rhodospirillum rubrum* während der Morphogenese des Photosynthese-Apparates, *Arch. Mikrobiol.* **66**:154–165.

Oelze, J., Biedermann, M., and Drews, G., 1969b, Die Morphogenese des Photosyntheseapparates von *Rhodospirillum rubrum.* I. Die Isolierung und Charakterisierung von zwei Membransystemen, *Biochim. Biophys. Acta* **173**:436.

Oelze, J., Schroeder, J., and Drews, G., 1970, Bacteriochlorophyll, fatty acid and protein synthesis in relation to thylakoid formation in mutant strains of *Rhodospirillum rubrum, J. Bacteriol.* **101**:669.

Okamura, M. Y., Steiner, L. A., and Feher, G., 1974, Characterization of reaction centers from photosynthetic bacteria. I. Subunit structure of the protein mediating the primary photochemistry in *Rhodopseudomonas spheroides* R-26, *Biochemistry* **13**:1394.

Olson, J. M., Filmer, D., Radloff, R., Romano, C. A., and Sybesma, C., 1963, The protein–chlorophyll-770 complex from green bacteria, in: *Bacterial Photosynthesis* (H.

Gest, A. San Pietro, and L. P. Vernon, eds.), pp. 423–431, Antioch Press, Yellow Springs, Ohio.

Reed, D. W., and Peters, G. A., 1972, Characterization of the pigments in reaction center preparations from *Rhodopseudomonas spheroides, J. Biol. Chem.* **247**:7148.

Schachman, H. K., Pardee, A. B., and Stanier, R. Y., 1952, Studies on the macromolecular organization of microbial cells, *Arch. Biochem. Biophys.* **38**:245.

Schick, J. S., and Drews, G., 1969, The morphogenesis of the bacterial photosynthetic apparatus. III. The features of a pheophytin–protein–carbohydrate complex excreted by mutant M46 of *Rhodospirillum rubrum, Biochim. Biophys. Acta* **183**:215.

Schmidt, G. L., and Kamen, M. D., 1971, Control of chlorophyll synthesis in *Chromatium vinosum, Arch. Mikrobiol.* **76**:51.

Schön, G., 1971, Der Einfluss der Kulturbedingungen auf den Nicotinamid-Adenin-Dinucleotid (phosphat)-Gehalt in Zellen von *Rhodospirillum rubrum, Arch. Mikrobiol.* **79**:147.

Schön, G., and Drews, G., 1968, Der Redoxzustand des NAD (P) und der Cytochrome b und c_2 in Abhängigkeit von pO_2 bei einigen Athiorhodaceae, *Arch. Mikrobiol.* **61**:317.

Schön, G., and Ladwig, R., 1970, Bacteriochlorophyllsynthese und Thylakoidmorphogenese in anaerobes Dunkelkultur von *Rhodosperillum rubrum, Arch. Mikrobiol.* **74**:356.

Schröder, J., and Drews, G., 1968, Quantitative Bestimmung des Fettsäuren von *Rhodospirillum rubrum* und *Rhodopseudomonas capsulata* während der Thylakoidmorphogenese, *Arch. Mikrobiol.* **64**:59.

Schröder, J., and Drews, G., 1969, Fettsäuregehalte in Lichtkulturen von *Rhodospirillum rubrum* während der Thylakoidmorphogenese, *Arch. Mikrobiol.* **69**:20.

Schröder, J., Biedermann, M., and Drews, G., 1969, Die Fettsäuren in ganzen Zellen, Thylakoiden, und Lipopolysacchariden von *Rhodospirillum rubrum* und *Rhodopseudomonas capsulata, Arch. Mikrobiol.* **66**:273.

Schwenker, U., and Gingrass, G., 1973, A carotoprotein from chromatophores of *Rhodospirillum rubrum, Biochem. Biophys. Res. Commun.* **51**:94.

Segen, B. J., and Gibson, K. D., 1971, Deficiencies of chromatophore proteins in some mutants of *Rhodopseudomonas spheroides* with altered pigments, *J. Bacteriol.* **105**:701.

Shaw, M. A., and Richards, W. R., 1971a, Evidence for the formation of membranous chromatophore precursor fractions in *Rhodopseudomonas spheroides, Biochem. Biophys. Res. Commun.* **45**:863.

Shaw, M. A., and Richards, W. R., 1971b, Studies on the morphogenesis of the photosynthetic apparatus in *Rhodopseudomonas spheroides*, in: *IInd International Congress on Photosynthesis, Stresa* (G. Forti, M. Avron, and A. Melandri, eds.), pp. 2733–2745, Dr. W. Junks, The Hague.

Singer, S. J., and Nicolson, H. M., 1972, The fluid mosaic model of the structure of cell membranes, *Science* **175**:720.

Sistrom, W. R., 1962a, Observations on the relationship between the formation of photopigments and the synthesis of protein in *Rhodopseudomonas spheroides*, *J. Gen. Microbiol.* **28**:599.

Sistrom, W. R., 1962b, The kinetics of the synthesis of photopigments in *Rhodopseudomonas spheroides*, *J. Gen. Microbiol.* **28**:607.

Sistrom, W. R., 1963, A note on the effect of inhibitors of electron transport and phosphorylation on photopigment synthesis in *Rhodopseudomonas spheroides*, in: *Bacterial Photosynthesis* (H. Gest, A. San Pietro, and L. P. Vernon, eds.), pp. 53–60, Antioch Press, Yellow Springs, Ohio.

Sistrom, W. R., 1966, The spectrum of bacteriochlorophyll *in vivo*: Observations on mutants of *Rhodopseudomonas spheroides* unable to grow photosynthetically, *Photochem. Photobiol.* **5**:845.

Sistrom, W. R., and Clayton, R. K., 1964, Studies on a mutant of *Rhodopseudomonas spheroides* unable to grow photosynthetically, *Biochim. Biophys. Acta* **88**:61.

Steiner, S., Sojka, G. A., Conti, S. F., Gest, H., and Lester, R. L., 1970, Modification of membrane composition in growing photosynthetic bacteria, *Biochim. Biophys. Acta* **203**:571.

Takemoto, J., 1974, Kinetics of photosynthetic membrane assembly in *Rhodopseudomonas spheroides*, *Arch. Biochem. Biophys.* **163**:515–520.

Takemoto, J., and Lascelles, J., 1973, The coupling between bacteriochlorophyll and membrane protein synthesis in *Rhodopseudomonas sphaeroides*, *Proc. Natl. Acad. Sci. U.S.A.* **70**:799.

Takemoto, J., and Lascelles, J., 1974, Function of membrane proteins coupled to bacteriochlorophyll synthesis; studies with wild type and mutant strains of *Rhodopseudomonas spheroides*, *Arch. Biochem. Biophys.* **163**:507.

Throm, E., Oelze, J., and Drews, G., 1970, The distribution of NADH oxidase in the membrane system of *Rhodospirillum rubrum*, *Arch. Mikrobiol.* **72**:361.

Tuboi, S., and Hayasaka, S., 1971, Control of δ-amino-levulinate synthetase activity in *Rhodopseudomonas spheroides*. I. Partial purification of the inactive form of fraction I, *Arch. Biochem. Biophys.* **146**:282.

Tuboi, S., and Hayasaka, S., 1972, Control of δ-amino-levulinate synthetase activity in *Rhodopseudomonas spheroides*. II. Requirement of a disulfide compound for the conversion of the inactive form of fraction I to the active form, *Arch. Biochem. Biophys.* **150**:690.

Tuboi, S., Kim, H. J., and Kikuchi, G., 1969, Occurrence of a specific and reversible inhibitor of δ-aminolevulinate synthetase in extracts of *Rhodopseudomonas spheroides*, *Arch. Biochem. Biophys.* **130**:92.

Tuboi, S., Kim, H. J., and Kikuchi, G., 1970, Differential induction of fraction I and fraction II of δ-aminolevulinate synthetase in *Rhodopseudomonas spheroides* under various incubation conditions, *Arch. Biochem. Biophys.* **138**:155.

Tuttle, A. L., and Gest, H., 1959, Subcellular particulate systems and the photochemical apparatus of *Rhodospirillum rubrum*, *Proc. Natl. Acad. Sci. U.S.A.* **45**:1261.

Uffen, R. L., and Wolfe, R. S., 1970, Anaerobic growth of purple nonsulfur bacteria under dark conditions, *J. Bacteriol.* **104**:462.

Wittenberg, T., and Sistrom, W. R., 1971, Mutant of *Rhodopseudomonas spheroides* unable to grow aerobically, *J. Bacteriol.* **106**:732.

Worden, P. B., and Sistrom, W. R., 1964, The preparation of and properties of bacterial chromatophore fractions, *J. Cell Biol.* **23**:135.

Wraight, C. A., Lueking, D. R., Fraley, R. T., and Kaplan, S., 1978, Synthesis of photopigments and electron transport components in synchronous phototrophic cultures of *Rhodopseudomonas sphaeroides*, *J. Biol. Chem.* **253**:465.

Yamashita, J., and Kamen, M. D., 1969, Observations on distribution of NADH oxidase in particles from dark-grown and light-grown *Rhodospirillum rubrum*, *Biochem. Biophys. Res. Commun.* **34**:418.

Zilinsky, J. W., Sojka, G. A., and Gest, H., 1971, Energy charge regulation in photosynthetic bacteria, *Biochem. Biophys. Res. Commun.* **42**:955.

CHAPTER 44

Control of Antenna Pigment Components

W. R. Sistrom

1. Introduction

The photosynthetically active pigments of photo-trophic bacteria occur in two functionally distinct forms. The first, reaction center bacteriochlorophyll (RC Bchl), is the only one that engages in productive photochemistry; the second form, the antenna pigments, absorbs radiant energy and transfers it to RC Bchl. The antenna pigments include both carotenoids and bacteriochlorophyll (Bchl).

Wassink *et al.* (1939) (see also Chapter 7) showed that the antenna Bchl's of species of both the Chromatiaceae and the Rhodospirillaceae have two noteworthy properties. The first is the complexity of the spectrum of intact cells or cell extracts in the region of Bchl absorption from about 780 to 900 nm. Depending on the species, the spectra are composed of two or three or even five distinct absorption maxima, even though only a single chemical species of Bchl is present (Wassink *et al.*, 1939; Vrendenberg and Amesz, 1966; Biebl and Drews, 1969). *Rhodo-spirillum rubrum* is an exception: the spectrum of Bchl has a single major absorption band at 885 nm (see, e.g., Biebl and Drews, 1969). The second un-usual characteristic noted by Wassink and co-workers is that the relative heights of these bands vary depending on the conditions under which the cells have been grown. Cohen-Bazire *et al.* (1957) confir-med this observation, and in addition showed that the ratio of carotenoid pigments to Bchl is dependent on growth conditions. Aagaard and Sistrom (1972)

showed more recently that the ratio of antenna Bchl to RC Bchl also depends on growth conditions, at least in *Rhodopseudomonas sphaeroides*.

These changes in properties should be distinguished from changes in the specific pigment content (Bchl or carotenoid per unit protein) of chromatophore material. The specific pigment content can be deter-mined only after careful purification of chromato-phore material from extraneous protein. Changes in the ratios of Bchl to carotenoids and of RC Bchl to total Bchl and changes in the spectrum of Bchl can, in principle, be determined from measurements on intact cells, although for technical reasons this is often difficult.

In this chapter, I shall summarize what is known about these changes in relative amounts of the various pigments and suggest a unified explanation for them. I shall deal only with variations in cultures that have reached steady-state levels of pigmentation in different light intensities or oxygen tensions. I shall not consider the kinetics of the changes after an alteration of light intensity or oxygen tension; these topics are dealt with in Chapter 43.

2. Observations

2.1. Spectrum of Bacteriochlorophyll *in Vivo*

Wassink's observations on the variability of the spectrum of Bchl in intact cells were limited to *Chromatium vinosum*. The work since then has extended these observations to various members of the

W. R. Sistrom · Department of Biology, University of Oregon, Eugene, Oregon 97403

Rhodospirillaceae (especially *Rp. sphaeroides*) and provided more quantitative data.

It is obvious simply from inspection of the numerous published spectra of extracts of *Rp. sphaeroides* with different specific Bchl contents that the shape of the spectrum between 790 and 900 nm is markedly dependent on the Bchl content of the cells (see, e.g., Cohen-Bazire *et al.*, 1957; Worden and Sistrom, 1964). It is also evident, as was pointed out by Wassink, that the spectra can be resolved into three bands with maxima at 800, 850 and about 875 nm. By assuming that the shape of the 850 and 875 nm bands is the same as that of the single symmetrical band of *Rs. rubrum*, it is possible (again following Wassink) to resolve these spectra into the component bands and assign values to the absorbance at each of the peak wavelengths. The results can be expressed as absorbance (OD) at each wavelength for a standard total Bchl concentration. The results of such an analysis are shown in Table 1. The obvious conclusions that can be drawn from these data are: (1) the 800 and 850 nm bands vary in concert (the ratio of absorbances at these wavelengths is constant); (2) there is an inverse relationship between the height of the 875 nm band relative to total Bchl and the specific Bchl content of the cells; and (3) there is a direct relationship between the height of the 800 or of the 850 nm band relative to total Bchl and the specific Bchl content. The last two conclusions are obviously not independent. For simplicity, I shall refer to the component of the spectrum with a maximum at 875 nm as B875, and, since they vary in parallel, to the 800 and 850 nm bands simply as B850. The designation B850 does not imply that both bands originate from the same molecule of Bchl.

Although there has been little comparative work done, it seems clear that several species show the same kind of change as does *Rp. sphaeroides* (*Rs.*

molischianum: Gibbs *et al.*, 1965; *Rp. gelatinosa* and *Rp. capsulata*: Barron, 1962; Sistrom, unpublished); other species, however, show a much less marked change in the spectrum (*Rhodomicrobium vannielii*: Barron, 1962). Raymond and Sistrom (1967) noted that the spectrum of *Ectothiorhodospira halophila* was the same in cultures grown at two different light intensities even though the spectrum is composed of at least three components.

A more extensive analysis of the sort exemplified in Table 1 was carried out for *Rp. sphaeroides* by Crounse *et al.* (1963b). Their data showed that the relative absorbance due to B875 decreased rapidly as the cellular specific Bchl content increased from about 0.5 to 2 or 3 μg Bchl/100 μg cell protein, and that the relative absorbance due to B850 increased rapidly over the same range of specific Bchl contents. The changes in the relative amounts of the two components became less marked at higher specific Bchl contents.

2.2. Ratio of Carotenoid Pigments to Bacteriochlorophyll

Essentially all the observations on variation in the relative amounts of carotenoids and Bchl have been made on strain Ga of *Rp. sphaeroides*. The reason is that the carotenoid composition of this mutant strain is very simple: it accumulates only neurosporene and chloroxanthin (Griffiths and Stanier, 1956). Since these pigments have the same spectrum and extinction coefficient and are easily and completely extracted into methanol, the carotenoid content can be determined accurately without difficulty. The complexity of the carotenoid compositions of most other phototrophic bacteria makes quantitative analyses difficult.

Some results for strain Ga of *Rp. sphaeroides*

Table 1. Antenna Bacteriochlorophyll Components and Specific Bacteriochlorophyll Content in *Rhodopseudomonas sphaeroides* Strain Ga[a]

Cellular specific Bchl content (μg Bchl/100 μg cell protein)	Absorbance at 1 mM total Bchl			Ratios of absorbances	
	875 nm	850 nm	800 nm	A_{850}/A_{800}	A_{850}/A_{875}
0.56	67	48	38	1.27	0.75
0.655	58.5	61.5	51.5	1.2	1.05
2.06	54	79	63	1.25	1.46
4.98	32	85	66	1.28	2.65

[a] Recalculated from data of Crounse *et al.* (1963b).

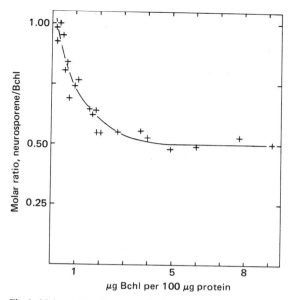

Fig. 1. Molar ratio of carotenoid pigments (neurosporene plus chloroxanthin) and specific Bchl content of *Rhodopseudomonas sphaeroides* strain Ga. The data were taken from Crounse *et al.* (1963*b*), Worden and Sistrom (1964), and Aagaard and Sistrom (unpublished). The molarity of the carotenoid pigments is based on the assumption that neurosporene and chloroxanthin are present in equal amounts.

2.3. Size of the Photosynthetic Unit

For our purposes, the size of the photosynthetic unit can be defined simply as the ratio of total Bchl (largely antenna Bchl) to RC protein. In *Rp. sphaeroides*, the latter can be determined most easily from the extent of the light-induced bleaching at 870 nm and the known differential extinction coefficient expressed in terms of the molarity of RC protein ($116\ mM^{-1}\ cm^{-1}$; Straley *et al.*, 1973). It should be borne in mind that although the results are expressed in terms of RC protein, this was not directly measured in the experiments discussed.

Aagaard and Sistrom (1972) determined the size of the photosynthetic unit in *Rp. sphaeroides* strain Ga grown in a variety of light intensities and oxygen tensions. They found that the size increases as the specific Bchl content of the cells increases. The size varied by a factor of about 10 for a 25-fold range of cellular specific Bchl contents.

It was shown that regardless of the specific Bchl content, the molar ratio of B875 to RC protein is constant and equal to about 25. The observed variation in the size of the photosynthetic unit is therefore entirely due to changes in the relative amount of B850. This idea was confirmed by the observations that in *Rs. rubrum* and in a carotenoidless mutant strain of *Rp. sphaeroides* (see below), in both of which B850 is absent and the antenna Bchl has a simple, single-banded spectrum, the size of the photosynthetic unit is very nearly invariant and is in the neighborhood of 30.

Lien *et al.* (1971) showed that the relative amounts of RC Bchl and of cytochrome c_2 (which can be taken as an index of the amount of RC Bchl) are much higher in an arsenate-resistant mutant strain of *Rp. capsulata* than in the wild type grown under similar conditions. The increased amount of RC Bchl is paralleled by an increase in the height of the absorption maximum corresponding to B875. The same workers (Lien *et al.*, 1973) obtained indirect evidence that the size of the photosynthetic unit in the same strain is higher in cells with lower specific Bchl contents.

Dierstein (1975) obtained results similar to those of Aagaard and Sistrom for *Rp. capsulata* growing semi-aerobically in continuous culture; under these conditions, the cellular specific Bchl content is directly related to the growth rate of the culture. For a fivefold change in specific Bchl content, there was a fourfold change in the size of the photosynthetic unit (Table 3).

It can be concluded from the data of Dierstein that the amount of RC protein per chromatophore protein is constant (Table 3). The data of Aagaard and Sistrom substantiate this conclusion, albeit rather weakly (Table 3).

collected from a number of sources are depicted in Fig. 1. The molar ratio of carotenoids (neurosporene + chloroxanthin) to Bchl decreases sharply with increasing specific Bchl content and tends to approach a limiting value of about 0.5. There is an obvious parallel here with the way the relative amount of B875 changes with specific Bchl content; we shall consider this point later. A similar pattern is apparent from the scanty data for *Rs. molischianum* shown in Table 2.

Table 2. Specific Bacteriochlorophyll Content and Carotenoid Composition of *Rhodospirillum molischianum*[a]

Bchl ($\mu g/100\ \mu g$ cell protein)	Molar ratio, Lycopene/Bchl
9.65	0.31
6.2	0.45
2.8	0.54
1.65	0.71

[a] Recalculated from data of Gibbs *et al.* (1965). In this reference, an incorrect value for the extinction coefficient of Bchl in methanol was used; the data have been recalculated using the correct value.

Table 3. Reaction Center Bacteriochlorophyll and Membrane Protein in
Rhodopseudomonas capsulata **and** *Rhodopseudomonas sphaeroides*

Bchl (μg/mg):			
Cell protein	Membrane protein	Molar ratio, total Bchl/RC Bchl	RC Bchl (μg/mg membrane protein)
Rhodopseudomonas capsulata[a]			
4.8	19.6	113	0.17
14.5	41.5	287	0.15
18.7	56.5	380	0.15
24.2	79.5	443	0.18
Rhodopseudomonas sphaeroides[b]			
5.2	13.2	72	0.19
7.2	18.6	77	0.24
12.0	38.5	100	0.38
20.5	54.0	222	0.24
52	122.0	330	0.37

[a] From Dierstein (1975).
[b] Recalculated from Aagaard and Sistrom (1972).

The data of Aagaard and Sistrom (1972) also showed that for both *Rs. rubrum* and *Rp. sphaeroides*, the amount of RC protein relative to total cell protein increases with increasing specific Bchl content. In the case of *Rp. sphaeroides*, the amount of RC Bchl per cell protein apparently reaches a limiting value at a specific Bchl content of about 3 nmol Bchl/100 μg cell protein. It follows from the constancy of the ratio of RC protein to total chromatophore protein that the amount of the latter relative to total cell protein must increase with increasing specific Bchl content of the cells. This was observed by Worden and Sistrom (1964) in *Rp. sphaeroides*. Taken together, the constant ratio of RC protein to chromatophore protein and the variable size of the photosynthetic unit imply that the specific Bchl content of purified chromatophore material increases with increasing cellular specific Bchl content. This was also observed by Worden and Sistrom (1964).

In the case of *Rs. rubrum*, it has been known for some time (Ketchum and Holt, 1970; Holt and Marr, 1965) that the Bchl content of purified chromatophore material is constant regardless of the cellular Bchl concentration. Since the size of the photosynthetic unit does not vary in this organism, the assumption that the amount of RC protein per chromatophore protein is constant is consistent with the invariant Bchl concentration of the chromatophore material.

It is clear that an organism that can vary the size of the photosynthetic unit can keep its RCs operating efficiently over a wide range of light intensities without overproduction of antenna pigments. Presumably, this is of selective advantage. More interesting and difficult to understand is the behavior of *Rs. rubrum*.

2.4. Mutations That Affect Antenna Bacteriochlorophyll

Two kinds of mutations alter the spectral characteristics of antenna Bchl: mutations that affect carotenoid biosynthesis and mutations that affect RC protein.

In mutant strains of *Rp. sphaeroides* in which carotenoid synthesis is blocked before the formation of neurosporene (see Chapter 39), the spectrum of Bchl is grossly altered (Griffiths and Stanier, 1956). The multibanded spectrum of the wild-type is replaced by one with a single, symmetrical major absorption band at about 870 nm and a small shoulder at 805 nm that is due to RC Bchl. Clearly, B850 is not present in these mutant strains. Whether the band at 870 nm is unmodified B875 is not entirely clear, primarily because of the difficulty in determining precisely the position of this band in the wild type. However, the fact that in a mutant of *Rs. rubrum* that is unable to synthesize unsaturated carotenoids (strain G9) the Bchl absorption maximum is shifted from 885 to

870 nm (Aagaard and Sistrom, 1972) suggests that the long-wavelength absorption band in carotenoidless mutant strains of *Rp. sphaeroides* is not due to unaltered B875.

It was tempting to conclude that the presence or absence of carotenoid pigments has a direct effect on the spectrum of Bchl. In fact, the author and colleagues succumbed to this temptation (Sistrom *et al.*, 1956). This simple explanation appears to be false for several reasons. First, phenocopies of the carotenoidless strains of both *Rp. gelatinosa* (Crounse *et al.*, 1963b) and *Rs. rubrum* (Nugent and Fuller, 1967) have normal Bchl spectra. Second, the parallelism already noted between increased amount of B850 and decreased relative carotenoid content is not what would be expected from this idea. The last, and most convincing, piece of evidence is provided by a mutant strain of *Rp. sphaeroides* isolated by Segen and Gibson (1971). Under semiaerobic conditions, this strain is brown rather than red as is the wild type. Although its carotenoid composition was not completely determined, it appeared that this strain accumulates the normal carotenoid, spheroidene, but is unable to oxygenate this pigment to spheroidenone, which gives semiaerobic cultures of the wild type their red color. Despite the presence of spheroidene, the spectrum of Bchl in this strain is similar to that in strains lacking colored carotenoids.

Although it has not been definitely proved, it seems likely that disruption of carotenoid biosynthesis and alteration of the spectrum of Bchl can result from a single mutation. This conclusion is based on the fact that these effects are correlated in independent isolates in several different species (Crounse *et al.*, 1963a; Sistrom, unpublished). The strain isolated by Segen and Gibson may be a double mutant.

A mutant strain of *Rp. sphaeroides* (strain PM-8) lacking RC Bchl was isolated by Sistrom *et al.* (1963) and characterized by Sistrom and Clayton (1964). Sistrom (1966) compared the spectra of Bchl in strain PM-8 and in the wild type (strain Ga), and found that the amount of B875 in strain PM-8 was considerably less than in strain Ga at similar specific Bchl contents. Furthermore, the spectrum of Bchl in a carotenoidless derivative of strain PM-8 is single-banded (as it is in carotenoidless but photosynthetically competent strains), but the peak is shifted from 870 to 863 nm. In photosynthetically competent revertants of the double mutant, the peak regains its 870 nm position.

In both kinds of mutation, the apparent pleiotropic effects are most reasonably explained by assuming that the primary defect in each case (blockage of carotenoid biosynthesis or absence of RC Bchl) is due to the absence or alteration of a specific protein, and that in the wild type, certain antenna Bchl molecules are bound to that protein.

3. Speculations

In this section, I shall outline a model that can account for the observations on *Rp. sphaeroides* discussed in this chapter. In brief, these observations are:

1. RC Bchl (and therefore RC protein) is a constant fraction of chromatophore protein.

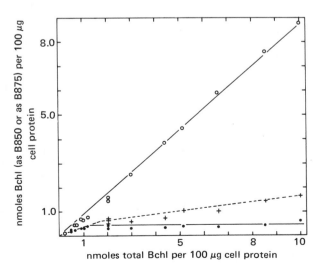

Fig. 2. Relative amounts of B850 and B875 and the cellular specific Bchl content of *Rhodopseudomonas sphaeroides* strain Ga. The data were taken from Aagaard and Sistrom (1972). The calculations are described in the text. Abscissa: Specific Bchl content of crude cell extracts. Ordinate: Bchl as either B850 (O——O) or B875 (+−−−+ and ●——●), μg/100 μg cell protein.

2. There are two species of antenna Bchl, B850 and B875; the synthesis of each is controlled independently.

3. The molecular ratio of B875 to RC protein is fixed and is equal to about 20–25.

4. The amounts of both B875 and carotenoids relative to total Bchl decrease steeply as the specific Bchl content increases in the range from about 0.2 to 2 or 3 nmol Bchl/100 μg cell protein; at larger specific Bchl contents, they decrease much less rapidly, if at all.

5. Mutant strains in which biosynthesis of carotenoids is blocked at (or before) the reaction leading to neurosporene do not have B850.

6. The spectral characteristics of B875 are altered in mutant strains lacking normal RC protein.

I shall assume that antenna pigments occur as protein complexes with fixed numbers of pigment molecules per protein. There is little direct evidence for this assumption in *Rp. sphaeroides*. It is in accord, however, with what is known about the structures of RC Bchl–protein complex (Chapter 19) and of the Bchl *a*–protein complex from *Chlorobium limicola* (Chapter 8). Various workers have identified what appear to be antenna Bchl–protein complexes from *Rp. sphaeroides* (see, e.g., Segen and Gibson, 1971; Clayton and Clayton, 1972; Takemoto and Lascelles, 1974). Hall *et al.* (1973) provided some evidence for the existence of two such pigment–protein complexes in wild-type *Rp. sphaeroides* that is in accord with the conclusion of Aagaard and Sistrom concerning the independence

of B850 and B875. The various Bchl–protein complexes are discussed fully in Chapter 7.

The second assumption of the model is that B850 and B875 are each associated with a fixed number of carotenoid molecules. I have already suggested that the absence of B850 in carotenoidless mutants can be explained on the supposition that the enzyme for the reaction leading to neurosporene is part of the B850 complex. This suggestion can be extended to the B875 complex; here, loss of enzymic activity results merely in a shift of the spectrum of the associated Bchl as is seen in *Rs. rubrum* and possibly in *Rp. sphaeroides*.

These assumptions require that the ratio of carotenoid to total Bchl be related in some fashion to the relative amounts of B850 and B875. Evidence of such a relationship is provided in Figs. 2 and 3.

These figures combine some of the data of Aagaard and Sistrom and some of the data from Fig. 1. Aagaard and Sistrom calculated the amount of B875 from analyses of spectra and an assumed extinction coefficient for the complex. The extinction coefficient was determined from the spectra of carotenoidless strains, in which there is only one major absorption peak. Aagaard and Sistrom employed two different modes of analysis that gave rather different results at high specific Bchl contents. In Fig. 2, both sets of values for B875 are shown expressed as nmol B875 Bchl/100 μg cell protein (Aagaard and Sistrom used the average of the two sets of values). The amount of B850 can be calculated simply, as the difference between total Bchl and B875 (neglecting the amount

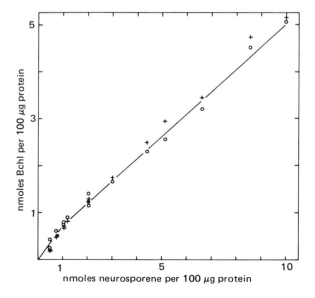

Fig. 3. Specific carotenoid content and specific Bchl content in extracts of *Rhodopseudomonas sphaeroides* strain Ga. The molarity of the carotenoids was calculated as in Fig. 1. (O) Data from Fig. 1; (+) data from Fig. 2 and assumption that Bchl/carotenoid is equal to 2 in B850 and 1 in B875.

of RC Bchl). The results of all these calculations are shown in Fig. 2. In the calculation, the average value of B875 at each specific Bchl content was used. This introduces no large error, since at specific Bchl contents, where the two B875 values are discordant, this component is at most only 20% of the total Bchl.

It can be seen that the amount of B850 increases more or less linearly with specific Bchl content. On the other hand, B875 increases at low specific Bchl contents and then either increases much less steeply or becomes constant, depending on which set of values for B875 is used. The important fact is that at small specific Bchl contents, B875 is a major fraction (up to 50%) of antenna Bchl, but at high specific Bchl contents is only 5–15% of the total.

The way in which carotenoid content relative to cell protein varies with specific Bchl content is shown in Fig. 3; it increases linearly or nearly so. This immediately suggests that either B875 has no carotenoid associated with it or the amounts of carotenoid associated with B875 and with B850 are nearly equal. From Fig. 1, it can be concluded that B875 must have carotenoid associated with it. This is because at high specific Bchl contents, where most of the Bchl is B850, the ratio of carotenoid to total Bchl approaches 0.5. On the other hand, at low specific Bchl contents, where the relative amount of B850 is much less, the ratio of carotenoid to total Bchl actually increases. This can occur only if B875 is associated with carotenoid (this conclusion depends, of course, on the assumption that the amount of carotenoids per antenna complex is constant).

Figure 3 shows the amount of carotenoid per cell protein calculated on the assumption that the Bchl/carotenoid ratio in B850 is 2 and in B875 is 1. The agreement is rather good.

According to this model, then, all the various qualitative changes in chromatophore properties associated with different cellular specific Bchl contents can be accounted for by variation of only two components: B850 and RC protein.

Note Added in Proof: Recently, Takemoto and Huang Kao (1977) have analyzed the amounts of RC protein and the protein believed to be associated with antenna Bchl by SDS–polyacrylamide gel electrophoresis of chromatophore material from *Rp. sphaeroides*. Their results extend the findings of Aagaard and Sistrom (1972) by demonstrating that the ratio of antenna Bchl protein to RC protein increases with increasing specific Bchl content of the cells.

4. References

Aagaard, J., and Sistrom, W. R., 1972, Control of synthesis of reaction center bacteriochlorophyll in photosynthetic bacteria, *Photochem. Photobiol.* **15**:209.

Barron, P. D., 1962, Chemical and physicochemical properties of bacterial chromatophore fractions, Ph.D. thesis, Harvard University, Cambridge, Massachusetts.

Biebl, H., and Drews, G., 1969, Das *in-Vivo*-Spektrum als taxonomisches Merkmal bei Untersuchungen zur Verbreitung von Athiorhodaceae, *Zentralbl. Bakteriol. Parasitenkd. Infektionskr. Hyg.* Abt. 2: **123**:425.

Clayton, R. K., and Clayton, B. J., 1972, Relations between pigments and proteins in the photosynthetic membranes of *Rhodopseudomonas* spheroides, *Biochim. Biophys. Acta* **283**:492.

Cohen-Bazire, G., Sistrom, W. R., and Stanier, R. Y., 1957, Kinetic studies of pigment synthesis by non-sulfur purple bacteria, *J. Cell. Comp. Physiol.* **49**:25.

Crounse, J. B., Feldman, R. P., and Clayton, R. K., 1963a, Accumulation of polyene precursors of neurosporene in mutant strains of *Rhodopseudomonas spheroides*, *Nature (London)* **198**:1227.

Crounse, J., Sistrom, W. R., and Nemser, S., 1963b, Carotenoid pigments and the *in vivo* spectrum of bacteriochlorophyll, *Photochem. Photobiol.* **2**:361.

Dierstein, R., 1975, Zur Regulation des Ausbildung des Photosyntheseapparates von *R. capsulata* in kontinuierlichen Kulturen, Dissertation, Universität Freiburg.

Gibbs, S. P., Sistrom, W. R., and Worden, P. B., 1965, The photosynthetic apparatus of *Rhodospirillum molischianum*, *J. Cell Biol.* **26**:395.

Griffiths, M., and Stanier, R. Y., 1956, Some mutational changes in the photosynthetic pigment system of *Rhodopseudomonas* spheroides, *J. Gen. Microbiol.* **14**:698.

Hall, R. L., Kung, M. C., Fu, M., Hales, B. J., and Loach, P. A., 1973, Comparison of phototrap complexes from chromatophores of *Rhodospirillum rubrum*, *Rhodopseudomonas* spheroides and the R-26 mutant of *Rhodopseudomonas* spheroides, *Photochem. Photobiol.* **18**:505.

Holt, S. C., and Marr, A. G., 1965, Effect of light intensity on the formation of intracytoplasmic membranes in *Rhodospirillum rubrum*, *J. Bacteriol.* **89**:1421.

Ketchum, P. A., and Holt, S. C., 1970, Isolation and characterization of the membranes from *Rhodospirillum rubrum*, *Biochim. Biophys. Acta* **196**:141.

Lien, S., San Pietro, A., and Gest, H., 1971, Mutational and physiological enhancement of photosynthetic energy conversion in *Rhodopseudomonas capsulata*, *Proc. Natl. Acad. Sci. U.S.A.* **68**:1912.

Lien, S., Gest, H., and San Pietro, A., 1973, Regulation of chlorophyll synthesis in photosynthetic bacteria, *Bioenergetics* **4**:423.

Nugent, N. A., and Fuller, R. C., 1967, Carotenoid biosynthesis in *Rhodospirillum rubrum*: Effect of pteridine inhibitor, *Science* **158**:922.

Raymond, J., and Sistrom, W. R., 1967, The isolation and preliminary characterization of a halophilic photosynthetic bacterium, *Arch. Mikrobiol.* **59**:255.

Segen, B. J., and Gibson, K. D., 1971, Deficiencies of chromatophore proteins in some mutants of *Rhodopseudomonas* spheroides with altered carotenoids, *J. Bacteriol.* **105**:701.

Sistrom, W. R., 1966, The spectrum of bacteriochlorophyll *in vivo*: Observations on mutants of *Rhodopseudomonas* spheroides unable to grow photosynthetically, *Photochem. Photobiol.* **5**:845.

Sistrom, W. R., and Clayton, R. K., 1964, Studies on a mutant of *Rhodopseudomonas* spheroides unable to grow photosynthetically, *Biochim. Biophys. Acta* **88**:61.

Sistrom, W. R., Griffiths, M., and Stanier, R. Y., 1956, The biology of a photosynthetic bacterium which lacks colored carotenoids, *J. Cell. Comp. Physiol.* **48**:473.

Sistrom, W. R., Ohlsson, B. M., and Crounse, J., 1963, Absence of light-induced absorbancy changes in a mutant of *Rhodopseudomonas spheroides* unable to grow photosynthetically, *Biochim. Biophys. Acta* **75**:285.

Straley, S. C., Parson, W. W., Mauzerall, D. C., and Clayton, R. K., 1973, Pigment content and molar extinction coefficients of photochemical reaction center from *Rhodopseudomonas spheroides*, *Biochim. Biophys. Acta* **305**:597.

Takemoto, J., and Lascelles, J., 1974, Function of membrane proteins coupled to bacteriochlorophyll synthesis, *Arch. Biochem. Biophys.* **163**:507.

Takemoto, J., and Huang Kao, M. Y. C., 1977, Effects of incident light levels on photosynthetic membrane polypeptide composition and assembly in *Rhodopseudomonas sphaeroides*, *J. Bacteriol.* **129**:1102.

Vrendenberg, W. J., and Amesz, J., 1966, Absorption characteristics of bacteriochlorophyll types in purple bacteria and efficiency of energy transfer between them, *Brookhaven Symp. Biol.*, No. 19, p. 49.

Wassink, E. C., Katz, E., and Dorrenstein, R., 1939, Infrared absorption spectra of various strains of purple bacteria, *Enzymologia* **7**:113.

Worden, P. B., and Sistrom, W. R., 1964, The preparation and properties of bacterial chromatophore fractions, *J. Cell Biol.* **23**:135.

CHAPTER 45

Respiration vs. Photosynthesis

Donald L. Keister

1. Introduction

Under natural conditions, photosynthetic micro-organisms receive light during only part of the solar day, and even if their growth is limited to this period, it is to be expected that they display some metabolism during dark periods. Under aerobic conditions, the purple bacteria, including the obligate anaerobes of the purple sulfur bacteria (Chromatiaceae), exhibit the ability to take up oxygen (Gaffron, 1933; Roelofsen, 1935; Nakamura, 1937). With, however, the one exception of *Thiocapsa roseopersicina*, which is discussed below, none of the Chromatiaceae can grow in the dark. On the other hand, several species of non-sulfur bacteria (Rhodospirillaceae) can grow in the dark aerobically. No respiration has been observed with the green bacteria with the exception of the unique *Chloroflexus aurantiacus* (see Chapter 9).

Ecologically, the ability to grow in the dark is of little importance, since these organisms are not found thriving under natural conditions unless light is periodically available to them. Thus, the aerobic growth is in essence a physiological potentiality that is demonstrable in the laboratory with pure cultures, but is without ecological significance. This potential enables us, however, to understand how the purple bacteria can survive in natural environments in which there is little or no light penetration.

Donald L. Keister · Charles F. Kettering Research Laboratory, Yellow Springs, Ohio 45387

2. Aerobic and Microaerophilic Growth

Eight species of *Rhodopseudomonas* and *Rhodospirillum*—*Rp. palustris*, *Rp. capsulata*, *Rp. sphaeroides*, *Rp. gelatinosa*, *Rp. acidophila*, *Rp. sulfidophila*, *Rs. rubrum*, and *Rs. tenue*—can grow aerobically in the dark at normal or slightly reduced oxygen partial pressures. In addition, Pfennig (1970) recently demonstrated that five other species that had previously been considered to be strict anaerobes can be grown in the dark under microaerophilic conditions: *Rp. viridis*, *Rs. fulvum*, *Rs. molischianum*, *Rs. photometricum*, and *Rhodomicrobium vannielii*. More surprisingly, he found that one species of purple sulfur bacteria, *Thiocapsia roseopersicina*, could also develop organotrophically under microaerophilic conditions in the dark. This species is the only member of the purple sulfur bacteria for which this physiological property has been demonstrated. The *Rhodospirillum* species grew only when fully pigmented and occurring 3–5 mm below the surface of agar cultures. *Rhodopseudomonas viridis* and *Rm. vannielii* grew pigmented below the surface and colorless up to the surface. Pfennig postulated that the two different modes of growth indicated the occurrence of two different oxidative energy-yielding mechanisms in the dark: (1) via the photosynthetic electron-transport chain in the case of the organisms that grew only when pigmented; and (2) via an oxidative electron-transport chain the the case of those organisms that grew colorless. This will be discussed in more detail below.

Some strains of the facultative anaerobes behave like true anaerobes when initially isolated, and can be grown only in the complete absence of oxygen. By repeated transfers, the physiological characteristics of these strains are modified so that ultimately they become able to develop aerobically (van Niel, 1941).

The media that will permit development in the dark are essentially the same as those in which growth can occur in the light. The growth factor requirements are unchanged. The major difference in metabolism is that under *anaerobic conditions* in the light, carbon dioxide serves as the hydrogen acceptor for substrate oxidation, and this reduced carbon is subsequently utilized in the production of cell materials. When oxygen is the hydrogen acceptor, a large proportion of the organic substrate is decomposed with the liberation of carbon dioxide, and the yield of cell material from the same amount of substrate is much smaller than when growth occurs anaerobically in the light.

3. Biochemical Basis of Obligate Phototrophy

Although several species of the Rhodospirillaceae have been considered to be obligate phototrophs, the demonstration by Pfennig (1970) that these species can grow in the dark under microaerophilic conditions removes them from this category. Only the Chromatiaceae and Chlorobiaceae are true obligate anaerobic phototrophs, but as noted above, one species of Chromatiaceae (*Thiocapsa roseopersicina*) and the green bacterium *Chl. aurantiacus* have been shown to grow in the dark. Further studies are therefore needed to determine whether other representatives of these groups also have this property.

Many species of the Chromatiaceae are found in nature in extremely turbid environments in which they must reside close to the surface to obtain light. Thus, they are frequently exposed to oxygen. Hurlbert (1967) found that with *Chromatium vinosum*, oxygen had no effect on the viability or motility of the cell for periods up to 21 hr. Thus, oxygen is bacteriostatic for *Chromatium*, not bactericidal as it is for some of the non-photosynthetic obligate anaerobes. The observation that motility was maintained in the presence of oxygen implies that energy metabolism was not grossly impaired, and yet *Chromatium* cannot grow in the dark under aerobic or anaerobic conditions. Why? (*Chromatium* cannot grow aerobically in the light either, as will be discussed later.)

There are potentially three classes of ATP-generating processes available to the Chromatiaceae: (1) aerobic respiration; (2) fermentation; and (3)

photophosphorylation. Let us examine the first two to see whether we can establish the basis for obligate phototrophy. As mentioned earlier, the Chromatiaceae have been shown to take up oxygen when supplied with substrate. The rates of respiration by the purple bacteria as a group are low. The rates of oxygen uptake by *Chr. vinosum* strain D are therefore low, but with either sulfide or thiosulfate, are only slightly less than the rate of oxygen uptake by *Rp. spheroides* with malate as substrate (Gibson, 1967), an organism that can grow aerobically. The rate of oxidation of malate by *Chr. vinosum* was about one-half that of thiosulfate. The rate of thiosulfate oxidation is as much as one-fifth that obtained with *Thiobacillus neapolitanus* (Gibson, 1967), a chemoautotroph that can grow with a doubling time of 1.5 hr. The pathway of thiosulfate oxidation may be similar (Smith, A. J., and Lascelles, 1966) in these organisms. The low rates of oxygen uptake, therefore, are not a sufficient explanation for the lack of aerobic growth.

Chromatiaceae do not have typical cytochrome-oxidase-type pigments, and oxidative phosphorylation has not been directly demonstrated in particles prepared from this class of bacteria. Gibson (1967), however, mentions that P/O ratios of about 1.0 were obtained in some experiments with whole cells in which intracellular ATP was measured, suggesting that *Chr. vinosum* may have some capacity for oxidative phosphorylation. This ratio is much lower than that obtained by Ramirez and Smith (1968) with whole cells of *Rs. rubrum* (P/O about 2.8), which can grow aerobically. A possible explanation for this low efficiency may be revealed by examining the current knowledge of the cytochrome region of the electron-transfer chain of *Chr. vinosum*. Olson and Chance (1960) and Morita *et al.* (1965) found two cytochrome components that were autooxidizable (see also Cusanovich *et al.*, 1968). The spectral characteristics of these pigments correspond closely to those of cytochromes isolated by Bartsch and Kamen (1960), which had midpoint potentials of about 0 V. The autooxidizable components become oxidized in starved-cell suspensions and reduced when substrate is added. In comparison with organisms that have typical electron-transport systems containing cytochromes *b*, *c* and *a* in which P/O ratios of 2 are found during electron transport between components with a redox potential of 0 V and oxygen, oxidation by components with 0-V midpoint potentials is inefficient in energy conversion.

These considerations led Gibson (1967) to conclude that no single factor could be pinpointed for the failure of *Chr. vinosum* to grow aerobically in the dark, but rather that a combination of factors including low rates of oxygen uptake and a low efficiency of aerobic phosphorylation must be responsible.

The second potential energy-generating process is anaerobic fermentation. van Gemerden (1968) demonstrated that in *Chromatium*, storage carbohydrate that was synthesized in the light was converted to poly-β-hydroxybutyric acid (PHB) in the dark. During this conversion, intracellular elemental sulfur was reduced to sulfide and CO_2 was released. van Gemerden postulated that storage carbohydrate was broken down to pyruvate via the Embden–Meyerhof pathway, followed by synthesis of PHB via acetyl-CoA and acetoacetyl-CoA. This conversion would yield 3 ATP per glycosyl unit. The PHB apparently was not metabolized further. This process could provide the cell with energy for maintenance and motility in the dark, but the energy yield is limited to the amount of storage carbohydrate available.

The failure of *Chromatium* to grow aerobically in the light can be explained by the effect of oxygen on bacteriochlorophyll (Bchl) synthesis. As Cohen-Bazire *et al.* (1957) found with the Rhodospirillaceae, oxygen inhibits Bchl synthesis in *Chromatium* (Hurlbert, 1967). This effect of oxygen is in itself sufficient to explain the obligate anaerobic phototrophic characteristics of *Chromatium*.

With *Rs. molischianum*, a Rhodospirillaceae that cannot grow for long in the presence of air, Sistrom (1965) found that the only immediate effect of oxygen was to inhibit Bchl and carotenoid synthesis. Growth as measured by an increase in optical density (unfortunately the viable count was not measured) and protein synthesis continued in the light until the inhibition of Bchl synthesis led to the inhibition of growth. With *Chr. vinosum*, Hurlbert (1967) found that cell division but not protein synthesis was inhibited immediately. Although oxygen had some effect on pyruvate and thiosulfate utilization, Hurlbert concluded that these effects were not striking enough to account for the immediate inhibition of cell division. He therefore suggested that oxygen may inhibit some essential cellular process such as DNA synthesis. Without specific data, it is difficult to comment on this suggestion.

There are, however, several hypotheses that could account for the oxygen sensitivity of obligately anaerobic bacteria. These were discussed in depth recently by Morris (1975) and will be summarized here:

1. Oxygen is a toxic agent. Since members of the purple bacteria that have been studied retain viability for extended periods in the presence of oxygen, this hypothesis does not apply.

2. The presence of oxygen prevents the attainment and maintenance of a low redox potential (E_h), which is required for growth. Under natural conditions, Chromatiaceae have been observed to grow at E_h values ranging from -230 to $+320$ mV (Baas-Becking and Wood, 1955). Other work has shown that the presence or absence of oxygen is seemingly more important than the actual E_h of the medium (Kondrat'eva, 1965). It is a fact that a change in the E_h of the media is characteristic of the growth of the purple bacteria, but this depends more on the substrate than the organism. These observations seem to exclude E_h as the dominant factor in growth.

3. Preferential reduction of oxygen consumes the cell's reducing power unproductively, so that insufficient remains for necessary biosyntheses. This hypothesis suggests that NAD(P)H is oxidized more rapidly by oxygen than it can be generated, thereby unbalancing the cells' pool of reducing power. No information is available on this in the Chromatiaceae, although the low P/O ratio found in whole cells (Gibson, 1967) indicates that this may be true.

4. Oxygen controls cellular activity by determining the intracellular concentration or state of a key metabolic regulator such as a redox couple liable to direct oxidation by oxygen or in equilibrium with another similar couple. This hypothesis was originally proposed to account for the control of Bchl synthesis by oxygen. Cohen-Bazire *et al.* (1957) and recently Davies *et al.* (1973) suggested that the redox state of some cellular component was the factor controlling Bchl synthesis, although several papers suggested that the ATP level was more important (Fanica-Gaignier, 1971; Schmidt and Kamen, 1971). In contrast, Oelze and Kamen (1971) could not demonstrate a regulating function of ATP during the transition from aerobic to anaerobic conditions. A detailed discussion of this hypothesis can be found in the preceding chapters of this book and in Gest (1972).

5. The cell contains key components bearing free —SH groups the oxidation of which by oxygen to the —S–S— form halts growth and metabolism. No studies are available on this potentiality in the Chromatiaceae.

In summary, the studies to date do not account for a specific mechanism to explain the immediate inhibition of growth by oxygen in the light observed with *Chr. vinosum*. In the dark, it is probably a combination of factors such as low respiratory rates coupled with a poor phosphorylation efficiency that prevents growth. Therefore, since oxygen inhibits Bchl synthesis, the Chromatiaceae are obligate anaerobic (or microaerophilic) phototrophs.

4. Effect of Light on Respiration: Interaction of the Photochemical and Respiratory Electron-Transport Chains

Before it was established that the purple bacteria do not evolve oxygen during photosynthesis, Nakamura (1937) observed that the rate of oxygen uptake by *Rp. palustris* was decreased by illumination. He interpreted this as evidence for the photochemical production of oxygen by the bacteria, not as a decreased oxygen consumption. Van Niel (1941) and many subsequent authors also observed this phenomenon in the nonsulfur bacteria. Complete suppression of oxygen uptake is frequently observed in *Rs. rubrum*, while only partial inhibition is usually observed in *Rp. sphaeroides* and *Rp. capsulata* (Clayton, 1955). I have occasionally observed only partial inhibition with *Rs. rubrum*. By treating the cells with short sequential periods of light and dark, however, complete inhibition can be attained (Keister, unpublished). This change from partial to complete inhibition is accompanied by an increase in the dark respiration, a phenomenon also observed by Fork and Goedheer (1964) and Ramírez and Smith (1968). The diminished utilization of oxygen in the light is not accompanied by an inhibition of growth, and growth continues essentially at the same rate in air as under anaerobiosis (Cohen-Bazire *et al.*, 1957). The utilization of substrates can even remain unchanged (van Niel, 1941; Morita, 1955), for van Niel demonstrated that acetate utilization was the same whether suspensions were illuminated in air or nitrogen, while oxygen uptake was completely suppressed in the former instance. This is not necessarily so, however, for Clayton (1955) found a different rate of succinate utilization in light and dark. Furthermore, Clayton demonstrated that the relative amount of substrate utilized via either respiration or photosynthesis depended on both the light intensity and the oxygen tension. The preferential mode of substrate utilization was through photosynthesis. Van Niel correctly interpreted these observations by suggesting that light was inhibiting respiration. Conclusive proof came in 1954 when Johnson and Brown (1954) demonstrated that $^{18}O/^{16}O$ ratios did not change during illumination of *Rs. rubrum*.

The light effect apparently was localized to the chromatophore when Katoh (1961) found that succinate oxidation was inhibited by light in cell-free extracts of *Rp. palustris*. [White and Vernon (1958) reported that high light intensities inhibited NADH oxidase activity in *Rs. rubrum* particles, but this was an irreversible phenomenon and is probably not related to the physiological effect of light.] In apparent contradiction, Horio and Kamen (1962*b*) and Kikuchi *et al.*

(1964) reported that light stimulated succinate oxidation in *Rs. rubrum* chromatophores, while L. Smith (1959) reported little effect of light. Somewhat later, my laboratory demonstrated that light inhibited NADH oxidation in *Rs. rubrum* in a reversible manner (Thore *et al.*, 1969), but stimulated succinate oxidation (Keister and Minton, 1971) as previously reported. Only Katoh's observation on the inhibition of succinate oxidation, which was done in *Rp. palustris*, remains unconfirmed.

Hence, there is no doubt that the diminished uptake of oxygen in the light is the result of the interaction between respiratory and photosynthetic processes. White and Vernon (1958) considered the possibility that this effect could be due to photophosphorylation. If oxidation is tightly coupled to phosphorylation as in mammalian mitochondria, the competition between photophosphorylation and oxidative phosphorylation for ADP and phosphate could explain the inhibition of respiration by light. In agreement with this idea, Ramírez and Smith (1968) found that 2-*n*-heptyl-4-hydroxyquinoline-*N*-oxide (HQNO) and carbonyl-cyanide trifluoromethoxyphenylhydrazine (FCCP) prevented the light-inhibition of respiration in whole cells of *Rs. rubrum*. HQNO and FCCP are well known as an inhibitor of photosynthetic electron transport and an uncoupler of phosphorylation, respectively, and therefore the finding of Ramírez and Smith supports this explanation of respiratory inhibition. Keister and Minton (1971) also observed the relief of inhibition of respiration by light by these compounds, but interpreted the results in a different way (see below).

A second possible mechanism of the light effect is that photosynthetic electron transport and oxidative electron transport compete for common hydrogen or electron donors (van Niel, 1941; Clayton, 1955; Morita, 1955; Horio and Kamen, 1962*b*; Fork and Goedheer, 1964), a concept that was discussed extensively by Horio and Kamen (1962*b*). From their work, they suggested that a single oxidation chain of electron carriers was adequate to account for both photosynthetic and respiratory electron transport. Ramírez and Smith (1968) and Welsch and Smith (1969), however, postulated two separate electron-transport chains as being more consistent with their observations on ATP synthesis in intact photosynthetically grown cells. The recent studies of Pfennig (1970) illustrating that the brown species of *Rhodospirillum*, i.e., *fulvum*, *molischianum*, and *photometricum*, which were previously considered to be obligate anaerobes, can grow in the dark provided the oxygen tension is low enough to permit pigment synthesis is not compatible with the idea of two separate

Table 1. Effect of Various Compounds on NADH Oxidation in the Light in Particles of Anaerobically Grown Cells[a]

Compound added	$-\Delta A_{340} \times 10$		Effect of light (% of dark)
	Dark	Light	
None	1.51	0.99	−35
Antimycin A (0.3 μM)	1.49	1.70	+14
Cl-CCP (3.0 μM)	1.89	1.51	−20
HHNQ (0.1 μM)	1.58	1.12	−29
Anti-A + Cl-CCP	1.91	2.86	+50
KCN (10 mM)	0.81	0.66	−19
KCN + anti-A	0.93	1.18	+27
KCN + Cl-CCP	0.84	0.75	−11
KCN + anti-A + Cl-CCP	0.81	2.99	+370
KCN + anti-A + Cl-CCP + HHNQ	0.82	1.17	+43

[a] From Thore *et al.* (1969). NADH oxidation was measured by the decrease in A_{340}. Cuvettes contained 17 μg Bchl/ml and were illuminated with red light of 1.2×10^5 ergs cm^{-2} sec^{-1} intensity.

electron-transport chains in these particular species. Pfennig (personal communication) supposes that the bacteria obtain ATP by oxidative phosphorylation via the photosynthetic electron-transport chain.

A third mechanism to account for the light inhibition of respiration resulted from the work of Thore *et al.* (1969) and Keister and Minton (1971). These authors postulated that a light-induced change in the oxidation–reduction state of a sensitive electron carrier causes the inhibition. Thore *et al.* (1969) found that light reversibly inhibited NADH oxidation in chromatophores prepared from light-grown cells of *Rs. rubrum*. The results in Table 1 show that the light-inhibition was only partially relieved by *m*-chlorocarbonyl cyanide phenylhydrazone (Cl-CCP) or antimycin A (or HQNO). The combination of these two compounds (an uncoupler of phosphorylation and an inhibitor of electron transport) not only prevented the

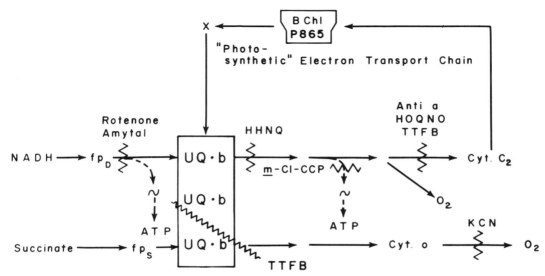

Fig. 1. Photosynthetic and respiratory electron-transport systems of *Rhodospirillum rubrum*.

Table 2. Effect of Light on the Oxidation of NADH by Fumarate[a]

Additions	$-\Delta A_{340} \times 10$		Inhibition by light (%)
	Dark	Light	
None	1.10	0.64	42
Cl-CCP (3 μM)	1.43	0.90	37
Antimycin A (0.3 μM)	1.05	0.98	7
Antimycin A + Cl-CCP	1.44	1.44	0

[a] From Keister and Minton (1971). NADH oxidation was measured anaerobically at pH 7.4. The reaction mixture contained 50 mM HEPES, 0.1 mM NADH, 1.7 mM fumarate, and 18 μg Bchl/ml. The red light intensity was 1.3×10^5 ergs cm^{-2} sec^{-1}.

light-inhibition, but also synergistically stimulated oxidation in the light. There was no synergistic stimulation in the dark. Furthermore, when cyanide was present to inhibit the dark oxidation, the combination of these inhibitors resulted in a fourfold stimulation of NADH oxidation in the light. The stimulation was reduced by another photosynthetic electron transport inhibitor, 2-hydroxy-3-(ω-cyclohexyloctyl)-1,4-naphthoquinone (HHNQ). This inhibitor totally in-

hibits photosynthetic phosphorylation (Keister and Yike, 1967), but has only a slight effect on NADH (or succinate) oxidation in the dark. Therefore, the light-stimulated pathway of oxygen uptake must occur via the photosynthetic pathway. These concepts are illustrated in Fig. 1. The site of the light-inhibition of NADH oxidation appears to be in the ubiquinone–cytochrome b segment of the electron-transport chain, for, as is shown in Table 2, light inhibited the oxida-

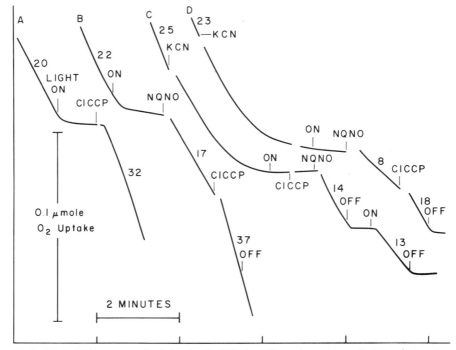

Fig. 2. Effect of light on respiration in intact cells of *Rhodospirillum rubrum*. Intact cells (10 mg/ml based on wet wt.) respiring on endogenous substrate were suspended in 50 mM HEPES, pH 7.4, and oxygen uptake was measured polarographically. Cl-CCP was 3 μM, NQNO was 2 μM, KCN was 1 mM, and the light intensity was 1×10^5 ergs cm^{-2} sec^{-1}. Reproduced from Keister and Minton (1971).

tion of NADH by fumarate under anaerobic conditions approximately the same amount as when oxygen was the oxidant. Antimycin A but not Cl–CCP prevented the inhibition.

It could be argued that these observations were made with chromatophores and therefore may or may not represent the physiological conditions of the intact cell. As shown in Fig. 2, however, the light-stimulated respiration observed in the presence of Cl-CCP and antimycin A in chromatophores can also be observed in whole cells [2-nonyl-4-hydroxyquinoline-*N*-oxide (NQNO) was used, for antimycin A was not effective in whole cells]. Further, this pathway is cyanide-insensitive (Fig. 2), and based on the known site of action of these inhibitors, must occur via the photosynthetic electron-transport chain. Succinate oxidation can also occur via the same cyanide-insensitive pathway (Keister and Minton, 1971). In contrast to the NADH oxidation, succinate oxidation is stimulated by light, and Cl-CCP and antimycin A have little effect.

The mechanism of these effects is not totally explainable with our present knowledge of the components of the electron-transport chain. The most likely explanation, however, would seem to be a light-induced change in the oxidation–reduction state of a sensitive carrier or carriers. Electron transport *in vitro* is quite sensitive to the redox state of the carriers, and either an overoxidizing or an overreducing environment inhibits photophosphorylation (Horio and Kamen, 1962a; Bose and Gest, 1963). The observation that the anaerobic oxidation of NADH by fumarate was inhibited by light to the same extent as the oxidation by oxygen seems to localize the site of inhibition to the dehydrogenase segment of the chain. To explain the light-stimulated oxidation of NADH in the presence of antimycin A and Cl-CCP, Thore *et al.* (1969) postulated that blocking electron transport with antimycin A caused the reduction of an auto-oxidizable carrier just prior to where cytochrome c_2 is shown in Fig. 1 and thereby stimulated electron flow to oxygen. Cl-CCP was required to relieve the rate-limiting phosphorylation site in this segment of the electron-transport chain. The attractiveness of this postulate was diminished when Keister and Minton (1971) found that light stimulated succinate oxidation; Cl-CCP and antimycin A had little effect. The rate of succinate oxidation, however, is about one-half that of NADH, and the potentials of NADH and succinate are quite different. Thus, the steady-state oxidation level of ubiquinone and cytochrome *b* is probably quite different in the presence of these reductants; even different pools of these carriers may be involved (Parson, 1967). It is therefore perhaps not surprising that light has different effects on the oxidation of these components.

To conclude, light inhibits NADH oxidation in chromatophores, and the site of the effect is probably in the dehydrogenase segment of the electron-transport chain. Under certain conditions in the light, oxidation of NADH and succinate occurs predominately via the photosynthetic system. This observation, coupled with the findings of Thore *et al.* (1969) that the only electron-transport-associated activity that increases during the transition of *Rs. rubrum* from light-growth to dark-growth is the terminal oxidase (probably cytochrome *o*), suggests that the dehydrogenase segment of the electron-transport chain is common to both the photosynthetic and the respiratory system. They diverge at the oxidase level. (A different opinion based on studies with intact cells can be found, however, in Chapter 33.) Whether the light effect observed in chromatophores is sufficient to account for the total inhibition of respiration observed in intact cells is debatable, and it may be that the removal of phosphate acceptors due to photophosphorylation also plays a role. Certainly, further studies are required to elucidate this interesting physiological effect of light.

5. References

Baas-Becking, L. G. M., and Wood, E. J., 1955, Biological processes in the estuarine environment. I and II. Ecology of the sulfur cycle, *Proc. K. Ned. Akad. Wet. Ser. B* **3**:160.

Bartsch, R. G., and Kamen, M. D., 1960, Isolation and properties of two soluble heme proteins in extracts of the photoanaerobe *Chromatium*, *J. Biol. Chem.* **235**:825.

Bose, S. K., and Gest, H., 1963, Bacterial photophosphorylation: Regulation by redox balance, *Proc. Natl. Acad. Sci. U.S.A.* **49**:337.

Clayton, R. K., 1955, Competition between light and dark metabolism in *Rhodospirillum rubrum*, *Arch. Mikrobiol.* **22**:195.

Cohen-Bazire, G., Sistrom, W. R., and Stanier, R. Y., 1957, Kinetic studies of pigment synthesis by non-sulfur purple bacteria, *J. Cell. Comp. Physiol.* **49**:25.

Cusanovich, M. A., Bartsch, R. G., and Kamen, M. D., 1968, Light-induced electron transport in *Chromatium* strain D. II. Light-induced absorbance changes in *Chromatium* chromatophores, *Biochim. Biophys. Acta* **153**:397.

Davies, R. C., Gorchein, A., Neuberger, A., Sandy, J. D., and Tait, G. H., 1973, Biosynthesis of bacteriochlorophyll, *Nature* (*London*) **245**:15.

Fanica-Gaignier, M., 1971, Control of bacteriochlorophyll synthesis by adenine nucleotides in *Rhodopseudomonas spheroides* Y, in: *Proceedings of the IInd International*

Congress on Photosynthesis Research (G. Forti, M. Avron, and A. Melandri, eds.), pp. 2721–2726, Dr. W. Junk, The Hague.

Fork, D. C., and Goedheer, J. C., 1964, Studies on light-induced inhibition of respiration in purple bacteria: Action spectra for *Rhodospirillum rubrum* and *Rhodopseudomonas spheroides, Biochim. Biophys. Acta* **79**:249.

Gaffron, H., 1933, Über die Stoffwechsel der Schwefelfreien Purpurbakterien, *Biochem. Z.* **260**:1.

van Gemerden, H., 1968, On the ATP generation by *Chromatium* in darkness, *Arch. Mikrobiol.* **64**:118.

Gest, .H., 1972, Energy conversion and generation of reducing power in bacterial photosynthesis, *Adv. Microb. Physiol.* **7**:243.

Gibson, J., 1967, Aerobic metabolism of *Chromatium* sp. strain D, *Arch. Mikrobiol.* **59**:104.

Horio, T., and Kamen, M. D., 1962a, Optimal oxidation–reduction potentials and endogenous co-factors in bacterial photophosphorylation, *Biochemistry* **1**:144.

Horio, T., and Kamen, M. D., 1962b, Observations on the respiratory system of *Rhodospirillum rubrum, Biochemistry* **1**:1141.

Hurlbert, R. E., 1967, Effect of oxygen on viability and substrate utilization in *Chromatium, J. Bacteriol.* **93**:1346.

Johnson, J. A., and Brown, A. H., 1954, The effect of light on the oxygen metabolism of *Rhodospirillum rubrum, Plant Physiol.* **29**:177.

Katoh, S., 1961, Inhibitory effect of light on oxygen-uptake by cell-free extracts and particulate fractions of *Rhodopseudomonas palustris, J. Biochem.* (*Tokyo*) **49**:126.

Keister, D. L., and Minton, M. J., 1971, Effect of light on respiration in *Rhodospirillum rubrum* chromatophores, in: *Energy Transduction in Respiration and Photosynthesis* (E. Quagliariello, S. Papa, and C. S. Rossi, eds.), pp. 375–384, Adriatica Editrice, Bari, Italy.

Keister, D. L., and Yike, N. J., 1967, Energy-linked reactions in photosynthetic bacteria. II. The energy-dependent reduction of NADP$^+$ by NADH in chromatophores of *Rhodospirillum rubrum, Biochemistry* **6**:3847.

Kikuchi, G., Yamada, H., and Sato, H., 1964, Light-stimulation of the respiratory activity of *Rhodospirillum rubrum* chromatophores, *Biochim. Biophys. Acta* **79**:446.

Kondrat'eva, E. N., 1965, *Photosynthetic Bacteria*, Israel Program for Scientific Translations, Jerusalem.

Morita, S., 1955, The effect of light on the metabolism of lactic acid by *Rhodopseudomonas palustris, J. Biochem.* (*Tokyo*) **42**:533.

Morita, S., Edwards, M., and Gibson, J., 1965, Influence of metabolic conditions on light-induced cytochrome changes in *Chromatium* D, *Biochim. Biophys. Acta* **109**:45.

Morris, J. G., 1975, The physiology of obligate anaerobiosis, *Adv. Microb. Physiol.* **12**:169.

Nakamura, H., 1937, Über die Photosynthese bei der schwefelfreien Purpurbakterie *Rhodobazillus palustris.* Beiträge zur Stoffwechselphysiologie der Purpurbakterie. I, *Acta Phytochim.* **9**:189.

Oelze, J., and Kamen, M. D., 1971, Adenosine triphosphate cellular levels in *Rhodospirillum rubrum* during transition from aerobic to anaerobic metabolism, *Biochim. Biophys. Acta* **234**:137.

Olson, J. M., and Chance, B., 1960, Oxidation–reduction reactions in the photosynthetic bacterium *Chromatium.* I. Absorption spectrum changes in whole cells, *Arch. Biochem. Biophys.* **88**:26.

Parson, W. W., 1967, Observations on the changes in ultraviolet absorbance caused by succinate and light in *Rhodospirillum rubrum, Biochim. Biophys. Acta* **143**:263.

Pfennig, N., 1970, Dark growth of phototrophic bacteria under microaerophilic conditions, *J. Gen. Microbiol.* **61**:ii.

Ramirez, J., and Smith, L., 1968, Synthesis of adenosine triphosphate in intact cells of *Rhodospirillum rubrum* and *Rhodopseudomonas sphaeroides* on oxygenation or illumination, *Biochim. Biophys. Acta* **153**:466.

Roelofsen, P. A., 1935, On photosynthesis of the Thiorhodaceae, Dissertation, Utrecht.

Schmidt, G. L., and Kamen, M. D., 1971, Control of chlorophyll synthesis in *Chromatium vinosum, Arch. Mikrobiol.* **76**:51.

Sistrom, W. R., 1965, Effect of oxygen on growth and the synthesis of bacteriochlorophyll in *Rhodospirillum molischianum, J. Bacteriol.* **89**:403.

Smith, A. J., and Lascelles, J., 1966, Thiosulfate metabolism and rhodanese in *Chromatium* sp. strain D, *J. Gen. Microbiol.* **42**:357.

Smith, L., 1959, Reactions of *Rhodospirillum rubrum* extract with cytochrome *c* and cytochrome c_2, *J. Biol. Chem.* **234**:1571.

Thore, A., Keister, D. L., and San Pietro, A., 1969, Studies on the respiratory system of aerobically (dark) and anaerobically (light) grown *Rhodospirillum rubrum, Arch. Mikrobiol.* **67**:378.

van Niel, C. B., 1941, The bacterial photosyntheses and their importance for the general problem of photosynthesis, *Adv. Enzymol.* **1**:263.

Welsch, F., and Smith, L., 1969, Kinetics of synthesis and utilization of adenosine triphosphate by intact cells of *Rhodospirillum rubrum, Biochemistry* **8**:3403.

White, F. G., and Vernon, L. P., 1958, Inhibition of reduced diphosphopyridine nucleotide oxidase activity of *Rhodospirillum rubrum* chromatophores upon illumination, *J. Biol. Chem.* **233**:217.

CHAPTER 46

Fermentative Metabolism and Growth of Photosynthetic Bacteria

R. L. Uffen

1. Introduction

Anoxygenic phototrophic bacteria represent a large group of microorganisms that share a common property: the ability to grow anaerobically in light and photosynthesize. In addition, some Rhodospirillaceae grow aerobically in darkness. As a result, cells are widely distributed in the natural environment occupying anaerobic, light and aerobic zones (Kondrat'eva, 1965; Pfennig, 1967). It was recognized some time ago, however, that when nonproliferating suspensions of Rhodospirillaceae or Chromatiaceae were placed into anaerobic, dark conditions, cells continued to metabolize. Yet, because anaerobic, dark growth could not be demonstrated, these reactions received only limited attention and, as a result, their significance was not fully appreciated. With the recent discovery that Rhodospirillaceae species indeed grow under proper anaerobic, dark conditions (Uffen and Wolfe, 1970), it seems important to discuss these data for the first time in a unified presentation.

This chapter is organized as follows: reactions demonstrated in cells after anaerobic transfer into dark, nongrowing conditions are discussed in Section 2.1. Some early studies of this type were also treated in a review by Gest (1951). Metabolic activities in Rhodospirillaceae growing under anaerobic, dark conditions are discussed in Section 2.2.

The literature surveyed in this discussion spans a period of 44 years, beginning with van Niel's publication in 1931 (van Niel, 1931) on enrichment and growth of Chromatiaceae and Chlorobiaceae.

2. Rhodospirillaceae

2.1. Anaerobic, Dark Metabolism

In 1930, it was generally agreed that Rhodospirillaceae metabolized a variety of organic substrates and grew anaerobically in light. The question whether cells could also grow anaerobically in darkness had not been resolved. Certain observations reported by Gaffron (1933) and Roelofsen (1934), however, suggested that such growth might occur, since light-grown cells formed both acid and gas products during anaerobic incubation in darkness. To learn more about this activity and perhaps to discover nutrient conditions to support anaerobic, dark growth, Gaffron and Roelofsen attempted to stimulate dark activity with a variety of organic supplements. Failure to detect significant increase in product formation, however, suggested that the "dark" reaction(s) functioned as a "peculiar" type of endogenous metabolism to provide maintenance energy for cells during dark periods (Muller, 1933; Roelofsen, 1934; van Niel, 1944). It appeared that the substrate or substrates catabolized were converted first into storage material

R. L. Uffen · Department of Microbiology, Michigan State University, East Lansing, Michigan 48824

during anaerobic, light-growth, and were then degraded when cells were placed in darkness. Additional evidence that reserve products were involved was suggested earlier by H. G. Derx in a personal communication to van Niel (1931). Using cultures enriched with photosynthetic bacteria, Derx noted development of iodine-staining granules in cells placed in sunlight during the day. The following morning, the granules were no longer detectable.

Shortly after the reports of Gaffron (1933) and Roelofsen (1934), Nakamura (1937b, 1939, 1941) observed that H_2 gas uptake or evolution could be stimulated by adding a variety of organic compounds to resting suspensions of *Rhodobacillus palustris* (*Rhodopseudomonas palustris*) incubated in anaerobic, dark conditions. Finally, about ten years later, Kohmiller and Gest (1951) reported that anaerobic, dark metabolism in *Rhodospirillum rubrum* cells could be stimulated similarly during incubation with sodium pyruvate.

In view of these results, after van Niel (1944) concluded from careful study that Rhodospirillaceae required light to grow anaerobically, it became increasingly difficult to explain why cells would not grow in darkness as well (Kohlmiller and Gest, 1951; Gest, 1951). Data obtained using resting cell suspensions of *Rs. rubrum* and *Rp. palustris* suggested that this growth might occur, and are discussed in separate sections below.

2.1.1. *Rhodospirillum rubrum*

In 1951, Kohlmiller and Gest (1951) published results of studies on anaerobic, dark metabolism in Rhodospirillaceae. Experiments were performed using resting suspensions of *Rs. rubrum* grown beforehand in a medium supplemented with malate and glutamate under either anaerobic, light conditions or aerobic, dark conditions. Although the ability of cells to "ferment" organic compounds such as lactate, succinate, fumarate, or malate was investigated, only pyruvate was used to any extent. When anaerobic, light-grown cells were suspended in phosphate buffer with sodium pyruvate and incubated anaerobically in darkness, cells evolved CO_2, trace amounts of H_2, and both acetic and propionic acids. Under these conditions, maximum amounts of pyruvate were fermented when CO_2 was allowed to accumulate in reaction vessels. Suspensions of cells grown aerobically in the dark fermented pyruvate similarly, except that substrate utilization was not CO_2-dependent and larger amounts of H_2 gas were formed. In addition, small but significant amounts of buryrate, valerate, and caproate were also detected. Based on these data, Kohlmil-

ler and Gest (1851) concluded that *Rs. rubrum* fermented sodium pyruvate anaerobically in the dark in a fashion similar to that of propionic acid bacteria.

Schön (1968) and Schön and Biedermann (1972, 1973) performed similar experiments using *Rs. rubrum* F1 grown aerobically in the dark and then "adapted" to anaerobic, dark conditions. In their studies, Schön and Biedermann noted that both acetate and propionate accumulated in anaerobic, dark reaction mixtures containing fructose or pyruvate. Moreover, use of ^{14}C-labeled fructose, pyruvate, or bicarbonate indicated that all three compounds were involved in formation of volatile acid products. In experiments with these anaerobic, dark-"adapted" cells, however, the amounts of acetate and propionate formed, particularly with fructose as substrate, were not always in agreement with theoretical values based on knowledge of fermentative metabolism in propionic acid bacteria (Wood, 1961). The authors proposed that the discrepancies might be explained by the substrate and culture conditions used before anaerobic, dark adaptation. Nevertheless, since dark-"adapted" cells formed variable amounts of propionate and since *Rs. rubrum* grown anaerobically in the dark produced only formic and acetic acids from sodium pyruvate (fructose did not support anaerobic, dark fermentative growth), a role for propionate formation in energy generation was unclear (Uffen, 1973a). From these results, it is possible that propionate functions more importantly as an intermediate during photosynthetic cell metabolism. If so, when cells are placed into anaerobic, dark conditions, formation of propionate might result during repression of photosynthetic enzyme system(s) in the early stages of transition to fermentative metabolism (Uffen, 1973a). Under these conditions, the appearance of propionate may also correlate with the transient increase in generation time of *Rs. rubrum* that was observed (Uffen, 1973b) when photosynthetically grown cells were transferred to low light or dark conditions.

When suspensions of anaerobic, light- or aerobic, dark-grown cells were incubated anaerobically in darkness with sodium pyruvate, not only acids but CO_2 plus trace amounts of H_2 gas were formed. With sodium formate, equimolar amounts of CO_2 and H_2 were produced. These results suggested that cells oxidized formate by formate hydrogen-lyase activity, and led to speculation that the volatile acid was an intermediate during degradation of sodium pyruvate (Kohlmiller and Gest, 1951; Bennett *et al.*, 1964; Schön and Biedermann, 1972, 1973). Additional evidence for this view was presented by Jungermann and Schön (1974). In their studies, suspensions of dark-"adapted" whole cells or high concentrations (8–

16 mg protein/ml) of cell-free extract (obtained from cells similarly adapted to darkness) appeared to degrade sodium [^{14}C]pyruvate with production of formate. Since activity was scarcely detectable using cells not treated in darkness beforehand, the authors concluded that the ability to produce the volatile acid by pyruvate formate-lyase activity was an important cellular event that occurred during the dark-adaptation period. From recent studies in our laboratory using fermenting *Rs. rubrum*, the reaction appears to be essential in anaerobic, dark-grown cells (Gorrell and Uffen, 1977). Nevertheless, since Jungermann and Schön (1974) did not measure formate–pyruvate exchange activity, characteristic of the lyase reaction, the functioning of alternate pyruvate-degrading pathway(s) during transition from light to dark anaerobic development could not be excluded. Specifically, pyruvate-ferredoxin oxidoreductase activity might occur in fermenting cells, even after long-term anaerobic dark growth, and also be important. Thus, small amounts of formate, measured in reactions using dark-"adapted" *Rs. rubrum* cells placed in nongrowing conditions, could be synthesized by reactions other than pyruvate formate-lyase using H_2 and CO_2 (Miller and Wolin, 1973; Thauer, 1972, 1973). The role of pyruvate-ferredoxin oxidoreductase during early stages of transition from light to dark, fermentative metabolism warrants further study.

The most important contribution of experiments on metabolic activity in suspensions of nonfermenting *Rs. rubrum* placed into darkness was the observation that light-independent reactions occurred. Once conditions were discovered to support anaerobic, dark *Rs. rubrum* growth (Uffen and Wolfe, 1970), however, it became apparent that energy-yielding fermentative activity could not easily be predicted from experiments with nonfermenting cells. For this reason, studies performed by Schön (1969, 1971) on changes in adenylate and pyridine nucleotide ratios in *Rs. rubrum* exposed to dark, nongrowing conditions will not be discussed. Although the author believes that these relationships may be important during transition to anaerobic, dark development, present speculation without data on changes in cells placed into strict anaerobic dark conditions, where growth can occur, might be misleading.

2.1.2. *Rhodopseudomonas palustris*

Nakamura (1937b, 1941) was first successful in stimulating anaerobic, dark metabolism in Rhodospirillaceae using photoheterotrophically grown *Rp. palustris*. In these studies, stimulation of gas uptake or evolution was measured manometrically. Addition of compounds such as formate, pyruvate, and glucose resulted in increased production of H_2 and CO_2 gases, whereas substrates such as malate, fumarate, L-asparagine, glyceraldehyde, and hexose diphosphate (fructose-1,6-diphosphate) stimulated H_2 uptake. Of particular interest, given the knowledge of anaerobic, dark metabolism in *Rs. rubrum* (Uffen, 1973a), was the ability of *Rp. palustris* to catabolize both formate and pyruvate in darkness.

From data obtained with these substrates, Nakamura (1937b, 1939, 1941) concluded that when photosynthetically grown *Rp. palustris* was placed into anaerobic, dark conditions, sodium pyruvate was degraded to acetate (which he did not report measuring), with formate as an intermediate in gas production. Subsequently, Yoch and Lindstrom (1969) were able to partially purify formate dehydrogenase from *Rp. palustris* after cells were grown photosynthetically in a medium supplemented with small amounts of formate. In their studies, enzymic degradation of formate occurred equally well in light or dark with NAD$^+$ as oxidant. Qadri and Hoare (1968) also measured formate dehydrogenase activity in photosynthetically grown cells. In their studies, hydrogenase present in cell extracts was observed to "couple" with formate dehydrogenase and oxidize formate to H_2 and CO_2. These observations, in support of Nakamura's earlier hypothesis, indicated that *Rp. palustris* developed formic hydrogenlyase activity. Whether cells metabolized sodium pyruvate, however, with formate as an intermediate, as proposed earlier by Nakamura, was unfortunately not established. In these studies, no mention of pyruvate formate-lyase activity was made (Yoch and Lindstrom, 1969; Qadri and Hoare, 1968).

On the contrary, in 1973, Qadri and Hoare (1973) suggested that *Rp. palustris* strain Q degraded pyruvate via a fermentative pyruvate decarboxylase reaction with production of CO_2, acetaldehyde, and acetoin. Similar pyruvate decarboxylase activity was observed in other species of Rhodospirillaceae. One of these species, *Rp. galatinosa*, was recently observed to grow in this laboratory in darkness under an argon atmosphere with sodium pyruvate (unpublished). Unfortunately, products formed during anaerobic dark growth have not been fully characterized, but unlike *Rs. rubrum*, cells did not produce H_2 gas.

As in studies using *Rs. rubrum*, experiments with anaerobic, light-grown *Rhodopseudomonas* spp. suggested that more than one pathway might function in cells to degrade sodium pyruvate. Unfortunately, since laboratory conditions have not been discovered to support rapid anaerobic, dark *Rp. palustris* growth, it is not known whether a pyruvate decarboxylase or pyruvate formate-lyase pathway or both can function to produce energy during anaerobic, dark cell development. Study of pyruvate degradation in anaerobic, dark-grown *Rp. gelatinosa*, an organism closely related to *Rp. palustris*, however, may provide an answer to this question.

2.2. Anaerobic, Dark Growth

2.2.1. *Rhodospirillum rubrum*

Several Rhodospirillaceae species grew slowly in darkness and developed pigmented colonies on the surface of solid medium prepared anaerobically by a modification of the Hungate technique (Uffen and Wolfe, 1970). These species included *Rp. palustris*, *Rp. sphaeroides*, *Rp. viridis*, and *Rs. rubrum*; *Rhomicrobium* species were not tested. Although cells developed slowly in the dark, with a generation time of about 5 days, addition of sodium pyruvate to anaerobic liquid medium stimulated *Rs. rubrum*, and cells required only about 8.6 hr to divide. Consequently, when anaerobic, light-grown *Rs. rubrum* S1 wild-type cells were transferred under strict anaerobic conditions into darkness, protein synthesis continued, cells exhibited longer generation times than light-grown cells, and after about 20 cell divisions, the concentration of bacteriochlorophyll *a* (Bchl *a*) decreased from 28 to about 14 nmol/mg protein. During continued long-term growth in darkness, Bchl *a* concentration slowly diminished to trace amounts ranging between 0.3 and 0.5 nmol/mg protein and generation times decreased (Uffen *et al.*, 1971; Uffen, 1973*a*). It was suggested that loss of pigmentation in anaerobic, dark cultures, however, occurred too slowly to be explained by simple dilution during growth, and cells either exercised inefficient repression or had become heterogenous for pigment production. The latter alternative appeared correct, since it was observed that cells obtained from these "nonpigmented" anaerobic, dark cultures developed a variety of different-colored colonies ranging from light pink to red during growth on solid medium in anaerobic bottles (Uffen and Wolfe, 1970) incubated without light. Subsequently, mutant phenotypes C and G1 were selected from dark-grown colonies on the basis of pigmentation. During dark growth, mutant C produced red-pigmented colonies and synthesized 5–10 nmol Bchl *a*/mg protein. Under similar anaerobic, dark growth conditions, mutant G1 formed light pink-colored colonies and produced only trace amounts of Bchl *a*. Growth properties, fine structure, light-induced oxidation reactions, and metabolism of sodium pyruvate by

mutant phenotypes were studied (Uffen *et al.*, 1971; Uffen, 1973*a*; Gorrell and Uffen, 1977).

(a) **Growth Properties.** Red-pigmented mutant C cells grew equally well aerobically or anaerobically in the dark or anaerobically in the light. Nonpigmented mutant G1, however, did not. While G1 grew both aerobically and anaerobically, most cells were sensitive to light. Only about 1 of 1.5×10^5 G1 cells grown anaerobically in the dark survived in anaerobic, light conditions and formed deep red-pigmented colonies characteristic of photosynthesizing *Rs. rubrum*. On the other hand, when G1 was grown first in aerobic, dark conditions and then transferred to anaerobic conditions in the light, the number of cells capable of growing and forming deep red-pigmented colonies increased 300-fold. Changes that occurred in dark, aerobic cultures of G1 that appeared to favor the development of light-*in*sensitive cells potentially capable of photosynthetic development in the light are a subject of continuing research. It is believed that "potentially photosynthetically competent" cells such as these and light-*in*sensitive, photosynthetically *in*competent cells recently isolated from anaerobic, dark G1 cultures (unpublished) will form the basis of a useful model system to study factors involved during the development of photosynthetic activity in *Rs. rubrum* exposed to light.

(b) **Light-Induced Oxidation Reactions.** Light-*minus*-dark difference spectra were obtained to assess the photosynthetic "competence" of anaerobic, dark-grown C and G1 (Uffen *et al.*, 1971). Although whole-cell suspensions of G1 exhibited light-induced "bleaching" of Bchl *a* centered at about 870 nm and a blue shift centered at 800 nm, characteristic of photosynthesizing *Rs. rubrum* cells, absorption changes were small and sluggish. The slow response of Bchl *a* in G1 to light and lack of detectable cytochrome oxidation reactions suggested that the trace amounts of photosynthetic pigment present were not directly involved during dark fermentative growth. In contrast, rapid light-induced oxidation reactions were measured in pigmented mutant C in midlogarithmic growth phase, but absorption changes were not typical of active photosynthesizing *Rs. rubrum*. When whole-cell suspensions of mutant C were used, light-*minus*-dark difference spectra in the near-IR spectral region were interpreted as a red shift of an absorption band at about 865 nm; a blue shift centered at 800 nm was not observed. Although similar difference spectra were obtained using *Rs. rubrum* in "late" stages of anaerobic, light growth (Fowler and Sybesma, 1970; Govindjee and Sybesma, 1970), dark-grown C cells were different and appeared to contain only the "*b*-type" cytochrome, or C428 pigment (Bartsch *et al.*,

1971; Uffen *et al.*, 1971). Light-induced oxidation of cytochrome c_2, associated with photosynthetic activity, was not detected. The function, if any, of unusual light-sensitive reactions in dark-grown C is not clear. More important, however, the growth properties of C suggested that the mutant was photosynthetically competent and cells formed deep red colonies on solid medium in light. When grown in the light, C also exhibited rapid, light-induced oxidation reactions similar to photosynthesizing wild-type *Rs. rubrum* S1.

From results using C cells, and the possibility of selecting additional pigment variants from anaerobic, dark cultures, it is believed that a study of fermentative-grown *Rs. rubrum* provides an opportunity to determine the quality of light that functions to derepress pigment synthesis and regulate change(s) in absorption characteristics in cells during development of photosynthetic competence.

(c) Internal Membrane Structure. In this section, I shall briefly discuss membrane structure(s) observed inside anaerobic, dark-grown *Rs. rubrum* C. For an

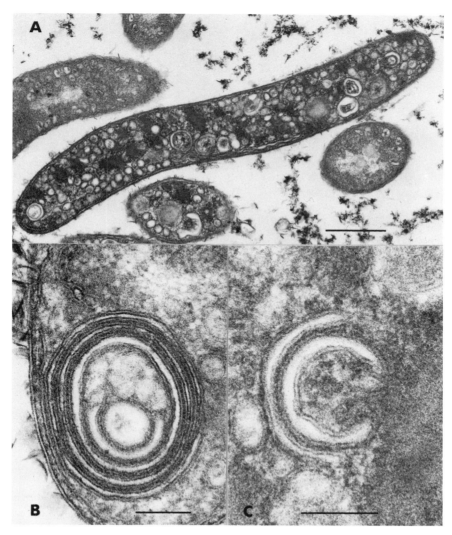

Fig. 1. Electron micrographs of thin sections of anaerobic, dark-grown *Rhodospirillum rubrum* mutant C. Cells fixed using glutaraldehyde (3% wt./vol.) followed by osmium tetroxide (Ryter–Kellenberger). The bars represent 0.5 μm (A) or 0.1 μm (B, C). Courtesy of H. S. Pankratz, Department of Microbiology, Michigan State University, East Lansing, Michigan.

extensive review of the synthesis and composition of photosynthetic membranes in anaerobic, light-grown cells, the reader should refer to Chapters 5 and 43.

In anaerobic, dark conditions, pigmented mutant C cells in midlogarithmic growth phase formed internal membrane structures resembling the chromatophores present in light-grown *Rs. rubrum*. In addition, other complex membranous bodies were observed (Fig. 1). Electron micrographs suggested that the structures developed from invagination of the cell membrane in a manner similar to chromatophore formation (Holt and Marr, 1965). During dark growth, however, spherical vesicles often became enlarged and, as a consequence,

probably collapsed into "goblet"-shaped structures filled with cytoplasmic material (Fig. 1C). Schön and Ladwig (1970) observed similar structures in non-proliferating *Rs. rubrum* cells in dark conditions. Electron micrographs of thin sections of highly developed bodies indicated an "onion"-like appearance (Fig. 1B). The structures appeared to contain a central "core" of coarse-textured ground substance, similar to cytoplasm, surrounded by alternating fine- and coarse-staining regions, each limited by a structured unit membrane. The organization, large size, and extent of these bodies in anaerobic, dark-grown C may provide an interesting system for study of membrane

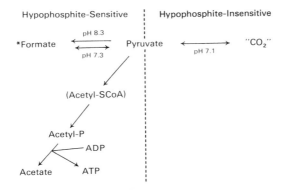

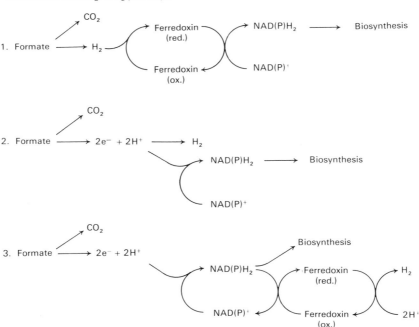

Fig. 2. Fermentation of sodium pyruvate by anaerobic, dark-grown *Rhodospirillum rubrum*.

growth and development. The question of how the complex membrane configurations develop, however, is unresolved. Membrane growth may occur at a point centered in the bowl of the "goblet" (Fig. 1C). On the basis of this model, growth would develop outward from the central core of the structure with surrounding membranous layers also continuing to grow to maintain a uniform spherical configuration. If, on the other hand, active membrane growth were concentrated at the "stem" of the goblet near the cell membrane, development would represent a repeating process of "engulfment" and layering by newly formed membrane material. Resolution between these alternatives is important to understanding chromatophore development, and would contribute to general knowledge about membrane synthesis.

(d) Pyruvate Metabolism. *Rhodospirillum rubrum* grew rapidly in complex (Uffen and Wolfe, 1970) or chemically defined* O_2-free media containing sodium pyruvate. Under dark conditions on complex medium (Uffen, 1973*a*), cells produced formate, acetate, and equimolar amounts of H_2 and CO_2 gases. Like most fermenting microorganisms, these cells incorporated only a small percentage of substrate into cell material and storage products, such as poly-β-hydroxybutyrate (Uffen *et al.*, 1971). The remaining 96% of the sodium pyruvate was utilized in energy metabolism and afforded about 1 mol ATP per mole consumed (Uffen, 1973*a*). Knowledge of fermentation products, energy-yield coefficients, and accumulation of poly-β-hydroxybutyrate inside cells during dark growth suggested that the energy substrate was degraded by the pyruvate formate-lyase or pyruvate-ferredoxin oxidoreductase pathway or both (Uffen *et al.*, 1971; Uffen, 1973*a*). A complete description of pyruvate-degrading reactions (summarized in Fig. 2) in anaerobic, dark-grown *Rs. rubrum* has been published (Gorrell and Uffen, 1977). The findings are discussed briefly below.

(1) Pyruvate-Ferredoxin Oxidoreductase. The presence of ferredoxins (Shanmugam *et al.*, 1972; Yoch *et al.*, 1975), the synthesis of poly-β-hydroxybutyrate (Bosshard-Heer and Bachofen, 1969), and the presence of pyruvate synthase activity in photosynthetically grown cells (Buchanan *et al.*, 1967) suggested that an energy-yielding pyruvate-ferredoxin oxidoreductase pathway might occur in *Rs. rubrum* during anaerobic, dark growth. In accord with these observations, both whole-cell suspensions and cell-free

* The medium was prepared as described earlier (Uffen *et al.*, 1971), but without yeast extract and peptone, and contained (g/100 ml distilled water): sodium pyruvate, 0.75; CaCl$_2$, 0.00075; with either L-alanine, 0.1; L-aspartic acid, 0.1; L-glutamic acid, 0.1; or NH$_4$Cl, 0.1.

extracts from anaerobic, dark-grown G1 and C catalyzed exchange of 1.23 and 1.70 μmol, respectively, of sodium [^{14}C]bicarbonate into sodium pyruvate per milligram protein after 30 min incubation. The reaction appeared to be specific for sodium bicarbonate and was not affected by addition of 9 mM sodium hypophosphite, a formate analogue and potent inhibitor of pyruvate formate lyase (Thauer *et al.*, 1972). More important, however, was the observation that after growth in complex sodium pyruvate medium with light, C cells exhibited similar low activity and exchanged 1.1 μmol sodium [^{14}C]bicarbonate into pyruvate per milligram protein after 30 min. These data indicated that bicarbonate exchange was not a unique property in fermenting cells. On the other hand, even though cells exhibited exchange activity, experiments performed using cell-free extracts from G1 suggested that pyruvate-ferredoxin oxidoreductase did not operate as an energy-yielding reaction during anaerobic, dark cell growth. Results of these studies, employing sodium hypophosphite where appropriate, are presented in Fig. 3, in which the amount of product formed per hour is indicated. Although maximum exchange of sodium [^{14}C]bicarbonate occurred in reactions at pH 6.8–7.1, no accumulation of acetyl-phosphate (Ac-P), the energy intermediate in the pathway, was detected in reactions performed over a wide range of pH values from 6 to 9.5. Consequently, the role of bicarbonate specific exchange activity present in anaerobic, dark-grown cells was not understood.

(2) Pyruvate Formate-Lyase. On the other hand, during anaerobic, dark growth with pyruvate, formate accumulated in spent fermentation medium, suggesting operation of a pyruvate formate-lyase reaction. As expected, suspensions of anaerobic, dark grown G1 and C cells suspended in solution at pH 7.0 catalyzed rapid exchange of sodium [^{14}C]formate into unlabeled sodium pyruvate. Values obtained using crude extracts from G1 in reactions performed over a wide range of pH values are presented in Fig. 3. Unlike exchange of sodium bicarbonate, however, the formate-specific reaction in dark, fermenting cells was strongly inhibited by sodium hypophosphite.

To determine whether pyruvate formate-lyase activity also occurred during photosynthetic development, measurement of sodium formate exchange was performed similarly, using only *Rs. rubrum* mutant C and wild-type S1 cells after light growth in complex medium containing sodium pyruvate, yeast extract, and peptone. Results showed that both C and S1 cells catalyzed exchange of 2.44 and 1.45 μmol sodium [^{14}C]formate mg protein after 30 min, respectively. Again, as in studies described using sodium [^{14}C]bi-

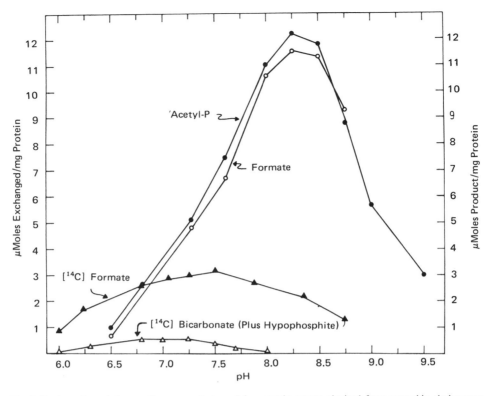

Fig. 3. Products formed from sodium pyruvate by cell-free protein extract obtained from anaerobic, dark-grown *Rhodospirillum rubrum* mutant G1. Reaction mixtures were incubated at 37°C under argon in darkness for 30 min. The amount of product formed per hour is shown. Products were assayed using the following procedures: sodium [^{14}C]pyruvate (Friedmann and Haugen, 1943), acetyl-P (Lipmann and Tuttle, 1945), and formate (Rabinowitz and Pricer, 1957).

carbonate, formate exchange did not appear to be an exclusive property of dark fermenting cells.* Additional experiments suggested that a formate-pyruvate-specific reaction was associated with energy generation.

Amounts of formate and Ac-P produced from pyruvate by crude extracts from anaerobic, dark-grown G1 were measured. The results are presented in Fig. 3. Unexpectedly, only small amounts of formate and Ac-P were detected in reactions performed at pH 7.0–7.5 in which ^{14}C-labeled formate was exchanged into pyruvate. Rather, maximum production of both formate and Ac-P occurred in reactions performed at

pH 8.3. At this pH value, both fermentation products were formed at a linear rate during 10 min incubation at 37°C. Similarly, in reactions performed at different pH values between 6.5 and 8.8, approximately equimolar amounts of both products were measured (Fig. 3). Like formate exchange activity, the pyruvate-degrading reaction was inhibited by 9 mM sodium hypophosphite. Production of Ac-P and formate suggested that *Rs. rubrum* metabolized sodium pyruvate by a pyruvate formate-lyase reaction similar to anaerobically grown *Escherichia coli* (Henning, 1963; Knappe *et al.*, 1974). In *Rs. rubrum*, regulation of the pathway might be influenced by pH changes inside cells during growth and development. Alternatively, different pyruvate-formate-specific reactions might be present.

Metabolic activities discussed to this point occurred not only in anaerobic, fermenting *Rs. rubrum*, but also in photosynthesizing cells. One pertinent difference between anaerobic, light- and dark-grown *Rs. rubrum*,

* These data appeared to contradict the suggestion by Jungermann and Schön (1974) that pyruvate formate-lyase activity occurred only in dark-"adapted" *Rs. rubrum* cells. This discrepancy, however, might be explained by our observation (Gorrell and Uffen, 1978) that pyruvate is required in the medium for *Rs. rubrum* to develop pyruvate formate-lyase during anaerobic, light growth.

however, was that fermenting cells did not require light to produce H_2 from sodium pyruvate (Gest, 1951, 1972). Light-*in*dependent hydrogenase activity is described in the next section.

(3) Hydrogen Production. During anaerobic, dark growth with sodium pyruvate, formate served as an intermediate in gas production. Evidence for this was obtained using suspensions of *Rs. rubrum* grown anaerobically in the dark. In these studies, it was observed that cells evolved equimolar amounts of CO_2 and H_2 from sodium formate or sodium pyruvate; yet if 2 mM sodium hypophosphite was added to reaction mixtures containing only sodium pyruvate, no gas was evolved. (In separate experiments, this concentration of sodium hypophosphite did not inhibit formate oxidation.) Furthermore, when cells were incubated with sodium [^{14}C]pyruvate, addition of unlabeled sodium formate caused a rapid decrease in the production of [^{14}C]CO_2.

Hydrogen gas formation by anaerobic, dark-grown *Rs. rubrum* appeared to be similar to reactions present in other anaerobic, heterotrophic microorganisms (Thauer *et al.*, 1972, 1974). Gas production from sodium formate was completely inhibited by carbon monoxide, and the oxidation reaction presumably did not require ATP. In contrast, production of H_2 gas by *Rs. rubrum* grown anaerobically in the light occurred by the CO-insensitive, ATP-dependent reaction associated with the nitrogenase system (Dalton and Mortenson, 1972; Dilworth *et al.*, 1965; Gest, 1972; Watt *et al.*, 1975). In this regard, anaerobic, light-grown *Rs. rubrum* actively reduced acetylene to ethylene, a measure of nitrogenase activity, while anaerobic, dark-grown cells did not.

Although *Rs. rubrum* produced H_2 from formate during anaerobic, dark growth with sodium pyruvate, the mechanism of formate oxidation is not yet known. In addition to the formation of H_2, formate may also be metabolized to provide reduced pyridine nucleotide for cell growth. Possible pathways are illustrated in Fig. 2. Formate could be oxidized directly to H_2 and CO_2 by formic hydrogen-lyase, as in *Escherichia coli* (Gest, 1951; Wood, 1961; Henning, 1963). Molecular H_2 evolved in this manner could then serve as substrate with ferredoxin present in cells (Shanmugam *et al.*, 1972; Yoch *et al.*, 1975) to reduce NAD(P)$^+$. On the other hand, cells might oxidize major amounts of formate to gas products, but formate dehydrogenase might also function to reduce required amounts of pyridine nucleotide directly. Finally, formate might be oxidized exclusively by NAD(P)-dependent formate dehydrogenase. Excess amounts of reduced pyridine nucleotide generated might then be oxidized with

release of H_2 gas through NAD(P)H-ferredoxin reductase (Jungermann *et al.*, 1971; Thauer *et al.*, 1971) in association with pyruvate-ferredoxin oxidoreductase exchange activity.*

As indicated above, numerous investigators have reported that photosynthetically grown *Rs. rubrum* evolved H_2 from certain substrates only in light (Gest, 1951, 1972). Under these conditions, light-dependent gas evolution was CO-*in*sensitive and occurred through the energy (ATP)-dependent dinitrogen-fixation pathway (Gest, 1972). Similar results were obtained in this laboratory using *Rs. rubrum* S1 exposed to sodium pyruvate after growth in chemically defined medium containing sodium salts of malate and glutamate (Kohlmiller and Gest, 1951). Quite unexpectedly, strain S1 grown similarly in the complex medium used to support anaerobic, dark growth of G1 or C produced only trace amounts of H_2 and exhibited low dinitrogen-fixation activity. During 30-min incubation with sodium pyruvate, these S1 cells reduced only 0.02 vs. 0.14 μmol acetylene/mg protein reduced by malate–glutamate-grown cells. Although changes that occurred in light-dependent H_2 metabolism in S1 during growth in complex medium were not understood, the results indicated that *Rs. rubrum* metabolized H_2 by two different mechanisms depending on availability of light during cell growth. While photosynthetically grown cells required light energy (or ATP), anaerobic, dark-grown cells could evolve H_2 equally well by light-independent (presumably also ATP-independent) reaction(s).

Since anaerobic, dark-grown mutant C, like G1, exhibited only CO-sensitive hydrogenase activity, the question arose whether similar reactions also occurred in anaerobic, light-grown C cells, or was H_2 produced only by ATP-dependent hydrogenase, as indicated in wild-type S1. To obtain answers to these question, *Rs. rubrum* mutant C was obtained from long-term anaerobic, light-grown cultures in which cells had undergone at least 500 generations. The results of these studies were surprising (Gorrell and Uffen, 1978).

---

* *Note Added in Proof.* Contrary to this speculation, attempts to detect the formation of NAD(P)H associated with formate oxidation were not successful, and no evidence for pyruvate-ferredoxin oxidoreductase activity in anaerobically dark-grown *Rs. rubrum* was obtained. Pyruvate:lipoate oxidoreductase was the only system identified which reacted with pyruvate to reduce NAD$^+$ in the dark-grown cells (Gorrell and Uffen, submitted). We now believe that pyruvate formate hydrogen-lyase (Gorrell and Uffen, 1977) and pyruvate:lipoate oxidoreductase operate in concert to form ATP and NADH, respectively, and support fermentative cell growth. Pyruvate:lipoate oxidoreductase was also recently demonstrated in photosynthetically grown *Rs. rubrum* (Lüderitz and Klemme, 1977).

In contrast to "wild-type" S1, anaerobic, light-grown C exhibited both light-*in*dependent and light-dependent hydrogenase activity. Evidence that light-*in*dependent hydrogenase activity occurred in mutant C was obtained by placing cells under CO in the light with sodium pyruvate. Under these incubation conditions, H_2 gas production decreased from 3.4 to 1.5 μmol/mg protein after 1 hr. In the same experiment, it appeared that H_2 gas formed from sodium pyruvate by anaerobic, light-grown mutant C under CO probably occurred through the light- (or ATP)-dependent nitrogenase system. As expected, when reaction mixtures incubated with CO in the light were transferred to darkness, no H_2 gas was produced. In addition, anaerobic, light-grown mutant C exhibited nitrogenase activity, measured by acetylene reduction, whereas dark-grown mutant C cells did not.

Discovery that both ATP-dependent and fermentative hydrogenase reactions occurred in anaerobic, light-grown *Rs. rubrum* mutant C was extremely important. To the author's knowledge, it provided the first evidence that "dark"- and light-dependent reactions with similar function can occur together in phototrophic bacteria and provided the basis for an experimental system to determine how light could influence and/or regulate the activity of the separate hydrogenase reactions. Furthermore, studies indicated that *Rs. rubrum* C and "wild-type" S1 were metabolically different.

From an ecological viewpoint, both light-*in*dependent and -dependent hydrogenase activity in *Rs. rubrum* C could provide cells with a selective advantage during growth under diurnal light conditions in nature. In this regard, it is interesting to speculate that "wild-type" *Rs. rubrum* S1 became altered during growth in continuous light conditions provided in most research laboratories. Consequently, "mutant" C might be more representative of naturally occurring *Rs. rubrum* cells. This idea might be easily tested by measuring hydrogen-producing reactions in newly isolated strains of *Rs. rubrum*.

In summary, it appeared that if photosynthetically grown wild-type *Rs. rubrum* was placed in strictly anaerobic, dark conditions, cells grew and often maintained pigments and membrane properties characteristic of photosynthesizing cells. Nevertheless, under anaerobic, dark conditions, *Rs. rubrum* differentiated and metabolized hydrogen, and presumably also pyridine nucleotide, similar to fermenting microorganisms. Once light-*in*dependent H_2 gas production became expressed during dark growth, the reaction appeared to occur in *Rs. rubrum* together with the analogous light-dependent reactions "normally" present in photosynthetically grown cells.

2.2.2. *Rhodopseudomonas* Species

As indicated earlier, while several *Rhodopseudomonas* species were observed to grow slowly in anaerobic, dark conditions on solid medium (Uffen and Wolfe, 1970), it was recently possible to obtain rapid anaerobic, dark growth in liquid using *Rs. gelatinosa*. Ability to grow anaerobically in darkness in complex liquid medium containing sodium pyruvate, peptone, and yeast extract appeared to be a general property of four strains tested in this laboratory. Preliminary studies suggested, however, that cells metabolized sodium pyruvate in a manner different from *Rs. rubrum*, since hydrogen was not produced during dark growth. In addition, three strains of a *Rhodopseudomonas* species that actively metabolized gelatin during photosynthetic growth exhibited the remarkable ability to grow methylotrophically under anaerobic, dark conditions with pure carbon monoxide (Uffen, 1976). This represents the first case of CO metabolism functioning for the cell as an important biological reaction.

Although anaerobic, dark growth occurred in a chemically defined mineral medium with CO as the only carbon and energy source, growth was stimulated by adding trypticase. Under these conditions, strain 1 cells had a generation time of 6.7 hr and reached a final concentration of $1–3 \times 10^9$ cells/ml of liquid medium. Resting suspensions of CO-grown cells metabolized about 6.7 μmol CO/mg protein in 1 hr and produced equimolar amounts of CO_2 and H_2 according to the equation: $CO + H_2O \rightarrow CO_2 + H_2$. Operation of this reaction was confirmed by using cell-free extracts of strain 1 cells grown anaerobically in the dark with CO. In accordance with the equation, when whole cells were suspended in tritium-labeled water and incubated with pure CO, $[^3H]H_2$ was produced at a linear rate with a constant specific activity.

Anaerobic, dark growth of strain 1 was CO-dependent. When cells were placed in liquid medium under argon, no growth occurred. Furthermore, no growth occurred when strain 1 cells were incubated anaerobically in the dark with methane, methanol, formaldehyde, formate, or a $H_2–CO_2$ gas mixture. As might be expected, when suspensions of CO-grown strain 1 were incubated in darkness with methane, methanol, formaldehyde, or formate, no oxidation activity, measured by H_2 gas formation, was detected.

Experiments suggested that the formation of CO-oxidizing activity occurred in cells during growth with CO. Attempts to demonstrate CO-oxidation using either aerobic dark or photosynthetically grown cells were unsuccessful. The interaction between light and

CO on cell metabolism, however, was complex. When CO-grown strain 1 was placed in the light with CO, cells grew and continued to metabolize the gas, as did anaerobic, dark grown cells. Furthermore, under these anaerobic conditions with CO, cells formed similar amounts of Bchl *a* in the light or in the dark (7.5 and 7.8 nmol/mg protein, respectively). These results suggested that CO-oxidizing activity in strain 1 was inducible and not sensitive to light. On the other hand, when CO-grown strain 1 was placed in the light (approximately 1300 lux) under argon, instead of CO, CO-oxidizing activity was lost, the concentration of Bchl *a* in the cells increased to 30 nmol/mg protein, and growth appeared to occur photosynthetically.

In accord with previous observations, transfer of photosynthesizing cells back into either light or dark conditions with CO should result in the reappearance of the CO-specific reactions. This, however, did not occur, and only after development of a complicated "training" procedure could CO-oxidation activity be regained in these cells (Dashkevicz and Uffen, unpublished). From these observations, it appeared that although strain 1 had the ability to metabolize and grow with CO, the activity was not readily re-expressed in the presence of CO after cells were grown with argon in the light. This curious situation could help explain our failure to detect CO-dependent, anaerobic, dark growth of certain laboratory strains of *Rp. gelatinosa*. Only *Rp. gelatinosa* ATCC 17011 developed CO-dependent growth, while ATCC strains 17013 and 17014 did not.

3. Chromatiaceae

3.1. Anaerobic, Dark Metabolism

The Chromatiaceae are strict anaerobes [except for *Thiocapsa roseopersicina*, which is reported to respire under microaerophilic conditions in the dark (Pfennig and Trüper, 1974)] that utilize either organic or inorganic compounds as reductants during photosynthetic development (van Niel, 1931, 1936). Since anaerobic, dark growth, however, has not been demonstrated in the laboratory, one may only speculate about this possibility based on rather fragmentary information in the literature. Like Rhodospirillaceae, when light-grown Chromatiaceae were transferred to darkness, the cells metabolized and produced acid and gas products. Moreover, under certain conditions, cells also appeared to reduce sulfur or nitrogen compounds. For this discussion, knowledge about these different activities is organized in separate sections below.

3.1.1. Acid Production

Results of experiments published by van Niel in 1931 showed that Chromatiaceae used either reduced sulfur or organic compounds as electron donors equally well during photosynthetic growth. On the basis of this discovery and knowledge of earlier reports suggesting that Chromatiaceae cells developed in enrichment cultures under dark conditions (for an excellent discussion of these early observations, refer to van Niel, 1931; Kondrat'eva, 1965), van Niel performed experiments to determine whether cells could also grow heterotrophically in darkness. Since growth failed to occur during dark incubation, aerobically or anaerobically, in medium supplemented with glucose, fructose, lactate, yeast extract, or mixtures of these compounds, it was concluded that Chromatiaceae could neither ferment organic compounds nor synthesize cell material in the absence of light energy.

Subsequently, however, both Gaffron (1934, 1935) and Roelofsen (1934) performed manometric studies with resting suspensions of "*Thiocystis*" species* from cultures grown photoautotrophically with sodium thiosulfate or sodium sulfide. Although Gaffron (1934, 1935) used cell suspensions obtained from enrichment cultures and Roelofsen (1934) performed experiments with pure cultures of strain 9 (*Chromatium vinosum* strain D), their results were similar. In both cases, photoautotrophically grown cell suspensions formed CO_2 and acids, primarily acetic acid (Hendley, 1955), during dark incubation under N_2 or N_2–CO_2 gas mixtures.

Subsequent attempts to determine the substrates used in the "fermentation" process were, however, inconclusive. Although Gaffron believed that cells oxidized organic compounds, such as malate and butyrate, to CO_2 and other acid products during dark incubation, his experimental procedures were complicated and made interpretation of data difficult. On the other hand, Roelofsen's attempts to stimulate CO_2 evolution with various organic substrates were unsuccessful. As a result, Roelofsen (1934) concluded that some reserve material deposited inside cells during growth in the light served as substrate for a light-independent "autofermentation" process to provide maintenance energy for cells in darkness. More recent studies performed by van Gemerden (1968*a,b*) supported this hypothesis. Van Gemerden observed that during photosynthetic development, *Chr. vinosum* strain D deposited polyglucose, which in the dark was metabolized by energy-yielding reactions to poly-β-hydroxybutyrate.

---

* The cells used by Gaffron and Roelofsen were later classified as *Chromatium vinosum* (Hendley, 1955; van Niel, 1936).

Although Roelofsen was unable to detect significant stimulation of CO_2 evolution from whole-cell suspensions of *Chr. vinosum* strain D in the presence of organic material, Nakamura (1939, 1941) reported that production of hydrogen gas could be stimulated under anaerobic, dark conditions using *Chr. minutissimum*. However, since Nakamura measured only gas uptake or evolution and did not analyze reaction mixtures for acid products, his data will be considered further in Section 3.1.2(c). Nevertheless, it appeared that "fermentative" activity in *Chr. minutissimum* could be stimulated; the failure by Roelofsen (1934) to obtain similar results with *Chr. vinosum* strain D by measuring CO_2 evolution may be explained by the oxidation state of compounds selected for use (e.g., fumarate and acetate) or species differences. The composite results of these studies indicated, however, that photoheterotrophically grown Chromatiaceae could excrete acid(s) during anaerobic, dark incubation. More recent evidence for one such acid-producing reaction was provided by Bennett and Fuller (1964) and Bennett *et al.* (1964). In studies with *Chr. vinosum* strain D, it was shown that during photoheterotrophic growth with L-malate, cells possessed a pyruvate-ferredoxin oxidoreductase pathway. More important, operation of the reaction in whole cells and cell-free extracts was light-*in*dependent and resulted in production of acetate plus equimolar amounts of H_2 and CO_2 gases. Gaffron (1935) observed a reaction in his studies that was perhaps similar, since incubation of resting-cell suspensions under CO, a potent inhibitor of ferredoxin-linked reactions (Thauer *et al.*, 1972, 1974), repressed acid production.

3.1.2. Metabolism of Inorganic Compounds

In addition to "fermentation" activity, it appeared that, in darkness, Chromatiaceae might also respire using oxidized sulfur or nitrogenous compounds as electron acceptors. A detailed account of sulfur metabolism in Chromatiaceae by Trüper is presented in Chapter 35.

(a) Sulfur Compounds. Gaffron (1934, 1935) reported that when resting suspensions of cells obtained from cultures enriched for Chromatiaceae were placed anaerobically in darkness, sulfate ion was reduced to H_2S. Furthermore, addition of glucose, hexosephosphate, butyrate, or malate to reaction mixtures stimulated both CO_2 and H_2S gas evolution. Gaffron interpreted these data to support the hypothesis that although certain degradation reactions occurred in darkness, in light, Chromatiaceae synthesized cell material autotrophically from CO_2 with H_2S as

reductant; direct incorporation of preformed organic compounds, as suggested by Muller (1933), did not occur. Other researchers, however, recognized that respiratory activity with sulfate might function to provide energy for cells during dark periods. Since Gaffron's experimental procedures were complicated, and the possibility existed that his enrichment cultures contained sulfate-reducing *Desulfovibrio* species (Roelofsen, 1934), van Niel (1936) performed experiments to reexamine this question using pure cultures of *Chr. vinosum* strain D (formerly strain 9 obtained from Roelofsen). Unlike Gaffron (1934, 1935), van Niel used "sulfur-free" cells grown photoheterotrophically with L-malate to eliminate possible H_2S production from sulfur stored inside cells and measured the H_2S purified from reaction mixtures after dark incubation with sulfate, thiosulfate, or sulfur. The results indicated that sulfur, not sulfate or thiosulfate, served best as substrate for H_2S production. Yet, since only very small amounts of gas were formed with organic compounds or under an H_2-atmosphere, it was concluded that the reduction reaction was not an important cell activity. In addition, after removal of H_2S, van Niel noted that significant amounts of iodine-reducing material remained in solution. These data and the observation that the iodine-reacting material disappeared when cell suspensions were returned again to light conditions led van Niel to suggest that in darkness, cells formed reduced material other than H_2S. From studies by van Gemerden (1968*a,b*) on dark metabolism of storage polymers formed in *Chr. vinosum* strain D during light growth, much larger amounts of H_2S would be expected to accumulate than van Niel recovered from reaction mixtures. It is quite possible that during sample preparation in van Niel's studies, quantities of O_2 were still present in cell suspensions at the beginning of experiments. In this event, although cells reduced sulfur in darkness, H_2S evolved could react chemically with residual O_2 to generate nonvolatile oxidized sulfur compounds and account for low amounts of sulfide purified from reactions (van Gemerden, personal communication). Unfortunately, the chemical nature of the iodine-reducing compound(s) reported by van Niel cannot be stated with certainty.

Nakamura (1939, 1941) performed manometric studies with *Chr. minutissimum* cells grown in light with Na_2S. Using these cells, he observed that during dark incubation, addition of sulfate or thiosulfate resulted in H_2 gas uptake. Under similar conditions, however, with pulverized sulfur, only small amounts of H_2 were used and at a very slow rate. The observation that larger amounts of H_2 were rapidly metabolized by cells exposed to sulfate or thiosulfate led Nakamura to believe that compounds were reduced to elemental sulfur and stored inside cells. Further reduction of sulfur and release of H_2S appeared as the rate-limiting step in the sequence of reduction reactions.

As a result of these studies, it apeared that *Chr. minutissimum* metabolized sulfur and could use H_2 as a reductant. The ability of *Chr. minutissimum* to reduce sulfate and thiosulfate should be examined further, however, since Nakamura observed only H_2 utilization and did not measure concentration changes in either sulfur reactants or product(s). This is especially important, since these reactions did not appear to occur in *Chr. vinosum* strain D (van Niel, 1936; Hendley, 1955).

(b) Nitrite and Nitrate. Nakamura (1939) further reported that hydrogen uptake by photoautotrophically grown *Chr. minutissimum* was stimulated not only by adding oxidized sulfur compounds to cell suspensions during dark incubation, but also by both potassium nitrite and potassium nitrate. [Gaffron (1933) and Nakamura (1937*a*, 1939) reported similar activity using Rhodospirillaceae species.*] It was proposed in these studies that nitrogenous compounds were reduced under H_2 to ammonia, but analysis of reaction mixtures was not reported. For this reason, although such reactions may represent an avenue for respiratory-energy generation during dark incubation, they should be examined in greater detail and products should be identified before a conclusion is arrived at.

(c) Hydrogen. As indicated in the previous discussion, hydrogen served as a substrate in certain light-independent reduction reactions involving sulfur compounds. Conversely, when photoheterotrophically grown cells were placed in darkness with reduced organic compounds, H_2 gas was evolved. *Chromatium minutissimum* (Nakamura, 1939, 1941) appeared to rapidly oxidize formate, pyruvate, glycerol, glycerol phosphate, or glucose; *Chr. vinosum* strain D evolved both H_2 and CO_2 during dark incubation with sodium pyruvate (Bennett and Fuller, 1964). These data, and the demonstration of pyruvate-ferredoxin oxidoreductase in *Chr. vinosum* strain D (Bennett *et al.*, 1964) and hydrogenase [recently purified from light-grown cells and characterized by Gilitz and Krasna (1975)], together with circumstantial evidence that *Chr. minutissimum* degraded formate and pyruvate by a pyruvate formate-lyase reaction (Nakamura, 1939, 1941), suggested that certain

Chromatiaceae might grow fermentatively in darkness as do Rhodospirillaceae species. Speculation about this possibility is presented in the next section.

3.2. Anaerobic, Dark Growth

Numerous reports on the distribution of Chromatiaceae in nature suggested that cells developed in darkness. Kondrat'eva (1965) provided an excellent summary of these data. One interesting citation discussed was a report by N. A. Volodin in 1935. Volodin noted that during oil-drilling operations in the Surakhany region in Russia, pink stratal waters were obtained from subterranean depths up to 1700 m. On examination, the water appeared to be enriched with "Chromatiaceae"-type cells, which suggested that development and growth might occur under these rather unusual natural conditions.

Even though anaerobic, dark growth of Chromatiaceae has not been accomplished in the laboratory, one might suggest on the basis of discussions presented, that cells respire anaerobically in darkness with sulfur. Hendley (1955) recognized this possibility, but was unable to demonstrate anaerobic growth of *Chr. vinosum* strain D after 1 week of dark incubation in anaerobic liquid medium containing colloidal sulfur, malate, pyruvate, and yeast extract. During the dark period, although cells remained motile and produced acetic acid, CO_2, and H_2S, growth occurred only when cultures were placed back into light. Similarly, van Gemerden (1968*a*,*b*) reported that *Chr. vinosum* strain D remained motile for long periods in darkness and metabolized poly-glucose, stored in cells during growth in the light together with sulfur (the terminal electron acceptor), to poly-β-hydroxybutyric acid. Under these dark conditions, however, reduction of sulfur to H_2S might proceed too slowly to provide adequate energy for cell growth to occur. If this idea is correct, the presence of sulfur compounds might dramatically influence dark metabolic activity in cells and function as a "preferred," but rate-limiting, respiratory oxidant. Evidence for this was provided by Roelofsen (1934), who observed that photoheterotrophically grown cells incubated in darkness evolved only H_2 gas in the absence of sulfur. Likewise, although *Chr. vinosum* strain D exhibited pyruvate-ferredoxin oxidoreductase (Bennett *et al.*, 1964) and hydrogenase (Gilitz and Krasna, 1975), in the presence of sulfur, H_2S, not H_2, was evolved during degradation of sodium pyruvate (Hendley, 1955). In the same studies, Hendley found that although CO blocked H_2 evolution, it did not effect H_2S formation from sulfur. As a consequence of these reports suggesting that sulfur

---

* It was brought to my attention that *Rs. rubrum* (Katoh, 1962) and *Rp. sphaeroides* (Satoh *et al.*, 1974) were reported to have denitrifying activity and to grown anaerobically in the dark with nitrate as electron acceptor (see Chapter 39).

occupies a complex role in "dark" cell metabolism, attempts to demonstrate dark fermentative growth of Chromatiaceae might be more rewarding under "sulfur-free" conditions with only small amounts of sulfoamino acids required for biosynthetic activity. In this regard, the supposition that Chromatiaceae cannot grow in darkness may not be adequately supported and should be tested further.*

4. Concluding Remarks

Anoxygenic phototrophic bacteria represent an ancient and remarkable group of microorganisms that experienced, and successfully competed at, every milestone in evolutionary history. This is most convincing in Rhodospirillaceae species, particularly *Rs. rubrum*, which preserved useful genetic information that could permit cells to grow under practically any condition in the natural aquatic environment by use of fermentative, photosynthetic, or respiratory reactions. In this laboratory, photosynthetic energy conversion is of special interest. Although information about fermentative growth is limited, studies performed to date indicate that anaerobic, dark-grown cells provide a unique model system to study regulation and development of photosynthetic capacity in cells exposed to radiant energy. With knowledge about metabolism in fermenting cells, it now appears possible to investigate how photosynthetic energy reaction(s) may develop, become integrated, and operate to regulate physiological activity in the biological system.

* After this chapter was prepared, *Thiocapsa roseopersicina* was reported to grow in the dark under anaerobic conditions with H_2 gas in a liquid medium supplemented with glucose and thiosulfate or sulfur. During anaerobic growth with glucose, pyruvate accumulated in the medium. Thiosulfate and sulfur appeared to be oxidized to sulfate, but trace amounts of H_2S were also produced (Krasil'nikova *et al.*, 1975).

5. References

Bartsch, R. G., Kakuno, T., Horio, T., and Kamen, M. D., 1971, Preparation and properties of *Rhodospirillum rubrum* cytochromes c_2, cc' and $b_{559.5}$, and flavin mononucleotide protein, *J. Biol. Chem.* **246**:4489.

Bennett, R., and Fuller, R. C., 1964, The pyruvate phosphoroclastic reaction in *Chromatium*: A probable role for ferredoxin in a photosynthetic bacterium, *Biochem. Biophys. Res. Commun.* **16**:300.

Bennett, R., Rigopoulos, N., and Fuller, R. C., 1964, The pyruvate phosphoroclastic reaction and light-dependent nitrogen fixation in bacterial photosynthesis, *Biochemistry* **52**:762.

Bosshard-Heer, E., and Bachofen, R., 1969, Synthese von Speicherstoffen aus Pyruvat durch *Rhodospirillum rubrum*, *Arch. Mikrobiol.* **65**:61.

Buchanan, B. B., Evans, M. C. W., and Arnon, D. I., 1967, Ferredoxin-dependent carbon assimilation in *Rhodospirillum rubrum*, *Arch. Mikrobiol.* **59**:32.

Dalton, H., and Mortenson, L. E., 1972, Dinitrogen (N_2) fixation (with a biochemical emphasis), *Bacteriol. Rev.* **36**:231.

Dilworth, M. J., Subramanian, D., Munson, T. O., and Burris, R. H., 1965, The adenosine triphosphate requirement for nitrogen fixation in cell-free extracts of *Clostridium pasteurianum*, *Biochim. Biophys. Acta* **99**:486.

Fowler, C. F., and Sybesma, C., 1970, Light- and chemically induced oxidation–reduction reactions in chromatophore fractions of *Rhodospirillum rubrum*, *Biochim. Biophys. Acta* **197**:276.

Friedemann, T. E., and Haugen, G. E., 1943, Pyruvic acid. II. The determination of keto acids in blood and urine, *J. Biol. Chem.* **147**:415.

Gaffron, H., 1933, Über den Stoffwechsel der schwefelfreien Purpurbakterian, *Biochem. Z.* **260**:1.

Gaffron, H., 1934, Über die Kohlensäureassimilation der roten Schwefelbacterien I, *Biochem. Z.* **269**:447.

Gaffron, H., 1935, Über die Kohlensäureassimilation der roten Schwefelbacterien II, *Biochem. Z.* **279**:1.

van Gemerden, H., 1968a, Utilization of reducing power in growing cultures of *Chromatium*, *Arch. Mikrobiol.* **64**:111.

van Gemerden, H., 1968b, On the ATP generation by *Chromatium* in darkness, *Arch. Mikrobiol.* **64**:118.

Gest, H., 1951, Metabolic patterns in photosynthetic bacteria, *Bacteriol. Rev.* **15**:183.

Gest, H., 1972, Energy conversion and generation of reducing power in bacterial photosynthesis, in: *Advances in Microbial Physiology* (A. H. Rose and D. W. Tempest, eds.), Vol. 7, pp. 243–282, Academic Press, New York.

Gilitz, P. H., and Krasna, A. I., 1975, Structural and catalytic properties of hydrogenase from *Chromatium*, *Biochemistry* **14**:2561.

Gorrell, T. E., and Uffen, R. L., 1977, Fermentative metabolism of pyruvate by *Rhodospirillum rubrum* after anaerobic growth in darkness, *J. Bacteriol.* **131**:533.

Gorrell, T. E., and Uffen, R. L., 1978, Light-dependent and light-independent production of hydrogen gas by photosynthesizing *Rhodospirillum rubrum* mutant C, *Photochem. Photobiol.* **27**:351.

Govindjee, R., and Sybesma, C., 1970, Light-induced reduction of pyridine nucleotide and its relation to light-induced electron transport in whole cells of *Rhodospirillum rubrum*, *Biochim. Biophys. Acta* **223**:251.

Hendley, D. D., 1955, Endogenous fermentation in Thiorhodaceae, *J. Bacteriol.* **70**:625.

Henning, U., 1963, Ein Regulationsmechanismus beim Abbau der Brenztraubensäure durch *Escherichia coli*, *Biochem. Z.* **337**:490.

Holt, S. C., and Marr, A. G., 1965, Isolation and purification of the intracytoplasmic membranes of *Rhodospirillum rubrum*, *J. Bacteriol.* **89**:1413.

Jungermann, K., and Schön, G., 1974, Pyruvate formate lyase in *Rhodospirillum rubrum* Ha adapted to anaerobic dark conditions, *Arch. Microbiol.* **99**:109.

Jungermann, K., Rupprecht, E., Ohrloff, C., Thauer, R., and Decker, K., 1971, Regulation of the reduced nicotinamide adenine dinucleotide–ferredoxin reductase system in *Clostridium kluyveri*, *J. Biol. Chem.* **246**:960.

Katoh, T., 1962, Nitrate reductase in the photosynthetic bacterium, *Rhodospirillum rubrum*: Adaptive formation of nitrate reductase, *Plant Cell Physiol.* **4**:199.

Knappe, J., Blaschkowski, H. P., Gröbner, P., and Schmitt, T., 1974, Pyruvate formate-lyase of *Escherichia coli*: The acetyl-enzyme intermediate, *Eur. J. Biochem.* **50**:253.

Kohlmiller, E. F., Jr., and Gest, H., 1951, A comparative study of the light and dark fermentations of organic acids by *Rhodospirillum rubrum*, *J. Bacteriol.* **61**:269.

Kondrat'eva, E. N., 1965, *Photosynthetic Bacteria*, Israel Program for Scientific Translations, Jerusalem, published for the U.S. Atomic Energy Commission and the National Science Foundation, Washington, D.C.

Krasil'nikova, E. N., Petushkova, Yu. P., and Kondrat'eva, E. N., 1975, Growth of the purple sulfur bacterium *Thiocapsa roseopersicina* in the dark under anaerobic conditions, *Mikrobiologiya* **44**:700.

Lipmann, R., and Tuttle, L. C., 1945, A specific micromethod for the determination of acyl-phosphates, *J. Biol. Chem.* **159**:21.

Lüderitz, R., and Klemme, J.-H., 1977, Isolierung und Charakterisierung eines membrangebundenen Pyruvat-dehydrogenase-Komplexes aus dem phototrophen Bakterium *Rhodospirillum rubrum*, *Z. Naturforsch.* **32c**:351.

Miller, T. L., and Wolin, M. J., 1973, Formation of hydrogen and formate by *Ruminococcus albus*, *J. Bacteriol.* **116**:836.

Muller, F. M., 1933, On the metabolism of the purple sulfur bacteria in organic media, *Arch. Mikrobiol.* **4**:131.

Nakamura, H., 1937a, Über die Photosynthese bei der schwefelfreien Purpurbakterie, *Rhodobacillus palustris*, *Acta Phytochim.* (Jpn.) **9**:189.

Nakamura, H., 1937b, Über das Vorkommen der Hydrogenylase in *Rhodobacillus palustris* und über ihre Rolle im Mechanismus der bakteriellen Photosynthese, *Acta. Photochim.* (Jpn.) **10**:211.

Nakamura, H., 1939, Weitere Untersuchungen über den Wasserstoffumsatz bei den Purpurbakterien, nebst einer Bemerkung über die gegenseitige Beziehung zwischen Thio- und Athiorhodaceen, *Acta Phytochim.* (Jpn.) **11**:109.

Nakamura, H., 1941, Weitere Untersuchungen über die bakterielle Photosynthese. Beiträge zur Stoffwechselphysiologie der Purpurbakterien, VI, *Acta Phytochim.* (Jpn.) **12**:43.

van Niel, C. B., 1931, On the morphology and physiology of the purple and green sulfur bacteria, *Arch. Mikrobiol.* **3**:1.

van Niel, C. B., 1936, On the metabolism of the Thiorhodaceae, *Arch. Mikrobiol.* **7**:323.

van Niel, C. B., 1944, The culture, general physiology, morphology, and classification of the non-sulfur purple and brown bacteria, *Bacteriol. Rev.* **8**:1.

Pfennig, N., 1967, Photosynthetic bacteria, *Annu. Rev. Microbiol.* **21**:285.

Pfennig, N., and Trüper, H. G., 1974, The phototrophic bacteria, in: *Bergey's Manual of Determinative Bacteriology* (R. E. Buchanan and N. E. Gibbons, ed.), 8th Ed., pp. 24–75, The Williams & Wilkins Co., Baltimore.

Qadri, S. M. H., and Hoare, D. S., 1968, Formic hydrogenlyase and the photoassimilation of formate by a strain of *Rhodopseudomonas palustris*, *J. Bacteriol.* **95**:2344.

Qadri, S. M. H., and Hoare, D. S., 1973, Pyruvic decarboxylase and acetoin formation in Athiorhodaceae, *Can. J. Microbiol.* **19**:1137.

Rabinowitz, J. D., and Pricer, W. E., Jr., 1957, An enzymatic method for the determination of formic acid, *J. Biol. Chem.* **229**:321.

Roelofsen, P. A., 1934, On the metabolism of the purple sulfur bacteria, *Proc. K. Ned. Akad. Wet.* **37**:660.

Satoh, T., Hoshino, Y., and Kitamura, H., 1974, Isolation of denitrifying photosynthetic bacteria, *Agric. Biol. Chem.* **38**:1749.

Schön, G., 1968, Fructoseverwertung und Bacteriochlorophyllsynthese in anaeroben Dunkel- und Lichtkulturen von *Rhodospirillum rubrum*, *Arch. Mikrobiol.* **63**:362.

Schön, G., 1969, Der Einfluss der Reservestoffe auf den ATP-Spiegel in Zellen von *Rhodospirillum rubrum* beim Übergang von aerober zu anaerober Dunkelkultur, *Arch. Mikrobiol.* **68**:40.

Schön, G., 1971, Der Einfluss der Kulturbedingungen auf den Nicotinamid-Adenin-Dinucleotid(phosphat)-Gehalt in Zellen von *Rhodospirillum rubrum*, *Arch. Mikrobiol.* **79**:147.

Schön, G., and Biedermann, M., 1972, Bildung flüchtiger Säuren bei der Vergärung von Pyruvat und Fructose in anaerober Dunkelkultur von *Rhodospirillum rubrum*, *Arch. Mikrobiol.* **85**:77.

Schön, G., and Biedermann, M., 1973, Growth and adaptive hydrogen production of *Rhodospirillum rubrum* (F₁) in anaerobic dark cultures, *Biochim. Biophys. Acta* **304**:65.

Schön, G., and Ladwig, R., 1970, Bacteriochlorophyllsynthese und Thylakoidmorphogenese in anaerober Dunkelkultur von *Rhodospirillum rubrum*, *Arch. Mikrobiol.* **74**:356.

Shanmugam, K. T., Buchanan, B. B., and Arnon, D. I., 1972, Ferredoxins in light- and dark-grown photosynthetic cells with special reference to *Rhodospirillum rubrum*, *Biochim. Biophys. Acta* **256**:477.

Thauer, R. K., 1972, CO_2 reduction to formate by NADPH. The initial step in the total synthesis of acetate from CO_2 in *Clostridium thermoaceticum*, *FEBS Lett.* **27**:111.

Thauer, R. K., 1973, CO_2 reduction to formate in *Clostridium acidiurici*, *J. Bacteriol.* **114**:443.

Thauer, R. K., Ruprecht, E., Ohrloff, C., Jungermann, K., and Decker, K., 1971, Regulation of the reduced nicotinamide adenine dinucleotide phosphate–ferredoxin reductase system in *Clostridium kluyveri*, *J. Biol. Chem.* **246**:954.

Thauer, R. K., Kirchiniawy, F. H., and Jungermann, K. A., 1972, Properties and function of the pyruvate-formate-lyase reaction in *Clostridia, Eur. J. Biochem.* **27**:282.

Thauer, R. K., Käufer, B., Zähringer, M., and Jungermann, K., 1974, The reaction of the iron–sulfur protein hydrogenase with carbon monoxide, *Eur. J. Biochem.* **42**:447.

Uffen, R. L., 1973*a*, Growth properties of *Rhodospirillum rubrum* mutants and fermentation of pyruvate in anaerobic, dark conditions, *J. Bacteriol.* **116**:874.

Uffen, R. L., 1973*b*, Effect of low-intensity light on growth response and bacteriochlorophyll concentration in *Rhodospirillum rubrum* mutant C, *J. Bacteriol.* **116**:1086.

Uffen, R. L., 1976, Anaerobic growth of a *Rhodopseudomonas* species in the dark with carbon monoxide as sole carbon and energy substrate, *Proc. Natl. Acad. Sci. U.S.A.* **73**:3298.

Uffen, R. L., and Wolfe, R. S., 1970, Anaerobic growth of purple nonsulfur bacteria under dark conditions, *J. Bacteriol.* **104**:462.

Uffen, R. L., Sybesma, C., and Wolfe, R. S., 1971, Mutants of *Rhodospirillum rubrum* obtained after long-term anaerobic, dark growth, *J. Bacteriol.* **108**:1348.

Watt, G. D., Bulen, W. A., Burns, A., and LaMont Hadfield, K., 1975, Stoichiometry, ATP/2e values, and energy requirements for reactions catalyzed by nitrogenase from *Azotobacter vinelandii, Biochemistry* **14**:4266.

Wood, W. A., 1961, Fermentation of carbohydrates, in: *The Bacteria* (I. C. Gunsalus and R. Y. Stanier, ed.), Vol. 2, pp. 59–149, Academic Press, New York.

Yoch, D. C., and Lindstrom, E. S., 1969, Nicotinamide adenine dinucleotide-dependent formate dehydrogenase from *Rhodopseudomonas palustris, Arch. Mikrobiol.* **67**:182.

Yoch, D. C., Arnon, D. I., and Sweeney, W. V., 1975, Characterization of two soluble ferredoxins as distinct from bound iron–sulfur proteins in the photosynthetic bacterium *Rhodospirillum rubrum, J. Biol. Chem.* **250**:8330.

CHAPTER 47

Genetics and Bacteriophage

Barry L. Marrs

1. Introduction

It is remarkable that the sex life of photosynthetic bacteria has retained an aura of mystery through an era of explicit bacterial sexuality. The genetics of bacteria and their viruses is a cornerstone of molecular biology, and the reasons this is so are the reasons for studying genetic systems and phage of the photosynthetic bacteria. Genetic exchange systems are necessary for a thorough analysis of mutant strains and for the construction of well-defined strains, both by the resolution of multiple mutations and by bringing together fully characterized mutations into one strain. Much of our understanding of bacteriochlorophyll (Bchl) and carotenoid biosynthesis relies on studies of mutant strains, and more recently, electron-transport systems have been constructively probed by biochemical analyses of mutant strains. Thus, the genetic capabilities mentioned above are relevant to many aspects of long-standing research interests involving photosynthetic bacteria. Furthermore, genetic exchange systems make possible the analysis of the arrangement of genes on chromosomes (mapping), which in turn is important for both understanding the regulation of gene expression and facilitating strain construction. The photosynthetic bacteria would represent an unrivaled model system for the study of the regulation of membrane formation

and differentiation if the tools for analysis of the regulation of gene expression could be fully developed for some member of this group of organisms. That genetics can be utilized in the ways mentioned above to accelerate progress in understanding biological systems is abundantly documented, but these capabilities have long remained unrealized potentials with regard to the photosynthetic bacteria.

Accordingly, the motivation for studying the virology of photosynthetic bacteria has most often been to obtain transducing phage for use as mediators of genetic exchange. In addition, specialized transducers have been sought because they provide a means of producing quantities of isolated bacterial genes that may be used to measure specific transcription products by RNA–DNA hybridization. Virology of the photosynthetic bacteria may also be expected to make contributions unrelated to genetics, since the interaction of phage and host cell is an intimate one, reflecting many aspects of cell physiology including surface changes, alterations in transcription and translation machinery, energy status, and others. Within the last few years, the molecular biology of photosynthetic bacteria has passed from its infancy into a promising adolescence. Phages have been isolated, including some temperate ones, and an unusual system of genetic exchange has been discovered in *Rhodopseudomonas capsulata*. These recent advances will be the subject of this chapter. Some aspects of this work were briefly reviewed by Wall and Gest (1974).

Barry L. Marrs · Department of Biochemistry, Saint Louis University School of Medicine, Saint Louis, Missouri 63104

2. Virology of Photosynthetic Bacteria

Phages have thus far been isolated for only three species of photosynthetic bacteria, all of the genus *Rhodopseudomonas*, namely, *Rp. palustris*, *Rp. sphaeroides*, and *Rp. capsulata*. It is interesting that the first viruses for this group of bacteria were not isolated until the late 1960's, although the lore of photosynthetic workers has it that many unsuccessful searches were conducted before that time. The most recent searches, by Gest and co-workers in Bloomington, Indiana, and Kaplan's group in Urbana, Illinois, have been the most fruitful, yielding large numbers of distinct phage types. One is tempted to speculate that a period of environmental enrichment, by means of laboratory drains, was necessary to raise host titers before phage titers reached easily detectable levels. These groups are now able, however, to isolate *Rhodopseudomonas* phages from a wide range of sources; thus, their discovery is more probably attributable to more thorough searches.

2.1. Summary of Phage Properties

The first virus reported for a photosynthetic bacterium was for *Rp. palustris* (Freund-Mölbert *et al.*, 1968; Bosecker *et al.*, 1972), but unfortunately this viral strain has been lost (Drews, personal communication). Six years later, three different laboratories described successful isolations of viruses for Rhodospirillaceae.

2.1.1. Phages of *Rhodopseudomonas sphaeroides*

RS1, a virulent, double-stranded-DNA-containing phage infecting *Rp. sphaeroides* 2.4.1, was isolated by Abeliovich and Kaplan (1974). This virus appears to be typical to those classified as "Group C" by Bradley (1967). It possesses a polyhedral head, approximately 65 nm in diameter, and a tail 60 nm long (Fig. 1a). The phage particle has a buoyant density in CsCl of 1.50 g/cm^3. The buoyant density and T_m of RS1 DNA correspond to a guanine plus cytosine content of about 45%, which is in marked contrast to that of the host (67%). The molecular weight was estimated by C_0t analysis and sucrose density-gradient sedimentation to be between 3.0 and 3.5 × 10^7. RS1 requires a divalent cation for adsorption, and adsorbs more readily to aerobically grown than to phototrophically grown cells. It grows only on *Rp. sphaeroides* strains 2.4.1 and "L," and appears to experience a form of host restriction on strain M29 : 5, which is a derivative of the wild-type strain 2.4.7. No plaques were formed on strains of *Rp. capsulata, Rp.*

palastris, Rp. gelatinosa, or *Rs. rubrum*. An unexpected finding was that the efficiency of productive infection was much lower in photosynthetically grown cells than in aerobically grown ones. Adsorbed phages could be detected as infectious centers with an efficiency of about 1 in aerobically grown cells, even at low multiplicities of infection (MOI = 0.01). On the other hand, only a small and variable fraction of the phages that adsorbed to photosynthetically grown cells resulted in productive infection. At MOI of greater than 10, however, all anaerobic cells give rise to infectious centers. It would be of interest to determine whether the efficiency of infection of photosynthetically grown cells is influenced by the intracellular membrane content, which could be conveniently altered by varying the light intensity during growth of the host culture.

Rϕ-1, discovered by Mural and Friedman (1974) in a search for temperate phages, also grows in *Rp. sphaeroides*. Fresh isolates of presumed rhodopseudomonads were grown in the presence of mitomycin C, and supernatants of those cultures were tested for lytic activity on other presumptive rhodopseudomonads. One strain (unnamed) gave rise to Rϕ-1, which lysed various rhodopseudomonad isolates, including strain 014, but was unable to form plaques on *Rp. sphaeroides* 2.4.1. Mutant derivatives of Rϕ-1 that could form plaques on strain 2.4.1 were isolated, but could no longer form plaques on the original host (strain 014). Apparently the authors intend to call both this presumed host range mutant and the original phages simply Rϕ-1, since no nomenclature is suggested for the host range mutant. A curious observation is that the original Rϕ-1 is chloroform-sensitive, whereas the "mutant" is chloroform-resistant. More rigorous demonstration of the identity of these two viruses seems necessary, since it is well established that superinfection by phage A of a strain lysogenic for phage B can result in production of either or both phages, or, if A and B are closely related, recombinants (Luria and Darnell, 1967). In this connection, it is interesting to note that Abeliovich and Kaplan reported that RS1 lysates of strain 2.4.1 contain about 0.1% of an unidentified phage that resembles Rϕ-1 in general morphology. Both have a polyhedral head and a long tail without visible tail fibers or sheaths, and they seem to belong in the morphological group designated "B" by Bradley (Figs. 1c,d). In any event, the form of Rϕ-1 that can infect strain 2.4.1 seems to be able to form a stable lysogen, since cells isolated from turbid plaques can be induced by UV irradiation to lyse and give rise to a burst of about 10 phages per cell. Preliminary tests for transduction with Rϕ-1 were negative; however, efforts continue to develop a useful phage-mediated gene transfer system with both Rϕ-1 and another temperate phage for *Rp. sphaeroides*, RS-2, isolated by Kaplan's group (personal communication). The latter phage is morphologically distinct from the others described for strain 2.4.1, resembling the coliphage T7. Lysates induced by this virus have a very low-level transducing activity for a variety of markers. Thus far, the highest frequency of transduction is approximately 6 transductants for a particular marker per 10^8 recipients. This is less than the frequency of spontaneous reversion for most mutations, and therefore transfer has been demonstrated for only a few markers that have unusually low reversion frequencies (Anderson and Kaplan, personal communication).

2.1.2. Phages of *Rhodopseudomonas capsulata*

Bacteriophage RC1 was isolated from sewage by Gest's group in Bloomington, Indiana (Schmidt *et al.*, 1974). The virus has a polyhedral head, a sheathed, contractile tail, and tail fibers associated with a base plate (Fig. 1b), and contains double-stranded DNA. It thus resembles the T-even coliphages and has the characteristics of Bradley's Group A. This virulent phage requires Ca^{2+} for adsorption and is specific for *Rp. capsulata*, infecting a wide range of independently isolated strains of this species (Wall *et al.*, 1975a). RC1 grows equally well in aerobically and photosynthetically grown hosts. Wall *et al.* (1975a) reported the isolation of 15 additional types of virulent phages for *Rp. capsulata* from Bloomington sewage. The phages were typed by analyzing the pattern of susceptibility of a collection of 33 different wild-type strains of *Rp. capsulata*, but no further characterization was reported, save that each type was tested for transducing activity with negative results. Wall and co-workers also screened their extensive wild-type *Rp. capsulata* collection for lysogens with negative results. Five of the wild-type strains proved to harbor phage, but in each case, the association between virus and host was not stable enough to be termed lysogeny, since between 20 and 80% of clones tested failed to give rise to phage. One of these carrier strains, B10, appears to carry two phage types, distinguishable by host range and plaque morphology. One of these phages, designated RC1001, produces tiny "asterisk-like" plaques (Wall *et al.*, 1975a), and sediments in the vicinity of 100 S in sucrose density gradients (Marrs, unpublished). These observations suggest that RC1001 may belong to one of the groups of tiny phages classified D, E, and F by Bradley.

2.2. Bioenergetics of Phage Infection

Schmidt *et al.* (1974) examined the bioenergetic aspects of bacteriophage RC1 replication. Their studies suggest that the nonsulfur purple photosynthetic bacteria share with the blue-green algae (Adolph and Haselkorn, 1972; Padan *et al.*, 1970) distinct advantages for studying the bioenergetics of phage development. The *in vivo* rate of ATP synthesis is particularly amenable to experimental manipulation, since it can be driven by light, and can thus be easily "programmed." The workers reported that continuous ATP regeneration throughout the latent period is essential for optimal phage replication. They also observed an intriguing threshold effect in the rate of ATP regeneration required for phage replication. Host cells containing different quantities of Bchl were obtained by varying the extent of aeration during dark growth. Samples of these cells were infected with phage, and the average burst sizes were determined. Phage development did not occur in cells with Bchl contents of 0.42 μg/mg dry wt. or less, although uninfected cells with lower Bchl contents were able to grow photosynthetically after a lag period. Cells with 0.62 μg/mg dry wt. gave rise to about 50% of the maximum number of phage per cell, and the burst size increased with increasing Bchl content above this threshold value, up to the maximum Bchl content tested, 2.64 μg/mg dry wt. Similar effects were demonstrated with photophosphorylation inhibitors and intermittent illumination (30 sec light, 30 sec dark). Schmidt *et al.* (1974) infer that there is a relatively high threshold rate of ATP regeneration necessary for orderly synthesis of phage components. They point out that since protein synthesis and Bchl synthesis stop about half-way through the 2-hr latent period, phage-infected cells cannot respond to energy stresses the way uninfected cells can, and thus the infected cell is essentially dependent on the state of the cells' energy-producing capacity at the time of infection.

3. Gene Transfer System of *Rhodopseudomonas capsulata*

3.1. Discovery

The *Rp. capsulata* genetic exchange process was discovered by screening fresh isolates of Rhodospirillaceae for recombination of antibiotic resistance markers (Marrs, 1974). Early experiments demonstrated that genetic information transfer was mediated by a nuclease-resistant vector found in cell-free filtrates. This vector, termed the *gene transfer agent* (GTA), is found in the medium of cultures entering the stationary phase of growth. GTA is adsorbed by recipient cells, which may then incorporate the genetic information carried, presumably with the concomitant loss of the corresponding resident genetic region of the recipient (Solioz *et al.*, 1975; Yen and Marrs, in prep.). Under the proper conditions, this process can occur at a frequency such that 0.1% of the recipient population will pick up and express a particular new genetic trait via GTA.

3.2. Physical Nature of the Gene Transfer Agent

GTA particles resemble small bacterial viruses. The accompanying electron micrograph (Fig. 1e) shows a 300 Å icosahedral head with short spikes and a tail of variable length. The tail is 50–60 Å in diameter, and it seems noncontractile. A stain-excluding "collar" is visible in "empty" particles. The identification of the particles seen under the electron microscope (EM particles) with GTA rests on the following evidence:

1. The EM particles are present in the same buoyant density fraction ($p = 1.34$ g/cc) of an RbCl isopycnic gradient as GTA activity.
2. The EM particles are of the approximate size expected from sedimentation analysis and from exclusion chromatography on controlled-pore-size glass beads (500 Å < GTA < 1500 Å) (Solioz, 1975).
3. The observed particle would be exected to carry a piece of DNA with the properties of GTA DNA, i.e., double-stranded, linear DNA of 3.6×10^6 daltons molecular weight, if one assumes that the EM particles should show the same DNA mass/head volume ratio as morphologically similar bacterial viruses. Furthermore, the family of morphologically similar bacterial viruses (Bradley's Group C) contain linear double-stranded DNA to the exclusion of single-stranded or circular forms.
4. An independent means of purification gave a collection of three kinds of particles visible in the EM, and among them was the particle in question.

The nature of the nucleic acid contained in GTA was studied by means of radioisotopic labeling, since only minute amounts of material were available. GTA was prepared (Solioz and Marrs, 1977) from a culture labeled with 5-methyl-[^{3}H]thymidine. The procedure employed gave a 4000-fold purification as measured by the increase in the ratio of gene transfer activity to TCA-precipitable counts per minute. The nucleic acid extracted from this preparation by phenol sedimented

as a single homogeneous 14 S band in both neutral and alkaline sucrose gradients. The labeled material was completely resistant to digestion by the single-strand-specific nuclease, S1, but on being heated and quickly cooled, it was converted into a form that was more than 90% acid-solubilized by the same enzyme. The labeled nucleic acid banded at 1.718 g/cc buoyant density in CsCl when native, and shifted to 1.734 g/cc after heat denaturation. The buoyant density in Cs_2SO_4 was 1.434 g/cc. These properties establish the nucleic acid contained in GTA as linear, double-stranded DNA with a molecular weight of about 3.6×10^6 daltons, and the same G+C content as *Rp. capsulata* (about 65%).

3.3. Physiology of the Gene Transfer Process

The modes of production, release, and uptake of GTA are not yet well understood. Most strains of *Rp. capsulata* isolated from nature can participate in the gene transfer process. Wall *et al.* (1975a) showed that of 33 strains examined: (1) at least half the strains examined can both donate and receive GTA; (2) two strains can only produce GTA; (3) eight strains can only receive GTA; and (4) six strains can neither produce nor receive. One strain, which produced GTA but could not receive genetic information from either itself or various other donors, was demonstrated to be defective in uptake of GTA (Marrs, unpublished). Strains that produced GTA were isolated from a wide range of geographic locations, including many areas of the United States, Puerto Rico and Germany, and thus the ability to produce GTA is widely distributed among *Rp. capsulata* strains in nature. Wall *et al.* (1975a) also screened a large number of other species of Rhodospirillaceae for interspecies GTA-mediated genetic exchange, but exchange occurred only among strains of *Rp. capsulata*. Twenty-three strains of *Rp. sphaeroides* have been screened for gene transfer to and from *Rp. capsulata* as well as in an intraspecific mode, all with negative results (P. Weaver, personal communication).

The production of GTA is greatly influenced by growth conditions, but a systematic study of these factors has not been undertaken. Relatively few media support the production of GTA, and even the peptone–yeast extract medium, which has been used in all the published GTA studies to date, shows variability with respect to yield when different lots of yeast extract are employed (Yen and Marrs, unpublished). Yen and Marrs (in prep.) have developed a synthetic medium that supports a high level of GTA production, but the critical differences between permissive and nonpermissive media have not been identified.

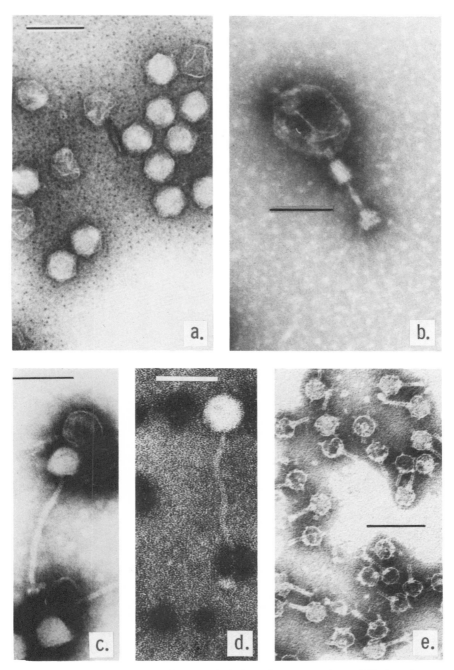

Fig. 1. Bacteriophages and gene transfer agent of *Rhodopseudomonas capsulata* and *Rhodopseudomonas sphaeroides*. Each bar is approximately 100 nm. (a) RS1, a virulent phage infecting *Rp. sphaeroides* (Abeliovich and Kaplan, 1974). (b) RC1, a virulent phage infecting *Rp. capsulata* (Schmidt *et al.*, 1974). (c) Rϕ-1, a temperate phage infecting *Rp. sphaeroids* (Mural and Friedman, 1974). (d) Unidentified particles found at a frequency of about 0.1% in purified preparations of RS1 (Abeliovich and Kaplan, 1974). (e) Gene transfer agent of *Rp. capsulata* (Yen and Marrs, unpublished).

Rapid growth is sometimes correlated with the production of high GTA titers, but this observation is of limited generality, since photosynthetic cultures produce about 100 times as much GTA as aerobic cultures, although the difference in growth rates may be negligible.

The time course of the appearance of GTA in a growing donor culture also varies with the composition of the medium; in general, however, GTA is released in one or more abrupt waves toward the end of exponential growth (Solioz et al., 1975). The turbidity of the culture does not decrease during GTA release, but it is possible that a small fraction of the donor population lyses to produce all the GTA observed. Yen and Marrs (in prep.) recently isolated mutant strains of Rp. capsulata that produce from 100 to 1000 times as much GTA per donor cell. These "overproducer" strains do appear to lyse during GTA production, and more work needs to be done to determine whether GTA release is always accompanied by cell lysis.

Free GTA disappears rapidly from a mixture of recipient cells and GTA in G-buffer (10 mM Tris, pH 7.8, 1 mM $MgCl_2$, 1 mM $CaCl_2$, 1 mM NaCl), and since the appearance of cells committed to expressing a GTA-derived phenotype is complementary to the disappearance of GTA, it seems reasonable to equate the disappearance with uptake (Solioz et al., 1975). Aerobically grown cells are less effective GTA recipients than cells from photosynthetic cultures, although it is not known whether the difference lies in uptake or expression. Nearly all the cells in a photosynthetic culture may be active recipients, as demonstrated by an analysis of the frequency of simultaneous independent gene transfers (Solioz et al., 1975). Gene transfer is inhibited by a wide variety of compounds and conditions, but in most instances, the affected step cannot yet be specified. Some of the more surprising inhibitory agents (>75% inhibition) are: glycerol (10%, vol/vol); sodium phosphate (10 mM, pH 7.8); NH_4Cl (100 mM); NaCl (50 mM), and peptone–yeast extract (the growth medium that gives the highest yields of GTA). NaCl and peptone–yeast extract appear to manifest their inhibitory effects on processes that normally occur after adsorption is complete. Sensitivities to KCN and anoxia demonstrate that energy metabolism is required during gene transfer.

After having considered the physical nature of the GTA and what is known of the genetic exchange process that is mediated by GTA, we may pose the question whether this system should be considered an example of generalized transduction, or whether it may be best regarded as a bacterial genetic exchange mechanism that does not involve bacteriophage. Although it is not possible to choose between these alternatives on the basis of existing data, it is instructive to set forth those features of the gene transfer system that distinguish it from previously described transduction systems:

1. No viral activity is found associated with gene transfer activity. Wall et al. (1975a) screened a large assortment of Rp. capsulata strains, but found none that would serve as an indicator for the hypothetical virus associated with GTA. One strain, B10, which produced GTA, was found to harbor phages capable of forming plaques on two indicator strains, but segregants of B10 that no longer produced detectable plaque-forming units were unchanged with respect to GTA production. Furthermore, some strains derived from B10 by mutagenesis (e.g., W16; Wall et al., 1975b) have lost the ability to produce GTA, but still produce phages (J. Wall, personal communications). There is no detectable killing of the recipient population on treatment with purified GTA (Marrs, unpublished). If there is a phage underlying GTA production, it would seem to be either defective or cryptic.

2. The majority of Rp. capsulata strains isolated from nature show gene transfer activity (Wall et al., 1975a). Thus, a phage involved in GTA production must be assumed to be extremely widespread, and since most Rp. capsulata strains are capable of acting as recipients for GTA, a wide host range would be implied.

3. We have been unable to transfer by "infection" with GTA the ability to produce GTA into recipient strains that do not spontaneously produce it (Yen and Marrs, in prep.).

4. The GTA particle is smaller in size than any known transducing phage, and it is smaller than any known phage of comparable morphological complexity. Reasoning by analogy with other phages, it would be remarkable if a particle with a head, head spikes, a collar, and a tail could maintain a nonlytic life cycle with such a small DNA complement. It might be argued, however, that the GTA observed in the electron microscope is not representative of the putative phage virion, but represents a defective particle.

Clearly, bacterial virology holds so many variations that one cannot determine whether or not a phage is involved in GTA production solely by comparing the present system to known transducing systems. We are currently examining GTA DNA to determine whether it contains reiterated DNA sequences that might indicate the presence of a hitherto undetected phage.

Lacking data to the contrary, it seems permissible to speculate that GTA may represent a "prephage" bacterial system, rather than a defective system evolved from a phage. We know little about the evolution of phages, and if a bacterial GTA-like system evolved in response to selective advantages conferred by recombination, it would then be easy to imagine the further evolution or escape of the system into the parasitic phages as we know them (Luria and Darnell, 1967).

3.4. Genetic Characteristics of *Rhodopseudomonas capsulata* Gene Transfer

Genetic markers acquired via gene transfer are eventually incorporated into a stably inherited form. Marrs (1978) found the same reversion frequency for a particular mutation (*trpA20*, inactivating tryptophan synthetase) in both the strain in which it originated and a strain to which it was transferred via GTA. This experiment also suggests that the GTA-borne marker replaces the resident marker, since a diploid would be expected to be tryptophan-independent (assuming that *trpA20* is recessive to wild type). Similarly, the gene transfer process is highly efficient in transferring markers conferring rifampicin resistance, a marker that is usually recessive. A collection of mutations blocking carotenoid biosynthesis at various stages has been studied, and all have been transferable even though they would be expected to be recessive.

If the *trpA20* marker were dominant, and genes acquired via GTA could not replace resident markers, no tryptophan-independent recombinants would be expected when a strain bearing *trpA20* was treated with wild-type GTA. Since a high frequency of tryptophan-independent recombinants are, in fact, recovered from this cross (Solioz *et al.*, 1975), we may conclude that the *trpA20* marker cannot be dominant unless GTA-borne genes *can* replace resident markers. Although none of the markers mentioned above has been demonstrated to be recessive by direct experiments, it seems unlikely that all the markers studied have been "unusual"; thus, taken together, the evidence supports the replacement of resident markers as the normal mode of incorporation of genetic information from GTA.

Two lines of evidence suggest that although the incoming genetic material may eventually be stably integrated into the recipient chromosome, it may exist for a few cell divisions in an unstable, nonreplicating state, and it may direct protein synthesis during that time. Both observations arise from experiments in which carotenoidless (blue-green) recipients are treated with GTA from a carotenoid-producing donor. Carotenoidless cells are sensitive to photooxidative killing (POK), whereas cells containing any colored carotenoids are not. Treatment of carotenoidless cells with GTA from carotenoid-producing donors results in an increase in survivors of subsequent POK, and one increased class of survivors remains genetically carotenoidless. When grown and retested for sensitivity to POK, they are indistinguishable from the original starting strain. Thus, they have been temporarily protected from POK by exposure to GTA. One hypothesis that could account for this observation is that some cells receive a wild-type copy of the gene that they lack, and that gene begins to function before it is integrated into the recipient chromosome. Thus, colored carotenoids may be produced even though the new genetic element may be lost through the lack of replication. An alternative hypothesis suggests that recombination of a newly acquired gene may occur in one chromosome of a cell with several chromosomes. If it began functioning before cell division partitioned its gene products into separate compartments, cells with carotenoidless chromosomes might receive enough mRNA and enzymes to enable them to synthesize sufficient carotenoids to survive POK. The latter hypothesis would predict that a mixed clone would result from this genetic event; i.e., the surviving cells should form a colony containing both carotenoidless and carotenoid-producing cells. Such colonies are observed (although they may arise by mechanisms other than that suggested), but pure blue-green surviving colonies are also observed, and we therefore favor the first hypothesis.

A second observation that suggests that the former hypothesis is correct concerns the appearance of sectored colonies. Colonies in which a donor phenotype appears as one-half or less of the colony are frequently observed, and these may arise in many ways. However, colonies in which the donor phenotype is present in more than half the cells probably arise by delayed recombination between chromosome and GTA elements. Another sectoring pattern that suggests the same phenomenon is one-half donor, one-half recombinant. Since both of these patterns are found, it seems likely that GTA DNA may remain in the cytoplasm in a nonreplicating but genetically competent state for several division cycles.

Two features of the gene transfer system facilitate strain construction. Many strains are self-fertile; i.e., they can receive and utilize genetic information from their own GTA. Second, strains may be used as recipients repeatedly without becoming "immune" to gene transfer, and thus, strains may be built up stepwise. The gentic linkages demonstrable in this system are somewhat weak for the purposes of strain construction. Markers separated by about four genes show only about 10% cotransfer, and cotransfer rapidly approaches zero for markers separated by more than about six genes (Yen and Marrs, 1976).

3.5. Mapping the Photosynthesis Region

Because cotransfer is limited to closely linked markers, this system is well suited for genetic mapping of regions about the size of an operon or smaller. Yen and Marrs (1976) succeeded in mapping a small region of the *Rp. capsulata* chromosome carrying five genes concerned with carotenoid biosynthesis and two genes for Bchl synthesis (Fig. 2). The mapping was accomplished by cotransfer analyses and ratio test crosses. Cotransfer data were converted to map distances by means of a new mapping function: cotransfer frequency $\phi = (1 - d)^2$ where d is the distance between two markers. This map function was shown to be appropriate by demonstrating that the distances generated by the function were additive. The map function is predicted by a model for gene transfer that assumes, among other things, that donor DNA is incorporated into the recipient chromosome by means of a pair of recombination events, one of which occurs at random between the incoming DNA and the homologous region of the host chromosome, while the second crossover occurs at or near the end of the donor molecule. If the model is correct, it gives physical meaning to genetically determined map distances, namely, that the DNA molecule carried by a GTA is 1 map unit long. This interpretation is supported by the observation that a typical "gene" in the mapped region seems to span somewhat less than 0.2 map unit, which suggests that GTA DNA carries about five genes. That genetic estimate is in excellent agreement with the physically determined size of 6 × 10³ base pairs, since an average gene is often taken as 10³ base pairs.

Cotransfer data for mapping the photopigment region were collected from a series of crosses in which the donors were mutationally blocked late in carotenoid biosynthesis (green or yellow mutants) and the recipients were either blocked early in carotenoid biosynthesis (blue-green mutants) or were blocked in Bchl synthesis (nonphotosynthetic). Transferants were selected for either resistance to POK (blue-green recipients) or ability to photosynthesize (nonphotosynthetic recipients), and cotransfers were scored by examining transferant colonies for carotenoid phenotypes.

A second type of genetic analysis was employed to determine whether pairs of markers were closely linked (e.g., in the same gene) or more widely separated on the chromosome. This procedure, termed the *ratio test*, can be used when both markers confer the same phenotype and that phenotype may be selected against. It was used to demonstrate that there are two separate genetic loci into which mutations causing the blue-green phenotype may fall.

The genes *crtB* and *crtE* are necessary for early steps in carotenoid biosynthesis, and thus mutations in either gene can cause the blue-green phenotype. Mutations in *crtC* or *crtD* cause the green phenotype, and mutations in the *crtA* locus cause a yellow phenotype. Chemical identification of the carotenoids accumulated in each case is in progress.

The *bchA* and *bchB* loci are involved in Bchl synthesis. Mutations in the former cause the accumulation of a presumed Bchl precursor that shows a 665 to 667 nm redmost absorption maximum, while *bchB* mutations result in accumulation of a 630 to 635 nm absorbing compound. Only one *bchB* mutation has been mapped, because it is difficult to use nonphotosynthetic donors or recipients in the gene

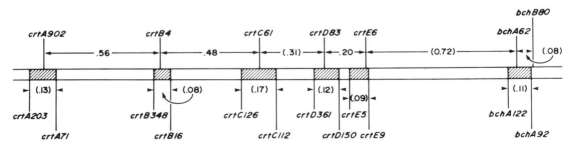

Fig. 2. Genetic map of the region of the *Rhodopseudomonas capsulata* genome devoted to carotenoid and Bchl synthesis. The numbers above the map represent distances, in map units, between specific markers in each gene. The numbers below the map give the distances between the (current) terminal markers in each gene, thus providing estimates of the minimum length of each gene. Distances obtained by subtraction are given in parentheses (Yen and Marrs, 1976).

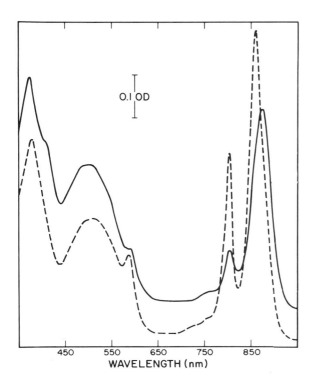

Fig. 3. Absorption spectra of wild-type and mutant strains of *Rhodopseudomonas capsulata*. Cells were grown in the dark with low aeration, and spectra were determined using centrifugal supernatants of cells disrupted by sonication. (– – –) Wild type; (——) carotenoid-producing mutant derived from crossing a blue-green recipient with wild-type GTA.

transfer system. The accumulated compounds have not been further characterized.

The observation that genes for carotenoid and Bchl biosynthesis are closely linked is exciting, because it suggests that certain steps in these pathways may be coordinately regulated at the transcriptional level. If this is indeed true, it would represent an unusual bacterial operon—a superoperon containing genes for the synthesis of two chemically unrelated compounds the synthesis of which are presumably coordinated because they function together.

An unexpected consequence of mapping the carotenoid region has been the resolution of the genotype of blue-green mutants into two contributing mutations. When any of the blue-green mutants thus far examined (seven) is treated with wild-type GTA, two types of colored-carotenoid-producing transferants are recovered: those with the wild-type Bchl absorption spectrum and those with an absorption spectrum in the IR resembling the blue-green parent (Fig. 3). This indicates that while one mutation results in the loss of colored carotenoids, a second mutation causes the characteristic alteration of the Bchl absorption, presumably reflecting a change in the environment of the bulk Bchl. If it is obligatory that these mutations accompany each other to produce viable blue-green strains, as preliminary evidence

indicates (Yen and Marrs, 1976), analysis of the nature of the change brought about by the second mutation may give new insight into the role of carotenoids in photosynthesis.

4. Prospects for Additional Genetic Exchange Systems

Photosynthetic bacterial research would benefit if genetic exchange systems for *Rp. sphaeroides* or *Rhodospirillum rubrum* could be discovered, since more is known about them than about *Rp. capsulata*. A few species of *Rhodopseudomonas* have been screened for a gene transfer system like that found for *Rp. capsulata*, but none has yet been found (Wall, personal communication). It seems likely that other photosynthetic species possess some means for genetic recombination that have been overlooked, and indeed other mechanisms for exchange in *Rp. capsulata* may exist. We may at least be confident that *Rp. capsulata* has the enzymes necessary to incorporate genetic information from an exogenous source.

The gene transfer system that we have studied in *Rp. capsulata* is useful for genetic analysis of relatively short regions of the chromosome, but systems that transfer longer tracts are needed. Furthermore, the current system does not appear to form stable merozygotes (partial diploids), so *cis–trans* analyses for defining cistrons are not easily performed.

As indicated earlier, the search for genetic exchange systems among photosynthetic bacteria does not take place in virgin territory. Serious efforts to find transduction or transformation have been conducted with marginal or negative results. The most promising area for future investigation may be conjugation. Sex factors that mobilize the host chromosome are known for many *Pseudomonas* species, and Olsen and Shipley (1973) demonstrated the transfer of a *Pseudomonas aeruginosa* R Factor, R1822, into both *Rp. sphaeroides* and *Rs. rubrum*. R1822 does not mobilize the *P. aeruginosa* chromosome, and it is not stably inherited by the photosynthetic bacteria, but the result suggests that some *Pseudomonas* sex factors might also have a host range that includes the photosynthetic pseudomonads. Furthermore, naturally occurring photosynthetic bacterial conjugation systems have not been sought on a large scale.

Kaplan and co-workers (personal communication) appear to have succeeded in transferring chloramphenicol resistance from *Escherichia coli* to *Rp. sphaeroides* using a P1 bacteriophage strain that carries the resistance marker. This finding opens up many possibilities, not the least of which is the development of a P1–*Rp. sphaeroides* system with which transduction may be carried out.

In conclusion, it seems safe to assume that the discovery of gene transfer in *Rp. capsulata* will accelerate the pace of research with the photosynthetic bacteria, and that equally useful additional genetic tools will be forthcoming.

Note Added in Proof: The transfer of chromosomal genes mediated by the promiscuous *Pseudomonas* sex factor plasmid R68.45 in *Rhodopseudomonas sphaeroides* has indeed been demonstrated by W. R. Sistrom (*J. Bacteriol.* **131**:526, 1977). This breakthrough should provide the long-awaited genetic system for *Rp. sphaeroides*, and point the way for establishing similar systems for other photosynthetic bacteria.

Acknowledgments

I thank Drs. D. Friedman, H. Gest, S. Kaplan, P. Weaver, and J. Wall for helpful discussions and sharing their unpublished observations, and A. Hu, R. LaMonica, J. Baldassare, and especially R. Narconis for their contributions to obtaining electron micrographs of the gene transfer agent.

Work in the author's laboratory was carried out with the excellent technical assistance of Sandra Bilyeu and was supported by grants from the National Institute of General Medical Sciences and the National Science Foundation.

5. References

Abeliovich, A., and Kaplan, S., 1974, Bacteriophages of *Rhodopseudomonas spheroides*: Isolation and characterization of a *Rhodopseudomonas spheroides* bacteriophage, *J. Virol.* **13**:1392.

Adolph, K. W., and Haselkorn, R., 1972, Photosynthesis and the development of blue-green algal virus N-1, *Virology* **47**:370.

Bosecker, K., Drews, G., and Tauschel, H. D., 1972, Untersuchungen zur Adsorption des Bacteriophagen Rpl an *Rhodopseudomonas palustris* 1e5, *Arch. Mikrobiol.* **87**:139.

Bradley, D. E., 1967, Ultrastructure of bacteriophages and bacteriocins, *Bacteriol. Rev.* **31**:230.

Freund-Mölbert, E., Drews, G., Bosecker, K., and Schubel, B., 1968, Morphologie and Wirtskreis eines neu isolierten *Rhodopseudomonas palustris* Phagen, *Arch. Mikrobiol.* **64**:1.

Luria, S. E., and Darnell, J. E., 1967, *General Virology*, John Wiley and Sons, New York.

Marrs, B. L., 1974, Genetic recombination in *Rhodopseudomonas capsulata*, *Proc. Natl. Acad. Sci. U.S.A.* **71**:971.

Marrs, B. L., 1978, Mutations and genetic manipulations as probes of bacterial photosynthesis, in: *Current Topics in Bioenergetics*, Vol. 8 (D. R. Sanadi and L. P. Vernon, eds.), pp. 261–294, Academic Press, New York (in press).

Mural, R. J., and Friedman, D. I., 1974, Isolation and characterization of a temperate bacteriophage specific for *Rhodopseudomonas spheroides*, *J. Virol.* **14**:1288.

Olsen, R. H., and Shipley, P., 1973, Host range and properties of the *Pseudomonas aeruginosa* R factor, R1822, *J. Bacteriol.* **113**:772.

Padan, E., Ginzburg, D., and Shilo, M., 1970, The reproductive cycle of cyanophage LPP1-G in *Plectonema boryanum* and its dependence on photosynthetic and respiratory systems, *Virology* **40**:514.

Schmidt, L. S., Yen, H. C., and Gest, H., 1974, Bioenergetic aspects of bacteriophage replication in the photosynthetic bacterium *Rhodopseudomonas capsulata*, *Arch. Biochem. Biophys.* **165**:229.

Solioz, M., 1975, The gene transfer agent of *Rhodopseudomonas capsulata*, Ph.D. thesis, Saint Louis University.

Solioz, M., and Marrs, B., 1977, The gene transfer agent of *Rhodopseudomonas capsulata*: Purification and characterization of its nucleic acid, *Arch. Biochem. Biophys.* **181**:300.

Solioz, M., Yen, H. C., and Marrs, B., 1975, Release and uptake of gene transfer agent by *Rhodopseudomonas capsulata*, *J. Bacteriol.* **123**:651.

Wall, J. D., and Gest, H., 1974, Prospects for the molecular biology of photosynthetic bacteria, in: *Proceedings of the Third International Congress on Photosynthesis* (M. Avron, ed.), pp. 1179–1188, Elsevier Scientific Publishing Co., Amsterdam.

Wall, J. D., Weaver, P. F., and Gest, H., 1975a, Gene transfer agents, bacteriophages, and bacteriocins of *Rhodopseudomonas capsulata, Arch. Microbiol.* **105**:217.

Wall, J. D., Weaver, P. F., and Gest, H., 1975b, Genetic transfer of nitrogenase-hydrogenase activity in *Rhodopseudomonas capsulata, Nature (London)* **258**:630.

Yen, H. C., and Marrs, B., 1976, A map of genes for carotenoid and bacteriochlorophyll biosynthesis in *Rhodopseudomonas capsulata, J. Bacteriol.* **126**:619.

CHAPTER 48

Ribosomes and RNA Metabolism

Ernest D. Gray

1. Introduction

Certain photosynthetic bacteria afford an unusual opportunity to study the molecular events involved in morphogenesis. The adaptation of these organisms to phototrophic growth involves the coordinated synthesis of the membranous intrusions of the photosynthetic apparatus, which can be easily induced by altering the growth conditions. The experimental system is a fascinating one that should lead to wider insights into molecular mechanisms involved in gene expression during morphogenesis.

An interest in the role of RNA in this process has led many investigators to an analysis and characterization of a variety of aspects of RNA metabolism. These studies have been carried out on a limited number of the Rhodospirillaceae, and while they have provided much insight into the molecular biology of these organisms, extrapolation to the rest of the photosynthetic bacteria must be done with much circumspection. The unusual rRNA complement of several of these organisms suggests that there may exist other differences in RNA metabolism. This chapter is an attempt to outline the present status of several lines of inquiry in this area; it is clearly incomplete, but it is indicative of the directions of research being pursued at present.

2. Ribosomes and Ribosomal RNA

2.1. Characteristics of Ribosomes and Ribosomal RNA

The ribosomes of those photosynthetic bacteria that have been examined have the general sedimentation characteristics common to most bacteria (Taylor and Storck, 1964). *Rhodospirillum rubrum*, *Rhodopseudomonas gelatinosa*, *Rp. sphaeroides*, and *Rp. palustris* all have ribosomes with sedimentation constants near 66 S. In *Rp. sphaeroides* (Friedman *et al.*, 1966) and *Rp. palustris* (Bhatnagar and Stachow, 1972), the functional ribosome is composed of a 29 S and a 45 S subunit that in *Rp. palustris* contain at least 23 and 28 proteins, respectively, similar to other bacteria (Nomura, 1970).

The majority of the Rhodospirillaceae that have been examined possess the usual complement of rRNA, although some interesting exceptions have been observed in which unusual rRNA species occur. Table 1 compares the molecular weights estimated by polyacrylamide gel electrophoresis of the major high-molecular-weight RNA species of many of the photosynthetic bacteria (Kaplan, personal communication). Most of these organisms contain a larger rRNA of approximately 1.0×10^6 daltons (1.0 M)* and a

* The rRNA species are referred to by their molecular weights estimated by electrophoretic mobility, since sedimentation analysis in most cases was not performed. rRNA species with molecular weights of 1.1×10^6, 0.93×10^6, 0.65×10^6, 0.53×10^6, and 0.40×10^6 are abbreviated as 1.1 M, 0.93 M, 0.65 M, 0.53 M, and 0.40 M RNAs.

Ernest D. Gray · Departments of Biochemistry and Pediatrics, University of Minnesota, Minneapolis, Minnesota 55414

Table 1. Ribosomal RNA of Photosynthetic Bacteria[a]

Bacterium	Source	Molecular weight $\times 10^6$	
Rp. sphaeroides (2.4.1)	Sistrom	0.54	0.43
Rp. capsulata (St. Louis)	Gest	0.62	0.48
Rp. gelatinosa	Pfennig	1.30	0.67
Rp. palustris (77/7)	Drews	1.20	0.63
Rp. acidophila	Pfennig	0.93	0.54
Rp. viridis	Pfennig	0.78	0.40
Rs. rubrum	Pfennig	1.04	0.59
Rs. molischianum	Pfennig	1.20	0.61
Rs. fulvum	Pfennig	1.15	0.53
Rs. photometricum	Pfennig	0.98	0.58
Rs. tenue	Pfennig	0.98	0.54
Rm. vannielii	Pfennig	1.15	0.59
Chr. vinosum	Pfennig	1.10	0.58
T. roseopersicina	Pfennig	1.00	0.53

Reference RNA

Esch. coli	—	1.08	0.57

[a] Molecular weights were estimated from electrophoretic mobilities on polyacrylamide gels. Cells were labeled with [³H]uracil and mixed with [¹⁴C]uracil-labeled *Esch. coli* before cell breakage in a French press in a buffer containing sodium dodecyl sulfate and ribonuclease inhibitors. Electrophoresis of lysates was carried out within 10 min after cell breakage (Kaplan, personal communication).

smaller rRNA of about 0.53×10^6 daltons (0.53 M) molecular weight. Obvious exceptions are *Rp. sphaeroides* and *Rp. capsulata*, both of which lack the 1.0 M RNA and in addition to a 0.53 M RNA possess an even smaller species of RNA. *Rhodopseudomonas viridis* possesses major RNA species of 0.78×10^6 daltons and 0.40×10^6 daltons, another example of an unusual rRNA complement. It would appear that the absence of the 1.0 M RNA is restricted to certain rhodopseudomonads.

The rRNA of *Rp. sphaeroides*, which has been most extensively studied, consists of two distinct 0.53 M species, one of which forms part of the 29 S ribosomal subunit; the other is from the 45 S ribosomal subunit, which also contains 0.40 M RNA (Marrs and Kaplan, 1971). The unusual absence of 1.0×10^6 dalton RNA in this organism was first reported by Lessie (1965a), but was attributed by others to the action of nucleases during isolation. Szilagyi (1968) and also Borda *et al.* (1969) demonstrated that an RNA that sedimented as 23 S in sucrose gradients could be isolated from *Rp. sphaeroides* under certain conditions. It was later shown (Marrs and Kaplan, 1971; Robinson and Sykes, 1973) that the 23 S RNA

demonstrated in these earlier reports was likely an aggregate of the 0.53 M RNA species that formed under the conditions of isolation and centrifugation. It is now well established that the observation of smaller rRNA species in these organisms is not artifactual, but represents an unusual processing sequence in the maturation of rRNA. It may be assumed that the rRNA species of *Rp. capsulata* and *Rp. viridis* are distributed in the ribosomes as in *Rp. sphaeroides*, but no evidence for this is presently available.

In addition to the higher-molecular-weight rRNA, 5 S rRNA species are a normal component of the ribosomes. In the Rhodospirillaceae, 5 S RNA has been demonstrated only in *Rp. sphaeroides* (Marrs and Kaplan, 1971), but is likely present in all species.

The 29 S ribosomal subunit of *Rp. sphaeroides* is very sensitive to the action of ribonuclease, which appears to lead to the disruption of the particle (Friedman *et al.*, 1966). This observation suggests that the rRNA must be ordered in the ribosome in such a way as to expose susceptible linkages to nuclease cleavage. Sensitivity of the particle to ribonuclease T_1 can be prevented by the binding of synthetic mRNA containing no susceptible bonds (polyU), indicating that the mRNA-binding site in this ribosomal subunit may involve rRNA. The dissociation of the 29 S ribosome after cleavage of the RNA emphasizes that this structure is much different from the 45 S subunit, which is stable under these conditions and functions even though its RNA is cleaved in the processing sequence. Robinson and Sykes (1973) showed that under the dissociating effect of EDTA, the 45 S ribosome differs from the corresponding *Escherichia coli* ribosomal subunit by splitting into two ribonucleoprotein particles, each containing one of the rRNA species. The presence in 50 S ribosomes from *Esch. coli* of an intact 1.1 M RNA species evidently prevents a similar disruption of the particle under such conditions. RNA in the larger subunit evidently also has a role in the stabilization of the particle, as it does in the smaller ribosome.

A recent report indicated that on transfer of *Rp. palustris* from anaerobic photosynthetic to aerobic dark conditions, a gradual loss of ribosomes occurs, followed by a reappearance at the time of resumption of growth (Mansour and Stachow, 1975). Such a finding is unusual, but may be related to the extended lag period of 100 hr after transfer to aerobic conditions. It has been well established that under certain nutritional deprivations, breakdown of bacterial ribosomes occurs (Horiuchi *et al.*, 1959; McCarthy, 1962; Ben-Hamida and Schlessinger, 1966), and a similar phenomenon may be involved with *Rp. palustris*, although the medium was adequate after growth resumed. It is difficult to understand how the transition to aerobic growth is accomplished without ribosomes and thus presumably without protein synthesis. A similar transition in *Rp. sphaeroides* results in only a slight lag in growth and no observable change in the cellular level of ribosomes (Gray, unpublished).

2.2. Ribosomal RNA Synthesis

2.2.1. Kinetics of Formation

The synthesis of rRNA in eukaryotes proceeds by the selective cleavage of a single large precursor molecule into the RNA species represented in the ribosomes (Attardi and Amaldi, 1970). A similar maturation sequence was recently demonstrated in *Esch. coli*, but with the difference that the large precursor transcript is observable only in mutant organisms lacking RNase III, the putative processing nuclease (Dunne and Studier, 1973; Nikolaev *et al.*, 1973). This precursor RNA, 1.8×10^6 daltons in size, is normally reduced before completion of transcription to smaller molecules that are further cleaved to the mature rRNAs. The products of this initial cleavage are precursors to the stable RNA species, and on further processing are reduced to these mature rRNAs (Adesnick and Levinthal, 1969; Pace *et al.*, 1970).

In *Rp. sphaeroides*, the first detectable rRNA species corresponded to the precursor rRNA molecules that are formed in *Esch. coli*. The rapidly formed high-molecular-weight RNA species were first observed by Lessie (1965b). Marrs and Kaplan (1971) demonstrated that a 23 S (1.1×10^6 daltons) RNA molecule appeared to be a precursor of the 0.53 M RNA of the 45 S ribosome, while an RNA of 0.65 M is likely the precursor of the 0.53 M RNA in the 29 S ribosomal subunit. We have examined in detail the kinetics of formation of the rRNA of *Rp. sphaeroides* using pulse-chase experiments (Tavernier, P. E., and Gray, E. D., in prep.) with analysis of the RNA products by electrophoresis on composite polyacrylamide–agarose gels (Dahlberg and Peacock, 1971). The spectrum of major RNA species observable during such a study includes RNAs with molecular weights of 1.1×10^6, 0.93×10^6, 0.65×10^6, 0.53×10^6, and 0.40×10^6 daltons. In anaerobic photosynthetically growing cells, the 1.1 M, 0.93 M, and 0.65 M RNA species follow a time course characteristic of precursor molecules, increasing before the uridine chase and decreasing afterward as radioactivity continues to accumulate in the stable 0.53 M and 0.40 M RNAs.

An analysis of these data together with the observations of Marrs and Kaplan (1971) led to the proposal of the processing scheme illustrated in Fig. 1.

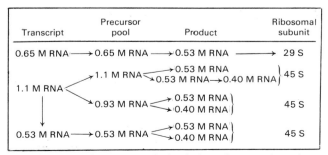

Fig. 1. Pathways of rRNA processing in *Rhodopseudomonas sphaeroides*.

The 0.65 M RNA transcripts enter a pool from which they are processed directly to the stable 0.53 M RNA of the 29 S ribosomal subunit. The processing of the RNA of the 45 S ribosomal subunit can apparently occur by several pathways. Some (approximately 50%) of the 1.1 M RNA transcripts are cleaved before the completion of transcription into two 0.53 M RNA molecules, analogously to the treatment of the entire ribosomal RNA gene transcript in *Esch. coli*. One of these 0.53 M RNA species represents the stable 0.53 M RNA, and the other is further reduced in size to the stable 0.40 M RNA. The remainder of the 1.1 M RNA transcripts are present as such and are processed by two pathways. Some are cleaved into 0.53 M RNA species, as in the sequence just described; the remainder are converted into 0.93 M RNA, which is subsequently converted to the mature RNA species. It appears that the same nucleolytic cleavages occur in both pathways, but in different temporal order.

The kinetics of rRNA formation in dark aerobically growing cultures of *Rp. sphaeroides* is basically similar to that in anaerobic photosynthetic cultures; however, a certain variation in processing of 1.1 M RNA occurs. The 1.1 M and 0.93 M RNA species in aerobic cultures follow a much different time course than in anaerobic photosynthetic cultures. There appears to be a rapid cleavage of 1.1 M to 0.93 M RNA species, so that the latter is in the majority throughout the experiment. The same cleavages occur during rRNA processing in both physiological states, but at different relative rates, perhaps due to an altered level of a processing nuclease. It is not known what relationship, if any, these differences may have to the adaptive formation of photosynthetic membranes.

2.2.2. Role of Protein Synthesis in Ribosomal RNA Maturation

In an attempt to study the role of methylation in RNA processing, a methionine-requiring mutant (AG-12) was isolated by nitrosoguanidine treatment of *Rp. sphaeroides* strain 2.4.1 (Gray, 1973). Following star-

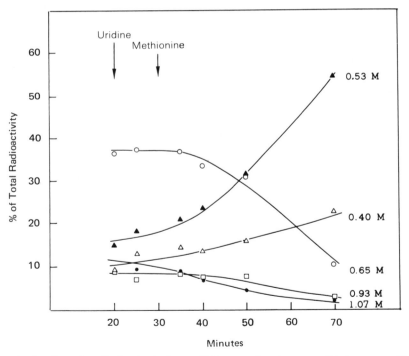

Fig. 2. Effect of methionine deprivation on RNA processing. An exponential-phase, anaerobic photosynthetically growing culture of *Rhodopseudomonas sphaeroides* strain AG-12 (met−) was harvested, resuspended in methionine-free medium, and incubated for 60 min in the light with N_2 gassing. [³H]Uridine was added at 0 time, followed at 20 min by 100-fold excess of unlabeled uridine and at 30 min by the addition of methionine. Samples were removed at various times, lysed with hot detergent, and phenol-extracted. The RNA extracts were electrophoresed on agarose–polyacrylamide composite gels, and the radioactivity was measured (Gray, 1973).

vation of this mutant organism for methionine under photosynthetic conditions, the rate of RNA synthesis gradually declined. The rRNA synthesized during methionine deprivation, however, was composed mainly of the precursor RNA species 1.1 M, 0.93 M, and 0.65 M RNA, with only low levels of stable 0.53 M and 0.40 M RNA present. Figure 2 illustrates the results of an experiment in which methionine-starved cells were allowed to incorporate [³H]uridine for 20 min, after which a chase of unlabeled uridine was added, followed 10 min later by methionine. The amounts of radioactivity in the various RNA species at the time of addition of unlabeled uridine indicate that only a low level of processing of precursor RNA into the stable forms has occurred. The uridine chase does not decrease the levels of precursor RNA; thus, they are evidently not turning over at any perceptible rate. On addition of methionine, however, there is a shift in the distribution of RNA species, with a continuing increase in the level of stable RNA accompanied by a decrease in the amounts of precursor forms. The results might suggest a close linkage between methylation of RNA and processing of precursors, but this is probably not the case. If chloramphenicol is added at the same time as methionine, no processing of the precursor RNA occurs. Furthermore, if chloramphenicol is added to a wild-type culture, the same accumulation of precursor RNA occurs as in the methionine-starved cells. Thus, the lack of protein synthesis is most likely responsible for the blocking in processing. The pool of ribosomal protein in *Esch. coli* is very small, and with restriction of protein synthesis, these proteins would soon be depleted (Schleif, 1968). If ribosomal assembly were necessary for precursor RNA processing, as appears to be the case in *Esch. coli* (Nierhaus *et al.*, 1973), then these RNA forms would accumulate in the absence of protein synthesis. The assembly of the ribosome may induce a conformation change in the RNA that is susceptible to nucleolytic cleavage.

2.3. Methylation of Ribosomal RNA

2.3.1. Distribution of Methylated Bases

Methylation is a posttranscriptional modification of RNA that likely plays a role in the specificity of nuclease action during the processing sequence. The maturation of rRNA in *Esch. coli* involves methylation of precursor molecules during ribosome assembly, although this need not be completed for formation of the stable rRNA species (Dahlberg and Peacock, 1971; Nierhaus *et al.*, 1973). The portions of the precursor that are removed during processing are

relatively methyl-deficient compared with the conserved sequences. The specificity of methylation of the rRNA of *Rp. sphaeroides* was studied by measuring the distribution of C[³H]₃ label in the various RNA species (Razel, A. J., and Gray, E. D., unpublished). The majority of methylated bases are present in the 0.53 M RNA of the 29 S ribosomal subunit. The remainder are distributed disproportionately among the RNA species of the 45 S subunit, with only 4% of the methyl groups present in 0.40 M RNA. The relative methyl deficiency of this latter RNA may be related to the unusual processing of *Rp. sphaeroides* rRNA. A more detailed examination of the distribution of methylation revealed that the rRNA of the 29 S ribosomal subunit contains 6,6-dimethyladenine as the principal methylated base with 2-methylguanine as the next most abundant species. Since the 0.40 M RNA contained such a small proportion of the methylated bases, total 45 S subunit RNA was analyzed, and a markedly different distribution than in the 29 S subunit RNA was observed. The major components are methylated adenine compounds with relatively little methylated guanine. The RNA of the 45 S ribosome contains a significant proportion of methylated pyrimidines, while these are vitually absent from the RNA of the 29 S ribosome. There are essentially no differences in the methylated base composition of rRNA isolated from anaerobic or aerobically grown cultures. Although the 0.40 M RNA contains only a small proportion of the methyl groups of the 45 S ribosome RNA, the distribution of the methylated bases is different than in the 0.53 M RNA. The 0.40 M RNA contains more methylated guanine, though the methylated adenines still represent the largest proportion of modified bases. The specificity of methylation that these results imply may be related to the processing sequence.

2.3.2. Kinetics of Ribosomal RNA Methylation

The kinetics of rRNA methylation in *Rp. sphaeroides* is quite different from that of RNA synthesis. The time course of incorporation of C[³H]₃-methionine into RNA is shown in Fig. 3. The amount of labeled methionine added to the culture was rapidly incorporated into protein as well as RNA, so the label in rRNA precursors was chased by endogenously synthesized methionine. Electrophoresis of deproteinized cell lysates reveals that only relatively small amounts of labeled methyl groups are incorporated into the 1.1 M and 0.93 M RNA. Most of the label appears in 0.65 M and 0.53 M RNA, with increasing amounts in 0.53 M RNA. In this particular experiment, little if any radioactivity is detectable in 0.40 M RNA, though in other studies, methylated

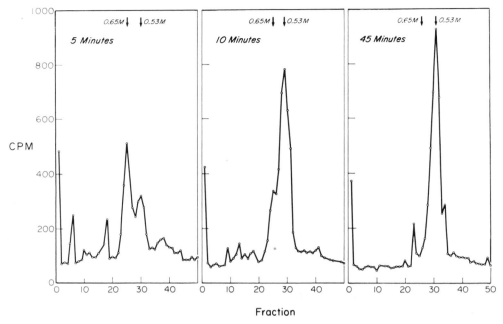

Fig. 3. Kinetics of rRNA methylation in *Rhodopseudomonas sphaeroides*. C[³H]₃-labeled methionine was added to an exponential-phase anaerobic photosynthetically growing culture of *Rp. sphaeroides* strain 2.4.1. Samples were removed at various times, and lysed, and the RNA purified by two successive phenol extractions was electrophoresed as described in Fig. 2.

bases have been observed in this RNA. The most abundant methylated RNA at early times of incorporation is the 0.65 M RNA, which is later converted to 0.53 M RNA. This observation is in consonance with the finding that the majority of methlated bases are present in the 0.53 M RNA from the 29 S ribosome. These results suggest that little methylation of 1.1 M and 0.93 M RNA occurs before they are subjected to nucleolytic cleavage. The early appearance of label in 0.53 M RNA may be due to the methylation of the 0.53 M RNA products of 1.1 M RNA that are formed prior to the completion of transcription.

3. Transfer RNA of *Rhodopseudomonas sphaeroides*

A role for transfer RNA (tRNA) in the regulation of protein synthesis has been suggested in such varied examples of differential gene expression as phage infection, sporulation in bacteria, and tumor formation and differentiation in higher oganisms (Littauer and Inouye, 1973). All these have been shown to be accompanied by qualitative or quantitative changes in isoaccepting tRNA that are proposed to reflect the activity of a regulatory mechanism.

The study of the adaptation of *Rp. sphaeroides*

from aerobic dark to photosynthetic growth led to an examination of the patterns of isoaccepting tRNA species in the differing physiological states (De Jesus and Gray, 1971). tRNAs from anaerobic and photosynthetic cultures were charged with amino acids using homologous preparations of amino acyl tRNA synthetases and chromatographed on benzoylated DEAE columns. This procedure demonstrated that as in other oganisms, the tRNAs consisted of multiple isoaccepting species. The profiles of isoaccepting aminoacyl tRNA species for most amino acids were identical from organisms grown in either physiological state. However, the phenylalanyl and tryptophanyl tRNAs from aerobic and anaerobic photosynthetic cultures both exhibited marked variations in their patterns of isoaccepting species. The phe-tRNA[phe] from photosynthetic cultures contains one major component, while "aerobic" phe-tRNA[phe] consists of three principal isoaccepting species. In different preparations of "aerobic" tRNA, the relative amounts of these vary, but all are present at a level at least equal to that of the major component of "photosynthetic" tRNA[phe]. These findings were recently confirmed in a study by Shepherd and Kaplan (1975), who observed similar variations in tRNA[phe] that were also independent of the source of aminoacyl tRNA synthetase. They observed that on transfer of aerobically growing cultures to anaerobic conditions, the pattern of phe-tRNA^phe rapidly assumed that characteristic of photosynthetic growth. The transition from anaerobic to aerobic growth was also accompanied by a change in distribution of phe-tRNA^phe species to those of tRNA from aerobic cultures. They concluded that the intracellular levels of tRNA^phe species are closely related to the biogenesis and maintenance of the photosynthetic organelles.

The tryptophanyl tRNA from both aerobic and anaerobic photosynthetic cultures consist of two major species, which vary considerably in their relative amounts in the two physiologic states. The ratio of major to minor species is 2.0 in "aerobic" trp-tRNA^trp and 8.0 in "anaerobic" trp-tRNA^trp. This difference is preserved when heterologous tRNA aminoacyl synthetase preparations are used to charge the tRNA from these sources. When tryptophanyl tRNA was isolated from cells incubated with labeled tryptophan, the same results were obtained on fractionation of the *in vivo* charged isoaccepting species. The difference in the proportions of isoaccepting tryptophanyl tRNA is thus not an artifact arising from *in vitro* charging. In *Esch. coli*, tRNA^trp can be converted to a form that differs in conformation and is inactive in protein synthesis (Gartland *et al.*, 1969). This conversion is believed to be related to the conditions of charging, and while a similar alteration in configuration cannot be excluded in tRNA^trp from *Rp. sphaeroides*, the correspondence of the *in vivo* isoaccepting tRNA pattern with that obtained in *in vitro* charged tRNA make it seem less likely. The conversion, if it occurs, might be related to the conditions of chromatography, but this has not yet been established.

The tRNA^phe and tRNA^trp variations observed in *Rp. sphaeroides* are independent of the source of aminoacyl synthetases, and thus may represent differential transcription of the tRNA genes in these differing morphogenetic states. Sporulation in *Bacillus subtilis* was shown by hybridization-competition analysis to result in such differential transcription (Jeng and Doi, 1975), and it was suggested that the novel tRNA species may be involved in translational regulation. In both sporulation and the adaptation of *Rp. sphaeroides* to photosynthetic growth, tRNAs could be important elements of regulatory systems.

Note Added in Proof: We have demonstrated that a precursor–product relationship exists between the two major tRNA^phe species (Razel and Gray, 1978). The tRNA^phe characteristic of anaerobic photosynthetically growing cultures is formed by post-transcriptional modification of the principal tRNA^phe of aerobic cultures. The conversion is accomplished by an enzyme activity formed during anaerobic photosynthetic growth.

4. RNA Synthesis during Formation of Photosynthetic Apparatus

The formation of photosynthetic apparatus in *Rp. sphaeroides* must involve the ordered formation of the variety of pigments, enzymes, and structural components of these organelles. This morphogenetic process must require protein and RNA synthesis (Bull and Lascelles, 1963; Sistrom, 1962; Higuchi *et al.*, 1965; Gray, 1967), since inhibitors of these cellular processes prevent chromatophore formation, as does the deprivation of adenine and histidine from mutants requiring these compounds.

The induction of formation of photosynthetic membranes by transfer of aerobic dark cultures of *Rp. sphaeroides* to anaerobic photosynthetic conditions results in a cessation of cell growth and division (Gray, 1967). The cellular levels of DNA, RNA, and protein remain constant, although there are considerable variations in the apparent rates of formation of macromolecules, as reflected in the rates of uptake of precursors. At the time of transfer to anaerobic light conditions, although the rate of accumulation of RNA is much restricted, there is formed an increased proportion of RNA that hybridizes to DNA. This increased level of what may be assumed to represent mRNA is maintained throughout the course of the transition to anaerobic photosynthetic growth.

The formation of bacteriochlorophyll (Bchl) and the photosynthetic apparatus can also be stimulated by a decrease in the light intensity incident on a culture of *Rp. sphaeroides* (Cohen-Bazire *et al.*, 1957). In cultures subjected to such a stepdown in light, it was found that an abrupt increase occurred in the proportion of RNA that was degraded following addition of proflavine, an inhibitor of RNA-chain elongation (Cost and Gray, 1967). Since this latter represents mainly mRNA, the increase in labile RNA is probably a reflection of an increased proportion of mRNA synthesis. The instantaneous rate of RNA synthesis on shift-down in light intensity undergoes a decrease in rate of the same order of magnitude as the decrease in the rate of RNA accumulation (see Section 5), indicating that rRNA synthesis is preferentially inhibited. The RNA synthesized following shift-down is therefore likely composed of an increased proportion of mRNA. It seems clear that transcriptional regulatory mechanisms are operative during this adaptation.

Witkin and Gibson (1972*b*) compared by hybridization-competition analysis the pulse-labeled RNA transcribed in aerobic and anaerobic photosynthetic cultures of *Rp. sphaeroides*. They were unable to detect any qualitative differences in RNA transcripts between the two physiologic states. In their comparisons, these authors employed not only chromosomal DNA but also the satellite DNA isolatable from this bacterium (Gibson and Niederman, 1970). No qualitative difference in RNA transcripts from aerobic or anaerobic photosynthetic cultures could be demonstrated in either of these species of DNA. Yamashita and Kamen (1968) made similar observations in *Rs. rubrum*, and also were unable to find any differences in the transcripts from aerobic or photosynthetic cultures.

The hybridization comparisons in all these experiments utilized conditions such that only differences among major species of mRNA would be detected. Since the level of mRNA specific for the formation of photosynthetic structures may vary considerably according to the growth conditions or even cell density (Witkin and Gibson, 1972b), this mRNA population may not represent a large enough proportion of the total mRNA population to be observable. A hybridization comparison utilizing RNA from cultures adaptively synthesizing photosynthetic membranes might allow the demonstration of mRNA specific to the adaptive process. Regulation of the formation of the photosynthetic apparatus by differential gene transcription is not ruled out by these results.

The possibility of translational regulation of photosynthetic organelle formation was suggested by Witkin and Gibson (1972a) and Yamashita and Kamen (1968). The former authors observed that there appears to be a higher rate of turnover of rapidly labeled RNA in anaerobic cultures of *Rp. sphaeroides* than in aerobic cultures. When the synthesis of RNA in this organism was inhibited by the addition of rifamycin, an inhibitor of RNA-chain initiation, a breakdown of rapidly labeled RNA occurred to a plateau level. The half-life of rapidly labeled RNA was determined to be 2 min and 4 min in anaerobic and aerobic cultures, respectively. The phenomenon cannot be ascribed to a variation in growth rate, since the doubling times under both growth conditions were identical. Transition of a culture from aerobic to anaerobic photosynthetic conditions or the reverse led to a similar transition of the half-life of rapidly labeled RNA. The stability of this RNA seems to be associated in some way with differential binding of the RNA to ribosomes.

The relationship of these observations to the mechanism of regulation of gene expression is not readily apparent, although there might be involvement

in control at the translational level. The variations in tRNA in the two physiological states previously described suggests that translational controls may be operational, but no direct evidence for this yet exists. The regulation of morphogenesis in the Rhodospirillaceae may involve both translational and transcriptional control elements. Some evidence exists for both of these, but a definition of differential transcription will probably require a genetic probe such as the DNA of a transducing phage carrying the "photosynthetic" genes.

5. Transcription and Translation during Synchronous Growth

The study of synchronous cultures of bacteria may provide insights into the properties of the organisms that cannot be deduced from random cultures. We have utilized synchronized cultures of *Rp. sphaeroides* to study the nature of transcription as reflected in the formation of enzymes during the formation of the photosynthetic organelles. While a full understanding of this remains elusive, a number of useful observations on the process were obtained (Ferretti and Gray, 1967, 1968).

The synchronization of *Rp. sphaeroides* was accomplished by the growth of cultures into controlled stationary phase followed by resuspension in fresh medium (Cutler and Evans, 1966). This procedure has the disadvantage that the synchronized organisms are not in a state of balanced growth, as are those cells synchronized by selection, but the procedure allows the synchronization of large cell populations. Synchronized photosynthetic anaerobic cultures of *Rp. sphaeroides* synthesize DNA, RNA, and protein, as indicated by the amounts of these macromolecules, more or less continuously throughout the cell cycle. The rate of uptake of labeled precursors of protein follows a similar increase continuous with growth, but the uptake of precursors into RNA and DNA is discontinuous with periodic increases in rate at specific points in the cell cycle. These discontinuities, in RNA labeling at least, are likely not due to variations in the rates of synthesis of the macromolecules, but to alterations in level of enzymes responsible for the uptake and conversion of precursors into the nucleotide triphosphates.

The discontinuous synthesis of enzymes is frequently observed in synchronized organisms, some enzymes being formed continuously but others produced in a periodic manner (Halvorson *et al.*, 1971). In *Rp. sphaeroides*, the initial enzymes of pyrrole biosynthesis were formed discontinuously at regular points in the cell cycle. In contrast to Bchl, which is formed continuously, succinate thiokinase [succinate:CoA ligase (GDP-forming) EC 6.2.1.4], δ-aminolevulinate (ALA) synthetase [succinyl CoA:glycine C succinyl transferase EC 2.3.1.37], and ALA dehydratase [5-aminolevulinate hydrolase EC 4.2.1.24] all underwent periodic increases in activity. Of two enzymes unrelated to Bchl formation, one, alkaline phosphatase [orthophosphoric monoester phosphohydrolase (alkaline optimum) EC 3.1.3.1] was formed periodically, and the other, ornithine transcarbamylase [carbamoyl phosphate:L-ornithine carbamoyl transferase EC 2.1.3.3], was synthesized in a more or less continuous manner. The variations in activity of ALA dehydratase, at least, were demonstrated not to be a result of feedback inhibition of activity and likely due to alterations in enzyme formation.

ALA synthetase and ALA dehydratase have been shown to undergo repression and derepression in response to environmental stimuli that regulate the formation of the photosynthetic apparatus (Lascelles, 1959, 1960). When a synchronous culture is subjected to a step-up in light intensity, the synchrony is maintained, but the synthesis of Bchl is abruptly halted, as is the formation of ALA synthetase and ALA dehydratase. This repression of enzyme formation is maintained until the light intensity is reduced to its previous level when there is a rapid increase in activity of both enzymes followed by a resumption of formation of Bchl. A similar result was observed when air was introduced into a synchronous photosynthetic culture. The synthesis of ALA synthetase and Bchl was halted and resumed only when the culture was sparged with N_2. At this point, there occurred an abrupt increase in ALA synthetase activity reaching the enzyme level of a control culture maintained under anaerobic conditions. In both the light shift and this latter experiment, the release of repression resulted in bursts of enzyme formation at a time in the cell cycle different from the usual point of discontinuous increases.

One explanation for discontinuous synthesis of enzymes is that the genome is transcribed in a regular and continuous sequence through the cell cycle (Halvorsen *et al.*, 1971). If this is the case in *Rp. sphaeroides*, the genome must also be available for transcription and translation at any time, since enzyme formation can apparently be derepressed at any time in the cell cycle. Another factor that may be involved in the periodicity of enzyme synthesis is oscillation in feedback repression circuits. In those systems studied, addition of repressor can alter the time in the cell cycle as well as the magnitude and frequency of oscillations in enzyme synthesis (Halvorsen *et al.*, 1971). Thus, sequential transcription and oscillatory repression may be combined to result in discontinuous enzyme formation. Analysis of differential transcription in synchronous cultures of *Rp. sphaeroides* to fully understand these interactions must await the isolation of a specific genetic probe.

6. Nucleotides and the Regulation of RNA Accumulation

6.1. ppGpp and RNA Formation

The adaptation of *Rp. sphaeroides* to growth at lowered light intensity represents a physiological response analogous to a nutritional shift-down in other bacteria (Eccleston and Gray, 1973). In both cases, the immediate response of the organisms is a sharp reduction in both the rate of RNA accumulation and the rate of growth (Fig. 4, insert); both gradually assume new values characteristic of the new medium or of the reduced light intensity (Lessie, 1965*b*). RNA accumulation during the stringent response to amino acid starvation is inhibited with similar abruptness (Edlin and Broda, 1968). All these cellular responses are accompanied by an abrupt increase in the levels of the guanine nucleotides: guanosine 5'-diphosphate 3'-diphosphate (ppGpp) and frequently guanosine 5'-triphosphate 3'-diphosphate (pppGpp) (Lazzarini *et al.*, 1971). These unusual nucleotides have been implicated as mediators in the regulation of rRNA formation as well as a number of other cellular processes (Gallant *et al.*, 1970). Since the adaptation to growth at lowered light intensity by *Rp. sphaeroides* involves a stimulation of formation of photosynthetic membranes, the possible role of these nucleotides in this process is of especial interest.

The decrease in rate of uptake of uracil and uridine after transfer of aerobic cultures to anaerobic conditions or following a light shift-down in anaerobic cultures (Gray, 1967; Cost and Gray, 1967) is similar to that observed in *Esch. coli* undergoing a stringent response to amino acid deprivation (Cashel, 1969). This effect in *Rp. sphaeroides* appears to be at least partially due to inhibition of transport of these compounds. The kinetics of this inhibition of transport is very similar to that of the abrupt increase in ppGpp levels (Eccleston and Gray, 1973). This nucleotide has been implicated as an inhibitor of base and nucleoside transport in *Esch. coli* (Hochstadt-Ozer and Cashel, 1972), and may exert a similar effect in *Rp. sphaeroides*. The variations in precursor uptake,

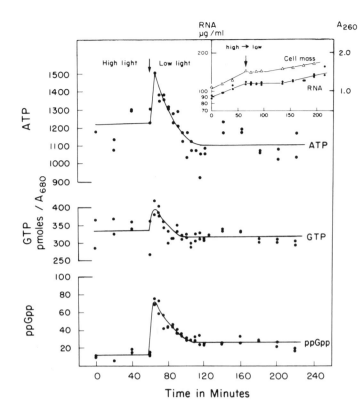

Fig. 4. Variations in nucleotide levels in *Rhodopseudomonas sphaeroides* during adaptation to lower light intensity. An exponential-phase anaerobic photosynthetically growing culture of *Rp. sphaeroides* strain 2.4.1 was incubated with $^{32}P_i$ for at least 2/3 generation to allow equilibration of nucleotide pools. The nucleotide levels were measured in samples removed at various times (Eccleston and Gray, 1973) before and after lowering the light intensity from 1100 to 240 foot candles.

however, are probably not related to the sudden decrease in rate of accumulation of RNA.

In *Esch. coli* undergoing the stringent response, the restriction of RNA formation is most notably reflected in the stable species, with a lesser effect on mRNA sythesis (Lazzarini and Dahlberg, 1971). The accumulation of rRNA in *Rp. sphaeroides* also ceases abruptly on light shift-down (Eccleston and Gray, 1973), although mRNA synthesis likely continues, since a greater proportion of the RNA that is formed is degraded following the addition of proflavin (Cost and Gray, 1967). Part of the increase in labile RNA may be due to accumulation of rRNA precursors consequent on a reduction in the rate of ribosomal assembly; these precursor RNA species, however, were not observed to turn over rapidly when they accumulated under conditions in which protein synthesis was prevented (Gray, 1973) (Fig. 2). Recently, measurements of the instantaneous rate of RNA synthesis utilizing the determination of precursor pool specific activity have indicated that the decrease in rate of RNA formation on light shift-down is of the same order as the decrease in rate of total cellular RNA accumulation (Eccleston, E. D., and Gray, E. D., unpublished). Since the cellular RNA is composed predominantly of rRNA, these results are further evidence that mainly rRNA synthesis is restricted. The synthesis of tRNA and 5 S RNA has not yet been studied in *Rp. sphaeroides* although in *Esch. coli* undergoing the stringent response tRNA is affected in a similar manner as the larger stable RNA species (Lazzarini and Dahlberg, 1971).

6.2. Nucleoside Triphosphate Levels and RNA Synthetic Rate

The regulation of rRNA synthesis by ppGpp in *Esch. coli* appears to involve an interaction of the nucleotide with components of the protein synthetic apparatus, elongation factors Tu and Ts, which in some way may alter the rate of initiation of RNA chains from rRNA cistrons (Travers, 1973). In addition to an apparent effect on rRNA-chain initiation, the accumulation of ppGpp may interfere with RNA synthesis by reducing the level of the precursor nucleoside triphosphates, particularly GTP (Gallant *et al.*, 1970; Edlin and Broda, 1968). In *Rp. sphaeroides*, however, this latter phenomenon does not appear to play a significant role in regulation of the rate of RNA synthesis. A study of the level of the purine nucleoside triphosphates during a shift-down in light intensity (Fig. 4) reveals instead an abrupt increase in their level coincident with the burst of accumulation of ppGpp. The inset illustrates the effect of the shift on RNA accumulation. The transient increase in the level of ATP and GTP might be due to the sudden decrease in the utilization of these compounds in RNA synthesis and would appear to be secondary to the actual inhibitory event. The levels of ATP and GTP at the reduced light intensity are only slightly lower than before the shift, the levels of UTP (Eccleston, E. D., and Gray, E. D., unpublished) following a similar time course. A limitation in level of substrates does not seem sufficient to account for the abrupt restriction in rate of RNA accumulation.

6.3. Nucleotides as Regulators of Formation of Photosynthetic Apparatus

The intracellular level of ATP has been proposed in several Rhodospirillaceae as a regulator of the synthesis of Bchl and the photosynthetic apparatus. Increased levels of ATP appear to reduce the rate of formation of Bchl (Zilinsky *et al.*, 1971; Fanica-Gaignier *et al.*, 1971). Miović and Gibson (1973), however, found no relationship between synthesis of the photosynthetic apparatus and ATP level or energy charge in *Chr. vinosum*. The results shown in Fig. 4 indicate that the pool size of ATP undergoes a sudden increase at the time of light shift-down and then returns to a lower level that is characteristic of the lowered light intensity, at most, 20% below the amount of ATP in the culture at the higher light intensity. Although the difference is modest, it is in the direction that would be predicted if ATP level acted as a regulator of Bchl formation. The initial increase in ATP level, while not observed by others due to their less frequent sampling, might suggest that if ATP is involved in this regulatory scheme, its role is a complex one.

The effects of ppGpp on RNA formation may not be restricted to those on rRNA. In *Esch. coli*, the *in vitro* expression of several operons is affected by this nucleotide, resulting in either a stimulation or an inhibition of enzyme formation, depending on the operon involved (Yang *et al.*, 1974). The abrupt increase in ppGpp levels following the induction of increased formation of photosynthetic apparatus by light shift-down suggested that the nucleotide might play a role in the morphogenetic sequence. A possibility existed that the transient in ppGpp level could simply be a consequence of the change in growth rate at shift-down. However, a shift from anaerobic to aerobic conditions in the light that results in an immediate cessation of Bchl formation does not significantly perturb the growth rate, but is accompanied by a decrease in the cellular level of ppGpp to approximately 50% of the preshift value (Eccleston, E. D., and Gray, E. D., unpublished). On return to anaerobic conditions, an abrupt increase in the amount of ppGpp to the preshift level occurs that is followed by a resumption of synthesis of Bchl. The extent of these changes in ppGpp is modest compared with the nearly sevenfold increase following a light shift-down, but they are compatible with a place for ppGpp in the regulation of Bchl synthesis.

Acknowledgments

The author is grateful to Dr. Sam Kaplan, Dr. Paul E. Tavernier, Eric D. Eccleston, and Anthony J. Razel for permission to include unpublished data. The research was supported by grants from the National Science Foundation and the Graduate School of the University of Minnesota. The author is also supported by N.I.H. Program Project Grant HE-06314.

7. References

Adesnick, M., and Levinthal, C., 1969, Synthesis and maturation of ribosomal RNA in *Escherichia coli*, *J. Mol. Biol.* **46**:281.

Attardi, G., and Amaldi, F., 1970, Structure and synthesis of ribosomal RNA, *Annu. Rev. Biochem.* **39**:183.

Ben-Hamida, F., and Schlessinger, D., 1966, Synthesis and breakdown of ribonucleic acid in *Escherichia coli* starving for nitrogen, *Biochim. Biophys. Acta* **119**:183.

Bhatnagar, Y. M., and Stachow, C. S., 1972, Ribosomal proteins of *Rhodopseudomonas palustris*, *J. Bacteriol.* **109**:1319.

Borda, L., Green, M. H., and Kamen, M. D., 1969, Sedimentation properties of ribonucleic acid from *Rhodopseudomonas sphaeroides*, *J. Gen. Microbiol.* **56**:345.

Bull. M. J., and Lascelles, J., 1963, The association of protein synthesis with the formation of pigments in some photosynthetic bacteria, *Biochem. J.* **87**:15.

Cashel, M., 1969, The control of ribonucleic acid synthesis in *Escherichia coli*. IV. Relevance of unusual phosphorylated compounds from amino acid starved stringent strains, *J. Biol. Chem.* **244**:3133.

Cohen-Bazire, G., Sistrom, W. R., and Stanier, R. Y., 1957, Kinetic studies of pigment synthesis by non-sulphur purple bacteria, *J. Cell. Comp. Physiol.* **49**:25.

Cost, H. B., and Gray, E. D., 1967, Rapidly labeled RNA synthesis during morphogenesis, *Biochim. Biophys. Acta* **138**:601.

Cutler, R. G., and Evans, J. E., 1966, Synchronization of bacteria by a stationary phase method, *J. Bacteriol.* **91**:469.

Dahlberg, A. E., and Peacock, A. C., 1971, Studies of 16 and 23 S ribosomal RNA of *Escherichia coli* using composite gel electrophoresis, *J. Mol. Biol.* **55**:61.

De Jesus, T. G. S., and Gray, E. D., 1971, Isoaccepting transfer RNA species in differing morphogenetic states of *Rhodopseudomonas sphaeroides*, *Biochim. Biophys. Acta* **254**:419.

Dunn, J. J., and Studier, F. W., 1973, T7 early RNAs and *Escherichia coli* ribosomal RNA are cut from large precursor RNAs *in vivo* by ribonuclease III, *Proc. Natl. Acad. Sci. U.S.A.* **70**:3296.

Eccleston, E. D., and Gray, E. D., 1973, Variations in ppGpp levels in *Rhodopseudomonas sphaeroides* during adaptation to decreased light intensity, *Biochem. Biophys. Res. Commun.* **54**:1370

Edlin, G., and Broda, P., 1968, Physiology and genetics of the "RNA control" locus in *Escherichia coli*, *Bacteriol. Rev.* **32**:206.

Fanica-Gaignier, M., Clement-Metral, J., and Kamen, M., 1971, Adenine nucleotide levels and photopigment synthesis in a growing photosynthetic bacterium, *Biochim. Biophys. Acta* **226**:135.

Ferretti, J. J., and Gray, E. D., 1967, Control of enzyme synthesis during adaptation in synchronously dividing populations of *Rhodopseudomonas sphaeroides*, *Biochem. Biophys. Res. Commun.* **29**:501.

Ferretti, J. J., and Gray, E. D., 1968, Enzyme and nucleic acid formation during synchronous growth of *Rhodopseudomonas sphaeroides*, *J. Bacteriol.* **95**:1400.

Friedman, D. I., Pollara, B., and Gray, E. D., 1966, Structural studies of the ribosomes of *Rhodopseudomonas sphaeroides*, *J. Mol. Biol.* **22**:53.

Gallant, J., Erlich, H., Hall, B., and Laffler, T., 1970, Analysis of the RC function, *Cold Spring Harbour Symp. Quant. Biol.* **35**:397.

Gartland, W. J., Ishida, T., Sueoka, N., and Nirenberg, M. W., 1969, Coding properties of two conformations of tryptophanyl tRNA in *Escherichia coli*, *J. Mol. Biol.* **44**:403.

Gibson, K. D., and Niederman, R. A., 1970, Characterization of two circular satellite species of deoxyribonucleic acid in *Rhodopseudomonas sphaeroides*, *Arch. Biochem. Biophys.* **141**:694.

Gray, E. D., 1967, Studies on the adaptive formation of photosynthetic structures in *Rhodopseudomonas sphaeroides*. I. Synthesis of macromolecules, *Biochim. Biophys. Acta* **138**:550.

Gray, E. D., 1973, Requirement of protein synthesis for maturation of ribosomal RNA in *Rhodopseudomonas sphaeroides*, *Biochim. Biophys. Acta* **331**:390.

Halvorsen, H. O., Carter, B. L. A., and Tauro, P., 1971, Synthesis of enzymes during the cell cycle, *Adv. Microb. Physiol.* **6**:47.

Higuchi, M., Goto, K. Fujimoto, M., Namiki, O., and Kikuchi, G., 1965, Effect of inhibitors of nucleic acid and protein synthesis on the induced synthesis of bacteriochlorophyll and Δ-aminolevulinic acid synthetase by *Rhodopseudomonas sphaeroides*, *Biochim. Biophys. Acta* **95**:94.

Hochstadt-Ozer, J., and Cashel, M., 1972, The regulation of purine utilization in bacteria. V. Inhibition of purine phosphoribosyl transferase activities and purine uptake in isolated membrane vesicles by guanosine tetraphosphate, *J. Biol. Chem.* **247**:7067.

Horiuchi, T., Horiuchi, S., and Mizuno, D., 1959, Degradation of ribonucleic acid in *Escherichia coli* in phosphorus-deficient culture, *Biochim. Biophys. Acta* **31**:570.

Jeng, Y.-H., and Doi, R. H., 1975, New transfer ribonucleic acid species during sporulation of *Bacillus subtilis*, *J. Bacteriol.* **121**:950.

Lascelles, J., 1959, Adaptation to form bacteriochlorophyll in *Rhodopseudomonas sphaeroides*: Changes in activity of enzymes concerned in pyrrole synthesis, *Biochem. J.* **72**:508.

Lascelles, J., 1960, The synthesis of enzymes concerned in bacteriochlorophyll formation in growing cultures of *Rhodopseudomonas sphaeroides*, *J. Gen. Microbiol.* **23**:487.

Lazzarini, R. A., and Dahlberg, A. E., 1971, The control of ribonucleic acid synthesis during amino acid deprivation in *Escherichia coli*, *J. Biol. Chem.* **246**:420.

Lazzarini, R. A., Cashel, M., and Gallant, J., 1971, On the regulation of guanosine tetraphosphate levels in stringent and relaxed strains of *Escherichia coli*, *J. Biol. Chem.* **246**:4381.

Lessie, T. G., 1965a, The atypical ribosomal RNA complement of *Rhodopseudomonas sphaeroides*, *J. Gen. Microbiol.* **39**:311.

Lessie, T. G., 1965b, RNA metabolism of *Rhodopseudomonas sphaeroides* during preferential photopigment synthesis, *J. Gen. Microbiol.* **41**:37.

Littauer, U. Z., and Inouye, H., 1973, Regulation of tRNA, *Annu. Rev. Biochem.* **42**:439.

Mansour, J. D., and Stachow, C. S., 1975, Structural changes in the ribosomes and ribosomal proteins of *Rhodopseudomonas palustris*, *Biochem. Biophys. Res. Comm.* **62**:276.

Marrs, B., and Kaplan, S., 1971, 23 S Precursor ribosomal RNA of *Rhodopseudomonas sphaeroides*, *J. Mol. Biol.* **49**:297.

McCarthy, B. J., 1962, The effects of magnesium starvation on the ribosome content of *Escherichia coli*, *Biochim. Biophys. Acta* **55**:880.

Miović, M. L., and Gibson, J., 1973, Nucleotide pools and adenylate energy charge in balanced and unbalanced growth of *Chromatium*, *J. Bacteriol.* **114**:86.

Nierhaus, K. H., Bordasch, K., and Homan, H. E., 1973, Ribosomal proteins. XLIII. *In vivo* assembly of ribosomal proteins, *J. Mol. Biol.* **74**:587.

Nikolaev, N., Silengo, L., and Schlessinger, D., 1973, Synthesis of a large precursor to ribosomal RNA in a mutant of *Escherichia coli*, *Proc. Natl. Acad. Sci. U.S.A.* **70**:3361.

Nomura, M., 1970, Bacterial ribosomes, *Bacteriol. Rev.* **34**:228.

Pace, B., Peterson, R. L., and Pace, N. R., 1970, Formation of all stable RNA species in *Escherichia coli* by posttranscriptional modification, *Proc. Natl. Acad. Sci. U.S.A.* **65**:1097.

Razel, A. J., and Gray, E. D., 1978, Interrelationships of isoacceptor phenylalanine tRNA species of *Rhodopseudomonas sphaeroides*, *J. Bacteriol.* **133** (in press).

Robinson, A., and Sykes, J., 1973, A comparison of the unfolding and dissociation of the large ribosome subunit from *Rhodopseudomonas sphaeroides* NCIB 8253 and *Escherichia coli* MRE600, *Biochem. J.* **133**:740.

Schleif, R. F., 1968, Origin of chloramphenicol particle protein, *J. Mol. Biol.* **37**:119.

Shepherd, W., and Kaplan. S., 1975, Variations in the amounts of three isoaccepting phenylalanyl tRNA species as a function of growth conditions in *Rhodopseudomonas sphaeroides*, *Abstr. Annu. Meet. Am. Soc. Microbiol.*, p. 102.

Sistrom, W. R., 1062, The kinetics of synthesis of photopigments in *Rhodopseudomonas sphaeroides*, *J. Gen. Microbiol.* **28**:607.

Szilagyi, J. F., 1968, 16S and 23S Components in the ribosomal ribonucleic acid of *Rhodopseudomonas sphaeroides*, *Biochem. J.* **109**:5.

Taylor, M. M., and Storck, R., 1964, Uniqueness of bacterial ribosomes, *Proc. Natl. Acad. Sci. U.S.A.* **52**:958.

Travers, A., 1973, Control of ribosomal RNA synthesis *in vitro*, *Nature (London)* **244**:15.

Witkin, S. S., and Gibson, K. D., 1972a, Changes in ribonucleic acid turnover during aerobic and anaerobic growth in *Rhodopseudomonas spheroides*, *J. Bacteriol.* **110**:677.

Witkin, S. S., and Gibson, K. D., 1972b, Ribonucleic acid from aerobically and anaerobically grown *Rhodopseudomonas sphaeroides*: Comparison by hybridization to chromosomal and satellite deoxyribonucleic acid, *J. Bacteriol.* **110**:684.

Yamashita, J., and Kamen, M., 1968, Observations on the nature of pulse-labeled RNAs from photosynthetically or hererotropically grown *Rhodospirillum rubrum*, *Biochim. Biophys. Acta* **161**:162.

Yang, H.-L., Zubay, G., Urm, E., Reiness, G., and Cashel, M., 1974, Effects of guanosine tetraphosphate, guanosine pentaphosphate and β-Δ methyl-lenyl-guanosine pentaphosphate on gene expression of *Escherichia coli in vitro*, *Proc. Natl. Acad. Sci. U.S.A.* **71**:63.

Zilinsky, J. W., Sojka, G. A., and Gest, H., 1971, Energy charge regulation in photosynthetic bacteria, *Biochem. Biophys. Res. Commun.* **42**:955.

CHAPTER 49

Phototaxis and Chemotaxis

W. R. Sistrom

1. Introduction

Photo- and chemotaxis in photosynthetic bacteria were reviewed more than 13 years ago by Clayton (1964). This review is occasioned not so much by new work in these areas since Clayton's review (of which there has been little), but by the progress that has been made in our understanding of how bacteria swim and of the mechanism of chemotaxis in other bacteria, especially *Escherichia coli*. I shall first summarize new information on the mechanics of swimming; I shall then review briefly the work on taxes in photosynthetic bacteria and discuss the salient features of chemotaxis in *Esch. coli*. Finally, I shall reassess the work on taxes in photosynthetic bacteria in the light of the newer findings.

2. Mechanics of Swimming

The fundamental cellular response of flagellate bacteria to a tactic stimulus is a modulation of the frequency of changes in swimming direction. The average rate of motion of a bacterium in a particular direction will be the greater, the smaller the frequency of changes away from that direction of swimming. Therefore, if cells change directions less frequently when swimming in a given direction than when swimming in a second direction, a population of cells will tend to move in the first direction. Some bacteria can change directions by simply backing up, in which case the new direction is approximately 180° away

from the old one. The majority of bacteria, however, change directions by interrupting smooth swimming with a brief period of thrashing about, after which smooth swimming is resumed. In this case, the new direction is largely independent of the original one. Clearly, an understanding of taxis requires an understanding of how bacteria swim, and especially of how they stop swimming and change directions. Since these were the topics of recent reviews by Berg (1975) and Adler (1975), in which references to the older literature can be found, I shall mention only the most significant findings.

The most notable result is the substantiation of the not infrequently stated but seldom believed idea that bacterial flagella are rigid, helical structures that function as Archimedean screws. The evidence comes largely from experiments in which cells of *Esch. coli* are tethered by attaching their flagella to the surface of a microscope slide by means of antibodies directed against the flagella. (It is just happenstance that the antibody sticks to the glass as well as reacts specifically with the flagella.) When this is done, the cells are seen to rotate. Perhaps the most telling experiment made use of a mutant strain of *Esch. coli* the flagella of which are "straight" and that is unable to swim, presumably because of the absence of the usual helical shape of the flagella. When cells of this mutant strain were tethered, they were observed to rotate more or less continuously (Silverman and Simon, 1974). Similar results were obtained with normally flagellated bacteria that had been grown in glucose so as to reduce the average number of flagella to about one per cell (Larsen *et al.*, 1974b). In both experiments, the observed rotation could not have been

W. R. Sistrom · Department of Biology, University of Oregon, Eugene, Oregon 97403

caused by other (nontethered) flagella turning the cell about the tethered flagellum as a pivot. It was concluded that the flagella can rotate relative to the cell and that their helical shape screws the cell through the medium. To allow direct observation of the rotation of the flagella, small latex beads were coated with antiflagella antibody and mixed with cells. Under the microscope, the beads appeared to be attached to the flagella (themselves invisible) at the rear of the cell and to rotate rapidly as the cell swam (Silverman and Simon, 1974).

Cells of *Esch. coli* are usually equipped with six to eight flagella randomly disposed over the surface of the cell; i.e., they are peritrichously flagellated. Pijper (see Weibull, 1960) and more recently Macnab and Koshland (1974) observed cells of other peritrichously flagellated bacteria with high-resolution, dark-field microscopy to visualize the bundles of flagella. The bundle projects from the rear of a swimming cell and has the appearance of a rotating corkscrew. In the past, the rotation was usually interpreted as due to propagation of helical waves along a stationary, cylindrical bundle of flagella (for a discussion, see Lowy and Spencer, 1968). It is now clear, however, that the bundle looks like a rotating corkscrew because it *is* a rotating corkscrew.

2.1. Change of Direction

In the absence of any chemotactic stimulus, the motion of *Esch. coli* consists of stretches or runs of smooth, fairly straight swimming interrupted by brief periods of rapid rotation of the cell (twiddles). After a twiddle, the direction of the next run of smooth swimming will, in general, be different from that of the previous run. The rate of rotation of the cell during a twiddle is greater than can be accounted for by Brownian motion (Berg and Brown, 1972). This fact and the observations by Macnab and Koshland (1974) that the bundles of flagella disappear during a twiddle and that individual flagella can be seen (albeit vaguely and fleetingly) suggest that twiddling is an active process driven by uncoordinated motion of flagella.

When cells are subjected to a sudden increase in concentration of a chemotactic attractant, the frequency of twiddling *decreases*; when they are subjected to a sudden decrease in the concentration of an attractant, the frequency of twiddling *increases* (Macnab and Koshland, 1972). The same response is seen when cells are swimming in a spatial gradient of attractant. The frequency of twiddling of cells swimming up the gradient is less than that of cells swimming down the gradient. This means, of course, that the average run length up the gradient is greater than that down the gradient. Berg and Brown (1972) found that

for cells of *Esch. coli* swimming up a serine gradient, the mean path between twiddles was about 30 μm, while for cells swimming down the same gradient, it was only 20 μm. As a result, the average rate of motion of a population of cells is greater up the gradient than down.

An analogous response is observed in cells tethered by antiflagella antibody. For example, in the absence of any stimulus, cells of *Esch. coli* rotate counterclockwise about 70% of the time; immediately after an increase in the concentration of an attractant, they rotate counterclockwise nearly 100% of the time. On the other hand, immediately after an increase in the concentration of a repellent (which should have the same effect as a decrease in the concentration of an attractant), the cells rotate counterclockwise only 4% of the time. It appears that in this case, clockwise rotation of tethered cells is correlated with twiddling and counterclockwise rotation with smooth swimming (runs).

Twiddling is not the only way in which cells can change the direction of swimming; indeed, it is probably restricted to peritrichously flagellated bacteria such as *Esch. coli*. Polarly flagellated rods change direction by an abrupt reversal that in most cases is followed a short time later by a second reversal. Since the two reversals are seldom 180° apart, this maneuver results in a new direction of swimming (Taylor and Koshland, 1974). This behavior is shown by *Chromatium okenii* and was originally observed by Buder (1915) after a sudden decrease in light intensity.

Many spirilla, including *Rhodospirillum rubrum*, can swim equally well in either direction; these cells change direction by a simple reversal that is not necessarily followed by a second reversal. It is likely that all these behaviors can be accounted for by reversals in the direction of rotation of rigid flagella.

2.2. Energy Source for Swimming

Larsen *et al.* (1974a) and Thipayathasana and Valentine (1974) showed that in *Esch. coli* and *Salmonella typhimurium*, the energy source for swimming comes directly from a high-energy intermediate in oxidative phosphorylation and not from ATP. This was demonstrated by the use of mutant strains that lack "coupling factor." Such strains can swim under aerobic conditions (in the presence of an oxidizable carbon source), but not under anaerobic conditions in which fermentation is the only source of ATP. The wildtype can swim under either aerobic or anaerobic conditions. In the latter situation, ATP derived from fermentation presumably can produce the high-energy intermediate because of the presence of the coupling factor. Furthermore, arsenate does not inhibit swimming so long as oxygen and an oxidizable carbon source are present, even though the intracellular concentration of ATP is reduced at least 300-fold.

There is no reason to believe that the same is not true of photosynthetic bacteria. Nultsch and Throm (1968) demonstrated that *Rs. rubrum* swims very slowly under anaerobic conditions in the dark and that ATP restores normal motility after many hours. This experiment does not show that ATP is the immediate source of energy for swimming, since the cells possessed coupling factor and could generate the high-energy intermediate from ATP. The inhibition of motility by dinitrophenol demonstrated by Throm (1968) also does not distinguish between ATP and a high-energy intermediate, since the uncoupler will reduce the concentration of both.

ATP is necessary, however, for chemotaxis in *Esch. coli*. The evidence for this is that arsenate reversibly inhibits chemotaxis, although, as we have just seen, it does not inhibit swimming (Larsen *et al.*, 1974*a*). There is a good deal of circumstantial evidence that *S*-adenosyl methionine (SAM) is needed for chemotaxis; the requirement for ATP in chemotaxis may simply reflect its role in the synthesis of SAM (Adler, 1975).

3. Taxes in Photosynthetic Bacteria

3.1. Phototaxis

Phototaxis is a motor response elicited by light. It is necessary at the outset to distinguish two possible responses to light. In the first, an organism moves so as to orient itself with respect to the direction of the light source in the absence of any gradient in the intensity of the light. Such a response is termed a *topophototaxis*; it is positive when the movement is toward and negative when the movement is away from the light source. Throm (1968) reported a weak topophototaxis in *Rs. rubrum*; there is no other report of topophototaxis in photosynthetic bacteria, and accordingly I shall not be concerned with it further. The second kind of response is elicited by a temporal change in light intensity without regard to the direction of the light source or to a spatial gradient in intensity. Such a response is called a *phobophototaxis*. The classic example of this response is the reversal of swimming direction when *Rs. rubrum* is subjected to a sudden decrease in light intensity. This behavior causes a population of the bacteria to accumulate in a spot of light: the cells experience a temporal decrease of intensity when they swim out of the spot and thus reverse and return to the spot; there is no change in direction when the cells swim into the spot. This is an example of a positive phobophototaxis.

Two techniques have been used to study phototaxis in photosynthetic bacteria. In the first, the accumulation of populations of cells in regions of different light intensities separated by a sharp boundary is observed. This method has been used extensively in determining action spectra and the effect of different intensities on phototaxis. It is most often used as a null method. The second technique is the observation under the microscope of responses of individual cells to temporal changes in light intensity. The change of intensity may be brought on by the experimenter or by the organism itself as it swims across a sharp intensity boundary. This method, although more tedious than the first, is less subject to artifacts and is considerably more informative. The errors and pitfalls of the two methods are discussed by Clayton (1964). Recent developments in instrumentation have reduced the tedium of observations on single cells (Berg, 1971; Lovely *et al.*, 1974); these newer methods have not, however, been applied to phototaxis.

3.1.1. Relationship of Phototaxis and Photosynthesis

Two lines of evidence suggest a close relationship of phototaxis and photometabolism. In the first place, in *Rs. rubrum*, the action spectrum for phototaxis is closely similar to that for photosynthesis, indicating that the two processes use the same receptor pigments (Manten, 1948; Clayton, 1953*a*). In the second place, there is a close parallelism between light saturation curves for photosynthesis and for phototaxis. These observations led Manten (1948) to propose that the phototactic response was occasioned by a transient decrease in the rate of photometabolism. A third piece of evidence is that by selection for nonphototactic mutant strains of *Rs. rubrum*, it is possible to isolate strains that are unable to photosynthesize (Weaver, 1971).

3.1.2. Quantitative Aspects of Phototaxis

Clayton (1953*b*) analyzed the response of individual cells of *Rs. rubrum* to variations in the strength, duration, shape, and frequency of pulses of light or darkness. The purpose of this research was to compare the response of the bacterium to that of nerve cells subjected to electrical stimuli. The general conclusion was that the behavior of *Rs. rubrum* was remarkably similar to the behavior of nerve cells, but on a considerably slower time scale. As examples, immediately following a response, a cell was refractory to a second stimulus; several subliminal stimuli could be summed to produced a response; and after a decrease of light intensity, cells showed a rhythmic reversal of swimming direction that could be suppressed by an increase in light intensity.

Another parallel between phototaxis of photosynthetic bacteria and sensory systems of higher organisms is adherence to Weber's law. This law states that the threshold stimulus (decrease of light intensity, in our case) is a constant fraction of the initial stimulus strength (light intensity before the decrease) over a fairly wide range of stimulus strengths. That photosynthetic bacteria behave in this fashion has been shown repeatedly (Schrammeck, 1934; Clayton, 1953a; Schlegel, 1956), although there is considerable variation in both the value of the threshold change of intensity and the range of intensities over which the threshold is a constant fraction of the background illumination. Thus, for *Rs. rubrum*, Clayton (1953a) found that a 2% decrease of intensity was sufficient to elicit a response over an approximately 100-fold range of intensities; on the other hand, Schrammeck found a threshold of 5% over a range of nearly 5000-fold of intensity. Schlegel (1956) extended these studies to *Chr. okenii, Thiospirillum jenese*, and *Chr. vinosum*. Since, as we shall see, chemotaxis in *Esch. coli* also adheres to Weber's law, I shall postpone discussion of the significance of this adherence to Section 4.

3.2. Chemotaxis

The only form of chemotaxis that has been analyzed in photosynthetic bacteria is the aerotaxis of *Rs. rubrum*; this bacterium can, of course, grow equally well aerobically in the dark or anaerobically in the light. In the dark or in dim light, *Rs. rubrum* is positively aerotactic; i.e., cells accumulate near a bubble of oxygen introduced into a cell suspension. A uniform field of bright light inhibits aerotaxis. Correspondingly, phototaxis is diminished in a well-aerated suspension (Clayton, 1955, 1958).

Various chemicals in addition to oxygen elicit tactic responses in *Rs. rubrum*. In general, carbon sources for growth are attractants. Inhibitors of photosynthesis (e.g., hydroxylamine) inhibit phototaxis, while inhibitors of respiration (e.g., cyanide) inhibit aerotaxis (Clayton, 1958).

3.3. The Clayton and the Links Hypotheses

On the basis of the then prevailing views of photo-metabolism and respiration in photosynthetic bacteria, Clayton (1958) analyzed his results on the interaction

of aerotaxis, chemotaxis, and phototaxis. He sought a single factor the cellular concentration of which would be similarly affected by decreases of light intensity, of oxygen tension, and of the concentration of carbon sources for growth, since the cells responded positively to all three of these stimuli. He concluded that "a compound whose rate of synthesis parallels the rate of . . . activities which support growth" satisfied the requirement. Most of Clayton's results (as well as those of other investigators) could then be accounted for by assuming that a tactic response was initiated by a decrease in the concentration of such a compound. The connection between the response of the flagella and the changing concentration of the compound was left unspecified.

Clayton pointed out that in the case of phototaxis, there was no reason to invoke an amplification of energy between stimulus and response and that there could therefore be a direct connection between the change in concentration and the motor response of the flagella. This led Clayton to adopt the much more concrete and simpler hypothesis of Links. This hypothesis states that a tactic response is initiated when the supply of energy to the flagellar motor apparatus decreases. The Links hypothesis says that all tactic responses are initiated by a decrease in the supply of energy and that there is no central processing of sensory information.

4. Chemotaxis in *Escherichia coli*

Since this topic was recently reviewed by Adler (1975) and by Berg (1975), it is necessary here only to mention those features that are of particular concern in attempting to understand phototaxis in photosynthetic bacteria. The tactic mechanism of bacteria is a remarkably subtle and sophisticated one, certainly much more so than envisioned by either the Clayton or the Links hypothesis, for the following reasons:

1. There is no necessary relationship between the effect of a compound on the gross metabolism of the cell and its ability to act as either an attractant or a repellant. For example, mutant strains that are unable to metabolize or even transport galactose respond chemotactically to the sugar. By itself, this result rules out both the Clayton and the Links hypotheses, in which tactic responses are assumed to be brought about by changes in the general metabolism of the cell.

2. Cells of *Esch. coli* are provided with a number of specific receptors for particular compounds or classes of compounds. The evidence for this is largely genetic, but partly behavioral and biochemical. The genetic evidence is the isolation of mutant strains that have

lost the ability to respond to certain substances, but not to others. On the basis of this sort of result, attractants and repellants can be grouped into classes; a single mutation eliminates the response to all members of a given class without affecting the responses to members of other classes. The behavioral evidence comes from competition experiments of the following sort: *Escherichia coli* is attracted to galactose, fucose, and serine (among many other substances). In the presence of a high concentration of galactose, cells do not respond to fucose, but do respond to serine. On this basis, galactose and fucose would be placed in one group, serine in a second. This classification of attractants and repellants agrees with the classification based on analyses of mutants.

In some cases, it has been possible to identify a specific protein with a particular receptor. These proteins function as binding proteins in certain transport systems; they can be separated from intact cells by gentle procedures and purified. The best studied of these proteins from the point of view of taxis is the galactose-binding protein. Certain mutant strains defective in galactose transport lack the galactose-binding protein; such strains are also defective in galactose chemotaxis. Furthermore, there is a fair correlation between the observed dissociation constants of the binding protein *in vitro* for various sugars and the threshold and saturating concentrations for chemotaxis with these same sugars.

Transport *per se* is not required for taxis, however, since mutants that are defective in a component of the galactose-transport system other than binding protein still respond chemotactically to galactose. Similarly, there are mutants that are specifically defective in galactose chemotaxis but that still possess the binding protein. The latter observation demonstrates that the specific chemical receptors have more than one component. The binding proteins can be considered as the specific sensors (the components that react directly with the detected substances) of the various receptor systems. Not all receptor systems utilize transport-related binding proteins.

3. Genetic evidence points to the existence of at least three components that are common to all chemotaxes. Strains that are defective in any one of these components are unable to respond to any chemotactic stimulus.

4. Mesibov *et al.* (1973) (see also Spudich and Koshland, 1975) showed that for two compounds, galactose and β-methylaspartate, chemotaxis adheres more or less to Weber's law. In this case, Weber's law takes the form that the smallest *percentage* change of concentration necessary to elicit a response is constant. The range of concentrations over which this is true is different for the two substances. As pointed out earlier, there is fair correlation between the concentration ranges and the binding constants of the binding proteins determined *in vitro*. A simple model, which can account for the sensitivity curves, says that a change in the fraction of binding protein to which the attractant is bound determines the tactic response.

5. *Escherichia coli* can detect spatial gradients such that the change in concentration along the length of the cell is only 1 part in 10^4. At concentrations of a few micromolar, this is a great deal smaller than the statistical fluctuations in the number of molecules in a volume of 0.1 μm^3 at each end of the cell, which is a reasonable size for a "sensing volume." It seems unlikely, therefore, that chemotaxis could be based on concentration differences over the length of the cell. Since cells swim at rates of the order of ten times their cell lengths per second, temporal comparison would be much easier (Macnab and Koshland, 1972). As has already been mentioned, it is known that *Esch. coli* does respond to temporal gradients in concentration of attractants or repellents. It had been known for many years, of course, that phototaxis is based on temporal changes in light intensity. A formal model for temporal comparison was given by Clayton (1953*b*); more recently, Macnab and Koshland (1972) suggested essentially the same model. In essence, these models say that the stimulus (change of concentration or of light intensity) initiates two reactions, one of which is fast and the other slow. The response occurs when the ratio of the extents to which the two reactions have proceeded attains some critical value.

6. There is circumstantial evidence implicating the cell membrane in the tactic response. One piece of this evidence comes from observations on bipolarly flagellated spiralla. When such cells are swimming, both bundles of flagella are directed toward the rear, the bundle at the front being turned back of the cell. Both bundles rotate in the same direction; the counter-rotation of the screw-shaped cell probably provides the main motive force. When the direction of swimming reverses, the rotation of the flagella reverses, and their orientations change so that the originally rearward-projecting tail bundle is now turned back over the new front end of the cell and the other bundle is now directed toward the rear (von Reichert, 1909).

Krieg *et al.* (1967) observed that *Spirillum volutans* can reverse its direction of swimming within at most about $\frac{1}{25}$ sec. This implies that both bundles of flagella reverse directions within this same time. Berg (1975) and others pointed out that this is too fast to be accounted for by diffusion of a low-molecular-weight molecule from one end of the cell to the other through the cytoplasm. Communication would be greatly speeded up by using the cell membrane either as a substrate for two-dimensional diffusion or as a transmitter of electrical impulses.

More direct evidence for a role of the membrane in chemotaxis was provided by Lofgren and Fox (1974). These workers compared the effect of temperature on motility and chemotaxis in cells of *Esch. coli* in which the major membrane fatty acid was either oleate or elaidate. The effect of temperature on motility was essentially the same for cells with oleate or with elaidate. On the other hand, there was a dramatic difference in the effect of temperature on chemotaxis of the two kinds of cells. The chemotactic response of cells containing elaidate fell precipitously between 32 and 28°C, but that of cells containing oleate was affected only slightly between 35 and 25°C. Independent evidence has shown that in cells containing elaidate, the membrane lipids are in an immobile state below approximately 30°C, while the lipids of cells containing oleate do not become immobile above 14°C. Thus, the Logfren–Fox experiment indicates both the importance of the cell membrane for chemotaxis and a requirement for a fluid lipid phase.

It would seem profitable to explore possible relationships between taxis and those electrochemical phenomena (pH gradient and membrane potential; see Chapter 26) in the membrane that are involved in energy transduction. This could involve the use of ionophorus antibiotics and other uncouplers, and of indicators such as the carotenoid absorption bandshift.

5. Reappraisal

It is clear that the Links hypothesis is no longer tenable, since its two postulates are not true. First, that all tactic responses involve a single intermediate of energy metabolism (ATP or something related to it) is disproved by the existence of chemotactic responses to substances that are not metabolized and by the loss of mutation of the ability to respond to a substance without loss of ability to metabolize the substance. Second, the proposal that a stimulus is perceived directly by the motor apparatus is disproved by the obvious genetic complexity of taxis in *Esch. coli*. It may be pointed out that the synchronous behavior of flagella of bipolar spirilla by itself militated against the Links hypothesis (Clayton, 1958).

In one fundamental respect, the phototactic (and presumably chemotactic) response of photosynthetic bacteria and the chemotactic response of *Esch. coli* are similar: both depend on temporal changes in stimulus intensity. As pointed out by Clayton (1953*b*)

and by Macnab and Koshland (1972), this implies that two reactions or processes with different rate constants are involved.

Because of lack of evidence, it is impossible to say whether taxis in the photosynthetic bacteria has other general features in common with taxis in *Esch. coli*. We have seen that the basic response in *Esch. coli* is a modulation of the frequency of changes in swimming direction (twiddles). Clearly, the reversal of swimming direction shown by *Rs. rubrum* following a decrease in light intensity corresponds to the increase in frequency of twiddling followed a decrease in attractant concentration seen in *Esch. coli*. However, there have been no observations that could answer the question whether an increase in light intensity decreases the frequency of reversal in *Rs. rubrum*. Clayton's observations on the "rhythmic" response of *Rs. rubrum* (Clayton, 1953*b*) are suggestive in this regard. He reported that after a step decrease in light intensity, cells reversed directions repeatedly for several seconds. An increase in light intensity inhibited this "rhythmic" response. Experiments to answer this question are certainly feasible.

Lack of direct evidence again makes it impossible to say whether or not photosynthetic bacteria have specific receptors for diverse stimuli. It is, of course, possible to update Clayton's proposal that aerotaxis and phototaxis in *Rs. rubrum* are initiated by a decrease in the rate of production of some metabolic intermediate by asserting the existence of a specific receptor for the intermediate.

It is perhaps more interesting to explore the possibility that some tactic responses may not involve specific receptors. In the first place, not all attractants (or repellents) for *Esch. coli* have been shown to have specific receptors. Most important, nothing is known concerning a specific receptor for oxygen in aerotaxis. Furthermore, Macnab and Koshland (1974) (see also Taylor and Koshland, 1975) found that *Sal. typhimurium* and *Esch. coli* increase their twiddling frequency when exposed to bright light; i.e., an increase in light intensity is a repellent. Unpublished action spectra are said to suggest the involvement of a flavin. It seems unlikely that these enteric bacteria would have developed specific receptors for light; more probably, the light may damage some component (e.g., a flavin) of a biochemical pathway. A part of the sensory machinery could recognize the damage either indirectly through a decrease in the output of the pathway or directly through physical interaction with the damaged component. An extension of the latter thought to the case of photo- and aerotaxis in *Rs. rubrum* leads to the following suggestion: A component of the tactic apparatus is in contact with a component of both the photochemical and the respiratory electron-transport chains and can recognize changes in the oxidation level of that component. A decrease in oxidation level (occasioned by a decrease in light intensity or in oxygen tension) would result in an increase in the frequency of reversal of swimming direction; an increase in oxidation level would result in the opposite change.

It is clear that the time is ripe for a further attack on the problem of taxes in the photosynthetic bacteria.

6. References

Adler, J., 1975, Chemotaxis in bacteria, *Annu. Rev. Biochem.* **44**:341.

Berg, H. C., 1971, How to track bacteria, *Rev. Sci. Instrum.* **42**:868.

Berg, H. C., 1975, Chemotaxis in bacteria, *Annu. Rev. Biophys. Bioeng.* **4**:119.

Berg, H. C., and Brown, D. A., 1972, Chemotaxis in *E. coli* analyzed by three-dimensional tracking, *Nature (London)* **239**:500.

Buder, J., 1915, Zur Kenntnis des *Thiospirillum jenense* und seiner Reaktionen an Lichtreize, *Jahrb. Wiss. Bot.* **56**:529.

Clayton, R. K., 1953a, Studies in the phototaxis of *Rhodospirillum rubrum.* I. Action spectrum, growth in green light and Weber law adherence, *Arch. Mikrobiol.* **19**:107.

Clayton, R. K., 1953b, Studies in the phototaxis of *Rhodospirillum rubrum.* III. Quantitative relations between stimulus and response, *Arch. Mikrobiol.* **19**:141.

Clayton, R. K., 1955, Tactic responses and metabolic activities in *Rhodospirillum rubrum,* *Arch. Mikrobiol.* **22**:204.

Clayton, R. K., 1958, On the interplay of environmental factors affecting taxis and motility in *Rhodospirillum rubrum,* *Arch. Mikrobiol.* **29**:189.

Clayton, R. K., 1964, Phototaxis in microorganisms, in: *Photophysiology,* Vol. 5 (A. Giese, ed.), pp. 51–77, Academic Press, New York.

Krieg, N. R., Tomelty, J. P., and Wells, J. S., Jr., 1967, Inhibition of flagellar coordination in *Spirillum volutanus,* *J. Bacteriol.* **94**:1431.

Larsen, S. H., Adler, J., Gargus, J. J., and Hogg, R. N., 1974a, Chemomechanical coupling without ATP: The source of energy for motility and chemotaxis in bacteria, *Proc. Natl. Acad. Sci. U.S.A.* **71**:1239.

Larsen, S. H., Reader, R. W., Kort, E. N., Tso, W.-W., and Adler, J., 1974b, Change in direction of flagella rotation is the basis of chemotactic response in *Escherichia coli,* *Nature (London)* **249**:74.

Lofgren, K. W., and Fox, C. F., 1974, Attractant-directed motility in *Escherichia coli*: Requirement for a fluid lipid phase, *J. Bacteriol.* **118**:1181.

Lovely, P., Dahlquist, F. W., Macnab, R., and Koshland, D. E., Jr., 1974, An instrument for recording motions of micro-organisms in chemical gradients, *Rev. Sci. Instrum.* **45**:683.

Lowy, J., and Spencer, M., 1968, Structure and function of bacterial flagella, *Symp. Soc. Exp. Biol.* **22**:215.

Macnab, R. M., and Koshland, D. E., Jr., 1972, The gradient-sensing mechanism in bacterial chemotaxis, *Proc. Natl. Acad. Sci. U.S.A.* **69**:2509.

Macnab, R. M., and Koshland, D. E., Jr., 1974, Bacterial motility and chemotaxis: Light induced tumbling response and visualization of individual flagella, *J. Mol. Biol.* **84**:399.

Manten, A., 1948, Phototaxis in the purple bacterium, *Rhodospirillum rubrum,* and the relation between phototaxis and photosynthesis, *Antonie van Loeuwenhoek; J. Microbiol. Serol.* **14**:65.

Mesibov, R., Orial, G. W., and Adler, J., 1973, The range of attractant concentrations for bacterial chemotaxis and the threshold and size of response over this range, *J. Gen. Physiol.* **62**:203.

Nultsch, W., and Throm, G., 1968, Equivalence of light and ATP in photokinesis of *Rhodospirillum rubrum, Nature (London)* **218**:697.

von Reichert, K., 1909, Über die Sichtbarmachung der Geisseln und die Geisselbewegung der Bakterien, *Zentralbl. Bakteriol. Abt. Orig.* **51**:14.

Schlegel, H.-G., 1956, Vergleichende Untersuchungen über die Lichtempfindlichkeit einiger Purpurbakterien, *Arch. Protistenkd.* **101**:69.

Schrammeck, J., 1934, Untersuchung über die Phototaxis der Purpurbakterien, *Beitr. Biol. Pflanz.* **22**:315.

Silverman, M., and Simon, M., 1974, Flagellar rotation and the mechanism of bacterial motility, *Nature (London)* **249**:73.

Spudich, J. L., and Koshland, D. E., Jr., 1975, Quantitation of the sensory response in bacterial chemotaxis, *Proc. Natl. Acad. Sci. U.S.A.* **72**:710.

Taylor, B. L., and Koshland, D. E., Jr., 1974, Reversal of flagellar rotation in monotrichous and peritrichous bacteria: Generation of changes in direction, *J. Bacteriol.* **119**:640.

Taylor, B. L., and Koshland, D. E., Jr., 1975, Intrinsic and extrinsic light responses of *Salmonella typhimurium* and *Escherichia coli, J. Bacteriol.* **123**:557.

Thipayathasana, P., and Valentine, R. C., 1974, The requirement for energy transducing ATPase for anaerobic motility in *Escherichia coli, Biochim. Biophys. Acta.* **347**:464.

Throm, G., 1968, Untersuchungen zum Reaktionmechanismus vom Phobotaxis und Kinesis an *Rhodospirillum rubrum, Arch. Protistenkd.* **110**:313.

Weaver, P., 1971, Temperature sensitive mutations of the photosynthetic apparatus of *Rhodospirillum rubrum, Proc. Natl. Acad. Sci. U.S.A.* **68**:136.

Weibull, C., 1960, Movement, in: *The Bacteria* (I. C. Gunsalus and R. Y. Stanier, eds.), pp. 153–205, Academic Press, New York.

CHAPTER 50

Quantum Efficiencies of Growth

Florian Göbel

1. Introduction

Phototrophic organisms are dependent on light (radiation energy) as their primary source of energy. With the introduction of the quantum theory, a quantitative correlation between light and photosynthesis became feasible. Warburg and Negelein (1922, 1923) were the first to apply the quantum concept to photosynthesis; they attempted to measure in unicellular green algae the optimal yield of light conversion in photosynthesis, i.e., the quantum efficiency of this process.

The early work was based on the assumption that photosynthesis in resting algal cells can be treated as a simple photochemical relationship between the numbers of quanta absorbed and the amount of a single product formed (e.g., O_2) or substrate consumed (e.g., CO_2) (see Kok, 1960). When pure cultures of phototrophic green and purple bacteria became available, these studies were extended to bacterial photosynthesis (see Larsen, 1953; Wiesner, 1966). The intention to measure quantum yields for individual photochemical reactions led to studies in which cell fragments rather than resting cells were used, e.g., chloroplast preparations for photoinduced $NADP^+$ reduction (Avron and Gozal, 1969) or the photoreduction of oxaloacetate (Heber and Kirk, 1975). The problem with all these studies is that the systems used were still so complex that many factors other than those actually considered could influence the quantum yield. Consequently, the measurements were finally limited

to the study of the quantum efficiency of the photooxidation of reaction center chlorophyll using reaction center preparations. Since this reaction is a primary photochemical process, the Stark–Einstein quantum equivalence law is strictly applicable (Duysens, 1952; Wraight and Clayton, 1973; Wraight et al., 1974). From a certain point of view, the measurement by Wraight and Clayton (1973) of a quantum efficiency of 1.02 ± 0.04 for the photooxidation of bacteriochlorophyll (Bchl) may therefore be considered as one definitive solution to the original problem posed by Warburg and Negelein to determine the optimal quantum efficiency of photosynthesis.

An alternative possibility is to extend the optimal-yield measurement to the entire growing organism. In this case, the number of quanta absorbed is related to the total biomass formed. With this approach, the efficiency of photosynthesis in different kinds of organisms and under various growth conditions can be determined comparatively, yielding biologically significant results.

For nonphototrophic growth, the well-known quotient, dry weight per mole of energy-yielding substrate consumed, Y_s, serves as the basis for a calculation of the energy requirement. For phototrophic growth, the dry-weight yield is related to quanta absorbed; thus, the quantum yield of growth corresponds to Y_s and is used similarly.

To study microbial growth under defined conditions, it is very useful to work with a chemostat. A chemostat, in contrast to a batch culture, provides continuous growth of a culture under steady-state

Florian Göbel · Institut für Mikrobiologie der Gesellschaft für Strahlen- und Umweltforschung mbH München in Göttingen, Göttingen, Germany

907

conditions; the growth rate of the culture is limited by the concentration of a required nutrient. To apply the chemostat technique to quantum requirement measurements, the quantum supply to the cells must be the growth-limiting factor. Light quanta are a very special "substrate." They exist only as a flux coming from an external radiation source; therefore, they cannot be distributed homogeneously in an absorbing cell suspension. Myers (1966) characterized the situation, when he discussed the light supply in algae cultures: "Faced with these complexities, we have considered it unlikely that a rigorous theoretical treatment can yet be made on dynamics of algal growth in relation to light as the most important controlling factor." In general, this statement is still valid today. Nevertheless, it is possible to calculate for flat culture vessels the mean irradiance, as the determining parameter for the quantum supply. I will give a short introduction to the special techniques used and the theory of light-limited continuous culture. I shall present results that show the relationship of quantum requirement to specific growth rate. The quantum requirement is in general a function of the substrates used (especially the carbon source). The correlation of quanta required and substrates used enables us, for example, to calculate energy requirement for nitrogen fixation.

2. Experimental System

2.1. Continuous-Culture Principle

I would like to characterize the principle of continuous culture only briefly, since the results presented below were obtained by use of the chemostat. For more detailed information, I recommend the following literature: Herbert (1961); Pfennig and Jannasch (1962); Noak (1968); Dean et al. (1972).

As the cell population grows, sterile medium is pumped in at a constant rate and the cell suspension is displaced into an overflow vessel at the same rate (see Figs. 1 and 3). The population would soon disappear through such a washout process if cell loss were not balanced by population increase through cell division. Washout and population growth may be described by simple equations. The ratio of the flow rate of medium, dV/dt, to the volume of the growth vessel, V, is the specific dilution rate, D:

$$D = \frac{(dV/dt)}{V}$$

D is a measure for the amount of volume change per unit of time; it is simultaneously a measure for cell loss through the rinsing-out process.

Under growth conditions, on the other hand, cell increase occurs that may be characterized thus:

$$\mu = \frac{dN}{dt \cdot N}$$

where μ is specific growth rate and N is the number or mass of cells.

The total variation in the number or mass of cells may be expressed by

$$\frac{dN}{dt} = \mu \cdot N - D \cdot N$$

Rinsing out and cell division work against each other. If both influences are the same, $\mu - D = 0$, then the number of cells in the culture remains constant.

$$\frac{dN}{dt} = N \cdot (\mu - D) = 0$$

Through this state of balance, it is possible to study the metabolic activity of a growing population without variations in the cell density. A continuous-culture system becomes a chemostat when a steady state is realized at constant flow rate.

As one can easily see, in such a steady state, the concentrations of substrates and metabolic products also do not vary. In static or batch culture, on the other hand, these quantities are continuously changing. For this reason, the continuous-culture system is particularly suited to the study of growth under defined and constant conditions. The results that will be presented herein were all arrived at through the use of steady states.

2.2. Scheme and Description of the Apparatus

Figure 1 shows a scheme of our experimental equipment. The culture vessel is formed as a flat section of a cylinder and is completely filled with cell suspension. The front side serves as an entrance window for radiation available for photosynthesis. The surface of the cylinder jacket is coated with a reflective layer to avoid loss of radiation. The rate of flow of medium into the culture vessel is controlled by the medium pump. We use a large area bolometer to measure the radiation. In Fig. 1, the bolometer is behind the culture vessel in order to measure the unabsorbed radiation that escapes from the rear window of the culture vessel. Figure 1 also shows the radiation source, consisting of a 2-kW incandescent lamp with iodine cycle and a set of absorption and interference filters. There are additional devices for temperature control, regulation of pH, intensive stirring of the culture, and aeration. It is essential that the entering and the transmitted radiation be measured, so that the quantum uptake of the

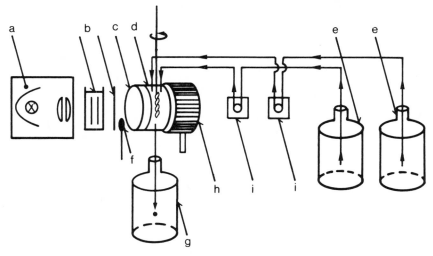

Fig. 1. Light chemostat. (a) Lamp housing; (b) water-cooled absorption and interference filters; (c) chamber for thermostatic circulation of water connected with an external thermostat; (d) culture vessel; (e) storage bottles for the sterile medium; (f) photoelectric element controlling the incident radiation; (g) collection reservoir; (h) bolometer for measuring the transmitted radiation; (i) medium pumps.

culture may be steadily monitored. The continuous emission spectrum of the incandescent lamp enables us, through the choice of suitable filters, to use monochromatic and polychromatic radiation in the wavelength range from 400 to 1200 nm. The medium is added in two separate, incomplete solutions, so that complications, such as the cells growing backward into the medium inlet tube, can be avoided.

2.3. Monochromatic Light

The incandescent lamp that we use emits polychromatic radiation and easily reaches the required radiation intensity. There are some reasons, however, to work with a filter monochromator. The scattering and absorption of light are easier to work with monochromatically because of their dependency on light wavelength. Radiation has to be expressed by photochemically relevant units, mol quanta or Einsteins. For this measurement, only the bolometrically measured radiation energy is required when using the quantum energy formula

$$E = h \cdot v = h \cdot \frac{c}{\lambda}$$

The dependence of photosynthetic performance on radiation wavelength can also be studied. It is possible, for instance, to compare the effectiveness of various photopigments using their different absorption wavelengths. It is also possible to carry out experiments with our equipment using polychromatic radiation. The bandwidths of the interference filters used in the

experiments discussed here are: 22 nm for the 522 nm filter and 19 nm for the 860 nm filter.

2.4. Choosing the Depth of the Culture Vessel

The thickness of the culture vessel is important, since thick layers can cause considerable loss of radiation intensity, thus creating dark areas through the cells' self-shadowing. Radiation weakening is necessary to allow for light-limited balance of growth, however, as will be described in more detail below. We worked with a 40-mm, and later a 28-mm, layer thickness. With a layer thickness of 28 mm and cell density of less than 0.1 mg dry wt./ml, the maximum loss of radiation intensity was 50%.

3. A Measure of Mean Irradiance

Since the quantum requirement for growth can be determined only when growth is limited by light, it is necessary to devise a suitable measure of availability of quanta in the culture vessel. The mean irradiance provides this measure.

3.1. Definition and Practical Determination of Mean Irradiance

If one considers that every ray of light penetrating into an absorbing medium is thereby weakened, so that cells that are in the center of a culture vessel receive less light than cells that are at the entry window, and those at the exit window receive even less, then the question becomes how to take this inequality into account? We have done this by using an average value of light intensity that takes into account both the weakening caused by absorption in the culture vessel and the directional distribution caused by scattering. This average value is called the *mean irradiance*, Ψ_m, and has units of Einsteins per second per square centimeter. It is given by

$$\Psi_m = \frac{\int_V \int_\omega B \cdot dV \cdot d\omega}{V}$$

where B, in Einst/sec · sr · cm², radiance, is the elementary flux of quanta; ω, in sr (steradians), is the solid angle through which flux radiates to the volume element; and V, in cm³, is the volume of the culture. For determination of Ψ_m, it is not necessary to develop mathematical functions for the distribution of light in the culture. A practical equation is (Göbel, 1969)

$$\Psi_m = \frac{\Delta\Phi \text{ cells}}{K_a \text{ cells} \cdot V \cdot 2.303}$$

where $\Delta\Phi$ cells is the radiant flux absorbed by the cells and K_a cells is the decadic pure absorption constant of the population.

Mean irradiance is proportional to the amount of radiation absorbed by the cells. Our goal, a measure for the light supply, as well as the availability of radiation for cell growth, is thus reached with Ψ_m. The equation for Ψ_m corresponds to the universally valid equation for the weakening of an elementary beam of light:

$$-dB = k_a \cdot B \cdot ds$$

$$B = -\frac{dB}{k_a \cdot ds}$$

where s is the path of radiation and k_a is the natural pure absorption constant.

With our apparatus, we can determine the absorbed flux of light, expressed by the following balance equation:

$$\Delta\Phi = \Phi_1 - (\Phi_2 + \Phi_3)$$

where $\Delta\Phi$ is the absorbed radiant flux, Φ_1 is the entering radiant flux, Φ_2 is the exciting radiant flux, and Φ_3 is the back-scattered radiant flux.

The pure absorption constant of the cell suspension, on the other hand, is difficult to determine. The practical problem in determining Ψ_m with the method shown here is the task of measuring the pure absorption of the scattering cell suspension.

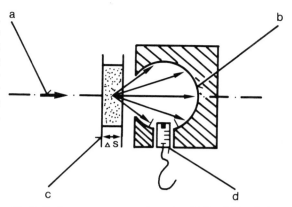

Fig. 2. Device to measure pure absorption. (a) Parallel irradiation; (b) integrating sphere for measurement of the scattered radiation; (c) cuvette filled with cell suspension; (d) photomultiplier.

3.2. Pure Absorption of Living Cell Suspensions

For determination of pure absorption, all loss of radiation caused by scattering must be excluded. To calculate the pure absorption constant of the cell suspension, k_a, we use Lambert's law:

$$\Phi_2/\Phi_1 = e^{-k_e\Delta s}$$

where Δs is the layer thickness, k_e is the extinction constant, and k_s is the apparent scattering constant. Since $k_e = k_a + k_s$, when k_s approaches zero, k_e becomes equal to k_a and

$$k_a = \frac{1}{\Delta s} \ln \frac{\Phi_1}{\Phi_2}$$

Figure 2 shows a diagram of the measuring system we use. As we could show through the comparison with pigment-free cells cultivated aerobically, it is possible, using small cuvettes of 1- to 2-mm thickness and an integrating sphere as the receiver for the diffuse transmitted radiation, to exclude loss of scattered radiation (Göbel, 1978). This is possible because of the predominantly forward direction of the singly scattered light of the bacteria (Bergter and Günther, 1961). Complications stemming from the sieve effect of the relatively large cells do not arise as long as a pure absorption constant for cell suspensions is employed.

4. Growth Kinetics in the Light Chemostat

The specific growth rate of photosynthetic bacteria is proportional to the intensity of irradiance so long as the concentrations of all necessary substrates are in excess. To achieve a light-limited steady state in a chemostat, the light-dependent growth rate has to balance the flow rate. Therefore, each value of specific flow rate demands a fixed mean irradiance, Ψ_m.

4.1. Establishment of Growth Balances by Changing Self-Shadowing

Figure 3 shows a diagram of the culture in which weakening of the entering irradiance by the cell suspension is indicated. The deeper layers, up to the exit window, are in increasingly deeper shadow. The degree of shadowing depends on the number of cells. In a chemostat, if the specific growth rate becomes greater than the dilution rate, the cell density and self-shading will increase and Ψ_m will consequently decrease; this will tend to reduce the specific growth rate until it is again equal to the dilution rate and the steady state is reestablished. Since Ψ_m depends on both incident irradiance and absorption (and scattering) of light in the growth vessel, we can anticipate that for a given specific growth rate, different incident irradiances will result in different cell densities such that the mean irradiance in the growth vessel (Ψ_m) is the same.

It is helpful, when starting a chemostat experiment, to allow the culture to grow to a high density without flow of medium, since balance is established rapidly during the following partial washing-out process.

The spatial variations in irradiance must be kept low, to avoid influences on cell growth. As a limiting case, the ideal homogeneous irradiation would be realized with a single layer of cells, so that the mean irradiance would equal the incident irradiance. But this would not allow the establishment of a steady state

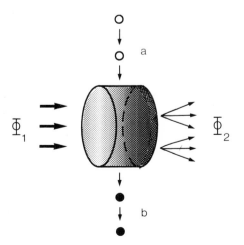

Fig. 3. Culture vessel. (a) Medium inflow; (b) cell-suspension outflow.

through self-shading, and therefore stands in contradiction to the principle of our light chemostat.

As a reciprocal measure of variations in irradiance in the culture vessel, the transmittance, δ, can be used. This is the ratio of exiting radiant flux, Φ_2, to entering radiant flux, Φ_1:

$$\delta = \frac{\Phi_2}{\Phi_1}$$

The transmittance becomes zero when the entering flux is fully absorbed, and it becomes unity when there is no absorption. Even in the wavelength region of maximal absorption by the pigments—minimal transmittance—we regularly reached $\delta > 0.4$. In our culture vessels with a depth of 2.8 cm, such high values of transmittance can be attained only with cell densities of less than 0.1 mg dry wt./ml.

To achieve a uniform supply of quanta to the cells, the cultures are stirred intensively.

As an important result, it should be borne in mind that the Ψ_m adjustment mechanism described above regulates by itself the growth rate in balance with the flow rate when incident irradiance is constant. This is true as long as growth is limited by light.

4.2. Transition from Light Limitation to Substrate Limitation of Growth

Figure 4 shows the results of steady-state experiments with *Rhodospirillum rubrum* strain Ha (Deutsche Sammlung von Mikroorganismen 107) at constant specific growth rate and increasing incident irradiance. Each experiment was carried out at a constant specific dilution rate and constant incident irradiance. At a specific dilution rate of 0.1/hr, one experiment takes a full week. An experiment is begun by inoculation of the medium in the thoroughly cleaned culture vessel. One day later the pumps are started and the flow of medium begins. One or two days after that, the cell concentration and exiting radiant flux have become constant. At least 30 hr later, the required measurements can be taken at the steady state. Entering and exiting radiant fluxes are measured with the bolometer. The back-scattered flux is measured by means of an integrating sphere, which is arranged between the monochromator of a photometer and a vessel filled with the cell suspension. We used a Zeiss DMR 21 photometer with a double-beam remission device. From the three radiant fluxes, one can calculate the quantum balance and the mean irradiance, Ψ_m. For the latter, the pure absorption constant, k_a, must also be known. For determinations of steady-state concentrations of dry cell material and

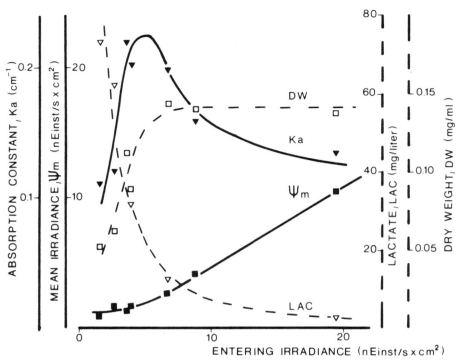

Fig. 4. Growth parameters as function of entering irradiance. Steady-state experiments with *Rhodospirillum rubrum* strain Ha. Wavelength of irradiance, 885 nm; constant specific dilution rate, 0.1 hr^{-1}.

substrate and product concentrations, samples are taken from the culture vessel. Before a new experiment is started, the culture vessel must be cleaned to remove slime and films that become attached to the walls. This requires special methods that are discussed later. Since each experiment takes such a long time, two chemostats were used simultaneously.

Dry weight per milliliter is used as a measure of cell concentration. The dry weight per milliliter increases with increasing incident irradiance until lactate, the only carbon source, becomes the limiting factor. The mean irradiance, Ψ_m, remains approximately constant while the dry weight per milliliter is increasing; Ψ_m increases proportionally with incident irradiance after the dry weight per milliliter has reached its maximum value, which is fixed by the concentration of lactate. We used 156 mg lactate/liter in the input medium; at this concentration, substrate becomes limiting at relatively low cell densities.

The absorption constant of the culture reaches a maximum at the incident irradiance corresponding to the shift from light limitation to substrate limitation. The increase in the absorption constant is due to the increase in cell density; the decrease is due to the decrease in pigment content of the cells, which reflects

a surplus of light when lactate is limiting. Pigment regulation, however, may not be effective enough to keep the quantum yield at its maximal level. I shall say that there is surplus light whenever the pigment content of the cells is lower than the maximum pigment content possible at a given specific growth rate.

4.3. Light Dependence and Light Saturation of Growth Rate

From what has been said so far, we can expect that the steady-state mean irradiance will increase with increasing specific growth rate. The results shown in Fig. 5 bear out this expectation. This figure shows the results of two series of experiments with *Rhodopseudomonas capsulata* strain Kb1 (DSM 155). One series used monochromatic light absorbed by Bchl (860 nm); the other used light absorbed primarily by the accessory carotenoid pigments (522 nm). For each specific growth rate, up to five experiments were done, each using a different incident light flux; the results were plotted as in Fig. 4. From these plots, the growth-limiting value of Ψ_m for each specific growth rate was determined; these are the results shown in Fig. 5.

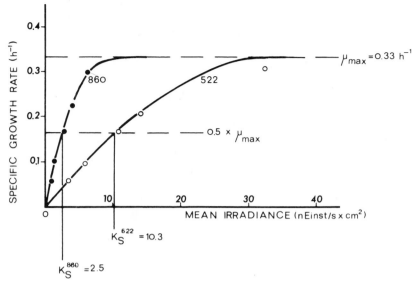

Fig. 5. Light dependence of specific growth rate. Steady-state experiments with *Rhodopseudomonas capsulata* strain Kb1. Wavelength of irradiation 522 nm for light-harvesting by carotenoids and 860 nm for light-harvesting by Bchl. μ_{max} was determined by washout experiments at saturating light intensities. Mineral medium (Pfennig, 1969) with lactate; temperature 30°C.

At any specific growth rate less than the maximum, the Ψ_m for light absorbed by the carotenoids is greater than that for light absorbed by Bchl. This matches the lower absorption power of the carotenoids as well as the higher quantum requirement for growth when the light is absorbed by these pigments rather than by Bchl.

The light dependence of the specific growth rate has the shape of the saturation function:

$$\mu = \mu_{max} \cdot \Psi_m / (K_s + \Psi_m)$$

where μ_{max} is the maximum specific growth rate and K_s is the saturation constant. This expression has the same form as the equation for substrate-limited growth rate formulated by Monod (1942). The shape of the saturation curve is characterized by the two parameters, μ_{max} and K_s. The maximum specific growth rate attainable in a particular medium is μ_{max}. Under the conditions used for the experiments shown in Fig. 5, we found, by separate washing-out experiments at saturating light intensities, that $\mu_{max} = 0.33$ hr^{-1}. This value was independent of the wavelength of the light. K_s is a reciprocal measure of the steepness of the saturation curve; it is defined as Ψ_m at $\mu = 0.5\, \mu_{max}$. We have determined that for $\lambda = 860$ nm, $K_s = 2.5$ nEinst·sec·cm^2, and for $\lambda = 522$ nm, $K_s = 10.3$ nEinst·sec·cm^2. Thus, with higher absorption power, there is a lower saturation constant and vice versa. The quantum yield measurements discussed in Section

7 were obtained from light-limited steady-state experiments. We found that the quantum yields, under these conditions, were dependent on the specific growth rate.

5. Light-Dependent Pigment Regulation and Growth Kinetics

The dependence on irradiance of Bchl *a* content for *Rs. rubrum* strain Ha is shown in Fig. 6. Pigmentation was measured in steady-state experiments from $\Psi_m = 1.8$ nEinst/sec·cm^2 up to 55.0 nEinst/sec·cm^2; all experiments were done at a specific growth rate of 0.1 hr^{-1}. Experiments with mean irradiance less than 1.8 nEinst/sec·cm^2 were not possible because this is the growth-limiting value. The dashed curve in Fig. 6 corresponds to the equation

$$P = P_{min} + \frac{K_s(P_{max} - P_{min})}{K_s + \Psi_m}$$

where P_{max} is the maximum pigment content, P_{min} is the minimum pigment content, and K_s is the saturation constant. The values of the three constants—$P_{max} = 25.14$ μg Bchl/mg dry wt., $P_{min} = 1.4$ μg Bchl/mg dry wt., and $K_s = 1.15$ nEinst/sec·cm^2—were determined from the experimental data. This equation was derived from the saturation function of light dependence of the specific growth rate. The derivation and the proof will be published elsewhere. For light saturation,

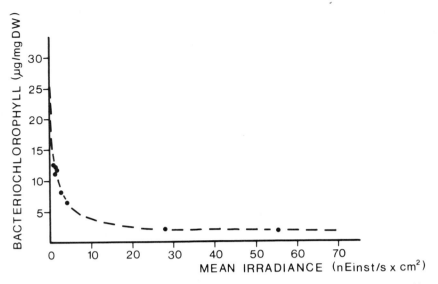

Fig. 6. Relationship of the Bchl content of cells to mean irradiance. Steady-state experiments with *Rhodospirillum rubrum* strain Ha. Wavelength of irradiation, 885 nm. (●) Experimental values; (————) saturation equation. See the text for further details.

$\Psi_m \gg K_s$, the equation gives $P = P_{min}$. For transition to darkness, $\Psi_m \to 0$, it gives $P = P_{max}$. This equation is of advantage compared with equations formulated earlier, e.g., Cohen-Bazire and Sistrom (1966):

$$C = K + A/J$$

where C is the pigment content, J is the irradiance, K is the minimum pigment content, and A is a constant. This equation gives an unlimited increase of pigment content for small values of irradiance.

The pigment contents of some organisms are compared in Table 1. Maximum values correspond to $\Psi \approx$ 40 nEinst/sec · cm²; minimum values, to $\Psi_m = 2$ nEinst/sec · cm² (these are not the true maximal and minimal values, but rather are merely the highest and lowest observed values). The absolute values as well as the extent of light-dependent regulation, i.e., the ratio of maximum to minimum pigment contents, are quite different for different organisms at equal irradiance. Maximal pigment contents should increase at irradiance less than $\Psi_m = 2$ nEinst/sec · cm². The ratio of the maximum to the minimum carotenoid content was always less than the ratio for Bchl content.

Table 1. Specific Bacteriochlorophyll Content of Phototrophically Grown Rhodospirillaceae

Organism	Mean irradiance (nEinst/sec · cm²)[a]	Bchl (µg/mg dry wt.)
Rs. rubrum strain Ha	2	12.6 (6.6)[b]
	40	1.9
Rp. acidophila strain 7050	2	21.5 (5.0)[b]
	40	4.3
Rs. tenue strain 3661	2	12.3 (3.7)[b]
	40	3.3
Rp. capsulata strain Kb1	2	20.0 (1.5)[b]
	40	13.0

[a] At wavelength corresponding to maximal absorption of Bchl (885 nm for *Rs. rubrum*, 860 nm for all others).

[b] Figures in parentheses are relative to Bchl content at 40 nEinst/sec · cm².

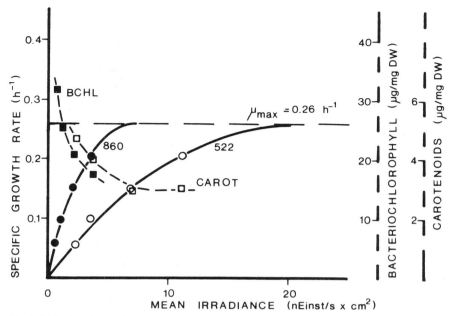

Fig. 7. Light dependence of specific growth rate and photopigments of cells. Steady-state experiments with *Rhodopseudomonas acidophila* strain 7050. Wavelength of irradiation 522 nm for light-harvesting by carotenoids and 860 nm for light-harvesting by Bchl. Carotenoid contents refer to the experiments at 522 nm; Bchl contents, to those at 860 nm. μ_{max} was determined by washout experiments at saturating light intensities. Mineral medium (Pfennig, 1969) with lactate; temperature 30°C.

In the state of light-limited growth, the three parameters mean irradiance, pigment content, and specific growth rate are in a fixed coordination. Each variation of the specific growth rate results in alterations of mean irradiance and pigment content. Figure 7 shows for *Rp. acidophila* strain 7050 the Bchl and carotenoid contents and the specific growth rate as dependent on mean irradiance. As in Fig. 5, two series of experiments are shown, one with quantum absorption by Bchl, the other with quantum absorption by the carotenoids. For an equal specific growth rate, 2.5 times stronger mean irradiance is required when carotenoids are the light-harvesting pigments.

To study growth kinetics, it is necessary to consider growth as dependent on quantum uptake rate as the only energy source. The quantum uptake rate per g cell dry weight, $\Delta\phi$, is proportional to the pigment content of the cells, P, and to mean irradiance, Ψ_m:

$$\Delta\phi = k_p \cdot P \cdot \Psi_m$$

where k_p is the absorption constant of the pigments. It can be seen from Fig. 7 that the pigment content of the cells decreases as the mean irradiance increases; therefore, the quantum uptake rate is not proportional to changes in irradiance. Since the quantum uptake rate equals energy supply, the reciprocal relationship between pigment content and irradiance reflects an adaptation of the cells to strongly varying light intensities. In this connection, the increased synthesis of photosynthetic membranes in dim light, i.e., under conditions of energy deficiency, is remarkable.

6. Adaptations of the Cell to Light Deficiency

To compare results of various experiments, quantum yield measurements must be carried out under conditions of light-limited growth. Light limitation means energy limitation. This expresses itself in the cells' increased content of light-harvesting pigments. This can also lead, however, to accumulation of reserve materials, to fermentation in the case of photoorganotrophic growth, and in the end to increased formation of slime resulting in growth on the walls of the culture vessel. Usually, from these phenomena, the dependence of photopigmentation on the intensity of light may be considered as one good example for physiologically significant regulations. If possible, formation of storage products, fermentation, and growth on the walls should be avoided, since they lead to particular difficulties in experimentation. For the survival of cells in the case of light deficiency, however, the transition to fermentation and the attachment of cells to places with particularly good light conditions are more important than the regulation of pigment content.

6.1. Production of Storage Material

The formation of storage material can always be expected when suitable substrates for the formation of the stored substances are in abundance, but where growth is limited by some other factor. In our experiments, low mean irradiance levels can induce the storage of reserve materials. Particularly with acetate, but also with malate, as the carbon source, we found a high content of poly-β-hydroxybutyric acid (PHB). Figure 8 shows values that were measured in a continuous culture of *Rs. rubrum* strain S1 (DSM 467, ATCC 11170) on acetate. The specific growth rate was constant, $\mu = 0.1$ hr^{-1}. With decreasing mean irradiance, the concentration of unused acetate increases, and parallel to that, the PHB content rises from 4 to as much as 70% dry wt. For the quantum requirement of dry weight formation, we found the relatively high value of 0.37 Einst/g dry wt. The dry weight consisted of 65% PHB. For this system, *Rs. rubrum* growing on acetate, it was not possible to determine separately the quantum requirement of PHB-free cells, because at light deficiency, PHB storage began before it was possible to measure the quantum requirement.

To avoid the complications due to storage of PHB in quantum yield determinations, it was necessary to select strains and substrates that would give minimal formation of PHB (see Table 2).

6.2. Transition to Fermentation

We found in experiments with *Rs. rubrum* strain Ha, with lactate as carbon source, that the yield decreases from 1.0 to 0.6 dry weight of cell material/lactate, when the transition from light saturation to light limitation is carried out. This transition is a step in the direction from anaerobic light to anaerobic dark conditions. It is known that Rhodospirillaceae with appropriate substrates, especially pyruvate, are capable of fermentative metabolism in anaerobic dark conditions (Kohlmiller and Gest 1951; Gürgün 1974).

In our case, additional fermentation experiments showed that *Rs. rubrum* strain Ha has little ability to decompose lactate and malate. The cells are not able, however, to grow with these substrates under anaerobic dark conditions. We found for malate that the assimilation rate returned to 4% during the change from light to darkness. The consumption of malate and lactate in darkness slowly died away, and after approximately 20 days was not detectable. We conclude that with light limitation, lactate, as well as malate, would serve not only as a source of carbon, but also as an energy substrate. For quantum yield

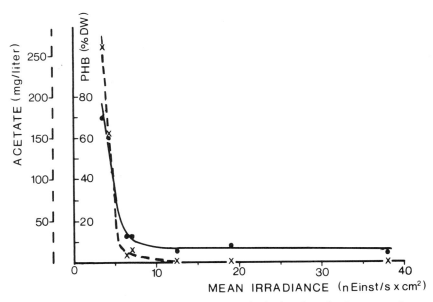

Fig. 8. Dependence of PHB content on acetate concentration in the culture. Steady-state experiments with *Rhodospirillum rubrum* strain S1.

Table 2. Screening of Bacterial Strains for Their Appropriateness for Use in Quantum Yield Measurements

	Phototrophic culture		Anaerobic dark culture		Appropriateness for quantum yield measurements
Organism and substrate	Relative growth rate[a]	pH optimum	Fermentation activity[b]	Storage material[c]	
Rp. acidophila strain 7050					
Malate	3	5.7	34%	+ +	
Lactate	3	5.8	2%	−	+ +
Acetate	No growth				
Rp. capsulata strain Kb1					
Malate	4	6.7	40%	−	
Lactate	4	6.7	13%	−	+
Acetate	5	>8	0%	−	+ +
Rp. palustris strain 6951					
Malate	2	6.7	13%	−	+
Lactate	2	6.7	20%	+	
Acetate	1	7.5	0%	+	

[a] By visual comparison of the turbidity increase.
[b] Percentage of carbon source decomposed after a dark period of 16 days. The dark period began after phototrophic precultivation. The concentration of carbon source when starting the dark period was equal to the dry weight concentration.
[c] By microscopic examination of cell inclusions.

measurements, therefore, we had to find a system in which the metabolism of the cells is in fact light-dependent. To achieve this, we studied the fermentation and decomposition of the carbon source under anaerobic dark conditions. The results of these experiments are presented in Table 2, which also contains a relative evaluation for the maximal growth rate.

6.3. Slime Production and Wall-Attached Growth

When phototrophic growth is continued over a long period of time, a layer of bacteria frequently develops on the walls of the culture vessel. When the culture is growing under light limitation, this occurs chiefly on the window where the light enters. With *Rp. capsulata*, for example, this phenomenon is linked with a slime capsule that is recognizable under the microscope. With *Rp. capsulata*, the tendency to wall-attached growth seems also to be dependent on the available carbon sources.

Growing cells that are not homogeneously distributed throughout the medium tend to hinder the establishment of balance between cell division and the rinsing-out process through a kind of continuous new inoculation. The total population then has no defined specific growth rate, and cell concentration in the outflow is no longer representative of the yield. Consequently, quantum yield measurements are also no longer possible. Therefore, after every experiment, i.e., about once a week, we cleaned our culture vessels with a surface-active cleaning agent to remove any adhering slime or other film that had formed (at a temperature of about 60°C, sodium laurylsulfate was found to be most effective). In especially stubborn cases, the attached cells were dissolved with the use of enzymes, the most effective of which were preparations of exoenzymes from myxobacteria that were kindly supplied by H. Hippe (1974). Because of the short growth time in batch cultures, we were unable to predict the tendency of organisms to form wall-attached growth.

7. Quantum Yield in Light-Limited Growth

The proportionality called quantum yield, Y_Q, may be calculated as the ratio between cell dry weight formed and the amount of quanta absorbed or as the ratio between specific growth rate, μ, and specific absorption rate of quanta, $\Delta\phi$, which is more appropriate to the continuous-culture principle:

$$Y_Q = \frac{\mu}{\Delta\phi}$$

Measurements of quantum yield require that light be the growth-limiting factor, where growth will increase or decrease with irradiance. By use of the light chemostat, we found that quantum yield is dependent on generation time. In recent work on ATP yield, corresponding findings were made. The absorption of quanta by Bchl or by carotenoids results in different quantum yields.

7.1. Dependence of Quantum Yield on Generation Time

In Fig. 9, two series of experiments are presented. Both were performed with *Rp. capsulata* strain Kb1 cultivated in monochromatic light. At $\lambda = 860$ nm, radiation is absorbed by Bchl; at $\lambda = 522$ nm, by the carotenoids. The generation time varies from 2.3 hr to 14 hr, respectively, and specific growth rates from 0.3 hr^{-1} to 0.05 hr^{-1}. For each specific growth rate, a

series of experiments was done at different incident radiant fluxes, and the results were plotted as in Fig. 4. From these plots, the light-limited levels of quantum yield and pigment content were determined; these values are shown in Figs. 9 and 10. In Fig. 9 the quantum yield increases with increasing specific growth rate; it reaches a maximum at $\mu = 0.1$ hr^{-1} and then falls to a value about one-half the maximum at higher growth rates. Fig. 10 shows the corresponding results with *Rp. acidophila* strain 7050.

An interpretation of the slope of the yield curves is possible, since the rise in the region of slow growth can be correlated with the influence of maintenance energy. Pirt (1965) gave a simple formula expressing the energy requirement for growth:

$$1/Y = 1/Y_{max} + m_e/\mu$$

This means that the energy requirement per unit of dry weight formed, $R = 1/Y$, consists of energy required for the formation of new cell material, $R_{min} = 1/Y_{max}$, and energy required for maintenance, m_e/μ; where m_e is the maintenance coefficient. The energy of maintenance does not contribute to formation of new cell material. The fraction of the total energy requirement

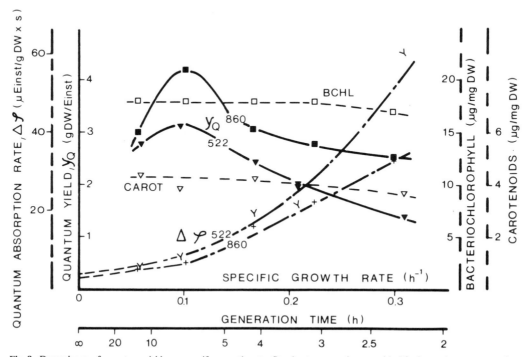

Fig. 9. Dependence of quantum yields on specific growth rate. Steady-state experiments with *Rhodopseudomonas capsulata* strain Kb1. Wavelength of irradiation 522 nm for light-harvesting by carotenoids and 860 nm for light-harvesting by Bchl. Carotenoid contents refer to the experiments with 522 nm; Bchl contents, to those with 860 nm. Mineral medium (Pfennig, 1969) with lactate; temperature 30°C.

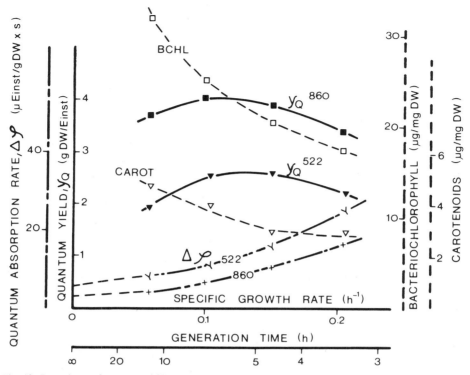

Fig. 10. Dependence of quantum yields on specific growth rate. Steady-state experiments with *Rhodopseudomonas acidophila* strain 7050. Wavelength of irradiation 522 nm for light-harvesting by carotenoids and 860 nm for light-harvesting by Bchl. Carotenoid contents refer to the experiments with 522 nm; Bchl contents, to those with 860 nm. Mineral medium (Pfennig, 1969) with lactate; temperature 30°C.

devoted to maintenance increases with increasing generation times, that is, with decreasing specific growth rate. According to the reciprocal of the Pirt formula:

$$Y = \frac{1}{1/Y_{max} + m_e/\mu}$$

we should get by plotting Y vs. μ a curve rising from zero and approximating Y_{max} asymptotically at high values of μ. In contrast to this shape, the curves Y_Q vs. μ in Figs. 9 and 10 decrease after reaching a maximum.

It has been found, however, that the growth yield per mole of ATP, Y_{ATP}, approaches a maximum value, Y_{ATP}^{max}, in accordance with Pirt's equation (de Vries *et al.*, 1970; Stouthamer and Bettenhaussen, 1973, 1975). This was established for anaerobic dark (fermentative) growth of bacteria, and suggests that Y_{ATP}^{max} and m_e are constant at all growth rates for a given bacterium in given conditions. If this is true for phototrophic bacteria, then the observed decrease in the quantum yield for growth at higher growth rates leads to the conclusion that at high growth rates, more

quanta are required for the formation of ATP than at low growth rates.

For discussing the influence of the pigment content of cells on quantum yields at different specific growth rates, the right viewpoint is given by the concept that the specific absorption rate, $\Delta\phi$ (see Section 5), not the mean irradiance, is the intrinsic limiting factor in light-dependent growth. This can be expressed by

$$\mu = Y_Q \cdot \Delta\phi$$

The quantum absorption rate per light-harvesting pigment molecule is the turnover number, $TN = \Delta\phi/P$, i.e., the quanta absorbed per pigment molecule per second. From these two equations, we get

$$TN = \mu/Y_Q \cdot P$$

We see from this equation that the turnover of quanta in the light-harvesting pigment molecules increases with increasing specific growth rate and with decreasing pigment content and decreasing quantum yield. If it is true that Y_{ATP}^{max} is constant, then the transfer of energy from the light-absorbing pigments to reaction center chlorophylls (RC Chl's) or the turnover of excitation states in the RCs becomes inefficient at higher growth rates.

The two organisms *Rp. capsulata* strain Kb1 (Fig. 9) and *Rp. acidophila* strain 7050 (Fig. 10) differ markedly in pigment regulation. The nearly constant pigment content of *Rp. capsulata* is correlated with a more rapid decrease of Y_Q at high specific growth rates and the active regulation of *Rp. acidophila* with a flat maximum of Y_Q.

With respect to the discussion above, not only the total amount of pigments but also the ratio of antenna to RC Chl's should influence the efficiency of absorbed light. In some cases, experimental data show variations of the ratio of antenna to RC pigments (Aagaard and Sistrom, 1972). In our laboratory, Lehmann (1974, unpublished) found the RC Chl per cell mass in *Rp. acidophila* strain 7050 to be nearly constant. When the total amount of pigment is reduced by high irradiance, this constancy leads to smaller photosynthetic units with less loss of quanta. This suggests that in the case of *Rp. acidophila* pigment, regulation controls not only the amount of pigments but also their organization, which may reduce the loss of quanta at high specific absorption rates.

Summarizing, one can say: At low specific growth rates and low irradiances, cells save quanta. In this region, the quantum yield increases to a maximum value as the fraction of energy devoted to maintenance decreases with increasing specific growth rate. At high specific growth rates and high irradiance, the quantum yield decreases. This seems to reflect an increased quantum requirement of photophosphorylation. Organisms vary considerably in their ability to adapt to low and high irradiance by variation of pigment content. At high irradiance, a small ratio of antenna to RC Chl's seems to result in more constant quantum yields.

7.2. Maintenance Coefficient and Quantum Yield Corrected for Maintenance

By extrapolation of the specific absorption rate in Figs. 9 and 10 to zero specific growth rate, the maintenance coefficient for quanta, m_q, can be determined. This is the amount of light energy required for maintenance without net growth. From the few suitable data, we determined as an average value for light absorbed by Bchl $m_q = 0.012$ Einst/g dry wt. $\cdot$ hr^{-1}. On the assumption that Pirt's formula

$$1/Y_Q = 1/Y_Q^{max} + m_q/\mu$$

is valid in the region of low specific growth rate up to the maximum value of Y_Q, we can calculate Y_Q^{max}. This is the quantum yield corrected for maintenance. For

light absorbed by Bchl, we found as an approximate value for both organisms $Y_Q^{max} = 8$ g dry wt./Einst. (For these approximate calculations, we included the data for 522 nm light reduced by the factor $Y_Q^{522\,nm}/Y_Q^{860\,nm}$; see the next section.)

7.3. Yield Differences with Bacteriochlorophylls and Carotenoids

With *Rp. capsulata* Kb1 and *Rp. acidophila* 7050, we have previously considered two corresponding series of experiments (Figs. 9 and 10), one using monochromatic irradiance absorbed only by Bchl at $\lambda = 860$ nm, the other using $\lambda = 522$ nm light absorbed almost entirely by carotenoids. In the light-harvesting system, the carotenoids are accessory pigments only, while the Bchl includes both light-harvesting Bchl and RC Bchl. We found that light absorbed only by the carotenoids satisfies the requirement for full growth; the same is true for absorption by Bchl. With both wavelengths, monochromatic irradiation yields the same μ_{max} as obtained with the full spectrum of a tungsten lamp.

In the experiments shown in Figs. 9 and 10 and in other similar experiments, the quantum yield is always higher when Bchl are absorbing. The mean values of the yield quotient are *Rp. capsulata Kb1*, $Y_Q^{522}/Y_Q^{860} = 0.76$; *Rp. acidophila* 7050, $Y_Q^{522}/Y_Q^{860} = 0.65$. This shows that the radiation with less quantum energy, $E = h \cdot c/\lambda$, absorbed by Bchl is more effective. We can interpret the yield quotient as efficiency of energy transfer from the carotenoids to Bchl or as the difference in the efficiencies of energy migration from antenna to RC pigments starting with two different components in the antenna system. This is in general agreement with determinations of excitation energy transfer by fluorescence yield measurements (Duysens, 1952, 1956).

8. Quantum Requirement of Photophosphorylation

There is no possibility of determining the rate of photophosphorylation directly during quantum yield measurements. It is possible, however, to calculate from Y_{ATP} values determined in other experiments and Y_Q values determined by us the quantum requirement of photophosphorylation, R_{ATP}^Q:

$$R_{ATP}^Q = Y_{ATP}/Y_Q$$

The values for Y_{ATP} presented in the literature differ considerably (Payne, 1970). The composition of the medium (especially the carbon and nitrogen sources) alters Y_{ATP}, and it was found to be dependent on specific growth rate (Stouthamer and Bettenhaussen, 1973; Stouthamer, 1973).

We shall assume that dry weight formation requires a similar phosphorylation rate under similar conditions of culture medium and at similar specific growth rates.

The quantum yields presented in Figs. 9 and 10 relate to growth on lactate in a mineral medium with NH_4^+ as nitrogen source. For this condition, we take from Stouthamer (personal communication) the values

$$Y_{ATP}^{max} = 8 \frac{\text{g dry wt.}}{\text{mol ATP}}, m_e = 4 \times 10^{-3} \frac{\text{mol ATP}}{\text{g dry wt.} \cdot \text{hr}}$$

Substituting these values in Pirt's formula

$$1/Y_{ATP} = 1/Y_{ATP}^{max} + m_e/\mu$$

we can calculate Y_{ATP} for each specific growth rate.

Table 3 gives values for R_{ATP}^Q calculated from Y_Q shown in Figs. 9 and 10 and Y_{ATP}, calculated as explained. The values refer to quantum absorption by Bchl. Considering Section 7.1, we used only quantum yields for specific growth rates up to the maximum of the Y_Q-curve.

As a mean value, we found

$$R_{ATP}^Q = 1.5 \text{ quanta/ATP}$$

This result suggests an intrinsic requirement of only 1 quantum per molecule of ATP. Because of the indirect determination, an intrinsic requirement of 2 quanta also seems to be possible.

9. Quantum Requirement Related to Substrates

The quantum requirement of growth for an organism is dependent on the composition of the medium; here, I report on the influence of different carbon and nitrogen sources. The quantum requirement of different strains growing in the same medium was found to be approximately constant.

9.1. Assimilation of CO_2 and Simple Organic Carbon Compounds

The media used were developed by Pfennig (1969). They consist of mineral salts and the organic carbon source. For autotrophic growth, the cultures were aerated with a mixture of 80% H_2 and 20% CO_2 without any carbon compounds in the medium other than bicarbonate. Nitrogen was available as ammonia.

Table 4 gives the minimal quantum requirement, R (mole quanta or Einsteins required per gram dry weight). For autotrophic growth of *Rp. acidophila* strain 10050 (DSM 145) with H_2 as electron donor, we measured $R = 0.69$ Einst/g dry wt. For the carbon content of dry cell material, we found 482 μg C/mg dry wt. This is somewhat less than the 526 μg C/mg dry wt. calculated from the formula $(C_5H_8O_2N)_n$ given by van Gemerden (1968) for the composition of photosynthetic bacteria. Using our values, we obtain 17.2 quanta/CO_2 fixed in cell material. Additionally, we found an extracellular carbon concentration in the culture solution (not bicarbonate) that constantly

Table 3. Quantum Requirement of Photophosphorylation[a]

μ (hr^{-1})	Y_Q $\left(\dfrac{\text{g dry wt.}}{\text{Einst}}\right)$	Y_{ATP} $\left(\dfrac{\text{g dry wt.}}{\text{mol ATP}}\right)$	R_{ATP}^Q $\left(\dfrac{\text{quanta}}{\text{ATP}}\right)$
	Rp. capsulata Kbl		
0.057	3.02	5.12	1.7
0.10	4.21	6.06	1.4
	Rp. acidophila 7050		
0.058	3.71	5.16	1.4
0.10	4.06	6.06	1.5
		Mean:	1.5

[a] Growth with lactate in mineral medium (Pfennig, 1969) with NH_4^+ as nitrogen source. Radiation absorbed by Bchl at wavelength 860 nm.

Table 4. Minimal Quantum Requirement as Related to Carbon Source[a]

Organism	Specific growth rate (hr^{-1})	Carbon source	Minimal quantum requirement (Einst/g dry wt.)
	Autotrophic growth[b]		
Rp. acidophila strain 10050	0.1	CO_2	0.69
	Heterotrophic growth[b]		
Rp. acidophila strain 10050	0.037	Methanol	0.77
Rs. rubrum strain S1	0.1	Acetate	0.37
Rs. rubrum strain Ha	0.1	Lactate	0.27
Rp. acidophila strain 7050	0.12	Lactate	0.244
Rp. capsulata strain Kb1	0.1	Lactate	0.238
Rp. acidophila strain 7050	0.15	Succinate	0.28

[a] Minimal values with respect to light-limited growth.

[b] Mineral medium (Pfennig, 1969) with ammonia as nitrogen source; H_2 as electron donor for autotrophic growth; temperature 30°C. Radiation absorbed by Bchl at wavelengths of 885 nm for *Rs. rubrum*, and at wavelength of 860 nm for all others.

equals 47% of the carbon fixed in cell material. Considering this to be fixed CO_2, we calculate 11.7 quanta/CO_2 fixed. With resting cells, quantum requirements of 4–9 quanta/CO_2 are given in the literature for different photosynthetic bacteria (see Larsen, 1953). From his experiments using static cultures of Chlorobiaceae, Larsen calculated 8–12 quanta/CO_2 with H_2 as the electron donor.

When methanol was used as a substrate, $R = 0.77$ Einst/g dry wt. (Table 4). This is more than the same organism needs for autotrophic growth. The photoassimilation of methanol by *Rp. acidophila* strain 10050 was described by the formula 2 $CH_3OH + CO_2 \rightarrow 3 CH_2O + H_2O$ (Quayle and Pfennig, 1975). For this process, no evidence was obtained for involvement of a reduced C_1-fixation sequence. Methanol is first oxidized and used as bicarbonate (Sahm *et al.*, 1976). A comparison of the maximal specific growth rate for *Rp. acidophila* strain 10050 growing autotrophically ($\mu_{max} = 0.16$ hr^{-1}) and growing on methanol ($\mu_{max} = 0.073$ hr^{-1}) indicates that methanol oxidation is the rate-limiting step. Thus, when the organism is grown on methanol, a higher R and lower μ_{max} are obtained.

Heterotrophic growth is given with the C_2-compound acetate. For R, we found half the amount needed for growth on methanol. For the C_3-compound lactate, R is also significantly smaller. In a

theoretical study on the amount of ATP required for synthesis of microbial cell material, Stouthamer (1973) calculated that 35% more ATP is needed for growth on acetate as compared with growth on lactate. This is almost the same ratio we get for quantum requirements of *Rs. rubrum*; 37% (Table 4). According to Stouthamer, the additional ATP is required for transport of carbon compounds as well as for the formation of glucose-6-phosphate, amino acids, and lipids.

For autotrophic growth (and for growth on acetate), quantum measurements are the only direct possibility for determination of the energy needed for synthesis of bacterial cell material; 86% more quanta are required for autotrophic growth as compared with growth on acetate.

With lactate as the carbon source we studied three organisms. The R's differ no more than $\pm 6\%$ from the mean value.

This comparison could give more precise information if the influence of maintenance energy were eliminated. To do this, more data about the dependency of R on specific growth rate are needed. We did not study this in all cases discussed above, but in Table 4 we compare only values belonging to the middle range of specific growth rate ($0.3 \cdot \mu_{max}$ to $0.6 \cdot \mu_{max}$), where the smallest values of R can be avoided.

9.2. Nitrogen Fixation and Assimilation of Various Nitrogen Sources

Most of the photosynthetic bacteria are able to fix nitrogen under anaerobic conditions. The quantum requirement of cells growing with ammonia rises dramatically when molecular nitrogen becomes the only nitrogen source. For this transition, Table 5 shows the increase of quantum requirement (R) for two organisms. The R becomes 57% higher for *Rp. capsulata* strain Kb1 and 88% higher for *Rp. acidophila* strain 7050. The nitrogen content of cell material was 81 μg N/mg dry wt. for *Rp. capsulata* strain Kb1 and 87 μg N/mg dry wt. for *Rp. acidophila* strain 7050. With these values, the number of additional quanta required per mole of N_2 fixed was calculated.

In Section 8, we showed that for the same organisms under the same growth conditions, 1.5 quanta/ATP are required when NH_4^+ is the nitrogen source. If this is valid during nitrogen fixation, then 31 ATP, in the case of *Rp. capsulata*, and 50 ATP in the case of *Rp. acidophila*, are required for fixation of one N_2. The maintenance coefficient will be much higher for nitrogen-fixing cells [as Dalton and Postgate (1969) found with *Azotobacter chroococcum*], and the increase of quantum absorption rate for nitrogen fixation will result in a decreasing efficiency of the light-harvesting system. Both influences will reduce the extremely high quantum as well as ATP requirements calculated above. Results of experiments with cell-free extracts (as well as those with whole cells, growing under anaerobic dark conditions) also yielded very high ATP requirements for nitrogen fixation. Postgate (1971) reported 30 ATP/N_2 from experiments in his laboratory with *Klebsiella pneumoniae* growing heterotrophically in the dark. His statement that "... the mysterious inefficiency of cell-free N_2 fixation appears to extend to intact anaerobes" seems also to be valid for nitrogen fixation of photosynthetic bacteria.

In Table 6, R is given for growth on various nitrogen sources. R decreases from the level of NH_4^+ assimilation only with yeast extract, which provides both nitrogen and carbon sources. In this case, the cells are able to assimilate most of the nitrogen compounds required (amino acids and nucleic acid bases) from the medium. In addition, yeast extract serves as an unknown mixture of carbon sources. This can be seen from the yield (g dry wt./g lactate), which rises from 0.65 to 8.7 (see Table 6).

With glutamate as the only nitrogen source, R

Table 5. Quantum Requirement of Nitrogen Fixation[a]

	Relative requirement[b]	$R_{N_2}^Q$ (quanta/molecule)	$R_{N_2}^{ATP}$
Rp. capsulata strain Kb1	1.57	47	31
Rp. acidophila strain 7050	1.88	75	50

[a] Mineral medium (Pfennig, 1969) with lactate; temperature 30°C. Radiation absorbed by Bchl at wavelength 860 nm.
[b] $R_{dry wt.}^Q$ on N_2 divided by $R_{dry wt.}^Q$ on NH_4^+.

Table 6. Quantum Requirement for Growth of *Rhodopseudomonas capsulata* Strain Kb1 on Various Nitrogen Sources[a]

Nitrogen source	$R_{dry wt.}^Q$ (Einst/g dry wt.)	Nitrogen content (μg N/mg dry wt.)	Yield (g dry wt./g lactate)
NH_4^+	0.24	100	0.65
N_2	0.37	81	0.57
NO_3^-	0.31	95	0.51
Glutamate	0.28	89	0.90
Yeast extract	0.17	123	8.7

[a] Mineral medium (Pfennig, 1969) with lactate as carbon source; specific growth rate, 0.1 hr^{-1}; temperature, 30°C; 860-nm light.

becomes 18% higher compared with NH_4^+. In a theoretical study, Stouthamer (1973) calculated a constant ATP requirement for microbial growth on glucose with NH_4^+ when all required amino acids were added. The small ATP requirement for amino acid formation from glucose and NH_4^+ and the ATP required for transport of the amino acids into the cells prevented a decrease in energy requirement when the amino acids were added. In our experiments with growth on lactate, amino acid formation is more expensive, and the addition of all amino acids should result in a lower ATP requirement. As shown by the nitrogen balance, determined from measurements of glutamate consumption and the nitrogen content of cell material, the cells use molecular nitrogen as a second nitrogen source. This additional fixation will cause the increased R. With glutamate, α-ketoglutarate is formed by transamination. α-Ketoglutarate and glutamate are additional carbon sources. Thus, the dry weight yield on lactate increases to 0.90.

For growth on nitrate, the R reaches 132% of its value for growth on NH_4^+; this is between the R for growth on NH_4^+ and on N_2. Probably the cells carry out a light-dependent, ferredoxin-mediated reduction of nitrate that increases the quantum requirement. The dry weight yield on lactate decreases with the oxidation level of the inorganic nitrogen source because more lactate is required as an electron donor for the reduction of the nitrogen sources to amino groups.

Table 6 also shows the different nitrogen contents of cell material. These values must be considered when energy requirements are related to the amount of nitrogen fixed in cell material.

ACKNOWLEDGMENTS

The organisms used in the studies presented in this chapter belong to the Rhodospirillaceae and were kindly provided by Norbert Pfennig, who has studied their nutrition and physiology extensively. I am grateful for his help concerning the biology of these organisms.

I want to acknowledge the Deutsche Forschungsgemeinschaft for financial support at the beginning of my work. Professor R. S. Wolfe is thanked for stylistic improvements and Hartmut Scheede for reliable technical assistance.

10. References

Aagaard, J., and Sistrom, W. R., 1972, Control of synthesis of reaction center bacteriochlorophyll in photosynthetic bacteria, *Photochem. Photobiol.* **15**:209.

Avron, M., and Gozal, B., 1969, Interaction between two photochemical systems in photoreactions of isolated chloroplasts, in: *Progress in Photosynthesis Research* (H. Metzner, ed.), Vol. III, pp. 1185–1196, Prof. Dr. Helmut Metzner, Tübingen.

Bergter, F., and Günther, H., 1961, Messung der Licht-Vorwärtsstreuung an synchronisierten Bakterienkulturen, *Monatsber. Dtsch. Akad. Wiss. Berlin* **3**:718.

Cohen-Bazire, G., and Sistrom, W. R., 1966, The procaryotic photosynthetic apparatus, in: *The Chlorophylls* (L. P. Vernon and G. R. Seely, ed.), pp. 311–342, Academic Press, New York.

Dalton, H., and Postgate, J. R., 1969, Growth and physiology of *Azotobacter chroococcum* in continuous culture, *J. Gen. Microbiol.* **56**:307.

Dean, A. C. R., Pirt, S. J., and Tempest, D. W., (eds.), 1972, *Environmental Control of Cell Synthesis and Function*, Academic Press, London and New York.

Duysens, L. N. M., 1952, Transfer of excitation energy in photosynthesis, Thesis, Utrecht, Netherlands.

Duysens, L. N. M., 1956, Energy transformations in photosynthesis, *Annu. Rev. Plant Physiol.* **7**:25.

van Gemerden, H., 1968, Utilization of reducing power in growing cultures of *Chromatium*, *Arch. Mikrobiol.* **64**:111.

Göbel, F., 1969, Measurements of the radiation balance in suspensions of photosynthetic bacteria, in: *Progress in Photosynthesis Research* (H. Metzner, ed.), Vol. II, pp. 1122–1127, Professor Dr. Helmut Metzner, Tübingen.

Göbel, F., 1978, Direct measurement of pure absorbance-spectra of living phototrophic microorganisms, *Biochim. Biophys. Acta* **538**:593.

Gürgün, V., 1974, Untersuchungen über den anaeroben Dunkelstoffwechsel einiger Arten der phototrophen Purpurbakterien, Dissertation, Göttingen.

Heber, U., and Kirk, M. R., 1975, Efficiency of coupling between phosphorylation and electron transport in intact chloroplasts, in: *Proceedings of the Third International Congress on Photosynthesis* (M. Avron, ed.), Vol. II, pp. 1041–1046, Elsevier Scientific Publishing Co., Amsterdam.

Herbert, D., 1961, A theoretical analysis of continuous culture systems, *Soc. Chem. Ind. Monogr.*, No. 12, pp. 21–53, London.

Hippe, H., 1974, Arbeiten und Ergebnisse, Deutsche Sammlung von Mikroorganismen, Arbeitsgruppe Göttingen, in: *Jahresbericht 1973*, Gesselschaft für Strahlen- und Umweltforschung mbH, München, Institut für Mikrobiologie, p. 52, Neuherberg bei München.

Kohlmiller, E. F., and Gest, H., 1951, A comparative study of the light and dark fermentations of organic acids by *Rhodospirillum rubrum*, *J. Bacteriol.* **61**:269.

Kok, B., 1960, Efficiency of photosynthesis, in: *Handbuch der Pflanzenphysiologie* (W. Ruhland, ed.), Vol. V, Part 1, pp. 566–633, Springer-Verlag, Berlin—Göttingen—Heidelberg.

Larsen, H., 1953, On the microbiology and biochemistry of the photosynthetic green and sulfur bacteria, Thesis, Trondheim, Norway.

Monod, J., 1942, Recherches sur la croissance des cultures bactériennes, Thèse, 2. Ed. Hermann, Paris, 1958.

Myers, J., 1966, On dynamics of algal growth, *J. Ferm. Technol.* **44**:344.

Noak, D., 1968, Biophysikalische Prinzipien der Populationsdynamik in der Mikrobiologie, in: *Fortschritte der experimentellen und theoretischen Biophysik* (W. Beier, ed.), Heft 8, VEB Georg Thieme, Leipzig.

Payne, W. J., 1970, Energy yields and growth of heterotrophs, *Annu. Rev. Microbiol.* **24**:17.

Pfennig, N., 1969, *Rhodopseudomonas acidophila*, sp. n., a new species of the budding purple nonsulfur bacteria, *J. Bacteriol.* **99**:597.

Pfennig, N., and Jannasch, H. W., 1962, Biologische Grundfragen bei der homokontinuierlichen Kultur von Mikroorganismen, *Ergeb. Biol.* **25**:93.

Pirt, S. J., 1965, The maintenance energy of bacteria in growing cultures, *Proc. R. Soc. London Ser. B* **163**:224.

Postgate, J., 1971, Fixation by free-living microbes, in: *The Chemistry and Biochemistry of Nitrogen Fixation* (J. R. Postgate, ed.), pp. 161–190, Plenum Press, London and New York.

Quayle, J. R., and Pfennig, N., 1975, Utilization of methanol by Rhodospirillaceae, *Arch. Microbiol.* **102**:193.

Sahm, H., Cox, R. B., and Quayle, J. R., 1976, Metabolism of methanol by *Rhodopseudomonas acidophila*, *J. Gen. Microbiol.* **94**:313.

Stouthamer, A. H., 1973, A theoretical study on the amount of ATP required for synthesis of microbial cell material, *Antonie van Leeuwenhoek; J. Microbiol. Serol.* **39**:545.

Stouthamer, A. H., and Bettenhaussen, C., 1973, Utilization of energy for growth and maintenance in continuous and batch cultures of microorganisms, *Biochim. Biophys. Acta* **301**:53.

Stouthamer, A. H., and Bettenhaussen, C. W., 1975, Determination of the efficiency of oxidative phosphorylation in continuous cultures of *Aerobacter aerogenes*, *Arch. Microbiol.* **102**:187.

de Vries, W., Kapteijn, W. M. C., van der Beek, E. G., and Stouthamer, A. H., 1970, Molar growth yields and fermentation balances of *Lactobacillus casei* L3 in batch cultures and continuous cultures, *J. Gen. Microbiol.* **63**:333.

Warburg, O., and Negelein, E., 1922, Über den Energieumsatz bei der Kohlensäureassimilation, *Z. Phys. Chem.* **106**:191.

Warburg, O., and Negelein, E., 1923, Über den Einfluss der Wellenlänge auf den Energieumsatz bei der Kohlensäureassimilation, *Z. Phys. Chem.* **106**:191.

Wiesner, W., 1966, Vergleichende Studien zum Quantenbedarf der Photoassimilation von Essigsäure durch photoheterotrophe Purpurbakterien und Grünalgen, *Ber. Dtsch. Bot. Gess.* **79**:58.

Wraight, C. A., and Clayton, R. K., 1973, The absolute quantum efficiency of bacteriochlorophyll photooxidation in reaction centres of *Rhodopseudomonas spheroides*, *Biochim. Biophys. Acta* **333**:246.

Wraight, C. A., Leigh, J. S., Dutton, P. L., and Clayton, R. K., 1974, The triplet state of reaction center bacteriochlorophyll: Determination of a relative quantum yield, *Biochim. Biophys. Acta* **333**:401.

APPENDIX

Lists of Mutant Strains

W. R. Sistrom

This appendix contains lists of mutant strains of *Rhodopseudomonas sphaeroides* (Table 1), *Rhodopseudomonas capsulata* (Table 2), *Rhodopseudomonas gelatinosa* and *Rhodopseudomonas palustris* (Table 3), and *Rhodospirillum rubrum* (Table 4). No attempt has been made to provide exhaustive lists; rather, the aim is to provide examples of the kinds of mutants that have been obtained and an introduction to the literature from which details of the procedures can be gotten.

Only strains that are defective in their energy metabolism or biosynthesis of photosynthetic pigments are included. Wild-type strains are listed for the sake of completeness. Auxotrophs and resistant strains are listed only if they are parents of other mutants.

Abbreviations and Notes

The abbreviations and notes below apply to all the tables.

Phenotype

accum., accumulates

Aer⁻, unable to grow aerobically

Aer⁻ (microaerophilic), grows aerobically only when cells are pigmented

ALA⁻, lacks δ-aminolevulinate synthase

Asn^R, resistant to arsenate

Bch⁻, defective in Bchl synthesis; the number in parentheses indicates the position (in nm) of the absorption maximum of the culture supernatant

Car⁻, defective in carotenoid biosynthesis; when known, the abnormal carotenoids accumulated are indicated. The letters in parentheses give a more precise description of the defect and phenotype: (BG) no colored carotenoids; (BG-G) blue-green under anaerobic conditions, green under semi-aerobic conditions; (G I) green, accumulates only neurosporene; (G II) green, accumulates neurosporene and chloroxanthin; (GR) green or blue-green under anaerobic conditions, red under semi-aerobic conditions—this is probably similar to the BG-G phenotype; (PY) pale yellow, accumulates spheroidene only; (Y) yellow, accumulates neurosporene, chloroxanthin, and spheroidene (no spheroidenone)

Car^ts, no colored carotenoids at restrictive temperature (temperature-sensitive)

GTA, gene transfer agent (see Chap. 47)

Met⁻, requires methionine

NADI⁻, lacks NADI oxidase

Ps⁻, no phototrophic growth

Ps^ts, no phototrophic growth at restrictive temperature (temperature-sensitive)

Rfm^R, resistant to rifampicin

Rxc⁻, lacks reaction center Bchl (no light-induced bleaching characteristic of oxidation of RC Bchl)

Ser⁻, requires serine

White, albino, no Bchl or carotenoids; these mutants can revert to wild-type and are therefore not simply double mutants (Bch⁻, Car⁻)

W. R. Sistrom · Department of Biology, University of Oregon, Eugene, Oregon 97403

Table 1. Mutant Strains of *Rhodopseudomonas sphaeroides*

No.	Strain	Phenotype	Parent	Method of isolation	Ref. Nos.[a]
1	ATH 2.4.1 (ATCC 17023)[b]	Wild-type			1
2	NCIB 8253[b]	Wild-type			2
3	CCl/R26	Car−(BG); no polyenes	ATH 2.4.1 (1)	UV	3
4	CCl/R34	Car−(BG); accum. phytoene	ATH 2.4.1 (1)	UV	3
5	CCl/R22	Car−(BG); accum. phytoene, phytofluene, and ζ-carotene	ATH 2.4.1 (1)	UV	3
6	AlOC	Car−(BG); accum. phytoene, phytofluene, and ζ-carotene	ATH 2.4.1 (1)	UV	3
7	uv-33	Car−(BG); accum. phytoene and chlorophyll *a* derivatives	ATH 2.4.1 (1)	UV	4, 5
8	Ga	Car−(G II)	ATH 2.4.1 (1)	Sp	1
9	M55[c]	Car−(G I)	ATH 2.4.1 (1)	UV	6
10	EMS 66[c]	Car−(G I)	ATH 2.4.1 (1)	EMS	6
11	PM-9	Car−(BG-G)	Ga (8)	UV	3
12	Brown[c]	Car−(Y)	ATH 2.4.1 (1)	UV	4
13	G[c]	Car−(G I)	NCIB 8253[d] (2)	NTG	7
14	BG[c]	Car−(BG)	NCIB 8253[d] (2)	NTG	7
15	B	Car−(PY?)	NCIB 8253[d] (2)	NTG	7
16	L-57[e]	White (Ps−)	NCIB 8253 (2)	Sp; repeated aerobic transfers	8, 9
17	6-6	Bch−; accum. porphobilinogen	NCIB 8253 (2)	NTG	10, 11
18	8-13	Bch−; lacks Mg-chelatase (?)	NCIB 8253 (2)	NTG	11, 12
19	Rm-7	Bch−; accum. Mg-protoporphyrin monomethyl ester	NCIB 8253 (2)	NTG	13
20	2-33R	Bch+; but accum. Mg-protoporphyrin monomethyl ester	2-33 (27) (2)	Sp; revertant of 2-33	9, 10
21	8-32	Bch−; accum. Mg-divinylpheoporphyrin a_5 (P631)	NCIB 8253 (2)	NTG	11, 14

No.	Strain	Phenotype	Parent	Mutagen	Ref.
22	8-29	Bch⁻; accum. 2-devinyl-2-α-hydroxy-ethylchlorophyllide (P661)	NCIB 8253 (2)	NTG	11, 14
23	8-47	Bch⁻; accum. 2-deacetyl-2-α-hydroxyethylbacteriochlorophyllide (P720)	NCIB 8253 (2)	NTG	11, 14
24	8-17	Bch⁻; accum. bacteriochlorophyllide (P770)	NCIB 8253 (2)	NTG	10, 14
25	H-5	ALA⁻	NCIB 8253 (2)	NTG; screen on ALA medium	15
26	6-6R	Bch⁺; but accum. porphobilinogen	6-6 (17)	Sp; revertant of 6-6	10
27	2-33	Bch⁻, Met⁻; accum. porphobilinogen and coproporphyrin III	NCIB 8253 (2)	NTG	9
28	L-57R	Makes Bchl and carotenoids under high aeration (control mutant?)	L-57 (16)	Sp; Ps⁺ revertant of L-57	16
29	TA-R	Makes Bchl and carotenoids under high aeration (control mutant?)	6-10 (31)	Sp; sel'n for Bchl synthesis in aerobic conditions	16
30	DW-R	Makes Bchl and carotenoids under high aeration (control mutant?)	NCIB 8253 (2)	NTG	16
31	6-10	Met⁻	NCIB 8253 (2)	NTG	16
32	113-20	Ps⁻ (Rxc⁻)	NCIB 8253 (2)	NTG, penicillin countersel'n	17
33	71-13	Ps⁻ (Rxc⁻)	NCIB 8253 (2)	NTG, penicillin countersel'n	17
34	71-20	Ps⁻ (Rxc⁻)	NCIB 8253 (2)	NTG, penicillin countersel'n	17
35	PM-8	Ps⁻ (Rxc⁻)	Ga (8)	UV	18
36	PM-8 bg15	Ps⁻ (Rxc⁻), Car⁻(BG)	PM-8 (35)	EMS	18
37	37ᶜ	Aer⁻ (microaerophilic)	Ga (8)	NTG	19
38	G1C	Car⁻(G1)	ATH 2.4.1 (1)ᶠ	NTG	20

[a] References: (1) Cohen-Bazire et al. (1956); (2) Lascelles (1956); (3) Crounse et al. (1963); (4) Griffiths and Stanier (1956); (5) Sistrom et al. (1956); (6) Sunada and Stanier (1968); (7) Segen and Gibson (1971); (8) Bull and Lascelles (1963); (9) Lascelles (1966); (10) Hatch and Lascelles (1972); (11) Lascelles and Altshuler (1967); (12) Lascelles and Hatch (1969); (13) J. Lascelles (unpublished); (14) Richards and Lascelles (1969); (15) Lascelles and Altshuler (1969); (16) Lascelles and Wertlieb (1971); (17) Takemoto and Lascelles (1974); (18) Sistrom and Clayton (1964); (19) Wittenberg and Sistrom (1971); (20) Crofts et al. (1974).

[b] NCIB (National Collection of Industrial Bacteria) strain 8253 is the same as ATCC (American Type Culture Collection) strain 17023.

[c] The strain has been lost.

[d] In the original description (reference 7), the wild-type parent was inadvertently designated NCIB 8327.

[e] This strain was called L-56 in reference 8; subsequent papers have referred to it as L-57.

[f] The parent of G1C was a derivative of ATH 2.4.1 resistant to nalidixic acid, streptomycin, and rifampicin; G1C has the same resistance pattern.

Table 2. Mutant Strains of *Rhodopseudomonas capsulata*

No.	Strain	Phenotype	Parent	Method of isolation	Ref. Nos.[a]
1	ATCC 23782	Wild-type, GTA recipient			1
2	B10	Wild-type, GTA recipient/donor			2
3	H9	Wild-type, GTA recipient/donor			2
4	37b4	Wild-type			3
5	Z-1	Asn^R	ATCC 23782 (1)	Sp	1
6	BH99	Rfm^R	H9 (3)	Sp	4
7	BB103	Sm^R	B10 (2)	Sp	4
8	M110	Asn^R, Car$^-$(BG)	Z-1 (5)	NTG	5
9	M1	Asn^R, Aer$^-$	Z-1 (5)	NTG	6
10	M2	Asn^R, Aer$^-$ (lacks succinic dehydrogenase)	Z-1 (5)	NTG	6
11	M3	Asn^R (lacks succinic dehydrogenase)	M2 (10)	NTG; sel'n for aerobic growth on malate	6
12	M4	Asn^R, NADI$^-$	Z-1 (5)	NTG	6
13	M5	Asn^R, Aer$^-$, NADI$^-$	Z-1 (5)	NTG	6
14	M6	Asn^R	M5 (13)	NTG; sel'n for aerobic growth	6
15	M7	Asn^R, NADI$^-$	M5 (13)	NTG; sel'n for aerobic growth	6
16	M106	Rfm^R, White	BH99 (6)	Sp	4
17	HH908	Rfm^R, Bch$^-$ (P630)	BH99 (6)	Sp	4
18	Y120	Car$^-$(BG), Bch$^-$(P670)	W6 (20)	NTG	4
19	HH910	Rfm^R, Bch$^-$ (P730)	BH99 (6)	Sp	4
20	W6	Car$^-$(BG)	B10 (2)	NTG	4, 7
21	MB1048	Sm^R, Car$^-$(BG)	BB103 (7)	NTG	4, 7
22	Y72	Sm^R, Car$^-$(G I)	BB103 (7)	NTG	4
23	Y83	Sm^R, Car$^-$(G II)	BB103 (7)	NTG	4
24	MB1044	Sm^R, Car$^-$(Y)	BB103 (7)	NTG	4
25	Y142	Sm^R, Ps$^-$(Rxc$^-$)	BB103 (7)	NTG	4
26	Y143	Sm^R, Car$^-$(PY)	BB103 (7)	NTG	4
27	L$_1$	Gain of ability to use glycerol	ATCC 23782 (1)	Sp; sel'n on glycerol	8
28	Serine-1	Ser$^-$ (lacks 3-phosphoglyceraldehyde dehydrogenase)	ATCC 23782 (1)	NTG	9
29	Serine-2	Ser$^-$ (lacks phosphoserine phosphatase)	ATCC 23782 (1)	NTG	9
30	Ala$^-$	Car$^-$(BG), Bch$^-$	37b4 (4)	NTG, penicillin countersel'n	10
31	Ala$^+$	Car$^-$(BG), Bch$^+$	Ala$^-$ (30)	NTG; sel'n for Ps$^+$ revertant	10

[a] References: (1) Zilensky *et al.* (1971); (2) Marrs (1974); (3) Biebl and Drews (1969); (4) Marrs (unpublished); (5) Solioz *et al.* (1975); (6) Marrs and Gest (1973); (7) Yen and Marrs (1976); (8) Lueking *et al.* (1973); (9) Beremand and Sojka (1975); (10) Drews *et al.* (1971).

Table 3. Mutant Strains of *Rhodopseudomonas gelatinosa* and *Rhodopseudomonas palustris*

No.	Strain	Phenotype	Parent	Method of isolation	Ref. Nos.[a]
		Rhodopseudomonas gelatinosa			
1	ATH 2.2.1	Wild-type			
2	TG-9	Wild-type			
3	NG 72	Car⁻(G II)	ATH 2.2.1 (1)	NTG	1
4	NG 78	Car⁻(G II)	ATH 2.2.1 (1)	NTG	1
5	NG 74	Car⁻(G I)	ATH 2.2.1 (1)	NTG	1
6	EM 1	Car⁻(BG)	TG-9 (2)	EMS	2
7	EM 7	Car⁻ (green, G I or G II?)	TG-9 (2)	EMS	2
		Rhodopseudomonas palustris			
1	Pale[b]	Low pigmentation	Wild-type	UV	3
2	Green-1	No Bchl, accum. protochlorophyll a[c]	Pale (1)	EMS	3
3	Green-2	No Bchl, accum. protochlorophyll a[c]	Pale (1)	NMU	3
4	Green-3	Bchl, accum. protochlorophyll a[c]	Pale (1)	NMU	3
5	Yellow	Bchl, accum. protochlorophyll a[c]	Pale (1)	NMU	3
6	Yellow-rose	Bchl, accum. pigment with λ_{max} (*in vivo*) of 667–669 nm	Pale (1)	NMU	3

[a] References: (1) Sunada and Stanier (1968); (2) R. K. Clayton (unpublished); (3) Fedenko *et al.* (1969).

[b] Editor's designation.

[c] Also contains protochlorophyllide. On the basis of observed color, it is assumed that carotenoid biosynthesis is defective.

Table 4. Mutant Strains of *Rhodospirillum rubrum*

No.	Strain	Phenotype	Parent	Method of isolation	Ref. Nos.[a]
1	ATH 1.1.1 (ATCC 11170)	Wild-type			
2	FR 1	Wild-type			
3	G-9	Car⁻(BG)	ATH 1.1.1		
4	VI	Car⁻(BG); accum. phytoene	FR1	NTG	1, 2
5	I3	Ps^{ts}	ATH 1.1.1	NTG; sel'n for temp.-sensitive phototaxis	3
6	B14	Car^{ts}	ATH 1.1.1	NTG	3
7	F24	Rxc⁻	S1	NTG	4, 5
8	F24.1	—[b]	F24	spont. revertant	5

[a] References: (1) Oelze and Golecki (1975); (2) Mandinas *et al.* (1974); (3) Weaver (1971); (4) del Valle-Tascón *et al.* (1975); (5) Picorel *et al.* (1977).
[b] The reaction center in F24.1 is apparently altered; there is little or no absorption at 800 nm and no light-induced absorbancy changes can be detected although ferri-cyanide-induced bleaching at 890 nm is observed.

Parent

The number in parentheses is the serial number of the parent in the table.

Method of Isolation

EMS, ethylmethane sulfonate
NMU, nitrosomethylurea
NTG, nitrosoguanidine
sel'n, selection
Sp, spontaneous
UV, ultraviolet radiation

References

These have been chosen to provide the best description of the strain; they are not necessarily the papers that describe the isolation of the strains.

Beremand, P., and Sojka, G. A., 1975, Mutant analysis of the primary route of serine and glycine biosynthesis in *Rhodopseudomonas capsulata, Abstr. Annu. Meet. Am. Soc. Microbiol.,* New York City, p. 181.

Biebl, H., and Drews, G., 1969, Das *in-Vivo*-Spektrum als taxonomisches Merkmal bei Untersuchungen zum Verbreitung von Athiorhodaceae, *Zentralbl. Bakteriol. Parasitenkd. Infektionskr. Hyg. Abt. 2* **123**:425.

Bull, M. J., and Lascelles, J., 1963, The association of protein synthesis with the formation of pigments in some photosynthetic bacteria, *Biochem. J.* **87**:15.

Cohen-Bazire, G., Sistrom, W. R., and Stanier, R. Y., 1956, Kinetic Studies of pigment synthesis by non-sulfur purple bacteria, *J. Cell. Comp. Physiol.* **49**:25.

Crofts, A. R., Prince, R. C., Holmes, N. G., and Crowther, D., 1974, Electrogenic electron transport and the carotenoid change in photosynthetic bacteria, in: *Proceedings of the Third International Congress on Photosynthesis* (M. Avron, ed.), p. 1131, Elsevier, Amsterdam.

Crounse, J. B., Feldman, R. P., and Clayton, R. K., 1963, Accumulation of polyene precursors of neurosporene in mutant strains of *Rhodopseudomonas spheroides, Nature* (*London*) **198**:1227.

del Valle-Tascón, S., Giménez-Gallego, G., and Ramírez, J. M., 1975, Light dependent ATP formation in a non-phototrophic mutant of *Rhodospirillum rubrum* deficient in oxygen photoreduction. *Biochem. Biophys. Res. Commun.* **66**:514.

Drews, G., Leutiger, I., and Ladwig, R., 1971, Production of protochlorophyll, protopheophytin and bacteriochlorophyll by the mutant Ala of *Rhodopseudomonas capsulata, Arch. Mikrobiol.* **76**:349.

Fedenko, E. P., Kondrat'eva, E. M., and Krasnovsky, A. A., 1969, Protochlorophyll mutants of *Rhodopseudomonas palustris, Nauchn. Dokl. Vyssh. Shk. Biol. Nauki* **8**:102.

Hatch, T. P., and Lascelles, J., 1972, Accumulation of porphobilinogen and other pyrroles by mutant and wild-type *Rhodopseudomonas spheroides, Arch. Biochem. Biophys.* **150**:147.

Lascelles, J., 1956, The synthesis of porphyrins and bacteriochlorophyll by cell suspensions of *Rhodopseudomonas sphaeroides, Biochem. J.* **62**:78.

Lascelles, J., 1966, The regulation of synthesis of iron and magensium tetrapyrroles: Observations with mutant strains of *Rhodopseudomonas spheroides, Biochem. J.* **100**:184.

Lascelles, J., and Altshuler, T., 1967, Some properties of mutant strains of *Rhodopseudomonas spheroides* which do not form bacteriochlorophyll, *Arch. Mikrobiol.* **59**:204.

Lascelles, J., and Altshuler, T., 1969, Mutant strains of *Rhodopseudomonas sphaeroides* lacking δ-aminolevulinate synthase: Growth, heme and bacteriochlorophyll synthesis, *J. Bacteriol.* **98**:721.

Lascelles, J., and Hatch, T. P., 1969, Bacteriochlorophyll and heme synthesis in *Rhodopseudomonas spheroides*: Possible role of heme in regulation of branched biosynthetic pathway, *J. Bacteriol.* **98**:712.

Lascelles, J., and Wertlieb, D., 1971, Mutant strains of *Rhodopseudomonas spheroides* which form photosynthetic pigments aerobically in the dark: Growth characteristics and enzymic activities, *Biochim. Biophys. Acta* **226**:328.

Lueking, D., Tokuhisa, D., and Sojka, G. A., 1973, Glycerol assimilation by a mutant of *Rhodopseudomonas capsulata, J. Bacteriol.* **115**:897.

Mandinas, B., Herber, R., and Villoutreix, J., 1974, Occurrence of *trans*-phytoene in micro-organisms grown in absence of carotenogenesis inhibitors, *Biochim. Biophys. Acta* **348**:357.

Marrs, B., 1974, Genetic recombination in *Rhodopseudomonas capsulata, Proc. Natl. Acad. Sci. U.S.A.* **71**:971.

Marrs, B., and Gest, H., 1973, Genetic mutations affecting the respiratory electron transport system of the photosynthetic bacterium *Rhodopseudomonas capsulata, J. Bacteriol.* **114**:1045.

Oelze, J., and Golecki, J. R., 1975, Properties of reaction center depleted membranes of *Rhodospirillum rubrum, Arch. Microbiol.* **102**:59.

Picorel, R., del Valle-Tascón, S., and Ramírez, J. M., 1977, Isolation of a photosynthetic strain of *Rhodospirillum rubrum* with an altered reaction center, *Arch. Biochem. Biophys.* **181**:665.

Richards, W. R., and Lascelles, J., 1969, The biosynthesis of bacteriochlorophyll. The characterization of latter stage intermediates from mutants of *Rhodopseudomonas spheroides, Biochemistry* **8**:3473.

Segen, B. J., and Gibson, K. D., 1971, Deficiencies of chromatophore proteins in some mutants of *Rhodopseudomonas spheroides* with altered carotenoids, *J. Bacteriol.* **105**:701.

Sistrom, W. R., and Clayton, R. K., 1964, Studies on a mutant of *Rhodopseudomonas sphaeroides* unable to grow photosynthetically, *Biochim. Biophys. Acta* **88**:61.

Sistrom, W. R., Griffiths, M., and Stanier, R. Y., 1956, A note on the porphyrins excreted by the blue-green mutant of *Rhodopseudomonas spheroides*, *J. Cell. Comp. Physiol.* **48**:459.

Solioz, M., Yen, H.-C., and Marrs, B., 1975, Release and uptake of gene transfer agent by *Rhodopseudomonas capsulata*, *J. Bacteriol.* **123**:651.

Sunada, K. V., and Stanier, R. Y., 1968, Observations on the pathway of carotenoid synthesis in *Rhodopseudomonas*, *Biochim. Biophys. Acta* **107**:38.

Takemoto, J., and Lascelles, J., 1974, Function of membrane proteins coupled to bacteriochlorophyll synthesis:

Studies with wild-type and mutant strains of *Rhodopseudomonas spheroides*, *Arch. Biochem. Biophys.* **163**:507.

Weaver, P., 1971, Temperature-sensitive mutations of the photosynthetic apparatus of *Rhodospirillum rubrum*, *Proc. Natl. Acad. Sci. U.S.A.* **68**:136.

Wittenberg, T., and Sistrom, W. R., 1971, Mutant of *Rhodopseudomonas spheroides* unable to grow aerobically, *J. Bacteriol.* **106**:732.

Yen, H.-C., and Marrs, B., 1976, Maps of genes for carotenoid and bacteriochlorophyll synthesis in *Rhodopseudomonas capsulata*, *J. Bacteriol.* **126**:619.

Zilensky, J. W., Sojka, G. A., and Gest, H., 1971, Energy charge regulation in photosynthetic bacteria, *Biochem. Biophys. Res. Commun.* **42**:955.

Index

76069

QR
88.5
P48

THE PHOTOSYNTHETIC BACTERIA

DATE DUE
